Y0-BWG-554

INDUSTRIAL ELECTRONICS REFERENCE BOOK

INDUSTRIAL ELECTRONICS REFERENCE BOOK

by

Electronics Engineers

of the

Westinghouse Electric Corporation

JOHN WILEY & SONS, INC., NEW YORK

CHAPMAN & HALL, LIMITED, LONDON

1948

SECOND PRINTING, JANUARY, 1949

PRINTED IN THE UNITED STATES OF AMERICA

FOREWORD

ELECTRONICS is a word which has been much used but also greatly abused in the past few years. From articles in the technical press, one might gain the erroneous impression that through electronics anything desired could be easily accomplished. It is the purpose of this book to give sufficient theoretical data and application information on electronic subjects so that, when an electronic device is designed or utilized, the reader may have an understanding of its possibilities as well as its limitations. Consequently, new applications or projects, carefully conceived, should have a better than even chance of giving the desired results, and electronics should take its place as a useful industrial tool.

In the fighting of World War II, electronics was an extensively utilized branch of the electrical art and as a result has been greatly glamorized. Electronics did much to speed the winning of the war. Its broad use brought many individuals into contact with electronic devices. Had these individuals lived a normal life, they would not have had to stake their lives on information obtained from or resultant operation of such devices. This factor advanced the use of electronics and proved that it could do things that were impossible by other methods.

The major problem in industry and a major purpose of this book, therefore, is to be sure that electronics as a tool is properly used. Our way of doing things should not be so rigid that change cannot be made, but any change should promise definite improvement over tried and proven methods. In the case where electronics provides a means of accomplishing something hitherto impossible, the only caution is to be sure that the application is economically sound and has a chance of surviving the trials of time.

A. C. MONTEITH, Manager
Headquarters Engineering Departments
Westinghouse Electric Corporation

April, 1948

PREFACE

SINCE 1940, widespread expansion has taken place in the field of electronic engineering. Marked advancement was made in the industrial phase of electronics with the result that this branch of engineering was directly related to the momentous production record of industry during this period. Developments took place so rapidly that it was almost impossible for one not highly skilled in such equipment to keep pace. The need for a digest embracing the application as well as design data for industrial electronic equipment under one cover was evident. The *Industrial Electronics Reference Book* was written to satisfy this need. This book has been prepared with the hope that the technical data and application information set forth will enable those concerned with the application and utilization of electronic apparatus to have a better understanding of the possibilities and limitations of such equipment.

The book, therefore, has been aimed at the practicing engineer who is faced with the problem of understanding the underlying principles as well as the scope and limiting factors of electronic apparatus as it is applied to industrial processes. It is felt that the book will also be of material value to graduates of engineering schools who are entering industry for the first time and who will be faced with the problem of approving or rejecting electronic equipment for use in industrial functions.

The *Industrial Electronics Reference Book* has been written by a group of engineers, each thoroughly versed in his particular branch of the subject. These particular authors were chosen because of their broad experience, deep interest, and close association, over a period of years, with that phase of the industrial electronics field about which they have written. The authors have been intimately associated with the rapidly expanding electronics field during the past decade and have imparted many valuable and lasting contributions to this branch of electrical engineering.

Chapters 1 through 3 cover the fundamental and theoretical concepts of basic laws governing the production and control of electrons. The kinetic theory of gases is reviewed. Modern views of atomic structure are discussed, and the electronic theory of solids is examined. The theory of electron emission is covered, and the types of emission and emitting surfaces are discussed. Theories covering the different types of emission, such as field emission, secondary emission, thermionic emission, are set forth. The many factors entering into the control of free electrons are pointed out. These cover space charge, electron motion in magnetic and electric fields, and forces on electrons in magnetic fields. Chapters 4 through 10 embrace the design, operation, and construction features of the different types of electronic tubes which are the fundamental part of any piece of electronic equipment. The design and operation of vacuum tubes, gas tubes, x-ray tubes, cathode-ray tubes as well as photoelectric devices and ultraviolet radiators are discussed. Electronic circuit components, such as the different resistance forms, capacitance and inductance, and also tuned circuits, filters, and transformers, are covered in Chapters 11 through 14, and the different circuits for rectifiers, amplifiers, oscillators, and various control circuits are presented in Chapters 15 through 18. The design and application factors relating to transmission lines and antennas, subjects which are becoming increasingly important in the industrial electronics field, are covered in Chapters 19 and 20.

The many different types of industrial equipment are reviewed in Chapters 21 through 34.

Power rectifiers and inverters, radio-frequency heating equipment, power-line carrier equipment, electronic instruments, such as electrometers, high-frequency ammeters, time-interval and speed meters, stroboscopes, balancing machines, electron microscopes, and many others, are explained. Industrial x-ray, motor control, precipitation equipment, regulation devices, resistance welding, and photoelectric equipment are discussed. Factors influencing design, application, and operation are emphasized. Limitations as well as advantages are set forth and discussed. Chapters 35 and 36 have been included to give the reader an indication of the factors involved in the care and maintenance of electronic tubes and electronic apparatus.

A list of references appears at the end of most of the chapters so that the reader may easily find additional information on any one of the specific subjects involved.

The Book Committee and the authors wish to acknowledge the work of G. P. Cardwell, D. R. Lynch, J. T. Carleton, W. Reid, and C. J. Madsen, who so ably assisted in the task of coordinating and following the book material through the production stage.

WESTINGHOUSE INDUSTRIAL ELECTRONICS BOOK COMMITTEE

Book Coordinator:
B. E. Rector, Electronics Engineer

S. L. Burgwin, Advisory Engineer
Electronic Control Engineering

R. N. Harmon, Manager of Engineering
Industrial Electronics Division

J. A. Hutcheson, Director
Research Laboratories

G. F. Jones, Electronics Engineer

D. D. Knowles, Manager of Electronics Engineering
Lamp Division

E. H. Vedder, Manager of Engineering
Electronic Control Division

Chairman:
F. R. Benedict, Manager
Industry Engineering Department

April, 1948

CONTENTS

1 Physical Background of Industrial Electronics 1
E. U. Condon and E. G. F. Arnott

2 Electron Emission 10
J. W. McNall

3 Control of Free Electrons 33
H. J. Dailey and J. H. Findlay

4 Electrical Conduction in Gases 47
D. E. Marshall

5 Vacuum Tubes 58
H. J. Dailey, J. H. Findlay, and E. C. Okress

6 Gas Tubes 99
D. E. Marshall

7 Photoelectric Devices 116
W. W. Watrous and D. E. Henry

8 Industrial X-Ray Tubes 138
Joseph Lempert

9 Cathode-Ray Tubes 162
P. E. Volz and P. E. Grandmont

10 Ultraviolet Radiators 171
H. C. Rentschler

11 Circuit Elements 176
D. G. Little

12 Tuned Circuits and Filters 188
Reuben Lee

13 Transformers 207
Reuben Lee

14 Vacuum Tubes as Circuit Elements 228
W. E. Shoupp

15 Rectifier Circuits 236
H. J. Bichsel

16 Amplifier Circuits 246
Reuben Lee

17 Circuits for Oscillators 261
E. U. Condon and William Altar

18 CIRCUITS FOR INDUSTRIAL CONTROL 278
H. J. Bichsel

19 TRANSMISSION LINES 290
E. U. Condon and William Altar

20 ANTENNAS 308
T. M. Bloomer

21 GENERAL REQUIREMENTS OF RECTIFIER APPLICATIONS 343
C. R. Marcum

22 MERCURY-ARC RECTIFIERS FOR POWER APPLICATION 348
G. F. Jones and C. R. Marcum

23 INVERTERS 361
J. L. Boyer

24 RADIO-FREQUENCY HEATING 375
D. Venable and T. P. Kinn

25 POWER-LINE CARRIER 442
F. S. Mabry

26 ELECTRONIC INSTRUMENTS 457
M. P. Vore

27 INDUSTRIAL X-RAY APPLICATIONS 503
D. E. Morgan and F. A. Trenkle

28 ELECTROSTATIC PRECIPITATION 526
E. H. R. Pegg

29 ELECTRONIC MOTOR CONTROL 533
K. P. Puchlowski

30 REGULATION 572
W. O. Osbon and V. B. Baker

31 RESISTANCE-WELDING CONTROL 599
C. B. Stadum, E. T. Hughes, F. R. Woodward, and H. J. Bichsel

32 INDUSTRIAL PHOTOELECTRIC CONTROL 628
V. B. Baker

33 APPLICATIONS OF ULTRAVIOLET RADIATIONS 641
H. C. Rentschler

34 RADAR: FUNDAMENTALS AND APPLICATIONS 645
Hugh Odishaw

35 CARE AND MAINTENANCE OF TUBES 659
C. J. Madsen

36 CARE AND MAINTENANCE OF ELECTRONIC APPARATUS 664
C. J. Madsen

INDEX 671

Chapter 1

PHYSICAL BACKGROUND OF INDUSTRIAL ELECTRONICS

E. U. Condon and E. G. F. Arnott

1·1 WHAT IS ELECTRONICS?

ELECTRONICS is that branch of physics or electrical engineering which deals with the phenomena of the flow of electric currents through vacuum, gases, and vapors. This chapter opens the subject with a review of the physical background of electronics. The processes by which electricity passes from one body to another through high vacuum or through gases are more numerous than the ways in which electricity passes through a wire, and these processes are more intimately bound up with the fundamental nature of matter. In fact the study of the processes which are put to work in this branch of electrical engineering has been one of the chief tools of the physicist in learning what we know today about electrons, protons, atomic structure, and related topics.

1·2 KINETIC THEORY OF GASES[1]

According to modern physics a gas is made up of a large number of molecules which occupy only a small fraction of the total space taken up by the gas. These molecules are independent units which move about with very little influence on each other except when they collide—like billiard balls on a billiard table. For many purposes it is not necessary to specify the internal structure of the molecules themselves.

When the gas is at a particular temperature and density, its molecules move at random. At any instant some are moving fast, others slowly and in different directions, and the particular behavior of any one changes rapidly as it collides with other molecules or with the walls of the containing vessel. The most outstanding mechanical property is that the gas exerts a pressure on the walls of the container, and this is interpreted to be due to the large number of molecular impacts against the wall. The average pressure on the walls is equal to two thirds of the total kinetic energy of translatory motion of the molecules contained in unit volume.

For gases at low enough pressures, and at high enough temperatures (far enough from the regime of the liquid state), the pressure is proportional to the density of the gas and to the absolute temperature, and inversely proportional to the molecular weight. This is expressed by

$$p = \rho \frac{RT}{M} \qquad (1\cdot1)$$

where T is the absolute temperature, ρ is the density, p is the pressure, M is the molecular weight, and R is a universal constant. In both engineering and physics in this field it is customary to use cgs units. That practice is followed here with convenient reference to other unit systems.

If t is the temperature on the Fahrenheit scale, then T, the absolute temperature in centigrade degrees, is

$$T = \tfrac{5}{9}t + 256$$

Density usually is given in grams per cubic centimeter. The density of air at 760 millimeters of mercury pressure and at 32 degrees Fahrenheit is 0.001293 grams per cubic centimeter —about 1/800 the density of water. Molecular weight is given on the usual chemical scale, which adopts 16 as the atomic weight of atomic oxygen. Then if p is measured in dynes per square centimeter, the gas constant R is 8.314×10^7 ergs per degree per mole.

A wide variety of pressure units is used in practice. These are related as follows.

Unit	Value in Dynes per Square Centimeter
Dynes per square centimeter	1
Micron = 10^{-3} mm Hg	1.333
Millimeters of mercury at 0°C	1.333×10^3
Inches of water at 4°C	2.491×10^3
Pounds per square inch	6.895×10^4
Bar	10^6
Atmosphere	1.0133×10^6

In practical electronics, pressures usually are expressed in millimeters of mercury, or in microns, but in theoretical work the dyne per square centimeter is more common.

All gases (at conditions far removed from the liquid state) at the same temperature and pressure contain the same number of molecules per unit volume. A quantity of material whose mass in grams is equal to the molecular weight is called a *mole*, and a mole contains the same number of molecules of any substance. This number is known as Avogadro's number and is equal to 6.0228×10^{23}. This is, for example, the number of molecules contained in 32 grams of oxygen gas, a little more than 1 ounce.

The molecules have a statistical distribution of speeds that is known as the Maxwell distribution law. It is easy to calculate the rms speed of the molecules by the formula

$$u = \sqrt{\frac{3p}{\rho}} = \sqrt{\frac{3RT}{M}} \qquad (1\cdot2)$$

where the units are as in equation 1·1. The velocity of a sound wave in a gas is given by

$$\sqrt{\frac{\gamma p}{\rho}} = \sqrt{\frac{\gamma RT}{M}} \qquad (1\cdot3)$$

where $1 < \gamma < 5/3$ for all gases, this being the ratio of specific heat at constant pressure to that at constant volume. Hence for all gases the rms speed of the molecules is somewhat greater than the velocity of sound in the same gas. Some typical values (at 15 degrees centigrade) are

Gas	u (in 10^3 cm/sec)
H_2	188.8
He	133.6
Ne	59.68
Air	49.82
Hg vapor	18.93

To convert to feet per second, multiply the listed values by 32.8, thus getting about 1630 feet per second for air, the velocity of sound being about 1100 feet per second.

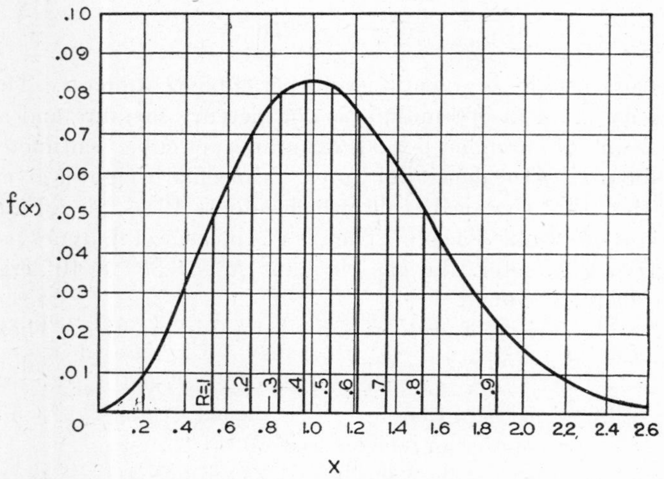

FIG. 1·1 Maxwellian distribution $f(x) = (4/\sqrt{\pi})x^2e^{-x^2}\,dx$ ($dx = 0.1$). Shows the distribution of the speeds among a group of molecules. The values of R are the fractions of molecules with speeds less than that determined by the corresponding value of x, where x is the ratio of the speed to the most probable speed.

The relative numbers of molecules having various speeds are indicated in Fig. 1·1, which shows that relatively few molecules go either very slowly or very fast compared with the rms speed.

Deflection of molecules by collision decreases the rate at which one gas diffuses into another in spite of the great molecular velocities previously mentioned. As a particular molecule moves about, it may have several collisions close together in space; then, again, it may go a relatively great distance before the next collision. These distances between collisions are called the free paths, and the average is called the mean free path. Evidently the mean free path varies inversely as the number of molecules in unit volume and inversely as the cross-sectional area of a molecule, for it is this target area that determines whether a collision takes place. This distribution of free paths is shown in Fig. 1·2.

For most purposes the mean free path l_m (in centimeters) may be calculated in terms of N, the number of molecules in unit volume (cubic centimeters), and the effective diameter d (in centimeters) of an equivalent spherical molecule by the Clausius formula:

$$l_m = \frac{1}{\sqrt{2}N\pi d^2} \qquad (1\cdot4)$$

Experimental information concerning l_m and d is obtained by developing theories about heat conduction, viscosity, and diffusion of gases. The rates of all these processes are determined primarily by the fact that molecular collisions interfere by scattering the molecules before they go far at ordinary pressures. For the study of electronics the details of this part of the theory are not especially important. Rather diverse kinds of measurements on gases are correlated in this way and lead to consistent experimental values of the collision diameters of molecules.

For ordinary purposes the following table of mean free paths in gases at 15 degrees centigrade and a pressure of

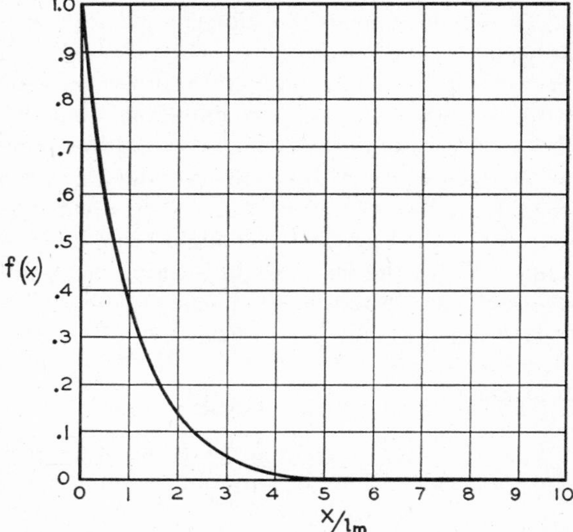

FIG. 1·2 Distribution of free paths $f(x) = e^{-x/l_m}$. Shows fraction of molecules which have gone a distance x without collision; x = length of free path; l_m = mean free path.

1 atmosphere, and the values of the effective molecular diameters as obtained from measurements on gas viscosity, will give all that is needed:

Gas		l_m (10^{-6} cm)	d (10^{-8} cm)
Hydrogen	H_2	11.77	2.74
Helium	He	18.62	2.18
Neon	Ne	13.22	2.59
Argon	A	6.66	3.64
Krypton	Kr	5.12	4.16
Xenon	X	3.76	4.85
Methane	CH_4	5.16	4.14
Nitrogen	N_2	6.28	3.75
Oxygen	O_2	6.79	3.61
Methyl bromide	CH_3Br	2.58	5.85
Mercury	Hg	8.32 at 219.4°C	4.26

In general the pressure in high-vacuum electronic devices is about 10^{-6} atmosphere or less, so the molecules go, on the average, several centimeters between collisions. In such devices the molecules usually go from wall to wall with little chance of striking another molecule. The mean free path of an electron moving among the molecules of a gas plays an important role in determining the way in which the electric discharge takes place in the gas and is $4\sqrt{2}$ times the mean free path for molecules.

1·3 THE ATOMIC NATURE OF ELECTRICITY [2]

In the older experiments on electrostatics and electric currents there is no indication that electric charge is anything but a continuous fluid—just as one does not recognize the molecular structure of water from the study of hydraulics. The existence of electrons and their properties, although interesting, therefore hardly plays a role in the classical field of electrical engineering. This is entirely different in electronics, which is founded on the detailed consideration of the elementary atomic units of charge.

The ways in which elementary substances combine to form chemical compounds can be explained by saying that matter consists of atoms of definite elements. Similarly, the first evidence that electric charge always occurs in definite units came with the discovery that, in electroplating, a definite amount of electric charge is needed to plate a gram mole of a univalent metal. This amount of charge is called the faraday and it is equal to 96,500 coulombs.

This is interpreted by saying that the univalent metal ions in solution, as for example each silver ion in silver nitrate solution, bear a definite electric charge such that the total charge carried by a gram mole of them, that is 6.0228×10^{23} of them, is equal to 96,500 coulombs. Therefore the charge on each of them is

$$e = \frac{96{,}500}{6.0228 \times 10^{23}} = 1.60 \times 10^{-19} \text{ coulomb} \qquad (1\cdot5)$$

It is generally believed *that all electric charge occurs in amounts that are multiples of this basic unit*, which is called the electronic charge.

Studies first made in the 1890's by J. J. Thomson on deflection of cathode-ray beams by electric and magnetic fields, in tubes that were primitive forerunners of the modern cathode-ray tube, indicated that such beams behaved like a stream of charged particles moving according to known mechanical laws. From these experiments the ratio of the charge to the mass of each particle could be determined. The modern accepted value for this quantity is

$$\frac{e}{m} = 1.76 \times 10^{8} \text{ coulombs per gram} \qquad (1\cdot6)$$

This is a large value compared with the same quantity for the lightest atom, hydrogen, for which

$$\left(\frac{e}{m}\right)_{\text{hydrogen}} = 0.958 \times 10^{5} \text{ coulombs per gram} \qquad (1\cdot7)$$

These results are interpreted by supposing the cathode-ray particles to have the same charge as the hydrogen ions involved in electrolysis, but to have a mass that is $1/1837$ of the mass of the hydrogen. These little negatively charged bits of matter are called *electrons*. According to modern views all matter is made up of atoms, which are themselves constructed of electrons and positively charged central nuclei that are relatively massive. Because of their small mass compared with the natural units of positive charge, electrons generally move much more rapidly in electric fields, and so usually electric currents are predominantly due to motions of these negative charges, although this is by no means always the case.

Positive ions in gas discharges have been studied, but never have shown particles of large e/m like that of electrons. The values are always comparable to the values of e/m appearing in electrolytic ions.

On the basis of this kind of evidence we conclude that negative electric charge comes in small units called electrons whose mass is about $1/1837$ that of hydrogen, the lightest atom known. Positive electric charge is always associated with masses of the same magnitudes as those of the atoms of matter, the magnitudes of the charge being the same as of the electron. (This statement overlooks the existence of the positron, a particle with positive charge and mass the same as the electron, which is observed only in experiments involving potentials of millions of volts or in nuclear transformations or cosmic-ray studies and which thus far has played no role in industrial electronics.)

1·4 THE STRUCTURE OF ATOMS [3]

The modern views of atomic structure are built on the ideas of Rutherford and Bohr, which were stated about 1912. According to their views, each atom consists of a central positively charged nucleus, which contains most of the mass of the atom, surrounded by enough electrons to make the atom electrically neutral. The atoms of the different chemical elements are distinguished principally by the number of such electrons and correspondingly by the magnitude of the positive charge on the nucleus.

The massive central nucleus is itself a composite structure made up of two kinds of particles. These are *protons*, which are simply the positively charged nuclei of ordinary hydrogen atoms, and *neutrons*. Neutrons are particles having almost the same mass as protons, but having no electric charge. The nature of the forces which hold these particles together within the nucleus is not fully known, and it forms one of the important problems of *nuclear physics*, which is the investigation of the properties of the atomic nucleus.[4]

Although changes in the nucleus can be produced artificially by special high-voltage bombardment with positive ions, and although they occur spontaneously in naturally radioactive materials, they have played no role up to now in industrial electronics and will not be discussed here. It will be noted in passing that one can produce artificially radioactive materials which are destined to play a great role in industrial chemistry and in medicine in the future.

The number of electrons in the atom is called its atomic number, and each chemical element is distinguished by a particular atomic number. These are given in Table 1·1 in which the known elements are arranged in a periodic table. The goal of atomic theory, which is well on its way to having been achieved, is to correlate all physical and chemical properties in terms of the detailed working out of this picture of atomic structure.

Table 1·1

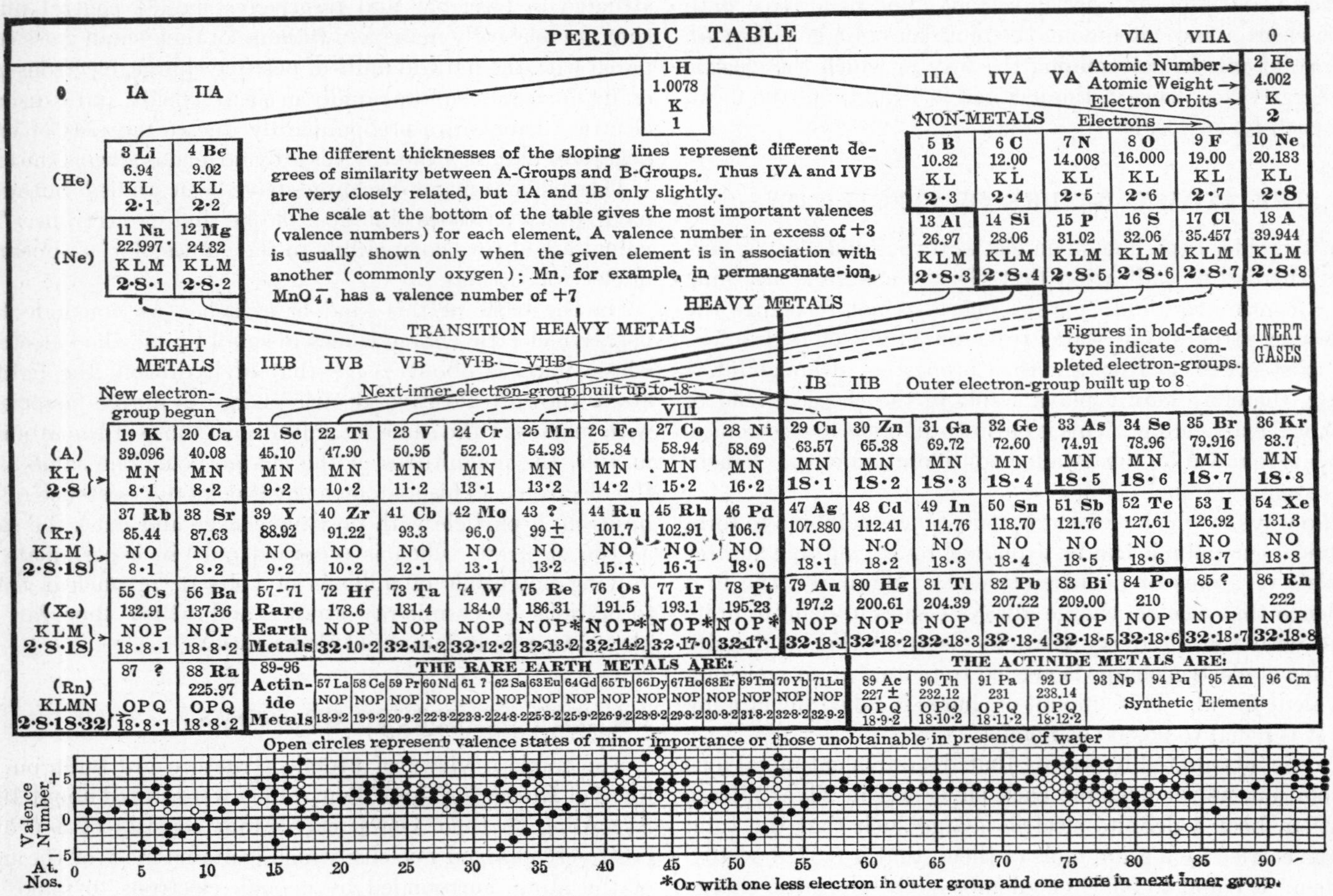

PERIODIC TABLE

The different thicknesses of the sloping lines represent different degrees of similarity between A-Groups and B-Groups. Thus IVA and IVB are very closely related, but 1A and 1B only slightly.

The scale at the bottom of the table gives the most important valences (valence numbers) for each element. A valence number in excess of +3 is usually shown only when the given element is in association with another (commonly oxygen); Mn, for example, in permanganate-ion, MnO_4^-, has a valence number of +7

Atomic Number → / Atomic Weight → / Electron Orbits → / Electrons →

LIGHT METALS — New electron-group begun. TRANSITION HEAVY METALS — Next-inner electron-group built up to 18. HEAVY METALS — Outer electron-group built up to 8. NON-METALS. INERT GASES.

Figures in bold-faced type indicate completed electron-groups.

0	IA	IIA	IIIB	IVB	VB	VIB	VIIB	VIII	VIII	VIII	IB	IIB	IIIA	IVA	VA	VIA	VIIA	0
									1 H 1.0078 K 1									2 He 4.002 K **2**
(He)	3 Li 6.94 KL **2**·1	4 Be 9.02 KL **2**·2											5 B 10.82 KL **2**·3	6 C 12.00 KL **2**·4	7 N 14.008 KL **2**·5	8 O 16.000 KL **2**·6	9 F 19.00 KL **2**·7	10 Ne 20.183 KL **2·8**
(Ne)	11 Na 22.997 KLM **2·8**·1	12 Mg 24.32 KLM **2·8**·2											13 Al 26.97 KLM **2·8**·3	14 Si 28.06 KLM **2·8**·4	15 P 31.02 KLM **2·8**·5	16 S 32.06 KLM **2·8**·6	17 Cl 35.457 KLM **2·8**·7	18 A 39.944 KLM **2·8·8**
(A) KL **2·8**	19 K 39.096 MN 8·1	20 Ca 40.08 MN 8·2	21 Sc 45.10 MN 9·2	22 Ti 47.90 MN 10·2	23 V 50.95 MN 11·2	24 Cr 52.01 MN 13·1	25 Mn 54.93 MN 13·2	26 Fe 55.84 MN 14·2	27 Co 58.94 MN 15·2	28 Ni 58.69 MN 16·2	29 Cu 63.57 MN **18**·1	30 Zn 65.38 MN **18**·2	31 Ga 69.72 MN **18**·3	32 Ge 72.60 MN **18**·4	33 As 74.91 MN **18**·5	34 Se 78.96 MN **18**·6	35 Br 79.916 MN **18**·7	36 Kr 83.7 MN **18·8**
(Kr) KLM **2·8·18**	37 Rb 85.44 NO 8·1	38 Sr 87.63 NO 8·2	39 Y 88.92 NO 9·2	40 Zr 91.22 NO 10·2	41 Cb 93.3 NO 12·1	42 Mo 96.0 NO 13·1	43 ? 99 ± NO 13·2	44 Ru 101.7 NO 15·1	45 Rh 102.91 NO 16·1	46 Pd 106.7 NO 18·0	47 Ag 107.880 NO 18·1	48 Cd 112.41 NO 18·2	49 In 114.76 NO 18·3	50 Sn 118.70 NO 18·4	51 Sb 121.76 NO 18·5	52 Te 127.61 NO 18·6	53 I 126.92 NO 18·7	54 Xe 131.3 NO 18·8
(Xe) KLM **2·8·18**	55 Cs 132.91 NOP 18·8·1	56 Ba 137.36 NOP 18·8·2	57–71 Rare Earth Metals	72 Hf 178.6 NOP **32**·10·2	73 Ta 181.4 NOP **32**·11·2	74 W 184.0 NOP **32**·12·2	75 Re 186.31 NOP* **32**·13·2	76 Os 191.5 NOP* **32**·14·2	77 Ir 193.1 NOP* **32**·17·0	78 Pt 195.23 NOP* **32**·17·1	79 Au 197.2 NOP **32**·18·1	80 Hg 200.61 NOP **32**·18·2	81 Tl 204.39 NOP **32**·18·3	82 Pb 207.22 NOP **32**·18·4	83 Bi 209.00 NOP **32**·18·5	84 Po 210 NOP **32**·18·6	85 ? NOP **32**·18·7	86 Rn 222 NOP **32·18·8**
(Rn) KLMN **2·8·18·32**	87 ? OPQ 18·8·1	88 Ra 225.97 OPQ 18·8·2	89-96 Actin- ide Metals															

THE RARE EARTH METALS ARE:

57 La	58 Ce	59 Pr	60 Nd	61 ?	62 Sa	63 Eu	64 Gd	65 Tb	66 Dy	67 Ho	68 Er	69 Tm	70 Yb	71 Lu
NOP	NOP	NOP	NOP	NOP	NOP	NOP	NOP	NOP	NOP	NOP	NOP	NOP	NOP	NOP
18·9·2	19·9·2	20·9·2	22·8·2	23·8·2	24·8·2	25·8·2	25·9·2	26·9·2	28·8·2	29·8·2	30·8·2	31·8·2	32·8·2	32·9·2

THE ACTINIDE METALS ARE:

89 Ac	90 Th	91 Pa	92 U	93 Np	94 Pu	95 Am	96 Cm
227 ±	232.12	231	238.14	Synthetic Elements			
OPQ	OPQ	OPQ	OPQ				
18·9·2	18·10·2	18·11·2	18·12·2				

Open circles represent valence states of minor importance or those unobtainable in presence of water

Valence Number: +5, 0, −5. At. Nos.: 0, 5, 10, 15, 20, 25, 30, 35, 40, 45, 50, 55, 60, 65, 70, 75, 80, 85, 90

*Or with one less electron in outer group and one more in next inner group.

Reprinted with permission from *Fundamental Chemistry*, 2nd ed., by H. G. Deming, published by John Wiley, 1947.

The atom is by no means a simple system in which the electrons move around the nucleus according to Newtonian laws of motion obeying the usual forces between electric charges. On the contrary it has been learned during the past 30 years that the behavior of electrons in atoms and molecules is governed by a different set of physical laws having some surprising characteristics to one who is versed in the mechanics of ordinary bodies.

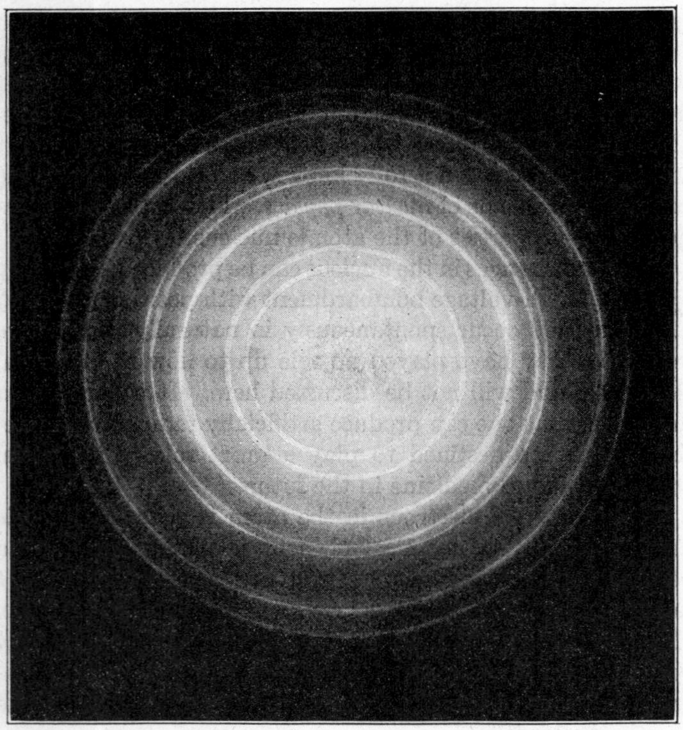

The mechanical principles applicable in atomic theory are known as *quantum mechanics*.[5] This forms an extensive part of modern physics. One of the most extraordinary things about a beam of electrons is that it behaves in some ways as if it were a wave motion instead of a stream of particles. In fact there is no proof that a beam of electrons is one or the other, for it has some of the attributes of a stream of particles and some of those of a wave motion. This peculiar duplex character is one of the most puzzling things in modern physics, and the nature of a cathode-ray beam is not completely understood in spite of the fact that enough is known to cover all practical cases likely to arise.

To be more explicit, a cathode-ray beam behaves like a stream of particles in the way it is deflected by electric or magnetic fields. In fact it was by such calculations that e/m was measured experimentally. But it behaves like a wave motion when it is scattered by going through a thin plate of crystalline material. Figure 1·3 shows the pattern produced by allowing electrons from an originally sharp beam to fall on a photographic plate after it has traversed a thin film of cobaltous oxide. The sharp rings in the picture show that the electrons are scattered in definitely determined directions. This phenomenon is known as *electron diffraction*.[6]

Fig 1·3 Electron diffraction photograph of cobaltous oxide.

Electron diffraction can be interpreted by assuming that if the electrons, regarded as moving particles with velocity v centimeters per second, are scattered by a crystal lattice they are scattered as if they were a wave motion of wavelength,

$$\lambda = \frac{h}{mv} \tag{1·8}$$

in which h is Planck's constant (6.62×10^{-27} erg-second) and m is the mass of the electron in grams. In a more practical and convenient form,

$$\lambda = \sqrt{\frac{150}{V}} \times 10^{-8} \text{ centimeter} \tag{1·9}$$

in which the electrons have the speed acquired by being accelerated from rest through a potential drop of V volts. Therefore the wavelength associated with electrons accelerated by 15 kilovolts is 10^{-9} centimeter, which is smaller than the separation of the regular layers of atoms in any crystal.

This is just one example of wave phenomena in the quantum mechanics of electronic motion, but it is an important one because it indicates that electronic behavior does not always follow classical mechanical laws.

1·5 ATOMIC ENERGY LEVELS [7]

The most outstanding characteristic of atoms and molecules is that they can exist only in certain sharply defined states of total energy. This is in sharp contradiction to ordinary mechanical systems. For example, a flywheel on a shaft may turn with any angular velocity, depending on starting conditions, and hence may store any amount of rotational kinetic energy.

But in the mechanics of atoms and molecules only particular energy values are allowed, and the system cannot exist in a state in which its total internal energy is different from one of these allowed values. The particular allowed values for an atom are a characteristic of that atom and have much to do with determining its other physical and chemical properties.

The state of least total energy for a particular atom is called its *normal state;* the higher energy levels are called *excited states.* The first state above the normal state is often called the *resonance level.*

These levels determine the wavelengths of light and x-rays which the atom is capable of emitting. If by some means an atom is in an excited state of which the total energy is W_1, it can in general make a transition to a state of lesser total energy W_2 and in so doing emit light of frequency ν (cycles per second) given by the relation

$$h\nu = W_1 - W_2 \tag{1·10}$$

where h is again Planck's constant. This relation affords a means of determining energy levels by analysis of the frequencies of the light in the spectrum of the material.

This existence of discrete energy levels for atoms is responsible for their behaving like elastic billiard balls under ordinary conditions in a gas. If an atom is in its normal state and collides rather gently with another atom, then, if the energy available is insufficient to put the atom in an excited state, it must leave the atom unaffected after the collision, since intermediate values are not allowed.

The same thing is true of excitation of atoms by electron impact. If an electron strikes an atom at a speed such that its energy is too small to raise the atom to the first excited level, then it is unable to give to the atom any energy (except for a small change in the translational speed of the atom as a whole due to its recoil, but this is small because the atom is so much more massive than the electron).

As an atom rises to states of higher and higher energy this usually corresponds to a sequence of states in which one of the electrons of the atom is less firmly bound to the atom. Finally there is a limiting state of energy corresponding to the complete removal from the atom of one of its electrons, leaving a singly charged positive ion, an atom with one electron less than the number needed to make it electrically neutral. Above this energy comes a sequence of levels corresponding to looser and looser binding of a second electron of the atom until finally there is a state corresponding to removal of two of the atomic electrons to give a doubly charged positive ion, and so on.

Just as emission of light is associated with a transition from a higher state to a lower state of energy, so also if an atom is in a lower state W_2 it can be raised to a higher state W_1 by absorption of light of frequency ν given by equation 1·10. If the frequency is high enough, the absorption of light may raise the atom to a state corresponding to ejection of one of the atomic electrons, leaving a positive ion. This process is known as photo-ionization and is important for the conduction of electricity in gases.

Yellow light has a wavelength of about 6000×10^{-8} centimeter and hence $\nu = 5 \times 10^{14}$ cycles per second, and so $h\nu = 3.38 \times 10^{-12}$ erg. This gives the magnitude of the difference in the energy levels of an atom involved in the emission of yellow light. A more convenient way of expressing the levels for electronics is to express their energy in terms of electron volts. An electron volt is the amount of energy acquired by an electron on being accelerated freely through a potential difference of 1 volt. This is 1.60×10^{-19} watt-second or 1.60×10^{-12} erg. Hence the energy difference involved in emission or absorption of yellow light is 2.06 electron volts.

An electron which has been accelerated through V volts (and so has V electron volts of energy) has $1.60 \times 10^{-12}V$ erg of energy and is therefore capable of raising an atom to a state from which, by dropping back to its normal state, it could emit radiation of wavelength λ such that $hc/\lambda = 1.60 \times 10^{-12}V$ (where $c = 3 \times 10^{10}$ centimeters per second), or $\lambda = 12{,}395/V$ (where λ is in units of 10^{-8} centimeter). Thus an atom must be excited by electrons having more than 1 volt of energy to emit visible radiation, and 10 volts suffices to make possible radiation in the deep ultraviolet. This relation holds also in the x-ray region and shows that thousands of volts are needed to get x-rays of wavelengths less than 10^{-8} centimeter.

Since in electronics the electron volt is the most convenient unit of energy, the ionization energy of an atom is sometimes called the ionization potential. The ionization potential of an atom or molecule is the voltage drop through which an electron must be freely accelerated to acquire enough energy to be able to ionize the atom or molecule by removal of one of its electrons. Generally the elements most electropositive in the chemical sense are the ones with the lowest ionization potential. The potentials corresponding to excitation of the atom to excited states are called the critical potentials of the atom.

It is important to know the critical potentials of the atoms and molecules involved in gaseous discharges. Some of the principal values (in volts) are:

Gas	Excitation	Ionization
Helium	19.73 volts	24.47 volts
Neon	16.60	21.47
Argon	11.57	15.69
Krypton	9.9	13.94
Xenon	8.3	12.08
Mercury	4.66	10.39
Sodium	2.09	5.12
Potassium	1.60	4.32
Hydrogen (molecular)	11.0	15.4

It is convenient to indicate energy levels of an atom by an *energy-level diagram* in which levels are indicated by horizontal lines with reference to a vertical scale of energy values.

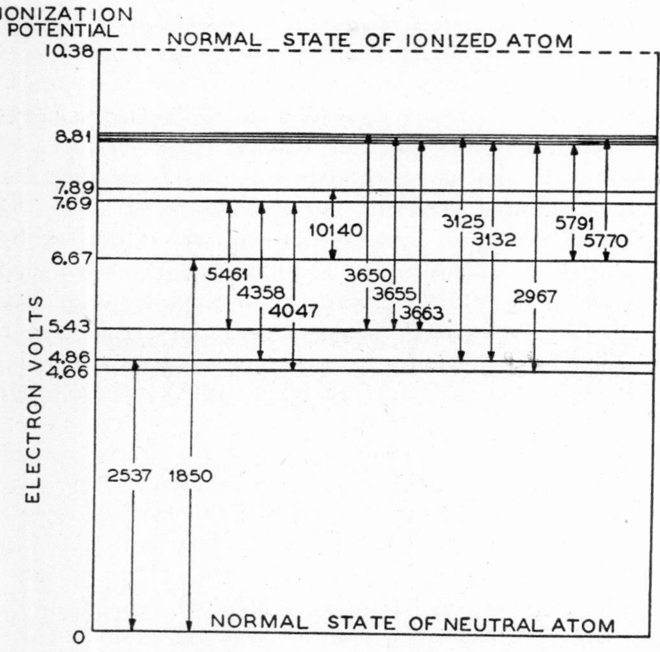

FIG. 1·4 Energy-level diagram of the lower states of mercury.

Then the different lines in the spectrum can be indicated by vertical lines showing energy levels between which the transition responsible for each line occurs. Figure 1·4 is an example of such a diagram. It is for mercury vapor, chosen because of the importance of mercury-vapor discharges in industrial electronics.

The atom is unlikely to make transitions from some of its excited states. These are called *metastable states.*[8] If the atom is put into a metastable state by any means it remains in such a state for a relatively long time until it gives its energy to some other atom or to the walls in a collision. Ordinary excited states of the atom do not last more than about 10^{-8} second before the atom returns to a lower state by emitting radiation. Another interesting phenomenon is that of *super-elastic* collisions: A previously excited atom struck by an electron may return to its normal state by giving up its excitation energy to the impacting electron, which therefore goes away with more energy than it originally had. These are important in some gas discharges.

Some gas atoms and molecules can attach an extra electron to themselves in a stable way, making *negative ions.*[9] Negative charge in this form behaves differently from free electrons. Materials known in chemistry as strongly electronegative do this most readily; for example, chlorine gas and carbon tetrachloride vapor. This accounts for the quenching action of such gases on corona discharge.

Some molecules can be dissociated into smaller fragments either with or without ionization. This is the basis of chemical reactions produced in certain discharges, such as the formation of ozone in a corona discharge in air.[10]

1·6 IONIZATION FUNCTIONS

In considering excitation and ionization of atoms or molecules by electron impact, a distinction must be made between whether a process can happen in the sense of there

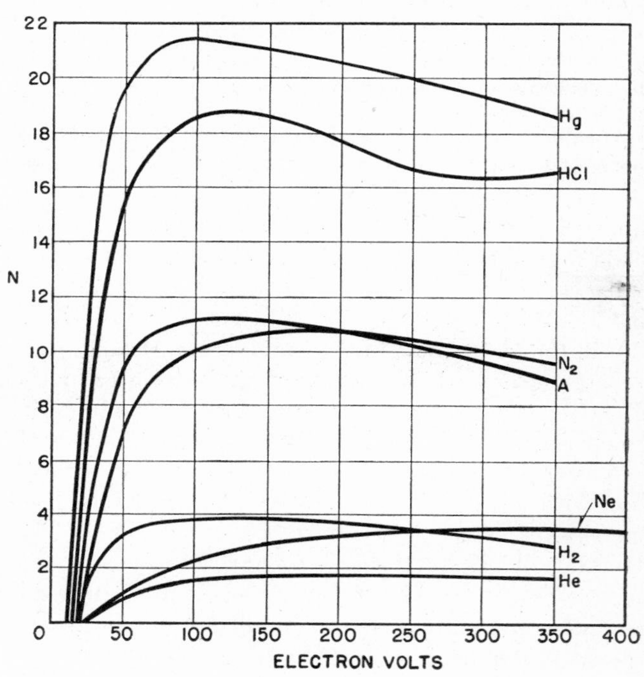

FIG. 1·5 Ionization functions for common atoms. Number of ions formed per centimeter path per millimeter pressure at 25°C. (Reprinted with permission from Compton and Van Voorhis, *Phys. Rev.*, Vol. 27, p. 724, 1926.)

being enough energy to permit it and the likelihood of its actually happening.

If an electron is moving with a certain energy V (measured in volts) through a gas containing N molecules per cubic centimeter, then the average distance it will go before experiencing a collision resulting in ionization of the atom is called the mean free path for ionization. This is given by

$$\frac{1}{NS}$$

where S (in square centimeters) is a quantity characteristic of the atom called its effective cross section for ionization. The larger S is, the more likely is the electron to strike one of the atoms. S depends on the energy of the electron and is zero for values of the energy less than the ionization potential. S as a function of energy is known as the *ionization function.* This has been determined experimentally for several gases, and the results are shown in Fig. 1·5. It is possible, though less likely, for an electron of sufficient energy

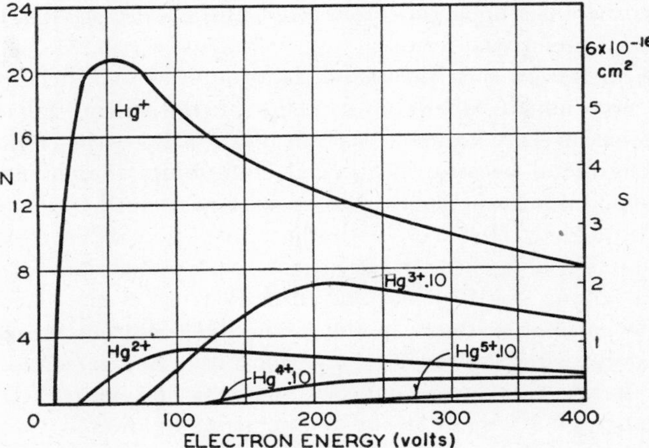

FIG. 1·6 Ionization functions for mercury. Number of singly, doubly, triply, quadruply, and quintuply charged mercury ions formed per electron per centimeter path per millimeter pressure at 0°C and effective cross-sectional area of the mercury atom. (Reprinted with permission from Walker Bleakney, *Phys. Rev.*, Vol. 35, p 145, Jan. 15, 1930.)

to knock off several electrons at one impact, forming a multiply charged positive ion. These processes have been particularly studied for mercury vapor by Bleakney whose results are shown in Fig. 1·6.

The general magnitude of the effective cross section is what would be expected for collision with an atom that is about 10^{-8} centimeter in diameter, which is the magnitude found from kinetic theory. S becomes very small for high-energy electrons. Although they have plenty of energy to ionize, they apparently move through the atom too fast to be effective in producing ionization. The penetrating power of high-energy electrons is foreshadowed in these results.

1·7 ELECTRONIC THEORY OF SOLIDS [11]

Quantum mechanical theory has also greatly extended our fundamental knowledge of the behavior of electrons in solid materials. The mathematical difficulties are very great, so this part of the subject is less developed than the theory for isolated atoms.

Electrically the most important distinction among solids is that between good conductors which are metals and good insulators which are non-metals. It is desirable to break the classification down further to include a class known as semi-conductors. These are materials like zinc oxide or cuprous oxide whose conductivity is much poorer than ordinary metals but much better than insulators. In contrast with metals the conductivity increases with increasing temperature. The conductivity of semi-conductors is believed to be due in all cases to the effect of impurities or slight departures of the composition from that of the chemical formula. The conductivity values are extremely sensitive to the purity and to the means of preparation of the material.

The non-conductors may be further classified into types as: (*a*) glasses or super-cooled liquids which do not have a well-defined crystal structure; (*b*) ionic crystals, like sodium chloride or magnesium oxide, distinguished by electrolytic conductivity at high temperatures; (*c*) valence crystals, like diamond and quartz, distinguished by great hardness, irregular cleavage, and very low conductivity; and (*d*) molecular crystals, such as the rare gases, hydrogen or methane, when cooled below the melting point.

The electrons in a metal are of two kinds, bound and free. The bound electrons are those which make up the inner shells of the atoms and exist in essentially the same states as when the atom is in the gaseous state. It is transitions between states of these tightly bound electrons that are involved in the emission and absorption of x-rays by matter. The fact that the x-ray spectra are very little changed by chemical combination or changes from gaseous to solid state shows that the inner electrons are not affected by such changes.

The outer electrons, which are the valence electrons when the atom is in chemical combination, are the ones which become the free electrons in the metal. Exact description of their motion is too difficult a problem to have been solved thus far. The general picture given by the theory is something like this:

Consider, for example, metallic copper which in the metallic state is regarded as having one free electron per atom. This means that there would be approximately 8.4×10^{22} free electrons per cubic centimeter. The atoms are arranged in a face-centered cubic lattice (Fig. 1·7) of which the edge of a basic cube is 3.609 angstroms, the distance apart of nearest neighbors being 2.55 angstroms. (An angstrom is 10^{-8} centimeter.) Any electron moves under the influence of

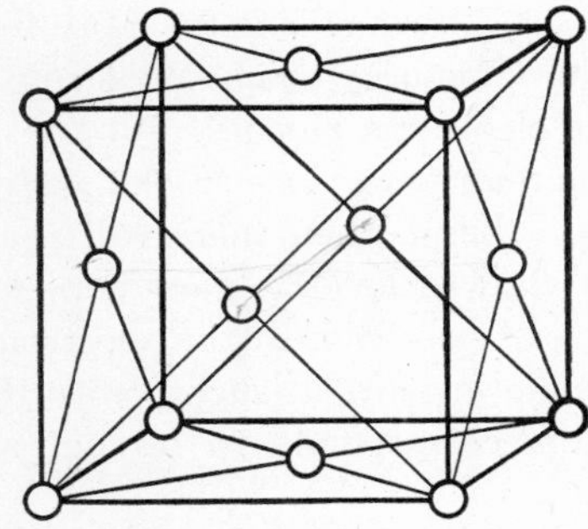

FIG. 1·7 Face-centered cubic lattice.

forces of attraction drawing it to the atoms and of forces of repulsion between itself and the other electrons. The complicated effect of the other electrons can be approximately replaced by a smoothed-out average field of force. Thus it is possible to think of each electron as moving in an average constant field. Instead of speaking of the force field directly it is more convenient to describe the field in terms of $V(x, y, z)$, the potential (volts) of this average electric field acting on the electrons. The potential energy of an electron in this field is then $-eV$, where $e = 1.60 \times 10^{-19}$ coulomb is the magnitude of the charge on the electron. Of course, V exists in three dimensions but its main qualitative properties are indicated in Fig. 1·8 in which the ordinates represent $-V$ which is the potential energy in electron volts. The abscissas indicate distance in the crystal in some plane which includes the nuclei of some of the atoms.

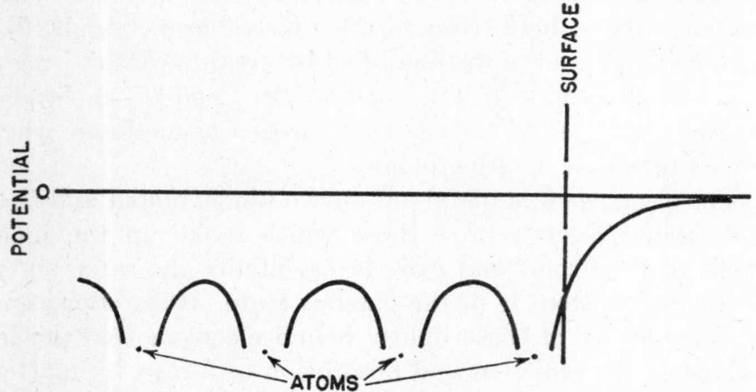

Fig. 1·8 Periodic potential energy of an electron in a metal lattice.

The deep negative valleys in the $-V$ plot are the result of the strong attraction of each nucleus for the electron. Where the curve is level between atoms, the opposite forces due to nuclei in opposite directions are balanced. At the right of the diagram conditions near the surface of the metal are shown. The electron is attracted to the metal as a whole so that the potential energy is higher outside the metal than inside.

According to quantum mechanics only certain total energies are allowed for the electrons, just as in the case of atoms. Here the theory gives the result that the allowed energy levels come in bands. Within a band the permitted levels lie so close together that they seem to form a practically continuous range of allowed values. Outside a band the levels are not allowed.

Another important fundamental rule is the exclusion principle. According to this, in any system in which there are several electrons, it is not possible for more than one electron to be in each allowed energy level. This rule is well founded in many parts of electronic physics.

If there are as many electrons in the system as there are allowed levels in a band, then there will necessarily be one electron in each allowed level. There corresponds a certain average motion of translation of an electron in each level, and the levels come in pairs in such a way that when every level is filled this corresponds to no net average motion of the electrons.

This is the situation which occurs in an insulator. The valence band of levels is completely filled so there is no net motion of the electrons. When a moderate electric field is applied, there is still no net motion of the electrons until the field becomes so great that electrons are caused to go into the allowed levels of a higher band. The higher band, being nearly empty, permits a net motion of electrons in the direction against the field. Some attempts to calculate the breakdown strength of insulators from this point of view have been successful enough to give confidence that the general interpretation is correct.

The concept of a semi-conductor is that this is a material which in an ideal state of purity would be an insulator, having a completely filled electron band. However, the electron band is disturbed by impurities (where slight departure from ideal chemical composition also counts as an impurity: Cu_2O with excess oxygen is therefore regarded as having oxygen impurity). Either this may act to trap some electrons near impurities, leaving the nominally full band not quite filled, or it may act to provide more than enough electrons to fill a band so that these go into the next higher band. In either case there is a departure from exact filledness of a band, and some conductivity results.

A metal, on the other hand, is a material in which there are not enough valence electrons to fill the valence band of allowed levels. Normally there is no average current flowing in the metal because in thermal equilibrium there tend to be equal numbers of electrons moving in each direction just as in a gas. However, with the application of an electric field a few more electrons go into states of motion against the field, giving rise to a net current. Because of the thermal agitation of the atoms in the lattice the electrons are continually being scattered by collisions, and this gives rise to the energy dissipation connected with the finite electrical resistance of a wire. As the temperature is lowered toward absolute zero this thermal agitation of the atoms is greatly reduced, and so the electrons are scattered less and the resistance is reduced.

At absolute zero of temperature the electrons will fill the levels of the allowed band up to some highest energy, as indicated in Fig. 1·9. All the levels lower than W_i are filled and all the levels above W_i are empty. At moderate temperatures this situation is altered. Owing to thermal motions some of the electrons occupy levels higher than W_i, and therefore some of the levels below W_i are not filled. The approximate range of energies over which this occurs is about kT on either side of W_i, where $k = 1.38 \times 10^{-16}$ erg per degree is the Boltzmann constant and T is the absolute temperature in centigrade degrees. Expressed in electron volts the Boltzmann constant is $k = 8.57 \times 10^{-5}$ electron volt per degree, so that at ordinary temperatures ($T = 300$ degrees) kT is about $1/40$ of an electron volt.

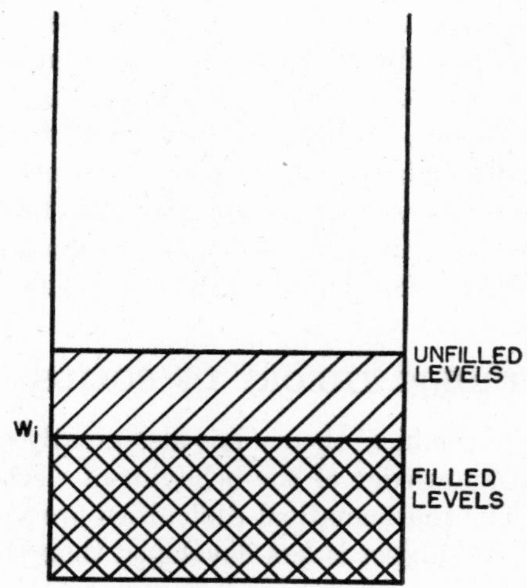

Fig. 1·9 Partly filled electron energy band in a metal.

The energy distribution for the free electrons in a metal is shown in Fig. 1·10 for both 0 degrees and 1500 degrees absolute. The shape of the distribution curve, even for the higher temperature, is very different from the Maxwellian distribution of Fig. 1·1, and this fact plays an important part in the proper explanation of the phenomena of electron emission.

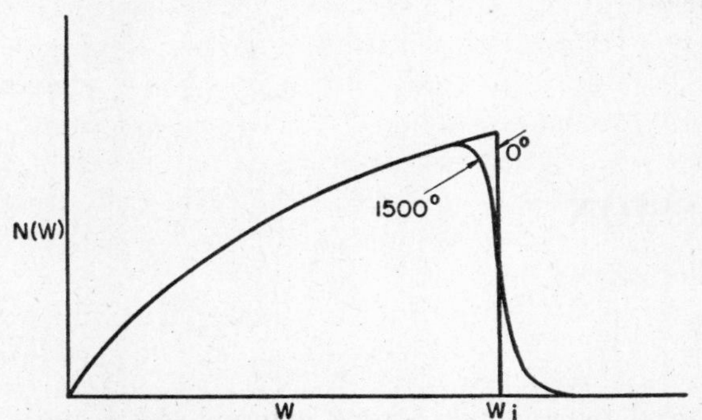

FIG. 1·10 Energy distribution of free electrons in a metal.

REFERENCES

1. *The Nature of a Gas*, L. B. Loeb, Wiley, 1931; *The Kinetic Theory of Gases*, E. H. Kennard, McGraw-Hill, 1938; *The Kinetic Theory of Gases*, L. B. Loeb, McGraw-Hill, 1927; *The Dynamical Theory of Gases*, J. H. Jeans, Cambridge Univ. Press, 1925.
2. *Electrons (+ and −), Protons, Photons, Neutrons and Cosmic Rays*, R. A. Millikan, Univ. of Chicago Press, 1935; *Introduction to Modern Physics*, F. K. Richtmyer, McGraw-Hill, 1928; *The Particles of Modern Physics*, J. D. Stranathan, Blakiston, 1942; *Atomic Structure*, L. B. Loeb, Wiley, 1938.
3. *The Structure of the Atom*, E. N. Andrade, Harcourt, Brace, 1923; *Atomic Physics*, Max Born, Stechert, 1937; *Atomic Structure*, L. B. Loeb, Wiley, 1938.
4. *Elements of Nuclear Physics*, Franco Rasetti, Prentice-Hall, 1936; *Nuclear Physics*, Norman Feather, Cambridge Univ. Press, 1936; *Applied Nuclear Physics*, E. Pollard and W. L. Davidson, Jr., Wiley, 1942.
5. *Quantum Mechanics*, E. U. Condon and P. M. Morse, McGraw-Hill, 1929; *Elements of Quantum Mechanics*, Saul Dushman, Wiley, 1938; *Introduction to Quantum Mechanics*, L. C. Pauling and E. B. Wilson, McGraw-Hill, 1935.
6. *Theory and Practice of Electron Diffraction*, G. P. Thomson and W. Cochrane, Macmillan, 1939.
7. *Atomic Structure and Atomic Spectra*, Gerhard Herzberg, Prentice-Hall, 1937; *The Theory of Atomic Spectra*, E. U. Condon and G. H. Shortley, Cambridge Univ. Press, 1935.
8. *Resonance Radiation and Excited Atoms*, A. C. G. Mitchell and M. W. Zemansky, Cambridge Univ. Press, 1934.
9. *Negative Ions*, H. S. W. Massey, Cambridge Univ. Press, 1938.
10. *Electrochemistry of Gases and Other Dielectrics*, G. Glockler and S. C. Lind, Wiley, 1939.
11. *The Modern Theory of Solids*, Frederick Seitz, McGraw-Hill, 1940.

Chapter 2

ELECTRON EMISSION

J. W. McNall

BECAUSE of its important duty, the electron source (usually referred to as the cathode) can be considered the heart of every electronic tube. The phenomenon of taking electrons from the external circuit and making them available as free electrons, to whatever vacuum, gas-, or vapor-filled region is desired, is defined as *electron emission.* In order to use electronic tubes intelligently, it is essential to be familiar with the characteristics of the various electron emitters and the theories of electron emitting processes because they determine many of the tube limitations and capabilities.

2·1 THEORY OF ELECTRON EMISSION

There are four general classifications of electron emitters: (1) thermionic, (2) field, (3) photoelectric, and (4) secondary emitters. Although these four types of electron emitters are radically different in their physical and operational characteristics, their mechanisms of emission are similar from the standpoint of the electron theory of metals and semiconductors.

As shown in Chapter 1, potential-energy distribution in a metal can be represented by Fig. 1·8. At a temperature of absolute zero, the maximum potential energy of an electron is W_i, the distribution of electrons having energies below W_i being as shown in Fig. 1·10. An electron outside the metal (at infinity) has potential energy W_a. The difference between W_a and W_i is known as the work function ϕ. It is different for different metals; hence it is a characteristic of the metal. The values of potential energy, kinetic energy, and work function usually are considered in terms of electron volts which are related to ergs by the expression

$$\text{One electron volt} = 1.60 \times 10^{-12} \text{ erg} \qquad (2\cdot1)$$

The work function therefore is the minimum energy that must be imparted to an electron so that it can be emitted from a metal at absolute zero. Although the metal contains many conduction electrons that are free to move throughout its interior, these electrons cannot escape from the metal at normal temperatures because as they arrive at the surface they are confronted by a potential barrier beyond which they cannot pass. To allow an appreciable number of electrons to overcome this potential boundary, sufficient energy must be imparted so that they can overcome the work function of the surface, or the height of the surface barrier must be reduced by applying strong external electrostatic fields to the surface. Thermionic, photoelectric, and secondary emission depend upon imparting energy to the electrons; field emission results from a reduction in the height of the surface barrier.

2·2 CONTACT DIFFERENCE OF POTENTIAL

In Section 1·7 it was shown that an electron has a lower potential energy between two nuclei than it would have in the absence of one of the nuclei. This results in a potential barrier at the surface as shown in Fig. 1·8. If a second metal is brought near the first, so that their surfaces are separated by only a few interatomic distances, the potential barriers of the two surfaces interact, and the resulting potential barrier between them is as shown in Fig. 2·1. If these two metals (A and B) are different, they have unequal work functions, ϕ_A and ϕ_B, and of course have potential barriers of different heights. If the two surfaces are brought into contact, as in Fig. 2·2, the W_i levels of the two metals adjust themselves to the same value, electrons flowing from the metal A, of lower work function, to the metal B, of higher

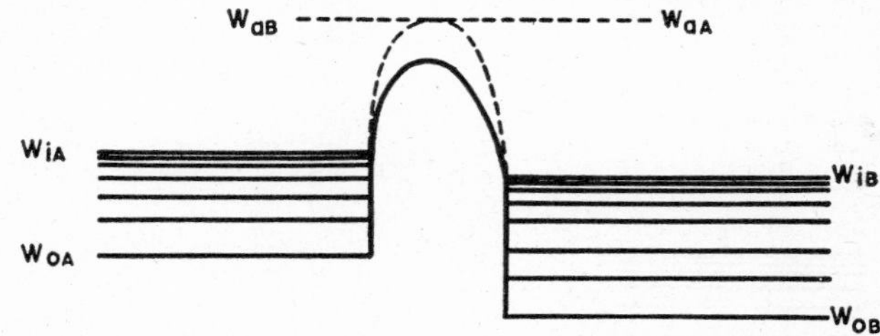

Fig. 2·1 Potential-energy diagram for two metal surfaces nearly in contact.

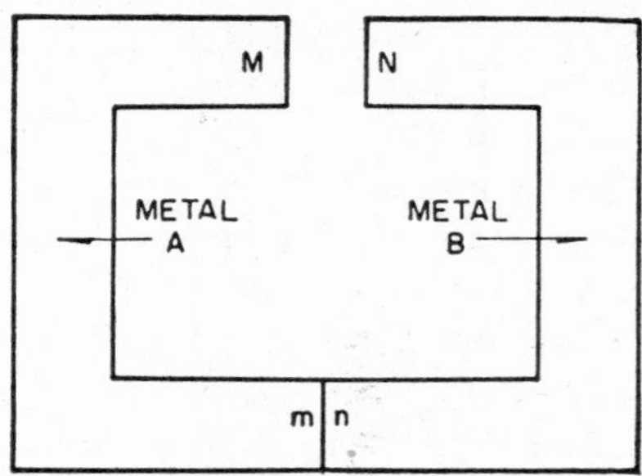

Fig. 2·2 Physical arrangement of two dissimilar metals for explaining contact difference of potential.

work function.[1,2] Figure 2·3 shows the potential-energy diagram for the two metals and their surfaces at the junction m-n. The potential barriers for both surfaces have disappeared. At the gap M-N between the other two ends of metals A and B (see Fig. 2·2), W_{iA} and W_{iB} still coincide, but W_{aA} does not coincide with W_{aB}, and the potential diagram between the two surfaces shown in Fig. 2·4 results.

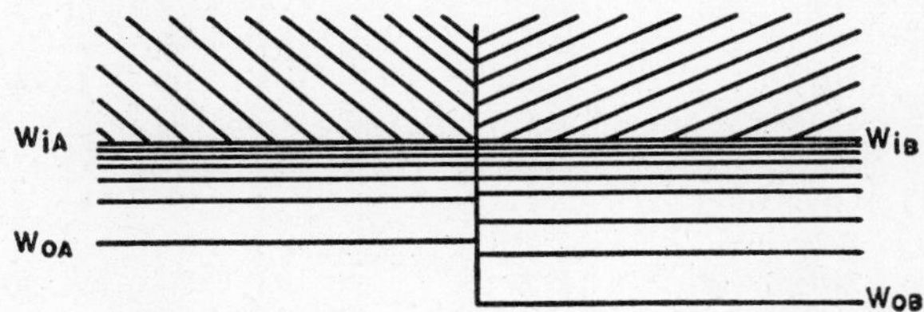

FIG. 2·3 Potential-energy diagram for two metal surfaces in contact (m-n junction of Fig. 2·2).

The dashed line joining W_{aA} and W_{aB} indicates an electrostatic field in the gap M-N. According to the principles of electrostatics, the field represented by the dashed line must be produced by surface charges. These charges were supplied to the surface by the flow of electrons from the metal of lower work function to the one of higher work function (A to B) when the contact was made between the two metals at m-n. At absolute zero, the difference in heights of the

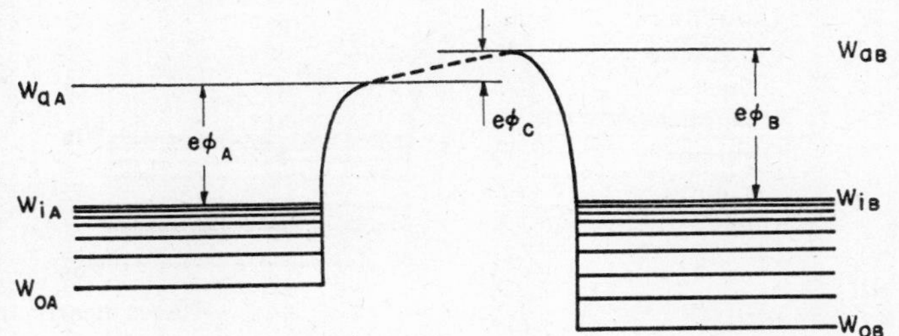

FIG. 2·4 Potential-energy diagram of gap (M-N of Fig. 2·2) between two dissimilar metals.

W_{aA} and W_{aB} levels, expressed in electron volts, is defined as the contact difference of potential ϕ_C. At absolute zero

$$\phi_C = \phi_B - \phi_A \tag{2·2}$$

At higher temperatures the contact difference of potential is only slightly different from its value at absolute zero.[2]

2·3 THERMIONIC EMISSION

The most generally used type of electron emitter is the thermionic cathode.[3] Thermionic emission results from the absorption of sufficient thermal energy by the conduction electrons to enable them to overcome the work function and leave the cathode surface. According to experiment, thermionic emission current increases with emitter temperature, the rate of increase being greater at higher temperatures, as shown in Fig. 2·5. To explain this important characteristic, only those electrons which have velocity components directed perpendicular to the surface of the metal need be considered. Imagine a system of rectangular co-ordinates in which the x-axis is normal to the surface. By using the Fermi-Dirac energy distribution,[4] an expression for the number of electrons having energies between W_x and $W_x + dW_x$ (associated with corresponding values of x-directed velocities) which pass a unit area in the y-z plane in unit time can be

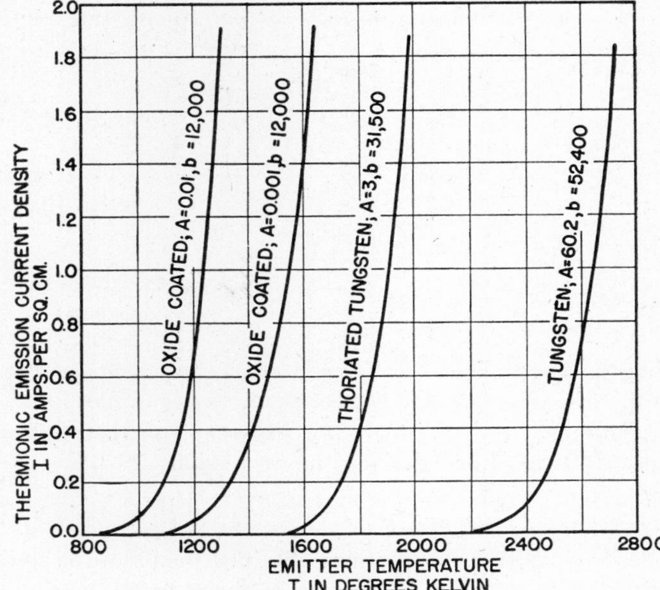

FIG. 2·5 Variation of thermionic emission with temperature. (Reprinted with permission from *Applied Electronics*, by the Electrical Engineering Staff, M.I.T., published by John Wiley, 1943.)

derived. If this number of electrons is designated by $N(W_x)\,dW_x$, then to a good approximation [5]

$$N(W_x)\,dW_x = \frac{4\pi m_e kT}{h^3}\,\epsilon^{-(W_x - W_i)/kT}\,dW_x \tag{2·3}$$

where m_e = the electronic mass
k = Boltzmann's constant
h = Planck's constant
T = the absolute temperature in centigrade degrees
ϵ = the Naperian base.

Figure 2·6 illustrates the variation of the distribution function $N(W_x)$ with associated energy W_x for temperatures of absolute zero and 1500 degrees Kelvin. To surmount the potential barrier and be emitted, however, the electron must have not only an x-directed velocity component, but also a

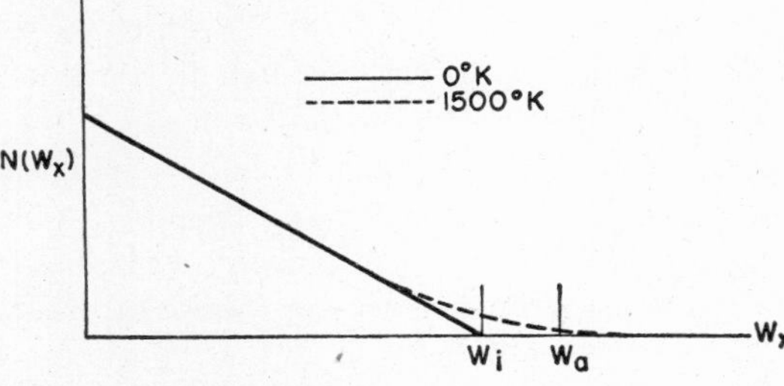

FIG. 2·6 Variation of the distribution function $N(W_x)$ with associated energy W_x.

kinetic energy associated with that velocity at least equivalent to the work function of the surface. By integrating equation 2·3 from $W_x = W_a$ to $W_x = \infty$ an expression for the number of emitted electrons is obtained. The value of this definite integral for 1500 degrees Kelvin is equal to the area under the dotted curve of Fig. 2·6 to the right of the ordinate $W_x = W_a$. Current carried by electrons is equal to the number of electrons emitted per second multiplied by the electronic charge; so the equation for the thermionic emission current density in amperes per square centimeter becomes

$$I = \frac{4\pi m_e k^2 e}{h^3} T^2 \epsilon^{-e\phi/kT} = AT^2 \epsilon^{-b/T} \qquad (2\cdot4)$$

where

$$A = \frac{4\pi m_e k^2 e}{h^3}$$

and

$$b = \frac{e\phi}{k}$$

Using the classical theory of free electrons, in which electrons in a metal are assumed to behave as an ideal gas and to have momenta which obey the Maxwell-Boltzmann distribution function, Richardson [6] derived the expression

$$I = aT^{1/2} \epsilon^{-e\phi/kT} \qquad (2\cdot5)$$

Richardson,[7] Wilson,[8] and others obtained the following similar equation

$$I = AT^2 \epsilon^{-e\phi/kT} \qquad (2\cdot6)$$

by basing their derivation on thermodynamic principles. Dushman [9] also derived an expression for A, sometimes called the thermionic emission constant, which is now known to have the value

$$A = \frac{4\pi m_e k^2 e}{h^3} \qquad (2\cdot7)$$

which is identical to the value of A obtained in the derivation of equation 2·4. Substituting values,

$$A = 120.4 \text{ amp/cm}^2 \text{ deg}^2 \qquad (2\cdot8)$$

Although careful measurements of thermionic emission from pure metals have been made by Langmuir,[10] Nottingham,[11] and others,[12] it is impossible to determine whether equation 2·5 or equation 2·6 is the more consistent with experimental results because the exponential term varies with T much more rapidly than T^2 or $T^{1/2}$. Such measurements, however, determine A to be about 60 for nearly all metals, the one unexplainable exception being platinum, which, according to DuBridge,[13] has a value of 17,000. A more recent determination by Whitney, however, resulted in a value of A of 32 for platinum. Experimental values of A and of the thermionic work function ϕ for several metals are shown in Table 2·1. Many explanations have been proposed to account for the fact that theory predicts a value of A about double that found from experiment. Some investigators [14] believe

Table 2·1 Experimental Value of the Work Function and the Thermionic Emission Constant for Several Metals

Atomic Number	Element	Symbol	Work Function (volts)	Thermionic Emission Constant A
3	Lithium	Li	2.1–2.9	
4	Beryllium	Be	3.16	
6	Carbon	C	2.5–4.7	5.93–30
11	Sodium	Na	1.9–2.46	
12	Magnesium	Mg	2.42	

Table 2·1 Experimental Value of the Work Function and the Thermionic Emission Constant for Several Metals (*Continued*)

Atomic Number	Element	Symbol	Work Function (volts)	Thermionic Emission Constant A
13	Aluminum	Al	2.5–3.6	
19	Potassium	K	1.76–2.25	
20	Calcium	Ca	2.24–3.2	60
26	Iron	Fe	4.72–4.77	
27	Cobalt	Co	4.12–4.28	
28	Nickel	Ni	5.01–5.03	1380–26.8
29	Copper	Cu	4.1–4.5	
30	Zinc	Zn	3.32–3.57	
32	Germanium	Ge	4.3	
37	Rubidium	Rb	1.8–2.19	
40	Zirconium	Zr	4.12	330
41	Columbium	Cb	3.96–4.01	57–37
42	Molybdenum	Mo	4.15	55–60
45	Rhodium	Rh	4.57–4.58	
46	Palladium	Pa	4.96–4.99	60
47	Silver	Ag	4.08–4.76	
48	Cadmium	Cd	4.0	
50	Tin	Sn	4.17–4.50	
55	Cesium	Cs	1.80–1.96	162
56	Barium	Ba	2.11	60
57	Lanthanum	La	3.3	
58	Cerium	Ce	2.6	
59	Praeseodymium	Pr	2.7	
60	Neodymium	Nd	3.3	
62	Samarium	Sa	3.2	
72	Hafnium	Hf	3.53	14.5
73	Tantalum	Ta	4.12	60
74	Tungsten	W	4.52	60
75	Rhenium	Re	5.1	200
76	Osmium	Os	4.7	
78	Platinum	Pt	5.32–6.27	32–17,000
79	Gold	Au	4.42–4.92	
80	Mercury	Hg	4.52	
82	Lead	Pb	3.5–4.1	
83	Bismuth	Bi	4.0–4.4	
90	Thorium	Th	3.3–3.6	70–60
	Compound	**Formula**		
	Calcium oxide	CaO	1.77	129–249
	Strontium oxide	SrO	1.27	4.1–258
	Barium oxide	BaO	0.99	2.9–272
	Combination			
	Barium, Calcium, Strontium oxides	(BaCaSr)O	1.24	0.0083
	Cesium on tungsten	W-Cs	1.36	3.2
	Barium on tungsten	W-Ba	1.56	1.5
	Barium on oxygen on tungsten	W-O-Ba	1.34	0.18
	Lanthanum on tungsten	W-La	2.71	8.0
	Cerium on tungsten	W-Ce	2.71	8.0
	Yttrium on tungsten	W-Yt	2.70	7.0
	Zirconium on tungsten	W-Zr	3.14	5.0
	Thorium on tungsten	W-Th	2.63	3.0
	Uranium on tungsten	W-U	2.84	3.2
	Thorium on molybdenum	Mo-Th	2.59	1.5

the discrepancy is attributable to approximations in deriving the thermionic equation; some [15, 16] contend that reflections of electrons at the surface of the metal result in a lower value of A; others [14] believe that neglect of the temperature coefficient of the work function may result in inaccuracy. Figure 2·7 shows a typical method of plotting

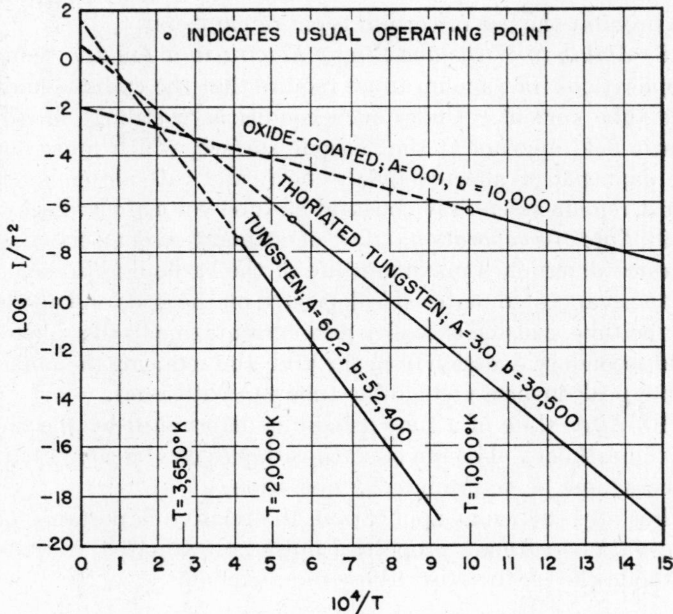

FIG. 2·7 Determination of the constants in Richardson's equation. The curves are plotted with I in amperes per square centimeter and T in degrees Kelvin. (Reprinted with permission from *Applied Electronics*, by the Electrical Engineering Staff, M.I.T., published by John Wiley, 1943.)

experimental data for determining A and ϕ. With A and ϕ known, the thermionic emission current at any value of temperature T can be computed.

Velocity Distribution of Thermionically Emitted Electrons

The motion of electrons in electrostatic and magnetic fields depends upon the initial velocities of those electrons, that is, the velocities of the electrons immediately after they have been emitted from the cathode. Velocities of escaping electrons are not zero, nor are they all equal, but instead they have a definite distribution which is Maxwellian.[16, 17, 21]

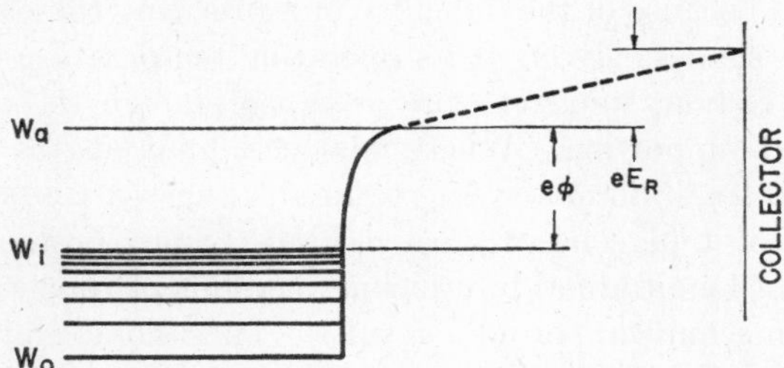

FIG. 2·8 Potential-energy diagram for retarding potential measurements.

The velocity distribution of escaping electrons can be determined experimentally by means of retarding potential measurements. A plane collecting electrode is placed in vacuum parallel to a plane cathode surface at a temperature T. As the potential of the collector is made increasingly negative with respect to the cathode, electron current to the collector (as indicated by a meter in series with the collector) decreases. For any given retarding potential E_R the only electrons reaching the collector are those with initial energies high enough to overcome the retarding field. In Fig. 2·8 the electron current reaching the collector consists of electrons which before emission had x-associated energies sufficiently high to overcome not only the work function of the surface but also the retarding field existing between the cathode and the collector. Therefore, if ϕ in equation 2·6 is replaced by $\phi + E_R$, the resulting expression gives collector current I_c as a function of retarding potential E_R:

$$I_c = AT^2\epsilon^{-[e(\phi+E_R)]/kT} = I\epsilon^{-(eE_R)/kT} \qquad (2\cdot9)$$

where E_R is the retarding potential, which has already been corrected for contact difference of potential between cathode and collector. This relation indicates that $\log_\epsilon I_c$ vs. E_R

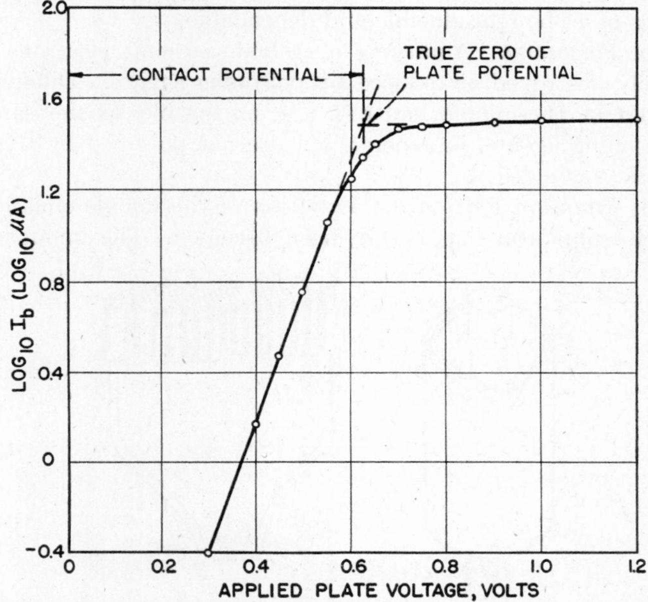

FIG. 2·9 Determination of contact difference of potential by retarding potential method. (Reprinted with permission from Millman and Seely, *Electronics*, 1941, p. 165.)

should result in a straight line with a slope $-e/kT$. Figure 2·9 indicates that the experimental data are consistent with the theory except for the higher currents, which have unexplainable deviations from theory. The failure of the "knee" of this curve to coincide with $E_R = 0$ is caused by contact difference of potential between cathode and collector. Its value in this case appears to be about -0.62 volt, indicating that the work function of the collector is 0.62 volt higher than that of the cathode. The average energy of the escaping electrons is $2kT$, which for tungsten at 2500 degrees Kelvin is 0.431 electron volt.

Characteristics of Thermionic Cathodes

Although the primary purpose of a cathode is to supply ample thermionic emission currents, the electron emitter must have other characteristics to satisfy the requirements of any particular application.[18, 19] The most important properties of thermionic emitters are these:

1. ***Preparation.*** It must be possible to fabricate the cathode of the desired material or materials in the required

form. This implies that the materials must be capable of being formed, spot-welded, or constructed by some other process. Preparation should be as simple as possible.

2. ***Strength.*** Cathode structures and materials must have sufficient mechanical strength at operating temperature to be self-supporting and capable of withstanding shock and vibration.

3. ***Activation.*** It is always preferable for a cathode to require a minimum of activation. In general, cathodes requiring a more involved activation schedule are the least reproducible in quality.

4. ***Reproducibility and Dependability.*** Successful operation of a tube usually depends upon its ability to give operation consistent with published tube characteristics and to continue to operate with unchanged electrical performance throughout its life. Since cathodes either directly or indirectly often determine or affect tube characteristics, cathodes must be reproducible and dependable.

5. ***Poisoning.*** Ability of a cathode to emit electrons is easily destroyed or reduced by poisoning agents. Different types of thermionic emitters are susceptible to the same poisoning agents, in general, but are not poisoned by them to the same degree.

6. ***Emission Efficiency.*** Variation of thermionic emission with temperature has already been discussed. The amount of cathode-heating power required to maintain an electron emitter at a temperature high enough to provide a desired thermionic emission current is important. The term *emission efficiency* and its value in milliamperes of emission current per watt of cathode heating power are convenient for comparing relative abilities of various cathodes. Emission efficiency of a cathode can be increased by providing it with heat shielding, which reflects or reradiates some of the heat back to the cathode. Examples of such shielding are shown in Fig. 2·10. In Fig. 2·10 (*a*), the shielding is in the form of a cylinder or cylinders of smooth refractory metal surrounding the heated cathode. In Fig. 2·10 (*b*) the cathode is designed so that the surfaces of the emitter shield each other. By such heat shielding, less cathode heating power is required to heat the cathode to its emitting temperature, and emission efficiency is increased. Since emission efficiency is so dependent upon heat shielding, values given later for various types of cathodes are for unshielded emitters.

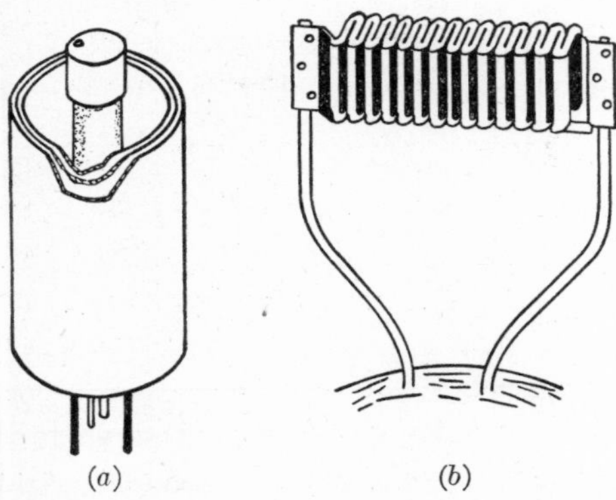

Fig. 2·10 Examples of (*a*) indirectly and (*b*) directly heated oxide-coated cathodes. (Reprinted with permission from *Applied Electronics*, by the Electrical Engineering Staff, M.I.T., published by John Wiley, 1943.)

7. ***Ability to Withstand Positive-Ion Bombardment.*** In tubes containing mercury vapor or inert gas the cathode is subject to appreciable positive-ion bombardment. Such a tube therefore requires a cathode capable of withstanding bombardment without damage. Even in vacuum tubes operating at high voltages, sufficient positive-ion bombardment of the cathode results from ionization of residual gas to prohibit the use of certain types of emitters.

8. ***Ability to Withstand Strong Electrostatic Fields.*** Some applications for vacuum tubes require that the cathode limit the tube current. Under such conditions a strong electrostatic field appears at the cathode surface, which must not be damaged or affected by the field, and cathode emission must remain essentially constant as tube voltage increases.

9. ***Rate of Evaporation.*** Rate of evaporation of an emitting surface determines maximum life of the cathode; also some of the evaporated material is deposited on the grids and anode of the tube, causing excessive grid currents due to thermionic and secondary emission from the grid, and reducing the maximum peak inverse voltage the tube can withstand.

10. ***Life.*** Life of a tube usually is determined by the life of the cathode. The life of some cathodes can be computed; others must be determined by operation.

For any particular application, the relative importance of the various cathode properties must be evaluated and the cathode chosen to satisfy those requirements.

Types of Thermionic Emitters

Pure Metals. From theory, and because emission varies with work function, it seems reasonable to expect that metals with lowest work function will be the most generally used. Actually, the best pure-metal thermionic emitters are tungsten, tantalum, and molybdenum, which have work functions of 4.52, 4.12, and 4.15 electron volts respectively. In spite of large work functions, these metals, particularly tungsten, are satisfactory thermionic emitters because their melting temperatures are so high and their evaporation rates so low that they can be operated at high enough temperatures to supply relatively large thermionic emission currents.

Because of its exceptionally high melting point, tungsten is prepared not by the usual metallurgical processes but by powder metallurgy. It is used as a cathode in the form of a straight or coiled filament. Other forms usually are impractical because of the difficulty of fabricating tungsten into complex shapes. Even at its operating temperature of 2500 degrees Kelvin, tungsten wire is strong enough to be essentially self-supporting. Where relatively long lengths of wire of rather small diameter are required, tungsten containing a fraction of 1 percent of silica is used. Silica promotes the growth of longitudinal interlocking crystals of tungsten, and greater mechanical stability results. Thermionic emission is obtained from tungsten merely by heating it to its emitting temperature. No activation process is necessary. Gases and other undesirable foreign matter in the metal or on its surface are rapidly eliminated at such high temperatures. In the manufacture of tubes, however, it is customary to heat tungsten cathodes to about 2800 degrees Kelvin for a short time to hasten degassing. Thermionic emission of a tungsten filament can be reproduced accurately from one cathode to another and can be depended upon to equal that calculated from equation 2·4, in which the accepted values of A and ϕ for tungsten are substituted. Although 4.52 electron volts is the accepted work function for tungsten, Nichols [20] discovered that the work function has different values for the different crystal faces of large single crystals of tungsten. Commercial tungsten wire is polycrystalline, so that variations in work function with crystal face average out, resulting in the accepted value of 4.52 electron volts for the work function and 60.2 for the thermionic-emission constant A.

Tungsten cathodes are poisoned [24] most easily by water vapor, but recover their emitting ability if the water vapor is removed by maintaining a better vacuum. They are not poisoned by mercury vapor or the inert gases, so they can be used in tubes containing those gases or vapors.

The thermionic emission efficiency of a tungsten cathode is extremely low at its usual operating temperature because of the high temperatures at which it must be operated to provide appreciable emission. This obviously is due to its relatively high work function. Emission efficiency of a cathode varies rapidly with its temperature, however, because electron emission is essentially an exponential function of temperature. Increase of emission efficiency with temperature of a straight tungsten wire is shown in Table 2·2. In spite of its high

Table 2·2 Emission Efficiency of a Straight Tungsten Wire

Temperature (degrees Kelvin)	2400	2450	2500	2550	2600	2650
Emission efficiency (milliamperes per watt)	2.01	2.95	4.27	6.04	8.56	11.9

operating temperature and correspondingly low emission efficiency, tungsten usually is used as the cathode material in vacuum tubes with operating voltages higher than a few thousand, because it can withstand the bombardment of high-energy positive ions resulting from ionization of the residual gas existing even in well-exhausted vacuum tubes. Cathodes other than those of the pure metal type are not satisfactory for high-voltage applications because their thermionic emission is reduced or destroyed by positive ion bombardment.

Because the surface of a tungsten cathode is relatively smooth, and because the surface atoms adhere strongly to underlying atoms, tungsten cathodes are unsurpassed in their ability to withstand strong electrostatic fields at their surfaces without damage and without emitting undesirably larger currents.

Although tungsten cathodes are usually operated at about 2500 degrees Kelvin, the rate of evaporation of the tungsten is sufficiently low at this temperature to have a reasonably long life. By use of the tables of Jones and Langmuir,[22] the life of tungsten filaments for various operating temperatures can be computed. The practical life of a filamentary pure-metal cathode heated by the passage of electrical current is considered to be ended when its cross-sectional area has decreased by 10 percent. A tungsten filament 0.010 inch in diameter operating at 2500 degrees Kelvin has an evaporation rate of 2.02×10^{-6} milligram per square centimeter-second and a life of 1690 hours.

Composite Surfaces. A cathode consisting of a layer of an electropositive metal on a metal base or on the oxidized surface of a metal is of the *composite-surface type*. In this group are such cathodes as thorium on tungsten, thorium on molybdenum, cesium on tungsten, cesium on oxygen on tungsten, barium on oxygen on tungsten, and various other combinations. Of these, thorium on tungsten is the most satisfactory as a thermionic source of electrons. Experiments by Langmuir [23] and others [25] have shown that the work functions of surfaces of cesium, barium, or strontium on tungsten or oxygenated tungsten are low (from 0.5 to 2.5 electron volts). The vapor pressure of the adsorbed atoms of cesium, barium, and strontium has been found to be so high, however, that those cathodes cannot be operated at sufficiently high temperatures to obtain thermionic current densities comparable with those obtainable from other types. Thorium-on-tungsten emitters therefore are the only composite-surface cathodes which are of general practical use.

Addition of silica to tungsten before sintering, swaging, and drawing has already been mentioned. Pure tungsten filaments in lamps have a tendency to grow large single crystals having dimensions comparable with the diameter of the wire. These crystals possess cleavage planes extending across the wire, so that one part of the crystal can slip with respect to the other and cause an "offset." Addition of a small percentage of thoria (thorium oxide) to the tungsten prevents the growth of such large single crystals, thereby increasing the life of the filament. In 1913 Langmuir and Rogers [26] discovered that the thermionic emission from such thoriated-tungsten filaments was much greater than that of pure tungsten.

Unlike tungsten, optimum thermionic emission from thoriated tungsten is obtained only after activation, which must be carried on in a good vacuum and consists of two parts:

1. The filament must be "flashed" for about 1 minute at a temperature of 2800 degrees Kelvin or higher.
2. The filament must be heated to a temperature between 2100 and 2400 degrees Kelvin for several minutes, after which it is operated at a temperature of 1900 to 2000 degrees Kelvin.

The most generally accepted theory of activation and electron emission of thoriated-tungsten filaments is that of Langmuir.[27, 28, 29, 30] According to this theory, flashing the thoriated-tungsten filament at a temperature of about 2800 degrees Kelvin cleans the surface and reduces some of the thoria to metallic thorium. The oxygen combines with tungsten atoms to form tungsten oxide, which has a sufficiently high vapor pressure at this temperature to volatilize and leave a tungsten filament containing thorium and unreduced thoria. Thorium atoms diffuse along the tungsten-crystal boundaries toward the surface where they evaporate and leave the surface relatively free of thorium. As a result of further heating at lower temperatures, some of the free thorium diffuses to the surface to build up a partial layer of thorium atoms.

To explain the characteristics of thoriated-tungsten filaments, four temperature ranges usually are considered. Below 1800 degrees the diffusion of thorium to the surface is not appreciable, and the emission immediately after flashing corresponds nearly to that of pure tungsten. When the flashed filament is heated to a temperature of 1900 to 2200 degrees Kelvin, the rate of diffusion of thorium to the surface is at first greater than its rate of evaporation, and a surface layer of polarized thorium atoms is formed gradually. Thorium evaporates more readily from a thorium surface than from a tungsten surface, so it is difficult to obtain more than a monomolecular layer of thorium on the surface except at lower temperatures. For a given temperature, the rate of diffusion of thorium to the surface is constant, but the rate of evaporation increases with the number of adsorbed thorium atoms. The final "fractional coverage" is reached when the rate of diffusion and the rate of evaporation are equal. Maximum emission is obtained when the work function of the filament surface is a minimum, so the effect of adsorbed atoms is to reduce the work function. This reduction is caused by the dipole moment of the adsorbed thorium atoms, which act as a double sheet of electric charge and polarize the surface. A simplified diagram of this polarized layer is shown in Fig. 2·11.

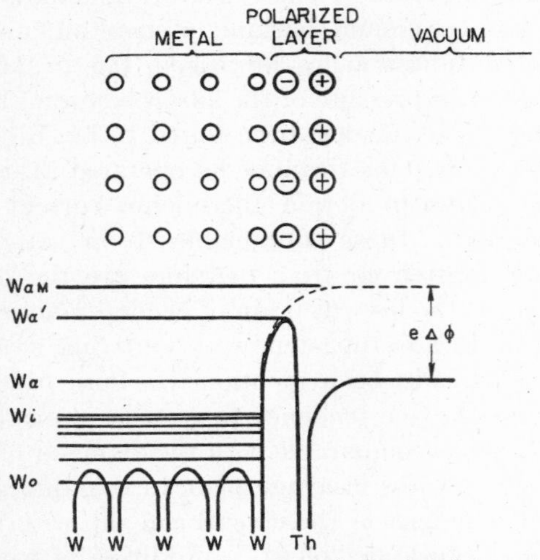

FIG. 2·11 Potential-energy diagram of thoriated-tungsten surface.

The work function is believed to decrease as the dipole moment per unit area of the surface increases. As the fractional coverage increases, the dipole moment at first increases but later decreases because of the mutual repelling action between thorium atoms. The fractional coverage resulting in maximum average dipole moment per unit area, and therefore in minimum work function, is about 70 percent, as shown in Fig. 2·12. The temperature corresponding to this optimum coverage is approximately 2100 degrees Kelvin. Also shown in Fig. 2·11 is a potential-energy diagram indicating the potential barrier of pure tungsten as a dashed line and that of tungsten having an adsorbed surface layer of thorium as a solid line. The decrease in work function caused by the polarized layer of thorium atoms is represented by the difference in heights of the potential barriers for tungsten and thoriated tungsten and is indicated by $\Delta\phi$. It seems that the decrease in work function should be the difference in height of the potential barriers for tungsten W_{aM} and of the potential maximum W'_a. This would be correct if all electrons having energies less than W'_a were reflected at the surface, only those having energies greater than W'_a being emitted. Fowler and Nordheim [31, 32] found that electrons in the metal having x-associated energies between W_i and W'_a have a finite probability of penetrating such a potential barrier and therefore of being emitted. By considering the electrons incident upon the barrier as electromagnetic waves having DeBroglie wavelengths given by

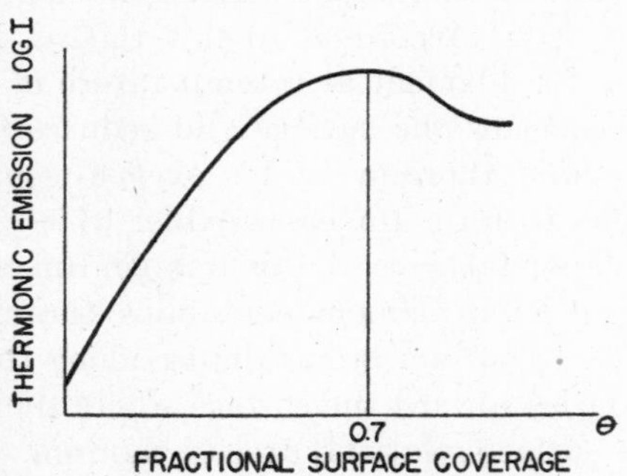

FIG. 2·12 Variation of thermionic emission from thoriated tungsten with the fraction of the surface covered by thorium.

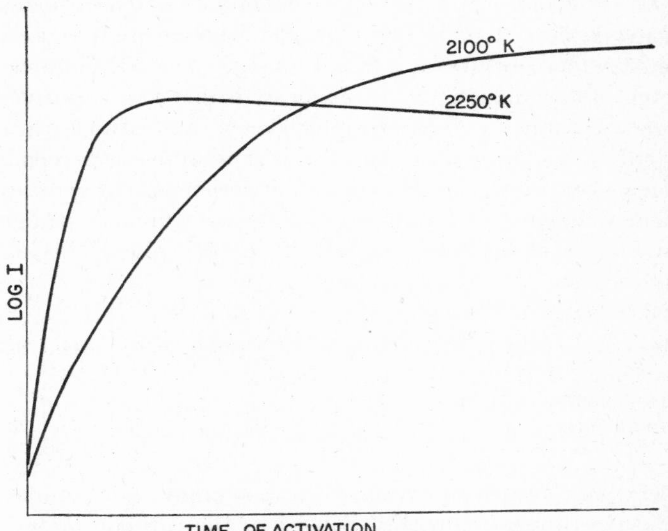

FIG. 2·13 Typical activation curves for thoriated-tungsten filaments.

$$\lambda = \frac{h}{m_e v} \qquad (2\cdot10)$$

where v is the velocity of the electron, Fowler and Nordheim were able to show that the mean transmission coefficient $\bar{D}(W)$ for electrons arriving at such a barrier was not zero. It is therefore consistent with theory to indicate the difference between the work function of thoriated tungsten and that of pure tungsten as shown in Fig. 2·11. $\bar{D}(W)$ is equal to $1 - \bar{r}$ where $\bar{r}$ is the mean reflection coefficient. The mean reflection coefficient modifies the value of A in equation 2·4, resulting in the more correct form for the thermionic emission current-density expression:

$$I = (1 - \bar{r})AT^2\epsilon^{-e\phi/kT} \qquad (2\cdot11)$$

The value of $(1 - \bar{r})A$ therefore is the quantity determined by experiment, the thermionic-emission constant A having the theoretical value given by equations 2·7 and 2·8. For thoriated tungsten with optimum fractional coverage, ϕ and A are generally accepted as 2.63 electron volts and 3.0 respectively. For activation other than the optimum, the work function is larger. Typical activation curves for thoriated tungsten are shown in Fig. 2·13. Each curve was obtained experimentally by maintaining the filament at a constant activation temperature, the thermionic emission being measured at 1560 degrees Kelvin at frequent intervals. Thermionic emission at 1560 degrees Kelvin is plotted in Fig. 2·13 as a function of the total time of activation. A low temperature was chosen so that the surface would remain unchanged during the test. Curves for activation temperatures of 2100 and 2250 degrees Kelvin are shown.

It is apparent that activation is more rapid at 2250 degrees Kelvin but has a final value much lower than for the optimum temperature of 2100 degrees Kelvin. The fractional coverage for the final state of activation at 2100 degrees Kelvin is probably about 70 percent; at 2250 degrees Kelvin it is appreciably less. The deactivation range (from 2300 to 2600 degrees Kelvin) for thoriated-tungsten filaments can best be explained by reference to Fig. 2·14, which is a qualitative sketch of the effect of operating a well-activated thoriated-tungsten filament at temperatures where deactivation occurs. Part *A* of the curve shows activation at 2100 degrees Kelvin (emission again measured at 1560 degrees Kelvin) to time t_1 where emission has almost reached its maximum value. If filament temperature is increased at time t_1 to 2400, 2500, or 2600 degrees Kelvin, the emission, still measured at about 1560 degrees Kelvin, decays as shown by curves *B*, *C*, or *D*, depending upon which deactivation temperature was used starting at time t_1.

It is clear, therefore, that the steady-state emission of a thoriated-tungsten filament is always greater than that for pure tungsten and often depends upon the state of activation of its surface as well as its operating temperature. Although the minimum work function surface is obtained at an operating temperature of 2100 degrees Kelvin, thorium evaporates rapidly at this temperature and filament life is shortened undesirably. Thoriated-tungsten filaments usually are operated at a temperature of about 1900 degrees Kelvin for satisfactorily long life. Even at this reduced temperature, the emission is about 10,000 times as large as for tungsten operated at this temperature, and is about five times as large as for tungsten at its normal operating temperature of 2500 degrees Kelvin. Table 2·3 lists values of thermionic emission and emission efficiencies of various types of emitters at several temperatures.

Table 2·3 Thermionic Emission and Emission Efficiencies of Various Emitters

Temperature (°K)	Electron Emission (amperes per square centimeter)			Emission Efficiency (milliamperes per watt)		
	2500	1900	1123	2500	1900	1123
Tungsten	0.26	2×10^{-4}	3×10^{-11}	4.	......	......
Thoriated tungsten	....	1.2	6×10^{-6}	..	25–100	......
Oxide-coated cathode	....	....	0.3	..	......	50–100

The ratio of emission efficiencies of thoriated-tungsten and tungsten filaments at their respective operating temperatures is about 25 to 1, as shown in this table. Increased emission efficiency and lower operating temperature of thoriated tungsten are important not only because of the saving in cathode-heating power, but also because the lower operating temperature reduces power-dissipation requirements of other electrodes in the tube which absorb some of the heat radiated by the cathode. Even at 1900 degrees Kelvin, the rate of evaporation of thorium from thoriated-tungsten cathodes is sufficiently high to result in contamination of other electrodes in the tube. Although this is not serious in some tubes, it is in other tubes in which grid currents must be kept small, and in which relatively strong electrostatic fields at the anode would cause field emission from thorium deposited on the anode.

The surface of a thoriated-tungsten cathode is damaged by positive-ion bombardment resulting from ionization of the residual gas in a vacuum tube if voltages in excess of several thousand volts are applied to the tube. Because a well-activated thoriated-tungsten cathode relies upon the existence of thorium atoms on its surface, it is quite sensitive to poisoning agents, particularly water vapor and oxygen. After the cathode is poisoned, all or most of its thermionic emission can be regained by removing the poisoning agent from the tube and repeating the usual activation schedule. During life the free thorium continues to diffuse to the surface and eventually evaporates until the supply of free thorium made available by the original flashing previous to activation is depleted. The cathode then can be reactivated in the usual manner several times during life until the supply of thoria is depleted.

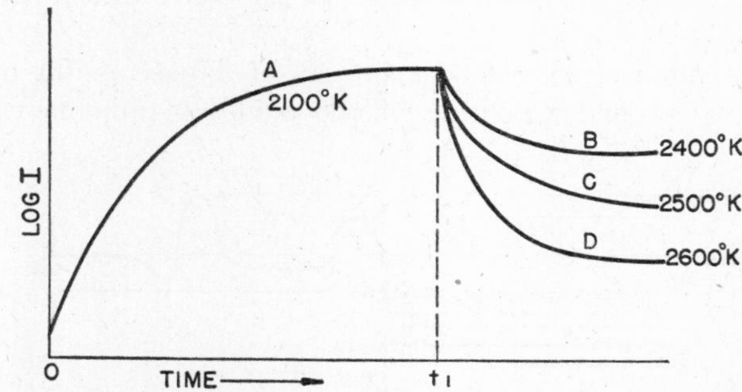

Fig. 2·14 Activation and deactivation curves for thoriated-tungsten filaments.

To increase their life, thoriated-tungsten filaments are often carbonized [33] by being heated between 1600 and 2000 degrees Kelvin in a hydrocarbon gas or vapor. The hydrocarbon is reduced and liberates carbon, which diffuses into the cathode. Usually carbonization is continued until the electrical resistivity of the filament is increased by 10 to 20 percent. Carbon in the cathode appears to reduce greatly the rate of evaporation of thorium. It is customary to operate carbonized thoriated-tungsten filaments at a temperature of 2000 to 2100 degrees Kelvin to take advantage of the increased thermionic emission, the life still remaining satisfactorily long. The higher operating temperature and the presence of the carbon are believed to make possible the reduction of thoria at operating temperatures so that thoriated-tungsten filaments can operate satisfactorily in vacuum tubes at somewhat higher voltages because the deactivation effect of the positive-ion bombardment is partially counteracted by the increased diffusion rate of thorium.

An electrostatic field at the cathode increases the electron emission from a thoriated-tungsten filament much more than from pure tungsten. This characteristic is objectionable for current-limiting applications.

Oxide-Coated Cathodes. The most often used and least understood type of thermionic emitter is the oxide-coated cathode. Its discovery dates back to 1903 when Wehnelt [34] noticed that a platinum ribbon coated with oxides of alkaline-earth metals was a copious emitter of electrons at relatively low temperatures (950 to 1200 degrees Kelvin). Since that time many investigators [48] have conducted innumerable experiments resulting in the formulation of several theories of the mechanism of thermionic emission from oxide-coated cathodes. In spite of the improvements resulting from such investigations, and from experience in manufacture, oxide-coated cathodes as they exist today leave much to be desired.

An oxide-coated cathode consists of a base metal upon which is applied a coating of the oxides of alkaline-earth

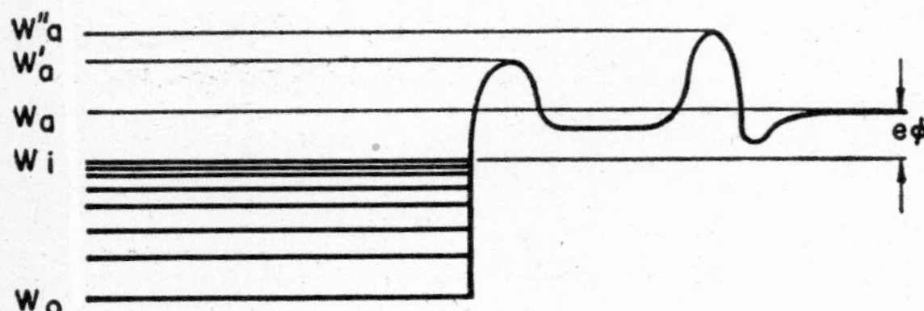

FIG. 2·15 Potential-energy diagram for an oxide-coated cathode.

metals. Nickel, Konal,* tungsten, and platinum have been used as base materials, although nickel and Konal are the most common. The coating consists of barium, strontium, and occasionally calcium oxides. In manufacture the coating usually is applied to the base metal not as oxides, which are chemically unstable in atmospheric air, but rather as carbonates, nitrates, oxalates, or hydroxides of alkaline earths. The most common method of applying the coating is spraying.[37] The desired mixture of alkaline-earth compounds is suspended in a finely divided state in an organic binder (such as nitrocellulose dissolved in amyl acetate), and the suspension is sprayed onto the base metal until the coating is from 10^{-2} to 10^{-3} centimeter thick. Sometimes the cathodes are coated by dipping the base metal in the desired suspension. A method used only occasionally is electrophoresis,[36] whereby the suspended particles of alkaline earths are deposited electrically on the base metal. Exact chemical composition of coating, type of binder, choice of base metal, and method of applying the coating to the base metal are determined by physical shape of the cathode, type of tube, and the experience of the manufacturer.

During exhaust of the tube, the binder usually is decomposed or volatilized in the baking process, the liberated gases and vapors being removed by vacuum pumps. The alkaline-earth compounds are then decomposed by sintering, which is a gradual heating of the cathode to a temperature of about 1300 degrees Kelvin. The products of sintering are the alkaline-earth oxides, and gases or volatile materials resulting from decomposition of the alkaline-earth compounds. Some investigators [38, 39, 40, 41] believe that, immediately after sintering, the cathode is in an activated state and that no further activation is necessary. This is correct, because the cathode is now capable of supplying large thermionic emission currents; however, drawing thermionic currents from the cathode by maintaining the other electrodes at a potential positive with respect to the cathode causes an increase of thermionic emission above that obtainable immediately after sintering.[42, 43, 44, 45] Increasing thermionic emission by drawing electron currents from the cathode is referred to as activation.[46]

For base metals containing reducing agents such as titanium, magnesium, or aluminum, a small part of the oxides is broken down at the sintering temperature to free barium and oxygen, the latter being pumped away. More free barium is formed during activation, the barium in either case being dispersed throughout the oxide coating, on its surface, and at the metal-oxide interface. Experimental evidence shows that activated cathodes are richer in free barium than unactivated ones, and the free-barium theory of the mechanism of emission from oxide cathodes has become generally accepted. According to this theory, the emission abilities of an oxide-coated cathode depend upon excess barium on the surface and throughout the coating. Many of the details of emission remain unexplained. One contention is that emission is caused by a monatomic layer of barium on the surface of the coating, another is that the emission is fostered by a thermionically emissive surface at the interface between the coating and the core metal,[41] and a third is that the emission is caused by both.

A qualitative potential-energy diagram for an oxide-coated cathode is shown in Fig. 2·15. This diagram is based on the supposition that there are two polarized layers of barium, one at the core-oxide interface and one at the surface of the oxide coating. As in thoriated-tungsten cathodes, electrons in the base metal need not have x-associated energies greater than W''_a to be emitted. Instead, electrons having x-associated energies greater than W_a but less than W''_a have greater than zero probabilities of penetrating the potential barriers. Electrons having x-associated energies between W''_a and W'_a are emitted by virtue of the transmission coefficient for the oxide surface barrier; those with energies between W'_a and W_a are emitted by virtue of the transmission coefficient for both barriers.

The experimentally determined value of the thermionic-emission constant is given by the quantity $(1 - \bar{r})A$, and the mean reflection coefficient $\bar{r}$ should be rather high because there are two potential barriers; therefore it seems that the experimentally determined thermionic-emission constant should be somewhat smaller than the constant for pure metals or even thoriated-tungsten cathodes. This is true. The best experimental values for the work function and the thermionic emission constant for oxide-coated cathodes are about 1.0 electron volt and 0.01 respectively. The low work function explains the large thermionic emission obtainable from oxide-coated cathodes.

An oxide coating has a low mechanical strength. The base metal must have sufficient strength, even at the operating temperature of oxide cathodes, to support itself and the coating. In many applications nickel is satisfactory as a base

* Konal is an alloy of nickel, cobalt, iron, and titanium (trademark registered, U. S. Patent Office).

metal. In others a stronger base metal is required, and for these Konal and tungsten are used as base materials because of their greater strength at the operating temperature. Some investigators believe that the use of Konal instead of nickel as a base metal results in better cathodes.[41]

One of the most serious difficulties in the manufacture of oxide-coated cathodes is their poor reproducibility. When hundreds or thousands of tubes containing oxide-coated cathodes are carefully prepared, sintered, and activated, large variations in cathode quality invariably occur. Cathodes usually are designed for much larger emission than is required, so that a reasonably large percentage of tubes have sufficient thermionic emission to satisfy required specifications. By refinements in methods and technique, as well as by strict control of purity of cathode materials, variations in cathode quality are reduced to a minimum.

Emission from a coated cathode is easily poisoned by water vapor, oxygen, or other electronegative gases.[47] This is consistent with the free-barium theory of emission, because electronegative elements such as oxygen combine chemically with the barium, thereby reducing the concentration of free barium in the cathode. A reducing gas such as hydrogen does not increase emission of a well-activated oxide cathode, but it at least partially restores emission of a cathode poisoned by oxygen. This also supports the free-barium theory.

The greatest advantage of coated cathodes is their ability to emit appreciable currents at lower temperatures than those required for tungsten or thoriated-tungsten cathodes. This results in a higher emission efficiency, which is important for many applications. Table 2·3 indicates that at a temperature of 1123 degrees Kelvin the thermionic emission from an oxide-coated cathode is 5×10^4 times as large as that from thoriated tungsten and 10^{10} times as large as that from tungsten at the same temperature. Emission from these three types of cathodes is of the same order of magnitude at their respective operating temperatures. Also shown in Table 2·3 is the emission efficiency of oxide-coated cathodes as compared with that of the other two types of cathodes. Figure 2·16 shows the variation of the thermionic emission with cathode-heating power for tungsten, thoriated tungsten, and oxide-coated cathodes.

Coated cathodes, like thoriated tungsten, are not suitable for use in high-voltage tubes because of the deactivating or destructive effect of high-energy positive-ion bombardment. In gas- or vapor-filled tubes, oxide cathodes are used almost exclusively. In such tubes the cathode is subject to positive-ion bombardment, but the energy of the ions is small, in the order of 10 to 20 volts. Several investigators have determined the maximum energy with which positive ions can bombard the cathode without causing deactivation or destruction.[49, 50]

If there is a strong electrostatic field at the surface of a coated cathode, emission currents[51, 52, 53] are much larger than those calculated from equation 2·4. This characteristic makes these cathodes unsuitable for current-limiting applications, but is a distinct advantage in gas- or vapor-filled tubes,[54] where the field is produced by a positive-ion space-charge sheath.

Experiments have shown[55, 56] that the vapor pressure of barium is much higher than that of its oxide or of strontium or its oxide. During operation barium evaporates from the cathode and condenses on other electrodes in the tube. The resulting contamination of the grids or anode of the tube is objectionable because of the higher grid currents and lower maximum permissible anode voltage.

After a coated cathode has been operated until its ability to emit electrons has almost ceased, some coating material still remains on the core metal. This is to be expected from the difference in the rates of evaporation of barium and those of barium oxide, strontium, and strontium oxide. As the cathode is operated the free barium evaporates until its

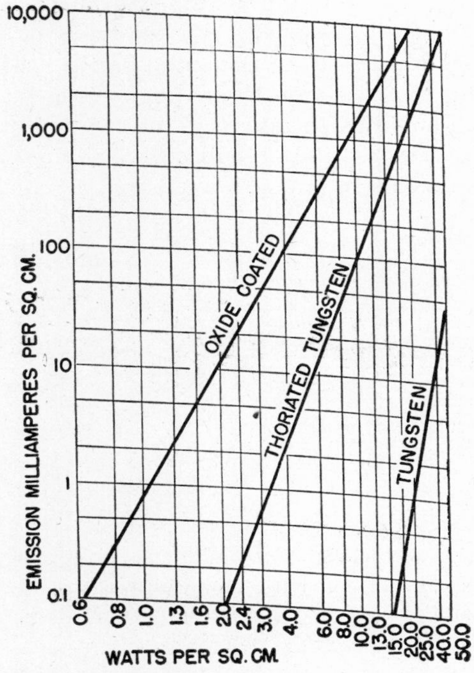

Fig. 2·16 Thermionic emission versus heater power for various cathode materials. (Reprinted with permission from *Physics of Electron Tubes*, by Koller, published by McGraw-Hill, 1934, p. 40.)

concentration in the coating and on the surface is reduced. Emission decreases accordingly in spite of the fact that much of the coating still remains. It is often possible to reactivate by the usual sintering and activation processes. This can be done only a limited number of times, after which it is impossible to regain the emission. In spite of the evaporation of barium, the life of an oxide-coated cathode is usually many thousand hours.

Many oxide-coated cathodes are heated to their operating temperatures by passing current through the base metal. Such cathodes are commonly referred to as *directly heated cathodes*. Because of their low operating temperature, coated cathodes can be heated indirectly, by radiation from a heater filament. Indirectly heated coated cathodes have come into quite general use. They can be made in shapes that are difficult or impossible to heat directly. One of their most important advantages over directly heated cathodes is their equipotential nature; in many applications the variation of potential along the length of a directly heated cathode is objectionable. Examples of directly and indirectly heated cathodes are shown in Figs. 2·10 (*b*) and 2·10 (*a*) respectively.

2·4 EMISSION IN THE PRESENCE OF A FIELD

Schottky Effect

Electron emission from a surface depends upon the height of the potential barrier; and reduction of that height results in increased emission. The larger thermionic emission from surfaces of low work function is an example of this effect. For a given surface the height of the potential barrier can be decreased by applying sufficiently strong electrostatic fields to the surface of the emitter.[57, 58, 59, 60, 61] By reducing the height of the potential barrier, the effective work function is reduced by the same amount, since the W_i level is unaffected by the application of external fields. To evaluate the decrease in effective work function requires a knowledge of the various forces which give rise to the potential barrier. The first retarding force on an electron attempting to escape is a *coulomb force* and is caused by the atoms in the crystal lattice. This type of force was discussed in Section 1·7 and was illustrated in Fig. 1·8. This retarding force decreases rapidly as the electron leaves the surface of the metal, and it is relatively weak when the electron is several atomic distances away. When the electron succeeds in getting several atomic distances away from the metal surface, the metal appears to the electron as a relatively homogeneous smooth surface rather than as an irregular surface of non-uniform charge caused by the discrete atoms of the crystal lattice. A second retarding force, commonly called the *image force,* is believed to act on an electron removed from the metal surface by a distance large compared with the atomic spacing in the crystal lattice. The image force can be considered as that between a charge $-e$ and a plane conducting surface a distance x away. Such a force is identical with that between two charges $-e$ and $+e$ separated by a distance $2x$, as shown in Fig. 2·17. The positive charge is thus the mirror image of the negative one, hence the name "image force." The expression for the image force is

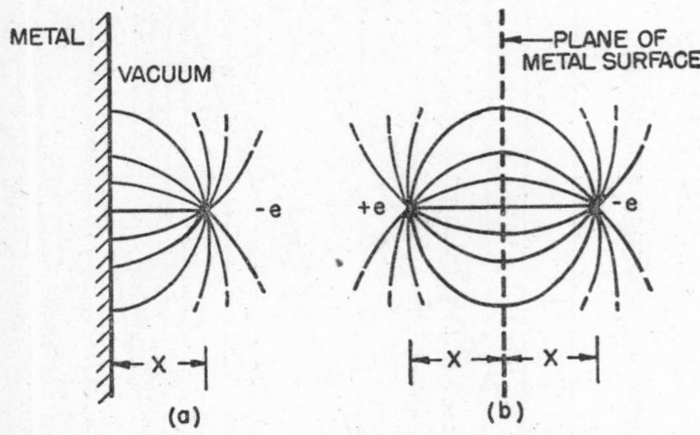

FIG. 2·17 Electric field between an electron and a metal and between an electron and its image. (Reprinted with permission from *Applied Electronics*, by the Electrical Engineering Staff, M.I.T., published by John Wiley, 1943.)

$$F = \frac{e^2}{4x^2} \qquad (2\cdot12)$$

which decreases as the distance increases, and increases to an infinite value as the separation x approaches zero. This equation is believed to be valid only when x is large compared with the atomic spacing. The potential barrier at the surface of a metal with no external field is caused by the combined coulomb and image forces and has the general appearance of curve 1, Fig. 2·18. Curve 2 of the same figure shows the potential barrier caused by the combined coulomb and image forces modified by the external accelerating field **E**. The maximum height of this barrier is lower than the W_a level of curve 1. If x_m is the distance of this potential maximum from the surface of the metal,[57, 59]

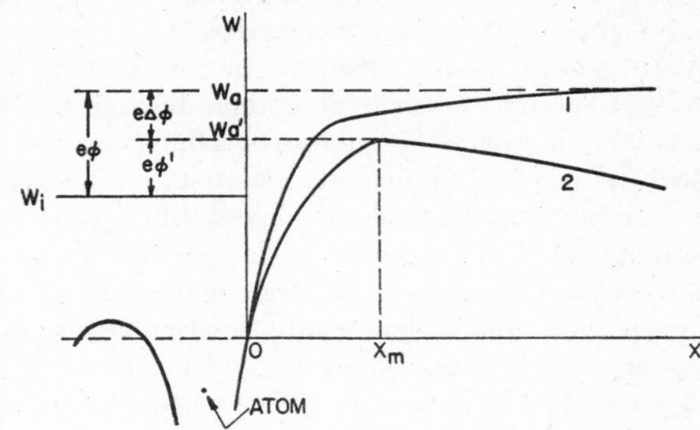

FIG. 2·18 Potential-energy diagram showing the effect of an electrostatic field on the height of the surface potential barrier. (Reprinted with permission from *Applied Electronics*, by the Electrical Engineering Staff, M.I.T., published by John Wiley, 1943.)

$$x_m = \frac{1}{2}\sqrt{\frac{e}{\mathbf{E}}} \qquad (2\cdot13)$$

and the decrease in the effective work function is

$$\Delta\phi = \phi - \phi' = \sqrt{e\mathbf{E}} \qquad (2\cdot14)$$

Electron emission in the presence of this applied field, therefore, is

$$I' = AT^2\epsilon^{-e\phi'/kT} = AT^2\epsilon^{-(e\phi - e\sqrt{e\mathbf{E}})/kT} \qquad (2\cdot15)$$

which, in terms of the zero field emission I, can be written

$$I' = AT^2\epsilon^{-e\phi/kT}\epsilon^{(e\sqrt{e\mathbf{E}})/kT} = I\epsilon^{(e\sqrt{e\mathbf{E}})/kT} \qquad (2\cdot16)$$

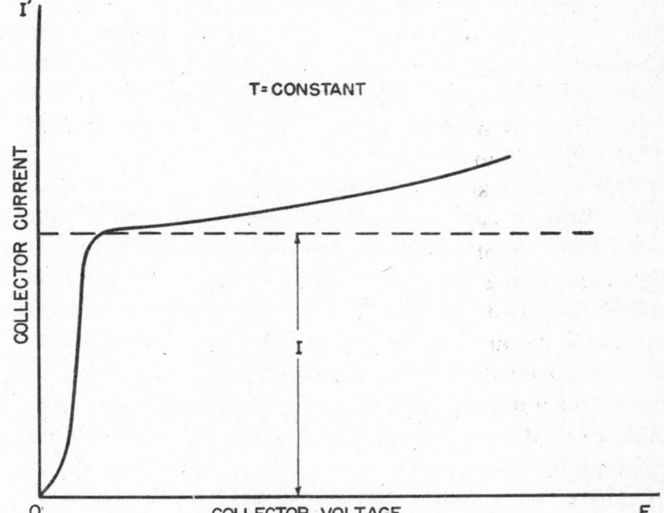

FIG. 2·19 Typical current-voltage characteristic showing poor saturation due to Schottky effect. (Reprinted with permission from *Applied Electronics*, by the Electrical Engineering Staff, M.I.T., published by John Wiley, 1943.)

This expression indicates that the electron emission from a thermionic emitting metal increases as the strength of the applied accelerating field increases. This effect, known as the Schottky [62] effect, explains the failure to obtain true emission saturation from a metal as the collecting voltage is increased, with emitter temperature remaining constant. A typical experimental result illustrating this is shown in Fig. 2·19. Equation 2·16 indicates that the logarithm of the current should be a linear function of the square root of the field; Fig. 2·20 shows that this is verified experimentally.

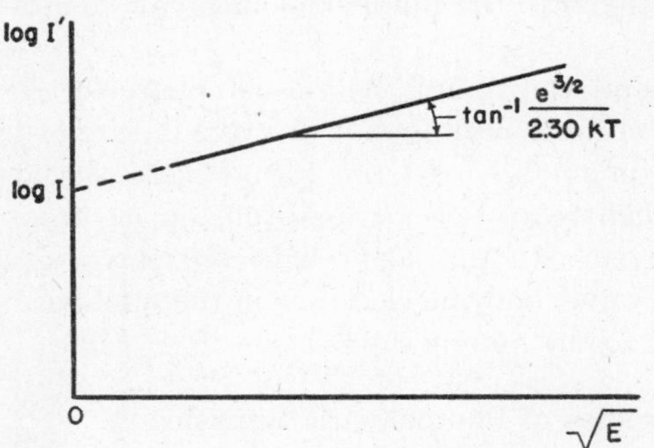

Fig. 2·20 Schottky plot for a pure-metal emitter illustrating extrapolation to zero field for determination of the true saturation current. (Reprinted with permission from *Applied Electronics*, by the Electrical Engineering Staff, M.I.T., published by John Wiley, 1943.)

The zero-field or true saturation emission current is indicated by the intercept in Fig. 2·20.

Both thoriated-tungsten and oxide-coated cathodes exhibit an "anomalous" Schottky effect.[63, 64] For stronger accelerating fields the Schottky effect is normal, because $\log_e I'$ is a linear function of $\mathbf{E}^{1/2}$ and the slope of the line is $e^{3/2}/kT$, as indicated in equation 2·16. For weaker fields the increase of the electron emission with field strength is many times larger than predicted by equation 2·16. This anomalous behavior, shown in Fig. 2·21, is believed to be caused by "patches" with different work functions on the surfaces of

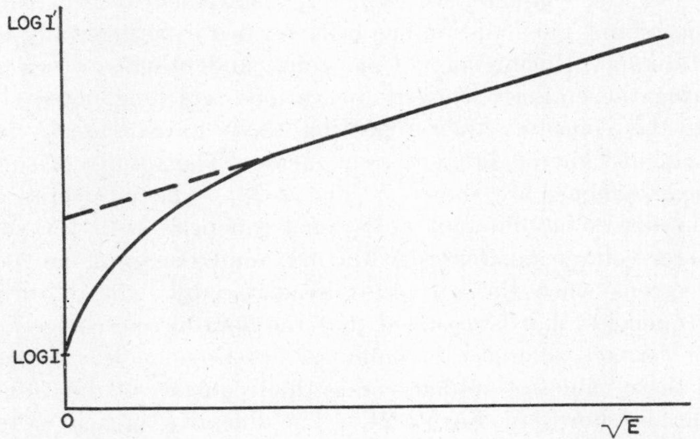

Fig. 2·21 Schottky plot for a composite or oxide-coated cathode illustrating anomalous Schottky effect. (Reprinted with permission from *Fundamentals of Engineering Electronics*, by W. G. Dow, published by John Wiley, 1943.)

thoriated-tungsten and oxide-coated cathodes. These patches cause local field effects which result in an abnormally large increase of electron emission for a small accelerating field. The effect of these local fields becomes negligible if a large field is applied to the cathode surface. Unusual field effects are displayed by oxide-coated cathodes because of the roughness of the surface.

In a vacuum tube an accelerating field at the cathode surface is produced by positive voltages on the grid or anode of the tube; in a gas-discharge tube, it is a result of the positive-ion space-charge sheath which forms outside the cathode surface.

Field Emission

As the potential gradient at the surface of a pure metal is increased, the potential barrier appears as shown qualitatively by curve A of Fig. 2·22. The barrier has been reduced in height, resulting in a large Schottky increase in electron emission, and it has been narrowed so that the barrier is penetrated by some electrons having energies less than W'_a.[64, 65, 66, 67] A small percentage of electrons in the Fermi band (energies less than W_i) also penetrate the narrow barrier at these field strengths.[31, 68, 69] As the accelerating field is increased, more of the emission is caused by penetration of electrons in the Fermi band.

At room temperature the thermionic and Schottky emission currents are negligible. If a sufficiently strong accelerating field is applied to the cathode surface, however, an

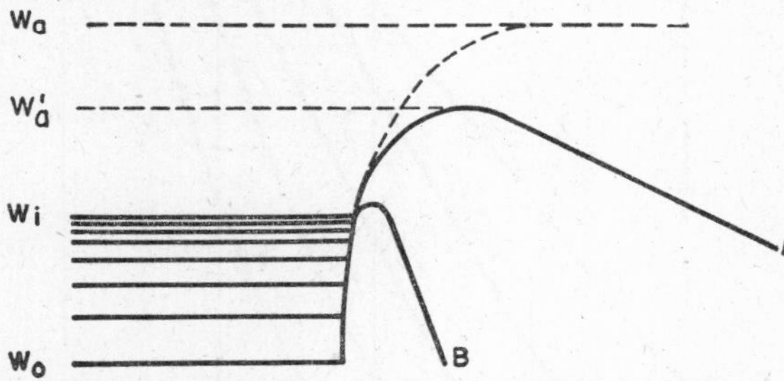

Fig. 2·22 Potential-energy diagram showing the effect of strong electrostatic fields on the height of the surface potential barrier.

appreciable number of electrons in the Fermi band can penetrate the resulting narrow barrier shown by curve B of Fig. 2·22. Electron emission caused only by the application of a strong accelerating field is called *field emission* or *autoelectronic emission*. Potential gradients of about 10^6 volts per centimeter can result in current densities of thousands of amperes per square centimeter.[70]

It is obvious that field emission is almost independent of temperature, but Schottky emission is not. From a wave-mechanical analysis, Fowler and Nordheim [31] arrived at a relation between field-emission current $I_\mathbf{E}$ from a pure metal and electric-field intensity $\mathbf{E}$,

$$I_\mathbf{E} = A_\mathbf{E}\mathbf{E}^2\epsilon^{-b_\mathbf{E}/\mathbf{E}} \tag{2·17}$$

where $A_\mathbf{E}$ and $b_\mathbf{E}$ are constants. This is surprisingly similar to the thermionic-emission relation, with temperature T replaced by the field strength $\mathbf{E}$. Although field-emission data are difficult to obtain experimentally, those which exist support the Fowler-Nordheim theory by displaying a linear relation between $\log I_\mathbf{E}$ and $1/\mathbf{E}$, and by being substantially independent of temperature.

2·5 PHOTOELECTRIC EMISSION

At a temperature of absolute zero, electrons must acquire an additional amount of energy at least equal to the work function of the surface in order to pass over the surface potential barrier. If the electrons acquire this energy by absorption of light energy, the resulting emission is called photoelectric emission.

In 1899, only a few years after the first indications of a photoelectric effect were observed by Hertz,[71] Elster and Geitel,[72] Lenard [73] and Thomson [74] discovered that kinetic energy of photoelectrically emitted electrons was independent of the intensity of the incident light, but that their number was proportional to it. To explain these results, Einstein [75] in 1905 suggested that light is composed of corpuscles each of which possesses an energy $h\nu$, where h is Planck's radiation constant and ν is the frequency of the

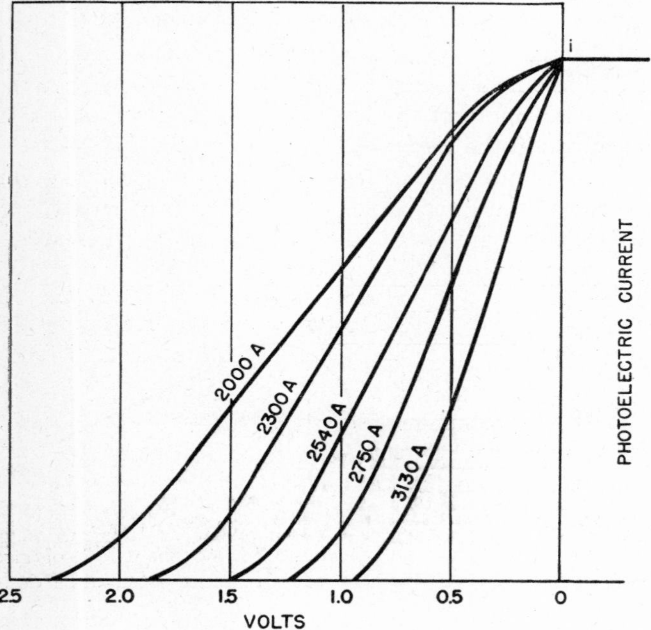

Fig. 2·23 Photoelectric current-voltage curves for aluminum. (Reprinted with permission from *Photoelectric Phenomena*, by A. L. Hughes and L. A. DuBridge, published by McGraw-Hill, 1932, p. 18.)

light. If it is assumed that this amount of energy is absorbed by an electron in the metal, the maximum energy of the emitted electron must be equal to the energy of the incident light quantum less the energy lost by the electron in overcoming the work function. Hence the maximum kinetic energy of a photoelectrically emitted electron is

$$W_{\max} = \tfrac{1}{2}mv^2 = h\nu - e\phi \tag{2·18}$$

This equation shows that, if the frequency of the incident radiation is such that the maximum energy of the photoelectrons is zero, the energy of the incident light quantum is exactly equal to $e\phi$. This minimum frequency for which electrons are photoelectrically emitted has become known as the *threshold frequency*, and the corresponding wavelength as the *threshold wavelength*. They are designated respectively as ν_0 and λ_0. Thus,

$$h\nu_0 = e\phi \quad \text{and} \quad \frac{hc}{\lambda_0} = e\phi \tag{2·19}$$

from which

$$\nu_0 = \frac{e\phi}{h} \quad \text{and} \quad \lambda_0 = \frac{hc}{e\phi} = \frac{12{,}395}{\phi} \tag{2·20}$$

where λ_0 is the threshold wavelength in angstrom units, ϕ is the work function in electron volts, and c is the velocity of light in centimeters per second.

It is particularly important that the *energy* of the individual quanta, not their number, determines whether photoelectrons are emitted from a given surface and also determines the maximum energy of the emitted photoelectrons. The number of photoelectrons emitted, however, is proportional to the number of incident quanta—the light intensity.

In general the many theories of photoelectric effect are based on either a corpuscular or wave nature of light and of electrons in metals. In the former case the action of an incident light corpuscle or quantum on electrons in metals is considered. In the latter light is treated as an electromagnetic wave, and the electrons in the metal are considered according to the Sommerfeld theory.[76, 77, 78, 11]

Characteristics of Photoelectric Emission [66]

Photoelectric surfaces are simple or composite in nature. Pure metals are examples of the simple type; the cesium-on-cesium oxide-on-silver surface is an example of the composite type. Most photoelectric properties depend upon the surface.

Three important characteristics of the photoelectric effect have already been mentioned, but are sufficiently important to be restated. They are:

1. The maximum energy of photoelectrically emitted electrons is a linear function of the frequency of the incident radiation.
2. The number of photoelectrically emitted electrons is proportional to the intensity of the incident light (number of quanta).
3. The energy of emitted electrons is independent of the intensity of the incident radiation.

The first characteristic is merely a statement of Einstein's equation. This relation has been verified experimentally by Millikan,[80] Richardson,[81] Compton,[81] and others, by measuring the emission current for various retarding potentials on the collector, then repeating these measurements for incident light of different frequencies. The results of such measurements are shown in Fig. 2·23. The intercepts of the curves for different radiation frequencies with the collector-voltage axis represent the maximum energy of emitted electrons when the surface is irradiated by light of those frequencies, if it is assumed that the appropriate correction for contact difference of potential has been made. A plot of these values of voltage versus the frequency of the radiation is shown in Fig. 2·24. The linearity of maximum electron energy with frequency is thus established, verifying Einstein's equation.

The proportionality between emission current and intensity of radiation has been investigated by Elster and Geitel,[72] Lenard,[73] Ladenburg,[82] and Richtmyer.[83] This law of photoelectricity holds over a wide range of light intensities.

Lenard, Ladenburg, Millikan, and others also proved experimentally that the maximum energy of emission of photoelectrons was independent of the intensity of the light. It was this fact which, being inconsistent with the original wave theory of light, led to the radically different conception of the photoelectric effect suggested by Einstein.

One of the most familiar characteristics of photoelectric emission is its spectral distribution curve,[66, 84] in which emission current is plotted as a function of the frequency or wavelength of the incident light. Such curves are important because they influence the choice of the proper photoelectric tube for a particular application. Spectral-distribution curves for the more important surfaces are given in Chapter 7.

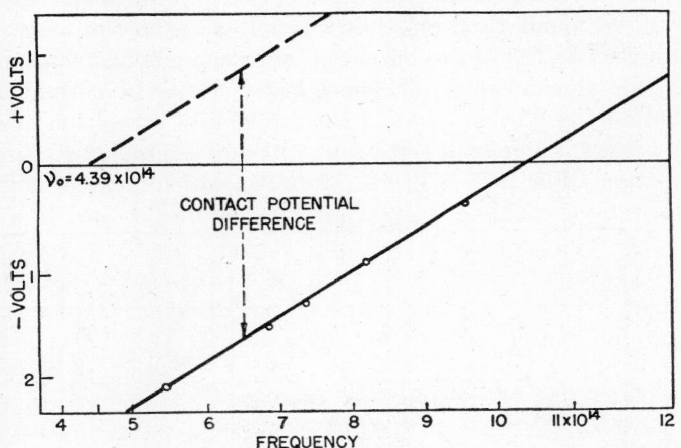

FIG. 2·24 Millikan's verification of Einstein's equation. (Reprinted with permission from *Photoelectric Phenomena*, by A. L. Hughes and L. A. DuBridge, published by McGraw-Hill, 1932, p. 21.)

Energy distribution of photoelectrons [66] is of little practical importance, but has contributed greatly to a better understanding of the photoelectric process. To determine the distribution experimentally, the retarding potential method is used. The resulting photoelectric current varies with collector voltage as shown by the solid line of Fig. 2·25. For any collector voltage, the collected current consists of electrons which have emission energies in excess of the retarding voltage on the collector (for no contact difference of potential), so the solid curve does not represent a true energy distribution but is an integrated distribution. The derivative of this curve indicates true energy distribution, shown by the dashed curve. This energy distribution depends upon the nature of the surface as indicated by Fig. 2·26, and the thickness as shown in Fig. 2·27. The total range of energies of photoelectrons for monochromatic incident light is from zero to a maximum determined from equation 2·18. The maximum obviously depends not only upon the frequency of the incident radiation but also upon the work function of the surface.

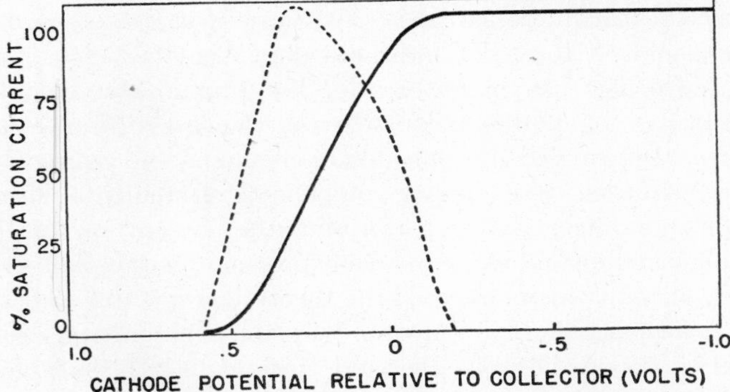

FIG. 2·25 Voltage-current curves for sodium-potassium alloy and energy distribution of photoelectrons. (Reprinted with permission from *Photoelectric Phenomena*, by A. L. Hughes and L. A. DuBridge, published by McGraw-Hill, 1932, p. 128.)

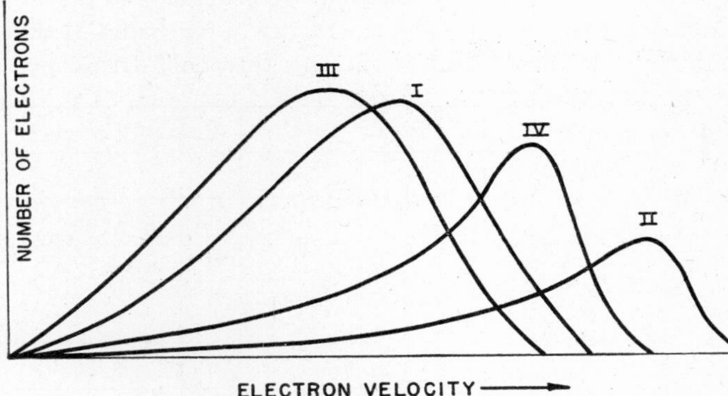

FIG. 2·26 Effect of gases on the velocity distribution of photoelectrons from platinum for 2536A.: I, fresh specimen; II, after 2 hr of heating in vacuum; III, after standing 15 hr in oxygen at 60 mm pressure; IV, after further heating in vacuum. (Reprinted with permission from *Photoelectric Phenomena*, by A. L. Hughes and L. A. DuBridge, published by McGraw-Hill, 1932, p. 127.)

Polarization of incident light has a large effect on spectral distribution, but no effect on energy distribution of photoelectrons.[66]

The effect of temperature on photoelectric emission has been investigated by Cardwell,[66] Winch,[85] and others. Large variations in temperature produce changes in the photoelectric surface, which in turn change its spectral distribution. Changes in crystal structure, composition, or even state (solid to liquid) can, of course, be produced by variations in temperature. A further temperature effect is caused by temperature variations of electron energies in the photoelectric cathode. This is particularly noticeable for frequencies near the threshold (Fig. 2·28), and makes accurate determinations of the photoelectric threshold difficult by observing the intercept in the spectral distribution. In developing a theory for temperature effect, Fowler [86] arrived

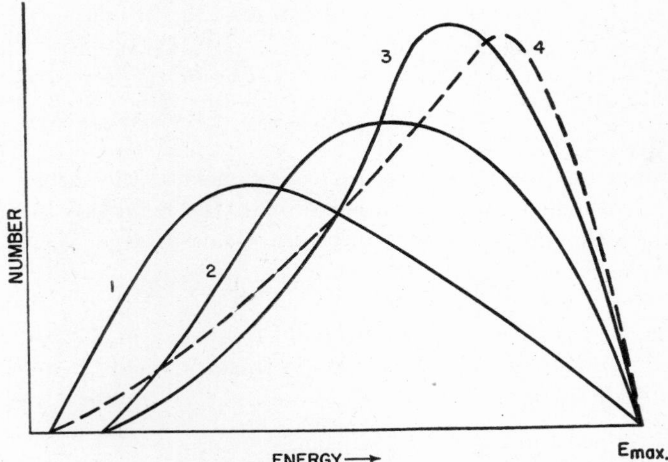

FIG. 2·27 Energy distribution of photoelectrons from silver: (1) thick layer; (2) film thickness, 3×10^{-6} cm; (3) 10^{-6} cm; (4) theoretical curve. (Reprinted with permission from *Photoelectric Phenomena*, by A. L. Hughes and L. A. DuBridge, published by McGraw-Hill, 1932, p. 231.)

at a new method for determining the threshold which compensates for the temperature of the cathode. This is described later.

The influence of electrostatic fields on photoelectric emission has not been thoroughly investigated, but the results of Lawrence and Linford [87] and the theory of Guth and Mullin [88] indicate that the reduction in the threshold frequency is

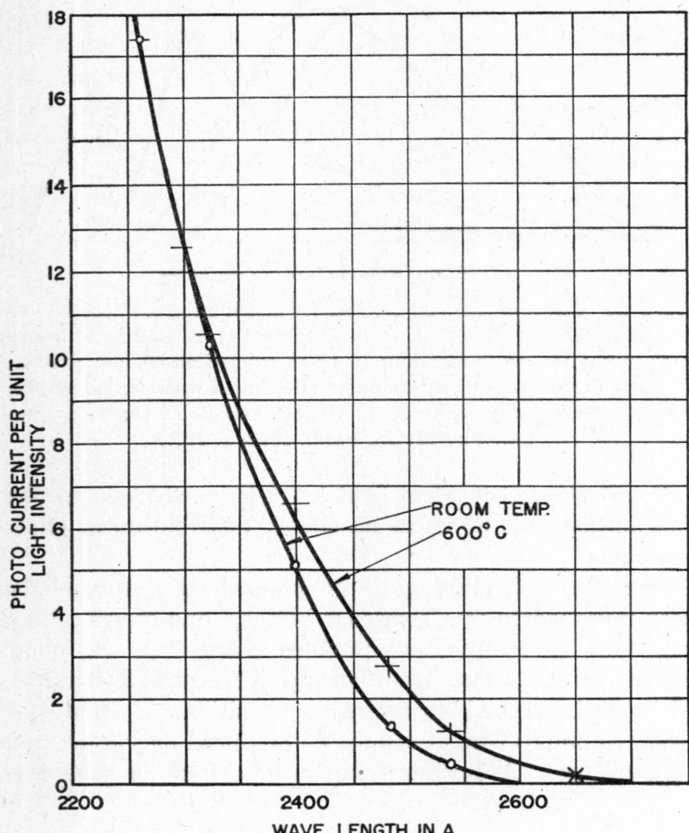

FIG. 2·28 Spectral distribution curves for silver at 20°C and 600°C. (Reprinted with permission from *Photoelectric Phenomena*, by A. L. Hughes and L. A. DuBridge, published by McGraw-Hill, 1932, p. 99.)

proportional to the square root of the field at the surface of the cathode. This is consistent with the Schottky theory and with equation 2·14, which can be rewritten

$$\Delta(h\nu_0) = h(\nu'_0 - \nu_0) = -e\sqrt{e\mathbf{E}} \qquad (2\cdot21)$$

where ν_0 is the zero-field threshold frequency and ν'_0 is the threshold frequency with an electrostatic field $\mathbf{E}$ at the cathode surface.

Theories of Photoelectric Emission

Many theories have been proposed,[66] each of which in its turn appeared to explain more of the characteristics of photoelectric emission, but no theory has satisfactorily explained all the observed phenomena quantitatively.

The first treatment of photoelectric emission based on the Sommerfeld theory of electrons in metals, the DeBroglie wavelength for electrons, and a wave-mechanical analysis of the interaction of electromagnetic radiation on electrons was that of Wentzel.[89] He considered electrons in the metal as standing waves, the wavelengths of which are determined by the DeBroglie relation (equation 2·10). These standing waves are perturbed by the incident radiation of frequency ν. After Schroedinger's wave equation and the boundary conditions have been satisfied, an analytical expression is obtained for the probability that the energy of the electron wave will change because of the perturbing effect of incident light. Wentzel found this probability small except for an energy change of $h\nu$, the Einstein concept of the energy of a quantum. A general expression for photoelectric emission current is then obtained by integrating the probability expression over all values of momenta consistent with the Fermi distribution function. For $(W_a - W_i) < h\nu < W_a$, which includes most

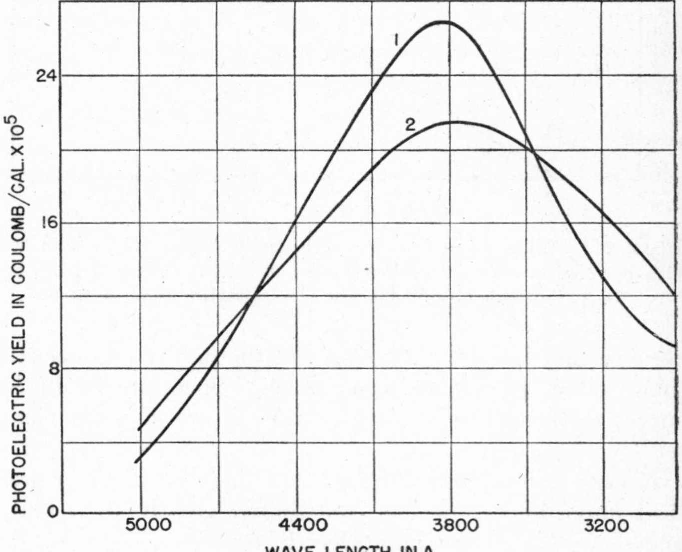

FIG. 2·29 Spectral-distribution curves for potassium: (1) theoretical curve; (2) experimental curve. (Reprinted with permission from *Photoelectric Phenomena*, by A. L. Hughes and L. A. DuBridge, published by McGraw-Hill, 1932, p. 232.)

experimental measurements, the photoelectric current is a function of both W_a and W_i instead of just $(W_a - W_i)$, the work function. This makes it possible for the theoretical spectral distribution to have a variety of shapes as well as thresholds. Houston [87] indicated that Wentzel's theory was in error because he integrated the probability over the momenta in all directions, whereas the integration should have been carried out only for momenta in the x-direction, that is, for electrons having components of momenta normal to the surface. After performing the integration in this manner, Houston arrived at an expression for the photoelectric current which results in the theoretical spectral distribution for potassium. Compared with the experimental distribution obtained by Fleischer and Teichmann,[90] this is in good agreement with experiment only for frequencies near the threshold ($\nu < 2\nu_0$).

Fröhlich [91] applied the method of Wentzel to a thin-film photoelectric surface with irregularities assumed to be small compared with the DeBroglie wavelength of the photoelectrons. The general shapes of the spectral-distribution curves obtained by Fröhlich indicate that the thin-film theory predicts spectral distributions that are of the correct shape qualitatively. The theory also qualitatively explained polarization selectivity. Reasonably good agreement was obtained between the energy distribution of photoelectrically emitted electrons as determined theoretically by Fröhlich and experimentally by Lukirsky and Prilezaev,[92] as shown in Fig. 2·27.

The theories and experimental results do not agree quantitatively on the photoelectric currents emitted for various frequencies of incident radiation. This agreement has been attained, however, by a theory of Tamm and Schubin.[93] They assume that the energy of the incident quanta must be absorbed by electrons bound in the metal or electrons held just outside the metal by the surface work-function field. They treat the quantum-electron interaction in a manner similar to that used by Wentzel. The spectral distribution resulting from their analysis agrees with the experimental results of Suhrmann and Theissing[94] both qualitatively and quantitatively, as shown in Fig. 2·29.

Some of these theories explain a single maximum in the spectral-distribution curve, but they do not show cause for a double maximum, as for a composite surface of cesium-

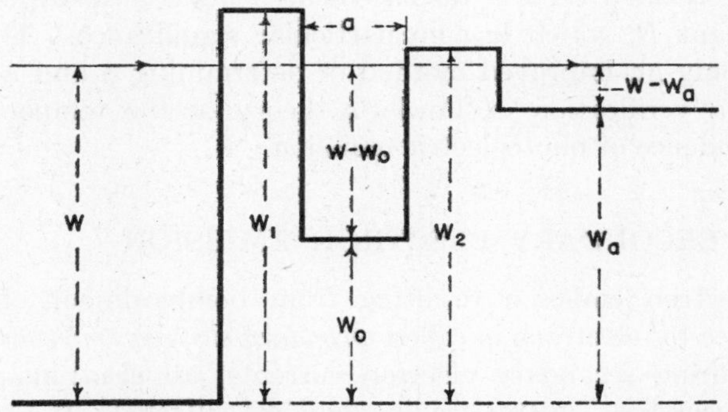

Fig. 2·30 Potential-energy diagram showing selective transmission of potential barrier. (Reprinted with permission from *Photoelectric Phenomena*, by A. L. Hughes and L. A. DuBridge, published by McGraw-Hill, 1932, p. 240.)

on-cesium oxide-on-silver. A theory of Campbell[95] and Fowler,[96] however, indicates the possibility of a second maximum for composite surfaces. Their explanation of this example of the selective photoelectric effect is based on a selective transmission of electrons through surface barriers characterized by a double hump, such as that shown in idealized form in Fig. 2·30. Using a wave-mechanical treatment, Fowler obtained an analytical expression for the transmission coefficient D for such a potential barrier. He found that $D = 0$ for $W < W_a$ and $D = 1$ for $W \gg W_1$. For intermediate values of energy D is small, but for certain critical values of W, D becomes significantly large. The values of W for which D exhibits maxima satisfy the relation

$$\frac{n\lambda}{2} = a \tag{2·22}$$

where

$$\lambda = \frac{h}{m_e v} = \frac{h}{\sqrt{2m_e(W - W_0)}}$$

and where a is the width of the potential minimum as shown in Fig. 2·30, n is an integer, and λ is the DeBroglie wavelength for an electron of energy $W - W_0$. The obvious explanation for the maximum values of D is that, for transmission, the potential minimum must have a width such that it is resonant for the frequency of the electron waves which penetrate the first barrier. This implies that standing waves are formed between the two potential maxima. Since photoelectric current is proportional to the transmission coefficient, and the energy W of the electron in the metal after absorption of the quantum is equal to its initial energy in the metal as determined by the Sommerfeld theory plus the absorbed energy $h\nu$, the spectral distribution should possess as many maxima as the transmission coefficient does for the range of energies used experimentally. For most composite surfaces, $W - W_0$ has values that restrict n of equation 2·22 to the single value of unity, although a few surfaces have sufficiently low values of W_0 to enable n to take on two values. In the former case a single maximum in the spectral distribution curve results; in the latter a second maximum appears. This theory thus can explain

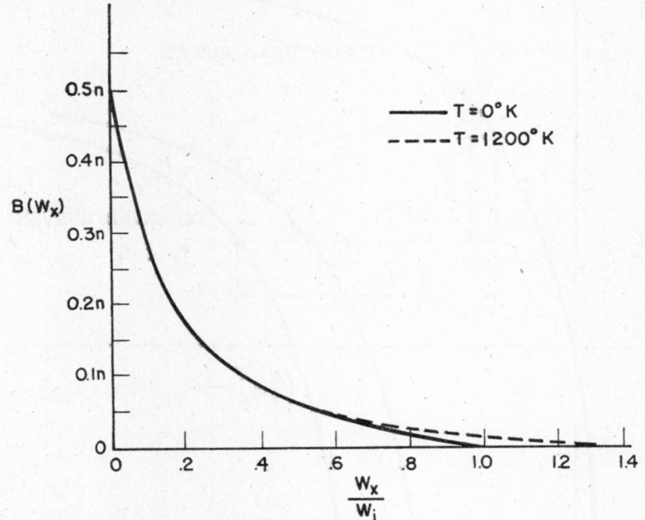

Fig. 2·31 Number of electrons per unit volume having x-associated energies at least equal to W_x/W_i.

qualitatively the general shape of spectral-distribution curves which display a selective photoelectric effect.

The effect of temperature on photoelectric emission[86] has already been mentioned, but no detailed account of it was given. Returning to the Sommerfeld theory of electrons in metals, it has been shown in Chapter 1 that the energy distribution of electrons in a metal at absolute zero and at 1500 degrees Kelvin is as shown in Fig. 1·10. At a temperature of absolute zero the kinetic energies of the electrons range from zero to a maximum W_i. At temperatures above absolute zero relatively few electrons possess kinetic energies greater than W_i, the number having energies less than W_i being diminished by the same amount. Since the x-associated energies instead of the total energies enable electrons to be emitted, an expression for the number of electrons per unit volume with x-associated energies between W_x and $W_x + dW_x$ can be obtained by integrating the Fermi-Dirac distribution function. By integrating the result thus obtained for absolute zero, the number $B(W_x)$ of electrons per unit volume at $T = 0$ which have x-associated energies at least equal to W_x is determined. The expression is

$$[B(W_x)]_{T=0} = \frac{n}{2}\left[1 - \frac{1}{2}\left(\frac{W_x}{W_i}\right)^{1/2}\left(3 - \frac{W_x}{W_i}\right)\right] \tag{2·23}$$

where n is the electron concentration. Equation 2·23, plotted in Fig. 2·31, indicates that the curve intercepts the

W_x-axis with zero slope at $W_x = W_i$. At temperatures above absolute zero, the curve approaches the W_x-axis asymptotically. If an electron of kinetic energy W_x absorbs the energy of a quantum $h\nu$ of incident light, its resulting energy W'_x is given by

$$W'_x = W_x + h\nu \tag{2·24}$$

Since the energy W'_x determines whether the electron possesses sufficient energy to pass over the surface potential

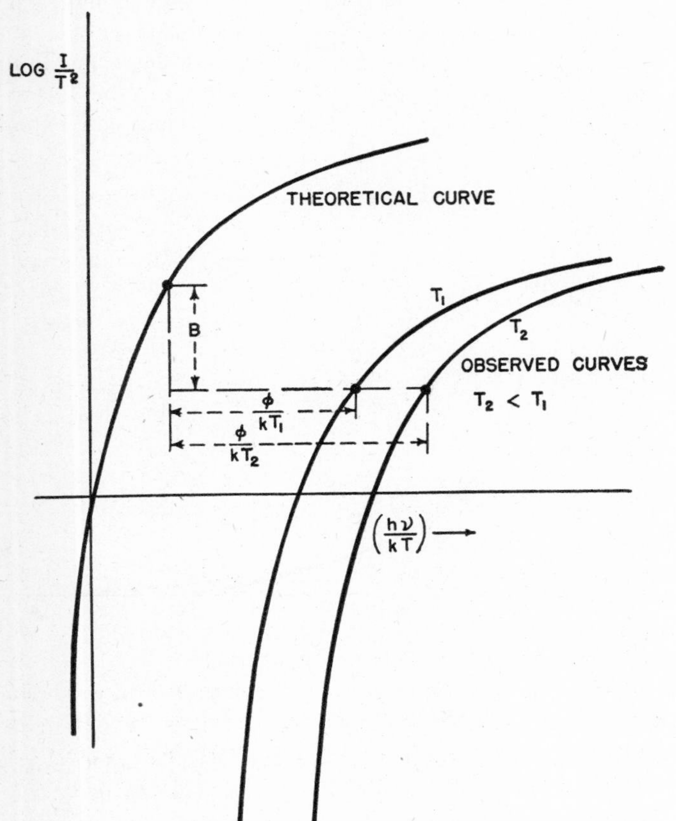

Fig. 2·32 Fowler's method of determining the true photoelectric threshold. (Reprinted with permission from *Photoelectric Phenomena*, by A. L. Hughes and L. A. DuBridge, published by McGraw-Hill, 1932, p. 244.)

barrier of height W_a, the photoelectric current must depend upon the initial energy distribution of the electrons in the metal, and it is directly proportional to the number of electrons having x-associated energies W'_x greater than W_a. The asymptotic nature of the curve shown in Fig. 2·31 for temperatures above absolute zero indicates that the intercept method of determining photoelectric threshold is inaccurate because the value thus obtained depends upon the temperature of the surface and the sensitivity of the measuring apparatus. As stated previously, Fowler's [86] theory of temperature effect provides a method of determining photoelectric threshold in which the temperature of the surface does not affect the result. Letting

$$x = \frac{h\nu - e\phi}{kT} = \frac{h(\nu - \nu_0)}{kT}$$

Fowler found that the relation between photoelectric current I, surface temperature T, and frequency ν of the incident radiation can be represented by an analytical expression of the form

$$\log \frac{I}{T^2} = B + \log f(x) \tag{2·25}$$

where

$$f(x) = \frac{\pi^2}{6} + \frac{x^2}{2} - \left(e^{-x} - \frac{e^{-2x}}{2^2} + \frac{e^{-3x}}{3^2} - \cdots\right) \quad \text{for } x \geqq 0$$

or

$$f(x) = \left(e^{x} - \frac{e^{2x}}{2^2} + \frac{e^{3x}}{3^2} - \cdots\right) \quad \text{for } x \leqq 0$$

and

$$B = \text{a constant}$$

The plot of log (I/T^2) versus $h\nu/kT$ as determined by equation 2·25 is shown in Fig. 2·32. Also indicated is the method of determining the work function and photoelectric threshold of a surface by observing the horizontal displacement of the observed curve required to "match" it to the theoretical one, the necessary vertical displacement being a measure of the constant B, which has no particular significance. This is not only an improved method of determining ϕ and ν_0 but also a verification of Fowler's theory of the temperature dependence of photoelectric emission.

2·6 SECONDARY ELECTRON EMISSION

Electron emission resulting from bombardment of the surface by electrons is called *secondary electron emission*. In measuring secondary electron currents, all electrons from the secondary emitter are unavoidably collected. If a material is bombarded by electrons, the electron current from it consists not only of emitted electrons but also of primary (or bombarding) electrons reflected from the surface. The distinction between reflected and true secondary electrons can be made on the basis of their energies. Secondary electrons are only those electrons actually emitted from the bombarded target. This distinction is important theoretically, but is of limited practical significance, because usually the number of electrons actually emitted from the target is much greater than the number reflected. The secondary emitting properties of a surface usually are described in terms of its yield δ, which is the average number of secondary electrons produced by one bombarding electron.

Theory of Secondary Electron Emission

The mechanism of secondary emission cannot be simply one of the impact of primary electrons with the free electrons of conduction of the metal. This would violate the law of conservation of momentum. More involved mechanisms, such as the interaction of primary electrons and the lattice, interaction between the primary electron and bound electrons of the lattice, and multiple collisions must be considered.

The theory of secondary electron emission leaves much to be desired. In fact, there are two principal theories which are basically different. One has been suggested by Bruining and others [97] and the other by Timofeev and others.[98] Although the theories are so different, they both explain secondary emission phenomena well enough so that it is impossible to decide between them.

The principal difficulty in setting up a theory is that the theory must explain not only the secondary emission of metals, but also the high secondary emission of composite surfaces. Fröhlich [99] in 1932 analyzed the secondary electron emission of metals by means of quantum mechanics, and arrived at a fairly satisfactory agreement with experimental results. His theory is not applicable to composite surfaces.

Bruining believes that secondary electrons are released by collision of primary electrons with bound electrons of the lattice. According to this theory, the low secondary electron emission of metals is caused by the large number of unoccupied energy bands in a metal. Semi-conductors have higher emission because they have no unoccupied energy bands. Hence, in a semi-conductor a secondary electron is less likely to lose its energy on its way to the surface, and therefore has a greater chance of being emitted.

The theory of Timofeev is based on the insulating qualities of a composite cathode. He asserts that the high secondary emission of composite cathodes is caused by field emission produced by positive ions left behind when secondary electrons are emitted. The positive ions, in a manner similar to the thin-film field emission of Malter [100] (discussed later), set up a high potential gradient which results in emission of electrons from within the semi-conductor. This effect is called an *inertia-less Malter effect* because there is no characteristic time delay in the rise and fall of the high secondary emission. The explanation of this is that no great charge need be built up and the positive ions which set up the high potential gradient recombine rapidly. Timofeev estimated the recombination time as 10^{-9} second, in which time an electron of one volt energy would travel 10^{-2} centimeter (which is about 10^5 times as great as the distance that must be traversed by a secondary electron). Hence, positive ions could cause the emission of numerous electrons before being neutralized by recombination. A secondary electron, on reaching the surface, is pulled away by the external electric field produced by the positively charged collector.

In brief, Timofeev believes that secondary emission of composite surfaces is composed of two distinct phenomena: ordinary secondary emission which is characteristic of metals with low coefficients of secondary emission ($\delta < 2$); and additional secondary emission produced by positive charges set up by the usual secondary electron emission, and which is composed of the field emission produced by the positive charges set up on the surface.

According to this theory the secondary emission of metals is low because of their high conductivity, which leads to rapid neutralization of the positive charge. A good secondary electron emitter is therefore a semi-conductor with metal molecules interspersed inside to provide centers for the production of field emission.

Yield and Types of Surfaces

Secondary electron emitters can be divided into several classes:

1. Metals.
2. Oxides of metals of high work function.
3. Compounds of metals.
4. Composite cathodes.
5. Insulators.
6. Thin-film field emitters.

Metals. All pure metals have low secondary emission ($\delta_{max} < 2$). In making measurements, the slightest traces of oxides and other impurities [101] must be excluded, for they can produce a four- or five-fold increase in secondary emission.

Secondary emission from surfaces is extremely sensitive to the slightest traces of impurities, so various experimenters disagree on the values of yield for surfaces. Probably the differences in yield are caused by variations in vacuum and in surface purity. For this reason it is difficult to make an accurate quantitative comparison between many of the published values of secondary-electron emission yield. The values given in this chapter are therefore subject to the same criticism and can be accepted only after the conditions under which they were obtained have been investigated.

Low work function does not necessarily mean high secondary emission, because secondary emission is not merely a matter of electron emission from the surface of a solid. It is a phenomenon which, in addition, involves excitation of secondary electrons by primary electrons, and the travel of secondary electrons to the surface of the emitter. Only when secondary electrons reach the surface of the metal does the phenomenon become one of electron emission from the surface. Hence, the structure of the emitter is of great importance.

In just one case is there a correspondence of secondary electron emission and work function. That is the secondary electron emission of metals under the bombardment of slow primary electrons [102] (10 to 15 volts). For this condition, the lower the work function is, the higher the secondary emission. However, maximum secondary emission in metals is at much higher voltages (500 to 600 volts), for which secondary emission does not correspond to work function.

Table 2.4 Correlation of the Work Function and Secondary Electron Emission Yield with Density of Several Elements

Metal	Density (g per cc)	Work Function (e.v.)	Maximum Yield (max)	Primary Electron Energy (e.v.)
Li	0.53	2.1–2.9	0.55	100
K	0.86	1.76–2.25	0.69	300
Rb	1.5	1.8–2.19	0.85	400
Mg	1.7	2.42	0.95	300
Be	1.8	3.16	0.90	200
Cs	1.9	1.80–1.96	0.90	400
Al	2.7	2.5–3.6	0.97	300
Ba	3.8	2.11	0.90	400
Sn	7.3	4.17–4.50	1.35	500
Fe	7.8	4.72–4.77	1.32	400
Ni	8.8	5.01–5.03	1.27	500
Cu	8.9	4.1–4.5	1.35	500
Mo	9.0	4.15	1.22	400
Bi	9.8	4.0–4.4	1.35	500
Ag	10.5	4.08–4.76	1.47	800
Pb	11.3	3.5–4.1	1.08	500
Ta	11.8	4.12	1.35	600
W	19.1	4.52	1.43	700
Au	19.3	4.42–4.92	1.47	700
Pt	21.4	5.32–6.27	1.78	700

If the density of the secondary-emitting metal is taken into account, a correlation between work function and secondary emission yield can be found. For metals of low density ($\rho < 4$ grams per cubic centimeter) and a low work function ($\phi < 4$ electron volts) the maximum secondary emission yield is less than unity, as shown in Table 2·4. For substances of higher density ($\rho > 7$ grams per cubic centimeter) and with a high work function ($\phi > 4$ electron volts) the maximum secondary emission yield is greater than unity. The lower yield for metals of low density and the higher yield for those of higher density can be explained on the basis of the difference in penetration of primary electrons. Primary electrons penetrate more deeply in materials of low density, resulting in a smaller yield because the secondary electrons are liberated at greater depths and suffer more energy loss by collision with atoms of the metal before they reach the surface.

Oxides of High-Work-Function Metals. Metals of high work function having an oxide layer on their surface have a low secondary emission yield, δ_{max} for copper oxide, molybdenum oxide, and silver oxide being 1.19, 1.33, and 1.18 respectively.

Compounds of Metals. Metallic compounds have secondary emission yields higher than the metals from which they were derived. Maximum yields range from 2 to 9. Yields for some metallic salts are given in Table 2·5.

TABLE 2·5 SECONDARY ELECTRON EMISSION YIELD FOR SEVERAL METALLIC COMPOUNDS

Metallic Compound		Maximum Yield
Barium oxide	BaO	4.8
Sodium chloride	NaCl	6.8
Lithium fluoride	LiF	5.6
Sodium fluoride	NaF	5.7
Copper fluoride	CuF_2	3.15
Barium fluoride	BaF_2	4.5
Potassium chloride	KCl	7.5
Rubidium chloride	RbCl	5.8
Cesium chloride	CsCl	6.5
Sodium bromide	NaBr	6.25
Sodium iodide	NaI	5.5
Potassium iodide	KI	5.6

Composite Cathodes. Still greater secondary emission yields are obtained from composite cathodes, properly activated.[103] Typical maximum yields are given in Table 2·6. For composite cathodes, the highest yield is obtained from cathodes which have built-in atoms of low-work-function metals.

Insulators. Though insulators may have some secondary emission,[104] the potential of the surface of the emitter always adjusts itself so that the net current to the emitter is zero. If the yield is less than one ($\delta < 1$), the emitter charges negatively until it reaches the cathode potential. If the secondary emission yield is greater than one ($\delta > 1$), the emitter becomes charged positively so that the secondary electron current is equal to the bombarding current, net current being zero. In vacuum tubes the glass walls become charged positively and sometimes cause high-voltage electron bombardment sufficient to melt the glass wall at the point. The maximum yields of quartz and glass are about 2.9 and 3.15.

TABLE 2·6 SECONDARY ELECTRON EMISSION YIELD FOR SEVERAL COMPOSITE CATHODES

Composite Cathode			Maximum Yield	Primary Electron Energy (e.v.)
Semi-conductor		Inbuilt atoms		
Molybdenum oxide	MoO_3	Pt	1.25	1200
Silver oxide	Ag_2O	Au	1.2	600
Silver oxide	Ag_2O	Ni	1.8	600
Silver oxide	Ag_2O	Cs	9.0	800
Silver sulphide	Ag_2S	Cs	17.5	400
Cesium oxide	Cs_2O	Au, Ag	4.5	800
Cesium oxide	Cs_2O	Ag	7.2	350
Beryllium oxide	BeO	Ni	12.0	750
Potassium iodide	KI	Cs	12.6	1800
Cesium chloride	CsCl	Cs	13.7	1700
Sodium chloride	NaCl	Cs	22.7	1400
Sodium chloride	NaCl	(Na)	9.7	700
Magnesium fluoride	MgF_2	(Mg)	4.1	800
Magnesium oxide	MgO	(Mg)	7.5	550
Barium oxide	BaO	(Ba)	6.5	500

Thin-Film Field Emitters. Thin-film field emission was first observed by Malter[105] and is therefore sometimes called the *Malter effect*. A typical surface showing this effect is made by forming a thin film of aluminum oxide (about 2000 angstrom units thick) on an aluminum base, and then covering with a layer of cesium. If this type of surface is bombarded by electrons, secondary emission yields are 1000 times the usual values. Further, the high secondary emission is not obtained immediately after the primary beam is turned on but rises slowly to the maximum. In addition, when the primary electron beam is turned off, the secondary emission does not immediately go to zero, but decreases slowly. Malter attributed the abnormally high secondary emission to field emission from the base metal. The high field is set up by positive charges left on the surface by emission of secondary electrons, acting through the thin film of aluminum oxide. The presence of positive charges on the surface has been demonstrated experimentally, thus supporting Malter's explanation of thin-film field emission, which is now well founded.

Characteristics of Secondary Electron Emission

Secondary electron emission varies not only with surface conditions and type of emitter, but also with other factors, such as the primary electron energy and the angle of incidence of primary electrons. The energy distribution is *one* of the characteristics of secondary electron emission.

Energy Distribution. Electrons from a surface bombarded by primary electrons can be collected by a more positive electrode in the vicinity. The electrons collected have the energy distribution shown in Fig. 2·33. Three types of electrons are to be distinguished:[106] I, elastically reflected electrons of the same energy as the primary electrons; II, inelastically reflected electrons; and III, true secondary electrons, which actually originate from within the emitter. The energy of the true secondary electrons, III, is low, generally below 10 electron volts.

Variation of Secondary Emission with Primary Energy. Secondary emission of solids varies with the energy of the primary electrons. For low energy, increased primary energy enables the bombarding electron to supply more energy; hence, it results in increased secondary emission. As

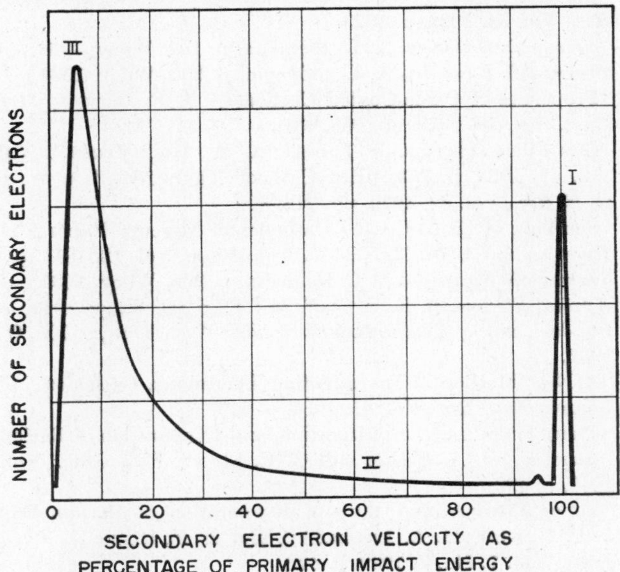

FIG. 2·33 Velocity distribution of secondary electrons for a primary electron impact energy of 155 volts. (Reprinted with permission from J. H. Owen Harries, "Secondary Electron Radiation," *Electronics*, Sept. 1944, p. 100.)

the energy of the primary electron is increased, however, the penetration increases, and secondary electrons are liberated at greater depths beneath the surface. The secondary electron then has farther to travel to get to the surface. Energy losses in collisions on the way to the surface reduce the energy of the secondary electron and therefore decrease secondary emission. Variation of the secondary emission of a metal with energy of the bombarding electrons is shown in Fig. 2·34. A broad maximum occurs at about 500 volts, and there is a slow decline thereafter.

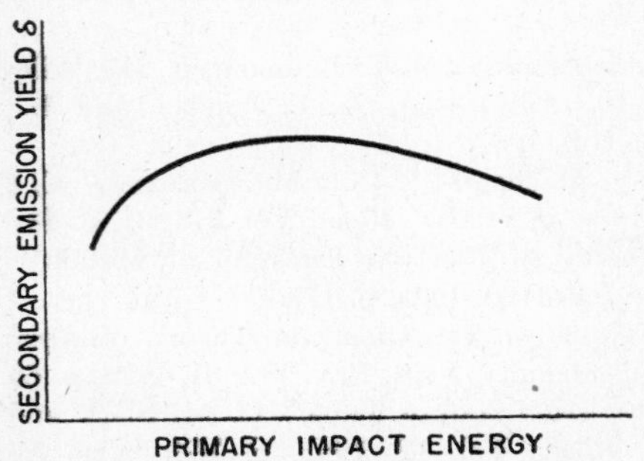

FIG. 2·34 Typical secondary emission yield-primary impact energy curve.

Variation of Secondary Emission with Primary Angle. Secondary emission of solids varies also with the angle of incidence of primary electron. It is maximum for oblique angles of incidence and minimum for normal incidence, because of the difference in penetration of primary electrons. Figure 2·35 indicates variation of secondary emission yield for various surfaces with angle of incidence of the primary beam.

Angular Distribution of Secondary Emission. Secondary electrons do not leave the surface in equal numbers at all angles. The angular distribution follows the cosine law. Maximum emission is normal to the surface and decreases as the angle to the normal increases.

Effect of Temperature. Secondary emission of metals seems to be almost independent of the temperature.[104] In the following metals it varies less than 4 percent over the tem-

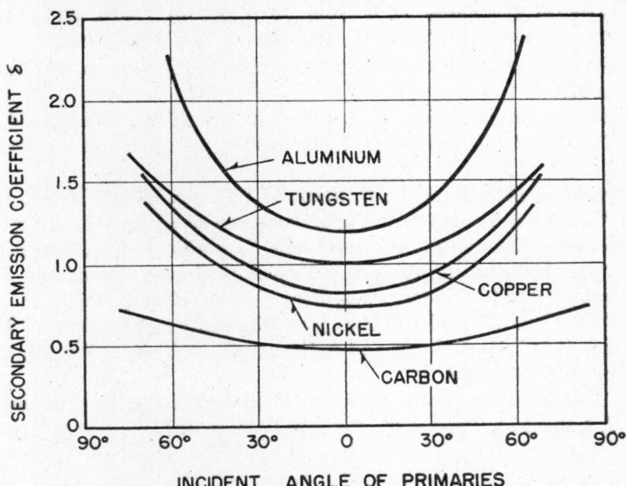

FIG. 2·35 Dependence of secondary emission yield upon the angle of incidence of the primary electrons. (Reprinted with permission from J. H. Owen Harries, "Secondary Electron Radiation, "*Electronics*, Sept. 1944, p. 105.)

perature ranges shown: tungsten, from 300 to 1700 degrees Kelvin; copper, from 300 to 750 degrees Kelvin; and molybdenum, from 300 to 1600 degrees Kelvin. Changes in secondary emission yield for iron, nickel, and cobalt correspond to the changes in crystal structure; however, these changes are less than 2 percent.

Secondary emission of composite cathodes varies with temperature, usually increasing with temperature. Increased secondary emission is explained by Bruining on the basis of the increased conductivity of the intermediate layer.

REFERENCES

1. *Fundamentals of Engineering Electronics*, W. G. Dow, p. 195, Wiley, 1937.
2. *Applied Electronics*, E. E. Staff, M.I.T., p. 73, Wiley, 1944.
3. "A Phenomena of the Edison Lamp," T. A. Edison, *Engineering*, Dec. 12, 1884, p. 553.
4. *Applied Electronics*, E. E. Staff, M.I.T., p. 68, Wiley, 1944.
5. *Applied Electronics*, E. E. Staff, M.I.T., p. 71, Wiley, 1944.
6. *Emission of Electricity from Hot Bodies*, O. W. Richardson, Longmans, Green, 2nd ed., 1921.
7. "Electron Theory of Matter," O. W. Richardson, *Phil. Mag.*, Vol. 23, 1912, pp. 601, 619; "Electron Theory of Thermo-Electricity," O. W. Richardson, *Phil. Mag.*, Vol. 24, 1912, p. 740; "Electron Theory and Gas Molecule Distribution in Field of Force," *Phil. Mag.*, Vol. 28, 1914, p. 633.
8. "Electrical Conductivity of Air and Salt Vapors," H. A. Wilson, *Trans. Roy. Soc. (London), Series A*, Vol. 197, 1901, p. 429; "Discharge of Electricity from Hot Platinum," H. A. Wilson, *Trans. Roy. Soc. (London) Series A*, Vol. 202, 1903, p. 258; "Discharge of Electricity by Hot Bodies," H. A. Wilson, *Trans. Roy. Soc. (London) Series A*, Vol. 24, 1912, p. 196.
9. "Thermionic Emission," S. Dushman, *Rev. Mod. Phys.*, Vol. 2, 1930, p. 381.

10. "Electrochemical Reactions of Tungsten, Thorium, Caesium, and Oxygen," I. Langmuir, *Ind. Eng. Chem.*, Vol. 22, 1930, p. 390.
11. "Thermionic Emission from Tungsten and Thoriated Tungsten Filaments," W. B. Nottingham, *Phys. Rev.*, Vol. 49, 1936, p. 78.
12. "Electron Emission from Tungsten, Molybdenum, and Tantalum," S. Dushman, H. N. Rowe, J. Ewald, and C. A. Kidner, *Phys. Rev.*, Vol. 25, 1925, p. 338.
13. *Photoelectric Phenomena*, A. L. Hughes and L. A. DuBridge, p. 56, McGraw-Hill, 1932; "Thermionic Emission of Platinum," L. V. Whitney, *Phys. Rev.*, Vol. 50, 1936, p. 1154.
14. *Applied Electronics*, E. E. Staff, M. I. T., pp. 77–78, Wiley, 1944.
15. *Thermionic Emission*, A. L. Reimann, p. 37, Wiley, 1934.
16. *Fundamentals of Engineering Electronics*, W. G. Dow, p. 236, Wiley, 1937.
17. *Thermionic Emission*, A. L. Reimann, pp. 41–50, Wiley, 1934.
18. *Physics of Electron Tubes*, L. R. Koller, pp. 19–60, McGraw-Hill, 1934.
19. *Applied Electronics*, E. E. Staff, M.I.T., pp. 84–94, Wiley, 1944.
20. "The Thermionic Constants of Tungsten as a Function of Crystallographic Direction," M. H. Nichols, *Phys. Rev.*, Vol. 57, 1940, p. 297.
21. "The Distribution of Initial Velocities among Thermionic Electrons," L. H. Germer, *Phys. Rev.*, Vol. 25, 1925, p. 795.
22. "The Characteristics of Tungsten Filaments as Functions of Temperature," H. A. Jones and I. Langmuir, *G. E. Rev.*, Vol. 30, 1927, pp. 310, 354, 408.
23. "The Electron Emission from Thoriated Tungsten Filaments," I. Langmuir, *Phys. Rev.*, Vol. 22, 1923, p. 357.
24. "The Effect of Space Charge and Residual Gases on Thermionic Currents in High Vacuum," I. Langmuir, *Phys. Rev.*, Vol. 2, 1913, p. 461.
25. *Thermionic Emission*, A. L. Reimann, pp. 103–187, Wiley, 1934.
26. "The Electron Emission from Tungsten Filaments containing Thorium," I. Langmuir and W. Rogers, *Phys. Rev.*, Vol. 4, 1914, p. 544.
27. "The Evaporation of Atoms, Ions, and Electrons from Caesium Films on Tungsten," I. Langmuir and J. B. Taylor, *Phys. Rev.*, Vol. 44, 1933, p. 423.
28. "Thoriated Tungsten Filaments," I. Langmuir, *J. Franklin Inst.*, Vol. 217, 1934, p. 543.
29. "Thermionic and Adsorption Characteristics of Thorium on Tungsten," D. H. Brattain and J. A. Becker, *Phys. Rev.*, Vol. 43, 1933, p. 428.
30. "Photoelectric and Thermionic Emission from Composite Surfaces," W. B. Nottingham, *Phys. Rev.*, Vol. 41, 1932, p. 793; "Influence of Accelerating Fields on the Photoelectric and Thermionic Work Function of Composite Surfaces," W. B. Nottingham, *Phys. Rev.*, Vol. 35, 1930, p. 1128.
31. "Electron Emission in Intense Electric Fields," R. H. Fowler and L. W. Nordheim, *Proc. Roy. Soc. (London), Series A*, Vol. 119, 1928, pp. 173–181.
32. "The Effect of the Image Force on the Emission and Reflection of Electrons by Metals," L. W. Nordheim, *Proc. Roy. Soc. (London), Series A*, Vol. 121, 1928, pp. 626–639.
33. *Physics of Electron Tubes*, L. R. Keller, p. 31, McGraw-Hill, 1934.
34. "On Cathode Rays at Glowing Cathodes," A. Wehnelt, *Deut. physik. Ges. Verhandl.*, Vol. 5, 1903, pp. 255, 423.
35. "Colloid Science and Radio Technique," A. Von Buzagh, *Kolloid-Z.*, Vol. 77, 1936, p. 172.
36. "Oxide Cathodes of Colloidal Structure," E. Patia and Z. Tomaschek, *Kolloid-Z.*, Vol. 74, 1936, and Vol. 75, 1936, p. 253 and p. 80 respectively.
37. "Oxide Cathodes," W. Statz, *Z. tech. Physik*, Vol. 8, 1927, p. 451.
38. "The Cause of Emission of Electrons from Hot Cathodes," A. Gehrts, *Deut. physik. Ges. Verhandl.*, Vol. 15, 1913, p. 1047.
39. "Variation with Temperature of the Work Function of Oxide-Coated Platinum," J. S. Glass, *Phys. Rev.*, Vol. 28, 1926, p. 521.
40. "Electron Emission from Oxide-Coated Filaments," L. R. Koller, *Phys. Rev.*, Vol. 26, 1925, p. 671.
41. "The Role of the Core Metal in Oxide-Coated Filaments," E. F. Lowry, *Phys. Rev.*, Vol. 35, 1930, p. 1367.
42. "Emission Work Function with Oxide Cathodes," H. Rothe, *Z. Physik*, Vol. 36, 1926, p. 737.
43. "On the Emission Mechanism of Oxide-Coated Cathodes," W. Espe, *Wiss. Veröffentl. Siemens-Werken*, Vol. 5 (III), 1926, pp. 29, 46.
44. "Thermionic Emission and Electrical Conductivity of Oxide Cathodes," A. L. Reimann and R. Murgoci, *Phil. Mag.*, Vol. 9, 1930, p. 440.
45. "Phenomena in Oxide-Coated Filaments," J. A. Becker, *Phys. Rev.*, Vol. 34, 1929, p. 1323.
46. *Thermionic Emission*, A. L. Reimann, p. 192, Wiley, 1934.
47. *Thermionic Emission*, A. L. Reimann, p. 202, Wiley, 1934.
48. "Properties of Oxide-Coated Cathodes," J. P. Blewett, *J. App. Phys.*, Vol. 10, 1939, pp. 668, 831.
49. "Gas Filled Thermionic Tubes," A. W. Hull, *Trans. A.I.E.E.*, Vol. 47, 1928, p. 753; "Hot Cathode Thyratrons," A. W. Hull, *G. E. Rev.*, Vol. 32, 1929, pp. 213, 390.
50. "Sparking of Oxide-Coated Cathodes in Mercury Vapor," D. D. Knowles and J. W. McNall, *J. App. Phys.*, Vol. 12, 1941, p. 149.
51. *Thermionic Emission*, A. L. Reimann, p. 205, Wiley, 1934.
52. *Applied Electronics*, E. E. Staff, M.I.T., p. 92, Wiley, 1944.
53. *Fundamentals of Engineering Electronics*, W. G. Dow, p. 205, Wiley, 1937.
54. "A New Method of Investigating Thermionic Cathodes," C. G. Found, *Phys. Rev.*, Vol. 45, 1934, p. 519.
55. "Thermionic Emission, Migration, and Evaporation of Barium on Tungsten," J. A. Becker and G. E. Moore, *Phil. Mag.*, Vol. 29, 1940, p. 129.
56. "Vapor Pressure and Rate of Evaporation of Barium Oxide," J. P. Blewett, H. A. Liebhafsky, and E. F. Hennelly, *J. Chem. Phys.*, Vol. 7, No. 7, 1939, p. 478.
57. *Applied Electronics*, E. E. Staff, M.I.T., p. 94, Wiley, 1944.
58. *Electronics*, J. Millman and S. Seely, p. 167, McGraw-Hill, 1941.
59. *Fundamentals of Engineering Electronics*, W. G. Dow, p. 192, Wiley, 1937.
60. *Thermionic Emission*, A. L. Reimann, p. 61, Wiley, 1934.
61. "On Electrical Fields Near Metallic Surfaces," J. A. Becker and D. W. Mueller, *Phys. Rev.*, Vol. 31, 1928, p. 431.
62. "Current Between an Incandescent Filament and Co-axial Cylinder," W. Schottky, *Physik. Z.*, Vol. 15, 1914, p. 624; "Emission of Electrons from an Incandescent Filament under the Action of a Retarding Potential," W. Schottky, *Ann. Physik*, Vol. 44, 1914, p. 1011; "Further Remarks on the Electronic Vapor Problem," W. Schottky, *Physik. Z.*, Vol. 20, 1919, p. 220; "Cold and Hot Electron Discharges," W. Schottky, *Z. Physik*, Vol. 14, 1923, p. 63.
63. "Schottky Effect and Contact Potential Measurements on Thoriated Tungsten Filaments," N. B. Reynolds, *Phys. Rev.*, Vol. 35, 1930, p. 158.
64. *Fundamentals of Engineering Electronics*, W. G. Dow, p. 204, Wiley, 1937.
65. *Thermionic Emission*, A. L. Reimann, p. 247, Wiley, 1934.
66. *Photoelectric Phenomena*, A. L. Hughes and L. A. DuBridge, McGraw-Hill, 1932.
67. "Quantum Mechanics of Collision Processes," E. U. Condon and P. M. Morse, *Rev. Mod. Phys.*, Vol. 3, 1931, p. 43.
68. "The Theory of Electron Emission of Metals," L. Nordheim, *Physik. Z.*, Vol. 30, 1929, p. 177.
69. "Three Notes on the Quantum Theory of Aperiodic Effects," J. R. Oppenheimer, *Phys. Rev.*, Vol. 31, 1928, p. 66.
70. "Field Emission X-Ray Tube," C. M. Slack and L. F. Ehrke, *J. App. Phys.*, Vol. 12, 1941, p. 165; "One Millionth-Second Radiography," C. M. Slack, *J. Photographic Soc. America*, Vol. 11, 1945, No. 7; "Field Emission Applied to Ultra-Speed X-ray Technique," C. M. Slack and E. R. Thilo, *Proc. 1st National Electronics Conference*, Oct. 1944.
71. "The Effect of Ultra Violet Light on the Electrical Discharge," H. Hertz, *Ann. Physik*, Vol. 31, 1887, p. 983.
72. "Note on the Dissipation of Negative Electricity through Sun and Daylight," J. Elster and H. Geitel, *Ann. Physik*, Vol. 38, 1889, pp. 40; "On the Use of Sodium Amalgams to Investigate Photoelectricity," J. Elster and H. Geitel, *Ann. Physik*, Vol. 41, 1890, p. 161; "Note on a New Type of Apparatus for the Demonstration of Photoelectric Discharge through Daylight," J. Elster and H. Geitel, *Ann. Physik*, Vol. 42, 1891, p. 564; "On the Com-

parison of Light Intensities by Means of Photoelectricity," J. Elster and H. Geitel, *Ann. Physik*, Vol. 48, 1892, p. 625.

73. "Production of Kathode Rays by Ultra-Violet Light," P. Lenard, *Sitzber. Akad. Wiss. Wien*, Vol. 108, 1899, p. 1649; "Production of Kathode Rays by Ultra-Violet Light," P. Lenard, *Ann. Physik*, Vol. 2, 1900, p. 359; "Light Electric Effect," P. Lenard, *Ann. Physik*, Vol. 8, 1902, p. 149.
74. "Masses of Ions," J. J. Thomson, *Phil. Mag.*, Vol. 48, 1899, p. 547.
75. "Generation and Transformation of Light," A. Einstein, *Ann. Physik*, Vol. 17, 1905, p. 132.
76. "Electron Theory of Metals on the Basis of the Fermi Statistics," A. Sommerfeld, *Z. Physik*, Vol. 47, 1928, p. 1.
77. "Statistical Theories of Matter, Radiation, and Electricity," K. K. Darrow, *Phys. Rev. Supp.*, Vol. 1, 1929, p. 90.
78. "The Electron Theory of Metals," R. H. Fowler, *Nature*, Vol. 126, 1930, p. 611.
79. *Physics of Electron Tubes*, L. R. Koller, Chapter XI, McGraw-Hill, 1934.
80. "A Direct Photoelectric Determination of Planck's h," R. A. Millikan, *Phys. Rev.*, Vol. 7, 1916, p. 363.
81. "Velocity of Electrons Liberated by Ultra-Violet Light," K. T. Compton, *Phil. Mag.*, Vol. 23, 1912, pp. 579, 594; "Theory of Photoelectric Action," O. W. Richardson, *Phil. Mag.*, Vol. 24, 1912, p. 570.
82. "Discharging Action of Ultra-Violet Light," E. Ladenburg, *Ann. Physik*, Vol. 12, 1903, p. 573.
83. "Dependence of Photoelectric Current on Light Intensity," F. K. Richtmyer, *Phys. Rev.*, Vol. 29, 1909, p. 71; "Photoelectric Effect with the Alkali Metals," F. K. Richtmyer, *Phys. Rev.*, Vol. 29, 1909, p. 404.
84. *Applied Electronics*, E. E. Staff, M.I.T., p. 104, Wiley, 1944.
85. "The Photoelectric Properties of Silver," R. P. Winch, *Phys. Rev.*, Vol. 37, 1931, p. 1269.
86. "The Analysis of Photoelectric Sensitivity Curves for Clean Metals at Various Temperatures," R. H. Fowler, *Phys. Rev.*, Vol. 38, 1931, p. 45.
87. "The Effect of Intense Electric Fields on the Photoelectric Properties of Metals," E. O. Lawrence and L. B. Linford, *Phys. Rev.*, Vol. 36, 1930, p. 482.
88. "Electron Emission of Metals in Electric Fields," E. Guth and C. J. Mullin, *Phys. Rev.*, Vol. 59, 1941, p. 867.
89. "Sommerfeld Theory," G. Wentzel, *Probleme der Modernen Physik*, p. 79.
90. "The Increase of the Photoelectric Effect of Potassium by Means of Hydrogen," R. Fleischer and H. Teichmann, *Z. Physik*, Vol. 61, 1930, p. 227.
91. "On the Photoeffect of Metals," H. Fröhlich, *Ann. Physik*, Vol. 7, 1930, p. 103; "The Photoelectric Effect in Thin Metallic Films," W. G. Penney, *Proc. Roy. Soc.* (*London*), *Series A*, Vol. 133, 1931, p. 407.
92. "On the Normal Photoeffect," P. Lukirsky and S. Prilezaev, *Z. Physik*, Vol. 49, 1928, p. 236.
93. "On the Theory of the Photoeffect of Metals," I. Tamm and S. Schubin, *Z. Physik*, Vol. 68, 1931, p. 97.
94. "On the Influence of Hydrogen on the Photoelectric Emission of Potassium," R. Suhrmann and H. Theissing, *Z. Physik*, Vol. 52, 1928, p. 453.
95. "Photoelectric Cells and Their Applications," N. R. Campbell, *Discussion of the Physical and Optical Societies of London*, 1930, p. 10.
96. "A Possible Explanation of the Selective Photoelectric Effect," R. H. Fowler, *Proc. Roy. Soc.* (*London*), *Series A*, Vol. 128, 1930, p. 123.
97. "Liberation Depth of Secondary Electrons," H. Bruining, *Physica*, Vol. 3, 1936, p. 1046; *The Secondary Electron Emission of Solid Bodies*, H. Bruining, Springer, Berlin, 1942; "Secondary Electron Emission, Part I, Metals," H. Bruining and J. de Boer, *Physica*, Vol. 5, 1938, pp. 17–30; "Absorption of Secondary Electrons, Part II," H. Bruining and J. de Boer, *Physica*, Vol. 5, 1938, pp. 901–912; "Slow Electron Bombardment, Part III," H. Bruining and J. de Boer, *Physica* Vol. 5, 1938, pp. 913–917; "Compounds with a High Capacity for Secondary Electron Emission, Part IV," H. Bruining and J. de Boer, *Physica*, Vol. 6, 1939, p. 823; "The Mechanism of Secondary Electron Emission, Part V," H. Bruining and J. de Boer, *Physica*, Vol. 6, 1939, p. 834; "Secondary Electron Emission from Metals with a Low Work Function," H. Bruining, *Physica*, Vol. 8, 1941, pp. 1161–1164; "The Secondary Electron Emission of Semi-Conductors, or Insulators," G. Maurer, *Z. Physik*, Vol. 118, 1941, p. 122; "On the Question of the Nature of Secondary Emission of Coated Cathodes," N. Morgulis, *J. Tech. Phys.* (*U.S.S.R.*), Vol. 9, 1939, p. 853.
98. "The Mechanism of Secondary Electron Emission from Composite Surfaces," P. Timofeev, *J. Tech. Phys. U.S.S.R.*, Vol. 10, 1940, p. 3; "The Secondary Electron Emission from Oxygen-Caesium Targets for Various Current Densities," P. Timofeev and J. Pjatnitsky, *J. Tech. Phys.* (*U.S.S.R*), Vol. 10, 1940, p. 39; "The Secondary Electron Emission from Pure and Caesium Treated Copper Oxide," A. Frimer, *J. Tech. Phys.* (*U.S.S.R.*), Vol. 10, 1940, p. 395; "Secondary Electron Emission from Thin Metal Films on Glass," P. W. Timofeev, A. Afanasjew, and A. Ignatow, *Tech. Phys. U.S.S.R.*, Vol. 3, No. 12, 1936, p. 1011; "Secondary Electron Emission from Oxidized Silver and Molybdenum Surfaces," A. W. Afanasjew and P. W. Timofeev, *Phys. Z. Sowjetunion*, Vol. 10, No. 6, 1936, p. 831.
99. "Theory of Secondary Electron Emission from Metals," H. Fröhlich, *Ann. Physik*, Vol. 13, 1932, p. 229.
100. "Anomalous Secondary Electron Emission," L. Malter, *Phys. Rev.*, Vol. 49, 1936, p. 478; "Field Enhanced Secondary Electron Emission," H. Nelson, *Phys. Rev.*, Vol. 57, 1940, p. 560; "Phenomena of Secondary Electron Emission," H. Nelson, *Phys. Rev.*, Vol. 55, 1939, p. 985.
101. "Secondary Electron Emission of Metals with a Low Work Function," H. Bruining and J. H. de Boer, *Physica*, Vol. 4, 1937, p. 473; "Secondary Electron Emission," H. Bruining and J. H. de Boer, *Physica*, Vol. 6, 1939, p. 941.
102. "Critical Potentials in Secondary Electron Emission from Iron, Nickel, and Molybdenum," R. L. Petry, *Phys. Rev.*, Vol. 26, 1925, p. 346; "Secondary Electron Emission from Metals in the Low Primary Energy Region," I. Gimpel and O. W. Richardson, *Proc. Roy. Soc.* (*London*), *Series A*, Vol. 182, Sept. 1943, pp. 17–47.
103. "Silver-Magnesium Alloy as a Secondary Electron Emitting Material," V. K. Zworykin, J. E. Ruedy, and E. W. Pike, *J. App. Phys.*, Vol. 12, 1941, p. 696.
104. "The Influence of Temperature on the Secondary Electron Emission of Metals," R. Kollath, *Ann. Physik*, Vol. 39, 1941, p. 19; "The Secondary Emission Factor of Electron Bombarded Insulators," H. Salow, *Z. tech. Physik*, Vol. 21, 1940, p. 8; "The Angular Dependence of Secondary Electron Emission from Insulators," H. Salow, *Physik. Z.*, Vol. 19, 1940, p. 434; "The Secondary Electron Emission of Pyrex-Glass," C. W. Mueller, *J. App. Phys.*, Vol. 16, 1945, p. 453.
105. "Thin Film Field Emission," L. Malter, *Phys. Rev.*, Vol. 50, 1936, p. 48; "Field Emission from Coated Cathodes by Electron Bombardment," H. Mahl, *Z. tech. Physik*, Vol. 18, 1937, p. 559; "Field Emission from Coated Cathodes by Electron Bombardment," H. Mahl, *Z. tech. Physik*, Vol. 19, 1938, p. 313.
106. "Secondary Electron Radiation," J. H. Owen Harries, *Electronics*, Sept. 1944, p. 100.
107. "Time Changes in Emission from Oxide-Coated Cathodes," J. P. Blewett, *Phys. Rev.*, Vol. 55, 1939, p. 713.
108. "Electron Theory of Metallic Conduction," O. W. Richardson, *Trans. Amer. Electrochem. Soc.*, Vol. 21, 1912, p. 69.
109. "The Effect of End Losses on Characteristics of Filaments of Tungsten and Other Materials," I. Langmuir, S. MacLane, and Katharine Blodgett, *Phys. Rev.*, Vol. 35, 1930, pp. 478, 503.
110. "Heat Conductivity of Tungsten and Cooling Effects of Leads upon Filaments at Low Temperatures," I. Langmuir and J. B. Taylor, *Phys. Rev.*, Vol. 50, 1936, pp. 68–87.
111. "The Thermionic Work Function of Tungsten," C. Davisson and L. H. Germer, *Phys. Rev.*, Vol. 20, 1922, p. 300.
112. *Electron Theory of Metals*, H. Fröhlich, Springer, Berlin, 1936.
113. "The Electronic Structure of Metals," J. C. Slater, *Rev. Mod. Phys.*, Vol. 6, 1934, p. 209.

114. "Modern Theory of Solids," F. Seitz and R. P. Johnson, *J. App. Phys.*, Vol. 8, 1936, pp. 84, 186, 246.
115. "A New Table of Values of the General Physical Constants," R. T. Birge, *Rev. Mod. Phys.*, Vol. 13, 1941, p. 233.
116. "The True Temperature Scale of an Oxide-Coated Filament," C. H. Prescott, Jr., and J. Morrison, *Rev. Sci. Instruments*, Vol. 10, 1939, p. 36.
117. "Emission from Oxide-Coated Cathodes," M. Benjamin and H. P. Rooksby, *Phil. Mag.*, Vol. 15, 1933, p. 810.
118. "Distribution of Autoelectronic Emission from Single Crystal Metal Points," M. Benjamin and R. O. Jenkins, *Proc. Roy. Soc. (London), Series A*, Vol. 176, Oct. 1940, pp. 262–279.
119. *Photocells and Their Application*, V. K. Zworykin and H. D. Wilson, 2nd ed., Wiley, 1934.
120. "Electrical Discharges in Gases," K. T. Compton and I. Langmuir, *Rev. Mod. Phys.*, Vol. 2, 1930, p. 123.
121. "Electrical Discharges in Gases," K. T. Compton and I. Langmuir, *Rev. Mod. Phys.*, Vol. 3, 1931, p. 191.
122. "Conductivity and Thermionic Emission of the Oxide Cathode," E. Nishibori and H. Kawamura, *Proc. Phys.-Math. Soc. Japan*, Vol. 22, 1940, p. 378.
123. "Conductivity and Thermionic Emission of the Oxide Cathode," E. Nishibori and H. Kawamura, *Proc. Phys.-Math. Soc. Japan*, Vol. 23, 1941, p. 37.
124. "Electron Emission and Adsorption Phenomena," J. H. De Boer, Macmillan, 1935.
125. "Secondary Electron Emission of Solid Bodies," R. Kollath, *Physik. Z.*, Vol. 38, 1937, p. 202.
126. "Secondary Electron Yield," F. Trey, *Physik. Z.*, Vol. 44, 1943, p. 38.
127. "Theory of Secondary Emission," D. E. Wooldridge, *Phys. Rev.*, Vol. 56, 1939, p. 562.
128. "Some High Temperature Properties of Niobium," A. L. Reimann and C. Kear-Grant, *Phil. Mag.*, Vol. 22, 1936, p. 34.
129. "Positive and Negative Thermionic Emission from Columbium," H. B. Wahlin and L. O. Sordahl, *Phys. Rev.*, Vol. 45, 1934, p. 886.

Chapter 3

CONTROL OF FREE ELECTRONS

H. J. Dailey and J. H. Findlay

AFTER release from a solid body, or emitting surface, an electron can be controlled and can be made to perform various functions. Since such an electron has a known electrostatic charge, the path it will follow can be accurately predicted when the difference of electric intensity can be calculated or measured.

This chapter deals with electrons in motion, the prediction of their behavior, and electrostatic means for controlling them after their release from the cathode.

3·1 POTENTIAL GRADIENT, ELECTROSTATIC POTENTIAL, ELECTRIC INTENSITY

The movement of an electron from one point to another in space is governed by the potential difference between these positions. The volts per centimeter between these two points is the *potential gradient.* The electric intensity (electrostatic field intensity) *at a point* is dimensionally and numerically equal to the potential gradient at the same point, although of opposite sign:

$$\mathbf{E} = -\frac{dV}{dx} \quad \text{electrostatic units} \qquad (3\cdot1)$$

conversely

$$V_2 - V_1 = -\int_1^2 \mathbf{E}\,dx \quad \text{electrostatic units} \qquad (3\cdot1a)$$

In equation 3·1, dx represents an incremental distance along the path of the electrostatic flux lines between points 1 and 2. Since **E** is a vector, it may be given in the rectangular co-ordinate system:

$$\mathbf{E}_x = -\frac{\partial V}{\partial x}, \quad \mathbf{E}_y = -\frac{\partial V}{\partial y}, \quad \mathbf{E}_z = -\frac{\partial V}{\partial z} \qquad (3\cdot1b)$$

The application of equation 3·1*b* to the electrostatic field between parallel plates usually involves only one dimension. For this special case, equation 3·1*b* becomes equation 3·1.

A graphical illustration of equation 3·1 is shown in Fig. 3·1. Distance between plates is represented to an arbitrary scale horizontally and the voltage between plates to an arbitrary scale vertically. The tangent of the angle between line AB and the horizontal at all points from $x = 0$ to $x = a$ represents the potential gradient at that point. The steeper the slope is, the greater the gradient and the greater the acceleration of an electron between the two plates. With no space charge between plates 1 and 2, the gradient at all points is equal to E/a, or as shown in Fig. 3·1 the gradient is also $(V_{x_2} - V_{x_1})/(x_2 - x_1)$.

The slope of line AB in Fig. 3·1 can be compared to a slope down which a ball is rolled. The steeper the slope the faster the ball will roll, and likewise the steeper the potential-gradient slope the faster will be the acceleration of electrons between plates 1 and 2.

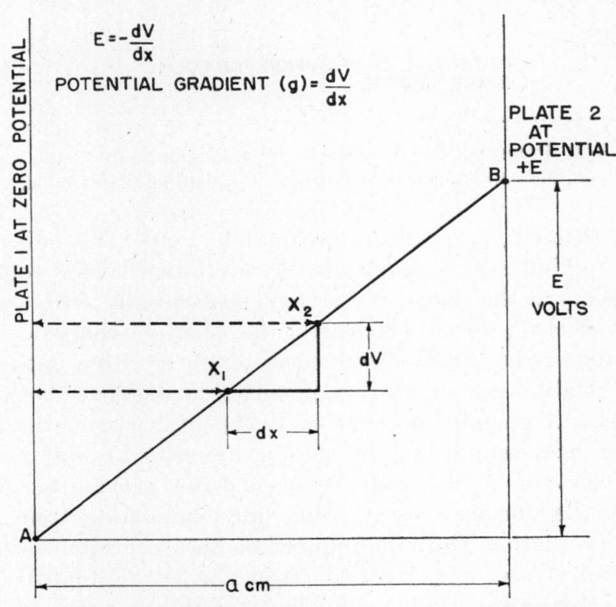

Fig. 3·1 Potential distribution in a space-charge-free parallel-plane diode.

3·2 SPACE CHARGE

If a space between parallel plates contains no space charge, the voltage gradient between those plates is constant. If the negative plate is an unlimited source of electrons and if a potential difference exists between the plates, the number of free electrons between the plates will increase until their number is limited by the potential difference between those plates. The voltage gradient is now no longer constant.

When electrons pass from the cathode to the anode, then the graphical representation shown in Fig. 3·1 changes to that shown in Fig. 3·2. As the electron density between plates 1 and 2 increases, **E** and the potential distribution change as the electron density increases from zero to the maximum that potential E can produce. The potential distribution then changes from that shown by the dotted line to that of the heavy line, and the current is said to be *space-charge limited.* Figure 3·2 shows that **E** varies over the distance from $x = 0$ to $x = a$. When x is approximately

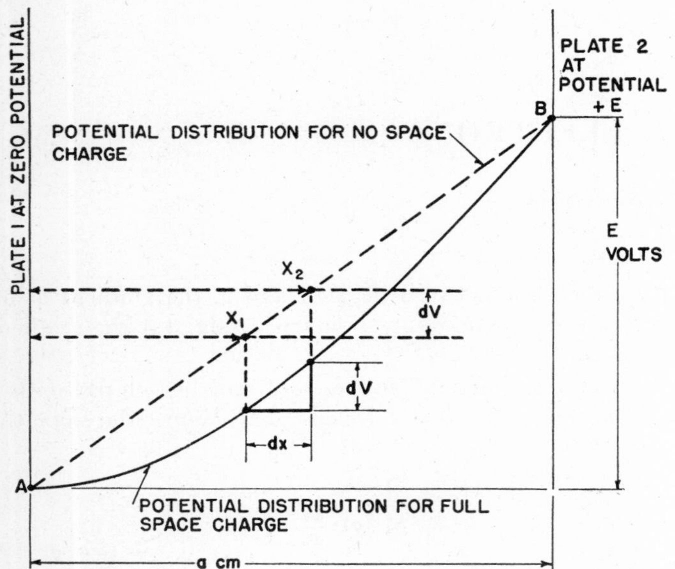

FIG. 3·2 Potential distribution in a space-charge-free and, under complete space-charge conditions, in a parallel-plane diode.

zero, then dV/dx is also approximately zero. This indicates that for full space charge the electric intensity (or voltage gradient) at the source of electrons is essentially zero.

As electron density increases, the negative charge of the electrons reduces the effectiveness of the positive potential E in establishing an effective potential in the space between plates. The number of electrostatic flux lines passing through a unit area establishes the electric intensity at that point. The electrons in the space between plates provide terminal points for the electrostatic flux lines emanating from the positive plate. Thus the number of lines per square centimeter decreases from right to left in Fig. 3·2, so that for full space charge essentially no flux lines terminate at the cathode surface. The velocity of an electron when it arrives at a positive electrode is the same for full space charge as for the space-charge-free condition. The time of flight of the electron from cathode to anode under similar conditions is different, as explained in Section 3·10.

3·3 POISSON'S AND LAPLACE'S EQUATIONS

Potential distribution diagrams such as those in Figs. 3·1 and 3·2 are most useful for simple geometric figures. Their usefulness may be expanded, however, to any configuration, if it is desirable to delineate the potential gradients graphically.

As explained in Section 3·2, the potential in-space varies as the charge density ρ. In three-dimensional co-ordinates, the space potential V and the charge density ρ are related thus:

$$\frac{\partial^2 V}{\partial x^2} + \frac{\partial^2 V}{\partial y^2} + \frac{\partial^2 V}{\partial z^2} = -4\pi\rho \quad \text{electrostatic units} \qquad (3\cdot2)$$

For one-dimensional configurations, equation 3·2 becomes

$$\frac{d^2 V}{dx^2} = -4\pi\rho \quad \text{electrostatic units} \qquad (3\cdot3)$$

For the cylindrical co-ordinate system equation 3·2 becomes

$$\frac{1}{r}\frac{\partial}{\partial r}\left(\frac{r\,\partial V}{\partial r}\right) + \frac{1}{r^2}\frac{\partial^2 V}{\partial \phi^2} + \frac{\partial^2 V}{\partial z^2} = -4\pi\rho \qquad (3\cdot4)$$

For the spherical co-ordinate system equation 3·4 becomes

$$\frac{1}{r^2}\frac{\partial}{\partial r}\left(\frac{r^2\,\partial V}{\partial r}\right) + \frac{1}{r^2 \sin\theta}\frac{\partial}{\partial\theta}\left(\sin\theta\frac{\partial V}{\partial\theta}\right) + \frac{1}{r^2\sin^2\theta}\frac{\partial^2 V}{\partial\phi^2} = -4\pi\rho \qquad (3\cdot5)$$

In the cylindrical co-ordinate system, the charge density and potential are usually symmetrical about the axis of the cylinder; thus equation 3·4 becomes

$$\frac{1}{r}\frac{d}{dr}\left(\frac{r\,dV}{dr}\right) = -4\pi\rho \qquad (3\cdot6)$$

Equation 3·6 may be converted to

$$\frac{d^2 V}{(d\log_\epsilon r)^2} = -2\rho' r \qquad (3\cdot6a)$$

where $\rho' = 2\pi r\rho$, the space charge density around the axis at radius r. Equation 3·2 is known as Poisson's equation. Equations 3·2 and 3·3 show that the rate of change of the gradient varies as 4π times the charge density ρ, and that the right side of equations 3·2 and 3·3 will be positive for electron densities and negative for positive-ion densities.

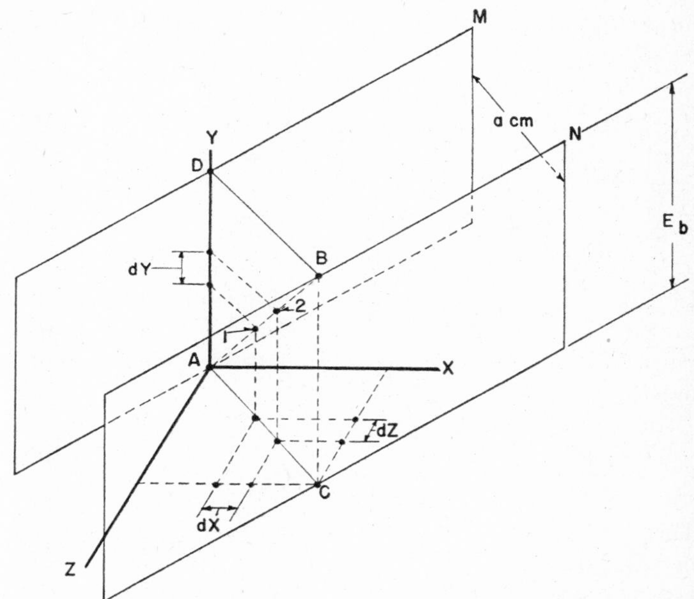

FIG. 3·3 Three-dimensional graphical representation of the potential distribution between parallel planes under space-charge-free conditions.

Calculations for electronic tubes usually involve space-charge-free conditions. Usually space-charge-free conditions are obtained by proper spacing of the elements and by applying potentials that cause a negative gradient at the cathode surface, thus preventing electron emission.

For space-charge-free conditions, equation 3·2 becomes

$$\frac{\partial^2 V}{\partial x^2} + \frac{\partial^2 V}{\partial y^2} + \frac{\partial^2 V}{\partial z^2} = 0 \qquad (3\cdot 6b)$$

This equation is known as the Laplace equation. For one-dimensional fields, it becomes

$$\frac{d^2 V}{dx^2} = 0 \qquad (3\cdot 6c)$$

In rectangular co-ordinates the electric intensity between parallel plates may be written

$$\frac{\partial V_x}{\partial x} + \frac{\partial V_y}{\partial y} + \frac{\partial V_z}{\partial z} = -\mathbf{E} \qquad (3\cdot 6d)$$

Equation 3·6d is illustrated in Fig. 3·3. Between plates M and N is a potential difference E, represented to an arbitrary scale vertically on plate N. Plate M contains the y-axis, and the x- and y-co-ordinates are laid out at equal angles from plate M.

3·4 CONSTANT-POTENTIAL LINE DISTRIBUTION

In Fig. 3·4 (a), curves 1 and 2 delineate potential change between cathode and anode along two paths, the upper curve between grid wires and the lower through grid wires. In Fig. 3·4 (b) are shown constant potential lines and electrostatic-flux lines under comparable conditions. At and near the cathode and anode surfaces the constant-potential lines are parallel to those surfaces. As the grid is approached,

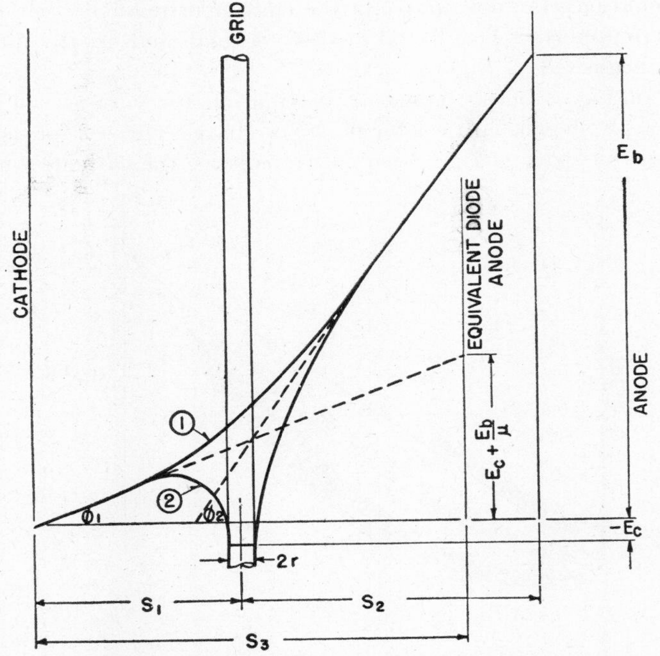

Fig. 3·4 (a) Potential distribution in a space-charge-free parallel-plane triode (and its equivalent diode).

the constant-potential lines deviate from straight lines because of the presence of the grid wires. The points at which the constant-potential lines deviate from straight lines coincide with the points on Fig. 3·4 (a) at which curves 1 and 2 begin to separate. Midway between grid wires there is a gradual transition from one gradient to another as shown by the gradual changing of the distance between constant-potential lines.

If a thin rubber membrane is stretched between two plane surfaces so the angle between the horizontal and the rubber

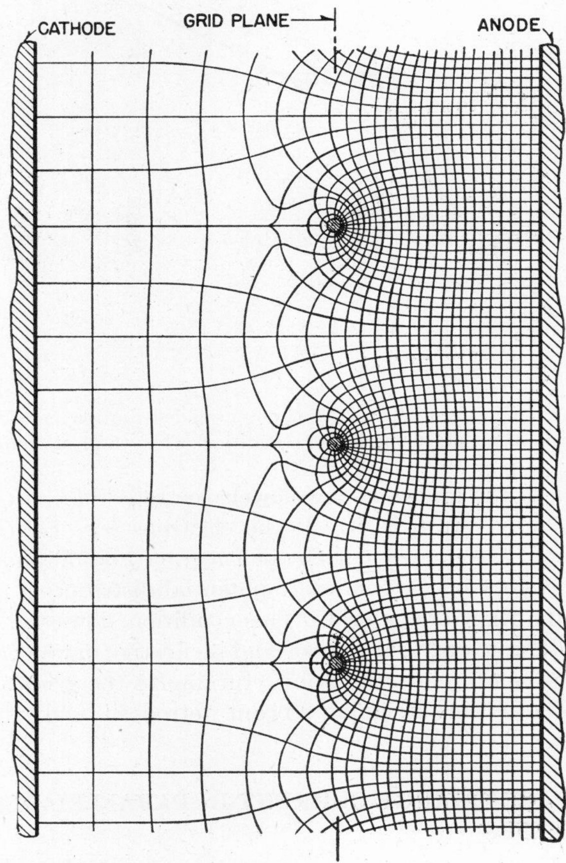

Fig. 3·4 (b) Space-charge-free constant potential line and electrostatic-flux line map of a triode.

membrane is comparable with the potential-gradient line in Fig. 3·1, and if rods are pressed downward on the rubber membrane corresponding to the grid wires, then the elevation contours on the rubber membrane correspond to the constant-potential lines in Fig. 3·4 (b). If the elevations are reversed, and the grid rods are pushed upward from below, small lead shot, released from the cathode side, may be used to study the travel of electrons through varying gradients. With the elevations reversed, the shot do not tend to strike the grid. This condition corresponds to a negative grid. If the grid rods are pushed downward, which for this mechanical demonstration corresponds to a positive grid, the shot now not only strike the grid but also remain in the depressions caused by the grid rods. This corresponds to electrons striking the grids and being taken into the grid-potential supply. Figure 3·4 (c) is a graphical illustration of the potential distribution in a space-charge-free triode with a positive grid. Curve 1 again represents potential distribution between grid wires and curve 2 through grid wires. Now the gradient dV/dx between grid and cathode is greater than between grid and anode. As the amplification factor μ is decreased (accomplished by increasing the spacing between grid wires), there is a greater divergence between curves 1 and 2, and conversely they approach each other as μ is increased (accomplished by decreasing the spacing between grid wires). The negative slope of curve 2 to the right of the grid wire indicates that electrons are slowed

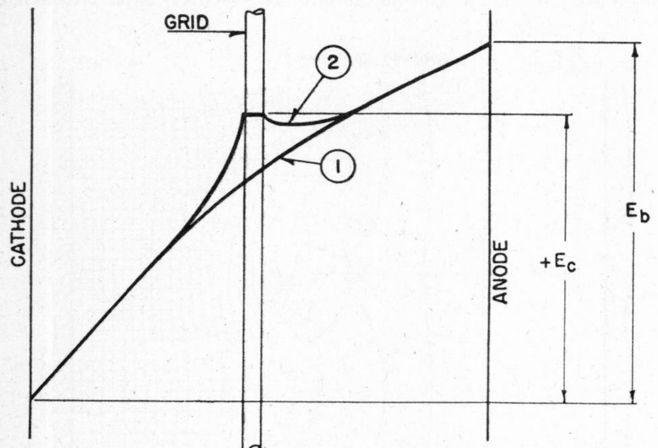

Fig. 3·4(c) Graphical illustration of potential distribution in a parallel-plane triode under space-charge-free conditions when the grid is positive.

adjacent to the grid, and the slower electrons that just miss the grid may return to it. If the electrons are "bunched" or beamed, a sufficient current density may be obtained to make the slope of the composite potential distribution negative to the right of the grid. This condition slows the electrons in their travel to the anode and is effective in preventing anode secondary electrons from returning to the grid. This effect is used in designing "beam tetrodes" and "beam pentodes."

3·5 SPACE-CHARGE CURRENT IN PARALLEL-PLANE DIODES

The electron current between two plane electrodes with complete space charge, when an ample supply of electrons is available, is limited by (1) area of electrodes and (2) the potential difference between the electrodes. The electron current $\mathbf{J}$ per unit area, the electron density ρ, and the electron velocity are related by

$$\mathbf{J} = \rho v \tag{3·7}$$

Electron velocity v is

$$v = \frac{(2eV)^{1/2}}{(m_e)^{1/2}} \tag{3·8}$$

where v = velocity an electron will have at any point which has a potential V in space or at an electrode at potential E

e = the charge on the electron

m_e = the mass of the electrons.

Combining equations 3·3, 3·7, and 3·8 and solving the resulting differential equation, we obtain

$$\mathbf{J} = 2.33 \times 10^{-6}\frac{E_b^{3/2}}{a^2} \tag{3·9}$$

where E_b = potential difference between cathode and anode

a = distance between cathode and anode in centimeters

$\mathbf{J}$ = the current density in amperes per square centimeter.

One step in the derivation of equation 3·9 is

$$\frac{1}{2}\left(\frac{dV}{dx}\right)^2 = \frac{8\pi\mathbf{J}}{\left(\frac{2e}{m_e}\right)^{1/2}}V^{1/2} + \mathbf{E}_0^2 \tag{3·10}$$

where $\mathbf{J}$ = current density

V = space potential

e = electron charge

m_e = electron mass

$\mathbf{E}_0$ = electric intensity at the cathode surface.

In Fig. 3·2, the heavy or space-charge line is tangential to the horizontal at plate 1. From equation 3·7,

$$\rho = \frac{\mathbf{J}}{v} \tag{3·11}$$

Combining equations 3·8 and 3·11,

$$\rho = \frac{\mathbf{J}}{\left(\frac{2eV}{m_e}\right)^{1/2}} \tag{3·12}$$

Since V approaches zero at the cathode and $\mathbf{J}$ is constant, then near the cathode ρ becomes large in comparison with ρ near the anode. With high electron density near the cathode, the anode voltage becomes less and less effective in releasing electrons from the cathode. The heavy line (condition of complete space-charge-limited current) in Fig. 3·2 shows that $\mathbf{E}$ approaches zero near the cathode.

The Langmuir-Childs equation (3·9) neglects the effect of initial electron velocities, that is, the velocity the electron would have at the surface of the emitter if no anode voltage were applied. This initial velocity is small and usually may be neglected.

In Fig. 3·5 the potential distribution for various conditions is shown: curve 1 for no space charge, curve 2 for normal full space charge when electrons leave the cathode with

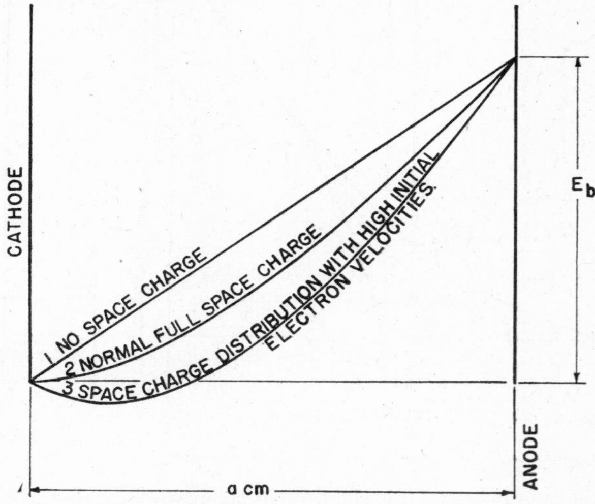

Fig. 3·5 Potential distribution in a parallel-plane diode showing the variations of potential with initial electron velocity.

essentially zero velocities, and curve 3 which shows a negative dV/dx if initial electron velocities are appreciably greater than zero. A complete mathematical solution has been developed for the potential distribution existing between parallel planes, whose initial electron velocities could be such that a virtual cathode is formed.[3, 4] Where dV/dx is negative near the cathode, anode voltage E_b has no effect on the number of electrons liberated. The current between the cathode on the left and the anode at the right is the same as if the source of electrons were at the point where $dV/dx = 0$ on curve 3 in Fig. 3·5; that is, that point for the anode is the source of electrons, or is the virtual cathode.

Initial electron velocities should be taken into account in dealing with close spacings and low voltages, but for most practical applications it may be neglected. In tubes with cylindrical elements, the effect of initial electron velocity is even less pronounced.

3·6 SPACE-CHARGE CURRENT IN CYLINDRICAL DIODES

Similarly to the current flow in parallel-plane structures, current density $\mathbf{J}'$ passing between the cathode and anode per centimeter of length, charge density ρ, and electron velocity v are related by

$$\mathbf{J}' = \rho' v \tag{3·13}$$

Combining equations 3·6*a*, 3·8, and 3·13, and solving the resulting differential equation, obtain

$$\mathbf{J}' = 14.65 \times 10^{-6} \frac{E_b^{3/2}}{r_a \beta^2} \tag{3·14}$$

where r_a is the radius of the anode
β^2 is a function of r_a/r_c with r_c being the cathode radius.

Values of β^2 may be obtained from Fig. 3·6. $\mathbf{J}'$ is the current per centimeter of length along the axis of the tube.

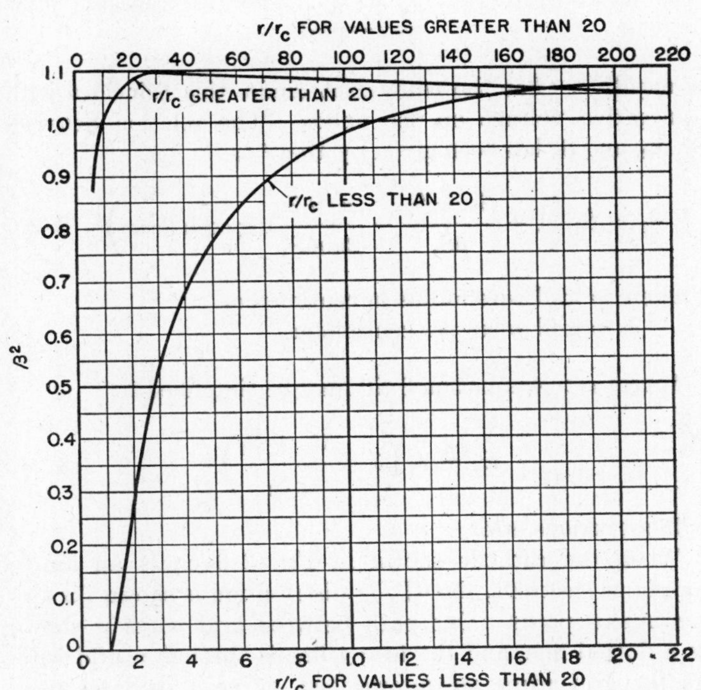

FIG. 3·6 Variations of β^2 with r/r_c. (Reprinted with permission from *Fundamentals of Engineering Electronics*, by W. G. Dow, published by John Wiley, 1937.)

From equation 3·14, it is evident that current $\mathbf{J}'$ varies nearly inversely with r_a, since β^2 contains r_a and is not a rapidly changing function. The lower curve in Fig. 3·7 indicates the potential distribution in a cylindrical diode with complete space charge.

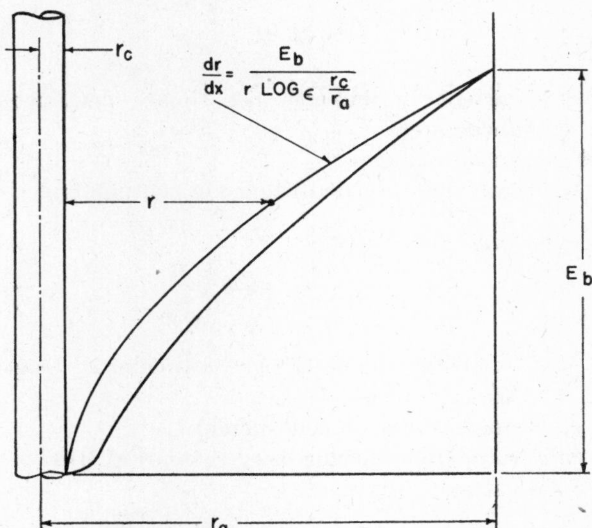

FIG. 3·7 Potential distribution in a cylindrical diode under space-charge-free and complete space-charge conditions.

3·7 POTENTIAL DISTRIBUTION IN SPACE-CHARGE-FREE DIODES

In Fig. 3·1, the space-charge-free potential distribution is shown. Since the potential varies inversely as x, dV/dx is constant from one plate to the other (if distortion of the field at the edges of the plates is neglected).

For cylindrical structures, the voltage gradient dV/dx is

$$\frac{dV}{dx} = \frac{E_b}{r \log_\epsilon \frac{r_a}{r_c}} \tag{3·15}$$

where r_a = anode radius
r_c = cathode radius
r = any distance from cathode to anode.

At the cathode surface, the gradient is maximum:

$$\left(\frac{dV}{dx}\right)_{\max} = \frac{E_b}{r_c \log_\epsilon \frac{r_a}{r_c}} \tag{3·16}$$

A potential-distribution curve for cylindrical diodes structures under space-charge-free conditions is the upper curve of Fig. 3·7. The minimum gradient is at the anode surface where

$$\left(\frac{dV}{dx}\right)_{\min} = \frac{E_b}{r_a \log_\epsilon \frac{r_a}{r_c}} \tag{3·17}$$

3·8 COMPARISON OF CURRENT FLOW IN PARALLEL PHASE AND CYLINDRICAL DIODES

An inspection of Figs. 3·2 and 3·7 shows that the voltage gradient near the cathode is greater for the cylindrical structure. For a given cathode-to-anode distance and at a given anode voltage a greater current will flow, or conversely for a given current less anode voltage will be required for the cylindrical structure.

For parallel structures:

$$\mathbf{J} = 2.33 \times 10^{-6} \frac{E_b^{3/2}}{a^2} \qquad (3\cdot 9)$$

where $\mathbf{J}$ = current in amperes per square centimeter of anode area
E_b = anode voltage
a = cathode-to-anode distance in centimeters.

For cylindrical structures:

$$\mathbf{J}' = 14.65 \times 10^{-6} \frac{E_b^{3/2}}{r_a\beta^2} \qquad (3\cdot 14)$$

where $\mathbf{J}'$ = current in amperes per centimeter of length of the structure
r_a = anode radius in centimeters
β^2 = a function varying as r_a/r_c (see Fig. 3·6)
r_c = cathode radius.

When the cathode-to-anode distances are the same, the current flow of the two structures may be compared. These distances are comparable when $r_a - r_c = a$.

Substituting $(r_a - r_c)$ for a and dividing equation 3·9 by equation 3·14,

$$\frac{\mathbf{J}}{\mathbf{J}'} = 0.159 \frac{r_a\beta^2}{(r_a - r_c)^2} \qquad (3\cdot 18)$$

This equation does not give a true comparison of current flow for the two structures as the effective anode areas are not equal. To correct for area the factor r_a is used:

$$\frac{\mathbf{J}}{\mathbf{J}'} = 0.159 \frac{r_a^2\beta^2}{(r_a - r_c)^2} \qquad (3\cdot 19)$$

When $r_a = 11$, $r_c = 1$, and $\beta^2 = 1$ (see Fig. 3·6),

$$\frac{\mathbf{J}}{\mathbf{J}'} = 0.604$$

Since $\mathbf{J}/\mathbf{J}'$ for a given cathode-to-anode distance is appreciably less than unity, cylindrical structures are preferable for lower anode voltage drop.

3·9 PARALLEL-PLANE TRIODES

If the electric intensities are known for a given tube structure, the electron behavior is known for that structure. In Fig. 3·4 (*a*) a plane structure is shown in which the grid wires are spaced relatively far apart. With the cathode cold and voltages $-E_c$ and E_b applied to the grid and anode respectively, the potential distribution is as shown by the heavy lines. Curve 1 is the distribution midway between grid wires, and curve 2 is through a grid wire. A diode may be constructed with spacing S_3 so that with an anode voltage $E_c + (E_b/\mu)$ applied, $\mathbf{E}(-dV/dx)$ at its cathode would have the same slope as shown in Fig. 3·4 (*a*) between cathode and grid. As μ (amplification factor) is increased, the angle between the potential line and the cathode decreases and curves 1 and 2 more nearly coincide, which indicates that the voltage gradient decreases at the cathode. As μ increases, the condition shown in Fig. 3·8 is reached. Here curves 1 and 2 approach each other, and curve 2 at the left of the grid has decreased so that dV/dx at the cathode is negative. In Fig. 3·4 (*a*) μ is low so that $E_c + (E_b/\mu)$ is positive even though E_c is negative; therefore dV/dx at the cathode is positive. In Fig. 3·8, where μ is high and E_c is negative, E_b/μ is less than E_c; therefore dV/dx at the cathode

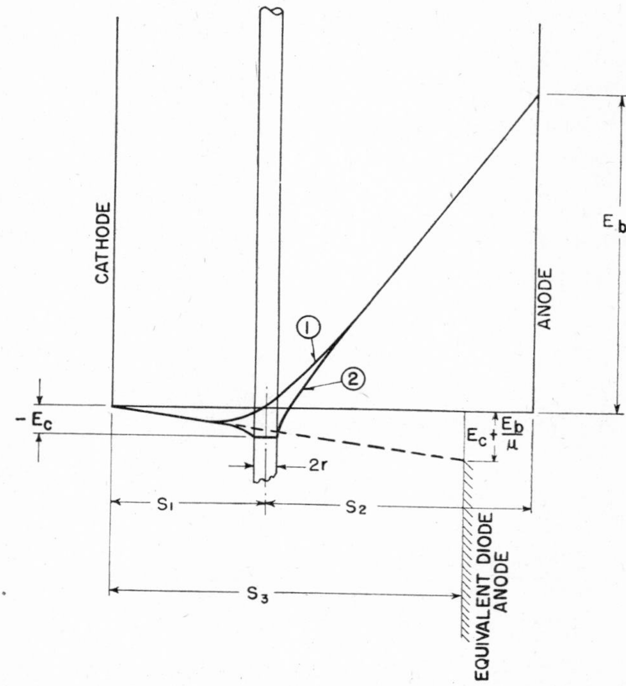

FIG. 3·8 Potential distribution in a space-charge-free parallel-plane high-μ triode where $E_c + (E_b/\mu) < 0$, with equivalent diode represented.

is negative. For the equivalent diode, this coincides with a negative voltage on the anode. The relation between S_1, S_2, and S_3 has been given by Dow [4] as

$$S_3 = S_1\left(1 + \frac{S_1 + S_2}{\mu S_1} - \frac{\mu + 2}{2\pi n\mu S_1}\log\cosh 2\pi nr\right) \qquad (3\cdot 20)$$

where r = grid wire radius in centimeters
n = grid wires per centimeter.

Where $\mu > 4$ equation 3·20 may be simplified to

$$S_3 = S_1\left(1 + \frac{S_1 + S_2}{\mu S_1}\right) \qquad (3\cdot 21)$$

without serious error.

When the cathode shown in Fig. 3·9 emits an ample supply of electrons, the dV/dx distribution is shown for two continuous paths. One path between grid wires is shown as a dotted line, and the heavy line is through a grid wire. As the grid wires are approached, curve 1 deviates from curve 2. Notice that dV/dx is positive for curve 1 to the left of the grid wire. For curve 2 it is positive, then zero, then negative. This means that no electrons will strike the grid under this condition, that is, when the grid is negative with respect to the cathode.

In Fig. 3·10 there is no anode current since dV/dx at the cathode is negative. Any electrons then leaving the cathode are returned to the cathode by a negative force.

An equation for current in parallel-plane triodes has been given by Fremlin [6] as

$$\mathbf{J} = \frac{2.34 \times 10^{-6}\left(E_c + \dfrac{E_b}{\mu}\right)^{3/2}}{\left(S_g^{4/3} + \dfrac{S_a^{4/3}}{\mu}\right)^{3/2}} \tag{3·22}$$

where $\mathbf{J}$ = amperes per unit area
S_a = cathode-to-anode distance in centimeters
S_g = cathode-to-grid distance in centimeters.

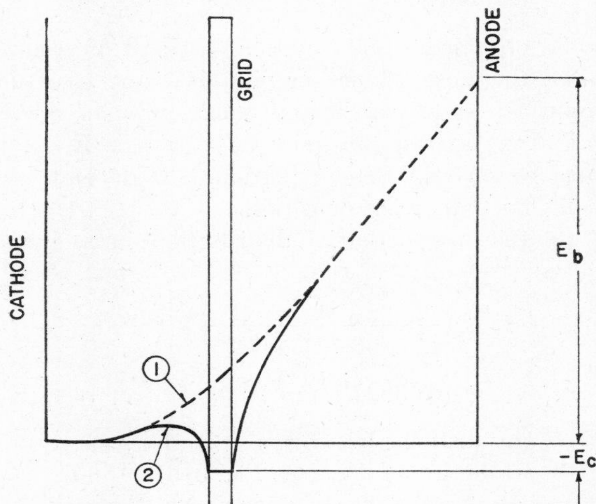

FIG. 3·9 Potential distribution in a parallel-plane triode under space-charge conditions where $E_c + (E_b/\mu) > 0$.

Equations 3·9 and 3·22 show the similarity of the equations of current in the diode and triode. The factor $E_c + (E_b/\mu)$ shows that for all values of μ this term may become zero for any value, plus or minus, of E_c or E_b. This indicates that for positive values of E_b the off-cathode gradient may be made negative (thus preventing emergence of electrons) by making E_c negative until $E_c + (E_b/\mu)$ is either zero or negative. At the point where the anode current just becomes zero, the ratio of anode voltage to grid voltage is termed the "cut-off" μ of the triode. The same effect may be accomplished for positive values of E_c by making E_b negative until $E_c + (E_b/\mu)$ equals 0 or is negative.

For current in cylindrical triodes, the current, voltage, and spacings are related by

$$\mathbf{J}' = \frac{14.65 \times 10^{-6}\left(E_c + \dfrac{E_b}{\mu}\right)^{3/2}}{\left[\dfrac{1}{\mu}(r_a\beta_{ca}{}^2)^{2/3} + (r_g\beta_{cg}{}^2)^{2/3}\right]^{3/2}} \tag{3·23}$$

$\mathbf{J}'$ = current density in amperes per unit length. Since the current density $\mathbf{J}'$ varies as the term $[E_c + (E_b/\mu)]^{3/2}$ for a given tube structure, this equation may be written

$$\mathbf{J}' = K\left(E_c + \frac{E_b}{\mu}\right)^{3/2} \tag{3·24}$$

The factor K is often called the "perviance" of a tube. For a cylindrical triode:

$$K = \frac{14.65 \times 10^{-6}}{\left[\dfrac{1}{\mu}(r_a\beta_{ca}{}^2)^{2/3} + (r_g\beta_{cg}{}^2)^{2/3}\right]^{3/2}} \tag{3·25}$$

In equations 3·23 and 3·25,

r_a = anode radius
r_g = grid radius
μ = amplification factor
$\beta_{ca}{}^2$ = a function of anode radius to cathode radius (see Fig. 3·6)
$\beta_{cg}{}^2$ = a function of grid radius to cathode radius (see Fig. 3·6).

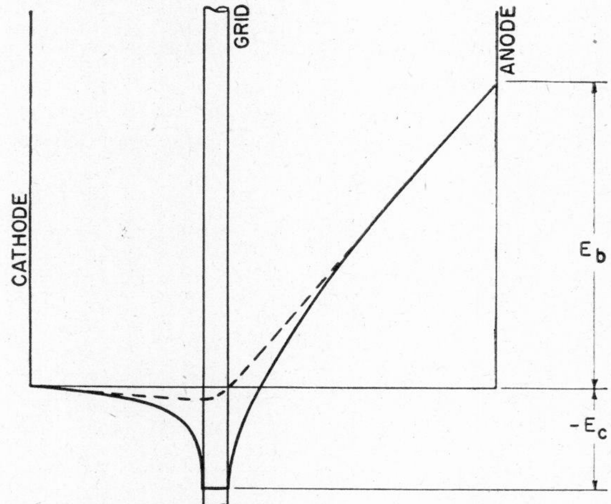

FIG. 3·10 Potential distribution in a parallel-plane low-μ triode under space-charge-free conditions where $E_c + (E_b/\mu) < 0$.

3·10 ELECTRON MOTION

Thus far electron motion has simply been called a "current" in a vacuum. This section deals more with the intrinsic properties of electrons when they are influenced by various electrostatic forces. Electron motion in an electrostatic field is so analogous to the motion of a projectile that the term electron ballistics is often used.

Where e is the electrostatic charge of the electron, m_e is the mass of the electron, and $\mathbf{E}$ is the volts per centimeter acting on the electron, the acceleration, velocity, and distance traveled are

$$a = \frac{d^2x}{dt^2} = \frac{\mathbf{E}e}{m_e}10^7 \text{ centimeters per second per second} \tag{3·26}$$

$$v = \frac{dx}{dt} = \frac{\mathbf{E}et}{m_e}10^7 \text{ centimeters per second} \tag{3·27}$$

$$x = \frac{\mathbf{E}et^2}{2m_e}10^7 + C \text{ centimeters} \tag{3·28}$$

$C = 0$ if initial velocity is zero. The 10^7 factor converts all units to the practical system.

In equation 3·26 no factor indicates initial velocity. This equation applies equally well to electrons starting from rest and to those which have initial velocity. At any time and at any place in their path a field of **E** (volts per centimeter) accelerates them by the degree determined by substituting that value of **E** in equation 3·26. Initial velocities must be added to determine final velocity. Equation 3·27 gives the velocity of the electron, starting from rest, at the end of t seconds in a constant field **E**. For a varying field, the velocity may be obtained by integration of equation 3·27, where the function representing the variation of **E** is substituted. In equation 3·28, x represents the distance an electron travels in t seconds in a constant field **E**. C is zero if the electron has no initial velocity. If C represents a finite velocity, Ct must be added to x.

In mechanics the comparable equations are

$$a = \frac{d^2x}{dt^2} = \frac{\mathbf{F}}{m} \tag{3·29}$$

where m = mass
$\mathbf{F}$ = force acting on mass m
a = acceleration.

By inspection, **F** corresponds to $\mathbf{E}e$ and m_e corresponds to mass m.

$$v = \frac{dx}{dt} = \frac{\mathbf{F}}{m}t \tag{3·30}$$

$$x = \frac{1}{2}\frac{\mathbf{F}}{m}t^2 \tag{3·31}$$

In Fig. 3·11, a unidirectional potential E_b with a superimposed potential $E_b \cos \omega t$ is shown. It is assumed that

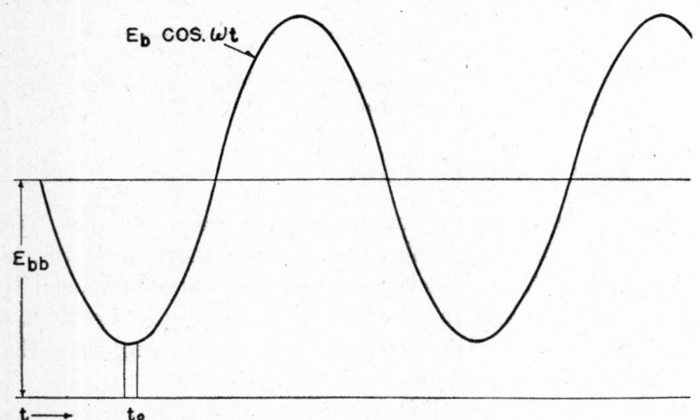

Fig. 3·11 Theoretical voltage diagram for E_b in a diode (or triode).

electrons are permitted to leave the cathode of a diode at time t_0 and when $\omega t = 0$ degrees. ($\omega = 2\pi f$ and f is in cycles per second.)

The acceleration in centimeters per second is

$$a = \frac{d^2x}{dt^2} = \frac{\mathbf{E}e}{m_e} - \frac{\mathbf{E}e \cos \omega t}{m_e} 10^7 \tag{3·32}$$

For small current densities $\mathbf{E} = E_b/s$, where E_b = unidirectional anode voltage and s = cathode-anode distance (for plane structures). For cylindrical structures $\mathbf{E} = E_b/[r \log \epsilon(r_a/r_c)]$ (see equation 3·15), and r is the radial distance at which **E** is to be determined.

For a cylindrical structure, successive calculations of a (from equation 3·32) must be based upon **E** in the region desired.

For high current densities, a different approach is necessary. Under complete space-charge-limited current conditions, the potential at various points between cathode and anode vary with distance and with current density.

For a parallel-plane structure, the potential at any point x between cathode and anode is

$$V = 3\left(\frac{3\pi^2 m_e \mathbf{J}^2}{2e}\right)^{1/3} x^{4/3} \tag{3·33}$$

From the solution of Poisson's equation (equation 3·2)

$$\frac{dV}{dx} = 878\mathbf{J}^{1/2}V^{1/4} \tag{3·34}$$

where $\mathbf{J}$ = amperes per square centimeter
V = potential in volts at point x centimeters from cathode.

J may be obtained from equation 3·9 and V from equation 3·33; therefore dV/dx for parallel plane structures at any point between cathode and anode may be calculated. Once dV/dx is known (and since $dV/dx = -\mathbf{E}$), equation 3·32 may be solved. Since equation 3·32 depends only on **E**, it holds for all current densities.

The solution for potential distribution in a cylindrical diode gives

$$V = \left(\frac{\mathbf{J}'}{2\pi \times 2.331 \times 10^{-6}}\right)^{2/3} (r\beta^2)^{2/3}$$
$$= (68{,}352\mathbf{J}')^{2/3}(r\beta^2)^{2/3} \tag{3·35}$$

where r = any distance between $r = r_c$ and $r = r_a$
β^2 = a function of r/r_c (see Fig. 3·6)
V = potential at the distance r from the cathode
$\mathbf{J}'$ = current density in amperes per centimeter of length and may be obtained by solving equation 3·14.

Since β^2 varies as r and r_c, dV/dr from equation 3·35 is not readily obtainable. It is a simpler approach to obtain dV/dr from several solutions of equation 3·35. The differences obtained from the solution of equation 3·35 for various values of r will be dV/dr or **E**. Therefore equation 3·32 may be solved for the cylindrical case.

Similarly the values of V may be applied in the solution of the equations for electron velocity and distance traveled:

$$v = \frac{dx}{dt} = \left(\frac{\mathbf{E}et}{m_e} - \frac{\mathbf{E}e \sin \omega t}{m_e}\right) \times 10^7 \tag{3·36}$$

$$x = \left(\frac{\mathbf{E}et^2}{2m_e} + \frac{\mathbf{E}e \cos \omega t}{\omega^2 m_e}\right) \times 10^7 \tag{3·37}$$

where $\mathbf{E}$ = voltage gradient
e = electron charge
m_e = mass of electron
$\omega = 2\pi f$ (f = frequency in cycles per second).

Equations 3·32, 3·36, and 3·37 apply to the solution of electron motion under the assumed condition given in Fig. 3·11.

The same basic equations, with modifications, may be applied to the solution of electron motion in triodes, tetrodes, or pentodes.

For low current densities the electric intensity between cathode and grid of a plane triode is, as an average (see Fig. 3·4),

$$\mathbf{E} = \frac{E_c + \dfrac{E_b}{\mu}}{S_3} \tag{3·38}$$

where S_3 is determinable in terms of tube geometry from equation 3·20. Since equation 3·38 determines the slope of the potential-distribution line and since S_1 (cathode-to-grid distance) is known for a given structure, the slope of the potential-distribution line between grid and anode is easily determinable. Assuming ϕ_1 is the angle between the potential-distribution line and the horizontal line (see Fig. 3·4 (*a*)), then

$$\mathbf{E}_{cg} = \tan \phi_1 = \frac{E_c + \dfrac{E_b}{\mu}}{S_3} \tag{3·39}$$

$$V_g = \tan \phi_1 S_1 \tag{3·40}$$

where V_g = effective potential at the grid plane
S_1 = cathode-to-grid distance in centimeters
$\mathbf{E}_{cg}$ = electric intensity between cathode and grid.

If ϕ_2 is the angle of the potential-distribution line between the grid anode and the horizontal,

$$\mathbf{E}_{ga} = \tan \phi_2 = \frac{E_b - \tan \phi_1 S_1}{S_2} \tag{3·41}$$

where S_2 = grid-to-anode distance
$\mathbf{E}_{ga}$ = electric intensity between grid and anode.

Thus equations 3·26, 3·27, and 3·28 may be solved for the plane triode. Equations 3·36 and 3·37 may not be solved

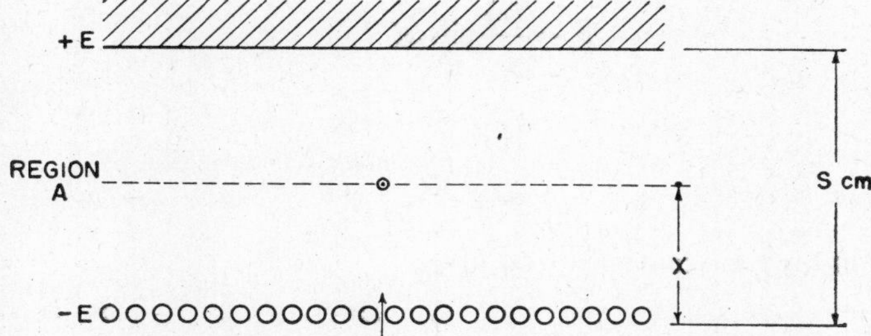

FIG. 3·12 Diagram representing an electron between grid and anode of a parallel plane (or cylindrical triode).

directly without assuming a value of t in equation 3·36 or a distance x or time t in equation 3·28.

The velocity of an electron after it has passed through a potential difference V is related to that potential by

$$v = \sqrt{\frac{2Ve}{m_e} 10^7} = 5.93 \times 10^7 (V)^{1/2} \tag{3·42}$$

For any point, V may be calculated and so may v. The time t may then be obtained by the use of equation 3·36. If an electron enters a region A as shown by the arrow in Fig. 3·12, its velocity at the point $x = S$ is obtainable from equation 3·41. This final velocity is independent of the velocity with which it entered the region.

From equations 3·36 to 3·41, the transit time required for an electron to travel from cathode to grid may be calculated for space-charge-free conditions.

The time required for an electron to travel from cathode to anode in a space-charge-free parallel-plane diode is

$$t = 3.37 \times 10^{-8} s (E_b)^{-1/2} \tag{3·43}$$

where s = separation of cathode and anode in centimeters
E_b = voltage applied to the anode.

Under full space charge conditions this time becomes

$$t = 5 \times 10^{-8} s (E_b)^{-1/2} \tag{3·44}$$

The ratio of equation 3·44 to equation 3·43 is 1.483. This indicates that it takes 48 percent longer for an electron to travel from the cathode to a point at potential V under full space-charge conditions than it does under essentially space-charge-free conditions. Therefore for plane or cylindrical structures, the transit time may be calculated under essentially space-charge-free conditions and converted to full space-charge conditions by the use of this factor.

3·11 FORCE ON AN ELECTRON IN A MAGNETIC FIELD

Since moving charges are an electric current, a force acts on an electron moving in a magnetic field. The force acting on a conductor of length l centimeters carrying a current of I amperes in a magnetic field of flux density $\mathbf{B}$ gauss is $\mathbf{B}l/10$ dynes provided the directions of I and $\mathbf{B}$ are perpendicular.

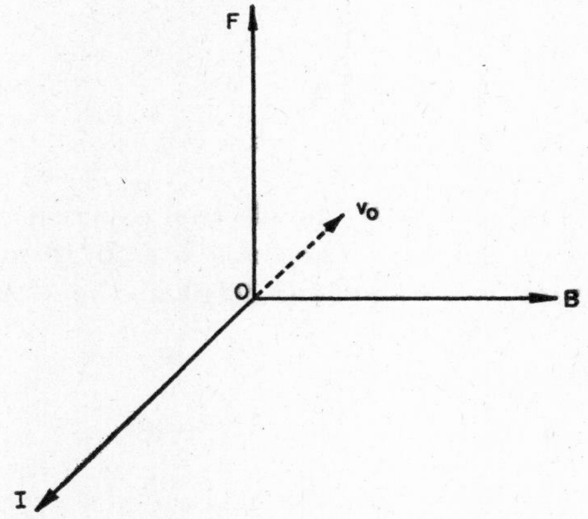

FIG. 3·13 Direction of the force on a conductor due to a magnetic field.

As shown in Fig. 3·13, the direction of this force $\mathbf{F}$ is perpendicular to the plane of I and $\mathbf{B}$ and has the direction of advance of a right-hand screw placed at O and rotated from I to $\mathbf{B}$. Since the conventional direction of current is the direction of flow of positive charge, the velocity of the electron, v, is shown opposite to I.

If a conductor of length l contains N electrons, and if the electron requires t seconds to travel the distance l, the current in the conductor is

$$I = \frac{Ne}{t} \tag{3·45}$$

where e is the charge on an electron in coulombs.

The force on the length l of conductor or also on the N electrons is

$$\frac{\mathbf{B}Il}{10} = \frac{\mathbf{B}Nel}{10t}$$

Since l/t is the velocity v of the electrons, the force per electron is

$$\mathbf{F} = \frac{\mathbf{B}ev}{10} \text{ dynes} \tag{3·46}$$

3·12 MOTION OF AN ELECTRON IN A MAGNETIC FIELD

Suppose that the magnetic field is along the $-Y$-axis as shown in Fig. 3·14 and that an electron starts from O with a velocity v_0 in the XY plane. The initial velocity v_0 has the component v_{0x} along the X-axis and v_{0y} along the Y-axis.

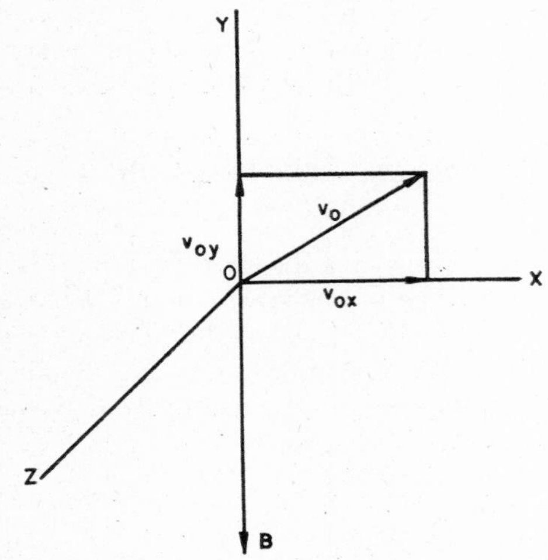

FIG. 3·14 Motion of an electron in a magnetic field.

Since v_y is parallel to $\mathbf{B}$, the electron experiences no force due to this component. The force is $ev_x\mathbf{B}/10$ in the $+Z$-direction owing to v_x and is $ev_z\mathbf{B}/10$ in the $-X$-direction owing to v_z.

Hence Newton's law gives

$$\begin{aligned} \mathbf{F}_x &= m_e \frac{dv_x}{dt} = -\frac{ev_z\mathbf{B}}{10} \\ \mathbf{F}_y &= m_e \frac{dv_y}{dt} = 0 \\ \mathbf{F}_z &= m_e \frac{dv_z}{dt} = \frac{ev_x\mathbf{B}}{10} \end{aligned} \tag{3·47}$$

where m_e is the mass of the electron in grams.

If $\omega = e\mathbf{B}/10m_e$, equations 3·47 become

$$\begin{aligned} \frac{dv_x}{dt} &= -\omega v_z \\ \frac{dv_y}{dt} &= 0 \\ \frac{dv_z}{dt} &= \omega v_x \end{aligned} \tag{3·48}$$

The second of equations 3·48 yields upon integration

$$\begin{aligned} v_y &= v_{0y} \\ y &= v_0 t \end{aligned} \tag{3·49}$$

if $y = 0$ when $t = 0$.

If the first of equations 3·48 is differentiated and a substitution is made for dv_z/dt from the third equation of this set,

$$\frac{d^2v_x}{dt^2} = -\frac{\omega dv_z}{dt} = -\omega^2 v_x$$

or

$$\frac{d^2v_x}{dt^2} + \omega^2 v_x = 0 \tag{3·50}$$

The general solution of equation 3·50 is

$$v_x = A \cos \omega t + C \sin \omega t \tag{3·51}$$

where A and C are constants. Since $v_z = 0$ when $t = 0$, from equation 3·48 $dv_x/dt = 0$ when $t = 0$. Also $v_x = v_{0x}$ when $t = 0$. Hence

$$A = v_{0x} \quad \text{and} \quad C = 0$$

Therefore

$$v_x = v_{0x} \cos \omega t \tag{3·52}$$

Equations 3·48 and 3·52 give

$$v_z = v_{0x} \sin \omega t \tag{3·53}$$

If we assume for the moment that $v_y = v_{0y} = 0$, then $v^2 = v_x^2 + v_z = v_{0x}^2$ or $v = v_{0x} =$ constant.

To determine the path of the electron when $v_y = 0$, integrate equations 3·52 and 3·53, using the initial conditions that $x = z = 0$, when $t = 0$. This yields

$$\begin{aligned} x &= \frac{v_{0x}}{\omega} \sin \omega t \\ z &= \frac{v_{0x}}{\omega}(1 - \cos \omega t) \end{aligned} \tag{3·54}$$

Combining equations 3·54 gives

$$\sin^2 \omega t + \cos^2 \omega t = 1 = \frac{x^2}{\left(\frac{v_{0x}}{\omega}\right)^2} + \left(\frac{z}{\frac{v_{0x}}{\omega}} - 1\right)^2$$

or

$$x^2 + \left(z - \frac{v_{0x}}{\omega}\right)^2 = \left(\frac{v_{0x}}{\omega}\right)^2 \tag{3·55}$$

which is the equation of a circle of radius

$$r = \frac{v_{0x}}{\omega} = \frac{10m_e v_{0x}}{e\mathbf{B}} \tag{3·56}$$

with center at $x = 0$, $z = v_{0x}/\omega$.

The angular velocity of the electron is

$$\frac{v_{0x}}{r} = \omega = \frac{e\mathbf{B}}{10m_e} \text{ radians per second} \tag{3·57}$$

The time for one revolution, or the period, is

$$T = \frac{2\pi}{\omega} = \frac{20\pi m_e}{e\mathbf{B}} \tag{3·58}$$

which becomes

$$T = \frac{3.57 \times 10^{-7}}{\mathbf{B}} \text{ second} \tag{3·59}$$

The period and angular velocity are independent of the speed or radius.

If $v_{0y} \neq 0$, the path is a helix of constant pitch:

$$p = v_{0y}T = \frac{2\pi v_{0y}}{\omega} \tag{3·60}$$

3·13 MOTION OF AN ELECTRON IN PARALLEL ELECTRIC AND MAGNETIC FIELDS

If the electric and magnetic fields are parallel and if the initial velocity of the electron is zero or parallel to the fields, the magnetic field has no effect on the electron. The electron moves in a direction parallel to the fields with a constant acceleration caused by the electric field.

If the electron has an initial velocity with a component perpendicular to the magnetic field, the path of the electron is a helix, as described in Section 3·12. The helix has a pitch which increases with time because of the acceleration produced by the electric field.

3·14 MOTION OF AN ELECTRON IN PERPENDICULAR ELECTRIC AND MAGNETIC FIELDS

The directions of the fields are as shown in Fig. 3·15.

The equations are the same as equations 3·47 except that the electric field $\mathbf{E}$ exerts an additional force $e\mathbf{E}\ 10^7$ volts per centimeter in the X-direction. The equations are

$$\mathbf{F}_x = m_e\frac{dv_x}{dt} = e\mathbf{E}10^7 - \frac{ev_z\mathbf{B}}{10}$$

$$\mathbf{F}_y = m_e\frac{dv_y}{dt} = 0 \tag{3·61}$$

$$\mathbf{F}_z = m_e\frac{dv_z}{dt} = \frac{ev_x\mathbf{B}}{10}$$

These reduce to

$$\frac{dv_x}{dt} = a - \omega v_z$$

$$\frac{dv_y}{dt} = 0 \tag{3·62}$$

$$\frac{dv_z}{dt} = \omega v_x$$

if

$$\omega = \frac{e\mathbf{B}}{10m_e} \quad \text{and} \quad a = \frac{e\mathbf{E}10^7}{m_e} \tag{3·63}$$

Suppose that the initial velocity of the electron is zero. Integration of the second of equations 3·62 yields

$$v_y = v_{0y} = 0 \tag{3·64}$$

If the first of equations 3·62 is differentiated and a substitution is made from the third equation for dv_z/dt,

$$\frac{d^2v_x}{dt^2} = -\omega\frac{dv_z}{dt} = -\omega^2 v_x \tag{3·65}$$

The general solution of this equation is

$$v_x = A\cos\omega t + C\sin\omega t \tag{3·66}$$

where A and C are constants. If the initial conditions that $v_{0x} = v_{0z} = 0$ and that $dv_x/dt = a$ when $t = 0$ are applied to this equation,

$$A = 0 \quad \text{and} \quad C = \frac{a}{\omega}$$

Therefore

$$v_x = \frac{a}{\omega}\sin\omega t$$

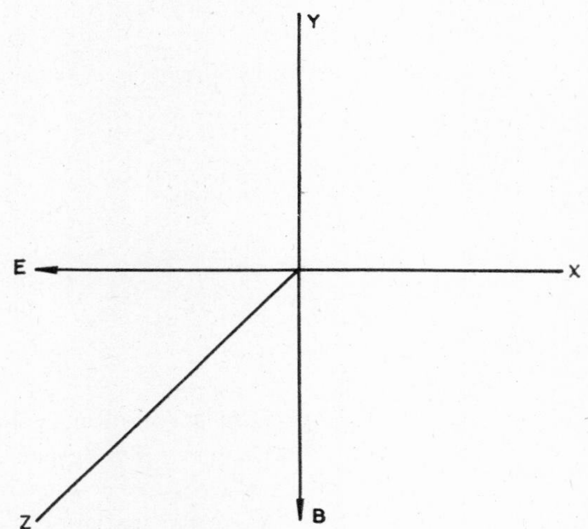

FIG. 3·15 Mutually perpendicular electric and magnetic fields.

The first of equations 3·62 gives

$$v_z = \frac{a}{\omega} - \frac{1}{\omega}\frac{dv_x}{dt}$$

Substitution for dv_x/dt obtained by differentiating the solution for v_x yields

$$v_z = \frac{a}{\omega} - \frac{a}{\omega}\cos\omega t$$

Thus

$$v_x = \frac{a}{\omega}\sin\omega t$$

$$v_y = 0 \tag{3·67}$$

$$v_z = \frac{a}{\omega} - \frac{a}{\omega}\cos\omega t$$

Integration of the first of equations 3·67 gives

$$x = -\frac{a}{\omega^2}\cos\omega t + D$$

Since $x = 0$ when $t = 0$, $D = a/\omega^2$. Similarly, integration of the third of equations 3·67 yields

$$z = \frac{at}{\omega} - \frac{a}{\omega^2}\sin\omega t + E$$

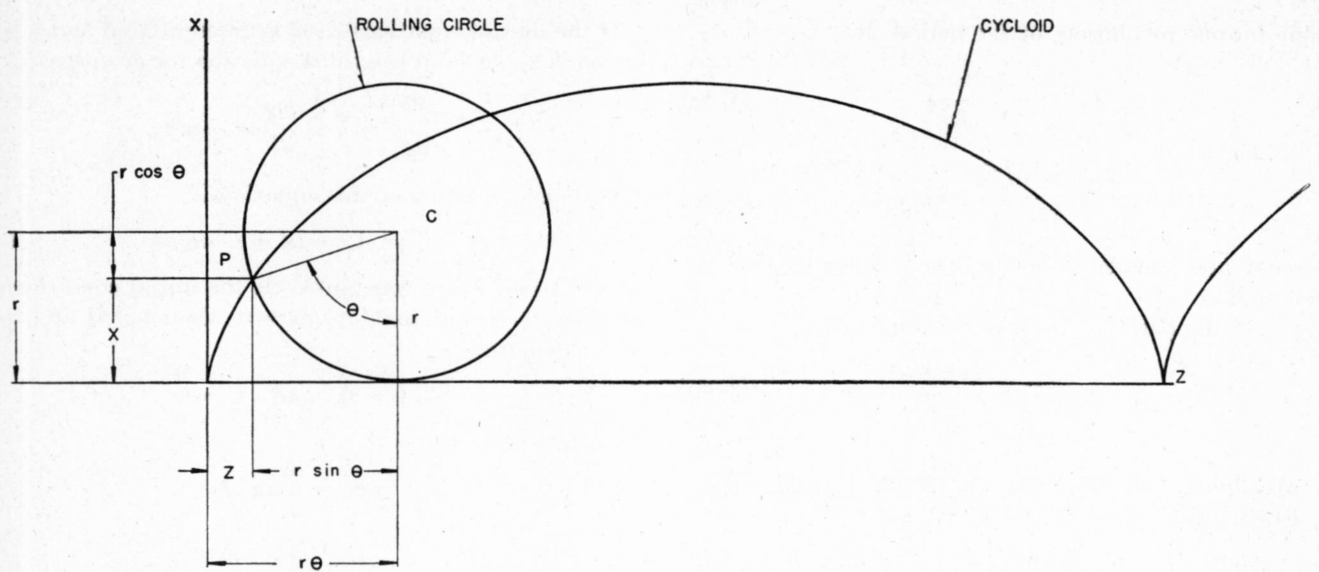

FIG. 3·16 Cycloidal motion of an electron in mutually perpendicular electric and magnetic fields.

Since $z = 0$ when $t = 0$, $E = 0$. Hence

$$\begin{aligned} x &= r(1 - \cos \omega t) \\ y &= 0 \\ z &= r(\omega t - \sin \omega t) \end{aligned} \tag{3·68}$$

where

$$r = \frac{a}{\omega^2} \tag{3·69}$$

Equations 3·68 are the equations of a common cycloid, the path generated by a point on the circumference of a circle of radius r rolling along the Z-axis. This is illustrated in Fig. 3·16, from which

$$\begin{aligned} x &= r(1 - \cos \theta) \\ z &= r(\theta - \sin \theta) \end{aligned} \tag{3·70}$$

which are the same as equations 3·68 if θ is equal to ωt. This shows that ω is the angular velocity of the rolling circle. The velocity of the center of the rolling circle in the Z-direction is

$$V = \frac{a}{\omega} = \frac{10^8 \mathbf{E}}{\mathbf{B}} \tag{3·71}$$

The equations for the path of an electron, if the initial velocities are not zero, may be determined in a similar manner. The only difference is in the determination of the constants of integration. The equations for the general case are

$$\begin{aligned} x &= \frac{v_{0x}}{\omega} \sin \omega t + \left(\frac{a}{\omega^2} - \frac{v_{0z}}{\omega}\right)(1 - \cos \omega t) \\ y &= v_{0y} t \\ z &= \frac{v_{0x}}{\omega}(1 - \cos \omega t) - \left(\frac{a}{\omega^2} - \frac{v_{0z}}{\omega}\right) \sin \omega t + \frac{a}{\omega} t \end{aligned} \tag{3·72}$$

3·15 MOTION OF AN ELECTRON IN ELECTRIC AND MAGNETIC FIELDS MAKING AN ARBITRARY ANGLE WITH EACH OTHER

This section deals with the case illustrated in Fig. 3·17. **B** lies along the $-Y$-axis. **E** lies in the XY-plane and makes an angle θ with the Y-axis. The results of Section 3·14 indicate that the equations for the path of the electron with zero initial velocity are

$$\begin{aligned} x &= \frac{a_x}{\omega^2}(1 - \cos \omega t) \\ y &= \tfrac{1}{2} a_y t^2 \\ z &= \frac{a_x t}{\omega} - \frac{a_x}{\omega} \sin \omega t \end{aligned} \tag{3·73}$$

where

$$\begin{aligned} a_x &= \frac{e\mathbf{E} \sin \theta 10^7}{m_e} \\ a_y &= \frac{e\mathbf{E} \cos \theta 10^7}{m_e} \end{aligned} \tag{3·74}$$

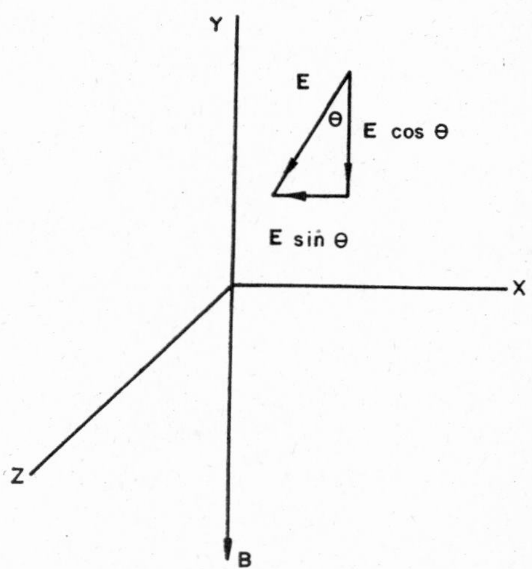

FIG. 3·17 Electric and magnetic fields making an arbitrary angle with each other.

Therefore, the motion is the same as though the fields were perpendicular to each other, except for the added acceleration along the Y-axis.

3·16 MOTION OF ELECTRONS IN A RADIAL ELECTRIC FIELD AND AN AXIAL MAGNETIC FIELD

Consider the case shown in Fig. 3·18. Here **B** is directed out of the paper. V_a is the potential difference between cathode and anode, and V is the potential of some point at distance r from the axis. Assume that electrons leave the cathode with zero velocity.

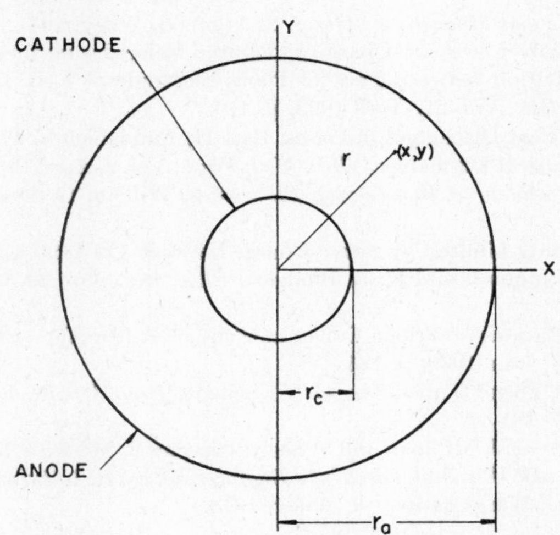

Fig. 3·18 Anode and cathode consisting of concentric cylinders.

The electric field alone accelerates an electron radially toward the anode, and the magnetic field bends the path of the electron away from its radial path.

The equations of motion are

$$m_e \frac{\partial^2 x}{\partial t^2} = -e \frac{\partial V}{\partial x} 10^7 - \frac{ev_y \mathbf{B}}{10}$$
$$m_e \frac{\partial^2 y}{\partial t^2} = e \frac{\partial V}{\partial y} 10^7 + \frac{ev_x \mathbf{B}}{10} \quad (3\cdot 75)$$

where $\partial V/\partial x$ and $\partial V/\partial y$ are the x and y components of the electric field and v_x and v_y are the x and y components of the velocity of the electron. Substituting,

$$v_x = \frac{\partial x}{dt} \quad \text{and} \quad v_y = \frac{\partial y}{dt}$$

and

$$\frac{\partial V}{\partial x} = \frac{x}{r}\frac{\partial V}{\partial r} \quad \text{and} \quad \frac{\partial V}{\partial y} = \frac{y}{r}\frac{\partial V}{\partial r}$$

in equations 3·75, and

$$m_e \frac{\partial^2 x}{\partial t^2} = -\frac{ex}{r}\frac{\partial V}{\partial r} 10^7 - \frac{e}{10}\frac{\partial y}{\partial t}\mathbf{B}$$
$$m_e \frac{\partial^2 y}{\partial t^2} = -\frac{ey}{r}\frac{\partial V}{\partial r} 10^7 + \frac{e}{10}\frac{\partial x}{\partial t}\mathbf{B} \quad (3\cdot 76)$$

By multiplying the first of equations 3·76 by y and the second by x and subtracting the first from the second,

$$m_e x \frac{\partial^2 y}{\partial t^2} - m_e y \frac{\partial^2 x}{\partial t^2} = \frac{\mathbf{B}e}{10} x \frac{\partial x}{\partial t} + \frac{\mathbf{B}e}{10} y \frac{\partial y}{\partial t} \quad (3\cdot 77)$$

Integrating equation 3·77,

$$m_e x \frac{\partial y}{\partial t} - m_e y \frac{\partial x}{\partial t} = \frac{\mathbf{B}ey^2}{20} + \frac{\mathbf{B}ex^2}{20} + Cm_e \quad (3\cdot 78)$$

where C is a constant of integration. Dividing both sides of equation 3·78 by m_e gives

$$x \frac{\partial y}{\partial t} - y \frac{\partial x}{\partial t} = \frac{\mathbf{B}e}{20m_e}(x^2 + y^2) + C \quad (3\cdot 79)$$

Substitution of

$$x = r\cos\theta, \; y = r\sin\theta, \; \text{and} \; \sin^2\theta + \cos^2\theta = 1 \quad (3\cdot 80)$$

into the left side of equation 3·79 gives

$$x \frac{\partial y}{\partial t} - y \frac{\partial x}{\partial t} = r^2 \frac{\partial \theta}{\partial t} \quad (3\cdot 81)$$

Substitution of

$$x^2 + y^2 = r^2$$

into the right side of equation 3·79 yields

$$\frac{\mathbf{B}e}{20m_e}(x^2 + y^2) + C = \frac{\mathbf{B}er^2}{20m_e} + C \quad (3\cdot 82)$$

Therefore,

$$r^2 \frac{\partial \theta}{\partial t} = \frac{\mathbf{B}er^2}{20m_e} + C \quad (3\cdot 83)$$

Using the initial conditions, $\partial\theta/\partial t = 0$ when $r = r_c$, the constant C can be evaluated, and

$$r^2 \frac{\partial \theta}{\partial t} - \frac{\mathbf{B}er^2}{20m_e} + \frac{\mathbf{B}er_c^2}{20m_e} = 0 \quad (3\cdot 84)$$

or

$$\frac{\partial \theta}{\partial t} = \frac{\mathbf{B}e}{20m_e}\left(1 - \frac{r_c^2}{r^2}\right) \quad (3\cdot 85)$$

Conservation of energy gives

$$\tfrac{1}{2} m_e v^2 = eV10^7 \quad (3\cdot 86)$$

also

$$v^2 = \left(\frac{\partial r}{\partial t}\right)^2 + r^2 \left(\frac{\partial \theta}{\partial t}\right)^2 \quad (3\cdot 87)$$

which with equation 3·86 gives

$$\left(\frac{\partial r}{\partial t}\right)^2 + r^2 \left(\frac{\partial \theta}{\partial t}\right)^2 = \frac{2eV10^7}{m_e} \quad (3\cdot 88)$$

Combining equations 3·88 and 3·85 to eliminate $\partial\theta/\partial t$,

$$\left(\frac{\partial r}{\partial t}\right)^2 + r^2 \left(\frac{\mathbf{B}e}{20m_e}\right)^2 \left(1 - \frac{r_c^2}{r^2}\right)^2 = \frac{2eV10^7}{m_e} \quad (3\cdot 89)$$

Equation 3·89 is a general equation describing the motion of an electron for any space-charge condition.

3·17 CUT-OFF CHARACTERISTICS OF A CYLINDRICAL MAGNETRON

At cut-off in a magnetron, all electrons are turned back toward the cathode just before they reach the anode. At cut-off the anode current is zero and $\partial r/\partial t$ in equation 3·89—the radial velocity—is zero for $r = r_a$, or more exactly for

r slightly less than r_a. If $\partial r/\partial t$ is set equal to zero for $r = r_a$ in equation 3·89,

$$\left(\frac{r_a \mathbf{B} e}{20 m_e}\right)^2 \left(1 - \frac{r_c^2}{r_a^2}\right)^2 = \frac{2eV10^7}{m_e} \qquad (3\cdot 90)$$

or

$$V = \frac{r_a^2 \mathbf{B}^2 e}{8 m_e 10^9} \left(1 - \frac{r_c^2}{r_a^2}\right)^2 \qquad (3\cdot 91)$$

which defines the cut-off voltage V.

Since $r_c \ll r_a$ in most cases,

$$V = \frac{r_a^2 \mathbf{B}^2 e}{8 m_e 10^9} \qquad (3\cdot 92)$$

or

$$\mathbf{B} = \frac{4 \times 10^4}{r_a} \sqrt{\frac{5 V m_e}{e}} \qquad (3\cdot 93)$$

3·18 APPLICATIONS

The equations derived in the foregoing sections apply as well to the motion of positive and negative ions as to the motion of electrons. In using the equations for ions, different values for mass must be used. The charge on ions is either positive or negative, and the magnitude of the charge is equal to the electronic charge or an integral multiple of it. Therefore, the equations can be applied to the motion of electrons or ions in cathode-ray tubes, mass spectrometers, cyclotrons, magnetrons, and other devices.

REFERENCES

1. *Electric and Magnetic Fields*, S. S. Attwood, Wiley, 1941.
2. "The Effect of Space-Charge and Initial Velocities on the Potential Distribution between Parallel Plane Electrodes," I. L. Langmuir, *Phys. Rev.*, Vol. 21, April 1923, p. 419.
3. "Electrical Discharges in Gases, Part II, Fundamental Phenomena in Electrical Discharges," *Rev. Mod. Phys.*, Vol. 2, April 1930, p. 191.
4. *Fundamentals of Engineering Electronics*, William G. Dow, Wiley, 1937.
5. "Currents Limited by Space-Charge between Co-Axial Cylinders," I. L. Langmuir and K. B. Blodgett, *Phys. Rev.*, Vol. 22, Oct. 1923, p. 347.
6. "Calculations of Triode Constants," by J. H. Fremlin, *Phil. Mag.*, Vol. 27, June 1939, p. 185.
7. "Beam Power Tubes," by O. H. Schade, *Proc. I.R.E.*, Feb. 1938, pp. 137–181.
8. *Electronics*, J. Millman and S. Seely, Chapter 2, McGraw-Hill, 1941.
9. *Hyper and Ultrahigh Frequency Engineering*, by R. I. Sarbacher and W. A. Edson, Chapter 16, Wiley, 1943.

Chapter 4

ELECTRICAL CONDUCTION IN GASES

D. E. Marshall

If a quantity of gas is included in a tube with a cathode which can emit sufficient electrons the tube has the following characteristics:

1. Anode-cathode voltage becomes almost independent of anode current, and in magnitude it is about equal to the ionization potential of the gas.
2. Grids are unable to modify anode-cathode current after conduction starts. Control consists of blocking conduction before starting.
3. The electrostatic field at the cathode is in the proper direction to assist in electron emission.

In a phototube, introduction of gas causes amplification of photoelectric currents; otherwise the current would be derived from cathode emission alone.

In a gas-filled tube with two or more electrodes having no independent electron emission, applied voltages of several hundred can cause discharges.

These phenomena form the basis for gas-tube engineering. Although the theories evolved to account for these effects are not in general completely satisfactory, they aid in understanding the phenomena.

4·1 IONIZATION

Ionization is one of the most important physical processes in gas discharges. To increase the current-carrying capacity of an electronic tube the electron space charge must be neutralized. This is done by gas ions. Their positive charges cancel the negative charges of the electrons in the anode-cathode space, and a slight preponderance of positive ions increases the field at the cathode and aids electron emission from the cathode.

Section 1·5 describes how an atom can be ionized by being struck by an electron with kinetic energy greater than the ionizing energy of the atom. This energy usually is given in terms of electron volts. For example, if the gas is mercury vapor the required ionizing energy is 10.4 electron volts. If the energy of the electron is greater than this amount an electron can be removed from the atom. The atom then becomes positively charged, and the original electron and the displaced electron are both free of the influence of the atom. After the collision the number of charged current carriers is increased by two. The additional electron can carry current by moving toward the anode, and the new positive ion can increase this current although it moves in the opposite direction.

It is not always necessary for the bombarding electron to have the full ionizing energy. Under some conditions an atom can be ionized by being struck by two electrons, if their total energy is greater than the ionizing energy of the atom. If the first electron excites the atom to a metastable state as described in Section 1·5 it remains in this state for a fairly long period, and the probability of its being hit again before it returns to its normal state is greater than if the initial excitation was not metastable. Excitation of an atom to the metastable state is rather rare, so, for this action which is known as "cumulative ionization" to become large, atom density and current density must be large.

The probability of ionization caused by collision of an atom and an electron is a function of the kinetic energy of the electron. If its energy is below the ionization potential the probability is zero; above this value the probability increases with the excess electron voltage. This is discussed in Section 1·6. For this reason if ionization is calculated from statistical data, the average electron energy per ion may be several times higher than the usual published figure obtained by detailed experimentation.

4·2 CURRENT CONDUCTION THROUGH IONIZED SPACE

Figure 4·1 shows some of the features of electrical conduction in a gas-filled tube. The body of the tube contains the plasma of the discharge. It is a region of low voltage gradient in which the densities of ions and electrons are nearly equal everywhere. Electrons move out of the cathode into the plasma and out of the plasma into the anode. This electron movement constitutes the only current-carrying medium at the anode. Positive ions are drawn to the cathode and to the walls of the tube. If the walls are insulated, an equal number of electrons also moves to the walls to balance the movement of positive ions. The movement of positive ions to the cathode forms a part of the current conduction at the cathode. The sum of positive-ion current and electron current at the cathode must be equal to the total current at the anode.

In the external circuit current is due to electron flow. Electrons can be considered to flow into the anode, through the external circuit, and into the cathode. As electrons flow from the external circuit into the tube through the cathode, some of them pass into the discharge directly because of thermal emission from the cathode or because of arc-spot emission. The other electrons are drawn out of the cathode by the ions which they neutralize. This is illustrated in Fig. 4·2. Gas ions are neutralized at the cathode surface, thus becoming neutral atoms again and diffusing back into the body of the tube.

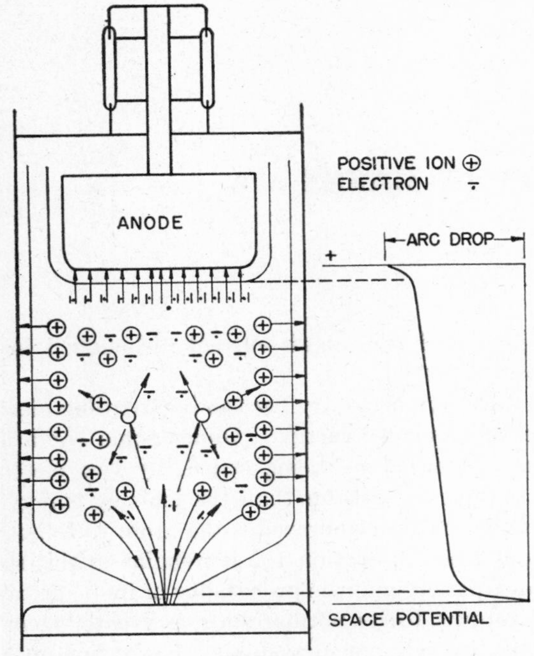

FIG. 4·1 Diagram of electrical phenomena in pool-type tubes.

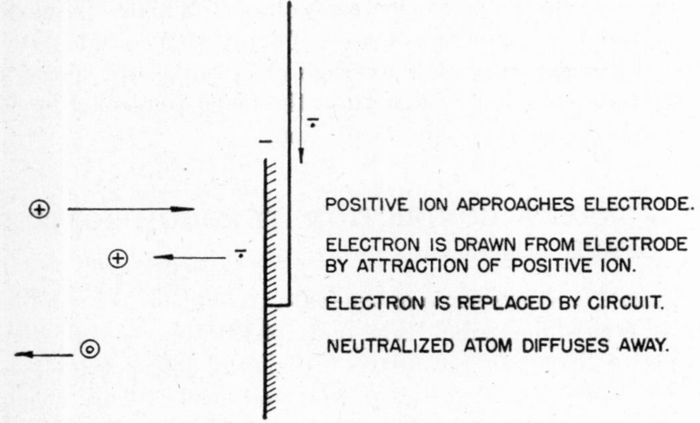

FIG. 4·2 Conduction to electrode by positive ions.

4·3 SPACE-CHARGE PHENOMENA

At the right in Fig. 4·1 is a space-potential diagram of the arc-discharge space. Voltage is distributed so that the field at the cathode is positive and the gradient through the plasma is small. A slight voltage rise at the anode is shown. The distribution of voltage is a function of the amount of unbalanced charge in the space. Complete calculation of voltage distribution would be complex,[17] but the phenomena can be illustrated by a theoretical case.

Assume that a given space contains both positive and negative charges. The total charge at any point is assumed to be uniform throughout the volume. This would be approximately true if the current were built up slowly from small values.

The relation between space potential and charge density is given by Poisson's equation (assuming one-dimensional case):

$$\frac{d^2V}{dx^2} = 4\pi e(\rho_e - \rho_p) \tag{4·1}$$

where ρ_e = electron-charge density
ρ_p = ion positive-charge density
e = electronic charge
V = space potential
x = distance from cathode to a point in space
d = cathode-to-anode distance.

Integrating this equation and setting $V = 0$ when $x = 0$,

$$V = \frac{V_a x}{d} + 2\pi e(\rho_e - \rho_p)(x^2 - dx) \tag{4·2}$$

where V_a equals the potential at the anode.

This is the general equation of a parabola. If the space charge is exactly neutralized, then

$$V = \frac{V_a x}{d} \tag{4·3}$$

and the voltage distribution is linear between anode and cathode, as shown in Fig. 4·3 by the line marked "no space charge." This line is the first term of equation 4·2. Space charge increases the space potential above the linear value if it is positive, and decreases it below the linear value if it is negative.

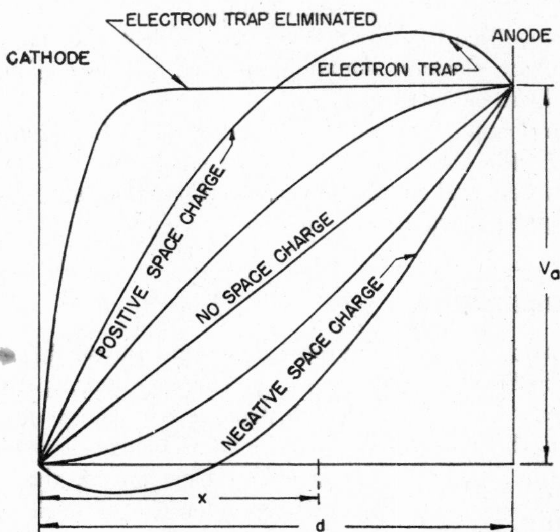

FIG. 4·3 Space-potential distribution with uniform charge distribution.

The field intensity at the cathode is

$$\left(\frac{dV}{dx}\right)_{x=0} = \frac{V_a}{d} - 2\pi e(\rho_e - \rho_p)d \tag{4·4}$$

If the space charge is positive the field intensity at the cathode is more positive than for a linear potential distribution. If the space charge is negative, as in high-vacuum tubes, the field intensity is more negative than for a linear potential distribution. If the space charge is sufficiently negative the field intensity at the cathode becomes negative.

The field intensity at the anode is

$$\left(\frac{dV}{dx}\right)_{x=d} = \frac{V_a}{d} + 2\pi e(\rho_e - \rho_p)d \qquad (4\cdot5)$$

This field intensity is negative if

$$(\rho_p - \rho_e) > \frac{V_a}{2\pi ed} \qquad (4\cdot6)$$

The parabolic curves in Fig. 4·3 show examples of all these variations.

The conclusions to be drawn from the above are that in general a negative space charge tends to limit current conduction through a tube, since the tendency is to decrease or reverse the field at the cathode, thus limiting electron flow from the cathode. A positive space charge tends to increase the field intensity at the cathode, thus drawing all emitted electrons into the discharge. For these reasons the current through a gas tube is limited only by the impedance of the external circuit.

This analysis can be used to trace the origin of the plasma. At the instant conduction starts, potential distribution is linear. Immediately positive ions are generated, and they leave the space more slowly than the electrons. The space charge then becomes positive, and a curve of space-charge distribution begins to assume the characteristic upward curvature. Soon the condition described by equation 4·6 is reached and a maximum appears in the potential-distribution curve, as shown in Fig. 4·3. Electrons moving from the cathode to anode then must gain momentum in traveling from the cathode to the potential maximum to enable them to continue their travel against the opposing field. This theory does not take into account any loss in momentum because of collisions of electrons with gas atoms. Collisions with gas atoms cause some of the electrons to lose momentum; consequently some of them are not able to move against the opposing field at the anode and are trapped. These trapped electrons neutralize some of the positive space charge and eliminate the potential maximum. As soon as the upward curvatures in the potential-distribution curve are eliminated the electrons can theoretically make their way to the anode. Potential distribution thus becomes constant at equilibrium through much of the discharge space. In this region a population of electrons, positive ions, and atoms exists. The field in this region is theoretically zero; however, losses to the tube walls require that a slight positive gradient exist.

As this phenomenon proceeds, the current in the external circuit increases. This current causes a voltage drop across the load impedance and reduces the voltage across the tube. The combination of establishing the plasma and increasing the current concentrates the tube voltage across a narrow region in the vicinity of the cathode and reduces the tube voltage to a minimum. Tube voltage reaches a lower limit somewhere near the ionizing potential of the gas. It must be large enough to accelerate the electrons and give them enough energy to insure a copious supply of positive ions.

4·4 THE PLASMA

The relatively field-free region described in Section 4·3 is known as the plasma. In this region the densities of positive ions and electrons are nearly equal. The atoms of gas are in violent motion. Their directions and speeds are completely at random. On the average the motion of all the atoms is zero, so that there is no mass movement of gas. Electrons entering the plasma from the cathode are accelerated by the cathode to plasma potential difference. They collide with atoms. If a collision ionizes an atom a large part of the energy of the electron is spent. The kinetic energy of the atom is not increased appreciably, because of the great difference in mass between the two colliding particles. After the collision the two resulting electrons divide the excess kinetic energy. They now have differing speeds in different directions. Collisions with atoms now become largely elastic because the electrons in general do not have enough energy to cause excitation or ionization in large quantity. These elastic collisions will result in further increases in the randomness of the electron velocities. Thus the electrons in the plasma have velocities similar to those of a gas, for they approach a Maxwellian velocity distribution (see Section 1·2). This distribution cannot be truly Maxwellian because electrons on the average must move toward the anode. This component of velocity toward the anode is known as the "drift velocity" and is superimposed on the random velocities. The electric forces accelerate the electrons to high velocities. The Maxwellian portion of their velocity distribution can be characterized by an equivalent temperature. Thus the term "electron temperature" is often used in discussing conditions in the plasma.

If a true equilibrium existed between electrons, ions, and atoms all three would eventually reach the same temperature. The interchange of energy between electrons, ions, and atoms is so small, however, that a long time would be required to establish equilibrium. During this time the ions and atoms would make many collisions with the tube envelope, and would transfer a portion of their energy to the tube walls. The energy gained by the ions and atoms from the electrons is lost therefore, and temperature equilibrium is never attained. The ions, being charged, gain some energy from the electric field. They are considered to have higher energy than the atoms, but are not usually considered to have Maxwellian distribution of velocities and therefore do not have true equivalent temperatures.

The motion of electrons constitutes a current, but because their average velocity is zero, the current is actually zero. The motion of electrons can be broken into positive and negative components and a current associated with each component. The relationship between electron temperature and electron current component is given by [3]

$$\mathbf{J}_e = e\rho_e\sqrt{\frac{kT_e}{2m_e}} \qquad (4\cdot7)$$

where $\mathbf{J}_e$ = random electron current density
k = Boltzmann constant
ρ_e = electron-charge density
T_e = absolute electron temperature
m_e = electron mass.

The electron current is large in comparison with the ion current. Although the ion energy cannot be characterized by a temperature, experimental evidence indicates that an equivalent temperature would be about half the electron temperature. The ratio of electrons and positive-ion current densities in the plasma then depends upon the ratio of the square root of their respective masses. For mercury this is 605.

If an insulated object is placed in the plasma it is bombarded by both electrons and ions. The transport of charge to the object causes it to be charged by the electron current and discharged by the positive-ion current. Since the electron current is greater, the object becomes charged negatively. Its potential becomes negative with respect to the space-potential level of the immediate surroundings, so approaching electrons of lower velocities are turned back until equal ion and electron currents strike the object. If the object is not insulated the difference in the two currents must be equal to the current drawn off by the object, and its potential adjusts itself accordingly.

4·5 VOLT-AMPERE CHARACTERISTIC OF ELECTRODES IN A PLASMA

If a small flat electrode is placed in the region of the plasma of a gas-discharge tube, and is provided with guard rings so that spacing and arrangement of the element do not affect the results, a volt-ampere characteristic such as that shown in Fig. 4·4 can be obtained. This is of value in studying the discharge and is characteristic of this phenomenon.[3]

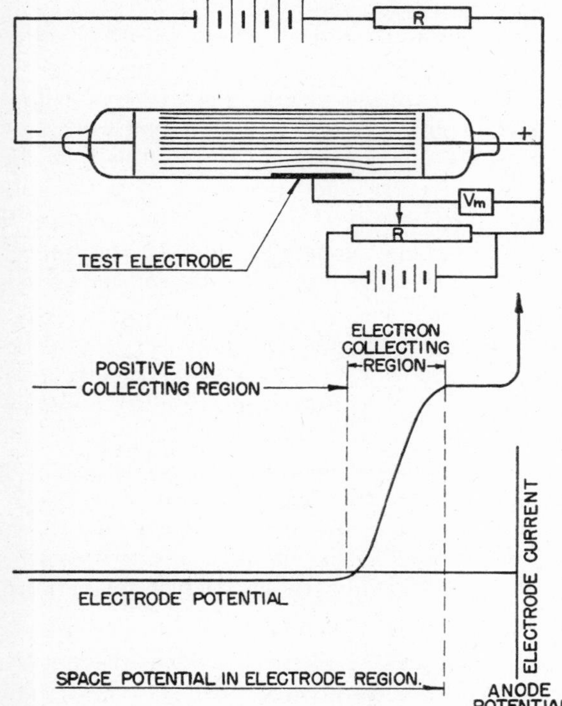

Fig. 4·4 Volt-ampere characteristics of a flat electrode in a plasma.

Figure 4·4 shows the circuit employed and the volt-ampere characteristic obtained by varying electrode potential with respect to one of the main discharge electrodes, in this case the anode. Starting at the region of high negative-electrode potential, a small current nearly independent of electrode potential is drawn from the electrode. Current at this point is due to positive ions. Electrons are repelled from the electrode by the negative potential. Ion current to the electrode is not balanced by electrons, and a space charge results. This space charge limits the distance into the plasma over which electrode potential is effective in accelerating ions to the electrode. The limit of the electrode's influence therefore can be considered as forming a "sheath" around the electrode. The positive-ion current is thus limited by the number of ions that strike this sheath. A change in electrode potential does not cause an increase of positive-ion current unless the outward movement of the sheath increases its area or in some way increases its current-collecting effectiveness.

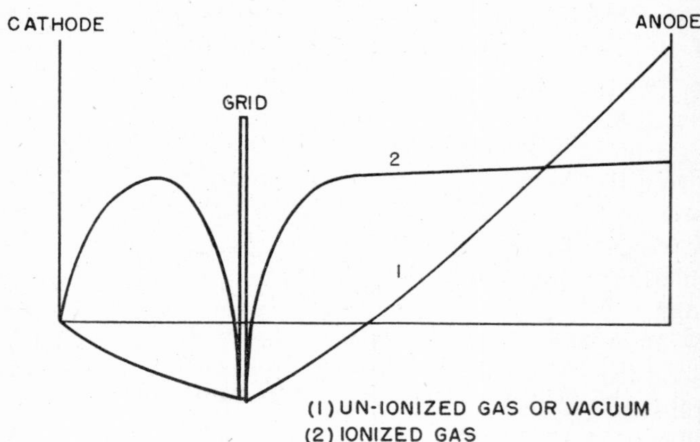

Fig. 4·5 Space-potential distribution in a gas-filled tube with grid.

As electrode potential is made more positive the sheath decreases in thickness and begins to collect electrons. Current is small at first because only high-velocity electrons can force their way against the opposing field. As electrode potential is made still more positive, large numbers of electrons of lower velocity are collected. When the electrode potential is equal to the space potential all electrons moving in the direction of the electrode strike it and are collected. A more positive electrode now cannot collect more electrons until its potential is equal to the anode potential. Up to this point a negative sheath surrounds the electrode, like the positive sheath mentioned before. When electrode potential becomes equal to the anode potential, the electrons moving through the sheath have enough energy to ionize the gas and break down the sheath. The electrode then draws current without limit and takes over the function of an anode.

This helps to explain the inability of a grid in a gas-filled tube to control magnitude of the cathode-anode current once the discharge has started. Figure 4·5 represents the potential distribution in a tube containing a grid. Before the gas is ionized the negative grid potential lowers the potential distribution and causes a negative field at the cathode which prevents electrons from leaving it. After the discharge has started the negative grid attracts positive ions whose space charge neutralizes the negative grid potential. Grid potential then has no effect at the cathode and cannot influence the current through the tube.

4·6 ELECTRON EMISSION AS GENERATED OR INCREASED BY POSITIVE-ION FIELDS

Increase of the field at the cathode by positive space charge is important in the theoretical explanation of electron emission from the cathode.[4, 5, 6] It is the only theory that gives a detailed and satisfying explanation of the action of the mercury-arc pool cathode. Unfortunately, quantitative applications of known theories do not fully support this view.[7]

As described in Section 2·1, electrons are normally held within the boundaries of metals by means of an electrostatic force of attraction to the atoms of the metal. The work necessary to overcome this force is known as the "work function," and for most metals it is between 4 and 5 electron volts. If a stronger field can be supplied externally to the metal so that the forces tending to hold the electron in the metal can be overcome by the external field, then the electrons can be removed from the metal by the action of this external field. If the field is relatively weak, only those electrons approaching the metal boundary with the highest velocities escape. As the strength of the field is increased the slower electrons also can escape; therefore the number of electrons emitted is a function of field strength. In the mercury arc the voltage difference from cathode to plasma is effective only in a thin space-charge sheath of positive ions, and fields of large magnitude can be computed from their effect at the cathode.

Since plasma potential is high with respect to cathode potential, the electrons gain their ionizing energy after traveling the distance between cathode and plasma. Many ions are generated then at the beginning of the plasma, and many of them are attracted to the cathode. Movement of ions into the cathode compares with the movement of electrons away from the cathode. Electrons are light and move at high velocity; their density in this space is low in comparison with that of ions. The relation [2] between ion-current density, cathode-plasma potential difference, and cathode-plasma distance is given by the space-charge formula used in vacuum-tube theory corrected for the larger mass of the ions:

$$\mathbf{J}_p = 2.331 \times (10)^{-6} \frac{V_{cp}^{3/2}}{d^2} \left(\frac{m_e}{m_p}\right)^{1/2} \qquad (4\cdot 8)$$

where

m_e = mass of electron
m_p = mass of positive ion
d = cathode-to-plasma distance in centimeters
V_{cp} = cathode-to-plasma potential difference
$\mathbf{J}_p$ = positive-ion current density.

The field at the cathode is given by

$$E_c = 5.7(10)^3 M^{1/4} V_{cp}^{1/4} \mathbf{J}_p^{1/2} \qquad (4\cdot 9)$$

where M = ion molecular weight.

To use these formulas qualitatively to explain phenomena at the cathode, the ratio $\mathbf{J}_p/\mathbf{J}_e$ of positive-ion current density to the electron-current density at the cathode must be estimated.

An estimate of this quantity can be made as follows: From experiments the cathode-plasma potential difference of a mercury arc is known to be somewhat less than the ionizing potential of mercury. The ionizing potential is 10.4 volts, and the cathode fall of potential is from 9.9 to 10 volts. The work-function energy of about 4.5 volts is supplied by the cathode field. The energy remaining, about 5.5 volts, is available for acceleration of electrons. This is sufficient for two-stage cumulative ionization but not for single-impact ionization.

If each two electrons emitted from the cathode produce one ion, and on the average every other new ion falls back into the cathode-fall region, the other ion with its electrons going forward into the plasma, then for each four electrons starting from the cathode one ion returns. The returning ion adds one unit of current to the four electron units, thus one fifth of the cathode current is ion current.

For mercury, at a cathode fall of 10 volts and $\mathbf{J}_p/\mathbf{J}_e = 0.2$, the field at the cathode surface is

$$E_c = 16{,}900\mathbf{J}_c^{1/2} \qquad (4\cdot 10)$$

where $\mathbf{J}_c$ = current density at the cathode.

The accepted experimental value for current density in the cathode spot is 4000 amperes per square centimeter. Using this value the field intensity at the cathode is estimated at 1,070,000 volts per centimeter.

This intensity usually is large enough to produce measurable emission. The only fields of practical importance are those large enough to produce electron currents equal to eight tenths of the total cathode current, or 3200 amperes per square centimeter. The Fowler-Nordheim formula,[8] based on wave mechanics, gives about the highest values for field-emission currents as a function of voltage. Computed by this formula, a field of about 42 million volts per centimeter is necessary to furnish the required electron-emission current.

This is a large discrepancy, and it suggests a re-examination of the assumptions made in the computations. The field as computated by the space-charge formula gives an average value because it assumes a uniform charge density, whereas the charges on ions are discontinuous. The surface of the mercury is assumed to be smooth, but it may be rough.

The Fowler-Nordheim formula assumes pure mercury. But experiments on field emission always show that impurities reduce the field required for emission. It might seem that the estimate of ion current at two tenths of the total tube current is too low. This formula gives a rapid variation of field current with field intensity, and at 30 million volts per centimeter the field-current yield would be only 10 amperes per centimeter. There does not seem to be any possibility of reducing the field intensity below this value. If a ratio of actual field to average field of 10 were assumed and all the cathode current were assumed to be ion current, a current density of 6600 amperes per square centimeter would be required at the cathode, according to this formula. Assuming that positive-ion current is equal to two tenths of the total current, the equation can balance only if the 10-to-1 field increase is taken and a cathode current of 96,000 amperes per square centimeter is assumed.

Even though field theory will not stand the test of quantitative analysis, it does offer a working hypothesis and is the only theory that gives a detailed analysis of the action of the mercury-pool cathode.

Conditions at the cathodes of thermionic tubes are different from those in pool tubes, because the field at the cathode is not responsible for the entire cathode emission.[9] The cathode emits electrons by thermal emission, and therefore cathode current is distributed uniformly over the cathode area. The action of the field at the cathode is similar but the intensity is smaller. The positive field at the cathode reduces the cathode work function, and the maximum emission is increased.

Schottky's equation [12] gives the ratio of increase of electron emission at the cathode caused by the action of the electrostatic field at the cathode surface. This ratio is

$$\frac{\mathbf{J}_e}{\mathbf{J}_{th}} = \epsilon^{4.4\sqrt{\mathbf{E}_c}/T} \qquad (4\cdot 11)$$

where $\mathbf{E}_c$ = electric-field intensity at the cathode
$\mathbf{J}_e$ = electric-emission density with help of field
$\mathbf{J}_{th}$ = thermal electron-emission density
T = absolute cathode temperature.

Assuming again that the cathode fall of potential is 10 volts and the ratio of positive-ion current to electron current at

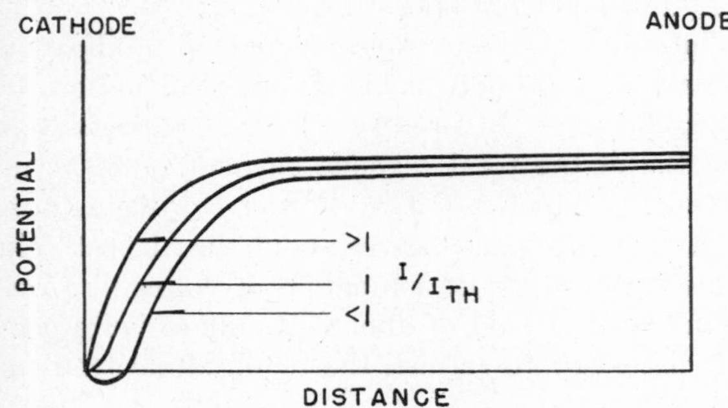

FIG. 4·6 Space-potential distribution diagram at the cathode of a gas-filled tube as a function of the ratio of cathode emission current to tube current.

the cathode is 0.2 for mercury vapor, a relation between the field at the cathode and the cathode current density is given by equation 4·10. Substituting from this equation into equation 4·11,

$$\frac{\mathbf{J}_e}{\mathbf{J}_{th}} = \epsilon^{570[(\mathbf{J}_c)^{1/4}/T]} \qquad (4\cdot 12)$$

Assuming for a coated cathode that T = 1100 degrees Kelvin, if $\mathbf{J}_c = 1$,

$$\frac{\mathbf{J}_e}{\mathbf{J}_{th}} = 1.68 \qquad (4\cdot 13)$$

if $\mathbf{J}_c = 10$,

$$\frac{\mathbf{J}_e}{\mathbf{J}_{th}} = 2.52 \qquad (4\cdot 14)$$

This derivation is given mainly to illustrate the physical theories involved, but the values in equations 4·13 and 4·14 seem to be of about the right order.

If electron emission from the cathode is greater than the circuit current, an electron space charge forms close to the cathode. This electron charge reduces the field intensity at the cathode and may actually reverse it. Thus for low circuit currents the electrons emitted by the cathode and entering the plasma are regulated by the electron space charge in front of the cathode. If circuit current is higher than the electron-emission current, this space charge vanishes and electron emission is increased by the positive field caused by the ions. Figure 4·6 illustrates the effect. If the circuit demand is too great the cathode current may concentrate to assist the field emission and cause the formation of an arc spot on the cathode. This is sometimes called "sparking."

4·7 PHENOMENA AT THE ANODE

The anode collects the electron current. Electrons strike the anode at the velocity imparted to them by the combination of potential difference in the anode-cathode space and the collisions with molecules in that space. This energy may amount to about 1 electron volt.

When an electron leaves the cathode it loses energy because of the image attraction of the emitting surface. The energy loss measured in electron volts is known as the "work function." It is about 5 electron volts. As an electron strikes the anode the reverse action takes place and the electron gains energy. This kinetic energy is released as the electron strikes the anode and appears as heat. The two effects are additive, so the energy lost at the anode is about 6 watts per ampere. Since the anode is in a vacuum the major dissipation of energy is by radiation. The radiation law is approximately

$$\text{Watts per square centimeter} = 5.7(10)^{-12}T^4 \qquad (4\cdot 15)$$

where T = anode temperature in degrees Kelvin.

Anode temperature should not exceed about 1000 degrees Kelvin. Therefore, the anode area is about

$$\frac{6}{5.7(10)^{-12} \times (10^3)^4} = 1 \text{ square centimeter per ampere} \qquad (4\cdot 16)$$

This figure is approximate, for it assumes black-body radiation and approximate values for loss factors.

The best material for anodes is graphite. It has high emissivity, it is refractory, and it is not injured by high temperature. Carbonized metal anodes are nearly as good and are often used in smaller tubes.

Arc-Back

Voltage rating is one of the least understood factors in tube design. Since losses in gas tubes are a function of the current, the power rating is limited by circuit voltage. It might seem feasible to design gas-tube circuits for high voltages, thereby increasing the power range of the tubes, but technical considerations limit circuit voltages. Technically the voltage rating is limited because of the tendency of the tube to "arc back." [15] This arc-back is a mysterious phenomenon. It is a spontaneous reversal of conduction in a tube. An arc spot appears on the anode, electrons are emitted from it, and reverse conduction follows. These are some of the causes of arc-back in pool and hot-cathode tubes:

1. Flakes of emission coating on the anode.
2. Flakes of insulating coating on the anode.

3. Improperly shielded junctions of dissimilar materials in the anode structure.
4. Impurities in the anode material.
5. Overheating of the tube.
6. Too rapid application of reverse voltage after conduction.
7. Poor design, causing localized fields.
8. Mercury on anodes.
9. High-velocity material from the cathode arc spot striking the anode.

Some features of design appear to minimize the tendency to arc back. None of these is fully successful. Smooth shaping of the anode, baffles interposed between anode and cathode to decrease deionization time or to reduce the transport of emissive coating, or mercury drops, and making field distribution uniform are some examples.

Arc-back follows no simple law. By careful attention to the design factors mentioned, the frequency of arc-back can be reduced but apparently cannot be eliminated. In designing circuits containing gas tubes some compensation for arc-back must be made by proper choice of circuit constants affecting the available arc-back current and by protective devices to protect apparatus and tubes.

4·8 DEIONIZATION

At the end of a conducting period, as anode current becomes zero, some of the ions remain in the anode-cathode space. In a thyratron these ions may cause the grid to be shielded by a positive-ion space charge for a time after conduction ends. If during this time anode voltage is reapplied, the thyratron may conduct even though the grid is at negative potential. In a rectifier this residual ionization may cause an increase in the frequency of arc-back.

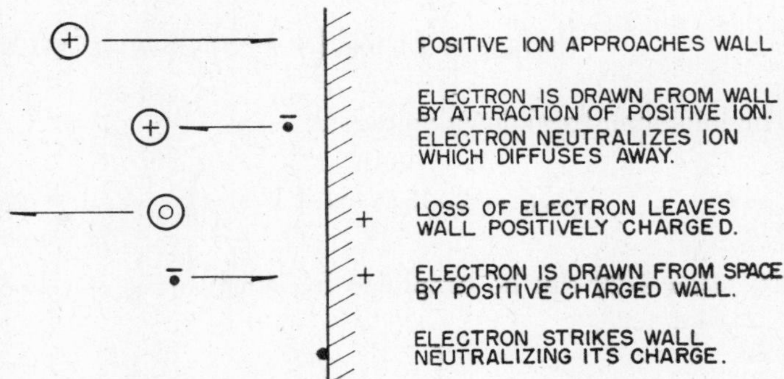

Fig. 4·7 Deionization process on walls of tube or on objects in ionized space.

Ions remaining at the end of the conducting period are in motion, and they find their way out of the space by diffusion through the gas. Experimentation indicates that ions and electrons in the space rarely recombine. The process probably is as follows: As shown in Fig. 4·7, the ions diffuse to some solid such as the tube walls. As an ion approaches the object, the electrostatic force between the object and the ion causes an electron to leave the wall and combine with the ion. This leaves the wall positively charged, and electrons are thereby attracted to the wall and restore it to its original potential. The neutralized ion or atom moves away from the tube wall to re-establish gas-pressure equilibrium. The deionizing process therefore requires enough time for the ions to move out of the anode-cathode space to the side walls. The time required can be computed by empirical formulas,[1] but little is known of the physics of the phenomenon.

4·9 AMPLIFICATION OF SMALL ELECTRON CURRENTS IN GASES

In many tubes the magnitude of the currents involved is too small to make space charge significant.[10, 11] Assume that a weak current of electrons is emitted by a cathode, say by photoelectric emission, and is conducted to an anode through a gas at a pressure high enough so that the anode-cathode distance includes many mean free paths. Assume also that the voltage supplied to the tube is relatively high so that the electrons can ionize the gas.

Electron current is increased throughout the tube, and therefore the electron current emitted from the cathode is amplified. The ratio of anode electron current I_{ea} to cathode electron current I_{ec} is

$$\frac{I_{ea}}{I_{ec}} = \epsilon^{\alpha d} \tag{4·17}$$

where α = the number of ionizing collisions per electron per centimeter in the direction of the field

d = the anode-cathode spacing in centimeters.

From the constant α the required average electron voltage per ionizing collision can be computed. The value so obtained is always higher than the ionizing potential of the atom, as explained in Section 4·1.

4·10 SELF-MAINTAINING DISCHARGE (BREAKDOWN)

Each electron leaving the cathode results in an amplified number reaching the anode. The additional electrons result from ionization of atoms, and an equal number of ions are formed which flow to the cathode.

The positive-ion current I_{pc} at the cathode is

$$I_{pc} = I_{ec}(\epsilon^{\alpha d} - 1) \tag{4·18}$$

In this type of discharge the cathode current is virtually all positive-ion current and the anode current is all electrons. Returning positive ions strike the cathode and are assumed to liberate additional electrons from the cathode by secondary emission. The number of electrons produced per positive ion is denoted by γ.

Current at the anode is

$$I_{ea} = \frac{I_{ec}\epsilon^{\alpha d}}{1 - \gamma(\epsilon^{\alpha d} - 1)} \tag{4·19}$$

The second term in the denominator is equal to the number of electrons produced by secondary emission at the cathode by the ions originally produced by one cathode electron. If this term becomes equal to 1, so that each electron leaving the cathode is replaced by another at the cathode, then the denominator is equal to zero, and the current increases theoretically to an infinite value. Actually it is limited by space charge or circuit constants before it can increase to any great value. The discharge is self-maintaining; that is, the discharge continues after the source of electron emission at the cathode is removed. The form of the discharge depends largely on external circuit constants. It is not necessary for electron emission to be supplied from the cathode. If sufficient voltage is applied to the tube this critical condition is maintained and the emission of but one electron from

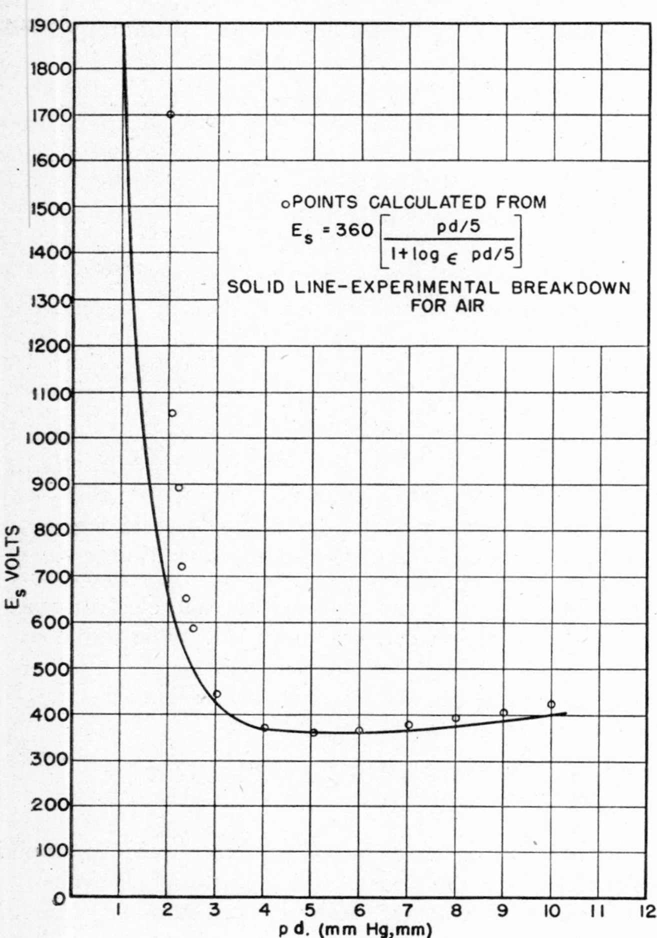

FIG. 4·8 Breakdown curve for air (dotted points calculated).

the cathode is enough to start the discharge. This single electron can be obtained by cosmic-ray bombardment, radioactivity, or other natural causes.

The theory can be carried on to derive the form of the breakdown voltage-pressure-distance curve. The constant α is a function of the anode-cathode voltage, spacing, pressure, and ionization potential of the gas. On the assumption that every electron gaining any energy greater than E_i electron volts produces 1 positive ion, the following expression applies for the constant α:

$$\alpha = \frac{1}{L} \epsilon^{-E_i d/EL} \tag{4·20}$$

where E_i = the effective ionizing potential
L = mean free path of electrons
d = anode-cathode spacing
E = anode-cathode voltage.

E_i is not the true ionizing potential of a single atom; it is an average which includes the efficiency of ionization.

If the denominator of equation 4·19 becomes zero the discharge becomes self-maintaining. Under these conditions

$$\epsilon^{\alpha d} = 1 + \frac{1}{\gamma} \tag{4·21}$$

If γ is assumed to be independent of pressure and field, and if the value for α at the breakdown voltage E_s (from equation 4·20) is substituted in equation 4·21, an expression for E_s can be obtained:[12]

$$E_s = \frac{E_i \dfrac{d}{L}}{\log_\epsilon \left[\dfrac{\dfrac{d}{L}}{\log_\epsilon \left(1 + \dfrac{1}{\gamma}\right)} \right]}$$

or

$$E_s = \frac{AE_i pd}{\log_\epsilon \left[\dfrac{Apd}{\log_\epsilon \left(1 + \dfrac{1}{\gamma}\right)} \right]} \tag{4·22}$$

where $A = 1/Lp$
p = pressure.

The sparking potential therefore depends on the product of pressure and distance. This is known as the Paschen law.

The curve has a minimum at

$$(pd)_{E_s \min} = 2.718 Lp \log_\epsilon \left(1 + \frac{1}{\gamma}\right) \tag{4·23}$$

and is asymptotic to the vertical line

$$(pd)_{E_s \infty} = Lp \log_\epsilon \left(1 + \frac{1}{\gamma}\right) \tag{4·24}$$

The minimum breakdown voltage is

$$E_{s \min} = 2.718 E_i \log_\epsilon \left(1 + \frac{1}{\gamma}\right) \tag{4·25}$$

Equation 4·22 can be simplified by substituting equations 4·23 and 4·25,

$$E_s = E_{s \min} \left[\frac{\dfrac{pd}{(pd)_{E_s \min}}}{1 + \log_\epsilon \dfrac{pd}{(pd)_{E_s \min}}} \right] \tag{4·26}$$

With a knowledge of the minimum breakdown voltage and the corresponding pd value, the theoretical E-pd curve can be laid out. This curve for air is shown in Fig. 4·8. It is adjusted to pass through the points $pd = 5$, $E_s = 360$. The curve rises more rapidly on both sides of the minimum than the experimental data given for air indicate it should. This is not surprising, because the computations are only approximate.

The shape of the curve with its high- and low-pressure regions is important in gas-tube engineering. To maintain high dielectric strength at low pressure it is necessary that the pressure be well below the minimum-pressure point. Dielectric strength is smaller for long distances than short distances. In the design of high-voltage tubes this must be considered, and long discharge paths must be avoided.

On the high-pressure side of the minimum the dielectric strength increases with path length.

4·11 GLOW DISCHARGE (COLD-CATHODE DISCHARGE)

The theory of breakdown shows that the lowest breakdown voltage of a discharge tube is obtained for a certain product of pressure and distance between electrodes. Assume now that at breakdown the circuit resistance is low enough to permit a large current and that *pd* is above the point at which breakdown voltage is minimum.[13]

Under these conditions space charges form, and they tend to increase the voltage gradient at the cathode. A plasma forms and in effect decreases the distance between cathode and anode. As this distance decreases, the voltage drop in this region also decreases until it reaches a minimum. The final regional voltage drop is called the "cathode fall" of potential. It is a constant determined by the gas used and the material of the cathode. Since cathode fall is constant and cathode-fall space is fixed by pressure, a space-charge region is set up at a certain voltage and distance. Such a condition can be maintained only if the current density has a definite value. The cathode region decreases in area until such a current density is reached. Under such conditions, cathode fall is independent of current. If the current through the tube is increased by changing the external circuit, the area of the discharge at the cathode increases. However, when the entire area of the cathode is covered, cathode fall is no longer constant but increases. Minimum cathode fall is about equal to minimum breakdown voltage. There is some difference because of distortion of the fields in the cathode-fall space by space charge.

For pressure higher than that required for minimum voltage breakdown the volt-ampere curve can be traced as follows; see Fig. 4·9 (*a*). At slightly above zero current the voltage is equal to the sparking potential. As the current is further increased the voltage falls until a complete plasma is built up. Cathode fall then becomes constant and tube voltage increases only slightly from plasma losses. When the cathode is completely covered, voltage rises with current. Behavior of such a tube is influenced greatly by the external circuit. If the circuit consists of a source of unidirectional voltage and a series resistor, the current characteristic can be drawn on the same diagram. Unless the resistance is very large when the source voltage is increased, the current changes suddenly immediately after breakdown. Tube voltage then falls.

If, conversely, the pressure is less than that required for minimum voltage breakdown, there is no possibility of reduction in cathode voltage drop because of a decrease of cathode-fall space. As current is increased and a plasma forms, the cathode-fall space necessarily is shorter and the cathode-fall voltage rises. The current, as before, does not cover the cathode at low currents. Cathode coverage increases as current increases until the cathode is covered. During this interval voltage rises with current. When the cathode is entirely covered the rate of rise of voltage with current increases. The volt-ampere curve to be expected is shown in Fig. 4·9 (*b*). Regardless of the size of the series resistance there is no sudden increase in current at breakdown.

During such a discharge patterns of light are seen in the tube and various dark spaces can be observed around the cathode. These are spaces in which the electrons have either too high or too low velocity for efficient ionization. The boundary of the cathode-fall space usually is visible

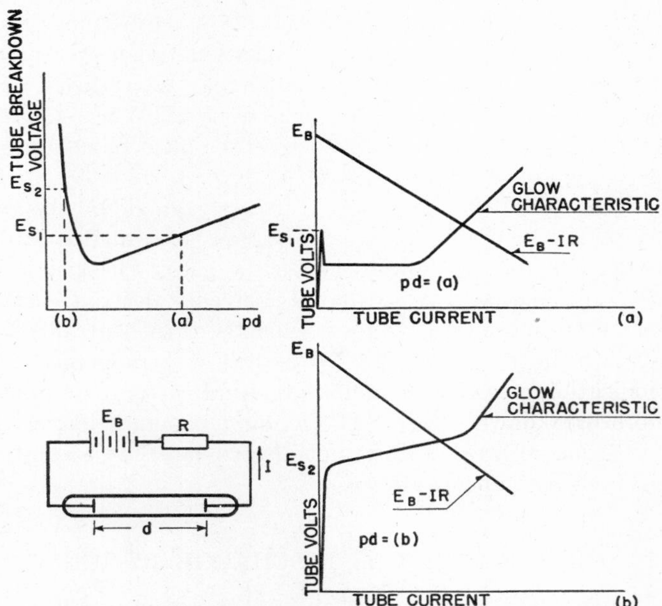

FIG. 4·9 Glow-discharge volt-ampere characteristic diagram.

because the plasma beyond has a uniform glow. Often striations (alternate transverse bands of light) appear. These are thought to result from alternate gain and loss of ionizing potential as the electrons speed up in the field and lose energy on ionization.

Breakdown voltage is a function of the effective ionization potential of the gas and secondary-emission efficiency of the cathode material. At high current, electron-emission current is increased by the heating effect of the ion bombardment, if the cathode has an emissive coating.

The condition for maintenance of a cold-cathode discharge has been described, but little has been said about how it is started. If the conditions are right for breakdown any disturbance in the gas, such as the emission of a photoelectron, is sufficient to start the discharge. A small quantity of a radium compound is often injected into the vacuum envelope to stabilize the value of the breakdown voltage. The radium causes the emission of enough electrons to start the discharge, if the voltage is above the breakdown value.

4·12 CONTROL OF GAS DISCHARGE BY ELECTROSTATIC FIELDS

If a grid is placed between an electron source and an anode, a potential on the grid can be of such value and polarity as to cancel or reverse the field from the anode, and thereby prevent electrons emitted from the cathode from leaving its surface. As the potential of the grid is made more positive the retarding field at the cathode is decreased, and electrons move from the cathode to the anode. These electrons ionize the gas. Ions are attracted to the negative grid, which at the same time repels the electrons. Ion current to the grid causes a positive space charge to form around the grid. The width of this space charge is limited, and it decreases as the positive-ion grid current increases. This positive space charge neutralizes the influence of the grid potential at distances greater than the width of the space charge. Upon further positive change in grid potential (and as soon as a sufficient portion of the anode-cathode space becomes grid-field-free), anode current and ionization increase rapidly. The grid potential becomes completely neutralized, and complete ionization or breakdown results. When conduction is not wanted the grid of such a tube must be held negative with respect to the cathode.

If the cathode is fully shielded from the anode by the grid structure then the field at the cathode is too small to cause enough electrons to flow to ionize the gas. The grid then must be made positive so that it increases electron current to generate more positive ions, which in turn neutralize the shielding effect of the grid. With such an arrangement the deionization rate is large and a large grid current is required before breakdown starts. The amount of ionization necessary to insure breakdown is not always constant, so such a tube is likely to be erratic.

4·13 CONTROL OF GAS DISCHARGE BY IGNITION

The theory of operation of an ignitor closely parallels the theory of the mercury-pool cathode, discussed in Section 4·6. Some computations [14] indicate that the passage of current down the ignitor (Fig. 4·10), concentrating at the mercury edge, can result in fields of about 10^6 volts per centimeter over atomic dimensions at the ignitor-mercury junction. This is comparable with the fields at the cathode originating from positive-ion space charge. It seems reasonable that the same phenomenon that maintains the arc with fields of

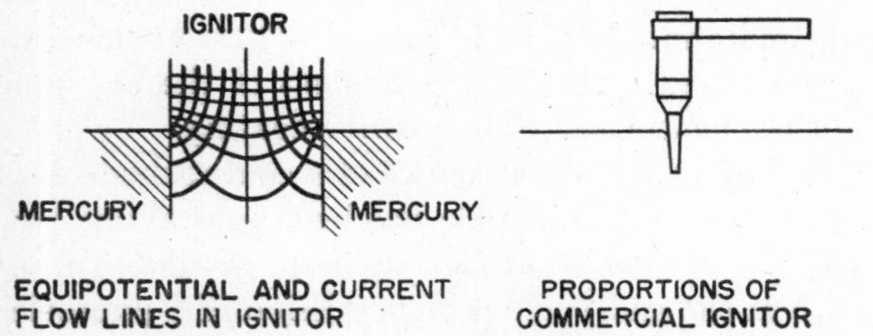

Fig. 4·10 Ignitor potential distribution and commercial form.

10^6 volts per centimeter in the one case would operate also in the second. Therefore, it can be assumed that the ignitor operates because of field emission near the mercury-ignitor junction.

Local heating by current through the ignitor probably causes a vapor pressure near the ignitor that is higher than the average pressure in the tube. Electrons are accelerated through the vapor by the voltage on the ignitor. After ionization the positive-ion space-charge field causes further emission of electrons at the cathode surface. The upper region of the ignitor acts as a collector for the first few electrons until the arc is transferred to the anode.

Oscillograms of ignitor firing voltage over a large number of cycles show that it has a variety of values, most of them between 50 and 150 percent of the most probable value.[18] Firing voltage varies, even under test conditions, with rate of rise of ignitor voltage, tube temperature, anode current, and other factors held constant as possible. That the surface of the mercury is in constant movement is undoubtedly of some importance in accounting for the variations. The

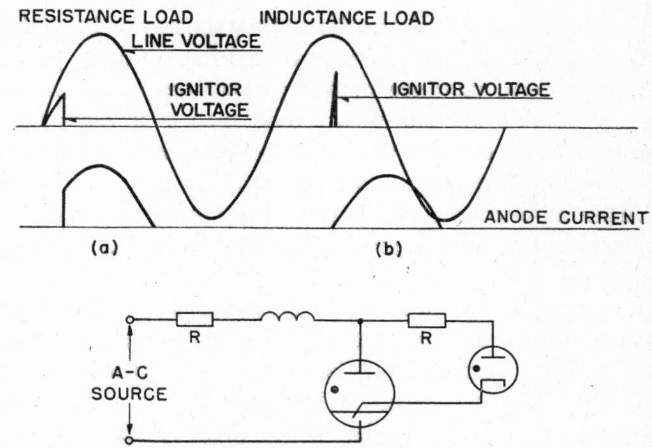

Fig. 4·11 Typical ignitor voltage wave forms as function of type of ignitron loading with anode firing.

meniscus at the ignitor-mercury junction makes an acute angle with the walls of the ignitor. The strong field and increased vapor pressure should be in this region. The movement of the mercury may cause fluctuations in this angle with consequent variations in field strength. In general, however, the variation seems to be characteristic of this type of phenomenon, as for example in arc-back.[15]

The duration of voltage for ignition is a function of the voltage itself. High voltages require less time. With extremely high voltage, ignitor operation in about 1 microsecond has been observed. At voltages of about 200, the time may be 100 microseconds. The resistance of the commonly used material decreases rapidly with temperature.

Volt-ampere curves of the dynamic ignitor characteristics are nearly straight lines over one cycle at a frequency of 60 cycles. The initial cycles show the highest resistance, which falls until equilibrium is reached. On intermittent duty the ignitor therefore may have a higher dynamic resistance than at 100 percent duty.

Certain types of circuits (Fig. 4·11) may cause current to pass through the ignitor for an appreciable time before current becomes high enough for firing. Under these circumstances resistive-energy loss heats the ignitor and reduces its resistance.

No fundamental properties of the ignitor serve as a measure of its quality. Current practice is to use a fixed standard testing circuit and base quality on comparative data. One such standard circuit consists of a half-wave rectifier with a resistive load and a 60-cycle circuit to pass current through the ignitron. The anode is connected to the ignitor through a phanotron. A peak voltmeter is used to measure maximum firing voltage and current at the ignitor with an alternating anode voltage of constant magnitude. These values must not be exceeded in more than a few cycles in a fixed time. If such a test is continuous, peak voltage and peak current indicate the quality of the ignitor. The firing voltage may be two or three times higher and the current two or three times lower if the tube is operated at low duty and with rapid application of voltage to the ignitor. This corresponds to firing the ignitor late in the a-c cycle on anode-firing applications.

4·14 CONTROL OF GAS DISCHARGE BY MAGNETIC FIELDS (PERMATRON)

A tube can be made similar to the thyratron, except that its control is magnetic rather than electrostatic.[16] In general, this tube has an anode and cathode similar to the usual thyratron elements, and an electron collector similar in appearance to the grid of a thyratron. As many parts as possible are non-magnetic.

A magnetic field transverse to the direction of electron flow from cathode to anode controls conduction. Such a combination of electrostatic and magnetic fields causes the electrons to move in cycloidal paths. If the magnetic field is great enough the electrons never advance far enough in the direction of the anode-cathode field to gain sufficient energy to ionize the gas. They finally strike the collector, which is connected to the cathode. If the magnetic field is weaker the electrons advance farther and are able to ionize the gas, and conduction starts. As in the thyratron, anode current must fall to zero before control can be regained.

REFERENCES

1. "Hot Cathode Thyratrons," A. W. Hull, *G.E. Rev.*, Part I, April 1929, pp. 213–223.
2. "Electrical Discharges in Gases," I. Langmuir and K. T. Compton, *Rev. Mod. Phys.*, Vol. 13, No. 2, April 1931, p. 213.
3. "Studies of Electric Discharges in Gases at Low Pressures," I. Langmuir and H. Mott Smith, *G.E. Rev.*, July 1924, pp. 449–455.
4. "Theory of the Electric Arc," K. T. Compton, *Phys. Rev.*, Vol. 21, 1923, p. 266.
5. "Positive Ion Currents in the Positive Column of the Mercury Arc," I. Langmuir, *G.E. Rev.*, Vol. 26, 1923, pp. 731–735.
6. "The Cathode Drop in an Electric Arc," S. S. MacKeown, *Phys. Rev.*, Vol. 34, 1929, pp. 611–614.
7. "The Cathode Fall of an Arc," R. C. Mason, *Phys. Rev.*, Vol. 38, 1931, pp. 427–440.
8. "Studies in the Emission of Electrons From Cold Metals," R. H. Fowler and L. Nordheim, *Proc. Roy. Soc. (London)*, *Series A*, Vol. 119, 1928, pp. 173–181; *Series A*, Vol. 121, 1928, pp. 626–639; *Series A*, Vol. 124, 1928, pp. 699–723.
9. "Methods of Investigating Thermionic Cathodes," C. G. Found, *Phys. Rev.*, Vol. 45, April 1934, pp. 519–527.
10. *Electricity in Gases*, V. S. Townsend, Oxford Univ. Press, 1915.
11. *Conduction of Electricity through Gases*, J. J. Thomson and G. P. Thomson, 1928–1933, Cambridge Univ. Press.
12. *Gaseous Conductors*, J. D. Cobine, McGraw-Hill, 1941.
13. *Conduction of Electricity in Gases*, J. Slepian, Westinghouse Electric Corporation, 1933.
14. "A New Method of Initiating the Cathode of an Arc," J. Slepian and L. R. Ludwig, *Trans. A.I.E.E.*, Vol. 52, June 1933, pp. 693–700.
15. "Back Fires in Mercury Arc Rectifiers," J. Slepian and L. R. Ludwig, *Trans. A.I.E.E.*, Vol. 51, March 1932, pp. 92–104.
16. "The Permatron," W. B. Overbeck, *Trans. A.I.E.E.*, Vol. 58, May 1939, pp. 224–227.
17. "A General Theory of the Plasma of an Arc," L. Tonks and I. Langmuir, *Phys. Rev.*, Vol. 34, 1929, pp. 876–922.
18. "Ignitor Characteristics," E. G. F. Arnott, *J. App. Phys.*, Vol. 12 September 1941, pp. 660–669.

Chapter 5

VACUUM TUBES

H. J. Dailey, J. H. Findlay, and E. C. Okress

WHEN the functions of each element of an electronic tube are well known, with the limitations of application, the tube may be used to obtain the best results.

In general, pliotrons are considered high-vacuum devices containing an electron source (cathode), control members (grid or grids), and a collector of electrons (anode or plate).

Cathodes in common use are, in their order of ability to withstand ion bombardment and high-voltage operation, (1) pure tungsten, (2) thoriated tungsten, (3) oxide-coated. Their emission efficiencies are in the reverse order.

Tube grids have definite limits of operation depending mainly upon the class of service, the anode voltage used, and the level of power operation.

Tube anodes have one primary function, that of collecting electrons passed through the various grids. The secondary function of the anode is that of dissipating the energy released at the anode surface by electrons impinging thereon. The tube rating sheets always contain the maximum voltage conditions, beyond which damage may often result.

This chapter will deal with the functions of the various tube elements and their relation to tube design, performance, and application.

5·1 VACUUM-TUBE CONSTRUCTION

Thermionic Emitters

Emission from all primary emitters is approximately the same per unit area at their normal operating temperatures, but emission efficiency of cathodes varies widely. Safe anode voltage varies with the type of emitter and, to a limited degree, with the conditions of operation.

Pure-Tungsten Emitters

Efficiency of pure-tungsten emitters varies between 4 and 10 milliamperes of emission per watt of energy expended in the filament. Permissible efficiency is dependent upon the diameter of the wire and the designed life. At a given temperature the evaporation rate of pure tungsten is constant per unit of area and is independent of wire diameter. Life of a pure-tungsten filament increases directly as the diameter is increased when the temperature is held constant. For a given life, filament temperature may be increased as the diameter is increased, and thus emission efficiency can be increased.

Emission is influenced little by ion bombardment; therefore tungsten filaments are used as electron sources in most high-voltage high-vacuum tubes.

Most pure-tungsten-filament tubes operate with high anode voltages, usually between 4000 and 20,000. Residual gas molecules are ionized by electron collisions, and the ions are accelerated at high velocities toward the cathode. Velocity at the cathode is

$$v = \sqrt{\frac{2\mathbf{E}e}{\frac{m_g}{m_e}}}\,10^7 \tag{5·1}$$

where $\mathbf{E}$ = average volts per centimeter between the point of ionization and the cathode

v = velocity in centimeters per second

e = electron charge in coulombs

m_e = electron mass in grams

m_g = ion mass in grams.

The designed life of tubes with pure-tungsten filaments for industrial service should be determined by the cost of power to excite the filament during its expected life and the cost per kilowatt-hour of a replacement tube. Let

p = filament power in kilowatts

n = hours of designed life of filament

a = cost of a new tube

w = cost of filament power in cents per kilowatthour

Then

$a + pnw$ = total tube operating cost during its life exclusive of anode power

When

$$a = pnw$$

the most economical combination is obtained. If the diameter of the filament wire and the current used to excite it are known, the temperature and predictable life are determinable.

Diameter of a tungsten filament (in mils) is related to the current required to heat it to a given temperature in degrees Kelvin by

$$d = A^{2/3}B' \tag{5·2}$$

where d = diameter in mils

A = filament current in amperes

B' = a factor that varies with temperature (see Fig. 5·1).

If wire diameter and the current required to heat the filament to its operating temperature are known, equation 5·2 can be solved for B', and the temperature can be determined from Fig. 5·1.

Once the operating temperature has been determined, electron emission, predictable life, variation of emission with life, and variation of filament current with life are determinable.

In Fig. 5·2 the intersection of the temperature and filament-diameter lines gives the emission in milliamperes per centimeter of filament length. This figure multiplied by the length of filament operating at full temperature gives the total emission. When the emission per centimeter of length has been found from Fig. 5·2, the filament life can be predicted from Fig. 5·3. The intersection of the line giving the emission in milliamperes per centimeter of length of the filament and the line for filament-wire diameter gives the predictable life in hours. A pure-tungsten filament is near the end of its useful life when the diameter has been reduced by tungsten evaporation to nine tenths of its original diameter. The end-of-life figures given in Fig. 5·3 are based on this observation.

By assuming various values for A in equation 5·2 and by using Figs. 5·2 and 5·3, the variation of life with filament current can be determined.

From Fig. 5·4 the variation with life of electron emission, filament current, wire diameter, temperature, and rate of evaporation are determinable for the value of the current A used in solving equation 5·2.

Where it is possible to precalculate the electron emission required for an application, the curves shown in Figs. 5·1, 5·2, 5·3, and 5·4 can be used to precalculate the filament current required to give that emission. Assume that six tenths of the initial tube emission is ample for a given requirement; it is then desirable to calculate filament current for seven tenths, to allow for line-voltage variations

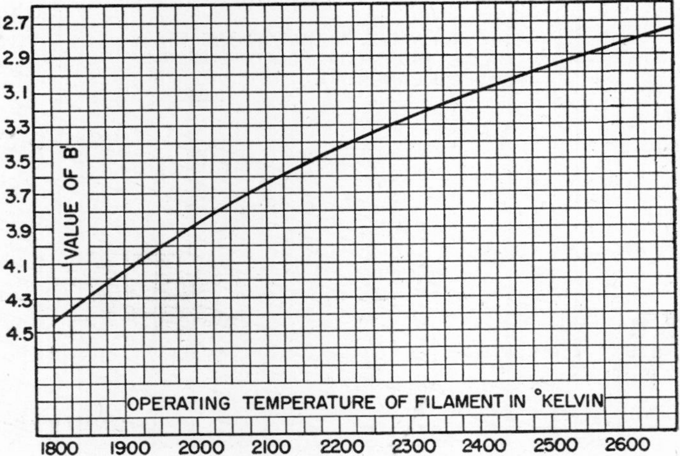

FIG. 5·1 Curve giving values of B' for equation 5·2.

and tungsten evaporation and to avoid frequent readjustments of filament voltage. Total required emission then is divided by the length of emitting filament to obtain the emission in milliamperes per centimeter. From Fig. 5·2 the required temperature is obtainable because the wire diameter is known. From Fig. 5·1, B' can be obtained for that temperature. Then, by substituting B' and the wire diameter in equation 5·2, the required current can be determined. The life in hours can be determined for that setting of filament current and the emission level has been assumed

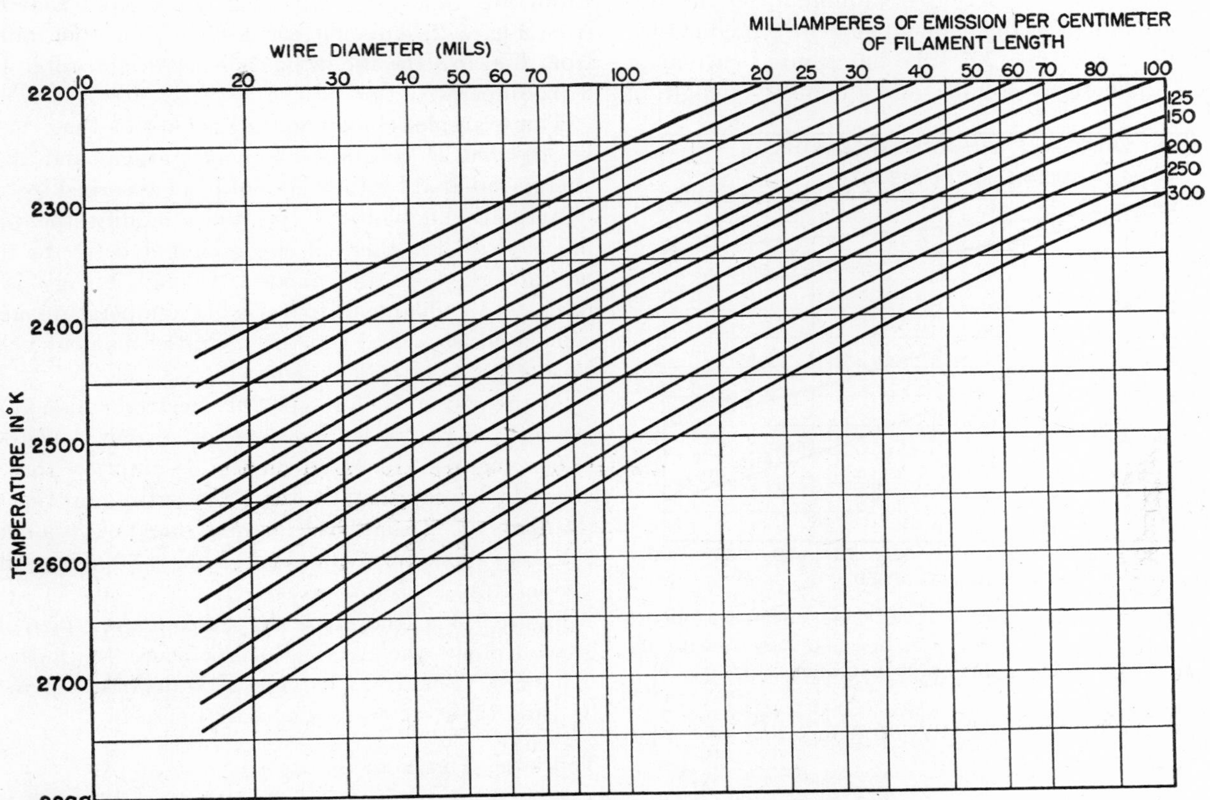

FIG. 5·2 Curves relating emission per unit length, wire diameter, and temperature for pure-tungsten filaments.

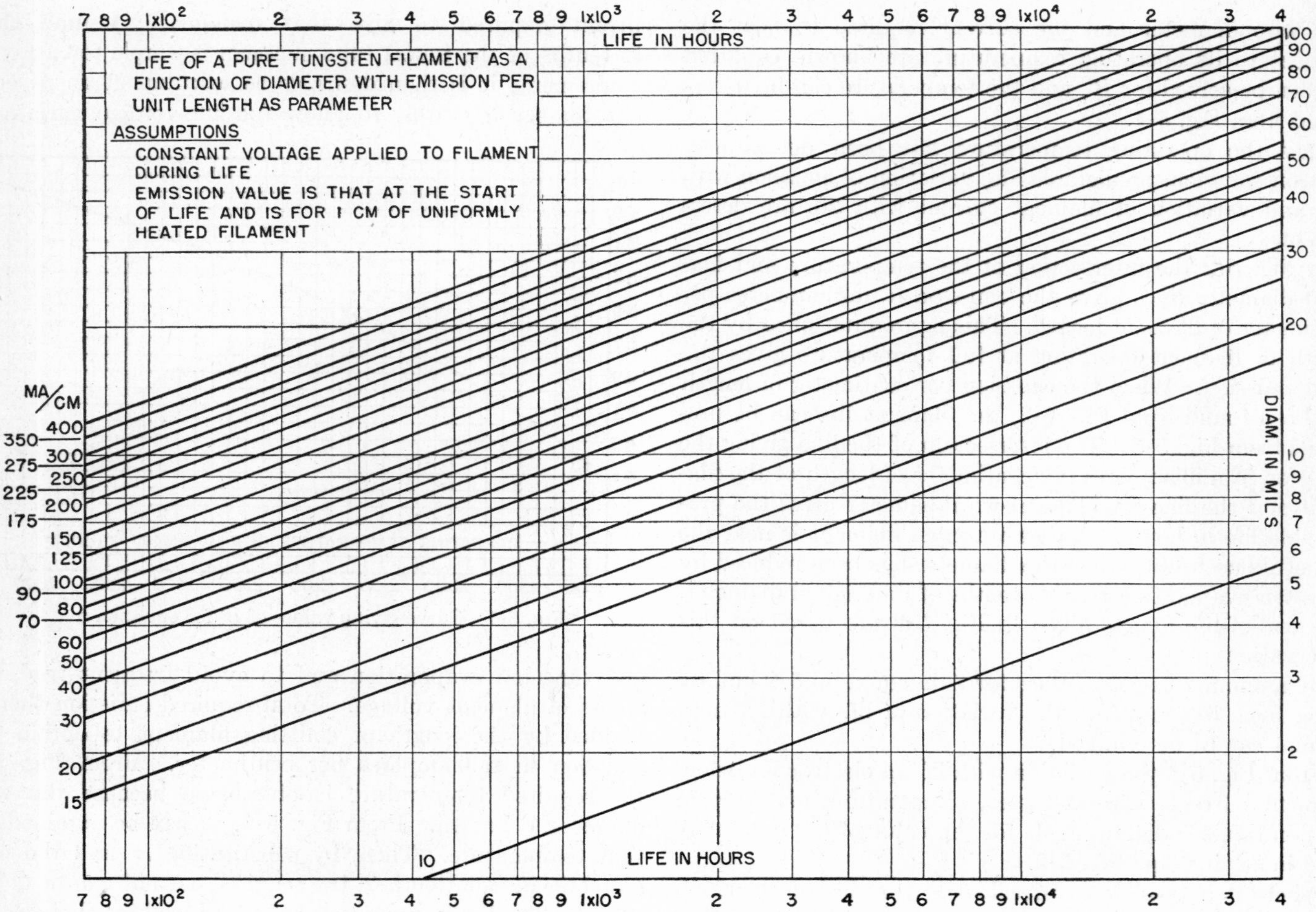

FIG. 5·3 Curves relating emission per unit length, diameter of wire, and life of pure-tungsten filaments.

to be 10 percent over the required minimum, so the frequency of readjustment of the filament current is predictable.

The filament of the type 207 tube has a rated current of 52 amperes, a diameter of 40 mils, and an effective length of

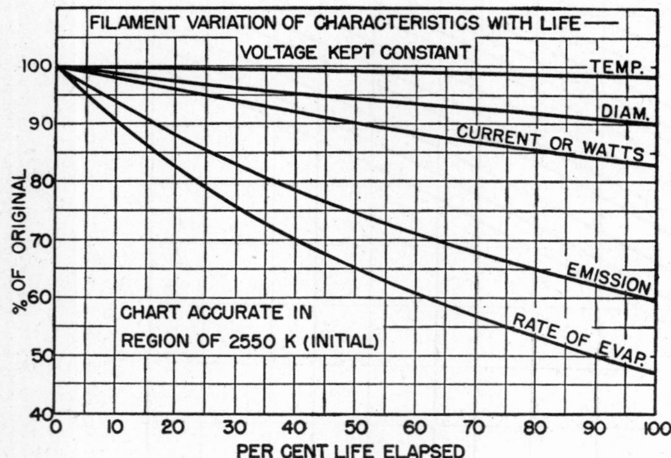

FIG. 5·4 Curves giving variation of temperature, diameter, watts, emission, and rate of evaporation of a pure-tungsten filament with life in terms of the original values at the beginning of life.

approximately 16 inches (40.6 centimeters). From equation 5·2,

$$B' = \frac{d}{A^{2/3}} = 2.87$$

From Fig. 5·1, the temperature in degrees Kelvin is 2590; from Fig. 5·2, emission per centimeter is 210 milliamperes; from Fig. 5·3 the life in hours is 4100, and total emission is 8·52 amperes.

This example is based on the assumption that: (1) the tube is operated at constant filament voltage, and (2) thermal energy radiated by the filament is all absorbed by the anode.

Actually line-voltage variations modify the predictable life, and all the thermal energy radiated by the filament is not absorbed by the anode. Enough energy is radiated back to the filament to raise its temperature above that calculated, so actual tube life at rated filament voltage may be less than that calculated.

At the end of life for type 207, electron emission is 5.1 amperes (from Fig. 5·4), and filament current is 43.16 amperes. (These figures are based on the assumption that tungsten vaporization is constant along the entire emitting length of the filament. Usually the filament does not vaporize evenly.) Filament diameter at the end of life is 36 mils, and filament temperature is 2543 degrees Kelvin.

Figure 5·4 is accurate in the temperature region of 2550 degrees Kelvin, the average temperature for most tungsten filaments. Between 2500 and 2600 degrees Kelvin the error in using these curves is less than 3 percent.

Thoriated-Tungsten Emitters

The electron-emission efficiency of a thoriated-tungsten cathode usually is from 10 to 50 times that of a pure-tungsten filament. The emission usually is between 100 and 200 milliamperes per watt expended in heating the cathode.

This type of primary emitter is susceptible to damage by positive-ion bombardment, especially if the positive ions are accelerated toward the cathode at high velocities. For this reason, tubes with thoriated-tungsten filaments usually are designed for operation at anode voltages less than 5000. Some tubes have been built for anode voltages higher than 5000, but in these the residual gas pressure is low, or the ratio of maximum emission current required for tube operation to the total emission the filament can supply is maintained at 0.25 or less.

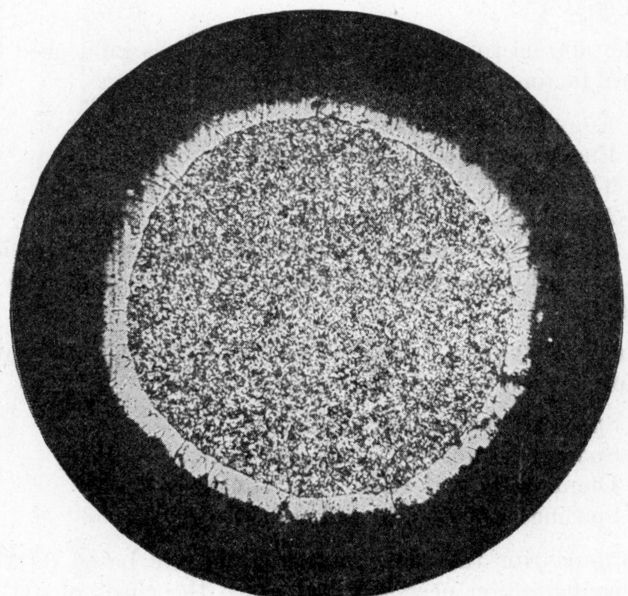

Fig. 5·5 Photomicrograph of a carburized thoriated-tungsten filament. The outer shell is composed of tungsten carbide crystals, and the center of thoriated tungsten. This filament is 0.010 in. in diameter.

If this ratio is maintained at 0.25 or less a dense cloud of electrons surrounds the cathode at all times. Positive ions must pass through this electron cloud to strike the filament. Many ions are neutralized; thus the number striking the cathode is reduced.

The ratio 0.25 may be exceeded if the anode voltage is not applied continuously. It may be increased to approximately 0.8 if the ratio of E_b on to E_b off is reduced to approximately 0.01 or less.

Carburization by heating the thoriated filament in a hydrocarbon atmosphere at a temperature between 1400 and 2400 degrees Kelvin gives it this high emission. When heated thus, free carbon is released from the hydrocarbon gas and is deposited on the filament surface. Subsequent heating in a vacuum causes a reaction between the tungsten and the carbon, forming tungsten carbide (W_2C). Figure 5·5 shows a photomicrograph of a carburized filament 0.010 inch in diameter.

Optimum operating temperature of such filaments is between 1950 and 2000 degrees Kelvin, with the permissible temperatures between 1900 and 2100 degrees. At 1900 degrees the life is longer but the usable emission is low; at 2100 degrees the available emission is much higher but the emission life is from about one-fourth to one-third the life for operation at temperatures between 1950 and 2000 degrees (if normal carburization is assumed).

The layer of tungsten carbide shown in Fig. 5·5 gradually changes to that shown in Fig. 5·6. At temperatures of 2100 degrees Kelvin and above this change is rapid. In this decarburized condition the filament loses its high emission efficiency and approximates the emission efficiency of pure tungsten at that temperature.

To determine operating temperature of a thoriated-tungsten filament, its diameter (or weight in milligrams per 200 millimeters of length) and the normal filament current per strand must be known.

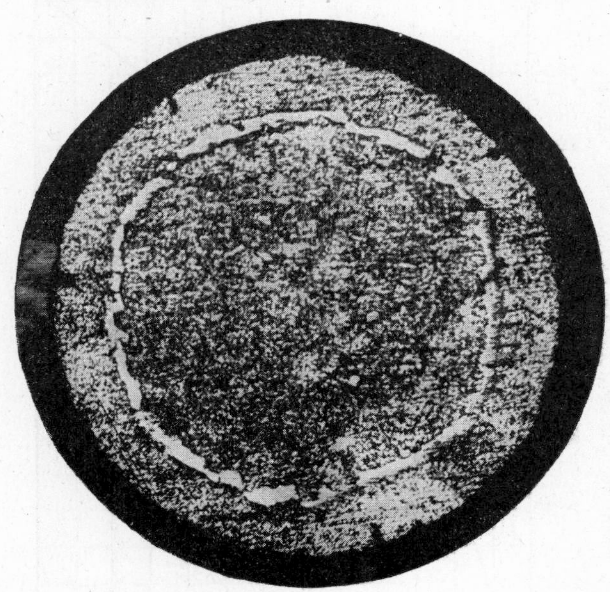

Fig. 5·6 Photomicrograph of a decarburized thoriated-tungsten filament. There are some remaining crystals of tungsten carbide remaining in the filament completely enclosed by the tungsten.

The diameter of the filament (in mils) is related to the weight by

$$d = \sqrt{\frac{W}{k}} \tag{5·3}$$

where d = diameter in mils
W = weight in milligrams per 200 millimeters
k = a conversion factor dependent upon the composition of the wire.

Values of k are:

	k	Density (g/cc)
Pure tungsten	1.931	19.06
Tungsten wire with 1% thoria	1.908	18.83
Tungsten wire with 1½% thoria	1.898	18.73
Tungsten wire with 2% thoria	1.875	18.47

Wire weight W is related to current, wire conversion composition factor k, and temperature factor B by

$$W = A^{4/3}kB \tag{5·4}$$

where W = wire weight in milligrams per 200 millimeters
A = heating current in amperes
k = a factor dependent upon the wire composition, as given for equation 5·3
B = a conversion factor varying with temperature (see Fig. 5·7).

Thus, if wire weight (or wire diameter) and filament current per strand are known, a solution of equation 5·4 for B and a reference to Fig. 5·7 determine the temperature. Should the temperature be approximately 2100 degrees Kelvin, it is too high for most applications. A temperature of approximately 1900° is satisfactory for low-load conditions.

If current per strand and wire diameter are known, equation 5·2 can be solved for B', and the temperature can be determined from Fig. 5·1. Figures 5·1 and 5·7 are applicable to both carburized thoriated tungsten and pure-tungsten filaments if the carburized filament has been carburized to a 20 percent increase in voltage at the rated filament current.

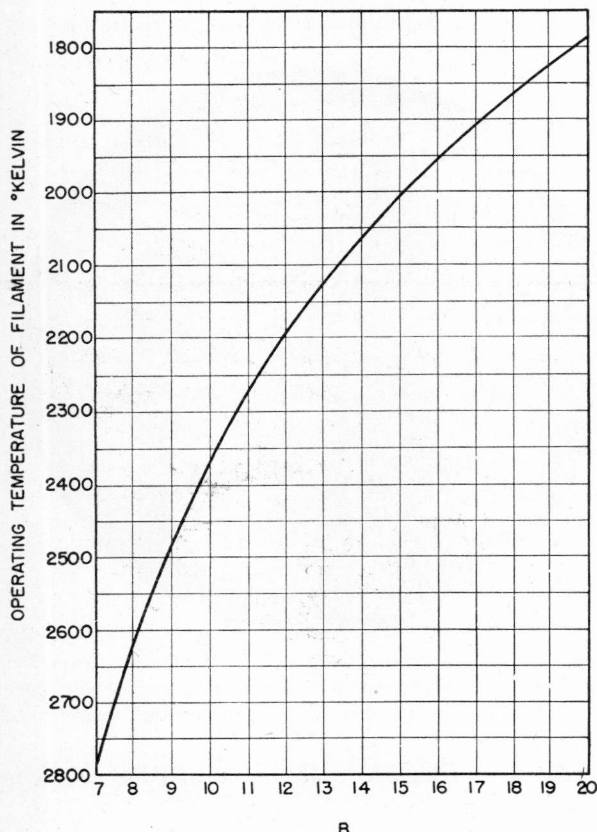

Fig. 5·7 Curve giving values of B in equation 5·4.

Oxide-Coated Cathodes

Oxide-coated filaments or cathodes usually are used only in high-vacuum tubes with anode potentials less than 1000 volts, especially if the electron-current density is relatively high. At low-current densities, or when anode voltage is applied intermittently, voltages of about 20,000 have been used successfully.

Anode voltage is limited apparently because this type of cathode cannot withstand bombardment by high-velocity ions (see equation 5·1) for long periods. If anode voltage is applied in pulses, it usually can be increased as the duration of the pulse decreases. Unless anode voltage is limited, craters form at the cathode surface and lead to the rapid destruction of the cathode. The time required for such craters to form determines the duration of the anode-voltage pulses.

Control, Screen, and Suppressor Grids

With an ideal grid the proper electrostatic potentials could be established at the desired points between cathode and anode, and no electrons would be intercepted. These electrostatic planes would be at constant potential.

An actual grid only approximates this ideal and has these disadvantages:

1. It intercepts electrons (if grid is at a positive potential).
2. It does not establish uniform electrostatic potentials; see Fig. 3·4 (*a*).

The normal function of the control grid is influenced by several factors, the most important of which are:

1. Amplification factor desired.
2. Energy radiated by the cathode.
3. The class of operation (class A, B, or C).
4. Applied anode voltage.
5. Secondary-emission characteristics of the grid material.
6. Energy dissipated in the grid in class B and class C operation.

The amplification factor (defined as $\partial E_b/\partial E_g$) is a function of:

1. Spacing between grid and anode.
2. Spacing between adjacent grid wires.
3. Diameter of grid wires.
4. Spacing between grid and cathode.

Formulas for the amplification factor are based on the grid configuration necessary to reduce the effect of anode voltage at the cathode to zero, that is, to make the factor $[E_g + (E_b/\mu)]$ equal to zero. (See equation 3·22.)

For plane structures:

$$\mu = \frac{2\pi ND - \log_\epsilon \cosh \pi Nd}{\log_\epsilon \coth \pi Nd} \tag{5·5}$$

$$\mu = \frac{2\pi D}{N \log_\epsilon \dfrac{1}{2 \sin \dfrac{\pi d}{2N}}} \tag{5·6}$$

where μ = amplification factor
N = pitch of grid winding
d = diameter of grid wire
D = distance from grid to anode.

Equation 5·6 is sufficiently accurate, when the grid-wire diameter is small in comparison with the center-line distance between adjacent grid wires. Equation 5·5 is accurate for all ratios of $d/(1/N)$. Neither equation is accurate if the cathode-to-grid distance is less than $1/N$.

For cylindrical structures:

$$\mu = \frac{2\pi N r_g \log_\epsilon \dfrac{r_a}{r_g} - \log_\epsilon \cosh \pi Nd}{\log_\epsilon \coth \pi Nd} \tag{5·7}$$

where μ = amplification factor
N = pitch of grid winding
d = diameter of grid wire
r_a = anode radius
r_g = grid radius to center of grid wires.

Formulas for μ usually are of the same general form with minor coefficient differences.

Part of the energy radiated by the cathode is intercepted by the grid, thus causing an increase in its operating temperature. If the grid is made with lateral turns on vertical supports, the supports are spaced midway between filament wires, which are vertical, so that supports intercept a minimum of radiated energy as well as a minimum number of electrons from the cathode.

The value of g should not exceed 200,000 volts per inch for most grid materials.

Secondary emission varies almost inversely with the diameter of the grid wire. Although large grid wire is desirable for mechanical strength, the use of grid wires larger than about 0.025 inch in diameter materially increases grid current in the positive-grid region (classes B and C); thus more power is required to drive the grid to the positive levels required for efficient class B or class C operation.

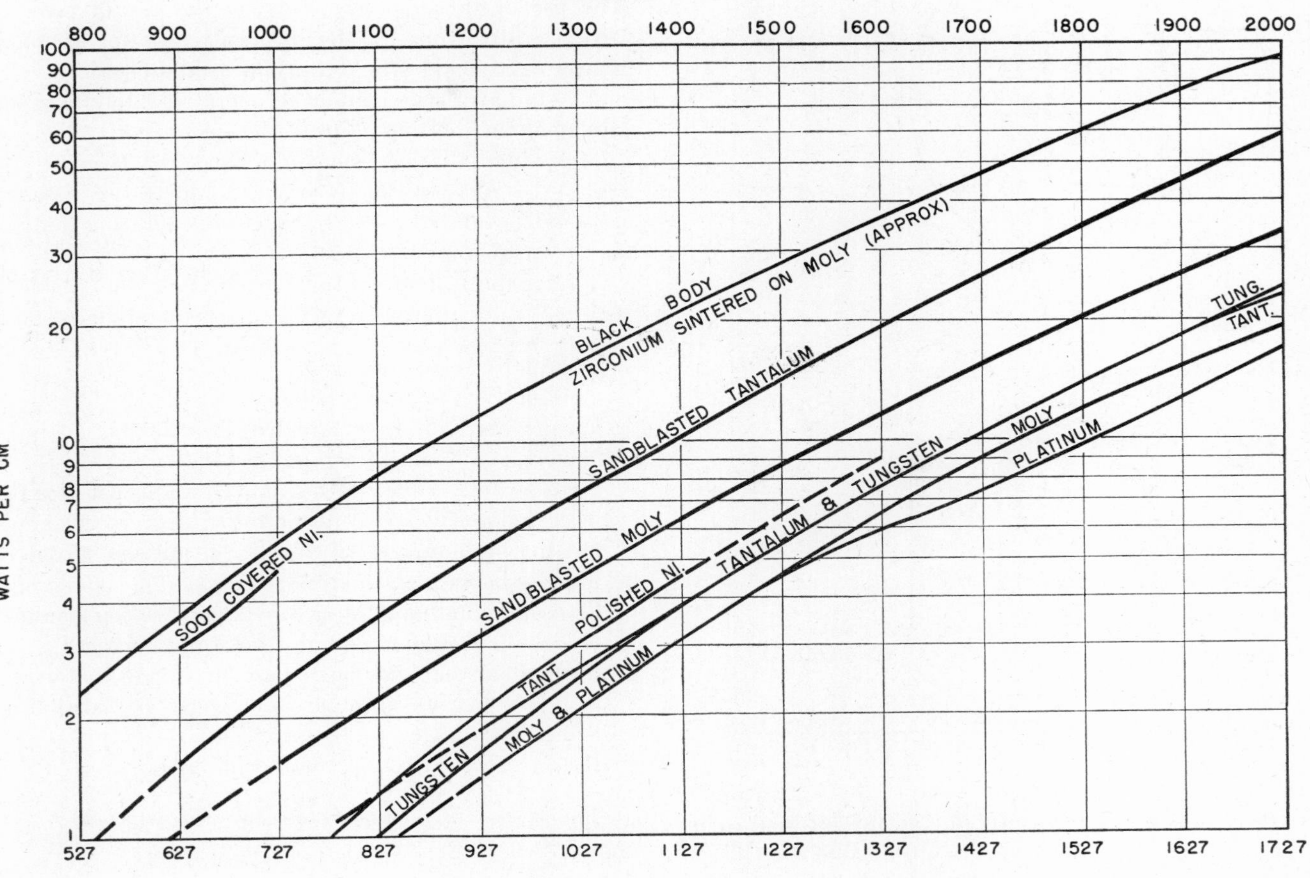

FIG. 5·8 Curves showing radiating properties of various anode materials, both smooth and sandblasted.

In class A service the grid is always negative with respect to the cathode; therefore, grid design is concerned only with the amplification factor desired and the energy radiated by cathode and anode.

If applied anode voltage is 10,000 volts or above, the radius of the grid wire should be as large as it can be made conveniently, to avoid field emission in any class of service. Voltage gradient g at the grid wire is, very nearly,

$$g = \frac{E_b}{r_a \log_\epsilon \dfrac{r_a}{r_g}} \cdot \frac{1}{nd} \qquad (5\cdot8)$$

where r_g = radius of a grid in inches
r_a = anode radius in inches
n = turns per inch of grid
d = diameter of grid wire in inches
E_b = anode voltage.

A compromise on grid design which combines all factors must then be used.

Anodes

The anode intercepts electrons that pass through the grids. Heat energy released by electrons striking the anode must be dissipated by the anode. This energy is the product of the instantaneous anode voltage and current averaged over a cycle or an interval of time.

The radiation-cooled anode is usually made from graphite, tantalum, molybdenum, nickel, or a combination of two or more of these materials. Radiation capabilities of these materials are shown in Fig. 5·8. Roughening the radiating surface by blasting it with carborundum grit or sand increases the energy-dissipating ability of the outer surfaces.

Tantalum not only is a radiator but also absorbs many gases that may be released during the life of the tube. Columbium and, to a striking degree, zirconium also are gas absorbers. Powdered zirconium is often used to coat graphite molybdenum or nickel anodes to act as a "getter" for gases released during the life of the tube.

If anode dissipation is greater than about 1 kilowatt, the anode is usually designed as a part of the vacuum envelope (external anode). The cooling medium is forced air or water, or a combination of the two.

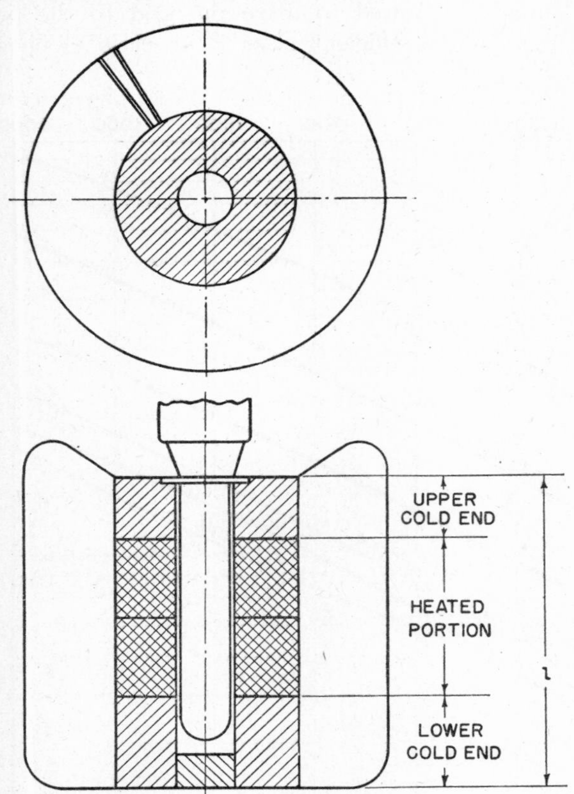

FIG. 5·9 Cross section of air cooler for tubes showing sections mentioned in the text.

A water jacket is designed for only slight separation between the outer surface of the anode and the internal surface of the water jacket. In any continuous system

$$va = \text{constant} \tag{5·9}$$

where v = velocity
a = cross-sectional area.

Therefore, reduction in cross section of water around the tube increases water velocity. This increased velocity is necessary to prevent formation of steam bubbles at the anode surface. If water velocity is not sufficiently high to carry steam bubbles away as rapidly as they are formed, the anode "hot-spots" because steam bubbles are comparatively good heat insulators. Water velocity must be high enough to produce a Reynolds number greater than 4000. Water flow then is turbulent, and cooling efficiency is increased.

The Reynolds number test is

$$Re = \frac{d\rho v}{\eta} \tag{5·10}$$

where d = equivalent duct diameter in centimeters (see equation 5·10d)
ρ = density
v = velocity in centimeters per second
η = viscosity.

Air cooling usually simplifies design and maintenance of radio-frequency equipment. Industrial equipment with air-cooled tubes can be moved from one power source to another; such equipment eliminates the problem created by a water supply containing soluble salts or alkalies.

With water-cooled external-anode tubes equipment can be designed more compactly, and energy dissipation usually can be increased without increasing anode size.

The radiator for a forced-air-cooled tube usually is soldered to a copper anode with tin, cadmium, or a soldering alloy. The use of these solders restricts the use of radiator core temperatures from 180 degrees centigrade for tin to approximately 220 degrees for cadmium or alloy solders.

The radiator core cannot be operated safely at a temperature near the melting point of the solder used. The temperature gradient through the solder used, the radiator core, and along the core to the point of temperature measurement must be accounted for in setting safe operating temperatures as shown by Fig. 5·9.

To compute temperature drop from the inside of the anode to the radiator core walls, these factors must be known:

Conductivity of copper
Conductivity of solder used
Energy dissipated per unit length in watts
Air velocity (centimeters per second) $= v$
Rate of heat transfer from fins to air in calories per second per square centimeter per degree centigrade $= h$
Heated perimeter of air ducts (centimeters) $= p$
Fin thickness (centimeters) $= \delta$
Cross-sectional area of air ducts (square centimeters) $= A$
Hydraulic radius of one air duct (centimeters) $= r_h$
Equivalent diameter of one air duct (centimeters) $= d$
Total cooling-fin area per unit length (square centimeters) $= s$
Rate of air flow (cubic feet per minute) $= q_a$

The necessary physical constants for air at 60°C are

$P = 1.06 \times 10^{-3}$ gram per cubic centimeter (air density)
$k_c = 6.25 \times 10^{-5}$ calorie per second per square centimeter per degree centigrade (thermal conductivity)
$C_p = 0.24$ calorie per gram per degree centigrade (specific heat)
$\eta = 1.95 \times 10^{-4}$ gram per second per centimeter (viscosity)

Average fin temperature above ambient is

$$T_{f\,\text{avg}} = \frac{m}{e_f} = \frac{m}{hs} \text{ degrees centigrade} \tag{5·10a}$$

where e_f = rate of heat transfer in calories per second per unit length and is equal to hs

$$h = 0.0225\,\frac{V^{0.8}}{d^{0.2}} \tag{5·10b}$$

m = heat generated per unit length in calories per second (5·10c)

$$d = 4r_h \tag{5·10d}$$

$$r_h = \frac{A}{p} \tag{5·10e}$$

The temperature is a maximum at the root of the fin which corresponds to the outer core temperature:

$$T_c = \frac{T_{f\,\text{avg}}}{\dfrac{\tanh Q_f}{Q_f}} \tag{5·10f}$$

$$Q_f = w\sqrt{\frac{2h}{\delta k_c}} \tag{5·10g}$$

where w = radial length of fin in centimeters
k_c = heat conductivity of copper
= 0.9 calorie per second per centimeter per degree centigrade.

To T_c must be added the temperature drop through the radiator core and the solder used to obtain the true anode temperature:

$$T_{\text{core drop}} = m \frac{1}{2\pi k_c} \log_\epsilon \frac{D_c}{d_c} \quad \text{degrees centigrade} \tag{5·10h}$$

where D_c = outer core diameter in centimeters
d_c = inner core diameter in centimeters.

$$T_{\text{solder drop}} = m \frac{t_s}{\pi d_c k_s} \quad \text{degrees centigrade} \tag{5·10i}$$

where t_s = solder thickness in centimeters
k_s = heat conductivity of solder (calories per second per square centimeter per degree centigrade)

Envelopes

The envelope is that portion of the tube which maintains the vacuum attained during the manufacturing process. Maintenance of vacuum requires not only a continuous glasswork but also a satisfactory metal-to-glass seal.

All tubes of the receiving type and some low-power transmitting tubes are insulated by soft glasses. The metal-glass seals are usually of an iron-core copper wire so proportioned that the combined expansion coefficient matches that of the glass used. Several patented soft-glass-sealing alloys are widely used.

Most transmitting tubes are made with hard-glass envelopes. Their metal-to-glass seals usually are of tungsten, molybdenum, or Kovar.* In hard-glass seals as well as soft-glass seals the seal material is oxidized and the glass is heated and pressed into contact with the oxide of the metal. The glass "wets" the oxide of the metal quite readily. If the oxide is too thick the oxide peels away from the metal. If the seal is overheated the oxide of the metal dissolves into the glass, leaving a weak seal. In external-anode tubes either the Housekeeper type of seal or a Kovar seal is usually used. In the Housekeeper seal (so named for its inventor) the copper is machined to a thin edge, usually about 0.0025 inch in thickness, is cleaned, oxidized, and sealed to a soft or hard glass. Even though an expansion coefficient difference of from 31×10^{-7} for type 707 glass to 150×10^{-7} for copper is often used, the flexing of the thin copper edge is sufficient for a satisfactory seal.

* A patented alloy of cobalt, iron, and nickel with an expansion coefficient of 45×10^{-7} centimeter per centimeter per degree centigrade.

Bases

A base may be the external terminal of the cathode, the grid or grids, the anode, or a combination of them.

Besides carrying the electrical connection to the proper electrode, the base radiates heat conducted to it and radiates contact losses between external connection and base. On high-voltage rectifier tubes the anode and cathode bases usually are rolled back at the edges to minimize sharp points from which cold-emission discharges may emanate.

If the tube is used at radio frequencies the base may be separated from the glass envelope by a low-loss ceramic material, which minimizes radio-frequency displacement currents between base and other elements.

Exhaust Procedures

Exhaust procedures for sealed-off tubes (tubes without a continuous pumping system) may be divided roughly into three groups:

Group A. High-production types which require high-speed exhaust equipment. In this group are all receiving types, as well as small transmitting types.

Group B. Semi-high-production types where several tubes are sealed to one pumping system.

Group C. Low-production types with one tube to each pumping system.

Group A is composed principally of the receiving types and transmitting types where the power input and voltages used are low. The degree of outgassing has only to be commensurate with the severity of usage. Once the tube is under vacuum the tube parts are heated and the gases are pumped out through the pump tubing (usually termed "tubulation"). This heating is accomplished by operating the cathode at a voltage somewhat above operating voltage, and by passing electron current between the cathode and other elements or heating the grids and anode by induction heating. Outgassing is accomplished by the first means plus the second or third, or by a combination of all three.

The cathodes of almost all receiving tubes are oxide-coated. The base material (usually nickel or nickel alloy) is coated with barium, strontium, and calcium carbonates. High-temperature flashing on exhaust changes the carbonates to the oxides. The residue gas, carbon dioxide, is removed by the pumping system.

$$BaCO_3,\ SrCO_3,\ CaCO_3 \rightarrow BaO,\ SrO,\ CaO + CO_2$$

This reaction usually takes places in an atmosphere containing some oxygen, so that free carbon is not liberated by the reaction. As the tube progresses to positions of higher vacuum on the exhaust machine, the elements are heated to higher temperatures. Usually at the last stage of the exhaust machine the "getter" is exploded. The getter is of barium, strontium, aluminum, magnesium, or some other metal which combines chemically with the residual gases in the tube as it vaporizes, to reduce gas pressure further. The getter also combines with gases released slowly from the elements as the tube is used.

Tubes in group B usually are sealed to separate risers from a common manifold which is evacuated by a common pumping system. All tubes on the manifold are given the same process simultaneously.

Glass absorbs water vapor, which is injurious to the cathode, so the tubes must be "baked" to remove it. For most tubes this baking is done at an oven temperature between 400 and 500 degrees centigrade, depending upon the type of glass used.

The cathode (usually filamentary in tubes of this exhaust class) operates at the highest temperature of any other element and is usually degassed first by flashing at about 2.3 times the normal filament voltage. The ratio of 2.3 applies only to tubes with thoriated-tungsten filaments. Tubes with pure-tungsten filaments, or kenetrons or pliotrons with oxide-coated filaments, are seldom exhausted by this exhaust procedure, except in experimental lots.

An induction-heating coil is placed around each tube, and the anode acts as a one-turn secondary of a radio-frequency transformer. Radio-frequency current induced in the anode heats it to the proper temperature for degassing. The grids also are heated to about the same temperature, except for the energy radiated from the open ends of the anode. Some manufacturers omit induction heating for tubes with tantalum anodes; instead the elements are heated by direct electron bombardment after the filament is outgassed.

After the elements have been degassed, the getter container (if one is used) is degassed by preheating to a temperature below that necessary to flash the getter, and flashing the getter is usually the final operation in the exhaust procedure.

For tubes in group C the exhaust procedure used is similar to that for group B, except that an individual pumping system ordinarily is used. The design of the pump and the pumping system controls the time required to obtain the desired pressure of 1×10^{-6} millimeter of mercury (or less) within the tube. Connections between the pump and the tube must be as large in diameter and as short as possible for high pumping speed. Figure 5·10 relates the variation of diameter and length of manifold to pumping speed.

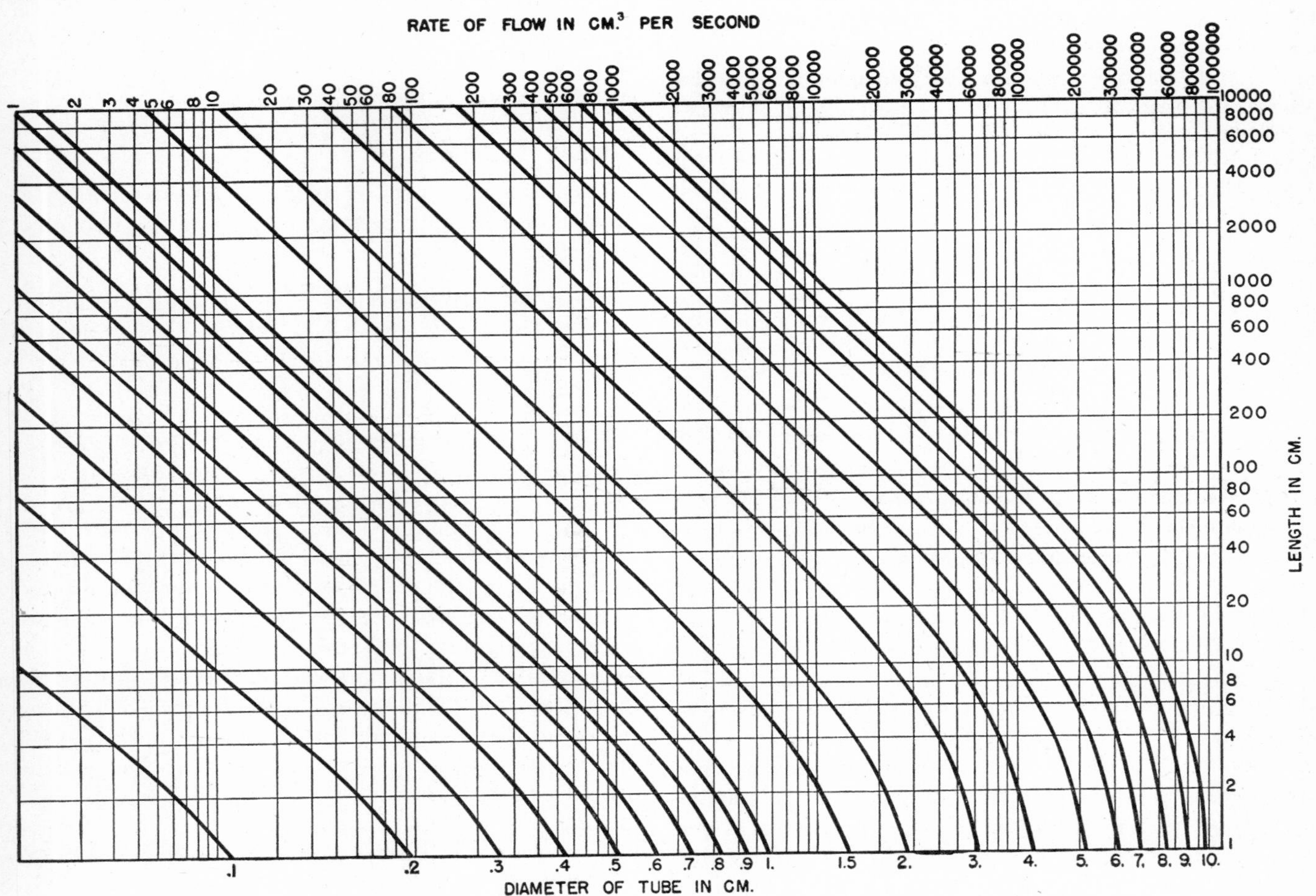

FIG. 5·10 Curves relating pumping speeds versus diameter and length of pumping apertures.

A vacuum of approximately 10^{-4} millimeter of mercury is usually desirable before the baking operation is performed, so that internal parts of the tube do not become oxidized. External parts are allowed to oxidize, except where copper-to-glass seals are used. Copper thickness at the glass seal is usually from 0.005 to 0.008 inch thick, and therefore must be protected from oxidation during baking and subsequent outgassing.

The tubes in this group usually have massive filaments containing large quantities of gas; therefore they are outgassed by increasing filament voltage (or current) in relatively slow steps. A large increase in filament temperature could release enough gas to contaminate the other elements of the tube or initiate an arc discharge between the filament terminals. Filament temperature is increased to about

2700 degrees Kelvin and the exhaust proceeds with the filament alone until gas pressure is approximately 10^{-6} millimeter of mercury.

Power required to exhaust tubes in this group usually varies between 1 and 8 kilovolts. For d-c power supplies, voltage is obtained more economically than current; therefore high voltage is usually used for "bombarding" supplies. In order to utilize high bombarding voltages the filament must be operated below normal operating voltage to limit electron emission to the necessary current; that is, tubes on exhaust must then be run with the filament operating "emission limited." Voltage from the bombarding supply is applied (usually to the grid), and filament current is increased until emission current multiplied by the unidirectional voltage applied gives the wattage required for exhaust.

As for filament outgassing, grid and anode outgassing must proceed progressively with low gas pressure in the tube, otherwise an arc discharge may destroy or damage the tube parts.

Grids of tubes in this exhaust group usually are made from tantalum, tungsten, or molybdenum. These refractory metals can be operated safely during exhaust at temperatures ranging from approximately 1500 degrees centigrade for molybdenum to 2200 degrees for tantalum and to 2600 degrees for tungsten. The energy dissipated in the grid, or grids, usually is sufficient to heat copper anodes to their outgassing temperature (750 to 850 degrees centigrade).

After exhaust the tube is sealed off from its pumping system. Gas and water vapor released from the glass during sealing off is removed by the pumping system. This operation does not increase residual gas pressure in the tube appreciably.

After the exhaust procedure, tube terminals are cleaned and the tube then is ready for testing.

Testing usually includes:

1. *Measurement of filament current at rated filament voltage (or vice versa).*

2. *Measurement of residual gas content.* This operation usually is performed by applying a positive potential to the anode and a sufficiently negative potential to the control grid to limit anode current to a given value. Usually the anode voltage and anode current are selected so that their product exceeds the rated anode dissipation by 30 to 50 percent. Passage of anode current ionizes residual gas.

Most gases form positive ions, so they are collected by the grid, the most negative element, and may be measured by an ammeter in the grid circuit. The ammeter does not measure ion current directly, but does measure electrons supplied by the grid-bias supply to the grid to neutralize ions collected by it.

3. *Measurement of amplification factor* (μ). Since μ is defined as

$$\mu = \frac{\partial E_b}{\partial E_g}$$

this measurement is made at a selected anode current, and E_b and E_g are varied to keep I_b constant. A bridge circuit sometimes is used for increased accuracy in making this measurement.

4. *Measurement of anode-current characteristics.* Once a circuit has been developed for use with a given tube type, the slope of the anode voltage-anode current curve must be maintained within a prescribed limit. This measurement normally is made with two different sets of grid and anode voltages, with limits applied to the anode current.

5. *Measurement of capacitance between tube elements.* Circuits for use at radio frequencies usually cover a range of frequencies, and tube capacitances partly determine the range of the variable capacitances or variable inductances in the circuit. If the designer knows the expected range of tube capacitance the circuits can be predetermined, and any tube within the prescribed limits can be operated within the desired frequency range.

6. *Measurement of cathode electron emission.* Tubes with oxide-coated and thoriated-tungsten cathodes must be capable of supplying peaks of emission required by both class B and class C operation. This measurement is made in either of two ways:

By method 1 (for thoriated-tungsten filaments only) the tube is placed in an oscillator, and the circuit is adjusted for maximum output. The load circuit contains an indicator of power output. Filament voltage is lowered until the power-output indicator shows a decrease of 10 percent. Filament voltage at this point must be less than nine tenths of the rated voltage.

By method 2 peak emission is measured directly by momentary application of a high positive potential to all grids and the anode simultaneously. A capacitance is charged from a d-c source and then is discharged through the tube under test. A non-inductive resistor of low value is connected in series with the tube. A cathode-ray tube is then connected across this resistor, and the voltage drop across it is read directly on the oscilloscope screen. The emission current is then calculated. If

s = sensitivity of the oscilloscope in volts per inch
d = observed oscilloscope deflection in inches
r = resistance of the resistor

emission current A in amperes is

$$A = \frac{sd}{r} \qquad (5\cdot 11)$$

The temperature of pure-tungsten filaments must be controlled accurately. If the filament is too hot at its rated voltage the life of the tube will be short, and if it is too cold the emission will be insufficient for proper circuit operation. Since emission is proportional to temperature, proper control of temperature is obtained by adjusting filament length. The filament wire is then processed to keep electron emission within appropriate limits.

5·2 VACUUM RECTIFIER TUBES

The vacuum rectifier tube (termed kenotron to differentiate from the gas-filled rectifier termed phanotron) basically consists of a source of electrons and a collector of electrons.

Electrons move from the cathode to the anode only if free electrons are available at the cathode and if the anode is positive with respect to the cathode. This electron movement can be regulated by controlling the number of electrons available and by adjusting the anode potential.

The voltage wave between cathode and anode in a half-wave rectifier circuit is shown in Fig. 5·11. The peak value

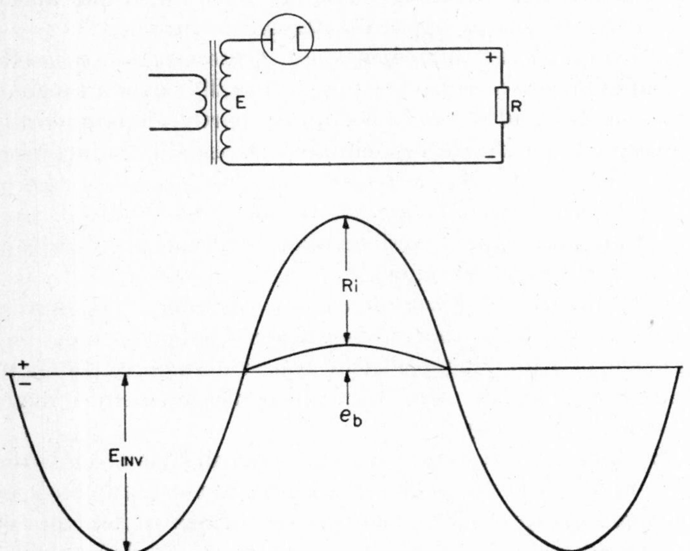

FIG. 5·11 Curve of voltage distribution and simplified circuit diagram of a kenotron rectifier circuit.

of rms voltage E is about equal to E_{inv}, if the transformer regulation is good. Tube voltage drop depends upon tube design as well as upon the value of R, for the voltage drop varies with the rectified tube current. When the tube is conducting,

$$R_i + e_b = 1.414E$$

The factors controlling voltage and current ratings of a kenotron are:

1. Average and peak current ratings are dependent primarily upon the number of free electrons available from the cathode. For pure-metal cathodes (such as tungsten or tantalum), peak current should not exceed the maximum electron emission (see curve 1 in Fig. 5·12). For maximum tube life, this peak-emission demand should not exceed approximately six tenths of the electron-emission capability of the tube (see Fig. 5·4).

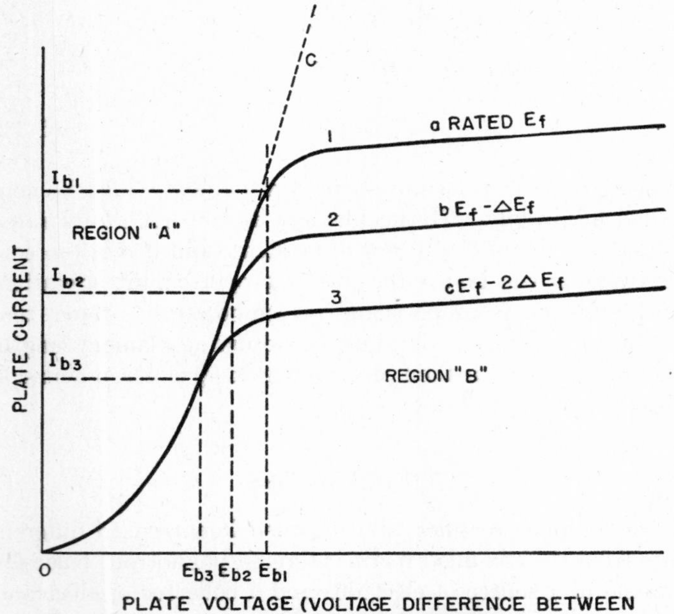

FIG. 5·12 Curves showing voltage distribution across a kenotron for various filament-voltage settings.

Figure 5·12 shows that, if the peak emission of the cathode is exceeded, the voltage drop in the tube increases rapidly. Anode dissipation ($I_{eff}E_{eff}$) then may exceed its maximum value, which may temporarily or permanently damage the tube.

If the peak-current rating is exceeded the forward voltage during the conducting half-cycle increases. This increased voltage increases the velocity of positive ions striking the cathode. With oxide-coated cathodes or thoriated-tungsten cathodes, this increased velocity of bombardment may impair electron emission of the cathode.

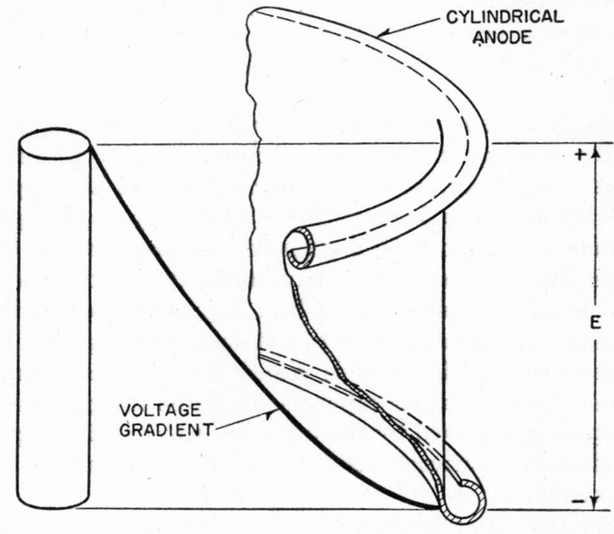

FIG. 5·13 Sectional sketch showing voltage gradient across a cylindrical kenotron during an inverse half-cycle.

2. Inverse-voltage ratings are dependent upon spacings between elements, contours of the elements, and external path between anode and cathode terminals. The design should limit the voltage gradient to less than 200,000 volts per inch for most surfaces.

Voltage across the tube elements is highest when the anode is negative, so rough surfaces, sharp points, and parts with small radii must be avoided on the anode to prevent electron-field emission.

In a parallel-plane diode (E_b+) with cold cathode, **E** is constant between cathode and anode. In a cylindrical diode (E_b+), **E** is greater near the cathode than near the anode; if the anode is negative with respect to the cathode, **E** near the anode is less than in a plane structure. If the anode were infinite in length the gradient would be as shown in Fig. 5·13, because the anode has a negative radius in calculating the gradient from anode to cathode. Rolling the anode edges with a large radius, as shown in Fig. 5·13, is a practical expedient toward extending the ends indefinitely.

If E_w is the work function of the anode material, ΔE_w or the decrease in work function caused by the electric field at the anode is

$$\Delta E_w = 300\sqrt{e\mathbf{E}10^7} \qquad (5\cdot 12)$$

where e = electron charge coulombs
$\mathbf{E}$ = electric-field intensity in volts per centimeter near the anode surface.

Element spacing and contours should be such that ΔE_w is as small as practical. External spacing between cathode and anode terminals must be such that there is no voltage breakdown along the external surface of the glass envelope even if it is coated with dust particles. Dust particles accumulate on the glass surface because of electrostatic charges.

Figure 5·12 is characteristic of all high-vacuum rectifiers with pure-tungsten cathodes. Curve 1 is a cross plot of voltage difference across the tube against current through the tube at the maximum rated filament voltage. Curve 2 is for a lowered filament voltage, and curve 3 for a still lower voltage.

With the filament voltage as in curve 1, the filament (or cathode) can supply the number of electrons required for voltage drop E_{b1} across the tube, thus giving the current I_{b1}. Likewise, if potential differences across the tube are E_{b2} and E_{b3}, the currents through the tube are I_{b2} and I_{b3}.

For a filament voltage E_f, I_{b1} represents the maximum electron-emission current that the filament can supply. If E_b is increased beyond E_{b1}, the tube is said to be running emission-limited. If filament voltage is E_f, and E_b is less than E_{b1}, the tube is said to be operating *space-charge-limited.* If the filament voltage is raised above E_f, the curve of anode current versus anode voltage follows the dotted curve C as predicted by equation 5·13.

The region of the chart to the left of the line marked E_{b1} is said to be the space-charge-limited region of the tube for that filament voltage. In this region the plate current follows Child's law:

$$\mathbf{J} = 2.331 \times 10^{-6} \frac{E^{3/2}}{s^2} \quad \text{(for flat structures)} \qquad (5\cdot 13)$$

where s = cathode-to-anode distance in centimeters
$\mathbf{J}$ = current density in amperes per square centimeter.

For filament voltages giving curves 2 and 3, the space-charge-limited regions are to the left of E_{b2} and E_{b3} respectively.

If the load on the rectifier causes a voltage drop greater than E_{b1}, E_{b2}, or E_{b3}, depending upon which filament voltage is used, the tube is then operating in the emission-limited region. If the filament could supply electrons in unlimited quantities, the anode voltage-anode current curve would continue as shown by C. Thus any excursion into the region to the right of curve O-C is into the emission-limited region B.

Curves a, b, c are essentially parallel and are inclined upward to the right. This increase of current with an increase of tube-voltage drop is caused by an effect first calculated by Schottky, hence called the Schottky effect. The effect can be calculated from Schottky's formula:

$$I = I_s\, \epsilon^{4.389\mathbf{E}^{1/2}/T} \qquad (5\cdot 14)$$

where I_s represents the current just to the left of the knee of the curve, and I is the current at the point being calculated and at the field intensity $\mathbf{E}$ with the filament at temperature T degrees Kelvin.

In the emission-limited region, $\mathbf{E}$ materially increases at the cathode. This increase effectively lowers the work function of the cathode surface, which in turn increases the emission current drawn from the cathode. With equations 5·14 and 5·12 the increase in emission current and decrease in work function respectively can be computed for the emission-limited region.

The power dissipated in the rectifier tube is the product of effective voltage across the tube and effective current through the tube during the conducting half-cycle. In the emission-limited region enough power may be dissipated in the tube to destroy or seriously damage it. The maximum current rating is controlled by the number of electrons available at the cathode and upon the dissipation capabilities of the anode.

Contours of the elements and spacings between them must be considered in determining inverse-voltage ratings. The voltage gradient at the surface of an element is inversely proportional to the square of the radius at the surface of that element. The radius must not be such that the gradient can cause field emission at the maximum inverse voltage.

Mechanical stresses must be considered also. The force which exists between rectifier elements is

$$F = \tfrac{1}{2}CE^2 \times 10^{-8} \qquad (5\cdot 15)$$

where C is the capacitance in farads and E is maximum inverse voltage. F is only numerically equal to the right-hand side of this equation.

Figure 5·14 shows the relative lengths of two rectifiers. Type WL-531 has an inverse rating of 50,000 volts with an external glass length of 4⅞ inches. The WL-616 has an inverse voltage rating of 150,000 volts and an external glass length of approximately 22 inches.

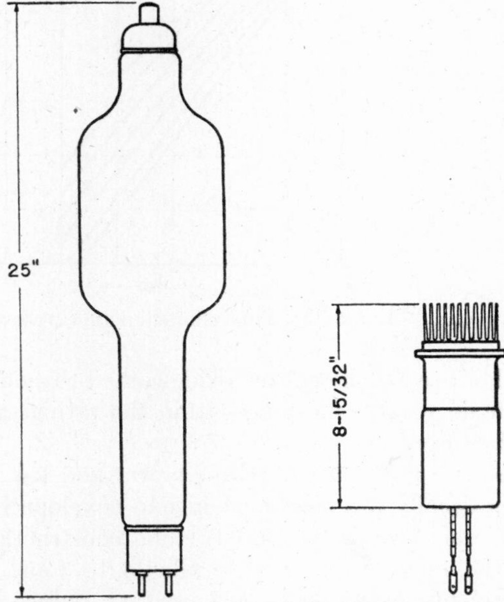

Fig. 5·14 Illustrating variations of tube length with inverse-voltage ratings.

If Fig. 5·12 were applicable to the WL-616, I_{b1} would be 750 milliamperes or less to avoid operating the tube emission-limited. Well-designed pure-tungsten-filament kenotrons have an excess of at least 40 percent in emission, because there is a decrease of emission during life. This excess provides ample emission at the end of tube life. Average plate current cannot usually be exceeded in any rectifier circuit without exceeding anode dissipation. If tubes with pure-tungsten cathodes are used at less than full anode current, less than the rated filament voltage may be used.

functions in the same envelope. Demodulation is basically the same function as that performed by power rectifiers.

The power level for demodulators is usually under 1.0 watt, the principal requirement being linearity of output to avoid audio-frequency distortion.

5·3 TRIODES

A triode is a three-electrode vacuum tube containing an anode, a cathode, and a control electrode, usually called a

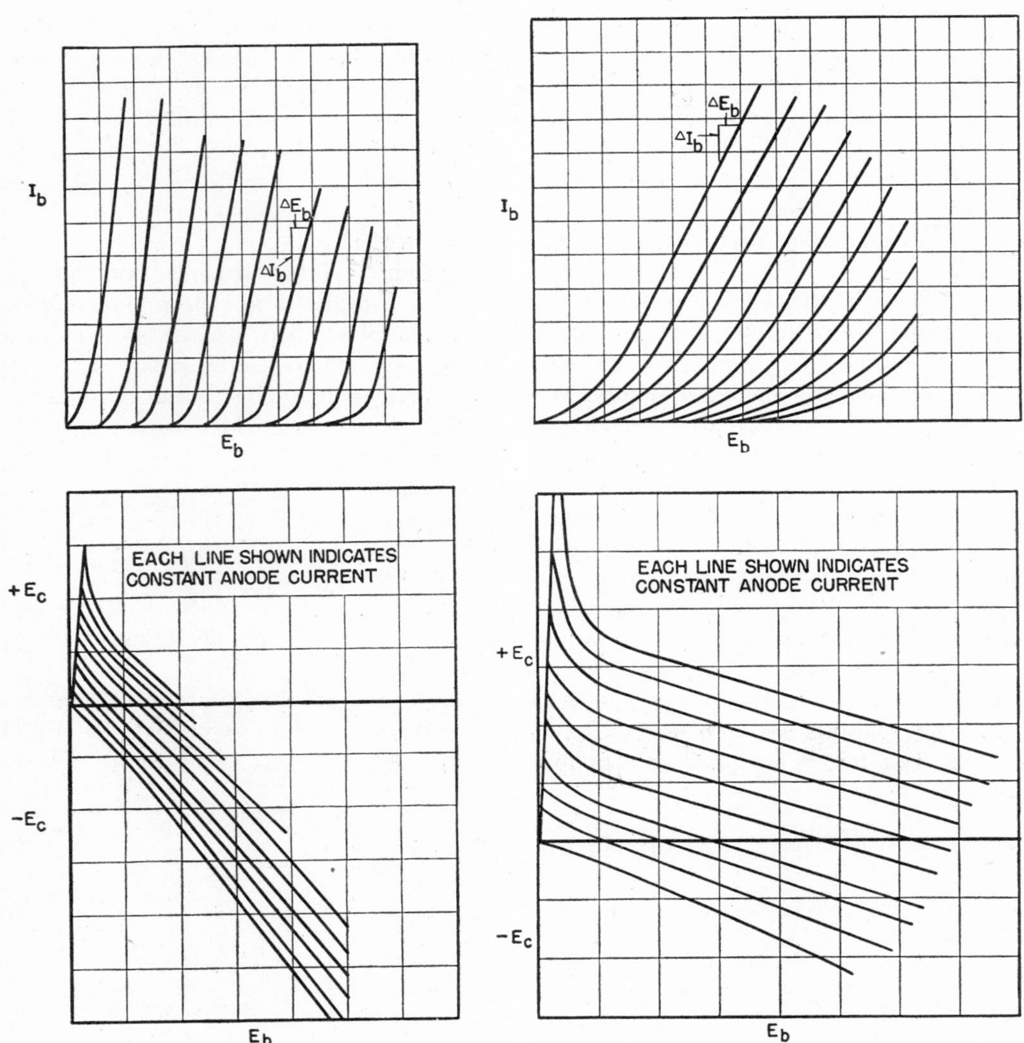

Fig. 5·15 Tube characteristic curves showing variation of shape of curves with amplification factor.

With thoriated-tungsten or oxide-coated-cathode kenotrons, cathode voltage must be within the ratings given by the manufacturer.

For full-wave rectification at low power and low voltage, two diodes usually are combined in one envelope by using common filament leads to simplify tube construction, thus reducing the number of tube terminals to four. Oxide-coated filaments or cathodes are used to reduce cathode power.

For demodulation or detection in radio receivers a single or double diode is commonly used. These diodes are often placed in the same envelope with other tube elements to save space or tube cost, thus combining two or more tube

grid. Physical size of triodes varies from the hearing-aid tube, approximately the diameter of a pencil and about 1¼ inches long, to demountable tubes about 10 inches in diameter and 6 feet long. Radio-frequency power outputs range from the few watts of the hearing aid and "acorn" types to approximately 500 kilowatts in some continuously pumped varieties.

Radio frequencies beyond 3000 megacycles can be generated by triodes, but the power outputs of triodes begin to decrease sharply between 50 and 100 megacycles (see Fig. 5·17).

Triodes usually are better suited to the generation of radio-frequency energy than are tetrodes or pentodes. Triode oscillator circuits and circuit adjustments, as well as their power supplies, are less complicated. The triode usually requires neutralization in radio-frequency amplifiers. Some types operated as grounded-grid radio-frequency amplifiers do not require neutralization.

For power-output levels up to approximately 2 kilowatts, the radiation-cooled glass-envelope tube is often used. For larger power levels water-cooled or forced-air-cooled-anode types are preferable. Radiation-cooled tubes have been built for power outputs of 5 kilowatts or greater, but external-anode types have had wider acceptance because of smaller size, greater ruggedness, and comparative freedom from trouble.

Plate-power conversion efficiencies of triodes usually are from 70 to 85 percent.

For general use, the amplification factors of triodes are between 12 and 50. Each extreme, as well as the center, of this range has advantages. If a tube with a high amplification factor is used as an oscillator, any condition that stops oscillation, such as a misadjusted circuit, may not cause damage to the tube. At zero grid bias the anode current may not be great enough to cause the anode to melt. Conditions that prevent oscillation are seldom found in well-designed equipment.

Tubes with high amplification factors have high dynamic anode resistances, thus matching better with high-impedance circuits. Conversely, the low dynamic anode resistance of tubes with low amplification factors favors their use for audio-frequency circuits in which amplification factors from 3.5 to 8 are desirable.

Figure 5·15 illustrates the transition in anode characteristics between amplification factors of 8 and 50. The WL-891 and WL-892 are identical in size, with the grids wound to give the required amplification factors.

Frequency limitations of triodes as oscillators are controlled primarily by two factors: (1) effective inductance of the grid, cathode, and anode leads; and (2) the time required for electrons to travel from cathode to anode.

Maximum frequency is controlled also by the capacitance between the three elements (Fig. 5·16). Inductances of the

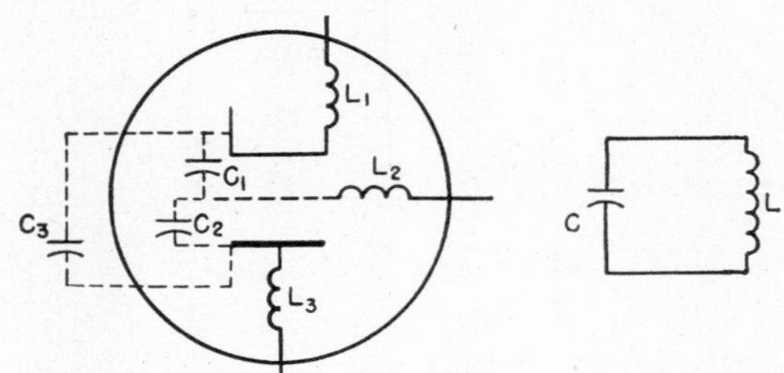

FIG. 5·16 Diagram illustrating tube-element capacitance and inductance and an equivalent circuit.

various leads are L_1, L_2, and L_3, and the tube capacitances are C_1, C_2, C_3. They can be combined by the following procedure to obtain C and L in Fig. 5·16, where

$$C = C_{pg} + \frac{C_{pk}C_{gk}}{C_{pk} + C_{gk}} \tag{5·16}$$

and L = La + Lg + inductance of short-circuiting strap between anode and grid.

Combined L and C substituted in the equation

$$f_m = \frac{1}{2\pi\sqrt{LC}} \tag{5·17}$$

give the highest frequency at which the tube will operate in conventional circuits.

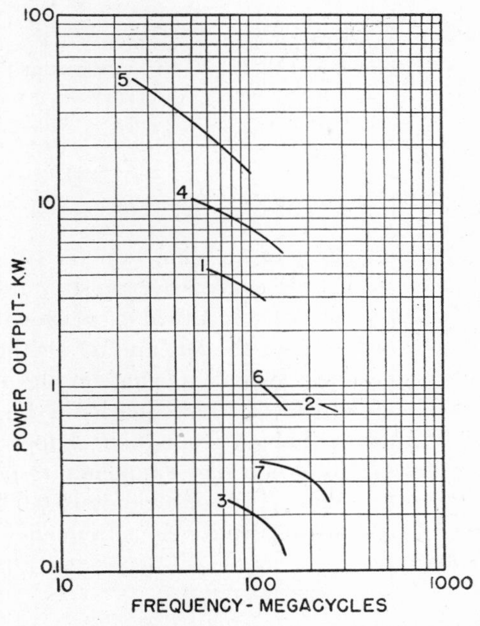

Curve	Tube Type Number	Number Elements
1	473	3
2	887–888	3
3	8001	3
4	889	3
5	880	3
6	827R	4
7	4–125A	4

FIG. 5·17 Curve showing output versus frequency of various triodes and tetrodes.

L can be approximated by taking the length of the tube elements plus the short-circuiting strap and substituting in the equation

$$L = 0.015D\left(2.3 \log_{10} \frac{8D}{d} - 2\right) \tag{5·18}$$

where L = inductance in microhenrys
D = mean diameter formed by loop of tube elements plus the short-circuiting strap in inches
d = average diameter of conductor in inches.

L is an approximation only if it is obtained from equation 5·18. Many tube elements and leads are made by flat ribbons with lower inductances than that given by equation 5·18. A tube with a short-circuiting strap constitutes an L-C circuit to which may be coupled a variable-frequency oscillator for determining the true f_m limitation from the physical limitation of tube size.

The foregoing does not account for electron transit time. If transit time determines the maximum frequency,

$$f_{\max} = \frac{3 \times 10^7}{\left(\frac{\mu}{E_b}\right)^{1/2} 3d_{kg} + \frac{2d_{ga}}{\mu + 1}} \tag{5·19}$$

where μ = amplification factor
d_{kg} = distance from grid to cathode in centimeters
d_{ga} = distance from grid to anode in centimeters
E_b = operating anode voltage
f_{max} = cycles per second.

Performance of a few popular triodes as indicated by equations 5·16, 5·17, 5·18, and 5·19 is as follows:

Type	Maximum Frequency from L and C (megacycles per second)	Maximum Frequency from Electron Transit (megacycles per second)
806	195	310
889	218	308
891	75	392
892	75	168

The predominant effect of amplification factor is evident in the frequency limitation caused by transit time of types 891 and 892. These tubes are identical except for the amplification factor, which is 8 for the 891 and 50 for the 892.

The relative effect of element spacing on the frequency limitation of triodes is evident from equation 5·19.

Lighthouse tubes (so named from their shape) use close grid-cathode spacing to obtain high-frequency response. In smaller tubes of this type d_{gk} is approximately 0.002 inch.

Figure 5·17 shows power output versus frequency response of several tube types.

5·4 TETRODES, PENTODES, AND BEAM-POWER TUBES

Tetrodes

One of the serious handicaps of triodes as radio-frequency amplifiers is the necessity for neutralizing the feedback (or regenerative) effect of the capacitance between the anode and the control grid. This handicap led to the introduction of a second grid between the anode and the control grid. The second grid is maintained at a positive potential with respect to the cathode and is by-passed by a low-impedance capacitor to the cathode. This screen grid reduces the effect that variations in anode potential can have on the grid potential; that is, the alternating component of the anode voltage which will exist between the anode and screen will have little or no effect in the region inside the screen. Since the screen removes the effect of the anode potential from inside the screen, the anode potential no longer has any material effect on anode current, and the screen- and control-grid potentials have almost complete control of the anode current.

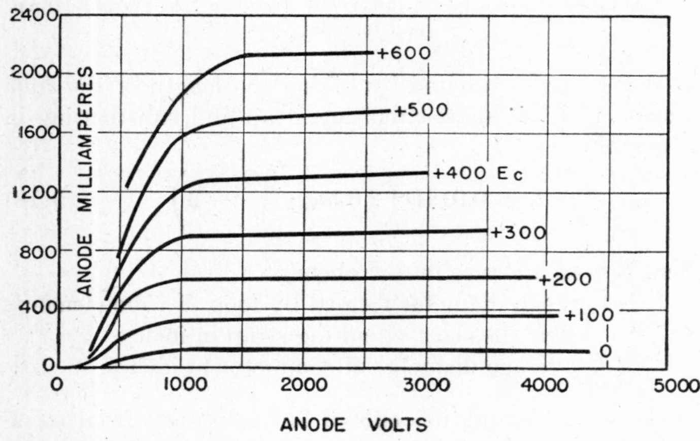

FIG. 5·18 Typical tetrode characteristic curves with screen voltage of +500 volts.

Current through a tetrode can be represented by

$$I_s = K\left(E_c + \frac{E_{c2}}{\mu_{cg}}\right)^{3/2} \tag{5·20}$$

where I_s = electron-emission current
K = a factor depending upon tube geometry (often termed perviance)
E_c = control-grid voltage
E_{c2} = screen-grid voltage
μ_{cg} = amplification factor of control grid to screen grid.

A more exact representation is

$$I_s = K\left(E_c + \frac{E_{c2}}{\mu_{cg}} + \frac{E_b}{\mu_{cg}\mu_{sg}}\right)^{3/2} \tag{5·21}$$

where E_b = anode voltage
μ_{sg} = amplification factor of screen grid to anode

and all other factors have the same significance as in equation 5·20.

In the region where anode voltage is equal to or less than screen voltage on the characteristic curves of a tetrode, the anode current may dip as shown in Fig. 5·18.

The dip is caused by secondary emission from the anode. If anode voltage is reduced so that it approximates the screen voltage, secondary electrons from the anode leave with enough energy to return to the screen, thus reducing net anode current and increasing screen current.

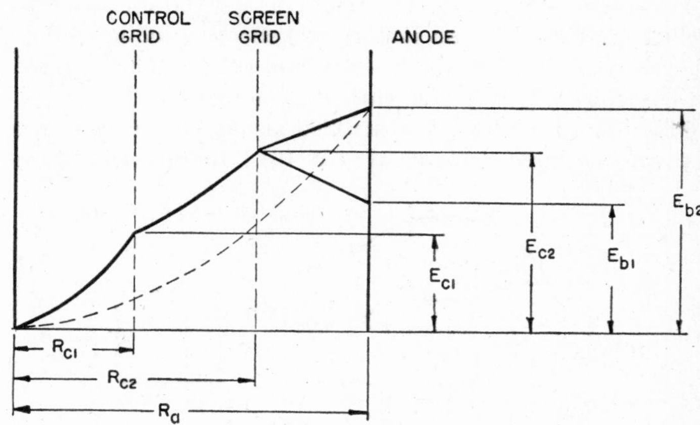

FIG. 5·19 Typical potential distribution in a plane tetrode for two values of anode potential. Dotted line shows typical potential distribution with no grids present.

If anode potential is reduced appreciably below screen-grid potential, electrons striking the anode do not usually have sufficient velocity to release secondary electrons; thus net anode current usually is greater under these conditions.

Figures 5·19 and 5·20 show potential distributions for plane and cylindrical tetrodes. For electrons traveling from cathode to anode, dV/dx is always positive when the anode voltage is E_{b2}. The magnitude of dV/dx in the screen-grid-to-anode region is controlled by $(E_b - E_{c2})/(r_a - r_{c2})$.

As this quantity approaches zero (that is, when $E_b = E_{c2}$), then $dV/dx = 0$, and the energy of secondary electrons is sufficient to cause them to return to the screen grid. Figure 5·21 shows that this condition is approached as E_b is reduced to some voltage between 700 and 800 (depending upon E_{c1}). This secondary-electron current to the screen grid approaches a maximum when $E_b = E_{c2}$ or E_b is slightly less than E_{c2}. If E_b is less than E_{c2}, then dv/dx is positive for secondary electrons leaving the anode; this quality aids their return to the screen grid. As E_b is reduced more, electrons from the cathode travel through a negative dV/dx between screen grid and anode, and their velocity is reduced so that their energy content on striking the anode is insufficient to release secondary electrons in the previous quantities.

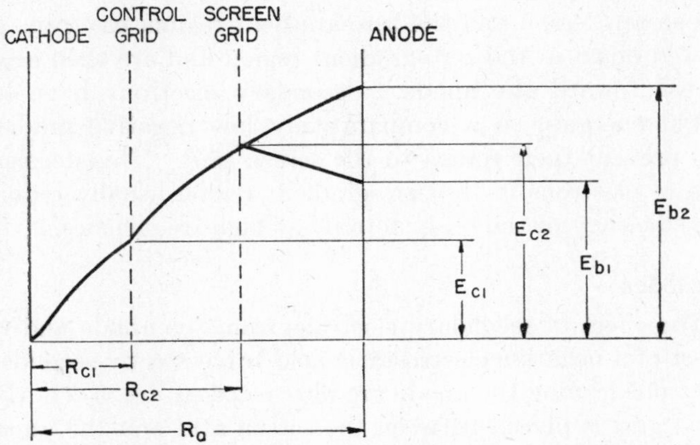

FIG. 5·20 Typical potential distribution for a cylindrical tetrode.

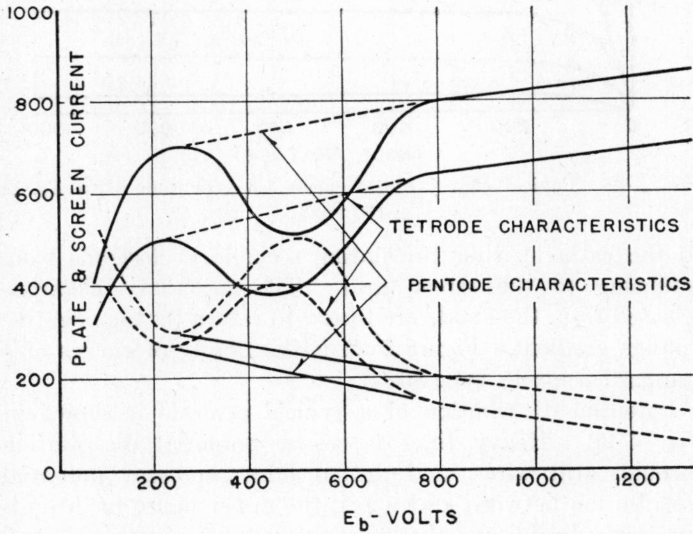

FIG. 5·21 Typical characteristic curves for tetrode and pentode for two values of screen voltage. Lower lines are for screen current and upper lines are for anode current.

This accounts for the increase in anode current for values of E_b appreciably less than E_{c2}. The equation

$$I_s = K\left(E_{c1} + \frac{E_{c2}}{\mu_{cg}} + \frac{E_{c3}}{\mu_{cg}\mu_{sg}} + \frac{E_b}{\mu_{cg}\mu_{sg}\mu_{supg}}\right)^{3/2} \qquad (5\cdot22)$$

determines maximum electron current from the cathode of a diode, triode, tetrode, or pentode by using applicable terms. Terms containing E_b may be dropped for tetrode or pentode calculations without serious error.

As shown in Fig. 5·18, the minimum anode-voltage excursion for satisfactory operation must be limited to that of the screen voltage multiplied by a factor between 1.1 and 1.25, depending upon the plate current dip at $E_b = E_{c2}$. This unfavorable characteristic has been utilized in some types of oscillators. Precalculation of tetrode performance follows the same procedure as that given for triodes later in this chapter.

Tetrodes are used in both radio- and audio-frequency circuits, although radio-frequency uses predominate. The tetrode offers high power gain and need not be neutralized. The low-power radio-frequency tetrode has been replaced by the pentode or beam tetrode for most purposes.

As an amplifier a tetrode has an advantage over a triode at high frequencies, because its anode voltage can lag grid voltage without seriously affecting plate efficiency. Transit time of electrons between screen and anode is the only transit-time effect in the tetrode that reduces its efficiency materially when it is used as an amplifier. Anode voltage is not affected by electrons between screen and cathode, so to the anode the screen is a source of electrons. Anode voltage therefore may lag grid voltage by the time necessary for electrons to travel from the cathode to the screen grid without an impairment in efficiency.

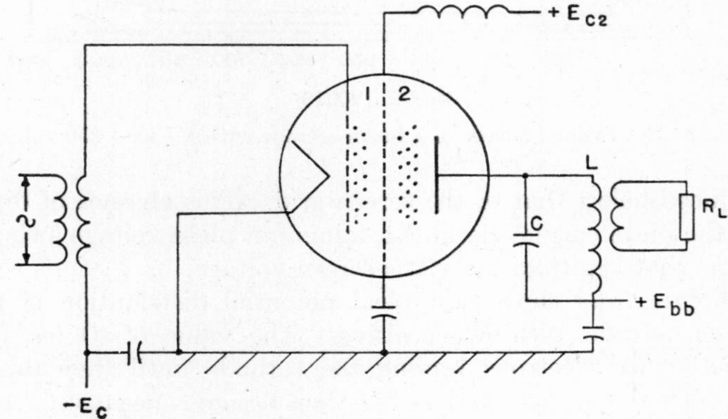

FIG. 5·22 Diagram illustrating the separation by the screen grid of a tetrode into two spheres of influence.

This effect can be further analyzed by an examination of Fig. 5·22. In the tube shown, the screen grid shields region 1 from the field set up within the tube by instantaneous anode voltage e_b. Conversely, electron charges in region 1 do not materially affect instantaneous anode potential e_b, for E_{c2} is held almost constant by the by-pass capacitor. As soon as the electron cloud moves from region 1 into region 2, then e_b is affected. Region 2 changes from an infinite d-c resistance to a finite resistance, and e_b is reduced in magnitude (the capacitance between anode and screen grid is reduced also), thus producing an oscillatory voltage in the anode inductance and capacitance. The time required for electrons to travel from the cathode to the screen grid thus does not affect the anode efficiency of the tetrode. Instantaneous anode voltage then may lag exciting voltages by the angle determined by the time required for the electrons to travel from cathode to screen grid without affecting anode efficiency.

Beam Tetrodes

The beam tetrode has two grids, and the configuration of the elements are such that the tube has electrical characteristics resembling both the pentode and the tetrode.

Figure 5·23 shows that the anode current of the beam tetrode does not decrease materially as the anode voltage is reduced below that of the screen grid. This absence of dip in the plate current permits a minimum plate-voltage swing somewhat less than the static screen voltage.

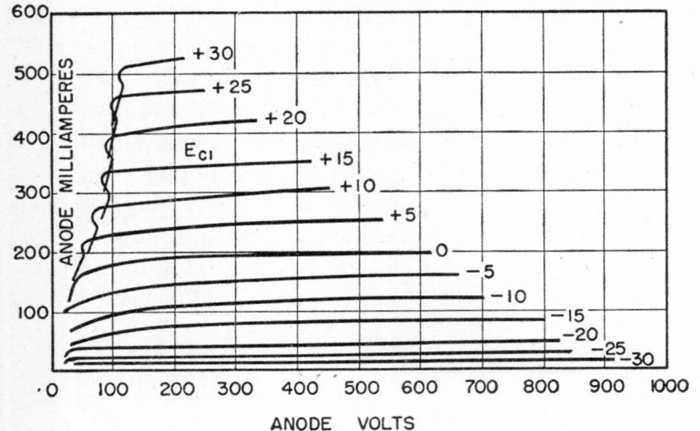

FIG. 5·23 Typical curves of a beam tetrode with E_{c2} at +250 volts.

Figure 5·24 shows a typical potential distribution of a beam tetrode with E_{c1} positive. The value of dV/dx is positive for electrons leaving the cathode until they pass through the screen grid; dV/dx then becomes negative. As a result the velocity of electrons is reduced and they are concentrated between screen and anode. This concentration is aided by the configuration of elements, which focuses the electrons along a relatively narrow beam and increases electron density between screen grid and anode. The concentration lowers the effective potential between screen and anode. At this point of minimum potential, dV/dx is zero. From the region of maximum electron concentration the electrons are accelerated toward the anode, thus forming another region in which dV/dx is positive for electrons traveling toward the anode. For secondary electrons from the anode, dV/dx is now negative, and these electrons are thus returned to the anode. A typical set of curves for a beam tetrode is shown in Fig. 5·23. There is no definite break in the anode current until the anode voltage is appreciably less than the screen voltage.

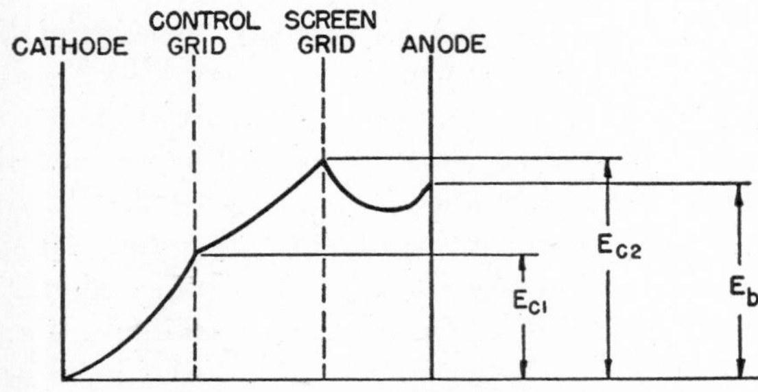

FIG. 5·24 Typical potential distribution in a beam tetrode.

Tube elements are arranged so that velocities of most of the electrons emitted are almost constant, and they arrive at a plane equidistant from the anode in essentially the same interval of time. The position of the electrons in space and their density establish a region in which voltage-space gradient is almost zero; therefore this space has the electrical effect of a suppressor grid without the grid being there. Since a zero gradient establishes a negative gradient between the screen region and the zero-gradient region, electrons are slowed down in the zero-gradient region and are then accelerated toward the anode. Secondary electrons have low initial energies, so a comparatively low negative gradient can prevent their return to the screen grid. This deceleration of electrons in the zero-gradient region usually reduces the efficiency of the beam tetrode at high frequencies.

Pentodes

The effect of secondary electrons from the anode and the effect of a negative electrostatic field traversed by secondary electrons leaving the anode are illustrated in Fig. 5·21. If a third grid is placed between the screen grid and the anode, the gradient between the third (suppressor) grid and the anode is made negative by connecting the suppressor grid to the cathode, thus preventing secondary electrons from returning to the screen grid. The secondary electrons released from the anode are forced to return by the negative voltage gradient. Figure 5·25 is the family of curves of a standard pentode, type 802.

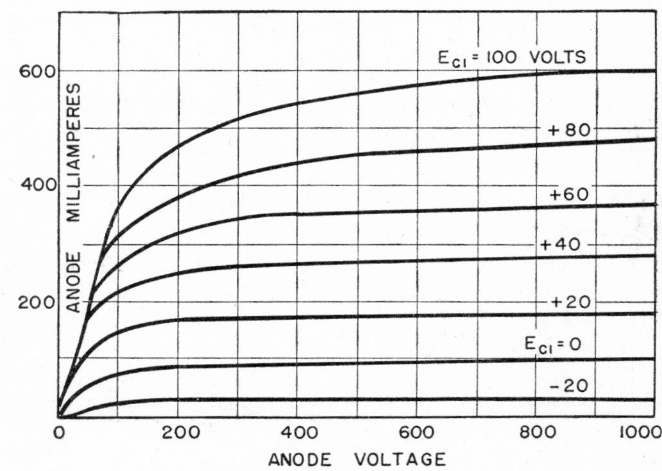

FIG. 5·25 Typical characteristic curves of a pentode with E_{c2} at +250 volts.

Potential distribution of a typical pentode is shown in Fig. 5·26. Heavy lines represent potential distribution through grid wires, and dotted lines represent potential distribution between grid wires; the upper figure for a positive control grid, and the lower figure for a negative control grid. In the upper figure the downward concavity of the potential line between the screen and suppressor grids is caused by electron concentration in this region.

The effective amplification factor of a pentode is the product of the amplification factors of all the grids, each referred to the previous grid or cathode. For power-output pentodes it usually is from 100 to 250, and for radio-frequency application it often is 1250 or higher.

The pentode was designed to prevent plate-current dip at low anode-voltage excursions, which means that the negative voltage gradient at the anode (for secondary electrons) must be great enough to prevent secondary electrons of highest velocity from passing through the suppressor grid and returning to the screen. For this reason the spacing between suppressor-grid wires must be close enough so that potential distribution between the grids is sufficient to prevent return of secondary electrons.

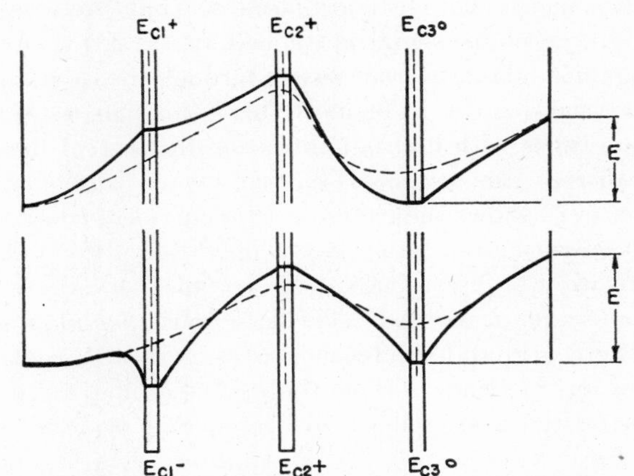

FIG. 5·26 Typical potential distribution in a pentode under space-charge-limited and space-charge-free conditions.

Figures 5·18 and 5·23 show a sharp break in the plate current of the beam tetrode at a low anode voltage and a gradual change of plate current at low anode voltages in a pentode. The beam tetrode depends upon electron density for secondary electron suppression. As anode voltage is reduced, the potential minimum between screen grid and anode gradually decreases, until at low anode voltages electron density decreases; a potential minimum is no longer established, and the sudden break occurs in the plate current. In a pentode the potential minimum is established by the suppressor, and thus there is no sharp disappearance of restraining force for secondary electrons.

5·5 MULTIPURPOSE TUBES

The pentagrid converter combines the function of a local oscillator and a first detector. The first two grids are connected to form a local oscillator, the third and fifth grids being connected to isolate the fourth grid from the effects of the oscillator section. Incoming signals are applied to the fourth grid. The circuit connected to the anode is tuned to the frequency difference of the incoming signal and the local oscillator.

Tube elements for rectification of the radio-frequency signal and amplification of the resultant audio-frequency signal are often combined into one envelope. This tube usually consists of either a single or double diode and a triode, tetrode, or pentode.

5·6 TUBE-DESIGN PARAMETERS

Designing of a high-vacuum pliotron is based on:

1. The type of service.
   a. Class A operation.
   b. Class B operation.
   c. Class C operation.
      (1) Minimum and maximum operating angles.
2. Anode power output required.
3. Maximum frequency of operation for full anode power output.
4. Modulation frequencies (if any).

Class A Operation

In typical class A service the tube is operated in the negative-grid region; therefore, there is no electron current to the grid. Tubes for this service require no special precautions regarding grid dissipation.

The power output P_o in class A service with a resistive load is

$$P_o = \frac{(E_{\max} - E_{\min})(I_{\max} - I_{\min})}{8} \tag{5·23}$$

Assume that a tube is to be designed for class A service and (1) that it is required to deliver 38 watts into a resistive load of 10,000 ohms and (2) that anode dissipation is to be 100 watts.

For a resistive load:

$$P_o = RI^2 \quad \text{or} \quad \frac{E_{\max}I_{\max}}{2} \tag{5·24}$$

where $I_{\text{rms}} = 61.7$ milliamperes
$I_{\max} = 87.4$ milliamperes
$E_{\max} = 870$ volts.

Total current change from peak to peak of the anode-voltage wave superimposed on the average anode voltage is 174.8 milliamperes or $2 \times I_{\max}$.

The direct anode current is equal to half the total anode current change when the anode current is reduced to zero on negative grid-voltage modulation excursions; therefore I_{d-c} may be assumed to be 87.4 milliamperes. Applied anode voltage then is

$$E = \frac{100}{0.0874} = 1145 \text{ volts} \tag{5·25}$$

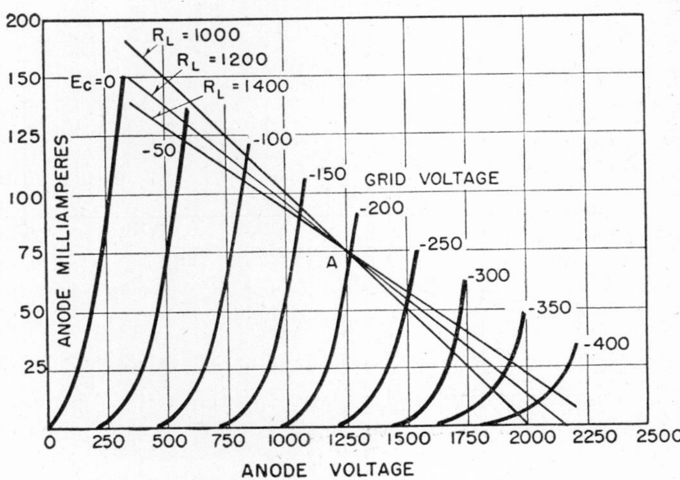

FIG. 5·27 Curves showing anode characteristics of a triode and operating lines for class A amplifier service.

Anode-voltage change must be 870 volts, so the minimum and maximum anode voltages are

$$e_{b\,\min} = 275 \text{ volts}$$

and

$$e_{b\,\max} = 2015 \text{ volts}$$

Anode current must be zero when $e_b = 2015$ volts, and 174.8 milliamperes when $e_b = 275$ volts. Grid voltage e_c may be zero but not positive under the latter conditions.

The relation

$$I_s = K\left(E_g + \frac{E_b}{\mu}\right)^{3/2} \qquad (5\cdot 26)$$

where I_s = emission current in amperes
E_g = grid voltage
E_b = anode voltage
μ = amplification factor
K = a constant depending upon tube geometry

shows that for a given value of K, I_s increases as μ decreases for any value of E_g. When $E_g = 0$, I_s is wholly dependent

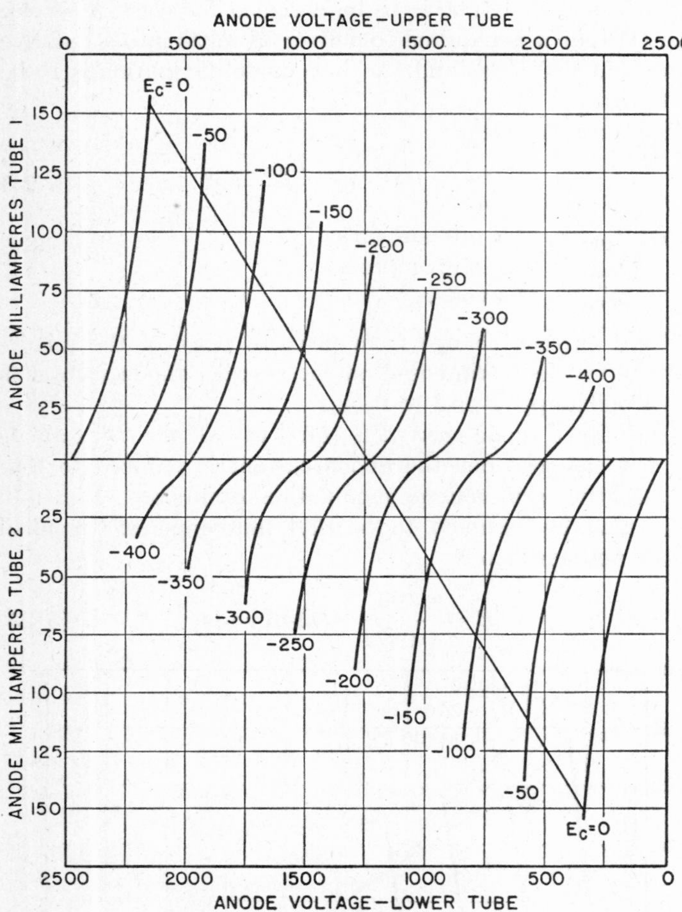

FIG. 5·28 Curves showing operating load line for two triodes operating in a push-pull class A amplifier.

upon E_b and μ. For most tube designs the lower practical limit for μ is approximately 3, with 5 preferred. Assuming $\mu = 5$, $K = 4.29 \times 10^{-5}$ to make $I_s = 174.8$ milliamperes when $E_b = 275$ volts and $E_g = 0$.

For $I_s = 0$,

$$-E_g = \tfrac{2175}{5} = -435 \text{ volts}$$

After substitution of the values calculated in equation 5·23, the power output is found to be 38 watts. Figures 5·27 and 5·28 give the characteristic curves of the type 845 tube, which is used primarily in class A service.

The peak I_s value of 174.8 milliamperes could be supplied with 2 watts of thoriated-tungsten-filament power, but, to satisfy the anode-area requirements of K, more filament power must be used to give sufficient effective cathode area and to have a practical cathode.

Class B Operation

In class B operation for either radio-frequency or audio-frequency service the grid is biased so that there is some anode current if no signal is applied to the grid. With a signal applied, anode current passes through the tube through 180 electrical degrees. For most tubes this bias is negative, but some tubes with high amplification factors can be operated with zero bias in class B service.

Figure 5·29 shows the characteristic curves of the WL-895 with an operating point for class B operation. Point A may be the focus of both grid and anode voltages for either audio- or radio-frequency service. Through point A will pass the axes of both grid and anode sine waves.

For radio-frequency service the grid is excited by a voltage driving the grid from A to C (one-half cycle) with no

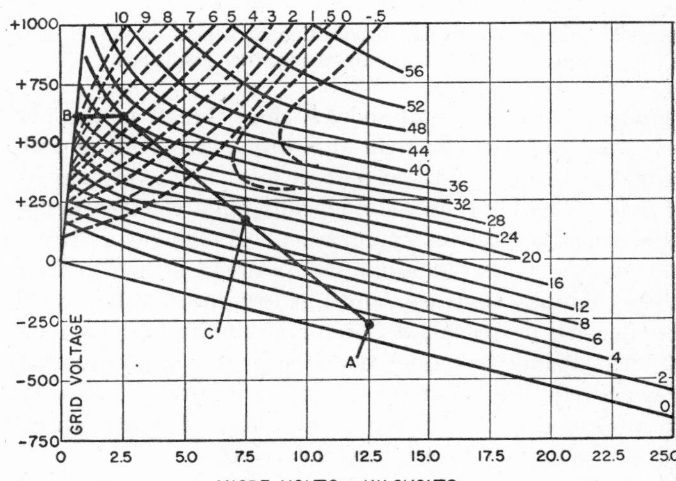

FIG. 5·29 Constant-current chart of the WL-895 with a class B r-f operating line shown. The operating line shown gives

	No Modulation	Full Modulation	
I_{bb}	3.87	10.1	amperes
I_g	0	0.935	ampere
P_o	15.5	82	kilowatts
P_g	0	812	watts

modulation voltages applied. Under 100 percent modulation the grid is driven to point B (one-half cycle). For the tube designer points A and B are of special interest.

In class B audio-frequency service the product of E_b and I_b at point A should be less than half of the permissible anode dissipation. With modulation to point B, power output is

$$P_o = \frac{34 \times 10{,}000}{2} = 170{,}000 \text{ watts}$$

of audio-power output for two tubes operated in push-pull, and anode dissipation is

$$P_a = 12{,}500 \times 10.1 - \frac{170{,}000}{2} = 41{,}300 \text{ watts}$$

thus indicating that an anode dissipation of approximately 40,000 watts is desirable. The required average dissipation is less. Average anode current for 100 percent modulation is 10.1 amperes, but the peak emission required from the cathode is 34 amperes of anode current plus 6 amperes of grid current. A pure-tungsten cathode to last 10,000 hours and cathode power of approximately 7800 watts are indicated.

I_s (space-charge current) of at least 40 amperes is required at point B in Fig. 5·29, so the tube must have a perviance factor

$$K = \frac{40}{\left(E_g + \frac{E_b}{\mu}\right)^{3/2}} \tag{5·27}$$

$$= \frac{40}{(630 + \frac{2500}{37})^{3/2}} = 2.18 \times 10^{-3}$$

for the type WL-895, which has an amplification factor of 37.0.

For cylindrical structures:

$$K = \frac{14.7 \times 10^{-6}}{\left[\frac{1}{\mu}(r_a\beta_{ca}{}^2)^{2/3} + (r_g\beta_{cg}{}^2)^{2/3}\right]^{3/2}} \text{ per inch of anode length} \tag{5·28}$$

where r_g = grid radius in inches
r_a = anode radius in inches
$\beta_{ca}{}^2$ = a function of the ratio of anode radius to cathode radius
$\beta_{cg}{}^2$ = a function of the ratio of grid radius to cathode radius; see Fig. 3·6 for the values of this function.

For parallel-plane structures:

$$K = \frac{2.34 \times 10^{-6}}{\left(l_g{}^{4/3} + \frac{1}{\mu}l_a{}^{4/3}\right)^{3/2}} \text{ per square inch of anode area} \tag{5·29}$$

where l_g = cathode-to-grid distance in inches
l_a = cathode-to-anode distance in inches.

In radio-frequency class B service the same analysis applies to tube requirements. The grid is excited to point C with no modulation, and to point B for full modulation. Power output at point B must be four times that at point C for minimum distortion.

The maximum practical limit on anode dissipation is approximately 2500 watts per square inch of anode area for water cooling. This limit is determined by the water pressure necessary to force the required quantity of water through the water jacket. For a given cross section of water wall there is also a minimum advisable anode dissipation, if water flow is reduced as anode dissipation is reduced. The minimum corresponds to the point at which water flow ceases to be turbulent. This point can be found by applying the formula

$$Re = \frac{D\rho V}{\eta} \tag{5·30}$$

where D = equivalent diameter of water-channel cross section in centimeters (it is equal to $4A/p$, where A is the area of water channel and p is the perimeter of water channel)
η = viscosity of cooling fluid
ρ = density of cooling fluid in grams per centimeter cube
V = velocity in centimeters per second.

If $Re > 4000$, the flow is turbulent. For Re of 4000 or below, the flow may be turbulent if the surfaces over which the water flows are rough.

For air-cooled external-anode tubes with radiators the practical limit on anode dissipation is controlled by several factors:

1. Maximum practical air velocity in centimeters per second. (At higher velocities noise is objectionable.)
2. Softening point of the solder used for joining the radiator to the anode.
3. Maximum safe temperature of the glass at the anode-envelope seal.

There is a minimum safe air velocity, if air velocity is reduced as anode dissipation is reduced. The Reynolds number applies to air velocity as well as to water velocity.

Under modulation as shown in Fig. 5·29 the power dissipated in the grid is 579 watts, approximately.

Figure 5·8 shows the heat radiated by various materials used in tube making. The effective radiating area of the grid necessary to limit primary emission from the grid can be approximated from the power-emissivity curve, rated cathode power, calculated grid dissipation, and thermionic emission of the material used.

Peak anode current is predicted from the type of service and the power output required. For class B service the peak anode current is 3.141 times the average anode current at 100 percent modulation. It is assumed that the grid intercepts the percentage of electrons determined by the ratio of grid cross section to total grid area. A positive grid voltage is assumed in precalculation, so the energy dissipated in the grid by electron bombardment is known for the most severe operating condition. The actual condition is often less severe.

Class C Operation

Class C operation is synonymous with class C amplifier or oscillator service. The same basic analysis of operation applies to plate-modulated class C service. In class C service anode current passes through the tube during less than 180 electrical degrees; thus, for a given power-output, peak anode current increases as the angle of operation decreases. In Fig. 5·30 the constant-current characteristic curves of the WL-895 with a performance analysis of that tube are given for an assumed set of conditions.

A-B is termed the operating line, and it represents 180 electrical degrees. The remaining 180 degrees would be described if a circle were described with A as a center and *A-B* as a radius. Figure 5·31 shows how grid- and anode-voltage vectors describe line *A-B-C*.

In any a-c circuit,

$$P = EI \cos\theta \tag{5·31}$$

where E = rms volts
I = rms amperes
P = power in watts
$\cos\theta$ = power factor.

When E and I are peak values, and θ is assumed to equal zero,

$$P = \frac{EI}{2} \tag{5·32}$$

From Fig. 5·32 peak anode current is rated to effective-load current for any angle of operation which may be obtained

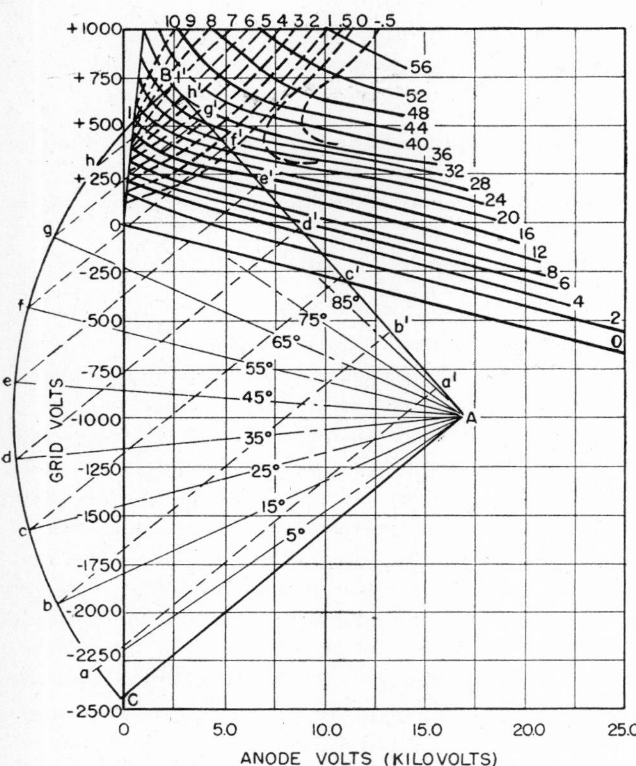

FIG. 5·30 Constant-current chart of the WL-895 with a class C amplifier or oscillator operating line shown. The operating line shown gives

Operating Degrees	Sin of Op. Degree		I_b at Angle Shown		Sin × Amperes	i_c at Angle Shown	Sin × Amperes
5	0.087	×	0	=	0		
15	0.258	×	0	=	0		
25	0.422	×	0.5	=	0.211		
35	0.573	×	5.0	=	2.865		
45	0.707	×	14.5	=	10.250		
55	0.819	×	24.5	=	20.065	0.75	0.614
65	0.906	×	33.0	=	29.898	3.50	3.170
75	0.966	×	34.5	=	33.327	5.80	5.600
85	0.996	×	35.0	=	34.860	6.80	6.772
			147.0		131.476	16.85	16.156

$$I_b = \frac{147}{18} = 8.167 \text{ amperes}$$

$$P_i = 17 \times 8.167 = 138.8 \text{ kilowatts}$$

$$P_o = \frac{E_o \Sigma i_b \sin \text{ amperes}}{18} = 105.8 \text{ kilowatts}$$

$$R_L = \frac{E_o^2}{2P_o} = 994 \text{ ohms}$$

$$I_c = \frac{16.85}{18} = 0.936 \text{ amperes}$$

$$P_g = \frac{E_g \Sigma i_g \sin \text{ amperes}}{18} = 1524 \text{ watts}$$

from equation 5·31, because the power desired and e_b may be assumed for the tube to be designed. e_b is 14.5 kilovolts for the conditions assumed in Fig. 5·30.

From Fig. 5·32 the ratio of peak tube current to effective-load current is 2.4 for an operating angle of 65 degrees (half the total operating angle), as shown in Fig. 5·30.

For the condition given,

$$I = \frac{105{,}800 \times 2}{14{,}500}$$

$$= 14.6 \text{ amperes effective}$$

Since the ratio is 2.4 the peak-tube current is 14.6 × 2.4 = 36.6 amperes, which coincides with point B in Fig. 5·30.

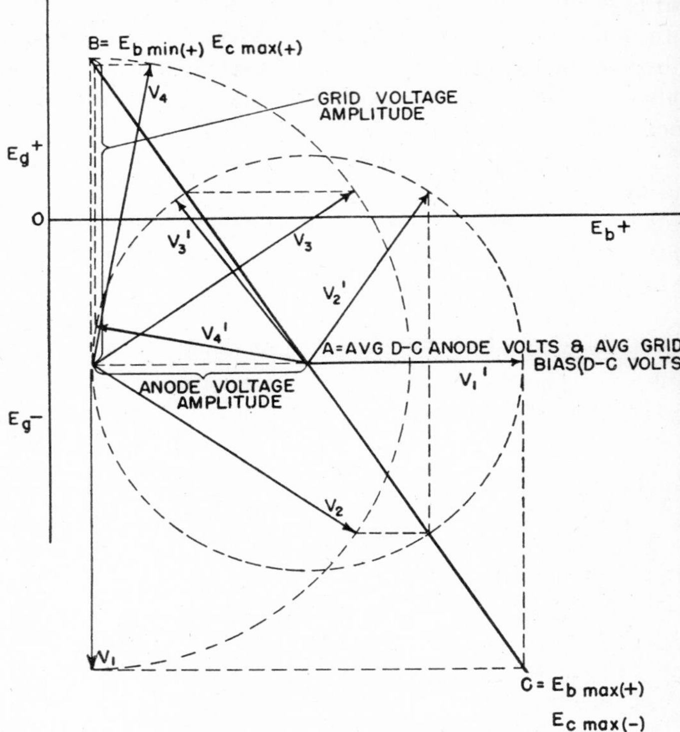

FIG. 5·31 Illustrating grid- and anode-voltage vector travel on a constant-current chart.

For this operating condition the cathode must provide minimum electron emission of 36 amperes of anode current plus 7 amperes of grid current throughout tube life (see

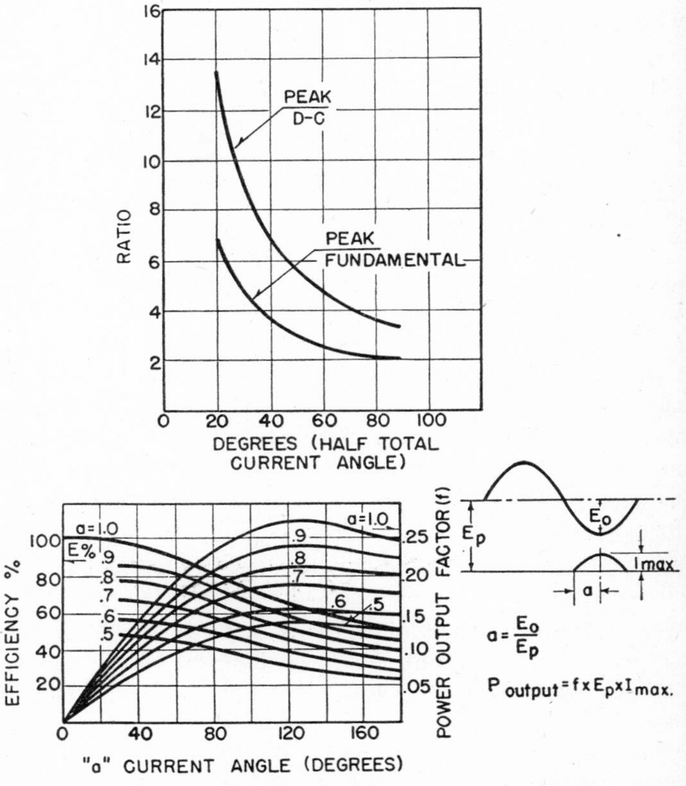

FIG. 5·32 Curves showing relation of peak-anode current to fundamental and average anode current for pliotrons operating in class C. The lower curves show the variation of efficiency and power output versus tube-anode-voltage operating angle.

Section 5·1). The anode must be capable of dissipating 138.8 − 105.8 or 33.0 kilowatts plus the cathode power plus grid losses of 1524 − 936 or 588 watts.

In addition, the grid and its insulation must be capable of withstanding the total grid-voltage swing from +700 volts to −2700 volts, and the anode must withstand a swing from +2500 volts to +31,500 volts.

Figure 5·30 shows that maximum negative grid voltage and maximum positive anode voltage occur simultaneously, so the total voltage between anode and grid can be 34,200 volts. With some types of operation even this value may be exceeded.

Plate-Modulated Class C Amplifier Service

A constant-current chart of the type WL-895 with two operating lines is shown in Fig. 5·33. Line *A-B* shows

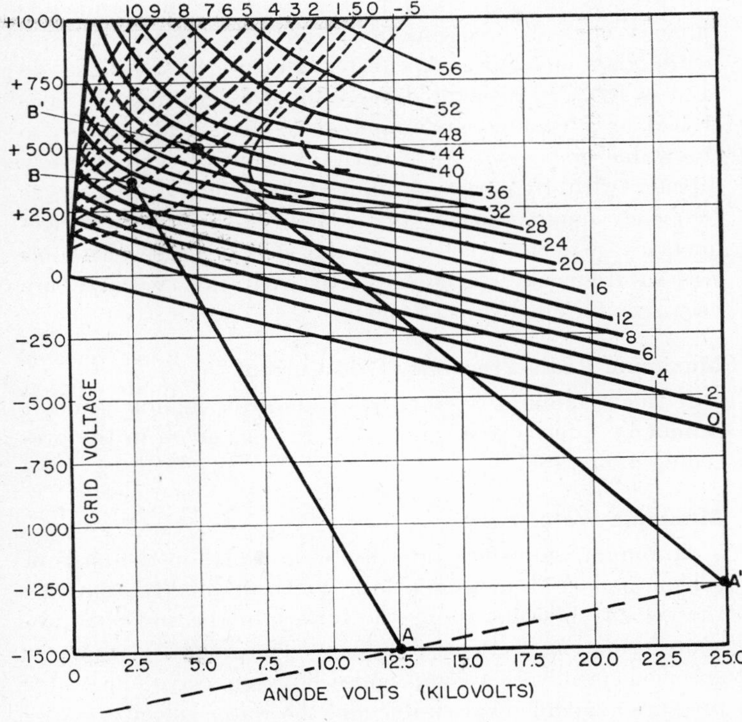

FIG. 5·33 Constant-current chart of a WL-895 with a typical operating line for 100 percent anode modulation.

normal conditions for carrier condition, that is, with no anode-modulating voltages applied. Line *A′B′* is for 100 percent anode modulation. For 100 percent modulation, applied anode voltage varies from zero to twice the anode voltage applied for the carrier condition.

For distortionless operation at maximum modulation the power output must be four times the power output under carrier conditions. The operating angle increases as modulation increases, so peak tube current required at maximum modulation is less than twice the peak anode current for carrier conditions. For design purposes a safety factor of two is often used.

For the conditions assumed, the anode voltage swings from +5000 volts to +45,000 volts at modulation peaks, but grid-voltage swings do not increase correspondingly.

The method of determining both peak and carrier power output is the same as for oscillator service.

5·7 TUBE-APPLICATION PARAMETERS

As a pliotron is sold, the tube designer specifies:

	Units
1. Amplification factor (μ)	. . .
2. Transconductance (g_m)	micromhos
3. Anode resistance (r_p)	ohms
4. Anode dissipation (maximum)	watts
5. Grid dissipation (seldom given)	watts
6. Maximum anode voltage	volts
7. Maximum negative grid voltage	volts
8. Maximum radio-frequency grid voltage	volts
9. Maximum radio-frequency grid current	amperes
10. Minimum water flow (if water-cooled)	gallons per minute
11. Minimum cubic feet of air per minute (if air-cooled)	
12. Frequency for maximum input (kilocycles or megacycles)	
13. Maximum input for classes *A*, *B*, *C*	
14. Miscellaneous items on tube operation	

Of these, items 1, 2, and 3 are beyond the control of the user; therefore items 4 through 14 have been termed tube-application parameters. The tube designer has taken into account all items before the tube-rating chart is published.

Anode Dissipation

On water-cooled tubes, anode dissipation is easy to determine. It can be measured by thermometers at both the inlet and outlet of the water jacket, plus a means of measuring the water flow.

$$\text{Anode dissipation} = (t_2 - t_1)g \times 0.264$$

where t_2 = outlet temperature in degrees centigrade
t_1 = inlet temperature in degrees centigrade
g = gallons per minute.

Anode dissipation varies with anode area, with amplification factor, and with type of service.

In forced-air-cooled types the dissipation can be measured by plotting a curve of air-temperature rise through the radiator against watts dissipated when the tube is operated statically; that is, with a negative grid bias to control the anode current to different values to give different points on a curve. If

t_1 = output air temperature with only the tube filament on
t_2 = output air temperature with W anode watts

then

$$\Delta t = (t_2 - t_1) = \text{temperature reading for each setting}$$

Δt gives the temperature for watts W.

If room temperature changes during calibration, then the change in degrees must be added to or subtracted from Δt for correct results. Plotting Δt versus W for several values of W gives a calibration curve for that radiator.

Once the calibration is made, the thermometer must be left in the same position. Temperature then can be observed with tube loaded, and anode dissipation can be determined from the calibration curve. Use of two or more thermometers (using average t_2) increases the accuracy of this measurement. Mercury thermometers must not be used for these measurements, for the radio-frequency field may destroy them.

For radiation-cooled tubes the same procedure can be followed. The thermometer (or thermometers) is placed at a given distance from the tube, and sufficient time is allowed for equilibrium temperature to be reached at each level of calibration. Tube and thermometer should be enclosed so that only normal air-convection currents pass by the tube and thermometers.

Grid Dissipation

Power that may be dissipated safely in the grid of a tube is not always published, because it is a measurement which cannot be made directly.

Ratings given by the tube manufacturer are based on calculations from the tube's characteristic curves and on performance measurements. If maximum radio-frequency grid amperes are specified, larger currents probably will cause overheating of the grid proper, connections to the grid, or both. Voltage in the grid circuit is

$$e_c = Z_c i_c \qquad (5\cdot33)$$

and Z_c is the impedance of the grid circuit at that frequency; therefore any increase of i_c increases e_c, which in turn may cause failure of the glass insulation between grid and anode or grid and filament.

Grid bias is the product of $R_c I_c$ (Fig. 5·30); therefore increasing either R_c or I_c increases the bias. With an increase of bias, the excursions of e_c about the bias axis must increase for proper tube operation, because e_c must make the same voltage excursion into the positive-grid region. Increasing either R_c or I_c so that the bias is beyond the maximum rating is undesirable.

If grid-excitation power is calculated from a constant-current chart modified so that all positive grid currents are assumed to be the same as they would be at that grid voltage on the diode line, then the power lost in the grid is approximately

$$P_{gh} = P_{gi} - P_c \qquad (5\cdot33a)$$

where P_{gh} = watts lost as heat in the grid itself
P_{gi} = total excitation watts
$P_c = RI_c^2$ watts lost in grid leak.

Maximum Anode Voltage

Maximum permissible anode voltage varies with the type of service for which the tube is to be used. In classes A, B, or C, it is usually about the same. In class C anode-modulated service the carrier condition is always under that of classes A, B, and C.

The limit on maximum plate voltage is the result of (1) flash arcs; (2) external glass-path length; (3) glass bombardment by electrons; and (4) practical limitations of power rectifiers, circuit components, required power, tube instability, and emission of x-rays.

Maximum ratings, as a rule, do not represent maximum capabilities. They more nearly represent the capabilities of the average tube of any given size and more nearly represent conditions under which average tubes will give long, satisfactory service.

Maximum Negative Grid Voltage

The peak radio-frequency voltage between grid and cathode is the sum of the average grid bias plus the maximum positive voltage excursion of the grid. The maximum negative excursion is this figure plus the negative grid bias. The insulation between grid and cathode, both internal and external, the amplification factor, and the shielding of the glass from stray electrons determines the maximum permissible negative grid voltage.

If grid bias is increased, the angle of current decreases, and for a given power output the demand on cathode emission increases. If this angle of flow is such that emission demand exceeds capabilities of the cathode, tube-voltage drop becomes excessive, thus increasing watts dissipated in the grid and anode.

Maximum Radio-Frequency Grid Voltage

The grid-voltage maximum is a function of permissible capacity currents (see equation 5·34) and the power losses in the glass envelope. A safe temperature for most glasses used is 180°C. Either the operating voltages must be controlled or a cooling medium used to limit temperature to this value or below. Although grid capacity currents vary directly with frequency, grid lead losses vary directly as frequency squared (neglecting skin-effect variation with frequency). Effective voltage at the grid decreases with increased frequency; thus increased excitation voltages are necessary for good tube efficiency.

Maximum Radio-Frequency Grid Current

If this maximum is supplied by the tube manufacturer it cannot be safely exceeded, for the reasons given in the preceding paragraph.

Minimum Water Flow

Minimum water-flow rates are established on the basis of calculations and measurements which indirectly establish the velocity of water along the surface of the anode to prevent formation of steam bubbles. The quantity of water specified usually is a compromise between maximum water pressure available or desirable and the water velocity. Adequate cooling can be obtained at higher water velocities and lower volume, but only at the expense of higher pressure.

Formation of mineral deposits sharply decreases the ability of the anode to dissipate energy. Water softeners help to prevent such deposits, and periodic removal of those that form is essential for trouble-free operation. Soft deposits can be removed safely with a stiff bristle brush, hard scale by the use of dilute hydrochloric acid.

Minimum Air Flow

The quantity of air required for forced-air-cooled tubes is specified for various types of service and often for various power levels. Because the tube radiator operates at high voltages, dust precipitation often partially blocks the air ducts.

It is desirable to inspect the radiator periodically to make sure air passages are not obstructed. An air vane operating a control relay is desirable to guard against tube failure in the event of blower failure or air-passage obstruction.

Frequency for Maximum Input

This limitation on the use of pliotrons is determined by (1) inductance of tube elements and leads; (2) electron transit time; and (3) capacitances between tube elements.

Item 2 is usually negligible up to approximately 50 megacycles, and it varies in importance with class of service, tube amplification factor, and the circuit used. Class B service and grounded-grid amplifier operation minimize the effect of transit time.

At radio frequencies interelectrode capacitances must be charged and discharged each cycle. Where

f = frequency in cycles per second
e_c = rms grid voltage
e_b = rms anode voltage
C_{gp} = grid-to-plate capacitance
C_{gk} = grid-to-cathode capacitance

for triodes the input charging current is

$$I_{\text{in}} = 2\pi f[e_c(C_{gp} + C_{gk}) + e_b C_{gp}] \tag{5·34}$$

and the output charging current is

$$I_{\text{out}} = 2\pi f C_{gp} e_b \tag{5·35}$$

for tetrodes:

$$I_{\text{in}} = 2\pi f(C_{gk} + C_{gs})e_c$$

$$I_{\text{out}} = 2\pi f(C_{ps})e_b$$

where C_{gs} = control-grid-to-screen-grid capacitance
C_{ps} = plate-to-screen-grid capacitance.

These currents must be carried by the element leads. If the inductance of the grid and anode leads is L_g and L_a, then the voltages along the grid and anode leads are

$$E_g = 2\pi f L_g I_{\text{in}} \tag{5·36}$$

$$E_b = 2\pi f L_a I_{\text{out}} \tag{5·37}$$

E_g and E_b must be in excess of the voltages required in the tube by the external radio-frequency circuits. By inspection, C_{in}, C_{out}, L_g, and L_a must be as low as possible for good high-frequency performance.

Maximum input ratings take into account:

1. The range of the ratio of maximum filament electron emission to average d-c anode current from approximately 3.14 for class B audio-frequency operation to approximately 10 in class C anode-modulated service.
2. Maximum anode dissipation.
3. External glass lengths.
4. Grid dissipation.
5. Maximum peak anode voltages.

Miscellaneous items for prolonging tube life:

1. Stay within manufacturer's ratings.
2. Operate pure-tungsten filaments as far under rated filament voltage as possible and yet maintain satisfactory operation.
3. Keep all contacts clean and tight.
4. Keep an ample supply of cooling medium flowing.
5. Observe manufacturer's recommendations regarding application and removal of voltages. For large tubes, the filament voltage should be applied and removed gradually or in steps to keep the instantaneous filament current (on starting) below 150 percent of maximum rating.
6. Do not subject the tube to sudden changes in temperature.

5·8 CONVENTIONAL TUBE DESIGN AS RELATED TO PERFORMANCE AT HIGH FREQUENCIES

Performance of a pliotron at high frequencies is controlled by:

1. Physical spacing of tube elements.
2. Physical lengths of active elements.
3. Inductance of tube elements.
4. Capacitances between elements.
5. Type of circuit in which the tube is used.
6. Class of service in which the tube is operated.

Performance of conventional tubes at high frequencies (if such performance is limited by electron transit time) is determined principally by the time required for electrons to move from the cathode to the grid plane.

In a triode this time is

$$t = \frac{5 \times 10^{-8}}{\left(E_g + \dfrac{E_b}{u}\right)^{1/2}} s \tag{5·38}$$

where s = cathode-to-grid distance in centimeters
E_g = control-grid voltage
E_b = anode voltage
μ = amplification factor.

For a tetrode equation 5·38 becomes

$$t = \frac{5 \times 10^{-8}}{\left(E_g + \dfrac{E_{sg}}{\mu_{cg}}\right)^{1/2}} s \tag{5·39}$$

where E_{sg} = screen-grid voltage
μ_{cg} = control-grid amplification factor.

Equation 5·39 neglects the effect of the anode voltage which, in tetrodes, usually is small.

By inspection of equations 5·38 and 5·39, s must be as small as possible for minimum electron-transit time between cathode and grid. At high frequencies tubes usually are used in circuits having distributed constants. If the tubes have active elements 1/16 wavelength or longer, the variations of radio-frequency voltage along the tube elements decrease over-all tube efficiency.

In the lighthouse tube the active tube elements are placed at the ends of distributed-constant circuits, and small spacings are used (usually 0.003 inch or less from cathode to control grid) to minimize the effects of limits 1 and 2. Disk electrode supports are used to minimize limit number 3.

Capacitances existing in the tube must be charged and discharged through the leads, so the inductances of such leads must be low to avoid loss of radio-frequency voltage at the tube element. Conversely, interelectrode capacitances must be as low as possible to minimize charging currents required for proper tube performance.

In a class A radio-frequency amplifier the control grid does not operate at a positive potential, therefore electron-transit time is greater than for either class B or class C service (see Fig. 5·17).

In a class B amplifier the grid goes positive earlier in the cycle than in any other service, and there is more time for electrons to be accelerated to the grid; consequently, tubes operated in this type of service operate better at high frequencies.

In class C amplifier service the grid is positive for less than 180 electrical degrees, so there is less time for electrons to be accelerated to the grid. Class C operation therefore is usually not so efficient as class B for high frequencies.

For either class B or class C service, operating the tube as a grounded-grid amplifier at high frequency may be desirable. Cathode-to-grid electron-transit time then is no longer of paramount importance, for anode voltage may lag grid voltage by the phase angle determined by cathode-to-grid transit time without appreciably affecting tube efficiency. This applies also to tetrodes, except that transit time between cathode and screen grid determines anode voltage lag. Figure 5·17 shows performance of several types of tubes for different classes of operation versus frequency.

5·9 THE KLYSTRON AND ITS RESONANT CAVITY [50, 51, 58, 60]

In Section 5·8 it was shown that electron-transit time limits the performance of conventional vacuum tubes at

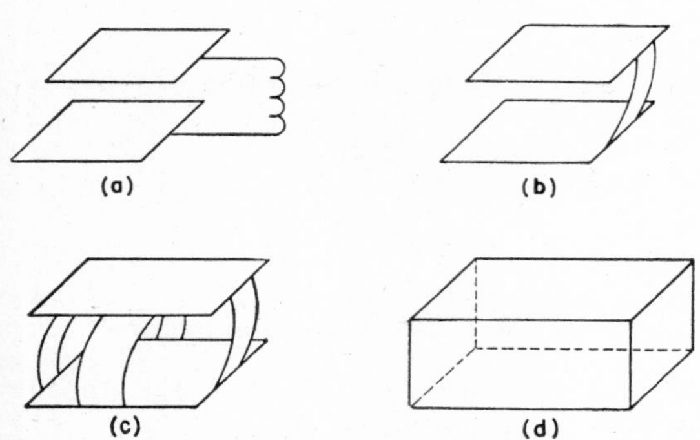

FIG. 5·34 Evolution of a resonant cavity from resonant circuit with lumped elements. (Reprinted with permission from *Fields and Waves in Modern Radio*, by Simon Ramo and John R. Whinnery, published by John Wiley, 1944.)

high frequencies. The rest of the chapter discusses tubes such as the klystron and the magnetron, in which this difficulty is overcome. Ordinary lumped circuit elements are not suitable as resonant circuits at extremely high frequencies, so these tubes use resonant cavities.

A conventional circuit with dimensions comparable to the wavelength at which it operates may lose too much energy by radiation; the resistance of wire circuits may become too high because of skin effect. These limitations immediately suggest that the circuit region be shielded and that current paths have as large an area as possible. The result is a hollow conducting box known as a cavity resonator with the electromagnetic energy confined inside. The inner walls of the box provide a comparatively large area for conducting current.

All mathematical analyses of cavity resonators are based on the solution of Maxwell's equations, subject to the boundary conditions that the tangential component of the electrical field and the normal component of the magnetic field vanish at the conducting walls, if as a first approximation the conductors are assumed to cause no losses. Because of the boundary conditions only certain discrete frequencies exist. These frequencies are determined by the size and shape of the enclosing surface.

A resonant cavity may be considered a logical evolution of lumped *L-C* circuits. A parallel resonant circuit is shown in Fig. 5·34 (*a*). If this circuit is used at high frequencies, C and L must be reduced. C can be reduced by moving the capacitor plates farther apart and decreasing their area. Inductance can be reduced to a single strap connected directly to the capacitor plates, as in Fig. 5·34 (*b*). Inductance can be decreased still more by placing four loops in parallel, as in Fig. 5·34 (*c*). Eventually this procedure produces the closed cavity shown in Fig. 5·34 (*d*).

Of course the extension of lumped-circuit theory to cavity resonators does not give a rigorous solution. The problem must be treated in terms of electromagnetic waves traveling in the cavity and being reflected from the walls. A standing-wave pattern is set up, with voltage nodes at the walls. The wavelength might be expected to have about the same dimensions as the cavity.

An extension of this reasoning and a more exact mathematical analysis gives

$$f = C\sqrt{\left(\frac{l}{2a}\right)^2 + \left(\frac{m}{2b}\right)^2 + \left(\frac{n}{2c}\right)^2} \tag{5·40}$$

and

$$\lambda = \frac{1}{\sqrt{\left(\frac{l}{2a}\right)^2 + \left(\frac{m}{2b}\right)^2 + \left(\frac{n}{2c}\right)^2}} \tag{5·41}$$

for a rectangular cavity having dimensions a, b, and c, where $l, m, n = 0, 1, 2, 3,$ and so on, but not more than one may equal zero. Also, l, m, and n are the number of half-wave variations of field along the x-, y-, and z-axes respectively. Each combination of l, m, and n produces resonance; hence an infinite number of oscillatory modes is possible. Two general classes of modes exist. Modes having components of electrical field in the direction of propagation but no components of magnetic field in the direction of propagation are known as E modes. Such modes do have components of magnetic field normal to the direction of propagation and, therefore, are known as transverse magnetic or TM modes. Modes which have components of magnetic field in the direction of propagation but no components of electric field in this direction are called H modes. These modes have components of electric field normal to the direction of propagation and are known as transverse electric or TE modes.

The longest resonant wavelength for a cube is that associated with the TE_{101} mode. This wavelength can be determined from equation 5·41 by putting $l = n = 1$, $m = 0$, and $a = b = c$. Then

$$\lambda = a\sqrt{2} \tag{5·42}$$

where a is the length of the cube. This shows that the longest wavelength is of about the same magnitude as the dimensions of the resonator.

Figure 5·35 shows the E lines in the xy plane for three transverse electric modes in a rectangular cavity. The

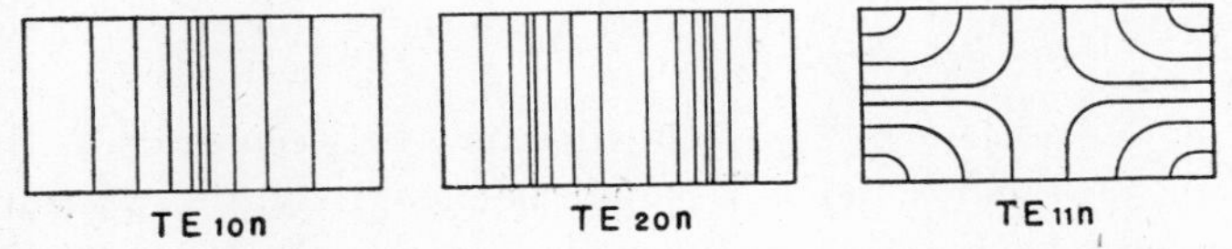

FIG. 5·35 Configurations of the electric field for transverse electric modes.

densities of the lines indicate the intensities of the fields. Figure 5·36 shows the H lines in the xy plane for two transverse magnetic modes.

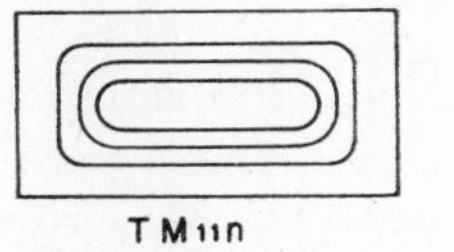

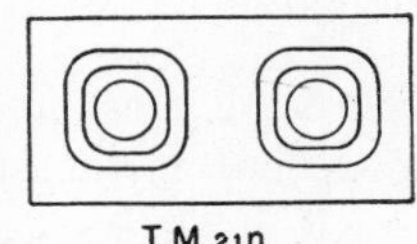

FIG. 5·36 Configurations of the magnetic field for transverse magnetic modes.

The Q or selectivity of any resonator can be defined as

$$Q = \pi \frac{\text{energy stored}}{\text{energy lost per half cycle}}$$

Because the walls of the cavity are not perfect conductors, the fields penetrate the inner surface slightly. This is known as skin effect. Consider that all the electromagnetic energy stored in the skin is lost as heat and is proportional to the volume penetrated, which for a cube is $6a^2\delta$, where a is the length of a side of the cube and δ is the skin depth. Assuming that the energy stored is proportional to the volume of space in the enclosure, which obviously is a^3,

$$Q = \frac{\pi}{6}\frac{a}{\delta}$$

Actually, for a cube

$$Q = \frac{a}{3\delta} \tag{5·43}$$

or, substituting for a from equation 5·42, Q of a cube for the mode of longest wavelength is

$$Q = \frac{\lambda}{3\sqrt{2}\delta} \tag{5·44}$$

Therefore, Q for a copper-cube resonator designed for $\lambda = 10$ centimeters is 18,800. This shows that cavity resonators have high values of Q.

In one type of resonator the region of maximum electric field appears across a short gap. This type of resonator can be used across two closely spaced grids of some high-frequency tubes.

One of the simplest resonators of this type is shown in Fig. 5·37. It consists of a short length of coaxial line short-circuited at one end and terminated at the other end by a short gap between two plates. If the gap is short compared to a wavelength it behaves like a lumped capacitance loading a length of transmission line. The equivalent circuit is shown in Fig. 5·38. For resonance the inductive reactance of the

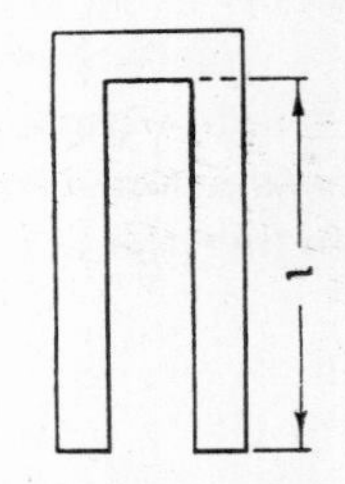

FIG. 5·37 Coaxial line resonator terminated by a short gap.

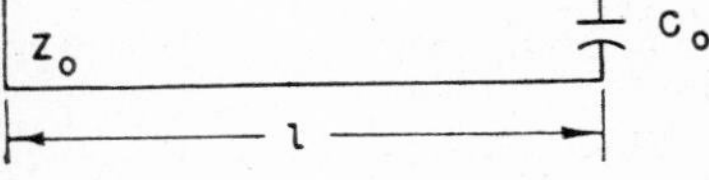

FIG. 5·38 Equivalent circuit for Fig. 5·37.

short-circuited length of transmission line must be equal in magnitude to the capacitive reactance of C_0. This gives

$$Z_0 = \tan \beta l = \frac{1}{\omega C_0} \tag{5·45}$$

where Z_0 is the characteristic impedance of the transmission line and $\beta = 2\pi/\lambda$. Equation 5·45 is only approximate. For a coaxial line made of non-magnetic conductors with air or vacuum as a dielectric,

$$Z_0 = 60 \log_\epsilon \left(\frac{r_0}{r_i}\right) \quad \text{ohms} \tag{5·46}$$

where r_0 and r_i are the radii of the outer and inner conductors, respectively.

5·10 GENERAL DESIGN OF KLYSTRON [53, 55]

The general design of a double-cavity klystron is shown in Fig. 5·39. It consists essentially of an electron gun, two

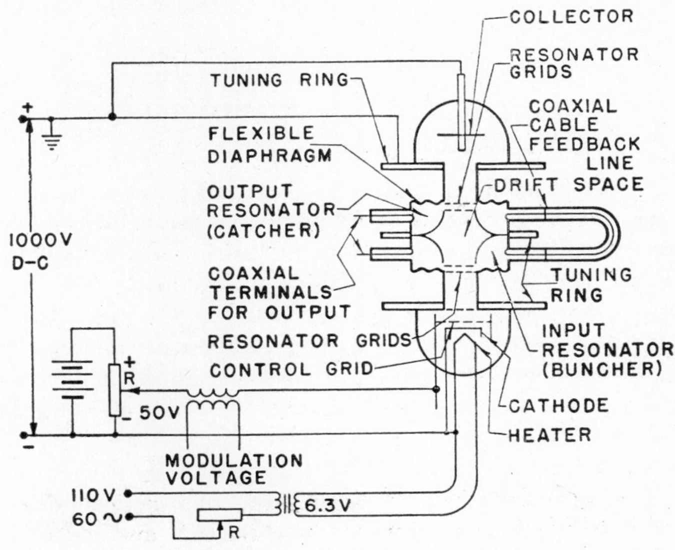

FIG. 5·39 Double-cavity klystron.

cavity resonators joined by a metal tube, and a collector electrode. The first cavity resonator is called the buncher, the second the catcher. The metal tube between resonators is the drift space. Resonators are tuned by varying their grid spacings by means of the flexible diaphragms. The two cavity resonators and the drift space are all at the same unidirectional potential, which usually is at ground. The cathode is at negative potential, and the collector usually is slightly positive.

Coupling to the resonators is through small wire loops connected to the coaxial terminals. The loops are placed so that a portion of the magnetic field surrounding the axis of the resonators passes through them.

5·11 VELOCITY MODULATION [53, 55]

To illustrate velocity modulation, Fig. 5·39 is stripped to its essentials as shown in Fig. 5·40. Electrons from the cathode enter the buncher with a velocity v_0 produced by the unidirectional voltage E_0. Assume an alternating voltage $E_1 \sin \omega t_1$ across the buncher grids. Assume also an alternating voltage $E_2 \cos (\omega t_2 - \alpha)$ across the catcher grids.

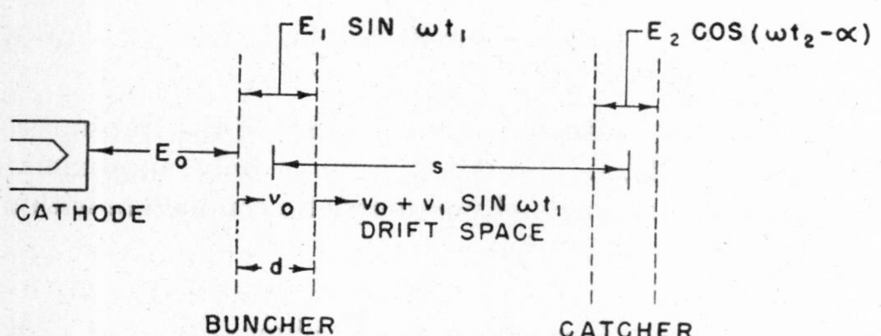

Fig. 5·40 Schematic diagram of klystron. (Reprinted with permission from David L. Webster, "Cathode Ray Bunching," *J. App. Phys.*, Vol. 10, p. 510.)

An electron passing the center of the buncher at the moment the electric field changes from opposing to aiding the forward motion of the electrons leaves the buncher with its speed unchanged. Another electron passing the center of the buncher a little earlier is decelerated, and an electron passing a little later is accelerated. This is called velocity modulation or variation.

5·12 CONDITIONS FOR MAXIMUM POWER CONVERSION [55, 56]

Velocity modulation is useless by itself for no output has been obtained. To get an output signal the velocity-modulated beam must be converted into an intensity-modulated beam. One way of doing this is by means of the drift space. While the electrons are making the relatively long flight in the field-free space of length s between buncher and catcher, the differences in speed cause electrons ahead and behind the one of unchanged speed to draw nearer to it. In other words, the electrons are bunched.

These bunches pass any fixed point with a frequency equal to that of the oscillations in the buncher so they can deliver power to the catcher as they pass through it and cause it to oscillate, provided the losses in the catcher are not too great. The wave form of the current in the bunches is not sinusoidal; therefore the catcher can oscillate at the same frequency as the buncher or at some multiple of it, depending upon the resonant frequency of the catcher. So that the bunches of electrons can deliver power to the catcher they must enter it as the electric field in it is in that phase which retards the motion of the electrons.

To deliver maximum power to the catcher, the following conditions must be satisfied:

1. The catcher must be placed where the electrons are in bunches of the best form; that is, the distance s in Fig. 5·40 must be properly chosen.
2. The catcher must oscillate in the correct phase.
3. Strength of the field in the catcher must be such as to reduce the velocity of the electrons in the center of the bunch to zero.
4. Distance d in Fig. 5·40 (distance between grids at the ends of the resonators) should be less than the distance traveled by an electron in one-half cycle. In other words, the proper relation is $d < v/2f = \lambda v/2c$, where v = velocity of the electron, f = frequency, λ = wavelength of the oscillations, and c = velocity of light.

5·13 MATHEMATICAL TREATMENT OF BUNCHING [52, 56]

The following treatment is only an approximation, for it is assumed that no debunching is caused by space charge, and that all electrons leave the cathode with zero velocity and move parallel to the axis of the structure. The grids of the buncher are assumed to define true equipotential planes, and no electrons are assumed to be lost from the beam by collision with grids or cavity walls.

In Fig. 5·40 consider an electron passing the center of the buncher at time t_1 and arriving at the center of the catcher at some later time t_2. The velocity of this electron after it leaves the buncher and enters the drift space is

$$v = v_0 + v_1 \sin \omega t_1 \tag{5·47}$$

Therefore

$$t_2 = t_1 + \frac{s}{v_0 + v_1 \sin \omega t_1} \tag{5·48}$$

or

$$t_2 = t_1 + \frac{s}{v_0}\left(1 + \frac{v_1}{v_0} \sin \omega t_1\right)^{-1} \tag{5·49}$$

or approximately

$$t_2 = t_1 + \frac{s}{v_0} - \frac{s v_1}{{v_0}^2} \sin \omega t_1 \tag{5·50}$$

since v_1 is small. Differentiating equation 5·50 with respect to t_1 gives

$$\frac{dt_2}{dt_1} = 1 - \frac{s v_1 \omega}{{v_0}^2} \cos \omega t_1 \tag{5·51}$$

Let i_0, i_1, and i_2 be the currents leaving the cathode, at the middle of the buncher, and the middle of the catcher, respectively. Now i_1 is virtually equal to i_0. Also the law of conservation of charge requires that

$$i_1(t_1)\, dt_1 = i_2(t_2)\, dt_2 \tag{5·52}$$

Hence

$$i_2(t_2) = \frac{i_1(t_1)}{\dfrac{dt_2}{dt_1}} = \frac{i_0}{\dfrac{dt_2}{dt_1}} \tag{5·53}$$

Equations 5·51 and 5·53 give

$$i_2(t_2) = \frac{i_0}{1 - \dfrac{sv_1\omega}{v_0^2}\cos\omega t_1} \tag{5·54}$$

which becomes

$$i_2(t_2) = i_0\left(1 + \frac{sv_1\omega}{v_0^2}\cos\omega t_1\right) \tag{5·55}$$

if $sv_1\omega/v_0^2 \ll 1$. The right-hand side of equation 5·55 is a sinusoidal function.

If $sv_1\omega/v_0^2 = 1$, from equation 5·54, current i_2 in the catcher is infinite once per cycle. When $sv_1\omega/v_0^2 > 1$, equation 5·54 appears to give a negative value for the current i_2 over part of each cycle. However, a negative current could not exist. Equation 5·51 explains the discrepancy, for dt_2/dt_1 also can be negative over part of each cycle for $sv_1\omega/v_0^2 > 1$. This merely means that fast electrons overtake slower ones which preceded them. Therefore, equation 5·54 can be used to calculate the magnitude of the current, which is infinite twice per cycle if $sv_1\omega/v_0^2 > 1$.

5·14 PHASE SHIFT IN THE KLYSTRON [52, 53, 57]

One type of phase shift is caused by time delay. A signal applied to the buncher affects the catcher only when the electrons controlled by the signal at the buncher reach the catcher. This time delay usually is long compared to a cycle. In calculating the time required for the electron to travel the distance s from buncher to catcher, the alternating voltage across the buncher grids can be ignored, and only the unidirectional voltage E_0 in Fig. 5·40 need be considered. Velocity of electrons in the drift space is

$$v_0 = 6 \times 10^7\sqrt{E_0} \text{ centimeter per second} \tag{5·56}$$

where E_0 is in practical volts.

The time required to cross the drift space is

$$T = \frac{s}{v_0} = \frac{s \times 10^{-7}}{6\sqrt{E_0}} \text{ second} \tag{5·57}$$

This can be converted to a phase angle θ by the equation

$$\theta = \omega T \tag{5·58}$$

or

$$\theta = \frac{\omega s \times 10^{-7}}{6\sqrt{E_0}} \text{ radian} \tag{5·59}$$

For a typical klystron, $s = 3$ centimeters and $f = 3 \times 10^9$ cycles per second. Therefore, if $E_0 = 900$ volts, T becomes 1.67×10^{-9} second, and θ becomes 10π radians or 5 cycles.

A bunch or current peak is formed around the electron which passed through the buncher at the time of zero voltage in the buncher. However, the current peak should pass through the catcher at the instant the voltage in the catcher is maximum; therefore, there should be a phase difference of 90 degrees between the buncher and catcher in addition to that caused by transit time.

5·15 THE KLYSTRON AS AN AMPLIFIER AND FREQUENCY MULTIPLIER [52, 53, 55]

The klystron serves as a power amplifier if power is fed into the buncher through its coaxial terminal. Amplified power can be utilized through the coaxial terminal connected to the catcher. Maximum theoretical electronic conversion efficiency of the klystron is 58 percent. Actually this is reduced, usually to less than 5 percent, because of electrons which leave the beam and strike the grids or resonator walls, debunching caused by space charge, and radio-frequency losses in the walls of the resonator.

If the degree of velocity modulation is increased by increasing buncher voltage or drift space, the intensity modulation in the catcher becomes less sinusoidal, and this deviation makes it possible for the klystron to serve as a frequency multiplier. With the klystron so used, the catcher is tuned to some harmonic of the buncher frequency.

5·16 THE KLYSTRON AS AN OSCILLATOR [52, 53, 55, 57, 59]

The klystron serves as an oscillator if power is fed back from the catcher to the buncher through a length of coaxial cable connected between the coaxial terminals coupled to the buncher and catcher.

In any oscillator the total phase shift around the loop must be $2\pi n$, where n may be any integer including zero. In most oscillators, phase shift is independent of applied voltage, but in the klystron the phase shift does depend upon the applied voltage. Hence only certain ranges of applied voltages produce oscillations.

The phase shift in the klystron is given by

$$2\pi n = \theta \pm \frac{\pi}{2} \tag{5·60}$$

where θ is defined in equation 5·59 and $\pi/2$ is the phase difference in radians between buncher and catcher voltage which may be either plus or minus, since buncher voltage may lead or lag catcher voltage. The actual value of this phase difference varies with the length of the feedback cable, the coupling loops, the resonators, the loading, and the frequency.

Equation 5·60 may be written

$$\frac{\theta}{2\pi} = n \pm \frac{1}{4} \tag{5·61}$$

From equation 5·58,

$$\frac{\theta}{2\pi} = fT = N \tag{5·62}$$

where N is the number of cycles during the transit from buncher to catcher. Equations 5·61 and 5·62 give

$$N = n \pm \tfrac{1}{4} \tag{5·63}$$

Equation 5·61 may be written

$$\frac{fs \times 10^{-7}}{6\sqrt{E_0}} = n \pm \frac{1}{4} \tag{5·64}$$

This shows that oscillation occurs only for those values of E_0 determined by equation 5·64 for integral values of n. Actually oscillation occurs over a small range of voltage on each side of E_0. Over this range the frequency varies according to equation 5·64; hence, frequency can be varied by changing E_0. In other words, electrical tuning is possible.

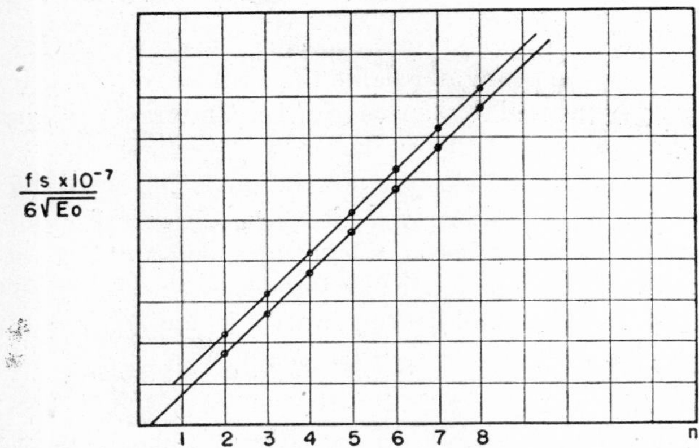

FIG. 5·41 Modes of oscillation of klystron. (Reprinted with permission from *Hyper and Ultra High Frequency Engineering*, by Robert I. Sarbacher and W. A. Edson, published by John Wiley, 1943.)

If the allowed values of the left-hand side of equation 5·64 are plotted against n the result is a straight line, as shown in Fig. 5·41. Two lines may be produced, because oscillation may occur at two different frequencies for any value of n.

Equation 5·64 indicates that the larger voltages are associated with the smaller values of n. Usually it is impossible to make the voltage large enough to operate at $n = 1$.

5·17 REFLEX KLYSTRON [52, 53, 59]

The reflex klystron shown in Fig. 5·42 has only one resonant cavity, which serves as both buncher and catcher. The electron beam leaving the cathode is velocity-modulated on

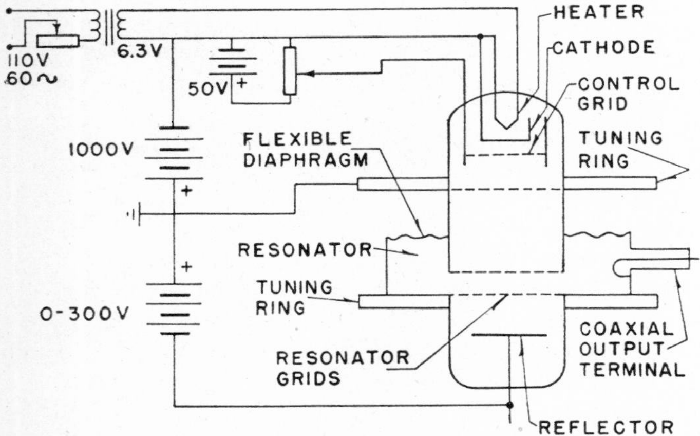

FIG. 5·42 Reflex klystron.

passing through the cavity. After leaving the resonant cavity the electrons move toward the reflector but are eventually turned back to the cavity resonator as the reflector is held at a negative potential. The path from the resonator toward the reflector and back again constitutes the drift space in which the electron beam becomes intensity-modulated. If voltages are properly adjusted, the bunched electrons return to the cavity resonator in the correct phase to deliver energy to the resonator, and the klystron oscillates.

As low-power local oscillators in superheterodyne receivers, reflex klystrons have a decided advantage over double-cavity klystrons, because varying the frequency of two high-Q resonators at exactly the same rate is difficult.

If reflector voltage in a reflex klystron is fixed, a series of accelerator voltages will produce oscillations as in a double-resonator klystron. Also several modes occur if accelerator voltage is fixed and reflector voltage is varied. Varying reflector voltage corresponds to changing the length of drift space. A change in frequency caused by a change in accelerator voltage can be compensated by a change in reflector voltage.

In a double-cavity klystron the slower electrons require a longer time to reach the catcher; therefore, electrons that pass through the buncher as the field is changing from deceleration to acceleration form the center of the bunch. In the reflex klystron the faster electrons travel farther and require a longer time to return to the resonator. Hence, the bunch is formed around the electrons that pass through the bunching stage as the field changes from acceleration to deceleration.

The electron bunch in the reflex klystron must pass through the cavity on returning from the reflector; at this time the electric field will decelerate the returning electrons or accelerate the electron beam going in the forward direction. Now the electron that forms the center of the bunch leaves the cavity as the electric field changes from acceleration to deceleration. Hence the equivalent of equation 5·63 for the reflex klystron is

$$N = n - \tfrac{1}{4} \tag{5·65}$$

In contrast to the double-cavity klystron, for any given value of n, oscillations occur at only one voltage.

5·18 SECONDARY-EMISSION REFLEX KLYSTRON

The reflex klystron oscillator in the 3000-megacycle region has an efficiency of about 1 percent and gives a power output of a few hundred milliwatts. Double-cavity klystrons are more efficient and give more power output, but are difficult to tune. Therefore, the reflex klystron is suitable only as a local oscillator, and the more efficient double-cavity tube has the disadvantage of being difficult to tune.

The reflex klystron is inefficient because of radio-frequency losses in the walls of the resonator, debunching of the electron beam, and interception of the beam by resonator walls and grids during both the forward and return path. Debunching is increased by an increase in beam-current density and drift distance. Since an electron beam of high current density is more divergent, an increase in current density also causes more interception. Therefore, it is difficult to increase power output of a reflex klystron beyond a certain point by simply increasing electron beam current. Furthermore, electronic emission of the thermionic cathode is limited.

Some reflex klystrons have reflector electrodes which contain good secondary emitters. In operating these tubes a positive voltage instead of the usual negative voltage is applied to the reflector. When a bunched electronic current strikes the secondary emitter a much greater bunched electronic current is emitted, returns to the resonator, and delivers a correspondingly greater amount of power to it. The secondary emitter in effect becomes a cathode which delivers large bunched currents to the resonator. Inefficiencies caused by debunching and divergence of the electronic beam from the secondary emitter are relatively low because the drift distance from the secondary emitter to the resonator is short. Some tubes of this type have efficiencies as high as 5 percent and power outputs up to 20 watts.

5·19 THE MAGNETRON OSCILLATOR

The multisegment magnetron oscillator [61] is discussed qualitatively in the light of modern knowledge. The present discussion deals exclusively with the traveling wave or rotating a-c field type of operation in multisegment magnetrons to which the classical literature refers as the "transit-time oscillation of higher (than the first) order" or "type B." The reader may consult the classical literature (collected by Gross [62] and Harvey [63]) for the remaining recognized types of oscillations, of which at least two are still acknowledged in split-anode or multisegment magnetrons. These remaining types are characterized, for instance, by the relative magnitude of the electron-transit time to the period of oscillations. The type in which the transit time is relatively negligible is labeled quasi-stationary, Habann, negative resistance, or dynatron type. The other type in which transit time is of paramount importance is labeled transit-time oscillation of the first order, type A, or electronic type. Although the overwhelming superiority of the split-anode or multisegment magnetrons compared with the non-split variety of Hull is such that negligible application is afforded the latter, nevertheless a substantial amount of theoretical investigation was done during the War on the non-split type by Brillouin in an effort to explore rigorously characteristics common to the multisegment magnetron.

Although valuable experimental data exist in the classical literature, the theoretical portion thereof is often inadequate or invalid, for the simple reason that the associated simplifying assumptions seriously limited the utility of the theoretical structure which was postulated to account for the mechanism of oscillation in magnetrons. Some obvious defects in the classical theories are the neglect of space charge and inadequate electron-trajectory determinations, although some valid theoretical concepts may be deduced without the inclusion of space charge, for example the Hull cut-off criteria.

Structurally, cylindrical magnetrons [61] are subdivided into single-anode and multisegment-anode types, each with a coaxial filament or a coaxial, unipotential, thermal or cold cathode, and an associated resonator system appropriate to the wavelength range. The region between the coaxial surfaces of the inner member (the cathode) and the outer member (the anode) is called the interaction space. The interaction space of the magnetron is permeated by a longitudinal, homogeneous, static magnetic field; a radial static electric field between cathode and anode; and oscillating electromagnetic fields. The resonator system, associated with the segments, is comprised of either lumped circuits,[62] lines,[62] or cavities,[64] depending upon the wavelength. For physical and electrical reasons the multicavity [61] magnetron is especially suitable for centimeter and lower decimeter wavelengths. The classical multisegment or split-anode [61] magnetron with associated resonant lines or lumped resonant circuits is more appropriate for the upper decimeter and meter wavelengths. An example of a multicavity magnetron appropriate for the centimeter and lower decimeter wavelength region is illustrated by Fig. 5·43. This example illustrates a development of the classical split-anode magnetron utilizing cavities in which mutual coupling between

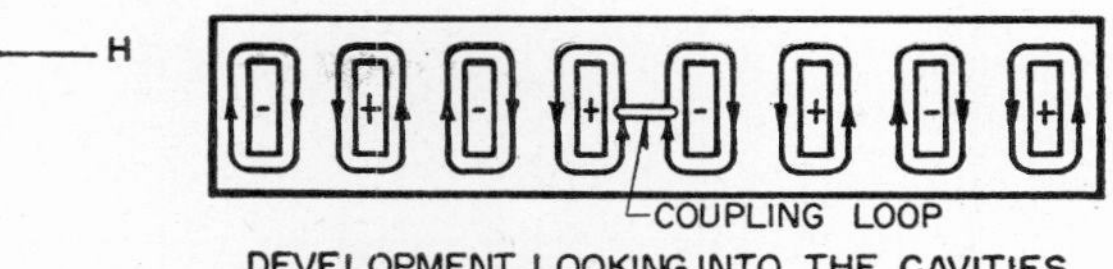

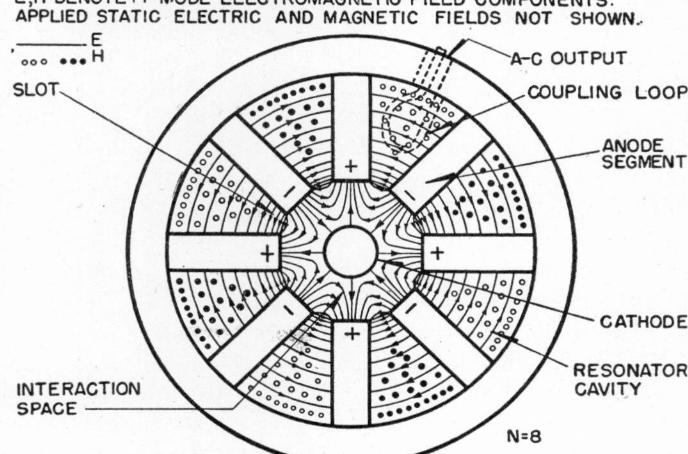

Fig. 5·43 Multicavity magnetron with pi-mode electromagnetic field distribution.

them is provided by the flux in the interaction space and the end chambers (the two termination regions) usually, though not always, provided at the ends of the cavities. By coupling the load to any one of the mutually coupled cavities a portion of the generated a-c energy can be extracted by means of a coupling loop as shown, or a resonant window can be used. By the latter arrangement efficient and convenient coupling can be made to a wave guide for very high power applications. The orientation and position of the coupling loop or resonant window may be readily derived by application of simple electromagnetic theory.

Usually the resonant system of the multicavity magnetron is comprised of either identical resonators or alternately identical resonators. The former is a symmetrical system such as illustrated by Fig. 5·43, whereas the latter is an asymmetrical system generally referred to as the "Rising Sun" anode on which only a relatively brief comment will be made. The asymmetrical system can be visualized with the aid of Fig. 5·43 by reducing the radial depth of alternate cavities by a constant factor so that two sets of symmetrical resonators constitute the anode. For efficient operation symmetrical magnetrons generally utilize anode straps,[61] whereas the asymmetrical magnetron inherently achieves the desired result without them. Of course, if the number of resonators or segments is sufficiently small, for example in the two or even four split-anode magnetron,[62] then efficient operation is achieved without them. One purpose of the asymmetrical system and the anode straps of the symmetrical system is to achieve adequate mode separation between the desired mode and its neighbors. An alternative to strapping of symmetrical magnetrons is to use certain higher modes; the results, however, are not favorable, especially with large number of resonators. Besides adequate mode separation it is essential to have efficient coupling between the electrons in the interaction space and the oscillating electromagnetic field.

The allowed modes of oscillation of the symmetrical cylindrical multicavity magnetron are grouped on a wavelength scale similar to band spectra. The group possessing the longest band wavelength is the one in general application for reasons shown later. This group is known as the fundamental multiplet and is characterized by variations of the oscillating electromagnetic field in the plane normal to the axis of the magnetron, for which edge effects are neglected. The modes are characterized by phase differences between alternating potentials on successive segments. Other groups have shorter band wavelengths, and their oscillating electromagnetic fields vary longitudinally as well as in the plane normal to the axis of the magnetron. Each component of the fundamental multiplet has a different wavelength, although the differences are relatively small and depend upon the effective inductance and capacity of the resonator and the number of coupled resonators. Relative intensities of the components in the fundamental multiplet are governed mainly by the coupling between the resonators and the electronic system via the interaction space. Other characteristics are the phase differences between segments and the conditions for sustenance of oscillations in the particular mode. All except one of the resonant modes in the fundamental multiplet, for which the number of component resonators, N, of the magnetron is even, are doubly degenerate when the resonant system is symmetrical and free of perturbation, whereas all the resonant modes in the fundamental group, for which N is odd, are doubly degenerate when the resonant system is symmetrical and free of perturbation. The term *doubly degenerate* mode refers to the number of linearly independent modes associated with the degenerate frequency, which in turn is defined as having more than one mode of oscillation but only one frequency. Such a mode is characterized by two identical opposite traveling waves around the anode. For a symmetrical multicavity magnetron with an even number of resonators there are, in the fundamental multiplet, $N - 1$ resonant modes though only $n = 1, 2, 3, \cdots N/2$ separate wavelengths, whereas for a symmetrical magnetron with odd N there are, in the fundamental multiplet, $N - 1$ resonant modes though only $n = 1, 2, \cdots (N-1)/2$ separate wavelengths. The non-degenerate mode is referred to as the $n = N/2$ or pi-mode.

Besides this non-degenerate mode of the symmetrical magnetron there is another which can be associated with the fundamental multiplet. This mode is referred to as the $n = 0$ or zero mode. The corresponding electromagnetic fields in the interaction space associated with these two modes is of the stationary type, though the zero-mode field is independent of the angular co-ordinate, whereas the pi-mode field is periodic in this respect. Although the zero mode has the desirable property of adequate mode separation from the adjacent modes, it possesses no obvious practical utility for the space-charge field does not interact favorably with it. Furthermore, the zero-mode field in the interaction space does not decrease from the anode toward the cathode as fast as the other modes of the fundamental multiplet. As a consequence, the desirable pi-mode field, which interacts satisfactorily with the space-charge field, may be interfered with by the zero mode under certain conditions so as to decrease the pi-mode conversion efficiency. Fortunately, the degree of zero-mode interference is usually nil in symmetrical magnetrons, especially if the cathode is sufficiently large with respect to the anode. The influence of the cathode diameter enters in another fashion, for as the number of segments increases the size of the cathode should increase also for efficient operation. The zero-mode interference in asymmetrical magnetrons is more serious than in symmetrical magnetrons, especially if the resonator asymmetry is large, or if the cathode is small with respect to the anode, or if both of these conditions exist. The degree of asymmetry denotes the degree of dissimilarity between any two adjacent resonators of the asymmetrical magnetron, for example, the ratio of the radial depth of two adjacent parallel-plane slot-type cavities. The surmounting of these difficulties has resulted in noteworthy performance of asymmetrical magnetrons, especially for relatively large numbers of cavities or resonators for which an unstrapped symmetrical magnetron would possess poor efficiency because of the inadequate mode separation between the desired pi-mode and its neighbors. In applications for which anode straps [61] are undesirable and a relatively large power output is desired, the asymmetrical magnetron is particularly adapted since efficient single-mode operation with a much longer anode and greater number of segments, that is, a large cathode, is possible than with a strapped symmetrical magnetron. It may be of interest to state briefly that the asymmetrical magnetron may be considered a composite of two sets of symmetrical resonators whose combined mode spectra comprise two groups of modes well separated from the pi-mode. The degree of isolation of the pi-mode from its neighboring modes is not quite so large as that obtainable with heavily strapped symmetrical magnetrons. However, this is not a significant disadvantage in view of the attractive properties pointed out previously, except perhaps for some very wide band tuning applications. The two groups of modes corresponding to the large and small resonator systems combine in and out of phase by pi-radians to form the complete spectrum of the asymmetrical magnetron. For example, consider an asymmetrical magnetron having a total number of resonators N, which is even; then the modes of the complete system as before are denoted by $n = 0, 1, 2, \cdots N/2$ or pi-mode. These are established by the in and out of phase relations between the corresponding components of the two groups of modes, $n_1[= 0, 1, 2, \cdots (N-2)/4]$ and $n_2[= 0, 1, 2, \cdots (N-2)/4$ modes. That is, the zero and pi-modes of the complete system are formed by the in phase and out of phase (by pi-radians) combination of the corresponding components, n_1, n_2, respectively. For the pi-mode all the identical large cavities are mutually in phase though out of phase with the identical small cavities. The mode separation properties of the asymmetrical magnetron may be realized from a consideration of a two-coupled cavity system which possesses two mode frequencies corresponding to resonators in and out of phase by pi-radians.

The higher mode numbers in the fundamental multiplet of the symmetrical multicavity magnetron are ideally found at shorter wavelengths. The $n = N/2$ or pi-mode of the fundamental multiplet is characterized by non-degenerate properties and highest conversion efficiency. It possesses a pi-phase difference of the alternating potential wave between adjacent segments, and the amplitude of the potential oscillations on all segments is the same for negligible zero-mode contamination, which is usual for a symmetric-resonator magnetron. A concept of n more general than the phase difference between adjacent segments is that it represents the number of times the rotating electromagnetic field or wave repeats around the anode. Perturbation or lack of symmetry in a symmetric-magnetron promotes non-degeneracy in the normal degenerate components of the $N - 1$ modes; that is, each has discrete wavelengths. A marked perturbation (different from that for variations in the radial extent of the interaction space) is observed in the mode distribution of the fundamental group, for practical end cavities and anode lengths. The order of the modes in the fundamental multiplet may be reversed or may even possess approximately a common wavelength if either the relative (with respect to wavelength) end cavity length or, for a sufficiently large end cavity, the relative anode length is inadequate.

5·20 STATIC CHARACTERISTICS OF THE MAGNETRON

Theoretically, for the static case of a magnetron with axial magnetic field H, there should be no anode current I for an anode voltage V less than the Hull [67] cut-off value V_c. Actually, at cut-off the anode current abruptly rises to a definite value for a given magnetic field, as illustrated by Fig. 5·44. The values of these anode currents at cut-off can be defined by an envelope, which is less than that prescribed by Langmuir [68] for space-charge-limited emission and no magnetic field. This envelope of currents at cut-off, known as the Allis curve, has been derived for the non-oscillatory case with space charge. As anode voltage increases above cut-off, anode current increases along the dashed line, the character of which is not yet well defined. Anode current for a given magnetic field continues to increase above cut-off with increase in anode voltage along the dashed line until at a sufficiently high voltage it blends with the Langmuir curve.

According to Langmuir's relation, which applies for no magnetic field, a definite anode voltage establishes a definite anode current. The rate of change of anode voltage, as a function of increasing magnetic field, is slow initially, then more rapid until the Allis curve is reached, beyond which the rate of change of anode voltage with magnetic field follows Hull's cut-off parabola independent of anode current, as illustrated by Fig. 5·45. The current contours do not quite merge with the cut-off parabola, if the ratio of the anode-to-cathode radius is less than a critical value ($\simeq 2$); instead they approach the parabola asymptotically at high fields.

In the static case ($V < V_c$) a cylindrical distribution of space charge immediately surrounds the cathode, which is referred to as the cathode sheath, under space-charge-limited conditions. In this thin region the magnetic field has a negligible influence on space charge, and the Langmuir space-charge relation is satisfied.

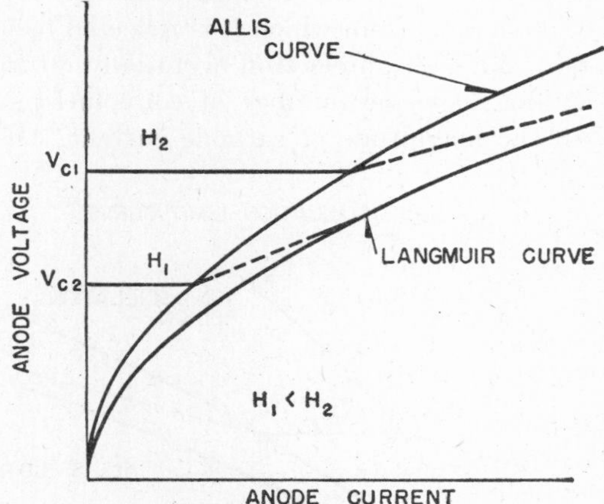

Fig. 5·44 Non-oscillation *V-I* curves.

From the Hamiltonian function for the electron in the combined fields in the interaction space of the magnetron, the angular momentum of the electron contributes to the electron's total potential energy. This energy varies radially in a periodic manner as the electron moves in cycloid-like convolutions from the cathode to and from the Hull cut-off boundary. Total potential energy is maximum wherever the radial component of the electron's velocity approaches zero. As a result of such radial retardations of the electron space-charge distribution is affected. Its density becomes maximum at the total-potential-energy maxima, which may be regarded as approximately equally spaced, thick, concentric sheaths of space charge surrounding the cathode. Their number depends upon cathode emission.

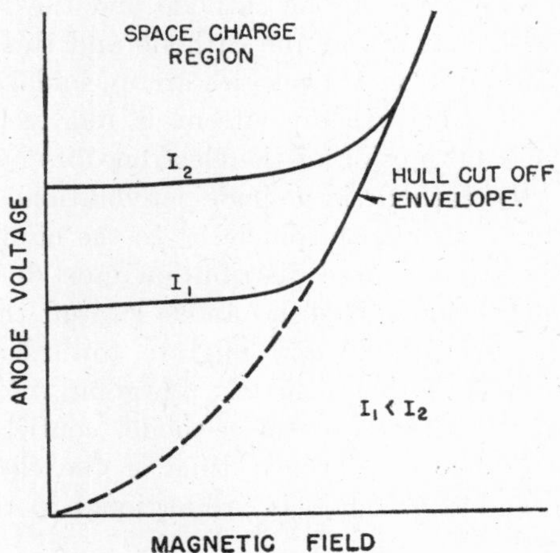

Fig. 5·45 Non-oscillation *V-H* curves.

Electrons emitted from the cathode, which is at a total-potential-energy maximum, are accelerated in the immediate region of the first total-potential-energy minimum, then are retarded, and subsequently stopped radially at the adjacent maximum, although they surmount this barrier because the maxima of total potential energy are approximately alike. In this manner, each electron is accelerated at each minimum and is radially retarded at each maximum. Electron motion is repeated at each space-charge sheath until it reaches the Hull cut-off boundary from which it is reflected back toward the cathode. Whether an electron eventually returns to the cathode within a specific number of convolutions depends partly upon the magnitude of cathode current. If cathode current is sufficiently high (the usual space-charge-limited case) the electron returns to the cathode within one convolution of its trajectory; that is, there is one minimum of total potential energy between the cathode and the Hull cut-off boundary, with maxima at the cathode and this boundary. Corresponding electron trajectories are cycloidal with cusps at the cathode. If cathode current is relatively low, the resultant cycloidal motion of the electron does not permit it to return to the cathode in a single convolution. The number of sheaths corresponds somewhat to the number of convolutions, for space-charge distribution must be consistent with electron motion. With no anode current the electrons which leave the cathode and migrate toward the cut-off boundary return to the cathode. Outgoing and incoming electrons, with reference to the cathode, contribute to the rotating space charge. Transit time of the electron from the cathode and back again is proportional to the number of these sheaths.

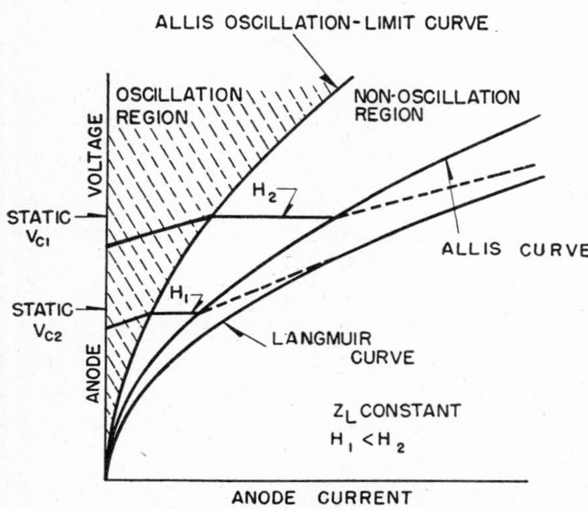

FIG. 5·46 Oscillation *V-I* curves.

In the region from cathode to cut-off boundary the potential function approximates a parabola, the degree of correspondence depending upon the number of space-charge sheaths, which in turn depends inversely upon cathode emission. In the region from the Hull cut-off boundary to the anode, in which there is no space charge, the potential is logarithmic. The potential barrier or cathode sheath assumes a suitable value to permit enough current to pass to satisfy the impressed boundary conditions or static anode voltage and magnetic field.

5·21 DYNAMIC CHARACTERISTICS OF THE SYMMETRIC MAGNETRON

Under oscillatory conditions anode current, for constant magnetic field H, begins at anode voltages below the Hull cut-off value V_c, but above a certain threshold, as illustrated by Fig. 5·46. This threshold requires that anode voltage be sufficient for a given magnetic field to allow the electrons to just reach the anode, that is, to maintain a finite alternating potential on the anode to sustain oscillations and permit anode current. Oscillation intensity for a given mode increases as anode voltage is increased above threshold value for a specific magnetic field. This process continues until oscillation intensity attains a maximum value, after which it decreases with further increase in anode voltage until oscillation ceases. The boundary including this limiting value is referred to as the Allis oscillation-limit curve. The reasons for this phenomenon will be evident later. Immediately after oscillation ceases, a static anode current ensues which eventually blends with the Langmuir curve, as illustrated by Fig. 5·46. It is evident from this figure that the oscillation region occupies a substantial portion of the area between the ordinate and the Allis curve.

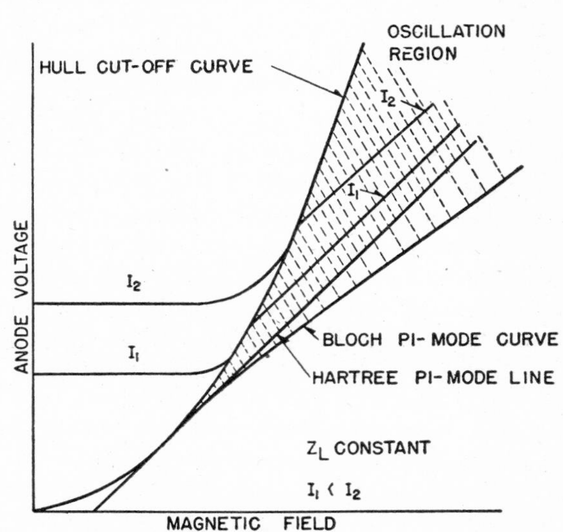

FIG. 5·47 Oscillation *V-H* curves.

Other features of the oscillation region are illustrated by Fig. 5·47, in which the region appears between the Hull cut-off curve and the threshold of oscillation represented by the Hartree pi-mode line and the Bloch pi-mode curve. At least the threshold criterion for the nth mode must be satisfied before oscillations in that mode can be initiated. The discrepancy in the Bloch and Hartree representations of the oscillation threshold lies in inadequacies in its theoretical interpretation. Hartree's criterion stipulates a minimum anode voltage for a given magnetic field which just permits electrons in the interaction space to reach the anode; based upon the premise that in the nth or pi-mode higher than the first term of the Fourier expansion of the a-c field around the anode are not relatively important. The Hartree criterion does not suppose a simple resonance between the electron and a-c field for mode selection but rather a modification involving the consideration that oscillations are derived from finite amplitudes of oscillation of anode potential. The Bloch criterion does not assume that the higher terms of the Fourier expansion of the a-c field in the vicinity of and around the anode are unimportant but that they represent significant contributions in enabling the electrons to reach the anode. The Hartree pi-mode line is a linear function of the applied static anode voltage and axial magnetic field, and it is tangent to the Hull cut-off curve. Bloch's criterion stipulates an anode voltage at which the electron is in resonance with the a-c field at the Hull [65] space-charge boundary at which the fundamental component of the a-c field is of paramount importance. Prior to Hartree and Bloch, Slater, neglecting space charge, proposed a purely resonance threshold criterion which stipulates an anode voltage for which electrons are in resonance with the a-c field at a plane halfway between the cathode and the anode. As illustrated in the figure, the Bloch pi-mode curve and the Hartree pi-mode line have a common tangent to the Hull cut-off curve. The Slater pi-mode line (not shown in Fig. 5·47, to avoid confusion) is displaced toward higher voltages than the corresponding Hartree pi-mode line; that is, for a common mode, the Slater mode line is above the Hartree mode line and intersects the Hull cut-off curve. Available experimental results involving pi-mode operation of symmetrical multicavity magnetrons apparently are more often in accord with the Bloch than with the Hartree threshold. The Slater, Hartree, and Bloch thresholds also retain their principal interrelationship for the lower order modes of the fundamental multiplet. The three threshold representations for each of the lower order modes lie between the Hull cut-off curve and the corresponding pi-mode thresholds, although each trio is at higher V-H values the lower the order of the mode. If no mode suppression is involved the various modes can be excited by suitable anode voltage and magnetic field. However, the magnetron generally is designed to operate in the most efficient mode, the pi-mode; other modes are suppressed by providing sufficient mode separation between the pi-mode and the others and attenuating the others by proper phasing. In this way single-mode operation is achieved at high efficiency over much of the oscillation region above the pi-mode threshold.

In the interaction space of the multisegment magnetron the electron motion consists of two independent components: a rotational component having a Larmor angular frequency $[\omega_H = (e/m)H]$ and a precessional or translational component of angular frequency $(\omega_p = E/Hr)$. The electron can impart a substantial portion of its total energy, either its angular kinetic or its potential energy, to the a-c field and hence to the resonant system. The proportion of the total energy transferred depends upon whether the field involved is a stationary field varying with time but independent of position, or a rotating field dependent upon both time and position. The first type of electromagnetic field obtains its energy from the electron's angular kinetic energy. The second type of field obtains its energy from the potential energy of the electron. Multisegment magnetrons can sustain the rotating type of field as well as the stationary field; the single-anode magnetron is capable of sustaining the stationary type only. The remaining discussion is restricted primarily to the steady-state properties of oscillation defined by the rotating electromagnetic or a-c field dependent upon both time and angular position.

A rotating charge distribution surrounds the cathode, which is uniform in angular distribution and is bounded radially by the Hull cut-off boundary. A perturbation, such as a finite voltage fluctuation on any anode segment, locally perturbs the charge distribution which, because of its rotation under the influence of the magnetic field, induces periodic voltage variations in the anode segment.[67] Oscillations cannot be expected to be initiated and sustained unless the anode voltage for a given magnetic field, although below cut-off ($V < V_c$), is above the mode threshold which, according to the Hartree criterion, will enable finite amplitudes of alternating potential on the anode segments to be established, and thus permit some electrons to reach the anode. Contributions to the a-c field in the interaction-space decrease from the anode to the cathode. The larger the mode number, the more rapid is this decrease; therefore if the Hull cut-off boundary is too far from the anode the electrons in the rotating space charge cannot get within sufficient influence of the desired field to be pulled out toward the anode. The decrease in field from anode to cathode is accentuated by space charge.

Of the several possible modes of oscillation initiated in the transient state by the electronic system only those which present favorable conditions for energy transfer from the electrons to the a-c field attain steady state. However, under some boundary conditions the only effective or predominant component wave corresponds to the pi-mode of oscillation. No consideration is given here to other circumstances under which the effective or predominant component may not be the pi-mode. The standing-wave pattern of the periodic potential on the anode segments, corresponding to the pi-mode, may be resolved into two equal and opposite rotating potential waves of which only the one rotating with the electrons appears as a constant field and can exert cumulative effects on the electron. This anode-potential wave corresponds to a rotating electromagnetic field around the cathode. An interaction ensues between the electric component of the rotating electromagnetic field and the electric component of the space charge under the action of the applied static electric and axial magnetic fields. Forces on the electron caused by the magnetic components of the rotating electromagnetic field and space charge may be neglected.

The electrons are characterized by less than the Larmor velocity, so the resultant phase relations between electrons and rotating field cause some electrons to be accelerated and force them to leave the field and return to the cathode. Others are decelerated, thus imparting energy to the field, and simultaneously are pulled out toward the anode. The acceleration process removes energy-absorbing electrons that decrease conversion efficiency. The decelerating process creates radial wave-like projections on the main cylindrical body of the rotating space charge at angular positions where the electrons travel in the same direction and at about the same angular velocity as the a-c field. Whether the tips of these waves reach the anode depends upon the magnitude of the rotating field. Figure 5·48 illustrates a case in which the field is of sufficient magnitude to accomplish this. Electrons in the projections need not be in exact resonance with the angular velocity of the field, as illustrated; a little variation below or above this value only modifies the symmetry of the waves. Furthermore, resonance pertains to average velocities of the electrons, so that the space-charge waves—not necessarily the electrons in them—are in resonance with the field.

Electrons travel directly across to the anode at cut-off ($V = V_c$). Below cut-off some electrons require one or more convolutions to reach the anode. Qualitatively, small anode current in a multicavity magnetron is synonymous with many convolutions of the electron trajectory or long transit time. Too small an anode current means too long a transit time, hence electrons normally near resonance may fall out and return to the cathode with substantial energy. Since a fair degree of resonance is required between electron motion and electromagnetic field in the interaction space, a few convolutions in a multicavity magnetron as a result of a-c field perturbation are beneficial. In fact, in split-plate magnetrons these several convolutions may be around the cathode. This provides enough time for electrons somewhat out of phase eventually to step in phase or resonance by the time they reach the space-charge boundary, and to establish favorable conditions under which energy may be efficiently imparted to the field. If sufficient time is not afforded, conversion efficiency is low or oscillations cannot even be sustained. In other words, too large an anode current caused by electrons reaching the anode too quickly causes a share of their energy to be imparted to the anode as heat. This resonance or phase condition is broad; that is, relatively large variations in operating conditions do not markedly influence operation. The main reason for close approach to resonance is to enable the electron to get to the anode more easily. Also for high conversion efficiency it is desirable that those electrons which reach the anode do so with a minimum of energy. Because of the rotating space charge many electrons which impart some of their energy to the field or are somewhat out of phase eventually drift into suitable regions in the interaction space so as to rehabilitate themselves and become useful for further work. If an out-of-phase electron enters an unfavorable region of the a-c field on emission from the cathode it is immediately returned to the cathode so that it absorbs a minimum of energy from the field.

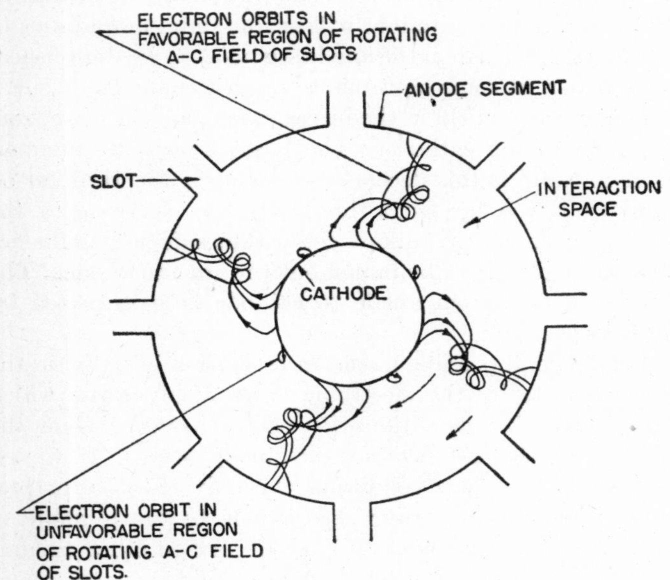

Fig. 5·48 Instantaneous representation of the rotating space charge for the pi-mode in a multicavity magnetron (periodic field not shown). (Reprinted by permission from British Committee on Valve Development Report, Mag. 41, Department of Scientific Research and Experiment.)

For maximum conversion efficiency the operating conditions (the applied static electric and magnetic field for given structural parameters) should be adjusted so that a maximum number of electrons arrive at the anode with a minimum of kinetic energy as a result of oscillations in a single mode. An additional advantage is gained by adjusting the magnetic field so that the angular velocity of the field is in resonance with the Larmor angular velocity. Fulfillment of these conditions permits an efficiency of 100 percent in the ideal case. Practically, efficiencies often are between 40 and 75 percent. At such efficiencies the a-c field amplitude is a substantial portion of the applied static electric field. Furthermore, to maintain high efficiency as the anode radius and number of segments increase it is necessary to increase the cathode radius proportionally.

The Hartree condition for oscillation can be clarified by a mechanical analogy provided that all forces on a particle in the chosen frame of reference appear conservative. To fulfill this prerequisite it is sufficient to rotate the frame of reference at the Larmor angular velocity. The particle leaves the cathode with an initial angular velocity $-\omega_H$ relative to the Larmor frame, which is greater than that of the potential-energy surface, which is assumed zero at the cathode. As the particle climbs on the surface toward the anode, under the action of gravity, its angular velocity decreases until it is equal to that of the rotating surface. During its climb the particle either gets through to the anode or falls back to the cathode, depending upon the form of the potential-energy surface between the cathode and anode. If the surface is not undulated around the angular co-ordinate the particle does not arrive at the anode unless the maximum extent of its orbit is greater than that prescribed by the equivalent Hull cut-off condition. If the surface is so undulated and is rotating, the particle leaving the cathode eventually can reach the anode. This rotating, rigid, undulated potential-energy surface corresponds to the rotating potential wave around the multisegment anode. On entering a trough of an undulation the particle can move farther toward the anode than if it were on a uniform surface, for it does not have to climb so high. Furthermore the rotating, undulated surface swings the particle around, eventually with its own angular velocity, and because of the imparted centrifugal force it reaches the anode. It could not do this on an unperturbed surface. The angular velocity of the rigid undulated surface, relative to the Larmor frame, must equal or exceed a certain minimum value if the particles are to reach the anode.

5·22 EQUIVALENT-CIRCUIT REPRESENTATION OF THE SYMMETRIC MAGNETRON

Two aids for interpretation of magnetron performance are the *equivalent-circuit* and *cavity-resonator* methods. The first method is based on the representation of the magnetron as an equivalent circuit with lumped parameters. The wavelength is assumed large compared with resonator elements. Even though this prerequisite is not followed precisely, useful approximations can still be derived. The cavity-resonator method,[68] based on the properties of the normal modes in a closed resonant system, is precise and yields more information than the other method without recourse to the artificial concept of an equivalent circuit. It is especially valuable if the wavelength is about equal to the resonator dimensions. The equivalent circuit method is more familiar and relatively simple, even though approximate, so it is considered here in preference to the cavity-resonator method for the symmetrical multicavity magnetron in the pi-mode of oscillation.

Either a series or a parallel lumped-circuit representation of the resonant system of the magnetron can be used. The final results are equivalent at the resonant frequency for one and the same mode of oscillation. For a multicavity magnetron, as illustrated by Fig. 5·43, the total equivalent lumped inductance L is equal to the reciprocal of the total number of individual identical circuits multiplied by the equivalent inductance of one of them. Total equivalent lumped capacity C is equal to the product of the total number of such identical resonators and the capacitance of one of them. Strap capacitance introduced by anode straps must be considered in C. Finally, the total equivalent resistance corresponding to resonator-system losses is $Q_u(C/L)^{1/2}$ involving the unloaded Q designated Q_u. Figure 5·49 illustrates the components of the resonator system of the magnetron and load. The coupling between resonant system LCR, the output device, and the base of the loop or iris at 3 is represented by an ideal transformer with turns ratio $1:1/L_m$, and the series inductances L_S where $L_S = L'[1 - (L_m^2/LL')]$, and L' represents the inductance of the loop. Other representations can be used for this coupling. The output device between the base of the loop at 3 and the transmission line at 2 can be represented by a passive linear non-dissipative four-terminal network[69] T. The oscillating space charge, looking to the left at 5 into the interaction space from the slots between the magnetron segments in an oscillating magnetron, is equivalent to an admittance, often referred to as an electron admittance. The conductance component of this admittance is negative and it more than offsets Q_u, so the magnetron can behave as a generator. The susceptance component of the admittance at 5 is responsible for the frequency variation discussed in Section 5·23. Determination of the manner in which the components of the admittance at 5 influences generated power and frequency of oscillation requires evaluation of the admittance, instantaneous currents flowing to the segments, at 5, and the alternating voltage between them. These factors are derived from data obtained by the familiar cold standing-wave measurements and impedance or Rieke diagrams. Standing-wave measurements yield

1. Cold impedance at 2, looking toward 5.
2. Transformation properties of output device T.
3. Impedance at 3, looking toward 5.
4. Unloaded Q, and in conjunction with the impedance or Rieke diagram at 2, the equivalent-circuit constants.

Conditions for deriving (1) from cold measurements require that the quiescent or cold magnetron act as load on the line, and that Z_L be replaced by a suitable signal generator. Then the position of minimum and standing-wave ratio in the line must be examined near resonance of the desired (pi) mode. The cold- and hot-magnetron frequencies are different because of space-charge and temperature effects. The transformation properties[70] of the coupling device T are determined from cold standing-wave measurements in lines connected to each side of the coupling device T. One end of one line is short-circuited and one end of the other line is connected to a signal generator. Once the transformation properties of the output device T are determined, impedances at 2 can be transformed to 3. With this information the performance characteristics of the hot magnetron, under applied boundary conditions at 5, can be evaluated.

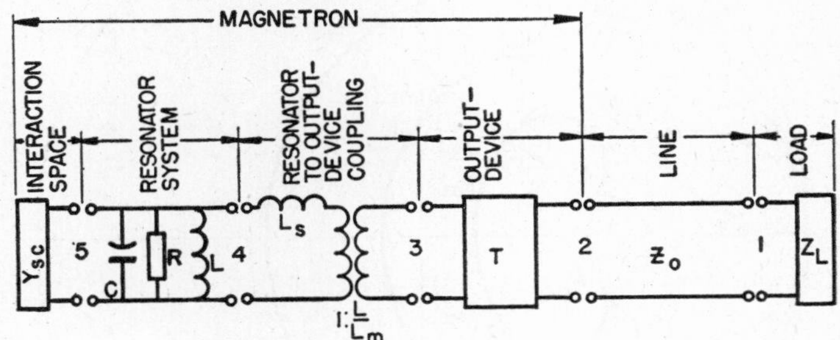

FIG. 5·49 Equivalent network representation.

Cold standing-wave measurements are relatively simple, as long as no resonances or frequency cut-offs are inherent in the output device near the magnetron mode under test. For reasonable accuracy the loaded Q of the magnetron should be relatively high.

5·23 OPERATIONAL DIAGRAMS FOR MAGNETRONS

Two principal types of characteristic diagrams, the impedance and the performance diagrams, are of importance to theoretical and practical aspects of magnetrons.

The admittance or Rieke diagram, as it is often called, can be prepared for either constant-anode voltage or current I with the magnetic field H constant in each type. The constant-current Rieke diagram, which is generally used, relates a-c power output P_o, frequency ν, and anode voltage V (optional) on a polar or admittance plane. The polar form of this diagram portrays operational characteristics in terms of the real component of the reflection coefficient $|\Gamma|$ as the radial co-ordinate. Phase angle ϕ of the voltage minimum with respect to an arbitrary reference plane on the magnetron, as 2 of Fig. 5·49, serves as an angular co-ordinate; 360 degrees of phase angle on the diagram corresponds to a half wavelength. The center of the polar diagram denotes matched-line conditions.

The polar form of the Rieke diagram can be transformed to an admittance plane or Smith chart.[71] Theoretically, the Rieke diagram on a Smith chart indicates that circles of constant-load conductance G_L should correspond approximately to contours of constant power if load susceptance has a negligible effect on output power. Actually power contours P are in accord with the predictions at relatively low reflection coefficients, illustrated by Fig. 5·50. Furthermore, circular arcs of constant-load susceptance B_L correspond approximately to contours of constant frequency. Although the susceptance curves, corresponding approximately to frequency contours, are shown symmetrical in Fig. 5·50, in the general case the line of convergence of these susceptance curves is inclined at some angle to the horizontal axis. Furthermore, in practice the power contours, corresponding approximately to the conductance contours, do not proceed to the point of infinite admittance, but rather close at some intermediate region as shown. The sink toward which frequency and power contours tend to converge is known as the *instability region* or *pulling sink* and is characterized by low-frequency stability of the magnetron.

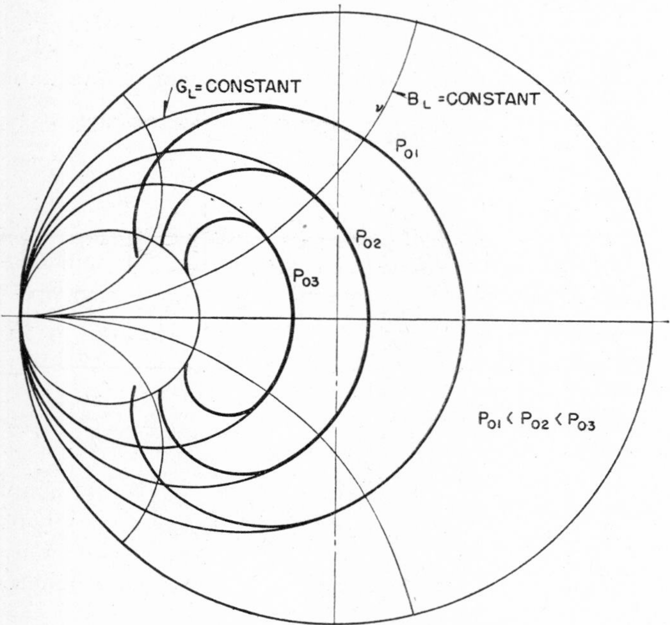

Fig. 5·50 The concept of Rieke diagrams.

Variation of frequency with the susceptance component of the load, while the conductance component remains constant, is termed *frequency pulling*. The degree of frequency pulling is referred to a specific load admittance or operating point on the Rieke diagram. It is specified as the maximum frequency variation across a circle corresponding to a specific (0.2) real reflection coefficient for all phase angles. The center of this circle corresponds to the chosen operating point. A practical example of a Rieke diagram for a two-cavity magnetron is illustrated by Fig. 5·51. Variations in the character of the Rieke diagrams are dependent upon the type of magnetron, output device, and operating conditions. Some interesting features of the Rieke diagrams are:

1. Data on the general frequency-stability characteristic of the magnetron over the indicated load-admittance range are provided. The extent and regions of spurious modes and instability caused by low values of Q_u of the magnetron as compared with the load also are revealed.

2. The proper load-impedance region is portrayed for optimum performance with regard to frequency pulling or stability, power output, and applied anode voltage for constant magnetic field, anode current, and nominal wavelength.

3. The degree to which matched-line operational conditions have been achieved, for chosen magnetron parameters and coupling devices, is disclosed.

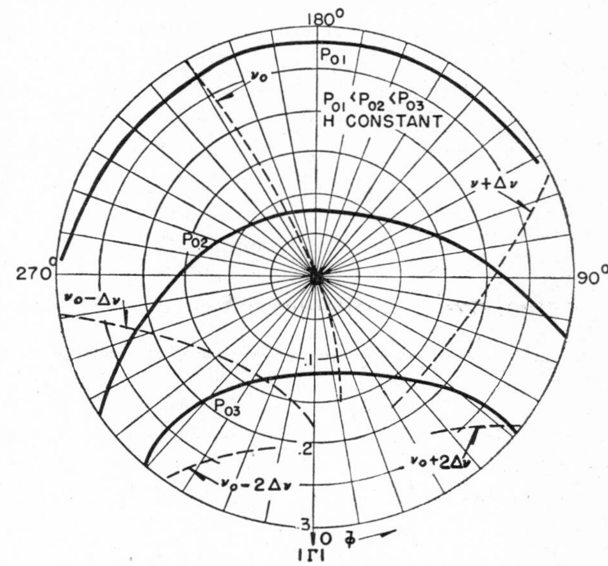

Fig. 5·51 A practical Rieke diagram (voltage contours not shown).

For a given operating point on the Rieke diagram, choice of which depends upon the particular application, further operational data can be obtained from performance diagrams. Such a diagram generally is prepared on a rectangular coordinate system with applied anode voltages as ordinate; resultant anode currents as abscissas, with magnetic field as parameter, relate efficiency, frequency (optional), and a-c power output at constant-load admittance. The performance diagram is confined within the indicated boundaries of the oscillation region of Figs. 5·46 and 5·47. A theoretical performance diagram for this region is illustrated

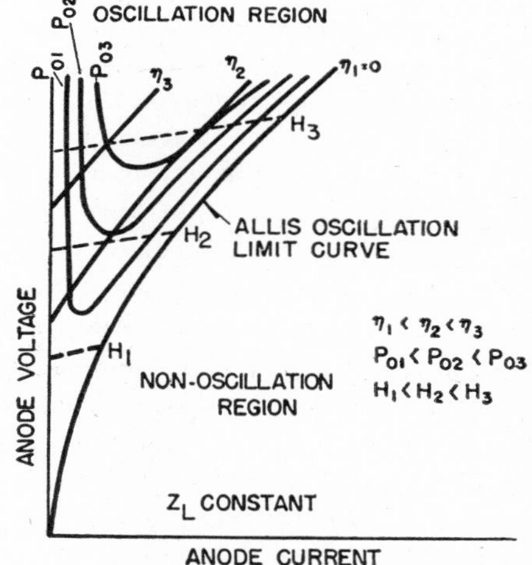

Fig. 5·52 The concept of performance diagrams.

by Fig. 5·52 for single-mode operation and constant-load admittance. The complete region of oscillations, as illustrated by Fig. 5·52, is not fully realized because of spurious modes caused by inadequate mode separation and mode coupling in the specific magnetron, by limited cathode emission, or by unfavorable relation between load and magnetron.

Variations in the character of performance diagrams depend upon the type of magnetron, load admittance, and operating conditions. Some interesting features of performance diagrams, such as the practical example for a two-cavity magnetron illustrated by Fig. 5·53 are:

1. An unstable or *mody* region at low currents.
2. A normal operational region. Small islands of instability of *modiness* may, on occasion, appear even in this region. However, the operating regions on Rieke and performance diagrams are chosen to avoid such areas as much as possible.

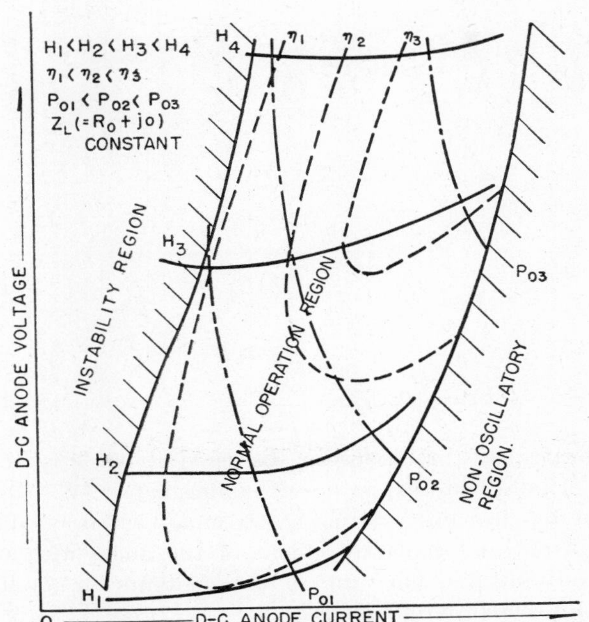

FIG. 5·53 A practical performance diagram (frequency contours not shown).

3. Non-oscillatory region at high currents. The high current boundary at the right in Fig. 5·53 is associated with cathode emission, inadequate oscillation build-up time under pulsed conditions, and electron-electromagnetic field synchronism (or phase-focusing) failure, which stipulates a maximum pi-mode d-c voltage at constant magnetic field.
4. Variation of the slope of the magnetic-field contours in the operational region. This phenomenon is also associated with cathode emission, and the slope is small if the current is space-charge-limited. The characteristics of the magnetic field contours illustrated in Fig. 5·53 are frequently exhibited by split-plate magnetrons with secondary-emission cold cathodes and is almost non-existent in most multicavity magnetrons with thermal cathodes.
5. Portrayal, by the operational region, of the broadness of the resonance condition (in applied anode voltage and magnetic field) for efficient energy conversion between space charge and resonant system and load.

The degree of frequency instability at any point in the performance diagram depends upon the degree of frequency variation with anode current, and is called *frequency pushing.*

5·24 SCALING OF MAGNETRONS

Analytical considerations and experimental data have justified voltage and wavelength scaling of magnetrons for equivalent operation. In voltage scaling, wavelength remains constant and the interaction space is varied. In wavelength scaling, the whole geometrical shape of the magnetron is maintained constant, while the whole structure is varied in the ratio of the new to the old wavelength. Emission, dissipation, and applied fields establish a limit to which scaling can be extended.

Magnetron scaling relations which must remain invariant for equivalent operation of two magnetrons having the same anode-to-cathode radii ratio σ, n (or N), and loaded Q, Q_L, are:

$$H\lambda$$

$$V\left(\frac{\lambda}{r_a}\right)^2$$

$$\frac{I\lambda^3}{r_a{}^2L}$$

V, I, L, r_a denote anode voltage, current, length, and radius, whereas H and λ denote magnetic field and wavelength, respectively.

These relations justify the types of scaling illustrated in the table on page 96. The prime denotes a scaled quantity. J, P, and G denote current density, power input or output, and parallel slot conductance, respectively.

A universal performance chart for any magnetron is obtained from Slater's reduced variables, wherein σ and n (or N) are not constrained, so that for the Hull curve,

$$\frac{V}{V_0} = \left(\frac{H}{H_0}\right)^2$$

for the Hartree line,

$$\frac{V}{V_0} = 2\frac{H}{H_0} - 1$$

for the Allis curve,

$$\frac{I}{I_0} = \left(\frac{V}{V_0}\right)^{3/2} = \left(\frac{H}{H_0}\right)^2$$

where

$$H_0 = 21.40 \frac{1}{n\lambda\left(1 - \frac{1}{\sigma^2}\right)} \quad \text{kilogauss}$$

$$V_0 = 9.997 \times 10^3 \left(\frac{r_a}{n\lambda}\right)^2 \quad \text{kilovolts}$$

are co-ordinates of the points of tangency of the Hartree line with the Hull curve for the n mode, and

$$I_0 = 2.096 \times 10^6 \frac{r_a{}^2L}{n^3\lambda^3} f(\sigma)\phi(\sigma) \quad \text{amperes}$$

represents the Allis current corresponding to V_0 and H_0, where

$$f(\sigma) \simeq 0.95 \text{ for } \sigma \text{ between 1.6 and 2.7}$$

$$\phi(\sigma) = \frac{1}{\left(1 - \frac{1}{\sigma^2}\right)^2 (\sigma + 1)}$$

Restricted Scaling

Ratio of Final to Initial Value of Parameter	Complete Wavelength Scaling (shape invariant) $\frac{\lambda'}{\lambda} = \frac{r'_a}{r_a}$		Special Wavelength Scaling (radial interaction space invariant) $\frac{r'_a}{r_a} = 1$		Voltage/Power Scaling (wavelength invariant) $\frac{\lambda'}{\lambda} = 1$	General
	$\frac{L'}{L} = 1$	$\frac{L'}{L} = \frac{\lambda'}{\lambda}$	$\frac{L'}{L} = 1$	$\frac{L'}{L} = \frac{\lambda'}{\lambda}$	$\frac{L'}{L} = 1$	
$\frac{H'}{H}$	$\frac{\lambda}{\lambda'}$	$\frac{\lambda}{\lambda'}$	$\frac{\lambda}{\lambda'}$	$\frac{\lambda}{\lambda'}$	1	$\frac{\lambda}{\lambda'}$
$\frac{V'}{V}$	1	1	$\left(\frac{\lambda}{\lambda'}\right)^2$	$\left(\frac{\lambda}{\lambda'}\right)^2$	$\left(\frac{r'_a}{r_a}\right)^2$	$\left(\frac{r'_a}{r_a}\right)^2\left(\frac{\lambda}{\lambda'}\right)^2$
$\frac{I'}{I}$	$\frac{\lambda}{\lambda'}$	1	$\left(\frac{\lambda}{\lambda'}\right)^3$	$\left(\frac{\lambda}{\lambda'}\right)^2$	$\left(\frac{r'_a}{r_a}\right)^2$	$\left(\frac{L'}{L}\right)\left(\frac{r'_a}{r_a}\right)^2\left(\frac{\lambda}{\lambda'}\right)^3$
$\frac{J'}{J}$	$\left(\frac{\lambda}{\lambda'}\right)^2$	$\left(\frac{\lambda}{\lambda'}\right)^2$	$\left(\frac{\lambda}{\lambda'}\right)^3$	$\left(\frac{\lambda}{\lambda'}\right)^3$	$\frac{r'_a}{r_a}$	$\left(\frac{r'_a}{r_a}\right)\left(\frac{\lambda}{\lambda'}\right)^3$
$\frac{P'}{P}$	$\frac{\lambda}{\lambda'}$	1	$\left(\frac{\lambda}{\lambda'}\right)^5$	$\left(\frac{\lambda}{\lambda'}\right)^4$	$\left(\frac{r'_a}{r_a}\right)^4$	$\left(\frac{L'}{L}\right)\left(\frac{r'_a}{r_a}\right)^4\left(\frac{\lambda}{\lambda'}\right)^5$
$\frac{G'}{G}$	$\frac{\lambda}{\lambda'}$	1	$\frac{\lambda}{\lambda'}$	1	1	$\left(\frac{L'}{L}\right)\left(\frac{\lambda}{\lambda'}\right)$

Hence these yield the general approximate scaling ratios in which the ratios of current, power, and conductance apply per unit length. The prime denotes the scaled quantity.

$$\frac{H'}{H} = \left(\frac{\lambda}{\lambda'}\right)$$

$$\frac{V'}{V} = \left(\frac{r'_a}{r_a}\right)^2\left(\frac{N}{N'}\right)\left(\frac{\lambda}{\lambda'}\right)^2\left[\frac{1-\dfrac{1}{\sigma'}}{1-\dfrac{1}{\sigma}}\right]$$

$$\frac{I'}{I} = \left(\frac{r'_a}{r_a}\right)^2\left(\frac{N}{N'}\right)\left(\frac{\lambda}{\lambda'}\right)^3$$

$$\frac{J'_c}{J_c} = \left(\frac{r'_a}{r_a}\right)\left(\frac{N}{N'}\right)\left(\frac{\lambda}{\lambda'}\right)^3\left(\frac{\sigma'}{\sigma}\right)$$

$$\frac{P'}{P} = \left(\frac{\lambda}{\lambda'}\right)^5\left(\frac{r'_a}{r_a}\right)^4\left(\frac{N}{N'}\right)^2\left[\frac{1-\dfrac{1}{\sigma'}}{1-\dfrac{1}{\sigma}}\right]$$

$$\frac{G'}{G} = \left[\frac{1-\dfrac{1}{\sigma}}{1-\dfrac{1}{\sigma'}}\right]\left(\frac{\lambda}{\lambda'}\right)$$

$$\frac{r'_c}{r_c} = \left(\frac{\sigma}{\sigma'}\right)\left(\frac{r'_a}{r_a}\right)$$

where J_c denotes the current density at the cathode.

In both the restricted and the general scaling tables above, end effects are neglected, and as a result the current per unit length of cathode is independent of length.

Another useful relationship, discussed above, involves the ratio of initial to final values of pulling figure PF, in terms of the matched line external Q, Q_E and frequency ν of the magnetron, and the complement of the angle with which the constant frequency and power contours on the Rieke diagram thereof intersects, sec α:

$$\frac{(PF)}{(PF)'} = M\frac{\nu}{\nu'}\frac{Q'_E}{Q_E}$$

$$M = \text{modulus} = \frac{\sec\alpha}{\sec\alpha'}$$

where Q_E is defined as 2π times the ratio of energy stored in the magnetron to the energy transmitted per cycle. In this respect, the ratio of initial to final pulling figures may be expressed in terms of the stored and transmitted energies.

5·25 SOME BASIC RELATIONS OF SYMMETRIC MAGNETRONS

The Fourier expansion of the a-c field around the anode in the interaction space of a magnetron may be resolved into waves of identical frequency of various forward and reverse angular velocities. The Fourier components which contribute to the a-c field in the interaction space may be represented by the relation

$$n = n_0 + \mu N$$

in which $\mu = 0, \pm 1, \pm 2, \cdots$, and negative integral values of n (that is, $\mu = -1, -2, \cdots$) denote wave rotation in reverse direction to the fundamental or electron motion. N, as before, denotes the number of resonators of the magnetron.

The fundamental is denoted by $|n| = n_0$ (that is, $\mu = 0$). The Hartree harmonics, on the other hand (not harmonics in the usual sense except when n is a factor of N), are denoted by $|n| \neq n_0$ (that is, $\mu \neq 0$). Now the absolute value of n, $|n|$, also denotes the periodicity of the wave of the mode around the anode, and n_0 denotes the Fourier component with a minimum value of n.

The most effective component n (in the Fourier expansion of the operating mode, n_0) of the a-c field in the interaction space, which assists electrons to the anode, for given anode voltage V, magnetic field H, wavelength λ, anode radius r_a, cathode radius r_c, and their ratio σ ($= r_a/r_c$), is obtained by assigning an integer, next in increasing sequence, to the integral part of Hartree's oscillation threshold:

$$n_x = \frac{21.4}{\lambda H} \frac{1}{\left(1 - \frac{1}{\sigma^2}\right) - \left[\left(1 - \frac{1}{\sigma^2}\right)^2 - 0.0454 \frac{V}{H^2 r_a^2}\right]^{1/2}}$$

where H and V are in kilogauss and kilovolts, and λ and r_a are expressed in centimeters.

Then the operating mode or mode of oscillation, n_0, is the minimum integer of the set

$$|n - \mu N|$$

for which the phase difference between adjacent segments is

$$\Delta\phi = \frac{2\pi n_0}{N} \qquad (\Delta\phi < \pi)$$

It is important to note that the most effective component n need not be the fundamental component of the a-c field in the interaction space of the magnetron.

Hull parabola:

$$V = 21.8 H^2 r_a^2 \left(1 - \frac{1}{\sigma^2}\right)^2$$

Slater resonance line:

$$V = \frac{944 r_a^2}{n\lambda}\left(1 - \frac{1}{\sigma^2}\right)\left(H - \frac{10.7}{n\lambda}\right)$$

Hartree line:

$$V = \frac{944 r_a^2}{n\lambda}\left[H\left(1 - \frac{1}{\sigma^2}\right) - \frac{10.7}{n\lambda}\right]$$

Bloch line:

$$V = \frac{944 r_c^2}{n\lambda}\left[\left(H + \frac{10.7}{n\lambda}\right)\log_e \sigma^2 - \frac{10.7}{n\lambda}\right]$$

The Bloch line is valid for

$$H > 32.1 \frac{1}{n\lambda\left(1 - \frac{1}{\sigma^2}\right)}$$

where V and H are in kilovolts and kilogauss, and λ, r_c, and r_a are expressed in centimeters.

REFERENCES

1. *Thermionic Emission*, T. J. Jones, Chemical Publishing Co., 1936.
2. *Physics of Electron Tubes*, L. R. Koller, McGraw-Hill, 1934.
3. "Comparative Analysis of Water Cooled Tubes as Class B Amplifiers," I. E. Mouromtseff and H. N. Kozanowski, *Proc. I.R.E.*, Vol. 23, No. 10, Oct. 1935, p. 1224.
4. *Communication Engineering*, W. L. Everitt, McGraw-Hill, 2nd ed., 1937.
5. "Analysis of Rectifier Operation," O. H. Schade, *Proc. I.R.E.*, Vol. 31, No. 7, July 1943.
6. "The Diode as a Half-Wave, Full-Wave, and Voltage Doubling Rectifier," N. H. Roberts, *Wireless Eng.*, Vol. 13, July 1936, pp. 351–362, and Aug. 1936, pp. 423–470.
7. "The Determination of Operating Data and Allowable Ratings of Vacuum Tube Rectifiers," J. C. Frommer, *Proc. I.R.E.*, Vol. 29, Sept. 1941, pp. 481–485.
8. *Radio Engineer's Handbook*, F. E. Terman, 1st ed., p. 589, McGraw-Hill, 1943.
9. *Fundamentals of Vacuum Tubes*, A. V. Eastman, McGraw-Hill, 1941.
10. "Vacuum Tubes of Small Dimensions for Use at Extremely High Frequencies," B. J. Thompson and G. M. Rose, Jr., *Proc. I.R.E.*, Vol. 4, No. 12, Dec. 1933, p. 1707.
11. "Disk-Seal Tubes," E. D. McArthur, *Electronics*, Feb. 1945, p. 98.
12. "Review of Demountable vs. Sealed-Off Power Tubes," I. E. Mouromtseff, H. J. Dailey, L. F. Werner, *Proc. I.R.E.*, Vol. 32, No. 11, Nov. 1944.
13. *Radio Engineer's Handbook*, Keith Henney, 3rd ed., McGraw-Hill, 1941.
14. "Grid Temperature as a Limiting Factor in Vacuum Tube Operation," I. E. Mouromtseff and H. N. Kozanowski, *Proc. I.R.E.*, Vol. 24, No. 3, 1936, p. 447.
15. "Grid Losses in Power Amplifiers," E. E. Spitzer, *Proc. I.R.E.*, Vol. 17, No. 6, 1929, p. 985.
16. "Determination of Grid Driving Power in Radio Frequency Amplifiers," H. P. Thomas, *Proc. I.R.E.*, Vol. 21, 1933, p. 1134.
17. "Modulating the Screen-Grid R.F. Amplifier," H. A. Robinson, *Proc. I.R.E.*, Vol. 20, No. 1, Jan. 1932, pp. 131–160.
18. "The Grid Coupled Dynatron," F. M. Gager, *Proc. I.R.E.*, Vol. 23, No. 9, Sept. 1935, p. 1048.
19. "A Quantitative Study of the Dynatron," F. M. Gager, *Proc. I.R.E.*, Vol. 23, No. 12, Dec. 1935, p. 1536.
20. *Theory of Multi-Electrode Vacuum Tubes*, H. A. Pidgeon, *Bell System Tech. J.*, Vol. 14, Jan. 1935, pp. 44–84.
21. "Power Output Characteristics of the Pentode," S. Ballentine and H. L. Cobb, *Proc. I.R.E.*, Vol. 18, No. 3, March 1930, p. 450.
22. "Beam Power Tubes," O. H. Schade, *Proc. I.R.E.*, Vol. 26, No. 2, Feb. 1938, p. 137.
23. "The Beam Power-Output Tube," J. F. Dreyer, *Electronics*, Vol. 9, No. 4, April 1936, p. 18.
24. "New Developments in Audio Power Tubes," R. S. Burnap, *RCA Rev.*, Vol. 1, No. 1, July 1936, p. 101.
25. *Radio Engineering*, F. E. Terman, 2nd ed., McGraw-Hill, 1937.
26. "Tubes as Power Oscillators," D. C. Prince, *Proc. I.R.E.*, Vol. 11, No. 10, Oct. 1923, pp. 11, 275, 405, 527.
27. "A New High Efficiency Power Amplifier for Modulated Waves," W. H. Doherty, *Proc. I.R.E.*, Vol. 24, No. 9, Sept. 1936.
28. "An Experimental 5 KW Doherty Amplifier," C. E. Strong and Samson, *Elec. Communication*, Vol. 16, No. 3, Jan. 1938.
29. "Calculation of Triode Constants," J. H. Fremlin, *Phil. Mag.*, Vol. 27, No. 185, June 1939.
30. "Formulas for the Amplification Constant for Three Element Tubes," F. B. Vodges and F. R. Elder, *Phys. Rev.*, Vol. 24, Dec. 1924, p. 683.
31. "Calculation of Characteristics and the Design of Triodes," Y. Kusonose, *Proc. I.R.E.*, Vol. 17, No. 10, Oct. 1929, pp. 1706–1749.
32. "Analysis of the Operation of Vacuum Tubes as Class C Amplifiers," I. E. Mouromtseff and H. N. Kozanowski, *Proc. I.R.E.*, Vol. 23, No. 7, July 1935.
33. "Calculation for Class A Amplifiers," E. K. Brown, *Radio Eng.*, Vol. 17, No. 1, Jan. 1937, p. 9.

34. "The Flash-Arc in High Power Valves," B. S. Bossling, *Proc. I.R.E., Wireless Section*, Vol. 7, No. 21, Sept. 1932, pp. 192–219.
35. "Rules for Prolonging Tube Life," H. J. Dailey, *Electronics*, Vol. 16, April 1943, p. 76.
36. "The Operating Characteristics of Power Tubes," E. L. Chafee, *J. App. Phys.*, Vol. 11, No. 6, 1938, p. 471.
37. "An Analysis of Class B and Class C Amplifiers," B. F. Miller, *Proc. I.R.E.*, Vol. 23, No. 5, 1935, p. 496.
38. "A Short Cut Method for Calculation of Harmonic Distortion in Wave Modulation," I. E. Mouromtseff and H. N. Kozanowski, *Proc. I.R.E.*, Vol. 22, No. 9, 1934, p. 1090.
39. "Graphical Harmonic Analysis," J. A. Hutcheson, *Electronics*, Vol. 9, No. 1, Jan. 1936, p. 16.
40. "A Simplified Harmonic Analysis," E. L. Chafee, *Rev. Sci. Instruments*, Vol. 7, 1936, p. 384; Vol. 8, 1937, p. 227.
41. "The Operation of Vacuum Tubes as Class B and Class C Amplifiers," C. E. Fay, *Proc. I.R.E.*, Vol. 11, No. 1, 1932, pp. 20, 548.
42. "Class B Amplifiers Considered from the Conventional Class A Standpoint," J. R. Nelson, *Proc. I.R.E.*, Vol. 21, No. 6, 1933, p. 858.
43. "Simplified Methods for Computing the Performance of Transmitting Tubes," W. G. Wagener, *Proc. I.R.E.*, Vol. 25, No. 1, 1937, p. 47.
44. "Optimum Operating Conditions for Class B Radio Frequency Amplifiers," W. L. Everitt, *Proc. I.R.E.*, Vol. 24, No. 2, 1936, p. 305.
45. "20 KW Tetrode for UHF Transmitters," A. V. Haeff, *Elec. Eng.*, Vol. 59, March 1940, p. 107.
46. *Communication Engineering*, W. L. Everitt, 2nd ed., McGraw-Hill, 1937.
47. "Space Current Flow in Vacuum Tube Structures," B. J. Thompson, *Proc. I.R.E.*, Vol. 31, No. 9, Sept. 1943, pp. 485–491.
48. "Skin Effect Formulas," J. R. Whinnery, *Electronics*, Vol. 15, Feb. 1942, p. 44.
49. "The Characteristics of Tungsten Filaments as Functions of Temperature," Dr. Howard A. Jones and Dr. Irving Langmuir, *G.E. Rev.*, June 1927, July 1927, Aug. 1927.
50. "A Type of Electrical Resonator," W. W. Hansen, *J. App. Phys.*, Vol. 9, Oct. 1938, p. 654.
51. *Fields and Waves in Modern Radio*, S. Ramo and J. R. Whinnery, Wiley, 1944.
52. *Hyper and Ultrahigh Frequency Engineering*, R. I. Sarbacher and W. A. Edson, Wiley, 1943.
53. *Klystron Technical Manual*, Sperry Gyroscope Co., 1944.
54. "Ultrahigh Frequency Generators," I. E. Mouromtseff, R. C. Retherford, and J. H. Findlay, *Electronics*, Vol. 15, April 1942, p. 45.
55. "A High Frequency Oscillator and Amplifier," R. H. Varian and S. F. Varian, *J. App. Phys.*, Vol. 10, May 1939, p. 321.
56. "Cathode-Ray Bunching," D. L. Webster, *J. App. Phys.*, Vol. 10, July 1939, p. 501.
57. "The Theory of Klystron Oscillations," D. L. Webster, *J. App. Phys.*, Vol. 10, Dec. 1939, p. 865.
58. "Principles of Microwave Radio," E. U. Condon, *Rev. Mod. Phys.*, Vol. 14, Oct. 1942, p. 341.
59. "Electronic Generation of Electromagnetic Oscillations," E. U. Condon, *J. App. Phys.*, Vol. 11, July 1940, p. 502.
60. "Forced Oscillations in Cavity Resonators," E. U. Condon, *J. App. Phys.*, Vol. 12, Feb. 1941, p. 129.
61. *Proposed Standards on Ultra-High-Frequency Electronics*, I.R.E. Sub-Committee on Advanced Developments, Definition of Terms, Section 4U1, 1945.
62. *Einführung Theorie und Technik der Dezimeterwellen*, O. Gross, Stechert, 1937.
63. *High Frequency Thermionic Tubes*, A. F. Harvey, Wiley, 1943.
64. "Obtaining Powerful Oscillations by the Magnetron in the Centimeter Waveband," N. F. Alekseev and D. D. Maliarov, *J. Tech. Phys. (U.S.S.R.)*, Vol. X, No. 15, 1940, p. 1299, Fig. 5.
65. "The Effect of a Uniform Magnetic Field on the Motion of Electrons between Coaxial Cylinders," A. W. Hull, *Phys. Rev.*, Vol. 18, 1921, p. 31.
66. "The Effect of Space Charge and Residual Gases on Thermionic Currents in High Vacuum," I. Langmuir, *Phys. Rev.*, Vol. 2, 1913, p. 450.
67. "The Induced Currents in Split Cylindrical Magnetrons by Moving Charges," A. Okazaki, *Electrotech. J. (Japan)*, April 1940, p. 83.
68. "Forced Oscillations in Cavity Resonators," E. U. Condon, *J. App. Phys.*, Vol. 12, Feb. 1941, p. 129.
69. "Matrix Theory of Four Terminal Networks," L. A. Pipes, *Phil. Mag.*, Vol. XXX, Serial 7, Nov. 1940, Section V, p. 384.
70. *Ibid.*, Section V, p. 384, equation 60.
71. "The Solution of Transmission Line Problems by the Use of the Circle Diagram of Impedance," W. Jackson and L. G. H. Huxley, *Institute of Electrical Engineering Journal*, Vol. 91, Part I, No. 44, Aug. 1944, p. 105.
72. "Practical Results from Theoretical Studies of Magnetrons," L. Brillouin, *Proc. Inst. Radio Eng.*, Vol. 32, April 1944, p. 216.
73. "The Magnetron as a Generator of Centimeter Waves," J. B. Fisk, H. D. Hagstrum, and P. L. Hartman, *Bell System Tech. J.*, Vol. XXV, April 1946, p. 167.
74. Improvements in High Frequency Electrical Oscillators, J. Sayers and C. S. Wright, British Patent Specification 588,916, June 6, 1947. This patent pertains to the invention of strapping of symmetrical magnetrons.
75. Improvements in High Frequency Electrical Oscillators, J. T. Randall, H. A. H. Boot, and C. S. Wright, British Patent Specification 588,185, May 16, 1947. This patent pertains to the invention of multicavity symmetrical magnetrons.
76. "Microwave Electronics," J. C. Slater, *Rev. Mod. Phys.*, Vol. 18, No. 4, Oct. 1946, p. 441.
77. "Resonant-Cavity Magnetron," J. W. Coltman, *Westinghouse Eng.*, Vol. 6, No. 6, Nov. 1946, p. 172.
78. "Reflex Oscillators," J. R. Pierce and W. G. Shepherd, *Bell System Tech. J.*, Vol. 26, No. 3, July 1947, p. 460.
79. "The Klystron-Radar Receiver Oscillator," S. Krasik, *Westinghouse Eng.*, Vol. 6, No. 6, Nov. 1946, p. 176.
80. *Klystron Tubes*, A. E. Harrison, McGraw-Hill, 1947.

Chapter 6

GAS TUBES

D. E. Marshall

BY definition, a vacuum tube containing sufficient gas to influence its operation is a gas tube. The influence of gas, in general, results from its ionization, so that positive charges are available to neutralize the electron space charge. Gas tubes, therefore, can conduct large currents at comparatively low voltage drop and therefore operate at high efficiency.

The term gas tube covers tubes containing gases that can condense within their operating temperature range, as well as tubes with gases which cannot so condense. Among the former are mercury-vapor tubes; the latter are xenon-, neon-, and argon-filled tubes.

Gas tubes range in size and capacity from large multianode pool rectifiers to small control thyratrons, or even smaller protective spark gaps of various types. Large tubes are applied in power rectification and inversion, in heavy-current resistance-welding control; smaller ones are used as rectifiers for d-c supply in radio receivers, as control relays, lock-in relays, and for other similar applications.

Detailed descriptions and ratings of commercial tubes are available in technical information sheets published by tube manufacturers.

6·1 GAS-TUBE OPERATION

The operation of a gas-filled tube is determined almost entirely by the circuit in which it is used. Since anode-cathode voltage is independent of grid voltage during conduction, the current through the tube after it has started can be changed only as determined by the voltages and impedances of the external circuit. The tube can be prevented from conducting in the forward direction, but after conduction starts it cannot be stopped or influenced by grid control. Conduction can be stopped only by reducing the current to zero long enough for the tube to deionize.

Control in D-C Circuits

In d-c circuits gas tubes can be used as sensitive indicating and lock-in relays, to discharge capacitors and for similar applications. Circuits using capacitors in such a manner that they can be discharged backward through the tubes, thus reducing the tube current to zero and stopping conduction, are used in inverter circuits, for example. By use of grid- or ignitor-controlled gas tubes in inverter circuits, alternating voltages controllable in magnitude and frequency can be obtained.

Control in A-C Circuits

In a-c circuits the current becomes zero twice each cycle, and conduction of the tube can start and stop because the current is stopped periodically by the external circuit. Starting time in the cycle can be controlled, so conduction can be made to start and stop periodically. The grid, therefore, does not modulate the current but does control the phase of starting the current through the tube.

Phase Control in A-C Power-Conversion Circuits

In polyphase rectifiers and inverter circuits, current through the tube continues until it is stopped by the action

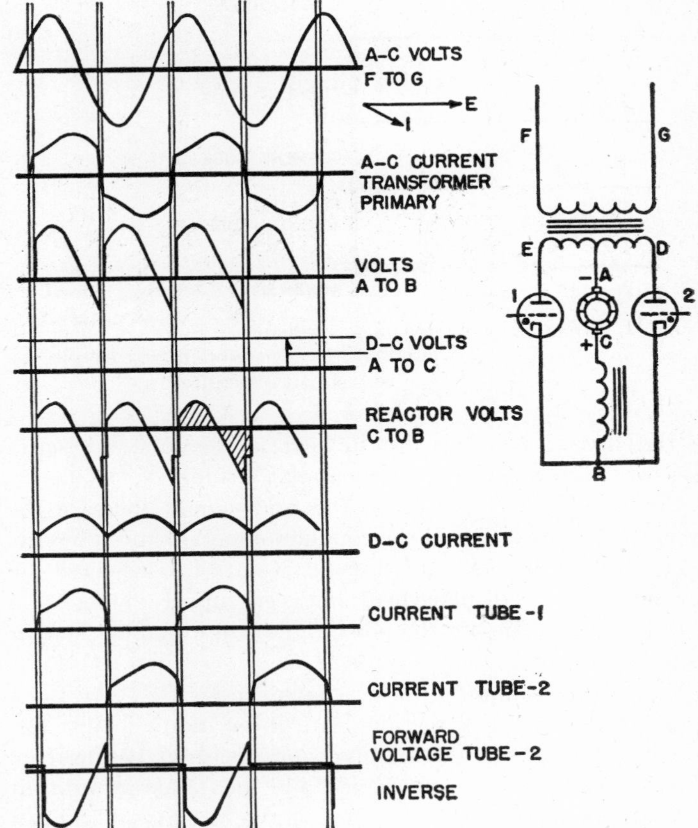

FIG. 6·1 Wave forms in rectification: single-phase full-wave operation, motor load. Derived on assumption of zero circuit resistance.

of firing the next tube in sequence at a definite phase of the next cycle; the time of conduction then becomes a predetermined fraction of a cycle. By controlling the phase of starting the tubes, the currents are of constant time duration, but are displaced in phase with respect to the alternating supply voltage. Control of the phase of firing changes the magnitude of the output power and reflects the change back into the supply circuit as a change in power factor.

When current is controlled in magnitude by the potential of a grid in a high-vacuum tube, the control is effected by increasing the voltage drop across the tube from cathode to anode. This increases the loss in the tube, and a large part of the circuit energy is lost in heat in the tube. A series resistor will control power in a similar fashion.

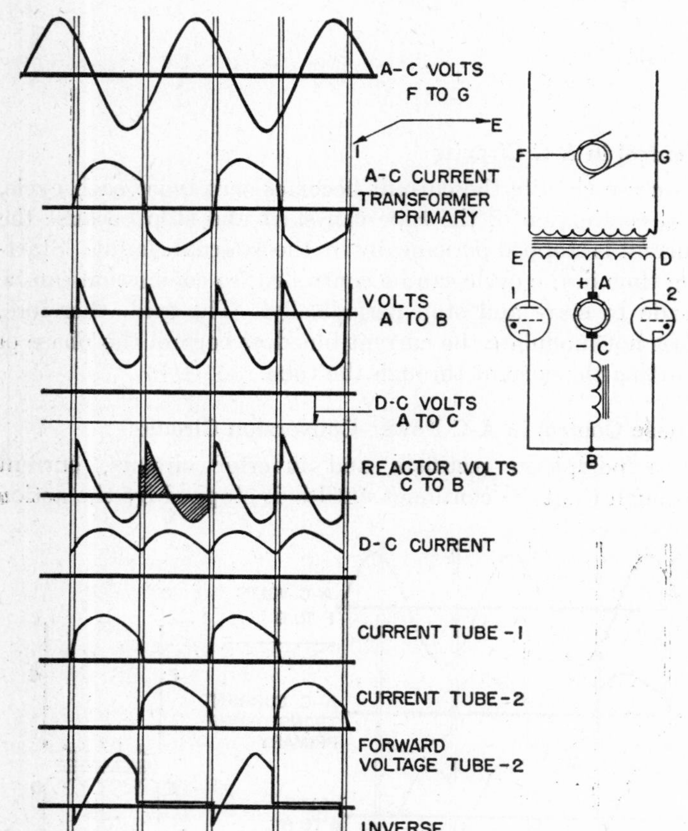

Fig. 6·2 Wave forms in inversion: single-phase full-wave operation, synchronous machine commutation, d-c generator drive. Derived on assumption of zero circuit resistance.

Conversely, in a gas tube, the grid cannot increase the anode-cathode voltage in the conduction period and, thereby, cannot increase the tube losses. Control is effected by causing a phase displacement between supply voltage and current, thus changing the power factor in a manner similar to the action of a series reactor in an a-c circuit.

The theoretical wave forms shown in Fig. 6·1 are for a single-phase full-wave rectifier with a d-c motor load. The wave forms of Fig. 6·2 for inversion are similarly derived. Their construction is simplified by the assumption that resistance is zero in all branches of the circuit. The load-smoothing inductance, however, is assumed to be finite rather than infinite, as it is usually assumed to be.

The shift in phase between input voltage and current mentioned is evident. The vector diagrams indicate the phase relations. Both rectifier and inverter have lagging wattless-power components, but the real power components are such as to feed power into the a-c supply system in inversion and into the d-c system in rectification. This change in phase is determined by the settings of the firing phase of the tubes.

Phase Control in A-C Welding Circuits

By using two gas tubes in inverse parallel, current can be passed in both directions through the combination. Switching action is obtained on a-c circuits with control of the current-starting time of each cycle.[1, 2, 3] If this combination is used to control current through an inductive load, the action of the tubes in stopping conduction at or near the zero-current point of the cycle interrupts the current when the stored magnetic energy is low, thus minimizing inductive disturbances.

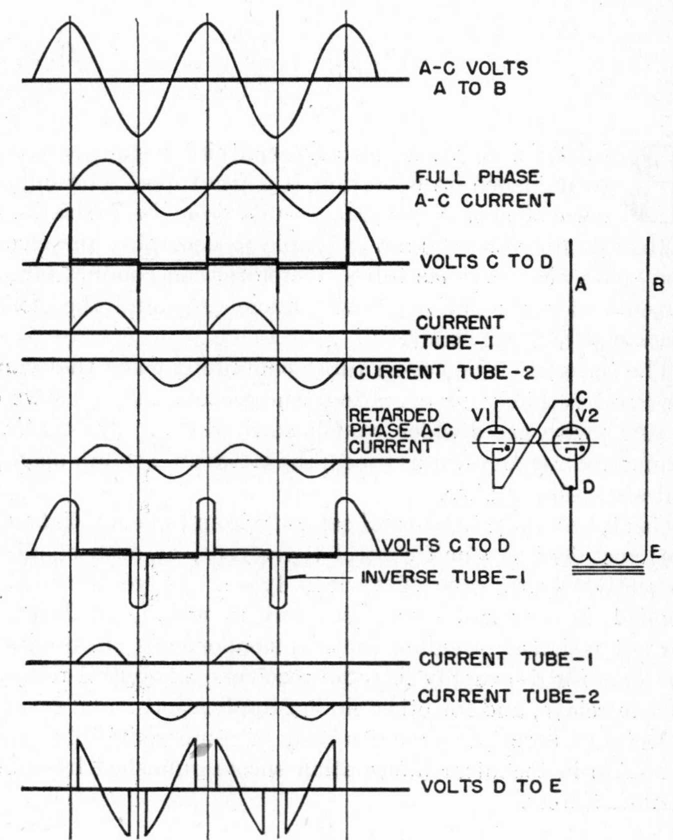

Fig. 6·3 Wave forms in a-c resistance welder control with and without phase control. Derived on assumption of zero-power-factor load.

Wave forms derived on the basis of zero-power-factor load are shown in Fig. 6·3. Two cases are shown, with and without phase delay in tube firing. Control efficiency is high, because current and voltage across the tubes, other than arc drop, are not simultaneous.

6·2 GAS-TUBE TECHNICAL DATA

Characteristics

Tube properties inherent in the design and not controllable by the user are called characteristics. For example, cathode-heater current at rated cathode voltage of a thyratron or the critical grid-voltage-control curve is a characteristic.

Figures for characteristics of a tube usually are given with plus and minus tolerances. It is not always stated whether these tolerances are for new tubes or whether they are valid throughout the guaranteed life of the tube. The present tendency, however, is to assign tolerances to cover expected deviations in the various characteristics throughout life. In addition to the deviations caused by use are the manufacturing tolerances and the effect of any unlisted parameters; for example, the effect of condensed mercury on the critical grid characteristic of a thyratron.

Often so-called typical characteristics are given in technical sheets. These usually show data for an average tube and do not show probable variation in characteristics in a manufacturing lot. They are of value in showing variation in characteristics as a function of various parameters.

Ratings

Lists of values limiting the amount that tubes can be stressed in various ways are called ratings. Ratings usually are given as maximum or peak values or maximum ranges; for example, the peak current rating of a thyratron or the maximum temperature range of a phanotron.

All values in one column of a rating sheet form a consistent system. Several columns of ratings may be needed for a given tube. For example, the demand-current rating of an ignitron can be 9600 amperes at 220 volts and 4800 amperes at 440 volts. Both rating groups are individually consistent, but 9600 amperes at 440 volts is not. The ratings do not apply individually but in groups.

Ideally it should be possible to assign to a tube ratings which would be independent of the type of duty for which it is used. This has not as yet been found feasible. For radio tubes typical operating conditions often are given. These list all circuit constants affecting tube operation. In general, industrial tubes are used for widely differing applications, so the rating system is used for most purposes instead of typical operating conditions.

To be more accurate, some types are rated for use in general classes of circuits; ignitron ratings for welding and rectifier circuits are in this class.

All ratings for gas- or vapor-filled tubes are on the "absolute maximum" basis, for no factors of safety are assumed.

Use of Technical Information

Rating information represents the manufacturer's contract with the user. By it, he agrees to furnish a device with renewals when necessary which will perform certain functions and will have a reasonable life. It need not reflect the actual capabilities of the devices, because the ratings may be purposely low to insure long life, or they may be tentative until field experience indicates that higher ratings are justified.

Tests made by the user indicating that a certain tube performs well at currents, voltages, or temperature beyond its ratings are not sufficient justification for use of the tube over its ratings. The tube may have short life if it is so used, or at some later time the manufacturer may redesign it in such a way that the capabilities of the tube more nearly coincide with the absolute maximum ratings.

The outline drawings shown may not be truly representative of the tube. The manufacturer may put extra width or length in the dimensions to insure sufficient room for future enlargement when tube design changes are made. An apparatus designer who designs his equipment around a sample may later find that replacement tubes are too large.

Variation in line voltage must be considered in designing circuits, because such ratings as peak inverse voltage and peak currents, as measured by a high-speed instrument such as an oscilloscope, should not be exceeded.

Short-circuit or arc-back currents should not exceed the surge-current ratings. Use of transformers of correct impedance or the use of additional resistors or reactors ordinarily satisfies this requirement.

Adequate ventilation must be provided for air-cooled mercury-vapor tubes to insure that the condensed mercury temperature does not exceed or fall below the rated value under conditions of the extreme range of ambient temperature. Temperature rise above ambient air temperature of a tube depends upon the natural convection of air. Obstructions to the natural flow of air are to be avoided.

Heat exchangers should be used for water-cooled tubes wherever water analysis indicates improper purity or cleanliness of the water. Parts of many metal tubes are made of stainless steel where cooling water touches them. High chlorine concentration and silt deposit in cooling water lead to early corrosion and failure of such tubes.

Circuits should be designed to operate properly by using tubes within the specified characteristic tolerances. As an example, circuits depending upon the accuracy of the thyratron grid-control characteristic do not represent good design. Control of thyratron grids should be effected by signals which pass through the control curves at the highest possible time rate so that temperature variations in the characteristic, variations in manufacture, and change in life do not unduly affect the phase of tube breakdown.

6·3 GENERAL GAS-TUBE RATINGS

Maximum Peak Current

Figures 6·1, 6·2, and 6·3 show that, even if equipment is operated continuously, the current conducted by the tubes is intermittent. In addition, the wave form of the currents varies in different circuits. The ratio of peak to average or rms currents, therefore, varies. For this reason, it is not sufficient to specify the current as measured by the electrical instrument in general use; oscilloscopic measurements or wave-form calculation must be used.

The maximum current a tube can conduct safely is limited by several factors. If it is a hot-cathode tube, the cathode is usually the limiting factor. If currents greater than the rated peak current are passed through such a tube, emission of the cathode may not be sufficient to furnish enough electrons to carry the current from the cathode. Electron emission then must be enhanced by the action of positive ions. To produce the required excess of ions, the arc drop of the tube increases. This voltage accelerates the ions moving back to the cathode, thus increasing their momentum. As these ions hit the cathode they damage it.

Cathodes of the pool type have high electron emission. Peak-current limits are applied to them to limit the probability of arc-back. The high current causes rapid evaporation of mercury from the cathode spot. This tends to increase vapor pressure and retard the diffusion rate. Deionizing time is increased. All these factors tend to cause the tube to lose control or arc-back. Therefore, the peak-current rating of a pool tube is controlled by the magnitude of other conditions, such as maximum inverse voltage. Rating tables for these tubes consequently have a series of peak-current ratings dependent upon average current rating and operating voltage.

Maximum Average Current

The average energy loss in any element of an electrical circuit is

$$P = \frac{1}{t}\int_0^t ei\,dt \tag{6·1}$$

where e = instantaneous voltage across the element
i = instantaneous current in amperes through the element
t = duration of current conduction
P = average energy loss in watts.

The anode-cathode fall of voltage in gas tubes is nearly independent of current. If it is assumed to be completely independent, it can be removed entirely from the integral as follows:

$$P = \frac{E}{t}\int_0^t i\,dt = EI_{d\text{-}c} \tag{6·2}$$

Thus, the power loss is equal to the product of the arc drop and the average current. This energy loss results in the heating of various parts of the tube. If the tube is not to be overheated, the maximum average current rating must not be exceeded.

Maximum Averaging Time

Gas-filled tubes are often operated intermittently, as in control circuits. Figure 6·4 illustrates such operation. The current is on for a short time, followed by a period of "off" time. Effective temperature of the tube increases and decreases. In such applications the average current conducted by the tube is controlled by the amount of the "off" time included in the quantity t in equation 6·2. If a large part of the "off" time is included, the average current corresponding to a given cycle is decreased. To limit tube heating, the time t over which the current is averaged in equation 6·2 must be limited to the manufacturer's rated maximum averaging time. If a rating is given as an average, the corresponding maximum averaging time should be stated.

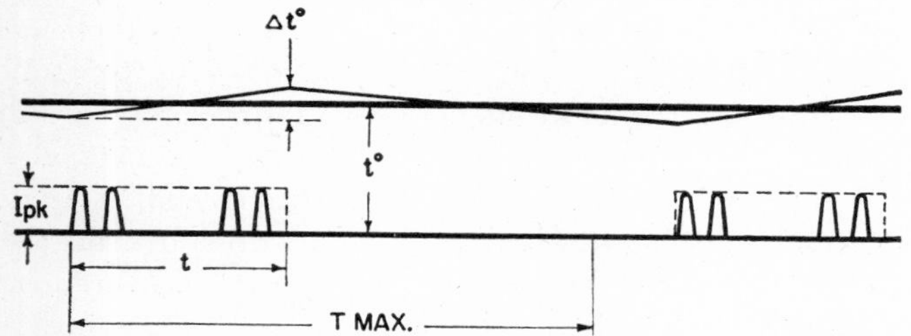

FIG. 6·4 Intermittent duty, maximum averaging time.

If the tube is operated within the rated maximum averaging time, the effective temperature of the tube will not vary more than a few percent on continuous duty.

The following discussion will illustrate the significance of the maximum averaging time concept. In Fig. 6·4, the variation in temperature of some critical part of the tube during the assumed intermittent-duty cycle is shown by the upper curve. The height of this curve represents equivalent temperature at any time. Average temperature is $t°$, as shown, and the variation in the temperature is $\Delta t°$. The temperature rise for comparatively short current duration is a function of the heat-storage and heat-dissipating capacity of the tube. This is expressed

$$\Delta t° \approx \frac{EI_{pk}t}{mc} \tag{6·3}$$

where E = arc drop
I_{pk} = peak current
t = periodic time of conduction
m = equivalent heat mass of tube
c = thermal capacity of unit mass.

$$t° \approx \frac{EI_a}{kA} \tag{6·4}$$

where I_a = average current
k = heat-loss constant
A = equivalent cooling area.

To prevent overheating the ratio $\Delta t°/t°$ should not exceed a certain value. Combining equations 6·3 and 6·4,

$$\frac{\Delta t°}{t°} \approx \frac{kAI_{pk}t}{mcI_a} \tag{6·5}$$

If maximum averaging time is to be considered, the basic equation used in applying tubes is

$$I_{pk}t \approx I_aT_m \tag{6·6}$$

where T_m = rated maximum averaging time.

Substituting,

$$\frac{\Delta t°}{t°} = \frac{kA}{mc}T_m \tag{6·7}$$

Thus, the value chosen for the maximum average time rating theoretically fixes the percentage of variation in tube temperature.

The equivalent temperature influencing tube operation cannot always be measured. Maximum averaging time usually is fixed by convention and proved by test. Once established, its use insures a nearly uniform factor of safety in the tube rating over a wide range of intermittent duty.

Maximum Surge Current

If the load circuit becomes short-circuited a large current passes through the tube until a circuit breaker opens or a fuse blows. If the impedance of the power source is small, large currents can pass through the tube. This may result in severe damage. The maximum surge-current rating indicates the regulation of the power source and speed of operation of protective devices necessary to protect the tube during a reasonable number of fault conditions. Short-circuit and arc-back tube currents are determined by the impedance of the circuits involved. They can be reduced below the surge-current rating by reactors or resistors in the anode leads or transformer primary or by choice of proper transformer impedance.

If rectifiers are operated in parallel with other rectifiers, or used to supply power for motors, the arc-back current may be increased by a current feedback from the d-c bus.[19, 20] As shown in Fig. 6·5, the reverse direct current from the motor increases arc-back or reverse surge current through the tube. Constants of the load circuit also are therefore important

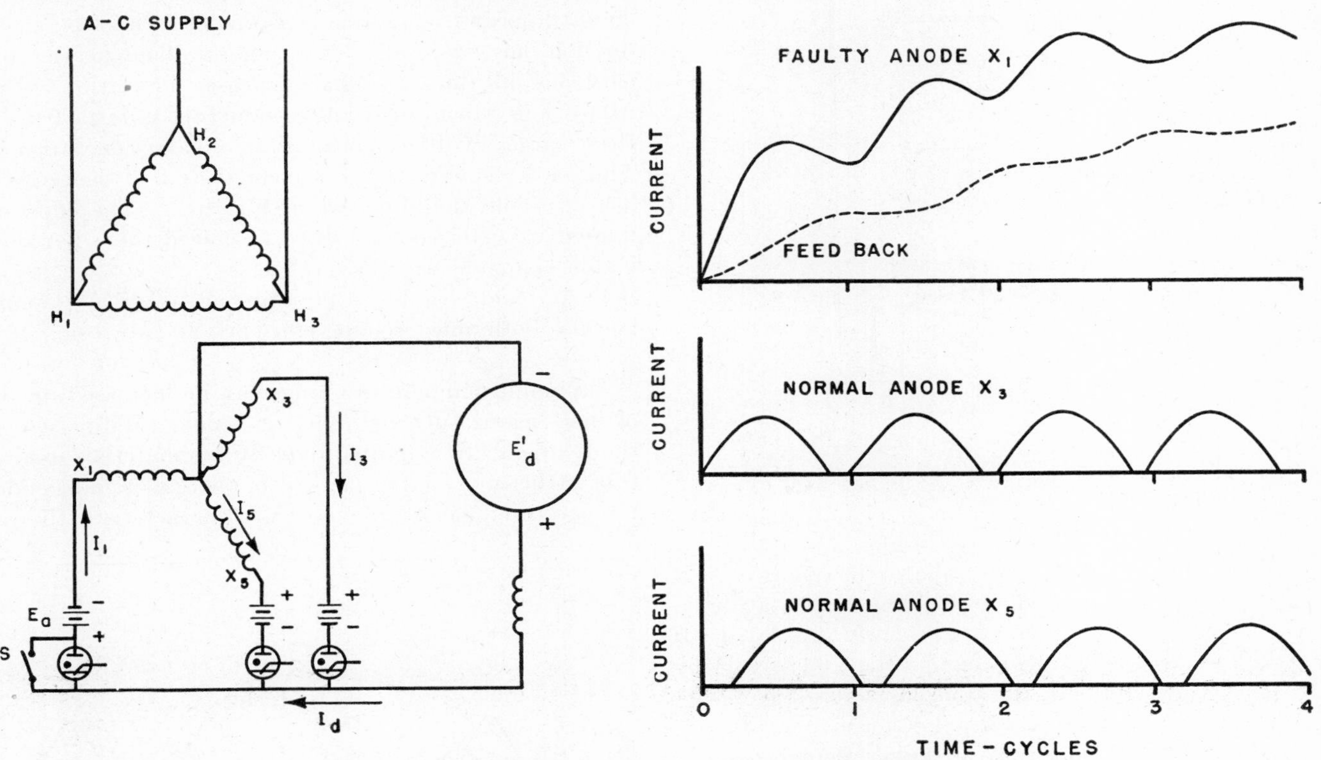

Fig. 6·5 Arc-back current in rectifier with current feedback from d-c side. (Reprinted with permission from Herskind and Kellogg, "Rectifier Fault Currents," *Trans. A.I.E.E.*, Vol. 64, pp. 145–150.)

and must be considered in addition to the transformer impedance.

Other factors result in a much lower surge-current rating than would be given if tube damage were the only consideration. There is some evidence to support the theory that arc-back may be influenced by the impedance of the power source. Therefore, reduced surge-current ratings are given for certain applications. This is based on commutation time of rectifier circuits and the rate of rise of current in arc-back. The first consideration limits ionization in the tube at the end of the conduction time, when inverse voltage is suddenly built up from anode to cathode. The second consideration is based on a belief that incipient arc-backs are more frequent than actual arc-backs. An arc-back may start but fail to develop because the time duration of the phenomena causing it is so short that the arc-back current cannot rise to a stable value before the cause disappears. If circuit regulation is decreased, then a larger number of incipient arc-backs can materialize into self-maintaining arcs.

Peak Inverse and Forward Voltage

Figure 6·1 shows that variation of the voltage across the tube is cyclic. During conduction the voltage is low but, when the current stops, the voltage in a rectifier circuit may suddenly increase to a comparatively high negative value, followed by a period of higher negative voltages. If the tube in Fig. 6·1 is grid-controlled, a period of positive voltage comes before conduction. In inversion service the conditions are reversed; immediately after conduction a negative peak may come as before, but it is followed by a cyclic period of positive voltage.

Maximum safe voltage in either the inverse or the forward direction is dependent upon the dielectric properties of the envelope, but more usually is limited by the tendency of the tube to arc back or lose grid control. Therefore, a given tube may have a series of peak-inverse-voltage or peak-forward-voltage ratings, depending upon peak and average currents and the temperature range of operation.

In tubes filled with gases such as xenon, the forward and inverse maximum voltage ratings take on an additional importance, because the gas pressure usually is as high as it can be made under the voltage ratings. Exceeding the voltage rating of a gas-filled tube may cause immediate formation of a glow discharge which causes loss of grid control; with mercury vapor it may cause an occasional arc-back.

Thermal Ratings

Mercury-vapor tubes of some types are rated in terms of the condensed mercury temperature range. This temperature should be measured in the region designated by the manufacturer at the point where the mercury actually condenses. This point usually is immediately above the base on glass tubes and on specially provided radiators on metal tubes, as shown in Fig. 6·6. These radiators are soldered to thin metal tubes which have low heat conductivity and in which the mercury condenses. Vapor pressure in the

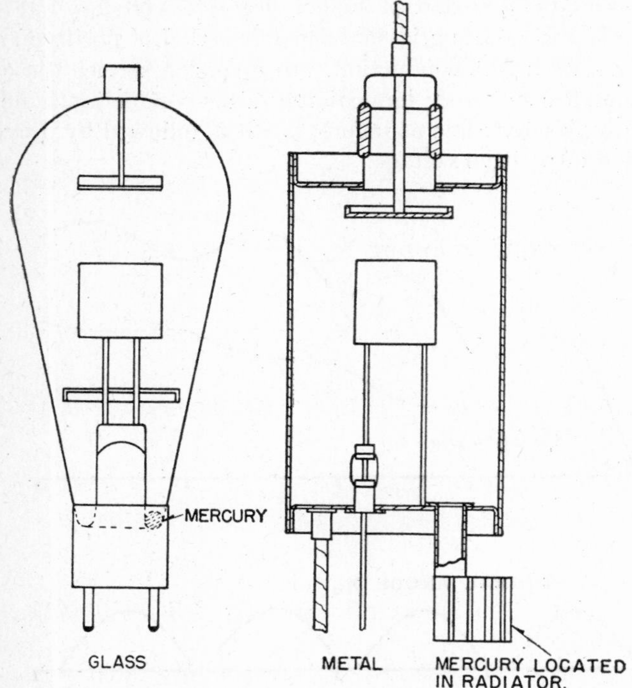

Fig. 6·6 Mercury condensing points in glass and metal tubes.

tube is closely controlled by the radiator temperature. If the mercury condenses in a region different from that designated, it is a sign that the tube is not properly ventilated. If the mercury condenses at a point high in the tube, it falls by gravity to the lower part of the tube; as a result it evaporates at a higher temperature than that measured. In this case, the mercury-vapor pressure is higher than necessary. Sometimes the falling mercury may cause internal short circuits.

The pressure and density of mercury vapor at room temperature is satisfactory for operation of ionic tubes, otherwise the tubes could only be operated inside heated or cooled enclosures. The range of temperature at which the pressure is suitable for operation is, however, undesirably narrow. A range of condensed-mercury temperature from 25° to 80°C is about the widest that can be tolerated.

Ventilation and cooling of mercury-vapor tubes is of the greatest importance. The low maximum mercury temperature of 80°C allows only a small temperature gradient compared to a water temperature of 40°C or an ambient air temperature of 60°C in enclosed apparatus. The temperature rise above ambient of a mercury tube of the hot-cathode air-cooled type under such conditions would have to be about 20°C, which is low for the amount of heat to be dissipated.

Water cooling is often used because the high specific heat and efficient heat transfer of water help to remove a large amount of heat without excessive temperature rise, and because the temperature of cooling water usually is lower than the ambient temperature of the air.

The temperature difference between the water and the surface from which heat is being removed depends upon the velocity of water. For this reason, it is necessary to give a water-cooled tube a rated minimum water flow. This flow must be independent of water temperature to insure that the operating temperature of the tube is not too high.

Exceeding the maximum temperature rating may cause arc-back, and reduction below the minimum may cause insufficient vapor density for proper conduction,[16] and may result in sudden stoppage of current. In inductive circuits this generates high-voltage oscillations and may cause insulation breakdown.[15] The high voltages caused by the entire circuit appear from anode to cathode of the tube after conduction has stopped. This high voltage accelerates the electrons in the tube and increases ionization efficiency. Many ions are formed and conduction is restarted. After the "surge" voltage dies away, conduction stops again. Thus, a cyclic variation in the conduction of the arc is established. This is illustrated in Fig. 6·7. The effect is not limited to water-cooled tubes; air-cooled tubes act similarly if tube temperature falls too low.

Purity and cleanliness of water are of great importance, because both affect cooling efficiency and corrosion of water jackets.

Maximum temperature range can be increased by the use of inert gases, such as argon or xenon. With these gases, the temperature is limited only by the materials used in the tube structure, for the density of the gas is independent of temperature. Gas, however, has a characteristic that makes

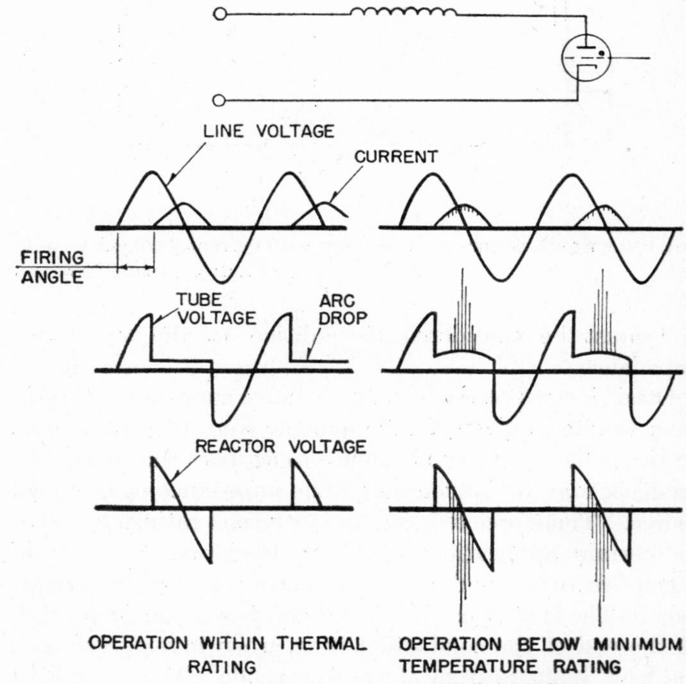

Fig. 6·7 Failure of mercury tube to conduct (surging) owing to low gas density.

its use difficult. It dissolves in the materials of which the tube is made. This is known as "clean up." As the tube operates, the gas becomes ionized and attaches itself to the tube parts. It can be recovered sometimes by heating the tube parts to drive off the gas.

For reasonable tube life, each gas tube usually is filled to a higher density than is necessary. As the gas "cleans up," a longer time interval is available before the gas density becomes too low for satisfactory operation. The use of this higher density limits the dielectric strength of the tube in the inverse direction, so gas-filled tubes usually are best adapted to low-voltage circuits. These limitations are not inherent and will be minimized as knowledge of the prevention of "clean up" increases. Mercury vapor also "cleans up"; but an excess of mercury can be provided, thus making vapor density a function of temperature, regardless of the amount combined with the tube parts.

6·4 ARC-DROP VOLTAGE

Anode-to-cathode voltage of a gas- or vapor-filled tube during conduction is known as arc drop. This drop usually

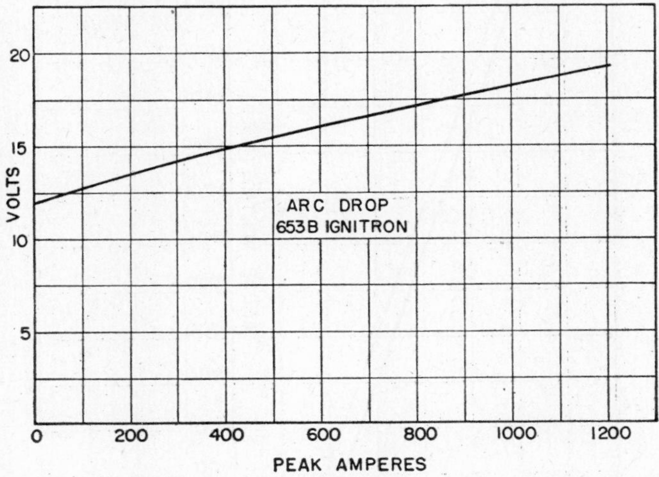

FIG. 6·8 Arc drop of typical ignitron as a function of peak current conducted.

is considered to be independent of the current conducted, but this is an approximation because experiment demonstrates that arc drop is a rising function of current. Figure 6·8 shows a typical arc-drop curve plotted as a function of peak current. Actually, arc drop varies with several other factors such as the vapor pressure. Usually the entire arc drop is insignificant in quality of circuit performance and is of interest only in computation of tube losses and over-all circuit efficiency. The arc-drop wave form as determined by an oscilloscope during tube operation is not flat, but is square in shape with a superimposed "hump" which is dependent upon instantaneous current.

Usually these minor effects are neglected and the average value of arc drop is determined by approximate integration of the arc-drop wave form for a circuit of the same general type in which the tube is most generally used. The arc-drop curves generally given, therefore, are not volt-ampere characteristics because they are not functions of instantaneous arc drop and conducted current, but are only averaged values corrected with respect to the peak current level at which the tube is to operate.

Because arc drop is a function of many variables, most tube-data sheets give only nominal values to indicate that the tube is of the low-loss type.

The arc drop of a hot-cathode tube differs in some respects from that of a tube of the pool type. Cathode emission is in general a function of the field at the cathode. If thermal emission is low, the current demanded by the circuit is obtained by an increase of voltage drop at the cathode. This drop added to the drop through the plasma results in increased arc drop. More variation is to be expected in arc drop of hot-cathode tubes than in tubes of the pool types, but it usually is a measure of tube quality and circuit efficiency rather than a characteristic influencing circuit operation.

A mercury arc is somewhat unstable at low currents, so the arc drop may rise rapidly at the end of a conduction cycle. In inductive circuits the voltage generated by the inductance may rise to high values. Resistors are often put in parallel with the arc or load to provide paths for current and reduce inductive voltage.

6·5 RATINGS AND CHARACTERISTICS OF HOT-CATHODE TUBES

Heater or Filament Voltage

The voltage at which the heater of an indirectly heated cathode or the filament of a direct heated cathode is operated is quite critical. Standard tolerance is ±5 percent. In some types of vacuum tubes, current can be limited by reducing the emission of the cathode. Such reduction causes the tube to operate at saturation with a high anode-cathode voltage. If gas pressure is low so that the gas is not ionized, such operation does not damage the cathode. In a gas tube the shortage of electron emission caused by low-temperature operation of the cathode is made up by ionic conduction, which increases cathode emission by field emission. The result is the same as discussed under Maximum Peak Current in Section 6·3. There is also a tendency for the discharge to form an arc on the cathode surface to increase the field emission. This formation of a concentrated arc is often called sparking, and it will materially damage the cathode.

Operation of the cathode at a temperature above the rated value shortens tube life because of the high evaporation rate of the oxide coating.

Heater or Filament Current

At rated heater or filament voltage the heater or filament current is a characteristic of the tube. It is given in the characteristic sheet to aid in the design of filament transformers. Many of these transformers are designed with high voltage regulation. Filament-current tolerance then is of importance in determining the constancy of the filament voltage with a given transformer tap setting when tubes are replaced in the socket.

The product of heater or filament voltage and current is the heating power in watts. It represents an added power loss and causes a loss of efficiency. In gas tubes this loss is to some extent under the control of the tube designer. The cathode can be designed so that the emission area is deeply recessed in the cathode structure, as illustrated in Fig. 6·9. The plasma extends into these cavities so that electrons emitted from the enclosed surfaces become available for eventual conduction to the anode, and at the same time the radiation-emitting area can be made quite small. The cathode structure can be surrounded by several sheets of

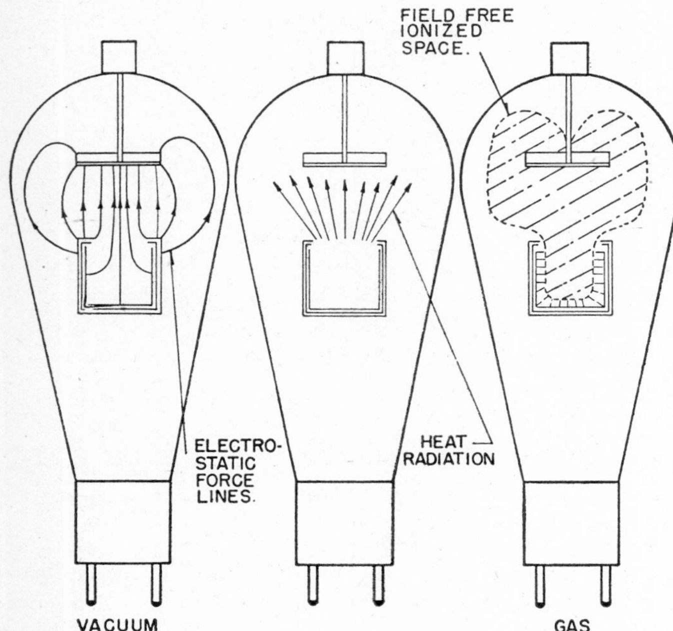

FIG. 6·9 Thermal efficiency of cavity cathodes in gas-filled tubes.

bright metal, as shown in Fig. 6·10, which act as radiation shields reflecting energy from the cathode and further reducing the heat losses.

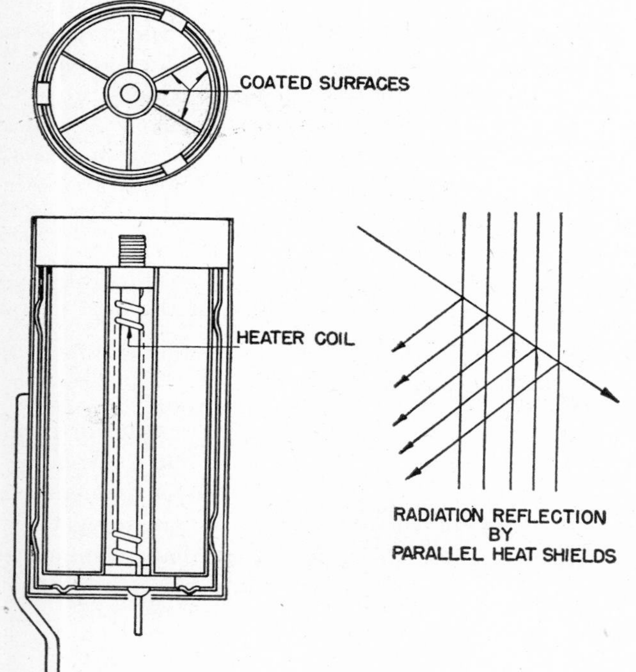

FIG. 6·10 Typical design of deep-cavity cathode showing reflector heat-conservation methods.

Cathode-Heating Time

As improvements in the cathode-heating efficiency are made, the ratio of energy input to cathode mass decreases. Total stored heat at a given temperature increases with mass. The time required to heat the cathode to a given temperature is a function of the heat mass and the input heating energy. As heating energy is reduced, time increases. Therefore, the tendency is to make the heating time as long as possible without sacrificing convenience in use. Cathode-heating time is a characteristic of the tube design and is shown in tube-data sheets. Failure to wait the specified time before starting conduction causes troubles similar to those resulting from reduced cathode-heater voltage.

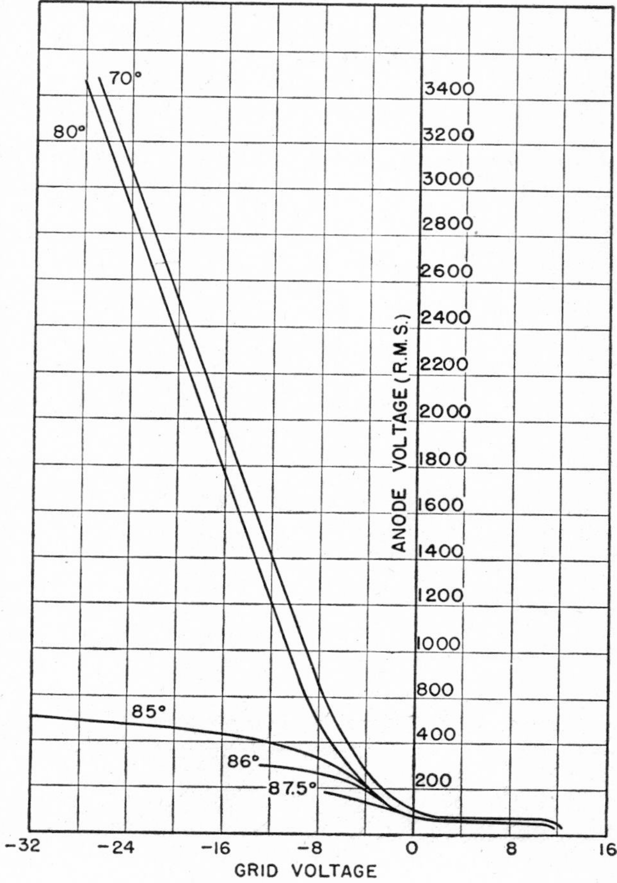

FIG. 6·11 Typical variation of thyratron critical grid-voltage curve as function of condensed-mercury temperature.

Heating time can be reduced somewhat by using light parts for cathodes. This may result in low surge-current ratings, because the light parts may not be able to carry the large currents without damage.

The use of a direct-heated shielded cathode has some advantage in reducing heating time. In an indirectly heated cathode the heater usually is a coil of bare-tungsten wire. The coil must be heated first, and the temperature rise of the emitting surfaces follows. In a directly heated shielded cathode, the emitting surfaces are heated first, thus saving some heating time.

In mercury tubes the cathode-heating time may be less than the time required to heat the tube to operating temperature. If the ambient temperature of the tube is less than the minimum condensed-mercury temperature rating, the tube is heated by energy loss from the cathode. This may require more time than to heat the cathode to its operating temperature.

6·6 RATINGS AND CHARACTERISTICS OF THYRATRONS

Thyratrons differ from other gas or vapor tubes by having one or more grids which control the starting of conduction of current through the tube.

Critical Grid Voltage

The blocking grid voltage of a thyratron is a function of anode voltage, tube temperature, or gas pressure, as shown in Fig. 6·11. Cathode emission also is important, as shown in Fig. 6·12, where emission is caused to vary experimentally by varying the cathode-heater voltage. In addition to these variations, a production variation in tube dimensions causes further variations in the grid-control curve.

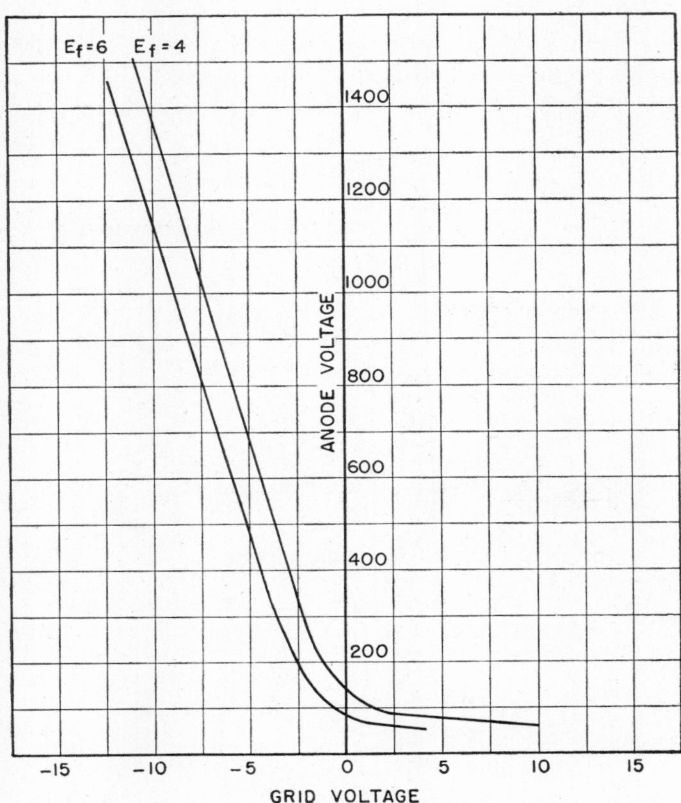

Fig. 6·12 Typical variation of thyratron critical grid-voltage curve as function of cathode-heater voltage.

These characteristics sometimes are depicted as broad regions, as shown in Fig. 6·13, instead of as sharply defined lines. Showing the characteristic in this way reduces the number of applications requiring high accuracy of the grid characteristic.

Shield-Grid Characteristics

Shield-grid or four-element thyratrons have two grids. The shield grid shields the control grid from the cathode and from the anode. By shielding the control grid from the cathode, less of the coating evaporated from the cathode is deposited on the grid. This coating on the grid promotes grid electron emission and must be minimized. By shielding the control grid from the anode, the anode-to-grid capacitance is reduced. This is important in reducing false operation upon sudden application of anode voltage (Fig. 6·14).

The critical curve of the control grid can be changed by changing the shield-grid-to-cathode potential. Usually the grid is connected to the cathode but, as shown in Fig. 6·15, a family of curves can be drawn to show the control-grid critical curve as a function of shield-grid potential. In this manner, a negative-control tube can be made to operate as a positive-control tube. If the potential of the shield grid is held negative so that the control-grid critical curve lies in a region of positive voltage of magnitude greater than the ionizing potential of the gas in the tube, the voltage necessary to cause the tube to start conducting may become considerably greater than the voltage to which the control grid must be reduced to prevent conduction. This is because the positive control grid draws a small current after the tube has started to conduct. This current is sufficient to keep the tube ionized. The potential of the control grid must be reduced below the ionizing potential to be completely certain that the current is stopped. A reduction to a value greater than the ionizing potential may at times be sufficient if the ionization becomes unstable. However, difficulty on some d-c control circuits may result.

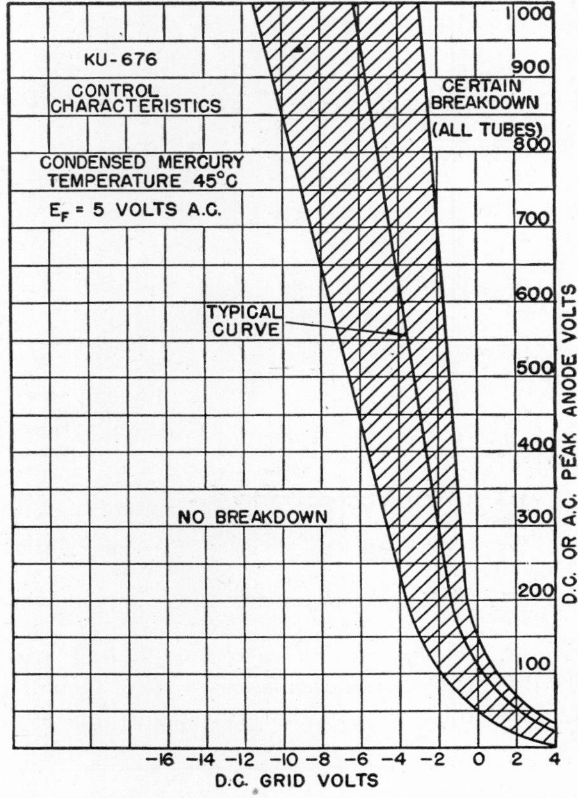

Fig. 6·13 Grid-control characteristic curve showing area of tolerance.

Grid Characteristics

Grid-current characteristics before conduction[4, 5] follow the course shown in Fig. 6·16. The dotted curve is based on the assumption that there is no grid electron emission or leakage current. The anode is assumed positive. Starting with a large negative grid potential, the grid current is at first zero; the tube is not conducting because the field at the grid is highly negative and prevents electrons from leaving the cathode. As the grid is made more positive, electrons begin to move to the anode from the cathode. They generate positive ions which are attracted to the negatively

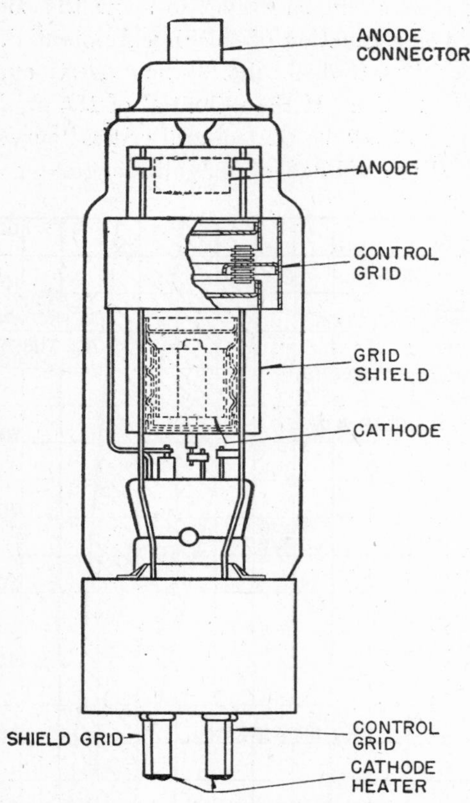

FIG. 6·14 Cross section of WL-672 shield-grid thyratron.

charged grid. Thus, the grid current is at first negative. As the grid is made more positive, electrons of high velocity begin to force their way to the grid against the opposing

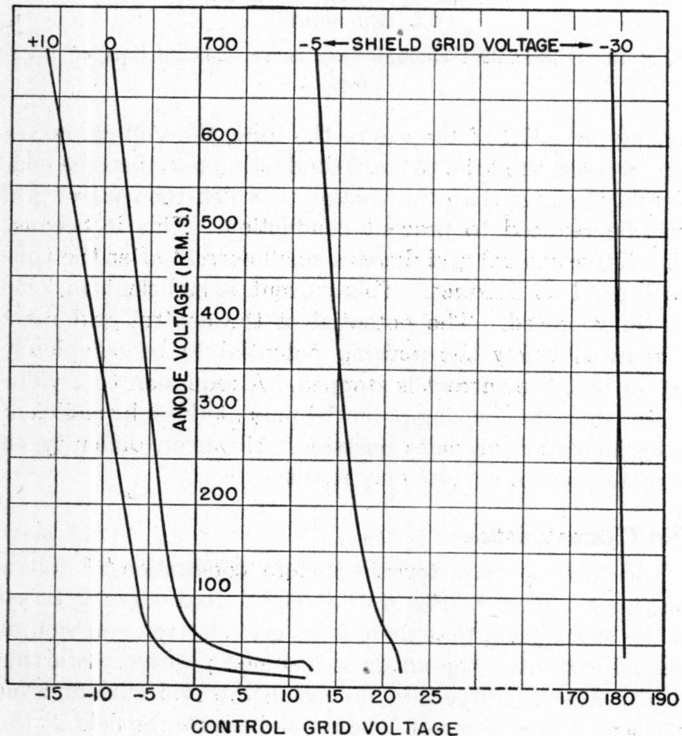

FIG. 6·15 Typical grid characteristics of a shield-grid thyratron.

potential, and the current at first decreases and then becomes positive as electron current increases. The grid is not always able to go through this complete cycle before the grid space charge becomes high enough to shield the grid, and tube conduction results.

The straight line in Fig. 6·16 is the regulation characteristic of a bias battery and series resistor. Voltage on the grid corresponds to the intersection of this line with the grid-current characteristic. Actual bias voltage ahead of the resistor at anode-to-cathode breakdown is a function of the size of the grid resistor. Grid current usually is increased by electron-emission current from the grid in addition to the currents previously described. This is shown by the line so marked in the diagram. Measurements of grid currents

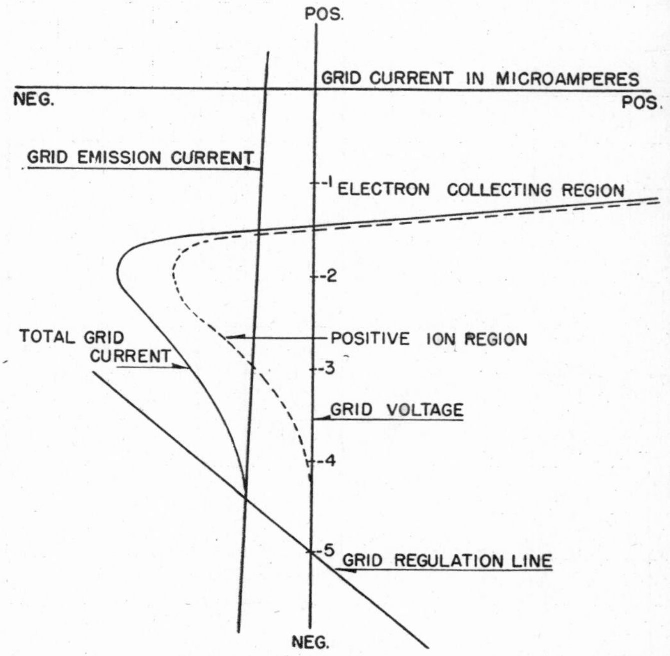

FIG. 6·16 Diagram of thyratron grid-current characteristic before conduction.

usually are taken by measuring the change in apparent breakdown voltage caused by the use of two values of grid resistors. A maximum value of this grid current before conduction is given in the manufacturer's tube-rating sheets under grid characteristics.

After the tube has begun to conduct, the grid current conforms to the phenomena described in Section 4·5. When the grid is negative, it collects ions, and when it is positive it collects electrons. The magnitudes of these currents are proportional to the anode currents. They follow the general course of the curves shown in Fig. 6·17.

Two grid-circuit regulation lines are shown, one corresponding to negative bias and the other to positive bias. Under either condition the actual grid potential as obtained from the intersection of these lines with the grid-current curves differs little from the cathode potential. During conduction the grid potential is slightly positive. If the anode current is small, or if grid resistance is low, the actual grid potential can become negative. Damage from ion bombardment of the grid coating may result. Some manufacturers rate the grid at a maximum of 10 volts negative after conduction to prevent such damage.

Grid Ratings

If grid resistance is too small, a heavy electron current may be drawn to the grid when the grid bias is positive. This causes heating of the grid, which is undesirable because its temperature should be kept as low as possible to minimize electron emission from the grid. In small tubes the current to the grid added to the anode current may overload the cathode. The ratings "maximum grid-current, peak positive, anode positive" and "maximum grid-current, average positive, anode positive" are used as guides in determining whether such overloading may occur.

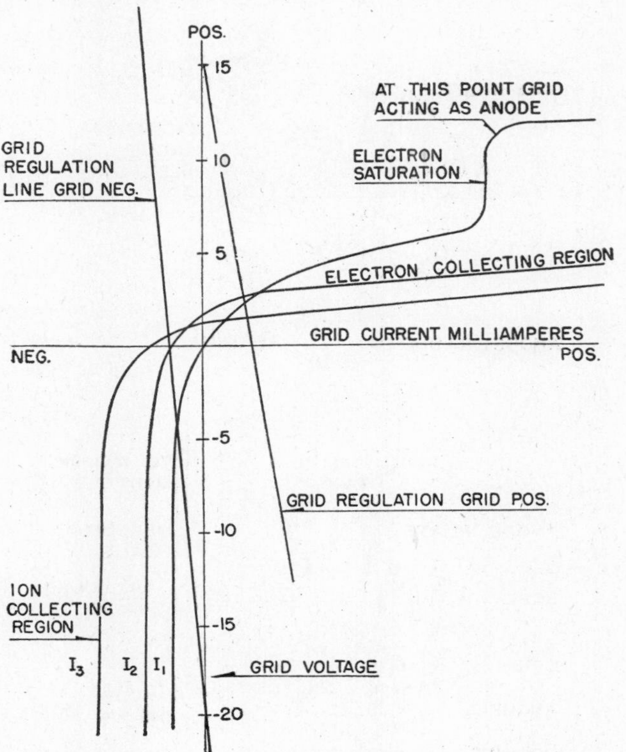

Fig. 6·17 Diagram of thyratron grid-current characteristic after conduction.

In high-voltage tubes, the control grid should not be more positive with respect to the cathode than the ionizing potential of the gas, as current drawn to the grid may cause the tube to arc back. This may not be too serious if grid-circuit resistance is high enough to limit the grid current to a small value. The rating applying here is "maximum grid-current, peak positive, anode negative."

If the grid voltage is negative when the anode is positive, the anode-to-grid voltage becomes the highest in the tube. Therefore, the probability of false discharge is increased. The rating covering this is "maximum grid voltage, peak negative, before conduction."

Ionization and Deionization Time

If grid potential is suddenly changed from negative to positive value greater than the critical voltage and held a few microseconds and then made negative again, ionization may not build up enough to cause the tube to conduct. The time required for the tube to ionize is called ionization time, and is a characteristic of the tube. It is a function of several variables, among which are anode voltage, grid voltage, tube temperature, and gas pressure.[6] Most rating sheets show only a nominal ionization time, usually 10 microseconds.

If the alternating supply frequency is increased too far, a gas or vapor tube which had previously operated correctly may fail to regain grid control between conducting cycles. This is because the anode-cathode space does not clear of ions before the tube is required to block. The time necessary is called the deionization time, and it depends upon diffusion of ions through the gas or vapor to the tube walls where they are neutralized after the conduction period. Tube-data sheets usually show only a nominal deionization time. The value usually given (except for specially designed tubes) is 1000 microseconds.

Grid-Capacitance Characteristics

Electrodes of all tubes are coupled electrostatically. One effect of this coupling is to make grid potential a function of anode potential. The change in grid voltage is determined by the potentiometer effect of the anode-grid and grid-cathode capacitances. If anode voltage is suddenly increased, the grid potential increases positively, tending to cause the tube to conduct. If the grid is properly biased, it is charged negatively, and this charge must be reversed before grid voltage can become positive with respect to the cathode.

A wave-form diagram of this is shown in Fig. 6·18. Here it is assumed that anode voltage is applied suddenly to a triode. The original grid voltage is negative because of the charge put on it by the bias battery. As anode voltage is applied suddenly, the charging current through the equivalent condensers is large. This current charges the upper capacitor and discharges the lower. If the lower capacitor is discharged and recharged oppositely, the grid potential becomes momentarily positive and the tube may start conducting without a definite grid signal.

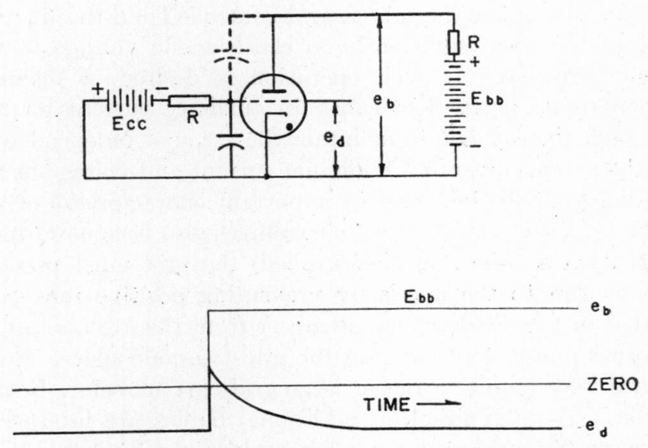

Fig. 6·18 Wave form of thyratron grid potential with suddenly applied anode potential.

The time the grid is positive is directly proportional to series grid resistance or in general to the impedance in the grid circuit. Even though the grid potential is positive during a disturbance, the tube will not fire provided that the time the grid is positive is less than the ionizing time corresponding to the anode potential, temperature, and the like. The grid circuit also serves as a wave trap for unwanted induced signals.

Shield-grid tubes have reduced anode-grid capacitance and thereby minimized danger of "shockover." However, they often have extremely sensitive grids and control-grid-to-cathode capacitors are required for the wave-trap function.

6·7 CHARACTERISTICS AND RATINGS OF MULTIANODE POOL-TYPE TUBES

Older types of tubes having pool cathodes were most often made in the multianode form, similar to that shown in Fig. 6·19.[7] Most of the ratings previously discussed apply to

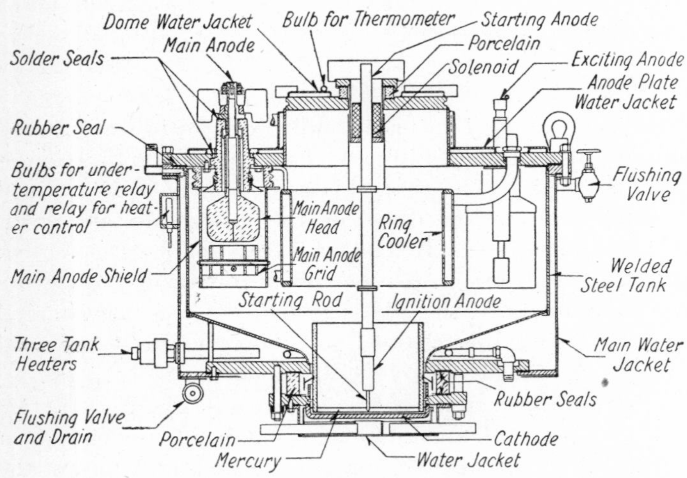

FIG. 6·19 Multianode mercury-arc rectifiers.

multianode tubes. The arc spot is maintained by a continuous current through the cathode. The direct output current on polyphase circuits does not stop even on resistance loads, unless the tube is grid-controlled and the starting point is delayed sufficiently to cause anode voltage to become negative. If such operation is desired, a separate circuit must be used to cause an auxiliary current to pass through the cathode to maintain the arc spot independently of the anode currents. Minimum current and voltage in the auxiliary circuit becomes an important characteristic of the tube. Maximum currents and voltages also become ratings.

If such a tube is grid-controlled, the grid must prevent conduction to the anode by preventing positive ions generated in the auxiliary anode arc or from the arc current to another phase from entering the grid-to-anode space. Such a grid may require a rather large grid current before breakdown. To minimize this, additional baffles are interposed between the grid and arc space to help deionize the space around the grid. The tube is designed with a fairly large distance between anode and cathode. Such distance and extra baffles cause additional arc drop. It is necessary to provide some means of initiating the arc and maintaining it. This is done by making and breaking a circuit to the mercury pool so that the interruption of the current causes breakdown. The arc may go out at times if, for example, the load current should intermittently stop. The initiating device must therefore be capable of restriking the arc automatically. Several methods of restriking are used. The usual method is to use a moving electrode controlled by a magnetic coil. Power required to operate the starting mechanism becomes a characteristic of the tube.

6·8 CHARACTERISTICS OF EXCITRONS

The Excitron is a single-anode tube of the pool type which requires a continuously operating auxiliary arc with characteristics similar to those mentioned in Section 6·7.[8] These auxiliary arcs may be of the full-wave a-c type or of the d-c type supplied from an auxiliary rectifier (Fig. 6·20).

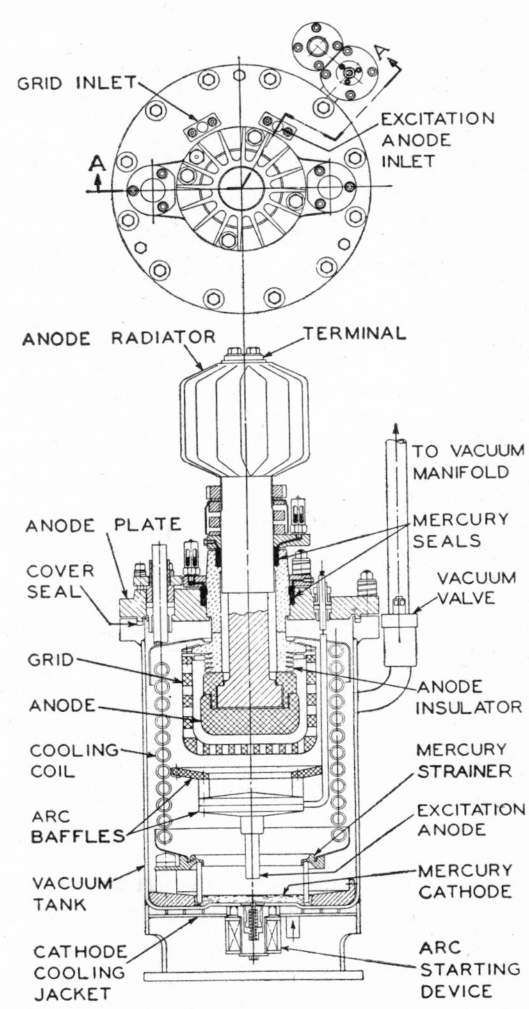

FIG. 6·20 Excitron. (Reprinted with permission from Winograd, "Development of the Excitron Type Rectifier," *Trans. A.I.E.E.*, Vol. 63, pp. 969, 978.)

The control function is obtained by a grid. The cathode must be separated from the tube walls to prevent the arc spot from moving to the edge of the mercury pool and attaching itself to the metal walls. This may be done by insulating the cathode from the tube walls or by interposing baffles.

Ratings and characteristics are similar to those described for tubes of the pool type.

6·9 RATINGS AND CHARACTERISTICS OF IGNITRON RECTIFIER TUBES

Ignitrons are used for two major purposes: power rectification and inversion (Figs. 6·1 and 6·2) and control switching of alternating currents as in resistance-welding control (Fig. 6·3). The first application requires continuous duty and has low probability of arc-back; the latter requires intermittent operation and conduction of large currents. In general, tubes are designed especially for each type of application.

Low-Voltage Ignitron Rectifiers

Rectifier ignitrons have deionization baffles between anode and cathode. These may or may not be connected to external circuits by sealed-in leads. The purpose of these parts is to shield the anode from the influence of the cathode by deflecting high-velocity particles away from the cathode and by providing surfaces at which the ions can combine with electrons and thereby be neutralized. A cross section of a sealed-off rectifier ignitron rated for d-c operation at 300 and 600 volts is shown in Fig. 6·21.

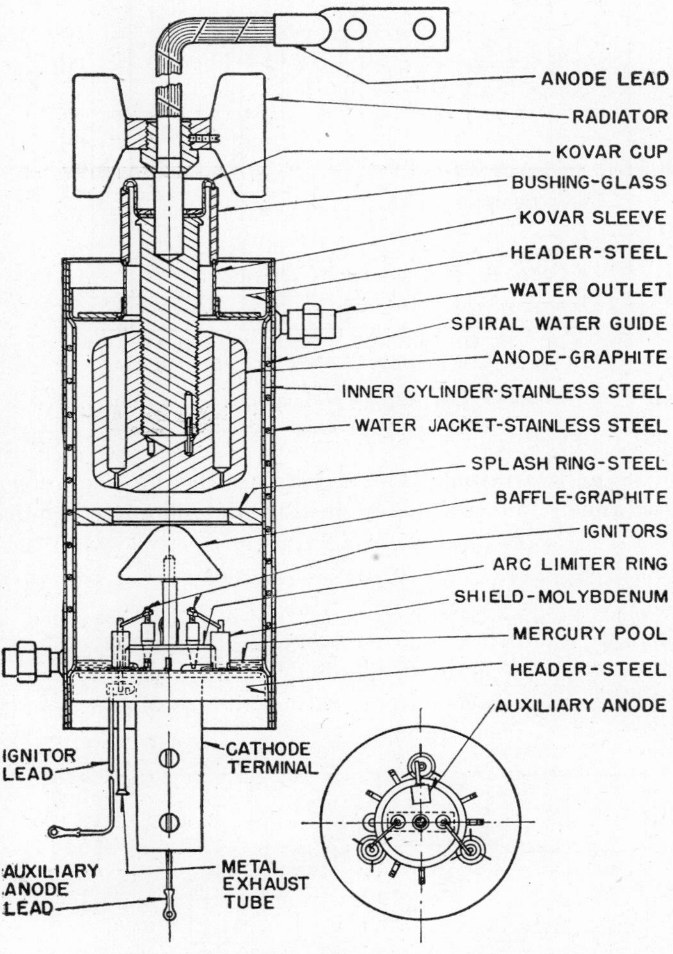

Fig. 6·21 Sealed ignitron for rectification service.

These tubes are rated at continuous duty, 150 percent of continuous duty for 2 hours, and 200 percent for 1 minute.

Current ratings depend upon tube voltage, being decreased as the voltage rating increases. Other ratings are similar to those for tubes previously described. In pulse-excitation circuits a small auxiliary anode lengthens the time during which the arc spot is available for anode pick-up. The original necessity for this extra firing time was on interphase-rectifier circuits with separate pulse ignitor excitation where the circuit was required to operate on either three or six phases. Ratings for this small anode are dependent upon the polarity of the main anode because the ability of one anode to resist arc-back depends upon the ionization by the other anode.

High-Voltage Ignitron Rectifiers

Ignitrons now are being developed for operation from 10 to 20 kilovolts. These high-voltage ratings are obtained by the addition of voltage-dividing grids which reduce the length of the conducting paths on which the large inverse voltages are impressed.[9]

Rectifier Ignitors

Rectifiers have two ignitors for long life. Either of the ignitors will operate the tube satisfactorily throughout its life, yet in rectification the duty is often continuous and requires a high percentage of duty on the ignitor. The use of two ignitors may be discontinued if experience proves it unnecessary.

Ignitors used in rectifier ignitrons have been developed for duty of this type with consideration for the variety of methods used in designing firing circuits. In general, continuous operating characteristics and long life are more important than intermittent operating characteristics.

6·10 RATINGS AND CHARACTERISTICS OF RESISTANCE-WELDER CONTROL IGNITRONS

Duty on an ignitron in resistance-welding control is illustrated in Fig. 6·3. Current is nearly sinusoidal, and the forward and inverse voltages are a function of the amount of phase delay. If there is no phase delay, the voltage across the tube after conduction is only that required to fire the ignitor of the succeeding tube, except of course at the end of a conduction period when the full line voltage must be blocked.

If the tube is operated with phase delay, conduction in each cycle is followed immediately by a sudden application of inverse voltage. The tube must block an equal forward voltage one-half cycle later.

If the impedance of the welder is constant, phase delay in firing causes a reduction in current, as shown in Fig. 6·3. This reduction in average current reduces heating in the tube.

Duration of inverse voltage increases when the phase is retarded; therefore arc-back or forward fire is more probable. Reduction in current reduces heating and increases time of application of inverse voltage. The two effects are to some degree compensating. The tube is performing its most difficult function on a given load impedance when the circuit is being operated with a delay in ignitor firing of about 10 degrees from full-cycle operations.

All ratings of welder-control ignitrons are made on the basis of zero phase retardation, or full-cycle conduction. Even though a welder may operate with considerable phase reduction, the tubes are applied as though the welder were to be operated with no phase retardation.

The use of baffles would hamper the passage of large intermittent currents in welder-control tubes. Arc-back is not so serious in this class of service for it does not cause the transformers to be short-circuited as in rectification service. For this reason welding-control tubes in general are made without baffles between cathode and anode (Fig. 6·22).

Ignitron-Welding Ratings

Sine-wave voltages and currents are assumed, and all ratings, where possible, are in rms values. The instantaneous rms current conducted by a pair of ignitrons connected in inverse parallel is called the demand current. The product

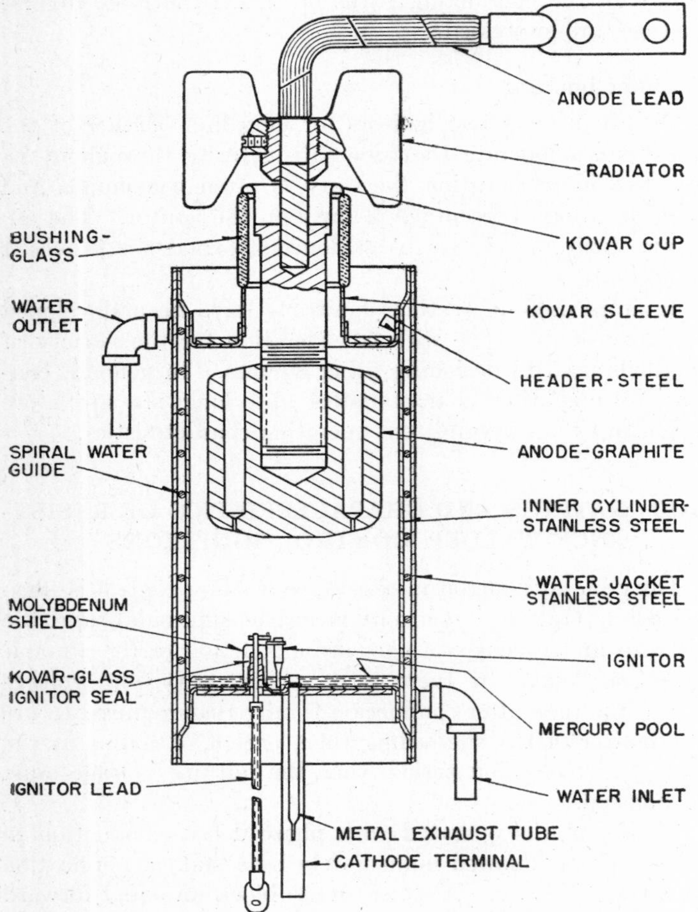

Fig. 6·22 Sealed ignitron for resistance-welder control.

of supply voltage and rms demand current is known as the kilovolt-ampere demand within the rated supply-voltage range.

A formal co-ordinated system of rating has been set up to insure that the line of tubes covers the field of application to the best advantage (Fig. 6·23). Maximum-demand kva rating increases by a factor of 2, and rated average current increases by a factor of 2.5 between successive sizes. The maximum average-current rating is valid up to a demand kva rating of one-third the maximum. At the maximum-demand kva rating the average current rating is decreased to 54 percent of its nominal value. Intermediate points are determined by logarithmic interpolation.

Curves showing demand current for given supply voltages as a function of the percentage of duty are given in tube-data sheets. These are chosen to be representative of 220- and 440-volt circuits, which are most often used. Examples are shown in Figs. 6·24 and 6·25.

Maximum averaging time also is a formalized rating. It is based on the rating of ½-second continuous duty at the

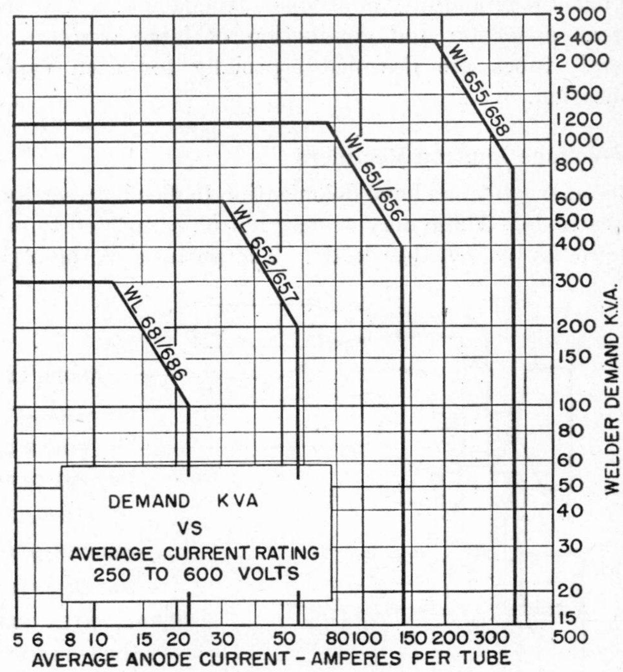

Fig. 6·23 Demand kva ratings of ignitrons for welding service as function of average current.

maximum kva rating. This determines the maximum averaging time. It can be determined by the following formula:

$$T_m = 0.225 \frac{I_D}{I_A} \quad \text{seconds} \tag{6·8}$$

where T_m = rated maximum averaging time
I_D = maximum-demand current rating
I_A = average-current rating at maximum-demand current rating.

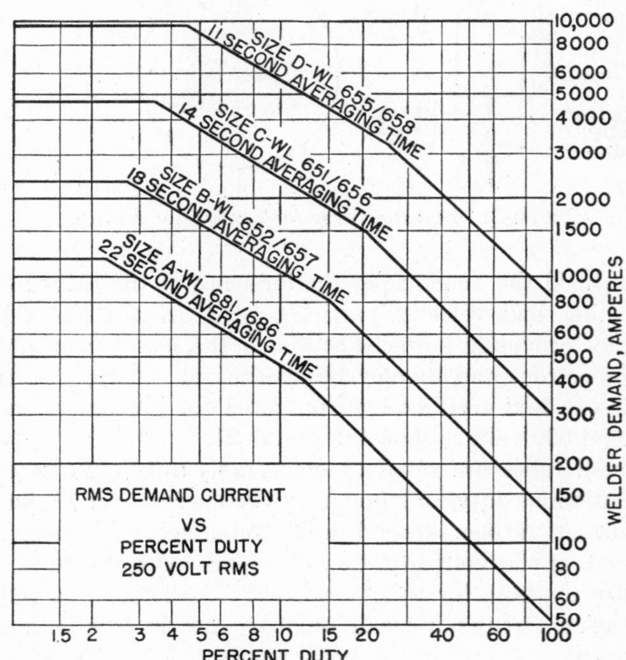

Fig. 6·24 Demand-current ratings of ignitrons for welding service as function of percent duty for 200- to 250-rms volt circuits.

The figures given in the rating sheet are "rounded off" values computed from this equation. Percentage of duty determined from the current-rating curves multiplied by the maximum averaging time is equal to the longest time of continuous operation allowed under the ratings at the given demand current. Such an impulse of current can be repeated at a period equal to the maximum-averaging-time rating.

Since maximum-demand current rating increases by a factor of 2 and the average-current rating increases by a factor of 2.5 between tube sizes, the rated maximum averaging time changes by a factor of 0.8 between sizes.

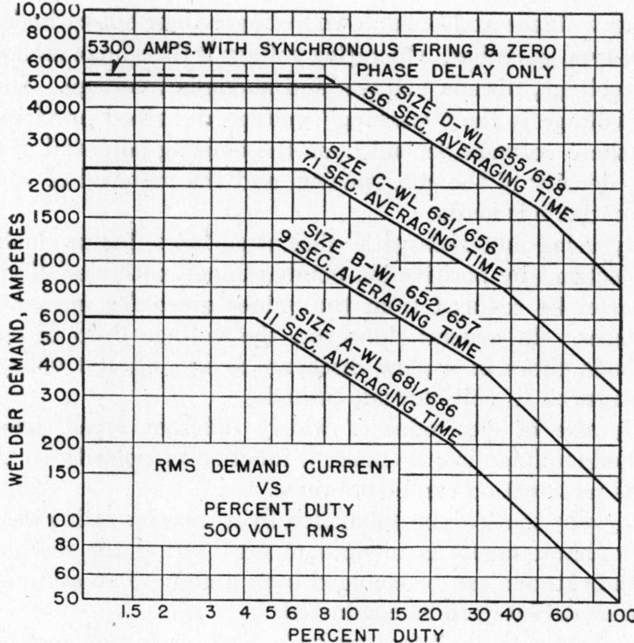

FIG. 6·25 Demand-current ratings of ignitrons for welding service as function of percent duty for 400- to 500-rms volt circuits.

On intermittent duty consisting of irregularly timed blocks of current of differing values of demand current, application of an ignitron must be based on the highest average current in any time period not longer than the maximum averaging time.

There is no restriction on how short a period of time may be used to compute average current per tube, but it must be taken at such a time in the welding cycle that the highest possible average current is obtained.

Welding-Control Ignitors

Ignitors used in welding-control ignitrons are developed chiefly for their characteristics in intermittent duty. Their ability to start the tube when it is cold is of primary importance. Usually ignitors in welding-control circuits are operated by so-called anode firing, as shown in Fig. 6·26. Ignitors are connected to the anode through a thyratron. When the ignitron begins to conduct, ignitron voltage falls and becomes too low to cause current through the thyratron in series with the ignitor. Current and voltage for operating the ignitor are drawn from the load. Therefore, operation of the ignitor depends upon the size of the load controlled by the ignitrons. As shown in Fig. 6·27 the current to such a load usually lags the voltage by an angle dependent upon load power factor. When the control thyratron operates, supply voltage is applied to the ignitor through the inductance of the load. Ignitor voltage and current are zero initially and will rise at a rate determined by load inductance, until current through the ignitor rises to the firing

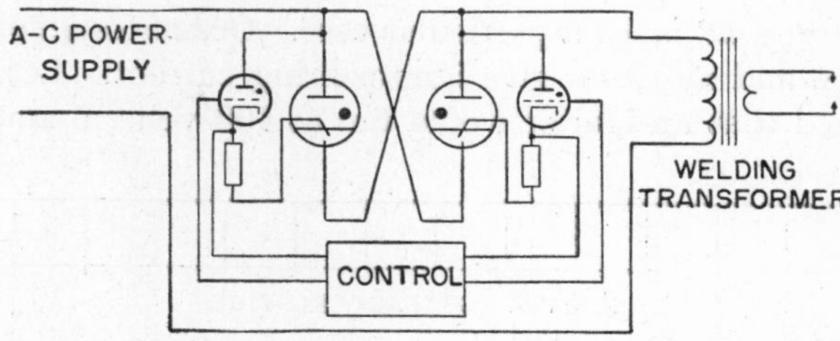

FIG. 6·26 Anode-firing connection of ignitrons in resistance-welder service.

value. Ignitor firing time, measured from the time at which the thyratron begins to conduct, varies with the circuit constants as well as with the ignitor characteristics. Below the firing point, current causes the ignitor to heat, and in general tends to increase firing current.

If the impedance of the load is low, the source voltage is impressed on the ignitor through the series-current-limiting resistor. In such a circuit (Fig. 6·26), current and voltage rise abruptly and continue until the ignitor fires. In this example the ignitor-firing time is clearly defined. In the rating sheet, "ignitor voltage, maximum positive peak

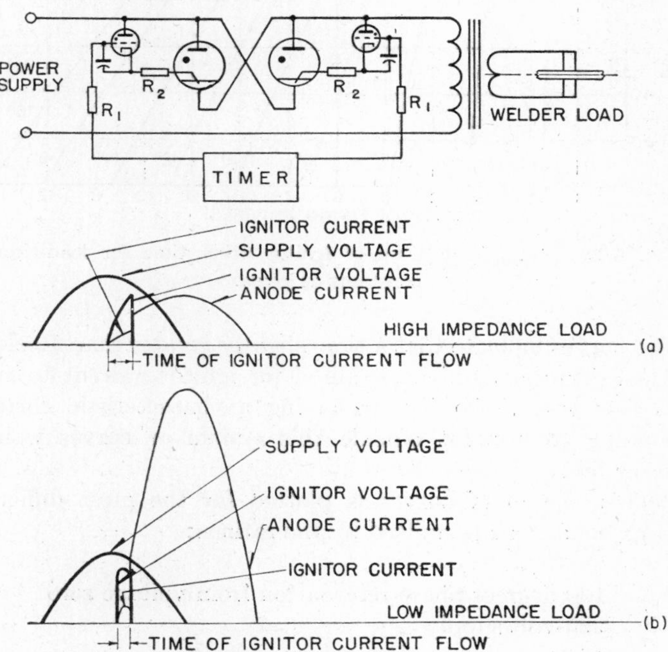

FIG. 6·27 Wave form of ignitor current and voltage as influenced by load impedance.

required" and the similar current characteristic indicate the values required. Applied constantly, they cause the ignitor to fire in less than "ignition time maximum." If the impedance of the load circuit is high, and anode firing is used, firing time may be longer than the indicated rating because the voltage applied to the ignitor is insufficient to fire the ignitor during much of the time.

The time required for ignitor current to rise to the firing value can be computed from the transient-current formulas for an *LR* circuit started at a time other than that of steady state. Ignitor impedance can be considered to be equal to a resistance given by the quotient of maximum firing voltage and current.

Figure 6·28 is for a particular case. Ordinates give firing voltage and abscissas give ignitor-firing current. Let it be assumed that an ignitor would fire at 100 volts, 6 amperes.

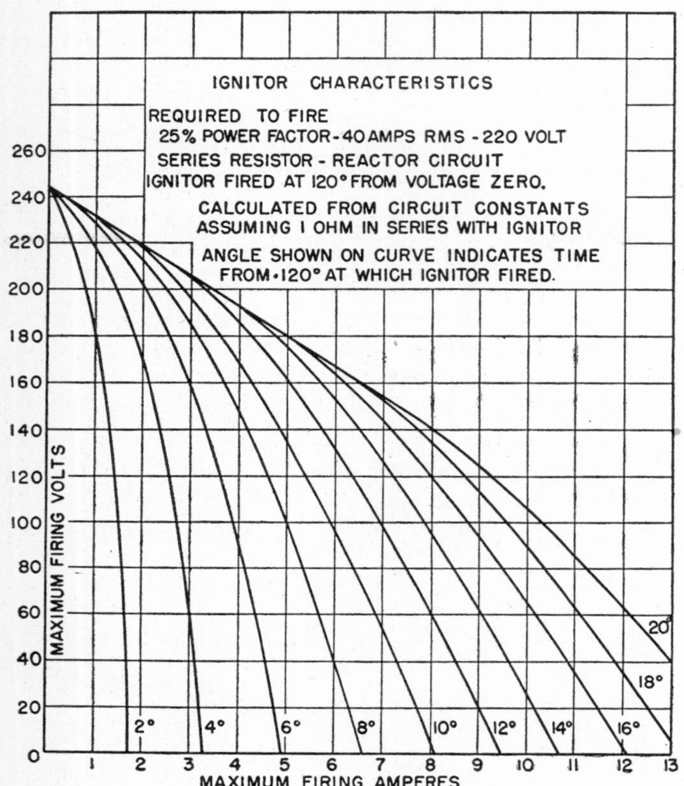

Fig. 6·28 Example diagram of ignitor firing time for load-circuit constants as given.

The curves indicate that a time slightly greater than 10 electrical degrees would be required for ignitor current to rise to this value. An ignitor having a characteristic corresponding to a point outside this system of curves would never fire.

This system of curves is plotted for the most difficult firing conditions permitted by the ratings:

120 degrees phase retardation from voltage zero
220-volt supply
40 amperes rms demand
0.25 power factor
1 ohm series ignitor resistance

Delay in firing does not represent a complete delay in load-current conduction, because the ignitor current passes through the load. Load current is somewhat reduced by the ignitor resistance until after the ignitor fires. Variation in ignitor characteristics can cause a slight variation in welder current.

6·11 CHARACTERISTICS OF COLD-CATHODE TUBES

The general theory of conduction between cold electrodes in a gas vapor is described in Section 4·10. Electron emission from the cathode is assumed to be due to positive-ion bombardment of the cathode. Other mechanisms—photoelectric emission from the cathode, and production of electrons in the gas by collision of positive ions and electrons—might be assumed, but the assumption of positive-ion bombardment is the most straightforward.

Cold-cathode tubes have higher cathode-to-anode operating voltages and breakdown voltages than hot-cathode or pool-type gas tubes. Operating voltage is the sum of cathode-fall voltage, plasma voltage, and anode-fall voltage. Starting voltage is the "sparking" voltage described previously. All these voltages depend upon the ionizing potential of the gas, density of the gas, spacing, and the material of which the cathode is made.

By using materials with efficient surfaces, for production of electrons by positive-ion bombardment, cathode-fall voltage can be reduced from the values given for pure-metal surfaces. In general, however, the voltage drop of cold-cathode tubes in a glow discharge is around 100 volts, in contrast to 10 volts for hot cathodes.

By use of electrodes of widely different areas, anode-cathode voltage can be made a function of polarity so that some rectification can be obtained.

Current that can be conducted by a tube operating on the glow characteristic is in milliamperes, but if the discharge at the cathode can be changed from a glow to an arc, much higher currents can be conducted.

In cold-cathode tubes cathode heating is not required. This is an advantage where the tube acts in a stand-by capacity, as in protective circuits. There is no delay in heating the cathode, as there is in hot-cathode tubes.

Cold-cathode tubes are often used as protective "gaps." These can operate as cold-cathode glow tubes with a breakdown voltage characteristic for initial breakdown of several hundred volts and a tube voltage drop of about 100 volts or, if heat-resistant electrodes are used, the current can be allowed to increase and form an arc, the voltage drop of which is from 10 to 15 volts. Some of these tubes do not conduct after the tube voltage falls below a certain "drop-out voltage." These tubes, in general, are of the glow-discharge type. Other types continue to conduct until the circuit is opened. These can be of the arc-discharge type.

The nearly constant anode-cathode operating voltage of the glow tube can be used for voltage regulation of small circuits. The tube is connected to the source through a resistor, and the load is connected across the tube. As source voltage varies, the current through the tube varies but the tube drop and the load voltage remain essentially constant.

Addition of a third electrode makes a glow-discharge tube controllable in a manner similar to the thyratron. The third electrode can be used as a "trigger" for a positive control device. When the glow is started by the third electrode the space ionizes, thus reducing breakdown voltage of the second electrode, which in turn breaks down. By proper choice of spacings of the various electrodes (even though the gas pressure in the bulb is the same for all electrodes) they can be made to have different breakdown voltages.

The materials of the various electrodes influence breakdown voltages and voltage drops. The areas of electrodes influence the volt-ampere characteristic of the electrode combination.

Characteristics of cold-cathode tubes may be complicated. For some tubes the grid-voltage anode-voltage control characteristic becomes a closed curve surrounding an area inside which the tube conducts no current and outside of which it does conduct. In general, ratings are similar to those for thyratrons.

REFERENCES

1. "A New Timer for Resistance Welding," R. N. Stoddard, *Trans. A.I.E.E.*, Vol. 63, 1944, p. 693.
2. "A High Power Welding Rectifier," D. Silverman and J. H. Cox, *Trans. A.I.E.E.*, Vol. 55, 1934, p. 1380.
3. "New Developments in Ignitron Welding Control," J. W. Dawson, *Trans. A.I.E.E.*, Vol. 55, 1936, p. 1371.
4. *Applied Electronics*, E. E. Staff, 1943 M.I.T., pp. 230–233, Wiley.
5. *Gaseous Conductors*, J. D. Cobine, 1941, p. 455, McGraw-Hill.
6. "Ionization Time of Thyratrons," A. E. Harrison, *Trans. A.I.E.E.*, Vol. 59, Discussion 1079, 1940, p. 747.
7. "Improvements in Mercury Arc Rectifiers," J. H. Cox, *Trans. A.I.E.E.*, Vol. 52, Dec. 1933, p. 1082.
8. "Development of Excitron Type Rectifier," H. Winograd, *Trans. A.I.E.E.*, Vol. 63, 1944, pp. 969–978.
9. "Pentode Ignitrons for Electronic Power Conversion," H. C. Steiner, J. L. Zehner, and H. E. Zuvers, *Trans. A.I.E.E.*, Vol. 63, 1944, p. 693.
10. "Sealed Off Ignitrons for Resistance Welding Control," D. Packard and J. H. Hutchings, *G.E. Rev.*, Vol. 40, 1937, p. 93.
11. "Mercury Arc Rectifiers and Ignitrons," J. H. Cox and D. E. Marshall, *Trans. Electrochem. Soc.*, Vol. LXXII, 1937, p. 183.
12. "Sealed Tube Ignitron Rectifiers," M. M. Morack and H. C. Steiner, *Trans. A.I.E.E.*, Vol. 61, 1942, p. 594.
13. "Ignitron Rectifiers in Industry," J. H. Cox and G. F. Jones, *Trans. A.I.E.E.*, Vol. 61, 1942, p. 713.
14. "Ignitrons for the Transportation Industry," J. H. Cox and G. F. Jones, *Trans. A.I.E.E.*, Vol. 58, 1939, p. 618.
15. "The Cause of High Voltage Surges in Rectifier Circuits," A. W. Hull and F. R. Elder, *J. App. Phys.*, Vol. 13, 1942, p. 372.
16. "The Theory of the Grid Glow Tube," D. D. Knowles, *Electric J.*, Vol. 27, 1930, pp. 116, 232.
17. "Grid Controlled Glow and Arc Discharge Tubes," D. D. Knowles and S. P. Sashoff, *Electronics*, Vol. 1, July 1930, p. 183.
18. "Cold Cathode Gas Filled Tubes as Circuit Elements," S. B. Ingram, *Trans. A.I.E.E.*, Vol. 58, 1939, p. 342.
19. "Arc Backs in Rectifier Circuits," R. D. Evans and A. J. Maslin, *Trans. A.I.E.E.*, Vol. 64, 1945, pp. 303–311.
20. "Rectifier Fault Currents," C. C. Herskind and H. L. Kellogg, *Trans. A.I.E.E.*, Vol. 64, 1945, pp. 145–150.

Chapter 7

PHOTOELECTRIC DEVICES

W. W. Watrous and D. E. Henry

PHOTOELECTRIC effect may be defined as an electrical change which is produced by the absorption of radiation by matter. This is a rather broad definition, and to make it less inclusive it is customary to include only electrical changes in which electrons are freed from atoms so that they constitute an electric current. The definition should be further modified to exclude radiant energy with wavelength longer than infrared.

Credit for the discovery of the photoelectric effect, upon which the modern phototube is based, goes to Heinrich Hertz.[1] * He made this discovery in 1887 while he was engaged in experimental verification of Maxwell's electromagnetic theory of light.

Hertz discovered that the length of a spark induced in an auxiliary micrometer gap was reduced if the light from the primary spark was prevented from falling on the gap. He traced this effect to ultraviolet generated in the discharge of the main gap, which fell on the electrodes of the second gap. The effect was increased if a large, clean plate was used as the negative electrode.

Hertz did not continue his investigation, but Wilhelm Hallwachs [2] took up the problem. He believed the effect was not dependent upon oscillatory spark discharges. Hallwachs' investigations showed that a clean zinc sphere connected to a gold-leaf electroscope and negatively charged would lose that charge if illuminated by light from an electric arc. He found further that light had no effect on a positively charged sphere, but that if the positively charged body was placed in the neighborhood of a negatively charged body which was irradiated by light, the positive body would lose its charge. Hallwachs concluded from these experiments that if a body is irradiated by ultraviolet radiation, negative electricity leaves the body and follows the electrostatic lines of force. In 1899 this negative electricity was identified by Lenard [3] and Thomson [4] as electrons. The effect discovered by Hallwachs is often known as the Hallwachs effect.

Several other investigators continued the study of the photoelectric effect. The most comprehensive work was accomplished by Elster and Geitel,[5] who first showed that certain alkali metals, particularly sodium and potassium, when amalgamated by mercury, become photoelectrically sensitive to visible light as well as to ultraviolet radiation. Because active surfaces such as sodium and potassium cannot be kept clean in air, they enclosed the sensitive surface in an evacuated glass bulb.[6, 7] This was the forerunner of the modern phototube. To Elster and Geitel must also go the credit for conceiving the first gas phototube. The same observers studied the effect of polarized light on a photo surface.

* Actually the discovery of the photovoltaic effect by Becquerel [61] in 1839 and of photoconduction by Smith [53] in 1873 precede the discovery of photoemission. Both these effects, although not truly "electronic" phenomena, are discussed in Sections 7·18 and 7·19.

7·1 FUNDAMENTAL LAWS

To understand the action of a photoelectric device, it is necessary first to become familiar with the fundamental laws of photoelectric emission. These laws show the relationship of the initial velocities, direction, and the number of electrons ejected from the photo surface to the variables of radiation intensity, frequency, and state of polarization.

First Law

The relation of photoelectric current to light intensity has been tested for the full range of light intensity. When corrections were made for errors in measurement, and for spurious effects in the particular surface employed, the following law was evolved: *Photoelectric current is directly proportional to the intensity of the incident light.* Figure 7·1

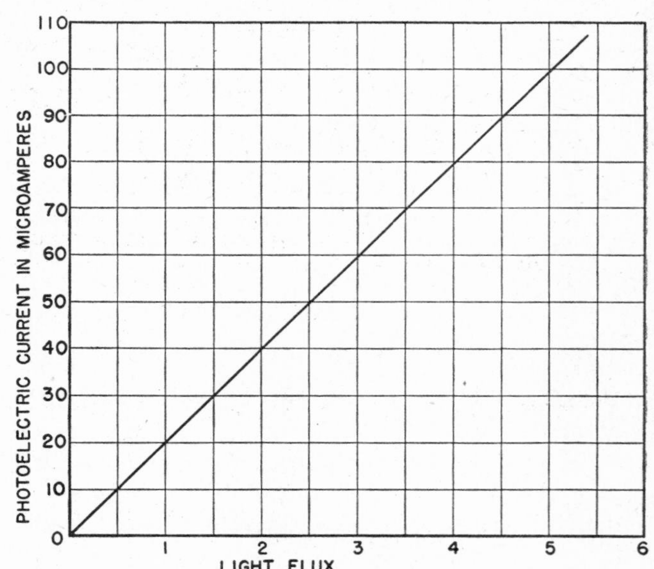

Fig. 7·1 Current-illumination curve for a typical vacuum phototube at voltage saturation.

shows a plot of the relation between photoelectric current and illumination.

Second Law

The second law of photoemission is based on the quantum theory of radiation. The quantum theory was first proposed by Max Planck [8] and later adopted by Einstein [9] to explain the connection between the energy of the emitted electron and the frequency of the radiation. This law was confirmed by several investigators,[10, 11, 12] and is expressed mathematically by equation 2·18. It is: *The maximum energy of the emitted photoelectrons is a linear function of the radiation frequency and is independent of the intensity of that radiation.*

Third Law

The third law of photoemission is derived from Einstein's original equation, and its expression is shown by equation 2·20. The third law may be stated: *The maximum wavelength at which photoemission occurs is dependent only upon the work function of the material of the photosensitive surface.*

Fourth Law

This law is concerned with the efficiency of emission. A spectral distribution curve, for example Fig. 7·11, plotted with absolute values as ordinates, shows the spectral sensitivity, or photoelectric yield for a particular surface over a wide band of wavelength. When the photoelectric current per unit radiant energy has been reduced to "electrons per absorbed quantum," it is known as the quantum yield, or quantum efficiency. This efficiency is a true measure of the performance of a photosensitive surface at a particular wavelength. In magnitude it varies from 10^{-3} percent to more than 30 percent. Quantum efficiency varies greatly with the type of surface, frequency of incident radiation, and state of polarization of the radiation. A theoretical maximum efficiency is reached when the energy in the electrons ejected from the cathode surface is equal to the energy of the absorbed radiation quanta.[13] Thus, true efficiency is given by the ratio

$$\frac{\text{Quantum yield at wavelength } \lambda}{\text{Maximum theoretical quantum yield}} = \frac{\theta_\lambda}{\theta_{\max}}$$

Figure 7·2 shows $\theta_\lambda/\theta_{\max}$ plotted as a function of wavelength for two types of surfaces. Surface 1 is a pure-metal surface, and surface 2 is a composite surface such as cesium-oxygen-silver. These curves show the great dependence of quantum efficiency on wavelength and surface.

All investigators [14, 15, 16] have found great differences in the amount of spectral selectivity depending upon the material of the surface, the amount of gas absorbed, the thickness of the sensitized layer, and the state of polarization of the incident light.

The fourth law may be stated: *Quantum efficiency depends upon the nature of the surface, on the wavelength of the incident light, and upon the state of polarization.* No detailed law can be postulated to show the exact relationship involved.

Fifth Law

Without a detailed discussion, the fifth law may be expressed: [17] *The distribution of electrons emerging from a metal surface is always symmetrical about the normal to the surface, regardless of the angle of incidence or plane of polarization of the incident light.*

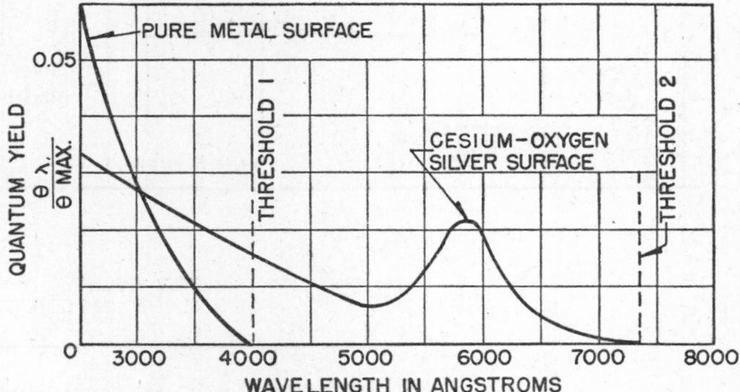

Fig. 7·2 Quantum yield-wavelength curve. (Reprinted with permission from *Photoelectric Cells*, by Campbell and Ritchie, published by Pittman and Sons, p. 28.)

7·2 RADIANT ENERGY

Radiation may be defined as energy traveling in the form of electromagnetic waves. Depending upon the source, the frequency of such radiation covers a band ranging from a few cycles per second to frequencies of some 10^{22} cycles per second. Frequencies radiated in the visible spectrum by a particular source determine its color. In addition, reflection, absorption, and transmission properties of the matter which receives the energy from the source are dependent upon the radiation frequency.

Figure 7·3 shows a chart which pictures the complete spectrum of radiant energy. A tremendous range of frequency is covered by this chart. Visible light is only a very small part of the whole. The portion of the spectrum of most interest in photoemission has been expanded to show the detail better. In general, the range of frequencies normally detected by photosensitive surfaces includes the visible band and parts of the infrared and ultraviolet bands. X-radiation also releases electrons from matter, but this effect is not usually included in the field of photoelectric devices.

The source of the radiation is important in the choice of the photosensitive surface. In accordance with the fourth law (Section 7·1), each type of surface has its own particular response characteristic which depends upon the frequency of the radiation. Therefore, to obtain maximum sensitivity, or to reduce the effect of unwanted radiation, the peak frequency response of the photoemissive surface should be, as nearly as possible, the same as the frequency of the source of radiant energy.

The Black Body

A black body, in the physical sense, is a body which is a temperature radiator, of uniform temperature, whose radiant flux in all parts of the spectrum is the maximum obtainable from any temperature radiator at the same temperature.

Any other material radiates less, at some or all wavelengths, than a black body. Such a radiator is called a black body because it absorbs all radiant energy that falls upon it.

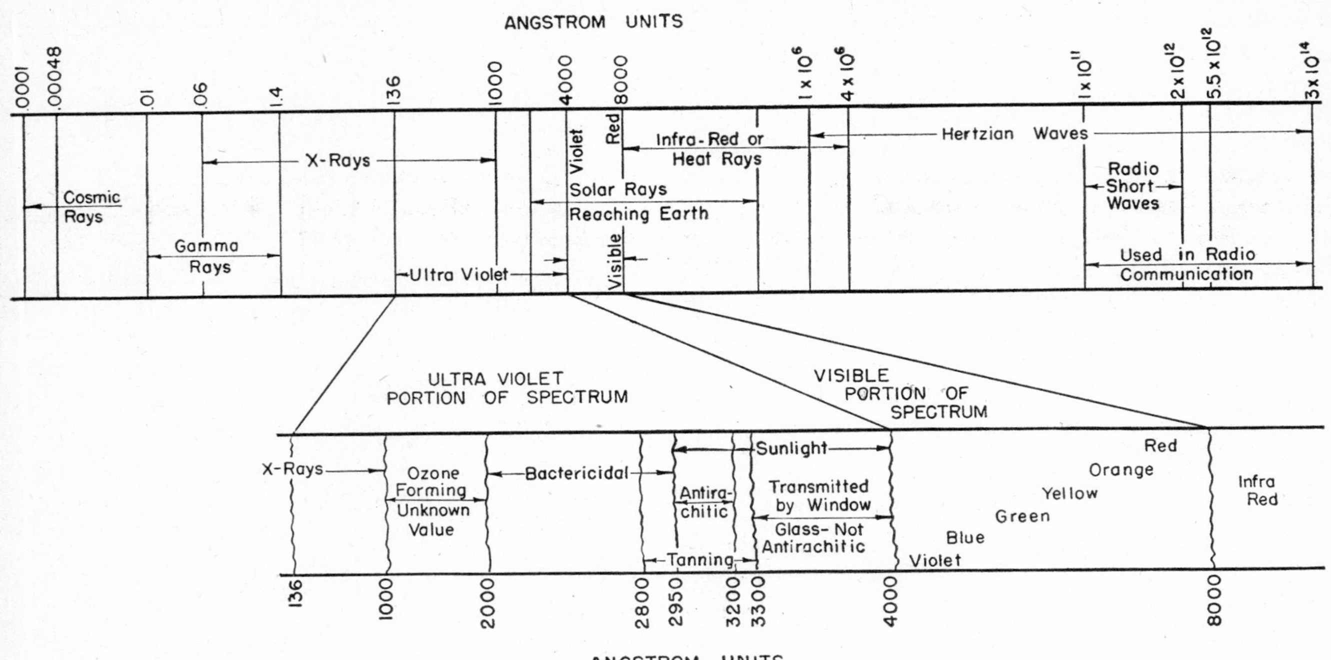

Fig. 7·3 Radiant-energy spectrum.

To determine how energy is radiated from a black body, it is necessary to know: (1) how the total energy radiated is related to the temperature of the body, and (2) how this energy is distributed over the radiant-energy spectrum.

The first question is answered by the Stefan-Boltzmann law:

$$W = \sigma T^4 \qquad (7\cdot1)$$

where W = radiant flux per unit area
T = temperature in degrees Kelvin
$\sigma = 5.7 \times 10^{-12}$ watts per cm^{-2} deg K^{-4} (Stefan-Boltzmann constant).

The distribution of radiant energy emitted by a black body was investigated by Wien [18] and Rayleigh and Jeans.[19] Both the Wien radiation law and the Rayleigh-Jeans radiation law were in error at certain wavelengths and temperatures. As a result, Max Planck [8] attacked the problem and, by the introduction of the quantum hypothesis, succeeded in developing his well-known law:

$$J_\lambda = \frac{c_1 \lambda^{-5}}{\epsilon^{c_2/\lambda T} - 1} \quad \text{microwatts per square centimeter per } 0.01\mu \text{ zone of spectrum} \qquad (7\cdot2)$$

where J_λ = spectral radiant intensity at wavelength λ for an 0.01μ zone of spectrum
T = temperature in degrees Kelvin
λ = wavelength in microns
$c_1 = 3.7403 \times 10^8$ microwatts per square centimeter per 0.01μ zone of spectrum
c_2 = 14,385 micron degree.

The values of the constants c_1 and c_2 will be found inconsistent, depending on the source.[20, 21, 22] Those given by Dr. Raymond T. Birge [23] are considered to be the latest available and are the values shown in equation 7·2.

If equation 7·2 is solved for a series of temperature values and the results plotted, the distribution curves shown in Fig. 7·4 are obtained. These curves show that, as temperature increases, the maximum radiant intensity occurs at successively shorter wavelengths. This shift can be expressed by a corollary to the Wien displacement law:[24]

$$\lambda_m = \frac{b}{T} \quad \text{microns} \qquad (7\cdot3)$$

where λ_m = wavelength in microns of maximum spectral response
T = temperatures in degrees Kelvin
b = 2897.

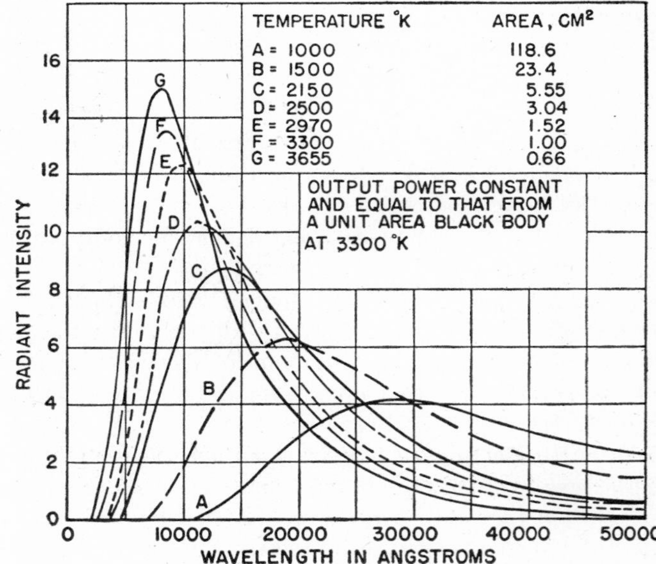

Fig. 7·4 Spectral distribution of the radiation from a black body as calculated by equation 7·2 for various temperatures. (Reprinted with permission from *Measurement of Radiant Energy*, by W. E. Forsythe, published by McGraw-Hill, 1937, p. 12.)

Points plotted from equation 7·3 will fit the curve drawn through the maximum-intensity points of the curves in Fig. 7·4. This formula is useful in determining either temperature or wavelength for a black body.

No general law has yet been written to describe radiation for other types of radiators. However, if the measured spectral distribution of any other kind of radiator is plotted on the same co-ordinates as a black body of the same temperature (Fig. 7·5), the ratio of corresponding ordinates for the two radiators is called spectral emissivity. This emissivity is valid for only one particular temperature and wavelength. If a table of spectral emissivity for a material is

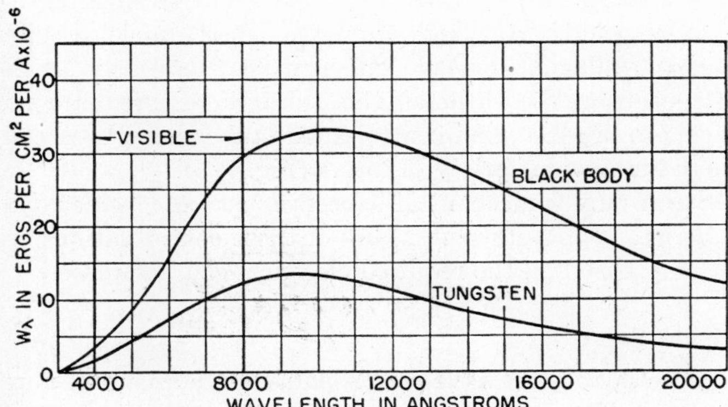

Fig. 7·5 Distribution of energy comparison between a black body and tungsten at a color temperature of 2870°K.

available, a spectral-distribution curve for a radiator other than a black body can be plotted for any definite set of conditions.

The black body is used as a standard for the measurement of radiant energy. Although no material is a true black body, artificial black bodies may be employed. The most common type of artificial black body is a cavity with uniformly heated opaque walls. A small hole is used for observation. The nearness of such an arrangement to a true black body is independent of the emissivity of the material of the cavity, except that for maximum accuracy the lower the emissivity the smaller the observation hole must be. From such a standard black body source, secondary standards may be made. One of the most common secondary standards, the standard used in the measurement of photoemissive tubes, is a tungsten filament at a color temperature of 2870 degrees Kelvin. At this temperature the spectral distribution of radiant energy in the visible band closely matches that of a black body of the same temperature.

Sources of Radiant Energy

Sources of radiant energy may be roughly classed as radiators of (1) continuous spectra and (2) discontinuous spectra, which may be either line or band spectra.

Some of the common sources [26, 27, 28] with their principal output characteristics are given in Table 7·1. An artificial black body is included for comparative purposes.

The radiant energy distribution, even in a continuous spectra light source, varies markedly with wavelength. In order to determine the relative proportion of energy in any given color band, measurement must be made with a spectrophotometer, unless actual spectral distribution curves are available.

Table 7·1 Sources of Luminous Energy

CONTINUOUS SPECTRA

Source	Approximate Wavelength Range
Sun	Short ultraviolet to long infrared
Black body (artificial)	Ultraviolet to long infrared
Incandescent lamp	Ultraviolet to long infrared
Carbon arc	Short ultraviolet to long infrared
Flame source	Visible through long infrared
Photoflash lamp	Ultraviolet to long infrared
Fluorescent lamp	Long ultraviolet to short infrared (disregards the line spectrum of mercury)
Super high-pressure mercury-arc lamp	Ultraviolet through long infrared

DISCONTINUOUS SPECTRA

Source (gas discharge)	Color of Principal Radiation
Helium	Yellow
Neon	Red
Argon	Blue
Sodium-vapor lamp	Yellow
Thallium	Green
Zinc arc	Red, blue
Mercury-arc lamp	Yellow, green, blue
Cadmium arc	Red, green, blue

Table 7·2 aids in determining representative wavelengths of various sources of radiation, particularly of color.

Table 7·2 Wavelengths of Various Radiations [25]

Source	Representative Wavelength (angstroms)
Shortest ultraviolet radiation	510
Shortest ultraviolet in the solar spectrum (limited by atmospheric absorption)	2,920
Lower limit of visible spectrum	3,900
Violet	4,100
Blue	4,700
Green	5,200
Maximum visual sensitivity	5,560
Yellow	5,800
Orange	6,000
Red	6,500
Upper limit of visible spectrum	8,100
Upper limit of solar spectrum	53,000
Infrared, shortest	8,100
Infrared, longest	3,140,000

(From "Wave Lengths of Various Radiations," *Handbook of Chemistry and Physics*, Vol. 24, p. 1993.)

Measurement of Radiant Energy

The measurement of radiant energy is considered beyond the scope of this chapter. However, it is believed that the inclusion of a few of the more important definitions of photometric quantities will materially assist in clarifying the nomenclature associated with the measurement of radiant energy. (See Section 7·20). In order to determine the magnitude of luminous flux, a quantity necessary for the measurement of phototube sensitivity, a simple formula for flux calculation is shown in Section 7·21.

The choice of units for the expression of the wavelength of radiant energy is usually a matter of convenience depending upon the particular portion of the radiant energy spectrum considered. To assist in more rapidly changing from one system of units to another, a conversion table (Table 7·3) is shown.

TABLE 7·3 WAVELENGTH UNITS

Name	Abbreviation	Value (millimeters)
Micron	μ	10^{-3}
Millimicron	$m\mu$	10^{-6}
Angstrom unit	A	10^{-7}
X-unit	XU	10^{-10}

7·3 CLASSIFICATION OF PHOTOSENSITIVE DEVICES

There are three general classes of devices the operation of which is based on the photoelectric effect. The first of these is the photoemissive cell, in which the absorption of radiation by the sensitive surface results in ejection of electrons into space. This type of device, formerly known as a photocell, is now called a *phototube*.

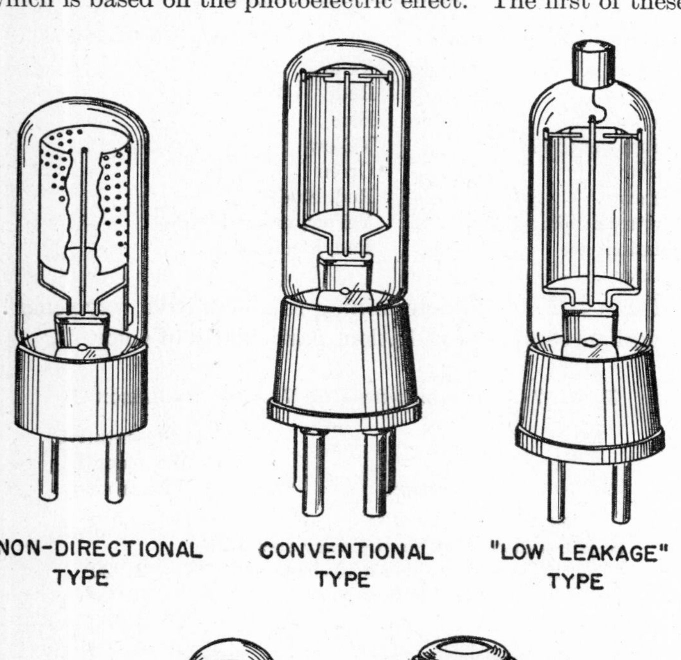

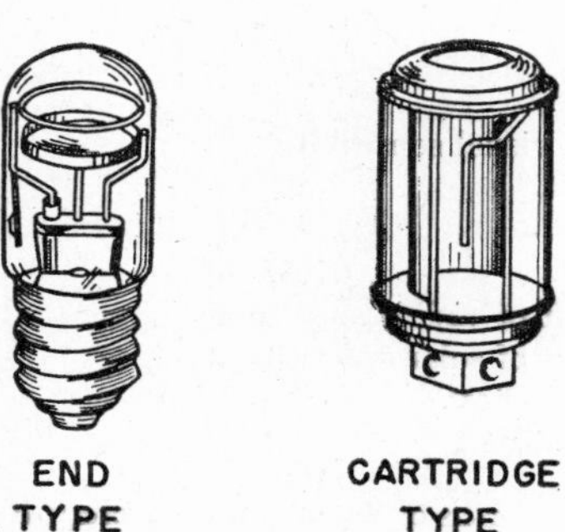

FIG. 7·6 Typical phototubes.

The second general class is the *photoconductive cell*, which is the earliest example of a photosensitive device. The principle of operation is that of actual change of electrical resistance under the influence of radiation.

The third class is the *photovoltaic cell*, which produces an electromotive force directly by absorbing radiation. This electromotive force usually is produced by photoelectrons released within the boundaries of the containing surfaces.

By the accepted definition of electronics, only the phototube is a truly electronic device. However, a brief description of photoconductive and photovoltaic cells is included here so that the reader may have as a basis for comparison the characteristics of all three classes of photoelectric devices

7·4 GENERAL DESIGN CONSIDERATIONS OF PHOTOTUBES

There are two general classes of phototubes: those with central anodes and those with central cathodes. Most phototubes are of the central-anode type. In such a tube the cathode can have a large electron-emitting surface, which greatly increases electron yield for a given intensity of illumination. Although use of a small anode requires increased voltage to provide sufficient field to attract all the electrons from the cathode, the voltages required are not difficult to produce, and central-anode tubes have been constructed in a wide variety of forms (Fig. 7·6).

Several purely mechanical aspects must be considered in the design of phototubes. Many of these considerations are primarily based on the required electrical characteristics.

Cathode Form

The cathode must have a reasonably large area and must be arranged to intercept the maximum quantity of light. Special cathode shapes such as spheres and vees have been employed to increase sensitivity by multiple reflections, but the results were disappointing. It is more practical to gather the light externally by the use of reflectors or lenses.

For some applications special structures such as the 360-degree photocathode surface and the "end on" type have been evolved. These forms are illustrated in Fig. 7·6.

Interelectrode Capacitance

Anode-cathode capacitance should be small to enable the tube to operate better at high modulated light frequencies.

Microphonics

Anode-cathode spacing should be rigidly fixed. Any motion of one electrode with respect to another, because of shock or vibration, causes fluctuation in output current and results in the phenomenon known as microphonics. This is important chiefly in such applications as the recording or reproduction of sound.

Leakage

Although this is an electrical phenomenon, the amount of leakage can be controlled by mechanical design. To minimize leakage, the anode and cathode connections are often brought out from the opposite ends of the tube. When the lead wires are not led from opposite ends of the tube, an internal sleeve (or sleeves) is usually fitted around the wires to increase the leakage path. This sleeve also reduces leakage directly by shielding the path between the anode and cathode leads from conducting materials which are sputtered or vaporized inside the tube during the evacuating process or during subsequent life. Proper choice of the sensitive surface material greatly aids in minimizing leakage, for some cathode materials vaporize and sputter less than others.

Glass Envelope

For visible or infrared radiation ordinary lime glass is satisfactory from the standpoint of transmission. At frequencies in the ultraviolet range, quartz or one of the new ultraviolet-transmitting glasses must be used. Sensitivity and high-frequency "cut-off" may be increased by using glass only a few mils thick for the "window." This is accomplished by drawing a large bubble inward in the bulb wall. As such a bubble stresses the glass in tension, it is sufficiently strong to withstand atmospheric pressure. See Fig. 7·9 (*b*).

7·5 PREPARATION OF PHOTOSENSITIVE SURFACES

Composite Surfaces

So many different photoemissive surfaces are used, each made in a different way, that it is impossible to describe all of them. At present the cesium-oxide-silver surface is one of the most common.

The cathode in such a tube is made of silver, or of other suitable materials silver plated. It is etched to increase its area, and then sealed into the bulb. The tube is evacuated and baked at a high temperature to remove foreign gases. Oxygen is then admitted, and a discharge current is passed between the electrodes to combine the oxygen with the silver surface. Depth of oxidation can be judged by observing the changes in color of the silver surface as oxidation proceeds. Cesium is next admitted, usually by vaporizing a cesium-bearing pellet. The cesium, when flashed, covers the interior of the bulb. By carefully controlled baking, the cesium is driven from the bulb walls and is deposited on the cathode. Excess cesium is absorbed by a layer of tin oxide, or similar material, usually painted on the back of the cathode. The tube is then "activated" by admitting an inert gas and passing controlled "shots" of current between the electrodes until the optimum depth of cesium layer is obtained. For maximum sensitivity, the cesium layer is only a few molecules thick.

Alterations in the process can produce phototubes markedly different in spectral response.

Pure-Metal Surfaces

Pure-metal surfaces are employed to obtain response to short wavelengths. One of the most satisfactory methods for the preparation of a pure-metal surface is cathodic sputtering. A rod, or wire, of the metal to be sputtered is mounted at the axis of a semi-cylindrical nickel plate.[29] Using the wire as the cathode and the plate as the anode, the wire is bombarded with positive ions by use of a d-c glow discharge in an inert gas such as argon. The contaminated surface of the wire is first sputtered over to the plate. Subsequent sputtering covers the contaminated surface of the plate and builds up a clean surface of sputtered material. A cylindrical shield which can be dropped into the neck of the bulb after the cell is completed protects the bulb from the sputtered material. When the sputtering is finished, the argon gas is removed with activated charcoal reduced to a low temperature in liquid air. The tube containing the charcoal is then sealed off from the bulb. Such technique is necessary as the sputtering process is lengthy and is normally done after the phototube has been removed from the vacuum pump. When used as a phototube, the plate becomes the cathode and the central wire forms the anode. Measurements of the photoelectric work function of sputtered-metal surfaces prepared by this method agree closely with the thermionic work functions of these same metals. The following are the measured threshold wavelengths and calculated photoelectric work functions for several sputtered-metal surfaces, compared with standard values of ϕ_T.

	Threshold	Work Function (ϕ_T)	Work Function (ϕ_P)
Tungsten	2750A *	4.52	4.50 volts
Molybdenum	2850A	4.15	4.34 volts
Tantalum	3022A *	4.12	4.10 volts

* Measured at the Research Laboratories of the Westinghouse Electric Corporation, Bloomfield, New Jersey.

Vaporization is another method commonly used for preparing surfaces of easily vaporized metals. As previously mentioned, this is the usual method for preparing cesium surfaces. It can be employed for such metals as magnesium,[34] calcium, barium, rubidium, sodium, and potassium also. All these metals are purified as much as possible before final deposition on the cathode. Methods of purification include reduction in the phototube envelope and multiple distillation outside the envelope.

7·6 TYPES OF PHOTOTUBE SURFACES

Probably all metals and insulators are photoemissive to some degree. Quantum efficiency and the spectral-selectivity-response curve depend almost entirely upon the materials and physical properties of the cathode surface. Equation 2·20, $\phi_P = 12{,}395/\lambda_0$ shows how the cut-off or threshold wavelength depends upon the work function of the cathode surface. For pure metal ϕ_P is related to the position of the element in the periodic table. In general, ϕ_P is small for the alkali group of metals and decreases with increasing atomic weight. Traces of surface impurities, either gaseous or metallic, alter the work function, change spectral response, and influence quantum yield, but controlled impurities on the cathode surface are beneficial, and can produce photoemissive surfaces with extremely high sensitivity and a wide range of spectral response.[30]

Table 7·4 shows photoelectric work functions for several common metals, grouped mostly according to their arrangement in the periodic table. Within each group the metals are tabulated in the order of their atomic number. Even with this arrangement metals in any one group do not all have consistently low or consistently high values of ϕ_P. This is because of the nature of the particular atom involved. Atoms with one or two electrons in remote outer orbits generally require comparatively little energy to release them. Such atomic structures make the best photoelectric emitters. The alkali metals and alkaline-earth metals are examples of atoms in which the electrons are loosely bound and are the materials most photoelectrically active to visible radiation. This is predicted by their low photoelectric work function. Figure 7·7 shows the shift in spectral response of alkali metals as their atomic numbers decrease.[31] In general, photoelectric work functions are almost identical with the corresponding thermionic work functions. Some investigators contend that except for experimental error they would be identical.

TABLE 7·4 WORK FUNCTIONS OF CLEAN-METAL SURFACES

Periodic Group	Atomic Number	Metal	Photoelectric Work Function (ϕ_p)
I	3	Li	(2.28)
	11	Na	(2.46)
	19	K	(2.24)
	37	Rb	(2.16–2.19)
	47	Ag	4.74(4.56–4.73) *
	55	Cs	(1.87–1.96)
	79	Au	4.90(4.73–4.82) *
II	12	Mg	(3.4) *
	20	Ca	(2.7) *
	30	Zn	3.32–3.57 *
	56	Ba	1.7 †
	80	Hg	(4.52)
III	13	Al	(2.5–3.6) *
IV	6	C	(4.7) *
	50	Sn	4.28–4.39
	82	Pb	(3.5–4.1) *
	90	Th	(3.3–3.6)
V	73	Ta	4.11(3.92–4.05) *
	83	Bi	4.0–4.4 *
VI	42	Mo	4.15(3.22) *
	74	W	4.52
VII	26	Fe	4.77
	27	Co	(4.12–4.28)
	28	Ni	5.01
	45	Rh	(4.57)
	46	Pd	4.97
	78	Pt	(6.30)

Values in parentheses are uncertain.
* Hughes and DuBridge, pp. 75 and 76 (1932).
† Zworykin and Wilson, p. 33 (1930).
All other values from Reiman, *Thermionic Emission*, John Wiley & Sons, Inc., p. 99.

From the photoelectric work function, the maximum wavelength to which the photosurface will respond can be predicted. It also suggests in a general fashion the relative sensitivity to be expected. The actual spectral-response curve cannot be precisely predicted, especially for composite-surface cathodes, because the shape of the curve depends on many variables, some of which are:

1. Surface materials employed.
2. Processing technique.
3. Thickness of sensitized surface.
4. Spectral emissivity of surface.
5. Spectral selectivity.
6. Glass-bulb transmission characteristic.

Pure metal surfaces usually have normal spectral-response curves and are characterized by extremely low electron emission because of their low quantum efficiencies. They show no fatigue effects.

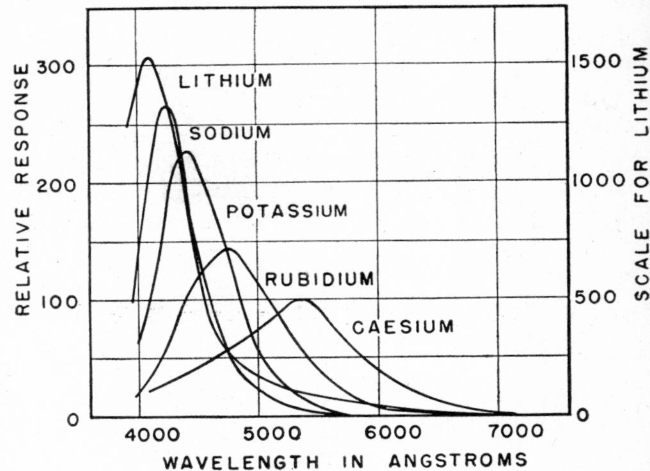

FIG. 7·7 Spectral distribution curves of the alkali metals. (Reprinted with permission from E. F. Seiler, "Color Sensitiveness of Photoelectric Cells," *Astrophys. J.*, Vol. 52, Oct. 1920, pp. 129–153.)

Ultraviolet Photosensitive Surfaces

Pure-metal photosensitive surfaces are employed chiefly for measurement of radiant energy, especially in the ultraviolet region. For such purposes the phototubes usually should have rather definite threshold frequencies. Many different pure metals have been used as cathode surfaces for such tubes. Some of these "pure-metal tubes" having threshold values in the ultraviolet are now commercially available.

Slight contamination of these surfaces markedly shifts the threshold frequency. Thin layers [32, 33] of alkali metals deposited on a metal base likewise result in a shift of the spectral response toward the infrared, with a decided increase in sensitivity.

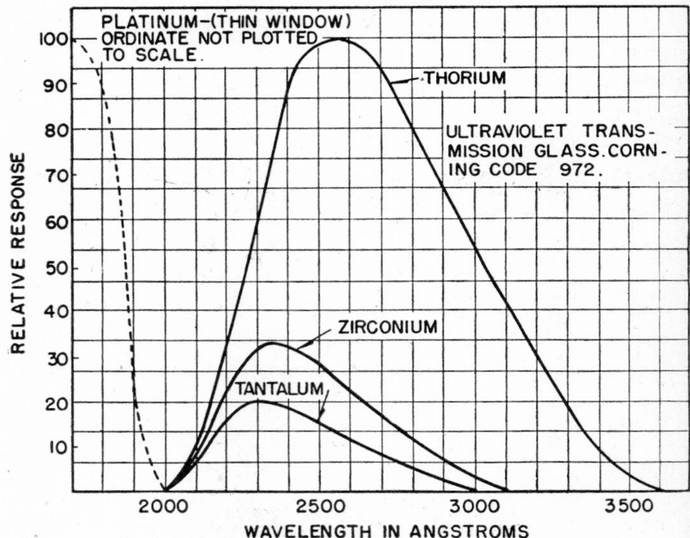

FIG. 7·8 Spectral distribution of commercial pure-metal photosurfaces using Corning code No. 972 glass envelopes.

In Fig. 7·8 the spectral-response curves of several commercial pure-metal cathode phototubes are plotted. The structure of these tubes is similar to that shown in Fig. 7·9 (*a*).

The thorium phototube with a sputtered-thorium surface has a threshold wavelength at about 3650 angstrom units. The peak in the response curve around 2550 angstrom units is due to the cut-off effect of the bulb, which is made of ultra-violet-transmitting glass. The curve would otherwise continue to rise until a maximum value, depending upon the quantum efficiency, was reached.

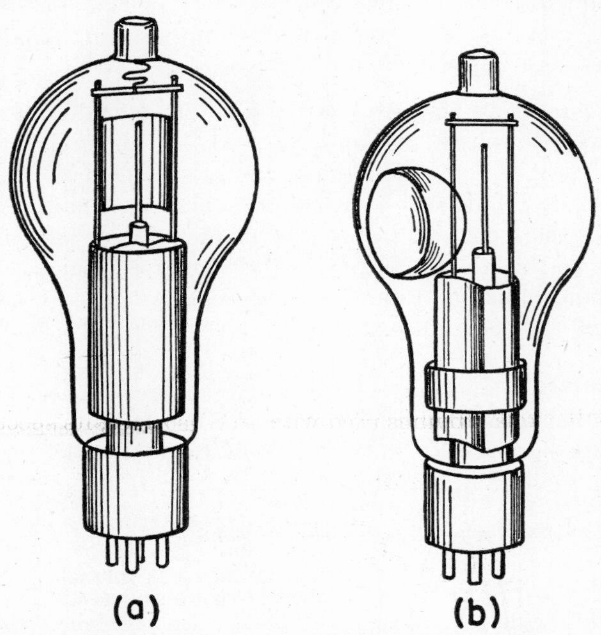

FIG. 7·9 (*a*) Ultraviolet phototube with thorium cathode. (*b*) Ultraviolet phototubes with a platinum cathode (thin window).

The sputtered-zirconium tube has a threshold value of 3150 angstrom units. This is the wavelength at which a reddening of the skin or the *erythemal* effect begins. The response curve reaches its peak at 2400 angstrom units, but with a suitable filter the sensitivity curve can be made to approximate the erythemal curve.

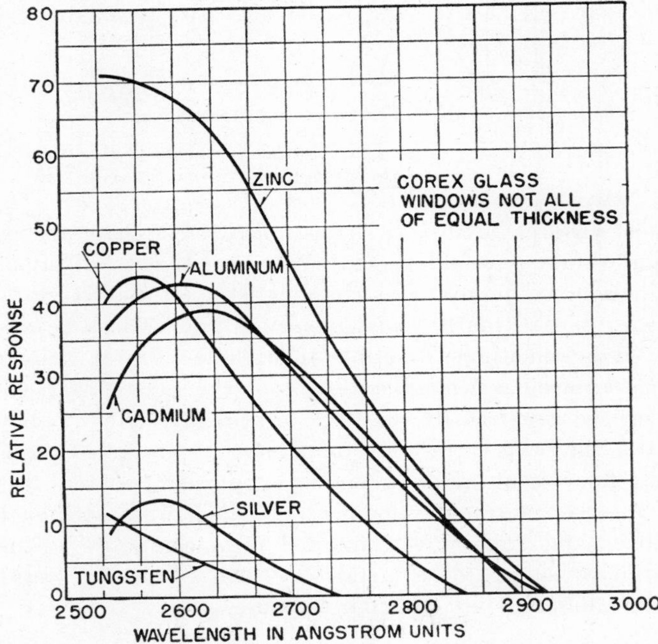

FIG. 7·10 Spectral distribution of a group of pure-metal photocathode surfaces using thin-window Corex D bulbs.

The sputtered-tantalum phototube has a spectral response similar enough to the bactericidal curve to be used to measure the "killing power" of ultraviolet lamps. The threshold is at 3022 angstrom units.

The platinum phototube, Fig. 7·9 (*b*), has a portion of the bulb in the form of a thin bubble or window, which transmits wavelengths below 1850 angstrom units. The threshold is at 2000 angstrom units. This permits measurements of ozone-producing rays, ultraviolet lamps, and arcs. The cathode of this tube is formed by sputtering a film of platinum onto the glass envelope through a hole in the shield.

There are several other types of ultraviolet photosensitive surfaces. One is composed of pure magnesium [34] and has a threshold at 3200 angstrom units. Another is a cadmium-magnesium alloy which has a threshold at about 3350 angstrom units and a peak at 2800 angstrom units.[35] Such surfaces are employed in phototubes to measure the erythemal radiations of the sun.

Figure 7·10 shows the spectral-response characteristic for a group of the more common metals from which photocathode surfaces have been prepared.

7·7 PHOTOSENSITIVE SURFACE DESIGNATIONS

In an attempt to standardize the common types of photosensitive surfaces, trade associations and engineering societies have assigned numbers to various surfaces which have common spectral-response curves. The illustrations of the various types of photosensitive surfaces shown in this chapter are typical of commercial surfaces. The surface number assigned to these typical surfaces in the text is not official,

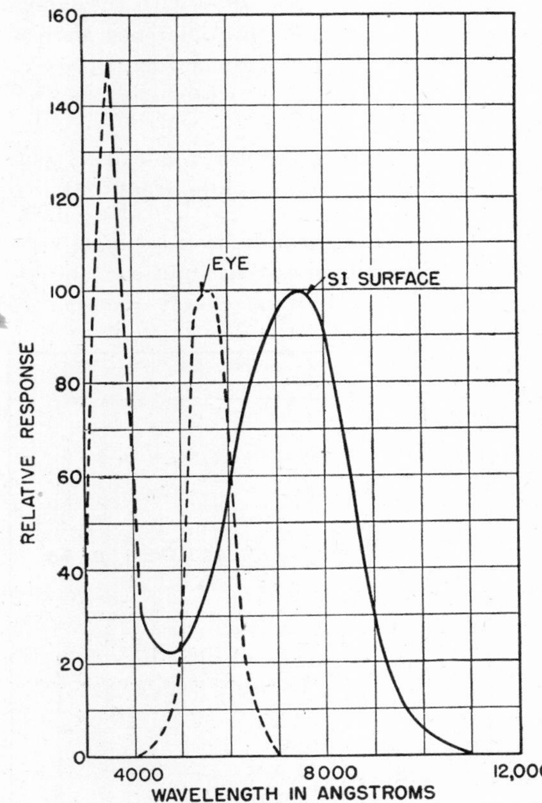

FIG. 7·11 Spectral-response characteristic of the S1 photosurface in a lime-glass bulb.

for these numbers have not been formally standardized. It is probable, however, that the standardized response curve will not be greatly different, so examples of surfaces corresponding as closely as possible with the proposed standards have been chosen. Surface numbers above S4 are used, but they are not included here because they have not been sufficiently standardized.

To obtain high sensitivity or a special spectral response characteristic, the composite type of surface is commonly used. In such surfaces, materials having large atomic radii and extremely low work functions can be used. One of the earliest composite surfaces and still one of the most common is the cesium-oxygen-silver surface. It is sensitive and much of its spectral-response curve is in the visible range.

S1 Surface

A form of the cesium-oxygen-silver type of surface is known as an S1 photosurface and is a good example of selective emission. The spectral-response curve of the S1 surface is shown in Fig. 7·11. Such a surface is sensitive in the red, and its emission is particularly high in the infrared region. A filter can be used to exclude response in the visible spectrum and permit only infrared sensitivity.

S2 Surface

The present S2 surface standard undoubtedly will be abandoned. Its spectral sensitivity is almost identical with the S1, and differences between tubes with the same type of surface may be greater than the difference shown in the standardized response curves for S1 and S2 surfaces.

S3 Surface

There has been some demand for a so-called "blue sensitive" tube. Such a characteristic is important for two reasons:

(1) Blue-sensitive tubes usually have a cut-off wavelength of approximately 8000 angstrom units. The spectral-response curve (Fig. 7·12) thus closely matches the visibility curve of the eye. This characteristic makes such a surface desirable for colorimetry.

(2) With low spectral response in the infrared, the output of a photosurface is almost independent of heat radiation. This is useful for applications that require masking the infrared response without the use of filters.

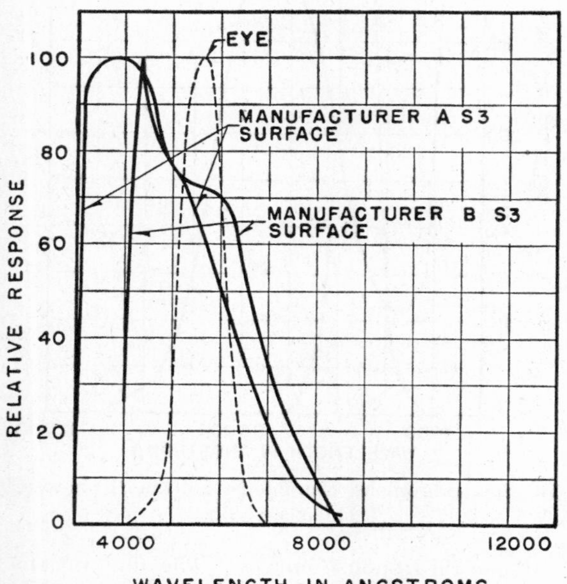

FIG. 7·12 Spectral-response characteristic of the S3 photosurface in a lime-glass bulb.

Blue-sensitive surfaces can be made in several ways. A composite surface of cesium-oxygen-silver with special treatment "peaks" in the blue region. Silver sputtered on cesium-oxygen-silver surfaces increases the blue response. Antimony-cesium deposits reported by Gorlich [36] show high sensitivity and maximum output in the blue-green region. The rubidium-rubidium oxide surface is probably the most widely used S3 type of surface.

S4 Surface

If a lime-glass bulb is used with an S4 surface, its spectral-response curve (Fig. 7·13) "peaks" in the blue-violet region.

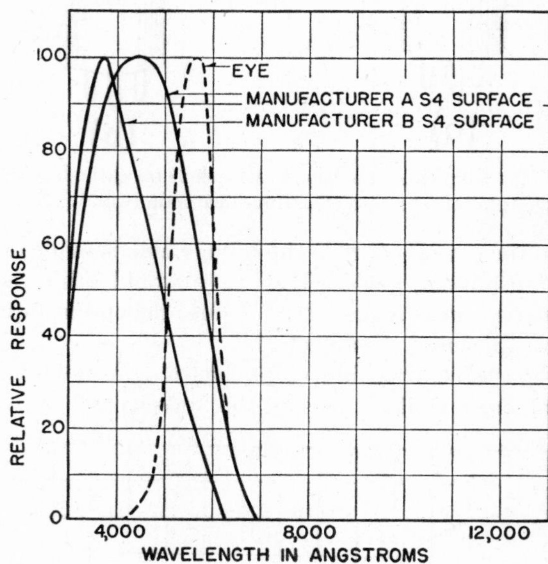

FIG. 7·13 Spectral-response characteristic of the S4 photosurface in a lime-glass bulb.

Such surfaces used with high-transmission glass, or with thin windows, would be extremely sensitive to ultraviolet radiation. The response characteristic of the S4 surface makes it particularly useful in certain applications because: (1) it is sometimes desirable to exclude infrared response (this surface has negligible response in the infrared); (2) such a surface is extremely sensitive to sunlight; and (3) phototubes using this surface give maximum response to radiation from fluorescent lamps or mercury-vapor light sources.

The S4 surface, development of which was reported by Glover and Janes,[37] is composed of a composite cesium-antimony surface on a metal base such as nickel. Cesium on magnesium gives a similar response.

Phototubes with spectral-response curves of almost any reasonable form can be made, but it may take years to develop some special surfaces so that they can be manufactured commercially. It is for this reason that attempts are being made to standardize only those surfaces for which there is a real requirement.

7·8 GAS PHOTOTUBES

During early experiments with photosensitive surfaces, someone noticed that in a given group of phototubes a few appeared to be unusually sensitive. This apparent high sensitivity was found to be caused by minute traces of gas left in the tubes by faulty pumping. Later, gas was deliberately admitted to increase sensitivity. This increase in sensitivity is known as gas amplification.

To avoid chemical reaction of the gas with the photosensitive surface, an inert gas is used as the filling medium. It can be any of the inert gases: xenon, krypton, helium, neon, or argon. Probably argon is most commonly used. Gas pressure has a marked effect on amplification, and there is an optimum pressure for maximum amplification. For practical reasons, this optimum pressure is seldom used. Curve 1 in Fig. 7·14 is the volt-ampere curve of a vacuum phototube for a definite value of illumination. This is assumed to be a tube of the central-anode type with a cylindrical cathode. At approximately 20 volts difference in potential, the output current of the tube saturates. At this anode voltage, the field is sufficient to insure that all electrons leaving the cathode are collected by the anode, and no further increase in anode voltage (within limits) has any effect on the phototube current.

Now let it be assumed that a small amount of gas has been added to the tube. The volt-ampere curve no longer shows saturation, but increases with anode voltage. (See curve 2, Fig. 7·14.) This increase is caused by an ionization current proportional to the photoelectric current. Amplification may be determined by the ratio of photoelectric current at some reference voltage (usually 90 volts) to the saturation current (usually measured at 15 volts). In Fig. 7·14 this is illustrated by the value I_b/I_s. This ratio is known as the *gas ratio* of a phototube and normally does not exceed 10. If one desires to compare two gas phototubes which have the same type of surface, but entirely different values of sensitivity, it is necessary to divide the sensitivity value by the gas ratio in order to determine the primary (vacuum) sensitivity. High primary sensitivity is the real measure of quality in a phototube. A gas ratio above 10 usually results in unstable output and a tendency to glow breakdown. Gas ratios for vacuum types are sometimes calculated as a measure of degree of vacuum.

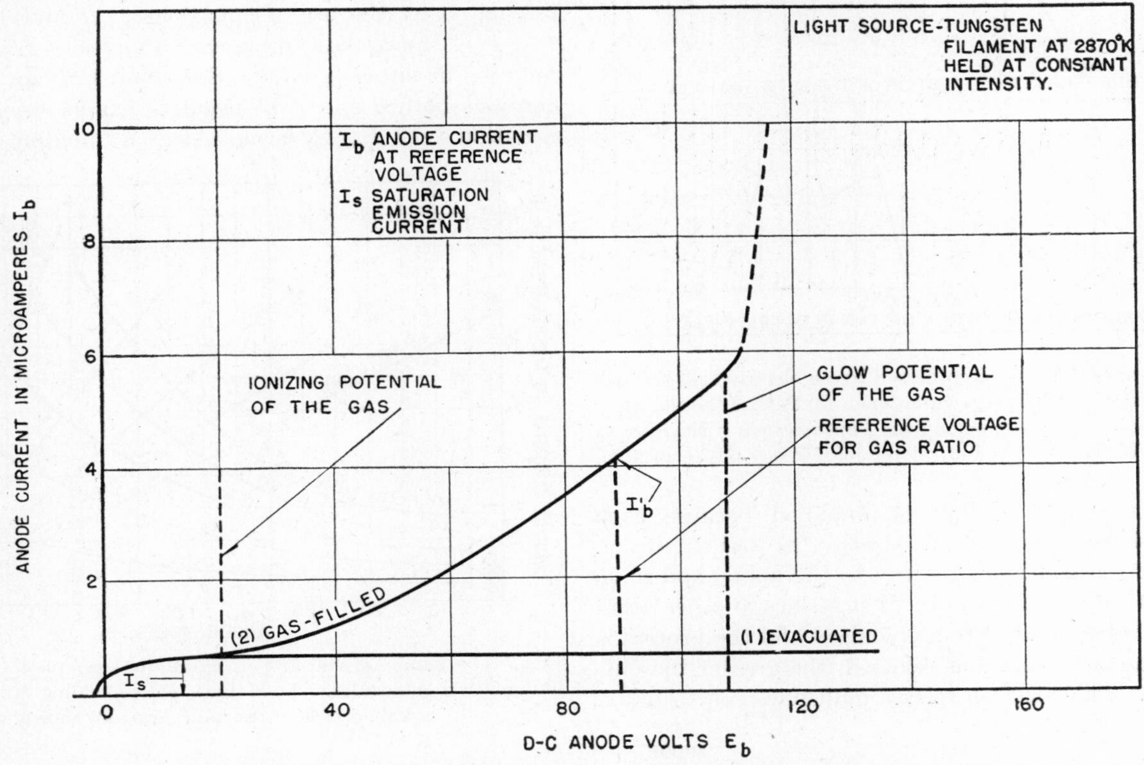

FIG. 7·14 Current-voltage characteristic for a typical gas and vacuum phototube.

The reasons for gas amplification are somewhat complicated, because increased electron emission can be attributed to any one of several possible sources. Among these are: (1) multiplication of electrons in space due to ionization; (2) production of electrons by positive-ion bombardment of the cathode; and (3) production of electrons by high fields at the cathode because of the accumulation of positive ions.

In Fig. 7·15 [38] is shown a possible path of an electron accelerated by an electric field. Whenever there is a collision between an atom and an electron, and the kinetic energy of the electron equals or exceeds the excitation potential of the atom, an inelastic collision may occur and the atom may become excited. If another electron of lower energy immediately strikes the excited atom, it may become ionized. If a collison with a normal atom occurs when the energy of the electron equals or exceeds the ionization potential, the atom may be ionized; that is, it loses an electron and becomes an ion. Atoms may also be ionized at an energy level less than that required for excitation if they are in the metastable state. In a phototube, however, the number of collisions with metastable atoms would be negligible. In the example shown in Fig. 7·15, ionization or excitation is indicated wherever the required kinetic energy is reached. Actually, in only a small percentage of these collisions does excitation or ionization occur.

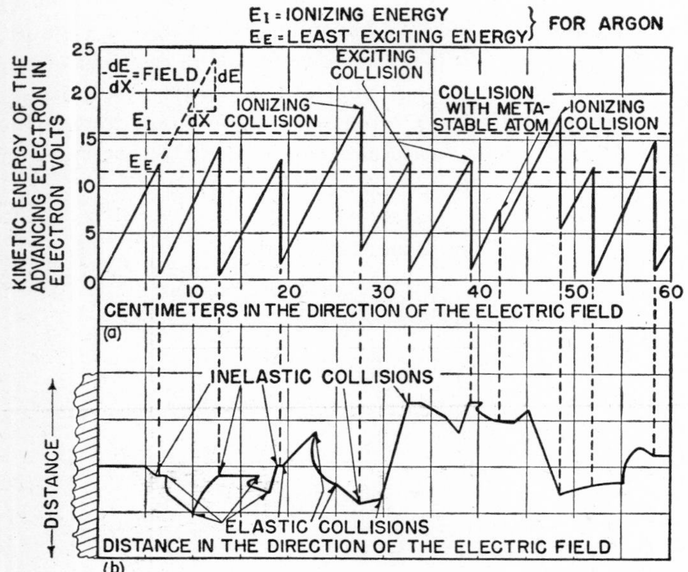

Fig. 7·15 Advance of an electron driven by an electric field through argon at low pressure. (*a*) Energy variations. (*b*) Path of the electron's motion. (Reprinted with permission from *Fundamentals of Engineering Electronics*, by W. G. Dow, published by John Wiley, 1930, p. 404.)

Initial photoelectric current is amplified because each ionizing collision produces a new electron as well as an ion. This new electron continues its way to the anode and in so doing collides with other atoms. If it strikes another atom with sufficient energy, the atom is ionized and the process is repeated. The more electrons released, the greater the current becomes. This electron release multiplies very rapidly.

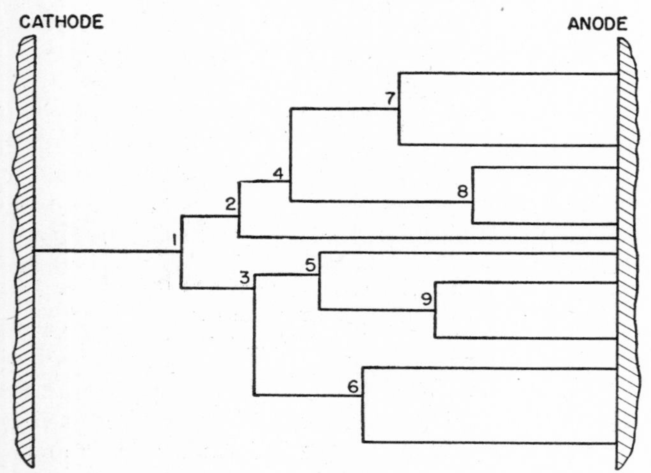

Fig. 7·16 Mechanism of gas amplification. One electron leaves the cathode, 10 electrons enter the anode. (Reprinted with permission from *Fundamentals of Engineering Electronics*, by William G. Dow, published by John Wiley, New York, 1930, p. 405.)

Figure 7·16 shows the possible yield from one electron leaving the cathode. Each ionizing collision has produced one additional electron, and the original electron starting from the cathode releases 10 electrons which reach the anode.

Figure 7·15 shows the advance of an electron at a relatively low pressure. The concentration of gas is such that there are not many collisions per volt, but there is opportunity for many electrons to acquire ionizing energy. If the pressure is too high, the electrons leaving the cathode have so many opportunities for collisions that they make elastic impacts or exciting collisions and rarely acquire enough energy for ionization. If gas concentration is too low, there are not enough collisions to cause any appreciable amount of ionization, and the phototube current approaches that of a vacuum tube.

Amplification for plane parallel-plate electrodes can be calculated by the use of equations 4·17 and 4·20. Equation 4·17 shows that amplification increases rapidly with an increase in either α or d. Equation 4·20 for α, however, contains both d and L as negative exponents, so that if in equation 4·17 I_{ea}/I_{ec} is plotted as a function of α a curve

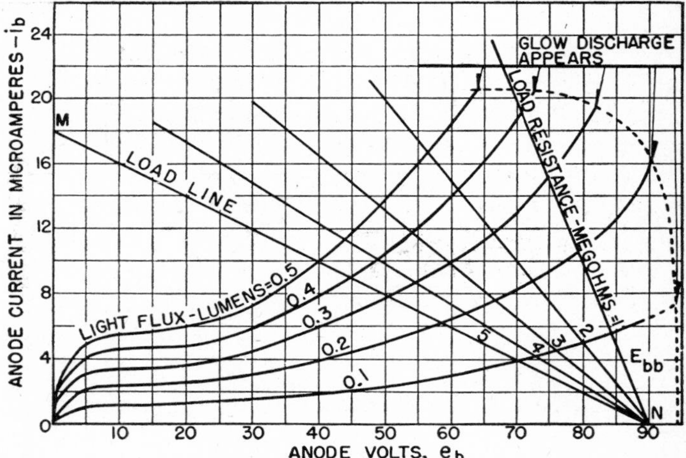

Fig. 7·17 Family of volt-ampere curves for a gas phototube at various illumination intensities. Load lines corresponding to a number of values of load resistance are also shown.

containing a maximum results, the ordinates depending upon the ratio d/L. To obtain the theoretical maximum of amplification, the conditions must be such as to make I_{ea}/I_{ec} a maximum.

However, if the gas amplification becomes too high, or if illumination is too great, a self-maintaining discharge results. Positive-ion generation then is so increased that the cathode is severely bombarded by positive ions, and the sensitive surface of almost any composite cathode would be destroyed. Normal gas ionization which produces gas amplification is not truly self-sustaining, and the current becomes almost negligible in the dark.

Curve 2 of Fig. 7·14 shows the volt-ampere curve of a gas tube for one value of illumination. If the volt-ampere curve is replotted for a series of illumination values, the result is a family of curves as shown in Fig. 7·17. These curves show several interesting things. As the illumination increases, gas amplification increases. Moreover, $\Delta e_b/\Delta i_b$ becomes smaller as illumination increases. If the anode voltage becomes higher than approximately 115, a glow discharge starts, regardless of the amount of illumination. At successively higher values of illumination, the voltage of transition to a glow is reduced until at approximately 75 volts a glow will not start regardless of the amount of illumination. It is for these reasons that a maximum peak-anode-voltage rating and a maximum anode-current rating are always shown in manufacturers' technical data. Limitation of anode current may, however, be made for other reasons.

On the curves of Fig. 7·17 is a series of load lines. To determine the position of these lines, assume a load resistance R_L. Let $\bar{r}_p = e_b/i_b$. This is the instantaneous tube resistance and is an inverse function of illumination. It is a continuously varying function so it is not directly determinable. In the circuit shown in Fig. 7·18,

$$E_{bb} = e_b + e_L \tag{7·4}$$

and

$$e_L = i_b R_L \tag{7·5}$$

From equations 7·4 and 7·5,

$$i_b = \frac{E_{bb} - e_b}{R_L} \tag{7·6}$$

Let $e_b = 0$; then $i_b = E_{bb}/R_L$. This is the co-ordinate for point M. Let $e_b = E_{bb}$; then $i_b = 0$. This is the co-ordinate for point N. The load line is drawn between the points M-N in Fig. 7·17. Current through the load may be determined graphically for any value of illumination or, if additional load lines are plotted, for any load resistance. If the light flux on the phototube is modulated, it is possible to calculate [39] the voltage output across load resistance R_L. The mean light intensity locates the quiescent operating point Q. Figure 7·19 shows the output voltage variation resulting from a sinusoidally modulated beam of light. If the light-flux variation is approximately linear, the corresponding phototube voltage output also varies linearly.

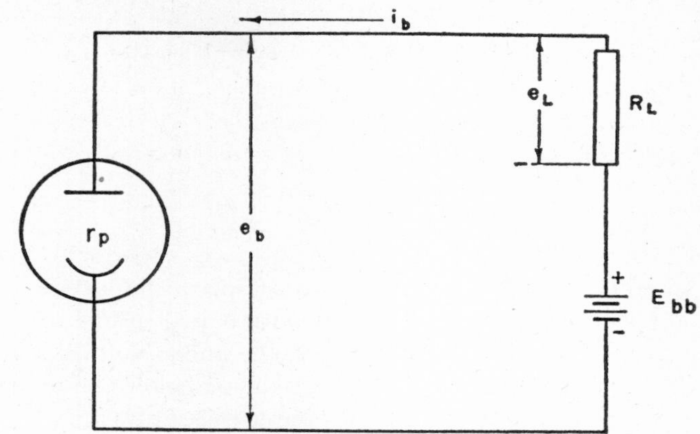

Fig. 7·18 Fundamental phototube circuit to illustrate use of load line.

The use of the family of volt-ampere curves has been illustrated for a gas phototube. The same general principles apply to phototubes of the vacuum type.

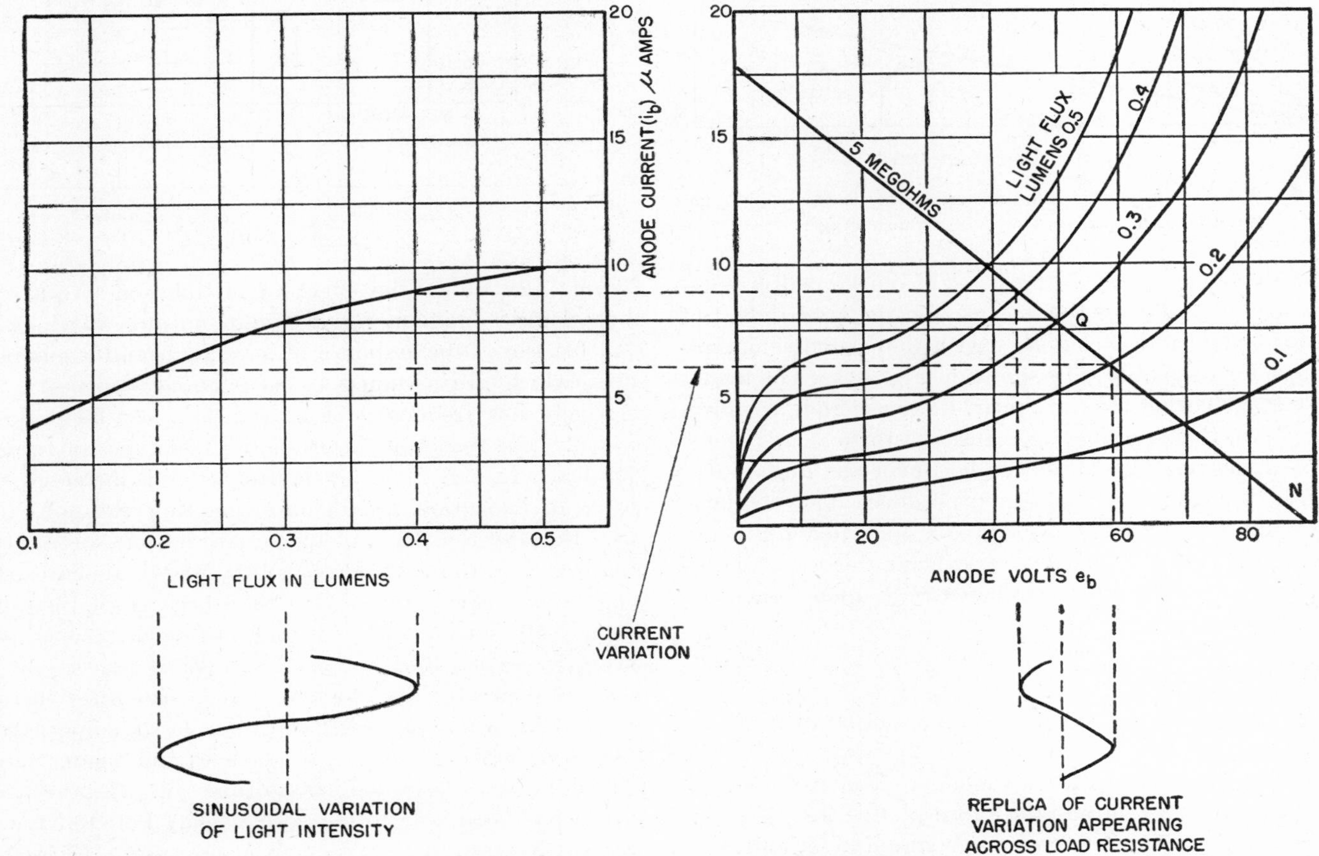

Fig. 7·19 Voltage-output variation resulting from a sinusoidally modulated beam of light. Circuit as in Fig. 7·18.

7·9 VACUUM VERSUS GAS PHOTOTUBES

The sensitivity of the gas phototube is so superior to the equivalent vacuum type that one may wonder why vacuum tubes are used at all. There are several good reasons for the continued use of vacuum phototubes. For some applications gas tubes are wholly unsuitable. A discussion of the more important disadvantages of a gas phototube follows.

Stability

The stability of a vacuum tube is always superior to that of an equivalent gas tube. This is of special importance in photometry or wherever constant output is of prime importance. To insure extreme stability, the anode voltage, even in the vacuum tube, sometimes is reduced below the ionization potential of possible gaseous impurities.

Linearity

As might be expected, the photoelectric current emitted by a gas phototube is not a true linear function of the illumination. Figure 7·20 illustrates this. The curve shown for

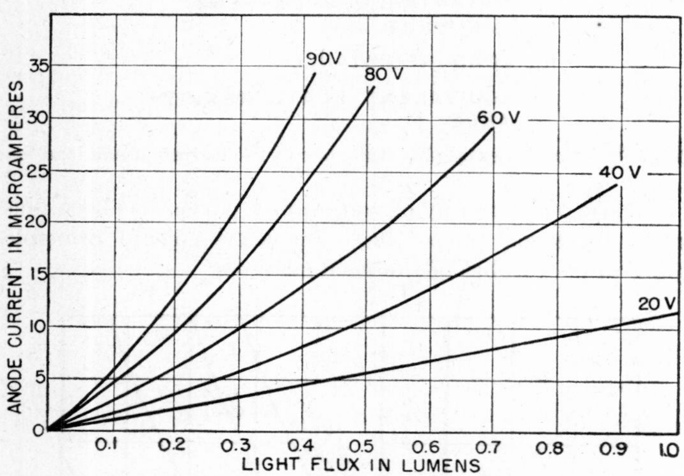

Fig. 7·20 Family of current-illumination characteristics for a gas phototube at various values of anode voltage.

the 20-volt anode supply is linear. This curve really represents vacuum-phototube conditions, for the probability of ionization is low and it compares with the current-illumination curve for a vacuum tube shown in Fig. 7·1. The higher anode-voltage curves are not linear, however, and, as shown by Fig. 7·19, this non-linearity causes distortion in the output. If a gas phototube is "worked" over the most linear portion of the range, the distortion produced does not limit response fidelity sufficiently to impair operation for many applications. The choice between a vacuum or a gas phototube, as regards linearity, depends entirely upon the conditions of application.

Dynamic Response

If a gas phototube is required to follow rapid fluctuations of light intensity, it has a non-linear dynamic sensitivity. The sensitivity decreases as the modulated light frequency increases. The most important cause of this loss of sensitivity is the finite time necessary for cumulative ionization of the gas, and for the diffusion of the ions to the bulb walls and cathode. The first and the lesser of the two effects is called ionization time, and the latter effect deionization time. As long as deionization persists, ions strike the cathode and new electrons are produced. This continues until equilibrium has been established; thus, photoemission current tends to lag any change in illumination. The time required for deionization varies from a few microseconds to several hundred microseconds, depending upon the construction of the tube, the nature of the gas, and the gas concentration.

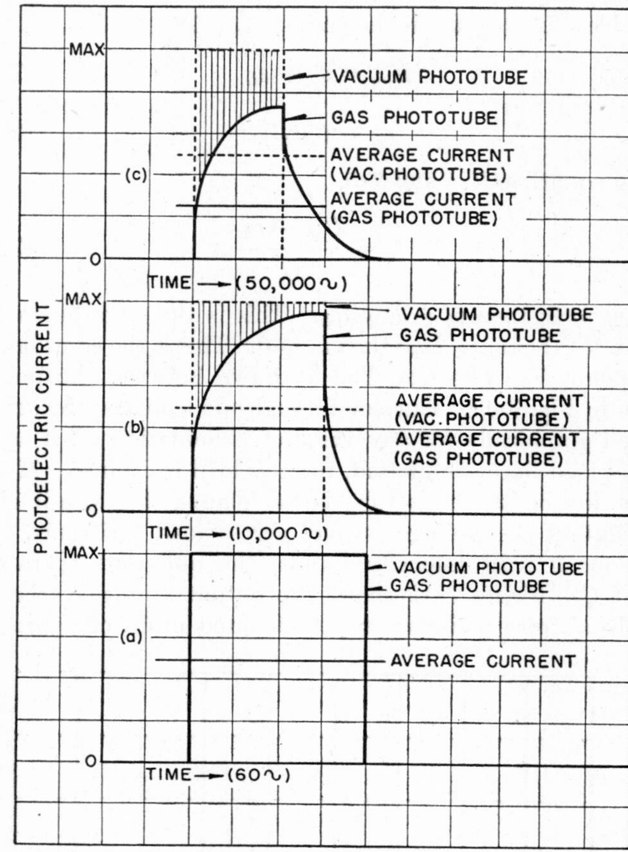

Fig. 7·21 Effect of light-modulation frequency on output current in a gas phototube.

Figure 7·21 shows the effect of modulation frequency on photoelectric emission for a square output wave. Figure 7·21 (*a*) shows the response of a vacuum and a gas phototube (which are assumed to have equal sensitivity) to a relatively low frequency of modulation, say 60 cycles per second. The response curves of the two types of tubes are identical. In Fig. 7·21 (*b*), the frequency is increased. The effects of ionization and deionization time are now apparent. Because the shape of the square output wave remains unchanged for the vacuum phototube, the shaded area shows the loss in current for the gas tube because of the lagging effect of the ions. At a still higher frequency, over 10,000 cycles per second for example, Fig. 7·21 (*c*), the loss in output is still greater and fidelity of response is seriously impaired. Figure 7·22 shows the relative dynamic response characteristic for a range of frequencies and anode voltages.

The vacuum phototube has no time lag. In other words, no one has been able to measure a delay between the incidence of light and the movement of electrons. Capacitance effects may appear to create a lag and can cause a falling frequency-response characteristic. These can be virtually eliminated by proper design of the associated circuit, however.

Glow Voltage

As indicated previously, a gas phototube breaks down with a destructive glow discharge if more than a specified voltage is applied across its elements. The vacuum tube is independent of such effects. This is important for some applications.

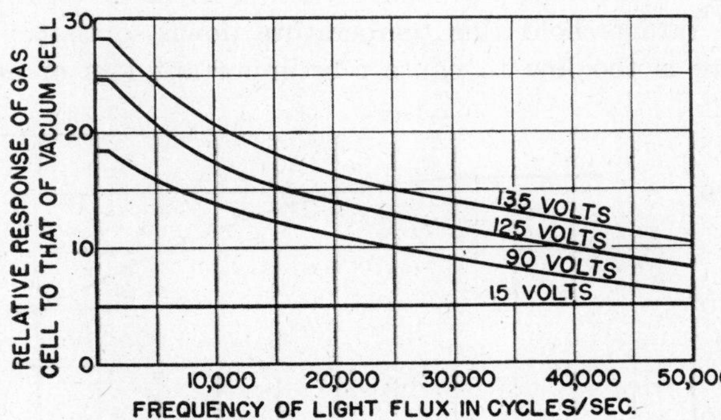

Fig. 7·22 Response-frequency curves for modulated light flux in a gas phototube. (Reprinted with permission from *Photocells and Their Application*, by V. K. Zworykin and E. D. Wilson, published by John Wiley, 1930, p. 81.)

7·10 SENSITIVITY

In any phototube, the introduction of gas increases sensitivity. A choice must be made between the undesirable characteristics of the gas tube and the lower sensitivity of the vacuum tube. It is possible to amplify the current of the vacuum phototube to increase its sensitivity and, if the light is modulated so that a-c amplifiers can be used, this is a practical method. If the light is unmodulated, or if the frequency of modulation is low, d-c amplifiers must be used. For low values of illumination, the use of a high-gain d-c amplifier is necessary. Such amplifiers are difficult to design and to operate from a stability basis. For this reason, highly sensitive phototubes have been developed. The present trend in phototube design appears to be toward highly sensitive vacuum phototubes. The more important facts about sensitivity are summarized in Table 7·5.

The sensitivity of phototubes having their response in the visible spectrum is called *luminous sensitivity*, and usually is defined as the ratio of microamperes per lumen of incident light. This measurement is made with a tungsten-filament light source at a color temperature of 2870 degrees Kelvin. Such measurements show only the integrated output and do not represent true sensitivity because: (1) distribution of energy in the source is a function of wavelength; and (2) the response of the photosurface may be most sensitive at a frequency beyond the output range of the source.

This difficulty has led to a much better method of determining sensitivity wherein sensitivity is defined as the ratio of photoelectric current to the energy in the received radiation for a definite wavelength band. This band is made as narrow as practical and usually is taken at the point of maximum sensitivity on the spectral-response curve. *Sensitivity* usually is expressed in microamperes per microwatt of radiant energy, and is valid not only in the visible spectrum, but also in the infrared and ultraviolet regions. When the value of sensitivity in $\mu A/\mu W$ at the point of maximum sensitivity and the relative spectral distribution curve of a tube are known, either the sensitivity for any wavelength zone or the total phototube sensitivity may be determined.

Table 7·5 General Dependence of Phototube Sensitivity upon Certain Variables

VARIABLE	SENSITIVITY	
	Gas Phototube	Vacuum Phototube
Type of surface	Very dependent.	Very dependent.
Intensity of illumination	Increases non-linearly with intensity (see Fig. 7·20).	Increases linearly with intensity (see Fig. 7·1).
Anode voltage	Increases with voltage.	Independent of voltage (above about 20 volts).
Wavelength	Very dependent; depends upon spectral-response curve.	Very dependent; depends upon spectral-response curve.
State of polarization	Dependent.	Dependent.
Modulation frequency	Decreases with increasing frequency of modulated light.	Independent of modulation frequency (capacity effects ignored).
Temperature	Independent of temperature (within ordinary ranges for pure-metal surfaces).*	Independent of temperature (within ordinary ranges for pure-metal surfaces).*

* See Fig. 7·23 for an example of the temperature effect on a composite surface phototube.

7·11 SECONDARY PHOTOTUBE CHARACTERISTICS

Several inherent characteristics of phototubes, usually of minor importance, must be considered in some applications. These characteristics depend upon such variables as type of tube (gas or vacuum), type of surface, mechanical construction, and operation history.

Dark Current

Dark current may be defined as the photoelectric current in a phototube that is not activated by radiant energy within the wavelength band to which the tube is considered sensitive. Dark current usually is caused by:

1. ***Thermionic Emission from the Cathode.*** Some emission is possible even though cathode temperature is far from that commonly considered necessary for thermionic emission.

2. ***Glow Currents.*** This is more noticeable in gas tubes, but it may occur in vacuum tubes because of residual atoms of gas in the tube. Glow current increases rapidly with anode voltage.

Both these effects produce small currents, usually only a small fraction of a microampere, so they are troublesome only in measuring light of extremely low intensity.

Leakage Current

This current is often loosely grouped with dark current. It is actually entirely different, although the result of masking flux change from weak values of illumination is the same. Leakage current is simply current through the ohmic resistance across the tube. It is almost always erratic. It can pass over the glass envelope, internally or externally, or across the base. Although not a characteristic function of a phototube, leakage in the tube socket or associated wiring has the same effect as tube leakage. Internal leakage is a property of tube design and can be controlled only by the manufacturer. Leading the anode (or cathode) connection from the top of the tube decreases both internal and external leakage. External leakage can be reduced by coating the surface of the glass envelope with certain non-hygroscopic waxes, such as sealing wax, ceresin, and picene, or silicones.

For extremely low leakage, double-end tubes plus wax are sometimes used. Though commercial phototubes have sufficiently low leakage for most applications without treatment, a film of moisture can reduce shunt resistance from 10^{12} megohms to only a few megohms. If high load resistances are employed, precautions against leakage are absolutely necessary.

Interelectrode Capacitance

Interelectrode capacitance causes loss of voltage or distortion if the intensity of illumination varies rapidly. At low frequencies of modulated light, with reasonably high values of illumination, the effect of capacitance is unnoticeable. At higher modulation frequencies, for weak illumination intensities, and for high values of load resistance, capacity becomes important. To light impulses of steep wave front or to high frequencies of modulated light, large interelectrode capacitance acts as though the terminals of the tube were shorted by a low, variable resistance. The effect of capacitance is as bad in a vacuum phototube as in a gas tube, except that it is lessened by the greater sensitivity of the gas type. Distortion caused by frequency variation can be minimized by amplifiers having rising response characteristics.

Fatigue

If a phototube, particularly one having a composite surface cathode, has been subjected to strong light, its sensitivity may decrease. Conversely during periods of inactivity, especially in the dark, its sensitivity increases. This is usually more noticeable in gas-filled tubes. The effect may be reduced by long "aging" by the manufacturer, and by the use of low values of light intensity by the user. It is well to arrange the optical system so that the light covers as much of the cathode surface as possible. Such procedure reduces the light-flux density on the cathode surface and prevents local changes in emission caused by fatigue or other effects from affecting photoelectric current as a whole. Several factors accelerate fatigue. Sensitivity of a poorly exhausted tube containing residual oxygen, water vapor, or other impurities may change rapidly because of contamination of the photosensitive surface. Pure-metal surfaces in a well-evacuated phototube show no fatigue. If a phototube is used with a filter, the filter itself may show fatigue at the shorter wavelengths.

Temperature

Normally photoemission is little affected by temperature. The magnitude of the temperature effect depends, to a large extent, upon the nature of the cathode surface. Some surfaces lose emission [40] at very low temperatures. The effect varies with the nature of the surface and the frequency of the light. Photosurfaces employing elements of high vapor pressure, such as cesium, are susceptible to temperatures above +100 degrees centigrade. If the maximum temperature rating is exceeded, even for a short time, there may be a permanent shift in sensitivity and spectral response. In extreme cases, if temperature is allowed to remain for a longer time at values in excess of rating, the tube may lose its sensitivity completely. The reason for this loss is that the high vapor-pressure materials on the cathode vaporize, leaving the cathode denuded of active surface. Manufacturers' ratings hold this temperature down so that vapor pressure is too low to cause any important loss of active

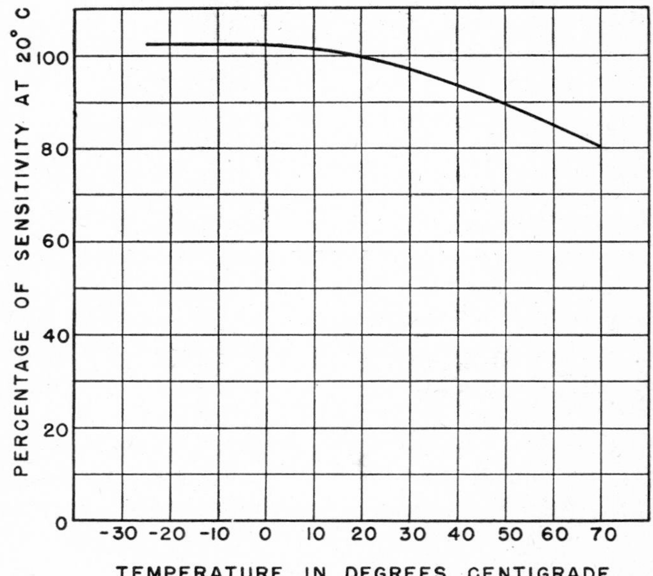

Fig. 7·23 Typical temperature characteristic of a cesium-oxygen-silver phototube.

surface. Other possible effects of operating at high temperature are the increase in internal leakage because of vaporized sensitive coating, and the increase in dark current. Figure 7·23 shows a typical temperature-characteristic curve for a cesium-oxygen-silver photosurface.

Maximum Current

Manufacturers' ratings give a maximum current for every type of phototube. The magnitude of this current depends upon the nature and area of the cathode surface. A vacuum tube usually can withstand higher current densities without damage than a gas tube, although excessive current densities are injurious in either type of tube. In the gas tube, glow discharge may start when the current rating is exceeded and cause rapid deterioration of the cathode surface.

7·12 MULTIPLIER PHOTOTUBE

One of the most promising phototubes is the electron-multiplier type of tube.[41, 42, 43] In this tube amplification is accomplished by focusing the electrons which are released from the cathode onto a series of secondary emissive surfaces called dynodes. (See Fig. 7·24.) This is accomplished by the special shaping of each dynode surface and by the application of ever-increasing positive increments of voltage from the photocathode to the anode. When suitable secondary emissive surfaces, such as silver-magnesium, or even the same material as that of the photocathode are used, yields from three to five electrons per primary electron per stage may be realized. If an average gain per stage per electron of 3.5 at 100 volts per stage, and 4.0 at 125 volts per stage is

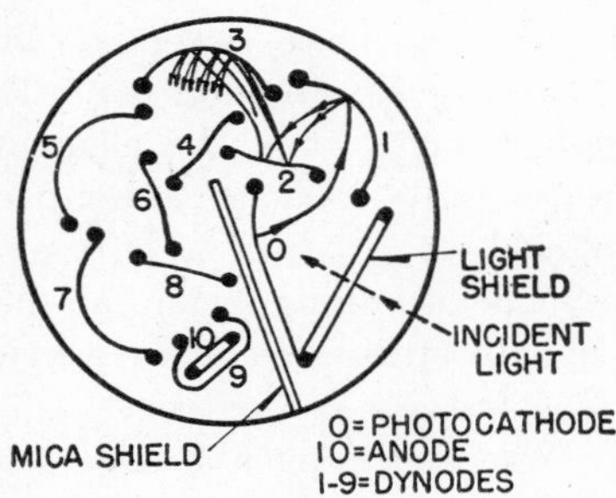

Fig. 7·24 Schematic diagram showing the electrode arrangement in a typical electron-multiplier phototube (RCA 931). (Reprinted with permission from "Technical Data Sheet for the 931A, Fig. 1," published by the R.C.A. Victor Division, 1943.)

assumed, gains of 80,000 and 260,000 respectively can be theoretically expected. The rapid change in gain with supply voltage, however, requires the use of a power source with extremely good voltage regulation. To hold the output of a nine-stage multiplier tube within 1 or 2 percent, the supply voltage must be held to 0.1 percent.

The spectral-response characteristic is determined solely by the characteristic of the cathode. The current-illumination curve is linear within the operating range. Operation beyond this range causes saturation in the output stage because of space charge.

Because there is no gas in the tube, the frequency response to modulated light is distorted only by tube capacitance.

Luminous sensitivity of a multiplier phototube to a light source of 2870 degrees Kelvin is about 2.0 amperes per lumen as compared with 15 to 45 microamperes per lumen for conventional phototubes.

One of the most important characteristics claimed for the multiplier tube is the reduction of signal-to-noise ratio, especially at low output. One reason for this reduction is that less gain is necessary in the amplifier for the same output voltage, so only the shot noise of photoemission, and not circuit noise, is significant.

Dark current multiplies in the same manner as photoemission current. The yield from the useful primary electrons is increased by the same exponent, so that the ratio of useful current to dark current is superior to that in a conventional tube. If primary dark current is reduced by the use of special surfaces, however, total dark current falls off rapidly so that the ratio of useful current to dark current increases sharply.

Construction of the type 931A phototube [44] (see Fig. 7·24) illustrates some of the methods used in overcoming certain design problems. When the multiplier tube is used dynamically, the anode potential and its field fluctuate continuously. This varying electric field interferes with electron focusing in the dynode region. To minimize this, dynode 9 is shaped so that it partially encloses the anode and shields it. The anode itself is constructed as a grid. This allows electrons from dynode 8 to pass through the anode to dynode 9, from which they are collected by the anode. Spacing between dynode 9 and the anode is such that almost all electrons released from dynode 9 are collected by the anode regardless of the instantaneous positive anode voltage. This feature permits the use of a wide range of load impedances.

A mica shield is interposed between the cathode surface and the anode to prevent ion feedback. Because of the high current density in the anode region, ions are likely to form from spurious gas in the tube. If these ions are not prevented from reaching the cathode surface, unwanted electron emission occurs and gas amplification results. This amplification, coupled with normal electron multiplication, causes complete loss of linearity in response, and may cause uncontrollable regeneration.

The light shield is a grill through which the light reaches the cathode surface. It is electrically connected to the cathode, and it serves as an electrostatic shield.

7·13 PHOTOGLOW TUBE

The photoglow tube [45] is a combination phototube and grid-glow tube. If the illumination on the sensitive surface of the gas-filled photoglow tube exceeds a critical value, the photoemissive current becomes unstable because of accumulative ionization, and the tube "fires." Current is limited only by the circuit resistance.

In construction the tube is relatively simple. It consists of a small central anode surrounded by a cathode of comparatively large area. This cathode has a surface, usually a pure metal, that withstands high current density without damage.

Gas pressure and spacing are so adjusted that, at normal working voltage and with no illumination, emissive current

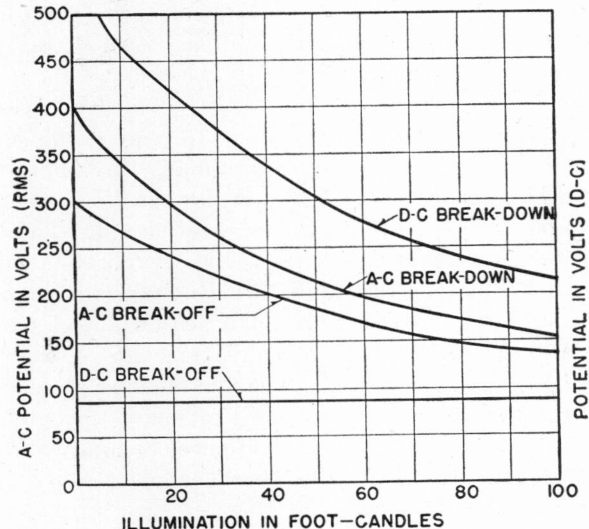

Fig. 7·25 A-c and d-c characteristics of a photoglow tube.

is negligible. If illumination for any fixed value of anode-cathode voltage exceeds the glow point, the tube characteristic undergoes a transition from gas amplification to glow discharge, as shown in Fig. 7·17. This results in a rapid current increase from perhaps microamperes to milliamperes.

Thus, a small change in illumination will trigger a current large enough to operate a relay directly. Although in a conventional phototube such a glow is injurious to the cathode, in a photoglow tube (because of the special cathode surface) the glow is not damaging, and it forms the actuating mechanism of the relay action.

The application of such a device, in its present stage of development, is that of a light-sensitive relay. It requires no stand-by power and can be operated with direct or alternating current. Figure 7·25 shows a typical operating characteristic. The "breakdown" curves represent the firing point of a typical tube at various anode voltages. The "break-off" curve indicates the value to which the anode potential must be reduced to stop the tube. With unidirectional supply voltage, the photoglow tube has a lock-in characteristic until the anode-cathode voltage is reduced below the drop-out point.

The photoglow tube now has little commercial application. Its main use is as a "watchdog" for commutator flashover in generating stations.

7·14 ICONOSCOPE †

The Iconoscope [46, 47] and tubes of its class at present represent the most advanced application of the photoelectric effect. The Iconoscope is a scanning tube in television applications. In common with any scanning device, it receives a visual image and transforms it into electrical energy. After amplification this energy can be transmitted to a distant point and reconverted into an image. Its chief advantage over other scanning devices is increased sensitivity, for the energy in each light element can be stored for the length of time taken by the scanning beam to repeat its trace.

The construction of the Iconoscope can best be visualized by considering a tube having a special photosensitive surface, an electron gun, and a suitable electron-beam-focusing mechanism. The electron gun and focusing mechanism are similar to the components of the ordinary cathode-ray tube. The special photosensitive surface is the heart of the device, and it is composed of a photosensitive mosaic structure having a large number of elements. In reality, the surface may be considered to be composed of many tiny phototubes. Each cathode element is connected through its capacitance to a common back plate, to the external circuit. The usual way of forming the mosaic surface is to evaporate a surface of silver on a sheet of thin mica with a metallically coated or graphitized back, or on a metal plate with an insulated surface. Another method is to spray silver oxide on the mica. This is subsequently reduced. The film of silver thus formed is not continuous, but is composed of many tiny globules. These are sensitized by the deposition of cesium, to form a cesium-oxygen-silver surface. The back plate is connected to one side of an amplifier.

The image of the subject to be scanned is focused on the mosaic surface. Because of the light incident on its surface, each elemental cathode emits electrons. The number of electrons emitted depends upon the intensity of illumination in one element of focused image. The result of this emission, because of the insulated cathode, is to charge the elemental cathode positively with respect to the back plate. This positive charge is balanced by an electronic space charge near the cathode surface. The strength of this field is proportional to the intensity of the illumination at the point considered.

The mosaic is now scanned rapidly by the electron beam from the gun. The negative field in front of the elemental cathode impedes the release of secondary electrons from the cathode surface. Thus, a smaller secondary electron current is produced by more intense illumination. This decrease in secondary-emission current when the elemental cathode is scanned, acting through its capacity to the back plate, is the mechanism employed to produce a current pulse in the external circuit. The scanning beam of electrons is not really small enough to strike just one cathode element. It has relatively large area in comparison with the size of a globule, and this large area prevents inequalities in electron emission because of unpreventable irregularities in the structure of the mosaic.

† Registered trademark, RCA Manufacturing Company.

7·15 ORTHICON ‡

The Orthicon [48, 49] is a modified form of Iconoscope in which the output-current curve, plotted as a function of the mosaic illumination, is a straight line. Its theoretical efficiency is 100 percent. The principal difference in operation is in the use of a low-velocity scanning beam, with which there is almost no secondary emission, and the electrons in the beam are themselves collected.

7·16 IMAGE ICONOSCOPE [50]

In this form of Iconoscope one photosensitive and one mosaic surface are employed. The image is focused first on a transparent photosurface. Electrons released from the back side of this surface are electromagnetically focused to form an electron image on a mosaic surface. This surface is composed of material, not photosensitive, but particularly effective in emitting secondary electrons. Because more intense fields can be used and because of greater secondary electron yield, its sensitivity is about ten times that of an ordinary Iconoscope.

7·17 IMAGE-DISSECTOR TUBE [51, 52]

The image-dissector tube is one of the early forms of electronic scanners. Instead of employing a mosaic surface, it uses an ordinary photoemissive surface. The subject to be scanned is focused on this surface. Electrons are emitted in numbers depending upon light intensities on the various elements of the focused picture. By use of magnetic fields, these electrons are made to pass down the tube, are focused, and are formed into an electronic image in the plane of an aperture, which is in front of a collecting anode. By additional sets of coils, the complete electron "picture" is scanned by moving it across the aperture at the usual scanning frequencies. Thus, the anode receives electron current directly proportional to the light intensity in the element of the original visual image. To increase electron yield, the col-

‡ RCA trademark.

lected electrons are "multiplied" by a series of secondary emissive surfaces.

7·18 PHOTOCONDUCTIVE CELLS

The drop in resistance of selenium exposed to light was first noticed in 1873,[53] fourteen years before Hertz discovered the photoelectric effect, but very little was done to find the fundamentals of this phenomenon until Gudden and Pohl [54] made a thorough investigation with a single crystal around 1925. They found that the primary effect was the liberation of electrons in direct proportion to the illumination, but that this electron current reduced the resistance of the selenium as a secondary effect and the secondary current masked the primary current.

The current through a selenium cell varies with the applied voltage, because the higher the voltage the larger the current

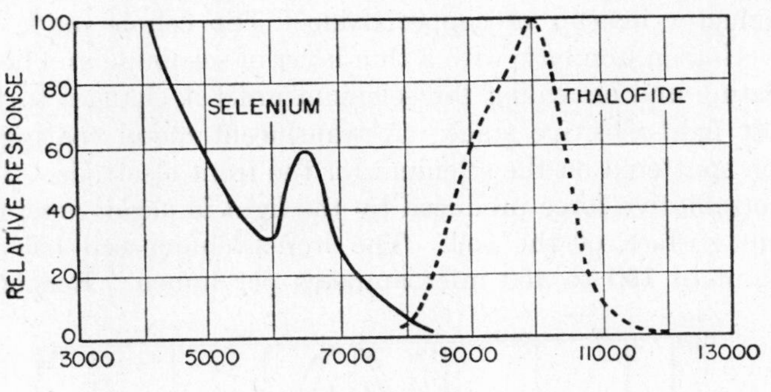

FIG. 7·26 Spectral-response curves for the selenium cell and the thalofide cell. (Reprinted by permission from *Photocells and Their Application*, by V. K. Zworykin and E. D. Wilson, published by John Wiley, 1930, p. 90.)

for the same change in resistance. The dark current also is greater for the higher voltage. Because too much voltage ruins the cell most cells are rated for use at 100 volts or less.

The current through the cell never reaches the full value immediately, but takes a little time to reach a peak. Afterward, with continued exposure, the current starts to drop off again. After the cell has been exposed to light, the dark resistance is lower than before, and it takes some time to return to the former resistance. If time between exposures is long enough to allow the cell to regain its dark resistance, the resistance varies approximately as the square root of the light intensity.[55] A rise in temperature decreases resistance of the cell and increases dark current. The one big disadvantage of the selenium cell (or bridge, as it is sometimes called) is its lack of linearity. There is, however, a rough relationship of output current to light intensity for short periods of exposure. The spectral sensitivity curve shows a response to the whole visible spectrum with a high sensitivity in the visible violet and a small peak in the red (Fig. 7·26).

The selenium cell is generally made by painting two electrodes with a metal paint in an interlocking grid arrangement on a piece of glass (Fig. 7·27) and covering the glass and grid with a thin layer of molten selenium. The resulting vitreous layer is then heat-treated until the selenium changes to the fine crystalline light-sensitive form.[56] Another method used in constructing the cells is to fill selenium in the spaces between metal plates arranged like a capacitor.[57] The object of either method is to get a lower resistance across the cell.

The selenium cell has been used for sound reproduction with a suitable capacitance-resistance coupling to the amplifier.[58] Its response to high frequencies is not so good as it is to low frequencies. In a relay circuit the current does not require amplification, because the output of these cells is from 500 to 1200 microamperes per lumen.

The thalofide cell [59] is similar to the selenium cell except that oxidized thalium sulphide is used instead of selenium for the light-sensitive material. The cell is generally used with infrared, because it has a peak sensitivity at 10,000 angstrom units (Fig. 7·26). A red bulb or an infrared-transmitting filter is used to protect it from the radiant energy of the blue end of the spectrum, which causes the cell to lose its sensitivity by changing the sulphide to sulphate. Low intensities of 0.5 foot-candle or less must be used, for higher intensities also cause deterioration.[60] The

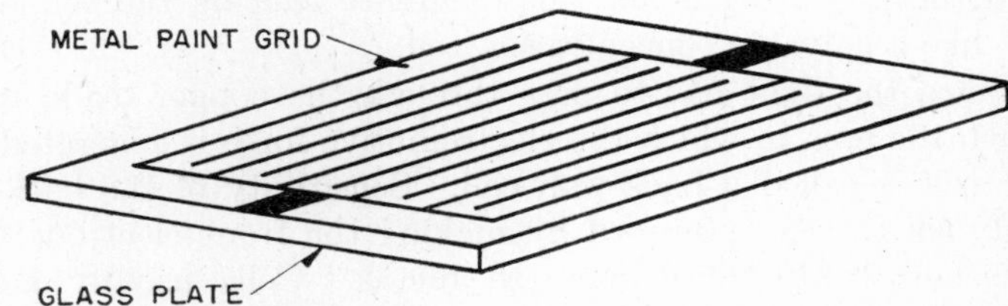

FIG. 7·27 Electrodes for selenium photoconductive cell.

cell has a much smaller time lag in reaching peak current than the selenium cell and very little lag in returning to its original dark resistance.

7·19 PHOTOVOLTAIC CELLS

Becquerel discovered in 1839 that an electromotive force was generated when he placed two similar electrodes in an electrolyte and allowed light to fall on one of them.[61] Later this production of electromotive force was noticed with crystals like those of zinc blende and galena. Several commercial photovoltaic cells, which used a copper oxide cathode in a liquid electrolyte, were produced around 1930. Although these cells had a current output of several hundred microamperes per lumen, they had the drawback of generating an electromotive force of 0.1 to 0.5 volts when they were not exposed to light. The copper oxide and iron-selenium semiconductor cells do not have this disadvantage.

Grondahl and Geiger [62] reported in 1927 that an electromotive force was developed when light was directed on the edge of a copper oxide rectifier plate. The first copper oxide cells were made by pressing a ring or screen of metal on the oxidized surface of the copper rectifier plate and connecting the screen and the copper plate to a galvanometer. The direction of current in the external circuit is from the copper to the copper oxide, as shown in Fig. 7·28. Schottky [63] investigated this photovoltaic effect of the cuprous oxide photocell and theorized that the electrons are liberated at the boundary of the copper and the copper oxide and pass through a barrier layer in one direction much more easily than in the other. By this theory the electrons would be expected to go in the same direction through a barrier or blocking layer as the electron current in the rectifier, but they go in the opposite direction. Although the idea of a unidirectional barrier has been discarded, some kind of barrier or insulating layer must retard the internal leakage to allow the electromotive force to develop between the

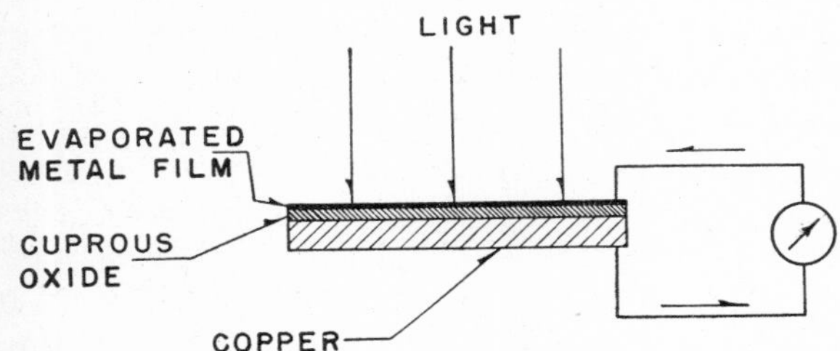

Fig. 7·28 Direction of electron current in a back-wall cell.

front and back electrodes of the cell. Lange [64] accounts for the electromotive force by the difference in the number of electrons released photoelectrically in the copper oxide and the insulating layer, this difference causing the cell to act like a galvanic concentration cell.

Since the light has to pass through the copper oxide to reach the area in which the electromotive force is generated, the cell is called a back-wall cell. Sensitivity of the back-wall cell can be improved by making the front electrode a translucent film of evaporated metal. Cells usually are made by oxidizing the copper at about 1000 degrees centigrade and then annealing around 600 degrees. The translucent front electrode—usually of gold, silver, or platinum—is then evaporated onto the copper oxide. The absorption of the shorter wavelengths of light by the copper oxide probably accounts for the back-wall cell having most of its sensitivity in the red and infrared. The response reaches a maximum between 6000 and 6500 angstrom units and decreases to a threshold at 66,000 angstrom units (Fig. 7·29). The output of the cell is only about 10 to 15 microamperes per lumen.

A much more sensitive cell can be made by cathodically sputtering the translucent metal electrode on the copper oxide.[65] Sputtering produces some change at the contact between the metal and the oxide, which causes the electromotive force to be generated at this juncture instead of at that between the copper and the oxide. Since the electrons are liberated at the front of the cell, it is called a front-wall

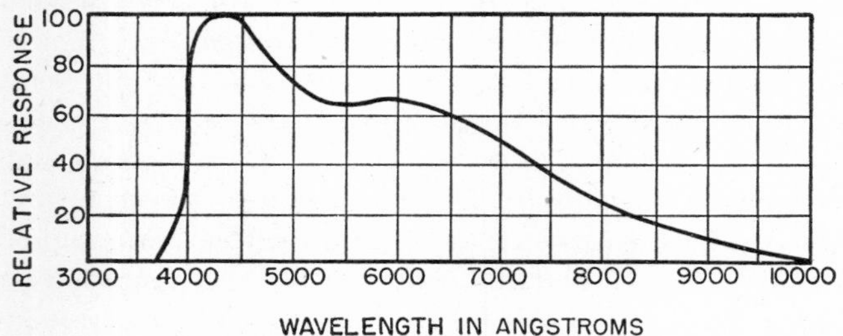

Fig. 7·29 Spectral-response curve for a copper oxide back-wall cell.

cell. The electromotive force is in the opposite direction to that of a back-wall cell; that is, current goes from the translucent film to the copper in the external circuit (Fig. 7·30). The output of this cell is from 100 to 150 microamperes per lumen. Peak sensitivity is at 5500 angstrom units, and the sensitivity covers the range from 4000 to 7000 angstrom units, as shown in Fig. 7·31. This is close to the visibility curve of the eye, which makes the cell useful as an illumination indicator. The current is large enough to use in a relay circuit without the use of batteries.

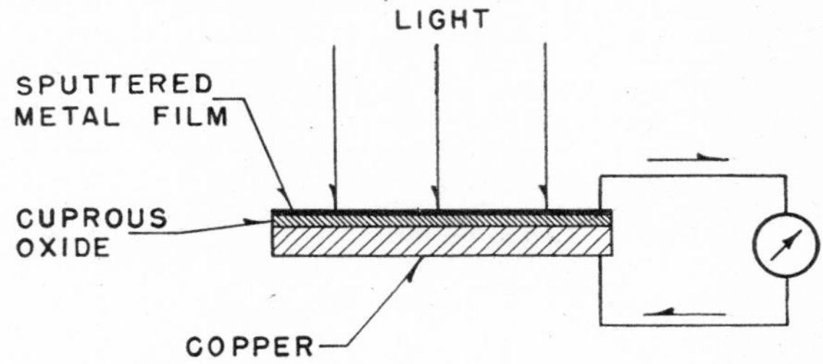

Fig. 7·30 Direction of electron current in a front-wall cell.

The iron-selenium cell was developed from the selenium rectifier. It is a front-wall cell with selenium as the semiconductor instead of copper oxide. The cell is made by covering an iron base with a thin layer of selenium, and heat-treating and annealing the selenium until it changes to the most light-sensitive stage. A translucent metal coating is then sputtered on the selenium for the front electrode. The electromotive force produced by the light is negative at the front surface of the cell. The iron-selenium cell has an output of 150 to 450 microamperes per lumen. Response

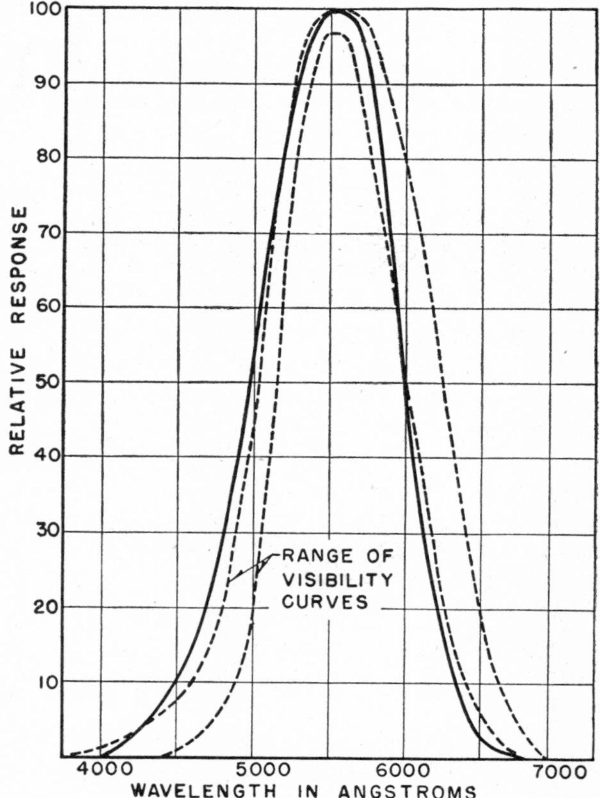

Fig. 7·31 Spectral-response curve for a front-wall copper oxide cell.

extends into the infrared and ultraviolet with a peak at 5550 angstrom units, and with the proper filter it closely follows the response of the eye (Fig. 7·32).

In all the semi-conductor photocells, current output is not quite proportional to light intensity. The current is nearly linear at medium intensities, if a low resistance is used in the external circuits as shown in Figs. 7·33 and 7·34. Linearity is less at higher intensities and also with higher external resistances, because of the internal leakage of the cells. The barrier or insulating layer is not perfect, and some of the electrons can leak back across this layer internally.

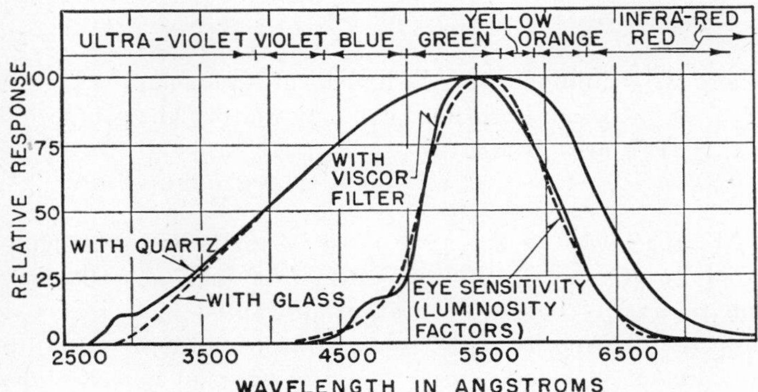

FIG. 7·32 Spectral-response curve of an iron-selenium cell. (Courtesy of Weston Electrical Instrument Corporation.)

Leakage becomes greater at high intensities because the large currents reduce the internal resistance of a cell. An increase in temperature also reduces resistance. Unless it is exposed to light, the cell has no output.

The open-circuit voltage produced by a semi-conductor cell is approximately proportional to the logarithm of light intensity.[66] The use of a large external resistance causes the current output to approach this same relation. To use this characteristic, photographic exposure meters are often made with a high resistance in the circuit.

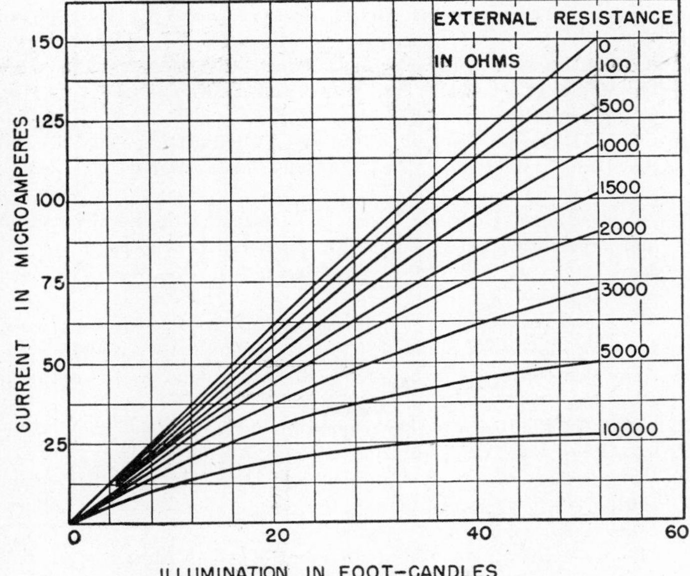

FIG. 7·33 Effect of illumination and external resistance on current output of a copper oxide front-wall cell.

Semi-conductor photocells have a capacitance formed by the insulating layer between the two electrodes. Spacing between electrodes is small, but the dielectric is poor, so the capacitance is about 0.1 microfarad per square centimeter. This capacitance causes lag in response and limits the use of the cell at high frequencies. The cells have a fatigue which is greater at higher intensities, but they generally regain sensitivity after a short time in the dark. The cells may lose sensitivity over long periods of time.

These photovoltaic cells produce enough electromotive force to be used directly in a meter circuit. Output can be increased with external voltage, but stability and linearity are reduced. The cells can also be used in an amplifier circuit, although the voltages generated are not large enough for best amplification. The main use of these cells is for illumination-intensity measurement.

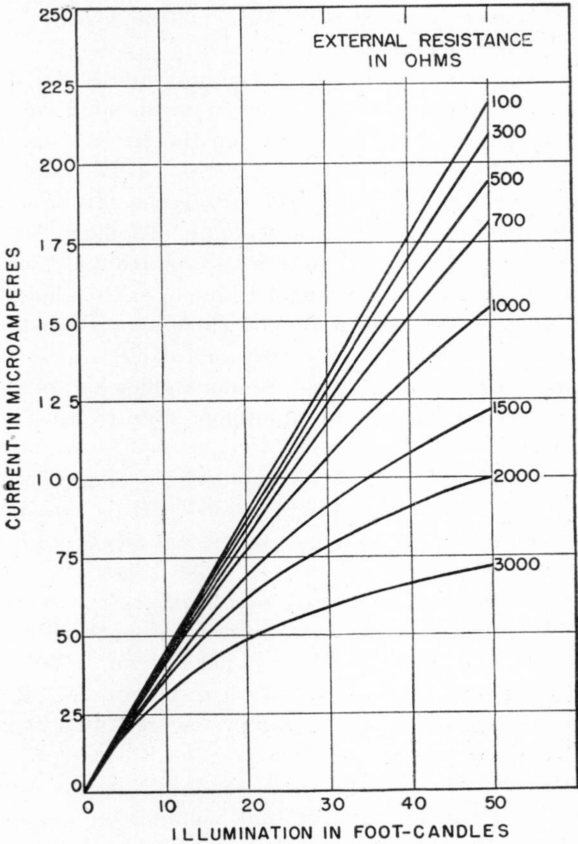

FIG. 7·34 Effect of illumination and external resistance on current output of an iron-selenium cell. (Courtesy of Weston Electrical Instrument Corporation.)

7·20 DEFINITION OF PHOTOMETRIC QUANTITIES

Light. For the purposes of illuminating engineering, light is radiant energy evaluated according to its capacity to produce visual sensation.

Luminous Flux (F). Luminous flux is the time rate of flow of light.

Lumen (*lm*). The unit of luminous flux. It is equal to the flux through a unit solid angle (steradian) from a uniform point source of 1 candle, or to the flux on a unit surface all points of which are at unit distance from a uniform point source of 1 candle.

Luminous Intensity ($I = dF/d\omega$). Luminous intensity of a source of light, in a given direction, is the solid-angular flux density in the direction in question. Hence, it is the luminous flux on a small surface normal to that direction, divided by the solid angle (in steradians) which the surface subtends at the source of light.

Candle (*c*). The candle is the unit of luminous intensity. The unit used in the United States is a specified fraction of the average horizontal candlepower of a group of 45 carbon-filament lamps preserved at the National Bureau of Standards, when the lamps are operated at specified voltages. This unit is identical, within the limits of uncertainty of measurement, with the International Candle established in 1909 by agreement between the national standardizing laboratories of France, Great Britain, and the United States, and adopted in 1921 by the International Commission on Illumination.

Candlepower ($I = dF/d\omega$; abbreviation, cp). Candlepower is luminous intensity expressed in candles.

Illumination ($E = dF/dA$). Illumination is the density of the luminous flux on a surface; it is the quotient of the flux by the area of the surface when the latter is uniformly illuminated.

Foot-Candle (ft-c). The foot-candle is the unit of illumination when the foot is taken as the unit of length. It is the illumination on a surface 1 square foot in area on which there is a uniformly distributed flux of 1 lumen, or the illumination produced at a surface all points of which are at a distance of 1 foot from a uniform point source of 1 candle.

Luminous Efficiency. The luminous efficiency of radiant energy is the ratio of the luminous flux to the radiant flux.

Radiant Energy (U). Radiant energy is energy traveling in the form of electromagnetic waves. It is measured in units of energy such as ergs, joules, calories, or kilowatt-hours.

Spectral Radiant Energy ($U_\lambda = dU/d\lambda$). Spectral radiant energy is radiant energy per unit wavelength interval at wavelength λ; for example, in ergs per micron.

Radiant Flux ($\phi = dU/dt$; alternate symbol, P). Radiant flux is the time rate of flow of radiant energy. It is expressed preferably in watts, or in ergs per second.

Radiant-Energy Density ($u = dU/dV$). Radiant-energy density is radiant energy per unit volume; for example, in ergs per cubic centimeter.

Radiant-Flux Density ($W = d\phi/dA$). Radiant-flux density at an element of surface is the ratio of radiant flux at that element of surface to the area of that element; for example, in watts per square centimeter. When referring to a source of radiant flux, this is also called radiancy.

Radiant Intensity ($J = d\phi/d\omega$). The radiant intensity of a source is the energy emitted per unit time, per unit solid angle about the direction considered; for example, in watts per steradian.

Spectral Radiant Intensity ($J_\lambda = dJ/d\lambda$). Spectral radiant intensity is radiant intensity per unit wavelength interval; for example, in watts per steradian or per micron.

Total Emissivity (ϵ_t). The total emissivity of an element of surface of a temperature radiator is the ratio of its radiant-flux density (radiancy) to that of a black body at the same temperature.

Spectral Emissivity (ϵ_λ). The spectral emissivity of an element of surface of a temperature radiator at any wavelength is the ratio of its radiant-flux density per unit wavelength interval (spectral radiancy), at that wavelength, to that of a black body at the same temperature.

7·21 FLUX CALCULATION

Photosensitivity to luminous flux is commonly measured in microamperes per lumen. To obtain a luminous flux of 1 lumen (or what is generally used, 0.1 lumen), the following formula may be used.

$$lm = \frac{CA}{d^2} \tag{7·7}$$

where lm = luminous flux in lumens at an aperture
C = mean horizontal candlepower of light source
A = area of aperture
d = distance between light source and aperture.

Measurements of this type must be made within a light-proof box having an interior surface of high absorptivity, and preferably baffled to prevent unwanted light reflected from the glass walls of the source appearing at the aperture (0.1 lumen is usually used for measurement purposes as 1 lumen may injure the photosensitive surface).

REFERENCES

1. "The Influence of Ultraviolet Light on Electric Discharge," H. Hertz, *Ann. Physik*, Vol. 31, 1886, pp. 983–1000.
2. "The Influence of Light on an Electrically Charged Body," W. Hallwachs, *Ann. Physik*, Vol. 33, 1888, pp. 301–312.
3. "Production of Cathode Rays by Ultraviolet Light," P. Lenard, *Ann. Physik*, Vol. 2, June 1900, pp. 359–375.
4. "Masses of Ions," J. J. Thomson, *Phil. Mag.*, Vol. 48, Dec. 1899, pp. 547–567.
5. "Notice on the Dispersion of Negative Electricity through Sun and Daylight," J. Elster and H. Geitel, *Ann. Physik*, Vol. 38, 1889, pp. 40–41.
6. "The Application of Sodium Amalgams to Electric Light Experiments," J. Elster and H. Geitel, *Ann. Physik*, Vol. 41, 1890, pp. 161–176.
7. "A New Form of Apparatus for Demonstrating Electrical Discharge by Means of Daylight," J. Elster and H. Geitel, *Ann. Physik*, Vol. 42, 1891, pp. 564–567.
8. "The Theory of the Law of Energy Distribution in the Normal Spectrum," M. Planck, *Deut. physik. Ges. Verhandl.*, Dec. 1900, p. 237.
9. "Concerning a Heuristic Viewpoint Dealing with the Production and Transformation of Light," A. Einstein, *Ann. Physik*, Vol. 17, July 1905, pp. 132–148; *Ann. Physik*, Vol. 20, May 11, 1906, pp. 199–206.
10. "Emission Velocities of Photo-Electrons," A. L. Hughes, *Trans. Roy. Soc. (London), Series A*, Vol. 212, Aug. 12, 1912, p 205.
11. "The Photoelectric Effect," O. W. Richardson and K. T. Compton, *Phil. Mag.*, Vol. 24, 1912, pp. 575–594.
12. "A Direct Photoelectric Determination of Planck's *h*," R. A. Millikan, *Phys. Rev.*, Vol. 7, March 1916, pp. 355–388.
13. *Photoelectric Cells*, N. R. Campbell and D. Ritchie, p. 29, Pitman, 1934.
14. *Ibid.*, p. 30.
15. "Photoelectric Effect from Thin Films of Alkali Metal on Silver," H. E. Ives and H. B. Briggs, *Phys. Rev.*, Vol. 38, Oct. 15, 1931, pp. 1477–1489.
16. "Photoelectric Effect on Platinum and Copper by Ultraviolet Light," R. Pohl, *Deut. physik. Ges. Verhandl.*, Vol. 10, 1909, pp. 339–359; "Photoelectric Effect on Mercury in Polarized Ultraviolet Light," *ibid.*, pp. 609–616; "Photoelectric Effect of Alkali Metals in Polarized Light and Its Dependence upon the Wavelength," *ibid.*, pp. 715–722.
17. "The Distribution in Direction of Photoelectrons from Alkali Metal Surfaces," T. C. Fry and H. E. Ives, *Phys. Rev.*, Vol. 32, July 1928, pp. 57–80.

18. "Division of Energy in Emission Spectrum of a Blackbody," W. Wien, *Ann. Physik*, Vol. 58, 1896, p. 662.
19. "On the Partition of Energy between Matter and Aether," J. H. Jeans, *Phil. Mag.*, Vol. 10, July 1905, pp. 91–97.
20. *Illuminating Engineering Nomenclature and Standards* (1942) 10.085.
21. *Measurement of Radiant Energy*, W. E. Forsythe, Chapter 1, p. 6, McGraw-Hill, 1937.
22. "The Blackbody," F. Benford, *G.E. Rev.*, Vol. 46, July 1943, pp. 377–382, August 1943, pp. 433–440.
23. "A New Table of Values of the General Physical Constants," Raymond T. Birge, *Rev. Mod. Phys.*, Vol. 13, 1941, p. 233.
24. "A New Relationship of the Radiation of a Blackbody to the Second Law," W. Wien, *Sitzber. Akad. Wiss. Berlin*, Feb. 9, 1893, pp. 55–62.
25. *Handbook of Chemistry and Physics*, Chemical Rubber Publishing Co., Vol. 24, p. 1939.
26. "Activating Light Sources for Luminescent Materials," E. W. Beggs, *J.O.S.A.*, Vol. 33, Feb. 1943, pp. 61–70.
27. "The Search for High Efficiency Light Sources," S. Dushman, *J.O.S.A.*, Vol. 27, Jan. 1937, pp. 1–24.
28. "Low Pressure Gaseous Discharge Lamps, Part I," S. Dushman, *Trans. A.I.E.E.*, Vol. 53, 1934, pp. 1204–1212, Part II, pp. 1283–1296.
29. "Photoelectric Emission from Different Metals," H. C. Rentschler, D. E. Henry, and K. O. Smith, *Rev. Sci. Instruments*, Vol. 3, Dec. 1932, p. 794.
30. "Effect of Oxygen upon the Photoelectric Thresholds of Metals," H. C. Rentschler and D. E. Henry, *J.O.S.A.*, Vol. 26, Jan. 1936, pp. 30–34.
31. "Color Sensitiveness of Photoelectric Cells," E. F. Seiler, *Astrophys. J.*, Vol. 52, Oct. 1920, pp. 129–153.
32. "Photoelectric Properties of Thin Films of Alkali Metals," H. E. Ives, *Astrophys. J.*, Vol. 60, Nov. 1942, pp. 209–230, Part II, *Astrophys. J.*, Vol. 64, Sept. 1926, pp. 128–135; "Electrical and Photoelectric Properties of Thin Films of Rubidium on Glass," H. E. Ives and A. L. Johnsrud, *Astrophys. J.*, Vol. 62, Dec. 1925, pp. 309–319; Abstract, *J.O.S.A.*, Vol. 12, May 1926, p. 486.
33. "Photoelectric Emission," H. C. Rentschler and D. E. Henry, *J. Franklin Inst.*, Vol. 223, Feb. 1937, pp. 135–145.
34. "A Photoelectric Cell for Measuring Ultra-Violet Solar and Sky Radiation on a Horizontal Plane," W. W. Coblentz and R. J. Cashman, *Am. Met. Soc. Bul.* 21, 1940, pp. 149–156.
35. "Cadmium Magnesium Alloy Photo-Tubes," L. R. Koller and A. H. Taylor, *J.O.S.A.*, Vol. 25, June 1935, p. 184.
36. "Sensitization of Transparent Compound Photocathodes," P. Gorlich, *Z. tech. Physik*, Vol. 18, Nov. 1937, pp. 460–462.
37. "A New High-Sensitivity Photosurface," A. M. Glover and R. B. Janes, *Electronics*, Vol. 13, Aug. 1940, pp. 26–27.
38. *Fundamentals of Engineering Electronics*, W. G. Dow, Chapter XVIII, p. 404, Wiley, 1937.
39. "Operating Characteristics in Photoelectric Tubes," G. F. Metcalf, *Proc. I.R.E.*, Vol. 17, Nov. 1929, pp. 2064–2071.
40. *Photoelectric Phenomena*, A. L. Hughes and L. A. DuBridge, Chapters 1–2, pp. 3–26, McGraw-Hill, 1932.
41. "The Secondary Emission Multiplier—A New Electronic Device," V. K. Zworykin, G. Morton, and L. Malter, *Proc. I.R.E.*, Vol. 24, March 1936, pp. 351–375.
42. "Electron Multiplier Design," J. R. Pierce, *Bell Lab. Rec.*, Vol. 16, May 1938, pp. 305–309.
43. "Recent Developments in Phototubes," R. B. Janes and A. M. Glover, *RCA Rev.*, Vol. 6, July 1941, pp. 43–54.
44. RCA Technical Data Bulletin for the RCA-931A.
45. D. D. Knowles, U. S. Patents 1,955,535 and 1,963,168.
46. "The Iconoscope—A Modern Version of the Electric Eye," V. K. Zworykin, *Proc. I.R.E.*, Vol. 22, Jan. 1934, pp. 16–32.
47. "Theory and Performance of the Iconoscope," V. K. Zworykin, G. A. Morton, and L. E. Flory, *Proc. I.R.E.*, Vol. 25, Aug. 1937, pp. 1071–1093.
48. "The Orthicon," *Electronics*, Vol. 12, July 1939, p. 11.
49. "Television Pick-Up Tubes using Low Velocity Electron Beam Scanning," H. Iams and A. Rose, *Proc. I.R.E.*, Vol. 27, Sept. 1939, pp. 547–555.
50. "The Image Iconoscope," H. Iams, G. A. Morton, and V. K. Zworykin, *Proc. I.R.E.*, Vol. 27, Sept. 1939, pp. 541–547.
51. "Television by Electron Image Scanning," P. T. Farnsworth, *J. Franklin Inst.*, Vol. 218, Oct. 1934, pp. 411–444.
52. "The Image Dissector," C. C. Larson and B. C. Gardner, *Electronics*, Vol. 12, Oct. 1939, p. 24.
53. "The Action of Light on Selenium," W. Smith, *J. Soc. Tel. Eng.*, Vol. 2, 1873, pp. 31–33 (letter).
54. "Photoelectric Conduction in Selenium," B. Gudden and R. Pohl, *Z. Physik*, Vol. 35, No. 4, 1925, pp. 243–259.
55. "On the Efficiency of Selenium as a Detector of Light," E. E. F. D'Albe, *Proc. Roy. Soc. (London), Series A*, Vol. 89, 1913, p. 75.
56. "Manufacture of Selenium Cells, Photoelectric Cells and Their Applications," C. E. S. Phillips, paper presented at Joint Meeting of Physics and Optical Society, June 4 and 5, 1930, *Trans. Opt. Soc. (London)*, Vol. 31, No. 4, 1929–1930, pp. 233–240.
57. "Selenium Cells," H. Thirring, *Proc. Phys. Soc. (London)*, Vol. 39, 1926, p. 97.
58. "Selenium Cells and Their Use in Sound Production Photocells—Their Applications," H. Thirring, *Trans. Opt. Soc. (London)*, Vol. 31, No. 4, 1929–1930, pp. 233–240.
59. "Thalofide Cell—A New Photo-Electric Substance," T. W. Case, *Phys. Rev.*, Vol. 15, No. 4, 1920, pp. 289–292.
60. "Photocells and Their Applications," T. W. Case, *Trans. Opt. Soc. (London)*, Vol. 31, No. 4, 1929–30, pp. 233–240.
61. "Report on the Effect of Electricity Produced by the Influence of Sun Rays," E. Becquerel, *Comptes rendues*, Vol. 9, 1839, pp. 561–567.
62. "Electronic Rectifier Is Suitable for Signalling Power Supply," L. O. Grondahl and P. H. Geiger, *Ry. Sig.*, Vol. 20, 1927, pp. 213–216.
63. "Cuprous Oxide Photoelectric Cells," W. Schottky, *Zeit. Techn. Phys.*, Vol. 11, No. 11, 1930; *Physik. Z.*, Vol. 31, 1930, pp. 917–925.
64. *Photoelements and Their Applications*, B. Lange, pp. 77–87, Reinhold, 1938.
65. "Barrier and Photoelectric Effects at the Boundary of Cuprous Oxide and Metal Films," E. Duhme and W. Schottky, *Naturwiss.*, Vol. 18, 1930, pp. 735–736.
66. *Photoelements and Their Applications*, B. Lange, p. 95, Reinhold, 1938.
67. *Photocells and Their Applications*, J. W. Anderson, Physical and Optical Societies, London, 1930.
68. *Photoelectric Cells*, N. R. Campbell and D. Ritchie, Pitman, London, 1934.
69. *Fundamentals of Engineering Electronics*, W. S. Dow, Wiley, 1937.
70. *Principles of Television Engineering*, D. G. Fink, McGraw-Hill, 1940.
71. *Measurement of Radiant Energy*, W. E. Forsythe, McGraw-Hill, 1937.
72. *Photoelectric Phenomena*, A. L. Hughes and L. A. DuBridge, McGraw-Hill, 1932.
73. *Industrial Electronics*, F. H. Gulliksen and E. H. Vedder, Wiley, 1935.
74. *Illuminating Engineering Standards*, Illuminating Engineering Society, New York City, A.S.A. Z7.1, 1942.
75. *Physics of Electron Tubes*, L. R. Koller, 2nd ed., McGraw-Hill, 1937.
76. *Photoelements and Their Applications*, B. Lange, Reinhold, 1938.
77. *Applied Electronics*, E. E. Staff, M.I.T., Wiley, 1943.
78. *Thermionic Emission*, A. L. Reiman, Wiley, 1934.
79. "Application of Conventional Vacuum Tubes in Unconventional Circuits," F. H. Shepard, *Proc. I.R.E.*, Vol. 24, No. 12, Dec. 1936.
80. *Conduction of Electricity in Gases, Lectures on*, J. Slepian, Westinghouse Elec. & Mfg. Co., East Pittsburgh 19, Pa.
81. *Photocells and Their Application*, V. Zworykin and E. Wilson, Wiley, 1930.

Chapter 8

INDUSTRIAL X-RAY TUBES

Joseph Lempert

THE discovery and use of x-rays by Roentgen in 1895 may be regarded as the first practical application of electronics. The importance of this discovery in introducing electronics is secondary only to the role x-rays played in startling the complacent nineteenth-century world with a phenomenon which could not be explained in terms of classical physics. X-rays have served as a key to modern quantum physics and all the phenomena associated with the modern concept of the atom. In addition, x-rays as used in radiology and x-ray diffraction are extremely valuable industrially. Finally, x-rays have filled a vital human need in medical radiology and therapy.

One of the reasons why x-rays were not discovered sooner is that special methods, such as light-sensitive emulsions and ionization chambers, are required for their detection. In addition, with the exception of natural radioactive processes, the combination of materials and circumstances necessary to produce x-rays do not occur in nature. Since x-rays result from the collisions of high-speed electrons with atoms, a source of such electrons is necessary. High voltages * are generally used for accelerating electrons. The process of x-ray production is confined to the evacuated chamber of an x-ray tube to avoid an excessive number of collisions between the electrons and the atoms of the gaseous medium, thus preventing arc-over between electrodes and maintaining electrical insulation. In addition, vacuum prevents oxidation of tube parts which operate at high temperature.

8·1 PHYSICAL PROPERTIES OF X-RAYS

Electromagnetic radiations of extremely short wavelength are produced by collision of high-velocity electrons and atoms. These radiations were called x-rays at the time of their discovery because of the mystery surrounding their origin. From numerous investigations x-rays are now known to be propagated at the speed of light; they follow the inverse square law, are unaffected by electric or magnetic fields, and can be reflected, diffracted, refracted, and polarized. X-rays can produce fluorescence and phosphorescence, and can modify, damage, or destroy living cells. Their ability to blacken sensitized film in proportion to their intensity, in addition to their property of penetrating solid matter, is the basis of radiography.

The quantity and intensity of x-rays can be measured in terms of their physical, chemical, or biological effects. The practical unit of x-ray quantity, the roentgen, is defined in terms of the ionizing properties of x-rays. One roentgen is that quantity of x-radiation which will ionize 1 cubic centimeter of air sufficiently at a temperature of zero degrees centigrade and an absolute pressure of 760 millimeters of mercury to produce 1 electrostatic unit of charge at saturated current. All secondary electrons within the volume must be fully utilized, and secondary radiation from the wall of the chamber must be avoided. The roentgen (or r unit) per second is the corresponding unit of intensity.

As indicated in Chapter 1, the energy of an x-ray photon of frequency ν is $h\nu$, where h is Planck's constant. To excite photons of frequency ν or energy $h\nu$, bombarding electrons must have kinetic energies equal to or greater than $h\nu$. It is thus evident that x-rays produced by collisions between electrons and the target of an x-ray tube have an upper frequency limit which is determined by the kinetic energy of the electrons as they impinge upon the target. The maximum frequency $\nu_{\max}$ that can be obtained by impressing E volts between the anode and cathode of an x-ray tube is determined by the following quantum equation:

$$\frac{Ee}{300} = h\nu_{\max} = \frac{1}{2}mv^2 \qquad (8\cdot1)$$

where m is electron mass, v is electron velocity, and e is electron charge. Thus, the highest frequency which can be produced by a 100-kilovolt potential across an x-ray tube is 2.4×10^{19} cycles per second. The energy associated with an x-ray quantum of this frequency is 1.6×10^{-7} erg, which by equation 8·1 is also the kinetic energy of a 100-kilovolt electron as it approaches the target. Since this energy is the maximum which can be transferred to a photon for this voltage, x-rays of higher frequencies can be produced only by increasing tube voltage.

Numerical values of the variables in equation 8·1 are plotted in Fig. 8·1 as a function of peak voltages. Curve A,

* Although high-speed electrons can be produced by magnetic fields and other methods, acceleration by high voltage is at present the simplest and most economical method of achieving a million electron volts or less.

which represents electron velocity calculated without a relativity correction, reaches the speed of light at 250 kilovolts. Actual electron velocities,[1] shown on curve *B*, never can attain the speed of light because of a relativity increase of mass with velocity which would result in infinite mass at the velocity of light.

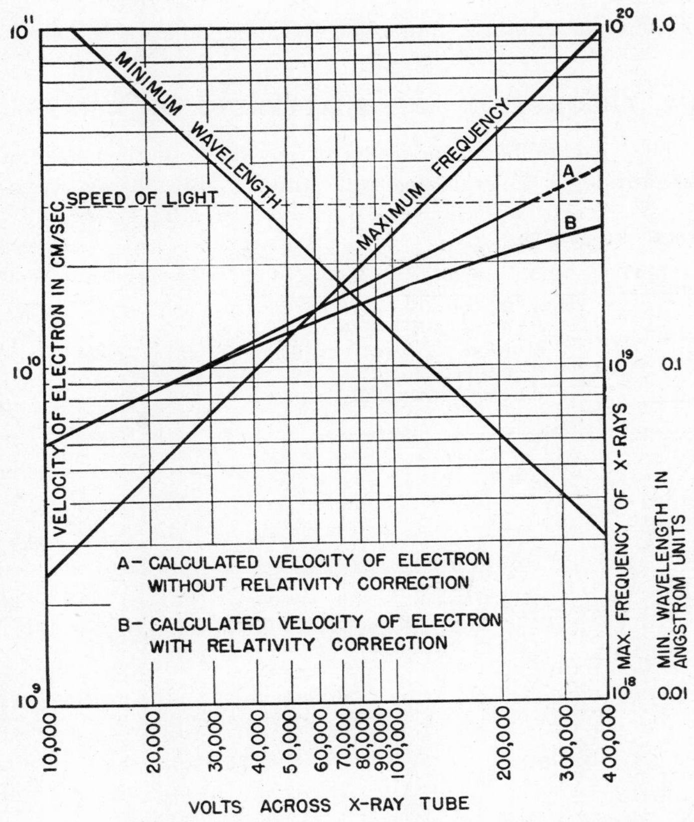

Fig. 8·1 Maximum frequency, minimum wavelength, and electron velocities as functions of impressed x-ray-tube voltage.

8·2 THE CONTINUOUS X-RAY SPECTRUM

The probability that the kinetic energy of an electron will be released in the form of high-frequency radiation upon collision with the target of an x-ray tube is small. It depends upon the atomic number of the target and the impressed tube voltage. The chance of transferring all the electron energy to a photon as the result of a single collision is even lower. There is a greater probability that a partial exchange of energy will occur; this means that a large number of the photons are of relatively low energy and long wavelength.

For a particular value of d-c tube voltage, a continuous spectrum of x-rays is produced, the maximum frequency of which is determined by equation 8·1. The character of the spectrum as determined by Ulrey [2] is shown in Fig. 8·2 in which monochromatic x-ray intensity is plotted as a function of wavelength. Thus, an x-ray tube operating at a given voltage produces a continuous band of wavelengths, the shortest being determined by the kinetic energy of the electrons as they impinge upon the target.

The maximum on the curve of relative intensity versus wavelength occurs at a wavelength approximately 1.5 times the maximum wavelength defined by equation 8·1. The frequency or wavelength characteristic of the continuous spectrum is independent of the material of the target. The total energy of the spectrum is approximately proportional to the atomic number of the target material, to the electron current, and to the square of the impressed tube voltage for constant tube current.

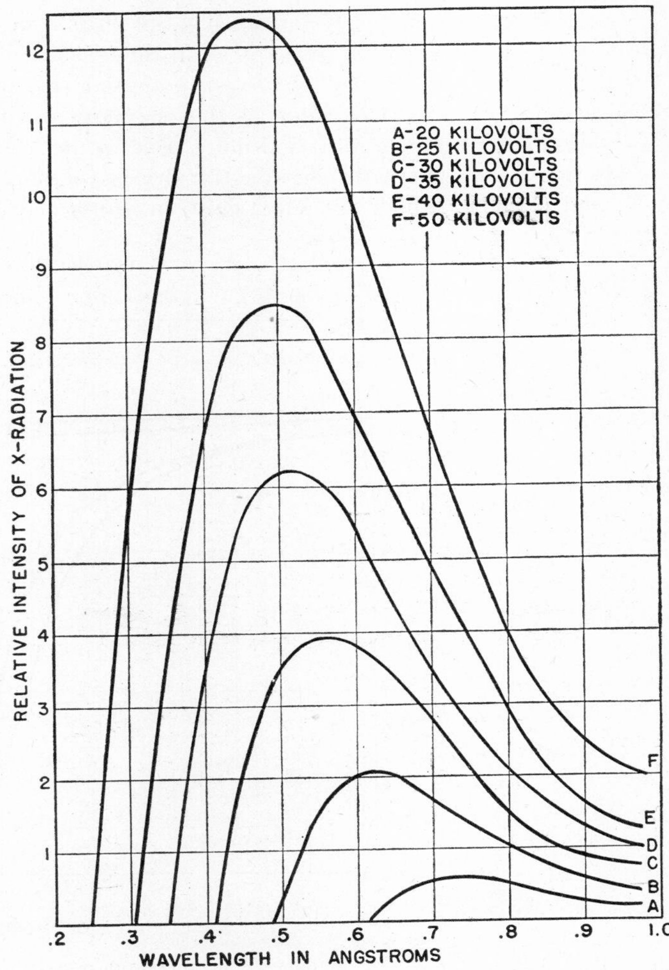

Fig. 8·2 Ulrey's curves for quality of x-radiation from a tungsten target for various values of tube voltage. The wavelength is indicated in units of 10^{-8} centimeter. (Reprinted with permission from C. T. Ulrey, "An Experimental Investigation of the Energy in the Continuous X-Ray Spectra of Certain Elements," *Phys. Rev.*, Vol. 11, 1918, p. 401.)

8·3 CHARACTERISTIC X-RAY LINES

In addition to the continuous x-ray spectrum which has frequency characteristics independent of the target material, some of the x-rays emitted have wavelengths characteristic of the target, provided certain energy conditions are fulfilled by the exciting electrons. These characteristic radiations owe their origins to electron transitions in the various orbits of the atom.

As described in Chapter 1, atomic electrons are represented in modern theory to occur in a succession of shells about the nucleus of the atom, each of these shells being composed of sub-shells or orbits. If an electron in a shell near the nucleus is removed by collisions with high-speed electrons or photons, an electron from an outer shell of the atom "falls" into the vacated position. If ΔW represents the difference in energy between the two levels, the frequency of the characteristic radiation is

$$\Delta W = h\nu \tag{8·2}$$

The so-called K shell is closest to the nucleus and is the shell of greatest stability and lowest potential energy. Radiation emitted because of electron transitions to this shell is referred to as characteristic K radiation. Subscripts indicate the shell sub-levels or orbits involved in the transitions. Since the K shell is the shell of lowest potential energy, ΔW is large. Thus, the frequency of the characteristic K radiation is higher and the radiation more penetrating than the radiation associated with electron transitions to the L, M, N, and O shells, which are outer electron orbits of the atom. Radiations produced by transitions to shells farther from the nucleus than the K shell accordingly are referred to as L, M, N, and O radiation, respectively, subscripts again being used to indicate the sub-levels.

The wavelength of the K_γ line, the x-ray line having the highest characteristic frequency for a given element, is indicated in Table 8·1. Minimum voltage required to transfer enough energy to an electron to excite the K_γ line of various elements is indicated in a separate column.

TABLE 8·1 CHARACTERISTIC EMISSION

Element	Atomic Number	Wavelength of K_γ Line	Equivalent Electron Volts
Aluminum	13	7.95	1,550
Chromium	24	2.06	6,000
Iron	26	1.74	7,100
Nickel	28	1.48	8,350
Copper	29	1.38	8,950
Molybdenum	42	0.619	20,000
Tungsten	74	0.179	69,000
Lead	82	0.141	87,500

The characteristic K lines of tungsten do not appear in Fig. 8·2 because the tube voltage is not high enough to excite the K lines, which occur at 0.179, 0.184, 0.185, 0.209, and 0.213 angstrom units. If tube voltage were increased to 150 kilovolts, the minimum wavelength of the continuous curve would shift to 0.082 angstrom units, and the K lines would appear as sharp lines or peaks rising from the continuous curve and reaching relatively high values of intensity.

8·4 ABSORPTION AND SCATTERING OF X-RAYS

The spacing between atoms of metals is of the order of magnitude of several angstrom units. Thus, x-ray wavelengths in Fig. 8·1 are small compared with lattice spacings. The probability of x-ray absorption by a solid decreases as the wavelength decreases. In other words, penetration of the rays increases with increasing tube voltage.

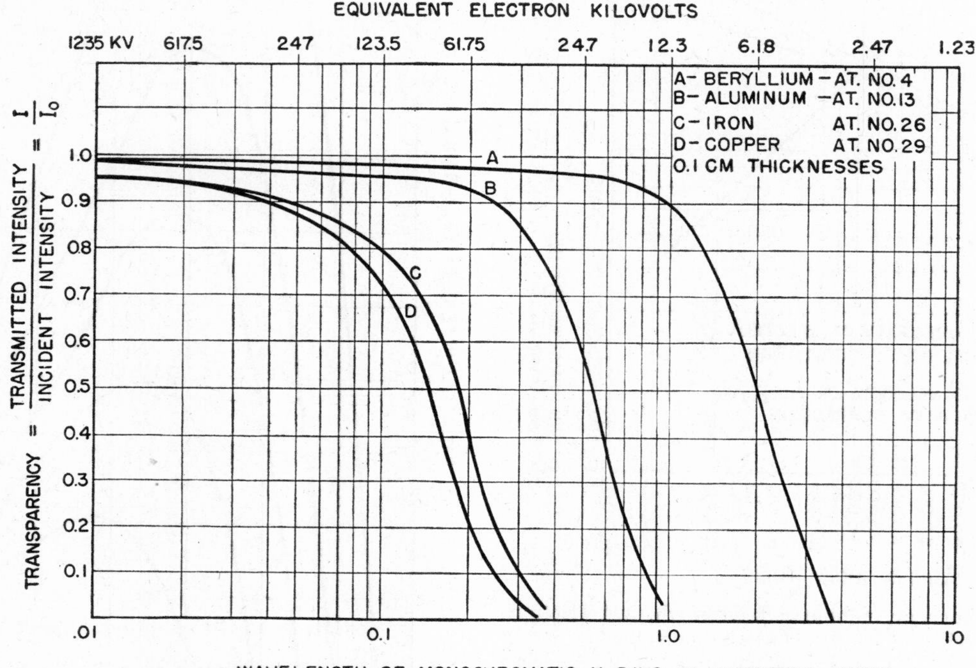

FIG. 8·3 X-ray transparency curve as a function of monochromatic wavelength for beryllium, aluminum, iron, and copper sheet.

In passing through material of thickness x, x-rays of intensity I are absorbed according to the following equation:

$$I = I_0 e^{-\mu x} = I_0 e^{(-\mu/\rho)\rho x} \tag{8·3}$$

where I is equal to the intensity of the emergent beam, μ is the absorption coefficient, and the absorption coefficient divided by the density ρ is defined as the mass absorption coefficient. The latter coefficient is commonly used because it is independent of the physical state of the material.

The mass absorption coefficient of a metal is a function of its atomic number and the wavelength of the incident radiation. Absorption involves removing x-ray photons from the primary beam, thus reducing the intensity of the transmitted beam. At usual radiographic voltages absorption is produced by the scattering of the beam by the material traversed and by photoelectric absorption of photons. The absorption coefficient μ/ρ may be expressed

$$\frac{\mu}{\rho} = \frac{\sigma}{\rho} + \frac{\tau}{\rho} \quad (8 \cdot 4)$$

where σ is the scattering coefficient, τ is the photoelectric absorption coefficient, and σ/ρ and τ/ρ are the mass-scattering and mass-photoelectric-absorption coefficients respectively.

Loss of radiation from the primary beam by scattering is a direct consequence of a change of photon direction by interaction of the x-ray beam with the atomic system of the material traversed. In general, scattering not only changes the direction of photons but also produces a degeneration of photon energy, resulting in increased wavelengths.

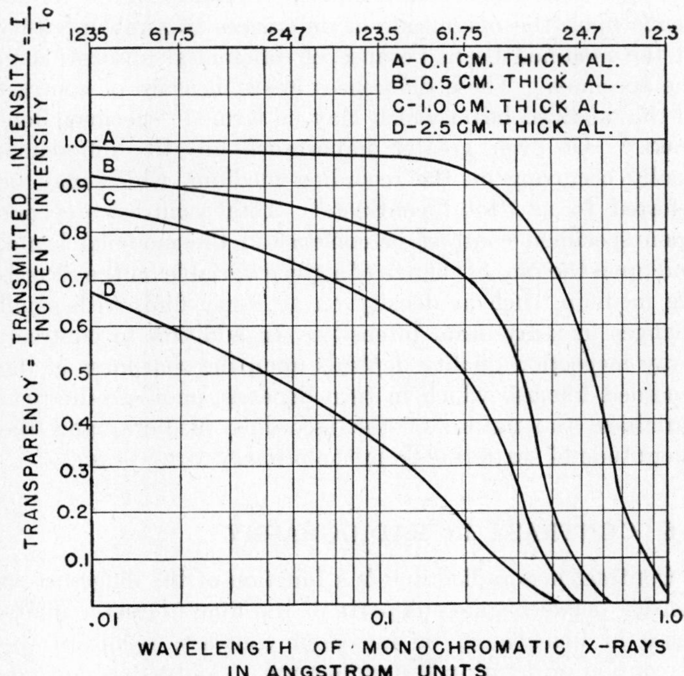

FIG. 8·4 X-ray transparency curve as a function of monochromatic wavelength for various thicknesses of aluminum sheet.

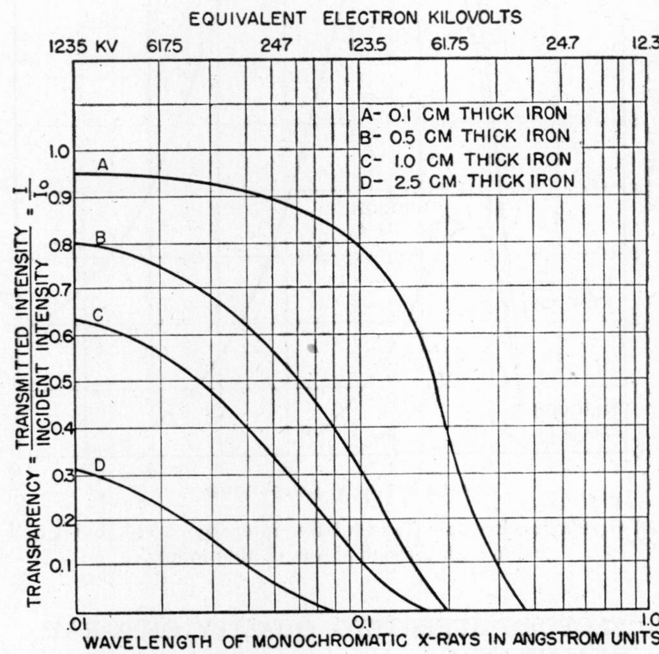

FIG. 8·5 X-ray transparency curve as a function of monochromatic wavelength for various thicknesses of iron sheet.

Photoelectric absorption is probable if the $h\nu$ energy of the photon is about the same as the energy required to remove an electron from a particular orbit in the atom, or to excite it to a higher level of potential energy within the atomic system. As would be expected, the critical absorption frequencies of a given element correspond closely to the characteristic K, L, M series of the material.

The photoelectric-absorption coefficient varies approximately as the cube of the wavelength within certain ranges of the spectrum. The scattering coefficient which, according to classical theory, should be a constant, actually varies with wavelength and atomic number.

The ratio I/I_0, as determined from various published data [3, 4, 5] on μ/ρ, is plotted in Fig. 8·3 for various metals as a function of the wavelength of homogeneous incident radiation. The minimum voltage capable of producing such monochromatic radiation according to equation 8·1 is indicated for reference only at the top of the figure. Actual impressed voltages would produce the continuous spectra of Fig. 8·2 and, if voltages were high enough, monochromatic lines characteristic of the material also would occur.

As shown in Fig. 8·3 materials of low atomic number, such as beryllium and aluminum, are much more transparent to x-radiation than are elements of higher atomic number, such as iron and copper. However, all materials are virtually opaque to radiations of very long wavelength, and the differences in the intensities of the transmitted beam become less as the wavelengths decrease in magnitude.

In the transparency curves for aluminum and iron (Figs. 8·4 and 8·5 respectively), increasing thickness of the specimen reduces the transmitted beam intensity. Thus, by traversing a non-homogeneous specimen, a reasonably uniform beam is modified so that it has intensity maxima closely conforming in size to the more x-ray-transparent areas of the specimen. The greater the difference in x-ray intensities and the larger the areas involved, the more obvious the lack of homogeneity of the specimen is to the eye when the image is viewed.

Actually the x-ray beam is heterogeneous. Since the long-wavelength portion of the beam does not have the penetrating ability of the short-wavelength fraction, more of the long-wavelength radiation is absorbed in passing through a given material, in this way tending to shift the maximum on the continuous x-ray curve to shorter wavelengths. The effect of various thicknesses of aluminum in modifying the quality of the 50-kilovolt curve of Fig. 8·2 is shown in Fig. 8·6. Wavelengths of the continuous spectrum are a function of the applied voltage. Thus a greater predominance of long-wavelength low-voltage radiation will occur if a constant-potential impressed voltage is replaced by half-wave or full-wave voltage of the same maximum value, because the average voltage is lower in both these instances.

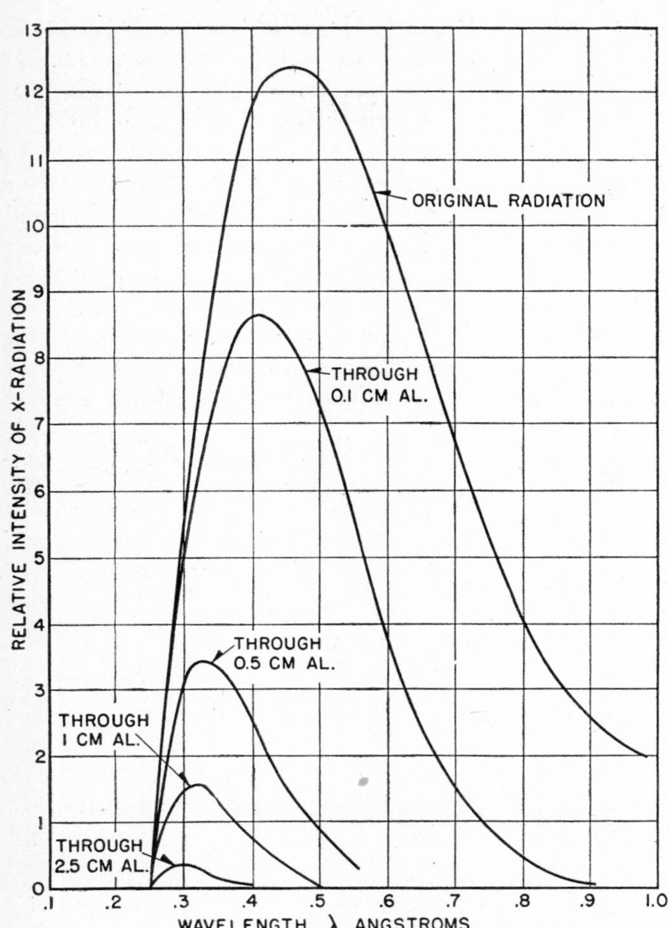

Fig. 8·6 Change in quality and relative intensity of 50-kilovolt radiation sent through sheets of aluminum.

8·5 FACTORS AFFECTING QUALITY OF X-RAY EXAMINATION

There are two stages in x-ray inspection: first, production of a non-homogeneous beam by differential absorption of the primary x-ray beam and, second, the recording of the intensity differences in the beam by conversion of the x-ray energy into a medium perceptible to the human senses. For simplicity, the x-ray beam, immediately after it has traversed a specimen, is called the "modified" x-ray beam.

The quantity of x-rays per second traversing a given specimen and reaching a recording medium, such as an x-ray film, can be adjusted by varying the peak voltage † across the tube, and the plate current through it. For a particular peak voltage the current through the tube determines the x-ray intensity at a given distance from the tube. Because x-ray intensity varies inversely as the square of the distance, the distance of a recording medium from the tube is one of the variables in x-ray inspection. The time of the x-ray exposure determines the dosage or total quantity of x-rays reaching a recording medium at a given distance from the tube. The product of kilovolts and milliamperes determines the power into the anode of the tube. Increasing the voltage and maintaining the product of kilovolts and milliamperes constant results in increased output, even though the power into the tube is constant, because the output from a tube is a function of applied voltage, and x-ray absorption by a specimen decreases with increasing tube voltage.

† Voltage across an x-ray tube is commonly expressed in peak kilovolts, and the current through it in terms of average direct current in milliamperes.

Several devices are used for indicating or recording intensity differences in an x-ray beam, among them x-ray sensitive film, fluoroscopic screens, ionization chambers, Geiger-Mueller tubes, or a combination of two or more of these devices.

The ability of the human senses to perceive, by means of some particular recording medium, a lack of homogeneity in the x-ray beam resulting from differential absorption in the specimen depends upon many variables. First, it depends upon the magnitude of differences in x-ray intensity in the modified beam because of differential absorption in the specimen. These differences are a measure of contrast in the modified beam which may be termed "specimen contrast." Of even greater importance are the differences finally occurring on the recording medium, which may be referred to as "total contrast." Total contrast depends upon specimen contrast or contrast in the modified beam, and upon the rate of change of a given variable on the recording medium, such as density on an x-ray film, with small changes in x-ray-beam intensity. In addition to contrast, x-ray inspection quality depends upon the sharpness of the recorded images, which in turn depends upon geometrical variables, the nature of the recording medium, and the magnitude of scattering from the primary x-ray beam.

8·6 CONTRAST IN RADIOGRAPHY

Contrast in a radiograph is a function of the difference in density between different parts of the film. Greater differences in blackening, or increased contrast, facilitate the perception of detail in the radiograph. Contrast is dependent largely upon the type of film employed and upon the factors which determine the amount and relative wavelength of the x-radiation used to make the exposure. As indicated in Section 8·5, radiographic contrast may be subdivided into film contrast and specimen contrast. Film contrast depends upon the film characteristics, and how the film is developed.

The general response characteristic of commercial film is indicated in Fig. 8·7. In this curve film density is plotted as a function of the logarithm of relative exposure. Film density is defined in terms of the amount of light the film transmits. If I_0 is the intensity of the light incident upon the film and I is the intensity of the light leaving the film, then film density D is defined as the log of the reciprocal of the transparency of the film or

$$D = \log \frac{1}{\dfrac{I}{I_0}} = \log \frac{I_0}{I} \qquad (8 \cdot 5)$$

The contrast of a given film is determined by the slope of the density-exposure curve. The steeper the curve, the greater are the differences in density between different parts of the radiograph. Thus, variation of primary beam intensity as produced by the specimen may be accentuated in terms of film density if a film of high contrast is used. The shoulder of the curve limits the contrast at high intensities. Such curves for some industrial films do not have shoulders within the range of usable densities.

In general, the quantity of x-rays reaching the film through the material being radiographed is adjusted so that the film densities fall on the portion of the curve with the steepest slope, where the contrast is greatest. The amount of radiation reaching the film through the specimen is controlled by adjusting plate current, time of exposure, voltage across the tube, and tube-film distance. A sample exposure chart for various sections of steel is shown in Fig. 8·8. The exposure factors are usually chosen to give a film density of about 1.0. Since the quality, or spectral distribution of x-ray intensity versus wavelength, depends upon the equipment used, technique charts must not only stipulate the film and the absence or presence of intensifying screens and filters, but also the circuit employed in generating the x-ray beam.

For a given material, specimen contrast is determined largely by tube voltage, and by the quality of x-radiation produced by the generator, which in turn depends upon the circuit employed and the amount of filtration in the primary x-ray beam. Figure 8·4 shows that monochromatic radiation which has traversed a specimen of aluminum would have far greater relative intensity differences at low voltages than at high voltages. In other words, the contrast on a film exposed to the emergent beam would be greater at low voltages because of greater specimen contrast.

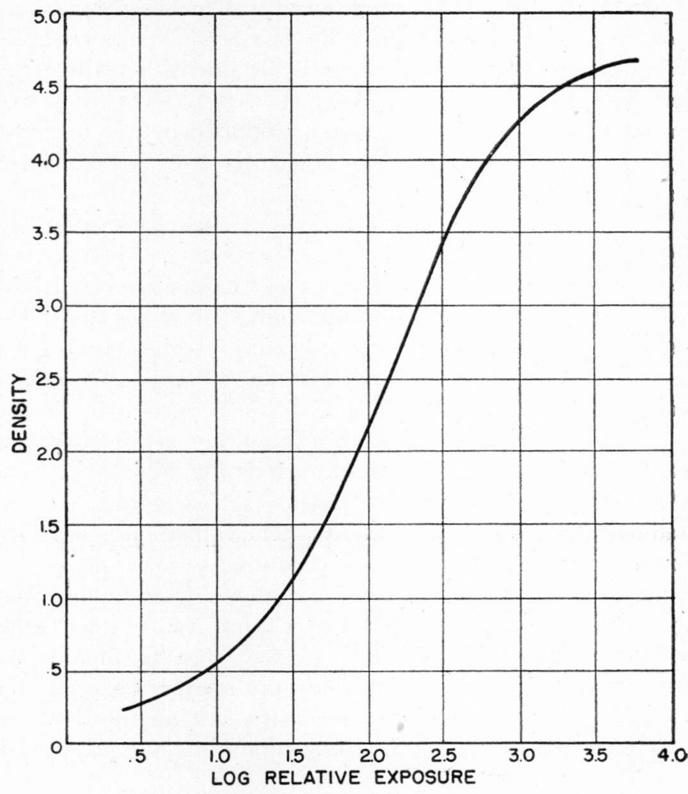

FIG. 8·7 Typical film density versus log-x-ray-exposure curve fo commercial x-ray film. New industrial films do not have shoulder.

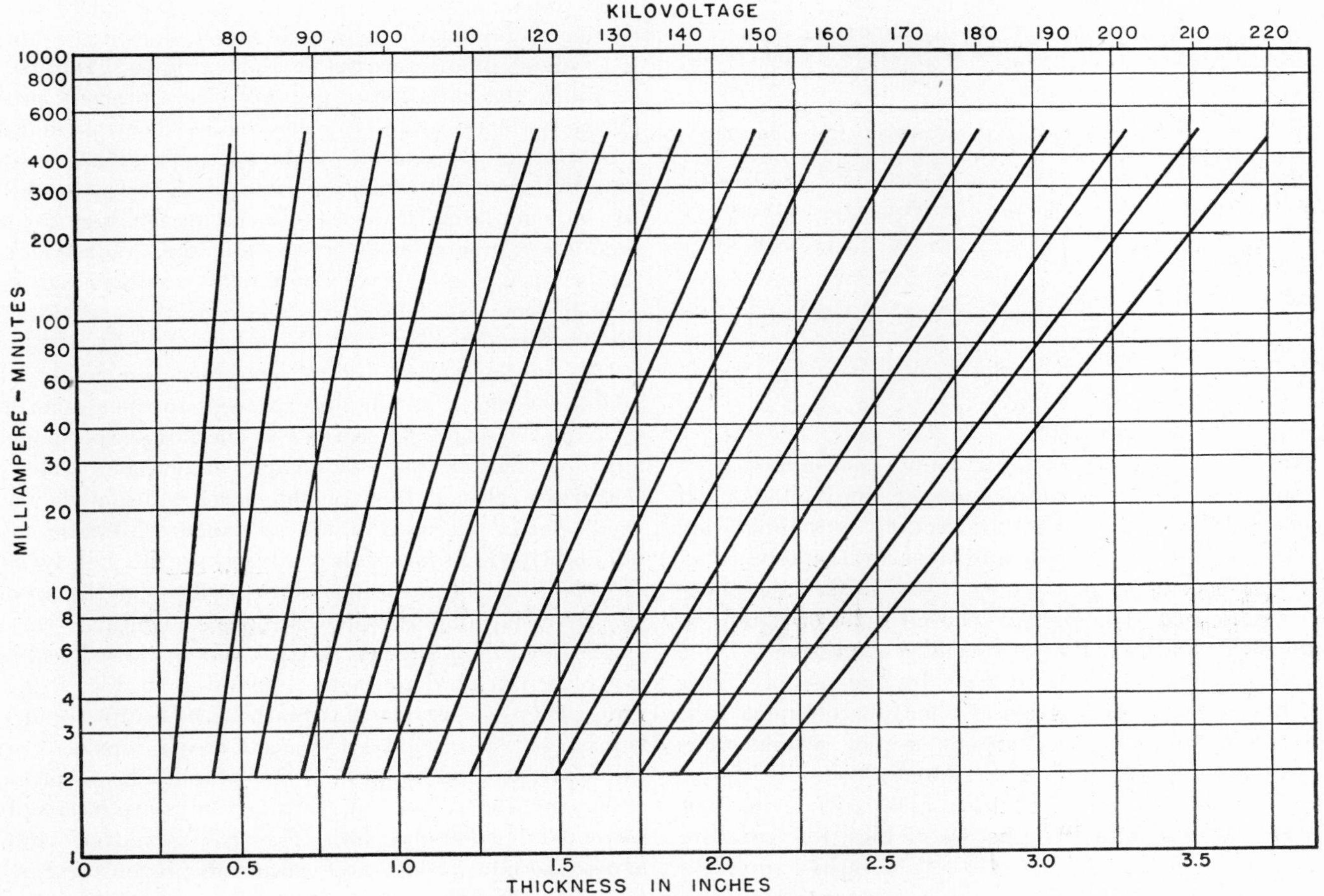

FIG. 8·8 Technique chart for rolled steel.

Monochromatic radiation is an abstract concept as far as most radiography is concerned. For actual impressed d-c voltages a continuous spectrum is obtained, instead of the monochromatic line indicated in Fig. 8·4. The predominance of lower-voltage radiation produces greater contrast than is indicated in the figure. Figures 8·3, 8·4, and 8·5 may therefore be considered to indicate minimum contrast.

Film contrast, as mentioned previously, is measured by the slope of the curve of Fig. 8·7. Specimen contrast may be expressed as the ratio of the x-ray intensities of various cross-sectional areas in the x-ray beam as it leaves the specimen. Since the intensity of the beam at a given tube voltage is inversely proportional to the milliampere-minutes required to produce a given density for a particular film, it is possible to obtain valuable information as to specimen contrast from a technique chart such as Fig. 8·8.

For this discussion, the technique factors in Fig. 8·8 are assumed to give a density of 1.0. The milliampere-minutes which, according to Fig. 8·8, are required to make exposures of film density equal to 1.0 for a steel specimen having sections 1.5 inches thick and 2.0 inches thick are listed in Table 8·2 for various voltages. It is evident that specimen contrast, as expressed by the ratio of x-ray intensity leaving the 1.5-inch section to the x-ray intensity leaving the 2.0-inch section, increases markedly as the tube voltage is decreased from 170 to 140 kilovolts.

TABLE 8·2 NUMERICAL EFFECT OF X-RAY TUBE KILOVOLTAGE ON CONTRAST AND LATITUDE

Kv	Thickness of Specimen (inches)	Milliampere-Minutes Required to Produce Film Density 1.0	Specimen Contrast or Ratio of X-Ray Intensity from 1.5-Inch Steel Section to Intensity from 2.0-Inch Steel Section for a Given Primary X-Ray Beam	Film Density Produced by 1.5-Inch Section when 2.0-Inch Section Is Held to Density of 1.0 Inch
170	1.5	2.2	10.9	3.3
	2.0	24		
150	1.5	11	19.1	3.8
	2.0	210		
140	1.5	35	28.6	4.3
	2.0	1000		

Usual radiographic techniques cannot accommodate a wide variation in specimen thickness at low voltages. If the technique is adapted to certain sections of the material, other sections may be underexposed or overexposed; that is, the film density may be too low, or so high that the radiograph cannot be viewed easily. The latitude, or ability to accommodate wide variation in specimen thickness, may be increased at the expense of contrast by the use of higher tube voltages. Greater latitude also may be obtained with special filters, and by simultaneous use of several films having different response characteristics.

The effect of voltage on contrast and latitude may be quantitatively determined by the use of Figs. 8·7 and 8·8. If the technique factors in Table 8·2 are adjusted to give a density of 1.0 for the 2.0-inch section of steel, the film densities produced by the greater x-ray intensity transmitted by the 1.5-inch section may be determined [6] directly from these two figures.

According to Table 8·2 the quantity of x-rays passing through the 1.5-inch section is 28.6, 19.1, and 10.9 times greater than the quantity passing through the 2-inch section for the 140-, 150-, and 170-kilovolt exposures respectively. From Fig. 8·7, the logarithm of the relative exposure required to produce a density of 1.0 for the film used is 1.4. Since the exposure for the 1.5-inch section is 28.6 times the exposure for the 2-inch section at 140 kilovolts, the log of relative exposure for the 1.5-inch section is equal to $1.4 + \log 28.6 = 2.86$. From Fig. 8·7, the density corresponding to the logarithm of relative exposure of 2.86 is 4.3. As indicated in Table 8·2, the densities which would be obtained through the 1.5-inch section for 150 and 170 kilovolts are 3.8 and 3.3 respectively, if the 2-inch section is held at a density of 1.0.

By using curves such as those in Figs. 8·7 and 8·8 it is possible to keep the maximum film densities low enough so that the radiographs can be viewed with ordinary illuminators. With a film viewer having greater illumination intensity some sections of film which might previously have been considered overexposed now can be examined. Recent film viewers are equipped with brilliant lamps in the center of the viewer, which permit viewing films with densities from 4 to 5.

8·7 SCATTERING IN RADIOGRAPHY

Scattered radiation contributes to the x-ray intensity leaving a specimen in a manner which does not conform to the radiographic properties of the specimen. It thus tends to nullify the variation in primary beam intensity produced by the specimen, and in this way reduces contrast and detail.

Scattering is produced by the specimen and by materials near the specimen which are exposed to the primary beam. In radiographing thick specimens, scattering frequently produces a greater effect at the film than radiation due to the primary beam. For example, the scattered radiation in radiographing ¾-inch steel sheets [7] is twice as intense as the primary radiation.

It is desirable to mask the specimen completely to prevent scattering from the film adjacent to the specimen, and to avoid overexposing sections of the film. Specimens usually are masked by x-ray-opaque diaphragms which limit the cross-sectional area of the primary beam to an area smaller than the area of the specimen. Metallic shot, or liquids with high x-ray absorptivity, are frequently used to cut down the beam at the outer surfaces of the specimen, thus reducing undercutting of the specimen and scattering.

Scattered radiation can be most effectively reduced by the use of a Bucky diaphragm, illustrated in Fig. 8·9. The grid, which consists of narrow strips of lead with an x-ray-transparent space between adjacent strips, is placed between the specimen and the film. The primary beam passes between the lead strips, but secondary rays are intercepted as indicated in the diagram. The grid is moved uniformly across the film during the exposure to prevent grid lines on the film. Radiographic sensitivity usually is increased by the use of a Bucky diaphragm.

If such a diaphragm is used, the x-ray intensity at the film is reduced, thus requiring an increase in exposure time which is frequently threefold or more. Moreover, use of the Bucky diaphragm requires an increase in distance between object and film, which reduces sharpness of the image on the film. Because of these disadvantages, and because

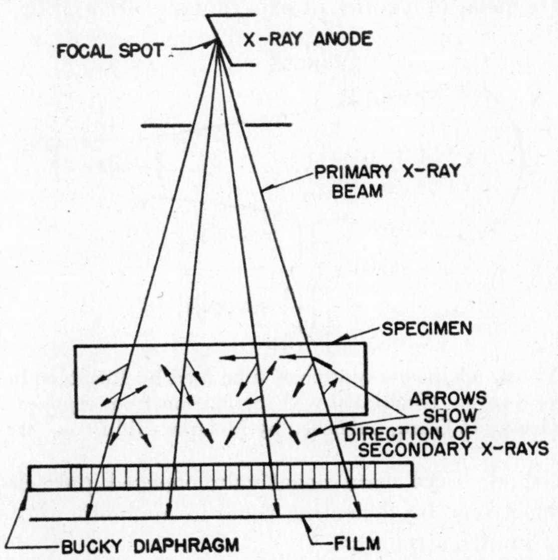

FIG. 8·9 Schematic diagram showing how primary rays pass through bucky to film while secondary scattered x-rays are intercepted.

the Bucky diaphragm cannot be made with the continuous travel required for long exposures, its industrial applications are limited.

Filters are widely used to reduce undesirable radiation at the film. Such filters have little effect on the quantity of radiation through the specimen, since only the radiation of short wavelengths, which is not absorbed appreciably by the filter, succeeds in penetrating the specimen. Filters absorb substantial quantities of low-voltage long-wavelength radiation which does not penetrate the specimen and which, unless filtered, causes high x-ray intensities at the film, with resulting scattering and undercutting.

Filters are frequently used in the primary beam between tube and specimen, and also between specimen and film. Filters placed near the tube can be smaller in size; and they lessen the probability that scattering from the filter proper will affect results. Filters between specimen and film have the advantage of reducing scattering originating in the specimen and the disadvantage of increasing specimen-film distance.

8·8 UNSHARPNESS ‡ IN RADIOGRAPHY

The finite size of focal spots in x-ray tubes produces radiographic images which are not completely sharp. The width of the unsharp region is a direct function of geometrical factors, and may be defined from the similar triangle relations in Fig. 8·10 (*a*) according to the following equation:

$$u_g = f \cdot \frac{d}{x} \tag{8·6}$$

‡ The term unsharpness is commonly used in place of the less negative expression sharpness, because it can be defined and measured.

where u_g is the so-called geometrical unsharpness, f is the focal spot width, d is the specimen-film distance, and x is the distance from the tube to the specimen. Increasing the film-specimen distance from d to d', as shown in Fig. 8·10 (*b*), decreases the width of the umbral region of the shadow—length bc in Fig. 8·10 (*a*)—to zero. For given values of x and d, the smallest cavity length l which will cast an umbral shadow is

$$l = f\frac{d}{d + x} \tag{8·7}$$

According to Ball and Draper,[8] the legibility of the image on the film depends upon total contrast ($D_i - D_b$) of Fig. 8·10 (*c*), which must exceed 0.02 density unit, and also upon the rate of change of film density with distance on the film $\Delta D/\Delta x$ which must exceed 0.2 density unit per inch. These factors tend to be higher if umbral shadows occur.

Motion of the focal spot, the specimen, or the film during the exposure affects sharpness adversely, producing a measurable value of so-called motion unsharpness. The focal spot and the film usually can be made stationary, but it may be convenient or necessary to have movement of the specimen. In Fig. 8·11, if v is the velocity of the edge of the specimen and t is the duration of the exposure, then motion unsharpness u_m caused by motion of the specimen parallel to the film is

$$u_m = \frac{vt(x + d)}{x} \tag{8·8}$$

In addition to geometric and motion unsharpness, there is an unsharpness associated with the type of film. Unless the radiograph is to be subsequently enlarged, film unsharpness is negligible, compared with the other types of unsharpness.

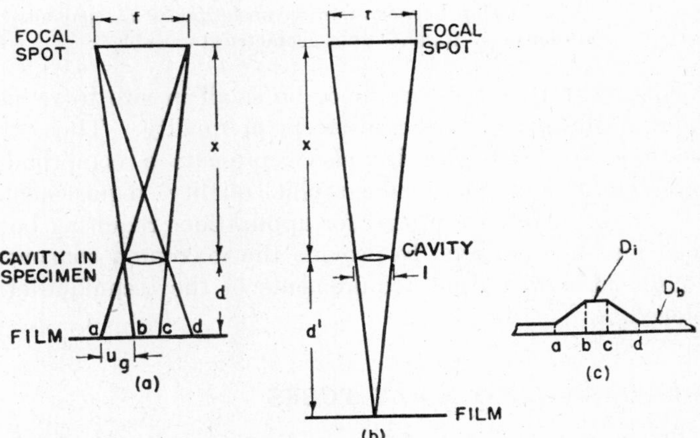

FIG. 8·10 (*a*) Relationship between geometric unsharpness and other geometric variables. Distances *ab* and *cd* are penumbral shadow, *bc* unbral shadow. (*b*) Geometric condition for the umbral shadow *bc* to equal zero. (*c*) Graphical representation of density on film for (*a*); D_i is image density and D_b is background density.

A more serious type of unsharpness is introduced by the use of intensifying screens, which are discussed in Chapter 27.

The cumulative or combined effect of the various types of unsharpness is referred to as total unsharpness. Total unsharpness limits the detail which can be seen in the radiograph. Effective total unsharpness, according to Warren,[9] is more closely approximated by the largest individual unsharpness than by the sum of the individual unsharpness values.

Radiographic techniques have conflicting requirements with respect to tube-film distance. Large tube-film distance produces sharp images and reduces distortion; but x-ray output is reduced according to the inverse square law. Practical tube-film distances represent compromises between output and radiographic quality requirements.

In Section 8·15 the factors affecting x-ray loading as a function of focal-spot size are discussed. Since loading into the focal spot determines output, the economical use of high output to produce more radiographs per unit time, and sometimes to minimize motion unsharpness, conflicts with the requirement that the focal spot be small in size to reduce total unsharpness in the interests of quality. Thus, the selection of focal-spot size also represents a compromise between output and radiographic quality requirements. Double focal tubes are used for applications requiring both high and low outputs, to provide the maximum sharpness consistent with output requirements of the technique employed.

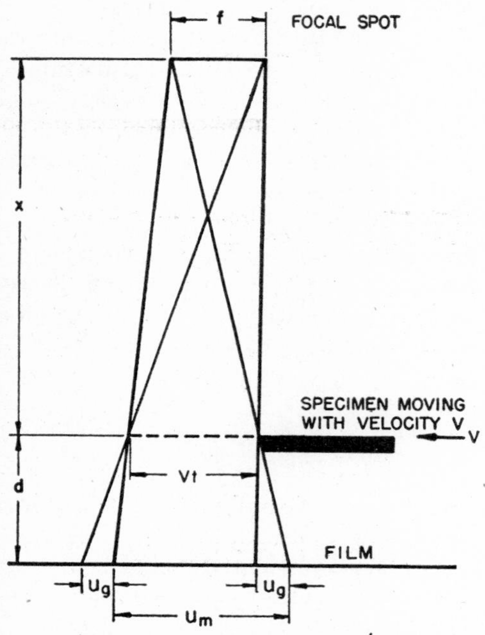

Fig. 8·11 Relationship between motion unsharpness u_m, geometrical unsharpness u_g, and other geometrical variables.

8·9 GAS-FILLED X-RAY TUBES

X-rays were discovered in 1895 by Roentgen, who used a gas discharge tube such as the one indicated in Fig. 8·12 (*a*). In this type of tube, gas molecules are ionized, or split into electrons and positive ions, upon application of high voltage. The positive ions then bombard the cathode under the influence of the electric field, producing electrons which are accelerated to the anode or anticathode, thus generating x-rays according to the principles indicated in Section 8·1. Actually in Roentgen's tube the electrons bombarded the glass at the anode end of the tube, producing x-rays and a glass fluorescence which was at first mistaken for the source of x-rays.

A similar type of tube had been made up in 1894 by Lenard in which the glass at the anode side of the tube was replaced by a thin window of aluminum. Lenard found that when "cathode rays," as the discharge was then called, were shot against the window, they penetrated it and excited a glow in the air immediately adjacent to the tube window. J. J. Thomson in a series of experiments, 1897–1898, proved that cathode rays were negatively charged particles, now called electrons, by deflecting them in magnetic and electric fields of known strengths.

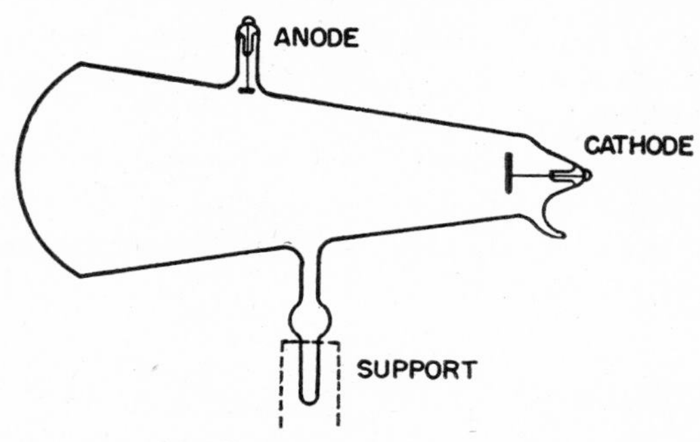

Fig. 8·12 (*a*) Cathode-ray type of tube used by Roentgen in the discovery of x-rays. (Reprinted with permission from *Applied X-Rays*, by George L. Clark, published by McGraw-Hill Book, 1940.)

As knowledge of the mechanism of x-ray production grew, the designs of the tubes were improved. The x-ray intensity from Roentgen's original tube was limited because the target was the glass wall of the tube and it was low in atomic number and had poor thermal properties. Later, gas-filled tubes were provided with metallic targets, as indicated in Fig. 8·12 (*b*), and also with concave cathodes to focus the cathode rays on the target.

A serious disadvantage of the gas-filled tubes was that the tube current was dependent upon tube voltage, and slight increases in tube voltage considerably augmented the ionization effects, thus causing more current to be drawn.

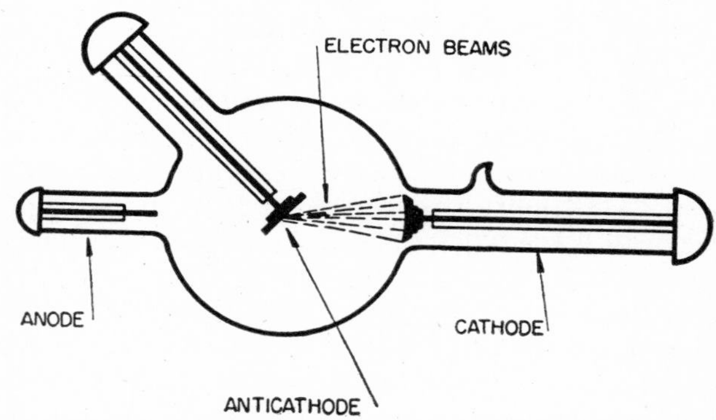

Fig. 8·12 (*b*) Gas-filled x-ray tube with metal target and concave cathode cup for focusing electron beam. (Reprinted with permission from *X-Rays*, by Kaye, published by Longmans, Green and Co., Ltd., 1923.)

An additional disadvantage of the gas-filled tubes arose from the critical nature of the pressure requirements. The normal operating pressure for these high-voltage discharge tubes was of the order of a number of microns. Too high a gas pressure in the tube would result in a "soft" tube, or one in which comparatively low values of tube voltage were required for the discharge. On the other hand, when the pressure in the tube became too low, higher voltages were required to produce the gas discharge, and the tube was said to become "hard."

Unfortunately, the gas pressure in such tubes decreased with continued life, owing to the "cleaning up" of gas caused by absorption on the walls of the tube. In order to maintain some degree of constancy of x-ray output during tube life, it was necessary to admit gas by diffusion into the tube by heating a small palladium tube attached to the bulb, or by passing a spark through a side tube containing a material capable of giving off gas.

Gas tubes are frequently used in precise x-ray diffraction work in order to avoid contamination of the target and, hence, of the spectral output of the tube by material evaporated from the filament.

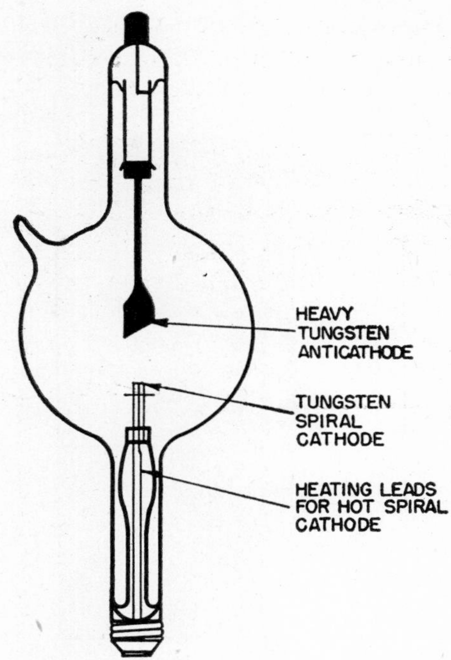

Fig. 8·14 Early type of hot-cathode high-vacuum tube. (Reprinted with permission from *X-Rays*, by Kaye, published by Longmans, Green and Co., Ltd., 1923.)

8·10 STANDARD TYPES OF X-RAY TUBES

Independence of tube current and tube voltage was the great advantage of the hot-filament vacuum tube developed by Coolidge. These tubes have long life, can be made smaller in size than earlier tubes, and can be operated directly from high-voltage transformers. Plate current is determined largely by the temperature of the tungsten filament. Typical characteristic curves of a hot-filament x-ray tube are shown in Fig. 8·13.

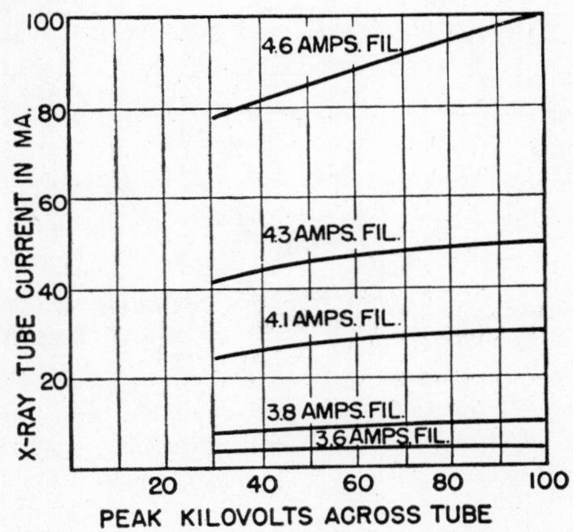

Fig. 8·13 X-ray-tube current as a function of full-wave peak kv across 100-kv x-ray tube for various values of filament current.

The early hot-filament x-ray tubes had tungsten targets, as illustrated in Fig. 8·14. Later tubes used copper-backed tungsten targets to conduct the heat through the envelope to an external radiator; this permitted the use of smaller bulbs. Protection against x-rays was obtained by the use of lead-glass shields outside the tubes. One of these is shown in Fig. 8·15. Other methods of protection against x-rays involved the use of lead-glass bulbs as part of the tube envelope; also metal center sections, which were lead-covered. More recently, so-called rayproof x-ray tubes have been produced. One of these is illustrated in Fig. 8·16. Such tubes are enclosed in insulating housings made of lead-impregnated plastic materials.

A dosage of x-rays not exceeding 0.1 r unit per day is considered safe. According to the Bureau of Standards,[10] a minimum thickness of 1.5 millimeters of lead or its equivalent is necessary for protection against 100-kilovolt-peak

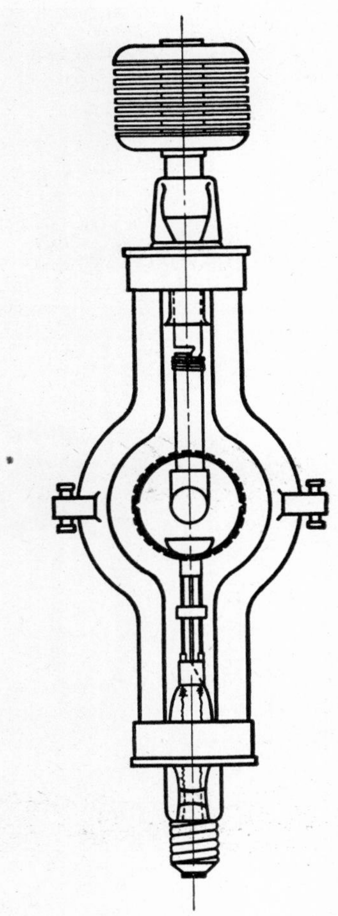

Fig. 8·15 Early type of copper-backed tungsten target x-ray tube with external lead-glass shield for x-ray protection. (Reprinted with permission from *The Science of Radiology*, by Glasser, published by Charles C. Thomas, Springfield, Ill.)

radiation. A minimum thickness equivalent to 2.5 millimeters of lead is considered necessary for 150-kilovolt-peak

FIG. 8·16 X-ray tube in non-shockproof rayproof housing.

radiation; 5.0 millimeters for 225-kilovolt-peak radiation. Required x-ray protection, indicated by Braestrup,[11] takes into account the continuous load on the tube and the dis-

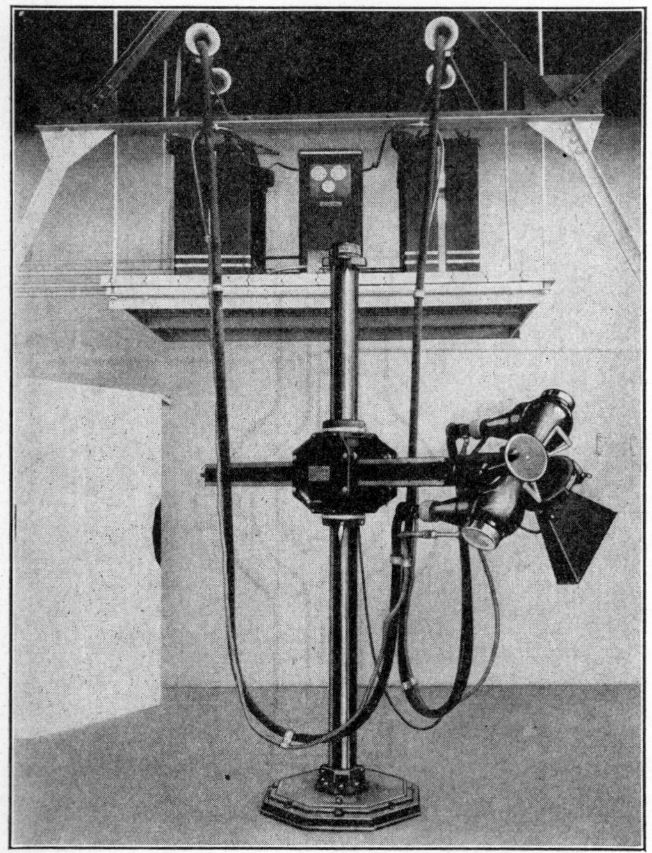

FIG. 8·17 Shockproof 220-kilovolt industrial x-ray unit with externally grounded high-voltage cables providing electrical connections to x-ray tube mounted in grounded head.

tance of operating personnel. Safety of personnel using x-ray equipment is increased by "shockproofing" the tube; that is, by mounting it in an oil-immersed grounded container or head, and making high-voltage connections to it through grounded cables. The x-ray-tubehead is generally equipped with a lead casing or an internal lead cylinder, or is manufactured of a heavy x-ray-absorbing material. Shockproofing the tube permits reducing its size, since the minimum spacing required to prevent sparkover is much smaller in oil than it is in air. A shockproof head connected to the transformer by internally grounded cables is shown in Fig. 8·17. An alternative method involves mounting both the tube and the high-voltage generating equipment in the same oil-filled container.

Most tubes manufactured now are of the shockproof type. Common oil-immersed tubes have maximum voltage ratings within the following ranges: 70–95, 100–125, 130–190, and 200–250 kilovolts.

Tubes with Maximum Ratings between 70 and 95 Kilovolts Peak

These tubes have copper-backed tungsten targets and are most frequently operated on self-rectified circuits in which alternating voltages are applied to the tube. Since electrons

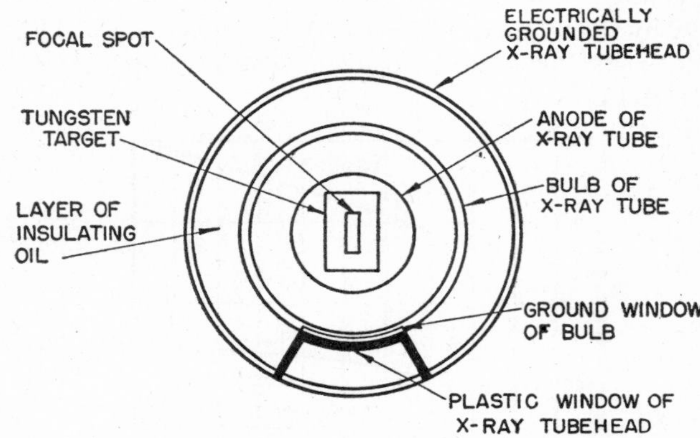

FIG. 8·18 Cross section looking toward anode through middle of x-ray tube. Relative positions of electrically grounded head, x-ray window, and insulating oil are indicated. X-ray protection is usually provided by a cylinder of lead (not shown) around tube.

can flow only from the cathode to the anode, current is unidirectional through the tube. The high-voltage transformer and the tube generally are mounted together in a small oil-filled portable container. An x-ray transparent material, usually a plastic, is mounted in the shockproof container to serve as the x-ray aperture. Such plastic windows are generally shaped as indicated in the cross-sectional sketch of a regular type of tubehead shown in Fig. 8·18. This arrangement permits maintaining electrical clearance between the tube and the grounded portion of the tubehead, at the same time minimizing the thickness of oil in the x-ray beam and thus minimizing absorption and scattering by the insulating oil. Tubes in this voltage range are used industrially for radiographing thin sections of materials of low atomic number, and for medical and dental radiography.

Tubes with Maximum Ratings between 100 and 125 Kilovolts Peak

Stationary-anode and rotating-anode tubes are used in this voltage range, usually with shockproof heads (see Fig.

8·18) connected to the generators by shockproof cables. The stationary-anode tubes have copper-backed tungsten targets, and the load is dissipated from the anode through a copper rod to an oil-immersed radiator. Rotating-anode tubes are manufactured both with copper-backed targets and with tungsten disk-type targets which dissipate load by radiation. Four rectifier full-wave generators and half-wave generators are most frequently used for tubes in this voltage range. Blowers and fans can be applied to the tube-heads to increase their continuous heat dissipation. Besides their industrial application, these tubes are used for medical diagnosis and for superficial therapy.

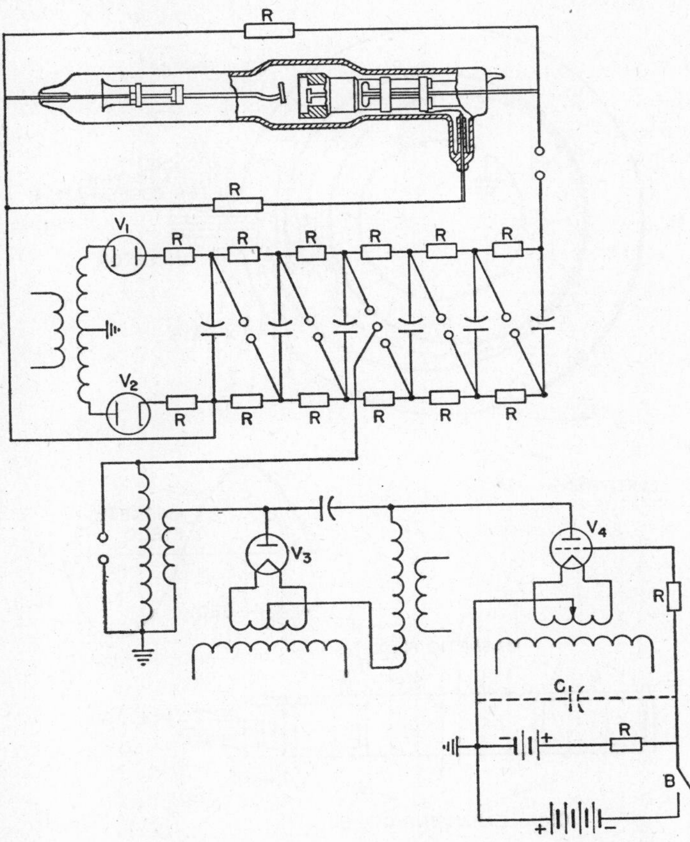

FIG. 8·19 High-speed x-ray tube capable of passing currents of several thousand amperes for periods of about 1 microsecond.

Tubes with Maximum Ratings between 130 and 190 Kilovolts Peak

These tubes have copper-backed tungsten targets. Forced-oil circulation is used sometimes to dissipate heat from the anode; or the oil in the tubehead may be cooled by water coils. These tubes are used most frequently with two rectifier half-wave and four rectifier full-wave generators. They are used for industrial radiography and intermediate therapy.

Tubes with Maximum Ratings between 200 and 250 Kilovolts Peak

These tubes generally have copper-backed tungsten targets. The anodes usually are cooled by forced-oil circulation, and the oil in turn is cooled by an external heat exchanger in which heat is transferred to a water-cooling system, or to a forced-air-cooling system. Such tubes are used with shockproof heads and interconnecting grounded cables, with constant potential supplies, with the Villard circuit mentioned in Chapter 27, and with half-wave generators. Frequently they are operated on a self-rectified basis in the same tank as the high-voltage transformer. They are used for industrial radiography and for deep therapy.

Air-insulated tubes are seldom operated much above 100 kilovolts for radiographic purposes. However, air-insulated tubes with large tungsten targets and large focal spots are used for intermediate and deep therapy at voltages as high as 200 kilovolts. The continuous rating of these tubes is frequently increased by the use of a blower on the bulb.

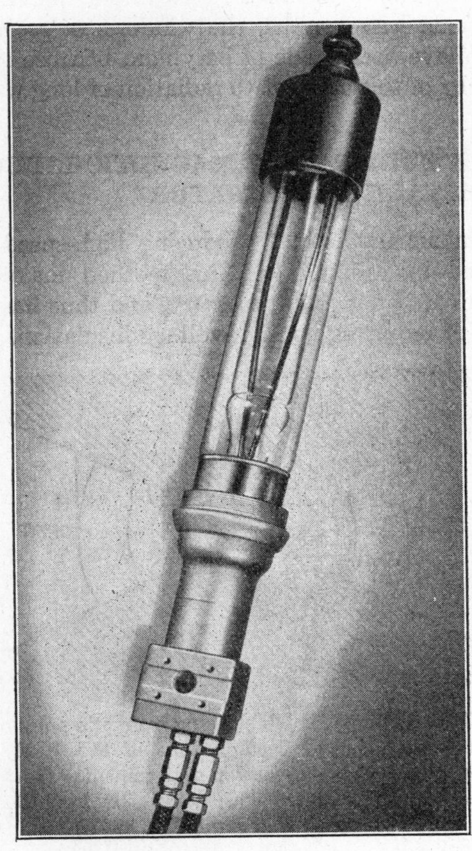

FIG. 8·20 Tube for x-ray diffraction work. (Courtesy of Machlett Laboratories, Inc.)

In some industrial applications it is necessary to make radiographs in times so short as to preclude conventional methods of timing exposures. To obtain useful x-ray intensities in such intervals, it is necessary that currents as large as 1000 amperes pass through the tube. The hot tungsten-filament tube cannot pass these currents because of the limitations imposed by space charge and safe operating temperatures of the filament. Slack and Ehrke [12] developed a 360-kilovolt cold-cathode tube with which an exposure of approximately 1 microsecond can be made. A sketch of this tube and the associated circuit is shown in Fig. 8·19. The circuit is arranged so that the capacitors are charged in parallel and discharged in series. The high voltage initially causes a discharge between the cold cathode and the focusing cup; this discharge releases metallic ions. Because of the high resistance in series with this circuit, the discharge transfers to the tungsten target of the tube, producing positive-ion bombardment of the cathode and bombarding the anode with high-speed electrons, which produce x-rays for the duration of the high-speed discharge.

Besides the tube types mentioned, sealed-off tubes are made for applications requiring 1000 and 2000 kilovolts.

A tube for diffraction work is shown in Fig. 8·20. Such tubes are used at voltages up to about 50 kilovolts with targets of special materials, such as copper, chromium, molybdenum, cobalt, and iron. Since the characteristic x-ray lines of the elements used as target materials are of relatively long wavelengths, the windows of recent diffraction tubes have been made of beryllium because of the high transparency of this element to radiation of long wavelength.

8·11 THE BETATRON, A MAGNETIC-INDUCTION ELECTRON ACCELERATOR

An ingenious method of producing high-speed electrons was devised by Kerst.[13, 14] This method makes a high-voltage source of power unnecessary, and thus has the great advantage of requiring no high-voltage insulation.

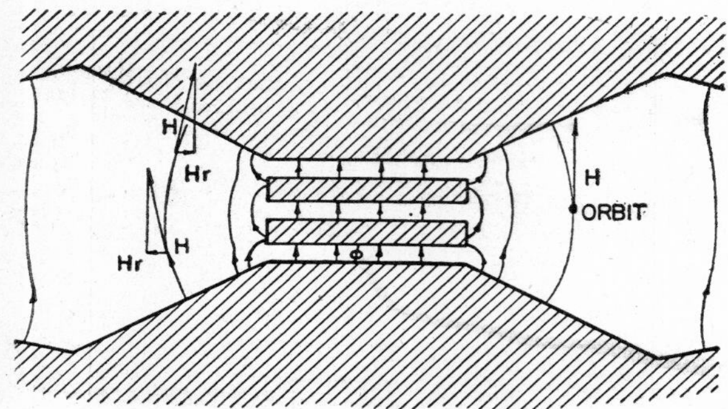

FIG. 8·21 Cross-sectional sketch showing magnetic poles of the betatron. The doughnut-shaped vacuum tube of Fig. 8·22 is inserted between these poles so that the electron path coincides with the electron orbit shown above. (Reprinted with permission from D. W. Kerst, "Betatron," *Radiology*, Feb. 1943.)

The principle of the method of accelerating electrons in the so-called magnetic-induction accelerator is not complicated. An electron of mass m and charge e, moving with velocity v in a magnetic field $\mathbf{H}$ perpendicular to its path, describes in the plane perpendicular to the magnetic field a circular motion. Its radius r may be expressed in electrostatic units by the following equation:

$$mv = \mathbf{H}\frac{re}{c} \tag{8·9}$$

where c is the speed of light.

As long as $\mathbf{H}$ is constant the electron moves in its orbit at a constant speed v. But if the magnetic flux ϕ enclosed by the orbit is increased with time, the electron is subject to a tangential electric field $E_\phi = \dfrac{1}{2\pi rc}\dfrac{d\phi}{dt}$ and is accelerated according to the equation:

$$eE_\phi = \frac{d(mv)}{dt} = \frac{e}{2\pi rc}\frac{d\phi}{dt} \tag{8·10}$$

If the magnetic field is chosen so that it increases proportionally with increase in momentum of the electron, r in equation 8·9 is a constant. By integrating equation 8·10 and equating with equation 8·9, Kerst has shown that a stable orbit r_0 is obtained, if the conditions of the following equation are satisfied:

$$\mathbf{H} = \frac{\phi}{2\pi r_0^2} \tag{8·11}$$

Thus the magnetic-field intensity at the orbit of the electron must be one-half the average intensity within the orbit. To satisfy this condition, the poles of the magnet used by Kerst, as illustrated in Fig. 8·21, are so shaped as to provide a

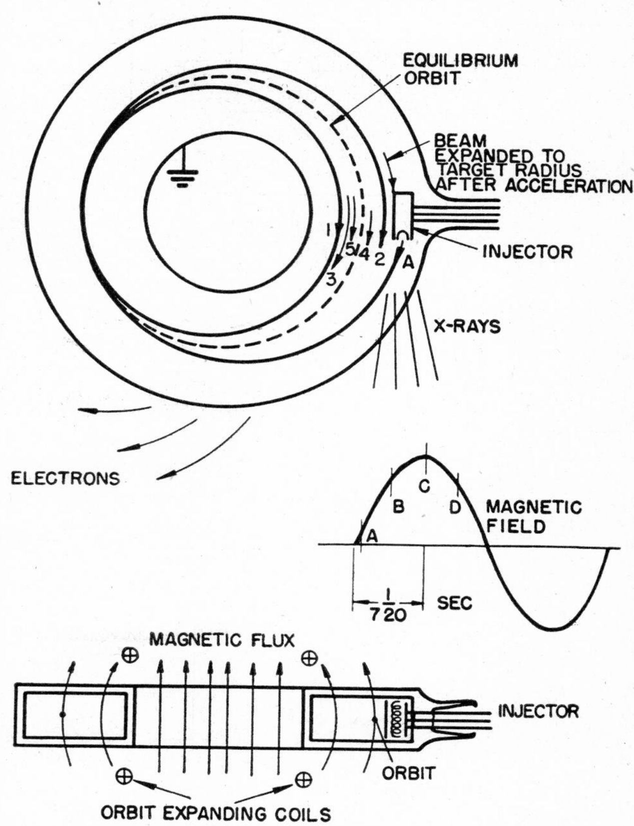

FIG. 8·22 Doughnut-shaped vacuum tube employed by Kerst in a 20-million-electron-volt betatron. Electrons are injected into the orbit at beginning of the cycle and are brought into contact with target at time C by expansion of the orbit. (Reprinted with permission from D. W. Kerst, "Betatron," *Radiology*, Feb. 1943.)

strong central magnetic field (and thus a high flux density) within the orbit, and a relatively weak magnetic field at the circumference of the orbit to maintain the electron within the r_0 orbit.

Stray electrons which are deflected from the equilibrium orbit by collisions with residual molecules in the vacuum can be made to go into a damped oscillation across the plane of the equilibrium orbit by the selection of a properly shaped magnetic field. As shown in Fig. 8·21, the magnetic lines of force must bulge outward between the poles of the magnet. Thus electrons with velocity components not parallel to the plane of the orbit are forced back toward this plane.

Kerst employed an alternating magnetic field of 600 cycles per second for his 2.5-million-volt betatron,[13] and a frequency of 180 cycles per second for the equipment with which he produced electrons having 20 million electron volts of energy.[14] The sequence of operation is shown on the magnetic-field curve of Fig. 8·22. Electrons are injected into the orbit at the beginning of the cycle at time A, and the orbit is expanded at time C, after the completion of the first quarter of a cycle. During this $\frac{1}{720}$-second period, the electrons rotate about the orbit, traveling a distance of some 100 miles and picking up a relatively small value of electron-volt energy with each revolution. X-rays are produced when the electrons in their expanded orbit spiral and hit the injector.

The energy attained by the electron as a result of the large number of revolutions while the magnetic flux is increasing is roughly the same as the voltage which would be generated by a transformer having the same number of turns and subject to similar magnetic conditions. Thus the betatron is similar to a transformer, except that high-voltage insulation is not required, nor is the special multisection x-ray tube insulated for the full output voltage.

The betatron, like the cyclotron, produces high-energy particles without the high-voltage source usually employed. In the cyclotron, however, the magnetic field is not a source of energy and is used only to cause the positive ions to revolve in spiral paths, and thus be subject to the accelerating effects of properly phased electric fields. Proper phasing is obtained by an increase in ion velocity to compensate for an increase in ion path per revolution. However, electrons because of their small mass rapidly approach the speed of light at high energies, and thus are unable to satisfy the condition of increasing velocity. In addition, requirements for high-frequency power would be excessive.

The betatron can be used as a source of high-energy electrons or x-rays. Baldwin and Koch [15] employed the high-voltage x-radiation from a 20-million-electron-volt betatron to make measurements on the quantum energies necessary to disintegrate the nuclei of various elements, as indicated by induced radioactivity.

8·12 FACTORS AFFECTING THE DESIGN OF X-RAY-TUBE ENVELOPES

Most x-ray tubes are diodes, and their construction and design, in relation to the number of elements, should be very simple. Much of the simplicity of construction is lost, however, in making the tube suitable for routine high-quality work on an economical basis. A cross-sectional sketch of a typical x-ray tube is shown in Fig. 8·23.

Two principal insulating requirements are imposed by the conditions of operation: first, the space between the electrodes must be insulating; and, second, the envelope must be able to withstand the maximum tube voltage without dielectric breakdown. For the space between anode and cathode to be insulating and yet permit electrons originating at the cathode to be accelerated to the anode, a low pressure must be maintained within the tube envelope. If a tube has too low a pressure to produce a gas discharge when a given voltage is impressed, it may (if the pressure is high enough) experience a gas discharge when the filament is energized, because of ionization of the gas by the electron current to the anode.

Insulating properties of the envelope are, of course, determined by the thickness, length, shape, and material of the envelope. A hard, highly resistant glass such as Pyrex is suitable, because it is relatively transparent to x-rays, is vacuum tight, and by the use of intermediate glasses can be sealed to the electrodes of the tube.

The length of air-insulated tubes is limited by the danger of external sparkover. Dimensions of shockproof tubes are

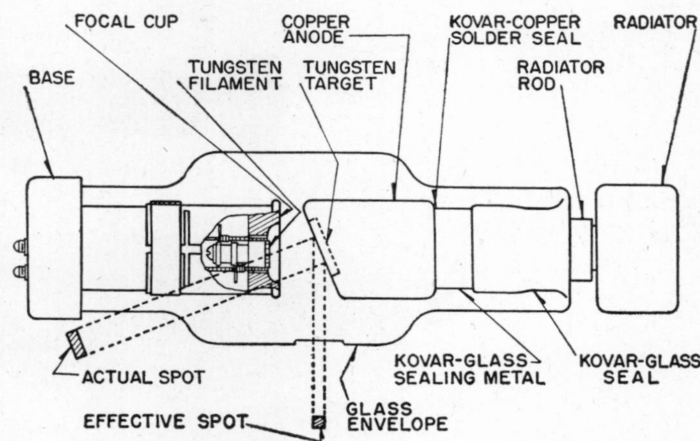

Fig. 8·23 Cross-sectional sketch of a typical x-ray tube.

not limited by external sparkover so much as by considerations of stable operation. Figure 8·24 indicates roughly glass-insulation lengths for oil-immersed and air-cooled tubes. Photographs of air- and oil-insulated tubes are shown in Fig. 8·25.

For stable operation, the anode-cathode spacing must be adjusted so that no cold-emission effects can occur. Single-section tubes rated at more than 400 kilovolts have serious design limitations imposed by the high fields between electrodes. Cold-emission discharges resulting from the high

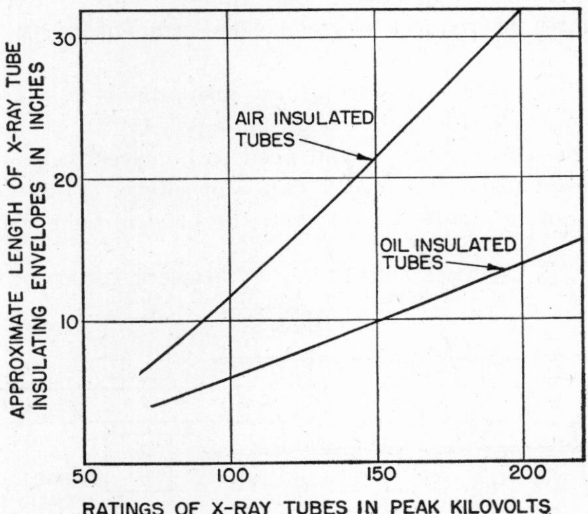

Fig. 8·24 Curves showing approximate glass-insulation over-all lengths for air- and oil-insulated single-section x-ray tubes.

fields tend to produce instability which may cause high-voltage oscillation, or which may affect voltage distribution on the bulb to produce electric stresses higher than the dielectric strength of a given thickness of envelope. The insulating member then is punctured, and the vacuum is destroyed.

The bulb tends to pick up negative charges during operation of the tube as a result of bombardment by electrons scattered from the anode. For high voltages, particularly if the tube is operated on a self-rectified basis, it is frequently desirable to use hooded-anode tubes. This construction eliminates or reduces electron bombardment of the glass wall.

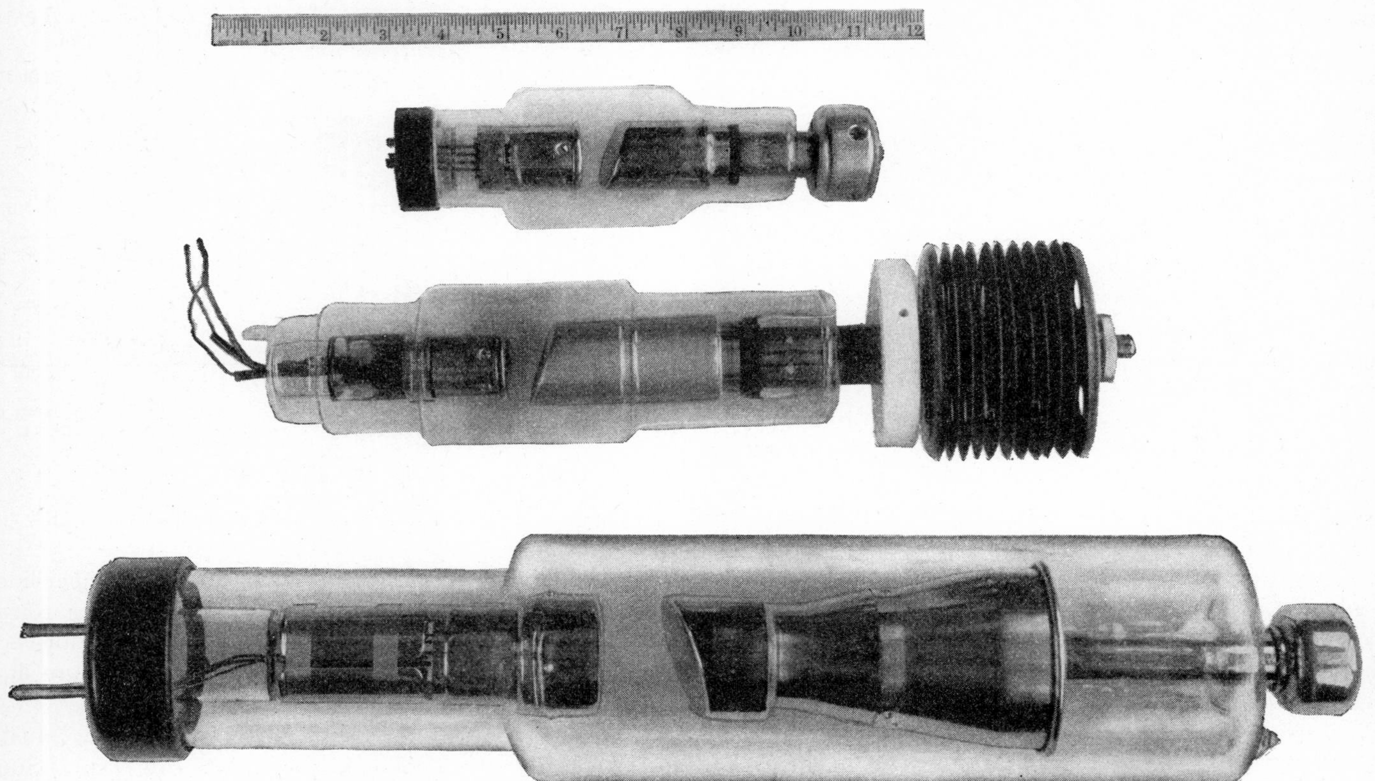

FIG. 8·25 Three modern types of x-ray tubes. The longer tube is a 250-kilovolt forced-oil-cooled industrial radiographic tube. The tube in the center is a 100-kilovolt air-insulated insert tube for use in rayproof housing. The short tube is a 100-kilovolt oil-immersed radiographic tube.

The dielectric strength of an insulator is a non-linear function of its thickness. High-voltage x-ray tubes designed as single-section units would need to be excessively long, if they were to have a proper factor of safety. High-voltage tubes can be made shorter, and the electric fields between electrodes and the electric stresses on the insulating envelope can be minimized by the use of the multisection principle. A typical multisection section is shown in Fig. 8·26. Since electron bombardment of the insulating members of these tubes produces serious consequences, a system of collimating electrodes is used to minimize this effect. Usually each section is tied to a definite potential by means of a potential dividing method such as a resistance divider. With appropriate spacings, a 1,000,000- or 2,000,000-volt tube, having lower electric fields between electrodes and less stress on the insulating envelope than some single-section tubes of much lower voltages, can be built.

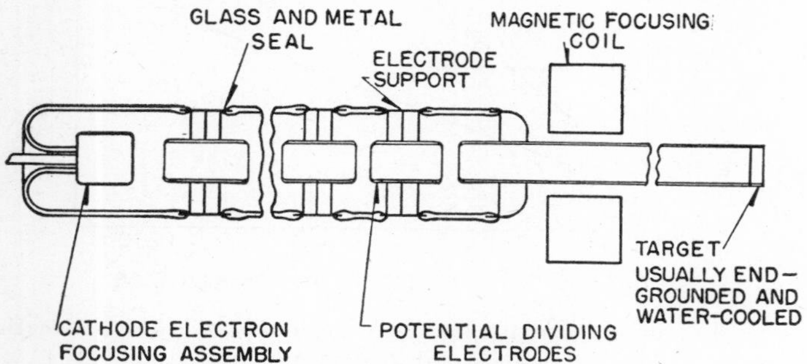

FIG. 8·26 Typical multisection x-ray tube designed to reduce high-voltage fields and electron bombardment of glass insulation.

To reduce absorption of x-rays by the glass walls of the tube, the useful beam frequently is brought out through a specially ground window in the bulb. The effect of variation in window thickness on the absorption of x-rays is shown in Fig. 8·27 as determined by Trout and Atlee.[16] According to equation 8·3 such curves should be straight lines when plotted on a semilogarithmic scale. The change of slope of the curves is caused by the change in quality of the beam as it traverses the material. As illustrated in the curves of Fig. 8·6, the effective wavelength of a beam becomes shorter as it traverses a given specimen. The shorter wavelength results in smaller values of the absorption coefficient and in a decrease in the slope of the curves as the thickness increases.

The quality of the beam therefore depends upon both the wave form of the impressed potential and the filtration in terms of the equivalent thickness of aluminum or copper. Thus, an x-ray tubehead will have an inherent filtration of x millimeters of aluminum, if this thickness will reduce the intensity of x-rays leaving the target by the same amount as the x-ray-absorbing materials ordinarily present in the system. For voltages in excess of 250 kilovolts, the inherent filtration is usually expressed in millimeters of copper.

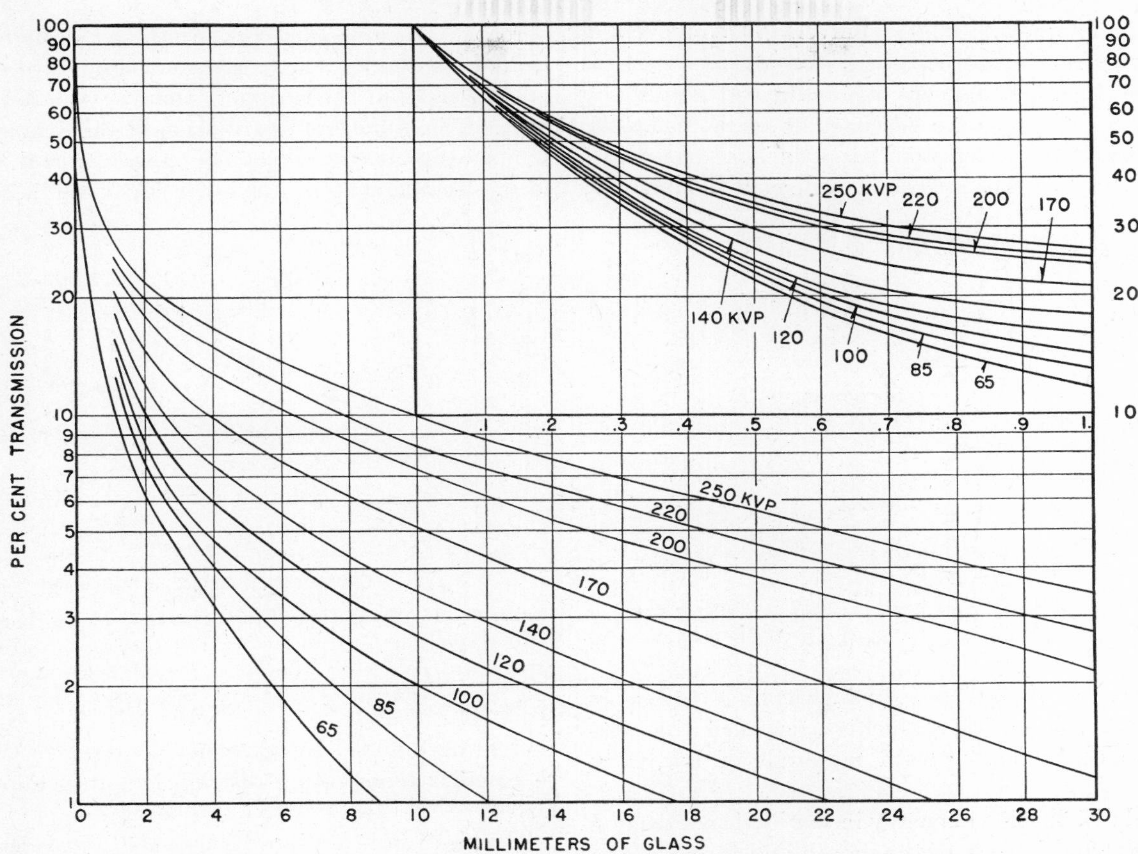

FIG. 8·27 Curves indicating the absorption of x-rays as a function of the thickness of Pyrex glass for various values of half-wave peak kilovolts. (Reprinted with permission from *Am. J. Roentgenology and Radium Therapy*, Vol. 30, No. 8, Aug. 1933, courtesy of Charles C. Thomas, publisher, Springfield, Ill.)

It is convenient to describe the quality of the beam from an x-ray generator in terms of the so-called half-value layer. Absorption curves like the ones shown in Fig. 8·27 are unique functions of the quality, or distribution of wavelengths in a beam. Thus, the thickness of copper or aluminum necessary to reduce the radiation intensity to one-half its original value provides a measure of the average or effective quality of the beam. The relation between the inherent filtration and the half-value layer is indicated in Fig. 8·28.

8·13 DESIGN OF ANODE AND CATHODE ASSEMBLIES

A radiographic tube usually is designed with its target at an angle of about 20 degrees with the direction of the useful beam. The projected area of the focal spot in the useful direction is smaller than the actual area of the focal spot on the target by the sine of 20 degrees or 0.342. This difference permits a substantial reduction in the effective size of the focal spot without affecting the x-ray output in the useful direction.

Although a smaller target angle would increase the ratio of actual spot size to projected spot size, this angle is seldom reduced beyond 15 degrees because the useful angle of the beam would become too small to permit the desired film coverage. In addition, x-rays emerging from the target at a small angle with respect to its plane surface are weaker in intensity because they travel a greater distance through the target material from their point of origin slightly beneath the surface of the target.

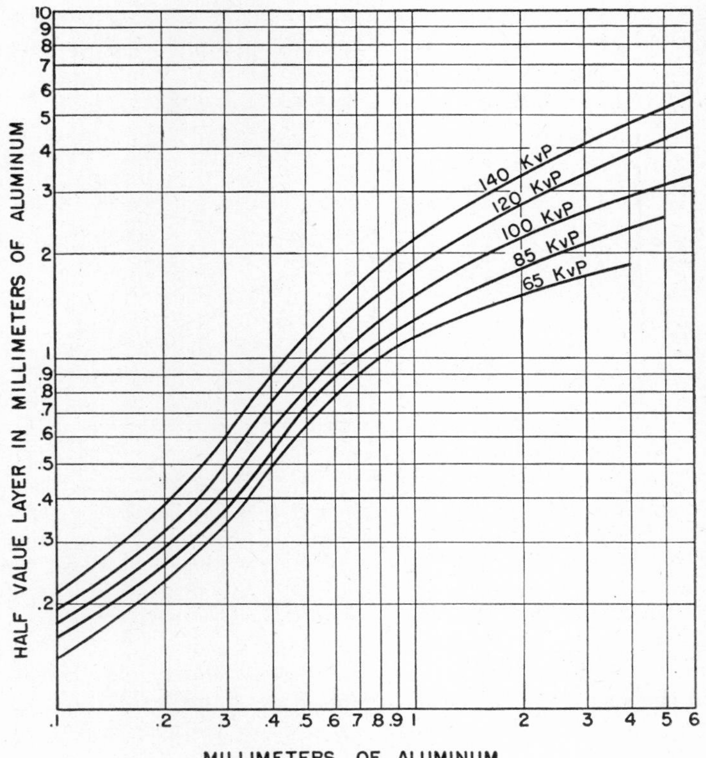

FIG. 8·28 Curves showing the relationship between the half-value layer in millimeters of aluminum and inherent filtration for various values of half-wave peak kilovolts. (Reprinted with permission from *Am. J. Roentgenology and Radium Therapy*, Vol. 47, No. 5, May 1942, courtesy of Charles C. Thomas, publisher, Springfield, Ill.)

Electrostatic focusing is almost universally employed in fixing the focal-spot size on the target. In the so-called line-focus tube the tungsten filament is mounted in a specially machined focal cup similar to the one shown in Fig. 8·23. Though the efficiency of tungsten filaments is low compared with the efficiency of thoriated and oxide-coated filaments,

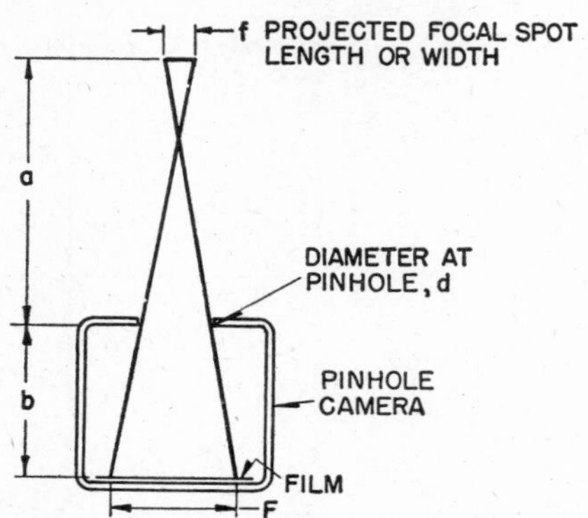

FIG. 8·29 Diagram showing geometrical arrangement of focal-spot camera. The diameter of pinhole d is usually 0.5 or 0.25 millimeter.

coated filaments are undesirable in high-voltage tubes because of cold-emission effects which occur when the electrodes become contaminated with materials of low work function.

The size and distribution of energy in the focal spot depend upon the widths and lengths of the mouth and throat of the cup, and upon the depth at which the filament is mounted. The energy distribution of a given focal spot may be determined by taking an x-ray picture of it, a pinhole camera being used as indicated in Fig. 8·29. The size of the projected focal spot f in terms of F, the apparent focal-spot size

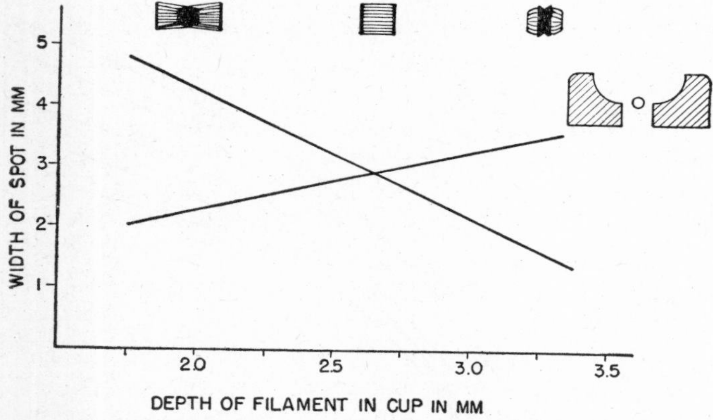

FIG. 8·30 (a) Separation of the pairs of lines forming the focal-spot pattern of line-focus tubes as a function of the filament mounting depth. The focal-spot patterns at several depths are indicated above the curves. (Reprinted with permission from N. C. Beese, "Focusing of Electrons in an X-Ray Tube," *Rev. Sci. Instruments*, July 1937.)

as measured on the film, and the other dimensions defined in the figure are given by the following equation:

$$f = \frac{a}{b}(F - d) - d \tag{8·12}$$

The distribution of energy in the focal spot as determined by Beese [17] is shown in Fig. 8·30 (a) as a function of depth of filament mounting. His data on the effect of varying the throat width appear in Fig. 8·30 (b). A uniform distribution of load on the focal spot is not desirable because cooling is more rapid at the periphery than at the center of the focal spot. Thus, uniform distribution of the load will produce a temperature maximum at the center, which will limit the load. Temperature at the outer edges will be relatively low.

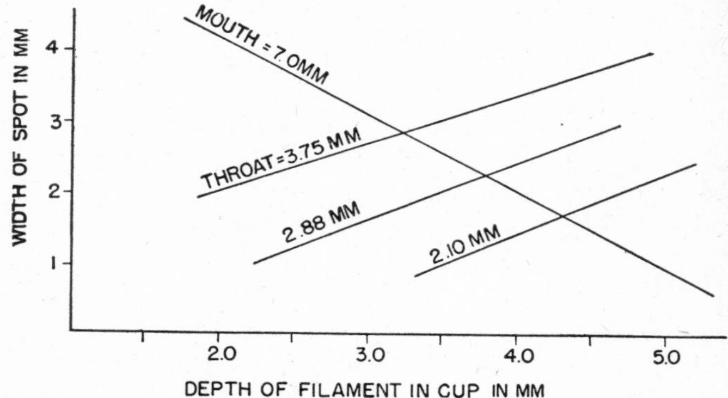

FIG. 8·30 (b) Separation of the pairs of focal-spot lines as a function of focal-spot depth for various values of focal-cup throat width. (Reprinted with permission from N. C. Beese, "Focusing of Electrons in an X-Ray Tube," *Rev. Sci. Instruments*, July 1937.)

Increasing the load energy along the edges of the spot tends to equalize temperature distribution, thus allowing greater loads.

The electron beam in a tube tends to broaden as it leaves the cathode because of the mutually repulsive forces exerted by like charges. However, magnetic forces of opposite direction,[18, 19, 20] which increase with beam velocity, are set up because of the flow of charge. Thus, as tube voltage is decreased, beam divergence increases. In addition, as the

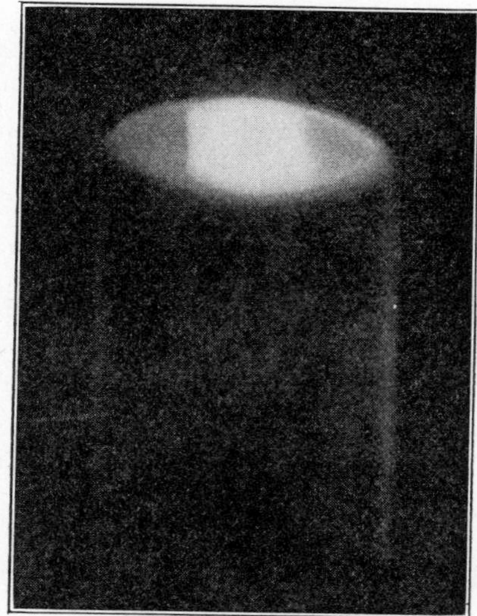

FIG. 8·31 Typical stem-radiation picture of copper-anode x-ray tube.

electron density in the beam is increased by increasing the tube current, the focal-spot size becomes larger. The change in focal-spot size with tube voltage and current, however, is relatively small within the limits established by tube ratings.

The focal spot is the primary source of the x-radiation, but other parts of the anode may provide appreciable radiation because of primary electrons reflected from the focal spot, and because of secondary and tertiary electrons leaving the focal-spot area at high speeds and swinging back to the anode under the influence of the high fields. Pinhole pictures indicating the relative intensity of the radiation leaving different portions of the anode are shown in Fig. 8·31.

The efficiency of production of x-rays is given approximately by the following equation:[21]

$$\frac{\text{X-ray energy}}{\text{Cathode-ray energy}} = KZE \qquad (8\cdot13)$$

where Z is the atomic number, E is the impressed voltage, and K is a constant. At 100 kilovolts the efficiency is between 0.5 and 1 percent for a tungsten target, assuming $K = 1.11 \times 10^{-9}$, as determined by Compton and Allison.[22] Remaining energy must be dissipated from the target as heat. For this reason, heat transfer from the anode is one of the important problems of x-ray-tube design.

Tungsten has been widely accepted as the target material because of its low vapor pressure, high melting point, high atomic number, and reasonably good thermal conductivity. In addition, it is economical, can be bonded with copper, and has satisfactory cold-emission properties.

Since a primary function of the anode in a tube with a copper-backed tungsten target is to conduct the heat through the envelope to the outside environment, the surface of the anode assembly must be regarded as part of the tube envelope, and the seal between the anode and the bulb must be vacuum-tight. To satisfy this requirement, a glass-to-metal seal like the one shown in Fig. 8·23 is made. The expansion coefficient of the glass-sealing metal and the glass must be very close through the range of temperatures encountered in processing the tube and in later service. Until recently, a feather-edged copper thimble was used to make the glass-to-metal seal. The thin-copper feather edge permitted differences in expansion between the copper and the glass, whereas a heavier edge would have caused the glass to crack. Unfortunately, the seal was difficult to fabricate because the feather edge was easily overheated during glassing operations. Kovar, an alloy developed by Scott,[23] matches the coefficient of expansion of some glasses closely throughout the useful temperature range. Since the glass used to seal Kovar does not exactly match the coefficient of expansion of Pyrex, a so-called graded seal made up of glasses with intermediate coefficients of expansion is used. The softening points of the intermediate glasses must be considered in selecting them, since it is desirable to have the glasses soften in the same temperature range.

In some medical applications high radiation intensity, a low degree of unsharpness, and short exposure time are required. To permit the desired loading of the tube it is necessary to rotate the anode, thus spreading the heat over a large area and confining the source of x-rays to a small effective area. Tube currents as large as 200 milliamperes are commonly used with a 1-millimeter effective focal spot; and currents as high as 500 milliamperes are used with a 2-millimeter effective focal spot. The usual exposure times are $\frac{1}{60}$, $\frac{1}{30}$, and $\frac{1}{20}$ second. Accurate timers are required to prevent overloading of the tube.

The anode, which is supported on ball bearings, is rotated at high speed by an oil-immersed induction motor, the coils of which are at ground potential and are mounted around the outside diameter of the anode portion of the bulb. The rotor of the induction motor is part of the anode assembly. Since these tubes must be outgassed on exhaust at temperatures somewhat higher than their normal loading, a good temperature-resistant vacuum lubricant must be used. In these tubes heat dissipation under continuous load creates a special design problem, for the bearings provide a poor thermal conducting path to the outside of the tube envelope, and they must be kept cool.

8·14 MANUFACTURE OF X-RAY TUBES

It is fairly easy to reduce the pressure in an x-ray tube to 10^{-6} or 10^{-7} millimeter of mercury. This pressure, if maintained, would insure the absence of unstable gas discharges during operation. Simple evacuation to a low pressure would not produce a stable tube. Liberation of gas by heating or by electron bombardment of the various internal surfaces of the tube during operation would quickly increase pressure enough to produce gas discharges upon application of high voltages. In addition, absorbed gas on the surfaces of the tube would cause cold emission, with consequent instability. Because of the importance of completely outgassing tube parts, x-ray tubes require careful manufacturing methods.

Tube parts must be highly polished to reduce electric fields. Most tube parts must be chemically cleaned, and some are heated by high-frequency induction in a vacuum bottle for preliminary outgassing before the exhaust operation, which permanently evacuates the tube by removing gas from both the tube volume and the tube walls. This so-called prevacuum treatment permits heating the tube parts to higher temperatures than can be tolerated during actual tube exhaust, since exhaust temperatures must be kept low enough so that vapor pressure deposits on the tube envelope do not impair insulation. Furthermore, the temperature which can be applied to parts sealed in the tube is limited by the proximity of parts having relatively low melting points, such as glass, or of metal parts having relatively high vapor pressures. Since many pure metals have low vapor pressures, whereas their oxides or other compounds have high vapor pressures, tube parts are frequently baked in hydrogen furnaces to reduce oxides. Hydrogen, which can easily be outgassed on exhaust, tends to replace other absorbed gases on these parts.

For convenience in sealing, the parts of a conventional tube are assembled as so-called cathode and anode assemblies. These assemblies are mounted on glass stems which have flare diameters closely conforming to the internal diameters of the ends of the envelope. The assemblies are sealed into the tube envelope. Nitrogen or a similar gas is usually flushed through the tube envelope during sealing to prevent oxidation of the tube parts. A tubulation which is later used to connect the tube to the exhaust system is also sealed to the tube. Immediately following sealing, the tube is connected to a preliminary exhaust system, is evacuated, and then is tipped off. This operation prevents the parts of the tube from oxidizing while they are cooling, keeps them hermetically sealed in an environment of low pressure until the tube is ready for exhaust, and permits testing the tube for "vacuum-tightness."

During exhaust the various parts of the tube are heated to temperatures that make outgassing reasonably complete. Usual exhaust schedules consist of cycles of baking, high-frequency-induction heat treatment, and high-voltage operation. During baking the entire tube is heated to a temperature limited only by the softening temperature of the glass bulb. High-frequency-induction heating frees the metal parts from gas, making subsequent operation of the tube on exhaust more stable. The temperature of the tube parts during high-voltage operation on exhaust is much higher than any temperature encountered during normal operation.

8·15 RATING OF X-RAY TUBES

As indicated in previous sections of this chapter, efficiency of x-ray production is rather low because of (1) the relatively low efficiency of conversion of electron energy into x-ray energy; (2) absorption of x-rays by the walls of the tube and by the oil and window of the tubehead; (3) absorption of x-rays by the specimen; (4) attenuation of x-rays according to the inverse square law; and (5) the low percentage of x-radiation which performs useful radiographic work on the film or recording medium.

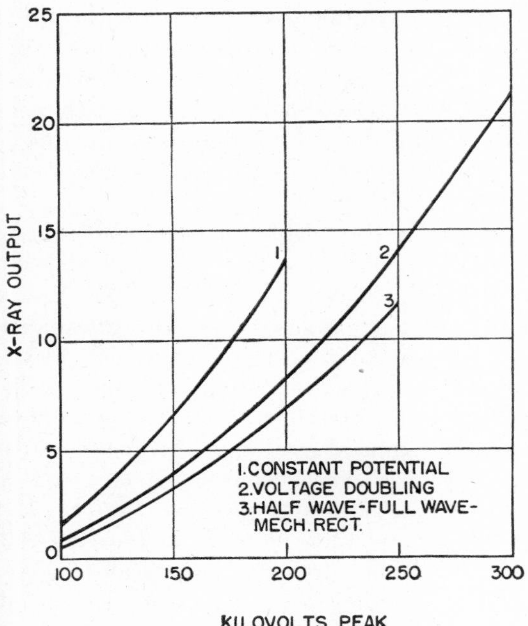

FIG. 8·32 X-ray output as function of peak kilovolts for various circuits. Tube has ¼-inch glass filter, and 1 millimeter of external copper filter. (Reprinted with permission from *Am. J. Roentgenology and Radium Therapy*, Vol. 30, Aug. 1933, courtesy of Charles C. Thomas, publisher, Springfield, Ill.)

Medical applications require high x-ray output for times which are short enough to minimize motion unsharpness. To keep the geometrical unsharpness low enough to permit the required diagnosis, tubes must have small focal spots. As the result of medical requirements, the safety factors indicated on short-time-rating charts are low compared with the factors of safety in other electronic tubes. Until recently, short-time-rating charts were provided only with tubes designed for medical use, with maximum voltages of 100 kilovolts or less. The rotating-anode tube was devised because, even with reduced factors of safety, the outputs of stationary-anode tubes were still too low to satisfy all the requirements for radiographing thick sections of the body.

An x-ray exposure involves impressing a given voltage across a tube, passing a given current through it by adjusting filament temperature, and controlling the time of exposure, usually by means of a timer in the primary circuit. Exposure factors for industrial radiographic work usually are shown on technique charts, such as Fig. 8·8, and are indicated in terms of peak voltage and the milliampere-seconds or milliampere-minutes required to produce a given film density. Because of the dependence of output upon wave form, such charts usually specify the particular circuit.

The wave form of a generator has an important influence on the quantity of x-rays, as shown in Fig. 8·32, where the roentgen-ray output of different circuits is indicated as a function of peak impressed-plate potentials. The higher x-ray output from constant-potential circuits is a direct consequence of the dependence of x-ray output upon voltage. Since a pulsating wave form impresses low potentials across the tube for substantial fractions of the cycle, the output is relatively lower.

Tube ratings, as expressed in terms of peak voltage and average direct current, depend upon the circuit employed. Thus a tube rated continuously at 200 kilovolts peak and 25 milliamperes for full-wave applications may have a rating of only 200 kilovolts and 18 milliamperes on constant potential. Power dissipated by the anode is the determining factor in ratings, and is essentially the same in both these examples.

Because of the low efficiency of x-ray production, almost all the energy going into the anode must be dissipated as heat. To achieve the desired x-ray outputs, energy input must be rather large. Energy input to the tube is limited, however, by the danger of melting or vaporizing some parts of the anode, or of exceeding safe operating temperatures for other parts of the tube. In oil-cooled tubes the external surfaces of the anode in contact with the oil must not become hot enough to deteriorate the oil, or "carbonize" it.

To prevent overheating of the anode assembly, the thermal path to convey heat energy from the anode to the external environment must be adequate, and loading limits must be established for all possible field conditions. A complete definition of ratings involves specifying the maximum average direct current in milliamperes as a function of the peak voltage across the tube and of exposure time. Limits must be assigned to the magnitude and frequency of successive exposures, and finally a continuous rating for the tube must be established. Such ratings, of course, vary with the circuit in which the tube is employed.

Limitations on Short-Time Loading of X-Ray Tubes

Short exposures are necessary to minimize motion unsharpness, to save time, and to increase the number of radiographs which can be taken in a given period of time. A conventional exposure chart for full-wave operation is shown in Fig. 8·33 for a tube having a projected focal-spot size of 4.2 square millimeters. Exposures up to several seconds have relatively low values of total energy, and they raise the average temperature of the anode system only slightly after the anode has come to equilibrium. However, this concentration of energy into a small area on the target during a short interval

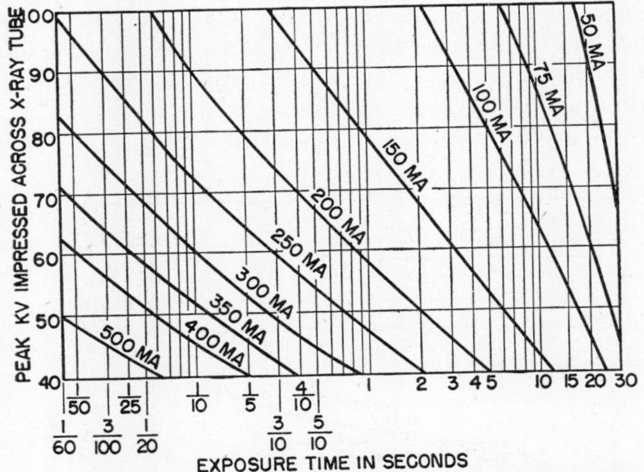

Fig. 8·33 Conventional type of rating chart for 100-kilovolt oil-immersed x-ray tube with projected focal-spot size of 4.2 square millimeters. Maximum values of peak kilovolt and d-c average milliamperes are indicated as a function of exposure time.

results in extremely high local temperatures of short duration. The limitation on ratings for short times results from the possibility of melting, cracking, and vaporizing the surface of the tungsten focal spot.

Temperature distribution from the surface of the tungsten target into the copper anode has been calculated by

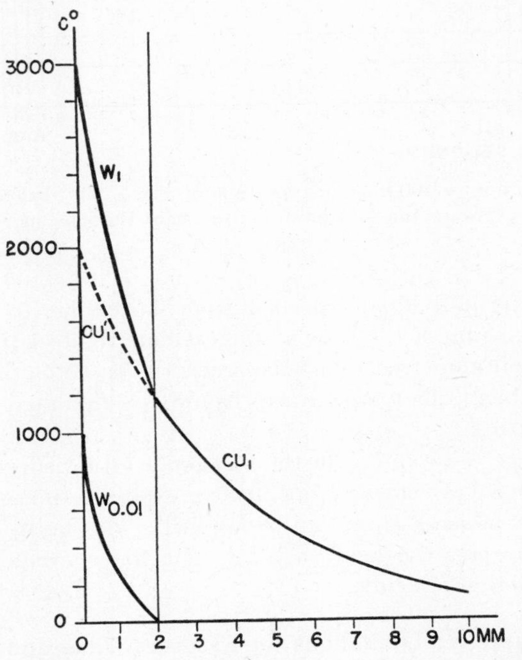

Fig. 8·34 Distribution of maximum temperatures for a copper-backed x-ray anode having a tungsten target 2 millimeters thick for loading of 200 watts per square millimeter and exposure times of 1 and 0.01 second. (Reprinted with permission from *X-Ray Research and Development*, Phillips Research Laboratories, p. 24.)

Bouwers,[24] who used simplifying approximations. Figure 8·34 indicates the maximum-temperature gradients he derived for a copper-backed tungsten target 2 millimeters thick for exposures of 1 second and 0.01 second. The curve W_1–Cu_1 indicates the temperature distribution for a load of 20,000 watts per square centimeter for 1 second. If the thickness of the target is reduced, the temperature of the surface of the tungsten also is reduced because copper has a higher thermal conductivity than tungsten. However, the maximum temperature of the copper is limited by its melting point at 1083 degrees centigrade; therefore a minimum thickness of tungsten must be maintained to prevent melting the copper adjacent to the tungsten. For exposures under 0.04 second, Bouwers has expressed the temperature of the target surface as

$$T = \frac{2Qa}{K\sqrt{\pi}}\sqrt{t} \qquad (8\cdot 14)$$

where Q is calories per square centimeter, K is the coefficient of thermal conductivity of the target material, t is the time

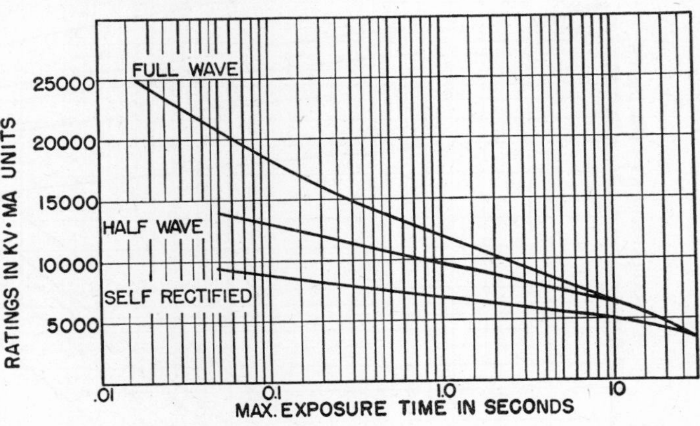

Fig. 8·35 Condensed rating chart for the tube in Fig. 8·33, indicating the products of the maximum values of peak kilovolts and d-c average milliamperes as a function of exposure time for full-wave, half-wave, and self-rectified types of operation.

in seconds, and a^2 is equal to the coefficient of thermal conductivity divided by the thermal capacity per cubic centimeter.

For a given exposure time the products of the maximum values of peak voltage and average direct current in milliamperes in the rating chart shown in Fig. 8·33 are equal to a constant. In other words, the energy applied to the anode during this time is the limiting factor in determining the rating. The conventional rating chart may be simplified by indicating only the fundamental information. Thus Fig. 8·35 indicates the product of peak voltage and average direct current in milliamperes (which is proportional to power) as a function of exposure time for full-wave, half-wave, and self-rectified operation. These curves, particularly for the short-exposure times, are parallel to the so-called melting curve for the tungsten focal spot in which the product of the peak voltage and average direct current in milliamperes required to melt the focal spot is plotted as a function of exposure time.

Although the average power on half-wave and self-rectified operation is approximately the same as that on full-wave operation for the same values of peak voltage and average direct current in milliamperes, the peak values of current are approximately twice as great for half-wave and self-rectified equipment as a result of the suppression of half a cycle. Thus the maximum of the cyclic variation of focal-spot temperature during an exposure at a given product of voltage and current is much higher for half-wave and self-rectified operation, and the half-wave and self-rectified ratings must be lowered with respect to full-wave ratings to prevent the greater intermittent peak loading from melting the spot.

Self-rectified or a-c operation applies a load to the target equivalent to the half-wave load but has an additional serious loading limitation. If tungsten is heated to incandescence its electron emission becomes appreciable. Since the inverse voltages in self-rectified operation would sweep electrons emitted by the anode to the cathode, the temperature of the focal spot must be limited to values which will not cause excessive electron emission. When back emission from the target is produced by a-c overloads, it frequently melts parts of the focal cup, or increases the temperature of the filament by electron bombardment, which in turn increases emission to the anode. This progressive increase in current will quickly destroy the tube if allowed to proceed.

As the exposure time in Fig. 8·35 is increased, the power rating of the tube decreases. However, total energy, which is proportional to the product of power and exposure time, becomes greater with increase in exposure time. For long exposures the rating is not limited by the possibility of overheating the focal spot, but by the heating of the entire anode by application of large values of energy long enough to allow dissipation of energy into the rear portion of the anode assembly. For long exposures the ratings are no longer limited by temperatures of the focal-spot region, and the kilovolt-milliampere power curves for full-wave, half-wave, and self-rectified operation merge into one curve.

The effect of focal-spot size in limiting power which can be applied for a given time (and hence the x-ray output which can be obtained from the tube during this time interval) is shown in Fig. 8·36. In this figure the full-wave kilovolt-milliampere values are indicated as functions of exposure time for various focal-spot sizes. Such curves define ratings for individual exposures from short times to the continuous rating of the tube at the extreme right of the chart. For an oil-immersed tube such a chart is based on maintaining ambient-oil temperature below a stipulated critical temperature.

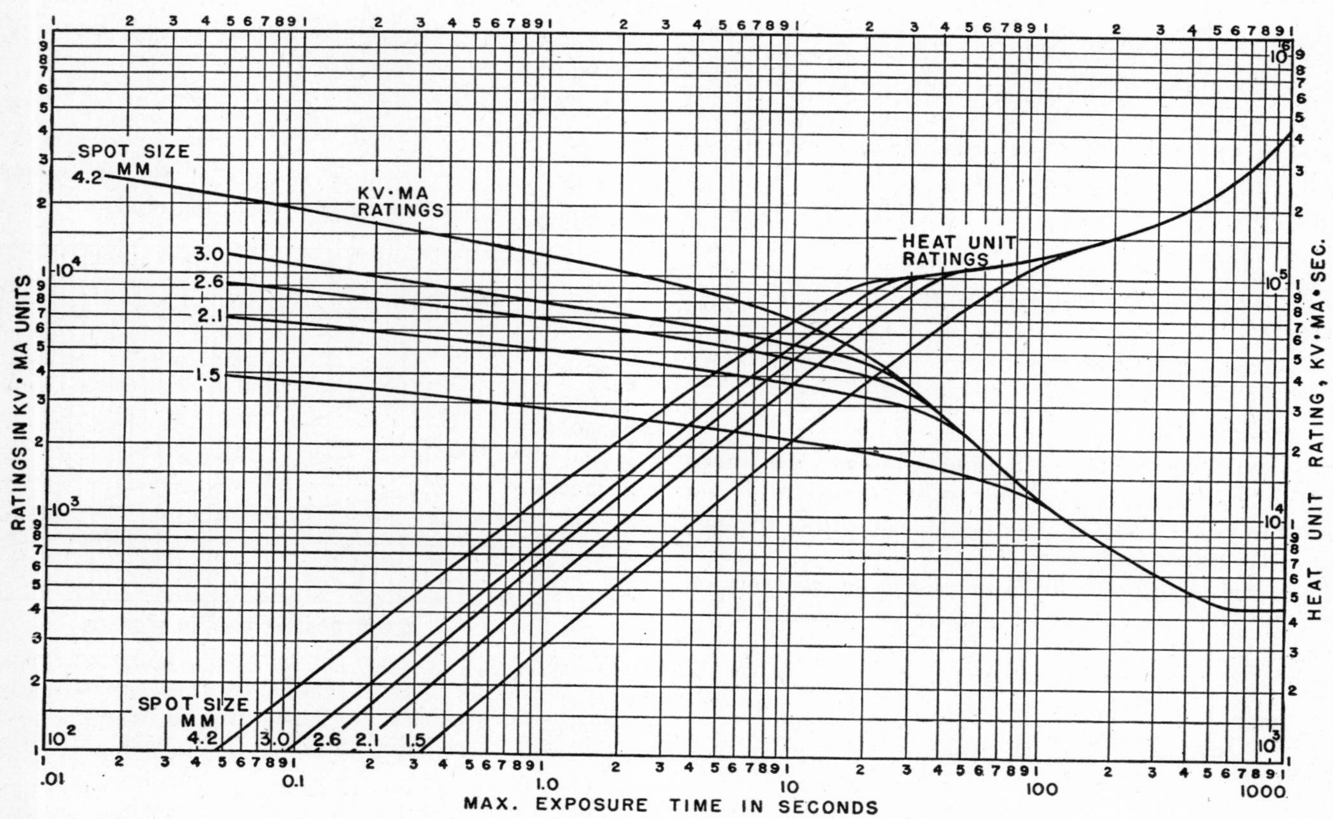

FIG. 8·36 Comprehensive full-wave rating chart showing maximum exposure factors for 100-kilovolt-type tube of Fig. 8·33. Kilovolt, milliampere, and heat-unit ratings are indicated for various focal-spot sizes as functions of exposure time for exposures from short times to the continuous rating of the tube.

In Figs. 8·35 and 8·36 the focal-spot-rating curves intersect a curve of sharper slope at long exposure times, further limiting the load which can be applied. This limiting curve depends upon the heat capacity and the continuous heat dissipation of the tube.

Heat Capacity; Limitations on Exposure Time and Length and Frequency of Successive Exposures

The rating of a tube may be expressed in terms of "heat units" as well as in kilovolt-milliampere units. Heat units are defined as the product of peak kilovolts, average direct current in milliamperes, and exposure time in seconds. The x-ray energy generated by a tube at a given voltage is directly proportional to the heat units of the exposure. The ratings indicated in Fig. 8·36 are shown also in terms of heat units.

Since the kilovolt-milliampere unit is proportional to power, the heat unit (which is the product of the kilovolt-milliampere or power unit) and time in seconds are proportional to energy. If Q_A is the energy applied to an anode during an exposure, then the average rise in temperature of the anode assembly may be expressed by the equation

$$Q_A = cm\Delta T \tag{8·15}$$

where c is the specific heat, m is the mass of the anode assembly, and ΔT is the average rise in temperature of the anode assembly. The energy or heat units applied to an anode as the result of a given exposure or series of exposures then determines the average temperature rise of the anode. If a particular temperature rise $\Delta T_{\max}$ is selected as the maximum allowable average temperature increase of the anode, by equation 8·15 a maximum energy input to the tube, which may be called $Q_{\max}$, also is fixed. $Q_{\max}$ in heat units is defined as the heat capacity of the tube. The temperature selected as the maximum average temperature at which the anode can be allowed to operate depends upon temperature limitations of the various parts of the anode and radiator assembly and upon considerations of stability during tube life.

The heat capacity of the tube limits the magnitude and frequency of successive exposures. Successive exposures may be applied to a tube within the limitations imposed by a short-time-exposure chart, such as Fig. 8·33, until the heat capacity of the tube is reached. A further limitation is imposed by the necessity of allowing high local temperatures at the focal spot to decrease by dissipation of heat into the rear portions of the anode. An interval of about 5 seconds between exposures usually is recommended. The heat capacity of a tube, in addition to limiting a series of short exposures, also limits long exposures. Figure 8·36 shows a plateau at 100,000 heat units on the curve of heat-unit rating versus time because the heat capacity of the tube described is 100,000 heat units.

Cooling Curve of an X-Ray Tube

The discussion of ratings thus far has not taken into consideration the fact that the energy going into the anode of the tube eventually is dissipated by the radiator as heat. If Q_t is the total energy applied during a given exposure, the distribution of energy a short time later may be expressed as

$$Q_t = Q_a + Q_d \tag{8·16}$$

where Q_a is the total heat energy in the anode, as indicated by the rise in its temperature, and Q_d is the total heat energy dissipated by the cooling system of the tube.

Single exposures in Fig. 8·36 lasting more than 30 seconds with the 4.2-square-millimeter focal spot are allowed to exceed the heat capacity of the tube. Heat units in excess of the 100,000-heat-unit capacity of the tube represent in these longer exposures the value of energy dissipated by the cooling system during the exposure. For example, according to Fig. 8·36, 150,000 heat units is a permissible loading for a 200-second exposure. During a 200-second exposure, 50,000 heat units is dissipated by the cooling system, and at the end of the exposure 100,000 heat units remains in the anode. If the tube is not further loaded, the remaining heat is dissipated at a rate dependent upon the cooling system.

A cooling chart, which indicates the rate of cooling as a function of the heat units in the tube, is shown in Fig. 8·37. It is evident that the rate of cooling should be greater at high values of heat units, since the anode temperature, according to equation 8·15, depends upon the heat-unit content of the anode, and since heat dissipation by the radiator is proportional to the temperature differential between the radiator and the ambient environment.

A cooling chart effectively increases the rating of the x-ray tube by providing a means for taking into account the heat units dissipated during a series of exposures. Suppose, for example, a series of exposures of 50,000 heat units each is proposed. What intervals between exposures must be maintained? Two 50,000-heat-unit exposures can be made immediately with the cold tube allowing only a 5-second interval to permit the focal-spot region to cool. The anode thus is brought to its maximum heat capacity. Figure 8·37 indicates that the tube must cool for approximately 2.3 minutes to allow the heat content of the anode to decrease to 50,000 heat units. Further successive 50,000-heat-unit exposures can be made without exceeding the heat capacity of the tube, if they are separated by 2.3-minute intervals. If a cooling interval of 13 minutes instead of 2.3 minutes is allowed after the tube has reached its 100,000-heat-unit capacity, two 50,000-heat-unit exposures can again be made with a 5-second interval between them. However, the heat input with such a combination of exposures would be approximately 100,000 heat units in 13 minutes, compared with a rate of 250,000 heat units during approximately a 12-minute interval for the succession of exposures first considered. The higher rate of heat input for the loading with the 2.3-minute cooling interval is a consequence of the higher rate of cooling at increased radiator temperatures.

Heating curves showing the heat-unit content of the anode as a function of exposure time are shown in Fig. 8·37 for various rates of heat input. For a given kilovolt-milliampere input and relatively long exposure times, the heat-unit content of the anode is somewhat lower than the heat-unit value of the exposure, because some heat is dissipated by the cooling system of the tube during the exposure.

Continuous Rating of an X-Ray Tube

The continuous rating is the maximum tube loading which can be dissipated continuously without heating any part of the tube above its critical temperature. This loading is represented in Fig. 8·37 by the heat-input curve (425 kilovolt-milliamperes) which brings the heat content of the anode to equilibrium at the full heat capacity of the tube. It is also represented in terms of heat units dissipated per unit time by the slope of the cooling curve at the full heat capacity of the tube.

Once equilibrium is established during continuous operation, all the energy passing through a given cross-sectional area in the thermal path between the focal spot and the thermal environment is equal to the thermal load transmitted through any other cross-sectional area in the thermal path. Since thermal loads are transmitted as the result of temperature gradients, the equalization of thermal load is accomplished by a gradual building up of temperature gradients through the thermal system of the anode until the energy outgo is just equal to the energy income. For example, the temperature of the oil-cooling or heat-transfer surface of a forced-oil-cooled tube rises upon application of the continuous rated load until it can just transmit to the circulating oil the load impressed upon the anode. The average temperature of the circulating oil rises until it is high enough head, and the external environment become high enough to dissipate the continuous loading.

The loading which the tubehead can dissipate is controlled by the design of the tubehead, and by whether a blower, a fan, or other heat exchange means is used. It depends also upon the maximum permissible temperature at which the tubehead can operate. The maximum permissible temperature depends principally upon the temperature-resistant

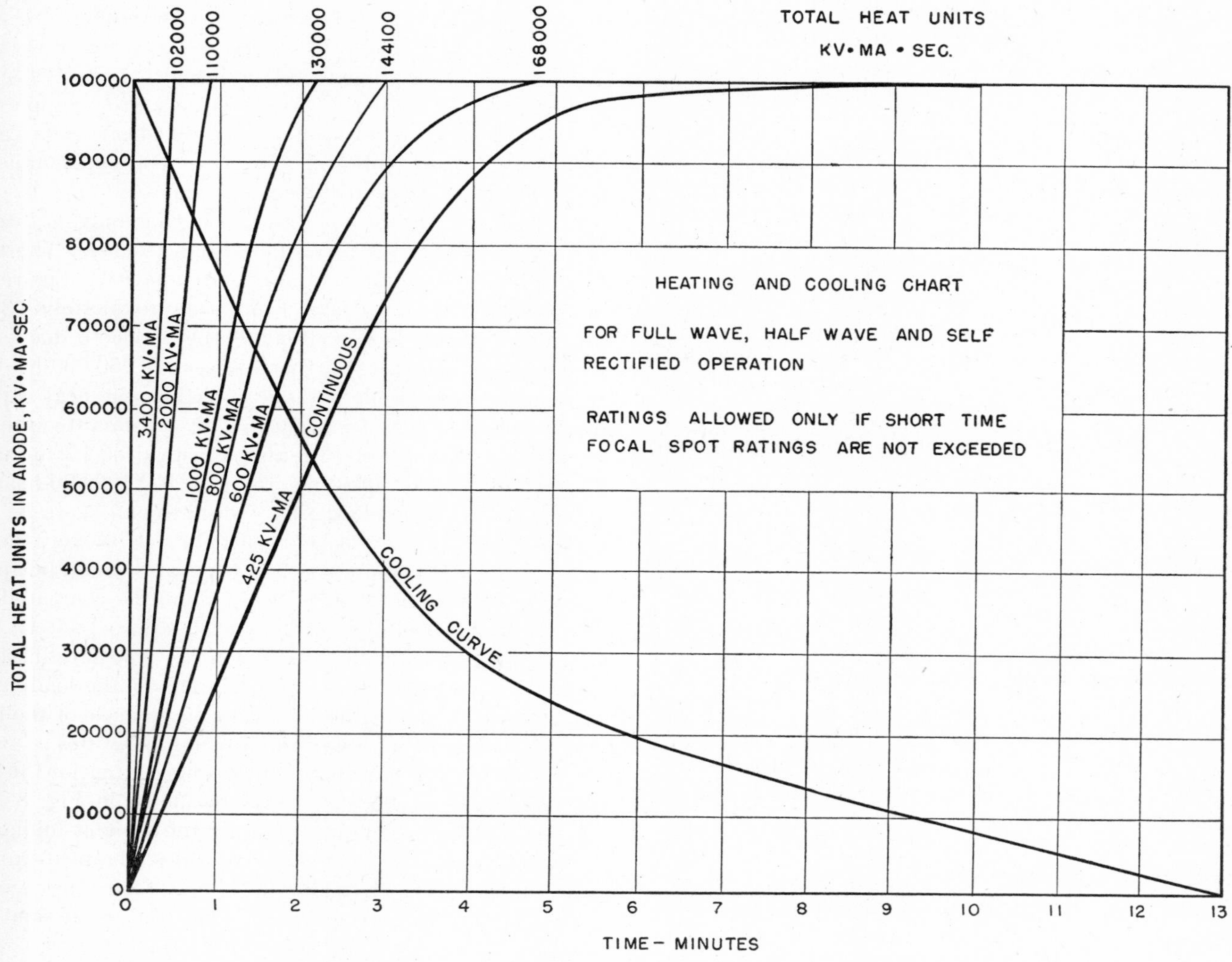

FIG. 8·37 Heating and cooling chart for oil-immersed x-ray tube of Fig. 8·33.

to transmit the energy being applied to the anode to an external heat exchanger, usually a water-cooled surface of some kind. At the same time the thermal gradients in the anode proper rise until they are high enough to compel the continuous passage of the load impressed upon the tube. At equilibrium the temperature of all portions of the tube should be below their critical values.

Ratings of X-Ray Tubes in Tubeheads

All the energy that goes into the anode eventually must be dissipated to the ambient environment. If the tube is oil-immersed, the load eventually must be dissipated to the oil and finally, through the walls of the tubehead, to the atmosphere. As the continuous rated load is applied to a tube, the temperature of the oil in the tubehead gradually rises until the thermal gradients between the oil, the tube- properties of the oil and of the insulation of the tubehead. The temperature-resistant property of the oil may be increased by a vacuum-treating process which frees it of dissolved gases and water vapor. Heat-expansion bellows are usually provided to allow for expansion of oil at high temperatures within the hermetically sealed tubehead.

The heat capacity of a tubehead may be defined as the number of heat units which will bring the tubehead to its maximum permissible temperature at the highest rate of loading permitted by the tube rating. Such a heat capacity should correspond closely to the heat capacity determined by heat-capacity equations, such as equation 8·15, the heat capacity of the tube, the head, and the oil being taken into consideration.

Cooling curves for tubeheads are similar to the cooling curve shown in Fig. 8·37, and they indicate the rate of loss of heat units as a function of time. The rate of heat-unit loss is maximum at the heat capacity of the tubehead, because of the greater dissipation of load at high temperatures, and it should correspond to the continuous rating of the tubehead.

REFERENCES

1. *X-ray Technology*, H. M. Terrill and C. T. Ulrey, Van Nostrand, 1930.
2. "An Experimental Investigation of the Energy in the Continuous X-Ray Spectra of Certain Elements," C. T. Ulrey, *Phys. Rev.*, Vol. 11, 1918, p. 401.
3. *Smithsonian Physical Tables*, Smithsonian Institution, 1921.
4. *Handbook of Chemistry and Physics*, C. D. Hodgman and H. N. Holmes, 25th ed., Chemical Rubber Publishing Co., 1940.
5. *Applied X-rays*, George L. Clark, McGraw-Hill, 1940.
6. "Estimating Radiographic Exposures for Multi-Thickness Specimens," H. E. Seemann and G. M. Corney, *Ind. Radiography*, Fall, 1943, pp. 33–38.
7. *Radiography of Materials*, Eastman Kodak Co., p. 19.
8. "An Investigation of the Apparatus Used in Radium Radiography," L. W. Ball and D. R. Draper, *ASTM Symposium on Radiography*, 1943, 187 pp.
9. "Roentgenographic Unsharpness of the Shadow of a Moving Object," S. R. Warren, *Radiology*, Vol. 34, June 1940, pp. 731–740.
10. *X-ray Protection*, National Bureau of Standards, HB20.
11. "Industrial Radiation Hazards," C. B. Braestrup, *Radiology*, Vol. 43, Sept. 1944, p. 286.
12. "Field Emission X-Ray Tube," C. M. Slack and L. F. Ehrke, *J. App. Phys.*, Vol. 12, Feb. 1941, p. 165.
13. "The Acceleration of Electrons by Magnetic Induction," D. W. Kerst, *Phys. Rev.*, Vol. 60, July 1, 1941, p. 47.
14. "A 20 Million-Induction-Volt Betatron or Induction Accelerator," D. W. Kerst, *Rev. Sci. Instruments*, Vol. 13, Sept. 1942, p. 387.
15. "Threshold Measurements on the Nuclear Photo-Effect," G. C. Baldwin and H. W. Koch, *Phys. Rev.*, Vol. 67, Jan. 1945, p. 1.
16. "Low Absorption Roentgen-Ray Measurements from 10 to 250 Kilovolts," E. D. Trout and Z. J. Atlee, *Am. J. Roentgenology and Radium Therapy*, Vol. XLVII, May 1942, p. 785.
17. "The Focusing of Electrons in an X-Ray Tube," N. C. Beese, *Rev. Sci. Instruments*, July 1937, Vol. 8, p. 258.
18. "The Dispersion of an Electron Beam," E. E. Watson, *Phil. Mag.*, Vol. 3, 1924, p. 849.
19. *Electron Optics, Theoretical and Practical*, L. M. Myers, Van Nostrand, 1939.
20. *Beitrog z. geometrischen electronen optik annalen der Physik*, M. Knoll and E. Ruska, 5 Folge, Band 12, p. 607, 1932.
21. "Energy of Roentgen Rays," R. T. Beatty, *Proc. Royal Soc. (London), Series A*, 89, 1913, p. 314.
22. *X-Rays in Theory and Experiments*, A. H. Compton and S. K. Allison, Van Nostrand, 1935.
23. "Expansion Properties of Low-Expansion Fe-Ni-Co Alloys," H. Scott, *Am. Inst. Mining Metallurgical Engrs., Tech. Pub.*, 318, 1930.
24. "Temperature Distribution on the Anode of an X-Ray Tube," A. Bouwers, Publication No. 332 *Z. tech. Physik*, 8271, 1927.
25. "Design and Application of X-Ray Tubes," Z. J. Atlee, *Electronics*, Oct. 1940, Vol. 13, No. 10, p. 26.
26. "The Reduction of Secondary Radiation and of Excessive Radiographic Contrast by Filtration," H. E. Seemann, *Proc. A.S.T.M.*, Vol. 40, 1940.
27. "Radiography and X-Ray Tube Design," J. Lempert, Vol. 16, No. 11, *Electronics*, Nov. 1943.
28. "Fluoroscopies and Fluoroscopy," W. E. Chamberlain, *Radiology*, Vol. 38, No. 4, April 1942, pp. 383–425.
29. "One Hundred Million-Volt-Induction Electron Accelerator," W. F. Westendorp and E. E. Charlton, *J. App. Phys.*, Vol. 16, No. 10, Oct. 1945.
30. "Trends in the Technique of Industrial Radiography," H. E. Seemann, *ASTM Bull*, March 1942.
31. "X-Rays," G. W. C. Kaye, Longmans, Green, 1923.
32. *The Science of Radiology*, O. Glasser, Thomas, 1933.
33. "A Study of Roentgen-Ray Distribution at 60–140 Kv P," Z. J. Atlee and E. D. Trout, *Radiology*, Vol. 40, April 1943, pp. 375–385.
34. "Standard Absorption Curves for Specifying the Quality of X-Radiation," L. S. Taylor and G. Singer, *Bur. Standards J. Research*, Vol. 12, April 1934, p. 666.
35. "The Comparison of High Voltage X-Ray Generators," L. S. Taylor and K. L. Tucker, *Bur. Standards J. Research*, Vol. 9, RP475, September 1932, pp. 333–352.
36. "Industrial Fluoroscopy of Light Materials," S. W. Smith, *Proceedings of the National Electronics Conference*, Vol. I, May 1945, pp. 430–446.
37. "Two Million-Volt Mobile X-ray Unit," E. E. Charlton and W. F. Westendorp, *Proceedings of the National Electronics Conference*, Vol. I, May 1945, pp. 413–429.
38. "Field Emission Applied to Ultra-Speed X-Ray Technique," C. M. Slack and E. R. Thilo, *Proceedings of the National Electronics Conference*, Vol. I, May 1945, pp. 447–456.
39. "Biological Effects of Ionizing Radiations," G. Failla, *J. App. Phys.*, Vol. 12, 1941, p. 279.

Chapter 9

CATHODE-RAY TUBES

P. E. Volz and P. E. Grandmont

A DEVICE using a beam of electrons (cathode rays) as a writing element is generally classified as a cathode-ray tube. The usual abbreviation, CR tube, is used throughout this chapter. There are two types of CR tubes, the sealed-off tube and the demountable tube, which have completely different applications. Components of a CR tube and its auxiliary equipment may be divided into three functional classifications:

1. Devices that produce, control, and focus a beam of electrons. These devices are generally called the electron gun. It is necessarily inside the evacuated envelope of the tube.
2. Devices that cause deflection of the electron beam. Either electrostatic or magnetostatic fields may be used, and the deflecting fields may be produced by devices inside or outside the evacuated envelope of the tube. Usually the beam can be deflected in two directions at right angles to each other.
3. The device (called a screen) that indicates the position and intensity of the beam. Some of the beam energy can be translated into visible light due to incandescence, fluorescence, or phosphorescence. The beam can produce a latent image on photographic film. In any case the beam must strike the screen material, so the screen is inside the evacuated envelope of the tube.

9·1 ELECTRON SOURCE

Thermionic emission, field emission, or emission caused by positive-ion bombardment may be used as the source of electrons. For thermionic emission either a directly heated cathode (filament) or an indirectly heated cathode (usually oxide-coated) may be used. To produce a focused beam of essentially circular cross section, an axially symmetric cathode surface is desirable. For this reason gun structures using thermionic emission have indirectly heated cathodes, the oxide-coated emitter being a surface of revolution. As the accelerating voltage used in the gun structures becomes very high there are two objections to the use of oxide-coated cathodes. First, a high field at the cathode may cause the oxide coating to flake off and, second, any residual gas in the tube becomes ionized and the resultant bombardment of the oxide coating by positive gas ions has a "poisoning" effect on the emitter as well as a mechanically destructive effect. For these reasons high-voltage CR-tube guns (say above 40 kilovolts of accelerating voltage) use other types of cathodes. As far as cathode life is concerned it would be possible to use pure-tungsten filaments; but, as previously mentioned, their use would introduce certain focusing difficulties. Cold cathodes therefore are widely used for these applications. Both field emission and positive ion bombardment play a part in the total electron emission from these cathodes. Emission is caused primarily by positive-ion bombardment. The cathode consists of a metal plug, usually aluminum, which has a low work function. The emitting surface of the cathode (surface facing the anode) is a surface of revolution. Gas pressure (usually about 10 microns) is used to control the discharge and hence the beam current. Since this type of cathode is used only in demountable CR tubes, which require their own vacuum system, internal gas pressure is easily controlled by means of an adjustable leak valve.

9·2 ELECTRON MOTION

A moving electron crossing the boundary between two spaces of different homogeneous potential in a path other than normal to the boundary leaves the boundary in a new direction. An electron moving in a magnetic field, but not parallel to it, also changes its direction. This change of direction is similar to the refraction of a light ray crossing the boundary between two media of different refractive index. Light is refracted at a finite number of refracting surfaces, but the refraction of electrons is a continuous process; that is, the refractive index changes continuously along the path of the electron.

9·3 ELECTRON PATH IN ELECTROSTATIC FIELD

Figure 9·1 represents the path of an electron in an electrostatic field. The x-axis is the boundary between two spaces of constant potential. Above the axis the potential is V_1, and below it is V_2. The path of the electron is shown as the lines v_1 and v_2 representing the electron velocity in the two constant potentials. In this case V_2 is assumed to be more positive than V_1 so that the electron is accelerated and v_2 is greater than v_1. The velocity component along the x-axis is tangent to the boundary and is the same in each space:

$$v_{1x} = v_{2x} \qquad (9\cdot1)$$

$$v_1 \sin \alpha_1 = v_2 \sin \alpha_2 \qquad (9\cdot2)$$

Rearranging,

$$\frac{\sin \alpha_1}{\sin \alpha_2} = \frac{v_2}{v_1} \tag{9·3}$$

The electron velocity is given by

$$v = \sqrt{\frac{2Ve}{m}}, \quad v = k\sqrt{V} \tag{9·4}$$

so that

$$\frac{\sin \alpha_1}{\sin \alpha_2} = \sqrt{\frac{V_2}{V_1}} \tag{9·5}$$

For any electrostatic field the electron trajectory could be found, to a fair degree of accuracy, by using a finite number

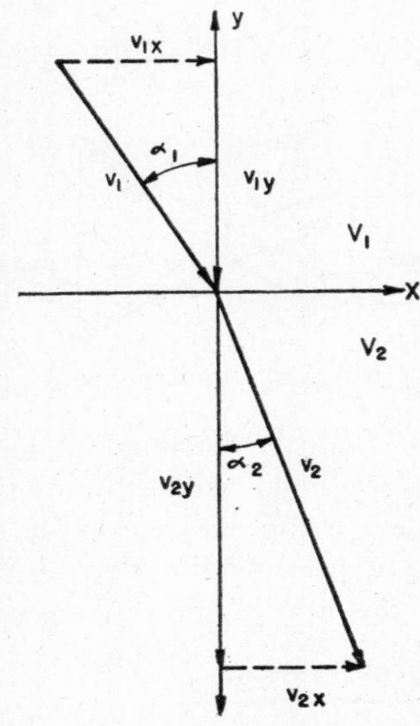

FIG. 9·1 Electron path in electrostatic field.

of equipotential lines and a repeated application of equation 9·5 at each boundary.

Figure 9·2 illustrates an electrostatic field in which the equipotential areas are parallel planes. Two electron trajectories are shown as 1 and 2 leaving point O. Let it be assumed that the potential becomes more positive from left to right.

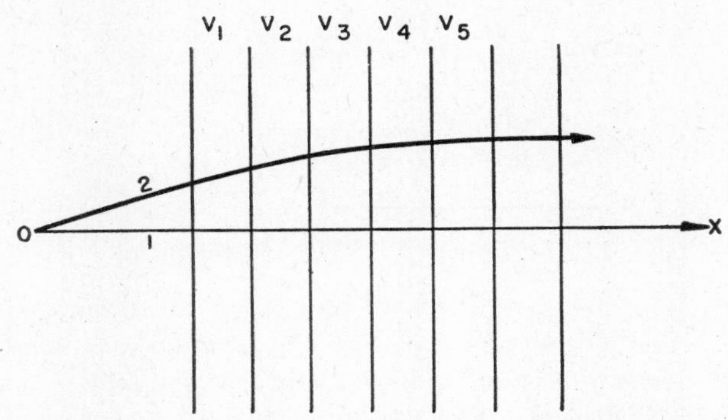

FIG. 9·2 Focusing field with equipotentials of parallel planes.

Electron trajectory 1 is a straight line normal to all equipotential planes. At each boundary trajectory 2 is bent toward the normal. For any number of equipotential planes a repeated application of equation 9·5 will show that trajectory 2 can only approach the normal, so that at any finite distance from O along the x-axis no image of O is formed. It is apparent that to form an image of point O at a finite distance along the axis, some of the equipotential planes must be curved surfaces, just as some of the light-refracting surfaces of a glass lens must be curved surfaces.

9·4 ELECTRON PATH IN MAGNETIC FIELD

An electron in motion passing into a homogeneous magnetic field has forces acting upon it, as described in Chapter 3. In Fig. 9·3 electron paths in such a field are shown. An end view of the field shows the electron paths to be circles, but because of the initial component of velocity in the x-direction (parallel to the field) the actual paths are helices. As shown

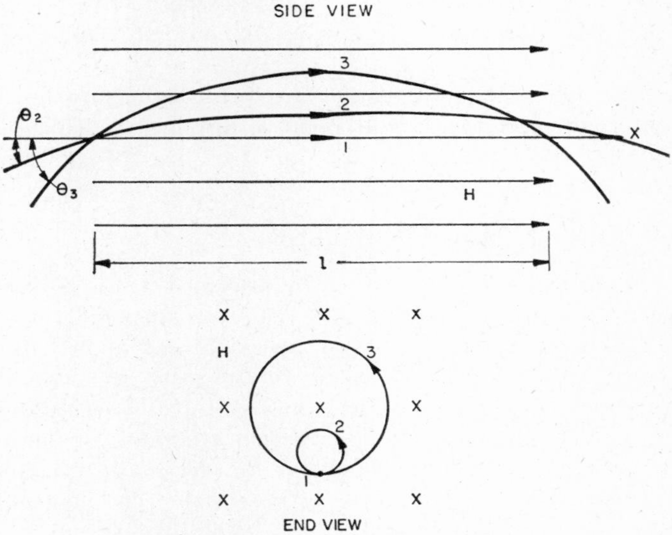

FIG. 9·3 Electron path in magnetostatic field.

in Chapter 3, the time for one revolution in the yz plane is independent of the velocity component in this plane, and is equal to

$$t = \frac{20\pi m}{He} \tag{9·6}$$

If the electrons entering this field all had the same linear velocity,

$$v_x = v \cos \theta \tag{9·7}$$

The length l (the distance from the entrance to the field to the point at which it again crosses the same zy co-ordinates) depends only upon the x component of electron velocity:

$$l = v_x t = v_x \frac{20\pi m}{He} \tag{9·8}$$

$$l = v \cos \theta \frac{20\pi m}{He} \tag{9·9}$$

The focal length of such a lens therefore is infinite and, as in the electrostatic lens, some of the field lines must be curved in order to have a lens of finite focal length.

9·5 SIMPLE ELECTROSTATIC LENS

The electrostatic lens with axial symmetry has equipotential planes which are surfaces of revolution. Because of this symmetry, the electron paths in a plane passing through the axis will explain the action.

Figure 9·4 is a plot of the equipotential surfaces and the electric field lines which are at all points perpendicular to the equipotential surfaces. The paths of three electrons are traced from object point a. Path 1 is on the axis and is a

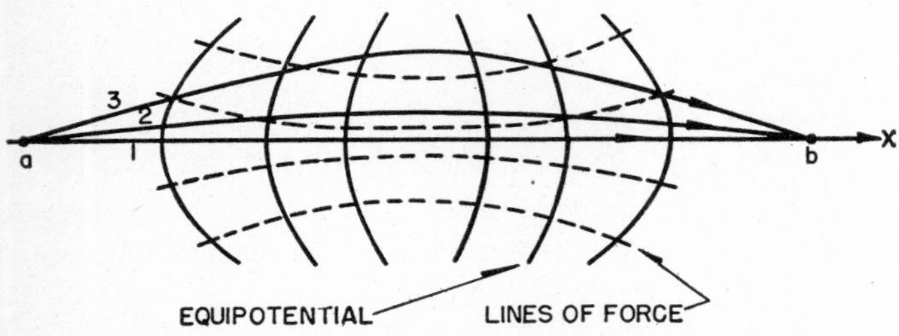

Fig. 9·4 Electrostatic electron lens.

straight line. Paths 2 and 3 are bent through different angles and meet the axis at point b, forming the image of point a.

9·6 SIMPLE MAGNETIC LENS

The magnetic lens is more easily discussed if it is reduced to a two-dimensional problem. Let it be assumed that a plane passes through the axis of an axially symmetric magnetic field, and that this plane rotates with the angular velocity of the electron under consideration. The electron path on this plane can be described in a manner similar to the electrostatic lens. In this plane the electron is refracted at boundaries of magnetic vector potential. The magnetic vector potential **A** is given by

$$\mathbf{H} = \text{curl}\,\mathbf{A} \qquad (9\cdot10)$$

For an axially symmetric field the magnetic potential is constant around a circle at right angles to the axis and with the axis as center. In the rotating plane lines of constant **A** may be drawn. The electron paths are shown in Fig. 9·5 for a magnetic lens. The paths of three electrons leaving the object point a are shown. Path 1 is a straight line, since

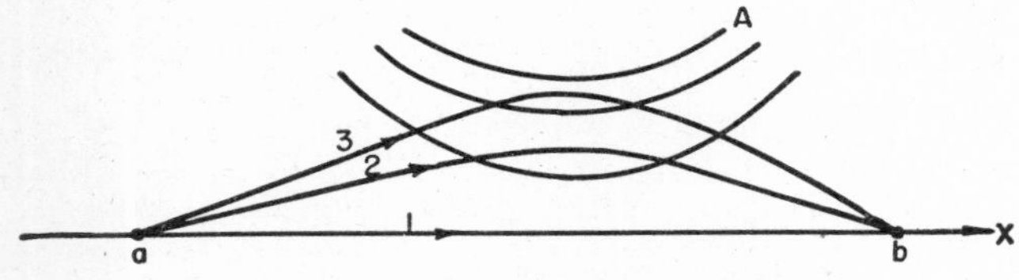

Fig. 9·5 Magnetostatic electron lens.

A is zero on the axis. Paths 2 and 3 are bent through different angles, and they meet the axis at point b and form the image of point a.

9·7 IMMERSION LENS

Figure 9·6 shows the cross section of an immersion lens. The cathode, control grid, and aperture are axially symmetric.

Since electrons are emitted from the cathode with different initial velocities they cross the lens axis at different distances from the cathode surface. Trajectories of several electrons are shown in Fig. 9·6. Although the electrons are not focused at one point on the lens axis they do pass through an area of minimum cross section. This area of minimum beam cross section is called the crossover region of the immersion lens. It is smaller in diameter than either the emitting cathode or the image of the cathode formed by this lens.

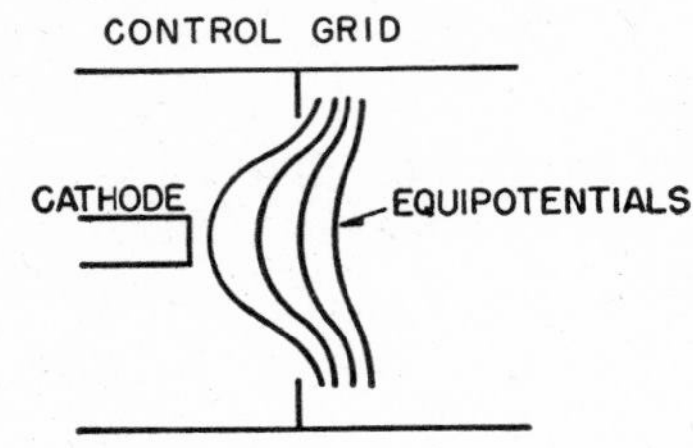

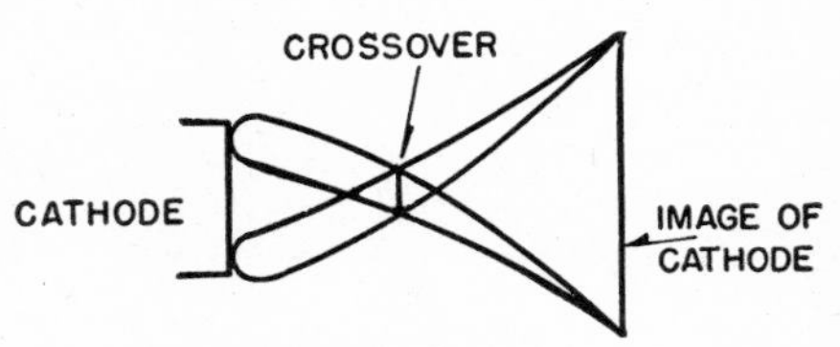

Fig. 9·6 Immersion lens.

For this reason it is usually the virtual image of the crossover region that is focused on the screen by another lens system, so that the spot size on the screen may be a minimum. In Fig. 9·7 the equipotential lines for an immersion lens are shown for two values of voltage between grid and cathode.

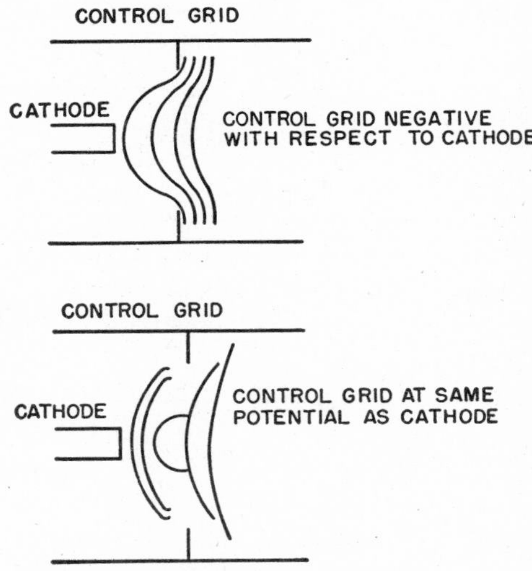

Fig. 9·7 Control of beam current by immersion lens.

Both the usable emitting area and the field at the cathode vary with this voltage; therefore the position and diameter of the crossover region vary with the grid voltage. Hence, if only the grid is modulated to vary beam current, some defocusing of the spot on the screen must result.

9·8 BIPOTENTIAL LENS

Figure 9·8 gives the equipotential plot for a bipotential lens formed of two cylinders. The focal length of this lens depends upon the geometry or ratio of cylinder diameters, and upon the ratio of voltages applied to the two cylinders.

To reduce distortion in the image produced by this lens, apertures are sometimes inserted in the cylinders. These apertures limit the beam angle admitted to the lens. Because the maximum displacement of electrons from the lens

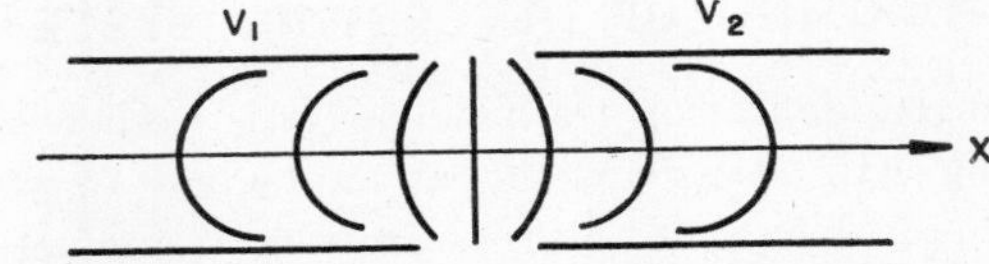

FIG. 9·8 Bipotential cylindrical electrostatic lens.

axis is reduced, some of the aberrations of the lens are reduced.

9·9 MAGNETIC LENS

Figure 9·9 shows two types of magnetic lenses. The first is a short air-core coil. To concentrate the field, coils are sometimes clad in soft iron, a small gap being left along the inside circumference of the iron casing. The magnetic lens rotates the image. Although this has no effect in focusing a

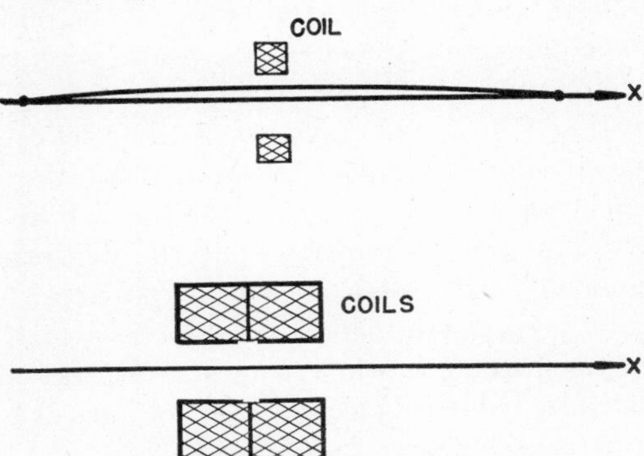

FIG. 9·9 Magnetic lenses.

circular object, it may be objectionable in other uses. A double coil may be used as shown in Fig. 9·9, where the fields are opposed and the rotation of the image by the first section is counterbalanced by an equal and opposite rotation of the second section of this iron-clad magnetic lens.

9·10 GAS FOCUSING

Electrons moving with sufficient velocity through a gas produce ionization. Slow-moving positive ions produce a focusing effect on the electrons, causing them to concentrate about the positive-ion core. Figure 9·10 shows two types of

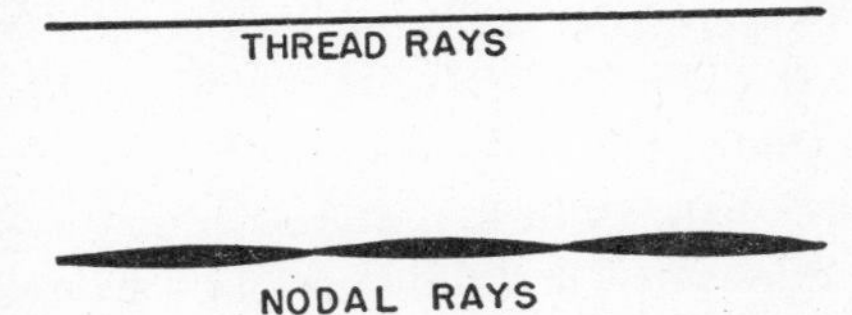

FIG. 9·10 Electron paths in gas focusing.

concentration. With nodal rays the lens effect produces successive images of the cathode. The thread rays resemble a narrow beam of light. The presence and type of focusing effect depend upon electron-beam density and velocity and the type and pressure of the gas. A finite time is required to reach equilibrium in which the rate at which positive ions are lost from the beam is equal to the ionization rate. Because of this time lag, gas-focused tubes are not suitable if rapid beam deflection is essential. Because of these difficulties, modern CR tubes are all of the high-vacuum type.

9·11 ELECTRON GUN

The device which forms and controls an electron beam is called an electron gun. Two types of electron guns are shown: the hot-cathode electrostatically focused type in Fig. 9·11, and the cold-cathode magnetically focused type

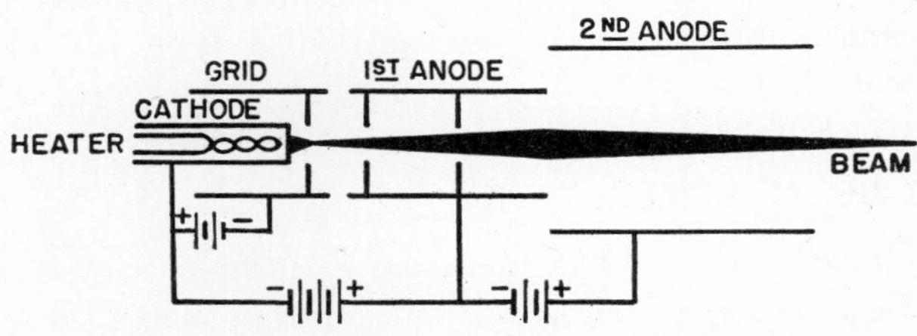

FIG. 9·11 Electron gun (used in sealed-off tube).

in Fig. 9·12. The electron gun shown in Fig. 9·11 uses an oxide-coated indirectly heated cathode as the electron source. The indirectly heated cathode is used because an axially symmetric emitting surface is desired. The filament is insulated from the cathode cup by means of a ceramic coating on the filament wire. It is desirable to use alternating current for heating the filament and, since varying magnetic fields in the region of the cathode would modulate the beam, the filament is constructed with little loop area. This is accomplished by first folding the coated filament wire into a hairpin and then coiling the hairpin. The cathode is usually a nickel cup, coated on the outside with the emissive mixture. The grid and cathode form an immersion lens, and the grid-cathode voltage is used to control the beam current. The first and second anodes form a bipotential lens which focuses the virtual image of the crossover region on the screen. The aberrations of the bipotential lens limit the size of the aperture which can be used. To restrict the angle of the beam admitted to this lens several apertures are added as shown. To obviate defocusing by modulation of grid voltage, either of two methods may be used. In one method the image of a small aperture may be focused on the screen. In the other

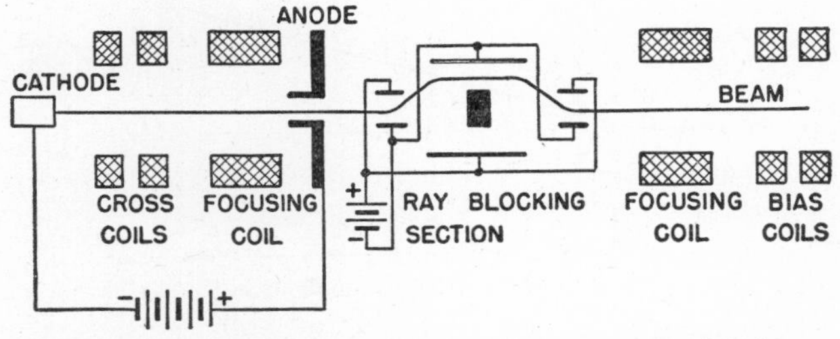

FIG. 9·12 Electron gun (used in demountable tube).

method a part of the modulating voltage is introduced between first and second anode and the focal length of the bipotential lens is changed to keep the crossover region in focus on the screen as the grid is modulated. This type of electron gun is used in many sealed-off CR tubes.

The electron gun shown in Fig. 9·12 uses a cold cathode because of the high accelerating voltage used. The cathode is an axially symmetric aluminum plug. Emission is caused by field emission and positive-ion bombardment. Focusing is done by means of magnetic fields, and the anode has a restricting aperture. As previously stated, gas pressure is varied to control beam current. This type of gun is used in demountable CR tubes requiring their own vacuum system, and gas pressure is adjusted by means of a leak valve. Instead of a grid, a ray-blocking section, which allows the beam to strike a target in the gun or to emerge from the gun, is used. In action this section is similar to a beam switch.

Such a large number of gun structures are in use that operating information on any particular type should be obtained from the CR tube manufacturer.

9·12 BEAM DEFLECTION

The beam must be deflected from its normal position if it is to be used as a writing element. Either electrostatic or magnetostatic fields may be used to deflect the beam.

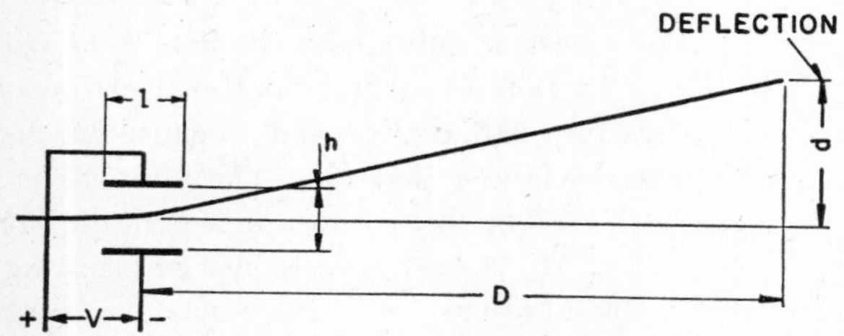

Fig. 9·13 Electrostatic deflection.

Electrostatic deflection is shown in Fig. 9·13. If it is assumed that there are no stray fields and no edge effects, deflection d is given by

$$d = \frac{1}{2}\frac{V}{E}\frac{l}{h}D \tag{9·11}$$

where V is the potential difference between the plates and E is the voltage through which the electron has fallen before entering the deflection system. Greater sensitivity may be obtained with curved plates, because they can be placed closer to the limiting beam positions, and hence produce a higher deflecting field for the same applied voltage.

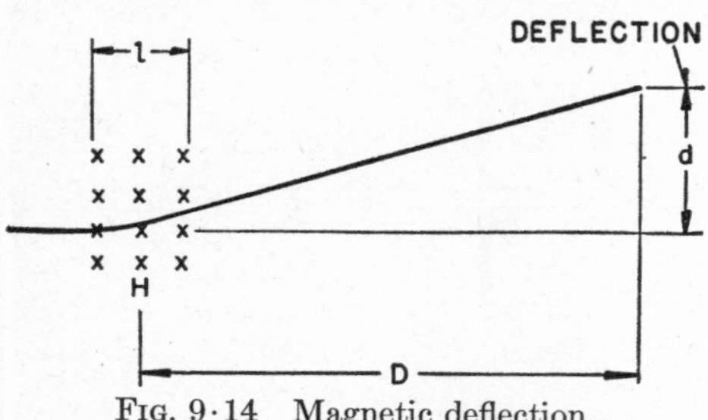

Fig. 9·14 Magnetic deflection.

Magnetic deflection is shown in Fig. 9·14. The magnetic field is constant and perpendicular to the plane of the paper over region l. The deflection d is given by

$$d = \frac{DHl}{\sqrt{2E\frac{e}{m}}} \tag{9·12}$$

where E is the potential difference through which the electron has fallen before entering the deflection system and H is the magnetic-field intensity.

9·13 DEFOCUSING OF DEFLECTING SYSTEMS

For magnetic deflection the field never is exactly uniform, and fringing will always occur at the edges of the field. Since the beam has a finite cross section, the effects of the deflecting field depend upon the position of the particular electron under consideration in the beam. With specially designed deflecting-field coils, defocusing can be made negligible.

Electrostatic deflection also may cause defocusing. For unbalanced deflection, with one deflection plate tied to the second anode and the potential of the other plate varied above and below this potential, the defocusing effect is large, and therefore the method is not generally satisfactory. For balanced deflection the potential midway between the deflecting plates is kept always at the beam potential. To produce deflection the potential of one plate is raised above beam potential, and the potential of the other plate is lowered by an equal amount below the beam potential. The resultant defocusing is the difference between the edge effects at the entrance and exit regions of the deflecting-plate system, which counteract each other so that the total defocusing effect of the balanced electrostatic system is small. This is based on the assumption that the deflecting plates are mechanically symmetrical. It is possible to design mechanically asymmetric deflecting systems which produce with unbalanced electrostatic voltages deflections similar to that produced by balanced electrostatic voltages in a mechanically symmetric system.

9·14 SCREENS

Translation of beam energy into a visible effect may be accomplished in several ways. A photosensitive surface such as a photographic film may be placed in the CR tube as a screen and be exposed to a trace of the beam. Removal of the film and subsequent development yield a permanent record of the intensity and position of the beam during the exposure. To allow insertion and removal of the film, the CR tube must be demountable. Such tubes are used for recording rapid non-recurrent electrical phenomena. Figure 9·15 is a photograph of such a tube with its associated equipment, showing the ports for insertion of the film drum or plates. For most uses an immediate visible effect on the screen is desired. This effect is produced by making the screen of a material that produces visible light when excited by the electron beam. The screen material may produce visible light by incandescence or luminescence. Emission of light from a material at low temperature is called luminescence. Luminescence of a material during excitation is called fluorescence; after removal of excitation it is called phosphorescence. Solid materials exhibiting fluorescence and phosphorescence are called phosphors. Some impurity, called an activator, is essential to the performance of some phosphors. Of the large number of phosphors known, zinc sulphide and zinc orthosilicate are widely used. Zinc orthosilicate is known as willemite. The decay of light after removal of the excitation is an important property of phosphors, for it determines what sort of modulation the light output of the phosphor can follow. Screen characteristics such as spectral characteristics of the emitted light and decay time (called persistence) may be varied by the composition of the phosphor. The glass surface on which the screen phosphor is placed is usually curved. An illuminated spot on the screen then directly illuminates the rest of the screen, causing background illumination. Some of the emitted light is reflected between the two glass surfaces, and halation is produced. Both effects limit the contrast which can be obtained. Figure 9·16 shows a CR tube with a 5-inch screen and all-electrostatic deflection. To improve the light contrast, tubes have been built with screens that can be viewed from the sides on which they are excited. The screen may then be made flat and suspended in the tube. Light output is up to 100 percent greater if the screen is viewed on the excitation side. However, because the gun must be mounted at an angle other than 90 degrees to the screen to allow direct viewing, complicated beam-deflection circuits result.

Fig. 9·15 Westinghouse CR high-voltage oscillograph.

For the production of large-screen pictures in television a small picture of high intensity might be formed in the CR tube and then projected on a large screen. If the required screen intensity is not too high, phosphors may be used for the screen material. As beam-current density is increased the light output is not increased proportionally; this increase indicates saturation. Furthermore, the high-energy beam has a destructive effect, and screen life is shortened.

If incandescence of the screen is used as the method of generating light, much higher beam intensities can be used. A thin metal-foil screen or a screen composed of lampblack has been proposed for this use. Various problems, such as the extremely thin foil needed for definition and the high beam powers needed, may prove such methods to be impractical.

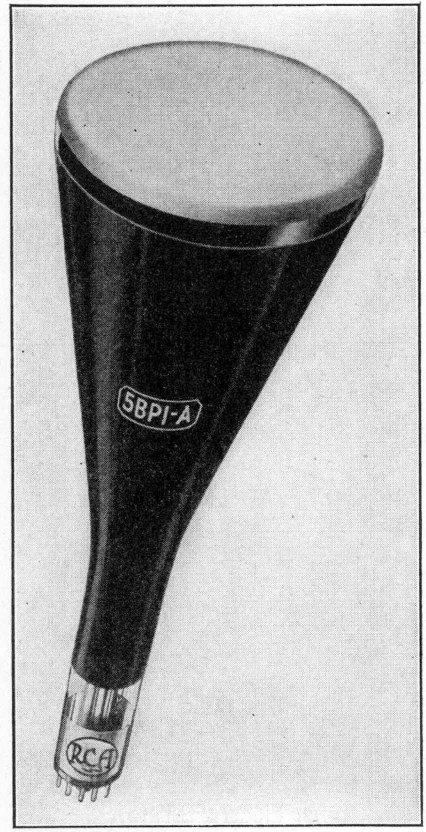

Fig. 9·16 CR tube with electrostatic deflection. (Courtesy of Radio Corporation of America.)

9·15 TRANSIT TIME

The deflecting force has been assumed to be constant as the electron traverses the length of the deflecting system. As the frequency applied to the deflecting system is increased the transit-time effect becomes noticeable, and deflection sensitivity decreases. It depends upon electron velocity and hence accelerating voltage, and upon the length of the deflecting system. The transit time effect must be considered only at higher frequencies, say from 10 to 20 megacycles upward.

9·16 RATINGS

High-vacuum CR tubes with translucent-phosphor screens are made in sizes ranging from 1 inch to over 20 inches in diameter and using from several hundred volts to over 10 kilovolts accelerating voltage. They may be constructed for all-electrostatic deflection, all-magnetic deflection, or a combination of both. Furthermore, the beam may be focused electrostatically or magnetically. Further variations include spectral characteristics and persistence of the screen. Manufacturers' ratings should be consulted to obtain such data as maximum safe accelerating voltage, various gun voltages to obtain focus, deflection sensitivity, and modulation characteristic.

9·17 APPLICATIONS; OSCILLOSCOPES AND OSCILLOGRAPHS

Since the electron beam can be readily deflected electrically and since it leaves a trace that persists a short time, it is well adapted to visual reproduction of the variations of electrical quantities; and since it is easy to convert other forms of energy to variations of an electrical current or field, the possible applications are numerous.

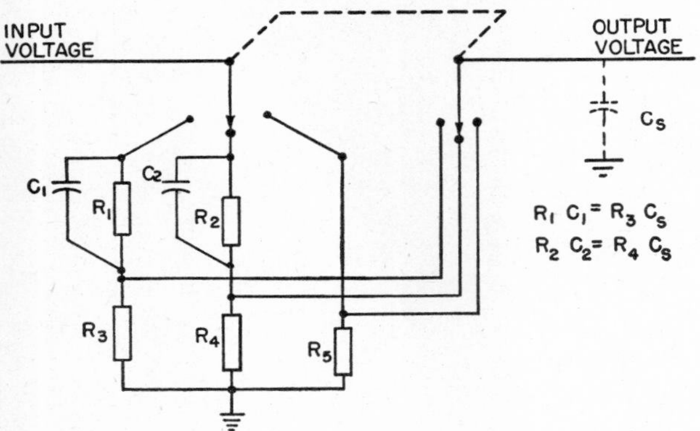

Fig. 9·17 Attenuator compensated for stray capacitance.

In such applications, electrostatic deflection of the beam is most common. The voltages required for 1 inch of deflection range from 50 to 700 volts, depending upon the accelerating voltage and other instrument constants. The voltages to be recorded range from millivolts to millions of volts. Hence, on the one side considerable amplification is often necessary, whereas on the other side high-ratio voltage dividers must be employed.

Since this amplification must be obtained without distortion of the input signal, the gain of the amplifiers must be constant, and the phase shift must be zero or proportional to frequency, over a suitable bandwidth. The bandwidth depends upon the frequency spectrum of the signal observed. Most commercial oscilloscope amplifiers have constant gain and linear phase shift from about 30 cycles per second to 50 or even several hundred kilocycles per second. For study of rapid signals of irregular wave shapes, a good response up to 5 megacycles or higher may be necessary.

For study of signals with d-c components, direct-coupled amplifiers are necessary, and they are provided in some commercial instruments.

The output voltage required of an amplifier is usually about that required for full-scale beam deflection. However, the input voltage applied may vary from a small fraction of a volt to several hundred volts; therefore suitable gain controls and attenuators must be provided, and they must have suitable amplitude and phase versus frequency response. This is usually obtained by the use of resistance-capacitance attenuators, in which the upper resistor of the divider is

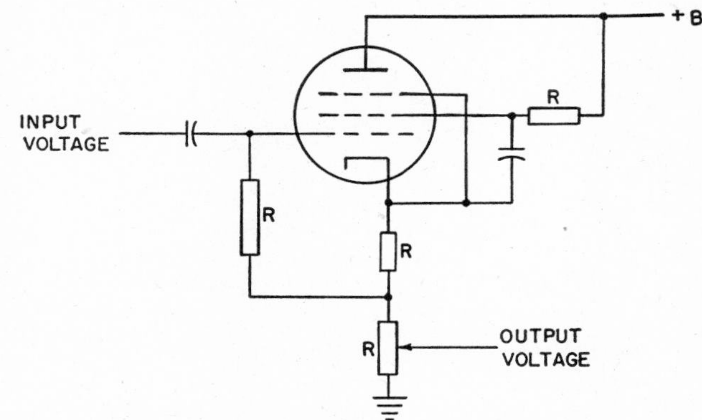

Fig. 9·18 Cathode-coupled input and gain-control stage.

shunted by a capacitor of such value as to compensate for the stray capacitance to ground from the switch arm by making the *RC* products equal. Such an attenuator is shown in Fig. 9·17.

For continuous gain control, a cathode-follower input stage can be used, with a comparatively low-resistance potentiometer for a load. Thus a high input impedance is presented, yet the resistance to ground from the gain-control contact is kept small, and thus the effect of stray capacitance to ground at higher frequencies is minimized. Such an input stage is shown in Fig. 9·18.

The CR tube is used mainly to study the variation of one quantity as a function of another. Both of these may be related signal voltages, but often it is desired to study the time variation of one signal only. For this purpose a sweep circuit is usually provided to deflect the beam from left to

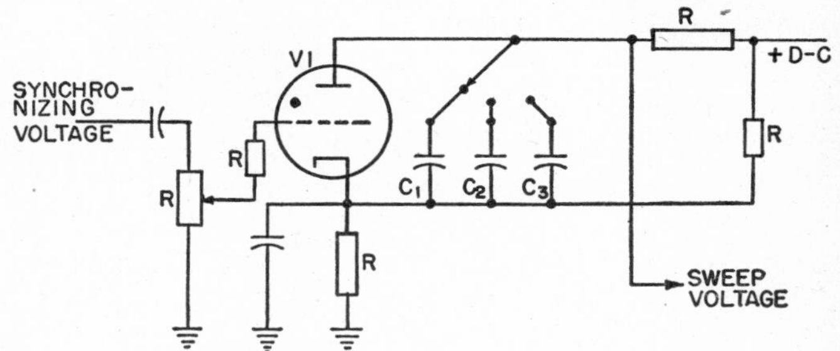

Fig. 9·19 Gas-tube time-base generator.

right, and the signal to be studied provides a vertical deflection. The time base should be adjustable in frequency, linear, and capable of being synchronized to the signal under study so as to give a stationary trace. A suitable time base may be obtained by means of a relaxation gas-tube oscillator of the type shown in Fig. 9·19. In this figure, capacitor C_1 is charged from a d-c source through resistor R until its voltage exceeds the breakdown voltage of thyratron $V1$. It then discharges rapidly through the thyratron. By adjusting the circuit so that only the initial part of the complete exponential charging curve is used, a linear forward sweep and rapid retrace can be obtained, as shown in Fig. 9·20. To lock in the sweep and signal under study, a small portion of the signal can be applied to the thyratron grid.

The gas-tube time base is useful at a sweep rate up to about 30 kilocycles. At higher frequencies the deionization time of the tube makes it useless. For higher sweep rates, some form of vacuum-tube sawtooth generator, such as that of Fig. 9·21 is necessary. Sweep is provided by the linear rise of voltage on C_2 as it charges at constant current through the pentode $V1$ from a d-c source. A rapid discharge takes place when the impulse shown drives the grid of triode $V2$ positive. Frequency is determined by the frequency of the impulse on the triode grid. This impulse must be derived from a suitable vacuum-tube impulse generator, and the sweep can be locked in by using some of the signal voltage to synchronize this oscillator.

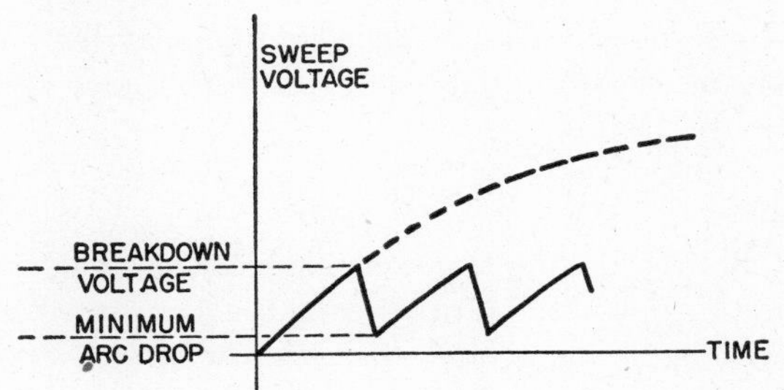

Fig. 9·20 Voltage output of gas-tube time-base generator.

The circuits described provide a continuously recurrent sawtooth, or linear, time base. If a very short impulse recurring at relatively long but regular intervals is to be studied, a sweep with a long rest period is desirable. For such purposes the vacuum-tube generator shown in Fig. 9·22 can be used. A d-c supply charges the capacitor C_2 through triode $V1$, also supplying current through pentode $V2$. Under this rest condition, the capacitor voltage is constant. The sweep is produced by applying a negative rectangular impulse to the triode grid. This impulse blocks the grid and allows the capacitor to discharge at constant current through pentode $V2$. The linear fall of potential at the anode of the pentode then gives the sweep. The rectangular tripping impulse can be applied at regular intervals or at will, one at a time, in which case a distinct single-sweep time base is obtained.

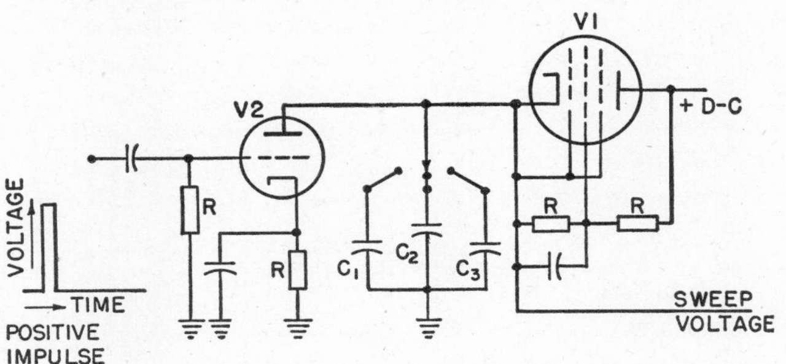

Fig. 9·21 Vacuum-tube time-base generator.

With all these methods the faint retrace as the beam returns from right to left is objectionable. In modern oscilloscopes it is eliminated by applying a "blanking" impulse to the CR-tube grid during the retrace. This impulse cuts the beam intensity to zero during the retrace.

For some applications a circular or spiral sweep may be useful. A circular sweep can be obtained by means of the phase-splitting circuit of Fig. 9·23. A sine wave is applied to a resistance-capacitance phase shifter, and the displaced voltages are applied to the CR-tube plates. If linear sawtooth modulation is applied to the input sine wave, a spiral sweep is produced. The linear increase must start at exactly the same point of the circular sweep at each cycle. The signal can be studied by using it to modulate the sweep voltage through a linear modulator. A radial deflection is produced.

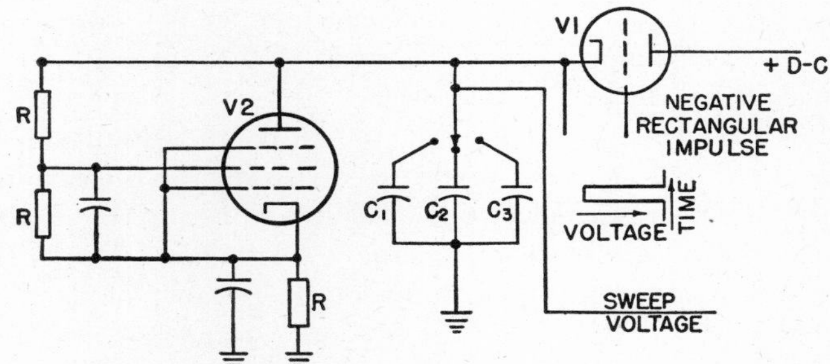

Fig. 9·22 Vacuum-tube time base with long rest period.

The CR tube is also used in plotting curves directly in Cartesian co-ordinates. For example, by use of a small alternator at the crankshaft and a pressure pick-up connected to the cylinder head, the pressure-volume diagram for a high-speed internal-combustion engine can be shown on the CR tube. It is necessary only to apply the alternator voltage, properly phased, to the horizontal plates, and the pressure pick-up voltage to the vertical plates.

The tube can even be used to plot families of curves by using grid control of the beam intensity. For example, unsynchronized sinusoids can be applied to the grid and anode of a vacuum tube, and thus cause its anode current to vary over the grid voltage-anode voltage plane. Then, if a voltage proportional to the anode current is used to excite a circuit producing short impulses for certain discrete values of its input voltage, these impulses can be used to brighten the CR-tube beam for definite values of the anode current. The beam would be cut off for all other values. Then, if the anode voltage of the tube is applied to the horizontal plates of the CR tube, and its grid voltage to the vertical plates, the dots produced on the screen show curves of con-

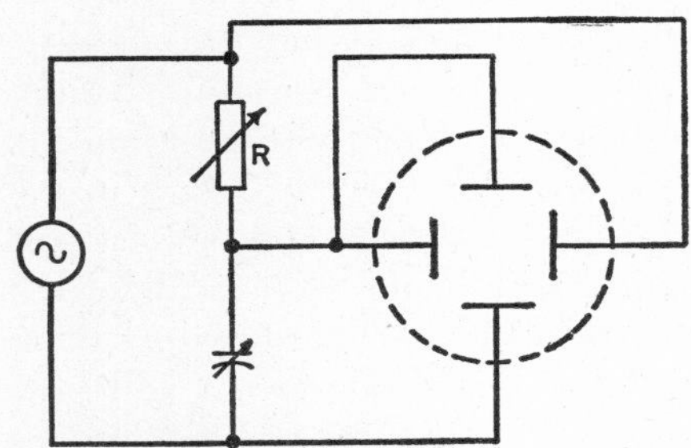

Fig. 9·23 Circuit for circular sweep on CR tube.

stant anode current on the grid voltage-anode voltage plane. Such use of the CR tube is sometimes referred to as "stroboscopic" beam modulation.

A variation of this technique is described in detail in a paper by Douma and Zjlstra.[4] Watson-Watt[5] seems to have been the first to use the "stroboscopic" control of cathode-ray beam intensity.

This last application is only one example of the use of beam-intensity modulation, or z-axis modulation. It can be used also to produce timing dots on a time plot of a signal. Thus, the exact duration of any phenomenon under study can be ascertained. Time intervals of 0.1 miscosecond or less can be measured in this way.

The CR tube can be used to study the simultaneous time variation of two quantities by using an "electronic switch," which is an electronic circuit connecting the oscilloscope input alternately to the two signals, switching from one to the other at a rapid rate. Because of the persistence of the CR-tube trace, both quantities are then plotted simultaneously.

CR tubes can be constructed with more than one beam for study of several simultaneous variables. One way of achieving this is to split a single beam vertically by a thin plate, the two beams being then deflected by a common set of horizontal plates, but by two separate sets of vertical plates.[6]

The trace on the screen of a sealed-off CR tube can be made to give a permanent record by photographing it directly. For such work a blue screen produces good results. In this way, records can be made of electrical transients and signals far beyond the range of the older string oscillograph. For most visual work, a green screen gives best results, because of the greater sensitivity of the eye to that color.

Although most commercial oscilloscopes use electrostatic deflection systems, it is possible to use magnetic deflection. In that case, the inductance of the deflection coils must be considered. To produce a sawtooth wave of current through the horizontal deflection coils, for example, a rectangular voltage must be used across them, and the resistance in the circuit must be low enough so that the current rises linearly over the sweep time. If coil resistance is appreciable, the applied voltage must rise slightly to maintain a linear current rise through the coils.

In general, magnetic deflection is not favored because of the varying impedance of the coils for different frequencies, and because of the phase shift produced by their inductance. They also require appreciable current and power from the deflection source. In large high-voltage tubes they have some advantages, for the voltages for electrostatic deflection then become quite large. Hence, magnetic deflection is common in large television CR tubes.

To power the amplifiers and sweep generators described, as well as to supply the high accelerating anode voltages needed on the CR tube, d-c supplies must be incorporated in an oscilloscope. These should be well filtered and regulated to insure stable amplifier gain and sweep frequency and to maintain constant deflection sensitivity and good focusing.

Since the cathode beam requires little current, radio-frequency supplies have been used for this purpose, the high potentials being obtained by resonance. Such supplies are small, they are easy to filter, and they remove most of the hazards associated with 60-cycle rectifiers at high voltage. They are especially adapted to use in home television receivers.

Television

The CR tube is important in television systems, where it provides the means of reproducing the transmitted image. The transmitted signal and the receiver circuits cause the CR beam to scan the screen from left to right. The beam starts at the upper left and, after reaching the bottom, returns to the upper left to scan between the lines previously traced. This "interlacing" reduces flicker on the screen. The transmitted image is reproduced by intensity modulation of the CR beam. During the retrace, when the beam travels from right to left, its intensity is cut to zero by a blanking signal on the grid. For television service, white screens are preferred to the green willemite screens generally used in oscilloscopes. Special CR tubes have been developed with red, blue, and green screens for an experimental "all-electronic" color television system.

REFERENCES

1. *Electron Optics in Television*, I. G. Maloff and D. W. Epstein, McGraw-Hill, 1938, pp. 190–195.
2. *Ibid.*, pp. 196–200.
3. *Ibid.*, pp. 208–210.
4. *Philips Tech. Rev.*, Vol. 4, Feb. 1939, p. 56.
5. *Application of the Cathode-Ray Oscillograph in Radio Research*, R. A. Watson-Watt, J. F. Herd, and L. H. Bainbridge-Bell, London, His Majesty's Stationery Office, 1933.
6. *Time Bases*, O. S. Puckle, Wiley, 1943, Appendix I, p. 141.
7. *Electron Optics*, L. M. Myers, Van Nostrand, 1939.
8. *Principles of Television Engineering*, D. G. Fink, McGraw-Hill, 1940.
9. *Electron Optics*, O. Klemperer, Cambridge Univ. Press, Macmillan Co., 1939.
10. *Television*, Vol. II, RCA Institutes Technical Press.
11. *Fluorescent Materials Used in Lighting and Elsewhere*, *U. S. Dept. of Interior*, I.C. 7276.
12. *Ultra-High-Frequency Techniques*, J. G. Brainerd, Glenn Koehler, H. J. Reich, and L. F. Woodruff, Van Nostrand, 1942.

Chapter 10

ULTRAVIOLET RADIATORS

H. C. Rentschler

VISIBLE radiation includes only about one octave of the complete electromagnetic spectrum (Fig. 7·3) and ranges in wavelength from 4000 to 8000 angstrom units. The angstrom unit is the customary unit of wavelength for this part of the spectrum, and is equal to 10^{-8} centimeter.

For convenience the visible spectrum is often broken up into sections identified by color or by wave bands: red to cover the range from about 8000 to 6500 angstrom units; violet from about 4500 to 4000; with orange, yellow, green, and blue intervening. The band of radiation extending from 4000 to about 2000 angstrom units is commonly designated as the ultraviolet region. To be more precise, the ultraviolet region extends to approximately 130 angstrom units. Air absorbs radiation below 2000 angstrom units so greatly that the region below 1700 angstrom units must be investigated in a vacuum or a gas such as hydrogen, which does not absorb the radiation. The region below 1700 angstrom units is known as the Schumann region of the ultraviolet.

Because these short waves are absorbed by air the practical uses of ultraviolet are limited to waves longer than 1700 angstrom units; therefore, the sources of radiation considered in this chapter are confined to the ultraviolet band from 4000 to 1700 angstrom units.

As with visible light, it is often convenient to subdivide ultraviolet into arbitrary regions: the "near" ultraviolet extending from 4000 to about 3200 angstrom units; the "middle" ultraviolet from 3200 to 2800; the "short" ultraviolet from 2800 to 2000; and a band covering radiations shorter than 2000 angstrom units. Radiations in the near ultraviolet produce fluorescence in certain minerals and organic compounds, and are used for many other applications. Middle ultraviolet causes sunburns and tanning, is effective for prevention and cure of rickets, and produces vitamin D. Short ultraviolet is bactericidal and also produces fluorescence in the minerals and compounds used in fluorescent lamps. Radiation shorter than 2000 angstrom units is largely absorbed by the oxygen in air; then the oxygen is converted into ozone.

10·1 PRODUCTION OF ULTRAVIOLET RADIATION

The intensity of radiation in the middle and short ultraviolet bands from incandescent filament lamps is too low to be of any practical use. Open arcs or spark discharges between metal electrodes or between carbon electrodes impregnated with metallic salts that produce spectral lines rich in ultraviolet have been used with some success. Such light sources are variable in output, and for most commercial applications they have been superseded by electrical discharges through mercury vapor in evacuated ultraviolet transmitting tubes containing inert gas or gas mixtures

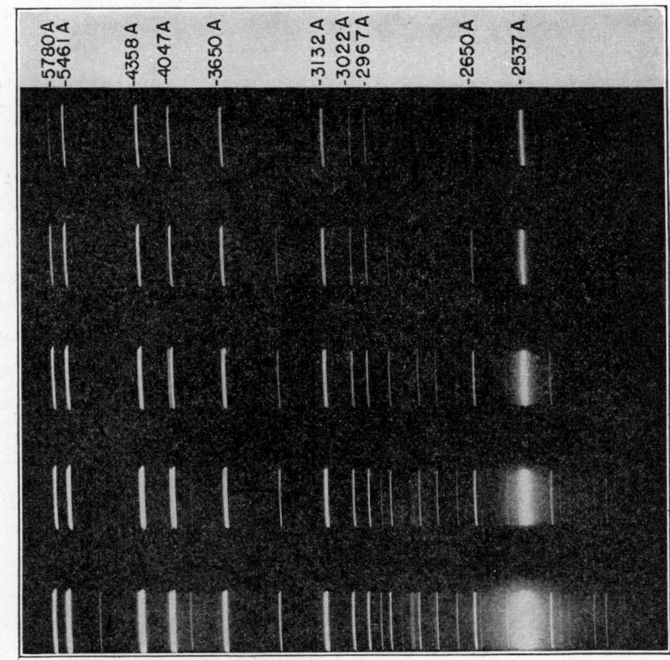

Fig. 10·1 Spectrograms of low-pressure mercury-vapor discharges in quartz tubes.

together with mercury. Such discharges through mercury vapor produce strong lines of radiation in the different ultraviolet regions (Fig. 10·1). The discharge also produces strong radiations between 1800 and 2000 angstrom units. These radiations are absorbed by air and by the emulsion of a standard photographic plate, and consequently are not shown in the spectrograms.

The spectrograms show that a large portion of the radiated energy is in the wave band around 2537 angstrom units. At a higher temperature the mercury-vapor pressure becomes appreciable and the 2537-angstrom radiation is absorbed. The degree of absorption depends upon the density of the mercury vapor, and the intensities of the other lines increase relative to the 2537-angstrom radiation. Table 10·1 shows the relative intensities of the different mercury lines: the intensity of the 3650-angstrom line is taken as 100; for discharges at pressures of a few microns of mercury, 200, 400, and 800 millimeters of mercury pressure respectively.

The relative intensities for 200, 400, and 800 millimeters are the values given by Lux and Pirani.[1] The values for pressure in microns are from measurements made on low-pressure discharges through a mixture of an inert gas with mercury vapor in a quartz tube.

TABLE 10·1 EFFECT OF VAPOR PRESSURE ON THE RELATIVE INTENSITIES OF MERCURY LINES

Wavelength		Pressure (millimeters)		
(angstroms)	Microns	200	400	800
2,537	6,000	38.0	29.0	15.7
2,650	180	21.3	22.4	24.4
2,804	6	10.2	11.3	12.7
2,967	31	15.2	14.8	16.2
3,022	15	29.6	31.4	33.0
3,341	7	7.4	8.1	8.7
3,650	100	100.0	100.0	100.0
4,050	118	45.0	36.0	32.0
4,358	336	68.2	55.3	49.5
5,461	197	80.9	68.2	64.2
5,770–90	42	71.3	75.6	79.1
10,150		32.8	33.4	37.1

10·2 LOW-PRESSURE ULTRAVIOLET LAMPS

Table 10·1 shows that the ultraviolet generated by a low-pressure discharge through mercury vapor is almost entirely confined to the resonance wave band around 2537 angstrom units. Gates[2] found that the bactericidal action of ultraviolet is limited to radiations shorter than 3000 angstrom units and is maximum at approximately 2660 angstrom units. Fluorescent powders such as zinc beryllium silicate, magnesium tungstate and cadmium borate are excited most efficiently by the resonance radiation of mercury. A low-pressure discharge through mercury vapor therefore provides a suitable means for generating bactericidal and fungicidal radiation or for exciting fluorescent powders used as coatings in fluorescent lamps.

Low-pressure bactericidal ultraviolet lamps are of either the cold-electrode or the hot-cathode type. In the cold-electrode type two hollow metal cups or cylindrical electrodes fastened to leads are sealed into the opposite ends of a long slender glass tube which transmits the ultraviolet. The lamp is exhausted and baked, and the electrodes are degassed. The tube is then filled with an inert gas or mixture of inert gases at a pressure of a few millimeters and a small amount of mercury; it is then sealed off. A small transformer supplies the discharge current through the lamp. The ultraviolet output and the efficiency of the lamp are controlled by several factors.

The Ultraviolet Transmission of the Glass

A special glass is required because ordinary lime or lead glasses are opaque to radiation at 2537 angstrom units. The transmission of these special glasses deteriorates rapidly during the first 100 or 200 hours, so that the ultraviolet output of a lamp drops as much as 15 to 20 percent in this time. Fused quartz has a higher transmission than any glass at this wavelength, but quartz transmits radiations shorter than 2000 angstrom units, and an undesirable production of ozone results. Ultraviolet lamps in special 96 percent silica glass produce about 40 percent more 2537 angstrom radiation than regular ultraviolet-transmitting glasses, with negligible solarization and almost no generation of ozone.

Length and Diameter of Lamp Tube

In a discharge lamp the power consumed consists of two parts: power lost at the electrodes (mainly because of cathode fall of potential at the negative electrode), which produces no ultraviolet or visible light; and power consumed in the discharge through the gas and vapor, which is determined by the current and the potential drop in the positive column. The cathode fall of potential for a cold electrode is relatively high, as a long slender tube is used to permit a low-current high-voltage discharge for efficient operation.

Electrodes

The power wasted at the electrodes (mainly at the cathode) varies with the size of the electrodes and the electrode material. Too small an electrode surface produces abnormal cathode drop and causes electrode disintegration. The cathode fall is lower for some metals than for others. Cathodes coated with alkaline-earth carbonates decomposed during exhaust have a low cathode fall but are unstable. Bare iron or other common metal electrodes have been used in some cold-cathode tubes (in sign lighting, for example) where efficiency is not too important. In bactericidal lamps, however, the electrode is of prime importance, and recent developments indicate that metallic thorium may improve electrodes as much as Vycor has improved glass. Although the cathode fall is somewhat higher than for alkaline-earth-coated electrodes, metallic thorium or its equivalent is more stable, and a discharge between such electrodes is constant during the life of a lamp.

Gas and Gas Pressure

Ultraviolet radiation from a discharge through an inert gas or a mixture of gases with a small amount of mercury vapor is produced almost entirely by the mercury vapor. The inert gas helps to start and maintain the discharge. To prevent clean-up of the inert gas by the discharge, argon or a mixture of inert gases at a pressure of about 20 millimeters, together with a small amount of mercury, is commonly used in cold-electrode lamps.

Lamp Temperature

At temperatures above 50 degrees centigrade the mercury-vapor density is sufficient to cause appreciable absorption of 2537-angstrom radiation. Efficient bactericidal lamps should operate at a relatively low temperature (only a few degrees above normal ambient temperature).

In a hot-cathode lamp the discharge tube is relatively short and has a larger diameter than the tube of the cold-electrode lamp. Tungsten filaments coated with alkaline-earth carbonates are sealed in the opposite ends of the tube. The lamp is exhausted and baked, and the coatings are decomposed. The tube then is filled with argon, at a pressure of about 4 millimeters, and a small amount of mercury. The filaments are connected in series through a bimetallic thermal relay which by its temperature lag allows the filaments to become preheated to electron-emitting temperature before it initiates the arc between the electrodes when the bimetallic relay breaks the connection between the filaments. A ballast in series with the lamp limits the current through the discharge, if the lamp is designed to operate on commercial line voltages. Bombardment of the filaments by the discharge keeps sections of the filament at emitting temperature during the operation.

The cold-electrode lamp has the advantage of instantaneous starting, it will operate at lower ambient temperature, it has longer life, it is more rugged, and its life is not affected by the number of times the lamp is turned on and off. The hot-cathode lamp operates on a lower voltage but, because of disintegration of the active material on the electrodes, it is susceptible to damage by frequent starting and line-voltage fluctuations. Its life is dependent upon the life of the electrodes and not upon the life of the container.

Four cold-electrode commercial ultraviolet lamps are shown in Fig. 10·2. They have discharge lengths between electrodes of 30, 20, 10, and 4 inches. Figure 10·3 shows similar hot-electrode ultraviolet lamps of 30-, 15-, and 8-watt sizes.

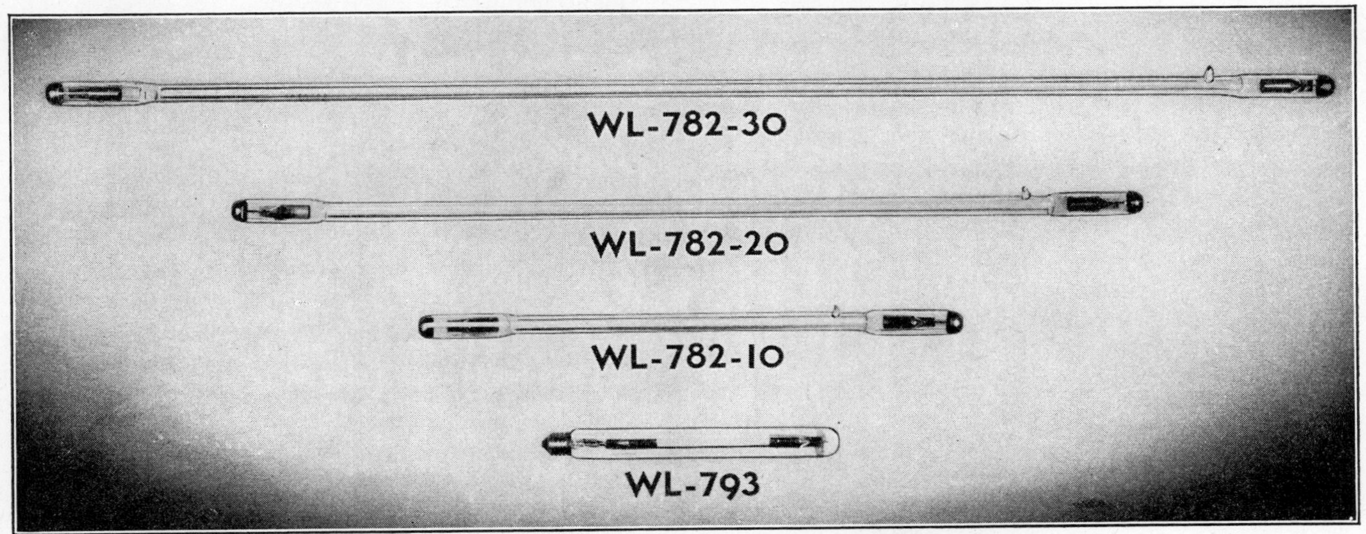

Fig. 10·2 Cold-electrode commercial ultraviolet lamps.

10·3 HIGH-PRESSURE ULTRAVIOLET LAMPS

Table 10·1 shows that a mercury discharge producing radiation efficiently in the near or middle ultraviolet band must operate at a high pressure, if the generated radiation is not to be confined to the mercury-resonance 2537-angstrom band. This type of discharge is readily obtained by using an inert gas with a limited amount of mercury to prevent the maximum pressure, with all the mercury vaporized, from rising above a fixed value.

When the voltage from a reactance transformer is applied to the electrodes a glow discharge is initiated between electrodes. They become heated by this discharge, and when they become thermionically emissive an arc is formed and the temperature and pressure rise until all the mercury is vaporized.

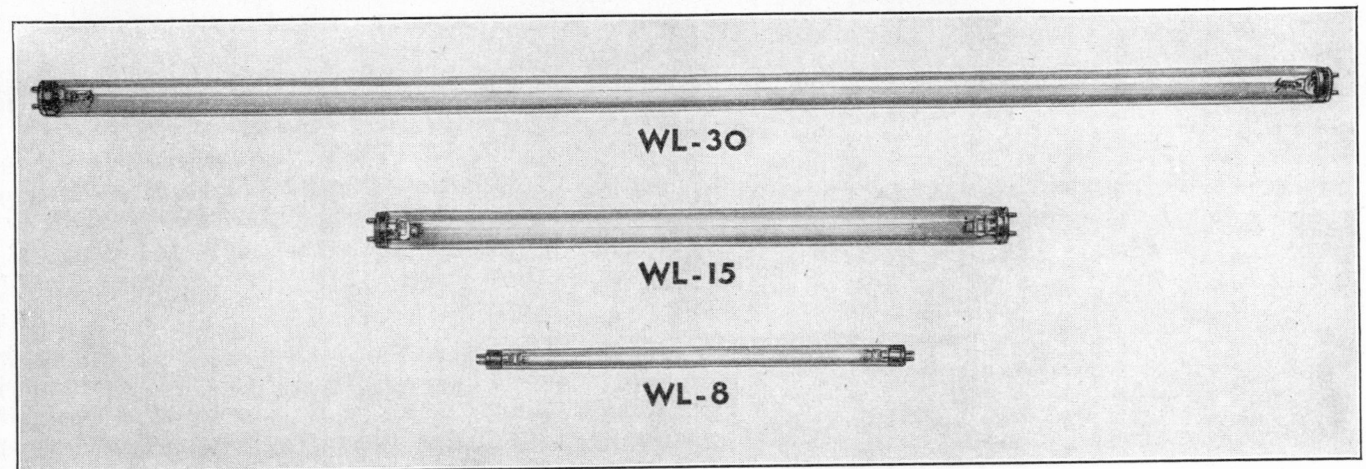

Fig. 10·3 Hot-electrode commercial ultraviolet lamps.

To protect these lamps against excessive and uneven cooling by air currents the lamp proper is generally sealed into an outer container. Because of the high pressure of the mercury vapor when the lamp is turned off, some time is necessary for the mercury to condense before it will start again.

Figure 10·4 shows a commercial high-intensity mercury-vapor lamp of the 400-watt size in glass for lighting. The lamp is enclosed in an outer tube to protect it from air currents. Such a lamp operates at a pressure of approximately 1 atmosphere.

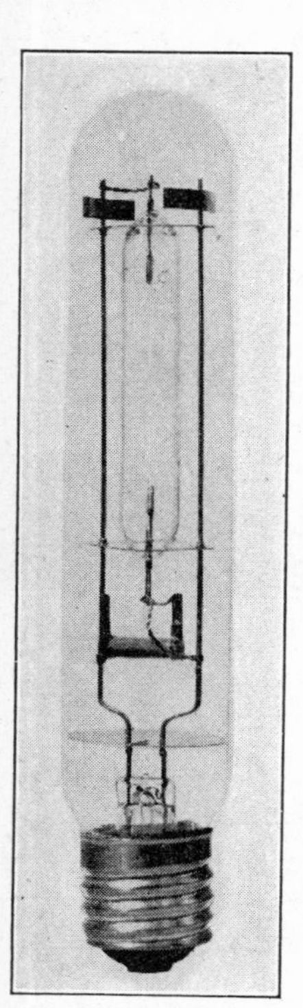

FIG. 10·4 400-Watt high-intensity mercury-vapor lamp.

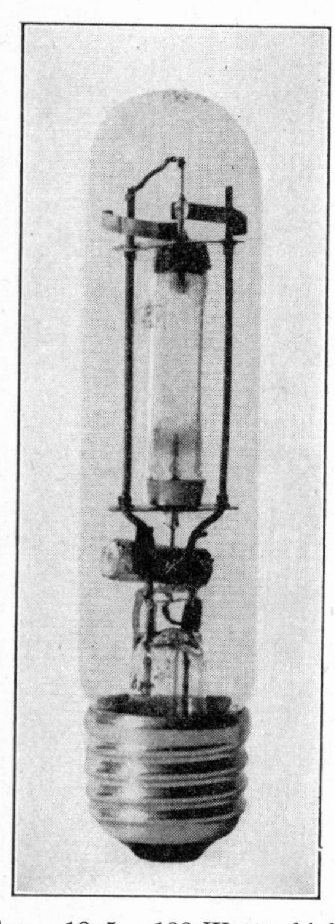

FIG. 10·5 100-Watt high-pressure quartz mercury-vapor lamp in outer bulb housing.

Figure 10·5 shows a small 100-watt high-intensity mercury-vapor lamp in quartz sealed into an outer bulb of glass. In this lamp the arc operates at a pressure of approximately 10 atmospheres.

These small quartz arcs in glass housings which transmit ultraviolet to approximately 2800 angstrom units are used as sun lamps. In a newer self-contained sun lamp the quartz-mercury tube, in series with a tungsten-filament ballast resistor, is mounted in a reflecting bulb so that both the ultraviolet and the heat from the filament pass through the end of the bulb. This lamp operates directly on a lighting circuit. Figure 10·6 shows such a self-contained commercial sun lamp.

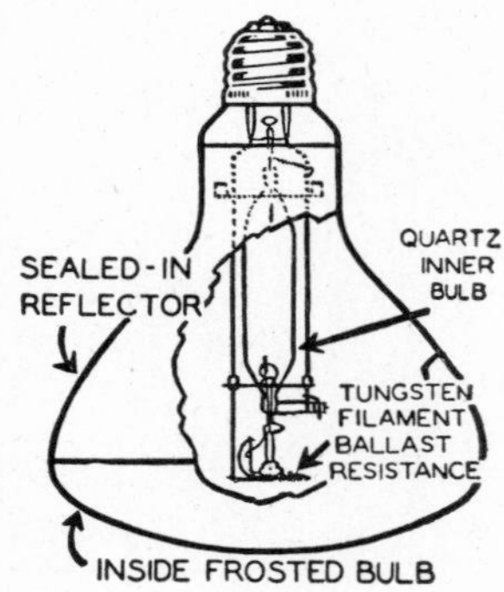

FIG. 10·6 Sun lamp.

Small high-pressure mercury lamps in quartz mounted in outer bulbs, which absorb visible light but transmit the near ultraviolet, are used for the so-called "black light" for producing fluorescent effects. Figure 10·7 shows a black-light lamp.

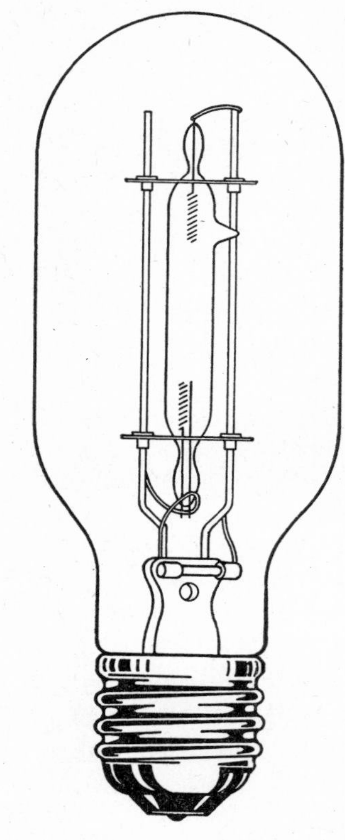

FIG. 10·7 100-Watt commercial black light (outer bulb is heat-resisting red-purple glass).

10·4 MEASUREMENT OF ULTRAVIOLET RADIATION

It is important to have a simple method of measuring the effective intensity and the amount of radiation used in a specific application. Since different wavelengths are not equally effective, a meter that evaluates the different wavelengths of radiation in proportion to their effectiveness is desirable.

The photoelectric threshold of a metal has a definite wavelength which is readily reproduced.[3] A specific use of ultraviolet radiation determines the longest wavelength that is effective. By proper choice of a phototube it is possible to measure only radiation shorter than the threshold wavelength of the phototube and, therefore, only that radiation which is effective for the application. Thus the phototube responds only to the ultraviolet which produces the reaction under investigation. As an example, the bactericidal action of ultraviolet is confined to radiation shorter than 3000 angstrom units. The photoelectric threshold of tantalum is approximately 3000 angstrom units. A tantalum phototube, therefore, measures only the bactericidal radiations, regardless of the source. The bactericidal action for radiations below 3000 angstrom units increases with decreasing wavelength, reaches a peak at 2660 angstrom units, and again diminishes for shorter wavelengths. The photoelectric response of tantalum increases for radiations shorter than 3000 angstrom units. By constructing the tantalum phototube in a glass of proper transmission the response of the phototube can be made to reach a peak close to 2660 angstrom units and again decreases for radiation of shorter wavelength. The response of this tube is therefore similar to the bactericidal action of the ultraviolet, and it may be used to evaluate the bactericidal activity for any exposure.

The sensitivity of the phototube is readily expressed in physical units by determining its response to a definite spectral line such as the 2537-angstrom radiation and by measuring the intensity of the same radiation with a thermopile. The effective intensity from a source of radiation as measured by such a phototube is thus expressed in the equivalent of 2537-angstrom radiation.

Other phototubes[4] are likewise suitable for measuring radiation in other bands of the ultraviolet region; for example, a zirconium phototube with a threshold at 3250 angstrom units for the erythemal region, or a platinum tube with a threshold of approximately 2000 angstrom units for ozone-producing radiation.

Since photoelectric currents are relatively weak a simple method of measuring these currents was devised.[5] A battery B charges a capacitor C_1 through a phototube $V1$ at a rate depending upon the intensity of the radiation falling on $V1$ and the response of the tube (Fig. 10·8). A special glow tube $V2$ with a thorium cathode and two anodes provides a means for discharging capacitor C_1 and automatically recording the number of times it is charged during an exposure. When capacitor C_1 is charged to the breakdown voltage

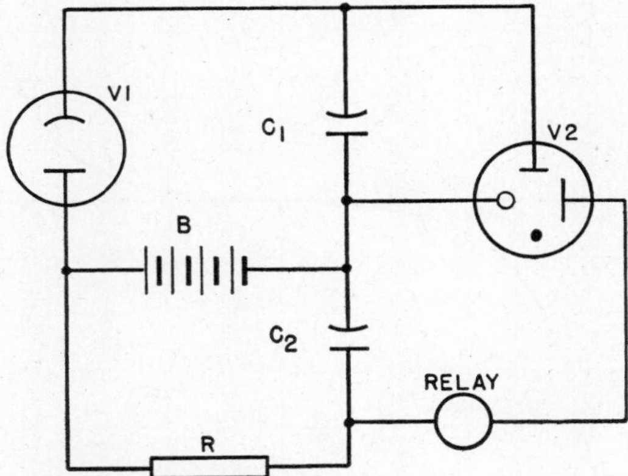

Fig. 10·8 Ultraviolet meter circuit.

between a and b, the starting anode and the cathode, a glow discharges C_1, thereby starting the main current. The larger capacitor C_2 charged through a high resistance R thus discharges through the relay, which operates the counter.

The number of discharges from a given exposure represents in arbitrary units the photoelectric current through the phototube $V1$ and therefore the integrated effective radiation to which the tube was exposed. The rate of charging and discharging capacitor C_1 represents the average intensity of the radiation.

The meter as described is calibrated so that each discharge is expressed in microwatt-seconds per square centimeter of energy of a given wavelength. For a tantalum tube the calibration expresses the bactericidal equivalent in microwatt-seconds per square centimeter of 2537-angstrom radiation.

Leakage in humid weather may cause serious errors unless special precautions are taken. This trouble is eliminated[6] by housing the capacitor and the glow tube in the neck of the cell and exhausting this compartment (Fig. 10·9). This

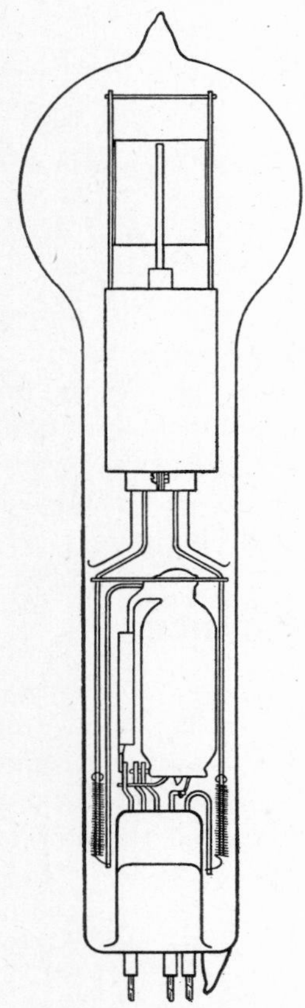

Fig. 10·9 Photoelectric cell.

meter with a tube suitable for the specific use affords an easy and reliable means for measuring effective radiation in the ultraviolet bands to which the phototube responds.

REFERENCES

1. *Tech.-Wiss. Abhandl. Osram-Konzern*, Vol. 3, p. 12, 1934.
2. "A Study of the Bactericidal Action of Ultraviolet Light," F. L. Gates, *J. Gen. Physiol.*, Vol. 13, 1929, pp. 231–260.
3. "Photoelectric Emission from Different Metals," H. C. Rentschler, D. E. Henry, and K. O. Smith, *Rev. Sci. Instruments*, Vol. 3, 1932, pp. 794–802.
4. "Lowering of the Photoelectric Work Function of Zirconium, Titanium, Thorium and Similar Metals by Dissolved Gases," H. C. Rentschler and D. E. Henry to appear in Vol. 87, *Electrochemical Society Journal*.
5. "An Ultraviolet Light Meter," H. C. Rentschler, *J. A.I.E.E.*, Vol. 49, 1930, pp. 113–115.
6. "Bactericidal Effect of Ultraviolet Radiation," H. C. Rentschler, R. Nagy, and G. Mouromtseff, *J. Bacteriology*, Vol. 41, 1941, pp. 745–774.

Chapter 11

CIRCUIT ELEMENTS

D. G. Little

THE three basic circuit elements are resistance, capacitance, and inductance. These elements never exist separately. Every electrical circuit always contains all three circuit elements.

Each circuit element has unique properties. Capacitance has the ability to store energy in the form of an electrostatic field in its dielectric material. An inductance stores energy in the form of a magnetic field surrounding it. Resistance cannot store energy but it always converts the electrical energy into some other form of energy, usually heat, light, or other electromagnetic radiations.

11·1 LINEAR RESISTANCE

The resistance of a current path may be considered as inversely proportional to the number of free electrons in that path. Metals such as silver or copper apparently have large quantities of free electrons; consequently, when an electromotive force is applied the flow of electrons is large and the resistance is said to be low. Insulators such as silk or porcelain have few free electrons and hence have a higher resistance. A perfect vacuum with no free electrons has an infinite resistance.

An electric current may be thought of as a migration or flow of free electrons through and between the atoms and molecules of the conductor. In practical terms a current of 1 ampere at a point in a circuit means that $6 \cdot 28$ times 10^{18} electrons flow past that point each second. At all temperatures except absolute zero the atoms of the conductor are agitated thermally, and they retard or resist the flow of free electrons. This resistance generally varies directly with absolute temperature. Thus at normal temperatures metallic conductors have specific resistances depending upon the material, but at absolute zero the resistance of all conductors approaches zero. Metallic conductors have constant resistance at constant temperature, or the voltage drop is directly proportional to the current. Such resistances are known as linear resistances because the value of the resistance is not affected by the current through it or the voltage across it, so long as temperature remains constant.

The unit of resistance is the ohm. The international ohm is that resistance offered to an unvarying electric current by a column of mercury at the temperature of melting ice, 14.4521 grams in mass, of a constant cross-sectional area, and of a length of 106.300 centimeters. A current of 1 ampere flowing through a resistance of 1 ohm causes a potential difference across the resistance of 1 volt.

11·2 NON-LINEAR RESISTANCE

Currents in other media, semi-metallic materials, high vacua, and ionized gases do not follow the linear law. In

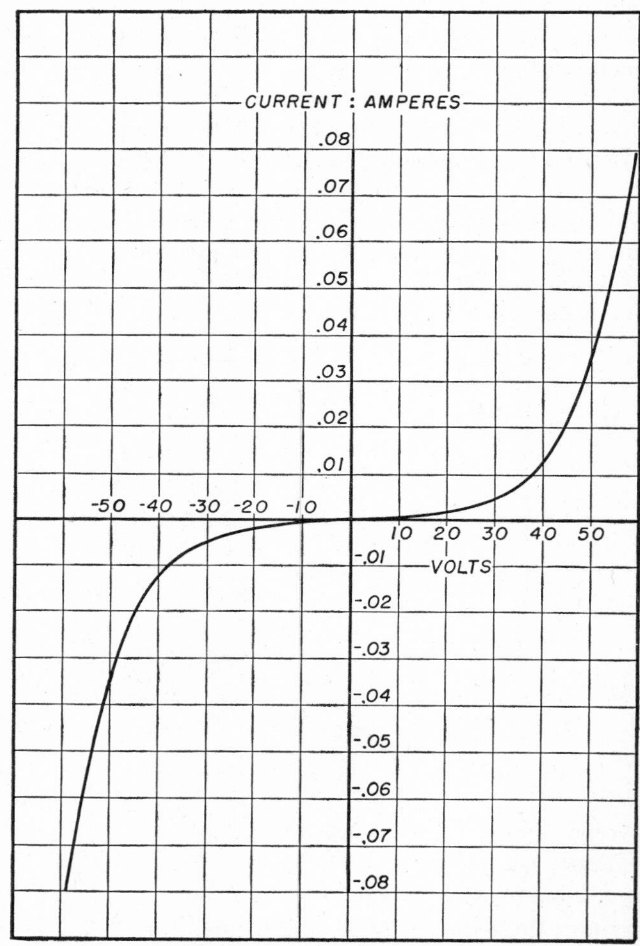

Fig. 11·1 Characteristics of typical Varistor.

high vacua the current varies as the three halves power of the voltage. In ionized gases the voltage drop is determined by the ionization potential of the gas, which depends upon the pressure and temperature and concentration of the gas, and this voltage drop is relatively independent of the current.

Other kinds of conductors are known generally as "Varistors." * In most of these the resistance decreases as the voltage increases for both directions of current. Varistors are often so made that their symmetrical characteristics (similar performance for either direction of current) are inherent in their construction. Other forms of Varistors are made from components (such as copper oxide rectifiers) in which the characteristics are not symmetrical. In this type, symmetry of performance is obtained by connecting two components in parallel and with opposite polarities connected together. The non-linear characteristics of Varistors are used in many applications, including voltage-surge protection, noise and overload limiting in telephone and other communication apparatus, voltage stabilization in the output of power supplies, frequency multiplication, and others. Figure 11·1 shows a characteristic curve for a typical Varistor.

Still other conductors are known as rectifiers (copper oxide, selenium, copper sulphide) because their resistances vary greatly with the direction of current but they are not linear resistances in either direction.

These metal or dry rectifiers are all based on the same fundamental principle. A thin barrier layer is used; on one side of the layer is a material containing a relatively large quantity of free electrons, and on the other side the material has relatively few free electrons. These devices therefore act as electronic valves and can be used as rectifiers to convert alternating current into direct current.

Figure 11·2 shows a characteristic curve for a typical dry rectifier.

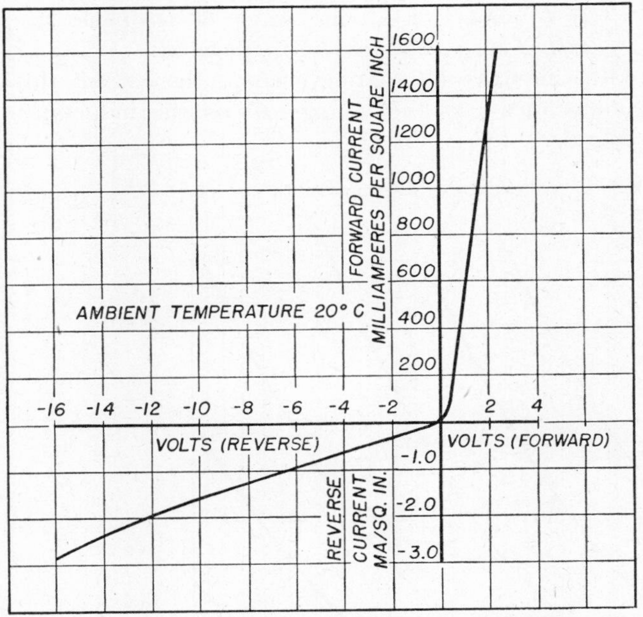

Fig. 11·2 Characteristics of typical copper oxide rectifier.

Figure 11·3 shows several commonly used types of resistors.

The resistance of a circuit thus follows various laws depending upon the medium through which the electrons flow, but electrical energy delivered to the resistance element always is converted into some other form of energy, usually heat, sometimes heat and light, and sometimes electromagnetic waves at frequencies lower than those of heat and light.

The effective resistance of several resistances in series is equal to the numerical sum of the individual resistances:

$$R = r_1 + r_2 + r_3 + \cdots \tag{11·1}$$

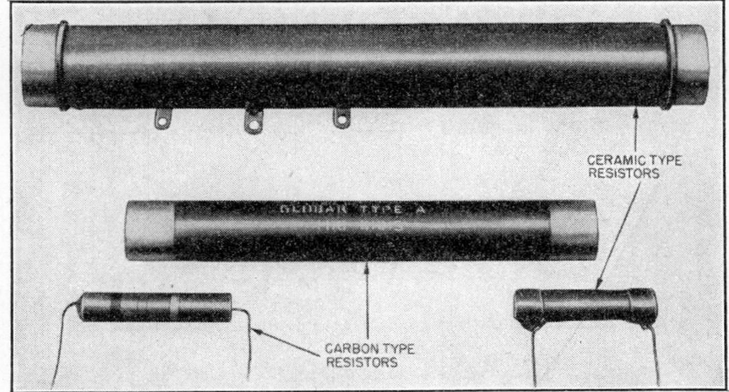

Fig. 11·3 Wire-wound and composition resistors.

The reciprocal of the effective resistance of several resistances connected in parallel is equal to the sum of the reciprocals of the individual resistances:

$$\frac{1}{R} = \frac{1}{r_1} + \frac{1}{r_2} + \frac{1}{r_3} + \cdots \tag{11·2}$$

11·3 OTHER RESISTANCE FORMS

Thus far resistance has been considered only for a continuous or steady-state direct current. With an alternating current the resistance acts differently, and it varies with the frequency of the current. Alternating current introduces "skin effect" and electromagnetic radiation, both of which increase the effective resistance of a conductor. Alternating current sets up an alternating magnetic field in and around the conductor. This field cuts the conductor and causes eddy currents which oppose the main current at the center of the conductor and add to it near the surface of the conductor. This effect increases with the frequency and the size of the conductor.

Figure 11·4 shows the increase in resistance for alternating currents as compared with the resistance to continuous direct currents expressed as a function of X where

$$X = \sqrt{\frac{8\pi f\mu}{R \times 10^9}} \tag{11·3}$$

f = frequency in cycles per second
μ = permeability of the conductor material
R = resistance per unit length of conductor—ohms per centimeter

The above relationship applies to round straight conductors in free space. If the conductor is adjacent to other conductors or wound in coil form, the a-c resistance will in general be increased.

Figure 11·5 shows the ratio of a-c resistance to d-c resistance of round, straight, isolated copper wire at various fre-

* Trademark of Bell Telephone Laboratories.

quencies. Several curves are drawn for various standard-gauge (B & S) wire sizes.

Electromagnetic radiation is difficult to explain in simple terms. Briefly, an alternating current in a conductor produces alternately magnetic and electric fields in the surrounding space. Because the velocity of propagation of these fields is finite, they do not have time to collapse or disappear as the current and voltage in the conductor passes through zero. Thus a certain amount of energy or field is left in space after each half cycle, and this energy is propagated further in space because the fields produced by succeeding cycles repel it. The higher the frequency of the current, the more energy is left in space and the greater the radiation. This loss of energy from the conductor makes its effective resistance higher; this increase is known as *radiation resistance*.

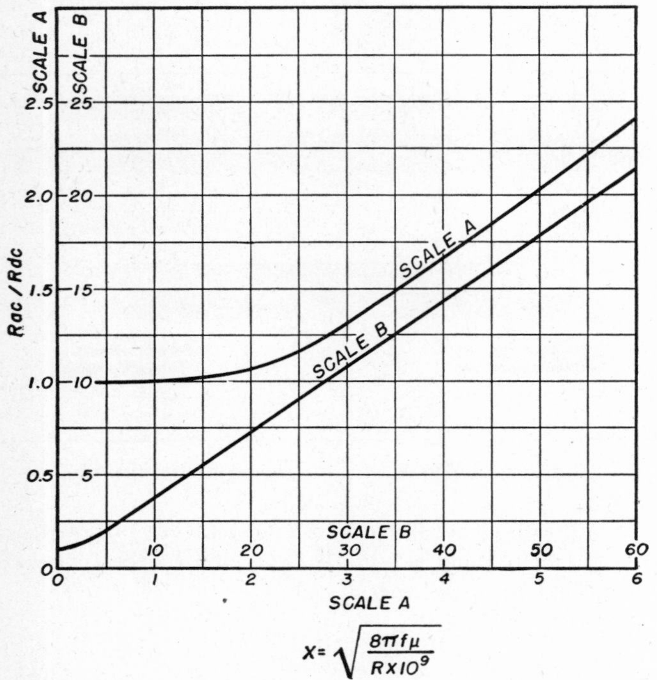

FIG. 11·4 Skin effect of round, straight, isolated conductors.

11·4 INDUCTANCE

The unit of inductance is the henry. If a potential of 1 volt is impressed across a circuit element and the current *changes* at a uniform rate of 1 ampere per second, then the inductance of the circuit element is 1 henry. Thus voltage across an inductance is proportional to the inductance multiplied by the rate of change of current. Inductance, because of its energy-storing characteristics, produces a voltage in such a direction as to keep the current constant. If the current increases, the inductance produces a voltage that opposes the current; if current decreases, the induced voltage tries to sustain the current. In mathematical form the expression for voltage across an inductance is

$$E_L = -L\frac{di}{dt} \tag{11·4}$$

Energy stored in an inductance at any instant is

$$\tfrac{1}{2}LI^2 \text{ joules or watt-seconds}$$

if current is in amperes and inductance is in henrys. In an a-c circuit the effect of an inductance appears as a reactance opposing the current. In mathematical form, the reactance $X_L = 2\pi fL$ ohms for a sine wave of frequency f, and $E_L = 2\pi fL \cdot I$ for a sine wave of frequency f.

With sine-wave alternating current, the current through an inductance lags the voltage across the inductance by 90 degrees.

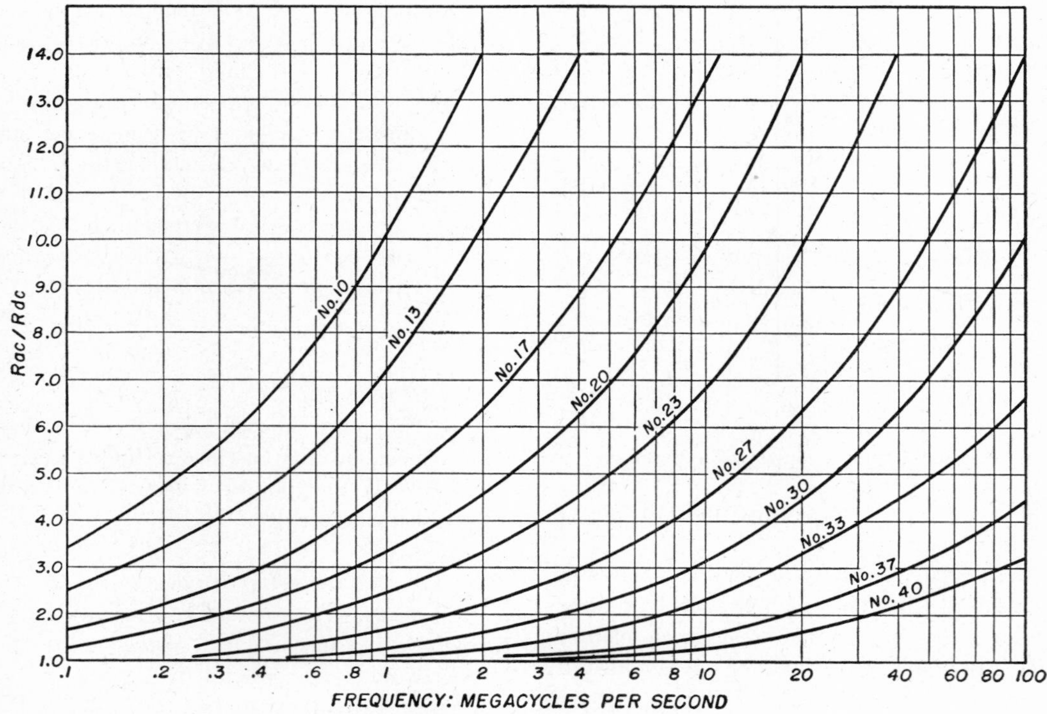

FIG. 11·5 Skin effect of round, copper wire.

The forms in which inductors are made vary with the amount of inductance desired and the voltage and current the inductors must withstand. Because a pure inductance (one having no resistance) would always return to the circuit all the energy stored in its field, it would have no losses; hence, as far as temperature rise is concerned it could be made extremely small in size for any amount of inductance. Practically, of course, resistance is always present in the conductor, and the watts lost as heat in the resistance must be dissipated in some manner. The over-all size of the unit is thus often dictated by the necessity for safely dissipating this heat energy.

FIG. 11·6 Large iron-core reactor.

FIG. 11·7 Typical small iron-core reactors.

Large values of inductance are used mostly in d-c circuits and low-frequency a-c circuits. They contain many turns of copper wire wound on a laminated iron core. The core, with a magnetic permeability many times that of air, increases the inductance of the coil. An air gap usually is placed in the core to prevent saturation of the iron. Figures 11·6 and 11·7 show some iron-core inductances. In the design of inductances for alternating current alone or alternating current superimposed on direct current, account must be taken of the eddy-current and hysteresis losses in the iron core as well as the I^2R loss in the copper, as all these losses appear as heat which must be dissipated.

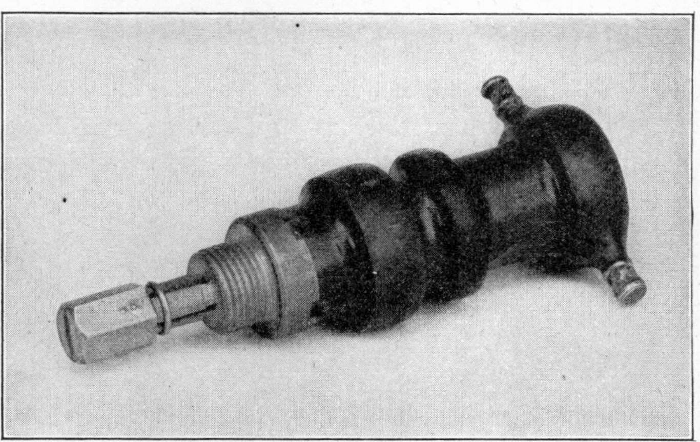

FIG. 11·8 Typical radio-receiver reactor.

Small inductances for radio-frequency alternating currents generally are single-layer or honeycomb-wound coils with finely laminated or powdered-iron cores or air cores. Figures 11·8 and 11·9 show typical radio-receiver and radio-transmitter inductances respectively. The inductance of a single-layer air-core coil is approximately

$$L = \frac{r^2 \times N^2}{9r + 10l} \qquad (11\cdot5)$$

FIG. 11·9 Typical radio-transmitter reactor.

where L = inductance in microhenrys
r = radius of the coil in inches
l = length of coil in inches
N = number of turns.

More accurately, $L = F \times N^2 \times d$ microhenrys where F is a rather involved function of the ratio d/l. Values of F usually are given in table or curve form in most electrical handbooks.

For very high frequencies small inductances are needed, and these usually are single-turn coils, hairpin-shaped conductors, or even straight conductors. Such an inductance is shown in Fig. 11·10.

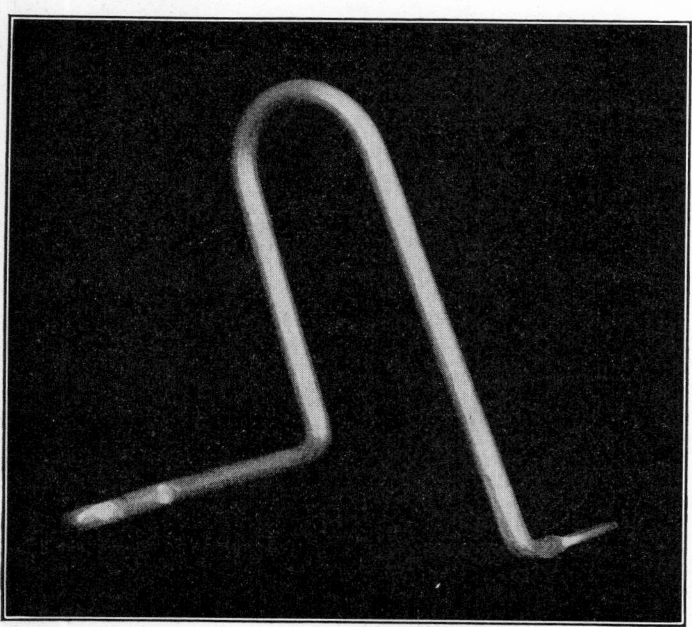

FIG. 11·10 Typical very high-frequency reactor.

The ideal inductance would have no effective resistance and no loss and would, therefore, be 100 percent efficient. Efficiency of an inductance is measured in terms of the ratio of the inductive reactance to the resistance at a given frequency. This ratio is known as Q. Q is equal to $2\pi fL/R$. The higher Q, the more efficient is the inductance design as far as heat loss is concerned.

11·5 CAPACITANCE

The unit of capacitance is the farad. A capacitor has a capacitance of 1 farad if 1 coulomb of stored energy causes a potential difference of 1 volt to exist across its terminals or electrodes.

Capacitance can be defined by the expression $C = Q/E$ farads, where Q is the electric charge in coulombs. Since 1 coulomb equals 1 ampere-second, $i = C(de/dt)$ amperes, and in a capacitor the energy storage $W = Ce^2/2$ joules or watt-seconds, if C is in farads and e is in volts.

Capacitance, because of its energy-storing characteristic, adds or subtracts a current in the circuit in such a direction as to try to maintain the voltage constant.

In an a-c circuit, the effect of a capacitance appears as a reactance, opposing the current. In mathematical form, its reactance $X_c = 1/2\pi fC$ for a sine wave of frequency f, and $E_c = I/2\pi fC$ for a sine wave of frequency f. With a sine wave alternating current, the current through a capacitance leads the voltage by 90 degrees.

A pure capacitance (one with no resistance) would return to the circuit all the energy stored in its electric field; hence, it would have no losses and would be 100 percent efficient. A capacitance in practical form has conductors to carry the current in and out of the plates surrounding the dielectric, and these conductors have resistance. Furthermore, the dielectric, unless it is a gas or a vacuum, has dielectric hysteresis. The amount of these losses determines how large the unit must be to dissipate these losses as heat.

Capacitors take various practical forms and utilize various kinds of dielectric—a vacuum, compressed gas, air, paper impregnated with oil or wax, mica, ceramics, and other kinds of insulating materials. Figures 11·11 and 11·13 show some of these types.

Dielectric materials have various values of specific inductive capacity ϵ, or dielectric constant. Air, vacuum, and compressed gases have a nominal ϵ of 1.0. Other materials have ϵ values higher than 1, usually between 2 and 10, but in some kinds of ceramics ϵ may be 10,000 or higher.

Another important property of dielectric materials is power factor. A perfect dielectric would have no leakage and no dielectric loss. The phase angle between current and voltage would be 90 degrees, and the power factor would be zero. Practical capacitors have power factors that vary between almost zero and 0.1 or greater. For high-frequency circuit capacitors, a figure known as loss factor is employed. This is the dielectric constant multiplied by the power factor. If an alternating voltage is impressed on a capacitor, the temperature rise of the dielectric varies as $f \times E \times$ loss factor. Table 11·1 shows ϵ, power factor, and loss factor for some common dielectric materials.

Oiled-paper capacitors are employed as filters to by-pass the a-c component in a d-c circuit; for example, the output of a rectifier. Here the requirements are high d-c resistance with less emphasis on low a-c loss because the a-c or ripple voltage is a small fraction of the direct voltage. Several

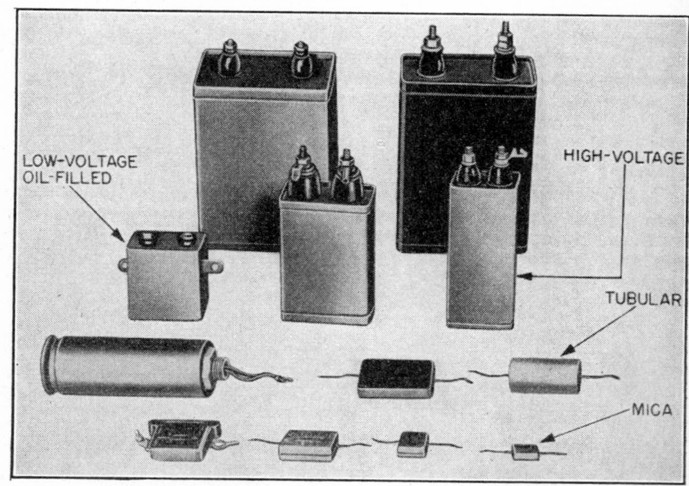

FIG. 11·11 Typical fixed capacitor units.

types of filter capacitors have almost constant capacitance, independent of temperature. The most constant type employs mineral oil as an impregnant. The capacitance is almost independent of ambient temperature, but the ϵ is low. Hence, the size and cost for a given capacitance and voltage rating are large. A second type uses castor oil in place of mineral oil. At a temperature of about -20 degrees centigrade, the capacitance starts to decrease until at -40 degrees centigrade the capacitance is only three quarters of its normal value. Castor-oil capacitors are smaller than mineral-oil capacitors of the same rating.

TABLE 11·1 PROPERTIES OF SOME COMMON DIELECTRICS

Material	ϵ	Percent P-F at 1 Mc	Loss Factor at 1 Mc
Cellulose acetate	6–8	3–6	18–48
Fiber	4–5	5.0	20–25
Glass, Crown	6.2	...	....
Glass, Flint	7.0	0.4	2.8
Glass, Pyrex	4.5	0.2	0.9
Methacrylic resin	2.8	2.0	5.6
Mica, clear India	7–7.3	0.02	0.14–0.146
Insanol	6–8	0.3	1.8–2.4
Phenol, black molded	5.5	3.5	19.25
Phenol, paper base	5.5	3.5	19.2
Phenol, cloth base	5.6	5.0	28.0
Porcelain, wet process	6.5–7	0.6	3.9–4.2
Porcelain, dry process	6.2–7.5	0.7	4.34–5.25
Quartz, fused	4.2	0.03	0.126
Rubber, braid	2–3	0.5–0.9	1.0–2.7
Styrene, polymerized	2.4–2.9	0.03	0.72–0.87

The most common type of oiled-paper capacitor employs a chlorinated diphenol synthetic oil as dielectric impregnant. The ϵ of this oil is fairly large and it ruptures at a rather high voltage, so that such capacitors are smaller and less costly per microfarad than mineral-oil or castor-oil units. Their capacitance, however, decreases about one quarter between 10 and -10 degrees centigrade. Figure 11·12 shows capacitance versus temperature curves for the three types of oiled-paper capacitors mentioned.

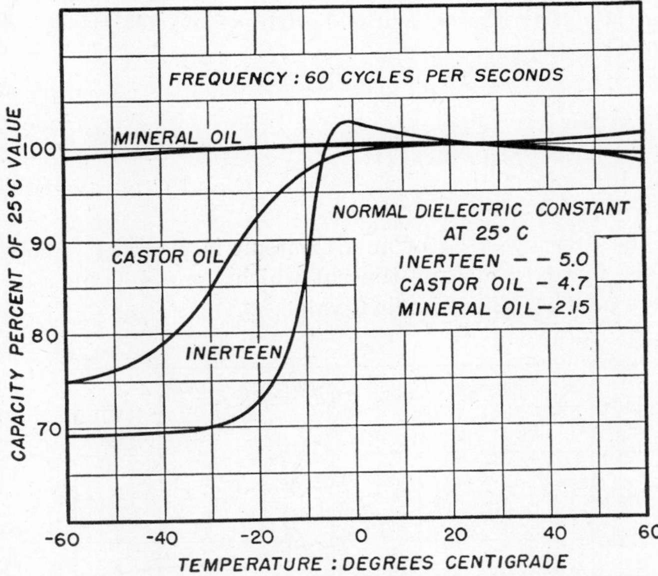

FIG. 11·12 Temperature-capacitance characteristics of liquid-filled paper capacitors.

If a large capacitance must be fitted into a small space, electrolytic capacitors are useful, provided some d-c leakage can be tolerated, as in power-supply units for home radio receivers. The active dielectric is a film of aluminum oxide which is formed by a current, from one sheet of aluminum foil to another, through the electrolyte between the plates. This film will stand only about 500 volts without puncture but, because of its thinness, an electrolytic capacitor as large as 1000 microfarads can be made to fit into a few cubic inches of space at a reasonable cost.

Capacitors are used in radio-frequency circuits for by-pass or d-c blocking, or to offset inductance in tuned circuits. The most common type probably is the variable air capacitor in series with a fixed coil to make a variable-frequency tuned circuit. See Fig. 11·13. Maximum capacitances of such units range from 5 to 5000 micromicrofarads, and their voltage ratings are from 200 to 20,000 volts.

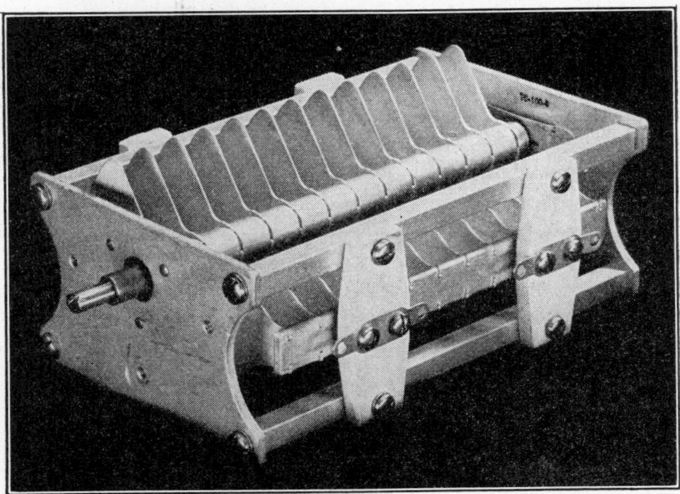

FIG. 11·13 Variable air capacitor.

For transmitters and oscillators, the compressed-gas capacitor is satisfactory. These are made with both fixed and variable capacitance and are similar to the variable air type except that the plates are enclosed in a pressure-tight case filled with an inert gas at 5 to 20 atmospheres pressure.

Vacuum capacitors also are used in transmitters. To obtain the benefits of close electrode spacing and high voltage breakdown, the vacuum must be nearly perfect, so the plates are sealed in a container, usually glass. Air, compressed-gas, and vacuum capacitors have almost no dielectric loss and therefore are very efficient.

If large capacitance is required for moderately high frequency, a low-loss liquid impregnant can be used for liquid-filled capacitors of relatively small size. The use of this impregnant results in capacitors that are smaller than equivalent air or gas-insulated units.

11·6 COMBINATIONS OF RESISTANCE AND INDUCTANCE

In d-c circuits where only steady-state conditions are important, capacitance and inductance have no effect on the operation of the circuit, and they may be omitted from the calculations. However, if starting, stopping, or transient conditions are to be considered even in a d-c circuit, all three circuit elements must be taken into account.

If a circuit consists of resistance and inductance in series and a constant voltage is impressed across them, the current starts from zero when the circuit is closed and rises to a constant value along the exponential curve:

$$i = \frac{E}{R}(1 - \epsilon^{-Rt/L}) \qquad (11\cdot6)$$

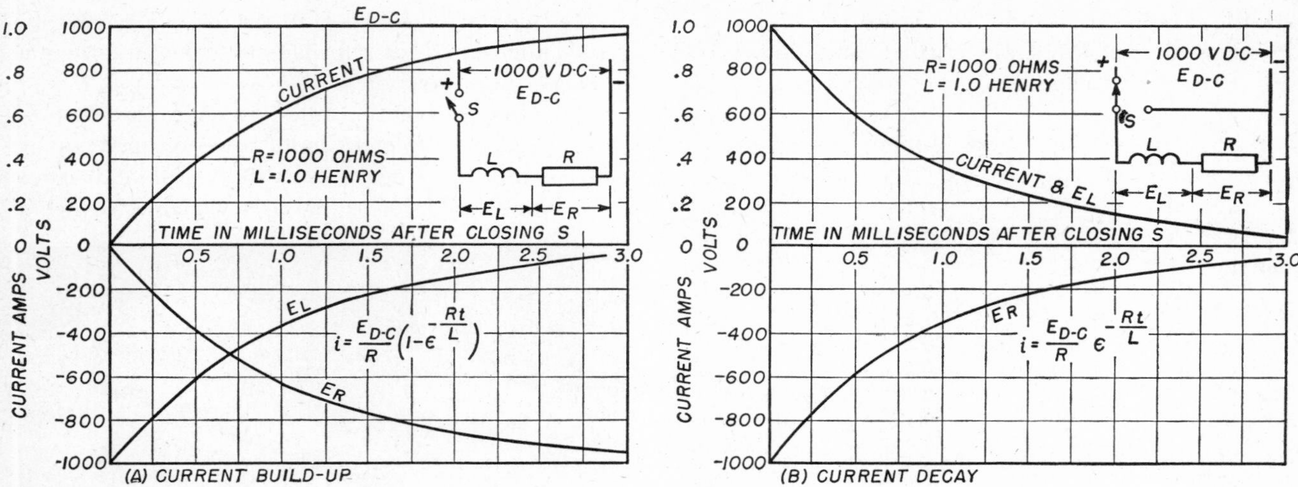

FIG. 11·14 D-c transient in RL series circuit.

If the source of voltage is removed without opening the circuits, the current falls to zero along a similar exponential curve:

$$i = \frac{E}{R}\epsilon^{-Rt/L} \tag{11·7}$$

Figure 11·14 shows current and voltage versus time. The instantaneous voltage across the resistor is iR, and across the inductance it is $L(di/dt)$; the two voltages added together equal the driving potential E. Thus at $t = 0$, $i = 0$, the voltage across the resistance is zero and across the inductance it is E. After steady current has been established, the rate of change of current is zero, and E equals IR across the resistance.

In an RL circuit the ratio L/R is known as the time constant. This is the time for the current to reach 0.632 of its final value. If L is in henrys and R is in ohms, the time constant is in seconds.

The impedance of a series RL circuit to a sine-wave alternating current of frequency f is

$$|Z| = \sqrt{R^2 + (2\pi fL)^2} \text{ ohms} \tag{11·8}$$

where R = total circuit resistance in ohms
L = circuit inductance in henrys.

11·7 COMBINATIONS OF RESISTANCE AND CAPACITANCE

A circuit of resistance and capacitance in series connected across a source of voltage E would cause a current to flow until the capacitor became charged to the voltage E following the exponential curve:

$$i = \frac{E}{R}\epsilon^{-t/RC} \tag{11·9}$$

At $t = 0$ all the voltage drop is across the resistor, at $t = \infty$ all the voltage drop is across the capacitor, and during the transient charging period the voltages across R and C add directly to equal voltage E. If E is removed without opening the circuit the capacitor discharges through the resistor following the curve:

$$i = -\frac{E}{R}\epsilon^{-t/RC}$$

and the sum of the voltages across R and C always equals zero (Fig. 11·15).

The time constant of an RC circuit is $T = RC$ (seconds, ohms, farads); it is the time at which the voltage across C has reached 0.632 of its final value.

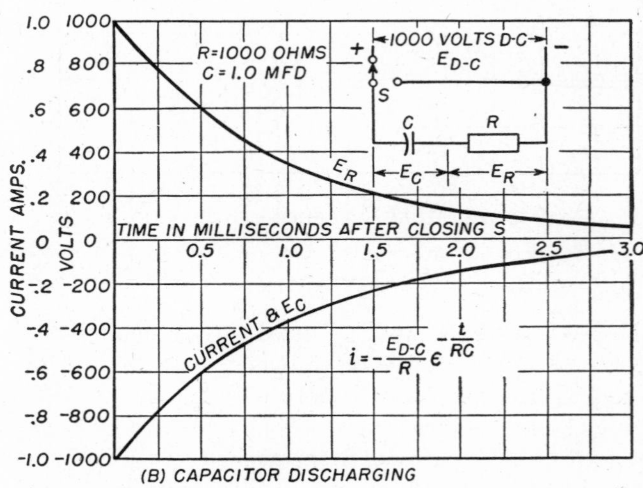

FIG. 11·15 D-c transient in RC series circuit.

The impedance of an RC series circuit to a sine-wave alternating voltage of frequency f is

$$|Z| = \sqrt{R^2 + \frac{1}{(2\pi f C)^2}} \quad \text{ohms} \tag{11·10}$$

11·8 COMBINATIONS OF INDUCTANCE, CAPACITANCE, AND RESISTANCE IN A-C CIRCUITS

Any network containing all the circuit components L, C, and R can, at any fixed frequency, be resolved into an equivalent circuit containing R and either C or L. If the reactances and resistances are expressed in complex form, the impedances can be combined in a manner directly analogous to the methods used for combining series and parallel resistances. As an example, consider the circuit as shown in Fig. 11·16. For a supply frequency f,

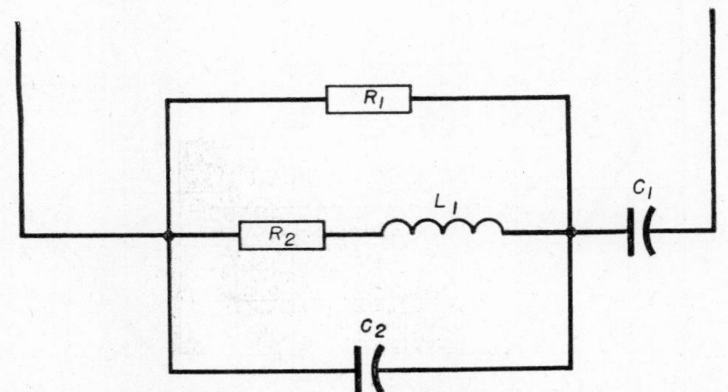

FIG. 11·16 Circuit diagram; generalized circuit of R, L, and C.

$$\begin{aligned} X_{C1} &= \frac{1}{2\pi f C_1} \quad (0 - j); \qquad X_{C2} = \frac{1}{2\pi f C_2} \quad (0 - j) \\ X_{L1} &= 2\pi f L \quad (0 + j) \\ R_1 &= R_1 \quad (1 + j0) \\ R_2 &= R_2 \quad (1 + j0) \end{aligned} \tag{11·11}$$

Then the total impedance of this circuit at frequency f is

$$Z_0 = X_{C1} + \frac{1}{\dfrac{1}{R_1} + \dfrac{1}{R_2 + X_{L1}} + \dfrac{1}{X_{C1}}} \tag{11·12}$$

This combination of impedances is similar to a combination of resistances connected as shown in Fig. 11·17.

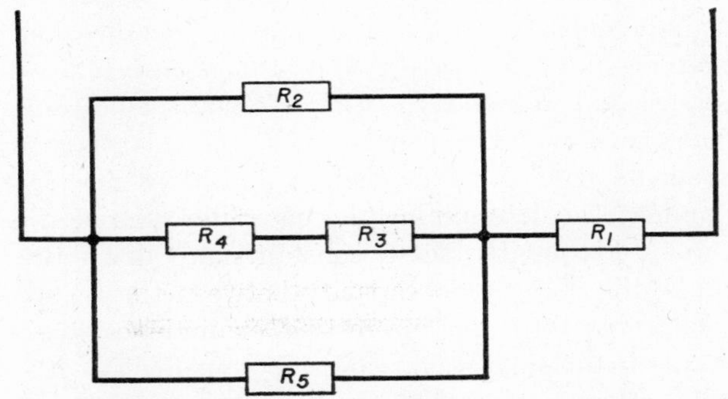

FIG. 11·17 Circuit diagram; generalized circuit of resistances.

The total equivalent resistance is

$$R_{\text{total}} = R_1 + \frac{1}{\dfrac{1}{R_2} + \dfrac{1}{R_3 + R_4} + \dfrac{1}{R_5}} \tag{11·13}$$

A combination of L, C, and R, in series, is commonly used in a-c circuits. Its distinguishing characteristic is that its impedance reaches a minimum at some frequency, usually called the "resonant" frequency. This circuit is thus "selective" with respect to frequency. Another characteristic of this circuit is that at or near the resonant frequency point it can be used as a voltage amplifier when the source of applied voltage has a relatively low impedance.

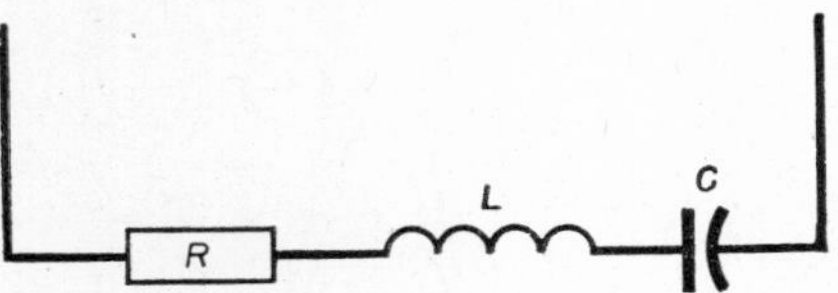

FIG. 11·18 Series circuit of R, L, and C.

For a circuit such as that shown in Fig. 11·18, impedance Z is

$$R + j\left(2\pi f L - \frac{1}{2\pi f C}\right)$$

at frequency f, and the magnitude $|Z|$ is

$$\sqrt{R^2 + \left(2\pi f L - \frac{1}{2\pi f C}\right)^2}$$

The circuit impedance Z has a minimum value equal to R, if f_0 is such that

$$2\pi f_0 L = \frac{1}{2\pi f_0 C} \tag{11·14}$$

where f_0 (resonant frequency) in cycles per second

$$= \frac{1}{2\pi\sqrt{L \cdot C}}$$

L = inductance in henrys
C = capacity in farads.

As frequency is changed above or below f_0, the circuit impedance increases. This rate of increase of impedance with change in frequency is a measure of the selectivity of the circuit. Selectivity increases as R decreases. Figure 11·19 shows the effect of changes in R on the shape of the resonance curve of a circuit containing L, C, and R in series, in which L is 25 microhenrys and C is 0.001013 microfarad.

Another important characteristic of such a circuit is its voltage-amplifying property at or near resonance. In Fig. 11·18 E_0 is the applied voltage. At resonance, E_C (across the capacitor) and E_L (across the inductance) are

$$E_L = E_C = E_0 \frac{X_L}{R} = E_0 \frac{X_C}{R} \tag{11·15}$$

Since the ratio X_L/R is defined as Q, then E_C and E_L equal E_0Q. Since Q can be made to be quite high, the voltages across the capacitance and inductance can be several times as large as the applied voltage. Series-tuned circuits in electronic devices are frequently designed with a high Q to obtain maximum selectivity and maximum voltage amplification.

Another common useful circuit is shown in Fig. 11·20. This kind of circuit is particularly useful when it is supplied from an a-c source with relatively high impedance. A vacuum-tube amplifier is such a source. The impedance Z_0 of such a circuit at a supply frequency f is

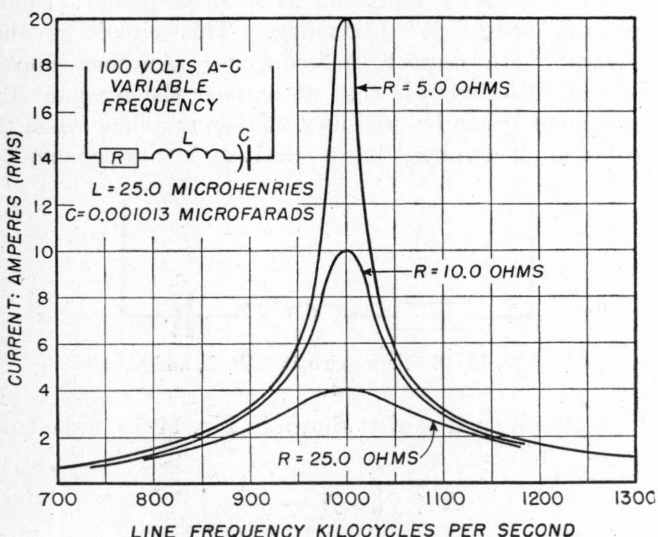

FIG. 11·19 Series resonance.

$$Z_0 = \frac{R + j2\pi f\{L[1 - (2\pi f)^2LC] - CR^2\}}{[1 - (2\pi f)^2LC]^2 + (2\pi fCR)^2} \qquad (11\cdot16)$$

Its magnitude is

$$|Z| = \sqrt{\frac{R^2 + (2\pi fL)^2}{[1 - (2\pi f)^2LC]^2 + (2\pi fCR)^2}} \qquad (11\cdot17)$$

At some one frequency f_0 the impedance of this circuit is a maximum, and for all other frequencies it is less.

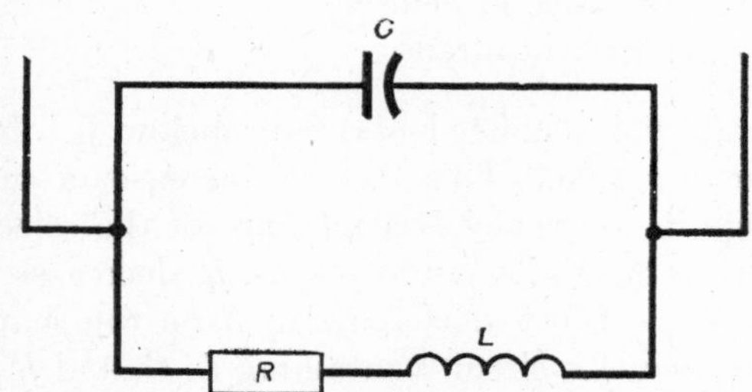

FIG. 11·20 Parallel circuit of R, L, and C.

For the usual parallel resonant circuit, the resistance is low compared with inductive and capacitive reactances, and the resonant frequency is approximately

$$f_0 = \frac{1}{2\pi\sqrt{LC}} \qquad (11\cdot18)$$

At resonance the impedance of such a circuit appears to be a pure resistance, and the total current is in phase with the applied voltage. If R is small with respect to X_L or X_C, the circuit impedance at resonance is

$$|Z_0| = \frac{L}{RC} \quad \text{ohms} \qquad (11\cdot19)$$

Figure 11·21 shows the variation in total current with frequency for a parallel circuit like that in Fig. 11·20 in which L is 25 microhenrys and C is 0.001013 microfarad. Curves are shown for several values of R.

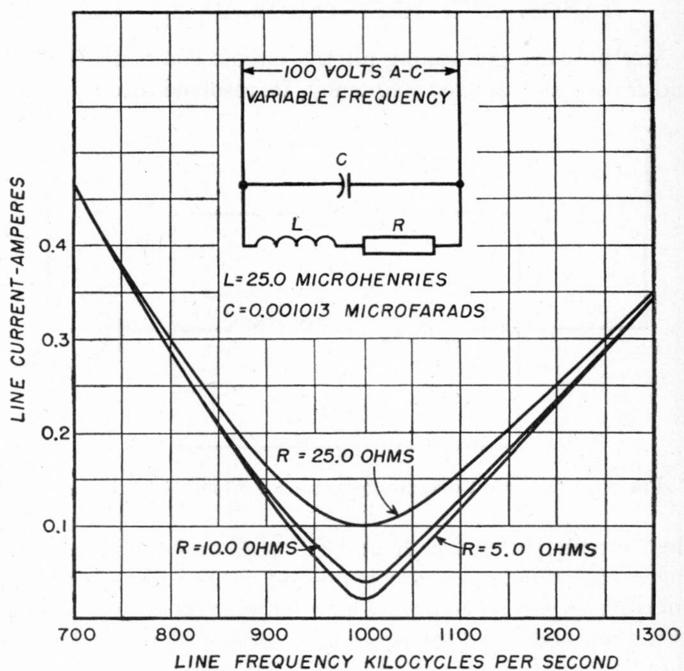

FIG. 11·21 Parallel resonance.

11·9 DISTRIBUTED CONSTANTS IN A-C CIRCUITS

Inductance, capacitance, and resistance may only be considered to be lumped constants as an approximation. It is sometimes necessary, owing to the circuit conditions or the operating condition, to take into consideration that these quantities actually are distributed over the circuit. An example of such a circuit is the transmission line. The characteristics of a circuit having distributed L, C, and R may be quite different from those of a similar circuit having lumped constants. Any circuit element always contains some inductance, some capacitance, and some resistance, regardless of its nominal function. Thus, a resistance has some inductance and capacitance, an inductance has some capacitance and resistance, and a capacitance has some resistance and inductance. Under certain conditions, usually at the higher frequencies, these secondary or residual components cannot be neglected, and they may actually predominate in determining the characteristics of the components. Frequently these constants are distributed; for example, the distributed capacity between turns and windings of a reactor. An interesting example of distributed constants is the single-layer coil shown in Fig. 11·22. Such a coil is often used to feed direct current to a plate or grid circuit of a vacuum-tube oscillator or amplifier. It presents high impedance to high-frequency voltages. On direct current and at low frequencies this coil acts as an inductance, but the capacitance between its turns combines with the inductance to change the resistance and impedance of the coil as the frequency is varied, as illustrated in Fig. 11·23. This impedance curve was obtained experimentally by connecting the coil across a source of known voltage at various frequencies and measuring current at the end of the coil. Radio-frequency milliammeters were connected in series at various points along the coil to measure current distribution along the coil. As frequency is increased from *a* to *b*, the coil is inductive and the reactance and impedance increase directly with frequency. At *b* the current in the capacitance between turns begins to be appreciable and, because it is 180 degrees out-of-phase with the inductive current, the total current decreases and thus makes the coil appear to have a higher impedance than that caused by the inductance. At *c* is what is known as "quarter-wave resonance" (distributed *LC* acts as a parallel-resonant circuit). At frequencies between *c* and *d* the current at the ends of the coil leads the voltage, the reactance is capacitive, and the impedance decreases with increasing frequency. At *d* is so-called "half-wave resonance," and the impedance is low. For further increases in frequency the impedance generally appears as either resistive or capacitive, rising to about the "three-quarter wave" resonant point and then decreasing again. The impedance continues to vary in a similar manner, going through minimum and maximum points as shown in Fig. 11·23. Resistance of the conductor and the dielectric loss between turns generally increase with frequency, and thus the resonance is limited until at very high frequencies the fluctuations become small or negligible. The design of a radio-frequency choke for high impedance over a wide range of frequencies is difficult. Two coils might be used, one similar to the coil just described in series with a second coil in which the high impedance point *c* would come at frequency *d*, and thus fill in the valley. However, caution

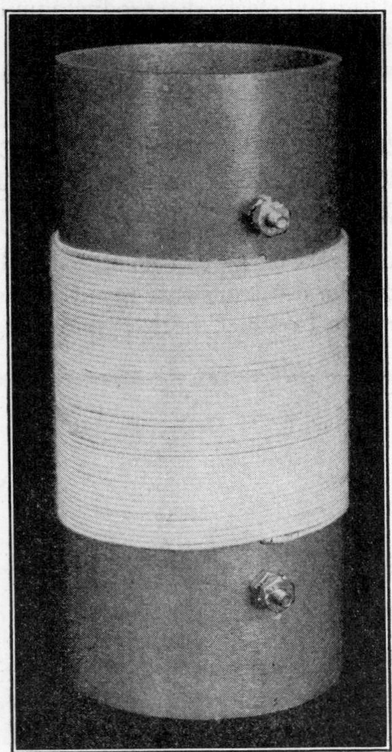

FIG. 11·22 Single-layer choke coil.

must be used, because at some frequency one coil is inductive and the other is capacitive, and a series-resonant circuit of low impedance at both frequencies results. One way of avoiding this is to design the coil for which point *c* is at the highest frequency to be covered, then connect several of these coils in series so that impedance is sufficiently large at the low-frequency end of the band.

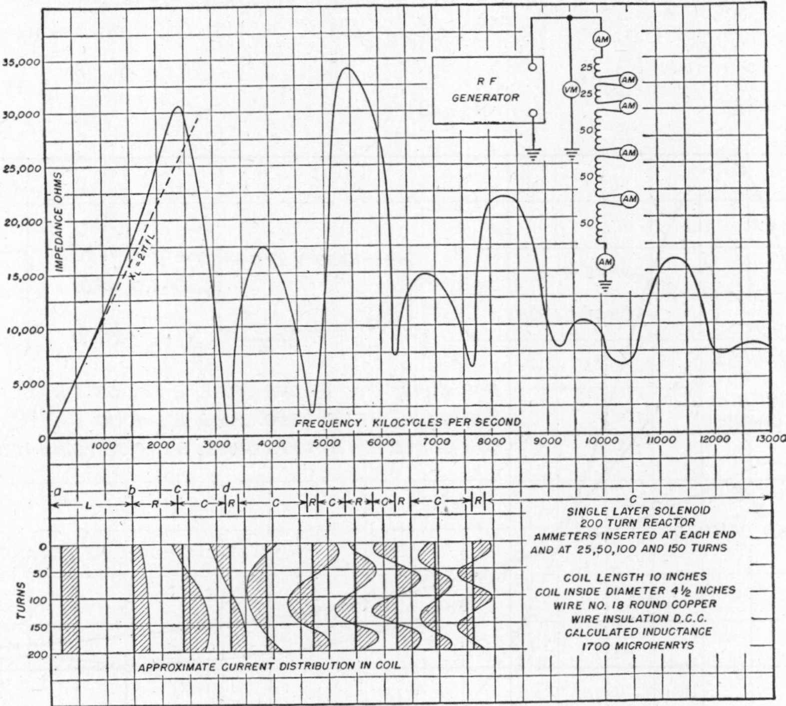

FIG. 11·23 Effect of distributed constants in a single-layer choke coil. (From *Electronic Transformers and Circuits* by Reuben Lee.)

The inductance of the leads of capacitors causes similar trouble, and in paper-foil capacitors the inductance of the many turns of the foil electrodes produces undesirable effects at high frequencies. The capacitance and inductance in resistors likewise can be detrimental at the higher frequencies.

11·10 CIRCUIT ELEMENTS OF INDUCTANCE, CAPACITANCE, AND RESISTANCE IN D-C CIRCUITS

Because electrical energy can be stored in both inductors and capacitors, the d-c transients resulting from changes of current in circuits in which both L and C are present are

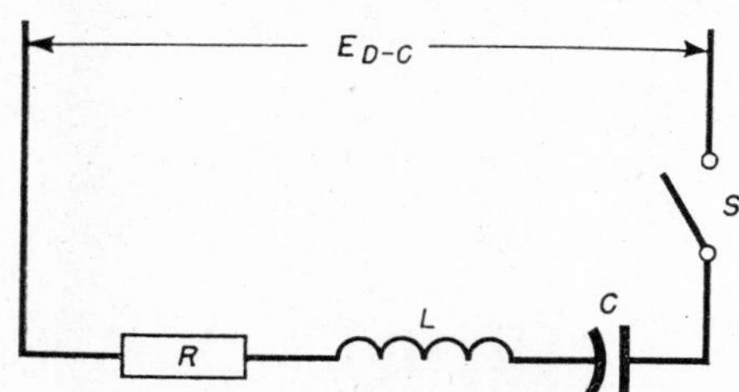

Fig. 11·24 Series circuit of R, L, and C.

frequently oscillatory and complex. In analyzing such transients the basic relation applying to instantaneous conditions must be used. For a reactor of inductance L,

$$e = -L\frac{di}{dt} \tag{11·20}$$

for a capacitor of capacitance C,

$$e = \frac{Q}{C} = \frac{\int_0^t i\,dt}{C} \tag{11·21}$$

for a resistor

$$e = Ri \tag{11·22}$$

From these relations the equations for transients can be established. The expressions usually are differential equations. Their solutions are complicated except in circuits of relatively simple form. Operational methods, either in the special form introduced by Heaviside or in the newer, more generalized forms, are useful in analyzing transients involving L, C, and R.

As an example, consider the sudden application of a unidirectional voltage to a circuit containing inductance, resistance, and capacitance in series as shown in Fig. 11·24. The current is the same in all three circuit elements and the sum of the voltage drops across them must be equal to the applied voltage. Therefore,

$$E_{d\text{-}c} = Ri + L\frac{di}{dt} + \frac{\int_0^t i\,dt}{C} \tag{11·23}$$

The solution of this equation is

$$i = \frac{E}{2bL}\,\epsilon^{-Rt/L}(\epsilon^{bt} - \epsilon^{-bt}) \tag{11·24}$$

where

$$b = \sqrt{\frac{R^2}{4L^2} - \frac{1}{LC}}$$

The general equation takes three forms according as b is real, imaginary, or zero.

If $R^2 > 4L/C$, then b is real, and

$$i = \frac{E}{bL}\,\epsilon^{-Rt/2L}\sinh bt \tag{11·25}$$

This is known as the damped or non-oscillatory condition, and it occurs where R is large.

If $R^2 < 4(L/C)$, then b is imaginary, and

$$i = \frac{E}{\beta L}\,\epsilon^{-Rt/2L}\sin \beta t \tag{11·26}$$

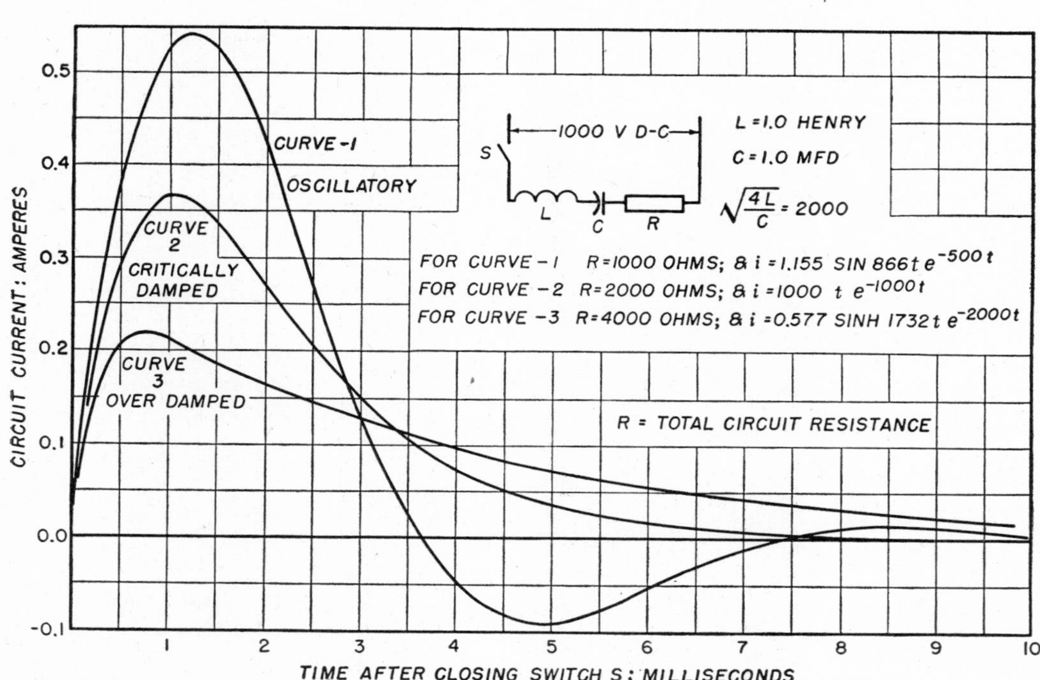

Fig. 11·25 D-c transients in L, C, and R series circuit.

where β is the numerical value of the imaginary b. This is known as the oscillatory condition, and it occurs for small values of R.

If $R^2 = 4L/C$, then $b = 0$, and

$$i = \frac{Et}{L} \epsilon^{-Rt/2L} \tag{11·27}$$

This is frequently referred to as the critically damped condition. The curves in Fig. 11·25 show current for a series circuit of L, C, and R, in which a steady unidirectional voltage is suddenly applied to the terminals. C is 1.0 microfarad and L is 1.0 microhenry.

For curve 1, R equals 1000 ohms, corresponding to an oscillatory condition since R is less than $\sqrt{4L/C}$. In curve 2, R equals $\sqrt{4L/C}$, corresponding to the critically damped condition. In curve 3, R is larger than $\sqrt{4L/C}$, and this corresponds to the damped condition.

Another common arrangement of L, C, and R in d-c circuits is in the form shown in Fig. 11·26. The voltage across C and R_2 is of interest. After the switch S is closed, the voltage across R_2 ultimately becomes equal to the applied voltage less the IR drop through resistor R_1. The voltage may oscillate above and below the final steady-state value with a magnitude and at a rate that depends on the values of L, C, R_1 and R_2.

With a steady source of voltage represented by E_1 and with R_1 representing the combined generator and reactor resistances, the voltage across load resistor R_2 at t seconds after closing switch S is

$$E_2 = E_1 R_2 \left[\frac{1}{R_1 + R_2} + 2A\epsilon^{-\alpha t} \cos(\omega t + \delta)\right] \tag{11·28}$$

where

$$\alpha + j\omega = \frac{K + \sqrt{K^2 - 4JS}}{2J} = P_1$$

and where

$$K = L + R_1 R_2 C$$

$$J = R_2 CL$$

$$S = R_1 + R_2$$

and

$$A\epsilon^{-j\delta} = \frac{P_1^2 LC + P_1 CR_1 + 1}{2JP_1^2 + KP_1}$$

This is applicable if $4JS$ is greater than K. Such a condition is usual for most filter circuits, and it indicates a transient oscillation. The effect is similar when the load is removed by opening the switch S.

Voltage across C rises to the no-load value after a transient oscillation. This voltage at t seconds after opening the switch S is

$$E_c = E_2 + E_1 \left[\frac{R_1}{R_1 + R_2} + 2A\epsilon^{-\alpha t} \cos(\omega t + \delta)\right] \tag{11·29}$$

where

$$\alpha + j\omega = \frac{K + \sqrt{K^2 - 4JS}}{2J} = P_1$$

$$K = L + R_1 R_2 C$$

$$J = R_2 CL$$

$$S = R_1 + R_2$$

$$A\epsilon^{-j\delta} = \frac{P_1 L + R_1}{2JP_1^2 + P_1 K}$$

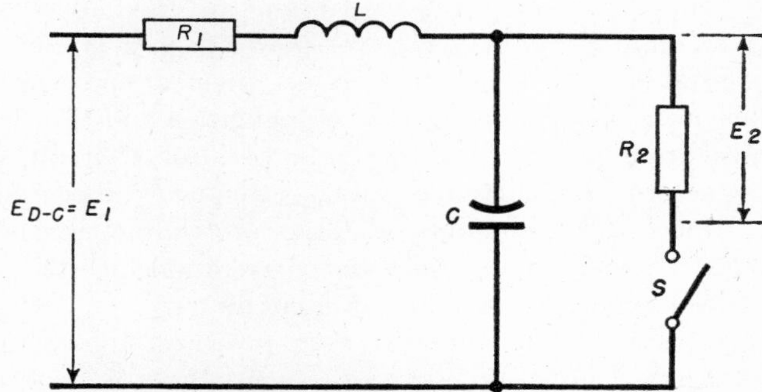

FIG. 11·26 Typical keying circuit.

As an approximation applying to usual filter circuits the maximum voltage deviation from the final value is given in percent by the expression

$$\text{Percent maximum deviation} = \frac{10^5}{R_2}\sqrt{\frac{L}{C}} \tag{11·30}$$

where R_2 is in ohms
L is in henrys
C is in microfarads.

As an example, if $L = 1.0$ henry, $C = 4.0$ microfarad, and $R_2 = 2000$ ohms, the percentage of deviation $= (10^5/2000)\sqrt{1/4} = 25$ percent. Because the resistance has a damping effect, the actual deviation is somewhat less than this approximate formula indicates. In the example just given, the actual deviation is about 22 percent. Likewise, as switch S opens, the voltage across the capacitor rises temporarily to a value above normal applied voltage. Approximate voltage rise in percent of normal can be determined by the expression used for the case of closing the switch.

Such a circuit is used in the telegraphic keying of radio transmitters. The transients of the radiated wave are affected by the constants of the filter circuits in the manner described.

Chapter 12

TUNED CIRCUITS AND FILTERS

Reuben Lee

IN general, tuned circuits and filters are used to obtain maximum current, voltage, or power from a system at desirable frequencies, or to reject them at undesirable frequencies. Incidental problems of regulation and transient response are important, but they arise because of circuit elements the primary function of which is frequency discrimination. Hence tuned circuits and filters are characterized by quantitative, and usually comparative, relations at different frequencies or in different parts of a circuit.

12·1 SERIES RESONANT CIRCUIT

One of the simplest tuned circuits is that shown in Fig. 12·1, in which resistance R, inductance L, and capacitance

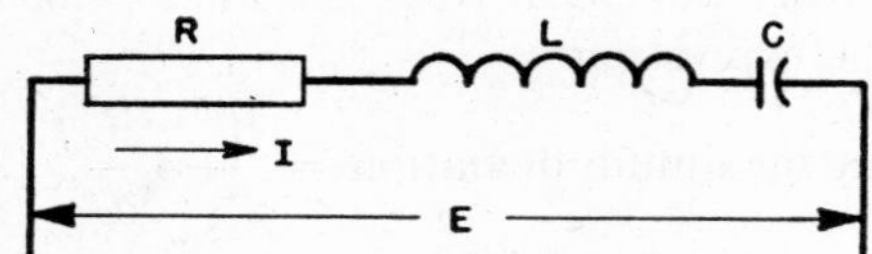

FIG. 12·1 Series RLC circuit.

C are connected in series across an alternating voltage E. If this voltage E is sinusoidal, the current I in the circuit also is sinusoidal. At some frequency the reactances of L and of C become equal, and the current in the circuit reaches a maximum and is equal to E/R. This condition is known as the tuned or resonant condition. It is achieved in one of three ways: (1) by varying the frequency of the applied voltage, (2) by varying L, or (3) by varying C. In Fig. 12·2 vector diagrams show the change in I for a fixed voltage E, inductance L, and resistance R, as capacitance C is increased beyond the point of maximum current. Voltages across the

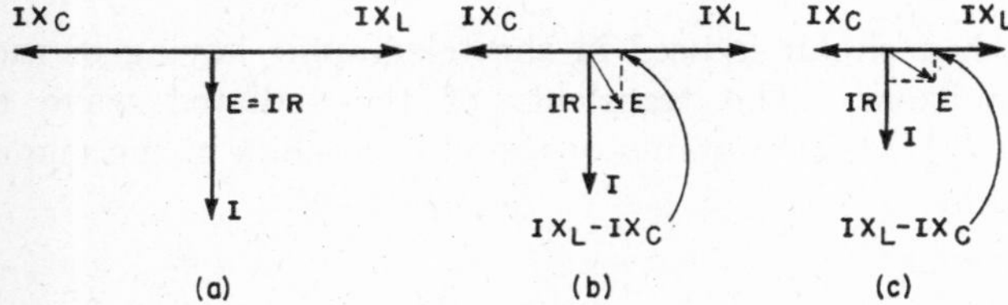

FIG. 12·2 Vector relations during detuning.

elements L and C become smaller as I decreases, that across C becoming smaller more rapidly. As this change is made, E begins to lead I. Therefore, the circuit acts as if it were an inductance in series with resistance R, the value of the inductive reactance being $X_L - X_C$. A similar change occurs if the capacitance is decreased on the other side of its resonant value. As the capacitance is decreased below the

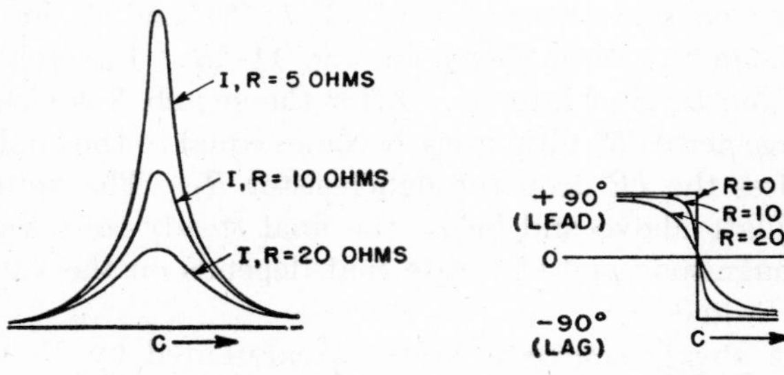

FIG. 12·3 Change in current as capacitance is varied.

FIG. 12·4 Change in current phase angle as capacitance is varied.

resonant value, the voltage lags the current, and thus the circuit acts as a capacitance in series with R. Figure 12·3 shows the change in current as capacitance is varied through this region. The influence of R in limiting the maximum current I is clearly shown. R may be an external resistance or only that of the coil, or both. Curves similar to those in Fig. 12·3 may be drawn if inductance or frequency are varied instead of capacitance (see Chapter 11).[1] Figure 12·4 shows how current I leads or lags E as the circuit is tuned through resonance.

For a given frequency, the current at zero capacitance is zero also, because circuit reactance is infinite at that value. At the other extreme, when C is infinite, circuit reactance is $R + j\omega L$. Thus the curves of Fig. 12·3 are not symmetrical at points widely separated from the resonant frequency, but are nearly symmetrical at resonance. Equations for impedance are given in Chapter 11.

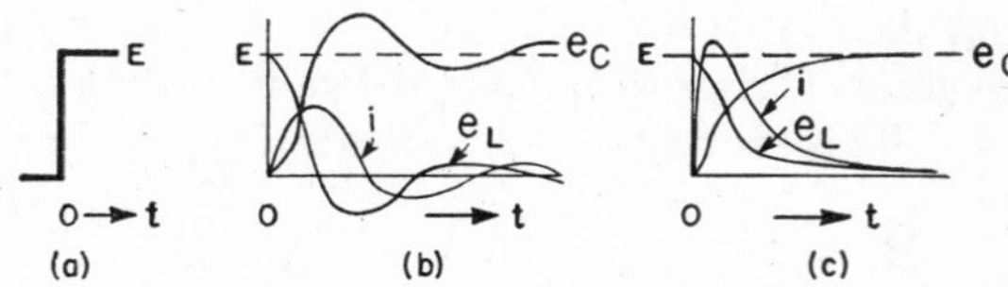

FIG. 12·5 (a) Step function. (b) Oscillatory circuit. (c) Non-oscillatory circuit.

If, instead of a sinusoidal voltage, a step function such as that in Fig. 12·5 (a) is applied, the resulting voltage and current depend upon whether the circuit is oscillatory. If the circuit is highly oscillatory, the voltage across C rises to a value nearly double that of the applied voltage E. The voltage across L becomes alternately positive and negative as it oscillates about zero.

Initial current is zero, and the current rises to a maximum at the instant voltage across L reaches zero. After that it oscillates about zero as an axis. Initially, all the voltage is across the inductor. Finally, it is all across the capacitor. These relations are illustrated for a moderately oscillatory circuit in Fig. 12·5 (*b*).[2]

If the circuit is non-oscillatory, the voltages and currents do not oscillate but become asymptotic to the value shown in Fig. 12·5 (*c*). The border line between the condition of oscillations and no oscillations is given by

$$R = 2\sqrt{\frac{L}{C}} \tag{12·1}$$

Frequency of oscillation is not exactly equal to $1/2\pi\sqrt{LC}$ but is

$$f_n = \frac{1}{2\pi}\sqrt{\frac{1}{LC} - \frac{R^2}{4L^2}} \tag{12·2}$$

There is always an interval of time from the initial application of the step function to that at which it reaches the value E. This interval can be regarded as a sort of phase shift, although a more accurate way of regarding it is in terms of time delay.

A comparison between the sinusoidal and step function performance can be obtained from Table 12·1.

Table 12·1 Performance of Resonant Circuits with Sinusoidal and Step-Function Voltages Applied

Quantity	Sinusoidal Voltage E	Step-Function Voltage E
Current	Maximum value E/R occurs at resonance.	Initially zero; maximum value E/Bt where B is the angular frequency of oscillations; final value zero.
Voltage across C	Approximately QE at resonance where $Q = \frac{\omega L}{R} = \frac{1}{\omega CR}$	Initially zero; may rise to maximum of $2E$ in oscillatory circuit; to E in non-oscillatory circuit; final value E.
Voltage across L	Approximately QE.	Initially E; may go negative to a maximum value of $-E$ in oscillatory circuit; final value zero.
Phase angle	Leading at $f <$ resonance; lagging at $f >$ resonance; zero at $f =$ resonance; approaches abrupt step as $Q \to \infty$	Time delay approximates a quarter cycle of natural frequency in oscillatory circuits; longer in non-oscillatory circuits.
Frequency	Resonance frequency $\frac{1}{2\pi\sqrt{LC}}$	Frequency of oscillations $\frac{1}{2\pi}\sqrt{\frac{1}{LC} - \frac{R^2}{4L^2}}$

12·2 PARALLEL RESONANT CIRCUIT

A typical parallel resonant circuit is shown in Fig. 12·6. Capacitor losses usually are negligible, and coil resistance is appreciable; this resistance may include the load resistance. If a sinusoidal voltage of constant frequency and amplitude is applied to this circuit, and if either C or L is varied, current in the variable branch changes slowly, but does not change in the other branch. The sum of these two currents, shown as I_Z in Fig. 12·6, approaches a minimum as resonance is reached, and this minimum is nearly zero if R is small.[3] In most electronic circuits, the impedance of the source is appreciable. For a constant voltage applied, for example, to the grid of a vacuum tube, the voltage on the resonant circuit is not constant but increases as the circuit is tuned. The voltage rises in somewhat the same way as the current does in Fig. 12·3. This is a result of the source impedance rather than a property of the parallel resonant circuit itself. This increase in voltage and the decrease in

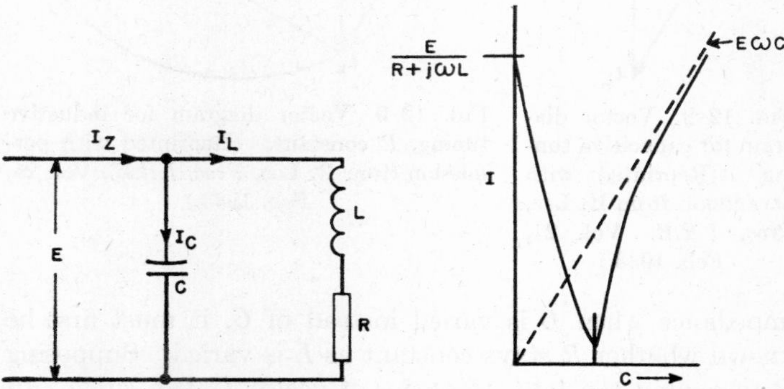

Fig. 12·6 Parallel resonant circuit.

Fig. 12·7 Current as parallel circuit is tuned through resonance.

current in Fig. 12·7 are asymmetrical, especially for low-Q circuits, as would be expected from the lack of symmetry in the circuit elements.

If the capacitance were reduced to zero, the circuit reactance would be $R + j\omega L$. If the capacitance became infinitely large, the circuit impedance would be zero. As capacitance is increased, the current in Fig. 12·7 becomes asymptotic to the line $E\omega C$. The general equation for impedance is given in Chapter 11.

In Fig. 12·8 is a vector diagram in full lines for the resonant condition of unity power factor. Total current I_Z entering the circuit is in phase with applied voltage E. Capacitor current I_C leads E by 90 degrees, and coil current I_L lags E by some angle θ. This angle is determined by the relation between R and X_L, where X_L is the coil reactance at the impressed frequency, or $2\pi fL$. It can be shown that

$$X_C = \frac{R^2 + {X_L}^2}{X_L} \tag{12·3}$$

and

$$Z = \frac{R^2 + {X_L}^2}{R} = \frac{L}{RC} \tag{12·4}$$

where Z = total circuit impedance. These are the mathematical relations between Z, X_C, X_L, and R for unity-power-factor resonance.

If C can be varied from the condition of unity power factor, and E is assumed to be constant, I_C varies accordingly, and I_L is not affected. Figure 12·8 shows that maximum impedance and unity power factor are obtained with the same setting of the capacitor, provided the circuit is tuned by varying C alone.

With inductance tuning the circuit relations may be different for the unity-power-factor and maximum-impedance conditions. To determine circuit relations for maximum

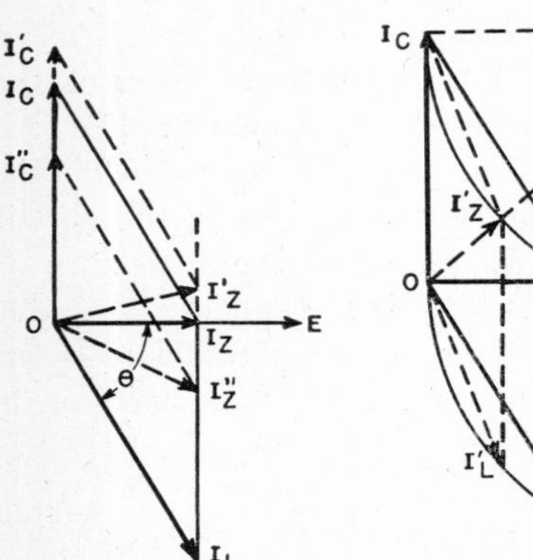

FIG. 12·8 Vector diagram for capacitive tuning. (Reprinted with permission from R. Lee, *Proc. I.R.E.* Vol. 21, Feb. 1933.)

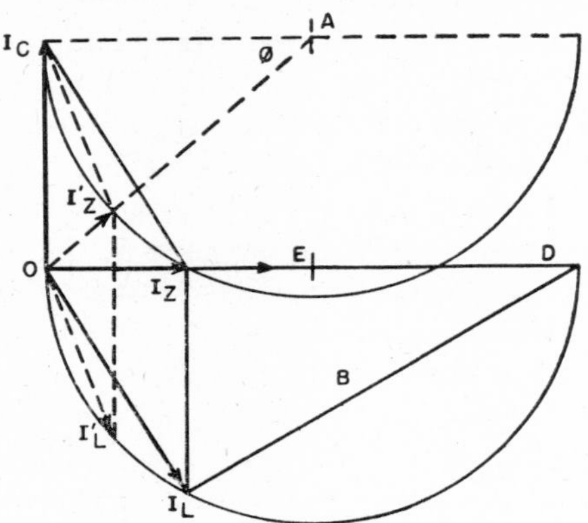

FIG. 12·9 Vector diagram for inductive tuning, R constant. (Reprinted with permission from R. Lee, *Proc. I.R.E.*, Vol. 21, Feb. 1933.)

impedance when L is varied instead of C, it must first be known whether R stays constant as L is varied. Supposing R to remain constant, the locus of vector I_L is a circle. In Fig. 12·9 the solid lines indicate the unity-power-factor condition. Let the circle be shifted vertically so that it passes through I_Z and becomes the circle with center A. The circuit relations for maximum impedance and inductive reactance are

$$Z = \frac{2X_C R}{\sqrt{X_C^2 + 4R^2} - X_C} \qquad (12\cdot5)$$

and

$$X_L = R\sqrt{\frac{2\sqrt{X_C^2 + 4R^2}}{\sqrt{X_C^2 + 4R^2} - X_C} - 1} \qquad (12\cdot6)$$

In practice R sometimes varies when L is varied. The unity-power-factor condition is shown again in Fig. 12·10 in solid lines. Considering X_L/R as a constant, the angle θ remains constant also. Impedance Z is maximum when the entering current is I_Z and

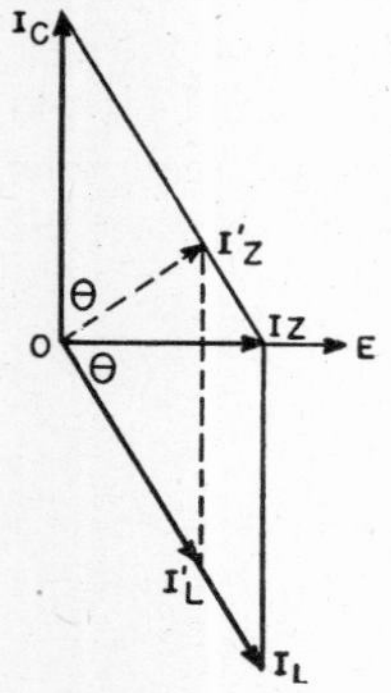

$$X_C = X_L \qquad (12\cdot7)$$

then

$$Z = \frac{X_L}{R}\sqrt{R^2 + X_L^2} \qquad (12\cdot8)$$

FIG. 12·10 Vector diagram for X_L/R a constant. (Reprinted with permission from R. Lee, *Proc. I.R.E.*, Vol. 21, Feb. 1933.)

Equations 12·7 and 12·8 are for maximum impedance when X_L/R is a constant and L is varied. The relation between X_C and X_L is the same as the relation in a series resonant circuit.

From vector diagrams similar to Figs. 12·9 and 12·10, a series of values is plotted in Figs. 12·11 and 12·12. These curves show I_L/I_C, X_L, and X_C at different values of Z and a given value of R (25 ohms) for unity power factor and maximum impedance. Figure 12·11 is for constant R, and Fig. 12·12 for X_L/R constant. For any other value of R, say R', abscissas and the reactance ordinates should be multiplied by $R'/25$ to make the curves

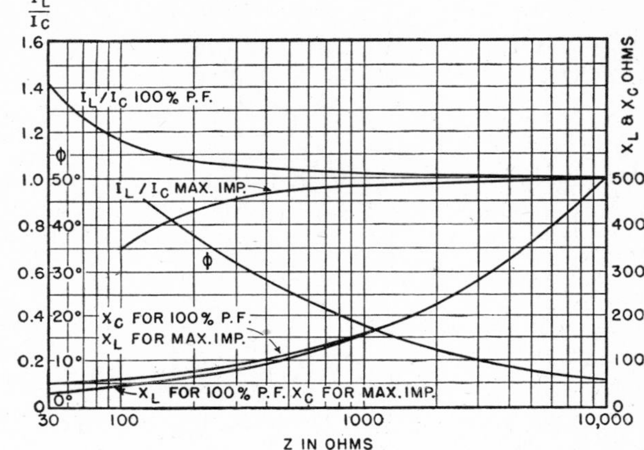

FIG. 12·11 Parallel resonance curves, R constant. (Reprinted with permission from R. Lee, *Proc. I.R.E.*, Vol. 21, Feb. 1933.)

applicable. I_L/I_C requires no scale alteration when these multiplications are made.

As Z, X_L, and X_C increase with respect to R, I_L/I_C becomes more nearly the same for the unity-power-factor

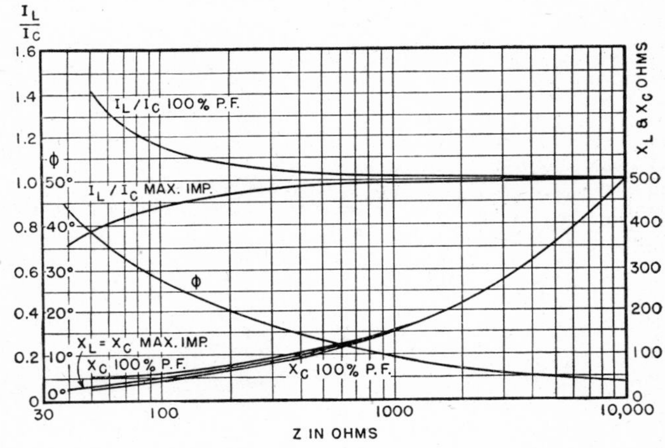

FIG. 12·12 Parallel resonance curves, X_L/R constant. (Reprinted with permission from R. Lee, *Proc. I.R.E.*, Vol. 21, Feb. 1933.)

and maximum-impedance conditions in both Figs. 12·11 and 12·12. As these conditions approach each other, the vector diagrams become less and less accurate, so that the algebraic relations become necessary for accurate calculations. The higher Z is relative to R, the higher is the ratio of volt-amperes in the circuit to the watts consumed in R. Therefore, at high-Q or volt-amperes-per-watt ratios, tuning for unity power factor becomes almost the same as for maximum impedance.

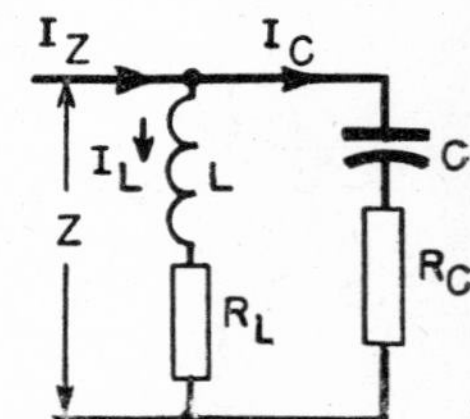

FIG. 12·13 Parallel circuit with resistance in both branches.

In some circuits the resistance of the capacitive branch is appreciable, but such circuits are rare. An unusual condition develops when resistance is purposely added to this branch, provided certain relative values are maintained. If in Fig. 12·13 the two resistors are of equal value, and if

$$R_c = R_L = \sqrt{\frac{L}{C}} \tag{12·9}$$

the impedance of the circuit is the same at all frequencies, and has unity power factor at all frequencies. A vector

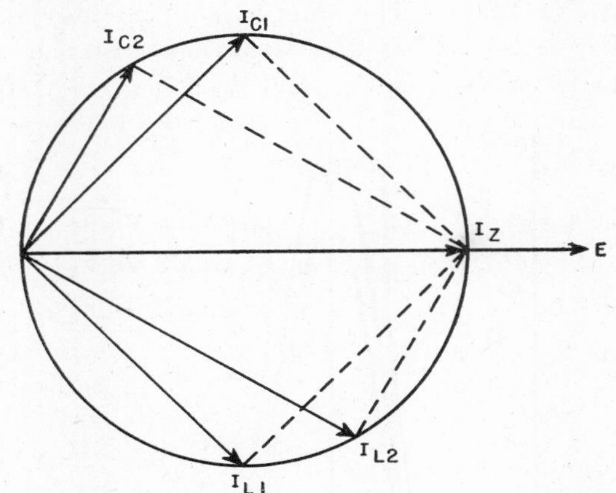

FIG. 12·14 Entering current I_Z in Fig. 12·13 is independent of frequency when $R_L = R_C = \sqrt{L/C}$. Relations for two frequencies.

diagram illustrating these properties at two frequencies is shown in Fig. 12·14. At the frequency for which I_C is the larger, it approximates I_Z in magnitude. This effect is sometimes used in networks to maintain constant impedance and power factor over a wide frequency band.

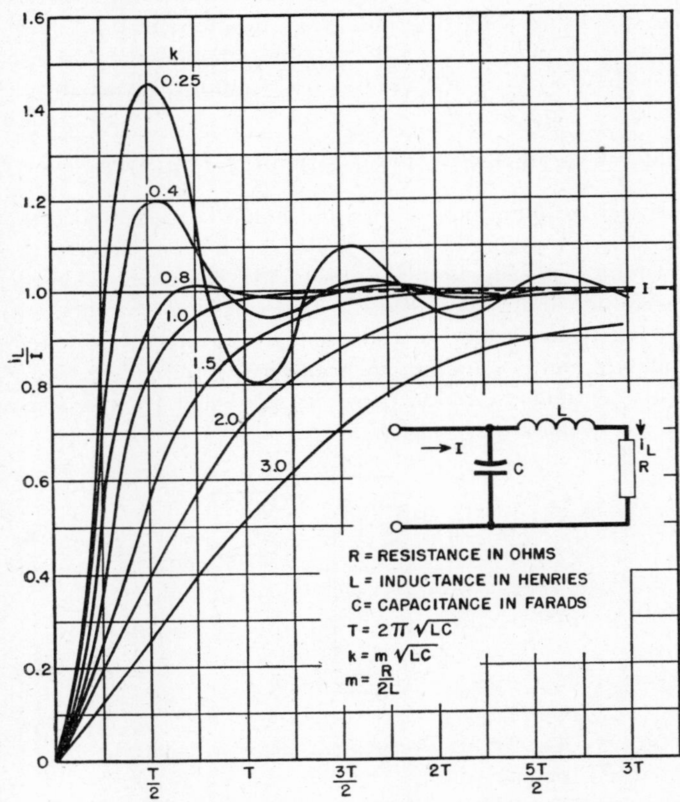

FIG. 12·15 Response of parallel circuit to suddenly impressed current. (From *Electronic Transformers and Circuits* by Reuben Lee.)

Into the parallel circuit of Fig. 12·6 may flow a suddenly impressed current, that is, a current having the same shape as the voltage of Fig. 12·5 (*a*). Anode current in a suddenly unblocked pentode amplifier has this shape. Current i_L in terms of the entering current I is plotted in Fig. 12·15 for several sets of values of circuit elements. Abscissas are not time but time constant, which is approximately the reciprocal of the angular frequency of the circuit.

I_Z lags E at frequencies below resonance and leads at frequencies higher than resonance. This situation is the reverse of that for a series resonant circuit. The phase angle is zero at resonance provided this resonance is taken as the unity-power-factor condition.

It is most feasible to compare the sinusoidal and step-function performance of a parallel circuit on a current basis. This is done in Table 12·2. In this table Q denotes the ratio of ωL to R, including load resistance. It is the Q of the coil only if the load resistance is zero.

TABLE 12·2 COMPARISON OF SINUSOIDAL AND STEP-FUNCTION CURRENTS IN A PARALLEL TUNED CIRCUIT

Quantity	Sinusoidal Current	Step-Function Current
1. Entering current I_Z	ERC/L for unity-power-factor condition (approximately true for maximum impedance for high Q), where $Q = \frac{\omega L}{R}$	Assumed constant after initial instant.
2. Current in inductive branch I_L	Approximately QI_Z.	Initially zero; rises to constant value I according to Fig. 12·15; may overshoot to $2I$ in highly oscillating circuit.
3. Current in capacitive branch I_C	Approximately QI_Z.	Initially I; at any instant $I - i_L$.
4. Phase angle	Lagging at $f <$ resonance; leading at $f >$ resonance; zero at f = resonance.	Time delay as shown in Fig. 12·15.
5. Frequency	Resonance frequency $\frac{1}{2\pi}\sqrt{\frac{1}{LC} - \frac{R^2}{L^2}}$	Frequency of oscillations $\frac{1}{2\pi}\sqrt{\frac{1}{LC} - \frac{R^2}{4L^2}}$

12·3 COUPLED CIRCUITS

The term coupling denotes voltage or power transfer from one part of a circuit to another. Coupling may be of several kinds, depending upon the type of circuit and the means used for transfer of power or voltage.[4] The term is rather loosely used, and may indicate an element across which the transfer is made, or may indicate the connecting means between different parts of a circuit. The discussion here is confined to inductive coupling. A discussion of one form of capacitance coupling is given in Section 12·4.

Consider first the circuit of Fig. 12·16, in which impedance Z_1 is complex and includes the inductance of the primary coil. Likewise, secondary impedance Z_2 is complex and includes the inductance of the secondary coil.

The equivalent impedance of the circuit of Fig. 12·16 when referred to the primary side is

$$Z' = Z_1 + \frac{X_M^2}{Z_2} \qquad (12\cdot10)$$

where $X_M = j\omega L_m$ ohms
ω = angular frequency of E_1
L_m = mutual inductance of the two coils in henrys.

The equivalent impedance referred to the secondary side is

$$Z'' = \frac{-jX_M}{Z_1 Z_2 + X_M^2} \qquad (12\cdot11)$$

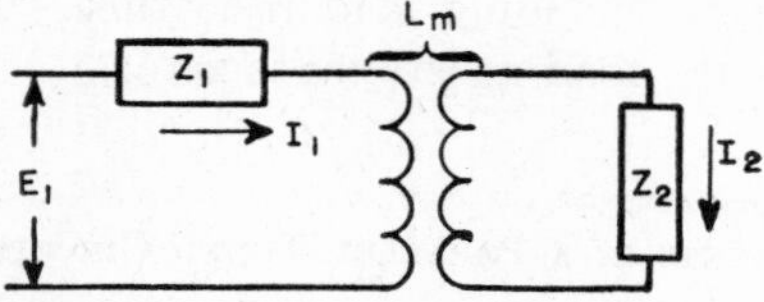

FIG. 12·16 Inductively coupled circuits (general case). (From *Electronic Transformers and Circuits* by Reuben Lee.)

If the primary resistance is zero, or virtually zero, and the secondary circuit is tuned to resonance, Z_2 is a pure resistance R. Under these conditions, equation 12·10 reduces to

$$R' = \frac{X_M^2}{R} \qquad (12\cdot12)$$

where R' is the equivalent resistance in the primary circuit.

Equation 12·12 gives the mutual inductance required for coupling a resistance R so it appears to be a resistance R' in the primary with a maximum power transfer between the two circuits.

The ratio of mutual inductance to the geometric mean of the primary and secondary self-inductance is known as the coefficient of coupling, or

$$k = \frac{L_m}{\sqrt{L_1 L_2}} \qquad (12\cdot13)$$

The value of k is never greater than unity even if coils are interleaved to the maximum possible extent. Values of k down to 0.01 or lower are common.[5]

A form of coupled circuit often used is shown in Fig. 12·17. A sinusoidal voltage E_1 is commonly impressed on the

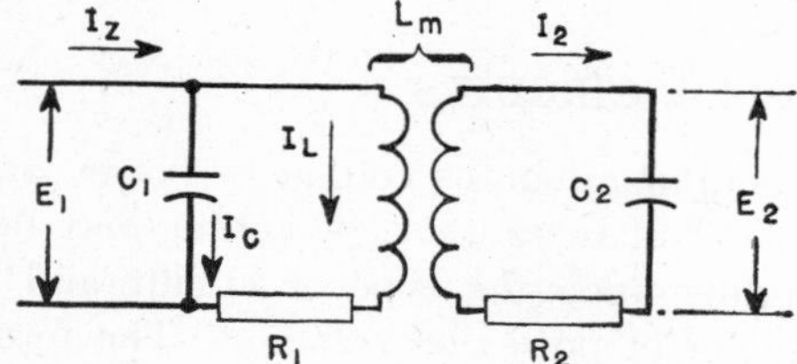

FIG. 12·17 Inductively coupled tuned circuits. (From *Electronic Transformers and Circuits* by Reuben Lee.)

primary circuit by a vacuum-tube amplifier. E_2 in this circuit depends upon impressed frequency, and is shown in Fig. 12·18 for resonance at three different values of coupling. If the value of coupling is such that

$$X_M = \sqrt{R_1 R_2} \qquad (12\cdot14)$$

the condition is similar to that of equation 12·12, in which maximum power or current is produced in the secondary circuit. Maximum current through C_2 gives maximum E_2. This coupling is known as the critical value. Smaller values of coupling give a smaller maximum of E_2. Values of coupling greater than critical result in a "double hump," as shown in Fig. 12·18. The amplitude of the resonant peaks and valleys, as well as the frequency distance between peaks, varies with circuit Q and coefficient of coupling k.

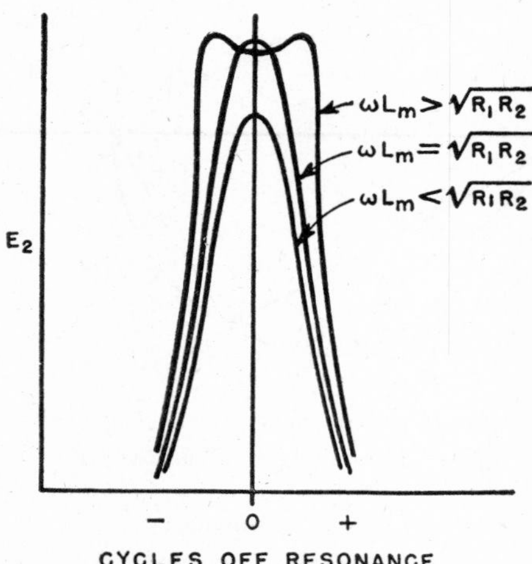

FIG. 12·18 Response curves for circuit of Fig. 12·17. (From *Electronic Transformers and Circuits* by Reuben Lee.)

The usefulness of curves such as the double-hump curve of Fig. 12·18 is that discrimination is afforded to adjacent frequencies and yet very little attenuation is offered to a band near the frequency normally corresponding to resonance.

12·4 POWER-AMPLIFIER OUTPUT CIRCUITS *

Power-amplifier output circuits should be designed so that the proper amount of power is delivered to the load with a minimum of distortion and harmonic content.[6] Consider the simplified circuit of Fig. 12·19, consisting of a tube with grid excitation from a preceding stage, an anode tank circuit C and L, and a coupling coil transferring the power from L to the antenna. Anode power is furnished from a source HV at a constant voltage E_b.

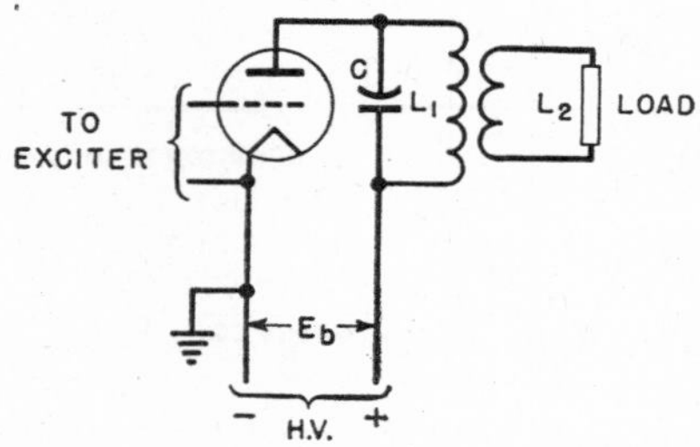

FIG. 12·19 Simple tuned amplifier circuit.

The tank circuit can be represented by Fig. 12·6, in which C and L correspond to similar quantities in Fig. 12·19, and R is the tank equivalent of load resistance plus tank-coil resistance. This circuit may be considered as having

* Sections 12·4, 12·5, and 12·6 are based on the author's "Design of Power Amplifier Output Circuits," *Radio Eng.*, July 1934.

an entering current I_Z, which is the a-c component of I_p, and which develops across the tank an effective alternating voltage E_p. The frequency at which this circuit operates is taken as the fundamental frequency unless otherwise stipulated.

The peak value of the alternating voltage is

$$E_{pk} = \frac{4E_b\eta}{\pi} \tag{12·15}$$

where η is anode efficiency, for values of $\eta < 0.785$.

The rms value of the tank voltage is $E_p = 0.707E_{pk}$. To obtain the currents and voltages here described, the fundamental component of I_Z must be in phase with E_p; that is, the tank circuit must be tuned to the unity-power-factor condition represented by the vector diagram of Fig. 12·8. This diagram shows that tank-capacitor current I_C leads I_Z and E_p by 90 degrees and that tank-coil current lags I_Z by an angle somewhat less than 90 degrees because of the resistance in the coil branch.

The product of E_p and I_C or I_L evidently may be (and in fact usually is) greater than the product of E_p and I_Z. The ratio K_T of these two products is the volt-amperes/watts or kva/kw ratio, the volt-amperes being the apparent power and the watts the true power, including losses, in the tank circuit.

If the combination of L and C in Fig. 12·19 were replaced by a simple resistance, the power amplifier would still function properly, and this would be the same as making volt-amperes/watts = 1. However, tube current I_p, which may have large harmonic components, would pass into this resistance unchanged, and the output of the amplifier would be high in harmonic content. But capacitor C in Fig. 12·19 allows the harmonic components to flow freely, while the coil L discriminates against them. Thus the effect of high K_T is to reduce harmonics; conversely, the lowest value of this ratio in Fig. 12·19 is fixed by the percentage of harmonics in the load circuit.

The upper limit of K_T is that of size or expense. This is true of single-frequency amplifiers but, if the impressed voltage is modulated, there is another limitation. Because of the time interval required to store or remove energy from the tank-circuit C and L, voltage E_p cannot be varied too rapidly, and the upper limit of K_T is determined by the highest modulation frequency. Thus the problem is to find the lower limit of K_T as determined by harmonic content and the upper limit from the highest modulation frequency.

12·5 HARMONIC CONTENT

Designate harmonic currents by primes (I'_p = any harmonic component of anode current, I'_L = any harmonic component of tank coil current), and let the symbols without primes refer to the fundamental component. The ratio I'_L/I_L is plotted for second and third harmonics in Fig. 12·20 for tank circuits working in an average class C amplifier. Higher harmonics have smaller amplitudes and are attenuated to a greater degree than the second or third, which are the only ones that usually need to be considered.

Two factors make Fig. 12·20 less accurate than could be desired. The first is distributed capacitance, which is not directly calculable. Its effect is to pass the higher harmonics between the turns of L_1 and over to L_2 more freely than Fig. 12·20 would indicate, especially at the higher frequencies. For this reason, a grounded static shield is sometimes placed between coils L_1 and L_2.

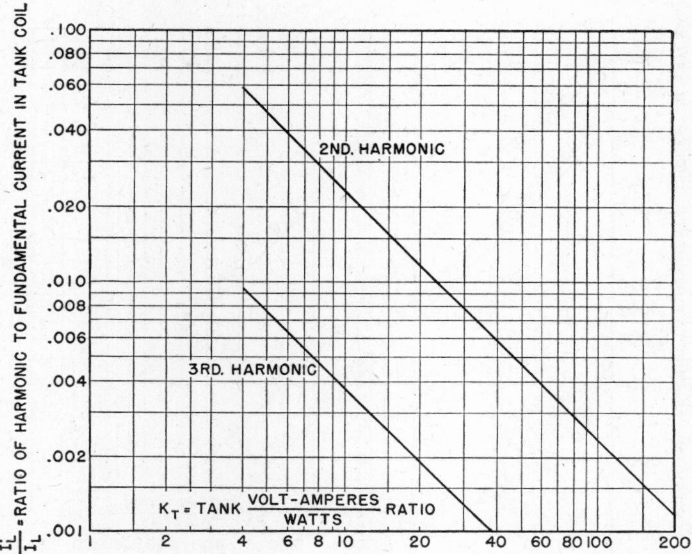

FIG. 12·20 Harmonic content of tank coil current for circuit shown in Fig. 12·19. (Reprinted with permission from R. Lee, *Radio Eng.*, Vol. 14, July 1934.)

The other factor not accounted for by Fig. 12·20 is possible load resonance at the harmonic frequency. If the load is an antenna which is resonant at some harmonic, the radiated field strength of this harmonic might be high, even though the harmonic component of antenna current were in accordance with Fig. 12·20. The antenna should be adjusted slightly to avoid such resonance, or the harmonic should be attenuated by some other means.

An antenna is rarely connected directly to the coil coupling as shown in Fig. 12·19. Usually some other apparatus is used to tune the antenna to resonance at the fundamental frequency. Nevertheless, a tank circuit is often coupled to a resistance in this way for other types of load.

Suppose a capacitive antenna is tuned to resonance by an inductance L_A. The output circuit could be shown as in Fig. 12·21, where L_A is the total inductance of the circuit,

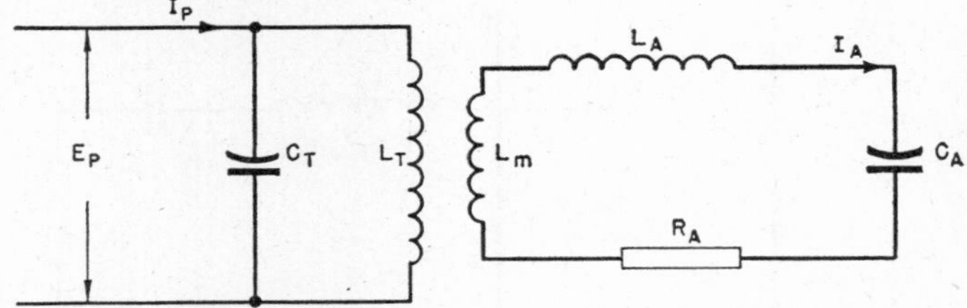

FIG. 12·21 Inductively coupled antenna circuit. (Reprinted with permission from R. Lee, *Radio Eng.*, Vol. 14, July 1934.)

C_A is the total capacitance of the circuit, and R_A is the sum of antenna and coil resistances. The antenna fundamental and harmonic currents, I_A and I'_A, respectively, are plotted in solid lines in Fig. 12·22 for the second and third harmonics between the harmonic current ratio I'_A/I_A and the product K_TK_A of the volt-ampere ratios in the tank and antenna circuits. Figure 12·22 is subject to the same general stipulations regarding tube current, capacitive coupling, and antenna impedance mentioned for Fig. 12·6.

The curves show the minimum ratio K_T that can be permitted in circuits like Figs. 12·21 and 12·6 for any ratio of

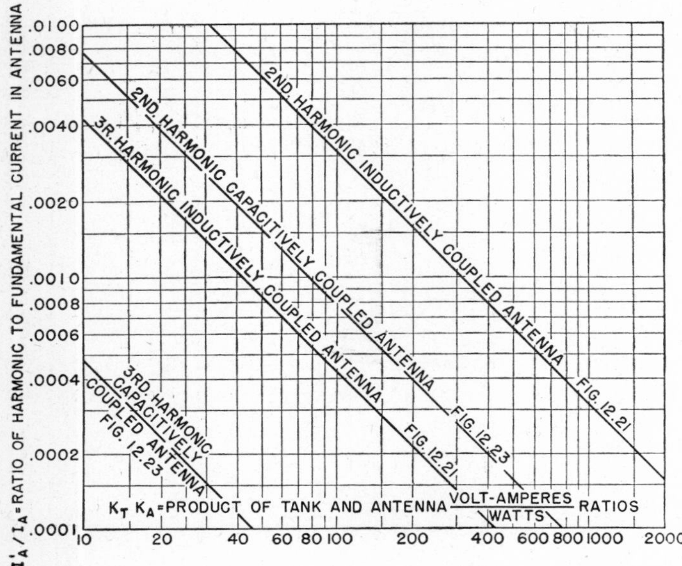

FIG. 12·22 Harmonic content of antenna current for circuits shown in Figs. 12·21 and 12·23. (Reprinted with permission from R. Lee, *Radio Eng.*, Vol. 14, July 1934.)

harmonic to fundamental antenna current. If this ratio cannot be made low enough without excessive size, other means can be used. One of the most common is connecting amplifier tubes in push-pull. By this means even harmonics are "balanced out" or reduced so that the harmonic of greatest magnitude is the third. Figures 12·20 and 12·22 show how harmonic current can be reduced in this way.

If push-pull is impractical, or if still further harmonic reduction is necessary, the capacitively coupled circuit of Fig. 12·23 can sometimes be used. The harmonic antenna-current ratio in this circuit is found from the broken lines of Fig. 12·22, if the capacitive reactance of capacitor C_c and the antenna resistance are low, as they usually are

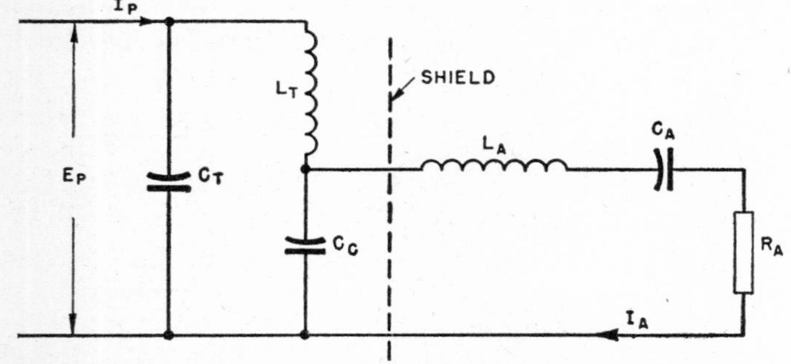

FIG. 12·23 Capacitively coupled antenna circuit. (Reprinted with permission from R. Lee, *Radio Eng.*, Vol. 14, July 1934.)

where this circuit is used. The effect of distributed capacitance in increasing harmonic currents is almost zero, particularly if the tank circuit is well shielded from the antenna circuit, as indicated in Fig. 12·23. Figure 12·22 shows that harmonic currents can be made very small with this circuit.

12·6 MODULATION

In a modulated amplifier, there is an upper limit of tank and antenna volt-amperes which must not be exceeded if distortion is to be avoided. This distortion is the result of applying and removing power from a circuit at a more rapid rate than the natural time of build-up and decay of the circuit. It imposes a limit upon the highest modulation frequency f_m as follows:

$$\frac{f}{f_m} = JK_A \qquad (12\cdot16)$$

where J is a constant depending upon the allowable audio distortion at f_m, and f is the operating carrier frequency. This equation neglects tank losses, and is accurate only for loosely coupled circuits, or for values of K_A and K_T greater

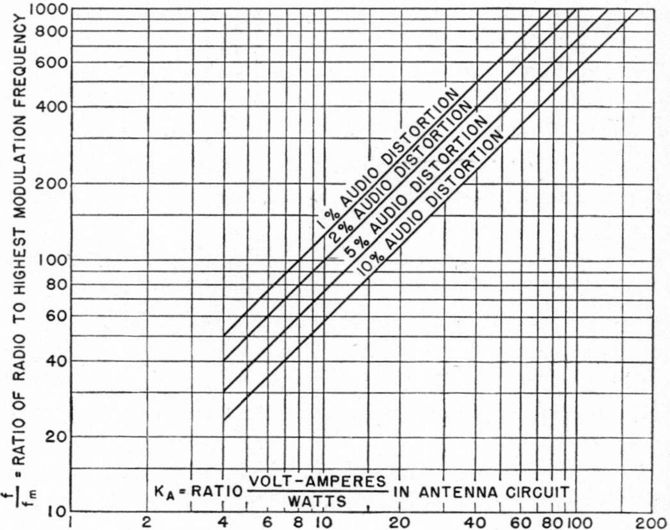

FIG. 12·24 Upper limit of modulation frequency. (Reprinted with permission from R. Lee, *Radio Eng.*, Vol. 14, July 1934.)

than 4. It is plotted in Fig. 12·24 for 1, 2, 5, and 10 percent audio distortion, for either inductive or capacitive antenna coupling. For the circuit of Fig. 12·19, the term J in equation 12·16 should have values one half of those just given, and K_A should be replaced by K_T.

12·7 TANK-CIRCUIT FREQUENCY RANGES

As long as the circuit is tuned to a single frequency, the variable part of capacitance or inductance required to tune the circuit is small, and the circuit performs approximately as desired. However, if a large frequency range is to be covered, changes in current may be too large. It is then necessary to limit the variation of tank current.

One common method of limiting it is to use tapped capacitors and coils. Capacitance values are chosen for each portion of the frequency range so that current remains fairly constant, and taps on the coil are likewise chosen for each of the portions of the frequency range. Exact tuning is then done by either a variable coil or a variable capacitor across the combination just mentioned. In Fig. 12·25 an upper and a lower current limit, I_1 and I_2 respectively, are shown as frequency is varied from f' to f^n. Corresponding to these frequencies are capacitances C', C'', and so on.

The total capacitance needed for the lowest frequency is

$$C' = \frac{I_2}{2\pi f' E} \qquad (12\cdot17)$$

where E is the alternating voltage across the tank circuit.

The minimum capacitance at the highest frequency f^n is

$$C^n = \frac{I_1}{2\pi f^n E} \qquad (12\cdot18)$$

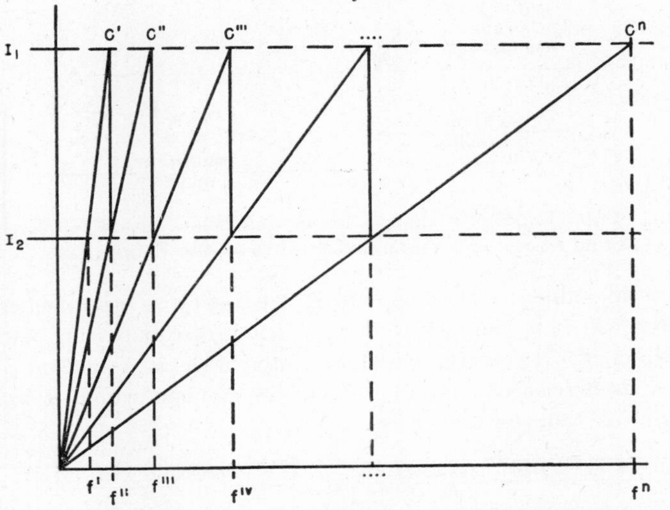

FIG. 12·25 Division of tank capacitance to limit tank-current variations.

12·8 PRINCIPLES OF WAVE FILTERS

Although tuned circuits are generally used to pass currents of certain frequencies at full amplitude and at the same time to suppress currents of other frequencies, the same thing can be accomplished by a type of network known as a wave filter.[7] In such a filter, the band of frequencies to be passed is known as the transmission band, and the band to be suppressed is known as the attenuation band. At some frequency, known as the cut-off frequency, the filter starts to attenuate. The transition between attenuation and transmission bands may be gradual or sharp, and the filter is said to have gradual or sharp cut-off accordingly.

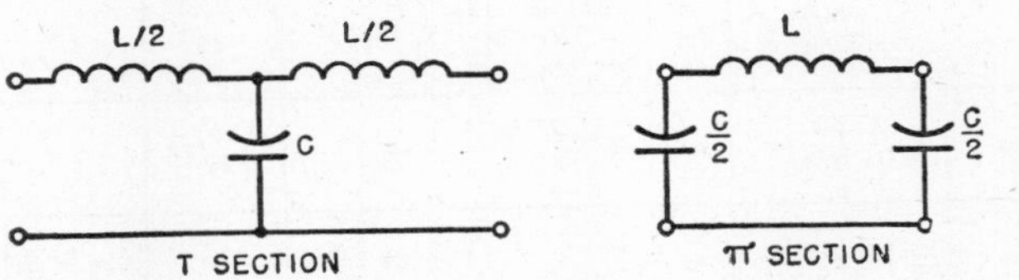

FIG. 12·26 Low-pass filter sections. (From *Electronic Transformers and Circuits* by Reuben Lee.)

In general, the elements of a wave filter are nearly pure reactances, to avoid introducing losses and attenuation in the transmission bands. For example, in the "low-pass" filter T-section shown in Fig. 12·26, the inductance arms shown as $L/2$ and the capacitance C are made with losses as low as possible. Capacitors used in filters have inherently low losses, but it is difficult to make the inductances with losses low enough. Values of Q ranging from 10 to 200 are common. The inductance and the frequency of transmission determine the value of Q. The relation of pure reactances in the transmission band is

$$0 > \frac{Z_1}{4Z_2} > -1 \qquad (12\cdot19)$$

where Z_1 is the reactance of the series arm and Z_2 is the reactance of the shunt arm. In Fig. 12·26, $Z_1 = 2\pi fL = 2\pi f(L/2 + L/2)$, and Z_2 = the reactance of C at transmission frequencies. The attenuation for sections of filter like Fig. 12·27 is shown in Fig. 12·27 for a pure-reactance

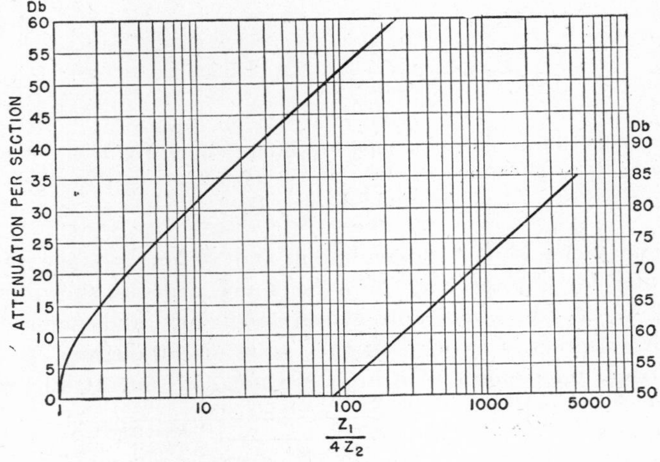

FIG. 12·27 Attenuation per section with pure reactance arms. (From *Electronic Transformers and Circuits* by Reuben Lee.)

network starting at the cut-off frequency. Attenuation is shown in decibels, and the abscissas are one quarter of the ratio of series to shunt reactances.

In the transmission band the sections of filter should be terminated in the proper impedance, for it delivers its full energy only into an impedance equal to its characteristic impedance. A properly terminated filter exhibits the same impedance at either end, if the opposite end is terminated in an impedance equal to its characteristic impedance. The impedance at any given point in the filter is called its image impedance; it is the same in either direction provided the source and sending-end impedances are equal. In general, image impedance is not the same for all points in the filter. For example, the impedance looking into the left or T-section of Fig. 12·26, if it is assumed that it is terminated properly, is not the same as that seen across capacitor C. For that reason another half series arm is added between C and the termination to keep proper impedance relations. Terminating sections at both the sending and receiving ends of a filter network are half sections; intermediate sections are full sections. A full T-section of the type shown in Fig. 12·26 includes an inductance L equal to $L/2 + L/2$. An image impedance at the input terminals of the T-section of Fig. 12·26 is known as the midseries impedance; that seen across capacitor C is known as the midshunt impedance.

Likewise, in the pi-section shown at the right of Fig. 12·26, the midshunt image impedance is at the input or output terminals. The midseries impedance is at a point in the middle of coil L. This section would terminate properly in its characteristic impedance at either end. Adjacent sections would have $C/2$ for the shunt arm, so that a full section would again be composed of a capacitor C and an inductance L. The choice of T- or pi-sections is determined by convenience in termination, or by the desired variation of image impedance with frequency.

If these precautions are not observed, reflections may cause a loss of power transfer in the transmission band.

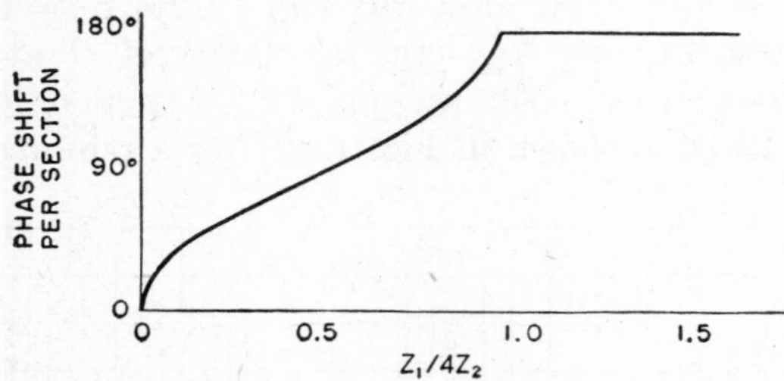

FIG. 12·28 Phase shift from input to output terminals of filter section as related to filter impedances. (From *Electronic Transformers and Circuits* by Reuben Lee.)

The phase shift between input and output voltage per section for a filter with pure reactance arms is shown in Fig. 12·28. It is zero only at zero frequency, and increases to 180 degrees at cut-off. It then stays at 180 degrees.

Image impedances of a filter are not constant at all frequencies, but change as shown in Fig. 12·29, depending upon whether midseries or midshunt values are taken. Midseries impedance decreases slowly from its nominal or charac-

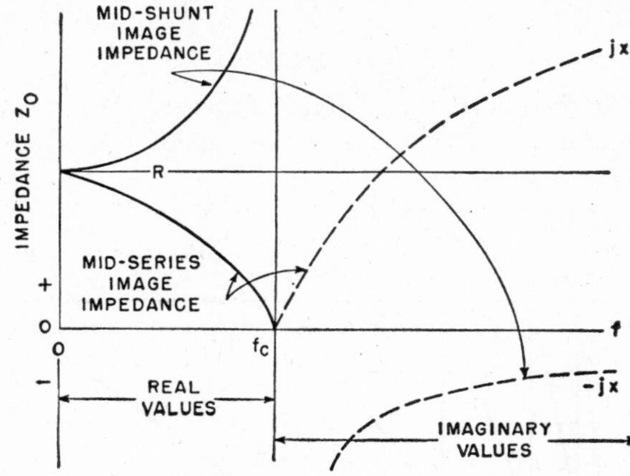

FIG. 12·29 Impedance changes above and below cut-off frequency. (From *Electronic Transformers and Circuits* by Reuben Lee.)

teristic value to zero as cut-off frequency is approached, after which it changes to a pure reactance increasing with frequency. Likewise, midshunt impedance is a resistance, and it increases to infinity at cut-off, beyond which it is a pure reactance of decreasing value.

NO.	TYPE	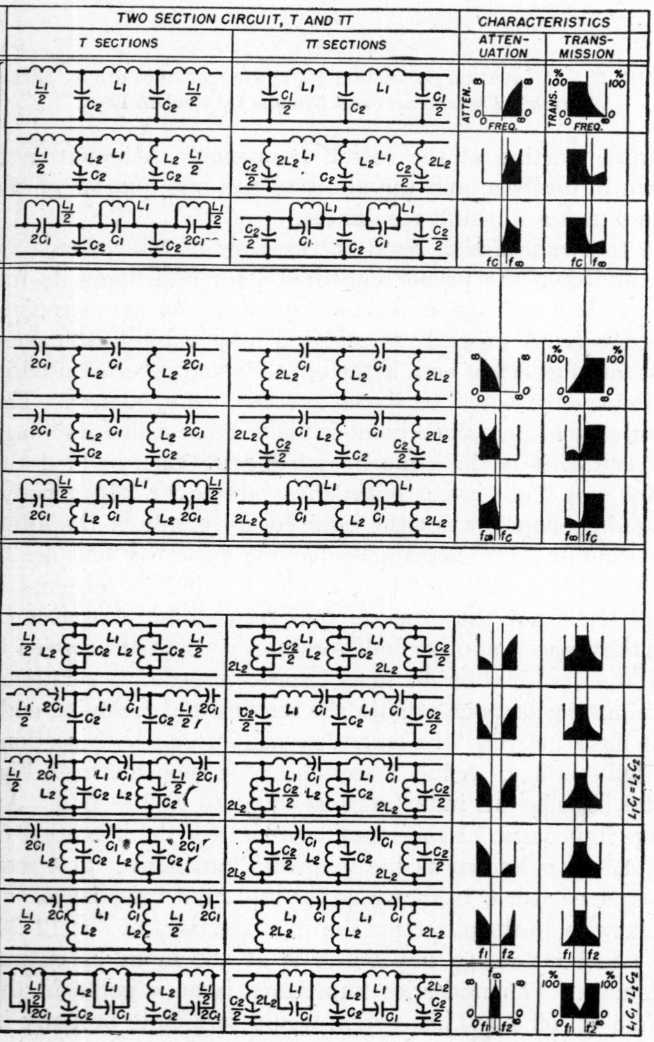 TWO SECTION CIRCUIT, T AND π: T SECTIONS	π SECTIONS	CHARACTERISTICS: ATTENUATION	TRANSMISSION	CUT OFF POINTS FROM FILTER CONSTANTS: $a = \frac{f_\infty}{f_c}$	f_c	f_∞	CHAR. IMPEDANCE AT ZERO $f = Z_0$	CONSTANTS FROM CUT OFF: L_1	L_2	C_1	C_2
1	LOW PASS					∞	$\frac{1}{\pi\sqrt{L_1C_2}}$	∞	$\sqrt{\frac{L_1}{C_2}}$	$L = \frac{Z_0}{\pi f_c}$			$C = \frac{1}{\pi f_c Z_0}$
2	LOW PASS					$\sqrt{1+\frac{L_1}{4L_2}}$	$\frac{1}{\pi\sqrt{C_2(L_1+4L_2)}}$	$\frac{1}{2\pi\sqrt{L_2C_2}}$	$\sqrt{\frac{L_1}{C_2}}$	$L_1 = mL$	$L_2 = \frac{1-m^2}{4m}L$		$C_2 = mC$
3	LOW PASS			f_c | f_∞	f_c | f_∞	$\sqrt{1+\frac{C_2}{4C_1}}$	$\frac{1}{\pi\sqrt{L_1(C_2+4C_1)}}$	$\frac{1}{2\pi\sqrt{L_1C_1}}$	$\sqrt{\frac{L_1}{C_2}}$	$L_1 = mL$		$C_1 = \frac{1-m^2}{4m}C$	$C_2 = mC$
						$a = \frac{f_c}{f_\infty}$			CHAR. IMPED. AT INFINITE FREQ. $= Z_0$				
4	HIGH PASS					∞	$\frac{1}{4\pi\sqrt{L_2C_1}}$	0	$\sqrt{\frac{L_2}{C_1}}$		$L = \frac{Z_0}{4\pi f_c}$	$C = \frac{1}{4\pi f_c Z_0}$	
5	HIGH PASS					$\sqrt{1+\frac{C_2}{4C_1}}$	$\frac{1}{4\pi}\sqrt{\frac{1}{L_2C_1}+\frac{4}{L_2C_2}}$	$\frac{1}{2\pi\sqrt{L_2C_2}}$	$\sqrt{\frac{L_2}{C_1}}$		$L_2 = \frac{L}{m}$	$C_1 = \frac{C}{m}$	$C_2 = \frac{4m}{1-m^2}C$
6	HIGH PASS			f_∞ | f_c	f_∞ | f_c	$\sqrt{1+\frac{L_1}{4L_2}}$	$\frac{1}{4\pi}\sqrt{\frac{1}{L_2C_2}+\frac{4}{L_1C_1}}$	$\frac{1}{2\pi\sqrt{L_1C_1}}$	$\sqrt{\frac{L_2}{C_1}}$	$L_1 = \frac{4m}{1-m^2}L$	$L_2 = \frac{L}{m}$	$C_1 = \frac{C}{m}$	
						LOWER FREQUENCY f_1	HIGHER FREQUENCY f_2	f_∞	CHAR. IMPED. Z_0 T SECTION AT $f_m = \sqrt{f_1 f_2}$	L_1	L_2	C_1	C_2
7	BAND PASS					$\frac{1}{2\pi\sqrt{L_2C_2}}$	$\frac{1}{2\pi}\sqrt{\frac{L_1+4L_2}{L_1L_2C_2}}$	∞	$\sqrt{\frac{L_1(L_1+4L_2)}{4L_2C_2}}$	$\frac{Z_0}{\pi(f_1+f_2)}$	$\frac{(f_2-f_1)Z_0}{4\pi f_1^2}$		$\frac{1}{\pi(f_2-f_1)Z_0}$
8	BAND PASS					$\frac{1}{2\pi\sqrt{L_1C_1}}$	$\frac{1}{2\pi}\sqrt{\frac{C_2+4C_1}{L_1C_1C_2}}$	∞	$\sqrt{\frac{L_1(C_2+4C_1)}{4C_1C_2}}-\sqrt{\frac{L_1}{4C_1}}$	$L_1 = \frac{Z_0}{\pi(f_2-f_1)}$		$C_1 = \frac{f_2-f_1}{4\pi f_1^2 Z_0}$	$C_2 = \frac{1}{\pi(f_1+f_2)Z_0}$
9	BAND PASS				$L_1C_1 = L_2C_2$	$\frac{1}{2\pi}\left[\sqrt{\frac{1}{L_1C_2}+\frac{1}{L_1C_1}}-\frac{1}{\sqrt{L_1C_2}}\right]$	$\frac{1}{2\pi}\left[\sqrt{\frac{1}{L_1C_2}+\frac{1}{L_1C_1}}+\frac{1}{\sqrt{L_1C_2}}\right]$	$0\ \infty$	$\sqrt{\frac{L_2}{C_1}} = \sqrt{\frac{L_1}{C_2}}$	$L_1 = \frac{Z_0}{\pi(f_2-f_1)}$	$L_2 = \frac{f_2-f_1}{4\pi f_1 f_2}Z_0$	$C_1 = \frac{f_2-f_1}{4\pi f_2 f_1 Z_0}$	$C_2 = \frac{1}{\pi(f_2-f_1)Z_0}$
10	BAND PASS					$\frac{1}{2\pi\sqrt{L_2(C_2+4C_1)}}$	$\frac{1}{2\pi\sqrt{L_2C_2}}$	0	$\sqrt{\frac{L_2}{C_1}\left[1-\frac{C_2}{4C_1}\right]}$		$L_2 = \frac{f_2-f_1}{4\pi f_1 f_2}Z_0$	$C_1 = \frac{f_1+f_2}{4\pi f_1 f_2 Z_0}$	$C_2 = \frac{f_1}{\pi f_2(f_2-f_1)Z_0}$
11	BAND PASS			f_1 | f_2	f_1 | f_2	$\frac{1}{2\pi\sqrt{C_1(L_1+4L_2)}}$	$\frac{1}{2\pi\sqrt{L_1C_1}}$	0	$\frac{\sqrt{L_1+4L_2}-\sqrt{L_1}}{2\sqrt{C_1}}$	$L_1 = \frac{f_1 Z_0}{\pi f_2(f_2-f_1)}$	$L_2 = \frac{(f_1+f_2)Z_0}{4\pi f_1 f_2}$	$C_1 = \frac{f_2-f_1}{4\pi f_1 f_2 Z_0}$	
12	BAND ELIM.			f_1 | f_2	f_1 | f_2; $L_1C_1 = L_2C_2$	$\frac{\sqrt{L_1C_2+16L_1C_1}-\sqrt{L_1C_1}}{8\pi L_1C_1}$	$\frac{\sqrt{L_1C_2+16L_1C_1}+L_1C_1}{8\pi L_1C_1}$	$\frac{1}{2\pi\sqrt{L_1C_1}}$	$\sqrt{\frac{L_1}{C_2}} = \sqrt{\frac{L_2}{C_1}}$	$L_1 = \frac{f_2-f_1}{\pi f_1 f_2}Z_0$	$L_2 = \frac{Z_0}{4\pi(f_2-f_1)}$	$C_1 = \frac{1}{4\pi(f_2-f_1)Z_0}$	$C_2 = \frac{f_2-f_1}{\pi f_1 f_2 Z_0}$

FIG. 12·30 Wave-filter characteristics. (From *Electronic Transformers and Circuits* by Reuben Lee.)

12·9 FILTER-DESIGN CHARTS

Filter design information and qualitative performance are shown in Fig. 12·30 for the more common types of filters. The configurations are half sections followed by whole sections and terminated by half sections. Cut-off frequency is designated as f_c, and the frequencies at which attenuation peaks are theoretically infinite are designated f_∞.

12·10 CONSTANT-*K* FILTERS

In a constant-*K* filter the product of the series and shunt impedances is a constant independent of frequency. That is,

$$Z_1 Z_2 = K^2 \tag{12·20}$$

where *K* is a constant. For pure reactance arms, this equation becomes

$$K = \sqrt{\frac{L}{C}} = R \tag{12·21}$$

Equation 12·21 expresses the characteristic impedance for a transmission line without losses and with unit inductance *L* and unit capacitance *C*. Therefore, *R* in this equation may be considered as the characteristic impedance of the filter. Constant-*K* filters are simple and are used widely. Filters 1, 4, 9, and 12 in Fig. 12·30 are of this type.

For constant-*K* filters equation 12·19 becomes

$$\left.\begin{aligned} \frac{-Z_1}{4Z_2} &= \frac{f^2}{f_c^2} \quad \text{for low-pass filters} \\ &= \frac{f_c^2}{f^2} \quad \text{for high-pass filters} \end{aligned}\right\} \tag{12·22}$$

This equation shows that the abscissas of Figs. 12·27 and 12·28 for constant-*K* filters are f^2/f_c^2 or f_c^2/f^2 for low- or high-pass filters.

Because the image impedance changes so much with frequency, it is difficult to terminate a constant-*K* filter for sharp cut-off. Reflection losses at or near cut-off are therefore large.

12·11 *m*-DERIVED FILTER

Filters 2, 3, 5, 6, 7, 8, 10, and 11 in Fig. 12·30 are *m*-derived filters. They have the same nominal image impedances as those of the constant-*K* prototype, but their attenuation, phase, and image impedance vary differently with frequency. The configurations of these filters are in general different from constant-*K* filters. The relations between the various elements of *m*-derived sections and their constant-*K* prototypes are shown in Fig. 12·30. The value of *m* is found from

$$m = \sqrt{1 - a^2} \tag{12·23}$$

where *a* is the ratio of cut-off frequency f_c to frequency of maximum attenuation f_∞ for low-pass filters, and the reciprocal of this ratio for high-pass filters. Variation of impedance is shown in Fig. 12·31 for several different values of *m*. For some filters the results are better, at least in the transmission band up to cut-off frequency f_c. In the attenuation band, attenuation becomes infinite for pure reactance elements at frequency f_∞, after which it drops below that of its constant-*K* prototype.

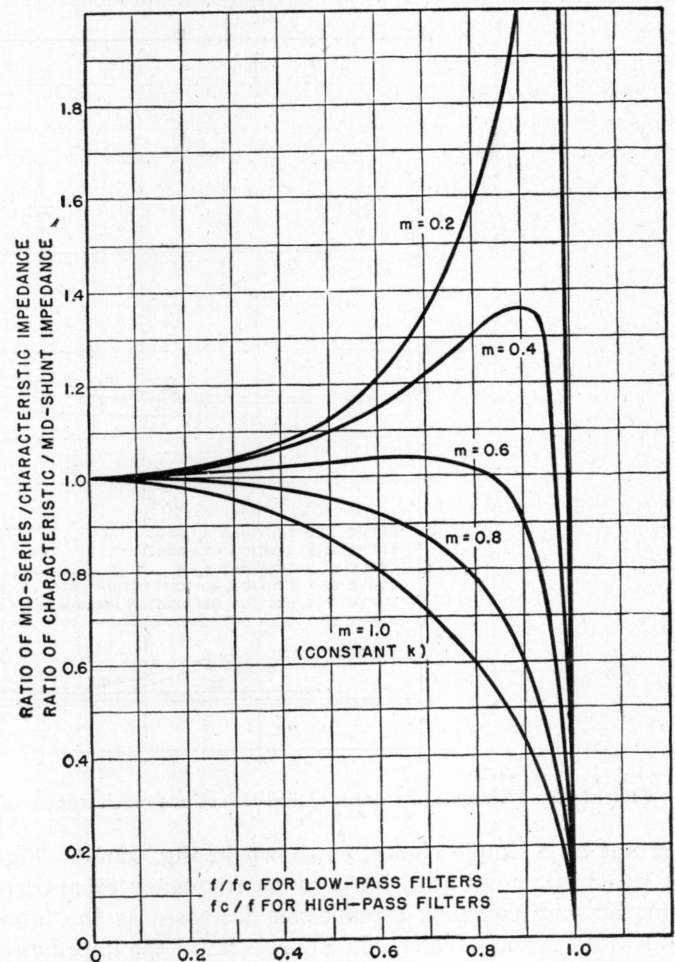

FIG. 12·31 Variation of image impedances in transmission band for *m*-derived filters. (From *Electronic Transformers and Circuits* by Reuben Lee.)

12·12 LIMITATIONS OF WAVE FILTERS

Several factors modify the performance of the wave filters shown in Figs. 12·27 and 12·28, especially in the cut-off region. One of these is reflection caused by mismatch of the characteristic impedance.[8,9] The load resistor usually is of constant value, whereas the image impedance usually drops to zero at cut-off. The resulting reflections cause a rounding of the attenuation curve in the cut-off region, as shown on curve *V* of Fig. 12·32. This is more pronounced in constant-*K* filters than in *m*-derived filters, because the impedance match is better in the *m*-derived type.

Another factor is the *Q* of the filter chokes. Curve *IV* of Fig. 12·32 shows typical rounding caused by this factor. Still another cause of the gradual slope of cut-off is the practice of inserting a resistor to simulate source impedance. Typical cases in which source and terminating resistances are equal have the effect shown by curves *I* and *IA* in Fig. 12·32. Correct prediction of filter response near cut-off requires a good deal of care. It cannot be taken from the usual attenuation charts.

In band-pass filters, the effects just noticed are present, with the additional complication of band width.[10] Band width must be chosen for high attenuation at unwanted frequencies and low attenuation at desired frequencies. This may not be a simple choice, as shown in Fig. 12·33. Figure 12·33 (*a*) shows how, for a given frequency separation from the midfrequency, attenuation decreases as the filter band width is made wider. In Fig. 12·33 (*b*) the impedance variation is shown to be much less with a wider band width. Therefore choosing a narrow band width attenuates frequencies in the transmission band because of reflections. The use of *m*-derived sections is of benefit in the latter case, as shown in Fig. 12·33 (*b*), but here again at attenuation frequencies beyond f_∞ the attenuation is often insufficient.

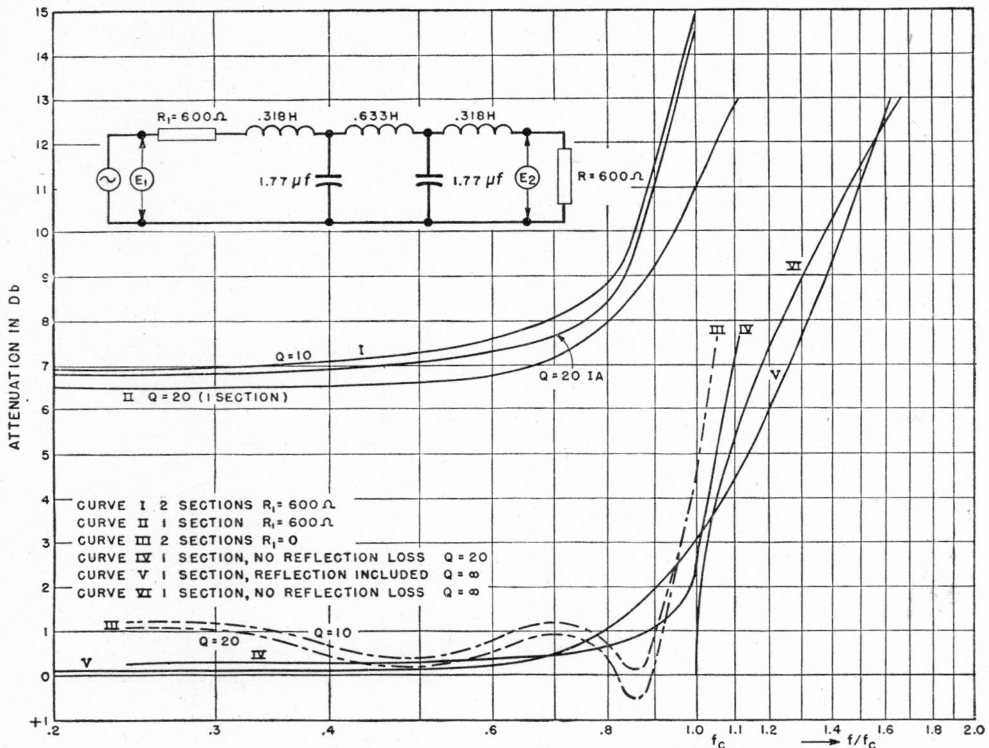

Fig. 12·32 Influence of reactor Q and reflections on sharpness of cut-off. (From *Electronic Transformers and Circuits* by Reuben Lee.)

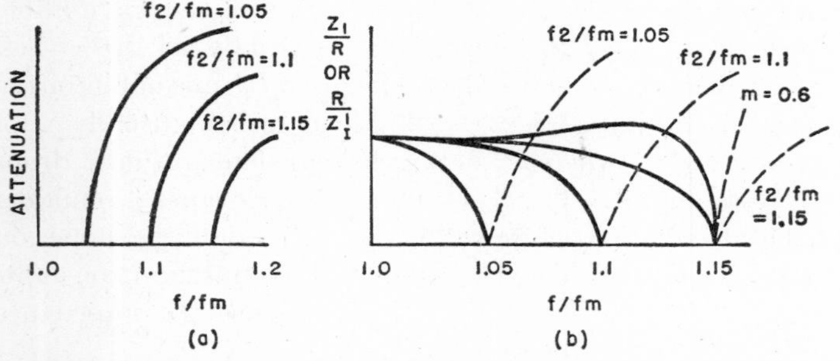

Fig. 12·33 (*a*) Attenuation and (*b*) impedance of band-pass filters. (From *Electronic Transformers and Circuits* by Reuben Lee.)

m-Derived terminating half sections for band-pass filter 9 in Fig. 12·30 are complicated, difficult to adjust, and seldom justified.

12·13 ARTIFICIAL LINES

Sometimes a certain amount of time delay must be interposed between one circuit and another; or, if a transmission line is not an exact multiple of 90 degrees long, some means must be found to increase its length to the next higher multiple of 90 degrees. For either of these purposes artificial lines are used. They may operate either at a single frequency or over a range of frequencies. They may be tapped for adjustment to suit any frequency in a given range, so that impedance and line length are correct. The configuration may be either T or π, high or low pass. Figure 12·34 shows these four combinations for any electrical length θ of line section. The line is assumed to be terminated in a pure resistance equal in value to the line characteristic impedance Z_0.

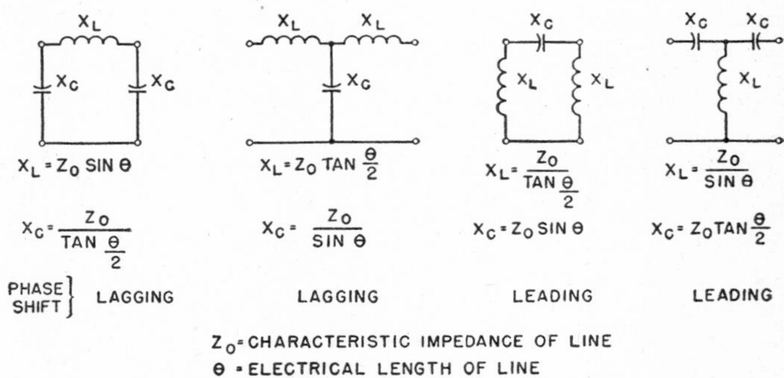

Fig. 12·34 Artificial line relations. (From *Electronic Transformers and Circuits* by Reuben Lee.)

12·14 D-C POWER-SUPPLY FILTERS

D-c power-supply filters have to pass only one frequency, which is zero. All ripple frequencies are in the attenuation band. Hence the whole wave-filter terminology of impedance-matching and cut-off frequencies may be discarded, and the analysis may be concentrated on attenuating properties. A rectifier filter should pass rectified direct current to the load without appreciable loss, but should attenuate the ripple so that it is not objectionable.

An example of the degree to which this filtering must be sometimes carried is the plate supply for a radio transmitter. From the microphone to the antenna of a high-power broadcast station, the power amplification may be 2×10^{15}. A ripple as great as 0.005 percent of the plate voltage at the microphone would produce a noise in the received wave loud enough to make the transmitted program inaudible.

Only rectifier voltages are considered here. D-c generators may be considered as special kinds of rectifiers. Different types of rectifiers have differing output-voltage waves, which affect filter design. Various common types of rectifiers are described in Chapter 15.

For a single-phase half-wave rectifier the rectified voltage across a resistive load R is shown in Fig. 12·35. It has a

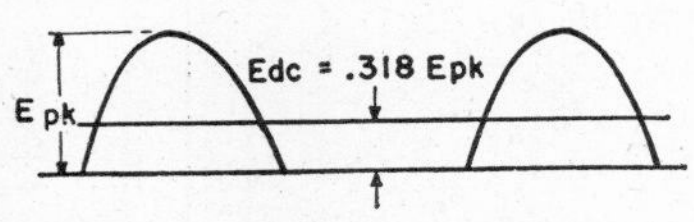

FIG. 12·35 Half-wave rectifier voltage.

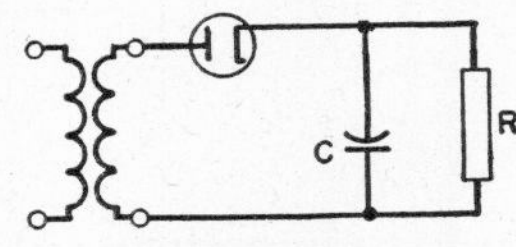

FIG. 12·36 Capacitor filter.

(From *Electronic Transformers and Circuits* by Reuben Lee.)

direct component with an average value of $0.45E_{a\text{-}c}$, and a series of alternating components. The fundamental alternating component has the same frequency as that of the supply voltage.

Single-phase half-wave rectifiers are used only if a low average value of load voltage and large variations in it are permissible. The chief advantage of this type of rectifier is its simplicity. A method of overcoming both its disadvantages is illustrated in Fig. 12·36, in which capacitor C shunts the load. With a suitable capacitor, $E_{d\text{-}c}$ sometimes can be increased to within a few percent of peak voltage E_{pk}. The principal disadvantage of this method is that the capacitor draws a large current during the charging interval Δt. This current is limited only by transformer and rectifier regulation; yet it must not be so large as to damage the rectifier. The higher the value of $E_{d\text{-}c}$ with respect to $E_{a\text{-}c}$ the larger is the charging current taken by C; consequently, if a smooth current wave is desired, some other method of filtering must be used.

After the limiting capacitor size has been reached, an inductive reactor may be employed. It may be placed on the rectifier side of the capacitor if load resistor R is high, or on the load side of the capacitor if R is low; see Fig. 12·37 (*a*) and (*b*). If the reactor is on the rectifier side,

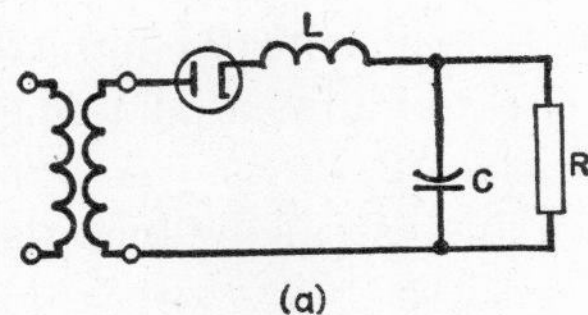

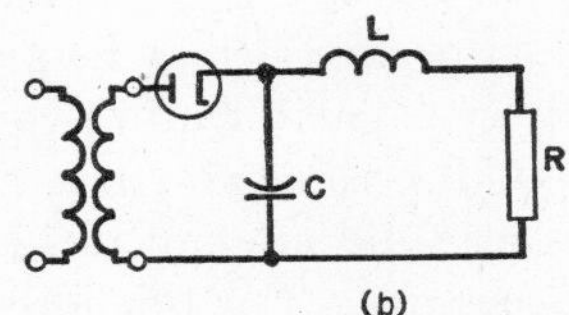

FIG. 12·37 (*a*) Inductor-input filter. (*b*) Capacitor-input filter. (From *Electronic Transformers and Circuits* by Reuben Lee.)

voltage $E_{d\text{-}c}$ has less than the average value $0.45E_{a\text{-}c}$, because the inductor delays build-up of current during the positive half cycle of voltage. Yet the inductor should have large reactance X_L, compared with the capacitive reactance X_C. If R is low, X_L should be large compared with R. In Fig. 12·37 (*a*) the ripple amplitude across R is $X_C/(X_L - X_C)$ times the amplitude generated by the rectifier, R being considered large compared with X_C. In Fig. 12·37 (*b*) the ripple amplitude across R is R/X_L times the value obtained with capacitor only. R is considered small compared with X_L.

High inductance is required for continuous current if the inductor is on the rectifier side of the capacitor in a half-wave rectifier circuit. Since current tends to flow only half the time, the rectified output is reduced accordingly.

Single-Phase Full-Wave Rectifier

In the single-phase full-wave rectifier, shown in Fig. 12·38, alternating components of the voltage have a fundamental frequency double that of the supply. Amplitudes of these

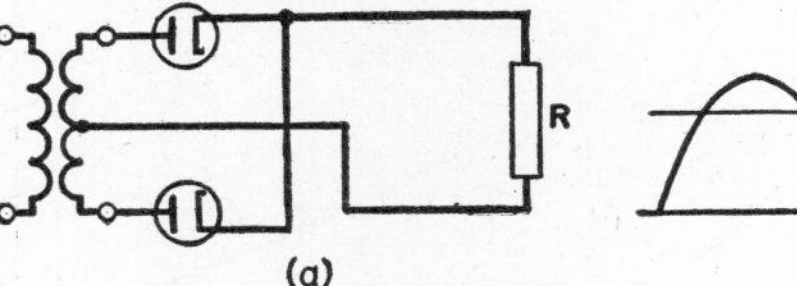

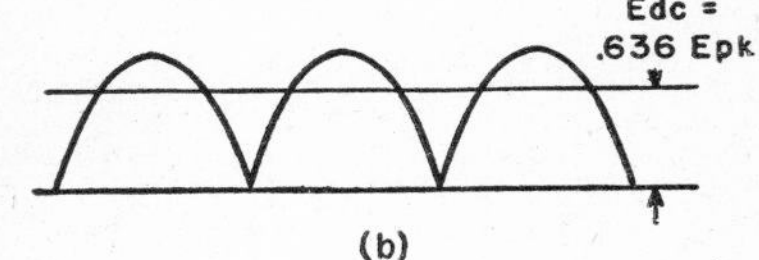

FIG. 12·38 (*a*) Single-phase full-wave rectifier. (*b*) Rectified voltage. (From *Electronic Transformers and Circuits* by Reuben Lee.)

components are much less than for the half-wave rectifier. At this higher frequency L and C are doubly effective; the smaller amplitude results in smaller percentage of ripple across the load. Current is continuous and $E_{d\text{-}c}$ has double the value it had in Fig. 12·35.

This rectifier uses only one half of the transformer winding at a time; that is, $E_{a\text{-}c}$ is only half the transformer secondary voltage. A circuit which utilizes the whole of this voltage in producing $E_{d\text{-}c}$ is shown in Fig. 12·39. The output-voltage relations are the same as those of Fig. 12·38. Although this circuit requires more rectifying devices, it eliminates the need for a transformer midtap.

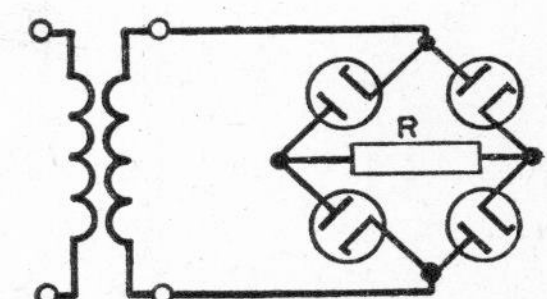

FIG. 12·39 Bridge rectifier. (From *Electronic Transformers and Circuits* by Reuben Lee.)

Polyphase Rectifiers

The effect of rectifying more than one phase is to superpose more voltages of the same peak value, but in different time relation to each other. Figures 12·40 (*a*) and 12·40 (*b*) compare the rectified output voltage for three-phase half-wave and full-wave rectifiers. These figures show that increasing the number of phases increases the value of $E_{d\text{-}c}$, increases the frequency of the alternating components, and

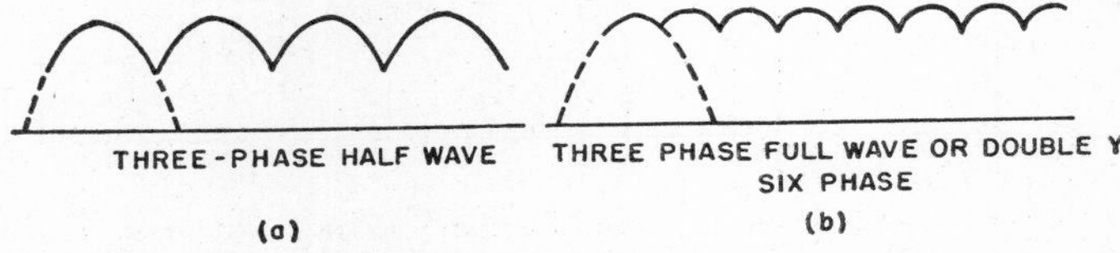

FIG. 12·40 Polyphase rectifier output waves. (From *Electronic Transformers and Circuits* by Reuben Lee.)

decreases the amplitude of these components. Ripple frequency is p times that of the unrectified alternating voltage, p being 1, 2, 3, and 6 for the respective waves. The frequency of any ripple harmonic is mp, where m is the order of the harmonic. Roughly, p represents the number of phases, provided that allowance is made for $p = 1$ and $p = 2$, and for the type of circuit with the higher values of p, as in Fig. 12·41.

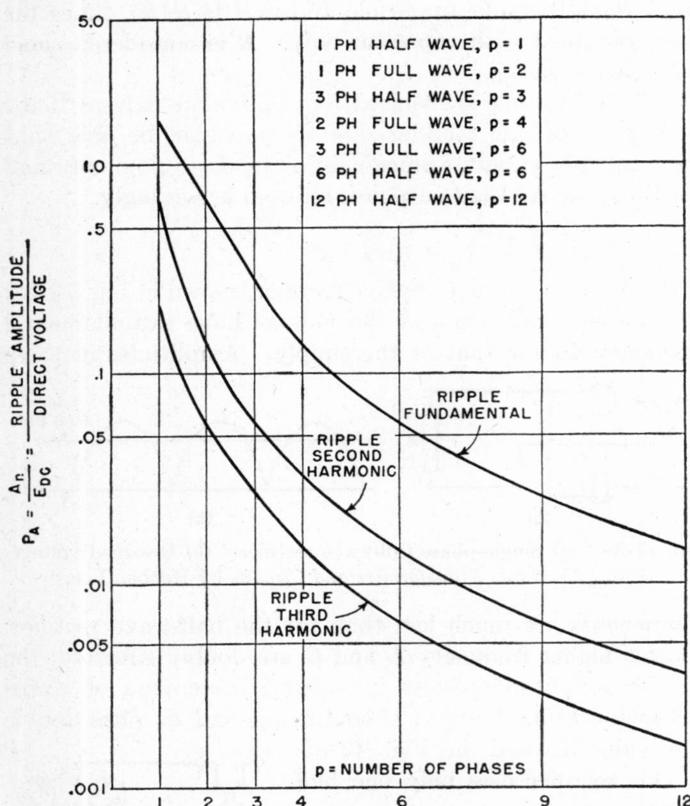

FIG. 12·41 Rectifier ripple voltage. (From *Electronic Transformers and Circuits* by Reuben Lee.)

Ripple amplitude is given in Fig. 12·41 for the ripple fundamental and second and third harmonics. In this curve, the ratio P_A of ripple amplitude to direct output voltage is plotted against the number of phases p. P_A diminishes a great deal for second and third harmonics. In general, if a filter effectively reduces the percentage of fundamental ripple across the load, the harmonics are negligible.

Multistage Filters

The filter shown in Fig. 12·37 (*a*) is known as an inductor-input filter. As far as filtering action is concerned, the

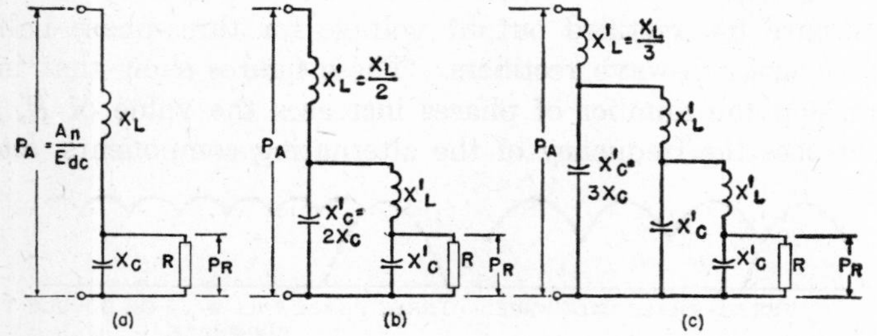

FIG. 12·42 Inductor-input filter circuits. (From *Electronic Transformers and Circuits* by Reuben Lee.)

rectifier is simply a source of non-sinusoidal alternating voltage across the filter. It is possible to replace the usual circuit representation by Fig. 12·42 (*a*). For any harmonic, say the nth, the voltage across the whole circuit is the harmonic amplitude A_n, and the voltage across the load is $P_R E_{d\text{-}c}$. P_R is the ripple allowable across the load, expressed as a fraction of the average voltage. Since load resistance R is large compared with X_C, the two voltages are almost in phase, and they bear the same ratio to each other as their respective reactances, or

$$\frac{P_A}{P_R} = \frac{X_L - X_C}{X_C} = \frac{X_L}{X_C} - 1 \qquad (12\cdot24)$$

The type of rectifier and the permissible amount of ripple in the load voltage determine the ratio of inductive to capacitive reactance.

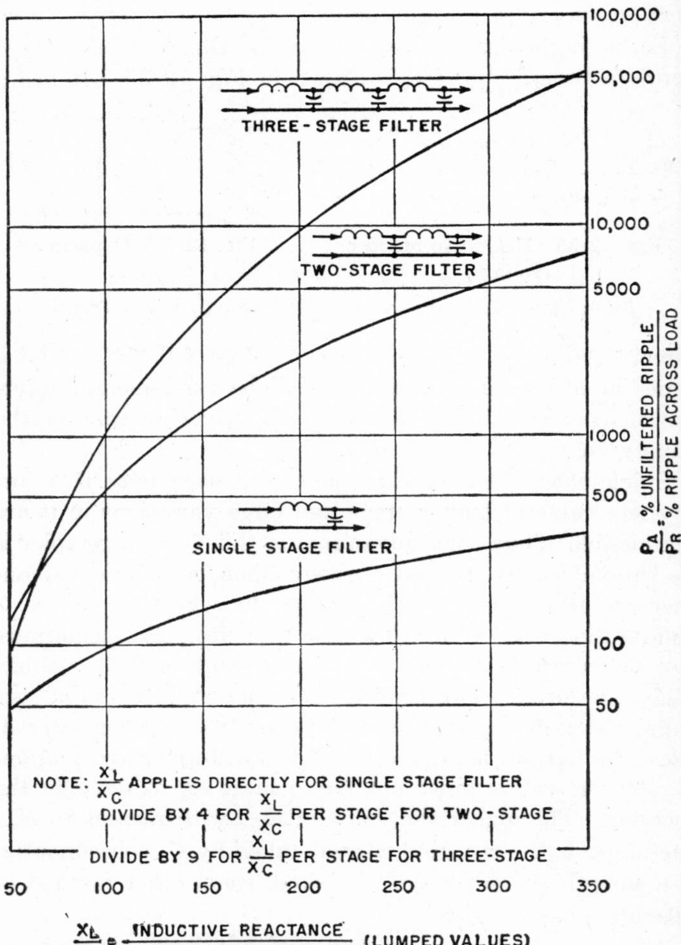

FIG. 12·43 Comparison of attenuation in 1, 2, and 3 filter stages. (From *Electronic Transformers and Circuits* by Reuben Lee.)

If P_R must be kept small, the single-stage filter of Fig. 12·42 (*a*) may require an abnormally large inductor and capacitor, and it may be preferable to split both the inductor and the capacitor into two separate equal units, and connect them like the two-stage filter of Fig. 12·42 (*b*). A much smaller total amount of inductance and of capacitance then is necessary. For this filter

$$\frac{P_A}{P_R} = \left(\frac{X'_L - X'_C}{X'_C}\right)^2 \qquad (12\cdot25)$$

X'_L and X'_C are the reactances of each inductor and capacitor in the circuit. Likewise, the three-stage filter of Fig.

12·42 (*c*) may be more practicable for still smaller values of P_R. In this filter,

$$\frac{P_A}{P_R} = \left(\frac{X'_L - X'_C}{X'_C}\right)^3 \tag{12·26}$$

and in general, for a *n*-stage filter,

$$\frac{P_A}{P_R} = \left(\frac{X'_L - X'_C}{X'_C}\right)^n \tag{12·27}$$

It is advantageous to use more than one stage only if the ratio P_A/P_R is high. That the gain from multistage filters is realized only for certain values of P_A/P_R is shown by filter, where otherwise more stages might be most economical.

Capacitor-Input Filters

It cannot always be assumed that transformer and rectifier voltage drops are negligible, if capacitor-input filters are used, because large peak currents are drawn by the capacitor during the charging interval. Such charging currents drawn through finite resistances affect both the output voltage and the ripple in a complicated manner, and a simple analysis such as that for inductor-input filters is impossible. Figure 12·44 is a plot of the ripple in the load of capacitor-input

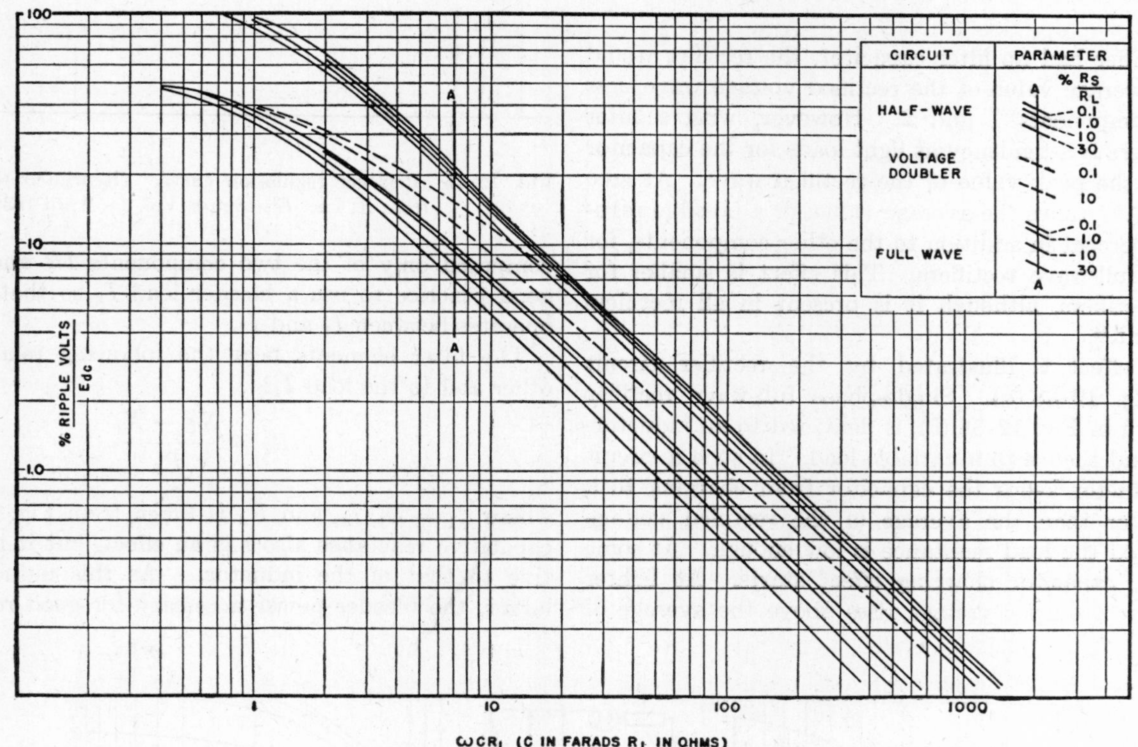

FIG. 12·44 Rms ripple voltage of capacitor-input circuits. (Reprinted with permission from O. H. Schade, *Proc. I.R.E.*, Vol. 31, July 1943.)

Fig. 12·43. The lower curve shows the relation between P_A/P_R and X_L/X_C for a single-stage filter. The second curve shows the increase in P_A/P_R gained by splitting the same X_L and X_C into a two-stage filter. As indicated in Fig. 12·42 (*b*), the inductance and capacitance have one-half their "lumped" value. The upper curve indicates the same increase for a three-stage filter, each inductor and capacitor of which have one third of their "lumped" or single-stage values. The attenuation in multistaging is enormous for high X_L/X_C. For lower values there may be a loss instead of a gain, as shown by the intersection of the two upper curves. These curves also intersect the lower curve if all are prolonged to the left. This condition may appear to be a puzzling one; but consider that, for $X_L/X_C = 50$ in the single-stage filter, the ratio is $\frac{1}{3}X_L/3X_C$ or 50/9 in the three-stage filter. The rather small advantage of the three-stage filter is not difficult to account for.

Other factors influence the selection of the number of filter stages. For example, modulation or keying may require that a definite size of filter capacitor be used across the load. Usually these conditions result in a single-stage

filters with various ratios of source resistance to load resistance, and for three types of rectifiers.[11,12] These curves are useful also if resistance is used in place of the inductance of an inductor-input filter. In Fig. 12·44 ω is 2π times the supply frequency, C is capacitance, R_L is the load resistance, and R_S is the source resistance.

If *L-C* filter stages follow a capacitor-input filter the ripple of the capacitor-input filter is reduced as in Fig. 12·43, except that P_A must be taken from Fig. 12·44. If an *R-C* filter stage follows any type of filter the ripple is reduced in the ratio R/X_C represented by the *R-C* stage.

12·15 D-C POWER-SUPPLY REGULATION

The regulation of a rectifier depends upon three distinct factors:

1. The d-c resistance or *IR* drop.
2. The commutation reactance or *IX* drop.
3. The capacitor charging effect.

The first factor can be reduced by the use of tubes, transformers, and inductors having low resistance.

Commutation reactance can be kept low by proper transformer design, particularly where the ratio of short-circuit current to normal load is high.[13] The amount of regulation caused by commutation reactance is

Percent commutation reactance drop

$$= \frac{100 I_{d\text{-}c} X}{2\sqrt{2} E_{a\text{-}c} \sin \frac{\pi}{p}} \qquad (12\cdot28)$$

where X = the leakage reactance per phase
$E_{a\text{-}c}$ = the rms value of the a-c phase voltage
p = number of phases in Fig. 12·41.

If the rectifier had no filter capacitor, the rectifier would deliver the average value of the rectified voltage wave, less regulation components 1 and 2. However, with a filter capacitor, there is a tendency at light loads for the capacitor to charge to the peak value of the rectified wave. At zero load, this is 1.57 times the average value, or a possible regulation of 57 percent in addition to the other components, for single-phase full-wave rectifiers. This effect is smaller for polyphase rectifiers, although it is present in all rectifiers to some extent.[14]

Capacitor effect is illustrated by the rectifier circuit shown in Fig. 12·38 (*a*). Single-phase full-wave rectifier output, shown in Fig. 12·38 (*b*), is delivered to an inductor-input filter and thence to a variable load. In such a circuit the filter inductor keeps the capacitor from charging to a voltage greater than the average of the rectified voltage wave, provided the load resistance is low enough. At some specific load, capacitor charging effect starts. At lighter loads, the direct output voltage rises above the average of the rectified wave, as shown by the typical regulation curve of Fig. 12·45.

Starting at zero load, the direct output voltage E_0 is 1.57 times the average of the rectified wave. As load increases, output voltage falls rapidly to E_1 as the current I_1 is reached. For any load greater than I_1, voltage drop is

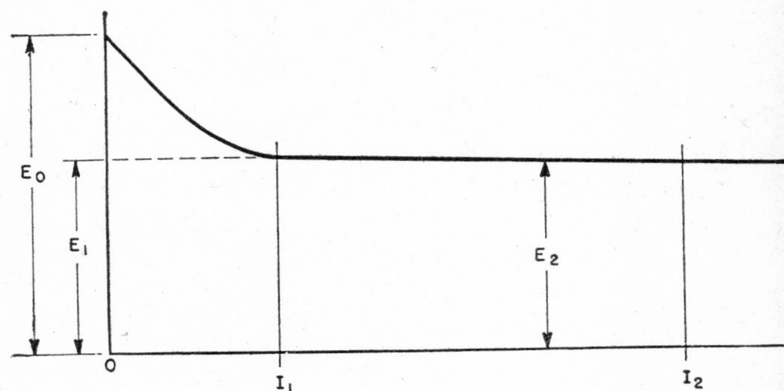

Fig. 12·45 Rectifier regulation curve. (Reprinted with permission from R. Lee, *Electronics*, Vol. 11, April 1938.)

composed only of the two components IR and IX. It is good practice to use a bleeder load I_1 so that the rectifier operates between I_1 and I_2.

The filter elements bear the following relation to each other and to the load I_1:

$$R_1 = \frac{X_L - X_c}{P_A} \qquad (12\cdot29)$$

where $R_1 = E_1/I_1$, and P_A is taken from Fig. 12·41. Here capacitive reactance also has an effect, but it is minor relative to that of the inductor. As this inductor becomes larger, the bleeder power necessary for good regulation can

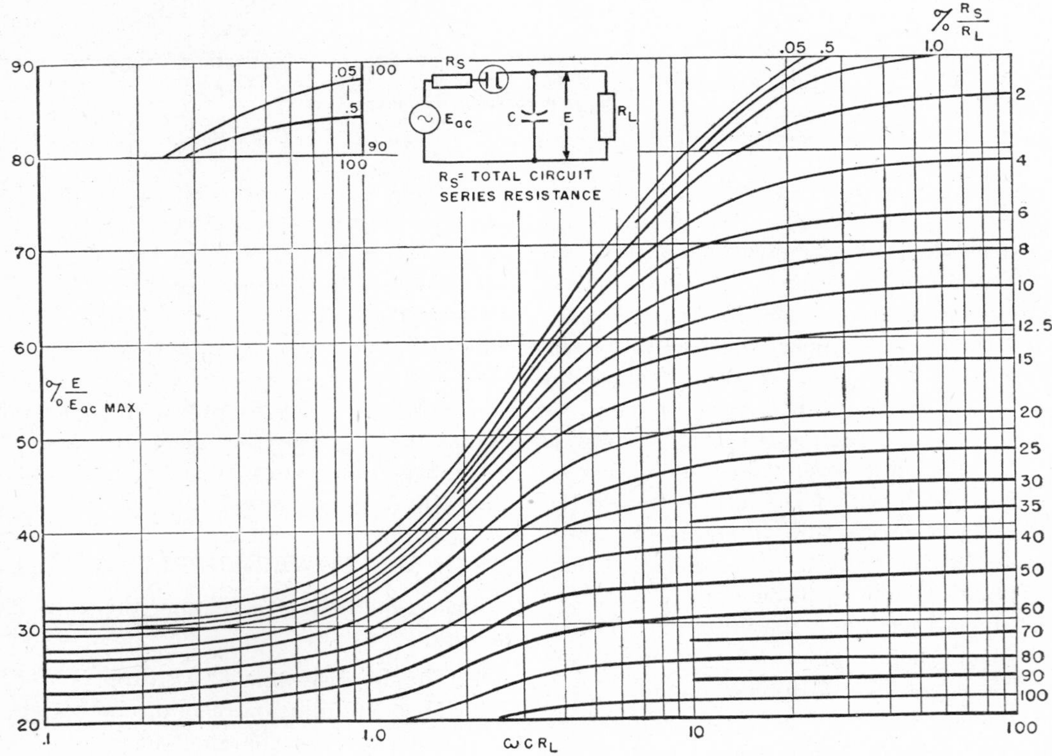

Fig. 12·46 Relation of peak sine volts to direct volts in half-wave capacitor-input circuit. (Reprinted with permission from O. H. Schade, *Proc. I.R.E.*, Vol. 31, July 1943.)

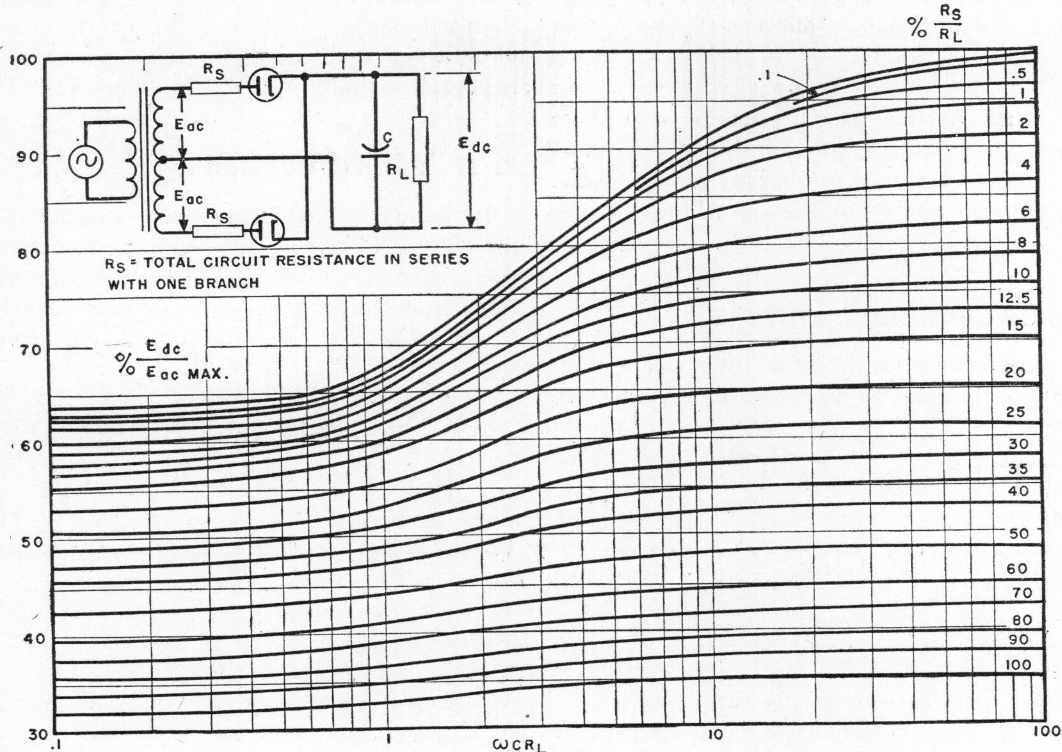

FIG. 12·47 Relation of peak sine volts to direct volts in full-wave capacitor-input circuit. (Reprinted with permission from O. H. Schade, *Proc. I.R.E.*, Vol. 31, July 1943.)

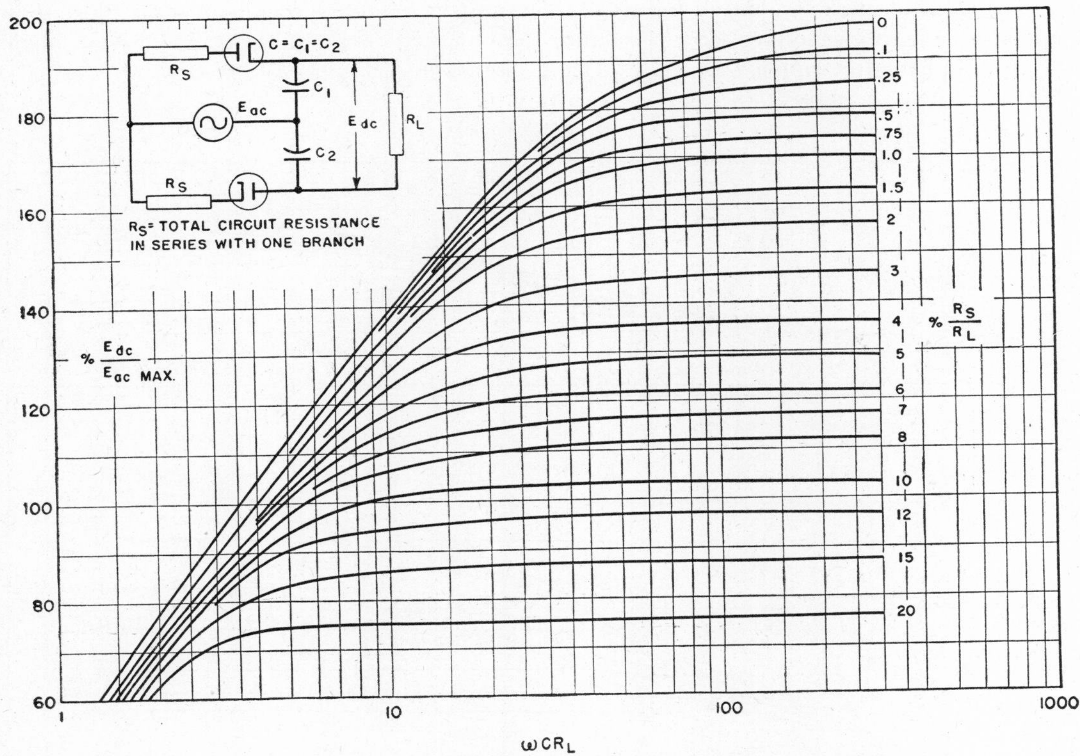

FIG. 12·48 Relation of peak sine volts to direct volts in voltage-doubling circuit. (Reprinted with permission from O. H. Schade, *Proc. I.R.E.*, July 1943.)

be made smaller. In polyphase rectifiers the rise in voltage is not so great because of the smaller difference between peak and average d-c output.

Capacitor-input filters have voltage curves shown in Figs. 12·46, 12·47, and 12·48. Over a range of load, these filters may give reasonably good regulation, but it is possible to get poor regulation at certain loads, as the curves show. Series resistance plays an important part in voltage regulation of this type of filter.

12·16 TUNED-POWER-SUPPLY FILTERS

Sometimes an inductor input filter is tuned as in Fig. 12·49. The addition of capacitor C_1 increases the effective

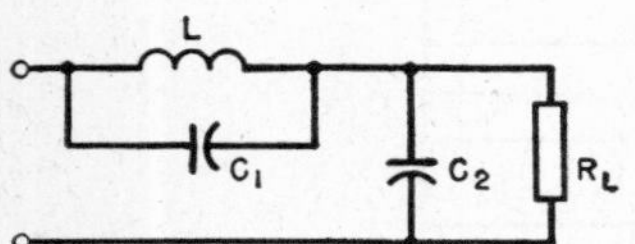

FIG. 12·49 Shunt-tuned power-supply filter.

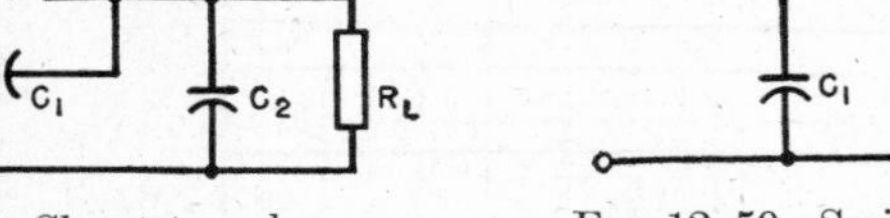

FIG. 12·50 Series-tuned power-supply filter.

(From *Electronic Transformers and Circuits* by Reuben Lee.)

reactance of the inductor at some frequency, usually the fundamental ripple frequency. Thus both regulation and ripple of this type of filter are improved. Tuning is not effective for the ripple harmonics, so that the use of high-Q filter inductors is unnecessary. The increase in effectiveness of the filter inductor is about three to one in this circuit.

Sometimes filters are tuned as in Fig. 12·50, with filter capacitor C_1 connected to a tap near the end of inductor L. The other filter capacitor C_2 is chosen to give series resonance across the load at the fundamental ripple frequency. The resulting ripple across load resistor R_L can be made lower than without the use of C_1. In this case, the ripple is attenuated more than in the usual inductor-input filter, but the regulation is not substantially different.

12·17 FILTER CURRENTS

If the inductance in an inductor-input filter were infinite, the current through it would remain constant. If commutation reactance were neglected, the current through each tube would be a square wave as shown by I_1 and I_2 of Fig. 12·51. The peak value of this current wave is the same as

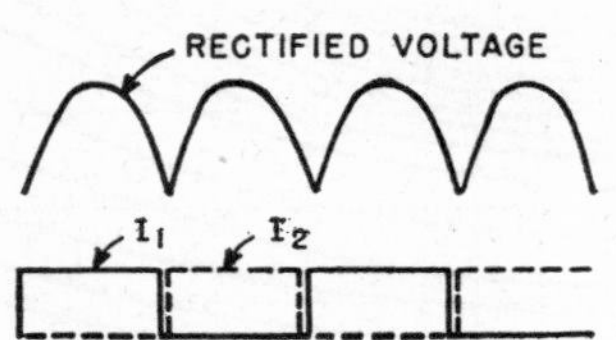

FIG. 12·51 Current waves in inductor-input filters.

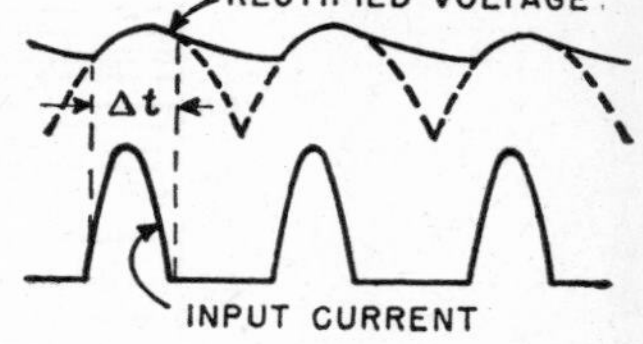

FIG. 12·52 Current waves in capacitor-input filters.

(From *Electronic Transformers and Circuits* by Reuben Lee.)

the d-c output of the rectifier, and the rms value is $0.707I_{d\text{-}c}$. With finite values of inductance, an appreciable amount of ripple current passes through the inductor and modulates I_1 and I_2, thus producing a larger rms value than before.

Capacitor-input filters draw current from the rectifier only during certain portions of the cycle Δt, as shown in Fig. 12·52. For a given average direct current, the peak and rms values of these current waves are much higher than for inductor-input filters. Values for single-phase rectifiers are given in Fig. 12·53.

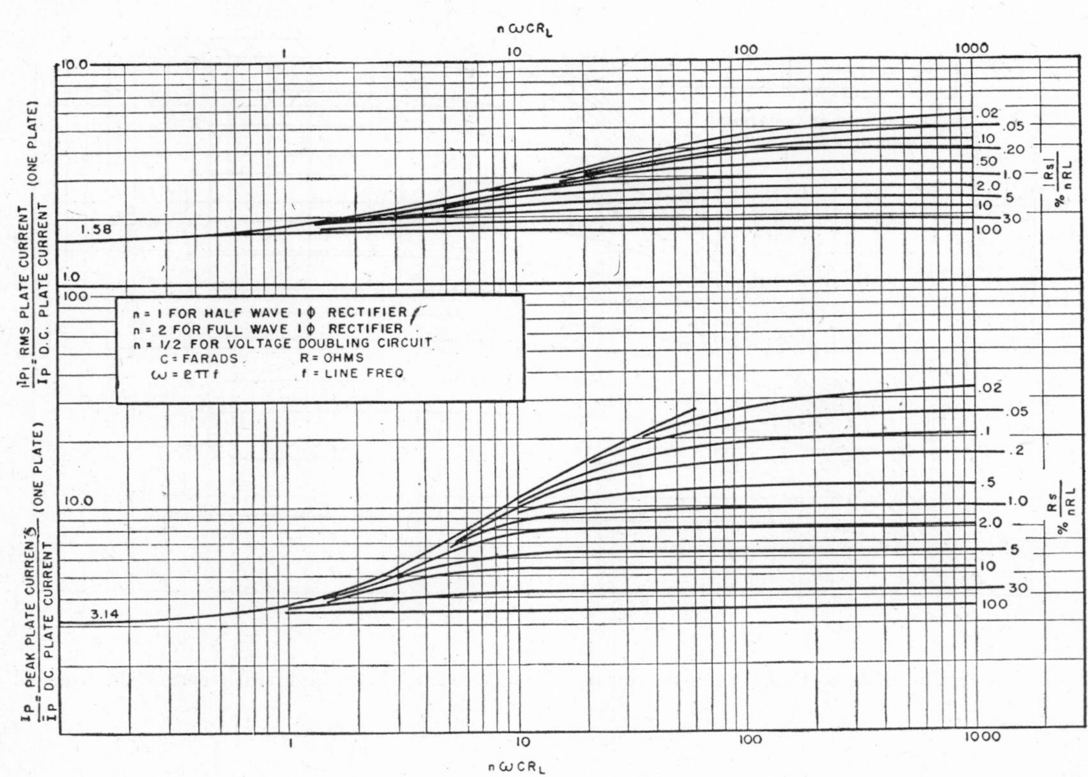

FIG. 12·53 Relation of peak, average, and rms diode current in capacitor-input circuits. (Reprinted with permission from O. H. Schade *Proc. I.R.E.*, Vol. 31, July 1943.)

In a shunt-tuned-power-supply filter such as that in Fig. 12·49, the current drawn from the rectifier is likely to be peaked because the two capacitors C_1 and C_2 are in series, without any intervening resistance or inductance

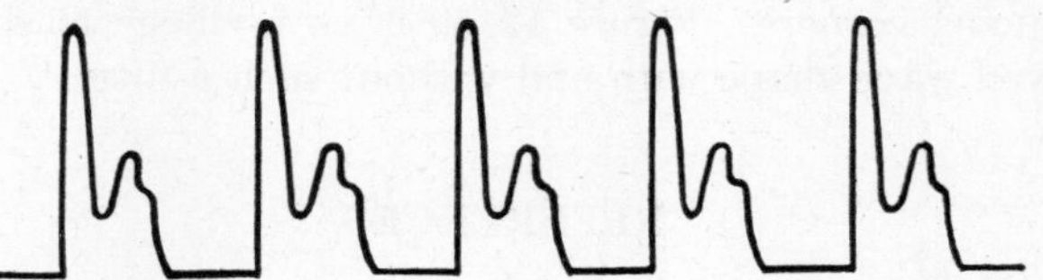

FIG. 12·54 Oscillogram of shunt-tuned input current. (From *Electronic Transformers and Circuits* by Reuben Lee.)

drop. This peak quickly subsides because of the influence of inductor L, and often an oscillation takes place on top of the current wave as shown in Fig. 12·54. The rectifier must be designed to withstand this peak current.

Currents in Figs. 12·51, 12·52, and 12·54 are reflected back into the a-c supply line, except that every other current wave is of reverse polarity. Small rectifiers have little effect on the power system, but large rectifiers may produce excessive interference in nearby telephone lines because of the

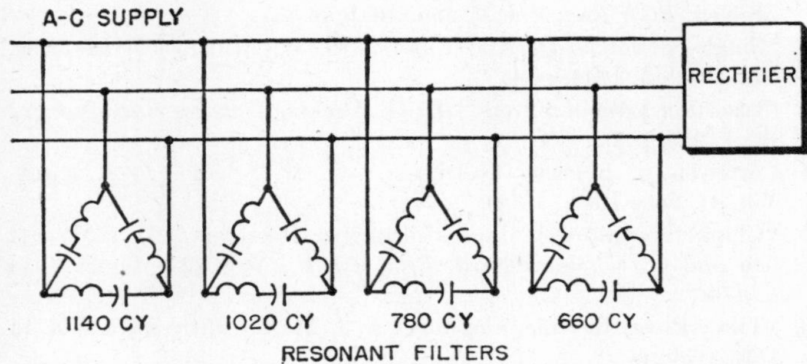

FIG. 12·55 A-c line filter for large power rectifier. (From *Electronic Transformers and Circuits* by Reuben Lee.)

large harmonic currents in rectifier loads. High commutation reactance reduces these line-current harmonics, but, since good regulation requires low commutation reactance, there is a limit to the control possible by this means. A-c line filters are used to attenuate line-current harmonics. A large rectifier with three-phase series resonant circuits designed to eliminate the 11th, 13th, 17th, and 19th harmonics of a 60-cycle system is shown in Fig. 12·55. Smaller rectifiers sometimes have filter sections such as those in Fig. 12·56;

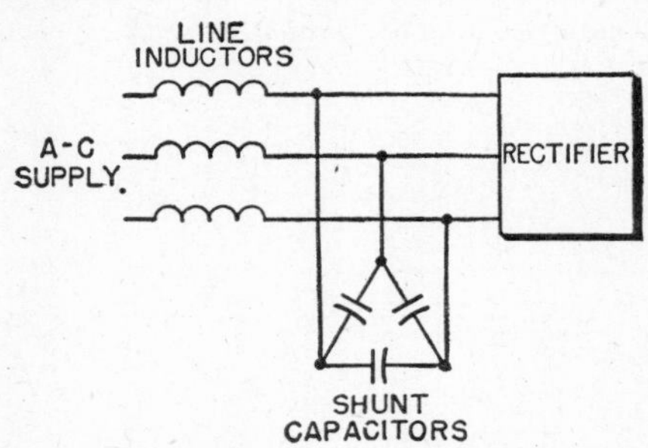

FIG. 12·56 A-c line filter for medium-sized power rectifier. (From *Electronic Transformers and Circuits* by Reuben Lee.)

these are rarely used in large installations because excessive voltage regulation is introduced by the line inductors.[15, 20]

Commutator motors and other noise-producing devices cause radio interference unless their output is filtered. Such filters usually use circuits tuned to radio frequencies.[16]

12·18 GRID-CONTROLLED-RECTIFIER FILTERS

In a grid-controlled rectifier the ripple, regulation, transients, and a-c harmonic currents exceed the values for corresponding rectifiers without grid control. In Fig. 12·57 values of ripple and the ratio $(X_L - X_C)/R_1$ for continuous current are plotted as functions of firing angle for a single-phase rectifier. The last curve is of importance, not only for its effect on good regulation, but also for proper rectifier performance.[17,19] Transients introduced by commutation

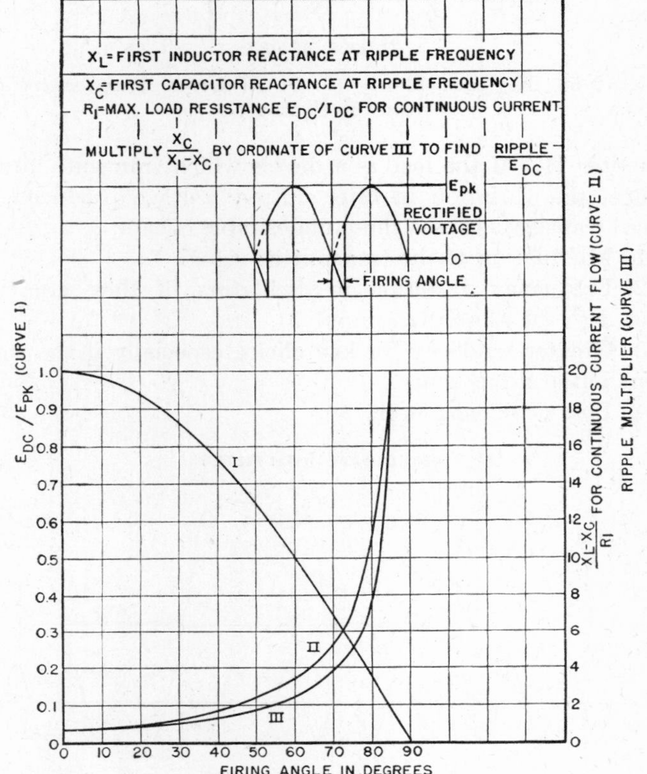

FIG. 12·57 Ripple and current continuity curves for single-phase full-wave grid-controlled rectifier. (From *Electronic Transformers and Circuits* by Reuben Lee.)

reactance are accentuated unless these values of filter inductors are adhered to. A-c line harmonics may increase several fold.

12·19 FILTER TRANSIENTS

If the current is cut off during each cycle, a transient current occurs each cycle. When power is first applied to the filter the transient may be smaller or larger than the cyclic transient, depending upon how fast the voltage builds up. In inductor-input filters the transient current can be approximated by the formula given in Section 12·1 for a step function applied to the simple series circuit, because the shunting effect of the load is slight in a well-proportioned filter. In capacitor-input filters, the same method can be used, but the leakage inductance of the anode transformer limits the transient current. The same is true of tuned-inductor filters of the kind shown in Fig. 12·49.

In some applications the load is varied or removed periodically. Examples of this are keyed or modulated amplifiers. Transients occur when the load is applied (key down) or removed (key up), causing a momentary drop or rise in plate voltage.[18] Formulas for these transients are given in

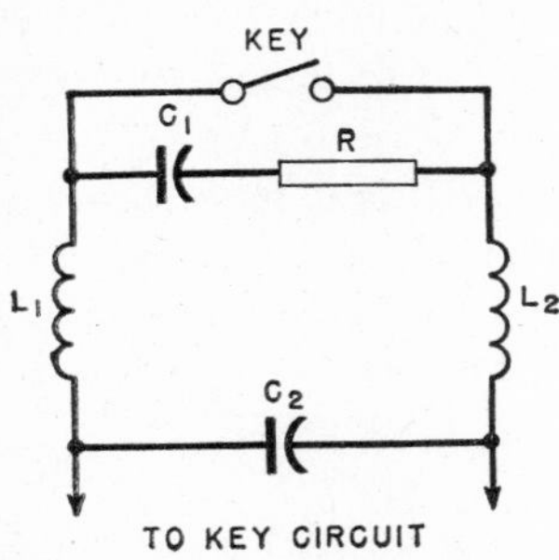

FIG. 12·58 Key-click filter. (From *Electronic Transformers and Circuits* by Reuben Lee.)

Chapter 11. If the load is a device which transmits intelligence, the variation in filter output voltage produced by these transients causes these undesirable effects:

1. Modulation of the transmitted signal.
2. Frequency variation in oscillators, if they are connected to the same plate supply.
3. Greater tendency for key clicks, especially if the transient initial dip is sharp.
4. Loss of signal power.

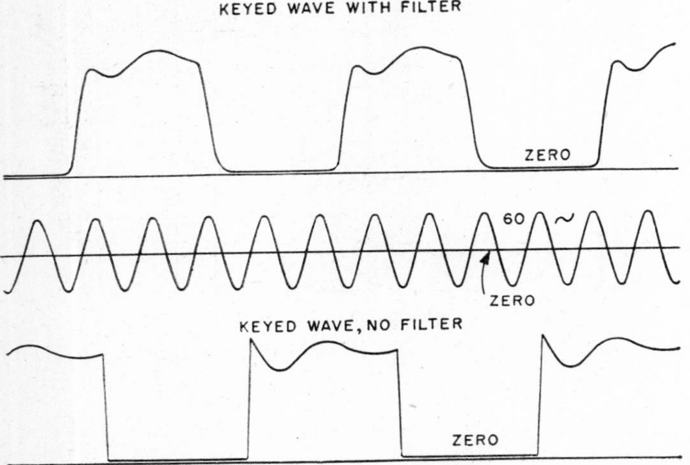

FIG. 12·59 Wave forms with and without key-click filter. (From *Electronic Transformers and Circuits* by Reuben Lee.)

Although the tendency for key clicks in the signal may be reduced by a suitably proportioned d-c supply filter, the clicks may not be entirely eliminated by it. If key-click elimination is necessary, some sort of key-click filter is used, of which Fig. 12·58 is an example. This filter rounds off the top and back of a wave and eliminates sharp, click-producing corners. Figure 12·59 is an oscillogram showing a keyed wave shape with and without such a filter.

REFERENCES

1. *Electrical Engineers' Handbook* Vol. V, H. Pender and K. McIlwain, Section 7, Wiley, 1936.
2. *Introduction to Electric Transients*, E. B. Kurtz and G. F. Corcoran, Wiley, 1935, Chapters III to V.
3. "A Practical Analysis of Parallel Resonance," R. Lee, *Proc. I.R.E.*, Vol. 21, Feb. 1933, p. 271.
4. *Electric Circuit Analysis*, M. G. Malti, Wiley, 1930, Chapter XI.
5. "Radio Instruments and Measurements," *National Bureau of Standards*, Circ. 74, Gov't. Printing Office, 1924, pp. 275–284.
6. "Design of Power Amplifier Output Circuits," R. Lee, *Radio Eng.*, Vol. 14, July 1934, p. 10.
7. *Transmission Networks and Wave Filters*, T. E. Shea, Van Nostrand, 1929.
8. "An Analysis of Constant *k* Low- and High-Pass Filters," O. S. Meixell, *RCA Rev.*, Vol. V, Jan. 1941, p. 337.
9. "Single-Section *m*-Derived Filters," C. W. Miller, *Wireless Eng.*, Vol. 21, Jan. 1944, p. 4.
10. "Insertion Loss of Filters," D. G. Tucker, *Wireless Eng.*, Vol. 22, Feb. 1945, p. 62.
11. "Analysis of Rectifier Operation," O. H. Schade, *Proc. I.R.E.*, Vol. 31, July 1943, p. 341.
12. "Characteristics of Voltage-Multiplying Rectifiers," D. L. Waidelich and C. L. Shackelford, *Proc. I.R.E.*, Vol. 32, August 1944, p. 470.
13. "Three Phase Rectifier Circuits," A. J. Maslin, *Electronics*, Vol. 11, Dec. 1936, p. 28.
14. "Solving a Rectifier Problem," R. Lee, *Electronics*, Vol. 11, April 1938, p. 39.
15. "Harmonics in A-C Circuits of Grid Controlled Rectifiers and Inverters," R. D. Evans and H. N. Muller, Jr., *Trans. A.I.E.E.*, Vol. 58, Supp., 1939, p. 861.
16. "Suppression of Radio Interference," C. V. Aggers and R. N. Stoddard, *Electric J.*, Vol. 31, August and Sept. 1934, p. 305.
17. "Filter Design for Grid Controlled Rectifiers," H. A. Thomas, *Electronics*, Vol. 19, Sept. 1944, p. 142.
18. "Radio-Telegraph Keying Transients," R. Lee, *Proc. I.R.E.*, Vol. 22, Feb. 1934, p. 213.
19. "Critical Inductance and Control Rectifiers," W. P. Overbeck, *Proc. I.R.E.*, Vol. 27, Oct. 1939, p. 655.
20. "Inductive Coordination Aspects of Rectifier Installations," A.I.E.E. Committee Report, *Trans. A.I.E.E.*, Vol. 65, July 1946, p. 417.

Chapter 13

TRANSFORMERS

Reuben Lee

IN weight, transformers used for electronic circuits range from a few ounces to many tons; ratings vary from zero to several million watts; and the frequency range extends from less than one cycle per second to several megacycles per second.

As might be expected, many different designs are used for these transformers, but one principle common to all is Faraday's law:

$$e = -N\frac{d\phi}{dt} \qquad (13\cdot1)$$

which means that the voltage induced in any coil of N turns is proportional to the number of turns and to the rate of change of magnetic flux in the coil. For a given voltage, if the rate of change of flux is small, the number of turns must be great; consequently, low-frequency transformers have many turns, and high-frequency transformers have few turns. As core materials are improved it becomes possible to operate transformers with higher and higher flux densities in the cores, and therefore fewer turns become necessary. Thus the upper limit of frequency is extended.

13·1 CONSTRUCTION AND SIZE

The majority of electronic transformers have one coil, and they use laminations of the shell type. Typical assemblies using laminations and type C cores are shown in Figs. 13·1 to 13·4. Laminations of the shell type are generally used for small transformers, and corresponding to them are assemblies with two core loops. It is simpler to assemble a single-core loop, and so a single core is often used for small transformers. For a 60-cycle supply the laminations are usually stacked to produce overlapping joints. This condition is approximated in the type C core with ground gap surfaces which fit closely together. To prevent saturation by direct current in the windings, either type of core can be used with core gaps: laminations by butt stacking without overlap, and type C cores by inserting the desired amount of gap insulation between the loops.

FIG. 13·1 Transformers with shell-lamination core.

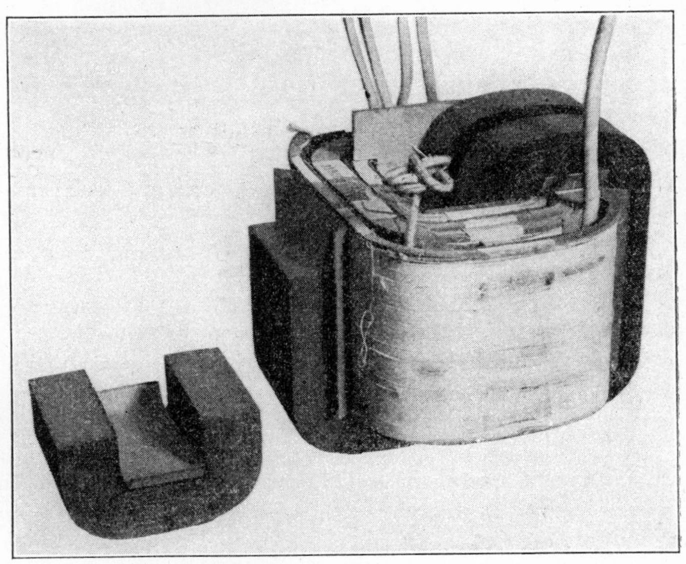

FIG. 13·2 Partly assembled transformer with type C core.

Both types of cores may be built into neat assemblies of the open type, such as those on the right of the panel in Fig. 13·5. For complete enclosure, assemblies like those in Fig. 13·6 are used.

The size of a power transformer is determined by one or more of several factors.

Rating

The product of volts and amperes fixes the number of turns and the cross section of wire in an appropriate core, and hence the total volume. The space for turns is reduced (hence size increases) if space must be provided for high-voltage insulation or spaces between sections of multisection windings. Sizes for low-voltage single-section 60-cycle enclosed transformers are given in Fig. 13·7.

Frequency

The size of a transformer is an inverse function of frequency. This relation holds within the range from 25 to

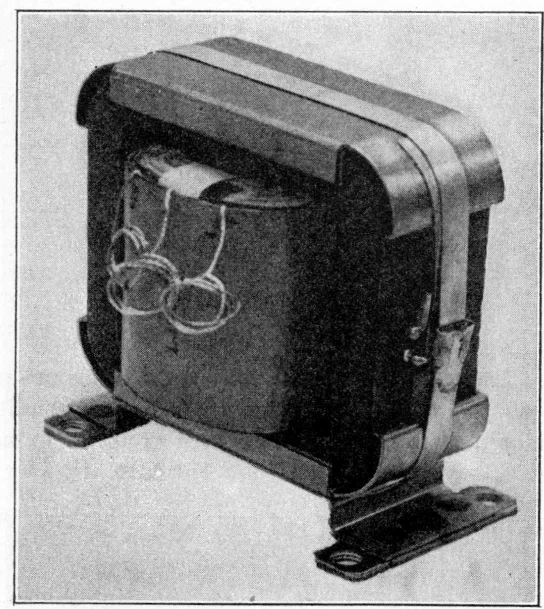

FIG. 13·3 Assembled cores and coil.

800 cycles. Typical combinations of core material and insulation are

Frequency	Strip Thickness	Gauss	Class of Insulation	Operating Temperature
60	0.014	15,000	A (organic)	95°C
400	0.005	12,500	B (glass and mica)	140°C
800	0.005	8,500	B	140°C

In very small units, these flux densities may be used with class A insulation throughout. The limit has not been reached in either flux density or operating temperature. The

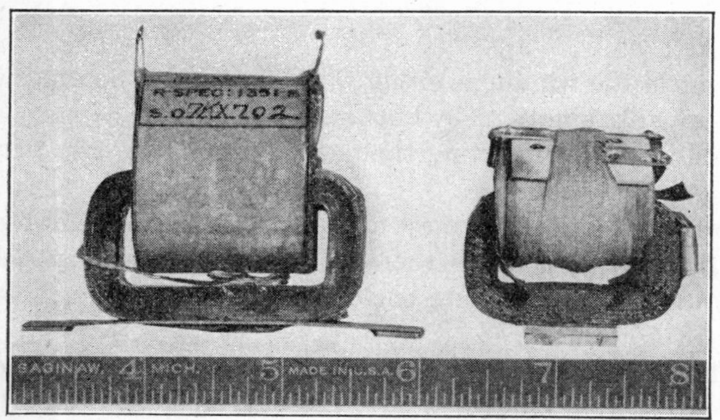

FIG. 13·4 Single-core, single-coil assembly.

necessity for small dimensions, especially in aircraft apparatus, requires use of materials at their fullest capabilities.

Insulation

As size decreases, the ability of a transformer to dissipate a given number of watts loss also decreases. Hence, it operates at a higher temperature. Transformers for 400- and 800-cycle power supplies can be made from 30 to 50 percent smaller with class B insulation than with class A insulation. Class B insulation, moreover, can better withstand extremes of ambient temperature, humidity, and altitude. Class B insulation is thus of special importance in aircraft apparatus. Usually at 60 cycles enough room is available to use class A

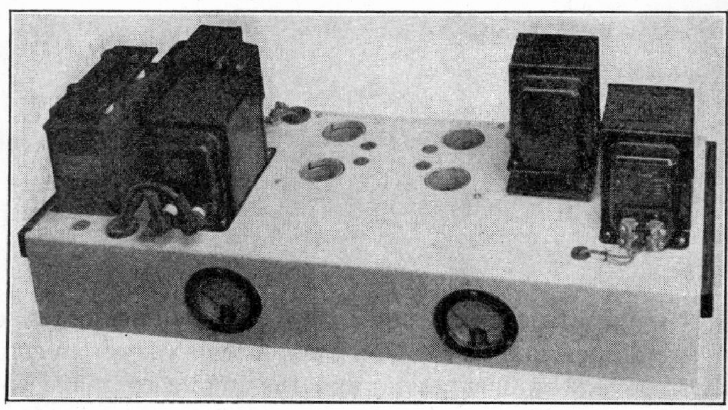

FIG. 13·5 Amplifier chassis mounting open-type transformers.

insulation, but mica is often used to reduce the size of high-voltage units.

The increase in operating temperature made possible by using class B insulation can be seen by referring to Fig. 13·8. The lower solid line represents the "eight-degree rule" for class A insulation (55 degrees centigrade temperature rise plus 10 degrees hot spot at 40 degrees ambient temperature) and is taken from data gathered over many years. The class B line has fewer data to support it. Life tests indicate that the average safe temperature rise could be 90 degrees at 40 degrees centigrade ambient temperature. This is shown by the dot-and-dash line.

Class B insulation can be worked much closer to the ultimate dielectric strength, but size then often depends upon creepage distance to the core. The use of filling compounds makes possible a substantial size reduction in high-voltage transformers by reducing creepage distance.

Varnishes used for impregnation of electrical coils have until lately been diluted by solvents to lower their viscosity and permit them to penetrate the windings. When the coils are baked the varnish dries, but it is left with tiny holes through which moisture can penetrate and in which corona

FIG. 13·6 Fully enclosed transformers.

may form. Eventually this corona destroys the insulation. It is therefore necessary to allow large clearances for high voltages, or to immerse the coils in oil. Either of these practices increases the size of a high-voltage transformer. New varnish such as Fosterite changes by heat polymerization from a liquid to a solid with only a slight change in volume. It has low viscosity, requires no solvent, and with care can be made to fill coil interstices completely. Because of this excellent filling, it is possible to reduce voltage clearances to a much smaller value than was formerly possible, and therefore to obtain a pronounced reduction in size.

Enclosure

With other factors such as frequency and grade of iron constant, large transformers dissipate less heat per unit of volume than do smaller ones, because dissipation area increases as the square of the equivalent spherical radius, whereas volume increases as its cube. Therefore large units commonly are of the open type; smaller units are totally enclosed. The apparent exception of 400- to 800-cycle transformers is because of the high power-dissipation rate with class B insulation. Totally enclosing these units might result in exploding the container. Where enclosure is feasible, it tends to increase size by limiting heat dissipation.

Regulation

Good voltage regulation, as required in some inductors and anode transformers, requires a larger cross section of copper than current-carrying capacity would normally dictate.

It is not good practice to apply so-called safety factors in specifying operating voltage, for it results in unduly bulky

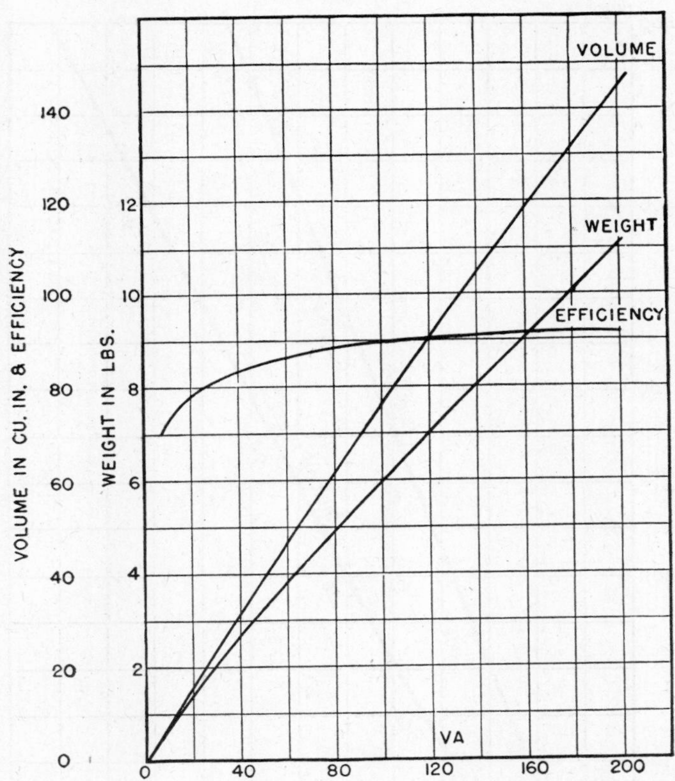

FIG. 13·7 Volt-amperes versus weight, volume, and efficiency. (From *Electronic Transformers and Circuits* by Reuben Lee.)

transformers. For class A insulation, the resistance of the insulation to corona over a long period is more important than the breakdown strength of the insulation in a 1-minute test. For example, a 20-mil thickness of treated cloth will withstand 10,000 volts for 1 minute, but corona starts at 1250 volts, and a higher voltage would puncture the insulation in a few weeks. It is much wiser to keep operating voltage from 20 to 30 percent below the corona limit than to guess at a percentage of the 1-minute breakdown voltage. It is important to make a distinction between test voltage and operating voltage. Operating voltage is the usual value specified.

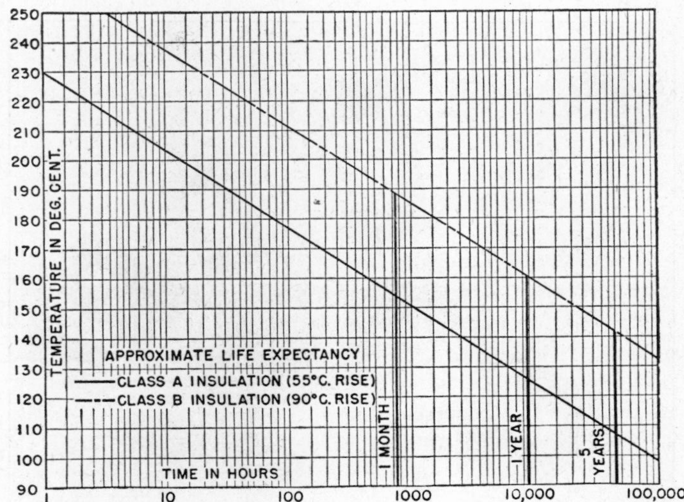

FIG. 13·8 Class A and class B insulation life. (Reprinted with permission from R. Lee, *Electronics*, Oct. 1939.)

13·2 CORE MATERIALS

A large variety of core steels are used in transformers for electronic circuits. The principal grades are listed in Table 13·1.

TABLE 13·1 CHARACTERISTICS AND USES OF THE PRINCIPAL CORE STEELS

Grade of Steel	Maximum Permeability (typical)	Saturation Flux Density (gauss)	Chief Uses
Silicon	10,000	12,000	Small power and voice-frequency audio transformers.
Hipersil	40,000	17,000	Larger sizes of power and wide-range audio transformers.
Hipernik	80,000	10,000	Small wide-range audio transformers; audio filter reactors.
Mumetal	200,000	6,000	Small wide-range audio transformers; audio filter reactors.
Conpernik	1,400	*	Linear and low-loss transformers.
Powdered iron	80	*	Low- and medium-frequency r-f transformers.

* Used for low-flux-density low-loss applications. As in power-transformer design, saturation flux densities must not be exceeded, because high exciting current produces high *IR* drops in the windings and high losses, and hence affects size.

Among the new steels is Hipersil, a steel in which the direction of grain orientation is controlled during the manufacturing process. Flux must follow this preferred or grain-oriented direction to utilize the full capabilities of the material. Grain-oriented cores are wound of strip in such a way that flux is in the lengthwise direction. Several cores are shown in Fig. 13·9.

Probably the most remarkable property of this material is its high saturation point. In Fig. 13·10 the comparison is given in terms of a hypothetical 60-cycle induction using high-grade conventional silicon steel. With this value as

FIG. 13·9 Hipersil cores.

100 percent, the induction with grain-oriented steel is 150 percent with no increase in magnetizing force. Another way of expressing this improvement is to say that the permeability of grain-oriented steel is higher at the maximum point (see Fig. 13·11). Iron loss is less than in silicon steel, as shown by Fig. 13·10.

Because of this increase in induction the core area can be smaller for the same magnetizing current; moreover, the mean length of turns can be reduced, so the amount of copper is reduced. For maximum benefit in power and distribution transformers the iron and copper losses are reproportioned. In small electronic transformers, the iron loss is

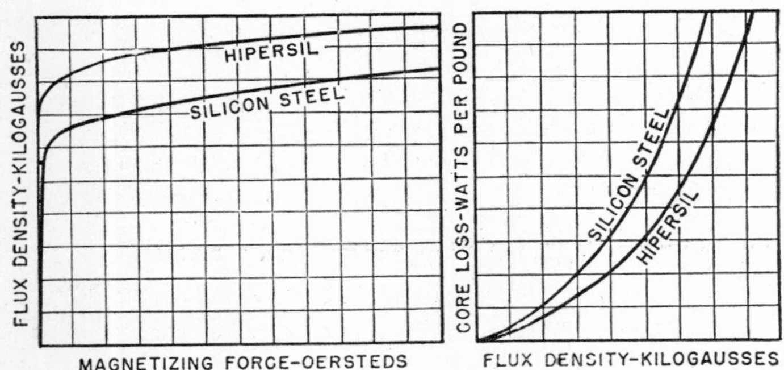

FIG. 13·10 Saturation curve and core loss of Hipersil versus silicon steel. (From *Electronic Transformers and Circuits* by Reuben Lee.)

usually a small part of the total loss, and the reduction in copper loss is of greater significance.

Inductors which carry direct current can be made smaller with grain-oriented than with ordinary silicon steel. At low voltage, where low induction is involved, grain-oriented steel has greater incremental permeability, and maintains it

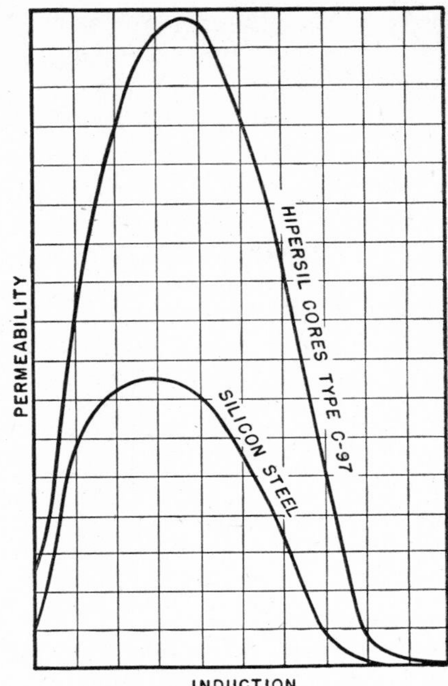

FIG. 13·11 Permeability of Hipersil and silicon steel.

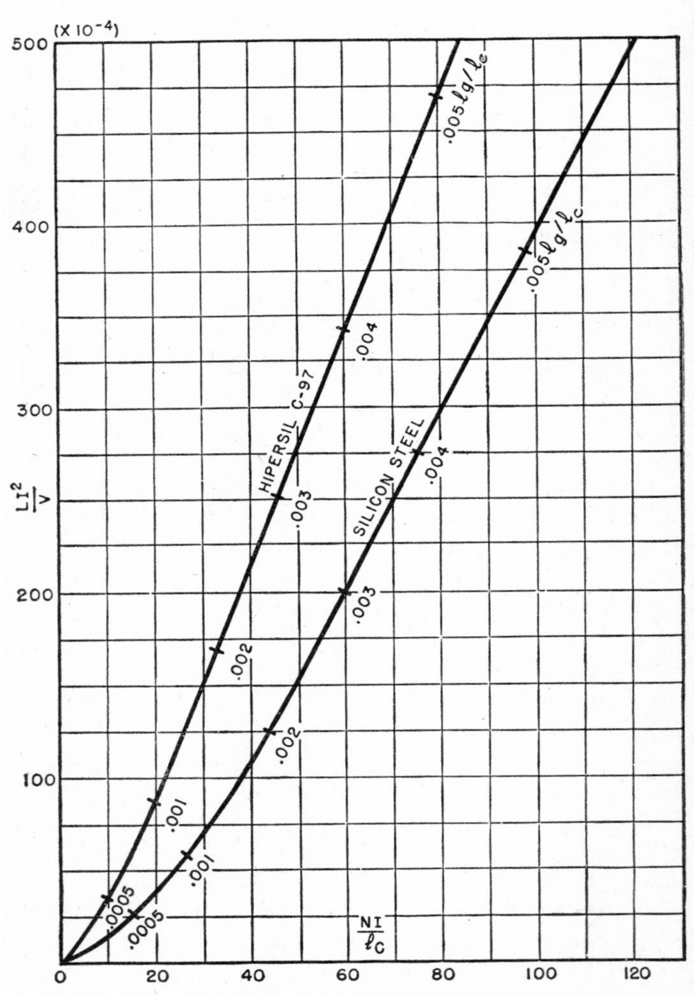

FIG. 13·12 Relation between energy per unit volume and ampere-turns per inch of core (l_g is the air gap, l_c the core length). (From *Electronic Transformers and Circuits* by Reuben Lee.)

at high flux densities. Consequently, a weight reduction of 50 percent is often feasible. Typical performance curves of Hipersil and silicon steel are shown in Fig. 13·12.

Grain-oriented steel does not replace high-nickel alloys for audio transformers. Such transformers usually work at low induction, and with little or no direct current. Some nickel-iron alloys have higher permeability at low flux density, so their use for this purpose continues.[1,2]

Hipersil can be used for transformers in various applications in the low and medium radio-frequency bands at power levels ranging up to hundreds of kilowatts. The same is true of video and pulse transformers, which may be regarded

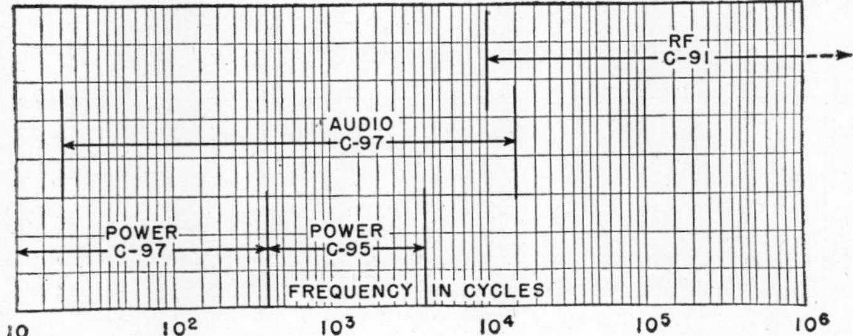

FIG. 13·13 Use of Hipersil in various frequency zones. (From *Electronic Transformers and Circuits* by Reuben Lee.)

as covering an extended frequency range down into the audio range and up into the medium radio-frequency range. Such transformers are grouped rather loosely as radio-frequency transformers in Fig. 13·13. In this figure the several classifications of radio-frequency, audio, and power transformers are shown with respect to their frequency ranges, and to the approximate gauge of the material. C-97 is 0.013 inch thick, C-95 is 0.005 inch thick, and C-91 is 0.002 inch thick.[3]

13·3 FILAMENT TRANSFORMERS

Low-voltage filament transformers are used for heating filaments of oscillator and amplifier tubes, at or near ground potential. Often the filament windings for the tubes of several stages are combined in one transformer. Sometimes several secondary windings are required. A transformer with five or six secondary windings is about half again as large and heavy as a unit with a single secondary winding. However, such a transformer is enough smaller than five or six separate units to warrant designing it specially for many applications.

Rectifier-tube filaments often operate at high direct voltages, and require windings with high-voltage insulation. It is usually not feasible to combine high-voltage windings with low-voltage windings, if the direct voltage is more than 3000, because of insulation difficulties, particularly in the lead joints. To reduce the length of filament leads and the voltage drop, rectifier-filament transformers are often located near the tubes, and the filaments are heated by separate transformers. In polyphase rectifiers all tube filaments are operated at high voltage, and some secondary windings may be combined.

Low-capacitance filament windings are sometimes required for high-frequency circuits. These are not particularly difficult to design, if the volt-ampere rating is small and the voltage is low. Air may occupy most of the space between windings as in Fig. 13·14. Larger ratings are more difficult,

FIG. 13·14 Low-capacity high-voltage filament transformer.

because the capacitance increases directly as the mean length of turns for a given spacing between windings.

Except for these differences the design of filament transformers does not differ much from the design of small 60-cycle

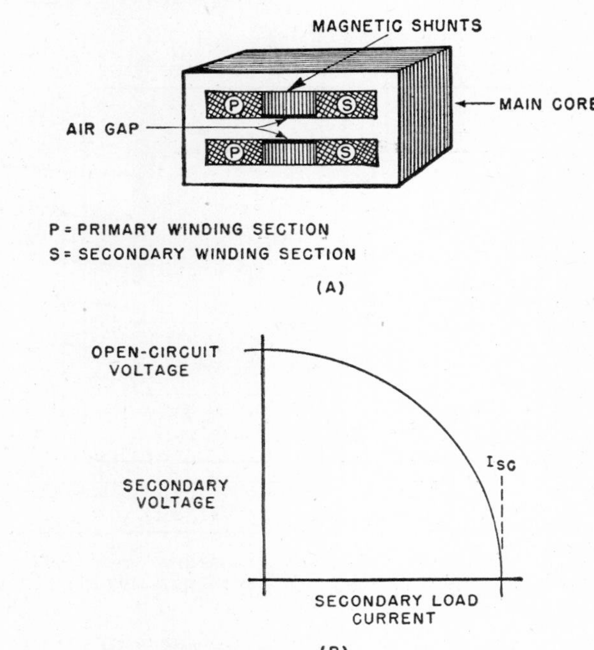

FIG. 13·15 (*A*) Current-limiting transformer. (*B*) Output voltage versus current curve. (From *Electronic Transformers and Circuits* by Reuben Lee.)

power transformers. The load is constant and of unity power factor. Leakage reactance is not important because it is almost 90 degrees out of phase with the load.[4]

Filaments of large vacuum tubes sometimes must be protected against the high initial current they would draw at rated filament voltage. This protection is secured by reducing starting voltage automatically through the use of a current-limiting transformer with magnetic shunts between primary and secondary windings. The shunts carry little flux at no load; as the load increases, the secondary forces more flux into the shunts until at current I_{sc} (Fig. 13·15) output voltage is zero. This same principle is used to limit current in high-impedance tube circuits.

13·4 ANODE TRANSFORMERS

Anode transformers may be classified as to type of rectifier and number of phases. Half-wave rectifiers carry unbalanced direct current, thus necessitating less alternating flux density, and hence larger transformers, than for full-wave rectifiers. A single-phase full-wave rectifier with two anodes has balanced direct current in the secondary, but because of the high secondary volt-amperes the transformer is larger for a given primary volt-ampere rating than a filament transformer normally would be. Bridge (four-anode) rectifiers have equal primary and secondary volt-amperes, as well as balanced direct current, and anode transformers for these rectifiers are smaller than for other types. Table 15·1 gives the volt-ampere ratings.[5]

Anode and filament windings are combined in small units for receivers, low-power transmitters, or bias supplies.

13·5 INDUCTORS

Inductors are used mainly to smooth out ripple voltage in d-c supplies, and therefore usually have air gaps in the cores

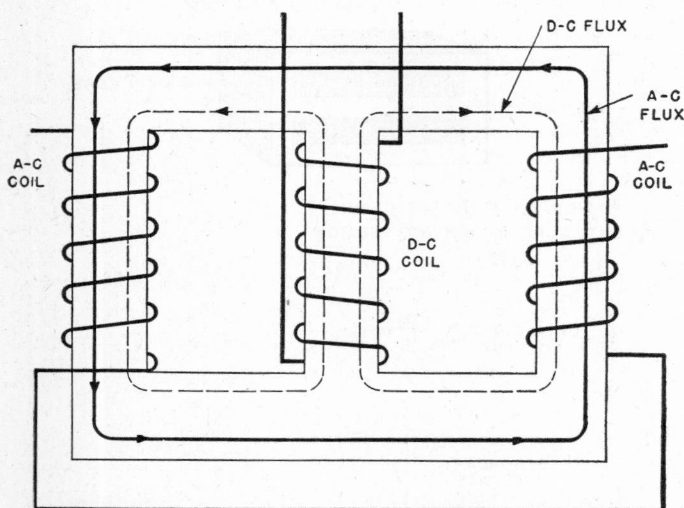

Fig. 13·16 Windings and core-flux paths in a saturable reactor. (From *Electronic Transformers and Circuits* by Reuben Lee.)

to prevent d-c saturation. The length of air gap, the size of the core, and the number of turns depend upon three interrelated factors: (1) inductance desired; (2) direct current in the winding; and (3) alternating volts across the winding.

Inductors in inductor-input filters of single-phase rectifiers are subject to the highest alternating voltage for a given unidirectional voltage. The inductance of this type of inductor influences:

1. Magnitude of ripple in the rectified output.
2. No-load to full-load regulation.
3. Transient voltage dip when load is suddenly applied, as in keyed loads.
4. Transient current through rectifier tubes when voltage is first applied to rectifier.
5. Peak current through tubes during each cycle.

It is therefore important that the inductance be right. Several of these properties can be improved by the use of

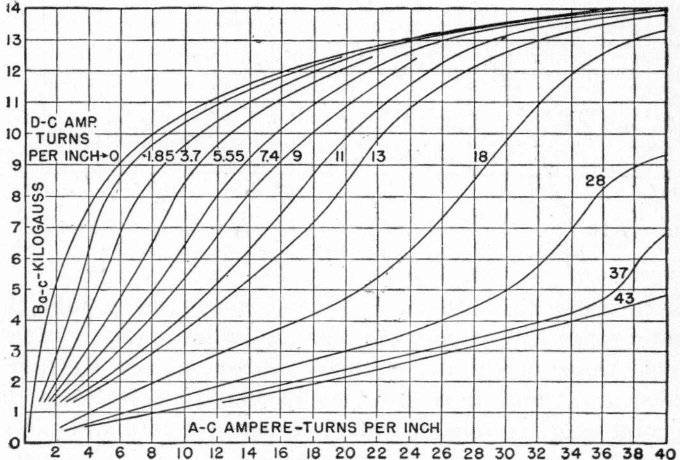

Fig. 13·17 Typical saturation curves for saturable reactor. (From *Electronic Transformers and Circuits* by Reuben Lee.)

swinging or tuned inductors. A swinging reactor is saturated at full load; therefore its inductance is lower at full load than at no load. The higher inductance at no load helps to decrease voltage regulation. The same result is obtained by shunt-tuning the inductor, but the inductance should be constant from no load to full load to preserve the tuned condition.

Insulation of an inductor differs with the type of rectifier and the way it is used in the circuit. Three-phase rectifiers, with low inherent ripple voltage, do not require as much insulation between turns and layers as single-phase rectifiers do. The size of an inductor for a given voltage and ratio of resistance to inductance is proportional to the energy content, or LI^2. Figure 13·12 shows the relation of size per unit volume to d-c magnetizing force for small a-c flux.[6,7]

If the function of the inductor is to control a-c circuits by means of large inductance changes, no air gap is used. An arrangement similar to that in Fig. 13·16 is often used to keep alternating voltages out of the d-c control circuits. Usually the alternating flux density changes widely with inductance changes. Figure 13·17 shows typical operating data for such inductors with cores made from silicon steel.[8]

In filter circuits another type of inductor is used. Air gaps are employed to increase the ratio of magnetizing current to loss current; thus Q or the ratio $\omega L/R$ is increased. For minimum loss, or maximum Q, there is an optimum length of air gap.

13·6 AUDIO-AMPLIFIER TRANSFORMERS

In audio work, the ratio of two voltages, E_1 and E_2, is usually stated in decibels (db) according to the definition

$$\text{db} = 20 \log_{10} \frac{E_1}{E_2} \tag{13·2}$$

Amplifier-voltage gain, transformer ratio, frequency response, and noise levels all may be expressed in decibels. If volume or voltage or power is given in decibels, it must be compared with a reference level; otherwise the term is meaningless.

Transmission lines at audio and higher frequencies have properties commonly ignored at 60 cycles. Such entities as line wavelength, characteristic impedance, and attenuation are important at audio frequencies. Of particular importance is impedance matching. If a transmission line has no attenuation its characteristic impedance is

$$Z_0 = \sqrt{\frac{L}{C}} \tag{13·3}$$

If such a line terminates in a pure resistance load equal in ohmic value to Z_0, all the power fed into the line appears in the load without attenuation or reflection. This process is called "matching" the impedance of the line. The notion has been extended to include the loading of vacuum tubes, but this is stretching the meaning of the term matching.

The major problem of audio-transformer design is obtaining good frequency response, or constant voltage ratio over the required audio-frequency range, when the transformer is operated with the audio apparatus for which it is intended. Several factors outside the transformer affect its performance:[9]

1. Impedance of the source of audio power.
2. Linearity of this impedance.
3. Impedance of the load fed by the transformer.
4. Audio frequencies involved.

In Fig. 13·18 the transformer that connects the audio source to its load may be represented by the diagram in Fig. 13·19 (*a*).

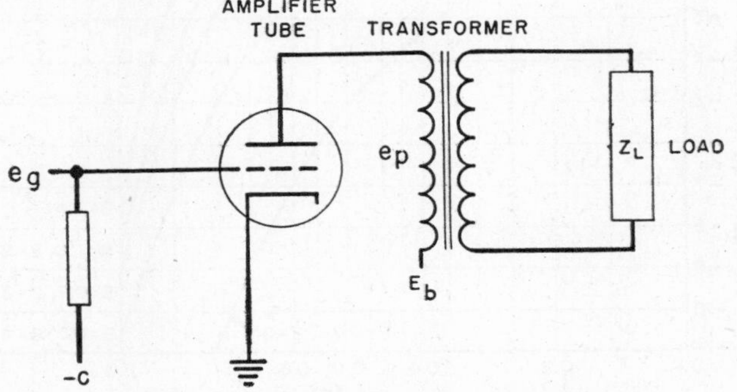

FIG. 13·18 Transformer-coupled amplifier.

13·7 LOW-FREQUENCY RESPONSE

At the low-frequency end of the audio range, leakage reactance is negligible. Resistance R_p may then be combined with Z_g to form R_1 for a pure-resistance source, and R_S combines with Z_L to form R_2 for a resistance load. At low audio frequencies both source and load are practically pure resistance, and the circuit shown in Fig. 13·19 (*b*) is the result. Here a^2 has been dropped; in other words, a transformer with a one-to-one ratio referred to the primary side is shown. X_N is the primary no-load reactance, or

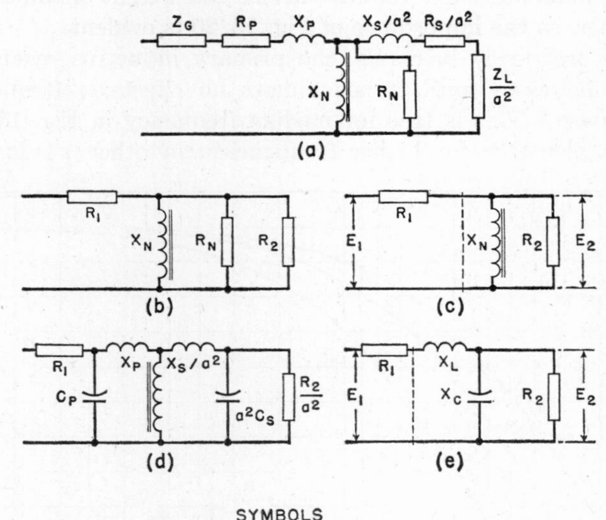

SYMBOLS

a = RATIO OF SEC. TO PRI. TURNS
C_P = PRI. WINDING CAPACITANCE
C_S = SEC. WINDING CAPACITANCE
$C_T = C_P + a^2 C_S$
f = ANY AUDIO FREQUENCY
f_r = RESONANCE FREQ. OF X_L & X_C
R_P = PRI. WINDING RESISTANCE
R_S = SEC. WINDING RESISTANCE
R_N = PRI. NO LOAD (CORE LOSS) EQUIVALENT RESISTANCE
X_N = PRI. OPEN CIRCUIT REACTANCE
X_P = PRI. LEAKAGE REACTANCE
X_S = SEC. LEAKAGE REACTANCE
$X_L = X_P + X_S/a^2$
X_C = TOTAL CAPACITY REACTANCE $= \frac{1}{2\pi f C_T}$
Z_G = SOURCE IMPEDANCE
Z_L = LOAD IMPEDANCE

FIG. 13·19 (*a*) Transformer equivalent circuit. (*b*) Low-frequency equivalent circuit. (*c*) Simplified low-frequency circuit. (*d*) High-frequency equivalent circuit. (*e*) Simplified high-frequency circuit. (Reprinted with permission from R. Lee, *Radio Eng.*, June 1937.)

$2\pi f$ times the primary open-circuit inductance (*OCL*) as measured at low frequencies.

If shunt resistance R_N is included in the load resistance R_2, the circuit becomes like that in Fig. 13·19 (*c*). Winding resistances are small compared with source and load resistances in well-designed transformers. Likewise, R_N is high compared with load resistance, especially if core material of good quality is used.

In Fig. 13·19 (*c*), R_1 represents the source impedance and R_2 the load impedance approximately. Three cases that deserve particular attention are: (*a*) $R_2 = R_1$; (*b*) $R_2 = 2R_1$; and (*c*) $R_2 = \infty$.

Of these, (*a*) corresponds to the usual line-matching transformer with source and load impedances equal; (*b*) is often recommended for maximum undistorted output of triodes; (*c*) is realized practically when the load is the grid circuit of a class A amplifier. For these cases, low-frequency response is plotted in Fig. 13·20 as "db down" from median. The median frequency is the geometric mean of the audio range, at which X_N/R_1 is large; this is true because of the high *OCL* of a good transformer. The equivalent voltage ratio E_2/E_1 has maxima of 0.5, 0.667, and 1.0 for cases (*a*), (*b*), and (*c*) respectively at the median frequency, or for

$X_N/R_1 = \infty$ in Fig. 13·20. Figure 13·20 is of direct use in determining the proper value of primary *OCL*. Permissible response deviation at the lowest audio frequency fixes X_N/R_1 from the curve which applies. This, combined with the source impedance, determines X_N. At the corresponding frequency, X_N represents a certain value of primary *OCL*. This inductance determines the size and weight of the transformer, so the importance of Fig. 13·20 is evident.

As frequency increases, the primary inductive reactance also increases until it has almost no effect on frequency response. This is true for median frequency in Fig. 13·20. It is also true for higher frequencies; in other words, the

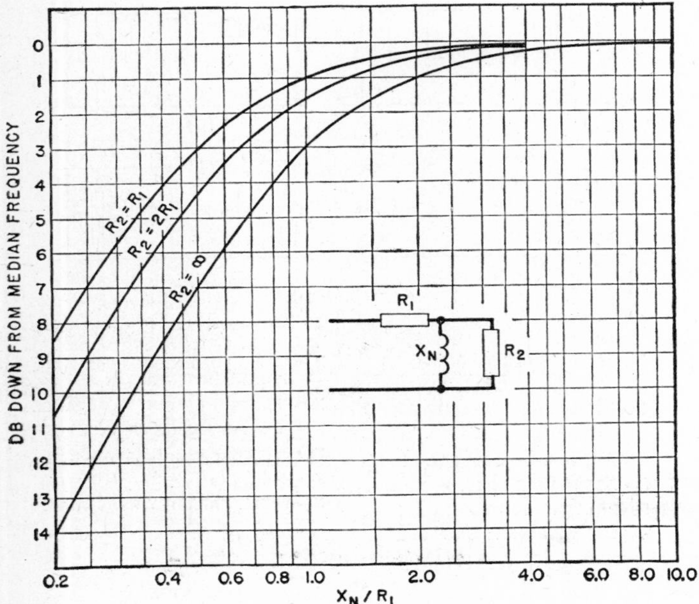

FIG. 13·20 Transformer characteristics at low frequencies. (From *Electronic Transformers and Circuits* by Reuben Lee.)

primary inductance has an influence only on the low-frequency end of the frequency-characteristic curve. A well-designed transformer has a uniform voltage ratio for a frequency range extending from the frequency at which X_N ceases to exert any appreciable influence upward to a zone designated as the high-frequency end of the transformer frequency range.

13·8 HIGH-FREQUENCY CHARACTERISTICS

The factors that influence high-frequency response are leakage inductance, winding capacitance, source impedance, and load impedance. Hence a new equivalent diagram, Fig. 13·19 (*d*), is necessary for the high-frequency end. Winding resistances are omitted or combined as in Fig. 13·19 (*b*). Winding capacitances are shown across the windings. If primary and secondary leakage inductances and capacitances are combined and X_N is omitted as if it were infinitely large, and a^2 is dropped as before, the circuit becomes that shown in Fig. 13·19 (*e*). X_L is the leakage reactance of both windings, X_C is the capacitive reactance of both windings, and R_2 is the load resistance, all referred to the primary side at a one-to-one ratio. X_L equals X_C at resonance frequency f_r.

Assign to the ratio X_C/R_1 a value B at frequency f_r. Then at any frequency f, $X_C/R_1 = Bf_r/f$. The three cases considered for low frequencies are plotted in Figs. 13·21, 13·22, and 13·23 with B as parameter. If X_C/R_1 has certain values

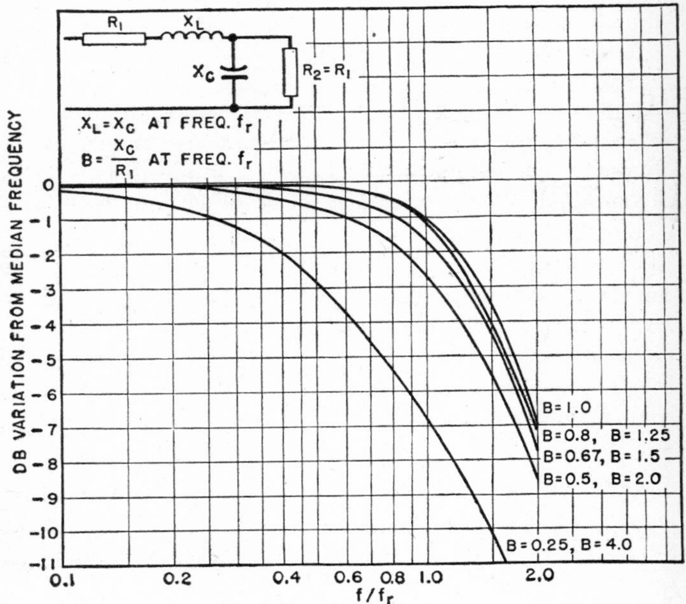

FIG. 13·21 Transformer characteristics at high frequencies (line matching). (From *Electronic Transformers and Circuits* by Reuben Lee.)

at frequency f_r, the audio-frequency characteristic is relatively flat up to frequencies approaching f_r. In particular, performance is good at $B = 1.0$ in all three figures.

In regarding leakage inductance and winding capacitance as "lumped" quantities, current distribution in the windings is assumed to be uniform throughout the range of frequencies

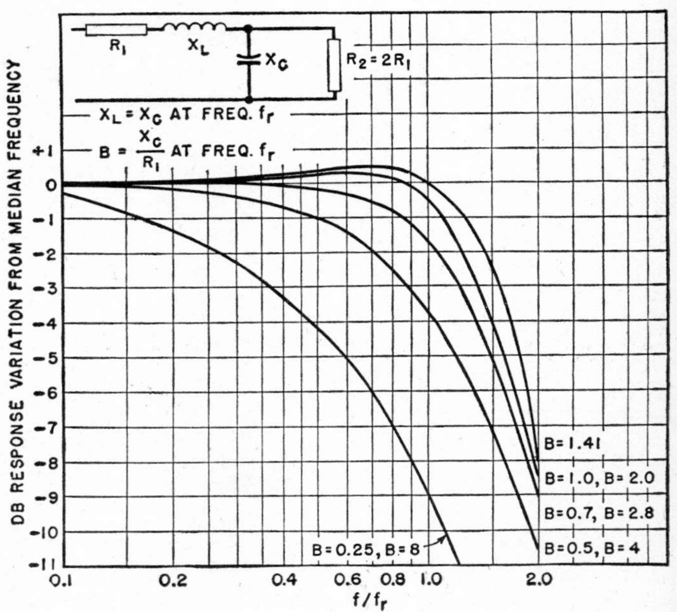

FIG. 13·22 Transformer characteristics at high frequencies (triode output). (Reprinted with permission from R. Lee, *Radio Eng.*, June 1937.)

considered. This assumption is valid up to the resonance frequency f_r. At frequencies higher than f_r the error in Figs. 13·21, 13·22, and 13·23 may be large. But good audio characteristics lie mainly below the frequency f_r, where the curves are correct within the assumed limits. The value of f_r should be such that the highest audio frequency to be

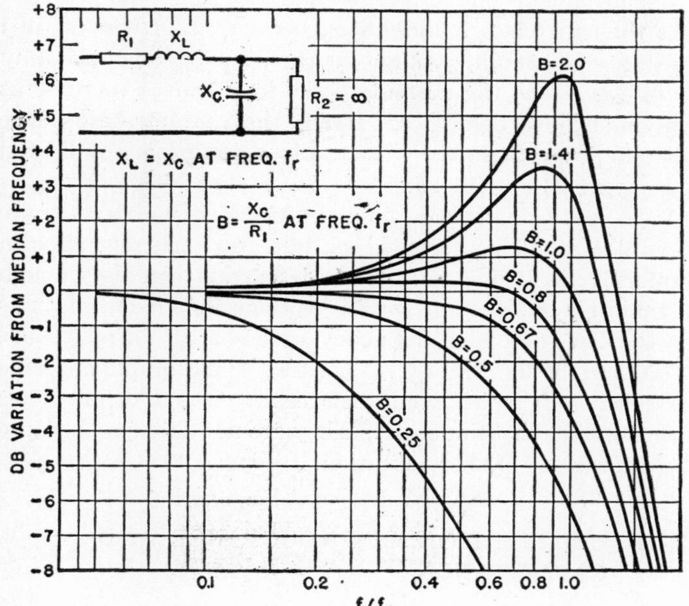

FIG. 13·23 Transformer characteristics at high frequencies (class A grid). (From *Electronic Transformers and Circuits* by Reuben Lee.)

covered lies on the flat part of the curve. X_C and f_r determine the values of winding capacitance and leakage inductance which must not be exceeded.

13·9 HARMONIC DISTORTION

Audio response may be good according to the curves of Sections 13·7 and 13·8, but at the same time the output may be badly distorted because of changes in load impedance.[10,*] This possibility is discussed below for the case in which the load impedance is twice the source impedance.

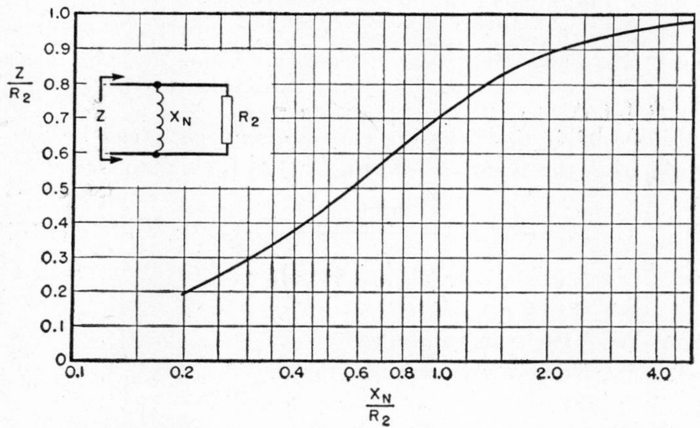

FIG. 13·24 Variation of load impedance with transformer characteristics at low frequencies. (Reprinted with permission from R. Lee, *Radio Eng.*, June 1937.)

Consider the load impedance to be equal to twice the source impedance, as distinct from the impedance of the entire circuit. The load impedance to the right of the dotted line of Figs. 13·19 (*c*) and 13·19 (*e*) is plotted in Figs. 13·24 and 13·25.

Figure 13·24 shows the change in load Z from its median-frequency value R_2 as frequency is lowered. Abscissas are necessarily X_N/R_2 instead of X_N/R_1 as in Fig. 13·20. Figure 13·25 shows the change in load impedance for several values of $D = X_C/R_2$. Impedance evidently varies widely from its median-frequency value, especially at lower values of D.

From Figs. 13·24 and 13·25, it is possible to compare the change in impedance with the previous frequency characteristics. Figure 13·20 shows that if the transformer characteristic is allowed to fall off 1.0 decibel at the lowest frequency, the corresponding X_N/R_1 is 1.3. This means that X_N/R_2 is 0.65. The corresponding load impedance in Fig. 13·24 is only 0.55 of its median-frequency value. Likewise,

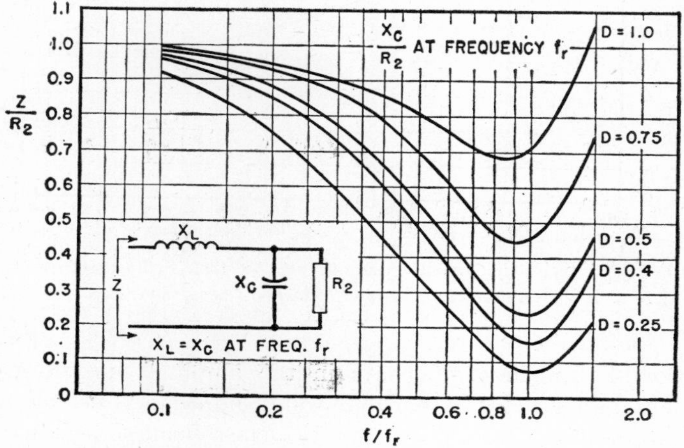

FIG. 13·25 Variation of load impedance with transformer characteristics at high frequencies. (Reprinted with permission from R. Lee, *Radio Eng.*, June 1937.)

for a 0.5-decibel drop of the frequency characteristic, the load impedance falls to 0.7 of R_2; whereas, for a load impedance of $0.9R_2$, the frequency characteristic can fall off only 0.1 decibel. It is evident that load impedance can vary widely, even with comparatively flat frequency characteristics.

At high audio frequencies the divergences are still greater. Suppose, for example, that a transformer has been designed so that X_C/R_1 is 1.0 at f_r (that is, $B = 1.0$ in Fig. 13·22). Suppose further that the highest audio frequency at which the transformer operates is $0.75f_r$. The transformer then has a relatively flat characteristic, with a slight rise near its upper limit of frequency. In Fig. 13·25 the curve corresponding to $B = 1.0$ is marked $D = 0.5$, for which at $0.75f_r$ the load impedance has dropped to 32 percent of R_2, an extremely poor match for the tube. For this reason, the distortion is large over a wide frequency range. It would be much better if B equaled 2.0 instead of 1.0, for the change in impedance is much less. Transformers with $B = 2.0$ have lower effective capacitance, but the leakage inductance may be proportionately greater than for transformers having $B = 1.0$.

Non-linearity of magnetizing current in the transformer produces distortion because the third harmonic in the magnetizing current is large if X_N/R_2 is small.[11]

* Section 13·9 is largely based on the author's "Distortion in High-Fidelity Audio Amplifiers," *Radio Eng.*, June 1937.

13·10 PUSH-PULL AMPLIFIERS: ANODE-CURRENT RISE

The analysis of single-side class A amplifiers in Section 13·9 applies to class A push-pull amplifiers, except that the

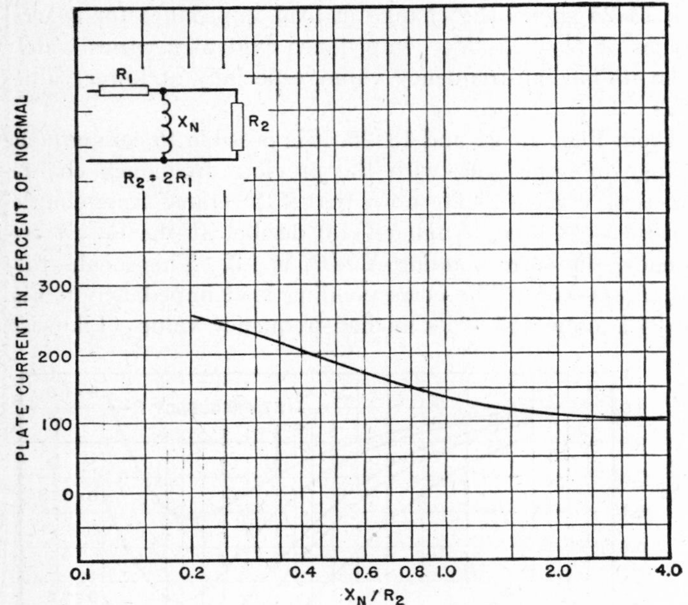

FIG. 13·26 Rise in anode current due to transformer impedance change at low frequencies. (From *Electronic Transformers and Circuits* by Reuben Lee.)

second-harmonic components in the amplifier output are due to unlike tubes, rather than to low-impedance distortion.

The internal tube resistance of a class B amplifier varies so much with the amount of signal voltage on the grids, power output, and anode voltage, that it is virtually impos-

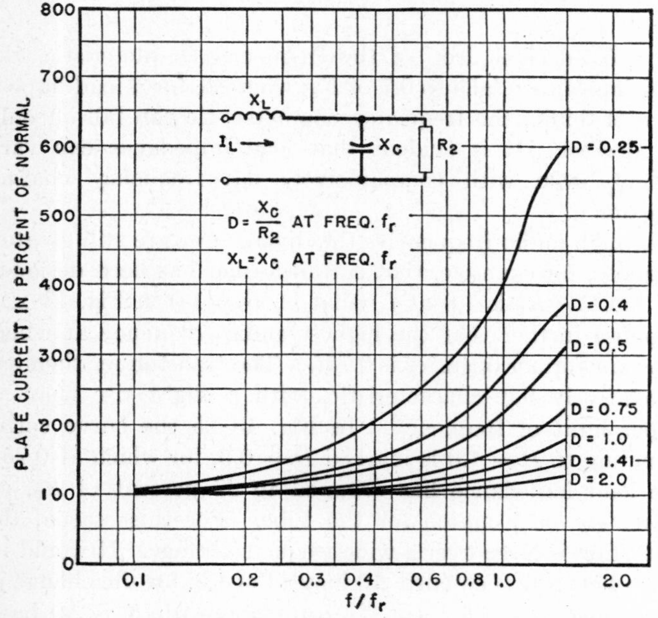

FIG. 13·27 Rise in anode current due to transformer impedance change at high frequencies. (From *Electronic Transformers and Circuits* by Reuben Lee.)

sible to draw curves similar to Fig. 13·22 for class B operation. One basis for designing class B transformers is to make sure that the load impedance, as determined from Figs. 13·24 and 13·25, does not fall below a given percentage of the load resistance R_2. This is discussed further in Section 13·13, as well as in Chapter 16.

Usually the decline in response is greater for class B amplifiers than for class A amplifiers, because the effect of tube resistance is greater. As an extreme, anode resistance may be so great that the current to the load cannot increase as load impedance decreases. Hence the characteristic curve falls off proportionately with the load impedance indicated in Figs. 13·24 and 13·25.

In a lightly loaded amplifier the frequency characteristic stays flat at higher frequencies, but anode current rises in proportion as it does so. If anode current can rise enough to maintain a constant output voltage, anode-current rise for the low- and high-frequency ranges are shown in Figs. 13·26 and 13·27. Many satisfactory audio amplifiers have anode currents which would be excessive if high or low notes were amplified continuously. They are not damaged because these tones are of short duration.

13·11 CALCULATION OF INDUCTANCE AND CAPACITANCE

At low frequencies the performance of a transformer-coupled amplifier depends upon open-circuit inductance of the transformer *OCL*, and at high frequencies upon the leak-

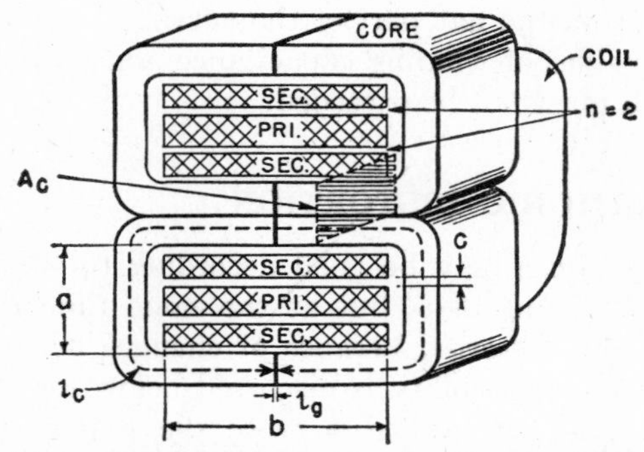

FIG. 13·28 Transformer coil sectionalizing. (From *Electronic Transformers and Circuits* by Reuben Lee.)

age inductance and winding capacitance. *OCL* can be computed from the formula (see Fig. 13·28 for dimensions):

$$OCL = \frac{3.2N^2A_c}{10^8\left(l_g + \dfrac{l_c}{\mu}\right)} \text{ henrys} \tag{13·4}$$

where N = turns in winding
A_c = core area in square inches
l_g = total length of air gap in inches
l_c = core length in inches
μ = permeability of core (if there is unbalanced direct current in the winding, this is the incremental permeability).

For windings of the concentric shell or core type, the total leakage inductance referred to any winding is

$$L = \frac{10.6N^2l(2nc + a)}{n^2b \times 10^9} \tag{13·5}$$

where N = turns in that winding
l = mean length of turn for whole coil
a = window opening height
b = winding width
c = insulation space
n = number of insulation spaces (number of primary-secondary interleavings).

Winding capacitance is not expressible in terms of a single formula. Effective winding capacitance is almost never measurable, because it depends upon the voltages at the various points of the winding. The major components of capacitance are from turn to turn, layer to layer, winding to winding, and windings to core. Also important are stray capacitance (including terminals, leads, and case), external capacitors, and vacuum-tube electrode capacitance. Layer-to-layer capacitance may be the major portion in high-voltage single-section windings, where thick winding insulation keeps the winding-to-winding and winding-to-core components small.

If a capacitance C with E_1 volts across it is to be referred to some other voltage E_2, the effective value at reference voltage E_2 is

$$C_e = C\frac{E_1^2}{E_2^2} \qquad (13\cdot6)$$

By use of equation 13·6 all capacitances in the transformer may be referred to the primary or secondary winding; the sum of these capacitances is the transformer capacitance used in various formulas and curves of the preceding sections.

In any element of the winding across which the voltage is substantially uniform throughout, the capacitance is

$$C = \frac{0.225A\epsilon}{t} \text{ micromicrofarads} \qquad (13\cdot7)$$

where A = area of winding element in square inches
ϵ = dielectric constant of insulation under winding
t = thickness in inches of insulation under winding.

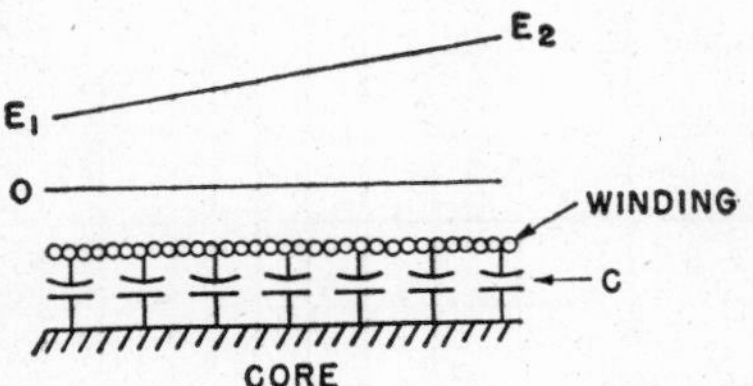

Fig. 13·29 Transformer winding with uniform voltage distribution. (From *Electronic Transformers and Circuits* by Reuben Lee.)

If the winding element has uniformly varying voltage across it, as in Fig. 13·29, the effective capacitance is

$$C_e = \frac{C(E_1^2 + E_2^2 + E_1E_2)}{3E^2} \qquad (13\cdot8)$$

where C = capacitance of winding element as found by equation 13·7
E_1 = minimum voltage across C
E_2 = maximum voltage across C
E = reference voltage for C_e.

If E_1 is zero and $E_2 = E$, equation 13·8 becomes

$$C_e = \frac{C}{3} \qquad (13\cdot9)$$

or the capacitance, say, to ground of a single-layer winding with its low-voltage end grounded is one third of the measured capacitance of the winding to ground.

Because effective capacitance is greatest at high voltages the capacitance in step-down transformers is mainly across the primary winding; in step-up transformers it is across the secondary winding.

13·12 PENTODE AMPLIFIERS

Pentodes are essentially constant-current devices. Load impedance is thus an indication of the output voltage, at

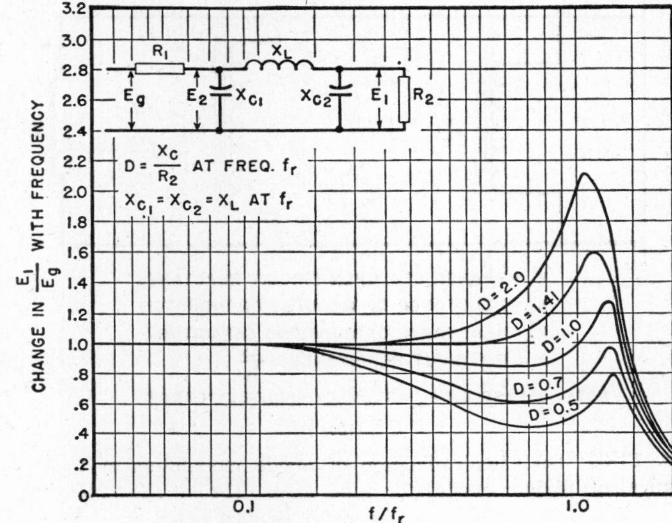

Fig. 13·30 Pentode frequency response with pi-filter output circuit. (From *Electronic Transformers and Circuits* by Reuben Lee.)

least for low frequencies. Response of low-frequency transformer-coupled pentode amplifiers can be taken from Fig. 13·24.

At high frequencies, leakage inductance of the transformer intervenes between the pentode and its load, so primary voltage and secondary or load voltage are not identical. In Fig. 13·30 the change of output voltage for a constant grid voltage at high frequencies is shown. In this figure the equivalent circuit is a pi-filter, which is more accurate for pentode transformers than the circuit of Fig. 13·19 (*e*). This is true because the load of a pentode is low in ohmic value compared to the tube resistance, and step-up transformers are not needed. Harmonic content of pentodes is high, especially in single-side amplifiers, and undue decrease of load impedance is undesirable.

13·13 MODULATION AND DRIVER TRANSFORMERS

In anode-modulated power amplifiers the audio power required to produce 100 percent modulation is one-half the power-amplifier input. Increased quality and reduction in size of components are achieved through the use of what may be called the pi-filter method.

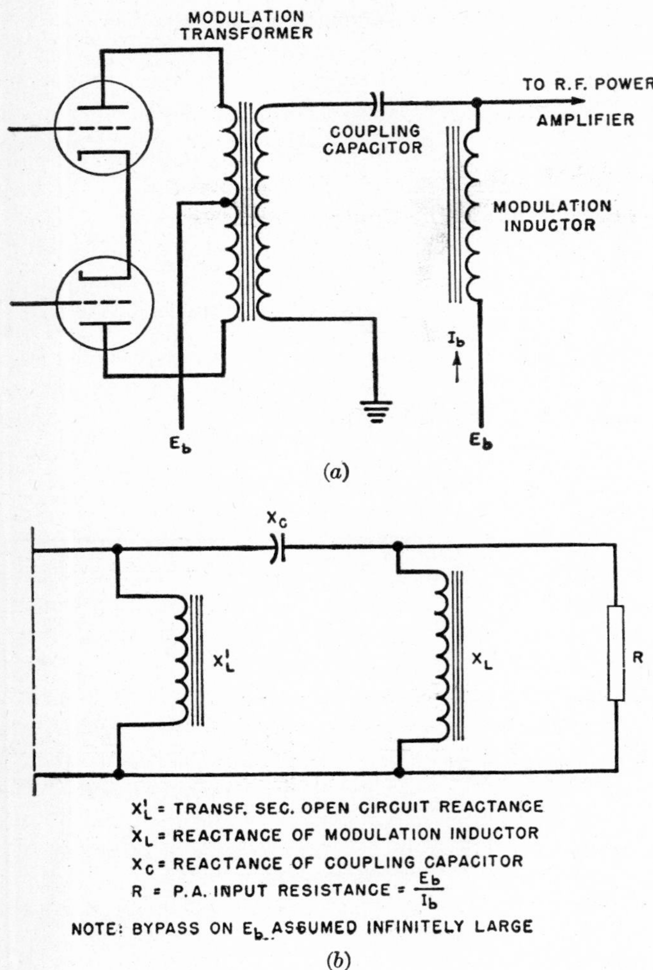

Fig. 13·31 (*a*) Circuit diagram of anode modulation system. (*b*) Equivalent pi-filter modulator tube load. (From *Electronic Transformers and Circuits* by Reuben Lee.)

For low audio frequencies, this method may be illustrated by means of the circuit diagram of Fig. 13·31 (*a*). The (usually class B) output-transformer open-circuit inductance, coupling capacitor, and modulation inductor combine to form a pi-section high-pass filter, Fig. 13·31 (*b*). The elements are proportioned for a characteristic impedance equal to the equivalent input resistance E_B/I_B of the modulated power amplifier.

Formerly these parts were made as large as was considered practical. Transformer-secondary and inductor reactances were each three or four times as large as the power-amplifier anode-input resistance, and coupling-capacitor reactance was a fraction of this resistance, at the lowest audio frequency. Advantages of the pi-filter are a reduction of the two inductive reactances to 1.41 the terminating load resistance, and increase in capacitive reactance to the same value, at a low audio frequency f_1.

Down to f_1 the filter maintains a tube load of almost 100 percent power factor, although the ohmic value rises to 190 percent of the terminating load resistance at f_1. The voltage required for constant output rises to 138 percent of normal. Partly compensating for this rise is the tendency of class B amplifier tubes to deliver higher voltages with higher tube-load impedances. Thus in a certain transmitter, the type 805 modulator tubes deliver 1035 peak volts per side into a normal tube load of 1860 ohms. At 30 cycles, the lowest audio frequency, load impedance rises to 3600 ohms, and the voltage required for full output is 1440. Plotting a 3600-ohm-load line on the tube curves shows that 1275 volts is delivered at 30 cycles, which is 1 decibel down from normal. To obtain the same frequency response with the older "brute force" method, at least twice the values of transformer and choke inductance and much more coupling capacitance would have been necessary.

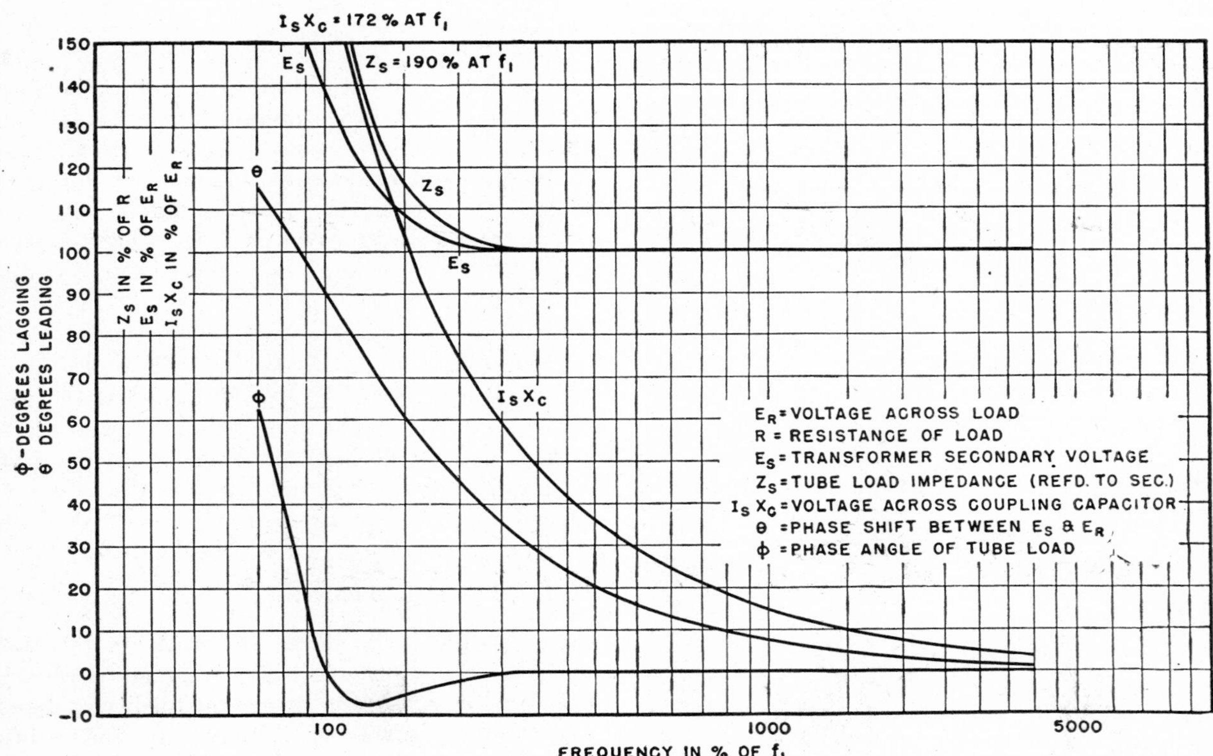

Fig. 13·32 Pi-filter basis of transformer design. (From *Electronic Transformers and Circuits* by Reuben Lee.)

These points are made clearer by Fig. 13·32 which shows how phase angle and phase shift, as well as load impedance, vary with frequency.

No matter what method of design is used, it is important that the modulator be loaded properly. If power-amplifier input should be interrupted while the modulator is fully excited, voltages on the various elements are likely to rise to dangerous values.

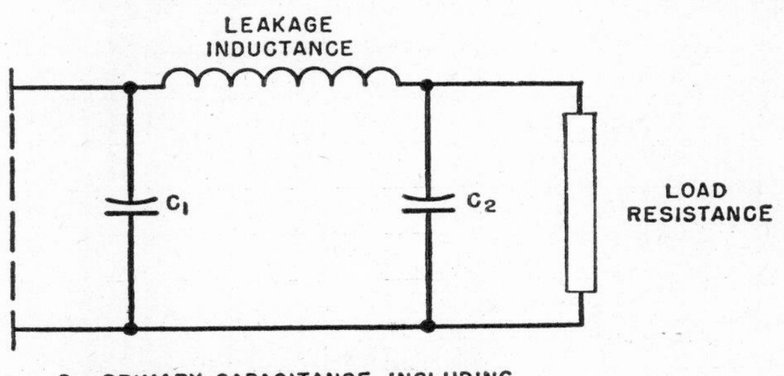

Fig. 13·33 Equivalent transformer diagram at high audio frequencies. (From *Electronic Transformers and Circuits* by Reuben Lee.)

The pi-filter method of design can be applied also at the higher audio frequencies.[12] Figure 13·33 shows how the usual winding capacitance and leakage inductance are arranged as a low-pass filter. Sometimes external L and C are added at the transformer terminals to produce the required proportions. In this way load impedance can be held constant and at low phase angle over a wide frequency range.

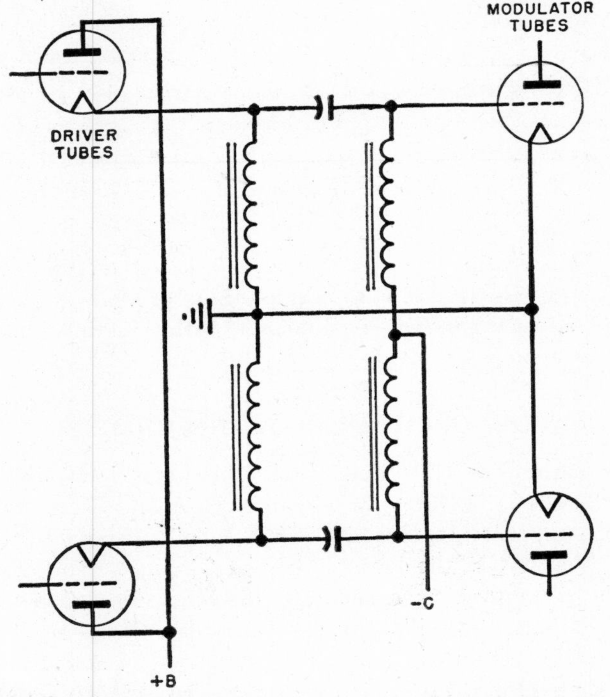

Fig. 13·34 Cathode-follower driver circuit. (Reprinted with permission from R. Lee, *Proc. I.R.E.*, April 1945.)

Requirements for class B modulator driver transformers are difficult to satisfy. Transformer load is non-linear, because grid current is far from sinusoidal. The driver tube must deliver these instantaneous peaks, although the average load is low; otherwise distortion appears in the modulator audio output and in the radio-frequency envelope. The grid current peaks are the equivalent of harmonic currents of higher order, and to insure their appearance in the modulator grid current requires an extension of the driver-transformer frequency range at both ends: on the high-frequency end because of the decreased leakage inductance necessary to allow the larger currents, and on the low-frequency end to prevent transformer magnetizing current from loading the driver tube so that it does not deliver peak grid power.

These conditions require transformers of exceptionally large size, so it is an advantage to dispense with driver transformers entirely. This is accomplished by the cathode-follower circuit, Fig. 13·34, which for a push-pull amplifier takes the form of a symmetrical pi-filter. The two audio-input inductors connect the driver-tube cathodes to ground, and carry their anode current. Coupling capacitors connect these inductors to the modulator-tube grid inductors, which carry modulator grid current. Sizes of inductors and coupling capacitors are chosen to give approximately constant impedance from the lowest audio frequency to the higher harmonics of the highest audio frequency.

The cathode-follower circuit is advantageous in another way. Leakage inductance in a driver transformer causes high-audio-frequency phase shift between driver and grid voltage. There is no such phase shift in the coupling-capacitor circuit. Since inverse feedback is often applied to audio amplifiers to reduce distortion the absence of phase shift is a great advantage.

13·14 SHIELDING

Gain of 80 to 100 decibels is often produced in high-gain amplifiers. In these amplifiers it is important that only the signal be amplified. Small amounts of extraneous voltage at the amplifier input may spoil the quality or even make unintelligible the received signal. One source of extraneous voltage or "hum" is in stray magnetic fields from power transformers in or near the amplifier. The stray fields enter the magnetic cores of input transformers and induce small voltages in the windings, which may be amplified to objectionably high values by the amplifier.

Several devices are used to reduce this hum pick-up:

1. The input transformer is located away from the power transformer.
2. The coil is oriented for minimum pick-up.
3. Magnetic shielding is employed.
4. Core-type construction is used.

The first expedient is limited by the space available for the amplifier, but since the field varies as the inverse cube of the distance from the source it is obviously helpful to locate the input transformer as far away from the power transformer as possible.

The second method is to shift the coil so that its axis is perpendicular to the field. It requires extra care in testing. Magnetic shielding is the "brute force" method of keeping out stray fields; core-type construction is effective and does not materially increase size. Of course, any of these methods increases manufacturing difficulties to a certain extent.

Magnetic shielding is ordinarily accomplished by a thick wall of ferrous metal or a series of thin, nesting boxes of high-permeability material encasing the windings and core of the input transformer. Neither type of shield is applied to the power transformer, because the flux lines originate at the power transformer and fan out in all directions from it. Much of the flux would strike the shield at right angles and pass through it. But the stray field near the input transformer is relatively uniform, and very few flux lines strike the shield at right angles. Thus more flux is by-passed by it.

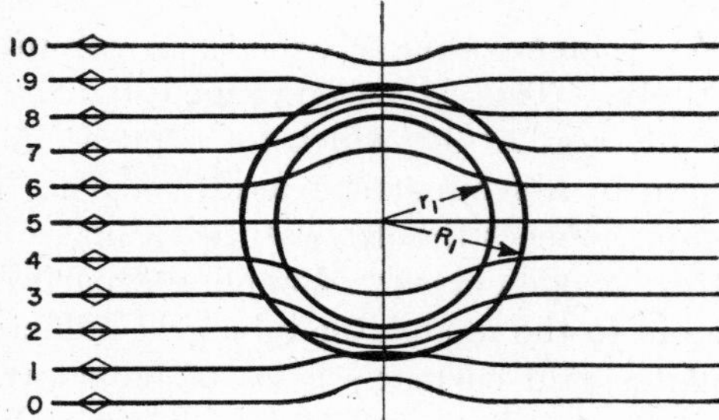

FIG. 13·35 Refraction of magnetic field by iron shield. (From *Electronic Transformers and Circuits* by Reuben Lee.)

The action of a thick shield in keeping stray flux out of its interior is roughly illustrated in Fig. 13·35.[13,14]

Multiple shields increase the action just mentioned, because eddy currents induced in the shields set up fluxes opposing the stray field. Sometimes alternate layers of copper and magnetic material are used for this purpose, when hum pick-up 50 or 60 decibels below the no-shield value is desired.

In core-type transformers the flux normally is in opposite directions in the two core legs, as shown in Fig. 13·36. A uniform external field, however, travels in the same direction in both legs, and induced voltages caused by it cancel each other in the two coils.

The relative effectiveness of these methods is shown in Fig. 13·37. The hum pick-up is given in decibels with zero decibel = 1.7 volts across 500 ohms. All curves are for 500-ohm windings working into their proper impedances, and with no orientation for minimizing hum. Using impedances much less than 500 ohms reduces the hum picked up. Orientation of coil position also reduces hum. For all types of units, there is a position of minimum hum. With the unshielded shell type the angle between the transformer coil and the field is almost 90 degrees and is extremely critical. With shielding, this angle is less critical, but the minimum amount of hum picked up is not noticeably reduced in this position. The core type is still less critical, especially with a shield. The minimum amount of hum picked up is from 10 to 20 decibels less than the shielded shell type in its minimum position.

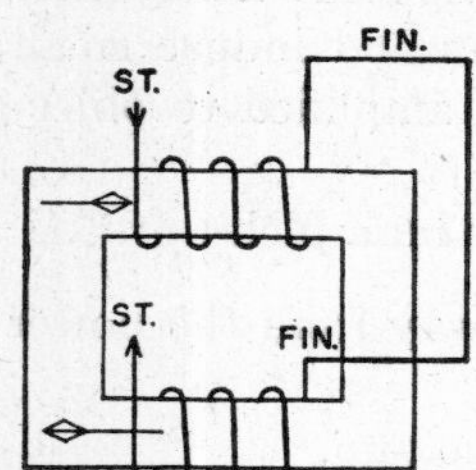

FIG. 13·36 Flux directions in a core-type transformer. (From *Electronic Transformers and Circuits* by Reuben Lee.)

Static shielding does not prevent normal voltage on a primary winding from being transferred inductively into a secondary winding. It is effective only against voltage transfer by capacitance between windings. High-frequency currents from vacuum-tube circuits are thus prevented from flowing back into the power circuits via filament and anode transformers. Without shielding, such currents may interfere with operation of receivers near-by. Likewise, voltages-to-ground on telephone lines are kept from interfering with normal voice-frequency voltages between lines. The extent of static shielding depends upon the amount of discrimination required. Usually a simple, thin, grounded layer of metal between windings is sufficient, with ends insulated to prevent a short circuit.

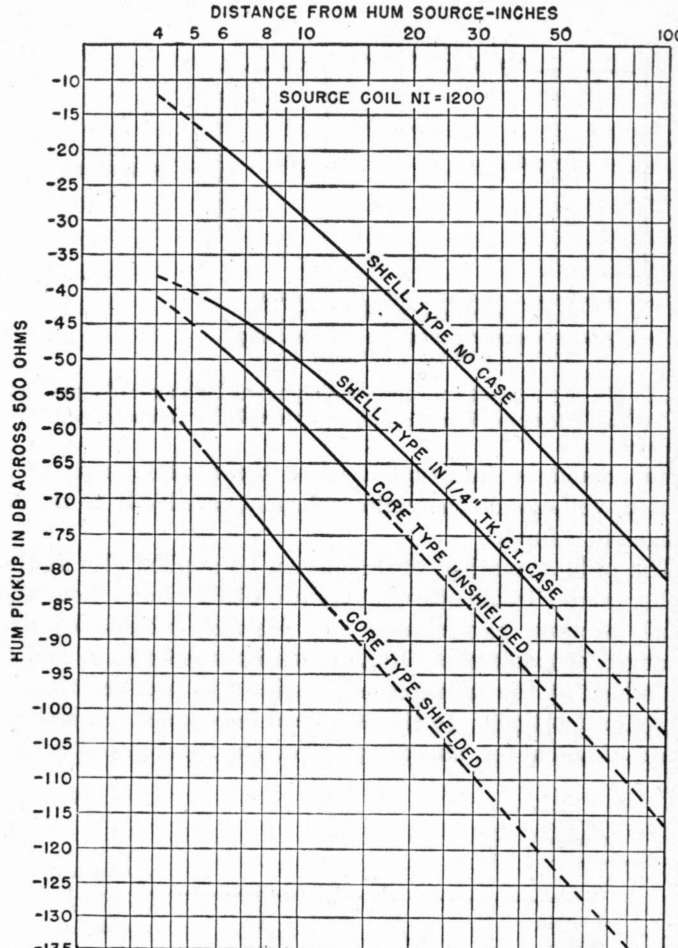

FIG. 13·37 Hum pick-up in input transformer. (From *Electronic Transformers and Circuits* by Reuben Lee.)

13·15 AUDIO-OSCILLATOR TRANSFORMERS

Transformer-coupled audio oscillators have circuits similar to that of Fig. 13·38. Transformer *OCL* and capacitor C_1 form a tank circuit, to which are coupled sufficient turns to drive the grid in the lower left-hand winding. The output circuit is coupled by a separate winding. For good wave shape in such an oscillator, triodes and class B operation are preferable. The ratio of turns between anode and grid circuits is determined by the voltage required for class B operation of the tube as if it were driven by a separate stage. Single tubes may be used, because the tank circuit maintains the sinusoidal wave shape over the half cycle during which the tube is not operating. Grid bias is obtained from the RC_2 circuit connected to the grid.

In such an oscillator, tube load equals transformer loss plus grid load plus output. In small oscillators, transformer loss may be an appreciable part of the total output. This loss consists of core, gap, and copper losses. Copper loss is large because of the relatively large tank current, and the wire size in the anode winding is larger than would be normally used for an ordinary amplifier. The gap is necessary to keep the inductance down to a value determined by the

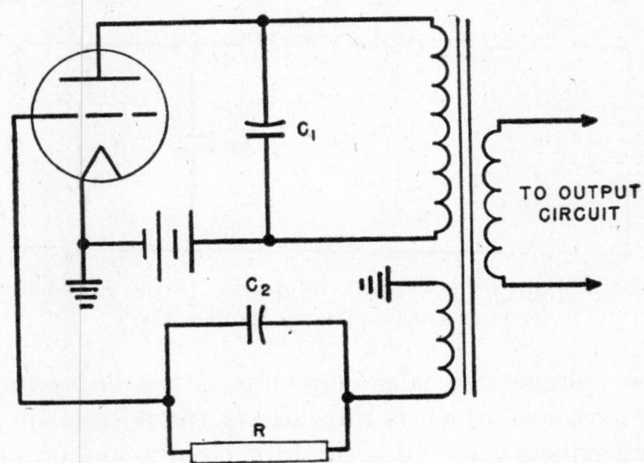

FIG. 13·38 Transformer-coupled audio oscillator. (From *Electronic Transformers and Circuits* by Reuben Lee.)

tank-circuit Q or volt-amperes. This value in turn is dictated by the required harmonic content, as mentioned in Chapter 12 for other types of tank circuits.

Class C audio oscillators are not often used for low-harmonic applications because of the difficulty of designing tank circuits with sufficiently high Q. Where large harmonic values can be tolerated, the transformer can be designed for low Q, but the wave form becomes non-sinusoidal. During the half cycle when the tube is operating, the wave has a roughly rectangular shape, and during the rest of the cycle it peaks sharply to a high amplitude in the opposite direction. Frequency of oscillation varies with changes in load; hence low-Q class C oscillators are to be avoided if good frequency stability is required.

13·16 POWER-LINE CARRIER TRANSFORMERS

In the carrier-frequency band of from 50 to 150 kilocycles, the same principles are involved as those discussed in preceding sections for audio transformers. The ratio of high to low frequencies is less, but at 50 kilocycles the curves for low-frequency operation apply, and they determine amplifier performance just as they do for 30-cycle audio operation. Likewise at 150 kilocycles the limiting factors of leakage inductance and winding capacitance must not interfere with proper operation.

In carrier-frequency transmitters, transformers are normally used for coupling between stages and for coupling the output stage to the line. They sometimes transform a large amount of carrier power. Spaced windings, used to reduce capacitance, usually have few turns, so that the transformer loss is mostly core loss. Two-mil Hipersil can be used advantageously in such transformers, because of its low losses and high permeability. In transmitter operation, class AB or class B amplifiers are commonly used, with or without modulation, which may be as high as 100 percent. In the receiver, input and interstage transformers are used for voltage gain or isolation purposes, and output transformers as described in Chapter 16. Transformers are used for line matching, especially where overhead lines are connected to underground cables. Line impedance changes abruptly, and transformers may be necessary for effective power transfer. Some of the transformers in Fig. 13·5 operate in the carrier band.

Transformers are used up to 500 kilocycles or higher. Capacitance limits the upper frequency at which transformers may be operated. In a tuned-circuit amplifier the tuning includes the incidental and tube capacitance as well as the tank-circuit capacitance. A transformer has no tuning to compensate for such capacitances.

13·17 AIR-CORE TRANSFORMERS

Some transformers widely used in radio-frequency circuits have no cores; others use small plugs of powdered iron. The main difference between such transformers and those with iron cores is that the iron core reduces exciting current required for inducing the secondary voltage to a small percentage of the load component of current. Air-core transformers are usually tuned to overcome the self-inductances of their respective windings.[15]

This type of transformer is seldom employed at frequencies below 50 kilocycles. If the wire is larger than 0.005 inch in diameter, it is usually subdivided into several strands in the type of cable known as *Litzendraht*, to reduce losses and thereby increase Q. Both the self-inductance and the mutual inductance of a coil can be increased by inserting a plug of powdered iron inside the coil tube. Such a coil is shown in Fig. 13·39, with the powdered-iron plug screened from view by the coil form. At the left end is shown the screw and lock by which inductance can be adjusted and maintained at a given value.

Powdered iron is available in several grades, from ordinary powdered iron to powdered nickel-alloy. The particles are

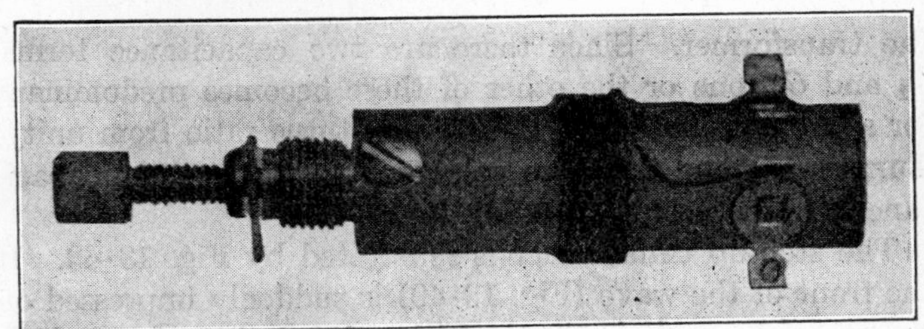

FIG. 13·39 Coil with variable powdered-iron-core plug. (From *Electronic Transformers and Circuits* by Reuben Lee.)

coated with insulating compound which separates them and reduces the permeability of the core to values ranging from 2 to 80, depending upon the grade of iron and the frequency.

13·18 VIDEO AND PULSE TRANSFORMERS

Square waves or pulses are used in the television and allied techniques to produce sharp definition of images or signals. A square or flat-topped pulse may be impressed upon the transformer by some source such as a vacuum tube, a transmission line, or even a switch and battery. Such a pulse is shown in Fig. 13·40, and a generalized circuit for the amplifier is shown in Fig. 13·41. As far as the transformer is concerned, the equivalent circuit for such an amplifier is given in Fig. 13·42. At least this is the circuit that applies to the front edge (a) of the pulse shown in Fig. 13·40 as rising abruptly from zero to some steady value E. This change is sudden, so that the transformer OCL can be considered as having infinite impedance to such a change, and is omitted in Fig. 13·42. Transformer leakage inductance, though, has

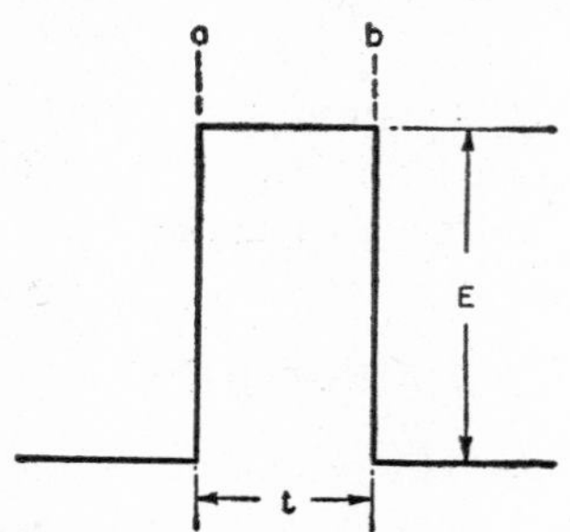

FIG. 13·40 Flat-topped pulse. (Reprinted with permission from R. Lee, *Electronics*, Aug. 1943.)

an appreciable influence and is shown as inductance L in Fig. 13·42. Resistor R_1 of Fig. 13·42 represents source impedance; transformer winding resistances are generally negligible compared to the source impedance. Winding capacitances are shown as C_1 and C_2 for the primary and secondary windings, respectively. The transformer load resistance, or the load resistance into which the amplifier works, is shown as R_2. All are referred to the same side of

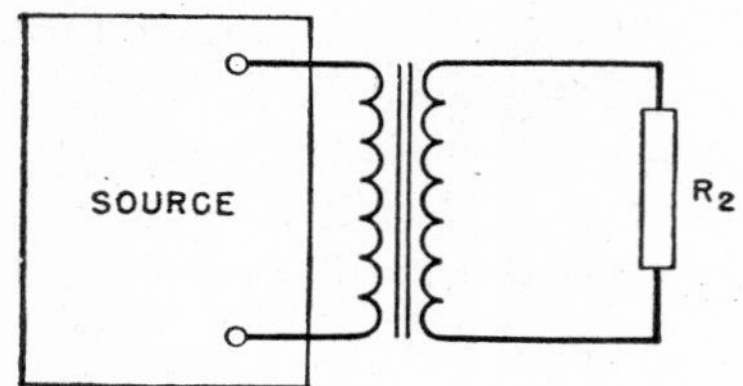

FIG. 13·41 Transformer coupling. (Reprinted with permission from R. Lee, *Electronics*, Aug. 1943.)

the transformer. Since there are two capacitance terms, C_1 and C_2, one or the other of these becomes predominant for any deviation of the transformer-turns ratio from unity. Turns ratio, and therefore voltage ratio, affect these capacitances, as discussed in Section 13·11.[16]

The step-up transformer is illustrated by Fig. 13·43. If the front of the wave (Fig. 13·40) is suddenly impressed on the transformer, it is simulated by the closing of switch S.

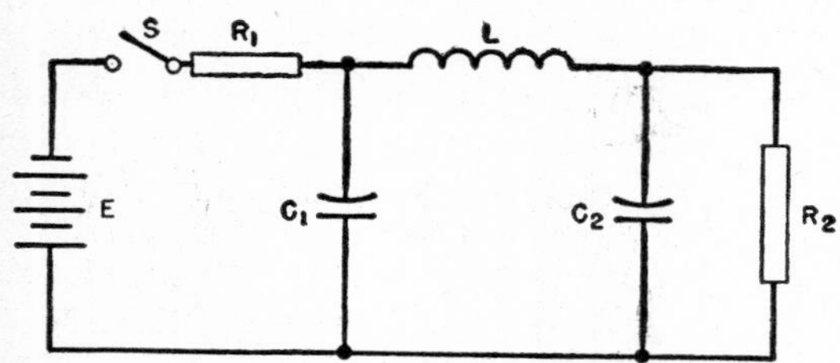

FIG. 13·42 Equivalent circuit. (Reprinted with permission from R. Lee, *Electronics*, Aug. 1943.)

At this initial instant, voltage e across R_2 is zero, and the current from battery E is also zero. Voltage e rises from its initial value of zero to its final steady value, $ER_2/(R_1 + R_2)$. Figure 13·44 shows the rate of rise of the transformed wave pulse for an amplifier with R_1 negligibly small. If the value E for the top of the pulse is multiplied by the ratio $R_2/(R_1 + R_2)$, the curves are reasonably accurate.

The scale of abscissas for these curves is not time, but percentage of the time constant T of the transformer. The equation for this time constant is given with the curves, and it is a function of leakage inductance and of capacitance C_2.

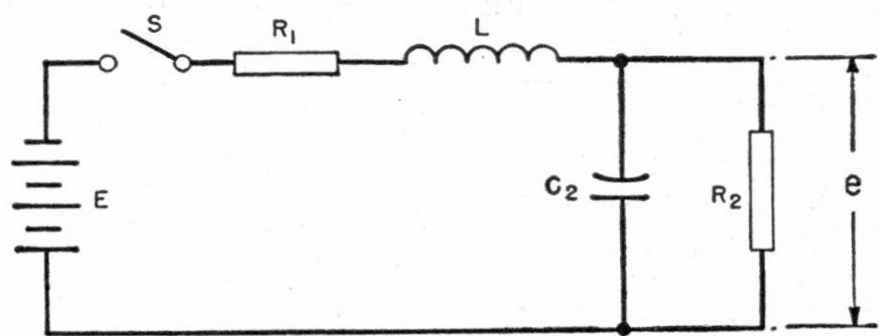

FIG. 13·43 Circuit for step-up transformer. (From *Electronic Transformers and Circuits* by Reuben Lee.)

Rate of voltage rise is governed to a marked extent by another factor k, which is the ratio of the decrement to the angular frequency for an oscillatory circuit, but retains the same form even if the circuit is not oscillatory. The relation of this factor k and the various constants of the transformer is given directly in Fig. 13·44. The greater the transformer leakage inductance and distributed capacitance, the slower the rate of rise is, although the effect of R_1 and R_2 is important, for they affect k. Voltage overshoot above the

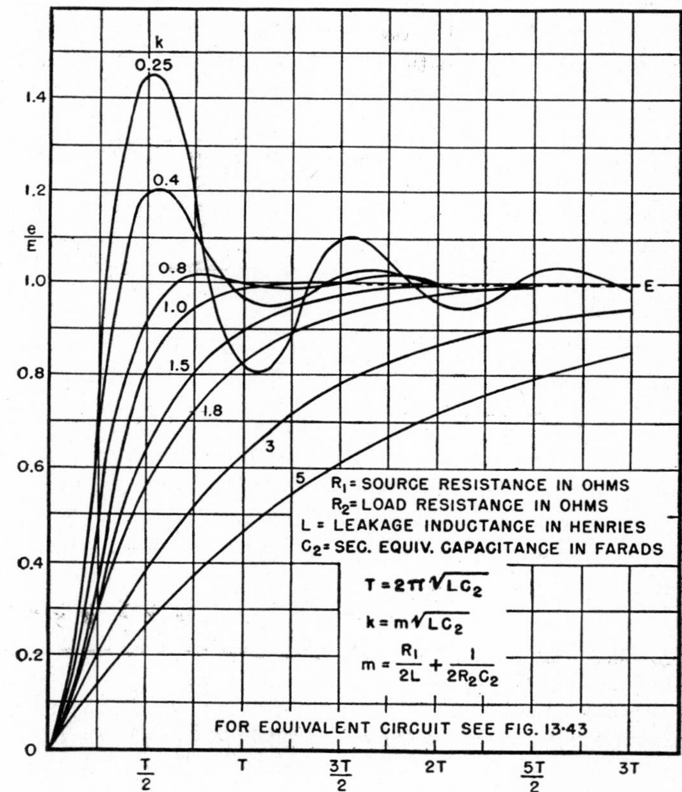

FIG. 13·44 Influence of transformer constants on front edge of pulse. (Reprinted with permission from R. Lee, *Electronics*, Aug. 1943.)

steady value E is given in Fig. 13·45 for oscillatory transformers. The time required to reach 95 percent of E is shown by Fig. 13·46.

Once the pulse top is reached, E is dependent upon the transformer OCL for its maintenance at this value. There is always a droop at the top of such a pulse. The equivalent circuit during this time is shown in Fig. 13·47. Here L is the *OCL* of the transformer, and R_1 and R_2 remain the same

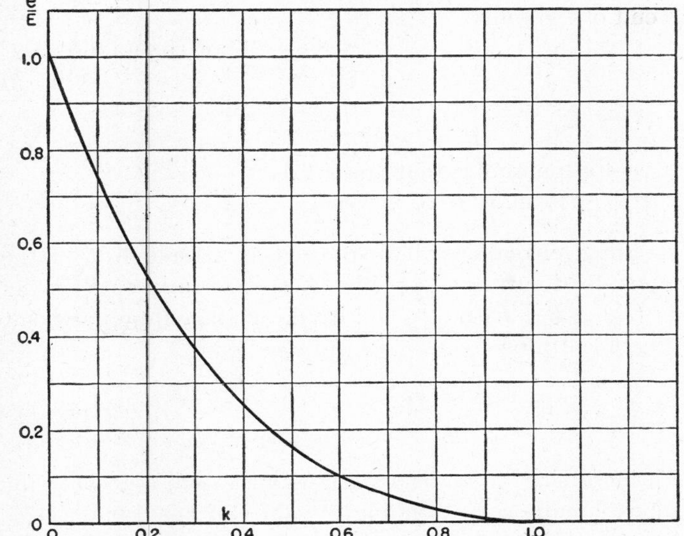

Fig. 13·45 Pulse-transformer overshoot voltage. (From *Electronic Transformers and Circuits* by Reuben Lee.)

as before. Since the rate of voltage change is relatively small during this period, capacitances C_1 and C_2 disappear. Also, since leakage inductance usually is small compared with the *OCL*, it is neglected. At the beginning of the pulse,

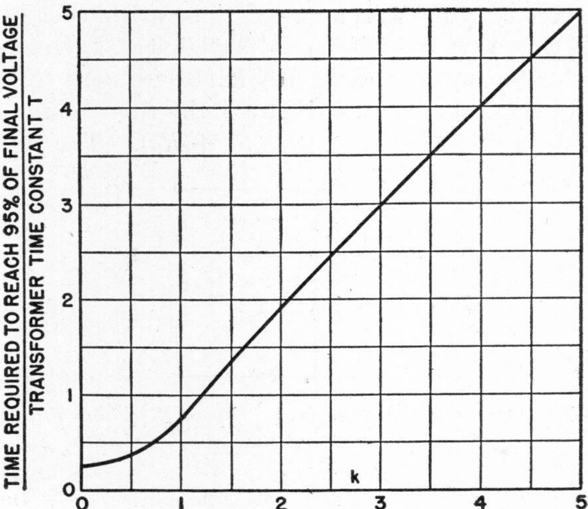

Fig. 13·46 Time required to reach 95 percent of final voltage. (From *Electronic Transformers and Circuits* by Reuben Lee.)

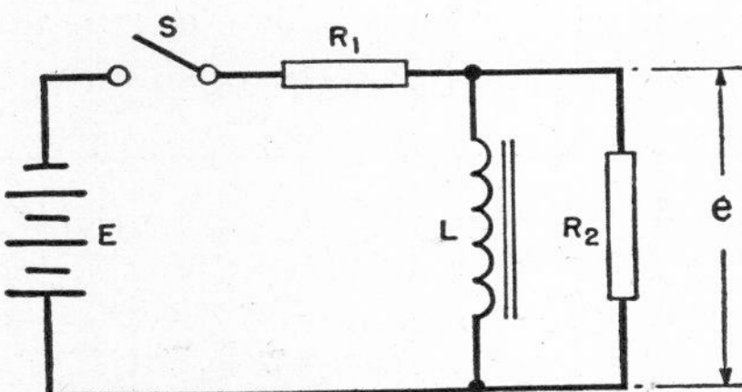

Fig. 13·47 Circuit for top of pulse. (From *Electronic Transformers and Circuits* by Reuben Lee.)

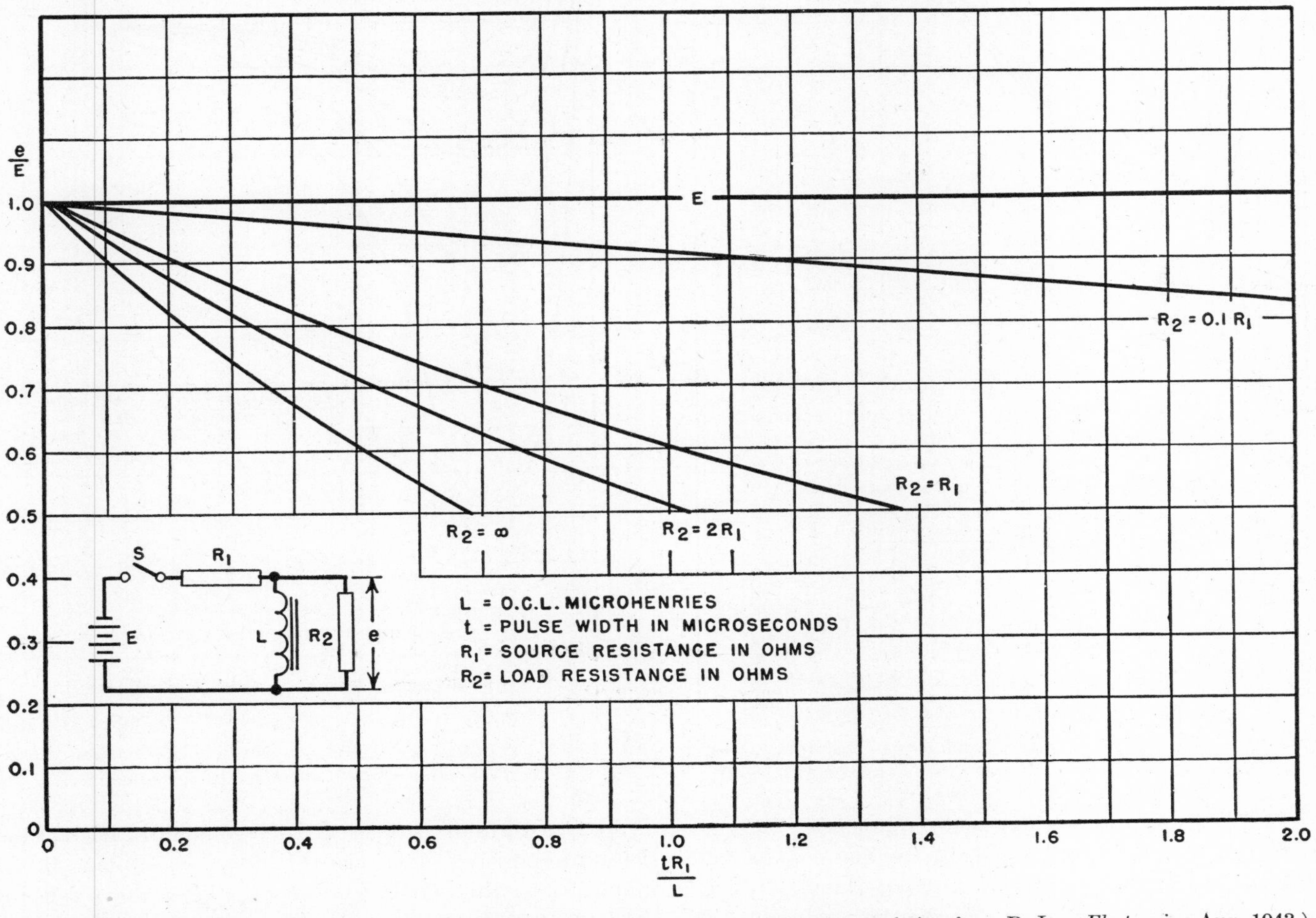

Fig. 13·48 Influence of transformer *OCL* on square-topped pulses. (Reprinted with permission from R. Lee, *Electronics*, Aug. 1943.)

the voltage e across R_2 is assumed to be at the steady value E, which is true only if R_1 is negligible. Therefore, the curves for the top of the wave, Fig. 13·48, need to be corrected in the same way as those for the front of the wave; e should be multiplied by $R_2/(R_1 + R_2)$. The abscissas

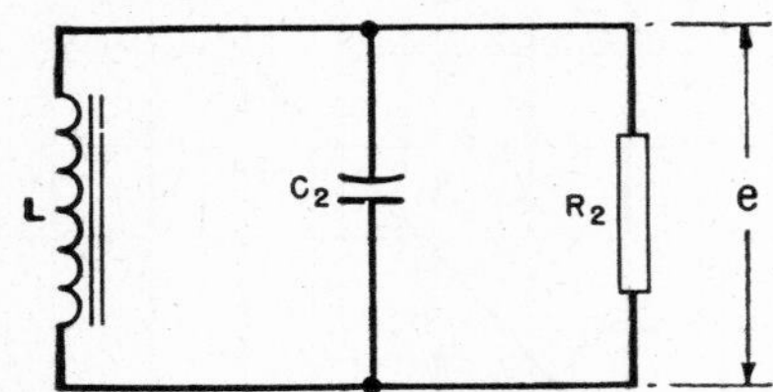

Fig. 13·49 Circuit for trailing edge. From *Electronic Transformers and Circuits* by Reuben Lee.)

are not time, but the product of time and R_1/L, the time being the duration of the pulse between points a and b in Fig. 13·40.

At instant b in Fig. 13·40, the switch S in Fig. 13·47 is assumed to be opened suddenly. The circuit now reverts to that of Fig. 13·49 in which L is the open-circuit inductance,

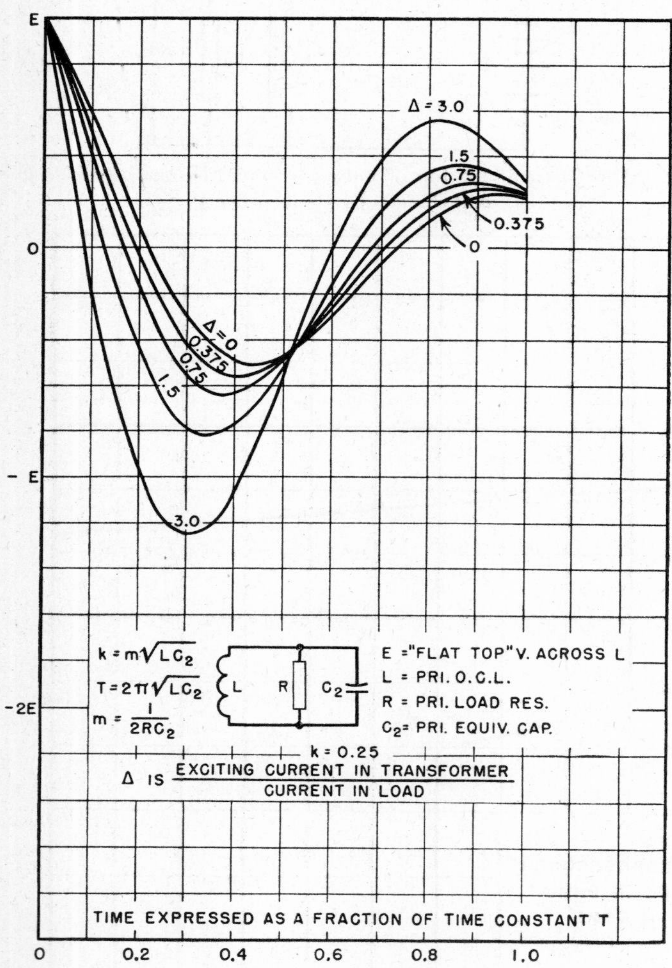

Fig. 13·50 Trailing edge of pulse, oscillatory transformer. (From *Electronic Transformers and Circuits* by Reuben Lee.)

but secondary capacitance C_2 is again appreciable. Figures 13·50, 13·51, and 13·52 illustrate the decline of pulse voltage. Abscissas are the time constant determined by OCL and C_2. Ordinates are given in terms of the original voltage E as if there were no droop at the top of the pulse. The curves apply even if there is droop, but the ordinates then should be multiplied by the fraction of E to which the voltage has fallen at the end of the pulse. The exciting current at the end of the pulse is

$$i_L = \frac{E}{mL}(1 - \epsilon^{-mt}) \qquad (13\cdot10)$$

where $m = R_1R_2/(R_1 + R_2)L$
t = pulse duration in seconds
L = primary OCL in henrys.

Exciting current can be expressed as a fraction Δ of the primary load current I, or $\Delta = i_L/I$. For any R_1/R_2 ratio, $\Delta = (R_1 + R_2)/R_1$ multiplied by the voltage droop at point b (Fig. 13·40), or

$$\frac{\text{Voltage at } b}{E} = 1 - \frac{R_1\Delta}{R_1 + R_2} \qquad (13\cdot11)$$

With increasing Δ the backswing is increased, especially for damped circuits corresponding to values of $k \geqq 1.0$.

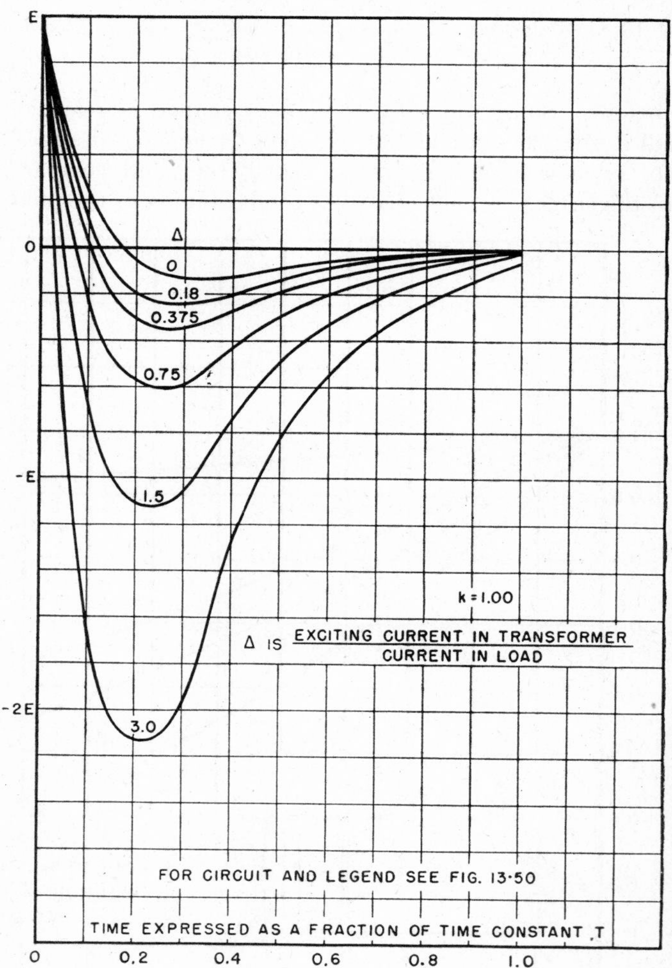

Fig. 13·51 Trailing edge of pulse, border-line case. (From *Electronic Transformers and Circuits* by Reuben Lee.)

In transformers with oscillatory constants the backswing becomes positive again on the first oscillation. In some applications this would appear as a false and undesirable indication of another pulse. The conditions for no oscillations are all included in the real values of the equivalent circuit angular frequency, or by the inequality

$$\frac{1}{4R_2{}^2C_2{}^2} > \frac{1}{LC_2}$$

or $\sqrt{L/C_2} > 2R_2$. Terms are defined in Fig. 13·50.

The quantity $\sqrt{L/C_2}$ may be regarded as the open-circuit impedance of the transformer. Its value must be more than twice the load resistance (on the basis of one-to-one ratio) to prevent oscillations after the trailing edge. This requires low distributed capacitance.

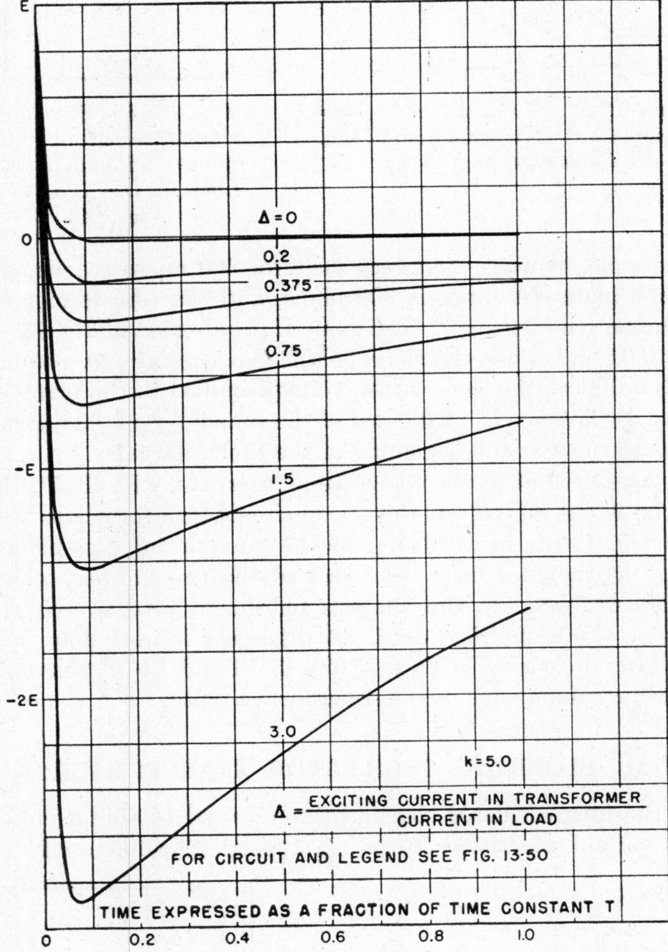

FIG. 13·52 Trailing edge of pulse, highly damped transformer. (From *Electronic Transformers and Circuits* by Reuben Lee.)

In a lightly loaded transformer Δ may be large. Besides the backswing voltage just mentioned, a resonance voltage may be produced by leakage inductance in combination with distributed capacitance. The circuit for this combination is shown in Fig. 13·43. Because L is small, the circuit is nearly always oscillatory. The oscillations are superposed on the backswing voltage, with a result similar to the oscillogram of Fig. 13·53.

Step-down transformers conform to the circuit of Fig. 13·54. The wave form of voltage rise is the same as for step-up transformers, except that the damping is less. The decrement, although still composed of two terms, has the resistances R_1 and R_2 in these two terms reversed with respect to the corresponding terms for the step-up transformer. Except for this, the front edge is little different in shape from that shown in Fig. 13·44. For most purposes, it can be regarded as being the same.

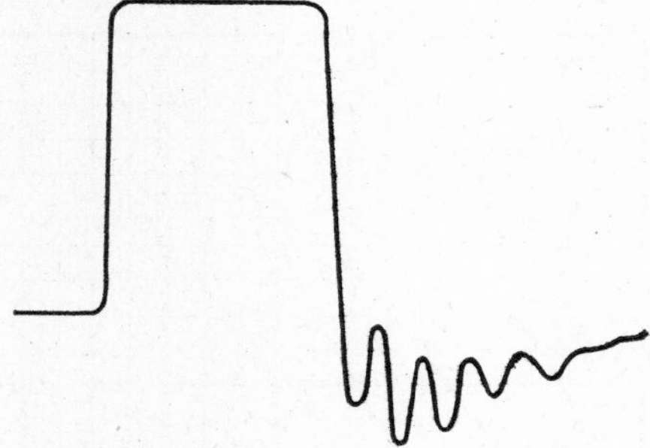

FIG. 13·53 Oscillogram of voltage pulse. (From *Electronic Transformers and Circuits* by Reuben Lee.)

Pentode amplifiers also can be represented by the circuit of Fig. 13·54. Here I is the current entering the primary winding from the tube, and it is constant over most of the voltage range. The front-edge response of these transformers is the same as that shown in Fig. 13·44 if the decrement is changed to $R_2/2L$. Load impedances may be nonlinear devices such as vacuum tubes. The voltage on such a load is often self-regulated to a constant flat-top value E. The current pulse rises after this value E is reached at a rate determined by leakage inductance and winding capacitance of the transformer and by the source and load impedances.

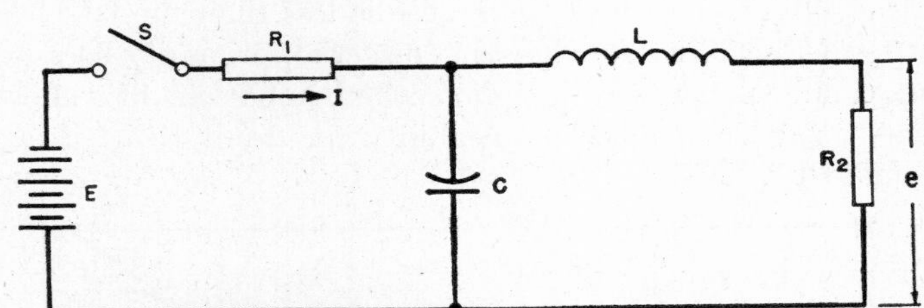

FIG. 13·54 Step-down-transformer equivalent circuits. (From *Electronic Transformers and Circuits* by Reuben Lee.)

13·19 PULSE-TRANSFORMER CORE MATERIAL

Typical curves of induction plotted against magnetizing force for pulse transformers are those of Fig. 13·55. Core induction builds up in the direction shown by the arrows. For a typical loop such as *obcd*, the slope of the loop (and hence permeability) rises gradually to the end of the pulse (point *b* in Fig. 13·40). Since magnetizing current starts decreasing at this point, magnetizing force H also starts decreasing. Current in the windings does not decay to zero immediately, but persists because of winding capacitance. Because the time is still increasing, permeability increases, and therefore induction B increases during a short interval after point *b*. The trailing edge of the pulse voltage soon reaches zero, and this corresponds to point *c* on the loop. At some interval later, the maximum backswing amplitude is reached, and this corresponds to point *d* on the loop. At this point the slope and the permeability are several times as large as at point *b*.

Inductions in pulse transformers may be low in small units where very little source power is available, or they may be high (several thousand gauss) in high-power units. This is true whether the pulse width is a few microseconds or 1000 microseconds.

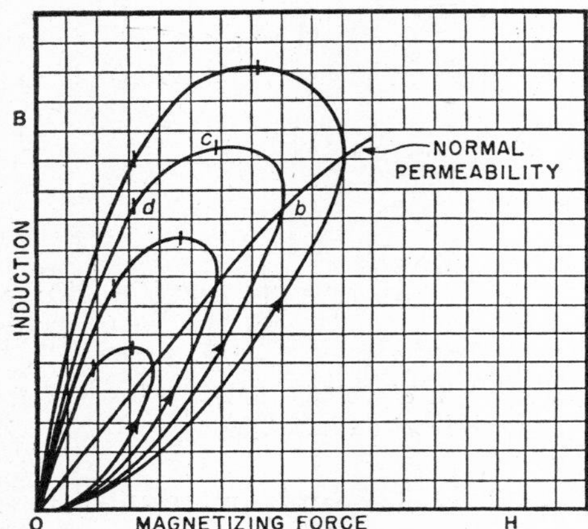

FIG. 13·55 Pulse *B-H* loops. (From *Electronic Transformers and Circuits* by Reuben Lee.)

For any number of pulses of varying amplitudes but of the same width, there are corresponding loops having respective amplitudes *c*. A curve drawn through point *b* of each loop is called the normal permeability curve, and this is ordinarily given as the permeability curve for the material. The permeability for a short-time pulse is less than the 60-cycle or d-c permeability for the same material, primarily because flux cannot penetrate throughout the laminations in a short time. Values of pulse permeability for 2-mil Hipersil are given in Fig. 13·56.

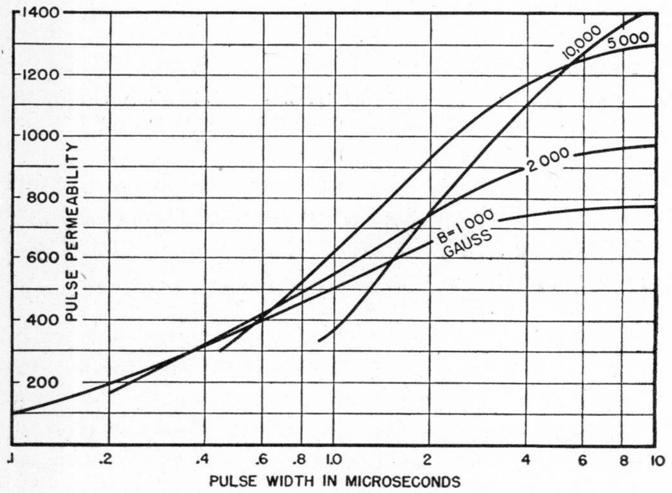

FIG. 13·56 Effective permeability versus pulse width. (From *Electronic Transformers and Circuits* by Reuben Lee.)

13·20 SAWTOOTH TRANSFORMERS

Probably the most common use of sawtooth transformers is to provide a linear sweep to elements of a cathode-ray oscilloscope. In such a circuit, the load on the transformer is negligible. Assume a linearly increasing voltage, as shown in Fig. 13·57, to be applied to the circuit of Fig. 13·58. Voltage *e* across inductance *L* has the same slope as the applied voltage times an exponential term, the value of which is determined by the resistance R_1 of the amplifier, the *OCL* of the transformer, and the time between the beginning and the end of the linear sweep. Under the conditions assumed, the value of the exponential for any interval of time can be taken from the curve marked $R_2 = \infty$ in Fig. 13·48. For

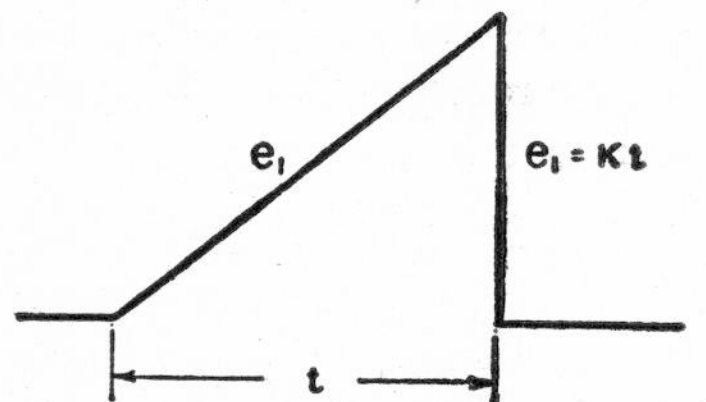

FIG. 13·57 Sawtooth wave. (Reprinted with permission from R. Lee, *Electronics*, Aug. 1943.)

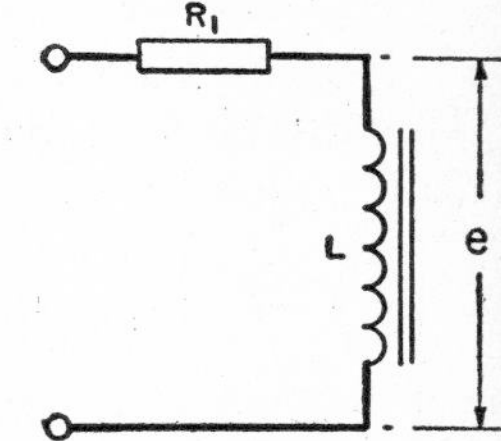

FIG. 13·58 Sawtooth transformer circuit. (Reprinted with permission from R. Lee, *Electronics*, Aug. 1943.)

example, suppose the sweep lasts for 500 microseconds, the tube plate resistance is 800 ohms, and the transformer inductance is 10 henrys. The value of the abscissa of Fig. 13·48 is 0.04 and, since the slope of this exponential curve equals its ordinate, the slope of the voltage applied to the plates of the oscilloscope is, at the end of the time interval, 96 percent of the slope at the beginning of the time interval.

Assume that at the end of time interval *t*, Fig. 13·57, the tube is cut off. Then the sweep-circuit transformer reverts to Fig. 13·49, in which C_2 has the same meaning as before, but R_2 includes only the losses of the transformer, which were neglected in the analysis for linearity of sweep. In other words, voltage does not disappear immediately, but follows the curves of Figs. 13·50, 13·51 and 13·52, the same as the trailing edge of the square-wave pulse.

13·21 BLOCKING OSCILLATOR TRANSFORMERS

Blocking oscillator transformers are used to obtain pulses at certain repetition rates. A typical blocking oscillator

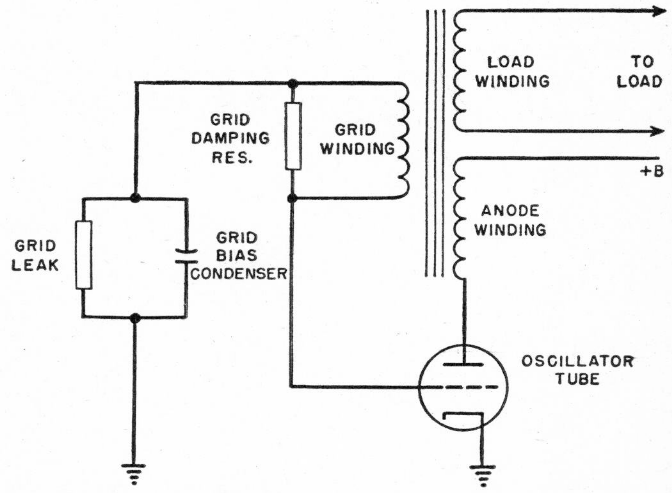

FIG. 13·59 Blocking oscillator. (From *Electronic Transformers and Circuits* by Reuben Lee.)

circuit is shown in Fig. 13·59. The grid is driven hard, and the magnitude of grid current usually is about equal to the magnitude of anode current. Hence the grid-winding turns are approximately the same as those of the anode winding.

Either a negative or a positive pulse voltage may be obtained. Instantaneous voltages and currents are shown in Fig. 13·60 for a load operating only on the positive pulse. The general shapes of these currents and voltages approximate those in a practical oscillator, except for superimposed ripples and oscillations, which are common. The next pulse

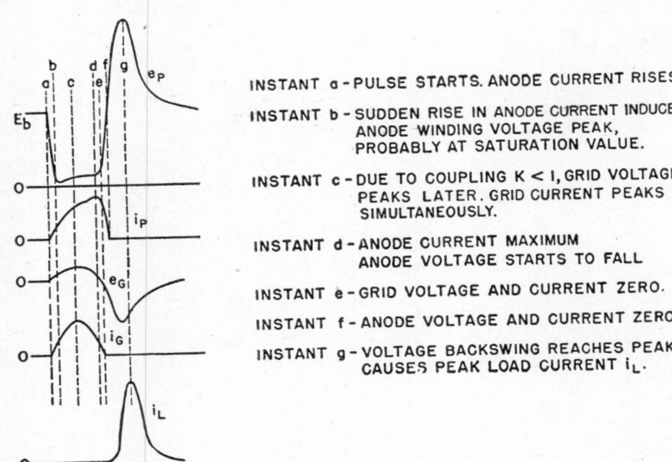

Fig. 13·60 Blocking-oscillator voltages and currents. (From *Electronic Transformers and Circuits* by Reuben Lee.)

occurs when the negative grid voltage has decreased to a point at which regeneration starts again. Hence the repetition rate depends upon grid bias R and C.

The negative pulse has a much squarer wave than the positive pulse, and consequently it is used where good wave shape is required. No matter how hard the grid is driven, anode resistance cannot be lowered below a certain value; hence a limit to the negative amplitude is formed. There is no such limit to the positive pulse, and this characteristic is used for voltage multiplication.

The slope of the front edge of the negative pulse is determined by the leakage inductance and capacitance, as in Fig. 13·44, with two exceptions: (1) the pulse is negative, and (2) the load is non-linear; hence there are no oscillations on the inverted top. The slope of this top can be computed from Fig. 13·48, provided the tube and load resistances are accurately known. Slope of the trailing edge can be found from Figs. 13·50, 13·51 and 13·52. If these curves are inverted, they show the amplitude and shape of the positive overshoot voltage.

Pulse width, shape, and amplitude also are affected by the ratio of grid turns to plate turns in the transformer; the situation parallels that of class C oscillators, as discussed in Section 13·15. Backswing pulse shapes usually are far from square. When the repetition rate is low and wave shape is unimportant, peaking transformers such as those described in Chapter 18 are used.

REFERENCES

1. "Magnetic Alloys of Iron, Nickel and Cobalt," G. W. Elmen, *Bell System Tech. J.*, Vol. 15, Jan. 1936, p. 113.
2. "Improvements in Communication Transformers," A. G. Ganz and A. G. Laird, *Bell System Tech. J.*, Vol. 15, Jan. 1936, p. 136.
3. "The Variation of the Magnetic Properties of Ferromagnetic Laminae with Frequency," C. Dannatt, *J.I.E.E.* (*London*), Vol. 79, Dec. 1936, p. 667.
4. *Principles of A-C Machinery*, R. R. Lawrence, McGraw-Hill, 1940, Chapters IX–XIX.
5. "Polyphase Rectification Special Connections," R. W. Armstrong, *Proc. I.R.E.*, Vol. 19, Jan. 1931, p. 78.
6. "Design of Reactances and Transformers which Carry Direct Current," C. R. Hanna, *J.A.I.E.E.*, Vol. 46, Feb. 1927, p. 128.
7. "Reactors in D-C Service," R. Lee, *Electronics*, Vol. 9, Sept. 1936, p. 18.
8. "Direct-Current Controlled Reactors," C. V. Aggers and W. E. Pakala, *Electric J.*, Vol. 34, Feb. 1937, p. 55.
9. "Design of Audio-Frequency Amplifier Circuits Using Transformers," Paul W. Klipsch, *Proc. I.R.E.*, Vol. 24, Feb. 1936, p. 219.
10. "Distortion in High Fidelity Audio Amplifiers," R. Lee, *Radio Eng.*, Vol. 17, June 1937, p. 16.
11. "Harmonic Distortion in Audio-Frequency Transformers," Norman Partridge, *Wireless Eng.*, Vol. 19, Sept., Oct., Nov., 1942.
12. "An Analysis of Distortion in Class B Audio Amplifiers," True McLean, *Proc. I.R.E.*, Vol. 24, March 1936, p. 487.
13. "Magnetic Shielding (Shielding of Magnetic Instruments from Steady Stray Fields)," S. L. Gokhale, *J.A.I.E.E.*, Vol. 48, Oct. 1929, p. 770.
14. "Magnetic Shielding of Transformers at Audio Frequencies," W. G. Gustafson, *Bell System Tech. J.*, Vol. 17, July 1938, p. 416.
15. *Radio Engineers' Handbook*, F. E. Terman, McGraw-Hill, 1943, Section 3.
16. "Iron-Core Components in Pulse Amplifiers," R. Lee, *Electronics*, Vol. 16, Aug. 1943, p. 115.

Chapter 14

VACUUM TUBES AS CIRCUIT ELEMENTS

W. E. Shoupp

THE OPERATION OF A VACUUM TUBE WITH LOAD

CHAPTER 5 considered characteristics of tubes that were not connected to any external circuit. Such a set of static characteristic curves (I_b versus E_b) is shown in Fig. 14·1 for a typical triode (type 2A3).

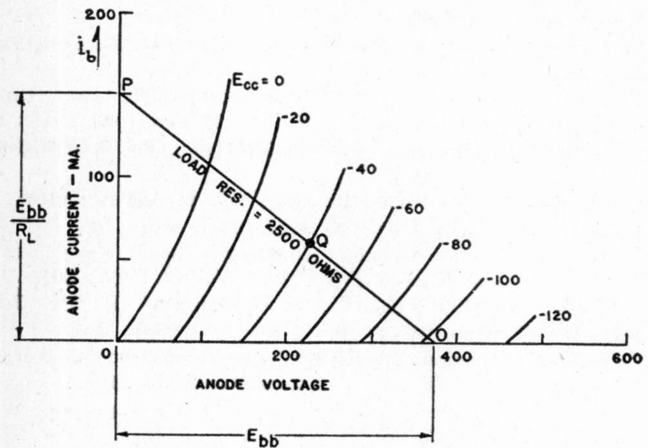

FIG. 14·1 Anode characteristics of 2A3 triode.

A *load line* has been added to the characteristic curves. This load line does not depend upon tube characteristics; it is a straight line with a slope $-1/R_L$, where R_L is the load

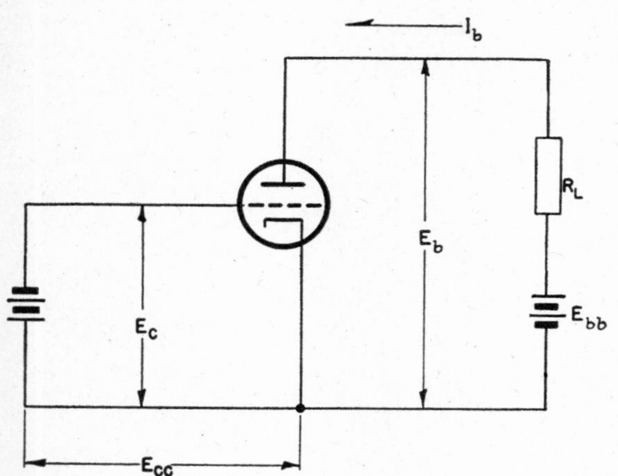

FIG. 14·2 Vacuum tube with resistance load R_L.

resistance in the anode circuit, as shown in Fig. 14·2. Voltage E_b is equal to anode-battery electromotive force E_{bb} minus I_bR_L, and is therefore represented by the straight line $I_b = E_{bb}/R_L - E_b/R_L$ in Fig. 14·1. This is a load line. The load line connects points P and O and, with the characteristic curves, determines the position of the operating point. P is the point at which no voltage appears across the tube; consequently, current has a value of E_{bb}/R_L at that point. Point O has a value equal to the B-supply voltage E_{bb} used, for that point corresponds to zero anode current, and the full B-supply voltage appears at the anode.

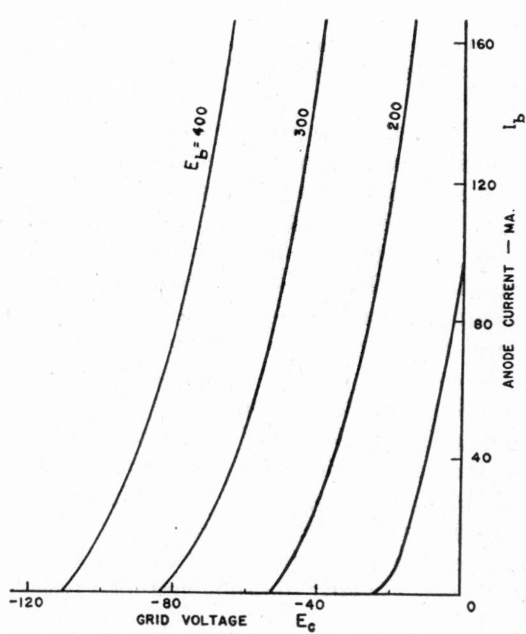

FIG. 14·3 Grid anode characteristics for 2A3 vacuum tube.

The *quiescent* or *operating point Q* lies on the load line (except for unbalanced push-pull amplifiers) and is the operating point for a d-c input to the grid of the tube. For the type 2A3 tube the recommended grid-bias voltage is $E_c = -40.0$ volts, for which Q is at the point indicated in Fig. 14·1. The load line is drawn for the recommended resistive load $R_L = 2500$ ohms; Q is then at the intersection of the recommended grid bias (-40.0 volts) with the load line.

If E_c is varied, the operating point moves up or down the load line corresponding to the value of E_c. For example, if E_c is -80 volts in Fig. 14·1, anode current is about 20 milliamperes; but if grid voltage is changed to zero, anode current rises to about 115 milliamperes.

For most applications curves of grid voltage E_c versus anode current I_b are more useful than the E_b-I_b curves given in Fig. 14·1. E_c-I_b curves, such as those shown in Fig. 14·3 for the type 2A3, are called transfer characteristics. They are static characteristics and have no relation to the load circuit.

The grid voltage-anode current characteristic with load, for a circuit such as that shown in Fig. 14·2, is called the *dynamic characteristic* and is given in curve A of Fig. 14·4, where it is superimposed upon the E_c-I_b curves of Fig. 14·3. The slope of the dynamic characteristic of curve A is less than that of the transfer-characteristic curves.

Flattening of the dynamic characteristic relative to the static characteristic is caused by the voltage drop I_bR_L across the load resistor. This voltage drop increases with current, so anode voltage E_b decreases with increasing current,

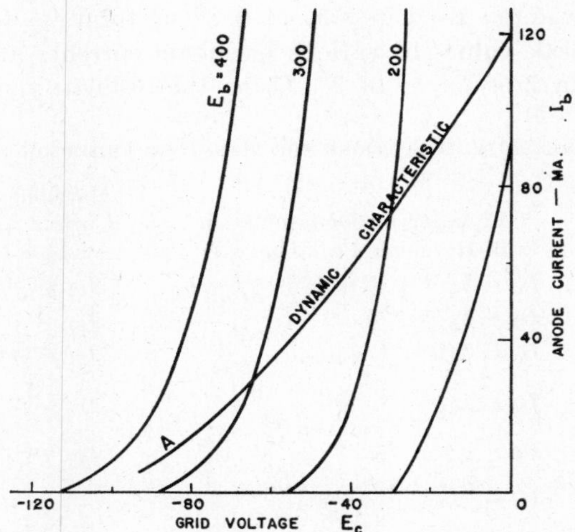

FIG. 14·4 Transfer characteristics with dynamic characteristic.

for the supply voltage E_{bb} remains constant. The slope of curve A (Fig. 14·4) is obtained by dividing the amplification factor by the total anode circuit resistance:

$$\text{Slope of } A = \frac{\mu}{r_p + R_L} \tag{14·1}$$

The static case corresponds to a load resistance R_L of zero; then the regular transfer-characteristic curves are as discussed previously, that is,

$$g_m = \frac{\mu}{r_p} = \frac{\text{amplification factor}}{\text{anode resistance}} \tag{14·2}$$

g_m is known as *transconductance* and is about 5250 micromhos for the type 2A3 triode. Amplification factor μ is the ratio of the change in anode voltage to the corresponding change in grid voltage in the opposite direction for constant anode current. With this background the characteristics of an amplifier can be analyzed. A method for determining the dynamic characteristic of the triode amplifier may now be developed, and furthermore it is possible to determine the distortion or harmonic content.

Consider again the circuit shown in Fig. 14·2, where a type 2A3 tube is operating into a resistive load. The recommended load for this tube from the tube handbook is $R_L = 2500$ ohms. Anode characteristics for the 2A3 tube are shown in Fig. 14·1. The static grid-anode characteristics are likewise given in Fig. 14·3. The first step is to determine the dynamic characteristic. The method is illustrated in Fig. 14·5, where the two sets of curves of Figs. 14·1 and

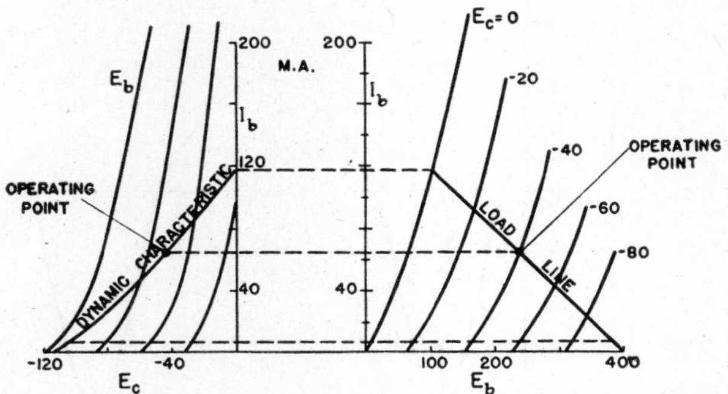

FIG. 14·5 Determination of dynamic characteristic for 2A3.

14·3 are drawn back to back. The points from the load line in the I_b-E_b curves are now connected to the various corresponding points in the I_b-E_c transfer-characteristic curves. These points determine the dynamic-characteristic curve as indicated.

14·1 DISTORTION IN THE TRIODE AMPLIFIER

In Fig. 14·6, the dynamic characteristic and the load line are shown with a sinusoidal attenuating voltage represented by $e_c = \sqrt{2}E_g \sin \omega t + E_c$ applied to the grid, where E_g is the effective value of the varying grid voltage. Wave forms W_2 and W_3 are determined graphically point by point.

A sine-wave input produces a sine-wave output only if the dynamic characteristic is a straight line. In the example in Fig. 14·6 the dynamic characteristic is noticeably curved;

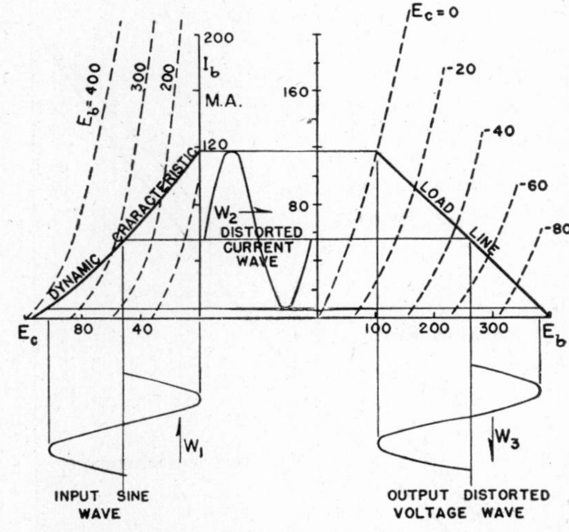

FIG. 14·6 Determination of output wave form for sine-wave input to 2A3.

therefore, output waves W_2 and W_3 are not sinusoidal. The load line in the I_b-E_b diagram is linear, so W_2 and W_3 have the same harmonic content. The distorted wave so produced may be represented by a Fourier series, and the relative values of the terms of the series denote harmonic content.

The output anode current wave W_2 usually is of particular interest in power-amplifier applications. It usually has a form somewhat as shown in Fig. 14·7.

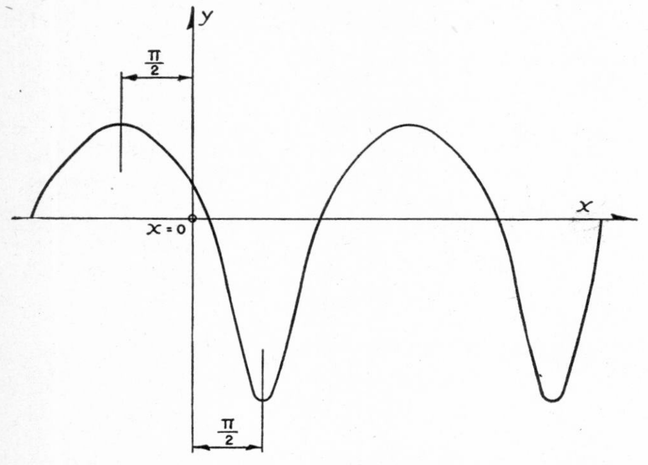

FIG. 14·7 Distorted wave.

This wave form obviously is not sinusoidal; however, it possesses mirror symmetry about the various maxima and minima. Provided that the origin ($x = 0$) is established at an ordinate displaced by $\pi/2$ from any maximum or minimum, the wave form may be represented by a Fourier series containing only sine terms of odd harmonics and cosine terms of even harmonics. Relative magnitudes of these harmonics are to be computed for a particular dynamic characteristic curve that describes the circuit. The method of determining the harmonic content up to the seventh is given here without analytical proof.[1] An analysis including seven harmonics usually is sufficient. The dynamic characteristic is first determined graphically and the ends A and B of the curve are connected by a straight line (Fig. 14·8). A different grid

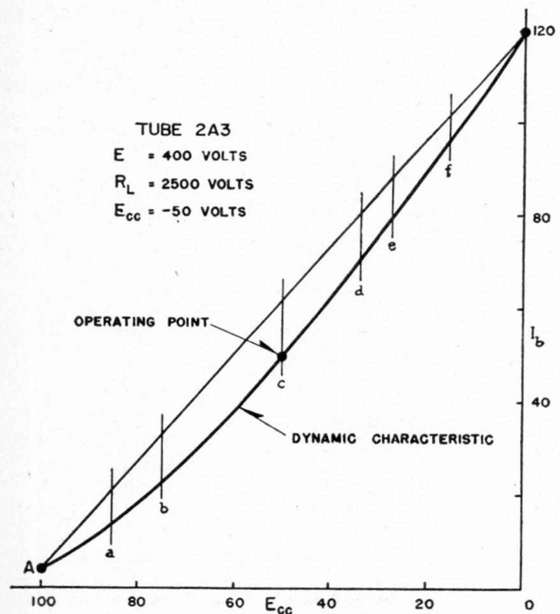

FIG. 14·8 Dynamic characteristic for harmonic determination.

bias is used in this example to cause sufficient distortion for illustrative purposes. Ordinarily a lower bias would be chosen to reduce harmonic content. Differences between the dynamic characteristic and the straight-line ideal characteristic are read off for six specific points: a, b, c, d, e, f. The abscissas to be used for the determination of six points are listed in Table 14·1, where E_{cc} is the value of E_c at the

TABLE 14·1 HARMONIC DIFFERENCE POINTS

Point	Abscissas (E_c)	E_c for 2A3	Ordinate Differences for Fig. 14·8
a	$1.707E_{cc}$	−85.1	−6.6
b	$1.50E_{cc}$	−75.0	−10.0
c	$1.00E_{cc}$	−50.0	−9.2
d	$0.700E_{cc}$	−35.0	−6.6
e	$0.500E_{cc}$	−25.0	−6.2
f	$0.293E_{cc}$	−14.6	−5.8

operating point. The third column of the table gives the abscissas, and the fourth column lists the ordinate differences for the example used in Fig. 14·8. The formulas for computing peak values of various harmonic currents are given in column 2 of Table 14·2. Using the ordinate differences

TABLE 14·2 EQUATIONS FOR HARMONIC DETERMINATION

Harmonic	Equation for Determination of Harmonic Peak Currents	Harmonic Peak Current for Fig. 14·8 (ma)
2	$I_{b2} = \frac{1}{3}(d + b) + \frac{1}{4}(c - f - a)$	$I_{b2} = -4.6$
3	$I_{b3} = \frac{1}{3}(d - b)$	$I_{b3} = +1.3$
4	$I_{b4} = \frac{1}{4}(c - f - a)$	$I_{b4} = +0.8$
5	$I_{b5} = \frac{1}{3}(d - b) + \dfrac{(a - f)}{2.83}$	$I_{b5} = +1.0$
6	$I_{b6} = c/2 - I_{02}$	$I_{b6} = +0.0$
7	$I_{b7} = \dfrac{g - 1.82I_{02} - 1.09I_{03}}{1.146} + \dfrac{0.655I_{b4} - 0.699I_{b5} - 0.751I_{b6}}{1.146}$	$I_{b7} = -0.76$

given in column 3 of Table 14·1, the harmonic currents determined for the 2A3 amplifier are given in column 3.

Minimum and maximum anode currents taken from the dynamic characteristic in Fig. 14·8 are 4.0 milliamperes and 120 milliamperes respectively. The fundamental peak current is

$$I_{b1} = \frac{I_a}{2} + I_{b3} - I_{b5} + I_{b7} \tag{14·3}$$

where I_a is the peak-to-peak current amplitude of the complex wave. For this specific example (from Fig. 14·8),

$$I_a = 120 - 4.0 = 116 \text{ milliamperes} \tag{14·4}$$

and, consequently,

$$I_{b1} = \tfrac{116}{2} + 1.3 - 1.0 - 0.76 = 57.5 \text{ milliamperes} \tag{14·5}$$

The percentage values of harmonic currents are the ratios of the values of harmonic current amplitudes to total current amplitudes. Distortions for this example are

Second harmonic	8.0 percent
Third harmonic	2.2 percent
Fourth harmonic	1.4 percent
Fifth harmonic	1.7 percent
Sixth harmonic	0.0 percent
Seventh harmonic	1.3 percent

Previous calculations have illustrated the determination of harmonic content in the 2A3 amplifier illustrated in Fig. 14·2. Formulas and methods for such determinations are applicable up to the eighth harmonic for amplifiers, modulators, and perfectly balanced push-pull amplifiers.

14·2 AMPLIFIER WITH REACTIVE LOAD

Thus far the load R_L has been assumed to be a pure resistance. If the load has a reactive component, the dynamic characteristic becomes an ellipse, provided there is no waveform distortion. The ratio of the minor to the major axis of the ellipse depends upon the magnitude of the reactive component relative to the resistive component of the load circuit. The shape of the dynamic characteristic[2] for various values of power factor is shown in Fig. 14·9. If the wave

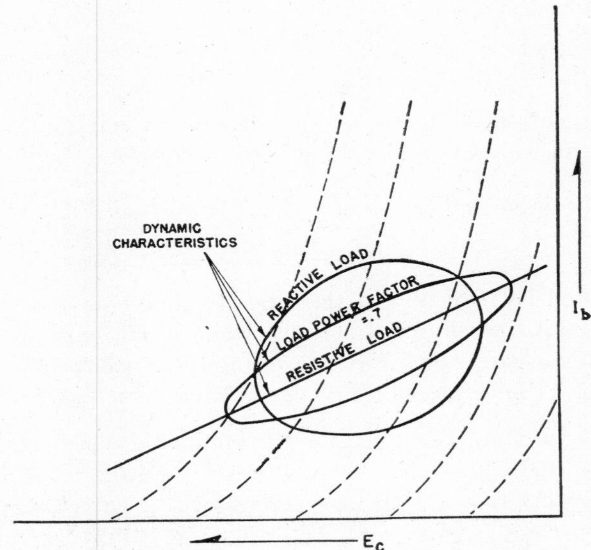

FIG. 14·9 Dynamic characteristics.

form is distorted, the ellipse is somewhat distorted. If the load has a reactive component, the analysis of the problem is so complicated that solutions are not generally attempted even for linear-tube characteristics. To obtain a reasonably accurate analytical solution the second and higher derivatives of tube characteristics must be known, because series representations are necessary. Because these derivatives are not generally available to the design engineer they must be obtained in the laboratory. Fortunately, solutions of problems involving reactive loads are seldom required, because final stages of a power amplifier usually work into a purely resistive load to extract maximum power from the amplifier. Reactance in the load usually decreases power factor.

Dynamic characteristic for reactive loads can be determined experimentally. It is usually preferable to use the experimental dynamic characteristic and from it determine the power output and harmonic distortion. An experimental arrangement that displays the dynamic characteristic on the screen of an oscilloscope is given in Fig. 14·10.

Alternating-voltage input is connected to the grid of the triode, which is in turn connected to the load circuit that may be reactive. Grid voltage is applied to the horizontal plates of the cathode-ray oscilloscope and anode current is applied to the vertical plates. The trace on the oscilloscope screen indicates the dynamic characteristic. The curve can

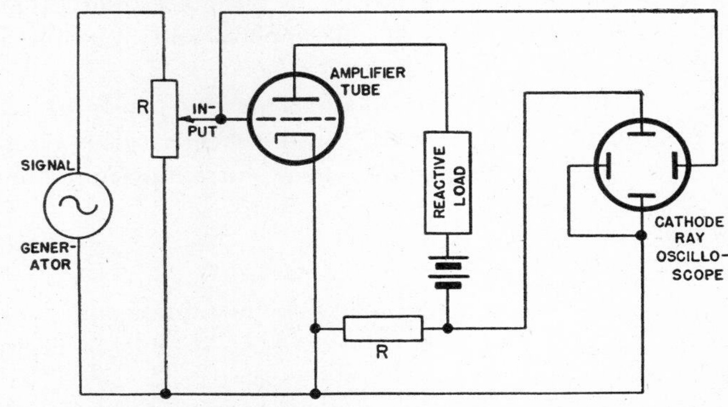

FIG. 14·10 Experimental arrangement for determining the dynamic characteristic.

be reproduced graphically from the oscilloscope screen and used to determine amplifier characteristics under operating conditions.

14·3 ANALYTICAL DESIGN EQUATIONS FOR TRIODE POWER AMPLIFIERS

The graphical method of determining amplifier characteristics, though quite accurate, frequently takes more work than is necessary, if only an approximate solution is required. The following approximate solutions are sufficiently accurate for most design purposes.[2] The assumed approximations are that the transfer characteristics are linear and parallel for currents above a certain minimum value I_O, and that the load is purely resistive. Under these conditions the dynamic characteristic is a straight line (Fig. 14·11).

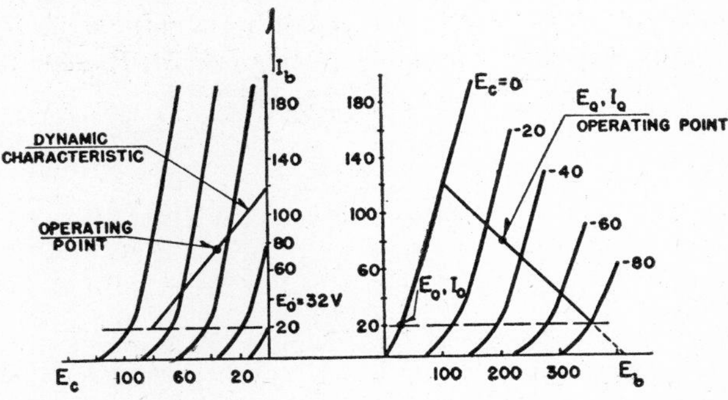

FIG. 14·11 Approximate characteristic curves for 2A3.

The following notation is used:

E_Q = anode voltage at operating point Q
E_O = anode voltage required to develop minimum allowed anode current I_O for zero grid bias
I_O = minimum allowed anode current
r_Q = anode resistance at operating point Q
R_L = load resistance
μ = amplification factor of tube
E_c = grid-bias voltage
E_g = peak signal voltage

Consider now that a sine-wave voltage input is connected to the grid of a triode biased with a negative voltage E_c, and that the tube is working into a resistive load R_L; Fig. 14·12 (*a*). The transfer characteristics are examined, and the region over which the characteristics are non-linear is marked off as shown in Fig. 14·11. Values of I_O and E_O are thereby determined to be 20 milliamperes and 32 volts respectively.

The electric field of the anode penetrates the grid and modifies the field at the cathode,[3] so the effective grid voltage is $e_c + (e_b/\mu)$. At the instant anode current is minimum,

$$\frac{E_O}{\mu} = e_c + \frac{e_b}{\mu} \tag{14·6}$$

Anode current I_O is minimum when instantaneous signal voltage e_g has a minimum value $-E_g$. At that time grid voltage is equal to twice bias voltage, that is,

$$e_c = -E_g - E_c = -2E_c \tag{14·7}$$

The triode amplifier in Fig. 14·12 (*a*) may be represented by the equivalent circuit shown in Fig. 14·12 (*b*), in which

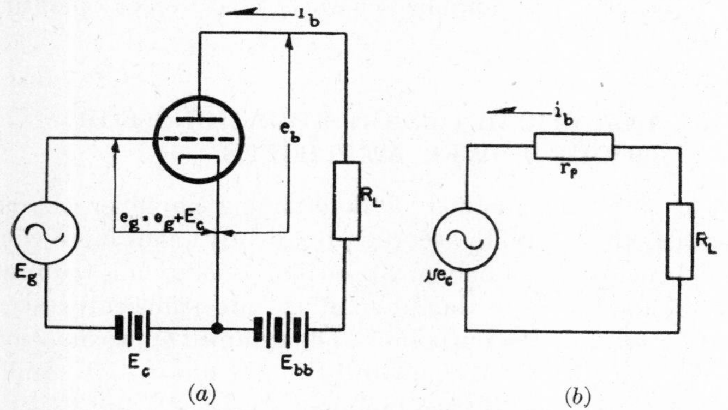

FIG. 14·12 (*a*) Amplifier circuit. (*b*) Equivalent circuit to 14·12 (*a*).

the tube is represented by a generator of output μe_c in series with a resistance equal to the anode resistance of the tube. When grid voltage is minimum in this circuit, the current is

$$i_b = \frac{\mu E_g}{R_L + r_p} \tag{14·8}$$

Corresponding voltage across the tube is

$$e_b = E_Q + \frac{\mu E_g R_L}{R_L + r_p} \tag{14·9}$$

Combining equations 14·6, 14·7, and 14·8 and solving for bias voltage E_c gives

$$E_c = \left(\frac{E_Q - E_O}{\mu}\right)\left(\frac{R_L + r_p}{R_L + 2r_p}\right) \tag{14·10}$$

If the grid bias is known, equation 14·10 may be used to determine the proper load resistance, since

$$R_L = r_p\left(\frac{\mu E_c}{E_Q - E_c(1 + \mu)} - 1\right) \tag{14·11}$$

Equation 14·10 expresses the proper grid bias for a triode amplifier.

Power output delivered to the load resistance is

$$\text{Power output} = \frac{1}{2}\left(\frac{\mu E_g}{R_L + r_p}\right)^2 R_L \tag{14·12}$$

Undistorted power output reaches its maximum as the crest signal voltage just equals the grid bias. Using the value of E_c from equation 14·10, equation 14·12 reduces to

Maximum power output

$$= \frac{1}{2}\frac{\mu^2 R_L}{(R_L + r_p)^2}\left(\frac{R_L + r_p}{R_L + 2r_p}\right)^2\left(\frac{E_Q - E_O}{\mu}\right)^2$$

$$= \frac{(E_Q - E_O)^2 R_L}{2(R_L + 2r_p)^2} \tag{14·13}$$

Anode voltage and current at the operating point (E_Q, I_Q) are, for linear characteristics, the average of the maximum and minimum instantaneous values; that is,

$$I_Q = \frac{I_{\max} + I_{\min}}{2}$$
$$E_Q = \frac{E_{\max} + E_{\min}}{2} \tag{14·14}$$

Power output in terms of the maximum, minimum, and operating currents and voltages then becomes

$$\text{Power output} = \tfrac{1}{2}(E_Q - E_{\min})(I_Q - I_{\min})$$
$$= \tfrac{1}{8}(E_{\max} - E_{\min})(I_{\max} - I_{\min}) \tag{14·15}$$

It should be remembered that the characteristic curves are assumed to be linear above a certain anode current $I_{\min}$. The following values for this example are obtained from Fig. 14·11 and the published data:

$E_{cc} = 40$ volts	$E_{\min} = 105$ volts	$\mu = 4.2$
$E_{\max} = 350$ volts	$I_q = 70$ ma	$I_{\min} = 20$ ma
$E_Q = 225$ volts	$I_{\max} = 120$ ma	$E_O = 32$ volts

Equation 14·15 gives for power output

$$\text{Power output} = \tfrac{1}{8}(350 - 105)(120 - 20) = 3.0 \text{ watts}$$

Anode efficiency is defined as the ratio of power output dissipated in the load resistance to the d-c power supplied to the anode circuit of the tube. Input power is $E_Q I_Q$, where I_Q is the direct current at the operating point. Using equation 14·12 for the output power gives

$$\text{Anode efficiency} = \frac{\dfrac{R_L E_g^2 \mu^2}{2(R_L + r_p)^2}}{E_Q I_Q} \tag{14·16}$$

or using equation 14·10 gives

$$\text{Anode efficiency} = \frac{R_L}{2(R_L + 2r_p)}\left(1 - \frac{I_{\min}}{I_Q}\right)\left(1 - \frac{E_O}{E_Q}\right) \tag{14·17}$$

Anode efficiency may be expressed also in terms of maximum and minimum voltages. When equation 14·15 is used, anode efficiency becomes

$$\text{Anode efficiency} = \frac{1}{2}\frac{(E_Q - E_{\min})(I_Q - I_{\min})}{E_Q I_Q} \tag{14·18}$$

or

$$\text{Anode efficiency} = \frac{(E_{\max} - E_{\min})(I_{\max} - I_{\min})}{8E_Q I_Q} \tag{14·19}$$

For the 2A3 tube (from equation 14·19),

$$\text{Anode efficiency} = \frac{(350 - 105)(120 - 20)}{8 \times 70 \times 225} = 19 \text{ percent}$$

Load resistance R_L required to give the maximum power output may be obtained by differentiating equation 14·13 with respect to R_L and equating this derivative to zero:

$$\frac{\partial P}{\partial R_L} = \frac{(E_Q - E_O)^2}{2} \frac{2r_p - R_L}{(R_L + 2r_p)^3} = 0 \qquad (14\cdot 20)$$

which gives

$$R_L = 2r_p \qquad (14\cdot 21)$$

Consequently, for maximum power output, load resistance must be twice the tube anode resistance. Using this condition reduces equation 14·10, for the proper grid bias, to

$$(\text{Proper grid bias for } R_L = 2r_p) = \frac{3}{4}\left(\frac{E_Q - E_O}{\mu}\right) \qquad (14\cdot 22)$$

Equation 14·13 reduces to

$$\text{Maximum power output} = \frac{(E_Q - E_O)^2}{16r_p} \qquad (14\cdot 23)$$

Experimental values of plate efficiency are from about 10 to 20 percent for most tubes used as amplifiers in which the grid is never driven positive.

The maximizing condition $R_L = 2r_p$ is frequently used in small power amplifiers that operate with anode voltages of about 300. For larger tubes, operation usually is limited by anode dissipation; consequently, load resistances are usually greater than $2r_p$, maximum anode dissipation determines proper grid bias, and the equations derived for $R_L = 2r_p$ do not apply.

A more rigorous analysis shows that equations 14·15 and 14·19 are applicable whether the dynamic characteristic is linear or not, provided that the output contains only even harmonics.

14·4 VACUUM TUBES AS VARIABLE-IMPEDANCE ELEMENTS

Vacuum tubes in simple circuits can be applied to the variation of magnitude and phase angle of impedance. Properties of circuits of this type are of particular interest to designers of oscillators, for circuits containing only resistance and capacitance can be made to act as variable inductances shunted by negative resistances. With such circuits it is comparatively simple to construct variable-frequency oscillators because variable resistors and capacitors are readily available. Commercial oscillators that operate on this resistance-tuned principle are available in the kilocycle range. Reactance tubes based on this fundamental principle are used in many frequency-modulated transmitters and receivers and in automatic-frequency-control circuits of ordinary receivers.

One basic circuit of the variable-impedance type [4] is shown in Fig. 14·13, in which a capacitance is connected from anode to grid. If an alternating voltage is impressed across terminals P and Q, the grid voltage of the triode is out of phase with the alternating anode voltage. Consequently, since anode current is controlled by grid voltage, anode current i_1 contains a component out of phase with the input voltage. Qualitatively the circuit acts like a reactance across P and Q. The magnitude of this reactance can be varied by changing the grid bias on the triode, so the circuit acts as a variable reactance. This simple explanation of the operation of this circuit is not sufficiently clear for design purposes, so a detailed analysis is needed.

From Fig. 14·13, when Kirchhoff's laws are applied,

$$i = i_1 + i_2 = \frac{e + \mu e_g}{r_p} + \frac{e}{\dfrac{1}{j\omega C} + R} \qquad (14\cdot 24)$$

and grid voltage is

$$e_g = \frac{eR}{R + \dfrac{1}{j\omega C}} \qquad (14\cdot 25)$$

Combining equations 14·24 and 14·25 gives as the equivalent admittance of the circuit

$$Y = \frac{i}{e} = \frac{R^2\omega^2C^2(1+\mu) + r_pR\omega^2C^2 + 1}{r_p(R^2\omega^2C^2 + 1)} + j\frac{(R\mu + r_p)(\omega C)}{r_p(R^2\omega^2C^2 + 1)} \qquad (14\cdot 26)$$

Inspection of equation 14·26 shows that the circuit acts as though it were a parallel combination of capacitive reactance X_e and resistance r_e where

$$X_e = -\frac{r_p(R^2\omega^2C + 1)}{(R\mu + r_p)\omega C} \qquad (14\cdot 27)$$

$$r_e = \frac{r_p(R^2\omega^2C^2 + 1)}{R^2\omega^2C^2(1+\mu) + r_pR\omega^2C^2 + 1} \qquad (14\cdot 28)$$

Since the transconductance of a triode is μ/r_p, the effective capacitance C_e across P and Q is

$$C_e = \frac{Cg_mR + C}{\omega^2R^2C^2 + 1} \qquad (14\cdot 29)$$

Usually the term g_mR is much greater than 1, so the second term in the numerator of equation 14·29 is negligible compared with the first. Therefore

$$C_e = \frac{RCg_m}{\omega^2R^2C^2 + 1} \qquad (14\cdot 30)$$

C_e is maximum when $R = 1/\omega C$, and

$$C_{e\,\max} = \frac{g_m}{2\omega} \qquad (14\cdot 31)$$

and the corresponding shunting resistance for this maximum is

$$r_{e\,\max} = \frac{2}{g_m} \qquad (14\cdot 32)$$

Now the value of the transconductance g_m of the triode may be varied by changing the grid bias; consequently, the circuit acts like a variable capacitance shunted by a variable resistance.

This basic circuit can be used as a circuit element in automatic tuning devices or as a variable capacitance for frequency modulation. In ordinary applications the circuit is used as the variable-capacitance element of a variable-frequency-oscillator tank circuit. In such a circuit it might appear from equation 14·31 that high values of C_e could be obtained by using a tube with large transconductance. However, from equations 14·31 and 14·32,

$$(r_e)_{\max} = \frac{1}{\omega(C_e)_{\max}} \tag{14·33}$$

Therefore, if r_e is to have a reasonable value, say 1000 ohms, $(C_e)_{\max}$ is about 0.16 microfarad at a frequency of 1000 cycles per second. This circuit therefore is not generally applicable if large effective capacitances must be produced, because the shunting resistance then becomes quite small.

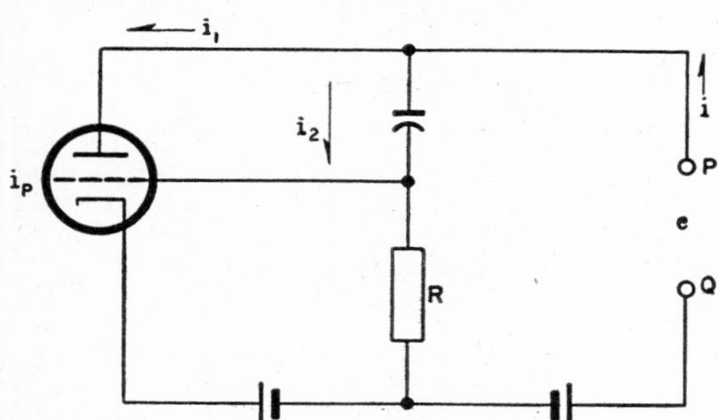

FIG. 14·13 Variable capacitance circuit.

The basic circuit (Fig. 14·13) can be modified to provide high shunting resistance, and it is possible sometimes to obtain capacitances of many microfarads.

14·5 VARIABLE-INDUCTANCE CIRCUIT

If capacitance C (Fig. 14·13) is replaced by an inductance L the effective reactance and resistance becomes

$$x_e = \frac{r_p(R^2 + \omega^2 L^2)}{\omega_L(r_p + \mu R)} \tag{14·34}$$

$$r_e = \frac{(R^2 + \omega^2 L^2) r_p}{Rr_p + R^2(1 + \mu) + \omega^2 L^2} \tag{14·35}$$

Consequently, by modifying parameters in Fig. 14·13 tube circuits can be made to act as variable-capacitance or variable-inductance circuits.

The tube shown in Fig. 14·13 is a triode, but any two elements or pairs of grids in multielectrode tubes may be used for the elements shown, if the control grid is provided with a proper bias. Connections for a pentode are shown in Fig. 14·14. The values of effective resistance r_e and reactance for the circuit in Fig. 14·14 become

$$x_e = \frac{-r_s(R^2\omega^2C^2 + 1)}{\omega C(\mu R + r_s)} \tag{14·36}$$

and

$$r_e = \frac{r_s(R^2\omega^2C^2 + 1)}{R^2\omega^2C^2(1 + \mu) + r_sR\omega^2C^2 + 1} \tag{14·37}$$

where r_s is screen resistance and μ is the screen-suppressor amplification factor. In this case μ is negative, and r_e becomes a negative resistance. Furthermore, μR may be greater than r_s; therefore the reactance x_e may be inductive, so that the circuit may act as a negative resistance shunted by an inductance. Sinusoidal oscillations [5] can be sustained with a circuit like that shown in Fig. 14·14 if a capacitance is connected across P and Q. A resistance-tuned oscillator can be constructed (Fig. 14·15) by connecting a capacitance C_1 and a resistance R_1 across P and Q, the resistor being necessary to provide screen-grid voltage.

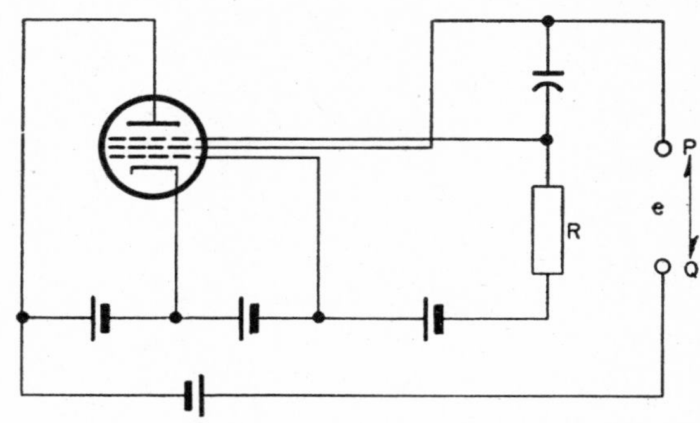

FIG. 14·14 Pentode impedance conversion circuit.

Oscillations are possible for the circuit in Fig. 14·15 if

$$|\mu| \geq \frac{R_1 + r_s}{R_1} + \frac{r_sC_1}{RC} + \frac{r_s}{R} \tag{14·38}$$

and the frequency of such an oscillator is given by

$$f = \frac{1}{2\pi}\sqrt{\frac{R_1 + r_s}{RR_1r_sCC_1}} \tag{14·39}$$

so the circuit parameters are designed to meet this condition.

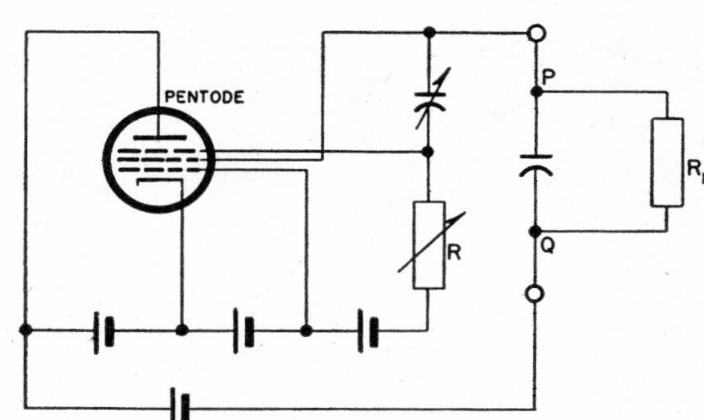

FIG. 14·15 Resistance-tuned oscillator.

14·6 THE VACUUM TUBE AS AN ELECTRONIC SWITCH

Tubes can be used also for rapid switching of electrical circuits. Applications of electronic circuits to high-power switching, such as in the control of welding circuits, are discussed in Chapter 31. One low-power switching application is of particular interest since it may be the basis of many unusual uses in the laboratory.

It is frequently desirable to observe simultaneously two separate phenomena on the same cathode-ray oscilloscope screen, such as voltage and current waves in a transformer.

Consequently, a switch operating at a rate faster than the oscillating frequency and the resolving time of the eye is required. The circuit described in this section acts as a single-pole double-throw switch that connects first one, then

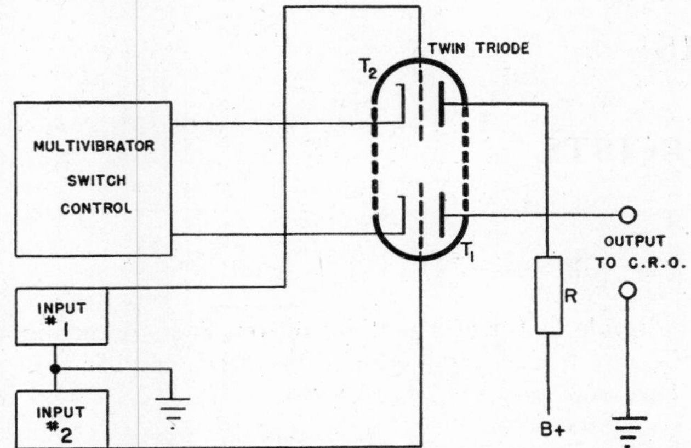

FIG. 14·16 Electronic switch.

another, input to the output terminals, which usually are connected to an oscilloscope for viewing the two separate wave disturbances. The output of a multivibrator oscillator as shown in Fig. 14·16 is so connected that it applies a high negative bias alternately to the cathodes of two amplifier tubes, 1 and 2, the output of which are connected together and to the output circuit,* which usually is a cathode-ray oscilloscope. The input grids of the two amplifiers are connected to the separate sources to be examined. The multivibrator switches off and on alternately, the separate amplifiers enabling the two input signals to enter the output circuit alternately. With such an arrangement a cathode-ray oscilloscope connected to the output of the device indicates both input signals alternately, but they are switched off and on at a rate as high as 100,000 times per second. To the eye both input signals 1 and 2 appear simultaneously on the oscilloscope screen. It is simpler to use such an instrument instead of two oscilloscopes, and the phase and amplitude of the two waves can be directly compared on the same viewing screen. It is likewise possible to cascade these devices to portray several wave patterns on the same oscilloscope screen.

REFERENCES

1. "Graphical Harmonic Analysis," J. A. Hutcheson, *Electronics*, Vol. 9, 1936, p. 16.
2. *Radio Engineering*, F. E. Terman, McGraw-Hill, 1937, p. 278.
3. *Radio Engineering*, F. E. Terman, McGraw-Hill, 1937, p. 117.
4. "The Use of Vacuum Tubes as Variable Impedance Elements," H. J. Reich, *Proc. I.R.E.*, Vol. 30, 1942, p. 288.
5. "Sinewaves in R.C. Oscillators," P. S. DeLaup, *Electronics*, 1941, p. 34.

* As in model 185A DuMont electronic switch.

Chapter 15

RECTIFIER CIRCUITS

H. J. Bichsel

A RECTIFIER is an asymetrical conducting device for producing unidirectional current from alternating current. Several different types of rectifiers now in use include the electronic, copper oxide, selenium, and mechanical rectifiers. The electronic rectifier has grown in popularity not only in the supply of small quantities of direct current for the operation of radio receivers, transmitters, and industrial electronic equipment, but also for supplying large amounts of direct current for the electric railways, mining equipment, electrochemical processes, and for general industrial use.

15·1 GENERAL CONSIDERATIONS

Basic Rectifier

A rectifier consists of a transformer and the rectifying element such as a tube or a copper oxide unit. The transformer is used to step the a-c supply voltage up or down to obtain the desired d-c output voltage from the rectifier. The transformer also isolates the rectifier output from the a-c supply. The rectifying element conducts current in only one direction and, thereby, produces a pulsating d-c voltage across any load connected to the rectifier output. A simple rectifier is shown in Fig. 15·1.

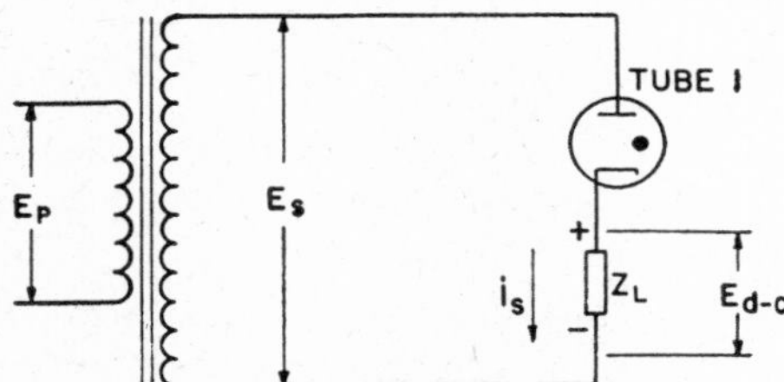

FIG. 15·1 Single-phase half-wave rectifier.

Ripple

Since the output voltage of a rectifier is pulsating direct current, the voltage can be analyzed by a Fourier series. Such an analysis shows that the output voltage consists of a d-c term and sinusoidal terms whose frequencies are multiples of the a-c supply frequency. The summation of the sinusoidal terms is the ripple voltage.

The percentage of ripple is the ratio of the rms value of the ripple voltage to the average value of the total voltage expressed in percent:

$$\text{Percent ripple} = \frac{E_r \times 100}{E_{d\text{-}c}} \tag{15·1}$$

The ripple factor of a rectifier output is also a commonly used term. It is defined as the ratio of the amplitude of the first harmonic of the ripple to the d-c output voltage or

$$\rho = \frac{E'_r}{E_{d\text{-}c}} \tag{15·2}$$

Loads

Rectifier loads may be almost every possible combination of resistance, inductance, and capacitance. However, the majority of industrial loads are combinations of resistance and inductance, and usually there is sufficient inductance to maintain the load current practically constant. Therefore the rectifier output current is assumed to be constant in the rectifier analysis in this chapter unless otherwise stated.

In some cases a resistance load is used. Here the wave form of the load current will be the same as the load voltage.

Capacitance loads are sometimes encountered when a capacitor filter is used. Rectifiers are sometimes used to charge large capacitors for storing electrical energy. An example of this is given in Section 31·31. If gaseous tubes are used in the rectifier, the capacitor charging current must be limited in some manner or the rectifier tubes will be damaged. This current can be limited by using resistance or inductance in series with the capacitor or by incorporating sufficient leakage reactance in the rectifier transformer to limit the rectifier short-circuit current to a value within the rating of the tubes.

Utility Factor

The utility factor of a rectifier transformer winding is the ratio of the d-c power carried by the winding to the volt-amperes of the winding. The higher the utility factor the more efficiently the winding is being used. The utility factor is a function of the wave form of the current and the ratio of the rms to the average voltage in the winding. Utility factor will be discussed in more detail for each type of rectifier in the following sections.

15·2 SINGLE-PHASE, HALF-WAVE RECTIFIER CIRCUIT

The single-phase, half-wave rectifier is the simplest possible rectifier circuit. As shown in Fig. 15·1, it consists of a transformer, a rectifying element, and a load. When the primary of the transformer is energized from an a-c source, an a-c voltage appears across its secondary. When the polarity of this voltage on the rectifying tube is such as to make the anode positive, the tube passes current through the load in the direction shown by the arrow. (The direction of conventional current flow is opposite to electron-current flow.) On the next half cycle, the anode of the tube is negative and, because the tube cannot pass current from cathode to anode (in the conventional sense) no current flows in the load. Thus the current through the load and the voltage across the load appear as shown in Fig. 15·2 when the load is resistive.

When the rectifying element is conducting, the wave shape of voltage across the load is the same as that of the secondary voltage of the rectifier transformer minus the

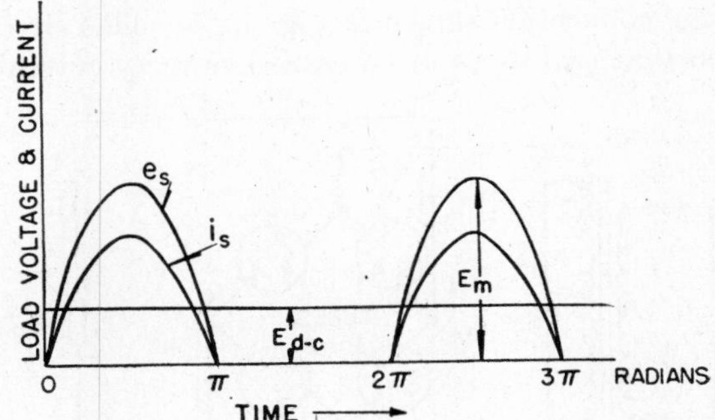

Fig. 15·2 Load voltage and current of a single-phase half-wave rectifier with resistance load.

drop or losses in the rectifying element. For simplicity, the rectifying elements are assumed to be ideal, having no losses. The drop through a phanatron, thyratron, or ignitron does not change appreciably with load; the tube drop is practically constant for a given type of tube. Tube drop in phanatrons and thyratrons may vary from 8 to 25 volts, whereas in ignitron tubes it ranges from 10 to 25 volts depending upon the size and design. The drop through vacuum tubes, copper oxide, and selenium rectifiers depends upon the load current, and it may vary quite widely as the type and size vary. The drop across these rectifiers can be determined from their voltage-current characteristics.

The ripple factor of the output voltage of the half-wave rectifier is given by

$$\rho = \frac{\pi}{2} = 1.57 \tag{15·3}$$

The d-c output voltage of the rectifier is

$$E_{d\text{-}c} = 0.318E_m = 0.45E_s \tag{15·4}$$

In the single-phase, half-wave rectifier, the total volt-amperes of the secondary winding for resistance load is

$$S = E_s \times I_s = \frac{E_{d\text{-}c}}{0.45} \times \frac{2 \times 1.11 I_{d\text{-}c}}{\sqrt{2}}$$

$$= 3.48E_{d\text{-}c}I_{d\text{-}c} \tag{15·5}$$

Inasmuch as current flows in a resistance load for half of each cycle, the utilization of the rectifier transformer is poor. The utility factor of the secondary is

$$\text{U.F.} = \frac{P}{S} = \frac{1}{3.48} = 0.287 \tag{15·6}$$

where $P = E_{d\text{-}c}I_{d\text{-}c}$ = d-c power output
S = total volt-amperes of secondaries.

The primary of the rectifier transformer must be designed to handle the load current plus the magnetizing current. Because the transformer is connected to an a-c supply, neither the primary current nor the voltage can have a d-c component. Therefore a current flows in the primary during the half cycle in which the tube is not conducting. This current has an average value that is equal and opposite to the half cycle of load current. Although the half cycle of magnetizing current does not have quite the same wave-shape as the half cycle of load current, satisfactory results can be obtained if the rms values of each half cycle are assumed equal. In this case the primary current is

$$I_p = I_s \frac{N_s}{N_p} = \frac{I_{d\text{-}c}}{0.45} \frac{N_s}{N_p} \tag{15·7}$$

where N_s = secondary turns
N_p = primary turns.

The primary volt-ampere rating is

$$S = E_p I_p = \frac{E_{d\text{-}c}}{0.45} \frac{N_p}{N_s} \times 2.22 I_{d\text{-}c} \frac{N_s}{N_p}$$

$$= 4.93E_{d\text{-}c}I_{d\text{-}c} \tag{15·8}$$

The primary utility factor is

$$\text{U.F.} = \frac{P}{S} = \frac{1}{4.93} = 0.203 \tag{15·9}$$

15·3 SINGLE-PHASE FULL-WAVE RECTIFIER

The next simplest type of rectifier is the single-phase full-wave type. It can be considered two half-wave rectifiers connected so that both halves of the applied sine wave of voltage are used. The electrical connections for this rectifier are given in Fig. 15·3. A center tap on the transformer secondary winding is required. Also, two tubes are required; although, in many cases, a double diode may be used.

When point A on the transformer becomes positive in respect to point B, tube 1 conducts from anode to cathode and through the load, as shown by i_{s1}. When point C becomes positive, tube 2 carries current in the direction shown by i_{s2}. The direction of current through the load is the same for both tubes. The wave shape of the load current is illustrated by Fig. 15·4.

The ripple factor for the output of the rectifier is

$$\rho = \frac{E'_r}{E_{d\text{-}c}} = \frac{\dfrac{4E_m}{3\pi}}{\dfrac{2}{\pi}E_m} = 0.667 \tag{15·10}$$

The d-c voltage output is

$$E_{d\text{-}c} = 0.636E_m = 0.90E_s \tag{15·11}$$

This is twice the d-c output of the half-wave rectifier.

Single-phase full-wave rectifiers are extensively used for power supplies, in radio equipment, instruments, and industrial electronic circuits. Filters can be designed to reduce the output ripple of the rectifier, thereby gaining smooth direct current for load currents ranging from a few milliamperes up to several hundred milliamperes.

The utilization of the rectifier transformer primary is much better in the full-wave rectifier than in the half-wave rectifier, because load current flows in the primary on both half cycles of the applied voltage wave. However, the utilization of each half of the secondary winding is low. For

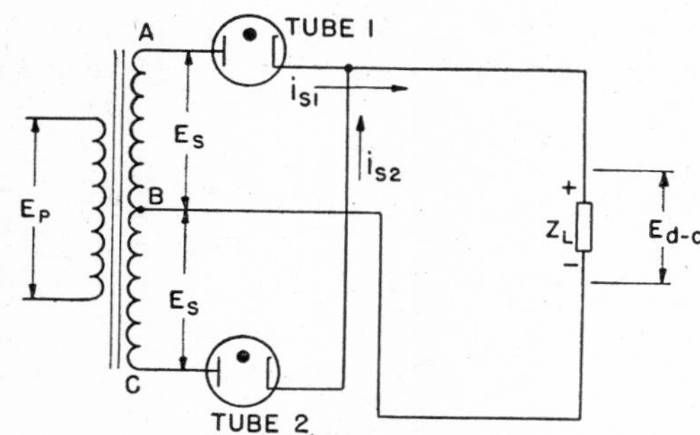

FIG. 15·3 Single-phase full-wave rectifier.

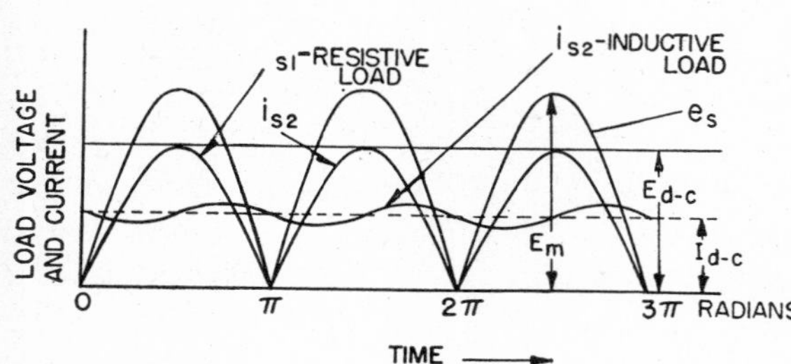

FIG. 15·4 Load voltage and current of a single-phase full-wave rectifier.

a single-phase full-wave rectifier, with resistance load, the total volt-amperes of the secondary is

$$S = 2\left(1.11E_{d\text{-}c} \times \frac{1.11I_{d\text{-}c}}{\sqrt{2}}\right) = 1.74E_{d\text{-}c}I_{d\text{-}c} \quad (15\cdot12)$$

For an inductive load, the current is practically constant. In this case the secondary volt-ampere rating of the transformer is

$$S = 2\left(1.11E_{d\text{-}c} \times \frac{I_{d\text{-}c}}{\sqrt{2}}\right) = 1.57E_{d\text{-}c}I_{d\text{-}c} \quad (15\cdot13)$$

The secondary utility factor of the transformer for resistance load is

$$\text{U.F.} = \frac{1}{1.74} = 0.584 \quad (15\cdot14)$$

For a reactive load

$$\text{U.F.} = \frac{1}{1.57} = 0.636 \quad (15\cdot15)$$

If the magnetizing current is neglected, the volt-ampere rating for the transformer primary on resistive load is

$$S = E_p \times I_p = 1.11E_{d\text{-}c}\frac{N_p}{N_s} \times 1.11I_{d\text{-}c}\frac{N_s}{N_p} \quad (15\cdot16)$$
$$= 1.24E_{d\text{-}c}I_{d\text{-}c}$$

The utility factor is then

$$\text{U.F.} = \frac{1}{1.24} = 0.81 \quad (15\cdot17)$$

The volt-ampere rating for the primary for inductive loads is

$$S = 1.11E_{d\text{-}c} \times I_{d\text{-}c} \quad (15\cdot18)$$

and

$$\text{U.F.} = \frac{1}{1.11} = 0.90 \quad (15\cdot19)$$

15·4 DOUBLE-WAY FULL-WAVE RECTIFIER

The need for the center tap on the transformer as used in the single-phase full-wave rectifier can be eliminated with the bridge connection shown in Fig. 15·5. This circuit is so arranged that two tubes in series always carry current. The

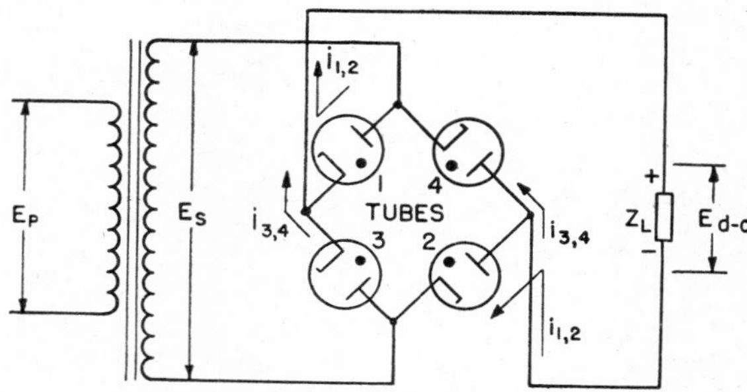

FIG. 15·5 Diametric double-way rectifier.

double-way rectifier is especially useful for high-voltage work inasmuch as the inverse voltage is divided between two tubes in series. However, the rectifier requires four tubes, and thus it is considerably more expensive than the single-phase full-wave rectifier. The rectifier transformer carries current during both half cycles instead of on alternate cycles as in the standard full-wave rectifier. Consequently the utilization of the transformer winding is high.

When the anode of tube 1 is positive, tube 1 conducts current, $i_{1,2}$, through the load, through tube 2 to the lower side of the rectifier transformer. On the next half cycle, the anodes of tubes 3 and 4 are positive, and current flows from the lower side of the transformer through tube 3, through the load and through tube 4 to the upper side of the transformer. The output voltage wave shape is the same as for the single-phase full-wave rectifier shown in Fig. 15·4.

This rectifier connection is frequently used with copper oxide rectifying elements because the circuit is easily made by applying the a-c voltage to the center taps of two parallel rectifier stacks. (The d-c output appears across the ends of the two stacks in parallel.) The connection can be easily visualized by replacing tubes 1 and 4 in Fig. 15·5 with a center-tapped stack and tubes 2 and 3 with another stack. The operation of the rectifier is the same as for the rectifier using tubes, except that the voltage drop is dependent upon the load current, whereas in thyratron tubes the rectifier voltage drop is practically constant.

15·5 VOLTAGE-DOUBLER RECTIFIER

In numerous cases d-c voltages in the neighborhood of 300 volts can be obtained directly from a 110-volt a-c line without a step-up transformer. This is desirable particularly in inexpensive equipment where the cost of a transformer might be prohibitive. Such a voltage can be obtained by using a voltage-doubler circuit similar to the one outlined in Fig. 15·6.

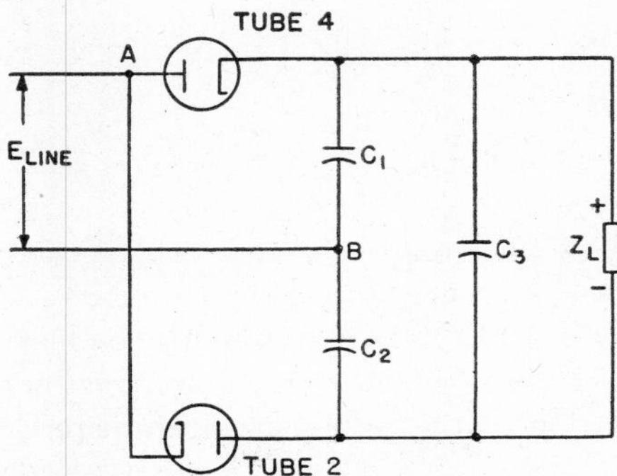

FIG. 15·6 Voltage-doubler circuit.

When point A is positive in respect to point B, tube 1 carries current and charges capacitor 1 to the peak of the line voltage, which in the case of 110-volt supply is

$$E_M = \sqrt{2} \times 110 = 155 \text{ volts} \qquad (15\cdot20)$$

On the next half cycle, when point B is positive in respect to point A, tube 2 conducts current and charges capacitor 2 to the peak of the a-c line voltage. The d-c output consists of the voltage across capacitors 1 and 2 in series; hence the d-c output voltage is 310 volts, if no load is assumed on the rectifier. In using a voltage-doubler rectifier it must be remembered that only a small load current should be drawn from it, otherwise excessive regulation results because capacitor 1 can be discharged by the load during the time capacitor 2 is charging and capacitor 2 can be discharged while capacitor 1 is charging. The result of this is apparent from Fig. 15·7.

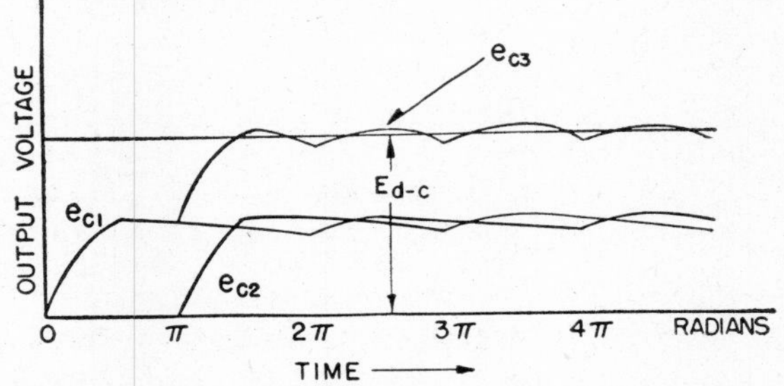

FIG. 15·7 Output voltage of a voltage-doubler circuit.

15·6 THREE-PHASE RECTIFIER

In the rectifier circuits previously discussed, the ripple factor of output voltage has been relatively high. Also the output current is limited because only one or two tubes are available to handle the load current. Where the percentage of ripple from the rectifier must be lower than that obtainable from a single-phase rectifier without a filter, or where the average current is greater than that supplied by one or two tubes, multiphase rectifiers may be used. Of the multiphase rectifiers, the three-phase rectifier is the simplest.

The transformer connections for this rectifier are usually a delta-connected primary and a wye-connected secondary. The wye-connected secondary is a necessity because one of the output leads of the rectifier must be connected to the neutral of the secondary winding. The diagram of the three-phase rectifier is given in Fig. 15·8.

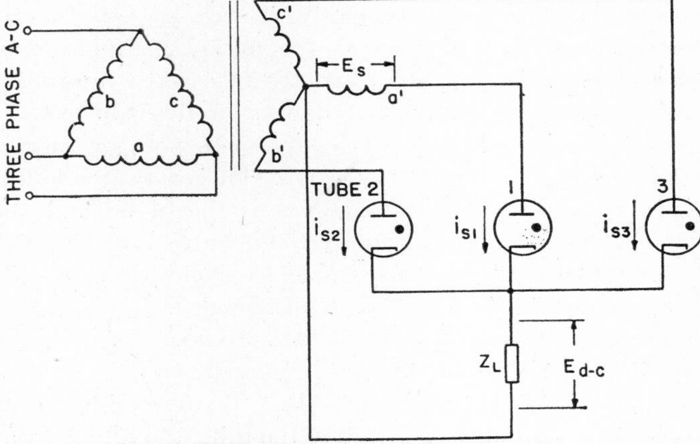

FIG. 15·8 Three-phase rectifier.

In analyzing the operation of this rectifier, it is assumed that the phase rotation of the secondary voltages is a', b', c', and also that phase a' is positive; therefore tube 1, which is connected to that phase, conducts current from anode to cathode and through the load back to the neutral of the transformer. Tube 1 continues to conduct current until the anode voltage of tube 2 becomes more positive than the anode voltage of tube 1. Tube 2 carries current from anode to cathode and through the load back to the neutral of the transformer, as did tube 1, and continues to carry current until the anode voltage of tube 3 has exceeded its anode voltage. Then the load current commutates from tube 2 to tube 3. Tube 3 likewise conducts current from its anode to its cathode and through the load back to the neutral of the transformer and continues to do so until the anode voltage of tube 1 exceeds the anode voltage of tube 3, thus causing the current to commutate to tube 1. Thereafter the general cycle of events described repeats through tubes 1, 2, and 3 as long as the anode voltage is applied and the load circuit is closed. The wave shape of the output voltage and currents of the rectifier is illustrated by Fig. 15·9.

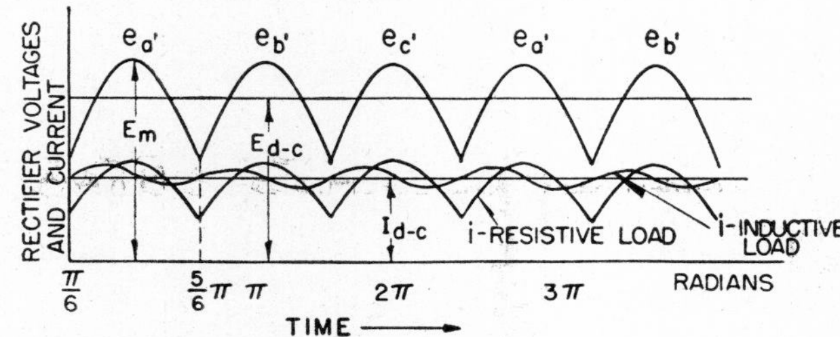

FIG. 15·9 Rectifier voltages and currents of a three-phase rectifier.

The d-c output voltage of the three-phase rectifier can be calculated by

$$E_{d\text{-}c} = \sqrt{2}E_s \frac{\sin \frac{\pi}{m}}{\frac{\pi}{m}} = 1.17E_s = 0.827E_m \qquad (15\cdot21)$$

where E_s = transformer secondary voltage, rms
E_m = transformer secondary voltage, maximum
m = number of secondary phases.

The ratio of d-c voltage to maximum voltage is considerably higher for the three-phase rectifier than for the single-phase rectifier discussed previously, as Fig. 15·9 shows. In general, this ratio increases as the number of phases increases; however, this is not strictly true when interphase transformers are used. The frequency of the ripple in the three-phase rectifier is 180 cycles for a 60-cycle a-c supply, or three times the supply frequency. In general for multiphase rectifiers the ripple frequency is the product of the number of phases and the supply frequency. For a double-way multiphase rectifier the ripple frequency is twice the number of phases times the supply frequency.

Each tube carries current for 120 degrees when the load is purely resistive (overlap caused by inductance in the transformer windings being neglected). Because each tube carries current for only one third of the time, the total output current of the rectifier can be made three times the average current rating of each tube, whereas in the single-phase full-wave rectifier, for example, the total current may be only double the average current rating of each tube. In other words, the current output from a rectifier can be increased directly as the number of phases is increased.

Three-phase rectifiers are used, for example, to furnish the d-c voltage for high-powered radio transmitting stations, and to supply the energy-storage resistance-welding machines, such as discussed in Chapter 31.

The volt-ampere rating of the three secondaries (total) in the three-phase rectifier on inductive load is

$$S = 3(E_s I_s) = 3\frac{E_{d\text{-}c} \times I_{d\text{-}c}}{1.17\sqrt{3}} = 1.481 E_{d\text{-}c} I_{d\text{-}c} \qquad (15\cdot 22)$$

and

$$\text{U.F.} = \frac{1}{1.481} = 0.675 \qquad (15\cdot 23)$$

This utility factor is one of the highest that can be obtained for a multiphase rectifier (other than double-way types) as indicated by Table 15·1. Therefore three-phase secondary connections are used extensively in power-rectifier circuits.

The calculation of the primary volt-ampere rating of the transformer is somewhat more complicated. Each secondary carries current for about 120 degrees each cycle; therefore the primary phase that supplies this secondary carries load current for 120 degrees. The magnitude of this current, as shown in Fig. 15·10, is

$$i_p = \frac{2}{3} I_{d\text{-}c} \frac{N_s}{N_p} \qquad (15\cdot 24)$$

During the remainder of the cycle the current in the primary is

$$i_p = \frac{1}{3} I_{d\text{-}c} \frac{N_s}{N_p} \qquad (15\cdot 25)$$

because the average current in the winding must be zero.

Table 15·1 Rectifier Circuit Relationships

Type of Rectifiers	Single-Phase Half-Wave	Single-Phase Full-Wave		Diametric Double-Way	Three-Phase Wye	Three-Phase Wye, Double-Way	Six-Phase Double Wye	Six-Phase
Load	Resistive	Resistive	Inductive	Inductive	Inductive	Inductive	Inductive	Inductive
$E_{d\text{-}c}$	$0.318E_m$ $0.45E_s$	$0.636E_m$ $0.90E_s$	$0.636E_m$ $0.90E_s$	$0.636E_m$ $0.90E_s$	$0.827E_m$ $1.17E_s$	$1.655E_m$ $2.34E_s$	$0.827E_m$ $1.17E_s$	$0.955E_m$ $1.35E_s$
Secondary phase current, rms	$1.57\ I_{d\text{-}c}$	$0.785I_{d\text{-}c}$	$0.707I_{d\text{-}c}$	$1.0I_{d\text{-}c}$	$0.577I_{d\text{-}c}$	$0.816I_{d\text{-}c}$	$0.289I_{d\text{-}c}$	$0.408I_{d\text{-}c}$
Secondary va $\div E_{d\text{-}c}I_{d\text{-}c}$	3.48	1.74	1.57	1.11	1.48	1.05	1.48	1.81
Secondary utility factor	0.287	0.584	0.636	0.90	0.675	0.95	0.675	0.552
Primary phase current, rms	$2.22I_{d\text{-}c}\frac{N_s}{N_p}$	$1.11I_{d\text{-}c}\frac{N_s}{N_p}$	$I_{d\text{-}c}\frac{N_s}{N_p}$	$I_{d\text{-}c}\frac{N_s}{N_p}$	$0.471I_{d\text{-}c}\frac{N_s}{N_p}$	$0.816I_{d\text{-}c}\frac{N_s}{N_p}$	$0.407I_{d\text{-}c}\frac{N_s}{N_p}$	$0.578I_{d\text{-}c}\frac{N_s}{N_p}$
Primary va $\div E_{d\text{-}c}I_{d\text{-}c}$	3.48*	1.24	1.11	1.11	1.21	1.05	1.05	1.28
Primary utility factor	0.287	0.81	0.90	0.90	0.827	0.95	0.95	0.780
Peak tube current	$3.14I_{d\text{-}c}$	$1.57I_{d\text{-}c}$	$I_{d\text{-}c}$	$I_{d\text{-}c}$	$I_{d\text{-}c}$	$I_{d\text{-}c}$	$0.5I_{d\text{-}c}$	$I_{d\text{-}c}$
Average tube current	$I_{d\text{-}c}$	$0.5I_{d\text{-}c}$	$0.5I_{d\text{-}c}$	$0.5I_{d\text{-}c}$	$0.33I_{d\text{-}c}$	$0.33I_{d\text{-}c}$	$0.167I_{d\text{-}c}$	$0.167I_{d\text{-}c}$
Peak inverse voltage	$3.14E_{d\text{-}c}$	$3.14E_{d\text{-}c}$	$3.14E_{d\text{-}c}$	$1.57E_{d\text{-}c}$	$2.09E_{d\text{-}c}$	$1.05E_{d\text{-}c}$	$2.09E_{d\text{-}c}$	$2.09E_{d\text{-}c}$
Ripple factor	1.57	0.667	0.667	0.667	0.25	0.057	0.057	0.057
Major ripple frequency	f	$2f$	$2f$	$2f$	$3f$	$6f$	$6f$	$6f$

E_s = transformer secondary voltage per leg, rms
E_m = transformer secondary voltage per leg, maximum
$E_{d\text{-}c}$ = load voltage, average
$I_{d\text{-}c}$ = load current, average
f = supply frequency, C.P.S.

The ratios given above neglect the voltage drop in tubes, transformer, and circuits and are given on the assumption that there is no counter-emf in the load and that the rectifier is phased fully forward. Primary volt-amperes do not include transformer losses.

* Transformer magnetizing current is extremely large for this connection and the primary volt-amperes may be as large as $5.0E_{d\text{-}c}I_{d\text{-}c}$ if the transformer losses are included.

The rms value of the primary current can be found by

$$I_p = \sqrt{\left(\frac{2}{3} I_{d\text{-}c} \frac{N_s}{N_p}\right)^2 \frac{1}{3} + \left(\frac{1}{3} I_{d\text{-}c} \frac{N_s}{N_p}\right)^2 \frac{2}{3}}$$

$$= \frac{\sqrt{2}}{3} I_{d\text{-}c} \frac{N_s}{N_p} = 0.471 I_{d\text{-}c} \frac{N_s}{N_p} \qquad (15\cdot 26)$$

The total volt-amperes of the primary is

$$S = 3(E_p \times I_p)$$

$$= 3\left(\frac{E_{d\text{-}c}\, N_p}{1.17\, N_s} \times 0.471 I_{d\text{-}c} \frac{N_s}{N_p}\right)$$

$$= 1.21 E_{d\text{-}c} I_{d\text{-}c} \qquad (15\cdot 27)$$

The utility factor is

$$\text{U.F.} = \frac{1}{1.21} = 0.827 \qquad (15\cdot 28)$$

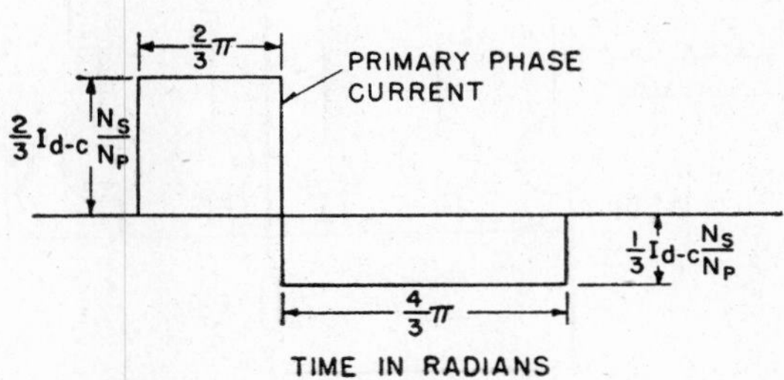

FIG. 15·10 Transformer primary and secondary currents in a three-phase rectifier.

15·7 THREE-PHASE WYE, DOUBLE-WAY RECTIFIER

The rectifier connection shown in Fig. 15·11 is widely used on high-voltage applications because the inverse voltage appears across two tubes in series. This means that the rectifier can operate at higher voltages without danger of arc-back (in the case of mercury-vapor tubes) or reverse conduction (when vacuum tubes are used). The output voltage wave shape and ripple factor for this rectifier are the same as for a six-phase rectifier. The ripple frequency is 360 cycles per second when operating from a 60-cycle supply. Consequently, comparatively small filters can be used to smooth the ripple to produce steady direct current. The output voltage and current wave shapes are shown in Fig. 15·12.

The operation of the rectifier can be understood by referring to Figs. 15·11 and 15·12. Assume that phase a' is positive and that tube 2 cannot start conduction until time $\pi/6$ is reached. The anode of tube 2 is positive, and the tube conducts current through the load providing that one of tubes 4, 5, or 6 can conduct. At time $\pi/6$, the cathode of tube 4, which is connected to phase b', is negative; therefore current flows through tube 2, through the load, and returns to the transformer through tube 4. The voltage appearing across the load while these tubes are conducting is

$$e = e_{a'} - e_{b'} \qquad (15\cdot 29)$$

where $e_{a'}$ = instantaneous voltage from phase a'.
$e_{b'}$ = instantaneous voltage from phase b'.

At time $\pi/4$, the voltage of phase c' becomes more negative than phase b', and the load current transfers from tube 4 to tube 6. Tube 6 continues to carry current until point $7\pi/6$ is reached. Then the current transfers to tube 5. Tube 2 carries current to point $5\pi/6$ where the voltage of phase b' becomes more positive than a' and the current

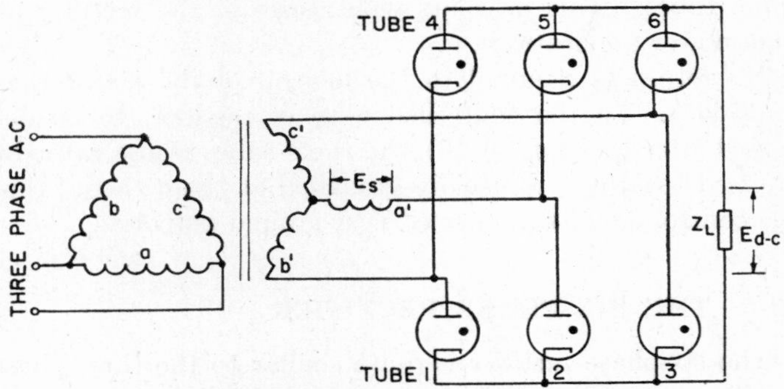

FIG. 15·11 Three-phase wye, double-way rectifier.

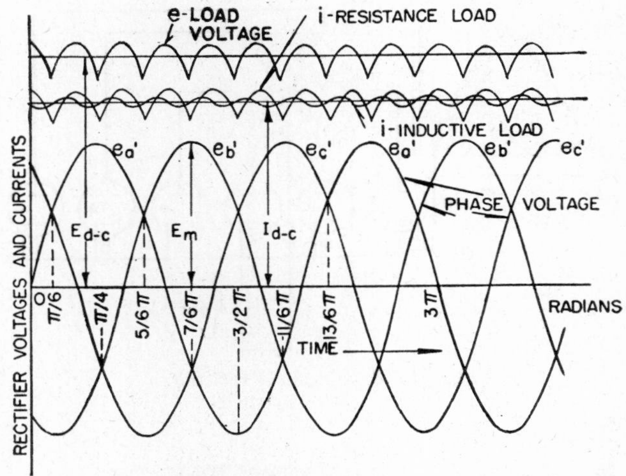

FIG. 15·12 Rectifier voltages and currents of a three-phase wye, double-way rectifier.

transfers to tube 1. Tube 1 conducts to point $3\pi/2$ where current transfers to tube 3, which conducts to point $13\pi/6$ where tube 2 again conducts, and the cycle of events is repeated. Thus each tube carries current for 120 degrees (assuming no overlap); a commutation occurs every 60 degrees, which results in a ripple frequency of

$$f_r = 6f \qquad (15\cdot 30)$$

where f = supply frequency.

The d-c output voltage is

$$E_{d\text{-}c} = \frac{2\sqrt{2}E_s \sin \dfrac{\pi}{m}}{\dfrac{\pi}{m}} = 2.34E_s = 1.655E_m \qquad (15\cdot 31)$$

where E_s = transformer secondary voltage, rms, line to neutral

E_m = transformer secondary voltage, maximum, line to neutral

m = number of phases.

The volt-ampere rating and utility factor of the transformer on this rectifier can be calculated in a manner similar to that used in the previous examples. The results are given in Table 15·1. The utility factors of both the primary and the secondary are higher than that of the three-phase rectifier. However, the rectifier uses six tubes and has the same average current rating as the three-phase rectifier using only three tubes. This prevents wide usage of the rectifier for high-power applications.

No return is required to the neutral of the transformer secondary. In the controlled type of rectifier, to be discussed later (Section 15·10), the three tubes whose cathodes are tied together are usually grounded at the cathode; thus the control circuits are placed near ground potential.

15·8 THE SIX-PHASE RECTIFIER

The six-phase rectifier is quite similar to the three-phase rectifier except that the number of tubes is increased to six, Fig. 15·13.

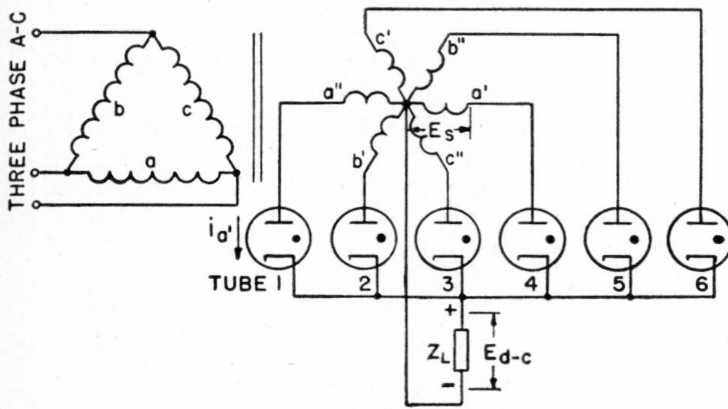

Fig. 15·13 Six-phase rectifier.

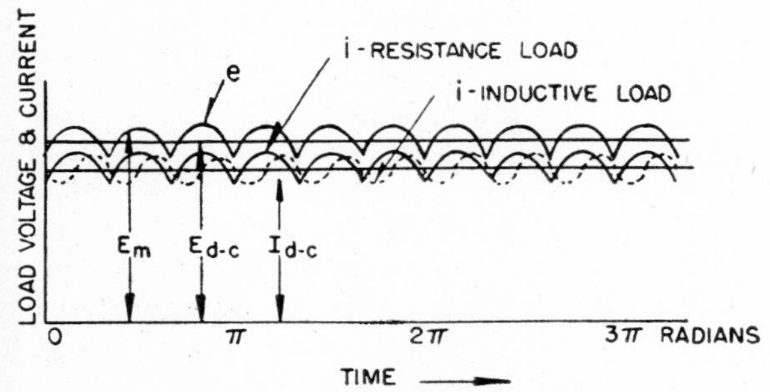

Fig. 15·14 Load voltage and current of a six-phase rectifier.

This rectifier is one of the types used for handling large amounts of power. Therefore, the tubes commonly used are ignitrons. These tubes, like other gaseous tubes, have voltage drops about 10 to 20 volts. They are efficient rectifiers.

The ripple output from the six-phase rectifier is relatively small, and the ratio of average voltage to peak voltage is high as shown by Table 15·1. The current and voltage wave shapes are shown in Fig. 15·14.

Disregarding commutation overlap, each tube conducts for 60 degrees or one sixth of the cycle; therefore the utility factor of the secondaries of the rectifier transformer is lower than that of the three-phase rectifier. However, the utility factor of the primaries is slightly higher than that of the three-phase rectifier because each of the primaries supplies two secondaries 180 degrees out of phase. Thus the magnetizing current for the transformer is reduced, and a lower volt-ampere rating and a higher utilization factor result. Nevertheless the utilization of the secondary is so poor that even this circuit is not used extensively for high-power rectifiers.

15·9 SIX-PHASE DOUBLE-WYE RECTIFIER

The six-phase double-wye rectifier is used as a basic circuit for high-power ignitron rectifiers because it produces a voltage with little ripple. The utilization of the rectifier transformer is high for both the primary and the secondaries.

The circuit for the six-phase double-wye rectifier is shown in Fig. 15·15. The rectifier transformer has one three-phase

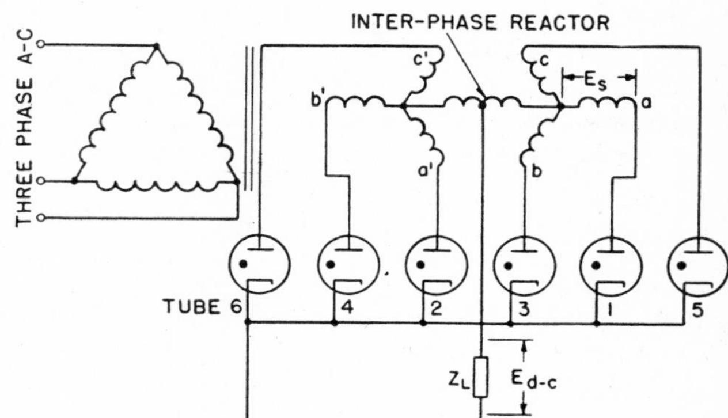

Fig. 15·15 Six-phase double-wye rectifier.

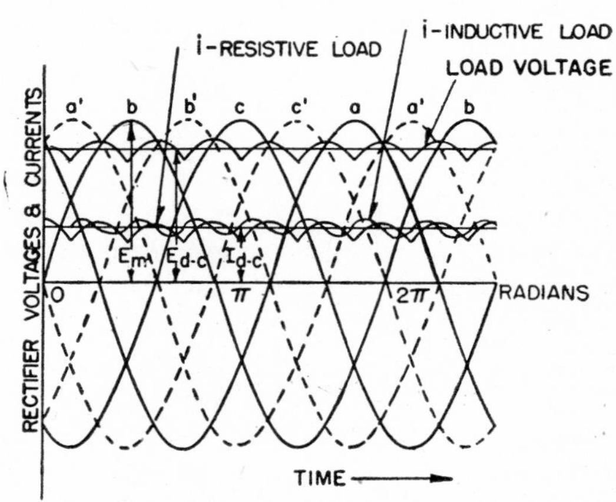

Fig. 15·16 Load voltages and currents of a six-phase double-wye rectifier.

primary. The secondary consists of two sets of three-phase wye-connected windings. The two sets of secondary windings are interconnected by means of an interphase reactor or balancing inductor. The negative return of the rectifier is connected to the center tap of this inductor. In operation, one phase in each set of secondaries is carrying current and the current is divided between the two secondaries by the interphase reactor. Hence the instantaneous output voltage is the average of the instantaneous voltages of the two secondaries which are conducting. The d-c voltage of the rectifier is

$$E_{d\text{-}c} = 1.17E_s = 0.827E_m \tag{15·32}$$

The wave forms of the output voltage and current for resistive and inductive loads are shown in Fig. 15·16. The ripple on the output voltage has a frequency of six times the supply frequency.

The volt-ampere rating of each set of secondaries of the rectifier transformer is

$$S = 3\left(\frac{1}{1.17}E_{d\text{-}c}\frac{I_{d\text{-}c}}{2\sqrt{3}}\right) = 0.74E_{d\text{-}c}I_{d\text{-}c} \tag{15·33}$$

The total volt-ampere rating of both secondaries is

$$S = 1.48E_{d\text{-}c}I_{d\text{-}c} \tag{15·34}$$

The utility factor of the secondaries is

$$\text{U.F.} = \frac{1}{1.48} = 0.675 \tag{15·35}$$

The total volt-ampere rating of the primary windings is found as follows. The primary current is a block with a magnitude of $\frac{1}{2}I_{d\text{-}c}(N_s/N_p)$ and $\frac{2}{3}\pi$ radians wide with a positive and negative block occurring each cycle. The rms value of this current wave is

$$I_{\text{pri.}} = \sqrt{\frac{1}{4}I_{d\text{-}c}^2\left(\frac{N_s}{N_p}\right)^2\frac{2}{3}} = \frac{1}{\sqrt{6}}I_{d\text{-}c}\frac{N_s}{N_p} \tag{15·36}$$

The total volt-ampere rating for the primary windings is

$$S = 3\left(\frac{1}{1.17}E_{d\text{-}c}\frac{N_p}{N_s} \times \frac{1}{\sqrt{6}}I_{d\text{-}c}\frac{N_s}{N_p}\right)$$

$$= 1.047E_{d\text{-}c}I_{d\text{-}c} \tag{15·37}$$

The utility factor of the primaries is

$$\text{U.F.} = \frac{1}{1.047} = 0.955 \tag{15·38}$$

By means of special transformer connections, more than two groups of three tubes can be operated to give a voltage wave shape equivalent to 12, 24, 36, or more phases. These rectifiers are discussed in Chapter 22.

15·10 GRID-CONTROLLED RECTIFIERS

The discussion has been limited to rectifiers using only diode tubes, which have no grids or control elements. All these rectifier circuits can be adapted for grid-controlled tubes so that complete control of the output voltage can be obtained by varying the voltage on the grids of the controlled tubes. Grid-controlled rectifiers are used when it is necessary to control the voltage or the current flowing into the load. Notable examples are the ignitron rectifiers discussed in Chapter 22, and the rectifiers used for energy-storage resistance-welding discussed in Chapter 31.

Several different methods are available for controlling the output of rectifiers. One of the most common is the phase-shift method in which the phase position of the voltages applied to the grids of the rectifier tubes is shifted. Details of phase-shifting circuits are given in Chapter 18.

A three-phase rectifier to which phase-shift control has been added is shown in Fig. 15·17. If the rectifier uses

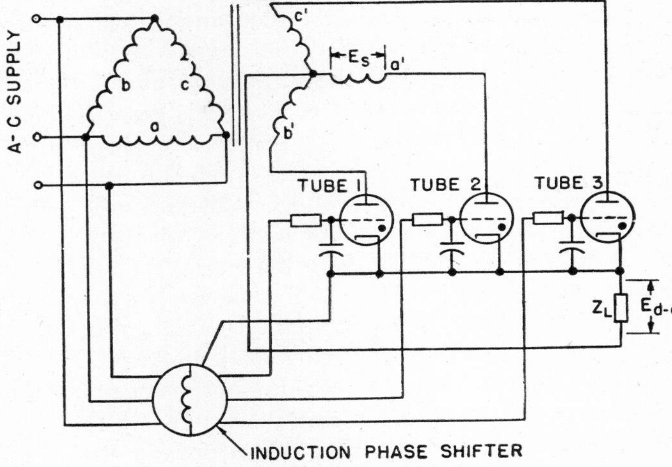

FIG. 15·17 Three-phase rectifier with phase shifter in the grid circuit.

thyratron tubes, the details of operation are as follows. Each thyratron tube is capable of carrying current only when its anode voltage is positive and its grid voltage is less negative than its critical grid voltage. By varying the phase position

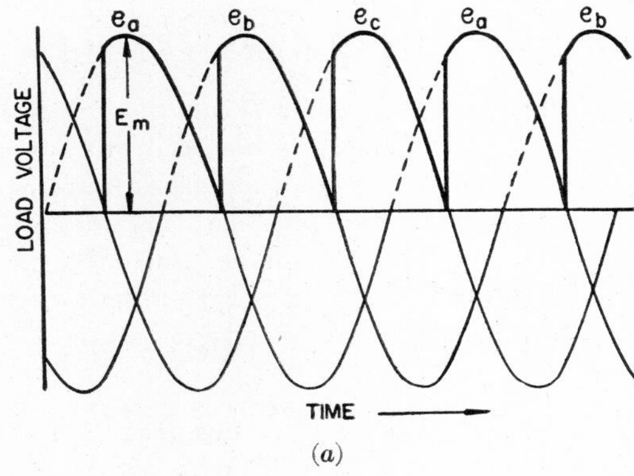

(a)

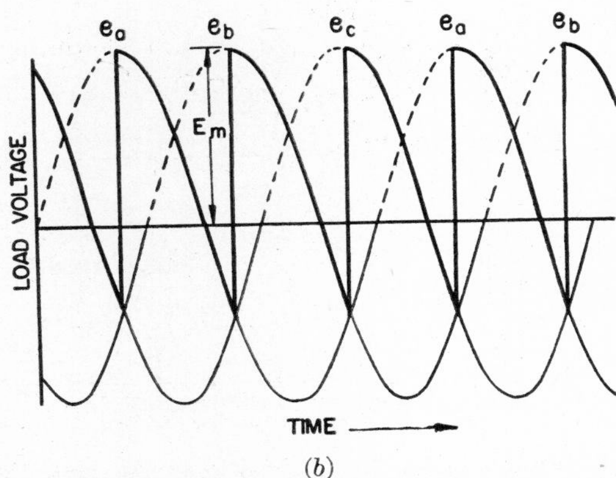

(b)

FIG. 15·18 Load voltage of a three-phase controlled rectifier. (a) Conduction delayed 30°. (b) Conduction delayed 60°.

of the grid voltages, the grid voltage can be made to become less negative than the critical characteristic at any desired point while the anode of the tube is positive. Once the thyratron is conducting, it continues to conduct until the voltage from anode to cathode falls below arc drop or the load current is commutated to the next tube. Figures 15·18 (*a*) and 15·18 (*b*) show the output voltage wave forms for several different phase positions of the grid voltage.

This method of controlling the outout voltage of rectifiers is usually applicable only when the output voltage is to be controlled manually (or electrically at a relatively slow rate). When it is necessary to vary the output voltage of a rectifier rapidly, the amplitude method of control is generally used. This type of control consists of a permanently phase-shifted a-c voltage plus a variable d-c voltage applied to the grids of the rectifier tubes. The a-c voltage is permanently phase-shifted by a definite amount for each rectifier tube. The amount of phase shift varies with the type of rectifier. For instance, to get complete control from zero to maximum voltage from a three-phase rectifier, the grid voltage must be shifted to 90 degrees lagging the anode voltage. The d-c grid voltage can be obtained from the output of a vacuum-tube control circuit. By changing the voltage on the input of the vacuum-tube control circuit the output can be changed rapidly, and thereby the output voltage of the rectifier is changed by making the grid voltage of the rectifier positive at different points. A circuit of this type is used for controlling the output of a rectifier for charging a large capacitor bank to a pre-set voltage in an energy-storage type of resistance welder discussed in Chapter 31.

15·11 CURRENT OVERLAP AND COMMUTATION

The reactance of the rectifier transformer has been assumed to be negligible. Also the load current is considered to transfer from one tube to another instantaneously at the point of commutation. In practice this ideal condition cannot be obtained. In many cases the effect of current overlap during the commutating period must be considered because it results in a decrease of output voltage.

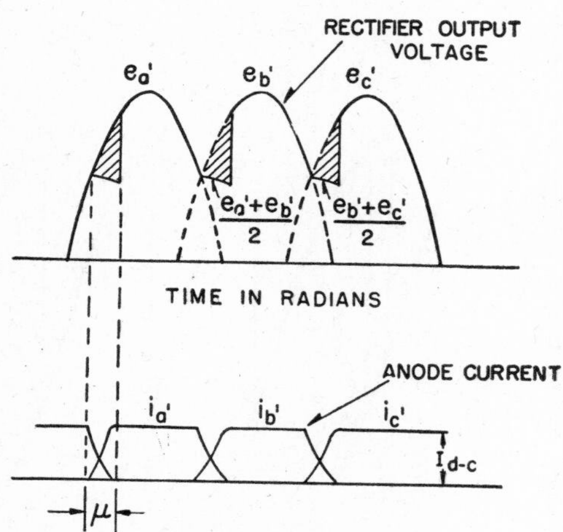

Fig. 15·19 Commutation and current overlap in a three-phase rectifier.

Figure 15·19 shows the phase voltages for two tubes in the three-phase rectifier of Fig. 15·8. The commutation period begins when $e_{b'}$ becomes more positive than $e_{a'}$, if it is assumed that phase a' has been carrying current $i_{a'}$. During the commutating period the rectifier output voltage is

$$e_r = \frac{e_{a'} + e_{b'}}{2} \tag{15·39}$$

Thus the output voltage of the rectifier is reduced by an amount proportional to the shaded area. To determine the voltage loss it is necessary to determine the amount of overlap. The voltage available for commutation is the difference between $e_{b'}$ and $e_{a'}$ (minus the arc drop of two tubes). The commutating voltage gives rise to a commutating current that flows around the transformer circuit closed by tubes 1 and 2, which conduct at the same time. The current is in such a direction as to increase i'_b and decrease i'_a. Because the rectifier tube cannot carry current in the reverse direction the commutating period ends when

$$i_{a'} = 0$$

or when the commutating current is equal to $I_{d\text{-}c}$.

The loss in voltage due to commutation is

$$E_c = \frac{m\sqrt{2}E_s}{2\pi} \sin\frac{\pi}{m}(1 - \cos\mu) \tag{15·40}$$

where m = number of phases
E_s = transformer secondary voltage
μ = commutation angle.

Also

$$E_c = \frac{mX_l I_{d\text{-}c}}{2\pi} \tag{15·41}$$

where m = number of phases
X_l = transformer reactance, line to neutral
$I_{d\text{-}c}$ = d-c load current.

These equations show that the loss in voltage due to commutation increases with the number of phases. It also increases as the commutating reactance and the load current are increased.

15·12 VOLTAGE REGULATION

The output voltage of the rectifier decreases as load current increases. This is caused by

1. Copper loss in the rectifier transformers.
2. Reactance in the rectifier transformer.
3. Voltage drop in the rectifying elements.

The reduction in output voltage due to copper loss in the transformer can be calculated by

$$E_r = \frac{P_r}{I_{d\text{-}c}} \tag{15·42}$$

The copper loss in watts, P_r, is obtained from design or test data on the transformer.

Voltage loss in the rectifying elements depends upon the type of rectifier used. If mercury-vapor tubes are used, the voltage drop will be practically constant and will be from

10 to 20 volts, depending upon the size and construction of the tubes. If vacuum tubes and copper oxide or selenium rectifiers are used, the voltage drop will vary with load current. The drop can be obtained from the volt-ampere characteristic of the rectifying element.

An expression for the decrease in voltage due to commutation is given in Section 15·11.

The complete equation for the d-c output voltage of a mercury-arc rectifier without phase control is

$$E_{d\text{-}c} = E_{d\text{-}o} - \frac{m}{2\pi} X I_{d\text{-}c} - \frac{P_r}{I_{d\text{-}c}} - E_0 \qquad (15\cdot43)$$

where $E_{d\text{-}o}$ = no-load output voltage
m = phases
X = commutating reactance
P_r = transformer copper loss in watts
E_0 = rectifier arc drop.

REFERENCES

1. *Alternating-Current Machines*, A. F. Puchstein and T. C. Lloyd, Wiley, 1942.
2. *Mercury Arc Power Rectifiers*, O. I. Marti and H. Winograd, McGraw-Hill, 1930.

Chapter 16

AMPLIFIER CIRCUITS

Reuben Lee

An amplifier is a device for increasing voltage, current, or power. The original wave form may or may not be maintained; the frequency usually is. An amplifier may be mechanical, electromechanical, electromagnetic, or electronic, or a combination of these. The electronic amplifier circuit usually consists of a vacuum tube in conjunction with capacitors, transformers, or resistors. Input voltage or current is impressed on some element of the tube, and higher voltage or current in the output circuit results.

16·1 AMPLIFIER POTENTIALS

Electronic amplifiers use tubes having three or more elements. In a triode the grid alters the voltage gradient between cathode and anode as shown in Fig. 16·1. It aids

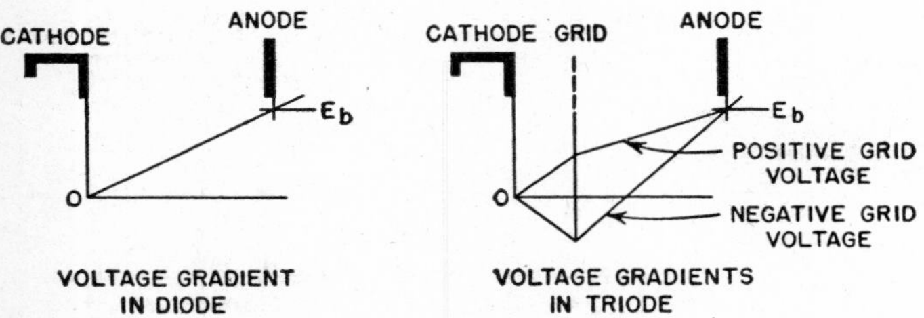

FIG. 16·1 Diode and triode voltage gradients. (From *Electronic Transformers and Circuits* by Reuben Lee.)

or opposes the movement of electrons from cathode to anode, depending upon whether it is positive or negative with respect to the cathode, which is shown at zero voltage in Fig. 16·1.

As grid voltage is made more and more negative, electron movement is diminished, and finally stops. At this point the anode current is zero; the condition is called *anode-current cut-off*.

If grid voltage is made more and more positive, eventually additional increase in grid voltage causes no increase of anode current. This is called *grid saturation.*

Tetrodes and pentodes have two and three grids respectively. Voltage gradient between cathode and anode is more complex than that indicated in Fig. 16·1. Advantages to be gained from additional grids are treated in Chapter 5.

Resistance-coupled amplifiers, as discussed in Chapter 14, are used mainly for voltage amplification. Large power output can be obtained only at low efficiency because of losses in the plate-circuit resistor.[1]

16·2 TRANSFORMER-COUPLED AMPLIFIERS

Amplifier circuits in which transformers are used can be represented by a circuit similar to Fig. 16·2 (*a*). A triode is shown with a voltage e_c impressed upon the grid, which is the grid bias (a constant negative direct voltage E_c), and a superimposed alternating voltage e_g. Anode voltage E_b is supplied through the primary of the transformer, across which appears an alternating voltage e_p. The secondary of the transformer is connected to a load Z_L. Under certain conditions this circuit may be simplified to that of Fig. 16·2 (*b*). A fictitious alternating voltage μe_g is impressed on the circuit, where μ is the tube amplification factor.

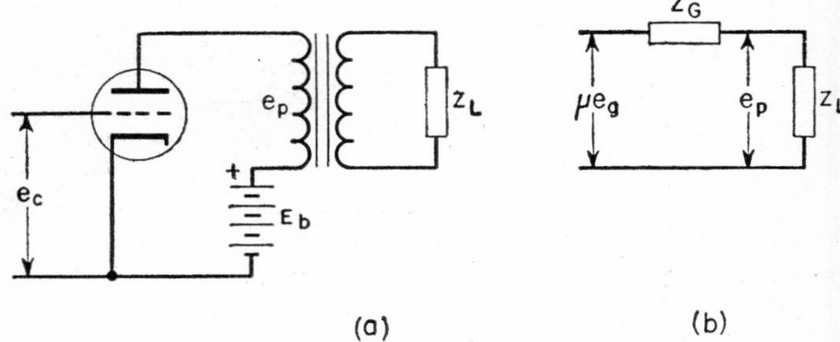

FIG. 16·2 (*a*) Transformer-coupled amplifier. (*b*) Equivalent circuit. (From *Electronic Transformers and Circuits* by Reuben Lee.)

Internal tube resistance Z_G is in series with Z_L, which is reflected by the transformer to the proper value in the primary circuit for tube operation. That is, Z_L in Fig. 16·2 (*b*) is equal to Z_L in Fig. 16·2 (*a*) only if the transformer has a one-to-one ratio. For any ratio of primary to secondary turns the two Z's are related by the square of the turns ratio. Winding resistances are regarded as zero, so that, with no grid signal, full voltage E_b appears on the plate of the vacuum tube.

Alternating voltage μe_g causes voltage e_p to appear across Z_L. Voltage e_p is not μ times e_g but

$$e_p = \mu e_g \frac{Z_L}{Z_G + Z_L} \tag{16·1}$$

Although transformer-coupled amplifiers are used sometimes for voltage amplification, they are used mostly where power output is required, and where good reproduction of the grid voltage is required in the plate circuit.

16·3 TUNED AMPLIFIERS

Figure 16·3 is the circuit for an amplifier in which output voltage appears across a parallel-tuned *tank* circuit. The circuit is shown coupled to a load Z_L. This type of amplifier is used for relatively large outputs, but voltage e_p is not

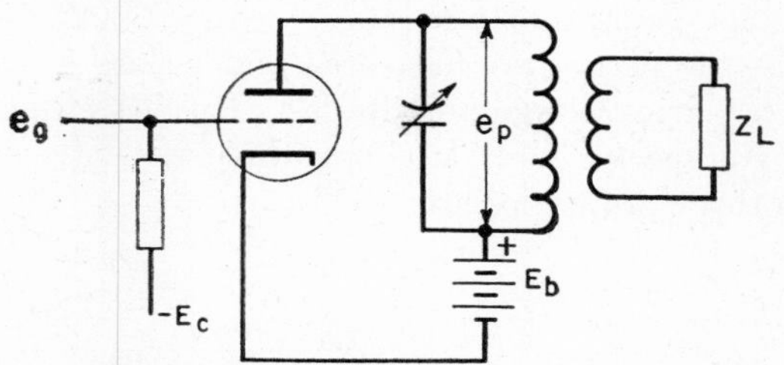

FIG. 16·3 Tuned amplifier. (From *Electronic Transformers and Circuits* by Reuben Lee.)

necessarily a good reproduction of e_g, and they are not related according to equation 16·1.

16·4 AMPLIFIER CLASSIFICATION

Amplifiers can be divided into classes according to their mode of operation. Following are standard definitions for these classes.[13]

Class A Amplifier

A class A amplifier is an amplifier in which the grid bias and alternating grid voltages are such that plate current in a specific tube flows at all times.

NOTE: To denote that grid current does not flow during any part of the input cycle, the subscript 1 may be added to the letter or letters of the class identification. The subscript 2 may be used to denote that grid current flows during some part of the cycle.

Class AB Amplifier

A class AB amplifier is an amplifier in which the grid bias and alternating grid voltages are such that plate current in a specific tube flows for appreciably more than half but less than the entire electrical cycle.

Class B Amplifier

A class B amplifier is an amplifier in which the grid bias is approximately equal to the cut-off value, so that the plate current is approximately zero when no exciting grid voltage is applied, and so that plate current in a specific tube flows for approximately one half of each cycle when an alternating grid voltage is applied.

Class C Amplifier

A class C amplifier is an amplifier in which the grid bias is appreciably greater than the cut-off value, so that the plate current in each tube is zero when no alternating grid voltage is applied, and so that plate current flows in a specific tube for appreciably less than one half of each cycle when an alternating grid voltage is applied.

Classes A, B, and C are illustrated in Fig. 16·4, in which the alternating components of plate current, plate voltage, grid voltage, and grid current are shown with the steady or average values, which are respectively I_b, E_b, E_c, and I_g. Relative plate and grid voltage amplitudes for these three types of amplifiers are shown in Fig. 16·4, and typical properties are summarized in Table 16·1.

TABLE 16·1 AMPLIFIER CLASSES

Amplifier Class	A	B	C
Anode efficiency			
(a) Theoretical maximum	50%	78.5% *	100%
(b) Practical value for low distortion	Up to 30%	40%–67% *	70%–85% †
Output proportional to	e_g^2	e_g^2	E_b^2 (grid saturated)
Grid current I_g	None	Small	Large (may $\approx I_b$)
Anode current I_b	Fairly constant	$e_g = 0$, I_b low $e_g = \max$, I_b high	$e_g = 0$, $I_b = 0$ $e_g = \max$, I_b high

* These values are for push-pull amplifiers.

† With a high-Q tank circuit, the efficiency depends upon excitation power.

Class A amplifiers have comparatively high no-signal anode current. Usually, the grid does not swing positive. Anode current remains comparatively constant when averaged over a whole a-c cycle. In a class B amplifier the grid is biased at a greater negative potential so that, with no signal, current is nearly cut off. Positive swings of grid voltage result in anode current, which causes a dip in residual voltage on the plate of the amplifier. Negative grid swings

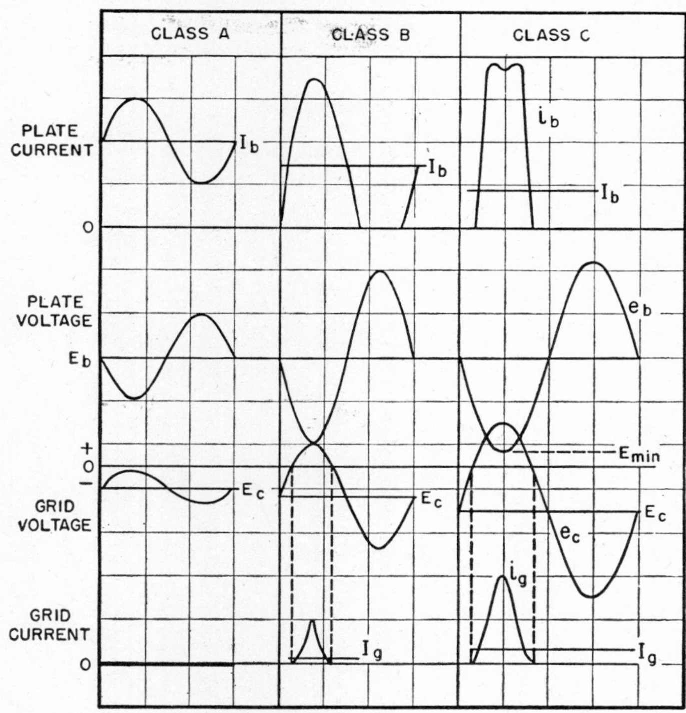

FIG. 16·4 Voltage and current relations in amplifiers. (From *Electronic Transformers and Circuits* by Reuben Lee.)

cause no plate current, but cause a positive plate-voltage swing. In a class C amplifier the grid is biased more negatively still, so plate current lasts less than half a cycle, and mostly when the plate voltage on the tube is at a relatively low value. Grid-current values are comparable with plate current. Output-voltage wave form is maintained by the tank circuit.

Operation may be improved by the use of two tubes connected in push-pull as shown in Fig. 16·5. This is the most common connection for class B amplifiers; it is also used in some class A amplifiers. Intermediate between class A and class B amplifiers are those known as class AB, with grid bias and efficiency intermediate between class A and class B. Such amplifiers are further subdivided into class AB_1 and class AB_2. A class AB_1 amplifier draws no grid current, but bias voltage is somewhat higher than the class A bias, and plate current may be discontinuous during the cycle when a grid signal is applied. Class AB_2 amplifiers draw grid current, but are not biased as close to cut-off as are class B amplifiers. Both class AB_1 and AB_2 are commonly used with the push-pull connection.

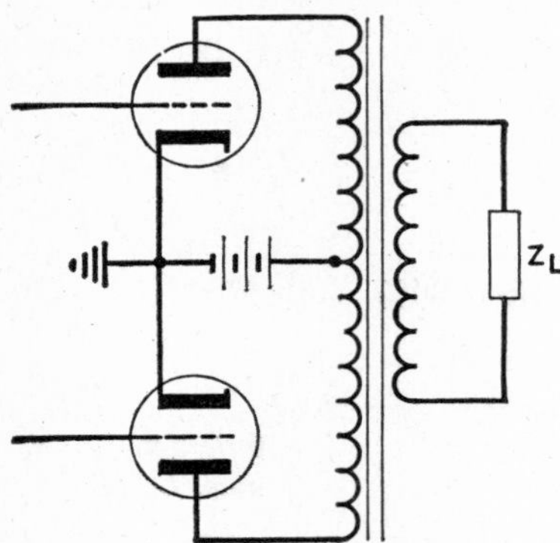

Fig. 16·5 Push-pull transformer-coupled amplifier. (From *Electronic Transformers and Circuits* by Reuben Lee.)

16·5 ANODE CHARACTERISTICS

Anode characteristics of vacuum tubes are commonly given in the form shown in Fig. 16·6, which applies specifically to the type 851 tube, a large air-cooled triode. Three common properties of a tube can be found from such characteristics: plate resistance r_p, amplification factor μ, and mutual conductance g_m. In Fig. 16·2 (*b*), plate resistance is the impedance Z_G, which represents the impedance of the source. It can be computed from the curves by measuring the change in plate voltage for a given change in plate current, with grid voltage constant. Amplification factor is the change in plate voltage for a given change in grid voltage, with plate current constant. Mutual conductance is the change of plate current for a given change in grid voltage, plate voltage being constant.

All three properties are measured under class A conditions at small a-c grid-voltage amplitudes. By measuring change in plate voltage for a given change in plate current in Fig. 16·6 the three properties are

$$r_p = \frac{\Delta e_p}{\Delta i_p} = \frac{1}{i - e \text{ slope}}$$

$$= \frac{1920 - 1770}{0.300 - 0.200} = 1500 \text{ ohms}$$

$$\mu = \frac{\Delta e'_p}{\Delta e_g} = \frac{1930 - 1530}{20} = 20$$

$$g_m = \frac{\Delta i'_p}{\Delta e_g} = \frac{0.510 - 0.200}{20}$$

$$= 0.0155 \text{ mhos} = 15{,}500 \text{ micromhos}$$

Two *e*'s and two *i*'s are involved. Which one to use for a given property must be determined by which of the three variables is held constant: grid voltage, plate current, or plate voltage. The values of r_p and μ agree closely with those given for the 851 tube in the manufacturer's handbook. The accuracy obtained here is the best that can be expected from vacuum-tube curves. Variations from one tube to another are much greater than the difference between calculated and published values.

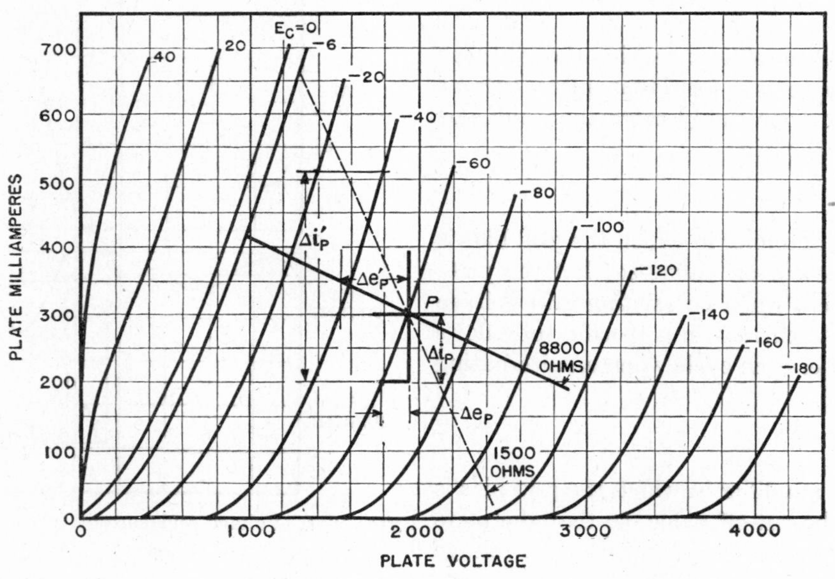

Fig. 16·6 Anode characteristics of 851 tubes. (From *Electronic Transformers and Circuits* by Reuben Lee.)

In a class A amplifier the static, or no-signal, condition of the 851 is $E_b = 1920$ volts, $E_c = -60$ volts, and $I_b = 300$ milliamperes. These values establish the operating point *P* in Fig. 16·6. A load of 8800 ohms is used for the first example. A maximum grid swing of 54 volts is assumed; that is, the grid swings from −6 to −114 volts on alternate half cycles. The 8800-ohm load line can be constructed in Fig. 16·6 as follows: 100 milliamperes plate-current swing times 8800 ohms gives 880 volts plate-voltage swing. Adding 100 milliamperes to 300 gives 400 plate milliamperes. Subtracting 880 volts from 1920 gives 1040 instantaneous plate voltage corresponding to the 100 milliamperes increase in plate current. This gives one point on the load line. Now a straight line is drawn between this point and the static point *P*. The ends of the load line are determined by the peak grid swings of −6 and −114. If now a sinusoidal grid voltage is assumed, plate current and plate voltage are sinusoidal also; that is, all points lie on the straight line representing an 8800-ohm load. This calculation may be checked by Table 16·2.

TABLE 16·2 851 TRIODE OPERATION WITH LOAD LINE = 8800 OHMS

E_b = 1920 volts Peak grid-voltage swing = 54 volts = e_g
E_c = −60 volts I_c = 0.300 ampere

θ (deg)	sin θ	e_g sin θ	e_c	e_b	i_b
0	0	0	−60	1920	0.300
30	0.5	27	−33	1440	0.355
60	0.866	47	−13	1090	0.395
90	1.0	54	−6	960	0.410
120	0.866	47	−13	1090	0.395
150	0.5	27	−33	1440	0.355
180	0	0	−60	1920	0.300
210	−0.5	−27	−87	2400	0.245
240	−0.866	−47	−107	2750	0.205
270	−1.0	−54	−114	2880	0.190
300	−0.866	−47	−107	2750	0.205
330	−0.5	−27	−87	2400	0.245
360	−0	0	−60	1920	0.300

Equal grid-voltage swings give equal plate-current and plate-voltage swings on this load line, and therefore the tube acts as a linear impedance. For such a load line, the circuit of Fig. 16·2 (*b*) correctly represents amplifier performance. In spite of the large voltage swing, power output is low because of the small plate-current swing. Peak a-c output is 0.11 times 960, or 105 watts. Multiplying each of these peak values by 0.707 to obtain rms values gives 52.5 watts average output from the tube. This is far below its capabilities. Power input is 300 milliamperes times 1920 volts, or 576 watts. Efficiency therefore is only 9.1 percent.

Next consider a 1500-ohm load line. Using the same peak-to-peak grid voltage, the positive voltage swing is 2400 − 1920 = 480 volts. Δi_p is 275 milliamperes, and $\Delta e_p \Delta i_p$ = 132 watts peak output. Negative Δe_p = 1920 − 1300 = 620 volts, Δi_p = 330 milliamperes, and $\Delta e_p \Delta i_p$ = 205 watts peak output. Average output then is (205 + 132)/4 = 84 watts, and efficiency is 14.6 percent. But the large difference in positive and negative plate-voltage swings causes harmonic distortion. Distortion is less with a 3000-ohm load or $2r_p$. This is the load generally given for triodes, and is known as the maximum undistorted power-output (UPO) load.

Suppose the tube of Fig. 16·6 were biased at −114 volts, the plate voltage raised to 2400, and the signal voltage doubled to 108 volts. Using the 1500-ohm load line, the positive plate current peak is 665 milliamperes and Δe_p = 1070. This gives a peak power of 710 watts or an average of 355. This power is for only one half cycle, because anode current is cut off in the other half cycle but, if for the second half cycle another tube is operating in the push-pull connection, then average power for the whole cycle will be 355 watts. Average current in one tube will be $665/\pi$ = 212 milliamperes, and in the two tubes 424 milliamperes. This is a power input of 1020 watts, so efficiency is 35 percent.

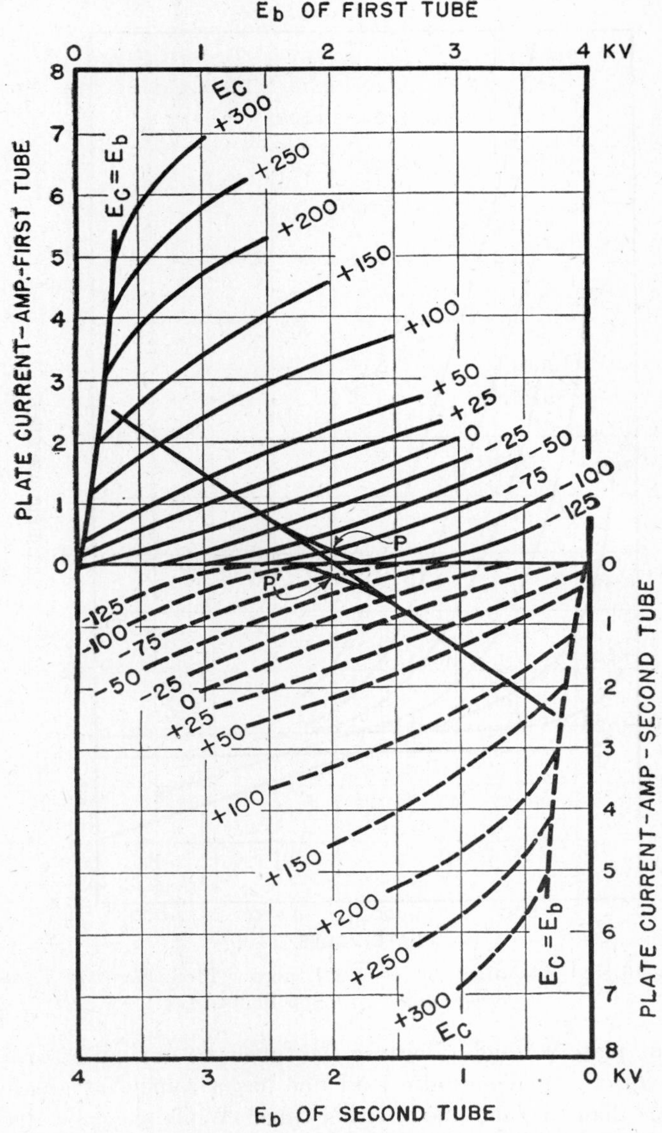

FIG. 16·7 Class B push-pull operation of 851 tubes. (From *Electronic Transformers and Circuits* by Reuben Lee.)

This efficiency and increased power input are still below the capabilities of the two tubes in push-pull class B, but they indicate how the efficiency of a pair of tubes can be increased by departing from class A conditions.

The turns ratio for a given load resistance is different for class A and class B amplifiers. For example, if Z_L in Fig. 16·5 is 1500 ohms, and this is to be reflected as 1500 ohms in the plate circuit of each tube in a class A amplifier, then the turns ratio is 1.41 step-down. This ratio changes the impedance to 3000 ohms for the whole primary winding. Since both tubes are working all the time, this is the same as 1500 ohms in each plate circuit. In a class B amplifier, only one tube works at a time. The turns ratio in this example from one half of the primary to the whole of the secondary is 1 to 1. The turns ratio for the whole primary to the secondary is 2 to 1.

Operation of class B amplifiers is easier to understand from inverted plate characteristics, plotted as in Fig. 16·7. A wider range of plate current is used. Plate voltage may swing as low as the "diode" line ($E_c = E_b$). Operating points P and P' are on the same vertical line, which represents the same plate voltage. Because the tubes are not quite biased to cut-off, the composite load line has a somewhat greater slope than the load line of either tube. While one tube gives up the load and the other accepts it, there is a short period each half cycle during which both tubes are active. The slope of the load line of the first tube decreases, and that of the second tube increases until it assumes the whole load. During this period tubes change from class B to class A operation and vice versa.

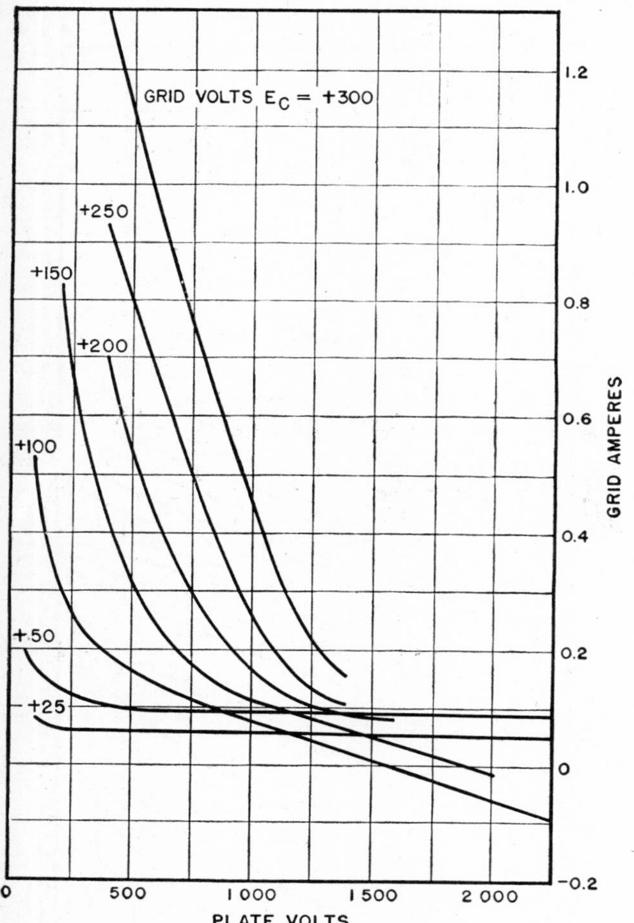

Fig. 16·8 Grid-current curves of 851 tubes. (From *Electronic Transformers and Circuits* by Reuben Lee.)

Plate-current swing is from zero to 2.5 amperes for each tube, and plate voltage swing is 1700 volts. Power output is 2.5 × (1700/2) = 2100 watts. Power input is approximately 2 × (2.5/π) amperes, or 1.59 amperes × 2000 volts = 3180 watts. Efficiency is 2100/3180 or 66 percent.

In a class C amplifier load lines like those for class B or class A operation are not feasible. The load changes from infinity at zero plate current to a low value at E_{min} in Fig. 16·4. On the assumption that the tank circuit maintains sinusoidal alternating plate voltage, and the alternating component of grid voltage is also sinusoidal, instantaneous plate currents can be determined from the plate-current characteristics, in a manner similar to that given for class A amplifiers. The same procedure can be used for finding grid current and grid power. See Table 16·3.

Table 16·3 851 Triode Class C Amplifier

E_b = 2000 volts e_g = 425 volts peak
E_c = −200 volts Data taken from Figs. 16·7 and 16·8.

θ (deg)	sin θ	$e_g \times$ sin θ	e_c	e_p	e_b	i_b	$e_p i_b$	i_g
0	0	0	−200	0	2,000	0	0	0
10	0.174	73	−127	−278	1,722	0	0	0
20	0.342	145	−55	−547	1,453	0.10	55	0
30	0.500	213	13	−800	1,200	0.85	680	0.02
40	0.643	273	73	−1,028	972	1.85	1,900	0.10
50	0.766	326	126	−1,225	775	2.60	3,180	0.14
60	0.866	368	168	−1,385	615	3.20	4,430	0.30
70	0.940	400	200	−1,505	495	3.85	5,790	0.55
80	0.985	419	219	−1,575	425	4.00	6,300	0.76
90	1.0	425	225	−1,600	400	4.00	6,400	0.80
						20.45	28,735	2.67

During the next 90 degrees the same figures may be put down in reverse order, but for negative e_g the output is zero. The table contains 10 ordinates, so averages are the sums divided by 20.

$$\frac{20.45}{20} = 1.023 \text{ amperes} = I_b$$

$$\text{Power input} = E_b I_b = 2{,}000 \times 1.023 = 2{,}045 \text{ watts}$$

$$\frac{28{,}735}{20} = 1{,}436 \text{ average watts output}$$

$$\text{Plate dissipation} = 2{,}045 - 1{,}436 = 609 \text{ watts}$$

$$\text{Plate efficiency} = \frac{1{,}436}{2{,}045} = 70.4 \text{ percent}$$

$$\text{Excitation power} = e_g\,(\text{max})\, I_g = 425 \times 0.134 = 57 \text{ watts}$$

Calculations for class C amplifiers are simplified by the use of constant-current curves, such as those mentioned in Chapter 5. Under the conditions of constant plate current, grid and plate a-c components of voltage are related linearly, so a straight line can be drawn to connect these two quantities on the constant-current characteristics. Instantaneous values of plate voltage and current and of grid voltage and current still must be tabulated, though, to determine output, input, and efficiency.[2] The value for E_{min} in Fig. 16·4 must be chosen carefully. The smaller E_{min}, the higher the plate efficiency, because virtually all the plate current is drawn at or near this plate voltage. However, if it is too low, the positive grid-voltage swing becomes higher than E_{min}, and the grid draws heavy current which subtracts from the plate current, thus causing a saddle in the top of the plate-current wave, as shown in Fig. 16·4. This not only decreases plate efficiency, but also increases the required grid excitation, which in turn requires more power from the driver. Ordinarily, E_{min} is from 10 to 25 percent of E_b.

In plate-modulated class C amplifiers, excitation must be large enough to cause grid saturation at 100 percent modulation; otherwise output would not be proportional to plate voltage, and the modulator would not modulate linearly.

16·6 TUBE CAPACITANCE

Capacitance between anode and cathode, grid and cathode, and grid and anode may be large enough in triode amplifiers to interfere with normal operation at radio frequencies, and in high-gain amplifiers at much lower frequencies. The input capacitance of a triode is given by this equation:[3]

$$C_{\text{input}} = C_{G\text{-}F} + (\alpha + 1)C_{G\text{-}P} \tag{16·2}$$

where $C_{G\text{-}F}$ = grid-to-cathode capacitance
$C_{G\text{-}P}$ = grid-to-anode capacitance
α = voltage gain of the stage.

Grid-to-anode capacitance in a stage with high voltage gain is likely to cause a large effective input capacitance, which may lower the impedance of the exciting amplifier and so decrease the overall gain, and may cause regeneration at high frequencies. In Fig. 16·9 the alternating component of voltage at point A is opposite in phase to that of the grid. Capacitance between anode and grid may be high enough to allow a large current and thus cause unwanted oscillations at frequencies determined by tube capacitance and other circuit elements. In tuned amplifiers of this kind, the grid-to-anode capacitance is neutralized by the addition of a variable capacitor C'. The tank coil has extra turns B-C in which a voltage is induced opposite in phase to that across A-B. With the proper setting of C' the grid can be adjusted to the same alternating potential as point B on the coil. Since point B is effectively by-passed for the operating frequency, feedback voltages cancel at the grid, and regeneration is prevented. C' is set by disconnecting anode supply E_b from the tube and applying excitation on the grid. If a sensitive radio-frequency ammeter is loosely coupled to the tank coil, and C' is varied until the ammeter reads zero, neutralization is correct.

The tetrode was developed with an additional grid between anode and control grid to reduce grid-to-anode capacitance. This additional grid is known as the *screen grid*, and it is operated at a positive potential with a-c by-pass to reduce grid-to-anode capacitance. The chief disadvantage of a screen-grid power tube is that anode-voltage swing is limited to the difference between anode voltage and screen voltage. The disadvantage is overcome by the addition of a third grid known as the *suppressor*, which removes this limitation and allows larger anode-voltage swings down to the diode line of the tube. Sometimes the third electrode is connected internally to the cathode. Similar characteristics are obtained with so-called *beam tubes*, which are tetrodes with special screen-grid spacings.

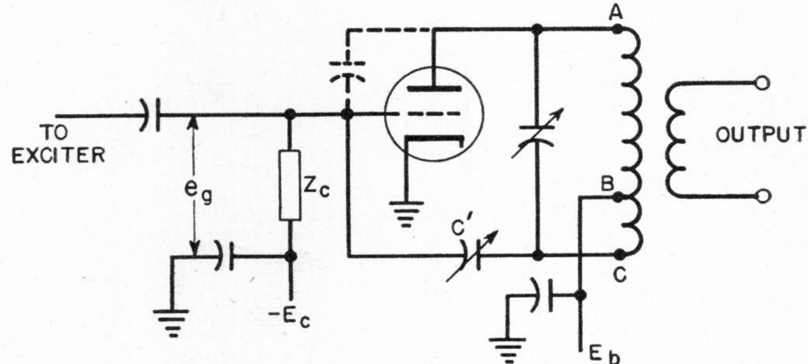

Fig. 16·9 Amplifier neutralization.

Figure 16·10 shows 6L6 beam-tube plate characteristics, with a typical load line of 2500 ohms. As a single-side amplifier, such a tube is likely to produce distortion because of the uneven spacing of constant-grid-voltage lines. Distortion is reduced in a push-pull amplifier, especially at high power output. Plate resistance r_p is high in pentodes and beam tubes—about 10 times the load resistance.

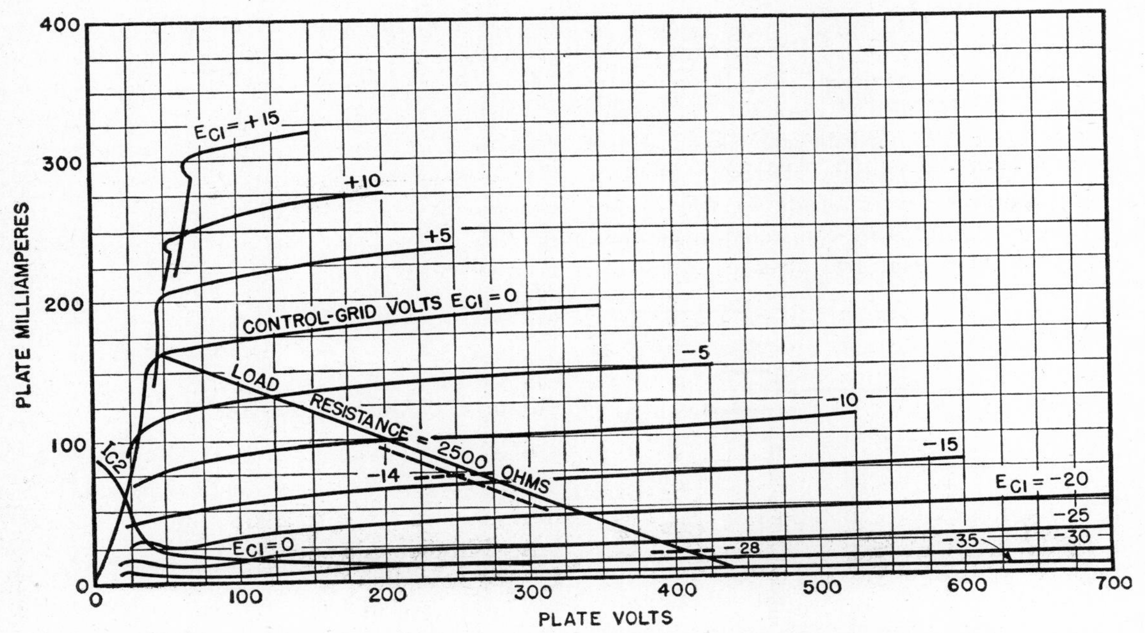

Fig. 16·10 Plate characteristics of 6L6 tubes. (From *Electronic Transformers and Circuits* by Reuben Lee.)

At very high radio frequencies, tube dimensions become too small for screen and suppressor grids, and triodes are used in grounded-grid circuits like that of Fig. 16·11. Excitation is applied in the cathode circuit, and at high frequencies this is a transmission line. With this circuit no neutralization is required, for the grid is already at ground potential. The plate efficiency apparently may exceed 100 percent because part of the output is fed directly from the input circuit.

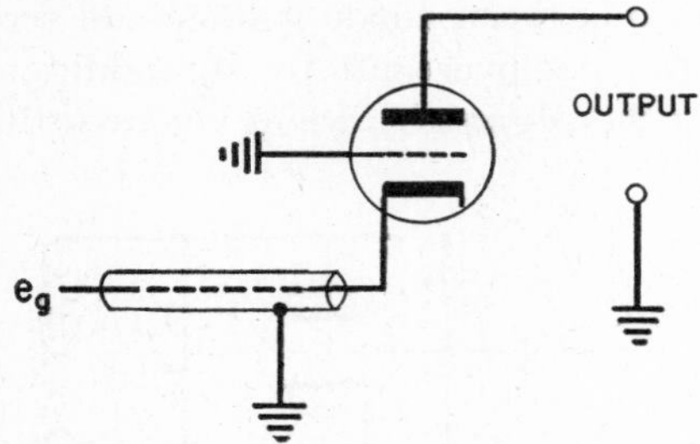

FIG. 16·11 Grounded-grid amplifier.

16·7 DIRECT COUPLING

In addition to the transformer coupling and resistance-capacitance coupling, direct coupling from one stage to another is used. Sometimes this is done inside the envelopes of small tubes. Figure 16·12 shows the internal connection of a 6N6 direct-coupled power-amplifier tube, in which the right-hand triode drives the left-hand triode. Direct-coupled amplifiers are used for amplifying direct currents, which cannot be amplified in circuits with transformers or capacitors. Stability in d-c amplifiers is attained only by careful control of the various grid and plate potentials,[4] because a change in them has the effect of a change in amplification. Direct-coupled amplifiers are used occasionally in a-c amplifiers at low frequencies to avoid the use of large capacitors or transformers.

16·8 AMPLIFIER STABILITY

In the construction of an amplifier, precautions must be observed to eliminate improper operation. Local or parasitic oscillations may appear at the natural resonant frequencies of circuit elements, connections, and tube electrodes. These oscillations can be detected and eliminated. Some of them can be avoided at the outset if short leads are used, especially in the grid circuit. If tubes are connected in parallel, resistors are used in the plate and grid leads to damp the

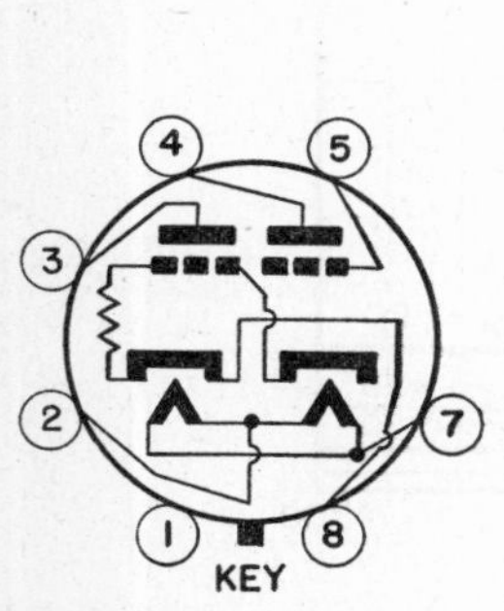

FIG. 16·12 Internal connection of a 6N6 direct-coupled amplifier tube.

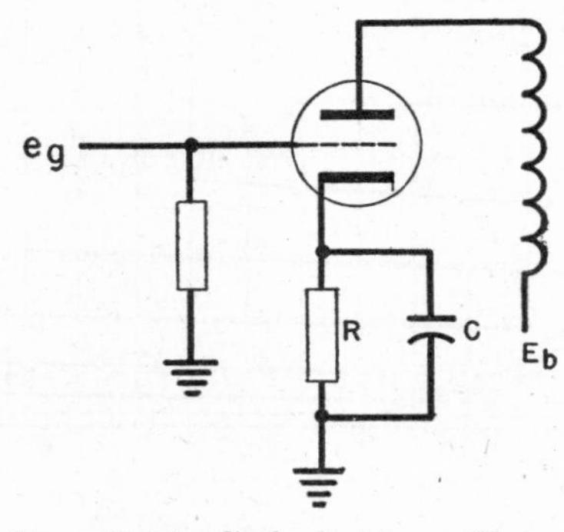

FIG. 16·13 Cathode bias. (From *Electronic Transformers and Circuits* by Reuben Lee.)

oscillations. Coils in circuits with widely different voltages are coupled loosely to prevent regeneration. In high-voltage circuits and circuits carrying large currents, shielding is used to prevent stray pick-up from one stage to another. If some circuit element in a push-pull amplifier is unbalanced, it may cause a push-push oscillation which can be eliminated by better balance, or by decoupling the tube plates at the unwanted frequency. Insufficient by-pass capacitance on plate or bias supplies may cause interstage coupling at low frequencies. The frequencies may be less than one cycle per second. This kind of instability is known as *motor-boating*.

If tubes are operated so that some electrode becomes a negative resistance during part of the cycle, oscillations may result which cannot be prevented except by avoiding the cause, or by some power-absorbing circuit which does not affect normal operation. Detection and elimination of this type of trouble require skill and judgment.

16·9 GRID BIAS

Grid-bias potential may be obtained from a separate bias rectifier, battery, or generator through a grid leak as in Fig. 16·9. This is called fixed bias. A class C amplifier grid can also be biased by its own rectifier action. Alternating voltage e_g from the exciter causes direct current in Z_C.

Polarity of the rectified voltage is such that the grid end of impedance Z_C is negative with respect to the other end. Sometimes combinations of fixed and grid-leak bias are used, especially in modulated class C amplifiers, because the two types of biasing arrangements produce opposing effects which cancel distortion of the modulation envelope.[2]

Bias can be obtained from a plate supply of the potentiometer type by connecting the potentiometer tap to ground, with one end of the potentiometer for plate voltage and the other end for grid voltage. The chief disadvantage is that part of the potentiometer must carry plate current. If the potentiometer resistance can be made low enough, this type of operation is practicable and saves a separate bias supply.

Cathode bias is shown in Fig. 16·13. Plate current through resistor R makes the cathode positive with respect to the grid and has the same effect as a negative voltage on the grid. R normally is shunted by capacitor C to by-pass variations in plate current and thus provide steady bias. This method is used widely with class A amplifiers in which there is no grid current and the bias cannot be developed across a grid leak.

16·10 INVERSE FEEDBACK

If part of the output is fed back to the input in such a way as to oppose it, the ripple, distortion, and frequency-response deviations in the output are reduced. Gain is reduced also, but with high-gain tubes an extra stage or two compensates for the reduction in gain caused by inverse feedback, and the improvement in performance usually justifies the extra stage or stages. In the amplifier of Fig. 16·14, a tapped resistor is shown across output voltage E_0, and a part of this output is fed back so that the input to the amplifier is

$$E_2 = E_1 - BE_0 \qquad (16\cdot3)$$

B is the portion of E_0 which is fed back. If α is the voltage amplification, if E_R and E_H are the ripple and distortion in the output without feedback, and if α', E'_R, and E'_H are the same properties with feedback, the following equations express E_0, when it is assumed that α, E_R, and E_H are independent:

Without feedback,

$$E_0 = \alpha E_2 + E_R + E_H \quad (16\cdot4)$$

With feedback,

$$E_0 = \alpha' E_1 + E'_R + E'_H \quad (16\cdot5)$$

From these equations it can be shown that [5]

$$\alpha' \approx \frac{1}{B} \quad (16\cdot6)$$

$$E'_R \approx \frac{E_R}{\alpha B} \quad (16\cdot7)$$

$$E'_H \approx \frac{E'_H}{\alpha B} \quad (16\cdot8)$$

With high-gain amplifiers and large amounts of feedback, the output ripple and harmonic distortion can be made

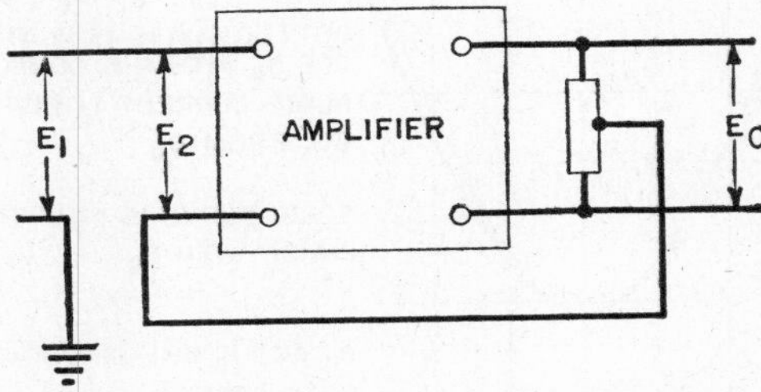

FIG. 16·14 Voltage feedback. (From *Electronic Transformers and Circuits* by Reuben Lee.)

small. Also, frequency response can be made flat, even with mediocre components. Inverse feedback is not used in class C amplifiers, because output and input are not linearly related.

Distributed capacitance and leakage inductance must be carefully matched in the inverse-feedback circuit so that the phase shift around the loop does not become 180 degrees. If it should come near this value, the inverse feedback turns into regenerative feedback and the amplifier becomes an oscillator at a frequency determined by the circuit constants.

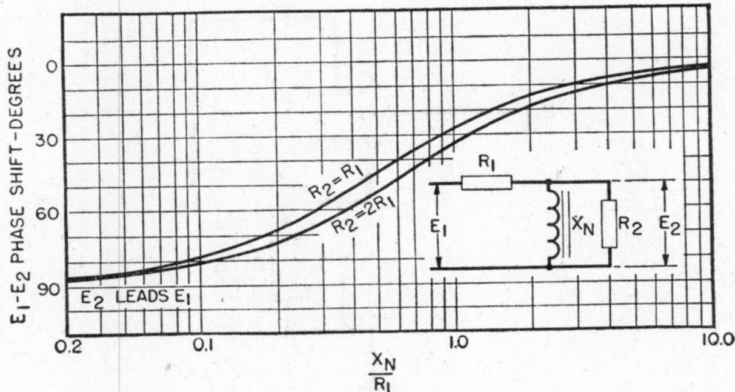

FIG. 16·15 Transformer-coupled amplifier low-frequency phase shift. (From *Electronic Transformers and Circuits* by Reuben Lee.)

To correct for distortion over a frequency range from 30 to 10,000 cycles, the amplifier should have low phase shift over a much wider range, say from 10 to 30,000 cycles. In the frequency interval from 10 to 30 cycles and from 10,000 to 30,000 cycles, both the amplifier and feedback circuits should taper off gradually to prevent oscillation.[6] Phase shift in transformer-coupled amplifiers is shown in Figs. 16·15 and 16·16 at low and high frequencies respectively. At high frequencies, 180-degree phase shift is possible; at low frequencies only 90 degrees is possible. In a resistance-coupled amplifier, phase shift is only 90 degrees at either low or high frequencies. Partly for this reason, partly because there is less capacitance in resistors than in transformers and good response is maintained at higher frequencies, it is in resistance-coupled amplifiers that inverse feedback is usually employed. However, if the distortion of a final stage is to be reduced, transformer coupling is involved. It is preferable to derive feedback voltage from the primary side of the output transformer. This is equivalent to tapping between R_1 and X_L in Fig. 16·16, where the phase shift is much less. The transformer must still present a load of fairly high impedance to

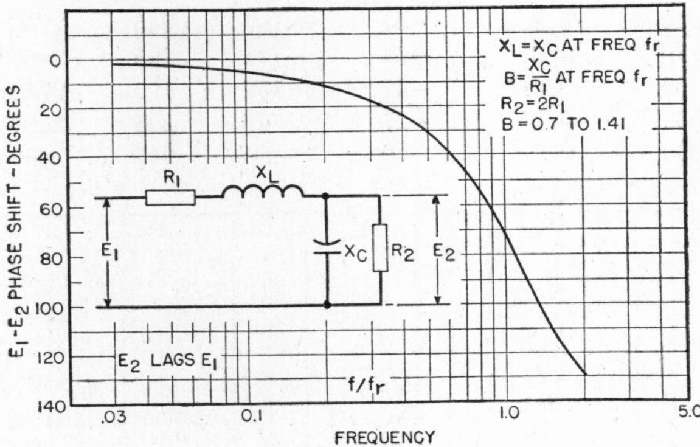

FIG. 16·16 Transformer-coupled amplifier high-frequency phase shift. (From *Electronic Transformers and Circuits* by Reuben Lee.)

the output tube throughout the marginal frequency intervals to permit gradual decrease of both amplification and feedback.

Current feedback is produced in the circuit of Fig. 16·13 by removing capacitor C. This introduces degeneration in the cathode-resistor circuit, and thus accomplishes the same thing as the bucking action of voltage feedback. It is also less affected by phase shift, and consequently is used in transformer-coupled amplifiers.

16·11 DISTORTION

The effect of low load impedance on distortion has already been mentioned, and so has the reduction of this distortion by means of feedback. Distortion is much greater with pentode or beam-tube amplifiers because of their non-uniform anode characteristics. See Fig. 16·10. The effect of a reactive load also is greater; it can be found for any amplifier by shifting plate current with respect to grid voltage for any angle of lag or lead. Table 16·4 gives a set of values for a load with a 30-degree phase angle. The load line becomes an ellipse, as shown in Fig. 16·17, and the wave form is distorted somewhat as in Fig. 16·18, which is typical of lightly loaded triodes. Distortion with heavier loads or with pentodes is much greater.

With a reactive load, efficiency and output are reduced, and plate dissipation is increased. In Fig. 16·17, the area of the ellipse represents power which diminishes the tube output and adds to plate dissipation. High open-circuit reactance in the transformer is necessary to prevent these undesirable effects.

TABLE 16·4 REACTIVE LOAD VOLTAGES AND CURRENTS

E_b = 1920 volts
Peak grid swing = 54 volts
E_c = −60 volts I_b = 0.300 ampere
Assume plate current lags e_g by 30 degrees.

θ	$\sin\theta$	$e_g \sin\theta$	e_c	i_p	e_p
0	0	0	−60	245	1850
30	0.5	27	−33	300	1400
60	0.866	47	−13	355	1080
90	1.0	54	−6	395	960
120	0.866	47	−13	410	1150
150	0.5	27	−33	395	1520
180	0	0	−60	355	2000
210	−0.5	−27	−87	300	2460
240	−0.866	−47	−107	245	2790
270	−1.0	−54	−114	205	2880
300	−0.866	−47	−107	190	2720
330	−0.5	−27	−87	205	2350
360	0	0	−60	245	1850

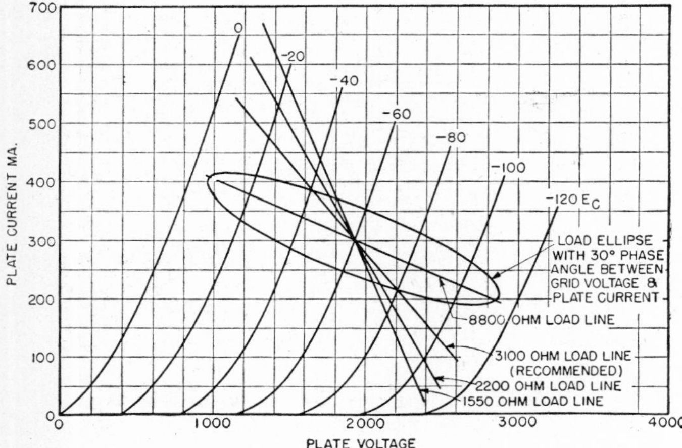

FIG. 16·17 851 triode with reactive load. (From *Electronic Transformers and Circuits* by Reuben Lee.)

If the amplifier is transformer coupled, the third harmonic in the magnetizing current causes distortion also. It can be found from the following formula: [10]

$$\frac{E_H}{E_f} = \frac{I_H R}{I_f X_N}\left(1 - \frac{R}{4X_N}\right) \tag{16·9}$$

where E_H = harmonic voltage amplitude
E_f = fundamental voltage amplitude
I_H = harmonic current
I_f = fundamental current
$R = R_1R_2/(R_1 + R_2)$.

R_1, R_2, and X_N are as shown in Fig. 16·15. Table 16·5 gives typical third- and fifth-harmonic currents for silicon-steel laminations, from which the corresponding distortion in equation 16·9 can be calculated. Since the fifth harmonic adds in quadrature to the third, it is safe to say that the major component of distortion in magnetizing current is third harmonic.

The effect of modulation on the distortion in tuned-output circuits is discussed in Chapter 12.

Harmonics in the output may be analyzed by the usual Fourier method or by simplifications of it.[11]

TABLE 16·5 HARMONIC COMPONENTS IN TYPICAL SILICON-STEEL MAGNETIZING CURRENT WITH ZERO-IMPEDANCE SOURCE

B_m (gauss)	Percentage of 3rd Harmonic	Percentage of 5th Harmonic
100	4	1
500	7	1.5
1,000	9	2.0
3,000	15	2.5
5,000	20	3.0
10,000	30	5.0

FIG. 16·18 Plate voltage wave forms with zero and 30° phase angle. (From *Electronic Transformers and Circuits* by Reuben Lee.)

16·12 CATHODE FOLLOWER

A special case of current feedback occurs in the circuit of Fig. 16·19, which is known as the cathode-follower circuit.[7] The anode is connected to the high-voltage supply E_b without any intervening impedance, so that for alternating currents it is virtually grounded. Grid voltage e_g must be large

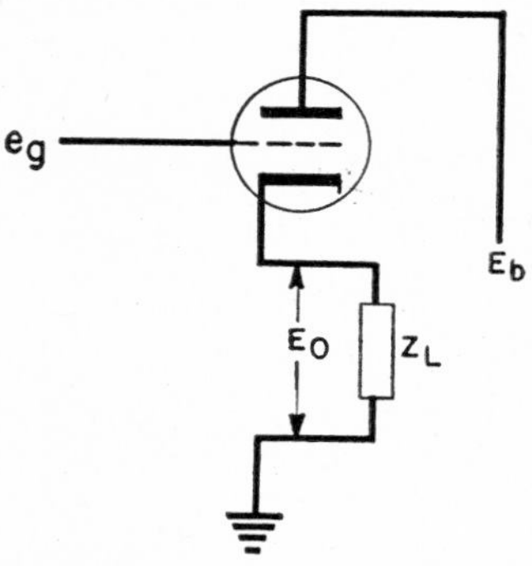

FIG. 16·19 Cathode follower. (From *Electronic Transformers and Circuits* by Reuben Lee.)

enough to include the output E_0 plus normal peak grid-to-cathode voltage. However, grid power is still the same as it would be if the cathode were grounded. This circuit is used if the output impedance Z_L is variable or of low power factor, so that full output from the tube would be difficult to produce in it. It has a low internal effective impedance as far as the output is concerned. It is approximately equal to the normal plate resistance r_p divided by the amplification factor of the tube. This is equivalent to saying that the effective internal impedance is approximately the reciprocal of the mutual conductance g_m for class A or class B amplifiers. The cathode follower has been used to drive grids in class B modulator tubes, which are highly variable loads. It produces nearly constant voltage output in such loads, and reduces distortion, but does not reduce plate dissipation with reactive loads.

16·13 MODULATION

Modulation is of two main types: *amplitude modulation* and *frequency modulation.*[8]

A carrier-frequency wave, the amplitude of which is varied or modulated at audio-frequency rates, has a shape similar to that shown in Fig. 16·20. The first few carrier-frequency cycles are shown unmodulated; the amplitude is then constant, and for this portion of the figure the wave, if it is of pure sine form, can be expressed by

$$e = E \sin \omega_c t \tag{16·10}$$

where E is the peak amplitude and ω_c is 2π times the carrier frequency. The modulated wave varies in amplitude above and below E. If this variation also is sinusoidal, the equation for the modulated wave is

$$e = E \sin \omega_c t(1 + M \sin \omega_a t) \tag{16·11}$$

ω_a being 2π times the modulating frequency, and M being the percentage of modulation or amplitude variation. This equation can be transformed into

$$e = E \sin \omega_c t + \frac{ME}{2} \cos (\omega_c - \omega_a)t - \frac{ME}{2} \cos (\omega_c + \omega_a)t \tag{16·12}$$

The right-hand terms signify, in order, the waves of carrier, lower-sideband, and upper-sideband frequencies.

Amplitude modulation usually is accomplished in one of two ways. In high-level modulation, the final carrier-amplifier plate voltage is connected with a modulation choke or transformer between the high-voltage plate supply and the power-amplifier input. In low-level modulation, some low-power driver stage is plate-modulated, and succeeding stages amplify the modulated wave to the desired output level. In all stages after the modulated stage, the amplifiers are operated class B with tuned-plate circuits. With zero modulation, the plate-voltage swing and efficiency are low, about 30 percent. At 100 percent modulation, efficiency increases to 60 percent because of the larger grid- and plate-voltage swings. Amplifier plate current remains almost constant. This system is still in use in low-power amplifiers, especially those for intermittent operation. Figure 16·21 shows a low-level modulator-amplifier chain.

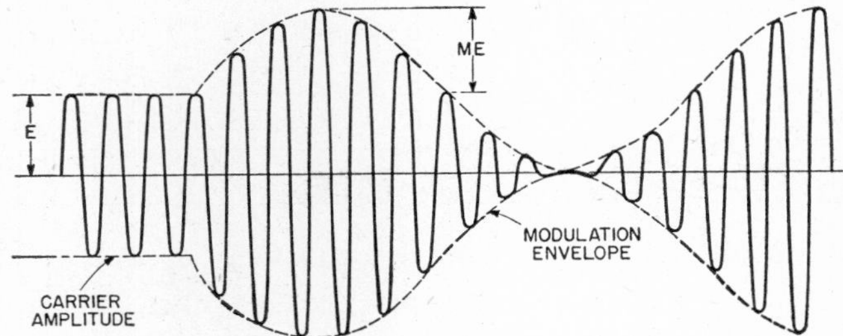

FIG. 16·20 Amplitude-modulated wave.

In high-level modulation, power-amplifier efficiency is high at all times. Modulator efficiency increases with percentage modulation as in Fig. 16·22. The main disadvantage is the large size of the high-power audio components. However, it is widely used in large amplifiers because of the power saving in continuous operation. Figure 16·23 is a diagram of an audio amplifier and modulator. The modulator has an output of 800 watts and is suitable for modulating a power-amplifier input of 1600 watts. The upper connection from the modulator output could be made to the lower end of the tank coil in Fig. 16·3. This amplifier has several

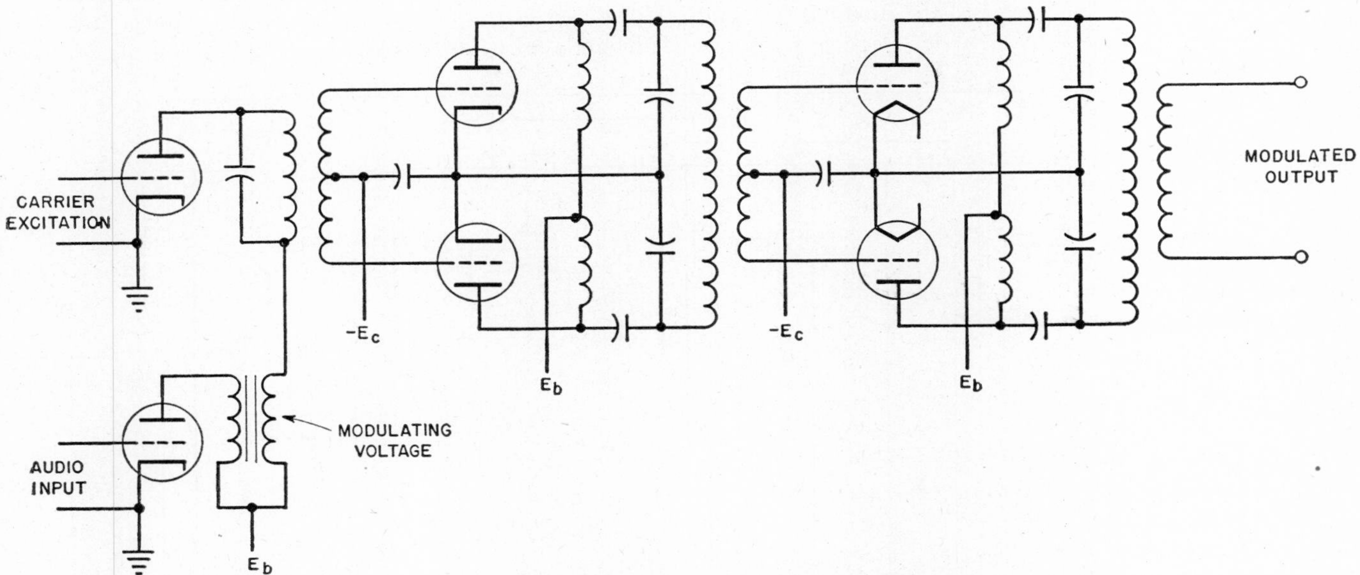

FIG. 16·21 Low-level modulation with class B linear carrier-frequency amplifier.

interesting features. Inverse feedback is applied from the plates of the modulators to the cathodes of the first amplifier tubes. All stages are resistance-coupled except the output.

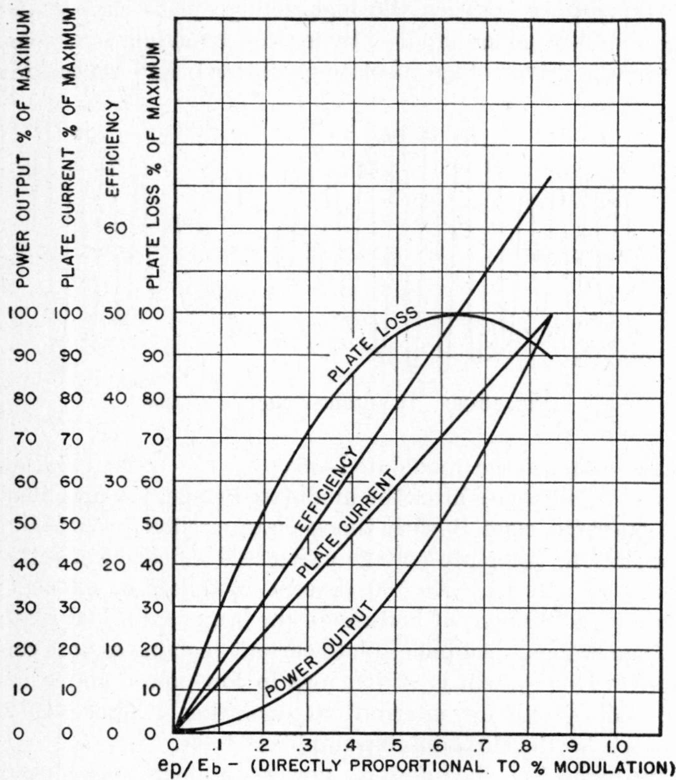

FIG. 16·22 Modulator properties vary with percentage modulation.

Current feedback is used in the first two stages, and a cathode follower is used to couple the driver to the modulator. The output transformer is shunted by a carrier-frequency capacitor C_1, and the choke *RFC* in the power-amplifier input lead is of carrier-frequency proportions also. These elements are added to obtain more uniform output impedance at high audio frequencies, the benefits of which are discussed in Chapter 13. Separate plate meters are provided for the tubes in the last two stages. Grid resistors are used for parasitic suppression in the last three stages.

Frequency modulation is produced by varying the frequency of the carrier-frequency oscillator in proportion to the amount of audio voltage applied, and at a rate proportional to the audio frequency. A frequency-modulated wave has constant amplitude but variable frequency like that shown in Fig. 16·24. It can be expressed by the following equation:

$$e = E \sin (\omega_c t + m_f \sin \omega_f t) \qquad (16\cdot 13)$$

where E = peak amplitude of carrier wave

$\omega_c = 2\pi \times$ carrier frequency

$\omega_f = 2\pi \times$ modulating frequency

$$m_f = \frac{\text{variation of carrier frequency}}{\text{modulating frequency}}$$

= modulation index.

For high fidelity this wave is composed of many frequency components of appreciable amplitude which require large sidebands compared with those of the amplitude-modulation system. For example, in broadcast frequency-modulated transmitters, the carrier frequency is about 100 megacycles and the sidebands are 75 kilocycles wide. The frequency-modulation system is therefore applicable mainly to very-high-frequency transmission.[9]

The useful part of a modulated wave is in the sidebands. If the carrier frequency is suppressed entirely, and only the sidebands are transmitted, it is still possible to transmit all

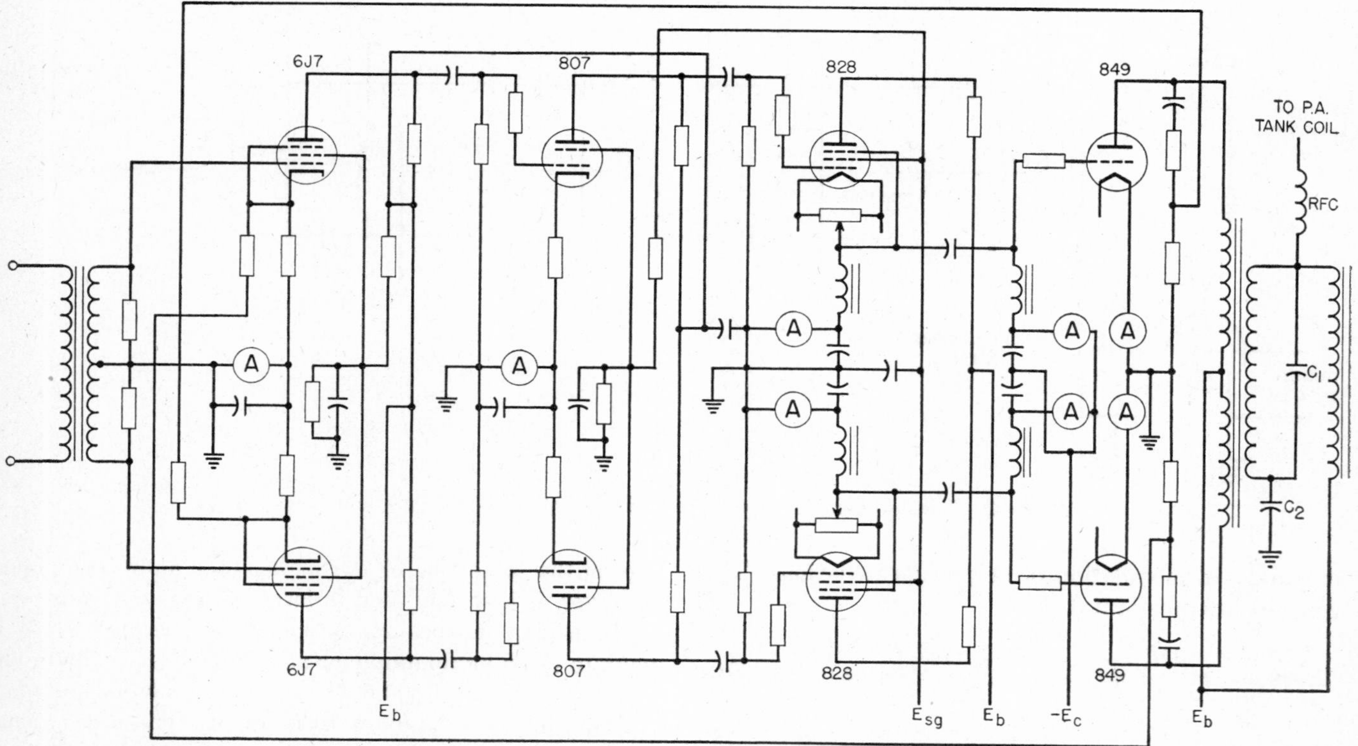

FIG. 16·23 Audio amplifier for high-level modulator. (From *Electronic Transformers and Circuits* by Reuben Lee.)

the communication, signal, or control for which modulation is used. If the upper sideband is separated from the lower, and only one sideband is transmitted, it is called single-sideband transmission. Amplification of single-sideband output results in higher power from the same output tubes for a given amount of intelligence transmitted.[9] Carrier voltage may be suppressed as shown in Fig. 16·25.

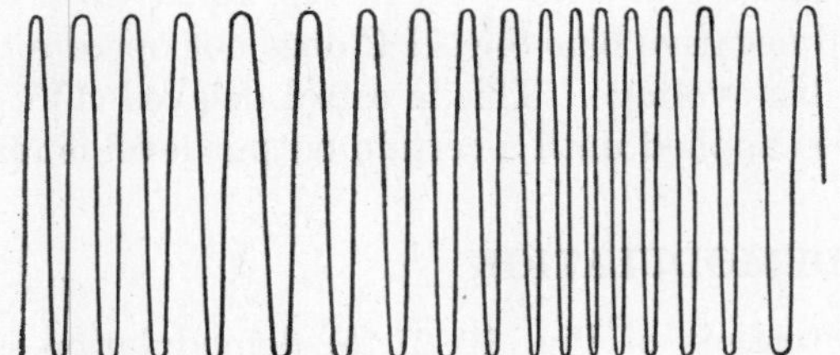

Fig. 16·24 Frequency-modulated wave.

A modulator using rectifying elements can be made as shown in Fig. 16·26. Such modulators ordinarily are employed at low to medium radio frequencies and at low power level; amplitude is increased by successive amplification. Rectifiers often are of the metal types. The carrier is suppressed in the circuit of Fig. 16·26, and double-sideband output is obtained. Single-sideband output can be obtained with special connections.[14]

If the audio or modulating frequency is close to the carrier frequency, tuned circuits do not respond properly unless they are tuned broadly. Distortion caused by power-amplifier output circuits is a function of the ratio of modulating frequency to carrier frequency. In an inductively coupled amplifier with tuning on both primary and secondary, the tuning curve is made flat on the top throughout the sideband range for good audio response. These subjects are discussed in Chapter 12.

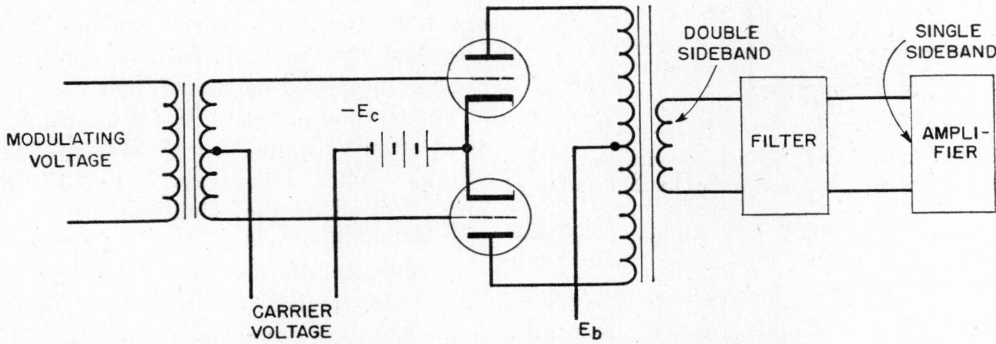

Fig. 16·25 Suppressed carrier and single-sideband output.

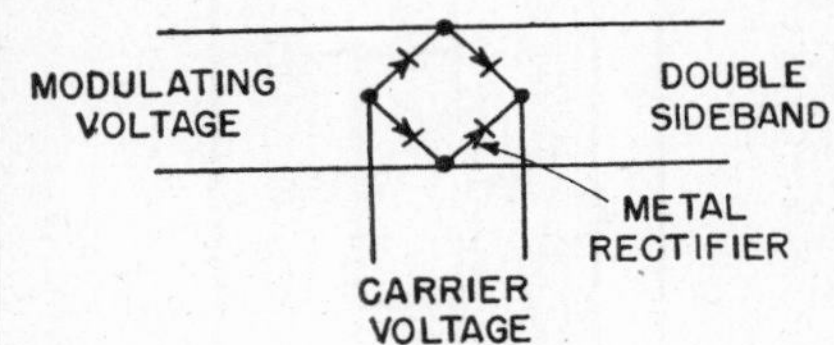

Fig. 16·26 Modulation with metal rectifier.

It is possible to amplitude-modulate by applying modulating voltage to the screen grid of a tetrode. The tube is operated at full output with no modulation. Modulation then produces a decrease in average power, because there is no plate-voltage increase with modulation and no increase in power output on modulation peaks. The same is true if the suppressor grid of a pentode is modulated. Both systems require relatively little power from the modulator, but they decrease the average power when modulation is applied. They are valuable if power requirements with modulation are small in comparison with the unmodulated carrier power.

16·14 AUTOMATIC GAIN CONTROL

Vacuum-tube amplification factor is not constant under all conditions of operation. For example, in the vicinity of the operating point P in Fig. 16·6, the amplification factor is constant. If the data are expanded to take into account high-current operation as in Fig. 16·7, the amplification factor in the region of high anode current and low anode voltage is no longer constant.

Some tubes are designed with large variations in amplification factor. These are known as variable-mu, remote cut-off, or super-control tubes. Their mutual conductance is highly variable with grid bias. Figure 16·27 (*a*) is the curve of mutual conductance for a tube of this kind. Such a characteristic can be used to reduce gain at high amplitudes and thus prevent overmodulation in audio systems. Figure 16·27 (*b*) shows a circuit that reduces gain automatically for excessive values of applied voltage e_g on the grid of the 6SK7 tube. This tube drives a 6L6 beam-power tube, the output of which is delivered through transformer T_1. On this transformer is an auxiliary winding S_2, which is connected to rectifier tube 6H6-1 and produces the rectified output across resistor R_2, having a negative potential at the point shown. With large signals the voltage rectified across R_2 is large, and it reduces the mutual conductance and plate-voltage swing of the 6SK7 tube. Nearly constant voltage is maintained in the 6L6 output.

If the power output of the 6L6 tube is delivered mainly into a linear a-c impedance, the slight additional load imposed by the gain control makes little difference. But, if all the output is delivered to rectifier loads, as it is in Fig. 16·27 (*b*), the non-linearity of both tube and load cause output distortion. This is true particularly of beam or pentode output tubes. The normal class A output of a 6L6 beam tube is 6 watts but, if output power is all rectified, only 50 milliwatts can be drawn without excessive distortion. Half-wave rectifiers and capacitor-input filter outputs are worst in this respect, because of the current discontinuities.

If the rectifier input of the automatic gain control is taken from a tuned amplifier, these difficulties decrease. The tuned-circuit capacitor readily supplies irregular current wave forms, provided the amplifier has sufficient power output available.

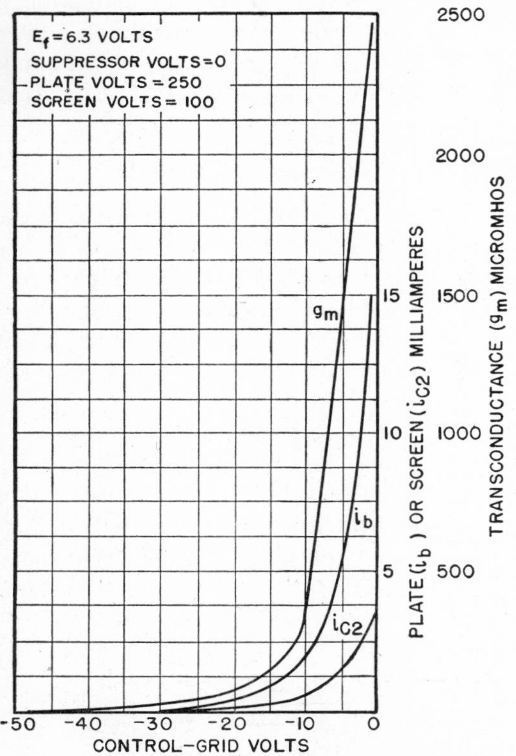

FIG. 16·27 (*a*) Variable-mu tube 6SK7 mutual conductance characteristics. (From *Electronic Transformers and Circuits* by Reuben Lee.)

Automatic volume control (*AVC*) in receivers is applied to either the radio-frequency or audio stages to maintain approximately constant volume in spite of fading or other causes of input-voltage variations. It is employed in audio amplifiers to maintain more uniform output volume with differing voice levels. In some amplifiers more than one stage is controlled, and the *AVC* action is amplified.

If the input grid resistor R_1 in Fig. 16·27 (*b*) is connected to a fixed negative bias the *AVC* does not operate below the value of bias voltage. This is called delayed *AVC*; with it, no *AVC* is applied until a certain output level is reached.

16·15 DEMODULATION

In the circuit of Fig. 16·27 (*b*) demodulation is accomplished by means of a diode 6H6-2. Each half of the radio-frequency cycle is rectified and filtered by L_1 and C_1, and the d-c power is absorbed in resistor R_3. Audio power is by-passed around R_3 by capacitor C_2, and audio voltage is impressed upon the primary of transformer T_2. If an amplitude-modulated wave is used in this amplifier, voltage e_g has the wave form of Fig. 16·20, and the output voltage across L_1 has the form shown in Fig. 16·28. The first few cycles are shown as full-wave rectified loops having a constant amplitude, that is, with no modulation. Audio output for this section of the wave is zero. Sinusoidal 100 percent modulation is shown in the rest of the figure. Average voltage left after the carrier-frequency half loops have been absorbed by the radio-frequency filter L_1C_1 is the audio voltage impressed on transformer T_2.

The method just described is known as *diode demodulation*. It is often accomplished by means of a single diode, with alternate lobes of the wave in Fig. 16·28 omitted. Triodes also are used for demodulation, with some amplification of the demodulated wave.

Similar circuits are used in power-line carrier receivers. The carrier frequency is from 50 to 150 kilocycles, and audio frequencies are employed for modulation. Transformer T_1 not only operates over the carrier-frequency range, but also

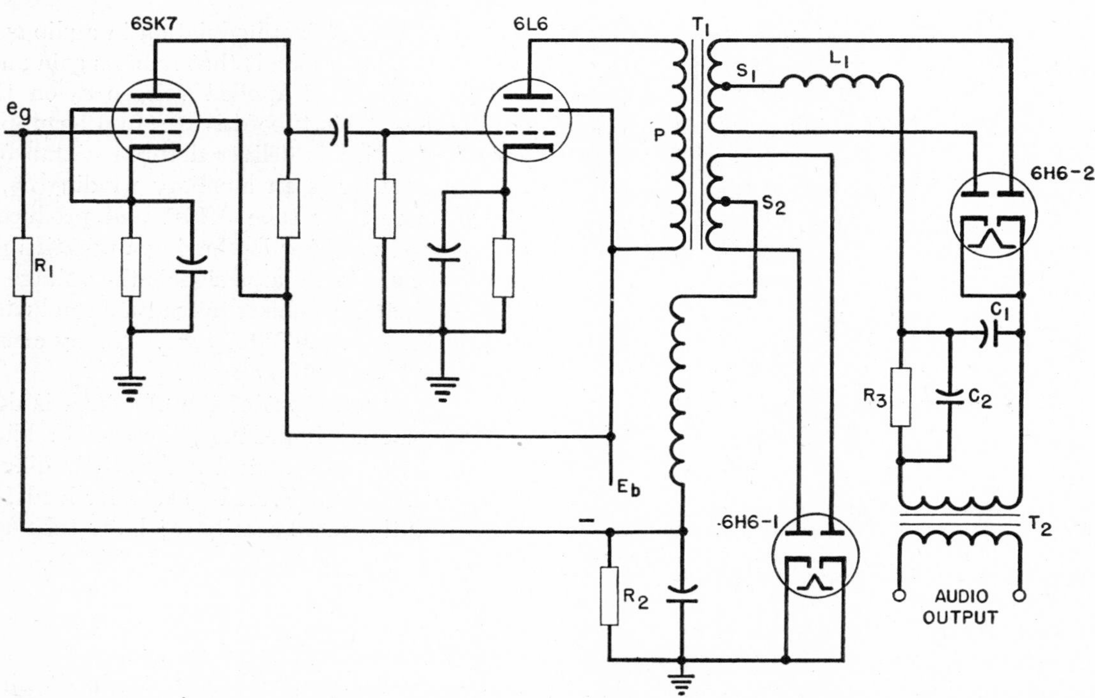

FIG. 16·27 (*b*) Automatic gain control and detector circuit. (From *Electronic Transformers and Circuits* by Reuben Lee.)

must deliver the correct amount of voltage to the automatic-gain-control tube 6H6-1 for proper *AVC* action, and the required output to the audio load without distortion. It must obtain these voltages from the nearly constant-current

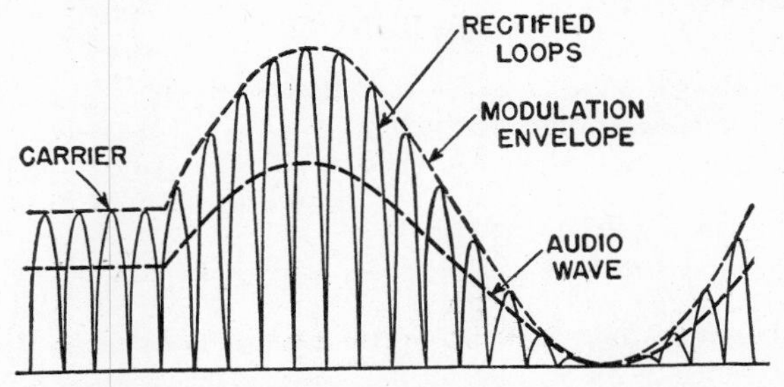

FIG. 16·28 Rectified amplitude-modulated wave. (From *Electronic Transformers and Circuits* by Reuben Lee.)

6L6 tube. The transformer ratio is determined by estimating the voltage swing produced by a square input-current wave; then dividing this by the voltage required to produce the necessary audio output after choke L_1 smooths the rectified lobes to the average values shown by the heavy dotted lines in Fig. 16·28.

Frequency-modulated waves can be demodulated by several means. One of the simplest is a mistuned pair of coupled resonant circuits. If the carrier frequency is on one of the "skirts," a change in frequency changes output-voltage amplitude, and subsequent amplification proceeds as in an amplitude-modulated receiver. This method is objectionable because the circuit is responsive to changes in signal amplitude and therefore to noise. If the reason for using frequency modulation is to obtain a large signal-to-noise ratio, this method is not used.

The circuit of Fig. 16·29 (*a*) can produce demodulation which is responsive to frequency-modulated waves but not

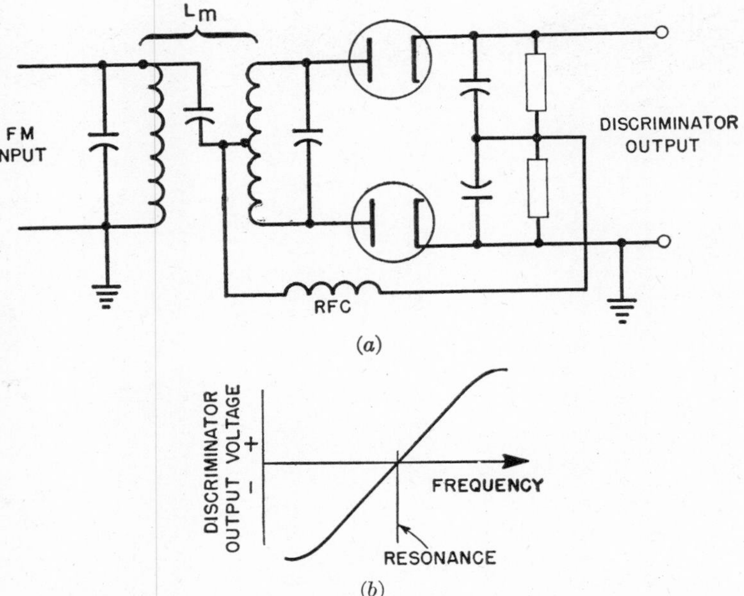

FIG. 16·29 F-m demodulator circuit and response.

to amplitude modulation. At resonant frequency each diode receives from the tuned secondary coil equal voltages which cancel in the output. At other frequencies the coil voltages are still equal, but are displaced in phase with respect to the input voltage. The diode voltages are then unequal, and the frequency-response curve is like that in Fig. 16·29 (*b*).

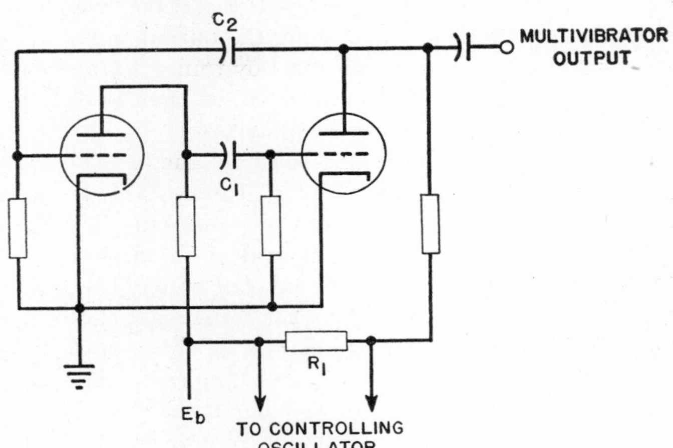

FIG. 16·30 Multivibrator circuit.

16·16 FREQUENCY MULTIPLICATION AND DIVISION

In Fig. 16·4 the saddle-backed current wave i_b in class C amplifiers was shown. If the grid is driven still harder, the

FIG. 16·31 Modulator for broadcast transmitter.

saddle may become very deep. If the tank circuit then is tuned to the second harmonic of the grid-circuit frequency, the output frequency is double the input frequency. Such a circuit is called a *frequency doubler*. Efficiency is less than it would be if the frequency were not doubled, because the plate-voltage and plate-current swings are less. Bias must be increased because of the increased excitation. Similarly, frequency may be tripled by tuning the output tank circuit to the third harmonic, or quadrupled by tuning to the fourth harmonic. Efficiency of the tripler is less than that of the doubler, and it is still less for the quadrupler.

Frequency division is accomplished by the means of the multivibrator [12] circuit shown in Fig. 16·30. This is really a resistance-coupled amplifier in which the output voltage from the plate is fed back to the grid of the first tube by means of capacitor C_2. This circuit can oscillate by itself, but the stability is poor and the wave form is approximately triangular. If sufficient voltage is injected across resistor R_1 from a controlling oscillator of stable frequency, the multivibrator can be made to lock in with it. If the injected voltage is decreased somewhat, the multivibrator output jumps to half the frequency of the injected voltage. Lowering the injected voltage still further increases the sub-multiple or ratio between the injected frequency and that of the multivibrator output. Division of frequency in a single multivibrator usually is not over 10 to 1. Further frequency division can be obtained by using one multivibrator output as the injected voltage for a second multivibrator, and so on. By this means two waves of widely different frequency can be compared. Since each multivibrator has a non-sinusoidal output, the output harmonic content is high, and frequency multiplication of any vibrator in the chain is possible. In one frequency-measuring system an oscillator operates at 100 kilocycles with an extremely stable frequency control. Its output is used to control a multivibrator delivering 10 kilocycles, and the output of the multivibrator then is multiplied to obtain 10-kilocycle intervals in the broadcast-frequency range for measuring broadcast-station operating frequencies.

REFERENCES

1. *Radiotron Designer's Handbook*, F. Langford-Smith, R.C.A. Mfg. Co., 1941, Parts 1 and 2.
2. "Analysis of the Operation of Vacuum Tubes as Class C Amplifiers," I. E. Mouromtseff and H. N. Kozanowski, *Proc. I.R.E.*, Vol. 23, July 1935, p. 752.
3. *Principles of Radio Communication*, J. H. Morecroft, Wiley, 1927, 2nd ed., p. 511.
4. "A High-Gain D-C Amplifier for Bio-Electric Recording," H. Goldberg, *Trans. A.I.E.E.*, Vol. 59, Jan. 1940, p. 60.
5. "Stabilized Feedback Amplifiers," H. S. Black, *Bell System Tech. J.*, Vol. 13, Jan. 1934, p. 1.
6. *Network Analysis and Feedback Amplifier Design*, H. W. Bode, Van Nostrand, 1945, Chapters XVI–XIX.
7. "Feedback," E. K. Sandeman, *Wireless Eng.*, Vol. XVII, Aug. 1940, p. 351.
8. *Communication Engineering*, W. L. Everitt, McGraw-Hill, 1937, 2nd ed., Chapters XV, XVI, XVII.
9. "A Comparison of the Amplitude Modulation, Frequency Modulation and Single Sideband Systems for Power-Line Carrier Transmission," R. C. Cheek, *Trans. A.I.E.E.*, Vol. 64, May 1945, p. 215.
10. "Harmonic Distortion in Audio Frequency Transformers," Norman Partridge, *Wireless Eng.*, Vol. 19, Sept., Oct., and Nov., 1942.
11. "Graphical Harmonic Analysis," J. A. Hutcheson, *Electronics*, Vol. 9, Jan. 1936, p. 16.
12. "The Adjustment of the Multivibrator for Frequency Division," V. J. Andrew, *Proc. I.R.E.*, Vol. 19, Nov. 1931, p. 1911.
13. *American Standard Definitions of Electrical Terms*, published by A.I.E.E., 1941, p. 234.
14. "A Simple Single-Sideband Carrier System," R. C. Cheek, *Westinghouse Eng.*, Vol. 5, Nov. 1945, p. 179.

Chapter 17

CIRCUITS FOR OSCILLATORS

E. U. Condon and William Altar

An electronic oscillator is a device employing electronic tubes and associated circuits fed from a power source, usually a constant-potential supply, and delivering a-c energy to a load. Many different forms of oscillators have been developed to take care of the varying performance requirements.

MAGNITUDE OF POWER OUTPUT

Power requirements range from a small fraction of a watt, as for laboratory testing oscillators or for local oscillators in heterodyne radio receivers, to power outputs of many kilowatts, as in radio-broadcast transmitters or in oscillators for induction heating.

MAGNITUDE OF FREQUENCY GENERATED

Electronic oscillators can be made to operate at extremely low frequencies (as low as a cycle per minute!) but there is seldom a need for these. For frequencies below about 10,000 cycles per second, electronic oscillators are widely used for low power applications (less than a hundred watts), for instance for measurement purposes. These are often called signal generators. For power applications, rotating machines (alternators) are generally more economical in this frequency range. Above 10 kilocycles per second to about 30 megacycles per second is the extended radio-frequency spectrum embracing the power applications of radio broadcasting and many industrial uses. Above 30 megacycles per second to about 300 megacycles per second is the so-called ultrahigh-frequency (uhf) region of the spectrum. This region is characterized by the increasing use of circuit elements with distributed constants, such as resonant transmission lines. Still higher frequencies having wave lengths less than 1 meter are referred to as microwaves. This region requires rather special techniques, which make use of such devices as cavity resonators, wave guides, and velocity-modulated electronic beams.

FREQUENCY STABILITY

Frequency stability is of great importance for radio broadcasting and for measurement work. In certain other applications, as in dielectric heating, it is desirable to have the frequency of the oscillator shift in accordance with the changing frequency characteristics of the load.

MODE OF OPERATION

A natural basis for classifying electronic oscillators is afforded by the operating principle peculiar to each class.

Feedback Oscillators

This class has the widest field of application, and accordingly receives most attention in this chapter. Feedback oscillators are essentially self-excited amplifiers. An amplifier takes a low voltage and current input and feeds a higher voltage, higher current output to a load. In a feedback oscillator this load contains a tank circuit which is sharply resonant to the desired frequency and also a working load, such as the antenna feed line in a radio transmitter, or the work in an induction heater. Coupled to the load in one of a variety of ways is the feedback circuit which leads back to provide the amplifier input.

Negative-Resistance Oscillators

These oscillators make use of some electronic device in which current decreases as the voltage drop increases. The long-familiar singing arc, which is of this type, operates on the basis of a falling characteristic for an arc maintained between solid electrodes at ordinary air pressure. Some triodes and tetrodes can be made to exhibit a negative-resistance characteristic when connected and operated in a particular way. The performance of certain oscillators using split-anode magnetrons is also based on this principle. In all these, the negative resistance neutralizes both the positive resistance of the load and the losses in the oscillator circuits, and thus changes the damping of oscillations associated with a positive resistance to an exponential build-up process of oscillations. This build-up is limited only by the non-linear character of the circuit, which makes the negative-resistance value a function of the operating level.

Relaxation Oscillators

No sharp dividing line distinguishes this class from the others. The term usually refers to oscillators in which the electrical condition changes abruptly and cyclically from one unstable condition to another; for example, a capacitor being charged through a resistor and shunted with a glow tube. As the voltage across the capacitor reaches the ignition voltage of the glow tube, the latter breaks down abruptly and discharges the condenser until its voltage is low enough to permit the glow tube to recover, whereupon a new cycle sets in with charging current flowing again to the capacitor.

The wave form of relaxation oscillators is usually far from sinusoidal, and the frequency is usually determined by elements with little inductive reactance. The "poor" wave form may be turned to good advantage in an important application of relaxation oscillators for frequency multiplication. Since the content of the output is rich in harmonic frequencies, it is only necessary to control accurately the repetition rate of the cycle in order that a suitable harmonic frequency may be obtained for use in frequency measurements or control.

Electron-Transit-Time Oscillators

These oscillators make use of velocity-modulation tubes such as single- and double-cavity klystrons and resonant-cavity magnetrons. Since their importance is exclusively in microwave applications, they are discussed in Chapter 5.

The theory of oscillator design offers mathematical difficulties owing to the non-linear characteristics of the tube. Approximations which assume linearity of these characteristics are necessarily inaccurate. Since a more thorough mathematical analysis is laborious without adding to an understanding of the simple underlying ideas, it has been found preferable to carry out oscillator design by combining a roughly approximate analytical procedure with a more detailed study making use of a "bread-board" model for testing purposes.

17·1 GENERAL PRINCIPLES OF FEEDBACK OSCILLATORS

Every feedback amplifier may be schematically represented as in Fig. 17·1. The tubes are in the amplifier unit,

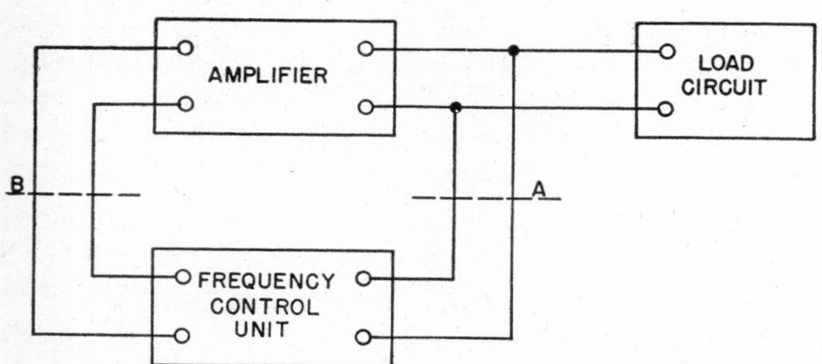

Fig. 17·1 Schematic block diagram of feedback amplifier.

the input terminals of which are at B. The amplifier output at A is divided between the load and the feedback circuit by which a fraction of the output is fed back through a frequency control unit to the amplifier input where it impresses input voltage. This general block diagram is more explicitly diagrammed for the simplest type of tuned-plate oscillator circuit (Fig. 17·2).

The exact analysis of any such oscillator circuit is too involved to be of much practical value. Usually, simplifying assumptions must be made which are too inaccurate to render useful quantitative information, but nevertheless help the general understanding by giving qualitative information.

The operating frequency and amplitude of an oscillator will adjust themselves so that the voltage output of the frequency control unit is exactly that needed at the input terminals of the amplifier in order to maintain the operating conditions unchanged. Voltages must be matched in both

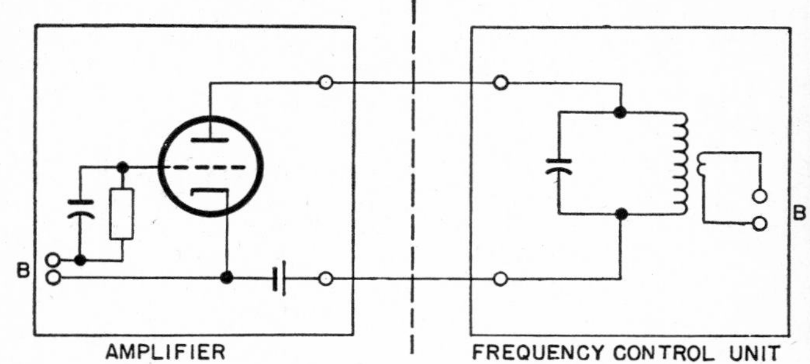

Fig. 17·2 Typical circuit of feedback amplifier.

amplitude and phase, so two requirements must be satisfied simultaneously. Horton [1] has given a useful graphical discussion of a typical case, on which the following is based.

Leaving wave form aside for the moment one assumes pure sine waves. The amplifier performance is characterized by a diagram as in Fig. 17·3 where the ratio of the input voltage V_B to the output voltage V_A is plotted against output voltage. The rise in the curve means that the tube does not amplify as well at higher output voltages and that the input voltage has to be raised more than in proportion to the desired increase of the output voltage. In general the curve

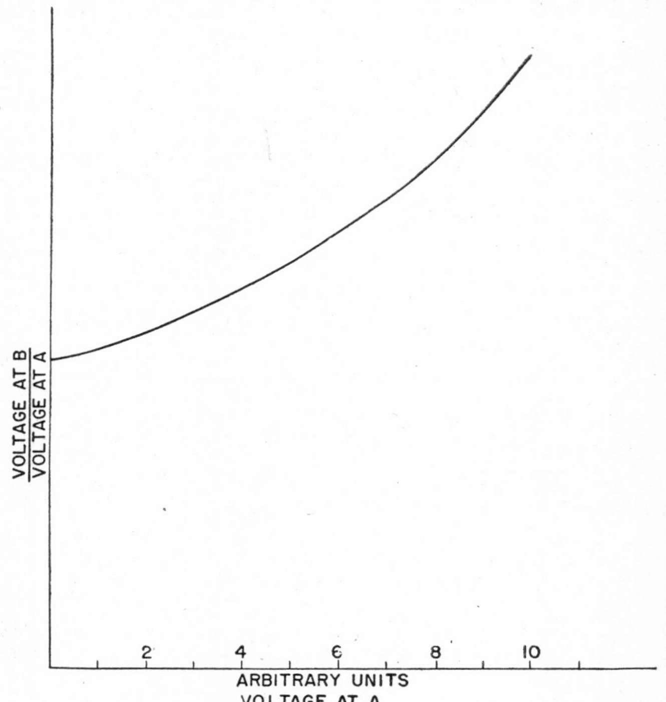

Fig. 17·3 Amplitude-response curve of amplifier. (Reprinted with permission from J. W. Horton, *Bell System Tech. J.*, Vol. 3, 1924.)

will shift somewhat with frequency so that a family of curves is needed to represent the amplifier gain under the various operating conditions. These curves may be re-plotted so that each curve shows the variation of the voltage ratio with frequency at a specified output voltage V_A, as shown in Fig. 17·5.

The frequency-control unit is a sharply resonant transmission system, and a plot of the voltage ratio V_B/V_A as a function of frequency is something like Fig. 17·4. In certain

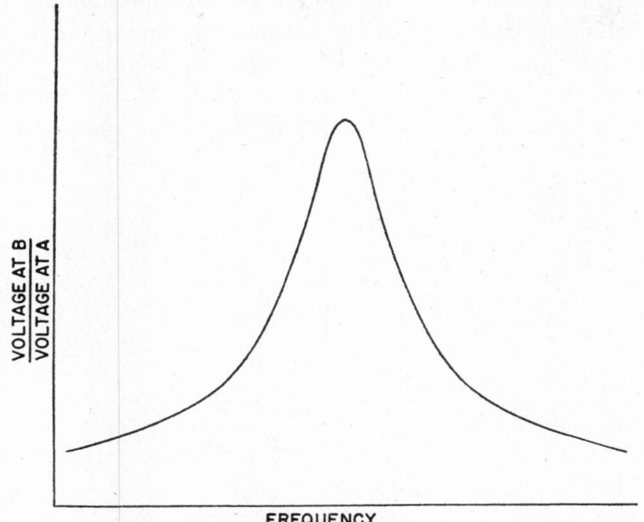

Fig. 17·4 Amplitude-response curve of frequency-control unit (high input level).

low-frequency applications employing inductors wound on iron cores, the effective resistance may increase with increasing voltage V_A. For such a circuit Fig. 17·4 should show a family of curves of the type shown, each for a different voltage V_A. These curves substantially merge into one at

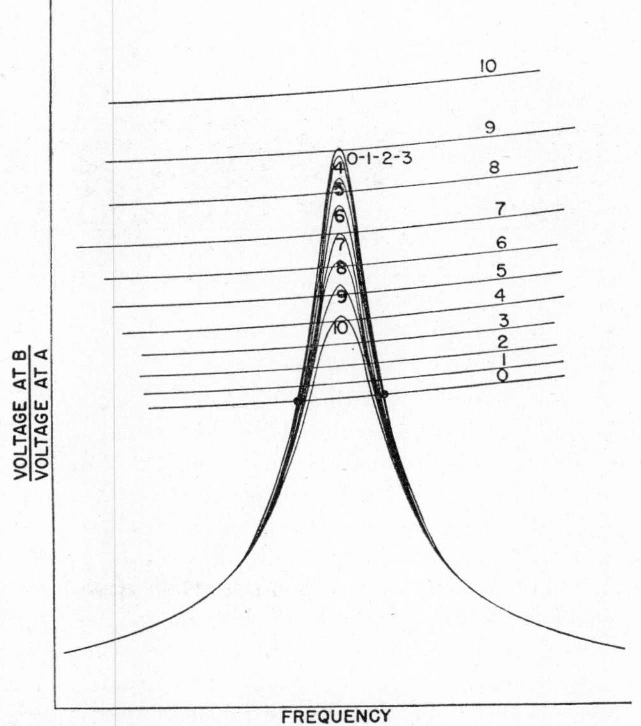

Fig. 17·5 Families of amplitude-response curves for amplifier and feedback circuits. (Reprinted with permission from J. W. Horton, *Bell System Tech. J.*, Vol. 3, 1924.)

frequencies far away from resonance but will show lower resonance peaks as the voltage V_A increases. This family of curves is also shown in Fig. 17·5 superposed on the family of amplifier response curves which are discussed in the preceding paragraph. Curves of each family are labeled in accordance with the V_A value, in arbitrary units. Consider the two characteristics labeled 0, indicating zero voltage at A. They intersect in two points marked with heavy dots, revealing two frequencies at which the voltage V_B matches around the feedback circuit in the limit of zero voltage at A. Similarly for $V_A = 1$ there are two possible frequencies, and so on. However, the curves for $V_A = 7$, as drawn in Fig. 17·5, are just tangent to each other, and thus give a single frequency, and at still higher voltages the corresponding curves no longer intersect. It is seen that the maximum voltage attainable at the oscillator output terminals can

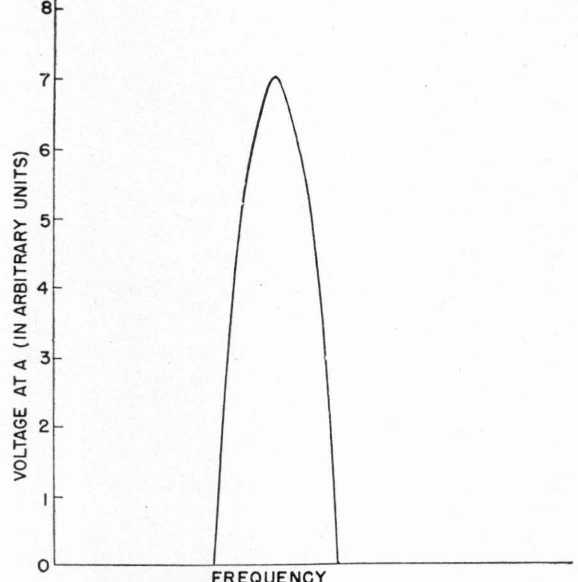

Fig. 17·6 Amplitude-balance curve for feedback oscillator.

at best reach 7 units, since the amplifier is unable at higher amplitudes to produce enough input voltage through the feedback circuit. Actually the oscillator will not necessarily operate at this voltage, though its operation must take place under conditions represented by one of its points of intersection between a pair of amplifier and feedback characteristics identically labeled. Thus is obtained the amplitude-balance curve of Fig. 17·6, representing the condition for matching voltage amplitudes in a diagram of output voltage versus frequency. As far as the amplitude match is concerned, any point of this curve would be acceptable as an operating point for the oscillator.

The correct operating point on the curve may be obtained by matching at B the phases of the feedback voltage and of the amplifier input voltage in complete analogy to the preceding consideration of the amplitudes. Figure 17·7 is the analogue for phases of Fig. 17·5 for amplitudes. As drawn, it indicates that the phase lag introduced by the amplifier is independent of voltage amplitudes, whereas the frequency-control unit shows a slight amplitude dependence in the sense which is consistent with the trend shown in Fig. 17·5, namely, greater proportionate loss at higher amplitudes. The intersection of corresponding pairs of curves again determines a relation between frequency and oscillator output voltage. This condition must be fulfilled in order that the feedback voltage be fed to the input terminals B in the correct phase. It is represented as a steep curve in Fig. 17·8, which also shows the amplitude balance curve of Fig. 17·6. It is seen that the phase balance curve rises vertically from zero and tilts toward the lower frequency side, in accordance with the sequence in which the feedback characteristics 0 to 10 of Fig. 17·7 are intersected by the amplifier characteristic. The operating point is that frequency and that voltage at A at which the amplitude-equality curve and the phase-equality curve intersect.

A quantitative evaluation of this graphical exposition will depend largely upon the selectivity of the feedback circuit,

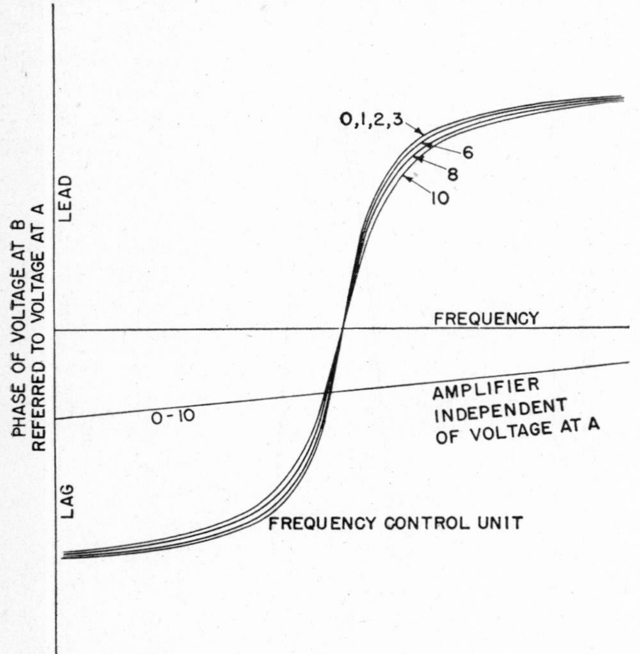

FIG. 17·7 Families of phase-response curves for amplifier and feedback circuits. (Reprinted with permission from J. W. Horton, *Bell System Tech. J.*, Vol. 3, 1924.)

that is, upon the so-called loaded Q of the frequency-control unit. By this is meant the circuit Q with the load attached. It is simply related to the half-power bandwidth, that is, to the frequency interval between half-power points which, on the voltage-ratio curve, represent about 71 percent of maximum voltage response. This bandwidth equals resonance frequency divided by the loaded Q of the circuit. Thus a convenient choice of units for the frequency scales laid out along the abscissas of Figs. 17·4 to 17·8 depends upon the Q value in each case. By opening up the scale in cases where the Q has a high value and crowding the frequency marks in other cases of low Q, one will end up with resonance curves of about identical appearance in the graphs. The amplifier characteristics generally show a much lesser dependance on frequency, and are nearly constant over the small frequency range when dealing with a sharply resonant feedback circuit. Nevertheless, it is this dependence, and the finite phase lag that is introduced by the amplifier, which make the resultant oscillator frequency differ slightly from the natural resonance frequency of the frequency-control unit.

That the oscillator is self-starting under the conditions assumed may be deduced from an inspection of Fig. 17·5 which shows that oscillations at a frequency near the resonance frequency of the feedback circuit can build up from a near-zero level. Small initial oscillations of frequencies covering the bandwidth of the unit invariably accompany the switching operation when the oscillator is turned on. Consider first an oscillation of a frequency belonging to one of the two heavily dotted intersection points of the curve pair labeled zero. In view of the amplitude balance existing for

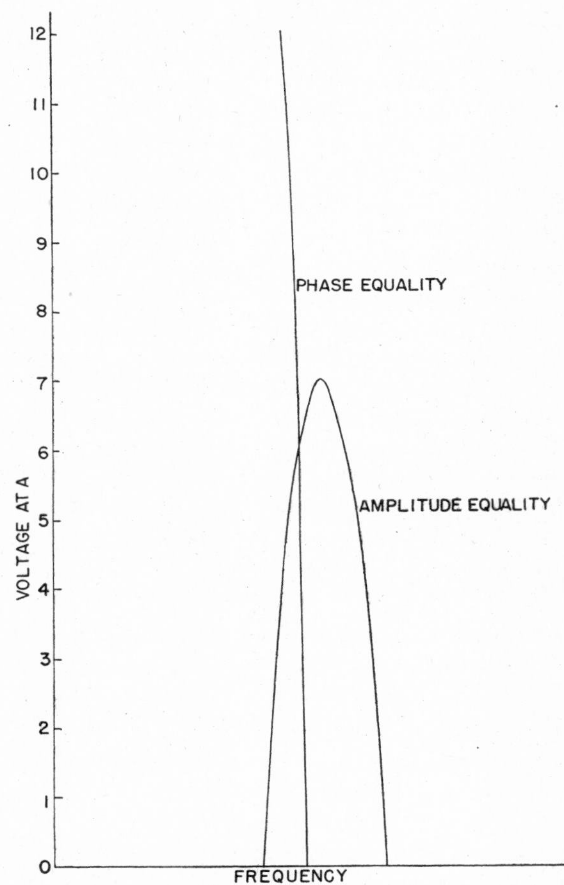

FIG. 17·8 Amplitude- and phase-balance curves for feedback oscillator. Intersection marks steady-state operation. (Reprinted with permission from J. W. Horton, *Bell System Tech. J.*, Vol. 3, 1924.)

this point the amplitude cannot build up to higher values, and in view of the phase unbalance its frequency cannot endure at its starting value; in other words, no build-up for oscillations of this frequency will occur. Next, examine a frequency near the resonance value of the feedback circuit. It will be seen that at the low amplitude level the gain factor V_A/V_B of the amplifier outweighs the attenuation factor of the feedback circuit by about 83 percent (or less for frequencies off the resonance point), so that the oscillation will build up rapidly. As its amplitude increases to a higher level, say 4, the factor which determines the build-up rate has been lowered from the initial 1.83 to 1.43. Finally at a level where the build-up curve intersects the amplitude-balance curve, this factor drops to unity, which marks the end of the build-up process. Actually the build-up does not take place at constant frequency, nor along the phase-equality curve as might be supposed. Rather, at every step during the build-up a relation between rate of frequency change and phase unbalance prevails which determines the process in detail.

The curves of Figs. 17·4 to 17·8 afford a convenient basis for discussing the effects of varied circuit parameters or load conditions. For example, suppose that the load is purely resistive and is increased, a change which has practically no effect on the phase-equality curve. The amplitude-equality curves of Figs. 17·6 and 17·8, on the other hand, will lose in height and in breadth, because the lowering of V_A attending the heavier load will tend to move the amplifier characteristics of Fig. 17·5 upward so that a pair of curves with lower label than before become tangent to each other. The breadth of the amplitude-equality curve at its base is also reduced. In view of the tilt of the phase-equality curve, the new intersection point lies at a lightly higher frequency and gives a lower operating voltage. As the load is increased further, oscillations will cease as soon as one of two things happen; either the amplitude-equality curve narrows down so that it no longer intersects the phase-equality curve, or the amplifier characteristics of Fig. 17·5 are raised to such an extent that even the pair of characteristic curves labeled zero no longer intersect. As a result of the first of these eventualities the oscillations will cease suddenly although they had existed at a finite power level for a slightly smaller load. The second process is characterized by a gradual dying-down to zero with increasing load.

17·2 CIRCUITS FOR FEEDBACK OSCILLATORS

Probably every conceivable combination of feedback elements has been tried at one time or another. Several arrangements have become well known and have been identified with names. Some of these circuits are given in Fig. 17·9. Although on casual inspection these circuits may look different, they have many points in common:

1. A parallel reactor-capacitor combination serves as the frequency-determining element. Occasionally, as in Fig. 17·9 (*g*), two such circuits tuned to the same frequency may be used.
2. A parallel resistor-capacitor combination connects the control grid to the rest of the circuit. This combination is roughly equivalent to the action of a C battery; its function is to make the grid negative with respect to the anode.
3. The control grid derives its radio-frequency potential from the anode circuit to which it is coupled by a feedback arrangement.
4. The main plate supply, indicated by the conventional battery symbol, is shunted by a capacitor of sufficient capacity to by-pass the plate supply so far as radio-frequency currents are concerned.

Power supplies for heating the cathode filaments are not indicated, and the load is not explicitly represented in the circuits. Loading may be accomplished in a variety of ways, usually by coupling either directly by tapping, or by mutual inductance, into the reactor of the basic *LC* combination of the tank circuit which determines the frequency.

In the design of oscillator circuits attention must be given to the steady components of current and voltage which in the various leads are superposed upon the radio-frequency currents and voltages. Frequently it is necessary to provide a path of low impedance for the radio currents between points which require different steady potential, or to keep radio currents from flowing through elements which have little resistance for steady currents. These two results can be accomplished by blocking capacitors and, respectively, by radio-frequency choke reactors. These elements add in a superficial way to the difference in appearance of the different types of feedback oscillators.

The seven types in Fig. 17·9 have the following advantages and disadvantages.

Hartley (a and b). This type requires but a single reactor with tapping. Series feed for the Hartley circuit has the disadvantage that the power supply terminals carry radio-frequency potential, and stray capacities from the power supply to ground will exercise a detuning effect. Also, the capacitance between cathode and power supply provides an undesired path for the radio-frequency currents, thus causing loss of radio-frequency power as well as detuning effects. In an improved version of the series-feed Hartley circuit a large capacitor is placed in series with the oscillating circuit coil next to the cathode connection, and the d-c plate supply is shunted around it. Shunt feed is the standard for Hartley circuits. It avoids the difficulties with the help of an extra choke.

Colpitts (c). The disadvantage of this type is that the ratio of grid voltage to plate voltage cannot be adjusted without disturbing the frequency adjustment.

Meissner (d). This type requires no tapping of the main tank reactor.

Tuned Plate (e). This is one of the most widely used types. Its disadvantage, that the tank circuit is at high potential, can be overcome by the use of a shunt feed as in (*b*).

Tuned Grid (f). This type oscillates only for a fairly narrow range of mutual inductance between tank circuit and plate circuit.

Tuned Plate—Tuned Grid (g). Feedback is accomplished through interelectrode capacitance in the tube. The feedback is relatively feeble so that a sharply resonant grid circuit is required to determine the operating frequency. In a widely used modification, this resonance is accomplished by placing a quartz crystal in the grid circuit and utilizing it as the frequency-determining element. To obtain oscillations, the plate circuit is tuned to a frequency higher than the characteristic crystal frequency, so that at the operating frequency the reactance in the plate circuit is inductive. This is referred to as the tuned grid-inductance loaded plate circuit.

17·3 DESIGN OF POWER OSCILLATORS

Feedback oscillators, like amplifiers, are classified as class A, class B, class C, according to the region of the characteristic curves in which the tube is operated. For most efficient power conversion, power oscillators are adjusted for class C operation, that is, with the grid biased sufficiently negative so that there is no plate current during a large portion of the cycle. Strictly, and particularly for low μ value

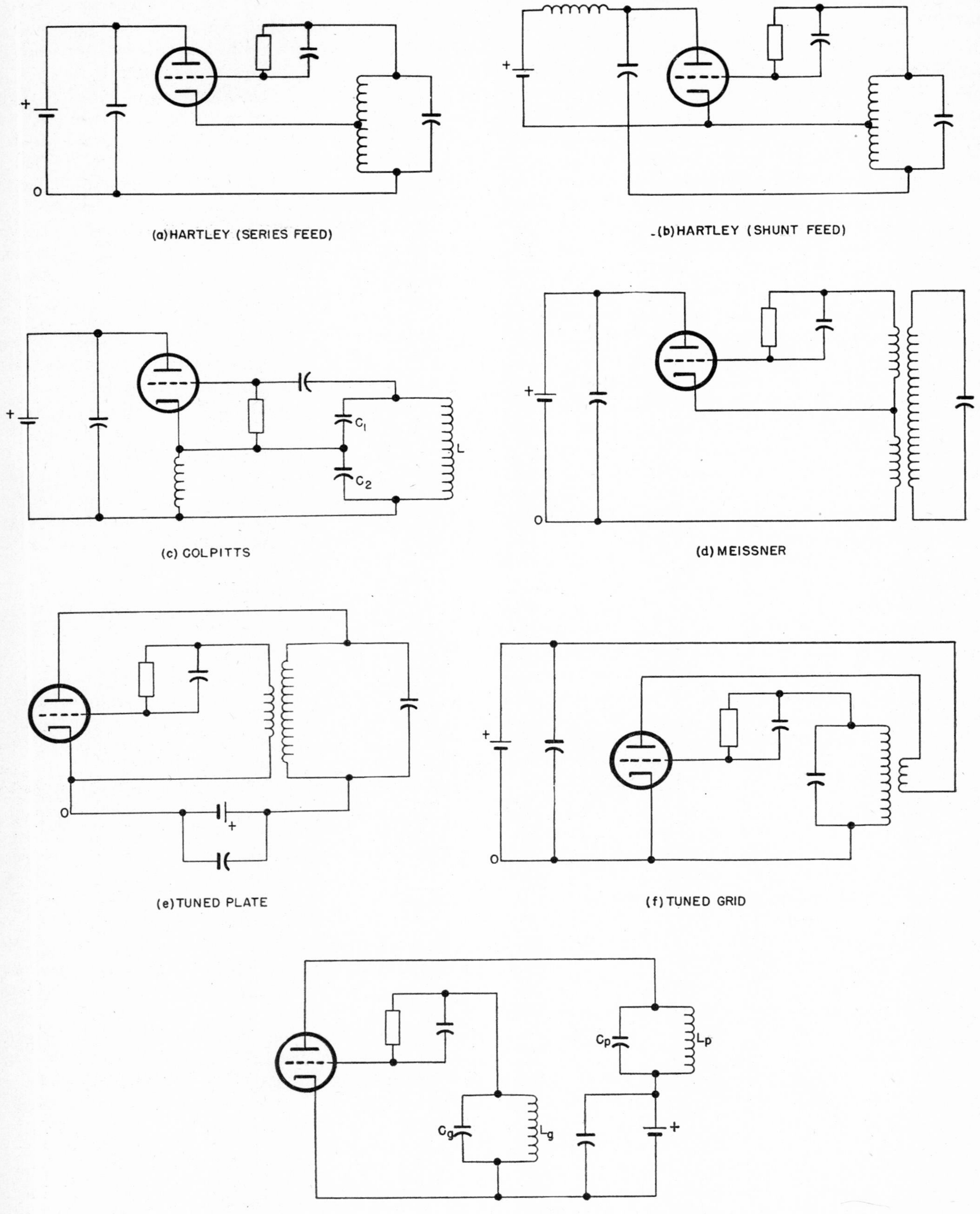

Fig. 17·9 Typical feedback oscillator circuits.

of the tube in question, cut-off for the plate current occurs at slightly negative grid potentials.

In class C operation with a constant negative bias, such as is derived from a C battery, the oscillator is not self-starting, because full negative bias does not permit enough plate current to build up the oscillations from low starting amplitudes. Therein lies the advantage of the grid-leak arrangement over a battery, because the grid bias is zero in the early build-up, or the operating point is well up along the curve with rather large plate current, so that the oscillations are able to start. The small grid current causes a voltage drop in the grid-leak resistor and so produces the necessary negative grid bias. As the average grid potential becomes more negative it reduces the plate current and amplifier action of the tube, and thus the amplitude of the oscillation is finally stabilized by the regulating action of the grid leak. The values R and C for the resistor and capacitor forming the grid leak are not critical. The time constant RC should be large compared with $1/2\pi f$, where f is the oscillation frequency. If R is too great, the grid bias builds up to such a large negative value that oscillations cease, to start again after a short delay while the capacitor in the grid circuit discharges through the resistor. The result is an intermittent oscillation, called "motorboating," and, incidentally, is a good illustration of relaxation oscillations.

To develop a class C power oscillator from which a load of given impedance is to be fed at a given power level, one must first adopt a working design for the tank circuit and its load coupling. According to circuit theory, the load may be represented by a reflected resistance in the tank circuit, that is, a series resistance which, at the operating frequency, is equivalent to the load circuit. This and the intrinsic resistance of the tank circuit are denoted, respectively, by R_s and R_1. The efficiency η_1 is that fraction of the total power generated by the tube, which is passed on to the load circuit. Evidently

$$\eta_1 = \frac{R_s}{R_1 + R_s} \tag{17·1}$$

Similarly if R_L is the resistance of the load and R_2 the resistance of the rest of the load circuit, then the efficiency of the load circuit

$$\eta_2 = \frac{R_L}{R_2 + R_L} \cdot \tag{17·2}$$

The over-all efficiency of the tank and load-coupling network is $\eta_1\eta_2$. Alternatively

$$\eta_1 = 1 - \frac{Q_{01L}}{Q_{01}} \tag{17·3}$$

$$\eta_2 = 1 - \frac{Q_{02L}}{Q_{02}} \tag{17·4}$$

where Q_{01L} belongs to the loaded tank circuit at resonance and Q_{01} to the unloaded tank circuit Q. Likewise Q_{02L} belongs to the coil in the loaded load circuit at resonance and Q_{02} to the unloaded load circuit.

Experience shows that the loaded Q of the tank circuit should not be less than about 12; and, as the unloaded Q can easily be made to exceed 100, η_1 can be made to exceed 90 percent. Similar large values are attainable for η_2.

If a loaded Q_0 of 12 is adopted for the tank circuit and $R_1 + R_s$ is known,

$$\omega L = Q_0(R_1 + R_s) \tag{17·5}$$

and hence L of the tank-circuit reactor can be computed. At a specified frequency this then determines C for the tank-circuit capacitor through $LC = 1/\omega^2$.

The plate current needed to maintain oscillations is about $1/Q_0$ of the magnitude of the current in the tank circuit as shown in the circuit and vector diagram of Fig. 17·10. Here

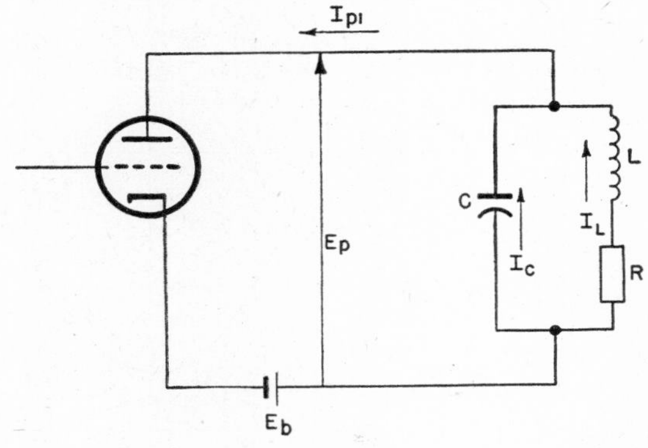

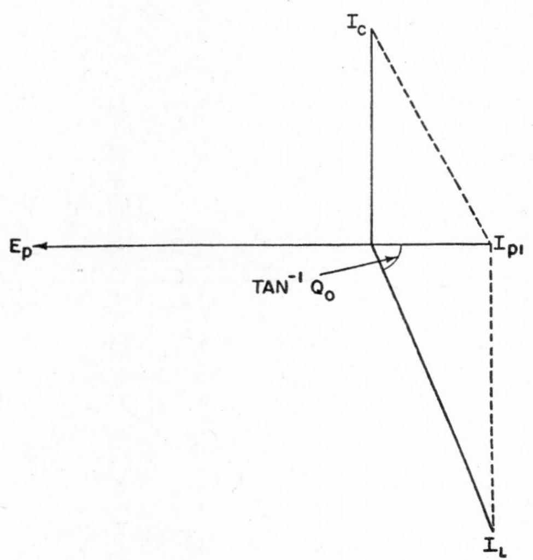

FIG. 17·10 Tank circuit of a feedback amplifier and its vector diagram.

the quantity I_{p1} is the amplitude of the fundamental-frequency Fourier component of the plate current. If P_L is the power to be supplied to the load and R_s is its effective series resistance in the plate circuit, the amplitude of the plate current fundamental is

$$I_{p1} = \frac{1}{Q_0}\sqrt{2\frac{P_L}{R_s}} \tag{17·6}$$

This carries the analysis back to the tube; therefore at this point it is necessary to consider some of the ideas underlying class C operation. Figure 17·11 shows the assumed sinusoidal voltages of plate and grid $e_p(t)$ and $e_g(t)$ for class C operation. The working part of the cycle is in the neighborhood of $\theta = 0$ when grid voltage is maximum and plate voltage is minimum. If $\theta = \omega t$,

$$e_p(\theta) = E_b(1 - a \cos \theta) \tag{17·7}$$

$$e_g(\theta) = E_c(1 - b \cos \theta) \tag{17·8}$$

where E_b is the voltage of the plate power supply and a is a coefficient less than unity; the oscillating amplitude is less than the average plate voltage, and consequently the plate is always positive relative to the cathode. The mean grid

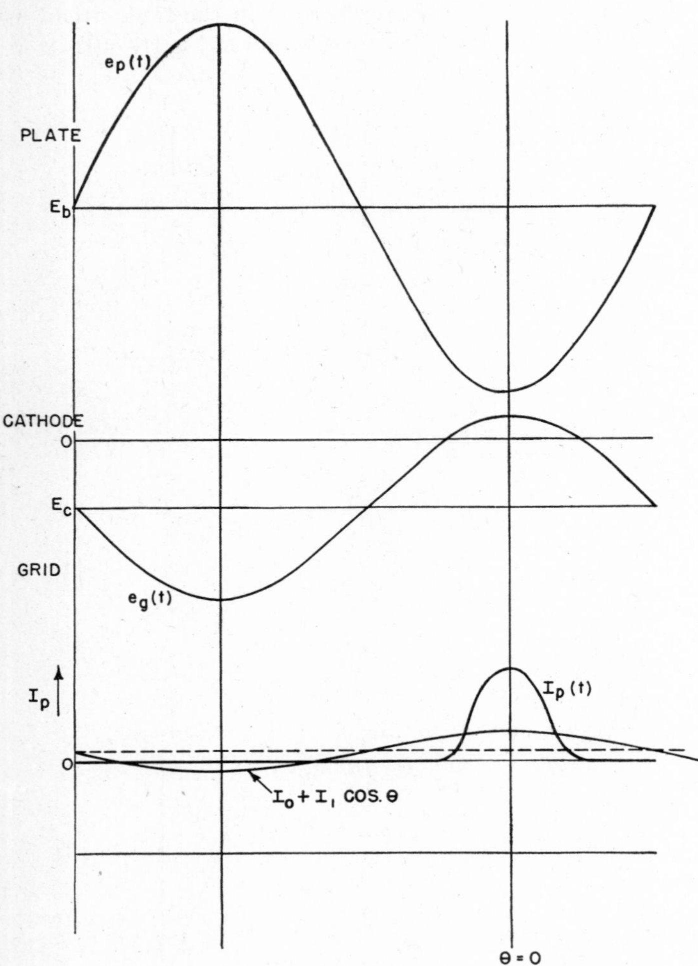

Fig. 17·11 Time curves of tube currents and voltages.

bias E_c is a negative number. The coefficient b is somewhat greater than unity; therefore the grid actually swings positive with respect to the cathode for an interval around $\theta = 0$.

During the part of the cycle when $e_g(\theta)$ is positive, there is plate current. Its amount can be found for a particular tube from its characteristic curve giving $i_p(e_g, e_p)$. A typical wave form for $i_p(\theta)$ is shown in Fig. 17·11, which shows it is far from sinusoidal.

The instantaneous power drawn at phase θ by the tube from the plate supply equals $E_b i_p(\theta)$ and the average power input is $E_b I_{p0}$, defining

$$I_{p0} = \frac{1}{2\pi} \int_0^{2\pi} i_p(\theta)\, d\theta$$

as the average plate current. Instantaneous power output to the tank circuit is $(aE_b \cos \theta) \times i_p(\theta)$, and average power output is

$$\frac{1}{2} aE_b I_{p1} = \frac{1}{2} \frac{I_{p1}}{I_m} \times a \times (E_b I_m) \tag{17·9}$$

where I_{p1} is the fundamental frequency component of $i_p(\theta)$ in phase with the alternating tank potential,

$$I_{p1} = \frac{1}{\pi} \int_0^{2\pi} i_p(\theta) \cos \theta\, d\theta$$

and I_m is the peak value of plate current.

The peak value of the plate current of a tube is limited by the electron emission of its cathode. For a given peak value, both I_{p0} and I_{p1}, and hence the power input and output, will vary in the same sense as the duration of the plate-current pulse. This duration in turn is essentially the time interval during which the grid has noticeably positive values. The biasing parameters E_c and b should then be chosen so that there is plate current during a moderately large portion of the cycle. Note, however, that $i_p(\theta)$ is most effective delivering power to the tank circuit when θ is zero; therefore a plate-current pulse of longer duration will deliver less than a proportionate amount of power yet will absorb full power from the plate supply. For better efficiency of power conversion, therefore, the grid should be biased to give a relatively short pulse.

Exact calculations have to be based on a determination of I_{p0} and I_{p1} from the wave shape $i_p(\theta)$ which in turn involves the tube characteristics and the assumed parameters E_b, E_c, a, and b. Results good enough for most design purposes may be obtained from approximate calculations by using some results of Terman and Ferns.[2] The tube characteristics may be represented approximately

$$i_p = k\left(e_g + \frac{e_p}{\mu}\right)^\alpha$$

where the exponent α is very nearly 1.5 for many tubes, and always between 1 and 2. It is understood that i_p is zero when $e_g + (e_p/\mu)$ is negative.

By substituting the assumed waves for e_p and e_g an approximate $i_p(\theta)$ is obtained. From this the maximum plate current I_{pm} as well as I_{p0} and I_{p1} can be computed in terms of the tube parameters, k, μ, and α, and of the voltage parameters E_b, E_c, a, and b. Curves of Fig. 17·12 provide a handy diagrammatic presentation of the results. Ordinates I_0/I_m and I_1/I_m are plotted against the total duration of the plate-current pulse, $\Delta\theta$ in degrees. Each curve is for a specific value of the exponent α. Obviously, the results are not too sensitive to this value over the practical range. When $\Delta\theta > 180$ degrees the parts of the plate current at the beginning and end of the pulse are harmful, tending to stop oscillations since $\cos \theta$ is negative. Nevertheless I_1/I_m does not have its maximum at $\Delta\theta = 180$ degrees because this harmful effect is small at first and is outweighed by the bigger average current in a wider pulse during the part of the cycle when $\cos \theta$ is positive.

The efficiency of the tube as a power convertor can be considered the product of two factors, η_p and η_g:

$$\eta_p = \frac{\text{total plate a-c power output}}{\text{input from plate supply}} = \frac{\dfrac{aE_bI_{p1}}{2}}{E_bI_{p0}} = \frac{aI_{p1}}{2I_{p0}}$$

$$\eta_g = 1 - \frac{\text{grid driving power}}{\text{total a-c plate output}}$$

For high efficiency a should be as near unity as possible; also the pulse duration should be short, as shown in Fig.

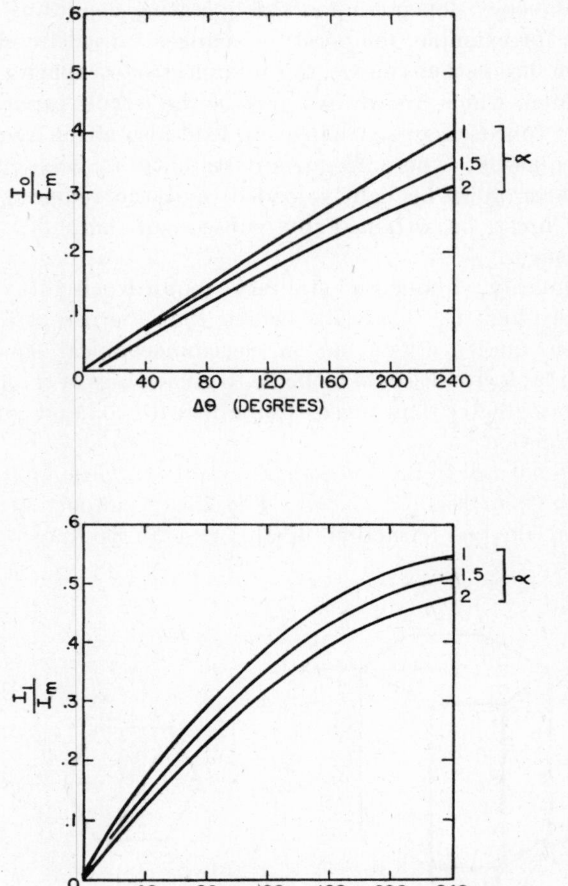

FIG. 17·12 Plot of the steady and the fundamental plate-current components as affected by biasing.

17·13, where the second factor of η_p is plotted against pulse duration from the values given in Fig. 17·12. The lower boundary of the shaded region is for $\alpha = 1$, the upper for $\alpha = 2$.

From the grid-current characteristic curves of the tube one calculates the average alternating grid driving power,

$$P_g = \frac{1}{2\pi}\int_0^{2\pi} [e_g(\theta) - E_c]i_g(\theta)\, d\theta$$

Grid current is always a pulse of short duration near the phase of maximum e_g; consequently if I_{g0} is the mean grid current as read by a d-c instrument in the grid line, the grid power is essentially

$$P_g = (1 - b)E_cI_{g0} \tag{17·10}$$

The mean grid current I_{g0} multiplied by the resistance of the grid leak R equals the mean grid-bias voltage E_c. On this basis one estimates the resistance needed for a particular value of E_c. In calculating η_g the losses in the feedback circuit and the grid power P_g needed by the tube are included in "grid-driving power."

It is considered good practice to design for $\Delta\theta$ about equal to 120 degrees, giving $I_{p1}/2I_{p0}$ somewhat above 0.9. Then if a also is made about equal to 0.9, $\eta_p \backsim 0.8$, making it possible to achieve an over-all efficiency of conversion of plate-supply power into a load power between 60 and 70 percent. Such a value of $\Delta\theta$ gives about 0.35 for I_1/I_m; therefore, with $a = 0.9$ the power output is $\frac{1}{2} \times 0.9 \times 0.35E_bI_m$, or about $E_bI_m/6.4$.

To return now to the design problem, a tube capable of giving the required I_{p1} must be used; that is, it must be capable of giving $I_m = I_{p1}/0.35$ or, say, $3I_{p1}$ and be capable

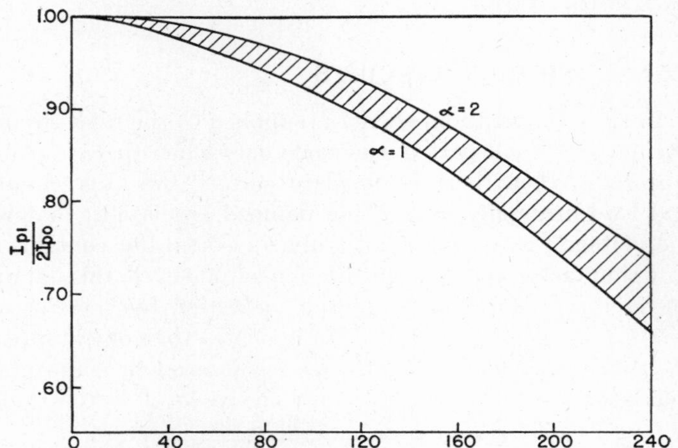

FIG. 17·13 Auxiliary plot for computing plate efficiencies.

of operating with sufficiently high plate voltage so that $\frac{1}{2}aE_bI_{p1}$ is equal to the required power output, including circuit losses and grid-driving power.

As an example, suppose 6 kilowatts is to be delivered to a 200-ohm load at a frequency of 200 kilocycles. The WL-892 pliotron is capable of giving 6.5 kilowatts at $E_b = 8$ kilovolts, and 10 kilowatts at $E_b = 10$ kilovolts. This seems to be a suitable tube. If $\eta_1\eta_2 = 0.8$ for the tank and load circuits, the tube power output required is 7.5 kilowatts; so, assuming $a = 0.9$ and $E_b = 10$ kilovolts, one will require $I_{p1} = 1.67$ amperes. For a loaded Q_0 of 12, this means that the current amplitude in the tank circuit is approximately $12 \times 1.67 = 20$ amperes; therefore the total effective resistance of the tank circuit should be $R = 37.5$ ohms. The inductive reactance of the reactor should be 390 ohms, giving

$$L = \frac{390}{1.25 \times 10^6} \text{ henrys} = 0.312 \text{ millihenry}$$

and the capacitance is

$$C = 2.06 \times 10^{-9} \text{ farad}$$

The load circuit needs to be coupled in such a way that the load impedance of 200 ohms is transformed to a little more than 30 ohms in the tank circuit.

Under these conditions, according to tube specifications, $E_c = -1300$ volts, and the average grid current is 0.18 ampere. Therefore the grid-leak resistor must be $R = 1300/0.18 = 7200$ ohms. The peak radio-frequency grid voltage is given as 2300, so the grid swings 1000 volts positive relative to the cathode. Therefore, the operating conditions probably refer to a smaller value of a than 0.9, perhaps more nearly 0.8, which calls for a slight revision of tank-circuit calculations.

A grid-driving power of 400 watts is required since the average grid current is 0.18 ampere and since it consists of short pulses triggered during the positive peaks of the grid voltage, that is, at 2300 volts. This addition to the load of the tube is not excessive, being less than 10 percent of the rated output. It is only necessary to design the feedback circuit so that it supplies the grid with the proper radio-frequency voltage. This circuit is designed by standard circuit calculations which will depend upon the type of feedback circuit used.

17·4 PUSH-PULL CIRCUITS

In class C operation, energy is supplied to the tank circuit during that small part of the cycle only when there is plate current. When there is no plate current the tank circuit and load are really undergoing damped oscillation; that is, the voltage wave form is not truly sinusoidal but contains a damping factor $\epsilon^{-\omega t/2Q_0}$. In the time of one cycle this damping factor produces an amplitude reduction by a factor of $\epsilon^{-\pi/Q_0}$, or 0.77 if $Q_0 = 12$. This indicates that operation at the minimum value of loaded Q_0 recommended in the preceding section is far from a smooth sine wave.

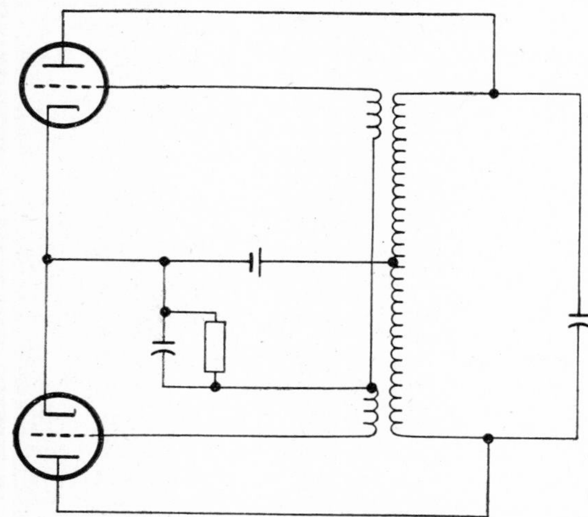

FIG. 17·14 Tuned-plate push-pull oscillator.

When the required power output is greater than can be obtained from one tube, two tubes may be used in some form of push-pull circuit. The tubes are connected in such a way that the tank circuit receives a pulse of plate current every half cycle, each tube providing only one pulse per cycle, but with pulses from the two tubes occurring with a half-cycle phase difference.

Such a circuit is shown in Fig. 17·14, which is the push-pull analogue of the tuned-plate circuit of Fig. 17·9 (*e*). Each of the feedback arrangements indicated in Fig. 17·9 can be extended to push-pull operation.

Push-pull operation gives a better wave form than single-tube operation at the same loaded Q_0, or alternatively permits operation with a lower loaded Q_0.

17·5 PIEZOELECTRIC CRYSTAL OSCILLATORS

If a high degree of frequency stability is required in an oscillator, the circuit should contain maximum attainable Q, and should be as free as possible from effects of temperature changes, mechanical shock, and so on. Much attention to detail is required in the design of the reactors and capacitors, a subject handled very thoroughly by Thomas.[3]

Frequency depends upon the operating conditions of the tube; for example, temperature changes caused by changed power dissipation change the interelectrode capacitances of the tube, which are always part of the circuit capacitances of the tuned circuit. Changes in load also affect frequency; consequently, where frequency stability is necessary the oscillator must be lightly loaded, and the required power level must be attained by subsequent amplification, if necessary.

Primarily, frequency stability requires circuits of extremely high Q. Cady discovered that the use of a piezoelectric quartz crystal as an electromechanical equivalent of the tank circuit could be made to give values of Q hundreds of times greater than those attainable with coil and capacitor construction.[4]

The natural form of a quartz crystal is a hexagonal prism, as shown in Fig. 17·15 (*a*). The Z-axis, important in the optical double refraction of the crystal, is known as its

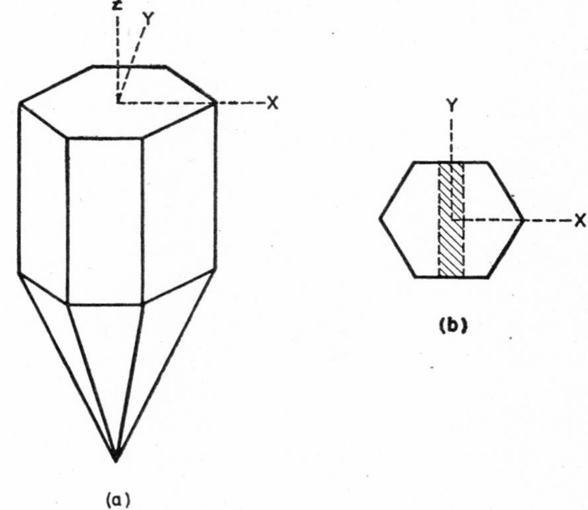

FIG. 17·15 X- and Y-cuts of piezoelectric quartz crystal.

optical axis. Any plane normal to the Z-axis contains three physically equivalent X-axes connecting the opposite corners of the hexagon, and three equvialent Y-axes normal to the faces of the hexagon, as shown in Fig. 17·15 (*b*). The X-axes are known as *electric* axes, and the Y-axes are known as *mechanical* axes (also sometimes called neutral axes). A flat section cut from a crystal, like the shaded part of Fig. 17·15 (*b*) is called an X-cut crystal (or Curie cut) if its plane faces are normal to an X-axis. Similarly one speaks of a Y-cut crystal if its plane faces are normal to a Y-axis.

If a mechanical tension or compression is applied to the

crystal along a Y-axis, the material becomes electrically polarized along an X-axis. That is, squeezing an X- or Y-cut crystal causes the appearance of charges on the crystal faces; conversely, if electrodes are attached to the crystal and a potential difference is applied, a mechanical stress is produced in the crystal. The charge per unit area on opposite faces is independent of the thickness of the X-cut crystal, and is directly proportional to the pressure (force per unit area). The relation is

$$q = d_{11}P \tag{17·11}$$

where d_{11} is known as a piezoelectric modulus, q is the charge per unit area in electrostatic units per square centimeter, and P is the pressure in dynes per square centimeter. For quartz the modulus d_{11} in these units is

$$d_{11} = 6.36 \times 10^{-8} \frac{\text{electrostatic units per square centimeter}}{\text{dynes per square centimeter}}$$

The same modulus measures the change in thickness of the plate which accompanies an applied potential, according to the relation

$$\delta = d_{11}E \tag{17·12}$$

where δ is the change in thickness in centimeters and E is the potential drop across the plate (X-cut) in statvolts. Hence a potential drop of 3000 volts across the thickness of an X-cut plate produces a thickness change of only 6.36×10^{-7} centimeter. Actually the piezoelectric effect is non-linear. The deformation is proportional to voltage up to about 2.5 kilovolts, but at 25 kilovolts it is about 30 percent less than the proportional value and seems to approach a limiting saturation value at 160 kilovolts.

Besides X-cut quartz many other ways of cutting the crystal are being used. Also, there are many other substances, especially tourmaline and Rochelle salt, with piezoelectric properties, but quartz has thus far found the widest field of application.

If an X-cut crystal is put between the plates of a parallel-plate capacitor to which an alternating voltage is applied, the piezoelectric deformation tends to excite the normal modes of free oscillation of the crystal. The simplest mode of oscillation of a rectangular plate, the length and breadth of which are large compared with the thickness d, is that in which the thickness contracts and expands in such a way that the displacement of a point at distance x from the central plane is proportional to sin $(\pi x/d)$. The frequency of such oscillations is approximately

$$f = \frac{1}{2d}\sqrt{\frac{C_{11}}{\rho}} \tag{17·13}$$

where $C_{11} = 8.55 \times 10^{11}$ dynes per square centimeter is the appropriate elastic modulus, and $\rho = 2.65$ grams per cubic centimeter is the density of quartz. Substituting these values,

$$f = \frac{283.9}{d} \frac{\text{kilocycles per second}}{\text{centimeters}}$$

that is, a crystal 1 millimeter thick oscillates at a frequency of 2.84 megacycles per second.

When such a crystal is placed between plane-parallel electrodes to which a sinusoidal voltage of nearly resonant frequency is applied, the crystal will respond with mechanical oscillations as well as with the usual capacitive displacement current. In addition to the dielectric capacitance representing energy stored in the electric field, the crystal has inductive reactance caused by the kinetic energy of the vibrating quartz; capacitive reactance caused by the elastic potential energy of the strained quartz; and resistance representing the dissipative effects of internal friction in the quartz, its rubbing against the supporting electrodes, and acoustic radiation of sound waves into the medium and the supports. Thus the mounted crystal represents an electromechanical system, a complete description of which must involve the mechanical material constants. Cady's analysis [5] has shown that the crystal, as far as the electrodes are concerned, is equivalent to the circuit of Fig. 17·16. According to this analysis, the crystal may be represented by an equivalent mechanical system of concentrated mass M, frictional resistance W, and stiffness including the effect of gap, G. The equation of motion of this system, for oscillations near resonance, is

$$M\frac{d^2x}{dt^2} + W\frac{dx}{dt} + G_x = kE_1 \sin \omega t \tag{17·14}$$

where the potential drop across the crystal, E_1 in statvolts, is related to E, the total voltage drop between electrodes, by

$$E_1 = \frac{d}{d + \epsilon w}E \tag{17·15}$$

and where x is the actual amplitude of motion of the surface of the crystal, and the piezoelectric constant is

$$k = \frac{2e_{11}A}{d + \epsilon w} \tag{17·16}$$

The equivalent mechanical constants are

$$M = \tfrac{1}{2}\rho A d \text{ grams} \tag{17·17}$$

$$W = \frac{\pi^2}{2}\rho F\frac{A}{d}\frac{\text{dynes}}{\text{centimeters per second}} \tag{17·18}$$

$$G = \frac{\pi^2}{2}C\frac{A}{d}\frac{\text{dynes}}{\text{centimeters}} \tag{17·19}$$

In these equations ρ is the density in grams per cubic centimeter, A is the area in square centimeters, and d is the thickness of the crystal in centimeters. The effective mass is only half the total mass, because the central layers do not take part in the motion at full amplitude. F is a measure of the internal viscosity of quartz. $F = 0.25$ has been reported as an empirical value. The elastic modulus c is a corrected value for C_{11} given by

$$c = C_{11} + \frac{4\pi e_{11}^2}{\epsilon} - \frac{32e_{11}^2 d}{\pi\epsilon(d + \epsilon w)} \tag{17·20}$$

Here $e_{11} = 4.77 \times 10^4$ is a piezoelectric modulus, ϵ is the dielectric constant (4.5), and w is the total air-gap thickness between the electrodes and the crystal.

In Fig. 17·16 capacitance C_1 is the capacitance of the quartz crystal computed with no regard to the piezoelectric property,

$$C_1 = \frac{\epsilon A}{4\pi d} (9 \times 10^{11})^{-1} \text{ farad} \qquad (17\cdot21)$$

and capacitance C_2 is the capacitance of the air gap,

$$C_2 = \frac{A}{4\pi w} (9 \times 10^{11})^{-1} \text{ farad} \qquad (17\cdot22)$$

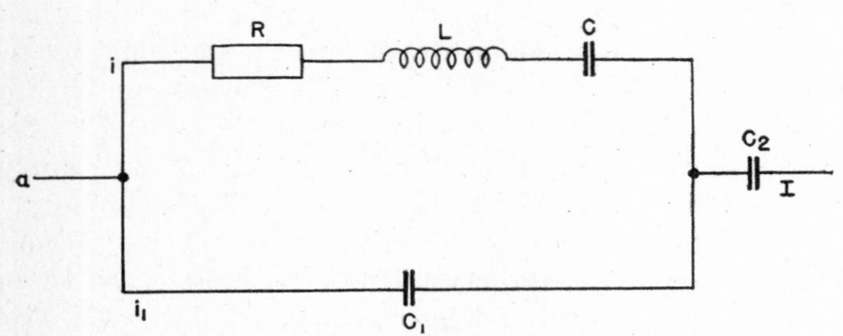

FIG. 17·16 Electric circuit equivalent of crystal with gap.

The equivalent electrical quantities are $L = M/k^2$, $R = W/k^2$, and $C = k^2/G$. Substitution of the parameter values for an X-cut quartz crystal gives

$$L = 130d \frac{(d + \epsilon w)^2}{A} \text{ henry}$$

$$C = 2.4 \times 10^{-15} \frac{Ad}{(d + \epsilon w)^2} \text{ farad}$$

$$R = 325 \frac{(d + \epsilon w)^2}{Ad} \text{ ohm}$$

$$\frac{L}{C} = 232 \frac{(d + \epsilon w)^2}{A} \text{ megohms}$$

$$Q = 715{,}000d$$

Actual values of Q for quartz are somewhat lower because of losses other than those due to internal friction. Van Dyke [6] found that a particular crystal mounted in air had $Q = 25{,}000$, but when the surface was etched and the crystal was mounted in vacuum, Q rose to 580,000.

These results indicate that the crystal is electrically equivalent to the usual resonant tank circuit and that very high Q values are obtainable. Many different circuits corresponding to different locations of the crystal elements in the grid or plate circuit and with different feedback arrangements have been tried.

This discussion has been limited to X-cut crystals, but there are many other ways of cutting the natural quartz, some of which are particularly designed for zero temperature coefficient of the frequency of vibration for a limited temperature range.

The practical upper frequency limit is around 50 megacycles, corresponding to a thickness for the X-cut crystal of only 0.057 millimeter. At higher frequencies the crystal would be too fragile both electrically and mechanically. It is possible, however, to drive thicker crystals at higher frequencies by harmonic excitation.

Quartz crystals have found extensive use, not only for maintaining accurate frequency control in oscillator circuits with electridal output, but also as the means for converting electrical power at supersonic frequencies into acoustic power. This application goes back to the first World War, when Langevin employed quartz crystals in an acoustic echo system for depth sounding and for the detection of such objects as submarines or icebergs. A modern device of this type has been described by Firestone,[7] who used a small quartz crystal operated in pulses at frequencies of several megacycles to locate defects in castings and forgings. These flaws in the material have the ability to reflect part of the supersonic energy back to the crystal. The latter, in a reversal of the mechanism whereby the mechanical waves were generated, responds to the mechanical stimulus of the echo with an electrical signal at the electrodes and so becomes the detector for the returning supersonic signal.

The many applications of intense supersonic radiation in fluids range from the preparation of colloids and emulsions to the killing of bacteria in milk and in other liquids. These applications have for their object a maximum density of supersonic energy in the fluid. Some experimenters have obtained sound-power densities in liquids of about 10 watts per square centimeter. Wood and Loomis,[8] who pioneered in the study of intense supersonic waves, achieved 35 watts per square centimeter. Sound-power density I (ergs per square centimeter second) is related to the pressure amplitude p (dynes per square centimeter), the density ρ (grams per cubic centimeter), and the velocity of sound v (centimeters per second) by the equation

$$I = \frac{p^2}{2\rho v} \qquad (17\cdot23)$$

from which the pressure amplitude is computed at 5 atmospheres in water when sound-power density is 10 watts per square centimeter. Hence cavitation and the formation of gas or vapor bubbles in the liquid set in at much lower power levels.

17·6 MAGNETOSTRICTION OSCILLATORS

Magnetostriction is another means by which the elastic vibrations of a solid medium can be coupled to an electric circuit, thus permitting the employment of such a resonant material as an oscillator tank circuit, and also providing an alternative means for exciting supersonic radiations at high power levels.

Magnetostriction is the change in length of a ferromagnetic rod when it is magnetized. Although the name suggests that the length always shortens, both elongation and contraction are observed, as shown in Fig. 17·17 for various ferromagnetic metals (*a*) and alloys (*b*). The effect is of the *same sign* for either sense of magnetization, and a cylindrical rod of ferromagnetic material, if in the alternating magnetic field of a coil, will pass through a complete cycle of magnetostrictive force during each half cycle of the alternating current. Therefore the current tends to excite resonant vibrations of the rod at twice its own frequency.

The inverse effect is also observed; that is, if a previously magnetized nickel rod is put under tension in the direction of its length, its magnetization is diminished; if the rod is compressed, the magnetization increases.

An electrical-circuit equivalent of a coil with a magnetic core, similar to that for quartz crystals, can be deduced. The fundamental frequency of a bar of length l having Young's modulus E and density ρ for longitudinal vibrations is

$$f = \left(\frac{1}{2}l\right)\sqrt{\frac{E}{\rho}} \qquad (17 \cdot 24)$$

For nickel the constants are such that a rod 10 centimeters in length has a fundamental frequency of 24.3 kilocycles per second.

Much of the study of magnetostriction oscillators was done by G. W. Pierce [9] and his co-workers. Generally these are

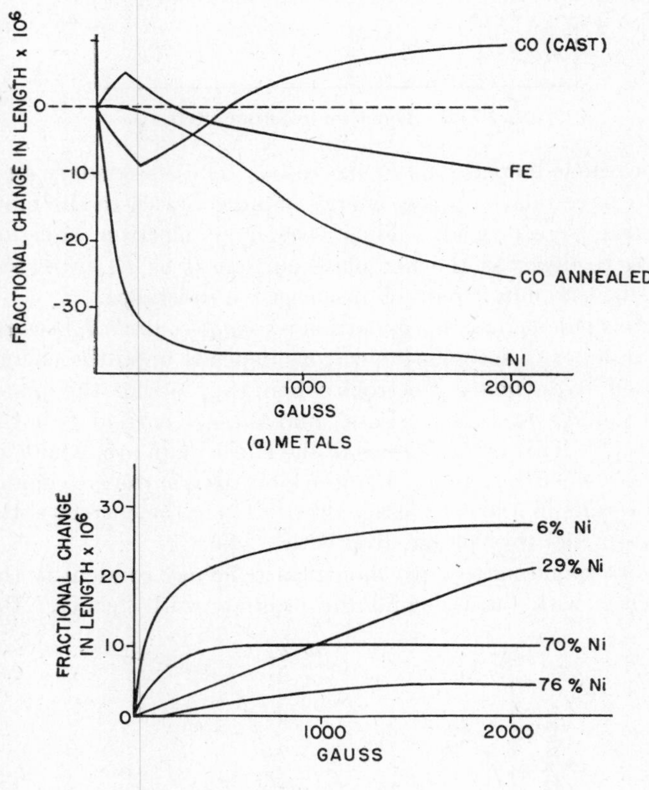

FIG. 17·17 Magnetostriction of ferromagnetic metals and alloys.

used for lower frequencies to a practical upper limit of about 70 kilocycles per second. As the most suitable material Pierce recommends Invar (36% nickel, 64% iron). Monel metals (68% nickel, 28% copper, with small amounts of iron, silicon, manganese, and carbon) give good results.

In some magnetostriction oscillators a d-c polarizing coil is used as well as the coil carrying the high-frequency current. This insures that the sense of magnetization of the rod is not reversed during the cycle, and the vibrations in the rod have the same frequency as the electrical oscillations.

Detailed description of a magnetostriction oscillator rated at about a kilowatt has been given by Salisbury and Porter.[10]

17·7 NEGATIVE-RESISTANCE OSCILLATORS

In a negative-resistance device an increase in current is accompanied by a decrease in the voltage drop across it. A good example is an ordinary arc between carbon electrodes at atmospheric pressure. The static characteristic of such an arc is well represented by an equation between the arc current i and its voltage drop e:

$$e = A + \frac{B}{i^n} \qquad (17 \cdot 25)$$

where A and B are empirical constants, and where the exponent n is usually close to unity.

Negative-resistance devices may be used to convert power from a supply of steady electromotive force to audio- or radio-frequency power. For an understanding of the performance of such elements under dynamic conditions, negative resistances may be divided into two classes: (1) if the voltage change lags the current change, as in an arc, the device is said to be *current-controlled;* and (2) if the current change lags behind the voltage change, as in the dynatron (discussed later), the device is said to be *voltage-controlled.* If the element could adjust itself to changing operating conditions instantaneously, or at any rate much faster than these changes, the relation between current and voltage at any instant would be represented by a point on the static characteristic, the curved line $e(i)$ in Fig. 17·18. Actually,

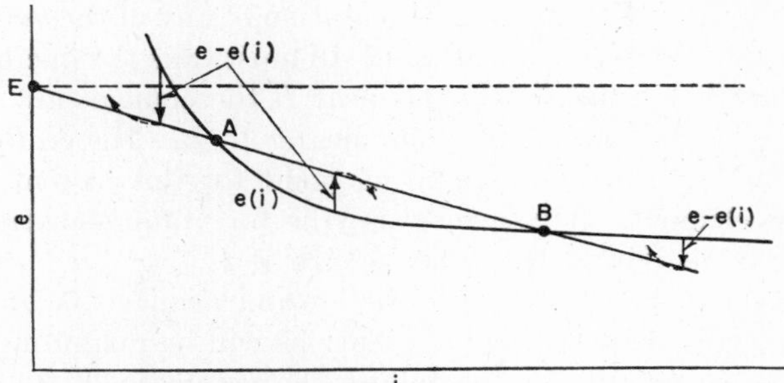

FIG. 17·18 Stable (B) and unstable (A) operation of current-controlled negative resistance with positive series resistance.

the changes of current and voltage can be so rapid that the operating point in the diagram follows a so-called dynamic characteristic instead. Generally, performance on a dynamic characteristic is described by an equation

$$e = e(i) + L(i)\frac{di}{dt} \qquad (17 \cdot 26)$$

Since the voltage change on a falling characteristic is of the opposite sense as the corresponding current change, $L(i)$ must have the positive sign for a current-controlled element, as in an inductance, and the negative sign for a voltage-controlled element. The only way for the element to operate at a point (e, i) not on the static characteristic $e(i)$ is in conformance with the last equation from which the proper rate of current change may be computed. If the voltage drop for a given current is instantaneously greater than that required by the static characteristic, then the current increases in a current-controlled element and decreases in a voltage-controlled element. Expressed differently, the operating point in the (e, i) plot will show a tendency toward clockwise rotation if the element is current-controlled, and counterclockwise if it is voltage controlled (Fig. 17·19).

As a simple consequence of this rule, the distinction between the two types of elements will be seen to affect the stability of operation even in the simple d-c series circuit shown in Fig. 17·20. The straight line (Fig. 17·18) represents the voltage

$$e = E - iR \tag{17·27}$$

which the battery can apply to the negative-resistance element at current i. Its two intersections A and B with the current-voltage characteristic of the element $e(i)$ are possible operating points. Which is stable? For a current-controlled element operating temporarily on the straight line and near A it will be seen that clockwise rotation of any of the vertical vectors $e - e(i)$ shown in Fig. 17·18 must take the operating point farther from A, whereas near B the changes are such as to move the operating point nearer to B. Hence B is a stable, A an unstable, operating point for the current-controlled element. The reverse is true for voltage-controlled elements, in view of the opposite sign of L.

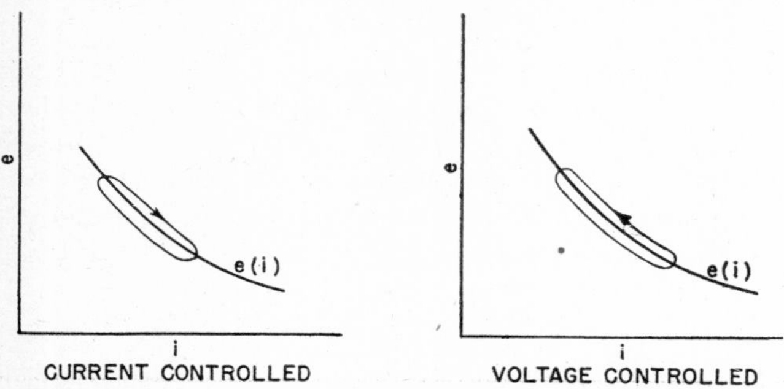

FIG. 17·19 Dynamic characteristics of current-controlled and voltage-controlled devices with falling characteristic.

If a shunt circuit with an inductive and a capacitive branch is connected across an arc, oscillations will be maintained in the circuit because of the falling characteristic of the arc. Energy transfer is from the arc to the oscillatory circuit, in accordance with the clockwise sense of the hysteresis loop in the (e, i) plot. The use of an arc for negative-resistance oscillators was historically important in the development of radio in the first decade of this century,[11] but arc oscillators have since been superseded by dynatron and other circuits employing electronic tubes.

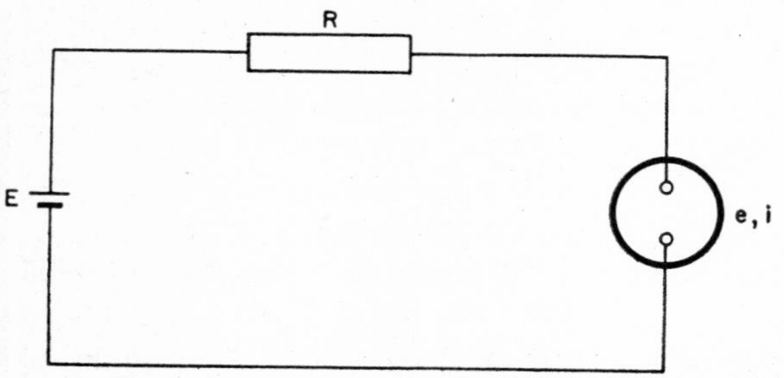

FIG. 17·20 Simple d-c series circuit of negative-resistance device.

The triode or tetrode, when used as a dynatron, gives a negative-resistance characteristic which is voltage controlled. It is not good practice to use a triode in this way, for the large positive potential needed at the grid may cause excessive drain on the plate-power supply, overheating of the grid, and excessive emission from the cathode, which is especially detrimental in tubes with oxide-coated cathodes. For this reason this discussion is confined to dynatron action of a tetrode. In Fig. 17·21 the screen-grid potential E_s is maintained at a fixed positive value of about 100 volts. The control grid potential E_c may be positive or negative as indicated. If a plate potential E_b somewhat lower than E_s is applied, electrons from the cathode (in quantities largely determined by E_c) go through the two grids and strike the plate. In so doing they release secondary electrons from the plate. Their number per unit of current incident from the cathode increases as E_b increases. These secondary electrons have relatively low energy and are drawn to the more positive screen grid. Thus secondary electron emission tends to decrease the net plate current i_b as E_b increases; that is, the plate circuit shows negative resistance.

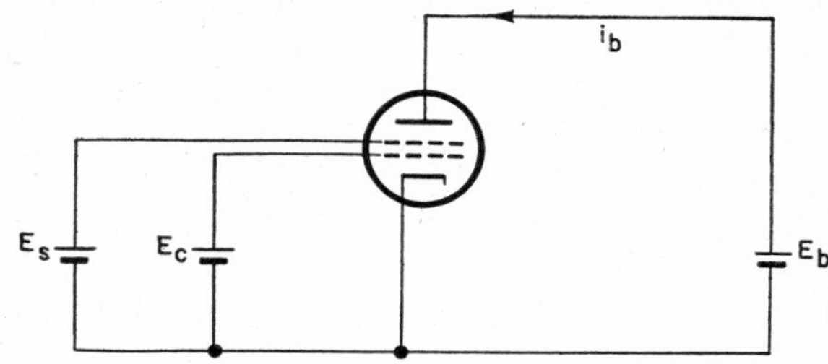

FIG. 17·21 Dynatron-connected tetrode.

It is evident that this operation is *voltage-controlled*, because the number of secondary electrons depends upon the energy carried by primary electrons when they strike the plate. The voltage E_b is the "cause" and the net current i_b is the "effect." This is the reverse of the effect in an arc, where an increase of current produces more ionization in the arc column and results in a lower voltage drop. There the current is the "cause" and the voltage drop is the "effect."

Corresponding to a psychological tendency to identify the abscissa with "cause" and the ordinate with "effect," the dynatron characteristic usually is plotted with E_b as abscissas and i_b as ordinates. In this "cause" versus "effect" plot, the oscillations of a voltage-controlled system are represented by hysteresis loops traversed in the clockwise sense, again indicating transfer of energy from the negative-resistance element to the (this time series-connected) load circuit.

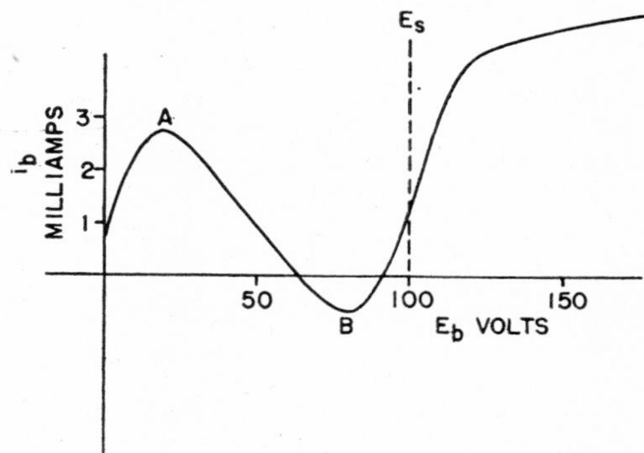

FIG. 17·22 Falling dynatron-characteristic plate current versus plate voltage.

An example is shown in Fig. 17·22. The negative-resistance portion is clearly evident. When $E_b > E_s$ the secondary electrons are presumably emitted from the plate even more copiously but are at once caught on the plate instead of the screen grid, which accounts for the rise to larger positive values of i_b as E_b becomes greater than E_s. Any such characteristic as Fig. 17·22 is for a particular value of the control-grid voltage E_c. As E_c is made more negative the maximum at A is diminished, since there are now fewer primary electrons. For the same reason there are fewer secondary electrons; therefore the negative conductance given by the general slope from A to B is diminished. Hence the magnitude of negative resistance may be controlled through E_c. Negative resistances of 10,000 to 20,000 ohms are typical in small tubes. The dynatron characteristic varies from one tube to another, nominally of the same type, because secondary emission is a property depending upon surface conditions of the plate and is therefore hard to control in tube manufacture.

A typical dynatron circuit is shown in Fig. 17·23. Here the control-grid potential is fixed by a grid-leak-biasing circuit. Alternatively, bias potential could be provided by a C battery.

If the characteristic of Fig. 17·21 and the voltages given in Fig. 17·22 were assumed, the steady average plate current would be close to zero. This is not always true; the operating point indicated happened to be one for which the current reduction by secondary emission just equaled the primary current to the plate.

In Fig. 17·23, if i_1 is the current through the inductive branch and i_2 the current through the capacitative branch, so that the plate current is $I = i_1 + i_2$, and E is the applied voltage (in this case 60 volts), then

$$L\frac{di_1}{dt} + Ri_1 + e = E$$

$$i_2 = -C\frac{de}{dt}$$

$$i_1 = I + C\frac{de}{dt}$$

where e is the voltage drop across the tube for current I. If I_0 and e_0 are the current and voltage of the steady operating point,

$$I(e) = I_0 + I'(e_0)\epsilon$$

approximately, where $\epsilon = e - e_0$ is the instantaneous voltage departure from the steady operating value. Hence, eliminating i_1 and i_2,

$$LC\frac{d^2\epsilon}{dt^2} + \left(L\frac{dI}{dt} + RC\frac{de}{dt}\right) + e + RI = E$$

and substituting the linear approximation for $I(e)$ gives

$$LC\frac{d^2\epsilon}{dt^2} + [LI'(e_0) + RC]\frac{d\epsilon}{dt} + [1 + RI'(e_0)]\epsilon = E - RI_0 - e_0 \quad (17\cdot28)$$

The steady values on the right must add up to zero; therefore

$$E - RI_0 = e_0$$

as discussed in relation to stability in the first part of this section, and the equation for the departure ϵ then is

$$LC\frac{d^2\epsilon}{dt^2} + [LI'(e_0) + RC]\frac{d\epsilon}{dt} + [1 + RI'(e_0)]\epsilon = 0 \quad (17\cdot29)$$

For this equation the usual solutions of the form $e^{\alpha t}$ represent oscillatory solutions growing in amplitude, provided

$$[LI'(e_0) + RC]\epsilon < 0$$

that is, $I'(e_0)$ must be negative, and

$$\frac{1}{|I'(e_0)|} < \frac{L}{RC}$$

In other words, the magnitude of the negative resistance must be less than L/RC so that oscillations will build up. As the amplitude of oscillation becomes finite, the tube begins

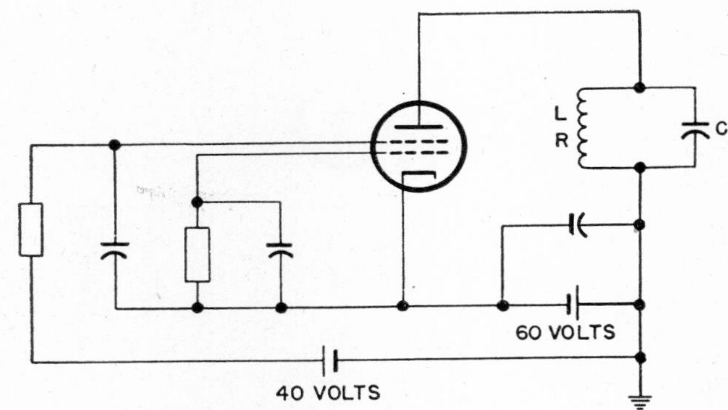

FIG. 17·23 Typical dynatron circuit.

to operate on a cycle involving the curved parts of the characteristic where $I'(e)$ is less than its value near the center of the operating range. This reduces the build-up rate of the amplitude until it comes to a finite limiting value, the magnitude of which is determined by the curvature of the characteristic rather than by its negative slope.

17·8 RELAXATION OSCILLATORS

In relaxation oscillators the frequency-determining element is intrinsically aperiodic, depending upon either the charging or discharging of a capacitor through a resistance, or the building up or decay of current in a reactor through a resistance. Because of the absence of a tank circuit in which stored energy is exchanged between an inductor and a capacitor, their wave form is generally "jerky," that is, far from sinusoidal.

Oscillators of this class are useful only where a special non-sinusoidal wave form is required. Sample applications are: (*a*) as a source rich in harmonics, for accurate frequency comparison between a standard-frequency oscillator and an unknown frequency which is greatly different in magnitude; (*b*) for generating "sawtooth" waves in which the voltage increases linearly with time up to a maximum and then snaps back to zero (this has an important application as the time base in the cathode-ray oscilloscope); (*c*) to give a periodic repetition of rectangular voltage pulses in various signaling devices, particularly in radar. As with oscillators of other types, relaxation oscillators have closely associated amplifiers of the same type. Because relaxation oscillators do not contain sharply tuned resonant circuits, they usually can be forced to operate at a desired frequency by introducing a signal voltage of that frequency into the circuit. This is important, for example, in the time base of a cathode-ray oscilloscope, which must be synchronized with the periodic phenomenon under observation to give a stationary pattern on the screen. The same is true of cathode-ray tubes in television receivers.

The simplest circuit of this type is indicated in Fig. 17·24. At first disregard the lead containing the coupling capacitor C_2. The tube is a cold-cathode gas diode. Such a tube does not pass appreciable current until the voltage drop across it is about 130 volts. Then a glow discharge continues until

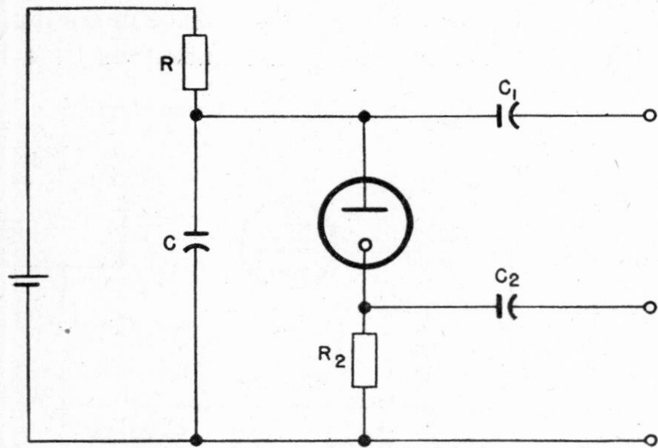

FIG. 17·24 Relaxation oscillator using a condenser charged cyclically through resistance R and discharged through cold-cathode gas diode.

the voltage falls to about 100 volts. Starting with capacitor C uncharged, the voltage V across C grows with time according to the law

$$V = E(1 - e^{-t/RC}) \tag{17·30}$$

until V reaches the breakdown voltage of the gas diode. When the discharge in the diode starts, it is effectively a low-resistance path across the capacitor; consequently the voltage of the latter drops abruptly to that value at which the discharge in the diode is extinguished. Then the potential across the capacitor again starts to build up until it reaches the ignition voltage of the gas diode, and so on. The result is a sawtooth wave form as in Fig. 17·25, in which the rising portions are not linear but are short sections of an exponential curve. The output of the oscillator may be taken off through the lead attached to the coupling capacitor C_1.

If it is assumed that V_1 is the extinction voltage and V_2 is the ignition voltage of the tube, and that the time of "flyback" (during which the capacitor is discharged through the diode) is negligible, then the period T of the oscillations is

$$T = RC \log \frac{E - V_1}{E - V_2} \tag{17·31}$$

To obtain a voltage rise after each discharge which is essentially linear with time, E must be large compared with V_1 and V_2. This requirement is an undesirable one, since it means that the supply must be several hundred volts to get an output voltage swing equal to only about 30 volts.

For a fixed choice of E and of V_1 and V_2 the period may be varied by varying R or C. To synchronize the oscillations with some periodic voltage under observation one introduces this voltage on the lead attached to coupling capacitor C_2 and tunes the circuit until its natural frequency is about the same as the voltage across R_2. The peaks of the signal voltage then determine the instants at which the voltage across the gas diode reaches its ignition value and trigger the discharge. Thus the relaxation oscillations are brought into synchronism with the signal voltage.

The voltage-rise curve is curved instead of linear because the voltage across resistor R diminishes as the capacitor becomes charged; therefore the current through it diminishes and the rate of voltage rise across the capacitor diminishes. Evidently a constant-current generator is needed, and it is approximately realized by using a high-voltage source in series with a high resistance.

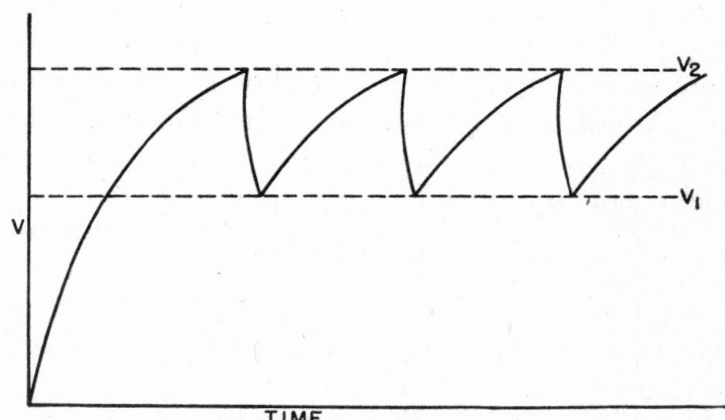

FIG. 17·25 Condenser voltage as a function of time: sawtooth wave form with exponential sections.

The simple circuit of Fig. 17·24 can be improved by replacing resistor R with a constant-current device such as a diode or a pentode working near plate-current saturation so that plate current does not change much with change in plate voltage. If R is the plate resistance of the tube (the rate of increase of plate voltage with plate current), then this is the effective value of R presented by the tube; and if I is the plate current, the effect is as though the voltage were IR, but in fact the plate voltage needed for the tube is much lower. Moreover, if a pentode is used in this way, the magnitude of R depends upon the voltages applied to its grids, thus affording a convenient possibility of remote control of frequency of relaxation oscillations.

Another drawback of the simple circuit of Fig. 17·24 is the relatively small difference between extinction and ignition voltage of simple gas diodes. This disadvantage was overcome by the use of the thyratron. The thyratron has a much greater separation between extinction and ignition voltage, which moreover is readily controllable.

Figure 17·26 is an adaptation of the circuit of Fig. 17·24 in which a thyratron is used in place of a gas diode and a pentode as the device which controls the charging rate of the capacitor.

The time constant R_3C_3 of the cathode-biasing circuit of the thyratron should be long compared with the longest period of oscillation. To prevent damage to the thyratron the protective resistor R_2 must be made large enough to limit the current through the tube to the safe rated value. With adjustable resistor R_1 the effective plate resistance of the pentode, and hence the frequency of oscillation of the circuit, can be varied.

These circuits ordinarily are considered to be limited to a maximum repetition rate of about 50 kilocycles per second. Gas thyratrons produce more stable operation than mercury-vapor thyratrons in such circuits.

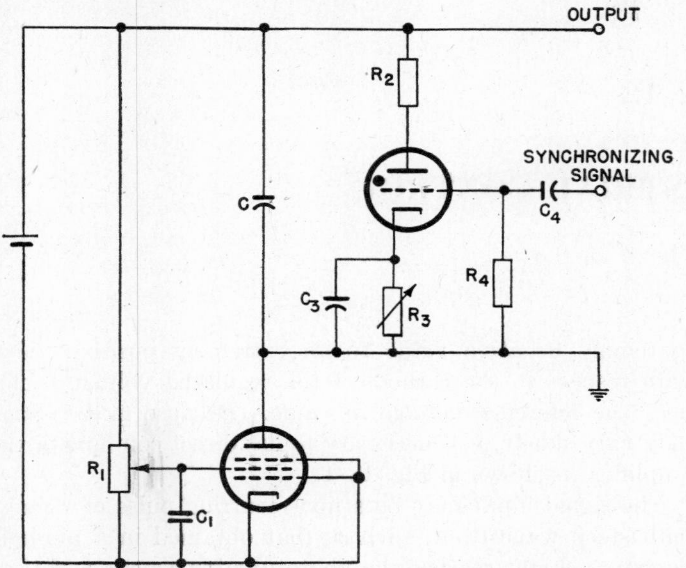

FIG. 17·26 Improved relaxation oscillator using discharge through thyratron and constant-current charging device.

The subject of time bases is treated thoroughly in a book by Puckle.[12]

The standard circuit for producing an output rich in harmonics is the multivibrator, so called because of the multitude of harmonics in its output. It is possible to work at least up to the 150th harmonic, obtaining enough signal for frequency comparisons. The multivibrator is simply a two-stage resistance-coupled amplifier in which the output is returned to the first tube as input, as indicated in Fig. 17·27.

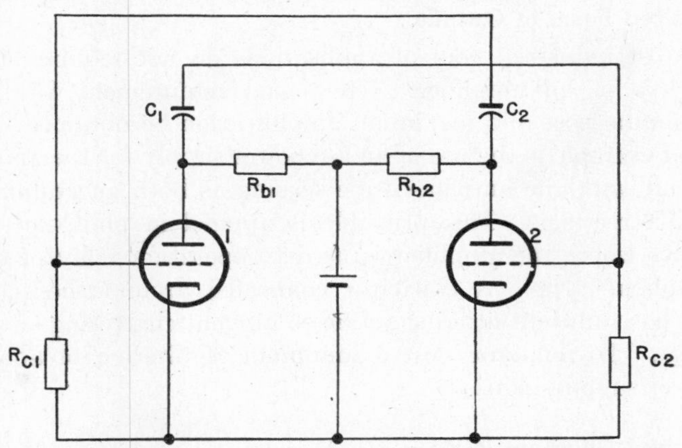

FIG. 17·27 Multivibrator circuit with two symmetrically connected tubes.

Figure 17·28 shows the various wave forms of a symmetrical multivibrator in which the tubes and corresponding resistors and capacitors are equivalent.

Frequency is determined mainly by the product R_cC but depends also upon other circuit parameters. Multivibrators will operate at frequencies up to about 100 kilocycles per second and as low as about 1 cycle per minute. If a signal voltage is injected into the circuit, the multivibrator may adjust itself to operation at a frequency which is an exact

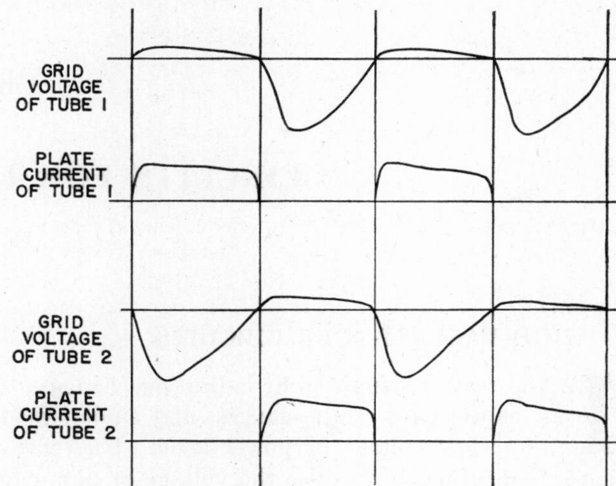

FIG. 17·28 Phases and wave forms in multivibrator tubes.

sub-multiple of the signal voltage. In this way the multivibrator finds important application in generation of sub-multiples as well as harmonic frequencies.[13]

REFERENCES

1. "Vacuum Tube Oscillators—A Graphical Method of Analysis," J. W. Horton, *Bell System Tech. J.*, Vol. 3, 1924, p. 508.
2. "The Calculation of Class C Amplifier and Harmonic Generator Performance of Screen Grid and Similar Tubes," F. E. Terman and J. H. Ferns, *Proc. I.R.E.*, Vol. 22, 1934, p. 359; "Calculation and Design of Class C Amplifiers," F. E. Terman and W. C. Roake, *Proc. I.R.E.*, Vol. 24, 1936, p. 620.
3. *Theory and Design of Valve Oscillators for Radio and Other Frequencies*, H. A. Thomas, Chapman and Hall, London, 1939.
4. *Quartz Oscillators and Their Applications*, P. Vigoreux, H. M. Stationery Office, London, 1939.
5. "The Piezoelectric Oscillator and the Effect of Electrode Spacing on Frequency," W. G. Cady, *Physics*, Vol. 7, 1936, p. 237.
6. "A Determination of Some of the Properties of the Piezo Electric-Quartz Resonator," K. S. Van Dyke, *Proc. I.R.E.*, Vol. 23, 1935, p. 386.
7. "Supersonic Reflectoscope for Interior Inspection," F. A. Firestone, *Metal Progress*, Vol. 48, No. 3, Sept. 1945, p. 505; "Supersonic Flaw Detector," R. B. De Lano, Jr., *Electronics*, Jan. 1946, p. 132.
8. "The Physical and Biological Effects of High-Frequency Sound-Waves of Great Intensity," R. W. Wood and A. L. Loomis, *Phil. Mag.*, Vol. 4, No. 7, 1927, p. 417.
9. "Magnetostriction Oscillators," G. W. Pierce, *Proc. Amer. Acad., Boston*, Vol. 63, 1928, p. 1.
10. "Efficient Piezoelectric Oscillator," W. W. Salisbury and C. W. Porter, *Rev. Sci. Instruments*, Vol. 10, 1939, p. 142.
11. *Wireless Telegraphy*, J. Zenneck, McGraw-Hill, 1915, Chapter IX.
12. *Time Bases*, O. S. Puckle, Wiley, 1945.
13. "A Convenient Method for Referring Secondary Frequency Standards to a Standard Time Interval," L. M. Hull and J. K. Clapp, *Proc. I.R.E.*, Vol. 17, 1929, p. 252; *Measurements in Radio Engineering*, F. E. Terman, McGraw-Hill, 1935, pp. 129–136.

Chapter 18

CIRCUITS FOR INDUSTRIAL CONTROL

H. J. Bichsel

18·1 HIGH-VACUUM-TUBE CIRCUITS

MANY control signals from indicating elements such as phototubes, strain gauges, and similar devices are at low voltage or power level. High-vacuum tubes are used primarily to raise the voltage or power level of the signal.

Designers tend to use thyratrons as amplifiers wherever possible because of their greater power capacity and their relay action; but thyratrons require large swings of grid voltage to bring about positive operation, and the signal may have to be amplified with high-vacuum tubes.

In comparison with radio applications, industrial applications of vacuum tubes are quite different. Industrial control applications seldom require radio frequency. The use of amplifier tubes in industrial circuits can best be studied by considering the type of signal input, the desired form of the output, and the desired amplifier characteristics.

Type of Signal

Three general types of signal are most frequently applied to an amplifier.

1. Direct current, slowly varying.
2. D-c bias with steep-wave-front pulse superimposed.
3. Periodic at audio frequency.

A slowly varying d-c signal is frequently obtained by balancing a regulated voltage against a reference voltage.

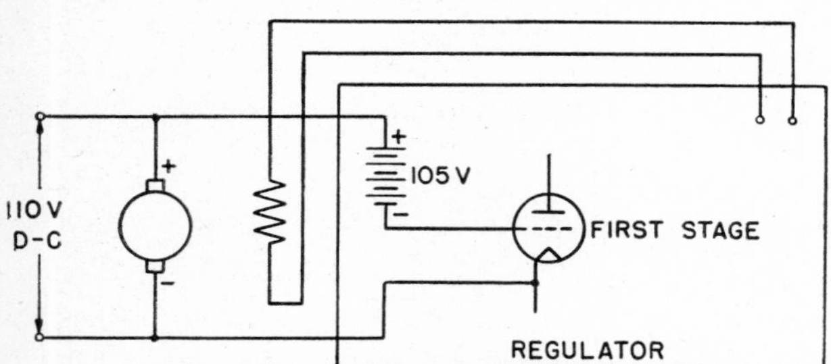

FIG. 18·1 Amplifier suitable for a slowly varying d-c signal.

An example of this is a direct voltage regulator with a constant battery-bias voltage connected to oppose the output of a d-c generator. The grid of the amplifier is actuated by the difference of the two voltages. Bias voltage might be 105 volts if the regulated generator voltage is to be 110. In an application of this kind, the grid may be swung from an extremely negative value to an extremely positive value with respect to the cathode if the regulated voltage is not near the reference voltage in value. Because grid voltage may vary slowly, it is necessary to use direct coupling of the amplifier, as shown in Fig. 18·1.

The signal input may be a non-recurring pulse of voltage with steep wave front, such as that obtained in a pin-hole detector. Light reaches the phototube for a period of time as small as one thousandth of a second while a small hole in a steel strip passes between the light source and the phototube. A capacitor may be used for coupling the phototube circuit to the first amplifier stage and for coupling successive stages of the amplifier.

Relatively few industrial control circuits have a periodic signal applied to the grid of a vacuum tube. One of the few examples is the amplification of the output of an a-c strain-gauge bridge to obtain sufficient power for automatic control. Audio-amplifier circuits similar to those used in radio are satisfactory.

Desired Form of Output

Most industrial control applications do not require high fidelity in an amplifier. The usual requirement is that minimum cost and maximum amplification be obtained. A good example is the use of an a-c power supply for the anode circuit with the amplifier tube serving as both an amplifier and a rectifier. This principle is applied in photoelectric relays where the amplifier is used to operate a relay of the telephone type. A capacitor connected around the relay coil has sufficient filtering action to prevent chattering of the relay. To minimize cost a minimum of filtering of anode power supplies is used.

Desired Characteristics

The desired characteristic of the amplifier often determines the details of design. The characteristic may vary, with the application, from that shown in curve A, Fig. 18·2, to that in curve B. In curve A, output of the amplifier is directly proportional to input signal. In curve B, no output is desired up to a certain input signal, at which point the output rises sharply to a maximum beyond which it does not increase in response to added input.

A good example of an application in which the characteristic of curve A would be desirable is an indicating arrangement in which a strain gauge must give a direct reading of the deflection of the member under test. The characteristic in curve B is regularly used in regulating systems such as a voltage regulator for which maximum sensitivity is desired over a range of one or two volts on each side of the regulated voltage. For example, the regulator would be designed so that no field current would be applied to the generator at 111 volts, and maximum field would be obtained at 109 volts.

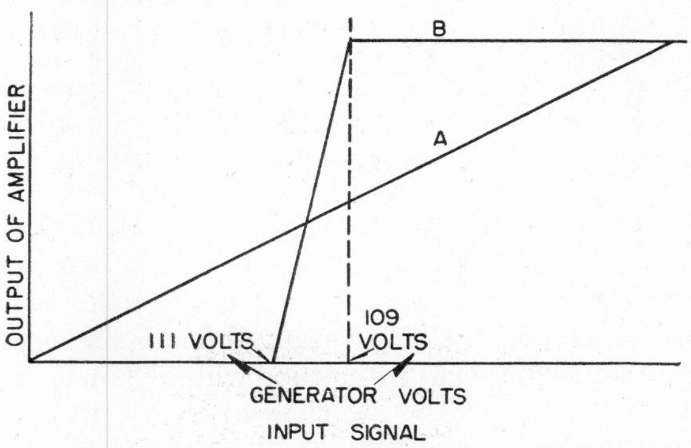

FIG. 18·2 Amplifier characteristics.

As a result, the amplifier has tremendous sensitivity so that balance at 110 volts is obtained with an accuracy of $\pm\frac{1}{4}$ volt.

18·2 BASIC PHANATRON CIRCUITS

The phanatron is most widely used as an uncontrolled rectifier tube (see Section 15·2), but in control circuits it is used also as an uncontrolled switch capable of "closing" whenever the anode voltage is positive and greater than the breakdown value for the tube (about 20 volts for most tubes). Two such applications of the phanatron are illustrated in Fig. 18·3. A capacitor C is originally charged with the polarity as shown. Contact M is open. To use capacitors with the smallest possible voltage ratings, capacitor voltage should not reverse during the discharge. The tube is used to prevent reversal of voltage. When contact M is closed, the capacitor is discharged into transformer T_1.

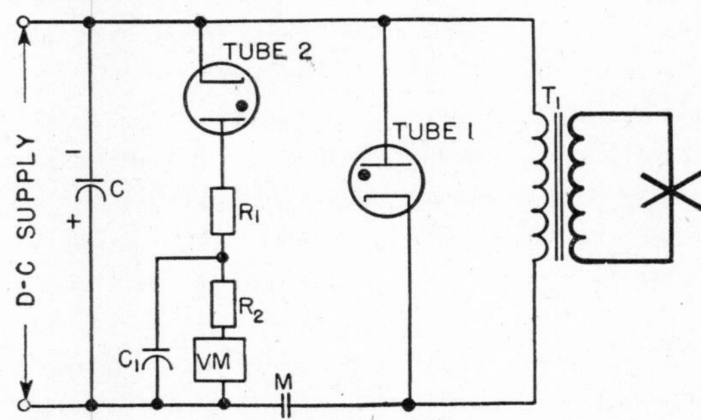

FIG. 18·3 Phanatrons used as uncontrolled switches.

Circuit resistance, inductance, and capacitance may be of such values that the voltage and current oscillate and cause the voltage across the transformer to reverse, making the anode of tube 1 become positive with respect to its cathode. It then conducts current and transfers the current from the capacitor circuit to the tube and minimizes reverse voltage on the capacitor. (The reverse voltage is equal to the arc drop of the tube.)

A second application of the phanatron as an uncontrolled switch is illustrated by tube 2 of Fig. 18·3. The maximum voltage across capacitor C is to be measured while the capacitor is being charged and discharged about 200 times per minute. This voltage can be measured by paralleling the voltmeter and its resistor with capacitor C_1 and charging the capacitor to the voltage across capacitor C by means of tube 2. If capacitor C_1 and resistor R_1 are small compared with capacitor C, the voltage across capacitor C_1 is equal to the voltage across capacitor C minus the arc drop of tube 2 as long as tube 2 conducts. If capacitor C is discharged tube 2 ceases to conduct because its cathode is more positive than its anode. Capacitor C_1 discharges slowly through the high resistance of the voltmeter until capacitor C again reaches nearly full charge. Tube 2 then conducts again and charges capacitor C_1 to the same voltage (minus arc drop) as capacitor C. Because of the charge on capacitor C_1, the voltmeter pointer falls only slightly during discharge of capacitor C; the voltmeter can quickly indicate the voltage on capacitor C when it is charged.

The voltage across a capacitor can be used for timing. If rapidly recurring intervals are to be timed, the capacitor

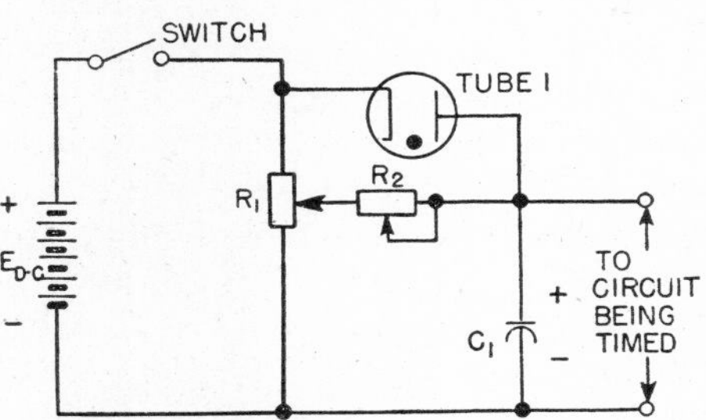

FIG. 18·4 Phanatron used as an uncontrolled switch in a timing circuit.

must be discharged quickly between intervals. A phanatron acting as an uncontrolled switch can accomplish this as shown in Fig. 18·4. When the switch is closed, capacitor C_1 is charged through resistors R_1 and R_2. R_1 is small and R_2 usually is comparatively large. Voltage across these resistors makes the anode of tube 1 negative and it cannot conduct. After the timing is finished capacitor C_1 is charged with the polarity shown. As the switch is opened the capacitor begins to discharge through R_1 and R_2, creating a voltage across them which initiates tube 1 and short-circuits R_2. The capacitor then discharges rapidly through R_1, which is small.

The phanatron may be used as a switch in any circuit in which the voltage reverses when the switching action is desired. Selection of the proper tube depends upon average and peak currents and inverse voltages of the circuits. Ignitrons may be used for currents too large for phanatrons, and a phanatron may be used to initiate the ignitron.

18·3 BASIC THYRATRON CIRCUITS

The thyratron is used extensively as a controlled tube to close d-c or a-c circuits at some predetermined instant, perhaps in response to an electrical impulse from the controlled circuit or from another circuit.

If the thyratron has a d-c supply, its basic circuit is shown in Fig. 18·5. Its components are similar to those used for vacuum-tube circuits: the tube, cathode-heater transformer, anode supply, load, grid-bias source, grid resistor, grid capacitor, and a switch.

The grid is biased negatively so that the thyratron does not conduct when the anode switch is closed (no input signal). The bias required depends upon the characteristics of the tube, type of load, magnitude of anode voltage, and the source of anode voltage. Minimum bias practical for most tubes is about 25 volts regardless of their characteristics or anode voltage. If the critical grid voltage is more than 15 volts negative, grid bias should be at least twice that value. These seemingly large values of bias are required to prevent false operation of tubes because of line-voltage

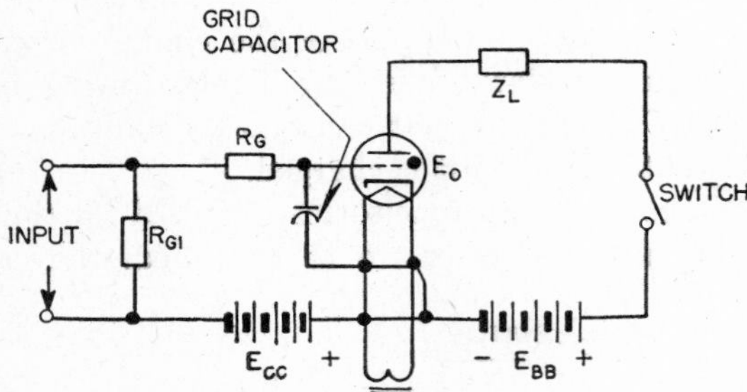

FIG. 18·5 Thyratron used as a controlled switch in a d-c circuit.

surges, temperature variations, and changes in characteristics caused by aging of the tube. If the applied negative grid voltage is too large, the tube may conduct from grid to anode, or it may arc back from grid to cathode. Maximum negative grid-voltage ratings usually are given in the rating sheets for thyratrons.

The grid resistor R_G serves three purposes: first, it limits the current in the grid circuit before and after the tube conducts; second, it partly filters line surges in the bias supply before they can reach the grid; third, it discharges from the grid capacitor any voltage accumulated on it because of surges in the anode circuit. To perform the first two functions the grid resistor should have a large resistance, but it should be small to perform the third function. Therefore the selection of the resistor is a compromise.

The grid capacitor is used mainly to prevent the tube from "shocking over" or conducting (even though proper negative bias is applied to the grid) if voltage is applied suddenly to the anode circuit. In some thyratrons the interelectrode capacitance from anode to grid is about as large as the capacitance from grid to cathode. If a steep wave front of voltage is applied between anode and cathode, the interelectrode capacitances charge rapidly and the grid voltage rises. If grid voltage rises to a value greater than the critical grid voltage, the tube conducts, provided the grid voltage is positive for a time long enough to ionize the gas in the tube. In effect the grid capacitor increases the grid-to-cathode capacitance of the tube to roughly 1000 times the grid-to-anode capacitance; therefore most of the voltage appears across the grid-to-anode capacitance when a steep wave front of voltage is applied, and the tube does not shock over.

The grid capacitor and resistor also filter out surges or disturbances that come in through the grid-bias supply or appear across the input resistor from another circuit. For some tubes, certain combinations of resistors and capacitors have become somewhat standardized. The resistance ranges from 10,000 ohms to about 1 megohm; capacitances may range from 0.0005 to 0.01 microfarad. Some typical values are:

Tube Type	Resistance (ohms)	Capacitance (microfarads)
RCA2050	20,000	0.001
KU627	100,000	0.002
WL672	100,000	0.002
WL677	100,000	0.01
WL414	50,000	0.01

In the anode circuit the load may be a combination of resistance and inductance, such as relay coil, or a combination of resistance and capacitance. Impedance of the load must be large enough to limit peak and average currents to the rated values; otherwise the tube will be damaged.

Thyratron Operation with Unidirectional Anode Voltage

When the switch in the anode circuit (Fig. 18·5) is closed, the tube does not conduct if the negative bias is larger than the critical value, and no signal is applied to the grid. To make the tube conduct, the grid voltage must be made more positive than the critical value. This can be done by supplying a voltage across the input resistor R_{G1} with the grid end positive. The current to produce this voltage may come from one of several different sources such as a vacuum-tube amplifier, another thyratron circuit, a phototube, an impulse transformer, a peaking transformer, or from another electrical device such as a motor or a generator.

Once the grid voltage becomes more positive than the critical value for at least 10 microseconds, the tube conducts and the tube drop decreases to about 15 volts. The voltage across the load therefore becomes

$$E_L = E_{BB} - E_0 \tag{18·1}$$

Impedance of the load must be large enough to limit tube current to its rated average value, or the tube will overheat and it may be destroyed or have its life shortened.

Once the tube has started to conduct, the grid is no longer effective in controlling current through the tube. To stop conduction reliably, anode voltage must be reduced to some value below the arc-drop voltage for an interval of time exceeding the deionization time of the tube. One of the easiest means of stopping the current is to open the switch in the anode circuit. Conduction can be stopped, however, by applying a voltage in series with the anode to make it negative with respect to the cathode. Such methods are used in inverters, trigger circuits, and other control circuits.

The d-c thyratron circuit is useful for applications that require "lock-in" operation, such as an overvoltage device that is reset manually. The d-c thyratron circuit is also useful if the thyratron is actuated by a single high-frequency impulse and if it then must conduct current for a long period of time.

Thyratron Operation with Alternating Anode Voltage

The basic circuit for operating the thyratron from an a-c source is shown in Fig. 18·6. The anode voltage becomes positive once each cycle. If the grid bias is less negative than the critical value while the anode is positive, the tube conducts until the end of the positive half cycle.

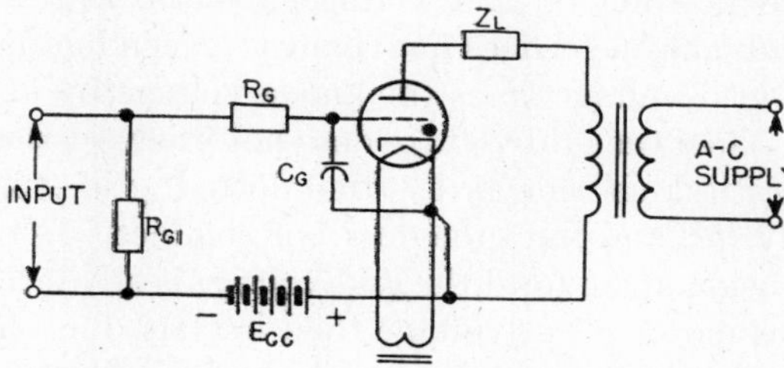

Fig. 18·6 Thyratron used as a controlled switch in an a-c circuit.

The method of obtaining the critical grid characteristic for alternating anode voltage is given in Fig. 18·7. If the grid voltage is more negative than the a-c critical grid charac-

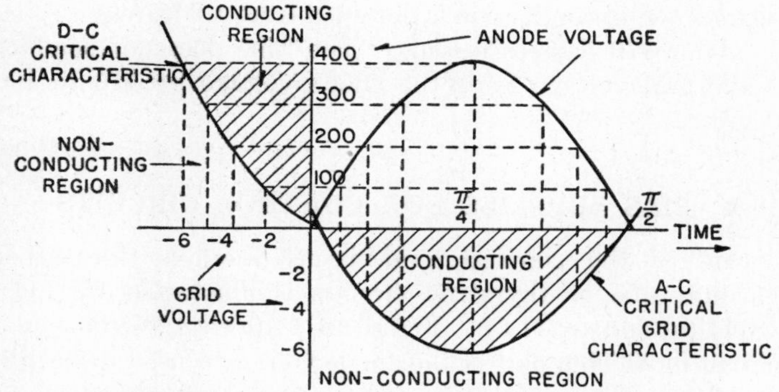

Fig. 18·7 Thyratron a-c critical grid characteristic.

teristic the tube does not start to conduct. As grid voltage becomes more positive than the critical value the tube starts to conduct.

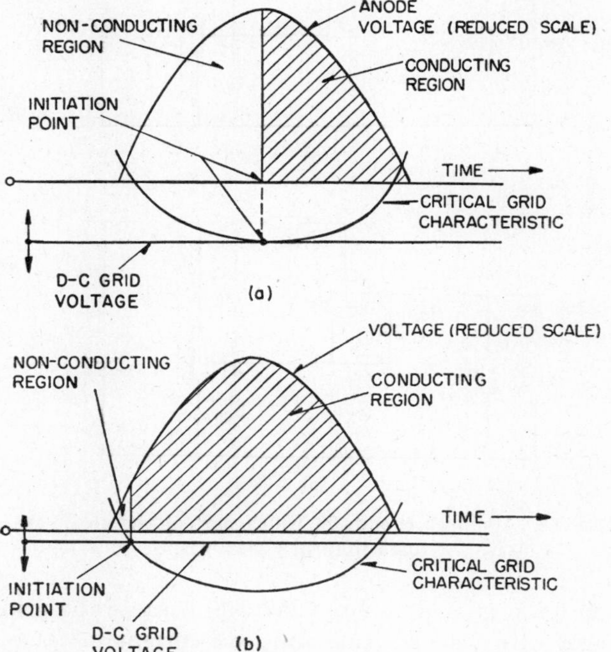

Fig. 18·8 Thyratron controlled by varying the d-c grid voltage.

Two common methods are used to control the initiation point. They are known as d-c amplitude control and a-c phase-shift control. Figures 18·8 (*a*) and 18·8 (*b*) illustrate d-c amplitude control. If the unidirectional grid voltage is more negative than the most negative point on the critical grid characteristics, the tube does not start conduction. As grid voltage is made less negative, the tube begins to conduct at the instant the voltage becomes less negative than the critical value. In Fig. 18·8 (*a*) the tube starts conduction at 90 degrees on the anode-voltage wave. This is the greatest possible initiation delay with unidirectional grid voltage because, as the grid voltage is made still more negative, it will not intersect the critical characteristic and the tube will

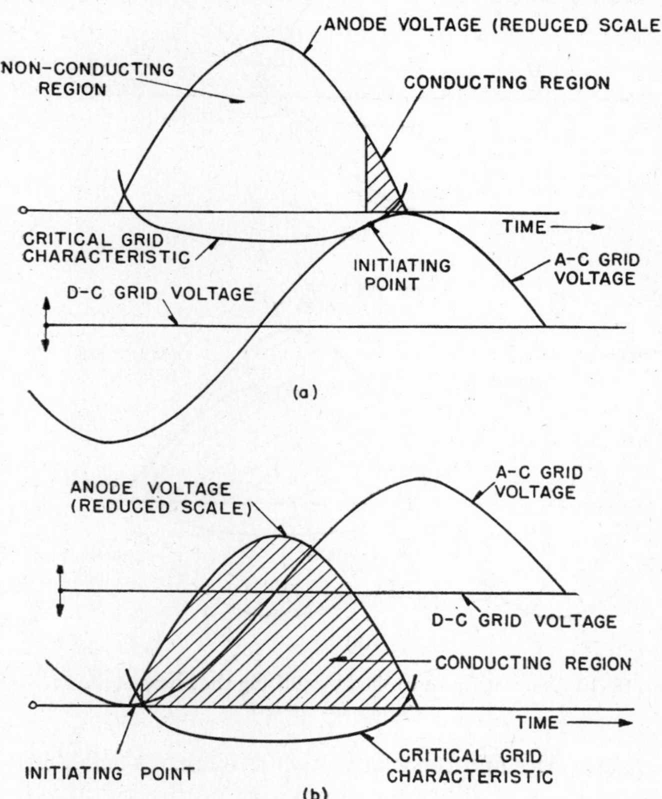

Fig. 18·9 Thyratron controlled by varying the d-c grid voltage on which there is superimposed a phase-shifted sinusoidal voltage.

not conduct at all. In Fig. 18·8 (*b*) the grid voltage is less negative, and the tube conducts nearly maximum current. To make the tube conduct maximum current it may be necessary to make the grid voltage slightly positive. Positive grid voltage measured inside the grid resistor should not exceed 8 or 10 volts, for an arc may occur between grid and cathode.

Output current cannot be closely controlled by the d-c amplitude method because the angle of intersection of the grid voltage and the critical characteristic is small. This means that a small change in amplitude of grid voltage causes a large change in the initiation point and consequently in tube current. Since small variations in the grid-bias voltage and in the critical grid characteristics cannot be prevented without undue expense, d-c amplitude control is seldom used. Usually a phase-shifted alternating voltage is superimposed on the unidirectional grid voltage as shown in Fig. 18·9 (*a*). The alternating grid voltage lags the anode voltage by 90 degrees; therefore its positive peak first intersects the critical characteristic near 180 degrees on the anode-voltage wave.

By varying the magnitude of the unidirectional grid voltage between positive and negative values equal to the peak value of the alternating grid voltage, tube current can be varied from zero to maximum as shown by Figs. 18·9 (*a*) and 18·9 (*b*). Tube current can be closely controlled because the angle of intersection between the critical grid characteristics and the grid voltage is large. A small change in magnitude of either the alternating or direct grid voltage has little effect on tube current, provided the alternating component has sufficient magnitude.

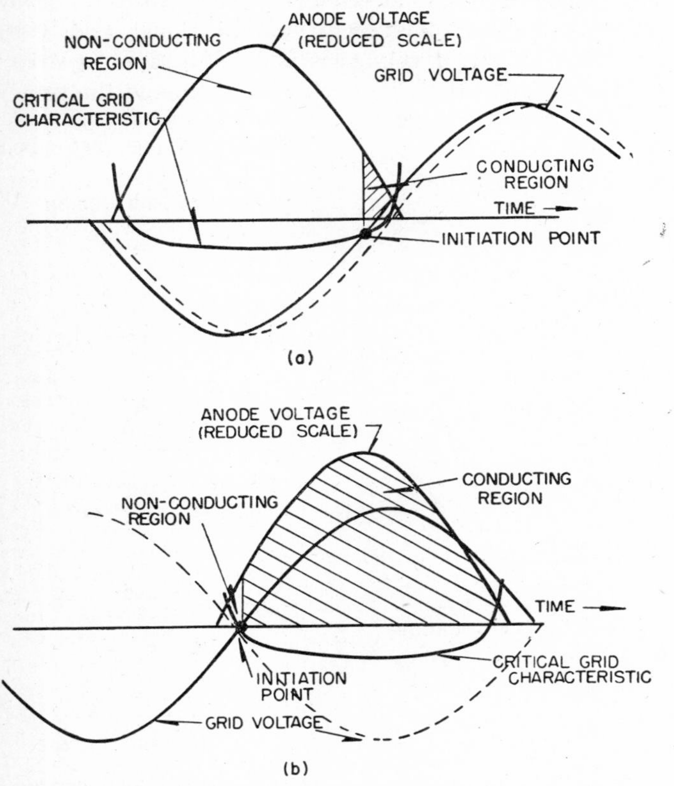

FIG. 18·10 Thyratron controlled by shifting the phase of the sinusoidal voltage on the grid.

Phase-shift control is accomplished by shifting the phase position of the grid voltage relative to the anode voltage of the tube. If grid voltage lags anode voltage by 160 degrees in phase position the tube conducts only for a small part of the half cycle, as shown by the solid grid-voltage wave of Fig. 18·10 (*a*). This figure also shows that the tube does not conduct if grid voltage lags anode voltage by 170 to 180 degrees (dotted grid-voltage wave) because the grid voltage does not intersect the critical grid characteristic in that region. The region of non-conduction depends upon tube characteristics and upon anode and grid voltages. In tubes without shield grids the region may be quite narrow; in tubes with shield grids it may be 30 or 40 degrees. In fact, the non-conduction region can be controlled by the shield-grid voltage.

As the phase position of grid voltage is advanced from that shown in Fig. 18·10 (*a*), the grid voltage intersects the critical grid voltage sooner in the half cycle, and the tube conducts for a longer time during each half cycle. Figure 18·10 (*b*) shows the initiation point and the conducting period when the grid voltage lags the anode voltage by 10 degrees (solid grid-voltage wave). Conduction is almost maximum. Maximum conduction is obtained by advancing the phase position of grid voltage about 5 degrees more. Further advancement does not change the conducting period until the grid voltage leads the anode voltage by about 170 degrees. Then the grid voltage does not intersect the critical characteristics, as shown by the dotted grid voltage in Fig. 18·10 (*b*), and the tube does not conduct. Conduction does not begin again until the grid voltage is advanced about 20 degrees more. Grid voltage then is phased as shown by the solid curve in Fig. 18·10 (*a*), and slight conduction begins.

It is evident that the conduction of a thyratron can be controlled continuously and smoothly from minimum to maximum, and off-on control can be obtained. Control is continuous if grid voltage lags anode voltage by 10 to 170 degrees. Full conduction is obtained from 10 degrees lagging to about 170 degrees leading. The tube does not conduct if the grid voltage leads the anode voltage by 170 to 190 degrees.

18·4 BIAS SUPPLIES FOR CONTROL CIRCUITS

One of the simplest grid-biasing methods for either vacuum-tube or thyratron circuits is illustrated by Figs. 18·11 (*a*) and 18·11 (*b*). Figure 18·11 (*a*) is a vacuum-tube circuit in which a part of the anode voltage is used to furnish grid bias. This circuit is used where output voltage need

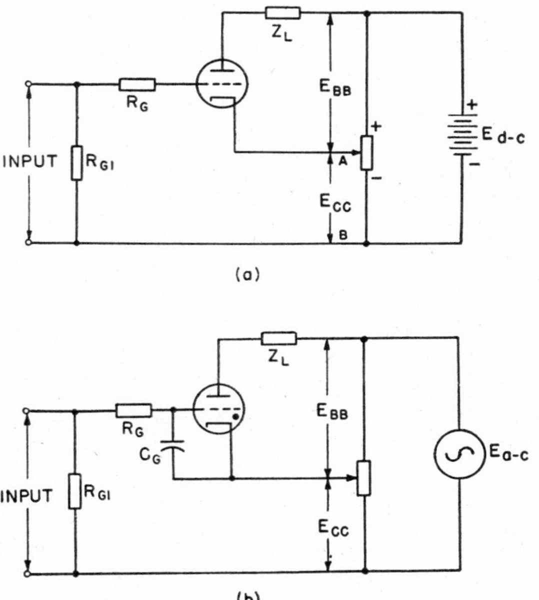

FIG. 18·11 Amplifier circuits in which part of the anode voltage is used to furnish grid bias.

not be proportional to input voltage, because the grid bias changes with load current. As load current increases, the voltage drop across the bleeder resistor from *A* to *B* increases. This increases the negative bias, which tends to decrease plate current. This effect can be minimized if the anode current is small in comparison with the power-supply bleeder current. If a thyratron were used in this circuit, the change of grid-bias with changing load current would have no influence, since grid voltage has little influence on tube current after the tube starts to conduct.

Figure 18·11 (*b*) involves a thyratron with a-c anode supply, part of which is used to supply grid bias. Here the change in grid bias because of changing tube current is not serious, for the tube is extinguished at the end of each positive half cycle of voltage. Bias voltage is not affected during the negative half cycle and the following positive half cycle until the tube is made to conduct by the grid-input signal.

In many applications ripple in the grid-bias voltage is not objectionable, and a half-wave rectifier and a capacitor-resistor filter, such as in Fig. 18·12 (*a*), can be used as a bias supply. The rectifier may be either a copper oxide rectifier or a vacuum tube. For voltages of less than 100, a copper oxide rectifier is probably less expensive, for no filament transformer is required. A capacitance of 2 microfarads and a resistance of 50,000 or 100,000 ohms make an adequate filter, because almost no current is drawn from the supply by the grid circuit.

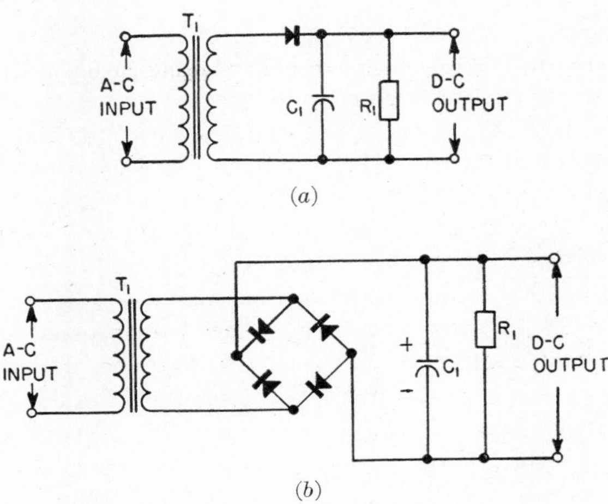

Fig. 18·12 Rectifiers suitable for furnishing grid bias.

For less ripple the full-wave copper oxide rectifier of Fig. 18·12 (*b*) may be employed. With the bridge type of rectifier a special center-tapped transformer is not needed. The filter in Fig. 18·12 (*b*) is the same as in Fig. 18·12 (*a*), but more complicated filters can be used if less ripple is desired.

18·5 PHASE-SHIFTING CIRCUITS

Many control circuits require voltages which are shifted in phase with respect to each other. In some, the phase shift can be fixed; in others, such as the voltage used on the grid of a thyratron to control output current, the phase shift must be variable. Resistance-capacitance and resistance-inductance combinations are used to phase-shift voltages.

Resistance-Inductance Phase Shifter

One of the most common phase-shifting circuits is the one in Fig. 18·13 (*a*). It consists of a transformer and a resistance-inductance combination. The transformer may be a step-up or step-down unit (with its secondary center-tapped), depending upon the magnitude of the supply voltage and the output voltage needed. Phase position of the output voltage can be shifted through more than 170 degrees by varying the resistance, if the resistance and inductance are properly chosen.

Figure 18·13 (*b*) is the vector diagram for the circuit of Fig. 18·13 (*a*). The reference vector is the transformer secondary voltage E, which is center-tapped. In the example shown by the solid lines the resistance is slightly smaller than the impedance of the inductor. Resistor voltage E_r is in phase with the current through the resistance and inductance. Voltage E_L across the inductance leads the current by 90 degrees. Output voltage is the vector voltage between the center tap and the junction of the resistor and inductor voltages. In the example shown by the dotted lines, the resistance has been increased, and the current lags the applied voltage by a smaller angle because the circuit is less inductive. The angle between resistor and reactor voltages is always 90 degrees, and the locus of a triangle, any two sides of which always intersect at right angles, is a semicircle, as shown. Output voltage E'_0 is more nearly in phase with E than the voltage E_0 of the first example.

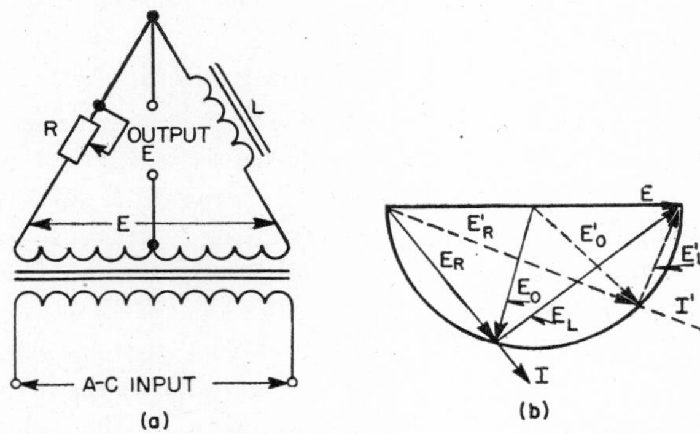

Fig. 18·13 Resistance-inductance phase-shifting circuit.

If the resistance is reduced to zero, the output voltage lags the applied voltage by 180 degrees. With maximum resistance in the circuit the output voltage is nearly in phase with the applied voltage; however, it always lags slightly because of the voltage drop across the inductor, even though the resistance is large compared with the impedance of the inductor.

In designing the circuit, minimum current through the resistance-reactance branch must be made large compared with the current drawn by the output circuit; otherwise, output voltage decreases as the phase-shifting resistance is increased. One practical circuit of this type uses a 1250-ohm resistor and a 0.7-henry inductor. This circuit is used as a phase-shifting circuit or "heat control" for a pair of thyratrons, which in turn initiate two ignitrons that control the primary current of a resistance-welding machine.

Resistance-Capacitance Phase-Shifting Circuit

Phase-shifted voltage can be obtained from a resistor-capacitor circuit like that in Fig. 18·14 (*a*). A phase shift of about 170 degrees can be obtained in this circuit by varying the resistance in the circuit. The solid-line vector shows voltage relations with small resistance in the circuit. Here output voltage E_0 leads the secondary voltage E by nearly 150 degrees. As resistance is increased output voltage becomes E'_0, which leads E by about 45 degrees.

This circuit is commonly used for a fixed phase shift, as in the grid circuit of a thyratron using amplitude control.

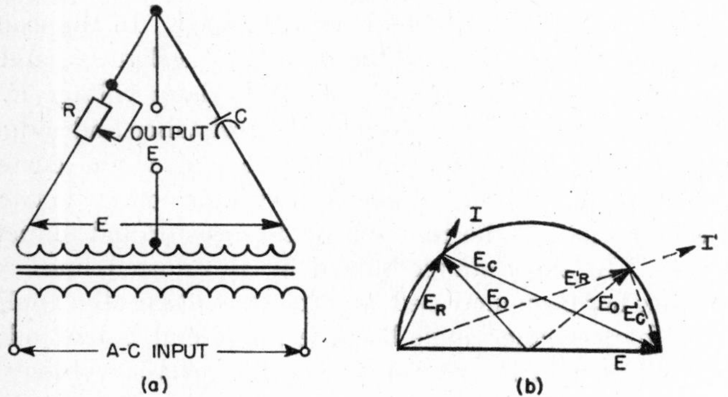

FIG. 18·14 Resistance-capacitance phase-shifting circuit.

Capacitor-Transformer Phase-Shifting Circuit

Another method of shifting phase position is illustrated in Fig. 18·15. Here a variable resistor is paralleled with a capacitor, and this combination is in series with the primary of a small transformer. An alternating voltage is applied to the input, and the phase-shifted voltage comes from the transformer secondary.

If the variable resistor is set to zero resistance, the primary of the transformer is then connected directly to the a-c supply, and the output is nearly in phase with the input. As resistance is increased the circuit becomes more capacitive, the current leads the applied voltage, and the phase position of the output voltage begins to advance. If the resistance is large compared with the capacitive reactance, output voltage can be made to lead the applied voltage by 90 degrees or more, depending upon the relative sizes of capacitive and inductive reactance of the circuit.

The shape of the output voltage wave from this circuit is distorted because of the voltage drop created by transformer magnetizing current in the series impedance. However, the circuit is extensively used with peaking transformers.

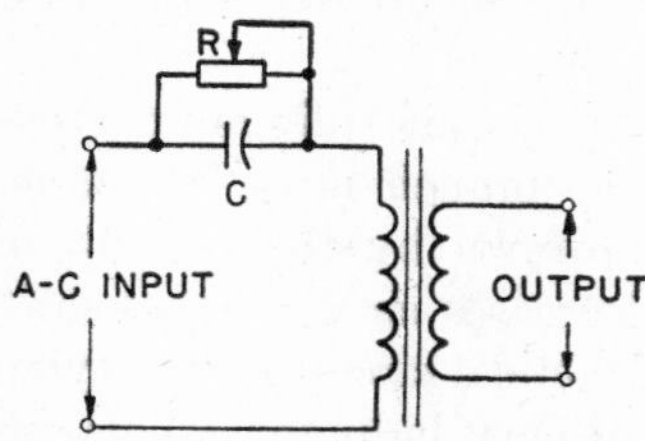

FIG. 18·15 Capacitor-transformer phase-shifting circuit.

Electrically Controlled Phase-Shifting Circuits

In some circuits phase shift is controlled by varying the resistance or inductance electrically. Resistance is varied by using the circuit shown in Fig. 18·16. The principle of this circuit is to reflect the resistance of the triode into the phase-shifting circuit through the full-wave rectifier and the transformer. The effective resistance of the vacuum tube can be varied by changing grid voltage, thus changing the reflected resistance and shifting the phase of the output voltage from the phase-shifting circuit.

Amount of phase shift is limited by the minimum resistance of the vacuum tube and its voltage and current ratings. In the circuit shown maximum shift is approximately 150 degrees, if the phase shifter has a high-impedance load.

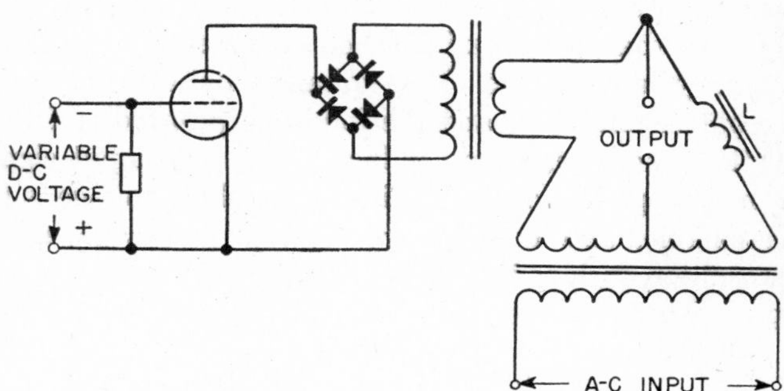

FIG. 18·16 Resistance-inductance phase-shifting circuit in which the resistance is varied electrically.

When the grid of the triode is least negative, the effective resistance of the tube is low. Output voltage lags input voltage by nearly 150 degrees. Output voltage is nearly in phase with input voltage when the grid voltage is highly negative and the tube's resistance is high.

Figure 18·17 shows a resistance-inductance phase-shifting circuit in which the resistance is fixed and the inductance is varied by using a saturable-core reactor. Phase shift is effected by varying the direct current in the saturating coil of the reactor, thereby changing the inductance of its a-c winding. Its reactance is maximum when voltage on the d-c coils is minimum, and it is minimum when voltage on the d-c coil is maximum or when the reactor is saturated. Output voltage of the phase shifter therefore is nearly in phase with applied voltage when voltage on the saturating coil is high, and it lags by almost 150 degrees when voltage on the d-c coil is zero.

Voltage for the saturating coil is adjusted by manual change of a potentiometer, or by a vacuum-tube system as

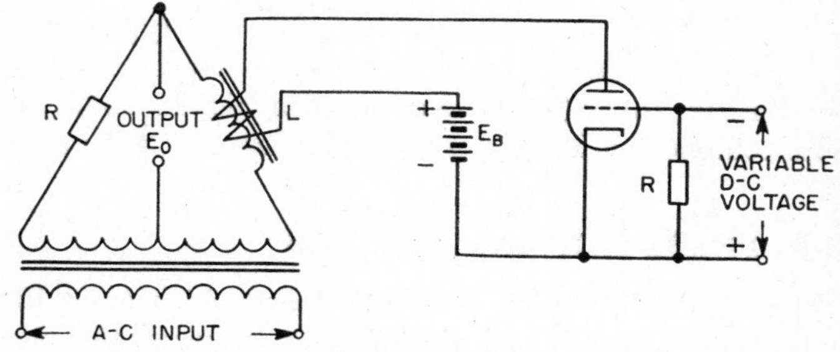

FIG. 18·17 Resistance-inductance phase-shifting circuit in which the inductance is varied electrically.

shown in Fig. 18·17. In the vacuum-tube system, output voltage is almost in phase with input voltage when the grid voltage on the tube is zero; input and output voltages are about 150 degrees out of phase when the grid voltage is highly negative. The control circuit can be electrically isolated from the phase-shifting circuit. This may be desirable if the phase-shifting circuit operates at high voltage, or if the control circuit is a complicated, directly coupled system.

18·6 IMPULSE-GENERATING AND PEAKING CIRCUITS

In many circuits using thyratron tubes, voltage peaks with steep wave fronts are needed for initiating conduction. Peaks are desirable for initiating conduction because they intersect the critical grid characteristic at an angle of about 90 degrees, thereby accurately controlling the time of initiation. Since the peak is narrow (10 electrical degrees), a large peak can be used to insure initiation without exceeding the grid current rating of the tube. If a peak is used, grid voltage does not remain positive when anode voltage is negative. The grid would be positive while the anode voltage was negative if a complete a-c wave were used to initiate conduction; Fig. 18·9 (*b*). This might cause the tube to

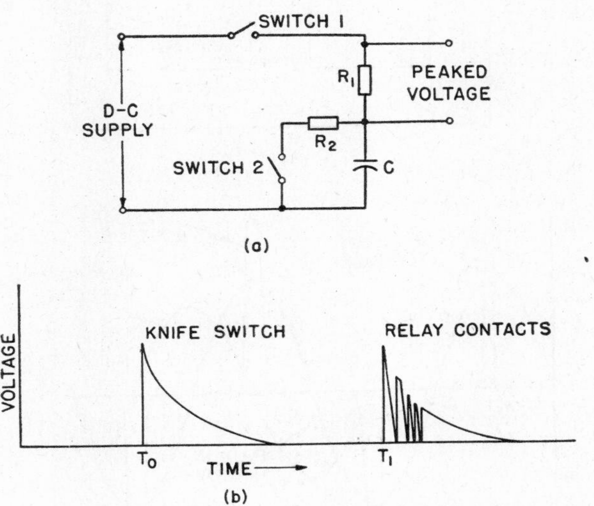

FIG. 18·18 Resistance-capacitance peaking circuit.

arc back because the grid keeps the gas in the tube ionized while the anode voltage is negative. Peaked voltages can be obtained from d-c resistive-inductive or resistive-capacitive circuits. Peaks can also be obtained by discharging a capacitor into a resistor or into the primary of the transformer. Other transformers, specially designed, can deliver peaked secondary voltages when they are supplied by an a-c source.

In the circuit of Fig. 18·18 (*a*) a single voltage peak is produced across resistor R_1 when switch 1 is closed, provided switch 2 is open. Figure 18·18 (*b*) shows the output voltage. Switch 1 was closed at time T_0. The magnitude of the peak is equal to the supply voltage (minus any voltage drop in the supply) since the voltage across the capacitor cannot change instantaneously. As the capacitor charges, the voltage across the resistor decreases exponentially to zero. The duration of the peak therefore depends upon the time constant of the circuit.

After the first peak the capacitor must be discharged before another peak of the same magnitude can be produced. The capacitor is discharged by closing switch 2. Resistor R_2 limits peak current through the switch to a safe value.

In commercial circuits the switch probably would be the contacts of a relay. A normally open contact would replace switch 1 and a normally closed contact would replace switch 2. Since the contacts of most relays bounce when they close, the peaked voltage may not be of the ideal shape shown by the first peak of Fig. 18·18 (*b*), but may be jagged and intermittent as shown by the second peak. If multiple peaks are objectionable, a thyratron can be connected in series with the relay contact to close the circuit. The relay contacts then are closed before the peak is desired, and the thyratron tube can be initiated after the contact has quit bouncing, to obtain the peak. The relay contacts must be opened between each peak to stop conduction of the thyratron because it operates on unidirectional voltage.

Figure 18·19 (*a*) illustrates a peaking circuit using a resistor-inductor combination. Either opening or closing the

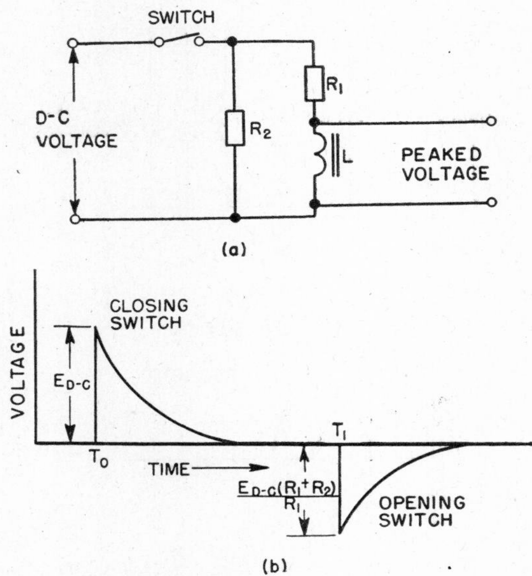

FIG. 18·19 Resistance-inductance peaking circuit.

switch produces a peak across the inductor. At the instant the switch is closed the current through resistor R_1 and the reactor is zero; therefore voltage drop across the resistor is zero and a voltage peak equal to the voltage of the d-c source appears across the inductor. Current through the inductor increases exponentially until it reaches a value determined by the voltage of the d-c source and the resistance. As the current increases, the voltage drop across resistor R_1 increases and the voltage across the inductor decreases, as shown in the first peak of Fig. 18·19 (*b*).

When the switch is opened at time T_1, the voltage peak is of opposite polarity. As the switch is opened reactor current begins to fall but the change in current induces a voltage of such polarity and magnitude as to keep the current constant. Current through resistors R_1 and R_2 decreases exponentially as the energy in the reactor is dissipated. This decrease is shown by the second peak of Fig. 18·19 (*b*). The peak voltage is the product of the original current and the resistance of the discharge path $(R_1 + R_2)$. The duration of the peak depends upon the time constant $L(R_1 + R_2)$.

If a relay contact is used in place of a manual initiating switch, a peak in both directions probably will be caused by the bouncing of the contacts as they close. This condition may not be objectionable; if it is, a single peak can be obtained by using a normally closed contact and opening the contact to obtain the peak. To obtain a second peak, the contact must be reclosed, and bouncing of contacts may produce another peak of the same polarity as the original peak. Unless the contact can be closed so that the peaks do not initiate conduction of the thyratron this circuit may not be satisfactory. A transformer can be used in place of the reactor to isolate the peaking circuit from the circuit in which the peak is used.

A suitable peak for initiating thyratrons can be produced by discharging a capacitor into a resistor or the primary of a transformer. Such a circuit is given in Fig. 18·20 (*a*). The wave shape of the peaked voltage depends upon the relative values of the capacitance, inductance, and resistance of the circuit. If a thyratron is used to close the circuit, the constants usually are selected so that the circuit is slightly oscillatory. The tube then is extinguished when the current reaches zero. The relation of circuit constants for this condition is

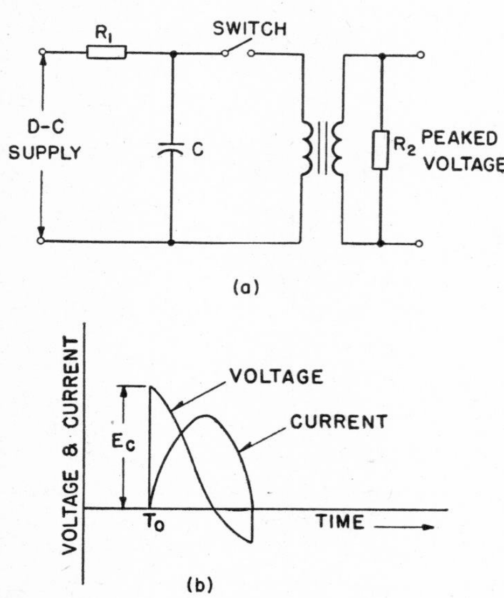

FIG. 18·20 Peaking circuit suitable for firing thyratrons.

$$R^2 < \frac{4L}{C} \qquad (18\cdot2)$$

Figure 18·20 (*b*) shows the wave shapes of the oscillatory voltage and current.

An A-C Peaking Transformer

Often a peak in every half cycle of an a-c wave is needed. Several different types of transformers have been designed to provide a narrow peak of output voltage on each half cycle. Figure 18·21 (*a*) shows the construction of one of these transformers. The magnetic core consists of three legs on one of which the primary is wound. The middle leg is a shunt with a small air gap in it, and the third leg consists of a few pieces of Hipernik over which the secondary is wound. Hipernik is a magnetic material which saturates very sharply at about 15 kilogauss.

With an alternating voltage applied to the primary, peaks of the type shown in Fig. 18·21 (*b*) are produced across the secondary. The peaking may be understood by considering the flux change in the Hipernik. Under steady-state conditions, the primary magnetizing current lags the applied voltage by 90 degrees, and the flux in the core is in phase with the magnetizing current. Starting at current zero, the flux in the secondary leg increases until the Hipernik saturates; the flux then travels through the shunt leg. Before the Hipernik saturates, the secondary voltage is of the same magnitude as it would be in a conventional transformer. After the Hipernik saturates, the secondary voltage quickly falls to zero, since $d\phi/dt = 0$, and remains small until the Hipernik desaturates. Then the flux changes from positive to negative when the current passes through zero on the next half cycle, at which time another peak in the opposite direction is generated.

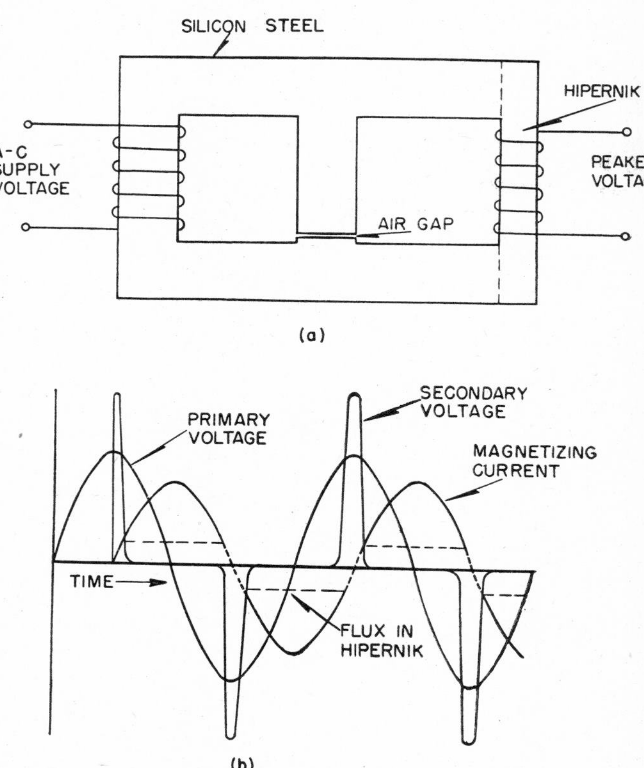

FIG. 18·21 Peaking transformer.

18·7 TIMING CIRCUITS

Timing circuits may use either vacuum tubes or thyratrons. The thyratron has gained popularity because of its characteristic of being either completely conducting or nonconducting.

Any timing circuit depends upon a consistent voltage with a known variation with time. Such voltages can be produced by resistor-capacitor or resistor-inductor circuits with d-c supplies. They are the same as those shown in Figs. 18·18 (*a*) and 18·19 (*a*), except that the capacitor voltage of Fig. 18·18 (*a*) and the resistor voltage of Fig. 18·19 (*a*) are used. The resistance-inductance timing circuit is seldom used because more power is required; therefore only the resistance-capacitance circuit of Fig. 18·22 (*a*) is considered here.

Either the capacitor charging time or discharging time can be used as the timing interval. The circuit of Fig. 18·22 (*a*) is designed to utilize charging time. As contact 1 is closed, capacitor C begins to charge from the d-c supply, provided contact 2 is open. Voltage across the capacitor terminals can be expressed

$$e_c = E(1 - \epsilon^{-t/R_1C}) \qquad (18\cdot3)$$

where E = voltage of d-c supply
R_1 = series resistance
C = capacitance in microfarads.

This equation is shown graphically in Fig. 18·22 (*b*.) The time constant of the circuit, T, is defined as the time required for the capacitor voltage to reach 63.2 percent of the source voltage, and it is expressed

$$T = RC \qquad (18\cdot4)$$

If the negative bias on the tube is 63.2 percent of E and the critical characteristic is assumed to be negligible, the time

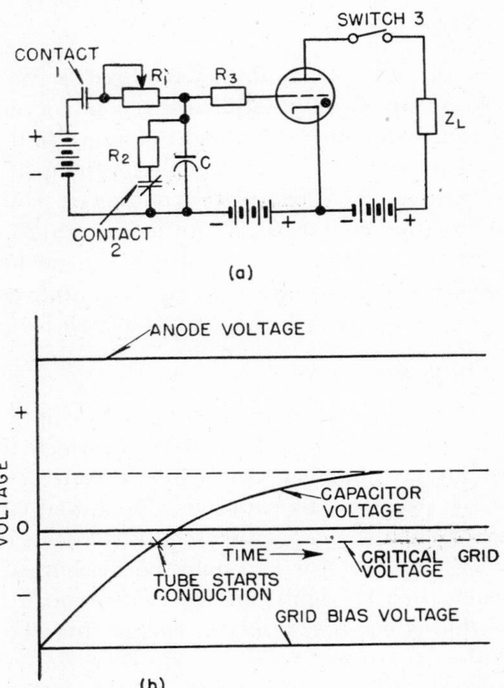

FIG. 18·22 Thyratron timing circuit utilizing capacitor charging time.

between the closing of contact 1 and the initiation of the tube is $T = RC$. Usually the capacitance is fixed and the resistance is variable, if variable timing is desired. Adjustment in timing is limited to about 20 to 1 and should be not more than 10 to 1 if an accuracy of ±1 to 2 percent in time is desired. If greater adjustment is required, a range switch can be used to increase the resistance or the capacitance.

Selection of proper values of timing resistance, voltage, and capacitance is important for accurate timing. Because capacitors are more expensive than resistors, designers tend to use a small capacitor and a large resistor. Although this practice produces the cheapest circuit, it may not be the best, because high-resistance potentiometers are short-lived, their resistance changes, and their operation is sometimes erratic when they become dusty or dirty. Moreover, inductive pick-up may produce a large voltage across a high impedance in the grid circuit. Generally it is good practice to use a variable resistance of less than 50,000 ohms, if possible. With this resistance, timing intervals of 1/60 to 1/2 second can be obtained accurately by using a 5- or 10-microfarad capacitor.

Timing voltage must be as large as the negative grid bias, or the thyratron will never fire. If the grid bias is about 63 percent of the timing voltage, the calculation of timing is simplified, and the timing voltage is changed rapidly enough to give accurate timing. Timing voltage should change at a rate of 200 volts per second at the operating point for a timing accuracy of ±1/120 second. With this precision a circuit in which the tube anode is operated from an a-c source can be timed to the nearest half cycle. Timers for a-c resistance-welding machines require timing of such precision.

After one timing operation has been completed, contact 1 is opened, contact 2 is closed, and the capacitor is discharged

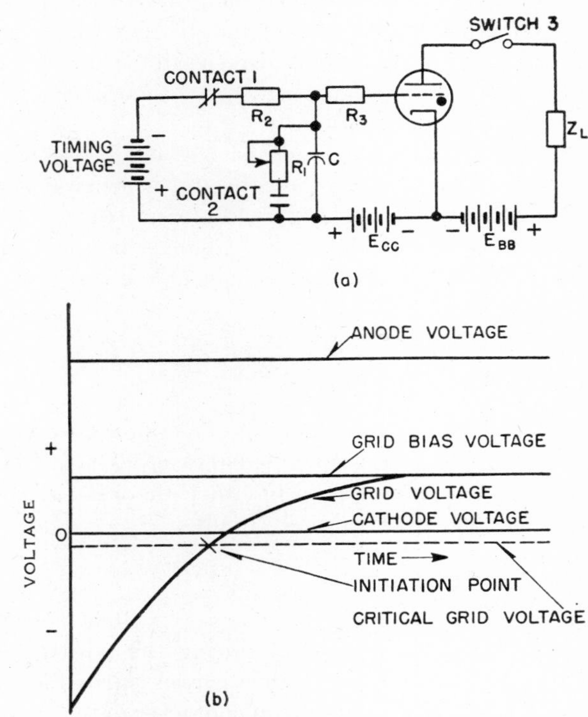

FIG. 18·23 Thyratron timing circuit utilizing capacitor discharging time.

through resistor R_2. This resistor may be only 50 or 100 ohms; therefore the discharge is rapid. The anode circuit of the thyratron is reset by opening switch 3. This operation is necessary only if unidirectional voltage is applied to the anode.

Capacitor-discharge time also may be used as the basis of a timing circuit, as shown in Fig. 18·23 (*a*). Here the capacitor is charged through contact 1 and resistor R_2 before the timing begins. A positive d-c bias is applied to the grid of the tube in series with the capacitor voltage, which is of such polarity as to make the grid negative.

Timing is started by opening contact 1 and closing contact 2 and allowing the capacitor to discharge through resistor R_1. The relative sizes of the resistance and capacitance determine the time required for the capacitor to discharge, just as these values determined the charging time of a resistor-capacitor combination. As the capacitor discharges, grid voltage becomes less and less negative, and the thyratron fires when the grid voltage becomes less negative than its critical grid characteristic, thus ending the timing period. See Fig. 18·23 (*b*). Before the circuit may be used again the timing capacitor must be charged by opening contact 2 and closing contact 1.

This circuit compares unfavorably with the capacitor-charge timing circuit in two ways: (1) more time must be allowed for the capacitor to be recharged than is required to discharge the capacitor for resetting the capacitor-charge circuit; (2) if the circuits have d-c supplies energized from an a-c source, the discharge timer is not so accurate as the capacitor-charge timer, because the discharge timer does not compensate for changes in line voltage which occur after the timing has started. The capacitor-charge timer compensates for changes in line voltage. If the line voltage decreases, the grid bias decreases and the d-c voltage for charging the capacitor decreases, and thereby the rate of increase of capacitor voltage is decreased. This tends to keep the timing interval constant. The same effect is evident for increase in line voltage.

The capacitor-discharge timer is not compensated for changes in line voltage, since the capacitor is charged before the timing interval starts and is disconnected from the charging source when timing starts. Changes in line voltage during the timing interval affect timing unless the grid-bias voltage is regulated.

18·8 IGNITRON CIRCUITS

The ignitron, like the thyratron, is a rectifier; however, it is used extensively as a switch capable of carrying large currents. The tube can be used in either a-c or d-c circuits, but is most frequently used in a-c circuits.

Operation in A-C Circuits

Two ignitrons are required for use as a switch in an a-c circuit. One tube carries current during the positive half cycle of line voltage, and the other carries current during the negative half cycle. An ignitron contactor in its simplest form is illustrated in Fig. 18·24. The ignitrons are connected in inverse parallel, and the ignitor circuits are connected in series through a control switch.

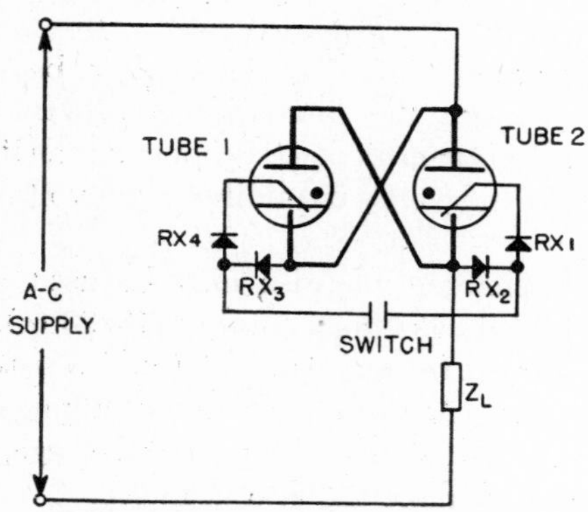

Fig. 18·24 Ignitron contactor.

To analyze the operation of the contactor, assume that the anode of tube 1 has just become positive and the anode of tube 2 is becoming negative as the starting switch is closed. Direction of ignitor current is from the positive line, through copper oxide rectifier 2, through the switch, through rectifier 4 and the ignitor of tube 1 to its cathode, to the negative line. When the anode voltage and ignitor current of tube 1 reach certain values, tube 1 conducts and causes the voltage across it to decrease to an arc drop of about 15 volts. Load current lasts for one-half cycle, or until it decreases to the minimum required to maintain the arc. If the control switch is still closed, conduction of tube 2 is initiated in the manner described for tube 1 and continues during the second half cycle. As long as the control switch is closed, the ignitron contactor conducts. Current and voltage across the load are as shown in Fig. 18·25. As soon as the switch is opened ignitor current can no longer continue, and the contactor ceases to conduct after completing the half cycle during which the control switch is open.

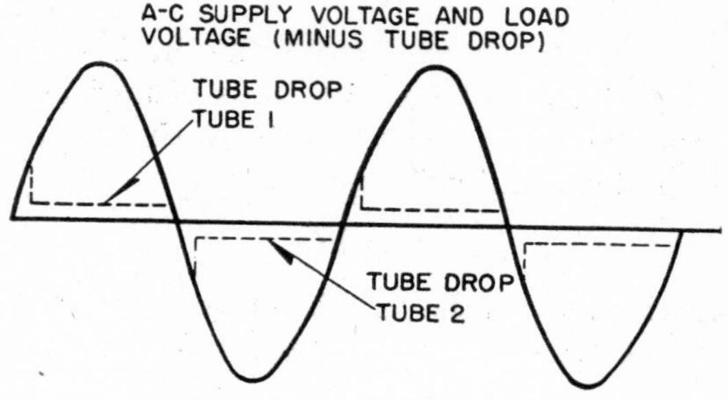

Fig. 18·25 Contactor voltages.

Operation in D-C Circuits

Ignitrons are used as switches in d-c circuits. Such an application is illustrated by Fig. 18·26, in which the tube is used to discharge a large capacitor into a load.

The switch from the d-c supply is closed, and capacitor *C* is charged. When the capacitor is fully charged, the switch is opened. The capacitor now may be discharged into the load by energizing the ignitron. In this circuit a thyratron energizes the ignitron. Ignition energy for the ignitron comes from capacitor *C*.

When the capacitor is to be discharged, a peak voltage is applied to the transformer in the grid circuit of tube 2 so that its grid voltage becomes positive, and the tube is made to conduct. When tube 2 conducts, capacitor *C* begins to discharge through the tube, the ignitor, and the cathode of tube 1 into the load. Current through the ignitor causes an arc between the ignitor and the mercury pool if the current is larger than a certain minimum value (about 40 amperes for most sealed-off ignitrons). This arc causes emission of electrons from the mercury pool, and the electrons are

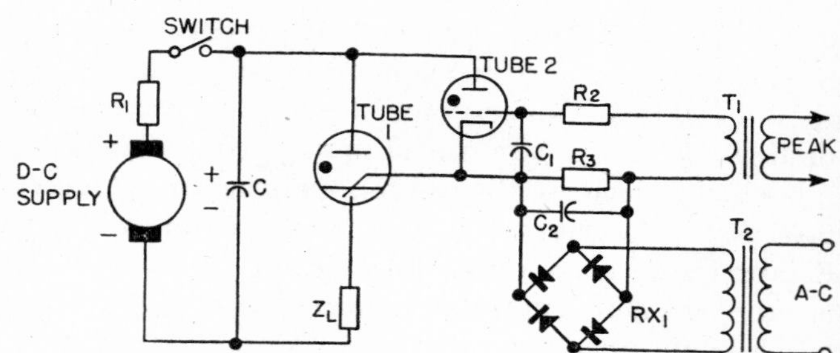

Fig. 18·26 Capacitor-discharge resistance-welding control circuit.

attracted toward the positive anode. Thus the gas in the tube is ionized and allowed to conduct the discharge current from the capacitor into the load. Once the ignitron is conducting, tube 2 is extinguished, because the arc voltage drop of tube 1 is less than the combined arc drop of tube 2 and the ignitor drop of tube 1.

After tube 1 conducts, current continues until the capacitor has discharged enough so that current through the tube approaches zero or becomes so small that the arc becomes unstable and goes out. If the arc continues for much longer than 0.1 second, the arc travels up the side wall of a metal ignitron unless a special ring is provided in the cathode to retain the arc on the mercury pool.

A circuit similar to that in Fig. 18·26 is used in capacitor-discharge resistance-welding controls.

The ignitron-initiating method shown in Fig. 18·26 is also applicable to ignitrons operating from a-c sources. Although this ignition system is one of the least complicated it has several disadvantages. Since ignitor current is carried by the load, the load impedance must be small enough to pass enough current to initiate the arc discharge. If the load is highly inductive, the ignitor may carry current for a large part of a half cycle before it reaches the proper value to ignite the tube. This may overheat the ignitor and destroy it. If the power factor of the load is high (from 90 to 100 percent) it is impossible to initiate conduction at the natural current zero because the supply voltage is too small to force sufficient current through the ignitor. For the same reason, initiation of conduction is not reliable on voltages less than 100 volts rms unless the power factor and impedance of the load are low. To overcome these limitations ignitrons are sometimes initiated by another system.

Figure 18·27 depicts the initiating system used on a low-voltage ignitron contactor similar to that of Fig. 18·24. The circuits involving tubes 3 and 4 are identical, so only one circuit need be analyzed. The initiating system consists of an a-c supply, a rectifier RX_1, capacitor C_1, and tube 3 with its grid bias and grid-impulse transformer. The winding of transformer T_3 in the rectifier circuit is phased so that capacitor C_1 is charged during the negative half cycle of voltage across tube 2. Resistor R_1 is chosen so that capacitor C_1 can be charged almost to the peak voltage of the transformer during one-half cycle. When the anode voltage of tube 2 becomes positive, tube 2 can be initiated by applying a voltage peak to the primary of transformer 1, to initiate tube 3 and discharge capacitor C_1 through the ignitor circuit of tube 2. Resistor R_2 must limit ignitor current to its maximum rated value. In some circuits a small air-core inductance is connected in series with the resistor to round off the current peak and increase its duration so that it lasts long enough to initiate ignitron conduction reliably. Conduction of tube 1 is initiated in the same manner by the initiating circuit connected to its ignitor. The point at which conduction in each ignitron is initiated in a particular half cycle is controlled by shifting the phase position of the peaks applied to transformers 1 and 2.

Because the energy for initiating conduction comes from the a-c supply during the negative half cycle and the ignitor

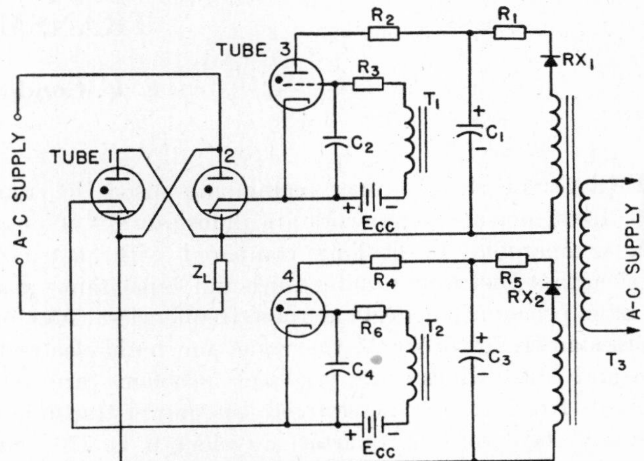

Fig. 18·27 Low-voltage ignitron contactor and initiating circuit.

current is not limited by the load impedance, conduction of the ignitrons can be initiated when their anode voltages exceed their breakdown voltages by only small amounts. This initiating circuit is more complicated and more expensive than the circuit shown in Fig. 18·26. Ignitron rectifiers, however, are initiated by circuits similar to the circuit in Fig. 18·27.

Circuit constants vary with the size of the ignitron. A small ignitron such as the WL-681/686 can be ignited by a capacitance of 20 microfarads charged to 300 or 400 volts with a resistance of 5 ohms for resistor 2.

REFERENCES

1. *Engineering Electronics*, Donald G. Fink, McGraw-Hill, 1941.
2. *Theory and Application of Electron Tubes*, Herbert J. Reich, McGraw-Hill, 1939.
3. *Electron Tubes in Industry*, Keith Henney, McGraw-Hill, 1934.
4. *Principles of Electronics*, Royce G. Kloeffler, Wiley, 1942.
5. *Industrial Electronics*, F. H. Gulliksen and E. H. Vedder, Wiley, 1935.
6. *Applied Electronics*, E.E. Staff, M.I.T., Wiley, 1943.

Chapter 19

TRANSMISSION LINES

E. U. Condon and William Altar

TRANSMISSION lines commonly used in radio-frequency electronic work are almost always of lengths comparable to, or long compared with, a quarter wavelength. Distributed inductance and capacitance in the line have pronounced effects at these frequencies. Although the fundamental theory is the same for both electrically long and short lines, the practical problems are much different from those encountered on power-transmission lines. At 60 cycles a quarter wavelength is 776 miles, which is considerably more than the longest transmission line in service.

19·1 BASIC EQUATIONS OF A TWO-CONDUCTOR LINE

Most types of two-conductor transmission lines, such as two parallel wires, a coaxial cable, or one wire parallel to a conducting earth, may be treated as equivalent to a circuit consisting of recurring sections of lumped elements as shown in Fig. 19·1. The inductors have self-inductance L_1 and

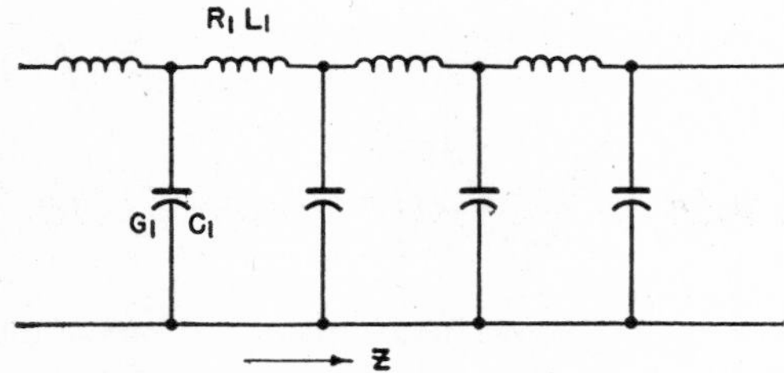

Fig. 19·1 Transmission line represented by network of recurrent elements.

resistance R_1. The capacitors have capacitance C_1 and conductance G_1. A large number N, say, of such meshes must be assumed per unit length of line so that NL_1 equals L, the inductance per unit length of line. Similarly, the total series resistance per unit length, the shunt conductance per unit length and the shunt capacitance per unit length, respectively, are denoted by R, G, and C.

If the upper conductor, at a distance z from an assumed pair of reference terminals and at time t, has a potential difference, or voltage, $E(z, t)$ against the opposite point on the lower conductor, and if $I(z, t)$ is the current (amperes) flowing toward positive z in the upper line and toward negative z in the lower line, E and I are related to each other and to the line parameters by differential equations:

$$\frac{\partial E}{\partial z} = -RI - L\frac{\partial I}{\partial t} \tag{19·1}$$

$$\frac{\partial I}{\partial z} = -GV - C\frac{\partial E}{\partial t} \tag{19·2}$$

According to the first equation, the line voltage changes along the line because of ohmic drop and also because of the back emf of self-induction. According to the second equation, the line current changes along the line because of ohmic shunt conductance and because of distributed capacitance.

19·2 PROPAGATION CONSTANT

While a transmission line has four characteristic parameters, R, L, C, and G, the manner in which E as well as I vary along the line for any given load is completely determined by one complex quantity γ, the so-called propagation constant of the line which is in turn a function of the characteristic line parameters.

The simplest case, a wave progressing without reflections along the line, can be discussed by assuming that the line extends homogeneously to $z = \infty$ and is fed at its input terminals, at $z = 0$, with current I_0 and voltage E_0 from a generator at frequency $\omega/2\pi$ cycles per second. Unless the line is free of loss, this wave must be damped as it progresses. Such a wave is expressed by the following solutions of equations 19·1 and 19·2:

$$E(z, t) = E_0\epsilon^{j\omega t - \gamma z} \tag{19·3}$$

$$I(z, t) = I_0\epsilon^{j\omega t - \gamma z} \tag{19·4}$$

On substituting these in equations 19·1 and 19·2 it is found that

$$\gamma E_0 = (R + j\omega L)I_0 \tag{19·5}$$

$$\gamma I_0 = (G + j\omega C)E_0 \tag{19·6}$$

from which the complex propagation constant becomes

$$\gamma = \sqrt{(R + j\omega L)(G + j\omega C)} \tag{19·7}$$

This shows that the propagation constant γ is a purely imaginary number for lines free of loss, and that its real part arises from the presence of one or both of the dissipation parameters R and G. Since practical lines have little dissipation, it is frequently desirable to introduce the quantity $k = -j\gamma$, which is real and positive if $R = G = 0$. In this case k bears a simple relation to the speed v with which the wave progresses along the line

$$k = \frac{\omega}{v} = \text{wave number in radians per unit line length} \qquad (19\cdot8)$$

by virtue of equations 19·3 and 19·4. At frequencies so high that only a negligible portion of magnetic flux penetrates into the conductors (small skin depth) $LC = 1/c^2$, where c is 3×10^{10} centimeters per second, the speed of light in free space. Equations 19·3 to 19·8 then describe a wave progressing with the speed of light. The penetration of magnetic flux into the conductors has the effect of slightly increasing L without affecting C, and thus of lowering the speed of propagation. Usually, however, this correction is practically negligible because most of the flux flows outside the conductors.

To demonstrate this explicitly, consider a transmission line of two parallel wires, having wires of radius r and a center-to-center distance $d \gg r$. The inductance per unit length is

$$L = 4 \times 10^{-9}\left(\log_\epsilon \frac{d}{r} + \mu\delta\right) \text{ henrys per centimeter} \qquad (19\cdot9)$$

the capacitance per unit length is

$$C = \left(36 \times 10^{11} \log \frac{d}{r}\right)^{-1} \text{ farad per centimeter} \qquad (19\cdot10)$$

and therefore

$$LC = \frac{1}{c^2}\left(1 + \frac{\mu\delta}{\log \frac{d}{r}}\right) \qquad (19\cdot11)$$

Here μ (usually equal to unity) is the magnetic permeability of the material of the wire, and δ is a correction factor which at low frequencies assumes the value 0.25 and at high frequencies sinks to zero. If f is the frequency and ρ the resistivity of the conductor material in microhm-centimeters, the parameter δ is a function of $x = 0.281\, r\sqrt{\mu f/\rho}$, and its variation with x is indicated in Table 19·1

Table 19·1

x	δ	x	δ	x	δ
0	0.250	4	0.171	10	0.070
1	0.249	5	0.139	20	0.035
2	0.240	6	0.116	50	0.014
3	0.211	7	0.100	100	0.007

For wires of unit permeability (copper, aluminum, and so on) and with a center-to-center distance 10 times the radius of either wire, the penetration of flux into the wires adds inductance sufficient to lower the speed of low-frequency waves to a value 5 percent below the speed of light. This statement is valid only if the line reactance ωL, even at the low frequencies considered, is big compared with the line resistance. This condition is fulfilled in ultrahigh-frequency and microwave practice where R and G are small compared with L and C, respectively. However, this is not true of the transmission lines common at lower radio and audio frequencies.

In terms of the previously defined skin-effect parameter x, the line resistance per unit length is

$$R = \frac{2\rho}{\pi r^2} f(x) \qquad (19\cdot12)$$

where $f(x)$ is given in Table 19·2.

Table 19·2

x	$f(x)$	x	$f(x)$	x	$f(x)$
0	1.0000	4	1.678	10	3.799
1	1.005	5	2.043	20	7.328
2	1.078	6	2.394	50	17.93
3	1.318	7	2.743	100	35.61

With the help of Table 19·2, the relative importance of line resistance and reactance can be discussed in terms of the ratio $Q = \omega L/R$. Generally, since the resistance rises quite slowly with the frequency and never faster than its square root, Q must rise at least as the square root of the frequency. For the two-conductor line

$$Q = \frac{x^2\left[\log_\epsilon \frac{d}{r} + \mu\delta(x)\right]}{2\mu f(x)} \qquad (19\cdot13)$$

The line will behave more like a resistance or else more like an inductive reactance, depending upon how much lower, or respectively higher, the operating frequency is compared with the frequency at which Q assumes the value unity. As an example, consider two copper wires of 0.1 centimeter radius 1 centimeter apart. Equation 19·13 then gives $Q = 1$ at approximately $f = 1700$ cycles per second, showing that the frequency region intermediate between the two extremes falls in the audio range.

A discussion of the relative importance of the distributed capacitance and the shunt conductance in equally general terms is not easy. The shunt conductance is not always continuously distributed, but it may be localized at insulator supports. These lumped contributions can often be ignored. Where an imperfect dielectric is used between conductors, the effective conductance may have noticeable effects. In such cases it is usually convenient to consider conductance in terms of a complex dielectric constant. Let C_0 be the capacitance per unit length of a line having the same geometry but with vacuum as the dielectric. Then for a medium of dielectric constant ϵ the capacitance per unit length is $C = C_0\epsilon$, and for an imperfect dielectric

$$\epsilon = \epsilon' - j\epsilon'' \qquad (19\cdot14)$$

where ϵ' and ϵ'' are characteristics of the dielectric material and usually vary with frequency. If the effective conductance arises entirely from dielectric loss, equation 19·6 takes the form

$$\gamma I_0 = (\omega\epsilon'' C_0 + j\omega\epsilon' C_0)E_0 \qquad (19\cdot15)$$

In other words the line behaves as if the conductance per unit length were $G = \epsilon''\omega C_0$. The ratio $G/\omega C$ of the quantities occurring in equation 19·7 thus reduces to ϵ''/ϵ', a quantity which is less than 0.01 for dielectric materials used in cable construction; therefore conductance losses are correspondingly small.

The propagation constant γ, a complex quantity in the presence of line loss, is resolved into its real and imaginary parts:

$$\gamma = \alpha + j\beta$$

where α and β are both positive quantities. A wave progressing in the z-direction may be expressed by substituting this into equations 19·3 and 19·4:

$$E(z, t) = E_0 \epsilon^{-\alpha z} \epsilon^{j(\omega t - \beta z)} \tag{19·16}$$

$$I(z, t) = I_0 \epsilon^{-\alpha z} \epsilon^{j(\omega t - \beta z)} \tag{19·17}$$

For lines with small loss, $\beta = \omega/v = 2\pi/\lambda$ denotes again the wave number in electric radians per unit length, as the quantity k did in the case of an ideal line. The so-called attenuation constant α measures the decrease in nepers of voltage and current amplitudes as the wave progresses. Since the average power crossing a plane z-constant is given by $P = (E_0 I_0/2)\epsilon^{-2\alpha z}$ for waves progressing in the positive z-direction, in a unit of distance the power level drops by a factor $\epsilon^{-2\alpha}$. A factor ϵ represents a change of power level by 4.34 decibels so that the attenuation of the power level is 8.68α decibels per unit length of line.

For the two-conductor line, by separation of equation 19·7 into its real and imaginary parts,

$$\alpha = \sqrt{\frac{\sqrt{(R^2 + \omega^2 L^2)(G^2 + \omega^2 C^2)} + RG - \omega^2 LC}{2}} \tag{19·18}$$

$$\beta = \sqrt{\frac{\sqrt{(R^2 + \omega^2 L^2)(G^2 + \omega^2 C^2)} - RG + \omega^2 LC}{2}} \tag{19·19}$$

If $R \ll \omega L$, $G \ll \omega C$, these formulas are well approximated by

$$\alpha = \frac{1}{2}\left(R\sqrt{\frac{C}{L}} + G\sqrt{\frac{L}{C}}\right) \tag{19·20}$$

$$\beta = \omega\sqrt{LC} + \frac{1}{4}\left(R\sqrt{\frac{C}{L}} - G\sqrt{\frac{L}{C}}\right) \tag{19·21}$$

The first of these shows that the attenuation of such lines does not change with frequency. The second relation shows that the phase velocity v at which the wave progresses may differ by small positive or negative amounts from the value $1/\sqrt{LC}$ obtained for an ideal line. The power transfer and the signal, on the other hand, proceed with the so-called group velocity u, given in this approximation by

$$\frac{1}{u} = \frac{\partial\beta}{\partial\omega} = \sqrt{LC}$$

$$u = \frac{1}{\sqrt{LC}} \tag{19·22}$$

which precludes any power or signal transfer at speeds greater than c.

At low frequencies this approximation cannot be used. In particular the attenuation as represented by equation 19·18 will vary with the frequency. The attenuation is independent of frequency only if $LG = RC$, and such a line is said to be *distortionless*, which would be very desirable in communication work. However, $RC \gg LG$ on parallel-wire transmission lines of ordinary dimensions and materials, and the lines distort badly. This condition could be remedied by using poorer insulation and thus increasing G, but only at the expense of large, if uniform, attenuation. An important discovery was made independently by Pupin and Campbell when they recognized that, by increasing the inductance of such lines, not only would the attenuation be made more nearly uniform but also it actually would be considerably reduced. This so-called "loading" increases the conductance loss and reduces in approximately reciprocal proportion the resistance loss in the line. Since the latter preponderates in the unloaded line, the net effect is a reduction of the total line losses by 50 to 60 percent or more. The added self-inductance does not need to be uniformly distributed along the line but can be arranged in lumped coils, provided only that the spacing between loading coils is small compared to one wavelength.

19·3 CHARACTERISTIC IMPEDANCE

In order that a wave may progress without reflections in accordance with equations 19·3 and 19·4, the ratio of voltage to current must be the same at all points z of the line. This ratio is called the characteristic impedance Z_0 of the line. It can be computed for $z = 0$ in terms of the characteristic line parameters R, L, G, C by eliminating γ from equations 19·5 and 19·6. The result is

$$Z_0 = \frac{E_0}{I_0} = \sqrt{\frac{R + j\omega L}{G + j\omega C}} \tag{19·23}$$

If a transmission line, instead of extending to $z = \infty$, is terminated in a load of impedance Z_0 no reflection will originate at the termination.

The characteristic impedance, generally a complex quantity, becomes real only if $LG = RC$, which is the condition for a distortionless line. Unless the characteristic impedance is real, voltage and current are not strictly in phase in a simple damped progressive wave.

The simplest example is a radio-frequency line with negligible losses, for which Z_0 becomes real and equal to $\sqrt{L/C}$ ohms. (Z_0 is expressed in ohms, not ohms per unit length, and refers to the impedance at input terminals of a line of infinite length.) The most important practical two-conductor lines are:

(*a*) *Parallel wires*, each of radius r, center-to-center distance d,

$$Z_0 = 120 \log_\epsilon \frac{d}{r} \text{ ohms} \tag{19·24}$$

(*b*) *Coaxial cable*, outer radius b, inner radius a, filled with a non-magnetic material of dielectric constant e,

$$Z_0 = \frac{60}{\sqrt{e}} \log_\epsilon \frac{b}{a} \text{ ohms} \tag{19·25}$$

(*c*) *Parallel flat strips,* separated by a distance d small compared with the width w, with space between them filled with a non-magnetic material of dielectric constant e,

$$Z_0 = \frac{120\pi d}{\sqrt{e}w} \quad \text{ohms} \tag{19·26}$$

These formulas show that the characteristic impedances are of the order of 50 to 500 ohms for lines of ordinary proportions.

A line for which G, but not R, is negligibly small is common in practice. The presence of copper loss has the obvious effect of attenuating the wave, since part of the power fed to the line is dissipated in the line resistance. Another and less obvious effect of the conductor resistance is that Z_0 is no longer a real quantity;

$$Z_0 = \sqrt{\frac{L}{C}\left(1 - \frac{j}{Q}\right)} \tag{19·27}$$

where again $Q = \omega L/R$. The form of this expression shows that the impedance of the infinite line at input terminals has a capacitively reactive component. The conductor resistance interferes with the transfer of energy along the line, and part of the energy stored capacitively in the electric field near the input terminals returns reactively to the generator instead of proceeding toward the load.

This effect assumes an extreme form at low frequencies where the inductive line reactance per unit length is small compared with the line resistance per unit length ($Q \ll 1$). The characteristic impedance then has a phase angle of approximately -45 degrees and a magnitude $\sqrt{R/\omega C}$ which increases indefinitely as frequency is diminished. The propagation constant becomes $\gamma = \sqrt{R\omega C}\epsilon^{-j(\pi/4)}$. The wave is therefore highly damped, and the reactive power equals the transmitted power at any line cross section. This extreme condition was a cause of considerable difficulty in early forms of submarine telegraph cables, until the remedy of magnetic loading to increase L was introduced.

To summarize: The two-conductor line has four parameters, the circuit constants R, L, C, and G. But the over-all behavior of the line is more aptly described in terms of two complex parameters, the characteristic impedance Z_0 and the propagation constant γ. It is well to point out that this description in terms of Z_0 and γ may also be extended to transmission lines other than two-conductor lines, such as wave guides (see Section 19·5), where it would be impossible to assign distributed circuit parameters or to describe the waves in terms of line voltage and conductor currents.

19·4 TRANSMISSION LINE WITH LOAD

If a transmission line of finite length is terminated in a load of impedance Z_L different from the characteristic line impedance, the electrical conditions on the line do not correspond simply to a single progressive wave traveling from generator to load. Instead a reflected wave is set up at the load and returned toward the generator. Superposition of the reflected wave and the incident wave produces standing waves along the line.

Generally, standing waves imply that the line is not used to full advantage because the voltage amplitudes in some portions, and the current amplitudes in other portions, of the line are much greater than the minima required to transmit the load power. This condition not only increases the line loss but also reduces the power-carrying capacity of the line, which is limited by the dielectric breakdown strength of the insulation. As an example, if the reflected wave has an amplitude 50 percent that of the oncoming wave, the line can carry only 33 percent of the power which it could supply to a non-reflecting load.

In 60-cycle power transmission even strong reflections at the load are not objectionable because the lines are too short electrically to exhibit much voltage or current variation with distance. It is then preferable to supply the variable loads from constant-voltage power sources and through transmission lines of characteristic impedance many times larger than the load impedances. In communication applications, on the other hand, one strives for a match between line impedance and line termination and controls the energy flow by varying both the signal voltage and the signal current in equal proportion.

Occasionally high standing-wave ratios do occur in high-frequency applications, for instance where short lengths of line are terminated in a short circuit to serve as circuit elements. Such line sections represent sharply resonant circuits, and at microwave and ultrahigh frequencies they are commonly used as oscillator tank circuits, instead of the lumped inductor-capacitor combinations (see Section 19·9).

In the presence of a reflected wave, the voltage-to-current ratio is no longer constant along the line but varies periodically with z. At the load terminals $z = a$, the ratio must assume the value required by the impedance of the load. This condition determines amplitude and phase of the reflected wave. If the voltage amplitudes and phases at $z = 0$ of the incident and the reflected waves are denoted by the a-c vectors E_i and E_r, then the voltage and the current at z are given by

$$E = (E_i\epsilon^{-\gamma z} + E_r\epsilon^{\gamma z})\epsilon^{j\omega t} \tag{19·28}$$

$$I = \frac{1}{Z_0}(E_i\epsilon^{-\gamma z} - E_r\epsilon^{\gamma z})\epsilon^{j\omega t} \tag{19·29}$$

The ratio of these can be made to assume any complex value by appropriate choice of the two complex numbers E_i and E_r. An equation for the ratio of these two wave amplitudes is obtained by setting the ratio of the expressions 19·28 and 19·29, taken at $z = a$, equal to the load impedance:

$$Z_L = Z_0\frac{E_i\epsilon^{-\gamma a} + E_r\epsilon^{\gamma a}}{E_i\epsilon^{-\gamma a} - E_r\epsilon^{\gamma a}} \tag{19·30}$$

The ratio of the reflected voltage to the incident voltage is called the complex reflection coefficient r. At a position z along the line,

$$r(z) = \frac{E_r\epsilon^{\gamma z}}{E_i\epsilon^{-\gamma z}} = r(a)\epsilon^{2\gamma(z-a)} \tag{19·31}$$

In particular, at the load terminals the reflection coefficient, by equation 19·30, becomes

$$r(a) = \frac{Z_L - Z_0}{Z_L + Z_0} \tag{19·32}$$

Evidently r depends only upon the ratio of Z_L to Z_0, and it is therefore convenient to choose the unit of impedance equal to Z_0.

The impedance at an arbitrary z is given by

$$Z = Z_0 \frac{1 + r(a)\epsilon^{2\gamma(z-a)}}{1 - r(a)\epsilon^{2\gamma(z-a)}} \tag{19·33}$$

or

$$Z = Z_0 \frac{1 + r(z)}{1 - r(z)} \tag{19·33a}$$

The last equation serves for computing input impedances Z for given load impedances Z_L. First one computes $r(a)$ from equation 19·32 and $r(z)$ from equation 19·31, which is then substituted in equation 19·33 (a).

19·5 PROPAGATION OF MODES OTHER THAN THE PRINCIPAL MODE

The definitions and statements of the past sections were presented with reference to a two-conductor line, and specifically to the principal mode on such a line. The principal mode is a process of wave propagation where at any line cross section the boundary of each conductor is an equipotential line, and where all currents are parallel to each other and to the direction of wave propagation. This permits one to define line voltage as the potential difference between conductors, and line current as the sum of all current elements, integrated over the section area of a conductor.

In all reference to impedance, propagation constant, line voltage and current, and in fact all through the customary presentation of circuit theory the assumption was implicit that the principal mode is the only one possible for the line or circuit under discussion. The need for explicit statements regarding the specific mode used does not arise in conventional power-, audio-, and radio-frequency applications because at these relatively low frequencies the principal mode happens to be the only one within the frequency band of transmission of lines and systems having conventional dimensions. Generally this is true as long as transverse line dimensions (conductor size and spacings) are small compared with the free-space wavelength of the wave.

In the microwave region with frequencies one hundred to ten thousand times higher than the highest radio frequencies and with wavelengths measured in centimeters, transverse line dimensions amount to considerable fractions of a wavelength, and the propagation of so-called wave-guide modes becomes a reality. In a coaxial line, for instance, the so-called cut-off wavelength for the lowest wave-guide mode equals roughly the arithmetic mean of the circumferences of outer and inner conductor, which means that the line operating at frequencies with free-space wavelengths shorter than the cut-off value will transmit at least one wave-guide mode in addition to the principal mode. A coaxial line with conductor diameters of 7 and 13 millimeters and operating in the 3 centimeter microwave range cannot be safely assumed to operate in the principal mode, and treatment by methods intended for a two-conductor line may lead to serious errors.

Operation in a wave-guide mode offers important advantages at microwave frequencies. Most important, such operation requires only one conductor, as a hollow metallic pipe of circular, rectangular, or other cross section, and thus avoids design complications which arise in coaxial lines with regard to a needed support for the inner conductor. Also, the power rating and line loss of a coaxial line are unfavorable if the line is small enough to meet the design requirements aimed at the exclusion of higher modes. Limitation for the power rating is the electric breakdown at the surface of the inner conductor where the electric gradient is relatively high. Also, the inner conductor causes high copper loss because of the concentration of current in a small sectional area. The supports for the inner conductor usually cause additional loss and increase the line attenuation.

Wave guides of various cross-sectional shapes, preferably rectangular, have become standard in microwave circuits. Just as in two-conductor lines, it is necessary to choose the wave-guide dimensions so that only one mode of propagation falls within the transmission range of the line. If transmission phenomena are restricted to one mode, transmission theory for a wave guide becomes formally identical, except for a few relatively unimportant details, with the theory of a two-conductor line operating in its principal mode. The differences concern numerical values for characteristic impedance and propagation constants, and a strong frequency dependence in guides of these two constants, which is absent in principal-mode propagation.

The electromagnetic fields associated with the principal mode in a homogeneous transmission line are characterized by the absence of longitudinal field components. For instance, in a coaxial line the electric lines of force are straight lines extending radially between the inner and outer metallic cylinder, and the magnetic field lines are circles concentric with the axis of the cylinders. It can be proved mathematically that the existence of a mode with transverse electric and magnetic fields (TEM mode) in a cylindric transmission line is contingent on the presence of two or more conductors. A TEM or principal mode cannot exist in a wave guide.

In a wave-guide mode either the electric field or the magnetic field, but not both, may be free of longitudinal field components, and every wave-guide mode can accordingly be classified as either a transverse electric (TE) or a transverse magnetic (TM) mode. The electromagnetic fields associated with a mode of either type can, for a hollow metallic cylinder of arbitrary cross section, be derived with the help of a two-dimensional key function.

Transverse Electric (TE) Modes

It can be verified by direct substitution that the following field vectors satisfy Maxwell's electromagnetic field equations everywhere inside a perfectly conducting hollow cylinder, as well as the boundary conditions at the metallic interface, and thus represent a transverse electric wave solution:

$$\begin{aligned} E_s &= k(u_z \times \text{grad } \psi)(\epsilon^{jk_z z} + r\epsilon^{-jk_z z})\epsilon^{-j\omega t} \\ E_z &= 0 \\ H_s &= -k_z \text{ grad } \psi(\epsilon^{jk_z z} - r\epsilon^{-jk_z z})\epsilon^{-j\omega t} \\ H_z &= jk_s^2\psi(\epsilon^{jk_z z} + r\epsilon^{-jk_z z})\epsilon^{-j\omega t} \end{aligned} \tag{19·34}$$

Here the two indices z and s refer to vector components respectively in the direction of propagation and in the sectional plane, and the quantity u_z denotes a unit vector in the direction of propagation. The key function ψ, a function of the two sectional variables x, y but not of z, is a solution of the two-dimensional wave equation:

$$(\Delta + k_s^2)\psi(x, y) = 0 \tag{19·35}$$

within the boundary and is subject to the boundary condition that at every point of the metallic boundary the gradient of ψ be tangent to the boundary:

$$\frac{\partial \psi}{\partial n} = 0 \tag{19·36}$$

This constitutes a so-called eigenvalue problem which means that solutions $\psi(x, y)$ will exist only for a discrete set of eigenvalues k_s. There are as many TE modes as there are solutions $k_{s,n}$ and ψ_n of the eigenvalue problem. Each eigenvalue $k_{s,n}$ determines the cut-off wavelength for the respective mode by virtue of the relation

$$\lambda_{\text{cut-off}} = \frac{2\pi}{k_{s,n}} \tag{19·37}$$

After k_s is known, k_z may be found by means of the relation

$$k_{s,n}^2 + k_{z,n}^2 = k^2 \tag{19·38}$$

An infinite set of real values $k_{s,n}$ usually exists, their magnitudes rising with the order number n of the mode. Thus in view of equation 19·38, only the lowest modes, if any, will have a k_z which is real, indicating wave propagation with the wave velocity (or phase velocity),

$$v = \frac{\omega}{\sqrt{k^2 - k_s^2}} \tag{19·39}$$

and the guide wavelength

$$\lambda_g = \frac{2\pi}{\sqrt{k^2 - k_s^2}} \tag{19·39a}$$

The higher modes will have a purely imaginary k_z, or a real propagation constant:

$$\gamma = \sqrt{k_s^2 - k^2} \tag{19·40}$$

indicating an exponential decrease for the field amplitudes in the positive or negative z-direction, instead of a sine variation which is characteristic of a propagating wave. The exponentially attenuated waves assume local importance in those portions of a line where the propagation of the dominant mode is impeded by an obstacle, such as a discontinuity in the line dimensions, a diaphragm or a bend or twist. The field disturbances of the dominant mode caused by such obstacles may well be thought of as a superposition of attenuated TE and TM modes of higher order, functioning as reservoirs for storage of energy during certain portions of each cycle; and the combined action of these modes may be expressed by the circuit equivalent of a reactive element shunted across the wave guide.[1]

Comparison of equation 19·39 with the corresponding equation for the principal mode in a two-conductor line (19·21) shows that the phase velocity for guide modes shows a strong frequency dependence increasing to infinite values as the frequency approaches the cut-off value, whereas a two-conductor line shows no frequency variation at all if perfectly conducting, and very little if losses are considered to be present. Whereas the phase velocity for propagation in a wave guide proceeds with a speed exceeding the speed of light c, the signals as well as power transfer proceed with the group velocity u which, in a wave guide, is given by the equation

$$u = \frac{c^2}{v} \tag{19·41}$$

This is always less than the speed of light, and it decreases toward zero as the frequency approaches its cut-off value.

The propagation constant for guide waves in the presence of guide loss can be computed with good accuracy by assuming that the field distribution as represented by equations 19·34 is not noticeably affected by the power dissipation; consequently the quantity β forming the imaginary part of the propagation constant remains unchanged. The real part α, expressing attenuation, can be computed in terms of the ψ-function by the following formula:

$$\alpha = \frac{1}{4}\sqrt{\frac{\rho}{\lambda_0}} \frac{k_s^2 \oint \psi^2 \, ds + \dfrac{k_z^2}{k_s^2} \oint (\nabla\psi)^2 \, ds}{kk_z \iint \psi^2 \, da}$$

nepers per centimeter (19·42)

The characteristic impedance of a wave guide is a somewhat ambiguous concept because it is founded on the concepts of line current and line voltage neither of which has a direct meaning when applied to wave guides. It is possible to define the characteristic guide impedance in such a manner that the net power transmitted to a non-reflecting load is correctly given by the conventional formula

$$P = \left(\frac{I^2}{2}\right) Z_0 \tag{19·43}$$

where I is the sum of all current components flowing in the positive z-direction, or toward the load. There are at any instant equal amounts of currents flowing in the opposite direction, but unlike in a two-conductor line they occupy parts of the same conductor as the outgoing current. On this basis one computes the total energy flow through the cross-sectional area A of the guide, and the current in the metal, in terms of the key function, and obtains for the characteristic impedance a general expression conforming with the definition of equation 19·43. For the lowest TE mode:

$$Z_0 = 120\pi \frac{k}{k_z} \frac{\iint |\nabla\psi|^2 \, da}{(\psi_{\max} - \psi_{\min})^2}$$

$$= 120\pi \frac{\lambda_g}{\lambda_0} \frac{(\text{mean square of grad } \psi)}{\dfrac{(\psi_{\max} - \psi_{\min})^2}{\text{area}}} \quad \text{ohms} \quad (19\cdot44)$$

The field distribution for a mode can be conveniently plotted, again with the help of the key or eigenfunction ψ. The electric lines of force associated with a TE mode are the lines ψ = constant, and the magnetic lines, projected into the sectional plane, are their orthogonal trajectories. Moreover, in order to map the electric field lines in the conventional manner which expresses the intensity of the field by the density of the lines mapped, it is only necessary to space the lines $\psi = c_1, \psi = c_2, \psi = c_3$, so that the chosen parameter values c_i come at equal intervals.

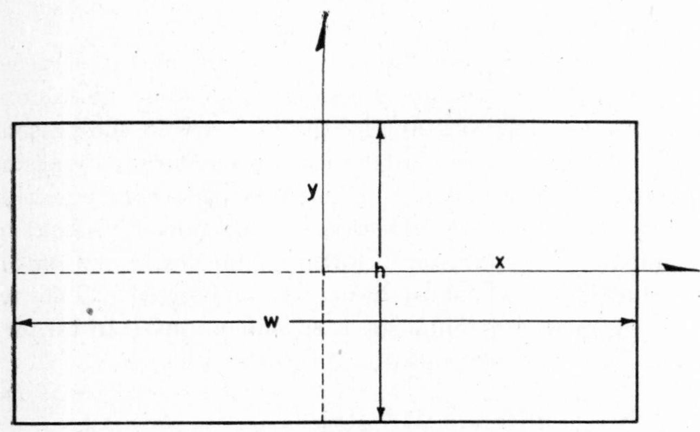

FIG. 19·2 Rectangular guide cross section.

As an illustration, the rectangular mode with cross section as shown in Fig. 19·2 has a twofold infinity of TE modes, $TE_{m,n}$ with corresponding key functions

$$\psi_{m,n}(x, y) = \cos m\pi \left(\frac{x}{w} + \frac{1}{2}\right) \cos n\pi \left(\frac{y}{h} + \frac{1}{2}\right)$$

This is in conformance with the boundary conditions, and yields for k_s and for the cut-off wavelength

$$k_s = \pi \sqrt{\left(\frac{m}{w}\right)^2 + \left(\frac{n}{h}\right)^2}$$

$$\lambda_{\text{cut-off}} = \frac{2}{\sqrt{\left(\dfrac{m}{w}\right)^2 + \left(\dfrac{n}{h}\right)^2}} \quad (19\cdot45)$$

The mode which is almost exclusively used in rectangular guides is the $TE_{1,0}$ mode the key function of which is

$$\psi_{1,0} = \sin \frac{\pi x}{w}$$

Its cut-off wavelength, according to equations 19·45, equals twice the guide width, and the guide wavelength varies with the free-space wavelength thus:

$$\lambda_g = \frac{\lambda_0}{\sqrt{1 - \left(\dfrac{\lambda_0}{2w}\right)^2}}$$

The attenuation due to copper loss in the guide walls, by equation 19·42, is

$$\alpha = \frac{1}{4}\sqrt{\frac{\rho}{\lambda_0}} \frac{\dfrac{(w + 2h)}{(\lambda_{\text{cut-off}})^2} + \dfrac{w}{(\lambda_g)^2}}{\dfrac{wh}{2\lambda_0\lambda_g}}$$

$$= \frac{\lambda_g}{\lambda_0}\sqrt{\frac{\rho}{\lambda_0}}\left(\frac{1}{2h} + \frac{\lambda_0^2}{4w^3}\right)$$

Considering the fact that the resistivity of copper, expressed in absolute units, is 6.0×10^{-8} centimeter, this gives for a three-centimeter wave guide of customary dimensions an attenuation of approximately 0.5×10^{-4} neper per centimeter. The power level is attenuated by approximately 1.32 decibel per hundred feet.

The characteristic impedance for the present case is, in accordance with equation 19·44,

$$Z_0 = 120\pi \frac{\lambda_g}{\lambda_0} \frac{k_s^2 \dfrac{wh}{2}}{[1 - (-1)]^2} = 15\pi^3 \frac{\dfrac{h}{w}}{\sqrt{1 - \left(\dfrac{\lambda_0}{2w}\right)^2}}$$

$$= \frac{463 \dfrac{h}{w}}{\sqrt{1 - \left(\dfrac{\lambda_0}{2w}\right)^2}} \quad \text{ohms} \quad (19\cdot46)$$

Although the characteristic impedance and the propagation constant as computed in the preceding paragraphs for a wave guide differ considerably from the corresponding quantities in a two-conductor line, there is a strong formal similarity between principal and wave-guide modes in all respects concerning the transmission process proper. This is borne out by an inspection of equations 19·34 as compared with the corresponding equations 19·28 and 19·29 for two-conductor lines, taken in conjunction with equation 19·31. The longitudinal component of the current flowing in the metallic boundary of a wave guide is proportional to the transverse, or sectional, component of the magnetic field adjacent to the boundary. Similarly, the electric field is a measure of the voltage at each section. Although it is true that the two sectional field components E_s and H_s vary within each line cross section, their ratio is constant according to equations 19·34 for an outgoing wave, and varies with varying reflection coefficient r exactly like the impedance of a two-conductor line. It is thus permissible to extend the transmission theory of two-conductor lines without formal changes to any wave-guide operating in a single TE mode. Of course, it is necessary to use the proper values of characteristic impedance and propagation constant at each frequency, both quantities now being frequency dependent.

Transverse Magnetic (TM) Modes

TM modes are less important in practical applications. Their mathematical theory is closely analogous to the one of TE modes. The field vectors are represented by the following solution of Maxwell's electromagnetic field equations:

$$\begin{aligned} H_s &= k(u_z \times \operatorname{grad} \chi)(\epsilon^{jk_z z} + r\epsilon^{-jk_z z})\epsilon^{-j\omega t} \\ H_z &= 0 \\ E_s &= k_z \operatorname{grad} \chi(\epsilon^{jk_z z} - r\epsilon^{-jk_z z})\epsilon^{-j\omega t} \\ E_z &= -jk_s^2\chi(\epsilon^{jk_z z} + r\epsilon^{-jk_z z})\epsilon^{-j\omega t} \end{aligned} \tag{19·47}$$

where the same notation is used as in equations 19·34. The key function $\chi(x, y)$ is a solution of the eigenvalue problem

$$(\Delta + k_s^2)\chi(x, y) = 0 \tag{19·48}$$

now subject to the condition that the value of $\dot{\chi}$ be zero all along the boundary of the guide. The cut-off wavelengths of the modes are again given by eigenvalues k_s; the phase velocity and the guide wavelength are computed by equations 19·38, 19·39, and 19·39*a* as before. The formula for attenuation of TM modes is simpler than for TE modes due to the absence of longitudinal magnetic field and hence of a transverse current in the boundary and associated copper loss. Instead of equation 19·42

$$\alpha_{\text{TM}} = \frac{1}{4}\sqrt{\frac{\rho}{\lambda_0}}\frac{k_z}{kk_s^2}\frac{\oint |\nabla\chi|^2\, ds}{\iint \chi^2\, da} \quad \text{nepers per centimeter} \tag{19·49}$$

The characteristic impedance for TM modes may be defined as for TE modes by equation 19·43. Taking the transverse E and H components from equations 19·47 for non-reflecting loads and computing net power and conductor current one has for the lowest TM mode

$$\begin{aligned} Z_0 &= 120\pi \frac{k_z}{k} \frac{\iint |\nabla\chi|^2\, da}{\left|\oint \nabla\chi \times ds\right|^2} \\ &= 120\pi \frac{\lambda_0}{\lambda_g} \frac{\iint \chi^2\, da}{k_z^2\left(\iint \chi\, da\right)^2} \quad \text{ohms} \end{aligned} \tag{19·50}$$

The line integral in equation 19·50 must be extended over the entire conductor boundary, while the mean values refer to the section area of the guide. Unlike in the TE modes, the current in TM modes is entirely longitudinal and, for the lowest TM mode, flows in the same direction all along the conductor boundary at a given line cross section. The seeming paradox of a net current without equal return current in another conductor, or in a different portion of the same conductor, resolves itself if attention is paid to the longitudinal electric field component which oscillates at a high frequency. The displacement current $\dot{D}_z/4\pi$ associated with this field component is electrodynamically equivalent to a conduction current; indeed, the total displacement current integrated over the section area of the guide is numerically equal and of opposite direction to the total line current obtained by the line integral over the entire conductor boundary.

As an illustration, the TM modes in a circular wave guide have the key function in polar coordinates r, φ:

$$\chi_{m,n}(r, \varphi) = J_m(k_{s;m,n}r) \cos m\varphi \quad m = 0, 1, 2 \cdots$$

subject to the boundary condition

$$J_m(k_{s;m,n}R) = 0$$

This means that the ratio of the guide radius to the cut-off wavelength equals the nth root of the Bessel function of order m, $J_m(x)$, divided by 2π.

For the lowest TM mode in a circular guide this root is $x_{0,1} = 2.40$ and

$$\lambda_{\text{cut-off}} = \frac{2\pi R}{2.40} = 2.62R$$

The magnetic lines are concentric circles. To represent the magnetic field in its proper strength they might be plotted with successive radii r given by

$$J_0\left(2.40\frac{r}{R}\right) = 0.0, 0.1, 0.2 \cdots 0.9 \text{ and } 1.00$$

Caution must be exercised in the practical use of formulas for characteristic impedance of wave-guide modes because of the arbitrariness of the defining equation 19·43. This merely reflects an analogous ambiguity regarding the impedance value which is to be associated with a given terminating load. If the terminating load is in the form of a wire connected across the height of a rectangular guide at its center, the wire resistance needed to absorb all power without reflection is clearly higher than for a vertical wire off-center. Again a different situation prevails if the terminating load is in the form of a resistive coating covering the whole section area of the guide; in this case it would be difficult to associate any simple resistance value with the coating. Generally speaking, formulas for characteristic impedance are useful only in connection with specified types of terminating loads. In the case of a resistive coating, the so-called wave impedance is more useful than the characteristic impedance. It is defined as the field ratio E_s/H_s, multiplied by 120π to convert to ohms. This ratio is constant over the guide section. A coating with a surface resistivity (ohms per square) equal to the wave impedance and backed by a metallic surface one-quarter wave farther from the generator will absorb all incident power without causing reflection, and thus represents a matched termination for a guide of arbitrary cross section. This point will be more fully discussed in Section 19·11.

19·6 MEASUREMENT AND EVALUATION OF STANDING-WAVE PATTERNS

At high frequencies where conventional methods of impedance measurement are impractical, analysis of the standing-wave pattern caused by reflections at the load affords a valuable method for determining the latter's circuit properties. This approach is of particular value in the case of wave-guide loads. In a wave guide, because of the absence of well-defined paths for the currents and terminal points for the voltages, the impedance concept loses much of its immediate significance, and it is preferable to specify such loads, not in terms of an absolute impedance but by its ratio to the characteristic guide impedance. This ratio can be directly computed from the measured value of the reflection coefficient, to which it is related by equation 19·33*a*. Indeed, it is often advisable to use the complex reflection coefficient itself, instead of an impedance, to characterize the load.

In carrying out standing-wave measurements along a transmission line, one must take precautions designed to exclude all modes of propagation other than the desired one. For instance, if the standing-wave detector is housed in a rectangular wave guide of transverse dimensions w and h, it is necessary that it be operated only at frequencies such that the free-space wavelength exceeds the quantity $2wh/\sqrt{w^2 + h^2}$ which is the cut-off wavelength for the $TE_{1,1}$ mode, the lowest mode to be excluded.

If the guide height equals about half the width, as is customary design practice, the $TE_{2,0}$ mode even has a slightly longer cut-off wavelength. Its symmetry character, however, is just opposite to that of the $TE_{1,0}$ mode; consequently its spurious excitation by that mode is much less likely.

If the transmission line is free of loss, the voltage and current distribution or, more generally, the electric and the magnetic fields are repeated periodically along the line in half-wavelength intervals. This is plausible since the phase of the incident wave is retarded by π radians per half wavelength in the direction toward the load, and the phase of the reflected wave is advanced an equal amount, adding up to a complete cycle of relative phase change between the two waves. The electric field, according to equation 19·28, varies as

$$E = E_i \epsilon^{j[\omega t - k(z-a)]}(1 + r\epsilon^{2jk(z-a)})$$

and its amplitude as

$$E = |E_i| \sqrt{1 + |r^2| + 2|r|\cos[2k(z-a) + \theta]} \qquad (19\cdot 51)$$

The voltage maxima are

$$|E_{\max}| = |E_i|(1 + |r|) \qquad (19\cdot 52)$$

and occur at places where the cosine term equals +1, that is

$$2k(z - a) + \theta = 2\pi n \qquad (n, \text{ an integer})$$

or

$$(z - a) = \frac{n\lambda}{2} - \frac{\theta}{4\pi}\lambda \qquad (19\cdot 53)$$

Similarly the voltage minima

$$|E_{\min}| = |E_i|(1 - |r|) \qquad (19\cdot 54)$$

are found at places where the cosine term equals −1, that is,

$$(z - a) = \left(n + \frac{1}{2}\right)\frac{\lambda}{2} - \frac{\theta\lambda}{4\pi} \qquad (19\cdot 55)$$

The square of the electric-field vector varies sinusoidally with the z-co-ordinate according to equation 19·51, which means that the standing-wave pattern becomes a sine curve when an indicator with quadratic response to the voltage or to the electric field is employed. This is the case in conventional microwave standing-wave detectors. At longer wavelengths diode voltmeters with linear response are often used, and they result in a more complicated pattern.

The complex reflection coefficient can readily be found from the known position and variation of the standing-wave pattern. The ratio of maximum to minimum voltage (or electric-field amplitude) determines the magnitude of the r-vector since from equations 19·52 and 19·54

$$\sigma \equiv \frac{|E_{\max}|}{|E_{\min}|} = \frac{1 + |r|}{1 - |r|} \qquad (19\cdot 56)$$

The quantity σ is known as the (amplitude) standing-wave ratio. When an indicator with quadratic response is employed, one measures the square of this quantity, σ^2, often referred to as the power standing-wave ratio.

The phase angle θ of the reflection coefficient is related to the positions of the maxima or minima. It must be borne in mind that the complex reflection coefficient is defined with reference to a certain line cross section ($z = a$ in the preceding), and that the load termination must be identified with this cross section. The physical dimensions of a load may be considerable compared with one wavelength. It is then necessary to select a reference cross section as the basis for evaluation, and the results are characteristic of the load as seen at the reference plane. If the reference plane is at $z = a$, the angle θ is proportional to the distance from that plane to the nearest maximum, to be counted positive in the direction toward the generator:

$$\theta = \frac{4\pi(a - z)}{\lambda} \text{ radians} = 720\frac{(a - z)}{\lambda} \text{ degrees} \qquad (19\cdot 57)$$

It is obviously permissible to disregard in θ terms which are an integral multiple of 360 degrees.

Standing-wave detectors for the analysis of standing-wave patterns make use of an electrostatic probe, that is, a thin wire protruding slightly through a longitudinal slot into the electric field of the transmission line. The probe response is rectified and measured by means of a sensitive d-c galvanometer. Two possible forms are schematically shown in Fig. 19·3. The first shows in cross section a standing-wave detector in a coaxial line, using a thermocouple meter for the r-f indicator. In the second, the standing-wave detector in a wave guide with rectangular cross section makes use of a rectifying crystal, for instance, a silicon crystal-tungsten combination. In each case the slot must be arranged at right angles to the tangential radio-frequency magnetic field so as not to interfere with the flow of currents in the conductor and not to give rise to radiation. The probe should not penetrate too far into the field space, otherwise it gives rise to reflections in addition to those caused by the load, affecting the magnitude of the probe response and thus distorting the standing-wave pattern.[2]

Probe and detector are mounted on a movable carriage so that they can be shifted to desired z-positions. These are read from a precise position scale the zero of which is preferably at the chosen reference cross section $z = a$.

It is not necessary to measure absolute voltages or absolute electric-field strengths since relative values only enter

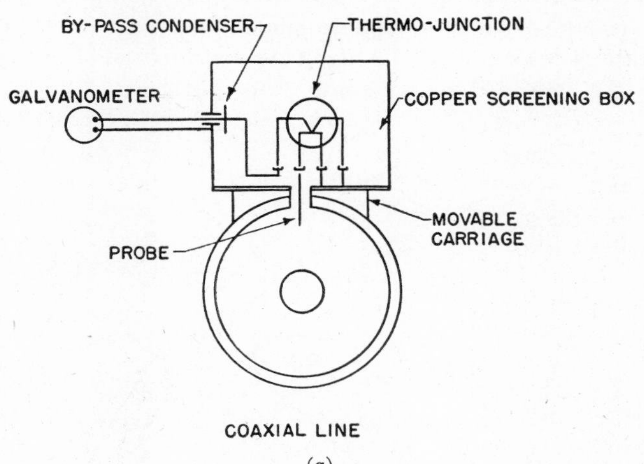

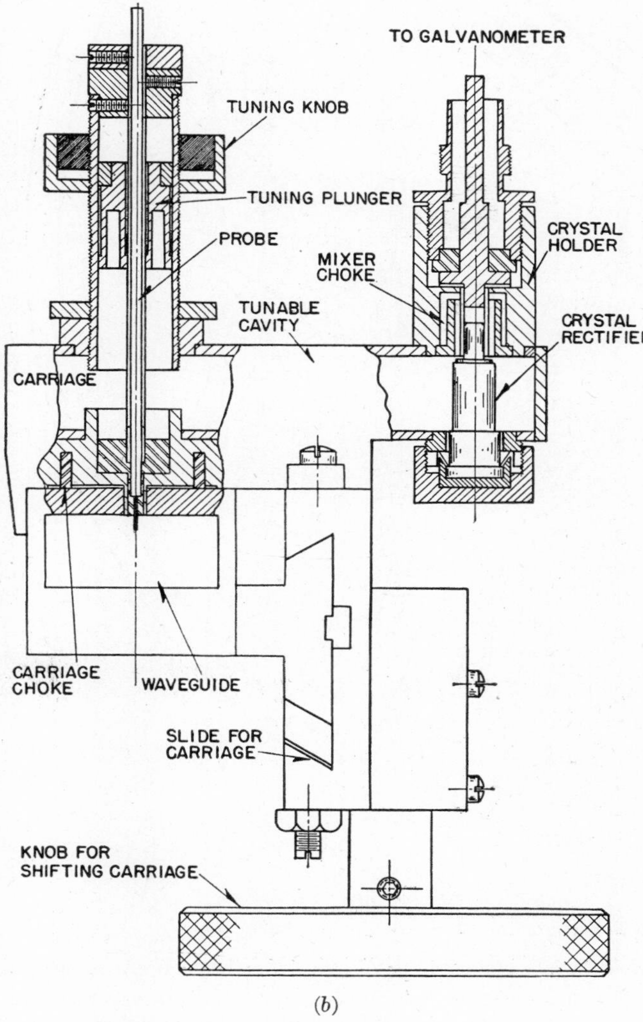

Fig. 19·3 Sectional view of standing-wave detector: (*a*) in coaxial line, using as an indicator a resistance wire in combination with a thermojunction; (*b*) in rectangular wave guide, using crystal detector.

into the computation of the reflection coefficient. Of course, it is important that the indicators be used only within the valid range of their assumed linear or quadratic response law.

19·7 CIRCULAR TRANSMISSION CHART

Most radio-frequency transmission-line calculations can be made with sufficient accuracy by graphical means. Of various charts devised for this purpose, one suggested by Smith [3] seems to be best suited, because all physical loads (complex impedance with positive real part) are represented on a finite region of the chart and also because changes of line length as well as shifts of the reference plane are represented by proportional displacements on the chart (Fig. 19·4).

In the Smith chart, a plot of contour lines of load resistance and load reactance has been drawn on the complex plane of reflection coefficients. These contour lines are useful in problems where it is desired to relate a given standing-wave ratio and position of the pattern to the load impedance, or conversely where one wants to find the reflection coefficient for a given load impedance. Many transmission-line problems can be solved without explicit reference to impedance or admittance, in which case the complex r-plane may be used without a plot of impedance or admittance contours. The properties of the chart are in fact best introduced without referring to impedance. In this form the chart consists simply of a circular line of demarkation, denoting infinite standing-wave ratio ($|r| = 1$). On the chart rim a position scale is laid out in fractional wavelengths, with one full revolution representing one-half wavelength. Pivoted at the center is a transparent ruler with a scale of amplitude standing-wave ratios, as shown in Fig. 19·4. To record the result of a standing-wave measurement in the chart, one swings the ruler into the angular position computed from the observed position of the electric maximum by equation 19·53. It is customary to measure the position of the electric minimum instead of the maximum because of the greater attainable accuracy. For this reason, the fractional wavelength scales given in the chart (Fig. 19·4) do not correspond to the angle θ but its supplement. The connotation "clockwise toward generator" refers to a change in the choice of reference plane; so, for a minimum on the generator side of the reference plane one uses that fractional wavelength scale which reads in the counterclockwise sense. The standing-wave pattern in Fig. 19·5 is represented in Fig. 19·4 by the position of the ruler shown and by the value $\sigma = 2.0$ indicated on the ruler scale; it corresponds to an angle $\theta = 0.058 \times 720$ degrees and to a value 0.308λ on the counterclockwise wavelength scale.

The center of the chart, for standing-wave ratio unity ($r = 0$), is indicative of a matched load ($Z = Z_0$). Points on the chart rim ($|r| =$ unity) are indicative of purely reactive loads completely reflecting the incident wave. Reactive loads can differ from each other in the phase relation between incident and reflected waves. For instance, if the transmission line is open-circuited at the reference plane $z = a$, the reflected wave and the incident wave are in phase synchronism at $z = a$, and the reflection coefficient is plus unity. This results in a maximum of the electric field at the open line end. If the line is short-circuited at $z = a$, the reflected and incident waves are in exact phase opposition ($r = -1$) at $z = a$, resulting in a node of the electric field. In the chart, these two cases are represented by vectors pointing to the right, and respectively to the left, from the origin, giving points as indicated in the chart (Fig. 19·6). If the load is a transmission line free of loss and open-circuited at a distance less than one-quarter wavelength from its input terminals, the representation in the chart is a point on the capacitive region of the chart where the stored energy is predominantly in the electric field. In a line measuring an exact integral multiple of quarter wavelengths the magnetic and electric fields store equal amounts of energy, indicating a state of resonance. The lag of the reflected relative to the

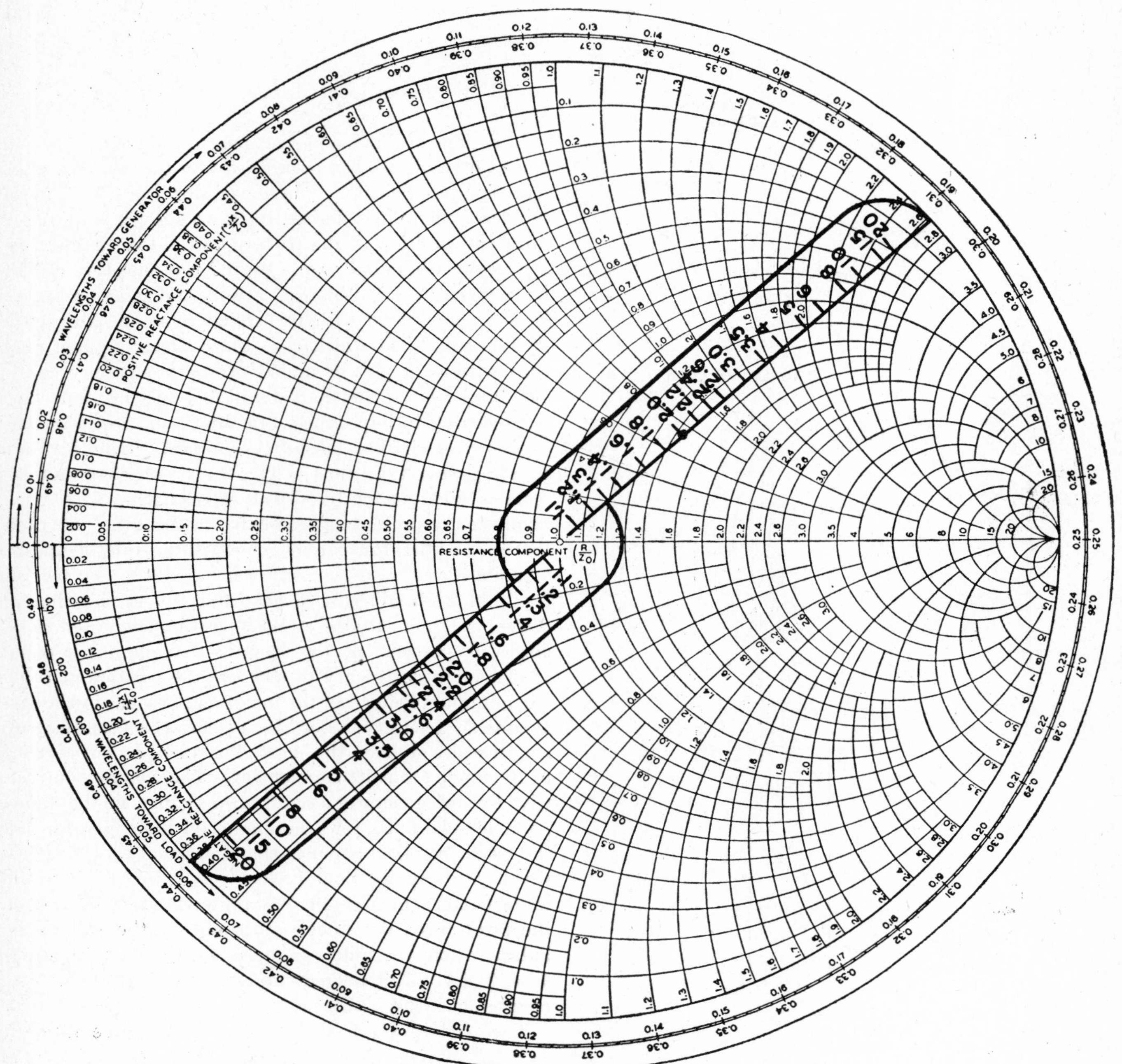

FIG. 19·4 Circular transmission chart with impedance contour lines.

lower half of the rim, because of a rule previously stated and also because the incident wave leads the reflected wave at the reference plane. To see this, one must note that the reflected wave at $z = a$ has a retarded phase compared with that at the open line end where both waves are in phase synchronism; and the incident wave there is again retarded relative to the incident wave at the reference plane.

Generally, points within the upper half of the chart circle represent inductive loads where the stored energy is predominantly in the magnetic fields. The lower half is the incident wave for capacitive loads is in agreement with customary sign conventions for reactance. If the termination is a small condenser C,

$$r = \frac{Z - Z_0}{Z + Z_0} = \frac{\dfrac{1}{j\omega C} - 1}{\dfrac{1}{j\omega C} + 1} \sim 1 - 2j\omega C \text{ (lagging!)}$$

To refer a given chart diagram to a new reference plane, all r-vectors must be rotated through angles proportional to the displacement, and in the clockwise sense if the new reference plane is nearer the generator. A full revolution in the chart corresponds to a displacement by a half wave.

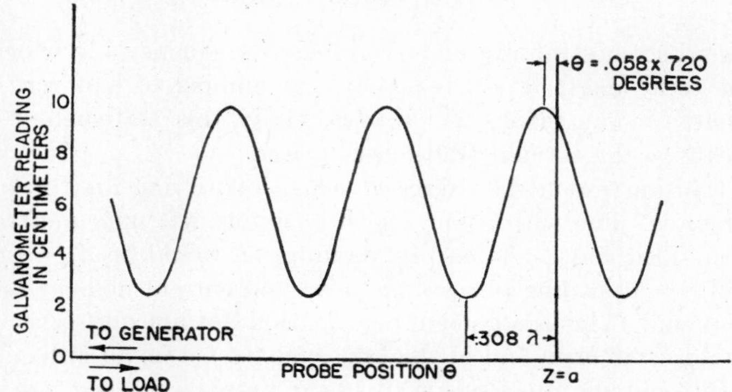

FIG. 19·5 Standing-wave pattern obtained with quadratic detector. Pattern corresponds to chart point in Fig. 19·4.

Contour lines of resistance and reactance (normalized by dividing through Z_0) can be mapped in the circular transmission chart, as shown in Fig. 19·4. These contour lines crowd toward the open-circuit point ($r = 1$) for large R and X values.

It is equally feasible to map, instead, the contour lines of conductance and susceptance on the same background as that of Fig. 19·6 (*a*). This will result in two families of

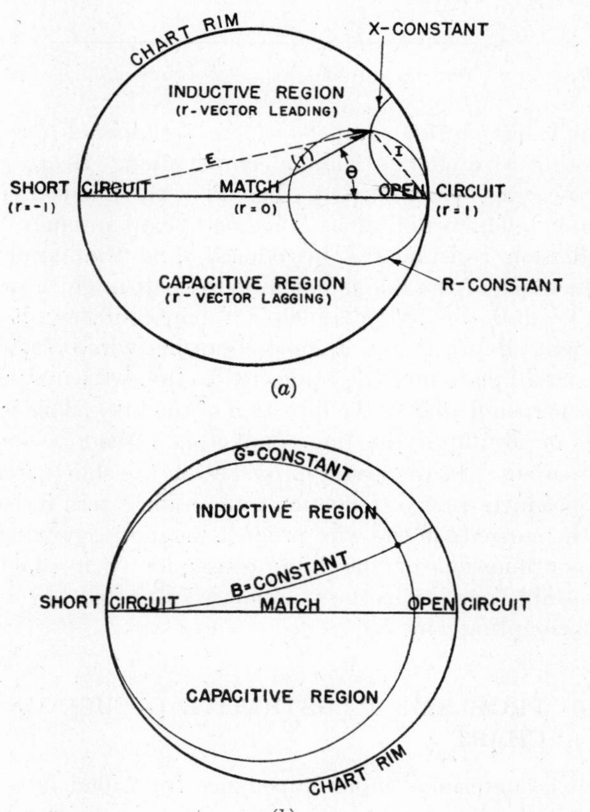

FIG. 19·6 Load regions in the circular chart: (*a*) with impedance contour lines; (*b*) with admittance contour lines.

circles much like the R and X contours, except that they appear rotated through 180 degrees; that is, they will crowd toward the short-circuit point ($r = -1$) for large Y values. It is expedient in this case to use Fig. 19·4 upside down for the contour lines but at the same time to retain the background unchanged, as in Fig. 19·6 (*b*).

The plot of load points in the circular transmission chart lends itself to a very handy diagrammatic representation of load voltage and load current (or transverse electric and magnetic fields). Since the electric vector at the reference plane, for incident amplitude unity, is given by $1 + r$, it will be seen that the chart vector pointing from the short-circuit point ($r_S = -1$) to the load point represents the electric field in magnitude and phase. The current, or the transverse magnetic field, are similarly represented by the chart vector pointing from the load point to the open-circuit point. The net power through the line is given by the product of these two chart vectors times the cosine of the angle enclosed, which is readily seen to be $1 - |r|^2$ for an incident wave of amplitude unity.

19·8 POWER TRANSFER AND STANDING-WAVE RATIO

More explicitly, the average power associated with the incident wave is $E_i^2/2Z_0$, and with the reflected wave it is $|r|^2(E_i^2/2Z_0)$, giving a net flow in the direction of the incident wave which is equal to the difference.

Since for small $|r|$ the power returned by the reflected wave is small like $|r|^2$, the presence of standing waves is a very sensitive test for reflected power. For example, if $|r|$ is 10 percent then the reflected power is only 1 percent of the incident power, yet the standing-wave ratio is $\sigma = 1.22$, or $\sigma^2 = 1.50$ for a standing-wave detector with quadratic response, which can be measured with great accuracy.

If the power rating of a line is limited by breakdown of insulation at a certain voltage E_B then the highest voltage occuring is $E_B = (1 + |r|)E_i$; therefore breakdown will occur when the line transmits power at the rate

$$P_B = \frac{E_B^2}{2Z_0}\frac{1 - |r|^2}{(1 + |r|)^2} = \frac{E_B^2}{2Z_0}\frac{1}{\sigma} \tag{19·58}$$

Maximum power therefore varies inversely as the amplitude standing-wave ratio; consequently careful attention must be paid to the matching of loads to the transmission lines in radio-frequency power circuits.

19·9 SHORT RESONANT LINES

If a two-conductor line is terminated in a short circuit ($Z_L = 0$) equation 19·32 yields the value $r = -1$. Reflection is 100 percent, and there is no net power flow along the line (neglecting its losses). The reflected wave cancels the voltage of the incident wave at the short-circuited end, thus producing a voltage node. Current amplitudes add, producing a current loop. If the line length is one-quarter wavelength then the other line end has a voltage maximum and zero current, giving infinite input impedance. If the line measures one-half wavelength there is a voltage node and a current loop at the input terminals, giving zero input impedance.

Generally, for a short-circuited two-conductor line of length a, the input impedance is

$$Z = jZ_0 \tan ka \tag{19·59}$$

Therefore any reactive value can be obtained by varying the length of the short-circuited line through a range of one-half wavelength. Similar statements hold for wave guides, but k in equation 19·59 must then be replaced by k_z.

It is instructive to check these statements against the procedure of the circular transmission chart. In the chart,

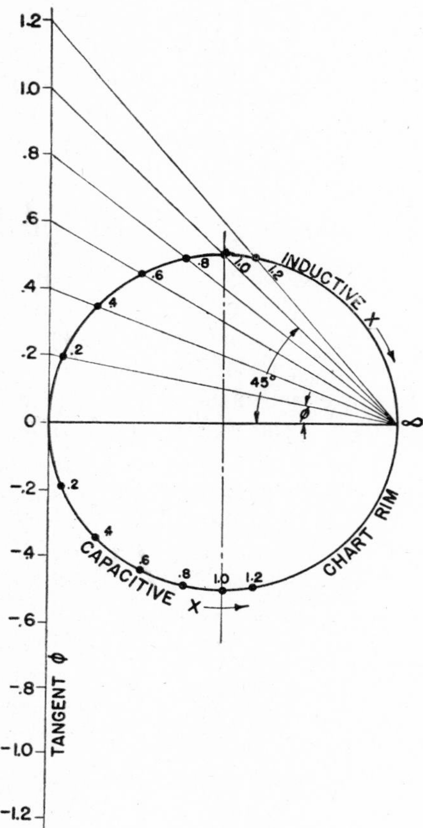

FIG. 19·7 Reactance labels along chart rim.

points on the rim are labeled with the proper reactance values by the construction of Fig. 19·7. The geometrical relations indicated in the figure lead directly to relation 19·59. It will also be noted that rotation of any point of Fig. 19·4 through 180 degrees will change its complex impedance label to its exact reciprocal value, substantiating the fact that a line of length $\lambda/4$ will make the input impedance equal to the reciprocal of the load impedance, counting the characteristic impedance as unity. In particular, a short circuit is transformed into an open circuit and vice versa, by interposing a one-quarter wavelength section.

Figure 19·8 represents a line of length $a + b$ short-circuited at each end and fed by a generator attached at a distance a from one end and b from the other. From equation 19·59 the input admittance at the generator is

$$Y = -j\frac{1}{Z_0}(\cot ka + \cot kb)$$

and therefore the input impedance at the generator is

$$Z = jZ_0 \frac{\sin ka \sin kb}{\sin k(a + b)} \tag{19·60}$$

This becomes infinite at any resonant frequency for which the total length $a + b$ is an integral number of half wavelengths. The reader will readily verify this statement by means of the circular transmission chart.

Caution should be exercised when identifying an "open-circuited" line with a line which is simply left unconnected at its load end. The natural assumption would be $Z_L = \infty$ and $r = 1$, leading to a voltage loop and current node at the load end. This is a good approximation but not quite exact, because the open end of the line tends to act as an antenna and to radiate into free space. As an antenna it has a radiation resistance terminating the line with an effective load even though no special circuit element is connected. The radiation resistance of the end of a parallel-wire line is large but not infinite, so r is nearly but not exactly equal to unity. Similar remarks hold for the unconnected end of a wave guide.

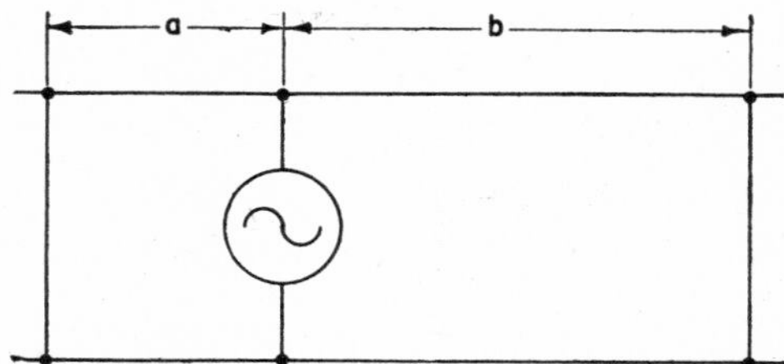

FIG. 19·8 Half-wave resonance on transmission-line section.

The behavior of a parallel-wire line terminated in a short-circuiting wire offers a similar complication. Because such a wire carries heavy current (the end of the line is a current loop), it tends to radiate as a magnetic loop antenna. Thus a radiation resistance is introduced, and the terminating conductance of the line is not infinite, as it might appear to be. To short-circuit a parallel-wire line accurately, it is not sufficient simply to use a short-circuiting wire. Instead, a large metal plate must be connected to the two wires with its plane perpendicular to the direction of the line. This is especially important if the line conductors extend beyond the short circuit. In the first approximation the short-circuiting wire produces perfect reflection like an ideal zero resistance, but the currents in the wire produce magnetic coupling with the continuation of the line; consequently a practically appreciable portion of the power is transmitted beyond the short-circuiting wire.

19·10 PROBLEMS ILLUSTRATIVE OF USE OF THE CHART

(*a*) To determine input impedance from load impedance and length of line. The load impedance is divided by the characteristic line impedance and thus reduced to natural units. The corresponding chart point is located with the help of the impedance contour lines. A change of reference plane from the load end to the input end is represented in the chart by a clockwise rotation of the load point through an angle representing the length of the line and read from the fractional wavelength scale at the rim of Fig. 19·4. The resulting point, referred to impedance contours, gives the desired input impedance.

(*b*) To determine the length of a two-conductor line of characteristic impedance 150 ohms short-circuited at one end and terminated with a 3-micromicrofarad capacitor so that the circuit will resonate at $f = 220$ megacycles.

At the given frequency, $1/j\omega C = -241j$ ohms $= (0 - 1.61j)Z_0$. On the circular chart this is represented by an angular reading of 0.338λ; and, since the line is to present zero impedance at the short-circuited end, the line length must be $(0.500 - 0.338)\lambda$, or 0.162λ. The wavelength at 220 megacycles is $\lambda = 136$ centimeters; therefore the line length required is 22.1 centimeters.

(*c*) To determine the input impedance with two or more circuit elements shunted across the line (Fig. 19·9). This problem requires the use of admittance contours in the chart.

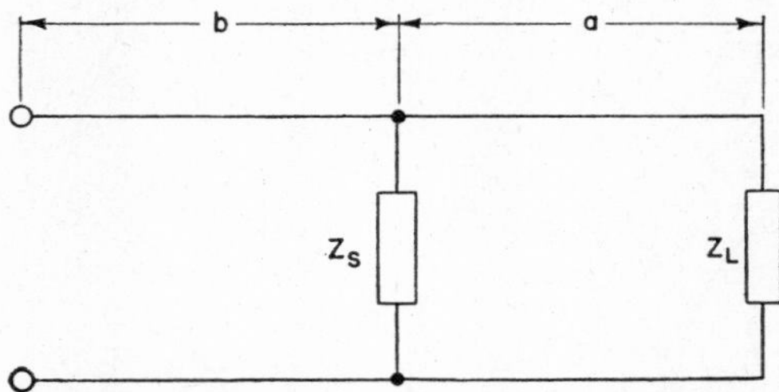

Fig. 19·9 Transmission line loaded on two terminal points.

Use the chart as in (*a*) to determine the line impedance as seen at a distance *a* from the load Z_L. Change to admittance; use the chart for finding the reciprocal value by rotating the point on a concentric circle through 180 degrees. Read the admittance value and add to it the admittance $1/Z_S$. The reciprocal of the sum is found by another rotation through 180 degrees, giving the line impedance as seen at Z_S. This forms the load impedance for the line section of length *b*, and the input impedance is found as in (*a*).

(*d*) Quarter-wave transformer. It is possible to match an arbitrary fixed load *Z* to a transmission line of characteristic impedance Z_0 by interposing a quarter-wave section of a line with different characteristic impedance Z'_0. For example, let it be required to match the load $(1.0 + 1.5j)Z_0$ to a coaxial line of characteristic impedance Z_0 by means of a quarter-wave transformer as shown in Fig. 19·10 (*a*). According to equation 19·24 the line section with larger inner conductor will have a smaller characteristic impedance Z'_0; therefore a given line impedance is represented by a larger fractional value with reference to Z'_0 than with reference to Z_0. If $Z'_0 = kZ_0$, change of the reference plane from a point (2) just outside to a point (3) just inside the quarter-wave transformer, Fig. 19·10 (*a*), must be accompanied by a change of the chart point to one bearing an impedance label that is larger by a factor $1/k$.

The quarter-wave transformer is placed relative to the load (1) so that the line impedance at (2) is a pure resistance of magnitude less than unity. According to the chart, Fig. 19·10 (*b*), a line length of 0.324λ will transform the given load $(1 + 1.5j)Z_0$ to a value $(0.25 + 0.0j)Z_0$ as seen at (2), and to $[(0.25 + 0.0j)/k]Z'_0$ as seen at (3). Traveling through one-quarter wavelength in the constricted section changes this to its reciprocal value $[k/(0.25 + 0.0j)]Z'_0$ as seen at (4). Transition to (5) requires multiplication with *k*; that is, $4.0k^2Z_0$ is the impedance seen at (5). Hence, in order to see a match at (5) and points to the left, it is necessary to make *k* equal to 0.50. For instance, if the ratio of outer to inner diameter in the line were 3.5, giving a characteristic impedance of 75 ohms, inspection of tables of the natural logarithms shows that this ratio must be reduced to 1.87 to give $k = 0.50$. In other words, the inner conductor must be increased by a factor $(b/a)^{1-k}$ to give the required transformation.

Note that the impedance seen at (2) would have to be a resistance larger than unity if the characteristic impedance

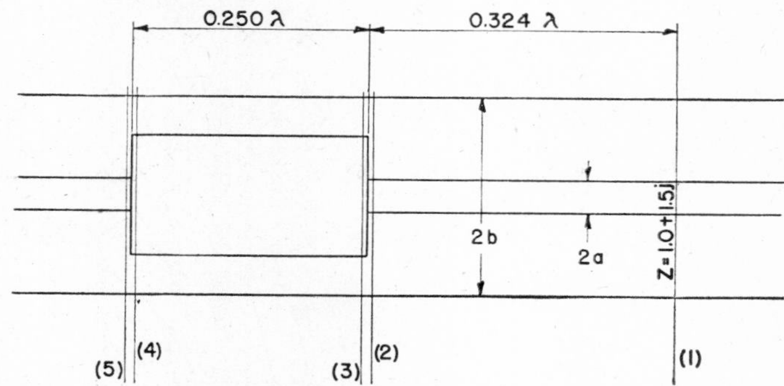

Fig. 19·10 (*a*) Quarter-wave transformer for matching an arbitrary fixed load, coaxial line arrangement for specified load.

inside the quarter-wave transformer had been larger than that of the line, instead of smaller as in the example.

(*e*) Load-to-line matching with three tuning stubs. A common device for matching arbitrary and variable loads to a line is the three-stub tuner consisting of three short lengths of transmission line connected in shunt across the main line at spacings of a quarter wavelength each and provided with adjustable short-circuiting bars. This device permits one to transform an arbitrary load Z_L to match the line Z_0 so that no standing waves will occur on the line left to the matcher, Fig. 19·11 (*a*).

Because there is a compensating circulation of reactive power between load and matching device, this part of the system has to be designed to withstand the attending higher voltage amplitudes.

For a discussion of the three-stub tuner one uses admittance contours in the chart because of the shunt connections. Let the load as seen at the first tuning stub (1) in Fig. 19·11 (*a*) be represented by the chart point (1*a*) in Fig. 19·11 (*b*). By properly placing the terminating bar of stub (1), a pure susceptance of any desired magnitude and sign (inductive for stubs shorter than one-quarter wavelength and capacitive for longer stubs) can be placed in parallel to the line admittance at (1). Chart point (1*a*) is moved to a new position (1*b*) somewhere on the circle of constant conductance through (1*a*). In particular one adjusts the admittance of the first stub so that (1*b*) also lies on the circle *C* of Fig. 19·11 (*b*). Moving on to stub (2) through one-quarter wavelength, circle *C* is transformed to *C′* by rotating it through 180 degrees around the chart center. *C′* will be seen to be a circle of constant conductance going through the chart center, hence it is possible to adjust the bar of the stub (2) so that the added line admittance due to the stub brings the chart point right into the center, indicating a match at point (2*b*). Thus a match has been obtained without using stub (3); in order not to spoil the match already obtained, stub (3) must be made one exact quarter wavelength long in order to present an open circuit (zero admittance) at point (3).

In Fig. 19·11 (*b*), a match was obtained by making stub (1) longer than $\lambda/4$ (capacitive), and stub (2) less than $\lambda/4$. There is another solution which would have required both stubs to be capacitive.

Closer examination shows that a similar procedure will work for all points (1*a*) outside of circle C' but for no point inside that circle. Thus the shaded region in Fig. 19·11 (*b*) indicates the region of the chart which can be matched by stubs (1) and (2). For points inside circle C' one has to make use of stub (3), but in this case one can do without stub (1). To see this, note that a load seen at stub (2) is certainly within the permissible region of stubs (2) and (3) if it was in the forbidden region of stubs (1) and (2). Thus, for any given load, one of the two pairs of stubs will always permit one to obtain a match.

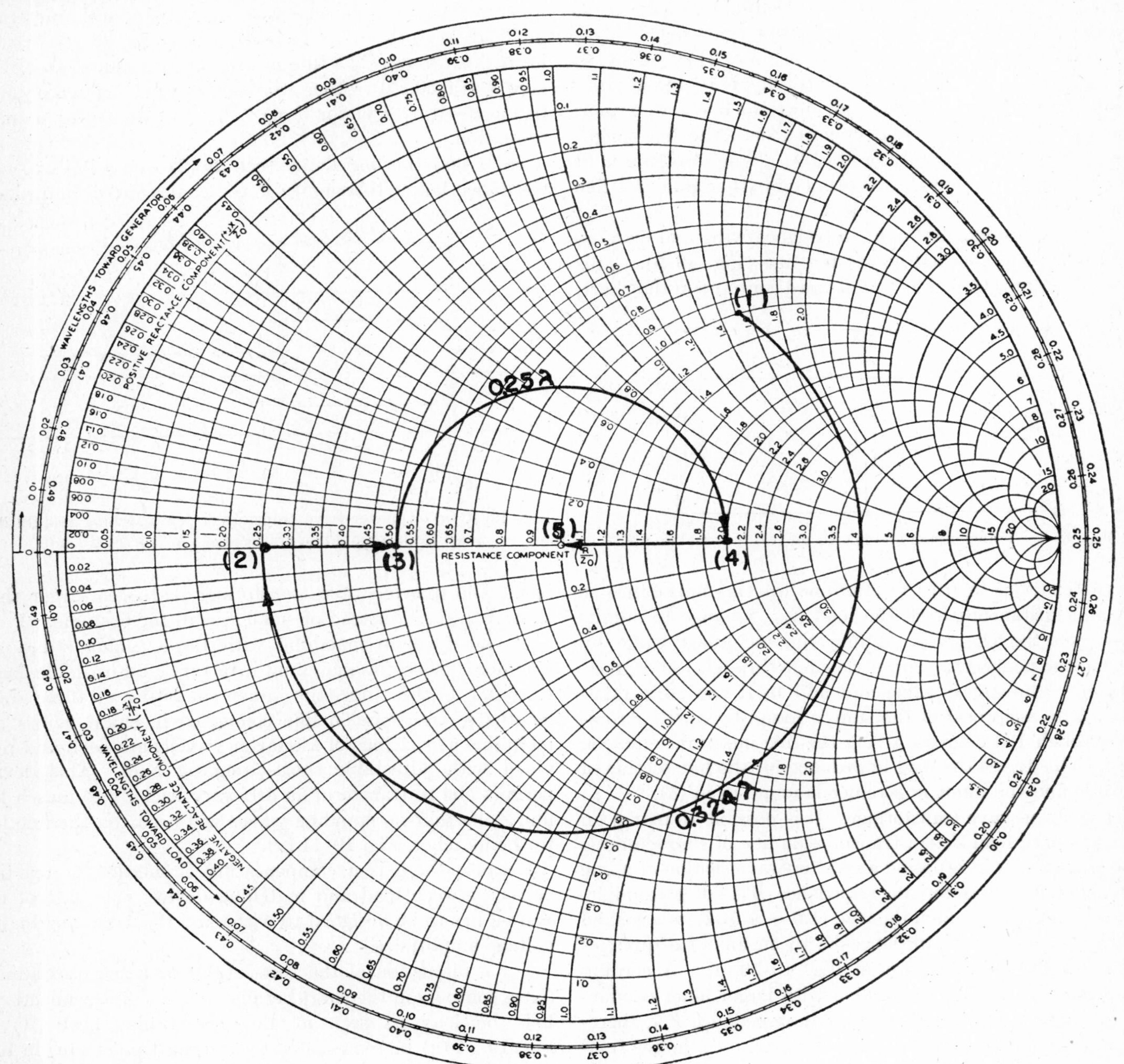

FIG. 19·10 (*b*) Dimensions of Fig. 19·10 (*a*) determined with the chart.

It should be pointed out that this result can be obtained even if the susceptance range of the three tuning elements is restricted to inductive (or else to capacitive) values. If the adjustable length of each stub is restricted to lengths between one-quarter and one-half wavelength so as to give only capacitive susceptances, the doubly shaded region in Fig. 19·11 (*b*) can be handled by stubs (1) and (2). Points outside the doubly shaded region can always be handled by stubs (2) and (3). This situation assumes practical significance in the case of three-screw tuners customary for matching loads to wave guides. A screw protruding into the electric field of a wave guide has a function very similar to that of a tuning stub but, owing to design conditions, it can usually add only capacitive but not inductive susceptances to

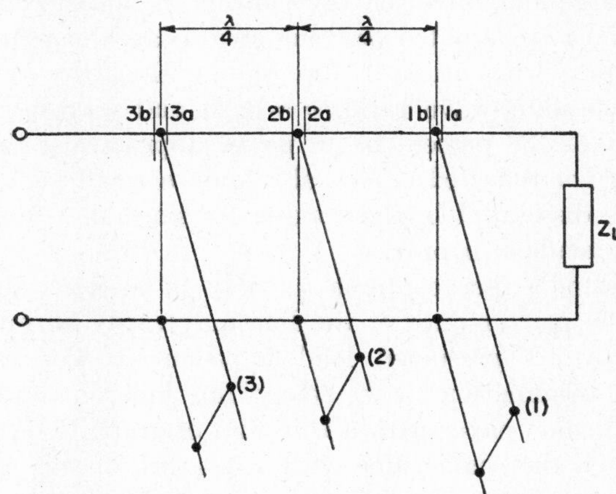

FIG. 19·11 (*a*) Three-stub tuner for matching arbitrarily variable loads, schematic view.

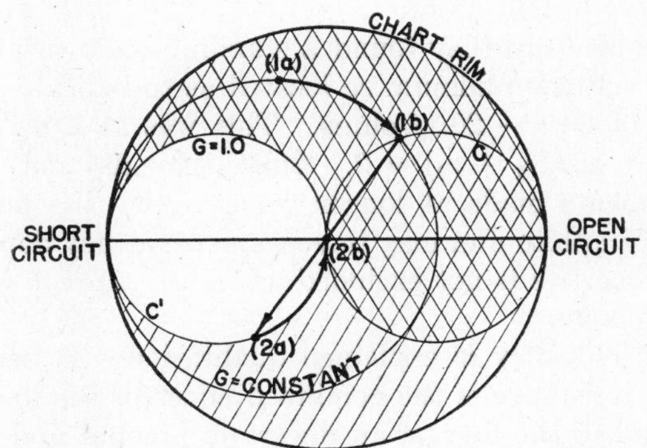

FIG. 19·11 (*b*) Solution obtained with the chart.

the line admittance. In spite of this restricted tuning range, the three screws are capable of matching an arbitrary load to the guide

19·11 TRANSMISSION THEORY, A COROLLARY OF FIELD THEORY

In the conventional presentation of its theory, a transmission line is usually presented as a limit of a circuit of recurring elements as in Fig. 19·1, and the performance is computed from the differential equations corresponding to this model. This is often referred to as the "engineering" form of the theory, suggesting a contrast with the "physicist's" approach by way of Maxwell's equations. In Section 19·5 it was shown how the "engineering" approach can be extended to wave guides in spite of the fact that these do not lend themselves to representation by equivalent circuits of recurring elements.

Practical transmission lines differ widely from their idealized models with strictly cylindric conductor configuration, in terms of which the theorems of transmission theory are usually demonstrated. The controversial element thereby injected into the subject is apt to create distrust in the "engineering" approach even where this treatment is entirely adequate. Fortunately an exposition of the limitations of the "engineering" approach is relatively simple.

Everyone believes, on the basis that no example to the contrary has ever been uncovered, that Maxwell's equations for the electromagnetic field are an adequate theoretical foundation for all work dealing with electromagnetic fields, at the frequencies which are employed in engineering. The theory of wave propagation along a line consisting of two parallel perfect conductors of arbitrary form, based on Maxwell's equations, shows that such waves exist with the electric and magnetic vectors **E** and **H** both accurately transverse to the axial direction of the line. Dielectric losses are easily taken into account because their distribution in space is proportional to the distribution of the electric energy.

However, ohmic losses in a material of imperfect conductors are not so rigorously dealt with because they require a slight departure from transversality of vectors **E** and **H**, thus causing a flow of energy into the conductors to provide for the dissipation in them. But that is a fine mathematical point with very little bearing on practical work. Fancy analysis ends by justifying the usual engineering method for lines made of good conductors such as are used in practice.

More difficult in general is the class of phenomena whose effects become pronounced at ultrahigh frequencies and beyond. This is a difficulty characteristic of all electric-circuit theory in which simple parameters like L and C are assigned to inductors and capacitors. At frequencies so high that the wavelength is not great compared with the size of the apparatus, the inductors have large distributed capacitance, and likewise capacitors show effects of inductive reactance. Sometimes such effects become pronounced at relatively low frequencies.

The same sort of difficulty exists in transmission-line theory. The relation to Maxwell's theory is well established only for smooth straight lines. But engineers are interested in lines with bends in them. Field theory deals with transverse waves only and may leave the impression that the whole calculation fails if the field is not transverse. Yet, if a line branches, as in stub-tuner arrangements, there must be some local distribution of fields near the branch where the fields are not transverse, because they cannot be transverse to both branches of the branched line at the same time. Similarly there are always local field irregularities wherever there are discontinuities in the line at an insulating support, where the spacing of a parallel wire line is changed, or where the sizes of the conductors in a coaxial cable are changed, for example.

All such irregularities cause local departures from the fields of the pure undisturbed transverse waves of the simple theory. As such they become localized storage places for electric or magnetic energy; in circuit theory these are represented by capacitors and inductors. The practical behavior of the irregularities is not difficult to understand, however difficult their calculation from their given geometry. In this respect the situation is like that in ordinary circuit theory—how seldom can the L's and C's of practical engineering inductors and capacitors be accurately calculated from their design dimensions! The shapes used in practice create difficult problems, and yet no one worries about any fundamental difficulty with the concepts of inductance and capacity on that account. Engineers use impedance bridges to measure L and C, and then confidently use the data in whatever circuit these circuit elements may be employed.

The same is true of the various disturbing circuit irregularities on actual two-conductor lines. There are always stretches where the line is regular. The irregularities are localized. Their general effect is to introduce reflections on an incident wave which can be studied by means of the standing-wave technique. By experimental study of the reflections a quantitative characterization of the effect of the irregularities can be given by describing them formally as shunt admittances.

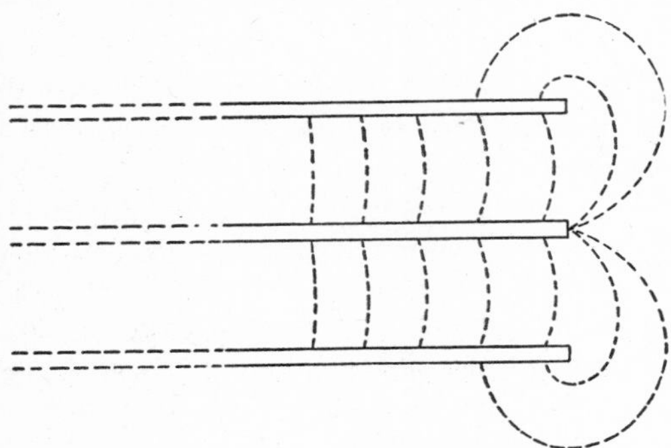

Fig. 19·12 Electric-field configuration at open end of coaxial line.

It is interesting to apply this point of view to some experimental data given by Jackson and Huxley.[4] They *measured* the nature of the termination of an open-ended coaxial line by observing the standing waves on such a line. The ratio of outer to inner conductor radii was 3.55, giving $Z_0 = 76$ ohms. They did this for various wavelengths expressed as ratio of wavelength to mean circumference of the two conductors, $\lambda/[\pi(r_1 + r_2)]$. When this ratio was 6.7, so that the wave was fairly long compared to transverse dimensions, they could not detect any shift in the voltage minimum away from a point $\lambda/4$ from the open end, and the standing-wave ratio was "large." In other words the behavior was as if $Z_L = \infty$, as is assumed in the simple "engineering" theory for an open line.

When the ratio of wavelength to mean circumference was brought down to 2.0, they observed that the position of the voltage node shifted slightly in such a way as to indicate an effective terminating impedance, $Z_L = 0 - 400j$ ohms. On a circular impedance chart it is hard to get an accurate value in this region, but nevertheless the load impedance obtained was clearly finite. When the ratio of wavelength to mean circumference was further dropped to 0.6, a standing-wave ratio of 4.25 was observed with a shift of the voltage minimum by 0.145λ, indicating an effective load impedance of $28 - 54j$ ohms. This is pretty far from being infinite!

The terminating impedance of the "engineering" approach supposedly represents the effect of departures of the actual field from the assumed transverse fields of a simple wave. Because the electrical field at the end is qualitatively like Fig. 19·12, there is some current at the end to carry charge to the outside of the outer conductor. In the laboratory it is easy to determine that the waves do run out and return along the outside of the outer conductor. More detailed experiments would probably show that the measured values are influenced by the positions of the supports of this outer conductor. These in turn could act as reflecting "terminations" of this outer line, which acts as a continuation of the line to cause variations in the effective values of Z_L. Since Z_L is rather small at the shortest wavelength, there is a rather large amount of energy pouring out of the open end which must be radiated or conducted away along the outer conductor. (Equation 19·56 shows that, for the observed standing-wave ratio, $\sigma = 4.25$ and correspondingly $|r|^2 = 0.62$, 38 percent of all power approaching the open end was not reflected.) Probably only a small part of this power is dissipated in the conductors since they were presumably made of copper.

The same paper contains a study of another situation where the performance of the line may easily be misrepresented by its presumed load impedance. To realize a matched termination for the coaxial line Jackson and Huxley obtained some paper with a thin film of graphite on it, and terminated the coaxial line with a flat disk of this material having a d-c resistance of 77 ohms—very close to the characteristic impedance of 76 ohms. Yet it is a fallacy to suppose that this termination fulfills physically the condition $Z_L = Z_0$ for absence of reflection.

Experimentally they found a standing-wave ratio of 3.57 and the voltage minimum so located as to indicate an effective Z_L of $23.1 - 20.1j$ ohms. This was at a wavelength such that $\lambda/\pi(r_1 + r_2) = 2$. Since they did not describe measurements made at longer wavelengths, the possibility that the graphite film may have some frequency dependence to its properties is not excluded, although a great variation seems unlikely.

The discrepancy is explained qualitatively as before, for the load resistance is the combined effect of the local fields at the end of the line, of an uncertain amount and difficult to calculate, together with the shunting effect of the graphite disk. The measured effective $1/Z_L$ is an admittance of $24.8 + 21.6j$ millimhos. If the graphite film retains its d-c value it accounts for 13.0 millimhos of admittance, leaving $11.8 + 21.6j$ to be accounted for by radiation and reactive properties of the local external fields just beyond the physical end of the line.

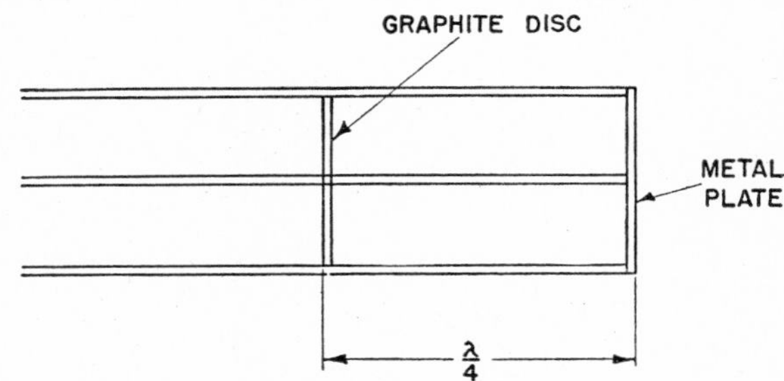

Fig. 19·13 Proper termination to obtain a matched load in coaxial line.

The analysis suggests the remedy: A line is properly terminated in a match only when Z_L really equals Z_0; therefore if the graphite resistance is to have the correct magnitude it must not be shunted, as a load, by any other part of the terminating resistance. This was accomplished as shown in Fig. 19·13. The line was continued beyond the graphite disk for $\lambda/4$ and terminated in a short-circuiting metal plate. The line beyond the graphite disk thus presents zero admittance at the graphite disk and so does not alter its shunt admittance. Therefore the complete termination satisfies the mathematical requirements of the theory. Experimentally a standing-wave ratio of 1.02 was found, and this ratio was equal to unity within the experimental accuracy.

The short-circuiting metal plate was made in the form of an adjustable piston, and it gave the best standing-wave ratio when it was only 0.88 quarter wavelength away from the graphite disk. It therefore was providing some inductive susceptance, which indicates that the graphite-on-paper disk was providing capacitive susceptance, as well as conductance to the total admittance reckoned at its location.

At the shortest wavelength, $\lambda/\pi(r_1 + r_2) = 0.6$, with the disk alone at the end of the line, the line behaved as if it were terminated with an admittance of $22.8 + 2.7j$ millimhos. Previously without the graphite disk the open end was found to be a termination of $7.6 + 14.6j$ millimhos. The difference, which is $15.2 - 11.9j$, is therefore the admittance contribution of the graphite disk. This time the disk appears to contribute a shunt inductive susceptance.

The experiments of Jackson and Huxley show that one cause of the seeming failure of the circuit theory of transmission is the improper identification of existing load conditions with their presumably equivalent circuit symbols. Another source of discrepancies is the existence of modes of propagation other than the one assumed, at either the ingoing or the outgoing reference terminals. With proper attention to both sources of error, it will invariably be found that one of the following three situations prevails.

(1) If the transmission system is a homogeneous line formed by cylindric conductor configurations of good conductivity, the transmission theory is mathematically equivalent to the field theory of one mode, the chosen mode of propagation. Caution is necessary in order that all other modes of propagation, and particularly radiation into free space, be excluded. At shorter wavelengths this means that the line must be properly dimensioned and must be completely enclosed by one of its conductors.

(2) If two, not necessarily identical, transmission lines lead into and out of an otherwise closed system of arbitrary internal conductor geometry, the system can be rigorously represented by an equivalent circuit of lumped elements, provided that the ingoing and outgoing transmission lines are designed to permit the existence of only one mode at each reference terminal (reference plane). No restriction is needed regarding the presence of higher modes inside the system except that their fields must not extend to the reference terminals.[5]

This is the usual situation in all engineering problems. The system is regarded as a transducer, or four-terminal network, the properties of which are empirically determined by three independent measurements such as open- and short-circuit tests at each specified frequency.

(3) If more than one mode of propagation can exist at either the input or the output terminals of the system, circuit concepts cannot safely be applied, and a more detailed analysis involving the electromagnetic fields is required. The most important illustration is the antenna problem; the outgoing power leaves the antenna not only in the form of dipole radiation (lowest TM mode of free space) but also in the form of higher modes as well. A simple and accurate circuit equivalent does not exist for situations in this category.

REFERENCES

1. "Equivalent Circuits for Discontinuities in Transmission Lines," J. R. Whinnery and H. W. Jamieson, *Proc. I.R.E.*, Vol. 32, 1944, p. 98.
2. "Probe Error in Standing-Wave Detectors," W. Altar, F. B. Marshall, and L. P. Hunter, *Proc. I.R.E.*, Vol. 34, Jan. 1946, p. 33.
3. "Transmission Line Calculator," P. H. Smith, *Electronics*, Vol. 12, 1939, p. 29; Vol. 17, 1944, p. 130.
4. "Solution of Transmission-Line Problems," Willis Jackson and L. G. H. Huxley, *J.I.E.E.* (London), Vol. 91, 1944, Part 3, p. 105.
5. "*Q*-Circles—A Means of Analysis of Resonant Microwave Systems, Part II," W. Altar, *Proc. I.R.E.*, Vol. 35, 1947, p. 478.

Chapter 20

ANTENNAS

T. M. Bloomer

An antenna is a system or structure for radiating or receiving radio waves, exclusive of the means for connecting its main portion with the associated apparatus. More simply, an antenna is a device for radiating or receiving electromagnetic waves.

Antennas are used for many purposes, such as point-to-point communications, broadcasting systems, direction-finding systems, systems of mensuration, transmission of power (on an experimental basis), radar, loran, and other systems.

Design is influenced by many factors. In broadcasting, for example, radiation should be concentrated in a given region for economic reasons. Sometimes, to minimize interference with stations on the same or nearly the same frequency, energy must be radiated in a specified direction at a level too low to cause interference. A receiving antenna may utilize directional characteristics to increase the voltage of the received signal applied to the terminals of the receiver, and to discriminate against noise from origins not in the same direction as the transmitter.

Another important factor is the ratio of the bandwidth to be radiated or received to the magnitude of the carrier frequency.

Antenna impedance should be in the range where matching networks are reasonably efficient, unless efficiency of the system is relatively unimportant.

The antenna must be designed so that it can be built from available material.

Specifications should describe:

1. Carrier frequency.
2. Power to be radiated.
3. Required field pattern.
4. Band width.
5. Permissible impedance range.
6. Maximum permissible dimensions.
7. Maximum weight allowable.
8. Any special features.

Some of these specifications are affected by economics. For example, it is sometimes less expensive to increase the power of the transmitter (or the gain of the receiver) than to increase antenna gain by decreasing the angle of radiation of the antenna.

There are two approaches to obtaining quantitative data for design:

1. *Trial-and-Error Methods.* Use of data from trial-and-error experimentation is a useful method of designing antennas. The data may be represented by charts, curves, or empirical equations fitted to experimental curves.
2. *Mathematical Computation.* The computation may be based on fundamental laws of physics and on assumptions. The accuracy of the conclusions depends upon the accuracy of the assumptions.

Frequently trial-and-error methods are used to correct for inaccuracies in computation or in basic assumptions.

20·1 MAXWELL'S EQUATIONS; THEORY OF RADIATION [1, 2]

From an academic standpoint it is interesting to know that Maxwell predicted electromagnetic waves mathematically before Hertz detected them experimentally. Maxwell's equations describe the laws of propagation of electromagnetic waves.

Ampere's Law

Ampere stated that the work done in carrying a unit magnetic pole in a closed path bounding a surface with a normal component of current I through the surface is $4\pi I$. In vector notation for three dimensions Ampere's law becomes

$$\text{curl}\,\mathbf{H} = 4\pi \frac{\mathbf{J}}{c} \tag{20·1}$$

where $\mathbf{H}$ = vector magnetic field (force on a unit pole) in oersteds

$\mathbf{J}$ = current density in statamperes per square centimeter

c = velocity of electromagnetic waves in centimeters per second (3×10^{10} in vacuum)

t = time in seconds.

Equation of Continuity

Maxwell extended Ampere's law by including the effects of capacitors in allowing for charges to pile up at points.

Now

$$\text{curl } \mathbf{H} = \frac{4\pi \mathbf{J} + \dfrac{\partial \mathbf{D}}{\partial t}}{c} \qquad (20\cdot2)$$

where **D** is the electric flux vector caused by the piling up of charges on capacitors in the system. Equation 20·2 is the first of the Maxwell equations.

Faraday's Law of Induction

By applying the Stokes' theorem to Faraday's law of induction, Maxwell obtained his second equation.

Faraday's law:

$$\oint \mathbf{E}\cdot d\mathbf{l} = -\frac{1}{c}\frac{\partial \Phi}{\partial t} = -\frac{1}{c}\frac{\partial}{\partial t}\int \mathbf{B}\cdot d\mathbf{A} \qquad (20\cdot3)$$

where $\mathbf{E}$ = vector electric field in statvolts per centimeter
$d\mathbf{l}$ = elemental length in centimeters
Φ = magnetic flux vector in maxwells
$\mathbf{B}$ = flux density vector in gauss
t = time in seconds
$d\mathbf{A}$ = elemental area in square centimeters.

In words, voltage induced in a loop is directly proportional to the time rate of change of magnetic flux through the loop.

Stokes' Theorem

$$\oint \mathbf{E}\cdot d\mathbf{l} = \int \text{curl } \mathbf{E}\cdot d\mathbf{A} \qquad (20\cdot4)$$

Stokes' theorem is virtually a definition of curl, for it expresses the conclusions to be drawn from the definition of curl. Substituting the right-hand part of equation 20·3 into the left-hand part of equation 20·4,

$$-\frac{1}{c}\frac{\partial}{\partial t}\int \mathbf{B}\cdot d\mathbf{A} = \int \text{curl } \mathbf{E}\cdot d\mathbf{A}$$

and

$$\text{curl } \mathbf{E} = -\frac{1}{c}\frac{\partial \mathbf{B}}{\partial t} \qquad (20\cdot5)$$

Equation 20·5 is Maxwell's second equation.

Maxwell's Equations

Restating, Maxwell's equations are:

$$\text{curl } \mathbf{H} = \frac{4\pi \mathbf{J} + \dfrac{\partial \mathbf{D}}{\partial t}}{c}$$

and

$$\text{curl } \mathbf{E} = -\frac{1}{c}\frac{\partial \mathbf{B}}{\partial t}$$

Add to these:

$$\mathbf{J} = \gamma \mathbf{E} \qquad (20\cdot6)$$

$$\text{div } \mathbf{B} = 0 \qquad (20\cdot7)$$

$$\text{div } \mathbf{D} = 4\pi\rho \qquad (20\cdot8)$$

where $\bar{\gamma}$ = admittance per unit volume in statmhos per cubic centimeter
ρ = charge density in statcoulombs per cubic centimeter

and other terms are as previously defined.

Maxwell's equations and equations 20·6, 20·7, and 20·8 are the basic equations of electromagnetic theory. They are given here, not for use, but to show that the natural laws involved are known. Equation 20·6 is a form of Ohm's law. Equation 20·7 is the mathematical way of saying that a line of magnetic flux is a closed loop. Equation 20·8 states that a line of electric flux must start and stop either on a charge or at infinity and that the electric flux diverging from a volume is equal to 4π times the charge within the volume.

It is known that there is a force on a unit magnetic pole in the region of a conductor carrying current. Measurements have been made to establish the magnitude and direction of the force. The relation has been extended to include the effects of capacitors. **H** is known in terms of **J** and **D**, and if **E** is changing, an **H** exists.

Faraday's law says that a changing magnetic field sets up a voltage in any circuit linked by the magnetic flux. Maxwell applied the Stokes' theorem and found that a changing magnetic field induces an electric field. The solutions to Maxwell's equations, particularly for free space, represent traveling waves, and Maxwell was thus able to show the existence of electromagnetic waves.

20·2 INDUCTION FIELD, RADIATION FIELD

Elemental Oscillating Doublet

Antennas can be analyzed by using an elemental oscillating doublet in free space, which is a wire of length l ($l \ll \lambda$),

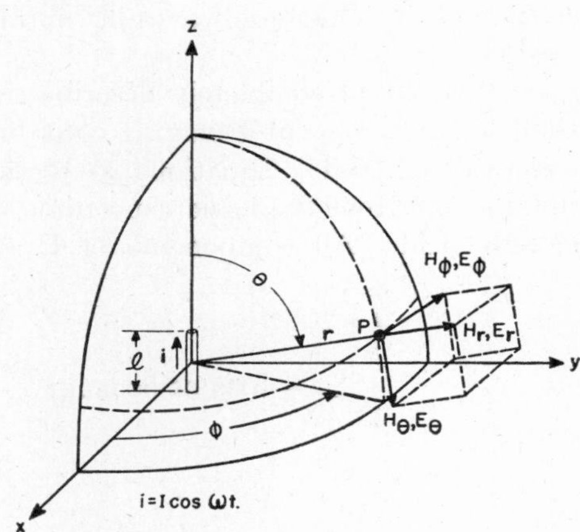

FIG. 20·1 Oscillating doublet in free space.

carrying a current of $I \cos \omega t$ as shown in Fig. 20·1 where

$\omega = 2\pi/\lambda$, the angular velocity in radians per second
λ = wavelength in centimeters
I = maximum current in statamperes
l = length in centimeters along the line $\theta = 0$
r = distance in centimeters from the midpoint of l to point P
θ = angle between x-axis and projection of r on plane $\theta = 90$ degrees
ϕ = angle to P measured from a line passing through midpoint of l and lying in the plane $\theta = 90$ degrees
t = time in seconds
$l \ll \lambda$.

From Maxwell's equations the components of the fields at any point P are

$$H_\phi = Il \sin\theta \left[-\frac{\cos\omega\left(t - \frac{r}{c}\right)}{c}\frac{1}{r^2} + \frac{\omega \sin\omega\left(t - \frac{r}{c}\right)}{c^2}\frac{1}{r} \right] \tag{20·9}$$

$$H_r = 0 \tag{20·10}$$

$$H_\theta = 0 \tag{20·11}$$

$$E_\phi = 0 \tag{20·12}$$

$$E_r = 2Il \cos\theta \left[\frac{\sin\omega\left(t - \frac{r}{c}\right)}{\omega}\frac{1}{r^3} + \frac{\cos\omega\left(t - \frac{r}{c}\right)}{c}\frac{1}{r^2} \right] \tag{20·13}$$

$$E_\theta = Il \sin\theta \left[\frac{\sin\omega\left(t - \frac{r}{c}\right)}{\omega}\frac{1}{r^3} + \frac{\cos\omega\left(t - \frac{r}{c}\right)}{c}\frac{1}{r^2} + \frac{\omega\sin\omega\left(t - \frac{r}{c}\right)}{c^2}\frac{1}{r} \right] \tag{20·14}$$

where H_ϕ, H_r, and H_θ are components of $\mathbf{H}$, the vector magnetic field intensity, in the ϕ-, r-, and θ-directions respectively, and are in oersteds; E_ϕ, E_r, and E_θ are components of $\mathbf{E}$, the vector electric-field intensity, in the ϕ-, r-, and θ-directions respectively, and are in statvolts per centimeter; c is the velocity of electromagnetic waves in centimeters per second.

Equations 20·9 to 20·14 completely describe the electric and magnetic fields at any point P having co-ordinates r, θ, and ϕ according to Fig. 20·1. Equations 20·10, 20·11, and 20·12 seem reasonable from the basic conception of a magnetic and electric field. All components of $\mathbf{H}$ except H_ϕ would be expected to be zero. E_ϕ also would be expected to be zero.

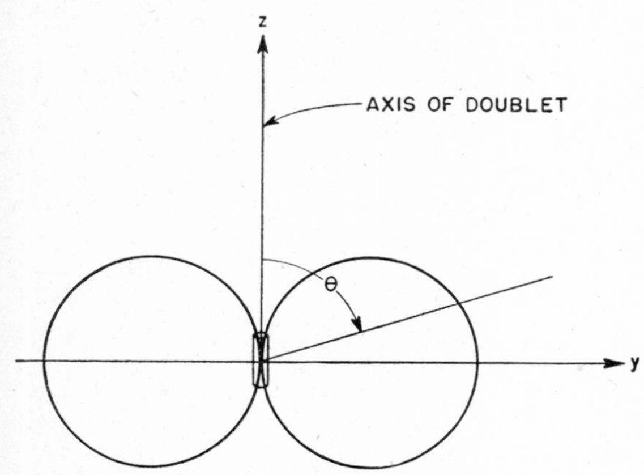

Fig. 20·2 Field pattern of an oscillating doublet.

Equation 20·9, describing H_ϕ, shows two components 90 degrees out of time phase for any given point in space. The first term in the brackets has a multiplier of $1/r^2$. The second term is directly proportional to $1/r$. The second term has its greatest significance for large values of r; and the first term predominates for small values of r. The term involving $1/r^2$ represents what is called the induction field, and the term involving $1/r$ represents the radiation field.

Equation 20·13, describing E_r, the radial electric field, is made up of two components 90 degrees out of time phase. The term involving $1/r^3$ represents the magnitude of the quasi-static field. The term involving $1/r^2$ represents the magnitude of the induction field as in equation 20·9. The induction portion of the electric field is 180 degrees out of time phase with the induction portion of the magnetic field. E_r has no radiation component.

Equation 20·14 describes E_θ, the electric-field intensity in a plane of the doublet. It has quasi-static, induction, and radiation terms. The quasi-static field is in time phase with the quasi-static field of E_r. The induction component of E_θ is also in time phase with the induction component of E_r. The radiation term of E_θ is in time phase with the radiation term of the magnetic field H_ϕ. When $r = \lambda/2\pi$ the induction and radiation fields of H_ϕ are equal, the quasi-static and induction terms of E_r are equal, and the quasi-static, induction, and radiation terms of E_θ are all equal.

Radiation Fields. For $r \gg \lambda$ we have the radiation fields:

$$H_\phi = Il \sin\theta \left[\frac{\omega\sin\omega\left(t - \frac{r}{c}\right)}{c^2}\frac{1}{r} \right] \tag{20·15}$$

$$E_r = 0$$

$$E_\theta = Il \sin\theta \left[\frac{\omega\sin\omega\left(t - \frac{r}{c}\right)}{c^2}\frac{1}{r} \right] \tag{20·16}$$

where the symbols are as identified for equations 20·9 to 20·14. Equations 20·15 and 20·16 are for traveling waves, as evidenced by the $t - (r/c)$ terms. The magnetic field is polarized so that it is perpendicular to the line of propagation and perpendicular to the line of the doublet. The electric field is polarized so that it is perpendicular to the line of propagation and in a plane of the doublet. Electric and magnetic polarizations are perpendicular to each other in space. They are in time phase and equal in magnitude if H_ϕ is in electromagnetic units and E_θ is in electrostatic units.

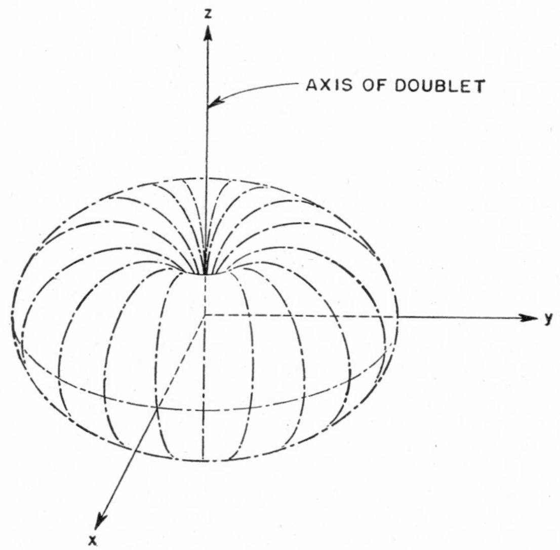

Fig. 20·3 Three-dimensional field pattern of an oscillating doublet.

Field Patterns. For a given r and ω the magnitude of the vectors is proportional to $\sin\theta$. A field pattern or radiation diagram of an antenna is the relation of the magnitude of its field to angular direction, always maintaining a constant distance from the antenna. Figure 20·2 shows the field pattern for the elemental oscillating doublet shown in Fig. 20·1. Equations 20·15 and 20·16 are functions of θ and not of ϕ. Hence for a three-dimensional representation of the pattern of an oscillating doublet, the pattern of Fig. 20·2 can be rotated about the line of the doublet ($\theta = 0$). (See Fig. 20·3.)

Equations 20·15 and 20·16 show that E_θ and H_ϕ for $r \gg \lambda$ vary inversely with distance.

Linear Antennas

Current Distribution. So far only an elemental oscillating doublet has been discussed. If the dimensions of an antenna in free space are known and the current distribution can be assumed with some accuracy, there is a good method of analyzing the field set up by an antenna. Consider, for example, an antenna in free space that is $\lambda/2$ long, where $\lambda = c/f$. Distribution of current on the antenna wire is not known in advance. If there were no radiation and the resistance of the antenna were negligible, the current distribution would be that of an open-circuited lossless transmission line. Current distribution then would be (Fig. 20·4)

$$I_z = I_0 \cos\frac{2\pi z}{\lambda}\cos\omega t \tag{20·17}$$

where I_z = peak current in statamperes at a distance of z centimeters from the center of the antenna

I_0 = peak current in statamperes at $z = 0$

and other terms are as previously defined. Figure 20·4 shows the magnitude of the current.

Field at a Remote Point. If the antenna of length $\lambda/2$ is considered to be made up of a large number of discrete elemental doublets joined end to end, each doublet having a length of Δz centimeters and a current of $I_0\cos(2\pi z_n/\lambda)\cos\omega t$ statamperes, with z_n representing the distance in centimeters from the center of the elemental doublet to the center of the over-all antenna, then by summation the field at any point P can be determined. Figure 20·4 depicts a discrete elemental doublet and its current.

The field at P is made up of the vector sum of the fields of all the discrete elemental dipoles in the over-all antenna.

For simplicity P is assumed to be so far away that lines from the end points of the antenna to P are effectively parallel, and the answers are not valid if P is closer. Angle θ_n between the axis of the antenna and the line joining doublet n to point P is almost equal to θ. Such a simplification is not required, but the resulting integration is far simpler.

Call r the distance in centimeters from the center of the antenna of Fig. 20·4 to point P; z_n is distance from the center of the antenna to the center of the nth elemental doublet in centimeters; Δz is the length of the doublet in centimeters; λ is the wavelength in centimeters; ω is angular velocity in radians per second; and t is time in seconds. Then $r - z_n\cos\theta$ is the distance from the nth discrete elemental doublet to point P.

H_ϕ resulting from doublet n carrying a current of $I_0\cos(2\pi z_n/\lambda)\cos\omega t$ statamperes is

$$H_{\phi n} = I_0\Delta z\cos\left(\frac{2\pi z_n}{\lambda}\right)\sin\theta\left[\frac{\omega}{c^2}\frac{1}{r - z_n\cos\theta}\sin\omega\left(t - \frac{r - z_n\cos\theta}{c}\right)\right]$$

If $z_n \gg r$ and $r - z_n \cong r$,

$$H_{\phi n} = I_0\Delta z\cos\left(\frac{2\pi z_n}{\lambda}\right)\sin\theta\left[\frac{\omega}{c^2}\frac{1}{r}\sin\omega\left(t - \frac{r - z_n\cos\theta}{c}\right)\right]$$

Similarly,

$$E_{\theta n} = I_0\Delta z\cos\left(\frac{2\pi z_n}{\lambda}\right)\sin\theta\left[\frac{\omega}{c^2}\frac{1}{r}\sin\omega\left(t - \frac{r - z_n\cos\theta}{c}\right)\right] \tag{20·18}$$

where $H_{\phi n}$ = ϕ component of vector magnetic field intensity in oersteds caused by doublet n

$E_{\theta n}$ = θ component of vector electric-field intensity in statvolts per centimeter caused by doublet n

and other terms are as previously defined.

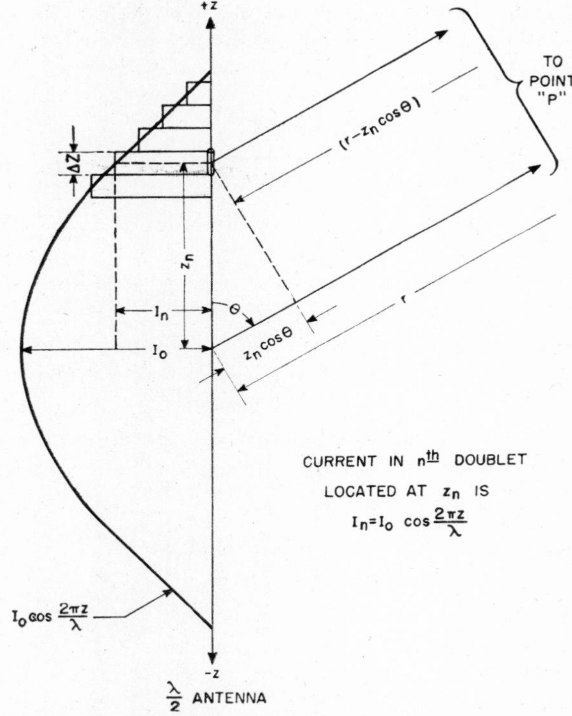

FIG. 20·4 Current distribution on a half-wave antenna in space.

Figure 20·5 shows the relative time phase and magnitude of the vector fields at P. Shrink Δz to 0 as a limit; then $z_n \to z$. The vector sum of all $H_{\phi n}$ elements at P is

$$H_\phi = I_0\frac{\omega}{c^2}\frac{1}{r}\sin\theta\sin\omega\left(t - \frac{r}{c}\right)\int_{-\lambda/4}^{+\lambda/4}\cos\frac{\pi z}{\lambda}\cos\frac{\omega z\cos\theta}{c}\,dz \tag{20·19}$$

Similarly,

$$E_\theta = I_0\frac{\omega}{c^2}\frac{1}{r}\sin\theta\sin\omega\left(t - \frac{r}{c}\right)\int_{-\lambda/4}^{+\lambda/4}\cos\frac{\pi z}{\lambda}\cos\frac{\omega z\cos\theta}{c}\,dz \tag{20·20}$$

Carrying out the integration,

$$H_\phi = 2\frac{I_0}{cr}\frac{\cos\left(\frac{\pi}{2}\cos\theta\right)}{\sin\theta}\sin\omega\left(t - \frac{r}{c}\right) \quad (20\cdot21)$$

and

$$E_\theta = 2\frac{I_0}{cr}\frac{\cos\left(\frac{\pi}{2}\cos\theta\right)}{\sin\theta}\sin\omega\left(t - \frac{r}{c}\right) \quad (20\cdot22)$$

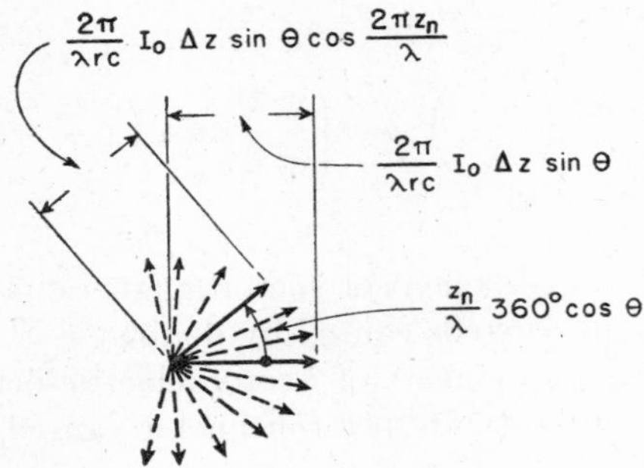

Fig. 20·5 Time vector components of H_ϕ or E_θ of an antenna element n at a remote point P.

The most used unit of signal strength is the strength of the electric field expressed in millivolts per meter. Equation 20·22 becomes

$$E_\theta = 37.3\frac{I_0}{r}\frac{\cos\left(\frac{\pi}{2}\cos\theta\right)}{\sin\theta}\sin\omega\left(t - \frac{r}{c}\right) \quad (20\cdot23)$$

where E_θ = electric field in the θ-direction at P in millivolts per meter (peak)

I_0 = current in amperes (peak) at a point $\lambda/4$ from the end of the antenna

r = distance in miles from the antenna to point P

c = velocity of electromagnetic waves (186,000 miles per second)

θ = angle between the axis of the antenna and the line joining the midpoint of the antenna to point P

and t and ω are as defined for equation 20·14.

Equation 20·23 may be written

$$E_\theta = K(\theta)37.3\frac{I_0}{r}\sin\omega\left(t - \frac{r}{c}\right)$$

where

$$K(\theta) = \frac{\cos\left(\frac{\pi}{2}\cos\theta\right)}{\sin\theta} \quad (20\cdot24)$$

$K(\theta)$ is the field pattern for a half-wave antenna in free space.

Equations 20·18 to 20·24 are for a half-wave antenna in free space, and are based upon r being much greater than $\lambda/2$ and upon the assumption of a current distribution on the antenna equivalent to that on a lossless line of the same length. Experience shows that the current distribution in antennas is not quite as assumed. Equations 20·21, 20·22, and 20·23 are accurate for $\theta > 15$ or 20 degrees. For $\theta < 15$ or 20 degrees the expressions are not accurate; here the error in assuming current distribution of a lossless line has its greatest effect. In most practical problems, however, it is sufficient to assume the current distribution of a lossless line.

General Considerations. Antennas of any length in free space may be analyzed in a manner similar to the method used for a half-wave antenna. The equation for intensity becomes more involved if the length of the antenna is not an integral number of half waves long. If the length of the antenna is an odd number of half wavelengths long,[3]

$$E_\theta = \frac{37.3}{r}I_0\frac{\cos\left(\frac{\pi l}{\lambda}\cos\theta\right)}{\sin\theta}\sin\omega\left(t - \frac{r}{c}\right) \quad (20\cdot25)$$

If the antenna is an even number of half wavelengths long,[3]

$$E_\theta = \frac{37.3}{r}I_0\frac{\sin\left(\frac{\pi l}{\lambda}\cos\theta\right)}{\sin\theta}\sin\omega\left(t - \frac{r}{c}\right) \quad (20\cdot26)$$

where l = length of the antenna in the same units as λ

λ = wavelength in the same units as l.

E_θ, I_0, θ, and r are as defined for equation 20·23, and t is as defined for equation 20·14.

20·3 IMAGES

In considering the effects of the ground on the radiated field at a point P in space, assume first that the earth is a perfect plane reflector, then at P there is a component of field from an elemental doublet along a direct path or ray

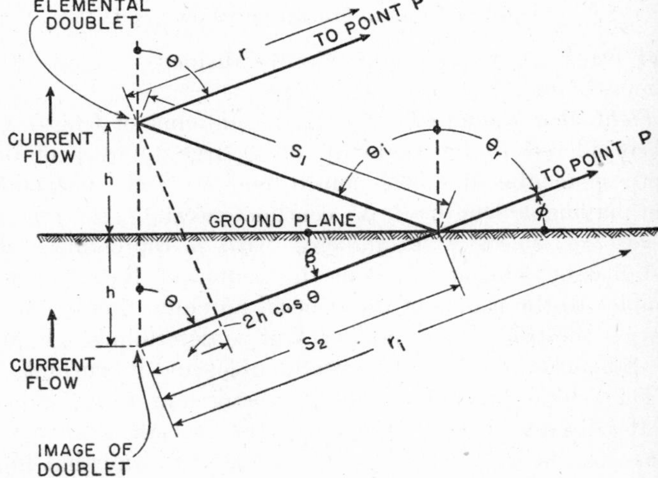

Fig. 20·6 Reflection by the use of images.

and another along a path involving reflection from the earth's surface. As shown in Fig. 20·6,

h = distance of the elemental doublet above ground in the same units as r

θ = angle from the axis of the elemental doublet made by r in going to point $P(r, \theta)$

r = distance from the center of the elemental doublet to point P in space

r_i = distance from the center of the image elemental doublet to point P

S_1 = distance from elemental doublet to the point of reflection
S_2 = distance from the image of the elemental doublet to the point of reflection
θ_i = angle of incidence
θ_r = angle of reflection

It can be shown that

$$\phi = \beta \qquad S_1 = S_2$$
$$\theta = \theta_i = \theta_r \qquad r_1 = 2h \cos \theta + r$$

Since $\phi = \beta$, the ray from the image elemental doublet to P coincides with the reflected ray from the elemental doublet after reflection. That $S_1 = S_2$ implies that the path lengths are the same. The earth usually can be considered to be a perfect plane reflector, and its effect can be replaced by the use of images. The analysis then proceeds as if the antenna and its image were in free space. The assumption of perfect reflection is adequate for most problems.

If the earth is not a perfect reflector, field strength at P can be calculated by using the reflection coefficient which is a function of the frequency of the wave and its polarization, the conductivity of the earth and its dielectric constant, and the angle of incidence. For any given problem the angle of incidence varies with the position of P.

If the earth is assumed to be a perfect plane reflector, the field of a *vertical* elemental doublet h units above the earth is made up at a point P in space of the usual field plus a field equal in magnitude but lagging in time phase by an angle of $360(2h/\lambda) \cos \theta$ degrees.

Vertical Radiators

Complex antennas, for example, vertical antennas [2] over a perfect earth, now can be analyzed. (Refer to Fig. 20·7.)

Current distribution is assumed to be sinusoidal, and the antenna is assumed to be driven from the end at the ground. Thereby the current at the top of the antenna must be zero, for there is no place for it to go. The equation for the field at a point P in space remote from a vertical antenna of height h is

$$E_\theta = \frac{37.3}{r \sin \theta} I_0[\cos (kh \cos \theta) - \cos kh] \sin \omega \left(t - \frac{r}{c}\right) \qquad (20 \cdot 27)$$

where E_θ = θ component of vector electric-field intensity in millivolts (peak) per meter at point P
I_0 = current in amperes (peak) at a current loop
r = distance in miles from the antenna to point P, provided $r \gg h$
θ = angle the ray to point P makes with the axis of the antenna
h = height of the antenna
k = $2\pi/\lambda$ where λ is the wavelength in the same units as h
c = velocity of electromagnetic waves in miles per second
ω = angular velocity in radians per second
t = time in seconds.

Equation 20·27 is valid also for an antenna of length $2h$, center fed, and in free space.

The vigorous field equations based on sinusoidal current distribution and a perfect earth at a point not remote from the antenna are useful in computing antenna impedances.[4]

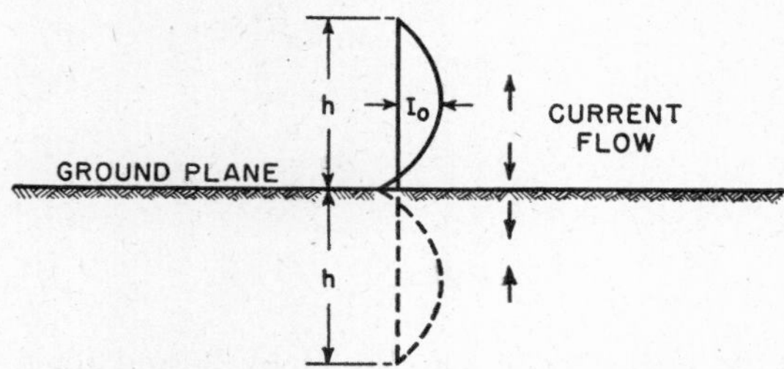

FIG. 20·7 Vertical antenna and its image.

20·4 SELF-IMPEDANCES

All the foregoing equations are in terms of the current at some point in the antenna system. Any designer, however, needs to know the power required to develop a given field at a given point. The impedance of the system at the feed point is important also.

Siegel and Labus analyzed antenna impedances effectively.[5] Their results apply to straight-line antennas of constant cross section fed at the center and positioned in free space, or to vertical antennas near the earth. Their results have been verified experimentally. Morrison and Smith further analyzed vertical antennas.[6]

Impedance of Center-Fed Linear Antennas in Space

Figure 20·8 (*b*) shows a linear antenna of length l fed at the center and having an impedance **Z** at the feed point, where

$$\mathbf{Z} = R + jX \qquad (20 \cdot 28)$$

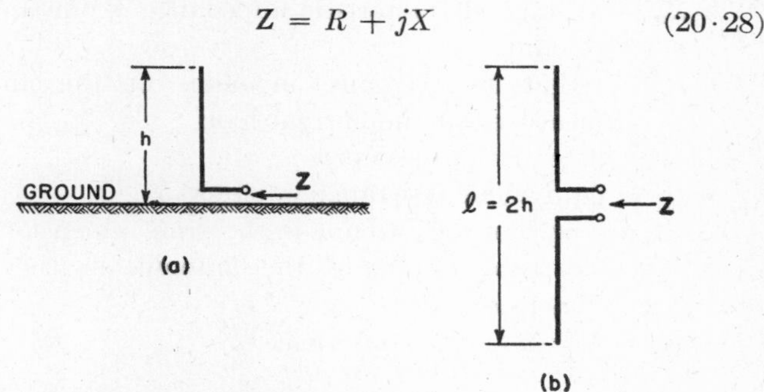

FIG. 20·8 Linear antennas.

Referring to Fig. 20·15,

$$R = Z_0 \frac{\sinh \alpha l - \dfrac{\alpha l}{4\pi \dfrac{h}{\lambda}} \sin 4\pi \dfrac{h}{\lambda}}{\cosh \alpha l - \cos 4\pi \dfrac{h}{\lambda}} \qquad (20 \cdot 29)$$

and from Fig. 20·16,

$$X = -Z_0 \frac{\sin 4\pi \dfrac{h}{\lambda} + \dfrac{\alpha l}{4\pi \dfrac{h}{\lambda}} \sinh \alpha l}{\cosh \alpha l - \cos 4\pi \dfrac{h}{\lambda}} \qquad (20 \cdot 30)$$

where

$$\alpha l = \frac{rl}{2Z_0} \qquad (20 \cdot 31)$$

Multiplying ordinates of Fig. 20·12 by 2,

$$rl = \frac{4R_0}{1 - \dfrac{\sin 4\pi \dfrac{h}{\lambda}}{4\pi \dfrac{h}{\lambda}}} \tag{20·32}$$

where

$$l = 2h \tag{20·33}$$

Multiplying ordinates of Fig. 20·11 by 2, and for $h \geqq 0.2\lambda$

$$R_0 \cong 30 \left[-\frac{\pi}{2} \sin 4\pi \frac{h}{\lambda} + \left(\log_e \frac{2h}{\lambda} + 1.722 \right) \cos 4\pi \frac{h}{\lambda} + 2\left(2.415 + \log_e \frac{2h}{\lambda} \right) \right] \tag{20·34a}$$

and for $h < 0.2\lambda$

$$R_0 \cong 320 \sin^4 \left(\pi \frac{h}{\lambda} \right) \tag{20·34b}$$

Likewise, by doubling ordinates of Figs. 20·13 and 20·14 and adding,

$$Z_0 = 120 \left[\log_e \frac{h}{\rho} - 1 - \frac{1}{2} \log_e \frac{2h}{\lambda} \right] \tag{20·35}$$

where $\mathbf{Z}$ = antenna impedance in ohms
R = resistive component of $\mathbf{Z}$ in ohms
X = reactive component of $\mathbf{Z}$ in ohms
Z_0 = average characteristic impedance of the antenna in ohms
R_0 = radiation resistance in ohms of the antenna measured at the current loop
αl = attenuation in nepers
rl = equivalent resistance in ohms
l = length of the antenna in the same dimensions as λ
ρ = electrical radius of the antenna in the same dimensions as λ
λ = wavelength in free space.

Equations 20·29 and 20·30 are obtained by considering the antenna and its image as a transmission line of constant characteristic impedance. The attenuation constant of the assumed transmission line is obtained by considering that the loop radiation resistance (based on sinusoidal current distribution) can be replaced by a resistance distributed equally along the antenna. From a first approximation of current distribution an attenuation constant αl has been established, from which a second approximation of the current distribution can be obtained.

Radiation Resistance. Radiation resistance is defined as

$$R_z = \frac{P_R}{I_z^2} \tag{20·36}$$

where R = radiation resistance in ohms referred to point z
P_R = radiated power in watts
I_z = current in amperes, rms, at point z on the antenna.

Radiation resistance is a function of current, which in turn is dependent upon position. If current at point z is known, and if the radiated power is known, the radiation resistance referred to point z of the antenna is defined by equation 20·36. Usually losses in the antenna conductors are so low that they can be neglected. Thereby the radiation resistance is almost the only resistance.

Base Impedance of Base-Insulated Vertical Radiators

Morrison and Smith [6] outlined the methods of calculating base impedances of single vertical towers. They investigated the application of the general method of Siegel and Labus to radiators used by broadcast stations; see Figs. 20·8 (*a*) and 20·9.

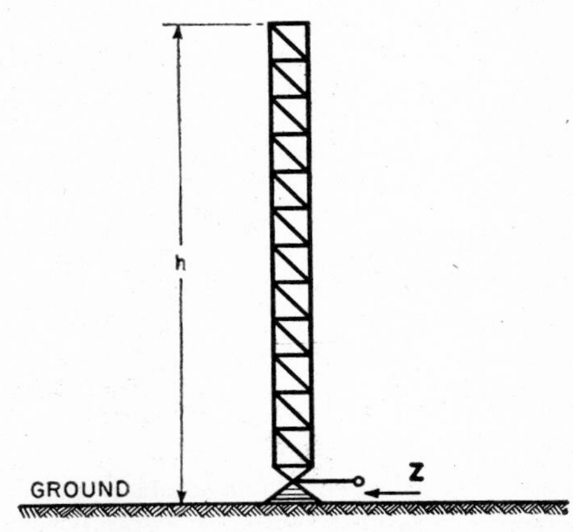

FIG. 20·9 Vertical base-insulated radiator of constant cross section.

$$\mathbf{Z} = R + jX \tag{20·37}$$

Referring to Fig. 20·15,

$$R = Z_0 \frac{\sinh \alpha l - \dfrac{\alpha l}{4\pi \dfrac{h}{\lambda}} \sin 4\pi \dfrac{h}{\lambda}}{\cosh \alpha l - \cos 4\pi \dfrac{h}{\lambda}} \tag{20·38}$$

and from Fig. 20·16,

$$X = -Z_0 \frac{\sin 4\pi \dfrac{h}{\lambda} + \dfrac{\alpha l}{4\pi \dfrac{h}{\lambda}} \sinh \alpha l}{\cosh \alpha l - \cos 4\pi \dfrac{h}{\lambda}} \tag{20·39}$$

where

$$\alpha l = \frac{rl}{2Z_0} \tag{20·40}$$

From Fig. 20·12,

$$rl = \frac{4R_0}{1 - \dfrac{\sin 4\pi \dfrac{h}{\lambda}}{4\pi \dfrac{h}{\lambda}}} \tag{20·41}$$

$$l = 2h \tag{20·42}$$

Referring to Fig. 20·11, for $h \geqq 0.2\lambda$

$$R_0 \cong 15\left[-\frac{\pi}{2}\sin 4\pi\frac{h}{\lambda} + \left(\log_e\frac{2h}{\lambda} + 1.722\right)\cos 4\pi\frac{h}{\lambda} + 2\left(2.415 + \log_e\frac{2h}{\lambda}\right)\right] \quad (20\cdot 43a)$$

for $h < 0.2\lambda$

$$R_0 \cong 160\sin^4\left(\pi\frac{h}{\lambda}\right) \quad (20\cdot 43b)$$

where

$$Z_0 = K_1 + K_2 \quad (20\cdot 44)$$

From Fig. 20·13,

$$K_1 = 60\log_e\frac{h}{\rho} \quad (20\cdot 45)$$

and from Fig. 20·14,

$$K_2 = -\left(60 + 30\log_e\frac{2h}{\lambda}\right) \quad (20\cdot 46)$$

Note that

$$\log_e x = 2.303\log_{10} x$$

where **Z** = base impedance in ohms
R = base resistance in ohms
X = base reactance in ohms
Z_0 = average characteristic impedance of the tower in ohms
R_0 = radiation resistance in ohms at the current loop
αl = attenuation constant in nepers
h = antenna height in the same dimensions as λ
rl = equivalent resistance in ohms
ρ = electrical radius of the antenna in the same dimensions as λ
λ = wavelength in free space.

A vertical antenna of height h, compared with a center-fed antenna of length $2h$ in free space has interesting characteristics. R_0 of equations 20·43*a* and 20·43*b* is half the magnitude of R_0 of equations 20·34*a* and 20·34*b*. The quantity rl of equation 20·41 also is half the magnitude of rl of equation 20·32. Z_0 of equation 20·44 is half the magnitude of Z_0 of equation 20·35. Therefore αl of equation 20·40 is numerically equal to αl of equation 20·31.

Similarly R and X of equations 20·38 and 20·39 are half the magnitudes of R and X of equations 20·29 and 20·30.

Modified Base Impedance. In practice **Z** of equation 20·37 should be modified. Because the earth is not perfect, the ground system contains an inductive reactance. In addition, a shunt capacitance should be included to make equations 20·38 and 20·39 agree with experience. (See Fig. 20·10.) L and C must be found by experience. The range of L is from 4 to 10 microhenrys, and the range of C is from 100 to 500 micromicrofarads, depending upon the dimensions of the tower and the characteristics of the ground system.

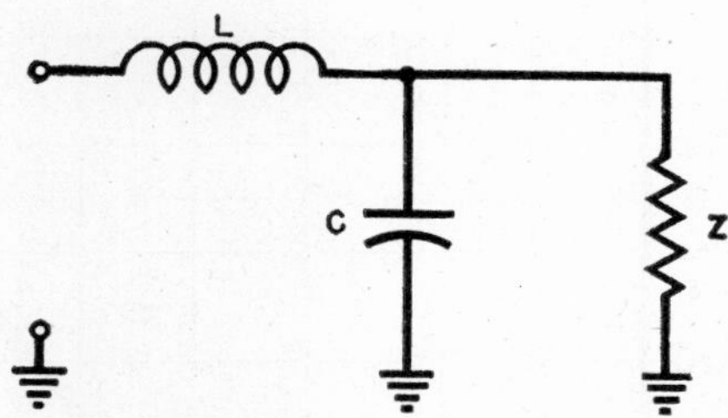

Fig. 20·10 Effects of base capacitance and ground circuit inductance.

Figures 20·11 to 20·16 are meant to assist in calculations of impedance. Figures 20·11 and 20·12 are similar to those of Siegel and Labus.[5] Figures 20·13 and 20·14 are similar to those of Morrison and Smith.[6]

Equations 20·44, 20·45, and 20·46, the equations for calculating the average characteristic impedance of the antenna, are valid only for towers of constant cross section. The series inductance of Fig. 20·10 can be neglected for most calculations. Shunt capacitance, however, is important and should be considered, particularly if $h \cong \lambda/2$ or is an integral multiple of $\lambda/2$. For slim masts 3 or 4 inches in diameter the base capacitance probably does not exceed 100 micromicrofarads. Normally for guyed towers of constant cross section, the capacitance will range from 150 to 300 micromicrofarads. For self-supporting towers with rather large dimensions the base capacitance might be from 400 to 500 micromicrofarads.

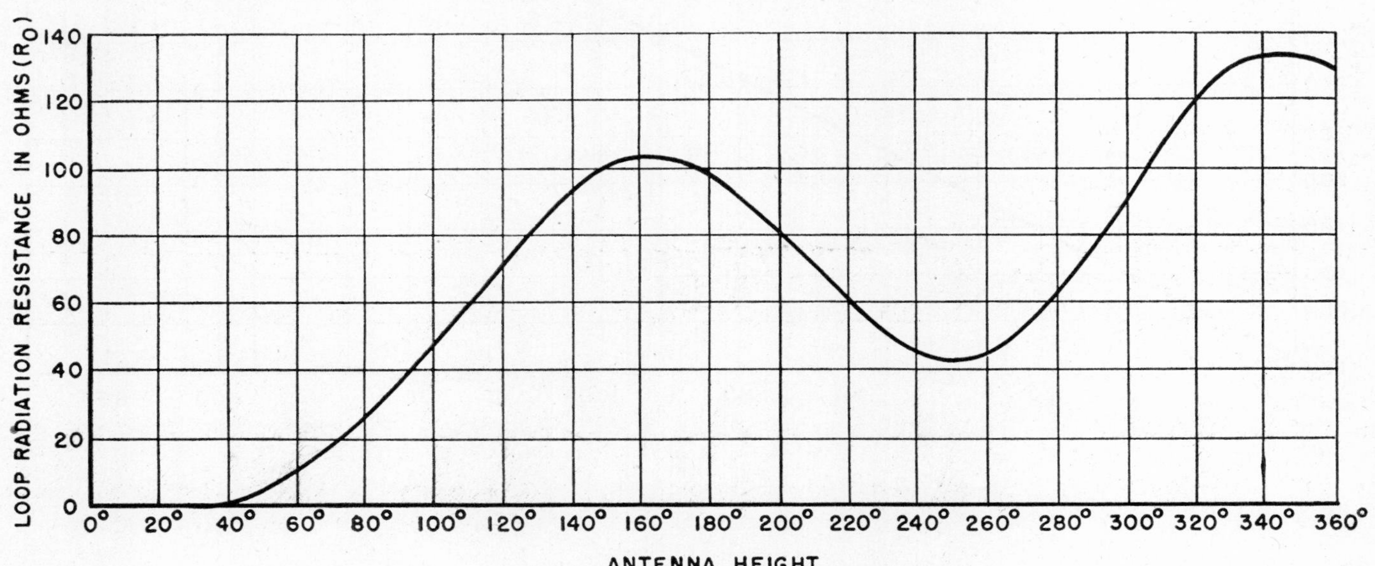

Fig. 20·11 Loop radiation resistance versus antenna height for a vertical antenna near ground.

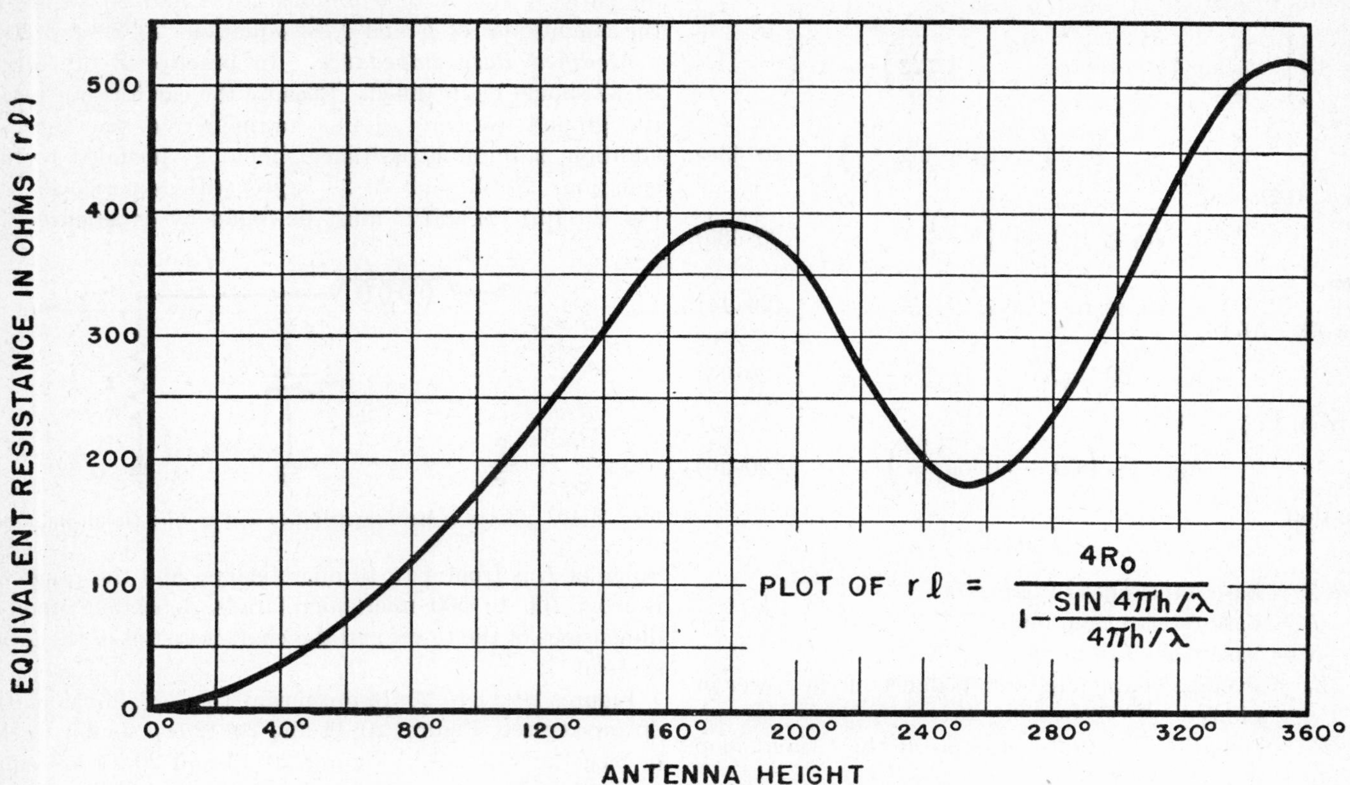

Fig. 20·12 Equivalent resistance of a vertical radiator near ground.

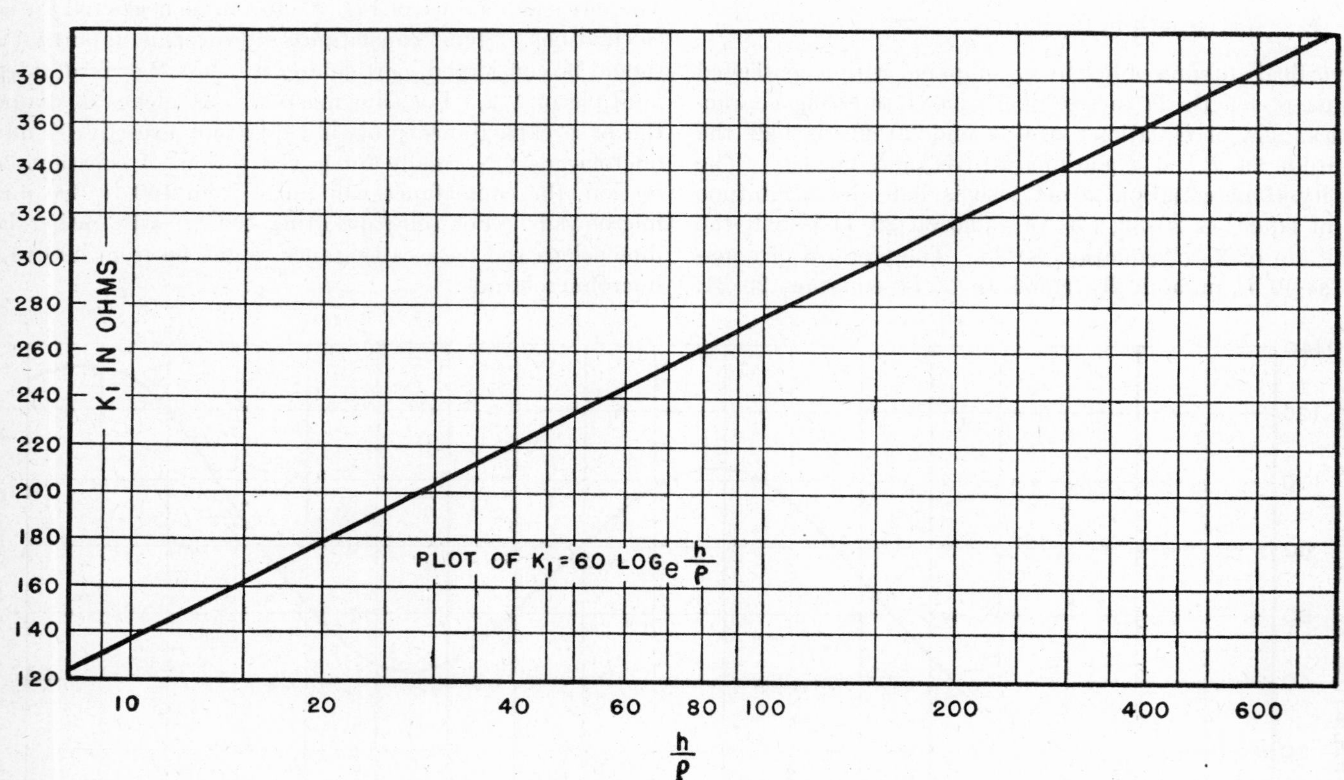

Fig. 20·13 Portion of characteristic impedance dependent upon slenderness ratio.

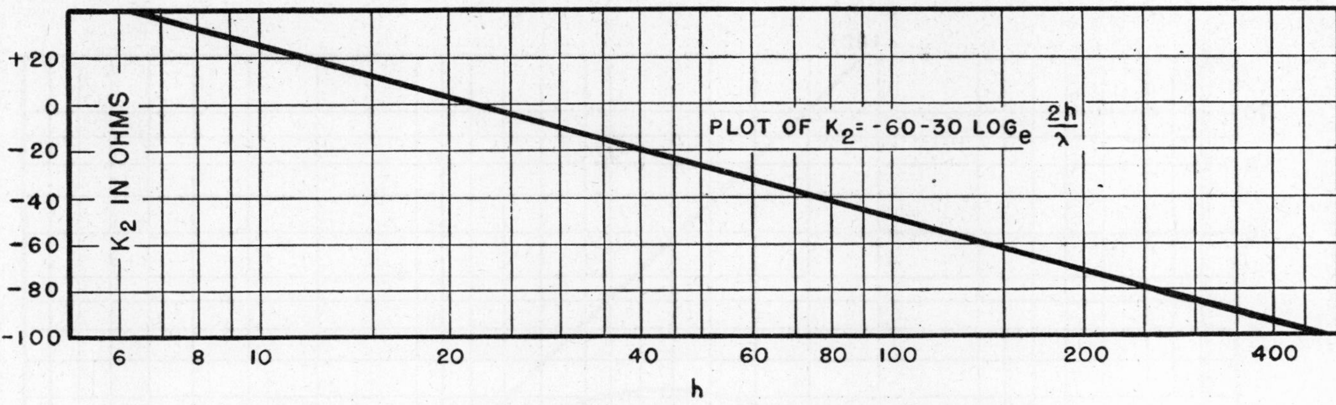

FIG. 20·14 Portion of characteristic impedance dependent upon antenna height.

To determine shunt capacitance for a given tower with a constant cross section the impedance of the tower at one frequency (preferably one that makes $h = \lambda/2$) must be known. From equations 20·37, 20·38, and 20·39, **Z** is calculated for the same frequency at which it was measured. The series inductance and shunt capacitance that will transform **Z** to the value measured should be computed. Series inductance and shunt capacitance can be determined more accurately from a polar plot of calculated and measured data for the tower.

Vertical towers of varying cross section can be best analyzed from actual data. However, a tower having a height shorter than $\lambda/3$ can be analyzed approximately for impedance by the methods outlined for towers of constant cross section. An electrical radius must be assumed, and the shunt capacitance should be assumed. The resulting predicted base impedance may be in error by 10 ohms in the resistive component and by from 20 to 30 ohms in the reactive component, both errors depending upon the height of the antenna.

Base Impedance of Top-Loaded Vertical Antennas

Loaded antennas (antennas having lumped constants in series or in shunt with sections of the radiator) are more difficult to analyze. Siegel analyzed capacitively end-loaded antennas.[7] The conclusions are given here (refer to Fig. 20·17).

The effective top-loading capacitance, if the top-loading is a disk, is

$$C = 10.8D \qquad (20\cdot47)$$

where C = effective top-loading capacitance in micromicrofarads

D = diameter of disk in feet.

The base impedance **Z** of the top-loaded (disk-loaded) vertical antenna is

$$\mathbf{Z} = R + jX$$

where

$$R \cong Z_0 \frac{\sinh \alpha l + \dfrac{\alpha l}{4\pi \frac{h}{\lambda}}\left[\sin 2\psi - \sin\left(4\pi \frac{h}{\lambda} + 2\psi\right)\right]}{\cosh \alpha l - \cos\left(4\pi \frac{h}{\lambda} + 2\psi\right)} \qquad (20\cdot48)$$

$$X \cong -Z_0 \frac{\sin\left(4\pi \frac{h}{\lambda} + 2\psi\right) + \dfrac{\alpha l}{4\pi \frac{h}{\lambda}} \sinh \alpha l}{\cosh \alpha l - \cos\left(4\pi \frac{h}{\lambda} + 2\psi\right)} \qquad (20\cdot49)$$

$$\psi = \tan^{-1}(Z_0 \omega C) \qquad (20\cdot50)$$

$$R_0 = 15\left\{-\left(2\,\mathrm{Si}\,4\pi\frac{h}{\lambda} - \mathrm{Si}\,8\pi\frac{h}{\lambda}\right)\sin\left(4\pi\frac{h}{\lambda} + 2\psi\right) + \left(1.722 + \log_e\frac{2h}{\lambda} + \mathrm{Ci}\,8\pi\frac{h}{\lambda} - 2\,\mathrm{Ci}\,4\pi\frac{h}{\lambda}\right)\cos\left(4\pi\frac{h}{\lambda} + 2\psi\right) + 2\left[2.415 + \log_e\frac{2h}{\lambda} - \mathrm{Ci}\,4\pi\frac{h}{\lambda} + \sin^2\psi\left(\frac{\sin 4\pi\frac{h}{\lambda}}{4\pi\frac{h}{\lambda}} - 1\right)\right]\right\} \qquad (20\cdot51)$$

or

$$R_0 \cong 15\left\{-\frac{\pi}{2}\sin\left(4\pi\frac{h}{\lambda} + 2\psi\right) + \left(1.722 + \log_e\frac{2h}{\lambda}\right)\cos\left(4\pi\frac{h}{\lambda} + 2\psi\right) + 2\left[2.415 + \log_e\frac{2h}{\lambda} + \sin^2\psi\left(\frac{\sin 4\pi\frac{h}{\lambda}}{4\pi\frac{h}{\lambda}} - 1\right)\right]\right\} \qquad (20\cdot52)$$

$$rl = \frac{4R_0}{1 + \dfrac{\sin 2\psi - \sin\left(4\pi\frac{h}{\lambda} + 2\psi\right)}{4\pi\frac{h}{\lambda}}} \qquad (20\cdot53)$$

$$\alpha l = \frac{rl}{2Z_0} \quad (\text{where } l = 2h) \qquad (20\cdot54)$$

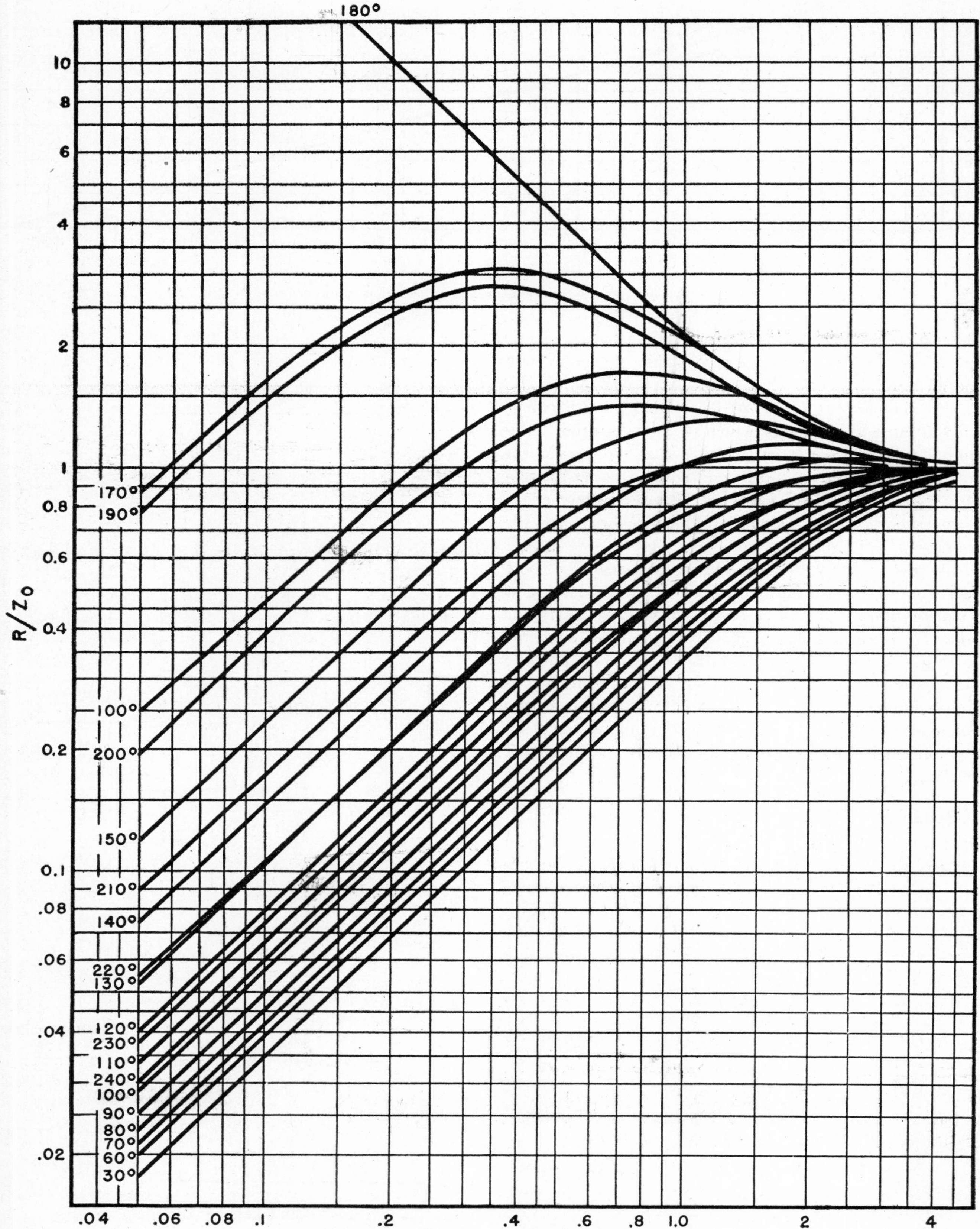

FIG. 20·15 Ratio of base resistance to characteristic impedance as a function of attenuation.

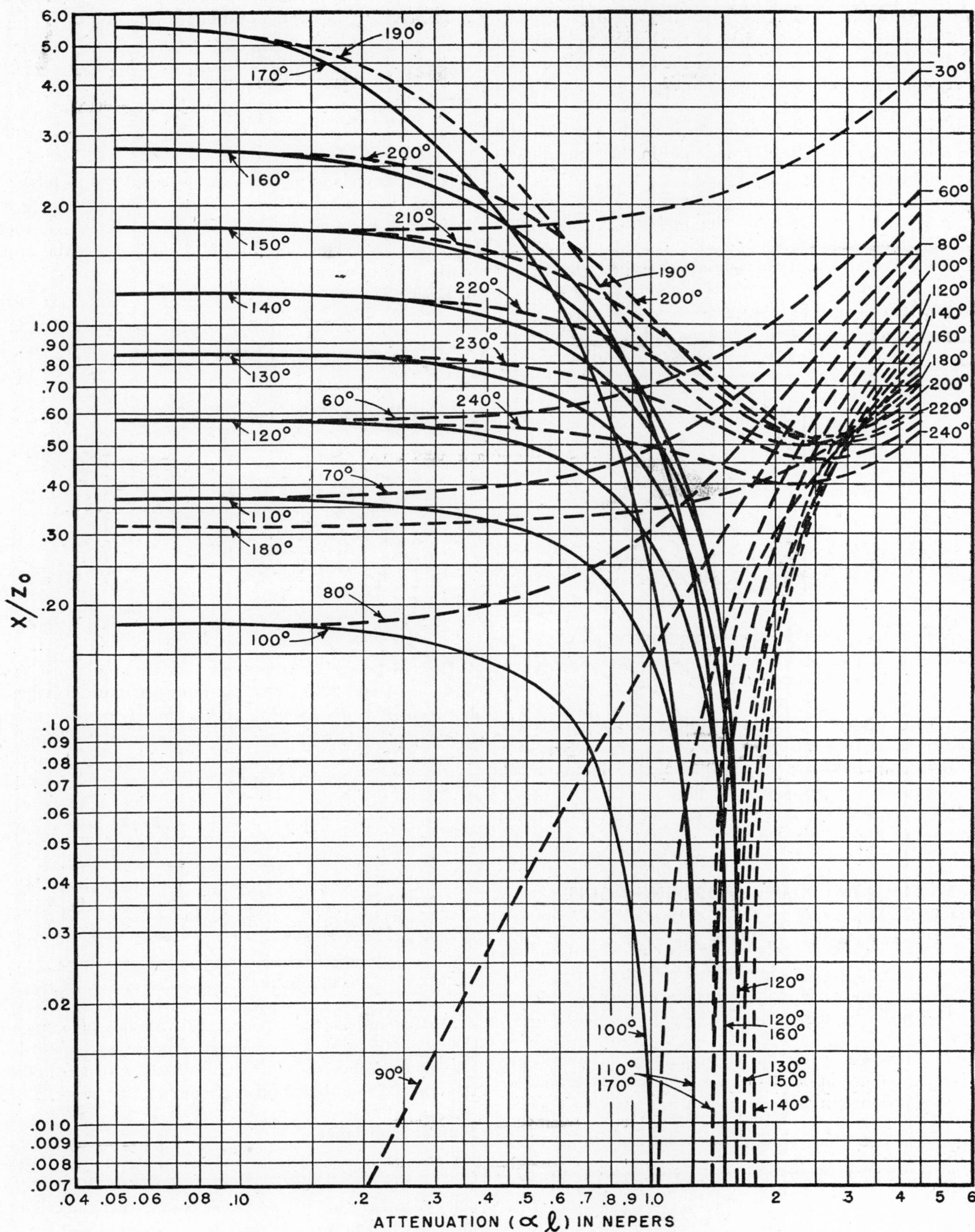

Fig. 20·16 Ratio of base reactance to characteristic impedance as a function of attenuation.

where $\mathbf{Z}$ = base impedance in ohms of the top-loaded antenna
R = base resistance in ohms of the top-loaded antenna
X = base reactance in ohms of the top-loaded antenna
Z_0 = average characteristic impedance in ohms from equation 20·44
R_0 = radiation resistance in ohms at the current loop
h = antenna height above ground in the same dimensions as λ
αl = attenuation constant in nepers
ρ = electrical radius of the antenna in the same dimensions as λ
ω = angular velocity in radians per second corresponding to λ
λ = wavelength in free space
ψ = number of electrical radians added to the antenna height by the use of the disk top-loading.

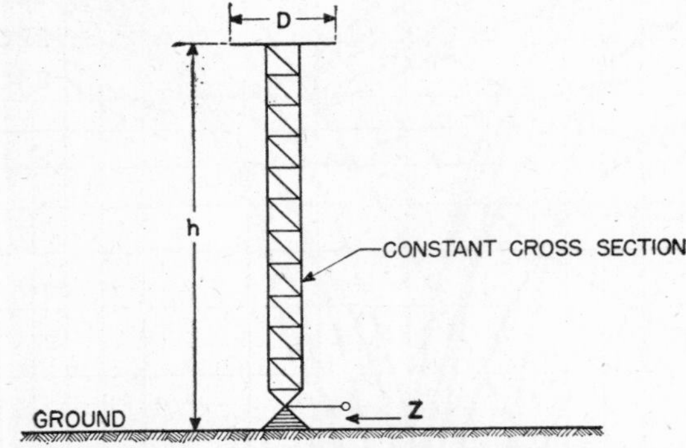

Fig. 20·17 Vertical, base-insulated, and top-loaded radiator of constant cross section.

Shunt capacitance of the top-loading can be effectively changed by insulating it from the tower and connecting to it by means of an inductor. By adjusting the inductor the effective capacitance can be set at will.

Effects of base capacitance and ground-lead inductance should modify $\mathbf{Z}$ as in Fig. 20·10.

20·5 MUTUAL IMPEDANCE, DRIVING-POINT IMPEDANCE

In a multiple-element antenna system a current in one element will set up voltages in all other elements. By definition, if a current in one circuit produces a voltage in another, a mutual impedance couples the two circuits (see Fig. 20·18).

$\mathbf{Z}_{12} = \mathbf{Z}_{21} = \mathbf{Z}_M$ = the mutual impedance in ohms
$\mathbf{Z}_{11}$ = impedance in ohms of mesh 1 with mesh 2 open
$\mathbf{Z}_{22}$ = impedance in ohms of mesh 2 with mesh 1 open

The equations describing circuit conditions are

$$\mathbf{E}_1 = \mathbf{I}_1(\mathbf{Z}_{11} + \mathbf{Z}_{12}) + (\mathbf{I}_1 - \mathbf{I}_2)(-\mathbf{Z}_{12})$$

$$\mathbf{E}_2 = (\mathbf{I}_2 - \mathbf{I}_1)(-\mathbf{Z}_{21}) + \mathbf{I}_2(\mathbf{Z}_{22} + \mathbf{Z}_{21})$$

These simplify to

$$\mathbf{E}_1 = \mathbf{I}_1\mathbf{Z}_{11} + \mathbf{I}_2\mathbf{Z}_{12} \quad (20\cdot55)$$

$$\mathbf{E}_2 = \mathbf{I}_1\mathbf{Z}_{21} + \mathbf{I}_2\mathbf{Z}_{22} \quad (20\cdot56)$$

The apparent or driving-point impedance of mesh 1 with a current of $\mathbf{I}_1$ in mesh 1 and of $\mathbf{I}_2$ in mesh 2 is

$$\mathbf{Z}'_1 = \frac{\mathbf{E}_1}{\mathbf{I}_1} = \mathbf{Z}_{11} + \frac{\mathbf{I}_2}{\mathbf{I}_1}\mathbf{Z}_{12} \quad (20\cdot57)$$

Similarly

$$\mathbf{Z}'_2 = \frac{\mathbf{I}_1}{\mathbf{I}_2}\mathbf{Z}_{21} + \mathbf{Z}_{22} \quad (20\cdot58)$$

The terminals in mesh 1 can be considered as terminals of element 1 of a two-element array. Similarly the terminals in mesh 2 become terminals of element 2. $\mathbf{Z}'_1$ is then the apparent impedance of element 1, and $\mathbf{Z}'_2$ is the apparent impedance of element 2.

With n elements involved

$$\begin{aligned} \mathbf{E}_1 &= \mathbf{I}_1\mathbf{Z}_{11} + \mathbf{I}_2\mathbf{Z}_{12} + \mathbf{I}_3\mathbf{Z}_{13} + \cdots + \mathbf{I}_n\mathbf{Z}_{1n} \\ \mathbf{E}_2 &= \mathbf{I}_1\mathbf{Z}_{21} + \mathbf{I}_2\mathbf{Z}_{22} + \mathbf{I}_3\mathbf{Z}_{23} + \cdots + \mathbf{I}_n\mathbf{Z}_{2n} \\ &\cdots\cdots\cdots \\ \mathbf{E}_n &= \mathbf{I}_1\mathbf{Z}_{n1} + \mathbf{I}_2\mathbf{Z}_{n2} + \mathbf{I}_3\mathbf{Z}_{n3} + \cdots + \mathbf{I}_n\mathbf{Z}_{nn} \end{aligned} \quad (20\cdot59)$$

$$\begin{aligned} \mathbf{Z}'_1 &= \mathbf{Z}_{11} + \frac{\mathbf{I}_2}{\mathbf{I}_1}\mathbf{Z}_{12} + \frac{\mathbf{I}_3}{\mathbf{I}_1}\mathbf{Z}_{13} + \cdots + \frac{\mathbf{I}_n}{\mathbf{I}_1}\mathbf{Z}_{1n} \\ \mathbf{Z}'_2 &= \frac{\mathbf{I}_1}{\mathbf{I}_2}\mathbf{Z}_{21} + \mathbf{Z}_{22} + \frac{\mathbf{I}_3}{\mathbf{I}_2}\mathbf{Z}_{23} + \cdots + \frac{\mathbf{I}_n}{\mathbf{I}_2}\mathbf{Z}_{2n} \\ &\cdots\cdots\cdots \\ \mathbf{Z}'_n &= \frac{\mathbf{I}_1}{\mathbf{I}_n}\mathbf{Z}_{n1} + \frac{\mathbf{I}_2}{\mathbf{I}_n}\mathbf{Z}_{n2} + \cdots + \frac{\mathbf{I}_{n-1}}{\mathbf{I}_n}\mathbf{Z}_{n,n-1} + \mathbf{Z}_{nn} \end{aligned} \quad (20\cdot60)$$

Here $\mathbf{Z}_{12}$ and $\mathbf{Z}_{21}$ represent mutual impedance between elements 1 and 2; $\mathbf{Z}_{13}$ and $\mathbf{Z}_{31}$ represent mutual impedances between elements 1 and 3; and so on. $\mathbf{I}_1$ represents current in element 1; $\mathbf{I}_2$ represents current in antenna 2; and so on. Currents should be specified for the particular points on the antennas for which the mutual impedances were evaluated.

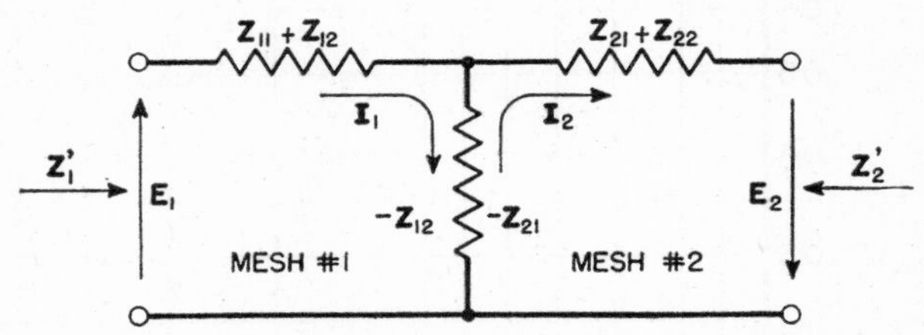

Fig. 20·18 Two-mesh network.

Mutual impedance between two meshes of a multimesh circuit is the ratio of the voltage in one mesh caused by current in the other mesh to that current with all meshes but the one last mentioned open circuited.

Mutual Impedance

Consider now a method of calculating mutual impedance between two antenna elements.[4] Refer to Fig. 20·18. A current of $\mathbf{I}_1$ in mesh 1 produces power (real and reactive) in mesh 1 when $\mathbf{I}_2$ is 0. If then $\mathbf{I}_2$ is permitted and $\mathbf{I}_1$ is maintained, the power in mesh 1 changes. This change is

$$\Delta\mathbf{P}_1 = \mathbf{I}_1\mathbf{I}_2\mathbf{Z}_{12} = \mathbf{I}_1\mathbf{I}_2\mathbf{Z}_{21}$$

for

$$\mathbf{Z}_{12} = \mathbf{Z}_{21}$$

Therefore

$$\mathbf{Z}_{12} = \frac{\Delta \mathbf{P}_1}{\mathbf{I}_1 \mathbf{I}_2}$$

or obviously

$$\mathbf{Z}_{12} = \frac{\Delta \mathbf{P}_1}{\mathbf{I}_1 \mathbf{I}_2} = \frac{\Delta \mathbf{P}_2}{\mathbf{I}_1 \mathbf{I}_2} = \frac{\Delta \mathbf{P}}{\mathbf{I}_1 \mathbf{I}_2} \qquad (20 \cdot 61)$$

Mutual impedance is referred to the points at which currents $\mathbf{I}_1$ and $\mathbf{I}_2$ are measured.

Equation 20·61 provides a means of obtaining mutual impedance. For any two elements, unity current of the same time phase at the current loops of both is assumed. Then by means of field equations and the integral of the Poynting vector (a power-density vector) $\Delta\mathbf{P}$ can be found. $\Delta\mathbf{P}$ is numerically equal to $\mathbf{Z}_M$ (mutual impedance referred to the current loops of the elements).

Consider elements 1 and 2:

$$\Delta \mathbf{P} = -\int_{z=0}^{z=a} \mathbf{E}_{z_2} \mathbf{I}_{z_1}\, dz \qquad (20 \cdot 62)$$

where $\mathbf{E}_{z_2}$ = component of electric field in volts per unit of z parallel to line of element 1 at a point z on element 1 caused by unit current at the current loop of element 2

$\mathbf{I}_{z_1}$ = current in amperes at a point z on element 1

a = length of element 1.

$\mathbf{E}_{z_2}$ now must be evaluated. Consider Figs. 20·19 (*a*) and 20·19 (*b*), which represent a center-fed antenna in space, of length $2h$, or a vertical antenna near the earth of height h. Current distribution I is represented by

$$I = I_0 \sin (h - z) \cos \omega t \qquad (20 \cdot 63)$$

The x and z components of the electric field at $P(x, z)$ caused by an antenna, are [4]

$$\mathbf{E}_x = j\frac{30}{x} I_0 \cos \omega t \left(\frac{z-h}{r_1} \epsilon^{-jkr_1} + \frac{z+h}{r_2} \epsilon^{-jkr_2} - 2\frac{z}{r_0} \epsilon^{-jkr_0} \cos kh \right) \qquad (20 \cdot 64)$$

$$\mathbf{E}_z = -j30 I_0 \cos \omega t \left(\frac{1}{r_1} \epsilon^{-jkr_1} + \frac{1}{r_2} \epsilon^{-jkr_2} - \frac{2}{r_0} \epsilon^{-jkr_0} \cos kh \right) \qquad (20 \cdot 65)$$

where $\mathbf{E}_x$ = electric field in x direction at point P in volts per unit of x

$\mathbf{E}_z$ = electric field in z direction at point P in volts per unit of r_0, r_1, or r_2

I_0 = loop current in amperes

z, r_0, r_1, r_2, h, and x are as shown in Figs. 20·19 (*a*) and 20·19 (*b*)

z, h, r_0, r_1, and r_2 should all be of the same dimensions in equation 20·64 for $\mathbf{E}_x$

r_0, r_1, r_2 should all be of the same dimensions in equation 20·65 for $\mathbf{E}_z$

ω = angular velocity in radians per second

k = that factor required to convert r_0, r_1, r_2, and h into radians corresponding to ω.

$\mathbf{E}_x$ and $\mathbf{E}_z$ are the only components of $\mathbf{E}$ in the case represented by Figs. 20·19 (*a*) and 20·19 (*b*).

For any case the component of $\mathbf{E}$ along another antenna can be calculated and substituted for $\mathbf{E}_{z_2}$ into equation 20·62 for $\Delta\mathbf{P}$. Integration usually is best performed by graphic methods. For loop currents of unity, $\Delta\mathbf{P}$ is numerically equal to the mutual impedance between the two elements. Figure 20·20 shows a plot of the magnitude of the mutual impedance versus separation for various heights of vertical-tower antennas. Figure 20·21 shows the phase relationship.

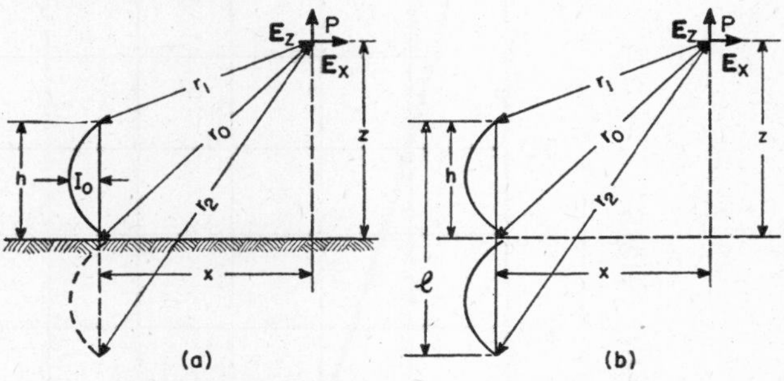

FIG. 20·19 Electric fields at a point near an antenna.

Driving-Point Impedance

Apparent impedances of individual elements of an antenna array can be computed if the elements are either center fed, linear, and in free space, or base fed, vertical, and adjacent to the earth.[8] The following information is required, and usually is determined by the radiation pattern required, physical requirements, and economics:

- Type of elements
- Element length (h or l)
- Electrical radius of each element
- Position of feed point of each element
- Geometric arrangements of elements
- Loop currents (relative magnitudes and phases)
- Frequency

With these data the mutual impedances referred to current loops (loop mutual impedances) are found by referring to charts showing mutual impedances, or by calculations as outlined in the preceding paragraphs. These loop mutual impedances are combined with the self-loop impedances by the mesh equations 20·60. The self-loop reactance is not yet known because it is affected by the attenuation which is, in turn, controlled by apparent loop resistance. However, by using the appropriate transmission line equations, 20·29 and 20·30 or 20·38 and 20·39, the feed-point resistance and the portion of feed-point reactance caused by the apparent loop resistance can be found for each element. The coupled loop reactance (that portion of the apparent loop reactance caused by current in other elements) should be modified by the ratio of apparent loop resistance to the apparent feed-point resistance with a change of the sign of the reactance once for each quarter wavelength between the loop and the feed point. So modified, the coupled reactance should be algebraically added to the already determined portion of apparent base reactance caused by the apparent loop resistance. The apparent feed-point impedance of each element is then known.

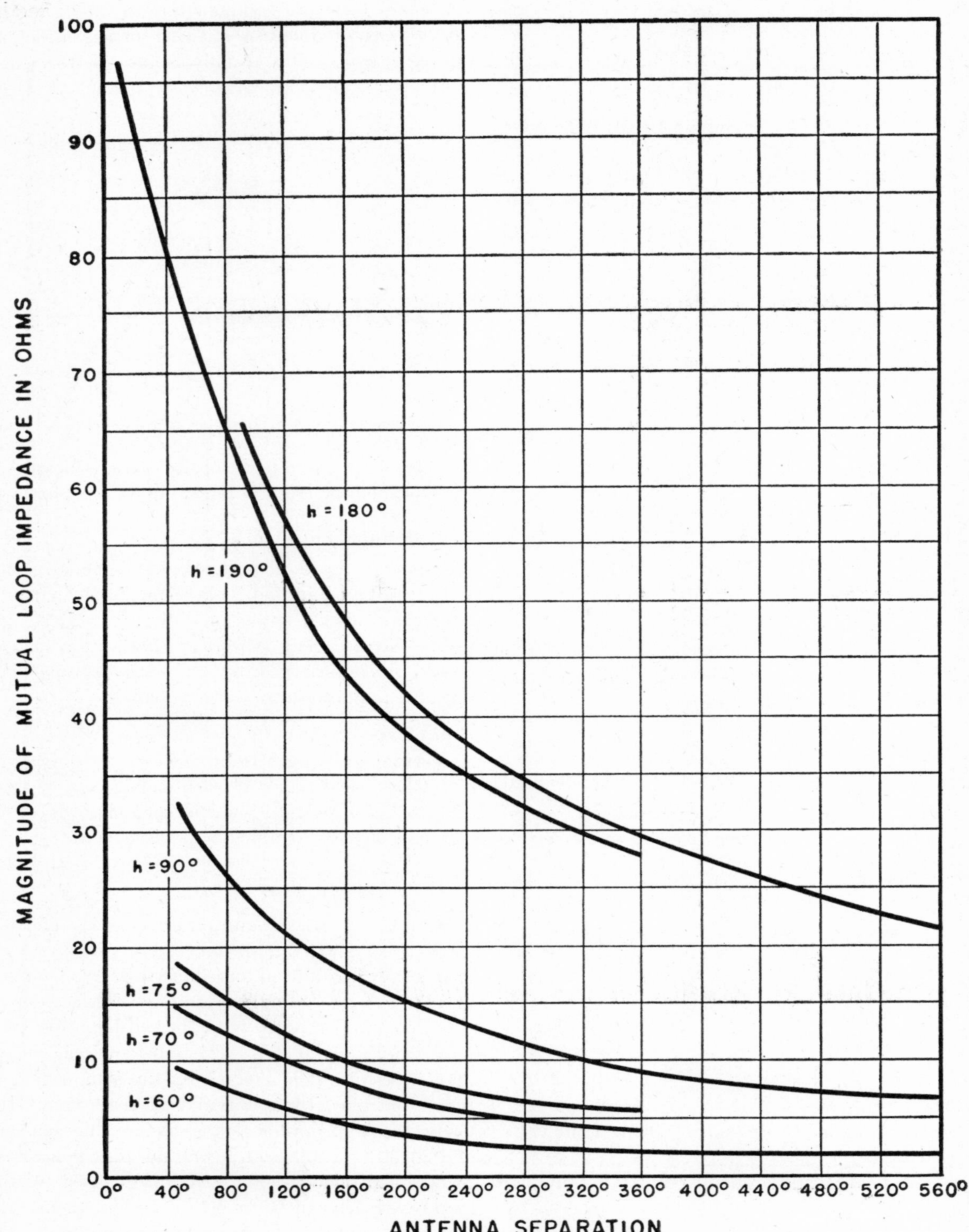

Fig. 20·20 Magnitude of mutual-loop impedance between two vertical radiators of the same height.

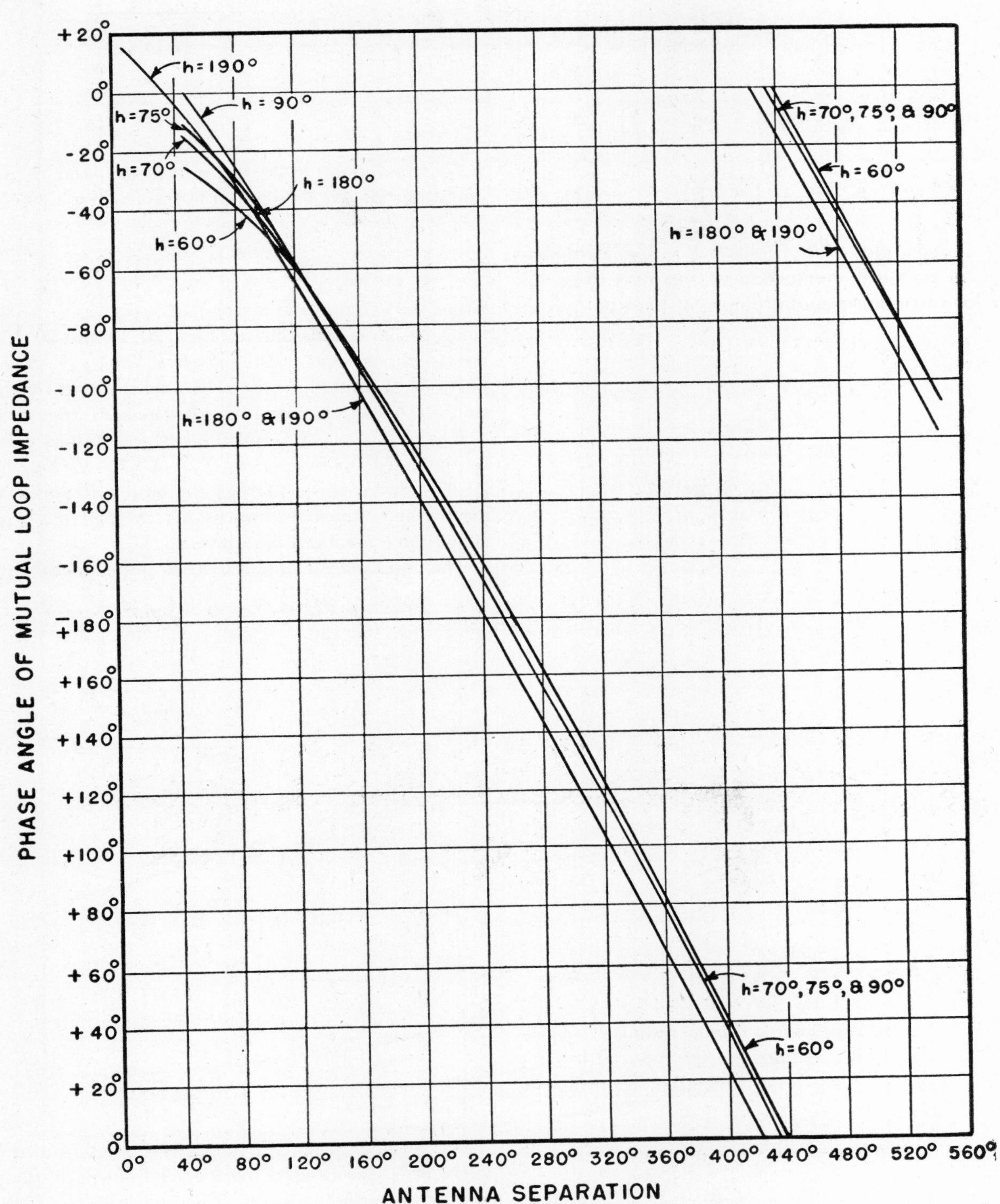

Fig. 20·21 Phase angle of mutual-loop impedance between two vertical radiators of the same height.

The procedure for vertical elements in an array is as follows. Obtain or calculate the self-loop resistances of each element R_{11}, R_{22}, R_{33}, $\cdots$, R_{nn} where the R_0 of antenna 1 is R_{11}, and so on. For vertical antennas of $h_n \geqq 0.2\lambda$

$$R_{nn} \cong 15\left[-\frac{\pi}{2}\sin\frac{4\pi h_n}{\lambda} + \left(\log_e\frac{2h_n}{\lambda} + 1.722\right)\cos 4\pi\frac{h_n}{\lambda} + 2\left(2.415 + \log_e\frac{2h_n}{\lambda}\right)\right] \qquad (20\cdot43a)$$

and for $h_n < 0.2\lambda$

$$R_{nn} \cong 160\sin^4\left(\frac{\pi h_n}{\lambda}\right) \qquad (20\cdot43b)$$

(See Fig. 20·11 for a plot of R_0 versus h/λ.) Using the data obtained, set up the self-loop impedances thus (X_{11}, X_{22}, $\cdots$, X_{nn} are self-loop reactances and are not yet known):

$$\begin{aligned} \mathbf{Z}_{11} &= R_{11} + jX_{11} \\ \mathbf{Z}_{22} &= R_{22} + jX_{22} \\ &\cdots\cdots\cdots \\ \mathbf{Z}_{nn} &= R_{nn} + jX_{nn} \end{aligned}$$

Obtain or calculate from equations 20·61 and 20·62 loop mutual impedance between all the elements obtaining (see Figs. 20·20 and 20·21):

$$\begin{aligned} &\mathbf{Z}_{12}, \mathbf{Z}_{13}, \cdots, \mathbf{Z}_{1n} \\ &\mathbf{Z}_{21}, \mathbf{Z}_{23}, \cdots, \mathbf{Z}_{2n} \\ &\cdots\cdots\cdots \\ &\mathbf{Z}_{n1}, \mathbf{Z}_{n2}, \cdots, \mathbf{Z}_{n,n-1} \end{aligned}$$

Solve for the apparent loop impedances by means of mesh equations 20·60:

$$\begin{aligned} \mathbf{Z}'_1 &= R_{11} + jX_{11} + \frac{\mathbf{I}_2}{\mathbf{I}_1}\mathbf{Z}_{12} + \frac{\mathbf{I}_3}{\mathbf{I}_1}\mathbf{Z}_{13} + \cdots + \frac{\mathbf{I}_n}{\mathbf{I}_1}\mathbf{Z}_{1n} \\ \mathbf{Z}'_2 &= \frac{\mathbf{I}_1}{\mathbf{I}_2}\mathbf{Z}_{21} + R_{22} + jX_{22} + \frac{\mathbf{I}_3}{\mathbf{I}_2}\mathbf{Z}_{23} + \cdots + \frac{\mathbf{I}_n}{\mathbf{I}_1}\mathbf{Z}_{2n} \qquad (20\cdot66) \\ &\cdots\cdots\cdots\cdots\cdots\cdots \\ \mathbf{Z}'_n &= \frac{\mathbf{I}_1}{\mathbf{I}_n}\mathbf{Z}_{n1} + \frac{\mathbf{I}_2}{\mathbf{I}_n}\mathbf{Z}_{n2} + \cdots + \frac{\mathbf{I}_{n-1}}{\mathbf{I}_n}\mathbf{Z}_{n,n-1} + R_{nn} + jX_{nn} \end{aligned}$$

Currents $\mathbf{I}_1$, $\mathbf{I}_2$, $\cdots$, are assumed, self-loop radiation resistance R_{11}, R_{22}, $\cdots$, and mutual impedances (referred to current loops) $\mathbf{Z}_{12}$, $\mathbf{Z}_{13}$, $\cdots$, have been calculated. Solution of equations 20·66 gives

$$\begin{aligned} \mathbf{Z}'_1 &= R_1 + j(X_{11} + X'_1) = R_1 + jX_1 \\ \mathbf{Z}'_2 &= R_2 + j(X_{22} + X'_2) = R_2 + jX_2 \qquad (20\cdot67) \\ &\cdots\cdots\cdots\cdots\cdots\cdots \\ \mathbf{Z}'_n &= R_n + j(X_{nn} + X'_n) = R_n + jX_n \end{aligned}$$

where X_{11}, X_{22}, X_{33}, $\cdots$, X_{nn} are not known because they are the components of the reactance determined by the amount of apparent loop resistance R_1, R_2, $\cdots$, R_n of each corresponding element. X_1, X_2, $\cdots$, X_n are the portions of apparent loop reactance caused by currents flowing in adjacent elements.

The next step is to convert the apparent loop radiation resistances to attenuation factors, for each tower, thus:

$$(rl)_n = \frac{4R_n}{1 - \dfrac{\sin 4\pi\dfrac{h_n}{\lambda}}{4\pi\dfrac{h_n}{\lambda}}} \qquad (20\cdot68)$$

R_n is from equation 20·67, and Z_0, K_1, and K_2 are given in equations 20·44, 20·45, and 20·46. Then

$$(\alpha l)_n = \frac{(rl)_n}{2Z_0} \qquad (20\cdot69)$$

From the attenuation factors $(\alpha l)_1$, $(\alpha l)_2$, $\cdots$, $(\alpha l)_n$ by equations 20·38 and 20·39 (see Figs. 20·15 and 20·16) calculate the apparent base resistances R_{1B}, R_{2B}, $\cdots$, R_{nB} and those portions of apparent base reactances caused by the attenuation factors as indicated by equations 20·38 and 20·39.

X_{1B}, X_{2B}, $\cdots$, X_{nB} are the base reactances corresponding to the self-loop reactances X_{11}, X_{22}, $\cdots$, X_{nn} of equation 20·67. So far those parts of the base reactances corresponding to the coupled reactances X'_1, X'_2, $\cdots$, X'_n of equation 20·67 have not been determined.

The equation for apparent base impedances states

$$\begin{aligned} \mathbf{Z}_{1B} &= R_{1B} + j(X_{1B} + X'_{1B}) \\ \mathbf{Z}_{2B} &= R_{2B} + j(X_{2B} + X'_{2B}) \qquad (20\cdot70) \\ &\cdots\cdots\cdots\cdots \\ \mathbf{Z}_{nB} &= R_{nB} + j(X_{nB} + X'_{nB}) \end{aligned}$$

where

$$\begin{aligned} X'_{1B} &= K_1X'_1 \\ X'_{2B} &= K_2X'_2 \qquad (20\cdot71) \\ &\cdots\cdots \\ X'_{nB} &= K_nX'_n \end{aligned}$$

and

$$\begin{aligned} K_1 &= \frac{R_{1B}}{R_1}(-1)^m \\ K_2 &= \frac{R_{2B}}{R_2}(-1)^m \qquad (20\cdot72) \\ &\cdots\cdots \\ K_n &= \frac{R_{nB}}{R_n}(-1)^m \end{aligned}$$

The actual signs of the K terms are determined by $(-1)^m$ where m is the integral number of free-space quarter wavelengths traversed in going from the loop to the base. For example, 1.7 quarter waves would have $m = 1$; 2 quarter waves would have $m = 2$.

The apparent base impedances should be modified as indicated by Fig. 20·10 and Section 20·4. For a specific example see Appendix I of this chapter.

For a more general solution similar methods may be used for any antenna system using linear elements, as long as they are fed at the center. By the use of images, any antenna element can be converted to a free-space equivalent, which

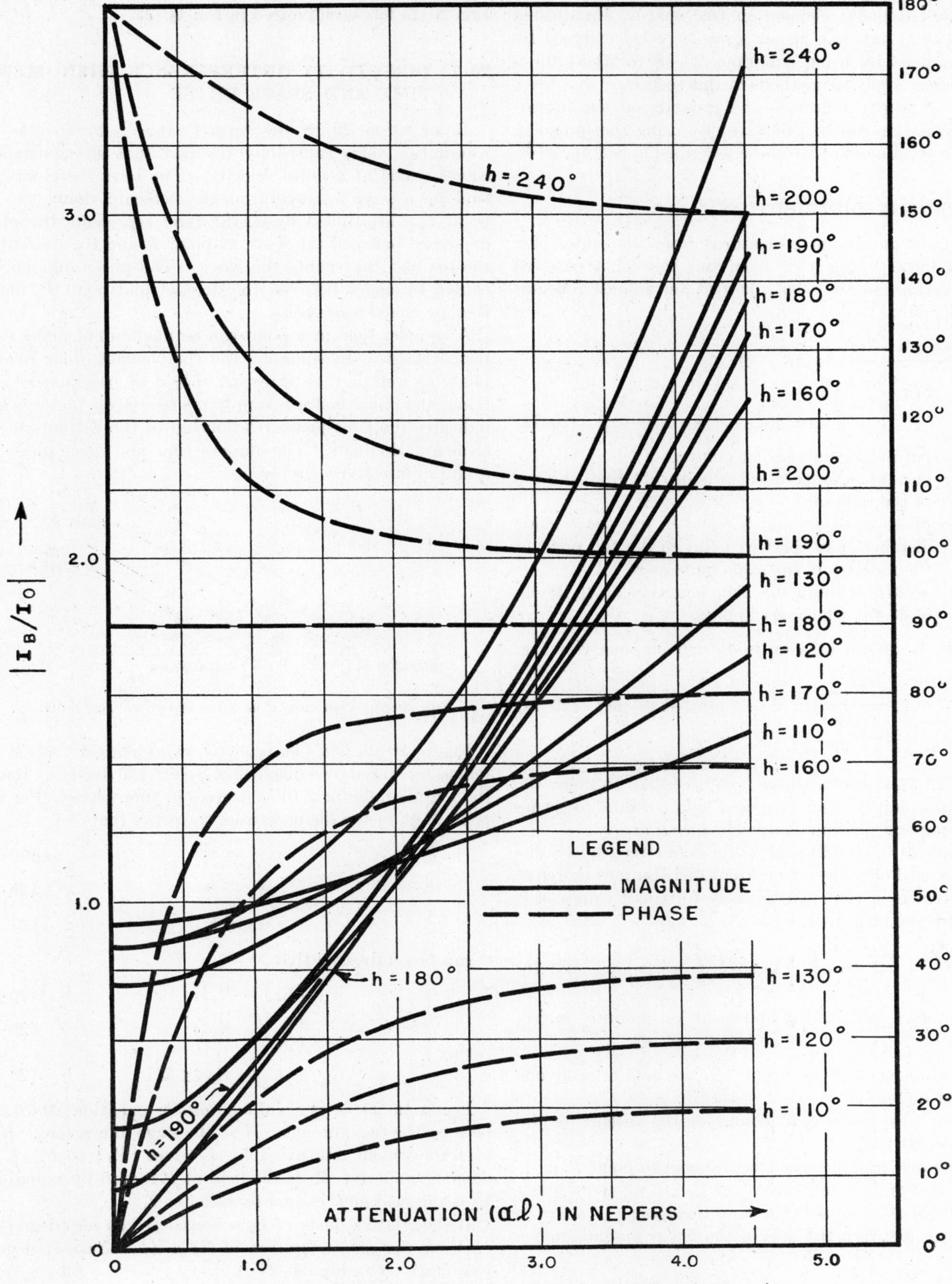

Fig. 20·22 Ratio of base current to loop current as a function of attenuation.

implies that actually a base-fed vertical element is equivalent to a center-fed linear element in free space. End-loaded (or top-loaded) antennas in an array may be analyzed by similar methods by using equations 20·48 to 20·54 for a start and then applying methods of this section.

Effects of nearby reflectors and grounds on the impedances of radiators can be calculated by using the foregoing methods in conjunction with those described in Section 20·3.

20·6 CURRENT DISTRIBUTION

The current at a point z measured from the center of a center-fed linear antenna or from the base of a base-fed radiator, vertical to and near a perfect and infinite reflector is (see Appendix II):[8, 9]

$$\mathbf{I}_z = \mathbf{I}_0 \frac{\cosh \alpha(h - z)}{\cosh \frac{\alpha\lambda}{4}} [\sin \beta(h - z) - j \tanh \alpha(h - z) \cos \beta(h - z)] \quad (20\cdot73)$$

where $\mathbf{I}_z$ = current in amperes, at point z

$\mathbf{I}_0$ = current in amperes, at a point $\lambda/4$ from the top of the radiator

β = phase constant and is equal to free-space value

h = distance from the feed point to the end point of the radiator in the same units as z

αh = $\alpha l/2$ (αl is obtainable from equation 20·31, 20·40, or 20·69, whichever applies)

If $z = 0$, $\mathbf{I}_B$ is the base or feed current, thus (see Fig. 20·22)·

$$\mathbf{I}_B = \mathbf{I}_0 \frac{\cosh \alpha h}{\cosh \frac{\alpha\lambda}{4}} [\sin \beta h - j \tanh \alpha h \cos \beta h] \quad (20\cdot74)$$

Equation 20·74 is useful where it is necessary to know the phase relationship between base and loop currents to set up the matching and phasing equipment for an array.

In Section 20·3, equation 20·27 for field strength at a distance was based on sinusoidal current distribution; therefore that equation is in error, particularly for small values of θ.

Vertical-plane field intensity is:[10]

$$E_\theta = j37.3 \frac{2\pi}{r} \epsilon^{-j\frac{2\pi r}{\lambda}} \sin \theta \int_{z=0}^{z=h} \mathbf{I}_z \cos\left(\frac{2\pi z}{\lambda} \cos \theta\right) dz \quad (20\cdot75)$$

for an antenna vertical to and near an infinite perfect reflector.

E_θ = θ component of the vector electric field in millivolts per meter

$\mathbf{I}_z$ = current in amperes at point z on the antenna (equation 20·73)

r = distance in miles from the antenna to point P

λ = wavelength in units of z

h = height of antenna in the same units as z

θ = angle between the axis of the linear antenna and a ray to P

Because the integral part of equation 20·75 is difficult to evaluate analytically, graphic methods are recommended.

In some arrays the phase relation between loop current and unshunted base current must be known; therefore equation 20·74 has been plotted in Fig. 20·22.

20·7 DIRECTIVITY (INTERFERENCE PHENOMENA), TIME AND SPACE PHASE

In equation 20·18 the term $[-(r - z_n \cos \theta)/c]$ is the retardation term caused by the fact that electromagnetic waves travel at a finite velocity. The term represents the time for a wave to travel from the elemental doublet to the point P. Multiplied by ω, the term represents the phase difference between the wave at point P and the current on the doublet that emits the wave. The phase difference is caused by space between the doublet and point P, and is thereby called *space phase*.

A complex field at a point can be made up of many component fields. In summing the components, their relative phase as well as their direction should be considered. The remaining phase factor is simply the *time-phase* factor, which indicates the time-phase relationship of the currents in the elemental doublets. The ideas of time and space phase are useful in directivity problems.

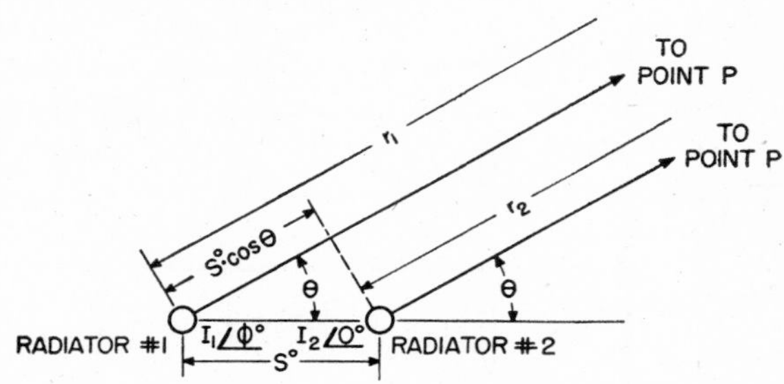

Fig. 20·23 Plan view of an array of two vertical radiators.

Figure 20·23 is a plan view of two identical vertical antennas separated by a distance of S electrical degrees. Radiator 1 leads radiator 2 by ϕ degrees in time phase. For any point P at such a distance from the array that

$$r_1 \cong r_2 \quad (20\cdot76)$$

$$r_1 \gg S \quad (20\cdot77)$$

$$r_2 \gg S \quad (20\cdot78)$$

it can be established that

$$|\mathbf{E}_1| = K_1 I_1 \quad (20\cdot79)$$

$$|\mathbf{E}_2| = K_2 I_2 \quad (20\cdot80)$$

with

$$K_1 \cong K_2 \cong K \quad (20\cdot81)$$

Equation 20·81 is valid because of equation 20·76 and because the two antennas are of the same dimensions. K_1 is a factor which, multiplied by I_1 gives $|\mathbf{E}_1|$ at P. K_2 is defined similarly. $\mathbf{E}_1$ is the field at P caused by radiator 1. $\mathbf{E}_2$ is the field at P caused by radiator 2.

If the relative phase of $\mathbf{E}_2$ is considered as zero degrees at P, then the phase position of $\mathbf{E}_1$ is $(\phi - S \cos \theta)$ degrees leading (refer to Fig. 20·24).

$$\mathbf{E}_2 = KI_2\underline{/0^\circ} \quad (20\cdot82)$$

$$\mathbf{E}_1 = KI_1\underline{/(\phi - S \cos \theta)^\circ} \quad (20\cdot83)$$

in which ϕ is the relative time phase and $S \cos \theta$ is the space phase. The sum of the fields is a simple vector summation:

$$\mathbf{E}_1 + \mathbf{E}_2 = KI_1[\cos(\phi - S\cos\theta) + j\sin(\phi - S\cos\theta)] + KI_2 \qquad (20\cdot84)$$

If I_1 should equal I_2,

$$\mathbf{E}_1 + \mathbf{E}_2 = 2KI_1 \cos\left(\frac{\phi}{2} - \frac{S}{2}\cos\theta\right) \Big/ \underline{\frac{\phi}{2} - \frac{S}{2}\cos\theta} \qquad (20\cdot85)$$

where K is not a function of θ because radiators 1 and 2 are vertical (therefore each one radiates uniformly for all θ's).

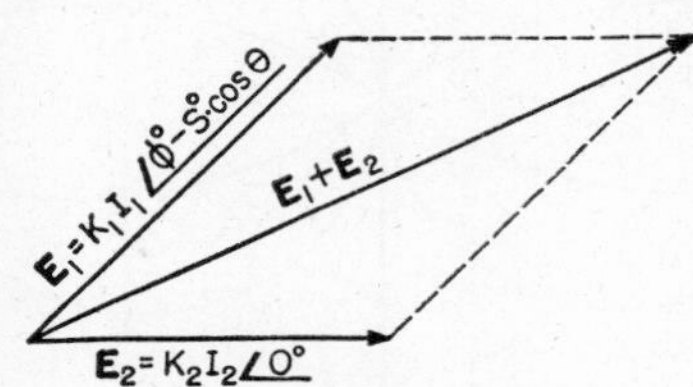

FIG. 20·24 Addition of vector electric fields at a point in space for array of Fig. 20·23.

The term $\cos[(\phi/2) - (S/2)\cos\theta]$ is the pattern term, for it indicates relative field in a horizontal plane of two identical radiators spaced S degrees, with radiator 1 having the same current as 2 except advanced ϕ degrees in time phase. Figure 20·25 shows patterns or plots of $\cos[(\phi/2) - (S/2)\cos\theta]$ using several values of ϕ and S.[4]

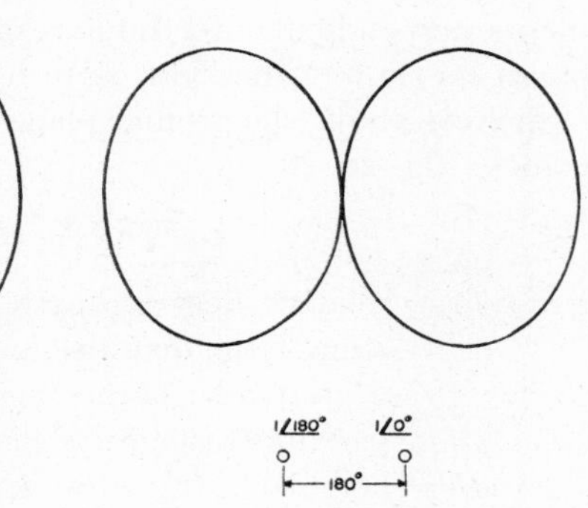
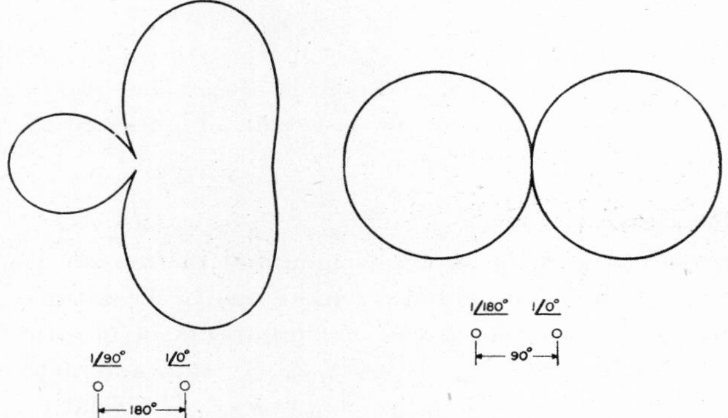

FIG. 20·25 Horizontal-radiation diagrams for two tower arrays having equal radiator currents.

For a more general case, suppose that radiators 1 and 2 are not vertical antennas but are any radiators, or an array of radiators, and are not identical, but have the same polarization. The points labeled 1 and 2 are the geometric centers of the radiators or arrays. Equation 20·81 is no longer valid, and in general

$$K_1 \neq K_2 \qquad (20\cdot86)$$

K_1 now represents the field pattern as a function of θ for antenna or array 1, and K_2 does the same for 2. If I_1 is the multiplier of K_1 required to represent the field at P from radiator or array 1, and if I_2 performs the same function for 2,

$$|\mathbf{E}_1| = K_1 I_1 \qquad (20\cdot87)$$

and

$$|\mathbf{E}_2| = K_2 I_2 \qquad (20\cdot88)$$

as in equations 20·79 and 20·80.

Total field at P becomes

$$\mathbf{E}_1 + \mathbf{E}_2 = K_1 I_1 \underline{/\phi - S\cos\theta} + K_2 I_2 \qquad (20\cdot89)$$

K_1 and K_2 can be functions of θ, and may even have differing relative phase positions which, in turn, may also be functions of θ. For antennas not having the same polarization, care must be taken to allow for the angle between the directions of the electric vectors as well as the time phases.

A similar analysis can be applied to systems that have more than two elements. The system of analysis can be extended to include three dimensions.[13]

20·8 THEOREM OF RECIPROCITY [11, 12]

Any four-terminal network composed of linear bilateral impedances can be duplicated, as far as external behavior at any one frequency is concerned, by an equivalent T network. For the circuit of Fig. 20·26

$$\frac{\mathbf{E}}{\mathbf{I}_2} = \frac{\mathbf{Z}_1\mathbf{Z}_2 + \mathbf{Z}_2\mathbf{Z}_3 + \mathbf{Z}_3\mathbf{Z}_1}{\mathbf{Z}_3} \qquad (20\cdot90)$$

Suppose the generator and ammeter are interchanged; this is the same as interchanging $\mathbf{Z}_1$ and $\mathbf{Z}_2$. Then

$$\frac{\mathbf{E}}{\mathbf{I}_1} = \frac{\mathbf{Z}_2\mathbf{Z}_1 + \mathbf{Z}_1\mathbf{Z}_3 + \mathbf{Z}_2\mathbf{Z}_3}{\mathbf{Z}_3} \qquad (20\cdot91)$$

which is the same as equation 20·90.

This ratio of input voltage to output current is called the transfer impedance. Generator and ammeter impedances are assumed to be zero, but it is necessary only to assume that they are equal.

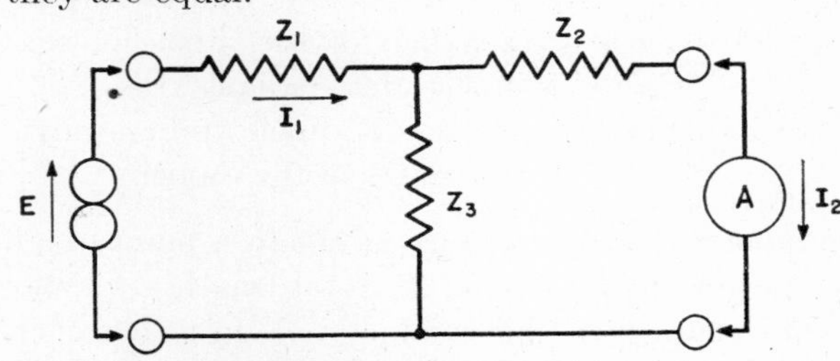

FIG. 20·26 Equivalent T network.

According to the reciprocity theorem in any four-terminal network composed of linear bilateral impedances the transfer impedance remains constant with an interchange of output and input terminals.

This means that if a voltage **E** applied to the terminals of a transmitting antenna produces a current of **I** amperes through the terminals of a receiving antenna, the voltage applied to the terminals of the receiving antenna would produce a current **I** through the terminals of the transmitting antenna, if the generator impedance were equal to the load impedance and if there were no non-linearities in the transmission medium.

The theorem is useful in theoretical and experimental analysis of antennas. For example, it might be more convenient in checking the pattern of a transmitting antenna to use it as a receiving antenna with a simple movable antenna as a transmitting antenna. Such a method gives valid results, but the orientation of the simple antenna should be considered so that polarization is not overlooked. Otherwise misleading results are obtained.

20·9 DIPOLES

One of the simplest and most widely used antennas is the dipole, an antenna composed of two straight conductors lying end to end along a straight line with a small space

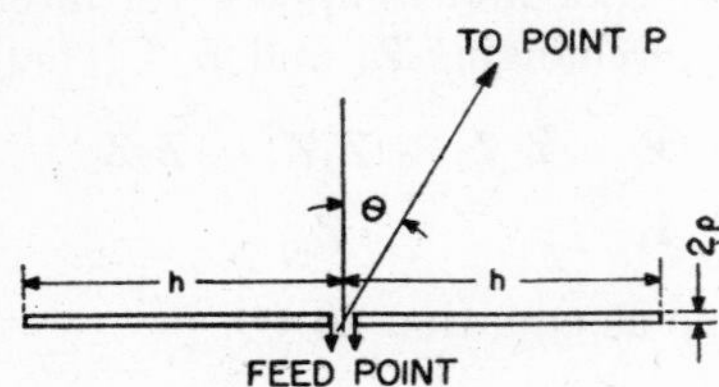

Fig. 20·27 Dipole antenna.

separating adjacent ends. The dipole is always fed at the center by means of a balanced feed system. Figures 20·27 and 20·8 (*b*) show typical dipoles.

Free-Space Dipoles

Radiated Field. If the dipole is mounted in free space, equation 20·27 gives the field at a remote point. Equations 20·64 and 20·65 give the x and z components of electric field close to the antenna. Figure 20·28 shows remote patterns for two different lengths of dipoles. The relative field pattern from a dipole in free space is

$$f(\theta) = \frac{\cos (kh \cos \theta) - \cos kh}{\sin \theta (1 - \cos kh)} \qquad (20\cdot92)$$

where $f(\theta)$ = relative field strength at any point P a constant large distance away from the dipole

kh = length of each half of the dipole in electrical degrees or in electrical radians

θ = angle between axis of dipole and a straight line to P from the center of the dipole.

Self-Impedance. Impedance at the feed point of a dipole in free space is a function of $2h$ (total length of dipole) and electrical radius ρ of the conductors. Impedance characteristics of dipoles were determined experimentally by Brown and Woodward.[14]

A method of calculating the free-space impedance of dipoles is presented in equations 20·28 to 20·35. For $h/\rho > 200$ the resistive component of the impedance of a half-wave dipole (kh = 90 degrees) is equal to 73.2 ohms.

Dipoles Near the Earth or Ground Plane

The field pattern and impedance of a dipole are affected by a nearby reflector. Effects of the ground plane can be calculated by the use of images (Section 20·3) and by incorporating impedance effects of the image according to the methods of Section 20·5.

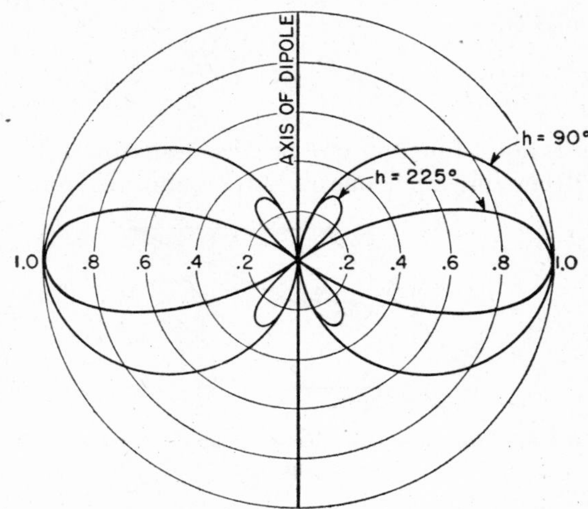

Fig. 20·28 Radiation diagram for dipoles in space.

Horizontal Dipole Near the Earth. The field pattern of a horizontal dipole near the earth is modified by its image. According to equation 20·92, for a dipole in free space the field is constant for a constant angle to the axis of the dipole and a constant distance away from it. If θ = 90 degrees (a plane perpendicular to the axis of the dipole and passing through its center) the field pattern is a circle; if the dipole is H degrees above the ground plane the remote field pattern becomes (Fig. 20·29)

$$f(\psi) = \cos (90^\circ - H^\circ \cos \psi) \qquad (20\cdot93)$$

where $f(\psi)$ = relative field strength at any point P a constant large distance away from the dipole, and in a plane perpendicular to the dipole and passing through its center

H = height of horizontal dipole above ground plane in electrical degrees

ψ = angle between a straight line from the center of the dipole to the point P and the vertical.

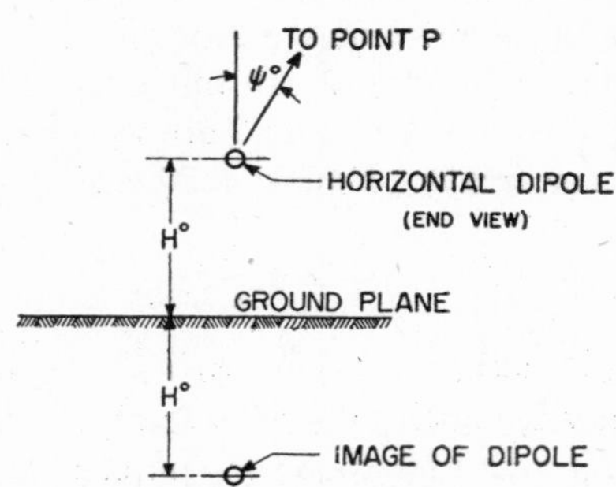

Fig. 20·29 Horizontal dipole H degrees above ground.

Figure 20·30 shows a plot of $f(\psi)$ against ψ for H = 90 degrees.

The driving-point impedance of a horizontal dipole is zero when it is a zero distance above the earth, and it varies as an ever-diminishing spiral that finally reaches free-space value

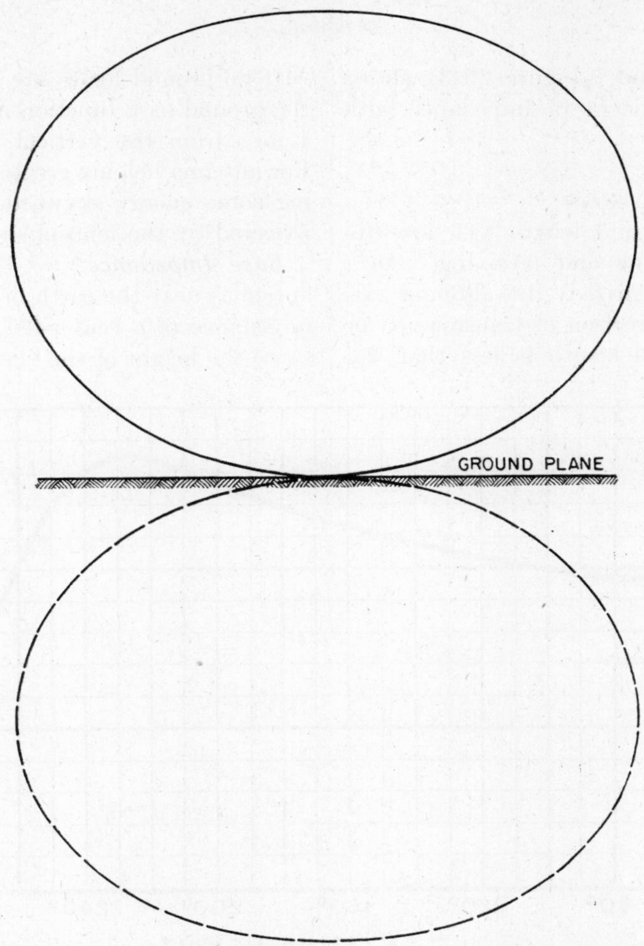

FIG. 20·30 Vertical-plane radiation diagram for a horizontal dipole 90 degrees above ground.

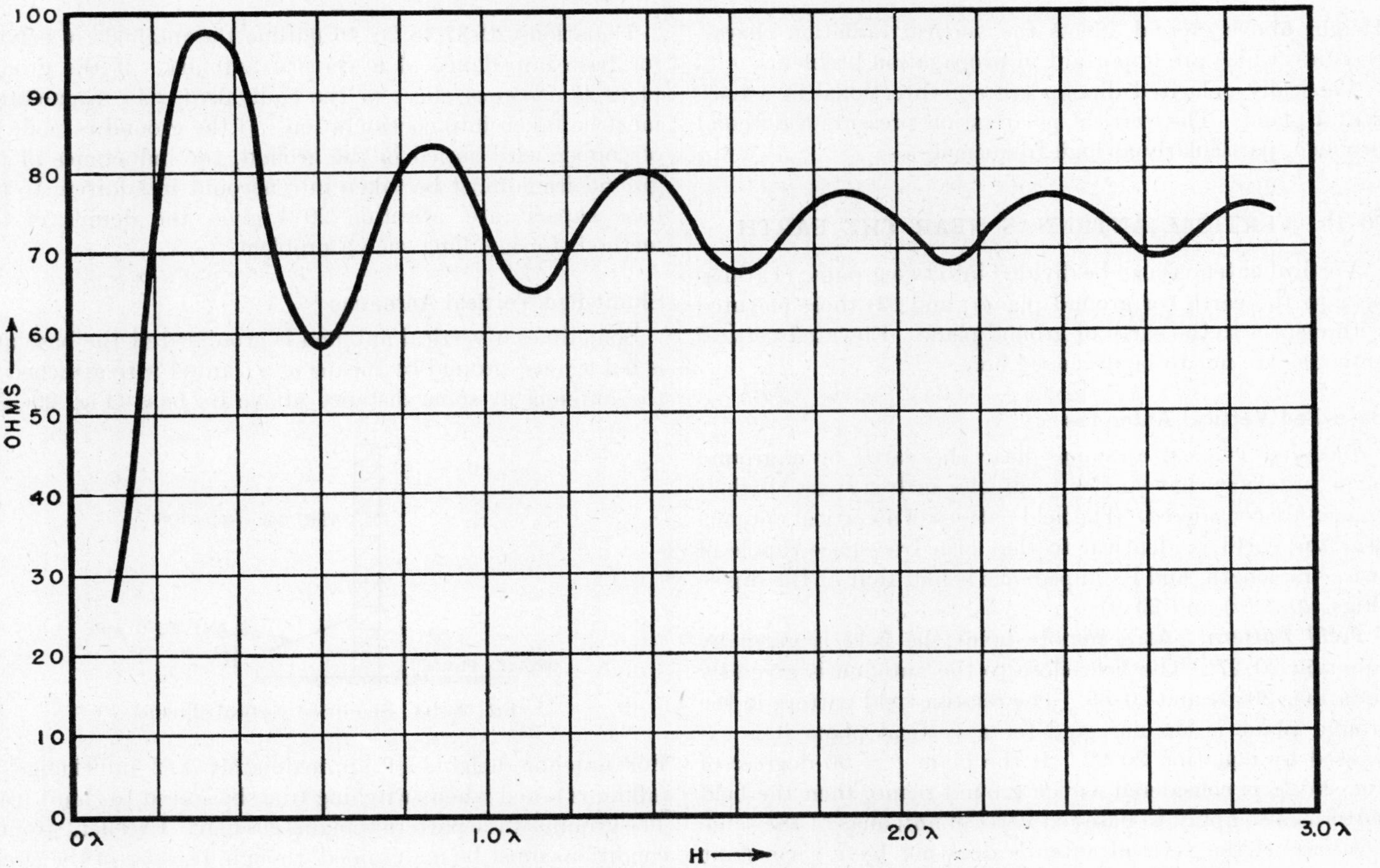

FIG. 20·31 Radiation resistance of a half-wave dipole as a function of height above a perfect earth.

for large heights above the ground. Figure 20·31 shows variation of the resistance component of impedance with spacing above ground.

General

Dipoles, particularly those of total length $\lambda/2$, are frequently used for both transmitting and receiving. As a horizontal dipole (horizontally polarized) the antenna axis should not point in the desired direction of transmission or reception, particularly if the total length is less than 2λ.

If no ground losses are assumed, the field strength along the ground as a function of antenna height at a distance of 1 mile from the vertical antenna is shown in Fig. 20·32. For antenna heights greater than $\lambda/2$, a lobe appears, directing some energy skyward. If that lobe is too large, it is reflected by the ionosphere back to ground to cause fading.

Base Impedance.[5, 6, 9, 10, 14] Base impedance of vertical antennas near the earth or ground plane is equal to half the impedance of a center-fed dipole of a total length equal to twice the height of the vertical antenna.

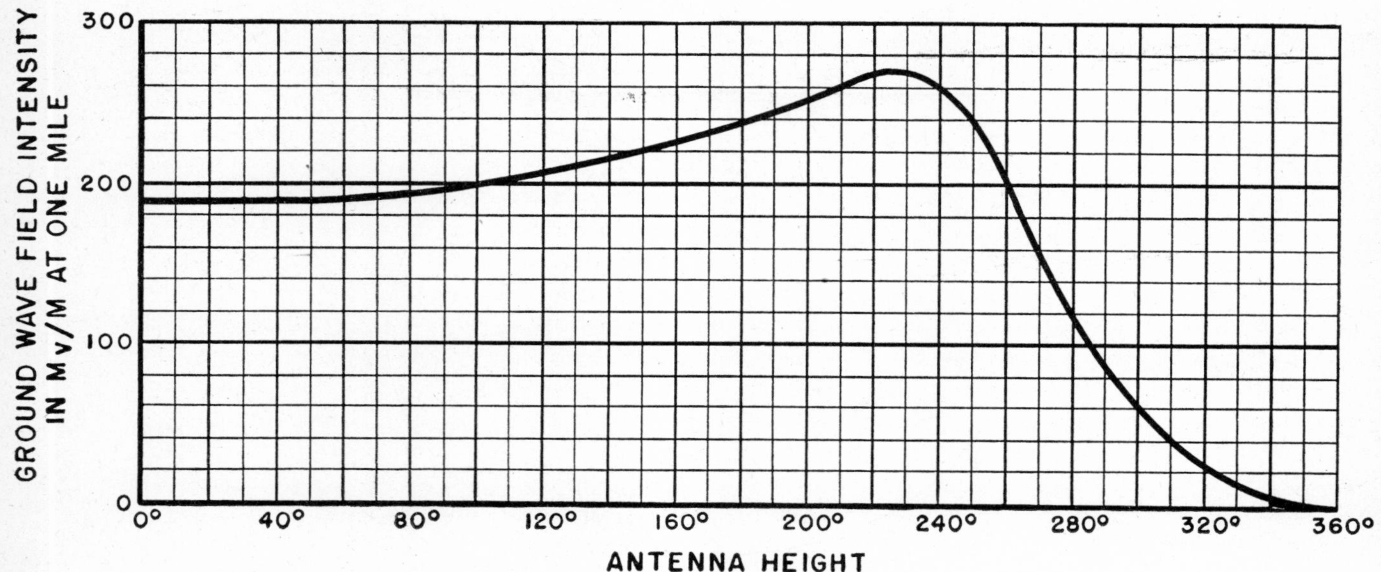

Fig. 20·32 Ground-wave field intensity as a function of antenna height for 1-kilowatt input.

Height above ground affects the vertical radiation characteristics, which are important in propagation problems.

Vertically polarized dipoles are non-directional in a horizontal plane. The vertical polarization presents a different problem, particularly at high frequencies.

20·10 VERTICAL ANTENNAS NEAR THE EARTH

Vertical antennas can be divided into two groups: (1) those near to the earth (or ground plane); and (2) those elevated with respect to the earth or ground plane. Elevated vertical antennas are not to be discussed here.

Base-Fed Vertical Antennas

Base-fed vertical antennas near the earth or a ground plane are somewhat similar to dipoles in free space, if their images are considered. The field pattern of a vertical antenna near the earth is identical to that of a free-space dipole of twice the length, and its impedance is half that of the dipole (Figs. 20·8 (*a*) and 20·9).

Field Pattern. At a remote point the field is given by equation 20·27. The field close to the antenna is given by equations 20·64 and 20·65. The relative field pattern in the ground plane is circular, and for a vertical plane it is expressed by equation 20·92. If the plane $\theta = 90$ degrees of Fig. 20·28 is considered as the ground plane, then the field patterns also apply to base-fed vertical antennas of the same h value. If the vertical antenna does not have a constant cross section, equation 20·92 and Fig. 20·28 no longer apply.

Equations 20·37 to 20·46 outline the methods of solving for base impedance of a vertical antenna. If the ground plane is a copper sheet or the equivalent, base capacitance must be taken into consideration. If the ground is made up of copper wire buried in the ground, the inductance of the ground leads must be taken into account in addition to the base capacitance. Section 20·4 gives the details of the methods for handling such a problem.

Shunt-Fed Vertical Antennas [6]

Sometimes a vertical antenna is grounded at the base and is fed against ground by means of a slanted wire attached to the antenna at some distance above its base (Fig. 20·33).

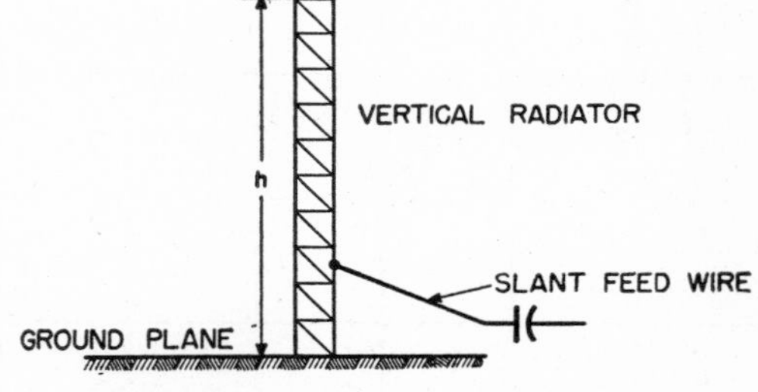

Fig. 20·33 Shunt-fed vertical antenna.

For antenna heights of approximately $\lambda/2$ no change in fading is noted when switching from base feed to shunt feed. The ground-wave pattern remains circular. Excellent ground conditions must be maintained, though, because of the circulating currents in the ground system.

Vertical Dipoles Near the Earth

Some vertical antennas are fed at their centers as dipoles. Such antennas can be analyzed by the general methods of Section 20·3.

Ground Systems [15]

In many vertical antenna systems, particularly those less than $\lambda/8$ in height, the ground losses are appreciable. About 120 ground radials of heavy-copper conductor, $\lambda/2$ long, spaced every 3 degrees, and at an approximate depth of 6 inches, are desirable. If ground loss is unusually large, a copper screen over the ground in the immediate vicinity of the radiator increases efficiency materially. For short-wave vertical antennas the ground should be of closely spaced conducting mesh or a sheet of conductor whose diameter is about $\lambda/2$.

20·11 STRAIGHT-WIRE ANTENNAS [3, 17, 18, 19]

Long-Wire Antennas

The field pattern of a long straight-wire antenna of integral multiples of $\lambda/2$ in length in free space is given by equations 20·25 and 20·26. The effects of a nearly plane earth on the field pattern of a straight-wire antenna can be computed by means of the principles outlined in Sections 20·3 and 20·7.

Radiation resistance referred to a current loop of such an antenna in free space is

$$R = 30\left[0.5772 + \log_e\left(4\pi\frac{l}{\lambda}\right) - \text{Ci}\left(4\pi\frac{l}{\lambda}\right)\right] \qquad (20\cdot94)$$

where R = radiation resistance in ohms at a current loop, and the other terms are as defined for equations 20·25 and 20·26.

The impedance of such an antenna near the earth is modified by its image according to the principles described in Section 20·5.

Wave Antenna

A wave antenna is a long-wire antenna terminated in its characteristic impedance at the far end. If its length is great, such an antenna is highly directional along its axis away from the generator. For a shorter antenna the major lobe divides into two beams. For a wire 10λ long the two beams lie about 15 degrees each side of the antenna axis. For a wire 2λ long the two beams are about 35 degrees on each side of the antenna axis.

V Antennas

A V antenna is composed of two horizontal wires disposed to form a V. The feed point is at the apex of the V. V antennas are of two types, terminated and resonant. A resonant V antenna can be designed by the use of equations 20·25 and 20·26. For equation 20·25 to become a maximum the smallest θ that makes $(\pi l/\lambda)\cos\theta$ go to an integral multiple of π should be determined. The numerator is then a maximum. The denominator is diminishing with θ at a slow rate compared with the change in the numerator with θ. If l/λ is large, the numerator has little effect except to cause equation 20·25 to maximize for a slightly smaller value of θ. The same applies to equation 20·26.

For example, suppose a V antenna is to be made from two wires, each 4λ long. The field pattern of a wire 4λ long is indicated in Fig. 20·34, which shows major lobes only. If

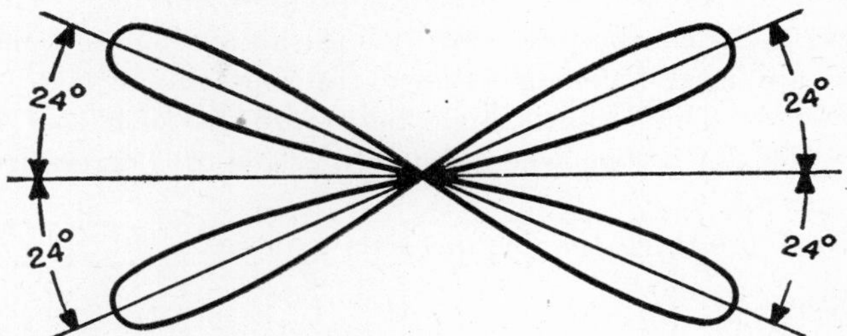

Fig. 20·34 Major radiation lobes of an antenna 4λ in length.

the two wires are placed at an angle of 48 degrees, the field is reinforced along the bisector of the angle. Figure 20·35 shows this.

A terminated V antenna is worked out in the same general manner as a wave antenna to give a unidirectional pattern instead of the bidirectional pattern of the resonant V shown in Fig. 20·35.

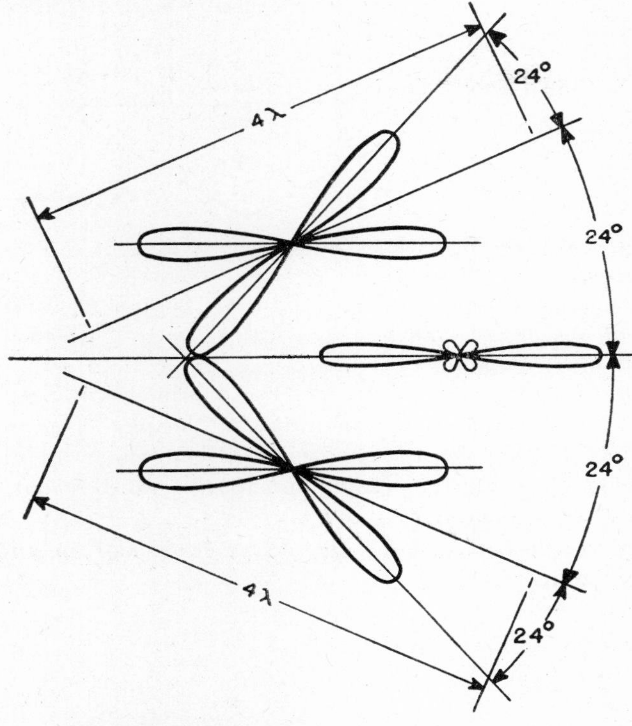

Fig. 20·35 Major radiation lobes of a V antenna.

Rhombic Antennas [19, 20, 21]

Two V antennas placed end to end to form a rhombus make up a rhombic antenna. There are two types of rhombic antennas, terminated and resonant. Figure 20·36 shows a terminated rhombic antenna.

Resonant. Resonant rhombic antennas, designed by the same methods as resonant V antennas, are bidirectional, as might be expected from the analysis of resonant V antennas. Input impedance of a rhombic antenna with legs more than a few wavelengths long is from 700 to 800 ohms, depending upon the wire size. The vertical pattern is a function of the height of the antenna above ground. It is also affected by the apex angles of the V antennas making up the rhombic antenna.

Terminated.[19] A terminated rhombic antenna is terminated in a resistor at the far end. The *ARRL Antenna Book*[19] describes clearly how to design a rhombic antenna. Large power gains can be produced by such an antenna. Usually they are used for long-distance transmission at high frequencies. The vertical angle best suited is such that the ray or beam is somewhere between 5 and 20 degrees from the ground.

There are three possible methods of design:

1. The ideal method.
2. The alignment method.
3. The compromise method.

The ideal method sets the parameters l, ϕ (see Fig. 20·36), and H (antenna height) for maximum power output along a line Δ degrees (the wave angle) from the ground.

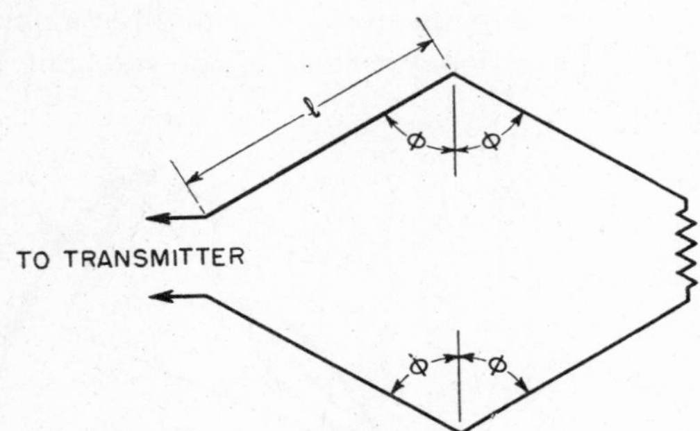

FIG. 20·36 Terminated rhombic antenna.

The alignment method sets the same parameters to centralize the major lobe of the vertical pattern about the desired wave angle Δ.

The compromise method sets the parameters if part of them are already set by conditions of the plot of ground, availability of poles for supporting the antenna, or other circumstances, for the compromise to give the best output for the wave angle Δ.

The design equations for the ideal method and the alignment method are[19]

$$H = \frac{\lambda}{4 \sin \Delta} \tag{20·95}$$

$$\phi = \sin^{-1} (\cos \Delta) \tag{20·96}$$

For the ideal method

$$l = \frac{\lambda}{2 \sin^2 \Delta} \tag{20·97}$$

For the alignment method

$$l = \frac{0.371}{\sin^2 \Delta} \tag{20·98}$$

where H = height of the antenna above the ground
ϕ = tilt angle (as shown in Fig. 20·36) in degrees
l = length of one leg in same units as H
λ = wavelength in same units as H
Δ = wave angle in degrees.

For terminating the rhombic antenna an iron-wire line can be used. Power up to 20 percent of the input may appear in the terminating resistor for a rhombic antenna with relatively short legs.

For example, a terminated rhombic antenna having H = 65 feet, l = 230 feet, and ϕ = 67.5 degrees shows an

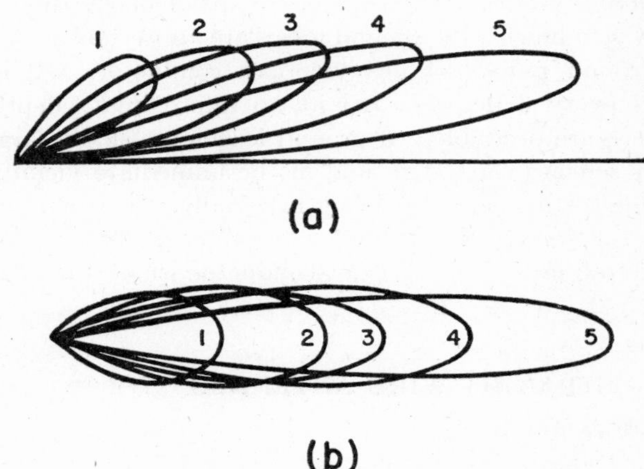

FIG. 20·37 Radiation diagrams of antenna 1 operating on 5 frequencies. (*a*) Vertical-radiation diagrams through the major lobes. (*b*) Radiation diagrams for a constant Δ through the maximum of (*a*).

ever-increasing sharpness and gain in the vertical plane and a constantly lowering beam angle for increasing frequency, as shown in Fig. 20·37 (*a*). Pattern 1 corresponds to 6140 kilocycles; pattern 2 to 9570; pattern 3 to 11,870; pattern 4 to 15,210; and pattern 5 to 21,540. Figure 20·37 (*b*) shows the pattern for a constant Δ through the maximum of each of the vertical patterns of Fig. 20·37 (*a*). With increasing frequencies the gain becomes greater and the beam becomes narrower. Figures 20·37 (*a*) and 20·37 (*b*) are the major lobes only, calculated according to the methods of Foster.[21]

Compare with this terminated rhombic antenna a similar antenna having l = 315 feet and ϕ = 70 degrees; both antennas operating on a frequency of 15,210 kilocycles, and at the same height above ground. Figure 20·38 (*a*) shows

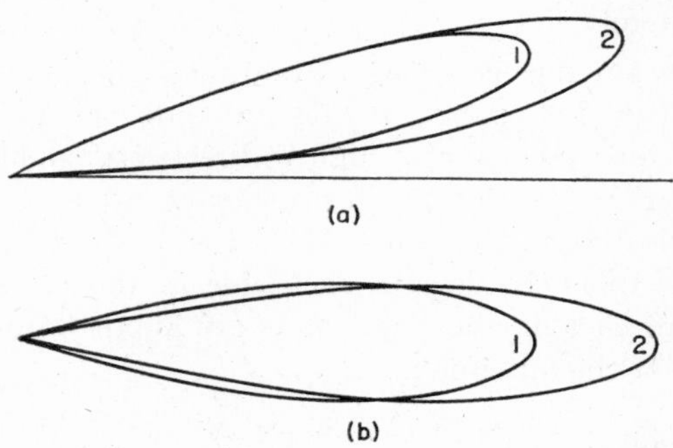

FIG. 20·38 Radiation diagrams of antennas 1 and 2 operating on 15,210 kilocycles. (*a*) Vertical-radiation diagrams throught he major lobes. (*b*) Radiation diagrams for a constant Δ through the maximums of (*a*).

the vertical pattern through the major lobes of antennas 1 and 2. Antenna 1 has its maximum in the vertical plane occurring for a slightly larger Δ than antenna 2. The major lobe of antenna 1 is slightly less than that of antenna 2. Figure 20·38 (*b*) shows the patterns for a constant Δ through the maxima of the vertical patterns of the two antennas.

20·12 LOADED ANTENNAS

Loaded antennas are used to provide a relatively high impedance for improved efficiency or to improve field pattern or signal strength in a specified direction.

A dipole can be loaded to increase its apparent length, or a vertical antenna can be loaded to increase its apparent height. For example, top-loading a vertical radiator to an apparent height of 190 degrees from a height of 150 degrees increases signal strength along the ground approximately 10 percent.

Loading may take two forms. Capacitance can be added to the free end or ends of the antenna, or the antenna can be insulated at various places, with reactances shunting the insulators.

Top-Loaded Vertical Antennas or End-Loaded Dipoles [7, 16]

Field patterns of top-loaded base-fed vertical antennas near the earth and end-loaded dipoles in free space are the same in shape and magnitude. The impedance of the vertical antenna is one-half that of an identically end-loaded identical dipole with a length twice that of the vertical antenna. The ground is assumed to be a perfect reflector. Figure 20·17 shows a top-loaded base-fed vertical antenna.

Radiated Field. The pattern along the ground (or, for end-loaded dipoles, in the plane perpendicular to the antenna axis and passing through the center of the antenna) is a maximum for an apparent height between 220 and 235 degrees. In other words, if any antenna with h from 90 to 200 degrees is top-loaded for maximum ground wave (or, for dipoles, the bisecting plane wave), the apparent height is in the range from 220 to 235 degrees.[16]

The vertical pattern of a vertical antenna is greatly affected by top-loading. The corresponding pattern of the end-loaded dipole is affected in the same manner.

The field pattern is [16]

$$f(\theta) = \frac{\cos \psi \cos (kh \cos \theta) - \cos \theta \sin \psi \sin (kh \cos \theta) - \cos (kh \cos \theta)}{\sin \theta [\cos \psi - \cos (kh + \psi)]} \qquad (20\cdot99)$$

where $f(\theta)$ = relative field strength at any point P a constant large distance from the antenna

ψ = amount of top or end loading (see equation 20·50) in radians

kh = height of vertical antenna or half the length of the dipole antenna in radians

θ = angle between the axis of the antenna and a straight line from the antenna to P.

Base Impedance. For center-fed dipoles the values for R and X of equations 20·48 and 20·49 should be doubled, and the effects of the ground should not be taken into consideration.

Sectionalized Antennas [16]

A sectionalized antenna is an antenna broken up by insulators, which are shunted by reactance. For a vertical radiator sectionalized near the top by an inductive reactance the vertical radiation characteristics and the effects on ground signal are approximately the same as for a top-loaded antenna of a height equal to the height of the sectionalizing insulators from the ground. The approximation becomes less and less valid for increasing heights of the section above the sectionalizing insulator.

20·13 LOOP ANTENNAS [11, 18]

Loop antennas usually are made up of one or more coils of wire in a single plane. They are of two general types: (1) those with a total peripheral length of conductor short compared with a wavelength; and (2) those with a total peripheral length of conductor an appreciable part of a wavelength.

Low-Frequency Loops

A low-frequency loop usually is constructed as shown in Fig. 20·39 (*a*) and is of the type whose total peripheral length of conductor is short compared with a wavelength.

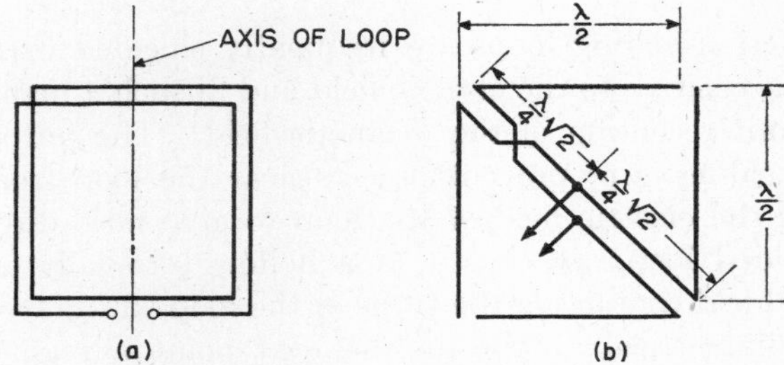

Fig. 20·39 Loop antennas. (*a*) Low-frequency. (*b*) High-frequency.

It is constructed of one or more turns and usually is symmetrical about an axis. Generally it is fed at a point on a conductor passing through the axis. Some loops are square, as in Fig. 20·39 (*a*); others are round or triangular.

Transmitting. Transmitting loops are seldom used because they have high Q and cannot be efficiently fed. If they are used for transmitting, the field strength in a plane perpendicular to the axis of the loop is expressed as [18]

$$E = 74.57 \frac{\pi^2 N}{r} \frac{A}{\lambda^2} I_s \cos \theta \qquad (20\cdot100)$$

where E = signal strength in millivolts per meter

r = miles from the loop to the point of measurement

θ = angle from the plane of the loop

I_s = current in the loop in amperes

N = turns in loop

A = area of loop in square units corresponding to the linear units of wavelength

λ = wavelength.

The field pattern of a loop in a plane that is the perpendicular bisector of the axis of the loop is

$$f(\theta) = \cos \theta \qquad (20\cdot101)$$

where $f(\theta)$ = relative signal strength for a constant distance from the loop

θ = angle from the plane of the loop, all points lying in a plane perpendicularly bisecting the axis of the loop.

Receiving. When used for receiving, the voltage induced in the loop is [11]

$$E' = 0.610\pi NE \frac{A}{\lambda} \cos\theta \qquad (20\cdot 102)$$

where E' = induced voltage in millivolts
N = turns in loop
A = area of loop in square feet orientated so that the axis of the loop is parallel to the polarization of the received signal
θ = angle between the plane of the loop and the direction of arrival of the signal
E = signal strength in millivolts per meter
λ = wavelength in feet.

Induced voltage is proportional to the cosine of the angle of arrival relative to the plane of the loop. If the loop is broadside toward the signal no voltage is induced. This principle is widely used in direction-finding equipment.

Direction-finding loops are frequently shielded to render them insensitive to the electric field and in such a manner as to permit reception of the magnetic field. The purpose of the shielding is to prevent objects near the loop from distorting the electric field of the loop to give false direction. A shielded loop can consist of a hollow toroidally shaped metal shield containing the turns of the loop within it. At a point diametrically opposite the feed point a segment of insulation is placed to prevent the shield from acting as a short-circuited turn of the loop.

Loop-receiving antennas are the best compromise for receiving antennas for low-frequency receivers, where space is limited.

High-Frequency Loops [31]

High-frequency loops are usually constructed in such a manner as to have a total peripheral length of conductor an appreciable part of a wavelength. Among the types in use for high-frequency broadcasting are the square loop and the circular antenna.

The square-loop antenna is constructed of four half-wave dipole radiators arranged to form a square, usually in a horizontal plane. The energy is fed to the loop in such a way that the loop currents are equal in magnitude and that the loop currents of adjacent dipoles have a relative time phase of 180 degrees. Figure 20·39 (*b*) shows a method of feeding a square loop.

A square loop has a gain of -1.4 decibels compared with a single $\lambda/2$ dipole. The field pattern is very similar to that of a vertical $\lambda/2$ dipole except that the vector electric field is horizontally polarized.

The circular antenna is a high-frequency loop antenna. It can be considered as a folded half-wave dipole, slightly shortened by capacitive end loading, bent to form a circle. The circular antenna is generally mounted in a horizontal plane.

The field pattern is more nearly circular than the field pattern of the square loop. The gain is about 1 decibel below that of a half-wave dipole.

20·14 DRIVEN DIRECTIVE ARRAYS [13, 18]

Unlike antenna elements in an array make a point-by-point summation necessary. If the elements are alike the radiation pattern becomes simple. Symmetry, both geometric and electric, results in still further simplicity. Southworth [13] describes complex arrays in some detail.

Array of Arrays

Figure 20·23 shows a plan view of an array of two vertical radiators spaced S electrical degrees apart, with antenna 1 having a relative current of $I_1\underline{/\phi^\circ}$ and antenna 2 having a current of $I_2\underline{/0^\circ}$. The resulting field is expressed by equation 20·84. For $I_1 = I_2$, equation 20·85 expresses the result. The pattern term of equation 20·85 is $[(\phi/2) - (S/2)\cos\theta]$. K is that number, which, multiplied by I_1, gives the absolute magnitude of the function. If antennas 1 and 2 were not identical antennas, but were identical arrays of antennas, K would include a constant term multiplied by the pattern term of the identical arrays.

Therefore in an array of arrays the resulting pattern is simply the product of the pattern (array term) of the composite array (assuming simple non-directional elements) and the pattern of the elemental arrays making up the major array. All elements must be alike.

For example, consider the system shown in Fig. 20·23 for $\phi = 0$, $I_1 = I_2$, and $S = 360$ degrees. Then the array term is $\cos(180^\circ \cos\theta)$. Suppose that antennas 1 and 2 are identical, that each is an array of two vertical antennas, and that the centers of the arrays correspond to the positions of antennas 1 and 2. Suppose that the elements of each elemental array are all in the same straight line, and that for each elemental array $\phi = 90$ degrees, $I_1 = I_2$, and $S = 100$ degrees. Then the relative pattern of each is expressed by $\cos(45^\circ - 50^\circ \cos\theta)$, and the pattern of the array of arrays is $\cos(45^\circ - 50^\circ\cos\theta)\cos(180^\circ\cos\theta)$. Computation of arrays can be long and tedious. Machine calculators have been devised to simplify the procedure.

Arrays of driven elements can take many configurations. For example, some arrays are made up of several rhombic or V antennas. For broadcast use, vertical elements in one line with a null and a major lobe appearing on opposite ends of the line are sometimes employed. Frequently broadside arrays are used with the major lobes on a line perpendicular to the line of antennas.

Turnstile Antennas.[22] High-frequency broadcast stations frequently use the *turnstile*, which is an antenna made up of two simple dipole radiators at right angles, each driven with the same current, but with 90 degrees time-phase difference (Fig. 20·40). The resultant pattern is almost circular where the patterns of the individual dipoles are figure eights, as can be seen if the figure eights are superposed at right angles and added vectorially.

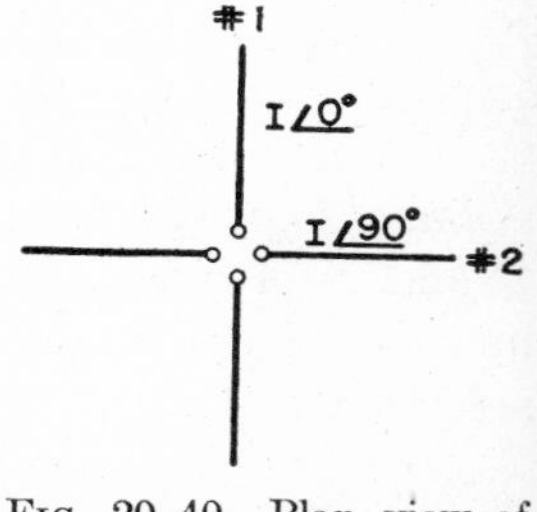

Fig. 20·40 Plan view of turnstile of dipoles.

Such antennas are stacked vertically above each other and fed in phase to form an array of turnstiles which has high vertical directivity and a circular horizontal pattern. Square loops and circular antennas are frequently stacked and fed in the same manner so as to concentrate the radiated energy in a horizontal plane.

20·15 ARRAYS USING REFLECTING SURFACES

Plane Reflectors

Antennas sometimes are placed in front of a flat-sheet reflector to obtain controlled directivity. A good example of this is shown in Section 20·9 by placing antennas having a circular pattern above ground (Fig. 20·29) and obtaining the general equation 20·93, a special case of which is shown in the field pattern of Fig. 20·30. However, directivity is controlled by the magnitude of H of Fig. 20·29.

Krause Antennas [23]

Krause antennas use flat-sheet reflectors intersecting at an angle as shown in Fig. 20·41 to direct the radiation of a

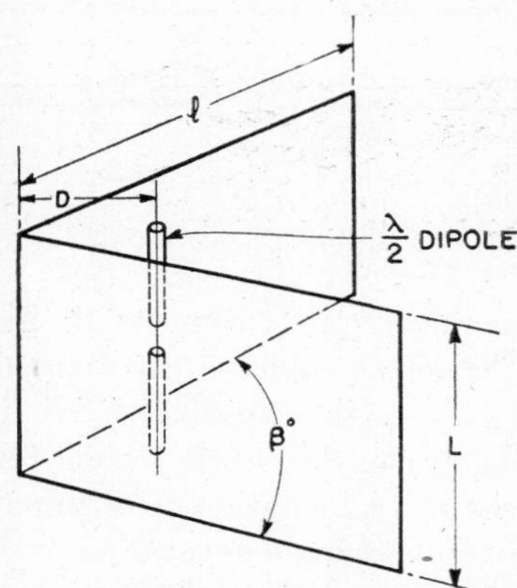

Fig. 20·41 Krause antenna.

simple dipole radiator in a given direction. The antenna can be completely analyzed by images, as discussed in Section 20·3, and the impedance can be analyzed by the methods of Section 20·5.

The gain from a Krause antenna ranges from 5 to 12 decibels over that of a dipole, depending upon the angle at which the reflecting surfaces intersect and the distance of the dipole from the apex of the angle. L should be at least $\lambda/2$, l should be at least 2.5λ, and D should be an eighth wave or greater. The sharper the angle β, the greater the gain. Sometimes rods parallel to the antenna are used in place of reflectors. They should be spaced closely.

Parabolic Reflectors

Two types of parabolic reflecting surfaces are used with antennas: parabolas of revolution and cylindrical parabolas. Parabolas of revolution are used for two-plane directivity. Cylindrical parabolas are used if the electromagnetic energy is to have one-plane directivity only.

Parabolic reflecting surfaces act on high-frequency radio waves much as they do for light in searchlights or reflecting telescopes. These are the major differences:

1. For parabolic reflectors of light the dimensions of the parabola are many times greater than the wavelength of light. For radio waves the dimensions may be from 10 to 20λ or greater.
2. Radio waves are diffracted more at the edges of the reflector than light would be from the same reflector.
3. There seems to be no good point source for radio waves. Therefore, it is difficult to get proper amplitude and phase illumination of the parabola for ideal functioning of the reflecting surfaces.

Sometimes it is necessary to shield the source of the radio waves from the main beam, because along the major axis of the beam the direct and reflected rays would almost cancel.

Power gains of several thousand may be obtained from well-designed parabolic reflecting surfaces used with an exciting antenna.

20·16 PARASITIC ARRAYS [4, 19]

A parasitic antenna or antenna element is not connected to the generator, but is close enough to a driven antenna to intercept energy, to set up currents in the parasitic element, and to affect the field of the driven antenna. Current in the parasitic element can be controlled by changing its distance from the driven antenna, by changing the diameter or length of the parasitic antenna, or by inserting reactance in it.

Usually parasitic antennas are used with half-wave driven antenna elements even in a complex array. The parasitic elements usually are about $\lambda/2$ in length. If such an element is placed ahead of the driven element in the main beam, it is called a *director*. If it is behind the driven element it is called a *reflector*. In any parasitic array the amplitude of current in the element must be balanced against the time-phase position of the current.

The gain of a sheet of dipole radiators in a broadside array can be improved by 4 decibels by adding parasitic reflectors.

Parasitic Antennas [19]

One type of antenna frequently used is a driven element with one or more parasitic elements, all parallel and mounted in the same plane (see Fig. 20·42). For a two-element beam (one driven element, and one director or reflector) the maximum forward gain compared with a half-wave dipole is 4 or

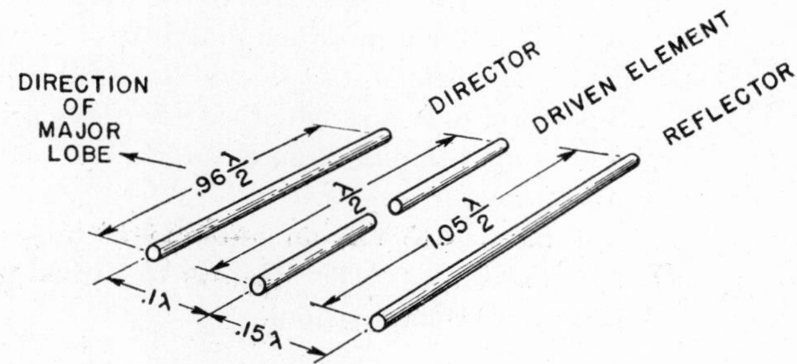

Fig. 20·42 Three-element parasitic array.

5 decibels and is 6 or 7 decibels for a three-element beam (one driven element and two parasitic elements). A four-element beam antenna has a slightly greater gain.

Radiation resistance for a two-element beam is about 15 ohms for $\lambda/10$ spacing between driven and parasitic element and ranges up to 60 ohms for $\lambda/4$ spacing. A three-element beam usually has a resistance from 8 to 10 ohms, and the driven element of a four-element beam has a resistance from 5 to 6 ohms. More elements have been used, but they become exceedingly critical. For multielement arrays the elements must be large to reduce power loss in the antenna, and an efficient feed system should be used to couple to the driven element.

Dimensions

It is best to space the first directors $\lambda/10$ in front of the driven element and any remaining $\lambda/10$ apart for maximum gain. The first reflector and any remaining should be spaced at 0.15λ intervals to the rear. Reflector spacing is not so critical as director spacing from the power-gain standpoint.

Proper lengths of the elements depend upon their spacing. For a three-element beam with $\lambda/10$ separation between director and driven element, and with 0.15λ separation between reflector and driven element, the director should be about 4 percent less than $\lambda/2$ in length and the reflector should be approximately 5 percent greater than $\lambda/2$. The length of the driven element is not critical but should be nearly $\lambda/2$.

In practice, the element lengths are set by trial and error. If gain is most important, length is not too critical, but if side and back lobes are to be minimized, lengths must be carefully adjusted.

A *Yagi antenna* is an array utilizing one driven element and several symmetrically placed (symmetrical about at least one axis) parasitic elements, all parallel to the driven element but not necessarily in the same plane. Section 20·5 describes a method for computing the driving-point impedance of the driven antenna in parasitic arrays as well as for driven arrays.

Brown [4] explains in detail the proper method of determining impedance of a driven antenna element with accompanying parasitic elements. Frequently a nearby obstruction adversely affects the performance of an antenna by acting as a parasitic element and causing a deterioration of pattern and a possible loss in efficiency.

20·17 BROAD-BAND SYSTEMS [14, 28, 29, 30]

Many systems require broad-band antennas. Television, for example, requires a good impedance match over a wide range of frequency. Sometimes it is desirable to have one antenna that will perform over a range of carrier frequencies. The major problem is maintaining a good impedance match to the line from the transmitter.

Usually the best plan is to use an antenna of inherent broad-band characteristics. Another plan is to utilize corrective networks in the coupling system.

Dipoles

Section 20·4 gives the theoretical aspects of broad-band dipoles, and Brown and Woodward [14] provide an experimental analysis of such dipoles. For center-fed half-wave dipoles a slenderness ratio (the ratio of one-half the total length to the diameter) of 20 or less produces a broad band. The shortening effect of such a slenderness ratio makes the apparent length approximately 270 degrees.

For a center-fed dipole changing the frequency to make the electrical length change from 160 to 180 degrees (based on free-space velocity) gives the following:

Slenderness Ratio	R(ohms)	X(ohms)
20	54 to 98	−7.6 to +50
10	62 to 112	−3.6 to +11
5	56 to 82	−13.0 to −34

Dipoles causing a 1.1 to 1 voltage standing-wave ratio on a matching line for a 1.5 percent bandwidth have been constructed.

Folded Dipoles [28]

A folded dipole can be made for broad-band use if the spacing between the conductors is made as wide as possible. Radiation of the folded dipole is similar to a conventional dipole for S, the center-to-center spacing between conductors

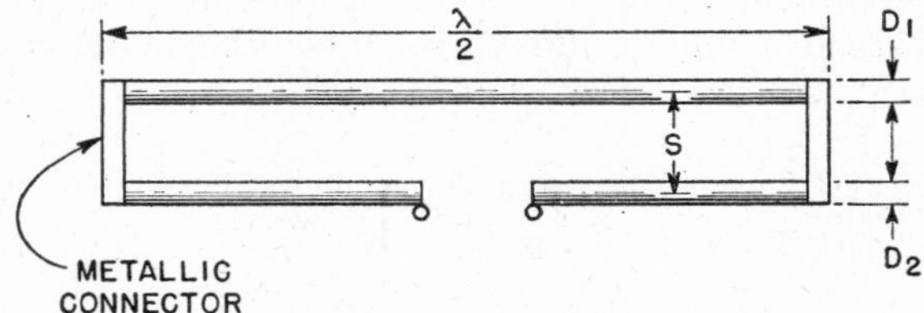

FIG. 20·43 Folded dipole.

(Fig. 20·43), less than $\lambda/16$. Bandwidth increases for increasing S. The larger D_1 and D_2, the diameters of the conductors, are, the greater the bandwidth, as in conventional dipoles. A folded dipole described by Carter [28] showed for a 10 percent bandwidth a variation of terminal impedance of ± 10 percent with a phase angle never greater than 20 degrees. Folded dipoles causing a 1.1 to 1 voltage standing-wave ratio on a matching line for a 3 percent bandwidth have been designed.

Folded dipoles, besides acting as dipoles, act as impedance transformers. If $D_1 = D_2$, the terminal impedance is four times that of a single dipole because the currents divide equally in the two conductors. If three conductors are placed in parallel and one is opened at the center, the impedance is nine times that of a single dipole. Impedance can be varied by changing the size of D_2, the diameter of the driven element, relative to the size of D_1, the diameter of the other element. For $D_2 > D_1$ the impedance is less than four times the impedance of a single dipole, and for $D_2 < D_1$ the impedance is greater than four times the impedance of a single dipole.

Antennas of Special Shapes

Other antennas having broad-band characteristics can be constructed. They are dipoles of cones placed apex to apex, dipoles of cones placed base to base, and dipoles of other shapes. Schelkunoff [29] has covered most of the theoretical aspects of radiators of such shapes.

For conical dipoles with apexes end to end, the length of each cone taken from apex to the circumference of the base should be about 0.365λ no matter what the apex angle is.[28]

If ψ of Fig. 20·44 is approximately 13 degrees the impedance is 380 ohms within ± 5 percent, and the phase angle is less than 10 degrees over a frequency band of 20 percent. Impedance varies as a function of ψ, the apex angle. For

$\psi = 5$ degrees the impedance is about 1000 ohms; for $\psi = 8$ degrees the impedance is nearly 700 ohms; for $\psi = 16$ degrees impedance is a little more than 200 ohms.

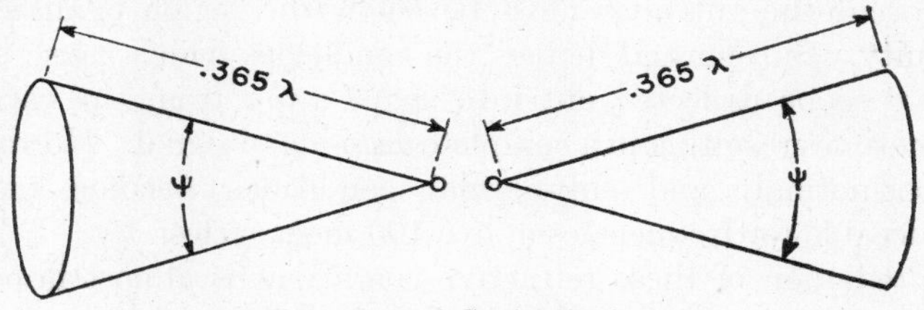

FIG. 20·44 Conic dipoles.

Circuit Design to Improve Bandwidth

One means of improving the bandwidth of an antenna is to place a similar element at an angle of 90 degrees to it in the same plane and feed it in phase quadrature. The common feed point has a broader bandwidth than either antenna alone. The quarter-wavelength difference in line lengths used to feed the elements changes the sign of the reactance of one antenna and cancels the unmodified reactance of the other antenna at the common feed point. Carter [28] shows data for a *turnstile* made up of folded dipoles. The impedance varies over a ±5 percent range at a phase angle always less than 5 degrees over at least 20 percent frequency band.

Similar methods can be used to help broaden the bands of complicated arrays, and some improvement can be made by impedance-matching methods combining fixed and distributed constants. In this way a system causing a two-to-one voltage standing-wave ratio (or less) over a frequency range of 50 percent was produced.[30]

20·18 PROPAGATION OF RADIO WAVES [18, 24, 25, 26, 27]

Prediction of the propagation of radio waves for the entire spectrum is exceedingly difficult. For frequencies up to 2 megacycles the factors affecting propagation are probably better known in a quantitative manner than for any other part of the spectrum. From 2 to 50 or 100 megacycles the condition of the ionosphere influences propagation. For frequencies above 100 megacycles the conditions of the lower atmosphere become increasingly important.

Maxwell's equations as shown in Section 20·1 express the laws of propagation of electromagnetic waves. For a complete analysis the theory of diffraction should also be known.

Factors Affecting Propagation

The following factors affect propagation:

1. Conductivity of earth along propagation path.
2. Dielectric constant of earth along propagation path.
3. Conducting layers called the ionosphere in the upper atmosphere.
4. Earth's magnetic field.
5. Refractive index of the atmosphere.
6. Absorption of the atmosphere.
7. Diffraction.
8. Reflection.

Not all factors are independent; reflection at the earth's surface involves earth conductivity and dielectric constant for the region of reflection.

The effect of a factor depends mostly upon frequency. Some factors vary on a daily, seasonal, and year-by-year cycle. The effectiveness and the height of the various layers of the ionosphere have been directly related to the sunspot cycle.

Propagation at Frequencies Below the Broadcast Band

Two modes of transmission are important below 550 kilocycles. A wave traveling along the ground is called a *ground wave.* It is attenuated by ground losses and the earth's curvature. Conductivity of the ground along the ground path is most important. The higher the frequency, the more important it is that the earth's conductivity be high for minimum signal attenuation. Another wave, called the *sky-wave,* is propagated by alternate reflections from the ionosphere and the earth. Factors affecting the sky-wave are ion density, height of the ionosphere, and reflection coefficient of the ground at points of reflection from the earth. Reflection coefficient of the ground for these frequencies is largely dependent upon conductivity.

Propagation for the Broadcast Band [27]

In the broadcast band, primary coverage as specified by the Federal Communications Commission is made up solely of ground wave. There is virtually no sky-wave propagation in the daytime. Therefore, primary coverage is determined by power, frequency, and ground constants.

At night the ionosphere contributes to sky-wave propagation. If the radiating system radiates an appreciable amount of energy at high angles measured from the ground, the sky-wave is reflected by the ionosphere into a region having appreciable ground-wave coverage. There it adds with its characteristic varying and uncertain phase position to the ground wave producing fading. The sky-wave produces secondary coverage beyond the fading region, but it is relatively unreliable.

Ground-wave propagation is good at the low-frequency end of the broadcast band, but deteriorates near the high-frequency end of the band. Conductivity of the ground is important also. For two ground conductivities the primary coverage (2-millivolt-per-meter contour) for a 50-kilowatt station would be

Frequency (kilocycles)	Radius (miles)	
	For Conductivity $\sigma = 1 \times 10^{-13}$ emu	For Conductivity $\sigma = 1 \times 10^{-14}$ emu
550	148	42.5
700	112	33.5
1000	74	25.0
1600	44	17.5

During the day coverage is completely by ground wave. At night a region of ground-wave coverage is surrounded by a fading zone or at least a zone of inadequate signal. The fading zone is surrounded by a ring of sky-wave coverage.

Propagation at Frequencies from 1600 to 30,000 Kilocycles

Ground-wave propagation disappears rapidly as frequency is increased above 1600 kilocycles. At 7000 kilocycles the maximum ground-wave coverage radius is from 4 to 5 miles. From the limits of the ground-wave coverage outward from the station is a region of no signal, which is called the *skip-distance region*. Beyond is the sky-wave coverage region which depends entirely upon the ionosphere.

From ionosphere soundings taken all over the world it is possible to predict the maximum usable frequency for communication between any two locations for several months in advance, and in a general way even for years in advance. Such predictions are based on the sunspot cycle, the season of the year, the time of day, and the geographical position of the points in question. Higher frequencies are best in the daytime, and lower frequencies are best at night. Reliable, world-wide communication is now maintained on frequencies as high as 25 megacycles.

Propagation at Frequencies above 30,000 Kilocycles [25, 26]

Long-distance transmission at frequencies above 30,000 kilocycles is unreliable. Such transmission depends upon sky-wave propagation. Line-of-sight propagation usually is the only reliable type.

There are two waves in line-of-sight propagation; one is the direct wave or ray which traverses a straight-line path from the transmitting antenna to the receiving antenna; the other ray is reflected from the ground to the receiving antenna (refer to Fig. 20·45). The direct ray and the reflected ray

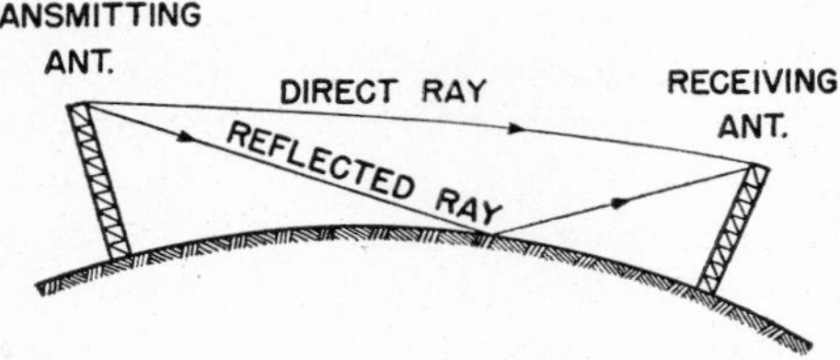

Fig. 20·45 Direct and reflected rays.

add vectorially at the receiving antenna. The phase of both rays depends upon the distance between the two antennas. In addition, the amplitude and phase of the reflected ray are dependent upon the reflection coefficient of the ground at the point of reflection. Reflection coefficient is a function of polarization, angle of incidence, conductivity, and dielectric constant. Conductivity is not important compared with the dielectric constant. For extremely high frequencies, conductivity may be ignored.

It is obvious that increasing antenna height will provide increasing coverage. Some diffraction around the earth increases propagation slightly beyond the line of sight. Under some conditions waves entering the troposphere are sent back to earth. Tropospheric waves generally are not constant. No ionospheric reflections are possible at frequencies higher than about 100 megacycles.

Normal conditions actually exist a small portion of the time. *Super-normal* and *sub-normal* conditions occur because of variations in the index of refraction of the earth's atmosphere for altitudes up to 1000 feet, and frequently there are sharp variations in the first 100 or 200 feet. Sometimes, depending upon the nature of the variation, the waves are *trapped* between a region of discontinuity in the index of refraction and the ground to produce almost free-space or super-normal fields far beyond the line of sight. For trapping to occur the antennas must be below the region of discontinuity. Sub-normal defines the conditions which cause the wave to be deflected out into space away from the earth, sometimes resulting in a complete *drop-out* of signal. Normal, super-normal, and sub-normal conditions become more noticeable for frequencies above 100 megacycles.

Prediction of these refractive conditions is almost impossible except in certain trade-wind regions, and refractive data for most paths is hard to obtain. Such refraction is influenced by temperature-height variations, relative-humidity variations, and variations of wind velocity with height.

Above 100 megacycles signal attenuation becomes more and more pronounced. Fog and dust particles serve to scatter and absorb the energy. Sometimes storm fronts reflect very high-frequency signals.

Weather, time of day, and conditions in the troposphere all affect propagation of high-frequency waves.

APPENDIX I [8]

An Example of Impedance and Phase-Shift Calculations for an Array Composed of Three Vertical Radiators

The antenna array of Fig. 20·46 has the field pattern of Fig. 20·47. To design equipment to feed power to the

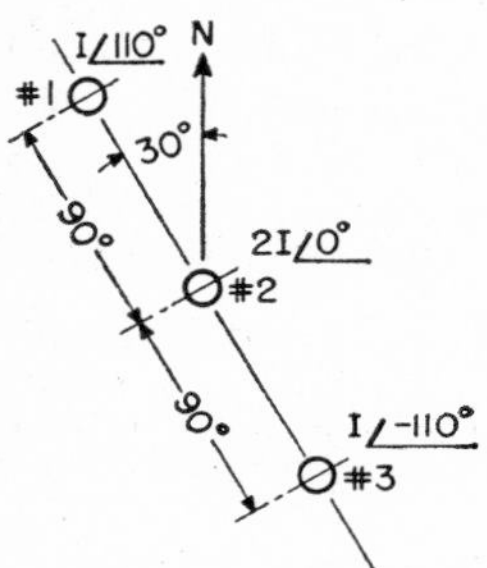

Fig. 20·46 Plan view of antenna array of three vertical radiators.

towers of such an array, the feed-point impedance of each tower and the phase relationship between the loop current and the current at the feed point for each tower must be known.

The following data apply to the array:

Radiated power	50 kilowatts
Frequency	1300 kilocycles (λ = 757 feet)
Towers	400 feet high (190°)
	6′ 6″ square cross section
	Vertical and base-insulated

Impedance of one tower (others not in place) is $140 - j203$ ohms at 1300 kilocycles

Spacing between towers 1 and 2 = 90° (189 feet)

Spacing between towers 2 and 3 = 90° (189 feet)

Spacing between towers 1 and 3 = 180° (378 feet)

Tower 1 relative current (loop) = $I\underline{/110^\circ}$

Tower 2 relative current (loop) = $2I\underline{/0^\circ}$

Tower 3 relative current (loop) = $I\underline{/-110^\circ}$

The first step is to determine the effects of the imperfect ground as shown by the circuit of Fig. 20·10. The modified base impedance of one tower alone is

$$\mathbf{Z}_{BM} = 140 - j203 \text{ ohms}$$

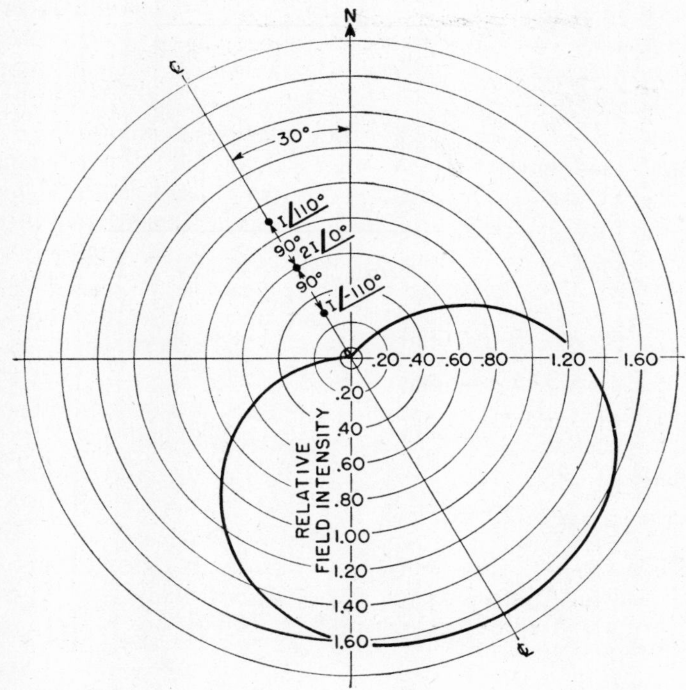

FIG. 20·47 Horizontal-plane radiation diagram of array of Fig. 20·46.

Calculate the base impedance (unmodified) by the use of equations 20·37 to 20·46. (Figs. 20·11 to 20·16 may be used in place of several of the equations.)

$$K_1 = 60 \log_e \frac{h}{\rho} \quad \text{(from equation 20·45)}$$

$$\text{Average } \rho = 3.93 \text{ feet}$$

$$h = 400 \text{ feet}$$

$$K_1 = 60 \log_e \frac{400}{3.93} = 60 \log_e 102 = 277.2$$

(refer to Fig. 20·13)

$$K_2 = -\left(60 + 30 \log_e \frac{2h}{\lambda}\right) \quad \text{(from equation 20·46)}$$

$$= -\left(60 + 30 \log_e \frac{800}{757}\right) = -61.7$$

(refer to Fig. 20·14)

$$Z_0 = K_1 + K_2 = 277.2 - 61.7 = 215.5 \text{ ohms}$$

(from equation 20·44)

For $h \geqq 0.2\lambda$

$$R_0 \cong 15\left[-\frac{\pi}{2}\sin 4\pi\frac{h}{\lambda} + \left(\log_e \frac{2h}{\lambda} + 1.722\right)\cos 4\pi\frac{h}{\lambda} + 2\left(2.415 + \log_e \frac{2h}{\lambda}\right)\right] \quad \text{(from equation 20·43}a\text{)}$$

$$\frac{4\pi h}{\lambda} \text{ radians} = 380°$$

$$\frac{2h}{\lambda} = 1.056$$

$$R_0 \cong 15\left[-\frac{\pi}{2}\sin 380° + (\log_e 1.056 + 1.722)\cos 380° + 2(2.415 + \log_e 1.056)\right] = 90.8 \text{ ohms}$$

(refer to Fig. 20·11)

$$rl = \frac{4R_0}{1 - \dfrac{\sin 4\pi \frac{h}{\lambda}}{4\pi \frac{h}{\lambda}}} \quad \text{(from equation 20·41)}$$

$$= \frac{4 \times 90.8}{1 - \dfrac{\sin 380°}{6.64}} = 383 \text{ ohms} \quad \text{(refer to Fig. 20·12)}$$

$$\alpha l = \frac{rl}{2Z_0} \quad \text{(from equation 20·40)}$$

$$= \frac{383}{2 \times 215.5} = 0.888 \text{ neper}$$

$$R = Z_0 \frac{\sinh \alpha l - \dfrac{\alpha l}{4\pi \frac{h}{\lambda}} \sin 4\pi\frac{h}{\lambda}}{\cosh \alpha l - \cos 4\pi \frac{h}{\lambda}} \quad \text{(from equation 20·38)}$$

$$X = -Z_0 \frac{\sin 4\pi\frac{h}{\lambda} + \dfrac{\alpha l}{4\pi \frac{h}{\lambda}} \sinh \alpha l}{\cosh \alpha l - \cos 4\pi \frac{h}{\lambda}} \quad \text{(from equation 20·39)}$$

$$R = 215.5 \frac{\sinh 0.888 - \dfrac{0.888}{6.64}\sin 380°}{\cosh 0.888 - \cos 380°} = 431 \text{ ohms}$$

(refer to Fig. 20·15)

$$X = -215.5 \frac{\sin 380° + \dfrac{0.888}{6.64}\sinh 0.888}{\cosh 0.888 - \cos 380°} = -209 \text{ ohms}$$

(refer to Fig. 20·16)

Then find L and C that, according to Fig. 20·10, will convert the unmodified base impedance $\mathbf{Z}_B$ to the modified base impedance $\mathbf{Z}_{BM}$, which is known to be 140 − j203 ohms.

The results are C = 275 micromicrofarads ($-jX_c = -j445$ ohms) and L = 3.0 microhenrys ($+jX_L = +j24.5$ ohms).

R_0 is 90.8 ohms for one of the towers alone; therefore

$$R_{11} = R_{22} = R_{33} = 90.8 \text{ ohms}$$

and

$$\mathbf{Z}_{11} = 90.8 + jX_{11}$$

$$\mathbf{Z}_{22} = 90.8 + jX_{22}$$

$$\mathbf{Z}_{33} = 90.8 + jX_{33}$$

X_{11}, X_{22}, X_{33} are self-loop reactances and are not yet known.

From Figs. 20·20 and 20·21, find

$\mathbf{Z}_{13} = \mathbf{Z}_{31} = 40.5\underline{/-122°}$ ohms (190° towers spaced 180°)

$\mathbf{Z}_{12} = \mathbf{Z}_{21} = 61.0\underline{/-40.8°}$ ohms (190° towers spaced 90°)

$\mathbf{Z}_{23} = \mathbf{Z}_{32} = 61.0\underline{/-40.8°}$ ohms (190° towers spaced 90°)

From the initial data for loop current

$$\mathbf{I}_1 = I\underline{/110°}$$
$$\mathbf{I}_2 = 2I\underline{/0°}$$
$$\mathbf{I}_3 = I\underline{/-110°}$$

Equation 20·66 applied to three towers becomes

$$\mathbf{Z'}_1 = R_{11} + jX_{11} + \frac{\mathbf{I}_2}{\mathbf{I}_1}\mathbf{Z}_{12} + \frac{\mathbf{I}_3}{\mathbf{I}_1}\mathbf{Z}_{13}$$

$$\mathbf{Z'}_2 = \frac{\mathbf{I}_1}{\mathbf{I}_2}\mathbf{Z}_{21} + R_{22} + jX_{22} + \frac{\mathbf{I}_3}{\mathbf{I}_2}\mathbf{Z}_{23}$$

$$\mathbf{Z'}_3 = \frac{\mathbf{I}_1}{\mathbf{I}_3}\mathbf{Z}_{31} + \frac{\mathbf{I}_2}{\mathbf{I}_3}\mathbf{Z}_{32} + R_{33} + jX_{33}$$

Substituting, the apparent loop impedances become, according to equation 20·67,

$$\begin{aligned}\mathbf{Z'}_1 &= 90.8 + jX_{11} + 2\underline{/-110°} \times 61\underline{/-40.8°} + 1\underline{/-220°} \\ &\quad \times 40.5\underline{/-122°} \\ &= 22.8 + j(X_{11} - 47.0) \text{ ohms} \\ &= R_1 + j(X_{11} + X'_1) \\ \mathbf{Z'}_2 &= 0.5\underline{/110°} \times 61\underline{/-40.8°} + 90.8 + jX_{22} + 0.5\underline{/-110°} \\ &\quad \times 61\underline{/-40.8°} \\ &= 75.0 + j(X_{22} + 13.6) \text{ ohms} \\ &= R_2 + j(X_{22} + X'_2) \\ \mathbf{Z'}_3 &= 1\underline{/220°} \times 40.5\underline{/-122°} + 2\underline{/110°} \times 61\underline{/-40.8°} \\ &\quad + 90.8 + jX_{33} \\ &= 128.5 + j(X_{33} + 154) \text{ ohms} \\ &= R_3 + j(X_{33} + X'_3)\end{aligned}$$

From equation 20·68,

$$(rl)_n = \frac{4R_{nn}}{1 - \dfrac{\sin 4\pi \dfrac{h_n}{\lambda}}{4\pi \dfrac{h_n}{\lambda}}}$$

therefore

$$(rl)_1 = 96.1$$
$$(rl)_2 = 316.5$$
$$(rl)_3 = 543$$

Z_0 for one tower is 215.5 ohms. Therefore, since the towers are identical,

$$Z_0 = Z_{01} = Z_{02} = Z_{03} = 215.5 \text{ ohms}$$

$(\alpha l)_n$ is determined from equation 20·69 thus:

$$(\alpha l)_n = \frac{(rl)_n}{2Z_0}$$

$$(\alpha l)_1 = 0.223 \text{ neper}$$
$$(\alpha l)_2 = 0.733 \text{ neper}$$
$$(\alpha l)_3 = 1.259 \text{ nepers}$$

From equation 20·38 and 20·39 or from Figs. 20·15 and 20·16,

$$R_{nB} = Z_0 \frac{\sinh (\alpha l)_n - \dfrac{(\alpha l)_n}{4\pi \dfrac{h_n}{\lambda}} \sin 4\pi \dfrac{h_n}{\lambda}}{\cosh (\alpha l)_n - \cos 4\pi \dfrac{h_n}{\lambda}}$$

$$X_{nB} = -Z_0 \frac{\sin \left(4\pi \dfrac{h_n}{\lambda}\right) + \dfrac{(\alpha l)_n}{4\pi \dfrac{h_n}{\lambda}} \sinh (\alpha l)_n}{\cosh (\alpha l)_n - \cos 4\pi \dfrac{h_n}{\lambda}}$$

$$R_{1B} = 535 \text{ ohms}; \quad X_{1B} = -884 \text{ ohms}$$
$$R_{2B} = 479 \text{ ohms}; \quad X_{2B} = -272 \text{ ohms}$$
$$R_{3B} = 345 \text{ ohms}; \quad X_{3B} = -145.5 \text{ ohms}$$

therefore

$$X'_{1B} = -\frac{535}{22.8} \times -47 = 1102 \text{ ohms}$$

$$X'_{2B} = -\frac{479}{75} \times 13.6 = -86.5 \text{ ohms}$$

$$X'_{3B} = -\frac{345}{128.5} \times 154 = -414 \text{ ohms}$$

From equation 20·67,

$$\mathbf{Z}_{nB} = R_{nB} + j(X_{nB} + X'_{nB})$$

Therefore the unmodified base impedances are

$$\mathbf{Z}_{1B} = 535 + j(-884 + 1102) = 535 + j218.5 = 578\underline{/22.2°}$$

$$\begin{aligned}\mathbf{Z}_{2B} &= 479 + j(-272 - 86.5) = 479 - j358.5 \\ &= 599\underline{/-36.9°}\end{aligned}$$

$$\begin{aligned}\mathbf{Z}_{3B} &= 345 + j(-145.5 - 414) = 345 - j559 \\ &= 657\underline{/-58.3°}\end{aligned}$$

These impedances are to be modified according to the circuit of Fig. 20·10 by a shunt capacitance of 275 micromicrofarads ($-j445$) and a series inductance of 3.0 microhenrys ($+j24.5$); therefore the modified impedances are

$$\begin{aligned}\mathbf{Z}_{1BM} &= 578\underline{/22.2°} \times \frac{445\underline{/-90°}}{584\underline{/-22.9°}} + j24.5 = 312 - j286 \\ &= 425\underline{/-42.4°} \text{ ohms}\end{aligned}$$

$$\begin{aligned}\mathbf{Z}_{2BM} &= 599\underline{/-36.9°} \times \frac{445\underline{/-90°}}{936\underline{/-59.2°}} + j24.5 = 108.3 \\ &\quad - j264 = 285\underline{/-67.7°} \text{ ohms}\end{aligned}$$

$$\begin{aligned}\mathbf{Z}_{3BM} &= 657\underline{/-58.3°} \times \frac{445\underline{/-90°}}{1062\underline{/-71.1°}} + j24.5 = 60.6 \\ &\quad - j243 = 251\underline{/-76.0°} \text{ ohms}\end{aligned}$$

The phase relations between the feed current and the loop current of each tower are valuable in designing equipment to feed such an array. Inspection of the foregoing equations shows that

Tower 1 modified feed current leads feed current by 67.1 degrees
Tower 2 modified feed current leads feed current by 30.8 degrees
Tower 3 modified feed current leads feed current by 18.9 degrees

By means of equation 20·74 and Fig. 20·22 the phase angle between feed current and loop current can be computed.

$$\mathbf{I}_{nB} = \mathbf{I}_{0n} \frac{\cosh (\alpha h)_n}{\cosh \frac{\alpha\lambda}{4}} [\sin (\beta h)_n - j \tanh (\alpha h)_n \cos (\beta h)_n]$$

where

$$(\alpha h)_n = \frac{(\alpha l)_n}{2}$$

Now

$$\mathbf{I}_{1B} = \mathbf{I}_1 \frac{\cosh 0.1115}{\cosh 0.0528} [\sin 190° - j \tanh 0.1115 \cos 190°]$$

$$= \mathbf{I}_1 \times 0.206\underline{/147.8°}$$

$$\mathbf{I}_{2B} = \mathbf{I}_2 \times 0.407\underline{/116.7°}$$

$$\mathbf{I}_{3B} = \mathbf{I}_3 \times 0.665\underline{/107.5°}$$

Therefore

Tower 1 feed current leads loop current by 147.8 degrees
Tower 2 feed current leads loop current by 116.7 degrees
Tower 3 feed current leads loop current by 107.5 degrees

and

Tower 1 modified feed current leads loop current by 147.8 + 67.1 = 214.9 degrees
Tower 2 modified feed current leads loop current by 116.6 + 30.8 = 147.4 degrees
Tower 3 modified feed current leads loop current by 107.5 + 18.9 = 126.4 degrees

The magnitude of base currents and the power to each tower can be computed thus:

$$\text{Total power} = 50 \text{ kilowatts}$$

$$R_1 = 22.8 \text{ ohms} \qquad \mathbf{I}_1 = I \underline{/\ 110°}$$

$$R_2 = 75.0 \text{ ohms} \qquad \mathbf{I}_2 = 2I \underline{/\ 0°}$$

$$R_3 = 128.5 \text{ ohms} \qquad \mathbf{I}_3 = I \underline{/-110°}$$

$$P_1 = 22.8I^2 \text{ watts} \qquad P_1 = 2{,}525 \text{ watts}$$

$$P_2 = 4 \times 75 \times I^2 \text{ watts} \qquad P_2 = 33{,}260 \text{ watts}$$

$$P_3 = 128.5I^2 \text{ watts} \qquad P_3 = 14{,}230 \text{ watts}$$

$$P_T = 451.3I^2 = 50{,}000 \text{ watts}$$

$$I = 10.53 \text{ amperes}$$

$$R_{1BM} = 312 \text{ ohms} \qquad I_{1BM} = \sqrt{\frac{2525}{312}} = 2.85 \text{ amperes}$$

$$R_{2BM} = 108.3 \text{ ohms} \qquad I_{2BM} = \sqrt{\frac{33{,}260}{108.3}} = 17.2 \text{ amperes}$$

$$R_{3BM} = 60.6 \text{ ohms} \qquad I_{3BM} = \sqrt{\frac{14{,}230}{60.6}} = 15.3 \text{ amperes}$$

where I_{1BM}, I_{2BM}, and I_{3BM} are the modified feed currents to each tower.

Complete information about the array is

$$\mathbf{Z}_{1BM} = 312 - j286 = 445\underline{/-42.4°} \text{ ohms;}$$

$$I_{1BM} = 2.85 \text{ amperes;} \quad P_1 = 2{,}525 \text{ watts}$$

Tower 1 modified feed current leads loop current by 214.9 degrees:

$$\mathbf{Z}_{2BM} = 108.3 - j264 = 285\underline{/-67.7°} \text{ ohms;}$$

$$I_{2BM} = 17.2 \text{ amperes;} \quad P_2 = 33{,}260 \text{ watts}$$

Tower 2 modified feed current leads loop current by 147.4 degrees:

$$\mathbf{Z}_{3BM} = 60.6 - j243 = 251\underline{/-76.0°} \text{ ohms;}$$

$$I_{3BM} = 15.3 \text{ amperes;} \quad P_3 = 14{,}230 \text{ watts}$$

Tower 3 modified feed current leads loop current by 126.4 degrees.

APPENDIX II

Analysis of Current Distribution on a Tower Radiator [9]

Given: a tower of height h, of ρ electrical radius, operated on a wavelength of λ. From equations 20·44 to 20·46 or equation 20·44 and Figs. 20·13 and 20·14, establish Z_0.

R_0 is established by equations 20·43*a*, 20·43*b*, or Fig. 20·11; rl is established by equation 20·41 or Fig. 20·12:

$$\alpha l = \frac{rl}{2Z_0}$$

$$l = 2h$$

$$\therefore \alpha h = \frac{\alpha l}{2}$$

If the tower is assumed to be a transmission line having a propagation constant $\boldsymbol{\gamma} = \alpha + j\beta$, where α is the attenuation per unit of h, and β is the phase shift per unit of h and having a $\mathbf{Z}_0 \cong Z_0\left(1 - j\frac{\alpha h}{\beta h}\right)$, the current distribution can be determined.

Basic transmission line equations state

$$\mathbf{I}_z = \frac{\mathbf{E}_R}{\mathbf{Z}_0} \sinh \boldsymbol{\gamma}(h - z)$$

where z is measured from the base and is in the same dimensions as h, and where $\mathbf{E}_R$ is the voltage at the top of the antenna where $z = h$.

$$\mathbf{I}_B = \frac{\mathbf{E}_R}{\mathbf{Z}_0} \sinh \boldsymbol{\gamma} h$$

$$\therefore \frac{\mathbf{E}_R}{\mathbf{Z}_0} = \frac{\mathbf{I}_B}{\sinh \boldsymbol{\gamma} h}$$

Substituting into the transmission-line equation for $\mathbf{I}_z$,

$$\mathbf{I}_z = \frac{\mathbf{I}_B}{\sinh \boldsymbol{\gamma} h} \sinh \boldsymbol{\gamma}(h - z)$$

But $\mathbf{I}_z = \mathbf{I}_0$ when $h - z = \lambda/4$. Therefore

$$\mathbf{I}_0 = \frac{\mathbf{I}_B}{\sinh \boldsymbol{\gamma} h} \sinh \boldsymbol{\gamma} \frac{\lambda}{4}$$

where $\mathbf{I}_0$ is the current at the loop nearest the top of the antenna.

$$\frac{\mathbf{I}_z}{\mathbf{I}_0} = \frac{\cancel{\mathbf{I}_B}}{\cancel{\sinh \boldsymbol{\gamma} h}} \sinh \boldsymbol{\gamma}(h - z) \frac{\cancel{\sinh \boldsymbol{\gamma} h}}{\cancel{\mathbf{I}_B} \sinh \boldsymbol{\gamma} \frac{\lambda}{4}} = \frac{\sinh \boldsymbol{\gamma}(h - z)}{\sinh \boldsymbol{\gamma} \frac{\lambda}{4}}$$

But since $\boldsymbol{\gamma} = \alpha + j\beta$,

$$\frac{\mathbf{I}_z}{\mathbf{I}_0} = \frac{\sinh \alpha(h-z) \cos \beta(h-z) + j \cosh \alpha(h-z) \sin \beta(h-z)}{0 \cdot \sinh \frac{\alpha\lambda}{4} + j \cosh \frac{\alpha\lambda}{4}}$$

$$\therefore \mathbf{I}_z = \mathbf{I}_0 \frac{\cosh \alpha(h-z)}{\cosh \frac{\alpha\lambda}{4}} [\sin \beta(h-z) - j \tanh \alpha(h-z) \cos \beta(h-z)]$$

which represents the current at any point z on the tower.

$$\mathbf{I}_B = \mathbf{I}_0 \frac{\cosh \alpha h}{\cosh \frac{\alpha\lambda}{4}} [\sin \beta h - j \tanh \alpha h \cos \beta h]$$

which represents base current (unmodified) in terms of loop current.

REFERENCES

1. *Fundamentals of Electric Waves*, H. H. Skilling, Wiley, 1942.
2. "Radiation and Radiating Systems," W. L. Everitt, a set of notes for classroom use at Ohio State University, 1938–1939.
3. "Development of Antennas," P. S. Carter, C. W. Hansell, and N. E. Lindenblad, *Proc. I.R.E.*, Vol. 19, No. 10, Oct. 1931, p. 1773.
4. "Directional Antennas," G. H. Brown, *Proc. I.R.E.*, Vol. 25, No. 1, Part II, Jan. 1937, p. 78.
5. "Scheinwiderstand von Antennen," E. Siegel and J. Labus, *Hochfrequenztechnik und Elektroakustik*, Vol. 43, No. 5, May 1934, pp. 166–172.
6. "The Shunt Excited Antenna," J. F. Morrison and P. H. Smith, *Proc. I.R.E.*, Vol. 25, No. 6, June 1937, pp. 673–696.
7. "Scheinwiderstand von beschwerten Antennen," E. Siegel, *Hochfrequenztechnik und Elektroakustik*, Vol. 43, No. 5, May 1934, pp. 172–176.
8. "Directional Antenna Systems," T. M. Bloomer, presented before the Baltimore Chapter of the Institute of Radio Engineers, Feb. 20, 1942.
9. "Broadcast Antennas, Some Notes on Their Design Computations," J. F. Morrison, *Pick-Ups*, July 1939, p. 12.
10. "Tower Antennas for Broadcast Use," H. E. Gihring and G. H. Brown, *Proc. I.R.E.*, Vol. 23, No. 4, Part II, April 1935, pp. 342–348.
11. *Communication Engineering*, W. L. Everitt, McGraw-Hill, 2nd ed., 1937.
12. "The Principle of Reciprocity in Antenna Theory," M. S. Neiman, *Proc. I.R.E.*, Vol. 31, No. 12, Dec. 1943, p. 666.
13. "Certain Factors Affecting the Gain of Directive Antennas," G. C. Southworth, *Proc. I.R.E.*, Vol. 31, No. 9, Sept. 1930, pp. 1502–1536.
14. "Experimentally Determined Impedance Characteristics of Cylindrical Antennas," G. H. Brown and O. M. Woodward, Jr., *Proc. I.R.E.*, Vol. 33, No. 4, April 1945, pp. 257–262.
15. "Ground Systems as a Factor in Antenna Efficiency," G. H. Brown, R. F. Lewis, and J. Epstein, *Proc. I.R.E.*, Vol. 25, No. 6, June 1937, pp. 753–787.
16. "A Critical Study of the Characteristics of Broadcast Antennas as Affected by Antenna Current Distribution," G. H. Brown, *Proc. I.R.E.*, Vol. 24, No. 1, Jan. 1936, pp. 48–81.
17. "Field Distribution and Radiation Resistance of a Straight Unloaded Antenna Radiator at One of Its Harmonics," S. A. Levin and C. J. Young, *Proc. I.R.E.*, Vol. 14, No. 5, Oct. 1926, p. 675.
18. *Radio Engineers' Handbook*, F. E. Terman, McGraw-Hill, 1943.
19. *The ARRL Antenna Book*, The American Radio Relay League, Inc., 3rd ed., 2nd printing, April 1945.
20. "Horizontal Rhombic Antennas," E. Bruce, A. C. Beck, and L. R. Lowry, *Proc. I.R.E.*, Vol. 23, No. 1, Jan. 1935, p. 24.
21. "Radiation from Rhombic Antennas," Donald Foster, *Proc. I.R.E.*, Vol. 25, No. 10, Oct. 1937, p. 1327.
22. "The Turnstile Antenna," G. H. Brown, *Electronics*, Vol. IX, April 1936, p. 14.
23. "The Corner-Reflector Antenna," J. D. Krause, *Proc. I.R.E.*, Vol. 28, No. 11, Nov. 1940, p. 513.
24. "Ground Wave Propagation," K. A. Norton, presented at the Fourth Annual Broadcast Engineering Conference, Feb. 10–21, 1941 (FCC 47475).
25. "Before the Federal Communications Commission Hearing in the Matter of Aural Broadcasting on Frequencies above 25,000 Kilocycles, March 18, 1940, Report by K. A. Norton on A Theory of Tropospheric Wave Propagation" (FCC 40003).
26. "Before the Federal Communications Commission, Television Hearing, January 15, 1940, Summary of Statement by K. A. Norton on Ultra-High Frequency Wave Propagation" (FCC 38521).
27. "Standards of Good Engineering Practice Concerning Standard Broadcast Stations (550–1600 kc.)," U. S. Govt. Printing Office, Washington, 1940.
28. "Simple Television Antennas," P. S. Carter, *RCA Rev.*, Vol. 4, Oct. 1939, pp. 168–185.
29. "Theory of Antennas of Arbitrary Size and Shape," S. A. Schelkunoff, *Proc. I.R.E.*, Vol. 29, No. 9, Sept. 1941, pp. 493–521.
30. "The Design of Broad-Band Aircraft-Antenna Systems," F. D. Bennett, P. D. Coleman, and A. S. Meier, *Proc. I.R.E.*, Vol. 33, No. 10, Oct. 1945, p. 671.
31. "FM Circular Antenna," M. W. Scheldorf, *G. E. Rev.*, March 1943, reprint by General Electric Co. GEA-4095.

Chapter 21

GENERAL REQUIREMENTS OF RECTIFIER APPLICATIONS

C. R. Marcum

THE American Institute of Electrical Engineers has proposed the following definitions of rectifying device, rectifier, and rectifier unit.

1. "A rectifying device is an elementary device, consisting of one anode and its cathode, which has the characteristic of conducting current effectively in only one direction."
2. "A rectifier is an integral assembly of one or more rectifying devices."
3. "A rectifier unit is an operative assembly consisting of the rectifier(s), the rectifier auxiliaries, the rectifier transformer equipment, and the essential switchgear."

Rectifying devices and rectifier units have been developed to meet the requirements of specific applications. The high cost of development and the economy of mass production necessitate that as few types of rectifying devices and rectifier units as possible be used to meet the requirements of all rectifier applications. For this reason, it is the usual practice to choose the rectifying device and the rectifier unit for an application from existing designs. Table 21·1 lists various types of rectifying devices and some of their most common applications.[1, 2, 3]

TABLE 21·1 ELECTRONIC RECTIFYING DEVICES AND THEIR APPLICATIONS

Tube Name	Type of Tube	Applications	Examples
Kenotron	Thermionic cathode, vacuum, diode	Low current, high voltage	X-ray, electrostatic precipitation
Rectigon (Tungar)	Thermionic cathode, gas, diode	Moderate current, low voltage	Battery charging
Phanotron	Thermionic cathode, gas, diode	Moderate current, up to 22,000 volts	Radio transmitters, industrial applications
Thyratron	Thermionic cathode, gas, triode or tetrode	Moderate current, up to 22,000 volts	Motor control, a-c voltage control, timing circuits, low-power d-c supply, electronic exciter
Mercury-arc rectifier (multianode)	Pool cathode, gas, continuously excited		Power rectifiers and inverters for industrial, electrochemical, and railway service
Excitron (single-anode)	Pool cathode, gas, continuously excited	High current, up to 20,000 volts	
Ignitron (single-anode)	Pool cathode, gas, synchronously excited		

21·1 THERMIONIC-CATHODE VACUUM TUBES

Vacuum tubes with thermionic cathodes have a high arc drop since there are few positive ions in the tube during conduction to neutralize the space charge caused by the large number of electrons flowing from cathode to anode. For this reason, vacuum tubes have a low efficiency. However, for the same reason, a grid may be used to change the characteristics of the tube during its conduction period. This type of rectifying device is usually used on low-power, low-current applications where control is more important than efficiency. High-vacuum tubes are built for hundreds of thousands of volts.

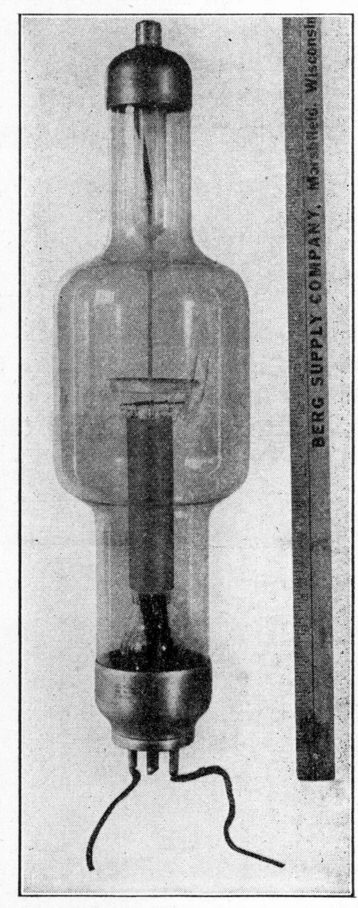

FIG. 21·1 Photograph of a kenotron rated 0.24 ampere average with a peak inverse voltage of 15,000 volts.

Kenotron

Kenotrons are high-vacuum, diode, rectifying devices used on high-voltage, low-current applications. See Fig. 21·1. Kenotrons are built in ratings ranging from 230,000 peak inverse volts at 30 milliamperes average anode current down to 1500 peak inverse volts at 3 milliamperes average anode current. The former is designed for current control by varying anode voltage. The latter is designed for current control by varying the filament voltage.

Kenotrons are used in power supplies for electrostatic precipitators, for high-voltage testing of materials, and for x-ray equipment. Since there is not the danger of glow discharge and possible breakdown that would be encountered with a gas tube and since the tungsten cathode can withstand the ion bombardment, the kenotron is best suited for such high-voltage applications. In these applications, the relatively high arc drop is not a disadvantage since it is such a small percentage of the d-c voltage. The arc drop due to space charge offers a partial, although not a complete, protection for the tube when applied on electrostatic precipitators where the accumulation of dust often causes short circuits on the rectifier.

The above-mentioned applications normally require a low-power rectifier; therefore, it is more economical to use the simplicity of the half-wave and full-wave rectifier circuits and provide filters than to go to the complexity of polyphase circuits.

At one time, it was common practice to use mechanical rectifiers for such high-voltage applications. Now, kenotron rectifiers are used because of improved wave form, increased efficiency, and decreased maintenance.

21·2 THERMIONIC-CATHODE GAS TUBES

Since the positive ions in a gas tube neutralize the space charge of the electrons, the tube has a low arc drop but is limited in current capacity by the electron-emitting ability of the cathode. A grid may be used to control the initiation of conduction, but the quantity of grid power required to control the tube becomes prohibitively large when there is ionized gas in the presence of the grid. The grid-controlled tube must be used only where circuit voltages allow sufficient time between conduction periods for the tube to deionize and the grid to regain control.

Rectigon (Tungar)

Rectigons are high-pressure-gas, diode, rectifying devices used on low-voltage applications requiring moderate currents.

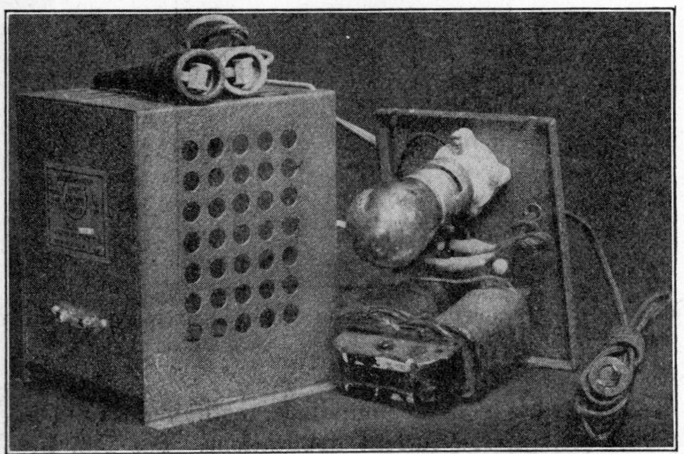

FIG. 21·2 Photograph of a Rectigon battery charger.

With a pressure of approximately 6 centimeters of argon or some other inert gas, the space charge is reduced to a minimum, and an arc drop as low as five volts results. The tubes are designed for a peak inverse voltage of approximately 375 volts. They have a tungsten filament which may be energized at the same time the anode is energized.

The main application of this tube is for battery-charging service. Since the tube does not require a time-delay relay for starting purposes and has a relatively high efficiency at low voltages, one or two tubes in a half-wave or full-wave rectifier circuit provide a very economical battery charger. See Fig. 21·2.

Phanatron

Phanatrons are low-pressure-gas, diode, rectifying devices used on applications requiring voltages up to 20,000 volts and moderate currents. Since the tube is filled with mercury vapor or an inert gas at a low pressure, the arc drop is low. If the tube is filled with neon, the arc drop is approximately 30 volts. If mercury vapor is used, the arc drop is approximately 13 volts. Phanatrons are manufactured with average current ratings ranging from 0.25 ampere to 10 amperes and peak inverse voltage ratings ranging from 1000 volts to 22,000 volts.

Phanatrons are used for d-c power supplies where no voltage control other than adjusting the transformer voltage is required. The most common applications are radio transmitters and industrial d-c power. Phanatrons are rugged tubes requiring little maintenance. The temperature at the base must be maintained between 25 and 70 degrees centigrade to obtain the desired mercury vapor pressure in the tube.

For low-power units, a half-wave or full-wave rectifier circuit is used. As the power ratings increase, the number of phases in the rectifier circuit is increased. For radio transmitters, it is common practice to use filters. Because of the mercury in the tube, it is necessary to have a time delay between the time of energizing the filament and energizing the anode. The arc drop remains practically constant with changing load. Since very high currents are encountered during short circuits, protection must be supplied to prevent damage to the coated cathode.

Thyratrons

Thyratrons are low-pressure-gas triodes or tetrodes capable of delivering moderate currents at peak inverse voltages up to 22,000 volts. See Fig. 21·3. Thyratrons are manufactured for average current ratings ranging from a few milliamperes to 12.5 amperes. Except for the availability of grid control, the application features of the thyratron are the same as those for the phanatron.

The high efficiency and range of ratings of thyratrons and the ability to control their initiation of conduction with very little power make them the best tube for control circuits and for d-c power supplies where voltage control is desirable.

With only a-c power available, a variable-speed d-c motor may be obtained by using thyratron rectifiers to supply the motor armature and field circuits.[4] See Fig. 21·4. Speeds below rated speed may be obtained by controlling the grids of the thyratron in the rectifier that supplies the armature. Speeds above rated speed may be obtained by controlling the grids of the field rectifier.

By using a saturating reactor with a thyratron rectifier supplying the d-c coil, stepless control of a-c voltages may be obtained. Two of the most common applications are electric-furnace temperature control and dimming of theater lights.[5] The a-c coil of the reactor is connected in series with the

Fig. 21·3 Photograph of a thyratron rated 2.5 amperes average with a peak inverse voltage of 1500 volts.

load. Either manually or automatically, the grids of the thyratrons are controlled to give the desired degree of saturation in the reactor. When the reactor is completely saturated, all the a-c voltage appears across the load. When the reactor is unsaturated, the a-c voltage divides between the

Fig. 21·4 Photograph of a Mot-o-trol.

reactor and the load. Temperature-indicating equipment that will deliver very little power may be used to control the input to a furnace by working on the grids of the thyratron. The ease of control of the thyratron rectifier has promoted its use as an electronic exciter for synchronous motors and generators.

Thyratrons have characteristics suitable for many types of timing circuits. A sensitive starting relay in d-c circuits, the control of stored energy welding circuits, and the timing of excitation circuits for ignitrons are some typical applications.[6]

21·3 POOL-CATHODE GAS TUBES

Pool-cathode gas tubes differ in characteristics from the thermionic cathode gas tube only in electron-emitting ability of the cathode. The only cathode in practical use today is the mercury cathode which can emit an unlimited number of electrons without damage to itself.

Pool-cathode tubes are used for the conversion of large blocks of power. Except where regeneration is required, mercury-arc rectifier units have become widely used at ratings from 75 up to 8500 kilowatts at d-c voltages from

Fig. 21·5 Photograph of a portable ignitron rectifier for mining service.

125 to 3000 volts. Tubes have been developed to withstand an inverse voltage of 22,000 volts.

Some of the most common uses of mercury-arc rectifiers are for rectification to provide power for trains, street railways, subways, mining equipment, manufacturing industries, and electrochemical processes.[7, 8, 9] See Fig. 21·5.

In electrochemical service, the rectifier has a load factor very near to unity. Mercury-arc rectifiers, in electrochemical service at voltages above 250 volts d-c, have the highest efficiency of any power-conversion equipment in common use today. For 250-volt service, at full load, the efficiency of the mercury-arc rectifier is about equal to that of a synchronous converter. Mercury-arc rectifiers are well suited for such constant load applications, since the best operation can be obtained when all the parts of the rectifier are operating at their optimum temperatures. This condition can be easily obtained on units operating with a constant load.

The largest rectifier installations in operation are in electrochemical service. On large installations, the rectifier load is usually a large portion of the a-c system capacity, resulting in telephone-interference problems. Such problems can be minimized by using a large number of phases in the rectifier transformers. (See Chapter 22.) Where many rectifiers are operated in parallel, the back feed through a rectifier that arcs back is so great that high-speed switchgear is necessary to protect the unit. The reduced noise and maintenance and the elimination of vibration problems, compared to those of other types of converters, make the mercury-arc rectifier very desirable.

Railroad, street railway, and mining service require that the rectifiers operate with a load factor less than unity and carry heavy overloads for short times. In such applications, the rectifier may operate a majority of the time at light loads. During rush hours, the load may be greater than full load with an occasional swing as high as 200 or 300 percent of full load. The ability of mercury-arc rectifiers to withstand heavy overloads makes them well suited for this service.

Since the noise of a mercury-arc rectifier is limited to the hum of the transformer, rectifiers are advantageous in street-railway applications where the sub-stations are often located in residential sections of the city. To obtain the advantages of the low maintenance requirements, many sub-stations are unattended and automatically controlled.

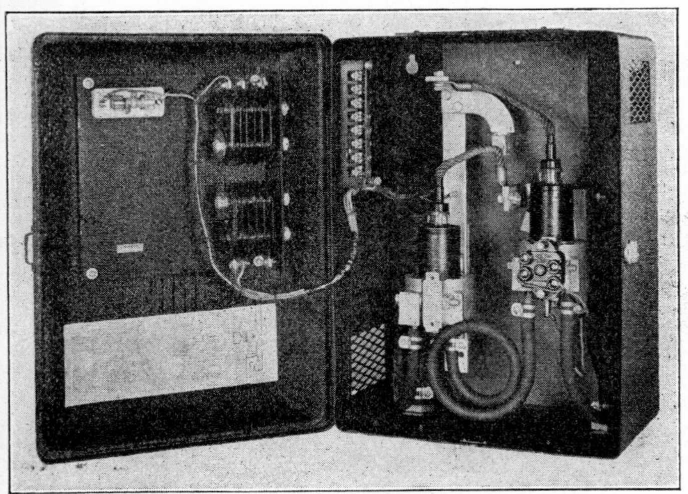

Fig. 21·6 Photograph of a Weld-o-trol using ignitrons.

If the rectifier is allowed to cool to a low temperature during light load periods and suddenly is required to deliver a heavy overload, it is possible that the ionization in the tube will be insufficient to carry the current. In such cases, the current flow will stop momentarily, causing a surge in the transformer winding. For such applications, it is necessary to have surge protective equipment for the transformer. It is common practice to limit the lowest temperature of the rectifier to such a value that surges do not occur. Where this is done, it may be possible for the tube cooling water to be considerably warmer than the surrounding air. During no-load periods, the anodes will become the coolest element in the tube and will collect mercury condensation. Mercury on the anode or insulating bushing may cause an arc-back. For these applications, anode heaters are required to maintain the anode temperature near or above the tube-wall temperature.

Mercury-arc rectifiers are used in many industrial plants and steel mills to provide d-c power for cranes, machine tools, auxiliary drives, and rolls. In industrial service, the rectifier normally operates with a low load factor with occasional moderate overloads. For conversion equipment which normally operates below full load, the mercury-arc rectifier has by far the highest efficiency. As the load drops below full load, the efficiency of rotating apparatus decreases very rapidly, while the rectifier efficiency drops only a few percent. In the normal industrial plant, there is no requirement for regenerative load, since the load generally consists of several small motors with the probability that only one regenerative motor will be on the d-c circuit at any one time.

Surge protective equipment is usually required on industrial applications. Single-anode pool-cathode tubes are widely used in the control of electric welders. See Fig. 21·6.

Multianode versus Single-Anode Rectifying Devices

The multianode mercury-arc rectifier has all its anodes in one vacuum envelope with a common cathode. The single-anode rectifier has each anode and its cathode in a separate vacuum envelope. Single-anode pool-cathode rectifying devices may be either the ignitron type or the excitron type. These two types differ in construction and method of operation, but not in application.

Since the multi-anode rectifier has all the anodes in one ionized space, suitable baffling and spacing must be maintained to deionize the space around each anode during the non-conducting period of that anode. Such baffling and spacing cause a high arc drop, usually of the order of 25 volts. The necessity for a common cathode eliminates the possibility of using such a rectifier in any circuit, such as a bridge circuit, that requires a separate cathode for each rectifier element. With only one vacuum envelope, a leak would cause a shutdown of the complete unit. This disadvantage is somewhat lessened with the use of sectional multianode rectifiers. A sectional rectifier is an assembly of two or more small multianode tanks. Such an arrangement gives more reliability, since only part of the capacity is lost if one tank fails. By making small units, the decreased spacing between anode and cathode results in a lower arc drop. The multi-anode has the advantage of a simplified excitation system. This advantage is noticeable in very small-capacity single-anode units where the excitation circuits become a sizable portion of the rectifier equipment.

With single-anode tubes, the space in the tube may be easily deionized during the non-conducting period. This deionization allows the designer to use less baffling and less spacing between the anode and the cathode, both of which reduce the arc drop. Since several vacuum envelopes are used in a rectifier unit, the maximum in reliability is obtained.

In the United States, the single-anode design has practically outmoded the multi-anode design. In Europe and Asia, multianodes only are manufactured; however, some single-anode designs have been bought from the United States, and plans are being made to manufacture single-anode rectifiers in several foreign countries.

Sealed versus Pumped Rectifiers

At the present time, sealed rectifiers are generally used for the lower ratings and pumped rectifiers are used for the larger ratings. The type of equipment recommended depends upon the application. The pumped rectifier has a higher first cost but has an indefinite life with proper maintenance. The sealed rectifiers have a lower first cost but the tubes are a replaceable item. The pumped rectifier has the disadvantage of periodic maintenance of the vacuum equipment, plus general conditioning of the tanks after long periods of operation. Capable maintenance personnel are required to recondition pumped rectifiers. Such services are available from the manufacturer. Since the vacuum of a pumped rectifier is not impaired by out-gassing at high overloads, these units may be designed for a higher short-time overload rating than sealed rectifiers for the same full-load rating.

REFERENCES

1. *Theory and Applications of Electron Tubes*, H. J. Reich, McGraw-Hill, 1939, Chapters 2, 13.
2. *Electron Tubes in Industry*, K. Henny, McGraw-Hill, 1934.
3. *Industrial Electronics*, F. H. Gulliksen and E. H. Vedder, Wiley, 1935.
4. "Theory of Rectifier D-C Motor Drive," E. H. Vedder and K. P. Puchlowski, *Trans. A.I.E.E.*, Vol. 62, 1943, pp. 863–870.
5. "Electronic Control for Resistance Furnaces," H. J. Hague, *Steel*, Aug. 14, 1944, p. 106.
6. "New Developments in Ignitron Welding Control," J. W. Dawson, *Trans. A.I.E.E.*, Vol. 55, 1936, p. 1371.
7. "Mercury Arc Rectifiers and Ignitrons," J. H. Cox and D. E. Marshall, *Trans. Electrochem. Soc.*, Vol. LXXII, 1937, p. 183.
8. "Development of Excitron Type Rectifier," H. Winograd, *Trans. A.I.E.E.*, Vol. 63, 1944, p. 969.
9. "Ignitron Rectifiers in Industry," J. H. Cox and G. F. Jones, *Trans. A.I.E.E.*, Vol. 61, 1942, p. 713.

Chapter 22

MERCURY-ARC RECTIFIERS FOR POWER APPLICATION

G. F. Jones and C. R. Marcum

22·1 RELATIONSHIP OF CHARACTERISTICS TO APPLICATION

Arc-Back

In a mercury-arc rectifier, the direction of current is from the anode to the electron-emitting cathode spot, and in that direction only. (Although current is in the direction of electron travel, it is universally accepted that the direction of current is in the opposite direction, from anode to cathode.) Current direction can be from cathode to anode if a cathode spot is formed on the anode. This condition is known as arc-back.

If the valve action of the rectifier fails because of the formation of a cathode spot on an anode, current from other anodes supplied by the same transformer is concentrated in the cathode spot, and thus the transformer secondary is short-circuited. Direct current from other conversion units in parallel with the rectifier passes from the positive bus, to the defective anode, and through its associated transformer winding to the negative bus, thus placing a short circuit on the d-c system.

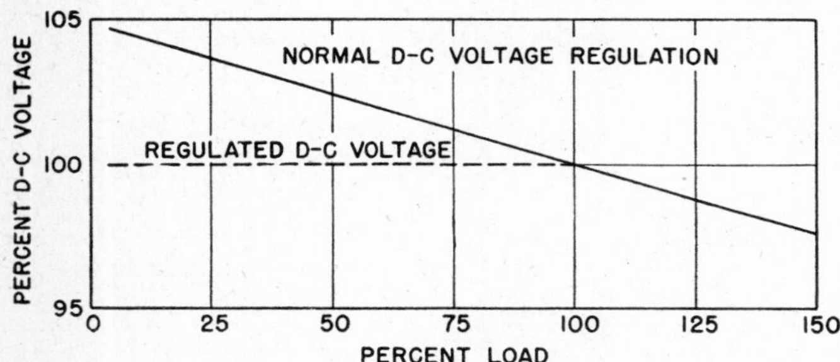

Fig. 22·1 Typical d-c voltage regulation curve.

Once started, an arc-back persists until power transfer from parallel conversion devices is interrupted, and until the a-c supply to the defective anode is opened or blocked by suitable biasing of control grids.

If the rectifier is protected by adequate switchgear, an arc-back does no damage. It simply causes a momentary interruption of power from the unit. Having removed the a-c and d-c faults, the switchgear can be reclosed immediately, or the grid biasing can be removed, and normal service is restored.

Most arc-backs are of a random nature. They are infrequent and are not an indication of faulty equipment or control. Frequent arc-back indicates some defect, and the rectifier unit should be taken out of service until the cause of arc-back is determined and corrected.

Cause of frequent arc-back usually is loss of normal vacuum, incorrect operation of the cooling system, or mechanical damage to internal parts.

Parallel Operation

The output-voltage-regulation curve of a rectifier unit supplied by constant alternating voltage is a straight line from light load through its overload rating. The slope of this curve varies, but in general it is between 5 and 6 percent from light load to full load. Rectifier units without output-voltage-control devices operate satisfactorily in parallel and share system demand in proportion to their ratings throughout the load range.

If the application requires a constant output voltage throughout the load range, a voltage regulator can be used. Cross-current compensation is required in combination with the constant-voltage characteristic of the regulating device to insure equal load division between parallel units. A rectifier on a given transformer tap and with fixed supply voltage cannot increase its output voltage above its normal regulation curve. The transformer must be designed for constant unidirectional voltage at the specified load and at the specified alternating input voltage. If the rectifier unit is loaded beyond this specified load, the voltage of the unit follows its natural regulation curve (Fig. 22·1).

Shunt-wound synchronous converters and unregulated rectifiers have approximately the same output-voltage characteristics from light load to full load. Beyond full load, however, the voltage characteristic of the synchronous converter drops off more rapidly. The voltage output of a rectifier unit or a synchronous converter is in direct proportion to the alternating input voltage.

A shunt-wound synchronous converter operates in parallel with a rectifier satisfactorily without regulating devices. Because the rectifier voltage in the overload ranges is somewhat higher than the corresponding voltage for the synchronous converter, the rectifier takes more than its proportion of the system load. This is satisfactory, because the rectifier has superior overload-carrying ability. If the converter is compound-wound, and if changing its output-voltage characteristic to approach that of a shunt machine is undesirable, the rectifier can be operated in parallel with it by use of a voltage regulator with cross-current compensation. Paralleling is simplified if the converter characteristic can be made to droop by strengthening the shunt field and weakening the compound field.

The voltage output of a synchronous motor-generator set is independent of the alternating voltage of the supply system. Voltage of the rectifier unit dips when the supply voltage dips; consequently paralleling may be more difficult. If supply voltage is constant, a rectifier unit can be operated in parallel with a shunt-wound d-c generator without voltage regulation. If the supply voltage fluctuates, the rectifier unit should be built so that it delivers the required voltage at full load when the alternating supply voltage is at its lowest value. A voltage regulator can then be used to control the output voltage of the rectifier so that it approximates that of the motor-generator set, either shunt- or compound-wound.

Regeneration

A rectifier unit arranged to convert alternating current to direct current will not invert direct current to alternating current unless the unit, with its control, is designed for inverter operation.

For some conversion applications some power may be returned to the d-c system by overhauling motors, and unless this power is absorbed the unidirectional voltage may rise beyond safe limits. If the normal load consists of many motors, power returned to the d-c system by an overhauling motor probably will be absorbed by other motors; however, if the load consists of a few motors, either for normal loading or during light-load periods, special provisions for absorbing regenerated power for short periods may be necessary. This is common with elevator motors, crane motors, reversing or rapid-slow-down motors, and some traction loads.

For these applications, the use of inverters to return power to the a-c system is not justified. Resistors to absorb the regenerated power are more economical. The resistors can be connected automatically by interlocks of the motor controls or by voltage indications of the d-c system. They can be disconnected automatically by timing devices or by relays responding to the system conditions and showing that regenerated power no longer need be absorbed.

Overload

A rectifier unit will carry its overload ratings in daily-duty cycle without any increase in maintenance cost. Overloading merely causes increased vaporization of mercury, which is condensed and returned to the cathode for reuse. The increased heat is dissipated by increasing the flow of cooling water through the rectifier or through the heat exchanger, depending upon the type of cooling system. If a recirculating cooling system is used with a water-to-air heat exchanger, the increased heat is dissipated by proper control of the fan.

The transformer is subject to additional heating on overloads. The application of overloads within the overload rating of the unit is limited by the ability of the transformer to return to full-load operating temperature between applications of successive overloads.

Current Relations

For a rectifier unit having a six-phase, double-wye transformer, the rms unidirectional pulsating current from each low-voltage terminal of the transformer is equal to the d-c cathode amperes multiplied by 0.289. For a quadruple-connected transformer, either six-phase parallel-double-wye or twelve-phase quadruple-zigzag, the current from each low-voltage terminal is equal to one-half the cathode current multiplied by 0.289. The current from each transformer secondary terminal for a six-phase star secondary is equal to the cathode current multiplied by 0.408.

The kilovolt-ampere input to the transformer primary is equal to the rectifier output in kilowatts plus the losses of the unit divided by the power factor of the unit.

Voltage Surges

Irregularities in the arc, which in turn cause rapid changes of current through the transformer windings, may cause voltage surges. Sudden changes in arc current, in microseconds, usually are caused by heavy loads applied suddenly when the cooling water is at a low temperature. Voltage surges can be suppressed at the source by capacitors in parallel with each arc path. Resistors usually are connected in series with the capacitors to control their rates of charge and discharge. They can be suppressed also by surge-arresting equipment connected across the transformer windings.

Noise and Vibration

Noise from a rectifier transformer is about the same as that from a standard power transformer of similar rating. The rectifier, except for one or two rotating auxiliaries, is static and noiseless. The rotary vacuum pump, if used, is small and its noise and vibration are negligible. A recirculating water-cooling system uses a motor-driven pump and, for installations using a water-to-air heat exchanger, a motor-driven fan with low noise level.

No special foundations are required for any of the equipment, the only requirement being that the foundations be reasonably level and strong enough to support the static weight of the equipment.

Harmonics

A mercury-arc rectifier unit distorts the wave shape of both the a-c supply system and the d-c system it supplies. This distortion depends upon the harmonics created by the rectifier, which in turn are fixed by the design of the rectifier unit and its mode of operation. Voltage-wave distortion of the a-c system is affected also by the impedance of the a-c system at the harmonic frequencies.

If the transformer is connected for six-phase operation, the harmonics in the d-c system are multiples of the sixth harmonic—sixth, twelfth, eighteenth, twenty-fourth, and so on.

If the transformer is connected for twelve-phase operation the number of harmonics theoretically is reduced by one half, the harmonics present being the twelfth, twenty-fourth, thirty-sixth, and so on. Actually, the odd multiples of the sixth harmonic are not eliminated, but are reduced in amplitude. Harmonics in the a-c supply system and unequal division of load between six-phase groups causes some of these odd harmonics to remain. They can be considered to be reduced to 25 percent of their values in a six-phase unit operating under the same system conditions.

The harmonics in the a-c system, for a six-phase unit, are the multiples of the base frequency plus and minus one: the fifth, seventh, eleventh, thirteenth, seventeenth, nineteenth, and so on. For a twelve-phase unit, the harmonics are the eleventh, thirteenth, twenty-third, twenty-fifth, thirty-fifth, thirty-seventh, and so on.

In a unit without output-voltage control the amplitude of the harmonics is fixed by the characteristics of the transformer, by the a-c supply system, and by the load, the amplitude of the harmonics increasing with load. If output voltage is controlled by delaying the point in the cycle at which the anodes pick up current, the number of harmonics remains the same but their amplitude is increased.

Transformers usually are connected for either six- or twelve-phase operation. A larger number of phases can be used, if several rectifiers are involved, by shifting the phase of the voltage supply to the units individually. By shifting the phase supply to two twelve-phase units so that they operate 15 degrees apart, over-all operation becomes twenty-four-phase, and thus the number of harmonics is decreased. For installations involving a large number of units, the phase supply to individual transformers can be shifted so that the number of phases becomes even larger. The number of phases obtainable in this manner is limited only by the number of rectifier units in the installation.

Harmonics caused by rectifier operation may cause telephone interference and may affect the heating of the a-c generator. For the average power application, the harmonics do not have any detrimental effect. With six or more phases, rectifier output will supply standard motors without changing their ratings, and will supply all types of industrial loads, in the same manner as rotating conversion equipment. For some applications, such as high-voltage plate supply for radio transmitters, and voltage supply for d-c motion-picture arc lights, filters must be used to reduce the harmonics to almost zero.

Telephone Interference

Amplitudes of the harmonics in both the a-c and the d-c systems are about inversely proportional to their harmonic frequencies. The exception to this general rule is that a harmonic can be amplified if the system is resonant to the frequency of that harmonic.

Harmonics cause telephone interference in proportion to their amplitudes and according to their position in the audio-frequency range. Phase control increases the severity of harmonic induction. Harmonics in the a-c and d-c systems do not in themselves constitute telephone interference. Such interference is caused by electromagnetic coupling or electrostatic exposure. Only a few rectifier installations cause telephone noise.

Experience with the large number of rectifier installations in operation has led to conclusions from which some general statements can be made.

If the a-c conductors are in cables, no telephone noise is caused in either cable or open-wire telephone lines. If the a-c system is made up of open wires and the telephone system is toll cable, no telephone noise is created. If the a-c system is made up of open wires and the telephone system is exchange cable, there may be some difficulty if the exposures are long and the systems are close together. If the a-c system and the telephone system are both made up of open wires, there probably will be some telephone noise if the systems are at highway separation or less, and if the exposure is for any considerable distance.

Exposures between telephone systems and d-c distribution systems follow the same general rule, there being little likelihood of telephone noise if the telephone system is in cable, and a possibility of telephone noise for open-wire telephone systems with long exposures and with the systems close together. Grounded d-c systems, notably for transportation service, generally have more severe exposures than do a-c circuits, because of greater coupling with the rail or ground return.

If the exposures with an a-c system are likely to cause telephone interference, the probability of noise decreases as the proportion of rectifier load carried by the a-c system decreases. That is, for a given exposure, the smaller the percentage of rectifier load to the total load on the system, the less chance for telephone interference.

The higher the number of phases the less is the likelihood of telephone interference, because many harmonics of low frequency and high amplitude in both the a-c and the d-c systems are eliminated or reduced.

Telephone interference can be reduced by transposing conductors of the telephone system, the power system, or both systems to reduce the coupling between circuits. Shielding or relocation of the circuits likewise may be effective, or the power system can be filtered to reduce the amplitude of the harmonics causing most interference.

Filters consist of one or more tuned circuits, each made up of a capacitor and a reactor. Each filter circuit is tuned to short-circuit the harmonic to be suppressed; the harmonic circulates locally through the filter and does not appear on the line. The number of harmonics to be suppressed determines the number of tuned circuits.

When a rectifier installation is to be made, the installation should be reported to the power company and the telephone company so that a preliminary study of possible telephone interference can be made. If no telephone interference is likely, the rectifier installation should be made without special interference-reducing apparatus. Tests after installation will show whether such measures are necessary. If the preliminary study shows the likelihood of telephone interference, estimates for interference-reducing devices can be obtained as recommended by the Edison Electric Institute.[1, 2]

Generator Heating

Harmonics cause some generator heating. Usually it is not serious, and it becomes important only if the generator carries a six-phase load consisting entirely of rectifiers, or if large amounts of phase control are used.

A generator of 0.80 power factor will carry a six-phase rectifier load, operating without phase control, without exceeding temperature guarantees, because the power factor of the load is greater than 0.90. A unity-power-factor generator requires a small percentage of derating for a six-phase rectifier load. If the generator carries load other than rectifiers, the heating problem is of less importance.

If the rectifier load consists of rectifier units of 12 or more phases without phase control, generator heating is not abnormal.

If the rectifier units are phase-controlled, harmonics are amplified and, if they form a large part of the generator load, heating becomes especially important.

22·2 ASSOCIATED APPARATUS

Transformers

Rectifier transformers can be connected many ways. The circuit selected depends upon the rating of the rectifier, the power source, and the purpose for which the unit is to be used. Usually, for power units, the transformer secondary is connected in double wye or in multiples or variations of the double-wye circuit. The resulting circuit gives the most

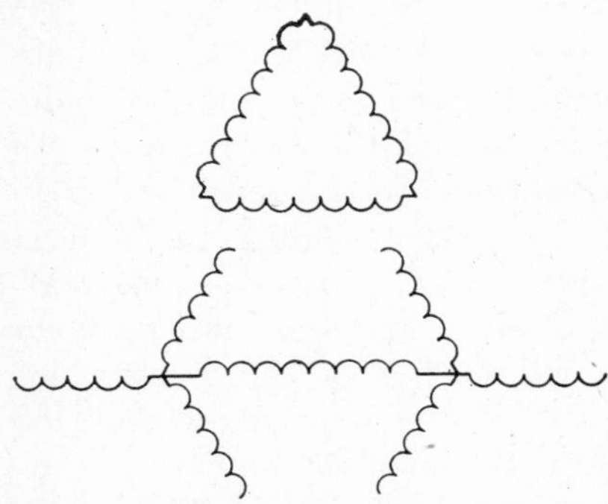

FIG. 22·2 Transformer connected delta, six-phase, double-wye.

economical transformer design and the optimum rating of the rectifier.

The circuit for a transformer having a delta-connected primary with a six-phase double-wye secondary is shown in Fig. 22·2.

A rectifier transformer differs from a normal power transformer in that the secondary windings do not carry current during some parts of the voltage cycle. The duration of the conducting periods is determined by the cyclic polarities of the transformer winding and by the valve action of the rectifier anodes. For this reason the primary and secondary windings have unequal capacities and the transformer is larger than a power transformer of the same input rating.

If a rectifier with six anodes is supplied by a six-phase double-wye transformer, each anode conducts current during 120 degrees of the cycle. Since two anodes carry current simultaneously, each anode carries half of the current output.

The neutrals of the two wye circuits are connected together through an interphase transformer. The midpoint of the interphase transformer is connected to the negative of the d-c system. A small pulsating excitation current circulates through the interphase transformer from the midpoint to the wye group having the highest instantaneous potential. Voltages in the wye groups are 60 degrees out of phase, and each wye group is at a higher potential than the other wye group three times each cycle. Excitation current, therefore, has a frequency three times that of the a-c supply system. This triple-frequency current is superposed on the load current, and it returns to the midpoint of the interphase transformer from the d-c system. The effect of this excita-

FIG. 22·3 Transformer connected delta, six-phase, parallel-double-wye.

tion current, in conjunction with the reactance of the interphase transformer, is to induce a voltage which opposes the voltage of the high-potential wye group and adds to the voltage of the low-potential wye group. These plus and minus voltages are equal to one-half the instantaneous voltage difference between wye groups, and their sum is always zero. The wye-group potentials with respect to the negative of the d-c system are therefore equalized; consequently, the

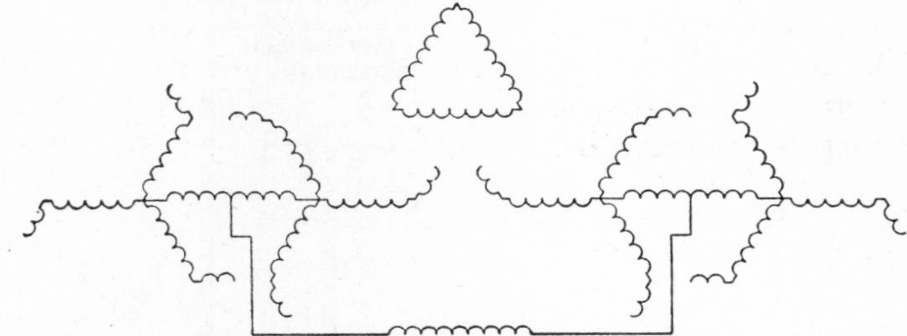

FIG. 22·4 Transformer connected delta, twelve-phase, quadruple-zigzag.

anode with highest instantaneous voltage in either wye group can carry current equally with the corresponding anode in the other wye group, even though they are out of phase.

The interphase transformer causes the wye groups to operate as independent three-phase units. This results in maximum utilization of the transformer, causes the anodes to operate under most favorable conditions, and increases the capacity of the rectifier to carry extreme overloads.

If the demand on the d-c system is below the interphase excitation requirement, the unit reverts to six-phase star operation, with a resultant zero-load voltage rise of 15 percent.

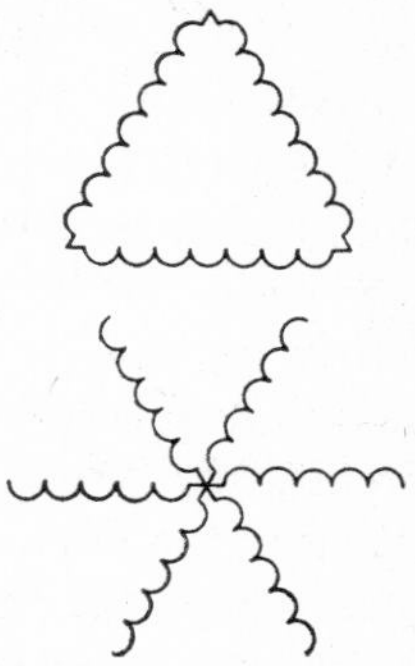

The interphase transformer is designed for a definite potential across its terminals. Phase control of the rectifier voltage increases this potential, and a larger interphase transformer is required. For this reason a rectifier transformer usually is limited to a specified percentage of output voltage reduction by phase control.

FIG. 22·5 Transformer connected delta, six-phase star.

Other circuits used for power-rectifier units are the six-phase parallel-double-wye and the twelve-phase quadruple-zigzag circuits, as shown in Figs. 22·3 and 22·4. These are modifications of the basic three-phase circuits and are used for large units. The third interphase transformer for the zigzag connection operates at 360 cycles, when it is supplied by a 60-cycle system, and displaces the neutrals of the other two interphase transformers which later are out of phase.

Figure 22·5 shows a six-phase star-connected transformer.

Switchgear

Switchgear for a rectifier unit is designed to start and stop the unit, and to give it adequate protection under abnormal conditions.

Varieties of switchgear range from manual devices to fully automatic apparatus which will start and stop the unit automatically according to the d-c system demand.

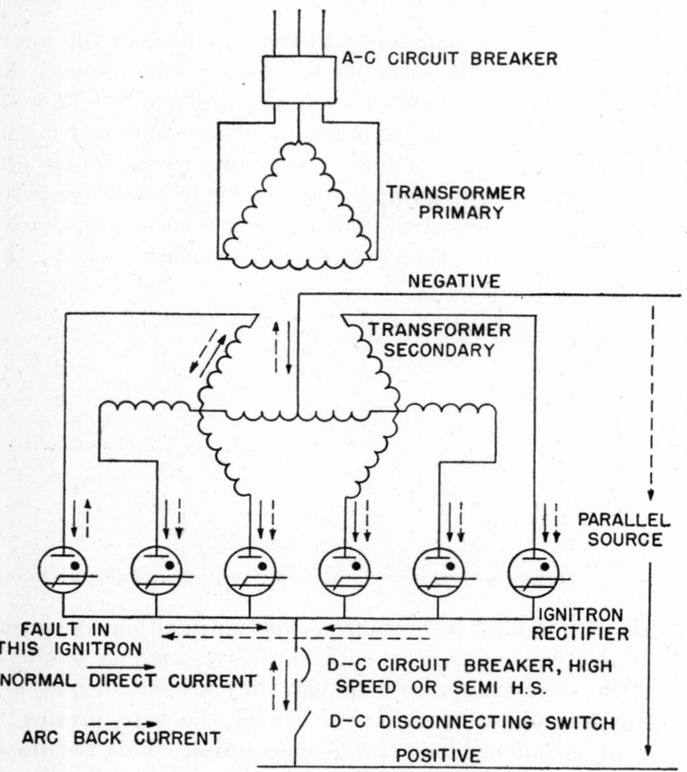

Fig. 22·6 Fundamental switchgear circuits.

The simplest switchgear arrangement consists of (1) an a-c circuit breaker between the transformer primary and the a-c supply system, and (2) a d-c breaker between the rectifier cathode and the d-c system, as shown in Fig. 22·6. The only operation normally necessary to start the unit is to close both breakers.

Selection of the a-c circuit breaker is determined by the voltage of the a-c system, the rating of the rectifier unit, and the kilovolt-amperes the system can deliver under short circuit. Selection of the d-c breaker is determined by the rating of the unit in kilowatts, its output voltage, and (if the unit is operated in parallel with other conversion units) by the capacity and type of units in parallel. Direct-current breakers may be semi-high-speed or high-speed cathode circuit breakers, or they may be multipole high-speed anode circuit breakers in combination with semi-high-speed cathode circuit breakers.

The function of the cathode circuit breaker is to open the d-c circuit at the time of normal shutdown or at the time of operation of some protective relay or by direct tripping at the time of an arc-back. The last-mentioned function is performed by a polarized overload trip which operates on reverse current only. Usually the cathode circuit breaker does not have a direct over-current trip since the unit can carry high overload swings without damage or additional maintenance. Overloads on d-c feeders are interrupted by feeder circuit breakers. If there are no d-c feeders, the cathode circuit breaker can perform the functions of a feeder as well as its normal functions.

During an arc-back the current from the normal anodes of the rectifier to the anode on which the cathode spot has formed can be interrupted by applying negative bias to the deionizing grids of all the anodes, by opening the a-c breaker, or by opening switchgear between the transformer secondary and the rectifier. Grid bias does not always suppress the current, and it must be backed up by the a-c switchgear in the event of failure. In general, the larger the rectifier unit the greater is the fault current and the greater is the stress on the apparatus involved in the fault. If the unit is rated at 2000 kilowatts or less, normal a-c switchgear operates fast enough to relieve the fault in time to prevent damage. For units larger than 2000 kilowatts, high-speed anode switchgear should be used to open the a-c fault.

At the time of arc-back there is a current from the d-c bus to the rectifier cathode, to the defective anode and through its associated transformer winding, to the negative terminal and back to the source. This reverse current from the d-c system appears only if there are other conversion units in parallel, or if the load is of the regenerating type. The stresses caused by reverse current depend upon the size of the rectifier unit and upon the amount of power which can be fed into the defective anode by the external source. If the rectifier unit is rated at 2000 kilowatts or less, and if all the conversion equipment connected to the d-c bus is less than approximately 2500 kilowatts, semi-high-speed cathode switchgear gives adequate protection. If the unit is paralleled with other conversion equipment, making the total capacity connected to the d-c bus more than 2500 kilowatts, stresses in the equipment are too great unless the reverse current is limited in time and magnitude by high-speed cathode switchgear. If the size of individual unit is larger than 2000 kilowatts, or if the total capacity in conversion units connected to the bus exceeds 6000 kilowatts, both a-c and d-c faults must be opened at high speed. Anode switchgear is for this dual function. Figure 22·7 shows the circuits of a rectifier unit with high-speed anode switchgear.

A pole of high-speed switchgear is connected in each anode circuit. Usually all the poles are operated by a common closing mechanism but are tripped individually by reverse current only. During an arc-back the current from the d-c system and from the other anodes of the rectifier to the defective anode is in the reverse direction. The circuit-breaker pole in the circuit of that anode opens at high speed, and the duration of the a-c fault and the magnitude of the direct current are limited. Opening the a-c circuit breaker, the cathode circuit breaker, or the other poles of the anode circuit breaker is unnecessary. For this arrangement of switchgear, the cathode circuit breaker is simply a load-break disconnecting switch operated for normal shutdown or by protective relays. It is not tripped by reverse current.

This type of protection not only removes a-c and d-c faults simultaneously at high speed but also permits the other anodes to deliver current to the d-c system as soon as the fault is removed. Service is not interrupted.

If a rotating conversion unit is operated in parallel with a rectifier unit on the d-c bus, high-speed switchgear usually is required to protect the commutator of the rotating machine during arc-back.

A rectifier is protected from overloads by means of inverse-time a-c overload relays and from a-c short circuit by instantaneous a-c overload relays. The transformer is protected from overheating by thermal relays. The rectifier is protected from operation at low vacuum and from abnormal water cooling by vacuum relays, thermal relays, and water-pressure relays. Additional protective devices such as transformer hot-spot relays, replica thermal relays, low-alternating-voltage relays, and misfire relays are available. Operation of the protective devices trips both a-c and d-c switchgear. If the sub-station is manually controlled, the unit is restored to service by the operator. If the station is automatically controlled, the switchgear is reclosed automatically or is locked out, depending upon the condition causing the shutdown. If the condition is self-correcting, the sub-station returns to service automatically after the condition becomes normal. If the condition is not self-correcting, the sub-station is locked out of service.

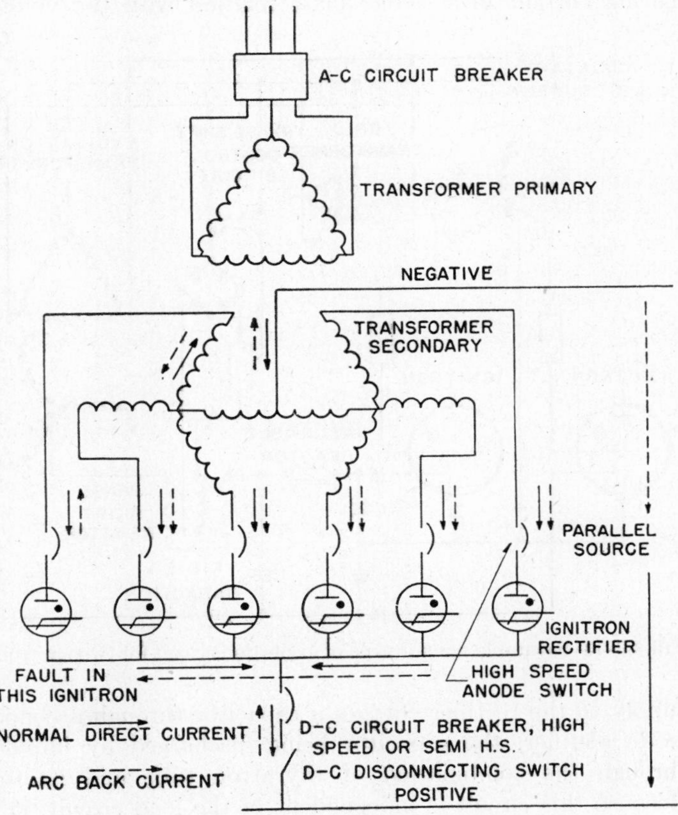

FIG. 22·7 High-speed anode switchgear circuits.

Excitation System

A mercury-arc rectifier must have an excitation system by means of which an electron-emitting spot can be formed at the cathode. If the rectifier assembly is made up of single-anode tubes, each must be provided with an excitation system. It may be a continuous-excitation arc or an intermittent-excitation arc, depending upon rectifier design.

In a continuously excited rectifier an arc between an auxiliary anode and the cathode is started by supplying a voltage between these two terminals from a separate excitation source. The source of voltage must have high regulation so that it can be short-circuited momentarily without drawing too much current, and voltage must be high enough to maintain a stable arc. The most common method of initiating the excitation arc is to depress a starting-anode rod into the mercury pool and then withdraw it, by spring action or by depressing a plunger in the mercury so that it projects a jet of mercury to a stationary starting anode. When the starting rod is withdrawn or when the mercury jet falls away from the anode, the arc is started. Either action is accomplished by an external solenoid actuating a movable armature inside the vacuum chamber. The wall of the chamber at this point is made of non-magnetic material.

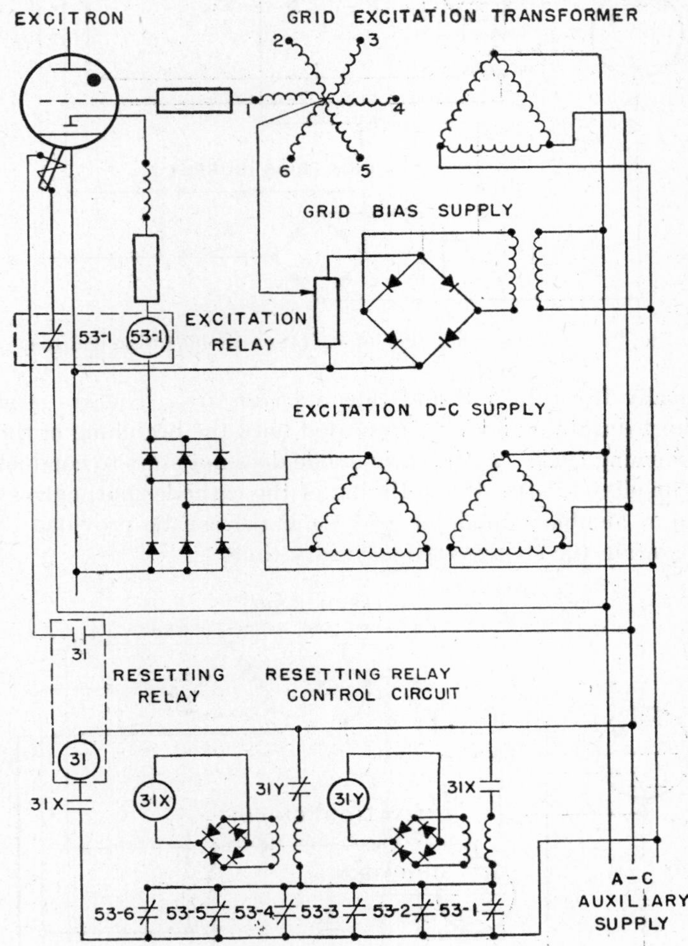

FIG. 22·8 Excitation circuit for an Excitron.

Either an a-c or a d-c excitation source can be used, but a d-c source is preferred because it assures correct polarity of the excitation arc, and because it requires fewer devices. Figure 22·8 shows an elementary circuit for a continuously excited rectifier.

In an intermittently excited rectifier, the excitation arc is started at the beginning of each cycle by applying timed impulses to a high-resistance rod permanently immersed in the mercury pool. As the current pulse passes from the high-resistance rod to the mercury, it vaporizes the mercury at the point of contact. This causes the mercury to separate

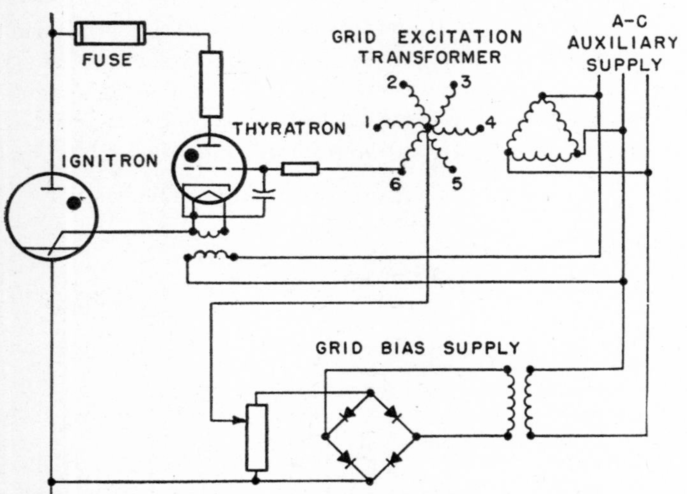

Fig. 22·9 Anode firing type of excitation circuit for an ignitron.

locally from the rod and thus start an arc. The arc is of short duration and is not repeated until the beginning of the following cycle. If the main anode does not start to conduct immediately after the initiation of the cathode spot, current to an auxiliary anode or grid (or both) can be provided to maintain the spot for the necessary time.

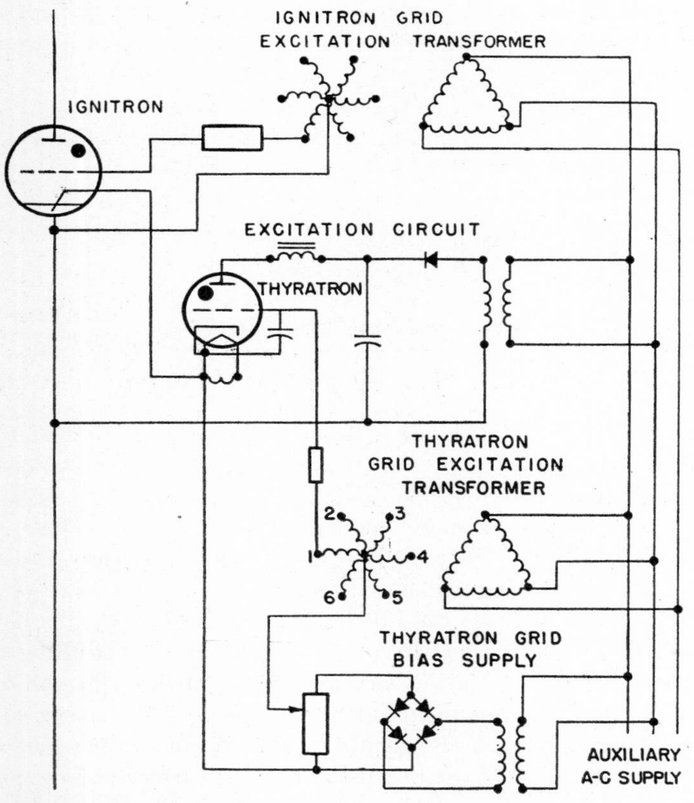

Fig. 22·10 Thyratron-capacitor type of excitation circuit for an ignitron.

The simplest form of circuit used for delivering timed impulses to the ignitor is shown in Fig. 22·9. It is known as anode firing. Load current passes through the ignitor and thyratron in series. Current magnitude is limited by a resistor. As soon as the cathode spot is formed, the load current is picked up by the main anode and the excitation circuit is short-circuited by the power arc in the tube. This extinguishes the excitation arc. The thyratron tube passes the positive impulse but blocks the negative impulse, so that only unidirectional pulses are supplied to the ignitor. It can be used also to time the impulse by means of its control grid. If the system load is lower than the excitation requirements, this excitation circuit becomes unstable, because the excitation current circulates through the load.

A similar circuit, except that it is independent of system load, is shown in Fig. 22·10. A separate source of alternating current in suitable phase relation with the voltage

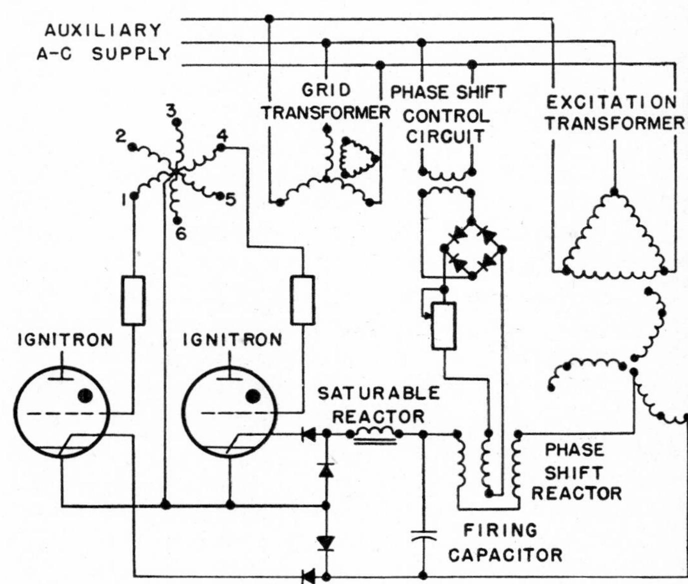

Fig. 22·11 Saturable-reactor type of excitation circuit for two ignitrons.

supply to the rectifier charges a capacitor through a copper oxide rectifier, the capacitor being discharged by impulse through the combination of thyratron tube and ignitor. Because this circuit is independent of the load circuit, it is stable at zero load on the rectifier.

A third excitation circuit, known as saturating reactor firing, is shown in Fig. 22·11. This circuit does not use thyratrons which have a limited life and require a cathode-heating time before they can be put into service. It uses a separate excitation source which charges a capacitor through a linear reactor instead of a dry rectifier. Capacitor charging, controlled by the reactor, is alternately at positive and negative polarity. The ignitor and a saturable reactor in series are connected in parallel with the capacitor. As capacitor voltage builds up, it is imposed across the saturable reactor and ignitor. When the reactor is saturated its inductance drops to almost zero and the capacitor is discharged through the ignitor. The saturable reactor is wound on a core of iron which saturates abruptly at a given point. The saturable reactor and capacitor form a tuned circuit that is permanently adjusted by the manufacturer. Copper oxide rectifiers direct the impulses to the proper ignitors. Both the positive and the negative capacitor charges can be used to supply impulses to ignitors 180 electrical degrees apart. This is the excitation circuit normally used for intermittently excited rectifier units.

Rectifier output voltage can be phase-controlled by varying the inductance in the capacitor-charging circuit, or by connecting a phase shifter between the primary of the excitation transformer and its a-c supply. The inductance in the capacitor circuit can be changed by use of an iron-core reactor

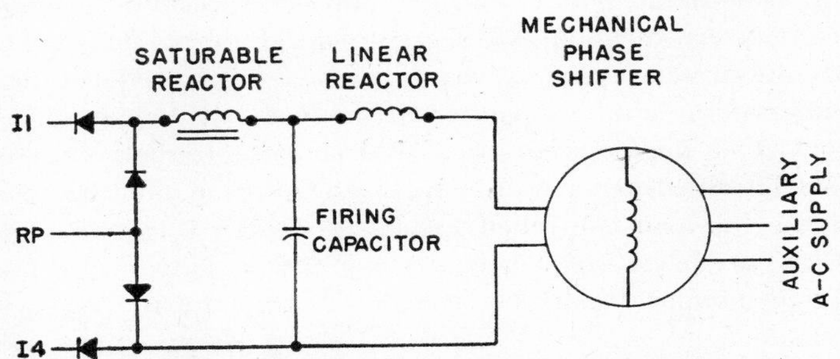

Fig. 22·12 Saturable-reactor type of excitation circuit with a mechanical phase shifter.

with a d-c saturating coil. Inductance of the reactor can be varied by controlling the amount of saturation. In this manner the phase position of the impulse supply to the ignitor can be shifted with respect to the anode voltage, thus changing the load-pick-up point of the anode in the cycle. With this excitation system the range of output-voltage reduction is limited to approximately 30 percent below normal.

The phase shifter between the primary of the excitation transformer and the alternating voltage source consists of a wound-rotor induction motor with a rotor that does not rotate, but which can be moved with respect to the stator. This position of the rotor usually is adjusted by means of a small driving motor with a high gear ratio. By the use of this circuit, shown in Fig. 22·12, the voltage output of the rectifier unit can be varied to zero.

Output voltage can be changed at high speed by using the variable-inductor circuit, but the speed of the phase-shifter circuit is somewhat slower, depending upon the speed with which the rotor of the phase shifter can be moved. A combination of the two circuits can be used for high-speed con-

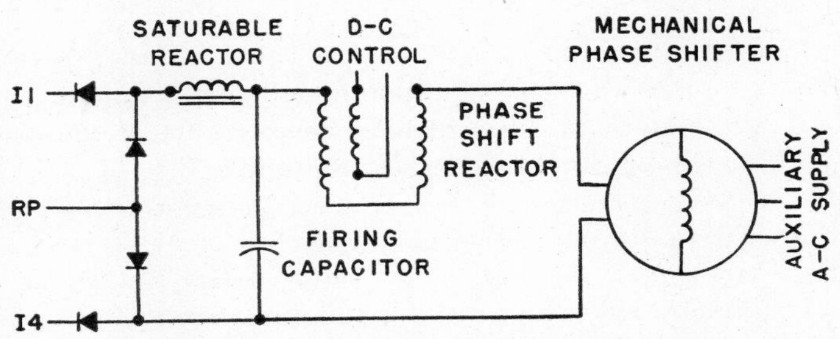

Fig. 22·13 Saturable-reactor type of excitation circuit with a phase-shifting reactor and a mechanical phase shifter.

trol for small voltage changes and slower control for large voltage changes. A combination circuit is shown in Fig. 22·13.

Cooling Equipment

The temperature of a mercury-vapor tube is important, because the molecular density of the vapor has an influence on the operating characteristics of the tube. The tube must have cooling and temperature-control apparatus to dissipate the losses in the arc at a temperature favorable for arc rectification. Temperature-control apparatus frequently includes

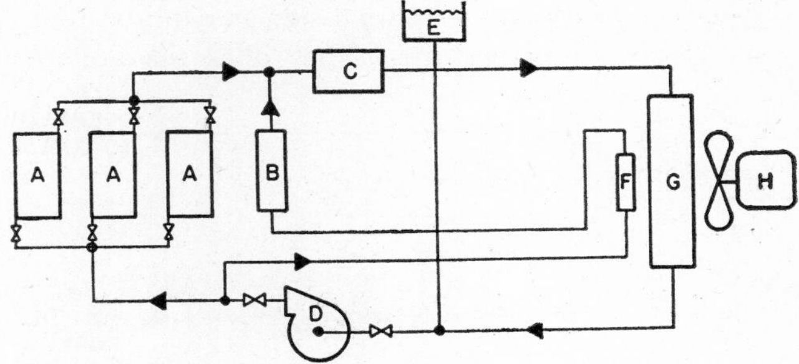

A- POOL CATHODE TUBE (PUMPED)
B- RECTIFIER VACUUM MANIFOLD AND MERCURY VAPOR PUMP COOLING COILS
C- HEATER TANK CONTAINING WATER HEATERS AND TEMPERATURE CONTROL THERMOSWITCHES
D- WATER PUMP
E- EXPANSION TANK
F- AUXILIARY WATER TO AIR RADIATOR
G- MAIN WATER TO AIR RADIATOR
H- FAN MOTOR

Fig. 22·14 Schematic diagram of a water-to-air cooling system.

heaters to be used at light load when the losses are too low for this purpose.

If a suitable water supply is not available for cooling steel-tank rectifiers, a water-to-air radiator with a forced-draft blower (water recirculating between radiator and tubes) is used. Standard water-cooled tubes can be used in this way. Such a system is shown in Fig. 22·14.

If temperature, pressure, and quality of cooling water are suitable, direct raw-water-cooling may be used. Figure 22·15 shows this system. The tubes are connected in series; therefore the minimum water rate for turbulent flow can be maintained and the full heat-absorbing capacity of the water can

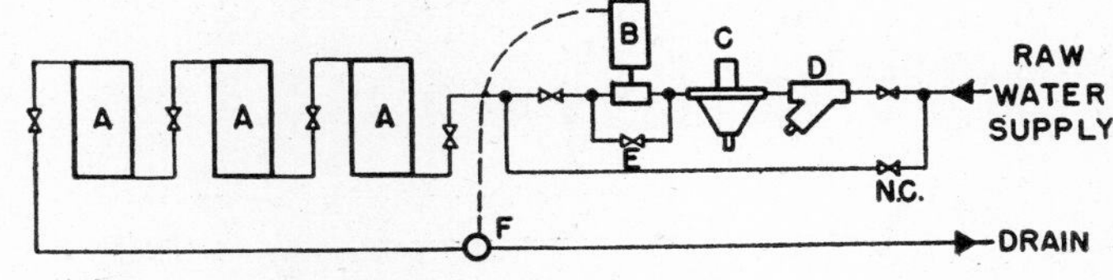

A- POOL CATHODE TUBE (SEALED)
B- TEMPERATURE REGULATING VALVE
C- PRESSURE REGULATING VALVE
D- STRAINER
E- BYPASS VALVE SET FOR MINIMUM ALLOWABLE WATER FLOW THROUGH TUBES
F- ACTUATING ELEMENT FOR TEMPERATURE REGULATING VALVE

Fig. 22·15 Schematic diagram of a direct raw-water-cooling system.

be used. The water must not contain an excessive amount of impurities that could cause corrosion or scaling. Temperature and pressure of the water must permit the tubes to operate at satisfactory temperatures with maximum or minimum load on the rectifier. This system is simple, but is used only on relatively inexpensive low-voltage tubes in which tube temperature and corrosion or scaling are not critical.

If the temperature and quality of cooling water are suitable but the pressure is insufficient, a recirculating raw-water system is used instead of direct raw-water cooling. The water is recirculated as shown in Fig. 22·16, and cold water is added at the pump-suction side as required to maintain the desired temperature. The tubes operate at the same temperature, with a constant water flow through them, and water heaters can be used in the system to avoid low temperatures.

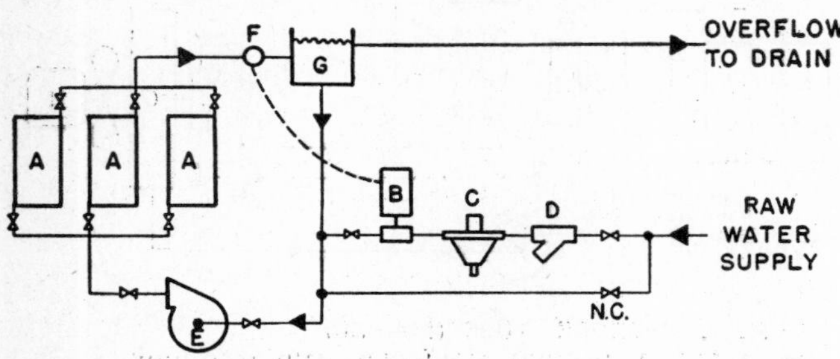

A- POOL CATHODE TUBE (SEALED)
B- TEMPERATURE REGULATING VALVE
C- PRESSURE REGULATING VALVE
D- STRAINER
E- WATER PUMP
F- ACTUATING ELEMENT FOR TEMPERATURE REGULATING VALVE
G- EXPANSION TANK

Fig. 22·16 Schematic diagram of a recirculating raw-water-cooling system.

If the quality of water is poor, or if tube temperature and corrosion or scaling are critical, a water-to-water system as shown in Fig. 22·17 is used. Water is circulated in a closed system through the tubes and one side of a water-to-water heat exchanger. Raw water is passed through the other side of the exchanger to maintain the desired temperature of recirculating water. Water heaters can be used, if they are needed.

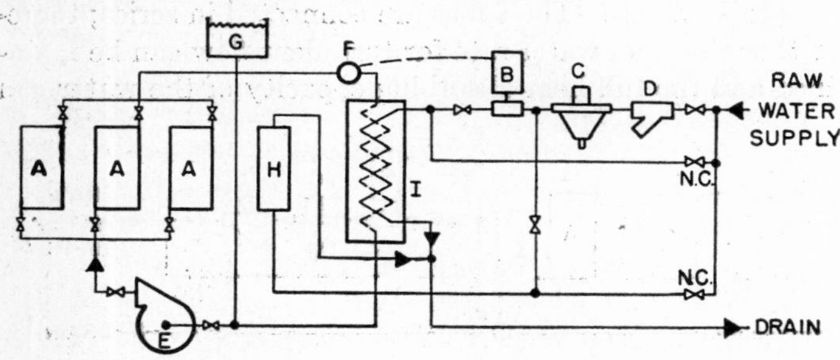

A- POOL CATHODE TUBE (PUMPED)
B- TEMPERATURE REGULATING VALVE
C- PRESSURE REGULATING VALVE
D- STRAINER
E- WATER PUMP
F- ACTUATING ELEMENT FOR TEMPERATURE REGULATING VALVE
G- EXPANSION TANK
H- RECTIFIER VACUUM MANIFOLD AND MERCURY VAPOR PUMP COOLING COILS
I- WATER TO WATER HEAT EXCHANGER

Fig. 22·17 Schematic diagram of a water-to-water cooling system.

Water-to-water and the water-to-air cooling systems are preferred, because the corrosiveness of water is reduced by recirculating in a closed system, which de-aerates it. In water systems which are not corrosion resisting, only distilled water made non-corrosive by treatment should be used in the recirculating system. The most successful treatment is (*a*) 0.5 percent sodium dichromate or (*b*) 0.1 percent sodium chromate. Borax is added to sodium dichromate to prevent the solution from becoming acid.

The temperature-control point for most rectifiers is at the water-discharge point. Most rectifiers operate best with a temperature of about 55 degrees centigrade at this point. Minimum temperatures vary widely. Ignitrons or other small-tank rectifiers operate satisfactorily at temperatures down to 5 degrees centigrade, but some of the larger grid-controlled rectifiers require temperatures within a few degrees of the full-load controlled value. Too high a temperature causes arc-backs frequently. At too low a temperature the mercury-vapor density becomes so low that the arc becomes unstable and tends to snap out. As a result surges are generated in the inductances of the circuit.

Rectifiers with vacuum equipment require cooling for the mercury-diffusion pumps. These pumps operate at temperatures as high as 50 degrees centigrade; consequently with water-to-air heat exchangers the main recirculating water can be used through an auxiliary radiator as shown in Fig. 22·14. Pumping speed increases as cooling-water temperature is decreased. For this reason, raw-water cooling is used if possible.

For the direct-raw-water and recirculating-raw-water cooling systems, computation of water consumption is relatively simple, if the heat transferred to the surrounding atmosphere (usually only a few percent of the total losses) is neglected. The method used is as follows:

$$H = \frac{E \times I_{d\text{-}c}}{1000} \qquad (22\cdot 1)$$

where H = heat to be dissipated in kilowatts
E = tube arc drop in volts
$I_{d\text{-}c}$ = d-c load current in amperes.

Then

$$g = \frac{3.8 \times H}{(t_0 - t_1)} \qquad (22\cdot 2)$$

where g = raw-water consumption in gallons per minute
t_0 = automatically controlled temperature of the discharge water in degrees centigrade
t_1 = temperature of raw water entering the rectifier in degrees centigrade.

A minimum raw-water rate for the direct-raw-water cooling system is set by the manufacturer to assure turbulent flow through the tubes.

In a water-to-water cooling system, different recirculating water rates are used for various tube sizes; heat-exchanger design varies with each manufacturer, so exact raw-water consumption must be obtained from the designer. For estimating, 0.35 gallon per minute per hundred d-c amperes for raw water at 25 degrees centigrade and 0.40 gallon per minute per hundred d-c amperes for raw water at 30 degrees centigrade, based on a discharge-water temperature of 55 degrees centigrade, can be used.

If continuity of operation is essential, additional apparatus and by-pass and isolating valves can be used to permit cleaning or repairing any device while the unit is in operation.

Vacuum Equipment

Operating gas density in a mercury-arc tube is controlled by the condensation temperature of mercury in the tube. Foreign gas pressure is controlled by vacuum pumps and indicated by vacuum gauges. Foreign gas pressure must be limited to about 1 micron for satisfactory arc operation, and to minimize chemical action of some of these gases on tube parts.

In recent years, it has become economically possible to manufacture permanently sealed tubes and replace them when, after a few years, the pressure becomes too high. The unlimited life of reparable, continuously pumped tubes make them popular. Vacuum equipment on a pumped rectifier must remove gases at a rate to balance remnant leakage.

Figure 22·18 shows a diagram of a typical pumping system. Vacuum valves are so placed that any tube, gauge, or pump can be removed without admitting air to the remaining tubes. These valves also permit the isolation of leaks in a relatively small part of the system, making them easier to find. Gases are pumped from the rectifier by a mercury-diffusion pump through a manifold cooled and baffled to avoid mercury transfer. The mercury pump discharges into an interstage reservoir. A mechanical pump of the vane type, discharging at atmospheric pressure, exhausts the interstage reservoir. This reservoir adds flexibility by permitting stoppage of the mechanical pump without stopping the pumping action. In some units a barometric seal consisting of a tube longer than barometric height dipping into a pool of mercury is incorporated in the interstage reservoir. This seal acts as a check valve if either the diffusion or the mechanical pump fails. A set of vacuum gauges completes the pumping system.

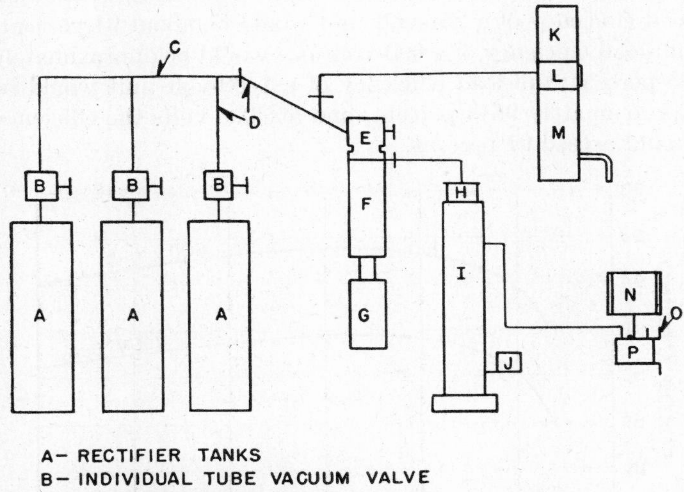

A– RECTIFIER TANKS
B– INDIVIDUAL TUBE VACUUM VALVE
C– VACUUM MANIFOLD
D– VACUUM PUMPING CONNECTION
E– HAND-OPERATE VACUUM VALVE
F– MERCURY DIFFUSION VACUUM PUMP
G– MERCURY PUMP HEATER
H– MERCURY TRAP
I– INTERSTAGE RESERVOIR WITH BAROMETRIC SEAL
J– PRESSURE INDICATING MANOMETER
K– HOT WIRE VACUUM GAUGE
L– THREE WAY VACUUM VALVE
M– McLEOD VACUUM GAUGE
N– ROTARY VACUUM PUMP MOTOR
O– DISCHARGE
P– ROTARY OIL-SEALED VACUUM PUMP

Fig. 22·18 Schematic diagram of a typical vacuum system for mercury-arc rectifiers.

The mercury-diffusion pump utilizes the same fluid as the rectifier, and a small transfer of fluid does not contaminate either device. The three-stage pump will pump against a back pressure as high as 25 millimeters of mercury.

Mechanical vacuum pumps differ in detail, but all use the principle of a rotary vane operating under oil to provide an oil seal. Various types of check valves are used to avoid admission of air or oil to the rectifier if the mechanical pump stops. The mechanical pump must pump at rates equal to the pumping speed of a mercury pump and must exhaust the interstage reservoir to a pressure less than the maximum discharge pressure of the mercury-diffusion pump.

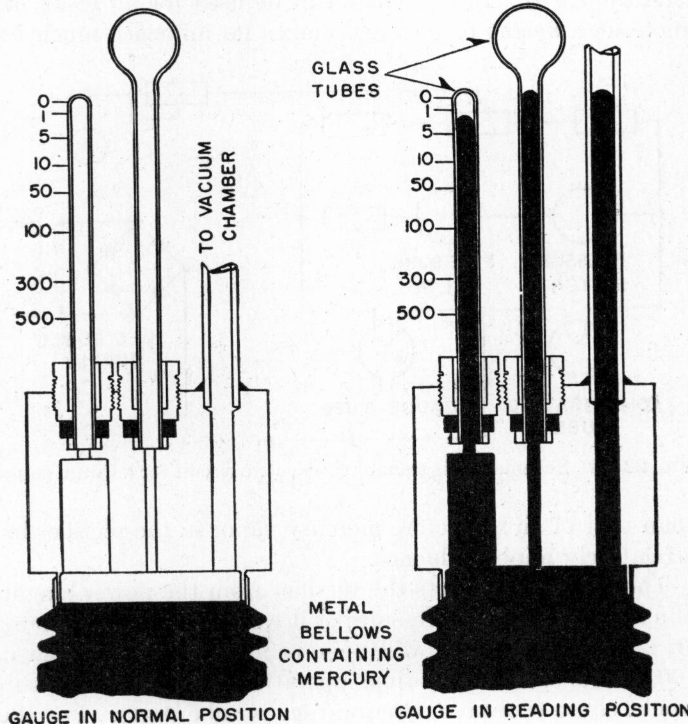

Fig. 22·19 Principle of a bellows type of McLeod gauge.

The most reliable vacuum gauge is the McLeod gauge, which takes a sample of the gas to be measured and compresses it by a specified amount to a readable value on a mercury column. Figure 22·19 illustrates this principle.

Since the McLeod gauge uses mercury as the operating medium, it measures only permanent gases; however, it operates on the sampling principle, and is not suitable for continuous reading or for unsupervised protective purposes. There are several forms of electric gauges, the principal types being the Pirani, thermocouple, and ionization gauges. For automatic indication and protection, the Pirani gauge is almost universally used for mercury-arc rectifiers because it is relatively simple and rugged; it has sufficient power to operate both an indicating meter and a contact-making relay, and it is most sensitive at a pressure of 1 micron.

The Pirani gauge operates on the principle that the thermal conductivity of a gas is proportional to its molecular density and inversely proportional to its molecular weight. In this gauge a tube with a filament is connected to the vacuum to be measured. A fixed current is passed through the filament. The temperature to which this current heats the filament is influenced by the pressure of the gas surrounding it. This temperature also influences the resistance of the filament. By connecting the filament in a bridge circuit with fixed resistances as shown in Fig. 22·20, the current through the meter circuit is related to the pressure surrounding the filament. The scale of the meter can be calibrated to read pressure, and a relay with the desired contact-making range is selected.

Although the Pirani gauge can be used for all gases, the molecular weight of mercury makes its influence much less than that of air; therefore mercury vapor in the rectifier has a relatively minor influence.

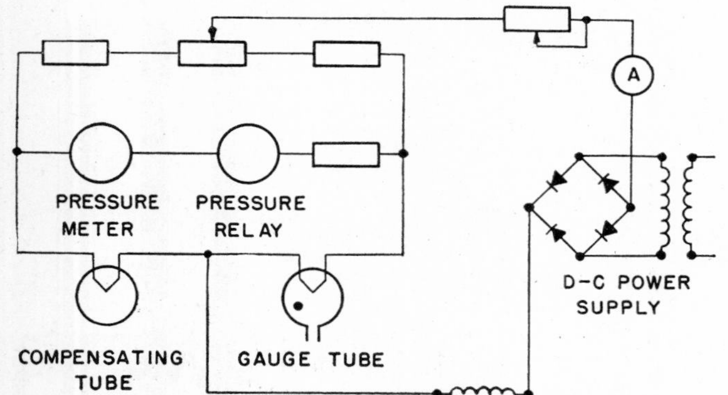

Fig. 22·20 Schematic diagram of the electric circuit for a Pirani gauge.

The relay disconnects the rectifier from the power circuits at a selected rectifier pressure; or it can be arranged to sound an alarm. A rectifier will operate at pressures as high as 100 microns, but long-time operation at such a pressure would cause excessive deterioration of tube parts. For this reason, the relay is usually set for some pressure less than 20 microns.

Efficiency

Losses in the component parts of a rectifier unit include transformer excitation and copper losses, rectifier arc-drop loss, and losses of the standard auxiliaries. Included with the auxiliary losses are the rectifier excitation loss, the loss in the vacuum-pumping system, and the loss in the continuously operating equipment of the rectifier-cooling system. A water-cooling system with no pumps or fans has no loss. If a recirculating cooling system is used, the losses in the pump-driving motors are included. Losses in auxiliaries such as fans, water heaters, and anode heaters are not included in efficiency calculations, because these devices operate intermittently, their operation depending upon load and ambient temperature.

The efficiency curve of a rectifier unit is almost flat from one-quarter load to full load. The smallness of no-load losses accounts for the high efficiency at light load. The arc loss is almost in proportion to the d-c load amperes; the arc voltage drop increases only slightly with load. The copper loss of the transformer, proportional to the square of the d-c load amperes, accounts for the decrease in efficiency at heavy loads, particularly in the overload region. A typical efficiency curve is shown in Fig. 22·21.

The loss in the arc is almost constant for a given ampere output, regardless of the output-voltage rating of the unit. Arc-voltage drop increases only slightly as the output voltage rating of the unit is increased. This loss is proportional to amperes and not kilowatts, so the arc loss of a low-voltage unit is a much higher percentage of the kilowatt rating than is the arc loss of a high-voltage unit. For example, the full-load efficiency of a 250-volt unit would be about 91 percent; full-load efficiency of a 600-volt unit would be approximately 95 percent; full-load efficiency of a 1500-volt unit would be approximately 96½ percent; and at 3000 volts the efficiency would exceed 97 percent.

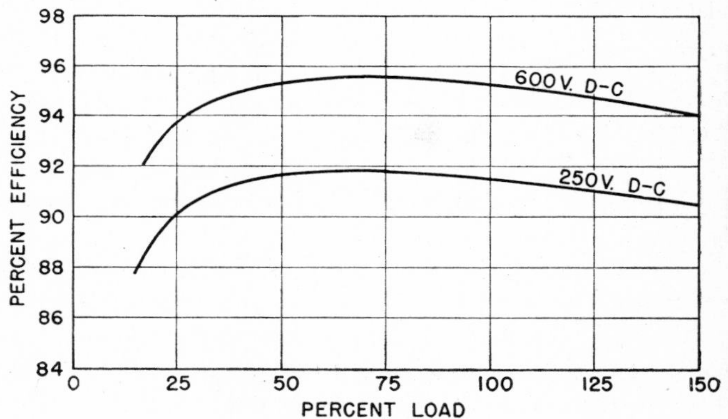

Fig. 22·21 Typical efficiency curves.

In general, the larger the rating of the rectifier unit, the higher its efficiency, particularly at smaller loads, because the no-load losses of the large unit are a small percentage of the unit rating as compared with smaller units.

Power Factor

The power factor of a rectifier unit is lagging. It is determined by transformer excitation; by harmonic components of the supply system; and by phase displacement (displacement factor) between supply voltage and current, which is caused by circuit reactance opposing transfer of current from one anode to another. At light loads the major influence on power factor is transformer excitation.

At normal loads power factor is influenced mostly by the reactances of the rectifier transformer and the a-c supply system. When current transfers from one anode to another, or commutates between rectifier elements, the time required for current transfer increases with the reactance of the total circuit. For a given circuit and transformer the time of transfer increases as the current transferred is increased. The effective combined reactance of the supply circuit and the transformer, which opposes transfer of current from one element to another, is known as commutating reactance. For given conditions, the higher the commutating reactance is, the greater the delay in transfer of current, the greater the lag of current with respect to voltage, and the lower the power factor. The time required to transfer current from one element to the next is known as the angle of overlap.

The harmonic content of the a-c system represents out-of-phase components which do not add to the active power but do add to the volt-amperes, and therefore influence power factor. The effect of harmonics on power factor is known as distortion factor.

In a twelve-phase unit the current is commutated between elements 12 times a cycle; for a six-phase unit, it is commutated between elements 6 times a cycle. Because all elements conduct for 120 degrees, the current commutated for each transfer of the twelve-phase unit is half that of a six-phase unit comparably rated. If both un ts had the same reactance, the displacement factor for the twelve-phase unit would be less than for the six-phase unit and would result in higher power factor. To maintain normal output-voltage regulation for a twelve-phase unit the reactance usually is double that of a six-phase unit. For this reason, the displacement factors of the two units usually are the same. The harmonics of a twelve-phase unit are much less than those of a six-phase unit, so the distortion factor for the twelve-phase unit is less. This accounts for the higher power factor of the twelve-phase unit. Typical power-factor curves for six and twelve-phase units are shown in Fig. 22·22.

The distortion factor of a twelve-phase unit is small, and any increase in the number of phases above 12 does not improve power factor appreciably.

Output-voltage control or phase control is accomplished by delaying the point in the cycle at which the anodes pick up current. This delay in pick-up adds to the angle of overlap, represents additional delay of current with respect to the supply voltage, and therefore reduces power factor. Reduction of power factor from the normal power factor of the unit is in direct proportion to the reduction of output voltage.

Regulation

Regulation guarantees are based on calculated values, the voltage drops of the various component parts of the rectifier unit being subtracted from the light-load voltage of the unit to determine output voltages at various loads. Voltage drop in the unit is made up of reactive drop in the transformer, copper drop in the transformer, and arc-drop in the rectifier.

For a six-phase double-wye transformer the formula for output voltage is

$$E_{d\text{-}c} = 1.17E_s - (0.2388IX) - \frac{P}{I} - E \qquad (22\cdot3)$$

where $E_{d\text{-}c}$ = output voltage at the specified load
E_s = anode-to-neutral no-load rms voltage
I = d-c ampere output of the rectifier unit
X = commutating reactance of transformer in ohms
P = copper loss of the transformer in watts
E = arc-voltage drop of the rectifier.

Voltage-regulation calculations are based on a sustained sinusoidal alternating voltage supply to the high-voltage terminals of the transformer. Regulation is influenced by the reactance of the a-c supply system. If the characteristics of the supply system are known, the reactance can be expressed in terms of transformer commutating reactance and can be included. Usually the reactance of the supply system is low compared with the reactance of the transformer, and has only a slight effect on regulation. The reactive effect of the supply system is somewhat greater than the voltage drop as measured by an a-c voltmeter because distortion of the voltage wave has an additional tendency to make the regulation curve steeper. If a system has unusually high reactance, the reactance of the system should be specified to the manufacturer.

The difference in arc-voltage drop between light load and full load usually is only 2 or 3 volts; therefore, it has little influence on regulation. This voltage drop is almost the same regardless of the output-voltage rating of the unit.

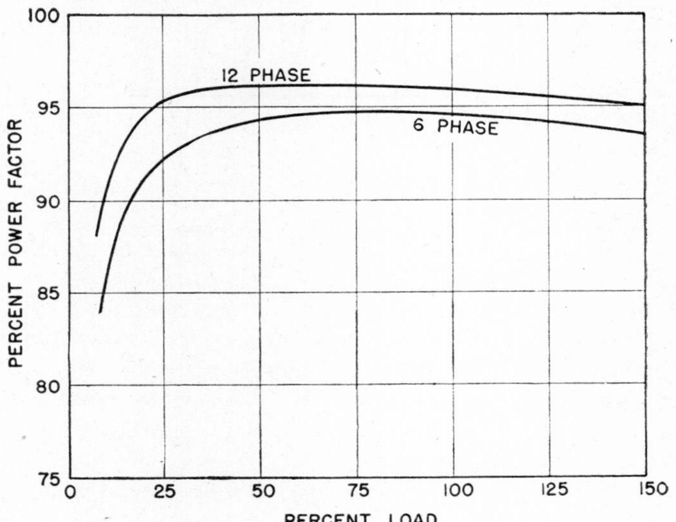

FIG. 22·22 Typical power-factor curves.

Usually the equivalent voltage drops in the component parts result in a regulation of 5 to 6 percent from full load to light load.

Output voltage of a rectifier unit with a twelve-phase quadruple zigzag transformer is

$$E_{d\text{-}c} = 1.17E_s - (0.1195IX) - \frac{P}{I} - E \qquad (22\cdot4)$$

This formula is similar to the formula for the six-phase double-wye unit, except that the effect of reactance is only one-half as much. The transformer usually is designed with a commutating reactance approximately double that of a six-phase double-wye transformer. The effect of the system reactance is, however, only one-half as much as when supplying a six-phase double-wye connected transformer.

Output voltage of a unit with a six-phase star transformer is

$$E_{d\text{-}c} = 1.35E_s - (0.95IX) - \frac{P}{I} - E \qquad (22\cdot5)$$

Here the effect of commutating reactance is greater than for either the six-phase double-wye or twelve-phase quadruple zig-zag units; regulation is greater, and is influenced more by the reactance of the a-c supply system.

Output voltage of a unit with a six-phase parallel double-wye transformer is the same as for a unit with a six-phase double-wye transformer.

If the load on the unit is lower than the excitation requirements of the interphase transformer, its operation will revert from six-phase, double-wye to six-phase, star. This reversion causes an output-voltage rise at zero load of from $1.17E_s$ to $1.35E_s$, a rise of 15 percent. The transition from double-wye to star operation is not sudden; the voltage rise increases as the load on the unit is decreased below the interphase excitation requirements. All rectifier units using interphase transformers have this light-load voltage rise, and special provisions may be required for its suppression. If the anticipated load on the unit is less than the excitation requirements of the interphase transformer, suppressing equipment should be provided. The simplest method is to load the unit permanently or intermittently by means of a resistor to provide an adequate load on the unit. The intermittent resistor can be controlled automatically so that it is connected into circuit only when it is required.

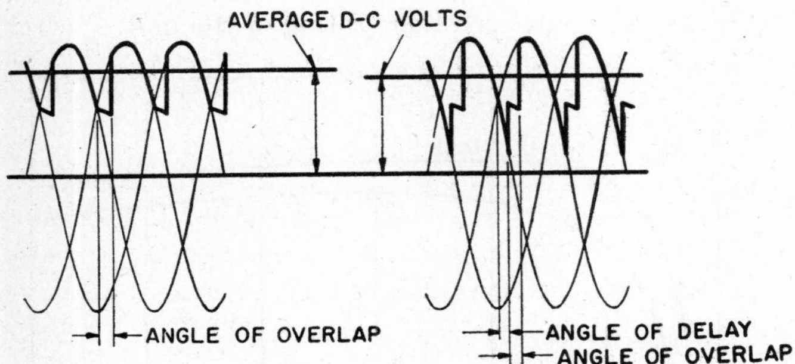

Fig. 22·23 D-c voltage reduction by phase control.

The usual interphase transformer is fully excited when the load on the rectifier unit is 0.5 percent of its full-load rating. If a large load is maintained, as in electrochemical loading, the interphase transformer may be designed for higher excitation, but it will not exceed 5.0 percent of the full-load rating of the unit.

Phase Control of Output Voltage

Voltage output of the unit at any given load bears a fixed relation to the alternating voltage. The only way in which output voltage can be increased is to change taps of the rectifier transformer, or to increase the a-c supply voltage. Output voltage can be decreased by delaying the point in the cycle at which the anodes begin to conduct current. This is accomplished by control grids in continuously excited rectifiers and by delaying the ignitron impulse in intermittently excited rectifiers. The effect of delayed pick-up of current by the anodes is shown in Fig. 22·23. Voltage output is the average of the voltage waves; because delay in pick-up results in a decreased average, output voltage is decreased proportionally. Although the average voltage is decreased, the peak voltage of the waves sometimes is maintained at the original value. This is important if stress, instead of the average output voltage, is to be reduced. Voltage stress is proportional to peak voltage, not to average voltage.

If the control equipment is set for a fixed angle of delay in pick-up, the regulation curve is parallel to the normal voltage-regulation curve, and is displaced from it by an amount proportional to the amount of fixed-excitation delay. Automatic regulators can vary the pick-up delay, depending upon the load on the unit, or can maintain a fixed voltage on the unit regardless of load.

Control of output voltage has an adverse effect upon the power factor of the unit and upon harmonics in both a-c and d-c systems.

Voltage-regulating equipment is not effective in suppressing light-load voltage rise, which must be suppressed in the same way as for a unit without voltage control.

REFERENCES

1. "Rectifier Wave Shape," Edison Electric Institute, 420 Lexington Ave., N. Y. C., Publication E1.
2. "Inductive Co-ordination Aspects of Rectifier Installations, an A.I.E.E. Subcommittee Report," *A.I.E.E. Tech. Paper* 46–105.

Chapter 23

INVERTERS

J. L. Boyer

THE conversion of d-c power to a-c power by electronic equipment can be accomplished by either an oscillator or an inverter. These two terms usually are applied to definite types of conversion equipment. A device using vacuum tubes or rectifying elements in which the flow of current through the elements is controlled continuously is called an *oscillator*. A device using gaseous tubes or rectifying elements in which only the start of current conduction can be controlled is called an *inverter*. The characteristics of these two types of converters are different, and therefore each has its own field of application. Only the inverter is considered in this chapter.

The inverter is applied where highly efficient static equipment is required and where the output a-c frequency is not above that at which gaseous tubes can be controlled.[1, 2] Ratings from a few watts to thousands of kilowatts can be obtained with the designs of gaseous tubes which are now available.[3] The normal lower frequency limit is in the low-power frequencies and can be as low as is practical for the transformer equipment. The tubes may have an extremely low-frequency limit, and some types can conduct a continuous current. The upper frequency limit is from about 2000 to 3000 cycles per second, which is about the practical frequency limit for which mercury-vapor gaseous tubes can now be designed to control the forward currents.

23·1 INVERTER OPERATION

If the angles of ignition retard of a rectifier are increased, the d-c voltage is reduced because the tubes conduct in periods during which the average of the working portion of the a-c voltages is lower. If the current is caused to continue flowing, the d-c terminal voltage can be reversed if the angles of retard are increased so the tubes conduct more of the time during the negative half of the a-c voltage wave than during the positive half. The a-c circuit receives energy if the current is caused to flow in the normal direction through the tubes by some voltage source in the d-c circuit. This type of operation with gaseous tubes is known as *inversion*.

The control characteristics of gaseous tubes under normal load conditions and with the usual control circuits are such that it is possible only to prevent the start of load current. For this reason if gaseous tubes are to control the transfer of energy from a d-c circuit to an a-c circuit, the power

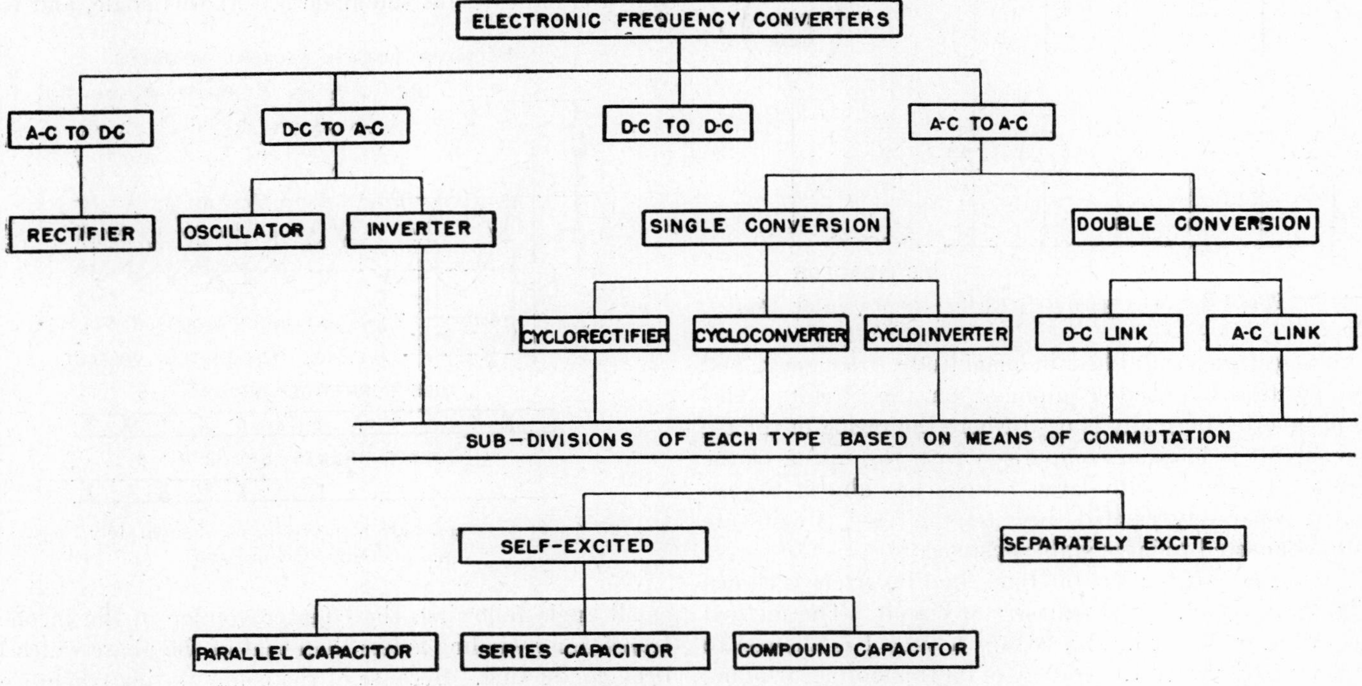

Fig. 23·1 Classification of electronic frequency converters.

circuit must have circuit voltages which can transfer the current from one tube to the next conducting tube.

A large number of circuits will meet the requirements for inverter operation, but all can be classified into the following types as determined by their operating conditions:

1. Separately excited inverter.
2. Self-excited inverter.
   a. Parallel capacitor
   b. Series capacitor
   c. Compound capacitor

The relation of these inverters to other types of electronic frequency converters is shown in Fig. 23·1. Since inverter operation is involved in some of the other types of electronic frequency converters, they are also considered here.

23·2 SEPARATELY EXCITED INVERTER

The determining characteristics of a separately excited inverter is that it supplies energy to an a-c circuit in which

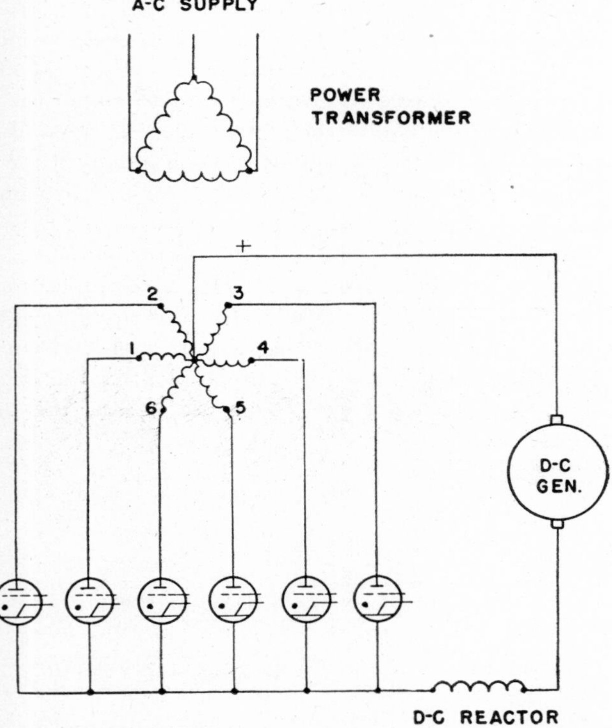

Fig. 23·2 Power circuit of separately excited six-phase star inverter.

the voltages are established in magnitude, frequency, and phase position by other equipment on the circuit. This type of inverter normally is used where the rating of the a-c power circuit is large in comparison with the rating of the inverter. Inverters with large ratings are almost always operated separately excited, because their a-c circuits are usually connected to large stable systems.[4, 5]

A typical circuit for separately excited inverters is shown in Fig. 23·2, which is a six-phase star circuit. The method of operation is shown in Fig. 23·3. If tube 1 is allowed to start conducting when the voltage of its transformer winding is opposed to the normal flow of current through the tube, the total direct current can be conducted through phase 1 because the impressed voltage of the d-c generator is greater than the transformer phase to neutral voltage. Energy is delivered to the transformer phase that is conducting. Current is prevented from flowing in the other windings by blocking the control electrodes of the other tubes.

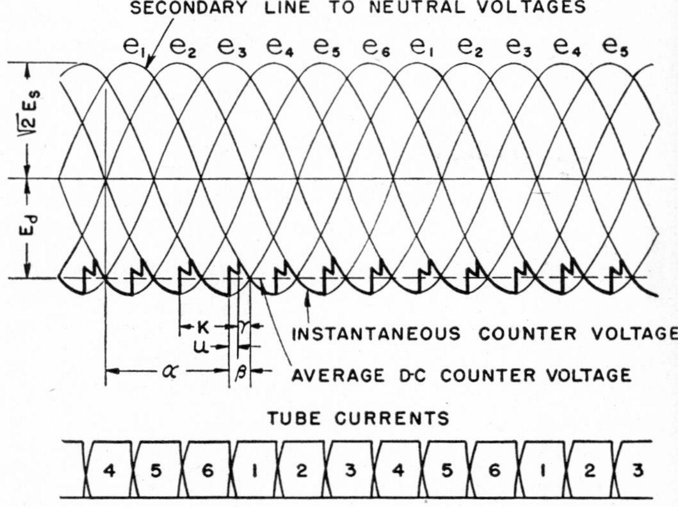

Fig. 23·3 Voltage and current wave forms of separately excited six-phase star inverter.

The usual inverter circuit carries a continuous direct current, because the inductance in the d-c circuit sustains current after the instantaneous counter voltage of the transformer winding is greater than the impressed voltage of the line. Current can be transferred from one transformer winding to another by releasing the second tube when the a-c voltage of its winding is less negative than that of the conducting tube. The angle during which this commutation can occur is shown in Fig. 23·3. The ignition angle of tube 2 is delayed from its zero-delay rectifier firing angle by about 150 degrees, which is indicated as the angle α. At this angle, and for a

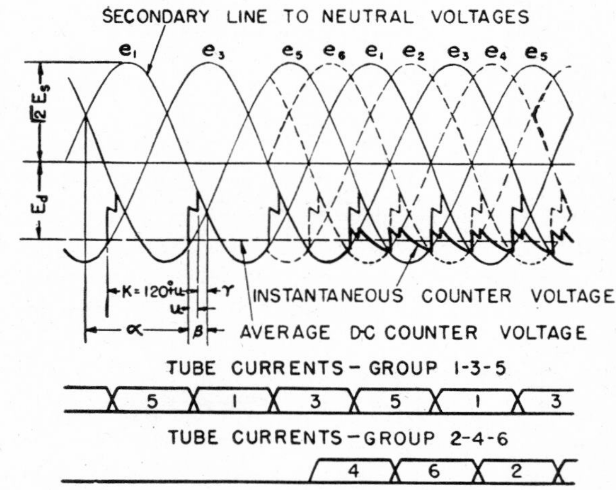

Fig. 23·4 Voltage and current wave forms of separately excited six-phase double-wye inverter.

small angle following, the counter voltage of the incoming transformer winding is less than that of the phase which has been conducting. Because of the commutating reactance of the transformer windings, a small angle is required to commutate the current from phase 1 to phase 2. The angle measured from the instant at which the current has been completely commutated to the instant at which the negative voltages of the outgoing and conducting phases are equal is the margin angle γ, and is the time available for the tube to deionize and gain control. Tube 2 conducts the current until it is time for tube 3 to start; then current is commutated from tube 2 to tube 3 by normal circuit voltages. The process continues until all tubes carry the current for slightly more than 60 degrees each cycle.

The operation of another important circuit for separately excited inverters is shown in Fig. 23·4. This circuit is the six-phase double wye, which is desirable for ignitron inverters of high capacity. (See Fig. 23·8.) Here this circuit has the same advantages as it does with a rectifier, and the reduction in peak ionization is especially important for inverter operation.

In large inverter installations in which the d-c voltage supply can be independent of all other equipment, the six-phase double-way circuit shown in Fig. 23·11 should be supplied. The transformer kva rating is lower, and under certain conditions the counter voltage drops to only half value, if there is a fault on an inverter tube. The circuit is used for high-voltage d-c circuits because the forward voltage on the tubes is only about half of that on single-way circuits.

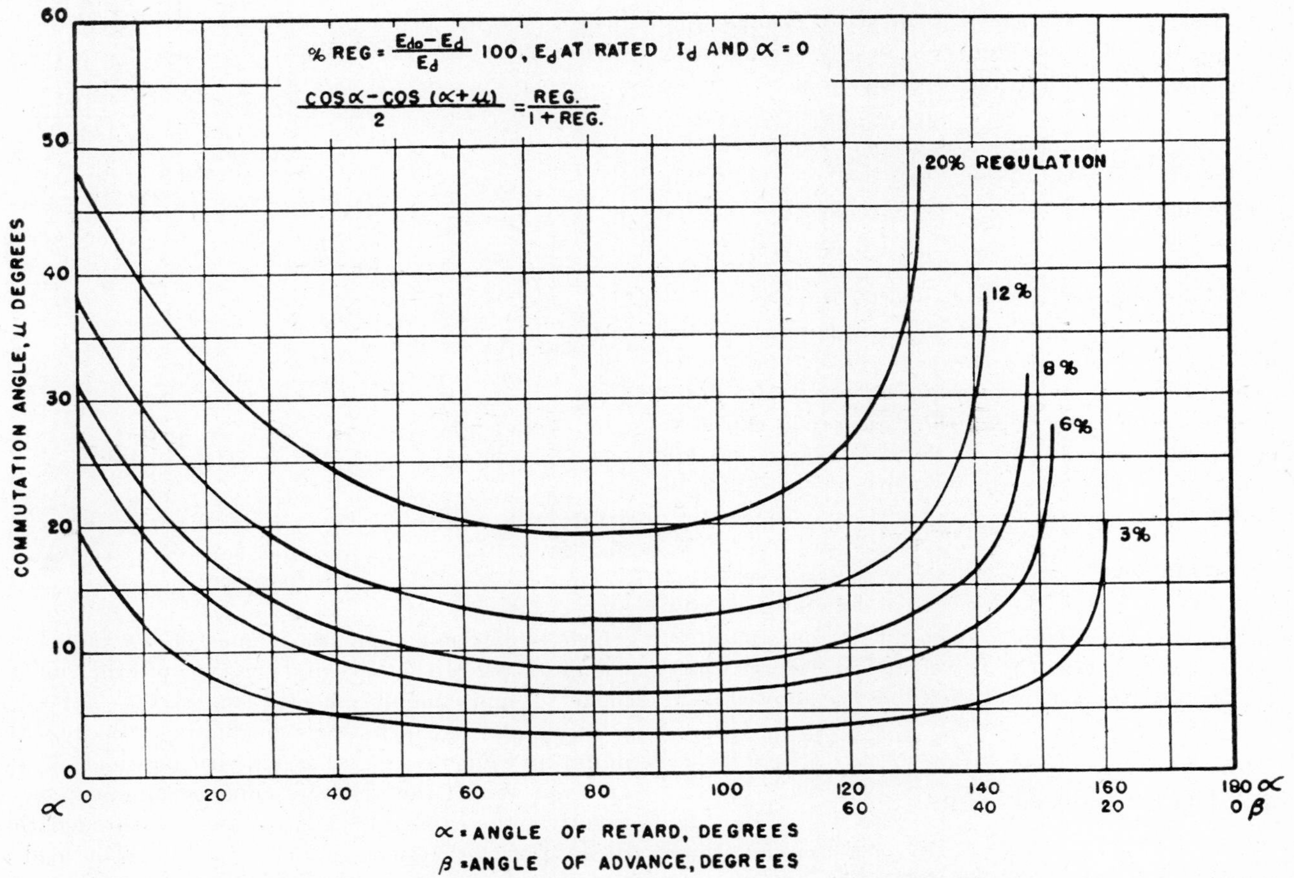

FIG. 23·5 Commutation angles for rectifier and inverter operation for several values of d-c voltage regulation. The commutation angles are independent of the number of phases.

Commutation Requirements

The major requirement that a separately excited inverter makes on the a-c circuit is that the a-c circuit receive a current which leads the voltage. This means that the inverter requires a lagging reactive current from the a-c circuit in order to commutate correctly. If the inverter is considered to be supplying the kilowatt load to some a-c equipment which also requires lagging current, the a-c circuit must furnish the lagging reactive current of both the inverter and its load. The magnitude of this reactive kva requirement can be determined separately, because the requirements of the load and the inverter are not related. The reactive kva requirements of the load depend upon the type of equipment and can be determined by normal methods. Reactive kva requirements of the inverter are dependent upon the angles of ignition advance of the tubes, the commutating reactance of the transformer, and the load current. Usually it is between 50 and 80 percent of the real power being converted. If circuit conditions are known, reactive kva can be calculated by the following formula:

$$\text{Reactive kva} = E_{d0} I_d \frac{2u - \sin 2\beta + \sin 2(\beta - u)}{4[\cos(\beta - u) - \cos\beta]} \qquad (23\cdot 1)$$

where E_{d0} = d-c voltage at no load operated as a rectifier with no ignition delay

I_d = d-c load current in average amperes

u = commutation angle in radians as an inverter

β = angle of advance of the inverter = $180° - \alpha$.

Conditions of operation of a separately excited inverter are determined by the ratio of the a-c voltage to the d-c impressed voltage, the load, the commutation angle, and the ignition angle. These quantities are interrelated and can be found by the following equation:

$$E_d = -\left\{E_{d0}\left[\cos\beta + \frac{\cos(\beta - u) - \cos\beta}{2}\right] + \frac{P_r}{I_d} + E_a\right\} \tag{23·2}$$

where E_d = average d-c voltage under load
P_r = power loss in transformer in watts
E_a = arc-drop voltage in tubes.

In the analysis of the separately excited inverter, the a-c voltage usually is considered to be constant in wave form and magnitude, and the d-c voltage is assumed to change with load. The angle of ignition advance is set by the operator or by some type of regulator, and with power inverters the advance angle is usually between 30 and 40 electrical degrees. The relation between the ignition angle and the commutation angle for various amounts of regulation for converters at full load is shown in Fig. 23·5.

Rating

The rating of a separately excited inverter usually is given in kilowatts, because it can convert only real power from the d-c circuit to the a-c circuit. However, since this type of inverter always has reactive kva requirements it also has a total kva rating, which determines the transformer capacity. The ranges of the a-c voltage and the d-c voltage and the overload limit have an important influence on the design of the over-all inverter system.

Frequency Limitations

A change in frequency of the a-c system does not have any great effect on the power-circuit operation of the separately excited inverter, so long as the power and excitation transformer equipment is suitable and the tubes are allowed sufficient time for deionization.

23.3 SELF-EXCITED INVERTERS

If an inverter is to supply power to an a-c circuit which does not have established voltages, a circuit which is self-excited must be used. This type of inverter is characterized by the fact that it has the determining influence on the wave form of the output a-c voltage. It may be of the parallel-capacitor type, the series-capacitor type, or a combination of the two types, depending upon the circuit arrangement used to obtain commutation. A requirement of a self-excited inverter is that it supplies a load current which, combined with the current required by the commutating capacitors, produces a resultant current which leads the output voltage. The different inverter circuits have a number of explanations of their method of operation, but fundamentally they are based on the principle of bringing the tube current to zero by some negative circuit voltages and blocking the tube by its control electrode before the circuit voltage reverses and becomes positive.[6, 7, 8, 9]

Parallel-Capacitor Inverter

A single-phase parallel-capacitor type of inverter is shown in Fig. 23·6. The operation of this circuit can be understood by assuming that tube 1 has started conducting. The full d-c voltage E_d will then be impressed on the winding 5 and the d-c inductance 8. At first, most of the voltage will be on the inductance, because the transformer is connected across the commutating capacitor 9, but as the capacitor becomes charged more voltage is impressed on the transformer. Finally almost the full d-c voltage is impressed on half of the transformer. The capacitor then is charged to twice the d-c voltage, because it is connected across the whole winding. The anode of tube 2 is positive with respect to its cathode by about double the d-c voltage, but it does not start to carry current because a cathode spot has not been established or the grid is held negative. In the first half of the cycle, load 10 receives a voltage the wave form of

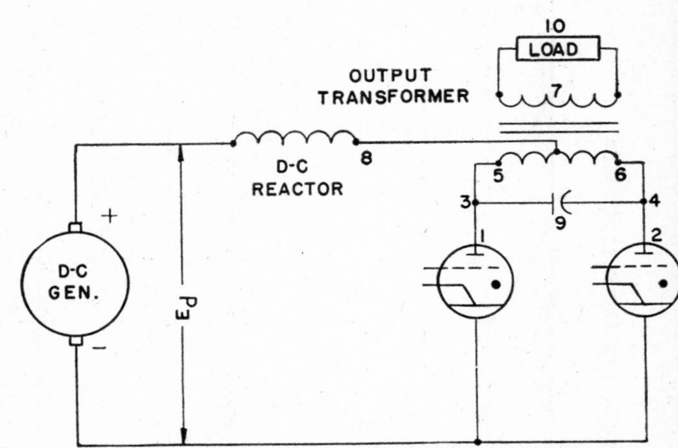

FIG. 23·6 Power circuit of parallel-capacitor, self-excited, single-phase inverter.

which is determined by the values of d-c inductance and commutating capacitance and the type of load, but it can be made to approximate half of a sine wave. Tube 2 is then released, and its anode starts conducting. Because the commutating capacitor was charged with terminals 3, negative, and 4, positive, the anode of tube 1 is immediately forced negative; the capacitor cannot discharge through the tubes because of the rectifier action of tube 1. The load current will then continue to flow through winding 5, through the capacitor, and then through tube 2 back to the input circuit as long as the load, the commutating reactances, and the commutating capacitor demand current flow in this direction. As the commutating capacitor becomes discharged and begins charging in the other direction the load current starts to flow in winding 6. The process then is repeated with tube 2, and the current later is commutated back to tube 1. The timing of the firing of the tube determines the output frequency.

Capacitor Requirements

The critical part of the self-excited inverter is the commutating capacitor. At a given load it determines the portion of the cycle which is allowed for the tubes to deionize. With a resistance load, low-output frequency, and low-power capacities a margin angle of about 18 degrees, which should be sufficient, can be obtained with a capacitor volt-ampere rating of about 60 percent of the load watts. High-capacity or high-frequency inverters may require a capacitor volt-ampere rating up to 100 percent of the load watts. A reasonably accurate method of estimating the capacitance required for lagging reactive load is to add the capacitance required to bring the load to unity displacement factor and estimate the remainder as a resistance-loaded inverter based on these percentages.

Applications

The parallel-capacitor type of inverter should be used for low-regulation a-c power supplies on which no heavy overloads are expected. The circuit elements can be selected so that almost any type of load can be supplied. This type of circuit has a definite overload limit. If it is exceeded, the inverter faults because two tubes start to conduct at the same time. This type of circuit is essentially a constant-voltage circuit.

Series-Capacitor Inverter

The series-capacitor inverter is shown in Fig. 23·7. This circuit has the advantage that it can supply a load with a widely variable impedance. A short-circuit current can be supplied, provided the tubes continue to control and do not

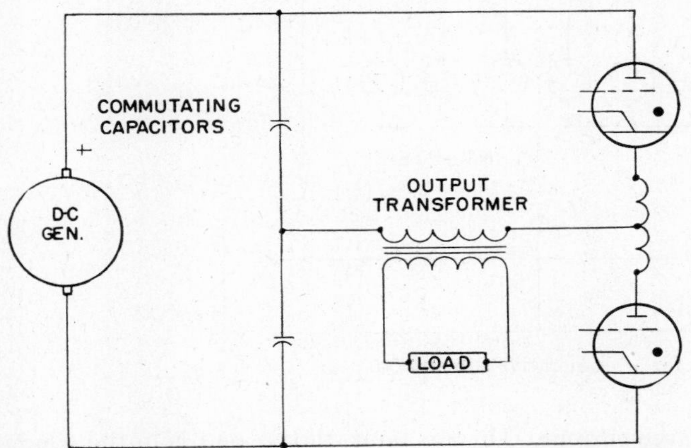

Fig. 23·7 Power circuit of series-capacitor self-excited single-phase inverter.

arc back. The circuit must carry some load at all times. Its voltage regulation always is rather high, and the voltage wave form changes considerably with load. This type of circuit is essentially a constant-current circuit.

Compound-Capacitor Inverter

By combining the parallel-capacitor and series-capacitor arrangements a compound-capacitor inverter circuit with a wide range of characteristics can be obtained. The parallel capacitor improves operation at light loads; the series capacitor makes it possible to operate at heavy loads. The wave form of the output voltage varies somewhat with load, but with the optimum capacitances it is approximately sinusoidal. The compound-capacitor inverter may become the most important type of self-excited inverter, because it has the best characteristics over a wide load range.

Special Circuit Conditions

Self-excited inverters should be avoided for some applications because of the possibility of obtaining circuit conditions that will prevent satisfactory operation. One of these is the starting of induction motors that require about the same rating as the inverter capacity. Resonance can be obtained which prevents the motor from reaching full speed, and can cause serious circuit disturbances. Special consideration should be given to self-excited inverters supplying loads with widely variable power-factor or variable-frequency requirements.

23·4 INVERTER FAULTS

If the load on an inverter is increased, a current will be reached at which the inverter suddenly fails and the d-c supply is short-circuited. The faults obtained on inverter operation are similar for all types of inverters. During correct operation, the blocked inverter tubes and the conducting transformer windings oppose the flow of forward current through them, and during a fault the counter voltage of the inverter falls to a value that averages zero over a cycle.

Arc-Through

The general term used for faults in which the average counter voltage falls to zero is the "arc-through," which is defined as the conduction of power current in the forward direction through a tube during a scheduled non-conducting period of the cycle. Arc-throughs have several causes and can be subdivided as (1) forward-fires, (2) misfires, and (3) incomplete commutations.

Forward-Fire. Failure of the tube to hold a forward voltage on the anode is known as a forward-fire. Failure is caused by the loss of control which may be due to the tube or its excitation circuit. If load current and voltage are at the limiting rating of the gaseous tubes there may be some loss of the control functions. In thyratrons this is caused by failure of the grid to prevent the start of load current. In ignitrons forward-fires are caused by the establishment of a cathode spot on the mercury surface or tube wall before the ignitor is fired. Control characteristics of thyratrons or ignitrons are partly determined by the residual ionization in the tube as the anode becomes positive. The amount of residual ionization is largely determined by load current, space between surfaces, and vapor temperatures.[10, 11] The excitation circuit can cause a tube to forward-fire by allowing the grid to become positive or by firing the ignitor during the non-conducting portion of the cycle. One important cause of the loss of grid control is a result of a sudden change in the main anode potential, which makes the grid positive for a short time because of the capacitance of the grid to the anode. Loss of control due to a change in grid potential is obviated by the use of external capacitors between the grids and cathode so that most of the surge voltage is impressed on the anode-to-grid capacitance.

Misfire. A misfire is a failure of a tube to start conducting current at its normal ignition angle. An arc-through is produced by a misfire because the current is not commutated from the conducting tube to the tube which has the misfire. The counter voltage in the transformer winding for the conducting tube reverses; then the full d-c voltage plus the a-c voltage tend to make the current increase. The misfire is caused by either the tube or its excitation circuit.

Incomplete Commutation. If the load on an inverter is increased until there are not sufficient volt-seconds during the advance angle to commutate the current from the transformer winding, the outgoing tube cannot gain control because the load current does not go to zero. The fault results from the fact that when the current is not completely commutated it commutates back to the original tube because the circuit voltage reverses. Current then continues in the first tube until the d-c source is interrupted or its voltage is removed. The incomplete-commutation type of fault usually is caused by the load circuit, but it may be caused by incorrect phasing of the excitation circuit. This type of fault is one of the main limitations of inverters, and the faults can occur regardless of the quality of the gaseous tubes that are used. However, the incomplete commutation fault merely produces a definite load limit for the equipment, and can be accurately predicted, so the inverter can be designed for any desired load.

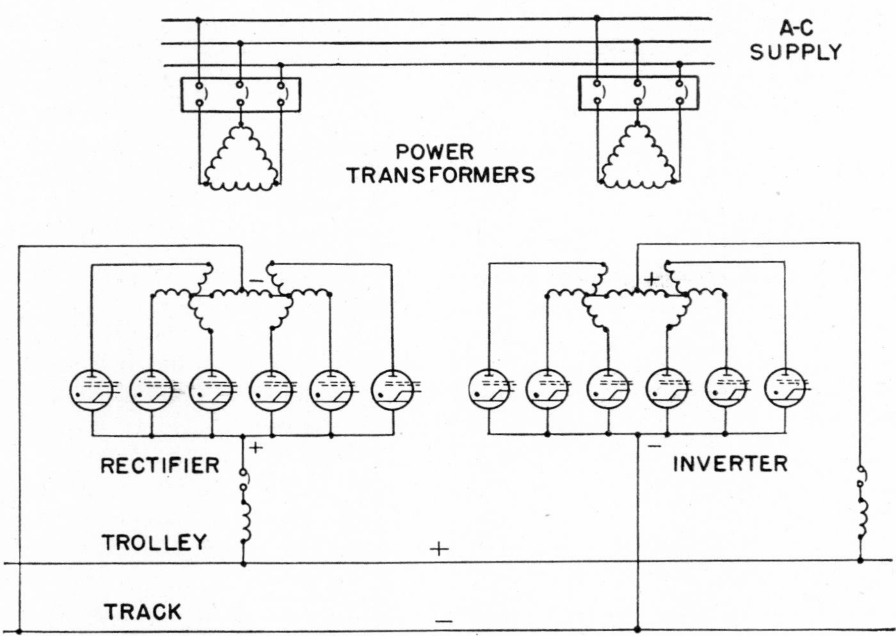

FIG. 23·8 Rectifier and inverter circuit for a regenerative sub-station.

Arc-Backs on Inverter Tubes

Arc-backs are sometimes found to occur in inverter operation, but they are not serious by themselves, since the circuit voltages do not maintain reverse current in the tubes. The arc-back may come at the end of the conduction period in the short angle when the anode is negative. The important disadvantage of the arc-back is that it usually becomes an arc-through because there is considerable ionization in the tube when the anode goes positive.

Clearing Inverter Faults

To clear an arc-through it is necessary only to remove the impressed d-c voltage, because no fault current is supplied by the other tubes in inverter operation provided they control correctly. Usually only one tube is involved; however, in some special circuits, such as the six-phase, double-wye, and the six-phase, double-way, a fault on one tube causes an overvoltage on the other group of tubes and may cause another fault.

An arc-through on an inverter does not always cause a complete circuit fault. Whether the fault is continuous is determined by the current in the faulted transformer phase at the times when current can be commutated to the next phase. The current tends to be commutated out of the faulted winding at the normal commutation angles, and if the current has increased only a slight amount this usually is possible. An inverter can be designed so that a single fault on the tube is corrected in one cycle, by using a large d-c inductance and exciting the tubes in order to commutate a large current. On this fault, the current is in the winding for one cycle, and when the second normal commutation angle, is reached the current is commutated and correct operation is started again.

23·5 APPLICATION OF INVERTERS FOR REGENERATIVE LOADS

If a large amount of energy must be returned by the load to the d-c circuit, the load cannot be supplied by rectifiers unless some provision is made to absorb the energy. It can be dissipated on resistors, or it can be returned to the a-c system by means of inversion. For railway substations in mountainous country, the rating of the inverter may have to be about the same as that of the rectifier. Since this type of converter is always connected to a stable a-c system, it usually is operated as a separately excited inverter.[12]

Several methods can be used to obtain regeneration. In one method the same unit can be changed from rectifier operation to inverter operation by switching the polarity of the d-c terminals and retarding the ignition angles of the tubes. Another method is to use one or more units to supply the rectifier load and one or more units operated as inverters to take regenerative loads. On heavy inverter overloads some of the units can be switched from rectifier to inverter operation to help carry the load, and when the regenerative load drops the units can be switched back to rectifier operation.

returning energy to the a-c circuit. The contactors are changed so that current can reverse, and the ignition angles are increased until inversion can take place. The magnitude of the regenerative current can be controlled by varying the firing angles of the inverter tubes. As the motor slows, the firing angles can be advanced and the motor speed can be reduced by a constant torque. The motor then can start in the opposite direction of rotation by rectifier action.

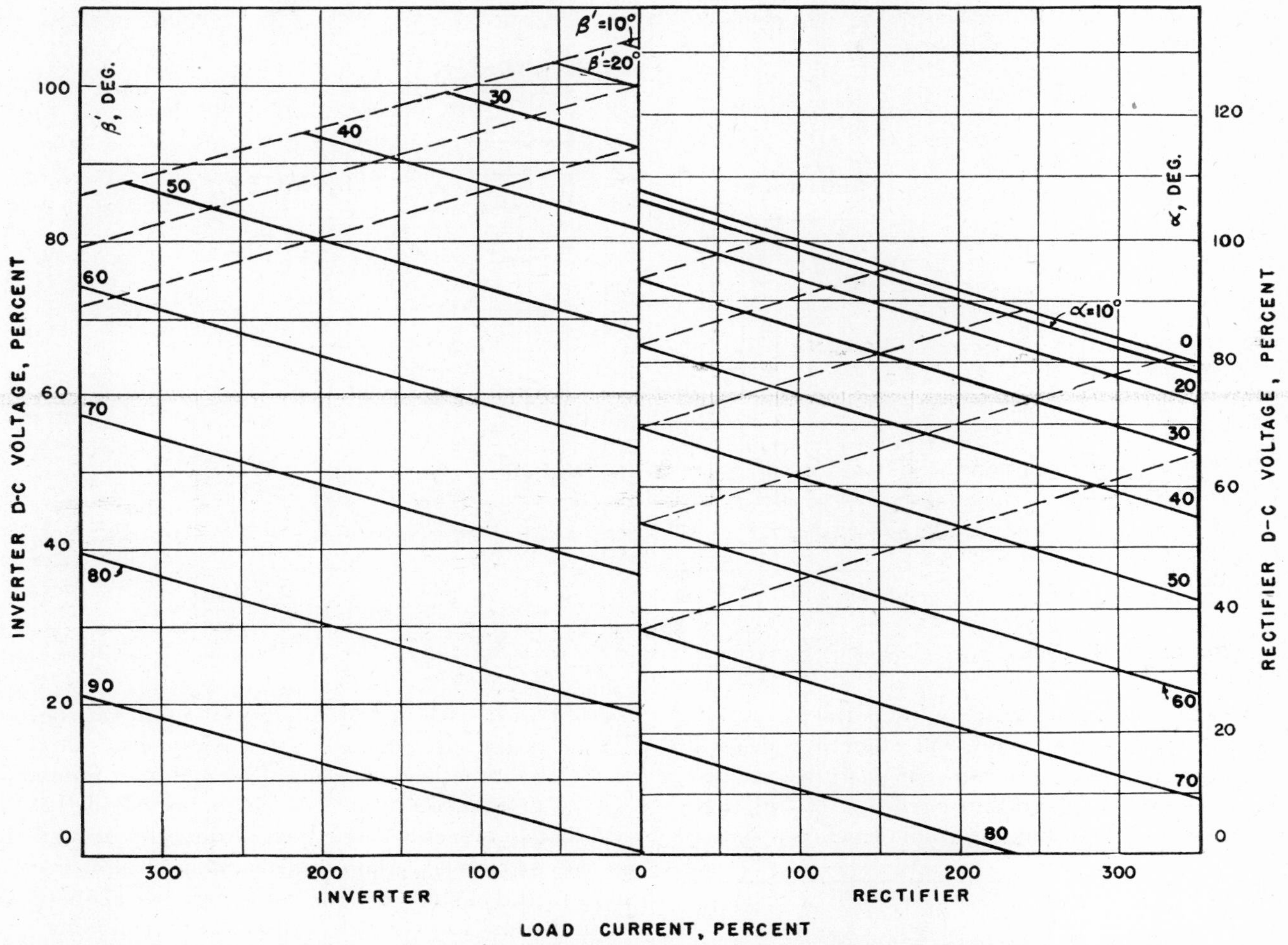

Fig. 23·9 Combined voltage-regulation characteristics of rectifier and inverter sub-station. Rectifier regulation is 8 percent, and inverter regulation is 6 percent with a higher transformer voltage.

If regeneration is required by a load, the only indication of the need for reversal of power transfer is a rise in the d-c voltage. The inverter counter voltage should be set by its ignition angles so that it starts to convert a regenerative load as soon as the rectifier voltage reaches its no-load value.

A typical rectifier and inverter sub-station for delivering or absorbing power from a d-c railroad trolley is shown in Fig. 23·8. The operating conditions which can be obtained are shown in Fig. 23·9. Because the a-c voltage of the transformer secondaries are higher on the inverter, small angles of ignition retard can be permitted on the rectifier and still allow sufficient margin angle on the inverter for operation with normal overloads.

Reversing D-C Motors

Reversing d-c motors is a special application for which inverter action may be required.[13] The action is the same as for regeneration, except that the reversal of the direction of power flow must take place over a wide range of d-c voltages. Ordinarily only one converter is used, as shown in Fig. 23·10. For sudden reversal, the inverter brakes the motor by

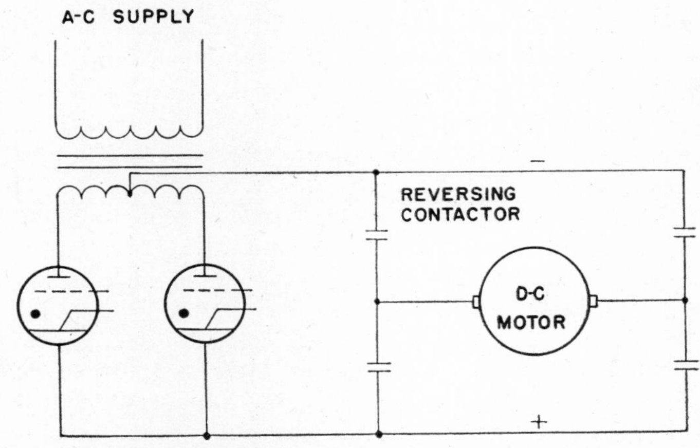

Fig. 23·10 Simplified circuit for single converter supplying d-c motor with contactors to allow regeneration on reversing.

23·6 RECTIFIER AND INVERTER FREQUENCY CHANGER

The application of inverters to frequency changing has become important in the past few years.[14, 15, 16] The electronic frequency changer converts power from a 60-cycle power system to a 25-cycle system for industrial plants. This operation requires a circuit that will allow a considerable variation in the frequency ratio without any great change in power flow. One of the important circuits with this characteristic is known as a d-c link electronic frequency changer.

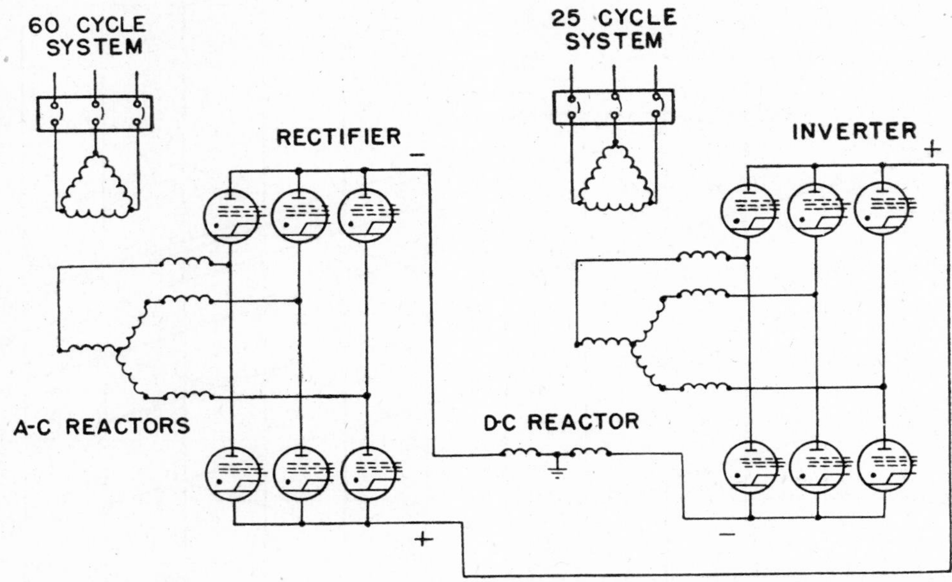

Fig. 23·11 D-c link electronic frequency converter for transfer of power between 60-cycle and 25-cycle power systems.

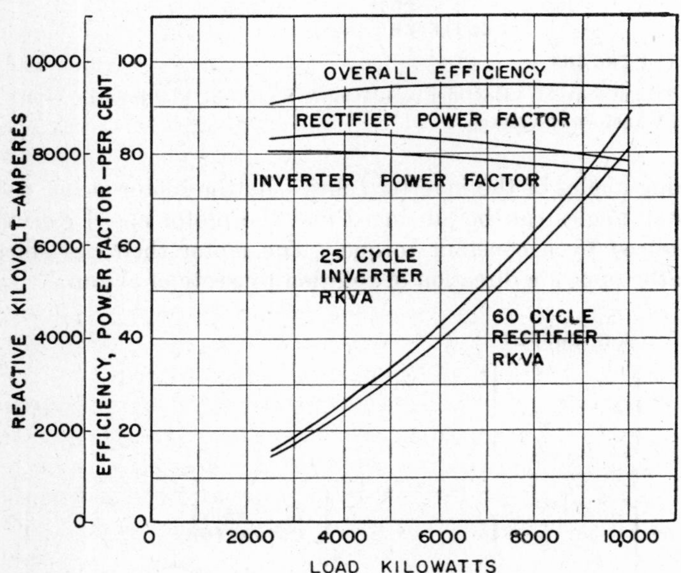

Fig. 23·12 Characteristics of a 10,000-kilowatt d-c link electronic frequency converter.

It consists of a rectifier connected to the a-c system supplying the power and an inverter connected to the a-c system receiving the power. (See Fig. 23·11.)

The magnitude of the d-c voltage is relatively independent of other equipment, and therefore the inverse or forward voltage on the tubes should be as high as possible to obtain a maximum efficiency. The upper limit in the d-c voltage is determined by the voltage limit of the tubes at which maximum power capacity can be obtained. The tubes usually are operated in a six-phase double-way circuit because the transformer kva is lower. For capacities greater than can be obtained with six tubes, several units can be operated with the d-c circuits in series; consequently all converters carry the same load currents. To maintain minimum voltage to ground, the units can be connected so that there is alternately a rectifier and then an inverter around the loop of the d-c circuit. Almost any number of units can be operated in the circuit, and the converters become less sensitive to individual inverter-tube faults as the number is increased.

Characteristics

The rectifier and inverter type of electronic frequency changer operates normally so that the two units are relatively independent except that the average d-c voltage and d-c current must be equal. The system frequencies are not related in any way, and the frequency ratio can vary over a large range as long as frequency limits of the excitation circuits and the transformer equipment are not exceeded on either unit. This characteristic and the fact that high efficiencies can be obtained are the outstanding advantages of this type of equipment. The performance of a typical 10,000-kilowatt electronic frequency changer is shown in Fig. 23·12. This unit was designed for conversion of power from a 60-cycle system to a 25-cycle system with provisions for reversing the direction of energy flow.

Applications for Electronic Frequency Changers

For large power ratings the electronic frequency changer has a higher initial cost than a synchronous-synchronous type of rotating frequency changer. The large number of parts and the complexity of the excitation and control circuits for the high-voltage ignitrons make the electronic frequency changer rather expensive. However, if there are some special requirements which cannot be easily accomplished by other types of frequency changers, the electronic type can usually be justified on its advantages of high efficiency and wide flexibility.

There are at present four installations of electronic frequency changers with rating above 5000 kilowatts. Three of the units are used for the conversion of 60-cycle power to 25-cycle power in industrial plants. The experience gained with these frequency changers indicates that reliable economical operation can be obtained in large electronic frequency changers. The prospects for improvement of this type of equipment in the future are very good, since as the ignitron tube design is improved it may be possible to reduce greatly the amount of excitation equipment.

23·7 SINGLE-CONVERSION ELECTRONIC FREQUENCY CHANGER

In the frequency changers which have been described, it has been necessary to change the a-c power of one frequency to d-c power and by inverter operation change the d-c power to the new frequency on the output a-c system. With single-conversion circuits, power is converted from one frequency directly to another frequency without any d-c link.[17, 18, 19]

The single-conversion type of electronic frequency changer can be subdivided into cycloconverters, cyclorectifiers, and cycloinverters. Application of these types is dependent upon frequency ratio and the need for reversibility of energy transfer. The cycloconverter is applied where energy must flow in either direction; the cyclorectifier is applied where energy flows from the higher frequency circuit to the lower frequency circuit; and the cycloinverter is applied where energy flows from the lower frequency circuit to the higher frequency circuit and under certain conditions for the opposite direction of power transfer. Each type can either be separately excited or self-excited on the power circuit. For separate excitation the higher frequency a-c circuit must have sufficient commutating kva capacity to transfer current from one tube to the next. Where there is no equipment on the higher frequency circuit to produce current commutation, or if the displacement angle of the current on the lower frequency circuit is not zero, the converter must be designed to operate self-excited.

23·8 CYCLOCONVERTER

An important type of single-conversion electronic frequency changer is the cycloconverter. When it is operated with a higher frequency circuit which has sufficient commutating capacity, the cycloconverter can supply almost any kind of lower frequency load. The direction of energy transfer can be reversed instantly without changes in the excitation circuit; the power factor can be either unity, leading, or lagging.

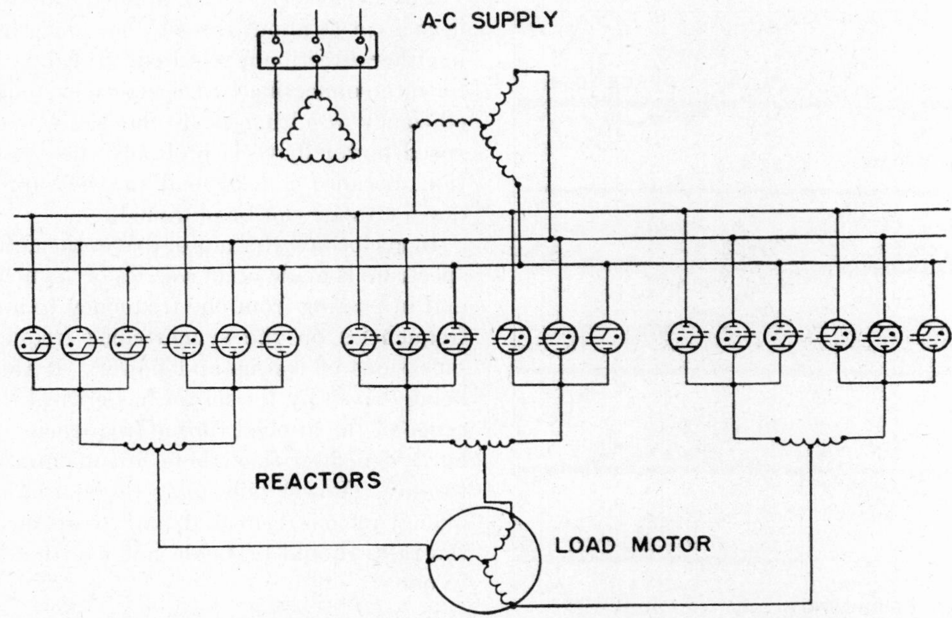

FIG. 23·13 Power circuit of a variable-speed synchronous motor using a cycloconverter type of single-conversion electronic frequency converter.

Circuit Operation

A typical cycloconverter circuit is shown in Fig. 23·13, where the load is a variable-speed synchronous or induction motor. Considering one phase of the output to the motor, one group of three tubes is connected as a three-phase converter to supply the positive current to the motor, and one group of three tubes is connected as a three-phase converter to supply the negative current for the motor. The tubes are controlled so that in one group several input frequency phase voltages combine to form half of the output voltage wave. Output frequency is determined by the number of input phases forming the envelope of half the output wave. This can be varied to produce a frequency that can be made to average any value between zero and some fraction to the input frequency.

There are many variations of the cycloconverter which can be applied to obtain different output voltage wave forms and special operating conditions. Its most important characteristic is that both positive and negative groups of tubes are controlled; therefore either group can conduct during any period. The group of tubes which conduct the current is determined by the circuit voltages. If the tubes conduct during periods when the higher frequency voltage is in the same direction as the current, the tubes then operating as rectifiers conduct current, and power flow is from the higher frequency circuit to the lower frequency circuit. If the tubes conduct when the higher frequency voltage is opposed to the current, the tubes then operating as inverters conduct the current, and the power flow is into the higher frequency circuit. The cylcoconverter can supply loads of any power factor because of its ability to permit the transfer of energy instantly in either direction.

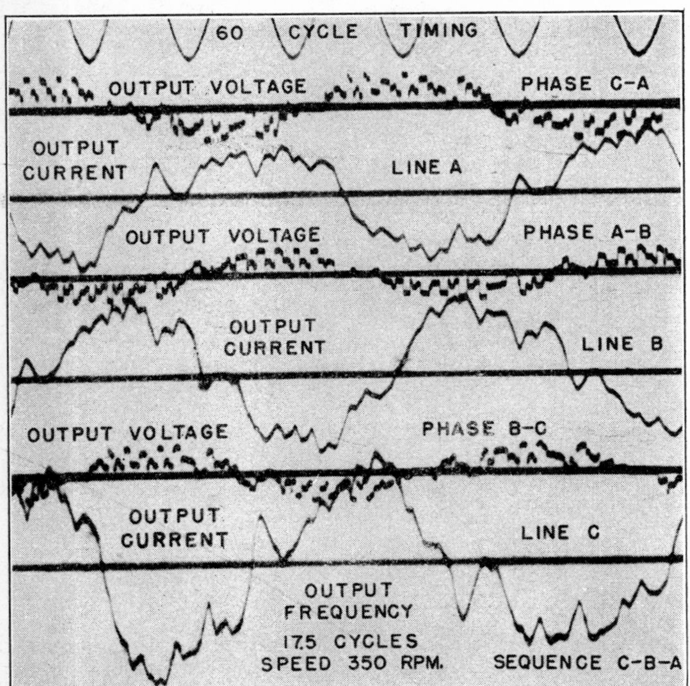

FIG. 23·14 Oscillogram of the voltage and current of a synchronous motor supplied by a cycloconverter.

The positive group of tubes must be separated from the negative group of tubes by a reactor, because there are instantaneous differences in the voltages between the two groups. One group of tubes is operating as a rectifier when the other group is operating as an inverter. During part of the cycle the instantaneous inverter counter voltage is less than the rectifier output voltage, and a pulse of current flows between the two groups. Because of this circulating current, a reactive current taken from the input circuit is not reflected in the output circuit, and therefore does not represent any real conversion of energy. The circulating current should be kept small by having sufficient reactance between the two tube groups, but this reactance should not be great enough to reduce appreciably the resultant power factor of the load.

An oscillogram of voltage and current wave forms of a typical three-phase to three-phase cycloconverter with a variable-speed synchronous motor load are shown in Fig. 23·14. The motor is rated at 250 horsepower at 400 revolutions per minute and 20 cycles input. The ignition points of the ignitrons which conduct the load current are determined by a low-energy commutator, on the shaft of the motor; therefore, the applied frequency cannot differ from the synchronous frequency of the motor. The speed of the motor and its direction of rotation are determined by the applied terminal voltage and the angular position of brushes with respect to the stator of the motor. The motor can be operated as a synchronous machine over its entire speed range, through zero frequency, with either direction of rotation, without changes in the power circuit.

Applications

The cycloconverter is applied where the frequency ratio is as great as about three to one and where power conversion in either direction is required. It can be used advantageously for interconnecting two electrical systems because of the high efficiency resulting from the single conversion. Variable-speed motor drive is probably the most important application, because a dead load can be supplied and a wide frequency range can be obtained.

In cycloconverter motor drives the actual output-frequency spectrum is made up of a series of "synchronous" frequencies, and in passing from one frequency to another there are beat frequencies on the synchronous frequencies. These cause pulsations of torque and power. If the mechanical system being driven by the motor has critical frequencies within the range of the applied torque frequencies, the equipment must be designed so that there are no unusually high amplified torques. Power pulsations become more serious as the frequency ratio is decreased, and to prevent serious disturbances the ratio should be made not less than two to one or three to one.

23·9 CYCLORECTIFIER

If the loads on a single-conversion electronic frequency changer do not require transfer of energy from the lower frequency circuit to the higher frequency circuit the simpler power and excitation circuits of the cyclorectifier can be applied.[20, 21] The lower frequency output of the cyclorectifier is obtained by controlling one group of tubes to supply a pulse of voltage and current during the first half cycle with a positive polarity, and controlling the other group of tubes to supply a pulse of voltage and current during the other half cycle with a negative polarity. Current in the group of tubes must have a displacement of nearly unity on the output voltage, otherwise current continues to be conducted in the last tube of each half cycle. If this current continues for more than one input phase, a circuit fault is produced.

If the load requires no displacement reactive kva, the cyclorectifier can be separately excited, that is, with no power-factor-correcting capacitors on the lower frequency output. If a load with lagging power factor is to be supplied, some provision must be made to bring the current on the cyclorectifier to unity displacement factor by either a rotating synchronous condenser or static capacitors. With either separate or self-excitation, the cyclorectifier is never able to convert energy from the lower frequency circuit to the higher frequency circuit.

The cyclorectifier type of a-c to a-c electronic frequency

changer has been applied to the conversion of 50-cycle power to 16⅔ cycle power for railway use in Europe.[22] In this type of service, conversion from three-phase to single-phase is required. Some form of energy-storage equipment, to provide a steady power flow on the three-phase side with pulsating power on the single-phase output, is desirable. A typical circuit with a multianode mercury-arc rectifier is shown in Fig. 23·15.

23·10 CYCLOINVERTER

If the transfer of energy is only from an a-c supply circuit to an excited a-c system, the cycloinverter type of operation should be selected. The supply-circuit frequency may be lower or higher than that of the system receiving power. The power circuit is about the same as for the cyclorectifier. To obtain a wide range in the ratio of the supply-circuit voltage to the receiving system voltage a reactor may be connected between the two groups of tubes in each supply-frequency phase, as in the cycloconverter.

The characteristics of the cycloinverter are determined by its excitation equipment. The tubes are controlled only by the system receiving power and are released only during the portion of the cycle in which a-c voltage is opposed to the flow of current. Each tube is thus released every cycle at an inverting angle. The impressed voltage of the power source then forces load current through the correct group of tubes against the counter voltage of the receiving system, causing energy to be transmitted to the receiving system. The conversion is entirely asynchronous, and the ignition angles are not modulated.

Separately excited cycloinverters have not been used commercially, but experimental units have been operated successfully. One application is with a wound-rotor induction motor to obtain variable speed. The cycloinverter is connected so that it returns energy from the rotor to the a-c input line. This type of variable-speed drive may have a low line power factor at high motor speeds.

23·11 SELF-EXCITED CYCLOINVERTER FOR INDUCTION HEATING

The most important application of the cycloinverter is for conversion of power from commercial frequencies to relatively high frequencies for induction heating.[23, 24] The cycloinverter is operated self-excited, usually by adding capacitors to the power-factor-correcting capacitors on the single-phase load. To obtain somewhat better control of the tube additional capacitors may be used on the primary of the higher frequency transformer.

Frequency Limits

The upper frequency limit of the cycloinverter is determined by the control characteristics of the gaseous tubes and the control circuits. Deionization time of the tubes depends upon spacing of tube electrodes, vapor density, current density, and some minor factors. The rate of decay of ionization can be partly controlled by special tube design, but it usually is between 50 and 100 microseconds. With the parallel-capacitor inverter the margin angle usually cannot be made greater than about 45 electrical degrees; therefore, the output frequency is limited to some value between 1000 and 2000 cycles per second. By reducing the tube load current below its rated capacity, operation at somewhat higher frequencies is possible, but the limit is about 3000 cycles per second.

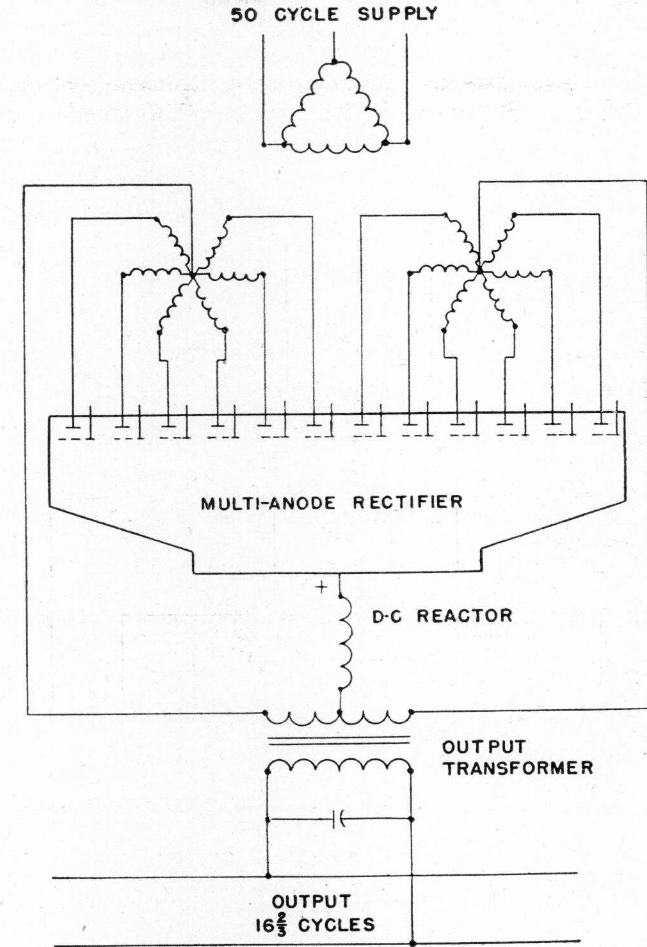

Fig. 23·15 Single-conversion electronic frequency converter circuit which has been used in railroad sub-stations for a-c electrification.

Circuits

One method of obtaining power at relatively high frequency by means of gaseous tubes is to invert d-c power by means of a two-tube self-excited inverter, but this method has the disadvantage that d-c power is required. With a cycloinverter circuit, power can be taken directly from the commercial power system. (See Fig. 23·16.) The load taken from the supply circuit is similar to that taken by a three-phase rectifier, because the d-c reactor usually has sufficient inductance to keep the high-frequency load current from being reflected into the input circuit.

In the cycloinverter the tubes operate in pairs on the input phases, so that each tube must be able to conduct during 120 electrical degrees of the input-frequency cycle plus the commutation angle for the input transformer. The current is commutated between the two tubes at the high-frequency rate by the parallel capacitor. After another phase of the input transformer becomes most positive, the pair of tubes on that phase start conducting current. The pairs of tubes continue to conduct the current and commutate to the next pair so that a constant current is maintained in the d-c circuit.

High-Frequency Ignitrons

To supply power ratings required for 1000- to 2000-cycle induction heating, melting, and forging, ignitrons or pool-cathode tubes are necessary. With ignitrons the cathode must not carry a continuous load arc or excitation arc, for the mercury pool is not insulated from the tank wall, and cathode spots forming on the wall of the tube may erode the steel wall. The cycloinverter type of circuit is used because it is not necessary to hold a continuous d-c arc. The ignitor can be fired on the input frequency with the grids controlling the output frequency.

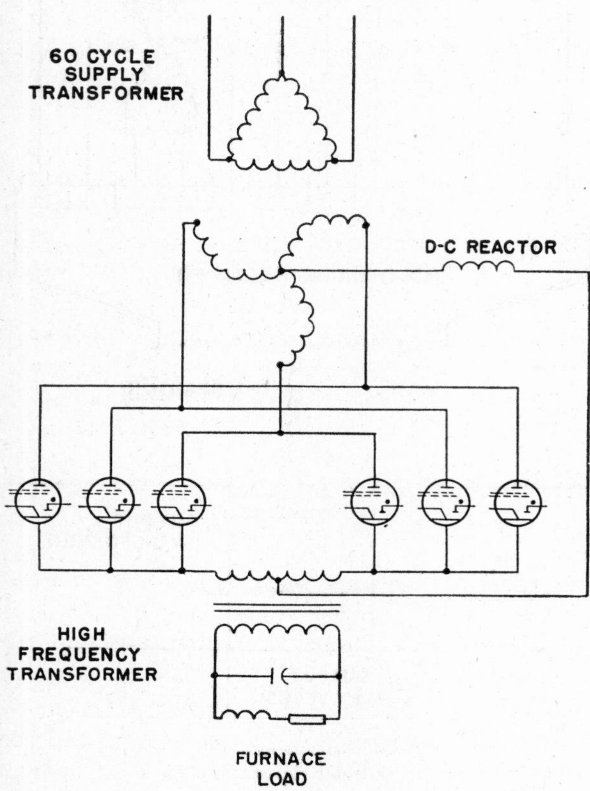

Fig. 23·16 Induction-heating cycloinverter power circuit using ignitron tubes.

For a given rating it is desirable to operate at as high a tube voltage as possible so that the tube current can be reduced and a shorter deionization time can be obtained. Several factors prevent the use of voltages above about 15,000 inverse or forward. Sputtering of the anode because of positive-ion bombardment of the anode surface is one of these. The best method of overcoming this is to design the tube with close spacings around the anode, and to use two or more graphite grids with small holes.

Operation Characteristics

The gaseous-tube cycloinverter can be used on almost any type of load if capacitors are employed to correct the displacement factor of the load current to slightly leading. With induction-heating furnaces the grids are best supplied from the high-frequency output. The frequency then is determined by the resonant frequency of the load circuit. With an induction furnace load, the resonant frequency changes as the temperature of the charge passes through the Curie point. If the frequency is to be held about constant, the number of capacitors can be changed to maintain approximately a constant resonant frequency; or, if close control is required, the grids of the gaseous tubes can be supplied by a separate vacuum-tube oscillator to establish the cycloinverter frequency.

23·12 D-C TRANSFORMER

Power can be transformed from a d-c circuit of one voltage to another d-c circuit with a higher or lower voltage by means of a frequency converter of the types described, except that the arrangement of the circuit components is changed. Instead of rectifying power from one a-c circuit, then inverting into another a-c circuit (as in the frequency changer), direct current is inverted to alternating current, and the required a-c voltage is then rectified to obtain the desired d-c voltage.

The practical need for the electronic d-c transformer has been limited, and no large power converters have been operated for any extended period of time. Low-power d-c transformers have been used for various purposes, but most of these were laboratory installations. Investigations have been made into the possibility of having electronic d-c transformers on railroad locomotives so that a high d-c voltage might be used on the trolley and a much lower d-c voltage on the motors, but the method is not attractive because of the large amount of conversion equipment required.

23·13 D-C TRANSMISSION

The practicability of high-voltage d-c power transmission is dependent upon the development of inverters of large capacity and the perfection of high-voltage mercury-arc tubes. Requirements for d-c transmission have been discussed for many years, and optimistic predictions have been made, but few have come true.[25] The most important technical obstacle has been the development of inverters.

The greatest advantages of d-c transmission are the increase in power which can be transmitted over a given line and the elimination of the stability problems. Power capacity can be increased, because for a given line insulation the d-c voltage can be equal to the crest value of the a-c voltage, and higher currents can be carried by some conductors, such as cables.

Constant-Current Circuit

Some of the early investigations of d-c transmission were based on d-c circuits with constant current and variable voltage to reduce disturbances caused by tube faults.[26] It now is agreed generally that the disadvantage of the large amount of reactor and capacitor equipment required for constant current is too great to be justified by the improvement in performance.

Constant-Voltage Circuit

Constant-voltage transmission equipment requires a circuit similar to the d-c link electronic frequency changer except that the rectifier is at the sending station of the transmission line, and the inverter is at the receiving station, and the two stations are several hundred miles apart. The circuits actually should contain many more converters in series, so that a d-c voltage of several hundred thousand volts can be obtained.[27] As many as 24 tubes in series may have to be used.

With large installations of rectifiers operating in parallel many phases are necessary to improve the wave forms of the currents in the a-c systems and to reduce the harmonics of voltage and current in the d-c circuit; otherwise telephone interference and other inductive co-ordination problems may result. The total effective number of phases should be at least 36 on a large installation. In many installations this may be a limiting factor which determines the maximum rating allowed for one rectifier or inverter circuit element.

Commutation on Receiving A-C System

One fundamental obstacle to the development of d-c transmission is the problem of obtaining commutation on the receiving end. Operation of the inverter is dependent upon a resultant transformer current leading the voltage by a displacement angle of about 30 or 40 degrees. If a lagging power-factor load is to be carried, capacitors or synchronous condensers must be provided to supply the lagging reactive kva required by the load and the inverter.

The receiving a-c system may not have any generating equipment of its own, and therefore it may be difficult to establish voltage patterns to supply a dead load. The best method of overcoming this difficulty is to start with light loading and have some of the inverters self-excited.

Forced commutation has been considered, but has not been worked out practically. Even the development of gaseous tubes with grids and grid circuits that could stop the current suddenly would not completely solve the problem, because capacitors will still be required to prevent abnormal circuit voltages. Special circuits to allow the inverter to carry a load with lagging power factor do not reduce the total amount of equipment required.

Application

There are certain power-transmission requirements which tend to make d-c transmission more favorable. Since the great saving with d-c transmission is in the construction cost of the line itself, the application must require great transmission distances and large blocks of power. Economic studies indicate that in the United States the power to be transmitted must be greater than 300,000 kilowatts at distances greater than 300 miles before d-c transmission can be expected to have any advantage over a-c transmission.[28] Conditions in Europe in 1939 indicated that for 400,000 kilowatts the distance beyond which d-c transmission would have been more economical was from 250 to 300 miles.[29] The d-c voltage to ground should be at least 300,000 volts.

If a typical d-c transmission line were to be constructed, on the basis of present economic relationships twelve converter units, each consisting of a six-phase double-way circuit with a d-c voltage rating of about 50,000 volts, might be required. The midpoint of the series could be grounded, and a transmission line with two conductors at 300,000 volts to ground and 600,000 volts between conductors would result. The conductor should have a current-carrying capacity of about 800 amperes. Such a transmission line would have a rating of 480,000 kilowatts.

A factor that tends to make d-c transmission more favorable is the interconnection of two existing power systems with different frequencies. The d-c transmission link would allow for frequency changing and make a flexible tie between the two systems.

Use of cable for the transmission line would make d-c transmission more desirable for shorter distances and at lower power. The higher cost of cable would limit these applications to transmission lines in which special climatic conditions, terrain, or strategic requirements must be overcome.

Faults

Faults on a d-c transmission line are of two types: those caused by short circuits on the line itself; those caused by the tubes. Most line short circuits would be caused by lightning, and with d-c transmission there would be little possibility of the arc clearing itself. There is some evidence to indicate that d-c transmission lines can have good arc-extinction characteristics.[30] The best results are obtained with an ungrounded midpoint, but the additional insulation required probably cannot be justified.

Faults caused by arc-backs of the rectifier tubes and arc-throughs of the inverter tubes are not so severe as a complete short circuit on the transmission line; however, with the large number of high-voltage tubes involved it will be necessary to obtain tubes that have a low fault rate. Ignitrons for inverse and forward voltages up to 20,000 volts have high arc-back rates compared with low-voltage tubes. Since there are no fundamental limitations on the design of high-voltage tubes, tube design will not be the type of limitation which would prevent the eventual perfection of d-c transmission.

In a d-c transmission system the faults would be cleared by means of arc-quenching of the tubes so that the fault would be cleared in a short time and power flow could be restored in a matter of a few cycles. Usually the fault can be cleared in such a short time that no disturbance can be noticed.

REFERENCES

1. "The Inverter," D. C. Prince, *G. E. Rev.*, Oct. 1925, Vol. XXVIII, No. 10, p. 676.
2. "The Ignitron Type of Inverter," C. F. Wagner and L. R. Ludwig, *Elec. Eng.*, Oct. 1934, pp. 1384–1388.
3. "History and Development of Electronic Power Converters," E. F. W. Alexanderson and E. L. Phillips, *Trans. A.I.E.E.*, Vol. 63, Sept. 1944, pp. 654–657.
4. "The Mercury Arc Rectifier Traction Sub-Station of the Electricity Supply Commission of the Natal and Reef Electrified System of the South African Railways," K. M. Cunliff, *Trans. South African Inst. Elec. Eng.*, March 1941, pp. 50–114.
5. "A 3000 KW Mobile Mutator Substation for Traction Service and Regenerative Braking Duty," E. Kern, *Brown Boveri Rev.*, May/June, 1943, p. 110.
6. "Series-Parallel Type Static Converters," C. A. Sabbah, *G. E. Rev.*, Part I, May 1931, pp. 238–301; Part II, Oct. 1931, pp. 580–589; Part III, Dec. 1931, pp. 738–744.

7. "The Parallel Type Inverter," F. N. Tempkins, *Trans. A.I.E.E.*, Sept. 1932, pp. 707–714.
8. "Parallel Inverter with Resistance Load," C. F. Wagner, *Elec. Eng.*, Nov. 1935, pp. 1227–1235.
9. "Parallel Inverter with Inductive Load," C. F. Wagner, *Elec. Eng.*, Sept. 1936, pp. 970–980.
10. "The Relation of Residual Ionization to Arc-Back in Thyratrons," K. H. Kingdon and E. J. Lawton, *G. E. Rev.*, Nov. 1939, Vol. 42, No. 11, pp. 474–478.
11. "Pentode Ignitrons for Electronic Power Converters," H. C. Steiner, J. L. Zehner, and H. E. Zuver, *Trans. A.I.E.E.*, Oct. 1944, pp. 693–697.
12. "High Voltage Ignitron Rectifiers and Inverters for Railroad Service," J. L. Boyer and C. G. Hagensick, *Trans. A.I.E.E.*, July 1946, pp. 463–470.
13. "Inverter Action on Reversing of Thyratron Motor Control," H. L. Palmer and H. H. Leigh, *Trans. A.I.E.E.*, April 1944, pp. 175–184.
14. "The Electronic Converter for Exchange of Power," F. W. Cramer, L. W. Morton, and A. G. Darling, *Trans. A.I.E.E.*, 1944, pp. 1059.
15. "Design of an Electronic Frequency Changer," C. H. Willis, R. W. Kuenning, E. F. Christensen, and B. D. Bedford, *Trans. A.I.E.E.*, Vol. 63, 1944, pp. 1070–1077.
16. "Switchgear and Control for an Electronic Power Converter," W. N. Gittings and A. W. Bateman, *Trans. A.I.E.E.*, Aug. 1944, pp. 585–589.
17. "The Mercury Arc Rectifier Applied to A-C Railway Electrification," O. K. Marti, *Trans. A.I.E.E.*, Sept. 1932, pp. 659–664.
18. "A Static Thermionic Tube Frequency Changer," A. Schmidt, Jr., and R. C. Griffith, *Elec. Eng.*, Oct. 1935, pp. 1063–1067.
19. "Network Coupling by Means of Static Electronic Frequency Changers," O. K. Marti, *Trans. A.I.E.E.*, Sept. 1940, pp. 495–502.
20. "A Study of the Thyratron Commutator Motor," C. H. Willis, *G. E. Rev.*, Feb. 1933, Vol. 36, No. 2, pp. 76–80.
21. "The Thyratron Motor," E. F. W. Alexanderson and A. H. Mittag, *Elec. Eng.*, Nov. 1934, pp. 1517–1523.
22. "Static Converter with Current and Voltage Smoothers for Flexibility Coupling a Three Phase 50 Cycle Network with a Single Phase $16\frac{2}{3}$ Cycle Network," C. Ehrensperger, *Brown Boveri Rev.*, Vol. 21, No. 6, June 1934, pp. 95–112.
23. "Mercury Arc Converter Masters Metal in Deep Heating," H. F. Storm, *Allis-Chalmers Elec. Rev.*, March 1946.
24. "Ignitron Converters for Induction Heating," R. J. Ballard and J. L. Boyer, *Proceedings of the National Electronics Conference*, Oct. 1946.
25. "Electronic Devices in D. C. Power Transmission," C. W. Stone, *Electronics*, March 1931, pp. 554–555.
26. "Constant-Current D. C. Transmission," C. H. Willis, B. D. Bedford, and F. R. Elder, *Elec. Eng.*, Jan. 1935, pp. 102–108.
27. "Transformer Connections for High Voltage D. C. Power Transmission," E. Kern, *Brown Boveri Rev.*, Oct. 1941, pp. 314–319.
28. "A Survey of Power Applications of Electronics," A. C. Monteith and C. F. Wagner, *Proceedings of the National Electronics Conference*, Oct. 1944, pp. 261–280.
29. "High Voltage D. C. Transmission," C. Ehrensperger, CIGRE Paper 103, June, 1946.
30. "The Forms in which the Earth-Fault Arc Occurs in a D-C Transmission Line with Insulated Middle Point," P. Weldvogel, *Brown Boveri Rev.*, Oct. 1941, pp. 299–303.

Chapter 24

RADIO-FREQUENCY HEATING *

D. Venable and T. P. Kinn

INDUCTION HEATING [1,2]

ROTATING machines supply induction heating power at frequencies up to 30,000 cycles per second and with power outputs up to several thousand kilowatts, with the exception that machines with frequencies above 10,000 cycles per second are limited in output. One rather new source of power for induction heating is the mercury-arc frequency changer, which produces frequencies as high as 3000 cycles per second. Its frequency depends upon the number of phases used and the deionization time of the mercury arc.

Spark-gap oscillators are used for frequencies from 20,000 to 400,000 cycles per second. They are used for heat treating, surface hardening, brazing, and soldering. Power outputs as high as 35 kilowatts are available commercially, but the spark-gap oscillator's efficiency varies from about 50 percent under normal operating conditions to about 15 percent under poor conditions, depending upon the type of load, the method of coupling to the load, and the conditions of the spark gap. Badly burned gaps very seriously decrease the oscillator's efficiency.

The vacuum-tube generator can be used for frequencies from about 100 kilocycles to several megacycles per second for induction heating,[3] with power output ranging from a few hundred watts to several hundred kilowatts. For thin casehardening, these higher frequencies are better than the lower frequencies produced by rotating machines. General induction heating for soldering, brazing, surface hardening, and annealing can be done with vacuum-tube generators ranging in frequency from 200 kilocycles to about 500 kilocycles. A few special induction-heating operations are done at about 15 megacycles per second.

DIELECTRIC HEATING [4]

Dielectric heating usually is done in the frequency range from 1 to 200 megacycles per second. Only low power output, perhaps several hundred watts, can be produced now in the upper frequency range. Most of the high-power generators now in use operate at frequencies from 1 to 50 megacycles per second. This high-frequency heating is useful in drying processes, such as the drying of rayon, wood, paper, and tobacco. It is used also in stress-annealing textile fibers, and gluing and bonding of woods and fabrics are accelerated by application of dielectric heat. In the plastic industries, dielectric heating is used to preheat plastic preforms, and to cure or set molded plastic products.

24·1 VACUUM-TUBE OSCILLATORS

The theory of operation of power oscillators has been discussed in Chapter 19, therefore no mention of operational theory is made here. Several types of oscillator circuits are in common use. The circuits are almost identical; the only prime difference is in the method of feedback used to drive the tube. Each circuit contains elements so constructed and connected that they can store energy. This is the "tank" of an oscillator consisting of a parallel tuned circuit.

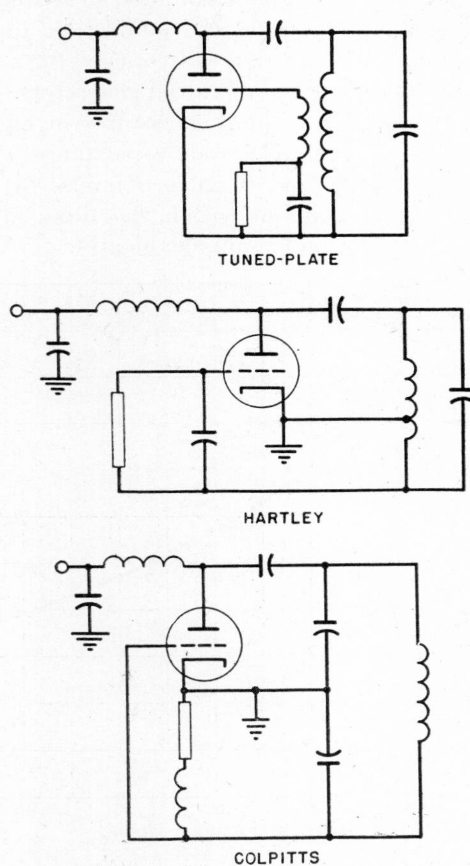

Fig. 24·1 Common oscillator circuits.

* The material in this chapter has been reproduced with permission from *Radio Frequency Heating Fundamentals and Applications—Instructor's Manual,* copyrighted by Westinghouse Electric Corporation, Pittsburgh, Pa., 1946.

Methods of feedback are shown in the diagrams of three circuits [5] in Fig. 24·1.

The diagram of a standard industrial 20-kilowatt 10-megacycle generator is shown in Fig. 24·2.

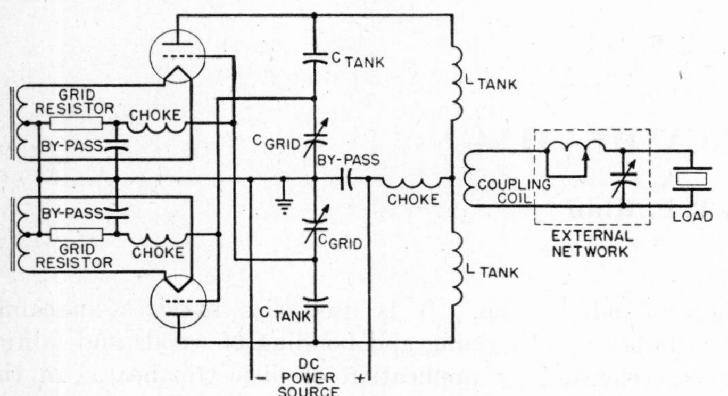

Fig. 24·2 Typical 20-kilowatt 10-megacycle generator circuit diagram.

Frequency and Power Rating

From the manufacturer's point of view there are basic limitations upon the frequency and power output of high-frequency electronic generators. Even vacuum tubes used in high-frequency generators are limited in frequency. For example, Fig. 24·3 illustrates the permissible input power and plate voltage as a function of frequency for a tube commonly used in high-frequency-heating generators.

Some of the frequency-limiting features in high-vacuum power tubes are: (1) interelectrode capacitance; (2) inductance of internal leads and filament strands; (3) dielectric losses in insulating material within the tube; (4) electron transit time; (5) electron loading of the grid.

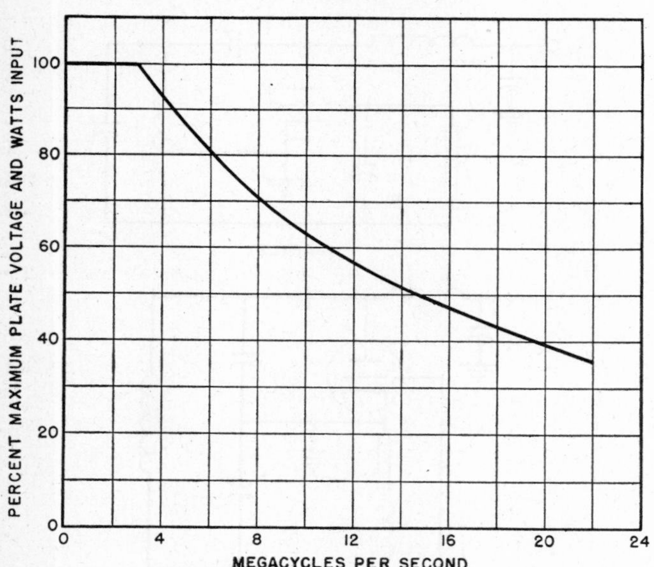

Fig. 24·3 Tube input versus frequency.

However, the vacuum tube is not the only component that is limited by frequency. Metering high-frequency currents is a serious problem in the megacycle range of frequencies. Commercial capacitors are frequency-limited. Like other electrical components at high frequencies, a capacitor becomes resonant at some particular frequency, and no longer acts as a capacitor but as a resistance. At even higher frequencies, capacitors may act inductively. Capacitors not specifically designed for high-frequency work may suffer voltage breakdown or overheating.

Another frequency limitation is caused by stray reactances. Stray inductance and capacitance may seem rather insignificant under d-c conditions, but at very high frequencies they possess reactances that very often impose an upper limit on the frequency at which they may be used.

Aside from the difficulties of generating high-frequency currents, the most important of the other problems that must be considered is the transmission of high-frequency energy. As wavelength becomes comparable to the distance of separation of the conductors of a transmission line, the theory of transmission lines generally used must be modified. In the decimeter and centimeter range of wavelengths, wave guides must be used as transmission lines. Coupling a load-impedance transforming network to such a line introduces more problems.

Symptoms of Various Loading Conditions

The vacuum tube as a generator of high-frequency currents, having a high internal impedance, should feed into a high-impedance load. Most loads in high-frequency induction heating have low resistive components, and their impedances are often relatively low. To overcome this disadvantage, and at the same time introduce a device which serves as a frequency-governing element for the generator, a parallel tuned circuit is utilized.[6] Some fundamental properties of a parallel tuned circuit are its current-amplifying properties and the flywheel effect of its circulating kva. This current-amplifying property may be interpreted as being an impedance-transformation property. The flywheel effect is that property which governs the frequency of the generator and its ability to maintain oscillations.

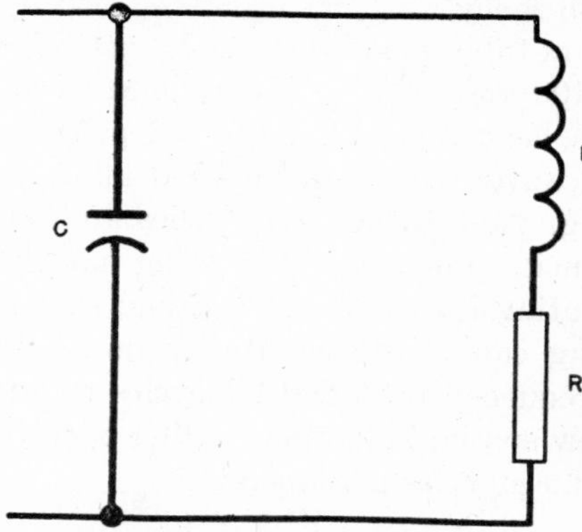

Fig. 24·4 Parallel tuned circuit.

At unity power factor, a parallel tuned circuit acts as a pure resistance much larger than the resistance in the circuit. This apparent high resistance can be shown, with the aid of Fig. 24·4, to be

$$R_a = \frac{R}{\omega^2 C^2 R^2 + (\omega^2 LC - 1)^2} = \frac{R^2 + \omega^2 L^2}{R} \qquad (24\cdot 1)$$

where R_a = apparent resistance of tuned circuit in ohms
R = resistance in ohms in series with inductance L
C = tuning capacitance in farads
L = inductance in henrys
$\omega = 2\pi \times$ frequency (cycles per second).

With little error for inductors with an over-all power factor less than 0.1, equation 24·1 can be reduced to

$$R_a = \frac{\omega^2 L^2}{R} \tag{24·2}$$

If the peak tank voltage is approximately 80 to 85 percent of the d-c plate voltage the oscillator tube operates most efficiently, and if circuit losses are kept at a minimum overall oscillator efficiency may be from 50 to 60 percent.

For a given tank Q, the ratio of circulating kva to kilowatts supplied to the tank circuit, the tube feeds into an apparent resistance as given in equation 24·1. For proper operation, this tank Q must be maintained as designed.

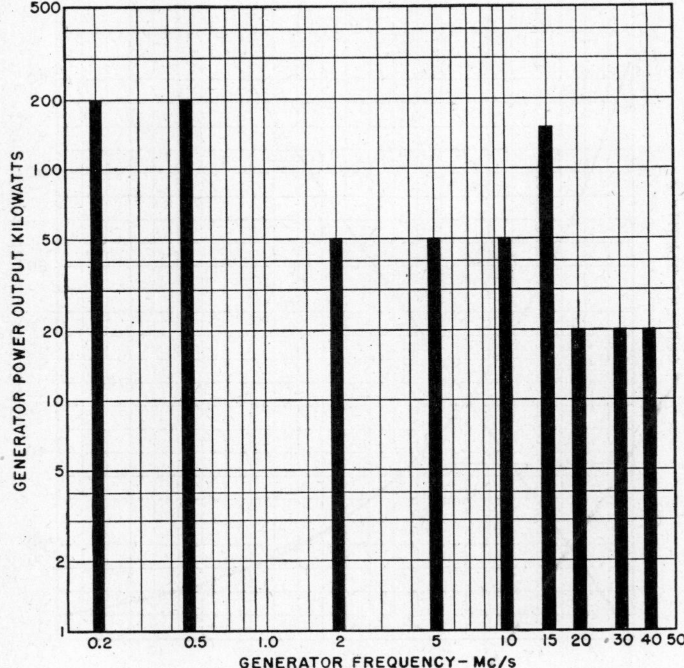

FIG. 24·5 Commercially available radio-frequency generators.

Therefore, if the tank voltage and Q are maintained as required, the tube supplies constant power to a constant load of R_a ohms.

It is well to note the conditions of operation that occur in an oscillator with changing load. If the resistive component R of the tank circuit is increased by increased coupling to a load, the tank Q is lowered, the apparent resistance into which the tube feeds is decreased, plate current increases, and grid and circulating tank currents decrease. Conversely, if the load is decoupled, reflected resistance R is decreased, tank Q is increased, apparent resistance into which the tube feeds is increased, plate current decreases, and grid and circulating tank current increase.

The tank Q is arbitrarily selected, but not without consideration. In general, loaded tank Q's range from 30 to 50 for commercial oscillators. Design and economic reasons govern the selection. Since the tank Q is the ratio of circulating kva to kilowatts supplied to the tank, a high Q implies high tank current. This current introduces circuit losses which must be dissipated satisfactorily or damage may result. Some tanks are water-cooled and usually are designed with a higher Q than air-cooled tanks intended to handle the same kilowatt rating. One of the economic problems facing the designer is the effect tank Q may have upon the over-all efficiency of the generator. Still another consideration is the application of the generator. This is both a design problem and an economic problem. The design problem is that of supplying power to a high-Q load. For example, if the load Q and the Q of the unloaded tank were comparable, then only about half of the oscillator's rated output could be supplied to the load, the remainder being lost in the oscillator tank and on the plate of the oscillator tube. The economic problem is evident.

For special cases, very high-Q tanks have been used and are now becoming commercially available. Some of these tank Q's range as high as 300 or 400 under load. Most of such tanks consist of pseudocavity resonators. In oscillators having high loaded tank Q's, the generator can be loaded with high-Q loads more satisfactorily, if it is assumed that the oscillator can be coupled to the load properly.

The rated power output of the set is not available to all types of loads. Loads of high Q limit the power that can be supplied to them.

Standard generators for high power are available as indicated in Fig. 24·5. This figure is not complete but is used to show the available power and frequency of present-day radio-frequency generators.

24·2 INDUCTION-HEATING THEORY

An electromotive force is induced in any conducting material in a changing electromagnetic field. If the conductor offers a complete path for the flow of current, the induced electromotive force produces a current along this path. In overcoming the resistance of the path work is done, and this work appears as heat.

Field Equations [7]

In attempting to explain the phenomenon of induction heating as a function of the induced voltages and currents, it seems in order to mention two very powerful tools. One of these is Ampere's law, which states that the work done in carrying a unit magnetic pole around a closed path across which an electric current is flowing is equal to 0.4π times the current. The other tool is that given by Faraday. Faraday's law states that, if the flux of induction through a closed path varies with time, an electromotive force equal to the time rate of decrease of flux is induced. These two laws lead to the so-called *field equations* which may be written in their vector form:

$$\begin{aligned} \text{curl } \mathbf{H} &= 0.4\pi \mathbf{J} \\ \text{curl } \mathbf{E} &= \mu 10^{-8} \frac{\partial \mathbf{H}}{\partial t} \end{aligned} \tag{24·3}$$

$\mathbf{H}$ is the vector magnetic field in oersteds, $\mathbf{J}$ is the vector current density in amperes per square centimeter, $\mathbf{E}$ is the vector electrical field in volts per centimeter and is assumed to be equal to the product of the resistivity ρ in microhm-centimeters, current density $\mathbf{J}$, and 10^{-6}. The permeability of the material is expressed as a numeric μ, and $\partial\mathbf{H}/\partial t$ is the time rate of change of the magnetic field $\mathbf{H}$.

An equation describing the distribution of current induced in a conducting medium by a harmonically changing magnetic field can be derived from these equations. This is the so-called *eddy-current equation* and is represented by the Laplacian

$$\nabla^2 \mathbf{J} = j\frac{8\pi^2\mu f}{\rho} \times 10^{-3}\mathbf{J} \qquad (24\cdot4)$$

in which the units have been previously defined. Most of the heat generated in a conducting body in a changing magnetic field is the heat generated by the flow of eddy currents; therefore only a small portion of the total heat is attributed to hysteresis.

Hysteresis Loss

Iron obeys the same laws of magnetic hysteresis at frequencies as high as 200 kilocycles per second as it does at commercial frequencies. The hysteresis-loss equation obtained empirically by Steinmetz is only an approximation for many steels:

$$P = \eta f B_m^{1.6} \qquad (24\cdot5)$$

in which η is a constant depending upon the steel and varying from 0.0002 for such magnetic materials as some special magnetic steels to 0.005 for soft steels or 0.075 for tool steels. For average steels η may be considered to be 0.005. The constant f is expressed in cycles per second, and B_m is the maximum flux density in gauss. If equation 24·5 is expressed in watts per cubic inch, the variables as defined being used,

$$P = 1.64\eta f B_m^{1.6} \times 10^{-6} \text{ watt per cubic inch} \qquad (24\cdot6)$$

This hysteresis loss may seem large, especially if the frequency is high, but the penetration of flux into the work usually is small. This shallow penetration is attributed to the shielding effect of eddy currents. Flux is assumed to vary linearly from a maximum at the surface to a depth δ'. Actually it varies exponentially, dropping off rapidly with distance from the surface. Hysteresis loss is small as long as the depth of penetration of flux remains small compared with the dimensions of the work that are perpendicular to the magnetic field.

Skin Effect [7]

Skin effect is the diminishing of current density from the surface to the interior of a conductor carrying alternating current. To visualize this effect mathematically in its simplest condition a harmonically varying magnetic field above an infinite conducting plane is assumed. The conditions thus imposed facilitate the use of equations 24·3 to examine the current distribution in the region at and below the surface of this infinite conducting plane.

The solution for an alternating magnetic field becomes

$$J_x = J_0 \epsilon^{-\alpha x} \epsilon^{j\alpha x} \qquad (24\cdot7)$$

where J_x represents current density at any point x units of distance below the surface of the plane, and J_0 is current density at the surface of the plane. The first exponential term represents the decrease of magnitude of the current density J_x as the distance from the surface increases. The second exponential term describes the phase relation between the current on the surface and the current at any distance x from the surface of the plane.

Depth of Penetration

Figure 24·6 illustrates the decrease in amplitude of current density as the distance from the surface increases. Where exponent αx becomes unity the current has fallen to $1/\epsilon$ or approximately 37 percent of the surface-current density. The depth to which this corresponds is defined as the *depth of penetration* of the current. This means that the

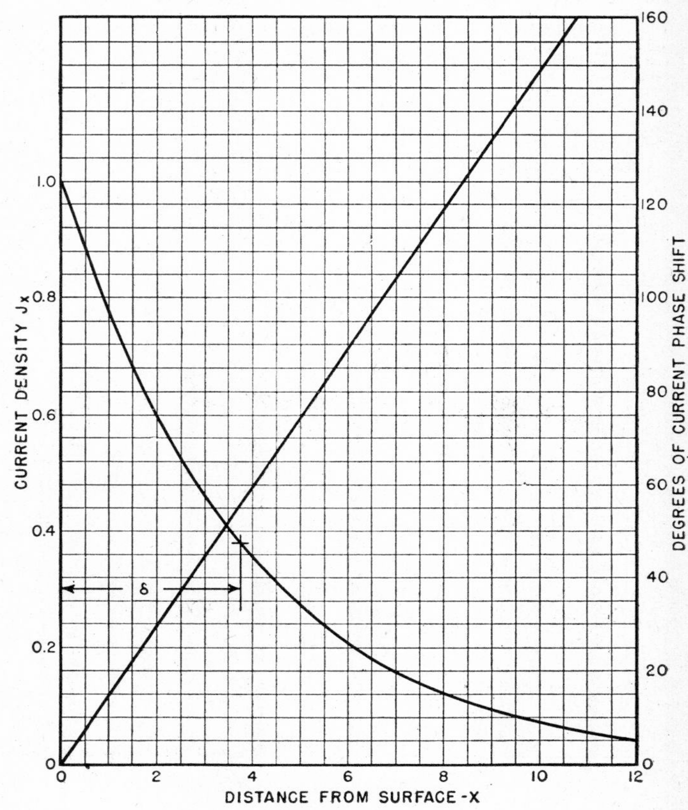

Fig. 24·6 Current decay and phase shift.

term α of the first exponent is equal to the reciprocal of the depth of penetration. That is, the depth of penetration may be defined by

$$\delta = 5.03\sqrt{\frac{\rho}{\mu f}} \text{ centimeters} \qquad (24\cdot8)$$

and

$$\delta = 1.98\sqrt{\frac{\rho}{\mu f}} \text{ inches} \qquad (24\cdot9)$$

where δ = depth of current penetration
ρ = resistivity in microhm-centimeters
μ = relative permeability
f = frequency in cycles per second.

The phase shift relative to the distance from the surface of the slab also is interesting. Figure 24·6 indicates the shift of phase of current as the distance x is changed.

When the depth below the surface becomes equal to δ the phase shift is equal to 1 radian, and Fig. 24·6 shows that at a distance from the surface little greater than δ there is a current component opposite in phase to that at the surface of the conductor. However, since there is so little current in the region $x > \delta$, this 180-degree phase shift has almost no effect.

The depth of penetration δ is that depth to which the total induced current could flow uniformly and produce the same heating effect as the current expressed by equations 24·4 and 24·7. Thus equations 24·8 and 24·9 define the effective cross section of a conductor of unit width carrying alternating currents with pronounced skin effect.

Heating Solid Cylinders [7, 8]

Assume that a cylinder is in a uniform electromagnetic field, with its axis parallel to the field, and consider an outer shell of radius r and thickness dr (Fig. 24·7). Let H be the

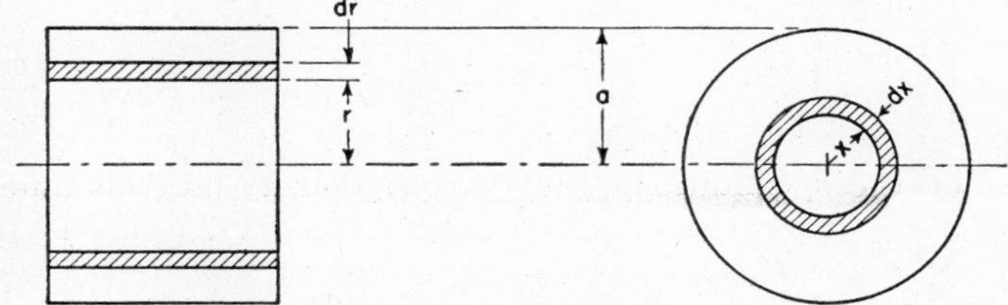

Fig. 24·7 Flux distribution in a solid conducting cylinder.

magnetizing force and J the current density at radius x (Fig. 24·7). The total flux within the cylinder of radius r is

$$\Phi = 2\pi\mu \int_0^r Hx\,dx \tag{24·10}$$

The permeability μ is assumed to be a constant.[9] This is not always true and is discussed later in more detail.

An electromotive force induced in the elementary shell is equal to the negative time rate of change of the flux Φ and is

$$dE = -2\pi\mu \times 10^{-8} \int_0^r x \frac{\partial H}{\partial t}\,dx \tag{24·11}$$

Therefore the current in a unit length of this elementary shell is

$$I = \frac{\mu}{r\rho} \times 10^{-2} \int_0^r x \frac{\partial H}{\partial t}\,dx \tag{24·12}$$

From equations 24·3 the current density in this shell is proportional to the negative time rate of change of magnetizing force H. Since sinusoidal variation of magnetizing force is assumed, a differential equation defining the flux distribution within the cylinder can be derived from the preceding equations. This equation is

$$\frac{d^2H}{dr^2} + \frac{1}{r}\frac{dH}{dr} - jm^2H = 0 \tag{24·13}$$

where

$$m^2 = \frac{8\pi^2\mu f}{\rho} \times 10^{-3}$$

The solution of this equation is used to define the current around the cylinder at any radius r. Thus the power generated in any elementary cylinder of radius r becomes simply

$$P = \frac{H^2m^2\rho M_1^2(mr)}{8\pi M_0^2(ma)} \tag{24·14}$$

where $M_1(mr)$ and $M_0(ma)$ are particular Bessel functions of the arguments mr and ma respectively. Now the total power generated in the cylinder can be determined by adding the power generated in each elementary shell from zero radius to radius $r = a$, the outside radius of the cylinder. To perform this addition exactly, equation 24·14 is integrated with respect to r between the limits $r = 0$ and $r = a$. This gives an equation for the power generated in a cylinder of unit length. In terms of power generated per unit volume, the integral of equation 24·14 becomes

$$P_v = \tfrac{1}{2}H_0^2\mu f G(ma) \times 10^{-7} \tag{24·15}$$

where

$$G(ma) = \frac{1}{ma}\left[\frac{ber(ma)ber'(ma) + bei(ma)bei'(ma)}{ber^2(ma) + bei^2(ma)}\right]$$

$$m^2 = \frac{8\pi^2\mu f}{\rho} \times 10^{-3}$$

and

$$H_0 = H\sqrt{2} = \text{peak magnetizing force (oersteds)}$$

The product ma has an important relation to the depth of penetration δ and the diameter of the cylinder d,

$$ma = \frac{d}{\delta\sqrt{2}} \tag{24·16}$$

where d and δ have the same dimensional units. The product ma is a pure numeric. Thus the power equation may be rewritten:

$$P_v = \frac{1}{2}H_0^2\mu f G\left(\frac{d}{\delta}\right) \times 10^{-7} \tag{24·17}$$

Figure 24·8 shows $G(d/\delta)$ as a function of d/δ. The function $G(d/\delta)$ is a maximum when $d/\delta \cong 3.5$. In this region the power absorbed by the cylinder is maximum for a given

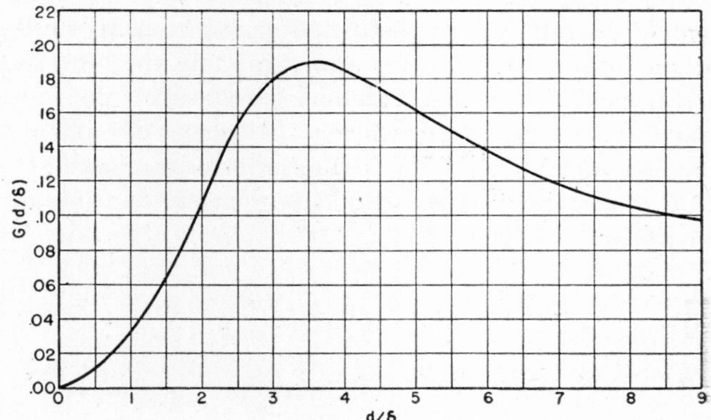

Fig. 24·8 $G(d/\delta)$ function curve.

frequency. For $d/\delta < 3.5$ the slope of the $G(d/\delta)$ curve is too steep for favorable energy absorption from the magnetic field. The frequency for $d/\delta = 3.5$ is not critical, since the curve in this region is not sharp, and any frequency for $d/\delta \geqq 3.5$ is satisfactory for heating cylinders, provided coil voltage and current limitations are not exceeded, and also provided stray reactances are not excessive.

For $d/\delta \geqq 14$, the function $G(d/\delta)$ is almost coincident with the curve $G(d/\delta) = \delta/d$. Then for practical purposes the power-density equation simplifies to

$$P_v = \frac{1}{2}\frac{\delta}{d}H_0^2\mu f \times 10^{-7} \text{ watt per cubic centimeter} \tag{24·18}$$

By expressing power density in watts per cubic inch and substituting in equation 24·18 the value of δ as previously defined in equation 24·9, the basic equation for inductive heating of cylinders is evolved:

$$P_v = 16.3 \times 10^{-7} H_0^2 \frac{\sqrt{\mu\rho f}}{d} \qquad (24 \cdot 19)$$

where P_v = power density in watts per cubic inch
H_0 = peak magnetizing force in oersteds
μ = permeability of cylinder
ρ = arithmetical mean resistivity of cylinder over the heating cycle in microhm-centimeters
f = frequency in cycles per second
d = diameter of cylinder in inches.

The permeability μ of magnetic materials is inversely proportional to magnetizing force, and for high magnetizing forces the slope of the *B-H* curve becomes almost unity, so the permeability changes from the surface of the load to the center. To determine effective permeability mathematically for a given load may be a difficult problem. Baker and Rosenberg [8] assign a value of permeability

$$\mu = \frac{1.8 B_M}{H_0} \qquad (24 \cdot 20)$$

which gives fairly accurate results for inductive heating of common low-carbon steels. B_M is the maximum flux density and is always above saturation, 16 kilogauss for most irons and steels. Therefore permeability becomes

$$\mu = \frac{28{,}800}{H_0} \qquad (24 \cdot 21)$$

When equation 24·21 is substituted in equation 24·19, there results a formula which gives approximately the power generated per cubic inch of a cylinder of iron. This power density is a fictitious quantity since actual power density at the surface is much higher than at the center of the load. However, P_v does represent an average power density throughout the cylinder. This equation is

$$P_v = 2.75 \times 10^{-4} H_0^{3/2} \frac{\sqrt{\rho f}}{d} \qquad (24 \cdot 22)$$

As mentioned previously, maximum power is absorbed from the magnetic field when $G(d/\delta)$ is maximum, or when $d/\delta \cong 3.5$, and it usually is undesirable to heat a cylinder inductively when $d/\delta < 3.5$. This establishes a minimum frequency limit. Let $d/\delta = 3.5$; then the permeability-frequency product becomes

$$\mu f_m = 48 \frac{\rho}{d^2} \qquad (24 \cdot 23)$$

where f_m is the minimum frequency in cycles per second, and the other quantities are as defined before. For the non-magnetic cylinder μ is unity.

If the cylinder is magnetic, minimum permissible frequency becomes

$$f_m = 0.5 \frac{\sqrt{P_v \rho}}{d} \qquad (24 \cdot 24)$$

Below this frequency operation is unstable. Maximum resistivity is chosen because then the greatest depth of penetration occurs.

Heating Hollow Cylinders

Because of skin effect or depth of current penetration, equations 24·19 and 24·22 should not be used for a hollow cylinder without considering its wall thickness. Consider a hollow cylinder of non-magnetic material (see Fig. 24·9).

Let the diameter and wall thickness of the cylinder be d and t inches respectively. Because the current density at the inside surface of the cylinder should be negligible, the depth of current penetration δ is arbitrarily chosen so that it will not be greater than 70.7 percent of the wall thickness.

$$\delta = \frac{t}{\sqrt{2}} = 1.98 \sqrt{\frac{\rho}{\mu f}} \quad \text{inches}$$

Under these conditions the current density at the inner surface of the cylinder falls to approximately 25 percent of its value at the outer surface.

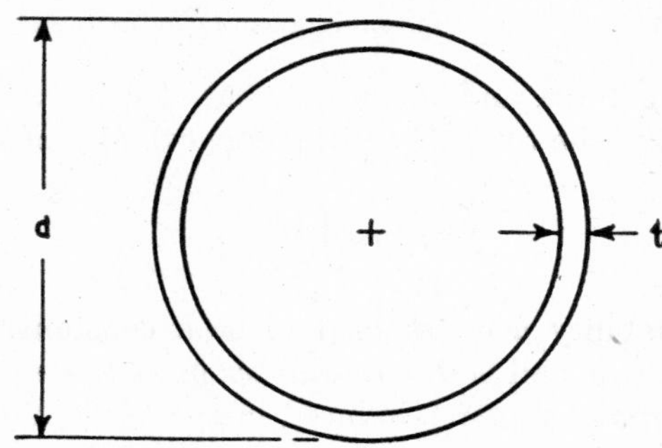

Fig. 24·9 Hollow conducting cylinder.

For non-magnetic loads the minimum frequency becomes

$$f_m = 7.85 \frac{\rho}{t^2} \quad \text{cycles per second} \qquad (24 \cdot 25)$$

For hollow magnetic cylinders this frequency is

$$f_m = 1.29 \sqrt{\frac{P_v d \rho}{t^3}} \quad \text{cycles per second} \qquad (24 \cdot 26)$$

Equations 24·17 to 24·26 have described a cylindrical load so that calculations can be made to find the required magnetizing force for a given heating task. In equations 24·17, 24·19 and 24·22 resistivity ρ is chosen to be the arithmetical mean resistivity of the load during the heating process. This is done to insure an average power input to a cylinder. However, ρ in equations 24·23 to 24·26 has been chosen to be a maximum to insure that the depth of penetration during the heating process never becomes too great for favorable power absorption.

If a hollow non-magnetic cylinder is to be heated at a lower frequency, and the depth of current penetration is much greater than the wall thickness, the cylinder can be analyzed as though it were a short-circuited single-turn secondary of an air-core transformer. The magnetizing force needed so that a hollow cylinder may absorb a specified amount of power (wall thickness is much less than the depth of current penetration) can be computed from this equation:

$$P_a = \frac{5.15f^2td^2\rho H_0^2 \times 10^{-9}}{\rho^2 + 1.6t^2d^2f^2 \times 10^{-2}} \tag{24·27}$$

where P_a is the required power density in watts per square inch, ρ is the arithmetical mean resistivity over the heating cycle in microhm-centimeters, and the other quantities are as previously defined.

Sometimes cylindrical hollow loads are to be heated at a low frequency; however, the required peak magnetizing force H_0 usually is high, which implies that large current is required to obtain large values of H_0. Large current usually causes large coil loss and therefore reduces coil efficiency.

Heating Slabs [9, 10]

Suitable equations for power absorbed and proper minimum frequency for strips or slabs can be derived from

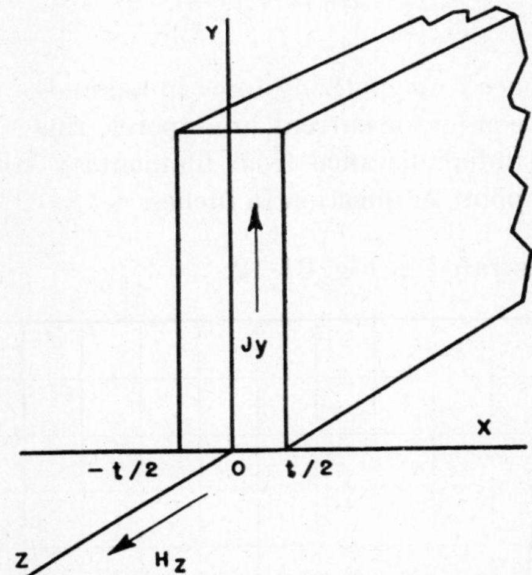

FIG. 24·10 Flux distribution in a solid conducting slab.

equations 24·3. If the depth of penetration of the magnetic flux, which obeys the same exponential law as the current penetration, is large in comparison with the thickness of the strip (that is, if flux density is fairly uniform throughout the cross section of the load) the calculation is the same as for transformer laminations.

The thickness is assumed to be much less than the width of the slab so that no appreciable power is absorbed by the edges. If an infinite-plane conductor is assumed to be placed in a uniform magnetic field H, currents are induced in the conductor according to equations 24·3. Figure 24·10 shows that equations 24·3 reduce to

$$\frac{dH_z}{dx} = -0.4\pi J_y \tag{24·28}$$

and

$$\frac{dE_y}{dx} = -\mu \times 10^{-8}\frac{\partial H_z}{\partial t} \tag{24·29}$$

By substituting $E_y = \rho J_y \times 10^{-6}$ in these equations and rearranging them, a single differential equation may be derived to define the magnetizing-force distribution within the slab:

$$\frac{d^2H_z}{dx^2} = \frac{4\pi\mu}{\rho} \times 10^{-3}\frac{\partial H_z}{\partial t} \tag{24·30}$$

From this equation, as from equation 24·13 for flux distribution within a cylinder, the magnetizing force can be determined at any point x within the material as a function of the magnetizing force at the surface. Thus distribution and magnitude of induced current densities can be found. An equation analogous to equation 24·15 can be derived for power density in watts per cubic centimeter as a function of surface magnetizing force, thickness of the strip, frequency, and electrical properties of the load material.

$$P_v = \tfrac{1}{2}H_0^2\mu f\, G(kt) \times 10^{-7} \tag{24·31}$$

where

$$G(kt) = \frac{1}{2kt}\left[\frac{\sinh kt - \sin kt}{\cosh kt + \cos kt}\right]$$

$$k^2 = \frac{4\pi^2\mu f}{\rho} \times 10^{-3}$$

$$H_0 = H\sqrt{2} = \text{peak magnetizing force in oersteds}$$

In equation 24·31, kt also bears a relation to the depth of penetration δ:

$$kt = \frac{t}{\delta} \tag{24·32}$$

where t = thickness of slab
δ = depth of penetration.

Both t and δ are in the same units; that is kt is also a pure numeric. Thus equation 24·31 becomes

$$P_v = \frac{1}{2}H_0^2\mu f\, G\left(\frac{t}{\delta}\right) \times 10^{-7} \tag{24·33}$$

Figure 24·11 shows $G(t/\delta)$ as a function of t/δ. The function $G(t/\delta)$ is a maximum when $t/\delta \cong 2.5$. In this vicinity the power absorbed is a maximum for a given frequency. As for a cylinder the slope of the $G(t/\delta)$ curve to the left of the maximum $G(t/\delta)$ is too steep for optimum

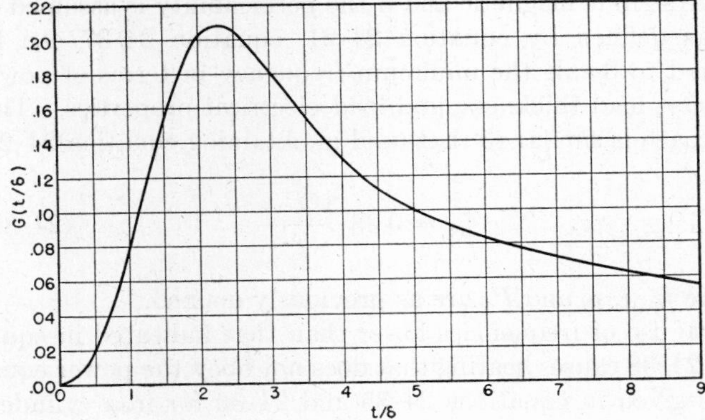

FIG. 24·11 $G(t/\delta)$ function curve.

heating. However, the $G(t/\delta)$ curve in the region of $t/\delta \geqq 2.5$ is satisfactory for heating slabs in an alternating magnetic field.

For $t/\delta > 10$, $G(t/\delta)$ may be replaced by the curve $G(t/\delta) = \delta/2t$, and the power-input equation reduces to

$$P_v = \frac{\delta}{4t}H_0^2\mu f \times 10^{-7} \tag{24·34}$$

To express power density in watts per cubic inch the right-hand side of equation 24·34 should be multiplied by

16.4. To reduce equation 24·34 even further, substitute the value $1.98\sqrt{\rho/\mu f}$ for δ given in equation 24·9. This becomes the fundamental equation for inductively heating a strip or slab:

$$P_v = 8.15 \times 10^{-7} H_0^2 \frac{\sqrt{\mu\rho f}}{t} \qquad (24\cdot35)$$

For non-magnetic material μ = unity.

If the value of μ defined by equation 24·21 is substituted in equation 24·35, the resultant equation describes the heating of magnetic steel slabs or sheet and is analogous to equation 24·22 for heating steel cylinders:[9]

$$P_v = 1.37 \times 10^{-4} H_0^{3/2} \frac{\sqrt{\rho f}}{t} \qquad (24\cdot36)$$

where P_v = power density in watts per cubic inch
H_0 = peak magnetizing force in oersteds
ρ = arithmetical mean resistivity of load over the heating cycle in microhm-centimeters
f = frequency in cycles per second
t = thickness of sheet or slab in inches.

Because of the limitations imposed upon the preceding equations if $t/\delta \geqq 2.5$, there is a minimum frequency range below which it is not advisable to apply the derived formulas. Therefore, let $t/\delta = 2.5$, with $G(t/\delta)$ a maximum. The permeability-frequency product now becomes

$$\mu f_m = 20.7 \frac{\rho}{t^2} \qquad (24\cdot37)$$

where f_m = minimum frequency in cycles per second
ρ = maximum resistivity in microhm-centimeters
t = thickness of strip in inches
μ = permeability.

If the strip is magnetic and if the permeability is assumed to be as defined by equation 24·21, equation 24·37 can be altered to define the minimum frequency in terms of power density, load thickness, and load electrical properties. This approach is similar to that used in obtaining equation 24·26.

$$f_m = 0.38 \frac{\sqrt{P_v \rho}}{t} \qquad (24\cdot38)$$

where f_m, t, ρ, and P_v are as previously defined.

The use of frequencies lower than that indicated in equation 24·38 causes heating that does not obey the power equations given in equations 24·35 and 24·36 for iron cylinders and slabs.

If power density is expressed in watts per square inch instead of watts per cubic inch, the resulting equations are even simpler than those defining the volume-power density. For the cylinder and slab the equations are the same. Thus for the non-magnetic load

$$P_a = 4.07 \times 10^{-7} H_0^2 \sqrt{\mu\rho f} \text{ watts per square inch} \qquad (24\cdot39)$$

and for the magnetic load

$$P_a = 6.87 \times 10^{-5} H_0^{3/2} \sqrt{\rho f} \text{ watts per square inch} \qquad (24\cdot40)$$

These equations are useful for loads with odd shapes.

The arithmetical mean resistivity over a given heating cycle is used in equations 24·33, 24·35, 24·36, 24·39, and 24·40; maximum resistivity is used in equations 24·37 and 24·38. The reasons are the same as for heating cylinders.

Proximity Heating

If the equations for power absorbed by the load are rearranged to give required peak magnetizing force in oersteds, the required coil turns and dimensions can be calculated. Now, however, suppose that the load cannot be heated by placing it in a solenoid; it must be heated by *proximity heating*, that is, by placing it in the field of a current-carrying conductor.

The field at any point near a filamentary current-carrying conductor not influenced by a conducting body is

$$H_0 = 1.11 \frac{I}{r} \qquad (24\cdot41)$$

where H_0 = peak magnetizing force in oersteds
I = conductor current in amperes, rms
r = radial distance from filamentary conductor to point in question in inches.

This is illustrated in Fig. 24·12.

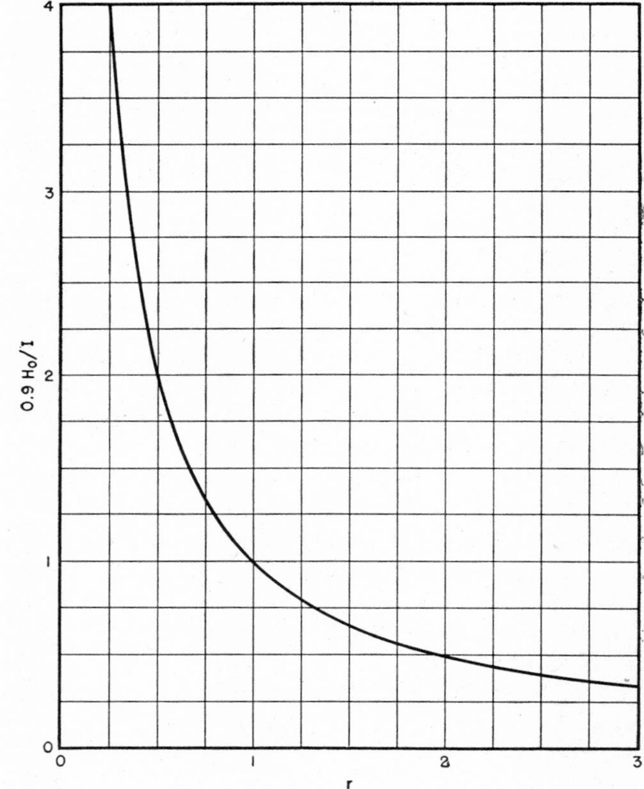

Fig. 24·12 Magnetic field at a point in space near a filamentary current-carrying conductor.

Now let an infinite slab of conducting material be brought into the field produced by the filamentary current-carrying conductor. An image effect is produced so that, if the depth of current penetration is small compared with other dimensions, the field component at the surface and parallel to the surface at any point a given distance r from the current-carrying conductor is

$$H_0 = 0.222 \frac{Ih}{h^2 + x^2} \qquad (24\cdot 42)$$

where h = effective distance from the filamentary conductor to surface of infinite slab in inches

x = distance along surface of slab normal to axis of filamentary conductor in inches

and H_0 and I are as defined before.

Figure 24·13 illustrates this magnetic field distribution as a function of x for various values of parameter h. For close

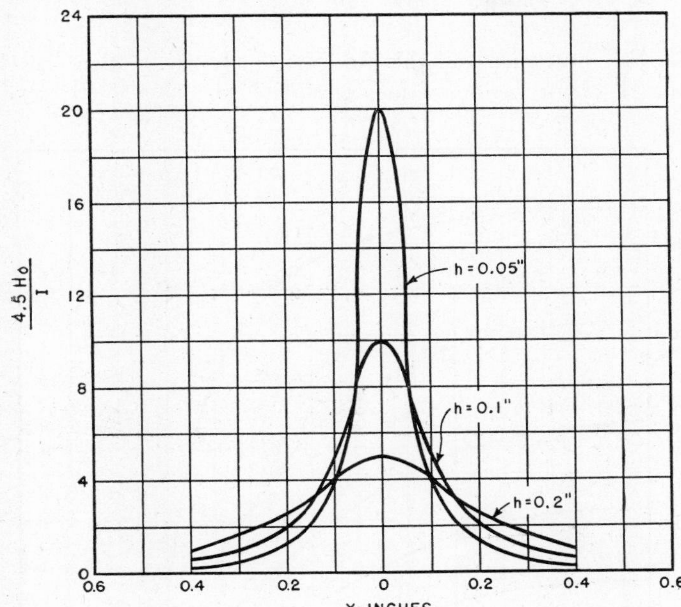

FIG. 24·13 Magnetic field at the surface of a plane conductor parallel to a filamentary current-carrying conductor.

coupling the magnetizing force becomes high just beneath the current-carrying conductor. The heating effect varies in a comparable manner, depending upon whether the load material is magnetic or non-magnetic.

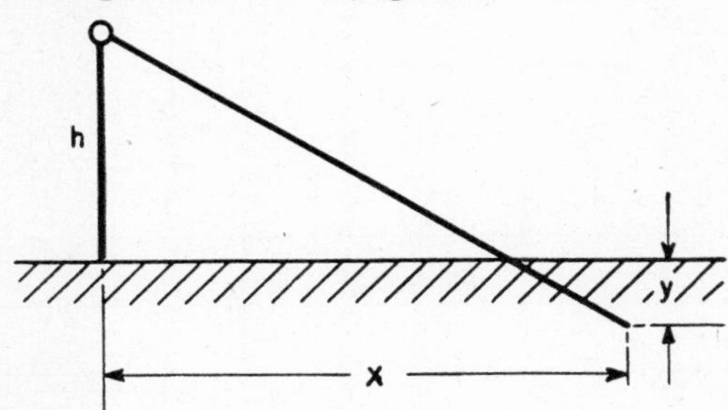

FIG. 24·14 Magnetic field beneath the surface of a plane conductor parallel to a filamentary current-carrying conductor.

With the aid of Fig. 24·14, current density J at any point in the infinite slab can be expressed by the Fourier integral [10]

$$J = \frac{I}{\pi} \int_0^\infty \{\sqrt{b^2 + 2j} - b\} e^{-hb - y\sqrt{b^2+2j}} \cos xb \, db$$

If the depth of current penetration is small, this equation reduces to

$$J = 0.222 \frac{Ih}{h^2 + x^2} \qquad (24\cdot 43)$$

where J is the surface current density, and I, h, and x are as previously defined.

A plot of this equation is identical with that of equation 24·42 for the flux distribution beneath the filamentary conductor.

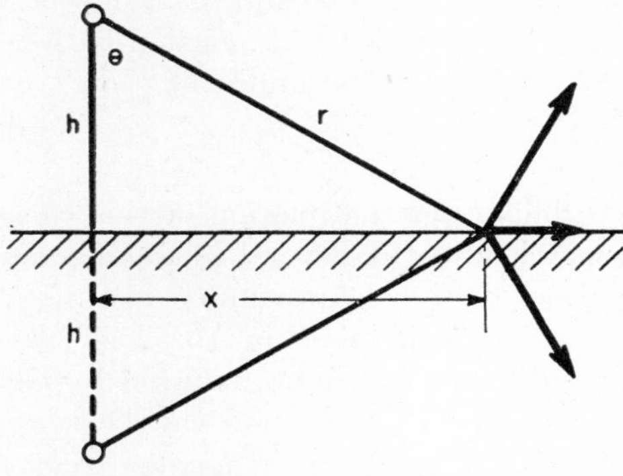

FIG. 24·15 Image effect of a current-carrying conductor.

From equation 24·43 (assuming that depth of current penetration in the infinite slab is δ) the power in watts absorbed by a slab of length l inches of the filamentary conductor becomes

$$P = 0.316 \times 10^{-7} I^2 \frac{l}{h} \sqrt{\mu \rho f} \qquad (24\cdot 44)$$

As mentioned previously, heating by proximity produces an image effect. In Fig. 24·15 the depth of penetration is

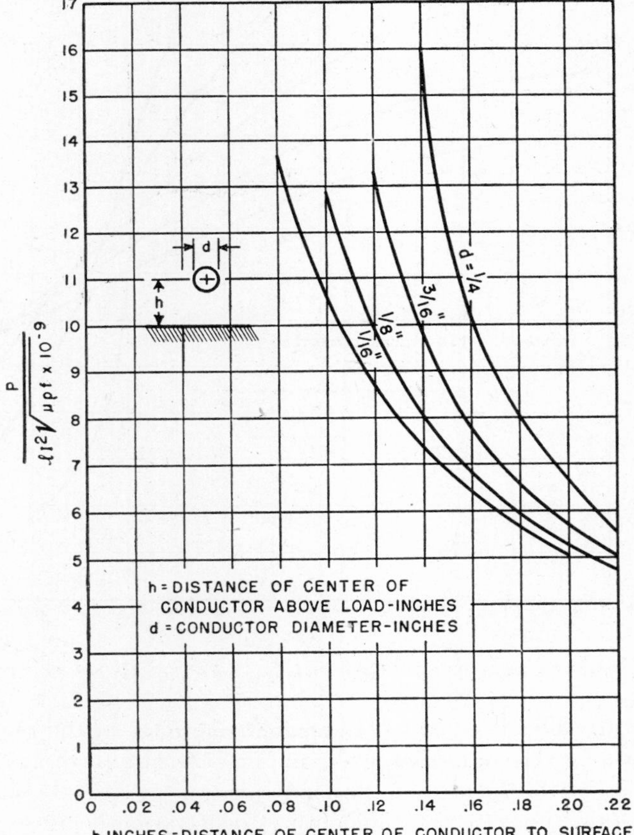

FIG. 24·16 Proximity heating.

assumed to be small compared with other dimensions. Current in the conductor above the surface of the load does not have its electrical center coincident with the geometrical center of the conductor. This non-coincidence implies that the effective height of the conductor above the surface of the load differs from the geometric height. Let the distance from the center of the current-carrying conductor be designated h_1, the effective height h, and the radius of the current-carrying conductor a. Then the relation between h and h_1 may be shown to be to a first approximation,

$$h^2 = h_1^2 - a^2 \qquad (24 \cdot 45)$$

Figure 24·16 illustrates the relation of this effective height upon proximity heating for several common conductor sizes.

This analysis applies to proximity heating with many conductors. Current induced in the load at any point becomes the sum of the currents induced by each current-carrying conductor. This addition can be simplified since the line integral of H around the magnetic circuit must equal $0.4\pi ni$. This is equal to the magnetizing force at the center of a long solenoid. However, the magnetizing force at any point on the surface of the load can be expressed as some constant times $0.4\pi ni$, the depth of penetration being small. Thus

$$H = 0.4\pi niK$$

If peak magnetizing force is desired and n is expressed in turns per inch,

$$H_0 = 0.7niK \text{ oersteds} \qquad (24 \cdot 46)$$

where H_0 = peak magnetizing force in oersteds
n = turns per inch
i = coil current in amperes

$$K = \frac{1}{\pi}\tan^{-1}\left[\frac{\frac{h}{l}}{\left(\frac{h}{l}\right)^2 + \left(\frac{x}{l}\right)^2 - 0.25}\right]$$

l = coil length in inches
h = distance between work and coil in inches
x = distance from center of coil along surface of slab in inches.

For the multiturn coil this function is shown in Figs. 24·17 and 24·18 for various values of h/l.

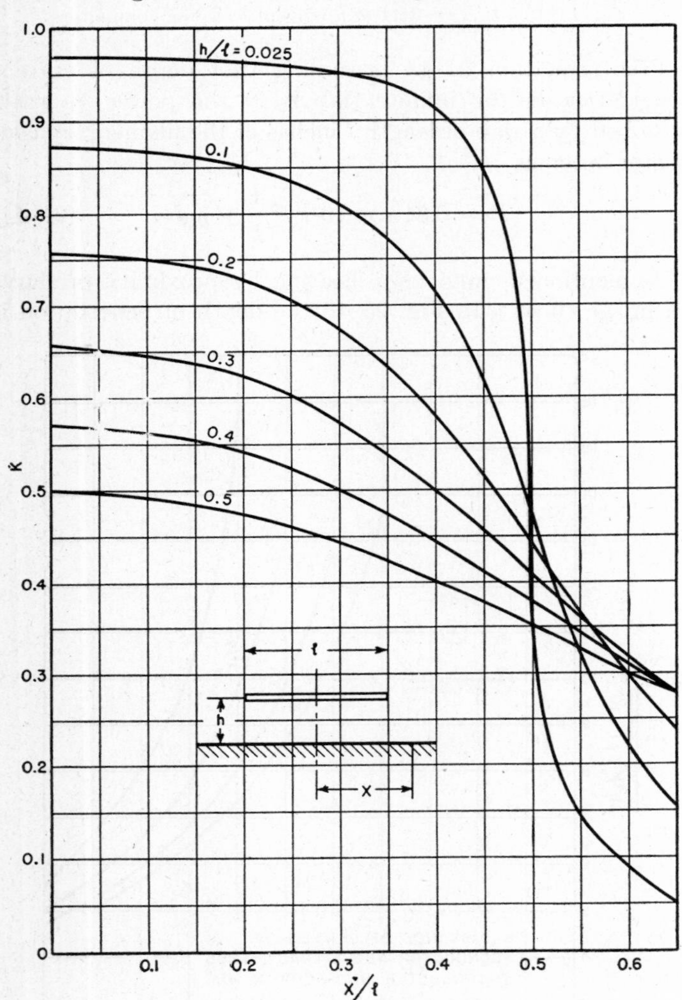

FIG. 24·17 Field distribution K factor.

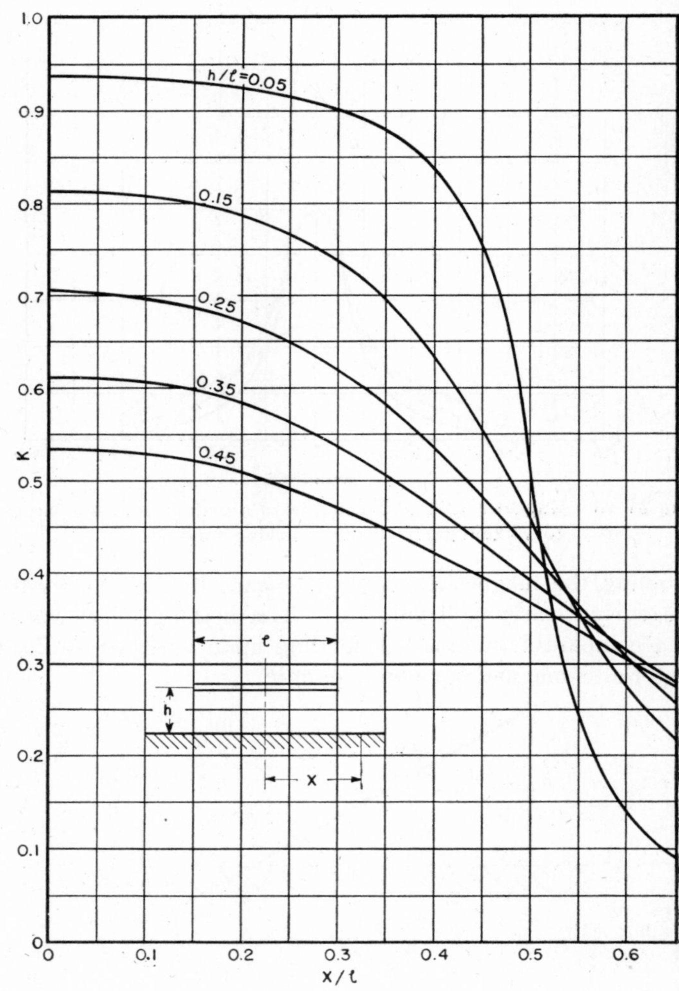

FIG. 24·18 Field distribution K factor.

Total Thermal Power

The total thermal power required for a given heating process may be considered to be the sum of several thermal requirements.

One of these requirements is the power required to raise the initial temperature of the load to the required temperature in the required length of time. This power P_1 is

$$P_1 = 17.6Mc\,\Delta t \qquad (24 \cdot 47)$$

where P_1 = thermal power in watts
M = pounds heated per minute
c = specific heat
Δt = temperature rise in degrees Fahrenheit.

The two variables in equation 24·47 that may need further consideration are M and c.

If the load properties are changed materially during heating, this change must be considered. An example is the vaporization of oil or water that may have been on or in the load. More important is the change in specific heat with a change in temperature. A good example is the radical change in the specific heat of iron as temperature is varied through the magnetic-transformation point. Another example, the more general case for non-magnetic loads, is the change in specific heat of brass as temperature is changed.

Radical changes in specific heat, as indicated by the specific-heat curve of iron in Fig. 24·19, must be accounted for by taking the average specific heat over the temperature range of the heating process. If the curve between the temperature limits does not approximate a straight line, numerical or graphical integration must be used to obtain the required average specific heat. This value is obtained by use of the heat-content curve for iron given in Section 24·12.

Besides the thermal power required to heat the load, the generator must supply all thermal losses: (1) radiation; (2) convection; and (3) conduction losses.

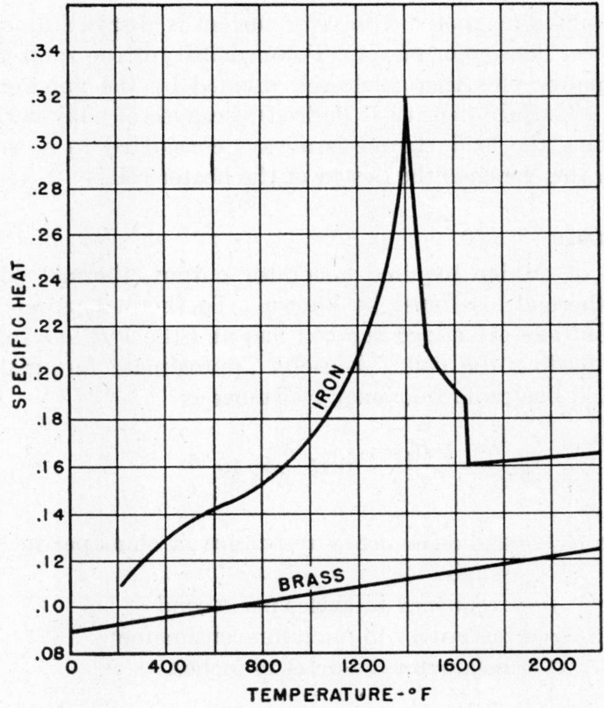

Fig. 24·19 Specific heat curve. (Reprinted with permission from *The Metal Iron*, by H. E. Cleaves and J. G. Thompson, published by McGraw-Hill, 1935.)

The first type of thermal loss P_2 is caused by radiant energy. If the area of the load is large this loss is important. Another important factor in radiation loss is the temperature. The following formula for radiation losses indicates that the loss is a function of the fourth power of temperature. Figure 24·20 shows radiation loss as a function of Fahrenheit temperature, based upon a 75-degree ambient temperature T_1.

$$P_2 = 37e\left[\left(\frac{T_2}{1000}\right)^4 - \left(\frac{T_1}{1000}\right)^4\right] \text{ watts per square inch} \tag{24·48}$$

in which the temperatures are Kelvin temperatures.

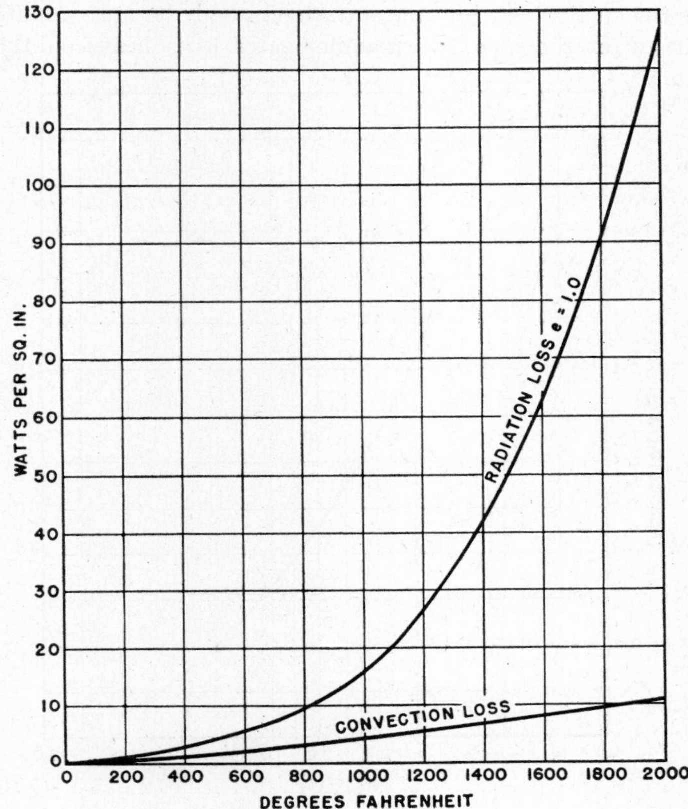

Fig. 24·20 Thermal-loss curve—radiation and convection.

The factor e is the *relative-emissivity factor.* Relative emissivity is the ratio of energy emitted per unit area of load surface to the energy emitted per unit area of a black body.

These are relative emissivities of the surfaces of various substances:

Aluminum, polished	0.10
Copper	0.15
Cast iron	0.25
Monel metal	0.43
Brass, polished	0.60
Copper, oxidized	0.60
Iron, black-plate	0.85

These values are approximate, for the condition of the surface of the material is not specifically defined, and the emissivity constant changes with variations in surface conditions. Emissivity of most surfaces approaches unity as temperature increases; that is, the emissivity of many surfaces approximates that of a black body at high temperature.

For many high-frequency heating jobs an average radiation loss may be used. This loss can be obtained with sufficient accuracy by graphical integration of the curve in Fig. 24·20 over the entire temperature range involved. Mathematical integration of equation 24·48 must be used if the upper and lower temperature limits do not fall on the curve of Fig. 24·20. If the power absorbed in watts per square inch is small compared with the maximum radiant-energy loss in watts per square inch, enough power must be absorbed from the magnetic field by the load to overcome load radiation loss at the highest temperature plus power to raise the temperature of the load to the required temperature in the desired length of time.

A second type of thermal loss is convection loss P_3, which is the energy lost from the surface of a body because minute air currents, created by expanding gases, carry heat from the load. It is expressed approximately in terms of Fahrenheit temperature rise Δt and area in square inches:

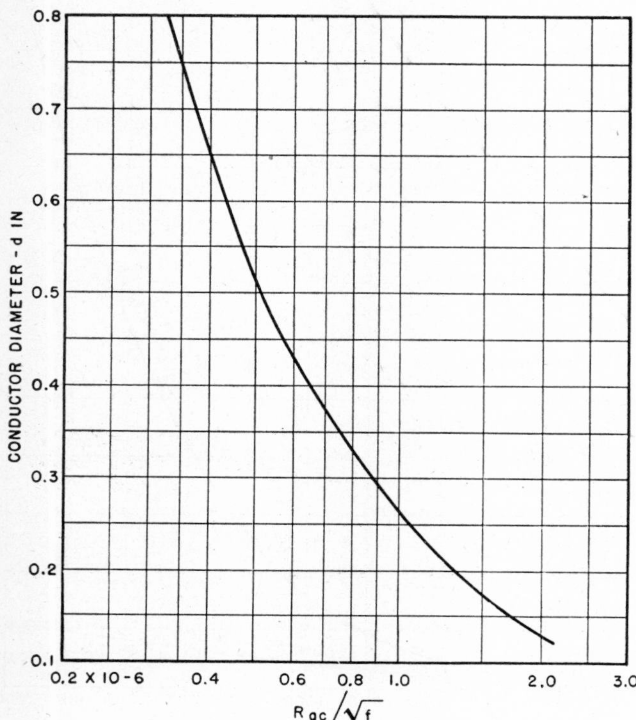

FIG. 24·21 Radio-frequency resistance of coil conductor of circular cross section.

$$P_3 = 4.66 \times 10^{-4} (\Delta t)^{1.33} \text{ watts per square inch} \qquad (24\cdot49)$$

If the process involves heating an appreciable area, convection loss is important. Figure 24·20 gives a convection-loss curve as a function of Fahrenheit temperature above 75 degrees.

Conduction losses P_4 sometimes are important.[11] A good example of this type of loss is the conduction of heat from the surface of a load piece being surface-hardened. Another example is the axial flow of heat from the end of a heated shaft.

The solution of heat-conduction problems is not simple. A general form of the heat-flow equation is

$$\frac{\partial \theta}{\partial t} = \frac{k}{sc}\left(\frac{\partial^2 \theta}{\partial x^2} + \frac{\partial^2 \theta}{\partial y^2} + \frac{\partial^2 \theta}{\partial z^2}\right) \qquad (24\cdot50)$$

Except in a few simple cases the conduction losses must be estimated.

When the depth of penetration is small, conduction loss may be considered comparable to the thermal case of constant-power application to the surface of a thick body. If constant power is applied to such a surface, then, as a function of surface-temperature rise, properties of the material, and time, the conduction loss in the body is

$$P_4 = 1.72\, \Delta t \sqrt{\frac{kcs}{m}} \qquad (24\cdot51)$$

where P_4 = watts per square inch conducted from the surface of an infinite slab
Δt = surface temperature rise in degrees Fahrenheit
k = average thermal conductivity [(calories per second per square centimeter)/(degrees centigrade per centimeter)] over Δt
c = average specific heat over Δt
s = average density in grams per cubic centimeter over Δt
m = heating time in minutes.

Equation 24·51 is rather limited. It does not apply to bodies of small dimensions, in which heat generated in one region affects surface temperature in another region in time m. From this equation the maximum time a given power density can be applied to a surface before the surface fuses can be estimated. It also tells the required power density needed to raise surface temperature Δt degrees Fahrenheit in m minutes, which is especially important in surface hardening.

Required magnetizing force or current is always a function of power density or power. Power density is the total thermal power plus thermal losses, divided by the volume (or area, if surface density is desired); because of eddy-current shielding the load volume is always considered to be solid. Upon this depends the design of the heater coil.

Coil Loss

To determine required generator output, thermal power and thermal loss must be known. To thermal power and thermal loss are added the coil loss and the I^2R loss in the lines feeding the coil. A useful approximate formula for coil and line radio-frequency resistance is

$$\frac{R_{ac}}{\sqrt{f}} = 0.15 \frac{\rho}{d} \times 10^{-6}$$

where R_{ac} = radio-frequency resistance in ohms per inch of conductor
f = frequency in cycles per second
ρ = resistivity in microhm-centimeters
d = conductor diameter in inches.

If a copper conductor is used for the coil, ρ is 1.724 microhm-centimeters, and

$$\frac{R_{ac}}{\sqrt{f}} = 0.26 \times \frac{10^{-6}}{d} \qquad (24\cdot52)$$

As an example, assume a length of copper conductor in the coil and transmission line 200 inches long, a coil-conductor diameter of ⅜ inch, and a frequency of 450 kilocycles. Then

$$R_{ac} = 0.465 \times 10^{-3} \text{ ohms per inch}$$

Total radio-frequency resistance is $200 \times 0.465 \times 10^{-3} = 0.093$ ohm. Equation 24·52 and the coil current define the coil loss by the formula

$$P_c = I^2 R_{ac} \tag{24·53}$$

where P_c = coil loss in watts
I = coil current in amperes, rms
R_{ac} = radio-frequency resistance of coil.

Equation 24·52 is plotted in Fig. 24·21, giving $R_{ac}/\sqrt{f}$ as a function of the coil-conductor diameter d. Multiplying $R_{ac}/\sqrt{f}$ by the square root of the frequency gives the radio-frequency resistance of the conductor in ohms per inch.

24·3 APPLICATION OF INDUCTION-HEATING THEORY

Regardless of the application, the number of ampere-turns to accomplish the job of heating must be determined. If the piece to be heated is symmetrical, such as a cylinder or slab, the problem is relatively simple, and the formulas of Section 24·2 are directly applicable. Inductive heating of almost all asymmetric loads may be resolved into equations 24·39 and 24·40 for power density per unit area.

As shown in Section 24·2 the frequency chosen for a particular application is important. There is a frequency range below which a load of given dimensions cannot be heated efficiently. For example, for a load of a given thickness, there is a minimum frequency below which small changes in the load characteristics cause large changes in absorbed power. Above this frequency small changes in load characteristics are not so noticeable. For this reason, the approximations of $G(d/\delta)$ and $G(t/\delta)$ curves were made for values of d/δ and t/δ, respectively, above certain established magnitudes. The formulas derived for loads of various shapes are not valid below this minimum frequency. The general equations are used if ratios of d/δ and t/δ are less than this established value.

Before ampere-turns or magnetizing force can be calculated, power density must be determined, as in Section 24·2. Sufficient power must be absorbed by the load to compensate for thermal losses such as radiation, convection, and conduction. Therefore the volume power density becomes

$$P_v = \frac{P}{\text{volume}} \text{ watts per cubic inch}$$

and the surface power density becomes

$$P_a = \frac{P}{\text{area}} \text{ watts per square inch}$$

where P is the total required thermal power in watts, including all thermal losses; the volume includes all space inside the coil occupied by the load, in cubic inches; and the area is the lateral surface of the load under coil, in square inches.

Magnetizing Force

The power-density formulas derived in Section 24·2 may be rearranged to indicate the peak magnetizing force required for a desired power absorption. These equations, rearranged, for the simplest cases, are

Non-magnetic load

$$H_0 = 1570 \left[\frac{P_a}{\sqrt{\rho f}}\right]^{0.5} \tag{24·54}$$

Magnetic load

$$H_0 = 598 \left[\frac{P_a}{\sqrt{\rho f}}\right]^{0.66} \tag{24·55}$$

where H_0 = peak magnetizing force in oersteds
P_a = surface-power density in watts per square inch
ρ = average resistivity in microhm-centimeters
f = frequency in cycles per second.

A similar expression can be obtained for the amperes required in a single current-carrying conductor inductively heating a load by proximity heating. This equation is

$$I = 5620 \left[\frac{hP}{l\sqrt{\mu\rho f}}\right]^{0.5} \tag{24·56}$$

where I = current in conductor in amperes, rms
P = total power required in watts
h = effective height above surface of load in inches (see equation 24·45)
l = length in inches of load being heated under current-carrying conductor
μ = relative permeability
ρ = average resistivity in microhm-centimeters
f = frequency in cycles per second.

A similar approach can be made with the aid of Figs. 24·17 and 24·18 for proximity heating with more than one conductor. The value of μ for magnetic loads is considered as that given by Rosenberg in equation 24·20.

Coil Design

Equations 24·54 to 24·56 indicate the magnetizing force required to produce a desired power density. Magnetizing force is a function of the ampere-turns per inch in a coil, and at a point on the surface of a conducting body it is a function of the square of the distance from a current-carrying conductor, if the depth of current penetration is small.

The coil conductor should have sufficient effective cross-sectional area to carry the required current safely and economically. Often it is impractical to use simple conductors with this current-carrying capacity. By water cooling, the resistivity of the coil conductor can be held to a minimum and at the same time the ampere-turns per inch can be increased greatly.

Coil-Loss Dissipation

If some allowable coil-temperature rise is selected, the power in kilowatts to heat Q gallons of water per minute Δt degrees Fahrenheit is

$$P = 0.147Q\,\Delta t \tag{24·57}$$

Equation 24·57 and Fig. 24·22 indicate theoretically the amount of power that may be dissipated in the coil current-carrying conductor if it is water-cooled tubing. This assumes 100 percent transfer efficiency of the energy in the coil conductor to the water. For practical purposes an assumed transfer efficiency of 50 percent is about right. Sometimes the coil I^2R is greater than the allowable coil dissipation. However, the coil can be fed with water from the center, instead of one end, and the flow of water into the coil can be doubled. If the water connection is made at any point above ground potential, it must be insulated from ground. Rubber hose, saran tubing, and porcelain coils have been used with satisfactory results.

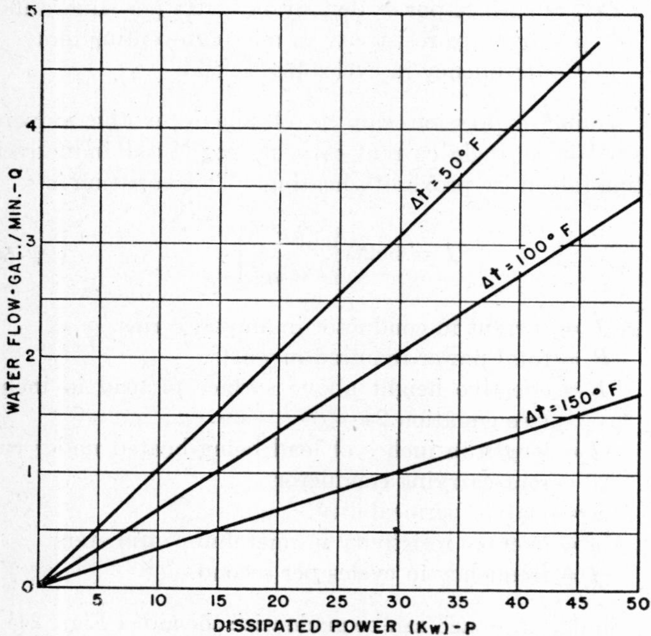

FIG. 24·22 Coil-cooling curves.

Ampere-Turns per Inch

Since the current-carrying capacity of a conductor is determined by its I^2R losses, the conductor must be capable of carrying the coil current without overheating. Equations 24·53 and 24·57 can be used if there is any uncertainty about current-carrying capacity of a given conductor. These equations define the effective cross section of the coil conductor as well as the necessary flow of water to dissipate I^2R losses. Once these quantities are established, the next step is to determine approximately (total coil voltage is considered later) the voltage per turn to provide sufficient spacing between turns to prevent possible arcing between turns. With long coils this computation is not difficult. Since the inductance of a long coil (a coil with cross-sectional dimensions small compared with its length) is approximately

$$L = 0.032N^2 \frac{A_c - A_w}{l} \text{ microhenrys} \qquad (24 \cdot 58)$$

where N = total number of turns in the coil
A_c = cross-sectional area of the coil without the load in square inches
A_w = cross-sectional area of the work (including hollow portion if work is hollow) within the coil in square inches
l = length of the coil in inches.

Then the voltage per turn may be expressed by

$$\frac{E}{N} = 0.2nfI_c(A_c - A_w) \times 10^{-6} \text{ volts, rms} \qquad (24 \cdot 59)$$

where n = turns per inch
f = frequency in cycles per second
I_c = coil current in amperes, rms.

Load Coupling

The coil must be designed so that there is no voltage breakdown between coil and load, or between turns of the coil. The points of greatest potential difference are at the coil terminals. If heating is to be uniform, the shape and location of the coil relative to the load must be considered. If the coil shape is changed to produce the desired uniformity of magnetic field, there is a change in its coupling (impedance). This change is important for constant-voltage generators such as motor-generator sets. For an oscillator, a load-impedance change usually can be compensated by changing the tank-coil reactance, since the greatest change in the load impedance is reactive. Thus the oscillator acts more nearly as a constant-current generator. If total tank reactance is held constant, the only load factors determining the magnitude of tank impedance into which the tube of the oscillator feeds are the resistance (variable with load coupling) reflected into the tank circuit and the load coil resistance (fixed for a given coil).

The impedance into which the usual oscillator feeds always acts as a pure resistance. This means that the tank-circuit impedance has no reactive component, and the power factor is unity. Thus the impedance of the tank circuit at resonance is [6]

$$R_a = \frac{R^2 + (\omega L)^2}{R} = \frac{R}{(R\omega C)^2 + (\omega^2 LC - 1)^2} \qquad (24 \cdot 60)$$

where R_a = apparent load resistance of the tube
R = sum of the resistances of tank coil, load coil, and reflected load resistance (tank-coil resistance usually is negligible)
ω = $2\pi f$ frequency in cycles per second
L = total tank inductance in henrys (includes inductance of load)
C = tuning shunt capacitance in farads.

Equation 24·60 becomes, as an approximation for a circuit Q of 10 or greater,

$$R_a = \frac{X^2}{R} \qquad (24 \cdot 61)$$

where $X = \omega L$ reactive ohms.

The last two equations indicate that the total series resistance in the tank circuit is critical. As shown by equations 24·42 and 24·46, the magnetizing force is proportional to the reciprocal of the square of the distance from the current-carrying conductor and proportional to the ampere-turns per inch. Now, since the power absorbed is a function of the magnetizing force at constant coil current, and a change in coupling changes the magnetizing force, the resistance reflected into the load coil from the work is changed, the resistance being simply the ratio of the power absorbed by the work to the square of the coil current. Thus, increasing the coil coupling to the work increases the resistance reflected into the load coil by the work; also, decreasing coil coupling decreases reflected resistance. An increase in coupling therefore decreases apparent resistance into which the generator feeds and thus an increase in generator current is demanded. The vacuum-tube generator, as well as the alternator, must feed into an optimum impedance so that the ratings of such generators are not exceeded when full power is demanded. The criterion in operating vacuum-tube generators is that the loaded tank Q, which is approximately equal to the ratio of the kva circulating in the tank circuit to the kilowatts supplied to the tank by the generator should be adjusted to the value stated by the manufacturer. Voltage and current ratings of high-frequency motor-generator sets define the minimum impedance, usually at unity power factor, into which this generator must feed.

Equations 24·60 and 24·61 indicate that the series resistance in the inductive branch of a parallel tuned circuit must be held close to a definite value for proper operation. Sometimes it is difficult to obtain enough reflected resistance (coil resistance held to a minimum) to lower the Q of the tank circuit enough to attain full output, especially with an oscillator. This is especially true of non-magnetic loads. Iron is usually the exception; however, when such material passes through the magnetic-transformation point it acts as a typical non-magnetic load. If high power densities are required, experience and theory show that, as far as the generator is concerned, the magnetic-transformation point is reached in short time. The short time is due to the induced currents flowing in the work to a depth δ as defined by equations 24·8 and 24·9, and a shallow surface layer is heated to temperatures above the magnetic-transformation point before the mass of the load has absorbed an appreciable amount of energy. This property is utilized in the practice of self-quenching, especially for surface hardening.

Heat Patterns and Coil Shapes

Heater coils are of various shapes and sizes, depending upon the application. Some consist of a single turn, some of many turns. A single-turn coil is shown in Figs. 24·23 and 24·24. The darkened area indicates qualitatively the heat pattern that may be expected. The magnetic field at a point on the surface of the load is a function of the square of the distance of the conductor from the load, and it also depends upon the shape and arrangement of the coil.

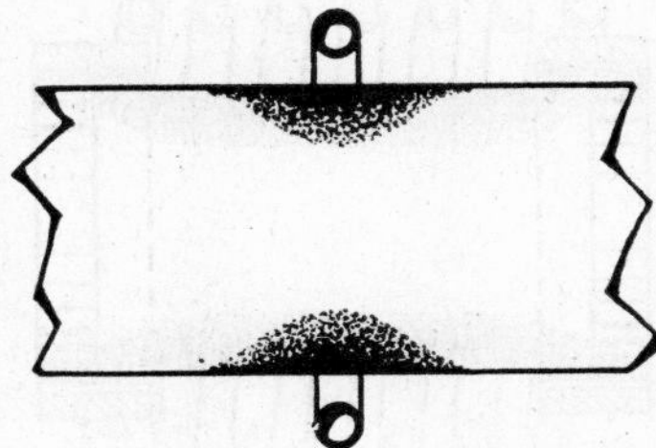

Fig. 24·23 Heat pattern with single-turn coil.

If the coil in Fig. 24·23 is coupled more closely to the load, the heating pattern becomes much deeper and narrower. If this were to be extended to multiturn coils, a "shadow" of the coil would appear as a heat pattern on the load. This shadow effect becomes more pronounced with multiturn coils of large pitch. This effect may be detrimental in hardening, but it may be overcome by rotating

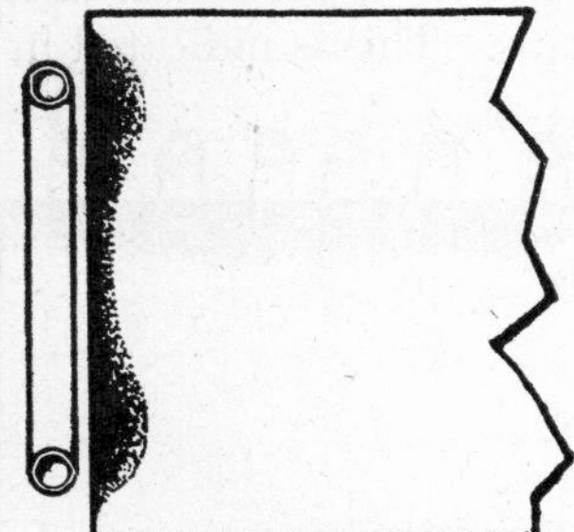

Fig. 24·24 Heat pattern with single-turn coil.

the coil or work about their common axis. Usually it is more convenient to rotate the work. Coil pitch ordinarily should be less than twice the coil-conductor diameter.

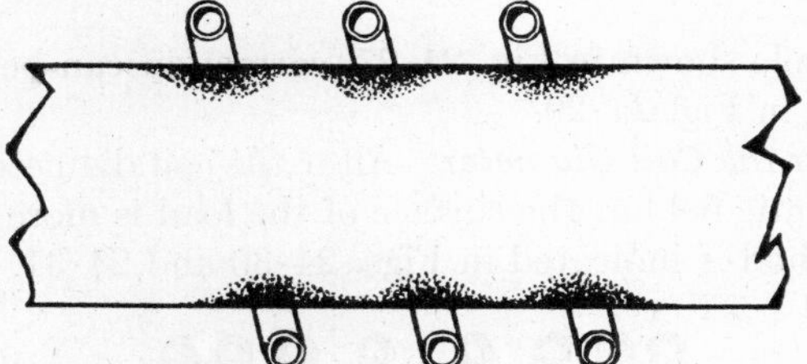

Fig. 24·25 Shadow effect in multiturn coil.

An example of the heat pattern for a closely coupled coil with large turn pitch is shown in Fig. 24·25. Assume that the pitch of the coil is small. Then the magnetic field along

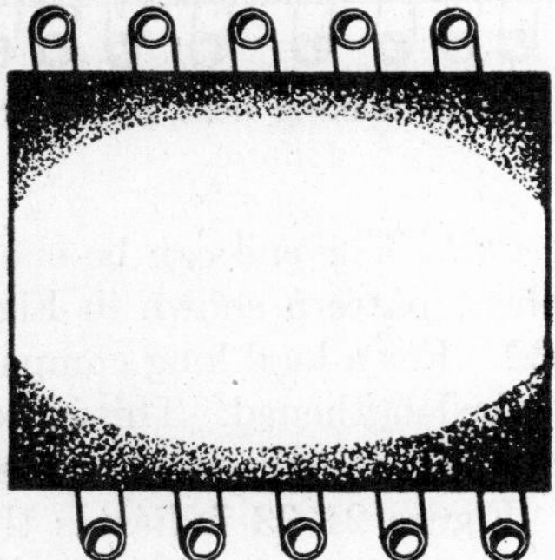

Fig. 24·26 Heat pattern in multiturn coil with short load.

the center of the coil is fairly uniform, dropping off rapidly near the ends of the coil. This change in the magnetic field is often serious. Consider, for example, a shaft of about the

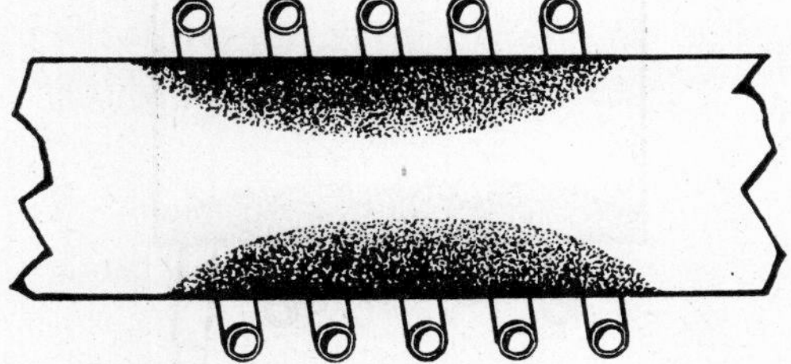

Fig. 24·27 Heat pattern in multiturn coil with long load.

same length as the coil. The expected heat pattern is indicated in Fig. 24·26. If the load is long compared with the coil length, the expected heat pattern is shown in Fig. 24·27. These patterns are qualitative only.

There are five corrective methods:

1. ***Change the Coil Pitch.*** For the example shown in Fig. 24·26, increase the pitch at the center of the coil relative to the pitch at the ends. This is indicated in Fig. 24·28. For the example shown in Fig. 24·27, correction can be made as indicated in Fig. 24·29.

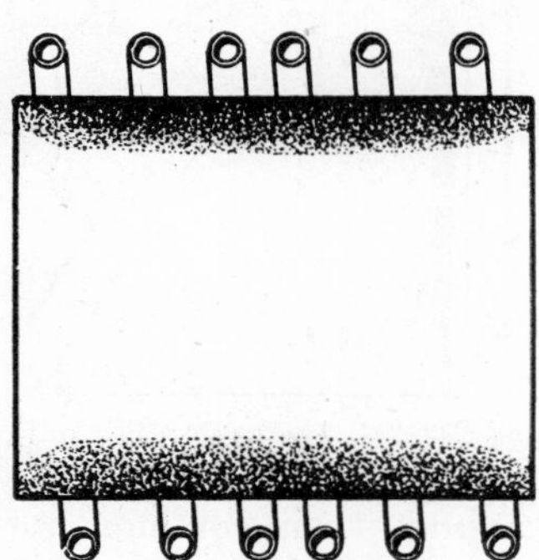

FIG. 24·28 A method of making heat pattern of Fig. 24·26 more uniform.

2. ***Alter the Coil Diameter.*** Alter the coil diameter so that the magnetic field at the surface of the load is more uniform. This method is indicated in Figs. 24·30 and 24·31.

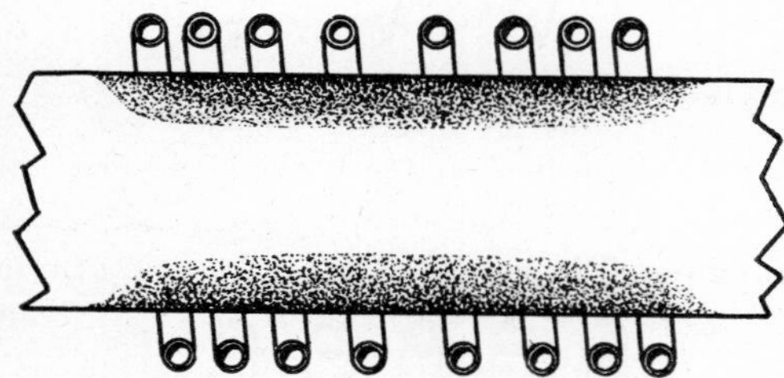

FIG. 24·29 A method of making heat pattern of Fig. 24·27 more uniform.

3. ***Shorten the Coil.*** The coil can be shortened to correct the non-uniform heat pattern shown in Fig. 24·26, as indicated in Fig. 24·32. For a load long compared with the coil length, the coil can be lengthened. This increases the amount of material being heated, but the region to be heated is heated more uniformly. Figure 24·33 indicates the expected heat pattern.

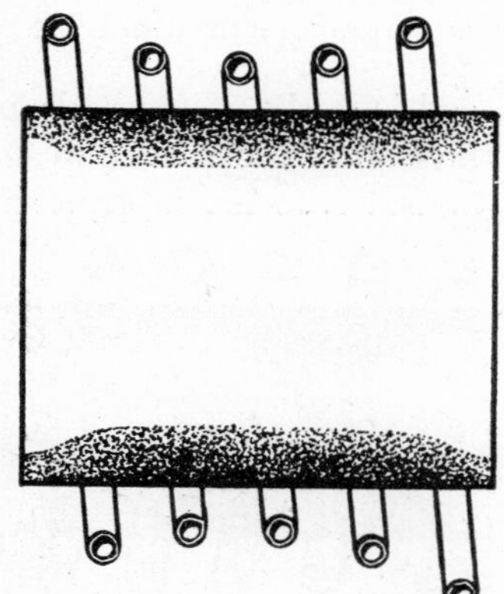

FIG. 24·30 A second method of correcting non-uniform heat pattern in Fig. 24·26.

4. ***Shield the Load Magnetically.*** Use a short-circuited turn to reduce the magnetic field. For the example shown in Fig. 24·26, this method is illustrated by Fig. 24·34.

5. ***Use Secondary Coupling.*** The heater coil consists of a split one-turn coil. The load coil is energized by inductive coupling from the generator. This method involves an added loss in the secondary coil (heater coil). By using copper for the secondary coil, and perhaps silver-plating it, the coil loss can be minimized. The secondary coil should be water-cooled. The combined coupling of the primary coil and the heater coil can be made close to unity. Thus the load is heated by a coil of low potential with a high current. The load coil can be coupled closer to the load because of its low voltage. Therefore, for heating loads of small cross-sectional area, this becomes a method to secure close

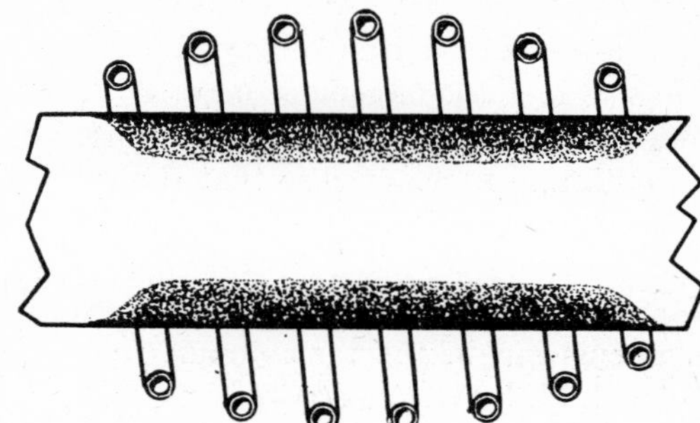

FIG. 24·31 A second method of correcting non-uniform heat pattern in Fig. 24·27.

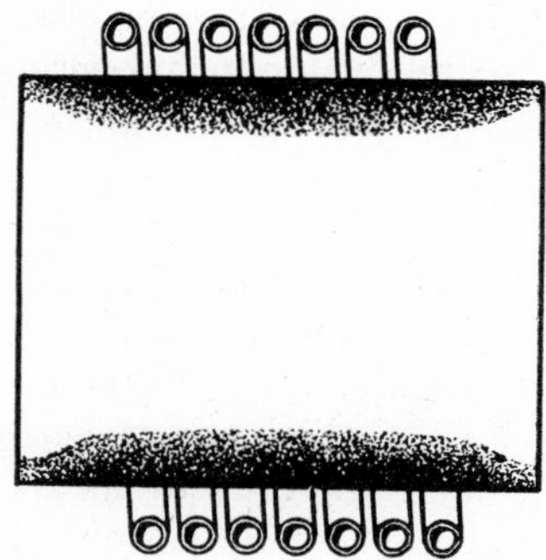

FIG. 24·32 A method of making more uniform heat patterns in short loads.

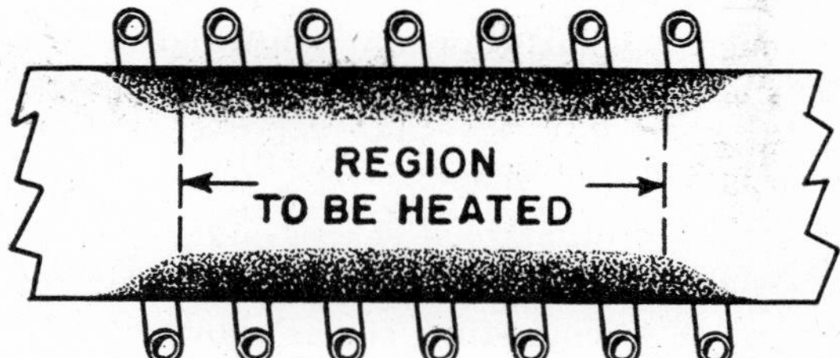

FIG. 24·33 A method of making more uniform heat patterns in long loads.

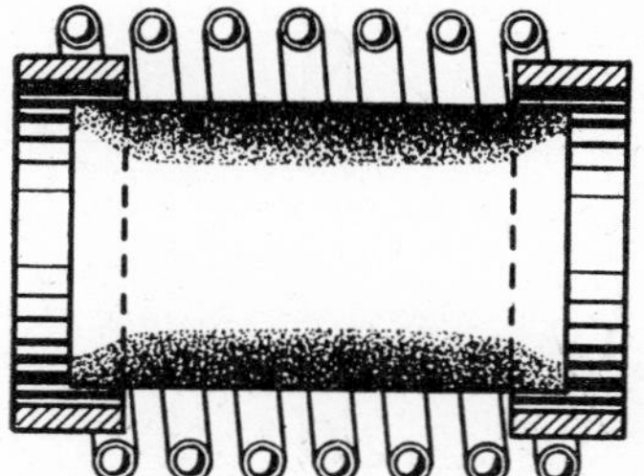

FIG. 24·34 Another method of producing more uniform heat patterns in short loads.

coupling. The fringing effect at the coil ends is reduced because of the greater uniformity of the magnetic field throughout the length of the secondary coil. This type of coil approximates the theoretical coil from which formulas are derived. Figure 24·35 shows an example of this method.

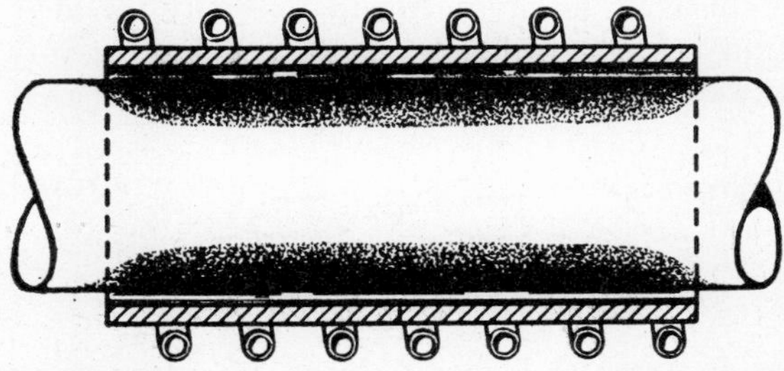

FIG. 24·35 Another method of producing more uniform heat patterns in long loads.

The size and shape of the piece to be heated suggests the shape of the coil. A tapered shaft or a bevel gear can be heated uniformly in a coil. Since the magnetizing force at the surface of the load is a function of the square of the distance of the load to the coil, the coil shape can be altered accordingly, as shown in Fig. 24·36. The taper of the coil is not the same as that of the shaft. The constant-pitch

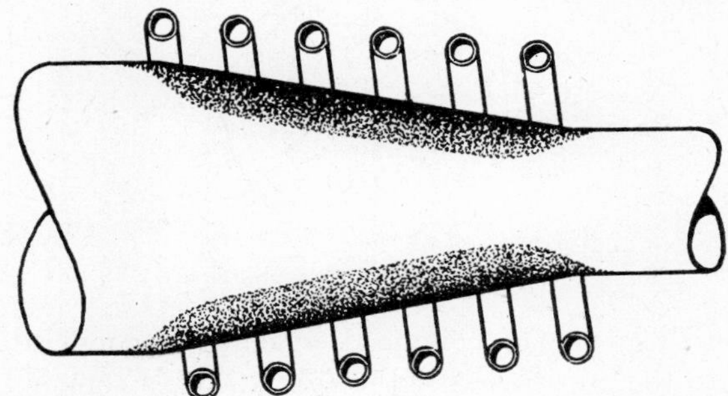

FIG. 24·36 A method of obtaining uniform heat patterns in tapered loads.

coil taper is made less than that of the shaft to obtain a uniform magnetic field at the surface of the taper. Another method is to change the pitch of the coil turns with the axis. This method is illustrated in Fig. 24·37. Here the coil is not tapered.

If corrective methods are not used, the heating pattern is not uniform. Figure 24·38 indicates the heating pattern

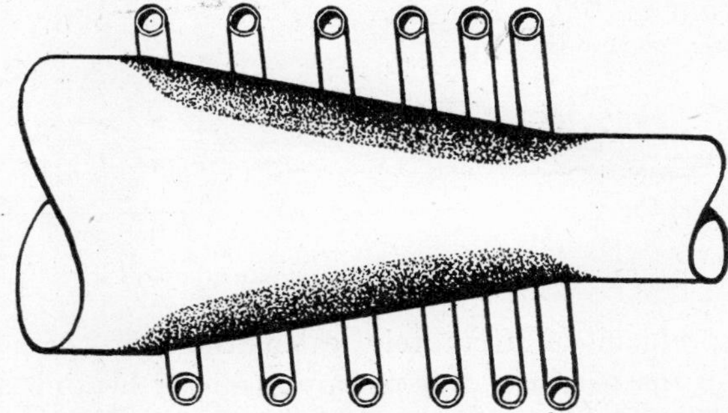

FIG. 24·37 Another method of obtaining uniform heat patterns in tapered loads.

that may be expected if the coil taper and shaft taper in Fig. 24·36 are equal. Figure 24·39 shows the expected heat pattern if a uniform pitch is used for the coil in Fig. 24·37. If, instead of using several turns, as shown in Figs. 24·36 and 24·37, one turn is used (the coil conductor having rectangular cross section), the method applied to Fig. 24·36

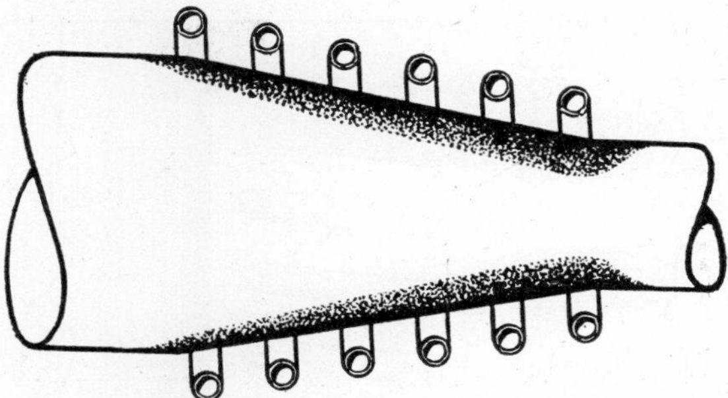

FIG. 24·38 Non-uniform heat pattern in tapered shaft.

should be used to obtain a uniform heat pattern. This is shown in Fig. 24·40.

The methods of obtaining uniform magnetizing force at the surface of the load, as shown in Figs. 24·36, 24·37, and 24·40, are useful if the load is long compared with the coil. If load length is about the same as coil length, the heat pat-

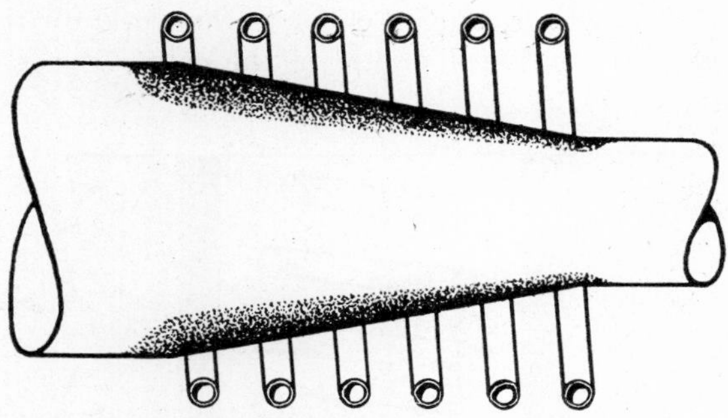

FIG. 24·39 Non-uniform heat pattern in tapered load.

tern is not uniform because of fringing effects. This condition may be partially corrected by the methods suggested for cylindrical loads.

With a given cylindrical conductor, the number of turns per inch may be increased by decreasing the pitch. Flattening the conductor decreases its current-carrying capacity,

FIG. 24·40 Another method of obtaining uniform heat patterns in tapered load.

but permits a smaller pitch. Since the current tends to concentrate about the inside of the coil, this decrease in pitch gives, in effect, a closer coupling to a load inside the coil.

Heating the wall of a cylindrical cavity with a solenoid inside it produces a marked fringing effect. Figure 24·41 shows the heat pattern of a regular solenoid concentric with

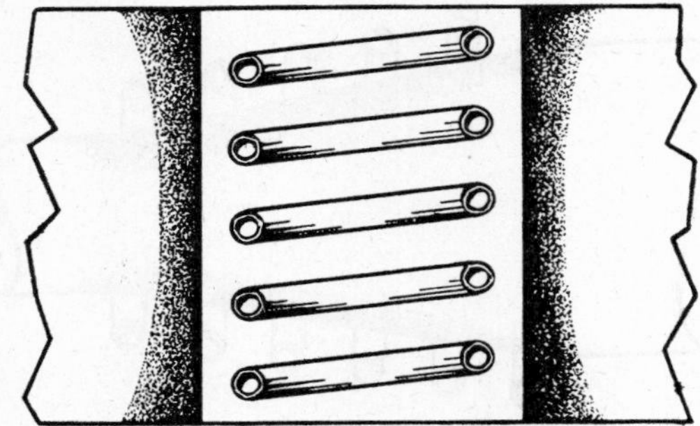

FIG. 24·41 Heat pattern expected when heating bores.

the cavity. A method of correction that lends itself to practical application is indicated in Fig. 24·42. Since the coil shape is altered the magnetizing force at the load surface is altered to produce uniform heating.

Instead of flattening a cylindrical-coil conductor to decrease the coil pitch, the conductor can be flattened to increase coil pitch; thus the coil is coupled closer to the work, for the current around the inside diameter of the coil is closer to the load surface.

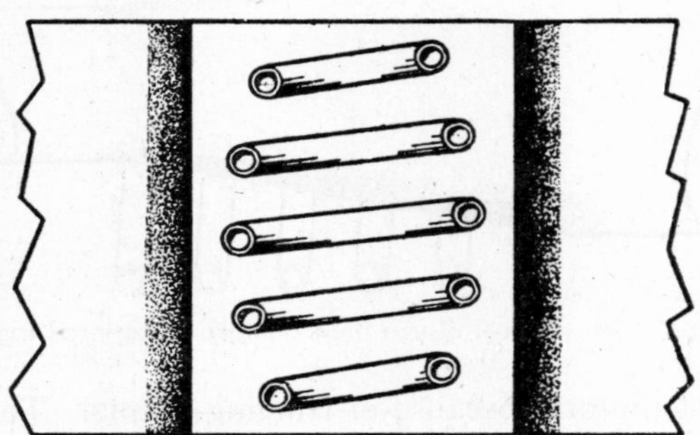

FIG. 24·42 Corrected heat pattern in bores.

One method of heating a cavity of small bore is to pass a single conductor through it. This becomes a form of proximity heating. Sometimes a hairpin coil consisting of several loops, the plane of each loop rotated relative to the plane of the other loops, is used. The work should be rotated continuously to minimize shadow effects.

Figure 24·43 illustrates the various types of coil conductors. Those in Figs. 24·43 (*a*) and 24·43 (*b*) are more commonly used than those in Figs. 24·43 (*c*) and 24·43 (*d*).

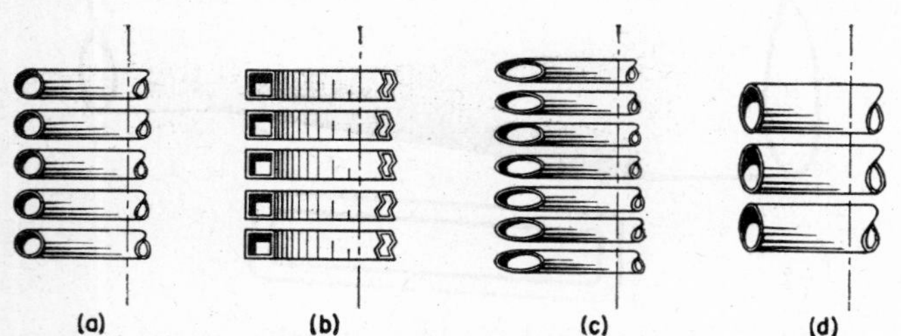

FIG. 24·43 Typical coil-conductor shapes.

If a surface is to be heated locally, a helix may be used, with the zone to be heated near the end of the helix. If space is limited, a pancake coil or a spiral coil may be used. Heating with a pancake coil is about three fourths as fast as heating shafts by surrounding coils.

The inductor block is a form of the single-turn coil. The coil is made of sheet copper, of thickness depending upon load requirements. The inductor block can be made to conform to the contour of the load, especially for gear teeth, and can be closely coupled. Load coupling at any given point may be made to give the desired heating at that point.

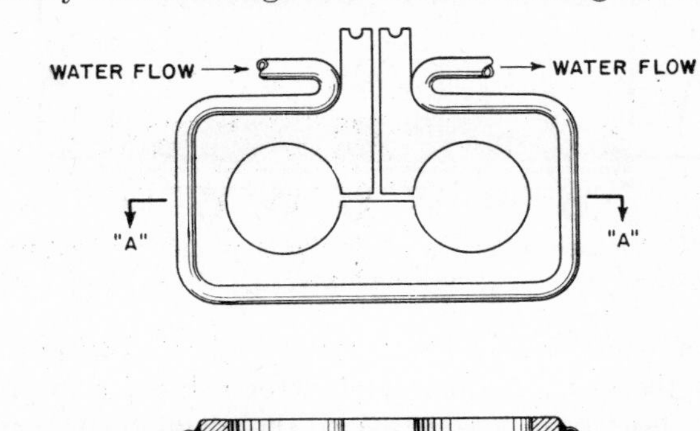

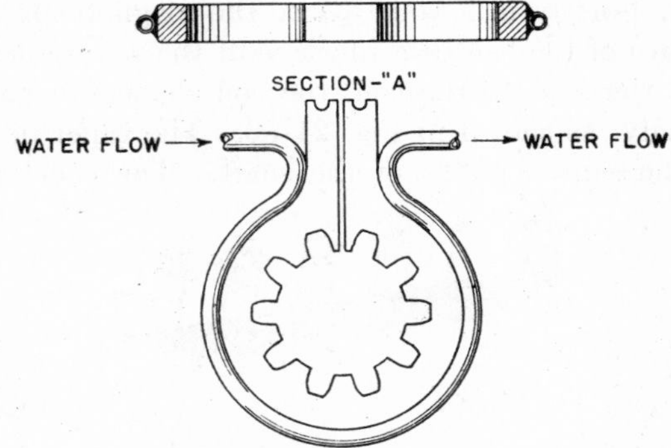

FIG. 24·44 Inductor blocks.

Because this type of coil needs cooling, a copper tube is brazed to the outer edge of the coil, and this copper tube is connected to a supply of cooling water. The coil terminals are connected to the power line or the terminals of the generator. Figure 24·44 shows two types of inductor blocks.

If a flat surface is to be heated, the spiral helix, a modified form of the pancake coil as shown in Fig. 24·45, is used. This is necessary because the field along the normal axis of the pancake is greater than at other points equidistant from the coil. It is much more difficult to couple a pancake coil to the load than to couple a coil encircling the load.

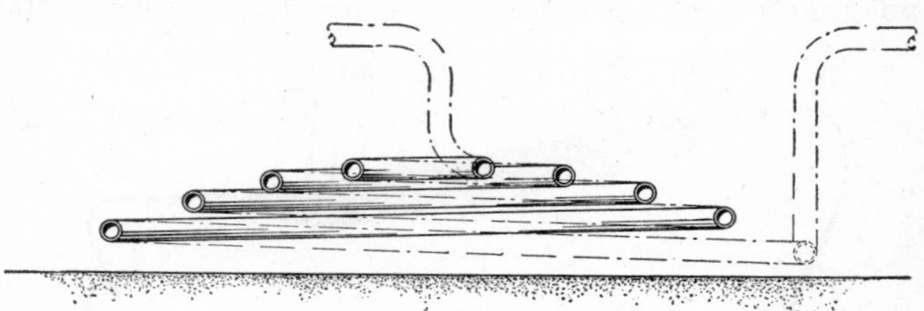

FIG. 24·45 Spiral helix or pancake coil.

The problem of accurately calculating the shape and ampere-turns per inch of a required coil for heating a particular load is very difficult. The magnetizing force at a given point along a current-carrying conductor is a rather simple function; however, if more than one conductor is used and the shape of the load is complex, the geometric function defining the magnetizing force at a given point for a given coil current is complicated.

For long solenoids the magnetizing force through a plane at the center and perpendicular to the axis is uniform. Therefore, for a coil with a length-to-diameter ratio of 10 or greater, the basic equation for the magnetizing force at the center of a long coil is sufficiently accurate:

$$H_0 = 0.7ni \tag{24·62}$$

where H_0 = peak magnetizing force in oersteds
n = coil turns per inch
i = coil current in amperes, rms.

For a coil with a length roughly equal to its diameter, equation 24·62 is not valid, but it can be used for an approximation. Trial and error must be used for coil design for irregularly shaped loads. Sometimes an electrolytic tank is used to determine the shape of a coil, especially for zonal heating.

Coil Voltage

Coil voltage with no load for a given current can be computed for several coil shapes. If a load is introduced inside a coil carrying high-frequency current, the coil voltage for a given current usually is reduced. Since voltage is the time rate of decrease of flux linking a circuit, the voltage of a long coil is

$$E_1 = 4.44fN[\phi] \times 10^{-8} \text{ volt} \tag{24·63}$$

where E_1 = induced coil voltage in volts, rms
f = frequency in cycles
N = total number of turns in coil
$[\phi]$ = numerical value of flux linking the coil.

The term $[\phi]$ is a complex quantity consisting of flux in the air and in the load. These two flux components are not in phase and must be added vectorially.

The complex form of the flux for a *non-magnetic* load is [9,12]

$$\phi = 6.45A_wH_0\left[\left(\frac{A_c}{A_w} + P - 1\right) - jQ\right] \text{ lines} \tag{24·64}$$

The absolute value of flux is

$$\phi = 6.45A_wH_0\left[\left(\frac{A_c}{A_w} + P - 1\right)^2 + Q^2\right]^{1/2} \text{ lines} \tag{24·65}$$

where ϕ = flux linking the coil
A_w = area of cross section of load in square inches
A_c = area of cross section of coil in square inches
P = factor proportional to the in-phase component of the flux in the load
Q = factor proportional to the out-of-phase component of the flux in the load.

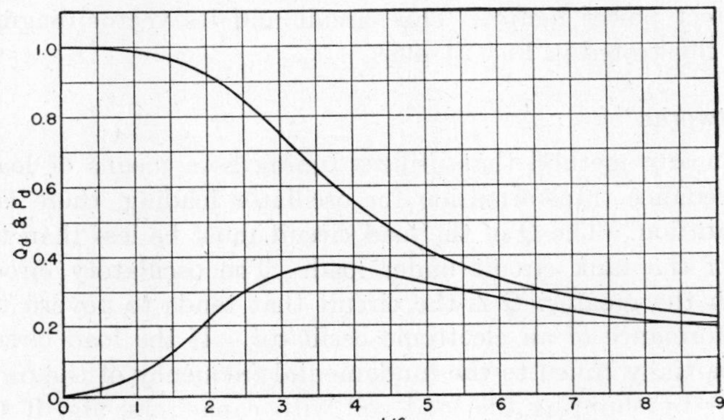

FIG. 24·46 P and Q functions for cylinders.

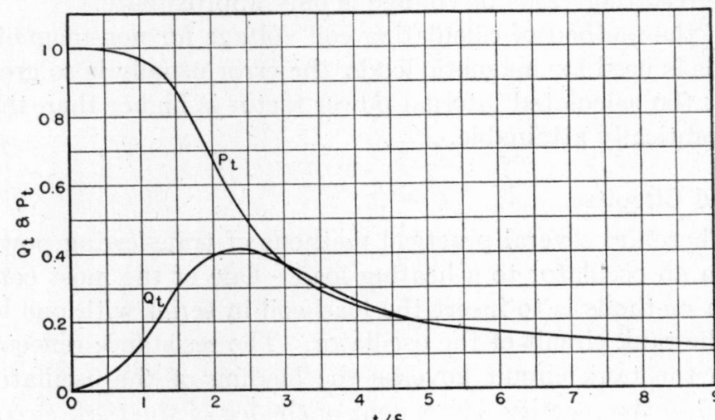

FIG. 24·47 P and Q functions for slabs.

Figures 24·46 and 24·47 show P and Q plotted as functions of d/δ and t/δ, respectively, for cylinders and slabs. For $d/\delta \geqq 14$

$$P_d = Q_d = \frac{2\delta}{d} \tag{24·66}$$

and for $t/\delta \geqq 10$

$$P_t = Q_t = \frac{\delta}{t} \tag{24·67}$$

where δ = depth of penetration in inches, as defined by equation 24·9
d = diameter of cylinder in inches
t = thickness of strip in inches.

At high frequencies and with loose coupling, P and Q become negligible.

The power factor of the coil, if coil I^2R loss and coil IR voltage drop are neglected, is the *internal power factor* defined by equation 24·68. This is the power factor of a lossless coil coupled to a load, and is known as the internal power factor of the coil and load.

$$\cos\theta_1 = \frac{P_T}{E_1I_c} \tag{24·68}$$

where $\cos\theta_1$ = internal power factor
P_T = total power absorbed by load in watts
E_1 = induced coil voltage in volts, rms (equation 24·63)
I_c = coil current in amperes, rms.

This power factor (equation 24·68) never can be greater than 0.707.

From Section 24·3 coil losses can be computed and, with the IR drop across the coil, the power factor of the coil and load may be expressed

$$\cos\theta = \frac{P_T + P_c}{E_cI_c} \tag{24·69}$$

where P_T = total power to load in watts
$\cos\theta$ = over-all power factor
P_c = coil I^2R loss in watts
E_c = total coil voltage in volts, rms
I_c = coil current in amperes, rms.

The product E_cI_c is defined by

$$E_cI_c = \sqrt{(E_1I_c \cos\theta_1 + I^2R)^2 + (E_1I_c \sin\theta_1 + I^2R)^2} \tag{24·70}$$

Coil voltage for heating magnetic loads can be estimated by computing the flux in the air space and adding to this the flux in the iron, if it is assumed that it penetrates uniformly to a depth δ below the surface of the iron load where

$$\delta = 1.16 \times 10^{-2} \sqrt{\frac{\rho H_0}{f}} \text{ inches} \tag{24·71}$$

in which H_0 is the peak magnetizing force in oersteds, and δ, ρ, and f are as defined in equation 24·9.

The flux in the iron is

$$\phi = 1.03 \times 10^5 A = 6.45BA \text{ lines} \tag{24·72}$$

where B is taken to be 16,000 gauss, $A = p\delta$ square inches, and p is the mean perimeter of the load in inches in a plane normal to the magnetizing force.

The total flux ϕ may then be substituted in equation 24·63. The resulting value of voltage is only approximate.

If the method of calculating coil voltage for non-magnetic loads is used for magnetic loads, the error usually is so great that the calculated internal power factor is higher than that theoretically attainable.

Load Circuits

There are several practical methods of transferring power from an oscillator to a heating load. One of the most common methods is to insert the load coil in series with one leg of the tank circuit of the oscillator. The resistance reflected into the tank circuit governs the loading of the oscillator, because the reactive component of the leg of the tank circuit in which the load is inserted is held constant, and only the load resistance appears to change. This circuit is illustrated in Fig. 24·48. Z_L, the load impedance, is equal to $R_s + jX_L$,

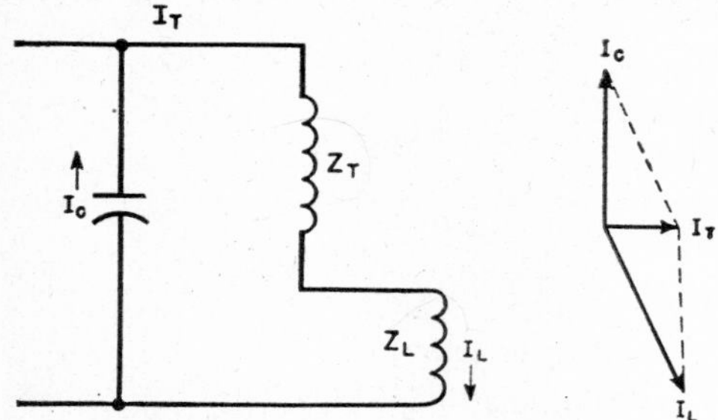

Fig. 24·48 Typical induction heating load circuit.

where R_s and X_L are the series resistive and reactive components respectively. The load-coil kva is necessarily less than that of the entire tank. The tank coil, represented by Z_T in Fig. 24·48, has little loss; consequently almost all the energy dissipated in the tank may be considered to be delivered to the load coil Z_L. Thus the Q of the tank is greater than that of the load coil. This is true also of an untuned transformer with its primary winding in series with the tank coil.

Power absorbed by the load coil and load is limited by the maximum tank current and the permissible voltage. For a given current, if it is assumed that the required coil voltage is available, the magnetizing force may be increased by decreasing coil pitch. This increase in magnetizing force increases the resistance reflected into the tank; the converse is true also. The pitch that can be used for a given coil current and coil dimensions cannot be reduced beyond a certain practical limit. The only way to increase magnetizing force further is to increase the coil current. With an oscillator as a generator (if coil voltage is not critical) this current sometimes is boosted by using shunting capacitors which partially

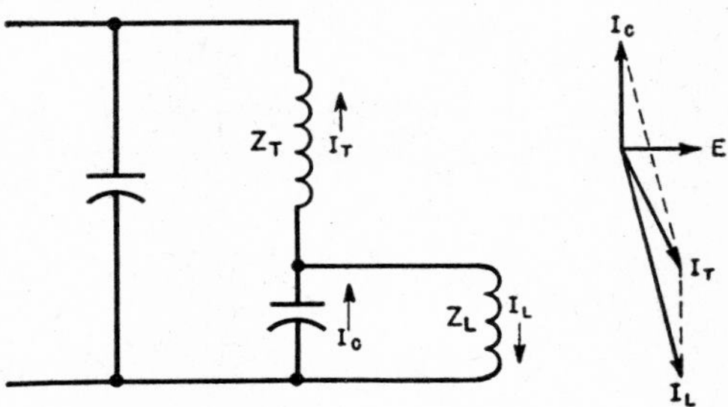

Fig. 24·49 Load-circuit and vector diagram for partial tuning.

correct power factor. This circuit and its vector diagram are illustrated in Fig. 24·49.

Networks [13]

In any method that utilizes tuning as a means of load-impedance transformation for oscillator loading, there is a limitation. The Q of the load circuit must be less than the Q of the tank circuit under load. The oscillatory circuit with the greatest Q is the circuit that tends to govern the performance of an electronic oscillator. If the load circuit is partially tuned to the fundamental frequency of the oscillator by shunting the load coil with capacitors, and if the Q of the load circuit is approximately equal that of the loaded tank, erratic operation can be expected. The frequency of the oscillator tends to change, depending upon load circuits and electrical constants of the tank. The result of this erratic action is improper valving action of the oscillator tube. This instability is invariably accompanied by increased anode dissipation and a substantial decrease in efficiency of the oscillator. Such excessive dissipation may damage the tube.

If the load coil is to be shunted by a capacitor, coil current is maximum for a given input current when the power factor of load coil and shunting capacitor combined is unity. Such a combination forms a parallel resonant circuit which acts as a high resistance. If this apparent resistance is greater than that into which the tube should feed, the oscillator no longer operates under optimum conditions. The Q of the load circuit under these conditions may be much greater than the tank-circuit Q. For these reasons only partial tuning should be used, if load-coil current is to be boosted by capacitive shunting. Tank current and coil current are approximately 180 degrees out of phase with the capacitor current, so these currents can be added algebraically (see Fig. 24·49 for vector diagram):

$$I_T = I_L - I_C \tag{24·73}$$

This method of boosting coil current (with an oscillator

as generator) is limited to the ratio of coil current to tank current of the order of 2.5 to 3.

If a motor generator is used as the source of power, complete correction of load power factor is desirable because high-frequency motor generators are usually rated for unity-power-factor loads. The capacitance required is

$$C = \frac{L}{Z_L^2} \tag{24·74}$$

where C = shunting capacitance in microfarads
L = loaded coil inductance in microhenrys
Z_L = impedance of loaded coil in ohms.

This equation suggests an expression that defines the impedance into which the generator feeds,

$$R_a = \frac{R_s^2 + X_L^2}{R_s} = \frac{Z_L^2}{R_s} \tag{24·75}$$

and

$$I_g = \frac{E_g R_s}{Z_L^2} = \frac{E_g R_s C}{L} \tag{24·76}$$

where R_a = apparent resistance of tuned load circuit in ohms
R_s = series resistance of loaded coil in ohms
X_L = series reactance of loaded coil in ohms
Z_L = absolute impedance of loaded coil in ohms
I_g = demanded generator current in amperes, rms
E_g = generator voltage in volts, rms
L = loaded load coil inductance in henrys
C = shunting capacitance in farads.

Figure 24·50 illustrates this principle.

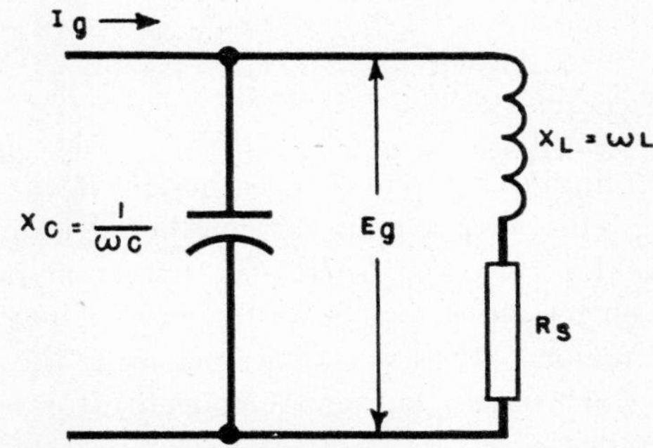

FIG. 24·50 Load circuit corrected for power factor.

Sometimes the coil voltage must be greater or less than the rated generator voltage because of difficulty in designing a coil to carry enough current to produce the required magnetic field. As a solution, networks can be used. In a T network

$$\frac{R_G}{R_s} = \frac{X_1 + X_3}{X_2 + X_3} \tag{24·77}$$

$$-R_G R_s = X_1X_2 + X_1X_3 + X_2X_3 \tag{24·78}$$

The network is assumed to consist of pure reactances. The asymmetric T network is shown in Fig. 24·51. R_G is the resistance the generator should feed. X_2 is the reactive component of the load, that is, the load-coil reactance plus the reactance reflected into the coil by the load. R_s is the coil resistance plus the resistance reflected into the coil by the load in the load coil.

For example, suppose that the generator voltage is 800 and that it becomes necessary to have 1600 volts across the load coil and at the same time boost the coil current. The load coil can be shunted by a capacitive reactance X_3, partially tuning the load coil. The choice of X_3 determines X_1 so that it can produce series resonance with X_2 and X_3 combined. A negative sign is employed for the capacitive reactance and a positive sign for the inductive reactance. To minimize network losses, it may be best to choose X_3 so that a capacitive reactance is required for X_1; however, the

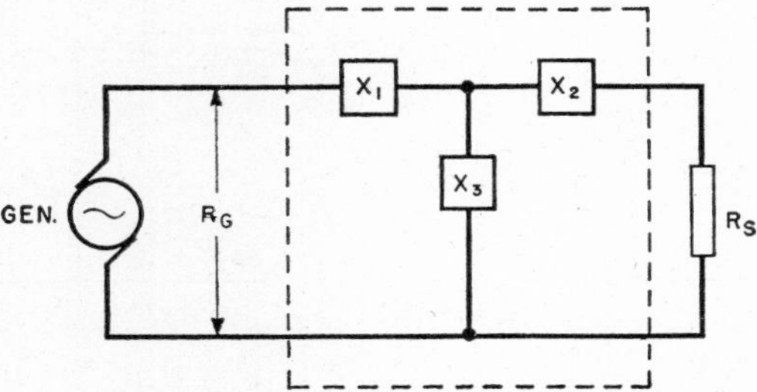

FIG. 24·51 Asymmetric T network.

cost of the capacitor must be considered, and so must the possible lower efficiency caused by losses in a network inductance. Networks carrying large current are expensive; therefore it is desirable if a coil can be designed such that the use of a network is avoided. If the required current is too large for a practical network, a transformer must be used.

Transformers and Current Concentrators [14]

For some induction heating, high power density is essential. With high-frequency generators only a small area can be heated at high power density. Zonal heat treatment is one application that often requires high power densities, especially in the field of surface hardening. Such types of applications demand very high current densities in the work pieces and in the heater coil. Seldom are the generator voltage and current exactly as required by the load impedance. Because such loads usually demand much more current than the generator can supply, a heater coil of only one or two turns, possibly three, must be used, depending upon the load requirements.

One method of transforming the load impedance into that required is to change the frequency. As frequency is increased, the power absorbed by the load at a given current increases. At high frequencies the power applied to a load is limited by high-voltage gradients between the work and the coil. This limitation is even more pronounced if the load is heated to red heat, which rapidly increases the ionization of the atmosphere between coil and work. Thus the coil voltage must be further reduced. If greater space is allowed between coil and work, the load-circuit impedance is increased and coil voltage is increased.

For loads 3 or 4 inches in diameter, the upper limit of frequency is about 1 megacycle per second. For smaller diameters, higher frequencies may be used. These are only practical limits; higher frequencies can be used, but greater precautions must be taken to insure proper insulation between coil and load. These precautions may not be economically justified.

Another method of increasing the power absorbed by the load is to increase load-coil current by shunting the load coil with power-factor-correction capacitors.

A third method is to change load impedance by a current transformer. Power-factor correction and the use of transformers are really in the same category. The transformer may be considered as a special case of either a T or a π network consisting of reactive elements. For a perfect match, one of the three reactive elements must bear an algebraic sign opposite to the signs of the remaining two elements. If a transformer is used, a perfect match cannot be made unless capacitive reactance is introduced in the appropriate place in the circuit.

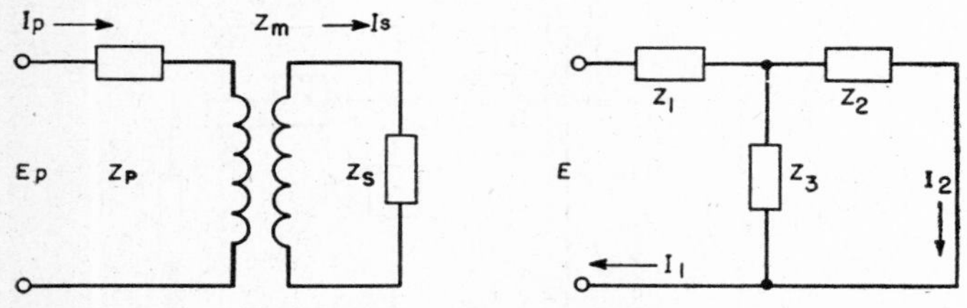

FIG. 24·52 Air-core transformer.

Figure 24·52 shows the transformer and its equivalent T network. The impedances involved are

$$Z_p = Z_1 + Z_3$$
$$Z_s = Z_2 + Z_3$$
$$Z_m = Z_3 \tag{24·79}$$

The driving-point impedance becomes

$$Z'_p = Z_1 + \frac{Z_2 Z_3}{Z_2 Z_3} = \frac{E_p}{I_p} \tag{24·80}$$

or in terms of the transformer impedances

$$Z'_p = Z_p - \frac{Z_m^{\ 2}}{Z_s} = \frac{E_p}{I_p} \tag{24·81}$$

where

$$Z_1 = Z_p + Z_m$$
$$Z_2 = Z_s + Z_m$$
$$Z_3 = -Z_m$$

These equations indicate modification of the impedance of the primary alone, which can be expressed by

$$\psi = \frac{Z_m^{\ 2}}{Z_s} \tag{24·82}$$

To obtain maximum regulation and efficiency from the transformer, a maximum of flux must interlink both its primary and its secondary. Thus the numerical value of ψ in equation 24·82 must be maximum; that is, mutual impedance must be maximum. Unity coefficient of coupling cannot be achieved in air-core transformers. As a means of compensating for this, the impedance match is best if the load impedance is equal to about 60 percent of the impedance of the secondary winding of the transformer. At best only about 25 percent of the primary kva is supplied to the load coil.

Such a transformer usually has a secondary winding consisting of a sheet of copper coaxial and concentric with the primary. This secondary sheet constitutes a single loop, sometimes two loops, mechanically in series, the load coil being connected to the secondary terminals with leads as short as possible. From 10 to 20 turns are used in the primary winding. Proper secondary impedance is secured by choice of dimensions of the secondary winding. The length of the transformer usually is a little greater than its diametric dimension. There are some advantages to using a secondary that encircles the primary, so that secondary leads can be kept short, and secondary impedance can be minimized. The transformer usually is encased in a nonferrous metallic shell filled with oil or pressurized gas for insulation.

Another type of high-frequency transformer is the spiral transformer, with its secondary turns in the interstices of the primary turns. It has a few distinct disadvantages. It is frequently limited by interwinding capacitance. Mechanically, the maintenance of proper interturn insulation and support is troublesome. The ratio of transformation is unity; that is, it is used only as an isolation transformer.

At frequencies above 10 megacycles the coaxial transformer is used. It is discussed in some detail in Section 24·8.

To reduce secondary impedance, thus developing higher ratios of current transformation, a unique transformer known as the current concentrator [14] has been developed. Its operation is based on the property of eddy currents around the "skin" of a conductor to shield the metal beneath the skin from the magnetic field.

A conductor in a magnetic field seems to press out the magnetic lines of force from the space these lines previously occupied.

$$\oint H\, dl = 0.4\pi ni \tag{24·83}$$

An analogy is the forcing out of water currents by a body in a stream of water. The concentrator consists merely of a primary winding and a one-turn secondary that forms the heater coil. This secondary is the conductor, immersed in a magnetic field. However, this field distortion or pressing out in a given region can cause an increase of magnetic field in another region. If this region including the increased field can be arbitrarily chosen, a concentrator is evolved. The field is decreased in the region occupied by the concentrator, but is highly concentrated in the region between the concentrator and the work.

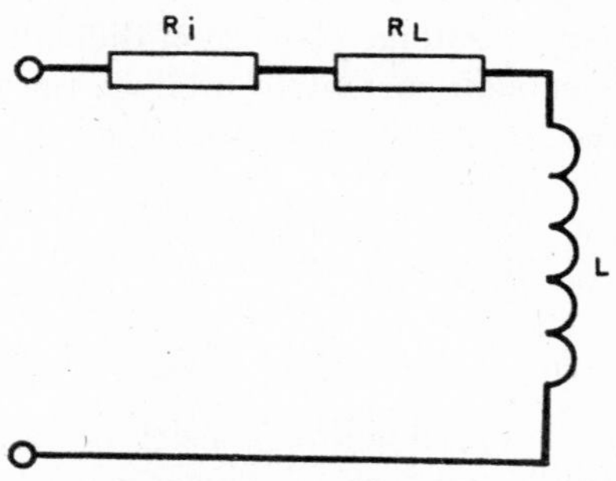

FIG. 24·53 Equivalent load-coil circuit.

If the concentrator had no resistance, it would have an efficiency of 100 percent. To minimize losses, the concentrator should be made of silver-plated copper. The equivalent circuit of a heater coil with load is shown in Fig. 24·53.

The efficiency of this circuit is

$$\eta_1 = \frac{R_L}{R_L + R_i} \tag{24·84}$$

where R_i is the coil resistance and R_L is effective reflected series resistance of the load. If a concentrator is used, there is an added resistance in the equivalent circuit shown in Fig. 24·54. Because the coupling is much closer with a

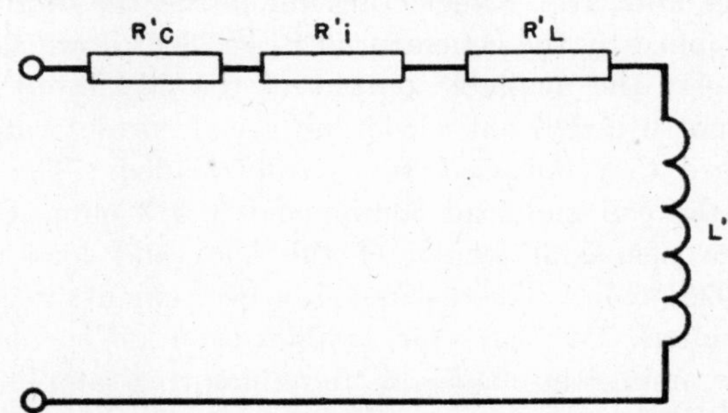

Fig. 24·54 Equivalent load-coil and current-concentrator circuit.

concentrator, the reflected load resistance sometimes is much larger so that

$$R'_L \gg R_L \tag{24·85}$$

The efficiency of this circuit is

$$\eta_2 = \frac{R'_L}{R'_L + R'_i + R'_c} \tag{24·86}$$

where R'_L is the reflected load resistance, R'_i is the primary-coil resistance, and R'_c is the reflected resistance of the concentrator. By proper design of the concentrator, this effective reflected concentrator resistance can be made small and the over-all efficiency η_2 can be made greater than η_1. Another inherent advantage of the concentrator is that less

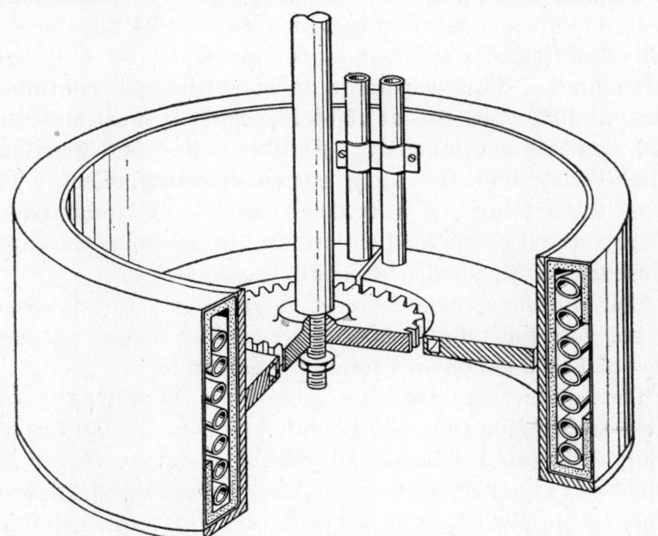

Fig. 24.55 External-current concentrator. (Reprinted with permission from G. Babat and M. Losinsky, *J. App. Phys.*, Vol. 11, p. 819, Dec. 1940.)

tuning capacitance is needed for the concentrator than if a simple multiturn coil is used. The reason for this is the reduction in cross section of the primary coil by the insertion of the concentrator, which reduces the total flux per unit current.

One type of concentrator is used for heating outer surfaces such as spur gear teeth; the other type is used for inside surfaces such as the surfaces of bores or internal ring gears. The internal concentrator is less efficient than the external type, and usually is less efficient than the comparable multiturn coil. Figures 24·55 and 24·56 show these two types of concentrators. The efficiency of the concentrator for external

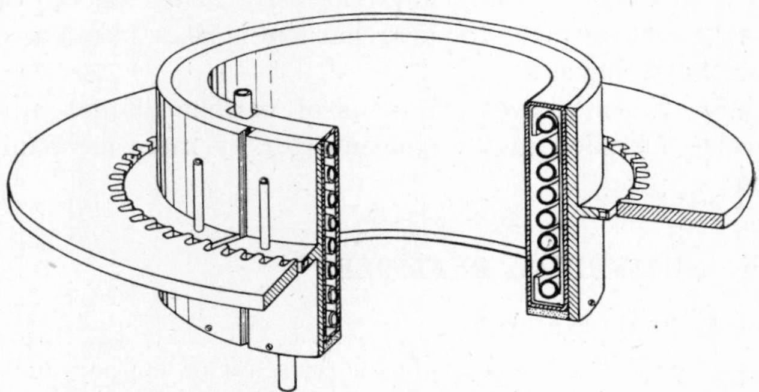

Fig. 24·56 Internal-current concentrator. (Reprinted with permission from G. Babat and M. Losinsky, *J. App. Phys.*, Vol. 11, p. 819, Dec. 1940.)

surfaces is almost independent of the radial dimension of the concentractor; at least efficiency is not so radically affected as is the efficiency of the concentractor for internal surfaces. An increase in the radial dimension of the concentrator for heating external surfaces drops the efficiency slightly, but an increase in the radial dimension of an internal concentrator radically decreases its efficiency. Figure 24·57 shows a curve of efficiency as a function of this radial dimension for a given thickness of concentrator ring. The curve is only an indication of how the efficiency of a partciular concen-

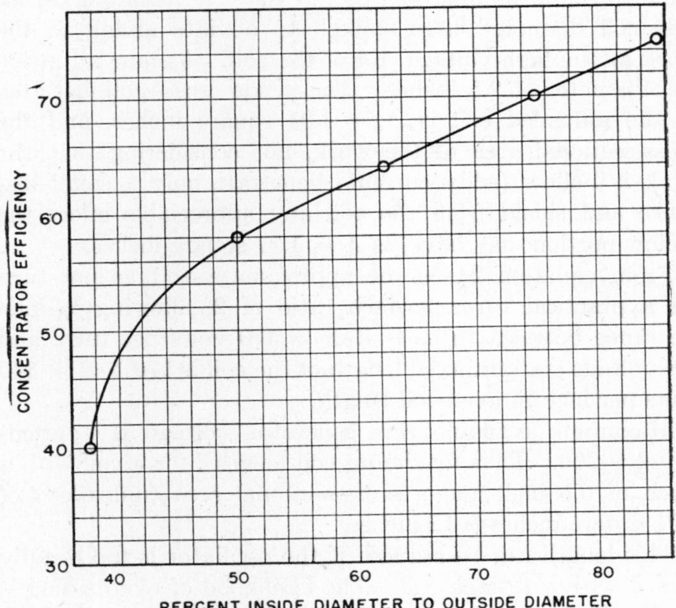

Fig. 24·57 Concentrator-coupling efficiency.

trator varies. For practical purposes the current tends to follow the path of least impedance, which is the path about the inside of the concentrator ring. The relative efficiency of a comparable concentrator for heating inside surfaces usually is low.

For hardening gears, the height of the heater coil must be

$$S = h - 2a \tag{24·87}$$

to obtain a uniform heat pattern, where h is the height of the load heated, and a is the space between the load and the coil. In gear hardening, the crest of the concentrator is made less than the face of the gear to prevent overheating of the rims of the teeth. Babat and Losinsky [14] developed a 200-kilowatt concentrator for gear hardening at a frequency of about 200 kilocycles.

Concentrator sleeves for loads of various diameters can be easily installed in the concentrator by mere mechanical changes.

24·4 NUMERICAL EXAMPLES

Non-Magnetic Materials

Assume that a hollow tube of an alloy of copper, nickel, and iron is to be heated from room temperature to 1800 degrees Fahrenheit. The outside diameter is 1.5 inches and the inside diameter is 1.25 inches. The problem is to determine the practical maximum production rate with a 50-kilowatt 450-kilocycle oscillator; tank current is 300 amperes rms and maximum rms terminal voltage is 4000.

According to equation 24·25, the minimum frequency is 85 kilocycles per second. Because a higher frequency is permissible, the use of 450 kilocycles is justified.

First consider the heater coil. A copper-tubing conductor with an outside diameter of $\frac{3}{8}$ inch is chosen; this conductor can carry 300 amperes continuously at 450 kilocycles, if water-cooled. Maximum rms coil voltage is limited to approximately 4000; consequently it seems best to determine the total number of turns in the coil, assuming an inside coil diameter large enough to obviate arcing to the work at the high-voltage end of the coil. Assume an inside coil diameter of 2.5 inches. The inside cross-sectional area of the unloaded coil is $A_c = 4.91$ square inches, and the cross-sectional area of the work, not considering that the work is hollow (eddy currents penetrate only a small distance and shield from the coil any appreciable effect the hollow portion may have) is $A = 1.77$ square inches.

From equation 24·59 the approximate voltage per turn for a practical turns-per-inch ratio of 2, allowing $\frac{1}{8}$-inch clearance between turns is $E/N = 170$ volts per turn, and the voltage per unit axial length of the coil is $170 \times 2 = 340$ volts per inch of axial-coil length.

Maximum available rms generator voltage is approximately 4000. Thus maximum coil length l for a coil with a pitch of 0.5 inch and a cross-sectional area (unloaded) of 4.91 square inches is 12 inches.

This length can be reduced if the oscillator is too heavily loaded with a 12-inch coil. The likelihood of overloading is small, because with non-magnetic material and such a coil the internal power factor (equation 24·68) is small.

From equation 24·52 or Fig. 24·21 the radio-frequency resistance per inch of conductor is $R = 0.465 \times 10^{-3}$ ohms.

The mean diameter of the coil is $2\frac{7}{8}$ inches, and total conductor length per turn is $3\frac{3}{8}$ inches. Total conductor length per axial inch of the coil is $6\frac{3}{4}$ inches. Thus the coil I^2R loss per axial inch is $P_c = 282$ watts, and the total coil I^2R loss at normal tank currents becomes $P_c = 3.4$ kilowatts. This coil loss contributes to some portion of the oscillator loading; however, energy absorbed by the tubing offers the greatest portion of the oscillator load. Power absorbed per unit length of load, according to equation 24·19, is $P_l = 2.06$ kilowatts ($\rho = 83$ microhm-centimeters and $H_0 = 420$ oersteds, from equation 24·62 being used). Thus total power absorbed by the load in the 12-inch coil is $P_t = 24.7$ kilowatts. Neglecting line losses, the total power to be supplied by the generator is $P_t = 28.1$ kilowatts.

Obviously the available output of the 50-kilowatt oscillator is not utilized, but a coil and a coil current can be so determined that the oscillator is fully loaded. The resistance of the coil and load combined is 0.313 ohm, and the resistance per axial length of the load and load coil is $R_l = 0.0261$ ohm. The discussion of load circuits in Section 24·3 pointed out that, for amplification of currents by partially tuning the load coil, the amount of amplification (if the high-frequency generator is an oscillator) should be limited to a maximum from 2.5 to 3. If the amplification is arbitrarily selected to be 2.4, the current in the load coil will be 720 amperes.

The use of 720 amperes of necessity limits coil length, if the same pitch and inside diameter are employed, to approximately 5 inches. More precisely, since combined load and load-coil resistances per axial inch are 0.0261 ohm the maximum coil length that can be used without demanding more than 50 kilowatts from the oscillator is 3.7 inches, or the total number of coil turns is 7.5.

The capacitance of the shunting capacitor used to boost the load coil current to 720 amperes must be determined. From equation 24·63 the rms coil voltage is $E_1 = 3200$, approximately. Therefore, the loaded coil reactance is $X_l = 4.5$ ohms, and the loaded coil inductance is $L = 1.6$ microhenrys.

By circuit analysis and equation 24·1 or 24·60, the load-coil shunting capacitance is computed to be $C = 0.046$ microfarad. This capacitor must withstand continuous duty at 4000 volts rms at 450 kilocycles. It must also carry 420 amperes continuously. Another value of capacitance would have had the same current-boosting effect. This other capacitance is $C = 0.13$ microfarad. To minimize the energy stored in the load circuit, and to minimize the capacitor current, the smaller capacitor is selected.

Coil efficiency is 88 percent; therefore about 6 kilowatts of coil loss must be dissipated by cooling water. A water flow of 1 or 2 gallons per minute is adequate.

The production rate obtainable from 44 kilowatts supplied to the tubing is 675 pounds per hour or 330 feet per hour. Expected line-to-load efficiency (at oscillator terminals) is about 60 percent, and an average of 84 kilowatts must be supplied by the power lines. This means that the kilowatt-hours per foot required to heat this tubing is 0.255.

This problem indicates the general solution for heating a non-magnetic material, and shows that, primarily because of a low internal power factor and limited output from the generator, some method of boosting coil current is necessary.

Heating Iron to Above Magnetic-Transformation Point

A manufacturer wishes to forge a low-carbon-steel rod $\frac{3}{8}$ inch in diameter. Production rate is 7 pounds per minute.

The process is to consist in feeding the rod to the induction-heating coils, heating from room temperature to 1850 degrees Fahrenheit, cutting lengths from the rod, and finally forging.

Assume a frequency of 450 kilocycles.

There are several approaches to this problem. One is to use a constant-pitch coil; another is to change coil pitch at a point where load temperature reaches the magnetic-transformation point.

The thermal power required, neglecting radiation and convection losses, is $P_i = 33.3$ kilowatts. Heating below the magnetic-transformation point is at an almost constant rate. Above the magnetic-transformation temperature the heating likewise is at an almost-constant rate, but not at the same rate as at the lower temperature. Proper manipulation of equation 24·48 gives a maximum radiation loss below the magnetic-transformation point of about 32 watts per square inch and an average of about 80 watts per square inch above the magnetic-transformation temperature to 1850 degrees Fahrenheit. The value of e, the relative emissivity, is about 0.8. The length of the coil, especially the section heating the load from the magnetic-transformation point to 1850 degrees Fahrenheit, is lengthened because of the extra power demanded by the load.

Because this heating process is continuous, the power required below the magnetic-transformation point and the power required above it may be considered individually. The two are considered together only for determining coil voltage and cooling, and if a single generator is used.

Neglecting radiation and convection losses, the power required to heat the work to the magnetic-transformation temperature of 1400 degrees Fahrenheit is $P = 24.7$ kilowatts. Because this particular part of the operation consists in heating the load from room temperature, the efficiency of power transfer from coil to load is expected to be rather high. Therefore, the use of a 50-kilowatt 450-kilocycle generator is satisfactory. From this generator the maximum available output voltage is 4000 and the tank current is 300 amperes.

Assume an inside coil diameter of 1.5 inches and ⅜-inch copper tubing as a coil conductor. From equation 24·59 the volts per turn is found to be approximately $E/N = 45$. This shows it is practical to use 1.6 turns per inch, and the voltage per turn then becomes 72 with ¼-inch air space between turns. Maximum length as defined by coil voltage per turn becomes 35 inches.

The power generated per axial inch of load (from equation 24·22) is 2.4 kilowatts.

The maximum radiation loss for the heat cycle below the magnetic-transformation point was given as 32 watts per square inch; therefore the maximum radiation loss per axial inch becomes 38 watts. This loss is almost negligible compared with the power absorption per axial inch, and may be neglected.

Coil resistance per axial inch (from equation 24·52) is $R = 4.55 \times 10^{-3}$ ohm, and the coil loss per inch of axial length is 400 watts. Therefore the total power per axial inch of coil to be supplied by the generator is $P/l = 2.8$ kilowatts. Then, since a total of 24.7 kilowatts is needed to heat the work to the magnetic-transformation point at the desired speed, the coil length for that heating is 10.25 inches. From this coil length above, the generator demand for heating to the magnetic-transformation point is $P = 28.8$ kilowatts. The voltage across this portion of the coil (from equation 24·63), neglecting any coil IR drop, is $E_1 = 1170$ volts.

Now let the load coil, using the same pitch and coil diameter, be extended so that the load in the non-magnetic state is heated as desired. Power absorbed by the load per axial inch (from equation 24·17) is $P = 420$ watts, if $\rho = 160$ microhm-centimeters and $H = 320$ peak oersteds. Since there is a maximum of 80 watts per square inch of the load or 95 watts per axial inch, the actual power absorbed per axial inch is reduced by 95 watts per axial inch. Therefore, $P = 325$ watts per axial inch.

From equation 24·47, the thermal power required to heat the work from 1400 to 1850 degrees Fahrenheit at the rate of 7 pounds per minute is 8.6 kilowatts. Therefore the coil length needed is 26.5 inches. Maximum coil length was determined previously as approximately 35 inches. The sum of the lengths of the two coil sections approaches 35 inches; therefore relative to coil length and coil voltage the total coil length is not excessive.

The coil loss of 400 watts per axial inch indicates that generator demand for heating the load from the magnetic-transformation point to 1850 degrees Fahrenheit is 19.3 kilowatts, and the total generator demand is 48.1 kilowatts.

Coil efficiency in heating the load to the magnetic-transformation point is approximately 86 percent, but the efficiency in heating from the magnetic-transformation point to 1850 degrees Fahrenheit is approximately 45 percent.

From equations 24·63 and 24·70 the over-all coil voltage is 4000. The kva required for heating a magnetic load, according to calculations, is 350 kva, and the calculated kva for a non-magnetic load is 810 kva.

As shown, the coil efficiency for the magnetic state is 86 percent, and for the non-magnetic state it is 45 percent. Thus coil losses become approximately 15 kilowatts.

To dissipate this energy, about 2 gallons of water per minute is necessary. This is shown by equation 24·57 or Fig. 24·22 if 50 percent of the energy is transferred to the water.

Magnetic Material

A manufacturer desires to heat a 10- by 0.01-inch low-carbon-steel strip from room temperature to 1000 degrees Fahrenheit at the rate of 15 feet per minute.

From equation 24·47, the useful power required is $P_1 = 12.7$ kilowatts. Thermal losses consist of radiation and convection losses; conduction loss performs useful work, and it is provided for in equation 24·47. Thus maximum radiation and convection losses become (Fig. 24·20) $P_2 = 220$ watts per inch of load length for a relative emissivity of 0.7, and $P_3 = 80$ watts per inch of load length.

Maximum loss is considered because, if this maximum loss is not supplied, the temperature of the surface of the steel strip will never reach 1000 degrees Fahrenheit. The total loss in watts then is taken to be 300 times the coil length in inches.

Thermally this loss appears to be suitable for a 20-kilowatt radio-frequency generator, and a 20-kilowatt 450-kilocycle generator is selected. The maximum terminal voltage is approximately 3000, and the current available at the terminals is 175 amperes. The required magnetizing force as a function of coil length can be determined from equation 24·36. This expression, when the total loss per inch of load length is considered, reduces to

$$\frac{12{,}700}{l} + 300 = 0.552 H_0^{3/2}$$

Required magnetizing forces for various lengths of coil, and the turns per inch needed to produce those magnetizing forces, are

l (inches)	H_0 (oersteds)	n (turns per inch)
5	200	1.65
10	135	1.1
15	110	0.9

An arbitrary selection of coil length can be made. Let a convenient turns ratio of one turn per inch be chosen; then the coil length must be approximately 12 inches.

Since the coil is rather short, the strip within the coil is not likely to vibrate appreciably. Therefore, the coil opening can be made small. But the strip may creep back and forth in the coil by as much as 2 inches. Thus, a coil opening of 1 by 12 inches can be chosen. Now the coil voltage, from equations 24·63 and 24·72, is $E_1 = 3200$. This voltage is a little higher than the maximum available voltage previously given. The difference is too small to be important, so it is neglected.

The coil resistance, from equation 24·52 or Fig. 24·21, is 0.11 ohm for ½-inch copper tubing; therefore coil loss becomes I^2R or 3.4 kilowatts.

The total kilowatts demanded by the load and coil is 19.7, which includes 3.6 kilowatts in thermal losses. The coil efficiency is

$$100 \frac{16.3}{19.7} = 83 \text{ percent}$$

Power cost per 1000 feet of strip is simply determined. Since the oscillator is expected to be 60 percent efficient, the power drain from the incoming line, neglecting stand-by operation such as filament excitation and air-blower power, for 19.7 kilowatts delivered to the load coil is 32.8 kilowatts. In 1 hour 900 feet of strip has been heated with an energy expenditure of 32.8 kilowatt-hours. Thus 36.5 kilowatt-hours are needed to heat 1000 feet of strip. At a power cost of $0.01 per kilowatt-hour, the hourly power cost, neglecting stand-by power, is $0.365.

24·5 INDUCTION SOLDERING, BRAZING, AND HEAT TREATING

Soft Soldering

Joining of metals by means of solders which melt at 361 to 621 degrees Fahrenheit is commonly termed *soft soldering.* Solders so used usually are binary alloys composed of tin and lead, or sometimes ternary alloys in which small amounts of antimony or cadmium are added to the tin and lead.

Tin-Lead Solders

Most of the tin-lead alloys change from a solid state to a plastic state, then to a liquid state, as they are heated. The temperature at which the alloy changes from the solid to the plastic state is called the *solidus point,* and the temperature at which the alloy becomes liquid is called the *liquidus point.* The solidus and liquidus points of various tin-lead alloys are given in Table 24·1. The complete tin-lead diagram is given in Fig. 24·58.

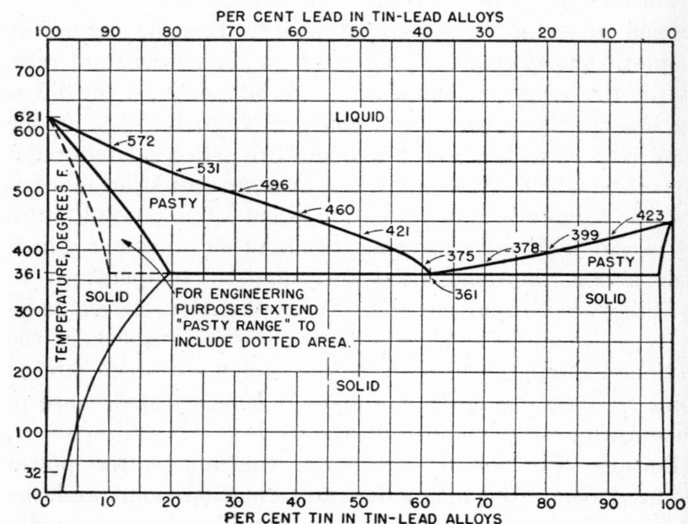

Fig. 24·58 Tin-lead diagram. (Reprinted with permission from A. Z. Mample, *Metals and Alloys,* Vol. 21, No. 3, p. 704, March 1945.)

Table 24·1 Critical Temperatures of Tin-Lead Alloys

Composition % Sn	% Pb	Solidus Point (°F)	Liquidus Point (°F)
0	100	621	621
5	95	556 (518) *	594
10	90	500 (361) *	572
15	85	428 (361) *	550
19.5	80.5	361	535
20	80	361	531
25	75	361	511
30	70	361	496
35	65	361	477
40	60	361	460
45	55	361	438
50	50	361	421
55	45	361	399
60	40	361	375
62	38	361	361
65	35	361	367
70	30	361	378
75	25	361	387
80	20	361	399
85	15	361	410
90	10	361	423
95	5	361	437
98	2	361	441
99	1	414	446
100	0	450	450

* For engineering purposes the solidus temperatures given in parentheses should be used.

The range of temperature in which the alloy is plastic decreases as the percentage of tin is increased until a critical composition of 62% tin and 38% lead is reached. This alloy, the *eutectic*, has no plastic state but changes suddenly from solid to liquid at a temperature of 361 degrees Fahrenheit. As Fig. 24·58 indicates, this eutectic temperature also is the solidus point for all tin-lead alloys containing from 19.5% to 98% tin. For engineering purposes the solidus point of alloys containing as little as 10% tin should be reduced to 361 degrees Fahrenheit. Both the solder and the surfaces to be soldered must be heated to a temperature above the liquidus point of the solder. At temperatures above the liquidus point, there may be some movement between the parts to be soldered without harmful effects. Such relative movement may be required to position the parts, but once the temperature drops below the liquidus point and the solder enters the plastic range, any movement between parts usually causes a defective joint. A defective joint of this type can be recognized by its crystalline appearance, but often it may pass a visual inspection. The elimination of movement is important both to the designer of parts to be soldered and to the designer of work-handling equipment for automatic soldering processes. Joints must be designed to be mechanically strong before soldering, or adequate cooling time must be provided without disturbing the soldered parts. Cooling time is proportional to the extent of the plastic range of the solders.

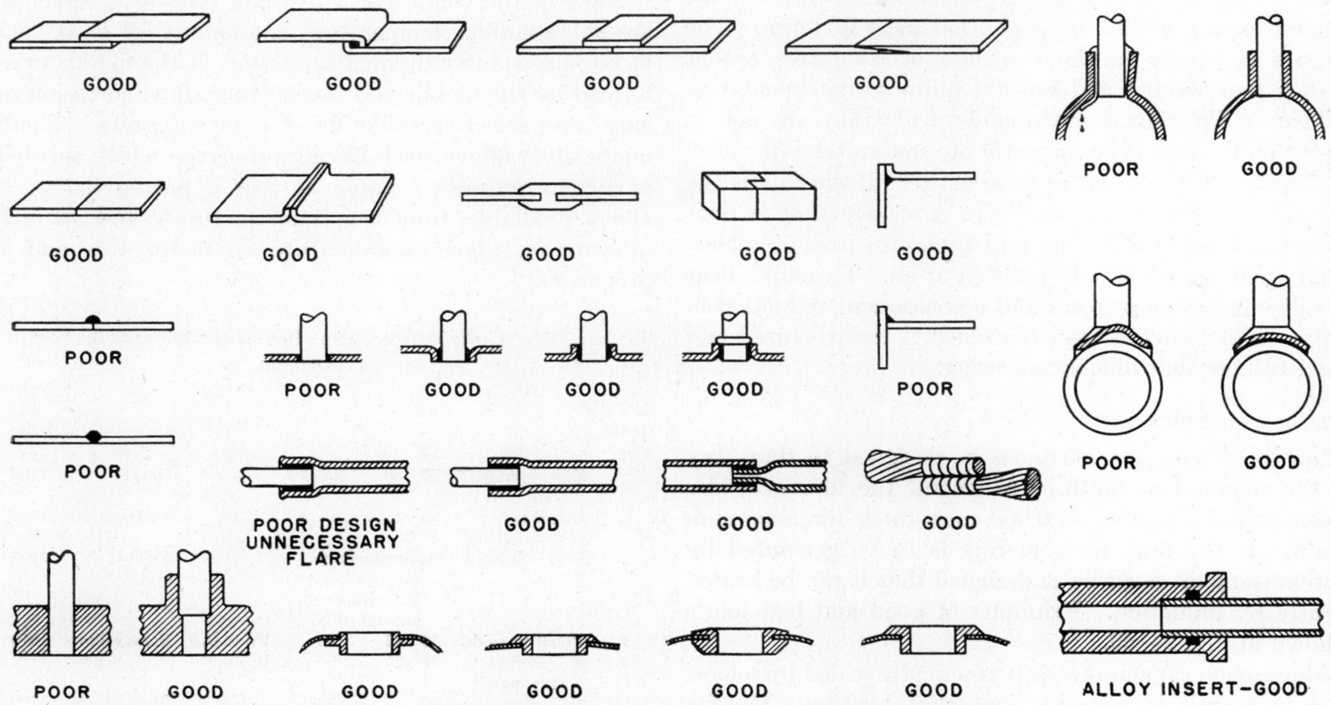

Fig. 24·59 Various grades of solder joints. (Courtesy of Handy and Harmon Company.)

High-Temperature Soft Solders

Frequently a soldered joint is required in apparatus subjected to high temperatures. Table 24·1 shows that low-tin solders have higher liquidus points than do the medium- to high-tin solders. Thus a low-tin solder might be assumed to be stronger than a higher tin alloy, such as the eutectic at high temperature. Such an assumption is erroneous. The solidus points of solders normally used are constant at the eutectic temperature of 361 degrees Fahrenheit. At temperatures above the solidus point solder is in its plastic state and has zero strength; hence nothing is to be gained by substituting a tin-poor solder for a higher tin solder for operation at high temperatures.

Special high-temperature solders are alloys of cadmium and zinc or silver; they have solidus and liquidus points high enough so that they can be used in selective soldering applications. Some of these high-temperature soft solders are listed in Table 24·2.

Table 24·2 High-Temperature Soft Solders

Composition		Solidus	Liquidus
% Cd	% Other metal	Point (°F)	Point (°F)
95	5 Ag	639	734
50	50 Sn	508	619
82.5	17.5 Sn	508	508

Fluxes for Soft Soldering

The heat required for soldering promotes the formation of oxides on the surfaces of the metals to be soldered. These oxides prevent solder from wetting the surfaces, and a poor joint is formed. A good flux, properly used, not only prevents oxidation, but also adds to the strength of the bond between solder and metal.

The requirements for a good flux are that it should dissolve oxides on the metal surfaces; prevent formation of oxides as the temperature is increased; increase the ability of solder to wet the metal surfaces by decreasing surface tension or at least preventing an increase in surface tension; promote alloying of the solder with the metal surfaces at temperatures below the melting point of the base metal; be stable at normal soldering temperatures; and leave a non-corrosive, non-conducting residue after soldering.

Several types of fluxes meet some or all of the requirements in varying degrees. Among these are acid or salt fluxes, paste fluxes, wax fluxes, and rosin. A salt-type flux consists of chlorides of ammonium, zinc, or aluminum. These are active and stable at soldering temperatures, and generally decrease the difficulty of soldering. They are corrosive, however, and leave a residue unsuitable for electrical work. Furthermore, they do not flow ahead of the solder, and are not suitable for sweating. Paste fluxes are solutions of salt fluxes in a grease or some other inert material. The properties are about the same as those of salt fluxes, except that they flow readily and are not quite so detrimental as salt fluxes. Wax fluxes are organic acids that are not so corrosive as the salt type, but still are not suitable for electrical work. They are not in general use. Rosin flux is the most nearly perfect flux. It leaves a residue that is non-corrosive and non-conducting and promotes ease of soldering and good joints, if it is properly used. Excessive heat causes disintegration of rosin and a consequent reduction in activity. If it is overheated too much it is carbonized and leaves a residue that hinders soldering.

Joints for Soft Soldering

Joints to be soft-soldered must be designed so that they have the required strength. In general the film of solder between surfaces should be 0.003 inch thick for maximum strength. If the heat for soldering is to be generated by induction the joint must be so designed that it can be heated efficiently by induction. Examples of good and bad joints are shown in Fig. 24·59.

Production on capacitor cans hermetically sealed by soldering can be greatly increased by induction heating. The fits in cans soldered by ordinary means are often poor and are not suitable for automatic soldering. A typical example is shown in Fig. 24·60 (*a*). In a can of this type solder flows over the bottom surface and may leave a gap in the seam.

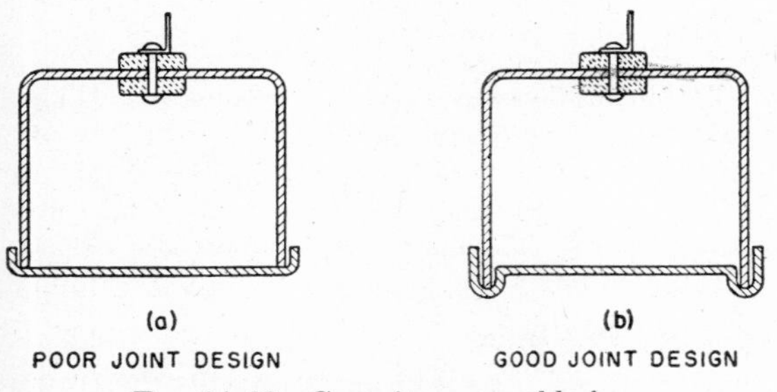

Fig. 24·60 Capacitor-can soldering.

A person using a soldering iron can add more solder to fill the gap, but automatic soldering involves the use of a fixed amount of pre-placed solder; therefore, if solder flows over a surface instead of around the seam the result is unsatisfactory. A joint designed as in Fig. 24·60 (*b*) provides a pool for the molten solder and prevents flow over adjacent surfaces. The amount of solder required to fill the pool and the seam can be accurately determined and placed in the joint before heating to reduce the probability of gaps.

Silver Brazing

As a result of the development of silver alloys that melt at temperatures as low as 1175 degrees Fahrenheit, parts formerly made of expensive castings or machinings have been largely replaced by fabrication. Silver brazing has made possible great savings in material, labor, and space requirements, and has produced products of increased strength and superior quality. Silver brazing is used also in the repair of broken parts.

Silver Solders

Some of the conditions governing the use of silver solder are color match, temperature characteristics of the metals to be joined, metallurgical properties of the metals or alloys formed by the metals and the brazing alloy, corrosion resistance, and selective soldering of adjacent parts. There are many silver alloys, each having properties which suit it for a particular purpose. Table 24·3 is a partial list of silver alloys available from one manufacturer, and it includes information about temperature characteristics and color properties.

Table 24·3 Composition and Physical Properties of Silver Solders

ASTM Grade No.	Name	Ag	Cu	Fe	Cd	Color	Melting Point (°F)	Flow Point (°F)
	Sil-Fos brazing alloy	15	80	(5% P)		Gray	1190	1300
	Easy-Flo brazing alloy	50	15.5	16.5	18	Yellow-white	1160	1175
	Easy-Flo No. 3	Approximately same as Easy-Flo with small amount of Ni				Yellow	1195	1270
1	TL silver solder	10	52	38	Up to 0.5	Brass yellow	1510	1600
2	AT special silver solder	20	25	35	Up to 0.5	Brass yellow	1430	1500
3	ATT silver solder	20	45	30	5	Brass yellow	1430	1500
4	DE silver solder	45	30	25		Yellow-white	1250	1370
5	ETX silver solder	50	34	16		Yellow-white	1280	1425
6	Easy silver solder	65	20	15		Silver white	1280	1325
7	Medium silver solder	70	20	10		Silver white	1335	1390
8	IT silver solder	80	16	4		Silver white	1360	1460

Fluxes for Silver Brazing

In silver brazing, oxides are readily formed on the metal surfaces. It is imperative that the surfaces to be joined be kept perfectly clean. A good flux should dissolve oxides on the metals, and should prevent further formation of oxides. The flux should be fluid at a temperature below the flow point of the brazing alloy and should flow thoroughly over the surfaces to be joined. Borax becomes fluid at about 1400 degrees Fahrenheit. It dissolves oxides, including chrome oxides, and is widely used in high-temperature brazing. The relatively high temperature at which it becomes active precludes its use with low-temperature brazing alloys. It can be used, but the major advantage of low-temperature alloys is lost because of the higher temperature required to keep the flux active. A paste flux of the fluoride type begins to fuse at 800 degrees Fahrenheit and becomes a thin colorless liquid at 1100 degrees. Furthermore, it remains active at temperatures up to 1600 degrees. It dissolves most of the oxides, including those of chromium. Thus it is useful in brazing steel, stainless steel, copper, brass, bronze, and nearly any other metal or alloy normally used. Only a small amount of flux is required. Excess flux can be removed by washing in hot water. Cleaning costs are thus eliminated or reduced.

Design of Joints for Silver Brazing

Silver alloys have tensile strengths from 40,000 to 60,000 pounds per square inch, and a good bond between the brazing alloy and the base metal may achieve a strength greater than that of the alloy itself, sometimes as much as 130,000 pounds per square inch.

Joints for silver brazing are of three general types: lap or shear, butt, and scarf. Lap or shear joints are preferable. In the design of a joint to be brazed, proper clearances must be maintained in the joint. Strength is maximum when the film in the joint is thin (from 0.001 to 0.006 inch). Clearances must be large enough so that uneven expansion does not make the joint so tight that the brazing alloy cannot penetrate it or so loose that capillary action cannot pull the alloy into the joint. If heat is to be generated rapidly in the material, as by induction heating, it is desirable to include in the joint a seat in which the brazing alloy can be placed prior to heating. In this way the proper amount of alloy is used, and the results are most consistent and economical. Examples of good and bad joints for silver brazing are shown in Fig. 24·59.

Jigs and Fixtures

Proper jigs help to make good brazed joints. The jigging may be simple, such as staking, riveting, spotwelding, or even spinning the parts together. For some jobs a more elaborate fixture may be necessary to position the parts and hold them securely. By eliminating motion between parts, the work can be handled after it is brazed but before it has cooled. This is of particular benefit in induction brazing, where much of the advantage of rapid heating would be nullified if it were necessary to provide for a long cooling period without moving the work.

Induction Soldering and Brazing

Because heat is generated within the material and can be confined to the area around the joint, brazing production can be increased by induction heating. This same localized heating allows one joint to be brazed near a previously brazed joint without melting the first joint. Furthermore, a brazed joint can be made near an area previously heat-treated, and it will leave the heat treatment unaffected. Since no oven or furnace is required, and no contact need be made with the work, induction heating is suited for production-line brazing. Heat or energy requirements for induction brazing usually are less than for other methods because of the localized heating effect. To realize these benefits heat must be generated faster than by other methods. Power required for heating materials to soldering or brazing temperatures can be calculated from requirements given in Section 24·2.

Heat Treating

High surface power densities can be produced with high-frequency induction heating that are not obtainable by any other method. High power densities permit heat treating surfaces without heating other parts of an object. Thus distortion is minimized. With such high surface power densities high surface hardness can be attained in low-carbon steels; SAE 1045 and carbon steels with higher carbon contents have been hardened satisfactorily to 60 to 65 Rockwell C, thus obviating special alloy steels for many industrial applications.

Metals in their solid state are composed of crystals formed by the atoms of the metal. Three types of crystalline

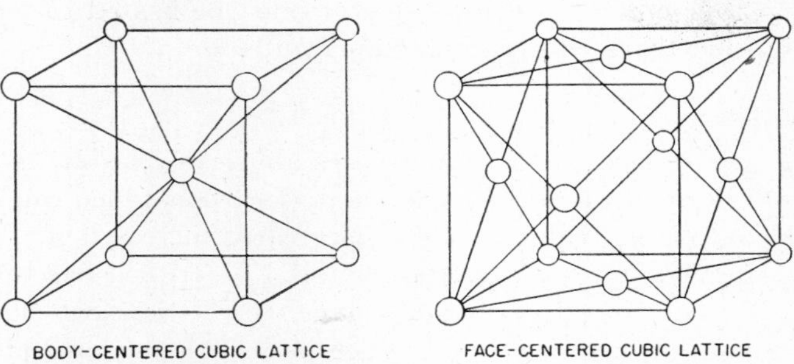

Fig. 24·61 Body-centered and face-centered cubic lattices.

structures are important in heat-treating processes. In heat treatment of steels only two of these crystals are important. These are the body-centered cubic lattice and the face-centered cubic lattice shown in Fig. 24·61. All steels possess one or the other or both of these crystals during heat treatment.

Atoms of iron have the property of holding atoms of carbon and atoms of other alloying materials in the interstices of the crystal entity to a degree determined by temperature and heat treatment. A partial iron-iron carbide diagram is shown in Fig. 24·62.

At about 1400 degrees Fahrenheit most carbon steels undergo a change of state from the body-centered to the face-centered crystalline structure. In the temperature

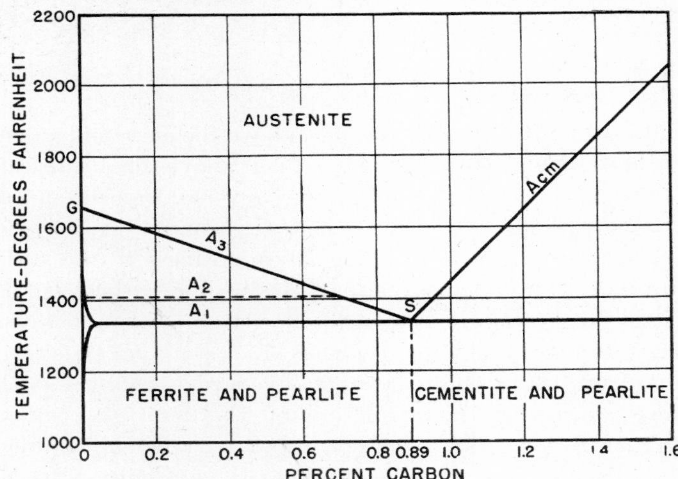

Fig. 24·62 Partial iron-iron carbide diagram.

region from room temperature to about 1340 degrees Fahrenheit steels normally exist in the body-centered crystalline state called *pearlite* and *ferrite*. The transformation does not generally occur at a specific temperature except at one point, the eutectoid point, designated by the letter S on the iron-iron carbide diagram of Fig. 24·62. At other points along the 1340-degree-Fahrenheit line, designated the A_1 critical, this crystalline change simply begins. For steels with carbon contents to the left of point S the transformation is completed at the line GS, which is termed the A_3 *critical*. In steels with carbon contents to the right of the eutectoid point S, but having less than 1.7 percent carbon, the transformation is completed at the A_{cm} *critical*. The particular state of steel, consisting of face-centered cubic crystals of iron in combination with carbon, above the transformation lines A_3 and A_{cm} is called *austenitic*. To harden, anneal, or normalize carbon steel it must be heated to temperatures above this critical temperature.

Hardening

If steel is heated to temperatures above the A_1 line, austenite begins to form. The amount of carbon which can be held within individual crystals is greatly increased in this transformed state. After the A_3 (or A_{cm}) critical has been reached all the steel crystals have been transformed to austenite. Because of excessive grain growth the hardening

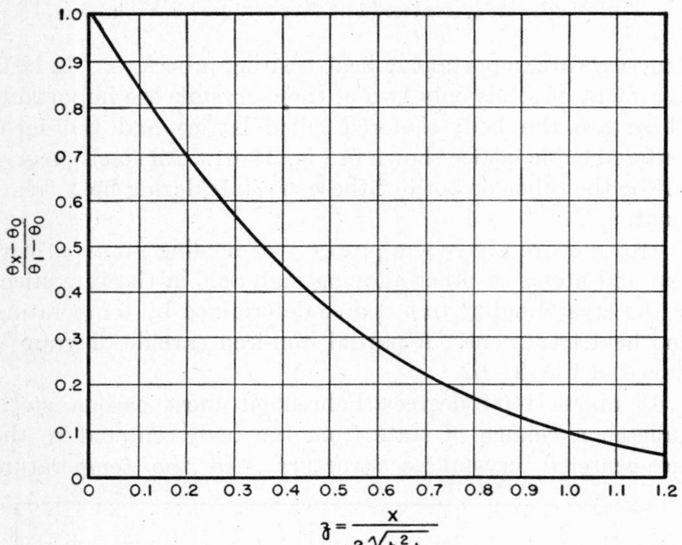

FIG. 24·63 Curve of equation 24·88.

temperature of high-carbon steels, above the eutectoid value, is held to within 100 degrees Fahrenheit above the A_1 critical. If austenite is cooled rapidly the reverse transformation process is prevented. Then a new combination of iron and carbon called *martensite*, which gives hardness to the steel, is formed. Sub-microscopic particles of iron carbide in the martensite seem to be responsible for this hardness, because they prevent the slipping of planes of crystals and thus tend to prevent distortion of the crystalline structure of the steel. Thus two principal factors determine whether a steel can be hardened. The first is that the steel must be transformed into austenite, and the second is that the austenite must be quenched rapidly enough to arrest the reverse transformation and cause the carbon to precipitate in the form of submicroscopic particles of iron carbide.

In furnace hardening the maximum temperature is not permitted to exceed 50 to 100 degrees Fahrenheit above the A_3 critical, because grain growth is excessive in steels subjected to higher temperatures. Furnace heat and induction heat are identical. However, this grain growth involves some time, and, because the desired temperatures can be attained by induction heating in a few seconds, higher temperatures are permissible with induction heating than with furnace heating. The time a piece of steel is exposed to hardening temperatures is roughly 1 hour per inch of minimum cross-sectional dimension for furnace heat. Heating plain carbon steels by induction usually takes only a few seconds. As soon as the hardening temperature has been reached the piece can be quenched. A little longer "soaking" time is needed for alloy steels, because the reactions in such steels are more sluggish. Even though heating time is increased for alloy steels, it still is measured in seconds.

Another advantage of induction heating is the ability to harden steels without protective atmosphere. Very little scale is developed in such short heating times.

Distortion is *not* eliminated by induction heating, but it is reduced.

In high-frequency induction hardening the surface of the steel usually is heated above the critical temperature, but the temperature of the interior is changed only little. In applying heat by the furnace method, the piece is immersed into a constant-temperature medium. Transient-temperature distribution under these conditions can be described mathematically. However, for induction heating, the body being heated cannot be considered as being immersed in a constant-temperature medium. Energy is applied at an almost constant rate. This may be termed the constant-power case as opposed to the constant-temperature case of furnace heating.

One of the few mathematical analyses of constant-power heating deals with temperature distribution in an infinitely thick slab. The depth of current penetration is assumed to be negligible. In terms of time, properties of the material, distance from the surface of the material, initial and final surface temperatures, and the constant power applied to the surface, this equation defines the transient temperature distribution:

$$\frac{\theta_x - \theta_0}{\theta_1 - \theta_0} = e^{-z^2} - z\sqrt{\pi}\left(1 - \frac{2}{\sqrt{\pi}}\int_0^z e^{-z^2}\,dz\right) \quad (24\cdot 88)$$

where θ_0 = initial body temperature in degrees centigrade
θ_1 = final surface temperature in degrees centigrade
θ_x = temperature in degrees centigrade at x centimeters from surface of slab after t seconds
$z = (x/2)\sqrt{\rho c/kt}$
x = distance in centimeters from surface of slab
ρ = density of slab material in grams per cubic centimeter
c = specific heat
k = thermal conductivity calories per second per square centimeter per degree centigrade per centimeter
t = heating time in seconds

$$\theta_1 - \theta_0 = \frac{P_A\sqrt{t}}{3.72\sqrt{k\rho c}}$$

P_a = watts per square centimeter.

To facilitate application, this equation is plotted in Fig. 24·63. This curve can be applied to the heating of cylinders or slabs where the temperature at the center is not appreciably affected during the heating cycle.

As an example equation 24·88 can be applied to hardening low-carbon steel, SAE 1045 being considered as having the minimum carbon content required for full hardness. The temperature-differential ratio can be plotted as a function of the distance from the surface in mils for various lengths of heating times in seconds, as in Fig. 24·64. Suppose that

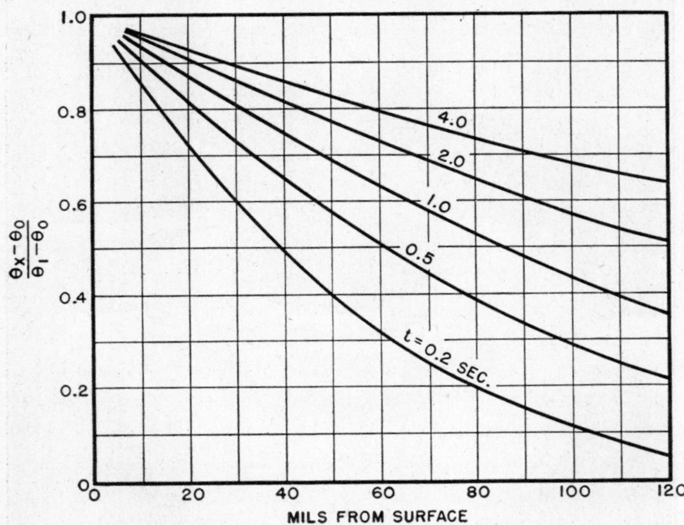

Fig. 24·64 Temperature differential ratio plotted as a function of distance from the surface for various heating times for steel.

the transformation critical for a particular type of steel is 1500 degrees Fahrenheit and that the desired case depth and heating time are 40 mils and 2 seconds respectively. Then the final surface temperature after two seconds can be determined from equation 24·88 or Fig. 24·64. This temperature is 1815 degrees Fahrenheit.

Power required to raise the temperature of the surface of an infinite slab $\theta_1 - \theta_0$ degrees is

$$P_a = 13.3\ \Delta t \sqrt{\frac{kc\rho}{t}} \tag{24·89}$$

where P_a = watts per square inch
Δt = surface temperature rise in degrees Fahrenheit
k = thermal conductivity in calories per second per square centimeter per degrees centigrade per centimeter
ρ = density in gram per cubic centimeter
c = specific heat
t = time in seconds.

Then the necessary power density becomes 6.3 kilowatts per square inch. So many variables affect the thermal and metallurgical properties of steels that unless these properties are known the curves can serve only as guides.

Some experimental work has been performed with rods of small diameter, but the results are not consistent and theoretically the smaller rods are heated throughout at almost uniform temperature. Therefore, heat treating such pieces does not involve temperature distribution and quenching but involves quenching only.

Figure 24·65 shows a section of a shaft hardened by high-frequency induction heating. This type of casehardening can be reproduced consistently, because of the precision with which such heating can be controlled. Gears may be surface-hardened with good contours. A sample contour is shown in Fig. 24·66.

Some surface-hardening applications require hardened cases from 0.020 to 0.060 inch thick. Lower frequencies

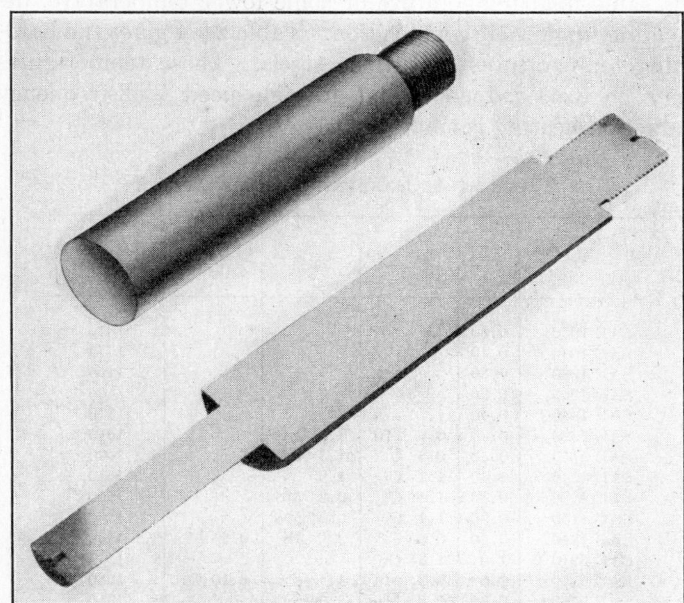

Fig. 24·65 Section of surface-hardened shaft.

supplied by rotating machines are most often used to produce case depths greater than 60 mils. For case-depth range of 20 to 60 mils the vacuum-tube radio-frequency generator has become firmly established. The frequency range usually employed for high-frequency induction hardening is between 100 and 500 kilocycles per second. Within this band of frequencies case depths as shallow as 10 or 15 mils can be produced consistently. Deeper cases can be obtained also, but it is generally more economical to use lower frequencies for deep hardening. If case depths of less than 10 mils are required, the megacycle frequency range may be employed.

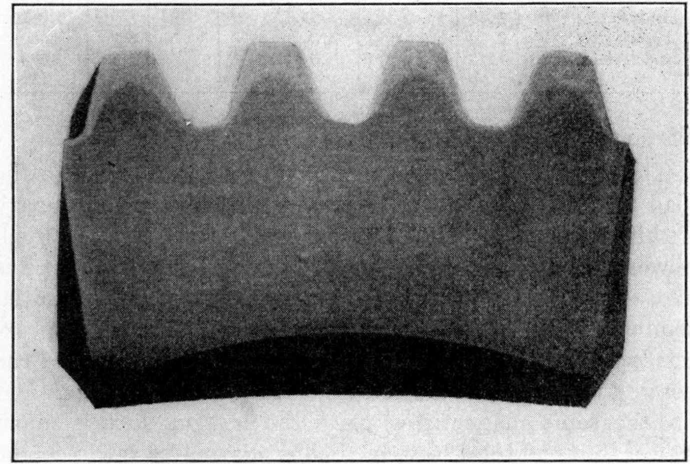

Fig. 24·66 Section of contour-hardened gear.

Few applications demand such thin cases, and often the use of such high frequencies is not economically justified.

Hardening temperatures of plain carbon steels can be determined from Fig. 24·62. The temperature may be raised 100 or 200 degrees Fahrenheit above the A_3 critical for low-carbon steels and above the A_1 critical for high-carbon steels. Special steels, such as tool steels, must be heated to temperatures close to the melting point of about 2300 degrees Fahrenheit. Other types of steel may require the same final temperature or some lower temperature, depending upon their composition. Table 24·4 gives the hardening temperatures of various steels. These temperatures may be exceeded somewhat if high-speed high-frequency induction heating is employed.

TABLE 24·4 HARDENING TEMPERATURES

	% C	% *X*	% *X*	% *X*	Temperature (°F)
SAE 1030	0.30				1510
SAE 1040	0.40				1532
SAE 1080	0.80				1495
SAE 1095	0.95				1460
SAE 1340	0.40	1.75 Mn			1510
SAE 2330	0.30	0.7 Mn	3.5 Ni		1510
SAE 3150	0.50	0.8 Cr	1.3 Ni		1490
SAE 3240	0.40	1.1 Cr	1.8 Ni		1510
SAE 4130	0.30	1.1 Cr	0.2 Mo		1560
SAE 4150	0.50	1.1 Cr	0.24 Mo		1560
SAE 4340	0.40	0.8 Cr	2 Ni	0.25 Mo	1560
SAE 5210 0	1.0	1.35 Cr			1545
SAE 6150	0.5	0.8 Mn	1 Cr	0.15 Va	1580
SAE 9260	0.60	0.8 Mn	2.0 Sn		1650

HIGHLY ALLOYED STEELS

No.	Composition						Temperature (°F)
	% C	% *X*	% Cr	% Mo	% *W*	% Va	
West 15	1.5		12.0	1		0.15	1850
West 21	1.0		5	1		0.3	1740
Cr-die steel	0.95		4				1560–1740
Tungsten die steel	1.30	0.45 S_1			5.5		1580
Medium carbon, medium molybdenum	0.75	0.74 Mn		4.0		0.2	1580
High-speed steel	0.70		4		18.0	1.1	2330
High-speed Mo steel	0.80		4	8.6	2.0	1.2	2190
Chisel steel	0.55	1.25 S_1	1.0		1.75		1650
Finishing steel	1.30		1.25		5.0		1520–1560
Hot-work steel	0.40		3.5		14.5	0.7	2280
Non-deforming steel	1.75		18.0				1830

Stationary Hardening

There are two principal methods of applying heat to the load by induction. One is to hold the piece stationary within the heater coil. During the heating interval high power densities are generated in the piece. The surface is exposed to high but almost constant power from the initial temperature to the magnetic-transformation point at approximately 1400 degrees Fahrenheit. For the remainder of the heating cycle the heater coil feeds a non-magnetic load, and for the same magnetizing force the heating rate is much slower, but the total heating time is still only a few seconds.

At the desired surface temperature, which is generally predetermined for a given generator output, the piece to be hardened is suddenly quenched. A laboratory set-up for stationary surface hardening is shown in Fig. 24·67. A more elaborate arrangement for hardening wrist pins at a high rate of production is shown in Fig. 24·68, and a production-line set-up for hardening a more complex structure is shown in Fig. 24·69. In this apparatus for hardening crankshafts the heated portions are quenched by means of a spray

FIG. 24·67 Laboratory set-up for stationary surface hardening.

emanating from the coil itself. This type of coil-quench combination is known as an integral-quench inductor coil.

If equation 24·89 is reconsidered it becomes evident that stationary hardening is limited to small surfaces. In gen-

FIG. 24·68 Production-line wrist-pin hardening.

eral about 10 kilowatts per square inch is used to produce average case depths. For small case depths the requirements defined by equation 24·89 may be as high as 30 or 40 kilowatts per square inch. Thus only a few square inches of load can be heated to hardening temperatures with a radio-frequency generator of average size, if stationary heating is employed.

Progressive Hardening

Progressive hardening can be used to harden large areas, such as a long section of a shaft. This method consists in

FIG. 24·69 Vertical crankshaft unit using a frequency of 9600 cycles. (Courtesy of Ohio Crankshaft Company.)

passing the work piece through the inductor coil at the required rate. As with stationary heating, a point on the surface is heated to the magnetic-transformation point at a relatively constant rate, and then the heating rate is reduced. By proper adjustment of the speed of travel of the work piece and the power input to the coil, cases ranging from 15 to 100 mils in depth can be produced consistently on work pieces of such shapes as shafts and boxes. Figure 24·70 illustrates the progressive hardening of a bar of square cross section. Immediately after the surface of the work piece is heated to hardening temperatures it is quenched.

Quenching

Hardness and depth of hardness depend upon the quench, once the temperature of the steel has been raised above the upper critical. The rate of quenching varies with the type and mass of steel and with the degree and depth of hardening. Usually only the surface of the metal is heated to the required temperature, and then the steel is quenched. During the time the steel is moved from the inductor coil to the quench, and even during the external quenching, the heated surface experiences some *self-quenching;* that is, some of the heat passes into the interior of the piece. Because of this the severity of the quench for induction hardening need not be so critical as the quench for furnace-heated pieces.

The type of quench employed depends upon the heating procedure, the steel, and the desired metallurgical results. One type is the submerged quench. The metal is dropped into a quench bath after the proper surface temperature has been attained. Severity of the quench depends upon the type of steel and the heating it has experienced. To attain full hardness, an SAE 1045 steel that has been deeply heated

FIG. 24·70 Progressive hardening, square shafting.

must be quenched in a submerged water spray or agitated brine. If the heating time had been less than 3 seconds, the quench could have been only agitated tap water.

The spray quench also is widely employed, most commonly with progressive hardening processes. A spray ring sprays water on the work piece as it leaves the work coil. For rapid cooling, water pressures from 40 to 100 pounds per square inch must be used, and the piece must be sprayed at the proper angle of incidence.

Self-quenching is employed for some particular applications such as the one illustrated in Fig. 24·71.

Water is generally used as a quench with induction heating. Cold agitated water produces rapid cooling; the submerged water spray gives the fastest obtainable rate of cooling. Temperature and gas content of the water affect the results. It is important also that surfaces being quenched be free of oil and oxide scale, which increase cooling time by their inherently lower thermal conductivity.

Oil probably is the next most commonly used quench. One advantage of oil is that the cooling rate below about 500 degrees Fahrenheit is slow, thus minimizing internal stresses and the possibility of quenching cracks.

The foregoing discussion applies primarily to high-frequency induction heating, but there is a wide overlapping of the application that can be accomplished by low-frequency (up to 10,000 cycles) and high-frequency (100 to 500 kilocycles) induction heating as applied to heat treating. Many furnace-heating applications also are accomplished by low- and high-frequency heating. For example, several deep-heating functions are accomplished by furnace heating as well as by low-frequency induction heating. Deep case-hardening is a good example; annealing of large diameter bars is another. Full hardness of a case is difficult to achieve in treating low-carbon-steel (SAE 1045) cylinders of large diameter if furnace heat is employed. The use of induction heating helps to achieve full hardness, the high frequencies being better for the shallow depths of hardness. Low- and high-frequency induction heating do not replace furnace heating.

Fig. 24·71 Self-quenching.

Annealing

Low frequencies have been most successfully employed for annealing large pieces (greater than about ½ inch in cross-sectional dimensions), both continuously and intermittently. Rods can be annealed continuously at low frequencies.

If material of small cross section (less than about 250 mils in diameter) is to be annealed inductively, high frequencies should be used. If such material is non-magnetic it usually is not practical to attempt to apply induction heating. The power factor of a load and heater coil is low. However, small iron wire has been annealed at temperatures at or below the magnetic-transformation temperatures.

Some of the advantages of induction heating are:

1. Heat can be generated by high-frequency heating at rates impossible by any other method; therefore heat treating can be performed at higher production rates by high-frequency heating than by other methods. It also permits obtaining hardness patterns heretofore not realized.
2. High-frequency heating provides accurate control of temperature; the application of power can be started and stopped instantaneously and varied at will. There is no time lag.
3. More uniform products are obtained because of the accurate control.
4. Heated materials are distorted less than by other methods. Scale is almost negligible.
5. Working conditions are improved. No continuously running furnaces are necessary. Almost all heat generated is in the material itself. All other heat is carried away by water cooling.
6. Somewhat higher hardness usually is obtained for a given surface temperature for steel products.
7. Since surface heating is employed for induction hardening, the metallurgical properties of the base metal are undisturbed.

24·6 DIELECTRIC-HEATING THEORY [26, 27, 28, 29]

Dielectric heating is the generation of heat by molecular action in dielectric material in an electrodynamic field. It is a means of distinguishing perfect dielectrics (those dielectrics having no losses when they are placed in a changing electric field) from imperfect dielectrics (those having losses or those that are heated when placed in a changing electric field). For dielectric heating, only imperfect dielectrics are considered. These losses are attributed to a property of dielectric material called *absorption*.

Properties of Imperfect Dielectrics [26, 30, 31, 32]

A few properties of dielectric materials, although not completely understood, give a simple method of computing dielectric losses under various conditions of voltage gradient and frequency.

In an electric field the normal structures of the atoms of a dielectric are deformed and rotated. This process is termed *polarization*,[28] and it is proportional to the field intensity or voltage gradient. The constant of proportionality obtained from the ratio of the polarization to the electric intensity is called *electric susceptibility*, and is almost constant for steady fields, but variable with changing fields. The electric displacement D is, by definition, equal to the sum of the electric field intensity E_1 and the polarization P; therefore

$$D = E_1 + P = (1 + \chi)E_1 \tag{24·90}$$

where

$$P = \chi E_1 \tag{24·91}$$

The relation between D and E then becomes

$$\frac{D}{E_1} = \epsilon \tag{24·92}$$

where $\epsilon = (1 + \chi)$ and is called the *dielectric constant* of the material. Since the electric susceptibility χ varies with frequency, depending on the material, the dielectric constant also varies with frequency.

In other words the dielectric constant at a given frequency represents the ratio of the electric displacement in a given dielectric to the displacement if the dielectric were a vacuum ($\epsilon = 1$). If the dielectric were a vacuum, the electric susceptibility would be zero, for there would be no polarization, and the dielectric constant would be unity.

The complement of the angle by which current through a dielectric is out of phase with the impressed voltage is called the *phase angle* of the dielectric. The dielectric phase angle is the complement of the power factor angle θ. If the dielectric phase angle is Φ,

$$\sin \Phi = \cos \theta \qquad (24\cdot93)$$

Usually in high-frequency heating the dielectric phase angle is small. Therefore, for angles as great as 30 degrees,

$$\sin \Phi = \Phi = \cos \theta \qquad (24\cdot94)$$

Another property, a function of the dielectric phase angle, is the dielectric *loss factor* ϵ''. This factor, by definition, is the product of the power factor Φ and the dielectric constant ϵ:

$$\epsilon'' = \Phi\epsilon \qquad (24\cdot95)$$

Power Absorption

With these equations, the relation of these properties to the familiar power equation

$$P = EI \cos \theta = EI\Phi \qquad (24\cdot96)$$

can be shown. If the load is in the form of parallel plates or a capacitor, then current can be expressed in terms of load geometry and load voltage:

$$I = \frac{E}{Z} \qquad (24\cdot97)$$

but the absolute value of Z is

$$Z = \sqrt{R^2 + X^2} \text{ ohms}$$

Since load impedance usually has a ratio of load reactance to load resistance greater than 10, the term Z can be reduced to

$$Z = X$$

Therefore

$$\text{Power} = \frac{E^2}{X} \Phi \text{ watts} \qquad (24\cdot98)$$

The reactance

$$X = \frac{1}{\omega c}$$

where

$$c = 0.224\epsilon \frac{A}{d} \times 10^{-12} \text{ farad}$$

all dimensions being in inches, and $\omega = 2\pi f$. Then the power equation becomes

$$\text{Power} = 1.41 f E^2 \epsilon'' \frac{A}{d} \times 10^{-12} \text{ watts} \qquad (24\cdot99)$$

Equation 24·99 can be reduced further if both the numerator and denominator are multiplied by d, the distance between the plates. Then the term dA reduces to the volume of the load, since the term A is the area of one electrode face. By rearranging terms and substituting the value of voltage gradient E_1 in place of E/d, the power equation takes its final form:

$$P_v = 1.41 f {E_1}^2 \epsilon'' \qquad (24\cdot100)$$

where P_v = power absorbed in watts per cubic inch
f = frequency in megacycles per second
E_1 = voltage gradient in kilovolts per inch, rms
ϵ'' = loss factor of dielectric material.

This shows that the power absorbed by an imperfect dielectric in an alternating electric field of E_1 kilovolts per inch and changing at the rate of f megacycles per second can be expressed simply as a function of measurable quantities.

The preceding discussion is based on the heating of homogeneous isotropic imperfect dielectric media completely filling the space between two parallel electrodes of A square inches each and separated by a distance of d inches. However, it is not necessary that the dielectric fill the space between the electrodes. Equation 24·100 is entirely independent of shape, size, or spacing of electrodes. The absorbed power is a function of the frequency of the alternating field, voltage gradient, and loss factor of the material. Shape, size, and spacing of electrodes for a given electrode voltage determine the voltage gradient in the work.

Power factor and dielectric constant are not entirely independent of frequency. Figure 24·72 indicates the dependence of loss factor, the product of dielectric constant and power factor, as a function of frequency for a particular dielectric. For the substance represented by Fig. 24·72 a

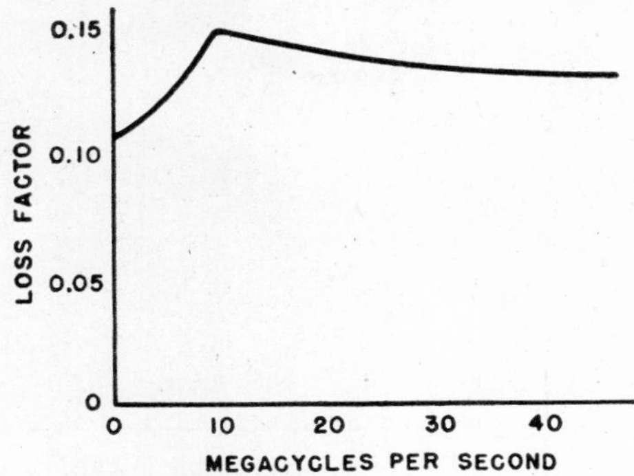

FIG. 24·72 Dielectric properties versus frequency.

small change in frequency does not appreciably change loss factor. These dielectric properties also change with moisture content, pressure, and temperature.

Measurement of Dielectric Properties [31, 32]

There are several methods of measuring power factor and dielectric constant at high frequencies. The prime difficulty is in measuring high-frequency currents and doing it rapidly. One method is known as the *susceptance-variation* method, which consists in measuring the properties of a series LCR circuit excited by a constant voltage alternating at the resonant frequency of the circuit. The initial circuit consists of an inductance L in series with two parallel capacitive reactances. One of these reactances is normally a sample holder (Fig. 24·73) containing a sample of the dielectric.

An instrument often used for measurement of these properties is the Q meter. This instrument consists of a variable-frequency oscillator used to excite the LCR circuit. Total inductance L and capacitance C can be varied to make this circuit resonant at a desired frequency. The output of the meter feeds an initial circuit as indicated in Fig. 24·74. Here voltage E_Q is the constant voltage applied across the series circuit. The inductance L_Q is the sum of the internal and external inductances. Capacitance C_Q is the internal capacitance of the instrument, which is variable. A sample holder contains a sample of the dielectric material. The capacitance of this holder with the sample is termed C_S.

The circuit without the sample in the sample holder has an inherent resistance designated R_Q, and the effective series resistance of the sample is designated R_S.

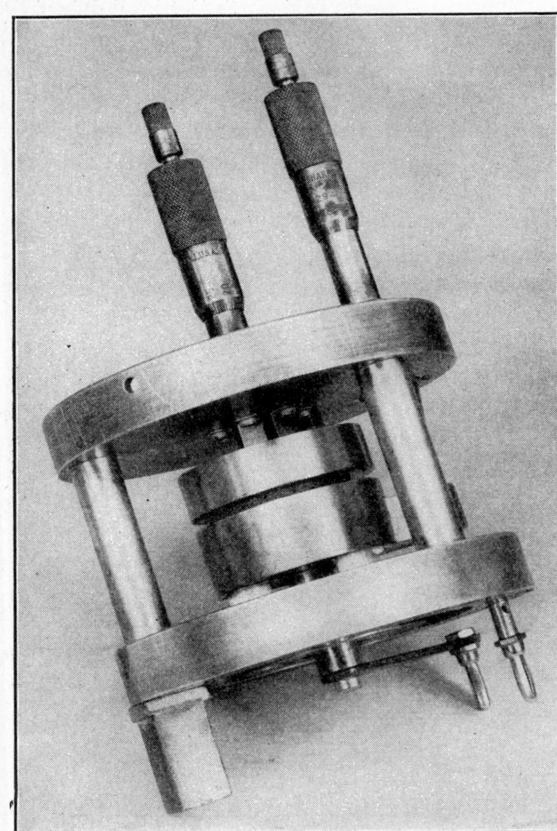

FIG. 24·73 Dielectric sample holder.

In taking these measurements the circuit Q and voltage with an input of E_Q volts are measured, with the sample holder containing a sample of the dielectric. Then the sample is removed and the sample holder is readjusted to the capacitance it had when it held the sample. Again circuit Q and voltage are measured at voltage E_Q. If instrument capacitance C_Q is now adjusted by both increasing and

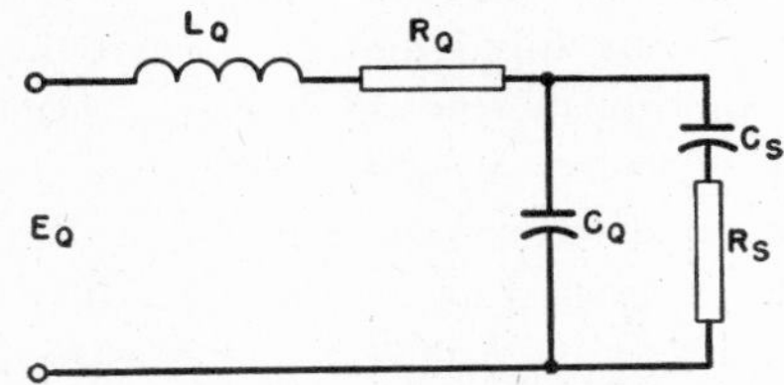

FIG. 24·74 Initial dielectric-property-measuring circuit.

decreasing the capacitance so that the Q meter is detuned to the half-power points, inherent circuit resistance can be determined. It is

$$R_Q = \frac{\Delta C}{\omega C} \tag{24·101}$$

where R_Q = inherent instrument resistance in ohms
C = total circuit capacitance in farads = $C_Q + C_S$
ΔC = average capacitance differential in farads needed to detune to half-power points on either side of the resonant peak
$\omega = 2\pi f$
f = frequency in cycles per second.

By combining the results of equation 24·101 with the observed capacitive-voltage readings, with and without the sample in the holder, the power factor of the dielectric in the holder becomes

$$\Phi = \frac{\Delta C}{C_S}\left(\frac{E_0 - E_1}{E_1}\right) \tag{24·102}$$

where E_0 = capacitive voltage with no sample in holder
E_1 = capacitive voltage with sample in holder
C_S = capacitance of sample holder with sample.

The dielectric constant can be determined in the initial procedure by equating the geometric capacitances of the sample holder with and without the samples. In all the operations with the sample held in the holder, the cross-sectional area of the sample is assumed to be equal to the area of one electrode face. Thus dielectric constant

$$\epsilon = \frac{d_0}{d_s} \tag{24·103}$$

where ϵ_0 = dielectric constant
d_0 = sample-holder electrode spacing with sample removed
d_s = sample-holder electrode spacing with sample in holder

and the capacitance of the sample holder with an electrode separation of d is equal to sample-holder capacitance with electrode separation d_s.

Power Requirements [33, 34]

As in induction heating the required power is that necessary to raise the temperature of the load to a given temperature in a specified length of time. Thermal losses such as conduction, convection, and radiation must be overcome by an additional supply of power.

The useful thermal requirements may be expressed as

$$P_1 = 17.6Mc\,\Delta t \text{ watts} \tag{24·104}$$

where P_1 = thermal power in watts
M = pounds of material heated per minute
c = specific heat of load material
Δ_t = load temperature rise in degrees Fahrenheit.

Equation 24·104 is one of the fundamental thermodynamic equations.

Radiation and convection losses are identical with those in induction heating, as given in Section 24·2. However, conduction loss is a little different. For conduction losses to exist there must be a thermal gradient at the boundary of the electrodes and the dielectric material. This gradient must have the proper algebraic sign. If the electrodes were properly heated from some external source the thermal gradient necessary for conduction losses would be absent and then there would be no conduction loss. Such a solution as this is not so simple, if the electrodes consist of cold, unheated press platens.

Figure 24·75 facilitates estimates of conduction losses into cold press platens. Ordinates represent ratios of conduction losses to useful power (equation 24·104); the abscissas represent a function of time, the properties of the load, and the thickness of the load. More specifically, the term B in Fig. 24·75 is

$$B = \frac{kt}{\rho c d^2} \tag{24·105}$$

where k = thermal conductivity in (calories per second per square centimeter) per (degrees centigrade per centimeter)
t = heating time in seconds
ρ = density of dielectric in grams per cubic centimeter
c = specific heat in calories per gram
d = thickness of load in inches.

One other power component is that of a change in state by the load or some portion of the load. An example is the power required to vaporize water. This power, termed P_5, is often as great as any other power term involved.

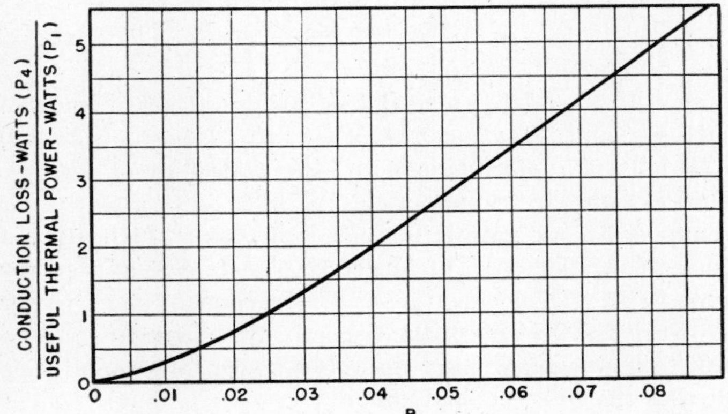

FIG. 24·75 Conduction-loss curve for dielectric heating in cold press.

Total load power then can be taken conservatively as

$$P_T = P_1 + P_2 + P_3 + P_4 + P_5 \tag{24·106}$$

where P_T = total power in watts
P_1 = useful thermal power in watts (equation 24·104)
P_2 = radiation loss in watts (equation 24·48)
P_3 = convection loss in watts (equation 24·49)
P_4 = conduction loss in watts (Fig. 24·75)
P_5 = power in watts to bring about a change of state.

I^2R losses in transmission lines and networks must be accounted for also and added to the thermal power of equation 24·104 to determine total generator power.

24·7 DIELECTRIC-HEATING APPLICATIONS

Power Density, Voltage, and Frequency

If the required total thermal power and the volume to be heated are known, the power density can be determined. This is the power per cubic inch of the volume, primarily influenced by the field between the electrodes, to heat the work at the desired rate.

Once volume power density is established, permissible voltage gradient and electrode voltage can be estimated. The limit of voltage gradient is indefinite, especially for radio-frequency voltages. Some dielectrics can withstand gradients of 3 to 5 kilovolts or more per inch. Other dielectrics break down under voltage gradients of several hundred volts per inch. A few of the factors influencing voltage breakdown are moisture content; the material with which the dielectric is impregnated; and mechanical forces, such as volume stresses, applied to the dielectric. Another factor that must be considered is possible voltage breakdown between electrodes at points other than through the dielectric material. Evolution of steam or some solvent vapor from the heated dielectric, as well as condensation of vapors on the electrodes, should be considered. Corona at high radio-frequency voltages may cause difficulty. Under normal conditions an upper limit of about 15 kilovolts is used for electrode voltages. Steam or solvent vapors near the electrodes reduce this limit. Radio-frequency voltages greater than 15 kilovolts may be used, but special precautions must be taken to eliminate corona and voltage breakdown. Corona discharge in air begins at about 75 kilovolts per inch voltage. The edges of straps must be beaded, or heavier conductors must be used, to obtain the edge radius necessary to eliminate breakdown and corona. Shorter conductors must be used. These changes affect the cost of apparatus, and usually decrease its flexibility.

Assuming that a voltage gradient and electrode voltage have been established, the required frequency for a given material is approximately

$$f = \frac{0.71P_v}{E_I{}^2\epsilon''} \text{ megacycles per second} \tag{24·107}$$

This formula defines the required frequency as a function of loss factor ϵ'' which is also a function of frequency. The only way of determining how ϵ'' varies with frequency is to measure power factor and loss factor over a frequency range that might be used for the particular heating process. This property of the dielectric may be fairly constant over a limited frequency range. Since most radio-frequency generators are designed for fixed frequencies, the approximate frequency indicated by equation 24·107 is close enough to determine the frequency of the generator. Once this frequency is established, the loss factor is thereby defined. If the chosen generator frequency is greater than the approximated frequency, the required voltage is decreased, assuming that loss factor and dielectric constant remain constant.

For a load consisting of a slab of dielectric material placed between two parallel electrodes (no air gap between load and electrodes) the voltage gradient in the work is the ratio of electrode voltage to distance of separation. Electrode voltage is the product of the voltage gradient and the distance between the electrodes, in this particular case.

An air gap between load material and electrodes may be necessary, or the electrodes may contain layers of material of different dielectric constants and power factors.

The voltage gradient across the kth layer is

$$E_k = \frac{E}{\epsilon_k \displaystyle\sum_{i=1}^{i=n} \frac{a_i}{\epsilon_i}} \tag{24·108}$$

where E_k = voltage gradient in kth layer in kilovolts per inch
E = total electrode voltage in kilovolts
ϵ_k = dielectric constant of the kth layer
a_i = thickness of ith layer in inches
ϵ_i = dielectric constant of ith layer.

For example, suppose parallel-plate electrodes are arranged with a half-inch air gap above a 2-inch load with a dielectric constant of 3. If equation 24·100 shows the required voltage gradient in the load to be 4 kilovolts per inch, equation 24·108 becomes

$$4 = \frac{E}{3(0.5 + \frac{2}{3})}$$

and the total electrode voltage becomes $E = 14$ kilovolts. If there had been no air gap, the electrode voltage would have been $E = 4 \times 2 = 8$ kilovolts; hence, the voltage across the air gap becomes 12 kilovolts per inch.

An analogous equation for voltage gradient across a cylindrical layer of dielectric between coaxial electrodes is

$$E_k = \frac{E \log_\epsilon \sigma_k}{\gamma_k \epsilon_k \sum_{i=1}^{i=n} \frac{\log_\epsilon \sigma_i}{\epsilon_i}} \text{ kilovolts per inch} \qquad (24 \cdot 109)$$

where E_k = voltage gradient in kth cylindrical dielectric layer in kilovolts per inch
E = total electrode voltage in kilovolts
σ_k = ratio of outer to inner radii of kth cylindrical dielectric layer
γ_k = difference in outer and inner radii of kth cylindrical layer in inches
ϵ_k = dielectric constant of kth cylindrical dielectric layer
σ_i = ratio of outer to inner radii of ith cylindrical dielectric layer
ϵ_i = dielectric constant of ith cylindrical dielectric layer.

If a load does not occupy the entire space between two parallel electrodes, the effective dielectric constant of the space between the electrodes is altered, and the effective power factor is changed. As illustrated in Fig. 24·76, the effective dielectric constant and power factor become

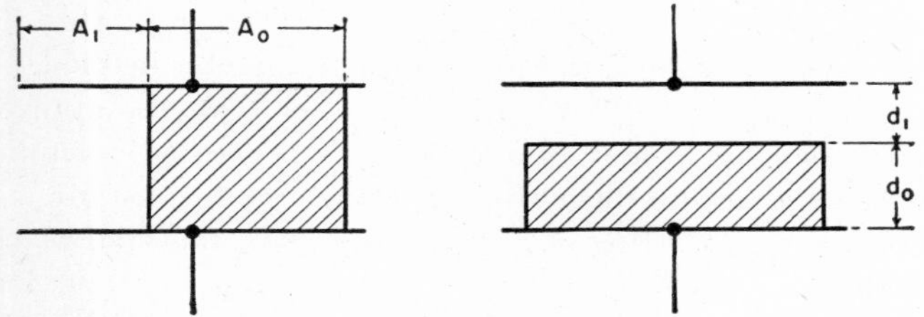

Fig. 24·76 Dielectric loads accompanied by air spaces.

$$\epsilon = \frac{\epsilon_0 + \frac{A_1}{A_0}}{1 + \frac{A_1}{A_0}} \qquad (24 \cdot 110)$$

$$\Phi = \frac{\Phi_0}{1 + \frac{A_1}{\epsilon_0 A_0}} \qquad (24 \cdot 111)$$

where ϵ = effective dielectric constant of volume between electrodes
Φ = effective power factor of volume between electrodes
ϵ_0 = dielectric constant of load material
Φ_0 = power factor of load material
A_0 = area of electrodes occupied by load
A_1 = area of electrodes not occupied by load.

If a series air gap is used the apparent changes in the dielectric constant and power factor are

$$\epsilon = \frac{1 + \frac{d_1}{d_0}}{\frac{1}{\epsilon_0} + \frac{d_1}{d_0}} \qquad (24 \cdot 112)$$

and

$$\Phi = \frac{\Phi_0}{1 + \epsilon_0 \frac{d_1}{d_0}} \qquad (24 \cdot 113)$$

where ϵ, Φ, ϵ_0, and Φ_0 are as defined before, d_0 is thickness of load layer, and d_1 = thickness of air gap. Equations 24·110 to 24·113 are applicable only to loads consisting of *isotropic homogeneous* dielectric materials.

Suppose an air gap of ½ inch is necessary in heating a 2-inch-thick dielectric of dielectric constant 4 and power factor 0.05. Then effective dielectric constant and power factor become 2.5 and 0.025 respectively, illustrating that a small series air gap greatly decreases power factor, and thus increases the Q of the load circuit.

Load Circuits

No matter how load elements and electrodes are arranged, load impedance may be viewed as a perfect capacitor shunted by a pure resistance, or in series with it. The equivalent series circuit is shown in Fig. 24·77; the shunt circuit in Fig. 24·78.

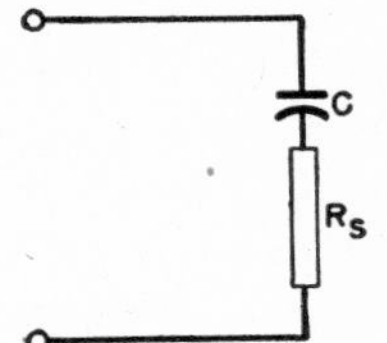

Fig. 24·77 Imperfect capacitor with equivalent series resistance.

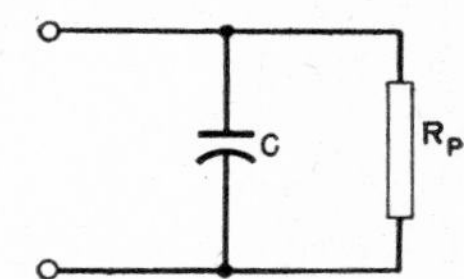

Fig. 24·78 Imperfect capacitor with equivalent parallel resistance.

The effective series resistance R_s of Fig. 24·77, with little error for high-Q loads, is

$$R_s = (X_c) \text{ (power factor) ohms} \qquad (24 \cdot 114)$$

and the effective shunt or parallel resistance R_p may be expressed

$$R_p = \frac{(X_c)}{\text{(power factor)}} \text{ ohms} \qquad (24 \cdot 115)$$

where the power factor is the dielectric power factor described in Section 24·6. If there are layers of different dielectrics between electrodes, the effective series resistance of the layer considered is indicated by equation 24·114, where X_c is the reactance of a capacitor having an electrode area equal that of the dielectric layer, and a separation equal to the thickness of the layer. Effective series resistance of several layers becomes

$$R_s = \sum_{i=1}^{i=n} X_1 \Phi_i \text{ ohms} \tag{24·116}$$

If there is an air gap between one electrode and the load, the effective series resistance of the load becomes

$$R_s = [(X_a)(\Phi_a) + (X_L)(\Phi_L)] \text{ ohms}$$

where X_a and X_L are reactances of air gap and load respectively, and Φ_a and Φ_L are the power factors of the air and the dielectric load respectively. Since Φ_a, for work below ionizing potentials, may be considered to be zero, the effective series resistance becomes

$$R_s = (X_L)(\Phi_L) \text{ ohms} \tag{24·117}$$

Suppose that load and electrodes have equal face areas of 100 square inches, that the load is two inches thick, and that there is a ½-inch air space between one electrode and the load. If dielectric constant is 3 and power factor is 0.05, the reactance of the load alone is

$$X_L = \frac{d \times 10^{12}}{1.41 feA} = 470 \text{ ohms}$$

at a frequency of 10 megacycles per second. Effective series resistance is $470 \times 0.05 = 23.5$ ohms. The reactance offered by the air gap alone is

$$X = \frac{d \times 10^{12}}{1.41 fA} = 350 \text{ ohms}$$

Because the power factor of air is zero, the series-resistive component of the air gap is zero. Figure 24·79 shows this circuit. The solution to the problem is similar if the load is considered to be a perfect capacitor shunted by a pure resistance.

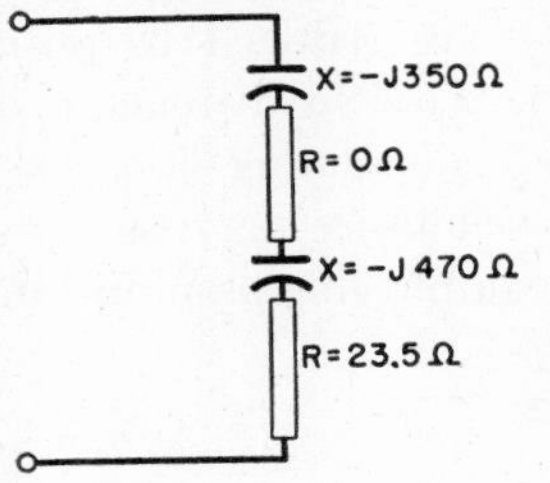

Fig. 24·79 Equivalent-load circuit utilizing series resistance.

The discussion of induction-heating loads pointed out that the load Q must be less than the Q of the tank circuit to supply available power to the load under some conditions. By introducing an air gap the effective reactance is increased, but the resistive component remains the same. This means that load Q is increased, because Q is the ratio of reactance to resistance or the ratio of kva to kilowatts for high-Q loads.

However, Q of the load without the air gap may be too great. To overcome this in extreme cases, the entire tank circuit is placed outside the oscillator. Figure 24·80 illustrates this method. The loops at each end constitute the inductance L of the tank circuit, and the capacitance of the two plane surfaces near the load material form the tank capacitance C. The Q of this LC circuit may be made high by increasing C and decreasing L. Loads with extremely low losses have been heated in this manner. Design and operating difficulties prohibit general use of this circuit.

As an example, a low-loss material with a Q of about 75 was heated by this method. It was almost impossible to heat this material in the usual manner with the available oscillator because the loaded-tank Q of the oscillator was about 30. When an air gap was used between the load and one electrode the load Q was further increased to about 120, or about four times the Q of the normally loaded tank. About 10 kilowatts or one-half the oscillator capacity was supplied the load at 75 percent of the maximum plate voltage. By this method the oscillator was loaded properly.

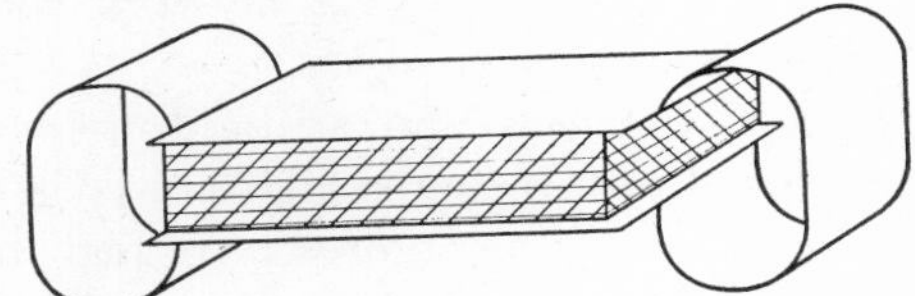

Fig. 24·80 High-Q tank for dielectric heating.

Another type of load circuit involves induction heating as well as dielectric heating. This method is used in the furniture industry for bonding wood. A conductor carrying high-frequency current passes close to the joint to be heated, as shown in Fig. 24·81. The electric and magnetic fields built up between the two conductors heat the dielectric material in the glue by dielectric heating, and induce currents in the glue while it is wet, to produce an additional I^2R loss.

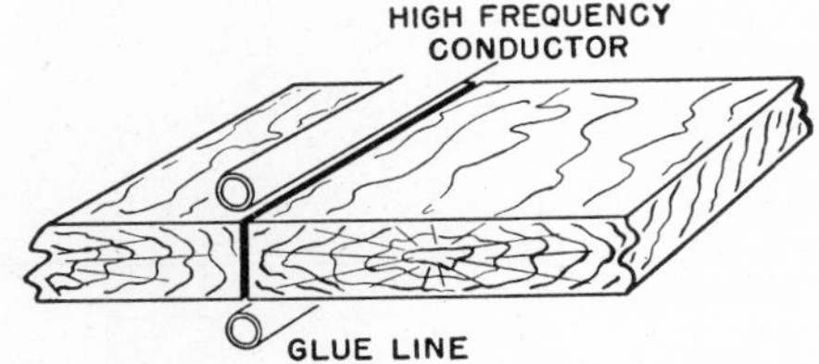

Fig. 24·81 Wood bonding by dielectric heating.

There are several methods of arranging the load to approximate a desired load impedance. An example of the lower limit is that of heating a single slab of a dielectric load. This load is simply two electrodes, one of which is grounded, the other at high potential.

An alternative is frequently used for heating two slabs of the dielectric simultaneously, as shown in Fig. 24·82. The two slabs of the same load material usually are identical in shape and size. This arrangement reduces the impedance to half that of a single slab. One advantage of this method, especially if presses are used, is the reduction of capacitive currents to extraneous grounds. Therefore the loaded-electrode ratio of kva to kilowatts approaches a minimum for a given load. There are numerous parallel-series circuits for loads; however, a minimum of circuit paths is most desirable.

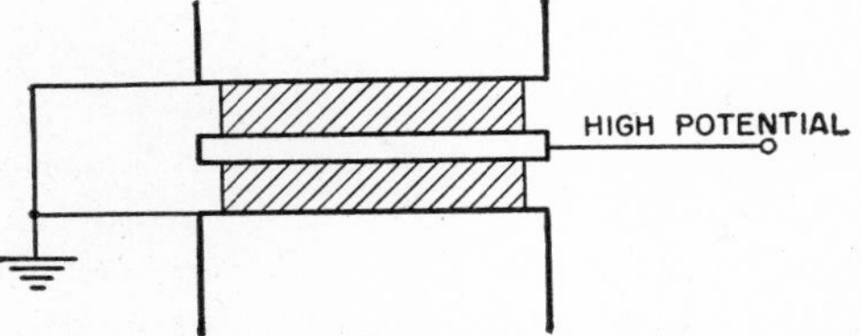

Fig. 24·82 Paralleled dielectric-heating load.

Matching Networks [35, 36, 37]

To transfer power to a load in the best manner, the load impedance must be transformed into an impedance which can utilize available generator current and voltage most favorably. At power frequencies, transformers are used often. Actually a transformer is a network consisting of reactive components with some losses.

At high audio-frequencies and at radio-frequencies, the use of an iron-core transformer for high power is limited by high core losses and stray reactances. At high frequencies air-core transformers are used more often. Their leakage reactance is large, which implies that regulation is poor. Power transformers for the megacycle range of frequencies are not generally used to couple dielectric-heating loads directly to a radio-frequency generator, but a generator can be coupled to a transmission line by means of an air-core transformer.

An impedance arrangement can be used in place of transformers as a means of transforming a load impedance. To minimize losses or increase efficiency of such a network, its elements are made as nearly pure reactance as is economically justified. The types of networks considered here are those reducible to the four-terminal networks shown in Fig. 24·83.

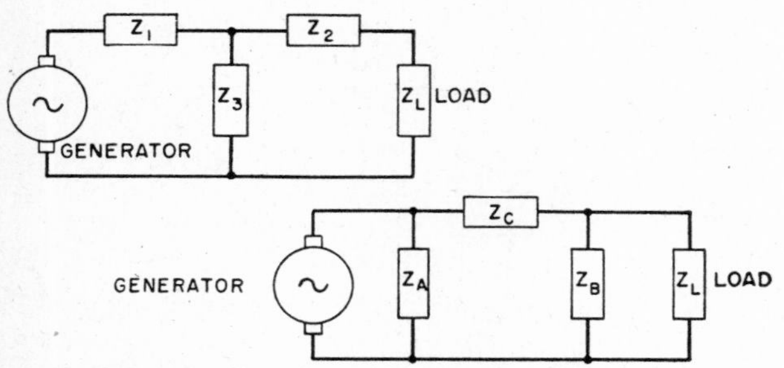

FIG. 24·83 Dielectric-heating networks.

The T network and the π network are electrically the same, if the components are chosen properly. The following equations indicate these relations:

$$Z_A = \frac{Z_Q}{Z_2}$$

$$Z_B = \frac{Z_Q}{Z_3} \qquad (24\cdot118)$$

$$Z_C = \frac{Z_Q}{Z_1}$$

where $Z_Q = Z_1Z_2 + Z_1Z_3 + Z_2Z_3$. By employing resonance, any desired load voltage and current can be produced within the limits of practical network components, voltage breakdown and corona, and the Q of the generator tank circuit if a vacuum-tube radio-frequency generator is used.

Equations 24·114 and 24·115 define the effective series and parallel load resistances. For convenience, the effective series resistance as defined by equation 24·114 is used throughout the remainder of this discussion, unless otherwise indicated, and is designated as R_s. Likewise the T network is used although there is no reason why the π network could not be used.

Both networks have three reactive elements. Networks with more than three elements usually can be reduced to equivalent three-element networks. These three elements form a network capable of performing three basic functions: transformation of resistance, transformation of reactance, and change of phase. Change of phase is least important; in dielectric-heating networks, load-impedance (resistance and reactance) transformation is the most necessary function. Load impedance usually is transformed from an impedance containing a reactive component as well as a resistive component to an impedance consisting of pure resistance.

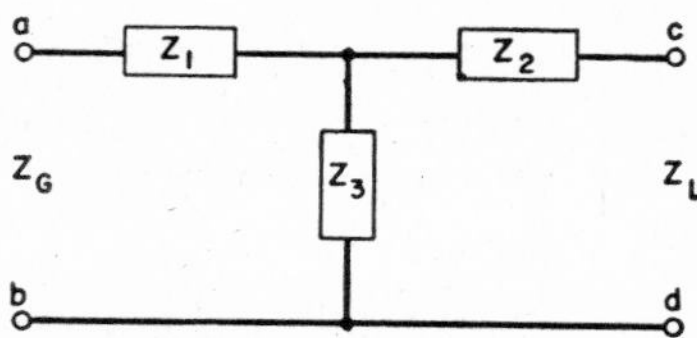

FIG. 24·84 T network with impedance elements.

In Fig. 24·84, Z_G represents the impedance into which load impedance Z_L is to be transformed by the network consisting of Z_1, Z_2, and Z_3. Then looking toward Z_L from Z_G, the combination of impedances in the network and the load should be equivalent to Z_G; also looking from Z_L toward Z_G the network combination should produce an equivalent impedance Z_L. Thus, looking in either direction from *a-b*, an impedance of Z_G should be seen; Z_L should be seen looking in either direction from *c-d*. This is known as matching impedances:

$$\frac{Z_G}{Z_L} = \frac{Z_1 + Z_3}{Z_2 + Z_3} \qquad (24\cdot119)$$

$$Z_GZ_L = Z_1Z_2 + Z_1Z_3 + Z_2Z_3 \qquad (24\cdot120)$$

These equations are necessary to determine the network impedance components if one of the network components, Z_G, and Z_L are known. Z_G and Z_L are reduced to pure resistances R_G and R_s respectively. Any load reactance may be considered to be a part (or all, depending upon the conditions) of Z_2, and any generator reactance may be considered to be a part of Z_1. Another simplifying assumption

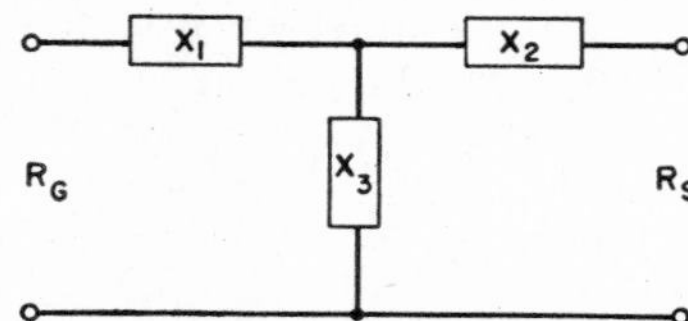

FIG. 24·85 T network with reactance elements.

is that network impedances may be considered as pure reactances, as illustrated in Fig. 24·85. Thus equations 24·119 and 24·120 reduce to

$$\frac{R_G}{R_s} = \frac{X_1 + X_3}{X_2 + X_3} \qquad (24\cdot121)$$

and

$$-R_GR_s = X_1X_2 + X_1X_3 + X_2X_3 \qquad (24\cdot122)$$

Equation 24·122 may indicate negative resistances; therefore such impedance transformation cannot be accomplished by reactive elements with identical algebraic signs. One capacitive (negative) reactance must be used if the remaining two elements are inductive (positive), or one inductive reactance must be used if the remaining two are capacitive.

One of the simplest networks is the single-element network, which is easily adjusted. Figure 24·86 represents the transformation from the series circuit to the equivalent

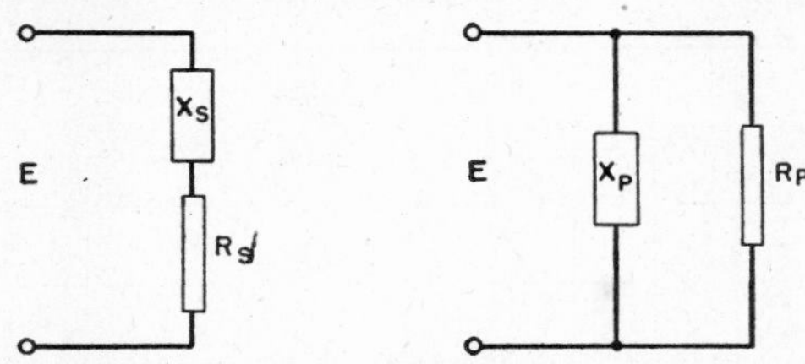

FIG. 24·86 Single-element network.

parallel circuit. The desired resistance R_p may be determined if the voltage to be applied across the load terminals to produce the required power is known. This voltage is

$$R_p = \frac{E^2}{P} \qquad (24\cdot123)$$

where E is the impressed voltage and P is the resultant power in watts; then from comparison of the two circuits in Fig. 24·86

$$R_p = \frac{R_s^2 + X_s^2}{R_s} \text{ ohms} \qquad (24\cdot124)$$

and

$$X_p = \frac{R_s^2 + X_s^2}{X_s} \text{ ohms} \qquad (24\cdot125)$$

This network is limited in its range of resistance transformation. Rearrange equations 24·124 and 24·125 for X_s as a function of R_p and R_s.

$$X_s = \pm\sqrt{R_sR_p - R_s^2} \text{ ohms} \qquad (24\cdot126)$$

If R_s is greater than R_p, equation 24·126 represents X_s as an imaginary number, which is impossible. The single-element network can be used only if the desired parallel resistance R_p is greater than the effective series-load resistance R_s. In equations 24·124 and 24·125 X_s represents load reactance as well as an additional series reactance and, if the additional series reactance has a resistive component, it is included in R_s.

In transforming R_s into R_p ($R_p > R_s$) the reactive component X_s is changed according to equation 24·125. For high-Q loads ($Q > 10$) the resistive component R_s is small compared with the reactive component X_s. Therefore equation 24·125 is slightly altered to $X_p \cong X_s$, which means that reactance X_s is shunted across the generator. Since this transformation does not eliminate the reactive component of the load, it cannot be used to match the characteristic impedance of a transmission line. However, if the load is coupled to the generator by an untuned transmission line, the single-element network may be used if the reactive component X_s ($\cong X_p$) does not alter the operating conditions by changing the frequency of the vacuum-tube radio-frequency generator. Usually some adjustment can be made on the tank-circuit reactances to correct for this additional reactance X_s ($\cong X_p$), so that the frequency of the generator remains constant and power can be delivered to the load limited by the transformation of resistance ($R_p > R_s$) and limited by the shunting reactances X_s ($\cong X_p$). X_p may be small compared with one leg of the tank circuit in the radio-frequency generator. This effect can be corrected to some extent, if X_p is a capacitive reactance, by removing the original tank capacitor from the circuit. This means that the load circuit serves as the tank-circuit capacitive leg. To carry this even further, suppose that the magnitude of X_p is much less than either branch of the tank circuit; without a different type of network full loading cannot be reached, even if the load Q is within the required upper limit. Frequency, as well as loading, is influenced, depending upon R_p and X_p.

If it is necessary to transform a load impedance with resistance R_s into an effective pure resistance R_G, greater than R_s, the single-element network is unsuitable; but the T (or π) network can be used. As a special case of the T network, and L network is used sometimes. This network, being a two-element network, can transform the load impedance into a pure resistance, so it is more flexible than the single-element network. The L network is most suitable if the ratio of transformation is 10 to 1 or greater (assuming transformation of resistances only, the load reactance being a portion of the network components).

Required shunting reactance X_3 can be determined by

$$\pm X_3 = \sqrt{R_GR_s} \text{ ohms} \qquad (24\cdot127)$$

indicating that this reactance may be either inductive (+) or capacitive (−). The remaining component X_2 must be numerically equal to X_3 and must bear the opposite algebraic sign; that is, if X_3 is capacitive, X_2 must be inductive, and vice versa. This network is shown in Fig. 24·87. The arrangement shown in Fig. 24·87 can be used for values of $R_G > R_s$ but, to reduce R_s, the network becomes like that shown in Fig. 24·88.

If load resistance must be changed by less than about 100 to 1, a T network is used. X_2 may be used to represent only load reactance, and this network becomes the familiar unsymmetrical T network. Substituting R_G, R_s, and X_2 in equations 24·121 and 24·122 produces two simultaneous equations from which X_1 and X_3 can be determined. These two newly formed equations reduce to a quadratic equation;

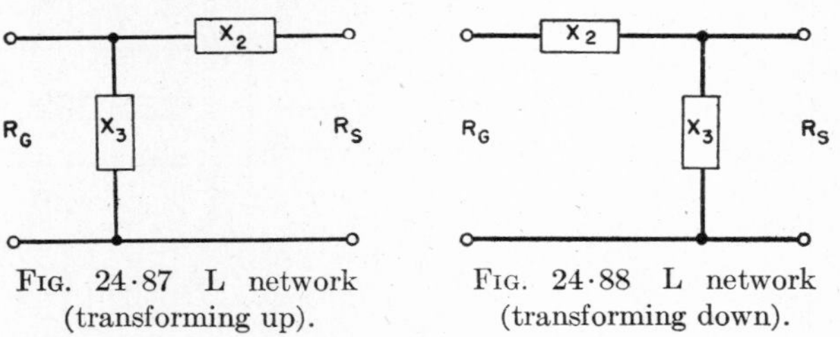

FIG. 24·87 L network (transforming up).

FIG. 24·88 L network (transforming down).

then there are two solutions, or two different sets of network components can be used.

If the required network components X_1 and X_3 are difficult to manufacture or are not economically justified, a suitable network may be found. One method is to change the effective load reactance by using a value of X_2 differing from the true load reactance. This change can be made by adding a series reactance. A variant of this method is to add a series reactance and to keep the values of all network components numerically equal. This is the symmetrical T network. This network can be solved easily by equations 24·111 and 24·122. From those equations X_3 is defined:

$$\pm X_3 = \sqrt{R_G R_s} \text{ ohms}$$

which is identical to equation 24·127.

The most common method of obtaining network reactances is to use coils and capacitors to give the desired inductive and capacitive reactances. If inductances larger than several microhenrys are needed, coils ordinarily are used. If a small inductance must be variable, its design is difficult, especially if the radio-frequency current is large. Because maximum network efficiency (minimum network losses) usually is desired, the choice of dimensions and conductor material becomes important. Power factor is the term associated with Q by

$$\text{Power factor} = \frac{1}{\sqrt{Q^2 + 1}} \qquad (24 \cdot 128)$$

With network coils a high Q is desirable so that the loss is minimum. Thus a coil must be designed for minimum resistance for a given inductive reactance.

As an example, suppose an L network is to be used with a load having $Q = 250$. If the shunting power-factor-correcting inductor with $Q = 250$ is used, the power absorbed by the load is roughly the same as the power absorbed by the network inductor; therefore power supplied by the generator must be approximately twice that absorbed by the load.

Resistance of the inductive-reactance coil sometimes can be reduced even more by using elements of short-circuited transmission lines as inductances. High Q can be produced with transmission-line elements or stubs. At low frequencies transmission-line stubs usually become too long, but at frequencies above 10 megacycles they are useful.

Figure 24·89 shows an actual installation using variable stubs to correct for changes in the load impedance during heating.

As loading is automatic for a specific load, constant radio-frequency power is applied to the load. If, instead of constant loading, a particular variation in loading is required,

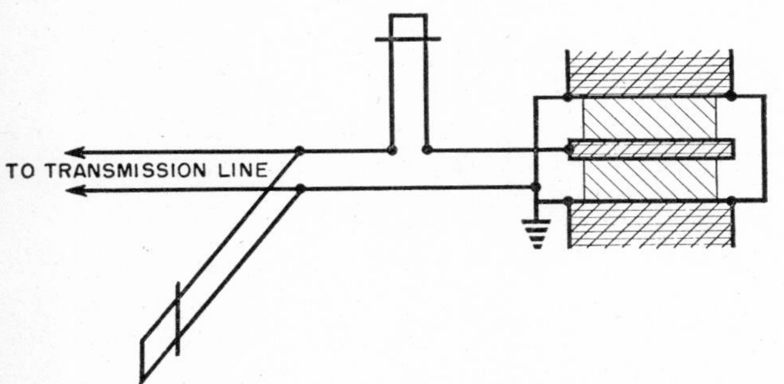

Fig. 24·89 Diagram of installation employing stubs.

it is feasible to build equipment to obtain it. The network shown in Fig. 24·89 is controlled so that loading varies as indicated in the curve of Fig. 24·90. The reason for this variation is indicated by the temperature curves, which are almost linear. This linearity means that constant power is used by the load, which in this case is desirable. Part of this power is produced by the radio-frequency generator; the remainder is generated by chemical reactions within this particular load.

Capacitors for networks may be air, gas, or vacuum capacitors, if low capacitive reactances are not needed. For capacitive reactances with lower losses, open-circuited coaxial transmission-line elements can be used. This type of reactive element usually is restricted to frequencies above about 20 megacycles.

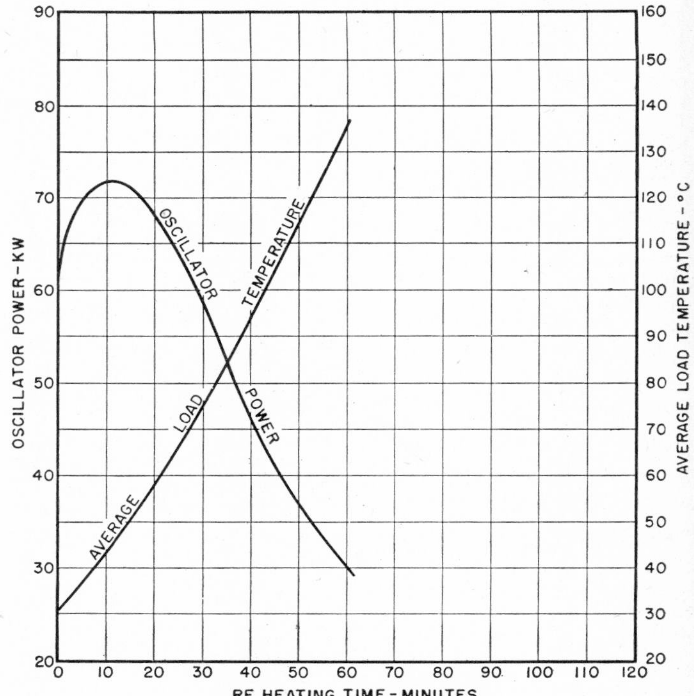

Fig. 24·90 Operational characteristics of installation.

Capacitors usually have extremely high Q, so their losses need cause no concern; but, if large currents are involved, the capacitor or parts of it may require cooling. This necessity is infrequent.

If the adjustments are critical and the network reactances are small, stray reactances of the leads of inductors and capacitors become important, especially at the higher frequencies. In some circuits the reactances of the leads in a branch of a network may be sufficient to serve as that leg of the network.

Because transmission-line stubs utilize standing waves to obtain low-loss reactances, voltage and current at various points along the length of the stub must be considered. Proper insulation is obtained by selection of the dimensions and spacing of the stub conductors. Current-carrying capacity also is a function of the dimensions. Loss in a transmission line is minimized by using a coaxial transmission line with a characteristic impedance of about 77 ohms. This fact applies also to stubs as reactance elements. If the stub is chosen for a particular characteristic impedance, the dimensions of the conductors are then defined.

Electrodes [38]

A changing electric field is needed to heat an imperfect dielectric. Assuming that uniformity of heating is desired, a specific field distribution, which implies unique shaped electrodes, is required. Electrodes may be considered approximately as transmission-line sections, devolving into simple capacitors if the linear dimensions are small enough to have negligible standing waves.

If a dielectric sheet is held stationary between two electrodes and parallel to them, the heating is not uniform if the potential varies from point to point along one electrode. Assume that a homogeneous isotropic dielectric is being heated. If the distance from the point of power application on the elctrode to another point is as great as one-eighth wavelength the difference in heating is 50 percent; if that distance is only one-sixteenth wavelength, the difference in heating becomes about 15 percent. Maximum electrical dimension from point of power application to any other point on the electrode therefore is limited usually to less than one-sixteenth wavelength.

As a guide, equation 24·136 indicates the effective electrical length of the electrode as a function of frequency and the dielectric constant of the material between the two electrodes.

The decrease in velocity of light traveling through a medium with an index of refraction η, is analogous to electromagnetic waves transmitted along the electrodes influenced by the dielectric with a dielectric constant ϵ. Two electromagnetic waves, one light and the other radio waves, are slowed according to the reciprocal of the index of refraction. The dielectric constant is equal to the square of the index of refraction.

If load requirements require electrodes longer than one-sixteenth wavelength, the effective electrical length can be changed to make the heating uniform. One method is to place inductive stubs about the edges of the electrodes, as in Fig. 24·91. The inductance of these stubs reduces the

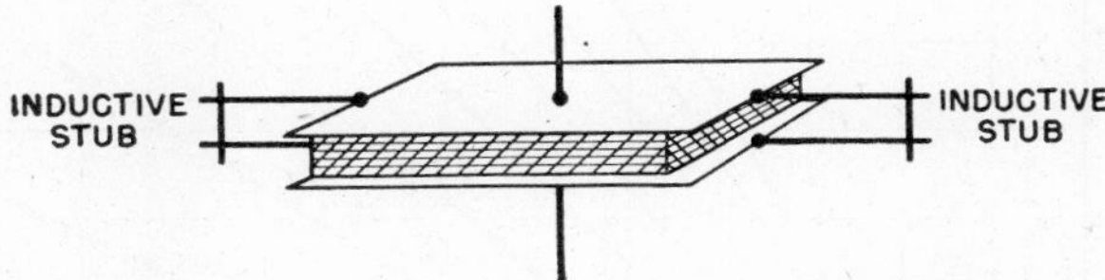

FIG. 24·91 Minimizing standing-wave effect by stubbing.

standing waves. In other words, stubs tend to cause the electrodes, as transmission lines, to be shortened electrically.

Electrically long electrodes may be required to build up voltage along the electrode, as in continuous heating of a material the power factor of which decreases during the heating process. To feed constant power per unit volume into the material as it passes between the electrodes, electrode voltage must be increased, and a constant dielectric constant assumed. Figure 24·92 illustrates the voltage distribution along an electrode fed from one end.

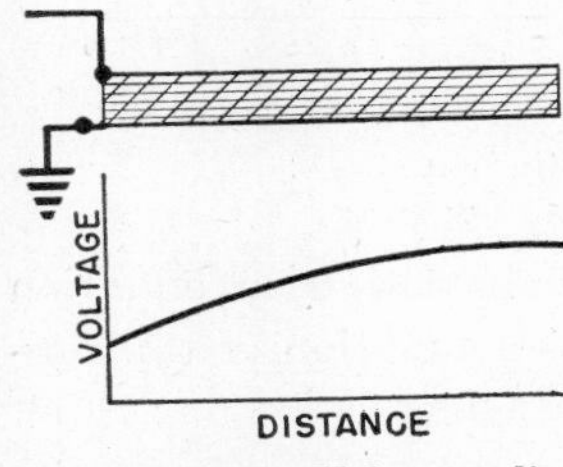

FIG. 24·92 Voltage distribution along narrow electrodes.

If a hollow dielectric cylinder is to be heated, the standing-wave phenomena previously discussed are not the only factors involved. The following equation defines voltage gradient in the space between the two electrodes:

$$E_r = \frac{E}{r \log_\epsilon \sigma}, \quad r_2 \geq r \geq r_1 \tag{24·129}$$

where E_r = voltage gradient in work at radius r in kilovolts per inch
E = electrode voltage in kilovolts
r = distance from axis in work in inches
σ = ratio of outside to inside radii, r_2 and r_1 inches respectively.

E_r decreases along an outward radial line and is independent of the dielectric constant, if the dielectric fills the space between the electrodes. For a required voltage gradient for a given power density, frequency, and loss factor, the necessary electrode voltage can be determined. Equation 24·129 indicates that voltage gradient changes from one cylinder to the other. This implies a difference in heating at the boundaries of the two cylinders and from point to point between the two cylinders in a radial direction. The ratio of the heating at the outer surface of the inner electrode to that at the inner surface of the outer electrode, neglecting thermal losses, is

$$\nu = \left(\frac{r_2}{r_1}\right)^2 \tag{24·130}$$

This equation indicates whether heating at the surface of the inner electrode will be excessive for a given electrode voltage, and how much reduction in electrode voltage is necessary.

The foregoing discussion need not be limited to the heating of concentric cylinders. For example, the difference in heating between two electrodes at a point where the electrodes make a bend of inside radius r_1 and outside radius r_2 can be determined, as Fig. 24·93 illustrates. The material

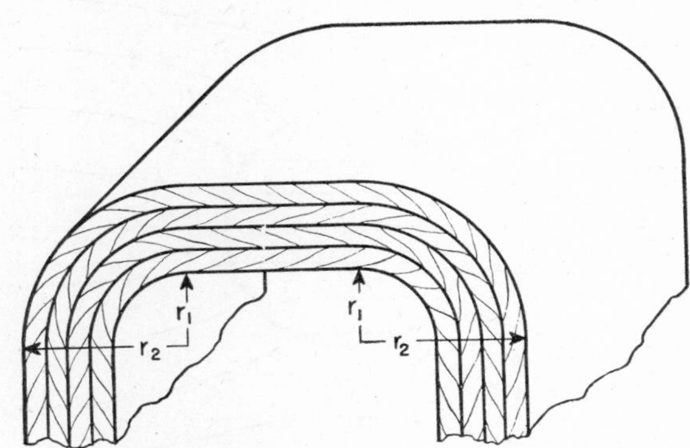

FIG. 24·93 Heating curved dielectrics.

between the electrodes is plywood. To get the required power density in the greater portion of the plywood, which lies between the flat uncurved portion of the electrodes, an electrode voltage of 3 kilovolts is required. Radii r_1 and r_2 are 2 and 3 inches respectively. Thus from equation 24·129 the voltage gradient at the surface of the electrode of radius r_1 is

$$E_r = \frac{3}{\log_\epsilon 1.5} = 3.7 \text{ kilovolts per inch}$$

and the voltage gradient at the surface of the electrode of radius r_2 is

$$E_r = \frac{3}{3 \times \log_\epsilon 1.5} = 2.48 \text{ kilovolts per inch}$$

The ratio of the heating rates can be determined by taking the square of the ratio of voltage gradients, or from equation 24·130:

$$\nu = \left(\frac{3}{2}\right)^2 = \left(\frac{3.7}{2.48}\right)^2 = 2.25$$

Thus the heating at the inner electrode is 2.25 times as great as that of the outer electrode and is

$$\left(\frac{3}{2.48}\right)^2 = 1.48$$

times as great as the heating to be expected between the flat uncurved portion of the electrodes. If r_1 and r_2 had been sharper, that is, the ratio of r_2 to r_1 greater, the difference of heating would have been even greater. If, however, the radii had been large, so that there was no appreciable curvature, then the formulas for plane electrodes would be applicable.

Sometimes a dielectric cylinder has dimensions that make it impossible to use electrodes at the ends. For such cylinders electrodes parallel to the axis may be used, if other conditions permit it. Such electrodes may be flat, but the heating is not uniform because of the non-uniformity of the voltage gradient.

If the heating is to be uniform, a uniform electric field must be established in the load. For this analysis the load is assumed to be a homogeneous isotropic imperfect dielectric cylinder, and the field is assumed not to be distorted by any conducting materials.

If a dielectric cylinder in a uniform electric field has its axis perpendicular to the electric field, the field is distorted

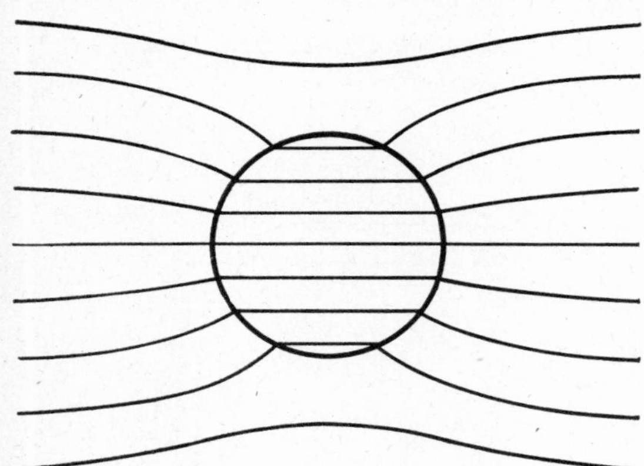

FIG. 24·94 Field distribution about cylindrical dielectric.

symmetrically, as indicated in Fig. 24·94. The field within the cylinder is uniform. (For proof of this see: Jeans, *Mathematical Theory of Electricity and Magnetism.*) To reproduce this condition the electrodes must occupy symmetrically equipotential planes.

Voltage gradient inside the dielectric cylinder is defined by

$$E_1 = \frac{2}{\epsilon + 1} E_0 \tag{24·131}$$

where E_1 = voltage gradient in the dielectric in kilovolts per inch
E_0 = voltage gradient in space before insertion of dielectric cylinder in kilovolts per inch
ϵ = dielectric constant of cylinder.

An equipotential plane for this particular case is defined by

$$E = E_0 \left(r - \frac{\epsilon - 1}{\epsilon + 1} \frac{b^2}{r}\right) \cos \theta \tag{24·132}$$

where E = electrode potential in kilovolts
E_0 = voltage gradient in space before insertion of dielectric cylinder in kilovolts per inch
r = radius vector with origin on the axis of the dielectric cylinder in inches
ϵ = dielectric constant of cylinder
b = radius of cylinder in inches
θ = angular displacement of radius vector and the direction of the field

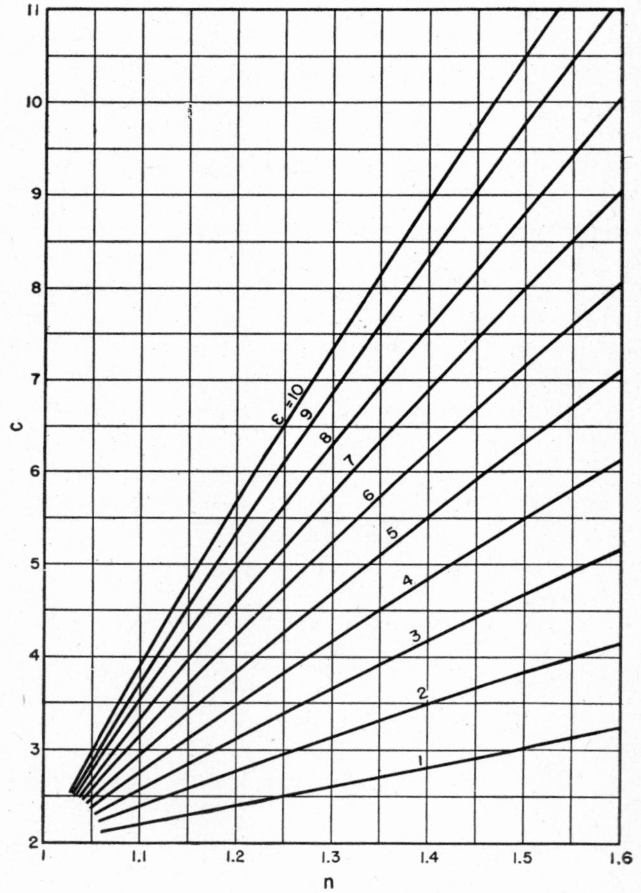

FIG. 24·95 Equipotential function curve.

From equations 24·131 and 24·132 can be derived an equation to define the shape of an electrode as a function of the conditions of the problem, neglecting end effects:

$$C = (\epsilon + 1)n = \frac{\epsilon - 1}{n} \tag{24·133}$$

where $C = (E/E_1)b \cos \theta$, the units of which have been previously defined

ϵ = dielectric constant of cylinder
n = ratio of radius vector r and the radius of the dielectric cylinder b.

Figure 24·95 shows a family of curves of C as a function of n for various dielectric constants from 1 to 10. These curves show a method of determining the shape of the electrodes.

If the voltage gradient in a solid cylinder is 1 kilovolt per inch (from equation 24·100), the radius and dielectric constant of the cylinder are 2.5 and 4 inches respectively. Electrodes must have a minimum space of ¼ inch from the load because of electrical and mechanical reasons. Thus $n = 1.1$. For $\theta = 0$ degrees, $C = 2.75$ (from Fig. 24·95). The constant C, however, is

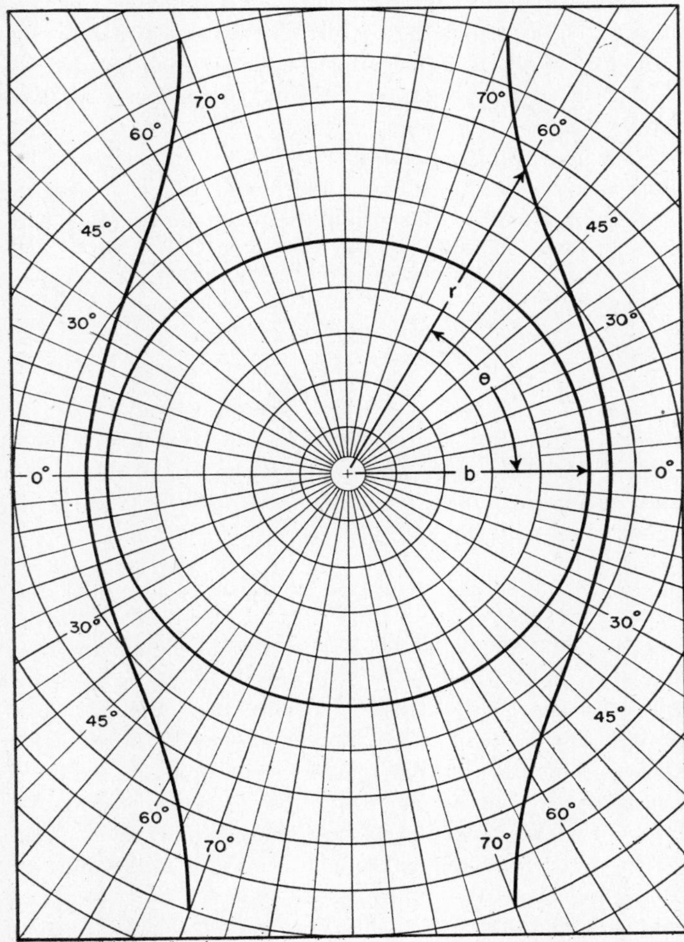

FIG. 24·96 Examples of electrodes for dielectric-heating cylindrical dielectric load.

$$C = \frac{E}{E_1 b \cos\theta}$$

When $\theta = 0$ degrees

$$C = \frac{E}{2.5}$$

therefore the electrode voltage becomes $E = 2.5 \times 2.75 = 6.88$ kilovolts. This gives a basis for determining the length of the radius vector $r = nb$ for various magnitudes of θ, so that an equipotential surface can be plotted.

The following tabulation gives r for values of θ equal to 0, 30, 45, 60, and 70 degrees. Those five angles are sufficient to determine practical electrode shapes.

θ	Cos θ	n	C	r
0	1.0	1.1	2.75	2.75
30	0.866	1.15	3.17	2.88
45	0.707	1.26	3.89	3.14
60	0.5	1.50	5.5	3.75
70	0.342	2.01	8.05	5.02

Figure 24·96 shows the actual shape of the electrodes. The use of polar-co-ordinate paper simplifies this work. The edges of the electrodes should be curled up to minimize flashover possibilities.

Many types of electrodes are not easily determined from a knowledge of electrical phenomena. Examples of such electrodes are shown in Fig. 24·97. Electrodes of similar shapes have been applied by numerous manufacturers. The principle is that the electric flux tends to take the path of least reluctance between electrodes. As denoted by the

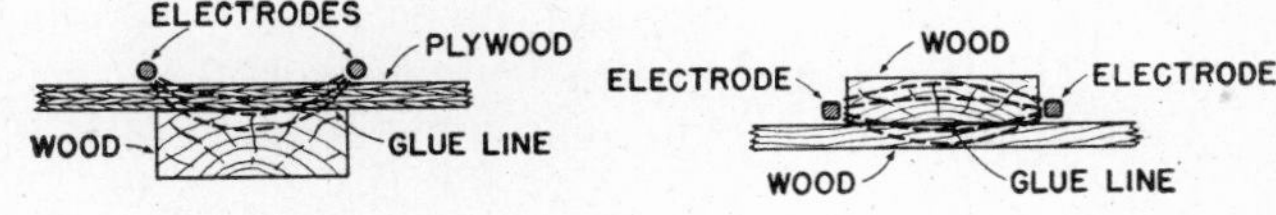

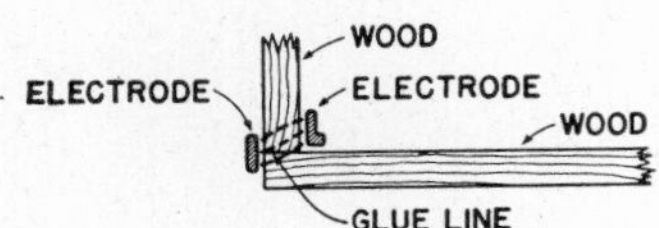

FIG. 24·97 Dielectric-heating electrodes.

dashed line in the examples shown in Fig. 24·97, the lines of flux may be expected to concentrate in the dielectric load. Determination of actual shape and position and voltage of a given electrode must be determined experimentally.

Transmission Lines [5, 13, 39]

The basic transmission-line equations are rewritten here for convenient reference:

$$E = E_L \cosh \gamma x + I_L Z_0 \sinh \gamma x \qquad (24\cdot 134)$$

$$I = I_L \cosh \gamma x + \frac{E}{Z_0} \sinh \gamma x \qquad (24\cdot 135)$$

where E = voltage across line at a distance x feet from the receiving end in volts
I = current through line at a distance x feet from the receiving end in amperes
E_L = receiving-end voltage in volts
I_L = receiving-end current in amperes
Z_0 = characteristic impedance of line in ohms
x = distance along line from receiving end in feet
γ = propagation constant, which is equal to the square root of the product of the impedance per foot and the admittance per foot. It is a complex quantity and is often written as $\gamma = \alpha + j\beta$, where α is the attenuation constant and β is the phase constant.

The real part of γ is the attenuation constant α, the function that determines the energy dissipated in the transmission line. The coaxial transmission line having minimum attenuation has a characteristic impedance of approximately 77 ohms.

The imaginary term $j\beta$ is the phase constant, which defines the velocity of propagation and wavelength along the transmission line:

$$\beta = \frac{2\pi}{\lambda} \quad \text{radians per unit length}$$

The approximate wavelength for lines is

$$\lambda = \frac{984}{f\sqrt{\epsilon}} \quad \text{feet} \tag{24·136}$$

where λ = wavelength along line in feet
f = frequency in megacycles per second
ϵ = dielectric constant of medium between the two transmission-line conductors.

Equation 24·136 applies to dielectric-heating electrodes to determine minimum frequency for given electrode dimensions, or to the minimum electrode dimension for a given frequency that provides the desired uniformity of heating.

The term

$$Z_0 = \sqrt{\frac{z}{y}} \tag{24·137}$$

where Z_0 = surge or characteristic impedance of the line in ohms
z = impedance per foot of the line in ohms
y = admittance per foot of the line in ohms.

In well-designed lines the attenuation constant α may be neglected in comparison with the phase constant β. By neglecting α, equation 24·137 may be reduced to

$$Z_0 = \sqrt{\frac{L}{C}} \quad \text{ohms} \tag{24·138}$$

where Z_0 = surge or characteristic impedance of the line in ohms
L = inductance per foot of line in henrys
C = capacitance per foot of line in farads.

A transmission line need not always be matched at the receiving end for the required transfer of power. If the voltage difference between the sending and receiving ends is not too great, and the line is less than a quarter wavelength long, a match of line and load may not be necessary. Usually a matching network is used for lines over an eighth wavelength long. Reflected waves can be eliminated by terminating the line in a resistance equal to the characteristic impedance Z_0 of the line. If the transmission line is properly terminated, the impedance at any point along the line therefore is that of a pure resistance equal to Z_0.

If the terminating end is mismatched, reflected waves, called *standing waves*, appear. Voltage and current at any point along the line become the sum of the principal and reflected waves of voltage and current. Excessive voltages and currents may result from a serious mismatch on lines a quarter wavelength long or greater.

Lines can be properly matched by use of a radio-frequency bridge. A perfect match throughout a heating cycle is rather rare, though, so transmission lines usually are designed to withstand abnormal voltages and currents.

At times standing waves are desirable, especially if transmission-line sections are used as impedances in networks. This point was mentioned in the discussion of stubs as network components.

The transmission-line equations indicate that high voltages and currents can be built up along a transmission line by proper termination of it. Excessive voltages may cause radiation, which may interfere with communication services, and may be objectionable by introducing an extraneous loss of energy. Radiating lines must be shielded as required by law. This trouble is more prevalent with open-wire lines than with concentric lines.

Reactances of open- and short-circuited transmission line sections or stubs now can be determined. If the receiving end is short-circuited the terminating voltage becomes zero. If the length of the line is l feet,

$$\frac{E}{I} = Z_{ss} = Z_0 \tanh(\alpha l + j\beta l) \tag{24.139}$$

$$= jZ_0 \tan \beta l \text{ ohms}, \quad \alpha \ll \beta$$

where Z_{ss} is the impedance at the sending end of the line of length l, α is neglected, and $\tanh j\beta l = j \tan \beta l$. A similar result is obtained for an open-circuited line. Terminating current is zero. Then

$$Z_{0s} = jZ_0 \cot \beta l \text{ ohms} \tag{24·140}$$

Figures 24·98 and 24·99 illustrate the change in input impedance as a function of the argument βl in electrical degrees for short-circuited line elements as well as for open-circuited line elements. By proper choice of the electrical length of the stub, the stub reactance can be made inductive or capacitive, whether the stub is open- or short-circuited. These curves are not valid as ratio of stub reactance to stub characteristic impedance approaches plus or minus infinity. For low-loss stubs they indicate the expected reactance.

For a given high Q a shunting inductive reactance of 55 ohms at 30 megacycles is required. Assume the characteristic impedance of the line to be 70 ohms. Then the stub reactance-to-characteristic-impedance ratio is 0.79. Figure 24·98 indicates that the electrical length of the stub for this ratio is 38 degrees. Converted to feet, the stub length becomes approximately 3.5 feet.

The curves of Fig. 24·99 can be used in the same manner to determine physical dimensions of open-circuited stubs.

Stubs, as well as transmission lines, must have adequate insulation and current-carrying capacity, for standing waves on stubs shift because of a changing load and oscillator frequency.

If equations 24·134 and 24·135 are rearranged to define the input impedance of the line as a function of the line properties and load impedance,

$$Z_s = Z_0 \frac{Z + Z_0 \tanh \gamma l}{Z_0 + Z \tanh \gamma l} \tag{24·141}$$

where Z_s = input impedance of line in ohms
Z_0 = characteristic impedance of line in ohms
Z = load impedance in ohms
γ = propagation constant in radians per foot
l = length of line in feet.

This equation implies a definite impedance transformation from the receiving end of the line to the sending or input end. It is used in the so-called coaxial line transformer. (Open-wire-line transformers may be used; however, the coaxial-line transformers have the advantage of minimizing radiation.) If the attenuation constant of the transmission line is small and can be neglected in comparison with the phase constant, equation 24·141 may be rewritten:

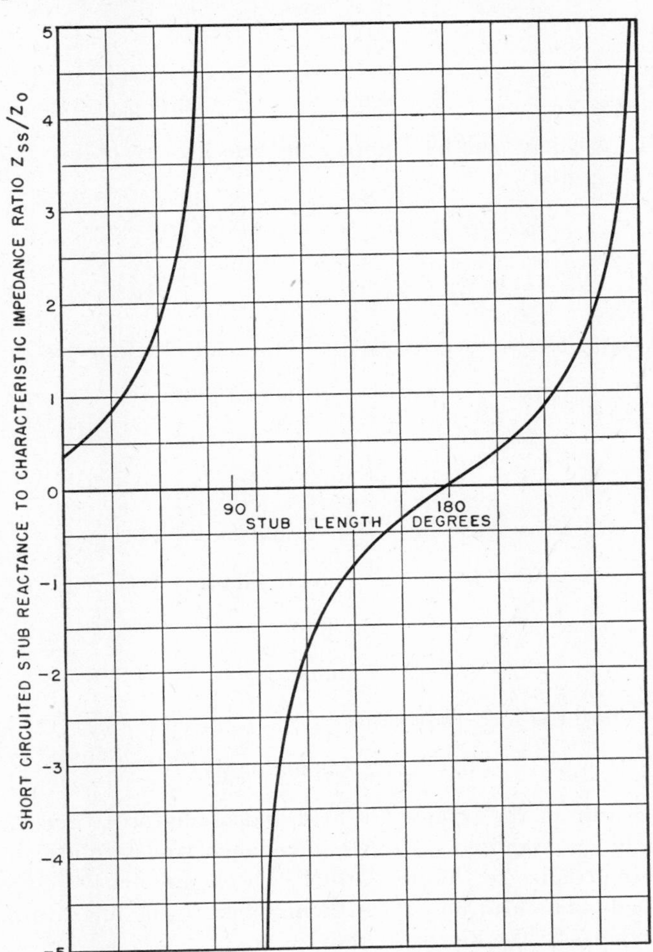

FIG. 24·98 Impedance of short-circuited transmission-line element.

$$Z_s = Z_0 \frac{ZZ_0(1 + \tan^2 \beta l) + j(Z_0^2 - Z^2) \tan \beta l}{Z_0^2 + Z^2 \tan^2 \beta l} \qquad (24\cdot142)$$

or

$$Z_s = Z_0 \frac{ZZ_0(1 + \cot^2 \beta l) + j(Z_0^2 - Z^2) \cot \beta l}{Z^2 + Z_0^2 \cot \beta l} \qquad (24\cdot143)$$

One of the simplest transformers is the quarter-wave transformer. For a particular length of transmission line and frequency, $\beta l = \pi/2$. Thus

$$Z_s = (R_s - jX_1)\left(\frac{Z_0}{Z_1}\right)^2 \qquad (24\cdot144)$$

where Z_s = sending-end impedance in ohms
Z_0 = characteristic impedance of the transmission line in ohms
Z_1 = absolute impedance of receiving end; also equal absolute impedance of load
R_s = series-resistive component of load impedance
X_1 = series-reactive component of load impedance. Negative algebraic sign indicates capacitive reactance, positive algebraic sign indicates inductive reactance.

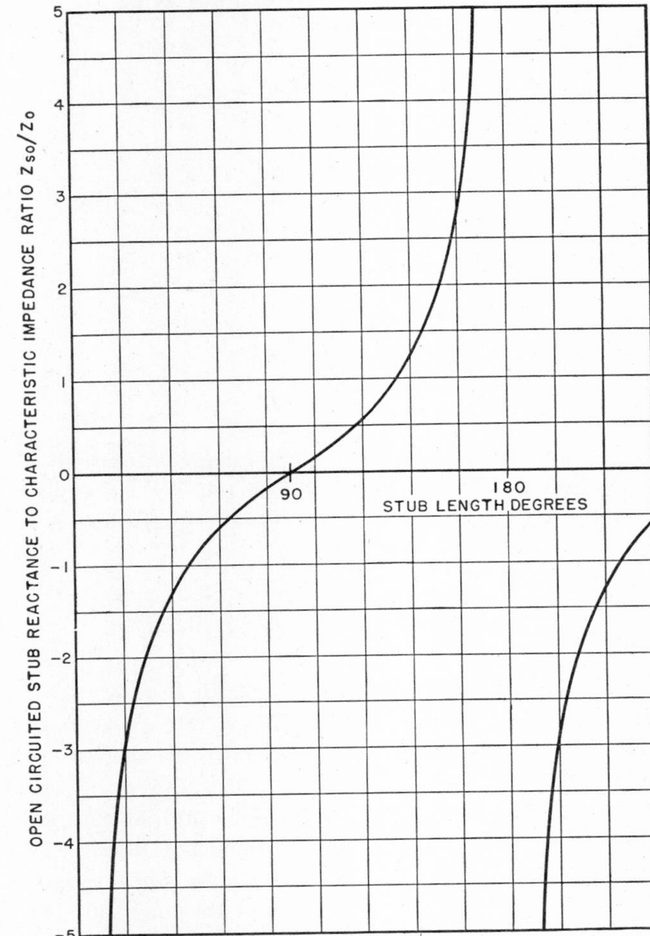

FIG. 24·99 Impedance of open-circuited transmission-line element.

The transformation does not correct for power factor, but if the load is inductive the sending-end impedance appears to be capacitive, and vice versa. Operation of such a transformer depends upon frequency of the generator; therefore any frequency change caused by the load must be corrected.

24·8 EXAMPLES OF DIELECTRIC HEATING

Three applications of dielectric heating are given here to emphasize the methods used to solve fundamental problems in dielectric heating. Factors other than those given may be imposed by a particular application; for example, limitations in heating a given product or mechanical limitations involving work-handling equipment.

Plastic Pre-Forms

A manufacturer of molded-plastic products desires to heat plastic pre-forms before molding. The production rate per press is four pre-forms every 30 seconds, and the required temperature rise is from 70 to 250 degrees Fahrenheit. Each pre-form has its own electrode assembly. The pre-form is 4 inches in diameter and 1 inch thick. The dielectric constant of the pre-form material is 7.0, and its power factor is 0.06. Weight per cubic inch of the plastic material is 0.05 pound, and its specific heat is 0.5. From equation 24·104 the required thermal power is

$$P_1 = 17.6Mc\,\Delta t \text{ watts}$$

$$M = \text{pounds per minute} = \frac{\text{volume} \times \text{density}}{\text{minute}}$$

$$= \frac{4 \times \pi \times 2^2 \times 1 \times 0.05}{0.5}$$

$$= 5.02 \text{ pounds per minute}$$

$$\Delta t = 250° - 70°$$

$$= 180°\text{F}$$

$$c = 0.5$$

$$P_1 = 17.6 \times 5.02 \times 0.5 \times 180$$

$$= 7900 \text{ watts}$$

Radiation loss for a black body at 250 degrees Fahrenheit is about 1 watt per square inch. If the relative emissivity is about 0.8, the radiation loss in watts per square inch would be 0.8. The total area is

$$4\pi\left[\frac{4^2}{4} \times 2 + 4\right] = 151 \text{ square inches}$$

Therefore the total radiated power at 250 degrees Fahrenheit is $P_2 = 151 \times 0.8 = 120$ watts. Convection loss at 250 degrees Fahrenheit is about 0.5 watt per square inch; therefore the convection loss $P_3 = 151 \times 0.5 = 75$ watts. Thermal loss is taken as the average of the sum of the radiation and convection losses between 70 and 250 degrees Fahrenheit and is $\frac{1}{2}(120 + 75) = 92.5$ watts, since radiation and convection losses at 70 degrees Fahrenheit are zero. Generator power, neglecting line and network losses, becomes $P = 7900 + 92.5 = 8000$ watts approximately, and the required thermal-power density becomes

$$P_v = \frac{8000}{50.2} \cong 160 \text{ watts per cubic inch}$$

Assume that an electrode voltage of 10 kilovolts can be safely obtained. The required frequency, from equation 24·107, is

$$f = \frac{0.71P_v}{E_1^2\epsilon''}$$

$$= \frac{0.71 \times 160}{100 \times 0.42}$$

$$= 2.65 \text{ megacycles per second}$$

and

$$E_1 = \tfrac{10}{1} = 10$$

$$= 10 \text{ kilovolts per inch}$$

A standard 5-megacycle 10-kilowatt oscillator is selected. Because electrode dimensions are small compared with one-sixteenth wavelength (equation 24·136), $\lambda = 75.4$ feet. This frequency is satisfactory for the size and shape of the load. With this choice of frequency, the required voltage gradient, from equation 24·100, is

$$E_1 = \frac{0.71P_v}{f\epsilon''}$$

$$= \frac{0.71 \times 160}{5 \times 0.42}$$

$$= 5.3 \text{ kilovolts per inch}$$

To calculate the load impedance the load reactance must be computed:

$$C = 0.224 \times 7 \times \frac{4\pi}{4} \times 4^2 \times 10^{-12}$$

$$= 79 \times 10^{-12} \text{ farad}$$

for four electrodes in parallel. The load reactance is then

$$X = -\frac{10^{12}}{2\pi \times 5 \times 79 \times 10^6}$$

$$= -403 \text{ ohms}$$

Effective series resistance, from equation 24·114, is

$$R_s = |X| \text{ (power factor)}$$

$$= 403 \times 0.06$$

$$= 24.2 \text{ ohms}$$

Therefore the load impedance is

$$Z = 24.2 - j403 \text{ ohms}$$

Because of the position of load and oscillator a transmission line is required. Proper impedance transformation of line impedance to the oscillator has been previously accomplished. The load now must be matched to the line. Surge impedance of the line is 70 ohms $= R_G$.

A T network is used with equations 24·121 and 24·122 to determine network components. The known values of reactance and resistance are: $R_G = 70$ ohms; $R_s = 24.2$ ohms; $X_2 = -403$ ohms; $-R_GR_s = X_1X_2 + X_1X_3 + X_2X_3$; and

$$\frac{R_G}{R_s} = \frac{X_1 + X_3}{X_2 + X_3}$$

By rearranging equations 24·121 and 24·122,

$$X_1^2 = R_G(R_s - R_G) + \frac{R_GX_2^2}{R_s}$$

$$X_1 = -684 \text{ ohms}, \quad X_3 = +250$$

$$X_1 = +684 \text{ ohms}, \quad X_3 = +972$$

Therefore the network may be either of two combinations. To minimize network I^2R losses the first combination is used ($X_1 = 684$ ohms with a load-shunting inductance having a reactance of 250 ohms).

Plywood Bonding

To increase production rate and improve the quality of his product, a manufacturer of plywood cabinets uses dielectric heating. He wants to bond plywood, in a specific shape indicated in Fig. 24·100.

The wood consists of five layers of density 0.0137 pound per cubic inch. Specific heat of the plywood is 0.378 calorie per gram per degree; dielectric constant and power factor are 3.2 and 0.04 respectively. A thermosetting glue which sets in 3 minutes at 250 degrees Fahrenheit is used. Production rate is to be 50 pieces per 8-hour day. If 8 minutes is chosen as the handling time plus the heating time to produce one plywood form, the required production rate can be achieved easily. Let the time to increase load temperature from room temperature (70 degrees Fahrenheit) to 250 degrees Fahrenheit final temperature be 3 minutes. Immediately after the load has been heated to 250 degrees, the oscillator can be switched to another press while the load previously heated is held at the required temperature for 3 minutes longer by the heated press.

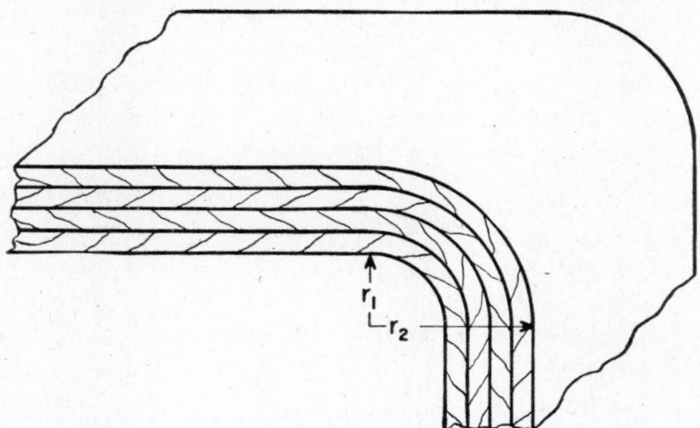

Fig. 24·100 Bonding plywood.

Required thermal power, from equation 24·104, is

$$P_1 = 17.6Mc\,\Delta t \text{ watts}$$

$$M = \text{volume} \times \frac{\text{density}}{\text{time}} = \text{pounds per minute}$$

$$= 530 \times \frac{0.0137}{3} = 2.43 \text{ pounds per minute}$$

$$c = 0.378$$

$$\Delta t = 250 - 70 = 180$$

$$P_1 = 17.6 \times 1.93 \times 0.378 \times 180$$

$$= 2900 \text{ watts}$$

Since the press is heated, the only thermal loss is in radiation and convection from the exposed edges, which have a total area of 73 square inches. If the emissivity of the wood is 0.8, radiation and convection losses per square inch are about 1.3 watts. Total thermal loss becomes 73 × 1.3 = 95 watts. Total generator power is approximately 3 kilowatts.

The load is heated with electrodes conforming to the shape of the load. Most of the heating is between the flat portions of the electrodes; therefore, average power density becomes

$$P_v = \tfrac{3000}{530} = 5.7 \text{ watts per cubic inch}$$

An electrode voltage of 2 kilovolts, or a voltage gradient of 4 kilovolts per inch, is allowable. Then the required frequency, from equation 24·107, is

$$f = \frac{0.71P_v}{E^2\epsilon''}$$

$$= 2 \text{ megacycles per second}$$

This frequency falls within a practical range. To reduce electrode voltage further, the frequency of the generator selected is 5 megacycles, a standard frequency. Electrode voltage becomes 1.25 kilovolts. This reduction of electrode voltage minimizes electrode insulation. The electrode dimensions are not excessive relative to the wavelength at 5 megacycles.

There is a difference in heating at the inner and outer radii of the electrodes at the bend in the piece. From equation 24·130 this difference is

$$\nu = \left(\frac{2}{1.5}\right)^2 = 1.77$$

times. To compare the heating at the bend to the heating under the flat, uncurved portion of the electrodes, the various voltage gradients must be known. From equation 24·129 the gradient at radius $r = 1.5$ when the electrode voltage is 1.25 kilovolts is

$$E = \frac{1.25}{1.5 \log_\epsilon 1.33}$$

$$= 2.95 \text{ kilovolts per inch}$$

Voltage gradient at radius $r = 2$ becomes

$$E_r = \frac{1.25}{2 \times \log_\epsilon 1.33}$$

$$= 2.22 \text{ kilovolts per inch}$$

The ratio of the square of the voltage gradients indicates relative heating. Therefore the ratio of the square of the voltage gradient at $r = 1.5$ to the square of the voltage gradient between the flat, uncurved portions of the electrodes gives one of the desired comparisons:

$$\nu = \left(\frac{2.95}{2.5}\right)^2 = 1.39$$

and as another comparison

$$\nu = \left(\frac{2.22}{2.5}\right)^2 = 0.79 \quad \text{for } r = 2$$

These ratios indicate 39 percent greater heating at the surface of the bend of the electrode having the smaller radius at the bend than the heating between the flat portions of the electrodes. Average heating in the bend, however, is almost equal to that throughout the remaining portion of the load. Because of this, and because the load is relatively thin, this difference in heating at the bend can be neglected.

Rayon Drying

A textile mill usually receives rayon in cakes at about 100 percent moisture content (dry-weight basis). This moisture content must be reduced to about 10 percent. Figures 24·101 and 24·102 show the dielectric properties of rayon at 200 degrees Fahrenheit as a function of moisture content.

As an example, a textile manufacturer desires to dry 50 1-pound cakes (dry weight) per hour. The initial and

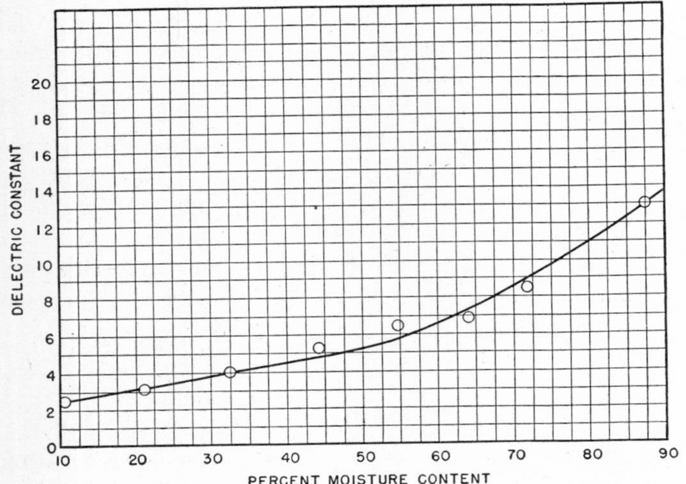

Fig. 24·101 Dielectric constant of rayon versus water content.

final moisture contents are to be 100 percent and 10 percent respectively. The total volume of rayon cakes heated per hour is to be 5000 cubic inches. Temperature rise is from 70 to 220 degrees Fahrenheit, or a 150-degree temperature increase. Two power components must be supplied by the radio-frequency power source. One of these is the power to heat the cakes from 70 to 220 degrees Fahrenheit. Another power requirement is that needed to evaporate moisture. For this particular arrangement about 5 percent of the total moisture is removed by the hot air. Thus 85 percent of the

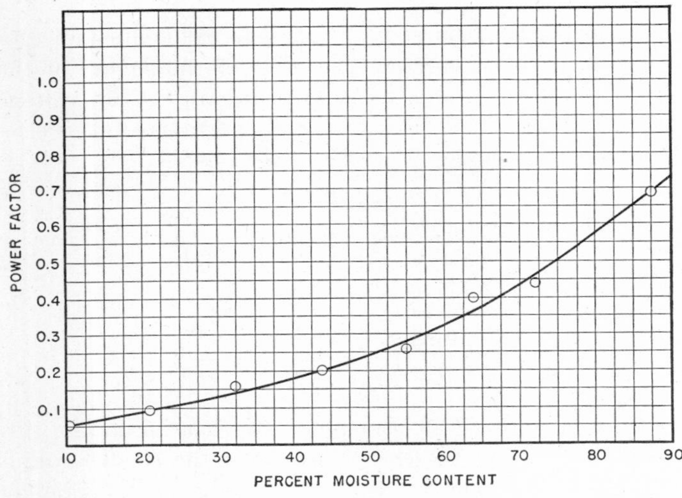

Fig. 24·102 Power factor of rayon versus water content.

moisture remains to be evaporated by radio-frequency energy.

Power required to heat the rayon from 70 to 220 degrees Fahrenheit can be further divided into that power necessary to heat the dry rayon and that power necessary to heat the moisture. The water is assumed to evaporate at 220 degrees Fahrenheit, 8 degrees Fahrenheit above boiling point for normal pressure because the recirculated hot air is under some pressure.

To minimize vapor pressure about the rayon cakes, hot air is recirculated at the rate of 2000 cubic feet per minute, and to dispose of the evaporated moisture (45 pounds per hour) hot air is exhausted at the rate of approximately 100 cubic feet per minute. This exhausted moist air must be replenished with cool dry air, which must be heated. There are also thermal losses through the dryer cabinet.

These power components can be supplied by a hot-air heater; the remainder, the power needed to heat the rayon from 70 to 220 degrees Fahrenheit and to vaporize 85 percent of the water, must be supplied by the generator. The power to heat the 50 cakes of rayon from 70 to 220 degrees Fahrenheit in 1 hour becomes P'_1, where

$$P'_1 = 17.6 \times \tfrac{50}{60} \times 150(1.00 + 0.35) = 3000 \text{ watts}$$

since the specific heat of the dry rayon is about 0.35 calorie per gram. The power required to vaporize 42.5 pounds (85 percent) of water per hour is

$$P''_1 = 17.6 \times \frac{42.5}{60} \times 970 = 12.000 \text{ watts}$$

Then the total power required becomes $P = P'_1 + P''_1 = 15{,}000$ watts.

The characteristic curves of the rayon shown in Figs. 24·101 and 24·102 indicate that the average load is relatively low, and consequently the line and network losses may be expected to be low unless the added air space in the electrodes increases the load reactance substantially; therefore a 20-kilowatt oscillator is recommended. As an approximation, a frequency of 10 megacycles per second is chosen; the necessary voltage gradient then becomes

$$E_1 = \left[\frac{0.71 \times P_v}{f\epsilon''}\right]^{1/2} \text{kilovolts}$$

Power density becomes

$$P_v = \frac{15{,}000}{5000} = 3.0 \text{ watts per inch}^3$$

where 50 rayon cakes are beneath the electrode at any instant during the drying process.

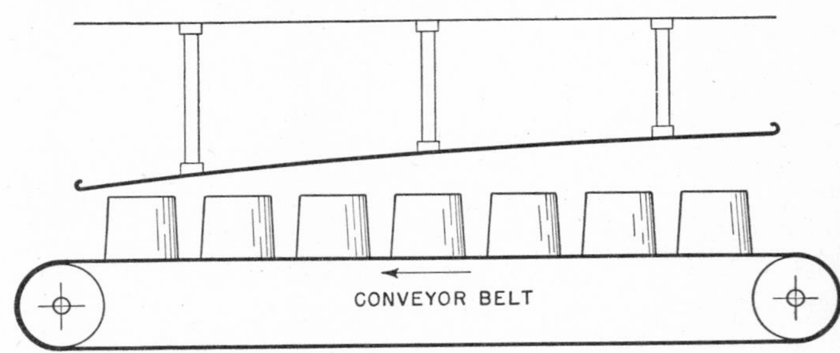

Fig. 24·103 Electrode shape for rayon heating.

Voltage gradient in the load beneath the electrode now can be determined, assuming constant power input, and assuming that the loss factor varies from one end of the electrode to the other as indicated in Figs. 24·101 and 24·102. Assuming an electrode voltage of 10 kilovolts, electrode shape and spacing are shown in Fig. 24·103.

The mean space between the work and the top electrode is 3 inches; therefore load reactance and power factor (or Q) can be determined approximately. These are such that load impedance is $Z = 2 - j108$ ohms, approximately. See Fig. 24·104 for this equivalent impedance.

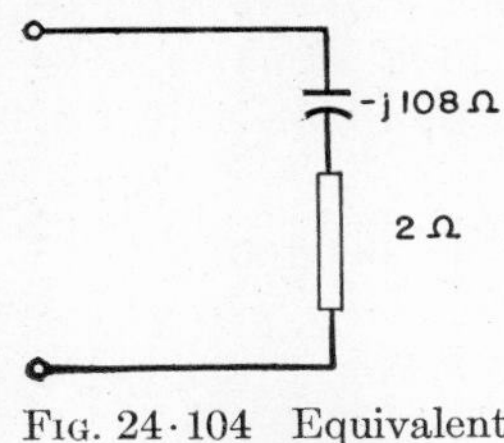

FIG. 24·104 Equivalent load impedance.

Average Q of the material only is 3.4, and the Q of the load circuit is 54. This high-load Q implies the use of high-Q reactance in a matching network. The required load current is $I = 90$ amperes (rms).

This drying process seems feasible with a 20-kilowatt 10-megacycle generator.

24·9 ECONOMICS OF HIGH-FREQUENCY HEATING

Induction heating is one of the more expensive methods of producing heat energy, but its high efficiency frequently affects the relatively high cost of the energy. In addition, it may not be possible to heat by conventional means a work piece that can be heated selectively by induction.

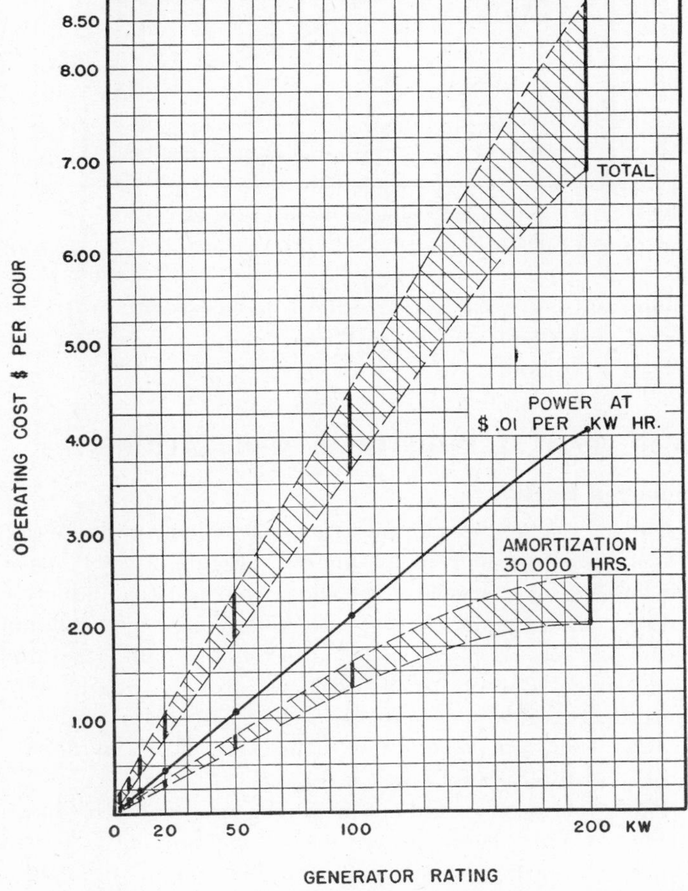

FIG. 24·105 Operating cost versus generator rating.

Electrical non-conductors usually are poor conductors of heat, and thick pieces are difficult to heat by conventional means. With dielectric methods, the heat is generated uniformly throughout the piece, if the piece is homogeneous. In some molding processes using thermosetting materials, rapid pre-heating is necessary, and dielectric heating is the only means of accomplishing it.

As the technical analysis of new applications often involves much time and expense, it is advisable to investigate the economics of the high-frequency application beforehand. The results of this investigation may be used as a guide to the maximum utilization of given facilities.

The curves of Fig. 24·105 show the operating costs in dollars per hour for various standard electronic generators. Total operating costs are broken down into several components, two of which are shown. Total cost and amortization cost are shown in bands of approximately 25 percent to allow for costs of accessories such as work-handling fixtures and mechanisms for normal applications. Input-energy cost is shown at $0.01 per kilowatt-hour; therefore, if the actual cost of energy differs from this, a correction can be made. If a generator must be operated at reduced load, the necessary correction can be made by subtracting from the total operating cost per hour the corresponding reduction of loading (energy cost), then adding 10 percent of the original energy cost to take care of the stand-by power consumption of the generator. For example, suppose a 200-kilowatt generator is to be used at 100-kilowatt output. Assuming a full-load operating cost of $6.90 per hour,

$$\$6.90 - (0.5 \times 4.07) + (0.1 \times 4.07) = \$5.28 \text{ per hour}$$

This is a fair approximation of the cost of operation between 50 and 90 percent of full load.

The curves of Fig. 24·105 are based on an assumed operational factor of 100 percent, which is seldom realized. Some of the other curves show the effects of various operational factors on cost. The curves of Fig. 24·105 represent operating costs of induction-heating generators of 5-, 10-, 20-, 50-, 100-, and 200-kilowatt ratings. These curves are also good for dielectric-heating generators of 2-, 5-, 10-, 20-, and 50-kilowatt ratings only. Operating cost for a 100-kilowatt dielectric-heating generator is slightly higher than that indicated on the curve for the 100-kilowatt induction-heating unit, and costs of work-handling equipment probably vary over wider limits than those indicated on the curve.

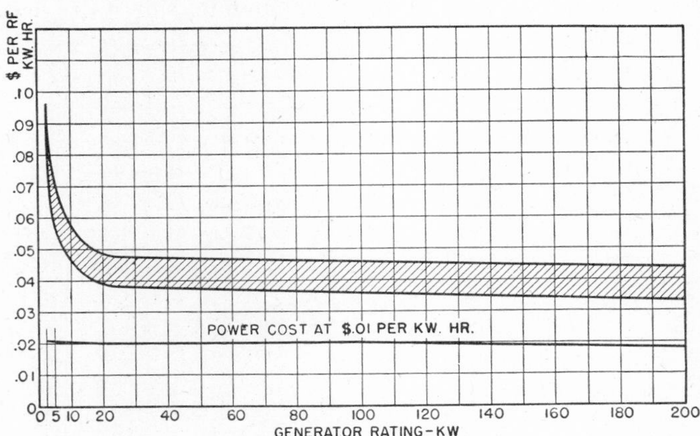

FIG. 24·106 Cost per kilowatt-hour versus generator rating.

Figure 24·106 shows costs per kilowatt-hour for various generator ratings. The component of input energy at $0.01 per kilowatt-hour is shown so that corrections can be applied if actual costs differ from the $0.01 rate. Cost per radio-frequency kilowatt-hour does not change appreciably for ratings above 20 kilowatts for a cost rate of $0.01 per kilowatt-hour. Usually ratings of 100 kilowatt or more are installed where energy can be purchased at rates lower than

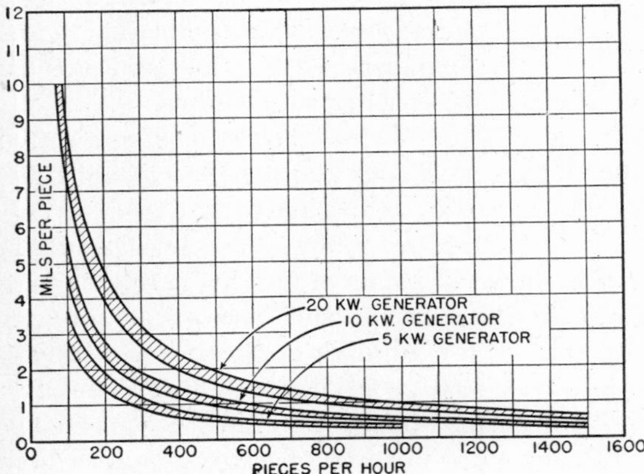

FIG. 24·107 Cost per piece versus pieces per hour.

$0.01 per kilowatt-hour; therefore, after applying a correction for a lower energy cost, the total operating cost of the larger generator is less than the values shown in Fig. 24·106.

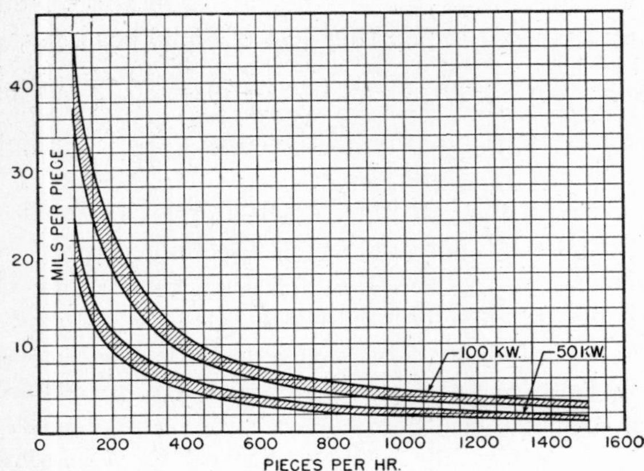

FIG. 24·108 Cost per piece versus pieces per hour.

Figures 24·107 and 24·108 show costs per piece versus pieces per hour for various sizes of generators from 5 to 100 kilowatts. After heating time and handling time have been

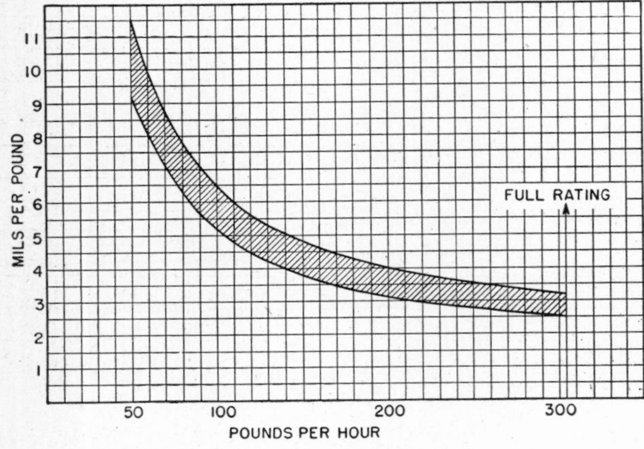

FIG. 24·109 Cost per pound versus pounds per hour.

determined, the minimum cost per piece can be read from the curves for the maximum production rate, or at 100 percent operational factor. The effect of any operational factor on the cost per piece can be seen from these curves.

Figure 24·109 shows the cost per pound of heating steel from 50 to 1400 degrees Fahrenheit with a 20-kilowatt generator, when the process is continuous. The effect of fractional production rates on the cost per pound can be determined from this curve.

Figure 24·110 shows the variation of cost per ton of flowing electrolytic tin-plated steel strip of a given size versus production rate, with a 600-kilowatt generator. The three

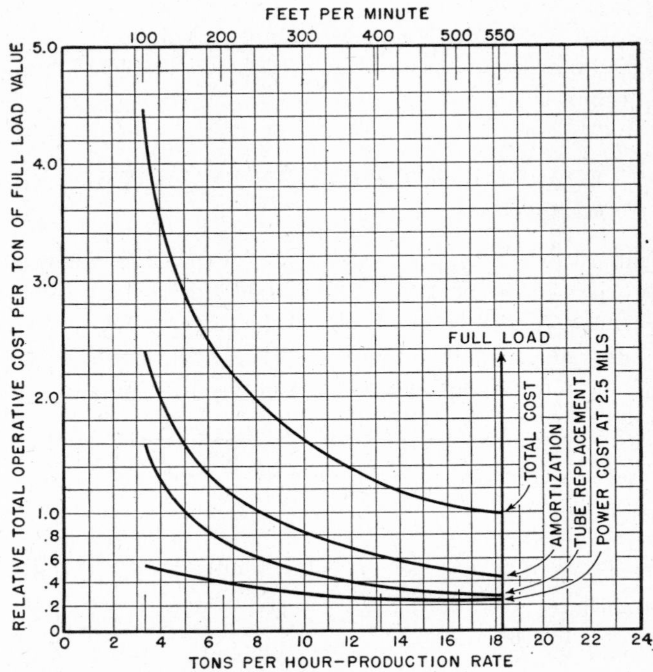

FIG. 24·110 Relative operative cost per ton versus tons per hour.

components of total cost are shown as amortization (36,000 hours), tube replacement cost, and energy cost at 2.5 mils per kilowatt-hour.

24·10 TYPICAL INDUSTRIAL APPLICATIONS

Induction Heating

One of the earliest applications of industrial induction heating was the melting of metals. Figure 24·111 shows the pouring of a special steel alloy from a 1-ton induction furnace. Furnaces of this type are widely used in the manufacture of special alloys such as nickel-chromium steel and some silver alloys. Such furnaces provide rapid, well-controlled, highly efficient heating. They also minimize product contamination. A close-up view of an induction-furnace installation is shown in Fig. 24·112.

The application of induction heating to forging is shown in Fig. 24·113. Here bars are heated inductively and then transferred to the nearby forging machine for the next operation. Figure 24·114 shows a similar induction furnace for heating 75-millimeter gun tubes.

A third application of induction heating is in soldering, brazing, and welding. This joining process is not necessarily limited to either high- or low-frequency heating. Figure 24·115 illustrates an application of low-frequency brazing. Here a current-boosting transformer feeds a single-turn inductor. The frequency is within the range of rotating machines. Another typical application is the brazing of capacitor cans. This is generally accomplished at a frequency of several hundred kilocycles per second. An example is shown in Fig. 24·116, where a single-turn hairpin heater coil is used. Radio-frequency current is secured by means of an air-core current transformer. To minimize lead impedance, flat-bus-type leads are used. Another brazing application is shown in Fig. 24·117, where brass flanges are brazed to the ends of a brass tube. As in Fig. 24·116, radio-frequency currents are produced by an air-core current transformer, which is directly in front of a generator.

Fig. 24·111 Induction melting of special steel alloy. (Courtesy of Ajax Electrothermic Corporation.)

Fig. 24·112 Induction-heating furnace installation. (Courtesy of Ajax Electrothermic Corporation.)

Zonal heating was a result of the development of induction heating. Low-frequency heating became the first employed for heat treating. Now the applications of machine-frequency heating are numerous and varied. One such application is the continuous heat treatment of bar stock. Intricate mechanisms handle the material being treated, as shown in Fig. 24·118. Even more interesting is the hardening of the bearing surfaces of crankshafts. This application requires an intricate machine with complex timing mechanisms. An operator may be seen inserting a crankshaft into such a machine in Fig. 24·119. Machine frequencies usually are used if deep casehardening is desired. An example is

Fig. 24·113 Induction-heating forge shop. (Courtesy of Ajax Electrothermic Corporation.)

Fig. 24·114 Induction-heating gun tube with 9600 cycles. (Courtesy of Ohio Crankshaft Company.)

Fig. 24·115 Low-frequency induction brazing. (Courtesy of Ajax Electrothermic Corporation.)

Fig. 24·116 High-frequency induction soldering of capacitor cans.

Fig. 24·117 Brazing flanges to ends of tube.

Fig. 24·118 Continuous heat treating of bar stock with rotating machine frequencies. (Courtesy of Ohio Crankshaft Company.)

Fig. 24·119 Crankshaft hardening with rotating machine frequencies. (Courtesy of Ohio Crankshaft Company.)

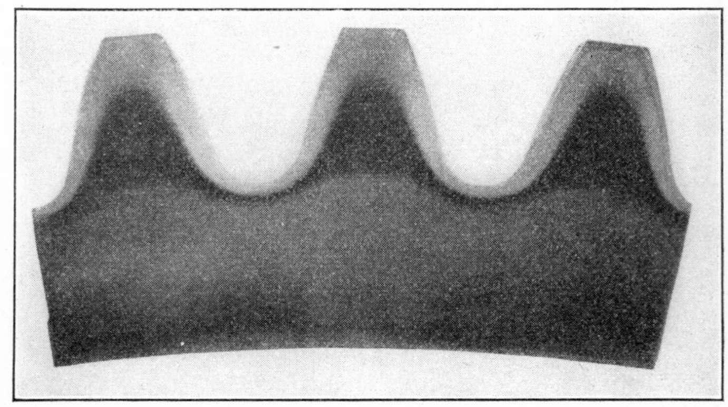

Fig. 24·120 A 30-inch-diameter, 6-inch-face, 1½-pitch sprocket contour-hardened using 9600 cycles. (Courtesy of Ohio Crankshaft Company.)

the contour-hardened sprocket of Fig. 24·120. Contour hardening is not limited to low frequencies. Figure 24·121 illustrates a gear contour-hardened at high frequency. The relation between case depth and frequency is not so simple as the depth-of-penetration formulas (equations 24·8 and

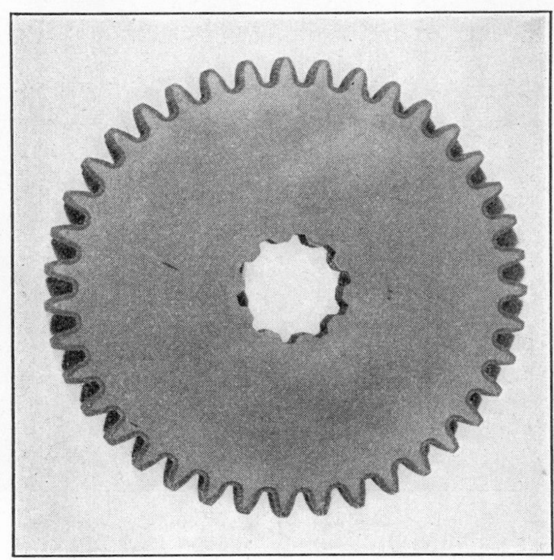

Fig. 24·121 Contour hardening of 4-inch-diameter spur gear with 450 kilocycles. Such results obtained if heating time is relatively short. Heat flow by conduction must be minimized. (Courtesy of Ohio Crankshaft Company.)

24·9) indicate. This relation is complicated, being a function of the electrical, physical, and thermal properties of the material as well as a function of heating time. Selection of the proper frequency is based upon experience combined with the depth-of-current-penetration equation and the dimensions of the piece to be heated. Another application is hardening wrist pins. A typical hardened wrist pin is shown in Fig. 24·122, where both transverse and longitudinal sections are shown. Work pieces of small cross section usually are best heated at higher frequencies.

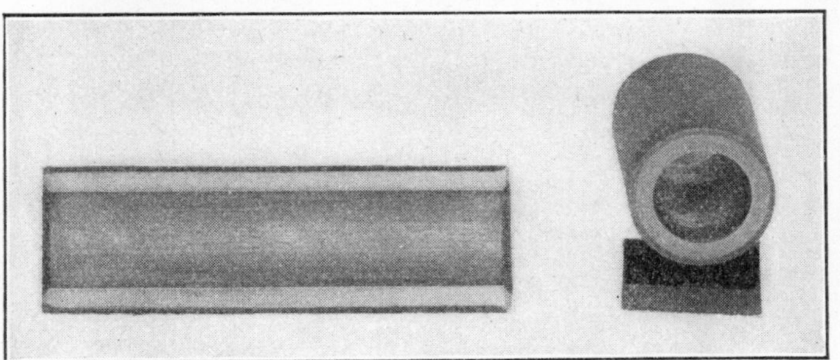

Fig. 24·122 Hardened wrist pins.

A common production-line application is the annealing of the ends of pins used as chain links. Figure 24·123 illustrates the mechanism for handling and processing these pins, and Fig. 24·124 illustrates the induction heating coils and their associated adjustment mechanisms. Here both ends of the pins are treated simultaneously.

Another application is tin fusion. Some tin lines fuse electro-plated tin on steel strip 30 inches wide and 10 mils thick at the rate of 1000 feet per minute. One typical tin line and heater coil are shown in Fig. 24·125.

Dielectric Heating

One of the most common dielectric-heating applications is the gluing or bonding of plywood. Plies of various shapes

Fig. 24·123 Annealing link pins. (Courtesy of Whitney Chain Company.)

and sizes can be heated quickly for bonding with thermosetting glues with few handling problems. Two differently shaped plywood loads are shown in Figs. 24·126 and 24·127. Figure 24·126 shows simultaneous bonding and forming of a plywood board of unusual shape. A simple pre-bonded plywood board is being placed in the press shown in Fig. 24·127.

Fig. 24·124 Annealing link pins. (Courtesy of Whitney Chain Company.)

Another popular use is heating plastic pre-forms. A good example of this is shown in Figs. 24·128, 24·129, and 24·130. In the first figure the plastic pre-forms are shown being placed between electrodes fed by a radio-frequency gen-

FIG. 24·125 Tin fusion.

FIG. 24·127 Plywood bonding.

FIG. 24·126 Plywood bonding.

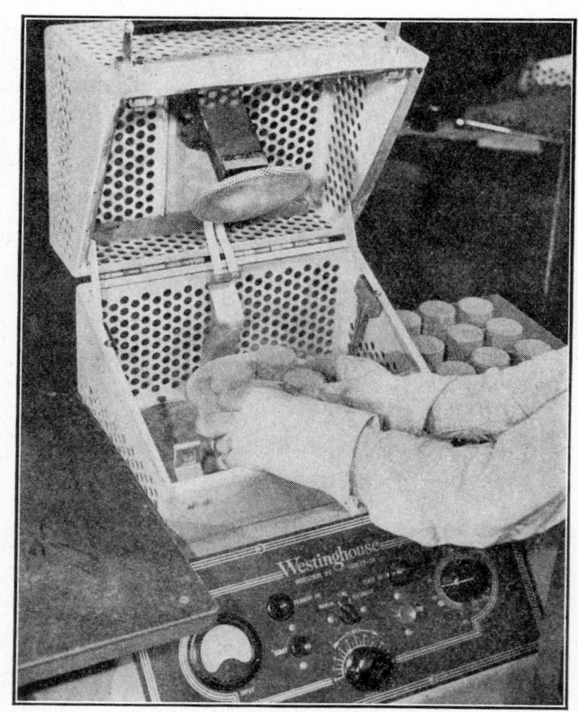

FIG. 24·128 Heating pre-forms. (Courtesy of Hemco Plastics Division of Bryant Electric Company.)

erator. The pre-forms, after heating, are inserted in the mold chamber as shown in Fig. 24·129. After pressure has been applied and the plastic has "set," the finished product is removed as indicated in Fig. 24·130.

Fig. 24·129 Placing pre-forms in press. (Courtesy of Hemco Plastics Division of Bryant Electric Company.)

The textile industry has other applications of dielectric heating. Some of these are twist setting and drying. Representative examples are given in Figs. 24·131 and 24·132. Figure 24·131 shows a rayon-drying process using a 20-kilowatt generator with accompanying oven. The conveyor carries rayon cakes beneath the high-voltage electrode, thence to the delivery end. Hot air is recirculated continuously, and moist air is exhausted to carry off water vapor. Figure 24·132 illustrates the heating of fibers for twist setting. The moisture content is maintained with the aid of the cellophane wrapping.

Fig. 24·130 Removing finished product from press. (Courtesy of Hemco Plastics Division of Bryant Electric Company.)

Fig. 24·131 Laboratory rayon dryer using 20-kilowatt radio-frequency generator.

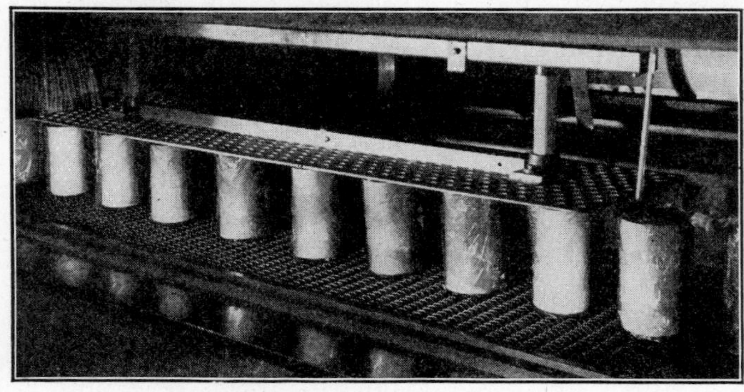

Fig. 24·132 Twist setting.

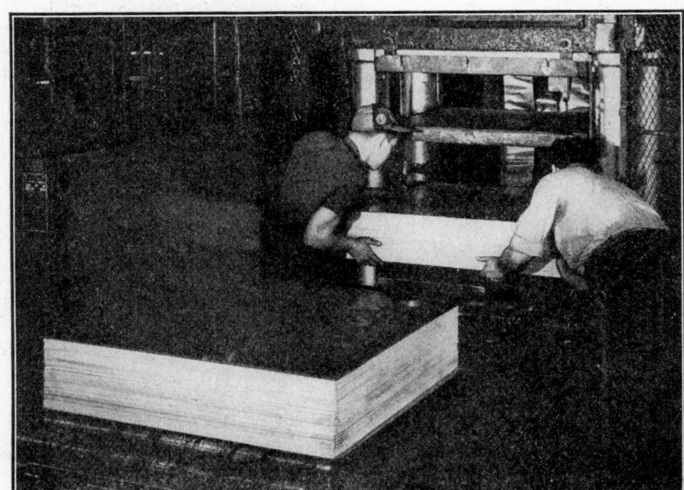

Fig. 24·133 Inserting raw Micarta boards in press to be cured by radio-frequency heating.

Fig. 24·134 Removing radio-frequency cured Micarta boards.

Fig. 24·135 Mixing latex. (Courtesy of Firestone Tire and Rubber Company.)

Fig. 24·136 Pouring latex into mold. (Courtesy of Firestone Tire and Rubber Company.)

Fig. 24·137 Dropping form into latex. (Courtesy of Firestone Tire and Rubber Company.)

Fig. 24·138 Latex is inserted in radio-frequency oven. (Courtesy of Firestone Tire and Rubber Company.)

Fig. 24·139 Cured mattress. (Courtesy of Firestone Tire and Rubber Company.)

FIG. 24·140 Removing mattress from mold. (Courtesy of Firestone Tire and Rubber Company.)

FIG. 24·142 Finished pre-washed mattress. (Courtesy of Firestone Tire and Rubber Company.)

In the heavy plastics industry one application is the curing of large plastic boards. Such an application is shown in Figs. 24·133 and 24·134. In the first is shown the loading of the press with raw material, and the second shows the removal of the cured boards. In this application the heating cycle has been reduced from 6 hours to 1 hour.

Another prominent application of dielectric heating is that of curing sponge rubber for mattresses. The process really begins with the proper mixing of the latex and the pouring of the mixture into the mattress mold as shown in Figs. 24·135, 24·136, and 24·137. The mold containing the uncured latex is then inserted into the radio-frequency

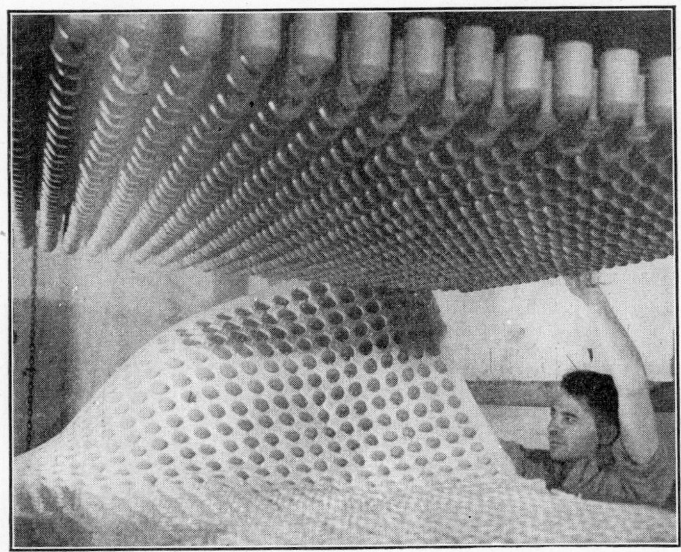

FIG. 24·141 Removing mattress from mold. (Courtesy of Firestone Tire and Rubber Company.)

furnace. Men are seen performing this operation in Fig. 24·138. A 100-kilowatt 13.6-megacycle generator can be seen in the air-conditioned room above the curing oven. After the proper curing time has elapsed the mattress is taken out of the oven and removed from the mold. The finished pre-washed mattress is shown in Fig. 24·142.

24·11 DATA AND FORMULAS

The tables and formulas in this section give the designer and operator of high-frequency heating apparatus a few of the necessary physical constants and formulas.

TABLE 24·5 PHYSICAL AND ELECTRICAL CONSTANTS OF METALS

(Temperature is 20°C unless specified otherwise)

Metal	Resistivity (microhm-centimeters)	Temperature Coefficient of Resistivity	Specific Gravity	Specific Heat	Thermal Conductivity
Aluminum	2.69	0.0042	2.70	0.214	0.48
Barium	9.8	0.0033	3.78	0.068	
Beryllium	18.5		1.8	0.397	0.385
Bismuth	120	0.004	9.8	0.029	0.019
Brass	(See Fig. 24·150.)		8.6	0.094	0.204
Bronze	17	0.0005	8.8	0.086	0.18
Carboloy	19.6			0.052	
Carbon	3500 at 0°C	−0.0009	3.52	0.165	
Calcium	4.59	0.00364(0–600°C)	1.54	0.157	
Chromium	13.1		6.92	0.11	0.165
Cobalt	9.7	0.0066 (0–100°C)	8.71	0.10	0.165
Constantan	42.4	Negligible, 0–100°C	8.88	0.098	0.054
Copper, annealed	1.724	0.0039	8.89	0.093	0.918
Copper, hard-drawn	1.77	0.0038	8.89	0.093	0.918
Gold	2.44	0.0034	19.3	0.032	0.70
Invar	71	0.002	8.0	0.12	0.026
Iron, pure	10	0.005	7.8	0.11	0.16
Iron, cast	79–104		7.03	0.12	0.11
Lead	22	0.0039	11.4	0.036	0.083
Magnesium	4.6	0.004	1.74	0.246	0.376
Manganese	5		7.42	0.121	
Manganin	44	0.00001	8.4	0.097	
Mercury	96	0.00089	13.6	0.033	0.148
Molybdenum	5.7	0.0047	9.1	0.065	0.35
Monel metal	42	0.002	8.9	0.127	0.06
Nichrome	100	0.0001	8.2		0.032
Nickel	7.8	0.0054	8.9	0.105	0.14
Palladium	11	0.0033	12.2	0.056	0.168
Phosphor-bronze	7.8	0.0018	8.9		
Platinum	10	0.003	21.4	0.035	0.167
Silver	1.59	0.0038	10.5	0.057	1.006
Sodium	4.3 at 0°C	0.0044	0.97	0.295	0.322
Steel, carbon *	(See added reference.)				
Tantalum	15.5	0.0033	16.6	0.036	0.130
Thorium	18	0.0021(20–1800°C)	11.5	0.027	
Tin	11.5	0.0042	7.3	0.054	0.155
Tungsten	5.5	0.0045	19	0.034	0.476
Zinc	5.8	0.0037	7.1	0.088	0.265
Graphite	800 at 0°C		2.25	0.17	0.057

* Resistivity of steels at 20°C may be approximated by $\rho = 11.25 + 4.5(c - 0.02)$ microhm-centimeter, where c = percent of carbon in steel. Temperature coefficient of resistance of carbon steel is taken to be 0.0043 microhm-centimeter per degree centigrade. Specific heat of carbon steels heated from room temperature to hardening temperatures may be taken to be 0.16. The foregoing properties of materials have been compiled from various handbooks. For more extensive data relative to alloys, refer to information published by the manufacturer of the alloy in question.

TABLE 24·6. CHARACTERISTICS OF DIELECTRIC MATERIALS

Name	Density		Therm. Cond.	Sp. Ht.	Soften-ing Temp.	1 Mc			5 Mc			10 Mc			15 Mc			30 Mc			Remarks
	gm/cc	Lb/in.^{3}				p-f	ϵ	ϵ''	p-f	ϵ	ϵ''	p-f	ϵ	ϵ''	p-f	ϵ	ϵ''	p-f	ϵ	ϵ''	
Amber	1.1	.04			250°C	.002	2.9	.0058													
Asbestos sheet	0.5	.0181		.195		.32	4.6	1.46	.21	4.1	.86	.15	3.9	.585	.12	3.8	.456	.08	3.3	.264	
Asbestos sheet impregnated with thermosetting plastic															.0015	1.022	.00153				
Beeswax, yellow	.96	.0348			60°C	.016	3.2	.0511													
Cellulose acetate	1.3 1.5	.047 .054	.0005	.45		.099	4.6	.455													
Cellulose nitrate	1.42	.053	.0003	.36	60–90°C	.085	6.18	.525													
Chocolate, cake form																		.0241	3.26	.0786	
Cottonseed																				.04	
Fiber-bone, vulcanized	1.3	.047	.0011			.05	5.75	.25–.27													
Fiberboard, grey																		.085	5.1	.434	
Fiber sheet, C-598																		.066	4.1	.27	
Fiber sheet, spauldite GB-1																		.034	4.43	.15	
Insanol (Mycalex)	3.5	.126	.0014			.001	7.5	.0075	.001	7.5	.0075	.001	7.5	.0075	.001	7.5	.0075	.0012	7.5	.0093	A very good insulator. Given for comparison.
Nylon				.55		.028	3.3	.0925	.024	3.3	.079	.024	3.3	.079	.024	3.25	.078	.022	3.2	.075	
Paper, corrugated asphalt impregnated									.022	1.42	.0315	.022	1.42	.0315	.020	1.42	.024	.020	1.4	.028	
Plexiglas									.022	2.78	.061							.017	2.78	.047	
Rubber, pure gum												.007	2.15	.015				.01	2.15	.0215	
95% uncured	2.4	.0866										.012	3.25	.039				.012	3.2	.0384	
95% cured	3.6	.13										.037	4	.148				.03	4	.12	
Medium grade, uncured	4.5	.162										.018	5	.090				.015	5	.075	
Medium grade, cured	4.3	.155										.10	6	.60				.082	5.8	.475	
Low-grade, uncured	4.6	.166										.02	6	.12				.018	5.9	.106	
Low-grade, cured	4.2	.152										.103	7.4	.764				.068	6.9	.47	
Reclaimed	5.6	.202										.03	4.6	.138				.025	4.5	.113	
Pure pale crepe												.004	2.1	.0084	.004	2.1	.0034	.005	2.2	.0105	
Buna S, pure												.006	2.2	.0132	.006	2.2	.0132	.007	2.2	.0154	
With carbon black																		.028	10.7	.29	
Synthetic, crumb																		.00281	1.67	.00467	
Shellac	1.01	.0364				.009	3.1	.028													
Soybean oil	.925	.0334		.5	−10°C							.098	1.9	.175	.088	2.3	.21	.07	2.5	.175	
Tobacco, green, uncured												.168	2.57	.432							
Twine, paper, spools, 20% moisture												.07	2.4	.16							
Wheel, grinding, 85% Alundum												.031	6.9	.214				.038	6.5	.247	
Grinding, 80% Alundum												.067	11.4	.765							
Wood, birch	.63	.0238	.0003			.0648	5.2	.337													Sp heat of wood = .226 + .000644 (T-32) where T is °F.
Maple	.62	.0224	.00038			.033	4.4	.145													
Oak	.69	.025	.00035			.0385	3.3	.127													
Plywood (fir)	.41	.0148	.00027									.035	3.1	.1085							
PLASTIC PRE-FORMS																					
Resinox 1337						.053	4	.212	.049	3.9	.19	.045	3.9	.175	.045	3.8	.171	.045	3.7	.166	
Resinox 3828	.80	.029							.06	4.4	.264	.05	4.4	.22	.047	4.4	.207	.052	4.4	.26	
Resinox 3942									.05	3.6	.18	.051	3.6	.183	.053	3.6	.19	.062	3.6	.223	
Resinox 6542									.039	2.65	.103	.04	2.65	.106	.041	2.65	.109	.044	2.7	.119	
Resinox 6565									.049	3.4	.166	.052	3.5	.182	.053	3.55	.188	.059	3.6	.212	
Resinox 6570	.55	.0199				.041	4.92	.20	.053	4.67	.25	.059	4.53	.27	.063	4.45	.28	.053	4.17	.22	
Resinox 6905									.046	2.95	.136	.046	2.95	.136	.047	2.95	.138	.048	2.95	.141	
Resinox 7013	.70	.0252				.014	3.79	.052	.010	3.73	.038	.0096	3.73	.036	.0092	3.73	.034	.0087	5.67	.032	
Resinox 7934						.0071	3.6	.0256	.0072	3.6	.0260	.0073	3.6	.0263	.0072	3.6	.0260	.0075	3.6	.0270	
Bakelite BM120 black	.5	.018		.35–.40		.034	4.11	.14	.044	3.95	.17	.052	3.84	.20	.053	3.77	.21	.047	3.6	.17	
BM261 black	.85–.90	.0316		.30														.052	4.7	.24	
BM1132 black	.11–.17	.005		.35–.40		.032	3.27	.11	.040	3.18	.13	.042	3.12	.13	.043	3.06	.13	.049	2.98	.15	
BM2498 black	.58–.62	.0216		.35–.40														.047	3.4	.16	
BM3200 black	.52–.56	.0195		.35–.40														.043	3.5	.15	
BM3510 black	.09–.12	.0036		.35–.40		.033	3.19	.11	.039	3.11	.12	.049	3.09	.15	.045	3.0	.14	.038	3	.11	
BM6102 black	.60–.62	.022		.35–.40														.044	3.5	.15	
BM6260 black	.40–.43	.015		.35–.40		.014	3.84	.05	.0099	3.8	.038	.0091	3.8	.035	.0081	3.77	.03	.0081	3.75	.03	Thermal conductivities of pre-form materials in the order of .0003 (cal/sec/sq cm)/(°C/cm)
BM8041 black	.60–.64	.024		.28–.32														.065	4.2	.28	
BM9131 black	.50–.55	.019		.28–.32														.013	2.9	.04	
BM9928 black	.68–.72	.0253		.28–.32														.062	4.3	.27	
BM13080 black	.27–.32	.0107		.35–.40														.062	3.5	.22	
BM13107 black	.81–.85	.03		.28–.32														.016	3.6	.06	
BM14111 tan	.12–.18	.0054		.35–.40														.054	2.5	.14	
BM16981 black	.78–.83	.029		.28–.32														.009	3.4	.03	
BM17100 black	.55–.59	.0206		.35–.40														.068	4.0	.28	
BM17610 black	.12–.16	.005		.28–.32														.065	3.1	.20	
Durez 75						.036	4.93	.18	.041	4.70	.19	.045	4.65	.21	.046	4.57	.21	.051	4.4	.22	
775-B						.026	4.31	.11	.029	4.26	.12	.034	4.20	.14	.035	4.13	.15	.041	4.01	.17	
760						.038	4.89	.19	.049	4.65	.23	.055	4.56	.25	.055	4.47	.25	.056	4.28	.24	
11540						.039	4.72	.19	.047	4.55	.21	.050	4.44	.22	.055	4.30	.24	.050	4.18	.21	
14-SB						.040	4.89	.19	.052	4.65	.24	.056	4.52	.25	.058	4.4	.26	.060	4.10	.25	
1863									.0057	2.98	.017							.0072	3.00	.0216	
Plaskon 3555						.035	4.28	.15	.036	4.12	.15	.037	4.03	.15	.035	3.96	.14	.035	3.89	.14	
Melmac 592						.032	4.73	.15	.033	4.55	.15	.034	4.52	.15	.035	4.45	.15	.035	4.38	.15	
Melmac 3020						.017	2.98	.051	.022	2.97	.066	.028	2.89	.080	.028	2.84	.0795	.0284	2.81	.080	
Melmac S6003									.016	3.25	.052	.016	3.25	.052	.016	3.25	.052	.016	3.25	.052	

Pre-form note:—Average specific heats for pre-forms are
General-purpose, wood-flow-filled, .35 to .40
Impact materials, cellulose-filled, .35 to .40
Heat-resistant, asbestos-filled, .28 to .32
Electrical materials, mica-filled, .28 to .32
Chemical-resistant materials, .28 to .32

General note: p-f = power factor
ϵ = dielectric constant
ϵ'' = loss factor

Pre-form note: Pre-form power factors given above are internal power factors for the pre-form. When using standard electrodes and cages, multiply by the following correction factor to correct for distributed capacity:

No air gap, .68
⅛″ air gap, .42

General note: correct all power factors for distributed capacity of electrodes and shielding.

General note: Thermal conductivity expressed as (cal/sec/sq cm)/(°C/cm)

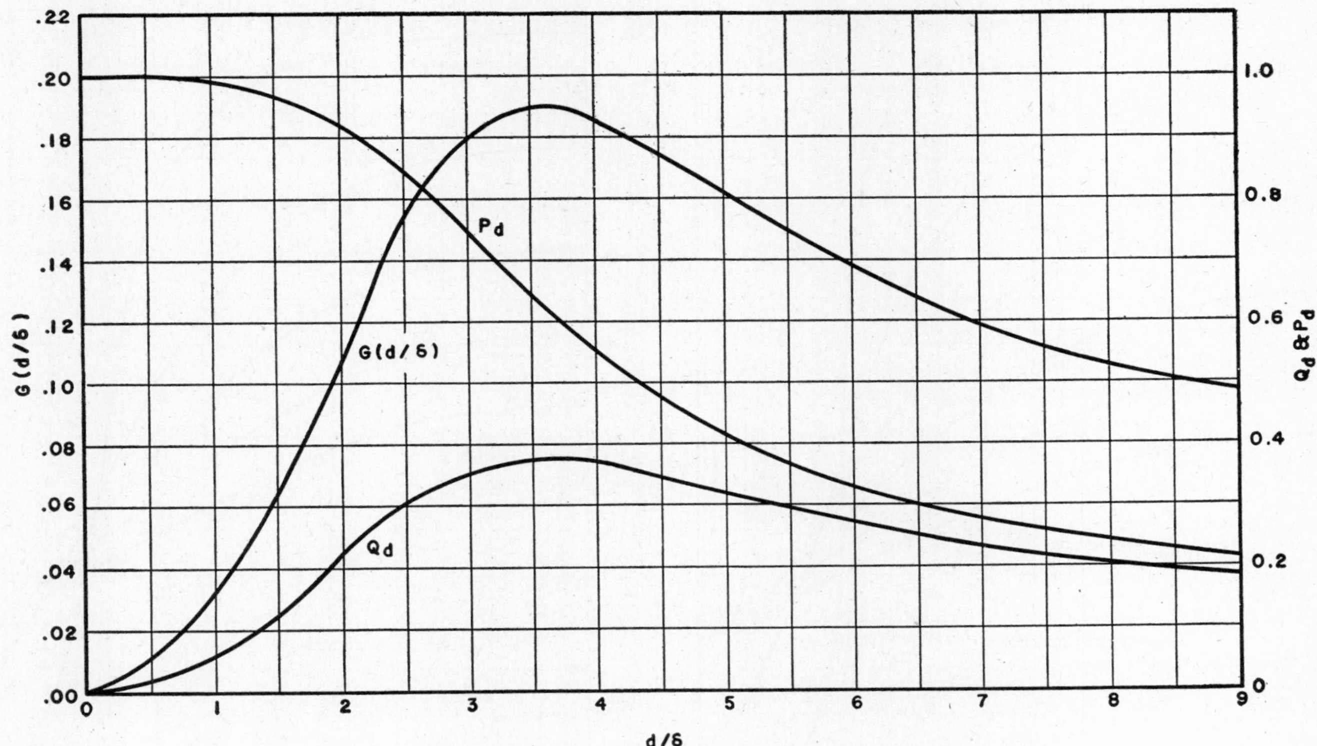

FIG. 24·143 *G*, *Q*, and *P* functions for cylinders.

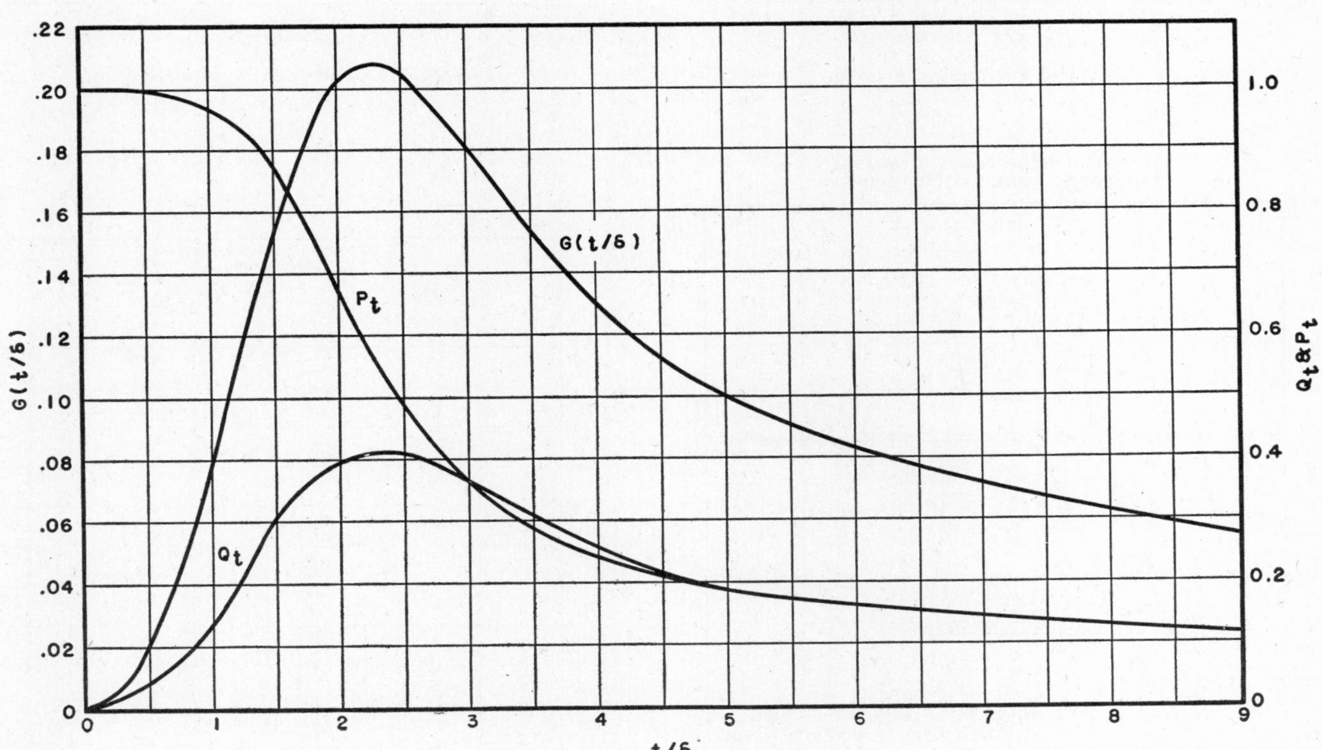

FIG. 24·144 *G*, *Q*, and *P* functions for slabs.

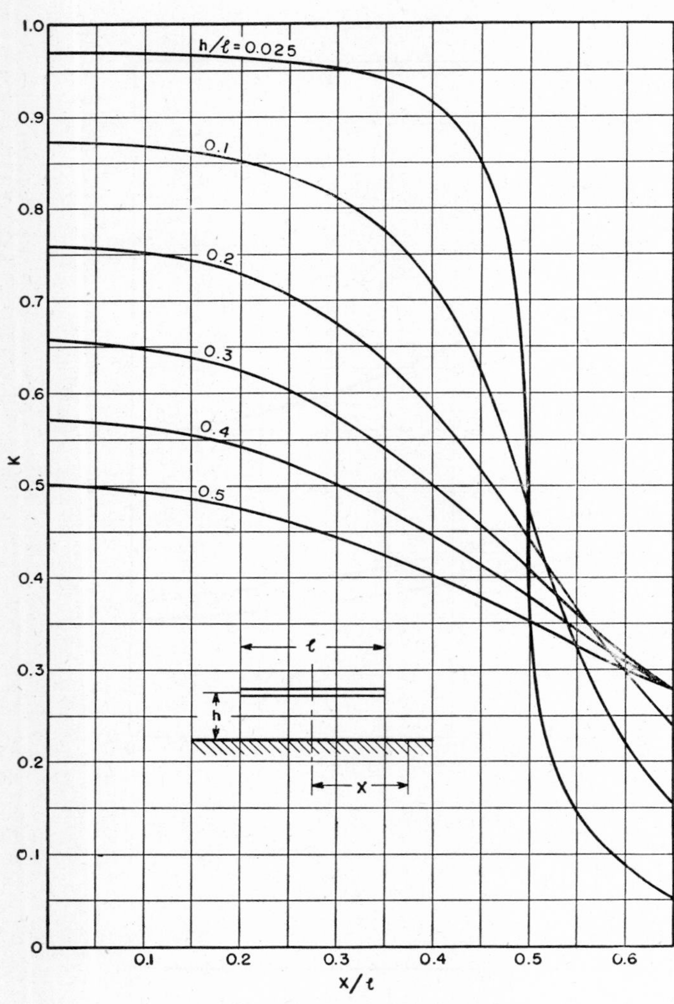

Fig. 24·145 Field distribution K factor.

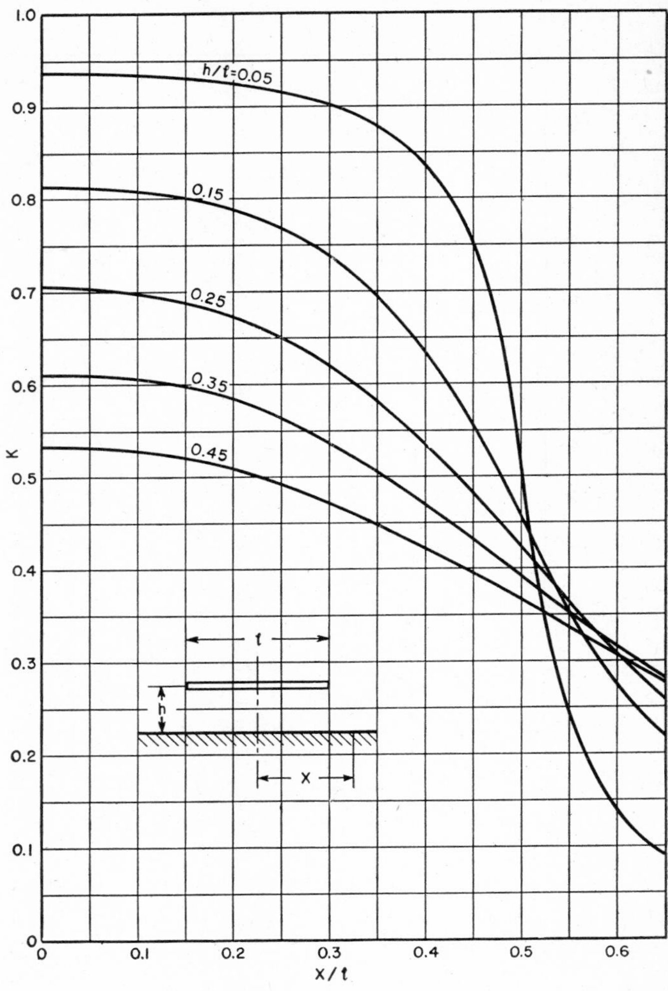

Fig. 24·146 Field distribution K factor.

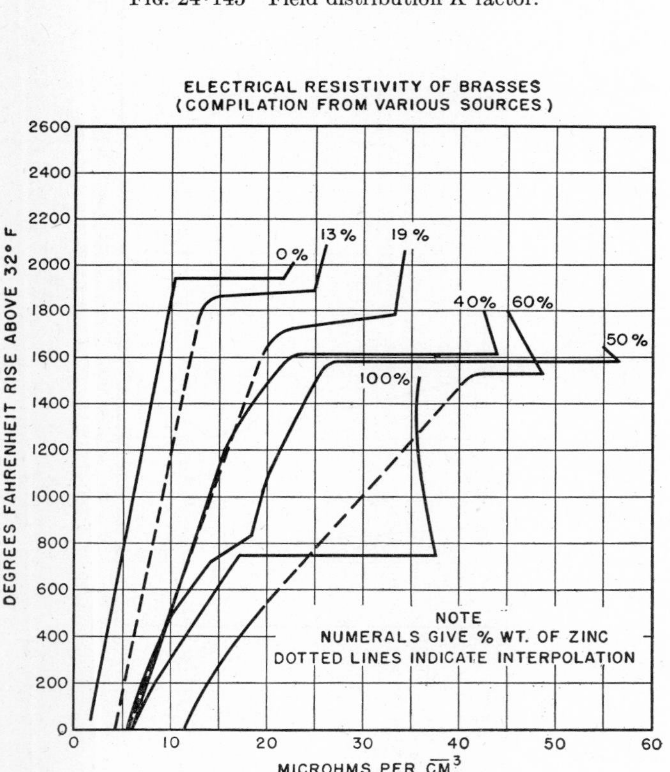

Fig. 24·147 Electrical resistivity of brasses.

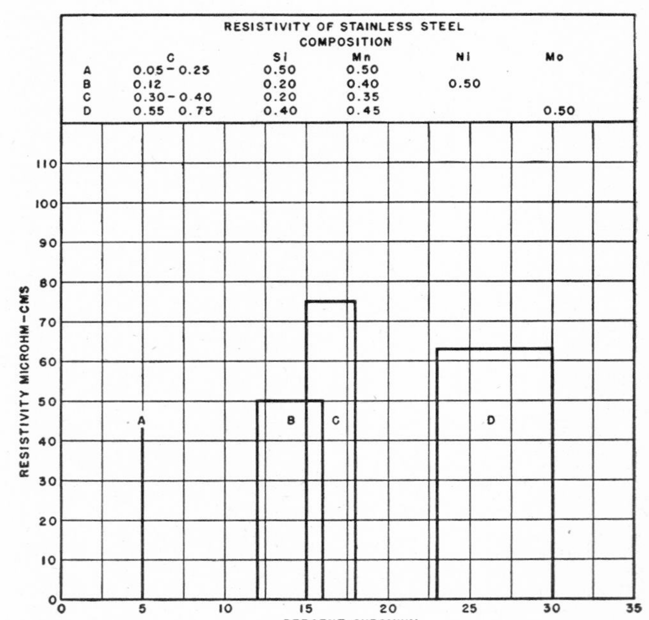

	C	Si	Mn	Ni	Mo
A	0.05 – 0.25	0.50	0.50		
B	0.12	0.20	0.40	0.50	
C	0.30 – 0.40	0.20	0.35		
D	0.55 0.75	0.40	0.45		0.50

Fig. 24·148 Electrical resistivity of stainless steels.

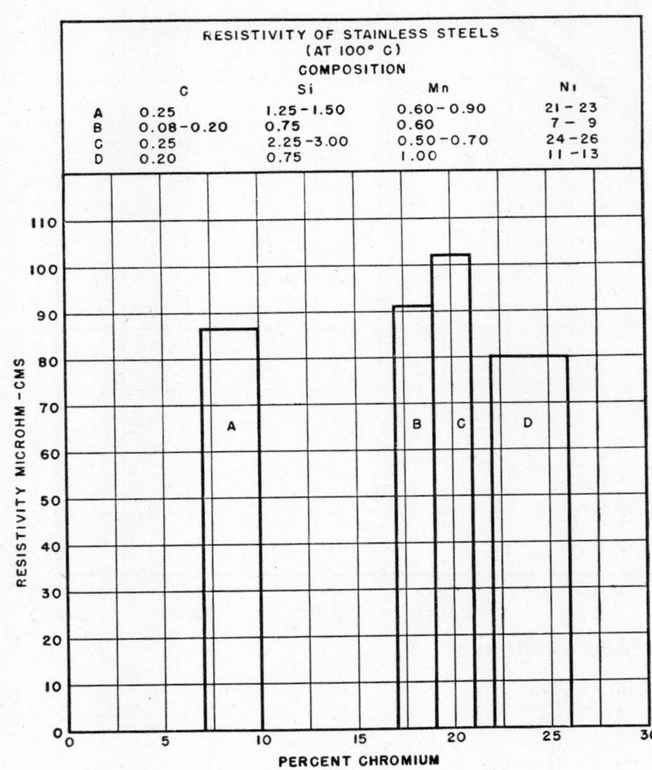

FIG. 24·149 Electrical resistivity of stainless steels.

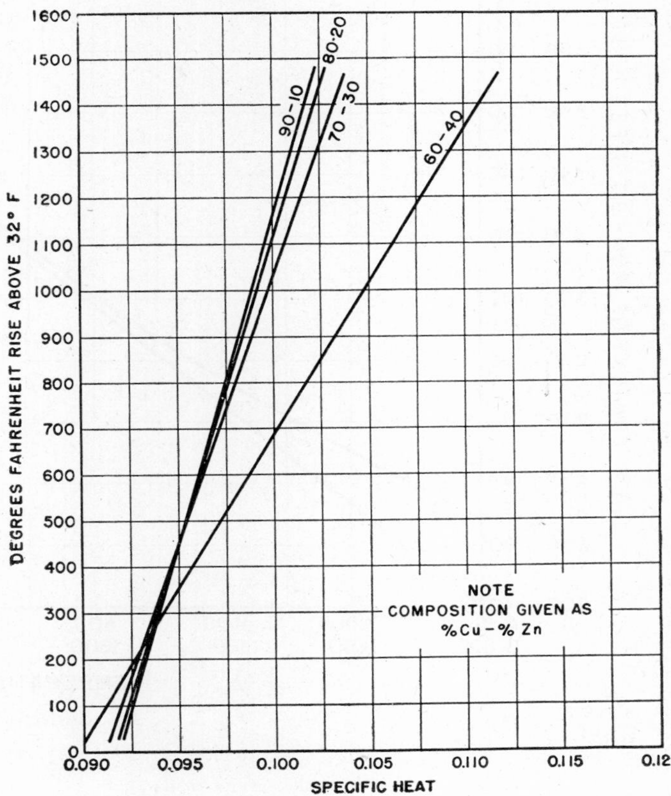

FIG. 24·150 Specific-heat curve of brasses.

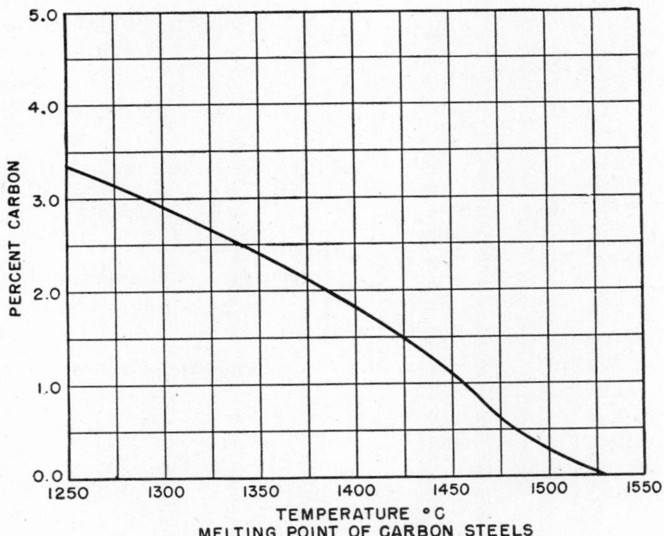

FIG. 24·151 Melting point of carbon steels.

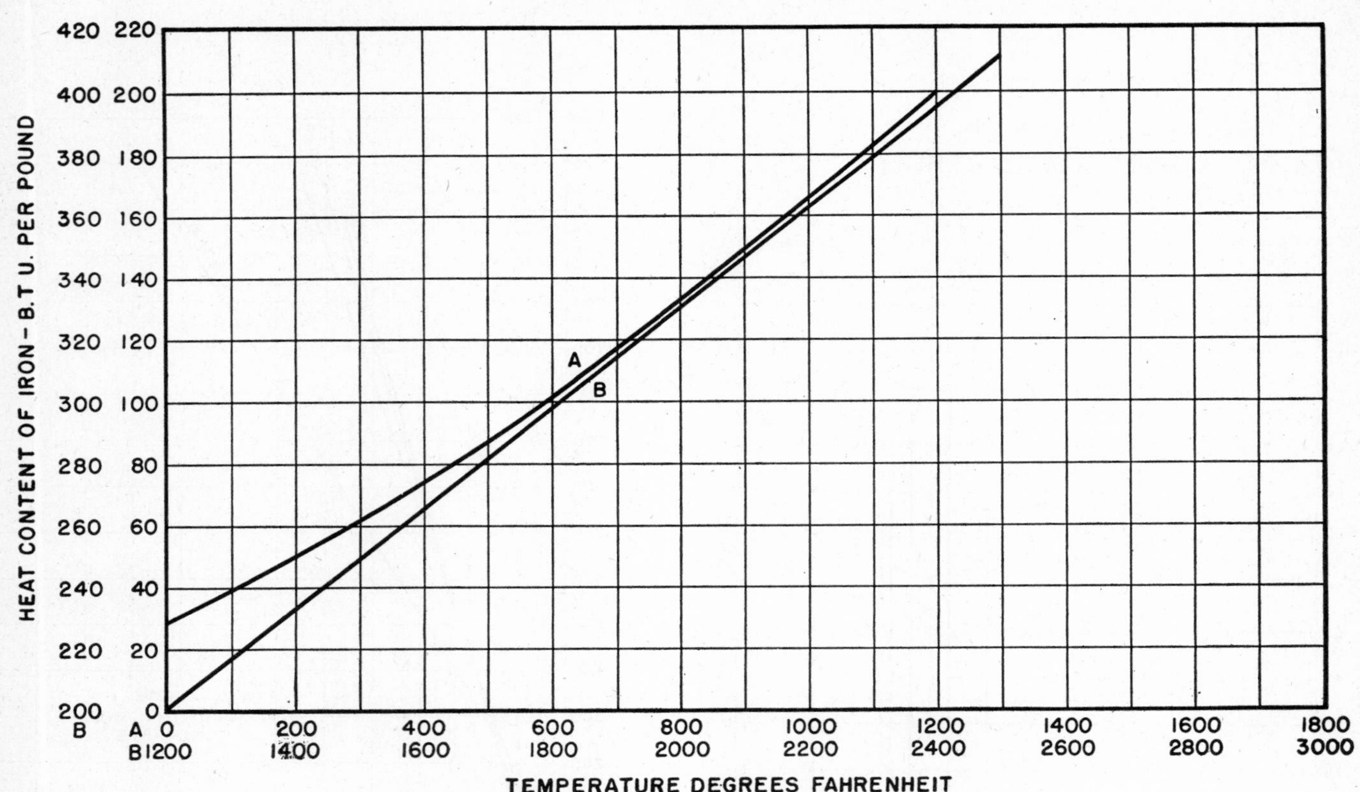

FIG. 24·152 Heat content of iron.

TABLE 24·7 CONVERSION FACTORS

To Convert	Into	Multiply By	Conversely Multiply By
Ampere-turns	gilberts	1.257	0.7958
Ampere-turns/in.	gilberts/cm	0.495	2.021
Btu	joules	1050	9.48×10^{-4}
Cal/sec/cm/°C	Btu/hr/ft/°F	0.00413	242
Cal/gm	Btu/lb	1.8	0.5555
Centigrade	Fahrenheit	$\frac{9}{5}$°C + 32	$\frac{5}{9}$(°F − 32)
Cu in.	cu cm	16.39	6.102×10^{-2}
Cu in.	gallons	4.329×10^{-3}	231
Erg	joule	10^{-7}	10^{7}
Gauss	lines/sq in.	6.452	0.1550
Grams/cu cm	lb/cu in.	0.03613	27.68
Grams	pounds	2.205×10^{-3}	453.6
Inches	centimeters	2.54	0.3937
Joules	calories	252	3.968×10^{-3}
Kilowatt-hours	Btu	3413	2.930×10^{-4}
Liters	cu cm	10^{3}	10^{-3}
Liters	gallons	0.26418	3.783
Sq in.	circular mils	1.273×10^{6}	7.854×10^{-7}
Sq in.	sq cm	6.452	0.155
Watts	Btu/min	0.05688	17.58
Watts	cal/sec	0.2389	4.185
Watts	ergs/sec	10^{7}	10^{-7}
Meters	feet	3.281	0.3048

Useful Formulas

Thermal. *Load-Thermal-Power Requirement.*

$$P_1 = 17.6Mc\,\Delta t \text{ watts}$$

where P_1 = load-thermal-power requirement in watts
M = pounds per minute of load material heated
c = specific heat of load material
Δt = temperature rise in degrees Fahrenheit.

Radiation Loss.

$$P_2 = 37e\left[\left(\frac{T_1}{1000}\right)^4 - \left(\frac{T_2}{1000}\right)^4\right]$$

where P_2 = radiation loss in watts per square inch
e = relative emissivity
T_1 = temperature of radiating surface in degrees Kelvin
T_2 = ambient temperature in degrees Kelvin
T = $\frac{5}{9}$ degrees Fahrenheit + 255.4 degrees Kelvin

Convection Loss.

$$P_3 = 4.66 \times 10^{-4}\,(\Delta t)^{1.33}$$

where P_3 = convection loss in watts per square inch
Δt = degrees rise in degrees Fahrenheit.

Alternating Current. *Coil Voltage.*

$$E = 2.86fNH_0 \times (A_c - A_L)10^7$$

where E = coil voltage in volts, rms
f = frequency in cycles per second
N = total number of turns in the coil
H_0 = peak magnetizing force in oersteds
A_c = cross-sectional area of coil in square inches
A_L = cross-sectional area of load in square inches.

This formula should not be used if the depth of penetration within the work is not small compared with the cross-sectional dimensions of the load. See the text for more accurate treatment.

Coil Current.

$$I = 1.43 \frac{H_0}{n}$$

where I = coil current in amperes, rms
H_0 = peak magnetizing force in oersteds
n = coil turns per inch.

Power-Factor Correction of Load Coil by Shunting Capacitor.

$$C = \frac{L}{Z^2}$$

where C = required capacitance in farads for complete power-factor correction
L = inductance of loaded coil in henrys
Z = absolute impedance of loaded coil.

Power-Factor Correction of Electrodes by Shunting Inductance.

$$L = CZ^2$$

where L = required inductance for complete power-factor correction in henrys
C = capacitance of loaded electrodes in farads
Z = absolute impedance of loaded electrodes.

Maximum Current Amplification in Parallel Resonant Circuit.

$$\frac{I_L}{I_g} = \frac{Z}{R} = Q \quad \text{when } Q > 10$$

where I_L = current in amperes in inductive leg
I_g = generator current in amperes
Z = impedance in ohms of inductive leg
R = resistance in ohms of inductive leg
Q = ratio of resistance in reactance in inductive leg.

T-Network Equations.

$$\frac{R_G}{R_s} = \frac{X_1 + X_3}{X_2 + X_3}$$

$$-R_G R_s = X_1X_2 + X_1X_3 + X_2X_3$$

where R = resistance in ohms
X = reactance in ohms.

Inductive reactance is denoted by positive algebraic sign; capacitive reactance by negative algebraic sign.

Coil Inductance for Solenoids.

$$L = \frac{r^2N^2}{9r + 10l}$$

where L = coil inductance in microhenrys
r = inside radius of coil in inches
N = number of coil turns
l = length of coil in inches.

This formula is accurate within 1 percent if $l > 0.8r$.

Coil Inductance for Long Coils.

$$L = \frac{0.032N^2(Ac - Aw)}{l}$$

where L = coil inductance in microhenrys
N = number of coil turns
l = coil length in inches
Ac = cross-sectional area of coil in square inches
Aw = cross-sectional area of work in square inches.

Capacitance of Parallel-Plate Electrodes with Parallel Layers of Different Dielectrics.

$$C = \frac{0.224A \times 10^{-12}}{\sum_{i=1}^{i=n} \frac{a_i}{\epsilon_i}}$$

where C = capacitance in farads
A = area of one electrode in square inches
a_i = thickness of ith dielectric layer
ϵ_i = dielectric constant of ith dielectric layer
n = number of layers of dielectric.

Voltage Gradient in Any Dielectric Layer in Parallel-Plate Electrodes with Parallel Layers of Different Dielectrics.

$$E_k = \frac{E}{\epsilon_k \sum_{i=1}^{i=n} \frac{a_i}{\epsilon_i}}$$

where E_k = voltage gradient in the kth layer in kilovolts per inch
E = electrode voltage in kilovolts
ϵ_k = dielectric constant of kth layer
a_i = thickness of ith layer in inches
ϵ_i = dielectric constant of ith layer
n = number of layers.

Capacitance per Foot of Coaxial Electrodes Having Coaxial Layers of Different Dielectrics.

$$C = \frac{16.9 \times 10^{-12}}{\sum_{i=1}^{i=n} \frac{\log_\epsilon \sigma_i}{\epsilon_i}}$$

where C = capacitance of electrodes in farads per linear foot
σ_i = ratio of outer to inner radii of the ith dielectric layer
ϵ_i = dielectric constant of ith layer
n = number of dielectric layers.

Voltage Gradient in Any Dielectric Layer in Coaxial Electrodes with Coaxial Layers of Different Dielectrics.

$$E_k = \frac{E \log_\epsilon \sigma_k}{\gamma_k \epsilon_k \sum_{i=1}^{i=n} \frac{\log_\epsilon \sigma}{\epsilon_i}}$$

where E_k = voltage gradient in kth layer in kilovolts per inch
E = electrode voltage in kilovolts
σ_k = ratio of outer to inner radii of kth dielectric layer
γ_k = difference in outer and inner radii of kth dielectric layer in inches
ϵ_k = dielectric constant of kth dielectric layer
σ_i = ratio of outer to inner radii of ith dielectric layer
ϵ_i = dielectric constant of ith dielectric layer.

Transmission-Line Equations, General.

$$E = E_L \cosh x\gamma + I_L Z_0 \sinh x\gamma$$

$$I = I_L \cosh x\gamma + \frac{E_L}{Z_0} \sinh x\gamma$$

where E = voltage across line at x feet from receiving end in volts
I = line current at x feet from receiving end in amperes
E_L = load (receiving-end) voltage in volts
I_L = load (receiving-end) current in amperes
Z_0 = characteristic impedance of line in ohms
x = distance along line in feet from receiving end
γ = propagation constant in complex radians per foot
$\gamma = \alpha + j\beta$
α = phase constant in radians per foot
β = attenuation constant in radians per foot.

Transmission-Line Input Impedance.

$$Z_s = Z_0 \frac{Z_L + Z_0 \tanh \gamma l}{Z_0 + Z_L \tanh \gamma l}$$

where Z_s = input impedance of transmission line in ohms
Z_0 = characteristic impedance of line in ohms
Z_L = load impedance at receiving end of line in ohms
γ = propagation constant in complex radians per foot
$\gamma = \alpha + j\beta$
α = attenuation constant in radians per foot
β = phase constant in radians per foot
l = length of transmission line in feet.

Power into a Dielectric.

$$P_v = 1.41 f E_1^2 \epsilon''$$

where P_v = power density in watts per cubic inch
f = frequency in megacycles per second
E_1 = voltage gradient in dielectric in kilovolts per inch
ϵ'' = loss factor of load material.

Wavelength along Electrodes or Transmission Lines.

$$\lambda = \frac{984}{f\sqrt{\epsilon}}$$

where λ = wavelength in feet
f = frequency in megacycles per second
ϵ = effective dielectric constant of region between electrodes.

Depth of Current Penetration.

$$\delta = 1.98 \sqrt{\frac{\rho}{\mu f}}$$

where δ = depth of current penetration in inches
ρ = resistivity of material in microhm-centimeters
f = frequency in cycles per second
μ = permeability of material
$\mu = 1.8(B_m/H_0)$ for magnetic materials
B_m = saturation flux density = 16,000 gauss
H_0 = peak magnetizing force in oersteds.

Power Input to Magnetic Material.

$$P_a = 4.1 H_0^2 \sqrt{\rho f} \times 10^{-7}$$

where P_a = power density in watts per square inch
H_0 = peak magnetizing force in oersteds
ρ = resistivity of work in microhm-centimeters
f = frequency in cycles per second.

Power Input into Non-Magnetic Material.*

$$P_a = 7 H_0^{3/2} \sqrt{\rho f} \times 10^{-5}$$

where P_a = power density in watts per square inch
H_0 = peak magnetizing force in oersteds
ρ = resistivity of work in microhm-centimeters
f = frequency in cycles per second

Power Input into a Surface by Single-Conductor Proximity Heating.

$$P_l = 0.316 \frac{I^2 \sqrt{\mu \rho f}}{h} \times 10^{-7}$$

where P_l = power absorbed in watts per inch of work and coil length
I = conductor current in amperes, rms
μ = permeability of work
$\mu = 1.8(B/H_0)$
B = saturation flux density = 16,000
H_0 = peak magnetizing force in oersteds at surface of work
ρ = resistivity of work in microhm-centimeters
f = frequency in cycles per second
h = effective height in inches of current-carrying conductor above surface of work = $[h_1^2 - a^2]^{1/2}$ for cylindrical current-carrying conductor
h_1 = height in inches of center of current-carrying conductor above surface of work
a = radius of current-carrying conductor.

REFERENCES

1. "Principles of Inductive Heating with High Frequency Currents," E. F. Northrup, *Trans. Amer. Electrochem. Soc.*, Vol. 35, 1919, p. 69.
2. "Inductive Heating," E. F. Northrup, *J. Franklin Inst.*, Vol. 201, Feb. 1926, pp. 221–244.
3. "Vacuum-Tube Radio Frequency Generator-Characteristics and Application to Induction Heating Problems," by T. P. Kinn, *Trans. A.I.E.E.*, Dec. Supp., 1944, p. 1290.
4. "Dielectric Heating Fundamentals," Douglas Venable, *Electronics*, Vol. 18, No. 11, 1945, pp. 120–124.
5. *Principles of Electricity and Electromagnetism*, G. P. Harnwell, McGraw-Hill, 1938.
6. *Principles of Radio Engineering*, R. S. Glasgow, McGraw-Hill, 1936.
7. *Bessel Functions for Engineers*, N. W. McLachlan, Oxford Univ. Press, 1934.
8. "Induction Heating of Moving Magnetic Strip," R. M. Baker, *Trans. A.I.E.E.*, Vol. 64, No. 4, 1945, pp. 184–189.
9. *Transient Electrical Phenomena and Oscillations*, C. P. Steinmetz, McGraw-Hill, 1920.
10. "Concentration of Heating Currents," E. Bennett, *Elec. Eng.*, Vol. 51, No. 8, Aug. 1932, p. 559.
11. "Temperature Distribution in Solids during Heating or Cooling," E. D. Williamson and L. H. Adams, *Phys. Rev.*, Vol. XIV, No. 2, 1919, p. 99.

* Formulas applicable only if depth of penetration is less than 1/5 diameter d of cylinder or 1/3 thickness t of slab.

12. "Heating of Non-Magnetic Electric Conductors by Magnetic Induction Longitudinal Flux," R. M. Baker, *Trans. A.I.E.E.*, Vol. 63, No. 6, 1944, pp. 273–278.
13. *Communication Engineering*, W. L. Everitt, McGraw-Hill, 1937.
14. "Concentrator of Eddy Currents for Zonal Heating of Steel Parts," G. Babat and M. Losinsky, *J. App. Phys.*, Vol. 11, Dec. 1940, p. 316.
15. "An Engineering Approach to Soldering with Tin-Lead Alloys," A. Z. Mample, *Metals and Alloys*, Vol. 21, No. 3, 4, March and April, 1945, pp. 702–707, 1000–1006.
16. "Induction Brazing and Soldering," H. U. Hjermstad, *The Iron Age*, Vol. 156, No. 6, Aug. 1945, pp. 56–60.
17. "Silver Alloy Brazing with Induction Heating," A. M. Setapen, *Trans. Amer. Electrochem. Soc.*, Vol. 86, 1944, pp. 277–297.
18. "Heat Treatment of Steel by High-Frequency Currents," by G. Babat and M. Losinsky, *J. I.E.E.* (*London*), Vol. 86, No. 518, 1940, pp. 161–168.
19. "Induction Hardening of Plain Carbon Steels," D. L. Martin and Florence E. Wiley, *Trans. Am. Soc. Metals*, Vol. XXXIV, 1945, p. 351.
20. *Alloys of Iron and Carbon*, Samuel Epstein, McGraw-Hill, 1936, Vol. I.
21. "Thin Case Hardening with Radio Frequency Energy," V. W. Sherman, *Aeron. Eng. Rev.*, Vol. 2, No. 5, 1943, pp. 7, 9, 11, 13, 15–16, 101.
22. "Inherent Characteristics of Induction Hardening," A. Tran and H. B. Osborn, Jr., *Am. Soc. Metals*, Preprint No. 49, 1940.
23. "Construction of Heating Coils for Induction Surface Hardening," G. Babat, *Heat Treating and Forging*, Vol. 27, 1942, p. 39.
24. *Mathematical Theory of the Conduction of Heat in Solids*, H. S. Carslaw, MacMillan, 1921, p. 581.
25. "The Theory of Quenching," J. H. Awbery, *Heat Treating and Forging*, Vol. 28, 1942.
26. *Lectures on Dielectric Theory and Insulation*, J. B. Whitehead, McGraw-Hill, 1928.
27. "Theory of Dielectric Constant and Energy Loss in Liquids and Solids," H. Fröhlich, *J. I.E.E.* (*London*), Vol. 91, Part 1, No. 48, Dec. 1944, p. 456.
28. *Principles of Electricity*, L. Page and N. Adams, Van Nostrand, 1931.
29. "Dielectric Heating," T. W. Dakin and R. W. Auxier, *Ind. Eng. Chem.*, Vol. 37, No. 3, March 1945, pp. 268–275.
30. "The Determination of Dielectric Properties at Very High Frequencies," J. G. Chafee, *Proc. I.R.E.*, Vol. 22, No. 8, Aug. 1934, pp. 1009–1020.
31. "The Measurement of the Permittivity and Power Factor of Dielectrics at Frequencies from 10^4 to 10^8 Cycles per Second," L. Hartshorn and W. H. Ward, *J. I.E.E.* (*London*), Vol. 79, No. 479, Nov. 1936, pp. 597–609.
32. *Radio Frequency Electrical Measurements*, Hugh A. Brown, McGraw-Hill, 1938.
33. "Heat Conduction Problems in Presses Used for Gluing Wood," G. H. Brown, *Proc. I.R.E.*, Vol. 31, No. 10, Oct. 1943, pp. 537–548.
34. "Thermal Conduction with Radio Heating," H. Herne, *Wireless Eng.*, Vol. 21, No. 251, Aug. 1944, pp. 377–382.
35. "L-Type Impedance Transforming Circuits," P. H. Smith, *Electronics*, Vol. 15, No. 3, March 1942, pp. 48–54, 125.
36. *Communication Circuits*, L. A. Ware and H. R. Reed, Wiley, 1942.
37. "Single- and Double-Stub Impedance Matching," A. H. Wing and J. Eisenstein, *J. App. Phys.*, Vol. 15, Aug. 1944, pp. 615–622.
38. *Static and Dynamic Electricity*, D. R. Smythe, McGraw-Hill, 1939.
39. "The Solution of Transmission Line Problems by Use of the Circle Diagram of Impedance," W. Jackson and L. G. H. Huxley, *J. I.E.E.* (*London*), Vol. 91, Part 3, 1944, p. 105.

Chapter 25

POWER-LINE CARRIER

F. S. Mabry

POWER-LINE carrier is the application of high-frequency carrier currents to power-transmission lines. These high-frequency currents are usually in the frequency band from 50 to 150 kilocycles. The term carrier implies that the high-frequency wave is modulated or varied proportionally by the signals to be transmitted. By using carriers of different frequencies many carrier circuits or channels may be established over a single wire or pair of wires. These channels, as they are most generally called, are used for telephone communication, relaying, telemetering, load control, supervisory control, and remote tripping. Reliability of these services is limited only by the reliability of the power line itself and of the carrier terminal apparatus.

Before the development of power-line carrier it was necessary to use telephone-wire lines for these functions. They were generally less satisfactory than power-line carrier for several reasons. Wind and sleet, which will completely carry away a telephone line, do no great harm to a power line because of its sturdy mechanical construction. Telephone lines owned by power companies generally parallel high-voltage power lines, and sometimes are subjected to noise and induction, which may cause improper functioning of the apparatus. It is difficult to protect apparatus and personnel against the hazard of high voltage in the telephone-wire line caused by induction or accidental contact between the telephone and the power lines. Because of the inherent advantages of using the power line itself as the conducting medium for a communication or control circuit the application of power-line carrier should be thoroughly understood and fully exploited.

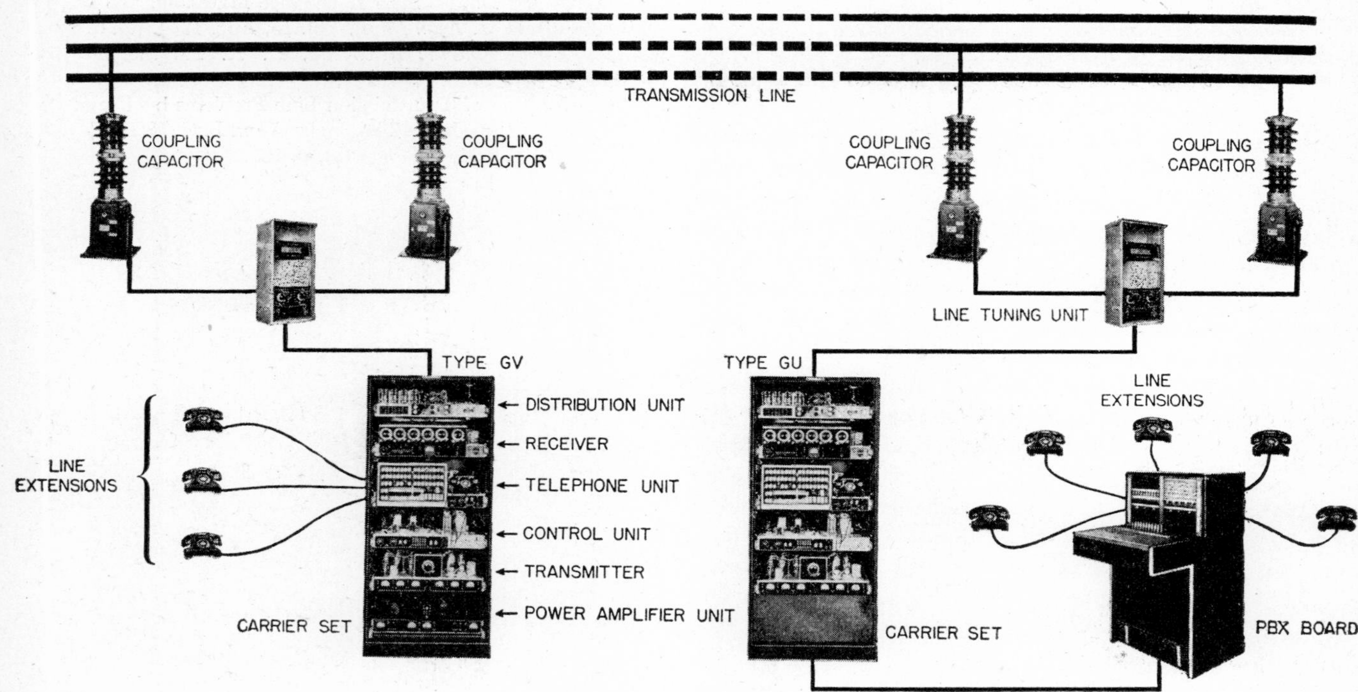

FIG. 25·1 A modern carrier-communication system.

25·1 CARRIER COMMUNICATION

Carrier-communication equipment is used both for dispatching services and for conducting the operations of a power system economically over long distances. The continuously available channel provides a ready means of transmitting intelligence to vital points, thereby permitting closer scheduling of power flow and prompt action to cope with an emergency.

The basic elements of a carrier-communication system are shown in Fig. 25·1. They consist of line-coupling capacitors, line-tuning units, carrier set, and hand telephones. This is a typical set-up for long-distance communication, and a rather elaborate carrier set is used. Phase-to-phase coupling minimizes line losses and noise, as explained later in this chapter.

The modulated carrier is fed into the transmission line through coupling capacitors especially designed to withstand line voltage under all weather conditions. The upper end of the capacitor connects directly to the line; the lower end connects to ground through a carrier frequency choke coil, or drain coil. The junction between the drain coil and the capacitor is connected to the line-tuning unit. The drain coil effectively grounds the bottom end of the coupling capacitors for the power-system frequency but has little or no effect on the carrier frequency. The line-tuning unit is an adjustable inductor used to neutralize the capacitive reactance of the coupling capacitor at the carrier frequency. Thus the carrier set, in effect, feeds directly into the transmission line.

The carrier set shown in Fig. 25·1 is of the automatic simplex type with selective ringing, and is arranged for several telephones or line extensions at both terminals. Any one of these extensions can be connected into a PBX board as shown at the right of Fig. 25·1 to further extend and multiply its use.

A carrier-communication system may be designed to use either one or two frequencies. If the same frequency is used for transmission in both directions (a single-frequency system) only one station can talk at a time without producing interference. The carrier must be turned on when the person speaks into the microphone and must be turned off again when he is through. This can be accomplished manually by a pushbutton or automatically by voice-operated electronic or mechanical relays. The set shown in Fig. 25·1 is a single-frequency electronically controlled set more commonly known as an automatic simplex set. The advantage of a single-frequency system is that any number of stations can be tuned to the same frequency and have communication among them just like a party line. The disadvantage is that when the line is busy no other station can interrupt the person using it because his receiver is turned off during his transmission. However, if the system is of the automatic type, he can be interrupted at any pause between words, for the electronic control changes from transmit to receive and vice versa in a few milliseconds.

In a two-frequency system the operator need not turn his transmitter off to listen to the other station; hence it is possible for the person listening to interrupt a transmission without waiting for a pause in the transmission. Two-frequency or duplex equipment is less expensive than automatic simplex, and is perhaps more satisfactory, since it is simpler to build and maintain. It is limited in application because only two stations can converse at once. If several stations are involved, half of them tuned to transmit on frequency *A* and the other half on frequency *B*, it is obvious that those tuned to frequency *A* cannot communicate with each other. Neither can those tuned to frequency *B*, but any *A*-frequency station can communicate with any *B*-frequency station. Another limitation of the two-frequency system is the number of frequencies required. Some systems have so many requirements for carrier channels that two frequencies cannot be spared for one channel.

A typical two-frequency duplex carrier set is shown in Fig. 25·2. Figure 25.3 shows a simplified schematic diagram of this equipment. This set has a carrier output of 25 watts, and its output frequencies are within the 50- to 150-kilocycle frequency band.

The transmitter employs one type 807 vacuum tube as a master oscillator. The oscillator drives six type 807 tubes in a push-pull-parallel radio-frequency power-amplifier stage. The output of the amplifier is coupled by the output transformer, *T*-2, to the coaxial cable. The coaxial cable terminates at the line-tuning unit which applies the radio-frequency output to the power line through the coupling capacitors. Grid-bias amplitude modulation is used. The bias on the amplifier tubes is obtained from a resistor in the common-cathode circuit. Radio-frequency excitation is controlled by adjusting the plate voltage on the master oscillator.

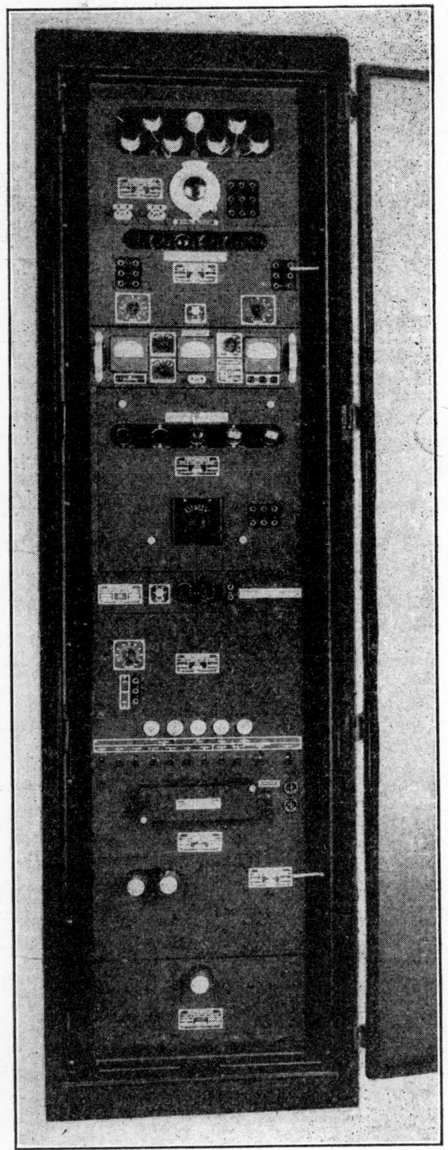

Fig. 25·2 Two-frequency duplex carrier-communication set.

The audio amplifier is of the constant-level type to prevent overmodulation of the carrier wave. A 40-decibel variation in the voice input level will produce a variation in the output of not over 6 decibels. This prevents distortion due to over-

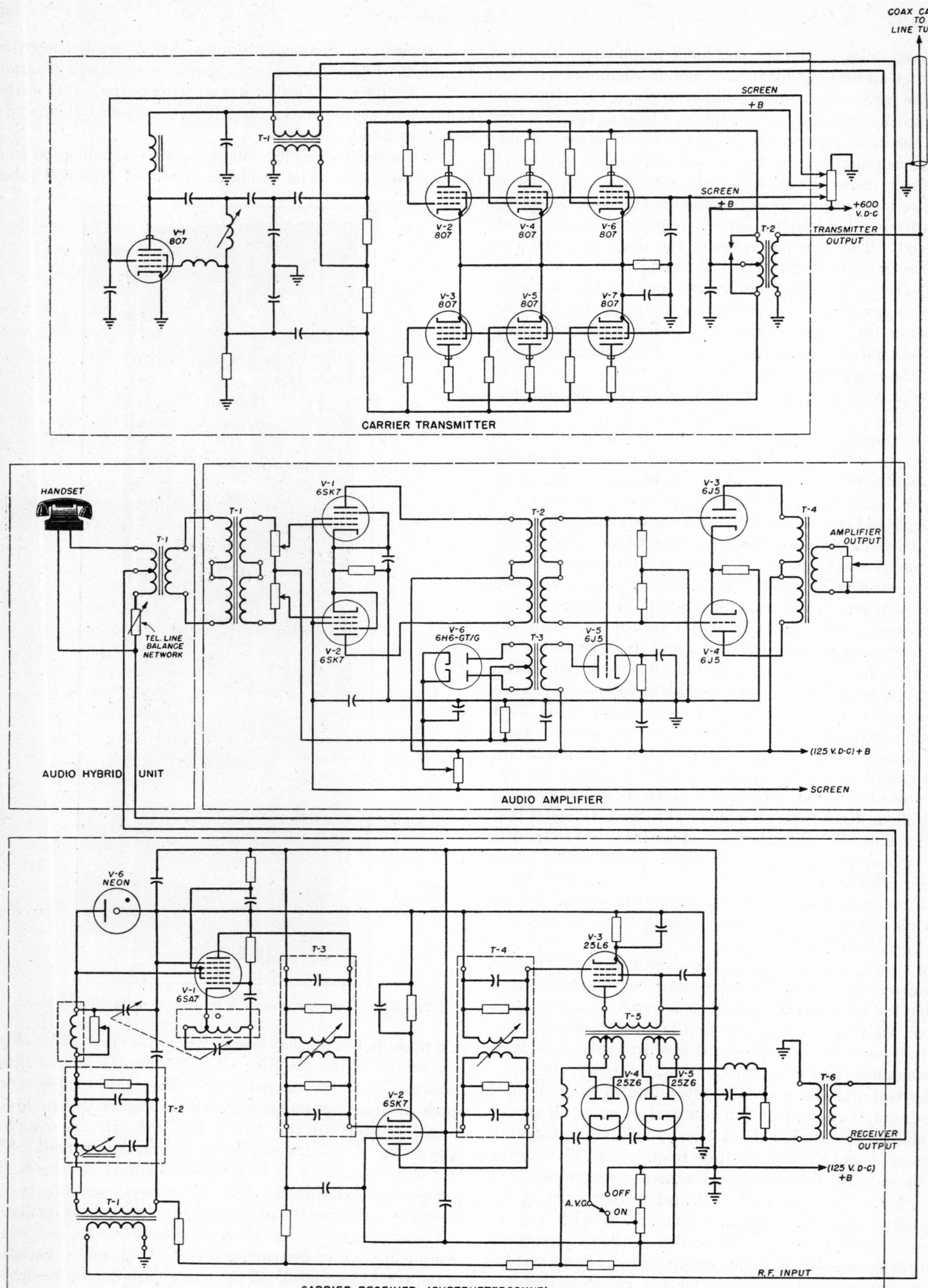

Fig. 25·3 Simplified schematic diagram of a two-frequency duplex carrier set.

modulation, and compensates for the difference in voice levels of different people. The input to the amplifier is obtained from the audio hybrid unit. This unit contains the necessary bridge circuits and balancing network to permit both the transmitter and the receiver to be connected to a two-wire telephone line. Without this balancing network, the output of the receiver would feed back into the audio amplifier and modulate the transmitter. If a similar condition existed at the other end of the carrier channel, regenerative "singing" would result and the channel would not be usable.

The receiver employs a superhetrodyne-type circuit with a 6SA7 mixer, a 6SK7 i-f amplifier, a 25L6 second i-f amplifier, and two 25Z6's as demodulators. The output of one 25Z6 is used for automatic volume control. The output of the other is applied to the telephone receiver. The automatic volume control will limit the variation in output to 6 decibels for a variation in input of 40 decibels. The over-all distortion of the transmitter and receiver combined is less than 10 percent.

The control relays and the calling circuits are not shown on Fig. 25·3. The set can use voice calling, automatic ring calling, code-bell calling, or dial selection by substituting panels containing appropriate calling circuits.

For voice calling, the receiver output is connected to loudspeakers at all stations, and the person calling announces the name of the person or place being called.

In automatic ring calling, the equipment rings a bell at the receiving station when the carrier is received and continues ringing until the call is answered.

Code-bell calling is similar to automatic ring calling except that a code button is provided so that the operator can break the call up into a definite code.

In dial-selection calling, the equipment functions like a dial-telephone system. The selector system will select the telephone set which was dialed, and only the bell on that set will ring.

25·2 TYPES OF MODULATION

The ultimate quality of transmission depends not only upon technical perfection of the apparatus but also upon the ratio of signal to noise in the telephone receiver or loudspeaker. The system of modulation has much to do with this ratio. Three systems are currently in use: amplitude modulation (AM), frequency modulation (FM), and single-sideband modulation (SSM).

Amplitude Modulation

Amplitude modulation is the oldest and by far the most widely used system. When a continuous wave of, say, 100,000 cycles is varied in amplitude at a 3000-cycle rate (which corresponds to the upper limit of voice frequencies) two new frequencies are produced: 100,000 + 3000 or 103,000 cycles; and 100,000 − 3000 or 97,000 cycles. The bandwidth must be 6000 cycles for good transmission and reception of a 3000-cycle tone. Since 3000 cycles is the highest voice frequency that must be transmitted for good intelligence, 6000 cycles represents the bandwidth required for an amplitude-modulated system. The random power noise in an AM receiver is directly proportional to bandwidth. The wider the bandwidth is, the greater the noise. This, of course, assumes that line noise presents a problem. Many carrier systems operate over short distances, and the received carrier energy is so far in excess of the noise energy that noise is not an important factor.

Frequency Modulation

In a frequency-modulated system the modulating wave varies the frequency of the carrier wave (instead of varying its amplitude) at the modulating frequency. Whenever the maximum change in carrier frequency is equal to the modulating frequency, the deviation ratio is said to be 1 to 1. FM broadcasting stations use a ratio of 5 to 1. For carrier communication a ratio of 1 to 1 is used because most power systems have the frequency band (50 to 150 kilocycles) divided up into 6000-cycle channels, and will permit no more than this per channel. An FM receiver is designed to be as sensitive as possible to frequency changes and as insensitive as possible to amplitude changes. The greater the deviation ratio of the modulation, the greater is the improvement in signal-to-noise ratio. For a deviation ratio of 1 to 1 an FM system has an improvement of 4.8 decibels in signal-to-noise ratio over an AM system having an equal amount of unmodulated carrier power. For an FM system to be successful the radio-frequency signal must be greater than the noise. Because of the action of the limiter tube in the receiver the output is nearly constant whether the input is signal, noise, or a combination of both. When no signal is being transmitted this noise output is highly objectionable; therefore all FM receivers employ a squelch circuit which shuts off the noise if no carrier is received. Thus if the signal level falls below the noise level, the output of the receiver becomes zero. In an AM system, communication continues even when the signal may be only one-fourth as strong as the noise.

Single-Sideband Modulation

Single-sideband modulation is similar to amplitude modulation except that the carrier and one of the two sidebands are eliminated. The remaining sideband requires only half the bandwidth of an AM system. For equal-peak transmitter-voltage outputs a single-sideband system has a signal-to-noise ratio 9 decibels greater than that of the AM system. In addition to the improvement in signal-to-noise ratios over amplitude and frequency modulation the single-sideband system utilizes a smaller portion of the frequency spectrum. In this way a great many more channels are made available for other uses or for future expansions.

25·3 CARRIER RELAYING

In the past, carrier relaying has usually been employed for one or more of the following reasons:

1. To improve stability.
2. To prevent relay operation when the system is out of synchronism.
3. To permit quick reclosing of breakers.
4. To increase system design flexibility.
5. To reduce shock to the system.

6. To prevent line-to-ground faults from becoming three-phase faults.
7. To improve ground relaying.
8. To provide a combination of services.

Today, carrier relaying is simply the logical choice for a network of high-voltage transmission lines, and many systems are changing whole groups of lines over to it. Carrier is the only form of pilot protection that is economical for lines longer than a few miles. Pilot protection is becoming more and more desirable because the first line of defense is practically independent of other lines; correct instantaneous relay operation does not hinge on adherence to a fixed system. Maintenance of lines does not interfere with correct relaying and additions, and new interconnections do not necessitate replanning of the whole relaying system. A fault on any line is cleared instantly at both ends without requiring time co-ordination.

Dependence is placed upon carrier transmission over the line only when the line is not faulted; therefore there is no cause for carrier failure due to a line fault. Line traps are used to prevent the transmitted carrier signal from being short-circuited by a nearby fault outside of the protected line.

If the carrier set is of the outdoor type, the line tuning unit is in the set, and only four physical components are involved. In fact, even with the carrier set indoors, if the distance to the coupling capacitor is not over a few hundred feet, the tuning unit can often be included in the set. Except when the distance is short, a matching transformer mounted as a part of the coupling capacitor is required.

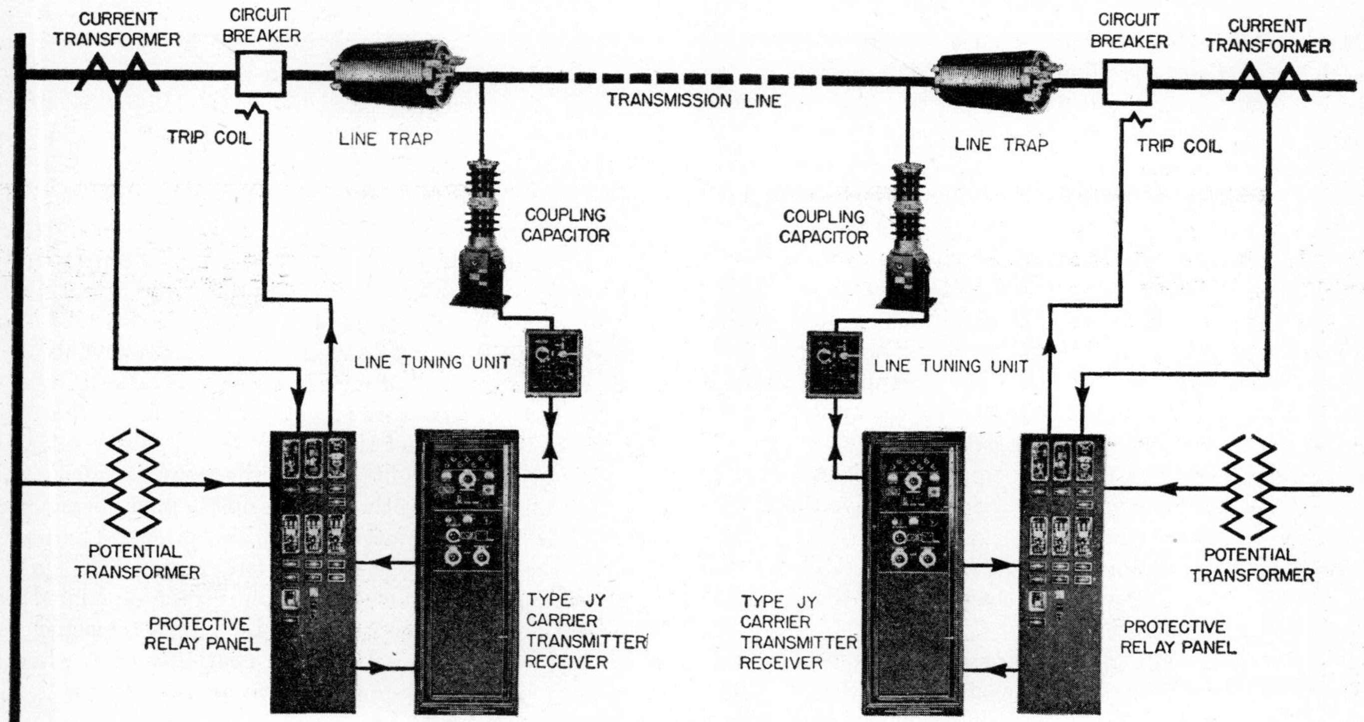

Fig. 25·4 Elements of a carrier relaying system.

The five basic elements of a carrier relay system are shown in Fig. 25·4. A line trap, a coupling capacitor, a line-tuning unit, a carrier set, and a protective relay panel are required at each end of the protected section of line.

The relays are connected to the line by current and potential transformers, and will detect the presence and the direction of a fault. Whenever the fault is toward the opposite end of the protected section the relay will trip the breaker unless restrained by the reception of carrier from the other end of the line. Reception of carrier indicates that the fault is beyond the end of the protected section and will be isolated by some other breaker. Thus a pair of equipments can detect almost instantly whether a fault is external or internal. If it is internal, no carrier is transmitted in either direction, and the relays trip the breakers as fast as possible. If it is external, carrier is transmitted over the line to prevent tripping.

The relay equipment shown in Fig. 25·4 is the high-speed distance carrier scheme. If the carrier is cut out of service the relays still provide the best type of non-carrier protection. More simplified schemes are often used. They do not give back-up protection, but they require only a single operating element for phase-to-ground carrier-relay protection. Currents at the two ends of the line are compared for fault discrimination, and hence potential transformers are not needed. This scheme is ideal where existing relays are adequate for back-up protection.

25·4 TELEMETERING

Carrier telemetering is the remote indication and metering of real and reactive power and other quantities. For effective operation of a power system the operator must know the power flow at important points of the system. Usually interconnecting lines are of prime importance since stability, heating, and contractual limits must not be exceeded. Proper operation of transmission systems is simplified by carrier telemetering, because it provides an accurate, prompt indication of power flow at all times.

The basic elements of a carrier telemetering system are

shown in Fig. 25·5. It is assumed that the power supplied from a generator into the system at one point is to be indicated at a remote load-dispatcher's office. The impulse-transmitting meter is similar to a watthour meter except that its contact-making device operates at a rate proportional to the speed of rotation of the disk. This contact keys the carrier transmitter at a corresponding rate and sends pulses of carrier current over the transmission line to the receiver at the load-dispatcher's office. In the receiver the carrier pulses are amplified and rectified. The resulting d-c pulses enter the impulse receiver, and they are converted into smooth direct current. This current actuates the indicating or recording instrument which is calibrated in units of the quantity being measured. Kilowatts, kilovolt-amperes, and other quantities can be transmitted in this way.

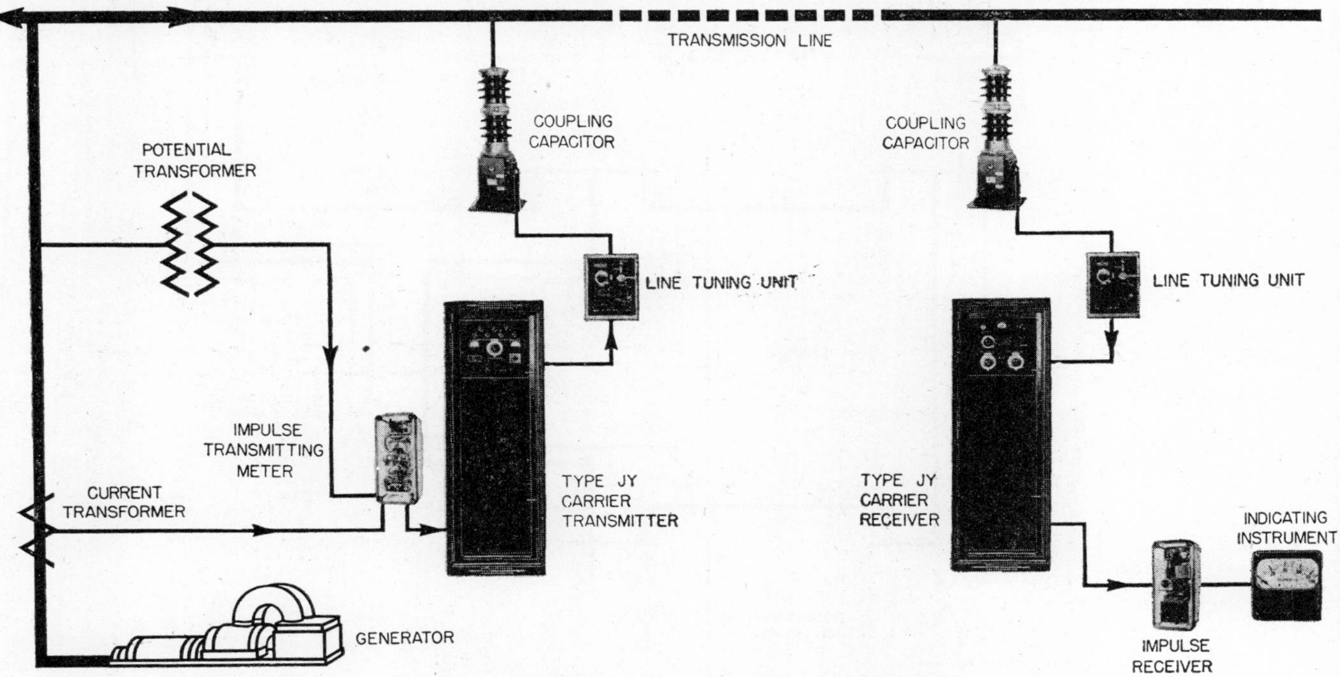

Fig. 25·5 Elements of a carrier telemetering system.

25·5 CARRIER LOAD CONTROL

Telemetering is frequently used with load-control equipment. Carrier load control is used for remote control of the load over a tie line or to control power output of a generator by means of signals sent over a power-line carrier channel. These signals may be initiated manually by the load dispatcher or automatically by load controllers located in his office.

A typical combined telemetering and load-control system is shown in Fig. 25·6. Two power systems *A* and *B* are assumed to be connected by two tie lines for which the billing points are at *M* and *N*. The dispatcher's office for system *B* is at point *O*. System *B* is extensive, having steam and water-power generation, but the steam generation is normally operated on a block-load basis, and any required load swings are handled by the hydroelectric station, which does not have enough water flow for continuous delivery of power.

The requisites for controlling the total load interchange over the tie lines between the two companies are: first, an indication at point *O* of the interchange power; and, second, a means of controlling the hydroelectric plant to maintain the desired tie-line loads. Power indications from *M* and *N* are brought to *O* by telemetering equipment. At point *O* these quantities are totalized and go into a single load controller. The load controller is essentially a load-recording instrument except that it has "raise" and "lower" contacts that operate if the total tie-line load is above or below the desired value. A load-control carrier channel is set up between *O* and the hydroelectric plant *H* over one of the power lines between these points. When the load controller at *O* indicates that more power should be generated by system *B* to increase power flow toward system *A*, a "raise" indication is sent over the carrier channel. For simplicity, assume that two carrier frequencies are used between *O* and *H* for load control. Actually, two audio tones over a single carrier channel probably would be used. One frequency would be used for increasing power, the other for decreasing power, and they are called "raise" and "lower" frequencies, respectively.

The load controller at *O* starts the carrier set, which sends out the raise frequency and transmits a signal with a duration proportional to the deviation of interchange power from that required. This signal is received at *H* and actuates the speed-changing motor on the governor of the waterwheel generator. There is a waiting period to give the telemetering equipment an opportunity to respond to this change; then if further correction is required another raise impulse is sent out, but if the load then is nearer to the desired value the duration of subsequent impulses becomes smaller until the system finally stabilizes with the desired load over the tie lines. For a lowering operation the process is reversed.

If several remote hydroelectric stations are used to control the load, the tie-line load controller is arranged to control one of the stations as described. Control of the remaining stations is by proportional-load controllers. The proportion of load to be carried by each of the waterpower stations is set up on proportional load controllers in the dispatcher's office. Telemetering channels from the hydraulic stations to the dispatcher's office indicate in each of the proportional load controllers the load on that particular station. If the load carried by a particular station differs from the proper proportion with respect to the master station, the proportional load controller sends out raise or lower impulses to increase or decrease the load of that station until the proper proportional load is established among all stations involved.

Carrier supervisory control is an ideal solution. Control of the Victorville and Silver Lake sub-stations out in the desert along the Boulder Dam to Los Angeles lines is a notable example of the application of carrier supervisory control.

The elements of a simple supervisory-control system are indicated in Fig. 25·7. In addition to the supervisory-control equipment at the dispatcher's office and at the sub-station end, a complete two-way carrier channel is required. This includes the transmitter and receiver at each end, the

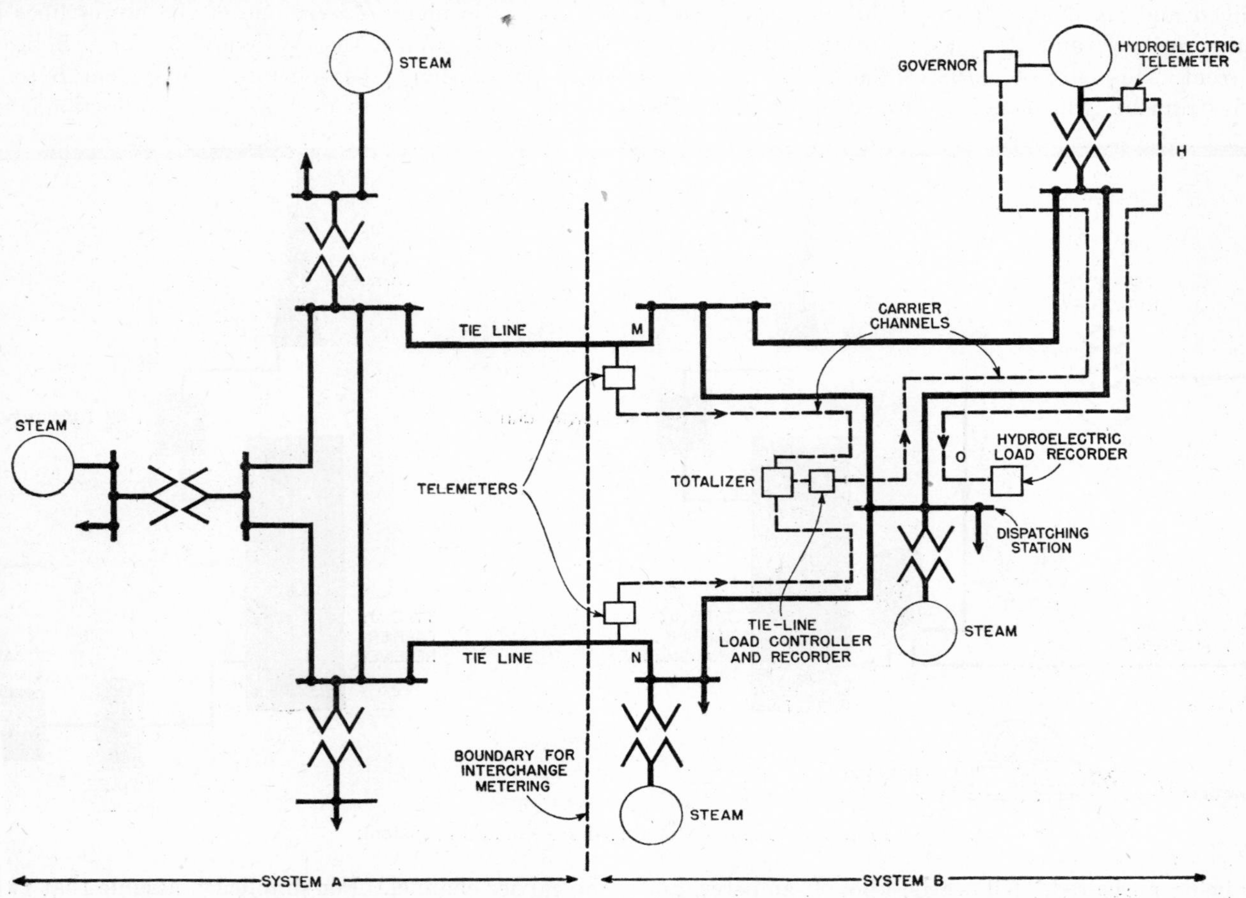

Fig. 25·6 Power-line carrier telemetering and load control.

If one of the hydroelectric plants contains several machines, all of which are to respond to the load-control signals, the usual procedure would be to put the primary control on one unit and use proportional load controllers that would redivide the load among other units.

25·6 CARRIER SUPERVISORY CONTROL

Supervisory control, as the name implies, is both control and supervision. It differs from ordinary remote control in that many devices or "points" are controlled and supervised by means of a single wire or carrier current. Carrier supervisory control differs little from supervisory control equipment operated over wires. The only difference is in the manner of sending indications of certain quantities.

With supervisory control it is unnecessary to staff isolated stations, and widely separated circuit breakers can be controlled by one centrally located operator. It would be impractical to use wire channels to control many tie switches and stations at isolated points along transmission lines.

line-tuning units, coupling capacitors, and any line traps that may be necessary.

Control and supervision of five circuit breakers from a remote point are shown in Fig. 25·7. Corresponding to each breaker is an escutcheon plate on the supervisor's equipment including indicating lights that show the position of the breaker and a control switch for its operation.

Any automatic or local operations of the breakers are reported over the carrier channel so that the operator has a continuous lamp indication of every device controlled.

Coded signals are sent over the line by telegraphic keying of the carrier channel. A contact in the dispatcher's equipment keys the transmitter to send out a pre-selected series of carrier pulses corresponding to the code used. The equipment is similar to that used in a dial-telephone system except that it is unnecessary for the operator to remember the number; he simply presses a button and the equipment "dials the number."

In addition to the control of circuit breakers, several metering or position indications can be carried over a supervisory-control channel, and the quantities can be controlled as they are measured. For example, amperes, volts, watts, power factor, reactive volt-amperes, or position of a device can be indicated and controlled over the carrier channel by using one supervisory point for each quantity.

After the operator has an indication of a particular quantity he can disconnect from that point by the master reset key; or, if there is no objection to continuous carrier on the circuit, any quantity can be indicated continuously. This continuous indication is convenient for the operator, for he can select a quantity that affords the chief basis for operation.

possible coincidence of a series of voice impulses with the code used for supervisory control.

25·7 BASIC APPLICATION PROBLEMS

The examples of carrier applications shown in Figs. 25·1, 25·4, 25·5, and 25·7 all have three essential elements in common, excluding the transmitting and receiving apparatus. These are the transmission line, the coupling capacitor, and the line-tuning unit. Auxiliary circuit elements often required are line traps and coaxial cable circuits.

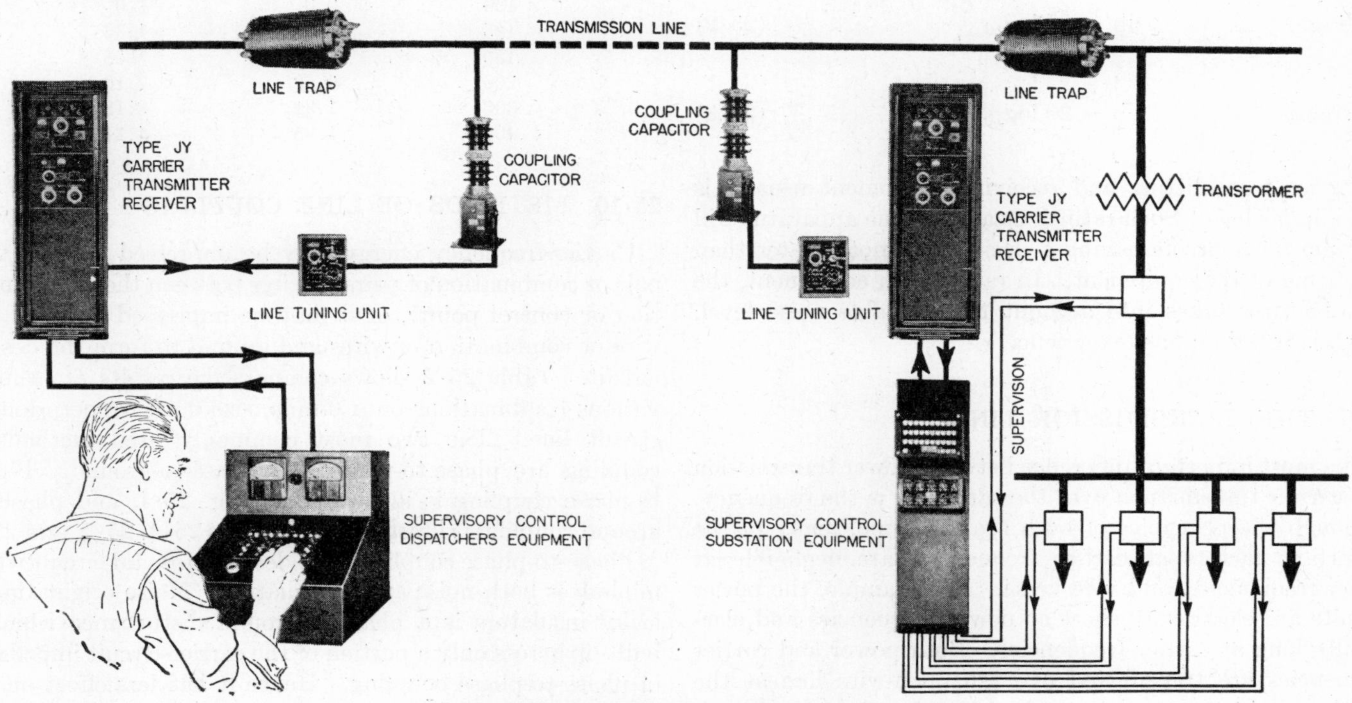

Fig. 25·7 Carrier supervisory control.

He can change positions to check other quantities and return to the most important quantity for continuous indication.

Demand and integrating metering also can be carried on over the carrier supervisory-control channel. During the demand interval the meter at the sending end stores up impulses proportional to the integrated kilowatthours. At the end of the demand interval the metering takes over the carrier channel for a short time and sends in rapid succession a number of pulses corresponding to the kilowatthours accumulated during the demand interval, such as 15 minutes. The receiving meter responds to the total number of impulses in each group. This response is a demand indication. It also runs the integrating meter ahead by an amount corresponding to the number of impulses in that group, and thus corrects the integrating meter each 15 minutes. In so doing, it utilizes the carrier channel only a brief part of the total time, and therefore the channel is available the remainder of the time for supervisory control.

If voice communication is to be used over the same carrier channel used for supervisory control, it is desirable to use one supervisory control point for ringing and to lock out the supervisory-control equipment during the telephone conversation. This method prevents misoperation because of a

All of these are parts of the carrier-frequency circuit, and much of the carrier-current energy is dissipated in them. The losses in all the circuit elements must be evaluated in order to determine how much power is required at the sending end of the line to produce a suitable signal at the receiving end of the line.

Whatever apparatus is attached to or inserted in the power circuit for carrier purposes must not appreciably alter the characteristics of the power circuit at 60-cycle or power frequency. It must merely adapt it to the carrier frequency and provide the necessary terminals for connecting the carrier energy.

25·8 TRANSMISSION LOSSES AND GAINS

In carrier transmission, it is convenient to consider the transmission characteristics of a system in terms of attenuation or the diminution of power along the line. The ratio between voltages, currents, or power at any two points is a measure of the attenuation of the circuit between these two points. It is not convenient in practice to express transmission losses or gains in terms of these ratios directly. Losses so expressed cannot be added to obtain total loss, but must be multiplied. When losses or gains of circuits or circuit elements are expressed in decibels (abbreviated db), they may be added directly to obtain total loss or gain of a system. The number of decibels gained or lost is equal to 10 times the logarithm (to the base 10) of the power ratio, or 20 times the logarithm of the voltage or current ratio. When voltage or current ratios are used to compute decibels the circuit impedance must be the same at the two reference points.

$$\text{(Power)} \qquad \text{db} = 10 \log_{10} \frac{P_1}{P_2} \tag{25·1}$$

$$\text{(Voltage)} \qquad \text{db} = 20 \log_{10} \frac{E_1}{E_2} \tag{25·2}$$

$$\text{(Current)} \qquad \text{db} = 20 \log_{10} \frac{I_1}{I_2} \tag{25·3}$$

Carrier transmitting and receiving equipment usually is rated in decibels. Such ratings mean that the apparatus will function on a circuit having a decibel loss not greater than the rating of the equipment. In rating such equipment, the manufacturer takes into account the probable noise level, which is based on past experience.

25·9 THE TRANSMISSION LINE

The most important difference between power transmission and carrier transmission over the same line is the frequency. Although the principles of both are the same, many of the important factors at carrier frequencies are negligible at power frequencies, and vice versa. For example, the power circuits are electrically short at power frequencies and electrically long at carrier frequencies. Both power and carrier frequencies are transmitted over an open-wire line at the speed of light, or approximately 186,000 miles per second. Since

$$\text{Wavelength} = \frac{\text{speed of propagation}}{\text{frequency}}$$

one wave at 60 cycles is 3100 miles long, whereas one wave at 60,000 cycles is only 3.1 miles long. This is why the average power line is said to be electrically long at carrier frequencies and short at power frequencies.

Another important difference between power transmission and carrier transmission is the relative efficiency. At power frequencies the impedance of a line as measured at the sending end is largely a function of terminal or load impedance. If line losses are too high, it is necessary merely to increase both load impedance and voltage, thereby reducing the current, to improve the line efficiency.

At carrier frequencies most of the carrier energy is dissipated in the line itself, and the load impedance has very little effect on the sending-end line current. This is especially true of long lines. The voltage and current relationship is fixed by the characteristic impedance of the line, and nothing practical can be done to change it. Efficiency reaches a maximum when the load impedance is equal to the surge impedance of the line. Table 25·1 shows the losses caused by mismatching impedances. Column 1 gives the value of the terminating impedance in percent of line impedance. Column 2 shows the loss if the terminating impedance is pure resistance. Column 3 shows the loss for a terminal impedance with a 45-degree phase angle.

TABLE 25·1 LOSS CAUSED BY MISMATCHING IMPEDANCES

Terminating Impedance in Percent of Line Impedance	Db Loss for Pure Resistance Termination	Db Loss for a Terminal with a 45° Phase Angle
25	1.95	2.0
50	0.52	1.0
75	0.1	0.88
100	0.0	1.0
125	0.06	1.32
150	0.19	1.58
200	0.52	2.16
300	1.32	3.18
400	1.95	4.14

25·10 METHODS OF LINE COUPLING

Carrier-frequency energy may be impressed between any pair or combination of pairs of wires between the communication or control points, or it may be impressed between any wire or combination of wires and ground to form the desired circuit. Table 25·2 shows comparative results of utilizing various combinations on a 6-mile, size 0000 copper, double-circuit line. The two most commonly used methods of coupling are phase-to-phase and phase-to-ground. Phase-to-phase coupling is illustrated in Fig. 25·1, and phase-to-ground coupling is illustrated in Figures 25·4, 25·5, and 25·7.

Phase-to-phase coupling provides an all-metal circuit which minimizes both noise and attenuation. Noise originating in leaky insulators is a phase-to-ground disturbance which is built up across only a portion of the carrier-circuit impedance in phase-to-phase coupling. Hence it has less effect on the carrier apparatus than the same noise would have on apparatus connected between the same phase and ground.

Noise components caused by lightning and static on the three phases of a line are generally in phase with each other, and tend to cancel one another in the coupling apparatus.

For these reasons phase-to-phase coupling generally is used for long lines on which noise and attenuation are important, and phase-to-ground coupling is used for short lines or wherever noise and attenuation are not important. Other special forms of coupling sometimes are used on double-circuit lines on which the carrier circuit must be maintained under various line-switching conditions.

25·11 CHARACTERISTIC IMPEDANCE

The characteristic impedance of a line is the impedance which would be measured at the sending-end terminals if the line were infinitely long. If a line is terminated in an impedance equal to its characteristic impedance, all the energy reaching the load will be absorbed by it, and none will be reflected. For all practical purposes the characteristic impedance of open-wire power lines at carrier frequencies is a pure resistance. The following formula may be used for calculating the characteristic impedance to within ±10 percent, if the lines are uniformly spaced and the distance between wires is small compared to the height above ground.

For two-conductor open-wire lines,

$$Z_0 = 276 \log_{10} \frac{D}{d} \tag{25·5}$$

where Z_0 = characteristic resistance in ohms
D = spacing between wires in inches
d = radius of conductors in inches.

For one-conductor and ground,

$$Z_0 = \frac{276 \log_{10} \frac{2D}{d}}{2} \tag{25·6}$$

where D = height above ground in inches
d = radius on conductors in inches.

The characteristic impedance of a line is important largely from the standpoint of the apparatus connected to it. Table 25·2 shows the results of measurements made on a three-phase, 6-mile-long, double-circuit line with size 0000 copper conductors. Note that the losses are the same for an all-copper circuit regardless of the combination of wires or the Z_0 of the combination. On the combinations using ground return the losses are least for the highest Z_0.

TABLE 25·2 CHARACTERISTIC IMPEDANCE AND LOSSES ON A DOUBLE-CIRCUIT LINE WITH 0000 COPPER CONDUCTORS FOR VARIOUS METHODS OF COUPLING

Coupling between	Frequency (kc)	Z_0	Loss (db per mile)
1 and 3	41	850	0.073
1 and 3	49.3	850	0.073
1, 4 and 3, 6	47.5	440	0.073
1, 4 and 3, 6	64.5	465	0.073
3 and ground	48	520	0.236
3 and ground	66	510	0.266
1 and 3 to ground	59	315	0.273
1 and 3 to ground	48	280	0.32
1, 3, 4, 6 to ground	47.5	170	0.353
1, 2, 3, 4, 5, 6 to ground	48.5	140	0.343

25·12 LINE LOSSES

Losses in an open-wire line depend upon many factors: frequency, conductor size, spacing, resistivity, radiation, leakage, unbalance, coupling to adjacent circuits, tap lines, and power apparatus connected to it. Since data on these various factors are seldom available it is more practical to measure them than to attempt to calculate them. Calculations based solely on line material, construction, and the frequency, other factors being neglected, give results inconsistent with test data and usually far too optimistic. Table 25·3 gives a summary of data on actual installations and tests, and is suitable for estimating purposes. The data are for phase-to-phase coupling. For phase-to-ground or line-to-ground coupling these values should be multiplied by a factor varying from 1 to 4 depending upon the length of line. The data in Table 25·2 were taken on a line 6 miles long. On a line 60 miles long the loss per mile was only 10 percent greater for phase-to-ground coupling than the phase-to-phase value.

TABLE 25·3 APPROXIMATE LOSSES IN DECIBELS PER MILE

(Phase-to-Phase Coupling)

System Kv	50 Kc	100 Kc	150 Kc
230	0.05	0.075	0.107
138	0.065	0.09	0.12
115	0.075	0.102	0.13
69	0.08	0.11	0.14
34.5	0.1	0.13	0.16
13.8	0.15	0.18	0.21

25·13 EFFECT OF TAP LINES

Any tap circuit connected to the carrier circuit will absorb energy from it. If both the main and the branch circuits were infinitely long, or if both terminated in resistances equal to their characteristic impedance, the energy would divide equally between them and the loss of half the energy would represent only 3 decibels. An additional loss of 0.52 decibel would occur as a result of the mismatch at the junction. This amount of loss usually can be tolerated. More often, however, a tap line is short, and it is terminated with a power transformer the impedance of which is very high at carrier frequencies. Under this condition the carrier energy reaching the end of the line is reflected back toward the source. If the line is of such a length that this reflected energy reaches the tap point 180 degrees out of phase with the original wave it will act as though the main line were short-circuited at the tap point (neglecting the attenuation of the tap line). This action occurs when the electrical length of the tap is ¼ wavelength long or any odd multiple of ¼ wavelength long (¼, ¾, 5/4, and so on). If the tap line is ½ wavelength long (or any multiple of half wave) the reflected wave is in phase with the original wave, the tap appears as a very high impedance, and very little energy is absorbed by it.

As an example, at 60,000 cycles, tap lines approximately 0.775, 2.352, 3.875, etc., miles long would absorb most of the carrier energy. Tap lines of 1.550, 3.10, 4.65 etc., miles long would have little effect. This is true only if the transformer impedance is infinitely high or if it has high pure resistance. It can be assumed that the impedance of a transformer is reactive at carrier frequencies; this assumption changes the mileage examples just given. If the transformer characteristics are unknown, the only safe way to determine the effect of the tap is to measure it.

25·14 THE COUPLING CAPACITOR

The most satisfactory means of coupling the carrier energy into the transmission line is to use coupling capacitors. A typical unit is shown in Fig. 25·8. The type shown is for rigid-base mounting although they are also made for flexible-suspension mounting. Antenna coupling has been used successfully as a substitute capacitor coupling but has several disadvantages. If more than one circuit is carried on a single line of towers the energy is coupled to all circuits, even though the energy is desired in only one of them. A break in the power line or antenna wire may cause the two to come in contact with the possibility of damage to the carrier apparatus. To prevent this, expensive protective apparatus is required.

Fig. 25·8 115-kilovolt coupling capacitor.

The modern coupling-capacitor unit also contains the necessary protective devices such as drain coil, grounding switch, and spark gap built into the base of the unit. The carrier apparatus is completely protected even from direct

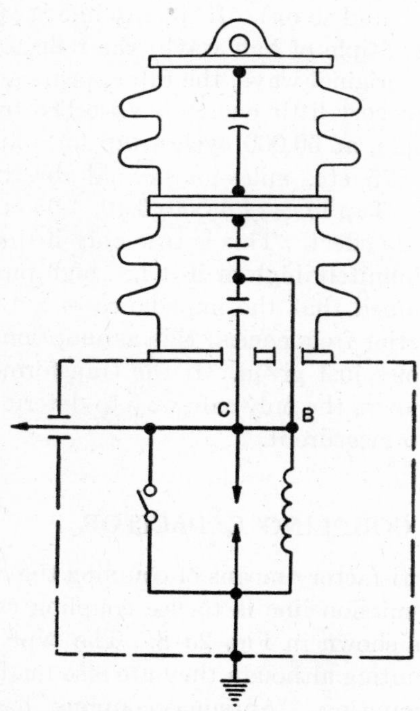

Fig. 25·9 Internal connections of 115-kilovolt coupling unit.

flashovers at the coupling unit. In order to insure this protection, the internal insulation strength of the capacitors is made greater than the external flashover voltage.

Figure 25·9 shows the internal connections of the unit illustrated in Fig. 25·8. For higher voltages additional standard sections are bolted to the top of the unit.

Coupling-capacitor units often are used with potential devices, the same unit functioning simultaneously as a carrier-coupling unit and a source of voltage for the potential device. A unit of this type is shown schematically in Fig. 25·10. As a potential device a single unit is capable of furnishing up to 150 watts of 60-cycle power depending upon the phase

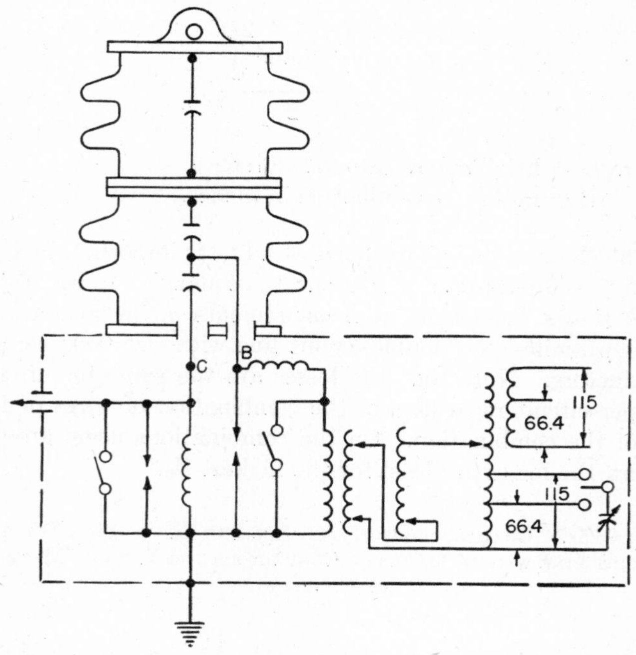

Fig. 25·10 115-kilovolt carrier-coupling unit for use with potential devices.

angle of the voltage in the load circuit in relation to the line voltage. Output is maximum when this phase difference is 90 degrees.

Capacitance and line-voltage ratings of coupling capacitors have been standardized. Table 25·4 shows the losses of standard units at various carrier frequencies. Obviously, the loss in the coupling unit is a small part of the total loss of the system.

Table 25·4 Standard Coupling Capacitors and Their Losses at 50, 100, and 150 Kilocycles

System Kv	Capacitance μf	Db Loss *		
		50 Kc	100 Kc	150 Kc
46	0.004	0.1	0.09	0.08
69	0.00275	0.15	0.14	0.12
92	0.002	0.2	0.175	0.16
115	0.00187	0.21	0.19	0.17
138	0.00137	0.3	0.26	0.22
161	0.00125	0.33	0.275	0.25
230	0.00094	0.4	0.36	0.32
287	0.00075	0.51	0.45	0.4

* Based on line-to-ground coupling Z_0 = 500 ohms.

25·15 THE LINE-TUNING UNIT

The line-tuning unit is simply an inductor used to neutralize the capacitive reactance of the coupling capacitor. The reactance X_c of the coupling capacitor is

$$X_c = \frac{1}{2\pi fc} \tag{25·7}$$

where X_c = capacitive reactance in ohms
f = frequency in cycles
c = capacitance in farads
π = 3.1416.

This reactance varies from 4240 ohms for a 0.00075-microfarad unit at 50 kilocycles to 266 ohms for a 0.004-microfarad

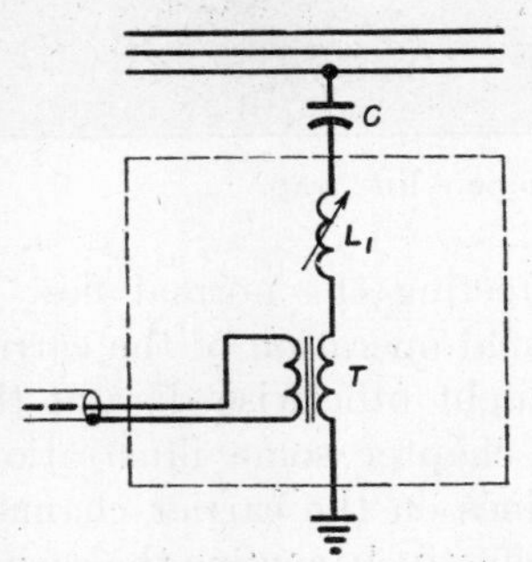

FIG. 25·11 Simple single-frequency tuning of single-coupling capacitor with coaxial-cable lead-in.

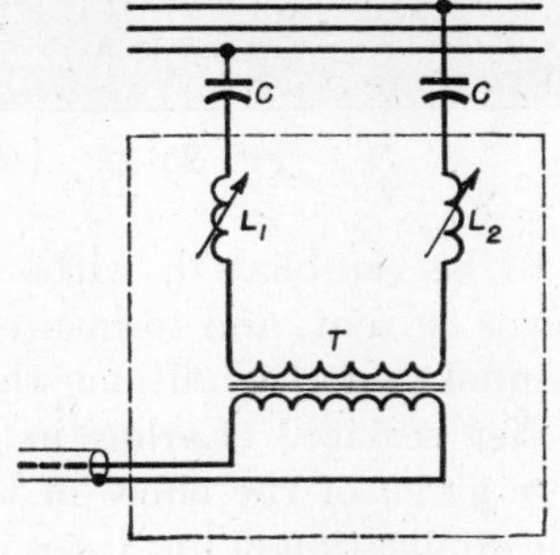

FIG. 25·12 Simple single-frequency tuning of two coupling capacitors (phase-to-phase or line-to-line coupling) with coaxial-cable lead-in.

unit at 150 kilocycles. These reactances correspond to inductances of 13.5 and 0.282 millihenrys, respectively.

Figures 25·11 to 25·15 inclusive illustrate some of the more common basic circuits of line-tuning units. In each of the single-frequency tuning circuits the inductance is adjusted for resonance with the coupling capacitor. Where the transformer T has appreciable leakage inductance it is considered part of the required tuning inductance.

In two-frequency circuits such as shown in Fig. 25·13, the inductor and capacitor which are connected in parallel (such

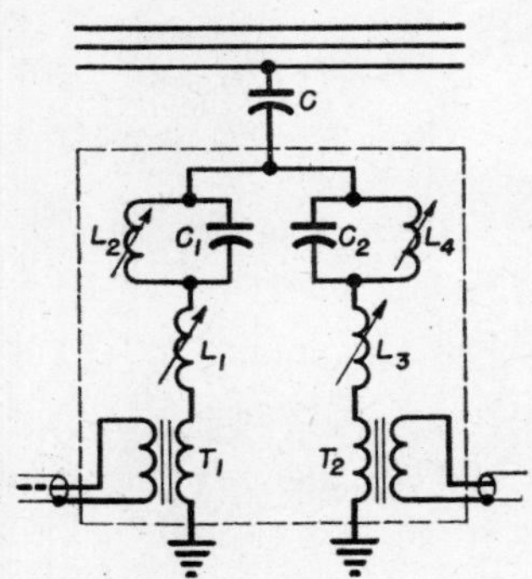

FIG. 25·13 Two-frequency tuning of a single-coupling capacitor with separate coaxial lead-ins for each frequency.

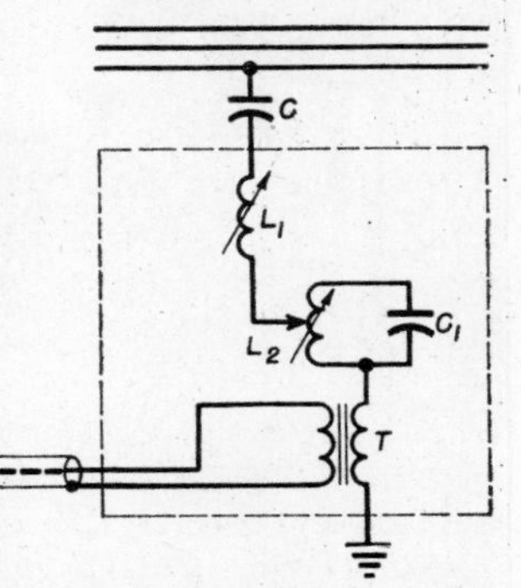

FIG. 25·14 Two-frequency tuning of a single-coupling capacitor with a single lead-in for both frequencies.

as L_2C_1) are tuned to the frequency not wanted in that branch of the circuit. This is the first step in the tuning-up procedure. The combination of L_1T_1C and the resultant reactance of L_2C_1 are then tuned to the frequency wanted in this branch. A similar procedure is followed for the other branch. At frequencies lower than that to which L_2C_1 is tuned the net reactance of these elements is inductive. If the desired frequency is close to the one not desired it is

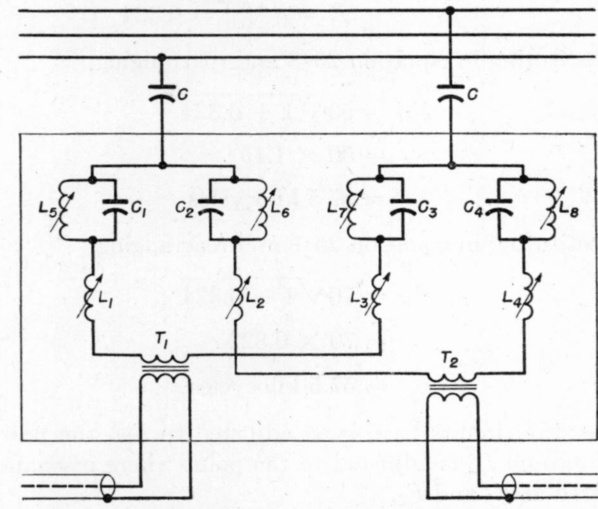

FIG. 25·15 Double-frequency tuning of two capacitors (phase-to-phase coupling) with separate coaxial lead-ins for each frequency.

necessary sometimes to insert an extra capacitor in series with L_1 to bring the system within the tuning range of L_1.

Figure 25·14 is a special type of circuit which is more difficult to adjust or tune since the elements individually are not resonated to either of the two frequencies to be used. Collectively they are, but individually they are not. Two-frequency tuning in this case is accomplished by a combination of series- and parallel-resonant circuits. Figure 25·16 illustrates the principles involved.

In Fig. 25·16, L and C are tuned to frequency F_M which is the geometric mean between F_1 and F_2, the two desired

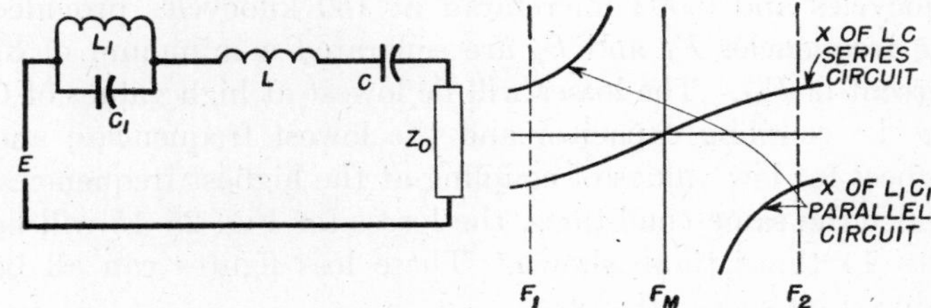

FIG. 25·16 Use of series and parallel resonance to produce two-frequency tuning.

frequencies. L_1C_1 is also tuned to frequency F_M. This frequency is found by use of the following relationship:

$$F_1 = \frac{F_M}{\sqrt{1+K}} \tag{25·8}$$

$$F_2 = \frac{F_M}{\sqrt{1-K}} \tag{25·9}$$

$$F_1\sqrt{1+K} = F_2\sqrt{1-K} \tag{25·10}$$

By substituting the desired frequencies for F_1 and F_2 in equation 25·10, a value of K is obtained. This value is then substituted in equation 25·8 or 25·9 to find F_M. As an example assume F_1 = 50 kilocycles and F_2 = 70 kilocycles.

$$50\sqrt{1+k} = 70\sqrt{1-K}$$

$$2500 + 2500K = 4900 - 4900K$$

$$7400K = 2400$$

$$K = \tfrac{2400}{7400} = 0.324$$

Substituting in equation 25·8 and rearranging,

$$F_M = 50\sqrt{1 + 0.324}$$

$$= 50 \times 1.15$$

$$= 57.5 \text{ kilocycles}$$

Substituting in equation 25·9 and rearranging,

$$F_M = 70\sqrt{1 - 0.324}$$

$$= 70 \times 0.822$$

$$= 57.5 \text{ kilocycles}$$

After the circuits have been adjusted to F_M, the position of the tap on L_2 is adjusted to the point where resonance is indicated at F_1 and F_2.

This principle applies to the method of tuning used in by-passes, line traps, line-tuning units, or wherever a combination of series and parallel resonance is used to produce two-frequency resonance.

Table 25·5 shows the losses encountered in a simple single-frequency tuning unit such as that in Fig. 25·11 when it is used to feed a line with a Z_0 of 500 ohms. For the circuit of Fig. 25·12 values shown in Table 25·5 should be multiplied by 2. General data on losses in two-frequency circuits cannot be tabulated because of the many possible combinations. Usually the losses are minimized when the two frequencies are separated as much as possible. For circuits like Fig. 25·13 the loss figures will vary from 4 to 7 times those shown in Table 25·5 when C_1 and C_2 are 0.004 microfarad at 50 kilocycles and 0.002 microfarad at 150 kilocycles provided the frequencies F_1 and F_2 are separated a minimum of 37 percent of F_M. The losses will be lowest at high values of C for the coupling capacitor and the lowest frequencies, and highest for low values of coupling at the highest frequencies. Under the same conditions, the losses for Fig. 25·15 will be 8 to 14 times those shown. These loss figures can all be reduced by special designs.

Table 25·5 Losses in Line-Tuning Units

System Kv	Capacitance μf	Db Loss *		
		50 Kc	100 Kc	150 Kc
46	0.004	0.16	0.15	.....
69	0.00275	0.2	0.17	0.17
92	0.002	0.22	0.19	0.19
115	0.00187	0.24	0.2	0.2
138	0.00137	0.26	0.225	0.215
161	0.00125	0.275	0.24	0.22
230	0.00094	0.32	0.27	0.26
287	0.00075	0.4	0.31	0.28

* Phase-to-ground coupling Z_0 = 500 ohms.

25·16 THE LINE TRAP

The primary function of a line trap is to prevent loss of the carrier-frequency energy in parts of the system outside the particular section of line over which the carrier channel is to be established, without affecting the normal flow of power current, and to insure normal operation of the carrier channel under conditions that might otherwise disrupt the carrier service. Earlier in this chapter some illustrations were given of the effect of tap lines on the carrier channel. The application of line traps will effectively confine the carrier frequency to its designated channel; permit transmission of carrier current when short circuits or grounds exist outside the section protected; conserve carrier energy by offering high impedance to its flow into other sections or branch lines; and allow normal operation with the power line intentionally grounded at the terminal stations.

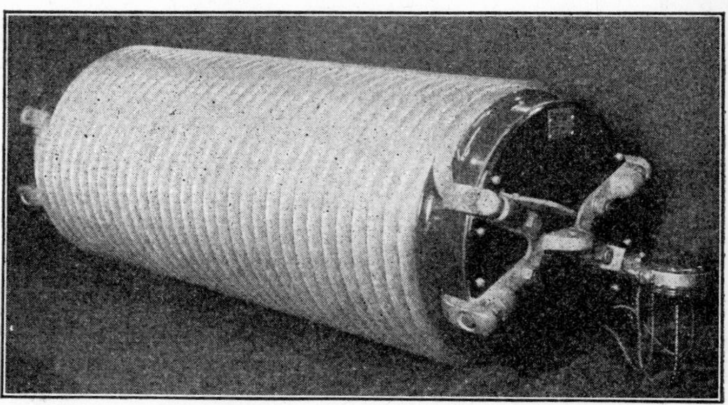

Fig. 25·17 A 400-ampere line trap.

Fig. 25·18 Installation of line traps and coupling units in a substation structure.

The line trap is simply a parallel-resonant circuit offering a high impedance over a band of frequencies centering around the resonant frequency. The resonant frequency must be the same frequency as that of the terminal carrier equipment.

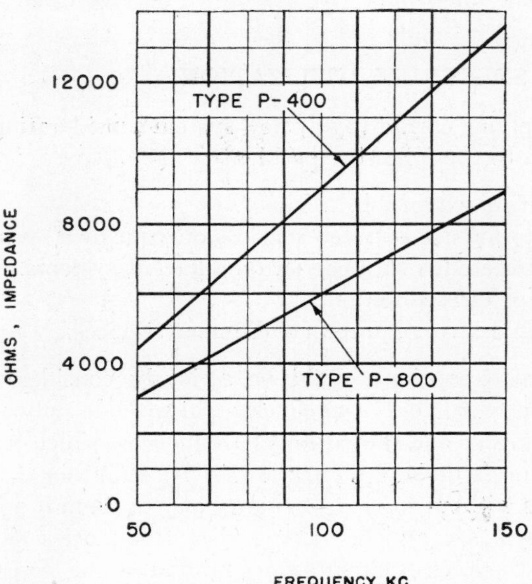

FIG. 25·19 Resonant impedance of type P line traps.

Line traps have been standardized at 400 and 800 amperes rating and are available for single- or double-frequency tuning. Figure 25·17 shows a 400-ampere trap, and Fig. 25·18 shows an installation of traps and coupling capacitors mounted in a sub-station structure.

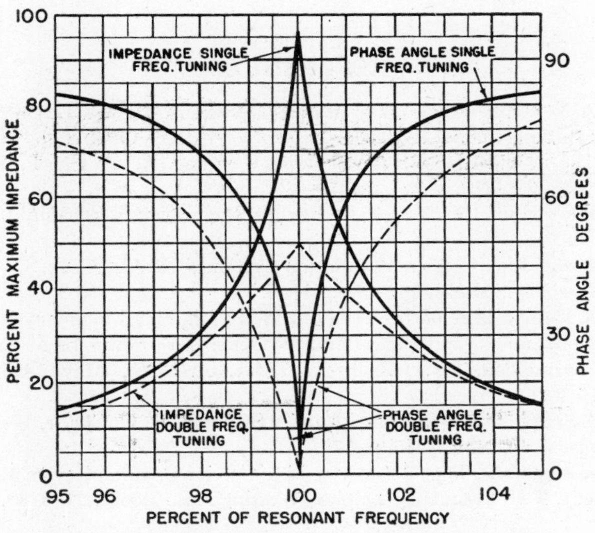

FIG. 25·20 Resonance curves of type P line traps.

The impedance which a line trap offers to the carrier is a pure resistance at the resonant frequency, and is

$$R_0 = \frac{(2\pi fL)^2}{R} \text{ ohms} \tag{25·11}$$

where R_0 = resonant impedance
π = 3.1416
f = frequency in cycles
L = inductance in henrys
R = radio-frequency resistance of coil alone.

The maximum values obtained commercially are shown in Fig. 25·19 for both 400- and 800-ampere traps. The difference in the two curves is due to the difference in the values

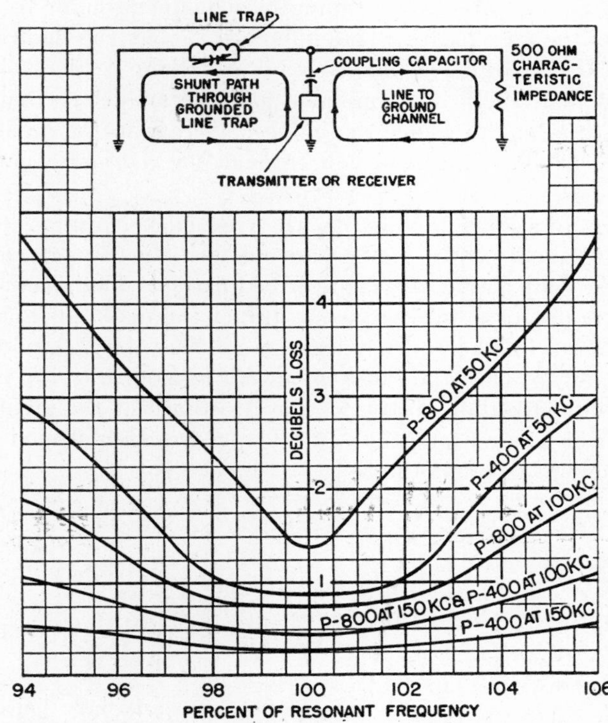

FIG. 25·21 Losses in type P single-frequency line traps when a line-to-ground carrier channel of 500 ohms characteristic impedance is short-circuited through line trap.

of L and R. Figure 25·20 shows how the impedance varies with frequency around the resonant frequency. Figures 25·21 and 25·22 show the loss to be expected in single- and double-frequency traps respectively.

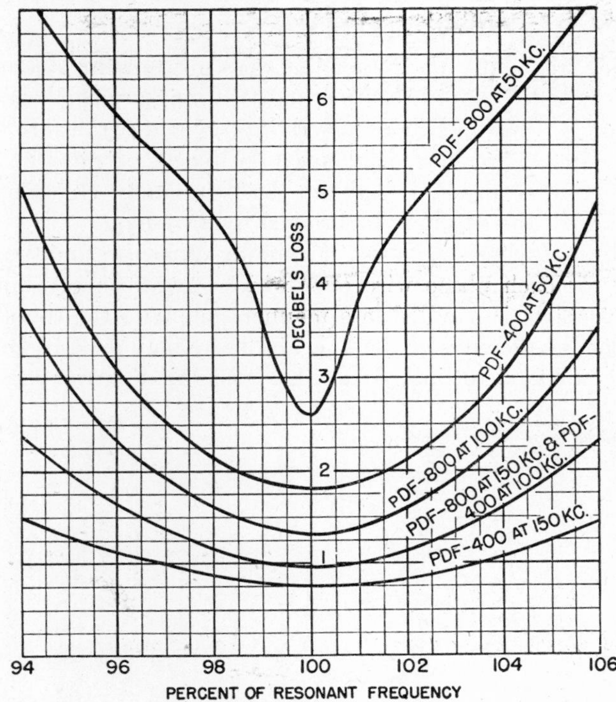

FIG. 25·22 Losses in type PDF double-frequency line traps when a line-to-ground carrier channel of 500 ohms characteristic impedance is short-circuited through line trap.

In order to make a line trap offer a high impedance at two frequencies an auxiliary resonant circuit (which does not carry power current) is coupled to the main trap. This produces two frequency tuning in a manner similar to that shown in Fig. 25·16. The difference is that the main trap coil is the parallel-resonant circuit (since the carrier voltage is applied across its terminals), and the auxiliary circuit is the series circuit (since the induced voltage in the auxiliary circuit causes current to flow through the coil and capacitor in series).

The procedure for figuring the frequency of tuning for a two-frequency trap is the same as for the two-frequency tuning unit of Fig. 25·14 described earlier. The maximum impedance of a two-frequency trap is figured the same as that for a single-frequency trap, except that the R term must contain not only the radio-frequency resistance of the main coil but also that portion of the resistance of the auxiliary coil which is reflected into it. The resistance reflected into one circuit from another is

$$R_1 = \frac{(2\pi f L_m)^2 R_2}{Z_2} \qquad (25\cdot 12)$$

where R_1 = reflected resistance
f = frequency in cycles
L_m = mutual inductance in henrys
Z_2 = net series impedance of auxiliary circuit, $\sqrt{(X_L - X_c)^2 + r_2^2}$

$$L_m = k\sqrt{L_1 L_2}$$

where k = value found from equation 25·1
L_1 = main trap inductance in henrys
L_2 = auxiliary circuit inductance in henrys.

25·17 BY-PASSES

Very frequently the power line does not present a continuous metallic circuit between the two points where carrier service is desired. There may be intervening transformers, open circuit breakers, disconnect switches, or possibly no connection of any kind between the two points. In most of these cases the gap may be bridged by a by-pass.

A by-pass consists of a set of coupling capacitors and tuning units for each phase wire over which carrier energy flows. The problem is handled in a manner similar to the coupling problem at the ends of the line, and there may even be traps in the line next to the by-passed point so that carrier service can be continuous even when the line is out and grounded. If the points where the coupling capacitors connect to the two lines are physically separated more than a hundred feet the line tuners may be connected together by coxial cable. Any of the line-tuning circuits shown may be used.

25·18 SPECTRUM UTILIZATION

In applying carrier to any new system consideration should be given to the following problems:

1. Future expansion.
2. Interference between services on your own system.
3. Interference with carrier on adjacent systems.
4. Type of modulation.
5. Signal level required for different services.

Future expansion should be seriously considered. The spectrum of 50 to 150 kilocycles will provide only so many channels, and one should select frequencies which will allow maximum frequency separation to be maintained between channels without having to change any of them if additional frequencies are added. The lowest frequencies should be used on the longest circuits since minimum line losses are encountered at these frequencies. Often the frequencies can be so grouped as to use common coupling capacitors for two or more frequencies. If the spectrum is crowded, single-sideband carrier may provide the necessary number of channels without interference. The length of a carrier-relaying circuit is usually shorter than a communication circuit. Therefore the higher frequencies should be used because attenuation will not be a problem.

REFERENCES

1. "A Versatile Power Line Carrier System," H. W. Lensner and J. B. Singel, A.I.E.E. Tech. Paper 44–34, Dec. 1943.
2. "The Application of Power Line Carrier," R. C. Cheek, *Power Plant Eng.*, Nov. 1945, pp. 76–81.
3. "Power Line Carrier," E. L. Harder, *Westinghouse Eng.*, July 1944, p. 98.
4. "Power Line Carrier Modulation Systems," R. C. Cheek, *Westinghouse Eng.*, March 1945, pp. 41–45.
5. "A Comparison of the Amplitude-Modulation, Frequency-Modulation, and Single-Side-Band Systems for Power-Line Carrier Transmission," R. C. Cheek, *Trans. A.I.E.E.*, May 1945, pp. 215–220.
6. "A Simple Single-Sideband Carrier System," R. C. Cheek, *Westinghouse Eng.*, Nov. 1945, p. 179.
7. Westinghouse Type JY Power Line Carrier Communication Equipment, Two-Frequency Duplex, Westinghouse Booklet B-3482.

Chapter 26

ELECTRONIC INSTRUMENTS

M. P. Vore

FOR measurement of simple single quantities such as potential, current, time, and frequency, mechanical and electromagnetic instruments are available. Within their operating ranges, these are accurate and reliable. Electronic instruments should be considered to supplement them, not replace them.

If the power available for operation of an instrument is minute, the amplifying properties of electronic tubes make it possible to design an instrument that does not unduly disturb the circuit to which it is connected. In other cases, because of the lack of inertia of electrons or the rectifying properties of electronic tubes, together with their small inductances and capacitances, electronic instruments can be made to operate at higher frequencies or over a broader range of frequencies than more conventional instruments. More complicated electronic instruments such as the electron microscope, mass spectrometer, and dynamic balancing machine exist because their functions are so fundamentally electronic that they cannot be performed by a non-electronic device, or because the use of electronics results in a less expensive, more flexible, or more easily operated unit.

Electronic instruments are widely used in all fields of research, in many types of testing, and in an increasing number of routine manufacturing operations. In this chapter the more common electronic instruments and their fundamental operating principles are described as an aid in applying the proper instrument to the job at hand. The reader who wishes to construct instruments and who needs detailed design information is urged to make use of the technical articles listed at the end of the chapter.

26·1 VACUUM-TUBE VOLTMETERS

For voltage measurements in d-c circuits or power-frequency a-c circuits that can furnish a moderate amount of power to an instrument, electromagnetic measuring instruments are preferable because they are simple, rugged, and accurate. There is, however, a wide field of usefulness for vacuum-tube voltmeters. In vacuum-tube voltmeters, the rectifying or amplifying properties of vacuum tubes are utilized to make voltage measurements without causing appreciable voltage drops in the circuits to which they are connected. Another advantage of the vacuum-tube voltmeter over the electromagnetic instrument is its ability to measure voltages at high frequencies or over a broad range of frequencies. The vacuum-tube voltmeter is especially useful also in the measurement of extremely small voltages. Not all vacuum-tube voltmeters possess these abilities equally.

Diode Voltmeters

Perhaps the simplest type of vacuum-tube voltmeter is the so-called diode voltmeter, the circuit of which is shown in Fig. 26·1. This type of voltmeter uses the rectifying properties of the diode to give a reading proportional to the average

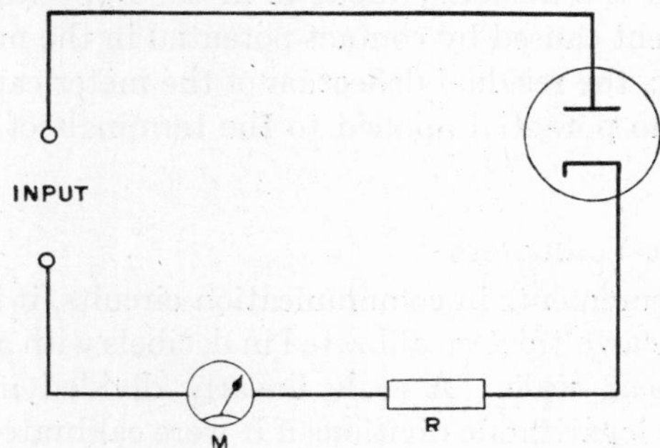

FIG. 26·1 Diode-voltmeter circuit.

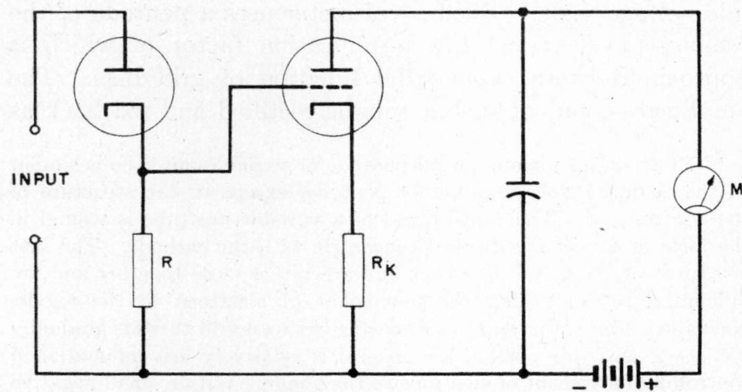

FIG. 26·2 Diode voltmeter with d-c amplifier.

value of alternating voltages over an extremely wide frequency range. If resistance R is large compared with the diode resistance, the reading of the meter is proportional to the average value of positive peaks of voltage.[1,2,3,4] For greater sensitivity, the simple voltmeter of Fig. 26·1 may be followed by a stage or more of d-c amplification as in Fig. 26·2.

If resistance R in Fig. 26·2 is replaced by a capacitance, the capacitance charges to the peak value of the signal voltage, and the reading of meter M is a function of the peak value of signal voltage instead of the average value. The capacitance usually is shunted by a high resistance to enable it to discharge when the input is removed. Inverse feedback in the stage of d-c amplification is provided by R_k which has the following beneficial effects: [5]

1. The reading of meter M is approximately independent of the constants of the tube used for d-c amplification.
2. The amplifier can handle large values of signal voltage because of cathode follower action.
3. If R_k is large, reading of meter M is proportional to signal input.
4. A multirange voltmeter can easily be made by arranging to vary R_k.

If the input terminals of the simple circuit of Fig. 26·1 are connected to a circuit, the contact potential between the dissimilar metals of the anode and cathode of the diode sends a current through the meter, even though no potential difference is applied to the input terminals. The residual deflection of the meter depends upon the resistance of the circuit to which it is connected. This difficulty can be overcome by the use of a second diode connected across the meter as shown in Fig. 26·15, so that current caused by the contact potential in the auxiliary diode is in the opposite direction to the current caused by contact potential in the main diode. In this way, the residual deflection of the meter can be made zero with no potential applied to the terminals of the diode voltmeter.

Logarithmic Voltmeters

For measurements in communication circuits, it is convenient to have a voltmeter calibrated in decibels with an approximately linear scale. A scale linearly divided in decibels would have logarithmic divisions if it were calibrated in volts. An amplifier with automatic gain control may be used as the basis of a voltmeter with a linear decibel scale or a logarithmic voltage scale.[6] Such a voltmeter uses a pentode of the remote-cut-off type,* the amplification factor of which is approximately an exponential function of grid bias. The amplified output of such a tube is rectified and fed back as grid bias in such a way as to decrease the gain of the tube. The resulting grid bias is almost a logarithmic function of the average input voltage to the tube. Figure 26·3 is a simplified circuit of such a voltmeter. The output of the amplifier is rectified by a diode. Rectified voltage appears across resistance $R1$ so that the grid is made more negative as the signal voltage increases. Capacitor C filters the rectified voltage. Bias voltage is registered by meter M, and this is nearly a logarithmic function of input voltage. To see why this is true, consider the expression for the output of the amplifier:

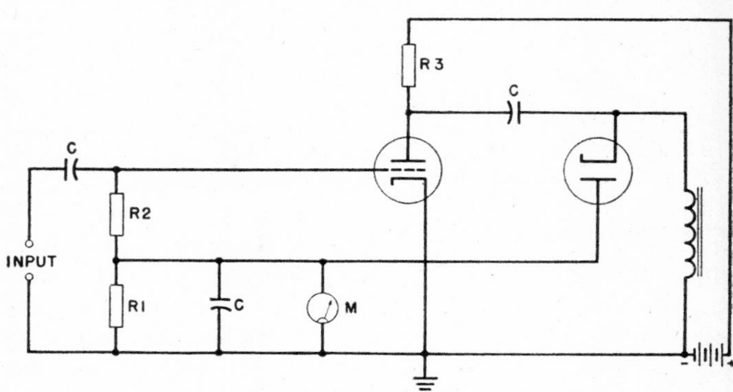

Fig. 26·3 Logarithmic voltmeter circuit.

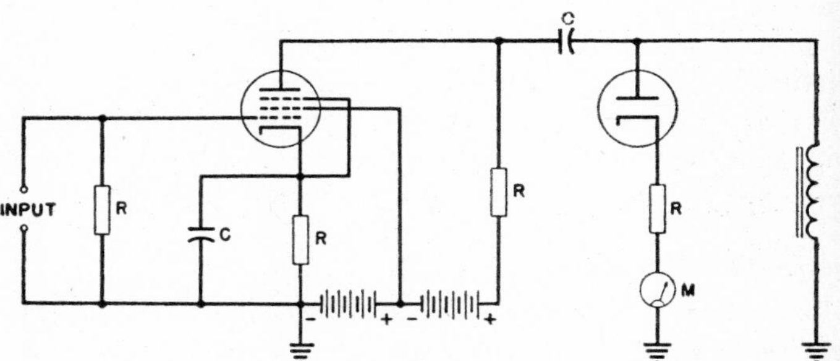

Fig. 26·4 Single-stage logarithmic voltmeter circuit.

$$e_0 = Ae_i\epsilon^{BE_c} \qquad (26\cdot1)$$

where e_0 = output voltage of the amplifier
e_i = input voltage to the amplifier
A, B = constants
E_c = grid bias applied to the amplifier
ϵ = base of natural logarithms.

Solving on the assumption that e_0 is a constant,

$$E_c = K - B \log_\epsilon e_i \qquad (26\cdot2)$$

Deflection of the meter is proportional to E_c, and this is theoretically a logarithmic function of the input to the amplifier. Actually, the assumption e_0 = constant is false, but if the amplifier is of high gain, e_0 is nearly constant, and the meter scale is nearly logarithmic.

A satisfactory logarithmic voltmeter can be made simply by rectifying the output of a remote-cut-off pentode amplifier and measuring the rectified d-c output.[7] A simple circuit for such a voltmeter is shown in Fig. 26·4. For a wider range of voltages, the two-stage circuit of Fig. 26·5 is sometimes used. For low input voltages, contribution to the meter current comes chiefly from the second stage. For higher inputs, the anode voltage of the second stage is zero

* The so-called remote-cut-off pentode or variable-mu tube is similar to the ordinary voltage-amplifier pentode, except in the structure of the control grid. The control grid of a variable-mu tube is wound in the form of a cylindrical spiral concentric with the cathode. The first few turns of the spiral near each end are rather close together and are influential in controlling the movement of electrons to the anode. Near the center of the spiral the spacing between grid turns is gradually increased, and this portion of the grid is relatively less influential in controlling movement of electrons to the anode. With a small negative bias applied to the grid, electrons move through all the spaces between its wires. For signals that are small compared with the grid bias, the amplification of the tube is comparable to that which would be obtained if the grid turns were uniformly spaced by the average spacing of the actual grid wires. If a relatively great negative bias is applied to the grid, the closely spaced end turns prevent movement of electrons through the spaces between them to the anode. All electrons that reach the anode do so by passing between the widely separated center turns. Amplification of the tube then is comparable to that which would be obtained from a tube with grid turns uniformly spaced by the distance between the center turns. This spacing is much larger than the average spacing, and the amplification of the tube is therefore less than when the negative bias is relatively small.

for almost half of the cycle, whereas the first stage continues to contribute to the meter current. For a suitably designed two-stage circuit, a linear calibration in decibels has been obtained from −40 to +20 decibels with a zero level of 1 volt.[7]

until the current through the diode, indicated by meter M, just ceases. Then the cathode potential indicated by voltmeter V is equal to the peak value of the alternating input voltage.

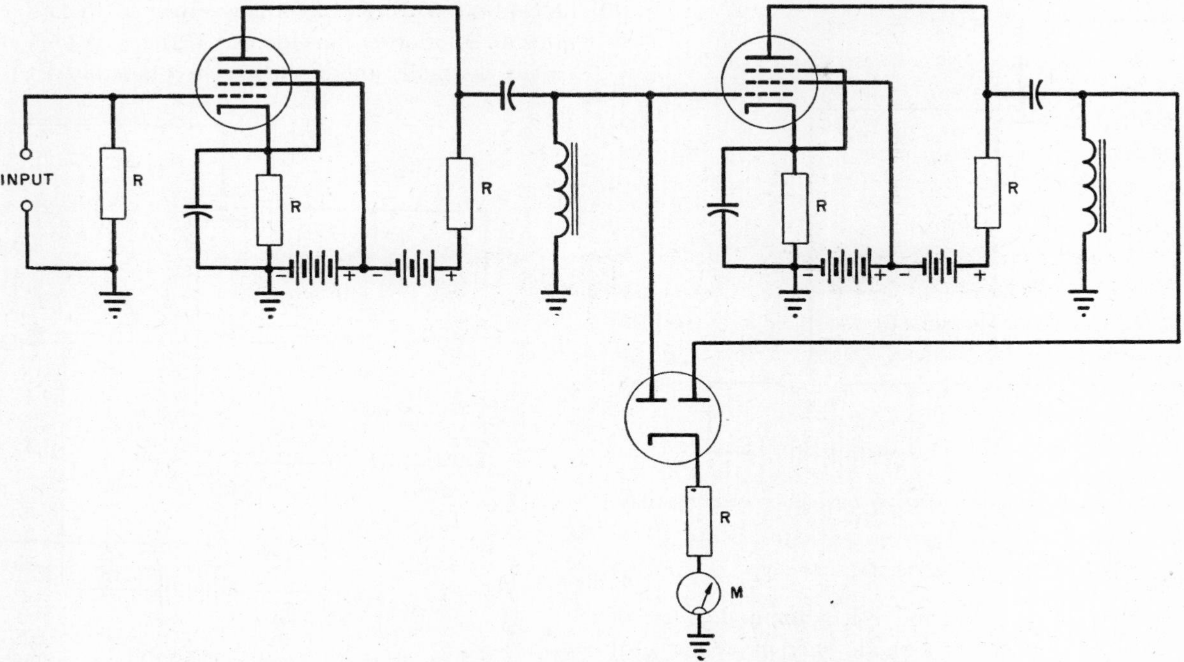

FIG. 26·5 Two-stage logarithmic voltmeter circuit.

Slide-Back Voltmeters

The slide-back voltmeter is an instrument for comparing the peak value of an unknown alternating voltage with a known unidirectional voltage. Numerous methods for doing this have been used.† [8, 9, 10, 11] A simple form of slide-back voltmeter which is useful for measuring fairly high voltages at high frequencies is shown in Fig. 26·6. The diode conducts

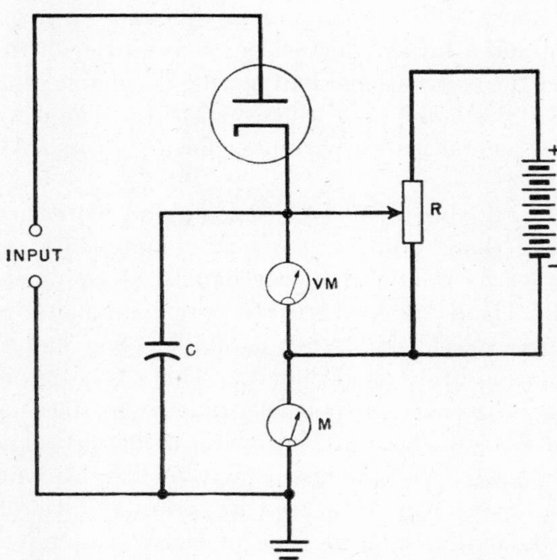

FIG. 26·6 Slide-back voltmeter circuit for high-voltage measurement.

only when the anode is positive with respect to the cathode. Cathode voltage is adjusted by means of a potentiometer

† R. A. Heising, U. S. Patent 1232919.

Transfer-Characteristic Voltmeters

Operation of transfer-characteristic voltmeters depends upon the relation between grid voltage and anode current of a vacuum tube. It has the disadvantage that the transfer characteristic is different for various individual tubes of the same type, and varies even for the same tube during its life. Such a meter, however, is simple to construct and to calibrate. Figure 26·7 is the circuit of the simplest type of transfer-characteristic voltmeter. The grid has a negative bias and

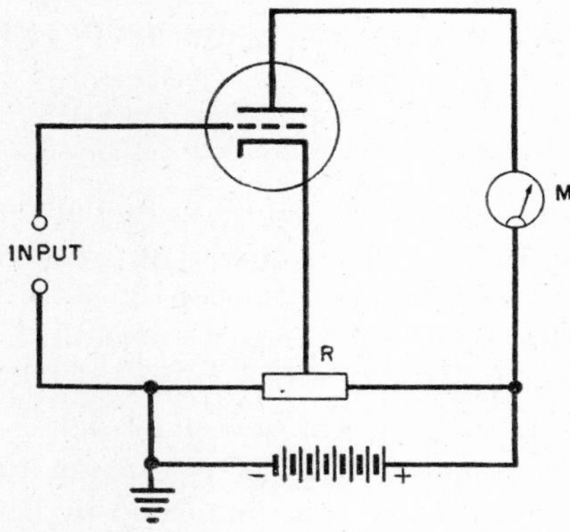

FIG. 26·7 Basic circuit of the transfer-characteristic voltmeter and the anode-detection voltmeter.

draws almost no current. Anode current is measured by meter M, and is a function of grid voltage. These are the rudiments of a voltmeter limited to the measurement of unidirectional voltages.

Another type of transfer-characteristic voltmeter is shown in Fig. 26·8. The grid voltage of a vacuum tube for zero anode current is related to the positive unidirectional or peak alternating voltage applied to the anode.[12] By adjusting the grid voltage by means of potentiometer R to make meter $M2$ read zero, the signal voltage can be read as a function of the reading of $M1$.

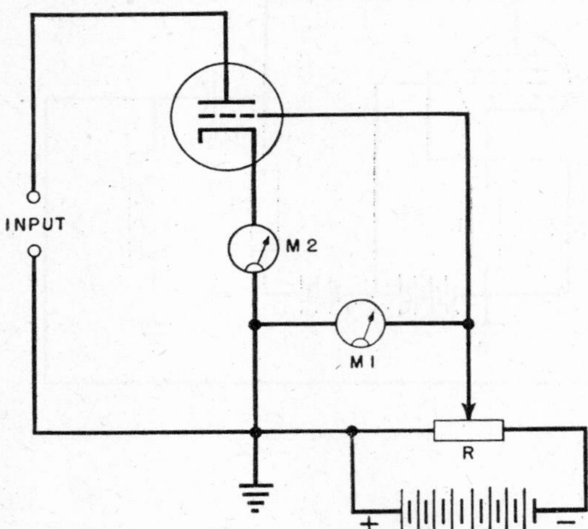

Fig. 26·8 Transfer-characteristic voltmeter for high-voltage measurement.

Anode-Detection Voltmeters

A common form of vacuum-tube voltmeter makes use of the non-linearity of the curve of anode current versus grid voltage.[13,14,15] Figure 26·7 shows the basic circuit of such a voltmeter. The change in reading of the anode-current meter depends upon the alternating-voltage input to the grid.

Let us assume that the grid bias is less than the cut-off value. For most tubes the instantaneous anode current is approximately a parabolic function of instantaneous grid voltage:

$$i_b = \frac{1}{r_1} e_c + \frac{1}{r_2^2} e_c^2 \qquad (26\cdot3)$$

where r_1, r_2 = constants. This equation shows that the change in average anode current is proportional to the sum of the squares of the peak values of the harmonics comprising the signal voltage, or, in other words, the rms signal voltage:

$$I_b = \frac{1}{2}\frac{1}{r_2^2}(E_1^2 + E_2^2 + E_3^2 + \cdots) \qquad (26\cdot4)$$

where E_1, E_2, etc., are peak values of fundamental and harmonics of the voltage to be measured. Equation 26·4 is valid if the assumption of equation 26·3 is valid. This assumption is not valid if the negative peak of the alternating signal voltage causes the grid bias to go below cut-off. It is approximately true, however, for a small signal voltage.[16]

Actually, the anode current of a vacuum tube contains terms of higher order in e_c than indicated by equation 26·3. For greater accuracy, the push-pull circuit of Fig. 26·9 may be used.[17] Current through the meter M is the sum of the average currents through $V1$ and $V2$. Since the grid voltage of $V2$ is 180 degrees out of phase with the grid voltage of $V1$, all the odd-order terms in the total anode current of the two tubes add to zero; consequently the fourth-order term is the first to introduce inaccuracy as a function of wave form. For triodes, this term usually is small compared with the second-order term.

Another type of anode-detection voltmeter can be made by biasing the tube to cut-off. With the tube so biased and with a high resistance in the anode circuit, the average anode current is nearly proportional to the average value of the positive loop of alternating grid voltage. Such a meter can be calibrated to read average voltage if the wave form contains no even-order harmonics. If the resistance in the anode circuit is small, anode current is more nearly proportional to the square of the grid voltage, and the meter can be calibrated to read rms volts if the wave form contains no even-order harmonics.

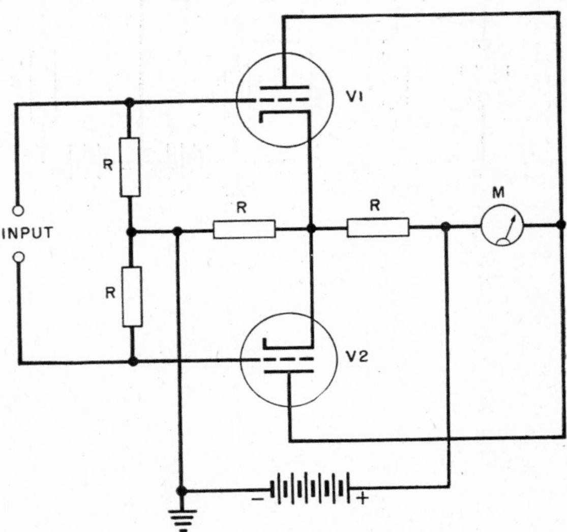

Fig. 26·9 Push-pull anode-detection voltmeter.

26·2 ELECTROMETERS

Conventional electromagnetic instruments can be used for measuring currents as small as about 10^{-6} ampere. Specially constructed sensitive instruments of this type measure currents to about 10^{-8} ampere. A modification, utilizing only one pivot and a rather short suspension in place of the other pivot and the conventional hair spring, is capable of measuring currents as small as 5×10^{-9} ampere. The d'Arsonval galvanometer measures currents down to approximately 10^{-10} ampere.

Sensitive double-pivot and single-pivot instruments require more than ordinary care and maintenance, and are relatively easily deranged by mechanical abuse or electrical overload. The d'Arsonval suspension galvanometer in addition requires special skill in use, careful leveling, and freedom from vibration in its environment. There is therefore some advantage in using electronic instruments for measurement of currents below about 10^{-6} ampere, if they avoid some of the objections to electromagnetic instruments. Amplification of vacuum tubes can be utilized to permit the use of a less sensitive and more rugged type of electromagnetic instrument than could ordinarily be connected directly to the circuit in which current is to be measured.

Electronic Microammeter

An amplifier with inverse feedback is the basis of one type of sensitive electronic ammeter. Figure 26·10 illustrates the basic circuit. An ordinary voltmeter is connected across the output terminals of the amplifier, and the entire output voltage is applied to the input terminals of the amplifier as inverse feedback. Under this condition, if the voltage gain

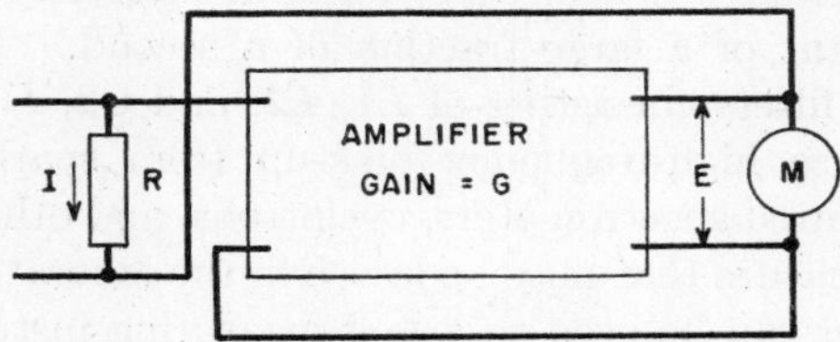

FIG. 26·10 Feedback amplifier for electrometer use.

of the amplifier without inverse feedback is G, the gain with inverse feedback is

$$G' = \frac{G}{1 + G} \tag{26·5}$$

If the gain without the feedback is much larger than unity, say 1000 or so, the gain with feedback is essentially unity. It also turns out that the stability of the amplifier, with respect to fluctuations in supply voltages, ambient temperature, and changes in tubes or tube constants, increases in almost direct proportion to the voltage gain without feedback. The change in reading of the voltmeter caused by an input current I is therefore

$$E = IR \tag{26·6}$$

$$I = \frac{E}{R} \tag{26·7}$$

The zero-signal reading of the voltmeter can be reduced to zero by sending current from an auxiliary battery through the meter in a direction opposite to that of the static component of amplifier anode current. Such a microammeter can be made as sensitive as desired by a proper choice of the amplifier-input resistor. Sensitivity of the instrument reaches its limit when the amplifier-input resistor is made so large that the potential drop across it produced by the normal grid current of the first-stage tube becomes perceptible in comparison with the potential drop caused by the current to be measured. If ordinary receiving tubes are used, this factor usually limits sensitivity to the measurement of current of 10^{-8} ampere and above.

Figure 26·11 is a simplified diagram of a microammeter which is simple and rugged.[18] For simplicity, the tubes are shown as triodes with indirectly heated cathodes, although actually they are pentodes of the battery-operated filament type. Several input resistors are provided to make it a multirange instrument. The amplifier is directly coupled and is therefore suited for d-c measurements. Gain of the amplifier without feedback is approximately 5000, and it therefore has great stability. A resistance-capacitance filter across the input resistor in the first stage by-passes a-c components of current that may be in the circuit in which measurements are being made, and prevents self-oscillation. The zero-signal reading of the output voltmeter is adjusted to zero by means of an auxiliary battery and potentiometer.

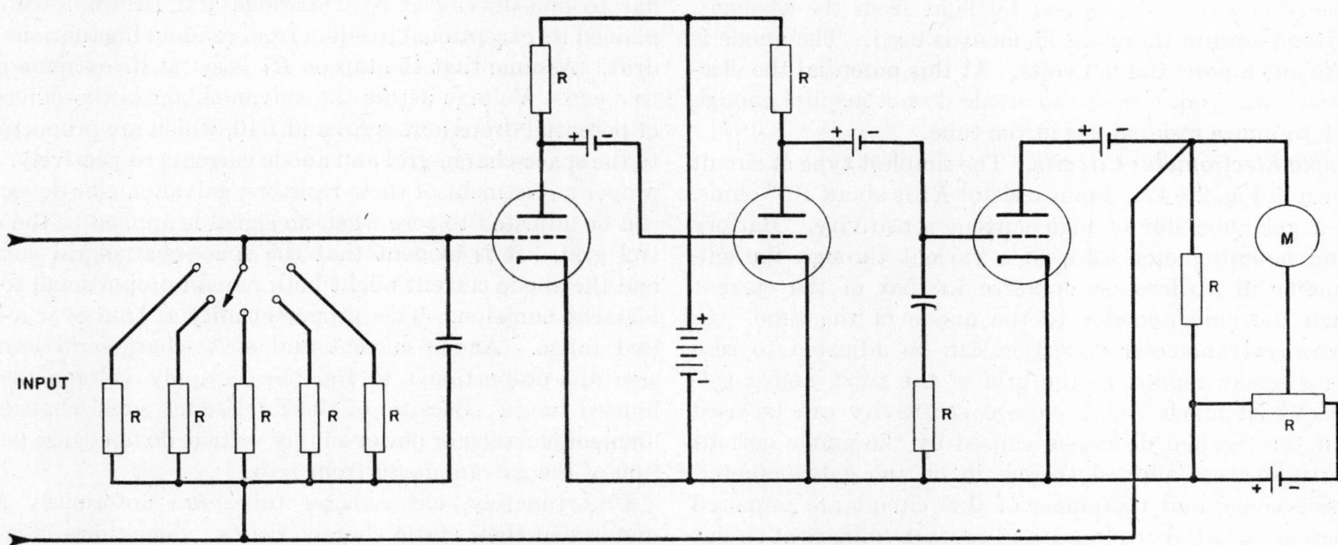

FIG. 26·11 Vacuum-tube microammeter.

Sensitive Electrometers

Measurement of direct currents of 10^{-15} or 10^{-16} ampere is sometimes necessary in research problems and in an increasing number of industrial applications. Measurement of ion currents in a mass spectrometer is one specific example.

Conventional d'Arsonval galvanometers are not practical for measuring currents much smaller than about 10^{-10} ampere, even if the power available in the circuit is unlimited, because of the difficulties of winding coils consisting of many turns of fine wire. To measure such small currents, a directly coupled amplifier must be used.

Electrometer Tubes. The first tube in such an amplifier must draw an extremely low grid current. Since 10^{-16} ampere is equivalent to about 640 electrons per second, and since most tubes have effective grid resistances of a few megohms, currents of such magnitude do not change the grid potential perceptibly.

The causes of grid current in vacuum tubes in class A operation are:[19]

1. Insulation leakage.
2. Occasional high-speed electrons which are emitted from the cathode and then hit the grid.

3. Positive-ion emission from the cathode.
4. Ionization of residual gas in the tube.
5. Electron emission from the grid, caused by x-rays generated by electron impact at the anode.
6. Photoelectric emission from the grid, caused by light from the filament.

Two tube types, the FP-54 and the D-96475, have been developed for electrometer use.[20,21] To minimize grid current, the grid lead is brought out through the top of the tube envelope on the end of a long stem, and all internal grid supports are made of quartz and have long leakage paths. In one of these tubes, no getter is used, partly because the getter materials emit photoelectrons and partly because some of the getter material may spatter onto the internal insulating surfaces when it is flashed. A 4-volt positive-space-charge grid is placed between cathode and control grid. The primary function of the space-charge grid is to overcome space charge, so that tube current is sufficient at low voltages. It also protects the control grid from positive ions and high-speed electrons from the filament. To avoid photoelectric emission from the grid caused by light from the filament, a low-temperature thoriated filament is used. The anode is operated at a potential 6.0 volts. At this potential the electrons crossing from cathode to anode do not acquire enough energy to ionize residual gas in the tube.

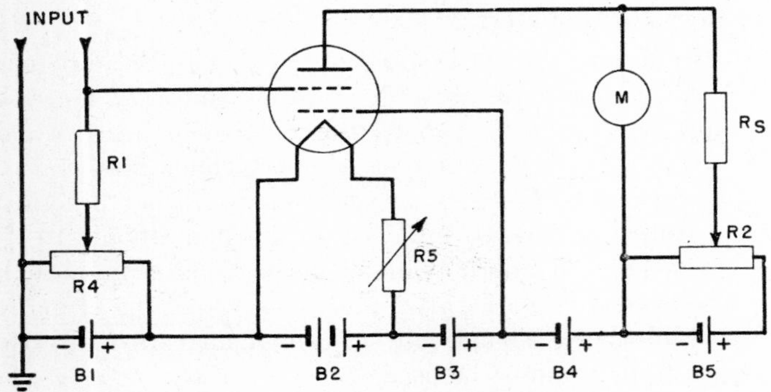

FIG. 26·12 Simple electrometer circuit.

Simple Electrometer Circuits. The simplest type of circuit is shown in Fig. 26·12. Input resistor $R1$ is about 10^{10} ohms. M is a galvanometer of high current sensitivity. Battery $B5$ and potentiometer $R2$ send a current through the galvanometer in a direction opposite to that of the current through the galvanometer to the anode of the tube. In this way galvanometer deflection can be adjusted to zero with no signal applied to the grid of the tube, and a galvanometer of much higher current sensitivity can be used than if the residual deflection caused by the anode current of the tube were allowed to remain on the galvanometer. The sensitivity and usefulness of this circuit are impaired by changes in battery voltages and contact voltages at circuit junctions, changes in filament emission with tube age, and thermal fluctuations in $R2$. All these changes tend to make unstable the balance point at which galvanometer deflection is zero with no signal applied to the tube.

Compensated Electrometer Circuits. Numerous bridge-type circuits, which are self-compensating for various causes of fluctuation, have been devised.[21–34] One of the best of these is shown in Fig. 26·13.[35] Resistors $R1$, $R2$, and the tube are mounted in an evacuated enclosure to prevent accumulation of atmospheric dust and moisture on the critical exterior insulating surfaces. Resistor $R1$, together with the input capacitance of the tube, forms a filter circuit with a time constant of a large fraction of a second. This filter, with other filters consisting of $L1$, $L2$, and $C2$, $L3$, $L4$, and $C3$, minimize high-frequency pick-up from spark spectrographs, infrared spectrometers, cyclotrons, and other apparatus of this nature that may be in use in the laboratory. This electrometer can be used as a null measuring instrument by impressing a bucking voltage from a potentiometer at point X. Resistor $R3$ and capacitor $C3$ filter surges caused by switching in the potentiometer. Although the grid would eventually lose by leakage through $R2$ any charge caused by a switching surge, the time constant of the grid capacitance

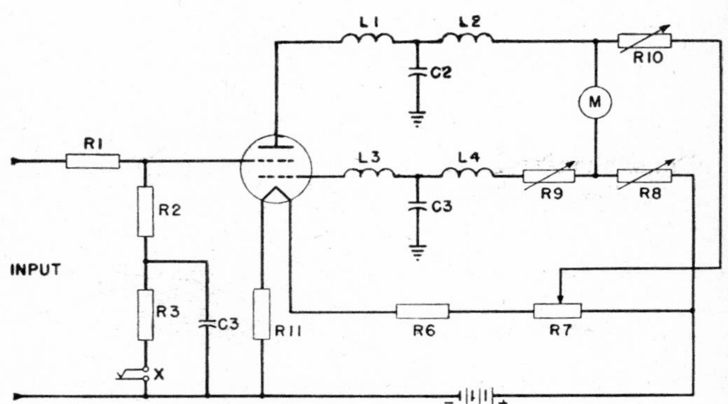

FIG. 26·13 Self-compensating electrometer circuit.

and $R2$ is so long (several seconds) that the wait would be annoying or even damaging with some types of measurements.

Theory of Compensated Circuits. These circuits are similar to one developed by DuBridge and Brown,[36] who explained its exceptional freedom from random fluctuations and drift. Assume that the tap on $R7$ is set at its extreme positive end. Voltage across the galvanometer is the difference of potential drops across $R8$ and $R10$, which are proportional to the space-charge-grid and anode currents respectively. By proper adjustment of these resistors, galvanometer deflection can be adjusted to zero when no signal is applied to the control grid. It is evident that the space-charge-grid current and the anode current might both remain proportional to the filament emission. This proportionality is true over a limited range. Anode current and space-charge-grid current also are proportional to the power-supply voltage over a limited range. Because of these relations, small changes in filament emission or power-supply voltage do not cause deviation of the galvanometer from zero.

Unfortunately, electrometer tubes are notoriously nonuniform in their static characteristics. Sometimes it is impossible to maintain stable operation and still impress on the electrodes those voltages which give maximum sensitivity. Moving the tap on $R7$ somewhat toward the negative end impresses a component of the power-supply voltage across the galvanometer in addition to the difference in potential drops across $R8$ and $R10$. This extra adjustment makes it possible to accommodate any electrometer tube that has so far been tried, without the necessity of rebuilding the circuit or suffering loss of sensitivity.

Measurement of High Resistance by Means of an Electrometer

Measurement of high resistances by conventional bridge circuits involves two difficulties. The first is that ordinary microammeters and galvanometers are not sufficiently sensitive to detect bridge balance accurately if the bridge arms are composed of high resistances. This could be overcome by use of a suitable amplifier between the bridge output and the detector, but the second and more serious difficulty would remain. The bridge must be composed of three resistors of accurately known value in addition to the unknown resistance. The magnitudes of the known resistors must be at least comparable with that of the unknown resistor. At least one of the known resistors must be adjustable, and its magnitude must be known accurately at every step in the adjustment. Such resistors usually are not available, or they may require excessive space.

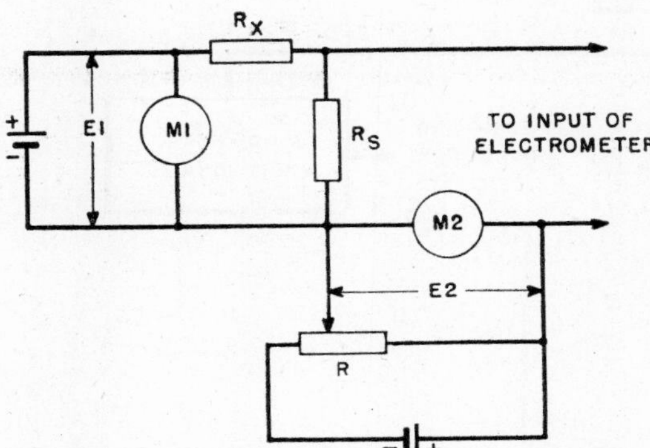

FIG. 26·14 Circuit for measuring high resistance.

A circuit for measuring high resistance, requiring two voltmeters, one standard fixed resistor, and an electrometer,[37] is illustrated in Fig. 26·14. R_x represents the unknown resistor, and R_s represents the standard resistor, which usually is also the input resistor of the electrometer. Resistance of the unknown is determined by adjusting the voltage $E2$ by means of potentiometer R until the output of the electrometer is the same as it was when both $E1$ and $E2$ were zero. The value of the unknown resistor then is

$$R_x = R_s\left(\frac{E1}{E2} - 1\right) \qquad (26\cdot8)$$

This method of measuring resistance is most often employed in determining the value of the type of resistor used in the input circuits of electrometers and related apparatus.

26·3 HIGH-FREQUENCY AMMETER

For measurement of moderate or high currents ranging from direct current through the audio range, a vacuum-tube instrument has little or no advantage over electromagnetic instruments. For measurement of high currents in the radio-frequency range, the circuit of Fig. 26·15 is useful. Coil $L1$ usually is a solenoid. It is placed near the conductor carrying the current to be measured, with its axis perpendicular to the direction of the current. Flux from the current to be measured links solenoid $L1$ and induces a voltage in it. The voltage is measured by a diode voltmeter. Tube $V1$ rectifies the current through meter M. Resistor $R1$ is used for calibration. Tube $V2$, because of the contact potential between its anode and cathode, sends current through the meter in a direction opposite to that of the meter current caused by the contact potential in tube $V1$. Resistor $R2$ controls this bucking current, and thus causes the meter to read zero when no voltage is being induced in solenoid $L1$. Since the voltage induced in $L1$ is proportional to the frequency of the current to be measured, this meter should be calibrated at the frequency at which it is to be used.

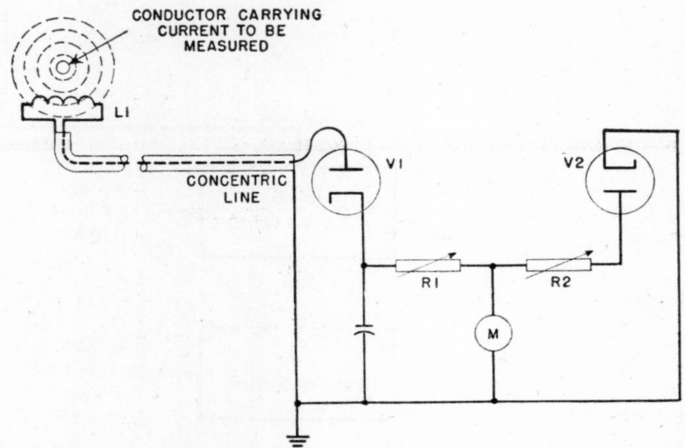

FIG. 26·15 High-frequency ammeter.

The principal application of this type of meter is in high-power broadcast transmitters and induction-heating equipment.

26·4 FREQUENCY METERS

Methods of measuring frequency fall into two classes: (1) direct measurement; and (2) comparison of the frequency to be measured with a known frequency standard. Among the direct measurements are the measurements of the charging current of a capacitor, use of bridge circuits, use of calibrated resonant circuits, and the use of Lecher-wire systems. Among the comparison methods of measuring frequency are the use of the cathode-ray oscilloscope and the heterodyne frequency meter. In general, the direct methods are quicker and require less equipment, but give less accuracy than the comparison methods.

Primary Frequency Standards

For frequency measurement by comparison, some easily available source of known frequencies must be available. Accuracy is limited only by the accuracy with which the frequency of the standard is known. Modern frequency standards are so accurate that frequency can be measured more accurately than any other electrical quantity.

Fundamental Standards. Fundamentally all frequency is measured by comparison with the frequency of rotation of the earth. This frequency is known accurately,[38] and it forms the primary frequency standard of 1 cycle per day. It is also far too low to be used directly for comparison with audio and radio frequencies ranging up to several million cycles per second; therefore primary frequency standards operating in a more usable frequency range are used to bridge the gap.

There are three possible primary standards of frequency. The most obvious is a precision clock or pendulum. The accuracy of such a device is high and could conceivably be made sufficiently accurate for the most precise measurements, but its low frequency of 1 cycle or so per second is not easy to utilize. All primary frequency standards now in use employ crystal-controlled master oscillators, which control the frequency of a chain of multivibrators.[39,40,41] Figure 26·16 is a block diagram of a typical primary standard. The synchronous clock is essentially a device for counting cycles. Comparison of the time as shown by this clock with sidereal time is a measure of the exact frequency of the crystal oscillator. Determination of the accuracy and constancy of frequencies is limited by the accuracy of determining sidereal time and by slight variations from day to day in the length of the sidereal day. For this reason frequency comparisons among primary frequency standards are more consistent within themselves than are the comparisons of these same frequency standards with sidereal time.

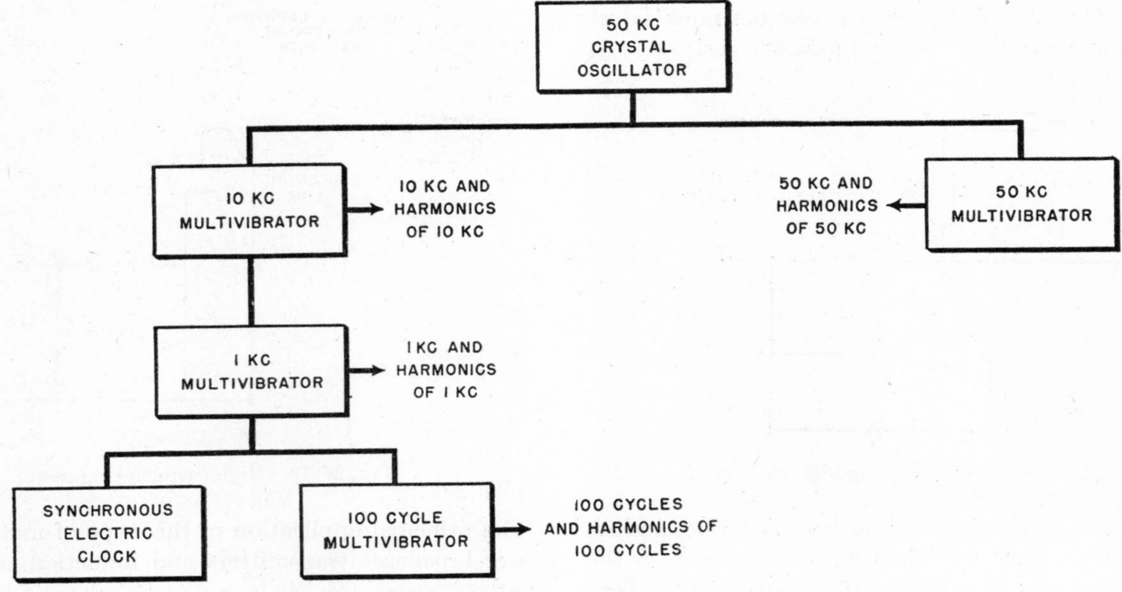

FIG. 26·16 Block diagram, primary frequency standard.

The output of a multivibrator is rich in harmonics. Harmonics of the order of the 300th, or even higher, are easily detectable. Thus, the frequency standard shown in Fig. 26·16 furnishes an immense number of frequencies spread over a broad spectrum. Each of these frequencies is known with the same accuracy as that of the crystal-controlled master oscillator. Thus, each frequency is available and useful for measurement purposes.

Construction. Such primary frequency standards are designed and constructed with care. The quartz crystal which determines the frequency of the master oscillator is carefully selected to have as low a temperature coefficient as possible and with the lowest possible decrement.[42] The crystal, master-oscillator tube, and other circuit constants in the master-oscillator circuit are enclosed in a temperature-controlled oven.[43,44] All coils in the master-oscillator circuit are wound on quartz forms in which grooves have been machined to insure accurate and stable spacing of turns. The temperature controlled enclosure in which the master oscillator is mounted is partially evacuated, and the pressure within this enclosure is carefully controlled. The entire master-oscillator circuit is mounted on springs in order to prevent shock excitation of the crystal or mechanical coupling between the master-oscillator crystal and any other sources of vibration or resonant systems. Finally, filament and power-supply voltages are carefully stabilized, and great care is taken to prevent power failure, which would stop the oscillator and necessitate a warm-up period before stability could again be attained.

Accuracy. A primary frequency standard such as this may be expected to maintain a frequency constant within one part in a million over a long period of time. If daily checks against astronomical time are made, the frequency can be maintained constant to within one part in 10 million. An error in frequency of one part in 10 million corresponds to a gain or loss of time of only about 2½ seconds per year. Figure 26·17 shows a primary frequency standard of the type just described. For reliability, two master oscillators were included in this installation. They can be seen under bell jars in the two ends of the cubicle. The control panel for the master oscillator is in the center.

Alternative Frequency Standards. An electrically driven tuning fork might be used in place of a crystal-controlled master oscillator for a primary frequency standard. Tuning forks have been used in this way, and their frequency stability can probably be made about as good as that of a quartz crystal.[45,46] Tuning-fork oscillators have not yet been so fully developed as crystal oscillators.

Standard-Frequency Sources. The National Bureau of Standards operates radio station WWV, which carries on scheduled broadcasts of standard-frequency signals. These signals have frequencies accurate to better than one part in

10 million.[47] Since these frequencies are easily available, they are useful in calibrating all kinds of frequency meters and secondary frequency standards. If the extreme accuracy of the signals from a primary frequency standard or from station WWV is not required, or if only a standard broadcast receiver is available, signals from the standard broadcast transmitter are useful as frequency standards. Federal Communications Commission regulations require the frequency of a standard broadcasting station to be within 50.0 cycles of the assigned frequency, and it is customary for the better class of stations to maintain their frequencies within 5 or 10 parts in a million of their assigned frequency.

Fig. 26·17 Primary frequency standard.

Comparison of Frequencies

Heterodyne Frequency Meter. To use a primary frequency standard for the measurement of frequency, it is necessary to be able to compare the unknown frequency with one of the known frequencies supplied by the primary standard. The most accurate method consists of the direct measurement of the difference between the unknown frequency and the nearest known frequency.[48] This involves the use of a radio receiver with an oscillating detector, a radio receiver with a non-oscillating detector, a heterodyne frequency meter, and an audio-frequency-measuring device. A heterodyne frequency meter is essentially a stable adjustable-frequency oscillator.

The procedure is as follows:[49]

1. The unknown frequency is tuned in on the receiver with an oscillating detector, and the detector is set to produce a beat note of some convenient frequency with the unknown frequency.
2. The output of the heterodyne frequency meter is then coupled to the receiver. The frequency meter is adjusted to zero beat with the unknown frequency.
3. The output of the heterodyne frequency meter is now loosely coupled to the input of a receiver with a non-oscillating detector. One of the frequencies in the 10-kilocycle sequence from the primary standard is coupled to the receiver. The receiver output contains a beat note that is the difference in frequency between that of the heterodyne frequency meter and that of the primary standard.
4. The audio-frequency beat note is then measured either directly or by comparison with a standard audio-frequency oscillator. The unknown frequency is the frequency from the primary standard, plus or minus the audio frequency. If slightly increasing the frequency of the heterodyne frequency meter increases the beat note, the audio frequency is to be added to the frequency from the standard. If increasing the frequency of the heterodyne frequency meter slightly decreases the audio beat note, the audio frequency is to be subtracted from the standard frequency.

A modification of this method employs several successive heterodynings to reduce the audio-beat note to a low frequency that can be measured to a fraction of a cycle.[50] This method can be even more accurate than the one described, but it requires elaborate equipment and therefore is not in general use.

The Cathode-Ray Oscilloscope. Two rather simple methods of comparing frequencies by means of a cathode-ray oscilloscope have been devised. One consists in impressing the known frequency on the horizontal deflecting plates of the oscilloscope tube and the unknown frequency on the vertical deflecting plates. The known frequency is adjusted until a stationary pattern appears on the oscilloscope screen. This pattern is known as a *Lissajous figure.* When the pattern is stationary, the ratio of the unknown frequency to the known frequency is the ratio of the number of times the Lissajous figure is tangent to a horizontal line to the number of times it is tangent to a vertical line. Figure 26·18 is a Lissajous figure obtained when the ratio of the frequency on the vertical deflecting plates to that on the horizontal deflecting plates is 3 to 2.

This method of comparing frequencies is extremely simple and is essentially as accurate as the adjustable known frequency. If large numbers are involved in the ratio of unknown to known frequency, however, and if either the known or unknown frequency contains appreciable amounts of higher harmonics, the Lissajous figure becomes too complicated to be analyzed easily.

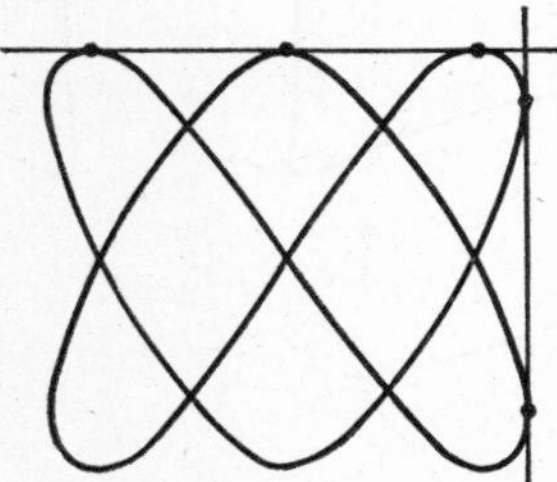

Fig. 26·18 Lissajous figure.

If the ratio of frequencies is large, say 10 or so, it is convenient to split the phase of the lower frequency by means of a resistor and capacitor. The two voltages with a 90-degree phase separation thus obtained are applied to the deflecting plates of an oscilloscope tube to produce a circle or broad ellipse on the screen. The higher frequency is superimposed on the positive potential applied to the accelerating anode of the oscilloscope tube. This causes the velocity of the electrons in the beam to be alternately increased and decreased from its mean value. If the electrons in the beam travel at less than their mean velocity, they remain in the field of the deflecting plates for a longer than normal period, and the diameter of the circle tends momentarily to become larger. Conversely, if the electrons travel at more than their mean velocity, the diameter of the circle tends to become less. The result is that a gear-shaped figure appears on the oscilloscope screen. The ratio of the higher frequency to the lower one is equal to the number of teeth on the gear divided by the number of times the spot traverses the complete circle before retracing a former path. Figure 26·19 shows a circuit for producing these gear-shaped figures and Fig. 26·20 shows some typical patterns.

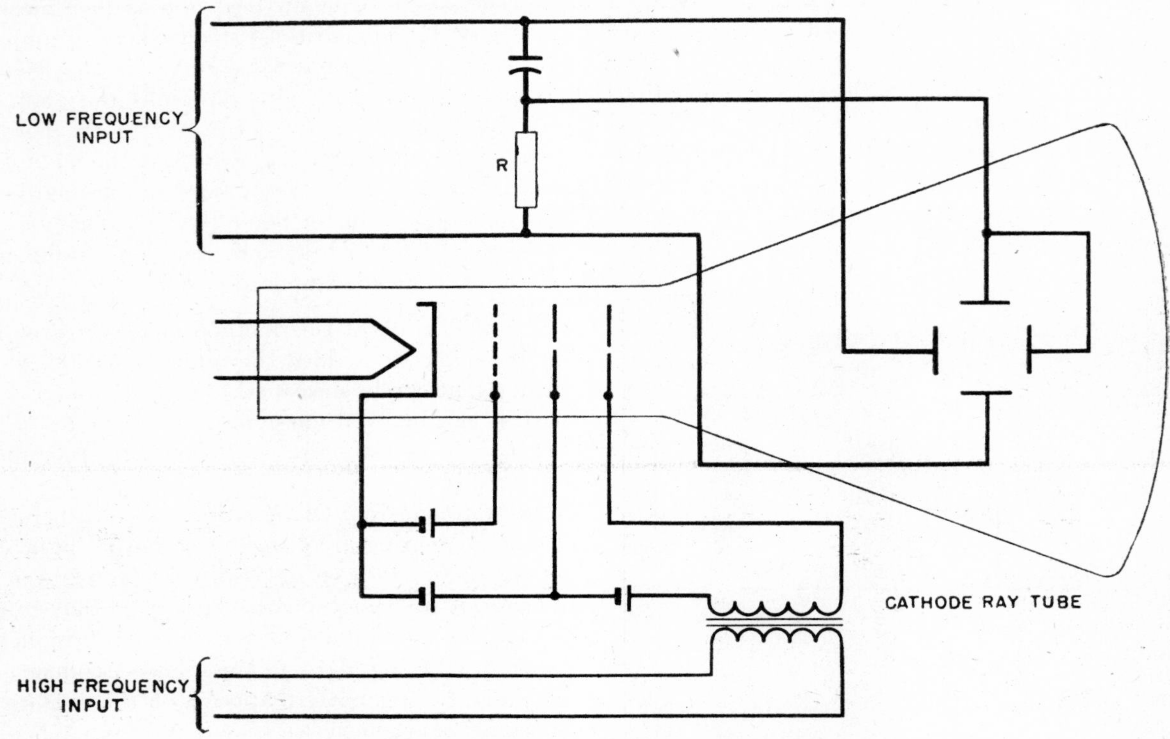

Fig. 26·19 Circuit for comparing frequencies.

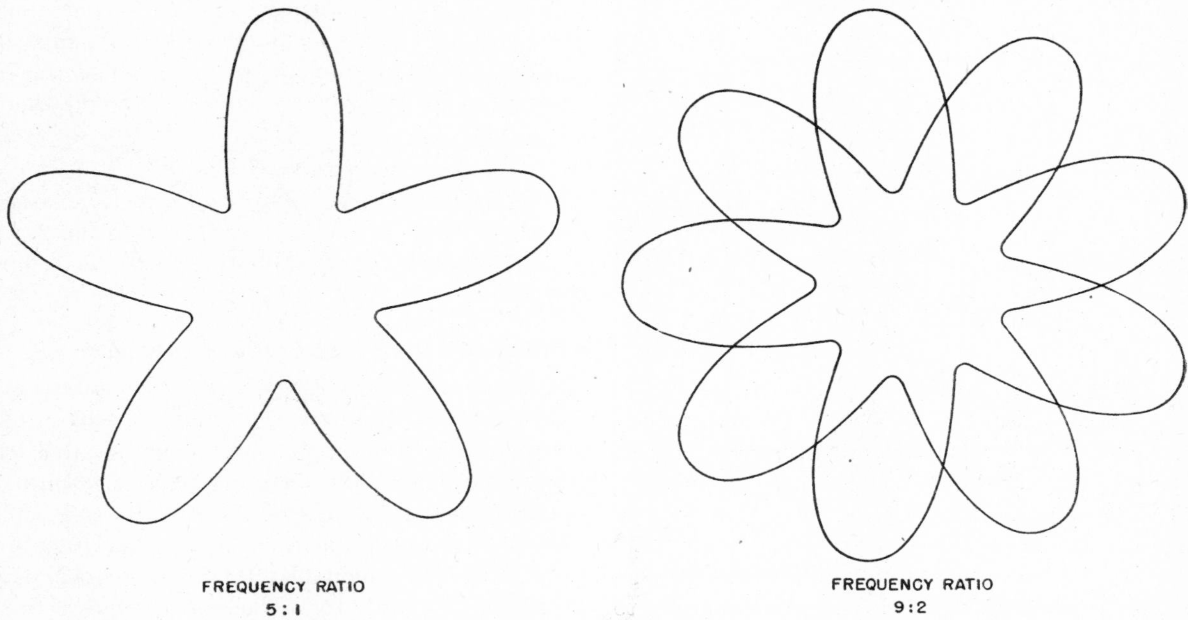

Fig. 26·20 Gear-shaped figures obtained with a cathode-ray oscilloscope.

Frequency Determination from Capacitor Charging Current

If a capacitor is charged from one circuit and discharged during alternate half cycles through a d-c meter, the average current through the meter is the product of capacitance, peak voltage to which it is charged, and applied frequency. Figure 26·21 illustrates such a circuit.[51] Tubes *V*3 and *V*4 are biased below cut-off by batteries *B*1 and *B*2 respectively. Grids of these tubes are excited in opposite phase by the secondary of transformer $T2$. During the half cycle that the grid of $V3$ is above cut-off, capacitor C charges to the voltage of $B3$. During the alternate half cycle, the grid of $V3$ is below cut-off and the grid of $V4$ is above cut-off; therefore capacitor C discharges through M and $V4$. The charge and discharge of capacitor C is complete only if the resistances of $V3$ and $V4$ are low while they conduct, and if C is small. In other words, the time constant of the circuit must be much less than the time of a half cycle. Tubes $V1$ and $V2$ act as amplifiers providing a large signal voltage to $V3$ and $V4$ to insure low tube resistance during the conducting part of the cycle. At frequencies below a few hundred cycles, a high-speed polarized relay can be used to perform the switching operation of $V3$ and $V4$.[52, 53]

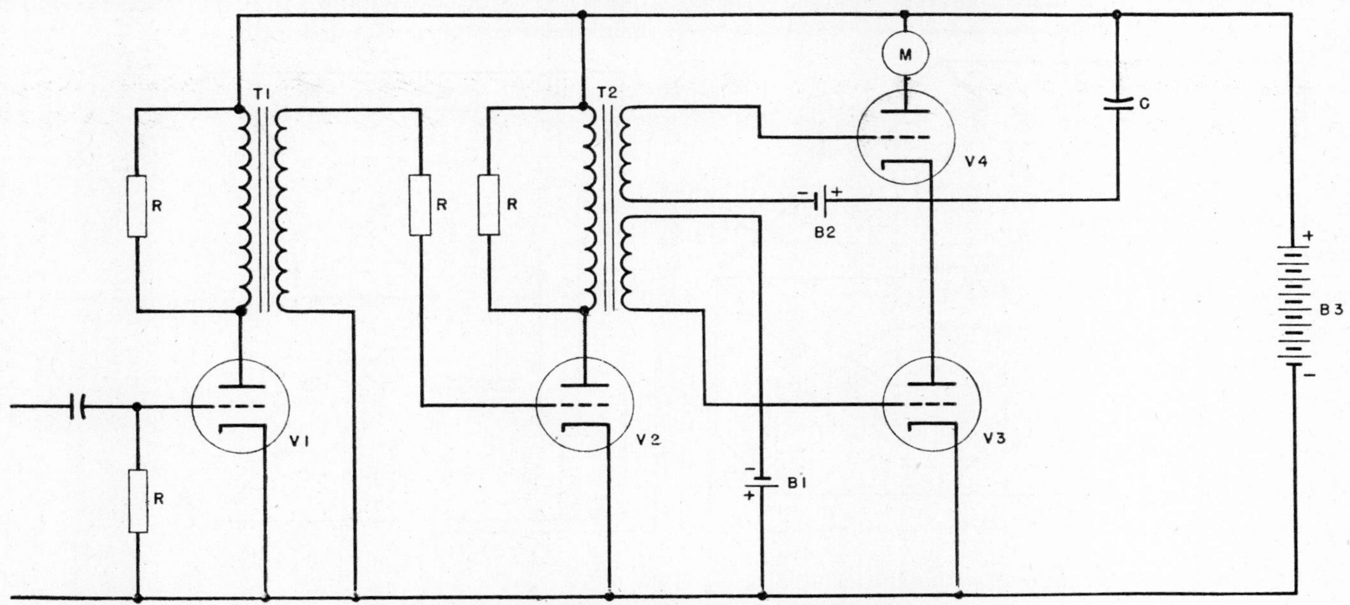

FIG. 26·21 Direct-reading frequency-meter circuit.

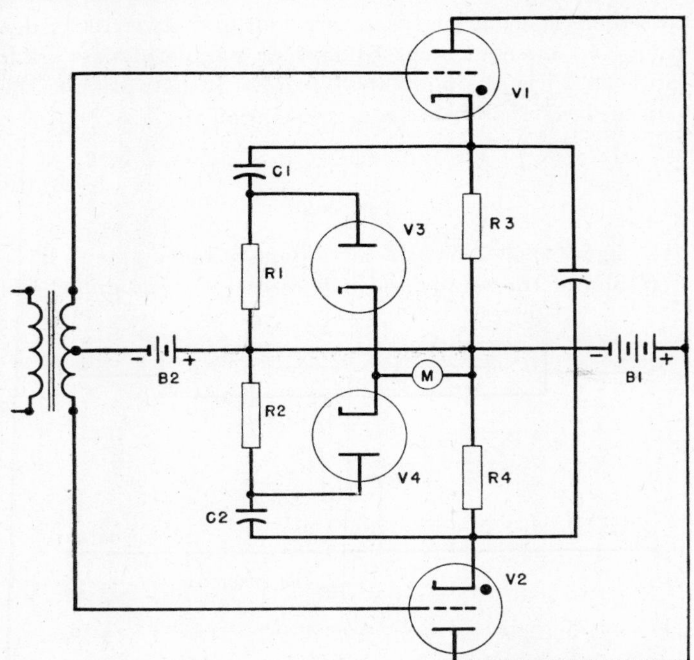

FIG. 26·22 Direct reading frequency meter circuit.

Figure 26·22 shows another method of accomplishing the same result.[39] Two gas-filled triodes are connected in a separately excited inverter circuit. Battery $B2$ biases the grids of thyratrons $V1$ and $V2$ to a value sufficient to prevent conduction. During the half cycle in which the grid of $V1$ is biased above the value required for conduction, capacitor $C1$ charges to the voltage of battery $B1$. A definite fraction of the charging current of these capacitors passes through diodes $V3$ and $V4$ and the microammeter. During the half cycle when $V1$ is not conducting, capacitor $C1$ discharges through resistors $R1$ and $R3$ while diode $V3$ prevents any reverse current from passing through the meter. In a similar manner, capacitor $C2$ discharges through $R2$ and $R4$ when thyratron $V2$ is not conducting.

Accuracy. Direct-reading frequency meters of the type just described can be made accurate to better than 1 percent, but probably not as good as 0.1 percent. Accuracy depends upon the constancy of the voltage of battery $B1$ and upon the resistance in the capacitor-charging and -discharging circuits being low enough for complete charge and discharge. Frequency meters of this type are useful in the audio-frequency range; that is, from about 20 cycles to a few thousand cycles.

Bridge Circuits for Measuring Frequency

The Wien Bridge. Bridge circuits fundamentally are not electronic, although electronic amplifiers are often used to aid in detecting balance.[54] Fundamentally, any bridge circuit for which frequency appears in one or both of the balance equations can be used for measuring frequency in terms of the known impedances of the bridge arms. One of the simplest and best of these circuits is the Wien bridge shown in Fig. 26·23. Provided that the relations between the circuit elements are as shown on the diagram and that R' resistance is zero, the balance equation of the bridge is

$$f = \frac{1}{2R_aC} \tag{26·9}$$

It is impossible practically to have the two ganged resistors R_a maintain strict equality throughout their range of adjustment. For this reason the small resistance R', which is a small percentage of R, is included to aid in obtaining a sharp balance.

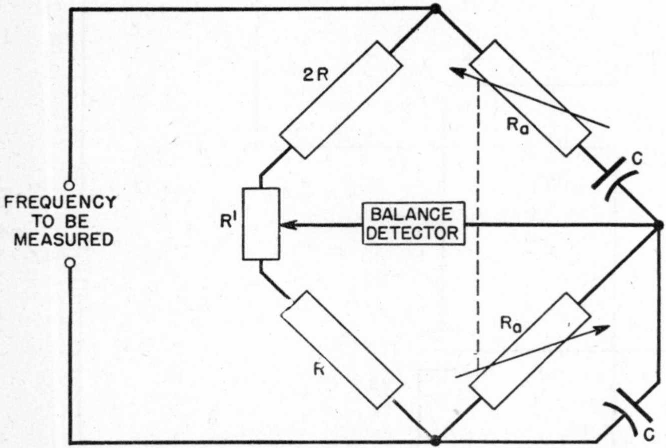

FIG. 26·23 Wien bridge circuit for frequency measurement.

Bridge Design. Several difficulties beset the user of a-c bridges. Inductive pick-up and capacitive pick-up are a serious source of trouble. Inductive pick-up can be reduced by using only resistors and capacitors in the bridge arms, if possible. Pick-up of any type is reduced by shielding the bridge and its circuit elements properly.[55-66] In making bridge measurements of frequency or impedance, the bridge seldom can be supplied with a source free of higher harmonics. When the bridge is balanced for the fundamental frequency it is unbalanced for the harmonics, and therefore the voltage across the balance detector is not zero even if the bridge is balanced. In the audio-frequency range, a good balance can be found despite this, if the balance detector consists of a set of head phones or some other device that gives an audible signal, because the human ear is sensitive to pitch and can discriminate accurately between harmonics and fundamentals. In the low-radio-frequency range, a cathode-ray oscilloscope gives a fairly accurate picture of the wave form of the voltage across the balance detector, which aids in obtaining a good balance.

Grounding of Bridges. In the upper audio-frequency range, stray capacitances from the bridge-circuit elements to ground can cause the balance point to be spurious or can prevent the existence of a balance point, especially if the impedances forming the bridge arms are rather high. The Wagner ground consists of a ground connection from a resistance connected across the supply line to the bridge. In using a bridge equipped with Wagner ground, as shown in Fig. 26·24, the balance detector is first connected between points 1 and 3 by means of switch $S1$ in the usual manner and the bridge is balanced as well as possible. Then one end of the balance detector is connected to ground by means of switch $S1$, and the Wagner ground tap is adjusted to give minimum response of the balance detector. The balance detector is then switched back to its normal position in the bridge and the bridge is rebalanced. By a series of such successive approximations, the true balance point is found. When the Wagner ground is adjusted in this manner and the bridge is balanced, both terminals of the balance detector are at ground potential; consequently there is no current through stray capacitance from points 1 or 3 of the bridge-to-ground. Stray capacitances to points 2 and 4 merely have their circuit completed through resistor R.

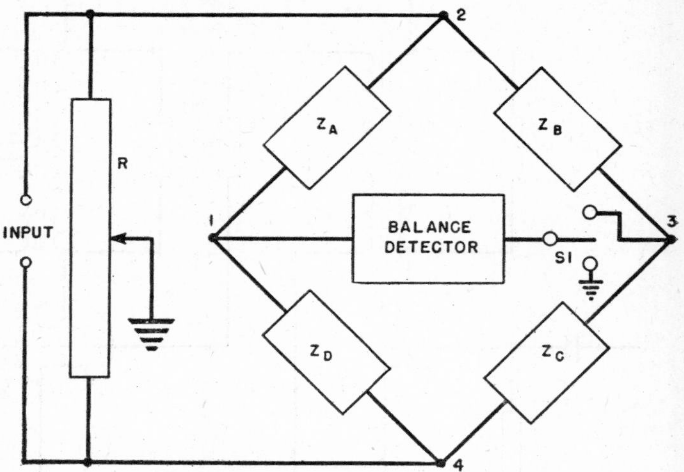

FIG. 26·24 Bridge circuit with Wagner ground.

T Networks for Frequency Measurement. A fundamental difficulty of bridge circuits for use with electronic apparatus and measurements is that either the source or the balance detector, or sometimes both, must operate at a potential above ground. This difficulty often can be overcome by the use of bridged or parallel T networks. These are three-terminal networks so constructed that, with certain relations between their elements, the transfer admittance is zero. Since these networks have only three terminals, both source and balance detector can have one terminal grounded. Sometimes there is a mathematical relationship between a bridge circuit and an equivalent bridged or paralleled network.[67] Figure 26·25 is the equivalent parallel T network for the Wien bridge,[68, 69] and the balance equation for it is

$$f = \frac{1}{\sqrt{8}\pi RC} \tag{26·10}$$

The figure shows that the advantage of being able to have a terminal of the source and balance detector grounded has

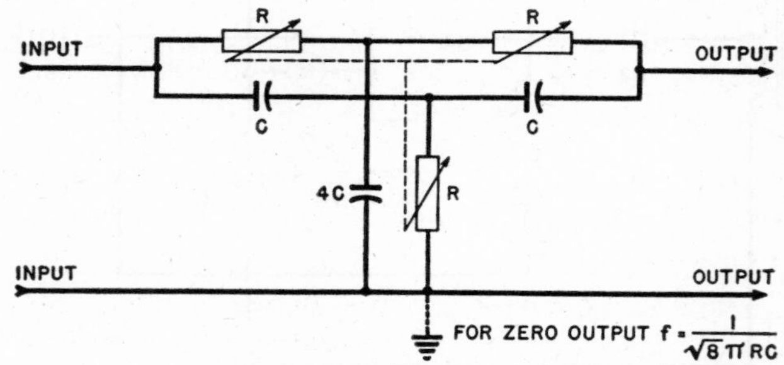

FIG. 26·25 Parallel T network equivalent of the Wien bridge.

been gained at the expense of varying three instead of two resistors simultaneously and keeping them all equal.

Resonant-Circuit Frequency Meters

For measuring impedances, the a-c bridges mentioned previously are used for frequencies ranging from the low audio-frequencies to radio-frequencies up to 60 megacycles, and even higher. For the measurement of frequency, however, bridge circuits become less useful than other means at frequencies above the audio range.

Construction. For quick measurement of frequencies from the audio range up to about 300 megacycles, if the accuracy need not be much better than 1 or 2 percent, calibrated resonant circuits are useful. A frequency meter employing calibrated resonant circuits consists essentially of an inductance, a variable capacitor, and a thermocouple milliammeter in a series circuit, as shown in Fig. 26·26. The inductance and capacitor are designed for the greatest possible stability of their constants with respect to temperature, age, and rough handling. The capacitor is carefully calibrated in terms of frequency. To cover a broad range of frequencies, several coils usually are employed. Often the coils are arranged to be easily interchangeable by means of plug-in connections.

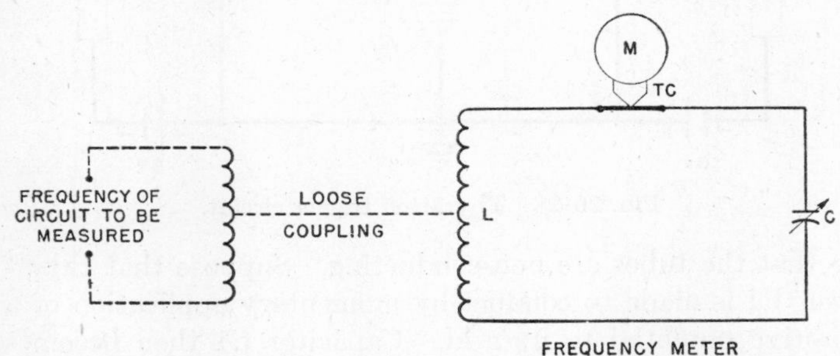

FIG. 26·26 Calebrated resonant circuit for frequency measurement.

Tuning for Maximum Secondary Current. There are two methods of using such a frequency meter. The first and more reliable method is to couple the coil L loosely with the circuit-carrying currents of the frequency to be measured and vary capacitance C slowly until resonance is indicated by a maximum reading of meter M. The frequency is then determined either from the capacitor dial directly or from a calibration curve of frequency versus capacitor-dial reading.

Tuning by Grid-Current Dip. The second method of use is applicable to small oscillators and amplifiers. Meter M and thermocouple TC may be omitted. Resonance of the coil and capacitor of the frequency meter are inferred from a pronounced decrease in grid current of the oscillator or amplifier, or a somewhat less pronounced increase in the anode current, at resonance.

Design Considerations. The effect of resistance of the meter and thermocouple on the natural frequency of the resonant circuit of the frequency meter is small and is accounted for in the calibration, but it is nevertheless important that this resistance be small to obtain a sharp resonance indication. A thermocouple of low resistance and with sufficient power output to operate a reasonably rugged milliammeter is likely to be delicate and easily damaged by overload or mechanical shock. A more rugged resonance indicator often used consists of a diode voltmeter connected across the capacitor. At high frequencies, where the internal capacitance of a diode masks its rectifying effect, a galena crystal is sometimes substituted for the diode.

Sources of Error. A resonant-circuit frequency meter must not be coupled too closely with the circuit under test. Close coupling causes several different effects depending upon the type of circuit to which the frequency meter is coupled and from what sort of source it draws its power.[70-72] These effects usually consist of one or more of the following: (1) broadening of the resonance peak; (2) error in calibration caused by reactance reflected into it from the primary circuit; and (3) change in the frequency of the current in the primary circuit because of reactance reflected into the primary circuit from the frequency meter.

Errors usually can be avoided by first tuning the frequency meter to approximate resonance and successively decreasing the coupling and retuning the frequency meter until the maximum reading of the resonance indicator is as small as can be conveniently distinguished.

Ultrahigh-Frequency Meters

At high frequencies the inductance and capacitance of the leads from coil to capacitor in the ordinary resonant-circuit frequency meter are comparable to those of the coils and capacitors themselves. This makes the calibration of the instrument unstable, and it cannot be tuned except over a limited frequency range.

Calibrated Transmission Lines. For measurement of frequencies from about 30 megacycles up to several hundred megacycles, the Lecher-wire system illustrated in Fig. 26·27 is useful.[73] The Lecher-wire system consists of a straight two-wire open-ended transmission line. Bridged across this line is a movable short-circuiting bar with a thermocouple instrument connected in parallel with a portion of the short-circuiting bar.

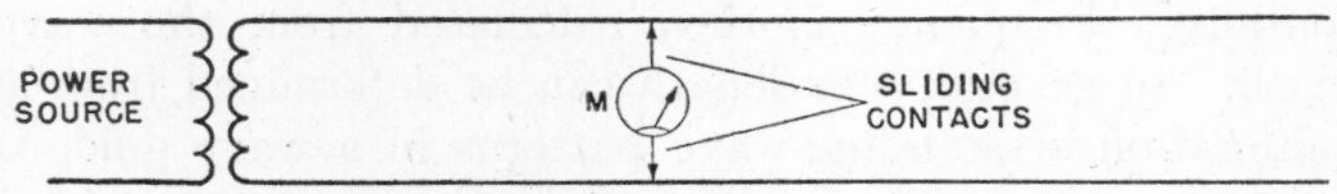

FIG. 26·27 Lecher-wire system for frequency measurement.

Theory. Current through the short-circuiting bar is maximum at positions along the line that causes the transmission line to be resonant at the frequency being measured. These positions are almost exactly a half wavelength apart. Frequency as related to wavelength is

$$f = \frac{300}{\lambda} \tag{26·11}$$

where f = frequency in megacycles

λ = wavelength in meters.

The procedure for measuring frequency is to slide the bar along until two maximum readings are found. The distance between the two positions of the bar that produce maximum readings, which is equal to a half wavelength, is then carefully measured, and the frequency is computed from equation 26·11.

Spurious Effects. Sometimes the maximum, instead of being fairly sharp, may be broad or have a double value. This broadness is caused by coupling between the open end of the line and the portion between the source and the short-circuiting bar. It may be avoided by placing additional bars across the open end of the line or by placing a shield (a large grounded metal disk with a hole in its center for the transmission line) around the portion of the line bearing the short-circuiting bar.[74]

Shortening the Line. The first current maximum can be shifted nearer to the input end of the line by connecting a variable capacitor across the line between the input end and the short-circuiting bar.[75, 76] This does not change the spacing of the maxima on the other side of the capacitor.

Precise Measurements. For ordinary measurements of frequency, equation 26·11 gives sufficiently accurate results. For more precise work, the frequency is given by the formula [77]

$$f = \frac{149.91}{s}(1 - \alpha) \qquad (26\cdot 12)$$

where s = distance in meters between positions of the short-circuiting bar for successive maxima

α = a correction factor, equal to about 0.001.

Resonant Cavities. One limitation of the Lecher-wire system is that, if the wires are to act like a transmission line instead of an antenna, the spacing between them must be rather small compared with a wavelength. It is almost impossible to construct a Lecher-wire system for measurement of frequencies of 3000 megacycles or above. Measurement of frequency in this range involves the use of resonant cavities and wave guides. As with Lecher-wire systems, wavelength is the quantity which is measured fundamentally. Frequency is then calculated from the wavelength. In general, wavelength can be determined from an examination of standing-wave patterns in a wave guide or cavity resonator. The wavelength in a resonator or guide is different from the wavelength in air, and depends upon the dimensions of the guide and its mode of oscillation.[78-85] Accuracy of calibrated, adjustable resonant cavities is comparable with that obtained with the common coil-and-capacitor frequency meter already described.

26·5 TIME-INTERVAL AND SPEED METERS

Circuits used for measuring time intervals can be adapted to measurement of speeds of moving bodies by measuring the time interval that elapses while the moving body travels between two points a known distance apart.

Methods of Measuring Time Intervals

Synchronous Clocks. Time intervals from a few seconds to a few minutes are usually measured by causing a synchronous clock or 60-cycle impulse counter [86] to start at the beginning of the time interval and to be de-energized at the end of the interval.

Change of Charge on Capacitor. Shorter time intervals ranging from a fraction of a second to a few seconds are measured by causing a constant current to start charging a capacitor at the beginning of the interval and cutting off the current at the end of the interval. The change of potential across the capacitor is measured and is a function of the time interval.

Cathode-Ray Oscilloscope. Measurement of intervals ranging from a few microseconds to a few milliseconds is most conveniently made by causing a small vertical pulse or pip to appear on an oscilloscope screen at the beginning and end of the time interval. The oscilloscope must be provided with a sweet circuit which carries the beam horizontally at a known constant speed during the time interval.

Design of Time-Interval Meters. Synchronous clocks are started and stopped, impulses are counted, and capacitors are charged, usually by a so-called "trigger circuit" or electronic switch. Some trigger circuits utilize thyratrons; others use vacuum tubes.

Thyratron Trigger Circuit. Figure 26·28 shows a trigger circuit using thyratrons.[87] The grids of both tubes are biased

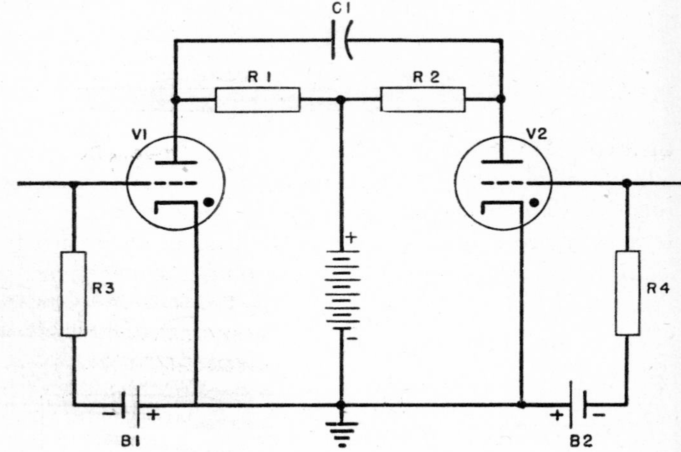

Fig. 26·28 Thyratron trigger circuit.

so that the tubes are non-conducting. Suppose that thyratron $V1$ is made to conduct by momentary application of a positive potential to its grid. Capacitor $C1$ then becomes charged, because of the potential drop in resistor $R1$. The plate of the capacitor adjacent to tube $V1$ is at the lower potential. If the grid of thyratron $V2$ now is made positive momentarily, this tube conducts. The potential at its anode drops suddenly from the supply voltage to the tube-drop voltage. Capacitor $C1$ cannot discharge and recharge to the opposite potential instantaneously. Therefore, the anode potential of $V1$ is momentarily reduced, and the tube stops conducting. Similarly, a positive pulse applied to the grid of $V1$ causes conduction to transfer from $V2$ to $V1$. The changing potentials at the anodes of the two tubes can be utilized to control other circuits.

Vacuum-Tube Trigger Circuit. Figure 26·29 shows one of the simpler and more reliable of the trigger circuits utilizing

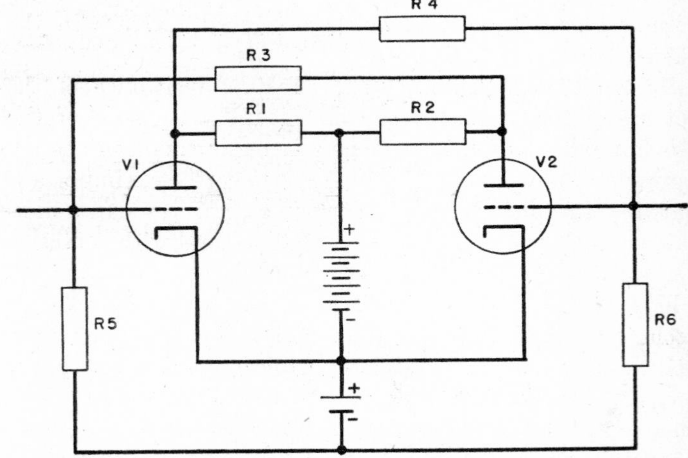

Fig. 26·29 Vacuum-tube trigger circuit.

vacuum tubes.[88] Anode and grid potentials and resistors are chosen in such a way that, when tube $V1$ carries anode current, the potential on the grid of $V2$ is below the cut-off value. If a positive potential is applied to the grid of $V2$, anode current starts. The potential at the anode of this tube decreases, and the decreased potential is transmitted to the grid of $V1$ through resistor $R3$. The decreased grid potential applied to $V1$ causes an increased anode potential of that tube, which is communicated to the grid of $V2$ through resistor $R4$, thereby further increasing anode current in tube $V2$ and decreasing the current in tube $V1$. In this way, conduction is transferred almost immediately from $V1$ to $V2$ if a positive potential is momentarily applied to the grid of $V2$. Similarly, conduction transfers back to $V1$ if a positive potential is applied to its grid, or if a negative potential is applied to the grid of $V2$.

Methods of Actuating Trigger Circuits. Signals for controlling trigger circuits can be supplied by contacts across resistors $R5$ and $R6$, by microphones across $R5$ and $R6$, by photoelectric cells in series with $R3$ and $R4$, or by a combination of these methods.

Figure 26·30 is the simplified circuit of a race timer.[89] Tubes $V1$ and $V2$ constitute a trigger circuit controlled by the microphone and by phototube $V3$. Tube $V4$ acts as a switch which controls a-c power to the impulse counter. The sound of the starter's gun is picked up by the microphone and stops conduction in $V2$. The change in anode potential of V2 raises the potential of the grid of V_4, allowing it to conduct every positive half cycle of the applied alternating anode voltage and thereby energizing the impulse counter. Interruption of light to the phototube at the end of the race stops conduction in phototube $V3$. Grid potential of $V1$ is thereby reduced below cut-off. The accompanying rise in its anode potential initiates conduction in $V2$. The decrease in anode potential of $V2$ is communicated to $V4$, biasing it below cut-off and stopping the impulse counter.

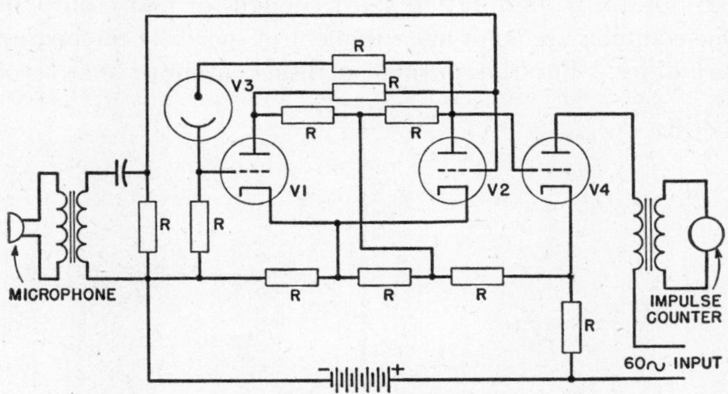

Fig. 26·30 Race timer circuit.

Figure 26·31 is a variation of this circuit for the measurement of shorter time intervals, and it is adapted to measurement of speed of moving bodies. When $V3$ conducts, the grid of $V5$ is biased below cut-off. Interruption of light to phototube $V1$ decreases the grid potential of $V3$ and stops conduction, raising the grid potential of $V2$ above the cut-off point, and capacitor $C1$ charges at essentially constant current through $V5$. Interruption of light to phototube $V2$ initiates the reverse cycle in the trigger circuit and stops the charging of capacitor $C1$. Potential across $C1$ is measured by a transfer characteristic voltmeter consisting of tube $V6$ and the milliammeter. The time interval between the interruption of light through the two phototubes depends upon the difference in milliammeter reading before and after the time interval. The instrument can be calibrated by interrupting light to the phototubes by a freely falling body or the bob of a pendulum of known length.

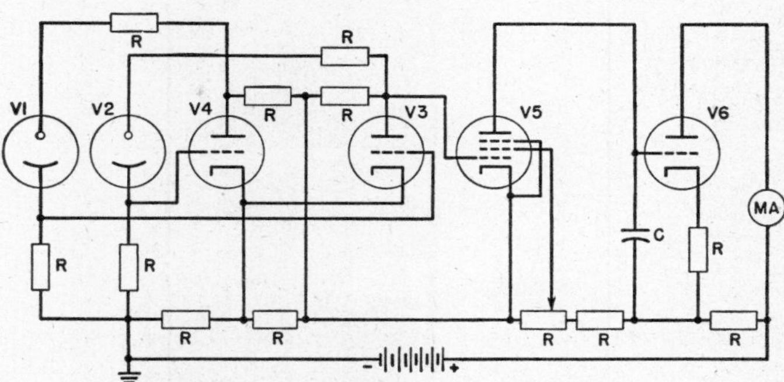

Fig. 26·31 Time-interval meter circuit.

26·6 STROBOSCOPES

A stroboscope is an instrument for studying periodic and rotary motion by rendering the moving body visible at only discrete points in its path. Usually the stroboscope consists of a source of brilliant flashes of light plus equipment which supplies power to the light source and controls the frequency of flashing.

Applications of the Stroboscope

Motion Study. The principal use of the stroboscope is in the study of rotary and reciprocating motions of high-speed machinery. Under ordinary steady light, bodies moving at high speed appear as a blur. When these motions are examined under light that flashes brilliantly approximately once per cycle of motion, the body is seen clearly and appears to be either stationary or moving slowly, depending upon whether the frequency at which the light flashes is exactly or approximately equal to the frequency of motion. Motion study of high-speed machinery under stroboscopic light gives information that cannot be obtained by running the machinery slowly, because centrifugal and inertia forces often cause distortion of moving bodies at high speeds. The same information can be secured from high-speed motion pictures. In addition, slow-motion pictures provide a permanent record from which accurate measurements can be obtained. The photographic equipment involved, however, is relatively expensive and difficult to set up. For most routine investigations, visual examination under stroboscopic light gives all the data needed.

Dynamic Balancing. In some dynamic balancing machines, the stroboscope is used to illuminate the work piece during its revolution when the point at which correction for unbalance is to be made passes under a pointer (see Section 26·8).

Measurement of Frequencies. The stroboscope can be used to measure speeds of rotation and frequencies of reciprocating motion. The advantage of the stroboscope over other instruments is that it imposes no mechanical load on the machinery on which measurements are being made. Thus, it does not disturb it from its normal speed or frequency of motion.

Stroboscope Design

Light Characteristics. The flashes of light should be as brilliant and as short as possible. The speed of the motion that can be studied is determined by the shortness of the light flashes. If the object under study moves appreciably during the flash, the observed image is somewhat blurred.

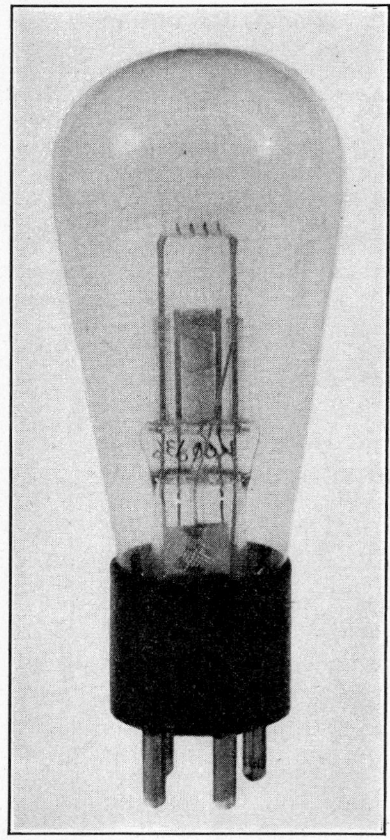

Fig. 26·32 Neon-filled thyratron.

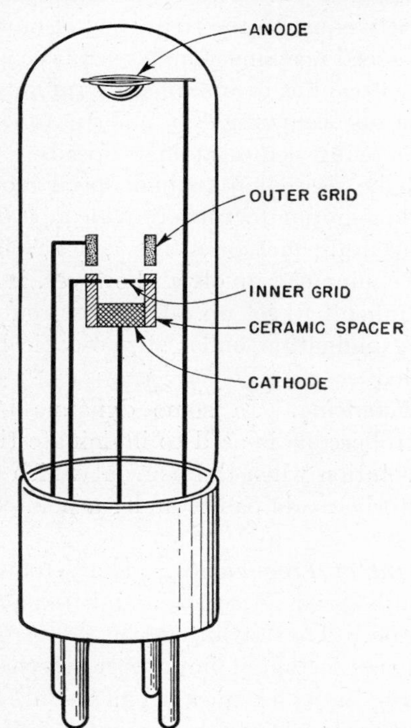

Fig. 26·33 Diagrammatic view of Strobotron tube.

The product of brilliance times the duration of the flash determines the level of stroboscopic illumination. This must be much greater than the level of general background illumination.

Frequency. The frequency of light flashes must be fairly stable and adjustable over a wide range. For some investigations, it is desirable to "freeze" the image by adjusting the frequency of flashes to correspond exactly with the frequency of motion. For other purposes it is better to cause the apparent motion to be much slower than that of the actual machine by adjusting the light to flash at a slightly lower frequency than that of the motion.

Effects of Stroboscopic Illumination

The apparent frequency of motion is the difference between the actual frequency of motion and the frequency of the stroboscopic light. If the light flashes at a slightly higher frequency than the motion, the apparent motion is at the difference frequency and in the reverse direction. Many "trick" effects are obtainable with a stroboscope.

Consider a rotating disk containing a set of twelve dots arranged in a circle about the center of rotation and a set of nine dots arranged in a concentric circle. Under stroboscopic light at various frequencies, the entire pattern may seem to remain stationary; both circles of dots may move in the direction of revolution at the same speed; both circles may move in a direction opposite to the revolution at the same speed; both circles may rotate in apparently opposite directions; or the circles may be made to appear to rotate in the same direction at different speeds.

Stroboscopic Light Sources and Control Circuits

Glow Lamp. The simplest stroboscopic light source is the neon glow lamp. It does not give particularly brilliant illumination, nor is its duration short enough for many purposes. One common use is for measurement of speeds of phonograph turntables. For this purpose a disk containing many con-

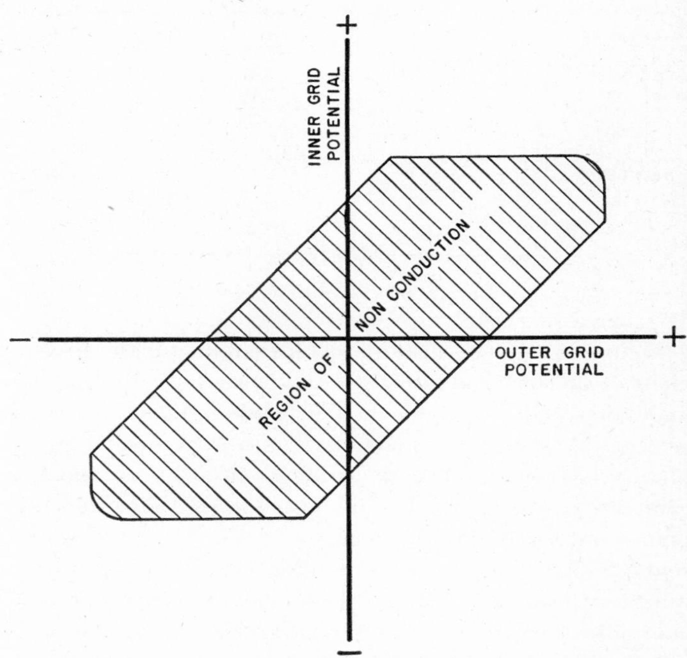

Fig. 26·34 Starting characteristic of the Strobotron.

centric circles of different numbers of dots is put on the turntable in place of the record, and is examined under light from the neon glow lamp energized from a 60-cycle power line. The rotational speed of the turntable is determined by observing which circle of dots appears stationary. This glow lamp can be used as a stroboscope at frequencies other than 60 cycles by using it in a simple relaxation oscillator circuit (see Chapter 17).

Neon-Filled Thyratron. A neon-filled thyratron is a simple and satisfactory stroboscopic light source. The type KU610 tube has been developed especially for this purpose and is widely used as a stroboscopic light source in dynamic balancing equipment. To obtain the greatest usable amount of light, the internal structure of the tube, as may be seen in Fig. 26·32, is slightly different from that of conventional thyratrons. The filament is in the upper end of the bulb.

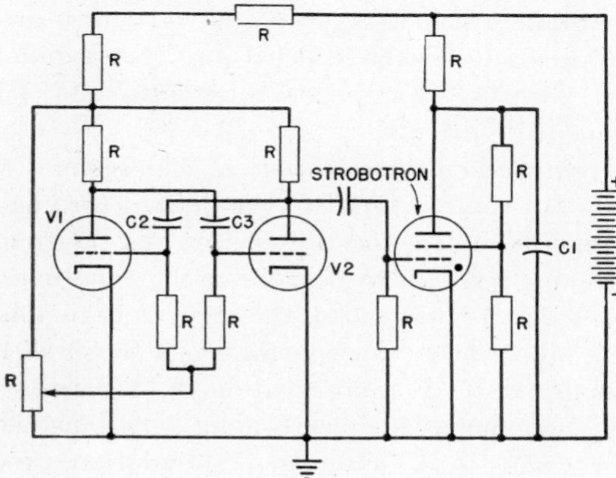

Fig. 26·35 Stroboscope circuit.

The anode, near the center of the bulb, is in the form of a small pin and is completely surrounded by the grid. The grid is a solid sheet-metal cylinder which contains one small hole in the end facing the filament. When the tube is conducting, a brilliant red glow fills the space between the filament and the hole in the grid. This tube is utilized in the circuit of Fig. 26·54. The action of this circuit is described in Section 26·8.

The Strobotron. One of the most popular stroboscopes is built around the Strobotron tube,[90, 91] a cold-cathode gas-discharging tube. Its internal structure is shown in Fig. 26·33. The cathode is cesium-coated. The two grids are used to start the discharge in the gas, which is neon and produces intense red light. The tube functions like a hot-cathode thyratron in that once the discharge is started the grids lose control and the discharge can be stopped only by removing the anode potential. The firing characteristics of this tube, however, are different from those of the conventional thyratron, as shown by Fig. 26·34. This is a plot of outer grid potential as abscissas and inner grid potential as ordinates. A region about the origin is shaded and bounded by a solid line within which conduction does not take place. If the potential of either the outer or the inner grid is increased or decreased to a point outside the shaded area, conduction starts and is extinguished only by removal of the anode potential.

Strobotron Firing Circuit. Equipped with a suitable reflector and firing circuit, this tube is a satisfactory stroboscope at flashing frequencies up to 150 cycles per second. Figure 26·35 is a simplified diagram of a stroboscope utilizing the Strobotron. Flashing is initiated by a multivibrator consisting of tubes $V1$ and $V2$. During the time the Strobotron is

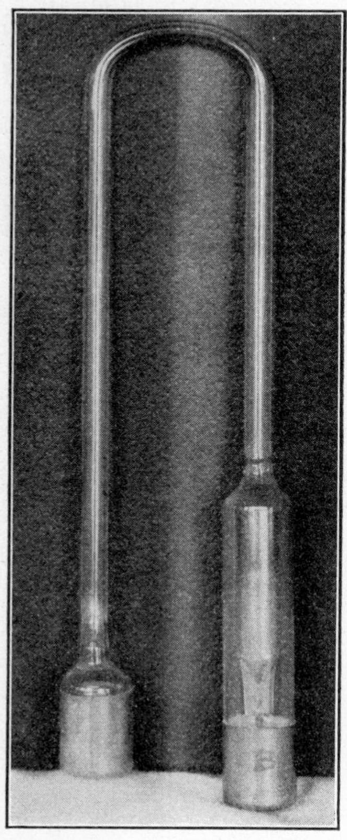

Fig. 26·36 Stroboglow lamp.

flashing, capacitor $C1$ discharges through it until the anode potential is too low to maintain conduction. The Strobotron then goes out and waits for the next impulse from the multivibrator. Frequency of the multivibrator is controlled by variation of the positive grid bias applied to it, and by the size of coupling capacitors $C2$ and $C3$.

Low-Pressure Arc. The most brilliant flashes of light of shortest duration are obtainable with a simple glass tube with electrodes sealed in the ends and filled with neon or mercury vapor at a reduced pressure. Figure 26·36 shows such a

Fig. 26·37 Stroboglow unit.

stroboscopic lamp. Both electrodes are cold, but the one intended to act as the cathode is made of a material that has a lower work function than the material of the anode.

Action of Arc Tubes. Flashing of this lamp is initiated by an electrical impulse of high amplitude and steep wave front applied to an electrode outside the glass envelope. This impulse ionizes some of the gas molecules inside of the tube and starts a discharge that is self-sustaining until the anode potential is removed. After removal of the anode potential, the return of the stroboscopic lamp to a non-conducting condition is more rapid than in any of the stroboscopic light sources previously described. This is because of its high ratio of surface to gas volume, and because ions recombine with electrons chiefly on surfaces. The stroboscopic tube illustrated can produce a 300,000-candlepower flash which lasts only 10 microseconds. Figure 26·37 is a view of a stroboscope designed around this lamp, and Fig. 26·38 is a simplified diagram of its circuit.

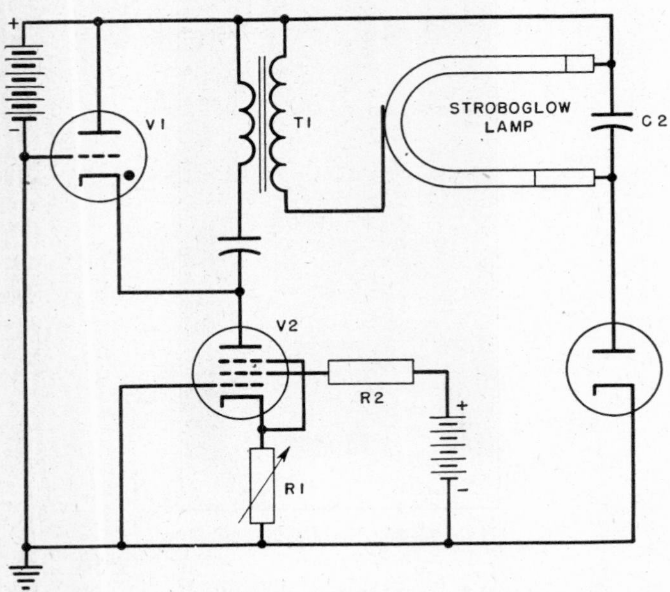

Fig. 26·38 Stroboglow circuit.

Arc-Tube Firing Circuit. In this circuit, thyratron $V1$, vacuum tube $V2$, resistor $R1$, and capacitor $C1$ function as they do in the sweep-circuit oscillator described in Chapter 9. When the capacitor has been charged to a potential at which thyratron $V1$ becomes conducting, it discharges through the primary winding of step-up transformer $T1$ and the thyratron. The resulting high potential suddenly applied to the electrode adjacent to the stroboscopic lamp produces enough ionization to start conduction in the lamp, which persists for a short time until discharge of capacitor $C2$ removes anode potential from the stroboscopic lamp. The frequency at which the lamp flashes is controlled by varying resistor $R1$, thus varying the charging rate of capacitor $C1$.

26·7 CATHODE-RAY OSCILLOSCOPE

The cathode-ray oscilloscope is probably the most versatile and most widely used of all electronic instruments. It is widely used in industrial measuring equipment and in testing equipment of various kinds.

Function of an Oscilloscope

The cathode-ray oscilloscope is a device for plotting any quantity that can be converted into an electrical potential on an axis of time or an axis of some other variable that can be converted into an electrical potential. Thus, it can be used for studying and measuring periodic and transient variations of electrical quantities such as voltage, current, and frequency; mechanical quantities such as displacement, speed, force, acceleration, and time; and optical quantities such as brightness and illumination.

The small inertia of an electron beam makes it ideal for converting high-frequency or rapid-transient phenomena into a visible indication or photographic record. The upper frequency limit of an oscilloscope is reached when the period of one cycle, or the duration of the transient being studied, approaches the length of time an electron remains the field of the deflecting plates of the oscilloscope tube.

Cathode-Ray Tube. Most cathode-ray tubes contain fluorescent screens for visual observation of the phenomenon under examination Photographic records are made by photographing the luminous trace that appears on the screen. A few demountable cathode-ray tubes have been so made that photographic plates can be inserted into the vacuum. Such tubes have been used for the study of transients of random and infrequent occurrence.

Fluorescent-Screen Characteristics. Fluorescent materials for cathode-ray-tube screens are classified according to persistence and color or spectral-energy distribution. The persistence of a screen is a measure of the time required for the light emission to decay after the electron beam has passed on to some other spot. For general use, a screen of medium persistence is used. For observation of transients or low-frequency phenomena, a screen of long persistence is preferable. For visual work, a screen material that produces a pleasing color is chosen; green is a favorite. For photographic work, blue screens are often used, because most film is more sensitive to radiations at the blue end of the spectrum. Excellent pictures can be taken, however, with green screens.

Special-Purpose Oscilloscope Tubes. Most cathode-ray tubes contain hot cathodes of the indirectly heated unipotential type. The upper frequency limit of a cathode-ray tube is extended by operating the tube at a high anode potential to increase the velocity of the electron beam. In tubes designed to operate at the highest anode potentials, cold cathodes are used. The electron beam forms some positive ions in the residual gases in the tube. These ions are accelerated toward the cathode by the anode potential. High-speed positive-ion bombardment would destroy the emitting properties of an oxide-coated or thoriated-tungsten cathode, and a pure tungsten cathode at emitting temperature gives off too much light. Tubes that are demountable so that photographic films or plates can be inserted into their envelopes usually contain cold cathodes which enable them to withstand the effects of positive-ion bombardment. The vacuum in such a tube is not usually as good as can be produced in a permanently sealed tube containing a getter.

Deflecting Amplifiers. The design of amplifiers to supply potentials to the deflecting electrodes of a cathode-ray tube is a painstaking job. For faithful reproduction of complex wave shapes, the gain of an amplifier should be almost uniform over a wide frequency range. Phase shift and non-uniform gain in an amplifier are related.[92] Usually phase shift appears before the gain has deviated enough to have a serious effect by itself. Therefore, the bandwidth for which the amplifier must have flat response is somewhat wider than might at first be thought necessary.

Amplifier Frequency Range. An amplifier for general use with cathode-ray tubes has a bandwidth of 100 kilocycles per second or so with a lower frequency limit of 5 to 10 cycles, although for special purposes an amplifier with a bandwidth or 2 megacycles may be used. Oscilloscope amplifiers usually are resistance-coupled. Gain is made uniform over the wide band by use of high-mutual-conductance amplifier tubes with inductive compensation in their anode circuits, and with inverse feedback. Good response at low frequencies requires large coupling capacitors with a minimum of dielectric leakage and absorption. To prevent "motorboating," the power supply usually must be carefully regulated and must present a low impedance to the amplifier. To obtain good response at high frequencies, stray capacitances to ground from high impedance circuits must be avoided. To prevent high-frequency oscillation, circuit components, wiring, and grounds must be laid out so that stray coupling between output and input is minimized.

Amplifier-Output Circuits. For cathode-ray tubes with screens smaller than about 3 inches in diameter, the amplifier output can be connected to one member of the pair of deflecting plates, the other member of the pair being grounded. For larger tubes, the amplifier must have a balanced output circuit so that the average potential of each pair of deflecting plates is constant; otherwise the electron beam does not remain in focus at the extremes of its deflection.

Attenuators. The attenuator used in the input circuit of the amplifier also can cause distortion unless a suitable variable capacitor is connected between the attenuator-input terminal and the adjustable center tap of the attenuator to compensate for stray amplifier-input capacitance from the attenuator center tap to ground.

Sweep Circuits. Most oscilloscopes for general use contain a time base or sweep circuit using thyratrons. Such circuits are satisfactory for maximum sweep frequencies from 20 to 50 kilocycles. At higher frequencies such circuits fail to function because the time required for a gas tube to deionize is comparable with the time of a sweep cycle. For high-sweep frequencies, vacuum-tube circuits are used. In such circuits, the thyratron usually is replaced by a biased multivibrator or similar trigger circuit, which is set off by the sweep circuit output voltage reaching a predetermined maximum value at the end of each cycle.[93]

Use of the Oscilloscope

Measurement of Amplifier Gain. An approximate measure of amplifier gain can be made as shown in Fig. 26·39. With the switch in position 1, the slider P of the potentiometer is adjusted to give a pattern of convenient height on the oscilloscope screen. Resistance from P to ground is measured, and is denoted by R_1. Next, with the switch in position 2, the slider P is adjusted to produce a pattern on the oscilloscope of the same height previously used. Resistance from P to ground is measured again, and is denoted by R_2. The voltage gain of the amplifier then is

$$\text{Gain} = \frac{R_1}{R_2} \tag{26·13}$$

Phase and Distortion Measurement. The oscilloscope is useful for detection and location of distortion and phase shift in amplifiers, transformers, and other three- or four-terminal networks. A signal generator supplying a sinusoidal voltage

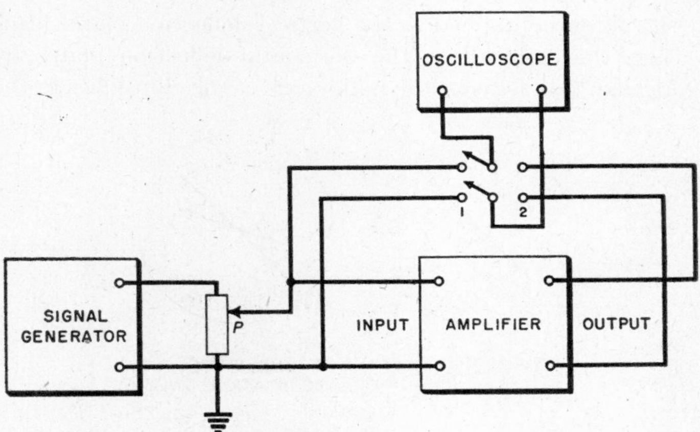

FIG. 26·39 Method of using oscilloscope to measure amplifier gain.

is connected to the input of the amplifier and to the horizontal-deflection circuit of the oscilloscope. The output terminals of the amplifier are connected to the vertical-deflection circuit of the oscilloscope, and the resulting pattern is observed on the screen. A straight line, Fig. 26·40 (*a*), indicates absence of phase shift and distortion. A line which is not straight, Fig. 26·40 (*b*), indicates distortion without

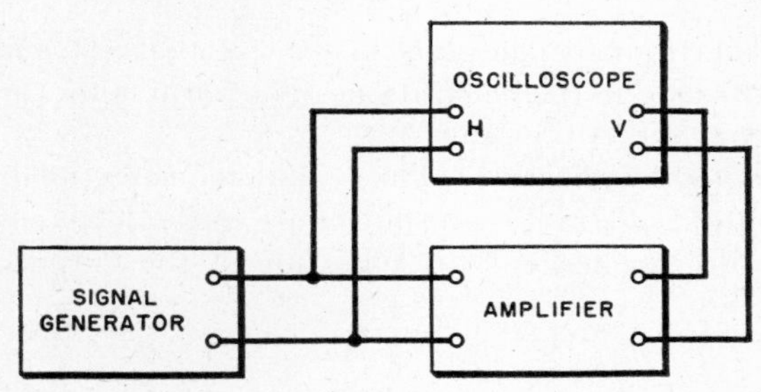

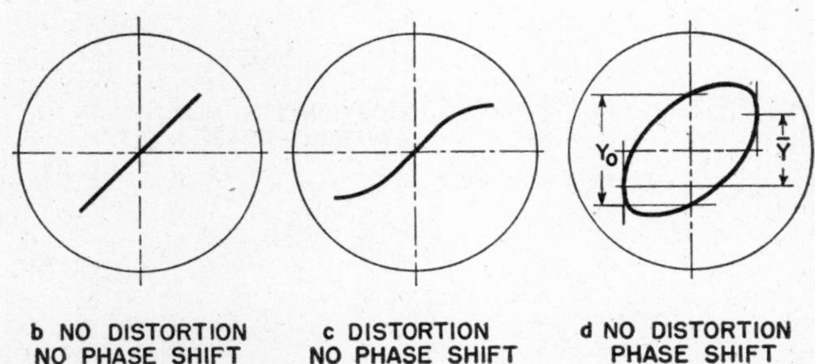

FIG. 26·40 Method of using oscilloscope to check distortion and phase shift.

phase shift. A smooth ellipse, Fig. 26·40 (*c*), indicates phase shift without distortion. The amount of phase shift is approximately

$$\tan\phi = \frac{Y_0}{Y} \tag{26·14}$$

where ϕ = phase angle by which the potential applied to the vertical-deflection plates leads or lags that applied to the horizontal-deflecting plates

Y_0, $\overline{Y}$ = pattern dimensions defined in Fig. 26·40.

If the potential applied to the vertical-deflection plates leads or lags that applied to the horizontal-deflection plates by more than 180 degrees, the major axis of the ellipse lies in the second and fourth quadrants instead of the first and third quadrants used in the illustration. The formula for the phase angle, however, is the same.

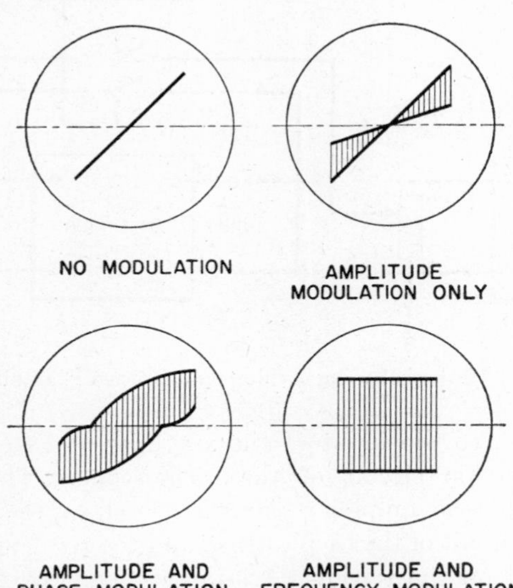

Fig. 26·41 Oscilloscope patterns for detection of phase and frequency modulation.

Modulation Indication. Phase and frequency modulation in the output of a radio transmitter are easily detected by the cathode-ray oscilloscope.[95] A portion of the output of the master oscillator of the transmitter is applied to the horizontal-deflection plates. Part of the transmitter output is fed to the vertical-deflection plates, and the phases of the two voltages are adjusted to produce a straight line on the oscilloscope screen when the transmitter is not modulated. Then, when the transmitter is modulated, the shape of the pattern on the screen indicates phase or frequency modulation as shown in Fig. 26·41.

Modulation Measurement. The degree of amplitude modulation may be measured, and distortion in the modulation may be detected if it is present.[94] A portion of the output of the modulating system supplied to the horizontal-deflection plates produces a pattern similar to one of those of Fig. 26·42. If no distortion is present, the pattern can be made to be bounded by straight lines by adjusting the phase of the audio voltage.

Frequency-Response Measurement. Frequency response of amplifiers, filters, and coupled tuned circuits such as those in radio receivers can be determined with a cathode-ray oscilloscope and an adjustable-frequency signal generator with a motor-driven tuning capacitor. Directly coupled to the shaft of the tuning capacitor is a potentiometer arranged for continuous rotation. The output of this potentiometer is fed to the horizontal-deflection plates of the oscilloscope so that for each position along the horizontal base line there is a corresponding frequency of the tuning capacitor. Output of the signal generator is fed into the circuit under test, and the output of the circuit is connected to the vertical-deflection plates of the oscilloscope. The luminous area on the screen is symmetrical about the horizontal base line, and the envelope of the upper half of the area is the frequency-response curve. The output of the circuit under test can be rectified and filtered; then only the frequency-response curve appears on the screen. Various modifications of this basic plan have been used. The equipment described by Hamburger [95] contains several refinements, including continuous calibration for amplitude and frequency.

Bridge-Balance Indication. If the output of a signal generator used to supply an a-c bridge contains harmonics, it is usually impossible to obtain a null or a sharp balance because a-c bridges can be balanced for only one frequency at a time. Iron-core components in the bridge circuit also produce harmonics which prevent sharp balance if a simple voltmeter or set of earphones is used as the balance detector. In using the oscilloscope as a balance detector the bridge output is applied to the vertical-deflection plates, and the ordinary linear sweep, synchronized with the input to the bridge, is

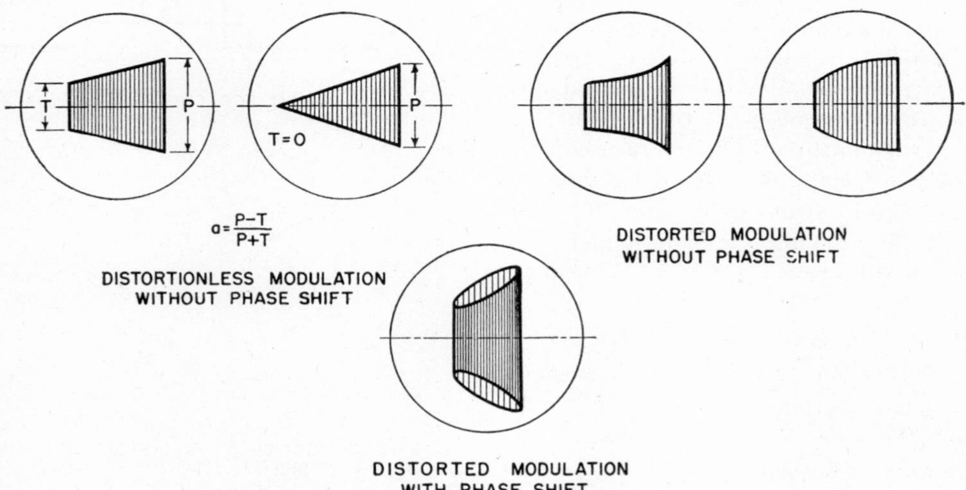

Fig. 26·42 Oscilloscope patterns for measurement of amplitude modulation.

applied to the horizontal-deflection plates. The oscilloscope indicates the wave shape of the bridge output on the screen, and makes it relatively easy to find the null point for the fundamental even with relatively large harmonic components.

Transmission Lines.[96] If a signal is fed into a transmission line, the receiving end of which is terminated by an impedance below the characteristic value of the line, the signal is reflected from the receiving end with a reversal of phase. If the line is terminated by an impedance above its characteristic value, the signal is reflected without reversal of phase. If the line is terminated in its characteristic impedance, all the energy of the signal is absorbed in the load and there is no reflection.

To determine whether a line is properly terminated, the vertical-deflection plates of the oscilloscope are connected across the sending end of the line. The horizontal-deflection plates are excited by the usual linear-sweep circuit. A generator of single pulses is connected to the sending end of the line. If a series of pulses, all in the same direction, appear on the screen, the line is terminated in too high an impedance. If the pulses are in alternating directions, the line is terminated in too low an impedance. If only a single pulse is seen, the line is terminated in its characteristic impedance.

Coil Testing. A rather interesting application of the cathode-ray oscilloscope is in the testing of electrical windings, such as motor coils, transformers, and relay coils.[97, 98] A capacitor charged to twice the test voltage is discharged periodically through the winding under test and a standard winding of identical design in series. The vertical-deflection plates of an oscilloscope are switched in synchronism with the discharge of the capacitor to record alternately the potential across the winding under test and the potential across the standard winding. The horizontal-deflection plates are supplied by a linear-sweep circuit synchronized with the discharge of the capacitor in such a way that the cathode-ray beam begins its journey from left to right at the instant of capacitor discharge. If both windings are identical, only a single trace, which shows the wave shape of the surge, shows on the cathode-ray screen. If the winding under test has one or more short-circuited turns or a ground fault, however, the wave shapes of the surges across the two windings are different, and because of persistence of the screen they appear as two distinct traces. Figure 26·43 is a photograph of equipment developed for this method of insulation testing.

Study of Transients. So far, most of the applications of the oscilloscope have been concerned with repetitive phenomena, but it is quite as useful for the study of transient phenomena. Transients may conveniently be divided into three classifications. The first consists of those under control of the investigator, which he can initiate at will. Switching surges are good examples. The second type consists of phenomena of random but frequent occurrence. Geiger-counter discharges caused by cosmic rays or other ionizing radiation are examples. The third type of transient consists of phenomena of random and infrequent occurrence, such as power-line surges caused by lightning.

Transients Initiated by Operator. The method of using an oscilloscope to study transients depends upon the nature of the transient according to the foregoing classes. Ordinarily the simplest way to study transients of the first class is to use an oscilloscope with a linear-sweep circuit so arranged that the transient is initiated once during every sweep cycle; that is, the sweep circuit is made to control the occurrence of the transient in such a way that it becomes a repetitive phenomenon while it is being studied.

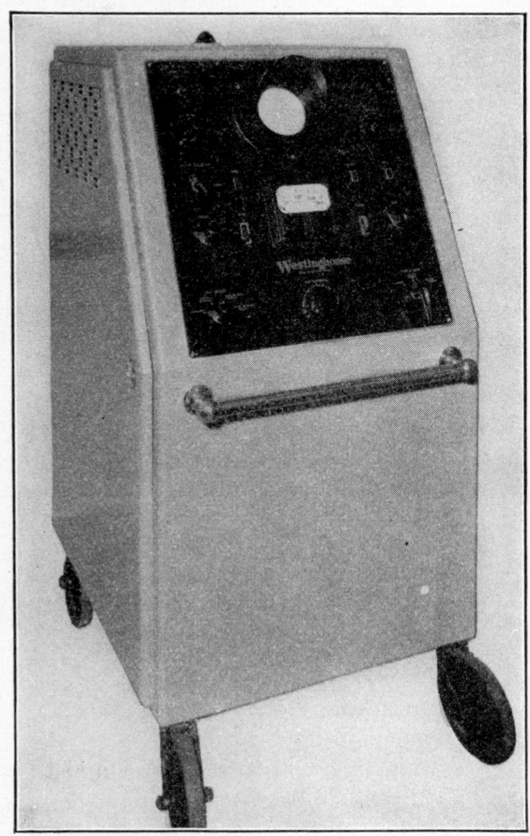

Fig. 26·43 Surge comparison tester.

Frequent Transients. For transients of the second class a linear-sweep circuit operating at any convenient frequency is used. A camera is set up before the oscilloscope, which may have a mask placed over the horizontal base line. The shutter of the camera is left open long enough to insure that a picture of the transient being studied is obtained, and is then closed.

Infrequent Transients. Study of transients of the third class by this method would result in long waits with the camera shutter open. By the time the transient occurred, the film would probably be ruined by stray light from the screen. For this type of transient a linear-sweep circuit that causes the cathode-ray beam to start across the screen when the transient occurs is used. Usually the beam is blocked off by a negative bias applied to the first grid of the cathode-ray tube. The bias is removed simultaneously as the sweep is initiated. The screen then is dark until the transient occurs, and the camera shutter can be left open for a long time without danger of ruining the film.

Time-Delay Circuits. Unfortunately, the transient voltage must rise from zero to some predetermined value to initiate the beam and its sweep. Often the time required for a signal to travel from the input terminals of the oscilloscope to the place at which it starts the sweep is greater than the time required for the same signal to travel through the amplifier applying it to the vertical-deflection plates. The first part of the transient then is not recorded on the screen, and is lost. If it is important that the first part of the transient be recorded, time delay can be introduced in the path traveled by the transient to the vertical-deflection plates. The simplest time-delay circuit consists of a length of properly terminated transmission line immediately following or preceding the vertical-deflection amplifier.[96]

26·8 BALANCING MACHINES

A body under the action of no forces other than that of gravity can rotate only about an axis through its center of gravity. Furthermore, it can rotate stably only about one of its three so-called principal axes of inertia.[99] For bodies of simple shapes such as parallelepipeds, ellipsoids, and cylinders the principal axes of inertia usually coincide with the geometric axes.

Unbalance in Rotating Bodies

If a rotating body is suspended in bearings, as most rotating bodies are, it is effectively constrained to rotation about the journal axis. If the journal axis does not coincide with a stable axis of rotation (principal axis of inertia) the constraints must exert forces to keep the body rotating about the journal axis. The projections of these forces on any plane alternate at a frequency equal to that of the frequency of rotation of the body.

Unbalance in a rotating member exerts rotating forces at the rotation frequency on the bearings. These forces are the resultants of the rotating centrifugal forces and couples exerted on the rotor by the unbalances. The forces are proportional to the product of the weight of the unbalance, its distance from the axis of rotation, and the square of the speed. These forces may rise to astounding values even for small unbalances. A practical rotor that might easily have a static unbalance of 0.1 ounce at a distance of 1 inch from the axis of rotation would weigh perhaps 2½ pounds. At a speed of 1800 revolutions per minute the total bearing load would be approximately double the static load. One serious effect of unbalance is that it shortens bearing life by increasing bearing load. This is particularly true in high-speed machinery.

Another bad effect of unbalance is the vibration it causes. Vibration may be transmitted to panel boards and cause damage to indicating instruments; it loosens nuts and screws, fatigues parts stressed by vibratory forces, and causes noise and annoyance to people in the vicinity. Today many small and medium-size motors operate at 3600 revolutions per minute, as do large power-plant turbines. Textile machinery operates at 8000 revolutions per minute, and marine-propulsion turbines operate at about 4000 revolutions per minute. Gas turbines are expected to run at 10,000 to 20,000 revolutions per minute, and precision grinding equipment and gyroscopic devices are operated at speeds approaching 60,000 revolutions per minute. The development of balancing equipment was brought about by this trend and has helped to continue it.

Balancing Machines

The Wattmeter Method of Vibration Measurement. To find the amount and the angular location for unbalance correction, it is necessary to measure both the amplitude and the phase of the vibration which the unbalance produces. One method of vibration measurement is the so-called wattmeter method illustrated in Fig. 26·44. A two-pole a-c generator is coupled or geared at a 1-to-1 ratio to the rotor to be balanced. It thus generates power at the same frequency as the rotation of the work piece. The output of this generator is fed to one coil of a wattmeter. A vibration pick-up which generates a voltage proportional to the amplitude of vibration, and with a fixed phase relationship to it,

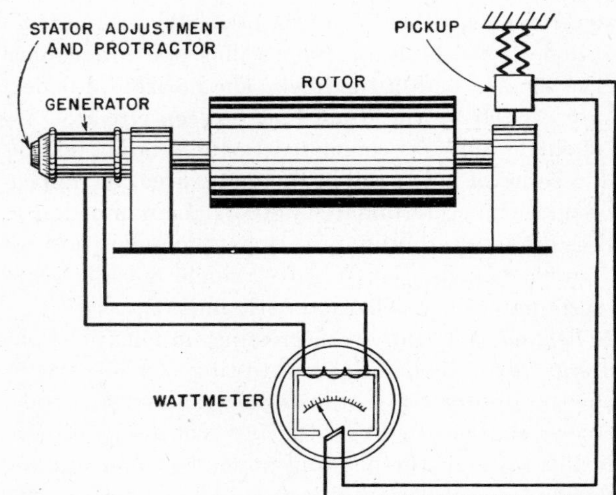

FIG. 26·44 Wattmeter method of vibration measurement.

has its output connected to the other coil of the wattmeter directly or through a suitable amplifier. Both generator current and pick-up voltage are of the same frequency; therefore the indication of the wattmeter is proportional to their product times the cosine of the phase angle between them. The generator stator is rotatable with respect to its frame and, by turning it, a position at which the phase angle is 90 degrees and the wattmeter reads zero can be found. Graduations on a protractor attached to the generator frame show the phase angle of the vibration with respect to the angular position of the work piece. Since, for balancing, only differences between phase angles of vibration are necessary, it is not necessary to know the radius of the rotor with respect to which phase is measured. This radius can be found from a previous calibration of the apparatus if it is desired.

Calculation of Unbalance Correction. For any rigid rotor the most general type of unbalance may be considered to be made up of two single weights, one in each of two arbitrarily chosen planes transverse to the axis of rotation.[100] The vibration at each bearing pedestal consists of contributions from each of the balancing planes. For an ideal mathematically linear system, such as that shown at the top of Fig. 26·45, the vibration at each bearing pedestal is simply the vector sum of the vibrations of that pedestal caused by unbalances in plane 1 and plane 2:

$$\dot{A} = \dot{k}_{A1}\dot{U}_1 + \dot{k}_{A2}\dot{U}_2 \qquad (26\cdot15)$$

$$\dot{B} = \dot{k}_{B1}\dot{U}_1 + \dot{k}_{B2}\dot{U}_2 \qquad (26\cdot16)$$

where $\dot{A}, \dot{B}$ = vibration (rotating vectors) at pedestals A and B respectively
$\dot{U}_1, \dot{U}_2$ = effective unbalances (rotating vectors) in planes 1 and 2 respectively
$\dot{k}_{A1}, \dot{k}_{A2}$ = mechanical admittances (complex quantities) of the dynamic system
$\dot{k}_{B1}, \dot{k}_{B2}$ = mechanical admittances (complex quantities) of the dynamic system.

These equations may be rearranged:

$$\dot{B} + \dot{\alpha}\dot{A} = \dot{\rho}\dot{U}_2 \tag{26·17}$$

and

$$\dot{A} + \dot{\gamma}\dot{B} = \dot{\sigma}\dot{U}_1 \tag{26·18}$$

where

$$\dot{\alpha} = -\frac{\dot{k}_{B1}}{\dot{k}_{A1}} \tag{26·19}$$

$$\dot{\gamma} = -\frac{\dot{k}_{A2}}{\dot{k}_{B2}} \tag{26·20}$$

$$\dot{\rho} = \dot{k}_{B2} + \dot{\alpha}\dot{k}_{A2} \tag{26·21}$$

$$\dot{\sigma} = \dot{k}_{A1} + \dot{\gamma}\dot{k}_{B1} \tag{26·22}$$

Writing the equations this way shows that linear combinations of the vibrations that are functions of the unbalance in either one of the planes separately can be made.

Quantities $\dot{\alpha}$, $\dot{\gamma}$, $\dot{\rho}$, and $\dot{\sigma}$ depend upon the mass and geometry of the rotor and on the spring constants of its supports. These constants usually cannot be computed conveniently, but they can be measured. First, the unbalanced rotor is run and the vibrations at the A and B pedestals are measured. Then a calibrating unbalance W_1 is placed in plane 1, and the vibrations at the two pedestals are measured again. Finally, a calibrating unbalance W_2 is placed in plane 2, and vibrations at the two pedestals are measured a third time.

The quantities $\dot{\alpha}$ and $\dot{\gamma}$ can be found from

$$\dot{\alpha} = -\frac{\dot{B}_1}{\dot{A}_1} \tag{26·23}$$

$$\dot{\gamma} = -\frac{\dot{A}_2}{\dot{B}_2} \tag{26·24}$$

where $\dot{A}_1, \dot{B}_1$ = change in vibrations at A and B pedestals caused by addition of calibration weight W_1
$\dot{A}_2, \dot{B}_2$ = change in vibrations at A and B pedestals caused by addition of calibration weight W_2.

To balance the rotor, it is not necessary to find $\dot{\rho}$ and $\dot{\sigma}$, but it is necessary to know U_1 and U_2 in terms of W_1 and W_2:

$$\dot{\beta} = \frac{\dot{B}_0 + \dot{\alpha}\dot{A}_0}{\dot{B}_2 + \dot{\alpha}\dot{A}_2} = -\frac{U_2}{W_2} \tag{26·25}$$

$$\dot{\delta} = \frac{\dot{A}_0 + \dot{\gamma}\dot{B}_0}{\dot{A}_1 + \dot{\gamma}\dot{B}_1} = -\frac{U_1}{W_1} \tag{26·26}$$

where $\dot{A}_0, \dot{B}_0$ = unbalance vibrations to be eliminated from A and B pedestals

and the required correction weights are

$$\dot{W}_{c1} = \dot{\delta}\dot{W}_1 \quad \text{in plane 1} \tag{26·27}$$

$$\dot{W}_{c2} = \dot{\beta}\dot{W}_2 \quad \text{in plane 2}$$

Figure 26·46 shows a set of equipment, developed for balancing large rotors in their own bearings, utilizing such unbalance calculations. The generator, which is coupled to the rotor and which supplies current to the field coil of the wattmeter, is shown on the left side of the picture. The vibration pick-up is on the right-hand side, and the switching and metering unit is in the case in the center. Figure 26·47 shows the equipment as used to balance a large alternator.

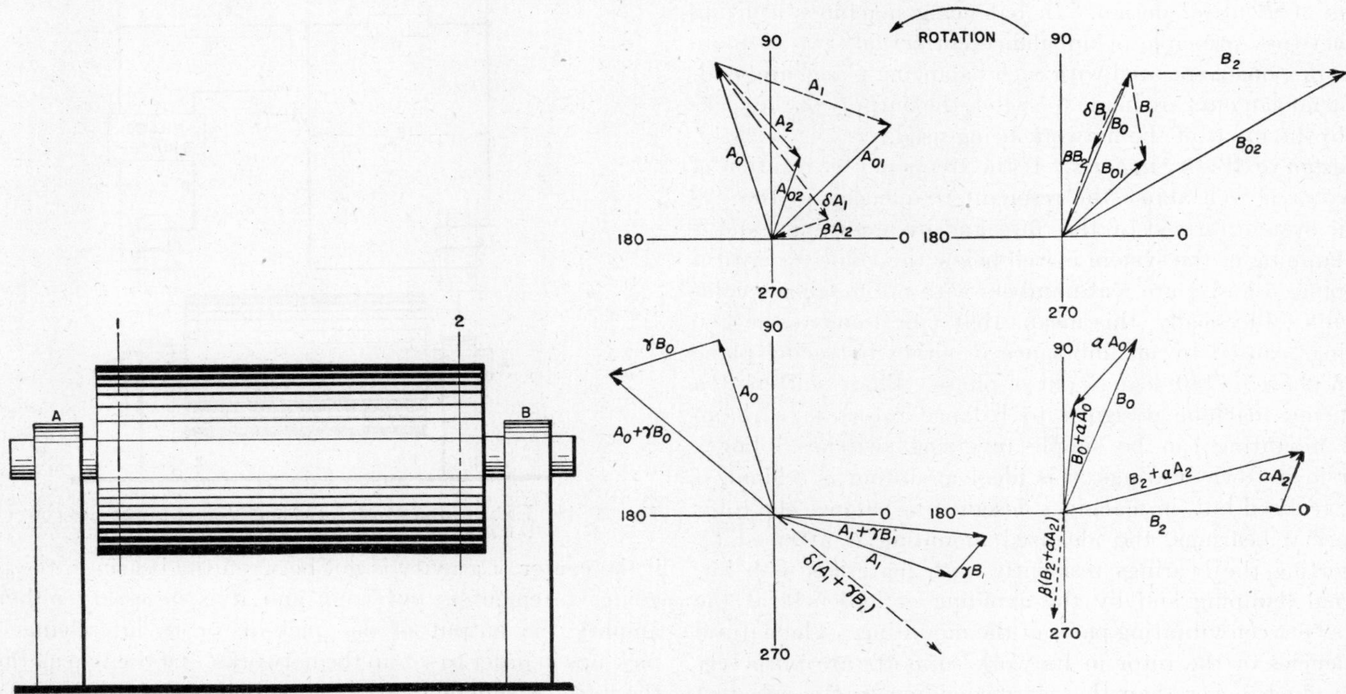

FIG. 26·45 Two-plane rotor balancing.

Electrical Networks for Unbalance Calculation. The calculation of unbalance can be made by rather simple electrical networks,[101] one of which is shown in Fig. 26·48. The vibration vector $\dot{A}_0 + \dot{\gamma}\dot{B}_0$ is proportional to the unbalance in plane 1 provided $\dot{\gamma}$ is chosen so that

$$\dot{A}_2 + \dot{\gamma}\dot{B}_2 = 0 \tag{26·28}$$

With the network method of unbalance calculation, two vibration pick-ups, the output voltages of which are proportional to vibration amplitude and in a fixed phase relation to

Fig. 26·46 Portable dynamic balancing equipment.

it, are used. The operation $\dot{\gamma}$ is performed on the voltage of one of the pick-ups by the combination of a phase shifter and an attenuator, and then the outputs of the pick-ups are added to produce a voltage proportional to unbalance in plane 1. The network is adjusted by introducing a calibrating weight in plane 2 of an otherwise perfectly balanced rotor and setting the attenuator and phase shifter in such a way that the output of the network is zero. Once the network is adjusted, it can be used without further adjustment for other rotors of identical design. In balancing machines utilizing the network principle of unbalance correction, two or more networks, one associated with each balancing plane, are used, and arrangements are made to switch the outputs of the pick-ups to the input of the network being used.

Design of Work Supports. If the frequency of rotation of the rotor is well above the resonant frequencies of the dynamic system formed by the rotor and its mountings, and if the damping of the system is well below the value for critical damping, $\dot{\alpha}$ and $\dot{\gamma}$ are real numbers with no imaginary components. Physically, this means that vibrations at the two bearings caused by an unbalance in either balancing plane are in phase or 180 degrees out of phase. Phase shifters in a balancing machine designed to balance rotors in such an ideal mounting can be simple reversing switches. For a rotor in its own bearings, this ideal mounting is seldom, if ever, realized but, in machines designed to balance the rotor in special bearings, the idealized mounting is attained by supporting the bearings resiliently with materials with low internal damping and by the avoiding of looseness at the joints between vibrating parts of the mounting. The natural frequencies of the rotor in its work supports are purposely made much lower than that corresponding to the speed at which balancing is to be done.

Network Design. The output voltage of the unbalance separation networks can be measured in amplitude and in phase by the wattmeter method. Figure 26·49 shows a simplified diagram of such a balancer. For simplicity, only one network is included. In the center of the network is a reversing switch *O-I*. The function of this switch is to interchange the vibration pick-ups. In the preceding discussion of unbalance calculation networks, it was assumed that the balancing planes lay on opposite sides of the center of gravity of the rotor and that the center of gravity was between the bearings. In that case, the voltage output of a pick-up caused by unbalance in one plane is larger if the unbalance is in the

Fig. 26·47 Balancing a large alternator rotor.

plane nearest the pick-up than if the unbalance is in the farther plane, and the network can be set up by attenuating the voltage of the pick-up nearest the balancing plane for which the network is to be set up. If both balancing planes are on the same side of the center of gravity, and sometimes

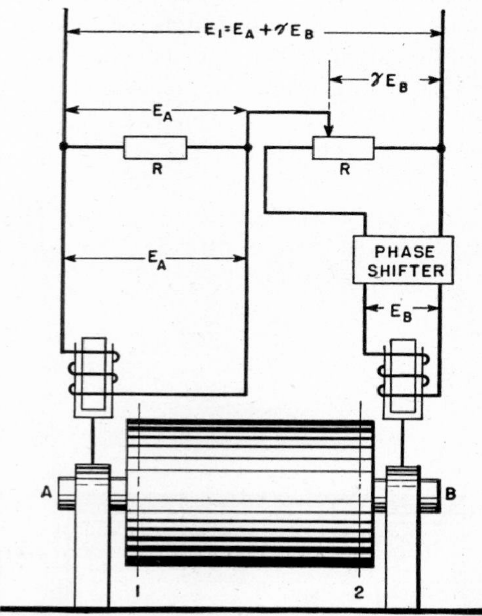

Fig. 26·48 Electrical network for calculation of unbalance correction.

if the center of gravity is not between the bearings, the foregoing statement is not true, and it is necessary either to amplify the output of one pick-up or to interchange the pick-ups in order to set up the networks. Of the two methods, the latter is simpler. To avoid starting out with a perfectly balanced rotor, a method of introducing voltages in series with the pick-ups in order to simulate a balanced rotor has been devised.‡

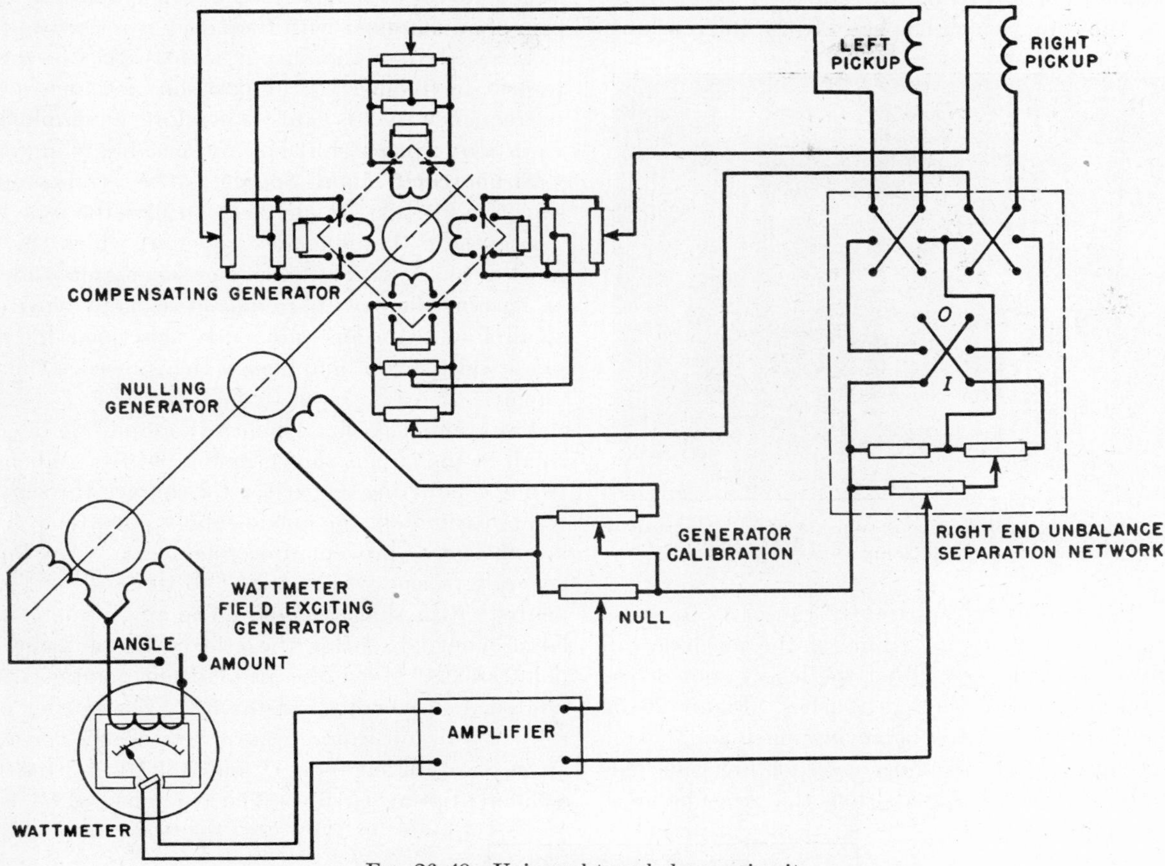

Fig. 26·49 Universal-type balancer circuit.

Compensation for Original Unbalance. The compensating generator supplies potentials in the pick-up circuits that simulate a balanced rotor so far as the wattmeter is concerned, and thus makes it possible to set up the networks without first balancing one rotor of each design by a cut-and-try process. Because the nulling generator introduces a potential in series with the output of the network, the meter reading can be adjusted to zero with the nulling potentiometer. The operator then can read the unbalances from the nulling potentiometer dials and need not remember the meter readings. Figure 26·50 is a view of such a machine. Figure 26·51 shows one of the largest balancing machines of this type ever built. A 50-ton marine-propulsion gear is shown in the work supports.

Fig. 26·50 Universal-type balancer. (Courtesy of Gisholt Machine Company.)

Stroboscopic Location of Unbalance. For balancing small rotors weighing from a few ounces to 300 pounds or so, a simpler method than the wattmeter method can be used. The amplitude of the output of the unbalance separation networks is measured by an ordinary d-c instrument after it

‡ J. G. Baker, U. S. Patent 2315478.

has been amplified and rectified. The phase angle of the network output is determined by causing it to initiate firing of a stroboscopic light source once per cycle. This stroboscopic light on the rotor, which has had identifying markings placed around its periphery, apparently causes it to stand still. A pointer attached to the frame of the machine can be positioned so that it is over either the heavy spot or the light spot of the rotor when the light flashes. Figure 26·52 is a simplified diagram of such a balancing machine.

Fig. 26·51 Balancer for 50-ton marine propulsion gear. (Courtesy of Gisholt Machine Company.)

Amplifier Design. In all machines used for precision balancing, the minute electrical signals from the pick-ups must be amplified. Exact measurement of the phase angle of the signal is as important as exact measurement of its amplitude. For this reason some special problems in amplifier design are created. The circuits usually are class A resistance-capacitance-coupled or transformer-coupled.

To achieve the necessary smallness of phase shift (most balancers are limited to 2 or 3 degrees of phase-angle error) a large amount of inverse feedback is employed. Variation of gain of an amplifier with frequency is related to the variation of phase shift of the output with respect to the input with respect to frequency.[92] Phase shift is much more sensitive to frequency than is gain. Therefore, an amplifier with small variation of phase shift has an exceedingly uniform gain.

Stroboscopic Light Source. The stroboscopic balancer shown in Figs. 26·52 and 26·53 utilizes the type KU610 tube. The cycle of the unbalance signal at which the light flashes must not vary with frequency or signal amplitude. Variation of flashing time with frequency is kept small by keeping small by the means previously described the variation of phase shift in the amplifier which precedes the stroboscope circuit.

Variation of flashing time with amplitude of signal is kept small in the circuit shown in Fig. 26·54. When the KU610 is not conducting, capacitor $C2$ charges through resistor $R5$ to approximately the anode-supply potential. The grid bias is sufficient to prevent tube conduction. The input signal is generally a sine wave applied to the grid of $V1$, which is a pentode with sharp cut-off. The amplitude of the signal is large enough to swing the grid of $V1$ far below the cut-off point and far above the grid-current point. Grid current is limited, however, by resistor $R2$. The output of $V1$ across $R3$ is a series of square waves with nearly vertical sides and approximately flat tops. Capacitor $C1$ and resistor $R4$ form a differentiating circuit. The reactance of $C1$ is large compared with the resistance of $R4$ at all frequencies in the operating range. This circuit causes the mathematical derivative of the output of $V1$ to appear across transformer $T1$ and on the grid of the KU610. The wave shape of this voltage is a series of sharp spikes in alternately negative and positive

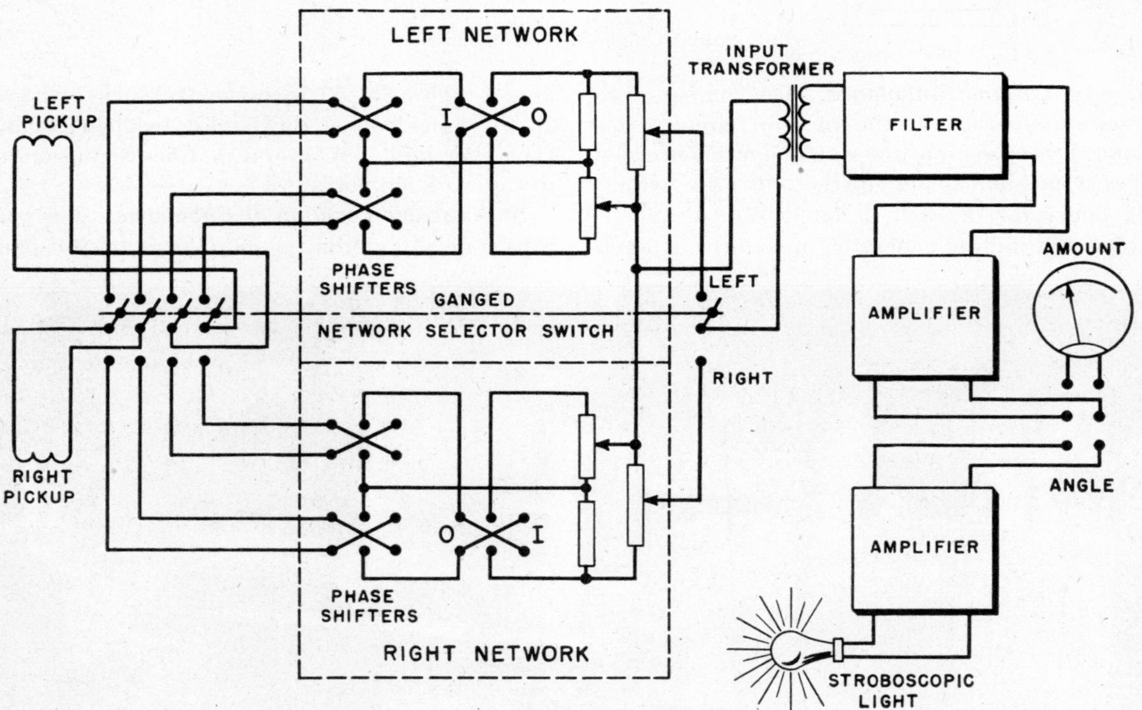

Fig. 26·52 Stroboscopic-type balancer circuit.

directions. Spikes in the positive direction, applied once per cycle of input signal to the grid of the KU610, cause capacitor $C2$ to discharge through the tube which emits light. If the potential across $C2$ and $L1$ becomes less than the ionization potential, the KU610 becomes non-conducting, and the grid regains control. The time during which the spike applied to the grid of the KU610 keeps it above the firing potential must be shorter than the time required for $C2$ to charge through $R5$. Otherwise, the KU610 ignites twice in quick succession per cycle of input and produces a blurred or multiple image. As applied to dynamic balancing machines, the formation of square waves by $V1$ and their subsequent differentiation assures that the KU610 will fire at the same

FIG. 26·53 Stroboscopic-type balancer. (Courtesy of Gisholt Machine Company.)

cyclic interval of input signal each time, regardless of variation in amplitude of the incoming signal.

Extraneous Vibrations. If the wattmeter scheme of vibration measurement is used, the wattmeter filters out the effects of all vibrations not caused by unbalance, that is, of frequencies differing from that of rotor rotation. In the stroboscopic balancer it is usually necessary to include a filter tunable to the frequency of rotation in the input circuit of the amplifier to avoid spurious indications caused by vibrations from nearby machinery or out-of-roundness of the rotor shaft. In the stroboscopic balancer of Figs. 26·56 and 26·53 the filter is shown as a narrow band-pass filter in the input to the amplifier. Such a filter can be sufficiently selective for all but the highest precision balancing.

Precision Balancing. During the last few years, there has been a need for equipment that will balance rotors so that their free vibration at running frequency will be less than 0.000002 inch. With few exceptions, rotors of this type run in ball bearings. The vibration introduced by ball bearings, although not at running frequency, is several hundred times as great as the allowable unbalance. The simple band-pass filter previously mentioned is not sufficiently selective for such balancing. Several methods of measuring unbalance vibration in the presence of relatively large amounts of disturbing vibration have been successfully applied.

Subtraction of Synchronous Signal from Unbalance Signal. One method of eliminating the effects of extraneous vibrations consists in adding to the unbalance signal from the vibration pick-up or unbalance-calculating network a reference voltage of the same frequency as the running frequency of the rotor and of adjustable phase with respect to the unbalance signal. Since the rotors to which this method is applied are almost always small, the reference voltage is generated by a photocell which faces a spot of contrasting brightness on the rotor. The resulting combination of the signal and the reference voltage is the algebraic sum of the signal caused by unbalance, the reference voltage, and signals of other frequencies caused by extraneous vibration. This combination is passed through several stages of filters and amplifiers, which attenuate the voltages caused by extraneous vibrations to a negligible amount, to an output meter. The output is the amplified algebraic sum of the reference voltage and the

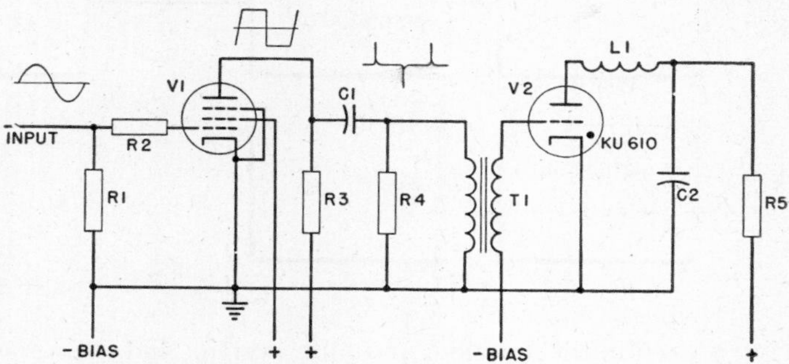

FIG. 26·54 Stroboscope circuit.

unbalance voltage. The phase of the reference voltage is then adjusted to make the output minimum, at which point the reference voltage is 180 degrees out of phase with the unbalance signal. The phase of the unbalance signal then can be obtained from the setting of the phase shifter associated with the reference voltage. Amplitude of the reference voltage is next adjusted until the output is zero, at which point the reference-voltage amplitude is equal to the unbalance-voltage amplitude. Reference-voltage amplitude is now measured by removing the pick-up signal from the input of the amplifier and filter so that the output-meter indication is due only to the reference voltage. Because of the extreme selectivity required of the filter system, it is not feasible to tune it to different frequencies for various work pieces. Any single machine is restricted to balancing rotors that run at a fixed speed.

Combination Wattmeter and Stroboscope Balancer. Another type of equipment for precision balancing is illustrated in the block diagram of Fig. 26·55. The signal from the unbalance-calculating network is fed through an amplifier and filter system to the potential coil of a wattmeter. The current coil of the wattmeter is excited by a reference voltage generated by a photocell, which faces a spot of contrasting brightness on the rotor to be balanced. If the phase shifter associated with the reference voltage is adjusted so that the wattmeter reads zero, the reference voltage is 90 degrees out of phase with the unbalance signal, and the stroboscope, which is also excited by the reference voltage, flashes each time the point from which weight must be removed from the rotor passes under a pointer. Next a phase shift of 90 degrees

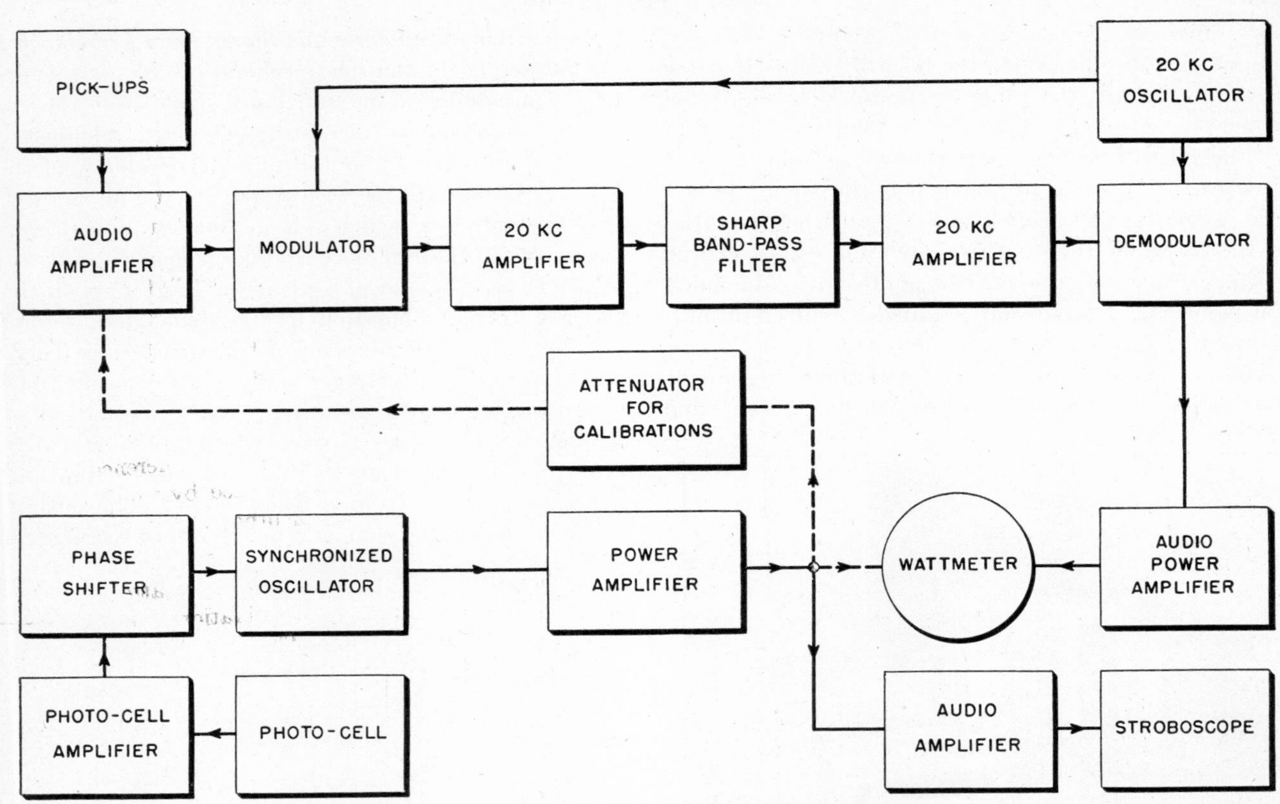

FIG. 26·55 Block diagram of microbalancer.

is introduced in the amplifier and filter system, and the wattmeter reads a maximum amount proportional to the amplitude of the unbalance signal.

Amplifier and Filter System. Theoretically, the use of the wattmeter eliminates the effects of extraneous signals on the indication of the meter. Actually, the amplitude of these signals may be so great that, if the entire signal from the unbalance-calculating network is amplified enough to produce a usable wattmeter deflection for small unbalances without filtering, the amplified extraneous signals caused by ball bearings and other sources might be so large that they would burn out the wattmeter coil. Filtering is achieved in the following way. The signal from the unbalance-calculating network modulates a signal generated by an oscillator at several thousand cycles per second. The modulated signal is then fed to a sharp band-pass filter permanently tuned to a fixed frequency. A filter can be built to operate at a fairly high fixed frequency with a pass band narrower than that of a filter tunable to the frequency of rotation of the work piece. The frequency of the oscillator is adjusted until the upper-side frequency corresponding to the oscillator frequency plus the frequency of the work piece is equal to the frequency to which the filter is tuned. The amplitude of this side frequency is a function of the amplitude of unbalance. Other side frequencies which are functions of extraneous vibrations are rejected by the filter. The output of the filter is amplified and demodulated. The output of the demodulator is equal in frequency to the frequency of rotation of the work piece and proportional in amplitude to the amount of unbalance. The output of the demodulator is again amplified and fed to the potential coil of the wattmeter. Figure 26·56 is a photograph of such a machine.

FIG. 26·56 Microbalancer. (Courtesy of Gisholt Machine Company.)

Extraneous Vibration Caused by Aerodynamic Disturbances. In balancing fans, superchargers, and propellers, a special problem arises. The balancing machine cannot distinguish between aerodynamic unbalance and mass unbalance because they both produce vibration at the rotational frequency. For small fans and impellers the problem is solved by shrouding them to minimize the aerodynamic effect and then proceeding to balance. For aircraft propellers the aerodynamic and mass unbalances must be corrected separately since the latter remains constant and the former varies with the altitude of the plane. A simple mass-unbalance correction that removed the total vibration would be correct only at one altitude. One solution to this problem § consists in mounting a large screen in the air stream of the propeller. The screen is pivoted about an axis perpendicular to the axis of rotation of the propeller. Vibration of the screen, measured by vibration pick-ups connected to a balancing machine in the usual manner, is a measure of aerodynamic unbalance only. This condition is corrected, and afterwards the mass unbalance is measured by another set of vibration pick-ups mounted along the shaft in the normal manner.

26·9 FATIGUE-TESTING EQUIPMENT

Fatigue of Metals

A metal repeatedly subjected to a stress somewhat below its ultimate strength eventually fails. The metal is said to become fatigued by the repeated application of stress. It is customary for the engineer designing a load-bearing structure to apply a so-called factor of safety; that is, he computes the stresses in the various members of the structure and then proportions the members so that they will support loads of a factor of safety times the maximum load to be expected without exceeding their ultimate strength.

Factor of Safety. The factor of safety, which at times has been called the factor of ignorance, is supposed to take care of abnormal load conditions and make allowance for the difference between the ultimate strength of the structure and the value of stress under which the structure would fail from fatigue. If, however, factors of safety considered proper by the bridge or elevator designer were to be applied to aircraft design, the plane would be so heavy that it would not be able to leave the ground. In many other types of equipment, the use of an excessive safety factor results in an impractical design.

Endurance Limit. For most metals at room temperatures a plot of stress against the number of cycles of stress until failure is of the shape of Fig. 26·57; that is, the curve starts

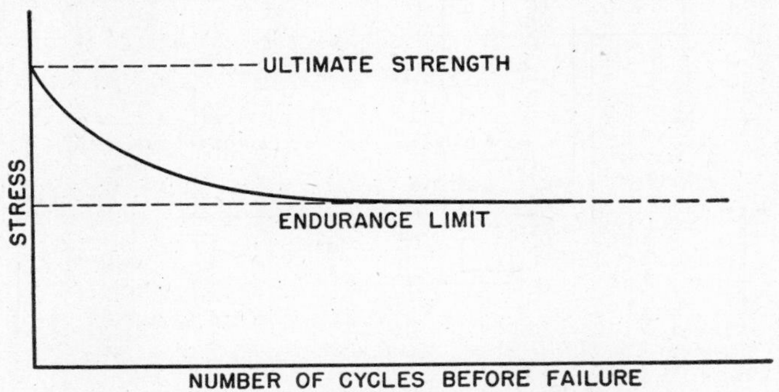

FIG. 26·57 Typical fatigue curve.

from the ultimate strength of the material and falls gradually, becoming asymptotic to a value of stress commonly called the endurance limit. A knowledge of the endurance limit is therefore of more practical value to the engineer than is the knowledge of ultimate strength. The curve falls close to the endurance limit only after a million or more cycles, depending upon the material.

§ Erle Martin and Murray C. Beebe, Jr., U. S. Patent 2343383.

Fatigue-Testing Equipment

To facilitate collection of fatigue data by applying repeated stresses at as high a frequency as possible, electronic equipment has been developed.‖ This apparatus consists essentially of the following components: a vibration motor applies an alternating force to the structure under test. The vibration motor draws its power from a vacuum-tube amplifier which contains, in addition to amplifier circuits, various control circuits.

Vibration Motor. The vibration motor is shown in partial cross section in Fig. 26·58 and pictorially in Fig. 26·59. The motor consists of a field structure containing a magnetizing coil which produces a strong unidirectional magnetic flux across the annular air gap between the top plate and the center core of the field structure. Mounted to the top plate are leaf springs which support a cast-aluminum spider. To the legs of the spider is attached a coil of copper strap, which is suspended in the air gap of the field structure. Current is brought to the drive coil through the springs that support the spider. The spider is steadied by a stud through its center. The lower end of the stud is positioned by leaf springs within the base of the motor, and the upper end of the stud projects through the spider for attachment of the test piece. Wound over the copper-strap portion of the drive coil is a winding of relatively small wire, known as the pick-up coil. The output of the pick-up coil is an alternating voltage induced by the movement of the drive coil in the magnetic field of the air gap.

Amplifiers. The amplifier, which supplies power to the vibration motor, is a conventional push-pull class AB driver stage capable of about 40 watts output, driving a class B push-pull amplifier of about 500 watts output to the vibration motor. The driver stage is excited either from an audio-frequency oscillator or from the pick-up coil on the moving element of the vibration motor.

Self-Excitation. If the test piece is driven at any one of its resonant frequencies, its inertia forces are balanced by its elastic forces. As a result large amplitudes of vibration of the test piece are produced at the expense of a relatively small vibratory force supplied by the vibration motor. When the pick-up coil of the vibration motor is connected with proper polarity to the input of the driver unit, the circuit oscillates at one of the resonant frequencies formed by the mechanical system of test piece plus the moving element of the vibration motor. The action is the same as that in the more conventional forms of feedback oscillators, except that the mechanically resonant test piece replaces the conventional electrically resonant tank circuit.

Elimination of Unwanted Modes of Vibration. Like any other oscillating circuit having more than one degree of freedom, oscillations tend to take place at the resonant frequency of the oscillating element for which the ratio of energy dissipated each cycle to the energy stored in the system is least.

‖ John A. Hutcheson: Vibration-Fatigue Apparatus, U. S. Patent 2300926 (1942); Apparatus for Vibration Testing, U. S. Patent 2326033 (1943).

For most mechanical specimens, oscillations take place most readily at the lowest natural mode of vibration. For fatigue testing, however, it is frequently desirable to vibrate the test piece at one of its higher natural modes of vibration. For this purpose, the filter (see Fig. 26·60) consisting of coil $L1$ and capacitor $C1$ is placed between the pick-up coil and the input to the driver unit, and is tuned to the frequency of the desired natural mode of vibration of the test piece. To produce sustained oscillations in any feedback oscillator, input power must be fed back in proper phase relationship to the output power. The filter permits this for only a small band of frequencies about the frequency to which it is tuned, and thus prevents oscillation at any frequency different from that to which the filter is tuned.

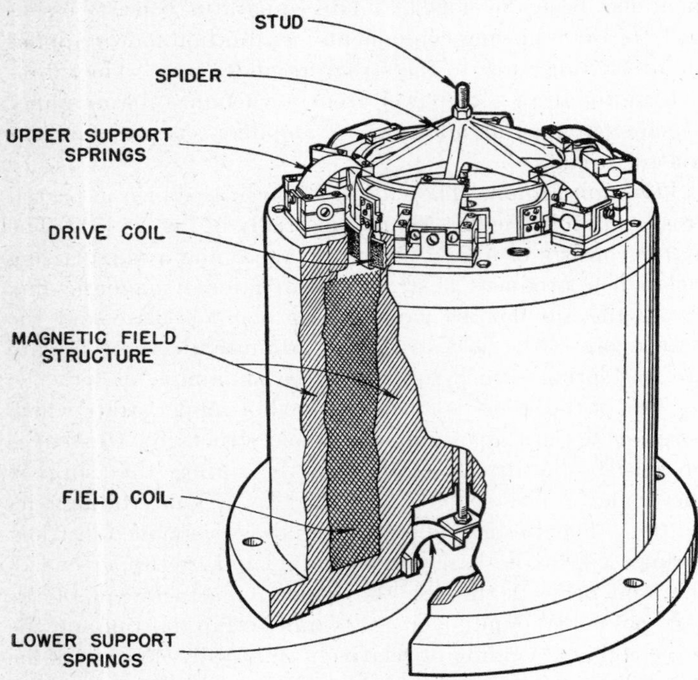

Fig. 26·58 Vibration motor.

Fig. 26·59 Vibration motor.

Control of Amplitude of Vibration. Usually a constant amplitude of vibration and stress in the specimen must be maintained, and the test must be stopped as soon as a crack forms. Figure 26·60, which is a simplified diagram of the equipment, shows how this is accomplished. Input voltage to the driver unit is applied to the diode voltmeter, consisting of tube $V5$ and $R3$, through transformer $T5$. The difference between the output of the diode voltmeter and a bias potential is applied to the grid of a triode tube $V6$ which forms one arm of a bridge circuit. The output terminals of the bridge circuit are connected to two relay coils and a milliammeter. If the amplitude of vibration of the test piece changes from the value at which the bridge was balanced, the bridge becomes unbalanced by a change in potential applied to the grid of tube $V6$. Output of the bridge energizes the coil of $K2$, which is a sensitive polarized relay. The contacts of $K2$ close in a direction that causes motor $B1$ to drive attenuator $R1$ in a direction that restores bridge balance and keeps amplitude of vibration of the test piece essentially constant.

Detection of Failure of Specimen. When the specimen starts to fail, its natural frequency usually changes appreciably. At the new frequency, attenuation in the filter is increased, and the bridge is thrown greatly out of balance, energizing relay $K1$ which has normally closed contacts that control the main power to the equipment. After relay $K1$ has been energized, the equipment is automatically shut down, and a clock used to time the test is stopped. This permits unattended operation. It also allows one to examine the specimen in which failure has started before the failure is so complete that it is impossible to determine how it started.

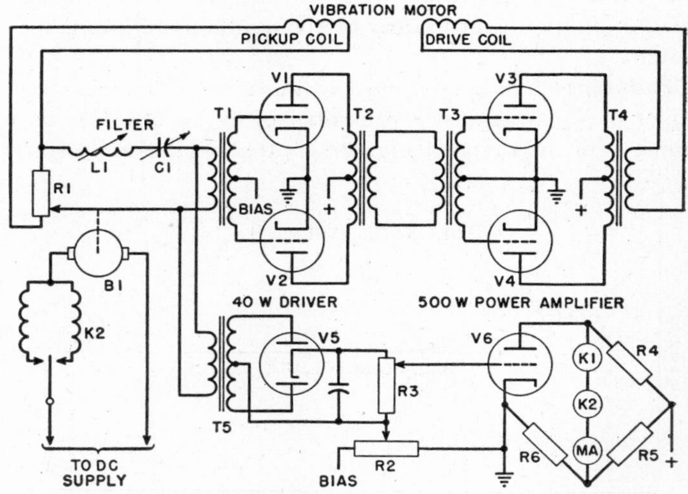

Fig. 26·60 Vibration-fatigue tester diagram.

Application of Fatigue-Testing Equipment

Determination of Natural Frequencies. In addition to fatigue testing this equipment can be used for determining the natural vibration frequencies of complicated structures and their vibration patterns at those frequencies. For determining natural frequencies and vibration patterns of structures, the driver unit is usually excited by an adjustable-frequency audio oscillator instead of the output of the vibration-motor pick-up coil.

Testing Non-Resonant Specimens. All but the flimsiest of specimens must be tested at their resonant frequencies. A vibration motor that would supply an alternating force large enough to stress a specimen beyond its endurance limit at any frequency other than its natural frequency would be impractically large. It is often desirable, however, to test a rather large structure at the natural frequency of one of its component parts, or to test at some arbitrary frequency at which a disturbing force is known to exist to determine

FIG. 26·61 Vibration motor equipped with resonator bar.

whether any of the smaller components have resonances that might contribute to fatigue failures. For such tests the test piece can be mounted resiliently so that the test piece and its mounting are resonant at a convenient frequency. Figure 26·61 is a photograph of a vibration motor with a bar mounted as a double cantilever to its top plate. The non-resonant test specimen (not shown on the photograph) is mounted on the resonator bar, and the vibrating system then becomes resonant with respect to the vibration motor. This system can be tuned by varying the length, width, or depth of the bar between its points of attachment.

Limitations of Fatigue-Testing Equipment. For any frequency of vibration there is an upper limit of amplitude above which the drive coil of the vibration motor cannot be operated without causing destructive stresses by its own inertia forces. Sometimes it is necessary to vibrate resonant test specimens at greater amplitudes than the vibration motor can endure. With certain types of specimens, this can be accomplished by interposing a coil spring between the drive-coil stud of the motor and the specimen. The specimen is vibrated above its resonant frequency; its inertia reactance at this frequency is cancelled by the elastic reactance of the spring.

26·10 THE MASS SPECTROMETER

A mass spectrometer is an instrument for determining relative amounts of atoms or molecules of different masses in a mixture. A mass spectrograph is an instrument for making precise determinations of the weights of atoms or molecules. Both instruments utilize the same fundamental principle of sorting ionized atoms and molecules of different weights by passing them through electric and magnetic fields. In construction, however, the design requirements result in two distinct types of instruments that have almost no superficial resemblance.

History of the Mass Spectrometer

The history of the development of the mass spectrometer has been excellently compiled by Jordan and Young.[102] The first mass spectrometer designed specifically for measuring relative proportions of ions of different masses was constructed by Dempster.[103] The arrangement of this spectrometer is shown in Fig. 26·62.

Other mass spectrometers of the same general type of design have been constructed by Bleakney,[104,105] Tate, Smith, and Vaughan,[149] Rittenberg, Keston, Rosebury, and Schoenheimer,[106] Tate and Smith,[107] Nier,[111] and Hipple.[108]

Principles of Design and Operation

The 180-Degree Mass Spectrometer. Dempster's mass spectrometer embodies the fundamental principles of all mass spectrometers regardless of variations in design and construction. Ions formed in the region $1C$ are accelerated through slit $S1$ by the positive potential on plate A into a region H shown shaded, where a magnetic field perpendicular to the plane of the paper exists. The stream of positive ions constitutes an electric current and the magnetic field exerts a force on the ions perpendicular both to the magnetic field and to the direction of travel of the ions. This causes the ions to travel in paths that are arcs of circles as shown by the solid and the dotted lines in the figure. The radius of curvature of the ion paths is

$$r = \frac{c}{H}\frac{V}{150}\frac{m}{e} \quad \text{centimeters} \qquad (26\cdot29)$$

where H = magnetic field in gauss
V = potential difference in volts by which the ions have been accelerated
e = charge on the ion in electrostatic units, that is, some multiple of the electronic charge of 4.8×10^{-10} electrostatic unit
m = mass of the ion in grams
c = the velocity of light = 3×10^{10} centimeters per second.

The action of the magnetic field, therefore, is to sort out ions of different masses (assuming that they are all singly ionized) into different paths. Only those of one mass follow the path with diameter equal to the distance from $S1$ to $S2$ and get through the slit $S2$ to collector C to be measured by an electrometer. The choice of the mass to be measured is made by adjusting either the ion-accelerating voltage V or the magnetic field H. All the ions of one mass that emerge from $S1$ within a small cone about the true axis of the slit converge to a focus on slit $S2$. This makes the requirements for collimating slits less severe and permits collection of a fairly large percentage of the total ion current.

Spectrometers with Sector-Shaped Magnets. A modified type of mass spectrometer in which the ions are deflected less than 180 degrees by the magnetic field is made possible by the fact that, if the central ray of a slightly divergent ion beam enters and leaves a sector-shaped magnetic field normal to its boundaries, the ions are brought to a focus on the intersection of the emergent central ray with a line through the ion source and the center of curvature of the central ray.[109] Figure 26·63 illustrates this. The resolution, or completeness of distinction between adjacent masses, is independent of the angle of deflection of the ions in the magnetic field, and is a function only of the radius of curvature in the field and of the widths of the defining slits.[110] The use of sector-shaped magnetic fields does not improve performance in any fundamental way, but it represents a great advance in the art. The advantages of this type of construction are that a sector-shaped field requires a much smaller magnet with less power to operate it and less equipment to control and regulate the field strength. The ion source and collector ends of the mass spectrometer are not between the magnet poles and are, therefore, much more accessible. The first mass spectrometer to use less than 180 degree deflection was constructed by Nier.[111] Others have been constructed by Coggeshall [112] and Hipple.[110]

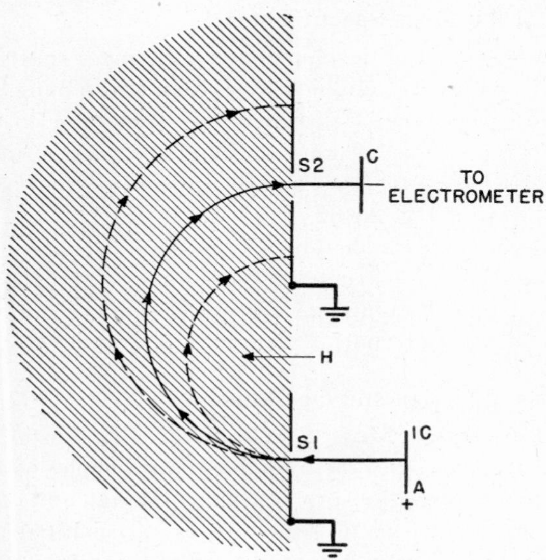

Fig. 26·62 Diagram of Dempster's mass spectrometer.

Comparison with Mass Spectrograph. Compared with a mass spectrograph, a mass spectrometer is a fairly small instrument. It supplies a relatively large ion current (from 10^{-15} to 10^{-10} ampere) to the collector. The mass spectrometer has sufficient resolution to separate one mass from another, but it is not suitable for precise determinations of mass. Determinations of the relative amounts of various masses can be made rapidly, for the ion current of each mass is collected separately, amplified, and measured by a galvanometer. In a mass spectrograph, the ion currents are too small to be amplified successfully; consequently they are detected by a photographic plate. Several masses are recorded simultaneously. Each mass is determined from the position of its image on the plate, but determination of its relative abundance can be made only by a measurement of the density of blackening produced.

Applications of the Mass Spectrometer

The mass spectrograph is a tool for the physicist engaged in research. The mass spectrometer was originally developed for determining the relative abundance of isotopes.[113–115] It became apparent that the mass spectrometer could perform other duties in pure and applied physics, chemistry, and biochemistry.

Fundamental Research. One of the earlier uses to which the mass spectrometer was put was the study of products of ionization and dissociation by electron impact.[116, 117, 146, 149]

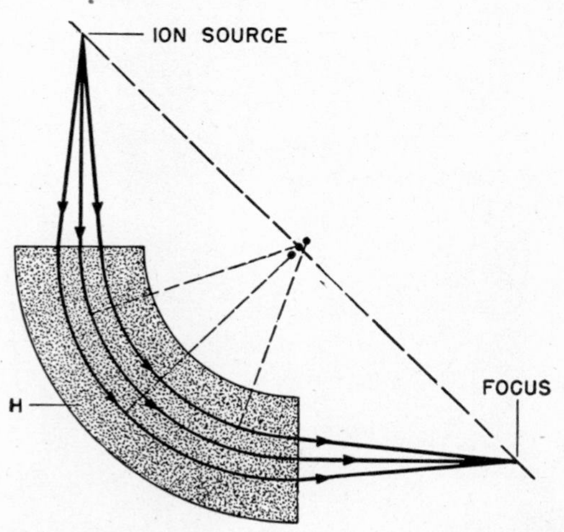

Fig. 26·63 The focusing property of a sector-shaped magnetic field.

Special types of mass spectrometers [117–123] have been made for separating and collecting minute quantities of pure isotopes for disintegration studies in nuclear physics and for tracer studies in biochemistry and medicine. Mass spectrometers are being used for detection of stable isotopes used as tracers in chemical and biochemical research.[106, 124–138] Investigations of this sort are producing important information not only in pure science but also in medicine.

Industrial Applications. One of the most promising uses for the mass spectrometer is in the industrial chemical laboratory for the analysis of complex organic gases.[112, 144, 145] The oil-refining industry has been making the greatest use of the mass spectrometer so far.

Petroleum Refining. Crude oils consist of a mixture of hydrocarbons ranging from the simple gas methane to highly complex molecules containing 45 or more carbon atoms. In general, the substances of lighter molecular weight are more volatile than the heavy compounds. These substances can be separated by distillation to yield illuminating gas, naphtha, gasoline, light and heavy lubricating oils, and residues of paraffin and asphaltic oils. Because the economic demand for petroleum products is not in the same proportion as their natural occurrence, the heavier molecules are cracked to increase the yield of the lighter gasoline and lubricating-oil constituents. The results of each step of the cracking process depend upon several variables. To assure the desired outcome, the various stages of the process must be monitored by chemical analysis. Conventional chemical analysis is sometimes too slow to be useful. Faster methods of analysis by means of infrared spectroscopy are used for some analyses, and the mass spectrometer provides a rapid method for others.

Synthetic-Rubber Manufacture. Manufacture of synthetic rubber consists essentially of the polymerization of unsaturated hydrocarbons. Like the petroleum-cracking process, it consists of many steps which need monitoring. The mass spectrometer, although not so extensively used in this field, partly because the industry is younger and partly because the applications are more difficult, may become tremendously useful.

Chemical Analysis with the Mass Spectrometer

Individual techniques and special problems often dictate departures from the general method of analysis given here, which is not always applicable; nor is it the only proper method. Suppose a mixture of oxygen and nitrogen is to be analyzed. The mass spectrometer is first calibrated with pure samples of these gases. Oxygen is first admitted to the ionization chamber of the mass spectrometer at a known low pressure through a flow-restricting device. A plot of ion current versus mass, which is a function of the magnetic field or ion-accelerating voltage (whichever is varied), is then made. The plot consists of several peaks, one corresponding to each mass. Because the flow through the flow-restricting device is not the same for all gases, and because the same ionizing electron current does not produce equal amounts of ions in all gases, equal peak heights do not necessarily mean equal partial pressures of the gases in a mixture. In passing through the flow-restricting device, however, and during the ionizing process, each gas behaves as though the other were not present; therefore, if the ionizing electron current and the energy of the electrons are held constant, the ratios of the peak heights are linear functions of the ratios of the numbers of molecules in the original mixture.

Calibration. For pure oxygen the peaks will correspond to the various isotopes and dissociation products of the molecule; a strong peak at mass 32 for the ionized molecule $O^{16}O^{16}$, with weaker peaks at masses 33 and 34 for the isotropic molecular ions $O^{16}O^{17}$ and $O^{17}O^{17}$ and weak peaks at masses 16 and 17 for the atomic ions O^{16} and O^{17}. Unless the natural relative abundances of the isotopes in any element have been purposely altered, and provided the temperature of the gas during ionization and the energy of the ionizing electrons are always kept the same, the ratios of all the peaks for any element to any peak of that element are always the same. Therefore the size of the most convenient peaks (usually the largest) of each element in a mixture may be taken as characteristic of the partial pressure of that element in the mixture. All the foregoing means that calibration for oxygen consists simply of determining the ratio of the peak height for $O^{16}O^{16}$ (mass 32) to the pressure of pure oxygen. The instrument is then calibrated in the same manner for nitrogen with pure nitrogen and, in general, it is calibrated with pure samples of each of the other gases to be analyzed.

Analysis. To determine relative amounts of oxygen and nitrogen in a mixture, which was the stated problem, the mixture is admitted at a known total pressure to the mass spectrometer. A plot of the peak heights is made and, from these data and the previous calibration data, the partial pressures of oxygen and nitrogen are obtained. The ratio of the partial pressure of each component to the total pressure is the mole fraction of that component in the mixture. This procedure is not restricted to two-component mixtures, but can in general be applied to mixtures of three or more components.

Hipple [141] gives the results of several, two-, three-, four-, and five-component analyses showing lack of agreement of from 0.2 to 5 percent between proportions in which the components were mixed. Hoover and Washburn [142] have reported results of several five-component analyses with comparable accuracies. Since it is important to know the properties of molecules to be analyzed with the mass spectrometer, a table has been prepared by Hipple [110] giving some references to molecules that have been studied with the mass spectrometer.[143–168]

Mass-Spectrometer Design

To obtain accurate results, the mass spectrometer must possess extraordinary stability and freedom from extraneous influences. A description of a typical mass spectrometer shows how some of the design problems are solved.

Tube. The heart of any mass spectrometer is the tube, for upon its design and construction depend the resolution, sensitivity, and reproducibility of the instrument. A 90-degree deflection of the ions was chosen because a smaller magnet is required than for the 180-degree type, and the length of path from ion source to collector is shorter than it would be with 60-degree ion deflection. This type of tube has the added advantage of being easy to mount, since only right angles are involved, and it is relatively rugged mechanically because its total length is short. The tube is shown pictorially in Figs. 26·64 and 26·65 and diagrammatically in Fig. 26·66. Electrons are generated by a hot filament and are collected by the anode 3. The sample is ionized in the box-like cavity between plates 2 and 4, and the ions are pushed out through the slit in plate 2 by a small positive potential applied to plate 4. They are then accelerated along the tube by the potential difference between plates 2 and 5.

Extraction of Ions. To approach the theoretical limit of resolution of the tube, the ion beam must be as nearly homogeneous in energy as possible. To secure a homogeneous energy distribution requires that ions emerge into the space between plates 2 and 5 with as small and as uniform a velocity as possible. This uniform velocity is achieved by keeping the "draw-out" potential between plates 2 and 4 small (1 or 2 volts) and by limiting the electron beam to a small dimension vertically to insure that all the ions originate in a nearly equipotential region. The electron beam is limited by the slits in plates 1 and 2 and by a magnetic field from a small auxiliary electromagnet that supplies a field in the direction of travel of the electrons.

Filament Structure. Thermal dissociation is minimized by an arrangement of the parts of the ionization chamber in such a way that the gas to be analyzed travels directly into the ionization chamber without first passing over the hot filament. Use of an oxide-coated or thoriated filament, which emits at a much lower temperature than pure tungsten, is undesirable because individual filaments vary in emissive characteristics. Even the same filament may act differently at different times. A pure-tungsten filament consuming as little power as possible overcomes this objectionable instability and does not heat the adjacent parts of the ionization chamber enough to cause a noticeable amount of thermal dissociation. Since the requirement of low power consumption results in a filament that is mechanically fragile and comparatively short-lived, it has been made removable by mounting it on a stem attached to a ground-glass seal. The side of plate 1 on which the filament is mounted is formed into a box-like space, and mating guides on the filament assembly insure self-alignment of the filament.

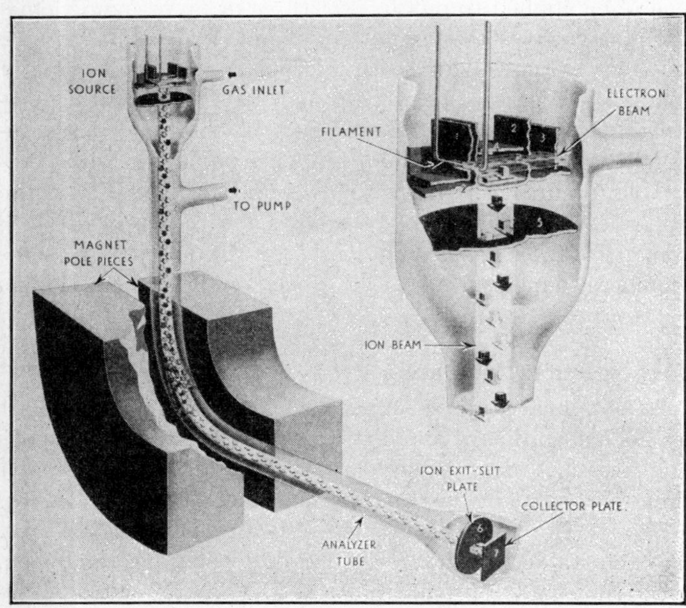

FIG. 26·64 Mass-spectrometer tube, pictorial diagram.

Shielding. To avoid distortion of the ion path by stray electric fields, a grounded metallic shield is placed inside the tube. Since both plate 5 and the exit slit are at ground potential, the ions travel in field-free space. The shielding within the tube is composed of a long narrow metal strip to which slightly overlapping rectangles of thin metal are spot-welded. As this structure is pulled into the tube, the rectangular strips bend to fit the interior contours snugly. This shield is visible in Fig. 26·65, inside the curved portion of the tube. Chem-

FIG. 26·65 Mass-spectrometer tube.

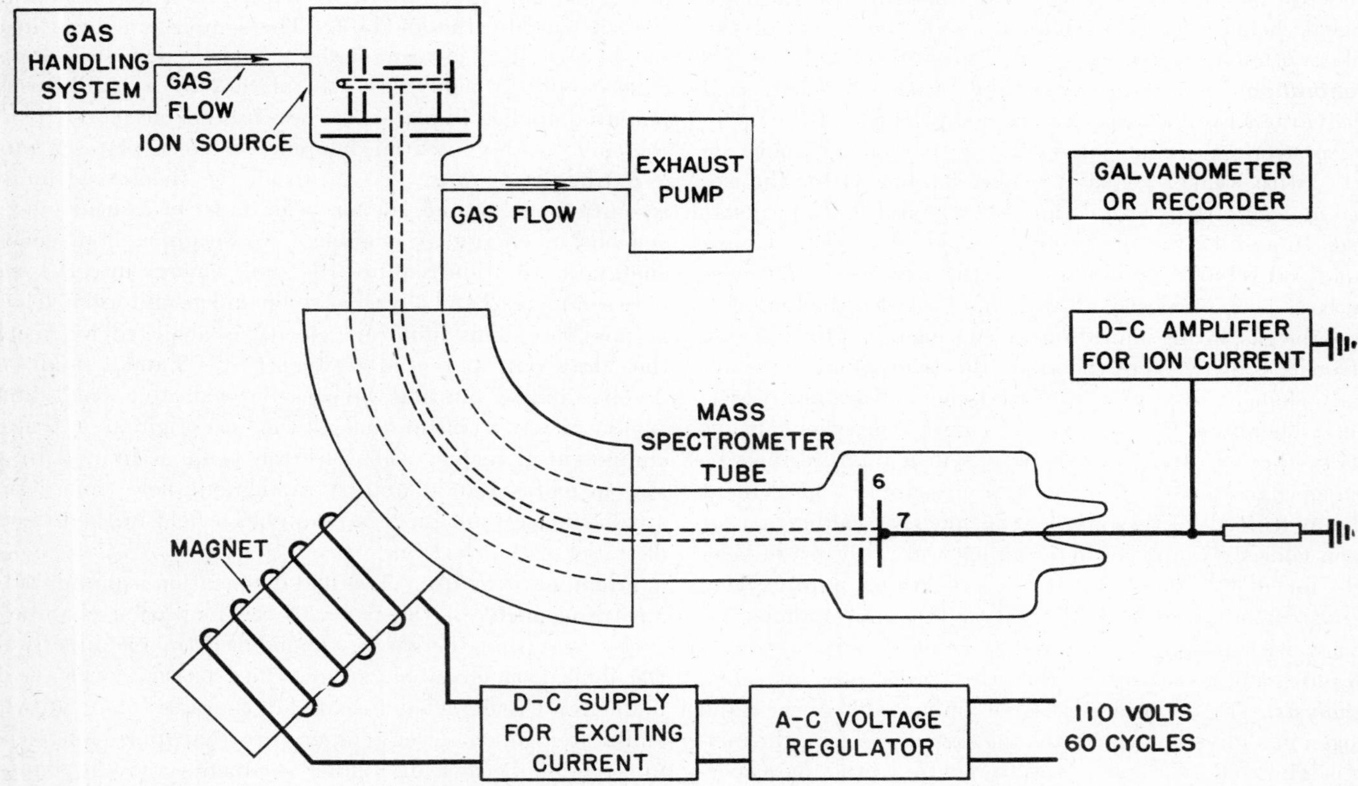

FIG. 26·66 (*a*) Block diagram of mass spectrometer.

ical methods of depositing thin films of metal on the interior walls of the glass tube have been tried with some measure of success.

Collection of Ions. Figure 26·66 shows the ion-collector plate 7 immediately behind the exit slit, plate 6. Actually there is an auxiliary guard ring between the exit slit and the ion collector. This guard ring is maintained at a potential slightly negative with respect to the ion collector to return any secondary electrons that might be emitted. Since secondary emission amounts to an effective multiplication of the ion current, it might seem a desirable thing, but it is too variable and unstable to be dependable for accurate measurements.

Gas Adsorption. To bake out adsorbed gases between analyses, resistance heaters are wound around the outside of the tube, and an internal winding to heat the ion-source parts is included. Some of the heavier unsaturated hydrocarbons, notably isoprene, become adsorbed on the internal metal parts of the tube during an analysis to such an extent that the resolution of the instrument is impaired, because these surface layers are insulating and acquire a charge which sets up disturbing electrostatic fields inside the tube. The formation of non-conducting surface layers was overcome by gold-plating all the metal parts inside the tube.

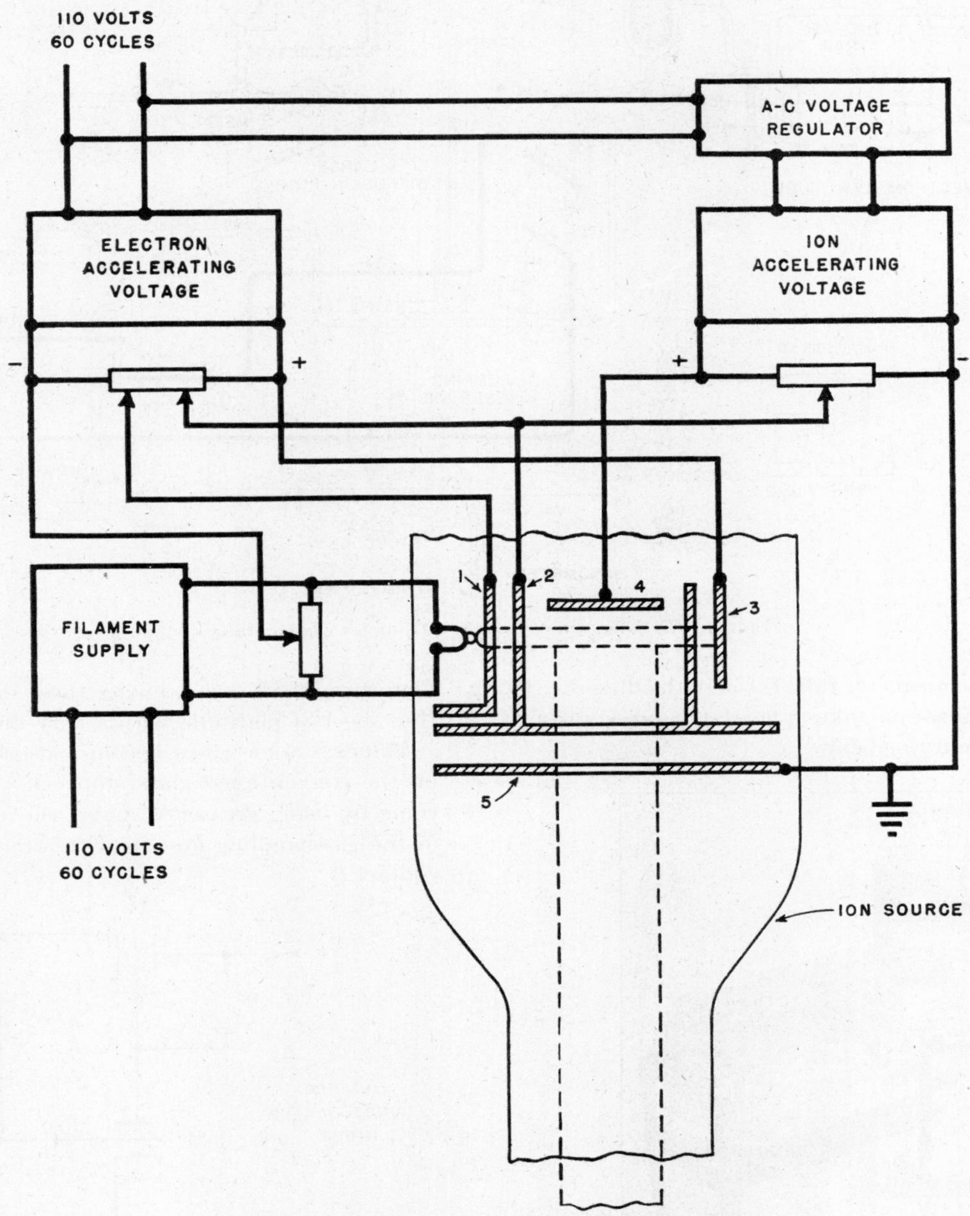

FIG. 26·66 (*b*) Block diagram of mass-spectrometer ion source.

Sample Introduction System. The gas-handling system for admitting samples and evacuating the mass spectrometer tube is shown in Figs. 26·67 and 26·68. Vacuum is produced and maintained by two conventional high-speed mercury diffusion pumps backed by a rotary forepump. The forepump is used also to evacuate the dome in which the electrometer is housed. Gas samples are brought to the instrument in a small flask. Then the sample is expanded into a larger reservoir, where its pressure is reduced to a few millimeters of mercury. The sample then leaks into the ionization cham-

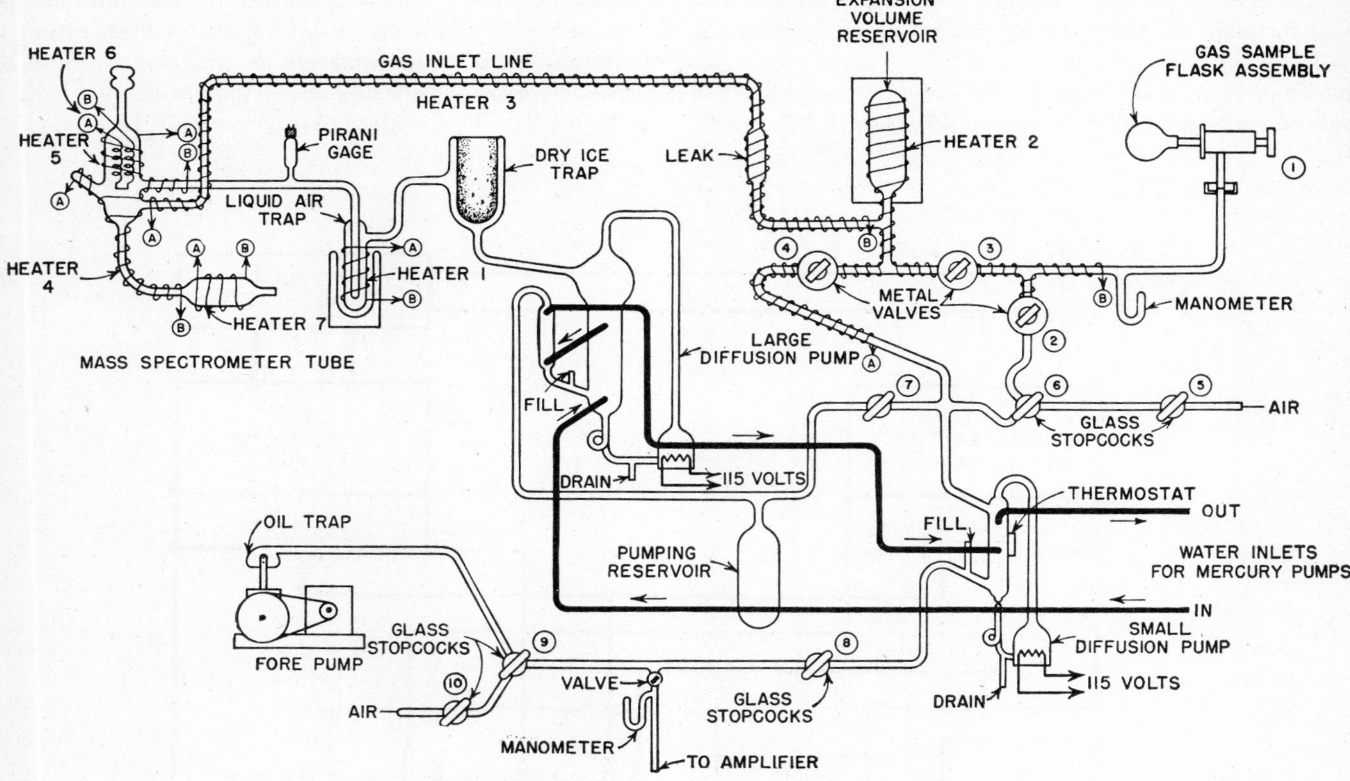

Fig. 26·67 Diagram of mass spectrometer gas-handling system.

ber of the mass-spectrometer tube through the flow-restricting device so that ionization takes place at a pressure of 10^{-4} millimeter of mercury or below.

Fig 26·68 Mass-spectrometer gas-handling system.

Flow-Restricting Device. The flow-restricting device must have a linear flow-pressure characteristic. The most satisfactory method of producing these devices is to puncture a thin sheet of platinum with a very fine needle.

Valves. Some gases become adsorbed on the grease used on the conventional glass stopcock. The difficulty is overcome by using greaseless metal vacuum valves at all points in the gas-handling system with which the sample comes in contact.

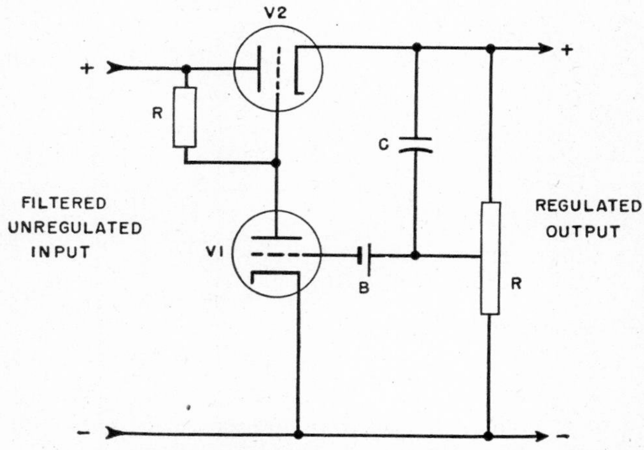

Fig. 26·69 D-c voltage-stabilizer diagram.

Electrometer. An electrometer of the type described in Section 26·2 and pictured in Fig. 26·13 is used to measure ion current.

Rectifiers. The unidirectional potentials necessary for accelerating the ions and for operating the magnet and the electrometer are furnished by conventional rectifiers. The alternating input voltage to the rectifiers is stabilized somewhat by a conventional saturable-reactor voltage stabilizer. After being filtered, the output of each rectifier is passed through an electronic voltage stabilizer of the type described by Hunt and Hickman [169] and by Neher and Pickering.[170] A generic circuit for this type of stabilizer is shown in Fig. 26·69. Vacuum tube $V1$ is biased by the difference in voltage between battery B and a portion of the output voltage. Variations in output voltage are amplified by this tube and applied to the grid of the regulator tube $V2$ in such a way as to correct the fluctuation. Capacitor C causes the regulator to function even more sensitively for ripple or transients than it does for slow changes in steady-state voltage. In other words, it prevents the regulator from hunting by enabling it to anticipate fluctuations. In some circuits of this type, glow tubes or voltage-regulator tubes are used instead of a battery to bias the amplifier tube. Although glow tubes are more convenient to use, they are too unstable with respect to age, temperature, and voltage changes for this application. With properly designed regulators fluctuations in output voltage can be made less than one part in several thousand during normal fluctuations in line voltage.

Magnet Power Supply. A slight additional amount of stability was obtained by energizing the magnet from the same rectifier that furnishes ion-accelerating voltage, because an increase in magnet current affects the ion beam in a way opposite that in which an increase in ion-acceleration voltage affects it. This stabilizing effect is not perfect, for deflection of the ion beam is proportional to the square of the magnetic field and only to the first power of accelerating potential.

have almost no effect on the number of ions formed. Figure 26·70 is a plot of an ionization-efficiency curve for nitrogen.

Filament-Emission Regulator. The number of ions formed is directly proportional to the number of electrons in the ionizing stream; consequently it is important to hold electron current constant. Since the filament requires a rather large current at low voltage, it is inconvenient to operate it from a battery. In fact, because active spots sometimes form on tungsten filaments, battery operation will not insure constancy of electron current. To overcome these objections,

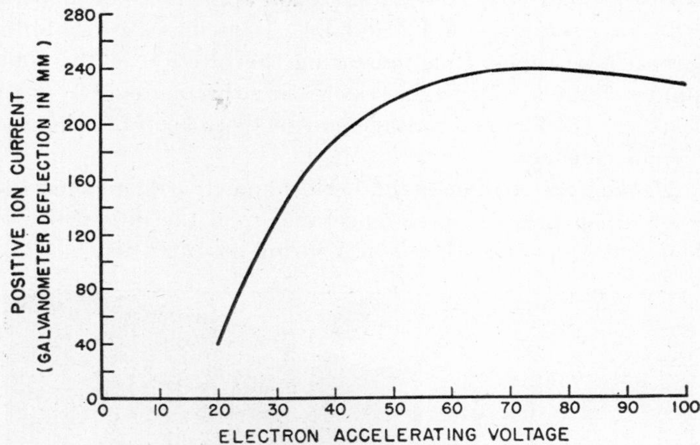

Fig. 26·70 Ionization efficiency of nitrogen.

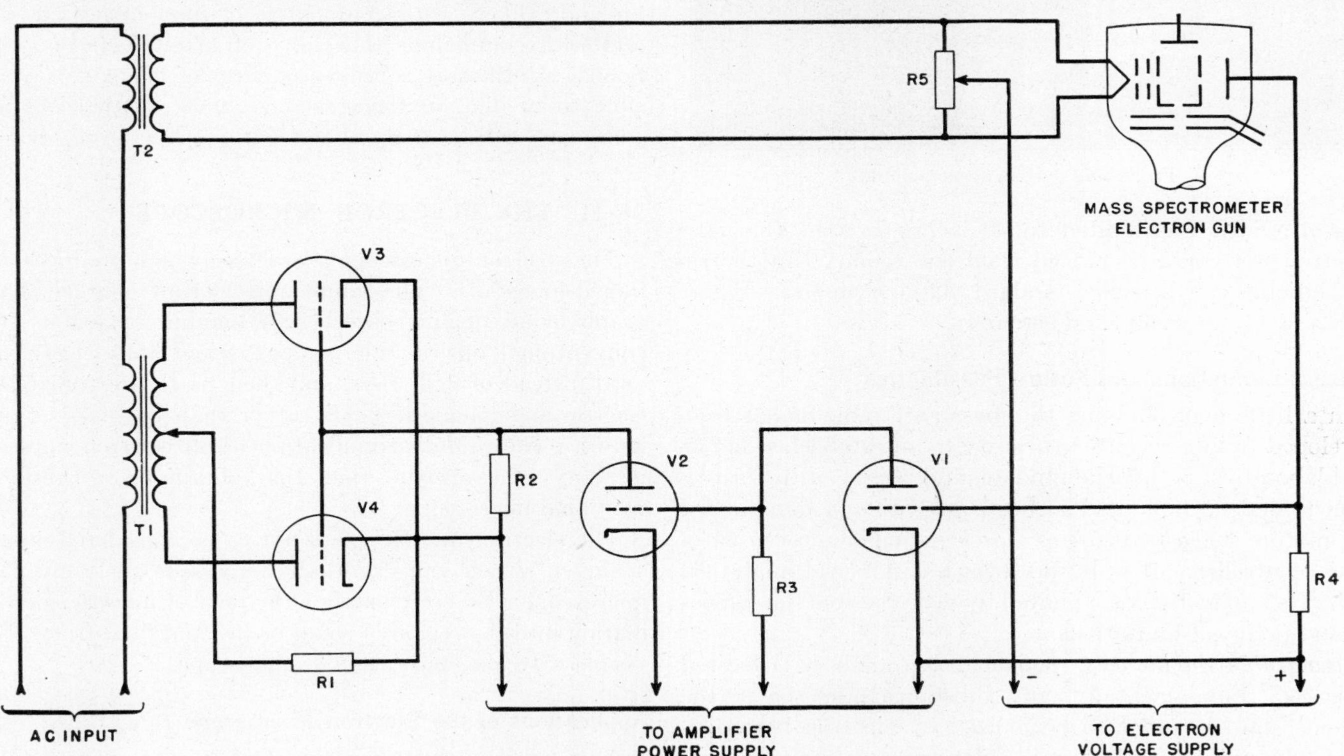

Fig. 26·71 Filament-emission regulator.

Electron Potential Supply. For supplying potential to the electron gun, a rectifier and filter with a two-stage cascaded glow-tube voltage regulator is satisfactory. It is customary to use electrons from 75 to 100 volts energy for ionizing. At these potentials, the ionization-efficiency curve of almost all gases is almost flat; therefore small changes in electron energy

the filament is operated on alternating current and is regulated so that the electron current collected by the anode is constant. Figure 26·71 is a simplified diagram of the circuit. Transformer $T1$ with vacuum tubes $V3$ and $V4$ forms a conventional grid-controlled rectifier feeding a load consisting of resistor $R1$. The primary of $T1$ is in series with the primary of the mass-spectrometer filament transformer $T2$ and forms a controlling adjustable impedance. Vacuum tubes $V1$ and $V2$ are pentodes but are shown as triodes. They are connected in a direct-coupled amplifier circuit. An increase in electron current after amplification appears as an increase in bias on the grids of $V3$ and $V4$. This increases the drop across these tubes, thus increasing the fraction of the total input voltage across $T1$ and decreasing that across $T2$. This decreases the filament voltage and tends to keep the electron current constant.

Mechanical Arrangement. Mechanical arrangement depends upon personal preference and upon the intended use. This instrument was built on a spring-mounted truck. The over-all width was limited to 30 inches so that the mass spectrometer could be moved from one room to another in the laboratory through standard 32-inch doors. Figure 26·72 shows the completed assembly.

FIG. 26·72 Mass spectrometer.

Present Limitations and Future Possibilities

In a little over 30 years the mass spectrometer has been developed from a scientific curiosity to an established industrial laboratory tool. The fundamental ability of the instrument to make a rapid and accurate analysis of a complicated gas mixture suggests its use as a process-line monitor or automatic controller. It is by no means a universal analytical tool. Its applications are limited by rather severe fundamental and practical limitations.

Analyzable Substances. Quantitative analysis is limited to gases. Fundamentally, any substance must be in the gaseous state in order to be ionized by electron bombardment. Liquids or solids might be vaporized before admitting them to the mass spectrometer tube. In fact, they are vaporized for determinations of mass. Unfortunately, the composition of the vapor from a liquid or solid is not representative of the composition of the liquid or solid until it is all vaporized. Sometimes it is feasible to vaporize completely a sample of mixtures of liquids with low boiling points.

Calibration. To analyze a sample, a pure sample of each component must be available so that the instrument can be calibrated for these components. If calibration data for all the components in a sample are available, it can theoretically be analyzed, both qualitatively and quantitatively; but, for most hydrocarbon analyses, these analyses involve time-consuming calculations. Such calculations can be made in a reasonable time if the identities of the components in the sample are known and only their relative proportions are to be determined.

Accuracy of Analysis. Accuracy depends greatly upon the number and type of components in the mixture. Accuracy of determination of peak height is fairly definite and constant. But, since molecules dissociate under electron bombardment, ions from two or more substances could contribute to a single peak. Accuracy then depends upon the relative sizes of the contributions to the peak being used for measurements. Masses of the molecular fragments range from the molecular weight of the molecule downward, and there is a unique peak for the heaviest molecule in the mixture. Accuracy is greater for analysis of a small amount of a heavy compound in the presence of lighter ones than for traces of light compounds in the presence of heavier ones.

Unsolved Problems. Some of the effects in the ionization chamber of a mass-spectrometer tube are not completely understood. They seem to consist of surface effects, chemical reactions on surfaces but not in bulk. Monomolecular films of some substances on the surfaces and edges of the slits that define the electron beam cause secondary emission of electrons, which upsets calibration. Generally these effects are more likely with compounds of high molecular weight and with compounds that have large numbers of multiple bonds. Until more is known of surface effects, it is not possible to predict on theoretical grounds whether any given compound can be analyzed with a mass spectrometer.

26·11 THE ELECTRON MICROSCOPE

The electron microscope is a device which produces an enlarged image of a transparent object or an enlarged shadowgraph of an opaque object in a manner analogous to the conventional optical microscope, except that electrons are used instead of light rays, and their paths are controlled by electric and magnetic fields rather than by glass or quartz lenses. The useful magnification of an electron microscope is many times greater than that obtainable with the conventional microscope.

The electron microscope is not to be considered as an instrument which will eventually supersede the conventional microscope; it merely extends the field of microscopic examination into the region of small bodies and fine structures not visible with the conventional microscope.

Applications of the Electron Microscope

Although the electron microscope is a comparatively new device, it has been applied to many problems in all branches of pure and applied science.

Medicine. One of the earliest uses of the electron microscope was in the field of medicine, where it has been used in the study of viruses. These substances cause several serious diseases, of which smallpox is one. Until examination with the electron microscope showed that they resembled living organisms, they had been very much a mystery, because they were too small to be seen under the conventional microscope or filtered out of liquid suspension. The electron microscope is providing data that will undoubtedly help to avoid and control virus diseases better than in the past.

Chemistry. The electron microscope has been used for the examination of large molecules, such as those of natural rubber and the various synthetic rubbers; it has thus contributed to some of the advances in synthetic-rubber technology. Other research has yielded information on the action of catalysts; on what gives paint its covering power; and on the factors that influence the setting and binding power of cements.

Engineering (Lubrication). In applied engineering, examinations of the surfaces of metals have given much valuable information on the mechanism of friction and lubrication. Studies such as these are increasingly important as the speeds of machinery are being increased.

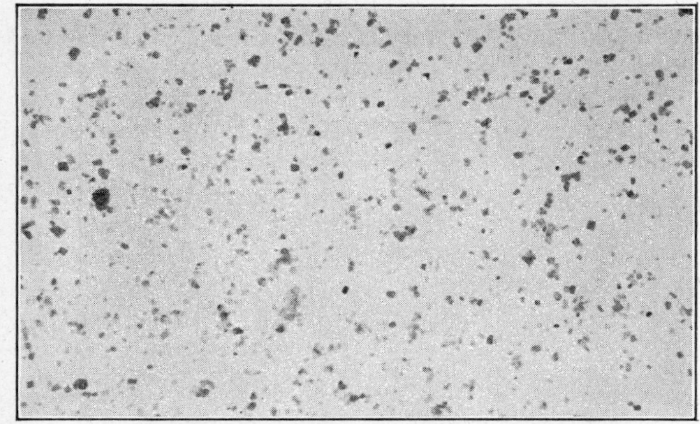

FIG. 26·73 Electron micrograph of particles of the pigment iron blue 13,000×. (Note that the shapes of the particles are apparent in this photograph, but in the optical micrograph, Fig. 26·74, the particles appear as diffuse circles.) (Reprinted with permission from Henry Green and Ernest F. Fullam, *J. App. Phys.*, Vol. 14, 1943, p. 336.)

Engineering (Corrosion). Some metals, for example stainless steels, protect themselves from corrosion by the formation of extremely thin oxide coatings on their surfaces. In the study of the protective nature of thin oxide films on metals, it is of interest to examine the detailed physical and chemical structure of the film. This is possible if the thin films are of the proper thickness for use in the electron microscope after they have been stripped from the metal or alloy. Oxide films of 100 to 500 angstrom units in thickness can be stripped by electrochemical or chemical means for examination by electron microscopy and diffraction.[171] Electron micrographs reveal the following information concerning the film: (1) particle size, (2) particle-size distribution, (3) particle shape, and (4) uniformity of the film. Electron-diffraction patterns are used to identify the composition of the films and to indicate the approximate particle size.

Principles of Operation

Magnification and Resolution. Two quantities describe the performance of any microscope: (1) magnification, and (2) resolving power. Magnification is numerically the ratio of the size of the image produced to the size of the object which the image represents. Theoretically, at least, there is no upper limit to the magnification with which a microscope can be built, but as magnification is increased beyond a certain point, no increase in the amount of detail or fine structure is observable in the image even though it continues to increase in gross dimensions. This occurs when the microscope has reached its limit of resolution. Resolution is usually expressed as the size of the smallest object that can be observed or the smallest distance by which two objects can be separated and still appear as two.

Limitation of Resolution by Diffraction. An object in the conventional microscope causes an image because the object intercepts some of the light, altering both its amount and the direction in which it travels after reaching the object. If the object is small enough to be comparable in size with a wavelength of light, the light waves merely close around it as waves of water close around a pile driven in a lake or bay and reach their destination unperturbed. The same pile,

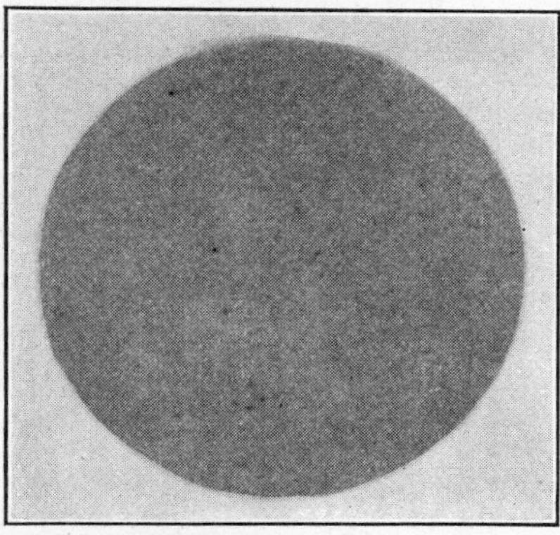

FIG. 26·74 Optical micrograph of particles of the pigment iron blue, 1250×. (Reprinted with permission from Henry Green and Ernest F. Fullam, *J. App. Phys.*, Vol. 14, 1943, p. 336.)

driven in a pond where there are small riplets of wavelengths short compared with the dimensions of the pile, casts a sharp shadow of calm water. In a similar manner, the size of an object observable in a microscope decreases as the wavelength of the medium used to illuminate it decreases.

The Quantum Theory. According to the quantum theory, the smallest possible quantity of electromagnetic energy is

$$W = h\nu \tag{26·30}$$

where h = (Planck's constant) 6.62×10^{-27} erg-second
ν = frequency (cycles per second).

This quantity of energy is indivisible. According to the theory of relativity, mass and energy are equivalent, and the units of mass and energy are related thus:

$$W = mc^2 \tag{26·31}$$

where m = mass in grams
c = velocity of light (3×10^{10} centimeters per second).

Considering light as a stream of corpuscles called photons, each containing one quantum of energy, and calculating the momentum (mc) of a light photon by combining equations 26·30 and 26·31, the result is

$$mc = \frac{h}{\lambda} \tag{26·32}$$

where λ = wavelength in centimeters. Actual measurements confirm this theory.

Electron Wavelength. Now consider a stream of electrons, not as discrete particles, but as electromagnetic wave phenomena. Rearranging equation 26·32 slightly and substituting the velocity of the electrons v for the velocity of light c produces the so-called de Broglie wavelength of an electron:

$$\lambda = \frac{h}{mv} \tag{26·33}$$

Measurement of the wavelength of electrons by diffraction experiments (similar to x-ray diffraction) confirms the wave character of electrons and give values of electron wavelength that are in agreement with equation 26·33. Rationalization of these two seemingly contradictory concepts, waves of energy versus particles of matter, is the concern of the philosophical physicist. In this chapter the correctness of both concepts is accepted, on the basis of the confirming experimental evidence, and in any given problem the concept that leads to the easiest solution is used.

Comparison of Light and Electron Wavelengths. The velocity acquired by an electron depends upon the potential difference through which it falls. Table 26·1 gives the de Broglie wavelengths of electrons that have been accelerated through various potential differences. For comparison, the wavelengths of various common types of radiant energy are listed in Table 26·2.

Table 26·1 De Broglie Wavelengths of Electrons

Potential Difference (volts)	Mass of Electron (grams)	Wavelength (centimeters)
1	9×10^{-28}	1.22×10^{-7}
100	9×10^{-28}	1.22×10^{-8}
10,000	9.14×10^{-28}	1.36×10^{-9}
1,000,000	10.43×10^{-28}	2.73×10^{-10}

Table 26·2 Wavelengths of Some Common Radiations

Radiation	Wavelength (centimeters)
Yellow light	5.8×10^{-5}
Blue light	4.4×10^{-5}
Ultraviolet	2.7×10^{-5}

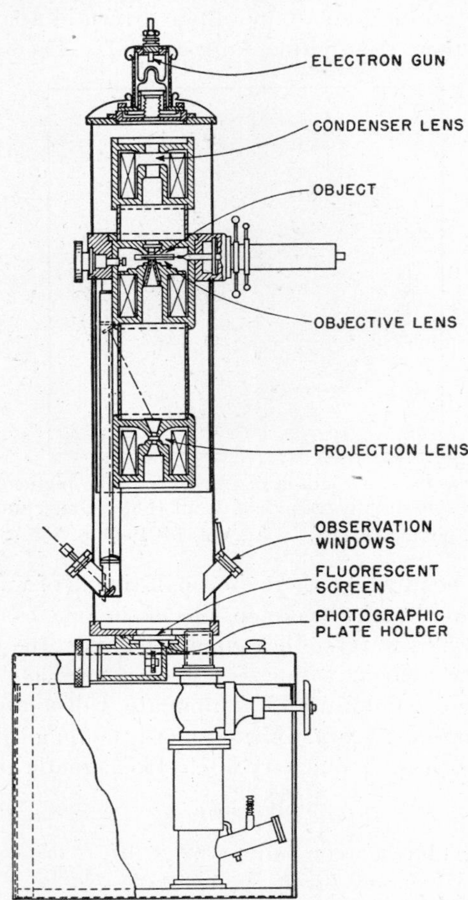

Fig. 26·75 Diagram of electron microscope. (Reprinted with permission from L. Marton, *Phys. Rev.*, Vol. 58, 1940, p. 58.)

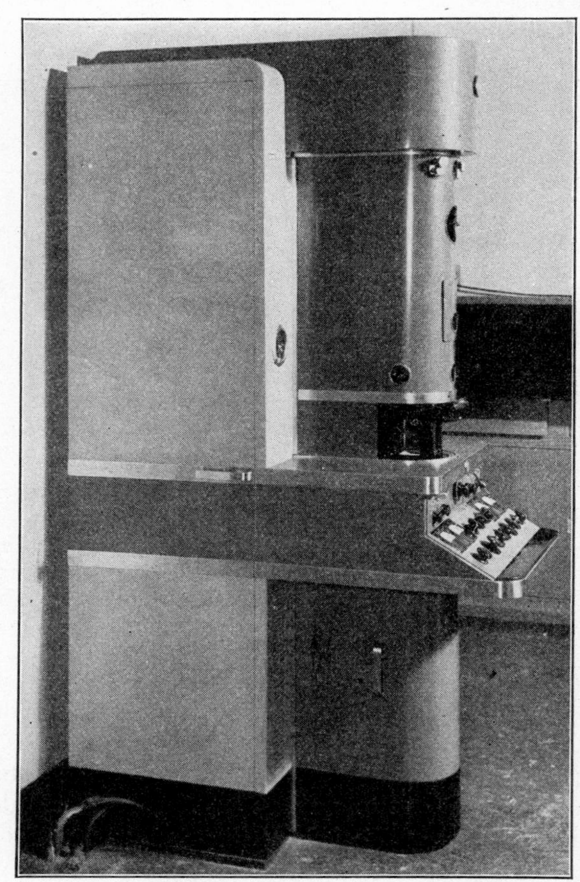

Fig. 26·76 Electron microscope. (Courtesy of Radio Corporation of America.)

History. Rudenberg [172] was aware that an electron stream could be guided by electric and magnetic fields as light rays are guided by lenses. In 1930, he conceived the idea of using electrons as the image-forming means in a microscope which, by virtue of the shorter wavelengths of electrons, should be capable of resolving much smaller objects than can be seen with a conventional microscope. Subsequently, patents ¶ disclosed the fundamental concepts of electron optics and the rudiments of the design of the electron microscope. Since that time, electron microscopes have been designed and built by various people.[173–179]

Construction. An electron microscope (Figs. 26·75 and 26·76) consists essentially of the following parts:

(*a*) *A Source of Electrons.* This usually consists of a hot filament surrounded by a smooth guard electrode containing

¶ Reinhold Rudenberg: U. S. Patent 2058914 (1936); U. S. Patent 2070319 (1937).

a small hole through which electrons emerge to be accelerated through the microscope.

(*b*) *A Condenser Lens.* The condenser lens may be either electrostatic or magnetic. This lens and all other lenses in the instrument usually are at ground potential; the cathode is at a large negative potential. Thus, in passing from the cathode to the anode aperture, electrons acquire a high velocity, and thereafter traverse the rest of the microscope at almost uniform speed. The condenser lens causes the divergent beam of electrons emerging from the hole in the anode to converge upon the object illuminating it, so to speak, with electrons.

(*c*) *An Object Mounting.* The object in a transmission-type electron microscope is always thin. Electrons pass through the object, which absorbs electrons, the number absorbed depending upon density and thickness of the object. The distribution over the surface of the object of the electrons that get through is what eventually causes the formation of an image. The object is placed on a thin organic or silica film, which in turn is suspended in a fine-mesh wire screen just beyond the condenser lens.

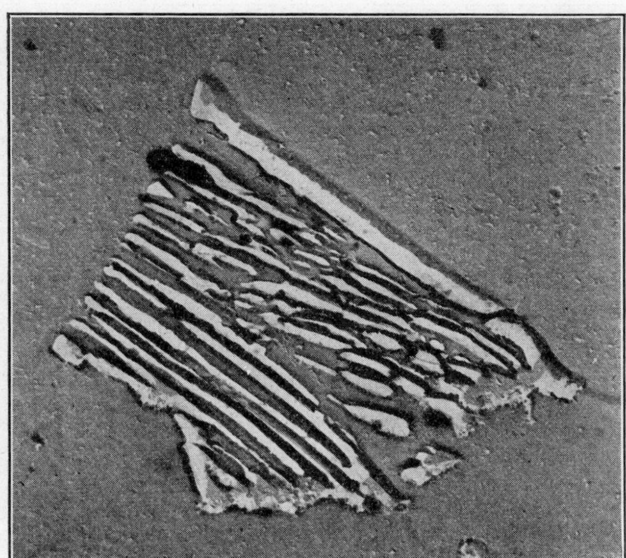

FIG. 26·77 A collodion replica, chromium shadowed, of a pearlitic inclusion in a low-carbon steel. (Courtesy of Radio Corporation of America.)

(*d*) *An Objective Lens.* The objective lens, which may be either electrostatic or magnetic, immediately follows the object. This lens forms an enlarged real or virtual electron image of the object.

(*e*) *A Projection Lens.* The projection lens, which may be either electrostatic or magnetic, magnifies the image formed by the objective lens and forms a real image.

(*f*) *An Observation Screen.* The image formed by the projection lens is cast upon an observation screen, which may be a photographic plate or a fluorescent screen for visual observation.

(*g*) *An Enclosure.* The enclosure surrounds the electron source, object, lenses, and viewing screen. A high vacuum is maintained within the enclosure, which is made of metal so that it also acts as a shield to prevent disturbance of the electron paths by external stray fields. In a practical electron microscope, an airlock is provided to seal off around the object mounting by vacuum-tight seals while the object is being mounted or dismounted.

Limitations and Techniques

Effect of Vacuum on the Specimen. Since the specimen is mounted in the evacuated enclosure of the electron microscope during examination, the examination of living organisms suspended in water is impossible. The passage of electrons through the object causes an appreciable generation of heat. This, coupled with the vacuum surrounding the object, is sufficient to destroy certain types of specimens, such as delicate plant and animal tissues. Many types of bacteria can be examined, however.

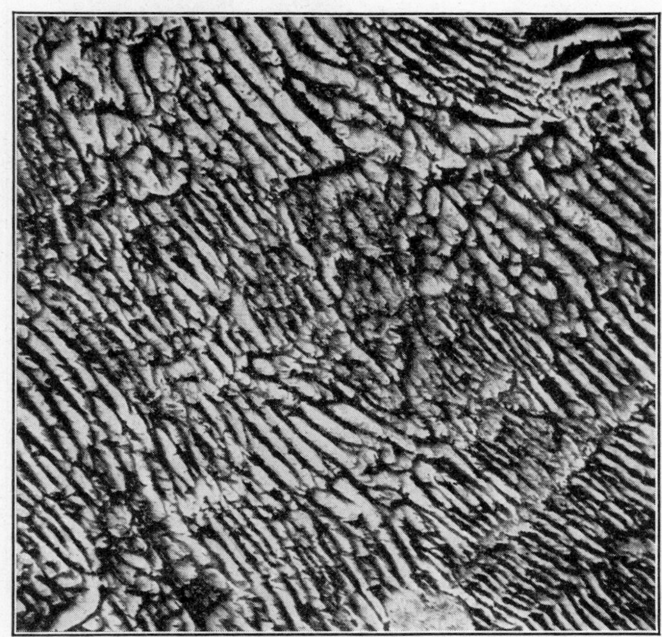

FIG. 26·78 Surface of 0.98 C hot-rolled steel with electrolytic polish and Nital etch. (Reprinted with permission from R. D. Heidenreich and V. G. Peck, *J. App. Phys.*, Vol. 14, 1943, p. 26.)

Specimen Thickness. The object in an electron microscope must be sufficiently thin to allow passage of electrons through it, if its internal structure is to be examined. A substance that is transparent to light is not necessarily transparent to electrons. The thickness of objects for electron-microscopic examination is limited to a fraction of a micron. The technique of handling such thin specimens has been the subject of much study. The most common practice is to float the specimen on water to prevent wrinkles. The fine-mesh object support is then inserted under the specimen, lifted out, and transferred to the microscope. The outlines of opaque specimens, such as smoke, bacteria, dust, and paint pigments, can be examined by dispersing them on a thin film of silica or organic material supported on the wire-mesh object mounting.

Replicas. For examination of the surfaces of opaque materials, such as metals, it is not so easy to arrange the electron microscope to illuminate the object by reflection as it is in the optical microscope. Consequently, the art of making thin replicas of surfaces has been extensively developed.[180–184] A replica consists of a thin film of collodion or special lacquer, which is deposited on the surface to be examined, allowed to dry, and later peeled off. The replica is then examined in the electron microscope in the same way that any thin transparent object is examined.

Measurement of Object Height. Because of the great depth of focus of the electron microscope, measurement of vertical distances in the object cannot usually be made by focusing successively on various planes through the object, as in a conventional microscope. Often pairs of stereoscopic pictures, made by successively photographing the object tilted first at one angle to the electron beam and then at a different angle, provide a means of measuring vertical distances. For some specimens of crystalline nature the two stereoscopic views of the specimen do not agree, and it is believed that the lack of agreement is due to reflection of electrons from the crystal faces into or out of the path of transmission through the microscope.[185] To overcome this apparent disparity a technique has been developed which involves depositing by evaporation a film of chromium over the specimen at an oblique angle to the film of collodion upon which it is mounted. In this way, protuberances on the object can be made to cast shadows, the length of which is from 5 to 10 times their height. The technique has been applied to the study of disease-producing viruses.[186, 187]

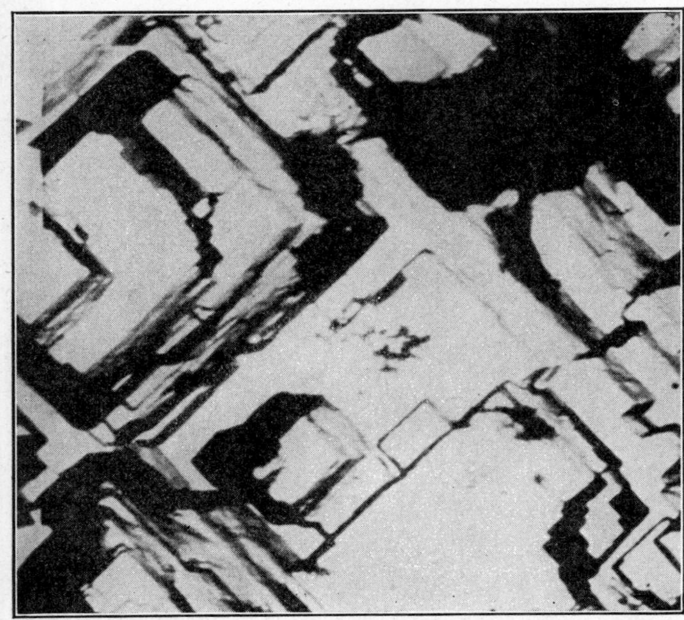

FIG. 26·79 Electron micrograph of oxide film from a deep-etched annealed aluminum sheet, 8000×. (Note the great depth of focus shown here compared with Fig. 26·80.) (Reprinted from F. Keller and A. H. Geisler, *J. App. Phys.*, Vol. 15, 1944, p. 700, with permission of Aluminum Company of America.)

Staining. In making some observations, particularly on plant and animal tissues, it is desirable to have a way of increasing the contrast between various parts of the tissue that have substantially the same transparency to electrons. Some substances, such as phosphotungstic acid, have high electron-scattering power.[188] Suitable stains combine selectively with various portions of the tissue and thus bring out the structure. This technique is analogous to the use of colored stains in conventional microscopy. Other substances containing atoms of large mass have been used as electron stains. Among them are osmic acid, phosphomolybdic acid, and silicotungstic acid.

Calibration. The proper method of calibrating an electron microscope for magnification is the subject of much discussion among electron microscopists. The problem consists in finding a dimensionally stable object small enough to be viewed under the electron microscope and at the same time large enough to be viewed and measured by means of a well-calibrated optical microscope. Some workers use the diameter of an exceedingly fine tungsten wire as the standard of length. One of the more successful calibration techniques consists of electron-microscopic examination of replicas of ruled diffraction gratings having as many as 30,000 lines per inch. The spacing of the grating lines can be determined accurately by spectroscopic means.

Limit of Resolution. Theoretically, the limit of resolution of the electron microscope is of the same order of magnitude as the wavelength of the electrons used.[189, 190] Practically, the resolution is limited by lens aberrations, unavoidable misalignment of the components of the microscope, and deposits on the objective pole-piece diaphragm to a value of 0.004 to 0.005 micron.[191] This represents a tremendous gain in resolving power over that obtainable with the optical microscope (0.25 micron) [192, 193] or with the ultraviolet microscope (0.11 micron).[194] The limit of resolution for the electron microscope really represents the size of an object that is just discernible. An object of this size appears as a diffuse circular dot. The object must be larger than this, if its shape is to be discernible.[195, 196]

FIG. 26·80 Light micrograph of oxide film from a deep-etched aluminum sheet, 500×. (Reprinted from F. Keller and A. H. Geisler, *J. App. Phys.*, Vol. 15, 1944, p. 700, with permission of Aluminum Company of America.)

High-Voltage Microscopes. The theoretical resolving power of the electron microscope increases with the electron-accelerating potential used. Most electron microscopes operate at electron energies between 30 and 100 kilovolts. Higher voltage microscopes have been built.[193] The higher electron velocity permits examination of thicker specimens, but the higher potential does not give any increase in the fundamental resolving power of the microscope.

Magnification. As a by-product of the high resolving power of the electron microscope, large useful magnifications can be used. Magnifications up to 25,000 times are produced by the microscope, and magnifications up to 100,000 times are possible by enlargement of the photograph of the image.

Depth of Focus. The lenses of an electron microscope, compared with the lenses of an optical microscope, possess long focal lengths and small apertures. These qualities give the electron microscope a relatively great depth of focus and make it possible to take stereoscopic pictures. To an observer accustomed to using an optical microscope, even the single visual images appear to possess a stereoscopic quality because there are few out-of-focus areas in the image.

An excellent bibliography on electron microscopes has been compiled which lists most of the sources of detailed information [197–199]

REFERENCES

1. "Crest Voltmeters," C. H. Sharp and E. D. Doyle, *Trans. A.I.E.E.*, Vol. 35, 1916, p. 99.
2. "The Crest Voltmeter," L. W. Chubb, *Trans. A.I.E.E.*, Vol. 35, 1916, p. 109.
3. "An Application of the Diode to the Measurement of A.C. Voltages," J. J. Taylor, *J. Sci. Instruments*, Vol. 3, 1925, p. 113.
4. "The Measurement of High Voltages with Special Reference to the Measurement of Peak Voltages," R. Davis, G. W. Bowdler, and W. G. Standring, *J. I.E.E. (London)*, Vol. 168, p. 1132.
5. W. N. Tuttle, *Gen. Radio Expts.*, Vol. 11, May 1937, p. 1.
6. "Variable-mu Tetrodes in Logarithmic Recording," S. Ballantine, *Electronics*, Jan. 1931, p. 472.
7. "A Vacuum-Tube Voltmeter with Logarithmic Response," F. V. Hunt, *Rev. Sci. Instruments*, Vol. 4, 1933, p. 672.
8. "Vacuum Tube Voltmeter of High Sensitivity," H. J. Reich, G. S. Marvin, and K. Stoll, *Electronics*, Sept. 1931, p. 109.
9. "Sharp Cutoff in Vacuum Tubes with Applications to Slide-Back Voltmeter," C. B. Aiken and L. C. Birdsall, *Trans. A.I.E.E.*, Vol. 57, 1938, p. 173.
10. "The Measurement of Peak Values of Alternating Currents and Voltages by Means of a Thyratron," E. Hughes, *J. Sci. Instruments*, Vol. 10, 1933, p. 180; "Peak Voltmeter," J. J. Ruiz, *Rev. Sci. Instruments*, Vol. 6, 1935, p. 169.
11. "Trigger Peak Voltmeter Using 'Hard' Valves," A. T. Starr, *Wireless Eng.*, Vol. 12, 1935, p. 601.
12. "*The Thermionic Vacuum Tube and Its Applications*," H. J. Van Der Bijl, McGraw-Hill, 1920, p. 369.
13. "The Thermionic Triode as Rectifier," E. B. Moullin and L. B. Turner, *J. I.E.E. (London)*, Vol. 60, 1922, p. 706.
14. "A Direct-Reading Thermionic Voltmeter and Its Applications," E. B. Moullin, *J. I.E.E. (London)*, Vol. 61, 1923, p. 295.
15. "A Sensitive Direct Reading Voltmeter and Ammeter for High Frequencies," E. B. Moullin, *Wireless World*, Vol. 10, 1922, p. 1.
16. "A Two-Range Vacuum Tube Voltmeter," C. M. Jansky and C. B. Feldman, *Trans. A.I.E.E.*, Vol. 47, 1928, p. 307.
17. "Double Tube Vacuum Tube Voltmeter," W. C. Michels, *Rev. Sci. Instruments*, Vol. 9, 1938, p. 10.
18. "An Improved Vacuum Tube Microammeter," A. W. Vance, *Rev. Sci. Instruments*, Vol. 7, 1936, p. 489.
19. "Measuring 1/100,000,000,000,000,000 of an Ampere," B. J. Thompson, *Electronics*, Vol. 1, 1930, p. 290.
20. "Mercury-Pool Cathode Arc Single-Tube Inverter," Y. Watanabe and R. J. Aoyama, *Inst. Elec. Eng. (Japan)*, Vol. 56, 1936, p. 983.
21. "Applications of Harmonic Commutation for Thyratron Inverters and Rectifiers," C. H. Willis, *Trans. A.I.E.E.*, Vol. 52, 1933, p. 701.
22. *Theory and Applications of Electron Tubes*, H. J. Reich, McGraw-Hill, 1939, Chapter 39.
23. "The Thermionic Amplification of Direct Current," P. A. MacDonald, *Physics*, Vol. 7, 1936, p. 265. (Includes bibliography of 29 items.)
24. "Thermionic Amplifiers for D.C.," F. Muller and W. Durichen, *Elektrotech.*, Vol. 42, 1936, p. 31.
25. "A Vacuum Tube Electrometer," H. Nelson, *Rev. Sci. Instruments*, Vol. 1, 1930, p. 281.
26. "A Low Grid Current Vacuum Tube," G. F. Metcalf and B. J. Thompson, *Phys. Rev.*, Vol. 36, 1930, p. 1489.
27. "Measurement of Radium Emanation Implants," P. A. MacDonald and E. M. Campbell, *Physics*, Vol. 6, 1935, p. 212.
28. "Direct-Current Amplifier Circuits for Use with the Electrometer Tube," D. B. Pennick, *Rev. Sci. Instruments*, Vol. 6, 1935, p. 115.
29. "Supersensitive Amplifier for Measuring Small Currents," F. J. Moles, *G.E. Rev.*, Vol. 36, 1933, p. 156.
30. "Amplification of Small Direct Currents," L. A. DuBridge, *Phys. Rev.*, Vol. 37, 1931, p. 392.
31. "An Improved Vacuum Tube Microammeter," A. W. Vance, *Rev. Sci. Instruments*, Vol. 7, 1936, p. 489.
32. "One-Tube Balanced Circuit for D-C Vacuum Tube Amplifiers of Very Small Currents," Walter Soller, *Rev. Sci. Instruments*, Vol. 3, 1932, p. 416.
33. "Improved Balanced Circuit for Use with Electrometer Tubes," L. A. Turner and C. O. Siegelin, *Rev. Sci. Instruments*, Vol. 4, 1933, p. 429.
34. "On Use of Vacuum Tube Electrometer with Extremely High Input Resistance," R. E. Burroughs and J. E. Furgeson, *Rev. Sci. Instruments*, Vol. 4, 1933, p. 406.
35. J. A. Hipple, unpublished data.
36. "Improved D.C. Amplifying Circuit," L. A. DuBridge and Hart Brown, *Rev. Sci. Instruments*, Vol. 4, 1933, p. 532.
37. "A Method for Measuring Very High Values of Resistance," G. M. Rose, Jr., *Rev. Sci. Instruments*, Vol. 2, 1931, p. 810.
38. "Time Service of U. S. Naval Observatory," J. F. Hellweg, *Trans. A.I.E.E.*, Vol. 51, 1932, p. 538.
39. *Measurements in Radio Engineering*, F. E. Terman, McGraw-Hill, 1935.
40. "A Convenient Method for Referring Secondary Frequency Standards to a Standard Time Interval," L. M. Hull and J. K. Clapp, *Proc. I.R.E.*, Vol. 17, Feb. 1929, pp. 252–271.
41. "'Universal' Frequency Standardization from a Single Frequency Standard," J. K. Clapp, *J. Opt. Soc. America*, Vol. 15, July 1927, p. 25.
42. "A High Precision Standard of Frequency," W. A. Marrison, *Proc. I.R.E.*, Vol. 17, 1929, p. 1103.
43. "Antenna-Measuring Equipment," J. K. Clapp, *Proc. I.R.E.*, Vol. 18, Dec. 1930.
44. "Thermostat Design for Frequency Standards," W. A. Marrison, *Proc. I.R.E.*, Vol. 16, 1928, p. 976.
45. "Frequency Measurement in Electrical Communication," J. A. Horton, N. H. Ricker, and W. A. Marrison, *Trans. A.I.E.E.*, Vol. 42, 1923, p. 730.
46. "A Precision Tuning Fork Frequency Standard," E. Norman, *Proc. I.R.E.*, Vol. 20, 1932, p. 1715.
47. "Monitoring Standard Frequency Emissions," E. G. Lapham, *Proc. I.R.E.*, Vol. 23, 1935, p. 719.
48. "Interpolation Methods for Use with Harmonic Frequency Standards," J. K. Clapp, *Proc. I.R.E.*, Vol. 18, 1930, p. 1575.
49. *Measurements in Radio Engineering*, F. E. Terman, McGraw-Hill, 1935.
50. "Device for Precise Measurement of High Frequencies," F. A. Polkinghorn and A. A. Roetken, *Proc. I.R.E.*, Vol. 19, 1931, p. 937.
51. "Direct-Reading Frequency Meter," F. Guarnaschelli and F. Vecchiacci, *Proc. I.R.E.*, Vol. 19, 1931, p. 659.
52. "A Precise and Rapid Method of Measuring Frequencies from Five to Two Hundred Cycles per Second," N. P. Case, *Proc. I.R.E.*, Vol. 18, 1930, p. 1586.
53. "A Direct-Reading Frequency Meter Suitable for High Speed Recording," F. V. Hunt, *Rev. Sci. Instruments*, Vol. 6, 1935, p. 43.
54. *Radio Engineers Handbook*, F. E. Terman, McGraw-Hill, 1943, Section 13.
55. "A Radio-Frequency Bridge for Impedance Measurements from 400 Kilocycles to 60 Megacycles," D. E. Sinclair, *Proc. I.R.E.*, Vol. 28, 1940, p. 497.
56. "A Shielded Bridge for Inductive Impedance," W. J. Shackleton, *Trans. A.I.E.E.*, Vol. 45, 1926, p. 1266.
57. "Measurements at Speech and Carrier Frequencies," W. J. Shackleton, *Bell System Tech. J.*, Vol. 6, 1927, p. 142.

58. "Shielding in High-Frequency Measurements," J. G. Ferguson, *Bell System Tech. J.*, Vol. 8, 1929, p. 560.
59. "Shielding in High-Frequency Measurements," J. G. Ferguson, *Trans. A.I.E.E.*, Vol. 48, 1929, p. 1286.
60. "Campbell-Shackelton Shielded Ratio Box," Leo Behr and A. J. Williams, *Proc. I.R.E.*, Vol. 20, 1932, p. 969.
61. "Resonance Bridge for Use at Frequencies up to 10 Megacycles per Second," C. L. Fortescue and G. Mole, *J. I.E.E. (London)*, Vol. 82, 1938, p. 687.
62. C. L. Fortescue and G. Mole, *J. I.E.E.*, *Wireless Section*, Vol. 13, 1938, p. 112.
63. "A 5-Megacycle Impedance Bridge," C. H. Young, *Bell Lab. Rec.*, Vol. 15, 1937, p. 261.
64. "Inductance and Capacitance Bridge," S. J. Zammataro, *Bell Lab. Rec.*, Vol. 16, 1938, p. 341.
65. "Radio Frequency Bridge for Impedance and Power Factor Measurements," D. W. Dye and T. I. Jones, *J. I.E.E. (London)*, Vol. 72, 1933, p. 169.
66. D. W. Dye and T. I. Jones, *J. I.E.E.*, *Wireless Section*, Vol. 8, 1933, p. 22.
67. "Bridged-T and Parallel-T Null Circuits for Measurements at Radio Frequencies," W. N. Tuttle. *Proc. I.R.E.*, Vol. 28, 1940, p. 23.
68. "New Type of Selective Circuit and Some Applications," H. H. Scott, *Proc. I.R.E.*, Vol. 26, 1938, p. 226.
69. R. F. Field, *Gen. Radio Expts.*, Vol. 6, 1931.
70. "Sensitive Frequency Meter for 30 to 340 Megacycle Range," E. L. Hall, *Electronics*, Vol. 14, 1941, p. 37.
71. *Radio Engineers Handbook*, F. E. Terman, McGraw-Hill, 1943, Section 3.
72. "Two-Mesh Tuned Coupled Circuit Filters," C. B. Aiken, *Proc. I.R.E.*, Vol. 25, 1937, p. 230.
73. "Single and Coupled-Circuit Systems," E. S. Purington, *Proc. I.R.E.*, Vol. 18, 1930, p. 983.
74. "A Method of Measuring Very Short Radio Wave Length and their Use in Frequency Standardization," F. N. Dunmore and F. H. Engel, *Proc. I.R.E.*, Vol. 11, 1923, p. 467.
75. "On a Double Hump Phenomenon of Current Through a Bridge Across Parallel Lines," Eijiro Takagishe, *Proc. I.R.E.*, Vol. 18, 1930, p. 513.
76. "Direct Radio Transmission on A Wave Length of 10 Meters," F. W. Dunmore and F. H. Engel, *Natl. Bur. Standards Sci. Paper*, 1923, No. 469, p. 1.
77. "Theory of Determination of Ultra-Radio Frequencies by Standing Waves on Wires," F. W. Dunmore and F. H. Engel, *Natl. Bur. Standards Sci. Paper*, 1924, No. 491, p. 487.
78. "Correction Factor for the Parallel Wire System Used in Absolute Radio Frequency Standardization," August Hund, *Proc. I.R.E.*, Vol. 12, 1924, p. 817.
79. "Some Fundamental Experiments with Wave Guides," G. C. Southworth, *Proc. I.R.E.*, Vol. 25, 1937, p. 807.
80. "Electromagnetic Waves in Hollow Metal Tubes of Rectangular Cross Section," L. J. Chu and W. L. Barrow, *Proc. I.R.E.*, Vol. 26, 1938, p. 1520.
81. "On the Passage of Electric Waves through Tubes or the Vibration of Dielectric Cylinders," Lord Rayleigh, *Phil. Mag.*, Vol. 43, 1897, p. 125.
82. "Transmission of Electromagnetic Waves in Hollow Tubes of Metal," W. L. Barrow, *Proc. I.R.E.*, Vol. 24, 1936, p. 1298.
83. "Hyper-Frequency Wave Guides—General Consideration and Experimental Results," G. C. Southworth, *Bell System Tech. J.*, Vol. 15, 1936, p. 284.
84. "Hyper-Frequency Wave Guides—Mathematical Theory," J. R. Carson, S. P. Mead, and S. A. Schelkunoff, *Bell System Tech. J.*, Vol. 15, 1936, p. 310.
85. *Ultra High Frequency Techniques*, J. G. Brainerd, Glenn Koehler, H. J. Reich, and L. F. Woodruff, Van Nostrand, 1942.
86. "An Impulse Counter, Tallying Device and Electric 'Stopwatch'—A Useful Laboratory Instrument," P. E. Klopsteg, *J. Opt. Sci. America and Rev. Sci. Inst.*, Vol. 19, 1929, p. 345.
87. "Hot Cathode Thyratrons," A. W. Hull, *G. E. Rev.*, Vol. 32, 1929, p. 399.
88. W. H. Eccles and F. W. Jordan, *Radio Rev.*, Vol. 1, 1919, p. 143.
89. "Vacuum Tube Relay and Race Timer," W. M. Roberds, *Rev. Sci. Instruments*, Vol. 2, 1931, p. 519.
90. "Cold Cathode Arc-Discharge Tube," K. J. Germeshausen and H. E. Edgerton, *Elec. Eng.*, Vol. 55, 1936, p. 790.
91. "The Strobotron," A. B. White, W. B. Nottingham, H. E. Edgerton, and K. J. Germeshausen, *Electronics*, March 1937, p. 18.
92. "Relations between Attenuation and Phase in Feed-Back Amplifier Design," H. W. Bode, *Bell System Tech. J.*, Vol. 19, 1940, p. 421.
93. *Time Bases*, O. S. Puckle, Wiley, 1944.
94. *Measurements in Radio Engineering*, F. E. Terman, McGraw-Hill, 1935.
95. "Calibrated Response Curve Tracer," G. L. Hamburger, *Wireless Eng.*, Vol. 22, 1945, p. 170.
96. "Cathode Ray Tubes and Their Applications," P. S. Christaldi, *Proc. I.R.E.*, Vol. 33, 1945, p. 373.
97. "Insulation Testing of Electric Windings," C. M. Foust and N. Rohots, *Elec. Eng.*, April 1943, p. 203.
98. "Winding Fault Detection and Location by Surge Comparison Testing," C. L. Moses and E. F. Harter, *A.I.E.E. Tech. Paper*, 1945, pp. 45–94.
99. *Theoretical Mechanics*, J. H. Jeans, Ginn, 1907.
100. *Mechanical Vibrations*, J. P. Den Hartog, McGraw-Hill, 1940.
101. "Balancing Rotors by Means of Electrical Networks," J. G. Baker and F. C. Rushing, *J. Franklin Inst.*, Vol. 222, 1936, p. 183.
102. "Short History of Isotopes and Measurement of Their Abundances," E. B. Jordan and Louis B. Young, *J. App. Phys.*, Vol. 13, 1942, p. 526.
103. "A New Method of Positive Ray Analysis," A. J. Dempster, *Phys. Rev.*, Vol. 11, 1918, p. 316.
104. "A New Method of Positive Ray Analysis and its Application to the Measurement of Ionization Potentials in Mercury Vapor," W. Bleakney, *Phys. Rev.*, Vol. 34, 1929, p. 157.
105. "The Ionization Potential of Molecular Hydrogen," W. Bleakney, *Phys. Rev.*, Vol. 40, 1932, p. 496.
106. "The Determination of Nitrogen Isotopes in Organic Compounds," D. Rittenberg, A. S. Keston, F. Rosebury, R. Schoenheimer, *J. Biol. Chem.*, Vol. 127, 1939, p. 291.
107. "Ionization Potentials and Probabilities for the Formation of Multiply Charged Ions in the Alkali Vapors and in Krypton and Xenon," J. T. Tate and P. T. Smith, *Phys. Rev.*, Vol. 46, 1934, p. 773.
08. "A Spherical Coil for a Mass Spectrometer," J. A. Hipple, *Phys. Rev.*, Vol. 55, 1939, p. 597.
109. "Shape of an Electron Beam in a Magnetic Field," N. F. Barber, *Proc. Leeds Phil. Soc.*, Vol. 2, 1933, p. 427.
110. "Gas Analysis with Mass Spectrometer," J. A. Hipple, *J. App. Phys.*, Vol. 13, 1942, p. 551.
111. "A Mass Spectrometer for Routine Isotope Abundance Measurements," A. O. Nier, *Rev. Sci. Instruments*, Vol. 11, 1940, p. 212.
112. N. D. Coggeshall, *Bull. Am. Phys. Soc.* (Baltimore Meeting, A37), 1942.
113. "Measurement of Relative Abundance with Mass Spectrometer," E. B. Jordan and N. D. Coggeshall, *J. App. Phys.*, Vol. 13, 1942, p. 539.
114. "A Mass-Spectrographic Study of the Isotopes of Hg, Xe, Kr, Be, I, As, and Cs," A. O. Nier, *Phys. Rev.*, Vol. 52, 1937, p. 933.
115. "The Iostopic Constitution of Strontium, Barium, Bismuth, Thallium and Mercury," A. O. Nier, *Phys. Rev.*, Vol. 54, 1938, p. 275; "The Relative Abundance of the Isotopes in Mn, Cb, Pd, Pt, Ir, Rh, and Co," M. B. Sampson and W. Bleakney, *Phys. Rev.*, Vol. 50, 1936, p. 732.
116. "Products and Processes of Ionization in Low Speed Electrons," H. D. Smyth, *Rev. Mod. Phys.*, Vol. 3, 1931, p. 347.
117. "A New Mass Spectrometer with Improved Focusing Properties," W. Bleakney and J. A. Hipple, *Phys. Rev.*, Vol. 53, 1938, p. 521.
118. "Nuclear Fission of Separated Uranium Isotopes," A. O. Nier, J. R. Dunning, and A. V. Grosse, *Phys. Rev.*, Vol. 57, 1940, p. 546 and p. 748.
119. "High-Intensity Mass-Spectrometer," W. R. Smythe, L. H. Rumbaugh, and S. S. West, *Phys. Rev.*, Vol. 45, 1934, p. 724.

120. "The Radioactive Isotope of Potassium," W. R. Smythe and A. Hemmendinger, *Phys. Rev.*, Vol. 51, 1937, p. 178.
121. "The Isolation of Lithium Isotopes with a Mass Spectrometer," L. H. Rumbaugh, *Phys. Rev.*, Vol. 49, 1936, p. 882.
122. "Separation of the Isotopes of Lithium and Some Nuclear Transformations Observed with them," M. L. Oliphant, E. S. Shire, and B. M. Crowther, *Proc. Roy. Soc.* (*London*), Vol. A146, 1934, p. 922.
123. "Relative Abundance of the Isotopes of Potassium and Rubidium," H. Bondy, G. Johannsen, and K. Popper, *Z. Physik*, Vol. 95, 1935, p. 46.
124. "Some Applications to Mass Spectrometric Analysis of Chemistry," D. Rittenberg, *J. App. Phys.*, Vol. 13, 1942, p. 561.
125. "A New Procedure for Quantitative Analysis by Isotope Dilution with Application to the Determination of Amino Acids and Fat Acids," D. Rittenberg and G. L. Foster, *J. Biol. Chem.*, Vol. 133, 1940, p. 737.
126. "The Glutamic Acid of Malignant Tumors," S. Graff, D. Rittenberg, and G. L. Foster, *J. Biol. Chem.*, Vol. 133, 1940, p. 745.
127. G. Hevesy, E. Hofer, and A. Krogh, *Skand. Arch. Physiol.*, Vol. 72, 1935, p. 199.
128. "Studies in Protein Metabolism," R. Schoenheimer, S. Ratner, and D. Rittenberg, *J. Biol. Chem.*, Vol. 130, 1939, p. 703.
129. "Oxygen Exchange Reactions of Organic Compounds and Water," M. Cohn and H. C. Urey, *J. Am. Chem. Soc.*, Vol. 60, 1938, p. 679.
130. "Heavy Oxygen (O^{18}) as a Tracer in the Study of Photosynthesis," S. Ruben, M. F. Randall, M. D. Kamen, and J. L. Hyde, *J. Am. Chem. Soc.*, Vol. 63, 1941, p. 877.
131. "The Kinetics of Isotopic Exchange between Carbon Dioxide, Bicarbonate Ion, Carbonate Ion, and Water," G. A. Mills and H. C. Urey, *J. Am. Chem. Soc.*, Vol. 62, 1940, p. 1019.
132. "The Utilization of Carbon Dioxide in the Dissimilation of Glycerol by Propionic Acid Bacteria," H. G. Wood, and C. H. Werkman, *Biochem. J.*, Vol. 30, 1936, p. 48.
133. "Biological Synthesis of Oxaloacetic Acid from Pyruvic Acid and Carbon Dioxide," H. A. Krebs and L. V. Eggleston, *Biochem. J.*, Vol. 34, 1940, p. 1383.
134. "The Utilization of Carbon Dioxide in the Synthesis of α-Ketoglutane Acid," E. A. Evans, Jr., and D. Slotkin, *J. Biol. Chem.*, Vol. 136, 1940, p. 301.
135. "Radioactive Carbon in the Study of Respiration in Heterotrophic Systems," S. Ruben and M. D. Kamen, *Proc. Nat. Acad. Sci. U.S.S.R.*, Vol. 26, 1940, p. 418.
136. "Heavy Carbon as a Tracer in Heterotrophic Carbon Dioxide Assimilation," H. G. Wood, C. H. Werkman, A. Hemingway, and A. O. Nier, *J. Biol. Chem.*, Vol. 139, 1940, p. 365.
137. "The Participation of Carbon Dioxide in the Carbohydrate Cycle," A. K. Solomon, B. Vennesland, F. W. Klemperer, J. M. Buchanan, and A. B. Hastings, *J. Biol. Chem.*, Vol. 140, 1941, p. 171.
138. "Reliability of Reactions Used to Locate Assimilated Carbon in Propionic Acid," H. G. Wood, C. H. Werkman, A. Hemingway, A. O. Nier, and C. G. Stuckwisch, *J. Am. Chem. Soc.*, Vol. 63, 1941, p. 2140.
139. "The Mass Spectrometer, Now an Industrial Tool," J. A. Hipple and H. E. Dralle, *Westinghouse Eng.*, Nov. 1943, p. 127.
140. Herbert Hoover, Jr., and Harold Washburn, *Calif. Oil World*, Vol. 34 1941, p. 21.
141. "Mass Spectrometer Aids Research," J. A. Hipple, *Electronics*, Nov. 1943, p. 120.
142. Herbert Hoover, Jr., and H. Washburn, *Proc. Calif. Natural Gas Assoc.*, 16th Annual Fall Meeting, 1941.
143. "Ionization Potentials and Probabilities for the Formation of Multiply Charged Ions in Helium, Neon, and Argon," W. Bleakney, *Phys. Rev.*, Vol. 36, 1930, p. 1303.
144. "Probability and Critical Potentials for the Formation of Multiply Charged Ions in Hg Vapor by Electron Impact," W. Bleakney, *Phys. Rev.*, Vol. 35, 1930, p. 139.
145. "The Ionization Potential of Molecular Hydrogen," W. Bleakney, *Phys. Rev.*, Vol. 40, 1932, p. 496.
146. "Ionization and Dissociation of Diatomic Molecules by Electron Impact," H. D. Hagstrum and J. T. Tate, *Phys. Rev.*, Vol. 59, 1941, p. 354.
147. "Mass Spectrograph Analysis of Bromine," J. P. Blewett, *Phys. Rev.*, Vol. 49, 1936, p. 900.
148. "The Ionization Processes of Iodine Interpreted by the Mass Spectrograph," T. R. Hogness and R. W. Harkness, *Phys. Rev.*, Vol. 32, 1928, p. 784.
149. "A Mass Spectrum Analysis of the Products of Ionization by Electron Impact in Nitrogen, Acetylene, Nitric Oxide, Cyanogen, and Carbon Monoxide," J. T. Tate, P. T. Smith, and A. L. Vaughan, *Phys. Rev.*, Vol. 48, 1935, p. 525.
150. "The Ionization of Nitrous Oxide and Nitrogen Dioxide by Electron Impact," H. D. Smyth and E. C. G. Stueckelberg, *Phys. Rev.*, Vol. 36, 1930, p. 478.
151. "The Ionization and Dissociation of Water Vapor and Ammonia by Electron Impact," M. M. Mann, A. Hustrulid, and J. T. Tate, *Phys. Rev.*, Vol. 58, 1940, p. 340.
152. "The Ionization of Carbon Dioxide by Electron Impact," H. D. Smyth and E. C. G. Stueckelberg, *Phys. Rev.*, Vol. 36, 1930, p. 472.
153. "Elementary Processes in Ion and Electron Collisions," H. von Kallmann and B. Rosen, *Z. Physik*, Vol. 61, 1930, p. 61.
154. "New Experiments on Ion and Electron Impact," E. von Friedlander, H. Kallmann, and B. Rosen, *Naturwiss.*, Vol. 19, 1931, p. 510.
155. "The Dissociation of HCN, C_2H_2, C_2N_2 and C_2H_4 by Electron Impact," A. Kusch, P. Hustrulid, and J. T. Tate, *Phys. Rev.*, Vol. 52, 1937, p. 843.
156. "A Mass Spectrograph Analysis of the Ions Produced in HCl under Electron Impact," A. O. Nier and E. E. Hanson, *Phys. Rev.*, Vol. 50, 1936, p. 722.
157. "Mass-Spectrograph Study of the Ionization and Dissociation by Electron Impact of Benzene and Carbon Bisulfide," E. G. Linder, *Phys. Rev.*, Vol. 41, 1932, p. 149; "Ionization of Carbon Disulfide by Electron Impact," H. D. Smyth and J. P. Blewett, *Phys. Rev.*, Vol. 46, 1934, p. 276.
158. "Ionization and Dissociation of Polyatomic Molecules by Electron Impact, I. Methane," L. G. Smith, *Phys. Rev.*, Vol. 51, 1937, p. 263.
159. "The Dissociation of Ethane by Electron Impact," J. A. Hipple, *Phys. Rev.*, Vol. 53, 1938, p. 530; "Ionization and Dissociation by Electron Impact: Normal Butane, Isobutane, and Ethane," D. P. Stevenson and J. A. Hipple, *J. Am. Chem. Soc.*, Vol. 64, 1942, p. 1588.
160. "Dissociation of Propane, Propylene, and Allene by Electron Impact," J. Delfosse and W. Bleakney, *Phys. Rev.*, Vol. 56, 1939, p. 256.
161. "Ionization and Dissociation by Electron Impact: Normal Butane, Isobutane, and Ethane," D. P. Stevenson and J. A. Hipple, *J. Am. Chem. Soc.*, Vol. 64, 1942, p. 1588.
162. "(CH_3Cl)," S. H. Bauer and T. R. Hogness, *J. Chem. Phys.*, Vol. 3, 1935, p. 687.
163. "Products of Ionization by Electron Impact in Methyl and Ethyl Alcohol," C. S. Cummings and W. Bleakney, *Phys. Rev.*, Vol. 58, 1940, p. 787.
164. "Analysis of B_2H_6 with a Mass Spectrometer," J. A. Hipple, *Phys. Rev.*, Vol. 57, 1940, p. 350A.
165. "The Dissociation of Benzene (C_6H_6), Pyridine (C_5H_5N), and Cyclohexane (C_6H_{12}) by Electron Impact," A. Hustrulid, P. Kusch, and J. T. Tate, *Phys. Rev.*, Vol. 54, 1938, p. 1037.
166. "A Modified Aston-Type Mass Spectrometer and Some Preliminary Results," D. D. Taylor, *Phys. Rev.*, Vol. 47, 1935, p. 666.
167. "Ionization and Dissociation of $CHBrF_2$ by Electron Impact," R. F. Baker and J. T. Tate, *Phys. Rev.*, Vol. 55, 1939, p. 236A.
168. "Ionization and Dissociation by Electron Impact in CCl_2F_2 and in CCl_4 Vapor," R. F. Baker and J. T. Tate, *Phys. Rev.*, Vol. 53, 1938, p. 683A.
169. "On Electronic Voltage Stabilizers," F. V. Hunt and R. W. Hickman, *Rev. Sci. Instruments*, Vol. 10, 1939, p. 6.
170. "Two Voltage Regulators," H. V. Neher and W. H. Pickering, *Rev. Sci. Instruments*, Vol. 10, 1939, p. 53.
171. E. A. Gulbranson, unpublished data.

172. "Early History of Electron Microscope," Reinhold Rudenberg, *J. App. Phys.*, Vol. 14, 1943, p. 434.
173. "A New Electron Microscope," L. C. Martin, R. V. Whelpton, and D. H. Parnum, *J. Sci. Instruments*, Vol. 14, 1937, p. 14.
174. "A Supermicroscope for Research Institutes," B. von Borries and E. Ruska, *Z. wiss. Mikroskop*, Vol. 56, 1939, p. 317.
175. "A New Simple Supermicroscope and Its Application in Bacteriology," E. Bruche and E. Haagen, *Naturwiss.*, Vol. 27, 1939, p. 809.
176. "A Report on the Development of the Electron Supermicroscope at Toronto," E. F. Burton, J. Hillier, and A. Prebus, *Phys. Rev.*, Vol. 56, 1939, p. 1171.
177. "New Electron Microscope," L. Marton, *Phys. Rev.*, Vol. 58, 1940, p. 57.
178. "Electron Microscope for Practical Laboratory Service," V. K. Zworykin, J. Hillier, and A. W. Vance, *Trans. A.I.E.E.*, Vol. 60, 1941, p. 157.
179. "Simplified Electron Microscope," C. H. Bachman and S. Ramo, *Phys. Rev.*, Vol. 62, 1942, p. 494.
180. "Electron Microscope Study of Surface Structure," R. D. Heidenreich and V. J. Peck, *Phys. Rev.*, Vol. 62, 1942, p. 292.
181. "Fine Structure of Metallic Surfaces with Electron Microscope," R. D. Heidenreich and V. G. Peck, *J. App. Phys.*, Vol. 14, 1943, p. 23.
182. "Surface Replicas for Use in Electron Microscope," V. J. Schaeffer and D. Harker, *J. App. Phys.*, Vol. 13, 1942, p. 427.
183. "New Methods for Preparing Surface Replicas for Microscopic Observation," V. J. Schaeffer, *Phys. Rev.*, Vol. 62, 1942, p. 495.
184. "Surface Studies with Electron Microscope," V. K. Zworykin and E. G. Ramberg, *J. App. Phys.*, Vol. 12, 1941, p. 642.
185. "Thickness of Electron Microscopic Objects," R. C. Williams and R. W. G. Wyckoff, *J. App. Phys.*, Vol. 15, 1944, p. 712.
186. "Electron Shadow Micrography of the Tobacco Mosaic Virus Protein," R. C. Williams and R. W. G. Wyckoff, *Science*, Vol. 101, 1945, p. 594.
187. "The Electron Micrography of Crystalline Plant Viruses," R. C. Williams and R. W. G. Wyckoff, *Science*, Vol. 102, 1945, p. 277.
188. "The Structure of Certain Muscle Fibrils as Revealed by the Use of Electron Stains," C. E. Hall, M. A. Jakus, and F. O. Schmitt, *J. App. Phys.*, Vol. 16, 1945, p. 459.
189. "Theoretical Resolving Power of the Electron Microscope," R. Rebsch, *Ann. Physik*, Vol. 31, 1938, p. 551.
190. "Limits of Resolving Power of the Electron Microscope," M. Ardenne, *Z. Phys.*, Vol. 108, 1938, p. 338.
191. "Some Applications of High Resolving Power of Electron Microscope," Henry Green and E. F. Fullam, *J. App. Phys.*, Vol. 14, 1943, p. 332.
192. *The Microscope*, Part II, Conrad Beck, Van Nostrand, 1924, p. 82.
193. "Preliminary Report on Development of 300 Kilovolt Magnetic Electron Microscope," V. K. Zworykin, J. Hillier, and A. W. Vance, *J. App. Phys.*, Vol. 12, 1941, p. 738.
194. "Photomicrography with Ultra-Violet Light," A. P. H. Trivelli, *Sci. Monlhly*, Vol. 33, 1931, p. 175.
195. "The Determination of the Shape and Size Distribution in Gold Colloids," B. von Borries and G. A. Kausche, *Kolloid Z.*, Vol. 90, 1940, p. 132.
196. "Applications of the Electron Microscope in Colloid Chemistry," L. Marton, *J. Phys. Chem.*, Vol. 46, 1942, p. 1023.
197. "A Bibliography of Electron Microscopy," C. Marton and Samuel Sass, *J. App. Phys.*, Vol. 14, 1943, p. 522.
198. "A Bibliography of Electron Microscopy II," C. Marton and S. Sass, *J. App. Phys.*, Vol. 15, 1944, p. 575.
199. "A Bibliography of Electron Microscopy III," C. Marton and S. Sass, *J. App. Phys.*, Vol. 16, 1945, p. 373.

Chapter 27

INDUSTRIAL X-RAY APPLICATIONS

D. E. Morgan and F. A. Trenkle

THE discovery of x-radiation by Professor W. C. Röntgen in 1895 touched off speculation as to its exact nature and that of the entire electromagnetic radiation spectrum. At present, the wave-mechanics theory seems to satisfy the observed phenomena, but in this chapter it is convenient for some purposes to ascribe continuous-wave characteristics to the radiation. For other purposes it is advantageous to consider the radiation as being composed of discrete bundles or photons of energy. This consideration need not lead to confusion, for the two concepts do not conflict in the applications considered here.

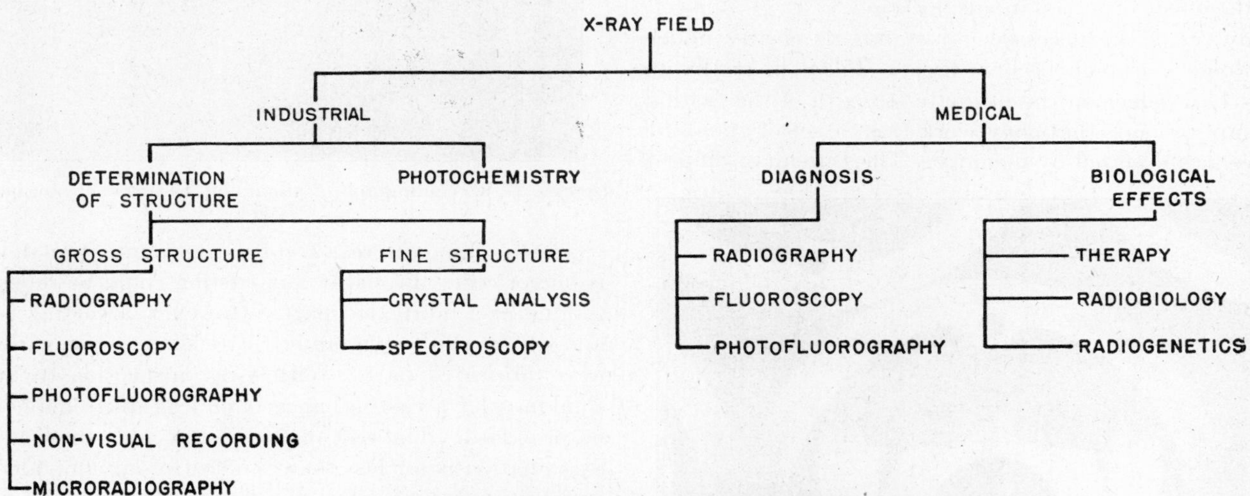

FIG. 27·1 Classification of the fields of use of x-ray apparatus.

In general, the commercially important x-ray applications can be divided into two classes, medical and industrial. Medical applications include diagnosis or treatment of human or animal ailments and studies of the effect of x-radiation on living organisms. Industrial applications necessarily include some which physicists or chemists may believe should not be in this category. Figure 27·1 shows that such an arbitrary division is not entirely adequate. For example, photochemistry and radiobiology overlap greatly even though they have been placed in different groups.

In this chapter, the discussion of industrial uses of x-radiation includes the objectives of use, the particular applications which have proved successful, and a description of the equipment used. At the end of the chapter is a general discussion of the objectives and the equipment used in medical work.

27·1 ECONOMICS OF GROSS-STRUCTURE EXAMINATION

Gross-structure examination is that use of x-radiation which reveals the homogeneity or non-homogeneity of materials when the conditions to be examined are larger than molecular size. For this application, x-ray equipment is but another inspection tool available to the manufacturer. The operator, like that of any other machine, must know the capabilities and limitations of the equipment in order to produce the best results.

The unique feature of an x-ray unit is that it enables the workman to "look inside" without destroying the parts being examined. It reveals internal structures in much the same way as light reveals the internal structure of transparent objects. It is no more difficult for the x-ray worker to locate a void in the center of a solid opaque piece than it is for the layman to locate an air hole in a piece of glass.

Metal-Casting Industries

Serious defects are common in metal castings. In Table 27·1 is a list of these defects.

Although it is not the purpose of this chapter to discuss the causes of these defects and the means of detecting them in the process of x-ray examination, some examples have been given in the illustrations to familiarize the reader with the appearance of the defect in a radiograph. (See Figs. 27·2, 27·3, and 27·4.)

TABLE 27·1 DEFECTS COMMON IN METAL CASTINGS

Gas cavities	Cracks
Blow holes	Hot tears
Porosity	Stress
Inclusions	Surface
Sand	Metal segregations
Slag or dross	Cold shut
Shrinkage	Miss run
Macroshrinkage	
Pipe	
Microshrinkage	
Porosity	

By x-ray examination the size and location of these defects can be determined. The nature of the defect, its extent, and its location are all significant in determining the ultimate strength of the part. A set of standards is necessary to judge the effect of the defect shown by the x-ray examination and, if no standards are available for the workman, he must make his own standards by testing several of the parts to destruction. From these test results a skilled workman is able to predict the effect of a flaw in any location.

Not always is the necessity for standards clearly understood. Some workmen become highly skilled in predicting the effect of a defect on the ultimate strength of the casting, which simply means that they work from a set of standards they have accumulated by memory. The Bureau of Ships of the Navy Department has published radiographic standards of steel castings, presenting radiographs of all the defects in their varying degrees. The various types of castings used by the Navy are classified into five groups, depending upon their use, and the standards indicate by reference to the radiograph the extent to which any defects may be present and still allow the castings to be acceptable.

FIG. 27·2 Radiograph of zinc-base die-casting showing blowholes.

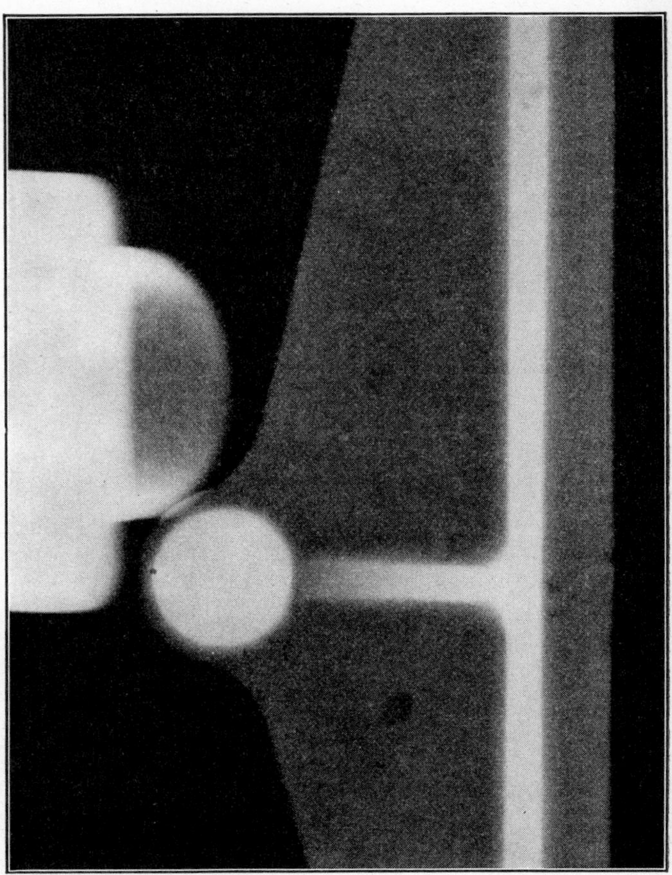

FIG. 27·3 Radiograph of aluminum castings showing porosity.

The fundamental reason for x-ray examination in foundries is one of cost. Probably any casting could be replaced by a forging or a fabricated part. However, a casting is cheaper and can be made in more intricate shapes than can a forging or a fabricated part. With x-ray inspection to insure the soundness of a casting, an economical and reliable part can be produced. The cost of the casting plus the cost of x-ray examination is far less than the cost of an equivalent forged or fabricated piece.

Welding Industries

Welds, like castings, may not be as sound as they appear on the surface. Table 27·2 gives a list of common defects in welded sections.

TABLE 27·2 DEFECTS COMMON IN WELDS

Cracks
  Shrinkage
  Stress
Gas cavities
  Blowholes
  Porosity
Inclusions
Lack of fusion

In the weld-shop as in the foundry, the x-ray machine lowers costs. An example is the manufacture of welded pressure vessels. The cost of welding plus x-ray inspection of these welds is far cheaper than any other means of manufacture. The x-ray machine has also been used as a means of education and certification of the welder. Before a welder is allowed to work on equipment for the Army or the Navy, a sample weld produced by him must be radiographed and approved. From a radiograph it is possible to point out to

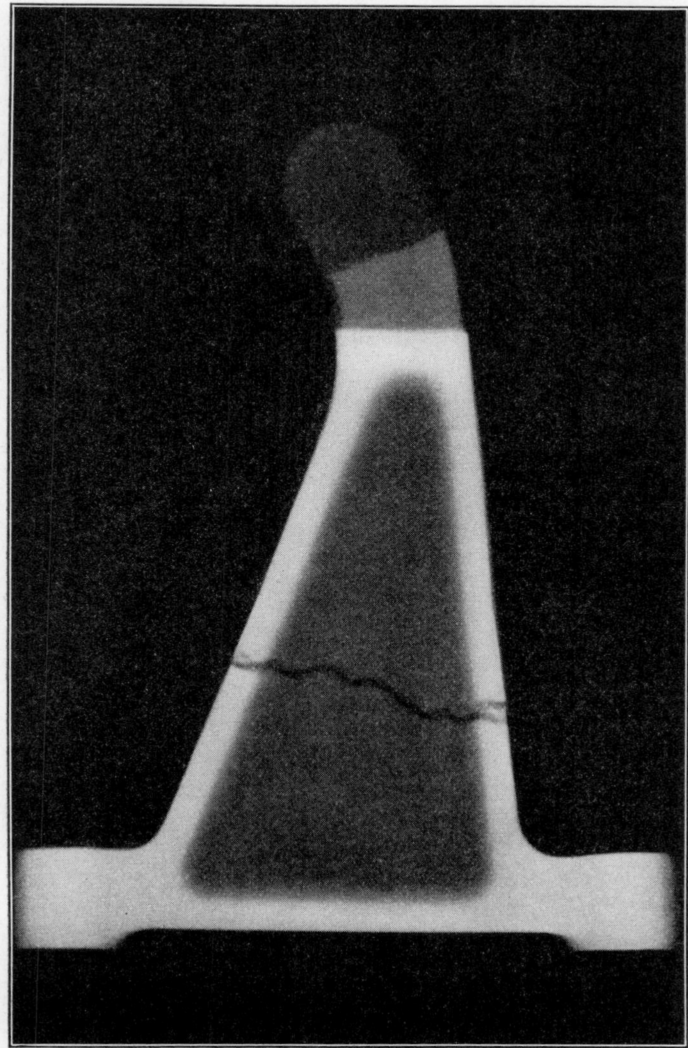

Fig. 27·4 Radiograph of aluminum casting showing stress crack. This crack was plainly visible on the surface but is pictured here because the loss in detail in reproduction necessitates accentuation of defects.

the workman the defects in his work and to suggest means for improving his welding technique.

Other Industries

In the plastics industry there is a constant need for x-ray examination. First, plastic parts must be sound. Sometimes metal inclusions are placed in a plastic part and, during the molding process, these pieces break or are displaced in position. By x-ray examination the extent of the damage can be determined immediately (Fig. 27·6).

Electrical manufacturers must be certain that electrical insulation is uniform, and that it does not include foreign material. With x-rays electrical parts such as bushings and insulators can be examined easily.

In the food industries, the possibility of including foreign and objectionable material with the food is always present. A reputable manufacturer must do all in his power to prevent inclusion of such material, and x-ray inspection is one way of doing it.

Fig. 27·5 Radiograph of butt weld of ½-inch-steel plates showing lack of fusion or insufficient penetration of weld leaving a line of unjoined metal at the butt.

A detailed list of this nature could go on and on, but it is possible here only to suggest general classifications of use; in other branches of engineering, the usefulness of this tool must be recognized as each problem arises.

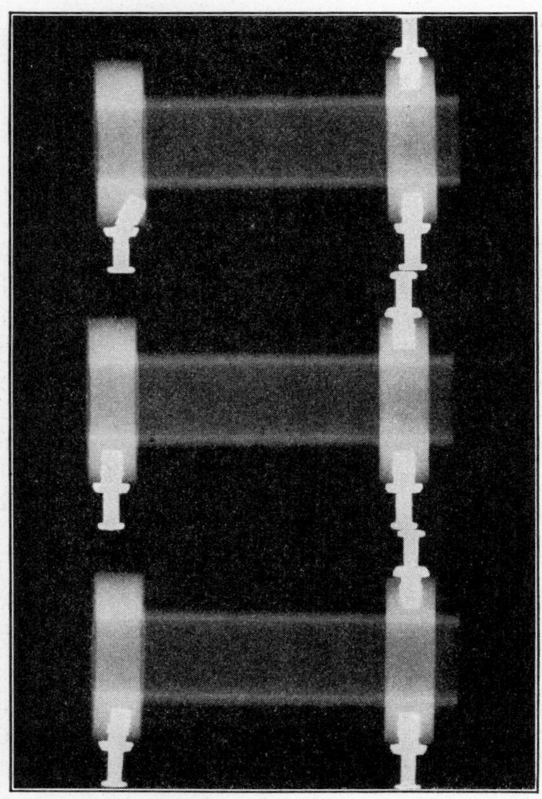

Fig. 27·6 Plastic-coil forms with metal inserts for electrical terminals. In the casting process, these terminals frequently broke just below the surface as shown by the ones bent in the radiograph.

Fundamentals of Application

There are fundamentally only three reasons for x-ray examination.

The first and most important reason for the use of industrial x-rays is that the cost of manufacture of a finished article may be lowered. With it, more economical processes of manufacture can be used, as in the casting and welding industries. Several examples will show possible additional savings.

Consider a machine-tool manufacturer. The bases for his machines probably are supplied by an outside foundry. Suppose that, when the first base was machined, a large flaw was found where a machined surface was called for on the drawing. Subsequent x-ray examination of all bases would allow the manufacturer to machine only those bases which were sound and to reject the remainder. In this way, he would save man-hours, machine-hours, and material, and thus be able to produce a given quantity of these bases with a minimum of machining time at a lower price. The saving in machining time far outweighed the cost of x-ray examination.

It is also possible to use x-ray examination in setting up a process. For example, a foundryman may find by first piece inspection that his casting technique is producing a bad part. X-ray examination shows what should be done to correct the difficulty. After the difficulty is corrected, it may be necessary to use x-rays only as an occasional spot check to make certain that the quality is uniform.

Fabrication problems often may be attacked in the same way. One electrical manufacturer was unable to pour a filling compound in a high-voltage bushing without leaving voids that caused electrical breakdown. With x-rays the results of various techniques of filling were studied and a satisfactory method was evolved.

The second reason for x-ray examination is that sometimes a valuable mechanism may be destroyed because of the failure of relatively inexpensive parts, or a costly delay in a large operation may result from the failure of an ordinarily unimportant part. If a flywheel on a reciprocating engine in a powerhouse fails, the engine itself is almost certain to be ruined. Failure of a gear driving a roll of a paper machine can cause a shutdown of days, which may cost the operator thousands of dollars in output and unprofitable labor. Failure of a small part in a printing press may mean that the entire edition of a newspaper is lost. So, even though x-ray examination of a particular part may raise its unit price, the most expensive part is more economical from an over-all viewpoint.

The third and last reason is that human life may depend upon the soundness of a part, or the exactness of internal material placement. Numerous examples of this may be found in the transportation industry.

27·2 X-RAY GENERATORS FOR GROSS-STRUCTURE EXAMINATION

Figure 27·7 shows the essential components of the x-ray-generating equipment. A control enables the operator to vary the voltage applied to the x-ray tube and to adjust the tube current by varying the filament heat. The x-ray transformer normally includes filament windings for both the high-voltage rectifier tubes and the x-ray tube as well as the high-voltage winding. An x-ray tube is needed, of course, for the actual generation of the radiation.

The connection from the x-ray tube to the high-voltage transformer is made by means of cables with rubber insulation and external metallic sheaths at ground potential. The x-ray tube is enclosed in a grounded metal housing and insulated by means of oil. By this arrangement, it is not possible for the operator to come into direct contact with the high-voltage circuit under normal operating conditions.

The x-ray tube can also be mounted in the same tank as the high-voltage transformer; therefore high-voltage cables are unnecessary. This type, known as a self-contained unit, has been adopted widely, but it has the obvious disadvantage of requiring a large mass to be moved when the position of the x-ray tube is to be changed. With equipment of low voltage and low current ratings, this kind of self-contained structure is not cumbersome, and is generally used.

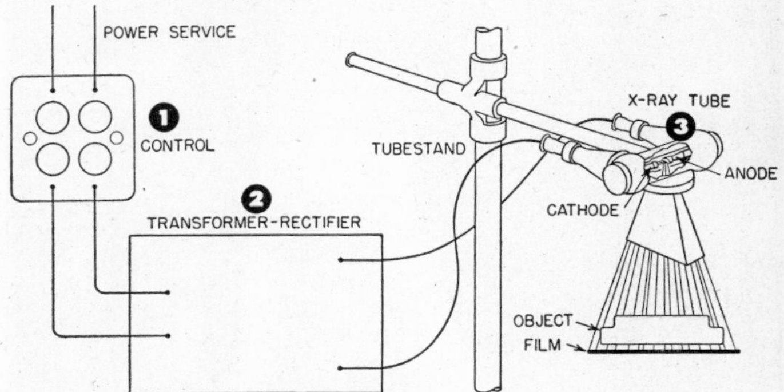

FIG. 27·7 Diagram showing the essential components of a simple x-ray generator of medium-voltage range.

The three primary functions of the control are simply to regulate the voltage applied to the tube, to regulate the current through it, and to control the duration of the x-ray exposure. In regulating the voltage applied to the tube there are several complications. It is necessary to assure the operator that the voltage he believes exists actually does exist at the tube. Such factors as transformer regulation and line regulation give different x-ray-tube voltages even though the control settings may not have been changed. The usual procedure is to use an autotransformer which is stepped to produce the necessary voltage division. If the voltage drop under load is known, the no-load setting is made higher by the amount of the voltage drop under load. Compensation for variations in incoming line voltage can be accomplished by the same means. The second function, that of varying the tube current, is fulfilled by the use of a rheostat or variable inductance in the tube-filament circuit. When the heat of the tube filament is varied, various x-ray tube currents are obtained. A milliammeter is usually in the control for indicating this current. The normal procedure is to measure the current through the transformer at the ground point and to use this as a measure of the tube current. Nearly all x-ray transformers are operated with their midpoints grounded because each half of the secondary circuit need only be insulated for one half of the voltage applied to the tube. The third important function of the control is to provide a means for accurate timing of the x-ray exposure. For this purpose, all that is necessary is a relay which removes the voltage from the tube after a predetermined time.

High-voltage equipment for x-ray work has quite different

requirements from those of high-voltage equipment used in power transmission and distribution. X-ray transformers are not subjected to the rough usage expected of other types of transformers. Voltage surges can be expected to be at a minimum, and the overload capacity can be almost zero, because the tube will fail long before the transformer has reached its limiting point. Furthermore, small space and light weight are of primary importance. This reduction in size is possible because only a limited current load is expected on these transformers, and because the insulation spacing may be reduced to a minimum. Because of the light current load reasonably small capacitors can be used as energy-storing devices, and voltage-doubling and similar circuits are practical.

X-ray generating equipment can be divided into five voltage classes; 0 to 50, 50 to 100, 100 to 150, 150 to 300, 300 kilovolts and above. This is a different division from that used in Chapter 8, and is chosen to fit the types of applications, not because of problems in equipment design. The voltage rating is always understood to be peak kilovolts, and the milliampere rating is always average.

0- to 50-Kilovolt Class

Equipment in this voltage class is generally used for microradiography and for diffraction work. Usually the control and transformer are in the same container and the transformer is "end-grounded." It is end-grounded because it is desirable to operate the anodes of tubes in this voltage class at ground potential so they may be cooled by tap water.

50- to 100-Kilovolt Class

A large portion of the radiographic work with light-alloy castings and with non-metallic material is done in the range of 50 to 100 kilovolts, but its use for steel and heavy metals is limited. Most of the equipment utilizes self-contained units, and the controls are usually of the simplest type. Equipment in this class is satisfactory for radiographic inspection of aluminum up to 2 inches thick.

100- to 150-Kilovolt Class

Most of the work done with equipment of this class is with light-steel sections, heavy light-alloy parts, and heavy plastic pieces. It can be used for radiographic inspection of 1 inch of steel, ¾ inch of brass, or from 5 to 6 inches of aluminum or magnesium. Half-wave rectified transformer circuits are used almost universally with cable connections to the x-ray tube. Figure 27·8 shows a modern unit rated at 150 kilovolts and 6 milliamperes. This unit uses a water-cooling coil inside the x-ray-tube head, with the heat being transferred from the tube to the insulating oil and then to the water in the cooling coils.

For this unit a new principle of controlling the tube current has been applied. The x-ray-tube current is carried by a bias resistor and thus controls the current through a high-vacuum tube. The current through this vacuum tube is then applied to the saturating coil of a saturable reactor. Coils on this saturable reactor also carry the filament current of the x-ray tube, which is alternating current. Any change in the current of the x-ray tube alters the bias of the vacuum tube, changes the saturating current for the saturable reactor, and varies the impedance in the filament circuit of the x-ray tube. The filament heat then changes because of the change in filament current and restores the x-ray-tube current to

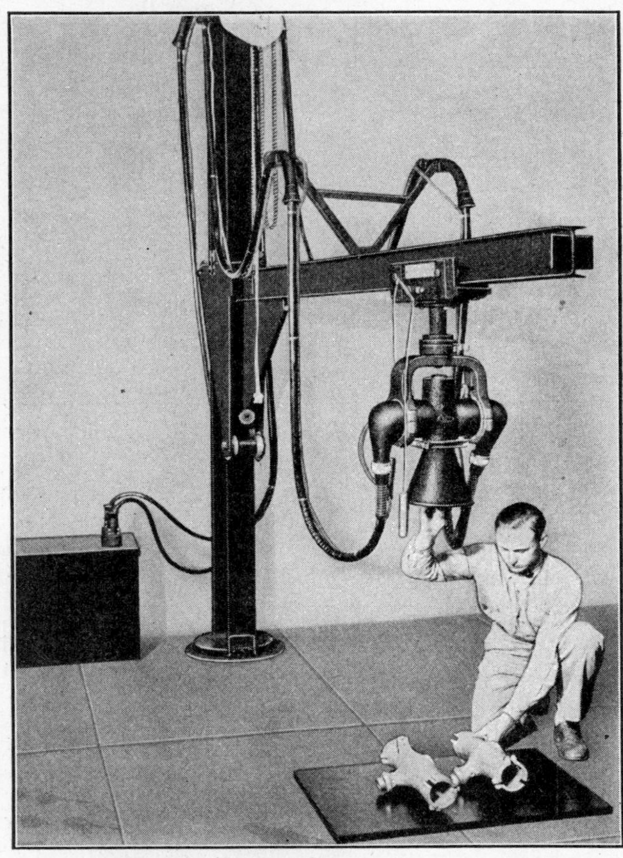

Fig. 27·8 A 150-kilovolt, 30-milliampere x-ray unit utilizing a jib crane mount for the x-ray tube. The control is outside of the room.

the proper value. This system enables the operator to make a current setting on a dial with absolute assurance that the current value he sets will be obtained and held during the entire exposure.

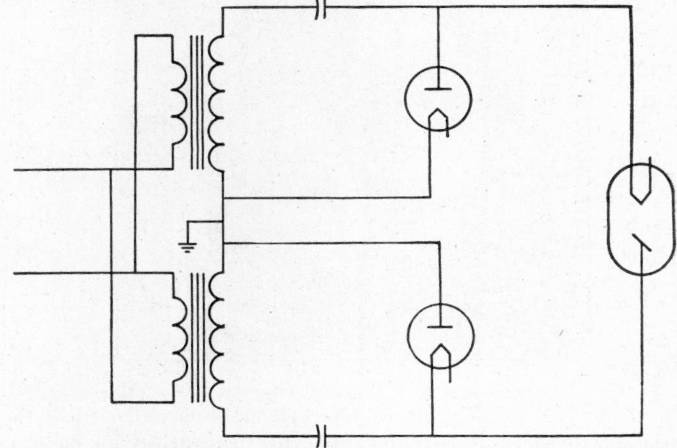

Fig. 27·9 Schematic wiring diagram showing one type of voltage-doubling circuit commonly used in x-ray units.

150- to 300-Kilovolt Class

In this category, steel up to 4 inches thick can be inspected. Generators of both the self-contained and cable-connected type are built in this voltage class. For these units, voltage-doubling circuits can be used economically.

Figure 27·9 shows a schematic diagram for a 220-kilovolt unit which utilizes a voltage-doubling circuit. Figure 27·10 shows a mobile 220-kilovolt unit incorporating a generator

Fig. 27·10 A 220-kilovolt, 25-milliampere mobile x-ray unit in position for radiography of a weld.

of this type. Sometimes it is more economical to transport the x-ray unit to the parts to be examined than to transport the parts to the x-ray unit. As discussed in Chapter 8, this unit utilizes oil cooling directly on the back of the anode of the tube. The oil serves as an insulator as well as a cooling agent and circulates through a heat exchanger with tap water as a primary coolant.

Fig. 27·11 Two-million-volt x-ray unit constructed at Massachusetts Institute of Technology. The outer shell has been removed in this photograph to disclose the equipotential rings and the driving motor in the base. The x-ray tube target is below the floor level. High-pressure gaseous insulation is used in this unit to reduce weight.

300 Kilovolts and Above

Some units rated at 400 kilovolts are extensions of the circuits used below 300 kilovolts. Most of such units have been of the self-contained type. They are satisfactory for radiography of steel up to about 4½ inches. However, the difficulties with secondary radiation become severe in this voltage range. As a result, emphasis has swung to units rated at 1000 kilovolts and above.

It is possible with 1000 kilovolts to radiograph about 5 or 6 inches of steel. Conventional high-voltage circuits are not applicable at these voltages as the weight of the equipment becomes excessive; therefore various unusual generating systems are utilized.

Since the current drain is relatively low for x-ray work, resonant circuits can be used for building up high voltages. To save weight, high-pressure gaseous insulation has been used in place of oil. One difficulty with this kind of unit is that the characteristics of a resonant circuit limit the range of voltage severely. This type of unit with a top rating of 1000 kilovolts operates only down to about 700 kilovolts,

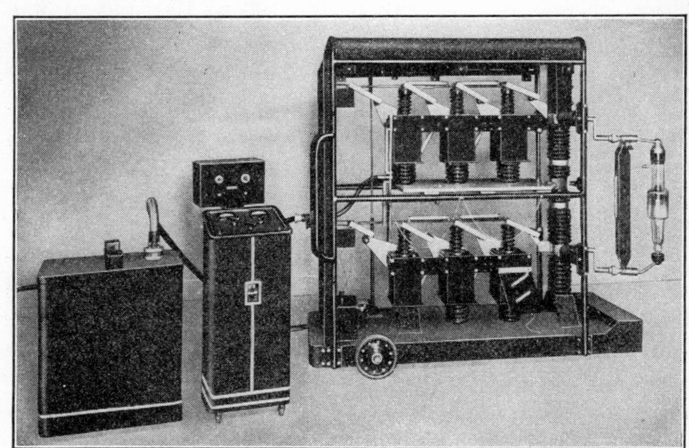

Fig. 27·12 Ultrahigh-speed unit for producing 1-microsecond exposures. At left is the high-voltage transformer supplying charging voltage to the surge generator. In the center is the control on which is mounted the interval timer for initiating firing of the surge generator at predetermined times after closing or opening a circuit. At right is the surge generator supplying power to the x-ray tube mounted on arms at extreme right.

and the output is much reduced at this voltage. However, units of this type with highest rating up to 2000 kilovolts are in successful operation.

The Van de Graaff principle of high-voltage generation has been used for equipment with a rating up to 5000 kilovolts. Constant potential is obtained with this type of equipment as compared to a pulsating potential for the resonant-transformer type of equipment. This unit overcomes the difficulty of the range of variation in allowable operating voltage, but the allowable current drain is lower. Figure 27·11 shows a unit of this type.

Ultrahigh-Speed Radiography

A radically different type of equipment has been made possible by the ultrahigh-speed x-ray tube described in Chapter 8. To supply the tube, a large quantity of power must be supplied for 1 or 2 microseconds. To obtain this power the x-ray tube is energized by a surge generator with a circuit of a modified Marx type to produce extremely short exposures. The exposures last approximately 1 microsecond; the tube current is approximately 2000 amperes; peak voltages up to 330 kilovolts are being successfully used. Such equipment is suitable for producing a radiograph through about 1 inch of steel, and will "stop" the motion of projectiles of even the highest velocity. Figure 27·12 shows one of these units ready for operation.

27·3 TECHNIQUE OF GROSS-STRUCTURE EXAMINATION

An x-ray field of uniform intensity is generated by an x-ray tube, and the radiation is allowed to pass through the object to be examined. Since the absorption of x-radiation depends upon the type and the amount of material which it must traverse, the x-ray field is transformed from one of uniform intensity to one in which the intensity varies according to the amount and type of material in the x-ray beam. The emergent field then has an intensity variation which parallels the variations in the object. The only difference in the various methods of inspection is in the means of transforming this x-ray pattern into either a visible pattern or a graph form.

In radiography, the recording medium is a film which is exposed for a period of time to the intensity of the x-ray pattern. Subsequent development and illumination of the film transfers the x-ray pattern into a visible image. In fluoroscopy, a visible image of the pattern is shown on a cardboard screen coated with small crystals of fluorescent material. The light pattern emitted from the surface of the screen is proportional at any point to the x-ray energy received; therefore, the light pattern follows the variation of the intensity of the x-ray pattern. Photofluorographic technique is an extension of fluoroscopic technique. A camera is set up so that the image on the fluorescent screen can be photographed on a film of small size. Non-visual recording systems produce directly a variation in an electrical quantity which can be used as a measure of x-ray field intensity. In microradiography, a film is made in much the same way as in radiographic work, but it is magnified many diameters for viewing.

Total radiation of an x-ray image comes from two sources: (1) primary radiation originating at the target of the tube, penetrating the object and forming the image; (2) secondary or scattered radiation which reaches the film and obscures the real image. Of these two sources, only the first is useful; the second is undesirable, and is to be eliminated if possible. In a satisfactory image little of the energy arriving at the recording medium is from secondary radiation.

Whenever x-radiation strikes any material, some of the primary radiation is changed into secondary radiation in much the same way as a beam of light is dispersed in a fog, as described in Chapter 8. Since the points of origin of secondary radiation are different from that of the x-ray-tube focal spot, any image produced by this radiation obscures the true image.

The secondary radiation is distributed spherically around the originating points. Therefore, it is possible to set up a grid of flat thin pieces of lead or other x-ray absorbing material aligned so that their thin surfaces lie in the direction of propagation of the primary beam. Since the strips of lead are thin, the primary beam is opposed by only a small amount of absorbing material. However, a large proportion of the secondary radiation will be stopped by this grid because of its different angular relationship with the surfaces of the lead strips. This grid or diaphragm is known as a Bucky diaphragm. Figure 27·13 illustrates the grid and the absorbing mechanism.

The statement that secondary radiation has approximately spherical distribution is true only at the lower voltages. At 1000 kilovolts, secondary radiation is predominantly in the forward direction; for this reason, results at this voltage are better than would be expected from experience with lower voltages. Obviously, if all the secondary radiation were propagated in exact line with the primary radiation, it would not be deleterious but advantageous, for it would intensify the image formed by the primary beam. At 1000 kilovolts,

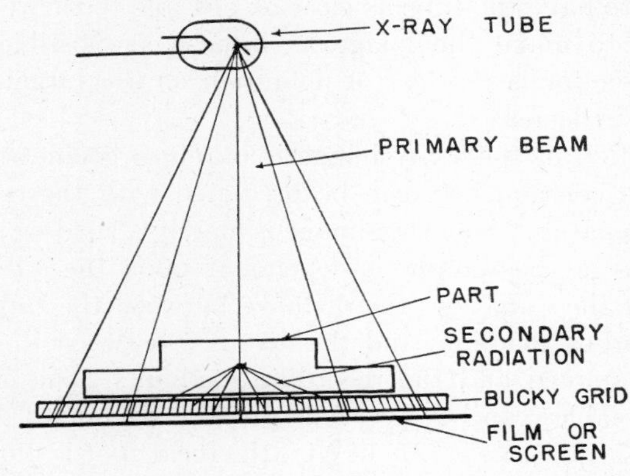

Fig. 27·13 Schematic diagram showing the mechanism of secondary radiation absorption by a Bucky grid.

the secondary distribution departs sufficiently from spherical distribution to show a marked advantage over the distribution at lower voltages.

Another means of reducing secondary radiation is based on the fact that secondary radiation is always of longer wavelength and is therefore less penetrating than the primary beam. A flat sheet of lead varying from 0.002 or 0.003 inch up to several millimeters in thickness, depending upon the x-ray-tube voltage, is placed between the object which generates these secondary radiations and the recording medium. Since secondary radiation is less penetrating than primary radiation, the percentage of secondary rays absorbed in the lead sheets is larger than that of the primary rays. After passing through the lead sheets, the x-ray field has a higher percentage of its intensity due to primary radiation and the relative intensity of the objectionable secondary radiation is diminished.

Three variables under the operator's control must be manipulated to produce the best image. These are (1) tube current, (2) tube voltage, and (3) distance between the x-ray-tube target and the recording medium.

Current through the tube is controlled by the filament heat and is indicated by an instrument on the control board. Primary radiation is directly proportional to this current; thus a tube current of 10 milliamperes produces twice as much radiation as a current of 5 milliamperes. Quality or penetrating power of the x-radiation is not directly affected by a change in tube current.

Voltage applied to the tube is regulated by taps on the autotransformer supplying primary voltage to the high-voltage transformer. Voltage influences both penetrating power (quality) of the radiation and also the amount (quantity) emitted. Because voltage influences both of these major characteristics of the beam, it is the most important variable at the command of the operator. It is also the most difficult to control, because the effect of voltage changes cannot be predicted except by practical knowledge.

The distance between the target of the tube and the recording medium, known as target-to-film distance in radiography and target-to-screen distance in fluoroscopy, influences the amount of radiation reaching the recording medium and also the definition of the x-ray image. The amount of radiation reaching the recording medium varies inversely as the square of the distance. Thus, it is nearly always advantageous to have the tube as close as possible to the recording medium to make the image as intense as possible. The limiting factor is the loss of definition as the target-to-film distance is decreased.

Definition of the x-ray image is purely a geometric function, and consequently can be predicted with the certainty of mathematics. Since the image is merely a shadow picture, sharpness of the shadow is dependent upon the size of the source of the radiation, the distance between the target and the recording medium, and the distance between the object being examined and the recording medium. The effect of these variables can be observed by the exaggerated drawing in Fig. 27·14. In Fig. 27·14 (*A*), the focal spot is illustrated as a point at *T*. The defect *O* in the specimen is close to the film *F*. A sharp image of the object is projected on the recording medium. The focal spot actually has a finite area, and the effect of substituting a large focal spot for the point source is illustrated in Fig. 27·14 (*B*). Each point in this large focal area projects its own shadow, and these shadows produce an indistinct image of the object. It is therefore advisable to make the focal spot as small as possible for the thickness of the material being examined.

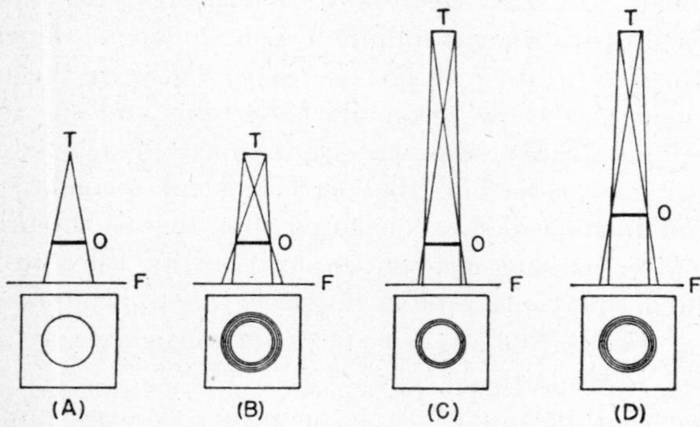

Fig. 27·14 The effect of the various factors controlling definition. The lowest figures represent a plan view of the image recorded on the film *F*. *T* represents the focal spot or x-ray source, and *O* the object.

The loss of definition caused by a large focal spot is lessened by using a greater target-to-film distance, as illustrated in Fig. 27·14 (*C*). It is advisable to use the greatest distance without decreasing the x-ray field intensity below a reasonable limit. Sharpness of the image is directly proportional to both focal spot size and target-to-film distance, and is dependent upon the distance between object and recording medium, as shown in Fig. 27·14 (*D*). Here the shadow is more diffused than in Fig. 27·14 (*C*), where the object was close to the recording medium. If the defects are suspected of being in one side of the specimen it is best to place that side as near the recording medium as possible. It is imperative to make the distance between the recording medium and the object being examined a minimum.

After the x-ray beam traverses the specimen its intensity varies over the entire area, paralleling the variation of part material and thickness. The degree of variation in intensity for different thicknesses is termed contrast. If the difference in intensity for a given change of thickness is great, the contrast is high. If the change in intensity for the same change of thickness is small, the contrast is low. Usually the means of recording the x-ray pattern increases the contrast in the x-ray beam. Increasing tube voltage decreases contrast, since the same increment of thickness absorbs less of the more penetrating radiation, and the difference in emergent radiation intensities is less. Since tube voltage is the only factor which influences penetrating power or quality of the beam, it is the only variable at the command of the operator by which contrast can be controlled.

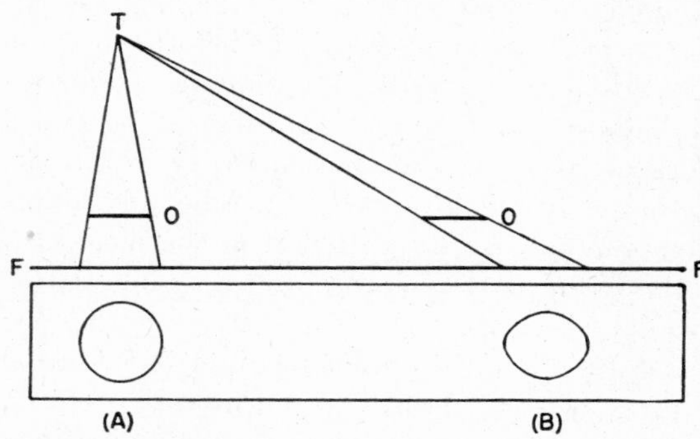

Fig. 27·15 The effect of the various factors controlling distortion and magnification. The lower figure represents a plan view of the image recorded by the film *F*. *T* represents a point source of x-radiation, and *O* the object.

The effect of contrast in the image is related intimately to the factor of definition. Obviously, a variation in thickness of material is more readily observable in a visible image if the variation causes a sudden or sharp change in x-ray-beam intensity, because the eye can detect sudden changes more easily than gradual changes. High contrast would be nearly worthless if each change of x-ray intensity had such poor definition at the point of change that it was indistinct. To produce a satisfactory image, contrast must be high and definition must be good.

The objective of high contrast must be tempered because of the frequent requirement that an x-ray image hold the changes in beam intensity to a usable range, even though the variation in thickness of the specimen is great. This factor is called latitude, and it refers to the variation in part thickness that may be allowed for usable maximum and minimum beam intensities. The choice of tube voltage, then, is always a compromise between the requirements of contrast and latitude.

Distortion and magnification in an image are also geometric functions, and are illustrated by Fig. 27·15. Image *A* is formed by having the tube centered over the object and the direction of the beam perpendicular to the recording plane. The image is circular in shape. Magnification is uniform and can be decreased by increasing the distance between the x-ray tube and recording plane. Image *B* is the projection of a beam which is not perpendicular to the film. The image is distorted, one axis being extended to make an oval shape, and the magnification is much greater than for image *A*.

Clearly the workman must be skilled in choosing voltage, current, and a set-up which will give the most intense image with good definition and acceptable contrast and latitude. In other words, he must choose these factors for maximum sensitivity over the entire area in which he is interested.

Strictly speaking, sensitivity is determined after the x-ray image has been recorded, but control of all the factors discussed in this section is essential for maximum sensitivity. Sensitivity is a measure of the success in detecting all defects in the part being examined. It usually is expressed as the ratio (in percent) of the thickness of the smallest detectable defect to the thickness of the base material. Rigid standards of sensitivity measurement have been set up. Loosely, a technique which can illustrate the presence of a void 0.015 inch thick in a specimen 1 inch thick has a sensitivity of 1.5 percent. With good laboratory procedures, it is possible to achieve a sensitivity of slightly less than 1 percent in radiography, from 5 to 15 percent in fluoroscopy, and from 3 to 7 percent in photofluorography. Sensitivity of non-visual recording is much better even than that of radiography, but progress is so rapid in this field that it is not wise to place a limit on the obtainable sensitivity. In practice sensitivity between 1 and 2 percent is obtained in radiography. The usefulness of x-ray examination as an inspection tool depends upon the sensitivity that must be obtained to make the anticipated defects legible in the resultant image.

Sensitivity is dependent upon all the factors affecting the x-ray image and all those involved in the recording system. To measure sensitivity it is customary to place additional thin pieces of the base material on the specimen. These telltales or penetrameters, as they are known, have been standardized.

The most common form of penetrameter (Fig. 27·16)[13] consists of separate telltales made of the same material as the object to be examined. For all materials except magnesium or magnesium alloys, the thickness of each penetrameter must be not greater than 2 percent of the thickness of the object, except sections less than ¼ inch thick, for which a penetrameter 0.005 inch thick is required. For magnesium and magnesium alloys, the thickness of each penetrameter must be not greater than 3 percent of the thickness of the object, except sections less than 0.17 inch, for which a penetrameter 0.005 inch thick is required. Drilled through each penetrameter are three holes with diameters two, three, and four times the thickness of the penetrameter. These penetrameters are always placed on the tube side of the specimen at the extremes of the field so that they are more difficult to detect than any defect in the specimen which would be closer to the recording plane and the center of the beam.

As pointed out in Chapter 8, the usual x-ray tube uses a line-focus principle, and the angle of the target in normal industrial work is 20 degrees. If an attempt is made to utilize radiation which comes from the tube anode at an angle of less than 5 degrees, the intensity decreases seriously because the radiation is not generated at the surface of the anode but may originate at a point several molecular layers below the surface of the target. Originating at this point, at grazing incidence to the target, the x-radiation must traverse some metallic material to reach the air and thus decreases in intensity.

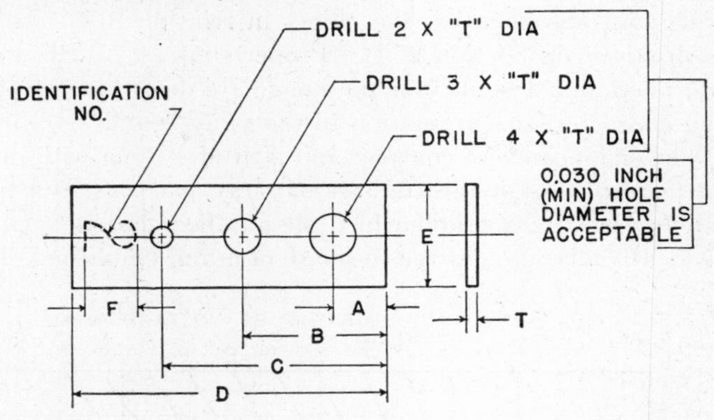

DIMENSION	SECTIONS UP TO AND INCLUDING 2.5 INCHES THICK	SECTIONS OVER 2.5 INCH THICK
A	1/4	1/2
B	11/16	1-1/16
C	1-1/16	1-9/16
D	1-1/2	2-1/4
E	1/2	1
F	1/4 (APPROX.)	3/8(APPROX)
T	SPECIFIED IN TEXT	

FIG. 27·16 Penetrameter dimensions. (Reprinted from Army-Navy Aeronautical Specification AN-1-26, Sept. 1944, with permission of the War and Navy Departments.)

27·4 INDUSTRIAL RADIOGRAPHY

The length of time radiation is allowed to strike the film is known as exposure time; it is recorded by an interlocked exposure timer on the control board, or by an independent timer. Although the incident intensity may be low, the film may be allowed to accumulate energy over a long period of time. This cumulative effect is one of the major advantages of this type of recording over the others. With it, low voltage can be used for increase in contrast, and long target-to-film distances can be used to increase definition.

Since total radiation reaching the film is directly proportional to both current and exposure time, these two variables are frequently combined into a product, milliampere-minutes. Ten milliampere-minutes may be produced by an exposure of 10 milliamperes for 1 minute or by 5 milliamperes for 2 minutes, and the effects on the film are the same. The choice of voltage, milliampere-minutes of exposure, and target-to-film distance can be determined from a technique chart. Figure 27·17 shows a typical technique chart.

Producing adequate film blackening in a radiograph is not the complete process of radiographic technique by any means; it is only the first step. The radiograph must possess other characteristics, but these must come before film blackening because without adequate film density the other characteristics could not be distinguished. Actually, film blackening is the means of bringing out the following necessary characteristics in a radiograph: (1) proper contrast and latitude; (2) maximum visibility of the image; (3) maximum definition; and (4) minimum distortion and magnification.

Maximum definition and minimum distortion and magnification are obtained by proper choice of target-to-film distance and placement of the object in relation to the beam, as discussed in Section 27·3. Proper contrast and latitude and maximum visibility of the image are determined by the film characteristics in relation to the x-ray beam.

The significance of contrast and latitude, along with other intricate phases of the transfer of x-ray patterns to a film, can be seen by a graph which shows the relation between incident radiation and the response of a film emulsion. This graph is known as an H & D curve (Hurter and Driffield) and relates film density to the logarithm of the exposure, as shown in Fig. 27·18.[7]

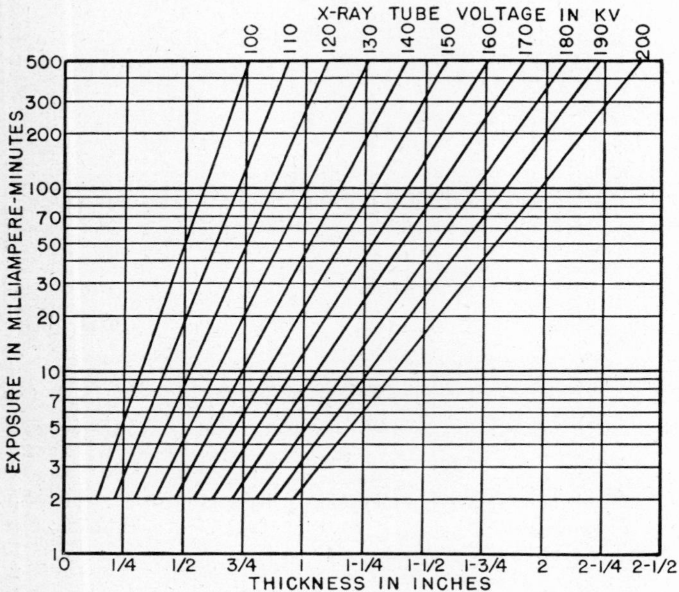

FIG. 27·17 Typical technique chart. This applies to rolled steel with no-screen film and 0.005-inch lead screens and a 36-inch target to film distance.

Density of a film is defined as the common logarithm of the ratio of the light incident on the film to the light transmitted through the film. A density of 1 indicates that one tenth of the incident light passes through the film. The useful range of the film is the straight-line portion of the curve, and ordinarily lies in a range of densities between 0.4 and 2.0 with ordinary viewing facilities, or to about 4.0 with high-intensity viewers.

An important phase of the transfer of radiation energy to a film is the speed of transfer; it is indicated by the position of the curve with respect to the vertical axis. Obviously, the film which gives a usable density with the least exposure is the fastest film. These curves are determined at definite voltages, and the slopes of the curves change with voltage, but the general relations remain the same.

Figure 27·19 is an interpretation of the curve shown in Fig. 27·18. The slope of the curve in Fig. 27·18 represents the contrast, since contrast in a film is really the rate of change of density with respect to exposure. The best density is that at which contrast is maximum if this point is within the usable density range. Figure 27·19 shows why the tendency in industrial radiography has been to use higher densities of 1.5 to 2.5. This limit is not because high contrast cannot be obtained at higher densities but because ordinary viewers cannot produce sufficient light intensity to penetrate the extremely high density.

Latitude is that characteristic of a film which determines the range of thickness of parts that can be examined on one film. If a part has a wide difference in thickness in its various sections, the latitude of the examining process must be sufficient to produce a usable density on the film at both the thinnest and the thickest sections. Thus a sacrifice must be made in contrast since, to obtain high contrast, a small change in part thickness must cause a large change in x-ray intensity. Both film characteristics and tube voltage affect latitude; therefore it is always necessary to consider the film in determining the proper voltage setting, not only from the viewpoint of proper exposure time determined by film speed, but also from the viewpoint of the latitude desired.

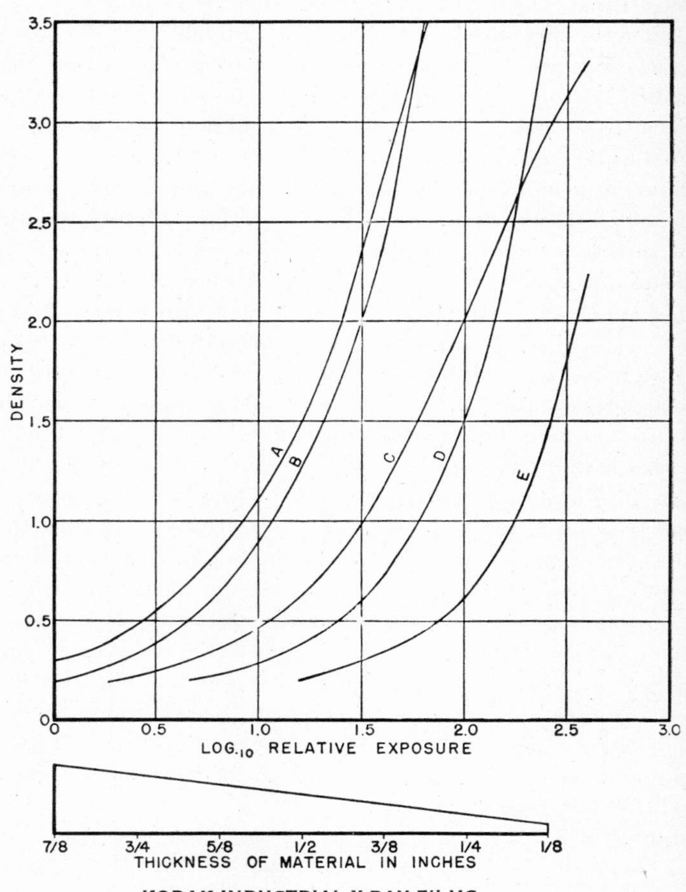

FIG. 27·18 Characteristic curves of films used in industrial radiography, for direct or metallic-screen x-ray exposures in the voltage range of 50 to 200 kilovolts. Curve *A* is for high-speed no-screen film, curve *B* for no-screen film, curve *C* for high-speed screen film, curve *D* for detail no-screen film, and curve *E* for fine-detail no-screen film. (Reprinted with permission from "Radiography of Materials," published by Eastman Kodak Company, 1943.)

Since the energy striking the film is directly proportional to exposure time, and since the absorption of x-radiation by materials is a logarithmic function for monochromatic radiation, it is possible to use a thickness of material instead of the logarithm of exposure on the H & D curves. This procedure has been followed in the bottom scale of Fig. 27·18; to assist in visualizing this scale, a thickness wedge is included. This conversion is true only for a monochromatic beam, but it is sufficiently close for approximate analysis.

Using the thickness-of-material scale as a guide, it is clear that, if the specimen is of uniform thickness, except for defects, it is desirable to use a film of high contrast because a small void in the material produces a noticeable increase in blackening in the image.

Such a high contrast characteristic could not be used if the specimen had several widely different thicknesses, because only one thickness would produce a legible density; thinner sections would appear black on the film and thicker sections would be clear areas. For such a specimen, a technique producing film of low contrast and wide latitude is indicated. Such technique does not produce great contrast in any section, and small changes of thickness, such as defects, may not be legible. If a radiograph with satisfactory sensitivity for each thickness of the specimen cannot be produced by one simple exposure, a technique giving higher contrast should be used, and the thinner section should be built up with additional material, or different exposures should be given to the individual sections.

Frequently calcium-tungstate intensifying screens are used to speed up the radiographic process. Calcium tungstate emits light under the influence of x-radiation, the amount of light emitted depending upon the intensity of the incident x-ray beam. These screens are made by coating pieces of cardboard with thin layers of fine calcium-tungstate crystals. One of these screens is placed on each side of the x-ray film, because x-ray film has a sensitive emulsion on both sides of the base. The total exposure of the film then comes from two sources: (1) absorption of x-radiation by the film itself; and (2) absorption of light emitted from the calcium-tungstate screens.

The intensification factor with calcium-tungstate screens is about 100 to 1, meaning that with other factors constant the x-ray exposure may be reduced by a factor of 100 if calcium-tungstate screens are used. Frequently, contrast can be increased by the use of intensifying screens. The screens have a direct effect in increasing contrast; moreover, since a satisfactory film can be produced with less exposure, film densities may be obtained with lower tube voltage and the same time of exposure. Reduction of voltage increases contrast. However, fine detail in the photographic image is never produced as well with screens as it is without them. The crystals cause fuzzy edges in the image because the entire crystal glows even though only a part of it may be irradiated. The screen "grain" then prints itself on the film and sharp contours become more diffuse.

Often visible in a radiograph is an under-cutting effect, if a thin section adjoins a thick section. The x-ray intensity reaching the film through the thick section is small compared with the intensity through the thin section. Film blackening then spreads underneath the thick section and produces an undercutting dark end effect which makes detection of defects in this edge almost impossible. Obviously, this undercutting happens most frequently at the outside edge of the part being examined. Two methods may be used to avoid it.

One is to shield the areas which normally would be subjected to radiations of high intensity. The shielding should follow closely the outline of the object or area under examination, and should reduce intensity through the thin areas to about the same as that through the thick areas. Lead sheet is frequently used but, if it cannot be made to fit the shape of the object, lead shot, copper shot, or lead powder can be used to mask the irregular shape or to fill large openings. These materials are sometimes formed into a workable paste by mixing them with wax. An alternative is to immerse the object in a water solution of a lead salt. If shielding is used in areas that are of no interest, radiopaque shielding is used to prevent creepage and fogging of the adjoining areas; in other cases, all the areas are of interest and the shielding is radiopaque only to the extent that it prevents complete blackening of the film.

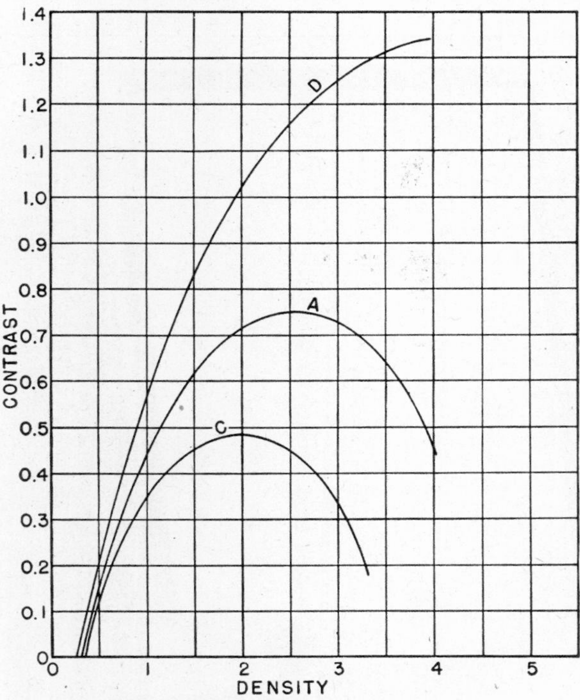

FIG. 27·19 Replot of the data given in Fig. 27·18 but using the slope of the curves, which is contrast, as abscissa. This points out the reason for the trend toward higher film densities because of the increase in contrast. The curves refer to the films given in Fig. 27·18.

A second means of reducing undesirable radiation is the use of filters. Filters of aluminum, copper, or lead are commonly used and, because of their uniform density and thickness, cast no image. A filter can be used between the x-ray tube and a specimen which does not completely fill the film area to reduce the background intensity of the unobstructed radiation without materially decreasing the intensity of the radiation of the real image. This is because the undesirable radiation along the unobstructed path contains much more radiation of long wavelengths than that which passes through the object before striking the film. Most of the darkening in the unobstructed areas is due to this long-wave radiation. A filter reduces this long-wave radiation without greatly affecting the short-wave radiation, thus reducing the difference in intensity between the x-ray along the unobstructed path and that passing through the part being examined.

Obviously the filter does absorb some radiation of all wavelengths, so exposure time or x-ray-tube current must be increased to compensate for the absorption of x-ray energy in the filter.

Usually it is preferable to place such a filter between the specimen and the film. In this position the filter serves the same purpose as before, and in addition eliminates some of the secondary radiation which originates in the object, as discussed in Section 27·3.

If the filter can be made of lead and can be placed in contact with the x-ray film, this method has an additional

Fig. 27·20 High-intensity illuminator using double illumination. A brilliant general field allows the illumination of the entire film. For very dense portions a high-intensity light covering a 4-inch circle in the center of the general field may be turned on.

advantage in that it causes intensification. The reason for the intensifying action is that some of the primary beam is absorbed in the lead screens, and the x-ray energy is transformed into heat, secondary radiation, photoelectrons, or recoil electrons. The energy transformed into heat has no effect on the film. Some of the secondary radiation strikes the film and produces exposure but, since the source of secondary radiation is so close to the x-ray film, the image produced by the primary beam is intensified. Since secondary radiation always has a longer wavelength than the primary radiation causing it, absorption of secondary radiation in the film is larger. Energy which is transformed into photoelectrons directed toward the x-ray film is almost completely absorbed by it because the photoelectrons have low penetrating power. This absorbed energy also causes exposure of the film. X-ray energy transformed into recoil electrons directed at the film also is almost completely absorbed by it. There is, then, a net gain in film exposure. It is reduced by the absorption of the primary beam by the lead screen, but its increase by the other factors far outweighs this decrease by absorption.

The film usually is sandwiched between two pieces of lead foil so that both intensify although only the one on the tube side of the film acts as a filter.

Lead-foil screens are much slower than calcium-tungstate intensifying screens. They have an intensifying factor of about 5, but this varies markedly with voltage. Lead-foil screens have almost no "grain" as calcium tungstate screens do; therefore lead-foil screens produce much better image visibility.

Some of the advantages of both lead-foil and calcium-tungstate screens can be obtained by using a lead-foil screen on the tube side of the film as a filter and a low-power intensifying agent and a calcium-tungstate screen on the far side as a high-power intensifier.

As mentioned in Section 27·3 a Bucky diaphragm, or grid, can be used to eliminate a large portion of the secondary radiation. Since the exposure is accumulated over a period of time, the Bucky Diaphragm can be moved in a direction perpendicular to the lines of the grid during the exposure to prevent formation of a sharp image of the grid. Because the various lines blur they are not detectable on the final film. This procedure is not often used industrially because of the difficulties of moving the diaphragm continuously during long exposures, and because lead-foil screens perform satisfactorily.

After the exposure has been made, the film is processed carefully so that the latent image is developed, fixed, washed, and dried. Although the processing of film is a simple routine, success is based on good technique. As in all chemical work, careful and orderly routines are important.

Fog on the film obscures the real image. A detailed discussion of the factors which produce fog and consequently decrease visibility of the image will not be given here. Some of the causes are:

1. Incorrect safe light in the darkroom.
2. Exposure of film to light.
3. Exposure to stray radiation.
4. Old film.
5. Oxidized developing solutions.
6. Warm developers.
7. Overdevelopment (to compensate for under exposure).
8. Fumes (lack of ventilation, especially when drying).

Radiographs must be viewed under the best possible conditions, so that everything in the radiograph is evident. Brilliant and uniformly distributed illumination should be used behind the film, and other illumination in the room should be minimized. If the film is smaller than the illuminator screen, or if small areas in a film are particularly important, the disturbing portion of the transmitted light from the illuminator should be masked.

Because the use of higher densities of film is becoming the practice in industrial radiography, it becomes desirable to have an illuminator with two light intensities. Usually a radiograph contains some dense areas which require strong illumination, but low-intensity illumination serves for the major portion of the radiograph. An illuminator for such radiographs can be built with a brilliant general illumination, which is satisfactory for a density of 2.0, and a high-intensity spot approximately 4 inches in diameter, which is satisfactory for densities up to 4.0. Figure 27·20 shows an illuminator of this type.

The ultrahigh-speed unit discussed in Section 27·2 is designed specifically for formation of a radiograph in the minimum time. To reduce the exposure time to about 1 microsecond, striking changes have to be made in the x-ray tube. The focal spot size must be increased to prevent destruction of the target by the tremendous short-time load. Total radiation emitted during this short exposure is less than is normally used for regular radiographic work where there is no limit on the time; therefore the output must be increased by raising the tube voltage above that which would be used for the same material thickness in conventional radiography. Shorter target-to-film distances are used also.

The target, which is two or three times as large as the conventional target, reduces definition; so does the use of shorter target-to-film distances. Higher voltage reduces contrast. An attempt to gain in definition by increasing target-to-film distance is accomplished by a loss in contrast, because tube voltage must be increased to hold the total radiation at the film constant. And, of course, the reverse is also true.

The ultrahigh-speed unit, then, cannot be considered as doing the same job as conventional radiographic equipment. If motion must be "stopped" it is an excellent tool; for this reason it is adaptable to problems in ballistics. Not only is it possible to radiograph projectiles in transit through gun barrels but also to radiograph the same projectiles where the smoke clouds or flash of light from the exploding powder makes high-speed photography impossible. Furthermore it is possible to take radiographs of exploding shells while the x-ray tube and film are protected by metal plates. Such radiographs are shown in Fig. 27·21.[15]

The nose of the shell has passed through the plate. There has been no shift of the detonating parts of the fuse.

The shell has started to swell in the region of the bourrelet. The booster has not yet detonated.

The shell body has swelled to almost twice normal diameter. The petals have constrained the bourrelet region of the shell from swelling as rapidly as the body of the shell. The base of the shell has swelled very little.

The shell has ruptured. The base of the shell is starting to mushroom out. The constraining effect of the petals is evident.

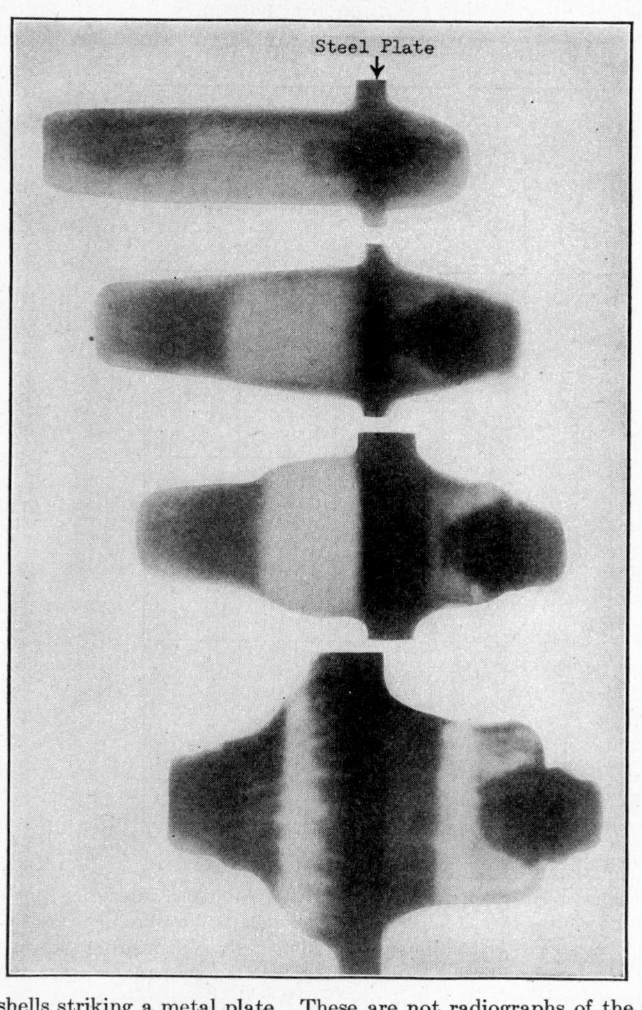

FIG. 27·21 A series of ultrahigh-speed radiographs of explosive shells striking a metal plate. These are not radiographs of the same shell but a series taken at different points in the explosion of a number of shells. From high-speed x-ray pictures like these taken under the direction of E. R. Thilo of Frankford Arsenal, ballistics experts learn many things about projectiles and armor plate.

In addition, high-speed mechanisms can be studied. The shell-ejecting mechanism of rifles has been studied in this way. The valve mechanism of high-speed compressors and motors could be so studied.

This equipment has a timer so that the exposure can be made at any pre-selected time after the opening or closing of a circuit. Therefore, it is possible to take a radiograph of an object at any time in the cycle of movement.

27·5 INDUSTRIAL FLUOROSCOPY

If viewing and inspection with a fluoroscopic screen is compared with radiography, there is an obvious saving in time, film cost, and processing equipment, but these advantages are offset by some disadvantages. In a radiograph, prolonging the exposure builds up sufficient x-ray energy to bring the film density to the desired level. Then, any amount of light necessary for viewing the image may be placed behind the film. In fluoroscopy, the screen brightness is a function only of instantaneous tube current and voltage. In other words, in radiography, energy can be stored over a long period of time; in fluoroscopic examination, screen brilliance depends only upon the power at that particular instant.

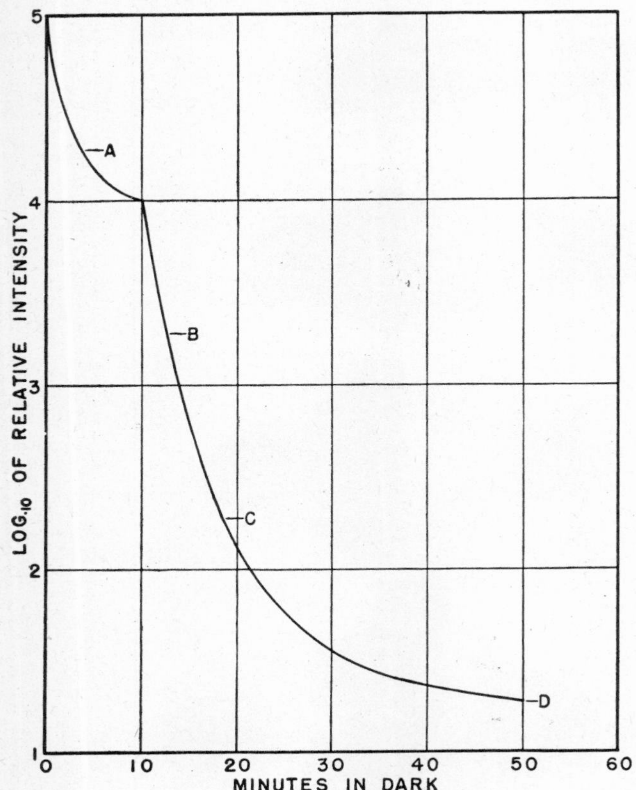

Fig. 27·22 Dark adaptation of the eye as a whole. First ten minutes is cone adaptation followed by rod adaptation. *A* indicates threshold at approximately 3 minutes after beginning dark adaptation. Increase of sensitivity is 10-fold, 100-fold, and 1000-fold respectively at points *B*, *C*, and *D*. (Reprinted with permission from W. E. Chamberlain, *Radiology*, Vol. 38, No. 4, p. 397, April 1942.)

Exposure time is no longer a variable, for no energy is stored in the fluoroscopic screen. Limitations of the output of x-ray tubes make fluoroscopic images relatively dim.

On the retina of the eye are two types of nerve endings. These are the recording medium for the light that reaches the retina. Each of these nerves carries an impulse to the brain, which incorporates all the impulses into a mass view at this point. These two types of nerve endings which perform different functions are known as rods and cones. At high levels of illumination, the cones are the dominant recording medium, and the rods play a relatively unimportant part. The cones are sensitive to slight differences in color or wavelength, and are the only nerve endings at the fovea or optical center of the eye.

At all other points on the retina are rod nerve endings which are used for vision at very low levels of illumination. Since the rods occur at positions away from the optical center of the eye, they are poor at viewing details. Everyone knows that a faint star in the heavens cannot be seen if the observer looks directly at the point where it should appear. However, if the observer looks slightly to the right or the left, the star becomes visible, since the image of the star is thrown onto an area on the retina at which rod vision becomes possible, and a small light intensity is picked up. These rods are not so sensitive to color or changes in light frequency but their acuity at low light levels is remarkable.

If the eye is allowed to rest in complete darkness, the rods become much more sensitive to low light intensities; consequently, in fluoroscopic work, a period of eye adaptation usually is allowed. This period is approximately one-half hour during which the observer is in complete darkness.

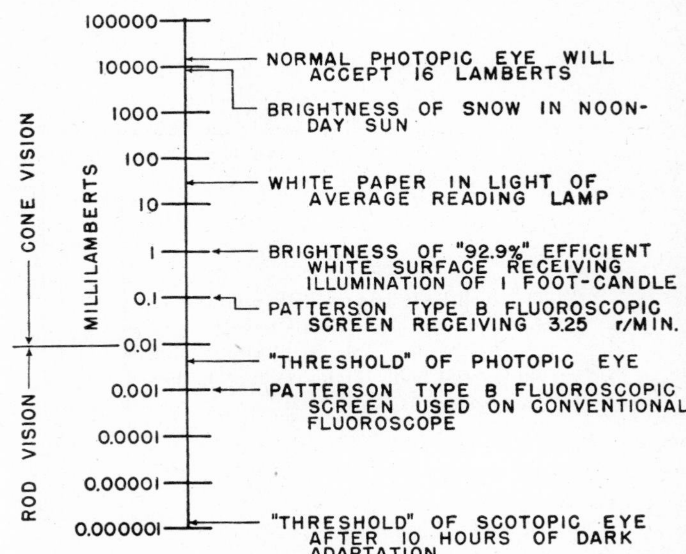

Fig. 27·23 The change-over between rod vision (low brightness levels) and cone vision (high brightness levels) is indicated as though it occurred suddenly at 0.01 millilambert. The transition actually takes place more gradually in the range between 0.01 and 0.1 millilambert. Even in the range between 1.0 and 0.1 millilambert where cone vision is operative, deterioration is visual acuity, and intensity discrimination with lowering of the brightness level becomes very noticeable. (Reprinted with permission from W. E. Chamberlain, *Radiology*, Vol. 38, No. 4, p. 385, April 1942.)

Figure 27·22 [17] shows a curve given by Chamberlain showing the speed of dark adaptation of the eye.

Figure 27·23 [17] shows the light intensities over which various nerve endings on the retina are used in normal sight. Shown in Fig. 27·24 [17] is the visual acuity at various illumination levels. Visual acuity is defined as the reciprocal of the angular distance which must separate two contours so that they may be recognized as discrete, the unit of separation being 1 minute of arc. At high levels of illumination the eye is a very sensitive device. However, at the low brightness levels, as in fluoroscopy, the eye becomes a poor instrument, and separation of contours must be large before the eye can discriminate between them. As a result the image may contain a great deal of detail which would be discernible at high levels of illumination, but the eye would not be able to detect it at the low levels obtainable from a fluoroscopic screen. This difficulty seems to have no simple solution. It is not economical to build x-ray tubes and generators to raise the brightness levels sufficiently to be of any value.

There are three ways to increase screen brightness: (1) an increase in voltage; (2) an increase in current; and (3) a decrease in target-to-screen distance. Raising the voltage increases screen brightness and increases the sensitivity of the human eye. However, at the same time it decreases contrast in the x-ray beam because of the smaller differential absorption of the beam generated at the higher voltage. Thus, a gain in one direction is neutralized by a loss in the other.

Reducing target-to-screen distance increases screen brightness according to the inverse square law. However, when this is done, sensitivity is reduced because of the loss of definition in the beam. In other words, the geometrical conditions change in order to decrease definition at the boundary of flaws or part contours.

Increasing tube current, of course, involves no direct loss of sensitivity, but several conditions limit the gain possible. Practical considerations limit the possible increases in tube current to factors of 2 or 3. This increase in tube current does not mean that visual acuity is increased by the same factor, as Fig. 27·24 shows clearly. A short portion of the curve may be considered to be a straight line. To double visual acuity from 0.6 to 1.2, screen brightness must be increased by a factor greater than 10, which is a practical impossibility at present.

Another complication is that an observer must be protected from the radiation. The usual method is to place lead glass, which is relatively opaque to x-radiation and relatively transparent to visible light, behind the fluorescent screen. Lead glass contains lead salts in solution. Its use introduces some loss, the chief loss being due to reflection of light at its surfaces.

Alternatively a mirror can be mounted so that it reflects the visible image at an angle to the direct x-ray beam. With this arrangement the observer can be protected with a thinner lead glass because he is no longer in the direct beam but is exposed only to scattered or secondary radiation. However, the observer is then necessarily at an effectively large distance from the fluoroscopic screen. The observer tends to view the screen at a distance of 6 to 10 inches, but he cannot if mirror viewing is used, and a net loss in sensitivity is the result.

In the voltage range from 30 to 300 kilovolts, observers have obtained sensitivities from 5 to 25 percent. Maximum sensitivity has been obtained only under ideal conditions and cannot be considered obtainable under average circumstances.

In many industries this range of sensitivity is acceptable. Consider a plastics manufacturer who wishes to use metallic inserts in his products. Relative x-ray opacities of the two materials are so far apart that a sensitivity from 15 to 20 percent is acceptable.

There is a second factor which limits the sensitivity of fluoroscopic equipment. In radiography, by proper choice of the x-ray film a light image can be produced with greater contrast than that of the original beam. So far, it has not been possible to produce fluoroscopic screens which have as high a contrast, if the term may be used loosely, as x-ray film. The screen maker must produce as high a light intensity as possible, which normally means a reduction in possible contrast, an increase in the screen's grain size, and a net loss in sensitivity. Economically it is not feasible to produce x-ray equipment with sufficiently high currents through the x-ray tube to enable the screen manufacturer to compromise with the requirement for high brightness levels.

One advantage of the fluoroscopic technique must not be overlooked. If the operator is allowed to move the piece as he views it, he has the equivalent of an infinite number of radiographs. If, for example, the object is to locate a crack in a part, it is almost impossible either by radiography or fluoroscopy to detect this crack as long as the x-ray beam is directed perpendicularly to the plane of the crack. However, if the part can be rotated so that the x-ray beam is directed at the crack longitudinally, the sensitivity for this particular view need not be extremely good to detect the flaw. It is unlikely that with one radiograph the crack would be so oriented as to be visible; with fluoroscopic examination the observer is almost certain to locate the defect if the part is moved about.

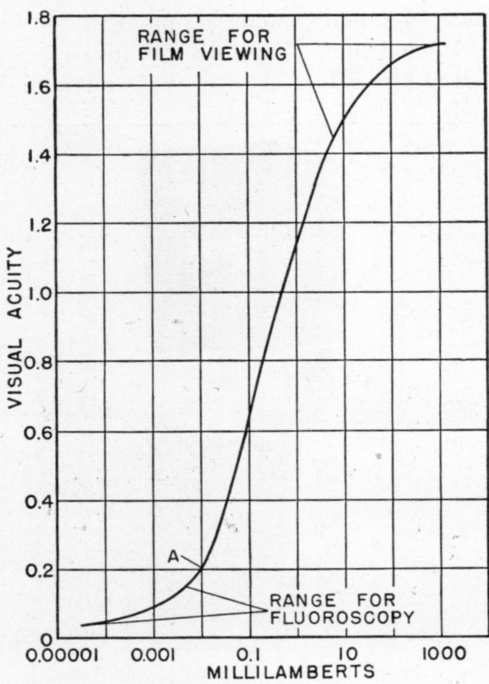

Fig. 27·24 The relationship of visual acuity to illumination. Visual acuity is defined as the reciprocal of the angular distance which must separate two contours in order that they may be recognized as discrete, the unit of separation being 1 minute of arc. This graph is after Hecht.[18] Point *A* indicates the approximate boundary between rod vision (low intensity) and cone vision (high intensity) but is open to question because the change-over does not occur suddenly. (Reprinted with permission from W. E. Chamberlain, *Radiology*, Vol. 38, No. 4, p. 405, April 1942.)

27·6 INDUSTRIAL PHOTOFLUOROGRAPHY

For many applications sensitivity as great as that given by radiography is not required, but the sensitivity of fluoroscopy is not sufficiently good. In this range a relatively new technique can be applied. It may be considered as an intermediate point between radiographic and fluoroscopic examination.

The new technique is photofluorography. It requires all the essential components of a fluoroscopic unit, except that the light image on the fluorescent screen is then photographed and the image is produced on a small film. In reducing the image to a small size, some sensitivity is lost even though the film is magnified for viewing.

Essential components of this system are an x-ray tube, a lead diaphragm for confining the beam to the prescribed area, a viewing field for the parts to be examined, a fluorescent screen, a lead-glass back (which prevents transmittal of x-radiation to the photographic film but passes light rays), and the lens system and camera to reproduce the fluorescent image on the small film. Figure 27·25 illustrates a unit of this type. Roll film can be used in the camera, and the film can be automatically advanced one frame at the time the exposure is terminated. Thus an unexposed film is in place, ready for each exposure.

FIG. 27·25 Photograph showing an industrial photofluorographic unit.

The unit has some of the advantages of radiography because the intensity of the image can be built up by prolonged exposure time. The resultant image may then be magnified and a sufficiently intense light source placed behind it so that viewing may be at a reasonable intensity level. In addition, a permanent record is obtained.

Obviously, the factors controlling quality of the image are similar to those in radiography or fluoroscopy, plus the additional factors of lens aberrations and the grain size of the recording film. Grain size of the film becomes important because of the reduction of the image and its magnification when viewed. Exposure times are necessarily longer than those used for direct radiography because of the less efficient fluoroscopic screen and the lens system. An exposure which would take approximately 1 minute for radiography would require 1¼ minutes with this type of equipment. Sensitivity is from 3 to 7 percent.

If a large quantity of work is to be done, the economy of this equipment, in many cases, justifies the decrease in sensitivity. It may be possible to take several views of a part and still have a more economical examination than with full-sized radiographs. Added to this is the additional ease of film handling. Systems have been worked out for processing large rolls of film quickly and easily, whereas processing individual flat films is long, tedious, and costly.

27·7 NON-VISUAL RECORDING

So far, only two means of recording the x-ray pattern have been suggested, film and fluorescent screen. But some characteristics of the x-ray beam besides its ability to affect a photographic film or a metallic salt crystal can be used for recording intensities.

Electrical Resistivity

X-radiation affects the electrical resistivity of various materials such as selenium. This change of resistivity can be correlated with x-ray intensity to give a method of recording intensity that is useful if high sensitivities are not required. For example, powder-filled blasting fuses must contain a continuous line of powder or the fuse fails. The x-ray intensity reaching the selenium cell is much different when the powder train is continuous and when the fuse contains no powder.

Ionization

X-radiation also has the power to ionize gases, and this process results in a current in that gas. The extent of ionization depends upon the x-ray intensity, and the degree of the ionization determines the current in the gas, since the number of ions determines the quantity of charge which can be conducted. The ionization method of recording x-ray intensity is a process of measuring electrical current or a quantity of charge. This method is used as the basic standard for the determination of x-ray intensity.

If an electric field is applied to a gas which is subject to ionization by x-rays, positively charged ions travel to the negative plate, and negatively charged ions travel to the positive plate. If the potential applied to the plates is zero, there is no resultant current; the ions simply recombine in the gas. As the potential is increased, the current for the same amount of radiation gradually increases until finally at some potential no further increases give an increase in current. This current value is called the saturation current.

The explanation for this characteristic is that at low potentials the ions tend to recombine. As potential is increased, the speed of the ions to the collecting plate is increased until finally at the potential causing saturation current there is no recombination, and all the ions produced travel to the collecting plates. If the potential is carried to an extreme the gas breaks down electrically and an arc is formed.

More practical equipment to measure the ionization produced by x-rays consists simply of a fine wire along the axis of a cylinder made of a good conductor of electricity. The interior of the cylinder is sealed at both ends and is filled with some gas, usually at atmospheric pressure.

The usual procedure is to charge this cylinder to a high potential. The insulation between the axial wire and the conducting cylinder must have high leakage resistance, because the capacitance in this cylinder is so small. The cylinder is then subjected to x-radiation, and the amount of charge which has leaked off through the ionized air is measured by means of an electrometer. Some instruments actually measure the ionization current produced and give a continuous reading.

If the voltage is carried higher, the current again increases after a plateau at the saturation value until electrical breakdown occurs between the two plates and an arc is formed. The explanation for this increase in current is that after a molecule is broken up into a heavy positively charged particle and a negatively charged electron at low potential the electron travels toward the positive plate and finally collides with an atom to which it becomes permanently attached, forming a negatively charged ion. If the electrical potential is sufficiently high, however, the electron may have sufficient energy at the time of collision with the neutral atom to cause a second ionization. This is ionization by collision.

Partial evacuation of the cylinder has two advantageous effects. First, the electrical potential required to cause an arc discharge between the plates becomes higher; that is, the dielectric strength is greater for a high vacuum than for atmospheric pressure. Second, the mean free path of the electron becomes longer, and a lower voltage between the plates can cause ionization by collision. Also, proper choice of the gas in the cylinder gives higher ionization currents.

Using this method, physicists have built equipment so sensitive that it can count single photons of x-ray energy. Both the proportional counter and the Geiger-Mueller counter can count this small energy level. The essential component of both these counters is a fine wire surrounded by a metallic cylinder Usually the assembly is enclosed in an evacuated glass envelope, and the amount and kind of residual gas is chosen with great care.

In the proportional counter, ionization by collision occurs close to the central wire. If a single photon of energy ionizes one molecule of gas, one electron is accelerated toward the central wire which carries a positive potential; on the way it collides with many particles. Not until this electron is close to the central wire is the electric field of sufficient intensity to accelerate the electron enough to cause ionization by collision. This single electron creates two, four, eight, and so on, causing a "Townsend avalanche." Thus, a tremendous current is produced by a single original ionization. If the photon has enough energy to increase the original ionization by two or three times, the current from the more numerous avalanches is proportionally greater. Therefore, the current caused by any one pulse is proportional to the energy of the original ionizing radiation. As soon as all the ionized particles in the avalanche are collected by the central wire and the enclosing cylinder, the discharge is extinguished.

If the potential on the central wire is increased, the violence of electron impacts is increased until finally the energy produced by the impacts generates short-wave radiations which ionize the gas at other points and initiate other avalanches. The process then would continue indefinitely, except that the positive ions, being heavy, do not move rapidly, and they collect in a sheath or cloud around the central wire and eventually diminish the electric field until ionization by collision ceases. Current and the duration of the discharge in the Geiger-Mueller tube do not depend upon the strength of the original ionizing radiation. Once the process is started the characteristics of the discharge are determined by the construction of the tube itself and the potential applied. Over a wide range of voltage the value of current and its duration are independent even of voltage variation.

Figure 27·26 illustrates the operation of these two devices. The sloping straight-line portion of the curve is the operating range for the proportional counter, and the plateau is for the Geiger-Mueller counter. Obviously the Geiger-Mueller counter is relatively insensitive to voltage changes, but the proportional counter is sensitive to them.

External circuits must be provided to record the number of pulses produced by these counters. Such circuits must be capable of counting pulses at the rate of thousands and even millions per second. However, the counting rate is not limited by external circuits but by the extinguishing time for the counters. Fast counters with counting rates of tens of thousands per second have been produced, and proportional counters have been produced with even higher counting rates, because their extinguishing time is inherently much shorter than that of the Geiger-Mueller counter.

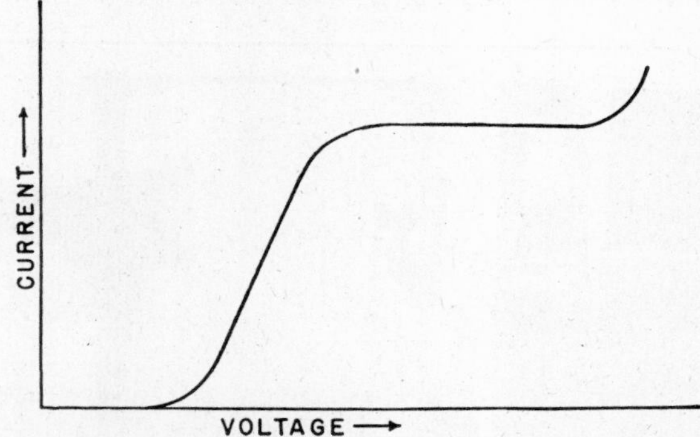

Fig. 27·26 Characteristics of a Geiger-Mueller tube. No absolute values of voltage or current are placed on the axes since they are very much a function of the tube design.

Since production of x-radiation and its absorption in gas is a random process, many individual counts must be made to ascertain the intensity accurately. Readings averaged for a total of 10,000 counts give the intensity with a probable error of 1 percent. To work to a probable error of 0.1 percent, 1,000,000 counts must be made.

If only one spot of predetermined location need be examined, this method of recording is excellent. It has been used, for example, in the examination of hand grenades to detect "duds." If there is no powder inside the hand-grenade shell, the difference in intensity between this grenade and others which are filled is detected by a Geiger-Mueller tube and the dud is rejected. It is possible also to scan an area with a small pencil of an x-ray beam, the intensity of which is recorded on a recording meter as the area is traversed. Fluctuations in the recording current trace the variation in density or thickness of the material. Scanning requires the highest counting rates, particularly if the flaws to be detected are small. The x-ray spot used to scan the area must have about the same dimensions as the flaw to be recorded. For large areas and small flaws the counting rate must be tremendous to make the inspection time of reasonable duration.

This means of recording is attractive because it can eliminate the use of the human eye as a recording medium. An entire inspection operation can be made mechanical and automatic; consequently the only attention necessary is that of maintenance and repair.

Multiplier Photocells

Electronic tubes utilizing electron-multiplier sections immediately following a photo-emissive material give tremendous gains in output currents compared with those available a few years ago. Tubes of this type have made it possible to work with extremely low light intensities and still produce reasonable current outputs. These electron-multiplier phototubes can be used with fluorescent screens to indicate x-ray intensity.

They are being used successfully at present for automatic control of exposure in medical radiography. For some uses this device may supersede the counters of both the proportional and Geiger-Mueller types, and may be used for many applications for which these counters are impractical.

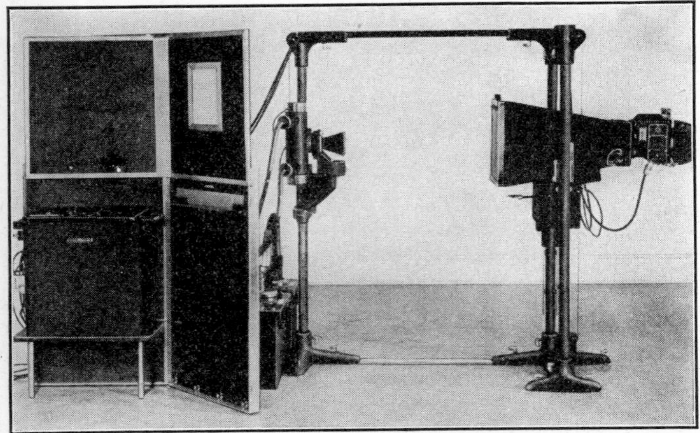

Fig. 27·27 A modern photofluorographic unit for producing chest x-ray images on 70-millimeter film. At left is the x-ray tube. In the large end of the hood at right is a fluorescent screen which is photographed by the camera attached to the small end of the hood at extreme right. The small box mounted on the bottom of the hood encompasses a multiplier-type phototube on which is focused by a lens system a portion of the light from the fluorescent screen. When the total amount of light received by the phototube reaches a value sufficient to produce a predetermined density on the film, the exposure is terminated by removing voltage from the x-ray tube.

A phototube of this type always has an inherent instability; first, because light striking a photosensitive surface causes the emission of electrons, which is an unpredictable and often erratic process. Second, this electron emission is followed by several stages of electron multiplication. The original electrons are accelerated by an electric field, and upon striking a sensitized surface they cause the emission of larger numbers of electrons. This also is an erratic process. Because one such tube uses nine stages of electron multiplication, the errors in it are raised to the ninth power. A drop of 2 percent in sensitivity of the electron emissive surfaces then reduces tube output by 17 percent.

Several schemes might be used to stabilize this tube, and some are successful now. One successful system consists of a standard light source connected so that it is lighted on alternate half cycles of voltage. The fluorescent screen is lighted on the other half cycles. During the time the tube is excited by the standard lamp its output is used to vary the accelerating voltage applied to the multiplier dynodes, and thus sensitivity is held constant.

In medical photofluorographic work this tube is used to control the length of the x-ray exposure to produce constant density on the film. Figure 27·27 shows a unit with this control. Since it is designed for the production of miniature films of the chest only, a portion of the light from a certain area of the fluorescent screen may be focused by a lens system on the phototube. This area of the screen always has the same part of the chest appearing on it. Then the current through the phototube charges a capacitor which fires a thyratron as a predetermined voltage is reached. The phototube holds the product of current and time constant. Since the current is proportional to light intensity on the fluorescent screen the total light emitted by the screen during one exposure is held constant, and the exposure is terminated at the proper time. Variations of as much as 10 percent in exposure time are not visible in the final films; therefore extremely accurate stabilization is not necessary. With this timer many patients of all sizes can be examined rapidly without making changes in x-ray control settings. The timer automatically controls exposure time to produce constant density on the films regardless of patient size.

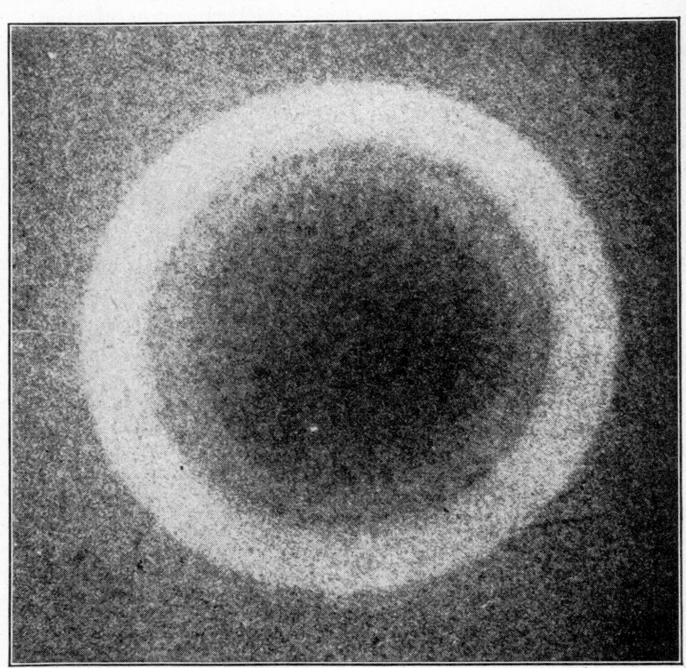

Fig. 27·28 Enlargement (10×) of sound 24-ST Alclad spotweld, 0.040 × 0.040 inch; radiographic tube; 20 kilovolts, 20 milliamperes; 78 seconds, 24-inch distance. (Reprinted with permission from Scott, Sutton, and Widmyer, *Welding J.* (*London*), *Research Supp.*, Nov. 1944.)

27·8 MICRORADIOGRAPHY

Improvements in spot-welding thin sheets of aluminum have enabled designers and producers of aircraft to substitute this quicker and neater method for riveted construction. X-ray equipment operating at 10 to 70 kilovolts is used to inspect these spot welds.[21]

The examination of spot welds in aluminum is one of many examples of a field which is being recognized as the large field of microradiography. Many metals can be reduced to thicknesses of 0.080 inch or less and examined by x-radiation. The image is produced on a film that can be enlarged and viewed so that the finest details of the structure of the thin sheet of metal are visible.

By this method the distribution of various component metals of an alloy can be determined. A piece of metal a few thousandths of an inch thick is placed in contact with fine-grain film. Over the top is placed a piece of lead foil with a central hole about ¼ inch in diameter. Ordinarily the metal to be examined is in contact with the film, and the light-tight envelope is placed around the outside. The x-ray exposure produces a small spot image on the film, which may be magnified by viewing through a microscope.

Because of the difference in absorption characteristics of the various metallic components of an alloy, the shadows in this small image clearly locate the component parts. It is possible to determine whether the various materials are in intimate mixture or whether there is some separation or segregation.

This microradiograph accomplishes somewhat the same results as a photomicrograph. However, the microradiograph shows the actual structure of the metal through a layer of finite thickness instead of the surface condition shown by a photomicrograph. Also, a large part of the long and tedious polishing and etching process necessary for the photomicrographic technique is eliminated.

It is also possible to filter the x-ray beam to obtain a monochromatic radiation of a previously chosen wavelength. Then, since the various metals have discontinuities in their characteristic absorption curves, the wavelength can be chosen in a range which gives a large difference in the absorption characteristics of the various constituents of the alloy.

This technique has become valuable in foundry control. Thin slices of defective parts of castings may be cut and examined by this method to give a more complete picture of the type and cause of the defects.

27·9 EXAMINATION OF FINE STRUCTURES

At the boundaries of light shadows cast by small obstacles, there are variations in the intensity of the light caused by the interference effect. If light is reflected from a surface containing ruled lines, with the spacing between them about equal to the wavelength of the light, incident white light produces a pattern of reflection which looks like a series of small rainbows. This pattern results from interference between the wavelengths of the original white light.

For years experimenters attempted to prove that x-radiation performed in the same way. None of these investigators was successful, however, because none was able to produce a diffraction grating with the ruled lines sufficiently close together.

In 1912, Professor M. Laue of Munich suggested that a crystal might serve as a solid diffraction grating for testing x-radiation. The orderly rows and columns of atoms in a crystal perform the same task as an optical diffraction grating, because the spacings between the atoms is roughly the same as the x-ray wavelength. The difference is that the x-ray diffraction grating is three-dimensional and the optical grating is two dimensional. In 1913, Friedrich and Knipping successfully tested Laue's theory and proved him to be correct.

The original analysis by Laue was complex; however, the Braggs were able to reduce it to a rather simple law:

$$n\lambda = 2d \sin \theta$$

where n = any integer
λ = wavelength of radiation
d = distance between crystal planes
θ = angle between crystal planes and radiation.

Figure 27·29 illustrates the meaning of the Bragg law. The incident radiation is represented by line A at an angle θ

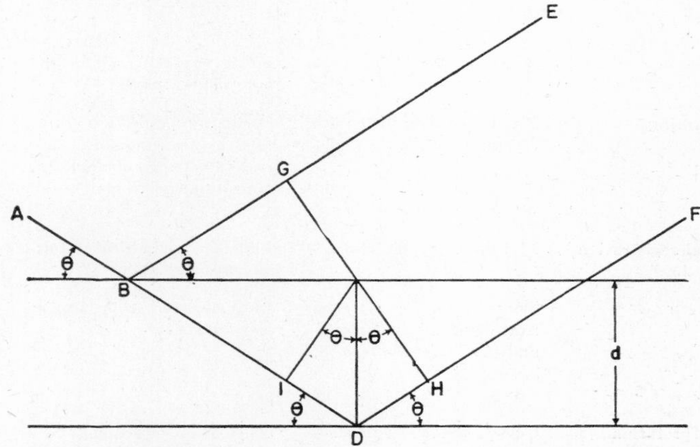

FIG. 27·29 Schematic diagram showing the derivation of the Bragg law.

to the surface of the crystal. At points B and D the x-ray beam strikes successive atoms. Reflection from atom B can be represented by line E, and the reflection from atom D by line F. Line GH is perpendicular to the path of the radiation E and F. If radiation at points G and H is to be in phase, line BG must be shorter than line BDH by one wavelength or an integral multiple of wavelengths. Line BDH is longer than line BG by a length equal to $2d \sin \theta$. Therefore, the condition that there shall be a reflected beam in the direction of these new lines is that the angle θ shall be governed by this equation.

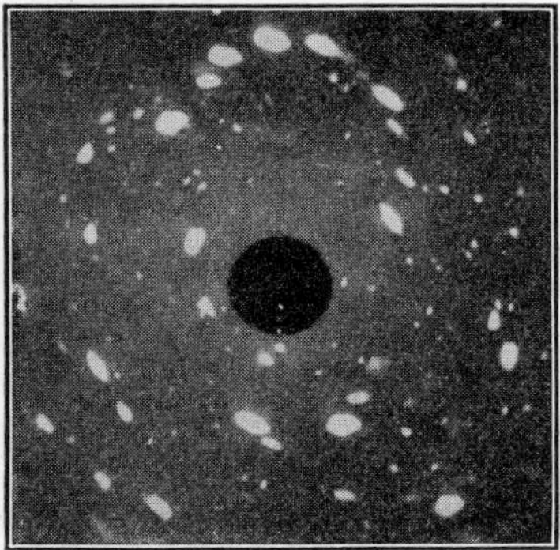

FIG. 27·30 Symmetrical Laue photograph of an iron crystal. (Reprinted with permission from *Applied X-Rays*, by G. L. Clark, published by McGraw-Hill, 1932.)

If the angle between the crystal and the radiation is changed, there will be a number of points where reflection will occur, corresponding to the angles at which the reflected radiations arrive in phase. If a film is exposed to the reflected radiation several dark spots will be produced on it if a pencil of the x-ray beam is used. If the wavelength of the radiation is known, the distance between the atoms of the crystal can be computed from the spacing of these spots or, if the distance between the atoms of the crystal is known, the wavelength of the radiation can be computed.

Many different techniques and types of equipment are used in various phases of this work. Table 27·3 gives in condensed form the salient points of the five most important techniques.

In general, the equipment utilizes an x-ray tube with its anode at ground potential, so the anode can be water-cooled and intense x-ray beams can be generated. Either a thin pencil or flat slit of beams is used. For some purposes the x-radiation as it comes from the tube can be used. Radiation coming from a tube covers a wide range of wavelengths and must be considered, by analogy with the visible spectrum, to be polychromatic. For other purposes monochromatic beams must be used, and they usually are produced by using the proper target material for the tube and by placing filters in the x-ray beam. Every metal has a characteristic radiation which is more intense than the over-all or general radiation level. By selecting a tube target, the characteristic wavelength of which is the wavelength required for the analysis, the emergent x-ray beam has an intense

TABLE 27·3 EXPERIMENTAL DIFFRACTION METHODS [9]

Method	Laue	Bragg	Rotating Crystal (Schiebold-Polanyi)	Powder (Hull-Debye-Scherrer)	Monochromatic Pin Hole (Fiber)
X-ray beam	Polychromatic	Monochromatic	Monochromatic	Monochromatic	Monochromatic
Beam definition	Pin hole	Slit	Pin hole usually; also wedge with knife edge on crystal face	Hull, slit; Debye-Scherrer pin hole	Pin hole
Specimen	Single crystal	Single crystal or film of oriented molecules	Single crystal	Powder or random aggregate	Any—fibers particularly
Mounting	Fixed according to definite crystal direction	Oscillation; reflection from face or transmission; successive settings of angle in ionization spectrometer	Rotated or oscillated over fixed angle around principal axis; mounting on goniometer head for proper orientation	Fixed, in small tubes, threads, wires, ribbons, etc.	Fixed with fiber axis oriented over pin hole and beam transmitted perpendicular
Usual registration	Flat photographic film	Flat or cylindrical film; ionization chamber	Flat or cylindrical film with crystal at center	Narrow film bent in arc with specimen at center	Flat film, perpendicular to beam
Pattern	Symmetrical spots each from different set of planes and particular value of d, λ, θ	Line spectrum from single set of planes	Diffraction spots lying on layer lines, parallel horizontal with cylindrical film, and hyperbolas with flat film	Line spectrum, each line corresponding to different sets of planes	For random aggregate, concentric rings or diagram 360 degrees in azimuth; for fibers, layer line pattern like rotating single crystal
Interpretation	Calculation of spacings according to spots; assignment of indices with assistance of stereographic or gnomonic projections; estimation of relative intensities	Calculation of spacings for set of planes involved in particular orientation ($n\lambda = 2d \sin \theta$) and thence from three experiments size of unit crystal cell; determination of missing orders, etc., in aiding structure factor	Measurement of identity period from layer lines by $I = n\lambda/\sin \mu_n$ and exact size of unit cell; straightforward indexing of diffraction spots on layer line; comparison with space-group criteria	Calculation of spacings for lines. In simpler cases assignment of indices from $\sin 2\theta$ data and unit-cell dimensions; measurement of line breadths for particle size	Same as for powders. Identity period along fiber axis from $I = n\lambda/\sin \mu_n$; orientation of planes in all fiber particles from positions of interference maxima on Debye-Scherrer rings
Chief uses	Symmetry, indices, and intensities for assignment of space-group. Practical determination of orientation as of large grains with respect to surface of sheet	Occasionally used to determine or check lattice spacing; for all films of long-chain carbon compounds	Commonly used for determining uniquely crystal structure and constitution where single crystals are available	Used in the great majority of cases (single crystals unavailable) for crystalline structure, allotropy, qualitative and quantitative analysis, purity, grain size from line breadths, etc.	Same for powder or random aggregates; determination of actual state of any specimen, such as degree of fibering, internal strain, etc., and the effect of any process such as working or heat treatment; thus for the control of industrial processes and as a method of specification
Modifications	Unsymmetrical patterns	....	Special goniometers (Weissenberg, etc.)	Bohlin-Westgren method has slit, flat sample, and film on same circumference permitting focus and rapid exposure by reducing absorption of rays in specimen to a minimum	Cylindrical film with axis perpendicular to beam or coaxial with beam; reflection from surface at fixed angle; back reflection

Reprinted with permission from *Applied X-Rays*, by G. L. Clark, published by McGraw-Hill, 1932.

Study of diffraction effects on crystals has shown that almost every solid substance has some crystal characteristics. There are only a few fundamental types of these crystal building blocks. For any compound or element there is a characteristic arrangement of atoms which is like that of no other material. The variables are the number and placement of the atoms in the fundamental crystals, the interatomic distances, and the crystal angles. By studying the diffraction pattern of a material its composition can be analyzed exactly. Card files giving the diffraction patterns for thousands of materials have been made available for use as standards in the identification of materials.[9]

component of the desired frequency along with a general radiation and additional characteristic radiations of lower intensity. By selecting filter materials with absorption characteristics favoring the transmission of the desired characteristic wavelengths, nearly all the radiation except the one wavelength desired can be eliminated.

It is also possible to detect the presence of residual strain in materials. Residual strain is necessarily accomplished by a shifting of atomic planes inside the crystals. Diffraction patterns of strained material show a characteristic difference from the normal diffraction pattern, and it is even possible to give an indication of the amount of strain.

With special techniques materials in powder form can be analyzed, even though large single crystals may not be obtained. In a powder many small crystals are arranged heterogeneously. If there is a preferred orientation of these crystals (if a majority are aligned in one direction) the diffraction pattern may show it also.

Preferred orientation is often important, since the characteristics of the material may depend upon the angle at which they are measured with respect to the direction of preferred crystal orientation. This fact is extremely valuable in studies of certain plastic materials.

27·10 PHOTOCHEMISTRY

Photochemistry is ordinarily outside the limits of the engineering field. It appeals to the theoretical physicist because it throws light on the interior atomic structure. However, for the engineer, only a general understanding of it is necessary. Physicists may say that the information in this section is not quite accurate, but it is sufficiently accurate for engineering purposes.

If an x-ray photon strikes an atom, the photon may be deflected and may change its direction without any effect on the energy content of either the photon or the atom. But if the atom absorbs the energy it causes extinction of the photon. The energy possessed by the photon may show up partially as heat generated within the atom itself. Another possibility is that one of the electrons close to the nucleus of the atom absorbs enough energy from the x-ray photon to escape the confines of the atom and become a free electron, leaving a space in the atomic structure which is filled by one of the electrons in the outer energy layer. As this electron falls into the space previously vacated, two things happen: first, a quantum of radiation with a wavelength characteristic of the particular atom is emitted; second, the atom becomes a positively charged ion by having lost the charge of one electron, and can then enter readily into chemical reactions.

27·11 MEDICAL X-RAYS

By far the largest application of x-ray equipment is in the medical profession. In general the equipment can be divided into two groups, diagnostic and therapeutic. Diagnostic equipment is that type of equipment which enables the doctor to view the internal parts of the human body and locate points of trouble. As in the industrial field, there are three general classes of techniques: radiography, fluoroscopy, and photofluorography. Radiotherapy is the treatment of human ailments with radiation.

27·12 MEDICAL RADIOGRAPHY

In medical radiography the object is to produce as clear a stationary view as possible of a part of the human body. Equipment must have a small focal spot, sufficient penetrating power, and a sufficient quantity of radiation to obtain a radiograph in a reasonably short time. Normally the voluntary and involuntary movement of body parts and organs must be considered in determining proper exposure time. This requirement has led to the development of equipment which generates a large quantity of radiation in a short time. As an example, the exposure for a chest radiograph must be of such short duration that the movement of the heart does not show on the finished radiograph. Exposures ranging from $\frac{1}{60}$ to $\frac{1}{10}$ second with tube currents of 200 to 500 milliamperes and voltages from 40 to 100 kilovolts are normal.

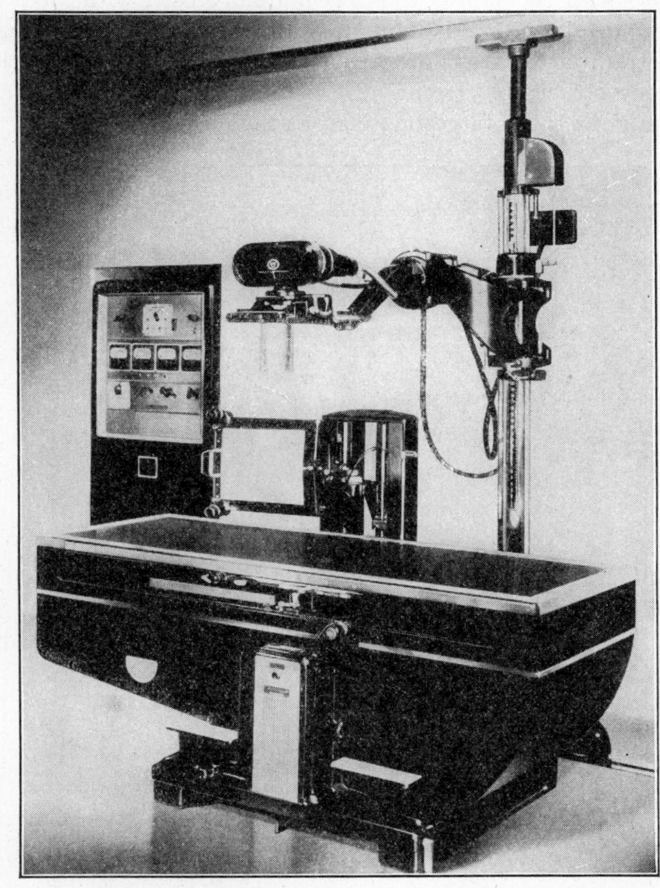

Fig. 27·31 A modern medical diagnostic unit using a rotating anode tube above the table for radiography and a stationary anode tube inside the table for fluoroscopy. The fluoroscopic screen is shown folded back out of the way leaving the unit set up for radiography. Table tilts to allow proper positioning of the patient.

Size of equipment must be minimized and must even be reduced below that acceptable for industrial equipment. Location of the equipment in office buildings and hospitals usually requires that it be small and light in weight. Furthermore, operation of the equipment by female technicians is common, and the equipment must be simple and light for their use.

In the most economical units, self-rectification is normally employed. In the more expensive equipment, bridge rectification is used. The only purpose in using circuits more complicated than those with self-rectification is to subject the x-ray tube to higher loadings. In industrial work, tube loading is limited by the amount of heat that can be conducted away from the tubehead by some cooling medium. However, with the short exposures normally used with medical equipment, loading is limited by the rate at which heat can be transmitted from the surface back into the body of the anode structure. If this rate of heat transfer is not sufficiently fast, the target material at the focal spot melts.

FIG. 27·32 A modern medical deep-therapy x-ray unit rated at 220 kilovolts constant potential.

In a self-rectified circuit the x-ray tube is never operated with the anode at greater than incandescent heat, and practically must never be allowed to approach this heat. As soon as the anode becomes incandescent it is a source of electrons, and the tube conducts a stream of electrons from the anode to the cathode structure and destroys itself. With a half-wave rectified circuit using one or two high-voltage rectifier tubes in series with the x-ray tube, the x-ray tube can be operated to the point at which the target reaches its melting point.

There is still further advantage in using a full-wave bridge-rectifier circuit. The x-ray output is a function of the average current through it. For the same average current, the peak current for a half-wave circuit must be twice that for a full-wave circuit. For the same rate of heat conduction into the anode body, the surface temperature at the target is higher at the current's peak for half-wave circuits than for full-wave circuits.

As an example of the performance that can be expected from an x-ray tube in the three types of circuits, consider the possible loading on an x-ray tube in a self-rectified circuit to be 100. Then the load the same tube can carry in a half-wave rectified circuit is about 200. The same tube operating in a full-wave rectified circuit can be loaded to about 250.

Short time exposures make possible the use of two devices rarely used in industrial radiography. The first is the rotating-anode x-ray tube. This tube is so designed that the electrons focus on the beveled edge of a large anode, several inches in diameter, which is rotated at high speed while an exposure is made. The tube can carry more load because the rotation of the anode spreads the generated heat over a large area.

The second device is the Bucky grid, or diaphragm, discussed in Sections 27·3 and 27·4. Because of the short exposure times, it is relatively simple to produce linear motion in one direction over the entire period of the exposure, and a marked improvement in the resulting radiographs is made possible.

27·13 MEDICAL FLUOROSCOPY

In fluoroscopy, moving or movable parts of the body that would otherwise be invisible, are studied. Fluoroscopic techniques are not as sensitive as those obtainable with radiography for the reasons outlined in the discussion of industrial fluoroscopy. It is valuable, though, because organic defects observable only during the movement of body organs can be seen, or internal changes can be observed during an induced movement of body parts or medical instruments otherwise invisible when in position. Frequently, radiopaque liquids are introduced into body cavities to outline contours otherwise invisible.

As a general rule, the same equipment is used as for radiography. The foregoing discussion of permissible x-ray-tube loading does not apply since the exposures for fluoroscopic work must be relatively long. Therefore, the requirements of the equipment for fluoroscopy are different from those of radiography, but it is usually economical to use one set of generating equipment which satisfies both requirements.

Limited patient tolerance to x-ray dosage requires a radiologist to work with low x-ray-tube currents and, consequently, fluoroscopic screen brightnesses lower than those obtainable in industrial work. The radiologist is careful that the patient is not given an excessive x-ray dose; he uses his x-ray equipment to cover as small a field as possible and keeps fluoroscopic exposures as short as possible.

27·14 MEDICAL PHOTOFLUOROGRAPHY

By routine x-ray examination, tuberculosis is detected early in its development while the disease may still be positively cured. The production of full-sized radiographs in the larger quantities needed for mass examination of this sort was prohibitively expensive until the photofluorographic type of equipment, discussed in Section 27·6, was adopted. With this equipment many people can be examined at a fraction of the cost and time required for full-sized radiographs. By photofluorography the suspicious cases can be detected, and the patients can return for full-sized radiographs and subsequent diagnosis.

27·15 THERAPY

The radiologist utilizes the fact that some abnormal growths are more sensitive to radiation than are normal tissues. For example, an equal quantity of radiation absorbed by a cancer growth and by normal tissue will eventually kill the cancer tissue, but the normal tissue will recover. Even various normal body tissues react differently to radiation.

The quality of the x-ray beam, as determined by tube voltage, is manipulated to concentrate as much of the energy as possible at the part of the body to be treated. For penetrating deep into the body tissues a high voltage is used, and for surface disorders a low voltage is needed. The patient is irradiated from many angles; the point to be treated is always concentrated on, but the patient is moved between treatments so that the radiation enters the body through different areas on the skin, and therefore the skin dose is kept as low as possible.

27·16 RADIOGENETICS

One of the first things the layman learns about x-radiation is that it has the power of producing sterility in humans, both male and female. This knowledge has caused some unnecessary fear, although great care should be exercised in this respect. Standards for protection have been set up which should be followed.

X-radiation affects the genes and chromosomes and changes their structure to produce mutations in the resultant offspring. Extensive experiments in radiogenetics have been carried on with insects and various other rapidly reproducing forms of life. Mutations have been produced, and have been inherited by subsequent generations from the time of irradiation. As far as is known, no effect of this nature has been observed in human beings.

REFERENCES

1. *Industrial Radiology*, Ancel St. John and Herbert R. Isenburger, Wiley, 1934.
2. "Radiographic Standards for Steel Castings," Bureau of Ships, Navy Dept., Washington, D. C., July 1, 1942.
3. *Engineering Radiography*, V. E. Pullin, G. Bell and Sons, Ltd., 1934.
4. *Industrial Radiology and Related Phenomena*, H. M. Muncheryan, Aircraft X-Ray Laboratories, 1934.
5. "Industrial Radiography with Radium," Technical Staff, Canadian Radium and Uranium Corp., 1942.
6. "Symposium on Radiography," American Society for Testing Materials, 1943.
7. "Radiography of Materials," Eastman Kodak Company, 1943.
8. *Radiologic Physics*, Charles Weyl, S. Reid Warren, Jr., and Dallett B. O'Neill, Charles C. Thomas, 1941.
9. *Applied X-Rays*, George L. Clark, McGraw-Hill, 1940.
10. *The Science of Radiology*, edited by Otto Glasser, Charles C. Thomas, 1933.
11. *Theoretical Principles of Roentgen Therapy*, Ernest A. Pohle and W. Edward Chamberlain, Lea and Febiger, 1938.
12. "Inspection of Metals," Harry B. Pulsifer, American Society for Metals, 1941.
13. "Inspection; Radiographic," Army-Navy Aeronautical Specification No. AN-1-26, Government Printing Office, Sept. 7, 1944.
14. *Radiology Physics*, John Kellock Robertson, Van Nostrand, 1941.
15. "Millionth of a Second X-Ray Snapshots," C. M. Slack, L. F. Ehrke and C. T. Zavales, *Westinghouse Eng.*, Vol. 5, No. 4, July 1945, p. 98.
16. "High Speed X-Rays," C. M. Slack, E. R. Thilo, and C. T. Zavales, *Electronic Industries*, Nov. 1944, p. 104.
17. "Fluoroscopes and Fluoroscopy," W. Edward Chamberlain, *Radiology*, Vol. 38, No. 4, April 1942, p. 383.
18. "Relation between Visual Acuity and Illumination," by Selig Hecht, *J. Gen. Physiol.*, Vol. 11, 1928, p. 255.
19. "Gamma-Ray Radiography and its Relation to X-Ray Radiography," N. L. Mochel, X-Ray Symposium of American Society for Testing Materials, 1936, p. 116.
20. "Applications of Geiger-Mueller Counters to Inspection with X-Rays and Gamma-Rays," Herbert Friedman, Herman F. Kaiser, and Arthur L. Christenson, *J. Am. Soc. Naval Eng.*, Vol. 54, 1942, p. 177.
21. "Radiography for Development and Control of Aluminum Alloy Spot Welding," G. W. Scott, Jr., L. F. Sutton, and J. H. Widmyer, *Welding J.* (*London*), *Research Supp.*, Nov. 1944.
22. "Technique and Applications of Industrial Microradiography," G. L. Clark and S. F. Gross, *Ind. Eng. Chem.*, *Anal. Ed.*, Vol. 14, No. 8, Aug. 15, 1942, p. 676.
23. "PE Tube X-Ray Timer," H. D. Moreland, *Electronic Industries*, Jan. 1945, p. 96.
24. "Safety Code for the Industrial Use of X-rays," American War Standard Z54.1–1946, American Standards Association.

Chapter 28

ELECTROSTATIC PRECIPITATION

E. H. R. Pegg

ELECTROSTATIC precipitation had its beginning in 1824 when Hohlfield observed that a corona discharge had a cleaning effect on dust-laden gases. Years later Sir Oliver Lodge and, later still, Dr. Cottrell developed it into a practical form and made successful installations for the precipitation of smelter fumes and other industrial dusts. It has been recognized for many years as the outstanding method of removing very fine solid or liquid particles from gases. This form of precipitator is limited in its uses, however, because its corona discharge generates so much ozone and oxides of nitrogen that the cleaned air is unsafe for human consumption. This result is due largely to the use of high direct potentials of 30,000 to 100,000 volts and high currents.

In 1936 Penney [1] announced the development of a new type of precipitator which was smaller, required less voltage and power, and generated so little ozone that air cleaned by it could be used for ventilation. Such a device, marketed under the trade name Precipitron, made possible some applications of electrostatic precipitation that were not previously practical, such as in the air-conditioning or ventilation systems of commercial and public buildings.

28·1 ANALYSIS OF ELECTROSTATIC PRECIPITATION

Fundamentally the process of electrostatic precipitation consists in first charging a dust particle and then separating it from a gas or air by means of its charge.

Four different things happen when air is passed through a precipitator:

1. The air is ionized.
2. The ions are caused to attach themselves to dust particles; thus they leave an accumulation of charge on each particle.
3. The charged particle is acted upon by the electrostatic field which carries it toward the electrode of opposite polarity.
4. Ozone and oxides of nitrogen are generated.

Separation of Particles from the Air

For reasons which will be apparent later let us skip over for a moment the charging of the particles and consider the separation of charged particles from the air. The motion of charged dust particles in an electric field can be calculated from Stokes's law. If we assume any given charge Q on the particle, a force QE will be acting to pull the particle toward the electrode of opposite polarity. The drift of the particle through the gas toward the electrode is opposed by viscous resistance of the gas. The velocity of drift, therefore, is

$$v_d = \frac{Q\mathbf{E}}{6\pi\mu r} \tag{28·1}$$

where v_d = velocity of drift in centimeters per second
Q = charge on the particle in electrostatic units
$\mathbf{E}$ = electrostatic field strength in statvolts per centimeter
μ = viscosity of the gas
r = radius of the particle in centimeters.

The drift velocity, therefore, is proportional to both the charge on the particle and the strength of the field.

Now consider a uniform electrostatic field between two oppositely charged electrodes of length l and spacing d, through which gas is flowing at a velocity v_g (see Fig. 28·1).

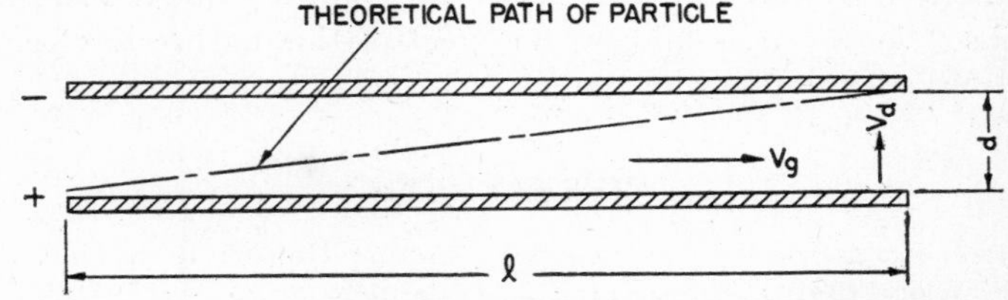

FIG. 28·1 The action of a charged dust particle in a uniform electrostatic field. (Reprinted with permission from G. W. Penney, *Elec. Eng.*, Jan. 1937.)

If a charged particle enters the field near the electrode of the same polarity, the time required to travel the distance l is

$$t_l = \frac{l}{v_g} \tag{28·2}$$

and the time required to travel the distance d is

$$t_d = \frac{d}{v_d} \tag{28·3}$$

In order that the particle shall not escape the field before it is deposited on the electrode, the time t_d must be equal to or less than t_l. Equating,

$$t_d = t_l = \frac{d}{v_d} = \frac{l}{v_g} \tag{28·4}$$

$$v_g = \frac{l\,v_d}{d} \tag{28·5}$$

To secure maximum capacity at lowest cost, v_g should be as large as possible and l as small as possible. Therefore, the velocity of drift v_d should be as high as possible. Since we have seen in equation 28·1 that v_d is proportional to the charge on the particle, it is important that this charge be high. Let us therefore go back and see how to obtain this charge.

Charging the Particle

There are two distinct methods by which a particle may acquire a charge: first, charging by an electrostatic field; and, second, charging by the molecular or heat motion of the ions or electrons.

Field Charging. When a high local electrostatic field is produced by impressing a high voltage on a fine wire, the voltage gradient near the wire is sufficient to produce ionization by collision and a self-maintaining discharge. In air the electrons quickly combine with air molecules to form negative ions; therefore, independent of the polarity of the wire, a unipolar ion stream results. If the wire is positive, the negative ions generated in the discharge are attracted back to the wire; the positive ions follow the lines of electrostatic force and stream toward the opposite or negative electrode.

Dust particles usually have a dielectric constant greater than 1 and distort the field (see Fig. 28·2), tending to concentrate the flux through the particle. The ions, then, following the lines of flux, are driven toward the particle, to which some of the ions attach themselves and cause the particle to become charged. Any charge on the particle exerts a repulsive force tending to repel ions. Therefore, as the charge accumulates, the lines of force are deflected away from the particle as in Fig. 28·2 (*b*). When this repulsive force is great enough to cancel the effect of the main electrostatic field and deflect all the lines of force away from the particle, no more ions can be brought to the particle by the field, and the particle then has maximum charge. This can be expressed by the equation [2]

$$n = \mathbf{E}\left(1 + 2\frac{\epsilon - 1}{\epsilon + 2}\right)\frac{r^2}{e} \qquad (28\cdot6)$$

where n = number of charges on the particle
$\mathbf{E}$ = electrostatic field strength
ϵ = dielectric constant of the particle
r = radius of particle (assumed spherical)
e = electronic charge.

Thus, for field charging, the maximum charge on a particle varies with the field strength $\mathbf{E}$. Since it is important that the charge on the particle be high, it is desirable to maintain the field strength $\mathbf{E}$ in the ionizer as high as possible.

Charging by Molecular or Heat Motion. In the field-charging method, the ions drift along the lines of force with a velocity equal to the product of their ion mobility and the field strength. The ion mobility in air is about 1.6 centimeters per second per volt per centimeter for positive ions. Thus, for a field strength of $\mathbf{E}$ = 1500 volts per centimeter, their velocity is about 2400 centimeters per second. This is only their drift velocity, not their actual velocity.

In addition to the drift velocity the ions move about at high velocity in random directions like gas molecules. This velocity for air at room temperature is about 10,000 centimeters per second. If there are N ions per cubic centimeter surrounding a dust particle, there will be a number of collisions of the ions with the particle, and some of the ions will diffuse to the particle, causing it to become charged. The greater the value of N is, the greater the number of collisions per unit of time and the greater the charge on the particle. The effect of this heat-motion charging becomes increasingly important as the particle size decreases. The mathematical

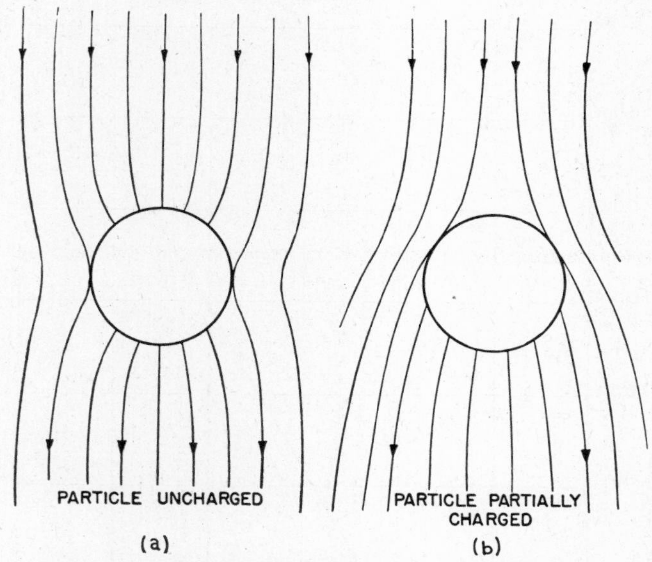

Fig. 28·2 Electrostatic field distorted by dust particle.

theory is too long to include here,[3,4,5] but Table 28·1 shows a few values of n obtained by this theory as compared with that given by equation 28·6. The particle tends to take the maximum number of charges, and it will be observed that, for very small particles of less than a micron, the charge due to heat motion is much greater.

Table 28·1 Theoretical Maximum Charge on a Dust Particle

Radius of Particle (centimeters)	Diameter (microns)	Dielectric Constant	Maximum Charge due to Field (Eq. 28·6, $\mathbf{E}$ = 5)	Maximum Charge due to Heat Motion
0.001	20	1	10,500	
0.001	20	4	21,000	2100
0.001	20	∞	31,500	
0.0001	2	1	105	
0.0001	2	4	210	207
0.0001	2	∞	315	
0.00001	0.2	1	1	
0.00001	0.2	4	2	19
0.00001	0.2	∞	3	

For large particles affected mainly by field charging, then, it is necessary to maintain the field strength as high as possible. For very small particles which depend mainly upon heat-motion charging, it is necessary to maintain a large value of N. The value of N is directly proportional to the current discharge per unit length of the electrode or wire. Furthermore, the current discharge per unit of wire is inversely proportional to the wire diameter for a given voltage and electrode spacing. Therefore, the finer the wire is, the greater the value of N and the greater the efficiency of precipitating small particles.

There are some practical limitations of wire size and current discharge. High currents from small wires tend to cause wire failures. Also, the generation of ozone in the glow discharge is linear with the discharge current as shown in Fig. 28·3. This figure also shows that for a given set of electrodes the rate of generation of ozone for negative ionization

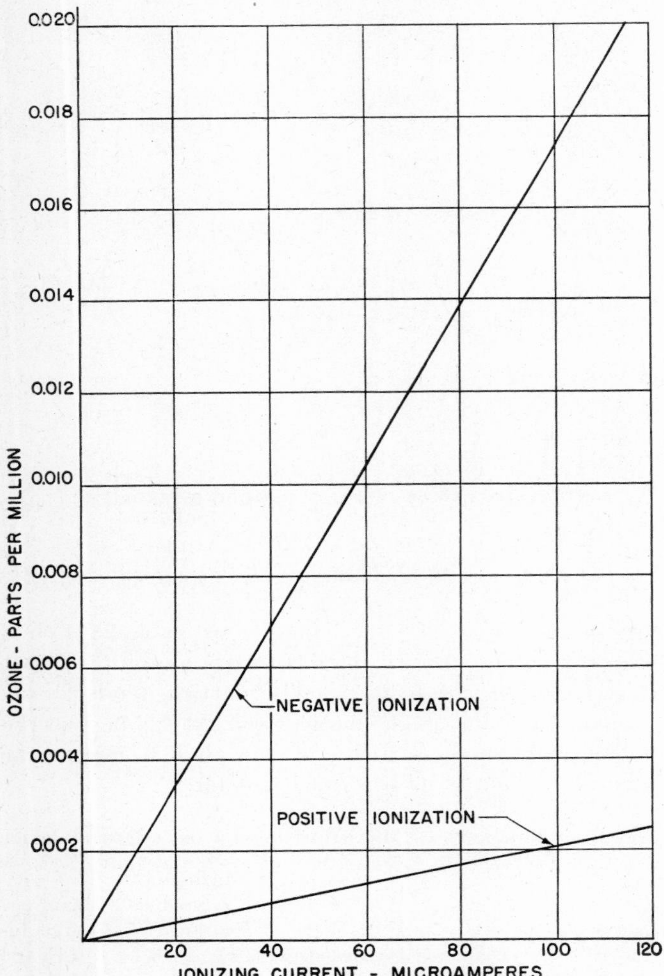

Fig. 28·3 Curves showing relation of ozone generation to ionizing current.

is about eight times that for positive ionization. For this reason positive ionization is used in the Precipitron even though negative ionization gives slightly better efficiency at the same voltage.

28·2 TWO-STAGE PRECIPITATOR

In the conventional precipitator the charging of the particles and their separation from the gas take place in one chamber with a discharging electrode throughout the length of travel of the gas (see Fig. 28·4). This method has been used for many years, and it still has a wide application in the precipitation of industrial dusts of high concentration.

In the two-stage precipitator the charging of the particles and their separation from the gas are accomplished in two separate chambers (see Fig. 28·5). This arrangement of separate ionization and parallel dust-collecting plates was proposed years ago. However, the combination which produces only a small amount of ozone was not discovered until 1933; therefore the use of this construction to obtain a more

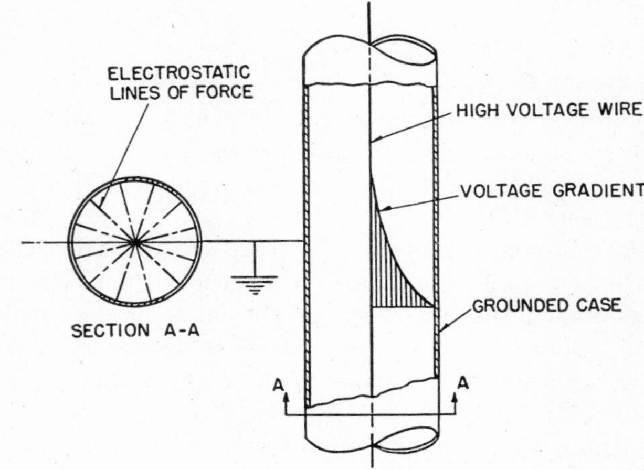

Fig. 28·4 Single-stage pipe-type precipitator.

compact and efficient precipitator for the removal of low concentrations of dust with a low generation of ozone and oxides of nitrogen is a relatively recent development. The charging unit is commonly called the ionizing unit, and the collecting plate assembly is commonly called a collector cell.

To limit the amount of ozone generated, it is essential that the air be exposed to ionization for a short time only. This is difficult to do in an ionizer such as that shown in Fig. 28·4, where the air flow is parallel to the wire. The shortest possible exposure, however, is obtained when the air flow is perpendicular to the ionizing wire, as in Fig. 28·5.

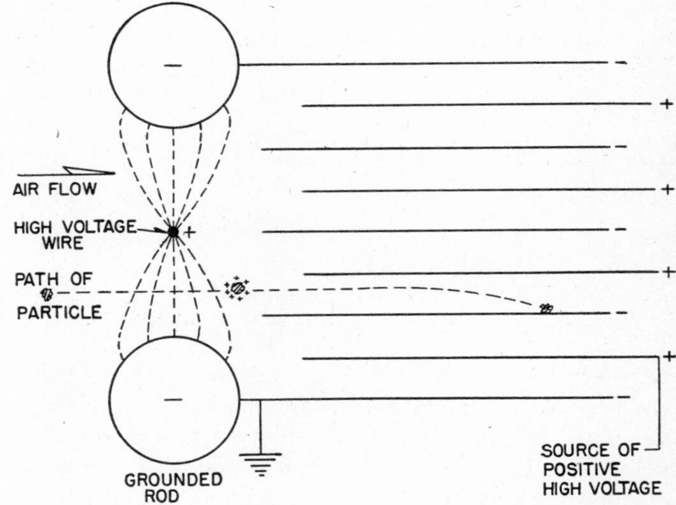

Fig. 28·5 Two-stage precipitator with parallel-plate collector.

A typical ionizing unit is shown in Fig. 28·6, where the high-voltage electrodes, or ionizing wires, are suspended between grounded tubular electrodes. For this type of ionizing unit, all other factors being constant, the ozone production is approximately proportional to the logarithm of the wire diameter. Therefore, from the ozone standpoint, the wire should be as small as possible within the limitations of mechanical strength and resistance to corrosion. With this as a starting point, the other design factors, such as spacing from the ionizing wire to the grounded electrode, diameter of the grounded electrode, voltage on the ionizing wire, and the velocity of air, must be selected from experimental data to charge the dust particles adequately and to keep the ozone generation at a harmless level. The ionizing unit shown in Fig. 28·6 employs wire of 0.0073-inch diameter, 1¼-inch wire-to-ground spacing, and 13,000 volts d-c potential on the ionizing wire. With this arrangement and an air velocity of 300 feet per minute, the particles are charged almost to a maximum, and the ozone concentration is in the order of 0.002 part per million of air.

Fig. 28·6 Typical ionizing unit for two-stage precipitator.

Fig. 28·7 Parallel-plate-type collector cell for two-stage precipitator.

The collector cell to go with this ionizing unit, and designed according to the principles of equations 28·1 to 28·5 inclusive, is shown in Fig. 28·7. This cell employs 6000 volts d-c potential on the charged plates, a spacing of 5⁄16 inch between plates of opposite polarity, and an effective plate length of 9 inches.

The power pack to supply 13,000 volts to the ionizers and 6000 volts to the collector cells has certain inherent design requirements. The d-c output must have less than 10 percent ripple, and must be approximately constant over a rather wide load range. It should be self-protecting against secondary short circuits. It must withstand instantaneous secondary short circuits, such as flashovers in the ionizer or cell without service interruption, but it should interrupt the service on a sustained secondary short circuit. It must have an indicating or alarm system for the indication of tube failure, short circuits, and normal conditions. It must be free from radio interference.

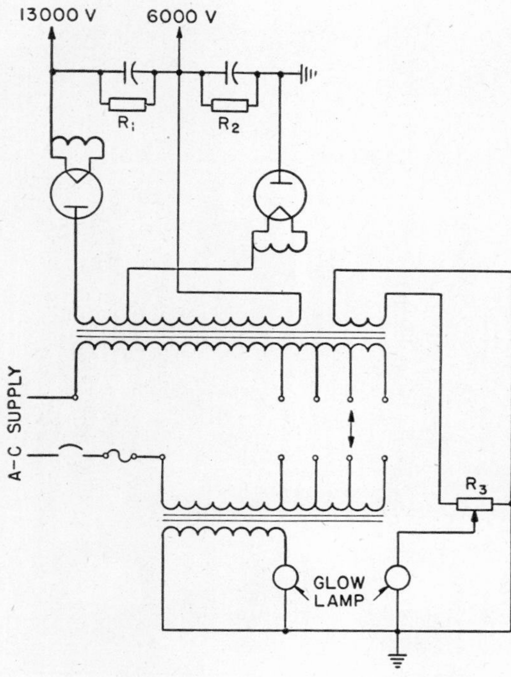

Fig. 28·8 Circuit diagram of power pack for the supply of d-c to two-stage precipitator.

The power pack employs a modified voltage-doubling circuit with two kenotron rectifiers, a high-reactance core-type transformer with magnetic shunt, and a two-section capacitor with built-in discharge resistors. The complete circuit of a typical power pack, including an indicating system, is shown in Fig. 28·8. Secondary load adjustment is obtained by means of primary transformer taps. Short circuits are indicated by a glow lamp energized by a current-to-voltage transformer in series with the transformer primary. For normal operation the lamp is not lighted, but on short circuit the voltage rises beyond the breakdown voltage and lights the lamp. Open circuits are indicated by a glow lamp energized by a tertiary winding on the same leg of the transformer as the secondary. For normal operation the lamp is not lighted, but on open circuit the tertiary voltage rises and lights the glow lamp. A thermal breaker in the primary provides protection against sustained short circuits, but has sufficient time lag to prevent service interruption by an instantaneous flashover. A photograph of this typical power pack is shown in Fig. 28·9.

28·3 EFFICIENCY OF A TWO-STAGE PRECIPITATOR

The equations derived on the basis of maximum charge indicate 100 percent efficiency up to a given critical velocity (equation 28·5) and a rapid decrease in efficiency above that velocity. For several reasons this abrupt transition is not obtained in practice. Particles vary greatly in size. All particles of a given size do not obtain the same charge because the dielectric constant varies, the field is not uniform over the ionizing region, and the shape of the particles varies. The equations governing the charging of small particles depend upon the probable movement of ions, and for small particles the number of ions is too small to give a uniform result. Figure 28·10 shows a typical curve obtained by test under laboratory conditions. This curve shows that the efficiency of a given unit can be increased to almost any desired value by merely reducing the velocity of gas through the precipitator.

Fig. 28·9 Power pack for two-stage precipitator.

The efficiency of a given unit varies widely with the method of test. Usually a weight test, based on the weight of dust removed from the gas, gives a high efficiency, since the larger particles are efficiently removed and constitute a large portion of the weight. A particle-count test, based on the number of particles, regardless of size, removed from the gas, gives a somewhat lower efficiency. With an air filter, for example, it is possible to obtain a weight efficiency of more than 95 percent, and yet a particle count test may show an efficiency of less than 20 percent.

Initially the two-stage precipitator was used for removing the dust, mainly black smoke particles, from ventilating air. In this application the primary interest is the tendency of the air to blacken wallpaper, draperies, and merchandise. This characteristic can be measured by drawing the air to be tested through a piece of white cloth or filter paper at a measured rate. This property led to the development of the blackness or discoloration test. In this test, samples of cleaned air and uncleaned air are drawn through filter papers at equal rates but for different lengths of time as required to obtain the same degree of discoloration on both filter papers. Thus the efficiency can be expressed in terms of the times required to obtain two samples of equal discoloration:

$$\text{Percent efficiency} = \left(1 - \frac{t_u}{t_c}\right) 100 \qquad (28 \cdot 7)$$

where t_u = time of uncleaned air sample
t_c = time of cleaned air sample.

The efficiency by blackness test approximates that by the particle-count test, and can be measured with very simple apparatus and in a short time compared with the complicated apparatus and long time required for a particle-count test. The efficiency curve of Fig. 28·10 was obtained by blackness tests on a precipitator operating on a mixture of smoke and air. A condition giving 95 percent efficiency based on this test would give well over 99 percent by the conventional weight test. Comparable efficiency can be obtained when the precipitator is operating on atmospheric air with no artificial dust added.

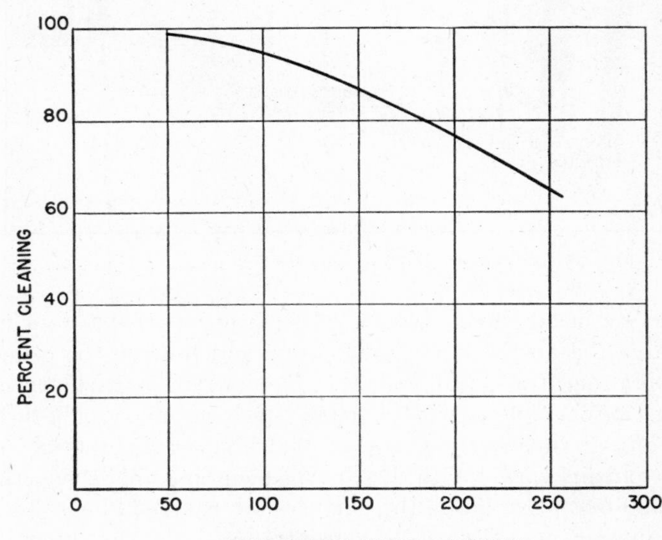

Fig. 28·10 Curve showing variation of cleaning efficiency with rate of air flow through the precipitator. (Reprinted with permission from G. W. Penney, *Elec. Eng.*, Jan. 1937.)

28·4 APPLICATION OF THE PRECIPITRON AIR CLEANER

As indicated in equation 28·5, the velocity of the gas v_g should be as large as possible for economic reasons. However, there is a practical limit to v_g. Once the dust particle has been deposited on the electrode it gives up its charge, and there is no force holding it on the electrode except its own adhesion. The velocity, therefore, must not be high enough to overcome this adhesive force and blow the particle off the electrode. Sometimes this limitation can be overcome by adding an adhesive coating to the electrode surface.

For ordinary purposes the Precipitron is rated at 90 percent efficiency by discoloration test when operated at an air velocity of 300 feet per minute. At 375 feet per minute it is rated at 85 percent efficiency.

At the 300-feet-per-minute rating the cell shown in Fig. 28·7 will clean 1800 cubic feet of air per minute.

For a given air-cleaning installation, the volume of air to be handled is first determined on the basis of ventilating and air-conditioning requirements. When this volume is established, it is divided by 1800 to determine the number of cells required. When this is determined the cells are arranged to fit the space available. Figure 28·11 is a typical installation showing how the cells are stacked up like boxes and bolted together to form a unit.

Fig. 28·11 A Precipitron installation showing the air-discharge side of the collector plates.

One feature of the Precipitron is that it has a very low resistance to air flow. Although this may be an advantage in some respects, it is a disadvantage in that it permits air to pass through too easily, and therefore is subjected to nonuniformity of velocity over the entire unit. For this reason, baffle plates are usually placed just ahead of the Precipitron to add the air resistance required to obtain uniform velocity distribution. The baffles usually take the form of roller-hung doors, as shown on Fig. 28·12, which may be rolled aside for accessibility to the Precipitron.

After a period of operation, usually several weeks, an accumulation of dirt is collected on the Precipitron and must be removed. Ordinarily, the dirt is removed by de-energizing the unit and washing off the accumulation with hot water. Drains in the floor or the bottom of the air duct carry the dirt and water into the sewer.

In some large installations, as in the ventilating system of a steel-mill motor room, the ventilating system cannot be shut down long enough to clean the Precipitron manually. In such places the Precipitron is sectionalized into several small groups so that one section at a time may be taken out of service for cleaning. In addition, automatic cleaning equipment is employed to clean each of the groups automatically at the press of a pushbutton. Such an installation is shown in Fig. 28·13.

Fig. 28·12 Air-intake side of a Precipitron installation with air-distribution baffles moved aside to show the ionizing units.

The Precipitron has been applied as the air-cleaning component of air-conditioning and ventilating systems in numerous commercial establishments such as office buildings, department stores, and restaurants. In such places the savings in redecoration costs, decreased soilage of merchandise, and increased comfort conditions are important.

In many industrial plants where a high degree of precision is required, as in the manufacture of aircraft engines, ball and roller bearings, bombsights, and other optical instruments, this precipitator has contributed greatly to both the quality and the quantity of the product. The textile industry is one in particular in which very large benefits are derived from the use of the Precipitron.

The cleaning of air for the ventilation of rotating electrical machinery and automatic switching equipment is another large field of application. Steel-mill motor rooms, synchronous condensers, and telephone exchanges, to name only a few, have all benefited from really clean air.

FIG. 28·13 A large Precipitron installation with an automatic washer. The washer can be seen behind the column, in position to wash cell group number 4.

FIG. 28·14 A domestic unit for application to warm-air heating systems. Unit has built-in washing system, which can be either manual or automatic, to dispose of the collected dirt.

Domestic units will be available soon for application to a central home warm-air heating system (see Fig. 28·14). Experimental units are now in operation on railroad cars which will make it possible in the future for passengers to smoke in any seat without annoyance to non-smoking passengers. New applications are constantly being suggested and explored as engineers become more aware of the contribution that electrostatically cleaned air can make to the comfort of the individual and to the improvement of many industrial processes.

REFERENCES

1. "A New Electrostatic Precipitator," G. W. Penney, *Elec. Eng.*, Jan. 1937, p. 128.
2. "Purification of Gases by Corona Discharge," Rudolph Von Ladenburg, *Ann. Physik*, Vol. 4, 1930, p. 863.
3. "Movement and Charge of Electrically Charged Particles in a Two Cylinder Capacitor," W. Deutsch, *Ann. Physik*, 1922, p. 335.
4. "The Mechanism of the Electrification of Small Particles in Clouds," P. Arendt and H. Kallman, *Z. Physik*, Vol. 35, 1926, p. 421.
5. "Charging of Particles Suspended in a Corona Discharge," H. Schweitzer, *Ann. Physik*, Vol. 4, 1930, p. 33.

Chapter 29

ELECTRONIC MOTOR CONTROL

K. P. Puchlowski

ELECTRONIC motor control represents one of the newest branches of industrial electronics. Although the first applications of electronic devices (tubes) to control the performance of an electric motor were made in 1930, those early attempts were not too successful from the practical point of view, and did not go much beyond the laboratory stage. However, recent developments and improvements in electronic motor-control systems have created an ever-increasing number of new practical possibilities, and have aroused considerable interest in industrial applications of electronic controls.

The term electronic motor control implies in its broader sense an electrical system consisting of an electric motor, combination of electronic elements (tubes), and a more or less complex network of control circuits. In recent years, however, the most significant engineering progress has been achieved in the field of electronic control of d-c motors, and at the present time these d-c motor systems have a much greater importance and practical usefulness than any of the existing a-c motor systems. Consequently, this chapter will be devoted exclusively to electronic control of d-c motors.

29·1 RECTIFIER DRIVES

Electronic d-c motor controls operate on the principle of rectification; that is, a rectifier converter is always used to provide the direct voltages necessary to drive and properly control a d-c motor. Thus, the term rectifier drive or electronic drive may be used to indicate that the motor is not only controlled electronically, but also that the power to it is supplied through electronic rectifying devices.

D-c motors with shunt excitation are particularly applicable as driving means for different industrial applications because of the well-known fact that speed of such motors can be readily varied by proper control of voltages applied to the armature and field windings.

It must be realized, however, that the advantage of speed-control characteristics of a d-c motor cannot be fully utilized unless an adjustable d-c voltage, to be applied to the armature, is available. On the one hand, the electric power available in about 90 percent of industrial establishments is of the a-c type, and cannot be used directly to drive a d-c motor. On the other hand, even if a d-c supply line were available, it would normally be of a constant-voltage type and, therefore, only field control by means of a rheostat could be accomplished. Speed control below the rated or "base" speed of the motor by means of a rheostat in the armature circuit is not acceptable in industrial applications on account of prohibitive losses, size and cost of the rheostat, and an excessively drooping speed-torque characteristic of the drive.

Thus, in order to take full advantage of a d-c motor characteristic, it is necessary to provide an adjustable-output voltage converter which would convert the a-c line voltage into an adjustable d-c voltage which is applied to the motor. The general trend is to increase the efficiency and speed of various industrial operations. This trend usually requires the application of individual converters to each driving motor in order to provide independent control especially adapted to individual conditions.

The very popular rotating converter consisting of an a-c motor driving a d-c generator, and known as the Ward-Leonard system, is one of the most widely used adjustable voltage supplies.

The rectifier converter, combined with a complete electronic control and regulating system, represents a new method of industrial control. Its prime advantage over the rotating type of converters is that only one rotating machine is used instead of the three, and often four, rotating machines, used in the other systems. In addition, the compactness of the electronic system, the flexibility for adaptation to various conditions, accuracy, and sensitivity are by no means of secondary importance.

Rectifier drives can be divided into two general classes:

1. Systems where the alternating voltage is varied either by means of a variable-ratio transformer, or by means of a saturable-core reactor, and then is rectified and applied to the armature of a d-c motor. Single-phase or polyphase rectifier systems may be used. The rectifying elements are either of the phanotron or rectigon-tube type or, sometimes, of the metallic type, such as copper oxide or selenium rectifiers.

Although drives of this group may have some advantages such as simplicity and low form factor of the armature current, they do not possess all the advantages of accuracy and flexibility of the drives of the second group.

2. Systems in which both rectification and control are performed electronically by means of controlled rectifier tubes

of the thyratron or ignitron type. These tubes handle the power and operate in conjunction with electronic control circuits using high-vacuum tubes of the so-called radio-receiver type.

This chapter will be mainly concerned with electronic systems of the second group since drives of that group have much greater flexibility and functional superiority. Their principle can be used not only for adjustable-speed systems, but also for controls in a virtually unlimited number of special applications requiring various automatic and regulating functions.

The electronic tubes used in industrial electronic systems can be classified in the following manner:

1. Mercury-vapor or inert-gas-filled, grid-controlled, hot-cathode rectifiers, commonly known as thyratrons.
2. Mercury-pool-type rectifiers with ignitor-controlled firing, known as ignitrons.
3. Mercury-vapor or inert-gas-filled, hot-cathode rectifier diodes, called phanotrons.
4. High-vacuum, low-power control tubes such as diodes, triodes, and pentodes of the so-called radio type.

Tubes of the first group are used as controlled power rectifiers for the armature and the field circuits of the motor. Ignitrons are used as controlled power rectifiers for the armature of larger motors (above 25 horsepower). Phanotrons are important as low-cost field-circuit rectifiers where no field control is required. Tubes of the fourth group and their allied circuits provide the necessary control of power rectifiers and in particular of those of the thyratron type. The principle of operation of electronic tubes of each of these groups has been described in Chapters 5 and 6.

29·2 APPLICATION CHARACTERISTICS OF D-C SHUNT MOTORS

In Section 29·1 the particular usefulness of operating characteristics of d-c shunt-wound motors in industry was specifically emphasized. The speed of a shunt-wound motor can be expressed by

$$n = \frac{E - IR}{C_1\phi} \tag{29·1}$$

where n = speed of motor
E = armature voltage
I = armature current
R = resistance of armature circuit
ϕ = operating magnetic flux
C_1 = coefficient of proportionality.

It is apparent from equation 29·1 that the speed of the motor can be controlled either by varying the armature voltage E or the operating flux ϕ, that is, by varying the field-excitation current of the motor.

The torque developed by a shunt-wound motor is directly proportional to the product of armature current and operating magnetic flux:

$$T = C_2 I\phi \tag{29·2}$$

The power developed by the motor is directly proportional to the product of angular speed and torque:

$$P = C_3 nT \tag{29·3}$$

Substituting the expression for the motor torque (equation 29·2) in equation 2·93, we obtain the power developed by the motor as:

$$P = C_4 nI\phi \tag{29·4}$$

or, considering equation 29·1,

$$P = C_5(E - IR)I \tag{29·5}$$

If the change in cooling conditions of the motor as well as the change in magnetic losses for different speeds is neglected, it may be assumed that the rated armature current (the armature current permissible from the point of view of motor-temperature rise) is the same for different speeds.

By analyzing equations 29·2 and 29·5 one can easily determine the manner in which speed control through variation of armature voltage and field excitation will affect the torque and the power which the motor is able to develop for a definite time without overheating.

Thus, for the armature voltage control, where the flux ϕ remains constant (neglecting armature reaction) and the speed of the motor varies in direct proportion to the voltage E (see equation 29·1), the available torque (equation 29·2) remains constant. The available power (equation 29·5) varies essentially in direct proportion to the armature voltage or speed (equation 29·4).

When the speed of the motor is controlled by means of the field excitation, that is, by means of varying the flux ϕ, with the armature voltage remaining constant, the available torque will change in direct proportion to the flux and, considering equation 29·1, the torque will vary in inverse proportion to the speed. The power which may be developed by the motor without overheating will remain constant, as equation 29·5 shows.

The so-called application characteristics of a shunt-wound d-c motor are shown in Fig. 29·1. This figure represents,

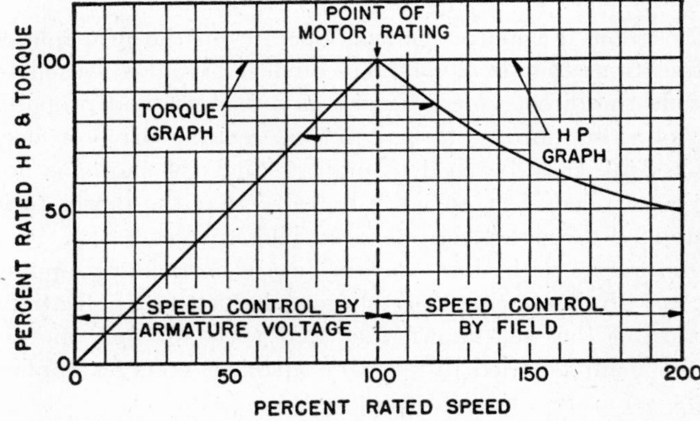

Fig. 29·1 Application characteristics of d-c shunt-wound motors.

as functions of speed by armature and field control, the horsepower and torque which can be obtained from a motor without overheating it, as previously analyzed. The proper interpretation of Fig. 29·1 is very important from the point of view of satisfactory selection of motors and controls in general. The armature-control range is often called the constant-torque range, and the field-control range the constant-horsepower range. Obviously, these terms refer to the torque or horsepower which can be developed by the motor at different speeds without overheating, and not to the torque or horsepower actually delivered by the motor. The latter will vary from application to application and will correspond to the requirements of the driven load. From a purely economical point of view, however, it is desirable to control the speed by armature voltage, for loads where torque remains more or less constant and independent of speed, and by field, where the torque required by the load changes approximately in inverse proportion to the speed, so that the required horsepower remains essentially constant.

29·3 CONTROLLED RECTIFICATION

Controlled rectification is one of the fundamental functions of any motor-control system whose operation is based on the principle of controlled angle of ignition of rectifier tubes of the thyratron or ignitron type. By controlling the angle of ignition it is possible to vary the armature voltage or the field current in some regular manner so as to obtain the desired characteristics of the drive.

Two basic circuits for controlled rectification are shown in Figs. 29·2 and 29·3. They both represent a symmetrical two-phase, half-wave, controlled rectifier (sometimes called a single-phase, full-wave rectifier). Figures 29·2 (*a*) and 29·3 (*a*) explain the principle of control of the angle of ignition for the two respective circuits.

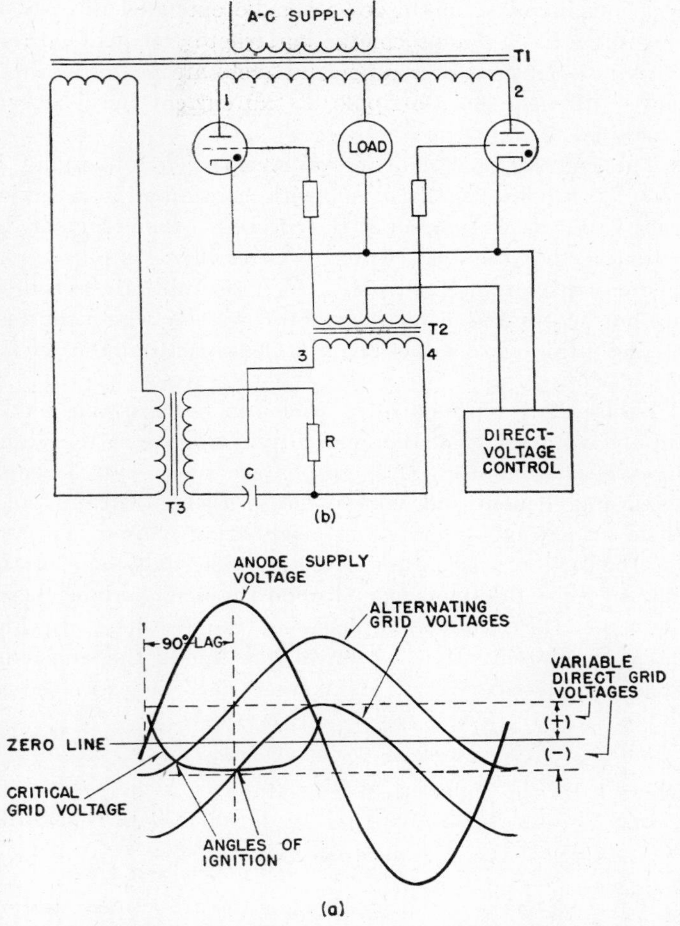

FIG. 29·2 Elementary circuit and principle of direct-voltage grid control of thyratron rectifiers.

In the circuit in Fig. 29·2 the control-grid voltage consists of an a-c and a d-c component. The a-c component is obtained from a conventional dephasing circuit comprising a transformer *T*3 with a center-tap winding, a resistor *R*, and a capacitor *C*. The alternating voltage between points 3 and 4, applied to the primary of the grid transformer, is permanently shifted in phase by about 90 degrees lag with respect to the anode-supply voltage, 1 and 2. The d-c component of grid voltage is introduced between the cathodes of rectifier tubes and the center tap of the secondary winding of the grid transformer *T*2. This d-c component is varied either manually or automatically in accordance with variations of some other quantity. In that manner, the a-c grid voltage component, which is superimposed upon the varying d-c component, may be shifted in the vertical direction, Fig. 29·2 (*a*), so that it will intersect the critical grid voltage of the thyratron at different points, and the angle of ignition corresponding to that point of intersection can thus be controlled, theoretically, from 0 to 180 degrees. It should be noted that the d-c control component of the grid voltage may assume negative as well as positive values, that is, below or above the cathode potential (see also Section 29·5).

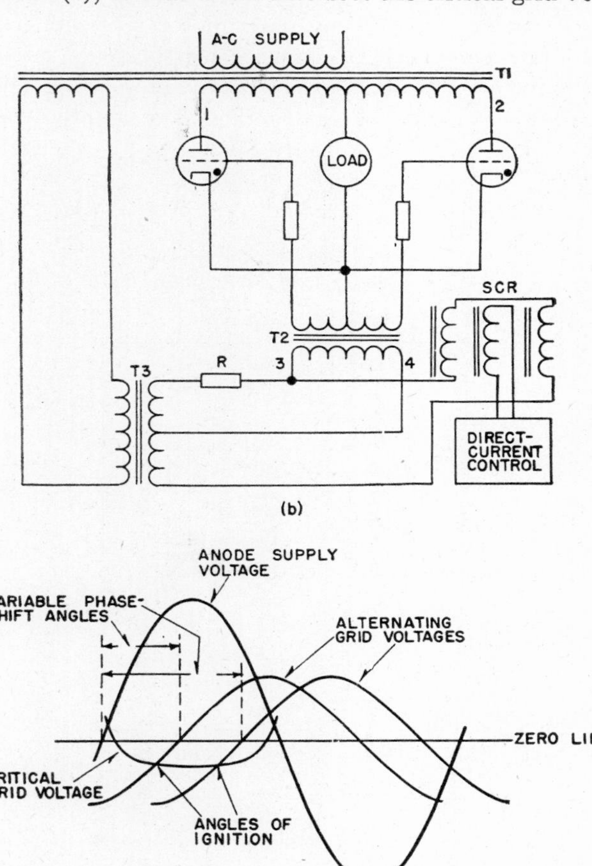

FIG. 29·3 Elementary circuit and principle of phase-shift control of thyratron rectifiers.

In the circuit in Fig. 29·3 the control-grid voltage has no d-c component, and is obtained directly from the grid transformer whose primary is connected to a dephasing circuit consisting of a center-tapped transformer $T3$, resistor R, and a saturable-core reactor SCR. The alternating voltage between points 3 and 4 is lagging in phase with respect to the anode-supply voltage by a variable angle. This phase angle depends upon the value of inductance of the saturable-core reactor, and can be controlled by varying the value of inductance through control of d-c current which saturates the core of the reactor. Thus, in the circuit in Fig. 29·3 the a-c grid voltage is shifted in the horizontal direction, parallel to the zero line, with respect to the anode-supply voltage and, in that manner, the angle of ignition of the armature rectifier tubes can be controlled manually or automatically by proper control of the d-c current in the saturable-core reactor.

The control method represented by circuit shown in Fig. 29·2 may be called "vertical" control of the angle of ignition of thyratrons, whereas the circuit in Fig. 29·3 represents the "horizontal" control.

The rotating armature of a d-c motor represents, under steady-state conditions, a circuit consisting of a resistance R, an inductance L, and an electromotive force E_g generated in the armature winding and acting as a counter voltage,

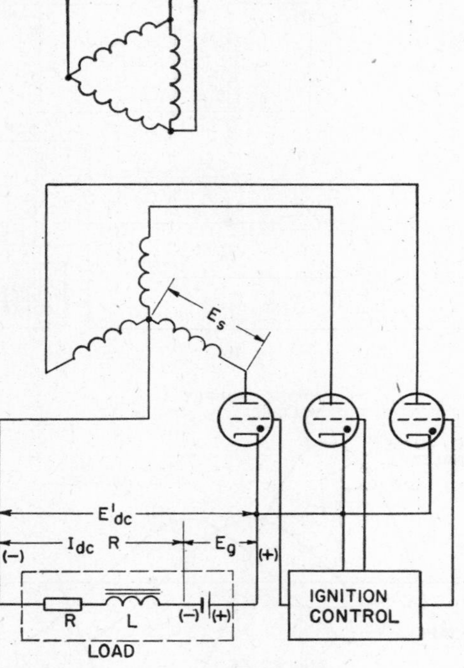

FIG. 29·4 Circuit of a three-phase thyratron rectifier for armature control, with equivalent load-circuit elements.

which tends to oppose the flow of current resulting from an external voltage applied to the armature terminals. This generated electromotive force is proportional to the speed of the motor, under the assumption of a constant operating flux. An equivalent circuit of a three-phase rectifier motor-armature system is shown in Fig. 29·4 where the generated electromotive force of the motor is represented by a battery, generating a voltage E_g, and the armature-winding resistance and inductance are represented by a resistor R and a reactor L respectively.

The principles and theory of controlled rectification are of fundamental importance in industrial control in general, and in motor control in particular. The relations involved should be fully understood before actual control circuits are discussed. Two cases of controlled rectification should be distinguished. The first one is the case of *discontinuous* load current where the current flowing in the load circuit, and supplied by a rectifier of a single-phase or a polyphase type, consists of discrete pulses with zero-current gaps between them. Discontinuous conduction can be obtained in any rectifier, regardless of the number of phases or the load-circuit inductance, if the angle of ignition is sufficiently delayed. The case of discontinuous conduction is of prime importance in electronic motor-control systems in general because the typical feature of these systems is a wide range of control of rectifier output voltages and, consequently, a wide range of control of the angle of ignition of rectifier tubes. Thus, the discontinuity of load current can always be encountered under certain operating conditions. From the theoretical point of view the case of discontinuous conduction is of fundamental importance since the relations and concepts involved in this case can be extended directly to the case of *continuous* conduction where the load current flows in a continuous manner and there are no zero-current gaps, although the a-c ripple of the current may be considerable.

The conventional analysis of rectifier circuits found in most textbooks usually deals with continuous conduction only, and often simplifies the problem even further by assuming that the load current of a rectifier is a pure direct current without any a-c ripple. Such assumptions, of course, are not acceptable in rectifier-motor systems where the case of discontinuous conduction is at least as common as the other one.

Figure 29·5 represents the time functions of the anode-supply voltage and of the load current flowing in the armature of a d-c motor. The graphs are referred to a single rectifying element and, consequently, a single current pulse is shown without reference to neighboring phases. In fact, for the case of discontinuous conduction the shape of the current pulse does not depend upon the neighboring phases, nor upon the number of phases of the rectifier. In Fig. 29·5, X represents the theoretical zero line of the sinusoidal anode-supply voltage. This voltage may be expressed as

$$e = \sqrt{2}E_s \sin x \qquad (29\cdot 6)$$

where E_s = rms value of anode voltage
$x = \omega t$ = variable time angle referred to the center of the system of co-ordinates.

Referring again to Fig. 29·5, E_0 is the arc-voltage drop of the rectifying element. This voltage drop may be assumed to be constant and equal to 15 to 20 volts. The opposing electromotive force generated in the armature is represented by E_g. The control-grid voltage and the critical grid voltage are not shown in Fig. 29·5. It is readily seen that a portion of the positive half cycle of the anode-supply voltage is used to overcome the sum of $E_0 + E_g = E_d$, and that only the portion of the anode voltage rising above the line X'' can produce any current flow in the armature circuit. Thus, the control-grid voltage and the critical grid voltage should be referred to the zero line X'', in the manner shown in Figs. 29·2 (*a*) and 29·3 (*a*).

Point N, corresponding to the time angle x_f, represents the point of ignition of the rectifying element, that is, the instant at which the rectifying element starts to conduct. The time angle x_f measured from the zero point O is called the *angle of ignition* of the rectifier. The time angle x_s corresponding to point R, at which the rectifier stops conducting, is called the *angle of extinction* of the rectifier.

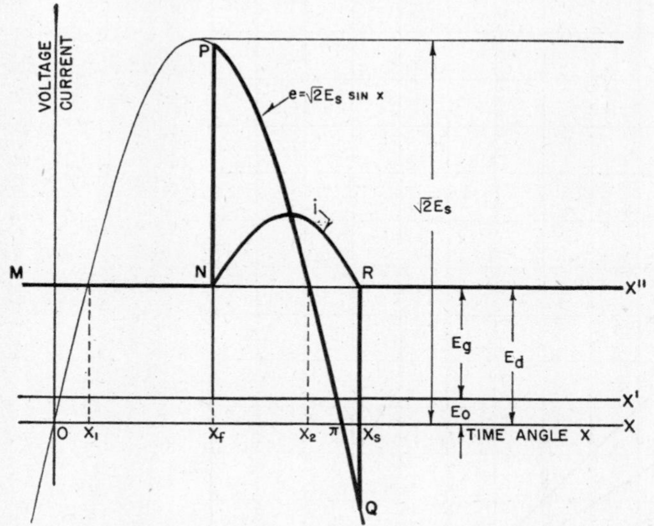

Fig. 29·5 Voltage and current time functions of a rectifier-armature system with a single rectifying element.

Thus, each rectifying element is conducting over the period $x_s - x_f$, and since the over-all cycle of the rectifying system is equal to $2\pi/p$, where p is the number of phases, it is apparent that the non-conductive period of the rectifier is equal to $(2\pi/p) - (x_s - x_f)$.

If line X' is regarded as representing the potential of the negative terminal of the motor armature, the line $MNPQRX''$, consisting partially of the zero line X'' and partially of the portion of the a-c anode-supply voltage, will represent the potential of the armature positive terminal, that is, the cathode potential of the rectifier (see Fig. 29·4). In other words, the line $MNPQRX''$ represents the time function of the voltage across the armature with respect to zero line X'.

The equation of the current pulse in the load circuit during the conductive period of the rectifying element is a combination of a sinusoidal and an exponential function of the time angle x:[2]

$$i = \frac{\sqrt{2}E_s}{R}\left\{\cos\theta \sin(x - \theta) - a + [a - \cos\theta \sin(x_f - \theta)]\epsilon^{-(x-x_f)/\tan\theta}\right\} \qquad (29\cdot7)$$

where θ = impedance angle of the load circuit:

$$\theta = \tan^{-1}\frac{\omega L}{R} \qquad (29\cdot8)$$

a = voltage coefficient, which also may be called speed coefficient:

$$a = \frac{E_g + E_0}{\sqrt{2}E_s} \qquad (29\cdot9)$$

$\omega = 2\pi f$ = angular line frequency
E_g = electromotive force generated in the armature
E_s = rms value of the anode-supply voltage
L = inductance of the load circuit
R = resistance of the load circuit
E_0 = rectifier arc-voltage drop.

The variable time angle x appearing as the independent variable in equation 29·7 is, of course, subject to limitation

$$x_f \leqq x \leqq x_s \qquad (29\cdot10)$$

since it follows from the definition of x_f and x_s that no current flows through the rectifying element for time angles outside the limits of expression 29·10. Graphs of equation 29·7 for different values of parameters a and x_f are shown in Fig. 29·6.

The voltage coefficient a has a particular significance. In the case of a rectifier-motor system, if the arc-voltage drop of the rectifier is neglected, and a constant operating flux in the motor is assumed, coefficient a will be directly proportional to the speed of the motor, as can be seen from equation 29·9. For that reason, a may be called the speed coefficient of the rectifier-motor system.

The fundamental relationship between the angle of ignition x_f and the angle of extinction x_s, for given values of

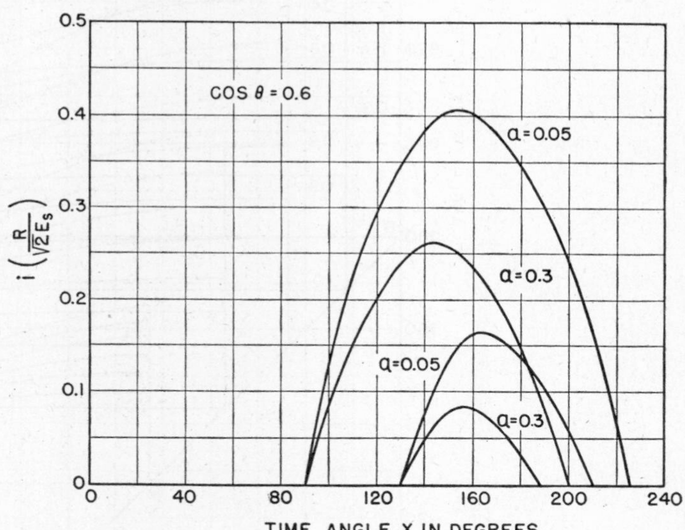

Fig. 29·6 Graphs of armature-current pulses plotted from equation 29·7.

the speed coefficient a and the impedance angle of the armature circuit θ, can be derived directly from equation 29·7 since for

$$x = x_s$$

$$i = 0$$

Thus, the relationship between x_f and x_s is

$$[a - \cos\theta \sin(x_s - \theta)]\epsilon^{x_s/\tan\theta} = [a - \cos\theta \sin(x_f - \theta)]\epsilon^{x_f/\tan\theta} \qquad (29\cdot11)$$

Equation 29·11 represents the angle of extinction x_s as an implicit function of the angle of ignition x_f. The solution for x_s of equation 29·11 is represented by a family of graphs in Fig. 29·7, where graphs of $x_s = f(x_f)$ are given for different values of the coefficient a and the impedance angle θ.[2,6] The value of x_s for given values of a and θ can be obtained from Fig. 29·7 directly by interpolation. The interpolation can be combined with any of the possible methods former voltage $\sqrt{2}E_s$, the value of coefficient a for a load consisting of a resistance and inductance only, such as the field winding of a motor, is of the order of 0.03, and the assumption of $a = 0$ is justified in many such cases.

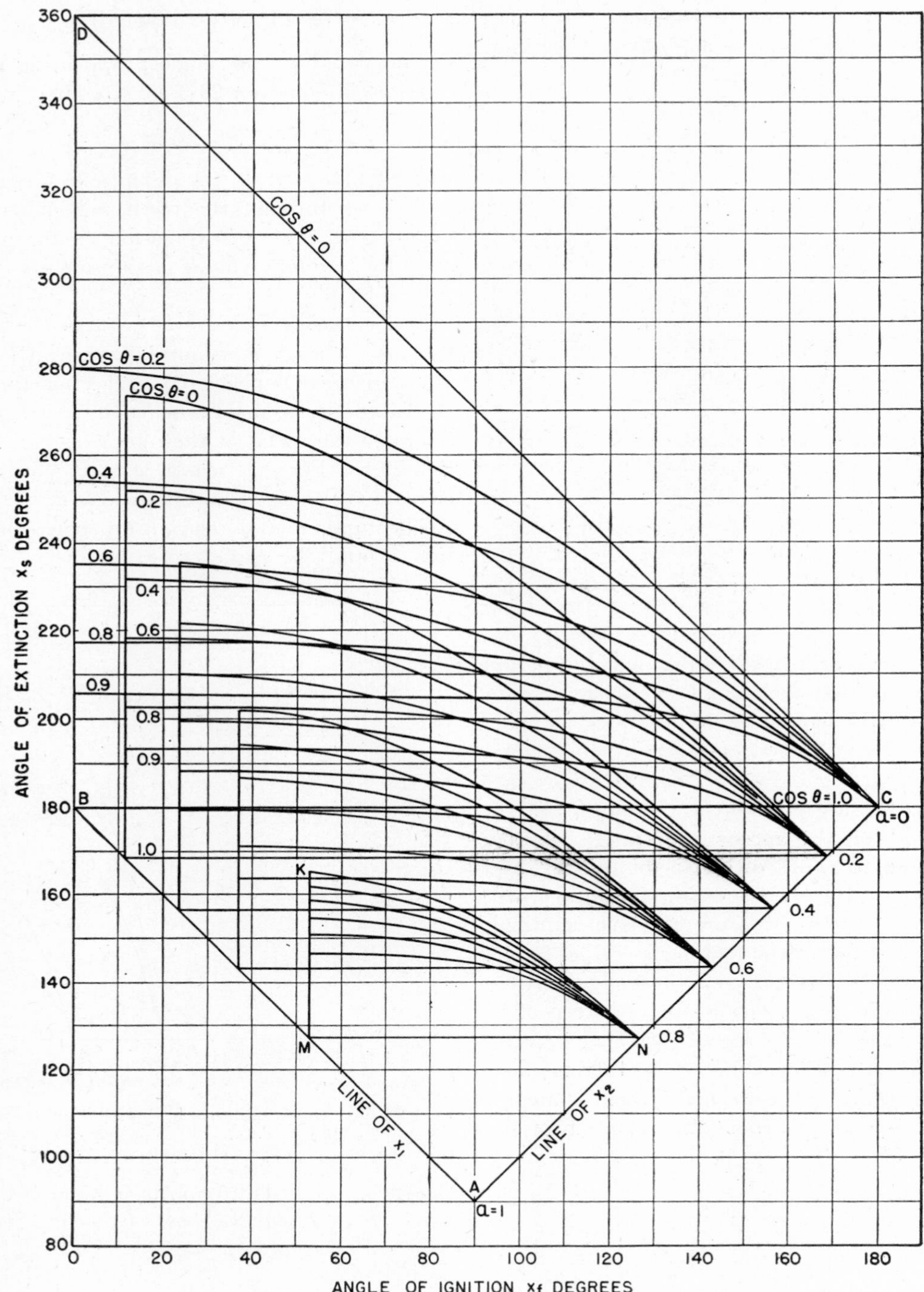

FIG. 29·7 Graphs of the angle of extinction x_s as a function of the angle of ignition x_f, for different parameters a and $\cos\theta$, plotted from equation 29·11.

of gradual approximations, if a greater degree of accuracy is required. Puchlowski has described one method of gradual approximations.[6] In the case where the load of the rectifier does not contain any generative element, that is, $E_g = 0$, and if the arc drop is neglected ($E_0 = 0$), then $a = 0$. The family of graphs for $a = 0$ is also shown in Fig. 29·7. Since E_0 is always low with respect to the amplitude of trans-

It can be easily understood that the angle of ignition x_f is subject to a very definite limitation:

$$x_1 < x_f < x_2 \tag{29·12}$$

where x_1 and x_2 denote particular border values of the angle of ignition, corresponding to the intersections of the zero line X'' with the graph of the anode-supply voltage (Fig.

29·5). The expressions for x_1 and x_2 can be readily obtained from the equation of the anode supply voltage (equation 29·6) referred to the zero line X'':

$$e_a = \sqrt{2}E_s \sin x - (E_g + E_0) \tag{29·13}$$

for the condition:

$$x = x_1$$

$$e_a = 0$$

and

$$x_2 = \pi - x_1$$

Thus, we obtain

$$x_1 = \sin^{-1} a, \quad 0 \leqq x_1 \leqq \frac{\pi}{2} \tag{29·14}$$

$$x_2 = \pi - \sin^{-1} a \tag{29·15}$$

The rectifier output voltage $E_{d\text{-}c}$ is defined as the average value of that portion of the transformer voltage which appears across the load during the conductive period of the rectifier. Obviously, the averaging is to be extended over the entire cycle of rectification, that is, over the period $2\pi/p$. It can be readily understood that, in the case of continuous conduction and also in the case of discontinuous conduction with a load consisting of R and L elements only (field winding of a motor), the rectifier output voltage can be identified with the voltage across the load terminals, as most of the textbooks on the subject of rectification assume. However, this assumption is not true when one has to deal with discontinuous conduction of the rectifier, combined with a load circuit containing an electromotive force E_g such as in the case of an armature circuit of a d-c motor.

Figure 29·5 shows that during the non-conductive period of the rectifier the voltage at the load terminals is equal to the electromotive force E_g generated in the load, whereas the rectifier output voltage during the same period is equal to zero. On the other hand, during the conductive period of the rectifier the load-terminal voltage is, of course, equal to the rectifier output voltage. Thus, it is immediately apparent that here the load-terminal voltage $E'_{d\text{-}c}$ is higher than the rectifier output voltage $E_{d\text{-}c}$.

In accordance with the previous definition of the rectifier output voltage, the expression for its average value can be derived directly from Fig. 29·5 by integrating the anode voltage function with respect to zero line X' over the period of conduction $x_s - x_f$, and averaging the result over the entire phase cycle $2\pi/p$:

$$E_{d\text{-}c} = \frac{1}{\frac{2\pi}{p}} \int_{x_f}^{x_s} (\sqrt{2}E_s \sin x - E_0)\, dx$$

$$E_{d\text{-}c} = \frac{pE_s}{\pi\sqrt{2}} [\cos x_f - \cos x_s - a_0(x_s - x_f)] \tag{29·16}$$

where

$$a_0 = \frac{E_0}{\sqrt{2}E_s} \tag{29·17}$$

and p = number of phases of the rectifier.

The average value of the armature voltage drop can be derived from Fig. 29·5 by integrating the voltage function with respect to zero line X'' over the conduction period of the rectifier, $x_s - x_f$, and averaging the result over the phase cycle $2\pi/p$:

$$I_{d\text{-}c}R = \frac{1}{\frac{2\pi}{p}} \int_{x_f}^{x_s} (\sqrt{2}E_s \sin x - E_d)\, dx$$

$$I_{d\text{-}c}R = \frac{pE_s}{\pi\sqrt{2}} [\cos x_f - \cos x_s - a(x_s - x_f)] \tag{29·18}$$

where

$$a = \frac{E_d}{\sqrt{2}E_s} = \frac{E_g + E_0}{\sqrt{2}E_s}$$

The average value of the direct voltage at the armature terminals $E'_{d\text{-}c}$ is, of course, equal to the sum of the counter electromotive force E_g and the armature voltage drop $I_{d\text{-}c}R$:

$$E'_{d\text{-}c} = E_g + I_{d\text{-}c}R$$

Thus, from equation 29·18

$$E'_{d\text{-}c} = \frac{pE_s}{\pi\sqrt{2}} \left[\cos x_f - \cos x_s - a(x_s - x_f) + a' \frac{2\pi}{p}\right] \tag{29·19}$$

where

$$a' = \frac{E_g}{\sqrt{2}E_s} \tag{29·20}$$

Equations 29·18 and 29·19 are general equations of controlled rectification, and, considering equation 29·23, their validity extends to all the cases of continuous and discontinuous conduction of the rectifier, and to any combination of R, L, and E_g in the load circuit. Most of the conventional forms of expressions for average values of load voltages and currents can be derived directly from equations 29·18 and 29·19.

When the motor is stalled, $E_g = 0$ and, in that case,

$$a' = 0 \quad \text{(see equation 29·20)}$$

Also, from equations 29·9 and 29·17

$$a = a_0 = \frac{E_0}{\sqrt{2}E_s}$$

Thus, it becomes apparent from equations 29·16, 29·18, and 29·19 that for a stalled motor

$$E_{d\text{-}c} = E'_{d\text{-}c} = I_{d\text{-}c}R$$

$$= \frac{pE_s}{\pi\sqrt{2}} [\cos x_f - \cos x_s - a_0(x_s - x_f)] \tag{29·21}$$

Equation 29·21 is valid for all cases of controlled rectification where the load circuit contains R and L elements only, such as, for example, the field winding of a d-c shunt-wound motor.

As mentioned previously, in many practical cases the arc-voltage drop of the rectifier E_0 can be neglected. Then

$$a_0 = 0 \quad \text{(see equation 29·17)}$$

and equation 29·21 for a stalled motor is further simplified:

$$E_{d\text{-}c} = E'_{d\text{-}c} = I_{d\text{-}c}R = \frac{pE_s}{\pi\sqrt{2}} (\cos x_f - \cos x_s) \tag{29·22}$$

In the case of *continuous conduction* of the rectifier (Fig. 29·8) the relationship between the angles of ignition and extinction (neglecting the rectifier-transformer leakage reactance) is

$$x_s = x_f + \frac{2\pi}{p} \tag{29·23}$$

By substituting equation 29·23 in equations 29·16, 29·18, and 29·19 we obtain, for the case of continuous conduction,

$$E_{d\text{-}c} = E'_{d\text{-}c} = \frac{pE_s}{\pi\sqrt{2}}\left[\cos x_f - \cos\left(x_f + \frac{2\pi}{p}\right) - a_0\frac{2\pi}{p}\right] \tag{29·24}$$

and

$$I_{d\text{-}c}R = \frac{pE_s}{\pi\sqrt{2}}\left[\cos x_f - \cos\left(x_f + \frac{2\pi}{p}\right) - a\frac{2\pi}{p}\right] \tag{29·25}$$

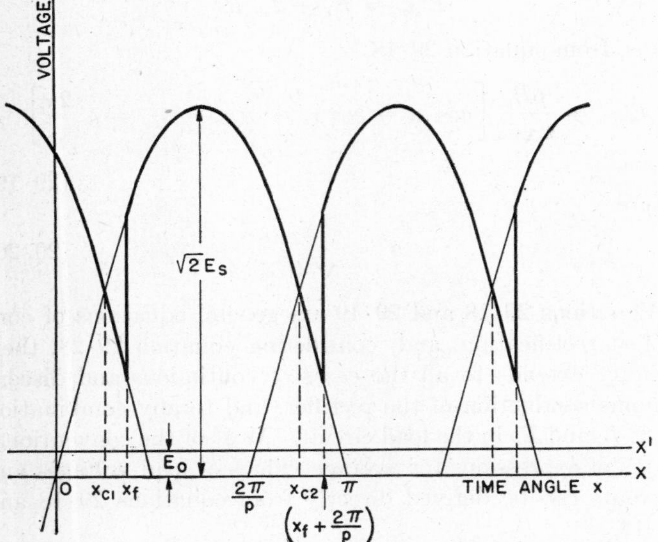

Fig. 29·8 Voltage wave for continuous conduction of a polyphase rectifier ($p = 3$).

Further, neglecting the rectifier-arc drop ($a_0 = 0$), and introducing the relationship between the angle of ignition x_f and the angle φ of the delay of ignition beyond the natural commutating point x_{c1}

$$\varphi = x_f - x_{c1} \quad \text{(see Fig. 29·8)}$$

$$\varphi = x_f - \frac{p\pi - 2\pi}{2p}$$

equation 29·24 for the rectifier output voltage can be directly transformed into the conventional form, valid only for continuous conduction and encountered in most standard textbooks:

$$E_{d\text{-}c} = \sqrt{2}E_s\frac{p}{\pi}\sin\frac{\pi}{p}\cos\varphi \tag{29·26}$$

Thus, the special form of equation 29·26 can be derived directly from the general equation 29·19. It should be pointed out that, in contrast with equation 29·19, coefficients a and a', which depend upon motor speed, do not appear in equation 29·24 and equation 29·26. Thus, for continuous conduction the armature voltage is not affected by the speed of the motor.

When the angle of ignition of a rectifying element of a polyphase rectifier is delayed beyond a certain critical value, the conduction of the rectifier will become discontinuous; that is, the current in the load will consist of discrete pulses, and there will be no commutation from one rectifying element to another.

The critical angle of ignition must satisfy the following equation:

$$x_{fc} = x_{sc} - \frac{2\pi}{p} \tag{29·27}$$

and the condition for continuous conduction is

$$x_f < (x_s) - \frac{2\pi}{p} \tag{29·28}$$

where (x_s) is the theoretical angle of extinction of a single rectifying element, obtained from Fig. 29·7. Discontinuous conduction is obtained when

$$x_f > x_s - \frac{2\pi}{p} \tag{29·29}$$

There always exists a maximum critical angle of ignition $x_{fc\,\max}$ such that, when the ignition is delayed beyond that maximum critical angle,

$$x_f > x_{fc\,\max} \tag{29·30}$$

the conduction will always be discontinuous, even for a purely inductive load ($\cos\theta = 0$). The maximum critical angle of ignition is a function of the motor speed and of the transformer voltage, both represented by the voltage coefficient a, and is expressed by the following formula

$$x_{fc\,\max} = \sin^{-1}\frac{\pi a}{p\sin\frac{\pi}{p}} - \frac{\pi}{p} \tag{29·31}$$

where

$$\frac{\pi}{p} < \sin^{-1}\frac{\pi a}{p\sin\frac{\pi}{p}} < \pi$$

and

$$p \neq 1$$

29·4 CHARACTERISTICS OF A RECTIFIER-MOTOR SYSTEM

In Section 29·3 general theoretical principles of controlled rectification were discussed in some detail, and a number of basic concepts and mathematical formulas was presented. In this section some of the characteristics of the rectifier-motor system will be discussed.

The graphical method of the qualitative analysis of characteristics will be used first, since it will help to clarify a number of concepts and explain the typical behavior of the system.

In Fig. 29·9 (*a*) are shown graphs of voltages and currents, represented as functions of the variable time angle x, for a motor whose armature is supplied by a symmetrical two-phase rectifier ($p = 2$, Fig. 29·2) with a delayed angle of ignition. Figure 29·9 (*b*) shows analogous conditions for a conventional d-c drive, and in Fig. 29·9 (*c*) rectifier-tube voltages (voltages appearing across the rectifying elements) are specifically emphasized. For the sake of simplification of the diagrams, the rectifier-tube arc-voltage drop has been neglected in Fig. 29·9.

In Figs. 29·9 (*a*) and 29·9 (*c*) are shown two sinusoidal anode-supply voltages for rectifier tubes 1 and 2 (see Fig. 29·2). For a symmetrical two-phase, half-wave rectifier (sometimes called single-phase full-wave rectifier, or bi-phase rectifier) these two voltage waves are displaced in phase with respect to each other by 180 degrees. It is assumed that the motor is running at constant speed; the counter electromotive force E_g generated in the armature winding is represented by line X'.

Tube 1 starts to conduct at point A, Fig. 29·9 (*a*), and at that instant the instantaneous voltage across the armature is represented by BG. From this point on, the armature voltage follows the anode-supply-voltage wave along the portion BE of the wave. At the instant when the rectifying element starts to conduct, the current is still zero, since the inductance of the armature winding prevents the current from rising immediately. If the instantaneous values of current are plotted in voltage scale as the iR drop with respect to zero line X', a pulse ACF will be obtained (see the equation of the current pulse, equation 29·7, as well as graphs of the current pulse, Fig. 29·6).

The iR-drop pulse will reach its peak at point C, where it intersects the anode-supply-voltage wave. At point D, where the a-c supply voltage intersects the line of E_g (line X'), there is no external voltage to cause the flow of current in the armature and, if the armature winding had a purely resistive character ($\cos \theta = 1$), the current would stop flowing at that point. In other words, in the idealized case of a non-inductive armature winding the armature current plotted in scale of iR drop would follow the shape of the a-c anode-supply voltage ABD. The inductance of the armature circuit, which prevents the current from rising sharply at the point of ignition A, also prevents the current from dying out at point D, where the external voltage causing the current flow is equal to zero. The electromotive force of inductance keeps the current flowing up to point F.

The general circuit equation

$$e = iR + L\frac{di}{dt} + E_g \tag{29·32}$$

can be interpreted graphically in the following manner [see Fig. 29·9 (*a*)].

At the point of ignition A the instantaneous supply voltage $e = BG$ consists of two components: component $AG = E_g$ balances the counter electromotive force E_g generated in the armature winding of the motor; component $BA = L(di/dt)$ constitutes the inductive voltage drop in the winding; that is, it balances the electromotive force of inductance $[-L(di/dt)]$ opposing the flow of current. The resistive voltage drop iR is equal to zero since $i = 0$. At point K, where the current reaches its maximum, $L(di/dt) = 0$, and the supply voltage $e = CL$ consists of two components $CK = iR$ and $KL = E_g$. Beyond point K, where the armature current starts to decrease, the electromotive force of inductance changes its sign and acts in the direction to maintain the flow of current in the circuit, so that, when the external voltage $(e - E_g)$ changes its sign at point D, the current is maintained by the electromotive force, $-L(di/dt)$. At point F, where the rectifying element 1 stops to conduct, the electromotive force of inductance is just equal to FE,

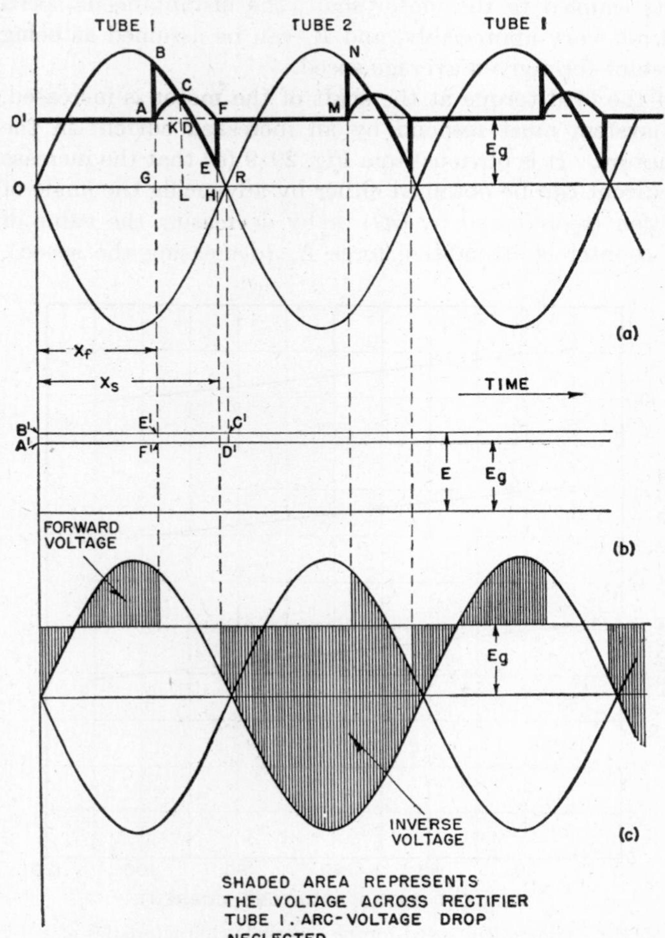

Fig. 29·9 Time functions of armature voltage, armature electromotive force, armature-voltage drop for: (*a*) two-phase symmetrical rectifier drive; (*b*) d-c drive; (*c*) time function of voltage across the rectifying element 1. Arc-voltage drop is neglected.

that is, to the difference of the counter electromotive force $FH = E_g$ and the supply voltage $e = EH$ (compare equation 29·32).

During the conductive period AF of the rectifying element, Fig. 29·9 (*a*), the voltage at the armature terminals will, of course, follow the anode-supply voltage along the portion BDE of the sinusoidal voltage wave. During the non-conductive period of the rectifier the voltage at the armature terminals will be equal to the electromotive force generated in the armature winding. Although this electromotive force cannot produce any current in the load circuit because of the rectifier, it will appear across the armature terminals and will fully affect the reading of any voltage-measuring instrument. Thus, the actual voltage existing at the armature terminals will follow the line $O'ABCEFMN \cdots$ with respect to the zero line X.

The average value of armature current, proportional to the average value of armature voltage drop (see equation 29·18), is directly proportional to the difference of area ABD

and area *DFE*, Fig. 29·9 (*a*). The instantaneous torque, proportional to the instantaneous value of the armature current, will, of course, follow the graph of the current and will have a pulsating character. However, owing to the moment of inertia of the motor armature and of all the other rotating parts coupled to the motor shaft, the instantaneous speed will not vary appreciably, and E_g can be assumed as being constant for a given average speed.

If the load torque at the shaft of the motor is increased, the system must respond by an increased current in the armature. It is obvious from Fig. 29·9 (*a*) that the increase in current can be obtained either by advancing the angle of ignition (represented by *OG*) or by decreasing the value of the counter electromotive force E_g (decreasing the speed).

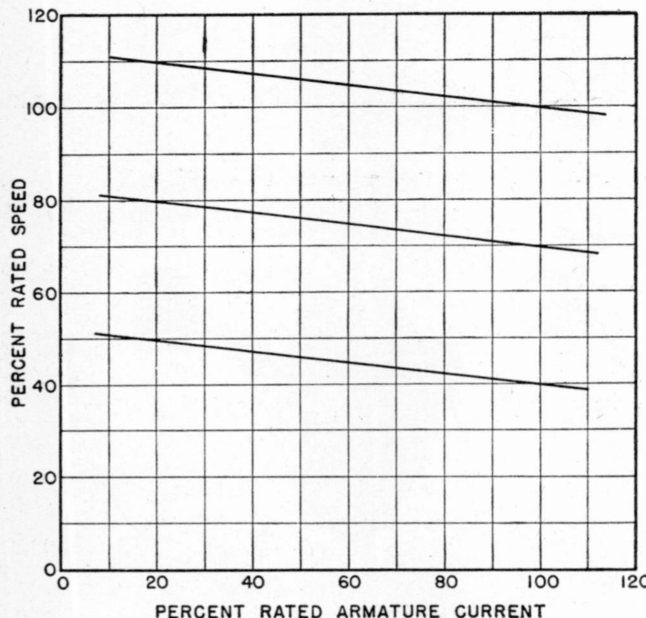

FIG. 29·10 Theoretical speed-torque (current) characteristics of a d-c shunt-wound motor drive. Constant field, armature reaction neglected.

If it is assumed that the angle of ignition (*OG*) remains unchanged, the increase in torque (or current) will result in a corresponding decrease in the counter electromotive force E_g. The same is, of course, true for a conventional d-c drive, whose armature voltage and counter electromotive force are shown in Fig. 29·9 (*b*), but the extent to which E_g (that is, speed) will have to decrease for a given increase in torque is different in each case.

Referring to Figs. 29·9 (*a*) and 29·9 (*b*), let it be assumed that in both cases the same average current flows in the armature circuit. Under these circumstances, the following equation for the average currents must be fulfilled:

$$\frac{\text{Area } ACF}{\pi} = \frac{\text{Area } A'B'C'D'}{\pi}$$

It is to be noted that in the case of a rectifier-motor system, Fig. 29·9 (*a*), the current flows during the portion *AF* of the cycle *OR* (that is, $x_s - x_f$), whereas for conventional direct-voltage drive, Fig. 29·9 (*b*), the current is uniformly distributed (in time) over the entire cycle ($B'C' = OR$). If in both cases the average value of armature current has to increase by the same amount, and the electromotive force has to decrease, it follows directly from comparison of Figs. 29·9 (*a*) and 29·9 (*b*) that E_g will have to decrease more in the case of a rectifier-motor system with discontinuous conduction, Fig. 29·9 (*a*), than in the case of a direct-voltage drive, Fig. 29·9 (*b*). This becomes immediately apparent because the same increase in armature current in both cases can be represented by the equal areas of two rectangles whose heights will represent the decrease in electromotive force, that is, in speed of the motor. In Fig. 29·9 (*a*) the base of the rectangle will be equal to the period of conduction $(x_s - x_f) < \pi$, and in Fig. 29·9 (*b*) the base of the rectangle will be, obviously, equal to π. Since the areas of the two rectangles are the same, and the base of the incremental rectangle in Fig. 29·9 (*b*) is greater, the height of the incremental rectangle for a rectifier-motor system will be greater than the height of the other rectangle. Consequently, the speed-torque characteristic of a rectifier drive with constant angle of ignition and discontinuous current flow in the armature will have considerably more droop than the speed-torque characteristic of a conventional direct-voltage drive.

A somewhat different approach to this problem can be obtained by analyzing equation 29·1:

$$n = \frac{E - IR}{C\phi}$$

Assuming a constant voltage E applied to the armature, and a constant operating flux ϕ, the speed of the motor will decrease with increasing current (torque) because of the armature voltage drop IR. Equation 29·1 is represented graphically in Fig. 29·10 for several different values of armature voltage.

Equation 29·1 also can be applied in the case of a rectifier drive under the assumption that both E and I are average values of periodical functions shown graphically in Fig. 29·9 (*a*). Yet there is a basic difference between the two cases because for a d-c motor system the voltage at the armature terminals does not depend upon the electromotive force E_g, speed, or load (disregarding the possible line voltage drop), whereas in the case of a rectifier drive with constant angle of ignition and discontinuous current flow, the armature voltage $E'_{d\text{-}c}$ depends upon the electromotive force E_g, speed n, and load current $I_{d\text{-}c}$ of the motor. This fact is clearly illustrated in Fig. 29·9 (*a*) where portions of armature voltage $O'A$, FM, $\cdots$, etc., over the non-conductive period of the rectifier are equal to the counter electromotive force of the motor. In that case E_g can be directly observed on the oscilloscope.

Referring again to equation 29·1, it becomes apparent that the droop of the speed-torque characteristic must be considerably greater for the rectifier drive with discontinuous armature current because not only is the armature voltage drop IR increasing with load, but there also the armature voltage is decreasing at the same time.

The effect of E_g on the armature voltage is also apparent from equation 29·19 where both a and a' are functions of E_g.

The speed-torque characteristics of a rectifier drive, for a given angle of ignition and a given impedance angle of the armature circuit, can be calculated from equations 29·18 and 29·25 and from the graphs in Fig. 29·7. Equations 29·18 and 29·25 can be rewritten as follows. For discontinuous conduction:

$$t_f = \frac{I_{d\text{-}c}R}{\sqrt{2}E_s} = \frac{p}{2\pi}[\cos x_f - \cos x_s - a(x_s - x_f)] \quad (29\cdot33)$$

For continuous conduction:

$$t_f = \frac{I_{d\text{-}c}R}{\sqrt{2}E_s} = \frac{p}{2\pi}\left[\cos x_f - \cos\left(x_f + \frac{2\pi}{p}\right) - a\frac{2\pi}{p}\right] \quad (29\cdot34)$$

The expression $t_f = I_{d\text{-}c}R/\sqrt{2}E_s$, representing the ratio of the armature voltage drop and the peak value of the rectifier-supply voltage, can be called the *torque factor* of the drive because it is directly proportional to the average armature current and to the average torque developed by the motor, if it is assumed that the operating field remains constant. This simplifying assumption is, of course, only approximately correct since the main flux will vary with load to a certain extent because of the armature reaction. The torque factor as well as the speed factor a, appearing in equations 29·33 and 29·34, is always less than unity:

$$0 < t_f < 1$$

$$0 < a < 1$$

As an example, let us assume the following operating data characterizing the conditions of the system:

Number of phases $p = 3$
Angle of ignition $x_f = 90°$
Load-impedance angle $\theta = \cos^{-1} 0.4$

The values of torque factor t_f for discontinuous conduction can be calculated, for given values of p, x_f, $\cos\theta$, and for different values of a, from equation 29·33 and the graphs in Fig. 29·7. It is to be noted that values of x_s appearing in equation 29·33 are taken from the graphs in Fig. 29·7 for given values of x_f, $\cos\theta$, and a. The calculated values of torque factor are assembled in Table 29·1. For continuous conduction of the rectifier the values of torque factor are calculated from equation 29·34. It is important to bear in mind that for this particular example the value of the critical angle of extinction is (equation 29·27)

$$x_{sc} = x_{fc} + \frac{2\pi}{p} = 210°$$

Discontinuous conduction takes place where the angle of extinction, obtained from Fig. 29·7, is less than $x_{sc} = 210$ degrees. Continuous conduction takes place where the theoretical angle of extinction (x_s), obtained for a single rectifying element (from Fig. 29·7), is greater than 210 degrees. The actual angle of extinction for continuous conduction is constant, and equal to $x_f + (2\pi/p) = 210$ degrees.

As stated previously, the coefficient a (equation 29·9) may be regarded as directly proportional to the speed of the motor, if the operating flux of the motor is assumed to be constant and if the arc-voltage drop of the rectifier is neglected. If the arc-voltage drop is taken into account, the coefficient a' (equation 29·20) is a better measure of the motor speed. Thus, although either of these coefficients may be called speed coefficient, a' represents the motor speed with a greater degree of accuracy. From equations 29·9, 29·17, and 29·20,

$$a = a' + a_0 \quad (29\cdot35)$$

$$a' = a - a_0 \quad (29\cdot36)$$

Assuming typical values for the arc-voltage drop and the rectifier-transformer phase voltage

$$E_0 = 15, \quad E_s = 350$$

we obtain from equation 29·17 a typical value of the coefficient a_0:

$$a_0 = \frac{15}{\sqrt{2} \times 350} = 0.0303$$

Values of the speed coefficient a', calculated from equation 29·36, also are given in Table 29·1.

TABLE 29·1 RESULTS OF CALCULATIONS OF ARMATURE VOLTAGE, ARMATURE CURRENT AND EMF FOR RECTIFIER-MOTOR SYSTEM, FIG. 29·4

$p = 3$, $x_f = 90°$, $\cos\theta = 0.4$, $E_s = 350$ volts, $a_0 = 0.0303$ critical angle of extinction $x_{sc} = 210°$

a	a'	x_s in degrees	t_f	v_a	
	$a' = \frac{E_g}{\sqrt{2}E_s}$ $a' = a - a_0$	Fig. 29·7	Eq. 29·33 Eq. 29·34	$v_a = \frac{E'_{d\text{-}c}}{\sqrt{2}E_s}$ $v_a = a' + t_f$	Conduction
1.0	0.9697	90	0.0000	0.9697	Discontinuous,
0.8	0.7697	151	0.0104	0.7801	$x_s < 210°$
0.7	0.6697	164	0.0268	0.6965	
0.6	0.5697	177	0.0418	0.6115	
0.5	0.4697	188	0.0644	0.5341	
0.4	0.3697	198.5	0.0908	0.4605	
0.3	0.2697	209	0.1197	0.3894	
0.25	0.2197	(214.5) *	0.1633	0.3830	Continuous,
0.2	0.1697	(218.5) *	0.2133	0.3830	$(x_s) > 210°$
0.1	0.0697	(228) *	0.3133	0.3830	
0.0303	0.0000	(236) *	0.3830	0.3830	

* Values of theoretical angle of extinction for continuous conduction $(x_s) > 210°$. Actual angle of extinction x_s for $(x_s) > 210°$ is constant and equal to $x_f + 2\pi/p = 210°$.

The speed-torque characteristic of the drive can be represented in terms of corresponding coefficients a' and t_f as $a' = f(t_f)$, and can be plotted directly from values of a' and t_f in Table 29·1. The graph of $a' = f(t_f)$ for $x_f = 90$ degrees, $\cos\theta = 0.4$, $p = 3$, and $a_0 = 0.0303$ is shown in Fig. 29·11.

The equation for the motor armature voltage of a rectifier drive was derived previously in Section 29·3 (equation 29·19). Now, a typical graph of the armature voltage as a function of load current or torque may be calculated and plotted on the basis of values of the speed coefficient a' and torque coefficient t_f contained in Table 29·1.

Dividing both sides of the equation

$$E'_{d\text{-}c} = E_g + I_{d\text{-}c}R \tag{29·37}$$

by the peak value of the rectifier transformer voltage $\sqrt{2}E_s$, we have

$$\frac{E'_{d\text{-}c}}{\sqrt{2}E_s} = \frac{E_g}{\sqrt{2}E_s} + \frac{I_{d\text{-}c}R}{\sqrt{2}E_s} \tag{29·38}$$

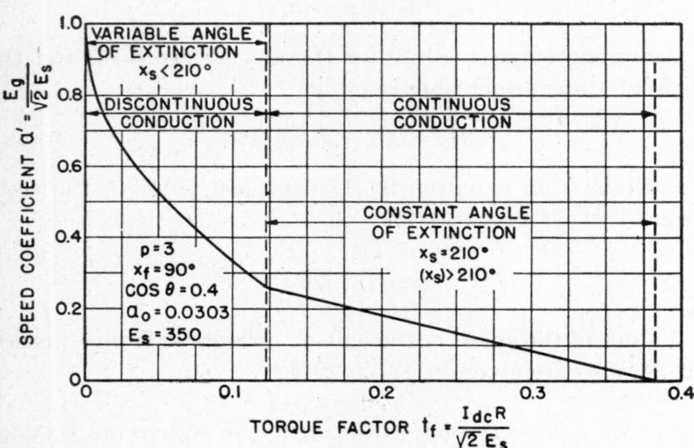

FIG. 29·11 Theoretical speed-torque characteristic, $a' = f(t_f)$, of a rectifier-motor system. Graph plotted from Table 29·1.

The ratio of the armature voltage $E'_{d\text{-}c}$ and the peak value of the rectifier-transformer phase voltage may be called the armature-voltage coefficient; it will be denoted by v_a. This coefficient, of course, is directly proportional to the armature voltage:

$$v_a = \frac{E'_{d\text{-}c}}{\sqrt{2}E_s} \tag{29·39}$$

Considering equations 29·20, 29·34, and 29·39, equation 29·38 can be represented as

$$v_a = a' + t_f \tag{29·40}$$

or

$$a' = v_a - t_f \tag{29·41}$$

It will be noted that values of coefficients a' and t_f for the particular case of $x_f = 90$ degrees, $\cos\theta = 0.4$, $p = 3$, and

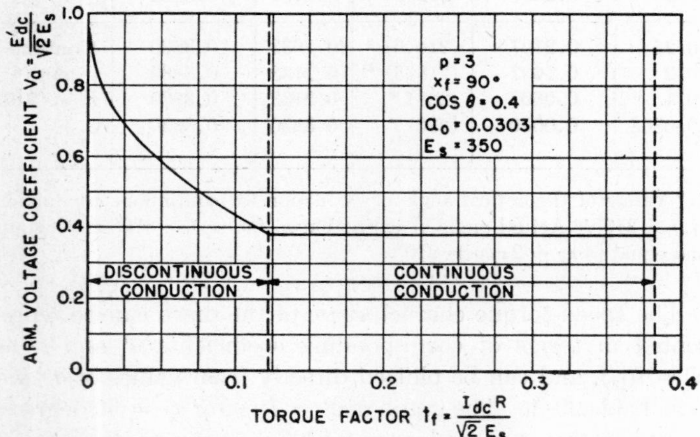

FIG. 29·12 Theoretical voltage-torque characteristic, $v_a = f(t_f)$, of a rectifier-motor system. Graph is plotted from Table 29·1 and corresponds to graph in Fig. 29·11.

$a_0 = 0.0303$ are given in Table 29·1; therefore the armature voltage coefficient v_a can be calculated directly from equation 29·40. Values of v_a also are given in Table 29·1. The graph of armature voltage versus torque, represented in terms of corresponding coefficients, $v_a = f(t_f)$, is shown in Fig. 29·12.

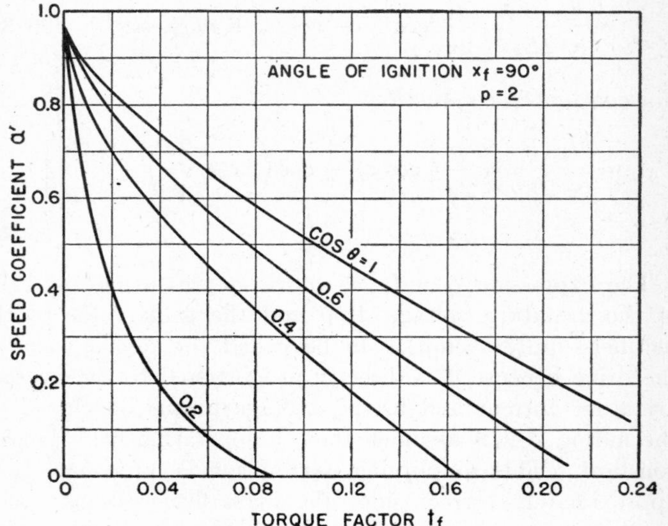

FIG. 29·13 Theoretical speed-torque characteristics of a rectifier-motor system, $a' = f(t_f)$, for $p = 2$, $x_f = 90°$, and four different load impedance angles.

Two families of graphs representing $a' = f(t_f)$ for $p = 2$, one for the angle of ignition of 90 degrees and different values of the load-impedance angle, the other one for the load-impedance angle $\theta = \cos^{-1} 0.4$ and different values of the angle of ignition, are shown in Figs. 29·13 and 29·14. These graphs are plotted on the basis of calculations similar to those given previously for the case $p = 3$, $x_f = 90$ degrees (see Table 29·1 and Fig. 29·11). The graphs can be used for estimating the speed of the motor for a given angle of ignition, armature-circuit constants, transformer voltage, and motor load.

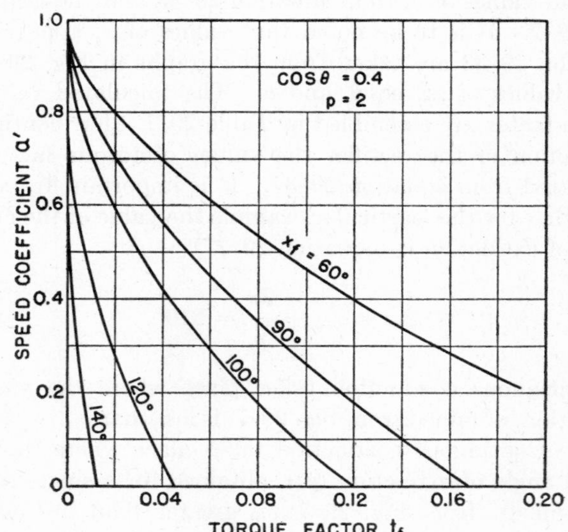

FIG. 29·14 Theoretical speed-torque characteristics of a rectifier-motor system, $a' = f(t_f)$, for $p = 2$, $\cos\theta = 0.4$, and five different angles of ignition.

If a more accurate calculation is required, a graph of $a' = f(t_f)$ can be plotted from several points calculated in the same manner as those for Table 29·1, on the basis of given values of the angle of ignition x_f, impedance angle of the armature circuit θ, and number of rectifier phases p.

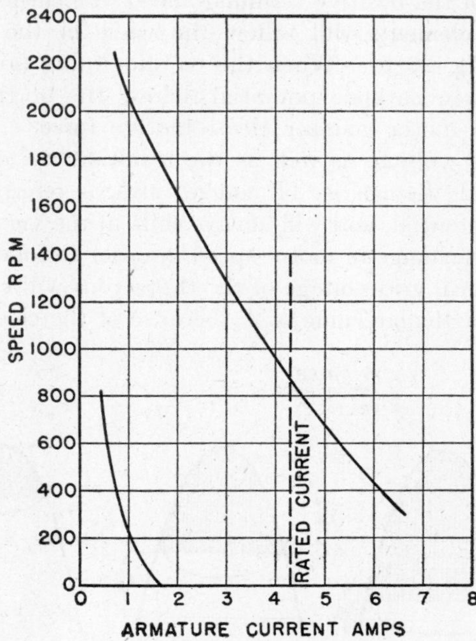

FIG. 29·15 Experimental speed-torque (current) characteristics of a rectifier-motor system for two different constant angles of ignition. $p = 2$. Motor rating: 1 hp, 1750 rpm, 230 volts. Constant field excitation.

Usually, four or five points will determine the curve adequately, and then the speed coefficient (hence the electromotive force and the speed of the motor) for any given load can be obtained directly from the graph.

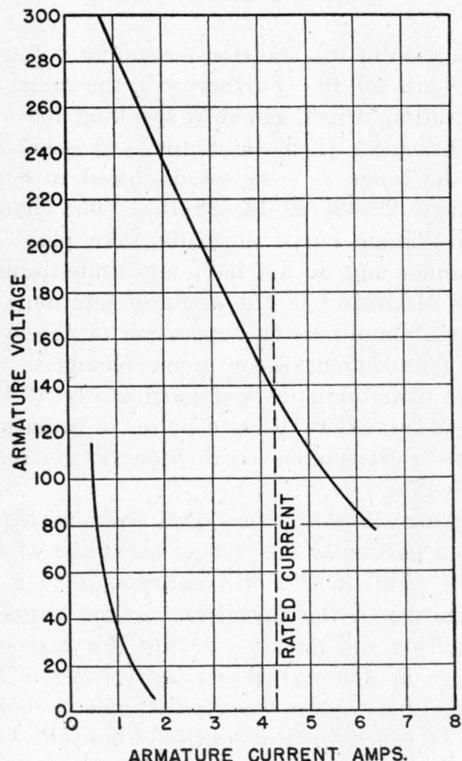

FIG. 29·16 Experimental voltage-torque (current) characteristics of a rectifier-motor system for two different constant angles of ignition. $p = 2$. Motor rating: 1 hp, 1750 rpm, 230 volts. Constant field excitation.

The experimental characteristics of a rectifier drive for $p = 2$ and a constant angle of ignition are shown in Figs. 29·15 and 29·16.

Analogous theoretical speed-torque characteristic for a conventional direct-voltage drive also can be represented in terms of speed and torque coefficients. Dividing both sides of the equation

$$E_g = E - IR$$

by the supply voltage which is equal to the armature voltage E,

$$\frac{E_g}{E} = 1 - \frac{IR}{E} \tag{29·42}$$

we can express equation 29·42 as

$$a_{d\text{-}c} = 1 - t_{f\,d\text{-}c} \tag{29·43}$$

where

$$a_{d\text{-}c} = \frac{E_g}{E} = \text{speed coefficient}$$

$$t_{f\,d\text{-}c} = \frac{IR}{E} = \text{torque coefficient}$$

It will be noted that equation 29·43 is analogous to equation 29·41. Since for a conventional d-c drive the armature voltage is equal to the supply voltage, the armature voltage coefficient $v_{a\,d\text{-}c}$ is equal to unity:

$$v_{a\,d\text{-}c} = 1$$

The graph of equation 29·43 is shown in Fig. 29·17.

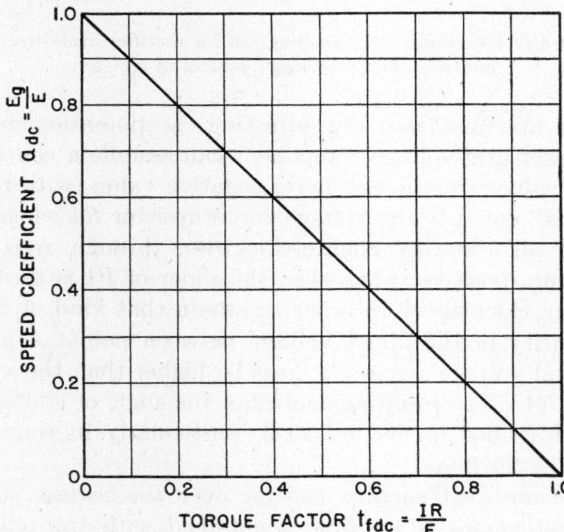

FIG. 29·17 Theoretical speed-torque characteristic, $a_{d\text{-}c} = f(t_{f\,d\text{-}c})$, of a d-c drive. Constant field; armature reaction neglected.

The speed and armature-voltage characteristics of rectifier drives will be discussed again in Sections 29·6 and 29·7 in connection with the description of various means and circuits for control.

29·5 ELEMENTARY SYSTEMS FOR ARMATURE CONTROL

An elementary scheme for armature-voltage control of a rectifier drive is shown in Fig. 29·18. It will be noted in

Fig. 29·18 that the power-rectifier circuit and the a-c grid-phase displacement circuit providing the alternating grid-voltage component are the same as those shown previously in Fig. 29·2. The variable d-c component of the thyratron grid circuit is applied between the common cathode connection of the two rectifier tubes (point A) and the center tap of the secondary winding of the control-grid transformer $T2$ (point B). The d-c grid component consists of two voltages opposing each other. One, across $R4$, is of constant magnitude and negative polarity with respect to the grids, the other one, of positive polarity, is obtained from a manually adjustable voltage divider (potentiometer) $P1$.

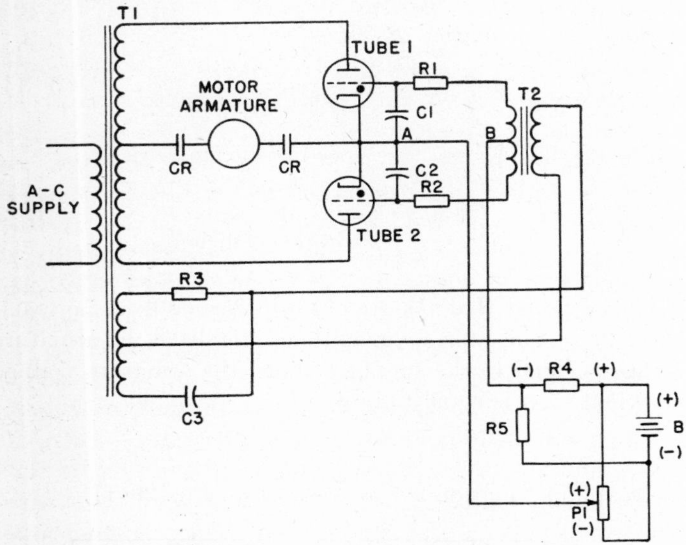

FIG. 29·18 Elementary circuit diagram for rectifier armature-voltage control. Direct-voltage control of ignition.

It is apparent that, by adjusting the potentiometer $P1$, the direct grid-voltage component can assume a continuous set of values from a maximum negative value (with respect to grids), equal to the voltage across resistor $R4$ when $P1$ is turned all the way counterclockwise, through zero, to a maximum positive value when the slider of $P1$ is turned all the way clockwise. In order to obtain that kind of change in polarity of the direct voltage between points A and B, the total voltage across $P1$ must be higher than the voltage across $R4$. The resulting control of the angle of ignition has been described in Section 29·3, particularly in connection with Fig. 29·2 (a).

The most satisfactory control over the widest possible range of angles of ignition is obtained with the following typical values of grid voltages:

Alternating grid voltage per tube, 25 to 35 volts rms
Phase displacement of the alternating grid voltage with respect to transformer voltage (lag), 90° to 120°
Constant d-c grid-voltage component, 40 to 55 volts
Maximum value of adjustable d-c component of opposite polarity, 80 to 120 volts

It is important to point out that when the motor is running (see Fig. 29·5) the center tap or the neutral point of the rectifier transformer can be assumed as being at constant potential represented by the zero line X' in Fig. 29·5. The negative terminal (brush) of the d-c motor, which in a normal drive system is connected to the neutral of the transformer, will be at the same potential and, consequently, the potential of the positive terminal and of the cathodes of the rectifying elements will follow the wave of the armature voltage (Fig. 29·5). When the rectifier tubes are not conducting, their cathode potential follows the horizontal line X'' of the motor counter electromotive force. The alternating grid voltage as well as the resultant d-c component of the grid voltage is, of course, always referred to the cathode potential, and will always shift in the vertical direction with a change in motor speed in order to follow E_g.

The critical grid voltage of the thyratrons will be different for each particular value of E_g because of the change in the

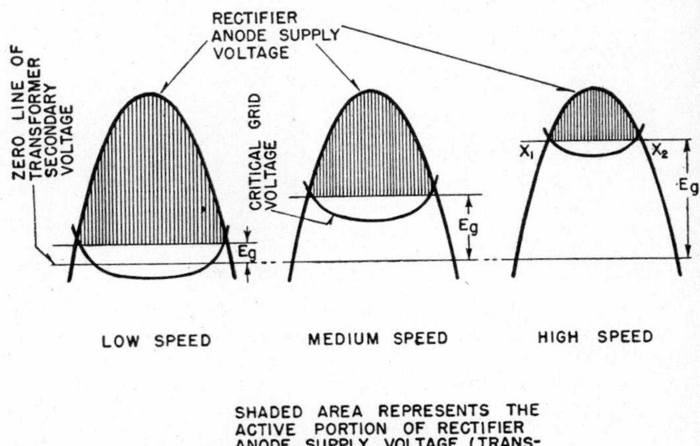

FIG. 29·19 Effect of motor electromotive force on the active portion of the rectifier-transformer voltage and on the critical grid-voltage characteristic of armature thyratrons.

"active" portion of the transformer-supply voltage, as illustrated in Figure 29·19. Furthermore, the range of possible angles of ignition, which, generally speaking, has a maximum of 0 to 180 degrees when the motor is at standstill, is narrowed to the range $x_2 - x_1$, as discussed in Section 29·3 (see equations 29·12, 29·14, 29·15). The changes in the critical-grid-voltage curve normally take place in a continuous manner and do not have any undesirable effect on the process of control of the angle of ignition. However, under certain conditions the operation of the system may be affected by a sudden discontinuous change in the critical grid-voltage characteristic, and particularly at the critical point of ignition, where discrete pulses of armature current join to give a continuous conduction of the rectifier (see equation 29·27).

If the change in the critical grid characteristic is disregarded, each particular position of the slider of the potentiometer $P1$ (Fig. 29·18) will correspond to a particular angle of ignition of the thyratron rectifier tubes, and this angle of ignition will remain constant (for a given position of the slider) for different speeds and torques of the motor. As mentioned previously, the type of control shown in Fig. 29·18 may be called vertical control of ignition.

Another type of elementary control, which can be called horizontal control, is shown in Fig. 29·20. The basic difference between the control system in Fig. 29·18 and the one in Fig. 29·20 is that the latter has no d-c component of the thyratron grid voltage, and the control of the angle of ignition is accomplished by means of a variable phase shift rather than by means of a vertical shift of the alternating grid-voltage wave. In the system shown in Fig. 29·20 the phase-shift circuit consists of a center-tapped transformer winding, a resistor $R3$, and a saturable-core reactor SCR.

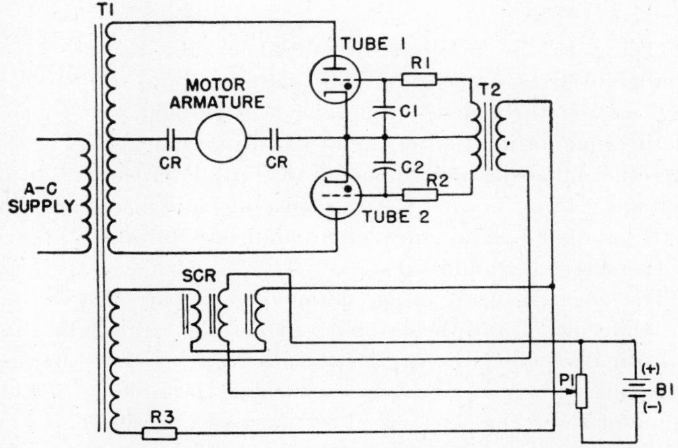

Fig. 29·20 Elementary circuit diagram for rectifier armature-voltage control. Phase-shift control of ignition.

The variable phase displacement can be obtained either by varying the resistance $R3$, or, as in Fig. 29·20, by varying the inductance of reactor SCR through the control of the direct saturating current by means of $P1$. The inductance can also be varied by changing the air gap in a plunger-type reactor. The plunger may be controlled manually or it may be coupled mechanically to some movable parts of the system.

In these two types of elementary control the angle of ignition for a particular setting of the control element remains approximately constant, and is not appreciably affected by the speed of the motor or the load current.

The speed and armature-voltage characteristics of the drives discussed in this section have been analyzed previously in Section 29·4 (Figs. 29·15 and 29·16). It is apparent that, although speed can be controlled by manual adjustment of the angle of ignition, a particular setting of the speed control element $P1$ does not correspond to any particular speed because the latter is influenced to a predominant extent by the load conditions of the motor. The excessively drooping speed-torque characteristics of these elementary drives make them unsuitable for most industrial applications, with the exception of a few special cases, where these "series" characteristics either are not objectionable or may even be advantageous, as for instance in some winding operations.

29·6 ARMATURE-VOLTAGE REGULATION

As pointed out in Section 29·4, there are two reasons for an excessively drooping speed-torque characteristic of an elementary rectifier drive when conduction is discontinuous. First, the armature-voltage drop, directly proportional to the torque, is the reason for drooping of the speed-torque characteristic, common to both rectifier and straight d-c drives. Second, in a rectifier drive, when the conduction is discontinuous, the armature voltage will decrease sharply with increasing torque. This is the primary reason for the excessive droop of the speed characteristic.

The speed regulation characteristic of a rectifier drive can be substantially improved by various special electronic control systems where the armature voltage is prevented from decreasing with the increase of torque by an automatic readjustment of the angle of ignition of the rectifier tubes, so that the angle of ignition x_f is advanced automatically for an increasing torque and delayed for a decreasing torque. In other words, an armature-voltage regulating system may be used to maintain the armature voltage constant, or practically constant, for any given setting of the control element (for instance potentiometer $P1$ in Figs. 29·18 and 29·20).

The diagram of one scheme of a rectifier drive with automatic armature-voltage regulation is shown in Fig. 29·21. In comparing the diagram in Fig. 29·21 with the elementary control diagram of Fig. 29·18 it will be noted that the control circuit, connected to points A and B in Fig. 29·18, is now connected to points A, B, and C in Fig. 29·21. Instead of a simple combination of two opposing direct voltages, we now have a more elaborate electronic control circuit centered around electronic tube 3. Tube 3 is a high-vacuum, high-mu triode or a pentode such as the 6J7 or 6SJ7.

The plate circuit of tube 3 starts at the positive side of the d-c plate-voltage supply, which for the sake of simplification is shown as obtained from battery $B1$. Then the plate circuit is followed through load resistor $R6$ to the plate of tube 3, through the tube to the cathode, and then to the negative side of $B1$. $B2$ supplies the voltage to the biasing resistor $R9$, and to the reference-voltage potentiometer $P1$. The grid circuit of tube 3 is followed from the cathode through $R9$, portion of $P1$, $R5$, grid resistor $R8$ to the grid.

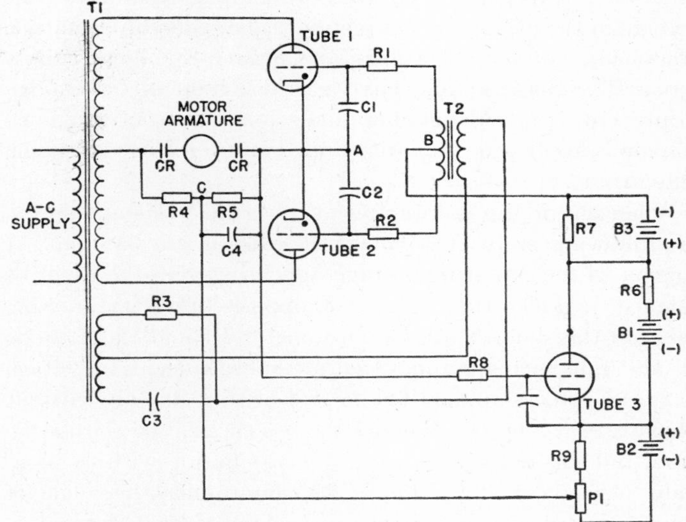

Fig. 29·21 Circuit diagram of an electronic drive of the regulator type for armature-voltage control and regulation.

The voltage across the armature is divided by means of a voltage divider consisting of resistors $R4$ and $R5$. The portion of armature voltage across $R5$ is introduced in the grid circuit of tube 3. The voltage across $R5$ is called the "indicating" voltage or the feedback voltage of the regulating system. Capacitor $C4$ reduces the a-c ripple of that portion of the armature voltage which is in the grid circuit of tube 3.

The third major control circuit is the grid circuit of the power thyratron tubes 1 and 2. The alternating-voltage components displaced in phase with respect to the alternating anode-supply voltages are obtained through the dephasing circuit *R*3-*C*3 and the grid transformer *T*2. These circuits have been described in Section 29·3 (see Fig. 29·2). The variable direct grid-voltage component between the cathode lead of thyratron tubes (point *A*) and the center tap of the secondary winding of the grid transformer *T*2 (point *B*) is developed in the following circuit: cathodes of tubes 1 and 2 (point *A*), resistor *R*5, portion of potentiometer *P*1, biasing resistor *R*9, tube 3 (cathode to plate), resistor *R*7, and center tap of transformer *T*2 (point *B*).

There are two principal direct-voltage components in the thyratron grid circuit corresponding to the voltages across *R*4 and across a portion of *P*1 in Fig. 29·18. These components are the variable voltage across tube 3, which has a positive polarity with respect to the grid, and the constant negative component supplied by the auxiliary voltage *B*3. The remaining two voltages, one across *R*5 and the other across the active portion of *P*1 and across *R*9, do not have any appreciable effect on the grids of rectifier tubes 1 and 2 since they are of opposing polarity and practically equal in magnitude.

The operation of control tube 3 is such that when the tube does not conduct any current the voltage across the tube is high, and a resultant positive direct grid-voltage component is applied to grids of rectifier tubes 1 and 2 so that, in accordance with previous discussions, the rectifier tubes will fire at the most advanced angle possible under the existing conditions, and the maximum possible voltage will be applied to the armature. On the other hand, when tube 3 is made to conduct full possible current, the voltage across it is very low (actually close to zero), and the resultant direct voltage in the grid circuit of rectifier tubes is highly negative, practically equal to the auxiliary negative voltage supply across *R*7. Because this negative grid voltage is just enough to prevent armature rectifier tubes 1 and 2 from firing, no voltage is applied to the armature when tube 3 is conducting full current.

When the motor is running at a certain constant speed, the following state of balance will exist in the system. A portion of the armature voltage appearing across *R*5 will be balanced against the voltage of opposite polarity existing between the slider of the potentiometer *P*1 and the cathode of tube 3, in such a manner that a resultant negative voltage of 2 or 3 volts is applied to the grid of tube 3. The current which tube 3 is conducting determines the voltage across the tube, and this voltage, in turn, in combination with the constant negative voltage across *R*7, determines the angle of ignition of tubes 1 and 2 and, hence, the voltage applied to the armature of the motor.

For a given setting of *P*1, the system shown in Fig. 29·21 will tend to maintain the armature voltage constant. Let us assume, for example, that owing to an increase in torque the armature voltage will start to decrease. Then the voltage across *R*5 will tend to decrease, and the resultant voltage applied to the grid of the high-vacuum amplifier tube 3 will become slightly more negative (note again that the voltage across *R*5 is positive and the voltage between the slider of *P*1 and the cathode of tube 3 is negative with respect to the grid of tube 3). Thus, the current of tube 3 will decrease and the voltage across that tube will increase so that the angle of ignition of tubes 1 and 2 will be immediately advanced, and will tend to prevent the armature voltage from decreasing. A similar action, but in the opposite direction, will take place when the armature voltage will tend to increase (for instance, on decreasing of motor torque).

Owing to the presence of amplifier tube 3, any small change in armature voltage will produce a powerful "restoring" action through the amplifier and through the control of the angle of ignition of the armature rectifier tubes in the direction opposing any change in magnitude of the armature voltage. Thus, normally, the whole system will remain in a state of rigid balance, provided conditions for the stability of the system are fulfilled.

The negative grid-voltage component of tube 3, between the slider of *P*1 and the cathode of the tube, represents what is normally called the "reference" voltage. Let us assume that this voltage is increased by turning the slider of potentiometer *P*1 in the clockwise direction (see Fig. 29·21). The resultant grid voltage of amplifier tube 3 will become more negative, the current of the tube will decrease, the voltage across the tube will increase, and the angle of ignition of rectifier tubes 1 and 2 will be advanced, so that a higher voltage will be applied to the armature, resulting in a correspondingly higher speed of the motor. If the slider of *P*1 is left in the new position, a new state of balance will be quickly established (time must be taken for the acceleration of the motor) at a higher armature voltage and higher motor speed. Thus, it can be readily seen that, by varying the reference voltage through the adjustment of *P*1, the speed of the motor can be arbitrarily controlled by varying its armature voltage.

The scheme shown in Fig. 29·21, and described in this section, is an example of a typical regulator. Although regulators in general are discussed in detail in Chapter 30, some of their basic features and characteristics must be mentioned here in view of their fundamental importance in most electronic motor-control systems.

A regulator in the strictly defined sense is a system (the term is sometimes applied to a portion of the system) where some physical quantity such as speed, voltage, current, or temperature, which is called "regulated quantity," is automatically maintained constant or is made to follow with a high degree of accuracy the variations of the reference quantity. Any regulator must have the following basic elements:

1. Indicator of the regulated quantity.
2. Reference quantity with which the regulated quantity is compared directly or indirectly.
3. An amplifying system necessary to provide satisfactory accuracy and sensitivity of the regulator.
4. Means to provide a "restoring force" resulting in the recovery of the system balance.

In addition, means for securing the stability of the system may be needed, such as different "damping" and "anticipatory" elements or devices. The system shown in Fig. 29·21 is a voltage regulator, and it has all the four basic elements.

In electrical systems the regulated quantity is usually translated into voltage, whose magnitude is proportional to the magnitude of the quantity to be regulated, since voltage can be most readily introduced or fed back into the amplifying system. In the regulating system previously described, the simple combination of resistors $R4$, $R5$, and capacitor $C4$ can be regarded as the indicator of the regulated quantity, that is, of the armature voltage. The reference quantity is the voltage provided between the slider of $P1$ and the cathode of the amplifier. This voltage is constant for a given setting of the potentiometer $P1$. The amplifying system is a very essential part of any regulator since without it proper accuracy could not be obtained, and in that sense the system would cease to be a regulator. In the system shown in Fig. 29·21 amplification of the system is provided by the amplifying properties of the high-vacuum tube 3, so that a slight change in the regulated quantity will cause a change in the voltage across the tube many times greater.

The restoring force is the "back-to-normal action" made sufficiently powerful through the previously discussed amplification. A slight deviation of the regulated quantity from its prescribed value will result in the appearance of the action of the system in the direction to prevent the deviation. Thus, in the system shown in Fig. 29·21, the restoring action consists of the variation of the voltage across tube 3, and the resulting changes in the angle of ignition of the armature-rectifier tubes acting in the direction to oppose any change in armature voltage.

It can be readily seen that a typical feature of a regulator is the cyclic interdependence of several quantities, the number of which may be two, three or more. Also, it should be noted that in a regulator system one may speak of the feedback of the regulated quantity in exactly the same sense as in electronic feedback amplifiers. Generally speaking, a regulator and a feedback amplifier are closely related, and very often sustained oscillations may develop in a high-gain regulating system. These oscillations, which are often called "hunting," are, of course, very undesirable phenomena, which must be eliminated by special stabilizing or anti-hunting means, usually required in most high-accuracy regulators. Some of these means will be described in Section 29·11.

29·7 ARMATURE-VOLTAGE-DROP COMPENSATION

In Section 29·6 an armature-voltage control and regulating system was discussed in some detail. The system shown in Fig. 29·21 represents a rectifier drive with armature control and, from the application point of view, is analogous to the well-known variable-voltage drive of the Ward-Leonard type.

However, in many industrial applications it is important that the drive provide a flat speed-torque characteristic; that is, that the speed of the motor remain essentially constant for different values of torque applied to the motor shaft. In other words, the system preferably should be a speed regulator which would automatically maintain a constant speed for varying loads. The system described in the preceding section is a voltage regulator, and although the armature voltage is maintained constant, the speed will normally decrease with increasing torque, owing to the armature IR drop, as in any conventional d-c drive with shunt-wound motors (see Fig. 29·17).

There are two ways to provide the desired constant-speed characteristic. The first is the speed-regulated drive where speed is maintained constant by a regulating system in which the speed is indicated directly by means of a rotating tachometer generator coupled mechanically to the motor. This type of drive has many advantages, the most important being the accuracy and consistency of operation. The main disadvantages are the necessity of providing a tachometer generator, and the problems of additional space as well as of mounting and coupling the tachometer generator.

The second method is the compensated drive where the armature voltage is regulated as described in Section 29·6 but, in addition, the angle of ignition is readjusted, and its readjustment depends upon the armature current, that is, upon the IR drop in the armature. The readjustment of the angle of ignition is such that with increasing current the angle of ignition of the main rectifier tubes is additionally advanced, and with decreasing current it is additionally delayed. This action, entirely automatic, can be superimposed upon the changes in the angle of ignition resulting from the armature-voltage regulation. The compensated type of drive permits the use of standard motors without additional shaft extensions and couplings for the tachometer generator. The speed-regulated drive will be discussed in Section 29·11. Here, circuits for IR-drop compensation will be described.

Figure 29·22 represents the basic diagram of one of the possible forms of the compensated system. It will be noted that the diagram is derived from the armature-voltage control and regulating scheme shown in Fig. 29·21 by the addition of another amplifier tube 4 and its allied circuits. In order to facilitate the comparison of both diagrams, the designations of circuit elements, common to the diagrams in Fig. 29·21 and Fig. 29·22, have been made the same. Tube 4 is a conventional high-vacuum triode. A new resistor $R10$ is introduced in series with the armature of the motor. This resistance is from two ohms to a fraction of an ohm, depending upon the motor current rating, and is used to indicate the load conditions of the motor. The resistor $R10$ is in the grid circuit of the IR-drop compensating tube 4. Other elements of that grid circuit consist of the anti-hunting control rheostat $P3$, the negative-bias resistor $R13$, and the grid resistor $R12$. The compensation-control potentiometer $P2$ and the resistor $R11$ form the load resistance of tube 4. A portion of the direct voltage across $P2$ is introduced into the grid circuit of the master control tube 3, so that this voltage tends to make the grid of tube 3 more negative.

Reference to the diagram in Fig. 29·22 will show that the resultant direct grid voltage of the master control tube 3 is a combination of three direct-voltage components:

1. The negative component obtained from the speed-control potentiometer $P1$ and biasing resistor $R9$.
2. The positive component across the armature-voltage feedback resistor $R5$.
3. The negative component across a portion of the IR-drop compensation control potentiometer $P2$.

The action of the IR-drop compensating circuit can be easily understood by analyzing first the control-grid circuit of tube 4 (Fig. 29·22), and then the control-grid circuit of tube 3. Referring again to Fig. 29·22, it will be seen that when the load torque applied to the shaft of the motor is increased and, consequently, the armature current is increased, the average voltage drop across $R10$ is increased, so that the control grid of tube 4 will become less negative. This condition will result in the increase of plate current of tube 4 and, hence, in the increase of voltage across $P2$. The control grid of master tube 3 will become more negative, and the plate current of tube 3 will be reduced. It will be recalled that the angle of ignition of armature thyratron tubes is

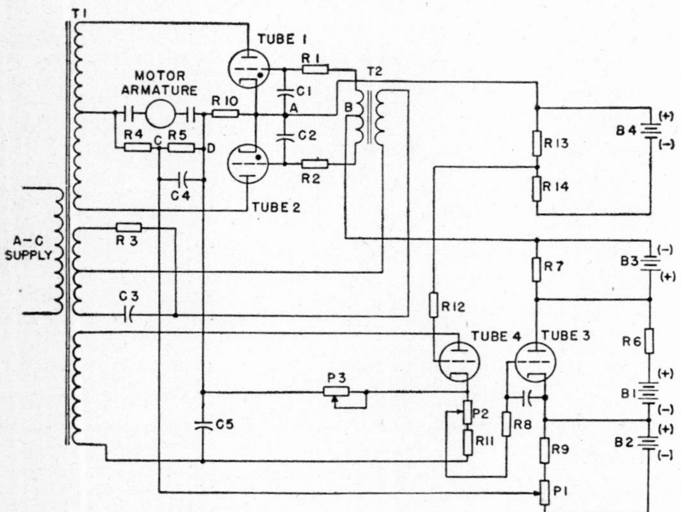

Fig. 29·22 Circuit diagram of an electronic compensated drive.

advanced when the plate current of the master control tube 3 is decreased, and the ignition is delayed when the plate current of that tube is increased. Thus, an increase in armature current will result in the reduction of the plate current of tube 3 and the advancement of the angle of ignition of the armature-rectifier tubes. The opposite is true when the armature current is reduced.

Since the above effect is caused by changes in armature current, and because it is added to the armature-voltage regulating action, which was described in detail in the previous section, it follows that the combination of the two effects will cause an increase in armature voltage with increasing armature current, and a decrease in voltage with decreasing current. If, by proper adjustment of the compensation-control potentiometer $P2$, the increase in armature voltage, for a given increase in armature current, is made to be approximately equal to the corresponding increase in the armature-voltage drop IR, then, from the expression

$$E_g = E - IR \tag{29·44}$$

it becomes apparent that for any change in torque, the two terms of the right side of equation 29·44 will be increased or decreased by the same amount, so that the electromotive force E_g will remain constant. Consequently, if the effect of armature reaction is neglected, the speed of the motor will remain constant and independent of the variations of torque.

It must be remembered, however, that it is not possible to make an adjustment of the rate of compensation by means of potentiometer $P2$, such that the condition

$$\Delta E = \Delta(IR) \tag{29·45}$$

be fulfilled for all possible loads and speeds of the motor. The reason is that the degree of compensation depends not only upon the setting of potentiometer $P2$, but also upon the operating angle of ignition and the magnitude of the electro-

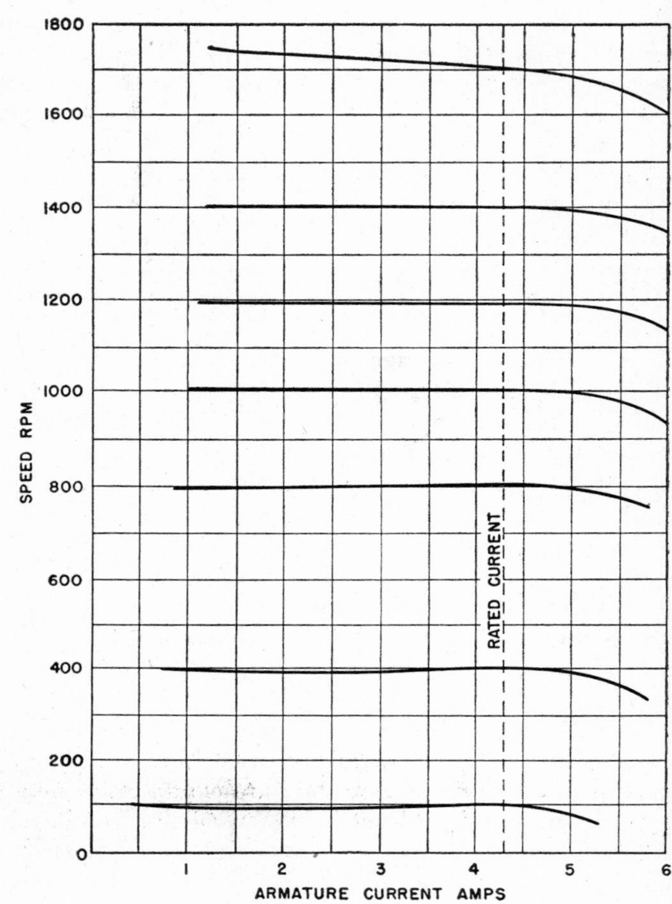

Fig. 29·23 Experimental speed-torque (current) characteristics of a compensated electronic drive. $p = 2$. Motor rating: 1 hp, 1750 rpm, 230 volts. Constant field excitation.

motive force, and those conditions will be different for different speeds and torques of the drive. Even if the condition for an ideal compensation (equation 29·45) is fulfilled for all operating speeds and loads, that is, if the electromotive force E_g for a given speed is actually kept constant, the speed may still vary because of the effect of armature reaction and the resultant variation of the operating flux in the motor. In fact, it will be recalled that the electromotive force of a shunt-wound motor is a true representation of its speed only under the simplifying assumption that the armature reaction is neglected and that the operating flux remains constant. Such a condition is actually never met in practice. Furthermore, the degree of compensation is affected by the characteristics of the armature circuit, mainly by the resistance and inductance of the armature winding, and a satisfactory adjustment of the compensation-control potentiometer for one motor may prove inadequate for another motor, even of a similar rating. Moreover, the resistance of the armature circuit is subject to some changes caused by variations in temperature of the winding, and even by variations in the brush drop.

In spite of all these difficulties, it is normally possible to obtain a satisfactory compensation at different speeds and loads for a given motor and for one optimum setting of the compensating control. As an example, it may be stated that, with a 10-to-1 speed range by armature-voltage control, 1 to 2 percent speed regulation from no load to full load at base speed and 5 to 7 percent at $\frac{1}{10}$ of the base speed is normal. With 20-to-1 speed range, from 8 to 12 percent speed regulation at $\frac{1}{20}$ of base speed may be expected.

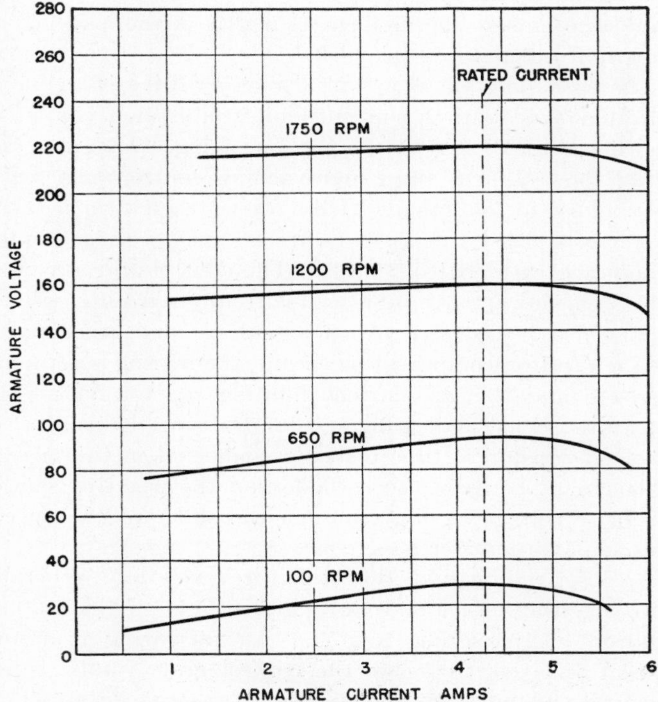

FIG. 29·24 Experimental graphs of armature voltage versus load current for an electronic compensated drive. $p = 2$. Motor rating: 1 hp, 1750 rpm, 230 volts. Constant field excitation.

These figures indicate that from the point of view of speed regulation the performance of a compensated electronic drive is much better than that of a conventional motor-generator-drive system where 8 to 12 percent regulation may be expected at base speed, with motor stalling at full rated torque when the no-load speed is reduced to about $\frac{1}{10}$ of base speed (100 percent regulation).

As an example of performance of a compensated rectifier drive, Fig. 29·23 represents a family of experimental speed-torque characteristics taken for a 1-horsepower, 1750 revolutions per minute compensated rectifier drive. In Fig. 29·24 are shown experimental characteristics of armature voltage as a function of armature current (torque) for different speeds. In connection with the experimental graphs shown in Figs. 29·23 and 29·24, several facts should be pointed out. The full rated field current was maintained throughout the test. The compensation-control potentiometer $P2$ was adjusted to an optimum setting to give a close speed regulation over a 20-to-1 speed range. The setting of $P2$ was the same for the entire range of speeds. From the graphs in Fig. 29·23 it may be seen that, within the load range from no-load current to full-load current of 4.3 amperes, the speed of the motor was kept within 4 percent for any speed of the 10-to-1 range. At $\frac{1}{19}$ of the base speed the regulation was about 7 percent. The increased drooping of the speed-torque characteristic for currents exceeding the rated value of 4.3 amperes was caused by the current-limiting effect of the drive. (Special current-limiting circuits, not included in Fig. 29·22, will be discussed in Section 29·8.) From graphs shown in Fig. 29·23 it is apparent that the motor was able to deliver full rated torque at a speed as low as $\frac{1}{19}$ of the rated speed, without any appreciable slow-down.

A peculiarity of voltage-current characteristics shown in Fig. 29·24 is the slope of curves which is greater for low speeds and smaller for higher speeds. This phenomenon can be explained by the effect of decreased system amplification at higher speeds. The fact that the armature-voltage curves in Fig. 29·24, which are not parallel to one another, result in the practically parallel speed curves shown in Fig. 29·23 can be explained by the influence of armature reaction in the motor. In fact, because the flux ϕ in equation 29·1 is actually a function of the load current I, the increment of armature voltage required to maintain constant speed in spite of IR drop decreases gradually with increasing speed.

29·8 CONTROL OF ACCELERATION

Electronic motor control of the rectifier type is particularly well suited for use as an automatic starter of the most elaborate type.

The three important factors to be considered during the process of acceleration of a motor are: (1) duration of the accelerating period; (2) safety of acceleration; and (3) magnitude of starting torque.

The term safety of acceleration implies such considerations as the magnitude of starting current and possible mechanical shocks, if the inrush current is too high. It is apparent that the three factors just mentioned are interrelated since the higher the current is, the higher the starting torque will be, and the shorter the duration of acceleration. In most industrial applications the duration of the accelerating period should be as short as possible, and a high starting torque is a definite advantage. However, the factor of safety puts a definite limitation on both time of acceleration and the magnitude of starting torque. In fact, the current during starting must not be too high; this requirement is imposed by both the motor and the control equipment, and sometimes also by the power-supply line. In conventional magnetic control systems the motor usually represents the limiting factor. The motor may be damaged by too high a starting current, and in particular the arcing on the commutator may be destructive to both the commutator and the brushes. In addition, the mechanical shock produced by a heavy starting current may be detrimental to the whole driving system, and it should be avoided or at least kept at a minimum. In electronic or rectifier drives, and particularly in those where thyratron tubes are used to transmit power to the armature, the rectifier tubes usually constitute the limiting factor because as a rule they have a very critical average- and peak-current rating which cannot be exceeded without risking permanent damage to the tubes. Thus, the factor of safety of acceleration assumes here an additional significance, and has to be given special consideration in the analysis and design of control of the whole process of acceleration.

There are two main types of automatically controlled acceleration: (1) time-delay acceleration; and (2) current-limit acceleration.

The time-delay acceleration is based on gradual increase of voltage applied to the armature of the motor, so that the speed of the motor and the generated electromotive force build up more or less simultaneously with the armature voltage, and in that manner heavy starting current is avoided. The basic feature of the time-delay acceleration is the lack of relationship between the build-up of the voltage and the speed, the voltage being increased gradually and independently of speed; that is, its instantaneous value during acceleration

$$e = f(t) \tag{29·46}$$

is solely a function of time t.

The chief disadvantage of the time-delay acceleration is the lack of automatic co-ordination between the behavior of the armature voltage during the acceleration, and the mechanical characteristic of the load, that is, the moment of inertia of all the rotating parts referred to the shaft of the motor. Thus, if the rise of the armature voltage is very slow and the WR^2 of the system is low, the current in the armature and the accelerating torque may be unnecessarily low, and the duration of acceleration will be unnecessarily long. With the exception of a few special cases where such conditions may be advantageous, this type of starting performance is inadequate for most industrial applications. On the other hand, if the rise of the armature voltage during starting is relatively fast and the WR^2 of the system is high, the safety of starting may be impaired, with current in the armature exceeding the permissible safe value. It is possible, of course, to design the time-delay-starting control system in such a manner that it will be adapted for a particular rotating system and give a satisfactory performance but, if the rotating system is changed, the difficulties as outlined above may be experienced.

The current-limit acceleration is a much more satisfactory and efficient type of controlled starting. Although that type of acceleration is also based on gradual increase of voltage applied to the armature, this voltage increase is not independent of the behavior of the motor, that is, its speed, starting current, torque, and so on. On the contrary, the armature voltage during the acceleration is influenced by the armature current, and is automatically reduced to prevent the starting current from exceeding a certain predetermined value. If the current limit is adjusted to a maximum safe value, the manner in which the armature voltage varies will be automatically adapted to the WR^2 of the load; therefore the acceleration will proceed in the most efficient manner and, for a given degree of safety, will be completed in the shortest possible time. The current-limit acceleration normally can be regarded as constant-torque acceleration because during a major portion of the accelerating period the armature current remains constant and equal to the predetermined current limit, whereas the field current is normally constant, at least until base speed is reached. Thus, the duration of the accelerating period can be expressed by the formula

$$t = \frac{0.00325n(WR^2)}{T} \tag{29·47}$$

where n = speed in revolutions per minute to which the motor is to accelerate
T = accelerating torque in pound-feet
t = time of acceleration in seconds

In formula 29·47, WR^2 in pound-feet square at the motor shaft is to include the moment of inertia of the load and of the armature of the motor itself.

The basic diagram of a typical rectifier drive of the compensated type with a current-limit-acceleration control is shown in Fig. 29·25. A comparison of Fig. 29·25 and Fig. 29·22 shows that a third high-vacuum control tube 5 has been added to the circuit. Tube 5 is a triode or a pentode characterized by a sharp cut-off. Its plate-voltage supply is common with that of tube 3. The control-grid circuit of tube 5 includes the motor load-current indicating resistor $R10$, which is also used as a load-current indicating means for the IR-drop compensating circuit, the biasing potentiometer $P4$, and the grid-current limiting resistor $R15$. The armature-voltage indicating resistor $R5$ is at the same time the load resistor of tube 5, being connected in the latter's plate circuit between the cathode and the negative side of the plate-voltage supply (through the slider of the speed-control potentiometer $P1$).

A closer analysis of the circuit reveals that, generally speaking, the action of tube 5 is just reverse that of the IR-drop compensating tube 4. An increasing armature current will tend to make the control grid of tube 5 less negative through the increase of the positive grid-voltage component across resistor $R10$. Tube 5 is normally cut off because of sufficiently high negative grid voltage obtained from biasing potentiometer $P4$. However, when the armature current reaches a certain critical value, tube 5 will start conducting, and its plate current, flowing through resistor $R5$, will make the grid voltage of the master control tube 3 less negative. Consequently, the plate current of tube 3 will increase, the angle of ignition of the main rectifier tubes 1 and 2 will be delayed, and such an adjustment of the armature voltage will result that the current will be prevented from exceeding a predetermined limit corresponding to the stalled torque of the motor.

It may be useful to repeat that within the normal operating range of armature currents, say from no-load to 150 percent of full-load current, tube 5 does not conduct any current being biased off by a high negative voltage from potentiometer $P4$. Thus, within that range the armature current affects solely the IR-drop compensating tube 4, and the compensating action takes place in the manner described in Section 29·7. If, however, the armature current rises above a certain predetermined value, tube 5, which is characterized by a sharp cut-off, will suddenly start to conduct. The amplifying action of tube 5 and its associated circuits is much stronger than that of compensating tube 4 and, when the current-limit tube conducts, the relative effect of the compensating tube is negligible. Because of the high amplification of tube 5, the slightest increase in armature current above a certain critical value results in a very strong action opposing that increase, and the maximum possible value which the armature current will reach will correspond to stalled torque of the motor. This maximum value of current (current limit) normally will be higher than the critical value of armature current corresponding to the cut-off point of tube 5.

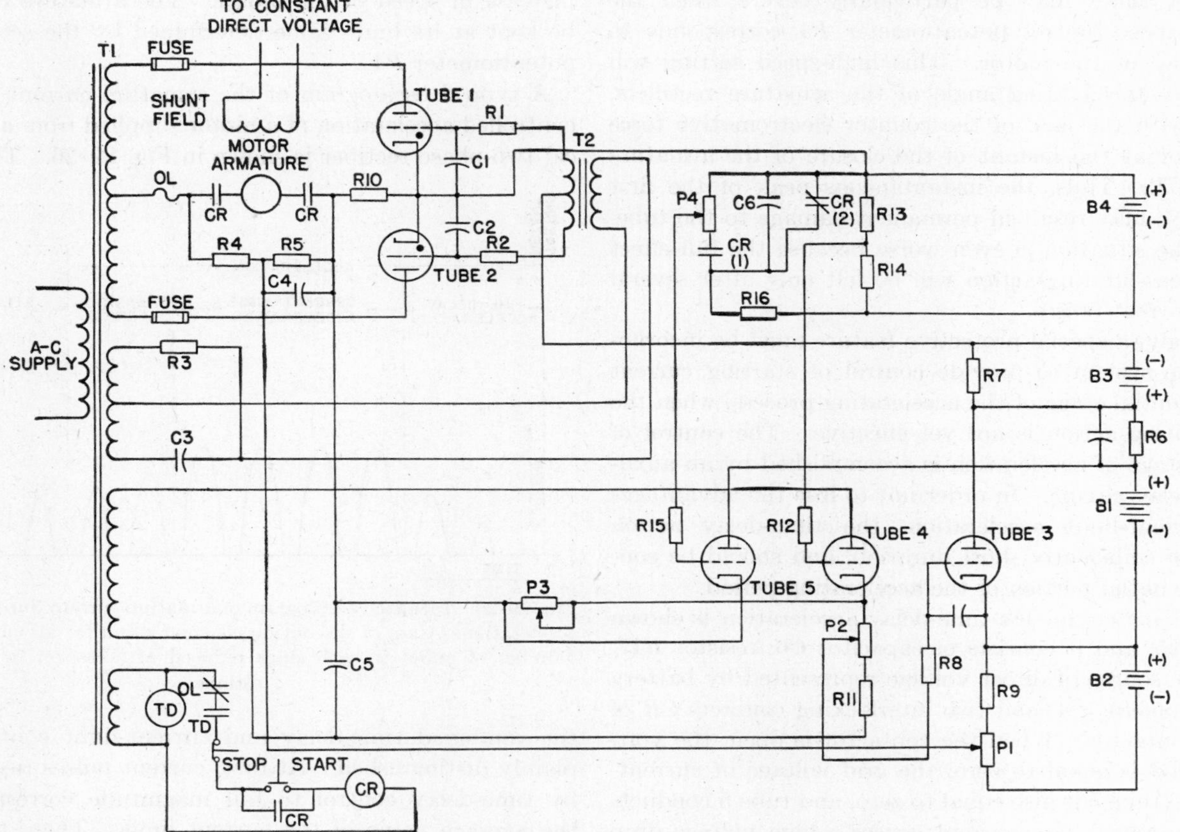

FIG. 29·25 Circuit diagram of an electronic adjustable-speed compensated drive with controlled acceleration.

neous value of anode current. If that value is exceeded, the emission of electrons from the cathode may momentarily become insufficient, and the positive-ion bombardment of the cathode may permanently damage or completely destroy the cathode-oxide coating. Thus, a thyratron may be destroyed by a single peak of current which exceeds the tube rating, even if the duration of the peak is as short as several microseconds. In previous discussions of the current-limit acceleration and current-limiting circuits, all the currents to which reference was made were average values. The

It was mentioned previously that the element of safety and of necessary protection, involved in the problem of automatic acceleration of a motor, refers not only to the motor and the line itself but, in electronic rectifier systems, also to the main rectifier tubes supplying the armature, which is particularly true if the rectifier tubes are of hot-cathode type, that is, thyratrons filled with either mercury vapor or inert gas. It must be remembered that there are two current ratings of those tubes, one the average-current rating, and the other the instantaneous peak-current rating. The effect of the rating of the tubes on the rating and type of the armature rectifier will be discussed in Section 29·14. The theoretical background of the instantaneous-peak value of armature current in a controlled rectifier will be discussed in Section 29·13. Here, the significance of the instantaneous-peak current in a rectifier tube will be briefly mentioned, however, because it involves important implications with relation to the process of acceleration of the motor.

It should be recalled that the peak-current rating of a thyratron tube refers to a maximum permissible instanta-

corresponding peak values of armature current are of particular importance from the point of view of the operation of tubes, and in some cases the instantaneous-peak current rather than the average current may constitute the actual determining factor of the power rating of the rectifier and the maximum permissible current limit.

Although the action of the current-limiting control normally reduces the average as well as the instantaneous-peak values of the current because of the existing definite relationship between the two, the peak-current conditions during the initial stage of starting are of a different nature. They involve the inability of the current-limiting circuits to exercise control over the peak current at the very beginning of the accelerating process. In fact, at the instant of starting, when the first armature-rectifier tube begins to conduct, the armature current as indicated by the voltage across the resistor $R10$ (Fig. 29·25) is zero and, consequently, there can be no current-limiting action by tube 5, and no effect on the angle of ignition of the rectifier tubes. Even if it is assumed that the current-limiting circuit can take full control in a short time corresponding to a fraction of a half cycle, its action on the control grid of the conducting thyratron would have no effect on the magnitude of the first pulse of current carried by the thyratron which started to conduct first. This condition is apparent from the fundamental principles of operation of grid-controlled rectifiers and, as a result, at least the first conductive half cycle in one of the tubes will be associated with an exceedingly high current peak, limited solely by the inductance and the resistance of the armature circuit. The instantaneous-peak currents in the rectifier tubes may be particularly severe when the setting of speed-control potentiometer *P*1 corresponds to a high speed of the motor. This high-speed setting will mean an advanced firing angle of the armature rectifiers, combined with the lack of the counter electromotive force of the motor at the instant of the closure of the armature contactor *CR*. Thus, the instantaneous peak of the first current pulse may result in permanent damage to the tube. Actually, the situation is even worse because the full effect of the current-limiting action will be felt only after several armature-current pulses.

Accordingly, a special protective feature must be incorporated in the system to provide control of starting current during the initial stage of the accelerating process, when the current-limiting action is not yet effective. The control of the initial stage of acceleration is accomplished by an auxiliary time-delay circuit. In order not to lose the advantages of the current-limit acceleration, the time-delay action should be of sufficiently short duration, and should be confined to the initial portion of the accelerating period.

The auxiliary circuit for time-delay acceleration is shown in Fig. 29·25, and it consists of capacitor *C*6, resistor *R*17, an auxiliary source of direct voltage represented by battery *B*4, potentiometer *P*4, and two interlocking contacts *CR* of the main contactor. When the contactor is open, the voltage across *P*4 is equal to zero, the grid voltage of current-limit control tube 5 is also equal to zero, and tube 5 conducts full plate current. This current causes a high voltage drop across resistor *R*5 and makes the grid voltage of master control tube 3 either zero or even positive, so that the latter will also conduct full plate current. The circuits are so designed that full plate current of master control tube 3, and the resulting low voltage across that tube, correspond to a complete cut-off of the main rectifier tubes.

As a result of this interlocking, the grid conditions of the main rectifiers at the instant of closure of the armature contactor are such that the rectifiers cannot conduct any current. The voltage across potentiometer *P*4, which provides the negative bias for current-limit tube 5, is zero when the contactor closes, and does not appear immediately after the closure of the contactor. In fact, the voltage across *P*4 is equal to the voltage across the starting capacitor *C*6, and the latter is charged gradually through the resistor *R*16. Thus, the negative biasing voltage of tube 5 gradually reappears, and the plate current of the latter decreases. As a result, the plate current of master control tube 3 also decreases, so that the angle of ignition of the rectifiers is gradually advanced, and they conduct a gradually increasing current. The current-limit tube, however, is not cut off completely by the reappearance of the negative biasing voltage across *P*4 as long as the acceleration is not completed. The time constant of the capacitor-charging circuit is normally relatively short, and the armature current, initially controlled by the time-delay auxiliary circuit *C*6-*R*16, soon increases to such a value that the voltage across armature series resistor *R*10 becomes sufficiently high to influence the conductivity of tube 5. In that manner, the current-limiting circuits take over the control of the angle of ignition, and prevent the firing angle from being advanced too rapidly so that the armature voltage will be increased in accordance with the increase in speed of the motor. The armature current will be kept at its limit value determined by the setting of the potentiometer *P*4.

A typical oscillogram of the armature current during the controlled acceleration of a motor supplied from a symmetrical two-phase rectifier is shown in Fig. 29·26. The effect of

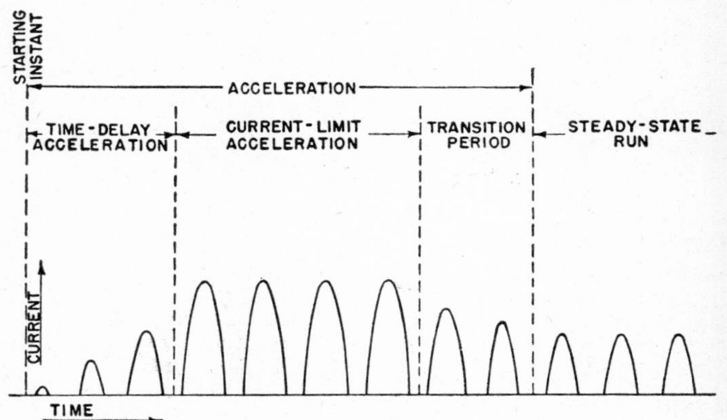

Fig. 29·26 Idealized oscillogram of armature current during controlled acceleration. Case of discontinuous conduction for all current values. Number of pulses in each stage reduced with respect to normal conditions.

the combined time-delay and current-limit control can be plainly distinguished. At first, current pulses rise gradually by time-delay control to full magnitude corresponding to the average value of the current limit. Then, the current does not rise any more but remains close to the limit value where it is controlled by the electronic current-limit system. Later, when the speed of the motor increases to a value approaching the steady-state speed (determined by the setting of *P*1 in Fig. 29·25), the current decreases to its normal operating value.

Referring again to Fig. 29·25, additional significance of the interlocking contacts of the auxiliary time-delay accelerating circuit should be noted. Contact *CR*(1) breaks the circuit of torque-control potentiometer *P*4 so that, when the main contactor is being opened, tubes 5 and 3 start to conduct full plate current, and the main rectifier tubes are cut off, even before the contactor has time to drop out completely. Consequently, any arcing which may have started across the main armature contacts on opening of the contactor will be suppressed immediately by the rectifier tubes. In order to eliminate any traces of arcing on the main contacts, an additional starting relay may be added. This relay would open the circuit of *P*4 a couple of cycles before the main contactor starts to drop out and, in that manner, the armature current will be actually interrupted by the main rectifiers, and not by the contactor. Contact *CR*(2) closes when the main contactor is deenergized, and capacitor $C6$ is discharged when the motor is being stopped. In that manner the starting circuit is reset for the next operation. It is obvious that the armature current-limiting control

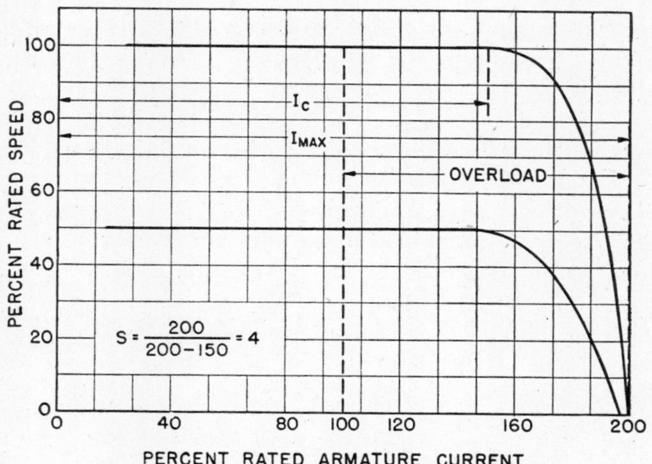

FIG. 29·27 Effect of current-limiting action on speed-torque characteristics. Illustration of current-limit "sharpness," $I_{max.} = 2I_r$, $I_c = 1.5I_r$.

operates not only during starting of the motor but also on overloads, when the armature current increases beyond its critical value.

The critical value of armature current I_c is defined as the average value of load current which corresponds to the cut-off point of current-limit tube 5. The current limit I_{max} is the maximum value of average load current obtainable under the given conditions, and corresponding to the stalled torque of the motor. The ratio of the current limit to the difference between the current limit and the critical value of armature current may be called the "sharpness" of the current-limiting action of the system:

$$s = \frac{I_{max}}{I_{max} - I_c} \tag{29·48}$$

where $I_{max} > I_c$. An ideal current-limiting system is characterized by $I_{max} = I_c$ or $s = \infty$.

The sharpness of the current limit has a great influence on the speed-torque characteristic of the drive. This becomes apparent if one considers that a low sharpness means that the current-limiting element (tube 5) will conduct over a wider range of load currents, and may lower the armature voltage with increasing current even for normal operating values of torque. For proper operation, the critical value of current should be higher than the rated current, $I_c > I_r$. If $I_c = 0$, tube 5 will conduct over the entire range of load currents, and the current-limit sharpness will be equal to unity. In that case, exceedingly drooping speed-torque characteristics will be obtained. Normally, a current-limit sharpness equal to 4 or higher may be considered satisfactory.

The current-limiting action on overloads and the significance of its sharpness are illustrated by means of two graphs in Fig. 29·27. The graphs represent normal speed-torque characteristics of the drive, drawn beyond the rated value of current and up to the stalling point of the motor, corresponding to the current limit of the system. The critical and limit values of armature current I_c and I_{max} are marked in Fig. 29·27, and their significance, as well as that of the current-limit sharpness, can be readily understood.

29·9 MODIFICATIONS OF BASIC CIRCUITS FOR COMPENSATED DRIVES

In Sections 29·6, 29·7, and 29·8 several basic networks for a controlled-rectifier drive were described, and their operation and significance were discussed in detail (see Figs. 29·21, 29·22, and 29·25). These networks represent controls for armature-voltage regulation (Section 29·6), armature-voltage-drop compensation (Section 29·7), armature-current limitation (Section 29·8), and time-delay acceleration (Section 29·8). There is, of course, a number of possible modifications of these circuits, and some of them may present different advantages over the others. Some of the most important modifications of particular practical usefulness will be briefly discussed in this section.

The diagram shown in Fig. 29·28 represents the basic power circuits for armature control as well as the phase-shift circuit of the type described in previous sections. The electronic-control network is shown symbolically. A comparison of Fig. 29·28 with the diagram described in the preceding section (Fig. 29·25), will show the following differences. The armature-current indicating resistor $R10$ shown in Fig. 29·25 is eliminated in Fig. 29·28, and the current indication is obtained by means of an auxiliary rectifier circuit consisting of transformer $T3$, double-anode rectifier tube 6, filtering capacitor $C7$, output resistors $R19$ and $R20$, and a protective resistor $R17$.

Transformer $T3$ is fundamentally a current transformer operated at secondary voltages much higher than is the normal practice. It has two primary windings, and a unidirectional pulsating current flows through each of these

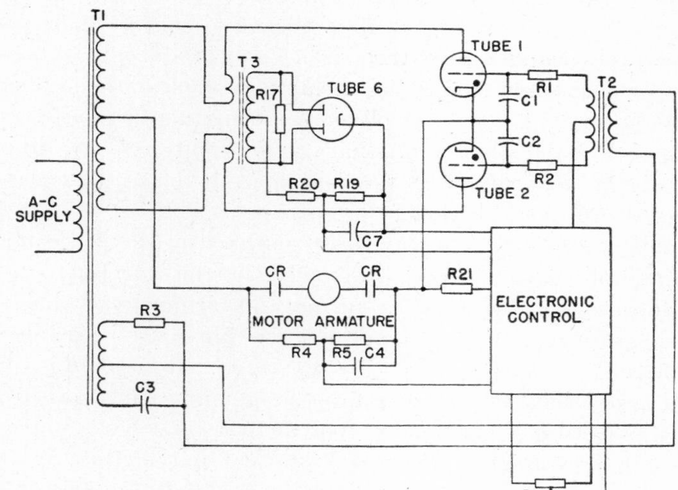

FIG. 29·28 Simplified circuit diagram of a rectifier drive with transformer-type load-current indication.

windings because the current is rectified by the main armature rectifiers, tube 1 and tube 2. The two primary windings are connected in such a manner that an alternating magnetic flux is produced in the core, so that, even with continuous current in the armature, saturation of the transformer is avoided. The equivalent load resistance of the transformer circuit is in general high, and the alternating voltage across the secondary winding may be of the order of several hundred volts. The secondary voltage is in general proportional to the primary current, that is, proportional to the load current of the motor. The alternating voltage across the secondary winding of the current transformer $T3$ is rectified in a conventional manner by means of a high-vacuum, double-anode rectifier tube 6. Because the rectified voltage contains an objectionable double-frequency ripple, the purpose of capacitor $C7$ is to reduce the effect of this ripple. A portion of the rectified voltage across resistor $R19$ is introduced into the electronic-control network, so that it forms the positive grid-voltage component of both IR-drop compensating tube and current-limiting tube (see Fig. 29·25).

Although the armature-current indicating circuit in Fig. 29·28 is more complex than the simple way of obtaining current indication through the series resistor $R10$ (Fig. 29·25), the current-transformer type of circuit has a number of advantages. Larger currents can be handled more readily through a current transformer than through a series resistor, whose resistance will have to become very small in order to avoid excessive power losses. The current-transformer circuit provides ample amplification, whereas the voltage drop across series resistor must be kept low in order not to dissipate too much power.

There is still another important advantage of the transformer-type current-indicating circuit. Figure 29·25 shows that resistor $R5$, which is part of the voltage divider across the armature and constitutes the armature-voltage feedback element, is also used as a load resistor for the current-limit tube 5. In order to obtain adequate amplification of the current-limit control circuit, the resistance of $R5$ must be at least 30,000 to 50,000 ohms. If it is lower, the sharpness of the current-limiting action will be impaired. On the other hand, it will be recalled that resistor $R5$ is in the grid circuit of the armature thyratron tubes. The grid current of thyratron tubes 1 and 2 flows through $R5$ and produces a voltage drop which tends to decrease the amplification of the system since it upsets the proper voltage distribution across resistors $R4$ and $R5$. This additional grid-current voltage drop should be kept as low as possible, and therefore the resistance of $R5$ should be low.

The contradictory requirements outlined in the preceding paragraph can be easily reconciled in the current-transformer type of the armature-current indicating circuit shown in Fig. 29·28. In this circuit the two previously described functions of resistor $R5$ can be separated. Resistor $R5$ will be used solely as an armature-voltage indicating resistor, and its resistance can be low, 1000 or 1500 ohms for example, so that the voltage drop across $R5$ caused by the grid current of tubes 1 and 2 will be negligible. On the other hand, the coupling resistor between circuits of the current-limit tube and the master control tube can now be moved out of the grid circuit of the main rectifier tubes, and its resistance can be made much higher, of the order of 100,000 to 250,000 ohms (resistor $R21$ in Fig. 29·28). In that way, the amplification of the current-limit control can be increased, and the sharpness of the current-limiting action improved.

Another important possible modification of the control network relates to the IR-drop compensating circuit. Theoretically, it is possible to obtain a voltage directly proportional to the electromotive force of a d-c motor by a proper combination of two voltages, one proportional to the armature voltage and the other proportional to the armature current. Essentially, the same thing has been accomplished in the circuit for a compensated drive described in Section 29·7 (see Fig. 29·22). In the simpler modified circuit, however, the IR-drop compensating tube is omitted, and the combination of the two voltages is fed directly into the grid circuit of master amplifier tube 3.

An electromotive-force indicating circuit for a shunt-wound motor with constant field excitation is shown in Fig. 29·29. From that figure, it is easy to calculate that the indicating voltage E_i is directly proportional to the electromotive force of the motor E_g, if the voltage across $R5$ is equal to

$$E_{R5} = \frac{r}{R + r} E_{d\text{-}c} \tag{29·49}$$

Then, the indicating voltage is:

$$E_i = \frac{r}{R + r} E_g \tag{29·50}$$

where $E_{d\text{-}c}$ = armature-supply voltage
E_g = electromotive force of the motor
r = resistance of the series current-indicating resistor
R = armature resistance.

Although the modified feedback circuit shown in Fig. 29·29, as well as the omission of tube 4 represents a simplification of the system, the practical usefulness of such a

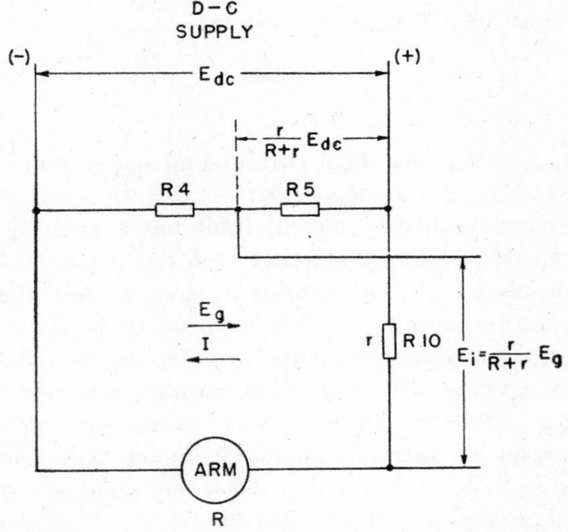

Fig. 29·29 Circuit for indication of the electromotive force of a d-c shunt-wound motor with constant excitation.

scheme is limited by several factors. For small motors of ½ horsepower and below the armature resistance is generally so high and the natural speed regulation of the motor so poor that, in order to obtain a constant-speed characteristic of the drive, an additional amplifier stage is required. This amplifier stage may be provided by an electronic tube in a circuit as shown in Fig. 29·25, or by the transformer-type current-indicating network, described previously in this section. For motors larger than 1 horsepower the stabilization of the system with indicating circuits as in Fig. 29·29 is, generally speaking, more difficult.

The current-transformer type of indicating feedback circuit (Fig. 29·28) can be used directly in combination with main amplifier tube 3, without the intermediate *IR*-drop compensating tube 4 (Fig. 29·25). The performance and stability of such a system are generally much better than those of the system using the indicating circuit shown in Fig. 29·29.

29·10 FIELD RECTIFIERS AND CONTROL

Most of the previous sections dealt exclusively with armature control of d-c motors, and in Sections 29·5, 29·6, 29·7, and 29·8 it was assumed that field excitation of the motor is supplied from a source of constant direct voltage. Normally, in electronic drives the field excitation is provided from a rectifier which may be of either a controllable or a non-controllable type.

In general-purpose electronic drives, field rectifiers and their controls do not present as many problems as armature circuits. In the first place, the currents to be handled by field rectifiers are in general much lower than armature currents. Secondly, the load circuit is of the high-inductance type, and thus provides an increased filtering action. Thirdly, since the speed range by field control is very rarely wider than 3 to 1, the range of field-current control usually does not exceed 5 to 1. The most important feature of field rectifier, however, is that the resistance of the load circuit, in contrast with that for armature rectifiers, is high, and is the natural limiting factor for the current. In armature rectifiers the situation is complicated by the presence of the electromotive force of the motor, which plays a predominant part in determining the behavior of the current, and calls for special means for current control (current limit).

In view of the above, three-phase rectification is normally not required for field-current supplies, mainly because of the relatively low currents involved, and the filtering action of the load. Furthermore, in most general-purpose electronic drives there is no particular need for field-current or field-voltage regulation or elaborate controls because in this case only a-c line-voltage variations are important as a factor affecting the speed of the motor through its field current. However, this effect is partly compensated by the simultaneous influence of armature controls. Only some special types of electronic drives may require an elaborate field control, particularly if the latter is used as part of the motor-speed regulating system.

A separate group of motor controls consists of rectifier field supplies (controlled rectifier exciters) and electronic regulators, combined with motor-generator sets for armature-voltage supply, rotating exciters, and the like. Although that kind of equipment may also be classed as electronic motor control, it is not covered in this chapter, which deals primarily with systems where both armature and field circuits of the motor are supplied from an electronic rectifier and controlled electronically. The electronic field-control equipment which operates in combination with a system of rotating machines, other than the driving motor itself, is discussed in Chapter 30.

Figure 29·30 represents a typical symmetrical two-phase phanotron-rectifier circuit for the field supply of a motor.

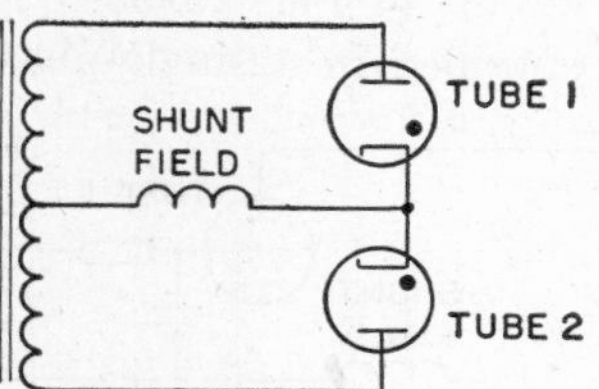

FIG. 29·30 Circuit diagram of a symmetrical two-phase field rectifier using two phanotron tubes.

The circuit shown in Fig. 29·31 is a single-phase, half-wave rectifier (tube 1) with a discharge tube 2 connected across the field winding. Here, advantage is taken of the high inductance of the field winding, so that the current from the transformer flowing through tube 1 during the positive half cycle is maintained by the high inductance of the field winding and flows through the discharge tube 2, after tube 1 has stopped conducting. In other words, the energy from the transformer, stored in the magnetic field during the conductive periods of tube 1, is being discharged through tube 2 during the non-conductive periods of tube 1. In this manner, a continuous flow of field current is obtained, and the a-c ripple is very moderate. Figures 29·30, 29·31, and 29·32 represent rectifiers for constant field. When the field of the motor is to be adjustable to provide speed control, controlled rectifiers, as shown either in Fig. 29·33 or in Fig. 29·34, are used. The circuit in Fig. 29·33 is a conventional symmetrical two-phase, half-wave controlled rectifier, and the circuit in Fig. 29·34 is a controlled rectifier of the single-phase discharge type, analogous to the one in Fig. 29·31, described previously.

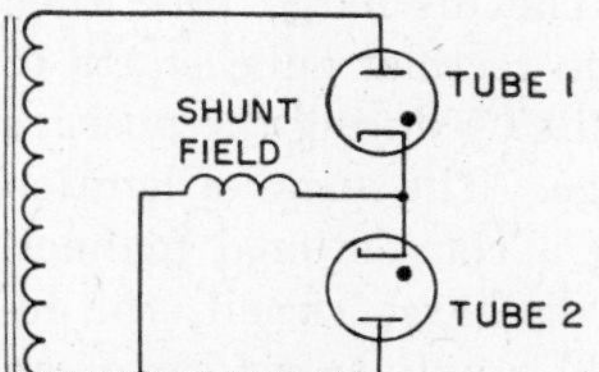

FIG. 29·31 Circuit diagram of a single-phase field rectifier using one phanotron as a rectifier tube and another phanotron as a field-discharge tube.

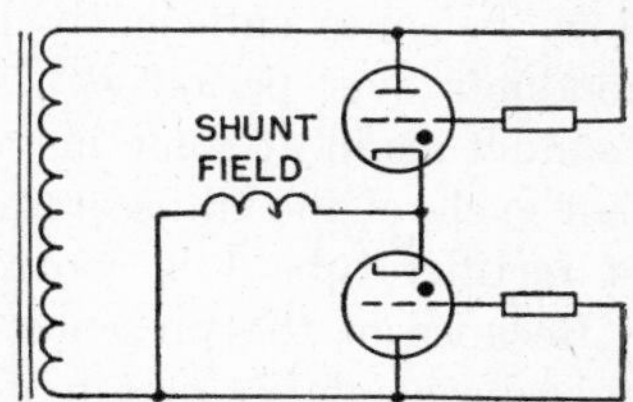

FIG. 29·32 Circuit diagram of a single-phase field rectifier with a discharge element. Two thyratrons are connected to perform as phanotrons.

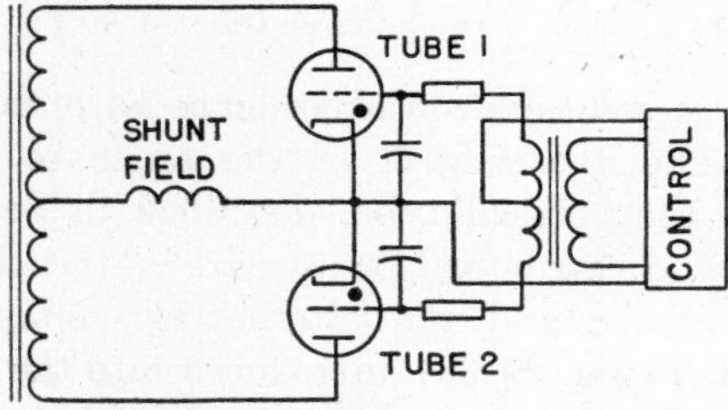

FIG. 29·33 Circuit diagram of a symmetrical two-phase, adjustable field rectifier.

The conventional rectifier circuit shown in Fig. 29·33 does not require any particular discussion since it has already been described in Sections 29·3 and 29·5. However, the discharge-type circuit in Fig. 29·34 calls for additional attention. It is apparent that this circuit is simpler than the conventional circuit in Fig. 29·33. The wave shape of the field current obtained in a single-phase rectifier system

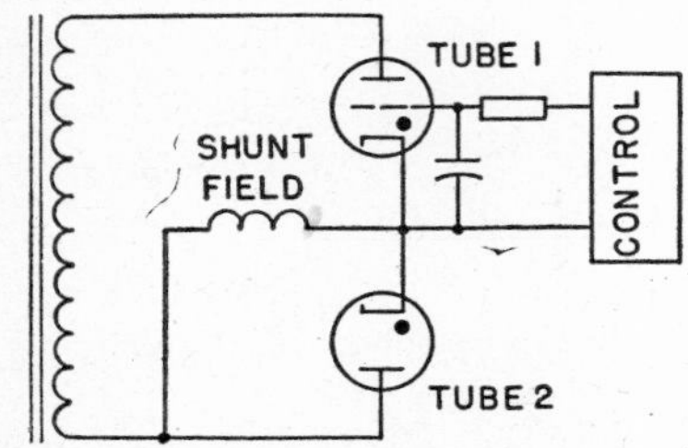

Fig. 29·34 Circuit diagram of a single-phase adjustable field rectifier with a field-discharge phanotron.

with a discharge element is shown in Fig. 29·35. It will be noted in Fig. 29·35 that rectifier tube 1 is conducting over a period AB, and that discharge tube 2 breaks down at point B, where the electromotive force generated in the field inductance and acting in the direction to sustain the current flow, becomes sufficiently high. The discharge tube then conducts over period BC, until the rectifier tube starts to conduct again at some instant (point C) during the positive half cycle of the transformer voltage. The angle of ignition of rectifier tube 1 is controlled in a conventional manner.

Because of the presence of the discharge circuit, the required secondary voltage of the field-supply transformer can be calculated from the simple formula for single-phase half-wave rectification:

$$E_s = 2.22(E_{d\text{-}c} + 15) \text{ volts} \tag{29·51}$$

where $E_{d\text{-}c}$ is the required maximum field voltage.

An example of a control circuit for the field rectifier of a general-purpose electronic drive is shown in Fig. 29·36.

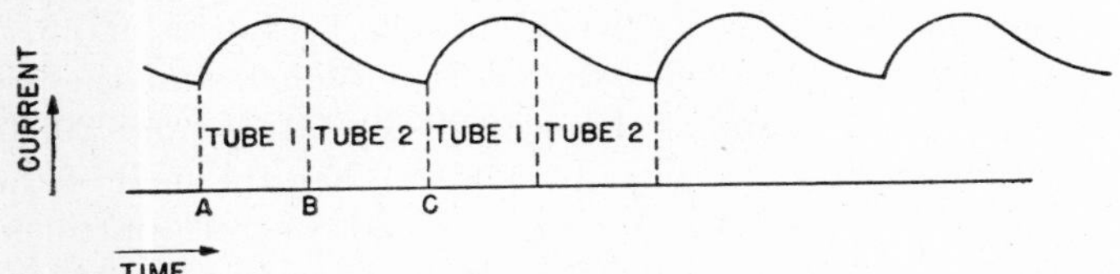

Fig. 29·35 Current wave shape in the inductive load circuit of a single-phase rectifier with discharge tube. (See Fig. 29·34.)

The rectifier is connected in accordance with the circuit in Fig. 29·34, and the control of the angle of ignition of tube 6 is of the vertical type, described in Section 29·5. Resistor $R3$ and capacitor $C3$ are part of the conventional phase-shift circuit, and the alternating voltage between points A and B (Fig. 29·36), introduced into the grid circuit of the field-supply thyratron, is permanently lagging in phase the alternating anode-supply voltage by about 90 degrees. In addition to that a-c component, an adjustable d-c component of the thyratron grid voltage is developed between points B and C. The d-c component consists of the following voltages (tracing the circuit in the direction from the cathode to the grid): negative voltages across $P5$ and $P1$ and positive voltages across $R25$, $P8$, and $P1A$.

The most important part of the circuit is the one consisting of potentiometers $P1$ and $P1A$. The former is the well-known armature-voltage speed-control potentiometer, whose significance was described in previous sections (see Fig. 29·25), and the latter is the field-current speed-control potentiometer. The sliders of the two speed-control potentiometers $P1$ and $P1A$ are coupled mechanically or mounted on the same shaft, and can be controlled simultaneously from the same control knob. Both potentiometers $P1$ and $P1A$ have center taps, and jumpers are connected across the right-hand half of $P1$ and the left-hand half of $P1A$. Since currents carried by sliders are negligible, it may be assumed that there is no voltage drop between any two points in the right-hand half of potentiometer $P1$ or in the left-hand half of potentiometer $P1A$. Thus, potentiometer

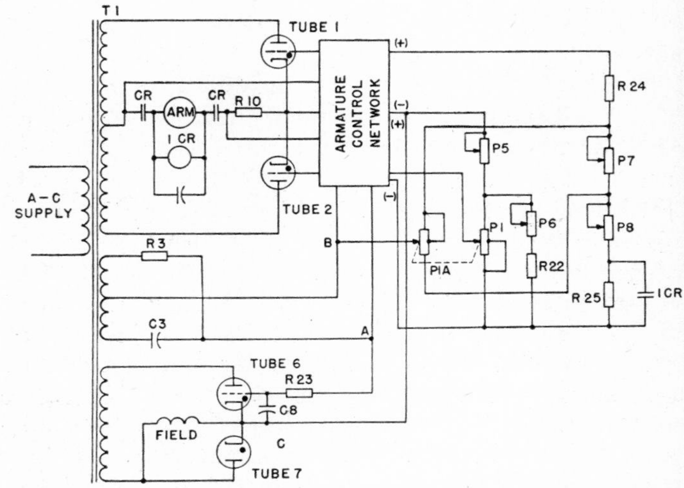

Fig. 29·36 Circuits for field control of a general-purpose rectifier drive.

$P1$ becomes ineffective when its slider is operated over its right-hand half, while the same applies to the left-hand half of potentiometer $P1A$. In this manner, the total range of the single speed-control dial is divided into halves. The first, from 0 to 50 percent rotation of the slider, is the armature-voltage speed-control range, and the second, from 50 to 100 percent rotation, is the field-current speed-control range. When the speed-control knob is turned from setting 0 to setting 50, the armature voltage is increased, and the speed increases from its minimum to its base or rated value. At the same time, the field current of the motor remains constant at its full rated value. When the knob is turned beyond setting 50, the armature voltage stays constant at its full value, the field is gradually weakened, and the speed will increase above its rated value to a maximum speed corresponding to setting 100 of the dial. The operation of potentiometer $P1A$ to produce weakening of the field current can be easily understood if it is noted that the voltage across that potentiometer represents the adjustable *positive* d-c component of the thyratron grid voltage. Consequently, the clockwise rotation of the slider of $P1A$ over the right-hand portion of the potentiometer will decrease the positive component of the thyratron grid voltage and reduce the output of the field rectifier (Fig. 29·36).

Two important requirements must be considered in order to obtain satisfactory field control. The first is that, when the motor is being started by pushbutton control, and the speed-control knob is turned beyond setting 50 (when the motor is to accelerate to a speed corresponding to a properly weakened field), the field current should be weakened only after the motor has accelerated to the base speed. In other words, the motor should always start with full field excitation in order to develop high starting torque, regardless of what the ultimate excitation of the field is to be. The second requirement is the proper range of control exercised by potentiometer *P1A*, so that the weakening of the field current will start as soon as the slider of *P1A* is moved beyond setting 50 in the clockwise direction, and the required top speed is attained for the extreme clockwise position of the slider.

These two requirements are fulfilled by proper distribution of direct voltages in the voltage divider consisting of *R*24, *P1A*, *P*8, and *R*25, as well as by means of a special starting relay 1*CR* (Fig. 29·36). The coil of relay 1*CR* is connected across the armature of the motor, and is so designed that the relay will become energized when the armature voltage is just a little less than its rated value, corresponding to the base speed of the motor. As will be recalled from Section 29·8 on controlled acceleration, the armature voltage during the initial stage of acceleration is low because of the action of the current-limit control, and it increases gradually as the speed of the motor builds up. By the time the armature voltage approaches its rated value, the motor has accelerated to approximately its base speed. Consequently, starting relay 1*CR* remains open until the motor reaches its base speed, and the closure of that relay, after the base speed is reached, is used as an indication that the control of the field current may now begin.

Figure 29·36 also reveals that the normally open contact of relay 1*CR* is connected across resistor *R*25. When relay 1*CR* is deenergized, the positive thyratron grid-voltage component across *R*25 is sufficiently high to make field thyratron tube 6 fire at a highly advanced angle, so that a full field current is obtained regardless of the setting of potentiometer *P1A* and rheostats *P*7 and *P*8. When, during the acceleration of the motor, the armature voltage is increased to its rated value, relay 1*CR* will close, and its contact will remove resistor *R*25 from the circuit. In this manner, the total positive component of the thyratron direct grid voltage will be reduced. If rheostats *P*7 and *P*8 are properly adjusted, the total positive direct-voltage component across *P1A* and *P*8 will be of such magnitude that, when the slider of *P1A* is in the middle position, the resultant direct-voltage component in the thyratron grid circuit will be positive and just about equal to the amplitude of the alternating-voltage component. These adjustments will assure a properly narrow "dead zone" of control; that is, the field current will be weakened as soon as the slider of *P1A* passes over the center tap if, of course, relay 1*CR* is closed at that time. Furthermore, the total positive direct-voltage component should be divided between *P1A* and *P*8 in such a manner that, with the slider of *P1A* in its extreme clockwise position, the positive direct-voltage component will be reduced, or even made sufficiently negative, to reduce the field current to a value corresponding to the required maximum speed of the motor. This division of voltages can be adjusted by means of rheostat *P*7. The two additional rheostats in Fig. 29·36, *P*5 and *P*6, are part of the armature-control system. Although these two elements are not shown in Fig. 29·25, they play a rather important part as minimum-speed and base-speed adjustments of armature control. In fact, *P*5 has exactly the same function as resistor *R*9 in Fig. 29·25. Making *R*9 adjustable permits the selection of the desirable minimum motor speed, corresponding to counterclockwise setting of armature speed-control potentiometer *P*1. On the other hand, the setting of rheostat *P*6 determines the voltage across armature speed-control potentiometer *P*1 and, consequently, the maximum armature voltage obtainable by adjustment of *P*1.

29·11 ELECTRONIC DRIVE OF TACHOMETER-GENERATOR TYPE

In preceding sections, circuits and characteristics of an electronic *IR*-drop compensated drive were described in considerable detail. In Section 29·7 it was also mentioned that a much higher accuracy of speed regulation can be obtained if a tachometer generator is used to indicate the speed of the motor directly, instead of the armature voltage-current indirect method of speed indication, which may be subject to various errors.

The main disadvantage of a tachometer-generator type of drive is the necessity to provide an additional small rotating machine and couple it to the shaft of the motor, sometimes through a speed-increasing gear. The application of a tachometer generator usually calls for a special motor with a double shaft extension. However, if a high degree of speed accuracy is required, and if the speed-control dial is to be calibrated, for example, in revolutions per minute or feet per minute, the tachometer-generator type of electronic drive is certainly the one to be used. The most important advantages of this type of drive are:

1. The accuracy of speed regulation with varying torques is of the order of ⅓ of 1 percent at base speed, and about 3 percent at $\frac{1}{10}$ of the base speed of the motor.
2. The consistency of speed setting is very good, and the speed-control dial may be calibrated with a high degree of accuracy.
3. The influence of line-voltage variations and of temperature changes of the motor armature and field windings is reduced to a minimum.
4. Circuits and initial adjustments of the drive are simplified.

Of all the types of tachometer generators the most suitable is the direct-voltage permanent-magnet type, since it does not require any separate excitation or additional amplification and rectification of its output voltage. Besides, tachometer generators of the permanent-magnet type usually have an excellent voltage-speed characteristic. Not only the output voltage is directly proportional to speed (for low output currents), but the characteristic is also practically independent of temperature, and in general is very consistent. The voltage output per unit speed is relatively high, and amounts to about 50 volts per 1000 revolutions per minute, as compared to 6 or 10 volts per 1000 revolutions per minute for a typical a-c tachometer generator.

A circuit diagram of a tachometer-generator type of electronic drive is shown in Fig. 29·37. Like all the preceding diagrams, the one in Fig. 29·37 is simplified to include solely the elements necessary for the proper explanation of the operation of the system. Accordingly, all auxiliary power supplies are represented by batteries, cathode heaters of all electron tubes are omitted, and some of the filtering and by-pass circuits are simplified. The system shown in Fig. 29·37 provides constant shunt-field excitation and all the basic features of armature control, that is, close speed regulation, wide stepless range of speed adjustment, and the current-limit control of starting and overloads.

speed indication is obtained directly from the tachometer generator, there is no need for any *IR*-drop compensation. Armature-voltage control and speed regulation are obtained through main control tube 6. The plate circuit of tube 6 comprises the d-c supply *B*1 and the load resistor *R*7. The grid circuit of tube 6 can be traced through *R*15, *P*1, *R*11, *R*12, *P*3, and *R*8. The grid-control circuit of armature thyratrons 1 and 2 includes an a-c component obtained from grid transformer *T*2. The primary of this transformer is connected to a phase-shift circuit *R*3-*C*3, and the alternating voltage applied to the grids of tubes 1 and 2 is lagging in phase the alternating anode-supply voltage by 90 to 115

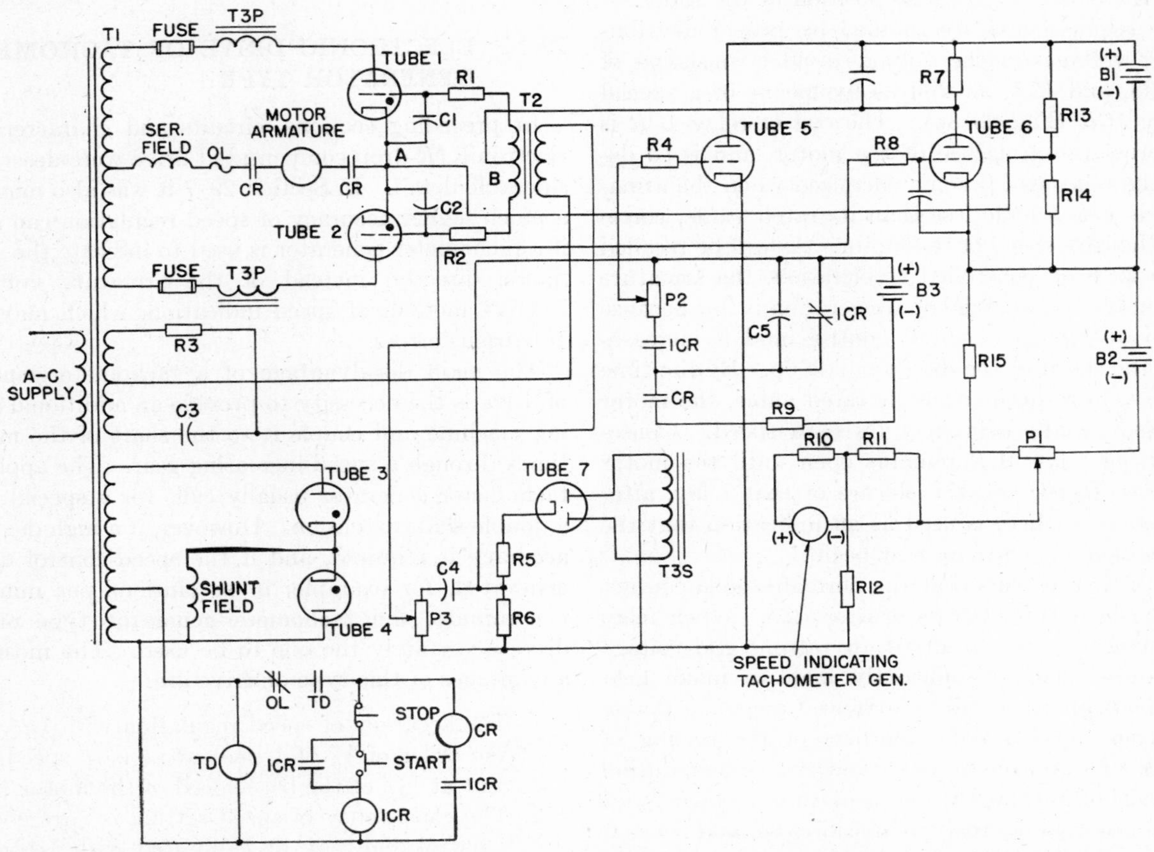

Fig. 29·37 Circuit diagram of a tachometer-generator type of electronic drive with controlled acceleration for high-accuracy armature control.

The system shown in Fig. 29·37 represents a typical speed regulator and has all the four regulator elements discussed in Section 29·6. The speed of the motor is indicated by the tachometer generator through its direct output voltage, proportional to the speed. This direct voltage, appearing across resistor *R*11, is compared with the adjustable standard reference voltage provided by potentiometer *P*1, and the difference of these two voltages is amplified by means of tube 6. The amplified changes in tachometer-generator voltage control the armature-supply thyratrons, thus forming the restoring force which opposes any change in motor speed caused by such conditions as changes in torque and temperature. Here, again, the closed cycle of interdependence of quantities, as discussed in Section 29·6, should be emphasized. This cycle is shown symbolically in Fig. 29·38.

With reference to diagram in Fig. 29·37, it will be noted that the armature-rectifier circuit is of the same conventional type as that described in earlier sections. Since the

degrees. In addition to the a-c component, the grid circuit of thyratrons 1 and 2 contains a variable d-c component developed between points *B* (center tap of the grid transformer) and *A* (common-cathode connection of the main rectifier tubes). This d-c component consists of a constant positive direct voltage across *R*13 and a variable negative direct voltage across *R*7. Since *R*7 is the load resistor in the plate circuit of main control tube 6, the firing angle of thyratrons 1 and 2 is delayed when the plate current of tube 6 is increased and, conversely, the firing angle of the thyratrons is advanced when the plate current of tube 6 is decreased. This portion of control is quite similar to that described in Section 29·6.

By referring again to the grid circuit of main control tube 6, it will be seen that the standard reference voltage of the regulating system is provided by d-c supply *B*2, and consists of the voltage across *R*15 and the adjustable voltage across a portion of *P*1. This reference voltage is negative with respect to the control grid of tube 6, and is opposed by the positive direct voltage across *R*11, which is proportional to the speed of the motor. The standard reference voltage is always higher by a few volts than the speed-indicating voltage and, as a result, a net negative voltage of several volts is applied to the grid of tube 6.

In accordance with previous considerations, a balance of the regulating system is obtained in such a manner that any change in the speed of the motor caused, for example, by a change in load torque will bring about a reaction of all elements of the closed cycle in the direction to oppose any change in the regulated quantity. Thus, for instance, a slight decrease in motor speed will result in a decrease of the speed-indicating voltage (*R*11), which will further cause a decrease in the plate current of tube 6. Consequently, the direct voltage across *R*7 will decrease, and the angle of ignition of the thyratron tubes will be advanced to cause an increase in the armature voltage, so as to oppose the change in motor speed. Because of the high amplification of the system, even a very slight change in motor speed will cause the appearance of a strong restoring force through the change in voltage across *R*7 and the change in the angle of ignition of thyratron tubes 1 and 2. As a result, the steady-state speed of the motor may deviate from the prescribed value, which corresponds to a given setting of speed-control unit *P*1, only to a very slight degree, which in terms of a percentage of speed may be called the error of the regulator.

It becomes apparent from Fig. 29·37 that a change in setting of reference-voltage potentiometer *P*1 will result in a new balance of the system. Thus, the increase in the reference voltage will cause a momentary decrease in the plate current of tube 6 and a decrease in voltage across *R*7, so that the angle of ignition of the thyratrons will be advanced, the armature voltage will be increased, and the motor will accelerate until a new balance is established. It should be remarked at this point that, because of high system amplification, the actual net difference of the reference voltage and the speed-indicating voltage will change in a hardly noticeable manner, so that the plate current of tube 6 will remain within the operating range of the amplifier.

Tube 5 and its allied circuits represent the armature current-limit control, whose principles have been discussed in Section 29·8. The armature-current indication is obtained by means of transformer *T*3, whose primary current windings are marked *T3P*, and whose secondary voltage winding is marked *T3S* (Fig. 29·37). The functions of this transformer have been described in more detail in Section 29·9. The direct voltage across *R*6, obtained from rectifier tube 7, is proportional to the armature current of the motor. In addition to the positive voltage across *R*6, the grid circuit of current-limit tube 5 contains an adjustable negative bias voltage provided by *P*2 and obtained from an auxiliary direct-voltage supply *B*3. *R*12 and *R*11 represent the coupling resistors between the plate circuit of current-limit tube 5 and the grid circuit of regulator tube 6. It will be remembered that, for normal operating values of the armature current, tube 5 is biased off by the negative voltage from *P*2 and does not conduct any current. When the armature current becomes high and exceeds a certain critical value, as during acceleration or overload of the motor, tube 5 will start to conduct and, as a result, the additional positive voltage appearing in the grid circuit of tube 6 across *R*12 and *R*11 will have a very strong delaying effect on the firing angle of the armature thyratrons. The operation of the current-limit circuits and the significance of the elements associated with *P*2 and *B*3 have been described previously.

Any high-gain regulating system may be subject to sustained oscillations, which sometimes are referred to as "hunting." Oscillations in a regulating system may develop because of various time delays caused by different inductances, capacitances, and the mechanical inertia of the system. If the amplification is high, the so-called overshooting will result, and sustained oscillations may develop. In order to

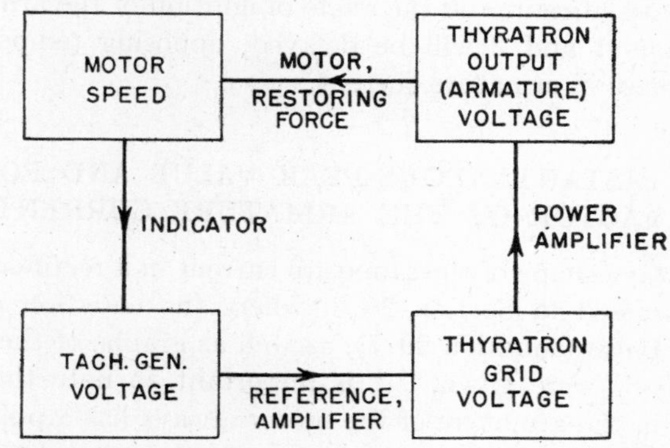

FIG. 29·38 Graphic representation of closed cycle of the speed regulator.

stabilize the system, and at the same time maintain a reasonably high amplification, an "anti-hunting" or damping arrangement must be used. A typical damping system should respond to the rate of change of one of the quantities involved, in such a manner that a fast change in that quantity will create a transient damping force acting in the direction opposite to that of the restoring force. The most important principle of the damping system is that the magnitude of the transient damping force is directly proportional to the *rate of change* of one of the quantities in the system, and not to the change itself. For this reason, the restoring force will be initially opposed by the transient damping force, and too abrupt changes in the system, which cause oscillations, will be prevented.

In Fig. 29·37 the anti-hunting circuit consists of current-indicating transformer *T*3, rectifier tube 7, resistors *R*5 and *R*6, capacitor *C*4, and potentiometer *P*3. It has been indicated previously that a direct voltage proportional to the armature current of the motor appears across *R*6. Capacitor *C*4 prevents any direct-voltage component from appearing across *P*3, that is, in the control-grid circuit of main amplifier tube 6. If, however, the system tends to oscillate, the armature current as well as the voltage across *R*6 will be subject to a fast change, and capacitor *C*4 will cause a transient current to flow in the circuit consisting of *C*4, *P*3, and *R*6. This transient current will be either a charging or a discharging current of capacitor *C*4, depending upon the sign of the rate of change of the voltage across *R*6. The magnitude of this transient current will depend upon the magnitude of the rate of change of the armature current, that is, of the voltage across *R*6. Thus, a transient damping voltage will appear across $P3$. A portion of potentiometer $P3$ is in the control-grid circuit of tube 6, and the transient damping voltage will act in such a way as to oppose a fast change in the armature current. As an example, let us consider the case where the armature current tends to increase at a high rate of change. The voltage across $R6$ will also increase in the same way, and a charging current of capacitor $C4$ will flow through $P3$, causing the appearance of a transient positive voltage in the grid circuit of tube 6. The magnitude of this grid voltage will depend upon the rate of change of the armature current and upon the setting of potentiometer $P3$, which thus may be used to adjust the amount of antihunting action in the system. A transient positive voltage in the grid circuit of tube 6 will cause a momentary increase in its plate current, and the angle of ignition of the armature thyratrons 1 and 2 will be delayed, opposing temporarily the increase in armature current.

29·12 INSTANTANEOUS-PEAK VALUE AND FORM FACTOR OF THE ARMATURE CURRENT

The wave shape of the armature current in a rectifier drive was discussed in Section 29·3, where the equation of the current pulse (equation 29·7), as well as graphs of the pulse (Fig. 29·6) were given. It is important to note that the current in the armature of the motor always has a pulsating character so that the a-c component of the current wave is very considerable, even when the individual pulses of current join and a continuous conduction of the rectifier is obtained.

From the analytical point of view, three concepts of current are to be considered, and each of these concepts plays an important part as a factor determining the operating conditions and the rating of the drive. These three concepts are: the average value, the instantaneous-peak value, and the rms value of the armature current. The definitions of the three values of current are well known from the general a-c theory, and apply here without modification. However, the significance of each of the three concepts, when applied to pulsating armature currents, should be analyzed in more detail.

In Fig. 29·39 are shown three typical cases of the armature-current wave shape. Figure 29·39 (*a*) represents a case of discontinuous current in the armature, where individual current pulses are distinct, and zero-current gaps exist between them. In Fig. 29·39 (*b*) a border case is shown, where the current pulses just join, but the instantaneous values of current in each pulse start from zero and drop to zero. Figure 29·39 (*c*) shows a typical case of continuous armature current, where commutation of load current from one tube to another takes place. The graphical significance of the average, instantaneous-peak, and rms values of current is indicated in the three parts of Fig. 29·39.

The average value of armature current is the d-c component of the periodical time function represented graphically in Fig. 29·39. This value has the greatest practical significance because it determines the average torque and the horsepower developed by the motor, and appears in all conventional calculations. In previous considerations, and in particular in Sections 29·2, 29·3, 29·4, 29·7, and 29·8, average values were usually meant wherever reference was made to armature current or voltage, and in Section 29·3 analytical methods for calculation of average values of currents and voltages for controlled rectification were discussed in detail.

The instantaneous-peak value of armature current is of vital interest from the point of view of behavior and proper selection of the main power-rectifier tubes. As already mentioned in Section 29·8, controlled-rectifier tubes of the

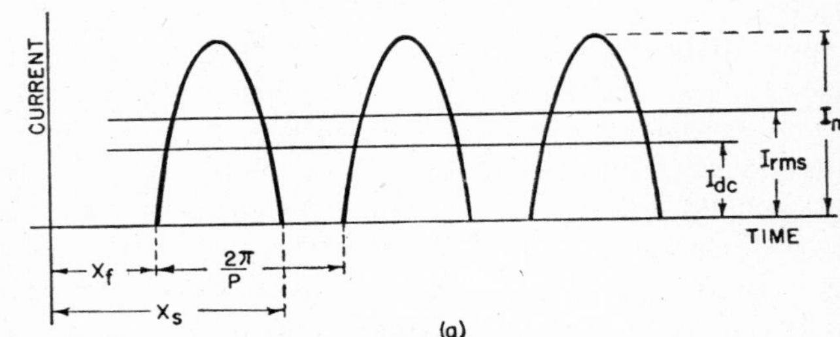

Fig. 29·39 (*a*) Typical wave shape of armature current for discontinuous conduction of the rectifier.

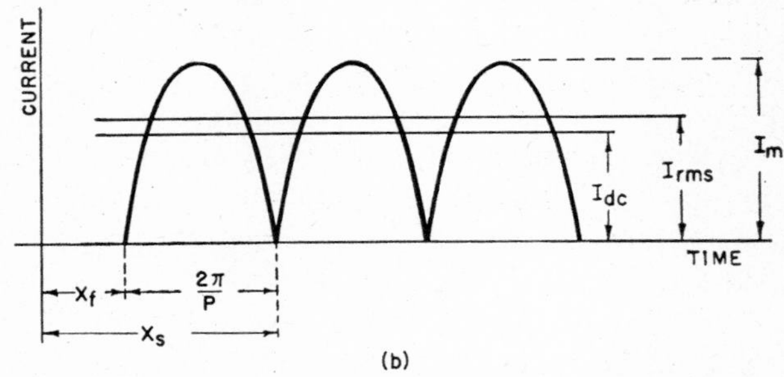

Fig. 29·39 (*b*) Wave shape of armature current for the critical case of conduction of the rectifier.

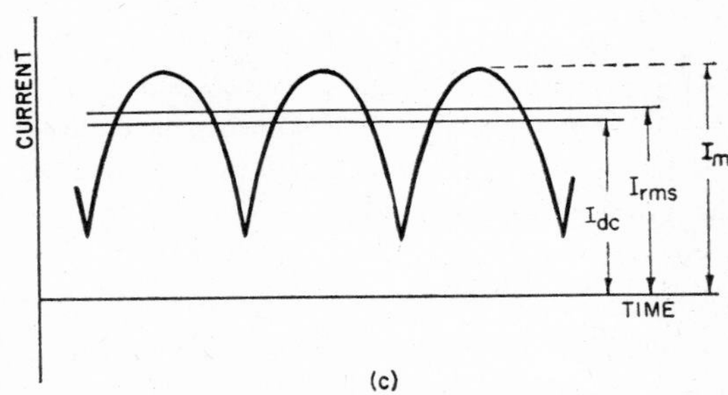

Fig. 29·39 (*c*) Typical wave shape of armature current for continuous conduction of the rectifier.

thyratron type are particularly sensitive to instantaneous-peak currents, and if the peak currents exceed the rated value for a given type of tube, the tube may be permanently damaged, even if the average-current rating of the tube has not been exceeded. Some thyratron tubes of the older type have a 6-to-1 or 8-to-1 ratio of instantaneous peak-current rating to average-current rating, but most of the modern thyratrons have a peak-to-average current ratio as high as 12 to 1.

In spite of such high peak-to-average current ratios, the condition in which the peak current in a tube exceeds the tube rating, although the average current is not excessive, can be created, particularly in polyphase, half-wave rectifier drives. It should be remembered that in half-wave polyphase rectifiers the average current in each rectifying element is obtained by dividing the average value of load current by the number of rectifying elements p:

$$I_{\text{tube}} = \frac{I_{\text{load}}}{p} \tag{29·52}$$

and, consequently, for a given load current the average current per tube can be reduced by increasing the number of phases p of the rectifier. The instantaneous-peak current, however, carried by each rectifying element is always equal to the instantaneous-peak current in the load circuit. Therefore, in polyphase rectifier drives the factor determining the rating of a given combination of rectifying elements is not only the average current-carrying capacity of each tube, but often their instantaneous peak-current rating.

In view of these considerations, it becomes apparent that calculations leading to the determination of the instantaneous-peak value of armature current, particularly in the more severe case of discontinuous conduction, are of considerable interest. Especially in thyratron drives, the instantaneous-peak current in tubes may be the factor determining the rating of the whole drive, since an excessive peak current is the most important single factor responsible for the permanent damage or even immediate destruction of the tubes.

In order to determine analytically the instantaneous-peak value of the armature-current pulse, the time angle corresponding to the maximum of the current-time function should be found first. By differentiating the equation of the current pulse (equation 29·7) with respect to the time angle x, and further equating the result to zero, the equation for the time angle x_m corresponding to the maximum value of the current pulse is obtained:

$$\cos(x_m - \theta)\epsilon^{x_m/\tan\theta} = \left[\frac{a}{\sin\theta} - \frac{\sin(x_f - \theta)}{\tan\theta}\right]\epsilon^{x_f/\tan\theta} \tag{29·53}$$

where the symbols are the same as those used in equation 29·7.

Equation 29·53 represents the relationship between the peak-current angle x_m and the angle of ignition x_f; that is, it represents x_m as an implicit function of x_f for different values of parameters a and θ. Although it is possible to calculate the value of x_m for each particular value of x_f, x_m cannot be represented by an explicit formula. The graphs of the function $x_m = f(x_f)$ for different values of parameters a and θ are shown in Fig. 29·40.

Considering the peak-current angle x_m as represented by equation 29·53, the expression for the peak value of the armature-current pulse can be obtained directly from equation 29·7:

$$I_m = \frac{\sqrt{2}E_s}{R}\{\cos\theta\sin(x_m - \theta) - a + [a - \cos\theta\sin(x_f - \theta)]\epsilon^{-(x_m - x_f)/\tan\theta}\} \tag{29·54}$$

where $x_m = f(x_f)$ may be calculated from equation 29·53 or estimated from Fig. 29·40.

It should be pointed out that, for the theoretical case of zero inductance in the load circuit ($\cos\theta = 1$), neither x_m nor I_m can be calculated from equations 29·53 and 29·54 because these equations assume an indeterminate form for $\theta = 0$. However, it is apparent from Fig. 29·5 that for a purely resistive load ($\theta = 0$)

$$\begin{aligned} x_m &= x_f \quad \text{when } x_f \geqq 90° \\ x_m &= 90° \quad \text{when } x_f < 90° \end{aligned} \tag{29·55}$$

Whereas equations 29·53 and 29·54 represent rigorous analytical relationships,[2] in some cases a simpler approximate formula for instantaneous-peak value of the load current for *discontinuous conduction* may prove more useful:

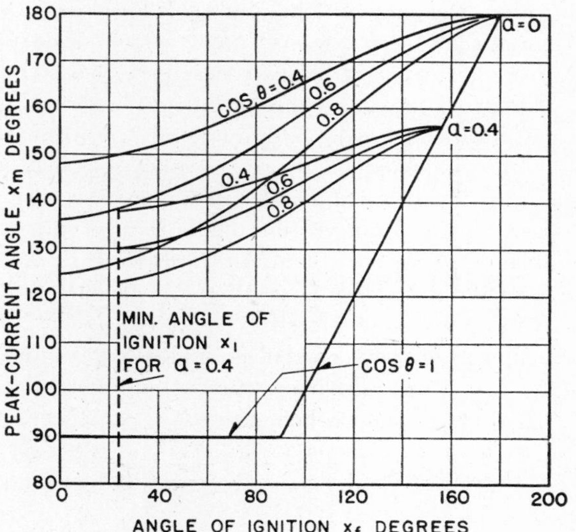

FIG. 29·40 Graphs of equation 29·53; $x_m = f(x_f)$.

$$I_m = \frac{\pi^2 I_{d\text{-}c}}{\sqrt{2\pi p(x_s - x_f)}} \tag{29·56}$$

where $I_{d\text{-}c}$ = average value of the load current
p = number of phases of the rectifier
x_f = angle of ignition in radians
x_s = angle of extinction in radians.

It will be recalled that the angle of extinction x_s can be calculated from equation 29·11 or estimated from Fig. 29·7.

The approximate formula for I_m, equation 29·56, can be derived under the simplifying assumption of a purely sinusoidal wave shape of each current pulse and a symmetrical location of its instantaneous peak. Although the above assumption, strictly speaking, is incorrect, the approximate formula 29·56 is useful for estimating purposes, particularly in the vicinity of the critical angle of ignition.

From the approximate formula for the instantaneous-peak current, equation 29·56, one can obtain directly the approximate expression for the so-called *peak coefficient* which is defined here as the ratio of peak and average currents, and represents a convenient way of describing the peak-current conditions in the load circuit of a rectifier:

$$f_m = \frac{I_m}{I_{d\text{-}c}} = \frac{\pi^2}{\sqrt{2\pi p(x_s - x_f)}} \tag{29·57}$$

The third significant value of the armature current is the rms value. The rms value of current is responsible for the losses in the armature winding of the motor and must be used in the well-known formula I^2R, which determines these losses. Thus, the rms value of current has a direct effect on the frame size and efficiency of the motor. In a conventional d-c drive, where a pure current flows through the brushes of the motor, the average current is equal to the rms value of current. In a rectifier drive, however, the presence of a considerable a-c component in the unidirectional armature current is responsible for the increase in the rms value, which is always higher here than the average value. Consequently, the efficiency of the motor is decreased so that very often the motor-frame size must be increased to dissipate additional losses, without an excessive rise of temperature. Moreover, the rms value of current is closely related to the instantaneous-peak value and, for the same average current, both increase or decrease simultaneously. Both the peak and the rms values of current have a very appreciable effect on the commutation of the motor, and it can be easily understood that the commutation becomes more difficult if, for a given average current, the peak current, as well as the rms current, is increased.

In rectifier-motor systems it is customary to deal with the *form factor* of the armature current, defined as the ratio of the rms value to the average value:

$$f_f = \frac{I_{\text{rms}}}{I_{d\text{-}c}} \tag{29·58}$$

Thus, the form factor is indicative of additional motor losses, as well as of the character of commutation for a given load current of the motor. The accurate analytical expression for the form factor is a rather involved one, and will not be given here. However, an approximate simplified formula for the form factor of the armature current for *discontinuous conduction*, based on the assumption of a purely sinusoidal symmetrical pulse, is given below, and may be used for estimating purposes:

$$f_f = \frac{\pi^2}{2\sqrt{\pi p(x_s - x_f)}} \tag{29·59}$$

For the border case, dividing discontinuous and continuous conduction of the rectifier, Fig. 29·39 (*b*), each rectifying element conducts over the entire phase cycle equal to $2\pi/p$; hence

$$x_s - x_f = \frac{2\pi}{p} \tag{29·60}$$

When equation 29·60 is substituted in equation 29·59, the form factor for the border case, Fig. 29·39 (*b*), becomes

$$f_{fb} = \frac{\pi}{2\sqrt{2}} = 1.11 \tag{29·61}$$

For the case of discontinuous conduction, Fig. 29·39 (*a*), the form factor is always greater than 1.11, and for the case of continuous conduction, Fig. 29·39 (*c*), the form factor of the load current is always less than 1.11:

$$\begin{gathered} f_{f\text{ disc.}} > 1.11 \\ 1 < f_{f\text{ cont.}} < 1.11 \end{gathered} \tag{29·62}$$

One will note that relationships 29·62 do not refer to the number of phases of the rectifier, motor speed, or load, and are based solely on the wave shape of the load current. Obviously, for the theoretical border case of a pure direct load current, the form factor is equal to unity.

It may be useful to analyze briefly the operating and design factors which have a direct effect on the wave shape of the armature current and, consequently, on the value of

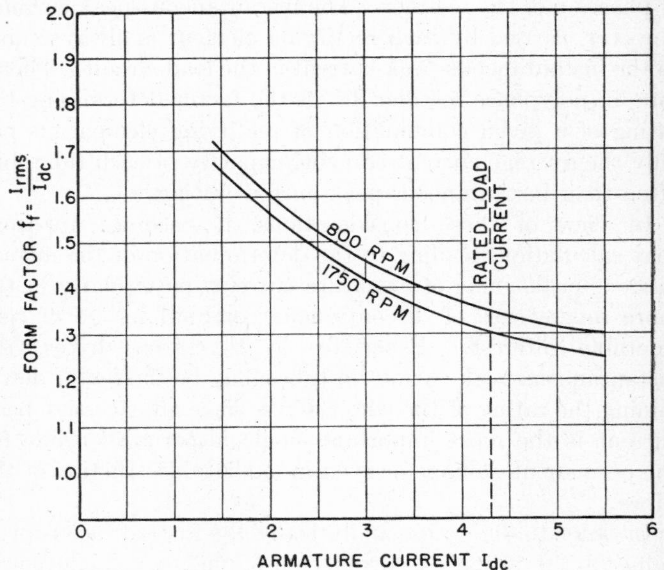

Fig. 29·41 Experimental graphs of form factor versus armature current for the rectifier drive: $p = 2$, $E_s = 400$ volts, 1 hp, 4.3 amperes.

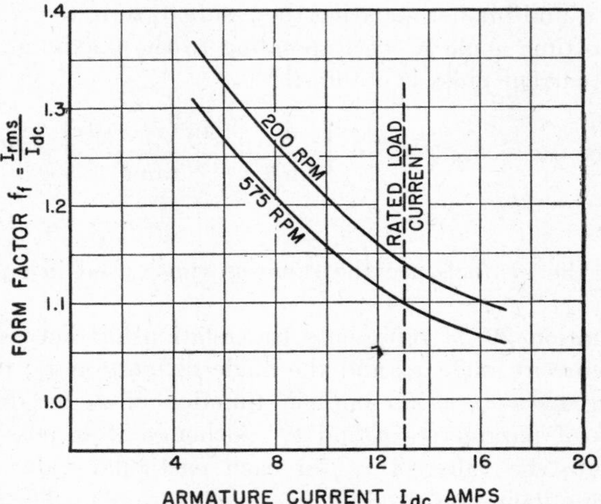

Fig. 29·42 Experimental graphs of form factor versus armature current for the rectifier drive: $p = 3$, $E_s = 400$ volts, 3 hp, 13 amperes.

the form factor. Among the most important ones are: (*a*) rectifier-transformer voltage, (*b*) number of phases of the rectifier, (*c*) impedance angle θ of the armature circuit, (*d*) load current in the armature, and (*e*) speed of the motor. Generally speaking, the value of the armature-current form factor increases when (*a*) is increasing, and it decreases when (*b*), (*c*), and (*d*) are increasing. The speed of the motor has relatively little effect on the form factor, although the latter has a general tendency to decrease with increasing motor speed. Several typical graphs of the armature-current form factor as a function of the average value of current, for different motor speeds, are shown in Figs. 29·41 and 29·42.

29·13 METHODS AND PRINCIPLES OF BRAKING

One of the important problems in any motor-control system is the problem of braking. Conventional mechanical brakes, magnetically operated, are sometimes used in electronic motor-control systems to bring the motor and the driven load to a quick stop, exactly as in any conventional drive.

Dynamic braking is one of the most widely used means of braking. It consists in connecting a resistor across the motor armature while the armature is disconnected from the supply terminals. This type of braking is often used in electronic motor-control systems. From the physical point of view, the process of dynamic braking is identical in any control system, and will not be discussed here in detail. It should be mentioned, however, that during dynamic braking the kinetic energy stored in the rotating system is transformed by the rotating armature into electric energy and dissipated in the dynamic-braking resistor. This type of braking has a number of disadvantages, the most important of which is the influence of the speed of the motor on the decelerating torque, which is high for high speeds and negligible for low speeds, where the electromotive force generated in the rotating machine is low. Thus, at low speeds, the motor will tend to coast, and a definite friction torque is required to bring the motor to a complete stop. Usually, this torque is provided by the natural friction of the system itself (bearings, windage, gears) but in some systems, particularly with high moment of inertia, this natural friction may be inadequate, and a magnetically operated mechanical brake must be used in combination with dynamic braking.

The third very important type of braking is regenerative braking. During regenerative braking the kinetic energy stored in the rotating mechanical system is transformed into electric energy in the armature of the rotating machine and delivered back to the supply line. Regenerative braking should not be confused with so-called "plugging," which is sometimes used in conventional d-c drives but never in electronic systems, where it may easily result in permanent damage to the power tubes. In conventional d-c motor-control systems plugging may be obtained by simply reversing the armature connections to the d-c supply terminals.

Figure 29·43 (*a*) shows the direction of applied voltage, armature current, and armature electromotive force during the normal run of the motor. Obviously, the armature current I has the same direction as the applied voltage E, and flows in opposition to the armature electromotive force E_g. This condition is also in agreement with the conventional direction of the flow and sign of electric power, that is, positive power is given away, negative power is absorbed. Thus, in Fig. 29·43 (*a*), power $+(EI)$ is given away by the d-c line, whereas power $-(E_gI)$ is absorbed by the rotating armature and transformed into mechanical energy.

Figure 29·43 (*b*) represents the conditions associated with "plugging," where the armature of a running motor was suddenly reconnected with respect to the supply line; that is, armature terminal $A2$, which was connected to $(-)$ of the line in Fig. 29·43 (*a*), is now connected to $(+)$ of the line, and terminal $A1$ is now connected to $(-)$ of the line. Since the motor armature and the associated mechanical parts continue to rotate in the same direction as in Fig. 29·43 (*a*) because of the moment of inertia of the system, the electromotive force generated in the armature still has temporarily the same direction as in Fig. 29·43 (*a*), that is, from $A2$ to $A1$.

Now it becomes apparent that the case of plugging shown in Fig. 29·43 (*b*) represents a "double" short circuit because not only is the line voltage E applied to a very low-resistance circuit without any opposing internal electromotive force, but also the electromotive force e_g developed in the armature

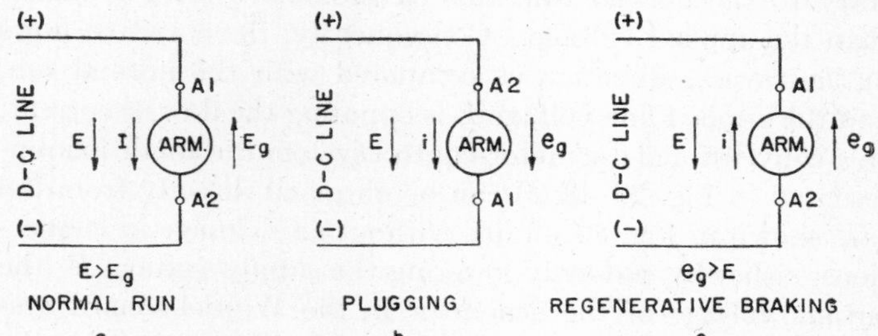

FIG. 29·43 Relative directions of line voltage, electromotive force, and armature current for (*a*) normal run, (*b*) plugging, and (*c*) regenerative braking of a d-c drive.

circuit acts in the same direction as the applied voltage, still increasing the current i. Thus, a very high short-circuit current will flow in the armature:

$$i = \frac{E + e_g}{R} \tag{29·63}$$

where R is the resistance of the armature circuit. Of course, e_g is a function of the motor speed or of time, since the speed of the motor is changing with time. The instantaneous electric power developed in the armature has a plus sign, and the electric energy developed in the armature during the braking period, and responsible for the braking action

$$W_b = \int_0^T e_g i\, dt = \int_0^T \frac{Ee_g + e_g^2}{R}\, dt \tag{29·64}$$

where T = total time of braking. If the friction of the system is neglected, the energy represented by equation 29·64 is equal to the kinetic energy stored in the mechanical system just before braking has started.

The total electric energy developed in the entire system during plugging to a stop in T seconds

$$W_s = \int_0^T (E + e_g) i\, dt = \int_0^T \frac{(E + e_g)^2}{R}\, dt \tag{29·65}$$

is dissipated in the form of heat, mainly in the resistance of the armature winding.

In analyzing that portion of electric power developed in the armature, which is responsible for the braking action during plugging,

$$P_b = e_g i = \frac{Ee_g + e_g^2}{R} \tag{29·66}$$

it will be seen that since both E and e_g (at the start of plugging) are high, and R is low, the power will be very high, and the braking action will be exceedingly sharp. At the same time, the short-circuit current i (equation 29·63) may be so high that, even for a short duration of braking, permanent damage to the system may result. Thus, it may be necessary to increase the resistance of the armature circuit R by introducing a resistor in series with the armature during plugging to reduce the magnitude of current as well as the shock of a very sharp braking.

Conditions for regenerative braking are represented in Fig. 29·43 (*c*). Here, as in a normal run, Fig. 29·43 (*a*), the electromotive force generated in the armature acts in the direction opposing the applied line voltage E but, contrary to the normal run, this electromotive force is higher than the applied voltage. Consequently, the current i flows in the reverse direction as compared with the normal run, and the applied line voltage E is opposing the flow of current. In a conventional d-c motor-control system the condition represented in Fig. 29·43 (*c*) can be obtained directly from the case shown in Fig. 29·43 (*a*), without any change in connections, either by suddenly lowering the supply voltage E (the output voltage of the generator in the Ward-Leonard system), or by increasing the electromotive force through the increase in the shunt-field excitation of the motor, if the latter was previously running at a weak field. The fundamental difference between the two cases, as shown in Figs. 29·43 (*b*) and 29·43 (*c*), is that for the regenerative braking the current in the armature flows in opposition to the applied voltage E. Consequently, the electric power at the supply terminals $-(Ei)$ has a negative sign, indicating that the line is absorbing power instead of delivering it as in the cases of both normal run and plugging. Thus, during regenerative braking the motor operates as a generator, and the kinetic energy stored in the rotating system is transformed in the motor into the electric energy and delivered to the line. In contrast with that, during plugging the motor operates as a combination of a generator, driven from the mechanical energy stored in the system, and a torque motor. During plugging, the electric energy generated by the motor, as well as the electric energy supplied to the motor from the line, both are dissipated in the form of heat in the resistance of the armature circuit.

The analysis of the general case of plugging and regeneration was necessary to stress the fundamental differences between the two in order to have the proper introduction into the more involved case of regeneration in electronic rectifier drives. First, it should be noted that plugging can be easily obtained in electronic drives in a manner very similar to that in conventional d-c systems by brute reversing of the armature contactor at the time the motor is running at high speed. However, this method of braking is never used in electronic thyratron drives since it invariably leads to currents of a magnitude destructive to electronic rectifiers.

With respect to regeneration, it must be borne in mind that in a rectifier drive it is not possible to obtain regenerative braking by simply lowering the output voltage of the rectifier (which here plays the part of the d-c supply line), or by increasing the electromotive force of the motor through the increase in its excitation, as is done normally in conventional d-c systems, Fig. 29·43 (*c*). The rectifier drives do not have the inherent ability to regenerate under those conditions, because regeneration requires a change in the direction of the armature-current flow with respect to the direction of the electromotive force. This change cannot be effected in a rectifier-motor system without reversing the polarity of armature electromotive force with respect to the rectifier system, since the current cannot flow in the reverse direction through the rectifying elements. This situation can be clearly seen in Fig. 29·44 which shows the diagram

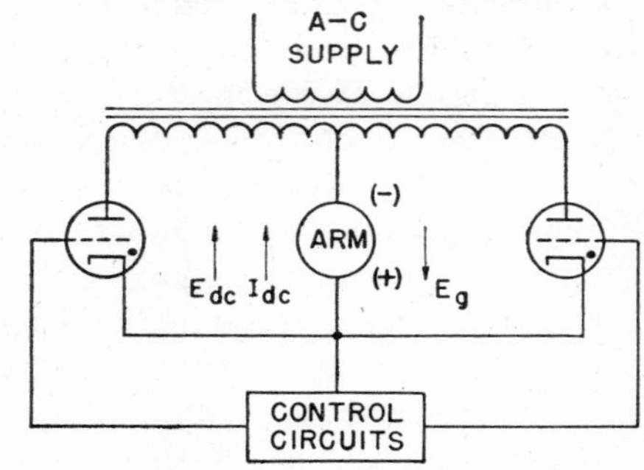

Fig. 29·44 Relative directions of rectifier output voltage $E_{d\text{-}c}$, motor electromotive force E_g, and armature current $I_{d\text{-}c}$ in a rectifier drive.

of a rectifier-motor system with polarities and directions of current and electromotive force indicated for the conditions of a normal run. If the motor electromotive force is increased, or the angle of ignition of the rectifying elements is delayed (this would correspond to lowering the d-c supply voltage), the current in the armature will not reverse its direction, and no regeneration will take place. The current in the armature and in the rectifier will simply drop to zero, the rectifier tubes will be momentarily cut off, and the motor will coast down to a lower speed determined by the new angle of ignition of the tubes, or by the new (increased) excitation of the motor. After the lower speed is reached, normal conduction will be resumed.

Regenerative braking in a rectifier drive can be obtained if three conditions are fulfilled simultaneously. First, as mentioned in the preceding paragraph, the polarity of the armature electromotive force must be reversed so that the rectifying system, shown in Fig. 29·44, will be transformed into an inverting system, shown in Fig. 29·45. In this

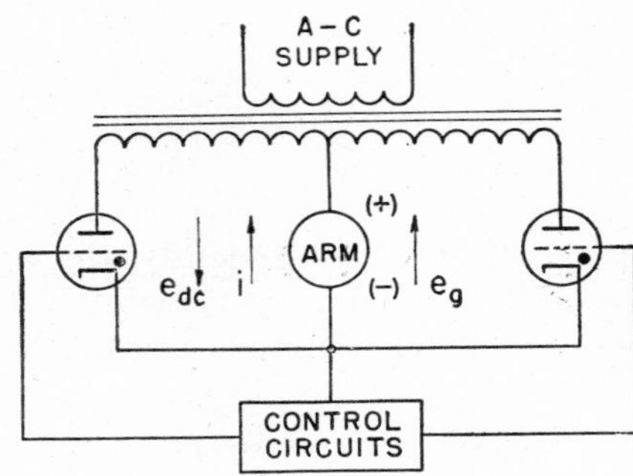

Fig. 29·45 Relative directions of inverter output voltage $e_{d\text{-}c}$, motor electromotive force e_g, and armature current i during regenerative braking.

manner, the current can change its direction with respect to the electromotive force in the armature, and still flow in the same and only possible direction through the rectifying elements. This condition can be accomplished by reversing either the armature contactor or the motor field. It must be borne in mind that the change in polarity of armature electromotive force is only one of the *three* necessary conditions for regeneration. If only this first condition is fulfilled, plugging and currents of destructive magnitude will result.

The second necessary condition for regenerative braking is the proper control of the angle of ignition of the rectifier tubes. Finally, the third and quite obvious condition for regeneration is the presence of a sufficiently high electromotive force of the motor, so that this electromotive force can be utilized as a current-producing voltage during the process of regeneration, when the motor operates as a generator of electric power. In other words, a source of energy is required in the load circuit. The fulfillment of the second condition, with the other two previously satisfied, normally leads to the problem of limiting the armature current, by means of controlled rectifier tubes, to a safe operating value. The mechanism of regeneration will be more fully understood in the following analysis of the process of inversion.

Figure 29·46 represents the sine wave of the rectifier-transformer voltage, combined with motor-generated voltages (electromotive forces) for both inversion and rectification, as well as a-c and d-c components of the grid voltages of thyratron rectifier tubes. Angles of ignition and armature-current pulses, during the process of motor reversal or stopping by regeneration, also are shown in Fig. 29·46. Let it be assumed that the fully excited motor is running at full speed in a given direction and that the main armature contactor is reversed, so that the polarity of the electromotive force generated in the armature becomes as shown in Fig. 29·45. Let it be further assumed that the grid-voltage supplies and controls for the thyratron tubes are of the same basic type as those described in Sections 29·5, 29·6, 29·7, 29·8 and Fig. 29·25, except the alternating grid-voltage component, obtained from the *R*3-*C*3 permanent phase-shift circuit, is made to lag the transformer voltage by about 120 degrees. Let it be assumed again, without going into details of circuit operation, that at the instant of closure of the reverse armature contactor (see Fig. 29·45) the direct component of the grid voltage is E_{c1}, and that the armature generated voltage (electromotive force) is E_{g1}, as shown in Fig. 29·46. Finally, let it be assumed for the purpose of simplification of Fig. 29·46 that the critical grid voltage of the thyratron rectifier tubes is zero, that is, that it coincides with the cathode potential of the tube.

Under these assumptions, and in view of the relative polarity of the electromotive force with respect to the transformer voltage (see Fig. 29·45), it is apparent that at the instant of closure of the reverse contactor (instant 1 in Fig. 29·46) the rectifier tubes will be cut off, and no current will flow in the circuit. At a certain subsequent instant 2 the generated voltage will become E_{g2}, the direct grid voltage E_{c2}, and the tubes will fire at the angle of ignition X_2. As shown in Fig. 29·46, the angle of ignition X_2 is equal to about 195 degrees. At the following instants 3, 4, 5, 6, and 7, the generated voltages are E_{g3}, E_{g4}, E_{g5}, E_{g6}, and E_{g7}, the direct grid voltages are E_{c3}, E_{c4}, E_{c5}, E_{c6}, and E_{c7}, and the angles of ignition are X_3, X_4, X_5, X_6, and X_7 respectively. At the instant 4, for example, the current in the armature will start at the angle of ignition X_4 equal to about 173 degrees, and will stop at the angle of extinction S_4. This current will cause an average armature-voltage drop proportional to the algebraic sum of the areas $(+)X_4MQ$ and $(-)QS_4T$:

$$IR \propto (\text{area } X_4MQ - \text{area } QS_4T) \qquad (29\cdot67)$$

This voltage drop is always positive.

The average value of the output voltage at the terminals of the transformer-rectifier system is proportional to the algebraic sum of areas $(+)NMP$ and $(-)PRT$:

$$E_{d\text{-}c} \propto (\text{area } NMP - \text{area } PRT) \qquad (29\cdot68)$$

Here, obviously, the area PRT is greater than the area NMP and, as a result, the average output voltage of the inverter is

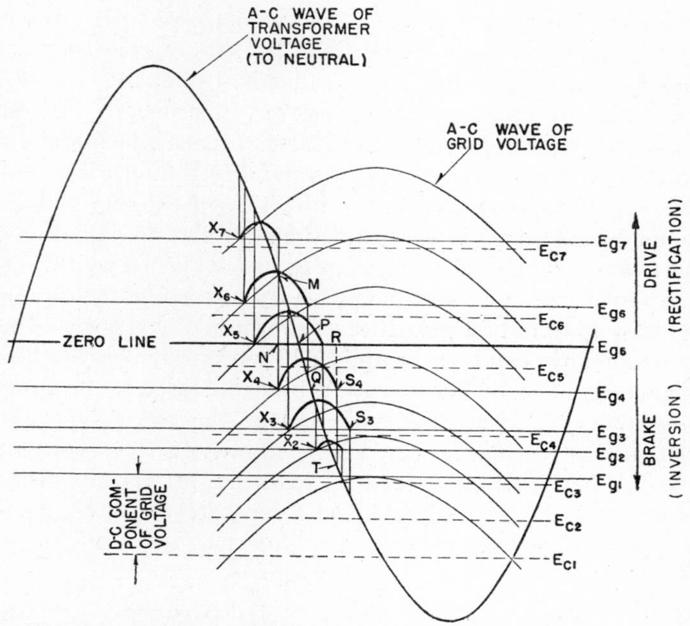

Fig. 29·46 Graphical explanation of the principle of regeneration in a rectifying-inverting system. (See Figs. 29·44, 29·45.)

negative, that is, opposing the flow of current. Also, it is apparent from Fig. 29·46 that for the instant 4 a greater portion of the armature current flows in opposition to the transformer voltage. This is consistent with the negative sign of the inverter output voltage represented for the instant 4 by equation 29·68, and with the direction of $e_{d\text{-}c}$ in Fig. 29·45.

The fact that the current flows in opposition to the transformer voltage means that during that period the power supplied by the transformer is negative; that is, the transformer becomes a receiver of electric energy supplied by the motor, operating now as a generator. Thus, a braking torque is developed in the motor, and this braking torque can be utilized for reversing, stopping, slow-down, or as a brake for overhauling loads. The significance of the automatic current-limiting action of the system, described in Section 29·8, should be specifically emphasized. The current-limiting action during inversion leads to such delays of firing angles of the main rectifier tubes that a net regenerative effect is obtained, as illustrated in Fig. 29·46.

During the process of braking there will be an instant when the average output voltage of the inverter becomes equal to zero and the process of regeneration stops. This will normally occur when the motor has already slowed down to a relatively low speed, so that its electromotive force is close to zero. From that point on, the transfer of energy will change its direction and will be from the transformer to the motor. The system will cease to be an inverter, and will become again a rectifier. The motor will accelerate in the reverse direction, unless disconnected from the system at zero speed (stopping), or unless its connections to the system are reversed to become again as shown in Fig. 29·44.

To reiterate the analogy between regeneration in a conventional d-c system, Fig. 29·43 (*c*), and that in a rectifier-motor system, it should be pointed out that in the latter case the relative directions of the line voltage E ($e_{d\text{-}c}$), armature current i, and electromotive force e_g are the same as shown in Fig. 29·43 (*c*). In conventional d-c systems, conditions as shown in Fig. 29·43 (*c*) can be obtained without any change in circuit connections by simply changing the relative magnitude of E and e_g, that is, by making $e_g > E$. In rectifier-motor systems, however, the conditions for regeneration, as shown in Fig. 29·43 (*c*), can be obtained only by a simultaneous change in the direction of the motor electromotive force, and in the direction of the line voltage E, as compared with their directions for the normal run of the motor. The first change can be accomplished by reversing the armature connections, or by reversing the motor field, so as to obtain polarities as shown in Fig. 29·45. The second change can be accomplished by properly delaying the angle of ignition of the thyratron tubes to obtain an average *negative* value (change in polarity) of the rectifier output voltage. A rectifier system with a negative output voltage, and an electromotive force in the load circuit with a polarity as shown in Fig. 29·45, is called a separately excited fixed-frequency inverter.

29·14 RATING, EFFICIENCY, AND POWER FACTOR

In electronic rectifier drives three main elements require a separate power rating. The electronic rectifier, the d-c motor, and the rectifier power transformer must be properly rated and combined in one system to give the most efficient and harmonious operation.

The electronic rectifier acts as a converter of electric energy and has a definite power rating. The rating of the rectifier is based on the fact that each electronic rectifying element has its very definite average current-carrying capacity and a definite instantaneous peak-current rating, which never should be exceeded. Although each electronic rectifier tube also has its forward and inverse peak-voltage rating, these have only an indirect effect on the power rating of the rectifier, since the rated direct voltage of the rectifier normally is chosen to correspond to the rated motor voltage. This voltage is usually 230 volts, but in some cases it may be 120, 350, 460 or 550 volts.

It has already been mentioned in preceding sections, and particularly in Section 29·12, that in polyphase half-wave thyratron rectifiers the instantaneous peak-current rating rather than the average current rating of each tube may become a factor in determining the over-all rating of the power rectifier. Obviously, the peak coefficient defined in Section 29·12 (equation 29·57) plays a very important part in that respect.

Usually, in assigning a power rating to a rectifier for armature power conversion, a somewhat different approach is used. Instead of assigning a proper power rating to each possible combination of rectifying elements on the basis of their individual current and voltage ratings, it is customary to start with standard horsepower ratings of motors, and select a proper rectifier for each of them. Considerations of cost, in addition to those of rating, play of course an important part in the selection of the most adequate combination of phases and types of electronic rectifying elements. Furthermore, the starting torque and the maximum current limit required also must be considered.

As an example, Table 29·2 shows the number of phases and the type of electronic tubes in armature-power rectifiers for standard horsepower ratings of the drive. The selection of rectifiers was based on standard 230-volt rating of d-c motors. It should be remarked that only half-wave polyphase rectifiers have been considered. Full-wave or bridge-type rectifiers are rarely used for power rectification in electronic drives, mainly because of the required double number of rectifying elements for the same rating, and poorer adaptability of these rectifiers to electronic control methods.

TABLE 29·2 EXAMPLE OF SELECTION OF THE TYPE OF RECTIFYING ELEMENTS AND THE NUMBER OF PHASES OF THE RECTIFIER FOR VARIOUS HP RATINGS OF 230-VOLT DRIVES

Hp	Rated Current	Current Limit	p	Average Current per Tube	Expected Inst. Peak Current	Type of Tubes	Average-Current Tube Rating	Peak-Current Tube Rating
1/3	1.5	3	2	0.75	6	3-C-23	1.5	6.0
1	4	8	2	2	16	672	3.0	30.0
1½	6	12	2	3	24	672	3.0	30.0
						or 624	6.4	77.0
2	8	16	2	4	32	624	6.4	77.0
3	12	24	3	3	48	624	6.4	77.0
5	20	40	3	7	80	414	12.5	100.0
7½	30	60	3	10	100	414	12.5	100.0
10	40	70	4	10	100	414	12.5	100.0

The influence of rectification on losses and frame sizes of d-c motors was discussed briefly in Section 29·12. Although d-c motors of *standard design* are normally used in electronic drives, and no special design modifications are required, the horsepower rating of the motor generally is affected by the wave shape of the armature load current. It will be recalled that the form factor of rectified armature current always is greater than unity, whereas standard d-c motors are rated on the basis of pure d-c supplies, which give the current form factor equal to unity. The copper losses in the armature winding are increased, with respect to normal d-c losses, in direct proportion to the square of the form factor. If, for instance, it is assumed that the form factor at full rated average armature current has a value of 1.25, then copper losses in the armature are increased 1.56 times, or by 56 percent. Magnetic losses in the armature, frame, and pole faces, as well as copper losses in the field of the motor, are normally increased too, but to a much lesser extent. Generally, an assumption of a total increase in losses of the motor of 50 to 60 percent is satisfactory for a rough estimation of the motor frame size. Conversely, an existing motor of a given horsepower rating often must be derated if it is to be used in rectifier drives; for example, a 1-horsepower motor will have to be derated to ½ or ¾ horsepower. A 2-horsepower motor will become a 1¼-horsepower motor, and a 3-horsepower motor will be rated 2 horsepower.

Sometimes, various means are used to reduce the armature-current form factor to such a low value (1.01 to 1.05) that no change in motor horsepower rating with respect to the conventional d-c case is required. Furthermore, the reduction of the form factor has a beneficial effect on the commutation of the motor, particularly at higher speeds. The reduction of the form factor can be accomplished by reducing the rectifier transformer voltage to a bare minimum, by increasing the number of phases of the rectifier, and by increasing the impedance angle of the load circuit. The latter can be accomplished by the addition of a reactor in series with the armature. Such an artificial increase in the impedance angle of the load circuit reduces the form factor, increases the horsepower rating of a given motor, and improves its commutation. However, the addition of a heavy and rather expensive reactor may offset the advantage of improved motor rating. Generally, the armature circuit reactors are used only in special cases, particularly where the form factor is to be reduced on account of an extremely difficult commutation, or to reduce the magnetic noise and possible vibrations caused by pulsating armature current.

The rating of the rectifier power transformer must be adapted to the horsepower rating of the drive. Normally, the determination of the kva rating of the polyphase power transformer is based on the continuous rating of the drive, without taking into account the starting and overload conditions. The kva of the secondary polyphase winding of the power transformer can be calculated from the formula

$$\text{Kva} = E_s I_{d\text{-}c} f_f \sqrt{p} \tag{29·69}$$

where E_s = rms value of the transformer secondary voltage to neutral

$I_{d\text{-}c}$ = rated value of motor armature current

f_f = form factor of the rectified armature current at the lowest operating speed and full rated torque

p = number of phases of the rectifier.

As an example, for a 3-horsepower electronic drive with $E'_{d\text{-}c} = 230$ volts, $I_{d\text{-}c} = 12$ amperes, $f_f = 1.2$, $p = 3$, $E_s = 300$ volts, the rating of the power transformer (from equation 29·69) will be 7.5 kva. For a rough estimation, the approximate kva of the power transformer for drives up to 15 horsepower can be obtained as

$$\text{Kva} = 2.5P \tag{29·70}$$

where P = horsepower rating of the drive.

In accordance with the general definition, the efficiency is the ratio of the output power of a system to the input power. For any converter of electric power into mechanical power, the efficiency can be represented by the general expression

$$\eta = \frac{P_{EL} - \Delta P}{P_{EL}} \tag{29·71}$$

where P_{EL} = total electric power delivered to the system from the line

ΔP = total power losses in the system.

In electronic rectifier drives the total losses ΔP have to include the total losses in the motor, losses for heating the cathodes of electron tubes, and in particular the cathodes of thyratron rectifiers, various losses in electronic control and other auxiliary circuits, arc-drop losses in power rectifier tubes and, finally, losses in the power transformer.

The efficiency of an electronic drive also can be expressed as

$$\eta = \frac{746P_M}{\sqrt{p_L} I_L E_L p_f} \tag{29·72}$$

where P_M = horsepower developed at the motor shaft

p_L = number of phases of the a-c supply line, normally 1 or 3

I_L = line current

E_L = line voltage

p_f = power factor at the line terminals.

Obviously, the efficiency of the system is a function of the torque developed by the motor, as well as of the motor speed, and increases with the increase in torque or speed.

In Fig. 29·47 are shown experimental graphs of the efficiency of an electronic drive as a function of the motor-armature current. With constant motor excitation, this current may be considered directly proportional to the motor torque. The electronic drive for which the experimental graphs in Fig. 29·47 were plotted is rated 1 horsepower, 230 volts, and uses a symmetrical two-phase, half-wave rectifier supplied from a single-phase, 220-volt, 60-cycle line.

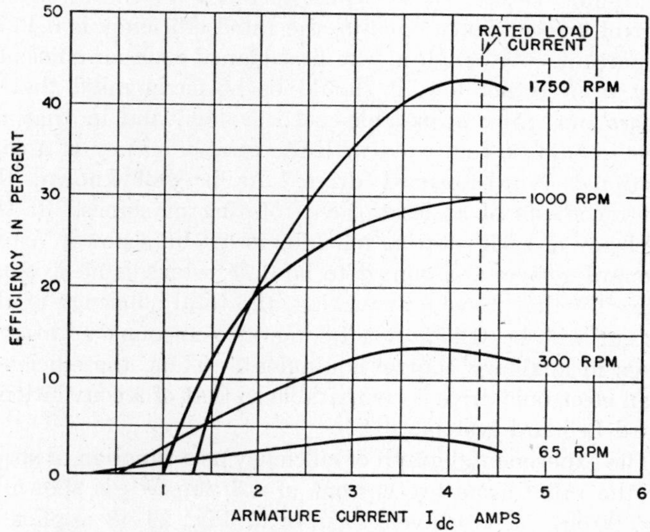

FIG. 29·47 Experimental graphs of efficiency of an adjustable-speed rectifier drive versus load current. ($p = 2$, $E_s = 400$ volts, 1 hp, 4.3 amperes.)

The base speed of the motor is 1750 revolutions per minute, and its rated direct armature current is 4.3 amperes. In Fig. 29·47 graphs of efficiency versus armature current are given for four different speeds. Full rated field excitation was maintained throughout the test. It will be noted that the efficiency is increasing with the increase in speed as well as in armature current. The efficiency becomes zero for no-load current of the motor since for this current no torque or power is developed at the shaft of the motor. The no-load current is a function of speed and it increases with the increase in speed because of the increase in magnetic and mechanical losses of the motor. This can be plainly seen in Fig. 29·47.

As should be expected, the efficiency at rated speed of 1750 revolutions per minute reaches its peak at about the rated value of armature current of 4.3 amperes. For this

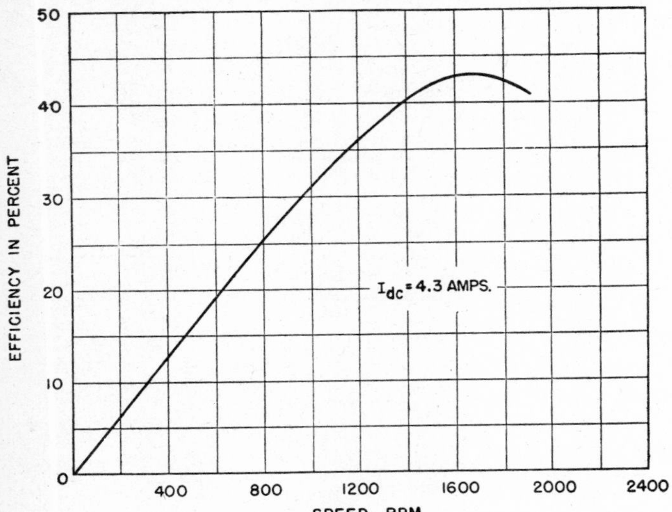

Fig. 29·48 Experimental graph of efficiency of an adjustable-speed rectifier drive versus speed, for constant armature current. (p = 2, E_s = 400 volts, 1 hp, 4.3 amperes.)

particular 1-horsepower, 1750-revolutions-per-minute, single-phase-line rectifier drive, which includes all the electronic control and auxiliary circuits, the rated efficiency is 0.43 or 43 percent. Although at the first glance such an efficiency may seem rather low, it should be borne in mind that it covers the entire adjustable-speed system, and in that respect should be compared with the total efficiency of a conventional Ward-Leonard drive. As is well known, the latter consists of at least three rotating machines. If the efficiency of each rotating machine in a 1-horsepower Ward-Leonard system is assumed to be 0.77, which is an average figure for that horsepower rating, the total efficiency of the system will be 0.456, not taking into account additional losses in auxiliary control equipment. Thus, the efficiency of an electronic drive is comparable to that of a conventional Ward-Leonard system.

The experimental graph of efficiency as a function of speed for the rated armature current of 4.3 amperes is shown in Fig. 29·48. The efficiency curve in Fig. 29·48 applies to the same drive as graphs in Fig. 29·47.

Electronic drives of higher horsepower rating, which use polyphase rectification, have, generally speaking, higher efficiency. As an example, the efficiency of a 10-horsepower drive with symmetrical four-phase rectification and a Scott-connected power transformer may be expected to be about 0.65.

The power factor at the a-c supply-line terminals of an electronic rectifier drive is of some interest, particularly wherever larger powers are involved. In general, the power factor is relatively low and, for higher powers, power-factor correcting means may be advisable in order to reduce the reactive power absorbed by the system. The main reason for a low power factor is the principle of control of the output voltage of the rectifier by delayed ignition of the rectifier tubes. The instant at which the current in each secondary phase of the transformer is allowed to start is arbitrarily controlled, and most often considerably delayed with respect to the voltage wave of that phase. This delay, of course, creates the effect of a highly inductive equivalent load circuit which may be thought of as being connected directly to the secondary terminals of the power transformer. Furthermore, it should be remembered that this inductive effect increases with increasing delay of ignition and, since that increase in delay corresponds to lower armature voltages and lower speeds of the motor, it follows that the power factor is directly affected by the speed. The lower the speed, the lower will be the power factor at the line terminals.

As an example, Fig. 29·49 represents the power factor for a 1-horsepower electronic drive as a function of the motor speed by armature voltage control. The load current of the drive was maintained at its rated value of 4.3 amperes. The graph in Fig. 29·49 applies to the same system as the graphs in Figs. 29·47 and 29·48.

Unlike efficiency, the power factor does not drop to zero at speeds approaching the standstill of the motor. In fact, even at standstill, where a very low voltage is applied to the armature and the angle of ignition of the rectifier tubes is extremely delayed, there are appreciable losses in the system, and these losses require the absorption of active power from the a-c line. The zero-speed losses include copper

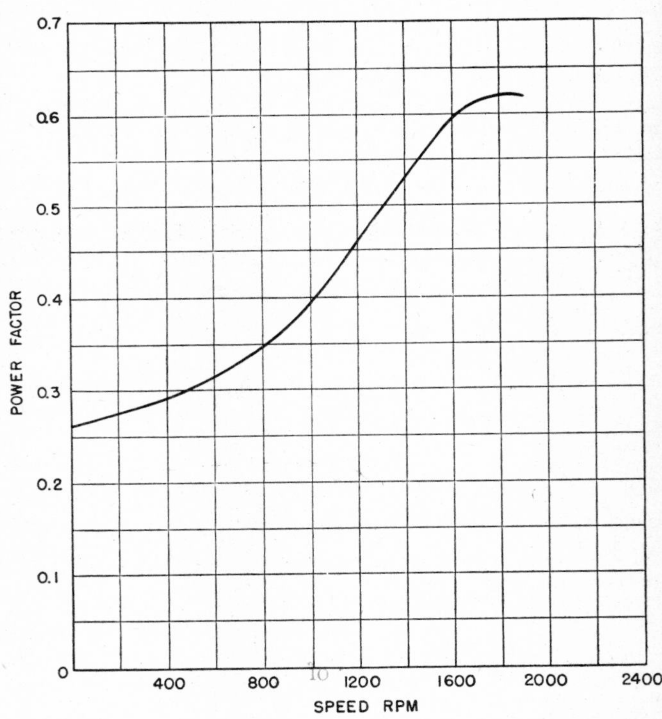

Fig. 29·49 Experimental graph of power factor at the line terminals of a rectifier drive versus speed, for constant load current of 4.3 amperes. (p = 2, E_s = 400 volts, 1 hp.)

losses in the motor caused by the full rated current in the armature, excitation losses of the motor, heater losses for the rectifier and other electron tubes, losses in auxiliary control circuits and in the power transformer. For speeds approaching zero the power factor is equal to about 0.26. At the rated speed of 1750 revolutions per minute, corresponding to the rated armature voltage of 230 volts, the power factor reaches its maximum of about 0.62.

Experimental graphs of the power factor plotted as a function of the armature current for different motor speeds are shown in Fig. 29·50. Generally speaking, the power factor does not change very appreciably with the load current. Particularly, for a speed of 1050 revolutions per

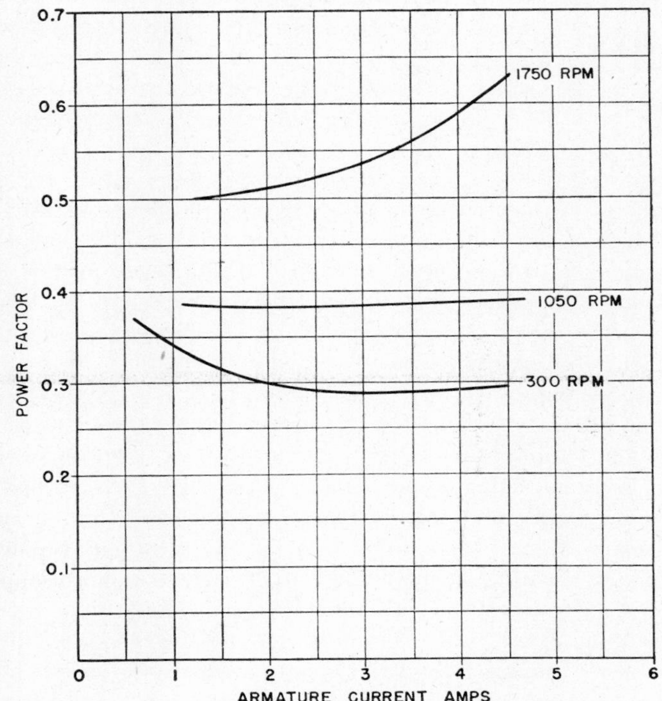

FIG. 29·50 Experimental graphs of power factor at the line terminals of a rectifier drive versus load current. ($p = 2$, $E_s = 400$ volts, 1 hp, 4.3 amperes.)

minute in Fig. 29·50 the power factor is practically independent of load, and is equal to about 0.384. At a full rated speed of 1750 revolutions per minute the power factor slowly rises with load from about 0.51 at no-load current to 0.62 at full rated load current.

The power factor is higher for larger polyphase systems, but even there it rarely exceeds 0.8 at rated armature voltage and rated load current.

29·15 CONCLUSIONS

In preceding sections, general principles of electronic motor control, and in particular of controlled rectification, were outlined. Several typical electronic control networks for rectifier-motor systems were described to illustrate possible solutions of some fundamental control problems. It should be remembered, however, that no attempt was made to cover all the recently developed circuits, some of which still remain undisclosed. In addition, the scope of this chapter could permit the inclusion of only a few of the existing circuits.

The control problems and the typical means for their solution, dealt with in this chapter, were approached from the point of view of an adjustable-speed electronic drive for general-purpose applications. In other words, specific application problems were not emphasized here, but only general elements of performance, such as speed control, speed-torque characteristics, control of acceleration, principles of rectification and inversion, were given special attention. Many problems of considerable interest either could not be mentioned at all, or had to be discussed very briefly.

General-purpose electronic drives and their main functions, which were given special attention in this chapter, represent only one group in the rapidly growing field of electronic motor control. In fact, the number of special electronic control systems and networks for rectifier drives is very great, and includes various electronic control and regulating systems for every conceivable application involving control of motors. To mention a few, hoist and crane drives, small traction drives, drives and regulators for various winding operations,[8] tension regulators, acceleration regulators, drives for stoker control, automatic-cycle drives for machine tools, and drives with spindle-load control for milling machines represent an active field for various electronic motor-control systems.

REFERENCES

1. "Thyratron Control of D-C Motors," G. W. Garman, *Trans. A.I.E.E.*, Vol. 57, 1938, pp. 335–342.
2. "Theory of Rectifier—D-C Motor Drive," E. H. Vedder and K. P. Puchlowski, *Trans. A.I.E.E.*, Vol. 62, 1943, pp. 863–870.
3. "Electronic Control of D-C Motors," K. P. Puchlowski, *Trans. A.I.E.E.*, Vol. 62, 1943, pp. 870–877.
4. "Thyratron Motor Control," E. E. Moyer and H. L. Palmer, *Trans. A.I.E.E.*, Vol. 62, 1943, pp. 706–712.
5. "Inverter Action on Reversing of Thyratron Motor Control," H. L. Palmer and H. H. Leigh, *Trans. A.I.E.E.*, Vol. 63, 1944, pp. 175–184.
6. "Voltage and Current Relations for Controlled Rectification with Inductive and Generative Loads," K. P. Puchlowski, *Trans. A.I.E.E.*, Vol. 64, 1945, pp. 255–260.
7. "Electronic Control of D-C Motors," E. E. Moyer, *Electronics*, 1943, pp. 98–103, 215–217.
8. "An Electronic Drive for Windup Reels," K. P. Puchlowski, *Trans. A.I.E.E.*, Vol. 65, 1946, pp. 585–591.

Chapter 30

REGULATION

W. O. Osbon and V. B. Baker

REGULATION is the automatic process of maintaining a designated quantity at a predetermined value or of varying that quantity according to some prearranged plan. More specifically, the process is one of matching the designated quantity to a reference or pattern in a manner to maintain the smallest possible difference or "error" between the two. Regulation of voltage on a d-c generator is a typical example. Here, all or a portion of the generated voltage is compared to a battery or other suitable constant reference voltage, and the difference between these voltages is amplified. A control device, operating on the amplified error signal, furnishes the required corrective means to make the error or difference voltage tend to approach zero.

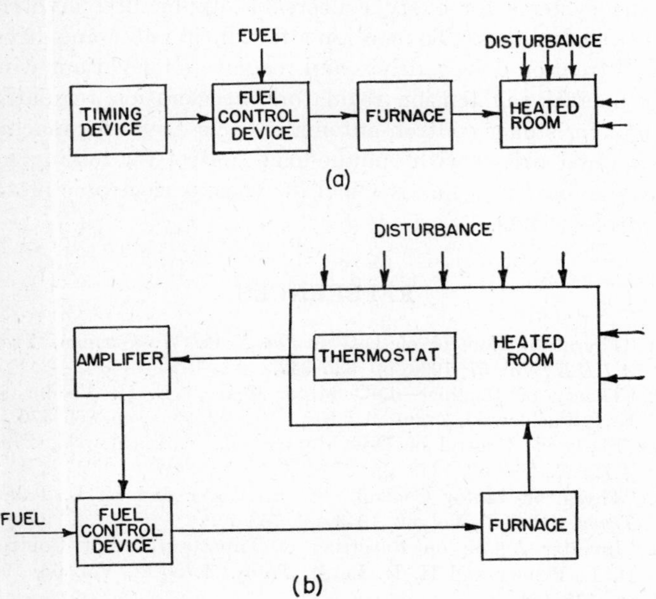

Fig. 30·1 Comparison of open-cycle and closed-cycle control systems.

Regulation is a closed-cycle process as contrasted to open-cycle control where the controlling element is in no way influenced by the controlled quantity. The fundamental difference between the two types of control is illustrated in the block diagrams of Fig. 30·1. Heat delivered to a room may be controlled as a function of time as in (*a*). Since the timing operation is not affected by room temperature, no compensation is made for temperature variation resulting from external disturbances. This process of delivering heat to the room is open-cycle. If, however, the timing device is replaced with a thermostat within the room and if the temperature measurement thus obtained is used to control fuel flow as in (*b*), the process becomes closed-cycle. Operation of each of the elements included in the closed system is influenced by some other element in the system and, in turn, affects the remaining components. This interdependence of elements forming a closed loop is characteristic of all regulated systems.

The process of regulating any quantity—voltage, current, position, temperature, velocity—requires the presence of certain basic elements which together constitute the regulated system. First, an error-detecting or -measuring device is necessary to provide a knowledge of the departure of the regulated quantity from the prescribed reference or pattern. Many such error detectors have been developed for a wide variety of applications.[1,2] In most systems it is desirable that the power required by the detecting device be as low as possible in order that its presence have little or no effect on the quantity being regulated except through the regulating process. Consequently, the power level of error intelligence is usually low, and some form of amplifier is required. The amplifier is the second basic element of a regulated system. Finally, there is the device in actual control of the regulated quantity which functions in accordance with the dictates of the amplified error signal to effect the desired regulation. In Fig. 30·1 (*b*) the error detector is the temperature-measuring element of the thermostat; the amplifier includes the thermostat contacts and the fuel control device, and the furnace is the output element which makes the required correction.

In the following discussions the term *regulator* is used to describe those elements of a closed-cycle regulated *system* which would otherwise not be present if the controlled quantity were not subjected to automatic regulation. In general, a regulator usually includes an error-measuring device and some form of error-signal amplifier to raise the power level. The error-measuring device includes both the reference quantity and the element for indicating the status of the regulated quantity. The results obtained with modern regulators would, in many cases, be impossible without suitable anti-hunting or stabilizing elements. By implication, these elements also are considered part of the regulator. In the system shown in Fig. 30·8 for regulation of motor speed, the reference battery, the speed-measuring tachometer, and the electronic device which supplies power to the exciter field are considered jointly as a speed regulator. None would be required in the absence of automatic regulation.

Regulating problems have been the concern of engineers from the very beginnings of industry, and the use of regulators is almost as old as the machines on which they work. Steam engines have long had mechanical speed governors. Hydraulic governors were the next important advance. With the development of electromechanical regulators many new quantities were brought under automatic control. In the earlier types, however, the requirements were often simple, and the regulators themselves rather crude. The development of electronics and its application to regulation problems brought about a tremendous change in the entire scope of automatic control. The field of application has been greatly expanded, and the high order of performance which is commonplace today was unthinkable a dozen or so years ago. Obviously, these results are a direct consequence of certain inherent characteristics of electronic devices that enable them to foreshadow their predecessors completely.

Electronic tubes are capable of providing with perfect fidelity the high amplifications which are requisite to high accuracy and high speed of response. Electronic amplifiers are uniquely free of time delays which are inherent in most other types of amplifiers such as those involving moving mechanical parts or magnetic fields. System stability, consequently, is usually more easily obtained with electronic amplifiers than with other types. The stability problem is further simplified by the ease with which stabilizing or antihunting elements may be incorporated in electronic amplifying circuits. Another important advantage of electronic devices in regulator applications is their very great power sensitivity. In many systems and processes the power levels available for actuating a regulator are well below the input power requirements of all other types of regulators except the electronic. Another advantage, perhaps most important from the standpoint of economy of manufacture, is the ease with which standard electronic components can be combined to form regulators for a wide variety of systems and processes.

All regulators may be classified according to the manner in which their corrective efforts vary with error. Most regulators can be classified into three fundamental groups: (1) the on-off or oscillating type, (2) the dead-zone type, and (3) the proportional type. Figure 30·2 shows a graphical representation of error versus corrective effort for each of the three types.

With the typical on-off type represented by Fig. 30·2 (*a*), any infinitesimal error in the regulated quantity produces a corrective effort of constant magnitude and of the proper sense to decrease the error. Since the corrective quantity theoretically can be either positive or negative with zero error, continuous oscillation about the nominal value of the regulated quantity is inherent with this type of regulator. The sudden appearance of maximum corrective effort, first in one direction and then in the other, which is characteristic of this type, may be conducive to rapid wear of moving parts.

In the typical dead-zone type represented by Fig. 30·2 (*b*), a zone of inaction exists in which the regulated quantity may vary between plus and minus error limits without initiating a correction. If, however, some disturbance causes the regulated quantity to exceed these limits of error, a corrective effort of constant magnitude and the proper sense to decrease the error is applied. With no time lag in the system, this correction exists as long as the error exceeds the dead-zone limits, and is removed when the deviation is again within these limits. In this type the regulating equipment remains at rest a major portion of the time, thus avoiding one of the principal objections to the on-off type.

Applications of the on-off and dead-zone types of regulator are usually limited to systems where the disturbances are relatively small and vary slowly with time. Both types are basically simple in design and operation.

As indicated in Fig. 30·2 (*c*), the corrective effort in the proportional type of regulator bears an essentially linear relationship to the error throughout the operating range of the regulator. The proportional regulator is ordinarily applied whenever rapid response and high accuracy are required.

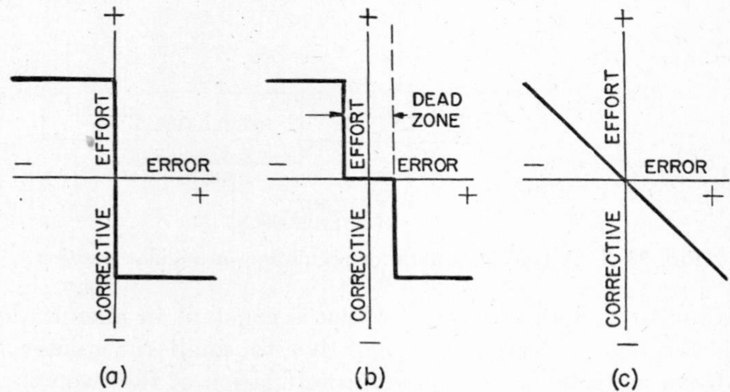

Fig. 30·2 Error versus corrective effort for the three fundamental types of regulators: (*a*) on-off type, (*b*) dead-zone type, (*c*) proportional type.

When properly designed it is capable of high steady-state and transient accuracy even with rapidly varying disturbances.

The three types of regulators are all susceptible to mathematical analysis, although the mathematical treatment of the on-off and dead-zone types [3,20] is not so straightforward as that of the proportional type. In the following sections the discussions of the on-off and dead-zone types is qualitative in character, whereas that of the proportional regulator is mathematical.

30·1 ON-OFF TYPE OF REGULATOR

An on-off type of regulator such as that shown schematically in Fig. 30·3 for the regulation of angular position is typical. The regulator element may be merely an error-controlled reversing contactor in the armature circuit of the d-c motor. Such a system includes inertia and some type of friction damping. To simplify the discussion, it will be assumed that the input angle is fixed or is changing very slowly. This is a reasonable assumption, since regulators of this type are seldom applied where conditions other than this exist.

Under ideal conditions, that is, if there were no time delay between the appearance of an error and the development of a corrective torque at the motor shaft, the system would oscillate at zero amplitude and infinite frequency. However, in any practical system there is always some time delay between these two events. In the system under discussion, this time lag is the time required to reverse the armature contactor and to build up the corrective torque after the reversal is completed. Its effect is to produce steady-state oscillations of a definite amplitude and finite frequency.

The system is extremely sensitive to time delay if the damping torque is viscous in character, that is, proportional to the velocity of the motor shaft. Under these conditions, the steady-state amplitude will be small only if the time delay is very short. On the other hand, if the damping torque

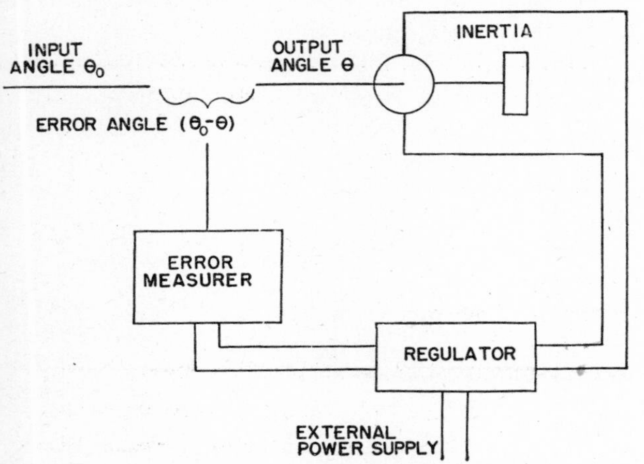

Fig. 30·3 A typical system for regulation of angular position.

is due to coulomb damping which is constant in magnitude, the system is relatively insensitive to small time delays.[3] Most practical systems have a combination of the two types of damping.

Figure 30·4 illustrates qualitatively the effects of time delay on the performance of a practical system. The system is assumed to be oscillating with an amplitude θ_1. Starting at point A, the positive corrective torque acts to decrease the negative error, and the output member accelerates. If there were no time delay in the system, the torque resulting from the initial error would reverse at point P, and the velocity of the motor shaft would start to decrease immediately. The presence of the time lag, however, permits the positive torque to remain in action beyond this point, and the system continues to accelerate until point B is reached, at which time the corrective torque reverses direction. At

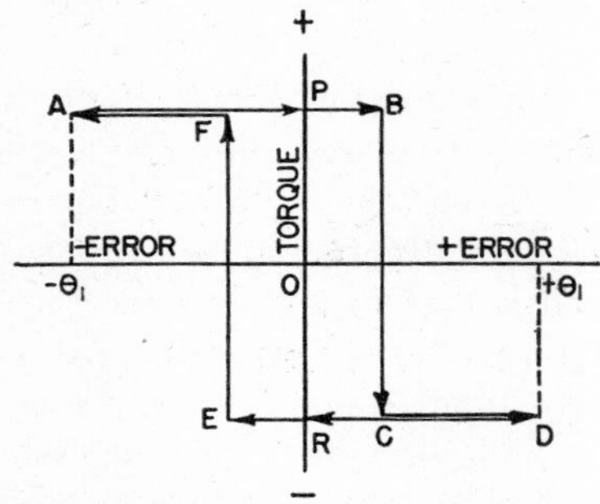

Fig. 30·4 Error versus corrective effort for a system with an on-off regulator and time delay.

point B the output velocity reaches a maximum value which is higher than the maximum velocity that would have occurred had the torque reversal taken place at point P with an initial error θ_1. Deceleration occurs between points C and D, and the velocity becomes zero and reverses at point D. During the remainder of the cycle from point D through points C, R, E, and F, back to A, the system follows the same sequence of events in the reverse direction. Because of the increased velocity at points B and E as a result of the system delay, an increment of energy is added to the system during each cycle of oscillation. This additional energy must be dissipated in the system, and the amplitude of oscillation, therefore, must continue to be θ_1, at which value the additional damping and friction loss corresponding to the higher peak output velocity exactly equals the energy added to the system as a result of the time delay. Thus, the system, when subjected to an initial error, instead of executing an oscillatory decay to zero amplitude as with zero time delay, will seek and find the amplitude θ_1 that fulfills the energy-balance requirement.

With a smaller time delay than that represented in Fig. 30·4, the regulator does the wrong thing for a shorter interval. The error angle at which torque reversal occurs is smaller, the system velocity is lower than that attained at point B, and the increment of energy added to the oscillation is smaller. Consequently, sufficient friction losses to equal this smaller block of energy will accrue with a maximum amplitude less than θ_1.

The performance of an on-off regulator system is, of course, influenced by factors other than time lag. The magnitude of the corrective torque produced at the motor shaft and the total inertia of the system also play important parts in the behavior of the system. Regardless of the type of damping, the amplitude of sustained oscillation is inversely proportional to the system inertia. That is, the oscillation amplitude of a high-inertia system is smaller than that of a low-inertia system, all other quantities being the same. In addition, the amplitude of steady-state oscillation is directly proportional to the magnitude of the corrective torque. For example, hunting will become more violent with an increase in torque at the motor shaft. Conversely, the amplitude of hunting will decrease if the corrective torque is lowered.

30·2 DEAD-ZONE REGULATOR

For purposes of comparison it is assumed that the on-off regulator of the system in the preceding section is replaced by a dead-zone regulator represented by Fig. 30·2 (*b*). In this case the regulator element of Fig. 30·3 may be essentially the same as before except that the corrective torque is not initiated until the error exceeds some predetermined value. As before, it is assumed that the input is either stationary or changing only very slowly.

Under ideal conditions, that is, if there is no time delay, it is apparent that the system cannot oscillate continuously. If the initial error exceeds the dead-zone limits, whether the system damping is purely viscous, purely coulomb, or a combination of both, there is an interval in each cycle when the only force acting on the system is the damping force. Thus, each time the regulated quantity passes through the dead zone, oscillatory energy is removed from the system by dissipation. The initial oscillations, therefore, decay in amplitude with each cycle until the error just fails to exceed the zone limit. Thereafter oscillations cease entirely.

From Fig. 30·2 (*b*) it is seen that the regulated quantity may vary within the dead-zone limits, and that as long as these limits are not exceeded no correction is initiated. The steady-state accuracy of this type of regulator is, therefore, defined by these limits since an error equal to one-half the width of the dead zone may exist indefinitely without initiating a correction.

The effect of time delay on the performance of this type of regulator can be seen from a consideration of Fig. 30·5. It is assumed that the system is subjected to an initial error $-\theta_1$. At A the initial output velocity is zero, and the positive restoring torque accelerates the system in the direction to decrease the error. At B the regulated quantity enters the dead zone and, in the absence of time lag, the restoring torque would fall to zero and the output velocity would start to decrease. However, with time delay, the torque persists for the duration of the lag. As a result, the velocity of the output member continues to increase from its value at B to some higher value at C inside the dead zone before the driving torque is actually removed. The additional energy represented by the increased velocity at C has been added to the system, and it must be dissipated in some manner.

During the ensuing period between D and F no corrective torque exists. The damping force alone acts on the system, removing some of the energy and thus decreasing the output velocity. At E in the cycle of oscillation, a reversed corrective torque would be applied under ideal conditions but, owing to time lag, angular displacement progresses to F outside the dead zone before this torque is applied. In the interval G–H, the restoring torque and the damping torque act in the same sense, both tending to decrease output velocity, which reaches zero at H. The reverse sequence then starts. An additional block of energy is again added to the oscillation as a result of time lag on re-entering the dead zone at I.

After the transient oscillatory period, the system may or may not oscillate continuously. Whether oscillations persist or not depends upon the control parameters. Considering only the effects of time lag, it is seen that this lag causes an addition of energy to the system on entering the dead zone, and permits the regulated quantity to deviate a finite amount beyond the zone limit when leaving the dead zone before the opposite restoring effort is applied. The combined effect is to aggravate the oscillation by raising the oscillatory energy level and causing the overshoot to be greater than it otherwise would be in the absence of lag. If the energy added to the system on entering the dead zone at B and I is greater than the energy losses which occur during the inactive periods represented by the intervals D–F and K–L of Fig. 30·5, continuous oscillation must exist. Since the additional energy imparted to the system is directly proportional to the time delay, it is evident that the shorter this delay the less will be the likelihood of sustained oscillation. In this connection, if the parameters of the system are such that continuous oscillation is possible, it is interesting to note that any disturbance must produce an initial error greater than a certain minimum value outside the dead zone before sustained oscillations occur. If the initial error is less than the critical value, the increment of energy added as a result of time lag will be less than the losses occurring within the inactive range, and the amplitude will decay until oscillation ceases entirely.

The width of the dead zone thus plays an important part in the performance of this type of regulator. From the standpoint of steady-state accuracy, it is desirable to keep the width small. However, if the width is small in comparison to the magnitude of the initial errors caused by the disturbances generally encountered in the application, oscillation beyond the dead-zone limits is likely to occur much of the time. The average accuracy thus obtained may not be as good as that obtained with a wider dead zone. Most designs are compromises between these two considerations.

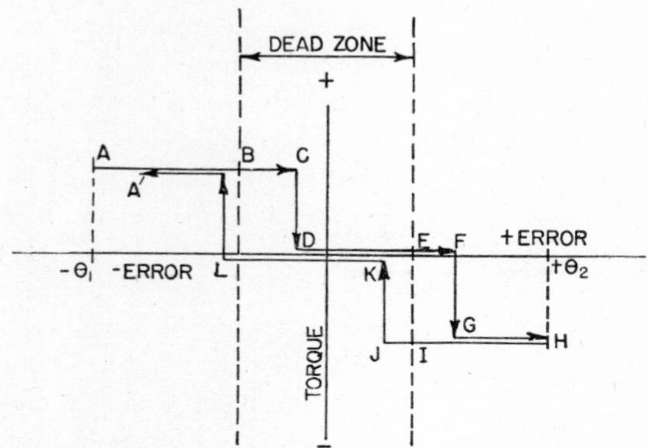

FIG. 30·5 Error versus corrective effort for a system with a dead-zone regulator and time delay.

As in the on-off regulator, the magnitude of the corrective torque, the nature and amount of the damping, and the system inertia are also important factors in the determination of the performance of the dead-zone regulator.

30·3 PROPORTIONAL REGULATORS

Proportional regulators are applied when high steady-state and transient accuracy and high speed of response are required. As pointed out above, proportional regulators are characterized by the fact that the corrective effort developed by the regulator due to an error is proportional to the magnitude of the error. For this reason the regulator and other elements of proportional systems have essentially linear characteristics over most of the operating range. The linear range of electron tubes is limited by the cut-off characteristics of their control elements, and the linear range of d-c rotating machines is limited by saturation of their magnetic circuits. Other types of control and power apparatus used in proportionally regulated systems have similar limitations to their ranges of linearity.

In practice each element of a proportional system is designed to have sufficient range to take care of the most severe disturbance to which the system will be subjected. These disturbances may be caused by changes in load on the system, changes in the reference quantity, temperature variations, voltage variations, or other similar effects. Although range limitations generally are troublesome in regulator design, they are frequently used to advantage by designing the system so that the cut-off characteristic of one element serves to protect other elements of the system.

The proportional type of regulator is more susceptible to accurate mathematical analysis than other types of regulators, and the theory of proportional regulators has reached a high state of development. The engineer who is familiar with the fundamentals of the theory and who has had some experience in their application can design complex servo systems with the assurance that the performance will be quite close to expectations.

Two general methods of attack are available for the theoretical investigation of regulated systems. The first of these is based on steady-state circuit theory. This method of analysis depends upon the fact that the performance of a servo system is completely defined by the steady-state frequency-response and phase-shift characteristics of the system. The theory of the method is quite complete,[4, 5, 6] and it has important advantages in regulator analysis, particularly for those systems which are subjected to disturbances of a periodic nature.

The other method of investigation is the analysis of the transient characteristics of the system. This method is based on the fact that the performance of a regulated system is also determined completely if its response to a suddenly applied disturbance of some specified form is known. If the system under investigation is to be subjected in operation to suddenly applied disturbances, the results of this method of analysis are more directly interpreted in terms of actual performance than those of the steady-state method.

It is probable that the suddenly applied type of system disturbance is more commonly encountered in industrial applications of regulators than the periodic disturbance. Consequently, the major part of the remainder of this chapter is devoted to a brief discussion of the transient theory of proportional regulators. The derivation of the differential equations of several typical regulators using d-c power machinery is given. The methods of solving the differential equations are indicated, and the devices used to insure system stability are described. Throughout the discussion the effect of machine characteristics on system performance is emphasized. Although much of the discussion is focused on specific systems used as examples, the analysis is presented in such a manner that it may be applied without difficulty either to more general systems or to other specific systems.

The work involved in making the more extensive investigations of the transient behavior of proportional regulator systems is immensely simplified by a powerful tool known as the mechanical transients analyzer.[7, 8] A complex regulator system can be reproduced on this device in the form of an analogous electrical circuit. The response of the system to any specified type of disturbance may then be determined for a wide variety of conditions in a small fraction of the time required to obtain the same data by computation.

30·4 GENERATOR-VOLTAGE REGULATOR

The generator-voltage regulator is one of the simplest types of proportional regulating systems incorporating rotating machines. A typical system is illustrated in Fig. 30·6. Here the field power of the main generator is supplied by an exciter which is in turn excited by the regulator amplifier. The input to the amplifier is the voltage error, which is the difference between the generator terminal voltage E and the reference voltage E_R. The latter voltage determines the value of the regulated voltage. The generator load is assumed to be resistive. In some applications the rotating exciter may be replaced by an electronic exciter consisting of a grid-controlled rectifier of the thyratron or ignitron type. If the main generator is an a-c machine its voltage must be rectified and filtered before it is matched against the reference voltage.

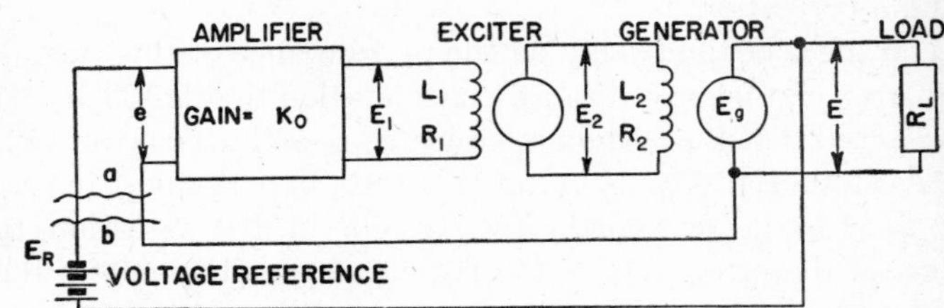

FIG. 30·6 A typical proportional voltage-regulated system.

Generator-Voltage Equation

If saturation is neglected the generated voltage may be written as

$$E_g = k_2\Phi_2 \tag{30·1}$$

where Φ_2 = the generator field flux
k_2 = a proportionality constant determined by the design constants and speed of the generator.

If eddy currents in the parts of the generator frame comprising the magnetic circuit are negligible, the flux during both steady-state and transient conditions is directly proportional to the generator field current i_2. Thus equation 30·1 may be written

$$E_g = K_2 i_2 \tag{30·2}$$

where K_2 is another proportionality constant determined by the design and speed of the machine. Moreover, the generator field current is related to the exciter voltage E_2 in accordance with the following equation:

$$E_2 = R_2 i_2 + L_2 \frac{di_2}{dt} \tag{30·3}$$

where R_2 = resistance of the generator field circuit
L_2 = inductance of the generator field circuit.

In operational notation [9] the derivative operator d/dt is replaced by p; and in operational calculus, which is commonly used in regulator theory, the operator p may be treated to a large extent as an ordinary variable. This property is extremely useful as it greatly simplifies the mathematical manipulations required in regulator analysis. Equation 30·3 may thus be solved for i_2 and written as follows:

$$i_2 = \frac{E_2}{R_2 + L_2 p} = \frac{E_2}{R_2(1 + T_2 p)} \tag{30·4}$$

where T_2 = the generator field-circuit time constant L_2/R_2. Substituting equation 30·4 in equation 30·2 gives

$$E_g = \frac{K_2}{R_2}\frac{E_2}{1 + T_2 p} \tag{30·5}$$

This is a differential equation in operational form expressing the generated voltage in terms of the field voltage and the circuit and machine constants.

Time Delay and Anticipatory Operators

It is well known that, when a varying voltage is applied to a generator field, the armature voltage lags behind the field voltage by an amount dependent upon the field-circuit time constant. This effect is illustrated in Fig. 30·7 for the case where a fixed voltage is applied suddenly to the field circuit at time $t = 0$. The initial rate of rise of armature voltage is such that if this rate were maintained the voltage would reach its ultimate value in time $T = L/R$. Actually in time T the voltage attains only $100[1 - (1/\epsilon)]$ or 63.2 percent of its final value. The differential equation 30·5 is a mathematical statement of this effect, and the curve of armature voltage in Fig. 30·7 (which will be recognized as the familiar exponential curve) is in fact a graph of the solution of equation 30·5 for the case of a suddenly applied field voltage. The factor in equation 30·5 which expresses this lagging time relation is $1/(1 + T_2p)$, and it is convenient, therefore, to consider

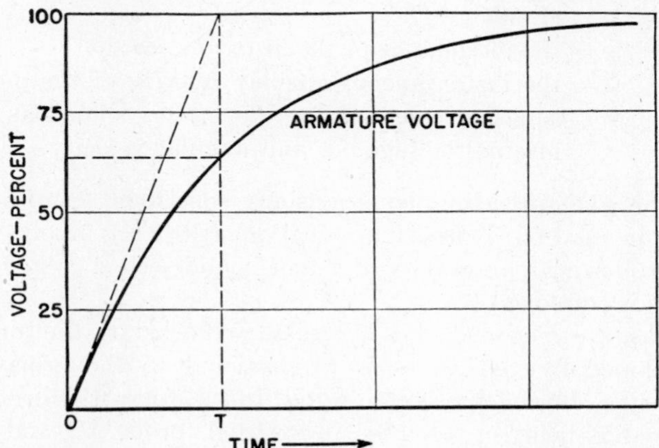

FIG. 30·7 Build-up of generator armature voltage after a sudden application of field voltage.

factors of the form $1 + Tp$ appearing in the denominator of an expression as *time-delay operators*. On the other hand, if equation 30·5 is rearranged to express the exciter voltage E_2 in terms of the generated voltage E_g the factor $1 + T_2p$ appears in the numerator of the equation:

$$E_2 = \frac{R_2}{K_2} E_g(1 + T_2p) \tag{30·6}$$

Since the generated voltage lags behind the exciter voltage when the latter is varying, the exciter voltage may be said to lead or anticipate the generated voltage. This effect is associated with the factor in the numerator of the form $1 + Tp$ which, consequently, is termed an *anticipatory operator*. Operators of these two types are fundamental to the theoretical treatment of regulator systems.

Derivation of System Equation

Referring again to Fig. 30·6, the generated voltage is equal to the sum of the load or terminal voltage and the internal generator drop. If armature circuit inductance is neglected the generated voltage is

$$E_g = E + i_aR_g \tag{30·7}$$

where i_a = the generator armature current
R_g = the generator armature resistance.

But, since $i_a = E/R_L$ where R_L is the load resistance, equation 30·7 may be written as

$$E_g = E\left(1 + \frac{R_g}{R_L}\right) = E\frac{R_T}{R_L} \tag{30·8}$$

Here R_T = the total generator armature circuit resistance $(R_L + R_g)$. Combining equations 30·8 and 30·5 gives the following expression for the terminal voltage in terms of the exciter voltage:

$$E = \frac{R_L}{R_T}\frac{K_2}{R_2}\frac{E_2}{1 + T_2p} \tag{30·9}$$

Now, by following the procedure used in the derivation of the generator-voltage equation 30·5, one may write a similar expression for the exciter voltage E_2 in terms of the amplifier output voltage E_1:

$$E_2 = \frac{K_1}{R_1}\frac{E_1}{1 + T_1p} \tag{30·10}$$

Here T_1 = the exciter field-circuit time constant L_1/R_1
R_1 = the total field-circuit resistance including the internal resistance of the amplifier-output stage
L_1 = the total field-circuit inductance.

If the amplifier is assumed to have no time delays its output voltage E_1 in terms of the input or error voltage e is

$$E_1 = K_0e = K_0(E_R - E) \tag{30·11}$$

where E_R = the reference voltage
E = the generator terminal voltage
K_0 = the voltage gain of the amplifier.

Combining equations 30·9, 30·10, and 30·11 to eliminate the voltages E_1 and E_2 gives

$$E = \frac{R_L}{R_T}\frac{K_0K_1K_2}{R_1R_2}\frac{E_R - E}{(1 + T_1p)(1 + T_2p)} \tag{30·12}$$

The quantity $K_0K_1K_2/R_1R_2$ is the product of the amplification factors of the various elements of the system, and may be replaced by a single quantity A', defined as the system amplification before the application of the load. The ratio R_L/R_T is the factor by which the system amplification is changed when the load is applied. Thus after the application of load one may write

$$E = A\frac{E_R - E}{(1 + T_1p)(1 + T_2p)} \tag{30·13}$$

where $A = A'R_L/R_T$.

The system amplification is the total amplification around the regulator "loop." In a physical system it can be measured by breaking the loop at some convenient point such as ab in Fig. 30·6 and determining the change in voltage at b due to a 1-volt change at a.

Equation 30·13 is the complete characteristic equation of the system of Fig. 30·6. If the system has more than two time delays the characteristic equation is

$$E = A\frac{E_R - E}{(1 + T_1p)(1 + T_2p)\cdots(1 + T_np)} \tag{30·14}$$

This is the general equation for a simple voltage regulator with any number of time delays and without special antihunting features. For a stable system this equation indicates that the generator voltage in the steady state (that is, when $p = 0$) is equal to the product of the system amplification and the voltage error ($E_R - E$), and that, immediately after a disturbance such as a load change, the generator voltage will approach its new steady-state value in a manner determined by the system time delays. It is obvious, therefore, that high steady-state accuracy depends upon high system amplification. High amplification also tends to minimize errors during transients, but these are further influenced to a great extent by the system time delays. In systems with two or more time delays, the tendency to oscillate or hunt is increased as the amplification is increased. This tendency and the anti-hunting arrangements used to minimize it are considered in later sections.

After it has been cleared of fractions and all terms containing E have been collected, equation 30·14 becomes

$$[(1 + T_1p)(1 + T_2p) \cdots (1 + T_np) + A]E = AE_R \qquad (30\cdot15)$$

and, when the indicated multiplications inside the brackets are performed, the equation takes the form

$$(a_np^n + a_{n-1}p^{n-1} + \cdots + a_2p^2 + a_1p + 1 + A)E = AE_R \qquad (30\cdot16)$$

The coefficients a_n, a_{n-1}, etc., in this equation are functions of the system time constants T_1, T_2, $\cdots$, T_n. The expression inside the brackets, when set equal to zero, is usually referred to as the auxiliary equation, and the right-hand side of the equation is the forcing function. This is the form of the characteristic equation generally used for determining system performance. Before demonstrating the treatment of such an equation, however, the derivation of the characteristic equations of typical speed and position regulators will be considered.

30·5 MOTOR-SPEED REGULATOR

Figure 30·8 illustrates a typical motor-speed-regulating system utilizing a tachometer generator to measure the motor speed and the familiar adjustable-voltage type of control.

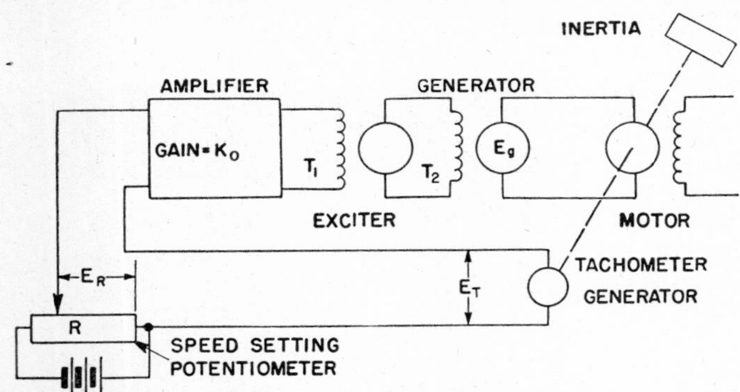

FIG. 30·8 A typical proportional speed-regulated system.

The motor field is excited from a constant voltage source, and it will be assumed that the field flux is constant or, in other words, that armature reaction is negligible. It will be observed that this system is essentially a voltage regulator, the regulated quantity being the voltage of the tachometer generator. However, since this voltage is directly proportional to tachometer speed, regulation of tachometer voltage is equivalent to the regulation of speed. Although Fig. 30·8 shows a d-c tachometer generator, a permanent-magnet a-c generator is frequently used, particularly when very precise regulation is required. In this case the outout of the a-c tachometer must usually be rectified and filtered before comparison with the reference voltage.

Motor-Speed Equation

In Fig. 30·8 the generated voltage E_g is made up of the impedance drop in the generator-motor armature circuit and the counter electromotive force or internal voltage of the motor:

$$E_g = i_a(R_a + pL_a) + G\Omega \qquad (30\cdot17)$$

where i_a = the armature current
R_a = the resistance of the generator-motor armature circuit
L_a = the inductance of the armature circuit
Ω = the instantaneous angular velocity of the motor
G = a design constant for the motor defined as the internal voltage per unit angular velocity.

In this and succeeding equations any consistent set of units may be used but, where numerical quantities are involved in the following paragraphs, the foot-pound-second system of units is employed.

Since the motor flux has been assumed constant the torque developed by the motor is proportional to the armature current. In the present example this torque will be considered as being opposed by an external torque M_L and the inertia reaction of the motor and load; thus

$$Bi_a = M_L + Ip\Omega \qquad (30\cdot18)$$

where Bi_a = the torque developed by the motor
B = a motor design constant expressing the torque per ampere
I = the total moment of inertia of the motor and its connected load
$p\Omega$ = the instantaneous acceleration.

The external torque M_L in the example being considered is assumed to be constant. Actually in some systems it will be zero, whereas in others there may be components of load torque which vary as some function of the speed or of time. A good example of a system with a variable load torque is the propeller drive for a wind tunnel where the load torque varies as the square of the speed and where there are usually no other external torques acting on the motor. We can express such a condition in equation 30·18 by replacing M_L with a term $M_0(\Omega/\Omega_0)^2$, where Ω_0 is the rated motor speed and M_0 is the load torque at speed Ω_0. This would lead, however, to a non-linear differential equation for the system, which is difficult to solve. Consequently, it is customary when analyzing systems of this sort to assume that the variations in speed are small; then the speed-torque characteristic of the load may be "linearized" at any particular operating point without serious error. For the general non-linear case the linear term representing the load torque to be substituted in equation 30·18 is the product of the instantaneous speed Ω and the slope of the speed-torque characteristic of the load at the steady-state operating speed. Thus in the case of the wind-tunnel propeller with a square-law characteristic the quantity $2M_1\Omega/\Omega_1$ would be used to represent the load torque in equation 30·18. Here Ω_1 is the steady-state operating speed and M_1 is the torque required by the propeller at this speed.

Equation 30·18 may be solved for i_a which is then substituted in equation 30·17 to give

$$BE_g = (M_L + Ip\Omega)(R_a + L_a p) + BG\Omega \qquad (30\cdot19)$$

This equation may be divided by R_a; and L_a/R_a, the armature circuit time constant, may be replaced by T_3 to obtain

$$\frac{BE_g}{R_a} - M_L = Ip\Omega(1 + T_3 p) + \frac{BG}{R_a}\Omega \qquad (30\cdot20)$$

In making the transition from equation 30·19 to equation 30·20 a term pM_LT_3 has been dropped because the derivative of the constant quantity M_LT_3 is zero. It is convenient now to divide equation 30·20 by the quantity

$$r_0 = \frac{BG}{R_a} \qquad (30\cdot21)$$

to obtain

$$\frac{E_g}{G} - \frac{M_L}{r_0} = \frac{I}{r_0}p\Omega(1 + T_3 p) + \Omega \qquad (30\cdot22)$$

This is the complete speed equation of the motor in terms of the generated voltage and the circuit and design constants of the machines and load.

Mechanical Time Constant

Equation 30·20 is a torque equation, and it is apparent from the last term of this expression that $r_0 = BG/R_a$ may be defined as the torque per unit speed. The accelerating torque is $Ip\Omega$, and it follows that the total inertia I may be defined as the torque per unit acceleration. Consequently, the ratio I/r_0 which appears in equation 30·22 has the dimension of time and is conveniently called the *mechanical time constant* T_0. This is an extremely useful quantity because it combines into a single term all the motor design factors which have an influence on system performance. The value of r_0 to be used in determining the mechanical time constant for a particular motor may be computed from equation 30·21 if the design constants G, B, and R_a are known; or it may be computed from the name plate data and certain measurable quantities as follows:

$$r_0 = \frac{M_0E_0}{2\pi nR_ai_0} \qquad (30\cdot23)$$

or

$$r_0 = \frac{550PE_0}{(2\pi n)^2R_ai_0} \qquad (30\cdot24)$$

where M_0 = full-load output torque in foot-pounds
n = rated speed in revolutions per second ($2\pi n$ = radians per second)
P = horsepower rating corresponding to torque M_0 and speed n
E_0 = rated voltage
i_0 = change in armature current due to application of load torque M_0
R_a = total armature circuit resistance including resistance of power source.

Since R_ai_0/E_0 represents the fractional speed drop due to torque M_0, equation 30·23 is equivalent to saying that the quantity r_0 is the reciprocal of the slope of the speed-torque curve where this curve is based on the assumption that speed droop is due to iR drop alone. Using the foot-pound-second system of units, the value of r_0 as computed in equations 30·23 and 30·24 is expressed in foot-pounds per radian per second.

For a particular motor, the value of the mechanical time constant $T_0 = I/r_0$ depends upon three factors: (*a*) the total inertia of the motor and its load, (*b*) the total armature circuit resistance, and (*c*) the motor field strength. Because it varies directly as the inertia, some control of its value may be exercised by adjustment of the load inertia. It also varies directly as the armature-circuit resistance since r_0 varies inversely as this resistance as indicated by equation 30·24. Finally it varies as the square of the motor field strength since, according to equation 30·24, r_0 varies inversely as the square of the speed, which in turn varies inversely as the motor field strength.

Derivation of System Equation

Following the methods used in the derivation of equation 30·12, the generated voltage E_g may be written as

$$E_g = \frac{K_0K_1K_2}{R_1R_2}\frac{E_R - E_T}{(1 + T_1p)(1 + T_2p)} \qquad (30\cdot25)$$

where E_R = the reference voltage
E_T = the tachometer voltage.

The other factors in equation 30·25 are the same as in equation 30·12. The quantity $E_R - E_T$ is proportional to the speed error, and if K_3 is a proportionality factor relating the tachometer voltage to motor speed,

$$E_R - E_T = K_3(\Omega_R - \Omega) \qquad (30\cdot26)$$

where $\Omega_R = E_R/K_3$ = a fictitious reference speed.

Equations 30·25 and 30·26 may be combined with 30·22 to give

$$\frac{K_0K_1K_2K_3}{GR_1R_2}\frac{\Omega_R - \Omega}{(1 + T_1p)(1 + T_2p)} - \frac{M_L}{r_0} = T_0p\Omega(1 + T_3p) + \Omega \qquad (30\cdot27)$$

This may be rewritten as

$$A\frac{\Omega_R - \Omega}{(1 + T_1p)(1 + T_2p)} - \frac{M_L}{r_0} = T_0p\Omega(1 + T_3p) + \Omega \qquad (30\cdot28)$$

where $A = K_0K_1K_2K_3/GR_1R_2$ = the regulator system amplification.

This is the complete characteristic equation of the speed regulator. As in the case of the voltage regulator the system amplification is the product of the amplification factors of the various elements of the system.

Equation 30·28 may be rearranged to give

$$\{[1 + T_0p(1 + T_3p)](1 + T_1p)(1 + T_2p) + A\}\Omega = A\Omega_R - \frac{M_L}{r_0} \qquad (30\cdot29)$$

Here also terms involving derivatives of constants have been dropped. If additional time delays are present they may be represented by corresponding additional factors in the left-hand part of the equation as in equation 30·15.

It will be observed that there is a marked similarity between this equation and equation 30·15. In fact, if the armature circuit inductance is negligible, that is, if $T_3 = 0$,* it will be noted that the left-hand side of the equation of a speed regulator having m electrical time delays is identical to that of a voltage regulator having $m + 1$ electrical time delays. After the indicated multiplications are performed, equation 30·29 can be put in the same form as equation 30·16.

If the regulator system is stable, equation 30·29 in the steady state ($p = 0$) reduces to

$$(A + 1)\Omega = A\Omega_R - \frac{M_L}{r_0} \tag{30·30}$$

When the load torque is zero we may say that Ω equals the no-load speed Ω_0. Then

$$\Omega_0 = \frac{A}{A + 1}\,\Omega_R \tag{30·31}$$

Now by definition $r_0 = M_L/s_L\Omega_0$, where s_L is the fractional speed drop and $s_L\Omega_0$ the actual speed drop which would occur as the result of the application of load M_L if the motor speed were unregulated. Therefore

$$\frac{M_L}{r_0} = s_L\Omega_0 = \frac{A}{A + 1}\,s_L\Omega_R \tag{30·32}$$

Substituting this in equation 30·30 and solving for Ω gives

$$\Omega = \frac{A}{A + 1}\,\Omega_R\left(1 - \frac{s_L}{A + 1}\right) \tag{30·33}$$

This equation indicates that the effect of the regulator is to reduce the steady-state fractional drop in speed caused by the application of a load by the factor $A + 1$.

Speed Measurement by Counter Electromotive Force

It is well known that the counter electromotive force or internal voltage of a motor is proportional to motor speed if the motor field strength is constant. Consequently the internal voltage is frequently used as a measure of speed for regulating purposes where precise regulation is not required.

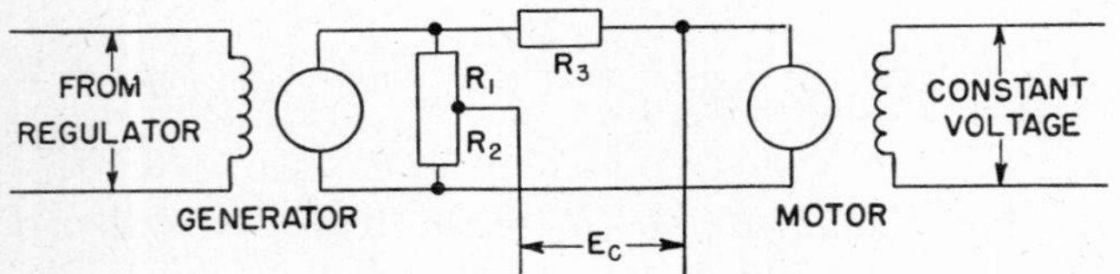

Fig. 30·9 Circuit for measurement of speed by counter-electromotive-force method.

Figure 30·9 shows a circuit commonly used to obtain a voltage proportional to the counter electromotive force. If the external series resistance R_3 is considered as part of the motor armature resistance, the voltage across R_1 is proportional to the motor terminal voltage. In addition, the voltage across R_3 is proportional to the iR drop in the motor. When the tap on the shunt resistor is properly adjusted the difference between these two voltages, E_c, therefore, can be made proportional to the internal voltage of the motor and may then be used instead of the tachometer voltage in the speed regulator. Because the motor field strength usually varies slightly with armature current due to armature reaction, and because the resistances may vary with temperature, the proportionality factor relating E_c and motor speed is not strictly constant. The result is that accuracy of regulation with this method of speed measurement is usually not as good as when a tachometer is used.

Speed Regulation by Motor Field Control

Another modification frequently used is to connect the motor armature across a constant voltage source and obtain regulation by motor field control. In this case the regulator furnishes either all or part of the motor field excitation, through an exciter if necessary. This system is less expensive than that illustrated in Fig. 30·8 because, if a suitable source of direct current is already available, it is unnecessary to furnish an a-c to d-c motor-generator set of sufficient capacity to supply the motor. From the standpoint of regulator performance, however, regulation by motor field control is often more difficult than with the adjustable-voltage system, particularly when a very wide range of operating speeds must be provided.

For a given armature voltage the speed of a motor varies *inversely* with field strength. Consequently an exact formulation of the system equation for this case results in a nonlinear differential equation. Because of this it is customary to linearize the system by assuming that variations in speed and field strength are small compared to their steady-state values; under this condition, second-order and higher order effects may be neglected. When this is done the system equation reduces to the same form as that for the adjustable-voltage system equation 30·29.

If it is necessary to vary the regulated speed of the motor there must be corresponding changes in the average field strength. It was pointed out previously that the mechanical time constant of a motor depends upon its field strength. Consequently the value of T_0 to be used in computing the characteristics of the system varies with the speed. Moreover, the motor-field time constant may vary as the regulated speed changes because of variations in the degree of saturation or because the field circuit resistance may be changed. Finally, the system amplification may also vary with the motor field strength. Thus for each different speed there is a corresponding value for each of these quantities, but as a part of the linearizing process it is customary to make the simplifying assumption that they are constant at any particular operating point.

30·6 POSITION REGULATOR

A typical position-regulated system is illustrated in Fig. 30·10. The system shown is similar to the speed-regulated system of Fig. 30·8 in that the adjustable-voltage d-c drive is used. The principal difference between the two systems is in the error-measuring and reference devices used. Whereas

* It will be recalled that this condition was also assumed for the voltage regulator.

the speed regulator utilized a tachometer generator to give a voltage proportional to angular velocity, the position system uses a device which gives a voltage proportional to angular displacement. The angular error-detection device shown in Fig. 30·10 consists of self-synchronous motors, commonly called synchros or selsyns, in the familiar "control transformer" connection. Any one of numerous other types of positional error-detection devices may of course be used equally as well.[1,2]

Derivation of System Equation

For purposes of analysis let us assume for the moment that there is a torsional spring member connected to the motor shaft as shown dotted in Fig. 30·10. This spring may be characterized by a spring constant or stiffness factor S_L which

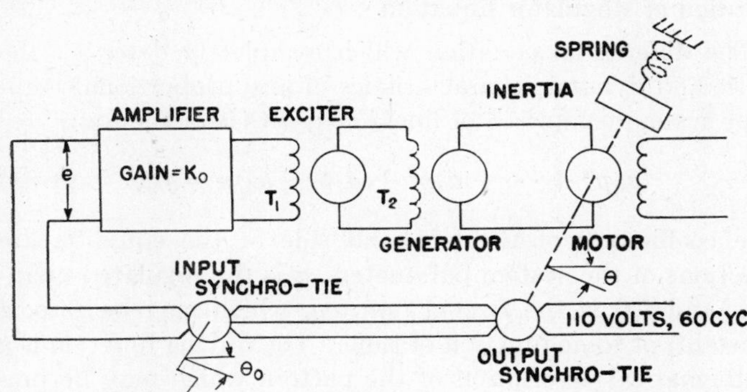

FIG. 30·10 A typical proportional position-regulated system.

is a measure of the torque per unit angular displacement θ of the spring. An example of a physical system having the equivalent of such a spring member is an automatic steering control for a ship where the water exerts a force on the rudder proportional to the angular displacement of the rudder. It will be assumed also that there is a component of load torque $r_Lp\theta$ proportional to output speed where r_L is the *damping coefficient* of the load.

As in equations 30·17 and 30·18 the following may be written for the position-regulated system:

$$E_g = i_aR_a(1 + T_3p) + Gp\theta \tag{30·34}$$

and

$$Bi_a = M_L + Ip^2\theta + r_Lp\theta + S_L\theta \tag{30·35}$$

where the instantaneous velocity and acceleration are now expressed in terms of the displacement angle θ as $p\theta$ and $p^2\theta$ respectively. These two equations may be combined by eliminating i_a between them; then, by making use of equation 30·21,

$$\frac{BE_g}{R_a} - M_L = (1 + T_3p)(Ip^2\theta + r_Lp\theta + S_L\theta) + r_0p\theta \tag{30·36}$$

Similarly to equations 30·25 and 30·26,

$$E_g = \frac{K_0K_1K_2}{R_1R_2}\frac{e}{(1 + T_1p)(1 + T_2p)} \tag{30·37}$$

and

$$e = K_4(\theta_0 - \theta) \tag{30·38}$$

where θ_0 = the input or reference angle
K_4 = a proportionality constant.

Substituting equations 30·37 and 30·38 in equation 30·36 gives

$$\frac{BK_0K_1K_2K_4}{R_aR_1R_2}\frac{\theta_0 - \theta}{(1 + T_1p)(1 + T_2p)} - M_L = (1 + T_3p)(Ip^2\theta + r_Lp\theta + S_L) + r_0p\theta \tag{30·39}$$

This is the torque equation for the complete position-regulated system. The first term of equation 30·39 is the torque developed by the motor as a result of regulator action and, since it is proportional to angular error $\theta_0 - \theta$, the factor $BK_0K_1K_2K_4/R_aR_1R_2$ has the dimensions of torque per unit angle or stiffness. Therefore this factor may be replaced by S_R which is then defined as the *regulator stiffness.*

Equation 30·39 may be rearranged as follows:

$$\{[(1 + T_3p)(Ip^2 + r_Lp + S_L) + r_0p](1 + T_1p)(1 + T_2p) + S_R\}\theta = S_R\theta_0 - M_L \tag{30·40}$$

This is a torque equation which can be converted to a position equation by dividing by S_L. If the ratio of the regulator stiffness to the load stiffness S_R/S_L is defined as the system amplification A, equation 30·40 may be put into the form

$$\left\{\left[(1 + T_3p)\left(\frac{I}{S_L}p^2 + \frac{r_L}{S_L}p + 1\right) + \frac{r_0}{S_L}p\right](1 + T_1p)(1 + T_2p) + A\right\}\theta = A\theta_0 - \frac{M_L}{S_L} \tag{30·41}$$

If additional electrical delays are present they may be represented in equations 30·40 and 30·41 by additional time-delay factors operating upon the quantity in the brackets.

As in the speed regulator analysis, let it again be assumed for the moment that the armature circuit inductance is negligible ($T_3 = 0$). Equation 30·41 may then be written

$$\left[\left(\frac{I}{S_L}p^2 + \frac{r_0 + r_L}{S_L}p + 1\right)(1 + T_1p)(1 + T_2p) + A\right]\theta = A\theta_0 - \frac{M_L}{S_L} \tag{30·42}$$

The ratio $(r_0 + r_L)/S_L$ has the dimensions of time and may be written as T''_0. Also,

$$\frac{I}{S_L} = \frac{I}{r_0 + r_L} \times \frac{r_0 + r_L}{S_L} = T'_0T''_0 \tag{30·43}$$

where $T'_0 = I/(r_0 + r_L)$ = the equivalent mechanical time constant of the motor and its load.

The quadratic factor within the brackets of equation 30·42 may then be written and factored as follows:

$$(T'_0T''_0p^2 + T''_0p + 1) = (1 + T'p)(1 + T''p) \tag{30·44}$$

where T' and T'' are either real or complex, depending upon the relative values of T'_0 and T''_0. If T' and T'' are real the factors of equation 30·44 may be considered as the equivalent time-delay factors of the motor and its connected load.

If these factors are put into equation 30·42 the result is

$$[(1 + T'p)(1 + T''p)(1 + T_1p)(1 + T_2p) + A]\theta = A\theta_0 - \frac{M_L}{S_0} \quad (30\cdot45)$$

Now if equation 30·45 is compared with equations 30·15 and 30·29, it is apparent that, when T_3 is assumed zero in each system, the auxiliary equation of a simple position-regulator system with a spring member at the output and having m electrical delays is identical with that of a simple speed-regulator system having $m + 1$ electrical delays, or with that of a simple voltage-regulator system with $m + 2$ electrical delays.

In many position-regulated systems there is no stiffness member in the output, and in some the damping torque (torque proportional to speed) introduced by the load is insignificant. When both of these conditions apply, equation 30·40 reduces to

$$\{[Ip^2(1 + T_3p) + r_0p](1 + T_1p)(1 + T_2p) + S_R\}\theta = S_R\theta_0 - M_L \quad (30\cdot46)$$

Equations 30·40 and 30·46 may be put into the form of equation 30·16 by performing the indicated multiplications.

In a system having a load containing a spring member, the steady-state angular position of the output, when the input velocity $p\theta_0 = 0$, is, according to equation 30·40,

$$\theta = \frac{S_R\theta_0 - M_L}{S_R + S_L} = \frac{A\theta_0 - \theta_L}{A + 1} \quad (30\cdot47)$$

where $\theta_L = M_L/S_L$.

The steady-state *error* for the same condition is

$$\theta_0 - \theta = \frac{\theta_0 + \theta_L}{A + 1} \quad (30\cdot48)$$

Thus, even when there is no external torque M_L, there cannot be perfect correspondence between the output and input positions except at $\theta_0 = 0$. It is obvious, however, that the error becomes smaller as the system amplification is increased. On the other hand, in the system having no output spring member, as represented by equation 30·46, the steady-state angular position of the output when $p\theta_0 = 0$ is

$$\theta = \theta_0 - \frac{M_L}{S_R} \quad (30\cdot49)$$

and the steady-state error is

$$\theta_0 - \theta = \frac{M_L}{S_R} \quad (30\cdot50)$$

In this case, in the absence of an external torque M_L, there is perfect correspondence in the steady-state condition between the output and input positions, and the error is zero. For both cases, however, there will be a steady-state error proportional to speed when the input angle is changing at some prescribed constant rate.[10]

30·7 REGULATOR CHARACTERISTICS

It has been demonstrated in the preceding sections that proportional voltage, speed, and position regulators may all be represented by differential equations of similar form but of varying degrees of complexity. Although only three specific examples have been worked out in detail, it should be evident that this statement may be extended to include *all* proportional systems regardless of the type of regulator or regulated quantity. It is feasible, therefore, to investigate the requirements for stability and the transient characteristics of all proportional regulators by considering the general differential equation which is applicable to all of them. In the following paragraphs this procedure is used in general, but in order to illustrate certain points it is convenient frequently to consider for purposes of clarity a specific type of system and to apply the conclusions of the investigation to more general types by inference.

Solution of Regulator Equation

The differential equation which completely describes the static and dynamic characteristics of any proportional regulator system comprised of linear elements is of the form

$$(a_np^n + \cdots + a_2p^2 + a_1p + a_0)x = f(t) \quad (30\cdot51)$$

The coefficients of the left-hand side of the equation are functions of the system parameters, x is the regulated quantity, and $f(t)$ is the *forcing function,* which may be zero, a constant, or some function of time. The forcing function is a mathematical description of the pattern which may be prescribed for the regulated quantity and of any external disturbing influences which may act to cause an error in the regulated quantity.

The *auxiliary equation* for the general proportional regulator is defined as that obtained by setting the left-hand side of equation 30·51 equal to zero. It is advantageous to divide the auxiliary equation by a_n:

$$p^n + b_np^{n-1} + \cdots + b_2p^2 + b_1p + b_0 = 0 \quad (30\cdot52)$$

Equation 30·52 in general is factorable[11,12,13] into a number of linear factors of the form $p + \alpha$ and quadratic factors of the form $p^2 + 2\beta p + \gamma$, where α, β, and γ are real numbers such that $\gamma > \beta^2$. Each linear factor of equation 30·52 has a root:

$$p = -\alpha \quad (30\cdot53)$$

and each quadratic factor has two complex roots:

$$p = -\beta \pm j\sqrt{\gamma - \beta^2} = -\beta \pm j\omega \quad (30\cdot54)$$

The roots expressed by equations 30·53 and 30·54 are obviously also roots of equation 30·52.

According to the elementary theory of differential equations[14] each of the linear factors of equation 30·52 represents an exponential variation with time of the regulated quantity x following a disturbance, of the form

$$x = C_1\epsilon^{-\alpha t} \quad (30\cdot55)$$

where $-\alpha$ = the root of the appropriate factor

C_1 = an arbitrary constant dependent upon the initial condition of the system.

Furthermore, each quadratic factor corresponds to an oscillatory variation of the regulated quantity represented by the expression

$$x = \epsilon^{-\beta t}(C_2 \cos \omega t + C_3 \sin \omega t) \qquad (30\cdot56)$$

where $\omega = \sqrt{\gamma - \beta^2}$ = the angular frequency of the oscillation

β = the damping factor.

The coefficients C_2 and C_3 are additional arbitrary constants determined by the initial conditions of the system. A differential equation of the nth order has a total of n arbitrary constants.

The complete solution of equation 30·51 will contain a *particular integral* determined by the forcing function $f(t)$ and the steady-state regulator characteristics, and the sum of all terms of the form of equations 30·55 and 30·56. The particular integral and the arbitrary constants may be evaluated, when necessary, by the classical methods of the calculus of differential equations [14] or by the more modern and frequently less laborious methods of operational calculus.[9,19]

System Stability and Damping

Fortunately, in most practical cases it is not necessary to determine completely the time variation of the regulated quantity, a process which frequently involves considerable computation. A determination of the roots (equations 30·53 and 30·54) of the auxiliary equation 30·52 usually will yield sufficient information concerning system performance to enable the engineer to decide if a certain regulator is suitable for a particular application.

Thus, from the roots of the linear factors (equation 30·53) it is possible to determine time $t_\alpha = 1/\alpha$ required after a disturbance for each of the exponential "responses" (equation 30·55) to acquire $100(1 - \epsilon^{-1})$ or 63 percent of its ultimate steady-state value.† In addition, from the roots of the quadratic factors (equation 30·54) one obtains the frequency $f = \omega/2\pi$ of each oscillatory "response" (equation 30·56), and also the rates of decay of the various oscillations as determined by the values of β. The factor $\epsilon^{-\beta}$ is the ratio of the oscillation amplitude at the end of any unit of time to that at the beginning of the same time unit. If β is a positive number the factor $\epsilon^{-\beta t}$ in equation 30·56 approaches zero as time increases, indicating that the corresponding oscillation amplitude decays with time. If one of the β's is negative the amplitude of the corresponding oscillation theoretically increases without limit, indicating that the regulator system is unstable. If $\beta = 0$ for one of the quadratic factors, the significance is that the oscillation amplitude will be maintained indefinitely at its initial value. This condition is the border line between stability and instability, and it represents an impractical operating condition.

Two oscillatory responses may have the same *time* rate of decay of oscillations, as indicated by equal values of β, but if the oscillation frequencies are different the response with the higher frequency oscillates through more cycles before a given fraction of the initial amplitude is reached. The higher frequency oscillation appears, therefore, to be relatively less well damped. For this reason the decay of oscillation amplitude *per cycle* has greater significance from an engineering point of view than the decay *per unit time*. The percentage amplitude decay in any time t is given by

$$\Delta = 100(1 - \epsilon^{-\beta t}) \qquad (30\cdot57)$$

and the decay *per unit time* is obtained by substituting $t = 1$ in equation 30·57:

$$\Delta_t = 100(1 - \epsilon^{-\beta}) \qquad (30\cdot58)$$

The decay *per cycle* is obtained by substituting the period of oscillation $t = 1/f$ in equation 30·57:

$$\Delta_f = 100(1 - \epsilon^{-\beta/f}) \qquad (30\cdot59)$$

In most regulator applications a rate of decay of 80 percent or more per cycle is considered satisfactory.

Stability Criteria

It has been indicated in the preceding paragraphs that an inspection of the roots of the auxiliary equation of a regulator system will reveal whether or not the system is stable. In many cases, however, it is desirable to know if a system is stable before the frequently long and involved calculations to determine the roots are performed. If the system is found to be unstable there is no need to solve for the roots. Time can usually be saved if some change in the system which will insure stability is sought before the roots are computed.

Routh's criteria [15] provide a very convenient means for checking the stability of a system if its auxiliary equation is known. According to Routh the first requirement for stability is that all coefficients of the auxiliary equation be positive. The other requirement is that a certain relationship, depending upon the order of the equation, exists between the various coefficients. For the lower order equations these relationships are as follows:

Cubic, $ap^3 + bp^2 + cp + d = 0$
  for stability $bc > ad$
Quartic, $ap^4 + bp^3 + cp^2 + dp + e = 0$
  for stability $bcd > b^2e + d^2a$
Quintic, $ap^5 + bp^4 + cp^3 + dp^2 + ep + f = 0$
  for stability $d(be - af)(bc - ad) > b(be - af)^2 + f(bc - ad)^2$

A system represented by a quadratic equation will of course be stable regardless of the value of the coefficients as long as the first criterion is satisfied. For equations of higher order than the fifth the expression relating the coefficients becomes so unwieldy as to be impractical. In these cases two alternatives are available: (1) a tabular method of setting up the stability criterion may be used (p. 129 [14]); (2) the order of the equation may be reduced by extracting one or more linear or quadratic factors so that one of the above relations is applicable.

If the left-hand side of the expression for stability *equals* the right-hand side, the system is on the border line between stability and instability, corresponding to $\beta = 0$ for one of the quadratic factors. A convenient means is thus provided by the stability criteria for determining the limiting values

† This assumes that the α's are positive, a necessary condition for stability. If one or more of the α's are negative the regulated quantity theoretically will increase indefinitely, and a "run-away" type of instability will result.

of some particular system parameter, such as the system amplification, between which the parameter must fall if the system is to be stable. The criterion of stability for the system can be set up with appropriate numerical values assigned to all quantities except the parameter in question. The inequality may then be solved for the unknown parameter to determine the range of values which yield a stable system. Values close to the limiting value are usually unsatisfactory operating values since they may result in a rate of decay of oscillations lower than the acceptable minimum of 80 percent per cycle.

30·8 EFFECT OF SYSTEM PARAMETERS ON REGULATOR PERFORMANCE

The essence of regulator engineering is the application of automatic controllers or regulators to conventional machines so that during disturbances the steady-state and transient errors remain within specified limits, and so that the system responds to changes in the reference quantity with acceptable rapidity. Since the machine characteristics have an important influence on regulator performance, they are of vital interest to the regulator designer. The design of the power machinery for a regulated system is usually based on the requirements of the application or, particularly when large sizes and powers are involved, on economical considerations. The result is that the constants of the machines seldom have values which a regulator engineer would select if he had a free choice. The regulator engineer too often must accept what is available in the way of machine characteristics, and then design around them a regulator which achieves the required results.

In this section the effects of the more important regulator and machine parameters on the characteristics and performance of simple regulated systems are discussed. In the following section some of the methods used by the regulator engineer to overcome certain limitations of machine design are described.

System Amplification

In the simple voltage-, speed-, and position-regulator systems described in Sections 30·4 to 30·6 the system amplification appears only in the constant term of the auxiliary equation. In more complex systems to be described later it may also affect other coefficients of the equation. In either case it has an important effect on the characteristics of the system. The most convenient way to demonstrate this effect is by means of an example, and for this purpose a simple two-delay regulator system will be considered. Armature circuit inductances will be neglected. As pointed out in the sections referred to, the system being considered may then be either a voltage regulator with two field circuit delays, a speed regulator with one field delay and output damping, or a position regulator with no electrical delays but with the output inertia, damping, and stiffness proportioned to give real values for the equivalent mechanical time constants (see equation 30·44).

The differential equation representing the system is

$$[(1 + t_1p)(1 + t_2p) + A]x = Ax_0 \qquad (30\cdot60)$$

In this equation t_1 and t_2 are the *effective* time constants of the system ‡ and x_0 is the reference quantity which in the following examples is assumed to be zero until $t = 0$, and constant thereafter. The roots of the auxiliary equation corresponding to equation 30·60 may be written as

$$p_1, p_2 = -\frac{Q+1}{2Qt_1} \pm \frac{1}{2Qt_1}\sqrt{(Q+1)^2 - 4Q(A+1)} \qquad (30\cdot61)$$

where $Q = t_2/t_1$.

The response of the system, as represented by the solution of equation 30·60, may be either overdamped, critically damped, or oscillatory, depending upon the nature of the roots p_1 and p_2 as determined by the radical in equation 30·61. If the roots are real and unequal the response is overdamped; if they are equal the response is critically damped; and if they are complex the response is oscillatory. For a given time-constant ratio Q it is apparent, therefore, that the nature of the response depends only upon the system amplification A. If x and px are both zero when $t = 0$, the complete solutions of equation 30·60 for the three cases are as follows.

Overdamped case, $(Q + 1)^2 > 4Q(A + 1)$:

$$\frac{x}{x_0} = \frac{A}{A+1}\left(1 - \frac{p_2}{p_2 - p_1}\epsilon^{p_1t} + \frac{p_1}{p_2 - p_1}\epsilon^{p_2t}\right) \qquad (30\cdot62)$$

where p_1 and p_2 are given by equation 30·61.

Critically damped case, $(Q + 1)^2 = 4Q(A + 1)$:

$$\frac{x}{x_0} = \frac{A}{A+1}[1 - (1 + p_1t)\epsilon^{p_1t}] \qquad (30\cdot63)$$

where $p_1 = (Q + 1)/2Qt_1$.

Oscillatory case, $(Q + 1)^2 < 4Q(A + 1)$:

$$\frac{x}{x_0} = \frac{A}{A+1}\left[1 - \epsilon^{-\beta t}\left(\cos\omega t + \frac{\beta}{\omega}\sin\omega t\right)\right] \qquad (30\cdot64)$$

where

$$\beta = \frac{Q+1}{2Qt_1} \qquad (30\cdot65)$$

$$\omega = \frac{1}{2Qt_1}\sqrt{4Q(A+1) - (Q+1)^2} \qquad (30\cdot66)$$

These equations are plotted in dimensionless form in Fig. 30·11 for various values of A and for $Q = N = 10$. The tendency toward more pronounced oscillations or hunting as the amplification is increased is clearly indicated. The curve for $A = 2.025$ corresponds to the critically damped condition.

‡ In the present discussion the effective time constants t_1 and t_2 equal the actual time constants T_1 and T_2, respectively, but they may have other values when anti-hunt circuits are applied to the system as described in the next section. The new terminology is introduced here because the equations which follow immediately are then directly applicable to later discussions. The factor Q will represent the ratio t_2/t_1, and N will represent T_2/T_1 to distinguish between the two ratios when they are unequal. Whenever the actual time constants are unequal, it is assumed that $T_2 > T_1$; likewise, $t_2 > t_1$.

An inspection of equations 30·65 and 30·66 reveals that for the system being considered the frequency of oscillation ω increases with the amplification factor, whereas the damping factor β is independent of the amplification. Consequently, in accordance with the discussion leading up to equations 30·58 and 30·59, the time rate of decay of oscillations of these systems is independent of system amplification, but the rate of decay per cycle decreases with increasing amplification. This effect is illustrated in Fig. 30·11, where all the oscillatory curves reach their steady-state values in approximately the same time. In more complex systems the damping factors, in general, also vary with amplification.

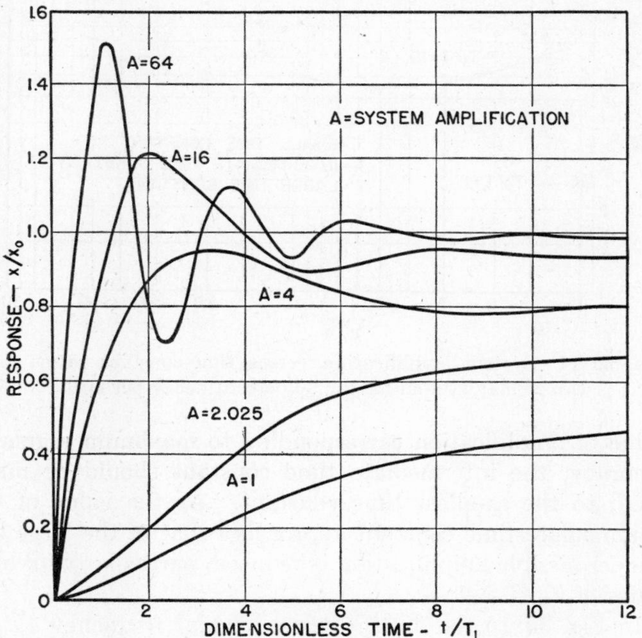

FIG. 30·11 Transient response of a two-delay system with $T_2/T_1 = 10$ after a sudden application of the reference quantity.

Figure 30·11 clearly illustrates the effect of system amplification on the steady-state error and the rapidity of response after a disturbance.

The minimum amplification of a regulator system is usually fixed by the steady-state accuracy requirements of the application. If oscillations of the system are highly damped, the speed of response of the system may be improved by increasing the amplification factor beyond the minimum value. The maximum usable value of amplification is limited only by hunting of the system.

System Time Constants; Two-Delay System

In most systems the amplification factor required to insure satisfactory accuracy is so high that it causes objectionable hunting unless the time constants of the system are modified in some way. Time constants of the field circuits of electrical machines can be reduced at the expense of greater power dissipation by the addition of resistance to the circuits.§ Armature-circuit time constants are reduced by the use of compensating or pole-face windings in the machines. The mechanical time constant of a motor may be increased by the addition of a flywheel or by a reduction in field strength. Where the load connected to a motor contributes appreciably to the total inertia, the mechanical time constant of the combination may be reduced by using reduction gearing between the motor and the load inertia. In addition to these direct methods of modifying the system time constants, various indirect or "synthetic" methods are also used in regulator practice. A discussion of these is reserved for a later section.

In the preceding paragraphs the characteristics of a two-delay system with fixed time constants were discussed. The effect of varying the relative values of the time constants in the same system will now be considered. For the oscillatory condition, which is of principal interest, the rate of decay per cycle is computed from equation 30·59; equations 30·65 and

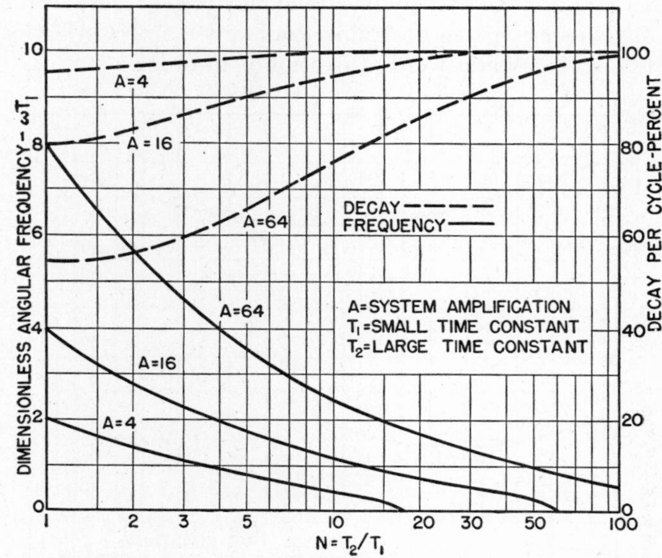

FIG. 30·12 Effect of time-constant ratio on frequency and decay per cycle of a two-delay system.

30·66 are used to determine the frequency $f = \omega/2\pi$ and the damping factor β. The dimensionless angular frequency $\omega t_1 (= \omega T_1)$ and the decay per cycle are plotted versus the time-constant ratio $Q = N$ in Fig. 30·12 for the several values of system amplification which yielded oscillatory responses in Fig. 30·11. It is apparent that the rate of decay is improved as Q is increased, the improvement being more pronounced for the higher amplifications. Also as Q is increased ωt_1 decreases until it reaches zero at the value of Q which produces critical damping.

The time-constant ratio may be increased either by increasing t_2 or by reducing t_1, and with respect to the improvement in rate of decay of oscillations it makes no difference which of the two time constants is changed. The speed of response of a system, however, is associated with the frequency of oscillation, the response being faster with a higher frequency. If the smaller time constant t_1 is held constant the actual system frequency decreases in the same manner as ωt_1 when Q is increased. On the other hand, if t_2 is held constant the system frequency will increase as Q is increased by reducing t_1. This is not immediately apparent from Fig. 30·12, but it becomes obvious when t_2/t_1 is substituted for Q in equation 30·66. Thus, when an improvement in damping must be obtained by a change in time-constant ratio,

§ This procedure also reduces the amplification of the machine but usually this can be compensated for by a corresponding increase in the amplification of some other element of the system.

it is preferable to make the change by decreasing the smaller time constant because this also increases the speed of response of the system. The superiority of a reduction in the smaller time constant is illustrated in Fig. 30·17. In this figure curve A represents the response of a two-delay system with a time-constant ratio of $Q = 10$. For curves B and C, Q has been increased to 100 by reducing the smaller delay to $t_1 = T_1/10$ for curve B, and by increasing the larger delay to $t_2 = 100T_1$ for curve C. The amplification A is 24.5 for all curves. The improvement of curve B over curve C is obvious.

System Time Constants; Three-Delay System

Because of the greater amount of work in solving higher order equations there may be a tendency to rely more heavily on Routh's criteria in designing the more complex systems.[16] In this connection a word of caution is in order.

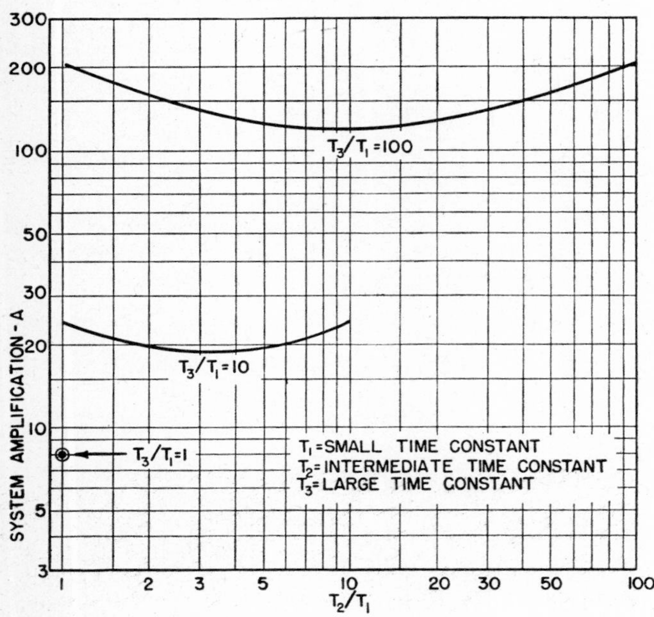

Fig. 30·13 System amplification versus time-constant ratios for a three-delay system at the limit of stability.

Figure 30·13 shows the system amplification corresponding to the border-line condition of stability for various time-constant ratios in a three-delay system. Here T_1 is the small time constant, T_2 is the intermediate time constant, and T_3 is the large time constant. These curves were computed from Routh's criterion for the cubic equation. They indicate that the amplification for the border-line condition is maximum when the intermediate time constant equals either of the other two time constants. Since a high stability limit is desirable, one is likely to conclude from these data that optimum regulator performance is obtained when T_2 equals either T_1 or T_3, and that it makes no difference which condition is chosen. Figures 30·14 and 30·15, which apply to a system having a rate of decay of 80 percent per cycle, show that such a conclusion is erroneous. A system operating at the stability border line is an impractical system, and the relation between permissible amplification and system time constants changes markedly when only practical systems are considered.

Figure 30·14 shows the value of amplification as a function of the time-constant ratios for a rate of decay of oscillations of 80 percent per cycle. To obtain the highest permissible

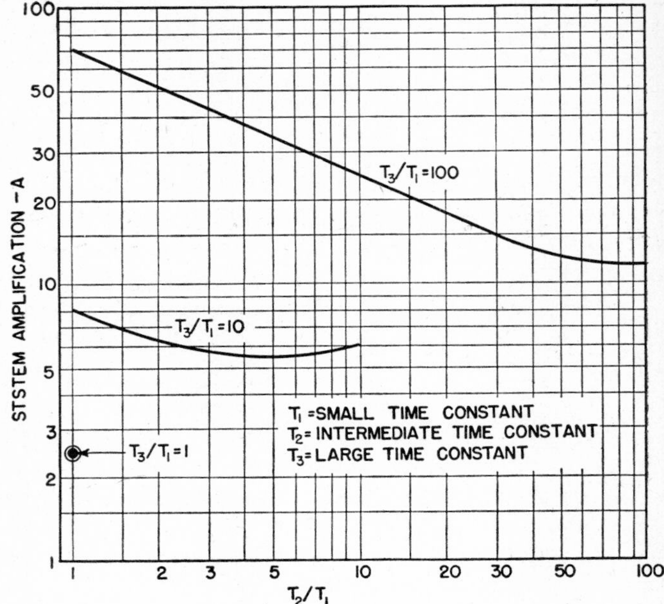

Fig. 30·14 System amplification versus time-constant ratios for a three-delay system having 80 percent decay per cycle.

value of amplification corresponding to maximum regulator accuracy, the intermediate time constant should be made equal to the smallest time constant. As the value of the intermediate time constant approaches that of the large one the permissible amplification is reduced, particularly for high ratios of T_3/T_1.

In Fig. 30·15 the dimensionless angular frequency ωT_1 is plotted against the time-constant ratios. Since a high

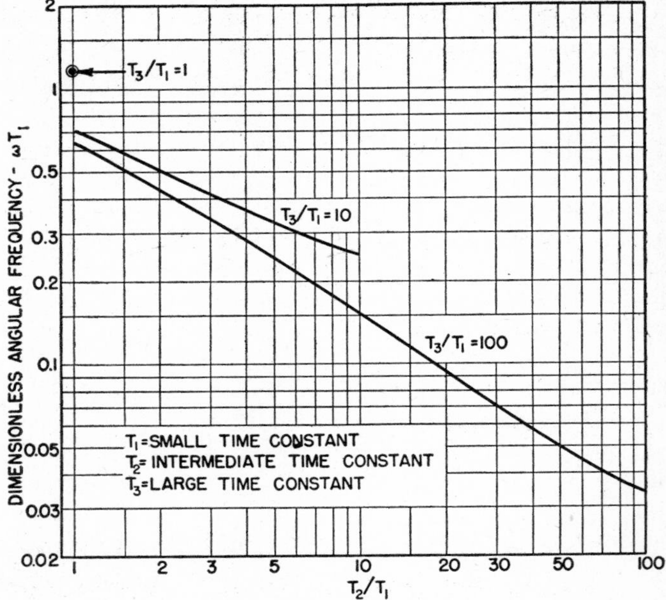

Fig. 30·15 Frequency versus time-constant ratios for a three-delay system having 80 percent decay per cycle.

regulator frequency is desirable for rapid system response, it is apparent here also that the optimum condition is realized when T_2 equals T_1.

In general it may be said that for optimum regulator characteristics the spread between the two largest system time constants should be as great as possible, and that all other delays should be made as small as possible compared to the second largest time constant.

30·9 ANTI-HUNTING CIRCUITS

It has been demonstrated that the ratios of the effective system time constants must exceed certain minimum values in order to achieve a specified accuracy of regulation with good stability and a high rate of response. It has also been pointed out that the design characteristics of machines used in regulator systems are usually dictated by economical considerations or by the requirements of the application. Because of this situation the inherent time constants of a system seldom meet the requirements for good regulation. For this reason it is often necessary to modify the time constants of the system by the indirect or "synthetic" methods mentioned previously. Such modifications are accomplished by means of stabilizing or "anti-hunting" circuits in the regulator. Anti-hunting circuits in general are of two basic types, *anticipator* circuits and *feedback* circuits. The two types are quite different in their effect on a system, and each type may take any one of several different forms. Frequently a single type of anti-hunting circuit is sufficient to stabilize a system, but in the more complex systems both basic types are commonly used. Anticipator circuits are sometimes called *lead* networks or *phase-advance* networks.

The use of an electronic regulator greatly simplifies the design and application of anti-hunting circuits. The high input impedance of electron tubes permits the use of small, high-impedance elements in the anti-hunting circuits. Consequently the circuits can be extremely flexible and easily adjusted. Moreover, the power drawn from the circuits to which they are connected can be negligible.

This section describes the effects of various simple anti-hunting circuit arrangements on system performance and also the more common forms of each type of circuit. Numerous variations of the basic types are possible.

Anticipator Circuit

Figure 30·16 shows a simple form of anticipator circuit consisting of a resistor-capacitor network connected between the error-measuring circuit and the regulator input. The

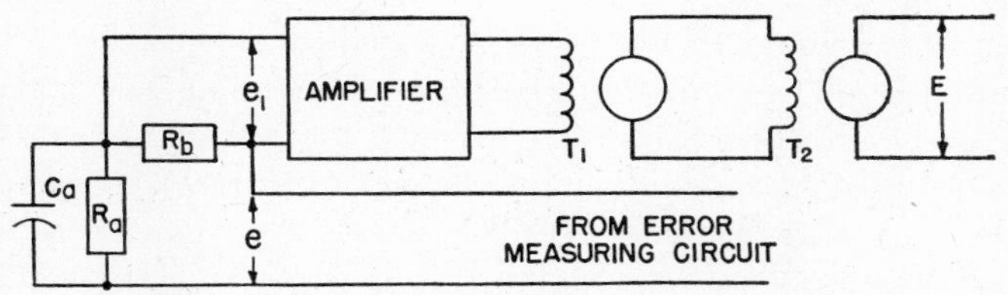

Fig. 30·16 A typical *RC* anticipator circuit.

regulator input voltage e_1 appearing across R_b may be considered as made up of two components, one due to the current flowing through R_a, and the other due to the current through C_a. The first component of voltage is proportional to the voltage e which is assumed here to be a d-c voltage proportional to error. Current flows through C_a only when the error voltage is changing and, if R_b is relatively small, the magnitude of this current is proportional to the rate of change of the error voltage. If the error voltage suddenly starts to change at a uniform rate, the rate component of input voltage appears immediately, whereas some time must elapse before the direct component reaches an appreciable magnitude. The input voltage e_1 may thus be said to *anticipate* the error voltage.

The effect of this anticipation on the regulator system is best demonstrated by setting down the equations for the circuit of Fig. 30·16. The input voltage may be expressed as

$$e_1 = \frac{1 + T_a p}{1 + kT_a p} ke \qquad (30\cdot67)$$

where T_a = the anticipator time constant R_aC_a

k = the anticipator ratio $R_b/(R_a + R_b)$.

Also

$$E = \frac{Ke_1}{(1 + T_1 p)(1 + T_2 p)} \qquad (30\cdot68)$$

or

$$E = \frac{1 + T_a p}{(1 + kT_a p)(1 + T_1 p)(1 + T_2 p)} kKe \qquad (30\cdot69)$$

where K = the steady-state amplification E/e_1. Now, if T_a is made equal to one of the time constants, the anticipatory operator in the numerator cancels one of the delay operators in the denominator. If T_a is made equal to T_1 the expression for E may be written

$$E = \frac{1}{(1 + kT_1 p)(1 + T_2 p)} kKe \qquad (30\cdot70)$$

The effective system time constants are now $t_1 = kT_1$ and $t_2 = T_2$, and the time-constant ratio is $Q = N/k$, where $N = T_2/T_1$. The anticipator circuit thus has increased the spread of the effective delays by *reducing the small one* by a factor k, and has thereby improved system performance. A subsidiary effect is a reduction in system amplification by the same factor k, but this effect can be compensated for by an increase in the gain of the amplifier.

The effect of an anticipator circuit with $k = 0.1$ on the response of a two-delay regulated system is illustrated in Fig. 30·17. Curve A represents the response of the system before the application of an anti-hunting circuit, and curve B is the response with the anticipator. The ratio of actual system time constants $N = T_2/T_1$ for the example is 10, making $Q = 100$. The system amplification is 24.5 for both cases. This value of amplification was chosen to give critical damping with the anticipator circuit. The improvement in speed of response and damping of case B is quite pronounced.

According to equation 30·61 the condition for critical damping is

$$A + 1 = \frac{(Q + 1)^2}{4Q} \qquad (30\cdot71)$$

For large values of Q, equation 30·71 reduces to the approximation

$$A \approx \frac{1}{4} Q = \frac{1}{4} \times \frac{N}{k} \qquad (30\cdot72)$$

It follows from equation 30·72 that the critically damped condition may be realized for high values of amplification if k is small. In practice, values of k may range from 0.5 to 0.05 or less.

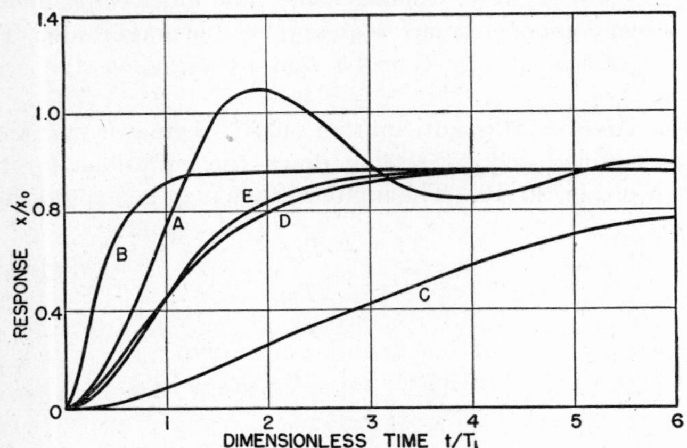

Fig. 30·17 Effect of anti-hunting circuits on the response of a two-delay system with $T_2/T_1 = 10$ and $A = 24.5$. Each anti-hunting circuit is adjusted for critical system damping: A, system without anti-hunting circuit; B, anticipator with 1 to 10 ratio; C, feedback around amplifier without delays; D, feedback around amplifier and one delay; E, feedback around amplifier and two delays.

Furthermore, according to equation 30·63, the rate of response of a critically damped 2-delay system depends upon

$$p_1 = \frac{Q+1}{2Qt_1} \tag{30·73}$$

When Q is large, it follows that

$$p_1 \approx \frac{1}{2t_1} \tag{30·74}$$

With an anticipator circuit, $t_1 = kT_1$ and $p_1 = 1/2kT_1$. Thus if $1/k$ and A are both increased in approximately the same proportion, the degree of damping remains essentially constant, but the speed of response, and hence the transient accuracy, of the system are improved. Owing to the higher amplification there is, of course, a corresponding improvement in steady-state accuracy also.

Feedback around Amplifier with No Delays

Figure 30·18 shows a feedback arrangement which, in comparison with the anticipator circuit, has a very interesting

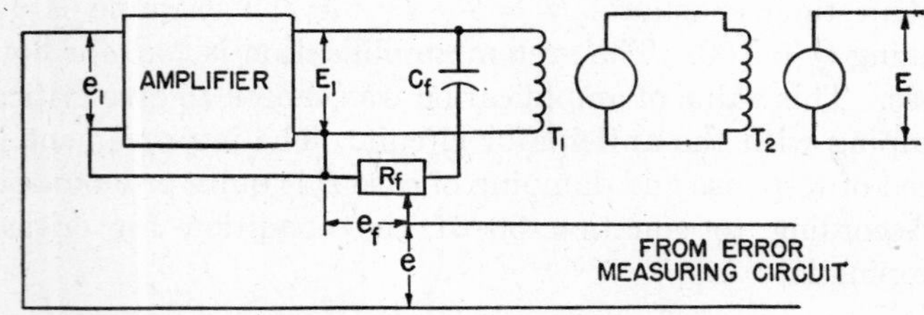

Fig. 30·18 Feedback around an amplifier without time delays.

feature. It is seldom used in practice, however, because it is inferior to other circuits described later. In Fig. 30·18 an RC circuit is connected between the output and the input of the regulator amplifier. It is assumed that there are no time delays in the amplifier. Changes in the amplifier output voltage are fed back to the input circuit in a degenerative sense. The feedback circuit prevents rapid variation of the output voltage and thus slows down the response of the amplifier.

For the circuit of Fig. 30·18,

$$e_f = \frac{k_f K_0 T_f p}{1 + T_f p} e_1 \tag{30·75}$$

where k_f = ratio of potentiometer R_f
K_0 = amplifier gain
T_f = feedback time constant $R_f C_f$.

The relation between the input circuit voltages is

$$e_1 = e - e_f \tag{30·76}$$

and, when this expression is combined with equation 30·75,

$$e_1 = \frac{1 + T_f p}{1 + K_f T_f p} e \tag{30·77}$$

where $K_f = k_f K_0 + 1$. This equation is identical in form with equation 30·67 for the anticipator, but in the latter case $k < 1$, whereas in the present instance $K_f > 1$. Substituting equation 30·77 in equation 30·68 gives

$$E = \frac{1 + T_f p}{(1 + T_1 p)(1 + T_2 p)(1 + K_f T_f p)} Ke \tag{30·78}$$

Now if $T_f = T_2$, the anticipatory operator $1 + T_f p$ cancels the delay operator $1 + T_2 p$, and equation 30·78 becomes

$$E = \frac{1}{(1 + T_1 p)(1 + K_f T_2 p)} Ke \tag{30·79}$$

The effective time constants of the system are now $t_1 = T_1$ and $t_2 = K_f T_2$. The feedback circuit thus effectively spreads the time constants *by increasing the large delay* by a factor K_f.

Curve C in Fig. 30·17 shows the result of the application of this feedback circuit to the two-delay regulated system considered previously. With an amplification of 24.5 the system is critically damped if $K_f = 10$, corresponding to $Q = 100$. The system response corresponding to curve C is just one-tenth as fast as that obtained with the anticipator circuit represented by curve B. This may be verified by reference to equation 30·73. For a fixed value of Q, p_1 depends only upon t_1 which, for the feedback circuit of the present example, is 10 times the value of t_1 obtained with the anticipator circuit.

If the amount of feedback is increased by increasing K_f the amplification corresponding to critical damping is also increased. However, according to equations 30·73 and 30·74 there is no corresponding improvement in the rate of response since the smaller effective time constant t_1 is unaffected by feedback. Curve C of Fig. 30·17 thus approximately represents the response of the system for any relatively large value of feedback provided that the amplification is adjusted in accordance with equation 30·71 to give critical damping for each value of feedback.

Feedback around Amplifier and One Delay

The feedback circuit shown in Fig. 30·19 is quite commonly used in stabilizing proportional regulator systems. This circuit differs from that of Fig. 30·18 in that the feedback circuit elements R_f and C_f are connected across the exciter armature rather than across the amplifier output. The feedback "loop," consisting of the feedback circuit, the amplifier imput circuit, the amplifier, and the exciter, thus includes the exciter time delay T_1. If the procedure used in deriving equation 30·78 is followed, a similar equation for the circuit of Fig. 30·19 may be written:

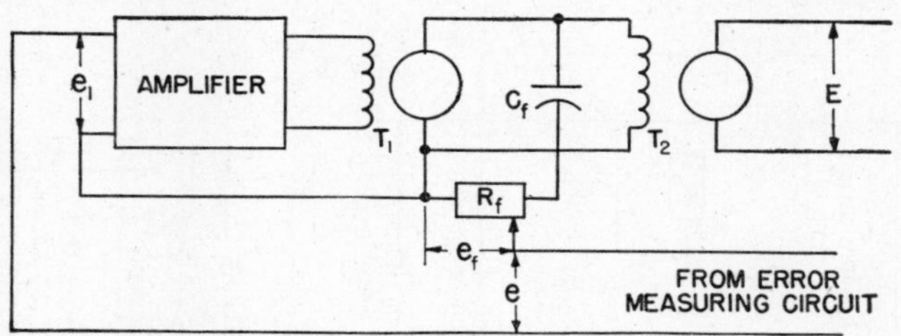

FIG. 30·19 Feedback around an amplifier and one time delay.

$$E = \frac{1 + T_f p}{(1 + T_2 p)[(1 + T_1 p)(1 + T_f p) + k_f K_0 K_e T_f p]} Ke \tag{30·80}$$

where K_e is the voltage gain of the exciter. If T_f is made equal to T_2, and if the multiplication inside the brackets is carried out, equation 30·80 becomes

$$E = \frac{1}{1 + (T_1 + K_f T_2)p + T_1 T_2 p^2} Ke \tag{30·81}$$

where for this case $K_f = k_f K_0 K_e + 1$. The denominator of this expression may be factored into two delay operators:

$$E = \frac{1}{\left(1 + \dfrac{T_1}{q} p\right)(1 + qT_2 p)} Ke \tag{30·82}$$

where K_f and q are related by the expression

$$1 + K_f N = \frac{1}{q} + qN \tag{30·83}$$

in which $N = T_2/T_1$.

According to equation 30·82 this feedback circuit in effect *modifies both time constants,* decreasing the smaller one by a factor q and increasing the larger by the same factor. If $q = \sqrt{10}$, corresponding to $K_f = 3.09$ when $N = 10$, the ratio of the effective time constants is $Q = 100$, the same as that of the two examples considered previously. Therefore, when this circuit is applied to the two-delay regulated system of the previous examples, the system again is critically damped when the amplification is 24.5. The response of the system is represented by curve D of Fig. 30·17. The superiority of this response over that obtained with the circuit of Fig. 30·18 clearly indicates why the latter circuit is infrequently used.

If the feedback is increased until $q = 10$, the effective time constants become $t_1 = T_1/10$ and $t_2 = 10T_2$. The time-constant ratio is then $Q = 1000$ for $N = 10$, and according to equation 30·71 the amplification may be increased to approximately 249 with critical damping. The new value of t_1 equals that obtained above with the anticipator. The speed of response with the increased feedback is thus represented approximately by curve B of Fig. 30·17. The steady-state accuracy is, of course, improved by a factor of about 10.

Feedback around Amplifier and Two Delays

The circuit shown in Fig. 30·20 is another commonly used feedback arrangement. In this circuit the feedback voltage is derived from a series resistor in the generator field circuit. Since the field current is in time phase with the generator voltage E, the feedback circuit may be connected across the generator armature with equivalent results. The field circuit connection shown in Fig. 30·20 must be used, however, when there are other direct connections between the generator-armature circuit and the error-measuring circuit, as in a voltage regulator for example. The feedback voltage e_f is delayed with respect to e_1 by both T_1 and T_2. There are consequently two delays in the feedback loop. Similar to equations 30·78 and 30·80,

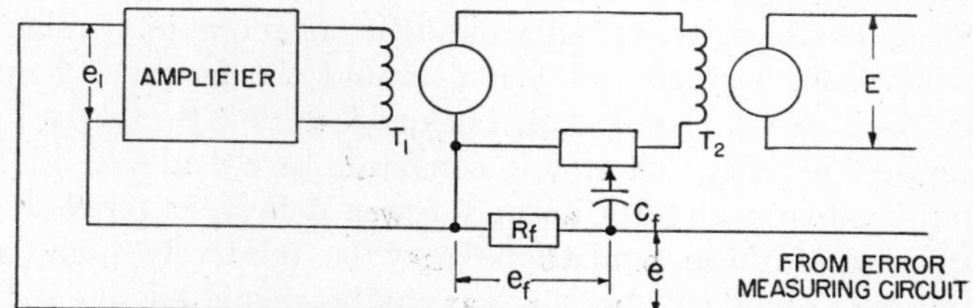

FIG. 30·20 Feedback around an amplifier and two time delays.

$$E = \frac{1 + T_f p}{(1 + T_1 p)(1 + T_2 p)(1 + T_f p) + K' T_f p} Ke \tag{30·84}$$

where K' is the gain around the feedback loop. It is not possible to cancel any factors in this equation.

When the feedback circuit is applied to a closed two-delay regulator system, the auxiliary equation becomes

$$(1 + T_1 p)(1 + T_2 p)(1 + T_f p) + (A + K')T_f p + A = 0 \tag{30·85}$$

This is a third-order equation which cannot be solved in general terms. Solutions can be obtained, however, if numerical values are assigned to the amplification factors and the time-constant ratios. Numerous combinations of values are possible. One condition of particular interest for comparison with the preceding examples is a combination of feedback-circuit constants which critically damps the system when $A = 24.5$ and $N = T_2/T_1 = 10$. One such combination gives an auxiliary equation having three equal roots. The values corresponding to this condition are

$$T_f = \frac{T_1}{6.58}$$

$$K' = 98.7$$

The complete solution of the system equation using these values for a suddenly applied constant reference quantity with the system initially at rest is

$$\frac{x}{x_0} = \frac{24.5}{25.5}\left[1 - \left(1 + 2.56\frac{t}{T_1} + 3.17\frac{t^2}{{T_1}^2}\right)\epsilon^{-2.56(t/T_1)}\right] \tag{30·86}$$

This equation is plotted as curve E in Fig. 30·17. The response is slightly better than that obtained with the circuit of Fig. 30·19.

With increased feedback and amplification the rate of response can be increased in this case also. For example, with $K' = 157.5$ and $A = 249$, the system is again critically damped and has an auxiliary equation with three real roots. The speed of response is approximately 3.3 times as fast as that represented by curve E of Fig. 30·17, or slightly faster than that indicated by curve B.

When feedback is applied around two or more delays there is a tendency toward oscillation around the feedback loop. Such oscillations usually can be made to decay rapidly by changing the feedback circuit constants or by adding additional feedback circuits around fewer delays.[17] Since the frequency of these oscillations may be relatively high, in some cases they will not appear in the regulated quantity with noticeable magnitude.

Alternative Anti-Hunting Circuits

The anti-hunting circuits described thus far have consisted solely of simple combinations of resistors and capacitors. Various other circuit arrangements are equally effective. Some of these are illustrated in Figs. 30·21, 30·22, and 30.23.

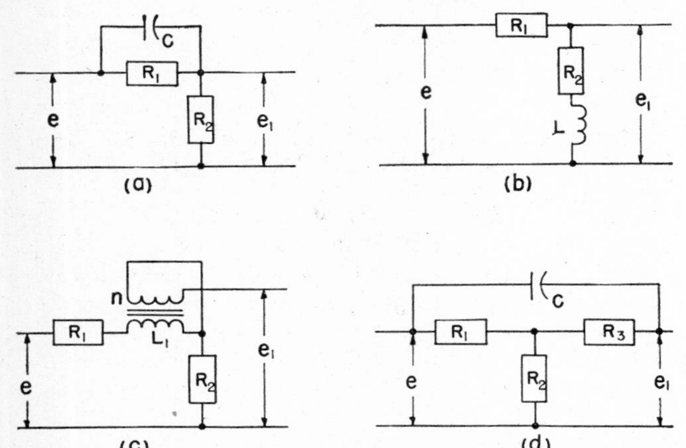

Fig. 30·21 Four common types of anticipator circuits for use with d-c error voltages.

Figure 30·21 shows the basic anticipator circuit (a), which has already been described, and three common variations. These anticipators are useful only with direct-current error voltages. The ratio of output voltage to input voltage for each circuit is of the form

$$\frac{e_1}{e} = k\frac{1 + T_a p}{1 + kT_a p}$$

where $k = R_2/(R_1 + R_2)$. The anticipation time T_a for each of the circuits is as follows:

(a) $$T_a = R_1 C$$

(b) $$T_a = \frac{L}{R_2}$$

(c) $$T_a = \frac{nL_1}{R_2}$$

where n = the transformer turns ratio

(d) $$T_a = \frac{R_1R_2 + R_1R_3 + R_2R_3}{R_2}C$$

It will be noted that when $R_3 = 0$, circuit d is identical with the basic circuit a. Circuit d is particularly useful where R_1 and R_2 must be relatively small for some reason. In this case circuit a requires a relatively large capacitor to obtain a given time constant, whereas with circuit b the capacitor can be small if R_3 is large.

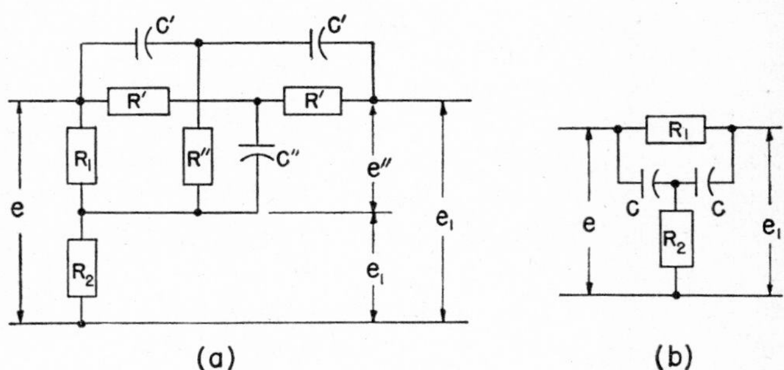

Fig. 30·22 Anticipator circuits for use with a-c error voltages.

The circuits of Fig. 30·22 are sometimes used as anticipators when the error signal is an alternating voltage of fixed frequency such as is obtained from the control-transformer error detector of Fig. 30·10. The elements in the parallel-T portion of circuit a in Fig. 30·22 are proportioned as follows:

$$R' = 2R''; \quad C' = \frac{C''}{2}; \quad R'C' = \frac{1}{\omega_0}$$

where ω_0 is the angular frequency of the applied alternating voltage. With these values the voltage output of the parallel-T circuit e'' is zero when the error voltage e is constant. When e is varying, the output voltage e'' is proportional to the rate of change of e. The voltage e' is directly proportional to the error voltage. Thus e_1 consists of one component of voltage proportional to error and another component proportional to rate of change of error. This is the characteristic feature of all anticipator circuits.

Circuit b of Fig. 30·22 is a simpler circuit with similar characteristics. The anticipation times obtainable with both circuits are relatively small, and consequently their application is limited to low-power regulator systems with small time delays. Moreover, because the anticipation time of these circuits is inversely proportional to the applied frequency, they are of little value for frequencies much above 60 cycles per second.

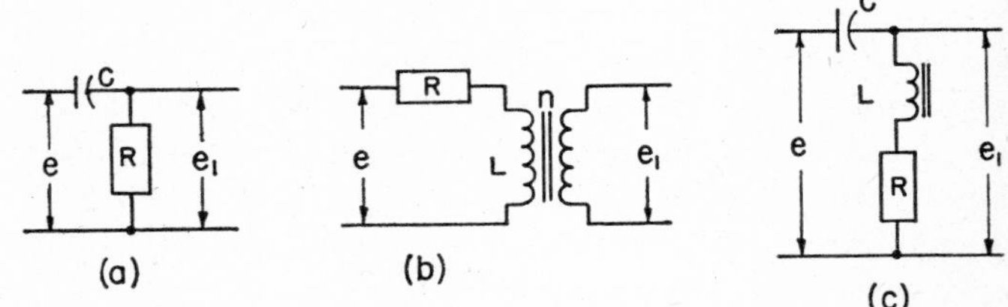

Fig. 30·23 Three common types of feedback circuits.

Figure 30·23 shows the basic feedback circuit a with two important modifications. The ratio of the output voltage e_1 to input voltage e for the various circuits is

(a) $$\frac{e_1}{e} = \frac{T_f p}{1 + T_f p}$$

where $T_f = RC$.

(*b*)
$$\frac{e_1}{e} = \frac{nT_fp}{1 + T_fp}$$

where $T_f = L/R$

$n =$ the transformer turns ratio.

(*c*)
$$\frac{e_1}{e} = \frac{T'_fp + T'_fT''_fp^2}{1 + T'_fp + T'_fT''_fp^2}$$

where $T'_f = RC$

$T''_f = L/R.$

Circuit *b* is especially useful where it is necessary to keep the error-measuring circuit electrically insulated from the circuit from which the feedback must be derived. The transformer primary is usually connected across a d-c source; therefore, the transformer must be so designed that its core is not saturated by the primary current.[18]

Circuit *c* is a variation of circuit *a*. With optimum circuit constants it can be adjusted to give almost as much stabilizing effect at hunting frequencies as circuit *a*, but with considerably less "holding-back" effect at rates of change lower than those corresponding to hunting frequencies. The rate of response of a system with circuit *c* is consequently somewhat faster than that with circuit *a* but the system damping is slightly worse.

Numerous arrangements other than those discussed in this section are also used in regulator practice for stabilization or for obtaining special performance characteristics. For example, in speed or position regulators of the type discussed above, a voltage proportional to the generator-motor armature current is sometimes fed back either directly or through modifying networks such as those shown in Fig. 30·23. Voltages proportional to various aspects of the output motion such as speed or acceleration may also be fed back. Frequently regenerative or positive feedback is used in small amounts to eliminate certain steady-state errors. A discussion of all the possible variations is beyond the scope of the present work; for further details the reader is referred to the list of references appended to this chapter and to other publications on the subject. All possible modifications are susceptible to analysis by means of the basic principles briefly outlined above.

30·10 APPLICATIONS

Electronic regulators are used in almost every industry for improving quality, increasing production, or for higher economy of operation. Certain processes are not possible at all without automatic regulation of some form. A brief review of even a reasonably complete list of typical applications cannot be included here. The several applications discussed illustrate the importance of electronic regulators as used in complex processes and indicate the proportions to which they have grown. However, their field of usefulness is being rapidly extended.

During World War II vast strides were made in regulator theory and design. The amount of development work for military purposes was tremendous. Most of this knowledge is applicable to the solution of peacetime regulator problems in industry. Much work is being done along these lines in the refinement of industrial systems and processes.

Multicolor Press Register Regulator

In multicolor printing the basic color impressions must be placed on top of one another with a high degree of accuracy; otherwise, the finished print has a blurred appearance displeasing to the reader. To provide continuous regulation of register to an accuracy of 0.003 inch or better on rotary-web presses operating as high as 1500 feet per minute, all-electronic regulators are used. Employing the photoelectric principle, these devices respond in the microsecond intervals available for error detection, and their fast action keeps the number of rejects to a minimum.

The application of one type of register regulator is shown in Fig. 30·24. The paper or web is fed through the press

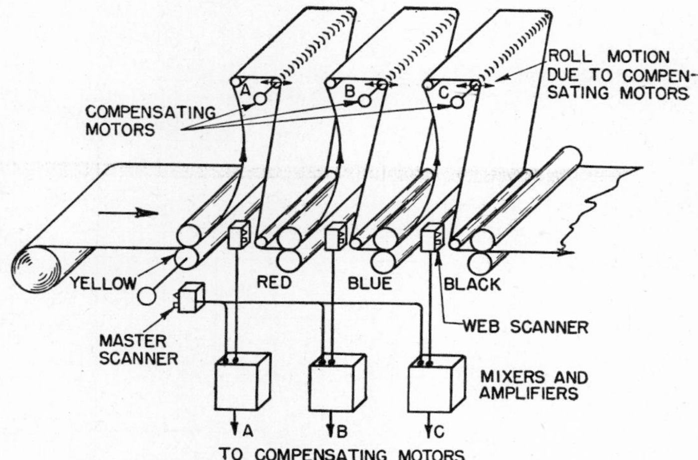

FIG. 30·24 Schematic regulator installation as applied to a multicolor printing press.

continuously. The first color impression is usually yellow, and when it is made a small yellow line is also printed near the margin of the web at 2½-inch intervals and at right angles to the direction of travel. These register marks are about 0.010 inch wide, from ⅜ to ½ inch long, and are the means of indicating the position of the web as it progresses through each suceeding printing stage.

On a lineshaft-driven press, rotation of a slotted disk is synchronized with the yellow impression cylinder. A master photoelectric scanner, observing the disk, provides electrical impulses which are in exact time phase with the register marks as they are printed. Close to each succeeding impression cylinder, identical photoelectric scanners observe the yellow register marks as they arrive at these locations and provide similar electrical impulses. Because all impression cylinders are driven from a common shaft, their positions relative to the shaft, and hence the impulses from the master scanner, are fixed. Thus, impulses from each of the web scanners are compared in time phase with the impulses from the master scanner. A difference between the two initiates a correction to bring them into exact coincidence.

The register regulator shown in Fig. 30·25 for performing this function is of the dead-zone type. The circuit element for comparing the time phase of impulses from the master scanner and a web scanner is the multivibrator trigger circuit comprised of tubes 1 and 2. Impulses from the master scanner are impressed on the grid of tube 1; impulses from a web scanner (for example, the scanner located at the red impression cylinder) are applied to the grid of tube 2. If the arrival of the register marks at this location is such that correct register of red on yellow occurs, impulses from the two scanners are impressed on the grids of tubes 1 and 2 at the same instant. The periods of maximum current in each tube are equal, and a square wave of equal positive and negative half cycles is obtained from their plate circuits similar to that shown in Fig. 30·26 (*a*).

occurring, the alternating components of bias voltages applied to the grids of the thyratrons from capacitors $C1$ and $C2$ are sufficiently small to permit conduction in both tubes. An alternating voltage is applied to the shunt field of the compensating motor. The net flux in the motor is therefore zero, and no rotation occurs.

The function of the duplex diode is to shift the cathode potentials of the thyratrons relative to their grids. This shift is such that the a-c voltage on their anodes is prevented from upsetting the bias voltage conditions of the tube whose instantaneous anode voltage is positive.

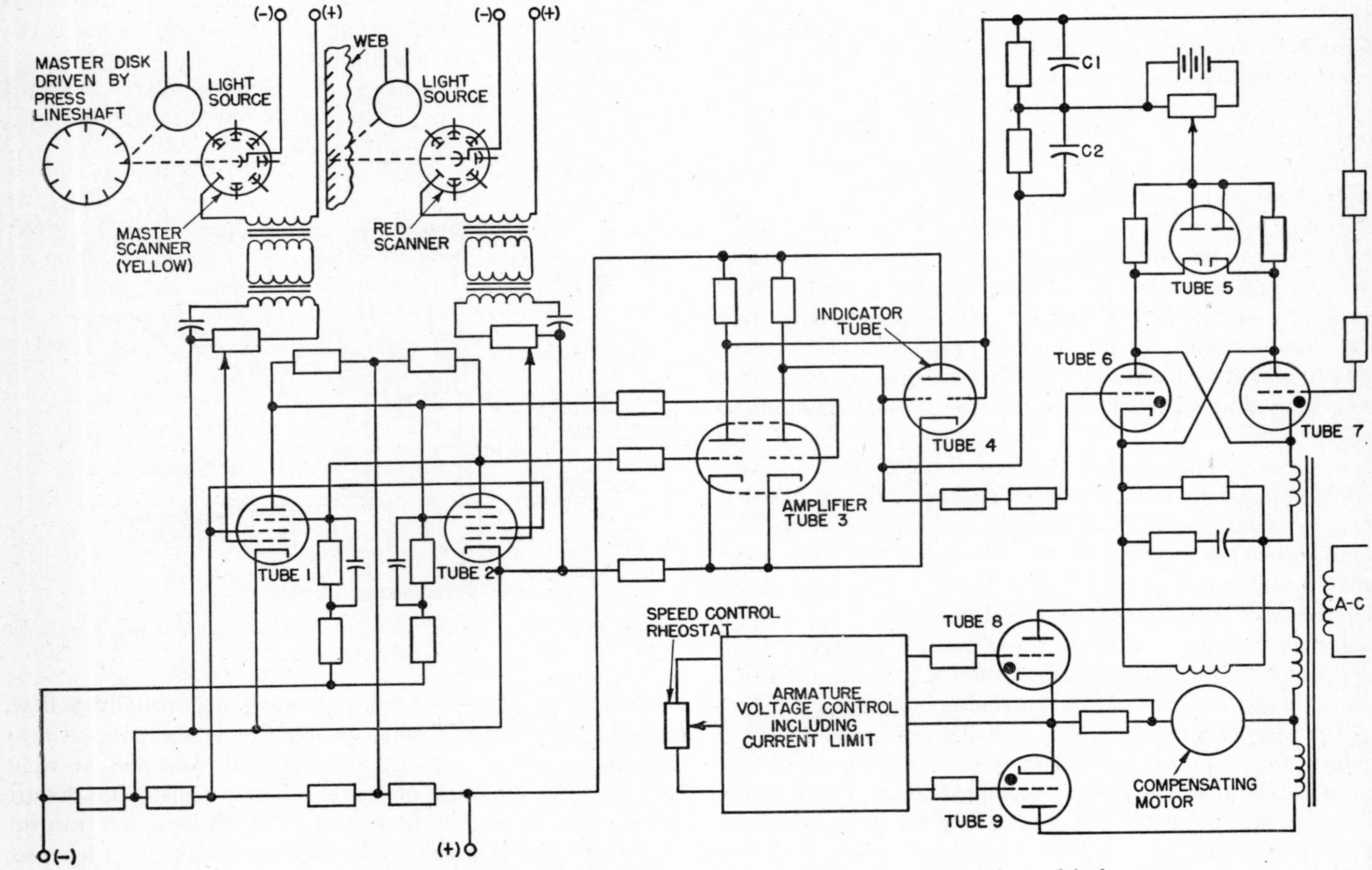

Fig. 30·25 Simplified schematic diagram of a register regulator for one section of a multicolor press.

The square-wave output voltage is applied to the grids of a duplex triode tube whose function is to decouple the trigger circuit element from the resistor-capacitor smoothing circuit.

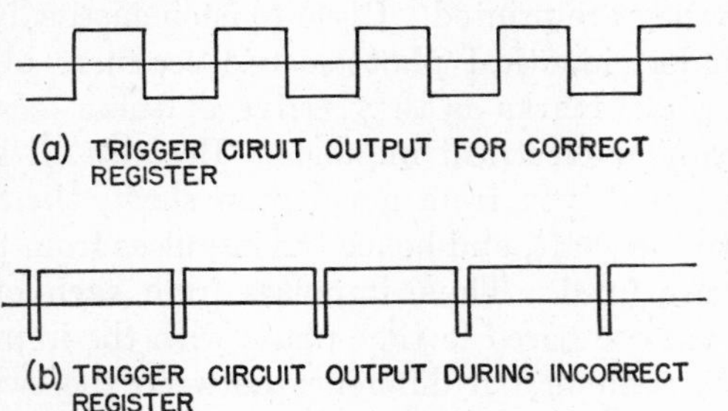

Fig. 30·26 Wave forms of trigger circuit output voltage.

The plate circuit of this tube also furnishes the necessary voltages for the indicator tube for visual indication of the conditions of register.

The voltages appearing across capacitors $C1$ and $C2$ form part of the bias voltages of two thyratron tubes connected in inverse-parallel. When correct register of red on yellow is

If web position at the red impression cylinder is incorrect for exact register of red on yellow, electrical impulses from the web scanner and the master scanner no longer coincide. One of the tubes in the trigger circuit passes maximum current for a much longer period than the other, and the wave obtained from their plate circuits is no longer symmetrical. Assuming that impulses from the web scanner lag those from the master scanner, plate current of tube 1 is maximum essentially all the time; maximum current exists in the plate circuit of tube 2 only during the short interval of lag between the two sets of impulses. The voltage wave is similar to that shown in Fig. 30·26 (*b*). The action of the resistor-capacitor smoothing circuit is to remove the short-time reversal from this wave entirely, resulting in a unidirectional voltage across capacitors $C1$ and $C2$. The magnitude of this voltage is sufficient to prevent one of the thyratrons from conducting. The other thyratron, conducting each time its anode is positive, applies rectified a-c voltage to the field of the compensating motor, and rotation is in the direction to decrease tension of the web between yellow and red impression cylinders. Correction ceases after the two sets of impulses again coincide, indicating exact register. If the web impulses lead the master impulses the opposite response of the system occurs, the motor rotating in the reverse direction to increase web tension. Changes in web tension are momentary, nominal tension being restored after a short time, as is evident from a consideration of the amount of paper entering and leaving a given loop.

FIG. 30·27 Complete assembly of regulators for a multicolor press for printing both sides of the paper simultaneously.

Armature power for the compensating motor is supplied by a grid-controlled thyratron rectifier of the full-wave type. Control tubes in this element of the system provide current limit operation of the motor. To obtain the maximum speed of correction consistent with stable operation, means for adjusting motor speed over a wide range are included.

Because this regulator is of the dead-zone type, the speed of the compensating motor is independent of the amount of position error existing in the web at the scanning location. The width of the dead zone is a function of the tube characteristics and other components of the trigger circuit, and is defined in units of time rather than distance. No adjustment is possible since the time characteristics of the trigger circuit are fixed by design. The dead zone permits the web to vary a small amount from the position for proper register without initiating a correction. Limits of variation are a function of web speed. Under the worst conditions, misregister resulting from the dead zone is within the acceptable limits of 0.003 inch.

The principal time delays of the system are the mechanical time constant of the compensating motor and its connected load, the equivalent time constant represented by the ratio of the length of paper between impression cylinders to the speed of the web, and the time delay in the motor shunt field. Since three major time delays are present, oscillation of the system can occur. However, the mechanical time constant and the field delay are short compared to the other time lag, and the system is inherently quite stable. Stability is further improved by limiting the speed of the compensating motor to a value no higher than the application requires.

The phototube used in each scanner is a photomultiplier type such as the 931A. Spectral response of this tube is maximum in the blue region; consequently, the thin yellow register mark, which to the eye is hard to distinguish on white paper, appears black to the photomultiplier tube, and thus gives maximum contrast. Vibration encountered on the press structure has little effect on the signal-to-noise ratio. Moreover, the high amplification obtained in the tube provides a strong output signal from the feeble light impulses.

Impedance-matching transformers in the scanners and the mixing panels provide high fidelity of impulse signal at the trigger circuits with separations of 100 feet or more between the two components.

Figure 30·27 shows a complete multicolor press register regulator assembly. Thyratron tubes are shown in only one circuit.

Speed Regulator for Wind-Tunnel Drive

To obtain high accuracy in wind-tunnel experiments, the speed of the propellers must be maintained within close limits. On some of the large modern drives, this problem is by no means simple. As the horsepower of the drive increases, the electrical equipment becomes more elaborate, and the problem of holding constant speed becomes increasingly difficult. In fact, wind-tunnel drives represent some of the most complex regulated systems yet encountered. For example, the mathematical expression describing the characteristics of the complete system shown in Fig. 30·28, including stiffnesses of the synchronous machines and the propeller shafts, is an eleventh-order differential equation. For practical purposes the equation can be reduced to a sixth- or seventh-order equation by neglecting certain effects which are relatively unimportant.

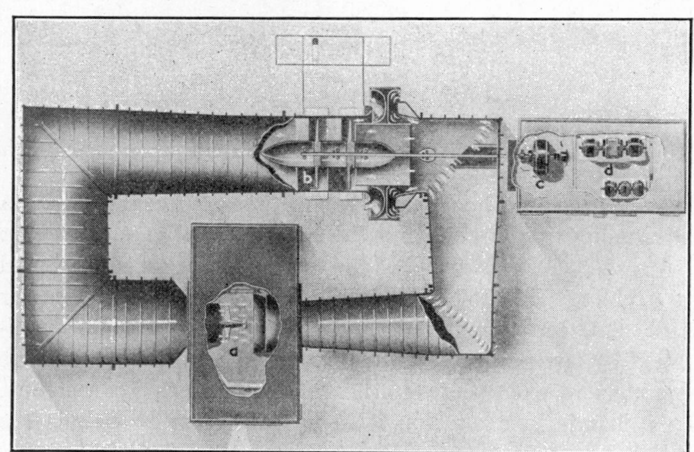

FIG. 30·28 A typical large wind tunnel.

A method for regulating the speed of the tunnel motor in a variable-frequency system is shown schematically in Fig. 30·29. The regulator is a proportional type, and it functions to maintain constant the frequency of the a-c supply to the synchronous motor driving the propeller.

Essential components of the regulating system are a speed-measuring pilot generator, a constant reference voltage, an electronic amplifier which furnishes power for two fields of a Rototrol supplying part of the excitation requirements of the d-c machine on the constant-speed motor-generator set, a current-limiting circuit, and the required stabilizing circuits.

A small change in speed of the tunnel motor is accompanied by a change in pilot generator voltage. The difference (or error) between this voltage and the reference voltage is amplified by the electronic element and the Rototrol to provide a change in excitation on the constant-speed d-c machine. This change in excitation is in the direction to reduce the initial change in speed.

The system is stabilized by using both anticipation and feedback. Without these important elements excessive overshooting and sustained oscillation would occur because of the many time delays in the system. Components of the anticipator circuit are resistors $R1$ and $R2$ and capacitor $C1$, all forming part of the error-measuring circuit. Resistor $R3$, potentiometer $P2$, and capacitor $C2$ constitute the feedback network. The feedback voltage is obtained from the Rototrol armature circuit. The voltage appearing across resistor $R3$ is a function of the rate of change of Rototrol voltage.

The output stage of the amplifier includes two power tubes of the vacuum type connected in push-pull. Each tube supplies excitation to one of the two opposing Rototrol fields. With zero voltage input to the amplifier, currents in the two fields are equal. Because the Rototrol has a 100 percent self-energizing field connected in its armature circuit, any value of steady-state excitation required by the auxiliary field of the constant-speed d-c machine can be maintained with zero error voltage. Any momentary error voltage other than zero causes the current in one Rototrol field to increase and the current in the other field to decrease. The action of the self-energizing field after the ensuing transient is to restore the balance of the Rototrol regulating fields by causing sufficient change in the auxiliary field winding of the constant-speed machine to produce the correction in speed necessary to return the error voltage to zero. The steady-state error voltage is thus maintained at zero throughout the entire regulator range.

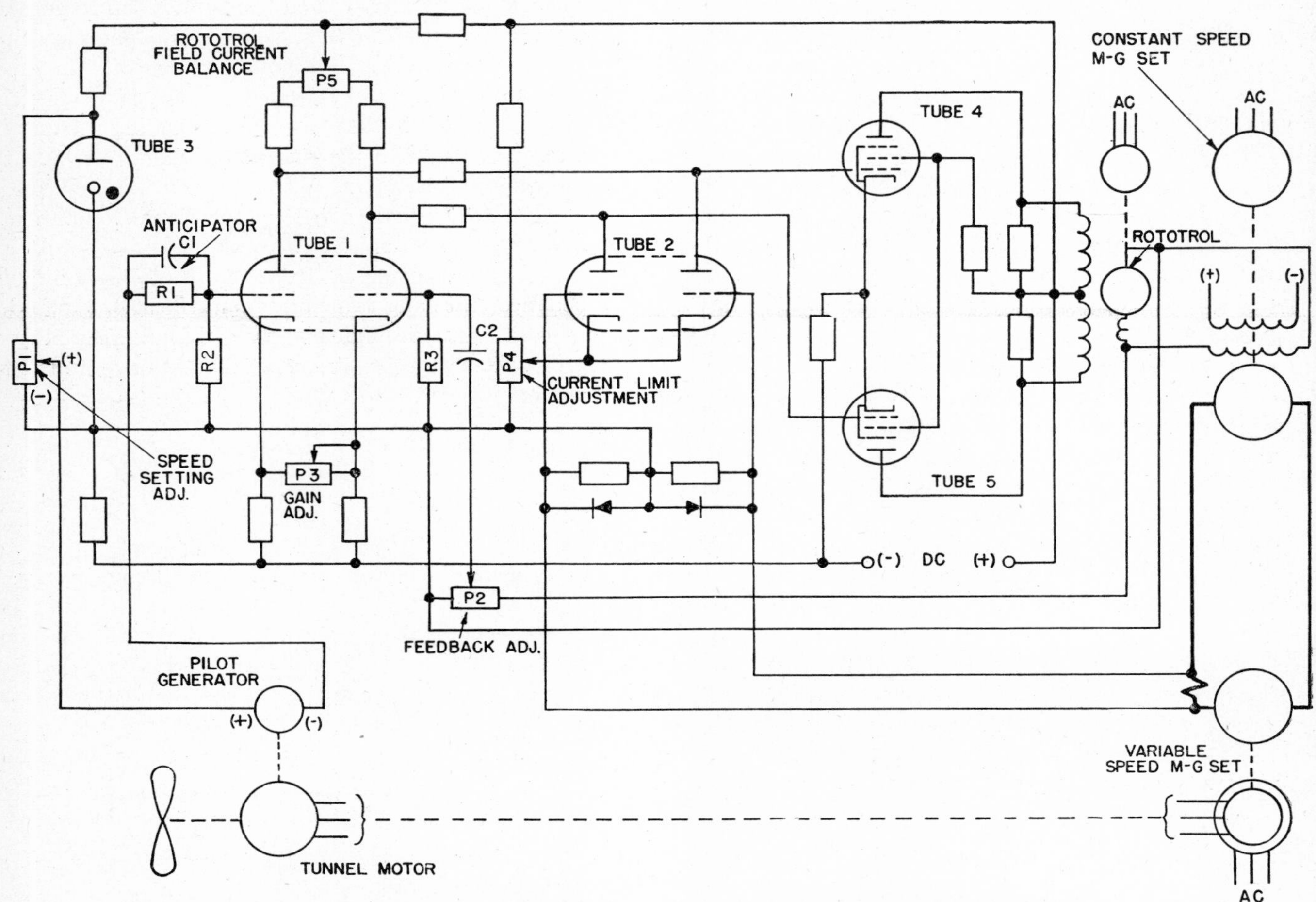

Fig. 30·29 Simplified schematic diagram of a typical wind-tunnel speed regulator.

In practice the tunnel motor is operated over a wide speed range for different test requirements. If a rapid change in speed is made, current flowing in the armature circuits of the main d-c machines could reach excessive values if provisions were not made to limit it. The current-limiting circuit shown in Fig. 30·29 performs this function. A current measurement is obtained from the voltage drop across the commutating field of one of the d-c machines. This voltage, applied to the rectox and resistor network, forms part of the bias voltages on the grids of tube 2, both elements of which are normally biased beyond cut-off. When the main armature current exceeds a value determined by the setting of potentiometer $P4$, the bias voltage on one of the elements of tube 2 becomes less than the cut-off value, and the resulting plate current alters the grid voltages of the power tubes. The original unbalance of the Rototrol fields causing the high current is thus modified in the direction to reduce the current in the main armature circuit. The rate of acceleration or deceleration of the system may thus be controlled at values which are safe for the electrical equipment.

Quite often electronic regulators are designed with insufficient consideration of the effects of leakage currents in component circuits, and erratic performance is thus encountered. The circuits including tubes 1 and 2 of Fig. 30·29 are novel in this respect in that essentially a common terminal exists between the error circuit, the feedback circuit, and the current-limit circuit. It will be noted that any paths where leakage currents may flow contain only low impedances. Thus voltage drops as a result of leakage currents either are not in critical portions of the circuit or are small enough to have a negligible effect on performance. This highly desirable feature is made possible by use of the phase-inverter circuit of tube 1. With this arrangement, the feedback voltage is introduced in the grid circuit of the second element. Leakage currents which may flow through the feedback circuit cannot affect the grid voltage of the first element of tube 1.

Amplifier sensitivity is controlled by a variable resistance linking the cathodes of tube 1. This resistance controls the amount of degeneration which occurs and thus determines the over-all gain.

FIG. 30·30 A propeller-drive installation in a large wind tunnel for testing full-size airplanes. Each of the six motors delivers a maximum of 6000 horsepower to a 40-foot-diameter propeller. The size of the equipment is emphasized by comparison with the man standing on the motor support in the lower center.

Figure 30·30 illustrates the enormous size of equipment which is precisely controlled by regulators of the type described above.

Speed Regulator for a Sectional Paper Machine

The various sections of paper-forming machines are frequently driven by individual motors. A typical machine with sectional drive is shown in Fig. 30·31. With this type of drive the relative speed of each section must be held within 0.1 percent or less to insure proper tension or "draw" of the web. Automatic regulation is the only solution because manual control is inadequate if not impossible. In addition to the high steady-state accuracy, the automatic regulating device must provide quick response of all sections including those having high inertia and long time delays, and must be suitable for 24-hour-a-day service without excessive maintenance. Recently, all-electronic regulators have been developed as a general improvement over previous methods. The new devices are adaptable to either single-generator or multiple-generator systems.

FIG. 30·31 A typical paper-forming machine with individual motors driving each section.

Figure 30·32 shows schematically one type of automatic regulator for use on a paper machine equipped with sectional drive. Each motorized section of the machine requiring individual "draw" or relative-speed control has its own electronic regulating elements. A separately driven master set acts as a reference to which the performance of each section of the machine is compared; the regulators function to maintain relative speeds constant.

In a complete regulated system, an a-c tachometer generator forming part of the master set furnishes a reference signal. A similar a-c tachometer, driven by the section motor to be regulated, supplies a signal whose magnitude is a function of section motor speed. After rectification these two signal voltages are filtered and combined in opposition to each other. Any difference voltage, indicating incorrect speed relation between the master and section motors, is amplified in a two-stage d-c amplifier whose output forms part of the grid bias circuit of a three-phase, half-wave thyratron rectifier. The resulting change in rectifier output, and hence shunt-field excitation of the section motor, causes the motor speed to change in the direction to decrease the difference voltage.

Accuracies of the order of 0.1 percent require high system amplification. The system has three principal time delays, namely, the time constant of the section-motor shunt field, the mechanical time constant, and the armature-circuit time constant. System stability is therefore obtained only through the use of suitable anti-hunting elements. Both anticipation and feedback are shown in Fig. 30·32. A voltage proportional to rate of change of error voltage supplements the normal grid voltage of the second stage of d-c amplification. This voltage is produced by the charging and discharging current of capacitor C flowing in resistor $R2$ during changes in plate current of the first amplifier stage.

Grid voltage on the second amplifier element is further modified by a feedback voltage proportional to rate of change of motor field current. This voltage is obtained through a damping transformer whose primary is excited in proportion to motor field current.

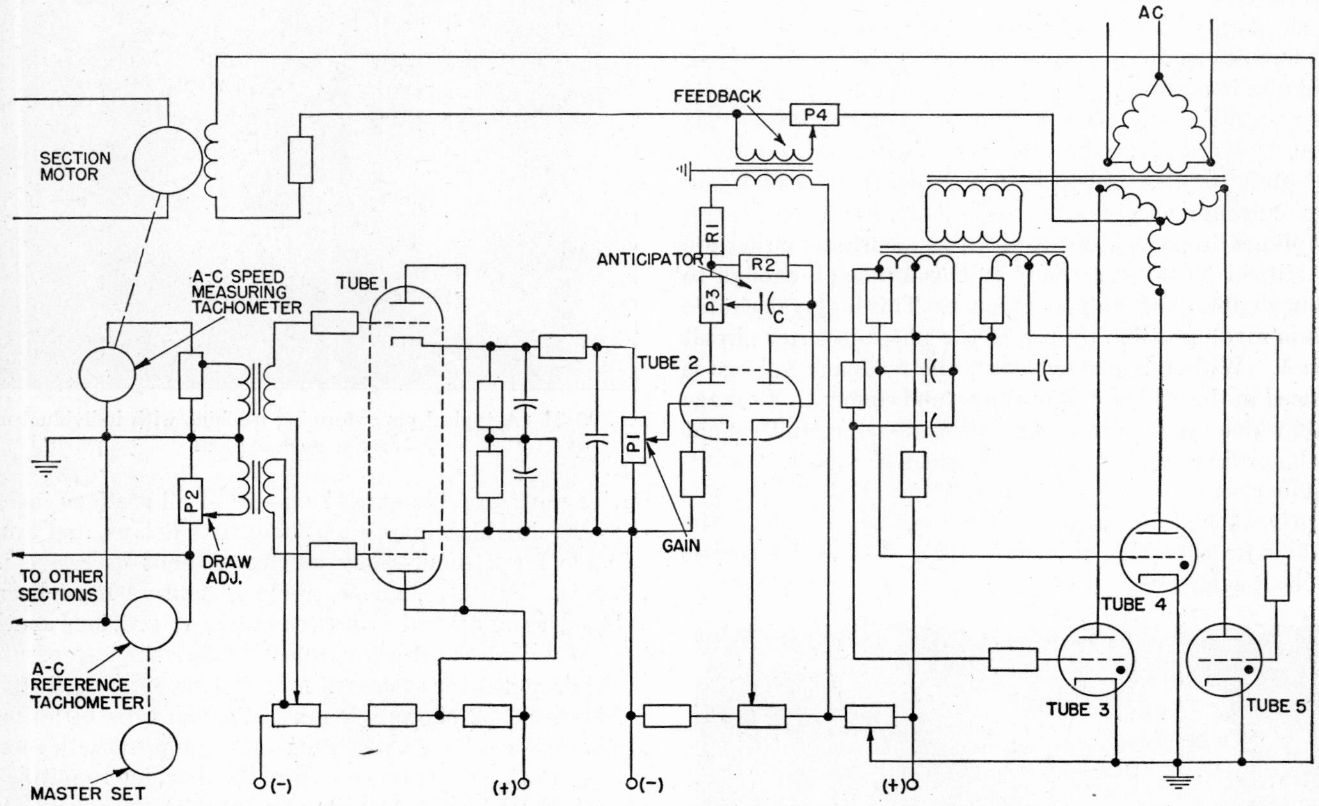

FIG. 30·32 Simplified circuit diagram of a regulator for accurately regulating the relative speed of one section of a paper machine.

The required occasional adjustment of draw between sections is obtained by speeding up or slowing down the section motor relative to the master set. These adjustments are accomplished by changing the relative magnitude of the master signal applied to the input terminals of the individual electronic units. For each master-signal input voltage there is a definite section tachometer speed which satisfies the requirements for zero difference voltage.

Electric-Arc Furnace Regulator

Power input to an electric-arc furnace such as that shown in Fig. 30·33 is controlled by adjustment of the position of the electrodes relative to the charge in the furnace. Frequent adjustment of position is necessary for several reasons. As the charge melts down, the electrodes must be lowered in order to maintain the arc. During operation the graphite electrodes are gradually consumed, and adjustment for this condition is necessary. Occasional short circuits caused by pieces of melting metal falling against the electrode necessitate rapid withdrawal. Electrode positioning is usually accomplished by a d-c motor operating through either mechanical or hydraulic elements.

Inasmuch as the voltage drop across the arc decreases as the current in the arc increases, a definite relation exists between the two quantities for every value of power input. To produce the required results, therefore, any automatic device for regulating the position of the electrodes must be responsive to both arc voltage and arc current in such a way that this relation is maintained. A simplified circuit of an electronic regulator for this purpose is shown in Fig. 30·34. Individual motor and control equipment is required for each electrode of the furnace. The control for only one electrode is shown.

The regulator operates by controlling the speed of the d-c motor over the required wide range by variation of the voltage applied to its armature, the shunt field excitation remaining constant. Armature-voltage supply is provided by two separate full-wave grid-controlled thyratron rectifiers. One of these rectifier elements runs the electrode motor in the direction to lower the electrode; the other, furnishing the opposite polarity to the motor terminals, raises the electrode.

Output voltage control of each rectifier is obtained in the conventional manner by a combination of d-c voltage and phase-shifted a-c voltage on the thyratron grids. The d-c component for each rectifier is furnished by a separate regulating tube whose output is determined by both arc voltage and arc current.

In Fig. 30·34 the grid voltage for tube 1 includes the voltage drops across resistors $R1$ and $R2$, each respectively proportional to arc current and arc voltage. Because these voltages oppose each other, their sum is zero for correct arc conditions. Any unbalance as a result of changing arc conditions alters the grid voltage of tube 1, and the change in its plate current flowing through potentiometer $P1$ modifies the d-c component of grid voltage on thyratrons 3 and 4. The same circuit conditions apply to tubes 2, 5, and 6 except that the polarities of the control voltages proportional to arc voltage and arc current are reversed.

When power is first applied to the furnace, maximum voltage appears across $R2$, causing the current in tube 1 to decrease and thereby making the grids of tubes 3 and 4 less negative. These thyratrons then conduct current, and the motor rotates in the direction to lower the electrode. Upon contact of the electrode with the metal in the furnace, the voltage across $R2$ drops to zero, and tubes 3 and 4 cease to conduct. When the second electrode strikes the metal a high current flows and a voltage appears across $R3$ which causes the output of tube 2 to decrease and permits thyratrons 5 and 6 to conduct. The motor then operates in the reverse direction, the electrode is raised, and the arc is established. The arc continues to lengthen until the required balance between potential and current control voltages is reached.

To preclude the possibility of both rectifiers attempting to supply power to the motor simultaneously, blocking resistors $R7$ and $R8$ are necessary. For example, assume that the rectifier for electrode lowering is conducting. The current flowing in the motor armature from A to B also flows through resistor $R7$, which is in the grid circuit of tube 2. The potential thus appearing across this resistor ultimately causes the grids of thyratrons 5 and 6 to become highly negative, and conduction by these tubes is thus impossible as long as thyratrons 3 and 4 are firing.

Speed of response of the system is practically instantaneous. There are no time delays within the electronic components of the regulator. Motor-armature inductance is very small. The only delay of any consequence is that repre-

Fig. 30·33 An electric-arc furnace showing the three electrodes controlled by the positioning motors.

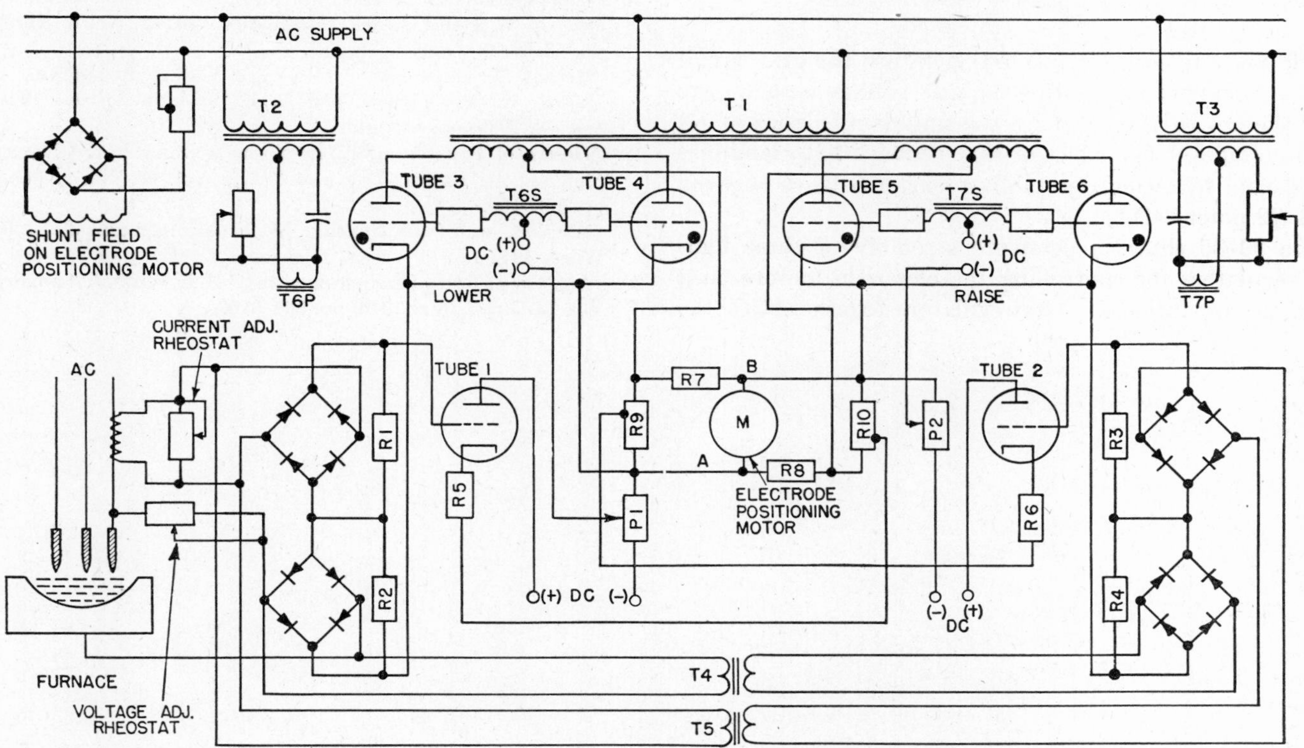

Fig. 30·34 Circuit diagram of a regulator for controlling power input to an electric-arc furnace by means of electrode positioning.

sented by the mechanical time constant. Consequently, the system is inherently stable. However, a refinement in performance is obtained by the action of resistors $R9$ and $R10$. If an unbalance in the potentials across $R1$ and $R2$ is assumed, the lowering rectifier will apply a definite voltage to the motor armature. However, a portion of this voltage is obtained from the drop across $R10$ and fed back into the grid circuit of tube 1. The polarity of this "speed" voltage is such as to oppose the action produced by the initial unbalance of $R1$ and $R2$ voltages. The effect of this feedback is to minimize the overshoot following re-establishment of balance between $R1$ and $R2$ potentials.

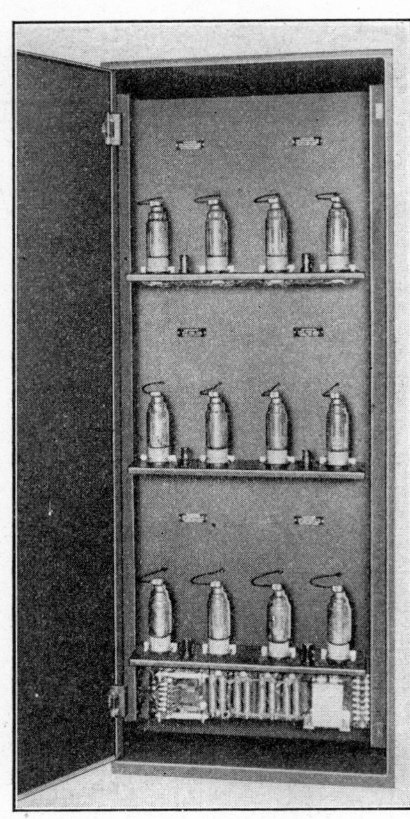

FIG. 30·35 Complete assembly of regulators for controlling the electrodes of a three-phase electric-arc furnace.

Figure 30·35 shows a complete assembly of three individual regulators for controlling the electrode motors on a three-phase arc furnace. The regulators together with the auxiliary control equipment are all mounted in a single steel enclosing cabinet for ease in assembly, test, and installation.

REFERENCES

1. "Forms and Principles of Servomechanisms," S. W. Herwald, *Westinghouse Eng.*, Vol. 6, No. 5, Sept. 1946, pp. 149–154.
2. *The Electronic Control Handbook*, Ralph R. Batcher and William Moulic, Caldwell-Clements, Inc., 1946.
3. "Theory of Servomechanisms," H. L. Hazen, *J. Franklin Inst.*, Vol. 218, No. 3, Sept. 1934, pp. 273–331.
4. *Network Analysis and Feed-back Amplifier Design*, H. W. Bode, Van Nostrand, 1945.
5. *Theory of Servomechanisms*, L. A. MacColl, Van Nostrand, 1945.
6. "Frequency Response of Automatic Control Systems," H. H. Harris, Jr., *Trans. A.I.E.E.*, Vol. 65, No. 8–9, Aug.-Sept., 1946, pp. 539–546.
7. "Electrical Analogy Methods Applied to Servomechanism Problems," G. D. McCann, S. W. Herwald, and H. S. Kirschbaum, *Trans. A.I.E.E.*, Vol. 65, No. 2, Feb. 1946, pp. 91–96.
8. "Dimensionless Analysis of Servomechanisms by Electrical Analogy," S. W. Herwald and G. D. McCann, *Trans. A.I.E.E.*, Vol. 65, No. 10, Oct. 1946, pp. 636–639.
9. *Operational Methods in Applied Mathematics*, H. S. Carslaw and J. C. Jaeger, Oxford University Press, 1941.
10. "Dynamic Behavior and Design of Servomechanism," Gordon S. Brown and A. C. Hall, *Trans. A.S.M.E.*, Vol. 68, No. 5, July 1946, pp. 503–522.
11. "Tracer-Controlled Position Regulator for Propeller Milling Machine," C. R. Hanna, W. O. Osbon, and R. A. Hartley, *Trans. A.I.E.E.*, Vol. 64, No. 4, April 1945, pp. 201–205.
12. "A Method of Successive Approximations of Evaluating the Real and Complex Roots of Cubic and Higher Order Equations," Shih-Nge Lin, *J. Math. Phys.*, Vol. 20, 1941, pp. 231–241.
13. "An Improvement on the G.C.D. Method for Complex Roots," F. L. Hitchcock, *J. Math. Phys.*, Vol. 23, No. 2, May 1944, pp. 69–74.
14. *Mathematics of Modern Engineering*, R. E. Doherty and E. G. Keller, Wiley, 1936.
15. *Advanced Rigid Dynamics*, E. J. Routh, MacMillan, 1892.
16. "Contributions to the Theory of Automatic Controllers and Followers," D. G. Prinz, *J. Sci. Instruments*, Vol. 12, No. 4, April 1944, pp. 53–64.
17. "Recent Developments in Generator Voltage Regulation," C. R. Hanna, K. A. Oplinger, and C. E. Valentine, *Trans. A.I.E.E.*, Vol. 58, 1939, pp. 838–844.
18. "Design of Reactors and Transformers which Carry Direct Current," C. R. Hanna, *Trans. A.I.E.E.*, Vol. 46, No. 2, Feb. 1927, pp. 155–160.
19. *Transients in Linear Systems*, M. F. Gardner and J. L. Barnes, Wiley, 1942.
20. "Analysis of Relay Servomechanisms," H. K. Weiss, *J. Aero. Sciences*, Vol. 13, No. 7, July 1946, pp. 364–376.

Chapter 31

RESISTANCE-WELDING CONTROL

C. B. Stadum, E. T. Hughes, F. R. Woodward, and H. J. Bichsel

IN order to visualize properly the field of electronic control for resistance welding, an understanding of the fundamental theory and application of the welding process is desirable. On some work electronic controls are not economically preferable but, in general, electronics has broadened the application and improved the quality of resistance welding to such a degree that it is considered indispensable to the process.

Resistance welding is the process of joining two or more metallic parts to form an assembly, the heat for the formation of the weld being created by the resistance of the parts and welding-machine electrodes to the passage of electric current supplied by the welding machine.

31·1 FUNDAMENTALS

Arrangement of the resistance-welding equipment components is illustrated in Fig. 31·1. It consists of: a rigid frame

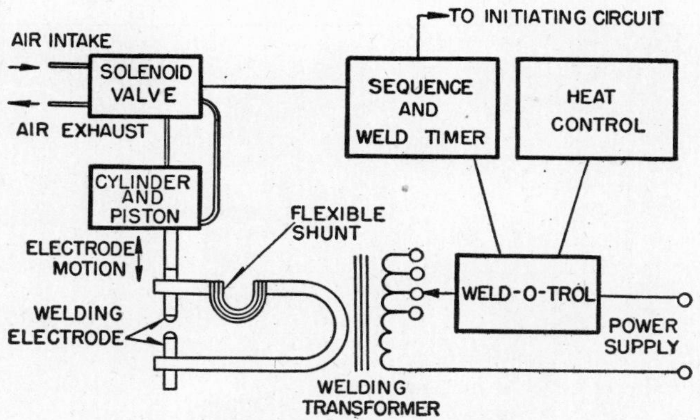

FIG. 31·1 Asynchronous welding-control system.

housing a high secondary-current welding transformer; the electrodes for contacting the work; the conductors connecting the electrodes with the welding transformer; means for exerting electrode pressure on the work; means of current regulation, either by regulation of transformer turns ratio as illustrated, or by electronic control; a contactor to interrupt the power to the welder transformer; and a timer which is capable of controlling the action of the contactor within the limits of accuracy required to produce the desired weld characteristics.

An illustration of a typical resistance-welding process, such as spot welding, is shown in Fig. 31·2, and it indicates the composite effects of heat generation and losses for the process. The heat being generated in these parts is calculated by the formula

$$W = I^2Rt \tag{31·1}$$

where W is the heat in watt-seconds, I is the current in amperes flowing through the weld, R is the resistance in ohms, and t is the time in seconds.

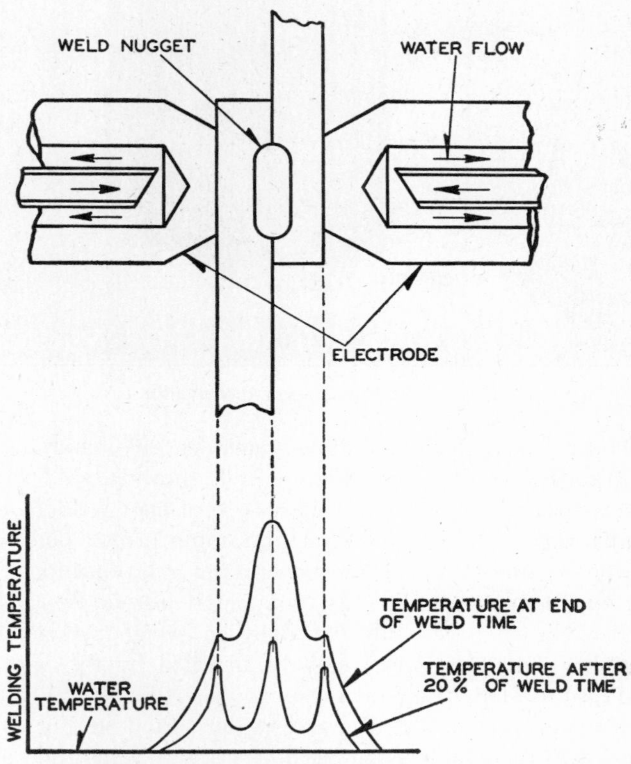

FIG. 31·2 Temperature distribution in a spot weld.

Not all the heat is generated at the proper point—at the juncture of the weldments (parts being welded). The flow of heat to or from this point, assisting or retarding the creation of the proper localized welding heat, is governed by the heat gradient established by the welding-current action on the various resistive components.

The factors affecting the amount of heat being produced by a given welding current for a unit of time are:

1. The electrical resistance of the materials being welded.
2. The electrical resistance of the electrode materials.

3. The contact resistance between weldments as determined by surface conditions, scale, welding pressure, and so on.

4. The contact resistance between the electrodes and weldments as determined by surface conditions, area of electrode contact, and welding pressure.

The weld heat losses are due to the removal of heat from the weld area by the electrodes, usually water-cooled, and by the material in the weldments themselves from the localized area where it is desired to produce a weld.

31·2 SPOT-WELDING MACHINES

Spot welding is that resistance-welding process which unites two or more sheets of metal, producing a lap joint by concentrating the welding current and pressure on the material by means of relatively small electrodes.

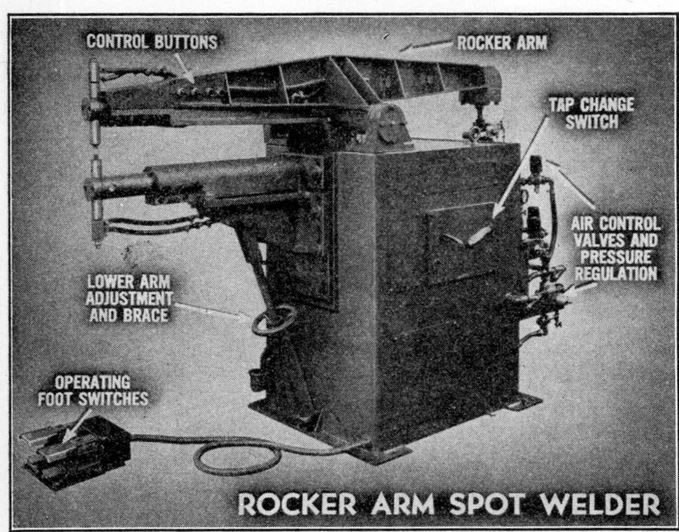

Fig. 31·3 Rocker-arm spot welder.

Many types of spot-welding machines are manufactured as standard equipment. Figure 31·3 shows a spot welder and associated equipment. This type of spot welder necessitates the use of devices for maintaining proper placement of the weldments to insure the proper relationship of the various components in the completed assembly. These devices must necessarily be light in weight since they, together with the work, must be handled by the operator and moved along the machine.

The portable or "gun" welder, illustrated in Fig. 31·4, consists of a portable assembly of electrodes, arms and means of applying pressure, and is connected to its transformer by a pair of heavy welding cables carrying the secondary or welding current. The welding electrodes can be moved around relatively independently of the heavy equipment such as the welder transformer and controls. This type of welding equipment is customarily used on assemblies which are difficult to handle at a stationary type of machine. Since the work and fixture remain stationary in this weld process, the locating devices may be as heavy and cumbersome as necessary in order to accomplish the desired location.

Another type of spot-welding equipment is that known as the "hydromatic" or "multispot" welder (Fig. 31·5), which operates a considerable number of electrodes, one for each required spot in the assembly, together with the necessary locating means. Here the electrodes operate in sequence by means of individual hydraulic pressure applications and weld power control. This type of equipment is necessarily constructed to do a particular welding job, and is special in that it must be rebuilt in order to be adapted to another assembly.

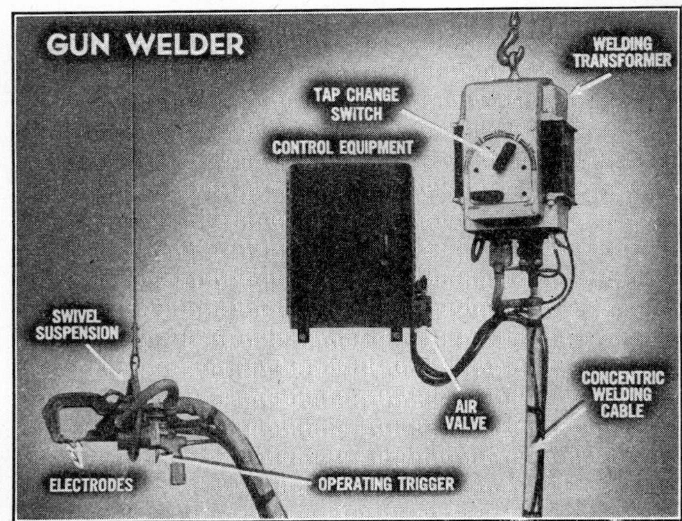

Fig. 31·4 Portable or gun welder.

31·3 SPOT-WELDING TECHNIQUE

Any combination of metals, similar or dissimilar, can be resistance-welded to achieve a bond. This bond may or may not be sufficient for the design requirements. In order to produce a strong weld, it is necessary to bring the contacting surfaces of the two weldments up to their respective welding temperatures at the same time and cause a bonding of a portion of each weldment in the formation of a "weld zone," "nugget," or "slug," terms typical of certain types of resistance welding. Thus, it is difficult to produce a satis-

Fig. 31·5 Hydromatic or multispot welder.

factory weld between two dissimilar metals which will not alloy, and any weld so produced—often termed a "stick" weld—will be nothing more than a surface molecular interlocking joint.

Materials can be better welded if the weldments are of nearly the same shape or thickness, if the materials have approximately the same electrical resistance and melting point, and if they will alloy properly. If extreme variations of electrical conductivity or weldment mass are present, it is possible to control to some degree the heat generated in and lost from the separate weldments by:

1. Use of an electrode of low conductivity against that material in which it is desired to increase the weld heat.
2. Use of a smaller electrode face to increase the heat in the desired part.
3. Use of short weld time, limiting all heat losses.

It is possible to approximate the required spot-welding set-up values from an examination of the physical characteristics of a given material. Thus, the proper electrode material is selected after consideration of the electrical conductivity, surface conditions, and yield strength of the weldment materials. Lower conductivity of the materials allows the usage of lower conductivity, harder electrodes since lower welding currents will be required. Poor surface conditions are compensated for by high-conductivity electrodes, because these softer electrodes will not "mushroom" excessively at the lower welding pressure required.

The duration of power application, or "weld time," required for the production of a spot weld having the desired characteristic of internal weld and external appearance, appears to be governed almost entirely by the thickness of the materials being welded.

The temperature gradients appearing across a spot weld at the completion of 20 percent of the weld time and at the end of the weld time are pictured in Fig. 31·2 according to the author's interpretation of this process. It can here be seen that the water cooling of the electrodes effectively limits the temperatures attained at the points of electrode contact with the weldments, and reduces adverse heat effects on these areas. The figure also indicates the considerable influence of contact resistance in the generation of heat during the first portion of the welding period. During the latter portion of the welding period, the weld nugget becomes thicker because the heat is generated by the body resistance of the material. Thus welds produced with longer "weld time" have a thicker weld nugget.

As a general rule, the weld time may be set up as equal to 150 cycles per inch thickness of the thinner sheet in contact with an electrode, if this thickness is 0.025 inch or less (one cycle—$\frac{1}{60}$ second). On heavier materials the time required is about 250 cycles per inch thickness. If the welding equipment is deficient in pressure or welding current, or has high mechanical inertia, it will be necessary to increase the weld time to get optimum results. Timing of less than five cycles should never be attempted without precise control equipment to eliminate inconsistencies.

The use of adequate electrode pressures results in the following effects contributing to good welding results:

1. Production of local intimate contact, concentrating current flow in the desired area.
2. Reduction in weld porosity caused by the low boiling point of certain elements.
3. Assistance in reducing cooling cracks by compensating for internal shrinkage with an external follow-up pressure.
4. Reduction of the tendency of the electrodes to alloy with the material being welded.

The electrode force or pressure requirement of the spot-weld process is higher for thicker materials, materials having higher yield strength, and those covered with scale, rust, or protective metallic coating. Thus, while 0.050-inch thickness

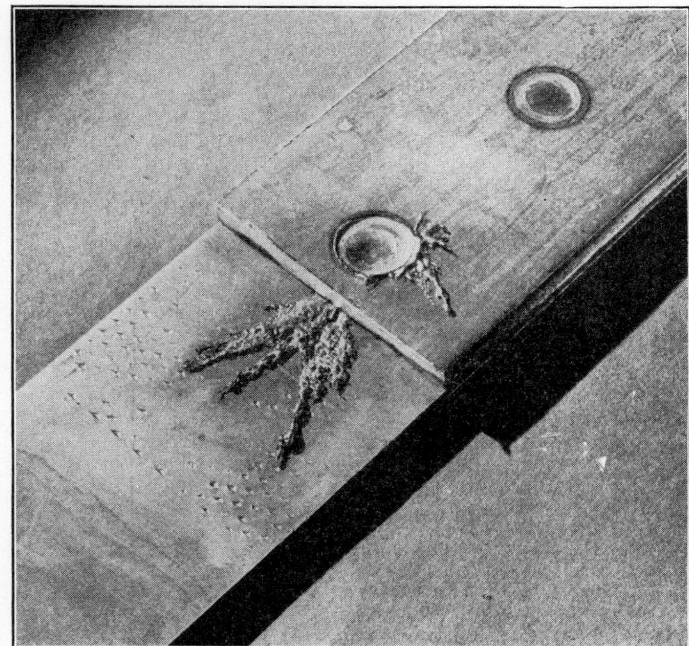

FIG. 31·6 Results of excessive current.

of hot-rolled low-carbon steel, pickled, can be welded readily with 600 pounds tip force, material of 0.100-inch thickness requires about 1200 pounds for optimum results, and a steel such as NAX 9115, having approximately twice the yield strength of mild steel, will require about 1200 pounds for best results on 0.050-inch thickness.

The current required for spot welding is determined somewhat by the preceding factors of electrode contour, weld time, and electrode force or weld pressure. These requirements are also based on the electrical conductivity, thermal conductivity, and melting point of the materials welded. Since the thermal conductivity is, to a degree, proportional to the electrical conductivity, its individual consideration can be neglected. The melting point will not necessarily be the actual temperature required for welding, but it is a very important characteristic in estimating weldability of dissimilar metal combinations.

Selection of proper spot-weld current values delivered consistently will result in:

1. The required design strength.
2. Good appearance with minimum tip indentation in the working or burning of the sheet surface.

3. A succession of spot welds consistent in strength.
4. A minimum of electrode maintenance.
5. A high rate of production.

Figure 31·6 illustrates the result of the application of an excessive amount of current in making a spot weld.

31·4 SPOT-WELDER SEQUENCE CONTROL

In practice the spot-welding process must embody a synchronization of the mechanical operation of the welding machine with the passage of welding current through the electrodes and materials, so that full electrode pressure may be attained prior to the time of application of the welding current, and that this pressure may be maintained until the weld "nugget" shall have cooled to a temperature where it regains a large portion of the inherent metal strength. The cooler, stronger material surrounding the weld nugget is thus forced by the welding pressure to move along with the inner plastic core of the weld as the latter shrinks on cooling, reducing any tendency toward formation of checks, cracks, or piping in the weld-nugget casting. Cracks in welds are therefore an indication of the possibility of an excessively short "hold" time after the end of welding current.

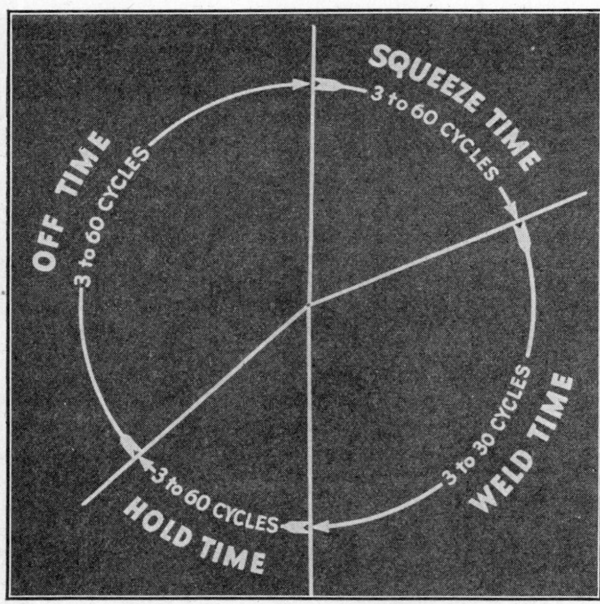

Fig. 31·7 Spot-weld timer functions.

In order to stabilize the welding conditions, it is desirable that the operations of weld timing and pressure application be accomplished automatically by means of some sort of sequence control. In one type of sequence timer controls the spot welder functions as illustrated in Fig. 31·7, wherein the start of operation is at the beginning of the "squeeze time" and is initiated by the operator's foot switch, pushbutton, or other means. A pre-set squeeze time allows the operation of the electrode pressure-actuating device and the attainment of the full weld pressure or electrode force before the initiation of the "weld time" wherein sufficient heat is generated in the part to produce the desired weld. The "hold time" then follows, allowing the weld to cool partially. During the "off time," the pressure-actuating device releases the weld pressure from the work, the electrodes separate, and the work is moved to the position where the next spot weld is desired. The sequence is then repeated, if the operator has not, in the meantime, released the pilot switch.

By means of automatic sequencing devices of this type, production speed may be increased and weld consistency improved to the point where excellent-quality welding may be obtained on the lighter gauges of sheet metal at operating speeds above 200 spots per minute.

Under some conditions the use of a "pulsation" weld time is desirable. This incorporates alternate periods of "heat" and "cool" times arranged over a total weld-interval time, as indicated in Fig. 31·8. This timing is particularly advantageous when relatively heavy materials, ⅛ inch in thickness and up, are being welded, especially if there are coatings of scale or rust on the surfaces. Pulsation welding is best applied on low-conductivity materials (less than 20 percent of copper, International Annealed Copper Standard), since heat losses become excessive on the higher conductivity materials with this process.

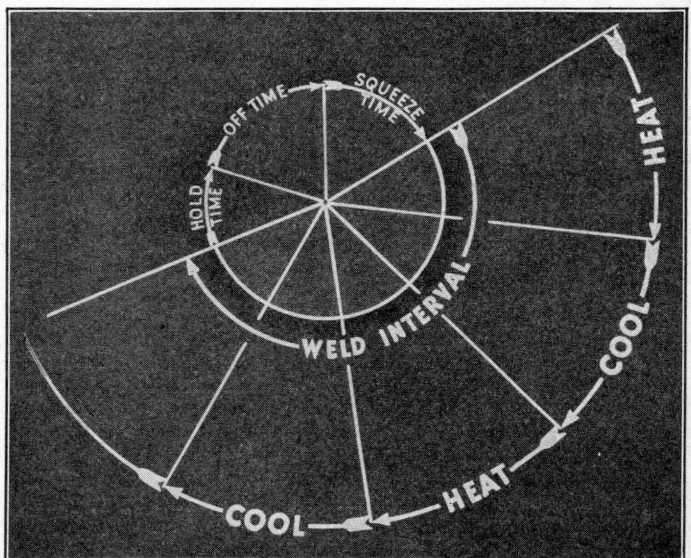

Fig. 31·8 Pulsation spot weld.

31·5 WELDING ELECTRODES

The electrodes of any resistance-welding machine are of prime importance from their influence on both production quality and operating cost. Electrode materials must carry large currents without burning or blistering with attendant work-marring effects, and must also have considerable mechanical strength to withstand the high welding pressures and repeated impacts on the material being processed. Since, in operation, the electrode face becomes quite hot, it is desirable that it have a high annealing temperature. An engineering compromise is usually necessary between hardness and conductivity since no electrode material has both maximum hardness and highest conductivity. Strong high-conductivity copper alloys are available, however, in many degrees of hardness.

The strength of the electrode material will be considerably greater and will reduce any tendency toward mushrooming, if the temperature is maintained at a low value. Low surface temperature of an electrode may be maintained by the following means:

1. Use of electrode materials having high electrical conductivity.
2. Use of materials having high thermal conductivity.
3. Use of low-temperature cooling water to other coolants.
4. Use of adequate amount of electrode coolant.
5. Maintenance of a clean, well-fitted joint between the electrode and the electrode holder.

The maintenance of a given contour or tip diameter during production is necessary to control the welding results obtained. If very consistent welding results are desired or if the materials being processed have critical welding requirements, it may be necessary to machine the electrodes periodically in order that the electrode diameter or contour may be re-restablished perfectly before it changes to a degree which may affect the weld heat.

There are two general types of electrode faces in favor at the present time for spot welding. The dome type has a spherical radius formed at its working face. The proper radius of this dome is greater for the spot welding of materials having a higher yield strength.

The flat type of electrode may be flat for the whole area of the electrode to reduce marking of the material, under which conditions its mating electrode must provide further localization of heat and pressure; or it may be flat for a reduced section at the tip of a truncated cone form, the sides tapering back to the full electrode section at an angle of 10 to 40 degrees from the face. With this type of electrode, the diameter of the contacting face is usually greater than $2t + 0.1$ inch, where t is the thickness of the contacting sheet of material.

The use of dome tips is becoming popular among production men because of the following advantages over other electrode shapes:

1. Ease of obtaining satisfactory electrode alignment in the machine.
2. Ease of maintenance of original contour with repeated cleaning or dressing.
3. Decrease in work marking with larger radius tips. Excellent results are obtained with electrodes having spherical radius faces of 4 to 10 inches.
4. More variation in other weld set-up factors permitted because of the increase in effective contact area with an increase in heat in the weld. This increase in area with indentation into the welded material acts effectively in reducing the tendency toward expulsion or "spitting" between the sheets being welded or between electrode and sheet, as illustrated in Fig. 31·6.

31·6 SEAM WELDING

The process of seam welding is very similar to that of spot welding except that the stud-type electrodes of spot welders are replaced with welding wheels which roll along the work. The rate of welding is usually from 2 to 100 feet per minute. At the higher speeds the current application is continuous, and each half cycle of alternating current produces a small weld overlapping the previously formed weld to form a continuous seam.

As indicated in Fig. 31·9, when best welding results are desired the welding power should be interrupted to provide "off" time for the cooling of each individual weld spot. One twelfth of an inch is a satisfactory spot spacing on No. 22 gauge materials to produce a pressure-tight seam weld. On heavier materials, however, the spot spacing must be greater in order to prevent excessive build-up of heat ahead of the welding wheel. Spot spacing for seam welding two pieces of ⅛-inch material should be about ⅙ of an inch. Seam welding of low-carbon steel can be accomplished very well with

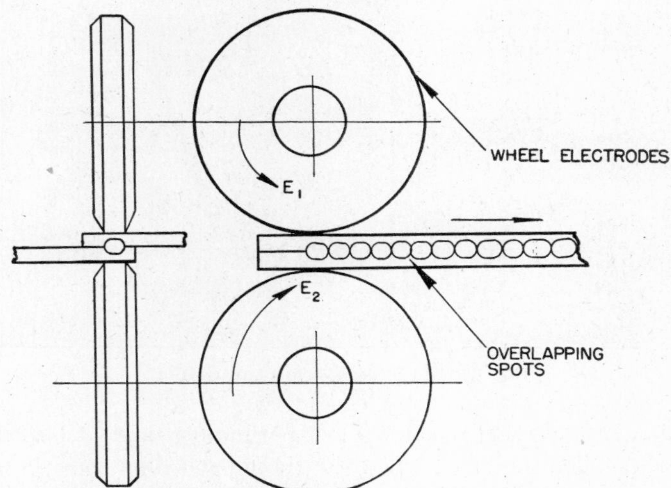

FIG. 31·9 Lap seam welder welding pair of tank stampings with a pressure-tight continuous seam.

"heat" and "cool" periods of equal duration, but materials having greater conductivity or less weldability require a relatively longer "cool" time. With a 60-cycle power supply the relationship of weld timing, wheel speed, and spots per inch may be indicated by the equation

$$t_w + t_c = \frac{300}{vN}$$

where t_w = weld time in cycles
t_c = cool time in cycles
v = velocity or speed of welding in feet per minute
N = spots produced per inch.

Most seam-welding machines weld two pieces of material whose edges overlap. Figure 31·9 illustrates a circumferential wheel arrangement of a lap seam welder welding a pair of tank stampings with a pressure-tight continuous seam. If the welding is not continuous, or the individual weld spots do not overlap, the weld is termed a roll spot or a stitch weld. Occasionally it is desirable to eliminate the double thickness of material at a joint by making a mash weld. This can be accomplished very well on the lighter gauges of material by lapping the material about 1½ times the sheet thickness and using broad flat welding wheels. The material must be clamped firmly in place so that the mashing action does not tend to shift the position of the sheets.

Some butt-seam or tube welders of a more special nature are used to produce a longitudinal butt joint in the fabrication of pipe or tubing. This tubing is usually progressively welded at high speed, continuous power application with no interruptions other than the inherent 60-cycle alternations being used.

A typical weld set-up for lap-seam-welding two pieces of 18-gauge low-carbon clean steel utilizes a pair of welding wheels 8 inches in diameter, having a wheel face width of 7/32 inch, and exerting a weld force or pressure of 750 pounds.

FIG. 31·10 Rocker-arm seam welder.

The welding-wheel speed is 9 feet per minute, and 11.1 spots per inch are produced when the timing schedule consists of two cycles "on time" and one cycle "off time," with the welding current 19,000 amperes. In this case, the necessity for power interruption at the rate of 1200 per minute requires that an electronic timer and contactor be employed, giving noiseless high-speed operation and insuring that successive half cycles passed to the welder transformer shall always be of opposite polarity. This is necessary for, if power were interrupted for an odd number of half cycles, producing successive current impulses of same polarity, the welder transformer would tend to saturate and draw excessive line current.

31·7 PROJECTION WELDING

Projection welding is a spot-welding process in which high localization of pressure and current is accomplished by means of irregularities or projections formed in one or both of the weldments. Since large flat electrodes can be used, electrode maintenance is considerably reduced.

Butt and Flash Welding

Butt and flash welding are those resistance-welding processes in which the butt ends of two pieces, such as rod or strip, are welded together. In a butt weld, the pieces are held together in firm contact while current heats the joint. After the heating of the joint, the two parts are forced together under extremely high pressure, and a solidly fused joint results. In a flash weld the two pieces are held together lightly, and a current is passed through the joint. Burning of the joint, accomplished by vigorous flashing during the passage of the current, results in eliminating the irregular surface of the joint and in heating the joint so that a strong weld is made when the two pieces are forced together under high pressures.

31·8 TIMING

The purpose of welding control is to:

1. Start and stop the welding current.
2. Control the amount of welding current.
3. Time the duration of the welding current.
4. Time the sequence of operation of the welding machine.

Timing the duration of welding current is a matter of controlling the time interval during which the switch or contactor shown in Fig. 31·1 is closed.

Duration of the welding current can be timed by the operator, who operates a pushbutton or manually operated switch. Welding time must be long, for an operator cannot consistently judge extremely small times. However, if welding current is of several seconds' duration, this method is satisfactory.

Some welding machines have motor-driven cams which raise and lower the heads of the welding machines. On this type of machine, an auxiliary cam times the duration of the welding current. The cam energizes a switch which, in turn, energizes a magnetic or electronic contactor. Varying welding time is obtained by varying the size of the cam. It is an awkward, slow process to adjust the cam to get variations in timing; therefore this system is not used except for high-production welding on which adjustments seldom are made.

For increased ease of timing adjustment, a pneumatic or tube-relay timer can be used. In the pneumatic type of timer air is allowed to escape through an orifice, the size of which can be varied. The time required for a given volume of air to pass through the orifice determines the length of welding time. Because the orifice opening can be varied by a

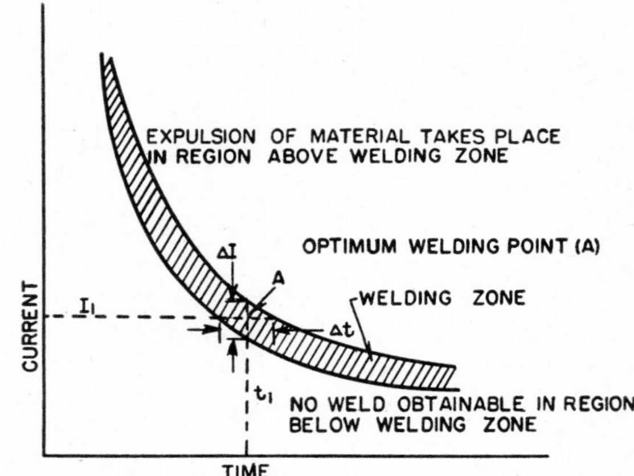

FIG. 31·11 Welding zone as a function of current and time.

needle valve, welding time can be adjusted easily. The tube-relay timer uses an electronic timing circuit to actuate a relay, which in turn controls the welding current. Pneumatic and tube-relay timers are equally accurate and easy to adjust; and although both are subject to a small amount of variation in timing, they are adequate for most timing functions. If no variation in time can be tolerated, the electronic timer is used. This and the tube-relay timer are discussed in detail later.

The importance of timing is illustrated by Fig. 31·11, which shows that a weld can be made with an infinite number of combinations of current and time. However, there is one combination of current and time which gives best welds. Other combinations may result in indentation of the weld or expulsion of the weld material. For example, in Fig. 31·11 the combination of I_1 and t_1 gives optimum welding conditions. At these values of current and time, there is no excessive material indentation and no expulsion. If welding time becomes greater than t_1 there is expulsion of material and indentation of the metal. If it becomes less than t_1 the metal does not weld.

In some materials such as mild steel I_1 and t_1 can vary somewhat (that is, ΔI and Δt can be large). For these materials a tube-relay welding timer is satisfactory. However, other materials such as aluminum have critical welding temperatures; the allowable variation in time and current is small, and the timer must be a precision timer. Equal in importance to the duration of welding current is the means of initiating current. One method of initiating may result in transient-free-welding currents; another method may cause transients.

31·9 TRANSIENTS AND THEIR INFLUENCE ON WELDING

In general, starting transients are obtained when the welding circuit is energized at an angle on the voltage wave which does not correspond to the natural power-factor angle. (The expression "energized at an angle on the voltage wave" is a convenient manner of expressing time.) That is, for a welding machine of a given power factor there is a definite phase relationship (angular displacement) between the steady-state current and voltage waves. If the circuit is energized at an angle other than the power-factor angle, an asymmetric current wave is produced. Figure 31·12 illustrates the effect of closing the circuit at different points. In Fig. 31·12 (*a*) the first half cycle of current is small and of short duration; the second half cycle is larger. This condition is caused by energizing the circuit at a point behind the natural power-factor angle of the welder. Figure 31·12 (*b*) shows the primary current when the circuit is energized at precisely the natural power-factor angle. No transients are present. Figure 31·12 (*c*) shows current for a weld energized at a point ahead of the natural power-factor angle; the first half cycle is much larger than the steady-state current, but the second half cycle is much smaller. In the cases shown in Figs. 31·12 (*a*) and 31·12 (*c*) the transient causing the variation diminishes gradually, and after a few cycles the current reaches a steady state. These figures illustrate maximum transients. The primary current may be anywhere between these two extremes, and never is consistent if a magnetic or electronic contactor is used to energize the circuit.

Figures 31·12 (*a*) to 31·12 (*c*) illustrate possible shapes of the primary current if the transformer does not saturate. Figure 31·12 (*d*) shows the current for a welding transformer that does saturate. It is similar to Fig. 31·12 (*c*), except for the high peak currents. With such a transformer the peak current may rise as high as 8 or 10 times the steady-state current.

These variations have much influence on welding and on the power system. Transients cause a large demand on the

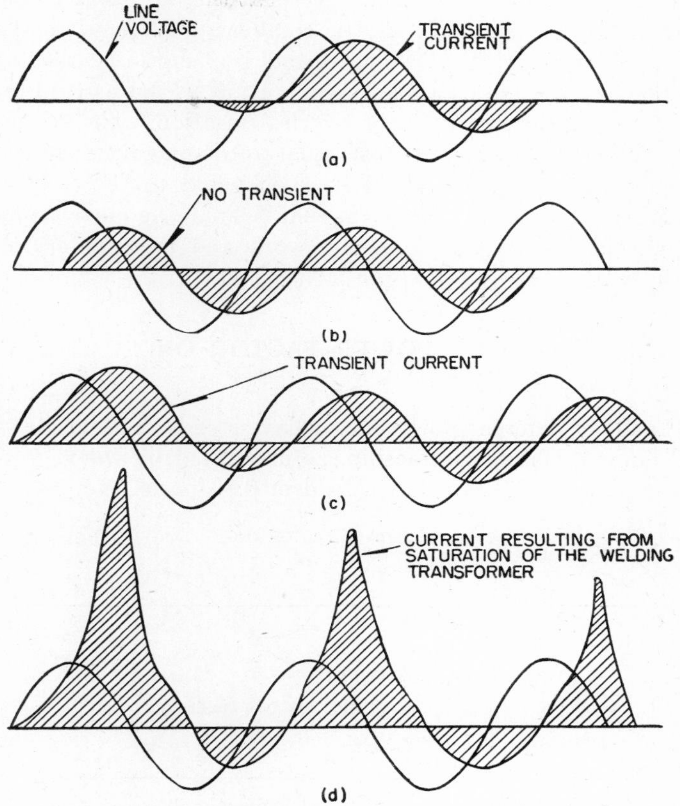

FIG. 31·12 Transients produced by closing the circuit at different points.

power line. This demand causes increased voltage drop in the line, and may cause light flicker. For this reason, a timer which eliminates starting transients is often used.

Besides light flicker, excessive transient currents in the power transformer, the power line, and the welding transformer may cause excessive heating of all three.

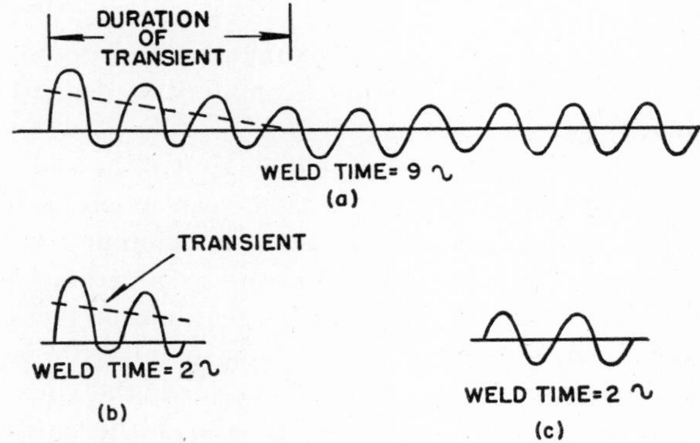

FIG. 31·13 Effects of transients for various weld times.

Figure 31·11 shows that, for consistent welds, both current and time must be relatively consistent. If welding time is short, the excessive energy caused by transients is sufficient to ruin a weld. Although current transients in the primary of the transformer are not directly transformed to the secondary, the excess energy delivered to the weld is roughly proportional to the transient. For simplicity only primary-current transients are shown in Fig. 31·12. Figure 31·13 (*a*) shows the effect of a transient for a weld time of 9 cycles. Duration of the transient is approximately 3 cycles. The increased current during the 4 cycles at the beginning of the weld has little influence on the total heat delivered to the weld. However, if the welding current is of short duration, as in Fig. 31·13 (*b*), the transient is important. Figure 31·13 (*b*) shows a transient in a weld of 2 cycles' duration. The magnitude of the transient is almost equal to the magnitude of the steady-state current. Total heat delivered to the weld is approximately twice as much as that desired, thereby causing expulsion of material from the work and marring of the surface.

31·10 EFFECT OF POWER FACTOR ON TRANSIENTS

Loads of different power factor have steady-state currents of different phase relationship, as illustrated by the oscillograms in Fig. 31·14. For a load of 0.80 power factor, the

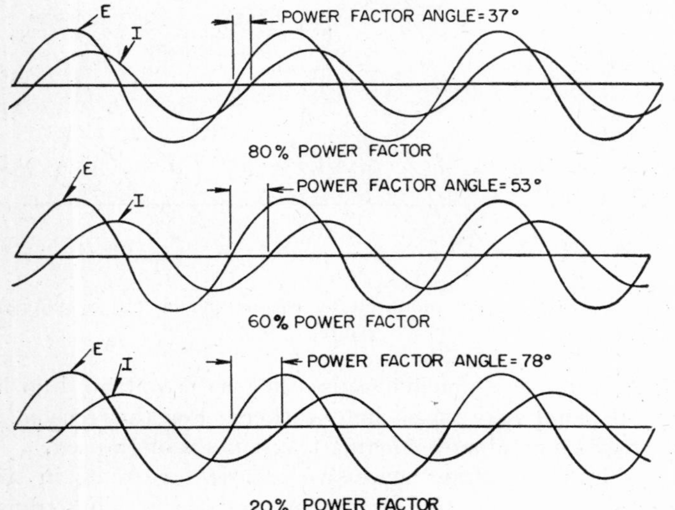

Fig. 31·14 Current and voltage phase relationships for various power-factor loads.

current lags the voltage by approximately 37 degrees. For a power factor of 0.60 it lags approximately 53 degrees, and for a power factor of 0.20 it lags 78 degrees. To start welding currents without transient the welding control must be able to start the welding current at a time corresponding to the natural current-zero point. This point is different for different power factors. Figure 31·15 shows oscillograms of the welding currents on machines of different power factors, if the machines are energized at the same point with respect to the voltage wave. Figure 31·15 (*a*) shows the current in a machine having a power factor of 0.50 if the current is energized at 60 degrees, which is the natural power-factor angle. There is no transient. Figure 31·15 (*b*) shows the current in a machine of 0.20 power factor also energized at 60 degrees, or in this case ahead of the natural power-factor angle. Figure 31·15 (*c*) shows the current in a machine of 0.80 power factor energized at 60 degrees, or after the natural power-factor angle. These figures show that the welding control must be adjustable to energize a welding current at the correct points corresponding to the different power factors of the machines, if transientless starting is necessary. Figure 31·16 shows the relation between power factor and energization point for transientless starting.

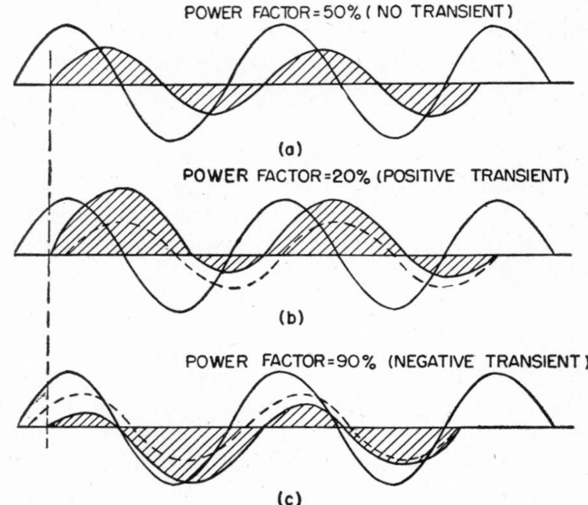

Fig. 31·15 Welding currents on machines of different power factor. Current is initiated at the same point with respect to the voltage wave.

31·11 SWITCHING OF WELDING CURRENT

Welding current can be switched synchronously or asynchronously. Asynchronous equipment includes either an electronic or a magnetic contactor. Synchronous equipment includes an electronic contactor in addition to auxiliary control circuits. An asynchronous contactor is one in which closing of the welding circuit is a random phenomenon; a synchronous contactor is one in which welding current is initiated at a point on the voltage wave corresponding to the natural power-factor angle of the welding machine. Thus, in a synchronous system no transients are present, whereas the asynchronous system may produce transients.

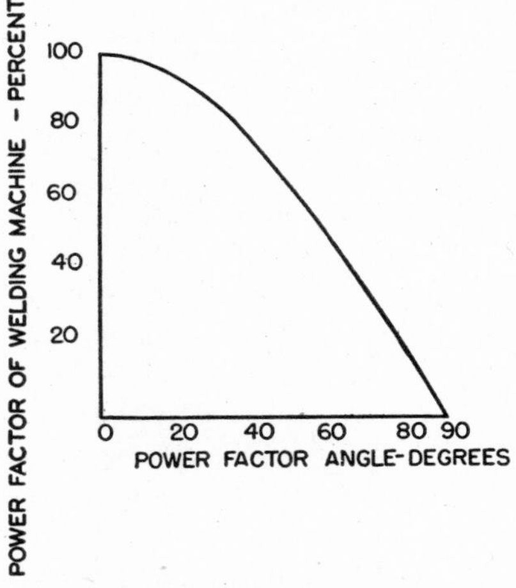

Fig. 31·16 Percent power factor as a function of power-factor angle.

31·12 ASYNCHRONOUS WELDING CONTROL

A complete asynchronous control system is shown in Fig. 31·1. It includes: (1) an electronic contactor; (2) a heat-control circuit to control the amount of current the contactor passes; and (3) a relay-tube type of sequence-weld timer to co-ordinate pressure, time, and current functions.

The electronic contactor is essentially a switch in series with the primary of the transformer. The heat control is a phase-shifting circuit used to control the length of time during each individual half cycle that the electronic contactor is effective. The sequence-weld timer controls the solenoid valve, which in turn controls the pressure device on the head of the welding machine. In addition to determining the pressure-time function required in the welding machine, it also determines the sequence and duration of the welding current. In general the operation is as follows. The operator presses the foot switch, which operates the sequence-weld timer. The sequence-weld timer in turn operates the pressure mechanism, lowering the welding electrodes on the work. After electrode pressure on the work is sufficient the sequence-weld timer energizes the electronic contactor, which allows current to pass for the length of time determined by the timer. After the weld has been made and has cooled sufficiently under pressure, the sequence-weld timer raises the head of the machine. If the initiating circuit is kept energized during the welding cycle, the sequence-weld timer repeats the entire sequence after the operator has had sufficient time to move his work in the throat of the welding machine.

31·13 THE ELECTRONIC CONTACTOR

The electronic contactor is equivalent to a single-pole single-throw electronic power switch. It consists of two igni-

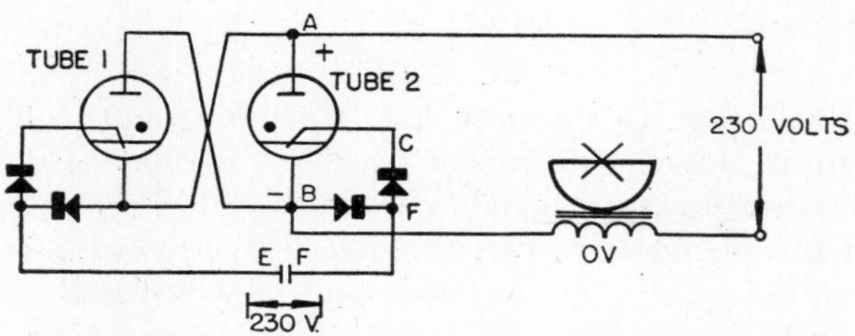

FIG. 31·17 Electronic contactor, ignitor circuit open.

trons in inverse-parallel; that is, the anode of one ignitron is connected to the cathode of the second tube, and vice versa. Alternating current passes through the primary of the welding transformer upon closing of the control circuit. Operation of the electronic contactor is illustrated by Figs. 31·17 to 31·19.

For an ignitron to pass current, (1) the anode of the tube must be positive relative to the cathode, and (2) a definite

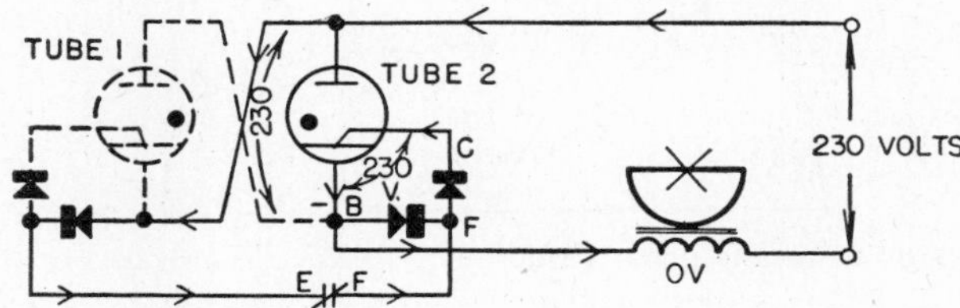

FIG. 31·18 Electronic contactor, ignitor current flowing.

igniter current must be passed (see Chapter 6). Suppose that in Fig. 31·17 the anode (lead *A*) of the ignitron is positive relative to the cathode. This tube is now ready to fire if sufficient igniter current is passed. However, since the igniter circuit is open, no igniter current can be passed but, as soon as the contacts between points *E* and *F* are closed, current can travel through the circuit designated by the arrows in Fig. 31·18. This current is sufficient to fire tube 2. After the tube fires, the voltage drop across it is approximately 15 volts (see Fig. 31·19). This reduction from full line voltage to 15

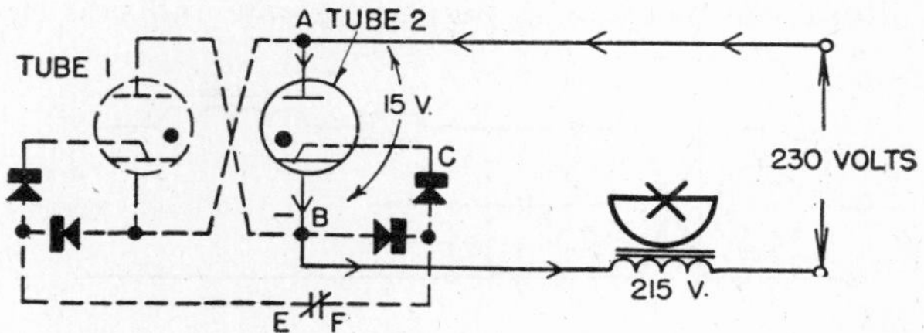

FIG. 31·19 Electronic contactor, tube 2 conducting.

volts causes the igniter current to fall to a fraction of its previous value. Current in the power circuit is now that shown by the arrows in Fig. 31·19. Since tube 2 carries current in one direction only, another ignitron must be connected in inverse parallel to conduct current in the opposite direction. In such a circuit, both halves of the a-c cycle are made effective, and an alternating voltage is applied to the welding transformer.

If the contacts between *E* and *F* are open, the igniter carries no current; hence there is no welding current. However, if the contacts are closed, igniter current and hence welding current will be carried by the circuit. By simply opening and closing the circuits the welding current can be controlled.

The purpose of the rectifier between points *F* and *C* is to prevent reverse current through the igniter of tube 2 whenever point *B* is positive relative to *A*. Reverse current through the igniter causes rapid deterioration of the igniter. The rectifier between points *B* and *F* allows igniter current for tube 1 to pass from *B* to *A* without passing through the igniter of tube 2. Although the primary purpose of the electronic contactor is to switch the welding currents on and off, it can also control the amount of current if a "heat control" is used with it.

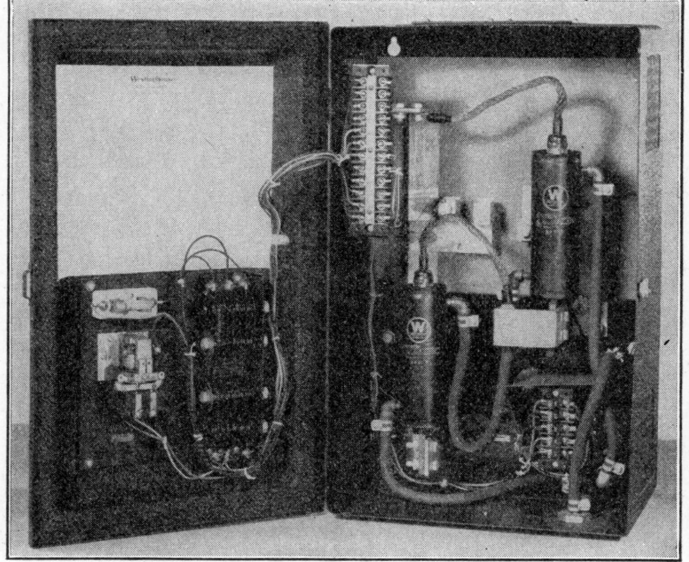

FIG. 31·20 Electronic contactor.

31·14 CONTROLLING CURRENT IN THE ELECTRONIC CONTACTOR

The principle of current control, or heat control as it is often called, is illustrated by Fig. 31·21. Assume that the load of two tubes in inverse parallel is inductive and that there is an alternating voltage on both the anode and grid circuits

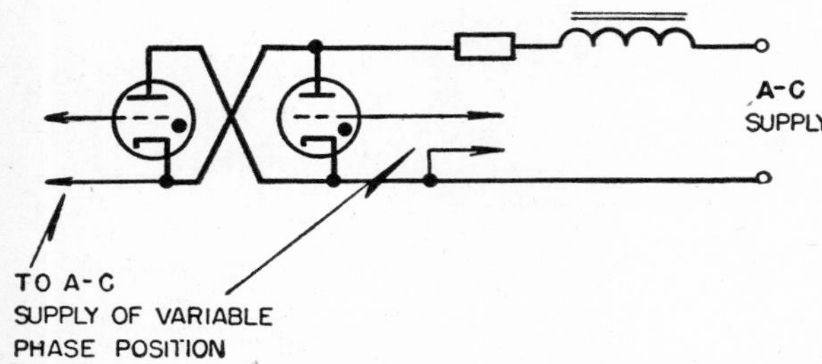

FIG. 31·21 Basic current control or heat-control circuit.

of the thyratrons. Assume also that a voltage which is out of phase with the line voltage in varying degree is impressed upon the grids of the thyratrons. For a thyratron to conduct, the anode voltage must be more positive than that of the cathode, and the grid voltage must be more positive than the critical grid voltage. (For this discussion the critical grid voltage is assumed to be zero.) Figure 31·22 (*a*) shows that in the region *A–B* the anode voltage is positive, and thereby the anode is in condition to conduct current. However, the grid voltage in this region is negative, and hence the tube

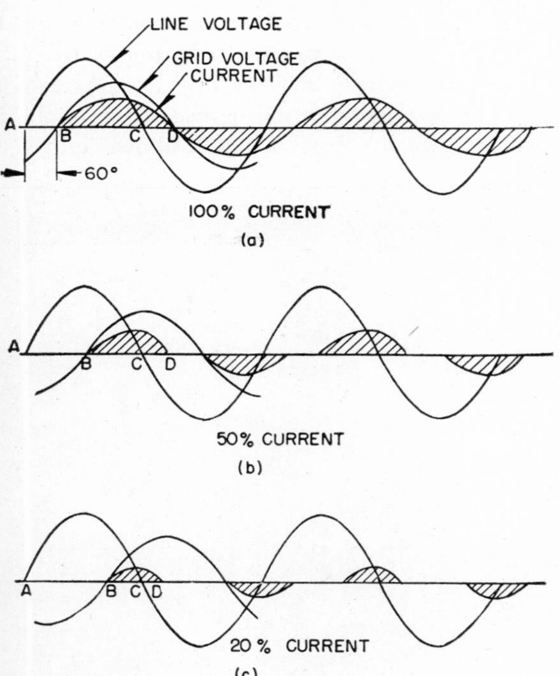

FIG. 31·22 Oscillograms showing the effect on the current of shifting the base of the grid voltage of the current-control circuit in Fig. 31·21.

cannot conduct. At point *B*, the anode voltage is positive, and the grid voltage is above the critical value; therefore the tube can conduct current. At this point the tube conducts or "fires." If the load controlled by the thyratron is resistive, conduction continues to point *C*; but if the load is inductive some energy is stored in the inductance and is released after the point corresponding to zero voltage. In other words, the current lags the voltage wave, and conduction stops at *D*.

If grid voltage is made to lag line voltage still more, the initiation point of the thyratron is further delayed, and a

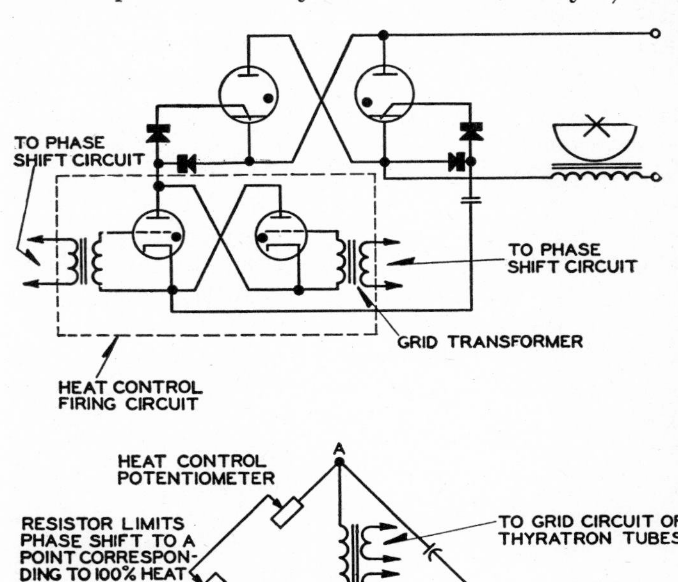

FIG. 31·23 Heat-control circuit.

smaller current, as shown in Fig. 31·22 (*b*) results. If grid voltage is phased still further back, the current carried by the thyratron is further reduced, as shown in Fig. 31·22 (*c*). If the phase angle of the grid voltage of the thyratron is controlled, the current the thyratron can carry during any given half cycle can be adjusted. This is the basis for the heat control on welding transformers.

The heat-control circuit is shown in Fig. 31·23. The thyratrons are used to control the instant of starting of the igniter current, which in turn controls the starting of the ignitron. The thyratron can be made to fire at any point on the voltage wave; hence it can be made to start the ignitron at that point. By controlling the phase relationship on the grid of

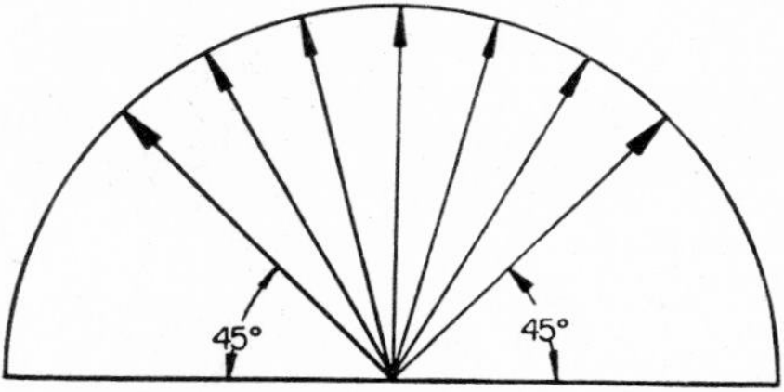

FIG. 31·24 Possible phase variations. Resistor shown in Fig. 31·23 is set for a minimum shift of 45 degrees.

the firing thyratron, the firing point of the ignitron can be controlled, and the current in the welding transformer can be varied. The welding currents are like those shown in Fig. 31·22.

The phase-shifting circuit described in Section 18·5 is often used in heat-control circuits but, in addition to the components shown, the secondaries of the transformer are connected into the grid circuits of the thyratrons. Since the phase position of the transformer primary is adjustable, the secondary phase position also varies, and thus the firing point of the thyratron is varied. The welding-control phase-shifting circuit also contains a variable resistance in series with the potentiometer to adjust the minimum phase shift.

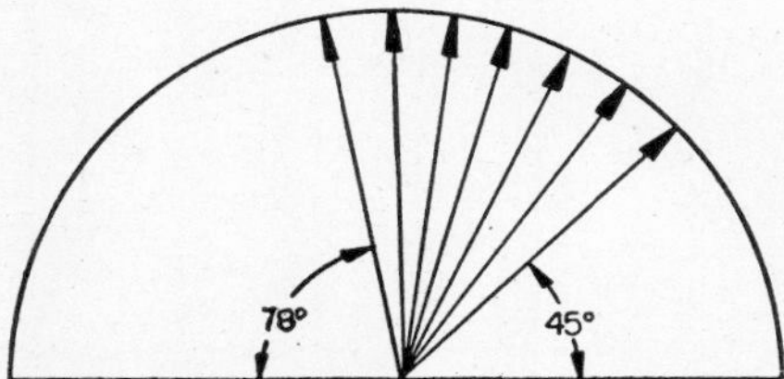

FIG. 31·25 Possible phase variations. The resistor shown in Fig. 31·23 is set for a minimum shift of 76 degrees.

Figures 31·24 and 31·25 show two possible combinations of phase variation for two corresponding adjustments of the resistor. In Fig. 31·24 the resistance is set for a minimum shift of 45 degrees. At this point welding current is maximum on a machine of 0.70 power factor. In Fig. 31·25 the resistance is set for a minimum phase shift of 78 degrees, at which point the current is maximum on a machine with a power-factor angle of 78 degrees (0.20 power factor). Regardless of the value of series resistance, maximum phase shift is approximately 135 degrees, and it is limited by capacitive reactance.

FIG. 31·26 Electronic heat control.

A maximum phase shift of 135 degrees gives approximately 20 percent of the maximum current, regardless of the power factor of the machine.

31·15 THE ASYNCHRONOUS SPOT-WELDING TIMER

In a welding transformer the duration of the current, as well as its magnitude, must be controlled. Figure 31·27 shows a simple type of welding timer. The block labeled $1TD$ is a tube-relay circuit which determines the time delay. When it is energized the timing mechanism is initiated. After a predetermined length of time (as determined by the electrical components in $1TD$) a relay in the circuit is energized. The relay contacts de-energize related control circuits. In the circuit shown in Fig. 31·27 relay $1CR$ is energized by closing the foot switch. The normally open contacts on $1CR$ energize the igniter circuit of the electronic contactor. When these contacts close, the electronic contactor passes welding current. After a given length of time determined by $1TD$, the relay in this circuit is energized. Its normally closed contacts, which are in series with the coil of $1CR$, open and de-energize it. De-energization of this relay opens the igniter circuit of the electronic contactor, and the welding current stops. By adjusting the time-delay circuit ($1TD$), welding time can be controlled.

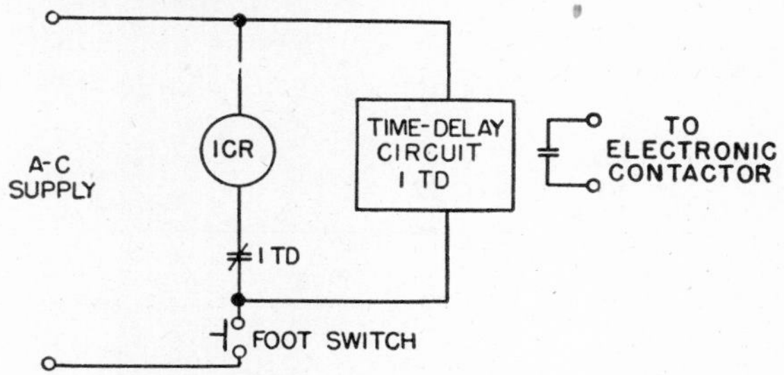

FIG. 31·27 Simple welding timer.

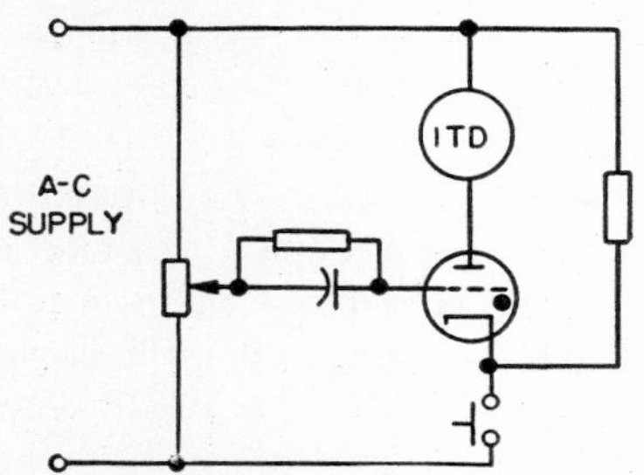

FIG. 31·28 Electronic time-delay relay.

Figure 31·28 is a diagram of the electronic time-delay relay. This circuit can be simplified (Figs. 31·29 and 31·30) to show the functions of each of the components. Figure 31·29, which is the equivalent circuit before the switch is closed, shows that the grid of the thyratron acts like an anode and conducts current in one direction only. Since the grid acts as an anode, it conducts only when it is positive

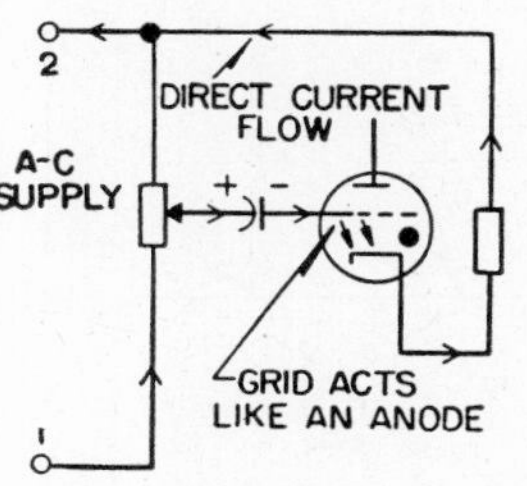

FIG. 31·29 Equivalent circuit of electronic time-delay relay shown in Fig. 31·28, switch open.

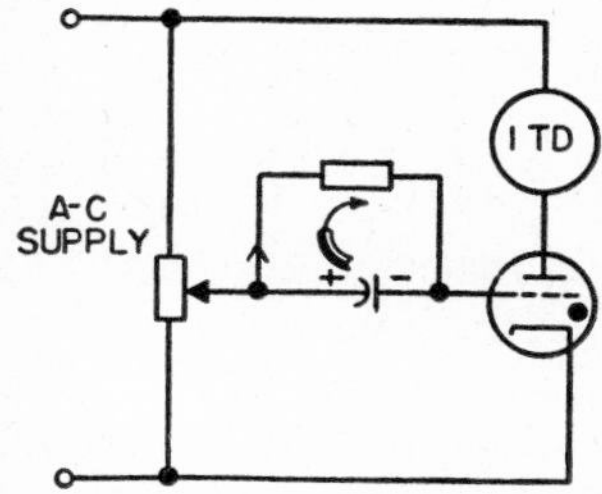

FIG. 31·30 Equivalent circuit of electronic time-delay relay shown in Fig. 31·28, switch closed.

relative to the cathode. Every time lead 1 becomes positive, current is carried through the circuit shown by the arrows. This unidirectional current charges the capacitor as shown by the oscillogram in Fig. 31·31. During the first positive half cycle after energization of the grid circuit, the capacitor receives most of its charge. During the next positive half cycle it receives a smaller portion of its charge. At the end

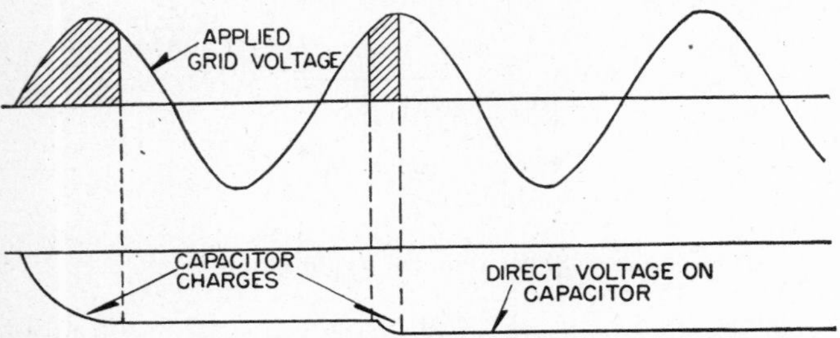

Fig. 31·31 Oscillogram of grid and grid-capacitor voltages of electronic time-delay relay, switch open.

of the second half cycle, the capacitor has been almost fully charged. Although a resistor is connected across the capacitor, its resistance is so high that leakage current during that portion of the cycle in which the grid is not charging is too small to be important. The capacitor can be charged to various voltages by varying the setting of the potentiometer. The closer the slider is to line 1, the larger is the voltage on the capacitor. The closer the point to line 2, the smaller the voltage on the capacitor. After the capacitor is charged, its negatively charged terminal is next to the grid.

After the capacitor is charged there is no further action until the foot switch is closed. Then the circuit is equivalent to that shown in Fig. 31·30. An anode voltage now is impressed on the thyratron, and it can fire when the grid becomes positive. However, because of the charge on the capacitor, the grid voltage is negative and the tube cannot fire. The length of time required for the charged capacitor to dissipate its energy through the resistor in parallel with it determines the time delay of the circuit. The sum of the unidirectional voltage and the alternating voltage, which

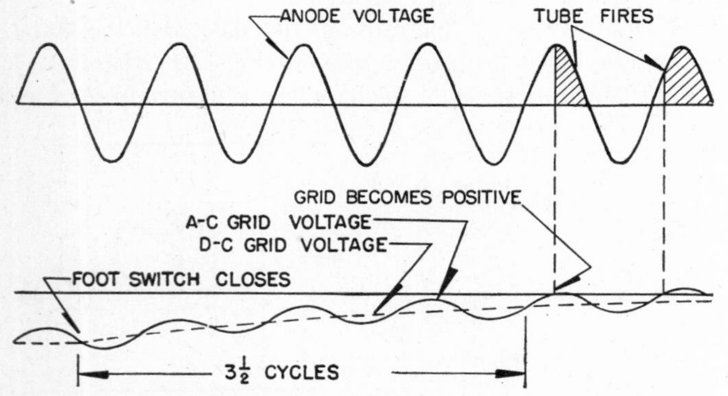

Fig. 31·32 Grid and anode voltages of the electronic time-delay relay, switch closed.

is also on the grid, gives the resultant voltage shown in Fig. 31·32. The grid is always negative within the zero-to-3½ cycle range. On the fourth cycle the grid voltage becomes positive. At this point the thyratron can fire and conduct current. A short time after it conducts, it energizes the control relay, which remains energized as long as the foot switch is closed. As soon as the foot switch is opened there is no longer any anode-cathode voltage on the tube conduction stops, and the relay is de-energized. At this time, the

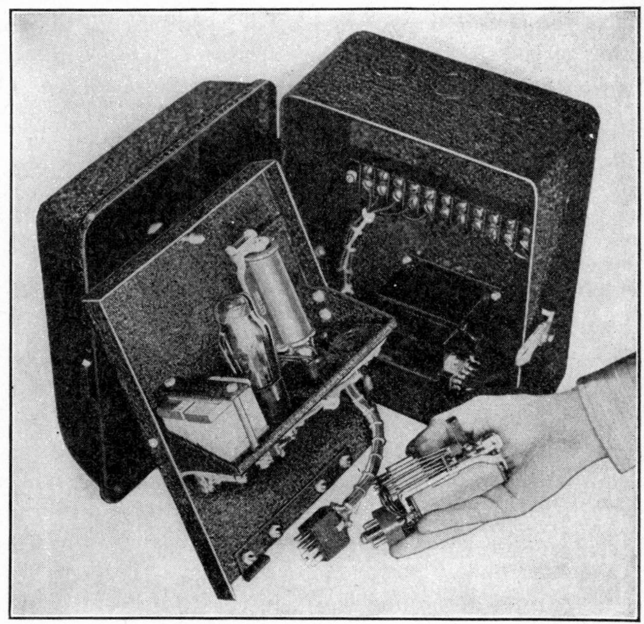

Fig. 31·33 Electronic welding timer.

capacitor again starts to charge. The time delay is the same when the foot switch is closed again.

31·16 THE ASYNCHRONOUS PULSATION SPOT-WELDING TIMER

For welding thick material, it is best to use a series of short-current pulses instead of one long-current pulse. The tube-relay timing circuit used to time the current impulses is shown in Fig. 31·34. The functions of this circuit can be divided into two parts, the "heat" or on time, and the "cool" or off time. Figure 31·35 shows the equivalent circuit during the heat time after the foot switch has been closed.

During this time an alternating voltage is impressed upon the grid of the tube. This voltage makes the grid sufficiently positive to energize relay $2TD$. This relay's contacts in the

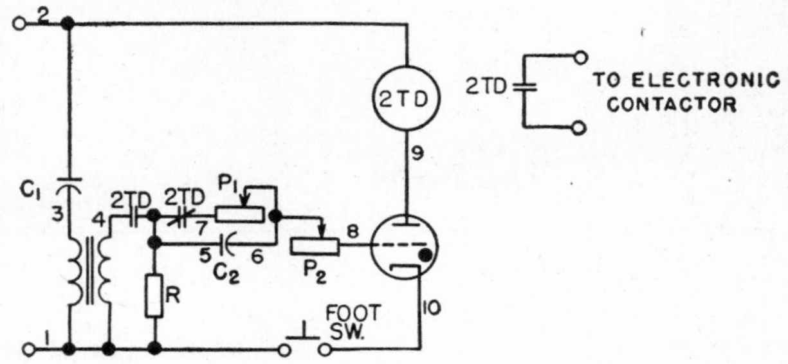

Fig. 31·34 Tube-relay pulsation timing circuit.

electronic-contactor circuit then energize the igniter circuit, and welding current is passed. A second action takes place when the tube fires. Since there is an alternating voltage on the grid of this tube, capacitor 2 is charged by the rectifying action of the grid of the tube. C_2 is charged by the unidirectional current. The time required to charge C_2 to a particular voltage depends upon the amount of resistance in series with C_2. The amount of resistance is variable and can be controlled at will by adjustment of the resistance in P_2.

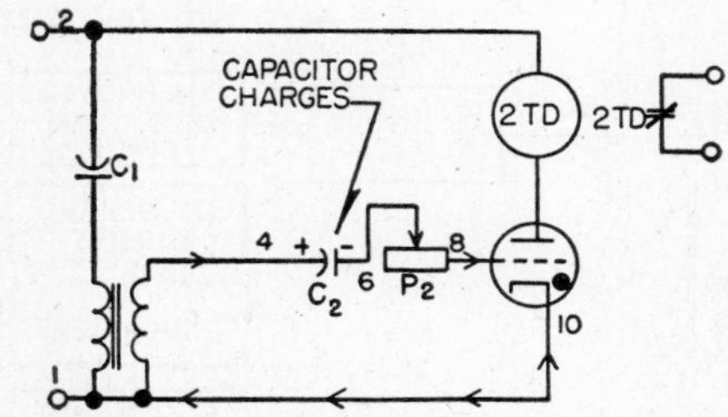

FIG. 31·35 Equivalent circuit tube-relay pulsation timer during "heat" time.

As the charging of this capacitor proceeds the alternating voltage on the grid drops lower and lower through the critical grid voltage of the thyratron. The tube then fires later and later in the positive half cycle (Fig. 31·37). The tube soon passes so little current that it is unable to keep $2TD$ energized, and the relay drops out. The alternating voltage on the grid of the tube not only lags the anode voltage but is also flat-topped; these characteristics are necessary to provide positive relay drop-out.

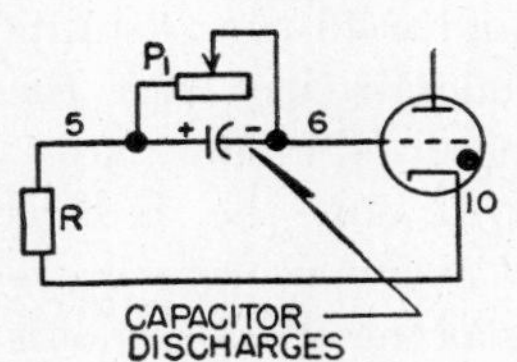

FIG. 31·36 Equivalent circuit tube, relay pulsation timer, during "cool" time.

After relay $2TD$ has dropped out and welding current has stopped, the equivalent circuit is as shown in Fig. 31·36. Contacts on $2TD$ between points 4 and 5 disconnect the alternating voltage from the grid circuit, leaving the capacitor voltage as the only voltage in the circuit. Since this grid voltage is negative. the tube cannot fire. The length of time required for this capacitor charge to leak off through potentiometer 1 depends upon the relative values of P_1 and C_2. The smaller the resistance of P_1 is, the shorter the time; the greater the resistance, the longer the time. After C_2 has dissipated most of its energy, the grid voltage is sufficiently close to zero to enable the tube to fire again. Relay $2TD$ then energizes and there is another heating period. This process of alternate heating and cooling periods continues until the circuit is de-energized by other timing devices.

31·17 TIMING THE SEQUENCE OF OPERATION

Among the other functions in a welding operation that must be timed is "squeeze" time, the period between the closing of the foot switch and the start of the welding current. Squeeze time is necessary so that the electrodes of the welding machine can be lowered and build up pressure on the work. Another function, "hold" time, is the period between the end of the weld current and the time at which the electrodes are raised. It must be timed to allow the weld to cool under pressure. A third function, "off" time, is the period between the raising and lowering of the electrodes. In this period, the operator moves his work to a new position. These functions are illustrated in Fig. 31·38.

31·18 SEQUENCE-WELD TIMER FOR SPOT WELDING

The NEMA type 3-B timer controls the squeeze, weld, hold, and off functions. The timing circuits are the same as those described in Section 31·15. Figures 31·38 and 31·39 show the operation. When the foot switch is closed, contactor $1CR$ is energized. This contactor has normally open

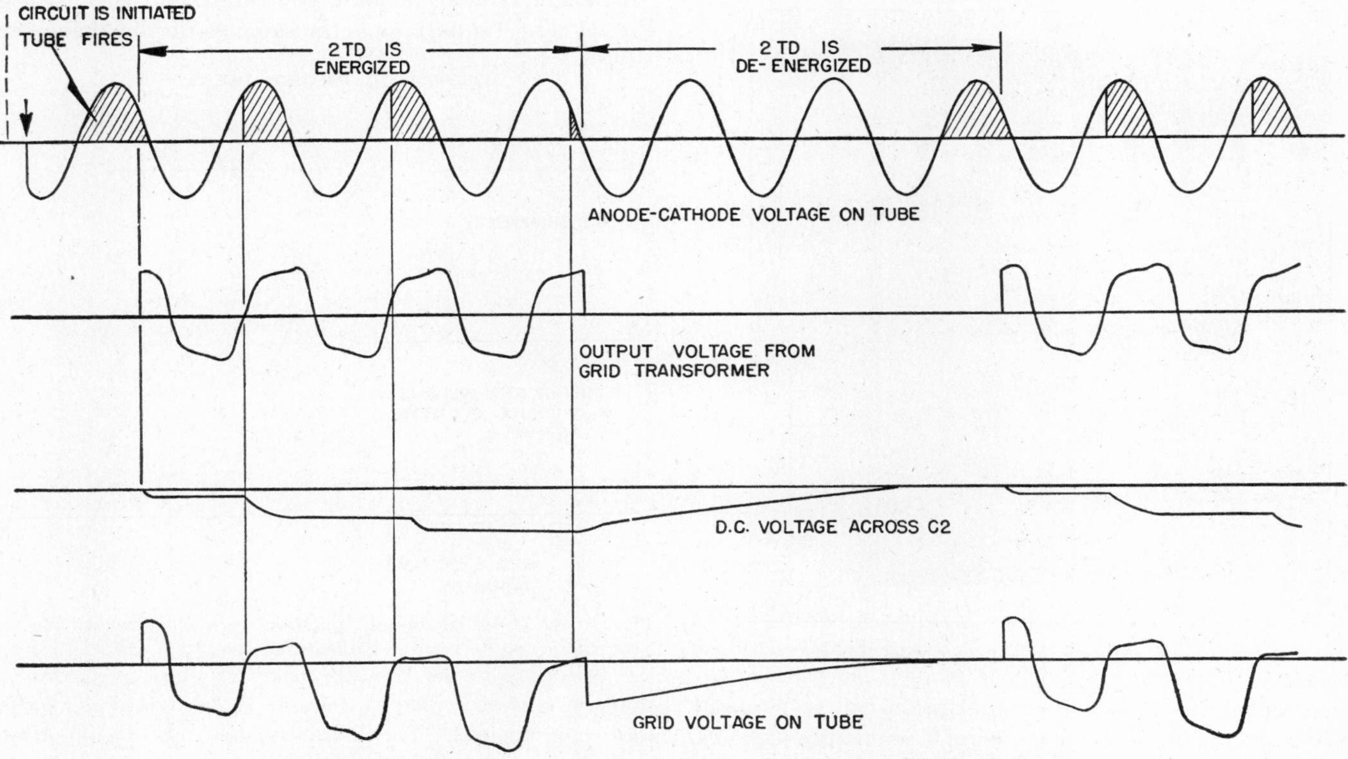

FIG. 31·37 Oscillogram of anode and grid voltages of the circuit shown in Fig. 31·34.

contacts which are used to energize a solenoid valve in the welding machine. Energization of this valve admits air into the upper chamber of a double-acting cylinder of the welding

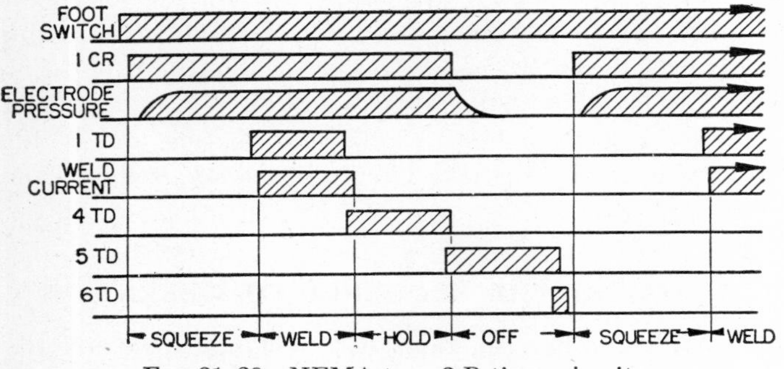

Fig. 31·38 NEMA type 3-B timer circuit.

machine. This lowers the electrode against the work piece and builds up pressure on it. At the time the foot switch energizes 1*CR*, the squeeze-time (1*TD*) circuit is energized.

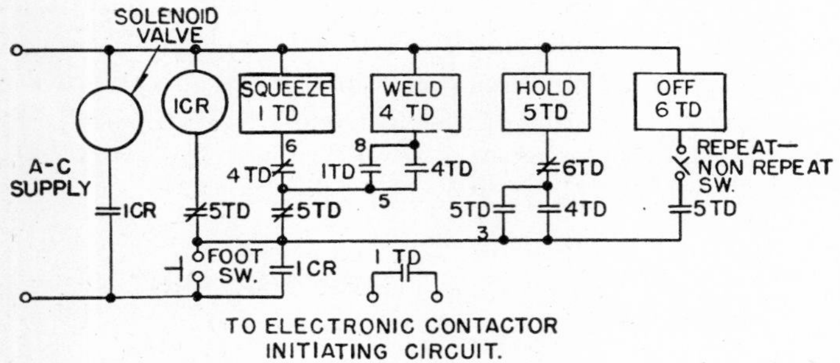

Fig. 31·39 Sequence of operation, NEMA type 3-B timer.

After this period has been timed, the relay in the squeeze-time delay circuit is energized. Its contacts between points 5 and 8 energize the weld-time circuit (4*TD*). After this period

Fig. 31·40 Automatic sequence-weld timer.

has been timed, the relay in the weld-time circuit is energized. Its back contacts between points 5 and 6 de-energize the 1*TD* circuit. De-energization of 1*TD* stops the welding current. As the 4*TD* relay operates it energizes the hold (5*TD*) circuit. After this time delay is over, relay 5*TD* is energized. Energization of this relay de-energizes 1*CR*, which de-energizes

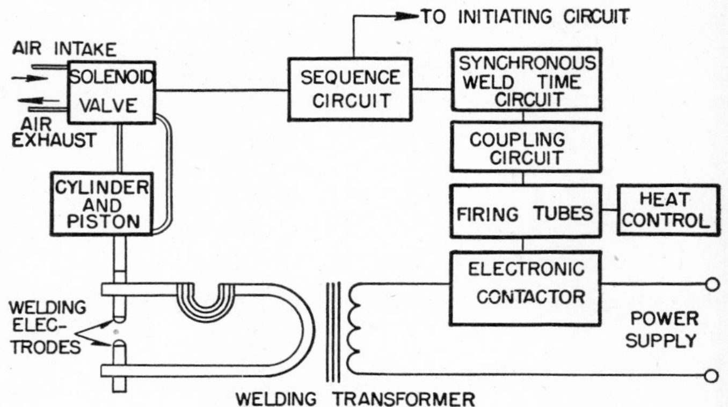

Fig. 31·41 Synchronous welding-control system.

the solenoid valve and raises the head of the welding machine. At the same time the squeeze and weld time circuits are de-energized and circuit 6*TD* is energized. After the 6*TD* time delay, the off-time relay is energized momentarily; thereby the hold-time circuit is tripped out, the 1*CR* is allowed to be energized, and thus the entire process is repeated. This cycle is continued as long as the operator holds the foot switch down. If the operator releases the foot switch while the timer is in the middle of its sequence, the timer nevertheless is through its sequence until the end of 5*TD* because of the lock-in provided by the 1*CR* contacts across the foot switch.

31·19 SYNCHRONOUS-PRECISION WELDING CONTROL

The synchronous welding control system is illustrated in Fig. 31·41. Its purpose is the same as that of the asynchro-

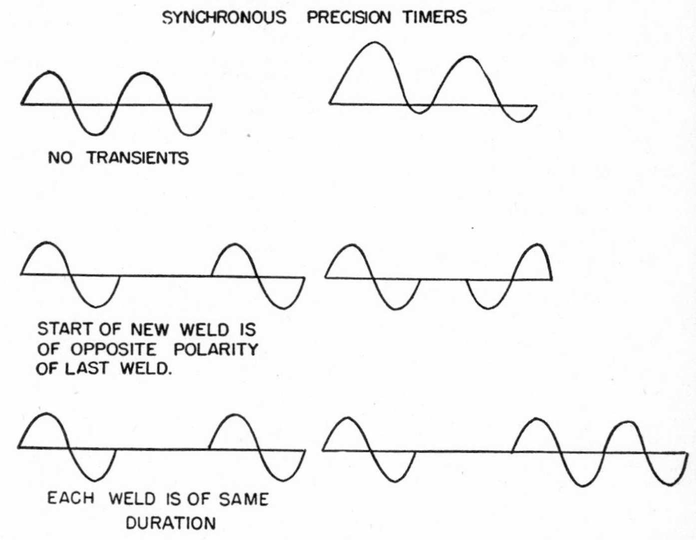

Fig. 31·42 Comparison of synchronous and asynchronous control features.

nous system described in Section 31·12, but it is necessarily more complicated. The distinctive features of a synchronous-precision control (see Fig. 31·42) in contrast to an asynchronous control are: (1) the welding contains no transients; (2) the start of a weld is of opposite polarity to the preceding weld to prevent the welding transformer from saturating as it would if two half cycles of current of the same polarity were passed; and (3) the weld time is precise.

31·20 ELECTRONIC CONTACTOR; FIRING MEANS FOR SYNCHRONOUS CONTROL

The electronic contactor and firing-tube circuit for synchronous control systems are shown in Fig. 31·43. The electronic contactor passes welding current as in the asynchronous system. The firing tubes control welding current and

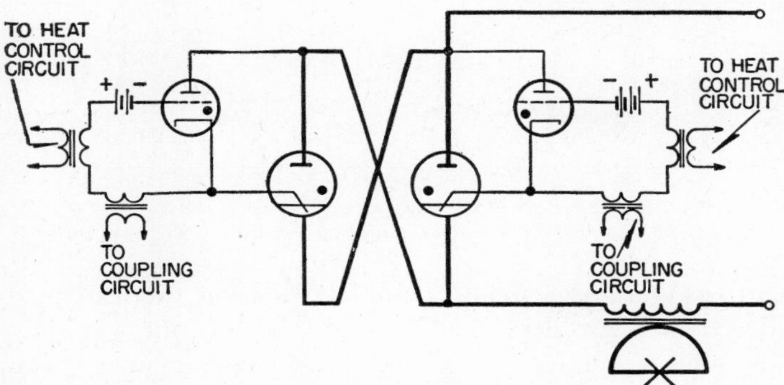

FIG. 31·43 Electronic contactor circuit for synchronous control systems.

turn the ignitrons on and off. The grid of each firing tube is controlled by three voltages: (1) a bias supply to keep the tubes normally de-energized; (2) a heat-control voltage to determine the instant in the one half cycle at which the tubes fire; and (3) a voltage from the coupling circuit to turn the firing tubes on and off. Figure 31·44 illustrates the various

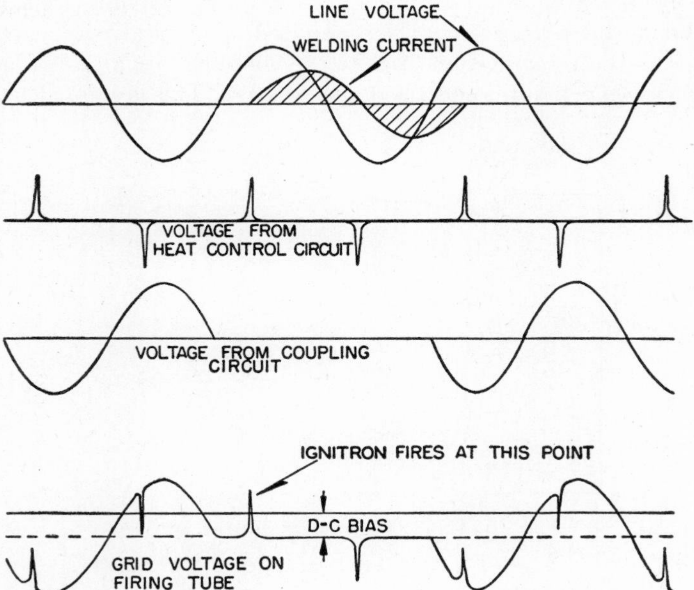

FIG. 31·44 Voltages in the electronic contactor circuit of Fig. 31·43.

voltages. The peaked voltage controls the welding current because it can be shifted in phase position. If it is phased forward, current is high; if it is phased back, current is low.

Voltage from the coupling circuit is an alternating voltage which appears when the electronic contactor is to be de-energized. When the contactor is to be energized, the coupling-circuit voltage is removed. Normally the firing tubes do not conduct because the sum of the alternating and unidirectional voltages prevent the grid from becoming positive when the anode is positive. If the coupling-circuit voltage is removed, the grid becomes positive, because the peaking voltage is greater than the bias, thus causing the thyratron to fire at a point corresponding to the phase position of the peak. This in turn fires the ignitrons, which pass welding current.

31·21 HEAT CONTROL FOR THE SYNCHRONOUS SYSTEM

The heat-control circuit (Fig. 31·45) is similar to the heat control described in Section 31·14. The phase-shifting cir-

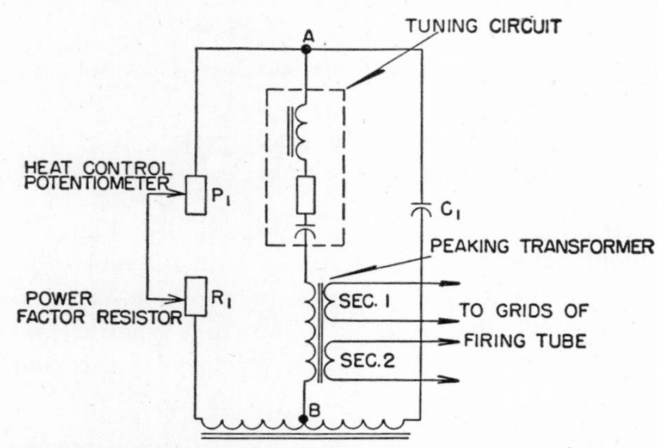

FIG. 31·45 Heat-control circuit for the synchronous control systems.

cuit is the same, except for the peaking transformer and the tuning circuit. The peaking transformer is of a special type, the secondary of which produces a peak when the primary alternating voltage is approximately at its maximum. The desired phase shift is from 30 to 135 degrees with respect to the line voltage. The phase-shifting circuit consisting of R_1, P_1, C_1 gives this amount of shift. However, the peaking transformer delivers peaks of voltage; therefore, when combined with the phase-shift circuit, the peaks actually shift

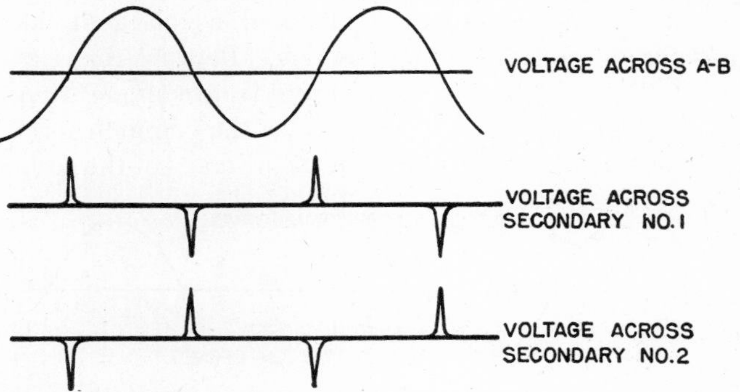

FIG. 31·46 Voltages in the heat-control circuit of Fig. 31·45.

from 120 to 225 degrees with respect to the line voltage. The tuning circuit in series with the peaking transformer phases the peaks so that they correspond with the desired range from 30 to 135 degrees. Figure 31·47 illustrates the shape of a welding current controlled by the heat-control peaks.

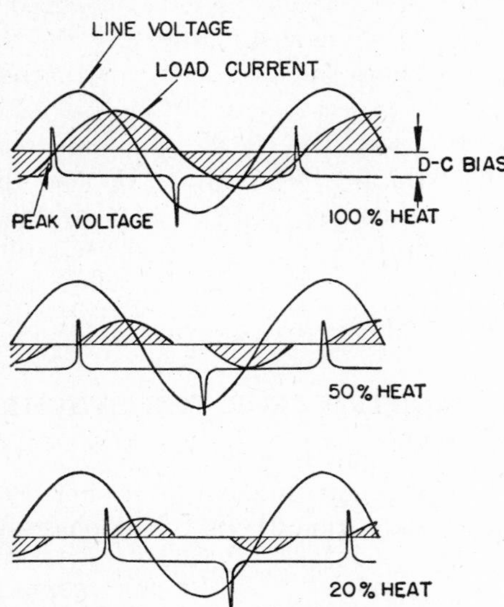

Fig. 31·47 Welding current controlled by heat-control peaks.

31·22 COUPLING CIRCUIT FOR THE SYNCHRONOUS SYSTEM

The coupling circuit is shown in Fig. 31·48. Figure 31·49 is the equivalent circuit with tubes 1 and 2 replaced by a switch, since two tubes in inverse parallel act like a single-pole a-c switch. Figure 31·49 shows that when the switch is open there is a voltage on the primary of the coupling

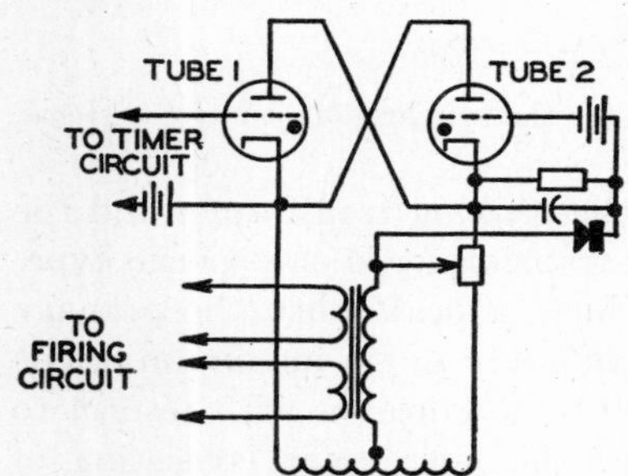

Fig. 31·48 Coupling circuit for synchronous control systems.

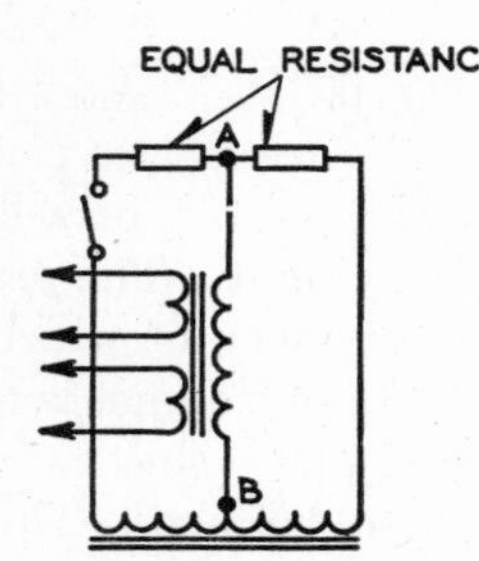

Fig. 31·49 Equivalent circuit for the coupling circuit of Fig. 31·48.

transformer. When the switch is closed, a voltage divider is placed across the transformer winding. Because the resistors in the voltage divider are equal, there is no voltage from the center tap of the power transformer to the common terminal of the resistors, and the voltage on the primary of the coupling

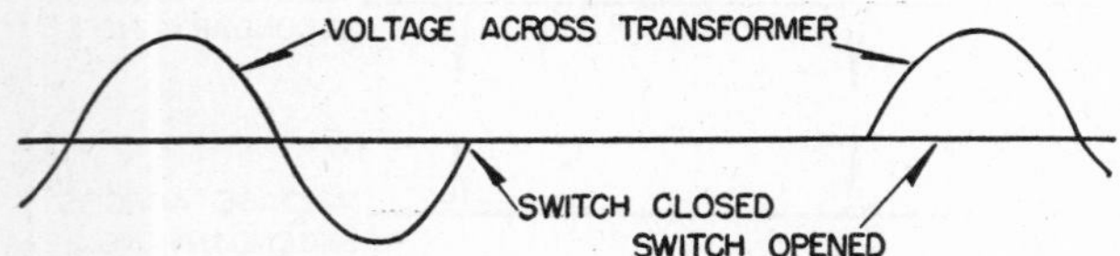

Fig. 31·50 Transformer voltage of coupling circuit of Fig. 31·48.

transformer is removed. The output voltage is the voltage on the grids of the firing tubes, as described in Section 31·20. When tube 1 in Fig. 31·48 (called the lead tube) is made conductive by having a positive voltage impressed on its grid, it fires; Fig. 31·51 (*a*). A voltage then is developed across the load resistor which is large enough to overcome the negative unidirectional bias on tube 2. Making the grid of tube 2 positive causes it to fire during the next half cycle.

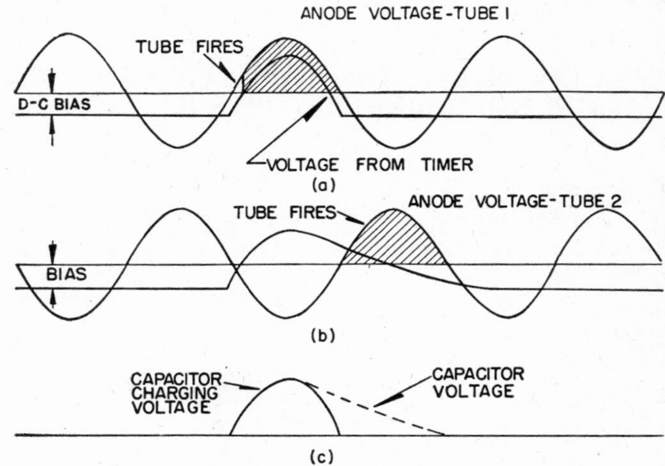

Fig. 31·51 Voltages in the coupling circuit of Fig. 31·48.

This is known as a follow action. Whenever the lead tube fires, tube 2 also fires or follows tube 1. In this way only full cycles of voltage are passed by the coupling current, and in turn only full cycles of welding current can pass. This is one of the requirements of a synchronous-precision weld timer.

31·23 SYNCHRONOUS-PRECISION SPOT TIMER

The synchronous-precision spot timer is shown in Fig. 31·52. The simplest equivalent circuit is shown in Fig. 31·53, in which each tube has been replaced by a switch. Before the beginning of the weld neither tube is energized, and no voltage appears across the resistor connected to the coupling circuit. When the timing circuit is energized, no voltage appears across the resistor connected to the coupling circuit. When the timing circuit is energized, tube 1 fires. This is equivalent

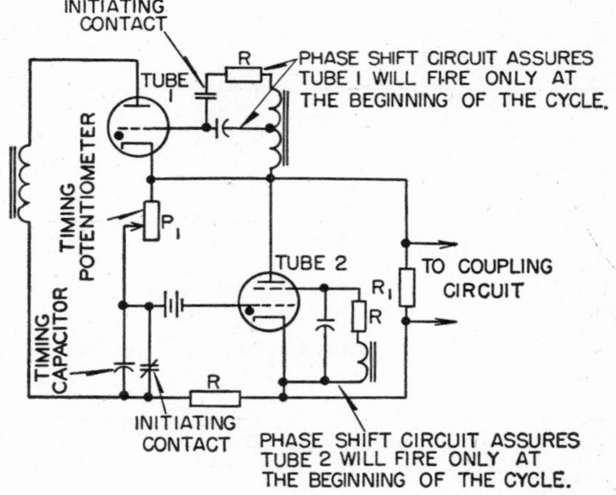

Fig. 31·52 Synchronous-precision spot-timer circuit.

to closing switch 1, and it causes voltage to appear across the resistor connected to the coupling circuit. This voltage causes the lead tube of the coupling circuit to fire, which in turn energizes the firing circuit. As tube 2 is fired, it short-circuits the resistor in the coupling-circuit grid, and the voltage disappears. The coupling circuit then ceases firing and the weld stops.

The oscillograms in Fig. 31·55 show the action of the spot-weld timer. Before the weld starts tube 1 is held non-

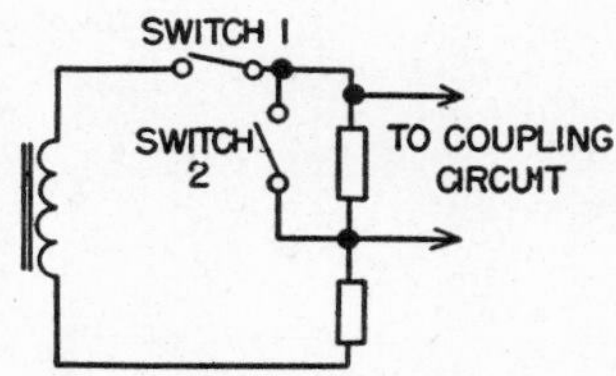

Fig. 31·53 Simplified equivalent circuit of synchronous precision spot-timer circuit of Fig. 31·52.

conductive because of a phase-shifted grid voltage. This voltage is of such phase position that it is always negative when the anode is positive, so the tube cannot fire. When the circuit is energized, the contact in the grid circuit changes the phase position of the grid voltage on tube 1, causing it

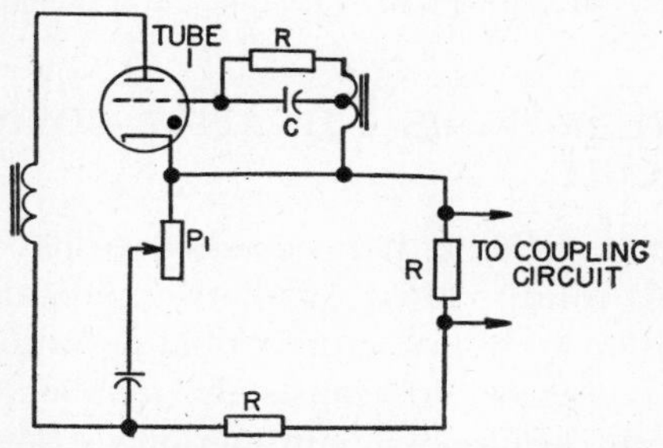

Fig. 31·54 Circuit of tube 1 of synchronous-precision spot-timer circuit of Fig. 31·52.

to fire within the first few degrees after the anode becomes positive. At this point tube 1 fires, energizes the coupling circuit, and causes another action shown in Fig. 31·54. Since tube 1 acts as a rectifier, it charges the capacitor in series with the potentiometer in its cathode circuit. The charging

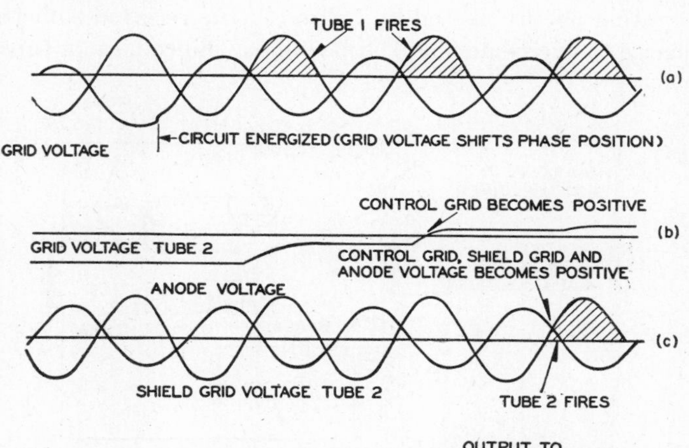

Fig. 31·55 Spot-weld-timer oscillogram.

can be seen in the oscillogram in Fig. 31·55 (*b*). Capacitor voltage overcomes the negative bias in the grid of tube 2; the shield-grid and anode voltages are positive, and tube 2 fires and stops the welding current. For tube 2 to fire, both its shield grid and its control grid must be positive when the anode is positive. The control grid is made positive by the timing capacitor. The shield-grid voltage is made positive by a phase-shifted alternating voltage, phased so that it becomes positive only during the first few degrees when the anode is positive. In this way only full half-cycle waves of voltage are impressed upon the grid of the lead coupling tubes.

The circuit is timed by varying resistance in the potentiometer in the cathode circuit of tube 1. The higher the

Fig. 31·56 Welding control unit.

resistance in series with the capacitor is, the longer the time required to charge the capacitor sufficiently to overcome the bias on the grid of tube 2. The smaller the resistor is, the shorter the time.

Proper choice of circuit components and the assurance that tubes 1 and 2 fire at a definite time in each individual half cycle make this timer a precise timer as well as a synchronous timer.

31·24 SYNCHRONOUS HEAT-COOL CIRCUIT

Figure 31·57 is a diagram of the synchronous heat-cool circuit. As tube 1 fires, it energizes the coupling circuit in the same manner as the starting tube in the spot timer. As it fires, there is a "heat" period. When it does not fire, there is a "cool" period.

When the circuit is energized, the charge on capacitor 1 overcomes bias 1 and causes tube 1 to fire. (The equivalent circuit then is as shown in Fig. 31·58.) When tube 1 fires, a voltage developed across R_1 energizes the coupling circuit. At the same time, capacitor C_2 starts to charge through the timing potentiometer. This process continues until the voltage on capacitor C_2 is sufficient to overcome bias 2. Tube

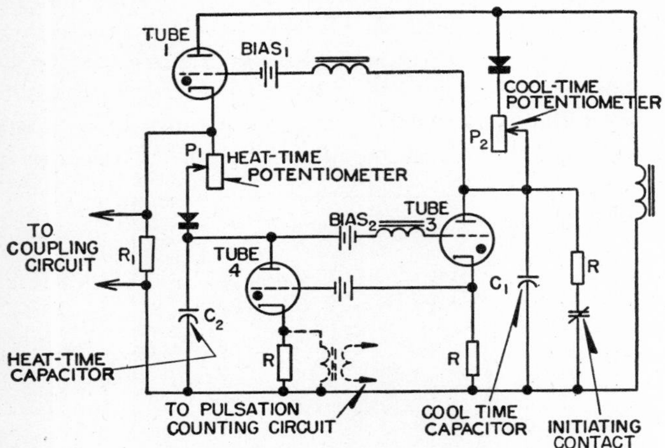

FIG. 31·57 Synchronous heat-cool circuit.

3 then fires. (The equivalent circuit is shown in Fig. 31·59.) When tube 3 fires it discharges capacitor C_1 and stops tube 1; then the weld is stopped. Voltage across R_2 then is sufficient to overcome bias 3. Tube 4 fires and discharges C_2. C_1

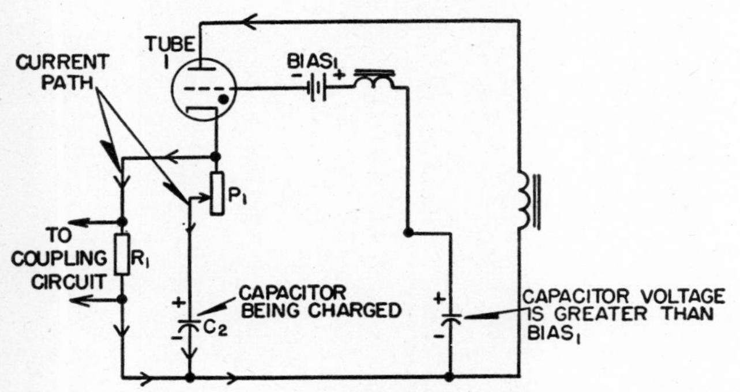

FIG. 31·58 Equivalent synchronous heat-cool circuit, tube 1 conducting.

again starts to charge through the copper oxide rectifier and potentiometer 2 (the "cool" time potentiometer). After this capacitor again has received enough charge to overcome bias 1, tube 1 fires again and energizes the coupling circuit.

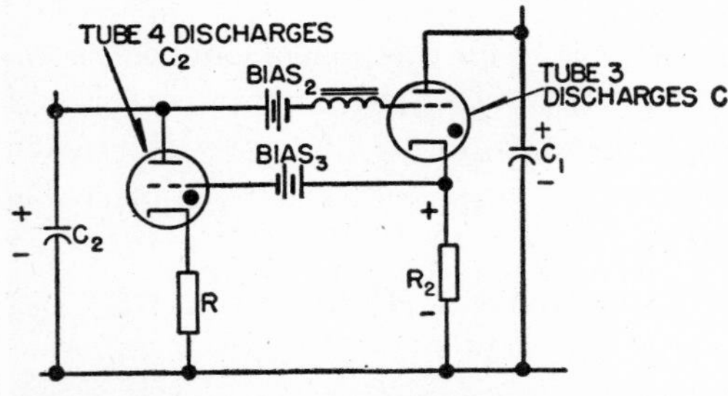

FIG. 31·59 Equivalent synchronous heat-cool circuit, tube 3 conducting.

This process is repeated until the circuit is turned off electronically, or until the initiating contactor is closed again.

To make the on and the off time synchronous, peaking transformers are adjusted so that they energize the tube at the zero voltage. Oscillograms of the operation are shown in Fig. 31·60.

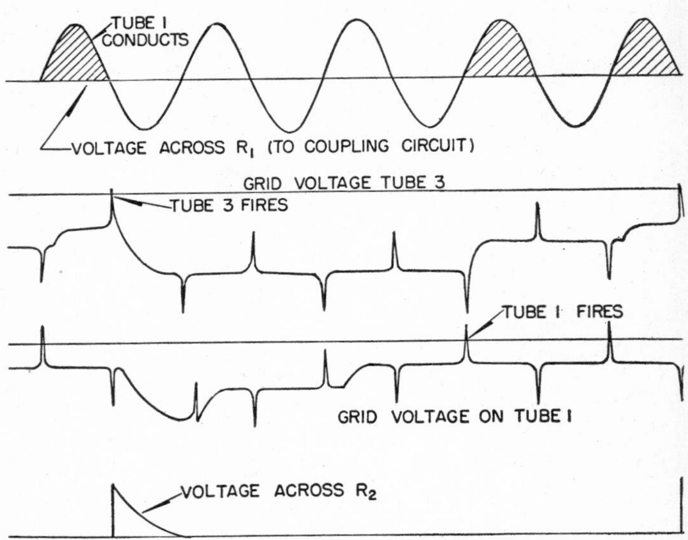

FIG. 31·60 Synchronous heat-cool-unit oscillogram.

31·25 SYNCHRONOUS PULSATION-COUNTING CIRCUIT

For pulsation welding, it is necessary to have a circuit to determine the total weld time either by measuring the total time interval or by counting individual pulsations. The circuit shown in Fig. 31·61 counts the number of pulsations produced in the heat-cool circuit and opens the circuit automatically after the last heat period. It takes its energization from the transformer shown by the dotted lines of Fig. 31·57. The outout of this transformer (shown by the oscillogram of Fig. 31·62) is an impulse which appears on the grid of tube 5 of Fig. 31·61. It is of sufficient magnitude to overcome the bias on tube 5 and cause this tube to fire each half cycle as the grid becomes positive. Making the grid positive allows tube 5 to conduct, which charges C_3 through the pulsation-counting potentiometer P_3. After C_3 has received sufficient charge to overcome bias 4, tube 6 fires. Since bias on tube 6

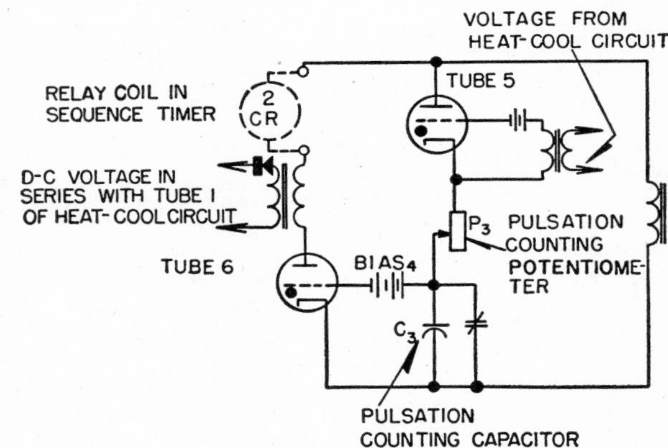

FIG. 31·61 Synchronous pulsation counting circuit.

remains positive, this tube operates every positive half cycle after that. The load of tube 6 is a transformer having a copper oxide rectifier in its secondary. Therefore, whenever tube 6 conducts, it produces a negative unidirectional bias which is in series with the grid of tube 1. The negative bias stops tube 1, which stops the weld.

P_3 can be adjusted to count from 1 to 15 pulsations regardless of the length of the heat or cool time. Oscillograms of the voltages and currents in the pulsation-counting circuit are shown in Fig. 31·62.

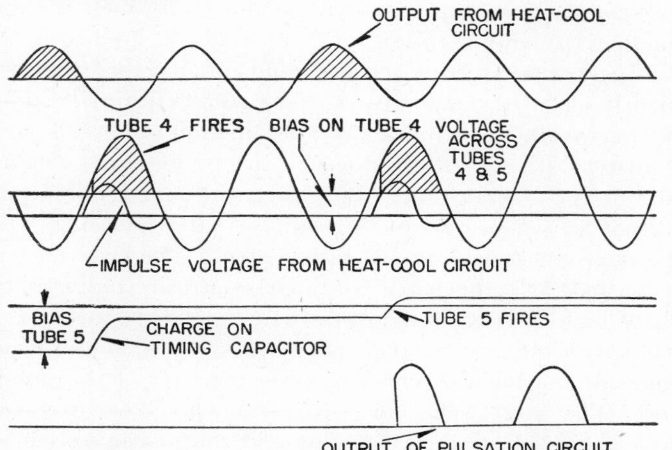

Fig. 31·62 Synchronous pulsation-counter oscillogram.

31·26 SEQUENCE TIMER FOR PULSATION SPOT TIMER

The pulsation spot timer can be used with sequence timer similar to that described in Section 31·18. The sequence timer co-ordinates electrode motion of the welding machine with the pulsation timer. When the operator wishes to make a weld, he presses the foot switch to lower the head of the welding machine. After a time interval (squeeze time) predetermined by the sequence timer, the pulsation spot timer operates. The pulsation timer determines the duration of both the heat and the cool periods in addition to counting out the total number of pulsations. After the total number of pulsations has been counted, operation again is transferred

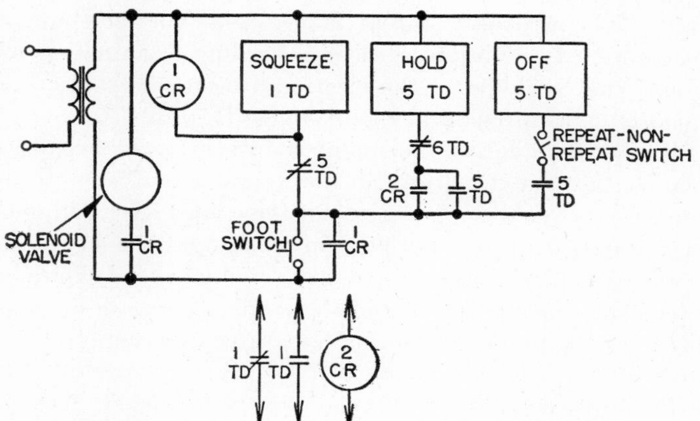

Fig. 31·63 Sequence timer circuit for pulsation spot timer.

to the sequence timer, which times the "hold" interval, the interval during which the weld is allowed to cool. After this timing function is over, the head of the welding machine again is raised, and the operator is allowed to move his work before the entire sequence is repeated. Operation of the sequence timer is illustrated in Fig. 31·64. The time-delay circuits shown as $1TD$, $5TD$, and $6TD$ are the same as those described in Section 31·15. When the operator's foot switch is closed, relay $1CR$ is energized, and in turn the solenoid valve is energized. This lowers the head of the welding machine. After a time determined by the $1TD$ circuit, relay $1TD$ is energized. Its external contacts energize the pulsation spot timer. The heat-cool circuit of the pulsation spot timer goes through its sequence and counts the total number of pulsations. After the pulsation-counting tube has operated, it energizes the hold ($5TD$) circuit. After a period predetermined by the $5TD$ potentiometer, the relay on $5TD$ de-energizes $1CR$ and the solenoid valve, and also energizes the "off" ($6TD$) circuit. The head of the welding machine rises while the operator moves his work. After $6TD$ has

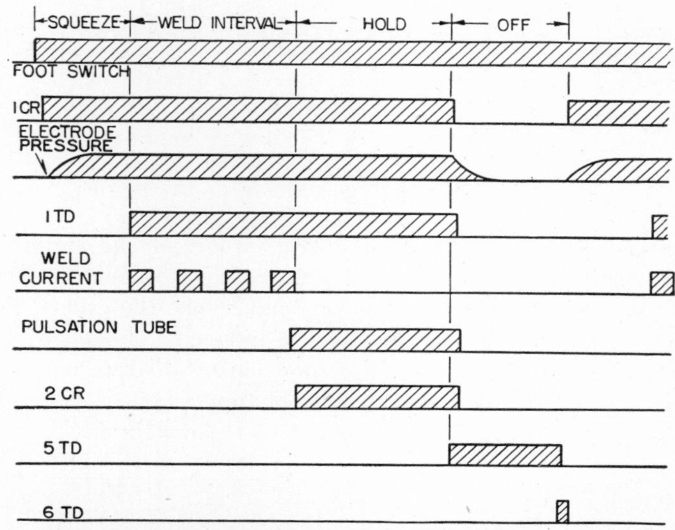

Fig. 31·64 Sequence of operations for pulsation spot timer.

completed its timing function it releases the hold ($5TD$) circuit, and the entire sequence is repeated. Operation of the relays and of the head of the welding machine is shown in Fig. 31·70.

31·27 ELECTRONIC SEAM TIMER

In making a seam weld, it is generally necessary to have interrupted spot welds, that is, alternate heat and cool periods. On a seam welder it is necessary only to lower the head of the welding machine on the work and to turn the welding current on and off accurately. The welding currents are of such high magnitude and the heat and cool periods are so short that a precision timer must be used. The control can be either totally electronic or semi-electronic. The electronic seam timer is the same as that used in the heat-cool circuit in the pulsation spot timer described in Section 31·25.

31·28 SEMI-ELECTRONIC SEAM TIMER

The semi-electronic seam timer is similar to the electronic seam timer. This timer not only uses the firing tube and coupling control circuit, but also has an electromechanical device as the source of the timing voltage. This device consists of a synchronous motor driving a disk with accurately spaced holes near the periphery. Pins placed in these holes pass through an air gap in the permanent magnet as the disk rotates. These pins change the reluctance of the magnetic circuit, and consequently the magnetic flux in the gap of the magnet is changed. This change of flux induces a voltage in a pick-up coil wound around the permanent magnet. The voltage is connected into the grid circuit of the coupling control lead tube. An oscillogram of it is shown in Fig. 31·65. At the beginning of every half cycle, the grid of the lead tube is made positive by the induced voltage on the pick-up coil. For every desired cycle of welding current, one pin is inserted. For every cool time, a pin is left out. Figure 31·66 shows the details of this type of timer.

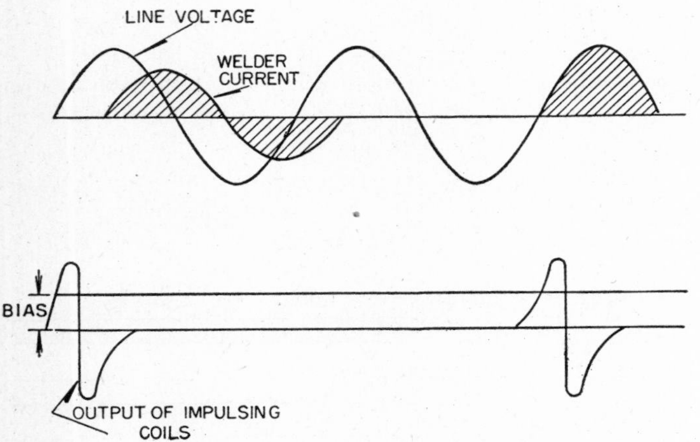

Fig. 31·65 Semi-electronic seam-timer oscillogram.

Fig. 31·66 Rotating disk and magnetic pick-up of semi-electronic seam timer.

31·29 ENERGY-STORAGE SYSTEMS; WELDING ALUMINUM

Energy-storage welding is resistance welding in which a device is used to store at low rate the energy to produce the welding current. Either mechanical or electrical devices may be used for energy storage. Mechanical devices, such as a flywheel on a motor generator, have not been widely used. Energy-storage welding in general is based on the use of a capacitor, a reactor, or a battery as the device in which energy is stored. The capacitor and reactor systems will be considered in detail.

Energy-storage systems have been developed in answer to problems in welding aluminum and magnesium. These two metals have high electrical conductivity, high thermal conductivity, rapid transition from solids to liquids as the melting temperature is reached, and affinity for copper at slightly elevated temperatures which results in rapid fouling of the welding electrodes. To obtain uniform welds with consistent strength it is necessary to provide large welding current for a short time. In general the magnitude of the current should be two to three times that required to weld the same thicknesses of low-carbon steel. The large current is needed to compensate for high electrical conductivity. These high currents would cause an excessive kva demand on the power supply with single-phase a-c systems, and in many applications would exceed the power capacity of the supply line. The short weld time is necessary to avoid excessive deformation of material near the weld zone. It also reduces the rate at which these metals alloy with the copper-alloy electrodes, because the maximum temperature at the welding electrodes can be reduced.

Energy-storage systems provide a means to obtain a large welding current with only a moderate demand on the power supply. Power-system loading can be further reduced by the use of multiphase rectifiers to provide d-c power. The weld zone can be heated quickly because the discharge current consists of a large initial peak which produces fusion. Timing circuits for the weld time are not required because the wave shape and magnitude of the welding current depend upon the amount of stored energy and the values of inductance and resistance in the welding transformer and welding machine. Magnitude of the welding current can be controlled closely, and can be made consistent in quality.

The higher weld currents require a corresponding increase in electrode pressure; therefore construction of the main frame and secondary throat is heavier than on standard machines. In addition, successful welding of aluminum requires that skidding of the electrodes be minimized; consequently extra mechanical rigidity must be provided.

Rapid softening of the metal results in rapid expansion and contraction of material in the weld zone. Therefore the pressure system to supply electrode force must have minimum inertia to provide quick follow-up pressure at the end of the weld. A typical solution is to use a spring or air bellows between the pressure piston and the welding electrode to cause the electrode to follow deformation of the material during the weld.

Pressure switches of a special type that can be operated from the bellows, spring, or other indicating devices are used to initiate the weld only after correct initial electrode force has been obtained. If this phase of the technique is neglected and follow-up pressure is not applied, cracks and porosity in the completed weld zone may lead to early failure of the weld through fatigue. Circuits and valves have been devised to apply electrode pressure much larger than welding pressure near the end of the weld. This is known as forging or double pressure.

From a metallurgical standpoint it is feasible to weld stainless steels, low-carbon steels, and the more weldable brasses in the thinner gauges with either the electromagnetic or the electrostatic system.

31·30 THE ELEMENTARY CAPACITOR-DISCHARGE CIRCUIT

The basic principle of the capacitor-discharge system can be best understood by considering the simplified circuit shown in Fig. 31·67.

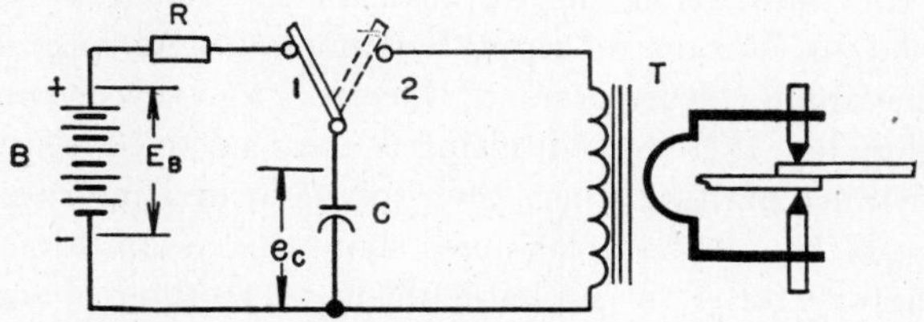

FIG. 31·67 Simplified capacitor-discharge circuit.

There are two fundamental steps in a complete weld. First, electrical energy is stored in capacitor C. Therefore, when the switch is closed to position 1, the capacitor (initially uncharged) is connected to the source of d-c power, which in this case is battery B. Current through resistor R and capacitor C decreases exponentially as energy is stored in the capacitor. Capacitor voltage e_c increases exponentially until it is equal to E_B, at which point maximum charge has been stored.

In step two, the weld is made when the switch is closed to position 2. The charged capacitor is then discharged into the primary winding of welding transformer T. The resultant discharge currents in both the primary and secondary circuits of the transformer depend upon the capacitance, the charge voltage of the capacitor bank, and the electrical characteristics of both the transformer and the secondary throat. The equivalent circuit is shown in Fig. 31·68. As in a-c resistance-

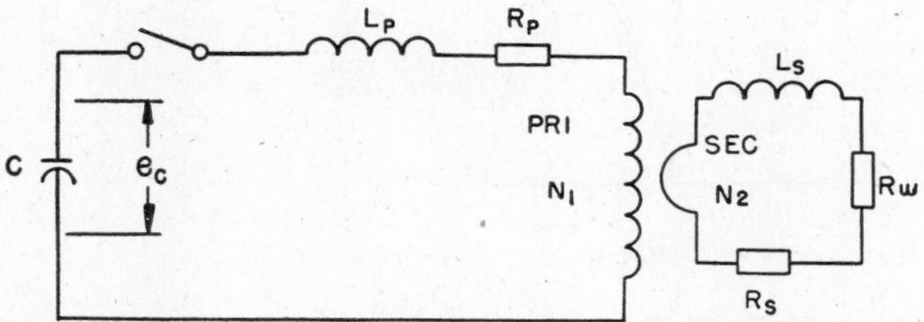

FIG. 31·68 Equivalent capacitor-discharge circuit.

welding machines, both inductance L_S and resistance R_S are present in the heavy secondary circuit which carries current to the electrodes from the transformer secondary. Resistance R_w in the work material is insignificant for aluminum but is larger than R_s for steel of any kind. These parameters are then reflected to the primary circuit, as the square of the transformer turns ratio, and along with capacitance C, determine whether the discharge current is oscillatory or non-oscillatory. In combination with the initial charge voltage E_C they also determine the magnitude of the currents in both circuits. Primary leakage reactance L_P of the welding transformer and primary resistance R_P usually are small, and have little influence on the discharge current. In general the circuit is designed to produce a damped oscillatory current upon discharge of the capacitor. With the weld completed, the cycle then can be repeated.

Figure 31·69 shows a block diagram of a complete capacitor-discharge system. The system consists of:

1. A capacitor bank in which the welding energy is stored.
2. A rectifier for converting a-c to d-c power for charging the capacitor.
3. A charge-control circuit for controlling the rectifier so that the capacitor voltage can be accurately regulated, and a "blocking" circuit which also works through the charge-control circuit to prevent the rectifier from conducting while the capacitor is being discharged.

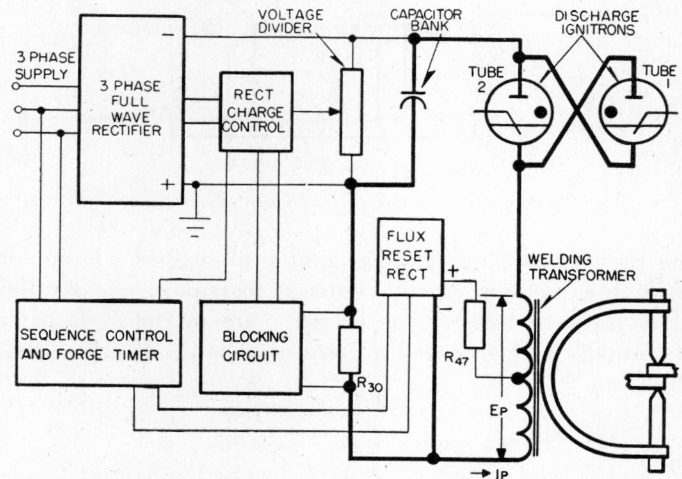

FIG. 31·69 Block diagram, capacitor-discharge system.

4. A discharge system to convert energy stored in the capacitor into welding current.
5. A flux-reset circuit to restore the flux of the welding transformer to the same value after each weld.
6. A sequencing system to co-ordinate the operation of control circuits with the mechanical operation of the welding machine.

31·31 RECTIFIER

The purpose of the rectifier is to provide d-c power to charge the capacitor bank in which the welding energy is stored.

Current in a capacitor being charged from a d-c source depends upon the applied voltage, the impedance in series with the capacitor, and the counter electromotive force of the capacitor. If capacitor voltage is zero, initial current is limited only by the applied voltage and the series impedance. If applied voltage is high and series impedance is low, initial current is high. Since the rectifier is the d-c source, a large initial current would result in a large kva demand from the power system. Therefore, the capacitor current at the start of charge must be limited if the demand on the power system is to be reasonable.

Capacitor current can be limited by resistance in series with the capacitor or by voltage regulation in the rectifier (for example, by building leakage reactance into the rectifier transformer). In most commercial systems the large power losses in a series resistor prohibit its use. Hence, current usually is limited by means of leakage reactance.

Under ideal conditions, capacitor current at the start of the charge is equal to the steady-state short-circuit current of the rectifier, in both average and peak values. However, when the rectifier circuit contains reactance, the rectifier tubes must be fired at the proper point in the cycle or an abnormally high transient current may be produced.

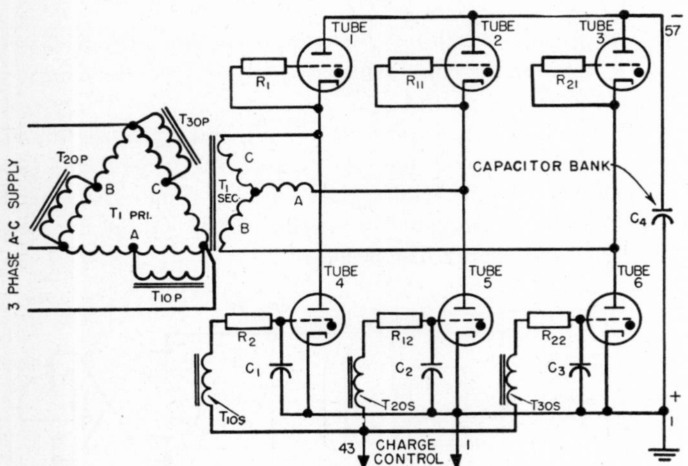

FIG. 31·70 Rectifier circuit, capacitor-discharge system.

The steady-state short-circuit current for which the rectifier is designed is determined by:

1. Maximum capacitance to be charged.
2. Maximum voltage to which the capacitors are to be charged.
3. No-load output voltage of the rectifier.
4. Time available for charging maximum capacitance to maximum voltage.

The relation between these factors is given by the equation

$$e_c = E(1 - \epsilon^{-t/RC}) \qquad (31\cdot2)$$

where e_c = capacitor voltage
E = d-c applied voltage
t = time in seconds
$R = E/I_{sc}$
I_{sc} = rectifier short-circuit current.

Rearranging and solving for the short-circuit current,

$$I_{sc} = \frac{EC}{t} \log_\epsilon \frac{1}{1 - \dfrac{e_c}{E}} \qquad (31\cdot3)$$

In this derivation the applied unidirectional voltage is assumed to contain no ripple component. As long as the rectifier ripple voltage is small, the error from this assumption is negligible. Reactance limiting is assumed to be equivalent to resistance limiting where the equivalent limiting resistance is $R = E/I_{sc}$. Laboratory tests show this assumption to be satisfactory for a three-phase full-wave rectifier charging a capacitor bank to one half (or less) of the no-load unidirectional voltage of the rectifier.

A three-phase full-wave rectifier, with a no-load unidirectional output voltage of approximately 6000, is used to charge the capacitor bank. The rectifier must charge 2640 microfarads to 2800 volts in ⅚ second. The calculated short-circuit current from equation 31·3 is 13 amperes, which agrees approximately with the values measured at the rectifier.

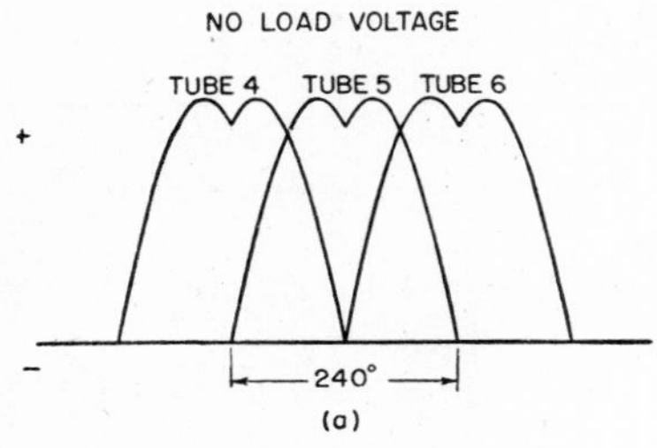

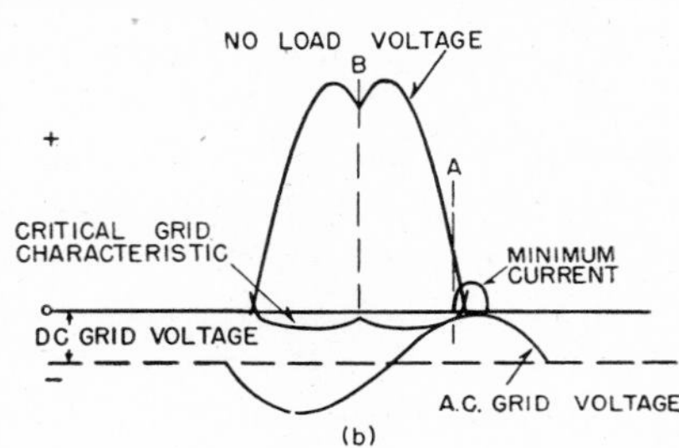

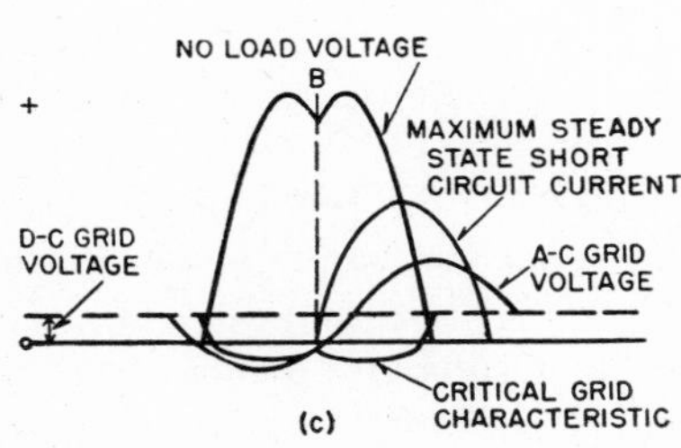

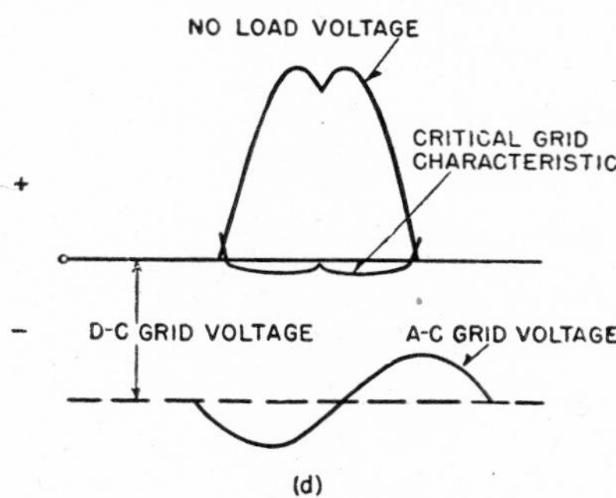

FIG. 31·71 Oscillograms showing the effect of changing grid bias on tubes 4, 5, and 6 of the rectifier in Fig. 31·70.

The diagram of the rectifier is given in Fig. 31·70. Tubes 4, 5, and 6 are grid-controlled thyratrons; tubes 1, 2, and 3 are used as phanotrons. Load current can be controlled by this circuit, because all of it must pass through the controlled tubes. Grid voltage on the controlled tubes consists of a fixed alternating voltage and a variable unidirectional voltage. The fixed alternating voltage is supplied by the secondaries of transformers with their primaries connected to the low-voltage winding of the rectifier transformer. The phase position of this alternating voltage is 120 degrees lagging the voltage of the rectifier transformer phase connected to the tube's anode. The variable unidirectional voltage is supplied by the charge-control system which is described in Section 31·32. By varying the magnitude of the unidirectional voltage, the rectifier current is controlled from maximum to minimum.

Figure 31·71 (*a*) shows the positive no-load voltage between anode and cathode of tubes 4, 5, and 6. Because of the large leakage reactance in the rectifier transformer, the voltage across each tube with the rectifier loaded is far different from that shown. However, no-load voltages are helpful in determining where a tube begins to conduct and where it stops conducting.

If grid voltage is made more positive than the critical value near point *A* in Fig. 31·71 (*b*), the rectifier conducts minimum current as shown. Current continues after the no-load voltage is zero, because of the leakage reactance in the transformer. The rectifier produces maximum steady-state short-circuit current when the voltages on the grids of the controlled tubes become more positive than the critical values at point *B* in Fig. 31·71 (*c*). However, the initiation point must be advanced as the counter electromotive force of the capacitor increases, so that maximum charging current can be maintained. This is accomplished by making the variable grid voltage more positive than that shown in Fig. 31·71 (*c*) in about 30 milliseconds after the start of charge. Each tube in the rectifier then fires as soon as its anode voltage exceeds its cathode voltage by the amount of its breakdown voltage, and the initiation point of the rectifier tubes automatically advances as the capacitor bank is charged. If the rectifier tubes are initiated before point *B* for the first several cycles after the capacitor starts to charge, the rectifier current contains a transient which may be as much as double the peak value of the steady-state short-circuit current. This transient may exceed the peak current rating of the rectifier tubes and may even cause an arc-back or loss of control in the rectifier. Transients can be avoided if the rectifier tubes are initiated later than point *B* at the start of charge and then gradually phased forward.

31·32 CHARGE-CONTROL CIRCUIT

The charge-control circuit regulates the voltage to which the capacitor bank charges, maintains the capacitor bank at that voltage until a welding discharge is desired, and blocks the rectifier during the discharge period. A diagram of such a circuit is shown in Fig. 31·72.

It is a special two-stage direct-coupled amplifier consisting of the amplifier tubes, their power supplies, and a voltage supply which serves as a reference in regulating the voltage of the capacitor bank.

The input to the charge control is a part of the voltage across resistor R_{27}. This voltage is proportional to that across the capacitor bank. The output of the charge-control circuit is the variable unidirectional voltage appearing between 1 and 43, and is applied to the grids of the controlled rectifier tubes as described in Section 31·31. This voltage varies from a positive value, equal to the peak of the alternating grid voltage on the rectifier, a negative value of approximately three times as much, as shown by Fig. 31·73. This figure is helpful in understanding the operation of the charge-control circuit. If it is assumed that the charge on the capacitor bank is zero, the grid of tube 10 is positive by the amount of the regulated reference voltage across potentiometer 6. The reference voltage selected by the potentiometer determines the voltage to which the capacitor bank charges. Since

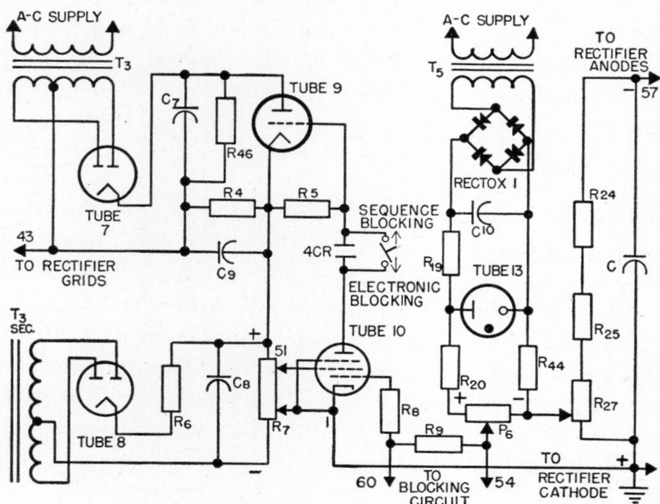

Fig. 31·72 Charge-control circuit.

the grid of tube 10 is positive, corresponding to point *A*, the tube conducts maximum anode current and produces sufficient voltage across its plate-load resistor to reduce the current through tube 9 almost to cut-off. Thus the voltage across resistor 4 is negligible. Output voltage of the charge control is the algebraic sum of the output voltage across resistor 4 and the voltage between points 51 and 1 on resistor 7. Therefore, lead 43 is positive with respect to lead 1. This positive voltage on the grids of the controlled rectifier tubes

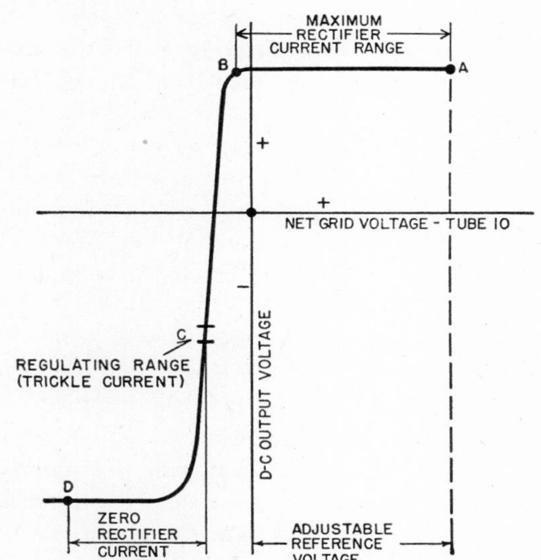

Fig. 31·73 Charge-control-circuit characteristics.

causes the rectifier to pass maximum current, thus charging the capacitor bank.

As the capacitor bank becomes charged, the voltage on the grid of tube 10 decreases. However, the output voltage of the charge control remains unchanged until the grid voltage of tube 10 reaches point *B*. Then a small decrease in the grid voltage of tube 10 causes the output voltage to swing negative to a value in region C. This decreases the rectifier current to a small value which is just enough to supply the bleeder current and the leakage of the capacitor bank. If this current is insufficient to maintain the capacitor voltage, the charge control causes the rectifier to increase its output until the capacitor voltage is maintained at the proper value. In this manner, capacitor voltage is regulated within 2 percent of the desired value until the capacitors are discharged to produce a weld. Voltage of the capacitor bank can be adjusted to any value between 1500 and 2800 by means of potentiometer P_6.

When the weld is made, the blocking circuit (described later in Section 31·33) decreases the grid voltage of tube 10 to point D in Fig. 31·73. D-c grid voltage of the rectifier tubes is now highly negative, as shown in Fig. 31·71 (*d*), and the rectifier is prevented from conducting current.

The grid voltage of tube 10 remains at point D until the discharge is completed; then the blocking voltage becomes zero, and grid voltage returns to point A. Output voltage of the charge control does not immediately become positive, however, because capacitor C_9 shunts resistor R_4. The discharge time constant of this resistor-capacitor combination is selected so that the output voltage of the charge control does not reach its full positive value for 30 milliseconds after the grid voltage of tube 10 returns to point A. When the output of the charge control reaches point C the rectifier begins to conduct. As this voltage becomes more positive, rectifier current gradually increases until it becomes maximum when the charge-control output voltage reaches point B. In this manner a transient in the charging current at the start of charge is avoided.

Not only must the charge control regulate the voltage of the capacitor bank at any desired value between 1500 and 2800 volts, but also it must bring the voltage to this value in the shortest possible time without undershooting or overshooting. Undershooting is caused by the charge control reducing the rectifier current before the desired voltage is reached. Reducing the rectifier current causes capacitor voltage to climb slowly to the correct value. Overshooting is caused by failure of the charge control to reduce the rectifier current soon enough, and the last half cycle of current conducted by the rectifier causes the capacitor voltage to become too great.

By properly selecting load resistances of tubes 9 and 10, the size of capacitor C_9, and the magnitude of the reference voltage, both overshooting and undershooting can be limited to less than 2 percent of the voltage to which the capacitor is being charged even at 200 operations per minute. Thus the charge control is capable of accurately controlling the voltage of the capacitor bank and the energy available for performing the weld.

31·33 BLOCKING CIRCUIT

The blocking circuit prevents the rectifier from passing current while the capacitor bank is discharging. If rectifier current is permitted during this time, tube 1 in Fig. 31·69 continues to conduct rectifier current through the primary of the welding transformer after the discharge is over. This prevents the capacitor bank from being recharged because it short circuits the capacitor.

A diagram of the blocking circuit is shown in Fig. 31·74. The blocking indication is the voltage across resistor R_{30} (also see Fig. 31·69) when the capacitor bank is discharging. This voltage is applied to the low-voltage winding of transformer T_8, which is designed for operation at a frequency of 1 cycle per second, so that it can transform the low-frequency voltages appearing across resistor 30 when certain combinations of capacitance and inductance are used in the discharge circuit. The high-voltage winding of transformer 8 is connected to the a-c terminals of a full-wave copper oxide rectifier, which rectifies the blocking signal. The voltage output

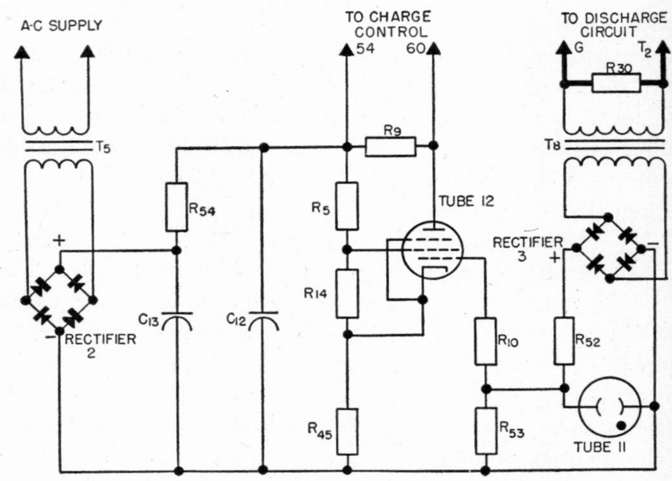

Fig. 31·74 Blocking circuit.

of the rectifier is impressed across neon tube 11, which limits the voltage applied to the grid of tube 12. The grid of this tube is normally biased just beyond cut-off. During the welding discharge the grid is driven positive, and the tube is made to conduct maximum current and thus produce maximum voltage across resistor R_9. This voltage is applied to the grid circuit of tube 10 in the charge-control circuit, making its grid negative and causing a highly negative direct voltage on the grids of the controlled rectifier tube and thus blocking the rectifier.

Blocking continues until the discharge current falls to about 2 amperes and the ignitron ceases to conduct. Since capacitor C_9 causes some delay in the charge-control circuit, the ignitron stops conducting before the rectifier can start again. Hence the blocking circuit accomplishes its purpose and allows the rectifier to begin charging the capacitor bank at the earliest possible moment after the discharge is over.

31·34 DISCHARGE CIRCUIT

Several different discharge circuits are used in capacitor-discharge controls. One is the full-cycle system. Another is the unidirectional or shunt-tube system.

The unidirectional system was the one originally used in all commercial capacitor-discharge systems. The full-cycle system is an improved system for high-speed operation.

Full-Cycle Discharge System

The full-cycle discharge circuit shown in Fig. 31·75 in heavy lines contains:

1. The capacitor bank.
2. The welding transformer and its secondary load circuit.
3. Two ignitrons and their associated thyratron firing tubes.
4. A flux-resetting rectifier.

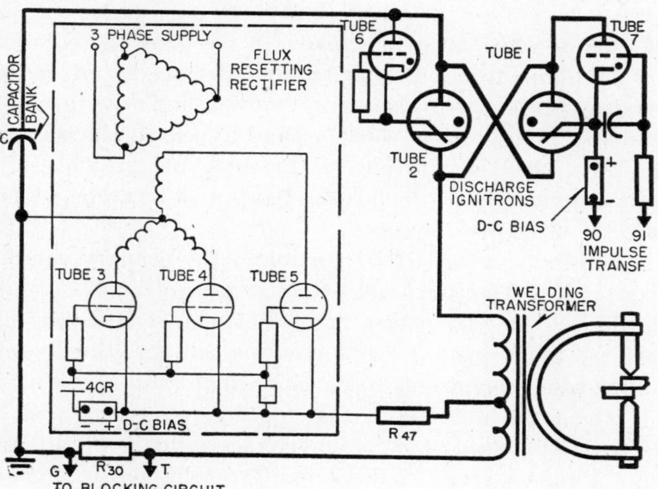

Fig. 31·75 Full-cycle discharge circuit.

The welder discharge circuit can be approximately represented by an equivalent inductance, resistance, and capacitance in series with a switch. This equivalent circuit is oscillatory for the usual values of resistance, inductance, and capacitance of a machine used for welding a wide range of thicknesses of aluminum.

In the full-cycle system tubes 1 and 2 are in inverse parallel. With proper control of the firing tubes, they permit a single positive and a single negative half cycle of current during discharge.

In the operation of the full-cycle discharge circuit the secondary of the welding transformer is closed and the flux-resetting rectifier is turned off by $4CR$ contacts just before

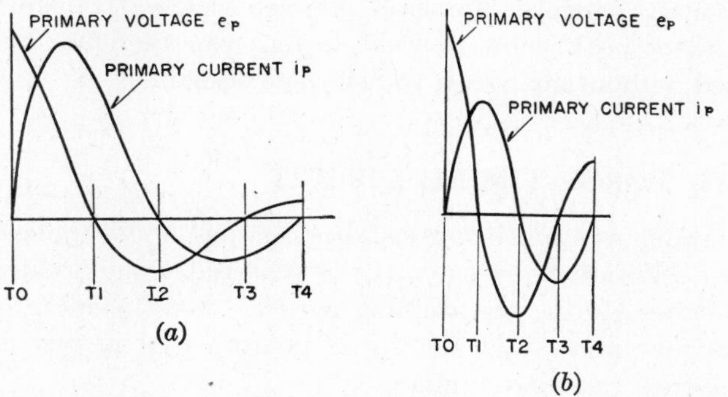

Fig. 31·76 Discharge currents and voltages for circuit of Fig. 31·75.

the weld is to be made. A single impulse of positive voltage is applied across leads 90–91 to the grid of firing tube 7 in Fig. 31·75 to initiate the discharge. Tube 1 is thereby initiated; the discharge current and voltage are shown by the oscillograms in Figs. 31·76 (*a*) and 31·76 (*b*).

Tube 1 conducts current from time T_0 to T_2 in Fig. 31·76 (*a*), and tube 2 begins to conduct at time T_2 and conducts current to the end of the discharge at T_4. Discharge current and voltage are displaced by a phase angle depending upon the circuit parameters. At the end of the discharge, the voltage of the capacitor bank has the same polarity as the initial charge voltage. This means that some of the original stored energy is returned to the capacitor bank for use on the next weld. It also means that the speed of the control is increased, because the charging time of the capacitor bank for subsequent operation is decreased.

The curve of flux versus magnetizing ampere-turns for a typical welding transformer is shown in Fig. 31·77. In this

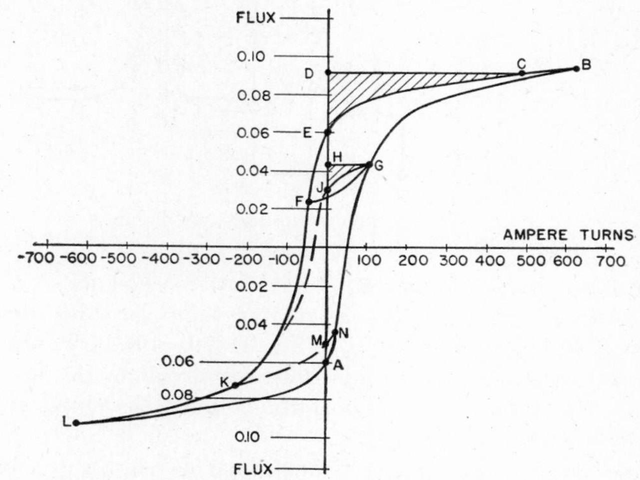

Fig. 31·77 Transformer flux, unidirectional-discharge circuit of Fig. 31·76.

figure the flux corresponds to the current and voltage oscillograms of Fig. 31·76 (*a*). Before the weld, it is assumed that the flux has been fixed at point K (the flux-reset point). During discharge the flux changes along the curve through points M, N, and G to reach a maximum B at time T_1 in Fig. 31·76 (*a*). From T_1 to T_3 the flux decreases along the curve through points C, E, and F. From T_3 to T_4 the flux is increased again, and the final value is reached at G at time T_4. Area GHJ is proportional to the residual energy in the transformer core at the end of the weld. The short-circuiting effect of the secondary circuit prevents any substantial decrease in this energy until the electrodes are opened. This energy must then be dissipated at the electrodes upon separation.

Unidirectional-Discharge System

The circuit for the unidirectional-discharge system shown in Fig. 31·78 consists of:

1. The capacitor bank.
2. The welding transformer and its secondary load circuit.
3. A series ignitron tube (or a magnetic contactor) to connect the capacitor to the primary of the welding transformer.
4. A shunt ignitron tube to prevent reversal of capacitor voltage and welding current.
5. Reversing contactors to change the direction of current through the transformer each operation to prevent saturation of the transformer.

If it is assumed that contactor A is closed and B is open, the operation of this circuit is as follows: Capacitor bank C has been charged to the proper voltage by the main rectifier. To start welding, the series ignitron, tube 1, is made to conduct. Current through the primary of the welding transformer rises rapidly, and capacitor voltage falls as shown by Fig. 31·79. Shortly after the current reaches its maximum, the capacitor voltage reverses (if the circuit is oscillatory).

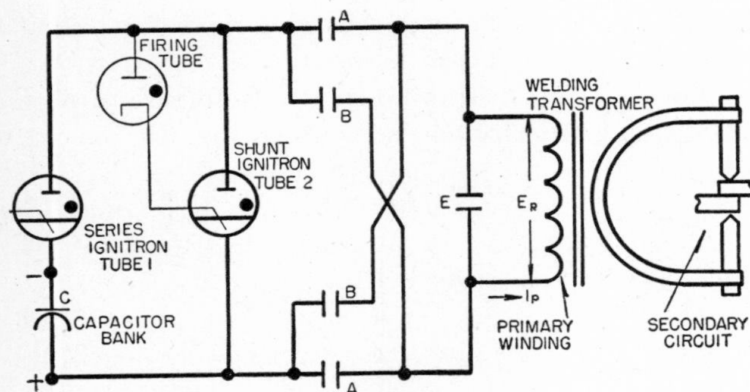

Fig. 31·78 Unidirectional-discharge circuit.

The shunt ignitron, tube 2, is made to conduct as soon as the reverse voltage becomes greater than the breakdown voltage of the tube at time T_2 in Fig. 31·79. Primary current then is transferred from the series-tube circuit to the shunt-tube circuit and decays exponentially until the shunt tube ceases to conduct.

The flux change in the transformer is shown by Fig. 31·77. At the start of discharge the flux is at a residual value at point A. As the discharge progresses the flux changes through N and G to point B at time T_1. While the shunt ignitron is conducting, the flux falls from B to C. The shaded area CDE represents the energy stored in the transformer iron at the end of discharge. It is much larger than the energy stored during a full-cycle discharge; therefore, a primary short-circuiting contactor (E in Fig. 31·78) must be

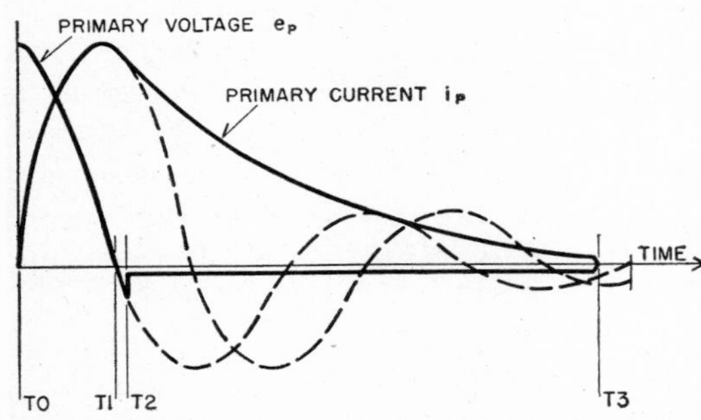

Fig. 31·79 Transformer current and voltages in the unidirectional-discharge circuit.

used to prevent sparking at the electrodes or a high surge voltage on the primary of the transformer when the electrodes are opened after a weld.

While the capacitor is recharging, contactor A is opened and B closes; thus the direction of the discharge current on the next weld is reversed. The contactors may be replaced by a flux-resetting rectifier such as that shown in Fig. 31·75. Elimination of the contactors is desirable for high-speed operation.

31·35 FLUX RESETTING

A flux-resetting rectifier supplies to parts of the primary winding of the welding transformer a current of opposite polarity to the first half cycle of discharge current. The resultant magnetomotive force applied to the transformer core resets the flux at the same value (K in Fig. 31·77) between welds. If the flux is not reset, the transformer iron tends to become saturated, and welding current is decreased.

Some important factors in economical rectifier design are:

1. The rectifier must be capable of resetting the core flux in a minimum time determined from the maximum desired operating speed. Any attempt to reset flux with a closed secondary circuit in as short a time as the discharge time requires a resetting current of the same magnitude as the weld current. Therefore, flux resetting must be accomplished while the electrodes are open.
2. Self-inductance of that portion of the primary winding carrying flux-resetting current must be known. Current-limiting resistor R_{47} must have a sufficient resistance to make the time constant of the circuit small, and thus permit the flux-resetting current to reach its final value quickly.
3. The type of rectifier used for flux resetting must be selected with regard to maximum ripple-voltage output under load. Ripple voltage in this circuit appears across primary and secondary windings of the welding transformer. If this voltage is too large, the secondary electrodes will spark if they are accidentally short-circuited by the operator as he inserts material in the machine for welding.
4. Ampere-turns supplied by the rectifier should be sufficient to establish the flux-reset point K of Fig. 31·77 at a value past the knee of the flux-ampere-turn curve to minimize variation in the magnetic characteristics of the iron in different transformers of the same rating.

The direct current required to reset the welding transformer flux therefore can be less than 3 amperes for nearly all transformers if these four factors are properly considered.

With a flux-resetting rectifier the initial value of flux K can be greater than the residual value obtained with the unidirectional-discharge circuit; therefore it is possible to obtain a greater change in flux from a given welding transformer. A greater peak value of weld current can therefore be obtained without change in transformer design.

31·36 FORGE-TIMING CIRCUIT

Forging, as applied to resistance welding, is the application of an additional pressure to the material being welded at some time during the welding period. Most welding engineers now agree that forging is instrumental in producing crack-free welds in aluminum.

One method of forging is to energize a solenoid air valve which acts to increase pressure on the welding electrodes. In any case, the objective is a rapid increase in pressure on the material at some time after the welding current has reached its peak. The instant of pressure application is adjustable at 8 to 192 milliseconds after the welding current is initiated. The forge-timing circuit provides this adjustable timing period.

Several different types of valves are used for forging, each having different characteristics. The minimum operating time of many valves is more than the minimum forge delay required. To obtain correct minimum forge delay with the slower valves, they must be energized before the welding current is initiated. For this purpose initiation of the welding current is delayed by means of a weld-delay circuit, as shown in Fig. 31·80. Forge-delay time is the sum of valve-delay time and valve-operating time minus weld-delay time.

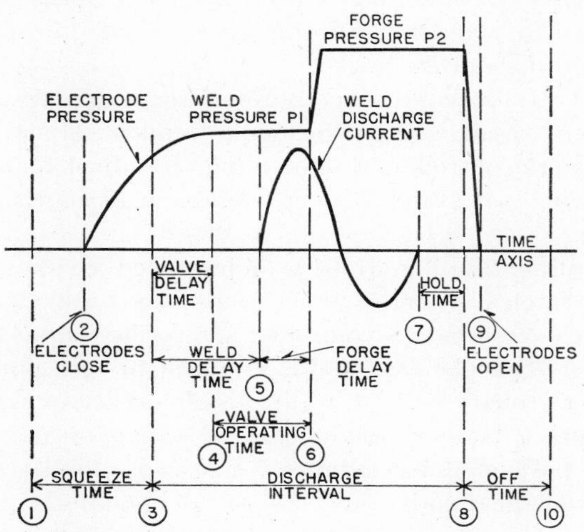

FIG. 31·80 Sequence of events, forging.

Several different methods can be used for energizing valves. Some can be energized by d-c sources; others must be energized by the discharge from a capacitor. The system in Fig. 31·81 uses a d-c supply for providing energy to operate the valve. The valve is energized by initiating tube 5.

Operation of the circuit is as follows: When a weld is to be made, relay $3CR$ is energized. As pressure on the electrodes reaches the proper value, relay $4CR$ is energized. Opening of a normally closed contact on relay $4CR$ creates an impulse on the grid of tube 4, which causes the tube to conduct current. This charges capacitors C_8 and C_9 through potentiometers P_1 and P_2 respectively. As the voltage on capacitor C_8 becomes great enough to overcome the negative bias across the part of resistor R_2 in the grid-cathode circuit of tube 5, the tube conducts and energizes the forging valve.

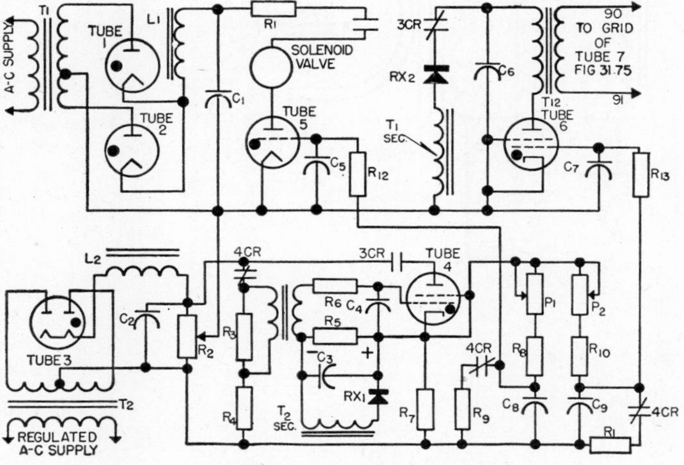

FIG. 31·81 Forge timing circuit.

Valve-delay time can be adjusted from 8 to 192 milliseconds by means of potentiometer P_1.

Operation of the weld-delay circuit is similar to that of the valve-delay circuit. As the voltage across capacitor C_9 overcomes the negative bias on tube 6, the tube conducts and discharges capacitor C_6 through the low-voltage winding of transformer T_{12}. This produces a voltage impulse across the high-voltage winding of the transformer, which is connected in the grid circuit of tube 7 in Fig. 31·75. The impulse causes the tube to conduct, which in turn initiates tube 1 through which the capacitor bank discharges into the welding transformer.

Weld-delay time is adjustable from 16 to 48 milliseconds by means of potentiometer P_2. Usually weld delay is set for 16 milliseconds, and forge delay is varied by adjusting only potentiometer 1 in the valve-delay circuit. In this manner, the forge delay can be varied from 8 to 192 milliseconds.

31·37 MAGNETIC-ENERGY STORAGE FOR RESISTANCE WELDING

The magnetic system is based on the principle that electrical energy can be stored in magnetic material, for example the iron core of a welding transformer, by passing a direct current through a winding or coil wound around this magnetic circuit. The amount of energy that can be stored depends upon both the volume of magnetic material and the maximum flux density in the magnetic material. A two-winding transformer or a single-winding reactor may be used for this purpose. The amount of energy that can be stored for a particular volume of iron can be increased if a small air gap is left in the core.

An elementary welder of this type could be constructed as shown in Fig. 31·82 with a battery, a welding transformer, a relay $1CR$ that can be adjusted to operate when its coil current reaches a pre-set value, a heavy contactor A, and an initiating switch.

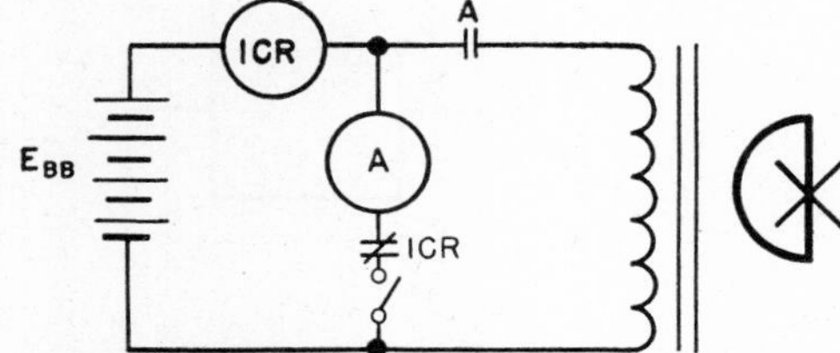

FIG. 31·82 Elementary magnetic-energy-storage system.

As the switch is closed, current passes from the battery through contact A and coil $1CR$ to the primary of the welding transformer.

This current is initially zero and does not increase instantaneously but rises exponentially because of the inductive effect of the welding transformer and its secondary circuit. A typical curve for charging current is shown in Fig. 31·83.

As primary current reaches a predetermined value relay $1CR$ operates and contactor A is de-energized. Contact A then interrupts the primary current. Rapidly decreasing primary current in the transformer induces a voltage momentarily across the secondary winding. As a result, secondary current increases to a peak value at the same time as primary current decreases to zero. Secondary current then decays exponentially, with the decay time-controlled by the resistance and inductance of the secondary circuit. The weld or secondary current is shown in Fig. 31·83.

Primary current must be interrupted quickly, or the energy stored in the transformer is dissipated in an arc across contact A. The rate at which current falls to zero in the primary determines the peak of secondary current in the welding transformer. The peak value can be controlled also by changing the value of primary current for which $1CR$ becomes energized. Interrupting at a smaller primary current decreases secondary current. This method is commonly used to control the magnitude of weld current. Resistance blocks in the secondary circuit can be changed to control the decaying portion of the secondary current wave.

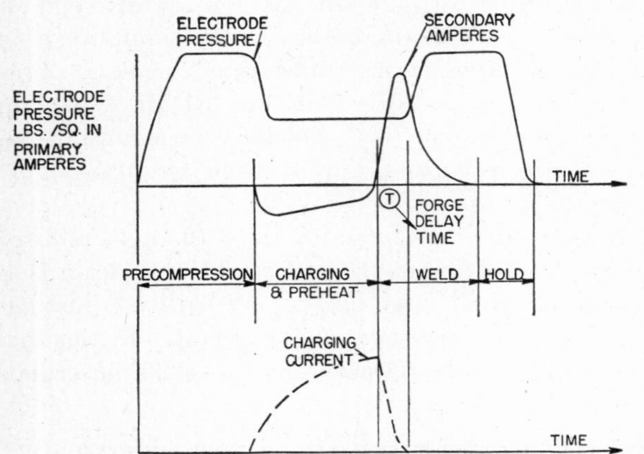

Fig. 31·83 Sequence of events, magnetic-energy-storage system.

If only a single contact were used to interrupt the primary current it would be impossible to control the magnitude of the secondary current within narrow limits, because it would be necessary to control the extinguishing of the arc across the interrupting contacts for each weld. Therefore, a series of four or eight contacts is used to insert resistance progressively in the primary circuit as shown in Fig. 31·84 (interrupter 1). By the time the first three contacts have opened, current has decreased to a small value for final interruption by contact 4. The drop-out or operating time of each contactor is carefully controlled by a resistor shunting each of the d-c operating coils A, B, C, and D.

Use of mechanical contactors to interrupt primary current has several disadvantages, including frequent adjustment of the resistors controlling drop-out time to adjust for wear of the moving parts and frequent replacement of the arcing contacts.

Interrupter 2 in Fig. 31·84 is an improved device. It contains a single contactor, but a capacitor bank of several hundred microfarads is connected across the contacts. The contactor is air-operated and is arranged to open quickly to interrupt primary current. There is almost no arcing across this contact because the uncharged capacitor provides a parallel path of low resistance. The capacitor is charged, and a voltage builds up across it to oppose the rectifier voltage. The whole action takes place quickly and forces the primary current to zero in the same time interval as with a mechanical contactor.

The commercial design of the electrical circuit for magnetic-energy storage is shown in Fig. 31·84. The battery is replaced by a low-voltage three-phase half-wave rectifier to convert three-phase a-c power to d-c power. The rectifier initiating thyratrons are phase-controlled to permit smooth adjustment of output voltage to any value between 80 and 155. In general, a lower unidirectional voltage is required for welding light gauges of aluminum to compensate for varia-

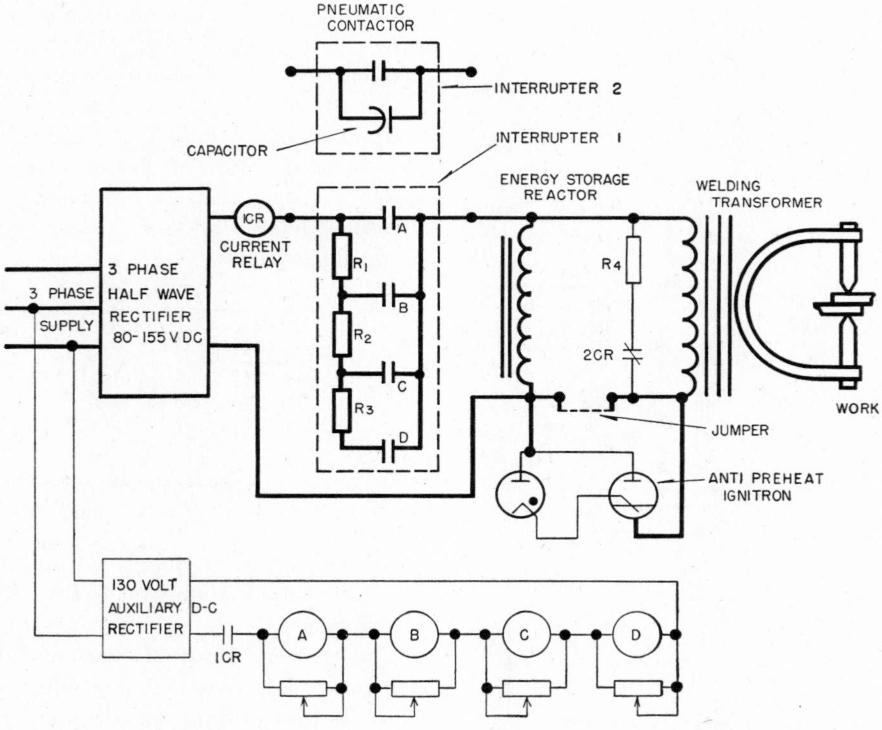

Fig. 31·84 Block diagram of magnetic-energy-storage welding system.

tions in operating time of the relay system used to control interruption of primary current. If primary charging current rises rapidly the variation in relay-operating time causes appreciable variation in the stored electrical energy at interruption; consequently weld current varies too much. The rate of rise of primary charging current is reduced by decreasing rectifier output voltage to obtain consistent amounts of energy for each weld.

In this system the material to be welded must be in place, with the welder secondary circuit closed, before energy storage or charging can begin, because there can be no elapsed time between the end of the charging period and the start of the weld or discharge period.

During the charging period, increasing primary current causes secondary current in the welder. This preheat current must be limited or the material will overheat and deform before the weld; therefore the rate of increase of primary current must be limited.

It is therefore sometimes necessary to use the reactor in Fig. 31·84 in parallel with the primary of the welding transformer to store energy in a reasonable length of time and avoid excessive preheat secondary current. Charging time on a standard machine varies from 0.05 to 0.5 second.

In welding thick materials the preheat current aids in making the weld by breaking down higher contact resistance between the two pieces of metal. In welding thin materials, the preheat current causes deformation of the metal even though a parallel storage reactor is used. This welder therefore contains an anti-preheat ignitron which functions when the jumper around it is removed. The tube blocks current in the welding transformer during the charging period. All energy is stored in the reactor. As the interrupter operates, voltage reverses across the reactor, and the ignitron conducts to transfer the stored magnetic energy from the reactor into the welding transformer.

A short-circuiting contactor *2CR* is required across the welder primary to prevent electrode sparking when the electrodes are opened immediately after a weld is made.

Sequence-timing circuits for the magnetic-energy storage system are similar to previously described circuits for energy-storage welders. However, one additional function is provided on magnetic energy storage welders. This is an additional electrode pressure referred to as "precompression" in Fig. 31·83. With the additional pressure the contact resistance is lowered slightly before the passage of welding current, since the material is forced into more intimate contact. The additional pressure aids in controlling weld consistency.

Sound welds can be obtained with this system over a wide range of materials with a slightly higher kva demand on the power system than for a comparable capacitor-discharge system. However, the reduction in kva over single-phase welders is substantial.

Chapter 32

INDUSTRIAL PHOTOELECTRIC CONTROL

V. B. Baker

APPLICATION of light-sensitive control is logically divided into two groups: (1) old jobs that can be done better or more economically by photoelectric control, and (2) functions which were formerly impossible. To replace some existing device, photoelectric control must be justified on the basis of economy or its ability to give better performance, a better product, or higher production. If a task can be performed in several ways, the merits of each solution must be considered carefully. It is not so difficult to justify the use of photoelectric control to perform a function which cannot be done satisfactorily in any other manner.

Any photoelectric device is vastly inferior to the human eye. Man has not yet made a device which can match certain abilities of the eye; for example, the eye does not have to be focused on a passing object, perhaps only a tiny speck, to be aware of its presence and direction of motion. No phototube has this ability. Much time and money have been spent on some photoelectric inspection jobs such as the determination of surface flaws in paper and tin plate or foreign material in milk and soft drink bottles, but success has been negligible. The solution of this type of problem may perhaps be found in the iconoscope and the methods of high-speed scanning used in television.

32·1 PHOTOTUBE AMPLIFIERS

Before phototubes can be put to work, their minute current output must be amplified. These currents may be only a fraction of a microampere, or as large as 20 or more microamperes. Generally, a given application requires a certain type of optical system to establish phototube illumination and, consequently, phototube current. The amplification required depends upon both phototube current and the current or power requirements of the device to be controlled. Besides amplifying the feeble currents, the amplifier serves as an impedance-matching element; the phototube is a high-impedance device, whereas the controlled device usually has a low-input impedance. Amplifiers have gains ranging from a few thousand to well over a hundred thousand. The controlled element usually is a bell or a light, a solenoid, a motor armature, or a generator field.

A vacuum-tube amplifier operates on voltage changes in the input circuit; consequently, phototube current must be converted into a proportional voltage prior to amplification. This is done by inserting impedance in series with the phototube. Such impedance usually consists of pure resistance or resistor-capacitor combinations; transformers are rarely used, because the required high-input impedance is difficult to obtain. Phototube load impedances generally range from 1 to 20 megohms. Such high impedance is required to produce reasonable voltage drops from the rather small currents passed by phototubes. In some special apparatus, for example, color matchers and gas detectors, load resistors may be several hundred megohms. Where such high resistances are used, extreme precautions against leakage currents must be taken.

Figure 32·1 shows a simple phototube amplifier using a vacuum tube with unidirectional potentials on anode and plate. With the phototube dark, it carries negligible current, and the amplifier tube is biased negatively by the voltage from potentiometer P. Plate current therefore is minimum or zero and the relay is de-energized. Increasing the light on the phototube causes it to pass current and develop a voltage drop across load resistor R. The voltage drop tends to make the grid potential less negative. If sufficient light strikes the phototube, the vacuum-tube plate current becomes large enough to operate the relay. Thus the relay is operated by varying the light between maximum and minimum limits as determined by the tube characteristic and the pick-up and drop-out currents of the relay. To energize the relay with a decrease in illumination, the positions of the phototube and the resistor R must be reversed in the circuit.

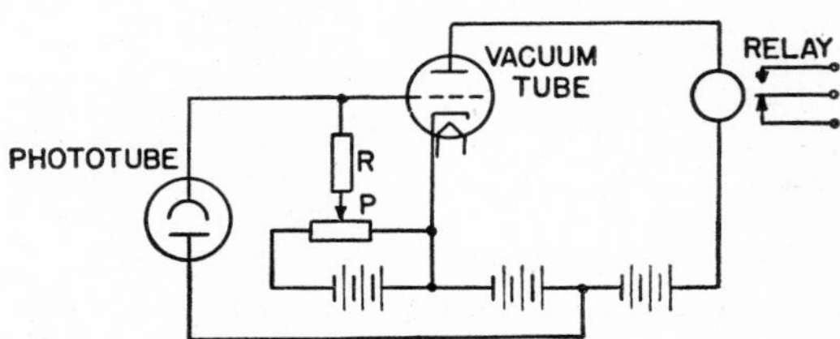

FIG. 32·1 Simple circuit of phototube amplifier using d-c voltages. Increase of light energizes relay.

A vacuum-tube amplifier which operates on alternating voltage is shown in Fig. 32·2 (*a*). The relay is energized by an increase in illumination, and circuit operation is similar to that of the amplifier in Fig. 32·1, except that the phototube and amplifier tube carry current only during the positive half cycle. Sensitivity and efficiency are therefore less than those of the d-c circuit. To prevent chattering of the relay, a capacitor can be shunted around its operating coil. Grid capacitor C prevents a time-phase difference between grid and plate voltages otherwise caused by phototube and interelectrode capacitances. The circuit arrangement shown in Fig. 32·2 (*b*) causes the relay to operate with a decrease in illumination.

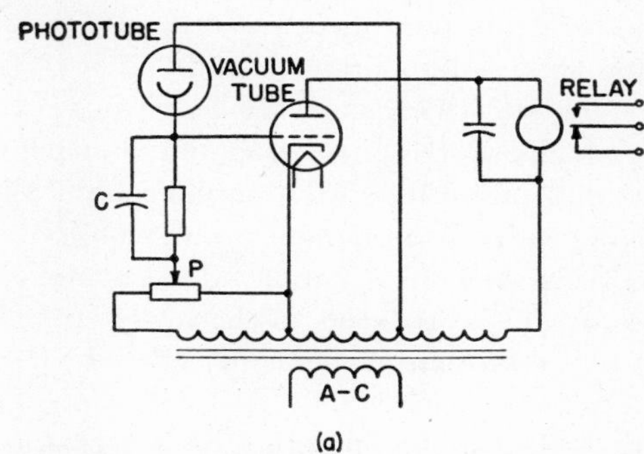

FIG. 32·2 (*a*) Phototube amplifier operated entirely on a-c voltages. Increasing light energizes relay.

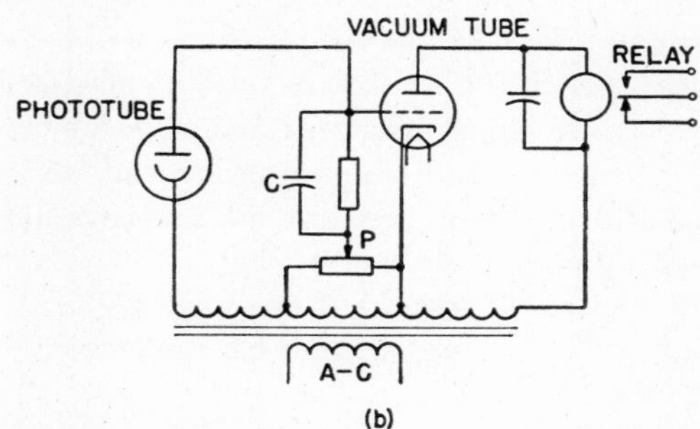

FIG. 32·2 (*b*) Circuit for energizing relay with a decrease in illumination.

Ordinary vacuum tubes can supply power for low-capacity relays only; for this reason, thyratrons or gas-filled tubes are extensively used in final stages of industrial phototube amplifiers. Because a thyratron grid loses control once current conduction has started, alternating anode voltage is used almost exclusively except where a lock-in feature is desired. A phototube circuit employing a single thyratron which actuates a relay directly is shown in Fig. 32·3. The grid

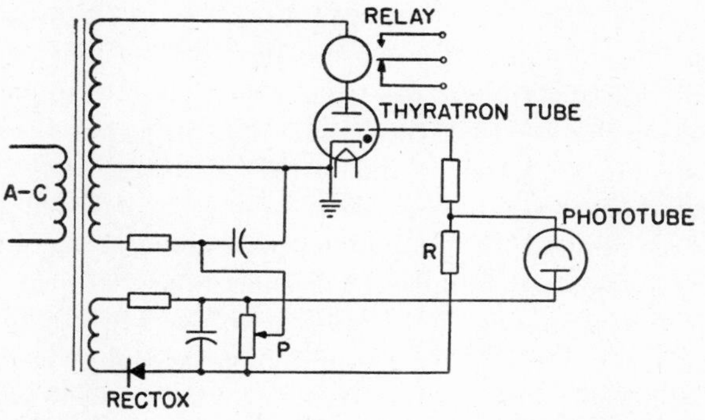

FIG. 32·3 Phototube amplifier using a single thyratron; operated on a-c voltages.

potential consists of a phase-shifted a-c component superimposed on a unidirectional voltage made up of a portion of the voltage across P and the drop across load resistor R. With no light on the phototube, the voltage across R is negligible, and the d-c bias prevents instantaneous grid voltage from intersecting the critical-grid-voltage characteristic of the tube; hence the thyratron remains non-conducting. With light, the phototube carries current, and the unidirectional voltage across R makes the total unidirectional component of grid voltage less negative. Instantaneous grid voltage thus becomes sufficiently less negative to initiate current conduction, and the relay becomes energized. Such a circuit requires an illumination level of around 5 foot-candles. Preceding the thyratron with a single vacuum-tube amplifier gives greater sensitivity.

Because bias voltages are difficult to maintain in d-c amplifiers, more than two stages of amplification are seldom used. Higher gains are obtained by using a-c amplifiers. If background light is troublesome, chopped or modulated light is used and the amplifier input circuit is arranged to respond

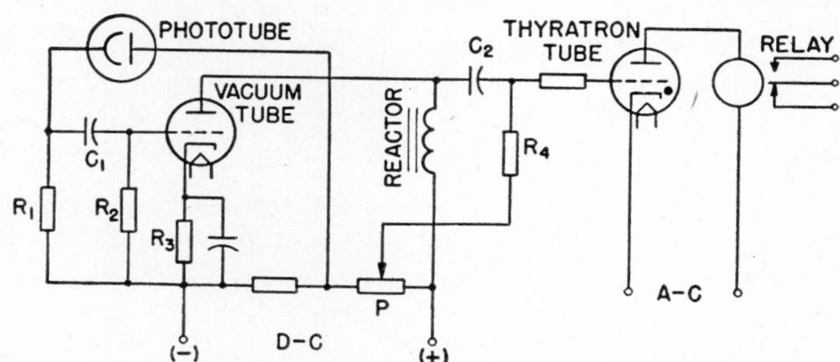

FIG. 32·4 Amplifier for use with modulated or chopped light.

to the rate of change of illumination instead of its total change. An a-c amplifier such as that shown in Fig. 32·4 then can be used for much higher amplification.

For steady illumination, the coupling capacitor C_1 is charged to the same voltage as the drop across R_1, and tube bias is determined solely by cathode resistor R_3. With modulated light on the phototube, the pulsating voltage across R_1 causes an alternating voltage across R_2, resulting from the charging and discharging currents of C_1. The frequency of this a-c component of bias voltage is the same as the light impulse frequency. This voltage is amplified in the vacuum tube circuit, and appears across the reactor in the plate circuit. Grid voltage for the thyratron is made up of a d-c component from potentiometer P and an a-c component across resistor R_4, which exists as long as modulated light strikes the phototube. Since the light frequency is greater than the frequency of the thyratron-anode supply, the relay always is energized except when the phototube is dark.

In many industrial applications a time delay in the photoelectric controller can be used to advantage. For example, in sequencing an industrial process or spacing objects on a conveyor belt, it may be desirable to initiate action only if the events or objects are out of order. The circuit shown in Fig. 32·5 incorporates time delay between light change and relay operation. With light on the phototube, thyratron A is non-conducting, and capacitor C is charged to the voltage between points M and N. The potential of the point D is therefore the same as the cathode of thyratron B, and thus it is permitted to conduct current and hold the relay energized. If phototube illumination is decreased sufficiently, thyratron A fires and effectively short-circuits the capacitor through current-limiting resistor R. The grid of thyratron B is immediately made highly negative, and the relay opens almost instantaneously with the light decrease. Restoring light to the phototube causes thyratron A to cease conducting, and the capacitor begins to charge at a rate determined by the total resistance in its charging circuit, of which potentiometer P is a portion. Thyratron B therefore is held nonconducting until the voltage drop caused by capacitor-charging current across the resistor and potentiometer between points D and N decreases enough to permit conduction.

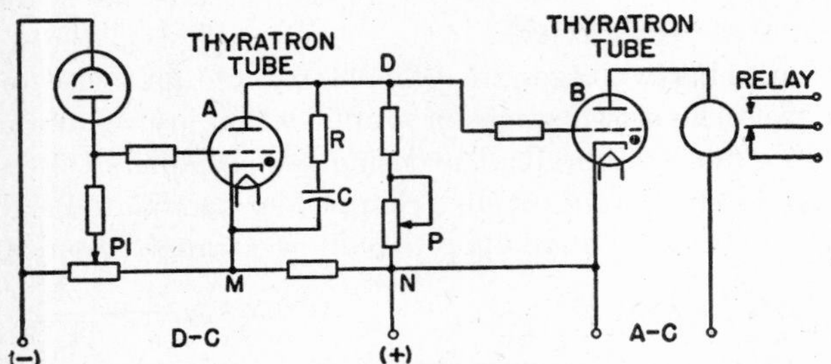

FIG. 32·5 Circuit incorporating a time delay between resumption of light and relay pick-up.

The relay does not become energized until a time interval following restoration of illumination. The length of this time delay is varied by adjusting the setting of P.

32·2 SENSITIVITY

Sensitivity may be defined in several ways. Absolute illumination requirements vary with amplification, type of phototube, and other factors, and may range from many foot-candles to a small fraction of a foot-candle. The change in phototube illumination necessary to initiate the control element may be as much as 100 percent or only a few percent. Color sensitivity depends upon the spectral response of the phototube itself, as discussed in Chapter 7. If the apparatus is responsive only to the rate of change of illumination instead of total change, another type of sensitivity must be considered.

With a phototube in the grid circuit of a small thyratron, absolute illumination requirements may be from 5 to 10 foot-candles with unidirectional anode voltage, and as high as 40 or 50 foot-candles with alternating voltage on the anode. With one stage of d-c amplification between phototube and thyratron, an illumination of about 1 foot-candle or less usually is sufficient. Many photoelectric relay applications, such as limit switching, counting, and sorting can be handled with this amount of phototube illumination. Additional amplification permits further reduction in light flux.

A photoelectric relay or control device can be made to operate satisfactorily with a change of only a few percent in light flux, but such a high degree of sensitivity is seldom needed in industrial equipment. The sensitivities usually are such that at least a 25 percent change is required. Such equipment can operate over long periods of time with a minimum of incorrect operation caused by spurious disturbances.

If the color of the controlling medium influences operation, the relative spectral-response characteristics of different phototubes must be considered. A phototube can be made with almost any desired spectral-response characteristic, but most commercial phototubes have color-response curves conforming to a limited number of standardized characteristics that fill most application needs. Since the output of ordinary incandescent lamps contains much red and infrared radiation, several commercial phototubes are designed for maximum sensitivity near the red end of the spectrum.

Use of infrared calls for careful attention to both absolute illumination level and phototube spectral response. Infrared filters commonly used to limit the amount of visible light may reduce the effectiveness of the light source from 50 to 90 percent. Light-source power, maximum spacing between light source and phototube, and amplifier gain all influence the selection of suitable equipment for a given job. A typical application using infrared is burglar-alarm systems, in which visible light should be reduced to the lowest practical limit; otherwise, the essential element of surprise may not be achieved.

Generally devices that respond to the rate of change of illumination are more sensitive (require a lower level of phototube illumination) than devices responsive only to total change. Phototube illumination of 0.02 or 0.03 foot-candle can be utilized with no more amplification than would normally be required with about 0.5 foot-candle and a phototube circuit responsive to total change of illumination. Devices responding to rate of change of illumination are well suited for outdoor applications where background illumination is ever-present and difficult to eliminate.

32·3 SPEED OF RESPONSE

Duration of light impulses may vary from 1 second or more to as little as 0.0005 second. If light-beam interruption lasts 0.1 second or longer, a small industrial relay has time to close, and a circuit responsive to absolute illumination (static control) can be used. If interruption is shorter, say 0.01 second, the controller is generally made responsive to rate of illumination change. Such circuits utilize the high-speed relay action of a thyratron to bridge the time gap between light impulse and relay closure. D-c anode voltage on the thyratron provides lock-in control, and manual or automatic resetting of the circuit is required after each operation.

On static control, the maximum number of operations per minute depends upon the closing time of the relay or other control device and the resetting time. For example, in a moderately high-speed counting application, the resetting time should be at least equal to the operating time. If the control device can be energized in 0.1 second, 0.2 second should be allowed for a complete cycle giving a maximum of 300 operations per minute. This maximum rate can be increased by arranging the thyratron to operate a high-speed magnetic counter directly. Some magnetic counters can operate as fast as 800 times a minute. Above this rate, all-electronic counters must be employed.

32·4 OPTICAL SYSTEMS

An optical system includes the light source and all optical parts up to the phototube cathode. The controlling medium itself sometimes must be considered part of the system.

Obviously, the needs of the application dictate the design

of this important component. The most elemental and most common optical system consists of a simple light source with the beam directly incident upon the phototube. The controlling medium causes partial or total interruption. There are thousands of applications of this type in scores of different functional arrangements. Limit-switch action on industrial machines and production lines, safety devices on machine tools and elevator doors, traffic control and automatic weighing systems (Fig. 32·6), and production-line counting and sorting of objects are but a few of the tasks performed by this simple arrangement. Optical equipment in more complicated apparatus such as sound motion pictures, facsimile transmission, and color matchers are extremely elaborate.

Fig. 32·6 Example of the application of simple photoelectric control for automatic weight checking. When weight is correct, light on the phototube located on the opposite side of the scale is blocked out by the pointer.

Three fundamental rules should be observed:

1. The light source should be capable of providing sufficient illumination for the worst operating conditions.
2. The controlling medium should be arranged to produce maximum absolute and percentage change in illumination on the phototube. A change of 50 percent or more is a good guide.
3. The light beam should cover as much area as possible of the phototube cathode to minimize local changes in cathode response.

Radiant energy from the light source may reach the phototube by direct transmission, or the rays may be bent by reflection or refraction. Direct transmission is used more extensively because it is simpler and meets the requirements of most applications. If reflected light is used, the optical system generally is arranged to work on diffused reflected light instead of spectrally reflected light. Tin-flow regulators in the steel industry and slitter and cutter register regulators in the paper and printing industries employ diffused reflected light extensively. The refractive index of a liquid or a translucent solid is sometimes used. Certain plastic materials can be used to channel light rays as desired. Light from a heated object is sometimes used to indicate temperature. Phototube excitation of this type is practicable only at temperatures above 1500 degrees Fahrenheit. Below this temperature, the intensity of the visible radiation is too low to be usable.

Because the success of any application depends upon design and care of the optical system, possible disturbing factors should be considered from the outset. Any one of such factors as supply-voltage fluctuations, extraneous phototube illumination, ambient temperature, dust, fumes, or humidity may directly cause or contribute to incorrect performance.

Line-voltage change has its greatest effect on the intensity of the light source. In the working range, light intensity of an ordinary incandescent lamp varies according to some exponential power greater than the cube of the applied voltage. For example, a 5 percent reduction in lamp voltage reduces light intensity about 16 percent (see Fig. 32·7). If a photoelectric control device is adjusted to operate with a change of only a few percent in luminous flux at the phototube, light-source voltage must be almost constant. Moreover, voltage changes on the electronic apparatus may initiate operation even with constant luminous flux.

Some supply-voltage variation usually can be tolerated because an illumination change of 50 percent or more can be obtained by suitable arrangement of the controlling medium.

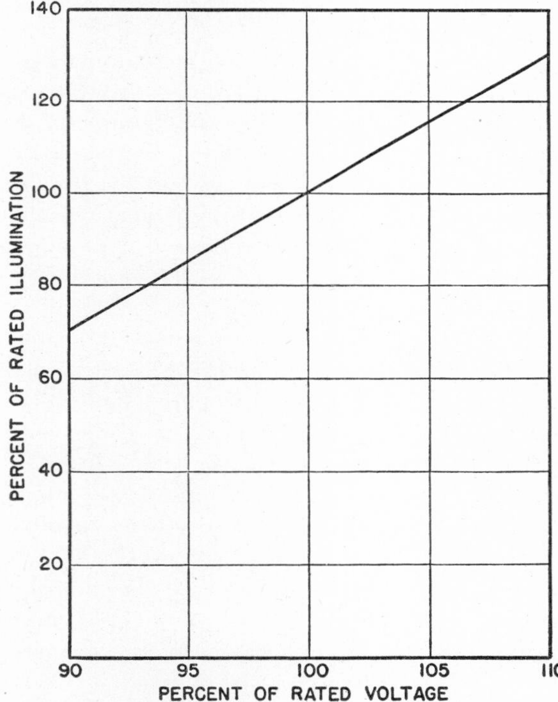

Fig. 32·7 Effect of variation of lamp voltage on light intensity obtained from ordinary incandescent lamps.

If available illumination change is much smaller, special precautions must be taken. If line-voltage changes in excess of a few percent are anticipated, a voltage-regulating transformer or other means must be used for providing constant voltage, especially at the light source. If the source of illumination is independent of line voltage and high sensitivity is needed, the electronic apparatus should be fed through voltage-regulating transformers or made independent of applied-voltage change.

A phototube cannot differentiate between light from the controlling source and all other extraneous or background illumination. It is imperative therefore that background light be considered in planning an installation. For example, if half the luminous flux at the phototube is extraneous light,

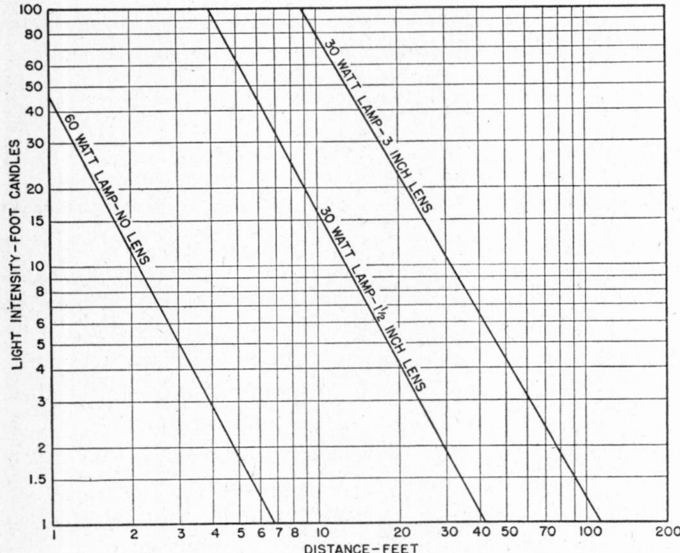

Fig. 32·8 Chart showing light intensity at various distances from the source under three different conditions.

total interruption of the controlled light beam may give only a 50 percent change at the phototube. Extraneous light always should be reduced to a practical minimum; the permissible amount of background light is determined entirely by the requirements of the particular application. Extraneous illumination can be reduced by shielding the phototube or by proper orientation of the apparatus. Its effects may be minimized by increasing the candlepower of the light source or by use of suitable filters.

Outdoors it is almost impossible to reduce extraneous illumination to the necessary minimum, if light source and phototube are far apart. Use of a modulated light beam (Fig. 32·10) and suitable amplifier (Fig. 32·4) renders ineffective the average illumination striking the phototube. The extraneous illumination is relatively constant, so operation of photoelectric equipment over great distances in broad daylight is possible.

Sensitivity of a phototube varies with the temperature of the cathode (see Fig. 7·23). Exceeding the maximum-temperature rating for an appreciable time may cause permanent reduction or complete loss of sensitivity. Maximum-temperature ratings of phototubes usually are 50 degrees centigrade or 100 degrees centigrade, depending upon the type; the more common phototubes used in industry are rated at 100 degrees centigrade. To insure reliability and long life they should not be subjected to ambient temperatures much in excess of 70 percent of their maximum ratings. If the illumination at the phototube emanates from a hot body, safeguards against radiant heat must be taken. The measured ambient temperature may be within reasonable limits, but radiant heat absorbed by the phototube cathode may heat it excessively. Suitable water filters therefore should be included in the optical system.

Dust, fumes, oil mist, or smoke may have an adverse effect on phototube illumination. Humidity and temperature variations may cause vapor condensation on lenses or other optical parts. Outdoors, fog and rain, and the normal effects of weather must be taken into account. The worst operating conditions must be anticipated, and the equipment must be planned accordingly.

Each mirror in the optical system should be assumed to absorb 50 percent of the light incident upon it. For example, two mirrors reduce the light intensity at the phototube to 25 percent of that obtainable from the source with no mirrors and the same beam length. Theoretically, the reduction per mirror is less than 50 percent, this being a practical figure based on operating experience. Mountings should be rigid; only a slight misalignment is greatly amplified by the beam arrangement.

Light Sources

Although many applications require special types of sources, a few general-purpose light sources satisfy the needs of the majority of uses. Of these there are three general types: (1) parallel beam for moderate distances; (2) concentrated beam for short distances; and (3) modulated beam for long distances or for use where the level of extraneous illumination is high.

The parallel-beam type usually consists of a low-voltage lamp at the focal point of a convex lens. Since the lamp filament has finite size, the light rays diverge slightly and the beam is not truly parallel. Effectiveness of lenses in collecting light from the source and projecting it in an approximately parallel beam is illustrated in Fig. 32·8. For example, a 60-watt lamp produces an illumination of 5 foot-candles at a distance of only 3 feet without lens; a 30-watt lamp and lens of 1½-inch diameter produces the same illumination at 18 feet, or at 46 feet with a 3-inch lens. This type of light source (Fig. 32·9) is used for most general-purpose applications where beam lengths may be only a few feet or as much as 30 or 40 feet.

Fig. 32·9 Simple light source with a 1½-inch-diameter lens. Thumb screw permits focal adjustment of lens for different operating condition.

Applications dealing with small objects or small displacements, or requiring a small, high-intensity spot of light generally require concentrated-beam light sources. Such a source usually consists of a low-voltage lamp placed outside the focal point of a convex lens to produce a converging beam. If the spot of light, which at the focal point is an image of the lamp filament, does not have the proper shape, apertures can be used to alter the spot. Beam lengths usually are only a few inches.

Fig. 32·10 A modulated light source. The 3-inch-diameter lens provides beam strength of suitable intensity for distances in excess of 200 feet.

For outdoor applications involving relatively long beams, chopped light affords distinct advantages. A typical modulated light source, self-contained in a weatherproof housing, is shown in Fig. 32·10. A small synchronous motor drives the perforated disk at 1800 revolutions per minute. Light from the low-voltage lamp behind the disk is interrupted between successive holes as the disk rotates, thus modulating the light beam 100 percent at a frequency of 540 cycles per second. Operating distances of 200 feet or more are possible with this type of light source.

Small low-voltage low-power lamps are used in each of the types of light sources discussed. The reason is that only a small portion of the image of the lamp filament at the phototube can be utilized; therefore color temperature of the filament is the sole factor in determining light intensity at the phototube. The filaments of most general-purpose lamps operate at about the same temperature; consequently little, of any, increase in intensity is gained by using a lamp of higher wattage. Lamps rated at 30 or 40 watts and 6 or 8 volts are most suitable for general use. There are exceptions, of course.

In the multicolor press-register regulators discussed later in this chapter, the lamp in each of the scanners is a special lamp of the projector type. Cutter-register regulators are usually equipped with this type of lamp. Lamps rich in ultraviolet, such as high-power mercury lamps, are used occasionally.

Phototubes

Although there are three types of light-sensitive cells, the phototube (photoemissive cell) is employed more widely in industrial apparatus than either the photovoltaic or the photoconductive cell. The phototube is fundamentally a high-impedance device and is therefore capable of developing large voltages across its load impedance. These voltages can be amplified, and this characteristic alone gives the phototube a flexibility and utility unmatched by the other two types of cells. Moreover, its high sensitivity, stability, reasonably wide ambient temperature range, and the positive control available over its spectral-response characteristics give it superiority.

In varying degrees, both the photovoltaic cell and the photoconductive cell lack these desirable characteristics; consequently, neither type has been used extensively in industrial apparatus. Photographic exposure meters, light-intensity meters, and a few industrial applications such as smoke alarms and illumination control constitute their greatest use.

Most phototubes used in industry are of the gas-filled type. The recommended maximum operating voltage for these tubes is around 90 volts. A voltage much higher may cause a glow discharge attended by damage to the cathode surface.

Phototubes are often used to differentiate between different colors, and the effects of their spectral characteristics are important. For example, a phototube having peak sensitivity in the red region of the spectrum gives only slightly different responses to a red object and a white object, although the eye indicates considerable difference. Blue or green, however, causes little response, and the maximum difference in response is between these colors and white. In the same way, a yellow object appears almost as black to a phototube having peak response in the blue region and low response elsewhere.

If plenty of space is available, the phototube is usually mounted in the amplifier cabinet, but it may be desirable to mount the phototube in its own housing, because of space or other limitations at the observation point. Maximum extension of a phototube from its amplifier depends upon the electrical characteristics of the connecting wires. With alternating voltage on the phototube the capacitance between wires determines the permissible extension. A 10-foot cable ordinarily used for such applications has a capacitive reactance of about 10 megohms, which is roughly the same as the phototube load impedance. This spurious impedance reduces the sensitivity of the equipment. Where greater extensions are required, a unidirectional potential must be applied to the phototube. Maximum extension then is limited only by the insulation resistance of the cable; however, cable capacitance in combination with the phototube load resistor forms an *RC* circuit which reduces the speed of response. Extensions above 50 feet should be avoided. Under any conditions, distance between phototube and amplifier should be as short as possible.

The cable connecting an extended phototube with its amplifier should be shielded, it should be of good grade, and it should not be spliced. BX or other ordinary rubber-covered wire should not be used. Anything which may tend to increase leakage currents, such as insulating tape (which usually is hygroscopic) or collection of dust, should be avoided. Conduit carrying phototube leads should be grounded and free from other power wires.

If moderate extension of the phototube is required, it may be advantageous to include in the phototube housing a single stage of amplification. The power level of the output circuit then can be more readily handled over long wires.

32·5 APPLICATIONS

Photoelectric controls have been used in so many different industries and in such a wide variety of forms that even a moderately complete description of them is impossible here. Specific applications have been selected to typify particular uses or to illustrate the adaptability of light-sensitive devices in solving special problems.

To simplify the application of photoelectric control manufacturers have developed many types of general-purpose photoelectric relays and control devices. These standardized units (Fig. 32·11) are adaptable to most of the more common

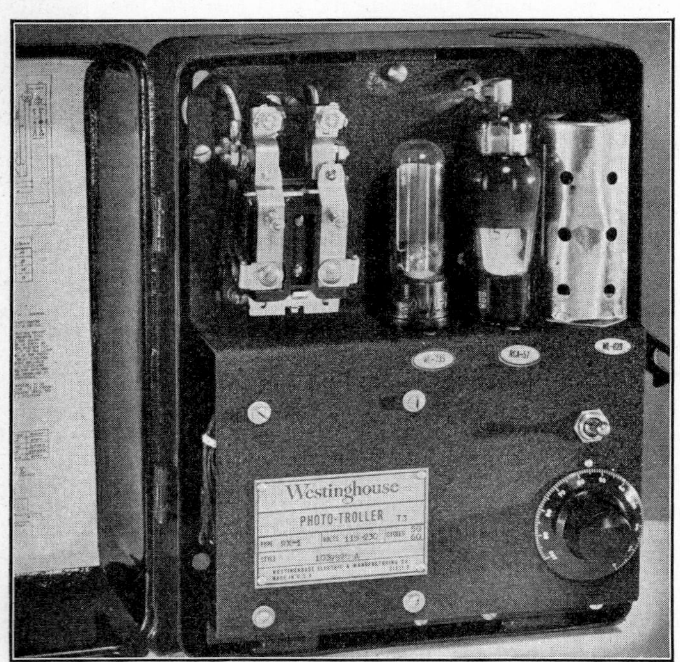

Fig. 32·11 General-purpose photoelectric relay.

uses; they also serve as basic components in more complicated control systems. Hence, special design is not necessary for most of the simpler applications.

General Control

On production lines including a painting operation, speed and efficiency have been improved by the addition of a simple "on-off" photoelectric control. A light beam spans the conveyor carrying the objects. Interruption of the beam by one of the objects initiates the paint sprayer, and the object is painted. Operation is timed so that painting is complete

Fig. 32·12 Photoelectric relay used for control of doors on a heat-treating oven.

when the sprayer is shut off automatically. No paint is wasted between objects, and the cycle proceeds at maximum speed.

The same simple arrangement can be used to open doors to a heat-treating furnace (Fig. 32·12), or to stop motion long enough to perform an operation on a piece. Similarly, objects can be counted, sorted, kept in fixed relation to one another, or made to initiate an operation for any number of special functions.

Interruption of a simple beam for limit-switching has countless uses. Liquid level in vats or tanks, as well as the level of material in bins, can be controlled. In rod and pipe mills of the steel industry, photoelectric limit switching stops the stock at a fixed position prior to cutting. Pieces of predetermined length are thus cut, and high-limit switch maintenance and loss are eliminated. Photoelectric equipment for elevator leveling stops the elevator accurately at each floor. These applications are typical; the possible uses number thousands.

Photoelectric controls have been used in various ways to initiate an operation in response to the temperature of a hot body. As an example, all sorts of small stock parts shaped by hot upsetting can be delivered from a forging heater to the upsetting machine at just the right working temperature. Stock parts usually are heated in a resistance-type forging heater. An electronic control with an infrared-sensitive phototube receives light emitted by the part being heated, and initiates rejection of the piece when it has reached a predetermined temperature as indicated by its color temperature. The operation is automatic, and stock pieces can be fed through the heater as fast as the process will permit with little or no attention by the operator.

The same type of control can be used for induction-heating applications, or with large resistance furnaces. Any operation, such as cutting off power, starting a quenching cycle,

Fig. 32·13 View showing the application of a pin-hole detector to a shearing line. The electronic-control cabinets are mounted adjacent to the light source located above the strip.

or initiating motion which can be accomplished by a small relay, can be performed by a simple, inexpensive light-sensitive relay.

A somewhat specialized sorting device, but one of the most successful, is the pin-hole detector used on shearing lines in the steel industry. In finishing mills a continuous trip of tin plate is unwound from a large roll and sheared into rectangular plates (Fig. 32·13). Prior to shearing, the pin-hole detector indicates the presence of a pin hole, and operates the classifier which sorts the sheets into two or more piles. Strip are connected in parallel, and each tube scans a portion of the sheet. Scanning zones are made to overlap by means of staggered cylindrical lenses. A light-tight seal is maintained between the sheet being inspected and the phototube housing; if the sheet contains no holes, no light enters the housing. Thus the most sensitive optical arrangement is provided.

Figure 32·14 shows interconnections of the various components. Phototubes are capacitively coupled to the grid of the 1603 vacuum tube, making the device responsive to rate of change of illumination rather than total change. Sudden appearance of a small quantity of light on any one of the paralleled phototubes causes the control-grid voltage of the 1603 to become more negative. This voltage change is amplified, and the potential of lead B in the final stage becomes highly positive with respect to the potential of lead D. Thyratron 1, the grid of which thus becomes positive with respect to its cathode, conducts current, and relay 1 closes to sound an alarm and initiate operation of the classifier. Since unidirectional voltage is applied to the anode of thyratron 1, it continues to conduct even after the initial impulse of light has passed and its grid potential has become negative again; however, when relay 1 closes, a set of back contacts opens and capacitor C_2, having been previously short-circuited, starts to charge. When the voltage across C_2 reaches a predetermined value, the time depending upon the setting of the potentiometer P, thyratron 2 fires and energizes relay 2. This opens the plate supply to thyratron 1, stopping its conduction. The alternating voltage applied to the anode of thyratron 2 causes it to cease conduction as soon as the back contacts on relay 1 close and short-circuit capacitor C_2 through the current-limiting discharge resistor R_s.

The same detector is used on trimming lines on which the strip is rewound into finished rolls instead of being cut into sheets. When it is so used, a device for marking the strip at the point where a hole is detected must be added. This

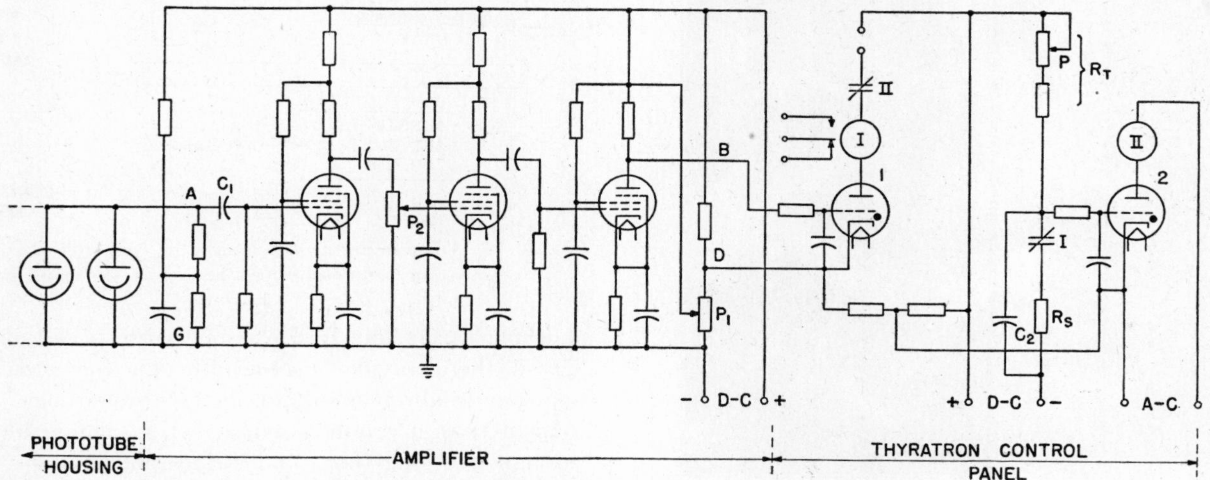

Fig. 32·14 Simplified circuit of the pin-hole detector showing elements of the phototube housing, the amplifier, and the thyratron control panel. Auxiliary contacts of relay I are for sounding alarm, operating classifier, and so on.

speeds in excess of 1000 feet per minute are in use, and the detector responds to holes as small as 1/100 inch in diameter.

Components of the detector include a light source, a phototube housing, a d-c amplifier, and a thyratron control panel. The light source consists of several low-voltage lamps or fluorescent lamps. Several phototubes mounted beneath it additional component includes two thyratrons in a full-wave rectifier circuit for energizing a relay and the marking solenoid, together with a vacuum tube acting as a variable-resistance element in a C-R timing circuit. A small pilot generator driven from the trimming line controls the timing sequence so that the marking solenoid remains energized long enough to make a mark of predetermined length on the sheet, regardless of the speed of the line. Action of the marking device is initiated when thyratron 1 in the control panel is made to conduct when a pin hole appears.

The pin-hole detector was designed primarily for metal strip but may be applied wherever opaque material is to be scanned. Even with opaque material and the tremendous change in illumination on the phototubes resulting from the appearance of even a very small hole, voltage amplifications as high as 250,000 may be necessary. In view of the difficulties inherent with gains higher than this, the device cannot be applied as a hole or flaw detector on translucent materials such as paper.

Many systems and machines for sorting have been devised. A photoelectric bean-sorting machine, manufactured by Electric Sorting Machine Company of Grand Rapids, Michigan, for example, examines each bean individually and separates the occasional off-color bean from the multitude of white beans. Sorting colored objects and production-line sorting for size or shape hold promise of raising efficiency or lowering cost in some industrial processes.[1,2]

Alarms and Safety Devices

Machines and people are protected by photoelectric alarm and safety devices. On punch presses and sheet-metal-bending machines, photoelectric controls render the operating control ineffective as long as the operator's hands are within danger zones. On many machines a simple light beam gives adequate coverage (Fig. 32·15); on others a curtain of light directly across the front of the operating position is necessary and, in some cases, two or three sides must be so equipped. All such controls are designed to fail safe. This attribute is essential since there is a human tendency, once confidence has been established, to rely heavily on protective devices, especially during periods of unusual fatigue. Failure of the light source or any element within the electronic component prevents normal operation.

Fig. 32·15 A light beam crisscrosses the front of this punch press to render controls inoperative as long as beam is interrupted by operator's hand.

Similarly, these devices are used on isolated d-c motors and generators for protection in case of flashover. Should a flashover occur, the control opens a circuit breaker or otherwise clears the machine of the cause. In plants processing inflammable material, and in warehouses, strategically located phototubes detect fire and start the sprinkling system or initiate other fire-control measures. Automatic sub-stations and switching vaults are often equipped with photoelectric control to sound an alarm at a central control point in case of fire in these remote installations. Elaborate photoelectric fire-detecting apparatus has been developed for use on ships. One such device[3] includes an air-sampling system which draws individual air samples from remote locations on the ship to a central indicating instrument on the bridge. Each air sample is passed in front of its own phototube. A fire is immediately indicated and located.

Photoelectric controls are used also for fire and smoke detection in the air ducts of forced ventilating and air-conditioning systems. Light passes through the duct and strikes a phototube. Smoke in the duct alters the quantity of light at the phototube and may initiate the control to close dampers or shut off ventilating fans in addition to sounding an alarm.

Photoelectric gas detectors give an on-the-spot indication of dangerous concentrations of harmful gases in industrial plants. Traces of carbon disulphide or mercury vapor smaller than one part in a million can be determined in a few seconds. One of these devices[4] is shown diagrammatically in Fig. 32·16. Operation is based on the principle that many gases,

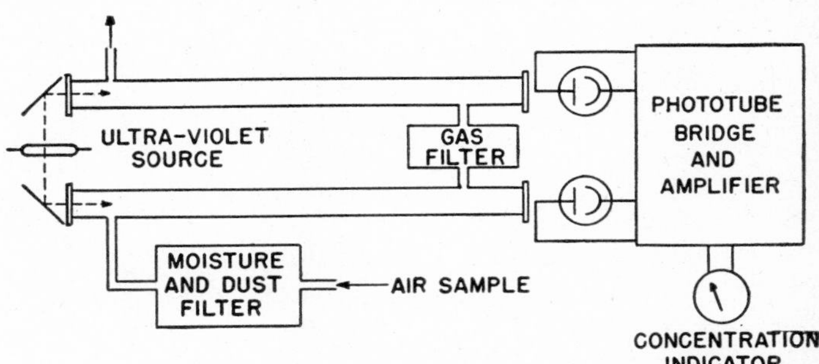

Fig. 32·16 Diagrammatic sketch showing elements of a gas detector for indicating the presence of a hazardous gas or vapor in air.

of which carbon disulphide is one, absorb certain bands of light in the ultraviolet region, although they show negligible tendency to absorb radiation in the visible range.

Light from a common source rich in ultraviolet is passed through two long tubes and strikes phototubes at the ends of the tubes. The phototubes are connected in a balanced bridge circuit (Fig. 32·17) so that unbalance in light on their cathodes is indicated on the microammeter. An air sample, free from dust and moisture, traverses the length of one of the tubes and passes through a filter to remove the hazardous gas. This air, free of carbon disulphide, continues through the length of the second tube and is expelled. Traces of carbon disulphide in the first tube unbalance the phototube bridge circuit and are indicated on the meter. The relative magnitude of this indication is a measure of the concentration.

A similar device[5] detects mercury vapor. These devices find their greatest application in the textile, chemical, glass, and other industries where the presence of a harmful gas in the air supply presents a health problem. Once the presence of a gas is known, adequate steps can be taken to keep the concentration below the toxic limit.

All sorts of burglar-alarm systems are based on the ability of the phototube to respond to infrared radiation. To be effective, visible light from the source must be negligible; consequently a powerful light source is needed if any appreciable distance is to be covered by the beam. Light beams crisscrossing a room or spanning windows protect against intruders by sounding alarms, turning on lights, or taking pictures of the intruder. Outdoor installations have been made to protect storage yards or other restricted areas.

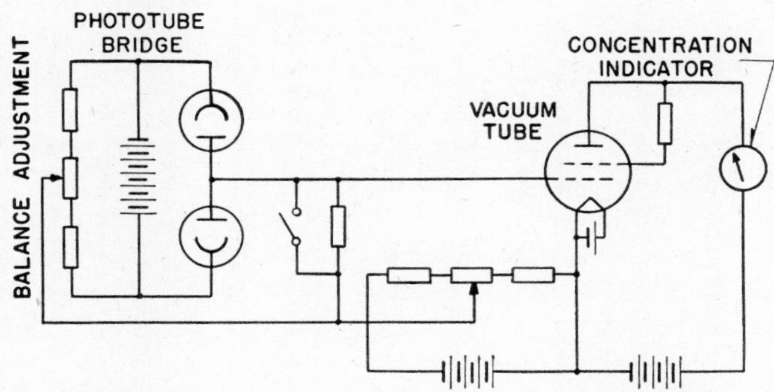

FGI. 32·17 Balanced phototube-bridge circuit gives a delicate indication of light unbalance due to presence of hazardous gas or vapor in one of the sampling tubes (FIG. 32·16).

Safety applications include photoelectric controls on elevator doors for passenger safety, in gas furnaces to close valves in case of pilot-flame failure, in plant traffic-control systems, on sewing machines operated by blind persons, and many others.

Measurement

Materials too wet or too weak for mechanical contact, or so hot that they cannot be approached, are easily checked for dimensions or weight with photoelectric measuring devices. In chemistry, variations in light transmitted through solutions or emitted by hot specimens yield information about chemical composition. Photoelectric densitometers have increased the speed of spectroscopic analyses tenfold and doubled the accuracy of conventional methods. Electronic color-matching devices capable of detecting color variations imperceptible to the human eye are used in the textile and paper industries.

Figure 32·18 shows the elements of a device for indicating the width of a traveling web of material such as paper, cloth, or tin plate.[6] No mechanical contact is necessary, and the optical system is designed to permit lateral "weaving" of the web within practical limits without affecting the accuracy of measurement. Light received by phototube 1 from its source is proportional to the masking effect of one edge of the web. Likewise, light on phototube 2 is proportional to the masking effect of the other edge of the web. For a given web width the light on each phototube is held almost constant by means of the current instruments which form part of the optical system. A small shift to the right, for example, decreases current in phototube 2 and increases in phototube 1. These unbalanced currents, after being suitably amplified, alter the positions of the moving coils and re-establish approximately equal illuminations on the phototubes. Total illumination on both phototubes is a measure of web width. If the web increases in width, total illumination decreases, and the sum of the current outputs from the two amplifiers to the width-indicating instrument is decreased. A wider web is thus indicated. The reverse is true with a decrease in web width.

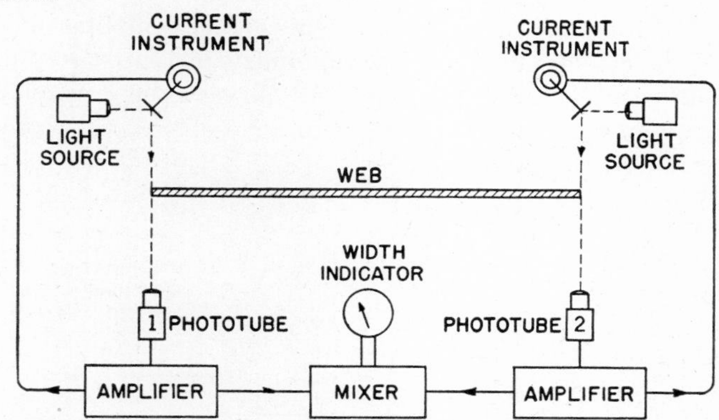

FIG. 32·18 Components of a device for measuring the width of a moving web or strip of material such as tin plate or rubber. Equipment functions to permit lateral shift of web.

Photoelectric instruments for indicating intensity of ultraviolet radiation have been available for several years; only recently has equipment been developed for measuring x-ray intensity.

In the irradiation of milk, ultraviolet radiation is directed on a thin film of the liquid for a specified length of time. Milk flow and radiation intensity must be held in fixed relation. For controlling the radiation, a meter which records ultraviolet intensity is used. The circuit of such a meter is shown in Fig. 10·8. The phototube is of the high-vacuum type, with maximum sensitivity in the ultraviolet. Radiation striking the phototube charges capacitor C_1 until the voltage across it breaks down the cold-cathode gas tube between the control anode and the cathode. The arc within the tube then transfers to the anode and cathode, and capacitor C_2 is discharged through a recorder coil replacing the relay shown in Fig. 10·8. Each discharge actuates a stylus which places a small dot on a revolving chart. Since the circuit is self-resetting (C_1 also discharges with the initial arc), the cycle repeats. Intensity of radiation is thus indicated by the frequency of marks on the chart.

Several methods of measuring x-ray intensity are in use, including the deionization or cloud chamber and the Geiger counter. These methods have some disadvantages which have limited their scope, particularly in industrial applications. Recently developed photoelectric equipment promises to overcome one of the chief difficulties—achieving speed of response. In some applications of x-rays, it is important to know the exact intensity of radiation during the exposure. This is essential to provide data for plotting intensity versus time when the time is short, or to permit making adjustments in intensity during exposure. Furthermore, an instrument of extremely short time lag may conceivably be used to determine the length of exposure when exposure time is a fraction of a second. The photoelectric instrument described here possesses almost instantaneous response; that is, the time lag between exposure and indication of the intensity of the exposure is in microseconds.

X-rays to be measured fall on a fluorescent screen, as shown in Fig. 32·19. Some of the radiation is thus converted into longer wavelengths near or in the visible range to which the phototube will respond. Consequently, phototube current can be used as a measure of x-ray intensity. Generally a multiplier phototube, such as the type 931, is used for initial amplification of the feeble currents. The amplifier itself may be conventional, but to minimize pick-up the input circuit usually includes a cathode follower tube. Amplifier output may be applied directly to an oscilloscope or other indicating or control device.

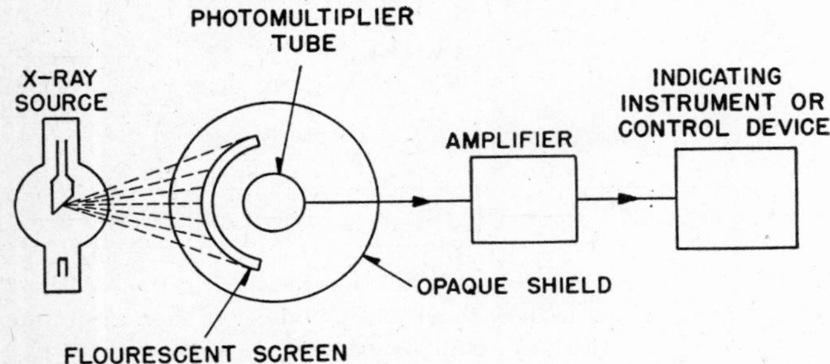

Fig. 32·19 Suggested arrangement of components for indication of x-ray intensity by photoelectric means.

The speed of response of this type of intensity indication opens new possibilities for such applications as high-speed inspection, and x-ray diffraction as well as x-ray emission and absorption of matter in chemical research. The high-energy level of x-rays suits them for chemical studies, and with a high-speed instrument for accurate intensity measurement these analyses can be made quickly and easily.

Regulators

A recent development is a photoelectric edge regulator for use on steel-mill trimming lines; it is applicable wherever the edge of a moving strip or web of material must be held in fixed relation to some part of the processing machine. Figure 32·20 shows diagrammatically the essential components as applied to a trimming line.

Steel strip is unwound from a roll, passed through a pair of speed-reference rolls, then through the trimmer, whence it is rewound on the reel. A scanner, which includes the phototube and a pre-amplifier, contains an aperture consisting of a slit approximately an inch long and 1/16 inch wide. Light from the source beneath the sheet excites the phototube in proportion to the position of the edge of the web in front of the aperture.

Phototube current when the aperture is half masked by the web establishes, after amplification, the necessary bias voltages on the power tubes of the amplifier to give equal currents in the two fields of the generator. Since these two fields are differential with respect to each other, no voltage is produced by the generator, the motor driving the hydraulic valve is at standstill, and no axial correction of the reel results. If the edge of the sheet moves, for example, outward from the page, phototube illumination decreases. Current in one of the generator fields decreases, and current in the other generator field increases. A voltage is thus applied to the armature of the d-c motor, and its rotation opens the hydraulic valve to produce an axial movement of the reel in the direction necessary to re-establish correct web position under the scanner. Initial movement of the web in the opposite direction reverses these reactions. The scanner moves

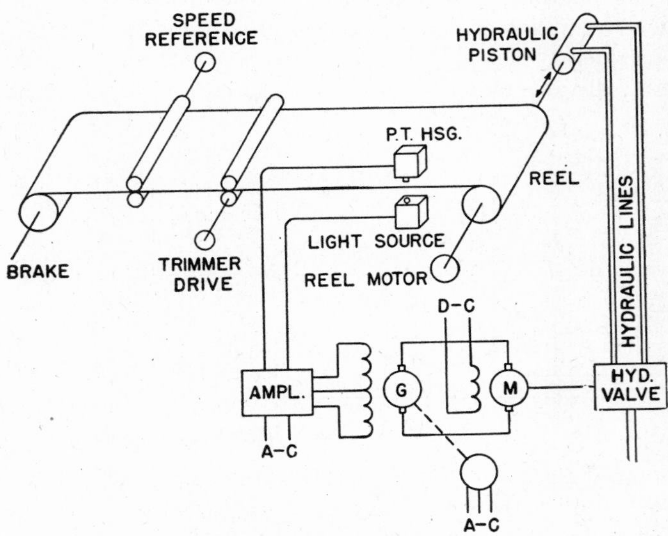

Fig. 32·20 Diagrammatic sketch of photoelectric edge control to provide an evenly wound coil.

with the reel, and the result is an evenly wound coil. Web position can be held within about 1/16 inch, provided the rate of change of web position with no corrective forces applied is not too great.

A device known as the side-register regulator performs a similar function, and can be used also to scan a printed line on a moving web of translucent material such as paper. A different scanning principle gives greater accuracy than the system just discussed. In the scanner shown in Fig. 32·21 a small metal disk containing four equally spaced lenses around its periphery is driven at 1800 revolutions per minute (on 60-cycle supply) by a small synchronous motor. The light source behind the disk projects a spot of light on the web, and the phototube receives diffused reflected light from this spot just as it crosses the line or edge of the web. Each time a spot passes over the edge or line, the phototube receives an impulse caused by the sudden change in reflected illumination. There are 4 × 1800 or 7200 such impulses per minute. The same number of half cycles occur per minute on the a-c supply.

These impulses are amplified and fed into a circuit containing two inversely connected thyratrons which function as half-wave rectifiers and supply the armature of a small d-c motor controlling the transverse position of the web. Since the thyratrons have alternating anode voltages of the same frequency as the synchronous-motor power supply, light impulse and phase angle of thyratron anode voltage have a fixed relation determined by web position. If the disk is properly positioned on the motor shaft, either thyratron may conduct current. Under these conditions essentially an alternating voltage is applied to the d-c motor armature, and there is no correction.

If the edge or line moves off the neutral position, the light impulse comes earlier or later in time phase relative to the thyratron anode voltage. Then only one of the thyratrons conducts current, and rectified alternating current is applied to the motor armature. The polarity of this voltage is such that the motor rotates in the direction necessary to correct for the displacement error of the web. Movement of the edge or line in the opposite direction causes the other thyratron to conduct, and motor rotation is reversed. For

FIG. 32·21 Photoelectric scanner of a slitter or side-register regulator showing synchronous motor, rotating disk with four lenses, and phototube housing.

reasonable rate of change of web position without correction, position accuracy of $\pm\frac{1}{64}$ inch may be obtained.

One of the most elaborate photoelectric systems for industrial use is the multicolor-press register regulator. A general discussion of the system has been given in Section 30·10.

In each of the web scanners (Fig. 32·22) a high-intensity light source illuminates the web at the point where the register marks pass. Lenses project an image of these marks on an aperture located in the optical path between the phototube and the web. When a register mark is directly on the optical center line the image completely covers the aperture. The sudden appearance of a mark at this point thus causes an abrupt change in illumination on the phototube cathode and an electrical impulse having a steep wavefront is produced. An optical amplification of 7 or 8 to 1 is used.

The register marks are about 0.01 inch wide and 0.5 inch long, and the aperture has the same relative shape. It is evident that the long dimensions of the marks and the aperture must be exactly parallel to each other; if they are not parallel the steepness of the impulse voltage wavefront is greatly reduced.

Register regulators of less complicated nature are used in the packaging industry. A photoelectric scanner observes register marks on printed wrapping material, and the electronic equipment maintains paper feed into the packaging machine in correct relation to the cutting cycle. These

FIG. 32·22 Web scanner used on multicolor press, showing light source, aperture, and photomultiplier tube all mounted in an explosion-resisting cabinet.

devices insure a complete, accurately placed design on each package.

Photoelectric regulators are maintaining loops in process lines of the steel, textile, and rubber industries (Fig. 32·23). Simple controls to maintain a loop of material within rather

FIG. 32·23 Loop between uncoiler on extreme right and 600-feet-per-minute shearing line on left. There are three phototubes below the inverted lighting function to maintain the proper loop length through the electronic control system contained in the pedestal-mounted cubicle mounted to the left. The operator inspects the upper side of the tin plate directly; the underside indirectly in the mirror.

wide limits have been applied extensively. Regulators that give corrections proportional to deviations for more exacting requirements have been successfully applied.

Electrolytically deposited tin on steel plate has a suede or matte appearance, whereas a mirror-like surface results after the tin has been heated to its melting point. This change is used in a photoelectric regulator applied to high-frequency induction heating. After tin has been electrolytically deposited on steel plate, the continuous strip passes through an induction-heating coil excited by a high-frequency oscillator. As the strip progresses through the coil, its temperature is raised until the tin melts. The melting zone within the coil is narrow, and its position is critical. If it is too near the exit end, melting is incomplete; if it is too far back in the coil, the temperature of the strip leaving the coil may be sufficiently high to discolor or burn the plate. The photoelectric regulator maintains this melting zone within fixed limits.

A phototube is arranged to observe diffused reflected light from the strip at the point where the flow line occurs. If melting occurs earlier in the coil, the mirror surface is under the phototube, and illumination decreases. The control then functions to lower the voltage applied to the oscillator and decrease its power output. A longer time then is required to heat the tin to the required temperature, and melting occurs later in the coil. If melting occurs near the leaving end of the coil, the matte surface is under the phototube, illumination level is high, and the control functions to raise the applied voltage to the oscillator and increase its output. This increases the temperature gradient along the coil and again brings the flow line within the dead zone of the regulating equipment.

REFERENCES

1. "Accurate Sorting of Colored Objects," Leland L. Antes, *Electronics*, Vol. 17, June 1944, p. 114.
2. "Photoelectric Dimension Gage," A. Edelman, *Electronic Industries*, Vol. 3, May 1944, p. 96.
3. "Detecting Fire at Sea," *Electronics*, Vol. 17, July 1944, p. 125.
4. "PE Tube Gas Detection," *Electronic Industries*, Vol. 3, March 1944, p. 108.
5. "Mercury Vapor Detector," *Electronic Industries*, Vol. 4, March 1945, p. 106.
6. "How to Measure Width of Moving Webs," E. H. Alexander, *G. E. Rev.*, Vol. 44, Nov. 1941.
7. *Electron Tubes in Industry*, Keith Henney, McGraw-Hill, 1937.
8. *Photocells and Their Applications*, V. K. Zworykin and E. D. Wilson, Wiley, 1930.
9. *Industrial Electronics*, F. H. Gulliksen and E. H. Vedder, Wiley, 1935.

Chapter 33

APPLICATIONS OF ULTRAVIOLET RADIATIONS

H. C. Rentschler

MANY physical, chemical, and biological reactions are influenced by light and particularly by ultraviolet radiation. In Chapter 10 lamps for producing radiation in the different regions of the ultraviolet spectrum are described. These lamps make possible numerous commercial uses of radiation which were heretofore feasible only to a limited extent. Applications may be classified as photochemical reactions, bactericidal and fungicidal actions, biological effects, and excitation of fluorescence and phosphorescence.

33·1 PHOTOCHEMICAL REACTIONS

Photochemical reactions are chemical changes brought about by radiant energy. One of the earliest of these reactions observed, about 1727, was the action of light on silver salts. This was the forerunner of photography.

Conversion of oxygen into ozone is a photochemical reaction which requires ultraviolet radiation shorter than about 2000 angstrom units. The near-ultraviolet band causes fading of colors in fabrics, and the radiation which produces the green color in plants is a maximum in the yellow portion of the spectrum. These examples show that different photochemical reactions are produced by different wave bands of radiation.

Popular interest in photochemical reactions was stimulated by the discovery that vitamin D is produced in some food products exposed to ultraviolet radiation, and that the prevention and cure of rickets are brought about by the use of foods rich in vitamin D or by direct exposure of the body to middle-ultraviolet radiation. Utilization of sunlight in promoting the growth of plants is one of the most important and mysterious photochemical reactions known.

Grotthus in 1817 concluded from his studies on the fading of alcoholic solutions of iron salts that chemical action is produced only by the absorbed radiation. All absorbed radiation does not cause photochemical changes. Sometimes absorbed energy is converted into heat and produces no photochemical reaction.

By an extensive study of the combination of hydrogen and chlorine under the action of light, Bunsen and Roscoe [1] proved experimentally that "photochemical action is proportional to the product of the intensity of the radiation and the time during which it acts." This is known as the Bunsen-Roscoe reciprocity law.

Many photochemical reactions produced by ultraviolet radiation and a theoretical discussion of these actions based on quantum theory are beyond the scope of this chapter, and the reader must refer to treatises on photochemical reactions.[2]

33·2 BACTERICIDAL AND FUNGICIDAL ACTIONS

The lethal action of sunlight on bacteria was demonstrated as early as 1877.[3] It has since been proven that spores as well as bacteria are destroyed by the sun's radiation.

In 1903 Barnard and Morgan [4] reported that bactericidal action is limited to wavelengths shorter than about 3000 angstrom units. In 1929 Gates [5] showed that the lethal action on bacteria increases from 3000 angstrom units to a maximum at approximately 2660 angstrom units, then decreases to a minimum at about 2370, and again rises for still shorter wavelengths.

The susceptibility of bacteria, spores, and other microorganisms to ultraviolet radiation under varying conditions has been widely investigated.[6,7,8,9,10,11]

In Chapter 10 a lamp that generates nearly all its radiation in the bactericidal region of the spectrum with practically no heat or other radiations has been described. Such a lamp, together with a simple ultraviolet meter for measuring the effective bactericidal ultraviolet radiation, has made possible the accumulation of data for the successful commercial applications of ultraviolet as a germicidal and fungicidal agent.[12,13] A few of the more important facts thus established are:

1. The Bunsen-Roscoe reciprocity law was verified over a range of intensities requiring from a few microseconds to many minutes for producing the same amount of radiation.
2. For low-intensity long-time exposures (from several minutes to 40 hours at room temperature) the effective bactericidal action is increased because of the difference in sensitivity to ultraviolet of a bacterium as it passes through its life span.
3. For weak exposures or intensities too low to produce lethal action, the rate at which bacteria develop is retarded by ultraviolet radiation.
4. An individual bacterium requires the same amount of radiant energy for destruction whether it is floating in air or is on the surface of the agar in a Petri plate.

5. The effective bactericidal intensity from an ultraviolet lamp at a distance of 1 meter perpendicular to the axis of the lamp is the equivalent of about 2 microwatts of 2537-angstrom radiation per square centimeter of surface for each watt consumed in the lamp.

6. Table 33·1 shows the approximate ultraviolet radiation required for destroying various percentages of the more common micro-organisms.

TABLE 33·1

Micro-Organism	Clicks on Standard Tantalum Cell Meter		
	80–90% Kill	90–95% Kill	95–100% Kill
Bacteria			
Streptococcus hemolyticus (alpha type)	15–18	18– 22	22– 25
Streptococcus hemolyticus (beta type)	15–23	23– 28	28– 35
Streptococcus lactis	25–30	30– 35	35– 40
Staphylococcus aureus	15–20	20– 24	24– 30
Staphylococcus albus	18–22	22– 25	25– 30
Neisseria catarrhalis	15–20	20– 25	25– 50
Micrococcus piltonensis	35–40	40– 50	50– 60
Micrococcus sphaeroides	50–60	60– 65	65– 75
Sarcina lutea	80–95	95–110	110–130
Corynebacterium diphtheriae	20–30	30– 40	40– 50
Shigella flexneri	15–20	20– 25	25– 30
Eberthella typhosa	15–20	20– 30	30– 35
Pseudomonas fluorescens	10–20	20– 25	25– 32
Escherichia coli	15–20	20– 25	25– 30
Proteus vulgaris	10–15	15– 20	20– 30
Serratia marcescens	15–20	20– 25	25– 30
Phytomonas tumefaciens	15–20	20– 25	25– 45
Bacillus anthracis	20–30	30– 50	50– 65
Bacillus fusiformis	10–15	15– 20	20– 40
Bacillus subtilis	25–35	35– 45	45– 55
Bacillus subtilis spores	40–50	50– 65	65–100
Spirillum rubrum	25–30	30– 35	35– 40
Yeasts			
Saccharomyces ellipsoideus (wine yeast)	20–30	30– 40	40– 60
Saccharomyces ellipsoideus (wild yeast from grapes)	40–50	50– 60	60– 80
Saccharomyces spp. (from orange juice)	25–30	30– 35	35– 50
Saccharomyces cerevisiae (molasses distillery yeast)	25–35	35– 40	40– 60
Brewers' yeast	12–15	15– 20	20– 30
Bakers' yeast	10–20	20– 25	25– 40
Common yeast cake	20–30	30– 40	40– 60
Molds			100% Kill
Penicillium roqueforti			150
Penicillium expansum			200
Penicillium digitatum			500
Aspergillus glaucus			350
Aspergillus flavus			500
Aspergillus niger			1500
Oospora lactis			80
Mucor racemosus			180
Rhizopus nigricans			1500

One click on standard meter = 220 microwatts × seconds per square centimeter.

Commercial applications of bactericidal radiation fall into three main groups: (1) destruction of micro-organisms floating in air, to prevent cross infection and contamination from air-borne micro-organisms; (2) destruction of bacteria and spores on the surface of solids; and (3) destruction of micro-organisms suspended in liquids.

Radiation for Destroying Air-Borne Organisms

Air-borne bacteria may be destroyed by exposure of the air to direct radiation from an ultraviolet source. This procedure is used over operating tables [14] in hospitals to prevent infections (see Fig. 33·1) and in laboratories during the transfer of sterile bacteriological media. Personnel in the rooms must be protected against excessive exposure by goggles and face masks.

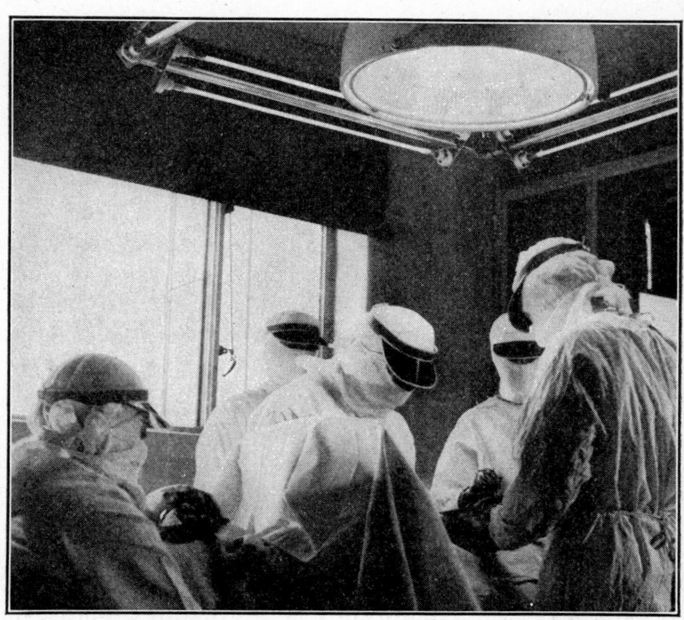

FIG. 33·1 Ultraviolet lamps over operating table at Duke Hospital, Durham, N. C.

A similar method is used extensively to prevent the spread of epidemics of respiratory diseases with poultry and animals, and also to protect perishable products against bacterial and mold spoilage.

By the use of semi-direct radiation the dangers from cross infection are minimized. Air in parts of a room not occupied is irradiated while the occupied space is shaded from the direct rays.[15] This type of radiation is particularly suitable in hospital wards and nurseries. Air-borne bacteria floating from one section of the room to another pass through a curtain of radiation; thus the possibility of cross infection is reduced. Figure 33·2 illustrates such an installation. In occupied rooms it is safe to irradiate only the upper air to avoid overexposure of the occupants. Bacteria are destroyed as they float into the irradiated space.[16] This method is suitable for offices, schools, or other places in which people congregate.

When ultraviolet lamps are placed in the ducts of an air-conditioning system [17] all the recirculated air is exposed to the radiation as it approaches and passes the lamps, and the chances of cross infection from one room to another on the same recirculating system are lessened. Figure 33·3 shows the arrangement of lamps for such an application.

Surface Irradiation

This application covers the destruction of bacteria and spores remaining on such surfaces as drinking glasses after they have been washed. The micro-organisms left in a thin film on the surface of glasses or those settling on them from the air are destroyed by direct exposure to ultraviolet.

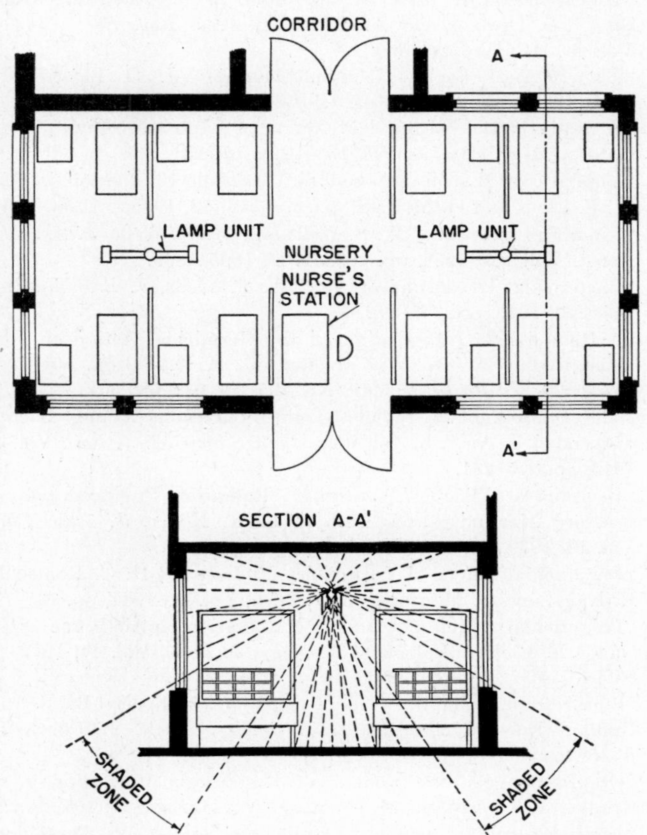

Fig. 33·2 Ultraviolet lamp units for air disinfection showing distribution of radiation with shaded zones.

Foods in refrigerators are subject to spoilage by bacterial and mold growth, most of which takes place on the surface. Such spoilage is greatly retarded by the use of ultraviolet radiation, which at the same time produces a small but controlled amount of ozone which destroys micro-organisms not directly exposed to the radiation. Objectionable odors are also materially reduced.

In the new method of meat tenderization, beef can be tenderized in a few days with the use of ultraviolet lamps. The process reduces surface spoilage by bacteria and mold and permits the aging of the beef in 36 to 48 hours at a temperature of 60 to 70 degrees Fahrenheit instead of the usual several weeks at a much lower temperature. The rate of tenderization is a function of temperature. By use of ultraviolet radiation the aging progresses more rapidly, and trimming losses are negligible.

Irradiation of Liquids

To destroy bacteria in a liquid the liquid must be transparent to ultraviolet radiation and must be subjected to a sufficiently high intensity for destruction of the organisms during the exposure. This method has been used for purifying water in swimming pools.

33·3 BIOLOGICAL EFFECTS

The beneficial and curative effects of ultraviolet irradiation have been demonstrated in cases of rickets and extrapulmonary tuberculosis. The healing of infected wounds is generally accelerated by ultraviolet radiation.

Mild exposures to special ultraviolet lamps such as sun lamps produce the same effect as sunlight bathing, but therapeutic irradiation of the body should be carried out under the supervision of a physician.

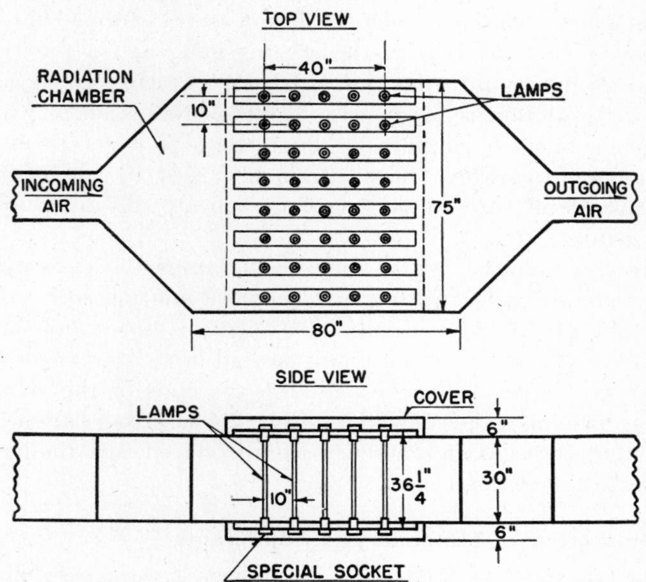

Fig. 33·3 Schematic arrangement of ultraviolet lamps in duct of an air-conditioning system.

Excessive exposure to ultraviolet radiation causes conjunctivitis and serious sunburn. The eyes always should be protected against direct exposure. Ordinary glasses are effective only against the radiation absorbed by the lenses, and additional protection is necessary to prevent radiation from reaching the eyes outside the glass rims. Moderate exposure to the middle-ultraviolet band of radiation produces tanning similar to that obtained from sunlight.

33·4 EXCITATION OF FLUORESCENCE AND PHOSPHORESCENCE

Many substances absorb radiant energy and transform a part of it into radiation of a different wavelength. The phenomenon is called fluorescence if emitted radiation lasts only while the exciting energy is applied, and phosphorescence when the emitted light continues for a time after the stimulus is removed. The wave band which stimulates fluorescence or phosphorescence and the color of the fluorescent or phosphorescent radiation depend upon the substance absorbing and transforming the radiation. Willemite, a zinc silicate, fluoresces to short ultraviolet radiation. Some specimens show phosphorescence with a brilliant green color.

Calcium tungstate crystals, similarly exposed, fluoresce blue. Zinc sulphide must be exposed to the near ultraviolet to cause efficient fluorescence.

The two most important uses of fluorescence and phosphorescence produced by ultraviolet radiation are in the commercial production of light, such as fluorescent lamps, and in the identification of materials and adulterants.

Fluorescent Lamps

In Chapter 10 an ultraviolet generator which produces ultraviolet mainly in the 2537-angstrom radiation band of mercury is described. If such a lamp is coated on the inside with a thin layer of zinc silicate, the ultraviolet radiation is transformed into a wide band in the green part of the spectrum. If this lamp is operated under optimum conditions a quantum of 2537-angstrom radiation is transformed into a quantum of radiation in the green region, and a practical transformation into light of nearly 50 percent of the generated radiation is produced. For the production of white light a blended mixture of fluorescent powders is used. Fluorescent powders most commonly used in commercial lamps are mixtures of zinc-beryllium silicate and magnesium tungstate.

In the manufacture of fluorescent lamps the glass tube generally is flushed with a nitrocellulose solution containing the phosphor in suspension. The tube is drained and dried at a low temperature and finally heated in air to remove the binder. Electrodes are sealed into the ends of the coated tube, which is then exhausted. Next, the electrodes are heat-treated, inert gas and mercury are introduced, and the lamp is sealed off and seasoned.

Identification of Materials and Adulterants

In the use of ultraviolet for identification purposes a filter which absorbs nearly all the visible light but transmits the ultraviolet is generally used over the ultraviolet light so that the specimen is not visible except for the portion which fluoresces. Materials are identified by their characteristic fluorescent colors. For example, an ore containing willemite (zinc silicate) fluoresces bright green. Mineral oils fluoresce and are readily distinguished from vegetable oils which do not fluoresce. This method of determining the presence or absence of certain substances is so universally applicable that its possibilities are innumerable. For details the reader is referred to a standard treatise [18] on the subject.

REFERENCES

1. "Photo-Chemical Investigation," R. Bunsen and H. Roscoe, *Poggendorff-Ann. Physik*, Vol. 96, 1855, p. 373.
2. *The Chemical Action of Ultraviolet Rays*, C. Ellis and A. A. Wells (revised and enlarged by F. F. Heyrother, published by Reinhold, 1941).
3. "Researches on the Effect of Light upon Bacteria and other Organisms," A. Downes and T. P. Blunt, *Proc. Roy. Soc.* (*London*), Vol. 26, 1877, pp. 488–500.
4. "The Physical Factors in Phototherapy," J. E. Barnard and H. DeR. Morgan, *Brit. Med. J.*, Vol. 2, 1903, pp. 1269–1271.
5. "A Study of the Bactericidal Action of Ultraviolet Light," F. L Gates, *J. Gen. Physiol.*, Vol. 13, 1929*a*, pp. 231–260.
6. "Viability of B. coli Exposed to Ultraviolet Radiation in Air," W. F. Wells and G. M. Fair, *Science*, Vol. 82, 1935, pp. 280–281.
7. "Air-Borne Infection," W. F. Wells and M. W. Wells, *J. Am. Med. Assoc.*, Vol. 107, 1936, pp. 1698–1703, 1805–1809.
8. "Bactericidal Irradiation of Air," W. F. Wells, *J. Franklin Inst.*, Vol. 229, 1940, pp. 347–372.
9. "A Radiometric Investigation of the Germicidal Action of Ultraviolet Radiation," W. W. Coblentz and H. R. Fulton, *Natl. Bur. Standards Sci. Paper*, Vol. 19, No. 495, pp. 641–680.
10. "The Efficacy of Ultraviolet Light Sources in Killing Bacteria Suspended in Air," B. Whistler, *Iowa State Coll. J. Sci.*, Vol. 14, 1940, pp. 215–231.
11. "Bactericidal Effect of Ultraviolet Radiation Produced by Low Pressure Mercury Vapor Lamp," L. R. Koller, *J. App. Phys.*, Vol. 10, 1939, pp. 624–630.
12. "Bactericidal Effect of Ultraviolet Radiation," H. C. Rentschler, R. Nagy, and G. Mouromtseff, *J. Bact.*, Vol. 41, 1941, pp. 745–774.
13. "Bactericidal Action of Ultraviolet Radiation on Air-Borne Organisms," H. C. Rentschler and R. Nagy, *J. Bact.*, Vol. 44, 1942, pp. 85–94.
14. "Sterilization of the Air in Operating Rooms by Special Bactericidal Radiant Energy," D. Hart, *J. Thoracic Surgery*, Vol. 6, No. 1, p. 4581.
15. "Observations on the Control of Respiratory Contagion in the Cradle," I. Rosenstern, *Aerobiology A.A.A.S. Publ.* 17, 1942, p. 242.
16. "Environmental Control of Epidemic Spread of Contagion," W. F. Wells, M. W. Wells, and T. S. Wilder, *Aerobiology A.A.A.S. Publ* 17, 1942, p. 206.
17. "Advantages of Bactericidal Ultraviolet Radiation in Air Conditioning Systems," H. C. Rentschler and R. Nagy, *A.S.H.V.E. J.*, Jan. 1940.
18. *Fluorescence Analysis in Ultraviolet Light*, J. A. Radley and J. Grant, Chapman and Hall, Ltd., 1939.

Chapter 34

RADAR: FUNDAMENTALS AND APPLICATIONS

Hugh Odishaw

THE principle of radar is the reflection of electromagnetic energy from an object in its path. The analogous behavior of sound and light waves affords a somewhat more familiar illustration of the phenomenon. Microwaves, radiations of wavelengths ranging from a few millimeters to several centimeters, behave much like light waves. Seeing depends upon reflection of light from the object to the eye. Detection of distant objects by radar likewise depends upon reflection of high-frequency radio energy.

This principle is useful because electromagnetic waves are propagated at a constant velocity of about 3×10^8 meters per second. Reflection of energy usually indicates the presence of a target within range of the system; and, because the velocity is constant, the target range can be measured by the lapse of time between transmission of a signal and reception of its reflection. Finally, electromagnetic waves travel in straight lines like light rays; therefore, the elevation and bearing of the target can be measured, because the angular position of the radar antenna with respect to some reference plane coincides with the angular position of the target. (Rotation and tilting of antennas is incidental, because the ratio of these velocities to the velocity of light may be considered to be zero.)

Although electromagnetic waves travel essentially in straight lines, shorter waves are preferable to longer ones for several reasons. Shorter waves can be more conveniently beamed. Beaming is a prerequisite for precise determination of bearing and elevation. Waves at broadcast frequencies would require a row of towering antennas stretched for miles for appreciable directional effects, but microwaves can be beamed with small, light reflectors. For this reason the trend has been more and more toward microwaves, and within this region toward wavelengths of a few centimeters. Another reason for using rather short waves is that they are less influenced by ground effects that account for the apparent curvature of commercial broadcast waves. This straight-line characteristic of the shorter waves is invaluable in obtaining accurate bearing and elevation data. Still another reason for preferring shorter waves is that for satisfactory reflection the ratio of wavelength to target dimensions should be small. Detection of planes, for example, requires relatively short waves for optimum reflection.

34·1 SIGNAL DIFFERENTIATION

One of the first problems in radar systems is the method of differentiating between the transmitted signal and the reflected signal. The same antenna is used for both transmission and reception because the synchronizing of two antennas is difficult; moreover, the receiving antenna would almost inevitably pick up some of the transmitted signal and, because the reflected signal is generally very weak, the problem of differentiation would remain. There are three methods of signal differentiation: continuous wave, frequency modulation, and pulse modulation.

In the continuous-wave system an unmodulated signal is transmitted continuously. If the target is in motion, the reflected energy undergoes a frequency shift (the Doppler effect), and this shift indicates the speed of the target relative to the detecting equipment. Stationary objects have no effect on the frequency of the signal, and the continuous-wave system is thus limited to detection of moving targets.

The frequency-modulation system is effective against stationary targets. The frequency of continuously transmitted energy is varied rapidly over a broad range of frequencies. In general, the reflected signal has a frequency different from that of the transmitted signal, and the frequency difference depends upon the travel time of the energy to and from the target. The two frequencies are combined in the receiver, and the beat frequency is detected. This frequency is a measure of travel time and hence of range. If the target or the detector is in motion, however, the Doppler effect introduces a frequency shift indistinguishable from the one depending upon the travel time of the signal, and the system is ineffective without compensating circuits.

In the third system, pulse modulation, the problem of single differentiation is avoided by having only one signal, either the transmitted or the reflected, at any moment. Short pulses of energy, from a fraction of a microsecond to several microseconds in duration, are radiated; they are followed by longer intervals of quiescence, usually from a few hundred to a few thousand microseconds in duration, during which reflections are received. If the range of the target is such that reflections return during a given period of transmission or multiples of it, differentiation usually is impossible, but these contingencies are not important.

34·2 PULSE-MODULATED MICROWAVE RADAR

A typical pulsed detection system may be divided into five major components: (1) control center, (2) transmitter, (3) radiator, (4) receiver, and (5) indicators. The control center establishes a pulse which initiates the transmission cycle and synchronizes the indicator sweeps. In response to this synchronizing pulse, the magnetron of the transmitter pulses, and a powerful, short burst of energy is fed to the radiator and is radiated into space. Reflected signals are intercepted by the antenna, detected and amplified in the receiver, and fed to the indicators where they appear visually. The function of a detection system is to obtain continuously one or more co-ordinates of a target with respect to the detector. Range is determined by time measurement; elevation and bearing by synchro-tie systems linked to the rotating radiator and to elements controlling the deflection circuits of the indicators.

The sequence of events in a representative microwave detection system is more complex than this. Figure 34·1 shows details of the components of such a system.

respect to the grounded anode. The magnetron (Figs. 34·2 and 34·3), consisting essentially of a cathode and a massive copper anode in a multiple resonant cavity lying in the field of a strong permanent magnet, emits a pulse of high power at high frequency during this microsecond.

Since the same antenna is used for both transmission and reception, the problem of signal diversion to the receiver during transmission arises. Branching of the transmitted energy is undesirable because of the resultant decrease in signal output (and hence in maximum range) and because of the saturation or destruction of the sensitive receiver. High receiver sensitivity is necessary because reflected signals are generally weak, as low in power as, say, 10^{-12} watt. If an appreciable portion of the powerful transmitted signal

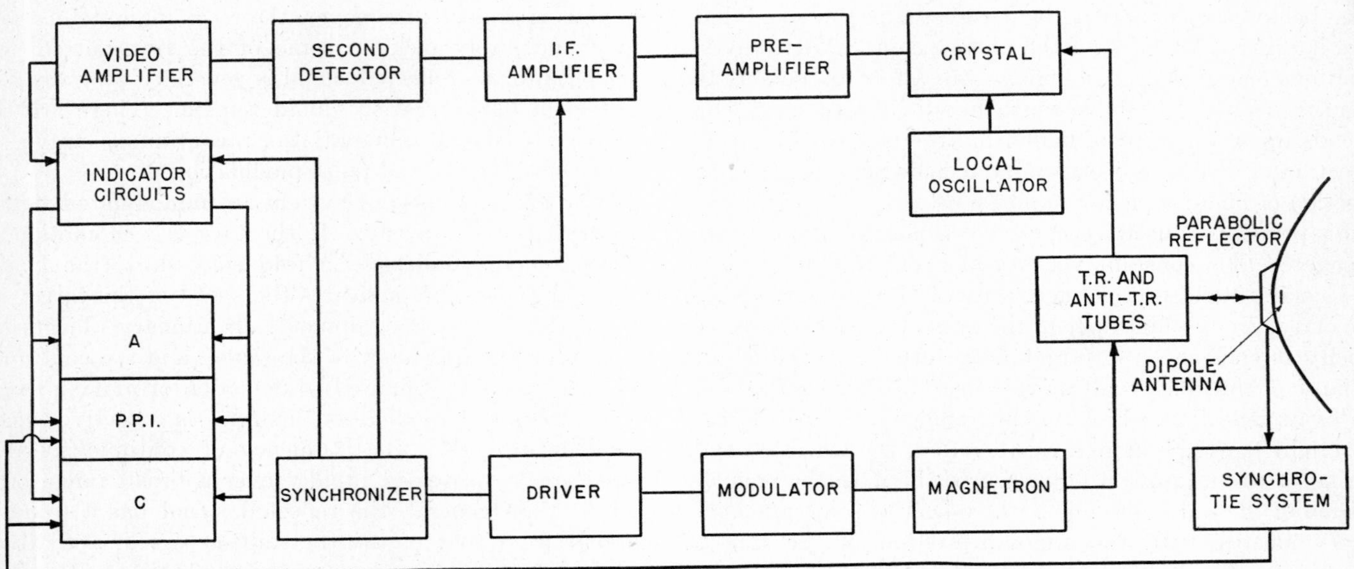

Fig. 34·1 Block diagram of a typical radar system.

The synchronizer unit has three functions. It creates a pulse that triggers the driver of the transmitter. The leading edge of this pulse establishes the timing of transmission and must be sharp and accurate. The pulse rises rapidly, then decays exponentially. At the same time, the synchronizer triggers the indicator circuits that control the indicator sweeps. Finally, the synchronizer creates the gate pulse which sensitizes the receiver. The purpose of the gate pulse is to energize the receiver throughout any portion of the cycle during which reflected signals return. Thus a particular range portion may be chosen for reception, excluding others. The width of this pulse is variable, and its position can be shifted from the minimum to the maximum range positions.

The synchronizing pulse from the control center triggers the driver, the output of which is a rectangular pulse of 1000 to 5000 watts. (All quantities given in this chapter are merely representative, and are used solely for illustrative purposes.) This pulse, in turn, initiates the modulator pulse, a rectangular pulse of high power (100,000 to 500,000 watts), lasting for a microsecond (a typical period). The modulator or keyer pulse is impressed between the anode and cathode of the magnetron, which is the high-frequency oscillator, with such a phase difference that the cathode is negative with

is permitted to enter the receiver, it may be burned out or saturated so that reflected signals cannot be detected. Conversely, a common antenna path can cause dissipation of reflected energy in the transmitter. For both these reasons, a method of opening and closing the paths to the receiver and transmitter is imperative.

The set of components designed for this function is known as the *duplexer*. It consists of two TR (transmit-receive) tubes, one for blocking the path to the receiver during transmission, the other for blocking the path to the transmitter during reception. The TR tube (Fig. 34·4) consists of a gas-filled copper cavity that resonates at the signal frequency, a pair of spark-gap electrodes, and a keep-alive electrode. The keep-alive electrode maintains a sufficient number of ions in the gas (a combination of hydrogen and water vapor) so that the gas ionizes completely and rapidly when the magnetron signal reaches the spark-gap electrodes. At this instant, free electrons in the gap between the two electrodes oscillate violently, and the electron density is so high that it produces a short-circuiting effect.

The tube intended to protect the receiver is called the TR tube; the tube preventing diversion of reflected energy to the transmitter is called the anti-TR tube. The TR tube fires almost instantly when the magnetron signal reaches the TR junction, short-circuiting the receiver from the antenna circuit. Most of the energy reaches the antenna, but a small part, representing the voltage the electrodes can withstand before firing, reaches the receiver, finally appearing on type A indicators as a zero-reference point.

The anti-TR tube is similar to the TR tube, but is backed by a short-circuiting plate. In conjunction with the associated wave-guide system it presents a very high impedance to energy traveling toward the transmitter.

Energy reaching the radiator must be beamed if target bearing and elevation are to be obtained; the sharper the beam, the more precise this information and the higher the discrimination between adjacent targets. A common radiator consists of a dipole antenna (although wave-guide horns can be used) and a parabolic reflector. Energy radiating back into the reflector is beamed. A reflecting shield in front of the dipole or another reflecting dipole directs the forward energy back to the reflector so that most of the energy is effectively beamed.

FIG. 34·2 Ten-centimeter magnetron.

Beam width depends upon wavelength of energy and diameter of reflector. Angular width of the beam is equal to the ratio of:

$$\theta = \frac{\lambda}{d} \tag{34·1}$$

or, in terms of radiator area, assuming circular aperture:

$$\theta = \frac{\lambda}{\dfrac{2\sqrt{A}}{\sqrt{\pi}}} \tag{34·2}$$

where θ = beam width in radians
λ = wavelength in meters
d = radiator diameter in meters
A = radiator area in square meters.

Angular width of the beam is fixed by the degree of discrimination and accuracy required of a given system. Once θ has been decided, λ and d may be chosen. Obviously, a variety of values provides the same ratio. Their choice depends upon numerous factors. The size of the target affects wavelength, which must be small in comparison with the dimensions of the targets; but attenuation at high frequencies tends to favor longer wavelengths for long-range systems. Nor can radiator size be chosen independently. Large radiators are suitable for permanent or semi-permanent ground installations; small ones are necessary on airplanes and light sea craft; intermediate sizes are practical on large ships.

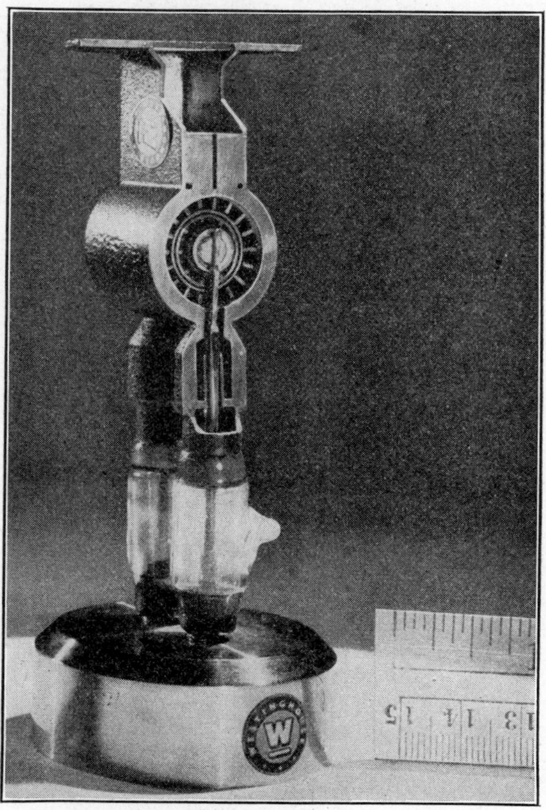

FIG. 34·3 Three-centimeter magnetron.

Usually the radiator of a detection system is rotated and tilted continuously, thus scanning a large volume of space surrounding the system. Bearing and elevation information depends upon co-ordinating radiator motion with indicator sweeps, usually by synchro-tie systems. One motor is activated by the rotary motion, another by tilting. Corresponding units rotate the magnetic-deflection coils or vary potentiometers of the indicators, establishing target position on the indicators. Target position at the instant of detection coincides with the angular position of the radiator.

The reflected signal enters the radiator and passes to the receiver. Attenuation is low because the anti-TR tube blocks the path to the transmitter. However, the signal usually is weak, requiring extensive amplification, and is high in frequency, requiring detection. Above 3 centimeters, tube detection is precluded because the transit time of electrons is of the order of the period of the high-frequency energy. The semi-conductor probe rectifier (Figs. 34·5 and 34·6) utilized for this purpose consists essentially of a "synthetic" crystal, a wire probe, and a containing porcelain cartridge. The crystal is silicon alloyed with small amounts of aluminum, boron, and beryllium. One surface of the

FIG. 34·4 TR tube.

crystal is nickel-plated to permit soldering; the other surface is ground and etched and is the working face of the crystal. The probe is a fine tungsten wire (a diameter of 0.005 inch

FIG. 34·5 Semi-conductor probe rectifier.

is common) tinned with an alloy of gold, platinum, and silver so that it can be soldered to the end piece.

Rectification depends upon the non-linear characteristics of the crystal (Fig. 34·7). The reflected signal and a signal of about the same frequency from the local oscillator (usually a klystron such as that shown in Fig. 34·8) are impressed on the crystal. For small current and voltage, the rectifying action of the crystal is expressed approximately by the equation

$$I = aV + bV^2 \tag{34·3}$$

where I is output current in amperes, V is total voltage, and a and b are constants depending upon the physical qualities

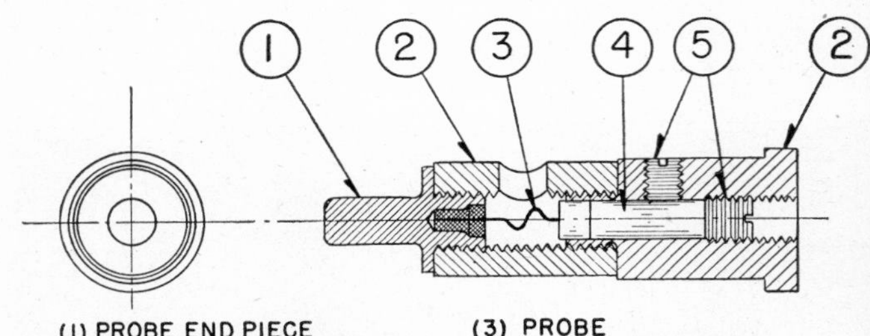

FIG. 34·6 Cross section drawing of the semi-conductor probe rectifier.

of the crystal. If $V_r \cos \omega_r t$ is the reflected signal voltage and $V_k \cos \omega_k t$ is the klystron voltage, the total voltage is the sum of these, and

$$I = \frac{b}{2}(V_r^2 + V_k^2) + aV_r \cos \omega_r t + \frac{bV_r^2}{2} \cos 2\omega_r t + aV_k \cos \omega_k t + \frac{bV_k^2}{2} \cos 2\omega_k t + bV_rV_k \cos (\omega_k + \omega_r)t + bV_rV_k \cos (\omega_r - \omega_k)t \tag{34·4}$$

Only the last component is desired. The first is a d-c component; the following five are components at the reflected

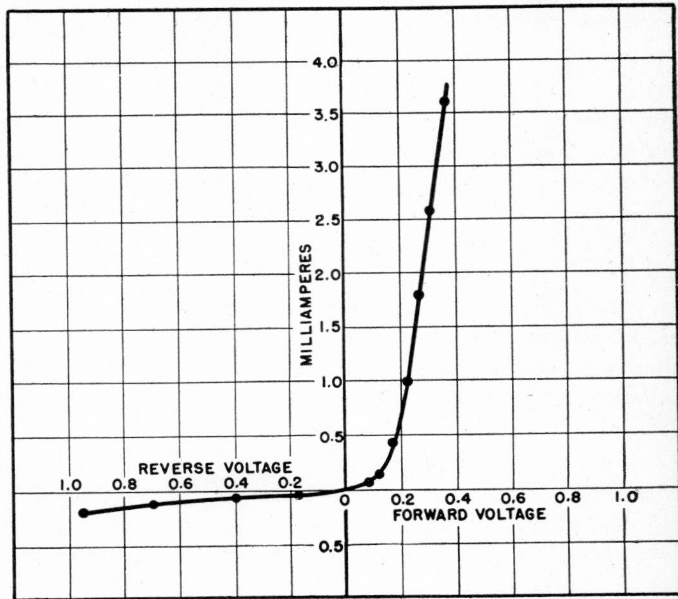

FIG. 34·7 Static characteristic curve for the semi-conductor probe rectifier.

and klystron-signal frequencies, at twice these frequencies, and at the sum of these frequencies. These components are by-passed to ground. The component representing a frequency equal to the difference in the two frequencies is at an intermediate frequency. Thus, if the reflected-signal frequency is 20,000 megacycles and the klystron frequency is 19,970 megacycles, the intermediate frequency is 30 megacycles.

Because the intermediate-frequency video amplifiers are tuned, the intermediate-frequency output of the mixer must be relatively stable. Stability of the intermediate-frequency signal depends upon stability of the transmitting magnetron and of the local oscillator. The magnetron frequency, in particular, is sensitive to changing impedance of the wave guide, antenna, and space complex. In addition, tuning the klystron is difficult, and once tuned its frequency inevitably shifts. Klystron frequency is sensitive also to voltage changes. Klystron voltage can be controlled by

Fig. 34·8 Reflex klystron.

feeding a small portion of its output to a mixer to secure an intermediate frequency signal and to a discriminator for a video signal. This signal is, in turn, used to control a thyratron which provides klystron control voltage. Drift of klystron frequency thus results in an automatic voltage correction tending to maintain the desired frequency. A simpler method is to use a standard-frequency cavity to control klystron frequency by providing correcting voltages. A small amount of the klystron output is fed to the resonant cavity; deviations from the intermediate frequency result in voltages which correct drift.

The intermediate-frequency signal proceeds through a series of amplifications. Because signal-to-noise ratio is critical, two stages of intermediate frequency amplification follow mixing, and these stages are closely coupled to the mixer. The signal then is conveyed to the receiver proper, in the console of the detection system, where it undergoes from five to seven additional stages of intermediate-frequency amplification. The signal frequency is lowered again, this time to a video frequency suitable for the indicators, by means of a simple diode detector. A few additional stages of amplification at the video level prepare the signal for visual presentation on the indicators.

Both intermediate-frequency and video amplifiers are tuned to the proper frequencies. For high discrimination the output signal must have sharp edges; therefore the intermediate-frequency and video amplifiers must be capable of passing certain frequency bands. Thus, if the magnetron pulse (which contains a band of frequencies) has a duration of 1.0 microsecond, the lowest frequency in the band is 500 kilocycles. For acceptable signal form, the amplifiers must be able to pass the third and fifth harmonics of this frequency, or 1.5 and 2.5 megacycles. This requirement means that the intermediate-frequency amplifier must be able to pass the upper and lower sidebands (from 3 to 5 megacycles wide), and the video stage must pass a band from 1.5 to 2.5 megacycles.

The transmitted signal is attenuated in proportion to the square of the distance from the transmitter, and the reflected signal is similarly attenuated as it travels away from the target, so the signal strength is variable. Targets near the detection system have powerful echoes; those far away, weak ones. Obviously the output of the receiver varies inversely as the fourth power of time for signals along any given line radiating from the system. To compensate for this undesirable variation, receiver gain must increase as the fourth power of time. One method of approximate compensation increases screen voltage to the third and fourth intermediate frequency amplification stages exponentially by means of cathode followers, varying the transconductance of those stages so that receiver output is uniform.

The receiver output is fed to the indicator, which consists of one or more cathode-ray tubes and the associated circuits forming, controlling, and modulating the electron beams. Magnetically deflected tubes are generally preferred, particularly for intensity modulation. They are shorter, do not need high deflection voltages, provide satisfactory focusing over the whole tube face even at high currents, and are more rugged mechanically than tubes with electrostatic deflection. The face of an indicator is constructed as flat as possible to increase the useful area and reduce parallax when scales are mounted on the tube faces.

34·3 INDICATORS

Indicators present visually the reflected signals so that the three target co-ordinates, range, bearing, and elevation, can be readily and continuously comprehended by the operator. Three co-ordinates are not usually presented on a single indicator; it is simpler to present range and bearing on one indicator, bearing and elevation on another, although the so-called G and H types of indicators present all three. In the type G indicator the tube is calibrated vertically in elevation and horizontally in bearing; the reflected signal appears as a bright spot; range is determined by "wings" extending from the spot; and the length of the wings is an inverse measure of range. In H indication, range appears vertically, bearing horizontally. Echoes appear as bright spots; elevation is approximated by the slope of short lines from the echo spots. The more common indicators are the A for range measurement; B for range and bearing; C for elevation and bearing; and PPI (plan position indicator) for range and bearing in a map-like fashion (Fig. 34·9). They may be classified as deflection or intensity-modulation indicators.

In deflection types the reflected signal deflects the cathode-ray beam. The most familiar of this group is the type A indicator, in which range is presented horizontally, signal intensity vertically. Sweep is triggered by a pulse from the synchronizing unit of the control center, the timing of which

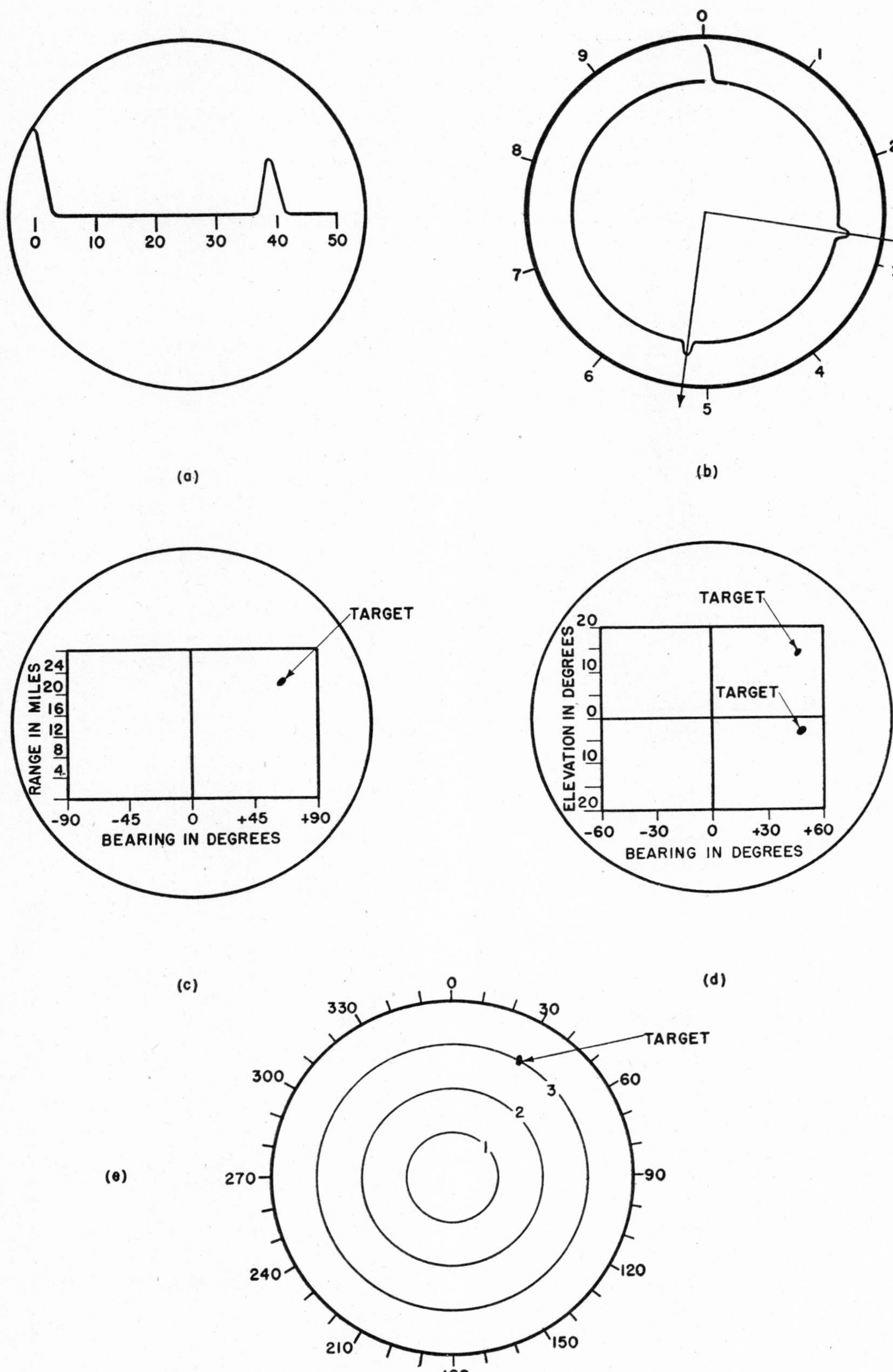

FIG. 34·9 Radar indicators: (*a*) A scope, (*b*) circular A scope, (*c*) B scope, (*d*) C scope, and (*e*) PPI.

coincides with the pulse which triggers the magnetron. The electron beam sweeps with constant intensity across the face of the tube in a straight line corresponding to the transmitted pulse cycle. If, for example, the pulse-repetition interval is 1000 microseconds, the trace represents 1000 microseconds. A small portion of the transmitted signal reaches the A scope, deflects the beam upward, and marks the zero-reference point. Reflected signals appear somewhere along the trace, deflecting the beam in the shape of an inverted V, called a "pip."

The A scope is calibrated in units of length, usually in miles. Calibration utilizes two facts: the known velocity of propagation (3×10^8 meters per second) and the round-trip path of the energy. Thus, if 100 microseconds elapse between the transmission of a signal and the reception of the echo, a total distance of 30,000 meters or 98,358 feet has been traversed, and the range to the target is 15,000 meters or 49,179 feet. Scales representing range on all indicators are calibrated directly in terms of straight-line target range. The A scope is widely used for tuning and adjusting the system because it shows the wave form of the signal and its intensity. Type A scopes are also used for accurate range measurement in systems employing other indications, or alone in systems supplying only range data.

A modified form of the A scope uses a circular sweep. One of the advantages of this modification is that flyback time causes no difficulty. In the ordinary A scope the electron beam must revert quickly from the end of the trace to its original position in time for the next cycle. In the circular A scope, the end of one sweep coincides with the beginning of the next. Still another advantage is greater precision because, for a given tube size, the trace can be about 3 times as long, and thus permit better scale calibration and easier interpretation.

Indicators of types B, C, and PPI use intensity-modulation, which is necessary if two or more co-ordinates are to be shown on a single scope. The electron beam sweeps across the face of the indicator tube in accordance with the coordinate data. Intensity of the beam is such that, in the absence of reflected signals, the tube is largely dark. Reflected signals are imposed on the tube so that beam intensity increases markedly, and a bright spot appears on the screen, which is coated with phosphors of long persistence.

Type C is the simplest of the intensity-modulated indicators. Information from the synchro-tie systems, which are linked to the moving radiator, is fed to the scope in such a manner that the electron beam moves horizontally to present bearing and vertically to present elevation. Target signals increase the intensity of the beam at points corresponding to their angular position in space with respect to the detection system, and a bright spot appears on the screen. Scale factors are generally linear. The length of the cycle of radiator motion in contrast to the relative shortness of the pulse cycle, however, results in a short scan line per pulse cycle. All noise is concentrated on the screen in this interval, and the signal-to-noise ratio is poor. Blanking circuits can be used to cut out all portions of the pulse cycle except that containing the reflected signal, but this requires a knowledge of the target range. Type C is thus generally limited to use with range indicators having good signal-to-noise ratios.

Type B is an intensity-modulated indicator in which the electron beam moves horizontally to present bearing, vertically to present range. Rectilinear relation of the co-ordinates results in a distorted map of the scanned region, because lines of equal bearing intervals are parallel vertical lines on the screen.

Next to the type A, the most widely used indicator is the PPI. Like type B, PPI is intensity-modulated and presents range and bearing. In PPI, however, the trace is swept from the center of the tube to the rim, each trace corresponding to the pulse cycle. Range is thus measured along radial lines. Rotation of the radiator drives the sweep through 360 degrees, instead of a maximum of about 180 degrees in type B. The 360-degree coverage of PPI and the undistorted (with horizontal scanning; other scanning introduces slant-range distortion) map-like presentation combine to make the PPI highly desirable, although type B gives greater range precision and angular resolution at short ranges. It is generally used with a type A scope, if accurate range measurements are necessary.

The map-like presentation of types B and PPI ensues from different reflectivities of different materials and varied angles of incidence. Water reflects comparatively little; ground, a fair amount; built-up areas and such targets as planes or ships, a good deal. Water thus appears dark or black on the screen; ground shows up fairly bright; built-up areas and metallic targets are bright.

Indicators are supervised by the control center, which contains the indicator circuits and the power supply. If the synchronizing unit of the control center triggers the modulator, a pulse is sent also to the indicator circuits, initiating the indicator sweeps. Indicator circuits include components for generating and amplifying the range sweeps, generating blanking pulses, electronic range markers and intensifier pulses, and for switching ranges. Electronic range markers appear in B as single markers and in PPI as a set of bright circles. Range switching is desirable if targets are close to the detection system, or if precise target information within a particular range bracket is needed. Under such conditions, the desired portion of the pulse cycle is expanded over the whole face of the indicator tube, other portions being blanked out.

34·4 TIME RELATIONS

Time relations of events in the representative detection system depend upon the synchronizing unit in the control center. Let it be assumed that the system transmits a 1.0-microsecond pulse and that the quiescent interval during which reflections return is 999 microseconds, making a pulse repetition interval of 1000 microseconds. The synchronizer delivers a sharp-edged pulse to the driver in the transmitter; the driver triggers the modulator with a low-power rectangular pulse; the modulator impresses a high-power rectangular pulse between the plate and the cathode of the magnetron; the magnetron oscillates at the designed frequency; and the pulse travels through the wave-guide system and is radiated by the reflector. There is virtually no time lag between responses of these various pulses, and all the pulses last for 1.0 microsecond.

In the meantime, the synchronizer delivers a triggering pulse to the indicator circuits to initiate the indicator sweeps. Simultaneously, the synchro-tie system rotates the magnetic deflection yokes controlling the angular position of the electron beam in the indicators in accordance with the motion of the radiator.

Type A indicators experience a deflection coinciding with the transmission of the magnetron pulse; this pip serves as the zero-range reference. Reflected signals return to the system some time during the quiescent interval. A gate pulse emanating from the synchronizer sensitizes the receiver for the reception of the echo. Deflection or intensity-modulated circuits impress the reflected signal on the indicators.

34·5 SCANNING METHODS

Effective search for a target requires that the space surrounding the detection system be scanned by the radiated beam. The method of scanning depends upon the function of the system. Ground and sea systems for detection of aircraft require rotation and tilting; systems for detection of surface craft need only horizontal rotation; aircraft searching for fixed surface targets require horizontal rotation at a fixed angle of tilt; aircraft searching for other planes require a forward beam rotated and tilted through rather small angles.

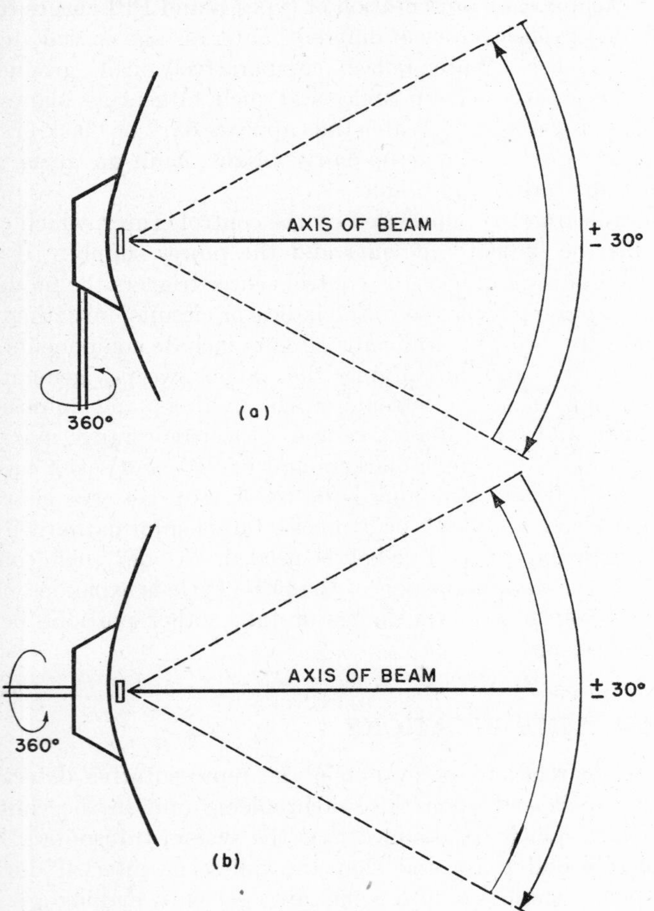

FIG. 34·10 Scanning methods: (a) helical and (b) spiral.

The principal methods of scanning are the helical and the spiral (Fig. 34·10). In the helical method the radiator is rotated constantly (about a vertical shaft) in the horizontal plane. On successive rotations, the radiator is tilted slightly. In this fashion a large volume of space is scanned continuously. A full 360 degrees of bearing is scanned, and the elevation angle depends upon the requirements of the system. In some ground systems the maximum angle above the horizontal is 60 degrees. Circular scanning is a special case of helical scanning, with a constant angle of tilts. Surface scanning at sea is circular scanning in which the angle of tilt is kept constant at about the horizontal level.

In spiral scanning the radiator is rotated rapidly about a horizontal axis by a spinning shaft. At the same time, the radiator is gradually tilted. The combination of motions causes the beam to sweep in a spiral. A special form of spiral scanning is conical scanning, in which the angle of tilt is fixed.

34·6 PULSE, POWER, AND RANGE

Pulse

In the design of a pulse-detection system the transmitted-pulse specification depends upon many interrelated factors such as wavelength, power limitations of magnetrons, and the design of sharp and short control pulses. Minimum and maximum range requirements suggest the value of the pulse and the length of the quiescent period. If, for example, it is tactically desirable to detect targets down to 150 meters, the pulse duration cannot exceed 1.0 microsecond, because the velocity of the energy is 300 meters per microsecond and the signal must travel 300 meters to and from the target. Similarly, if a maximum range of 75,000 meters is necessary, the time from the leading edge of one pulse to the leading edge of the next must be at least 500 microseconds. Pulse duration also affects discrimination. It has already been pointed out that the width of the beam determines angular discrimination. Thus, a beam 6 degrees in width covers a length of 104.82 meters at a range of 1000 meters. Targets separated by less than this distance merge into a single object, and for better angular discrimination a narrower beam width must be used.

Pulse width determines radial discrimination. If the pulse lasts 1.0 microsecond, for example, and two targets are separated by 150 meters, the energy reflected from the leading edge of the pulse striking the farther target returns to a position corresponding to the nearer one as the trailing edge of the pulse reaches the nearer one. As reflection ceases from the nearer target, it begins from the farther one, and the two targets are blurred into one. For better radial discrimination, then, shorter pulses are desirable.

Maximum range, however, depends upon total energy radiated. With given magnetrons a longer pulse means greater energy radiation and greater range.

The pulse is ordinarily described in terms of width w (Fig. 34·11) and repetition frequency f_r. Pulse-repetition interval has already been defined as the period between the leading edge of one pulse and that of the next; in other words, the repetition interval i is the sum of pulse width w

and the quiescent interval (during which the transmitter is quiescent) q:

$$i = w + q \tag{34·5}$$

Repetition frequency f_r obviously is the number of such repetition intervals or cycles per second and is equal to the reciprocal of the repetition interval:

$$f_r = \frac{1}{i} \tag{34·6}$$

Ratio of pulse width to repetition interval is the ratio of time during which the transmitter is on to the time it is silent. It is known as the duty cycle c:

$$c = \frac{w}{i} \tag{34·7}$$

The duty cycle is significant because it determines average transmitter power. Average power (Fig. 34·11) of rectangu-

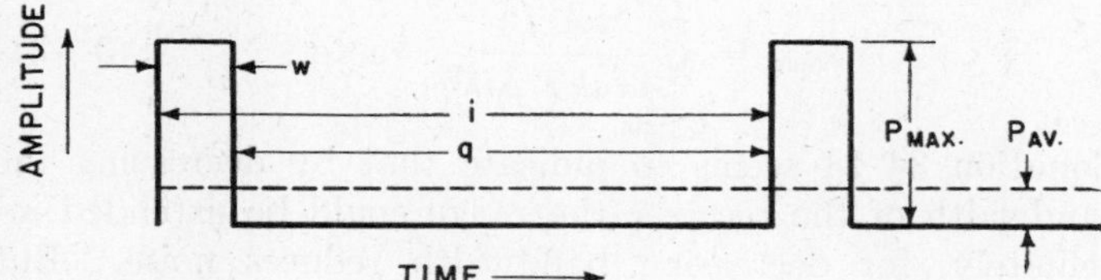

FIG. 34·11 Amplitude versus time of transmitted signal.

lar pulses, it is evident, equals the product of maximum power and pulse width divided by pulse-repetition interval:

$$P_{\text{ave}} = \frac{P_{\text{max}}w}{i} \tag{34·8}$$

$$P_{\text{ave}} = P_{\text{max}}c \tag{34·9}$$

Sometimes average power is expressed in terms of repetition frequency and pulse width. Using equations 34·6, 34·7, and 34·9,

$$P_{\text{ave}} = wf_rP_{\text{max}} \tag{34·10}$$

Thus, if a system has a pulse width of 1.0 microsecond and a repetition interval of 1000 microseconds, the repetition frequency is 0.001 cycle per microsecond or 1000 cycles per second; the duty cycle is 0.001 microsecond. If the magnetron has a peak output of 100 kilowatts, average power is 100 watts.

Power

The relation of power to range involves consideration of power density in the region of the target, power reflected from the target, reflected power density in the region of the detector, and reflected power intercepted by the radiator.

If energy were radiated from a point source, radiation would be uniform in all directions. It is assumed that the point is at the center of a sphere having a radius r. Since all the power must pass through the sphere, power density at the surface of the sphere is total power divided by the area of the sphere:

$$D = \frac{P_r}{4\pi r^2} \quad \text{watts per square meter} \tag{34·11}$$

where D = power density in watts per square meter
P_r = transmitted power in watts
r = distance from the point source in meters.

A point source radiating 100 kilowatts thus would produce a power density of 8×10^{-3} watt per square meter at a distance of 1000 meters.

With a dipole and reflector, however, the uniform spherical distribution of a point source is concentrated within a beam. The concentration is called the gain of the radiator, and results from the characteristics of the dipole and the reflector. A dipole has a gain of $3/2$ in the direction of maximum radiation; a paraboloid has a gain of $8\pi A/3\lambda^2$, where A is aperture area of the paraboloid in square meters and λ is wavelength in meters. Total radiator gain G is the product of these factors:

$$G = \frac{3}{2}\frac{8\pi A}{3\lambda^2}$$

$$G = \frac{4\pi A}{\lambda^2} \tag{34·12}$$

For a reflector with an area of 1.0 square meter, radiating energy at a wavelength of 10 centimeters, the gain would be 1256.

Power density P'_r created by a radiator having a gain G is the product of the gain and the density of a uniformly radiating source of the power,

$$P'_r = GD$$

$$= \frac{AP_r}{r^2\lambda^2} \quad \text{watts per square meter} \tag{34·13}$$

Total power reflected at random by the target is equal to the product of power density in the region of the target and the scattering cross-sectional area of the target. Effective scattering area of a target is a complicated function depending upon target area, shape, material, configuration, and wavelength of incident energy. It is determined experimentally. If ρ represents this reflecting area, scattered power P_s is the product of ρ and power density at a distance r from the source of energy:

$$P_s = \rho P'_r$$

$$= \frac{\rho AP_r}{r^2\lambda^2} \tag{34·14}$$

Under the previously given conditions a scattering area of 1.0 square meter would give a reflected power of 10 watts.

Total power reflected by the target is scattered in a random fashion. The target may be considered as the source of P_s watts, and for simplicity point radiation is assumed. Reflected power density P'_s at any distance r from the target source is

$$P'_s = \frac{P_s}{4\pi r^2} \tag{34·15}$$

$$= \frac{\rho P_r}{4\pi r^4\lambda^2} \tag{34·16}$$

For ρ = 1 square meter, A = 1 square meter, P_r = 100 kilowatts, r = 1000 meters, and λ = 10 centimeters, the reflected power density is 8.2×10^{-7} watt per square meter.

Power intercepted by the radiator P_i is simply the product of aperture area A and reflected power density:

$$P_i = AP'_s$$

$$= \frac{\rho A^2 P_r}{4\pi r^4 \lambda^2} \text{ watts} \qquad (34\cdot17)$$

If $A = 1$ square meter, 8.2×10^{-7} watt enters the receiver.

Solving equation 34·17 for range indicates the relation of range and power gives

$$r = \sqrt[4]{\frac{\rho A^2 P_r}{4\pi P_i \lambda^2}} \text{ meters} \qquad (34\cdot18)$$

This equation reveals that range varies directly with the fourth root of transmitted power. Doubling transmitted power increases the range only by the factor 1.19. Doubling the range would require 16 times as much transmitted power.

Range depends also upon the area of the radiator and the wavelength of transmitted energy. Increasing area or decreasing wavelength increases range. Reflector size is determined by allowable size and weight; wavelength depends upon availability of high-power oscillators at various frequencies, and on beam-width requirements. Equation 34·1 fixes beam width in terms of radiator diameter d and wave length λ:

$$\theta = \frac{\lambda}{d} \text{ radians} \qquad (34\cdot1)$$

In terms of radiator area,

$$\theta = \frac{\lambda}{2\sqrt{A}} = \frac{\lambda\sqrt{\pi}}{2\sqrt{A}} \qquad (34\cdot2)$$

For a given beam width, slight increases in wavelength permit large increases in area. Thus, if large radiators are practical, range can be increased without sacrificing beam definition by using longer waves and much larger radiators.

Maximum range can be obtained by substituting for P_i in equation 34·18 the experimentally determined minimum power P_{min} to which the receiver will respond satisfactorily:

$$r_{max} = \sqrt[4]{\frac{\rho A^2 P_r}{4\pi P_{min} \lambda^2}} \text{ meters} \qquad (34\cdot19)$$

The expression is based on reflections from targets in free space. Minimum-response power determines maximum range. If it is assumed that P_{min} has been established by measuring maximum range at which intelligible reflections are received,

$$P_{min} = \frac{\rho A^2 P_r}{4\pi {r_{max}}^4 \lambda^2} \qquad (34\cdot20)$$

Offhand, it appears that any signal, no matter how minute, can be amplified, detected, and made useful. Even a receiver with no power loss, however, is limited in its signal response by the thermal noise of tubes and circuits. For this reason, a maximum signal-to-noise ratio is desired.

The theoretical lower limit set by thermal noise is

$$P_{th} = kT\,\Delta f \text{ watts} \qquad (34\cdot21)$$

where P_{th} = thermal noise power
k = Boltzmann's constant (1.37×10^{-23} watt-second per degree)
T = absolute temperature in degrees
Δf = receiver bandwidth.

In practice, minimum receiver power must be from 5 to 100 times greater than the theoretical value:

$$P_{min} = nP_{th} \text{ watts} \qquad (34\cdot22)$$

where n is the factor compensating for non-thermal noise and power losses. Substituting equation 34·22 in equation 34·19,

$$r_{max} = \sqrt[4]{\frac{\rho A^2 P_r}{4\pi n P_{th} \lambda^2}} \text{ meters} \qquad (34\cdot23)$$

or, incorporating equation 34·21,

$$r_{max} = \sqrt[4]{\frac{\rho A^2 P_r}{4\pi nkT\,\Delta f \lambda^2}} \text{ meters} \qquad (34\cdot24)$$

Equation 34·24 seems to indicate that by decreasing the bandwidth of the receiver the range could be extended indefinitely, for decreasing bandwidth reduces noise. But equation 34·24 is not universally applicable, because decreasing the bandwidth excludes portions of the signal at a frequency including an extended bandwidth. For each different pulse duration w some value of Δf gives a maximum signal-to-noise ratio. This is approximately

$$\Delta f = \frac{1.2}{w} \qquad (34\cdot25)$$

Maximum range can be expressed in terms of average power and the pulse repetition frequency, for

$$P_r = P_{max} = \frac{P_{ave}}{wf_r} \text{ watts} \qquad (34\cdot26)$$

Combining equations 34·25 in 34·26,

$$P_r = \frac{P_{ave}\,\Delta f}{1.5 f_r} \text{ watts} \qquad (34\cdot27)$$

and substituting in equation 34·24,

$$r_{max} = \sqrt[4]{\frac{\rho A^2 P_{ave}}{6\pi nkTf_r \lambda^2}} \text{ meters} \qquad (34\cdot28)$$

The form of the equation for range falsely suggests that range is proportional to power because such quantities as P_r and P_{ave} appear in it. Actually, range is determined by the *energy per pulse*. It does not matter whether 100 kilowatts of peak power is delivered for one microsecond or 40 kilowatts for 2.5 microseconds; the energy is 100,000 watt-microseconds in either case. Rewriting equation 34·26,

$$P_{ave} = P_r w f_r \qquad (34\cdot29)$$

where $P_r = P_{max}$; substituting in equation 34·28,

$$r_{max} = \sqrt[4]{\frac{\rho A^2 w P_r}{6\pi nkT \lambda^2}} \text{ meters} \qquad (34\cdot30)$$

For a particular system,

$$\sqrt[4]{\frac{\rho A^2}{6\pi nkT\lambda^2}} = C \tag{34·31}$$

where C is a constant, and

$$r_{\max} = C\sqrt[4]{wP_r}$$

$$r_{\max} = C\sqrt[4]{\text{peak power} \times \text{pulse width}}$$

$$r_{\max} = C\sqrt[4]{\text{energy per pulse}} \tag{34·32}$$

For a particular radar set, w is fixed by the minimum range which must be controlled. This suggests that peak power rather than energy affects range. However, w is determined by a compromise. Peak power is limited by the transmitter tubes. For detection at close range, w must be small; but small w means a small amount of energy and reduces maximum detection range. If w is large, energy output is large and maximum detection range is greater; but large w means that detection at close range is sacrificed. In the end, a compromise must be made between maximum and minimum detection ranges, and this means a compromise in the value of w for a given peak power.

In this analysis the peak signal amplitude is assumed to be as large as possible relative to the noise. But signal and noise may also be distinguished through the differences in their wave forms. If the wave form of the signal is apparent (as in type A indicators), the bandwidth may be greater than $1.2/w$. As bandwidth increases, noise variations become more uniform and finer, and the signal becomes more rectangular. Type A indicators reproduce the form of the signal, so this method of signal-noise differentiation is feasible.

PPI and some other indicators do not show wave forms; only a spot appears on the screen. Because the effect of better wave form is hidden, peak signal amplitude alone establishes signal ascendency over noise, and a bandwidth of $1.5/w$ is customary. Energy per pulse determines the ratio of signal to noise. If the signal is to predominate, energy per pulse must be large. For detection in regions close to the radar unit, a short pulse is necessary; that is, peak power of the transmitter must be large.

34·7 POSITION-FIXING RADAR

Neither the true principle of reflection nor the microwave frequencies usually associated with radar are found in position-fixing systems.

The first beacon bombing systems were developed by the British. They depended on measurement of aircraft range from two ground beacons. Ground equipment consists of pulsed transmitters and receivers; the air-borne unit consists of a receiver that automatically triggers a transmitter on reception of the beacon pulses. In effect, the air-borne system simulates a target of exceedingly high reflectivity. Position of the plane is determined by measuring the range of the aircraft from each of the two beacons. Knowing the ranges, the ground controller can establish the position of the plane as the intersection of two circles having as their radii the indicated ranges (Fig. 34·12).

This system requires ground controllers, and only one plane or formation at a time can be controlled. Another system, called Gee-H, in which the aircraft interrogates the beacons, circumvents this limitation. The aircraft sends out a pulse which triggers the ground beacons. A cathode-ray indicator is used for measurement of ranges, which is done on the plane, because the position of the beacons is known and the problem is simply the measurement of travel time of the energy as in radar. The beacons serve, in a sense, as known points of unusually high reflectivity.

The Gee-H system is an extension of the Gee navigational system. Both the American Loran and the British Gee systems depend upon the measurement of time differences. Both use frequencies in the kilocycle band, and both employ pulsed ground beacons and simple radar-type receivers

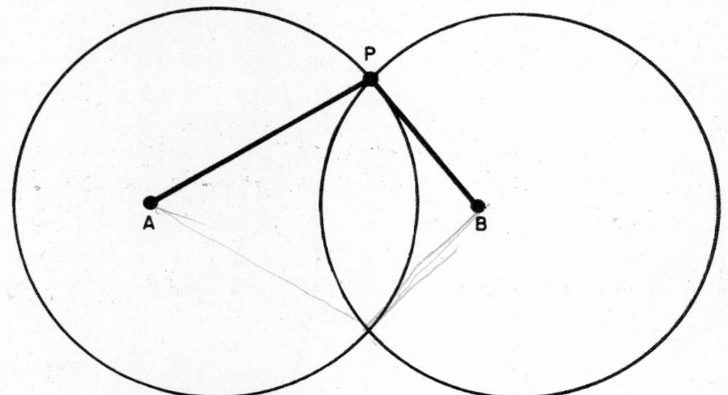

Fig. 34·12 Beacon method of position determination by range measurement.

in aircraft. Gee, however, uses ground waves only and is limited to line-of-sight ranges. Maximum range of any line-of-sight system is given approximately by

$$R_{\max} = \sqrt{2h} \tag{34·33}$$

where $R_{\max}$ is in miles if altitude h is in feet. Heights of both plane and beacon are included in h although the height of the beacon is usually insignificant in comparison with the altitude of the plane. Thus, if beacon height is assumed to be negligible, a plane at 20,000 feet can receive signals from a beacon 200 miles away.

The lower frequency Loran system uses sky-waves in addition to ground waves. Loran day coverage over water is 700 nautical miles; over land, about 300 statute miles. At night, Heaviside reflections extend this range to about 1400 miles over both land and sea.

Consider two ground beacons M and S and the airplane P in Fig. 34·13. Assume that both beacons are transmitting simultaneously. The two signals travel away from their respective beacons at the constant velocity of light, and equal increments of distance are covered in equal increments of time. Plane P intercepts both signals, and the indicator displays them along a line (much as in type A indication). If the plane is equidistant from the beacons, the two signals are superimposed, for they arrive at the same time. If, however, the plane is nearer to M than to S, there is a difference in travel time between the two signals, and this difference corresponds to the space interval between the two signals on the screen. Measurement of this calibrated interval, therefore, reveals the time difference between the two signals.

Now the locus of a point, the difference of whose distances from two points is a constant, is a hyperbola; therefore, for any position of the plane with respect to the beacons (within their range), the time difference specifies a position on a hyperbola. Thus, a time difference of 30 microseconds indicates that plane P is somewhere along hyperbola A-A'. Reception of similar signals from another pair of beacons determines another hyperbola, and the position of the plane is at the intersection of these two hyperbolas.

It is possible that the plane is somewhere along a hyperbola having its focus at S instead of M. This ambiguity is avoided

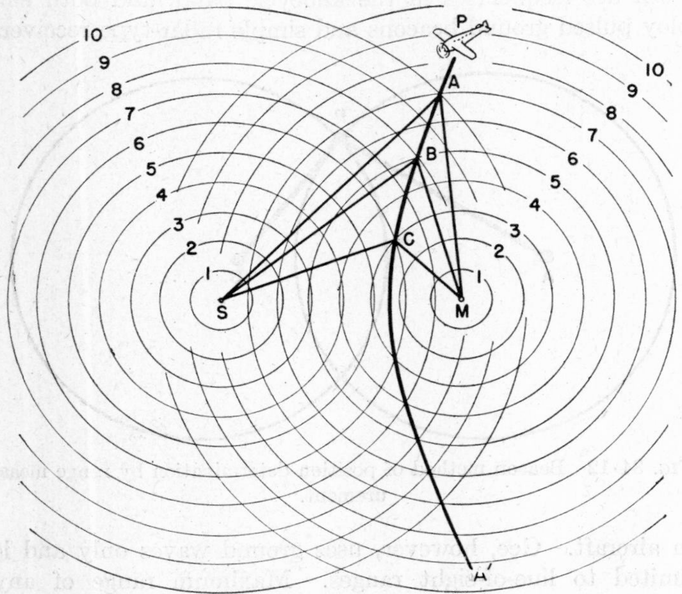

FIG. 34·13 Beacon method of position determination by measurement of time differences between signals from pairs of stations.

by proper synchronization of the two beacons. The "master" M beacon triggers prior to the "slave" S, and this pre-established delay is calculated so that, within beacon range, one time-difference reading indicates only one hyperbola on the navigational chart.

34·8 APPLICATIONS OF RADAR

An immediate application of radar lies in marine navigation. Relatively low-power systems with PPI presentation are adequate for both navigation and anti-collision purposes. For air-borne functions, however, such systems are not likely to be preferred. For air navigation precision systems of the position-fixing type, like Loran, offer accuracy, simplicity, and an advantage in size and weight. Ground radar systems, moreover, can simplify traffic control and blind landings.

Marine Radar

A marine-radar system already in commercial use consists of an antenna pedestal and an indicator console. Expert operators are not required, because the apparatus has been greatly simplified. Designed for navigation and anti-collision purposes, the system uses PPI presentation (Fig. 34·14).

The upper half of the antenna pedestal contains the cut paraboloidal antenna, the a-c driving motor and associated gears, and the synchro-tie motor. The lower section encloses the high-voltage power supply, the modulator, and the radio-frequency head. The console (Fig. 34·15), mounted in the wheelhouse or bridge, encloses the low-voltage power supply, the intermediate- and video-frequency amplifiers, the PPI indicator and associated circuits, and all controls.

Auxiliary equipment consists of a motor-generator set, remote PPI indicators, and an azimuth stabilizer for true bearing. The motor-generator set is necessary only if the vessel's primary power supply does not provide the 115-volt single-phase 60-cycle energy required. As many as three remote PPI indicators can be linked to the main indicator. Finally, vessels equipped with a standard gyrocompass can use azimuth stabilization of the PPI presentation for true bearing.

Components of marine radar are shown in Fig. 34·16 (*a*). Of particular interest are the modulator, Fig. 34·16 (*b*), and the radio-frequency head, Fig. 34·16 (*c*). To secure the proper sharp signal from the magnetron, an accurately formed high-power signal must be impressed between the cathode and the anode of the magnetron. The high-voltage power supply delivers 3000 volts to the modulator. A charging inductor and a resonant pulse network consisting of capacitors and inductors increase this voltage to 5100. Because such a voltage would oscillate in amplitude, becoming gradually damped, a hold-off diode is inserted between the charging inductor and the pulse network to prevent a reversal in current direction and thus maintain the voltage at 5100.

Accurate timing of pulses, necessary in pulse modulation, is achieved by use of a sine-wave oscillator, a blocking oscillator, and a thyratron. The blocking oscillator eliminates negative portions of the accurately timed signals from the sine-wave oscillator and converts the positive portion into a sharp pulse that rises sharply and decays exponentially. This pulse is applied to the grid of the thyratron, the anode and cathode of which are linked to the pulse network of the modulator and the pulse transformer in the radio-frequency head. The thyratron fires regularly on receiving the positive pulse, discharging the pulse network through the primary of the pulse transformer. The impedance of the transformer is equal to that of the network; therefore the 5100 volts are divided between the two, and a negative pulse of 2550 volts is applied to the transformer winding.

The modulator provides a high-voltage pulse of 0.4 microsecond duration, repeated at the rate of 2000 times per second. This signal is stepped up to approximately minus 12,500 volts by the pulse transformer, and is then applied to the cathode of the magnetron, driving it negative with respect to the anode and causing the tube to oscillate. The magnetron is of the 3-centimeter type, giving a signal frequency of about 10,000 megacycles, and has a peak power output of more than 15 kilowatts.

Wave guides conduct the energy to the radiator. Instead of a dipole antenna, a horn-type wave guide feeds the energy to the truncated paraboloid that radiates the pulse in a vertical fan pattern, 2 degrees wide horizontally and approximately 15 degrees vertically. The radiator is rotated continuously at 12 revolutions per minute by a small 115-volt a-c motor.

A synthetic crystal and a klystron local oscillator are used for detection; the superheterodyne principle is utilized. Intermediate- and video-frequency amplifiers are tuned to accept certain bandwidths, and the signal from the mixer must remain within these ranges. To compensate for magnetron variation, an automatic frequency control (AFC) system maintains the klystron at such a frequency that the intermediate frequency signal remains at 60 megacycles. A dynamic method of control, relying on a crystal mixer, amplifiers, discriminator, control tube, and sweep generator, performs this function by altering the frequency-sensitive voltage of the reflex klystron.

The AFC crystal is coupled directionally to the output wave guide so that only a small portion of the transmitted signal is intercepted and almost none of the reflected signal is absorbed. This signal is mixed with a portion of the klystron output. The intermediate frequency signal is amplified and applied to the discriminator (a capacitor-inductor circuit), which is highly sensitive to frequency. The resulting signal is amplified and fed to the control tube that actuates the sweep generator used to vary continuously the klystron voltage over a predetermined range. The net effect of the AFC is to keep the klystron frequency at a value that ensures a 60-megacycle intermediate-frequency signal output from the radio-frequency receiver.

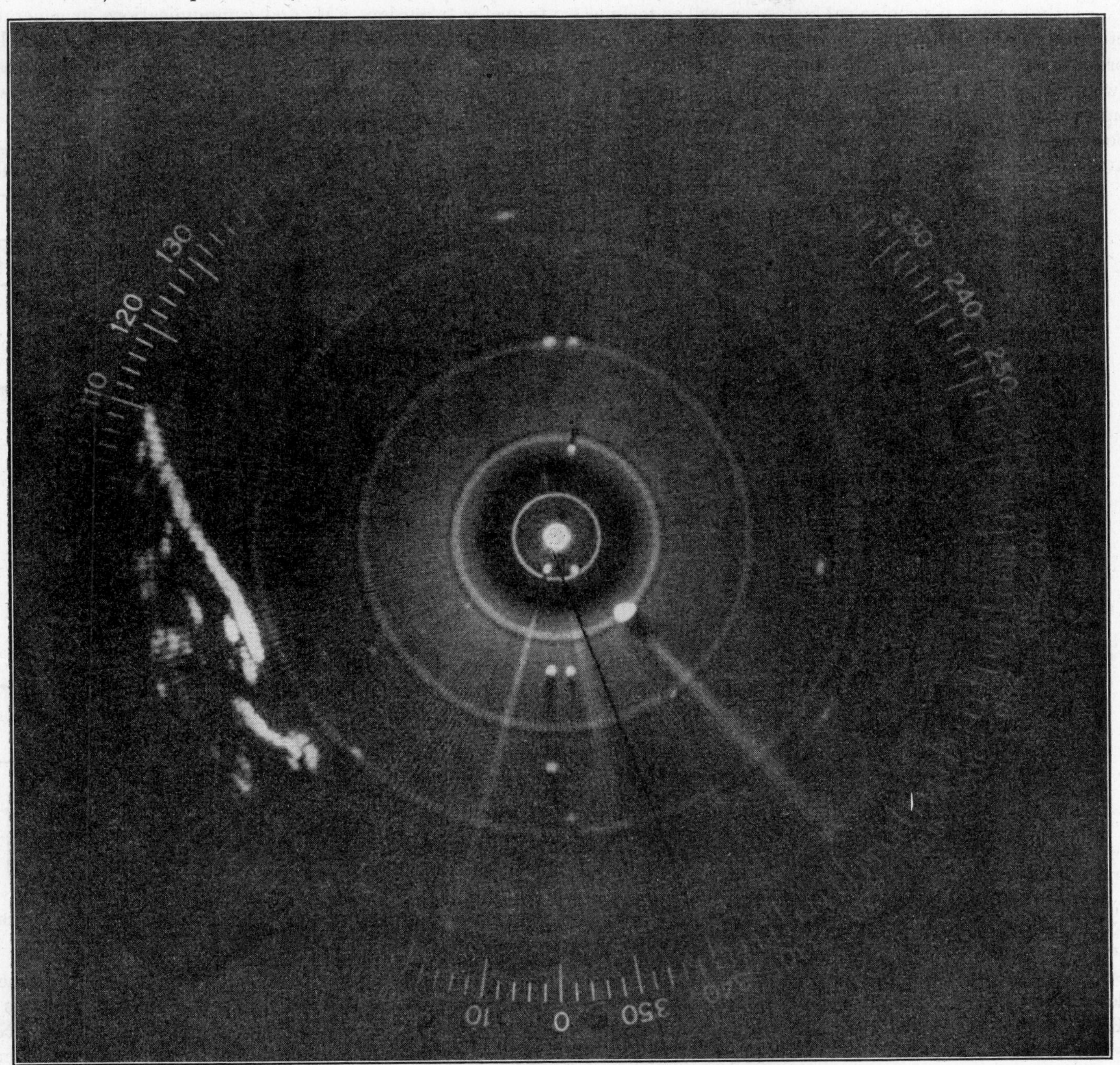

Fig. 34·14 PPI presentation of marine radar permitting blind navigation and furnishing anti-collision protection.

Maximum range is 32 miles. This range depends upon the amount of power transmitted and upon the height of the antenna. In general, ship-borne radar systems are limited in range by antenna elevation. Minimum range depends upon pulse width. The 0.4-microsecond pulse permits detection of objects as close as 100 yards.

Aircraft Systems

The nature of radar systems for air-borne use is more complex, partly because psychological factors as well as technical factors are involved, and partly because the possibilities are more extensive. It appears that Loran or a related system is well suited for transoceanic and continental navigation. Loran can serve just as well for domestic flights, but there is a possibility that other systems will assume this role. Ground search and control systems may direct planes over ultrahigh-frequency communication channels; beacon systems, like Loran but short-ranged and relying on range measurement (in contrast to Loran's time measurements) of coded signals, may replace radio beacons domestically; and ground systems transmitting televised-like displays to aircraft offer another possibility, but the choice and development of the most suitable system of this type are in the experimental stage.

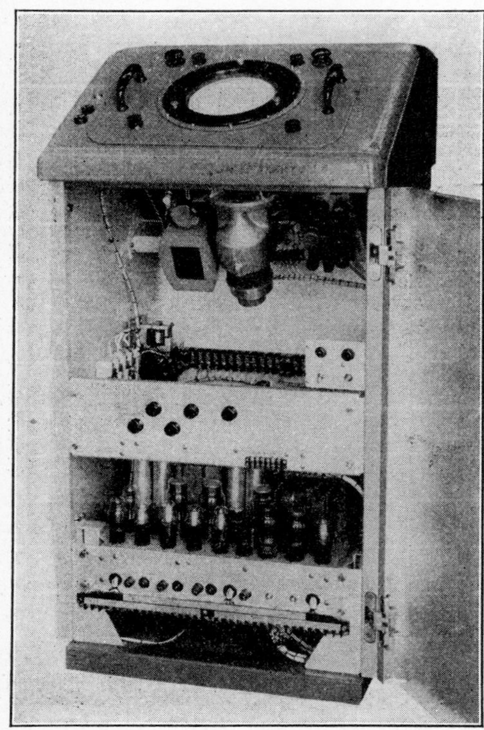

Fig. 34·15 Marine-radar console.

Radar offers still another aid to flying: blind landing. Systems of the GCA (ground-controlled approach) type, combining search and control with precision range and elevation functions, can guide planes to obscured landing fields. Some beacon systems also offer such possibilities, and the eventual practical form of blind-landing equipment also remains to be decided.

One of the short-range position-fixing systems called Shoran promises to revolutionize cartography. It can establish the position of ground points with greater accuracy than ground crews using conventional surveying instruments.

General Applications

The ultimate significance of radar lies in its general effect on the whole field of electronics. Such radar components as the magnetron, TR tube, resnatron, and semi-conductor rectifier indicate that the components necessary for utilization of higher frequencies are available. At the same time, extension of the radio-frequency spectrum during the last few years has been great. The value of this expanded spectrum for television, FM, and communications in general is self-evident; its value for such specific applications as dielectric and induction heating remains a subject of future investigation. These fields have profited by the techniques as well as the components developed in radar.

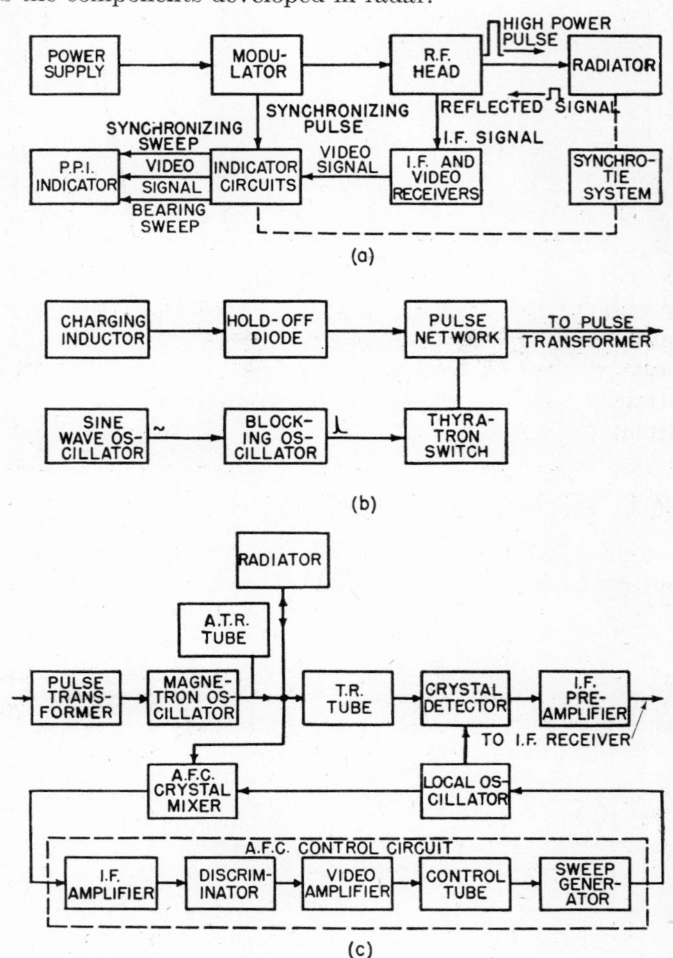

Fig. 34·16 Westinghouse marine radar: (*a*) system block diagram, (*b*) modulator, (*c*) radio-frequency head.

The development of miniature tubes began a movement that radar rapidly extended. In systems using hundreds of tubes, space became critical, and new designs resulted. Not only were small components designed and made, but a new concept of circuits was developed. The concept of tubes as short-lived devices was discarded, and tubes as durable as other circuit components are now available. Not only are these tubes smaller, but also whole circuits can be included within the evacuated envelope. From an industrial point of view, electronic devices hitherto prohibitive in size and weight will be practical. From a scientific point of view, accurate instrumentation, impractical in some fields hitherto, is feasible.

One of the greatest contributions of radar lies in its accurate measurement of short intervals of time. Timing circuits capable of measuring thousandths of a microsecond are already in existence.

Chapter 35

CARE AND MAINTENANCE OF TUBES

C. J. Madsen

INDUSTRIAL electronic apparatus usually includes circuits and components selected for their inherent stability and durability, because the apparatus may be required to operate over long periods without expert attention. Some components and circuits nevertheless require special care in installation and proper maintenance if peak efficiency and uninterrupted operation are to be attained. Extended tube life is often a direct result of such attention.

35·1 EXTENDING TUBE LIFE

Following a few general rules usually contributes to longer life and more reliable operation. They are:

1. Use the tube or device within its ratings.
2. Operate tubes at proper cathode voltages as recommended by manufacturers. Measure this voltage regularly with an accurate voltmeter.
3. Apply cathode voltages in the manner specified by the manufacturer.
4. Keep the tube clean, particularly its contact terminals, glass envelope, and anode (if water-cooled).
5. Maintain an ample supply of cooling medium, if it is required.
6. Cathode and anode voltages should not be applied simultaneously except as permitted by manufacturer's ratings.
7. Handle tubes with care. Avoid sudden mechanical or thermal shocks.
8. Store spare tubes carefully. Some tubes require special care as to position during storage.

Overload Capacity

Few types of tubes have appreciable overload capacity. The term overload refers to operation in which the interelement space current or potential (such as anode current or anode voltage) exceeds the tube rating. The average rating usually applies, although certain tubes have peak ratings which apply in special modes of operation.

An overload of short duration in a hot-cathode tube may not destroy the tube. Well-designed apparatus usually includes protective overload devices to protect the tubes and equipment under anode-current overloads. Occasionally an overload persists long enough to heat tube elements to a temperature sufficient to liberate occluded residual gases, and these gases may make the tube inoperative. If elements are not destroyed it is often possible to restore the tube to almost normal performance. The reconditioning procedure is controlled entirely by the type of cathode material.

Cathode Operation

The importance of operating the cathode at the proper voltage cannot be overemphasized. Many types of tubes operate satisfactorily with line voltage variations greater than 5 percent. Such variation causes trouble in some apparatus, and it always has an adverse effect on tube life. The type of cathode material determines the liberties that may be taken or precautions that must be exercised. Cathode materials can be divided into four general classes:

1. Pure-tungsten cathodes.
2. Thoriated-tungsten cathodes.
3. Oxide-coated cathodes, directly and indirectly heated.
4. Mercury-pool cathodes.

Pure-Tungsten Cathodes. Pure-tungsten cathodes usually are designed for operation at approximately 2550 degrees Kelvin. If the normal temperature can be reduced by only 50 degrees, tube life can be theoretically doubled. Conversely, if temperature is increased, useful life is shortened. The most common cause of failure is cathode burn-out, and precautions which minimize abuse of this element will pay dividends in trouble-free service. Tungsten cathodes are used in large pliotrons and kenotrons, tubes of the high-vacuum type.

If the tube is lightly loaded, its life can be increased by operation of the cathode at reduced voltage. A 5 percent reduction in cathode voltage results in an increase in theoretical life of 100 percent. The curves shown in Fig. 35·1 show expected life and emission expressed in percentages versus cathode voltage. Hence the lowest cathode voltage that produces satisfactory performance is the voltage that gives the longest tube life. Such adjustments should be made by an experienced electronic service man.

Allowance must be made for variations in cathode voltage caused by line-voltage fluctuations. Adequate emission must be available for satisfactory performance while line voltage is at its lowest excursion.

If line voltage varies widely the average cathode voltage must be somewhat higher than if the power voltage variations are small. Hence tube life is shortened. For example, suppose the emission from a tungsten-cathode tube at 94 percent of normal voltage is adequate. Add 1 percent for safety factor. Line-voltage fluctuations are ±5 percent. To assure the 95 percent value during periods of low line voltage, the cathode must be adjusted to an average value of 100 percent. This adjustment means that the cathode voltage will be 105 percent some of the time. As shown by Fig. 35·1 the theoretical life will be materially less than if line-voltage variations are only ±2 percent. If line-voltage varies only ±2 percent the average voltage can be adjusted to 97 percent, with maximum variations being between 95 and 99 percent of normal. Resulting life is influenced further by the relative time the voltage is at various values.

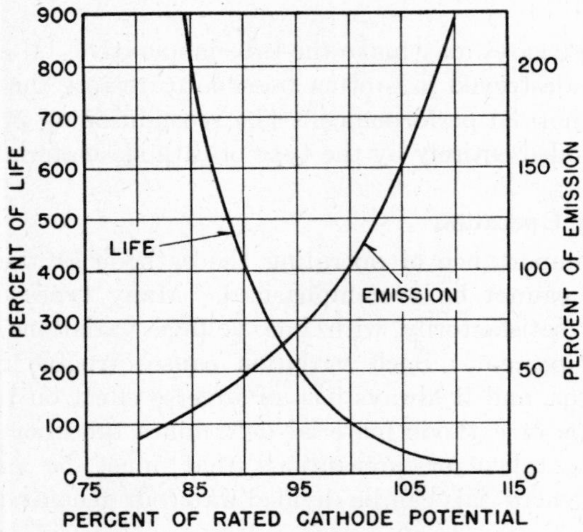

FIG. 35·1 Life and emission as affected by cathode potential on tubes with pure-tungsten cathodes. Values based on rated cathode voltage as 100 percent.

One method of determining the minimum safe value is to operate the device at normal load and normal cathode voltage, then reduce cathode voltage very gradually until anode current, grid current, output, anode temperature, or cooling-water temperature changes. Such a test should be conducted by an experienced man as the first indication may be minute.

This point represents the lowest cathode voltage that can be tolerated under all conditions of line voltage. The final adjustment is then determined by the expected variations from the line voltage used at the time of the test for minimum cathode voltage. The final adjusted value for the given load condition is always higher than this minimum cathode voltage, as at least 1 percent safety factor should be added to the line voltage correction.

In a water-cooled tube, steam bubbles may form at the anode surface if the cathode voltage is reduced below this minimum. The distinctive sound produced in the water jacket serves as an additional check. Such operation for any appreciable length of time will damage or destroy the tube.

Another factor important to cathode life in large tubes is the manner in which cathode voltage is applied. Stresses in the cathode caused by sudden application or removal of the voltage can cause premature failure. Stresses at starting are particularly severe, and most tube manufacturers limit starting current to 1.5 times the rated current. The usual way of limiting starting current is by means of current-limiting transformers, but it can be done by starting devices employing resistance, reactance, or variable voltage.

Anode voltage may be applied to tungsten-cathode tubes before the cathode has reached normal operating temperature without deleterious effects.

Thoriated-Tungsten Cathodes. Tubes employing thoriated-tungsten cathodes are usually of the high-vacuum type such as pliotrons and kenotrons. Tubes using thoriated-tungsten cathodes usually are designed for a cathode temperature of approximately 1900 degrees Kelvin. If a thoriated-tungsten cathode is operated at a voltage below normal it may lose emission rapidly, because thorium is not carried to the cathode surface fast enough to maintain normal emission. If the cathode is operated at a voltage above normal, the tungsten carbide on the cathode surface is reduced to pure tungsten at an excessive rate. This process affects the emission characteristic of the cathode. Useful life of the thoriated-cathode usually ends before burn-out, and is marked by loss of emission. Thoriated cathode tubes should be operated at rated cathode voltage.

Loss of emission may be caused by cathode "poisoning" by gases liberated from one of the elements during a short overload. The gas is adsorbed to some degree by the cathode material, which changes its emission characteristic. If the amount of gas liberated or the degree of poisoning is not excessive the tube may be restored by following the procedure outlined in Section 35·4.

Because cathode current in thoriated-tungsten tubes usually is relatively low, starting stresses are negligible. The time required for the cathode to reach normal temperature is short; cathode and anode voltages may be applied simultaneously without adverse effects on tube life.

Oxide-Coated Cathodes. Cathodes of the oxide-coated type operate at temperatures much lower than do those of tungsten or thoriated tungsten. Correct color temperature varies somewhat with the coating material, but is usually from dull red to cherry red. Phanotrons and thyratrons, tubes of the gas-filled or mercury-vapor type, commonly have oxide-coated cathodes. For maximum tube life, rated cathode voltage must be maintained at all times within 5 percent. If a coated cathode is operated below its rated voltage, the active material may be destroyed by ion bombardment, particularly in tubes employing high anode voltages. If cathode voltage is too high the emission surface is effused faster than normal, and in extreme cases may be transferred to some degree to other elements of the tube. Useful life of oxide-coated cathodes usually is completed before burn-out. In the indirectly heated type useful life usually ends without failure of the heater element. Loss of emission is the common end of life. Such tubes cannot be restored by "rejuvenating" processes.

Tubes employing high anode voltages must have their cathodes at normal temperature before anode voltage is applied. This is particularly true of gas-filled or mercury-vapor tubes such as phanotrons or thyratrons, and the apparatus using them ordinarily includes a time-delay device to prevent application of anode potential until the cathode has been heated to operating temperature.

Hot-cathode mercury-vapor tubes require additional precautions in mounting and initial operation. They should always be mounted with the base down. A new tube must be operated with only cathode or heater voltage applied for a period of 15 to 30 minutes to insure proper distribution of mercury before anode voltage is applied. The tube should not be inverted, in order that the amount of mercury on the upper portion of the envelope or on the elements may be minimized.

Mercury-Pool Cathodes. Mercury-pool cathodes are not subject to the limitations of hot cathodes because emission is initiated and controlled by factors other than cathode temperature; hence the cathode is not affected by overloads or voltage conditions. The ignitron is a common tube of this type. Electrodes or other devices for starting emission are subject to some limitations, usually current. Tolerance of these circuits usually is so wide that they function under conditions far beyond the limits of the associated apparatus. In addition the design of the "firing" circuit can be made almost completely self-protecting. The manufacturer's recommendation for mounting must be followed if reliable performance is to be expected.

Since mercury-pool cathodes are not subject to burn-out or loss of cathode emission, their normal life is measured in years. Most common causes of failure are rough handling, overloads, and failure of firing electrodes or circuits.

Regulating Cathode Voltages. If supply voltage varies widely it is often economical to use an automatic voltage-stabilizing device to maintain cathode voltages within reasonably close limits. For low power the static type of voltage-stabilizing transformer is satisfactory. If more than 5 kva is involved, the induction type of regulator with automatic control usually is preferable.

35·2 TUBE MAINTENANCE

Cleanliness

Cleanliness of vacuum tubes is particularly important in sensitive or high-impedance devices and in those employing high voltages or high frequencies.

Corrosion, dirt, or lint on the tube contacts may affect performance and tube life adversely in several ways. Cathode contacts usually carry larger currents than any other tube contacts. Hence a high resistance at this point reduces cathode voltage, and local heating produces additional oxidation, which in turn increases contact resistance. The effect is cumulative, it causes erratic operation, and it may produce tube failure. Tube terminals or contacts may be cleaned by occasional polishing with No. 00000 sandpaper or crocus cloth, followed by wiping with a cloth dampened with alcohol. Terminals should be kept tight, but mechanical strains on the glass-to-metal seals of the tube should be avoided. Spring clips used for connections to anodes or grids of some tubes can be cleaned in the same way.

Glass surfaces should be cleaned regularly with a cloth dampened in alcohol. The cleaning should not be done while the tube is still hot. Dirt on the glass may not affect performance, but in high-voltage or high-frequency applications it may lead to localized heating of the glass and subsequent tube failure.

In water-cooled pliotron tubes where water touches the anode, it may become coated with a scale deposit unless distilled water is used. This scale is mainly lime, similar to boiler scale, although it varies with the local water supply. Scale should be removed frequently, because it reduces the heat transfer from the anode. If allowed to accumulate indiscriminately, it can cause tube failure. Scale can be removed by immersing the anode in a 10 to 25 percent solution of hydrochloric acid. It may be removed by careful use of a scraper also.

Cooling

Air-Cooled Tubes. Little need be done to assure proper cooling of tubes cooled by natural ventilation, if the equipment was designed to provide proper ventilation. Such tubes handle relatively low power and the amount of heat produced is small. Ventilating openings in the cabinet housing the tubes should not be obstructed.

For tubes using forced ventilation, an adequate supply of air is important. Some blowers have air filters at the intakes to reduce the amount of dust drawn in. Failure to clean or replace this filter regularly results in restricted cooling and may cause failure of the tube. The intervals between cleaning or replacement depend upon local conditions. A thermometer mounted on the radiator or near the envelope of the tube will give a clue to the reduction in ventilation, particularly if daily readings are recorded and compared with the ambient temperature surrounding the equipment. Intake and exhaust openings must not be blocked.

The amount of cooling air for phanotrons and thyratrons employing mercury vapor is not particularly critical but such tubes are critical to temperature. Ventilation, whether natural or forced, usually is arranged to maintain a portion of the envelope just above the base at a temperature within the optimum operating range. Some designs provide for heating the tube compartment, and if tubes of this type are operated where temperatures fall below 50 degrees Fahrenheit, heating may be necessary. At the other extreme, high ambient temperatures may cause arc-backs. This upper limit varies with tube type and voltage conditions, so reference to the manufacturer's rating is necessary. In extreme cases a means of cooling the air applied to mercury-vapor tubes must be provided.

A small area of the envelope near the base must be kept at a temperature within the limits dictated by the operating conditions of the tube in the circuit, if reliable performance is to be maintained.

Water-Cooled Tubes. Water-cooled tubes depend upon an adequate flow of water to carry away heat fast enough to maintain the cooled parts at a safe operating temperature. The recommended flow as specified by the tube manufacturer should be maintained at all times when the tube is in operation. Inadequate flow of water at high temperature may cause formation of steam bubbles at the anode surface of large pliotrons where the water is in direct contact with it. This can contribute to premature tube failure. Lack of adequate cooling in ignitrons may result in frequent arc-backs.

By electrolysis and scale formation, hard water may cause a gradual constriction of some part of the water system. Hence water flow and plumbing fittings must be inspected regularly. The fittings on either end of an insulating section of hose or ceramic water coil or column are particularly subject to corrosion or electrolysis unless they have protective "targets." Targets should be checked periodically and replaced when they have disintegrated.

Cooling-water temperature is important. The tube manufacturer's rating sheet should be consulted to be sure operation is within safe limits.

In many ignitrons and similar tubes the shell or metal envelope is water-cooled. Since the shell usually is at cathode potential the use of isolating water columns (lengths of rubber or insulating hose) is sometimes necessary. Fittings and tube are subject to corrosion, scale, and colloidal deposits if impure water is used. Water connections and cooling jackets should be flushed out at three-month intervals with high-pressure water to maintain proper cooling efficiency. If scale deposits are found, a cleaning compound such as Oakite or equivalent material is used. After cleaning, the system must be flushed thoroughly.

The most reliable source of clean cooling water should be selected. If it contains scale or colloidal materials, a filter or conditioner should be installed. The conditioner or filter should not introduce ion-producing chemicals, because excessive electrolysis will result.

If scale and corrosion cause much trouble a closed system, which utilizes distilled water in the tube circuit and employs a heat exchanger to carry away heat, should be installed. The heat exchanger may be of the water-to-water or water-to-air type, depending upon local conditions and the power involved.

Water should not be permitted to drip or spray on the glass portions of tube envelopes, particularly when the tubes are in operation. The sudden thermal shock almost invariably results in tube failure.

35·3 HANDLING AND STORAGE

Glass tubes, which are recognized as being fragile, are seldom exposed to the rough treatment often given metal tubes. Yet the internal structure of a metal tube is similar to that of a glass tube, and it can be irreparably damaged by mechanical shocks. A safe rule is to handle all tubes as though they were precision instruments. Mechanical shocks may displace internal elements of the tube as well as shatter the glass envelope or insulating members supporting the elements. Thermal shock may be equally damaging to the glass portion of the tube. A hot tube should not be placed against a cold metal body or in contact with cold liquids. A cracked glass envelope or seal may result from such carelessness.

New tubes should be tested at the first opportunity in equipment similar to that in which they will be used. They should be unpacked carefully and tested as follows:

Kenotrons with Tungsten Cathodes. This tube is operated with cathode voltage only for 10 minutes at normal cathode voltage. On high-cathode-current tubes, low initial cathode voltage should be applied and gradually increased to normal to limit the cathode-current surge. Reduced anode voltage is applied and the tube is operated for an additional 10 minutes. Then anode voltage is increased to normal and the tube is operated for 10 minutes again. If the tube shows signs of gas, the conditioning procedure given in Section 35·4 should be followed.

The tube then should be returned to its shipping carton, or to a storage cupboard with some provision for protecting tubes against mechanical damage. Tubes of this type may be stored in any position. Spare tubes should be tested every 3 months, the same procedure being followed.

Kenotrons with Thoriated-Tungsten Cathodes. The tube is operated with cathode only for 1 minute at normal cathode voltage, and then for 5 minutes with anode voltage applied. Tubes of this type should be stored as indicated previously.

Test the spare-tube stock periodically to assure that no defective tubes are carried in stock.

Pliotrons with Tungsten Cathodes. The tube is operated with the cathode only (with application of reduced voltage or limiting of current by a starting circuit), at rated voltage for 10 minutes. Cooling must be normal during the test. Reduced anode voltage is applied approximately half of normal, and the tube is operated for an additional 10 minutes. Anode voltage is increased to normal, and the tube is operated for 10 minutes. If any tubes show signs of gas, they are conditioned as described in Section 35·4.

The tube then should be returned to its shipping carton or a suitable storage cupboard. Tubes of this type occupy less space and are less subject to mechanical damage if they are stored vertically, although the position in storage does not affect tube operation. Spare tubes should be tested every 3 months.

Pliotrons with Thoriated-Tungsten or Oxide-Coated Cathodes. The tube is operated for 1 minute with normal cathode current only. Normal anode voltage is applied for 5 minutes. Then the tube is returned to the shipping carton or storage cupboard. Tubes of this type may be stored in any position. Spare tubes should be tested every 3 months.

Phanotrons and Thyratrons. The tube is operated with only the cathode energized normally for a period of 10 to 15 minutes or until the mercury disappears from the elements and the upper portion of the tube envelope. Gas-filled tubes need be operated for only a minute. Normal anode voltage is applied, and the tube is operated for an additional 10 minutes. If the tube arcs back, the cathode only is operated for an additional 5 minutes, and the temperature of the condensed mercury is checked by measuring bulb temperature just above the base. If it is within normal limits (see manufacturer's rating sheet), which are usually between 60 and 122 degrees Fahrenheit, anode voltage is reapplied. Operation should be normal if the tube is not defective. If the tube appears to be normal, operation is continued for 5 minutes. When the tube is being removed from its socket, *every precaution is taken to prevent mercury from splattering on the upper portion of the bulb or electrodes. The tube is stored in an upright position.* Tubes so handled can be placed in active service with only the normal cathode-heating delay, and need not be reconditioned.

Spare tubes should be checked every 3 months. Initial cathode-heating time may be reduced on recheck to the normal time delay prior to anode-voltage application.

Gas-filled tubes can be stored in any position without the necessity of conditioning prior to service.

Ignitrons. The tube is placed in a vertical position with the cathode pool at the bottom. With an ohmmeter, test lamp, or test meter, a test is made from anode to the metal envelope for a short circuit. If a short circuit is indicated, it can be eliminated by tilting the tube about 45 degrees and tapping the tube with the bare hand. When the tube is returned to its normal position, a check to see that the short circuit has been relieved is made. The procedure is repeated if necessary. The igniter circuit is checked for continuity. The tube is handled and stored in an upright position.

Phototubes. The tube is operated for 2 hours. Characteristics of a phototube which has not been in service for a month or more may change slightly, but will return to normal with 1 or 2 hours of operation at normal voltage and illumination.

Some ultraviolet-sensitive phototubes have shields within the tubes. If the shield is not at the bottom of the tube when it is unpacked, the tube base is held downward and tapped gently until the shield slides down. It may then be tested in the normal manner.

Phototubes can be stored in their shipping cartons, and should not be exposed to excessive light, such as direct sunlight, for long periods.

35·4 RESTORATION OF INOPERATIVE TUBES

Kenotrons and pliotrons which may have become inoperative by a temporary overload and the liberation of a small amount of gas sometimes can be restored to useful service. Occasionally tubes which have been stored for long periods also show indications of gas.

A tube containing a small amount of gas produces a faint bluish glow filling almost all the envelope when anode potential is applied. This is often accompanied by erratic anode current, which rises rapidly and trips the overload relays. The slight bluish iridescence on the inner surface of a glass envelope should not be confused with a glow caused by gas. Some glass, under electron bombardment, produces a faint bluish glow which emanates from the inner surface of the glass envelope only. Sometimes the quantity of gas is so minute that the tube may appear normal until a transient voltage produces ionization and the overload relay is tripped. In this case the glow is momentary and can be seen only by careful observation.

If the tube contains a large amount of gas, the glow is pink or bluish pink. Such tubes seldom can be restored, although it is worth attempting for large tubes.

In tungsten- or thoriated-tungsten-cathode tubes a cloud of white fumes within the envelope on application of cathode voltage is an indication that the tube is filled with air. A cracked seal or envelope is probably responsible, and the tube cannot be salvaged.

Restoring Tungsten-Cathode Tubes. Kenotrons and pliotrons respond to the same general treatment, and if the amount of gas is not excessive restoration is almost certain. Pliotrons should be operated as oscillators during restoration, if possible.

Apply normal cathode voltage for a period of 15 to 30 minutes. Cooling should be normal.

Apply about one fourth of the normal anode voltage and load the tube lightly. If gas is indicated at this voltage, the tube should be operated with cathode only for an additional 30 minutes and anode voltage applied. It is operated at this anode voltage for 15 minutes. Anode voltage is increased to approximately one-half normal and the tube is operated for 15 minutes. During the next 2 hours the anode voltage is gradually increased to normal. If at any stage of this seasoning process gas is indicated, the anode voltage is reduced slightly, and the process is resumed until normal anode voltage can be applied without any indication of gas or instability.

Restoring Thoriated-Tungsten-Cathode Tubes. Tubes which apparently have lost emission because of an overload of short duration sometimes can be restored by operation of the cathode only at 120 percent of normal voltage for a period of 10 minutes, with no anode voltage applied. Cathode voltage is returned to normal for 10 minutes, and then anode voltage is applied. If anode current is approximately the same as that of a normal tube in the same circuit, the tube probably will give many additional hours of service. If emission is still low, the tube is operated with twice the normal cathode voltage for 10 seconds. Cathode voltage is reduced to normal for 10 minutes, and then anode voltage is applied. Operation is checked as before. If emission is still low the tube is probably beyond recovery.

REFERENCES

1. Westinghouse Electronic Tube Manual.
2. RCA Tube Handbook.
3. G.E. Electronic Tube Data.
4. "Rules for Prolonging Tube Life," Hampton J. Dailey, *Electronics*, April 1943, pp. 76–78.
5. "Factors Determining Industrial Tube Life," John F. Dryer, Jr. *Electronic Industries*, Dec. 1945, p. 94.

Chapter 36

CARE AND MAINTENANCE OF ELECTRONIC APPARATUS

C. J. Madsen

ELECTRONIC equipment for industrial service normally is designed to operate for extended periods without attention of skilled personnel. Protective devices and special circuit designs minimize interruptions or serious failures. Some components and circuits require special care in installation and special maintenance, though all electronic equipment requires reasonable care in installation and periodic maintenance if peak efficiency and trouble-free operation are to be attained.

36·1 INSTALLATION

Before installation of electronic equipment, the manufacturer's instructions should be read thoroughly. If the installation is made properly, hours of maintenance work and "trouble shooting" may be saved.

The following check list should be helpful:

1. Select a dry, clean location for the equipment.
2. Provide adequate clearances for routine servicing.
3. See that line voltage and frequency agree with the nameplate ratings of the equipment. Be sure the line has adequate current-carrying capacity, and that voltage variations with load are less than 5 percent.
4. Sufficient cooling media of ample purity must be available. This is extremely important in electronic apparatus employing water as the coolant.
5. Provide a good ground connection on sensitive or high-frequency apparatus.
6. Follow the manufacturer's instructions for installation.
7. Before applying power, check transformer taps (when provided). Check the equipment for loose connections or other damage.
8. Check safety devices or safety interlocks for proper operation.
9. Follow the manufacturer's instructions on initial adjustments.
10. Follow the manufacturer's operating instructions.
11. Avoid overloading. With the exception of equipment employing ignitrons or their equivalent, overloads reduce tube life and produce unsatisfactory operation.
12. Establish a maintenance and service schedule.

Location

Location of the equipment must, of course, be convenient for the function to be performed. Usually several choices are possible; a position which is dry and clean and fulfills the initial requirement is preferable. This choice is particularly important for highly sensitive electronic apparatus employing high amplification. Equipment employing high voltages, such as an industrial x-ray unit or radio-frequency generators, provide more hours of operation with less maintenance if environment is dry and clean.

In equipment employing phanotrons or thyratrons, ambient temperature at location is important unless some means is provided to maintain the tube-base temperature between the limits specified by the manufacturer. The common limits for such tubes are from 15 to 50 degrees centigrade (59 to 122 degrees Fahrenheit). The ambient temperatures which result in the above tube base or mercury temperature are usually from 10 to 15 degrees less than the figures given, because some heat is generated within the unit and tube, and this heat raises the temperature a few degrees above the outside room temperature. In other words, room temperature should fall between the limits of 50 and 110 degrees Fahrenheit unless some form of temperature control is provided for the equipment.

Provision for access to service doors, shields, or panels is important.

Power Line Voltage

Many types of electronic apparatus function satisfactorily with line-voltage variations greater than plus or minus 5 percent, but tube life usually is affected adversely. The criteria for using automatic voltage regulation are not sharply defined. Consultation with the manufacturer is desirable, particularly if variations in line voltage or phase balance exceed plus or minus 10 percent. The cost of improper operation with some processes may far outweigh the cost of automatic voltage regulation; or wide voltage variation may shorten tube life, in which case a voltage regulator on the cathode supply only may be justified.

Cooling Media

Purity of cooling water is important. If its specific resistance is greater than 100,000 ohms per cubic centimeter, the water is satisfactory; if less than 10,000 ohms per cubic centimeter, the use of a closed system with a water-to-water or water-to-air heat exchanger should be considered. In such a system distilled water or its equivalent is circulated through the equipment, then through the heat exchanger or radiator, and finally to a storage tank for recirculation.

The effect of impure water on tube life is described in Section 35·2. In addition, other serious results may be brought about by impure water. Precipitation in the piping may restrict water flow dangerously. Electrolysis is frequent; in severe cases the water line has been corroded so that replacement is necessary after only 3 or 4 years. This action does not however usually extend more than 25 or 30 feet from the electronic unit and is most severe if high unidirectional voltages are employed.

If the air is humid and the cooling water is cold, condensation accumulates on the surfaces of all pipes, tube jackets, and other parts carrying the water. This condensation may decrease surface leakage resistance, or drops of water may fall on some electrical component and cause erratic operation or failure. Some means of controlling temperature of incoming water to keep it above the dew point is then necessary. Control is rather easy in a closed cooling system, but in a system employing tap water and draining the exhaust water into the sewer, it is difficult, hence the importance of a dry location.

In air-cooled equipment, the air should be as free of dust and chemical vapors as possible. Dust does not affect operation immediately, but if it is allowed to accumulate indefinitely it causes trouble by reducing surface insulation resistance or increasing contact resistance. Some chemical vapors cause corrosion and increase contact resistance.

Temperature of cooling air is important in equipment employing mercury-vapor tubes. If it falls below 50 degrees Fahrenheit, the tube may fail to ionize. An automatic auxiliary heater is often built into equipment which may be required to operate at low temperatures. If the cooling air is at a temperature above 120 degrees Fahrenheit, arc-backs may interrupt operation. Artificial cooling is necessary where high ambient temperatures cannot be avoided.

Ground Connections

The importance of good ground connections cannot be overemphasized. Dependence upon conduit, power line, BX cable, and gas pipes for ground connections should be avoided. A short length of No. 8 or 10 wire should be run to the nearest water pipe, or to a driven ground. In radio-frequency generating equipment, the ground lead becomes most important. The generator and the work circuit should be connected together, if they are separate units, and grounded by means of wide copper strap. The thickness of this strap need be sufficient only for freedom from mechanical damage, but the width should be from 2 to 6 inches, depending upon the length of the lead. If such equipment is two or more stories above ground, a large copper or tin sheet common to both generator and work circuit should be placed under the equipment and grounded at several points with copper strap. Grounds then can be connected to this ground sheet. In dielectric heating particularly, the length and inductance of ground leads are important. At 27 megacycles (a common frequency), a quarter wavelength is approximately 9 feet. A lead this long may be at ground potential at one end and several hundred volts above ground at the other. If necessary to run such a long lead between units and ground, the ground sheet is recommended. Several copper straps separated a foot or more may be used to simulate a solid sheet and will reduce the effective inductance of the ground leads.

Manufacturer's Instructions

The manufacturer probably knows his particular equipment—its virtues and limitations—better than anyone, hence his recommendations should be followed for best results. The instruction book should be read thoroughly before installation is started. The equipment should be checked as it is unpacked, for parts or connections loose or broken in transit. Installation wiring must be checked before power is applied. Cathode potentials should be checked and adjusted if necessary before anode potentials are applied. In some equipment employing rectifier tubes, the rectifier anode connection may be removed from the tube cap if a separate switch is not provided to control application of anode voltage. When cathode potentials are being measured the circuit diagram should be checked to be sure dangerous potentials do not exist between cathode circuit and ground, even with anode leads removed. Mercury-vapor tubes should be conditioned by 15 minutes of operation with cathode only energized.

Initial adjustments should be made carefully. Taps are frequently supplied on transformers to permit line-voltage adjustment. Cathode voltages should be set to the specified values by means of an accurate voltmeter. If an a-c d-c circuit checker is employed, the calibration in the ranges used for cathode potentials should be checked against a reliable dynamometer or induction type of instrument.

When adjustments have been completed, the operation as checked against typical operation or instrument readings (usually supplied in instruction books) should be close to the values given. Any wide deviation should be thoroughly investigated.

36·2 OPERATION

A routine servicing or maintenance schedule is probably the most valuable insurance for dependable operation. The proper period of such a schedule varies with the type of apparatus. Initially the schedule should call for more frequent inspection than seems necessary. As experience is gained the period between service inspections may be increased.

Examinations must be made periodically to determine that enough of the cooling medium is being supplied, that water or air flow is normal and unobstructed and is at the proper temperature.

A daily log of meter readings is helpful in checking operation. Variations in adjustment or operation should be noted in this log to facilitate servicing.

Overloading of equipment should be avoided in normal operation. Most tubes, particularly those of the hot-cathode type, have little overload capacity; hence overloading causes short tube life and questionable performance.

36·3 MAINTENANCE

Maintenance is primarily insurance for continued reliable operation of equipment. Work should be done when the equipment is not in service, so that it can be done carefully and according to a plan.

It is necessary to know that all power has been removed from the equipment before maintenance work is undertaken. Capacitors should be discharged before they are touched with the bare hand. Equipment should be cool, for resistors or vacuum-tube envelopes may operate at a temperature sufficient to produce painful burns. Often the involuntary reaction from an accidental burn results in serious damage from dropped tools or displacement of components.

Tools and Instruments

The following tools and instruments are useful:

1. High-resistance volt-ohmmeter.
2. Several ranges of d-c ammeters or an instrument with several ranges from 10 milliamperes to 10 amperes.
3. One-inch paint brush.
4. Dental mirror, non-magnifying.
5. Lint-free dust cloth.
6. Fine-cut file.
7. Burnishing tool.
8. Several sizes of screw drivers and an offset screw driver.
9. Set of open-end or box wrenches up to half-inch size, or a small adjustable wrench.
10. Special tools as specified by manufacturer of equipment.

Basic Maintenance Operations

Maintenance work can be divided into five phases:

1. Inspection.
2. Cleaning.
3. Repair or service.
4. Lubrication.
5. Adjustment.

Inspection. Inspection is probably the most important phase of effective maintenance. Minor defects or deviations, if discovered early, can be corrected before major difficulties arise. Personnel must be completely familiar with the equipment when it is functioning normally. Inspection should be made by physical contact as well as by vision.

A visual check for overheated parts should be made by looking for signs of discoloration, blistering, leakage of insulating compounds, oxidation of metal contact surfaces, or malformation of components. Temperature of components (except resistors and vacuum tubes) can be checked by touch. Transformers, capacitors, motor bearings, and wiring should not be too hot to touch comfortably with the palm of the hand. If temperatures are in doubt, a thermometer should be used to ascertain actual temperature. The maximum normal temperature of any component (except tubes and resistors and copper leads carrying high radio-frequency currents) should not exceed 90 degrees Fahrenheit above ambient room temperature.

Electrical connections should be checked for temperature and looseness. Jerking or pulling wiring to check connections should be avoided, for either will eventually break connections. Wiring and components should be in original positions.

If the equipment is subject to vibration, however light, mounting screws must be checked for looseness. Freedom and resiliency of shock mountings or other cushioning devices must also be checked.

Check for leaks in water-cooling systems and clogged filters in air-cooled equipment.

Tubes should not be removed from the sockets in routine inspection and cleaning unless evidence of trouble is discovered. Unnecessary handling of tubes imposes additional hazards which are normally unwarranted.

Equipment should be inspected for cracked or chipped insulators and resistors; leaking capacitors, transformers, or reactors; cracked instrument cases or glass; and faulty indicator lamps.

Cleaning. For routine cleaning of tubes, the instructions of Section 35·2 should be followed. All components, especially insulators, must be wiped off with a clean cloth. Cleaning fluid is useful in removing oil-laden dust from insulators and other components, but the fluid residue should be wiped clean with a dry cloth. The small paint brush is useful in removing dust from corners and cavities. During cleaning any evidence of oil, water, or compound leakage should be checked.

A leaking capacitor should be replaced, if possible. An oil capacitor may be temporarily repaired by removing the capacitor and soldering the leaking seam or joint. Such a repair should be considered temporary, although it may be permanent if it is made before appreciable oil has escaped. The unit should be replaced as soon as possible, however, because the internal condition of the unit cannot be determined.

Transformers or other components that leak compound should be checked during operation for excessive temperature rise by mounting a thermometer in contact with the unit. The thermometer should be placed so that it can be observed during operation, if possible. If overheating is indicated, the cause should be determined and corrected.

Repair or Service. Loose connections should be tightened. The use of pliers to tighten nuts or screws should be avoided. A tool of the proper size should be used and the stripping of threads should be avoided. Screws which clamp porcelain, ceramic, or glass insulation or components should be tightened just to the point where the component is held firmly.

Cracked, chipped, or broken insulator bushings or cases should be replaced as soon as possible.

Relay and contactor contacts should be checked for burning and pitting or evidence of poor contact. Many types of contacts are employed in electronic equipment, and manufacturers usually specify special care which may be required. In the absence of specific instructions, it is usually safe to employ the burnishing tool on small contacts to remove small pits or slight roughness. If the burning or pit is more than surface deep, a fine-cut file may be used, and care must be exercised to maintain the original contour of the contact.

Silver contacts often show signs of discoloration. If they are not pitted or rough, only slight burnishing is necessary. The discoloration is probably caused by a thin layer of silver oxide (tarnishing) which is a good conductor of electricity. This film does not affect proper relay contact.

Excessive filing should be avoided because some contacts have only a thin layer of silver or other contact alloy. The final smoothing should be with the burnishing tool. If the contact is badly deteriorated, replace it with a new contact. The use of sandpaper or crocus cloth on contacts is to be avoided if it is not specified by the manufacturer, because the abrasive elements become lodged in the metal and prevent proper contact.

After the contacts are burnished or filed, the relay or contactor should be operated by hand to check for positive contact. The armature or operating mechanism should move freely without binding or dragging. Contacts must have proper alignment and correct spacing. Contact springs should be in good condition.

Contact arms must be adjusted if they have been displaced during cleaning. Correct setting and proper adjustment for contactors and relays usually are described in the instruction book for the equipment.

Switch contacts which are totally enclosed are difficult to check and service. A check on the physical operation of the switch, in which toggle action, freedom of movement, and amount of spring tension are noted, may indicate indefinite or questionable operation.

Open switches can be checked and their contacts can be serviced by a technique similar to that employed on relay contacts. Rotary-blade switches should be checked for proper contact without excessive wear of the stationary contacts. The manufacturer's instructions for adjustment should be followed if operation is improper.

Rheostats in which contact is made directly on the resistance winding should be cleaned only. Rheostats with contacts of the button type may be filed or burnished as necessary to clear burning or pitting of contact surfaces.

Rotating machines such as blowers and pumps should be checked for excessive vibration, brush wear, and condition of commutator or slip rings.

Lubrication. Oil should be used sparingly. Excessive use of oil leads to accumulation of dust, and the resulting contamination leads to early failure. Relay bearings should be oiled sparingly or not at all. Many contactor and relay manufacturers recommend no oil. Shaft bearings for switches, rheostats, and similar mechanical components may be lubricated with a drop of light machine oil. Switch contacts with evidence of excessive wear may be lubricated with a thin film of pure petroleum jelly. The excess should be completely removed.

Ball bearings in variable capacitors should not be lubricated, and only a drop of oil should be applied to sleeve bearings if there is evidence of binding or dryness.

Bearings in rotating units should be greased or oiled in accordance with the manufacturer's instructions.

Adjustments. When the other phases of maintenance have been completed, a check on all adjustments should be made, including:

1. Relay and contactor adjustments for proper clearance, contact, or timing.
2. Safety interlocks for proper operation.
3. Fuses for proper rating.
4. Indicator lamps for proper indication.
5. Tube-cathode voltages. If more than 5 percent from correct value, re-adjust transformer taps, rheostat, or other control device. Make allowance for line-voltage variations. See Section 35·1.
6. Complete final operating adjustments as outlined in the instruction book for the equipment. Final instrument readings, dial adjustments, and performance should agree with the values obtained under previous normal operating conditions.

36·4 TROUBLE SHOOTING

Location of causes for improper operation is usually not particularly difficult if the basic operation of the equipment is understood. Frequently repairs can be made without elaborate instruments or tools; however, a thorough knowledge of the circuits and a few necessary instruments are of great assistance in locating any source of trouble.

A volt-ohmmeter with a-c and d-c voltage scales and a-c and d-c current scales is a particularly useful instrument. The voltmeter should have a resistance of at least 1000 ohms per volt so that circuit conditions are not disturbed by connection of the instrument. In some circuits an instrument of higher resistance is desirable, and for such circuits a vacuum-tube voltmeter or cathode-ray oscilloscope is useful.

Study the schematic and wiring diagrams of the equipment. If possible, the operation of the equipment should be observed, and the circuits or phase of operation in which normal functioning fails should be noted.

A careful study of the schematic diagram then often discloses possible causes of improper operation. The wiring diagram shows connections between components and assists in determining which leads must be disconnected for subsequent tests. These components can be checked, or the circuit functions can be isolated and checked for trouble. The volt-ohmmeter is useful in checking both circuit components and measurement of voltages in the circuit. *High voltages may appear at unexpected points in the circuits of inoperative equipment.*

Some manufacturers supply charts or marked photographs in their instruction books, giving normal voltage or current at various points in the circuit. Such data are useful in tracing or isolating faulty operation.

If a tube appears to be the possible source of trouble, a replacement should be tried, or the suspected tube should be tested. The replacement method is probably the most useful method of checking industrial tubes, because few satisfactory testers are available. If the replacement tube does not correct the trouble, the original tube may be returned to its socket and the other components in the circuit under suspicion should be checked. Where possible cathode and anode voltages should be measured.

Resistance of all transformers and reactor windings can be readily checked with an ohmmeter. Resistors, as well as open or short circuits in the wiring, can be checked in the same manner. Resistance readings for resistor units should be compared with the values given on the diagram or in the electrical parts list usually provided in the instruction book for the equipment. One must be sure that the readings taken really represent the resistance of the component under test.

Parallel paths often make it necessary to disconnect the wires to the component to avoid erroneous results. A capacitor may be checked for failure caused by a short circuit or open circuit by utilization of the high-resistance range on the ohmmeter. To check for an open circuit, the ohmmeter is connected momentarily to the isolated terminals of the unit. A momentary "flick" of the needle may be observed. If not, the ohmmeter leads are quickly reversed and their contacts touched to the terminals of the capacitor. A reverse "flick" of the meter probably will be observed at the instant of contact if the capacitor is not defective. This method is not reliable on units of low capacitance, but is useful in checking units of 0.01 microfarad and larger. The resistance of a capacitor usually is well above 1 megohm. If a lower reading is obtained, a test should be made on an identical unit elsewhere in the equipment. The resistance of electrolytic capacitors cannot be reliably checked with the ohmmeter because they may have low leakage resistances. An electrolytic capacitor can be disconnected from the circuit momentarily, or a replacement can be made. If operation is changed materially by disconnection of the capacitor, a replacement should be tried for a positive indication of a defective unit.

A few items of general trouble in nearly all types of electronic equipment are given in the following list with possible causes and suggested corrective measures.

COMPLETE EQUIPMENT INOPERATIVE

Causes	Corrective Measures
Line switch may be open.	Close line switch.
Fuses may be "blown."	Replace fuse.
Failure of power-supply line or low-line voltage.	Check with voltmeter for proper line voltage.

POWER SUPPLY AVAILABLE BUT CATHODES FAIL TO LIGHT

Causes	Corrective Measures
Tube type may be such that no illumination is visible. Thyratrons and phanotrons often have shielded or enclosed cathodes.	Such tubes feel warm if cathode is energized. Be sure anode volts have not been applied when touching either glass or metal tubes.
Water supply may be shut off.	Turn on.
Water interlock may be inoperative.	Check.
Forced-air supply may be cut off.	Check air supply.
Air interlock may be inoperative.	Check.
Cathode contactor may fail to close.	Check interlocks in coil circuit and contactor coil.
Open connection in primary or secondary of cathode circuit.	Check wiring and transformer windings.
Tube cathode may be open-circuited.	Check visually or with ohmmeter. Replace tube if inoperative.
Incorrect power-line voltage or frequency	Check nameplate data against power line voltage and frequency rating.
Control circuit may be open, missing connection or broken lead.	Check fuses and circuit with the wiring and interconnection diagram.
Tubes may be in wrong socket.	Check location of tubes with wiring diagram.
Tube may be inoperative.	Try replacement tube.

POWER SUPPLY AVAILABLE, TUBE CATHODES ENERGIZED, BUT ANODE VOLTAGE NOT APPLIED

Causes	Corrective Measures
Time delay relay may not have functioned.	Wait for completion of timing cycle.
Anode-power-control switch may not be closed.	Push or close anode-power-control switch.
Equipment may be connected for remote control.	Check position of auxiliary control switches for proper position for operation intended.
Anode-control contactor may not be energized.	Check door and other protective and safety interlocks. Check control wiring and interconnections against wiring and interconnection diagrams. Check for broken connections in the contactor coil circuit. Check overload relays. Check voltage across coil and, if present, replace contactor coil.

TUBE CATHODES ENERGIZED, ANODE CONTROL CONTACTOR CLOSED, BUT ANODE VOLTAGE NOT APPLIED

Causes	Corrective Measures
Anode supply circuit open.	Check anode power-supply circuit for open switch or blown fuses and correct by closing switch or replacing fuse.
Rectifier tubes not firing (in equipment using rectifier tubes for d-c anode supply).	Check air temperature near tube and provide heat to supply temperature specified in tube instructions. Check cathode voltage and correct if low, using extreme caution as high voltage to ground may exist.

FREQUENT UNSCHEDULED INTERRUPTIONS

Causes	Corrective Measures
Loose connection or broken lead.	Check for loose connection. If lead broken due to excessive vibration replace with flexible lead.
Equipment may be overloaded, tripping overload protective devices.	Reduce loading. Check adjustment of protective device.
Line voltage may be fluctuating excessively.	Check line voltage with recording voltmeter. Install voltage regulator. Adjust transformer taps (if any) for average voltage.
Air or water flow may be erratic, tripping protective devices.	Correct water or air flow. Clean or replace filters. Check adjustment of protective devices.
Arc-backs in phanotron or thyratron tubes.	Be sure tube was properly conditioned as in Section 35·3. Provide forced-air cooling to reduce mercury-vapor temperature to value specified by manufacturer. Check applied anode voltage and, if above rating of tube, adjust taps on anode supply transformer.

OPERATION SATISFACTORY BUT TUBE LIFE REDUCED

Causes	Corrective Measures
Incorrect cathode voltage.	Check and correct if found more than 5 percent from rating. In checking cathode voltages, use care. Some cathodes may be at high voltage above ground.
Wide fluctuation in line voltage.	Install voltage-regulating transformer.
Improper care and handling.	See Sections 35·1, 35·2, and 35·3 for tube care and maintenance.
Excessive vibration or shock.	Provide shock mounting and connect to socket and tube with extra flexible leads.
Incorrect cooling.	Check air or water flow and temperature.
Overloading.	Check adjustments for correct loading. Check anode voltage adjustment or taps.
Intermittent operation.	For intermittent use, tube life may be improved by leaving cathodes energized during idle periods of less than 30 minutes.

OVERHEATED COMPONENTS

Causes	Corrective Measures
Inadequate cooling.	Check and clean air filter. Remove obstructions blocking ventilating louvers in cabinet. Check blower-motor operation (if used).
Overload.	Determine cause of overload and correct.
Defective component.	Replace.

REFERENCES

1. "Hints on Electrical Maintenance," Westinghouse Electric Corp. Section 21.
2. "Installation, Maintenance and Servicing of Electronic Motor Control," B. J. Dalton, *Electrical Contracting*, Dec. 1944, pp. 124–132; Jan. 1945, pp. 116–126.
3. "How to Maintain Electric Equipment," *General Electric Manual*, pp. 90–91.

INDEX

Absorption, of x-rays, 6, 140, 152, 509
 of yellow light, 5
Acceleration, control of, in motor, 551
Acids for electron scattering, 498
Adjustments in maintenance, 667
Admittance, electron, 93
Advance angle, in inverter, 393
Aerodynamic unbalance, 484
Air-borne radar system, 658
Air-cooled tube maintenance, 661
Air-core transformer, 221, 395
Air navigation by radar, 656
Alarms, photoelectric, 636
Alkali metals, 18, 122
Aluminum welding, 618
Ampere, definition of, 176
Ampere's law, 308, 377
Amplification, of electron current in gases, 53
 of regulator system, 577, 581, 584
Amplification factor, definition of, 62, 248
Amplifier, bias for, 252; *see also* Grid bias
 classes, A, 247, 249, 255
  AB, 247, 249
  B, 247, 249
  C, 79, 247, 249, 252, 255
 constant output, 458
 definition of, 246
 direct-coupled, 252
 distortion in, 253
 equivalent circuit of triode, 232
 feedback in, 252
 gain, measurement of, 475
 grounded grid of, 251
 inductive compensation of, 475
 klystron as, 85
 "motorboating" in, 252
 neutralization of, 251
 pentode, 191, 217, 225
 phase shift, measurement of, 475
  in transformer-coupled, 253
 for phototube, 628
 for power, design of, 192
 for power-line carrier transformer, 221
 push-pull, 216, 248
 with reactive load, 231
 for regulator, advantages of electronic type, 573, 587
 stability, 252
 transformer-coupled, 213, 246
 transformer for, 213, 217
 triode power, 229, 231
 tuned, 247
 wave forms in, 247, 254
Amplitude modulation, measurement of, 476
Angstrom unit, definition of, 7, 181
Angular position regulator, 574
Annealing by high frequency heating, 408
Anode, *see* Tubes, anodes of
Antenna, definition of, 308
Antenna arrays, 321, 326, 334
Antennas, apparent feed-point resistance of, 321
 apparent impedance of, 320, 324
Antennas (*Continued*)
 apparent loop resistance of, 321, 324
 attenuation constant of, 314, 317, 324
 base capacitance of, 315, 317
 base fed, 313, 320, 324, 330, 333, 338, 341
 base impedance of, 314, 317, 320, 324, 330
 broad-band, 336
 capacitance-loaded, 333
 characteristic impedance of, 313, 324, 339, 341
 circular, 334
 current distribution in, 311, 321, 325, 341
 dipole, 328, 330, 333, 336
 director, 335
 driven element, 335
 driving-point impedance of, 320, 324, 328
 electrical radius of, 314, 317, 320
 end-loaded, 317, 326
 equivalent resistance of, 314, 324, 339, 341
 feed-point resistance of, 321
 feed-point reactance of, 321
 field intensity of, 311, 321, 326, 330, 333
 field pattern of, 311, 326, 339
 folded dipole, 336
 ground loss of, 331
 ground systems for, 331
 half-wave, 311, 328
 image of, 312, 328
 Krause, 335
 linear, 311
 loaded, 317, 320, 326, 333
 long-wire, 331
 loop, 333
 losses in, 314
 multiple-element, 320, 326, 335
 mutual impedance of, 320
 parabolic, 335
 radiation resistance of, 314, 320, 329, 331
 receiving, 328, 330, 334
 reciprocity theorem applied to, 327
 reflector, 335
 rhombic, 331
 sectionalized, 333
 self-impedance of, 313, 328
 self-loop impedance of, 321, 324
 self-supporting towers, 315
 shunt capacitance, 315, 317
 square loop, 334
 top-loaded, 317, 320, 326
 tower for, 314, 317, 321, 324
 transmission line equations for, 313, 317, 321
 turnstile, 334, 337
 V type, 331
 vertical, 313, 317, 321, 324, 326, 330, 333
 Yagi, 336
Anticipator circuits, 587, 590
Anti-hunting circuits, 587
Anti-TR tube, 646
Arc back, 52, 348, 366
Arc-discharge, falling characteristics of, 273
Arc drop, 105
Arc-furnace regulator, 596
Arc-through, in inverter, 365
Artificial line, 198
Atom, adsorbed on metals, 15
 discussion of, 3
 ionization potentials, 6
Atomic energy levels, 5
Atomic number, 3
Atomic structure, 3
Atomic theory, 3
Atomic weight, determination of, 487
Attenuation, in antennas, 314, 317, 324
 in carrier transmission, 449
 in wave filters, 192, 195
Attenuation constant, for TE wave-guide mode, 295
 for TM wave-guide mode, 297
 for two-conductor line, 292
Attenuator, compensated for stray capacitance, 168
Austenite, 404
Auto-electronic emission, 21
Automatic gain control, 257
Average voltage, measurement of, 457, 460
Avogadro's number, 1

Bactericidal ultraviolet lamp, 172, 641
Balancing machines, 478
Bandwidth, filter, 198
Barrier, potential, 10, 16, 18, 25
Barrier layer, 133
Battery charger, rectifying device for, 344
Beacon bombing systems, 655
Beam tubes, 74, 251, 253, 257
Beam width, radar, 647
Bessel's equation, 479
Betatron, 150
Bias, 233, 246, 252, 268, 282
Bipotential lens, 164
Black body, 117
"Black-light" lamp, 174
Boltzmann's constant, 8, 11
Bombardment, positive-ion, of thermionic cathodes, 14
  of thoriated-tungsten surfaces, 17
Bores, high-frequency heating of, 392
Bragg's law, 521
Braking of motor, 565
Brazing by high-frequency heating, 402
Bridge-balance detectors, 468, 476
Bridge-type rectifier, 199
Broadcast, standard, accuracy of frequency, 465
Bucky diaphragm, 144, 509, 514, 524
Bureau of Standards, frequency broadcasts, 464
By-pass capacitor, 219, 251

Candle, definition of, 136
Candlepower, definition of, 136
Capacitance, calculation of, in transformers 217
 definition and description of, 180
 filter, 195, 199, 202
 in pulse transformers, 222

Capacitance (*Continued*)
in transformer winding, 213, 217, 219, 221
in vacuum tubes, 251
Capacitive coupling, 194, 218
Capacitive current, 190, 193
Capacitor charging current, 199
Capacitor-discharge welders, 618–625
Capacitor effect, 202
Capacitor input filter, 199, 201
Capacitors, charging time for welding, 620
frequency limitation in high-frequency generators, 376
inductance of leads in, 186
line coupling, 452
tuning of, 189, 192, 194, 204
types of, 180
Carrier communication, discussion of, 442
Carrier load control, 447
Carrier relaying, 445
Carrier supervisory control, 448
Cathode bias, 252
Cathode follower, 219, 254
Cathode operation, for reduced maintenance, 659, 669
Cathode-ray beam modulation, 170
Cathode-ray oscilloscope, *see also* Cathode-ray tube
amplifiers in, 474
applications, 465, 475, 477
Cathode-ray tube, *see also* Cathode-ray oscilloscope
cathodes, 162
components of, 162
curve plotting of, 169
deflection systems, 166
defocusing, 166
electron guns, 165
ratings, 168
screens, 166, 474
time base for, 276, 475
transit-time effect, 167
for x-rays, 146
Cathodes, *see* Tubes, cathodes of
Cavity, *see* Resonant cavity
Cells, cuprous oxide, 133
iron-selenium, 134
photoconductive, 120, 133, 633
photoelectric, *see* Phototubes
photovoltaic, 120, 133, 633
selenium, 133
thalofide, 133
Characteristic line impedance, for TE wave-guide mode, 295
for TM wave-guide mode, 297
for two-conductor line, 292, 451
Charge-control circuit, for capacitor-discharge welders, 621
Charge-to-mass ratio, 3
Child's law, 69
Circuit elements, 176
Clock, synchronous, 464, 470
Closed-cycle control, 572
Coaxial line, modes of operation, 294
Coaxial line transformer, 421
Coils for high-frequency heating, 387
Cold-cathode oscilloscope tube, 474
Cold-cathode thyratron, 473
Cold-cathode tubes, characteristics of, 55, 114
Collisions, super-elastic, 6
Colpitts oscillator circuit, 265, 375
Combinations of inductance, capacitance, and resistance, 181, 183, 186
Commutation, in inverter, 362
Commutation angle, rectifier and inverter, 363
Commutation reactance, in rectifiers, 202, 205
Conduction loss, in high-frequency heating, 386
Conductors, description of, 7, 176
Cone vision, 516
Contact difference of potential, 10, 458
Convection loss, high-frequency heating, 386, 438
Coolidge tube, for x-rays, 147
Cooling media, maintenance of, 664
Corona, cleaning effect on gases, 526
in transformers, 208
Corpuscular theory of light, 22
Coulomb force, 20
Counter, *see also* Timer, Timing circuits
impulse-type, 470
for welding, 616
Coupled circuits, 191, 194
Coupling, close, effect of, 469
coefficient of, 192, 227
to load in high-frequency heating, 388
Critical potentials of atoms and molecules, 6
Crystal analysis, *see* Diffraction
Crystals, galena, use of, 469
ionic, 7
molecular, 7
piezoelectric, Curie cut, 270
quartz, 464
valence, 7
Cuprous oxide cell, 133
Current concentrator for high-frequency heating, 395
Current-limit control, motor acceleration, 552
Current-limit regulator, circuit for, 594
Cut-off, definition of, 246
Cycloconverter, 361, 369
Cycloinverter, 361, 371
Cyclorectifier, 361, 370
Cyclotron, 151
Cylinders, high-frequency heating of, 379

Damping coefficient, 579, 581
Damping of regulator system, 583, 587
Damping transformer, 590
D-c motor control, advantages of, 533; *see also* Motors
D-c transformer, 372
D-c transmission, 372
De Broglie wavelength of electrons, 16, 496
Dead-zone regulator, 573, 591
Decibel, definition of, 213, 450
Deionization, 53
Demodulation, 258
Density, gas, 1
Dielectric constant, complex representation, 291
definition of, 408
effective, in high-frequency heating, 412
Dielectric heating, *see* High-frequency heating
Dielectric loss factor, 409
Dielectric materials, properties of, 180, 434
Dielectric phase angle, 409
Dielectric properties, measurement of, 409
Differentiating circuit, 482
Diffraction, electron, 4
x-ray, 521
Diffraction grating, 521
Diode, for modulation, 258
parallel plane, 38
potential distribution in, 33, 37
space-charge current in, 36
voltage gradient in, 256
Diode voltmeter, 457
Discoloration test, 530
Discriminator, 259
Distortion, *see also* Harmonic distortion, Harmonics
in amplifier, 229, 253
detection of, 475
of x-ray image, 510
Distortionless line, 292
Distributed constants in a-c circuits, 184
Doublet, elemental oscillating, 309, 311, 326
Duplexer, 646
Dushman's thermionic-emission equation, 12
Dust particles, in electrostatic field, 527
separation from gases, 526
Duty cycle of radar transmitter, 653
Dynamic braking, 565
Dynamic characteristics of tubes, 229
Dynatron circuit, 273

Eccles-Jordan trigger circuit, 470
Eddy-current equation, 378
Edge regulation, photoelectric control for, 638
Einstein's law of photoelectric emission, 22
Electric and magnetic field, *see* Magnetic and electric field
Electric-arc furnace regulator, 536
Electrodes, for high-frequency heating, 416
for welding, 602
Electromagnetic field equations, 377
Electromagnetic radiation, 178
Electromagnetic wave nature of electrons, 16
Electromagnetic waves, beaming of, 645
Electrometers, 460, 462
Electron, definition of, 3
Electron admittance, 93
Electron beam, deflection of, 166, 168
defocusing of, 166
divergence in x-ray tube, 154
formation of, 162, 165
in gas, 165
recording, 166
Electron diffraction, 4
Electron-diffraction pattern, use of, 495
Electron emission, auto-electronic, 21
definition of, 10
efficiency of, 14, 17
from energy viewpoint, 9
field, 10, 21
in gases, 51
photoelectric, *see* Photoelectric emission
saturation, 20
secondary, characteristics of, 28
of composite surfaces, 27
definition of, 26
in dynatrons, 274
from grids, 14
effect of impurities, 27
from insulators, 28
of metals, 27
in reflex klystron, 86
effect of temperature, 29
theories of, 26
variation with primary energy, 29
theories of, 10
thermionic, discussion of, 11
from grids, 14
from metals, 12
from thoriated tungsten, 15
from tungsten, 59
types of, 10
Electron-emission current, thermionic, 11

Electron emitters, 27
Electron focusing, in betatron, 150
in a gas, 165
in x-ray tubes, 152
Electron gun, for cathode-ray tubes, 162, 165
Electron impact, 5
Electron lens, 163
Electron micrograph, use of, 495
Electron microscope, 494–499
Electron motion, in constant field, 39, 58
in varying field, 40
Electron path, in electrostatic or magnetostatic field, 162
Electron stain, 498
Electron transit time, *see* Transit time
Electron volt, definition of, 5, 10
Electron wavelength, 496
Electronic apparatus, maintenance of, 665
Electronic charge, 3
Electronic contactor, for welding, 607, 613
Electronic drive, *see also* Motors
compensated, 549
current-limit action in, 552, 555
efficiency of, 569
inversion in, 566
power factor, 570
selection of tubes for, 568
with tachometer generator, 549
voltage-regulated, 547
Electronic microammeter, 460
Electronic switch, 234
Electronics, definition of, 1
Electrons, angular velocity of, in magnetic field, 42
average energy of escape, 13
bound, 7
bunching of, in klystron, 84
conduction of, 10
cycloidal path of, in magnetic and electric fields, 44
distribution of, in metals, 10
electromagnetic-wave nature of, 16
energy distribution of, 9
force on, in magnetic field, 41
free, 7, 10, 12
helical path of, in magnetic field, 43
high-speed, for x-ray, 150
initial velocities of, 13
interaction of, 91
mass of, 11
mean free path of, 3
motion of, in magnetic and electric fields, 43
path of, in magnetic field, 42
potential energy of, 7
reflected, 26
secondary, 28
in solids, 7
thermionic, velocity distribution, 13
transit time of, 82; *see also* Transit time
valence of, 7
wavelength of, 5, 16
Electrophoresis, 18
Electrostatic lens, 163
Electrostatic precipitation, *see also* Precipitator
rectifying device for, 343
Emission, electron, *see* Electron emission
photoelectric, *see* Photoelectric emission
secondary electron, *see* Electron emission, secondary
thermionic, *see* Electron emission, thermionic
of x-rays, 139
Emission equation, 12
Emission of radiation, 5
Emission saturation, 30
Emissivity, definition of, 136, 385
Emitters, electron, 10
Energy, atomic levels, 5
average, of escaping electrons, 13
conversion of, in a resistance, 177
excited state of, 5
of ionization, 5
normal state of, 5
potential diagram for metal surfaces, 10
stored in a capacitance, 176, 180
stored in an inductance, 176, 178
Energy band, 5
Energy distribution, of electrons, 9
Fermi-Dirac, 11
of photoelectrons, 23
Energy levels, 5
Energy-level diagram, 6
Energy-storage welders, 618–625
Evacuation of tubes, 65
Excitation system, for mercury-arc rectifier, 353
Excitron, 110, 343, 345, 348
Exclusion principle, 8

Farad, definition of, 207
Faraday's law, 207, 377
of induction, 309
Fatigue-testing equipment, 485
Federal Communications Commission, broadcast-frequency standard, 465
Feedback, in amplifiers, 252
inverse, in d-c amplification, 458
in microammeter circuit, 460
in regulators, 587
Feedback oscillators, 261, 265, 270
Fermi-Dirac distribution, 11
Ferrite, 403
Field control of motor, 557
Field emission, 21
Filaments, tungsten, thoriated tungsten, 15
Film, characteristics of, 512
for x-ray, log-density curve, 143
Film fog, 514
Film viewers, x-ray, 144
Filters, a-c line, 205
attenuation in, 195, 199, 201
band pass, 196, 198
capacitor-input, 257
capacitor type, 195, 199, 201
characteristics of, 195, 197
constant-*K* type, 197
current wave form of, 188, 204
cut-off frequency of, 195, 197
for d-c power supplies, 198
design charts for, 196
high-pass, 196
inductor-input, 195, 202, 205, 212
infrared for phototubes, 630
for key-click elimination, 206
limitations of, 197
low-pass, 196
m-derived type, 197
multistage, 200
phase shift in, 196, 218
pi-section, 195, 218
radio-frequency, 258
radiography application of, 145
for rectifiers, 198, 201, 205
termination, 195
transients in, 205
Filters (*Continued*)
T-section type, 195
tuned-power-supply type, 204
for x-rays, 513
Flaw detection, supersonic, 272
Fluorescent lamp, ultraviolet, 644
Fluorescent screen, characteristics of, 474
Fluoroscopic screen, 516
Fluoroscopy, eye adaption for, 516
industrial, 503, 515
medical, 503, 524
Flux, luminous, calculation of, 136
for soldering with high-frequency heating, 402
Flux-resetting rectifier, 624
Foot-candle, definition of, 136
Forge-timing circuit, capacitor-discharge welder, 624
Forging pressure, 618
Forward fire of inverter, 365
Fosterite, 208
Fowler-Nordheim formula, 51
Frequency, control of, in radar systems, 649 657
cut-off in phototubes, 121
measurement of, with T network, 468
multiplication, 259
in power-line carrier, 442
processional, 91
resonant, determination of, 486
supersonic, 272
threshold, photoelectric, 22
translational, 91
as affecting tube design, 81
Frequency changer, 368
Frequency meters, 463, 468
Frequency modulation, 256, 476, 645
Frequency response, measurement of, 476
relation of phase shift to, 474, 482
Frequency stability in oscillators, 270, 272
Frequency standards, sources of, 464

Gain in amplifier, measurement of, 475
phase shift relation to, 474, 482
Gain control, automatic, 458
for low-impedance cathode follower, 168
Galena crystal, 469
Gas, conduction of, 47
current in ionized, 176
density of, 1
electron focusing in, 165
pressure of, 1
residual, ionization of, 17
temperature of, 1
x-ray ionization of, 518
Gas detector, photoelectric, 636
Gas ratio, definition of, 125
Gas tube, anode phenomena in, 52
arc-back in, 52, 103
arc drop in, 105
breakdown in, 53
cathode, indirect heater, 106
pool-tube theory, 51
cathode field, 51
characteristics of, 47, 100
clean up of, 104
control of discharge, 55, 57
deionization in, 53
discharge, self-maintaining, 53
electron emission in, 51
grid of, 50

Gas tube (*Continued*)
grid space-potential diagram, 50
inverse voltage in, 103
ionization of, 47
phase control of, 99
phototubes, 125
positive-ion space-charge sheath in, 19
potential distribution in, 48, 52
pressure-density curve, 54
ratings of, 101
as rectifying devices, 344
Schottky effect in, 52
space charge in, 48
"sparking" at cathode of, 105
thermionic emission in, 52
wave forms, a-c control, 100
Gases, kinetic theory of, 1
in mercury discharge lamp, 172
Gee-H system, position-fixing radar, 655
Geiger-Mueller counter, 519
Generators, for high frequency heating, 375
Generator-voltage regulator, 576
Glass, for transmitting ultraviolet radiation, 172
Glow discharge, 55
Glow lamp, neon, 472
Grid bias, computation of, in triode amplifier, 233
for control circuits, 282
definition of, 246
in oscillator design, 268
types of, 252
Grid dissipation, 80
Grid saturation, 246
Grids, *see* Tubes, grids of
Ground, conductivity of, 313
connections, 665
Wagner, 468
Ground-controlled approach (GCA), 658
Ground fault, detection of, 477
Group velocity, in transmission lines, 292
in waveguides, 294

Hallwacko effect, 116
Hardening by high-frequency heating, 404, 406
Harmonic distortion, 215, 217, 219, 221, 229, 251, 259; *see also* Distortion, Harmonics
Harmonics, *see also* Distortion, Harmonic distortion
a-c line current, 205
in coupled circuits, 193
effect on bridge balance, 468
Hartree, 97
multivibrator generation of, 277
ripple, 200
effect on voltmeter readings, 460
Hartley oscillator, 265, 375
Hartree line, 90, 95, 97
Heat control, for welding, 608, 613
Heat treating by high-frequency heating, 403
Henry, definition of, 178
High-frequency heating, applications, dielectric, 375, 411, 429
induction, 375, 387, 426
of bores, 392
brazing by, 403
coils for, 386–393, 438
conduction loss, 386
convection loss, 386, 438
conversion factors for, 438
cycloinverter for, 371

High-frequency heating (*Continued*)
of cylinders, 379
dielectric constant in, effective, 412
dielectric-heating theory, 408
dielectric loss factor in, 409
dielectric materials, table of characteristics, 434
dielectric phase angle in, 409
discussion of, general, 375
economics of, 425
electrodes for, 416
examples of, numerical, for magnetic and non-magnetic materials, 398
for plastic preforms, 421
for plywood bonding, 423
for rayon drying, 423
flux for soldering, 402
formulas for, general, 438
frequency selection, 376, 411
hardening by, 404, 406
heat treating of metals, 403
induction-heating theory, 377
inductor blocks for, 392
load circuits, 376, 394, 412
load coupling, 388
loss factor of dielectrics, 434
magnetic field distribution, 383
magnetic shielding, 390
matching networks, 414
networks involved in, 394, 414
oscillators, tubes for, 376
penetration of current, 378
power factor involved, 393, 395, 409, 412, 416
power of generators available, 377
proximity, 377
quenching, 407
radiation loss, 385, 438
shadow effect, 389
shielding, magnetic, 390
skin effect, 378
of slabs, 381
soldering by, 400
spark-gap oscillators, 375
thermal conductivity, table for metals and dielectrics, 433
thermal power requirements, 384, 438
Hot-cathode tubes, characteristics of, 105
Hull, cut-off curve or envelope, 89
magnetron, 87
space-charge boundary, 91, 95, 97
Hunting of regulators, 573, 583
Hysteresis loss, 378

Iconoscope, description of, 132
Ignitors, 56, 111
Ignitron, *see also* Mercury-arc rectifier, Mercury vapor tubes, Excitron
anode, auxiliary, 111
application of, 343, 345, 534
care of, 661
initiating circuits, 289
for motor control, 534
ratings of, 110
for resistance welding, 111, 288, 607
test of, 663
voltage of, 111
Image-dissector tube, 132
Image force, 20
Image Iconoscope, 132
Images, of antennas, 312
Immersion lens, 164
Impact, electron, 5

Impedance, input of transmission line, 294
mutual, 320
surge, of transmission lines, 419
transfer, 327
Impedance chart, 299
Impedance-matching transformer, 213
Impulse circuits, 285; *see also* Timer, Timing circuits
Impulse counter, 470
Indicators, radar, 649
Inductance, definition and description of, 178
Inductance, capacitance, and resistance, *see* Combinations of
Induction field, 309
Induction heating, *see* High-frequency heating
Inductor blocks for high-frequency heating, 392
Inductors, 212
Inherent filtration in x-ray tubes, 152
Inspection, in maintenance, 666
by x-rays, 142
Instruments, for maintenance, 666
Insulators, 7, 28
Intensifying screens, 513
Inverse feedback, *see* Feedback
Inverse square law, for x-rays, 138, 142, 146, 156
Inverse voltage, in gas tubes, 103
Inversion, in rectifier drives, 566
wave forms, 100
Inverter, advance angle, 363
applications, 365
arc-back, 366
arc-through, 365
commutation, 362
definition of, 361
faults, 365
forward-fire, 365
frequency limitations, 361, 364
for frequency meter, 467
ignition angle, 362
margin angle, 363
misfire, 365
parallel-capacitor type, 364
ratings of, 364
regeneration with, 366
self-excited, 364
separately excited, 362
series-capacitor type, 364
theory of operation, 361
Ionic crystals, 7
Ionization, by collision, 519
cross section for, in a gas, 7
cumulative, 47
in gas tubes, 17, 47
mean free path for, 6
of mercury vapor, 7
metastable, 47
photo-, 5
probability of, 47
by x-rays, 518
Ionization energy, 5
Ionization function, 7
Ionization potential, 5
Ionized space, conduction through, 47
Ionizing power, of x-rays, 518
Ionosphere, 337
Ions, focus of in mass spectrometer, 487
motion of, in magnetic and electric field, 46
negative, 6
positive, 3, 5
Iris, resonant, 87
Iron-iron carbide diagram, 403

Iron-selenium cell, 134
Isotopes, in mass spectrometer, 488

Kenotron, 343, 662
Klystron, as amplifier, 85
  description of, 83
  frequency control of, 648, 657
Konal, 18

Langmuir curve, 89
Langmuir-Child equation, 36
Laplace's equation, 35
Laplacian, 378
Larmor angular frequency, 91
Lattice, body-centered cubic, 403
  face-centered cubic, 7, 403
Lecher wire, 469
Lenard tube, for x-rays, 146
Lens, electron, 163, 497
Line traps, in power-line carrier, 454
Line-tuning units, 454
Light, absorption of, 5
  corpuscular theory of, 22
  definition of, 135
  wavelength of, 135
Lissajous figure, use in cathode-ray oscilloscope, 465
Load controller, for power-line carrier, 447
Load coupling in high-frequency heating, 388
Logarithmic voltmeters, 458
Loop regulation, photoelectric control for, 639
Loran, 655, 658
Loss factor, in dielectric materials, 180, 434
Lubrication of electronic apparatus, 667
Lumen, definition of, 135
Luminous efficiency, definition of, 135

Machine time constants, 576, 580, 584
Magnetic and electric field, electron motion in, 43
  ion motion in, 46
Magnetic-energy storage welders, 618, 625
Magnetic field, electron action in, 41, 163
Magnetic field distribution in high-frequency heating, 383
Magnetic lens, 164
Magnetic shielding in high-frequency heating, 390
Magnetostriction oscillators, 272
Magnetron oscillator, cut-off characteristics of cylindrical type, 45
  discussion of, 87, 646
  equivalent-circuit representations, 92
  operational diagrams for, 93
  scaling of, 95
  static characteristics of, 89
  symmetric, dynamic characteristics, 90
Maintenance, air-cooled tubes, 661
  discussion, general, 664
  effects of temperature, 661, 665
  testing involved, 667
  water-cooled tubes, 661
Malter effect, 27
Marine radar, 656
Mass spectrometer, 487–494
Matched termination of transmission line, reasons for, 293
  by three-screw tuner, 305
  by three-stub tuner, 303
Matching networks for high-frequency heating, 414
Maxwell-Boltzmann distribution functions, 12
Maxwellian distribution of velocities, 2, 13, 49
Maxwell's equations, 308
Mean free path, of electron, 3
  for ionization, 7
  of molecule, 2
Mechanical time constant in regulation systems, 579, 585
Meissner oscillator, 265
Mercury-arc rectifier, *see also* Excitron, Ignitron, Rectifier
  application of, 343, 345
  arc-back, 348
  control and protection, 352
  cooling systems, 355
  d-c voltage, characteristics and control, 359
  efficiency, 358
  excitation systems, 353
  harmonics, 349
  overload ability, 349
  paralleling of units, 348
  power factor, 358
  ratio, a-c to d-c current, 349
  reverse power absorption, 349
  transformer connections, 351
  vacuum systems, 357
  voltage surges of, 104, 349
Mercury vapor, discharge, relative intensities at different pressures, 172
  energy-level diagram, 6
  ionization of, 7
  spectrogram of discharge, 171
Mercury-vapor tubes, cathode operation, 659, 662
  conditioning of, 662
  cooling of, 104, 662
  operating temperature, 660, 662, 664
  voltage surges, 104, 349
Metals, adsorbed atoms of, 15
  alkaline-earth, 18, 122
  characteristics, table of, 433
  crystalline structure of, 403
  fatigue-testing of, 485
  radiograph of castings of, 504
Metastable states of energy, 6
Metering, demand over carrier channel, 449
Microammeter, 460
Microradiography, 503, 507, 509, 520
Microscope, electron, 494–499
Microwave radar, 645
Misfire of inverter, 365
Modulation, in amplifiers, 79, 195, 217, 255
  amplitude, 445, 476
  frequency, 445, 476
  grid-bias amplitude, 443
  measurement of, 476
  single-sideband, 445
  transformer, 217
  types of, 255, 445
Mole, definition of, 1
Mole fraction, 489
Molecular crystals, 7
Molecular diameter, 6
Molecular velocities, 2
Molecular weight, 1
Molecule, mean free path of, 2
Monochromatic radiation, x-rays, 144
Motion study, high-speed machinery, 471
Motor speed regulator, 578
"Motorboating," 252, 475
Motors (supplied by rectifier), acceleration control, 551
  armature current, average value of, 539
Motors (*Continued*)
    continuous, 536, 540, 562
    discontinuous, 536–539, 562
    form factor of, 564, 568
    peak value of, 562
    rms value of, 564
    wave shape of, 537, 554, 562
  armature-current pulse, 537
  armature of, equivalent circuit for, 536
    inductive voltage drop in, 541
    terminal voltage of, 539
    voltage, regulated for, 547
    voltage drop in, 539, 540–542
  armature-voltage control, 546
  armature-voltage drop, calculations of, 539
  graphical analysis of characteristics of, 540
  ratings of, 534
  speed control of, 534
  speed-torque characteristics, 542–545
  torque factor, 543
  tubes, for control of, 534, 551, 562, 568
  voltage-torque characteristics, 544
Mot-o-trol, 345
Multianode pool-type tubes, 110
Multicolor press register regulator, 591
Multiplier phototube, 130
Multivibrator, 259, 277, 464, 473
Mutual conductance, 248, 255, 257
Mutual inductance, 192, 221

Negative ions, 6
Negative-resistance oscillators, 261, 273
Neon glow lamp, 472
Neon thyratron, 473
Networks, anticipator, 587, 590
  anti-hunting, 587
  bridged T, 468
  equivalent T for, 327
  in high-frequency heating, 394, 414
  mesh equations for, 320, 324
  phase-shift measurement in, 475
  reciprocity theorem for, 327
  unbalance calculation in, 480
Neutralization of amplifier, 251
Neutron, 3
Nucleus, discussion of, 3
  disintegration of by bombardment, 151

Ohm, definition of, 176
Open-cycle control, 572
Optical systems, for phototubes, 630
Optics, electron, 496
Orthicon, description of, 132
Oscillations, build-up of, 263, 275
  conditions for, 224
  parasitic, 252
  in regulator systems, 573, 583
Oscillators, circuits for, 261–277
  class C operation, 265, 267, 270
  classification of, 261
  Colpitts, 265, 375
  crystal-controlled, 464
  definition of, 361
  feedback, 261, 265, 270
  grid bias in, 268
  Hartley, 265, 375
  for high-frequency heating, 373
  klystron as, 85
  magnetron as, *see* Magnetron oscillator
  magnetostriction, 273
  Meissner, 265
  negative resistance, 261, 273

Oscillators (*Continued*)
operating efficiency of, 267
piezoelectric crystal for, 270
for power, 265–270
push-pull circuit for, 270
relaxation, 261, 275
resistance-tuned, 234
spark-gap, 375
transformer-coupled, blocking, 226
class B, 220
class C, 221, 227
transit time in, 262
tuned-grid type, 265
tuned-plate type, 265, 270, 375
Oscilloscope, *see* Cathode-ray oscilloscope
Overheating of electronic apparatus, 666, 669
Overload capacity of tubes, 659, 664
Oxide-coated cathodes, 18, 62, 65
Ozone, generation of in precipitator, 528

Pancake coil for high-frequency heating, 392
Paper-mill speed regulator, 595
Parallel-tuned circuit, 184, 376, 395
Parasitic oscillations, 252
Peak voltage, measurement of, 457, 459
Peaking circuits, 285; *see also* Transformers, for peaking
Penetration of current in high-frequency heating, 378
Pentagrid converter, 75
Pentode, 74, 458
Pentode amplifier, 191, 217, 225
Periodic table, 4
Permatron, 57
Permeability, effective in high-frequency heating, 380
table of, for principal core steels, 209
Persistence, oscilloscope screen, 474
Phanatron, 279, 343, 534, 557, 660, 662
Phase angle, measurement of, 478, 482
in resistance-welding control, 608
in transformers, 218
in tuned circuits, 188
Phase control, in a-c power conversion, 99
Phase modulation, detection of, 476
Phase shift, in amplifiers, 253, 474, 478
in artificial lines, 198
circuits for, 283
in filters, 196
in klystron, 85
measurement of, 475
relation of gain to, 474, 482
Phosphors, 166
Photocell, definition of, 120
Photochemical reaction, ultraviolet effects, 641
Photochemistry, 503, 523
Photoconductive cell, 120, 133, 633
Photoelectric absorption of x-rays, 141
Photoelectric alarms and safety devices, 636
Photoelectric cell, *see* Phototubes
Photoelectric control, 628
Photoelectric controller, 634
Photoelectric current, dynamic response, 118
linearity of, 116, 128
saturation, 125
stability, 128
Photoelectric edge control, 638
Photoelectric effect, definition and fundamental laws, 116
Photoelectric emission, characteristics of, 22
direction of, 117
Einstein's law for, 22
Photoelectric emission (*Continued*)
electron velocity in, 117
effect of electrostatic fields on, 24
fatigue, 122, 130
fundamental laws of, 116
effect of light intensity on, 22
maximum energy of, 117
maximum wavelength of, 117
spectral distribution curve, 23
effect of temperature on, 130
theories of, 24
time lag in, 128
Photoelectric gas detector, 636
Photoelectric measurement, 637
Photoelectric optical systems, 631
Photoelectric register regulator, 639
Photoelectric regulation, 638
Photoelectric response, 123
Photoelectric tin reflow regulator, 640
Photoelectric tubes, *see* Phototubes
Photoelectric width control, 637
Photoelectric work function, 22, 121
Photofluorography, 503, 509, 517, 520, 524
Photoglow tube, 131
Photo-ionization, definition of, 5
Photometric quantities, definitions of, 135
Photomultiplier tube, 520
Photosensitive devices, classification of, 120
Photosensitive surface, as affected by current density, 130
designation of, 123
ultraviolet, 122
Phototubes, amplifier for, 628
applications, 634–640
background illumination, 632
care of, 663
compared to human eye, 628
current-voltage characteristics, 125
definition of, 120
fatigue of, 122, 130
frequency response, 128
gas-filled, 125, 633
gas ratio, 125
glow voltage in, 125, 128
installation of, 633
interelectrode capacitance, 120, 130
leakage in, 120, 129
light sources for, 632
load impedance for, 628
load line of, 127
maximum ambient temperature, 632
maximum operating voltage, 633
microphonics in, 120
for multicolor press, 592
multiple reflections in, 120
as multiplier, 131, 520
optical systems for, 630
ratings of, 126, 130
saturation current, 125
sensitivity, 129, 131, 630, 633
spectral characteristics, 629, 633
storage of, 663
surface preparation, 121
thermionic currents in, 129
in thyratron-tube circuit, 629
time lag in, 128
types of, 120
for ultraviolet-radiation measurement, 174
in vacuum-tube circuit, 628
vacuum-type, 128
in x-ray applications, 520
Photovoltaic cell, 120, 133, 633
Photovoltaic effect, 133
Piezoelectric crystal oscillators, 270
Pilot protection, 446
Pin-hole detector, 635
Planck's constant, 5, 138, 495
Planck's law, 118
Plasma, description of, 49
Plastic pre-forms, high-frequency heating of, 421
Plate-modulated class C amplifier, 79
Pliotron, 75, 79, 662
Plugging, for braking, 565
Plywood bonding by high-frequency heating, 423
"Poisoning" of cathodes, 660
Poisson's equation, 34
Polarization, definition of, 408
Pool-type tubes, 48, 51, 345
Position-fixing radar, 655
Position regulator, 574, 580, 591
Positive ion, bombardment of thermionic cathode, 14
conduction to electrode, 48
current, 51
definition of, 5
sheath, 50
space-charge sheath, 19, 21
Positron, 3
Potential, contact difference of, 10
ionization, 5
Potential gradient, definition of, 33
Power factor, of dielectric materials, 180
in high-frequency heating, 393, 395, 409, 412, 416
in rectification, 100
effect on transients in welding, 606
Power-line carrier, discussion of, 442
losses in, coupling capacitor, 452
line, 451
transmission line, compared to power-transmission line, 450
Power transfer, maximum in inductively coupled circuits, 192
Precipitator, single-stage, 528
two stage, collector cell for, 529
description of, 528
efficiency of, 530
ionizing unit for, 529
power pack for, 529
rectifying device for, 343
Precipitron, applications of, 530
automatic washer for, 528
theory of, 526
Pressure-density curve for gas tubes, 54
Pressure units, table of, 1
Printing press register regulator, 591
Progressive hardening by high-frequency heating, 407
Projection welding, 604
Propagation constant, definition of, 290
of TE mode, 295
of TM mode, 297
of two-conductor line, 290
Propagation of radio waves, 337
Proportional regulator, 573, 575, 582
Protons, 3
Proximity heating, 382
Push-pull oscillator, 270

Q, of crystals, 272
formula for, 180
of resonant cavity, 83

Quantum mechanics, 4
Quantum theory, 495
Quantum yield, 117
Quarter-wave resonance, 185
Quarter-wave transformer, 303
Quenching, in high-frequency heating, 407

Race timer, 471
Radar, anti-TR tube, 646
applications of, 656
beam width, 647
continuous wave, 645
discrimination, 645, 652
duplexer, 646
duty cycle, 653
frequency-modulation, 645
indicators, 649
operation of typical microwave system, 645
position-fixing, 655
power involved, 652
principle of, 645
pulse modulation, 645
pulse width, 652
radiator gain, 653
range, 652, 654
reflecting target area, 653
repetition frequency and interval, 653
scanning, 652
semi-conductor probe rectifier, 647
signal differentiation, 645
signal rectification, 648
signal-to-noise ratio, 654
synchronizer, 646
TR tube, 646
Radiant energy, definition of, 117, 136
measurement of, 119
sources of, 119
spectrum, 118
Radiant-energy density, definition of, 136
Radiant flux, definition of, 136
Radiant-flux density, definition of, 136
Radiant intensity, definition of, 136
Radiation, electromagnetic, 178
loss in high-frequency heating, 385, 438
ultraviolet, 5, 171, 174, 642, 644
x-ray, 5
Radiation constant, Planck's, 22
Radiation field, antennas, 310
Radio interference, 205
Radioactive materials, 3
Radiogenetics, 503, 525
Radiography, *see also* X-rays, X-ray tubes
Bucky diaphragm, use in, 144
characteristics of, 511
contrast in, 142
filters, use in, 145
industrial, 505, 511
latitude of exposures in, 144
log-density curves, 143
masking of specimens in, 144
medical, 503, 523
quality of, 142, 146
scattering in, 140, 144
ultrahigh-speed, 508, 515
Rayon drying by high-frequency heating, 423
Rectifier, *see also* Mercury-arc rectifier, Ignitron, Excitron, Pool-type tubes
angle of ignition and extinction, 537
applications, 238, 241
arc drop, 536
balancing inductors, 242
blocking circuit for, in welding, 622
Rectifier (*Continued*)
capacitor-input filter, 201
characteristics of typical dry-type, 177
commutation, 244
commutation angle, 363
commutation reactance, 202, 205
conduction of, continuous, 536, 540, 562
criterion for, 540
discontinuous, 536–539, 562
control of, 535
copper oxide type, 177
current capacity, 240
definition of, 236, 343
for electrochemical service, 345
Excitron as, *see* Excitron
filters for, 199, 212
for flux-resetting in welding, 624
full-wave, 199, 203, 212
grid-controlled, 205, 243, 343, 620
half-wave, 199, 202, 204, 212
ignitron as, *see* Ignitron
for industrial service, 346
load resistance of, 201
for motor control, 535; *see also* Motors
overload capacity, 348
phase control, 99
phases, effect on ripple, 199
power factor, 100
power losses in, 244
for railway and mining service, 346
ratings for motor control, 562, 568
regulation, 201, 205
for resistance welding, 619, 624
ripple voltage, 199, 205, 212
selenium type, 177
series-resistance, 200, 203
single-phase, 199, 202, 204, 212, 237
six-phase double-wye, 241
theory of operation, 236, 241
three-phase, 46, 203, 239
voltage regulation, 239
wave forms, 99
for x-ray equipment, 524, 529
Rectifier device, comparison of multianode and single-anode mercury-arc type, 346
comparison of sealed and pumped mercury-arc types, 343
definition of, 343
pool-cathode gas type, 345
for radio transmitters, 344
thermionic-cathode gas type, 344
thermionic-cathode vacuum type, 343
Rectigon, 343
Reflected waves, due to loads, 293
Reflection coefficient, definition of, 293
relation to standing-wave ratio, 298
Reflex klystron, 86
Regeneration with inverters, 366
Regenerative braking, 565
Register regulator, 591, 639
Regulation, closed-cycle, 572
of generator voltage, 576
of motor speed, 551, 559, 578
Regulation system stability, Routh's criteria, 583
Regulator, angular frequency of, 583
of angular position, 574
anticipator circuits for, 587
anti-hunting circuits for, 587
applications, 591
for arc furnace, 596
characteristics, 582
Regulator (*Continued*)
components of, 572
current-limit circuit for, 594
damping in, 584, 587
dead-zone type, 573
error detectors, 572, 581
feedback circuits, 587
hunting, 561, 573, 583
leakage currents, 595
multicolor press register, 591, 639
on-off type, 573
oscillations, 573, 584
photoelectric-edge type, 638
principle of, 548
proportional type, 573, 575
register type, 591, 639
response, 583
for sectional paper mill, 595
stability, 583
stiffness, 581
types of, 573
wind-tunnel speed type, 593
Regulator-system amplification, 579
Regulator-system damping, 549, 583
Regulator-system equation, 577, 581, 584
Relative emissivity, definition and table, 385
Relativity, theory of, 495
Relaxation oscillators, 261, 275
Replicas, of surfaces, 497
Resistance, high, measurement of, 463
at high frequencies, 386
increase in, due to skin effect, 177
linear and other forms, 176
radiation, 178
specific, 176
Resistance and capacitance, *see* Combinations of
Resistance and inductance, *see* Combinations of
Resistance welding, *see also* Welding
control for, 599
ignitrons for, 112, 607
Resistor, typical wire-wound and composition, 177
Resolution, of electron microscope, 495, 498
of mass spectrometer, 488
Resonance, frequency for, determination of, 486
of parallel circuits, 184
response curve for inductively coupled circuit, 192
of series circuit, 183
in short transmission lines, 301
Resonance level, 5
Resonant cavity, in klystron, 82
Q of, 83
wavelength measurement in, 470
Resonant circuits, 188, 508
Resonant iris, 87
Resonant window, 87
Resonator bar, in fatigue testing, 487
Restoration of inoperative tubes, 663
Richardson's thermionic-emission equation, 12
Rieke diagram, 93
Ripple, *see* Harmonics, Harmonic distortion
Rms voltage, measurement of, 460
Rod vision, 516
Roentgen, definition of, 138
Rotation, measurement of speed of, 471
Rotational fluoroscopy, 517
Rototrol, 594
Routh's stability criteria for regulator systems, 583

Safety devices, photoelectric, 636
Saturation, electron emission, 20
of grid, 246, 250
photoelectric current, 125
Scaling of magnetrons, 95
Scanning, radar, 652
Scattering of x-rays, 140, 144
Schottky effect, 20, 69, 152
Schumann region of the ultraviolet, 171
Screens, calcium tungstate, 513
for cathode-ray tubes, 162, 166
Seals, for vacuum tubes, 65
Seam winding, 603, 617
Secondary electron emission, *see* Emission, secondary electron
Secondary electron emitters, types of, 27
Secondary emission yield, 26
Secondary radiation, 508
Selenium, x-ray effect on, 518
Selenium cell, 133
Semi-conductors, 7
Sensitivity, *see also* Spectral sensitivity
in *RLC* circuit, 183
in x-ray inspection, 511, 517
Series resonant circuits, 188
Shadow effect in high-frequency heating, 389
Sheath, positive-ion, 50
Shield, magnetic, 207
Shield grid of thyratron, 107
Shielding, magnetic, in high-frequency heating, 390
in transformers, 219
from x-rays, 513
Short circuit in coils, detection of, 477
Single-sideband transmission, 256
Skin effect, 177, 378
Slabs, high-frequency heating of, 381
Slater's reduced variables, 95
Slide-back voltmeter, 459
Smith chart, circular, 299
Soldering by high-frequency heating, 400
Solids, electron theory of, 7
Sound-wave velocity, 2
Space charge, current calculation of, 34, 36, 69
in gas tubes, 48
potential variation with, 34
rotating, 91
Space-flow current, 69
Spark-gap oscillator for high-frequency heating, 375
"Sparking" at cathode of gas tubes, 105
Specific gravity, tables of, 433
Specific inductive capacity, table, 180
Spectral distribution, metal surfaces, S1, S2, S3, S4, 122
photoelectric emission, curve for, 23
Spectral emissivity, definition of, 119, 136
Spectral radial energy, definition of, 136
Spectral response, 122
Spectral sensitivity, 123, 133
Spectrogram of mercury-vapor discharge, 171
Spectrometer, 487–494
Speed measurement, by counter electromotive force, 580
by tachometer generator, 578
Speed regulation, 578
Spot welding, 600, 609
Stabilizing circuits, for rectifiers, 587
Stain, electron, 498
Standing waves, discussion of, 293
on electrodes in high-frequency heating, 417
Standing-wave detector, 298
Standing-wave pattern, analysis of, 301
Standing-wave ratio, in transmission line, 293, 297
Stefan-Boltzmann law, 118
Steinmetz, hysteresis-loss equation, 378
Stokes' theorem, 309
Stroboglow, 473
Stroboscopes, 471, 481
Strobotron, 473
Stub tuner for transmission lines, 303
Sun lamp, 174
Super-elastic collisions, 6
Supersonic flaw detection, 272
Supersonic frequency generation, 272
Supervisory control, power-line carrier, 448
Surface tension, effect on soldering by high-frequency heating, 401
Susceptance variation, method of measuring dielectric properties, 409
Susceptibility, definition of, 408
Sweep generator, 168

T network, 468
Tachometer-generator speed measurement, 578
Tachometer-generator type of electronic drive, 559
TE modes in wave guides, 294
Telegraph cable, submarine, 292
Telemetering, 446
Telephone interference, 220
Telephone line, inductive loading, 292
TEM modes in transmission lines, 294
Tetrode, beam type, 74, 251
characteristics of, 72
Thalofide cell, 133
Thermal conductivity, table, 433
Thermal noise, 654
Thermal power in high-frequency heating, 384
Thermionic cathode, *see* Tubes, cathodes of
Thermionic emission, *see* Electron emission, thermionic
Thoriated tungsten, 15, 17, 60, 660
Thorium, rate of diffusion, 16
Three-stub tuner, 303
Threshold frequency, photoelectric, 22
Thyratron, applications of, 279, 343
basic circuits, 280
characteristics of, 107, 281, 546
cold-cathode, 473
neon-filled, 473
for phototube amplifier, 629
as rectifier for motor control, 535
in relaxation oscillator, 276
shield grid of, 107
"shockover" in, 280
storage of, 662
Time-base generator, 168
Time constant, definition of, 182
Time constants in regulator systems, of armature circuits, 579, 584
general, 584, 587
of generator field, 576, 585
mechanical, 579, 585
of motor field, 580
Time delay, in photoelectric controller, 629
in regulator systems, 574, 585
Time-delay circuits, 477, 609, 629
Time-delay control, motor acceleration, 552, 554
Time intervals, method of measuring, 470
Timer, calibration of, 471
race, 471
Timing circuits, 286, 613, 616
Tin-lead alloys, 400
Tin reflowing, photoelectric control for, 640
TM modes in wave guides, 294, 297
Tools for maintenance, 666
Townsend avalanche, 519
TR tube, 646
Transconductance, definition of, 229
Transformers, air-core type, 221, 395
for amplifiers, class B, 216, 218
equivalent circuit, 213, 221, 226
frequency response of, 213, 218
high-frequency, 221
power input, 218
transformer-coupled, 213, 221, 225, 246, 253
turns ratio in, 213, 249
anode type, for rectifiers, 212
audio-amplifier type, 213
audio-oscillator type, 220
backswing voltage, 224, 227
blocking-oscillator type, 226
breakdown voltage, 209
capacitance, of air-core type, 221
calculation of, 217
effective, 215, 217
of filament type, 211
of pulse type, 222
of winding, 217
coil, mean length of turn, 210, 217
coil impregnation, 208
coil interleaving, 217
coil orientation, 219
coil section, 216
coil treatment, 208
core, area of, 210, 216
length of, 210, 216
losses in, 210, 220
material used, 209
permeability of, 226
saturation of, 207, 209, 217
core gap, 207, 210, 212, 216, 220
core type, 207, 216, 219
corona in, 208
creepage distance, 208
damping, 590
d-c, 372
distortion in, *see* Harmonics, Harmonic distortion
driver type, 219
eddy currents in, 220
enclosure of, 209
equivalent circuit of, 213, 218
exciting current, 209, 221, 224
feedback in, 590
filament type, 211
flux in, time variation of, 207
flux density, *see* Transformers, induction
flux path, 212, 220
frequency characteristics, 213
gap, *see* Transformers, core gap
harmonic distortion in, *see* Harmonic distortion
heat dissipation in, 209
for high-frequency heating, 395
hum reduction, 219
hysteresis loop, 226
impedance, open-circuit, 225
for impedance matching, 213
incremental permeability, 210, 216
induced voltage, 207, 219
inductance of, calculation of, 217
formulas for, 216

Transformers, inductance of (*Continued*)
leakage, 214, 216, 219, 221
mutual, 221
open-circuit (*OCL*), 213, 216, 218, 220, 222, 226
reactor, 212, 218
induction (*B*) curve, 210
insulation, 208, 211, 217
lamination, 207, 221, 226
leakage flux, 219
leakage reactance, 212
Litzendraht cable for, 221
losses, 210, 213, 220
magnetic field of, 220
magnetic shunt, 212
magnetizing current, 215, 219, 225, 254
magnetizing force (*H*), 210, 212, 225
modulation type, 217
non-linear load of, 219, 221, 225, 227
for peaking, 227, 286, 613
in pentode amplifiers, 217
in power-line carrier amplifier, 221
pulse type, 221
in push-pull amplifiers, 216
ratings of, 207
rectangular pulse, 222
for rectifier drives, 569
regulation of voltage, 201, 209, 212
saturation of, 207, 210
sawtooth type, 226
shell type, 207, 216
shielding of, 219
size of, 208, 219
step-down type, 217, 225
step-up type, 217, 221
time constant in pulse type, 222, 224
for transmission line, quarter-wave, 303
varnish used in construction of, 208
video type, 221
for welding, 623
windings of, capacitance of, 214, 221
concentric, 214, 221
multi-, 207, 211
reactance of, 214
resistance of, 213, 222
single-layer, 217
Transients, damping of, in *RLC* circuits, 182, 186
in filters, 205
in keying, 187
in regulators, 582, 588
study of, with cathode-ray oscilloscope, 477
in welding, 605
Transients analyzer, 576
Transit time, effect on cathode-ray tubes, 167
in conventional tubes, 82
of electrons, 82
in oscillators, 262
Transmission, of direct current, 372
Transmission chart, circular, 299
Transmission coefficient of potential barrier for electrons, 16, 25
Transmission line termination, discussion of, 293, 297, 303, 305
determining of, 477
Transmission lines, characteristic impedance of, 292
for direct current, 372
distortionless, 292
in general, 290
general theory as related to field theory, 305
magnetic loading, 292
Transmission lines (*Continued*)
matching load to, 420
for power-line carrier, 450
power transfer and power rating of, 301
propagation constant of, 290, 419
resonating sections of, 301
Smith chart for, 299, 302
standing waves on, 297, 301, 420
Transparency curves (x-rays), 140–142
Transverse electric (TE) and transverse magnetic (TM) modes of propagation, 294, 305
Trigger circuit, Eccles-Jordan, 470
Triode, characteristics of, 248, 250
description of, 70
maximum frequency of, 71
potential distribution in, 35
for power amplifiers, 231
voltage gradient, 246
"Trouble shooting," 667
Tube types, anti-TR, 646
beam-power, 74, 251
cold-cathode, 14, 55, 473
Coolidge, x-ray, 147
diode, for modulation, 258
parallel-plane, 38
potential distribution in, 33, 37
space-charge current in, 36
voltage gradient in, 246
ignitron, *see* Ignitron
kenotron, 67, 343, 662
Lenard, for x-ray, 146
pentagrid converter, 75
pentodes, 74, 458
phanatron, application, 279, 343
care of, 662
phototube, *see* Phototubes
pliotron, 75, 79, 662
power tubes, 75, 79
rectigon, 343
TR, 646
tetrode, 72, 74, 251
thyratron, application of, 279, 343
basic circuits, 280
characteristics of, 107, 281, 344, 546
cold-cathode, 473
neon-filled, 473
for phototube amplifier, 629
as rectifier in motor control, 535
in relaxation oscillator, 276
shield grid of, 107
"shockover" in, 280
storage of, 662
triode, characteristics of, 248, 250
description of, 70
maximum frequency of, 71
potential distribution in, 35
for power amplifiers, 231
voltage gradient, 246
tungar, 343
Tubes, air-cooled, 661
amplification factor of, 62
anode radiators, design of, 64
anodes of, characteristics of, 248
cooling of, 64
current calculation, 39
dissipation at, 79
efficiency of, 232
function of, 63
materials for, 63
radiator, design for, 64
rotating for x-ray tube, 524
Tubes (*Continued*)
capacitance of, 251
cathodes of, in cathode-ray tubes, 162
oxide-coated, 18, 62, 65, 660
"poisoning" of, 660
pool type, 345
of pure metal, Schottky effect, 20
thermionic, 13, 343
of thoriated tungsten, 15, 17, 60, 660
of tungsten, 58, 659
classes of operation, 75
conductance, mutual, 248, 255, 257
construction of, general, 58
dynamic characteristics, 229
envelope of, 65
evacuation of, 65
frequencies involved, 81, 376
grid dissipation, 80
grid saturation, definition of, 246
grids of, control of, 62
screen, 62, 251, 257
suppressor, 62, 251, 257
handling of, 662
life of, prolonging, 659
load lines, graphical construction, 228, 249
mutual conductance in, 248
operating points of, 228
overload capacity of, 659
"poisoning" of cathode, 660
quiescent point of, 228
reactance of, 233
restoration of, 663
screens for cathode-ray tubes, 162, 166
seals for, 65
short life, causes of, 659, 669
storage of, 662
temperature effect on, 659, 662, 665, 668
testing of, 67, 662
transconductance of, definition of, 229
as variable-impedance elements, 233
voltage ratings, 80, 659, 664
water-cooled, 661
x-ray, *see* X-ray tubes
Tuned grid oscillator, 265
Tuned plate oscillator, 265, 270, 375
Tungsten, emission from, 59
as x-ray target, 155
Tungsten filaments, 15
Tuning fork, electrically driven, 464

Ultraviolet, photosensitive surfaces, 122
Ultraviolet lamp, 172
Ultraviolet meter, 175
Ultraviolet radiation, 5, 171, 174, 642, 644

Vacuum-tube voltmeters, 457
Valence band, 8
Valence crystals, 7
Valence electrons, 7
Van de Graff generator, 508
Varistor, 176
Vibration, 471, 478, 486
Viewers, high-intensity, 512, 514
Visual acuity, 516
Volt, electron, definition of, 10
Voltage doubler, 203, 239, 507
Voltage regulation, of generator, 576
of rectifier, 239
Voltmeters, 457

Wagner ground, 468
Water-cooling system, maintenance of, 661, 664

Wattmeter, use of, in vibration measurement, 478
Wave guide, circular, 297
  cut-off wavelength, 294
  rectangular, 296
Wave velocity, 290, 295
Wavelength, cut-off, 294
  of electrons, De Broglie, 16, 496
Wavelength conversion table, 120
Weight, molecular, 1
Welders, certification of, 505
Welding, butt and flash, 604
  capacitor-discharge system, 618–625
  electrodes for, 602
  fundamentals, 601
  ignitrons for, 111
  projection, 604
  seam, 603
  sequence control for, 602
  spot, 600, 609
  time, 601, 603
  timers for, 609, 614, 617
  timing, purpose of, 604
  transients in, 605
Width measurement, photoelectric control for, 637
Wien's law, 118
Wind-tunnel speed regulation, 593
Work function, definition of, 10
  electrostatic field effect on, 20
  photoelectric, 22
  table of, 12, 121
  temperature coefficient of, 13
  thoriated-tungsten surface, 16
  variation of, 69

X-ray generators, 506, 523
X-ray image, magnification of, 510
X-ray therapy, 503, 525
X-ray tube heads, 148, 160
X-ray tubes, *see also* X-rays, Radiography
  air-insulated, 146, 151
  anode, rotating, 148, 155
  beam divergence, 154
  characteristic curve of, 147
  cold-cathode type, 149
  Coolidge type, 147
  cooling of, 149, 156, 160
  design of, 151, 153
  diffraction type, 147, 150
  double-focus type, 146
  early types, 146
  efficiency of target material, 155
  exhaust of, 155
  focusing in, 152, 154
  gas-filled, 146
  high-speed type, 149
  Kovar seals for, 155
  Lenard type, 146
  manufacture of, 155
  medical applications of, 156
  multisection type, 152
  oil-immersed type, 148, 151, 156
  operating characteristics of, 147
  ratings of, 148, 156
  rayproof type, 148, 152
  size of, 151
  target angle, 153
  tungsten targets, 155
  windows, 148, 152
X-rays, *see also* X-ray tubes, Radiography
  absorption of, 6, 140, 152, 509
  applications of, 503, 523
  beam quality, 152
X-rays (*Continued*)
  Bucky diaphragm, 144
  control of, 506
  discovery of, 138, 146
  efficiency of production, 155
  emission characteristics, 139
  film control, 142
  film-density characteristics, 142
  frequency of, 138
  half-value layer, 153
  inherent filtration, 152
  intensity measurement, photoelectric control for, 637
  inverse square law for, 138, 142, 146, 156
  K_γ line of elements, 140
  in medical profession, 523
  method of producing, 138, 146
  monochromatic radiation, 144
  photoelectric absorption of, 140
  photons, 138
  phototube control of, 520
  pin-hole camera, 154
  portable radiographic generator for, 148
  properties of, 138
  protection against, 147
  radiation, 5
  roentgen unit, 138
  scattering of, 140, 144
  sensitivity, 511
  specimen contrast, 142
  spectrum, 139
  transparency curves, 140
  wavelengths of, 5, 139

Yield, secondary emission, definition of, 26

Zero field emission, 20